Plates, plumes, and planetary processes

Edited by

Gillian R. Foulger
Department of Earth Sciences
Durham University
Durham, DH1 3LE
UK

Donna M. Jurdy
Department of Geological Sciences
Northwestern University
Evanston, Illinois 60208
USA

THE
GEOLOGICAL
SOCIETY
OF AMERICA®

Special Paper 430

3300 Penrose Place, P.O. Box 9140 ▪ Boulder, Colorado 80301-9140, USA

2007

Published by The Geological Society of America, Inc.
3300 Penrose Place, P.O. Box 9140, Boulder, Colorado 80301-9140, USA
www.geosociety.org

Printed in U.S.A.

GSA Books Science Editor: Marion E. Bickford

Library of Congress Cataloging-in-Publication Data
Plates, plumes, and planetary processes / edited by Gillian R. Foulger, Donna M. Jurdy.
    p.   cm.—(Special paper ; 430)
    Includes bibliographical references and index.
    ISBN 978-0-8137-2430-0 (pbk. : alk. paper)
    1. Mantle plumes. 2. Volcanic plumes. 3. Volcanism. 4. Plate tectonics. 5. Seismology. I. Foulger, Gillian R., 1952–   II. Jurdy, Donna M., 1946–
QE527.7.P53 2007
551.1′16—dc22

2007009601

Cover: "Do we exist or not?"—The deep mantle plume's view of his/her predicament. Adapted from "Plumacy reprise," by P.R. Vogt and J.C. Holden, Chapter 45, Figure 2.

10 9 8 7 6 5 4 3 2 1

Contents

Preface vii

Scope of the Volume, the Review Process, and Acknowledgments ix

Plates and Plumes

   1. The "plate" model for the genesis of melting anomalies 1
      Gillian R. Foulger

   2. Origins of the plume hypothesis and some of its implications 29
      Norman H. Sleep

   3. The eclogite engine: Chemical geodynamics as a Galileo thermometer 47
      Don L. Anderson

   4. Plate velocities in the hotspot reference frame 65
      W. Jason Morgan and Jason Phipps Morgan

Mantle Convection and Seismology

   5. Implications of lower-mantle structural heterogeneity for the existence and nature of
      whole-mantle plumes 79
      Edward J. Garnero, Thorne Lay, and Allen McNamara

   6. The structure of thermal plumes and geophysical observations 103
      Scott D. King and Hannah L. Redmond

   7. Seismic observations of transition-zone discontinuities beneath hotspot locations 121
      Arwen Deuss

   8. Lower-mantle material properties and convection models of multiscale plumes 137
      Ctirad Matyska and David A. Yuen

   9. Global plume-fed asthenosphere flow—I: Motivation and model development 165
      Michiko Yamamoto, Jason Phipps Morgan, and W. Jason Morgan

   10. Global plume-fed asthenosphere flow—II: Application to the geochemical segmentation
      of mid-ocean ridges 189
      Michiko Yamamoto, Jason Phipps Morgan, and W. Jason Morgan

   11. The Hawaiian SWELL pilot experiment—Evidence for lithosphere rejuvenation from
      ocean bottom surface wave data 209
      Gabi Laske, Jason Phipps Morgan, and John A. Orcutt

Heat and Temperature

12. *Crystallization temperatures of tholeiite parental liquids: Implications for the existence of thermally driven mantle plumes* 235
   Trevor J. Falloon, David H. Green, and Leonid V. Danyushevsky

13. *Potential effects of hydrothermal circulation and magmatism on heatflow at hotspot swells* 261
   Carol A. Stein and Richard P. Von Herzen

14. *Crustal geotherm in southern Deccan basalt province, India: The Moho is as cold as adjoining cratons* 275
   P. Senthil Kumar, Rajeev Menon, and G. Koti Reddy

Geochronology, Hotspot Fixity, and Reference Frames

15. *A quantitative tool for detecting alteration in undisturbed rocks and minerals—I: Water, chemical weathering, and atmospheric argon* 285
   Ajoy K. Baksi

16. *A quantitative tool for detecting alteration in undisturbed rocks and minerals—II: Application to argon ages related to hotspots* 305
   Ajoy K. Baksi

17. *Divergence between paleomagnetic and hotspot-model–predicted polar wander for the Pacific plate with implications for hotspot fixity* 335
   William W. Sager

18. *Global kinematics in deep versus shallow hotspot reference frames* 359
   Marco Cuffaro and Carlo Doglioni

19. *Ridge-crossing seamount chains: A nonthermal approach* 375
   Erin K. Beutel and Don L. Anderson

Oceanic Melting Anomalies

20. *The OIB paradox* 387
   J. Godfrey Fitton

21. *ΔNb and the role of magma mixing at the East Pacific Rise and Iceland* 413
   James H. Natland

22. *Speculations on Cretaceous tectonic history of the northwest Pacific and a tectonic origin for the Hawaii hotspot* 451
   Ian O. Norton

23. *A plate model for Jurassic to Recent intraplate volcanism in the Pacific Ocean basin* 471
   Alan D. Smith

24. *Propagation of the Hawaiian-Emperor volcano chain by Pacific plate cooling stress* 497
   William D. Stuart, Gillian R. Foulger, and Michael Barall

25. *Geophysical characterization of mantle melting anomalies: A crustal view* 507
   Valentí Sallarès and Alcinoe Calahorrano

26. *The North Atlantic Igneous Province: A review of models for its formation* 525
   Romain Meyer, Jolante van Wijk, and Laurent Gernigon

27. *Origin of the Bermuda volcanoes and the Bermuda Rise: History, observations, models,
    and puzzles* 553
    Peter R. Vogt and Wu-Yeol Jung

Continental Melting Anomalies

28. **Lithospheric control of Gondwana breakup: Implications of a trans-Gondwana
    icosahedral fracture system** 593
    James W. Sears

29. **The origin of post-Paleozoic magmatism in eastern Paraguay** 603
    Piero Comin-Chiaramonti, Andrea Marzoli, Celso de Barros Gomes, Anderson Milan,
    Claudio Riccomini, Victor Fernandez Velázquez, Marta M.S. Mantovani, Paul Renne,
    Colombo Celso Gaeta Tassinari, and Paulo Marcos Vasconcelos

30. **The origin of the Columbia River flood basalt province: Plume versus nonplume models** 635
    Peter R. Hooper, Victor E. Camp, Stephen P. Reidel, and Martin E. Ross

31. **Evaluation of different models for the origin of the Siberian Traps** 669
    Alexei V. Ivanov

32. **Eastern Anatolia: A hotspot in a collision zone without a mantle plume** 693
    Mehmet Keskin

33. **Phantom plumes in Europe and the circum-Mediterranean region** 723
    Michele Lustrino and Eugenio Carminati

34. **Mechanisms of crustal growth in large igneous provinces: The north Atlantic province
    as a case study** 747
    Laurent Geoffroy, Charles Aubourg, Jean-Paul Callot, and Jean-Alix Barrat

35. **K-T magmatism and basin tectonism in western Rajasthan, India, results from
    extensional tectonics and not from Réunion plume activity** 775
    Kamal K. Sharma

36. **Plume-related regional prevolcanic uplift in the Deccan Traps: Absence of evidence,
    evidence of absence** 785
    H.C. Sheth

37. **Nd and Sr isotope systematics and geochemistry of a plume-related Early Cretaceous
    alkaline-mafic-ultramafic igneous complex from Jasra, Shillong plateau,
    northeastern India** 815
    Rajesh K. Srivastava and Anup K. Sinha

38. **A bimodal large igneous province and the plume debate: The Paleoproterozoic
    Dongargarh Group, central India** 831
    Sarajit Sensarma

39. **Thick, high-velocity crust in the Emeishan large igneous province, southwestern China:
    Evidence for crustal growth by magmatic underplating or intraplating** 841
    Yi-Gang Xu and Bin He

Planetary Evolution

40. **The coronae of Venus: Impact, plume, or other origin?** 859
    Donna M. Jurdy and Paul R. Stoddard

41. *An alternative Venus* 879
    Warren B. Hamilton

42. *Interaction between local magma ocean evolution and mantle dynamics on Mars* 913
    Chris C. Reese, Viatcheslav S. Solomatov, and Christopher P. Orth

Education

43. *The mantle plume debate in undergraduate geoscience education: Overview, history,
    and recommendations* 933
    Brennan T. Jordan

Platonics and Plumacy

44. *Graphic solutions to problems of plumacy* 945
    John C. Holden and Peter R. Vogt

45. *Plumacy reprise* 955
    Peter R. Vogt and John C. Holden

Index 975

Preface

The turn of the twenty-first century ushered in an upsurge of questioning of then-current views of the links between mantle convection and surface kinematics and volcanism. Since the early 1970s the mantle plume hypothesis had been the most popular explanation for volcanism apparently not explained by plate tectonics, including both large-volume magmatism and volcanism in regions distant from plate boundaries.

In 1963 J. Tuzo Wilson attributed the Hawaiian and some parallel Pacific island chains to volcanism above rising currents of convection cells in the upper mantle. This hypothesis was developed and expanded by W. Jason Morgan in 1971, who proposed a mechanism for the formation and sustaining of "hotspots" as sources of anomalous surface volcanism. Morgan's plume hypothesis envisaged a cylindrical column of hot material rising from the deep mantle to continually feed the surface volcanism. Furthermore, he proposed approximately twenty likely locations for such plumes, namely, Hawaii, Iceland, Yellowstone, the Azores, Easter Island, the Galápagos, Réunion, and about twelve others. Plume theory provided an elegant explanation for time-progressive volcanic chains and relative fixity among melting anomalies on diverse plates. It also offered a reference frame for kinematics, appealing because in plate tectonics, no continent or boundary type can be assumed to be fixed. Morgan's original plume hypothesis makes definite, testable predictions; starting with an observed hotspot track on one plate, and the known relative motion between two plates, predicted tracks on other plates are uniquely calculable.

Major later expansions of the hypothesis included the proposals that plume development involves a plume-head/plume-tail sequence, represented by the emplacement of large igneous provinces followed by time-progressive volcanic chains, and that lower-mantle and core-mantle–boundary geochemical tracers may be identified in hotspot basalts. Many more plumes than the original twenty or so were proposed, popular lists typically containing fifty to one hundred candidates. Planetary scientists extended the hypothesis to other terrestrial planets, and applied it to explain features interpreted as volcanic on Venus and Mars.

Nevertheless, many scientists are disappointed at what they consider to be the limited ability of the hypothesis to predict observations. Skeptics point out, for example, that the association of large igneous provinces with time-progressive volcanic chains is the exception rather than the rule, that many chains are not time progressive, that hotspots apparently are not relatively fixed, and

that the interpretation of some geochemical species as being of deep-lower-mantle origin is not safe. They decry what they see as too little criticism of the plume explanation for volcanism and note instances of discrepancies being "explained" by samples not yet observed, or proposed local variations in the nature of the presumed plume.

But what are the alternatives? The challenge of developing alternatives is seen by many as a fresh, win-win avenue of investigation. If alternatives can be developed and tested, they might supplant the plume hypothesis and bring about arguably the most significant paradigm shift in Earth science since the advent of plate tectonics. Conversely, if they are found to be unsustainable, they will serve to strengthen the plume hypothesis.

In an effort to develop alternatives, the Geological Society of America sponsored the Penrose Conference *Plume IV: Beyond the Plume Hypothesis,* in Hveragerdi, Iceland, 25–29 August 2003. At this meeting, more than sixty scientists from twelve nations brainstormed the problem, the result of which was the book *Plates, plumes, and paradigms* (P^3), published as GSA Special Paper 388, in the fall of 2005. One of the many positive consequences of the publication of P^3 was an upsurge of enthusiasm among plume-advocate scientists to engage in debate. As a result, the American Geophysical Union sponsored the Chapman Conference *The Great Plume Debate,* at Ft. William, Scotland, 28 August–1 September 2005. There, the idea for the present book, *Plates, plumes, and planetary processes* (P^4), to include both plume-advocate and alternative-advocate points of view, was conceived.

P^4 is an expression of the struggle of the Earth science community to deal with the formidable challenge presented by the current debate about the existence of mantle plumes and the nature of hotspot sources. Its contents represent well the current state of thinking. During the past four years, the debate has grown from embryonic beginnings in the private correspondence of a few to become a global, popular, cross-disciplinary subject. The present volume marks an important landmark in this development: It is the first compendium in which a balance of plume-advocate and alternative-advocate chapters was actively sought, and which focuses on challenge and debate, rather than relying on assumption of preferred theories.

Many facets of the current debate are reflected in P^4. It is clear that a broad spectrum of opinion exists, ranging from the view that the available evidence essentially rules plumes out to the view that it requires them. Most investigators occupy the middle ground. To perceive Earth scientists to be divided into two separate camps is to distort reality.

Virtually every branch of Earth science contributes relevant data, a fact that is both a major strength of the subject and also a formidable challenge to practitioners. How can an individual scientist develop an informed opinion when it is first necessary to understand and judge evidence from almost all the major subdisciplines of geology, geophysics, and geochemistry? A common temptation in the past has been to acknowledge the shortcomings in one's own data while trusting that data interpretations from other (less familiar) subdisciplines are decisive. When faced with the requirement for a seemingly impossible breadth of knowledge and understanding, the temptation to accept the judgment of perceived magisters becomes strong. We hope that the broad subject matter offered in P^4 will serve to reduce such temptation.

Much progress has been achieved since the 2003 Penrose Conference. Work has become focused on discriminants that have the greatest potential to rule certain models out. Heat, temperature, mechanisms for producing the observed magmatic volumes and eruption rates, and the nature and location of the magma sources have all received close attention. Hand in hand with these efforts, the uniqueness of common interpretive schemes for many kinds of data, for example, seismic and geochemical, have been revisited.

Alternative theories have matured and strengthened and are now clearly divided into two categories. One, the plate model, attributes melting anomalies to processes related to plate tectonics. The occurrence of magmatism is attributed to permissive extensional stresses in the lithosphere, which result directly or indirectly from plate tectonic processes. Intraplate magmatism is an expression of the nonrigidity of the plates. Variable magmatic volumes are attributed to variable source fusibility, inhomogeneity being maintained by plate tectonic processes. Bolide impacts comprise the second alternative theory. It is applied to Venus and Mars, and its ability to explain the formation of large igneous provinces on Earth is being explored.

The debate has sometimes been inaccurately portrayed as being emotionally charged to the extent that objective judgment has been suspended. We direct readers interested in the human aspects of the subject to the insightful analysis chapters that comprise the final section of P^4.

A particularly pleasing aspect of the debate is that many younger scientists are becoming involved and contributing fresh and innovative new ideas, novel investigative approaches, and cross-disciplinary working groups. The debate is now commonly taught in undergraduate and graduate classrooms, guaranteeing a healthy supply of enquiring new minds to the subject in the future. The origin of melting anomalies is a subject that most professional Earth scientists would today agree is, to a greater or lesser extent, undecided—the time is not yet ripe for widespread agreement on any one theory to be reached. Thus, the redoubling of effort among the experienced, and the influx of more fresh minds to the subject, are needed and welcomed. We, the editors, hope that the present book will both equip and encourage scientists at all levels, and in many fields, in their pursuit of this subject.

Gillian R. Foulger and Donna M. Jurdy
26 February 2007

Scope of the Volume, the Review Process, and Acknowledgments

The chapters in this book are grouped into sections: Plates and Plumes; Mantle Convection and Seismology; Heat and Temperature; Geochronology, Hotspot Fixity, and Reference Frames; Oceanic Melting Anomalies; Continental Melting Anomalies; Planetary Evolution; Education; and Platonics and Plumacy. Many of the chapters are followed by a block of discussion.

Each chapter was formally reviewed by two or more external reviewers and by both of the volume editors. The review process for chapters authored by one of the editors was handled by the other editor. We sought reviewers expected to advocate contrasting viewpoints, to expose authors to the most challenging criticisms possible, and to encourage the confrontation of difficulties.

We extend our grateful thanks to all the reviewers. They included Walter Zuern, Jerry Winterer, Marge Wilson, Howard Wilshire, Marie-Claude Williamson, Mike Widdowson, Mike Walter, Peter Vogt, Claudio Vita-Finzi, Surendra Verma, Jolante van Wijk, Arie van den Berg, Bob Tilling, John Tarduno, Christopher Talbot, William Stuart, Ellen Stofan, Paul Stoddard, Joann Stock, Bernhard Steinberger, Seth Stein, Norman Sleep, Tom Sisson, Hetu Sheth, Alison Shaw, Dave Scholl, Andy Saunders, Will Sager, Lars Rupke, Sergio Rocchi, Keith Putirka, Angelo Peccerillo, Manoj Pandit, Colin Pain, Jim Ogg, Jim Natland, David Naar, Dietmar Mueller, Tim Minshull, Romain Meyer, Greg McHone, Joe McCall, Stephen Marshak, John Mahoney, Phil Leat, Art Lachenbruch, Scott King, Walter Kiefer, Richard Ketcham, Jeff Karson, Bruce Julian, Garrett Ito, Peter Hooper, Anne Hofmeister, Albrecht Hofmann, Kaj Hoernle, Jon Hernlund, Dion Heinz, Chris Hawkesworth, Rob Harris, Warren Hamilton, Stuart Hall, Tristram Hales, Yanni Gunnell, Jeff Gu, David Green, Richard Gordon, Godfrey Fitton, Cinzia Farnetani, Trevor Falloon, Richard Ernst, Linda Elkins-Tanton, Zhijun Du, Carlo Doglioni, Bob Detrick, Bob Christiansen, Françoise Chalot-Prat, Paterno Castillo, Hans-Peter Bunge, Erin Beutel, Keith Bell, Alexander Basilevsky, Ajoy Baksi, Jayne Aubele, Donald Argus, Don Anderson, Ercan Aldanmaz, and an additional twenty anonymous reviewers. Of the 107 reviewers, 81% waived their anonymity.

Chapters were posted on the World Wide Web as they were accepted, and formal discussion of them was open to the public from November 2006 until the end of January 2007. Contributions were posted rapidly on the Web as they were submitted. The final products are printed herein, following the relevant chapters. To the editors' knowledge, this is the first time such an exercise has been conducted. Our experience was that it was extremely successful. Response was vigorous and many e-mails of appreciation were received from both participants and observers.

We gratefully thank the Geological Society of America for facilitating both P^3 and this, its daughter volume, P^4.

The Geological Society of America
Special Paper 430
2007

The "plate" model for the genesis of melting anomalies

Gillian R. Foulger*
Department of Earth Sciences, Durham University, Durham DH1 3LE, UK

ABSTRACT

The plate tectonic processes, or "plate," model for the genesis of melting anomalies ("hotspots") attributes them to shallow-sourced phenomena related to plate tectonics. It postulates that volcanism occurs where the lithosphere is in extension, and that the volume of melt produced is related primarily to the fertility of the source material tapped. This model is supported in general by the observation that most present-day "hotspots" erupt either on or near spreading ridges or in continental rift zones and intraplate regions observed or predicted to be extending. Ocean island basalt-like geochemistry is evidence for source fertility at productive melting anomalies. Plate tectonics involves a rich diversity of processes, and as a result, the plate model is in harmony with many characteristics of the global melting-anomaly constellation that have tended to be underemphasized. The melting anomalies that have been classified as "hotspots" and "hotspot tracks" exhibit extreme variability. This variability suggests that a "one size fits all" model to explain them, such as the classical plume model, is inappropriate, and that local context is important. Associated vertical motion may comprise pre-, peri-, or post-emplacement uplift or subsidence. The total volume erupted ranges from trivial in the case of minor seamount chains to ~10^8 km^3 for the proposed composite Ontong Java–Manihiki–Hikurangi plateau. Time progressions along chains may be extremely regular or absent. Several avenues of testing of the hypothesis are being explored and are stimulating an unprecedented and healthy degree of critical debate regarding the results. Determining seismologically the physical conditions beneath melting anomalies is challenging because of problems of resolution and interpretation of velocity anomalies in terms of medium properties. Petrological approaches to determining source temperature and composition are controversial and still under development. Modeling the heat budget in large igneous provinces requires knowledge of the volume and time scale of emplacement, which is often poorly known. Although ocean island basalt–type geochemistry is generally agreed to be derived from recycled near-surface materials, the specifics are still disputed. Examples are discussed from the Atlantic and Pacific oceans, which show much commonality. Each ocean hosts a single, currently forming, major tholeiitic province (Iceland and Hawaii). Both of these comprise large igneous provinces that are forming late in the sequences of associated volcanism rather than at their beginnings. Each ocean contains several melting anomalies on or near spreading ridges, both time- and non-time-progressive linear volcanic chains of various lengths, and regions of scattered volcanism several hundred kilometers broad. Many continental large igneous provinces lie on the edges of continents and clearly formed in association with continental breakup. Other volcanism is associated with extension in rift valleys, back-arc regions, or above sites

*E-mails: g.r.foulger@durham.ac.uk, gfoulger@usgs.gov.

Foulger, G.R., 2007, The "plate" model for the genesis of melting anomalies, *in* Foulger, G.R., and Jurdy, D.M., eds., Plates, plumes, and planetary processes: Geological Society of America Special Paper 430, p. 1–28, doi: 10.1130/2007.2430(01). For permission to copy, contact editing@geosociety.org. ©2007 The Geological Society of America. All rights reserved.

of slab tearing or break-off. Specific plate models have been developed for some melting anomalies, but others still await detailed application of the theory. The subject is currently developing rapidly and poses a rich array of crucial but challenging questions that need to be addressed.

Keywords: plate, plume, hotspot, Iceland, Hawaii

INTRODUCTION

It is common to hear, when debates are held regarding the origin of melting anomalies, "But what alternatives are there to the plume hypothesis?" This question may even be posed immediately after alternatives have just been described. The objective of this article is to lay out the alternative known as the "plate model," so that future work may build on what has already been achieved rather than comprise reiterations of what has been done in the past. This article does not seek to describe in detail the plume model, nor to compare the plate and plume models, though a few passing remarks on these issues are necessary.

It is common also to hear the statement that no viable alternative to the plume model exists. What is "viable" is a matter of debate. At present no model, plume or otherwise, is without unresolved issues, and it would be premature to accept any without questioning its fundamental validity and subjecting it to rigorous tests (Foulger et al., 2005c; Foulger, 2006a).

The concept of a global "hotspot" phenomenon emerged shortly after plate tectonic theory had been established (Anderson and Natland, 2005; Glen, 2005). Plate tectonics provided an elegant explanation for much of Earth's volcanism, and in so doing brought into focus the existence of many exceptions. Volcanic regions away from spreading plate boundaries or subduction zones, and high-volume on-ridge volcanism were considered to be unexplained by plate tectonics. The term "hotspot" was coined by Wilson (1963), who suggested that the time-progressive Hawaiian Island chain resulted from motion of the Pacific lithosphere over a hot region in the mantle beneath. The concept was extended by Morgan (1971), who suggested that a global constellation of approximately twenty such "hotspots" exists on Earth. He further postulated that they were fixed relative to one another and that this could be explained if they were fueled by source material delivered in hot plumes that rose by virtue of their thermal buoyancy from the deep mantle. The deep mantle was thought to be convecting only slowly and therefore able to provide a stable reference frame.

The model was further developed by laboratory and numerical convection modeling, which suggested that plumes rise from a thermal boundary layer at the core-mantle boundary and comprise bulbous heads followed by narrow tails (Campbell and Griffiths, 1990). Geochemical study of basalts from ocean islands thought to be "hotspots" revealed compositions distinct from mid-ocean ridge basalt (MORB), in particular, being more enriched in incompatible elements and containing high $^3He/^4He$ ratios (Schilling, 1973; Kellogg and Wasserburg, 1990). Such basalts are called "ocean-island basalts" (OIB). Because it was assumed that "hotspots" are fueled by plumes from the deep mantle, their geochemical characteristics were thought to reveal lower mantle characteristics. The numbers of postulated "hotspots" and plumes rose to approximately fifty or more on many lists (Fig. 1).

The main features and predicted observables of the original classical plume model were recently laid out by Campbell (2006) in a clear and concise summary. These traits are a long period of precursory uplift followed by large igneous province (LIP) eruption, subsequent dwindling of the magmatic rate as the plume head exhausts itself and is replaced by plume tail volcanism, high temperatures, and a conduit extending from the surface to the core-mantle boundary. Fixity relative to other "hotspots" and OIB geochemistry are also expected. High $^3He/^4He$ ratios were predicted following their initial observation in some basalts from Hawaii and Iceland, although this feature is not intrinsic to the original plume hypothesis.

Skepticism about the plume model that has arisen since the beginning of the twenty-first century has resulted largely from a growing awareness that most of these basic predictions are not fulfilled at most "hotspots." Indeed, no single volcanic system on Earth unequivocally exhibits all the required characteristics listed above (Courtillot et al., 2003; Anderson, 2005b), a full set of which can only be garnered by combining the observations from more than one locality. Even for Hawaii, the type example of a plume, evidence is absent for a precursory LIP with uplift (Shapiro et al., 2006), the locus of volcanism has not remained fixed relative to the paleomagnetic pole (Tarduno and Cottrell, 1997), volcanism is increasing and not dwindling, and as yet a conduit extending from the surface to the core-mantle boundary has not been imaged seismically (Montelli et al., 2004). These difficulties have traditionally been dealt with in the contemporary (in contrast to the classical) plume context by ad hoc adaptations of the model to suit each locality. These adaptations include suggestions that predicted but unobserved features exist but have not yet been discovered, are fundamentally undetectable, or are due to the long-distance lateral transport of melt in the mantle from a distant plume (e.g., Ebinger and Sleep, 1998; Niu et al., 2002). More sophisticated models (e.g., of uplift and mantle convection) have also suggested that some of the original predictions were too simplistic, to the point of negating the very features that the original model was invented to explain (e.g., Steinberger et al., 2004).

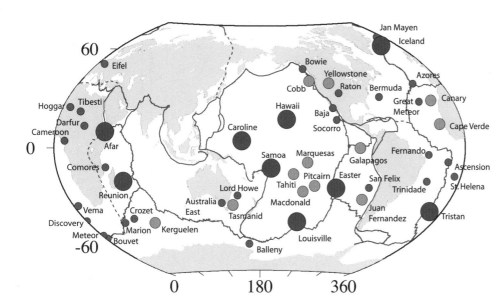

Figure 1. Map showing global "hotspots" compiled using the list of forty-nine of Courtillot et al. (2003). (See commentary on this list by Anderson, 2005b). Red, yellow, and green dots indicate "hotspots" proposed to be sourced at the core-mantle boundary, the base of the upper mantle, and within the lithosphere, respectively. Thin blue lines indicate plate boundaries. For additional maps and relevant resources, see Sandwell et al. (2005).

This approach is unacceptable to many scientists, who have in the past few years worked to develop alternative models. The most widely applicable of these alternatives attributes melting anomalies on Earth's surface to shallow, plate tectonic–related processes (Anderson, 2000, 2001, 2005a; Foulger and Natland, 2003; Foulger, 2004; Foulger et al., 2005c). Fundamentally, this model is based on processes associated with Earth's top thermal boundary layer—the surface—where heat is transferred from the solid Earth to the atmosphere. This alternative model is referred to here as the "plate model." It is the inverse of the plume model, which attributes "hotspots" to processes associated with Earth's bottom thermal boundary layer, the core-mantle boundary. Extraterrestrial models, involving meteorite impacts, form a separate category of mechanism and will not be dealt with in this article.

Much has already been achieved by a few researchers in a short time and has, most critically, highlighted problems that had gone largely unacknowledged in recent years. In this article, I summarize the basic fundamentals of the plate model, outline some key observations that have inspired its development, and briefly suggest what may be the most critical research goals for the future.

THE PLATE MODEL

Statement of the Model

The plate model attributes melting anomalies on the surface of Earth to shallow-based processes related to plate tectonics (Anderson, 2001). Simply put, it suggests that where the lithosphere is in extension, permissive volcanism will occur (Favela and Anderson, 1999; Natland and Winterer, 2005). The total magmatic volume is related to the fertility of the source material beneath, and the magmatic rate is related to factors that include the rate of extension, thickness of the lithosphere, fertility, temperature, and the availability of pre-existing melt.

Source fertility varies as a result of the dehomogenizing effects of processes related to plate tectonics, which recycle surface material and thereby refertilize the shallow mantle. At localities where extension occurs over refractory, infertile mantle, little melt will be produced. Where the source is fusible, fertile, and possibly partially molten to begin with, large volumes of melt will be emplaced.

Heat plays a role through being the ultimate driver of plate tectonics, and related temperature variations in the mantle and lithosphere will influence the volume of melt produced. Nevertheless, the plate model for the genesis of melting anomalies is essentially an athermal, top-driven one (Anderson, 2001). Heat is transported via processes including mantle convection, melt advection at mid-ocean ridges, slab subduction, gravitational instabilities, and delamination. Thus temperature and composition vary laterally as a result of plate tectonics. However, melting anomalies are primarily related to the structures and processes associated with plate tectonics, which are close to the surface thermal boundary layer, and not to point-source influxes of heat from the deep mantle that are essentially independent of plate tectonics.

Stress

The plate model suggests that the locations of melting anomalies are controlled by stress (Anderson, 2002). Indicators of extensional stress in the lithosphere include spreading ridges in the oceans, rift zones on land, and magmatism itself. The plate model predicts that surface melting anomalies will be preferentially located at or near these features. This prediction is largely borne out by the observation that approximately one-third of all melting anomalies are on or close to mid-ocean ridges (Fig. 1). Of these, several are at localities where extension is exceptionally large as a result of ridge-ridge-ridge triple junctions (e.g., the Easter, Afar, Bouvet, and the Azores melting anomalies). The Easter and

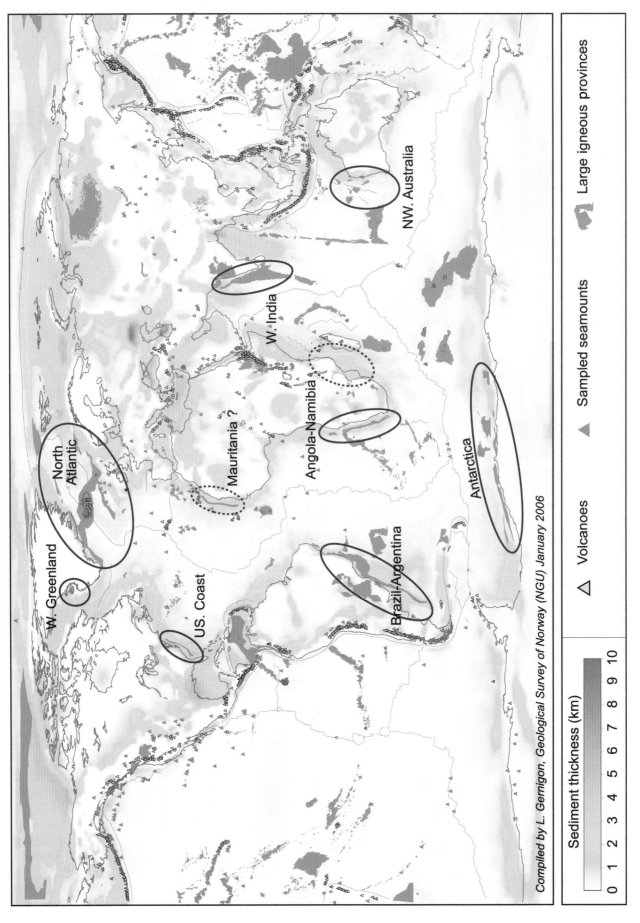

Figure 2. Map of global large igneous provinces. B&R—Basin and Range province; CRB—Columbia River basalts; EAR—East African rift; HLP—High Lava Plains; Y-SRP—Yellowstone–Snake River Plain. Base map by Laurent Gernigon, Geological Survey of Norway, January 2006. Data after Coffin and Eldholm (1994), Simkin and Siebert (1994), and Laske and Masters (1997). Coastline from USGS. Isochrons after Müller et al. (1997).

Iceland melting anomalies are also associated with oceanic microplates (Schilling et al., 1985a; Foulger, 2006b). Oceanic LIPs that formed at or near ridges include Shatsky Rise, which formed at a ridge-ridge-ridge triple junction (Sager et al., 1999; Sager, 2005) and the Iceland plateau, which is currently being created (Fig. 2). It is speculated that the Ontong Java plateau also formed near a ridge (Larson, 1997), and that it comprises only part of a much larger plateau that includes the Manihiki and Hikurangi plateaus (Taylor, 2006), but these suggestions remain to be confirmed.

Most melting anomalies in continental regions occur within or on the edges of extending regions or rift zones or at lithospheric discontinuities that predate the extension and magmatism. The extending Basin and Range province in the western United States is associated with widespread volcanism, including the persistent Yellowstone anomaly that lies on its northern edge. The East African rift is the location of much volcanism, including Afar at its northern end. Cameroon volcanism occurs on the southern flank of the parallel Benue Trough (Fitton and James, 1986). Many LIPs that now lie partially on continental lithosphere erupted where rifting led to continental breakup (e.g., the North Atlantic Igneous province, the Central Atlantic Magmatic province, and the Deccan Traps). Others formed as continents converged (e.g., Silver et al., 2006).

Some low-volume oceanic melting anomalies have been attributed to intraplate extension (e.g., due to thermal contraction; Sandwell et al., 1995; Lynch, 1999; Smith, 2003; Sandwell and Fialko, 2004). Global stress field maps show that volcanic regions occur in areas of tensile stress (Lithgow-Bertelloni and Richards, 1995; Lithgow-Bertelloni and Guynn, 2004). Calculations of the stress field from cooling for the whole Pacific plate predict large extensional stresses—in particular, near Samoa, Easter, and Louisville—and contraction normal to the orientation of the Hawaiian chain (Lithgow-Bertelloni and Guynn, 2004; Stuart et al., this volume). Globally, plate-circuit-closure calculations suggest that the total amount of thermal contraction occurring in the interior of oceanic plates is equivalent to a slow-spreading ridge (Gordon and Royer, 2005).

The question arises whether the presence of volcanism per se may be interpreted as evidence for extension (Favela and Anderson, 1999). Volcanic chains or lineations are expected to develop along extensional structures, such as fissures, faults, or cracks, and their orientation may thus be interpreted as indicating instantaneous orientation of extensional stress. Point sources of volcanism may be interpreted as localized extensional stress, and migration of the locus of volcanism with time may be interpreted as migration of the locus of extension. The building of volcanic edifices may also modulate stress locally and influence the location of subsequent eruptions (Hieronymus and Bercovici, 1999, 2000).

Source Inhomogeneity

Plate tectonic processes that create a heterogeneous mantle include melt extraction at mid-ocean ridges and subduction of oceanic lithosphere at trenches. Fertile and enriched material is continually advected from below into both the oceanic and continental mantle lithosphere, causing metasomatism and gravitational instability of over-thickened crust and lithosphere. Detachment recycles this material back into the mantle. Inhomogeneities are distributed throughout the shallow mantle by convection. As a result of these processes, the mantle is not uniformly comprised of peridotite or pyrolite but may vary radically in fertility from place to place (Meibom and Anderson, 2004; Anderson, 2005a; Fig. 3).

At mid-ocean ridges, plate separation proceeds hand in hand with the extraction of the most fusible 10–20% of the mantle source beneath, which is intruded and erupted to form new ocean crust. During the oceanic lithosphere's journey across the ocean, its thickness continually increases as it cools, asthenospheric mantle is plated onto its bottom, it is metasomatized, and sediments accumulate on its top. The slab that is re-injected into the mantle at subduction zones is thus a complex lithological package that includes the various rock types that make up the crustal section and the mantle lithosphere. The basaltic and gabbroic portions become transformed to eclogite at depths of ~50–60 km. At subduction zones, the downgoing slab experiences geochemical alteration as a result of dehydration and selective extraction of components into the mantle wedge. At

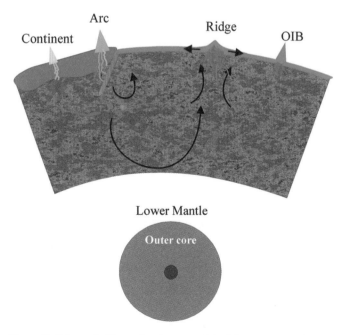

Figure 3. Schematic illustration (not to scale) of the upper mantle as a heterogeneous assemblage of depleted, infertile residues (bluish colors) and enriched, fertile, subducted oceanic crust, lithosphere, and sediments (reddish colors). The heterogeneities are statistical in nature and have wide ranges in shape, size, melting point, age, and origin (from Meibom and Anderson, 2004).

greater depths, different parts of the slab may become separated and recycled in the mantle separately.

Where the lithosphere becomes thickened (e.g., at island arcs and in collision belts), eclogitization of the lower crust and mantle lithosphere may occur, increasing its density and causing it to become negatively buoyant (Kay and Kay, 1993). A Rayleigh-Taylor (gravitational) instability may then develop, causing the mantle lithosphere and possibly also parts of the lower crust to detach and become recycled in the upper mantle (Tanton and Hager, 2000; Elkins-Tanton, 2005; Fig. 4). The continental mantle lithosphere thus recycled may be very old and may have been refertilized by metasomatism. Other mechanisms for detaching and recycling the lower continental crust into the mantle have also been suggested (Anderson, this volume).

To what depths are these materials recycled? Whereas lithospheric detachment events may be episodic, slab subduction is widespread and its locality well known. Seismology and mineral physics suggests that the depths to which slabs sink are variable. A high-velocity body that is apparently continuous from the transition zone to the core-mantle boundary has been detected seismically beneath North America and is commonly interpreted as a subducted slab (e.g., Grand, 1994). However, this scenario is the exception rather than the rule, and volume considerations cast doubt on whether a slab interpretation for this entire structure is plausible. Other down-going slabs clearly do not penetrate into the lower mantle (e.g., in the western Pacific; Ritsema et al., 1999; Gorbatov et al., 2000). It is likely that much material, both subducted and detached, is recycled in the upper mantle. Likely depths of density equilibration have been suggested by Anderson (2005a).

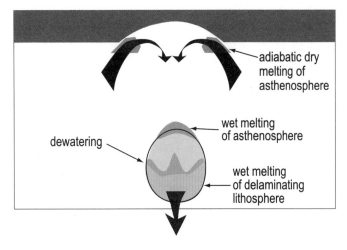

Figure 4. Schematic model for magma production during detachment of the mantle lithosphere as a result of gravitational instability. As the lithosphere sinks, asthenosphere is sucked into the resulting evacuated dome and may melt adiabatically. If the instability is hydrous, it may dewater as it sinks and heats, triggering melting of the mantle or of itself (from Elkins-Tanton, 2005).

OBSERVATIONS

Vertical Motion

The pattern of vertical motion associated with melt extraction is potentially a strong diagnostic of the source process. Differential vertical motion is expected to accompany transient volcanism, in particular LIP emplacement, and to scale with the volume of eruption. It can be studied using sedimentology on both terrestrial and marine sediments deposited prior to, during, and following eruption, and geomorphology. Of particular interest is comparing observations with the predictions of the classical, simple plume model that widespread regional domal uplift precedes LIP emplacement by 10–20 Ma (Campbell, 2006). This emplacement is expected to be followed by post-eruption subsidence as the LIP is carried by plate motion away from the hot plume (Griffiths and Campbell, 1991; Clift, 2005). Lithospheric detachment models, in contrast, predict that surface subsidence is associated with detachment of the dense, thickened lower lithosphere and precedes the LIP volcanism that occurs as mantle material from greater depths flows in to replace the lost lithosphere (Kay and Kay, 1993; Elkins-Tanton, 2005).

The picture that emerges from observations is variable, with precursory or syn-eruption uplift associated with some LIPs but not others. The Siberian Traps and the proposed composite Ontong Java–Manihiki–Hikurangi plateau represent the world's largest continental and oceanic LIPs, respectively, and yet neither is associated with major precursory uplift (Neal et al., 1997; Czamanske, 1998; Ito and Clift, 1998; Ingle and Coffin, 2004; Ivanov, this volume). The Siberian Traps have a volume of some 4×10^6 km^3 (Masaitis, 1983) and were erupted at ca. 200 Ma onto old continental lithosphere with a thickness of 100–200 km. They overlie the largest coal resource in the world and have been the subject of intensive research, including the drilling of hundreds of boreholes. The geology and stratigraphy of the region are thus known in great detail. The results show that the region subsided rapidly and continually prior to and during LIP emplacement, and that this subsidence was substantial and widespread (Czamanske, 1998; Ivanov, this volume).

The proposed composite Ontong Java–Manihiki–Hikurangi plateau has a combined volume of ~10^8 km^3 (Taylor, 2006). A plume origin for such a vast LIP would predict pre-eruption uplift of several kilometers. However, paleodepth estimates from volatiles in glass samples and from the study of sediments immediately overlying the crust from several ocean drill holes give elevations ranging from approximately sea level down to ~2.5 km below sea level at the time of eruption (Michael, 1999; Ingle and Coffin, 2004; Roberge et al., 2005). Subsidence subsequent to its formation has also been studied using oceanic sediments and was found to be slower than for normal oceanic lithosphere, consistent with an underlying mantle that is cooler than average, not warmer (Clift, 2005).

The pattern of vertical motion accompanying emplacement of the North Atlantic Igneous province is more complicated.

Volcanism occurred in two phases. The first occurred at 61–59 Ma and was distributed in a northwest-trending belt passing through northern Britain and east and west Greenland. The second phase occurred at 54 Ma, was associated with the opening of the North Atlantic, and was distributed in a NNE-trending belt along the newly formed passive margins. The first phase was preceded by limited uplift in the vicinity of some igneous centers. Kilometer-scale regional uplift did not occur until the second stage. It was contemporaneous with the second stage of volcanism and did not significantly predate it, as is expected for an impinging plume (Maclennan and Jones, 2006). It was closely associated with the new continental margins and comprised several episodes of uplift and magmatism. The main phase of uplift occurred in a short burst that may have lasted only a few tens of thousands of years. The pattern of subsequent subsidence permits only moderate temperature anomalies of no more than ~100 °C down to ~100 km depth and precludes very high temperature or very thick thermal sources (Clift, 2005).

Simpler, precursory uplift has been reported to be associated with the ca. 250-Ma Emeishan LIP in southwest China. There, variations in thickness of sedimentary rocks underlying the basalts suggest kilometer-scale domal uplift throughout a region ~900 km wide (He et al., 2003; Xu and He, this volume). Even though other cases have been reported (Rainbird and Ernst, 2001), this very recently reported result has been cited as the best example of domal uplift associated with LIP emplacement (Campbell, 2006).

Brief mention may be made of other LIPs. There is no unequivocal evidence for regular, regional uplift prior to emplacement of the Deccan Traps, which erupted at ca. 67 Ma and had a volume of ~1.5×10^6 km³ before erosion. Instead there is evidence that such uplift did not occur (Sheth, 2006a, this volume). There is evidence of subsidence prior to emplacement of the Columbia River basalts and of uplift immediately afterward (Hales et al., 2005), though this sequence of events is contested (Hooper et al., this volume). Study of sedimentary cores from many oceanic plateaus has produced little evidence for post-emplacement subsidence (e.g., at Shatsky Rise, Kerguelen plateau, Walvis Ridge, Ninetyeast Ridge; Clift, 2005).

The picture that emerges in general is one of considerable complexity. At some well-studied LIPs, notably the world's largest continental and oceanic ones, either subsidence or no uplift is observed prior to emplacement. Precursory, kilometer-scale uplift is not observed at most volcanic rifted margins. Uplift is associated with LIP emplacement at some localities, but it may not occur prior to eruption, and its pattern may be complex. Most oceanic plateaus subsided little, if at all, following volcanism, suggesting only minor temperature anomalies, if any, in the mantle when they formed.

In view of these variable observations, it is interesting to note the recent work of Burov and Guillou-Frottier (2005b), who modeled the uplift predicted to accompany the influx of hot source material using realistic continental lithosphere rheologies. They modeled lithosphere that included elastic-brittle-ductile properties and a stratified structure. Their results suggest that the simple classical expectation of regional domal uplift is oversimplified, and a more complex pattern might occur (Fig. 5). This work reopens the question regarding the vertical motions expected to accompany all source models. It suggests that a reassessment of predictions is required, and brings into question whether the occurrence of precursory domal uplift may be used to test the plume hypothesis.

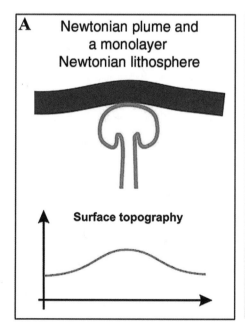

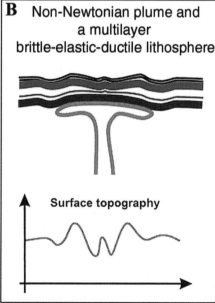

Figure 5. Sketches of plume-lithosphere interactions. (A) A "conventional" plume impinges on a single-layer viscous lithosphere, resulting in a single, long-wavelength topographic signature. (B) When realistic lithosphere rheology (brittle-elastic-ductile) and multilayer structure are considered, several wavelengths of surface topographic undulations are expected and both local uplifts and subsidences are predicted (from Burov and Guillou-Frottier, 2005a).

Volume, Rate, and Chronology of Volcanism

The total volume of magma and rate and chronology of emplacement at melting anomalies all vary throughout their entire feasible ranges. The high-volume end of the spectrum is represented by LIPs, the largest of which is the ~10^8 km^3 proposed composite Ontong Java–Manihiki–Hikurangi plateau (Taylor, 2006). Determining the lowest-volume LIP is, however, a matter of definition. LIPs were originally defined as having an areal extent of >0.1×10^6 km^2 (Coffin and Eldholm, 1994) but Sheth (2007a), as part of a proposed classification scheme for LIPs, recently suggested that 0.05×10^6 km^2 is a more appropriate lower limit (see Sheth, 2006b, and linked Web pages for a substantial discussion; Bryan and Ernst, 2007). Such debate hints that the sizes of volcanic provinces may form a continuous spectrum and invites the question of whether there is a natural break in the size distribution that might form the basis for theories of genesis.

At the low end of the volume scale are such island chains as the Samoa, Tahiti, Marquesas, Pitcairn, and Louisville chains (Fig. 1). The eruption rates of these chains throughout their lifetimes have been much less than 0.01 km^3/yr. In some cases, chains are discontinuous, and the eruption rate was almost zero for long periods (e.g., for the youngest ca. 30 Ma of the Louisville chain). In these cases, extrapolations or interpolations are speculative and model-dependent, and it is valid to question whether the volcanism results, in some sense, from a single "source."

Magmatic rates also vary widely. Thick sequences of lavas with similar ages are observed at many LIPs, leading to the view that they are emplaced very rapidly at rates that may exceed 1 km^3/yr and be sustained for one or more millions of years. However, age dating is insufficiently accurate to provide a precise figure, and where much of the LIP volume is inaccessible to sampling, such an assumption may not be supported by observational evidence. For example, it is commonly assumed that the Ontong Java plateau was emplaced rapidly, perhaps over just a few million years, giving a sustained magmatic rate of up to several km^3/yr (Fitton et al., 2004). However, this rate is conjectural, as the thick lower crust cannot be sampled. Ito and Clift (1998) suggested that the slow postemplacement subsidence observed there might be explained if the lower crust continued to be thickened by intrusions from a plume tail over a period of several tens of million years, but this seems unlikely, if only because the plateau drifted by ~2000 km between 120 and 90 Ma (Kroenke et al., 2004).

The very high eruption rates thought to occur at LIPs may be compared with the much lower rates at island chains. The magmatic rate along the Emperor and Hawaiian chains is shown in Figure 6. No pre-chain LIP is credibly associated with the Emperor chain. Its older end abuts the Aleutian Trench, but thick oceanic plateaus are probably not completely subductable (Cloos, 1993). There is no evidence for recycled subducted Emperor chain material in the lavas of Kamchatka (Shapiro et al., 2006).

The magmatic rate during emplacement of the Emperor and Hawaiian chains was modest for most of their history (Fig. 6).

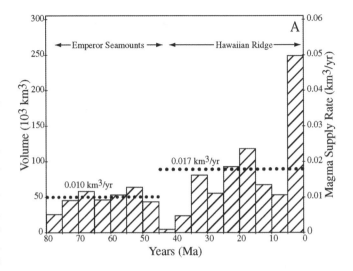

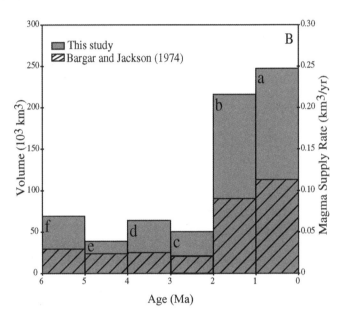

Figure 6. Volumes and magma supply rates for the Emperor and Hawaiian volcanic chains. (A) Histogram along the Emperor seamounts and Hawaiian ridge at 5-Ma intervals (from Bargar and Jackson, 1974). Dotted lines mark the long-term average magma supply rates. (B) Histogram of the Hawaiian Islands at 1-Ma intervals. Gray bars show volumes and magma supply rates calculated by Robinson and Eakins (2006), who take into account extra volume contained by subsidence of the Cretaceous sea floor. Hatched bars show the results from Bargar and Jackson (1974).

It was ~0.01 km^3/yr for much of the lifetime of the Emperor chain. This rate is approximately that of a single mid-ocean-ridge spreading segment. The volume rate declined to almost zero at the time of the bend (ca. 47 Ma; Sharp and Clague, 2006), and grew to somewhat higher rates, or up to ~0.17 km^3/yr for much of the Hawaiian chain. Very recently however, the magmatic production rate underwent a rapid increase, and is currently ~0.25 km^3/yr (Robinson and Eakins, 2006), which is of the order of the magmatic rate inferred for some LIPs. The young, high-volume end of the Hawaiian chain that formed over the past ca. 5 Ma has an area of ~0.15×10^6 km^2 and a volume of ~0.6×10^6 km^3 (Joel Robinson, personal commun., 2006) which means it is a LIP, even according to the higher-end classification scheme of Coffin and Eldholm (1994). The Hawaiian archipelago thus comprises a LIP forming at the end of a time-progressive chain rather than at its beginning.

The Iceland region is another locality where magmatic rate has increased with time rather than dwindled. Magmatism in excess of local ocean-crust formation has occurred there for the ~54-Ma duration of the opening of the north Atlantic. Following continental breakup, the excess magmatic rate was less than ~0.1 km^3/yr for the first ~10 Ma, during which time the Iceland-Faeroe Ridge was built, with its modest north-south extent of ~100 km (Fig. 7). Since ca. 44 Ma, however, the Icelandic Volcanic plateau has been built. It has a north-south extent of up to ~600 km and a crustal thickness of up to ~20 km greater than

the crust on the Reykjanes Ridge. It must thus have been built at an excess rate of ~0.25 km^3/yr, approximately the same as at present-day Hawaii. The Iceland plateau presently covers an area of ~0.3×10^6 km^2 and thus has an excess volume of ~5×10^6 km^3. It therefore also comprises a LIP that is forming late in the sequence of assumed associated magmatism rather than at the beginning.

The formation of a LIP followed by continuation of volcanism to form a time-progressive volcanic chain is a rare exception rather than the general rule (e.g., Anderson, 2005b; Beutel and Anderson, this volume). No trace of such associations exist, for example, for the Siberian Traps, the Ontong Java plateau, Afar, Hawaii, Samoa, or Louisville. Of the forty-nine "hotspots" cataloged by Courtillot et al. (2003) only thirteen associations of LIPs and volcanic chains are listed (Crozet, Easter, Fernando, Galapagos, Iceland, Kerguelen, Macdonald, Marion, Marquesas, Réunion, Tristan, Vema, and Yellowstone), and only two of these are cited as being unequivocal (Réunion/Deccan

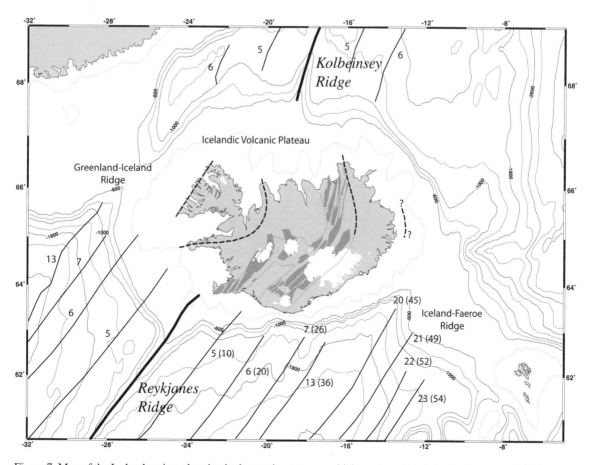

Figure 7. Map of the Iceland region, showing bathymetric contours, which are thought to indicate the extent of thick crust in the region. The island of Iceland makes up the innermost, subaerial part of the broader, mostly submarine, Icelandic volcanic plateau. Oceanic magnetic anomalies (Nunns, 1983) are labeled with anomaly number. Approximate ages in Ma are shown in parentheses after the anomaly number on the eastern flank of the Reykjanes Ridge. Thick black lines—axes of the Reykjanes and Kolbeinsey ridges; thin lines on land—outlines of the neovolcanic zones; gray—spreading segments; white—glaciers; dashed lines—extinct rift zones (two in west Iceland and two in east Iceland).

and Tristan/Parana). Both of these are contested by other authors, however (Burke, 1996; Fairhead and Wilson, 2005; Sheth, 2005).

Structure of the Mantle from Seismology

Seismology is able to probe the mantle in a focused way that is not possible using any other method. Many different approaches are available that vary in target and resolution. Of particular interest is the question of what structures underlie currently active melting anomalies, whether they are thermal, and whether they are continuous between the surface and the deep mantle.

Teleseismic tomography uses regional seismometer networks and can resolve structure on a scale of 50–100 km down to depths roughly equal to the breadth of the network. This breadth is typically a few hundred kilometers. Such experiments have been conducted to study the Iceland, Yellowstone, and Eifel melting anomalies (Iyer et al., 1981; Ritter et al., 2000; Foulger et al., 2001; Christiansen et al., 2002). The results in all these cases are consistent with the presence of underlying low-velocity bodies that are confined to the upper mantle. No teleseismic tomography experiment has yet been conducted on a scale that enables structure through and beneath the mantle transition zone, 410–660 km deep, to be well imaged. Thus, the continuity of structures between the upper and lower mantle has not yet been studied using this method.

Global tomography provides a continuous image of the mantle but at the much lower resolution of 500–1000 km (e.g., Ritsema et al., 1999). The results show a sharp contrast in structural character between the upper and lower mantles. Variations in the spherical harmonic power of velocity throughout the mantle shows that the Earth is characterized by strong heterogeneity in the longer-wavelength components in the upper mantle but weak heterogeneity in the lower mantle (Gu et al., 2001). At individual melting anomalies (e.g., Iceland, Yellowstone, Eifel), global tomography generally shows that the underlying low-velocity structures are confined to the upper mantle. Recent reports of low-velocity anomalies beneath some "hotspots" traversing the entire mantle based on finite-frequency tomography (e.g., Montelli et al., 2004) have been seriously challenged and remain to be confirmed (van der Hilst and de Hoop, 2005).

To study in detail possible continuity of structures through the transition zone, receiver functions have been applied to determine topography on the 410-km and 660-km discontinuities that comprise its upper and lower boundaries, respectively. These discontinuities result from mineralogical phase changes, and their depths are predicted to change by measurable amounts in the presence of temperature or compositional anomalies (Bina and Helffrich, 1994; Presnall, 1995; Fig. 8). The results in general show broad global correlations, with a thin transition zone beneath oceans and a thicker one beneath continents, but within individual provinces, there is little correlation between the topography on the two discontinuities (Gu and Dziewonski,

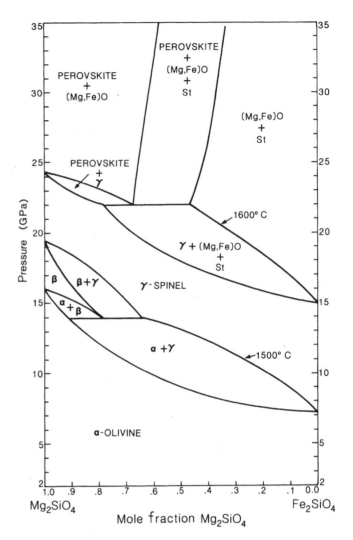

Figure 8. Minerological phase changes in the transition zone (from Presnall, 1995). This figure assumes a homogeneous mantle. For a discussion of the role of eclogite in the transition zone, see Anderson (this volume). Abbreviation: st—stishovite.

2001). This pattern is also observed at melting anomalies (e.g., Yellowstone, Iceland, Eifel). There, deflections on the 410-km discontinuity are observed but no corresponding deflections on the 660-km discontinuity (Dueker and Sheehan, 1997; Grunewald et al., 2001; Du et al., 2006).

Global tomography suffers from formidable problems of resolution and coverage, which become increasingly severe with depth, and there are many regions of the mantle devoid of seismic rays. Repeatability among different models is also a problem, with proposed lists of deeply sourced "hotspots" often exhibiting little overlap (Anderson, 2005b; Table 1). Nevertheless, the greatest barrier to using seismology to study melting anomalies is the inherent ambiguity in the physical interpretation of velocity anomalies. Seismic velocity varies because of variations in composition, mineralogical phase, anisotropy, the pres-

TABLE 1. HOTSPOTS REPORTED TO BE UNDERLAIN BY SEISMIC ANOMALIES TRAVERSING THE UPPER MANTLE ONLY AND THE WHOLE MANTLE, AND HOTSPOTS DEFINED AS ARISING FROM D″

Hotspot name	Traversing the upper mantle only*	Traversing the whole mantle[†]	Defined as arising from the core-mantle boundary[§]
Afar	X		X
Hawaii	X		X
Bowie	X		
Samoa	X	X	
Macdonald	X		
Easter	X	X	X
Louisville	X		X
Iceland	X		X
Tahiti		X	
Azores		X	
Canaries		X	
Ascension		X	
Réunion			X
Tristan			X

Note: Only Easter is common to all three lists.
*Traversing the upper mantle only—the deepest detected (Ritsema and Allen, 2003).
[†]From Montelli et al. (2004).
[§]Arising from D″, according to Courtillot et al. (2003).

ate state of lithospheric stress. Study of temperature variations in the presumed source region is thus important for determining the source process. This study is not an easy task, however. The interpretation of seismic velocity variations is ambiguous. Surface heatflow is insensitive to mantle temperature because thermal conduction in rocks is slow, and ground-water circulation may complicate matters. Thus, petrological methods may be important. However, in between melt formation and eruption on the surface, basalts experience a complex history, and as a result, the field of geothermometry is still controversial and developing rapidly.

Because the absolute potential temperature of the mantle is poorly known, temperature differences between volcanic provinces are generally studied. Specifically, melting anomalies are compared with mid-ocean ridges, which are assumed to represent average background mantle temperature. The most robust approach is to study olivine control lines. When a primary mantle-derived melt cools in a crustal magma chamber, its composition initially follows a liquid line of descent controlled by the precipitation of olivine only, during which time its MgO contents decrease. If samples of the instantaneous liquids formed during this cooling trajectory are preserved and erupted, they may be sampled. The original composition of the parental/primary liquid may then be back-calculated. Olivine is incrementally added (i.e., the reverse of crystal fractionation) to an evolved liquid composition (which itself lies on the olivine-only crystallization path) to obtain equilibrium with the most mag-

ence of partial melt and volatiles, and temperature (Anderson, 1989; Fig. 9). Interpreting anomalies is thus highly ambiguous, and nowhere is this difficulty better illustrated than in the case of the Pacific and south Atlantic "superplumes." These are low-velocity bodies several thousand kilometers broad that extend from the core-mantle boundary to the shallow mantle. It is widely assumed that they are thermal in origin and fuel surface melting anomalies either directly or indirectly. However, Trampert et al. (2004) were able to separate out the thermal and compositional components using bulk sound wave and shear wave velocities obtained from normal modes. They found that the "superplumes" are largely compositional in origin and anomalously dense, not thermally buoyant. Very recent high-pressure experiments on the silicate phases that exist at core-mantle boundary depths suggest that the low-velocity anomalies in the D″ layer are also dense and not thermal in origin (Mao et al., 2005).

In addition to having significant bearing on geodynamic models of the mantle, these results are an important cautionary tale concerning interpretation of seismic velocity anomalies. They also highlight the general scientific rule that it is dangerous to interpret observations under the assumption that one out of many possible interpretations is correct.

Temperature and Heat

The extraction of large volumes of melt requires either high temperature, a fusible source, or both, along with an appropri-

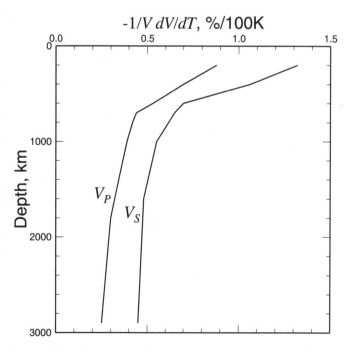

Figure 9. Sensitivity of seismic wavespeeds to temperature at different depths in the Earth's mantle (after Karato, 1993). These functions depend on composition and mineralogy and are uncertain by perhaps 30% (reproduced from Julian, 2005).

nesian olivine phenocryst observed, using an olivine geothermometer (Green and Falloon, 2005). The composition of the
parental liquid can then be compared with experimental mantle
melt compositions to deduce a pressure and temperature of mantle equilibration and a mantle potential temperature. It is necessary to assume a number of parameters for this calculation,
including source composition, degree of partial melting, pressure of initial melting, latent heat of fusion, and heat capacities.

The method has been applied to Hawaii, which is the only
locality where high-MgO basaltic glasses have been found
(Clague et al., 1991). Attempts have been made to apply it to
other regions (e.g., Iceland), but the only high-MgO basalts
found there are cumulate rocks that cannot safely be assumed
to represent liquid compositions. Putirka (2005), using an olivine geothermometer described in the same paper, reported a
temperature of ~245 ± 52 °C hotter for Hawaii compared with
mid-ocean ridges for olivine crystallization temperatures of
parental liquids. Based on this result, Putirka (2005) postulated
that there is a difference in mantle potential temperature of
~250 °C between the Hawaiian and MORB sources. Green et al.
(2001), Green and Falloon (2005), and Falloon et al. (this volume), however, report very similar temperatures for the olivine
crystallization temperatures of parental liquids for basalts from
Hawaii, Iceland, and Réunion and a subset of MORBs. They followed a similar approach but used the olivine geothermometer
of Ford et al. (1983). That geothermometer uses partition coefficients more appropriate for olivine crystallization at crustal
pressures than does any other olivine geothermometer. The
parental liquid compositions calculated by Green et al. (2001),
Green and Falloon (2005), and Falloon et al. (this volume) show
very similar pressures and temperatures of mantle equilibration
when compared with the relevant experimental data. They thus
conclude that it is unlikely that significant differences exist between the potential temperatures of the mantle sources for
Hawaiian and the hottest MORB parental liquid compositions.
The differences between the results of Putirka (2005) and Green
and Falloon (2005) using the same approach can be explained
entirely by the differences in performance of the olivine geothermometers used (Falloon et al., 2007a).

The question of heat is separate from that of temperature.
The advection of large volumes of melt to the near-surface requires a mechanism for extracting large amounts of heat from
the mantle. The heat required consists of the specific heat needed
to raise the source material to its solidus, plus latent heat of melting. The amount of melt that can be produced by thermal upwellings was modeled numerically by Cordery et al. (1997) and
Leitch et al. (1997). The results showed that a pyrolite upwelling
is incapable of generating any melt at all for reasonable temperature anomalies and lithosphere ages greater than a few million
years. Melting of entrained eclogite, which is more fusible, was
therefore modeled. Figure 10 shows the volumes of melt produced assuming that all the melt came from the fusible eclogite
component and the latent heat was obtained from conduction
from the surrounding material, not supplied by decompression.

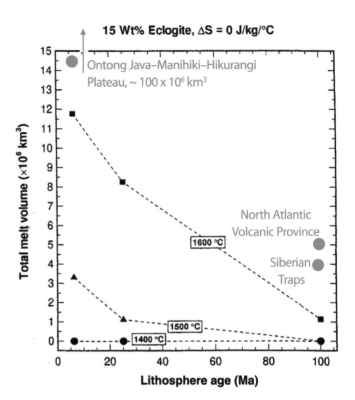

Figure 10. Volumes of melt calculated using finite-element modeling of
plumes rising from the core-mantle boundary, plotted against lithosphere age, which is a proxy for lithosphere thickness. Numbers indicate temperature anomaly, assuming a normal mantle potential
temperature of 1300 °C. It is assumed that all the melt comes from
fusible eclogite entrained in the plume, and that the latent heat is obtained by conduction from the surrounding material. It is not supplied
by decompression, and thus may be set to be zero in the calculations
(adapted from Cordery et al., 1997). This model cannot account for the
volumes of magma observed in the largest of the large igneous
provinces for reasonable temperature anomalies.

Only small or moderate LIPs erupting through thin lithosphere
could be simulated, but not LIPs such as the Siberian Traps,
which are large and erupted through thick lithosphere. The lithosphere beneath Hawaii is ca. 100 Ma old, and thermal upwelling
models can only simulate magmatic rates <<0.1 km³/yr for reasonable temperatures.

In the case of near-ridge melting anomalies, the question
arises whether the excess melting can be modeled by isentropic
upwelling of fusible eclogite at otherwise normal parts of the
ridge system. Such a mechanism has been suggested to account
for the large volumes of melt produced at Iceland, the Ontong
Java plateau, and other oceanic plateaus in the Atlantic and Indian oceans (Yaxley and Green, 1998; Anderson, 2005a; Foulger and Anderson, 2005; Korenaga, 2005). At present the energy
required to melt the relevant minerals at the appropriate temperatures and pressures is not known sufficiently well to be able
to answer this question. It has also been suggested that subducted ocean crust or fusible, metasomatized lithospheric mantle may warm by conduction of heat from its surroundings in the

upper mantle, causing it to rise and melt in a runaway fashion (Anderson, this volume). Melt accumulating from long-term warming of fusible material in the mantle by conduction may also pond and erupt on a much shorter time scale than that of accumulation (Silver et al., 2006). These ideas are currently speculative, however. At present no fully developed, robust numerical model currently exists that can explain the melt volumes and magmatic production rates thought to occur at large LIPs erupted through thick lithosphere.

Geochemistry

Significant geochemical differences are observed among many OIBs and MORBs (Hofmann and White, 1982). In particular the former include samples that are relatively enriched in incompatible elements, such as U and Th, and in light rare earth elements relative to heavy ones. Radiogenic isotope ratios (e.g., $^{87}Sr/^{86}Sr$) in MORB are indicative of past melting events that caused relative depletion in the parent elements compared to the source of OIB. OIBs sometimes have high ratios of primordial noble gas isotopes relative to radiogenic ones.

These characteristics of OIB represent end-members, however, and a range of compositions may be observed that overlap those of MORB (e.g., at Iceland). Furthermore, all the characteristics listed above may not be found in every sample. For example, in basalts from Baffin Island, the ratio of $^3He/^4He$ correlates with depletion in incompatible elements (Stuart et al., 2003). The observed geochemical signatures cannot be explained by simple binary mixing of two sources.

A remarkable aspect of OIB geochemistry is that similar compositions are seen in a wide variety of settings, including oceanic melting anomalies and low-volume magmas erupted in continental rifts, where a deep thermal origin is implausible (Natland and Winterer, 2005; Fitton, this volume). Eruption of similar compositions may furthermore continue for tens or even hundreds of millions of years at the same locality. For example, in the Scottish Midland Valley, OIB-like volcanism persisted for ~60 Ma, during which time the lithosphere is thought to have drifted ~15° relative to the deep mantle (Fitton, this volume). A fixed, focused source for the magmas observed is thus unlikely, and a source that is either ubiquitous (e.g., Anderson, 1995) or carried with the plate seems to be required.

The most likely source of OIB signatures is recycled near-surface materials (Hofmann and White, 1982). Throughout its history, the Earth has become chemically stratified by partial melting that preferentially extracts elements incompatible in mantle phases and transports them upward. The mantle has been extensively processed in this way throughout geological time. The source of OIB must be either some region of the mantle containing material relatively little processed by melt extraction during Earth history or recycled near-surface material. There is little geochemical support for the former scenario, which is at odds with the expectation that the mantle went through an early, largely molten stage (e.g., Anderson, 1989).

Candidate near-surface material that may be recycled into the mantle includes oceanic crust, along with overlying pelagic and terrigenous sediments and metasomatized mantle lithosphere. Ocean crust is reintroduced into the mantle at subduction zones. Metasomatized arc or continental lithosphere and lower crust may become detached during orogenic events and recycled back into the convecting upper mantle (e.g., McKenzie and O'Nions, 1995).

Chauvel and Hemond (2000), Breddam (2002), and Foulger et al. (2005b) have advocated a source for Icelandic tholeiitic basalts in discrete lithologies from subducted oceanic crust, in particular of Iapetus origin. Green et al. (1967) long ago pointed out the difficulty of producing from this source the silica-undersaturated basalts observed at many ocean islands. More recently, Niu and O'Hara (2003) objected to oceanic crust as a source, citing the inability of remelted oceanic crust to produce high-MgO parent melts, the isotopic depletion of ancient oceanic crust, and the depletion of water-soluble incompatible elements that it is expected to have undergone during subduction. Niu and O'Hara (2003) suggest that deeper portions of recycled oceanic lithosphere are a more likely source for OIB.

Pilet et al. (2002, 2004) recently proposed an origin in variably metasomatized mantle lithosphere by partial melting of metasomatic veins plus the enclosing lithospheric mantle. This model could generate isotopic and trace element ratio variations similar to those observed both in oceanic islands and continental rift zones. It would be applicable to both continental and oceanic lithosphere and might thus explain the ubiquity of OIB.

Helium isotope ratios make up a special category of geochemical observations because, whereas it is generally acknowledged that other geochemical characteristics are not diagnostic of source depth, high $^3He/^4He$ is widely assumed to indicate a deep lower-mantle origin. In some samples from melting anomalies (e.g., Iceland, Hawaii), values of $^3He/^4He$ as high as ~50 Ra (the atmospheric value of 1.38×10^{-6}) are measured. This ratio may be compared with values of 6–10 times the atmospheric value, which is the maximum typically found in samples from mid-ocean ridges. The high values have traditionally been interpreted as resulting from high levels of the primordial isotope ^3He, considered to be stored in a near-primordial lower-mantle region (e.g., Kellogg and Wasserburg, 1990). An alternative interpretation is that they result from low levels of ^4He, a radiogenic decay product of U and Th (Fig. 11). In this model, He is stored in a low-U+Th matrix (e.g., dunite cumulates), and ancient, high values of $^3He/^4He$ are preserved for long periods with little change (Anderson, 1998a,b; Natland, 2003; Meibom et al., 2005). Dunite cumulates occur in the lower parts of oceanic crust, and this model thus suggests that high $^3He/^4He$ ratios may be explained by recycled subducted oceanic crustal lithologies in the same way as other OIB characteristics. The Ne isotope ratios found in some OIBs are low in ^{21}Ne compared with those measured in MORBs. ^{21}Ne is produced by the nucleogenic decay of ^{24}Mg and ^{18}O when irradiated by U and Th decay products. It is thus produced at the same rate as ^4He,

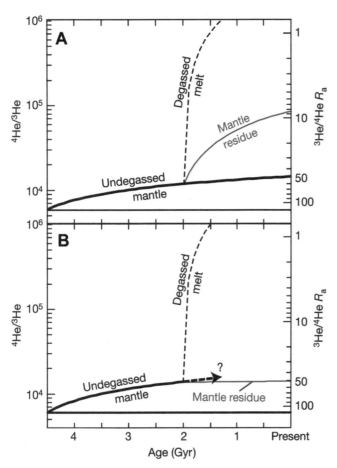

Figure 11. Scenarios for ³He/⁴He isotope ratios for the lower-mantle (plume) and upper-mantle (plate) models (from Parman et al., 2005). Age is plotted versus He isotope ratio. Both (A) and (B) start with an initial ³He/⁴He of 120 Ra (the atmospheric value of 1.38×10^6) at 4.5 Ga. (A) the standard (plume) model attributes high ³He/⁴He to high residual ³He from a little-degassed reservoir in the lower mantle. Undegassed mantle (thick black line), thought to comprise part or all of the lower mantle, evolves to a present-day value of ~50 Ra and is assumed to be sampled by ocean island basalt (OIB). A melting event at 2 Ga increases the $^{238}U/^3He$ of the melt relative to the undegassed mantle such that it evolves to a present-day value of 8 Ra (gray line). The standard model suggests that this is the mid-ocean ridge basalt (MORB) source. If the melt approaches the surface and degasses, $^{238}U/^3He$ increases and crustal materials made from such melts will rapidly evolve very low ³He/⁴He ratios (dashed line). (B) Alternative model in which He is more compatible than U + Th (line styles the same as in panel A). Here the highest ³He/⁴He OIBs are melts of mantle residue. Melting at 2 Ga produces a mantle residue with a $^{238}U/^3He$ one-tenth that of the undegassed mantle, which evolves to a present-day ³He/⁴He of 50 (gray line). As in (A), the degassed melt rapidly evolves low ³He/⁴He. Simply put, the plume model in (A) suggests that high ³He/⁴He arises from a primordial source with high absolute concentrations of ³He whereas the model in (B) suggests that high ³He/⁴He arises from a source depleted in U + Th by an earlier melting event and is thus low in ⁴He. Such a model was originally proposed by Anderson (1998a,b).

which suggests that the neon isotope ratios observed in OIB may be explained in the same way as the helium isotope ratios.

The implication for mantle dynamics of OIB geochemical signatures then revolves around the depths to which recycled near-surface materials are circulated, which involves considerations of density. Basalt transforms to the denser rock eclogite at depths of 50–60 km, which is expected to encourage sinking (O'Hara, 1975). The depth at which it reaches neutral buoyancy is disputed, however. Deep-source models cite tomographic evidence to propose that downgoing subducted slabs reach the core-mantle boundary and are recycled back to the surface from there in plumes (e.g., Kellogg and Wasserburg, 1990). The plate model proposes that neutral buoyancy for much recycled material is reached in the upper mantle, where it reheats by thermal conduction from the surrounding mantle (e.g., Anderson, 2006). The time scale of slab reheating is much less than the age of the plate on subduction, and because eclogite is more fusible than peridotite, its temperature may approach or even exceed its solidus in the upper mantle, providing potentially highly productive source material to supply melting anomalies.

EXAMPLES

Atlantic Ocean

The primary melting anomalies in the Atlantic Ocean include Iceland, the Azores, Bermuda, the Canary Islands, the Cape Verde Islands, the Ascension-Cameroon system, and Tristan (Figs. 1 and 12). All are very different in character. Plate models have been suggested for the Iceland and Tristan systems. The Bermuda melting anomaly is discussed in detail by Vogt and Jung (this volume). Plate models for other Atlantic melting anomalies have yet to be developed.

The melting anomaly associated with Iceland is by far the most voluminous in the Atlantic Ocean and the only one to produce large quantities of tholeiite. The magmatic rate has varied irregularly over time. High-volume volcanism along a ~2500-km zone of rifting accompanied the breakup of Laurasia and the opening of the north Atlantic Ocean at ca. 54 Ma. This volcanic activity subsequently dwindled as normal-thickness ocean crust began to form, but the magmatic rate continued at a high level along a ~100-km-long portion of the Mid-Atlantic Ridge, building the Iceland-Faeroe Ridge (Fig. 7). Starting at ca. 44 Ma, the part of the ridge that produced excess volcanism greatly lengthened to attain a present-day north-south extent of ~600 km. If the whole of the ~30-km-thick seismic crust is melt (an assumption that is questioned; Björnsson et al., 2005; Foulger, 2005) the present-day excess melt production rate is ~0.25 km³/yr. However, apart from precursory volcanism at 61–59 Ma, magmatism has always been centered on the Mid-Atlantic Ridge, and there is little evidence for a time-progressive volcanic track (Foulger, 2003, 2006b; Foulger and Anderson, 2005; Lundin and Doré, 2005). The area of the Icelandic platform is $~0.3 \times 10^6$ km², so

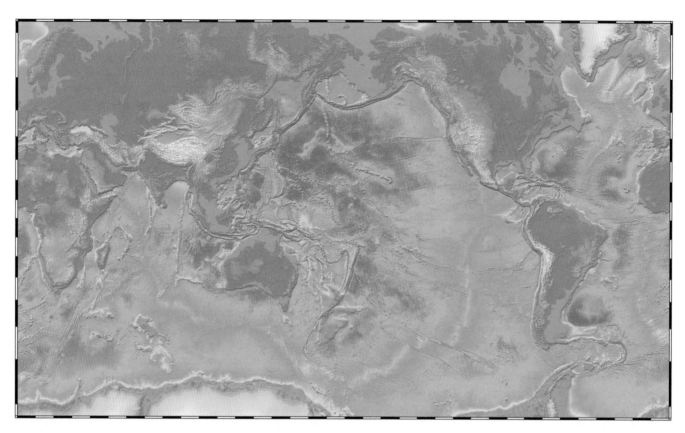

Figure 12. Topography of the Earth based on a global compilation of land data (GTOPO30, http://edcdaac.usgs.gov/gtopo30/README.html) and ocean data from a combination of sparse ship soundings and marine gravity anomalies derived from satellite altimetry (Smith and Sandwell, 1997).

it qualifies as a LIP (Coffin and Eldholm, 1994; Sheth, 2006b, 2007a).

Temperature estimates for the mantle beneath Iceland have been obtained using seismology, petrology, bathymetry, vertical motion, and heatflow (see Foulger et al., 2005a, for a review). All either require or are consistent with temperature anomalies no greater than 50–100 °C compared with the mean for mid-ocean ridges (Foulger et al., 2005a). Seismic experiments show that the whole north Atlantic is underlain by a low-velocity zone that extends from the surface to the base of the upper mantle (Ritsema et al., 1999). Its strength is consistent with either a temperature anomaly of ~100 °C throughout most of its volume or a fraction of a percent of partial melt, which might result from the presence of a small amount of carbonate (Presnall and Gudfinnsson, 2005). Detailed tomography and receiver function studies agree that the anomaly terminates in the transition zone (Ritsema et al., 1999; Foulger et al., 2001; Montelli et al., 2004; Du et al., 2006).

These results, coupled with observed tectonic correlations, have inspired development of a plate-based model for Iceland. The Greenland-Iceland-Faeroe Ridge formed colinearly with a northwest-trending zone of precursory magmatism that erupted at 61–59 Ma. This trend is also colinear with the trend of the western frontal thrust of the Caledonian suture where it runs off-shore in east Greenland and re-emerges on land in Britain. At this latitude, the Mid-Atlantic Ridge, newly formed at ca. 54 Ma, crossed the outer limit of the Caledonian suture. Here, the trend of the Mid-Atlantic Ridge also changes radically, from ~N35°E to N20°E. Tectonic style has been persistently locally complex in the Iceland region, involving migrating ridges, parallel-pair spreading, and both oceanic and continental microplate formation since the earliest opening of the ocean, when a transform fault existed at this latitude (Nunns, 1983; Foulger, 2006b).

The plate model relates excess magmatism in the Iceland region to processes related to extension (Foulger, 2002; Foulger and Anderson, 2005; Foulger et al., 2005a,b). The volcanism that accompanied the early opening of the north Atlantic was clearly associated with continental breakup (van Wijk et al., 2004). The Iceland melting anomaly comprises a portion of the spreading plate boundary where the mantle beneath is unusually fertile as a result of entrained eclogite from slabs trapped in the Caledonian suture when it closed at ca. 400 Ma. This entrained eclogite is now dispersed locally in the mantle below where the suture formerly lay. Although north Atlantic mantle in general may be somewhat warmer than the global average, the excess magmatism is attributed essentially to isentropic upwelling of mantle with enhanced fertility and not to high temperature. This model naturally explains the co-location of the region with the Caledonian western frontal thrust, the persistence of the melting anomaly at the Mid-Atlantic Ridge, and the evidence for normal temperature or only small temperature anomalies. In this model, the mantle seismic anomaly represents a thickening of the ubiquitous global low-velocity zone, which is thought to be caused by small degrees of partial melt induced by volatiles, in particular CO_2 (Presnall and Gudfinnsson, 2005; Dasgupta and Hirschmann, 2006). It may also be related to a high eclogite content (Foulger and Anderson, 2005; Foulger et al., 2005b). Enhanced volatile content is observed in erupted lavas and is expected in a mantle made unusually fertile by recycled near-surface material. The unstable tectonics in the Iceland region is clearly a continuation of behavior that dates from the earliest opening of the ocean, when the newly formed ridge crossed a major structural divide (Skogseid et al., 2000).

The Azores melting anomaly lies at a ridge-ridge-ridge triple junction, where the Azores ultra-slow spreading ridge meets the Mid-Atlantic Ridge. The Azores plateau, which is underlain by crust ~10–14 km thick, has been emplaced since ca. 20 Ma by magmatism along all three ridge branches. The Azores branch is deduced from kinematic models, earthquake focal mechanisms, and bathymetry to be an oblique ridge, spreading at a rate of 3–4 mm/yr (Lourenco et al., 1998). As is also the case with Iceland, its geochemical footprint along the Mid-Atlantic Ridge is asymmetric, extending to 1000 km south of the Azores but only to 250 km to the north. Moreira et al. (1999) suggested that a recycled oceanic crustal component originating in subcontinental lithosphere delaminated during opening of the north Atlantic can explain the Pb isotope systematics observed there, where $^{203}Pb/^{204}Pb$ is up to 11.3. The discovery of zircons dated at 330 Ma and 1.6 Ga in the Kane fracture zone ~1500 km south of the Azores lends support to this hypothesis (Pilot et al., 1998). A promising plate model for the Azores melting anomaly would then involve enhanced magmatism resulting from source fertility, peaking at the triple junction as a result of locally enhanced extension and focused flow (Georgen and Lin, 2002).

Magmatism in the Ascension-Cameroon region is complex. The Cameroon line has erupted along much of its length between ca. 130 Ma and the present, and no regular time progression is observed (Fitton and Dunlop, 1985; Fitton, 1987; Ernst and Buchan, 2002). It has been suggested that it is related to reactivation of the Central African Shear Zone (Moreau et al., 1987). Volcanism continues to the southwest as the St. Helena seamount chain, for which time-progressive ages are reported (O'Connor et al., 1999). Extrapolation of this age progression would suggest a present-day location for the locus of melt extraction ~500 km west of St. Helena. No current volcanism is observed there, however. Young volcanism in the region is, instead, scattered over many seamounts distributed throughout a region some hundreds of kilometers broad, including Ascension Island, St. Helena, and the Circe seamount (O'Connor et al., 1999). It has been suggested that a nearby geochemical anomaly observed on the Mid-Atlantic Ridge results from lateral flow from this melting anomaly (Schilling et al., 1985b). However, gravity data suggest a crustal thickness anomaly of only ~4 km in the Ascension region, with gravity variations resulting from crustal thickness variations and not variations in mantle temperature (Bruguier et al., 2003). In view of this suite of observations, a likely plate model would involve diffuse permissive volcanism as a result of variable extensional stress in the African plate. The mantle beneath the Atlantic is variable in fertility (the "blob" model; e.g., Bruguier et al., 2003; Vogt and Jung, 2005). Variations in the African plate stress field may be linked to tectonic evolution of the northern collision boundary with Europe, and old sutures and fault zones may be preferentially activated (Fairhead and Wilson, 2005).

A similar plate model has been proposed for the Tristan melting anomaly (Fairhead and Wilson, 2005). Like the Ascension-Cameroon province, this region also comprises a southwest-orientated zone of seamount volcanism—the Walvis Ridge. There are few reliable radiometric dates (Baksi, 1999), but the youngest volcanism is at the southwestern end, some hundreds of kilometers east of the Mid-Atlantic Ridge. Like the Ascension-Cameroon province, it, too, is scattered over a region several hundred kilometers broad. In contrast with that province, however, the Walvis Ridge is mirrored by similar volcanism west of the Mid-Atlantic Ridge, the conjugate Rio Grande Rise on the South American plate. These two ridges exhibit entirely different morphologies, however.

Fairhead and Wilson (2005) combined satellite gravity and plate motion data from GPS (Global Positioning System) with extensive information constraining the tectonic history of Africa and South America. They concluded that changes in stratigraphy in African rift basins reflected variations in the state of stress of the African plate resulting from changes in distant plate motions (e.g., India-Eurasia and Africa-Europe collisions). They proposed a plate model whereby periodic stress release permitted episodic excess volcanism near the Mid-Atlantic Ridge and on flanking aseismic ridges. These include the Walvis Ridge and the Rio Grande Rise, and volcanism occurs along shear, wrench, and extensional deformation structures. Additional observations in support of this model are reported from seismic reflection studies of the Cabo Frio area, part of the Rio Grande Rise immediately offshore Brazil (Oreiro et al., 2006).

The Atlantic Ocean contains a number of other, smaller-scale melting anomalies, some of which comprise short archipelagos or zones of seamounts. These include the Canary Islands, the Cape Verde Islands, the Discovery seamount chain, and various seamounts in the Azores region. There is a paucity of reliable dates, but volcanism may be locally short- or long-lived. Such volcanism fits best a model of varying stress and source fusibility and fits poorly a model involving large thermal diapirs.

Pacific Ocean

The most remarkable currently active melting anomaly in the Pacific Ocean is the Hawaiian archipelago at the southeast end of the Emperor and Hawaiian volcanic chains (Fig. 12). This system is unique on Earth. No other currently active volcanic system exhibits the same combination of longevity, extreme regularity of time progression, remoteness from plate boundaries and continental edges, and a high present-day production rate of tholeiitic basalt. The most widely suggested nonthermal model for the Emperor and Hawaiian volcanic chains is based on a propagating crack. This model was originally suggested by Dana (1849), who observed that the islands aged to the northwest and suggested that they formed over a major fissure zone caused by cooling and shrinkage of the outer layers of the Earth. This inspired insight forms the basis for some of the most recent efforts to explain linear volcanic chains in the Pacific (e.g., Sandwell and Fialko, 2004).

Models to explain Emperor and Hawaiian volcanism must account for a number of observations, including:

1. The Emperor chain started on or near a spreading ridge, and there is no evidence that it was preceded by LIP eruption (Shapiro et al., 2006; Norton, this volume);
2. Both chains are regularly time progressive (Sharp and Clague, 2006);
3. With respect to the geomagnetic pole, the locus of melt extraction propagated south by ~800 km during emplacement of the Emperor chain, but was stationary during emplacement of the Hawaiian chain (Tarduno and Cottrell, 1997);
4. The direction of propagation changed by ~60° at ca. 47 Ma, at the time of the "bend" (Sharp and Clague, 2006). No corresponding change in the direction of Pacific plate motion occurred at this time (Norton, 1995);
5. The melt extraction rate has increased by an order of magnitude during the past ~2 m.y. (Robinson and Eakins, 2006), resulting in LIP emplacement following the development of the time-progressive chains rather than preceding it (Fig. 6). This change in magmatic rate has apparently been accompanied by a very recent change in propagation direction from approximately southeast to approximately SSE;
6. The younger half of the Hawaiian chain is surrounded by a bathymetric swell (Fig. 12).

A great deal of research on the Hawaiian Islands has been directed at probing the underlying mantle structure. However, work is hampered by the small size of the "Big" Island and the large percentage of the Hawaiian archipelago that is under water. These factors limit the aperture of land seismic networks and geochemical sampling.

The temperature of the mantle beneath Hawaii has been investigated using sea-bottom heatflow measurements and petrology. Heatflow measurements across the Hawaiian swell do not detect a heatflow anomaly of the kind expected for a thermal source (von Herzen et al., 1989), although recently it has been suggested that hydrothermal circulations mask expected anomalies (Harris and McNutt, 2005). Estimates from picrite glass samples of the mantle temperature anomaly vary from zero to ~250 °C (Green and Falloon, 2005; Putirka, 2005; Falloon et al., this volume).

The geochemistry of the lavas is variable, exhibiting both enriched and depleted signatures consistent with components of continental and marine sediments and oceanic crust, and having variable isotopic ratios. The geochemistry varies geographically where the Hawaiian and Emperor chains cross fracture zones (e.g., the Mendocino, Murray, and Molakai fracture zones; Basu and Faggart, 1996). Variations with time area also reported; for example, Mukhopadhyay et al. (2003) describe changes in $^3He/^4He$ of up to 8 Ra during a single century in Kauai volcano. Observations suggest that different volcanoes are fed by different magma sources, suggesting a chemically heterogeneous source. This possibility is in keeping with recent work that suggests the present-day source contains up to 30% of recycled crust (e.g., Sobolev et al., 2005). The volume of magma erupted at Hawaii cannot currently be explained by any model, plume or plate, without appealing to a source much more fertile than lherzolite (Cordery et al., 1997; Leitch et al., 1997). The maximum $^3He/^4He$ observed at Hawaii is 35 Ra in samples from Loihi (Graham, 2002). Calculations based on a lherzolite source suggest that the lavas were last in equilibrium at ~90–100 km depth, corresponding to the estimated base of the Cretaceous lithosphere on which Hawaii is emplaced.

Teleseismic tomography conducted using a network on the Big Island found little significant structure in the top ~100 km (Ellsworth, 1977). Larger-scale teleseismic tomography involving a ~600-km-long array of sensors on several islands found low-velocity anomalies beneath the islands of Maui and Molokai, 250 km northwest of the Big Island, but no low-wavespeed anomaly beneath the Big Island itself down to the maximum depth of good resolution there at ~150 km (Wolfe et al., 2002). Ray coverage throughout the Pacific Ocean is patchy, and whole-mantle tomography can only image very large-scale features (Julian, 2005). It reveals continuous low-velocity material between the surface in the Hawaiian region and the core-mantle boundary beneath a large swath of the Earth ranging from the New Hebrides and Samoa, throughout the south Pacific, and north along the East Pacific Rise, depending on the line of cross-section selected (Ritsema, 2005). This global-scale anomaly is associated with the "Pacific superplume," which has recently been shown to be largely compositional in origin (Trampert et al., 2004). Finite-frequency tomography reveals low velocities beneath the Hawaiian region that die out in the mid-mantle (Montelli et al., 2004). Transition zone thickness beneath the Hawaiian region is ~229 km (Gu and Dziewonski, 2001), which is ~13 km thinner than the global average of 242 km but typical for the central Pacific as a whole.

Only a little work has been done to develop a numerical plate model for Hawaii involving a propagating crack (Shaw, 1973; Shaw and Jackson, 1973; Turcotte and Oxburgh, 1973; Turcotte, 1974; Jackson and Shaw, 1975; Jackson et al., 1975; Lithgow-Bertelloni and Guynn, 2004; Natland and Winterer, 2005). Stuart et al. (this volume) calculated the stress field that would result from cooling of the Pacific plate and found that the orientation of stress in the region of Hawaii was optimal for the southeastward propagation of a crack tip. It seems likely that the major changes in magmatic rate, from near zero at the time of the bend to the present-day maximum, must reflect variations in source composition rather than variations solely in temperature. A stress-based model could also naturally explain the time-progressive chains; variable propagation rates; and the sharpness of the bend, which argues against a cause in slowly varying thermal convection structures.

Favela and Anderson (1999) suggested that Hawaiian volcanism initiated as a result of a change in stress in the plate. Smith (this volume) suggests that the cause of this stress change was a nearly simultaneous change in the western part of the Pacific plate from being bounded by the Kula-Pacific and North New Guinea–Pacific ridges to being subducted beneath the Aleutian and Izu-Bonin-Mariana arcs. The volume of the bathymetric swell is proportional to the volume of magma emplaced on the surface and may be explained by buoyant residuum left over from melt extraction from the mantle (Phipps Morgan et al., 1995). Notwithstanding these speculations, a crack model for Hawaiian volcanism urgently needs to be tested, ideally by measuring extension normal to the chain. However, detecting extension in the deep ocean that might be on the order of 1 mm/yr and total no more than 1 km beneath a lava pile ~25 km thick is currently technologically challenging.

Aside from the Emperor-Hawaiian system, there is great diversity in the nature of melting anomalies in the Pacific. Low-volume seamounts of alkalic OIB are widely scattered over the ocean floor, and most do not plausibly form discrete chains. However, linear arrays of seamounts are common, but chronological progressions range from being regular (e.g., the Louisville chain; Watts et al., 1988) to highly variable (e.g., the Austral chain; McNutt et al., 1997). Many chains are of little significance volumetrically, are associated with essentially none of the characteristics expected of plumes, and may readily be explained by thermal contraction and cracking of the lithosphere (Sandwell et al., 1995; Sandwell and Fialko, 2004). Volumetrically larger melting anomalies do exist, for example, Samoa and the Galapagos, but neither of these is associated with time progression of volcanism. Dates from the Galapagos suggest a broad, heterogeneous melting anomaly. Volcanism in the Samoa region is located at the northern end of the Tonga Trench, where major extension is predicted by models of Pacific lithosphere cooling (Stuart et al., this volume).

Continents

Magmatism on the continents resembles that of the oceans in that a broad spectrum of volume, period of emplacement, time progressiveness, and relationship with tectonic events and features are exhibited. The volumes of magma observed range from very small to LIP-sized, with volumes up to ~4×10^6 km^3 in the case of the Siberian Traps (Fig. 2). Much of the Siberian Traps is thought to have erupted within a few million years (Kamo et al., 2003; Ivanov, this volume). However, this intensity may be compared with volcanism in the Basin and Range province, western United States, where widespread, relatively low-volume eruptions have continued since ca. 20 Ma to the present day and are sourced beneath a region ~1000 km broad.

Basaltic continental magmas typically have OIB-like compositions and range from alkalic to tholeiitic types. Low-volume alkalic OIB-like basalts are widespread in many rift settings, where they are presumed to be associated with the rifting process (e.g., Hooper et al., this volume). High-volume basaltic LIPs generally also have OIB-like compositions, although the Siberian Traps is an exception (Fitton, this volume). It has recently been recognized that silicic provinces may be sufficiently large to be classified as LIPs, and their duration of emplacement may be up to 40 Ma (Bryan et al., 2002). LIPs may be dominantly basaltic or silicic, though at least one province with subequal proportions of each is known (Sensarma et al., 2004). Large-volume silicic melts must come from remelting the continental crust. These observations need to be accounted for in models proposed to explain the origin of continental volcanism and the time scales and mechanisms of heat supply.

The history of vertical motion associated with the largest LIPs also varies widely, and includes cases of subsidence (e.g., the Siberian Traps, the Columbia River basalts; Czamanske, 1998; Hales et al., 2005), lack of uplift (e.g., the Deccan Traps; Sheth, this volume), and uplift (e.g., the Emeishan basalts; Rainbird and Ernst, 2001; He et al., 2003; Xu and He, this volume). The chronological relationship between vertical motion and eruption is also variable. For example, uplift may precede volcanism (e.g., the Emeishan basalts), accompany it (e.g., the North Atlantic Igneous province; Maclennan and Jones, 2006) or postdate it (e.g., the Whitsunday province; Bryan et al., 2002).

Major time-progressive trails of volcanism are rare, the best example being the Yellowstone–Snake River Plain chain of silicic calderas that formed from ca. 17 Ma to the present (Christiansen et al., 2002). This chain is aligned with the plate direction but is paired with the High Lava Plains time-progressive volcanic track, which youngs in the opposite direction (Jordan, 2005).

Volcanism is often associated with significant tectonic structures or events. Many continental LIPs are coastal, erupted when continents broke up (e.g., the North Atlantic Igneous Province, the Central Atlantic Magmatic province, the Deccan Traps), and are clearly associated with the rifting process. Van Wijk et al. (2001) modeled decompression melting and found that the volumes observed at many volcanic margins can be accounted for simply by isentropic upwelling of asthenosphere in response to continental rifting. Volcanism commonly occurs in continental rifts that have not developed to the point of continental breakup (e.g., the east African rift) and may be particularly intense at triple junctions (e.g., Afar). Well-documented pre-emplacement subsidence has led to the suggestion that some major continental LIPs erupt in response to the development of gravitational instability and detachment of the mantle lithosphere and possibly part of the lower crust (Tanton and Hager, 2000; Hales et al., 2005; Lustrino, 2005). Back-arc and slab deformation and fragmentation processes have been suggested to have triggered volcanism in such regions as Mexico, Turkey, the Basin and Range province, and east Asia (e.g., Ferrari, 2004; Keskin, this volume).

DISCUSSION

The primary objective of this chapter is to present the rationale behind the plate model for the genesis of melting anomalies. Although some observations are consistent with the classical plume model originally defined by Morgan (1971) and recently reiterated by Campbell (2006), many features of volcanic regions are not naturally expected for a classical plume source, and the predictions of that model are often not fulfilled. These failures have, in recent years, inspired the quest for a theory that may explain the observations more wholly and have greater powers of prediction.

The term "plume" is used flexibly to describe a wide variety of phenomena, some of which are remote from the original model of Morgan (1971). Some, ironically, fall within the plate model. These include thermal uprisings in the mantle resulting from continental insulation (Burov and Guillou-Frottier, 2005b), conductive and radiogenic warming of eclogite (Davies and Bunge, 2006), and "lithospheric plumes" (Courtillot et al., 2003). Such broad and undefined usage of the term can clearly, in itself, be a barrier to progress.

To be viable, a model must be consistent with the holistic observations, including field geology, tectonics, igneous petrology, geochronology, geochemistry, and geophysics. This requirement presents a formidable challenge to researchers, given the extraordinary degree of specialization that now exists in all branches of modern Earth science. Integrated science is, however, imperative for progress in this highly cross-disciplinary field.

Although melting anomalies or "hotspots" have traditionally been considered a single phenomenological class, a general review reveals extreme variation in almost every respect. This variability includes the melt volume, pattern of vertical motion, chronology of eruption, petrology, underlying mantle structure, and suite of characteristics at a single locality. Under such circumstances, no single, simple generic model is possible for every "hotspot" that does not require much ad hoc special pleading. Melting anomalies are united essentially only by the commonality of unusual melt extraction. Even in this respect, there is a continuous spectrum in the volume produced—from extraordinarily large LIPs to trivial eruptions—and in composition. A natural bimodalism might be expected if fundamentally different genesis processes are at work, but no such bimodalism has been identified to date. A unifying model for anomalous vol-

canism is required that can account for the full spectrum of size, longevity, composition, and spatial and temporal distributions of volcanism.

Two LIPs are currently in the process of formation, Iceland (with a present-day area and volume of ~0.3×10^6 km^2 and ~5×10^6 km^3, respectively) and Hawaii (with a present-day area and volume of ~0.15×10^6 km^2 and ~0.6×10^6 km^3, respectively). Both are forming late in the sequence of associated volcanism. In the context of the plume model, these systems amount to plume-head volcanism occurring after plume-tail volcanism and would be interpreted as indicating a pulse in the temperature or flow of the plume tail. In the context of plate interpretations, an explanation would be expected in variations of extension rate or the fertility of the source material being tapped. Melting anomalies that exhibit LIP volcanism followed by low-volume, time-progressive volcanic chain formation, as predicted by the plume model, are at best very rare, and perhaps nonexistent.

There is no unequivocal evidence for systematically elevated temperature at LIPs or "hotspots" in general. Petrological approaches using olivine control lines are still contested. The traditional assumption that high-forsterite olivines and picrite cumulates require high temperature is now recognized to be incorrect (Green et al., 2001; Green and Falloon, 2005; Presnall and Gudfinnsson, 2005; Falloon et al., 2007a, this volume). The eruption of large volumes of tholeiitic magma at LIPs requires the removal of a large quantity of heat from the mantle, but direct evidence that this removal is engineered by temperatures that are hundreds of degrees above those beneath normal mid-ocean ridges remains elusive. However, the OIB geochemical signature of "hotspot" lavas is prima facie evidence for fusible material in the source, and the percentage in recent Hawaiian and Icelandic lavas may be large (e.g., Foulger et al., 2005b; Sobolev et al., 2005).

OIB geochemistry is usually but not always associated with melting anomalies. It is observed in eruptive sequences with a wide range of tectonic settings, distributions, volumes, and longevities. OIB-like eruptive sequences may appear to be stationary in some frame and to erupt independently of plate motion, giving rise to time-progressive sequences. At the other extreme, they may appear to travel with the plate. The generally agreed-upon association of OIB geochemistry with recycled, near-surface materials naturally suggests shallow-based models, though the precise sources remain unclear. The depth of origin of the noble gas signatures associated with OIB remains controversial, with recent work tending to support shallow sources (Stuart et al., 2003; Parman et al., 2005; Fig. 11).

Usually, the differences between the volcanism in the fast-spreading Pacific Ocean and the slow-spreading Atlantic Ocean are most strongly emphasized. However, there is considerable commonality. Each ocean contains a single major tholeiitic province (Hawaii and Iceland) and several melting anomalies associated with the ridge system. These include the Cobb, Galapagos, Easter, and Louisville anomalies in the Pacific and Iceland, the Azores, Ascension, Tristan, and the Bouvet triple junction in the Atlantic. In both oceans, there are long, linear seamount chains, both time-progressive and nonprogressive, along with many shorter, low-volume chains and regions, where volcanism is scattered over broad areas.

The greatest challenge remains to explain quantitatively the largest melt volumes and volume rates that the Earth is capable of producing at the surface. Small volumes of alkalic lava, assumed to be formed by small-percentage melting, are a common observation in rifts and near extensional faulting and are assumed to be related to the lithosphere extension process. However, very large eruptive rates of tholeiitic basalt, which is thought to require a high degree of melting of a peridotite source or almost complete melting of an eclogite source, have yet to be fully explained. Quantitative modeling of decompression melting in plumes, even if containing fusible material, cannot reproduce, at reasonable temperatures, the volumes of the largest LIPs, where the lithosphere has significant thickness (Cordery et al., 1997). Decompression accompanying lithospheric rifting can explain the volcanic margins associated with continental breakup (van Wijk et al., 2001) and LIPs formed at mid-ocean ridges (Korenaga, 2005). The large temperature anomalies that have been suggested, but so far defy observation, are not required.

The problem for both thermal and athermal models then lies in explaining significant eruptive volumes at locations where the lithosphere is thick and large-scale extension is not observed. Where surface extension is minor, or where upwellings stall at the base of thick lithosphere, it is not clear how isentropic decompression can provide the energy required to melt large volumes of source material. The eruption of pre-existing, ponded melt has been suggested (e.g., Silver et al., 2006), but work still needs to be done to explain how extremely large volumes of melt can accumulate and be retained in the mantle for long periods. There are, however, few regions of this sort. Hawaii is the only presently active example, and has only become so in the past 2 Ma. Older examples include LIPs that erupted in the interiors of continents, far from rifted margins, such as the Siberian Traps (Ivanov, this volume).

The plate model for the source of melting anomalies is at an embryonic stage of development, and much work remains to be done before its full potential can be assessed. Urgent avenues of investigation include:

1. Mapping and modeling variations in stress in the lithosphere, both locally and on regional and plate-wide scales;
2. Numerical modeling of the melt volumes produced in different extensional settings;
3. Quantifying the volumes of melt produced by isentropic upwelling of nonperidotic source material;
4. Modeling convection;
5. Solving the paradox of the origin of OIB;
6. Studying in detail the history of vertical motion associated with major LIPs and comparing with the predictions of plate models;

7. Understanding the mechanism that produces very large melt volumes and eruption rates;
8. Determining the continuity of seismic structure through the transition zone beneath active melting anomalies;
9. Understanding how to interpret seismic anomalies;
10. Resolving the present controversy regarding the source temperature of "hotspot" lavas;
11. Investigating the relationship between extensional faulting and related volcanism; and
12. Erecting and testing plate models for melting anomalies that have not yet been subject to this type of scrutiny.

ACKNOWLEDGMENTS

I thank Don Anderson, Richard Ernst, Trevor Falloon, Godfrey Fitton, Donna Jurdy, John Mahoney, Dean Presnall, Joel Robinson, David Sandwell, Hetu Sheth, Norman Sleep, and Alan Smith for helpful comments.

REFERENCES CITED

Anderson, D.L., 1989, Theory of the Earth: Boston, Blackwell Scientific Publications, 366 p.

Anderson, D.L., 1995, Lithosphere, asthenosphere and perisphere: Reviews of Geophysics, v. 33, p. 125–149, doi: 10.1029/94RG02785.

Anderson, D.L., 1998a, The helium paradoxes: Proceedings of the National Academy of Sciences, USA, v. 95, p. 4822–4827, doi: 10.1073/pnas.95.9 .4822.

Anderson, D.L., 1998b, A model to explain the various paradoxes associated with mantle noble gas geochemistry: Proceedings of the National Academy of Sciences, USA, v. 95, p. 9087–9092, doi: 10.1073/pnas.95.16.9087.

Anderson, D.L., 2000, The thermal state of the upper mantle; No role for mantle plumes: Geophysical Research Letters, v. 27, p. 3623–3626, doi: 10.1029/2000GL011533.

Anderson, D.L., 2001, Top-down tectonics?: Science, v. 293, p. 2016–2018, doi: 10.1126/science.1065448.

Anderson, D.L., 2002, Occam's razor; Simplicity, complexity and global geodynamics: Proceedings of the American Philosophical Society, v. 146, p. 56–76.

Anderson, D.L., 2005a, Large igneous provinces, delamination, and fertile mantle: Elements, v. 1, p. 271–275.

Anderson, D.L., 2005b, Scoring hotspots: The plume and plate paradigms, *in* Foulger, G.R., et al., eds., Plates, plumes, and paradigms: Boulder, Colorado, Geological Society of America Special Paper 430, p. 31–54.

Anderson, D.L., 2006, Speculations on the nature and cause of mantle heterogeneity: Tectonophysics, v. 416, p. 7–22, doi: 10.1016/j.tecto.2005.07.011.

Anderson, D.L., 2007 (this volume), The Eclogite engine: Chemical geodynamics as a Galileo thermometer, in Foulger, G.R., and Jurdy, D.M., eds., Plates, plumes, and planetary processes: Boulder, Colorado, Geological Society of America Special Paper 430, doi: 10.1130/2007.2430(03).

Anderson, D.L., and Natland, J.H., 2005, A brief history of the plume hypothesis and its competitors: Concept and controversy, *in* Foulger, G.R., et al., eds., Plates, plumes, and paradigms: Boulder, Colorado, Geological Society of America Special Paper 388, p. 119–145.

Baksi, A.K., 1999, Reevaluation of plate motion models based on hotspot tracks in the Atlantic and Indian oceans: Journal of Geology, v. 107, p. 13–26, doi: 10.1086/314329.

Bargar, K.E., and Jackson, E.D., 1974, Calculated volumes of individual shield volcanoes along the Hawaiian-Emperor chain: U.S. Geological Survey Journal of Research, v. 2, p. 545–550.

Basu, A.R., and Faggart, B.E., 1996, Temporal variation in the Hawaiian mantle plume: The Lanai anomaly, the Molokai fracture zone and a seawater-altered lithospheric component in Hawaiian volcanism, *in* Basu, A., and Hart, S., eds., Earth processes: Reading the isotopic code: Washington, D.C., American Geophysical Union Geophysical Monograph 95, p. 149–159.

Beutel, E.K., and Anderson, D.L., 2007 (this volume), Ridge-crossing seamount chains: A nonthermal approach, *in* Foulger, G.R., and Jurdy, D.M., eds., Plates, plumes, and planetary processes: Boulder, Colorado, Geological Society of America Special Paper 430, doi: 10.1130/2007.2430(19).

Bina, C., and Helffrich, G., 1994, Phase transition Clapyron slopes and transition zone seismic discontinuity topography: Journal of Geophysical Research, v. 99, p. 15,853–15,860, doi: 10.1029/94JB00462.

Björnsson, A., Eysteinsson, H., and Beblo, M., 2005, Crustal formation and magma genesis beneath Iceland: Magnetotelluric constraints, *in* Foulger, G.R., et al., eds., Plates, plumes, and paradigms: Boulder, Colorado, Geological Society of America Special Paper 388, p. 665–686.

Breddam, K., 2002, Kistufell: Primitive melt from the Iceland mantle plume: Journal of Petrology, v. 43, p. 345–373, doi: 10.1093/petrology/43.2.345.

Bruguier, N.J., Minshull, T.A., and Brozena, J.M., 2003, Morphology and tectonics of the mid-Atlantic ridge, 7°–12°S: Journal of Geophysical Research, v. 108, p. 2093, doi: 10.1029/2001JB001172.

Bryan, S.E., and Ernst, R.E., 2007, Revised definition of large igneous province (LIP): Earth Science Reviews, in press.

Bryan, S.E., Riley, T.R., Jerram, D.A., Leat, P.T., and Stephens, C.J., 2002, Silicic volcanism: An under-valued component of large igneous provinces and volcanic rifted margins, *in* Menzies, M.A., et al., eds., Magmatic rifted margins: Boulder, Colorado, Geological Society of America Special Paper 362, p. 99–120.

Burke, K., 1996, The African plate: South African Journal of Geology, v. 99, p. 341–409.

Burov, E., and Guillou-Frottier, L., 2005a, Modeling plume head–continental lithosphere interaction using a tectonically realistic lithosphere: http://www.mantleplumes.org/LithUplift.html.

Burov, E., and Guillou-Frottier, L., 2005b, The plume head–continental lithosphere interaction using a tectonically realistic formulation for the lithosphere: Geophysical Journal International, v. 161, p. 469–490, doi: 10.1111/j.1365-246X.2005.02588.x.

Campbell, I.H., 2006, Large igneous provinces and the mantle plume hypothesis: Elements, v. 1, p. 265–269.

Campbell, I.H., and Griffiths, R.W., 1990, Implications of mantle plume structure for the evolution of flood basalts: Earth and Planetary Science Letters, v. 99, p. 79–93, doi: 10.1016/0012-821X(90)90072-6.

Chauvel, C., and Hemond, C., 2000, Melting of a complete section of recycled oceanic crust: Trace element and Pb isotopic evidence from Iceland: Geochemistry, Geophysics, Geosystems, v. 1, p. 1999GC000002.

Christiansen, R.L., Foulger, G.R., and Evans, J.R., 2002, Upper mantle origin of the Yellowstone hotspot: Geological Society of America Bulletin, v. 114, p. 1245–1256, doi: 10.1130/0016-7606(2002)114<1245:UMOOTY>2.0.CO;2.

Clague, D.A., Weber, W.S., and Dixon, J.E., 1991, Picritic glasses from Hawaii: Nature, v. 353, p. 553–556, doi: 10.1038/353553a0.

Clift, P.D., 2005, Sedimentary evidence for moderate mantle temperature anomalies associated with hotspot volcanism, *in* Foulger, G.R., et al., eds., Plates, plumes, and paradigms: Boulder, Colorado, Geological Society of America Special Paper 388, p. 279–288.

Cloos, M., 1993, Lithospheric buoyancy and collisional orogenesis: Subduction of oceanic plateaus, continental margins, island arcs, spreading ridges, and seamounts: Geological Society of America Bulletin, v. 105, p. 715–737, doi: 10.1130/0016-7606(1993)105<0715:LBACOS>2.3.CO;2.

Coffin, M.F., and Eldholm, O., 1994, Large igneous provinces: Crustal structure, dimensions and external consequences: Reviews of Geophysics, v. 32, p. 1–36, doi: 10.1029/93RG02508.

Cordery, M.J., Davies, G.F., and Campbell, I.H., 1997, Genesis of flood basalts

from eclogite-bearing mantle plumes: Journal of Geophysical Research, v. 102, p. 20,179–20,197, doi: 10.1029/97JB00648.

Courtillot, V., Davaillie, A., Besse, J., and Stock, J., 2003, Three distinct types of hotspots in the Earth's mantle: Earth and Planetary Science Letters, v. 205, p. 295–308, doi: 10.1016/S0012-821X(02)01048-8.

Czamanske, G.K., 1998, Demise of the Siberian plume: Paleogeographic and paleotectonic reconstruction from the prevolcanic and volcanic records, north-central Siberia: International Geology Review, v. 40, p. 95–115.

Dana, J.D., 1849, Geology, *in* Wilkes, C., ed., United States exploring expedition, volume 10, with atlas: Philadelphia and New York, C. Sherman and Putnam, 756 p.

Dasgupta, R., and Hirschmann, M.M., 2006, Melting in the Earth's deep upper mantle caused by carbon dioxide: Nature, v. 440, p. 659–662, doi: 10.1038/nature04612.

Davies, J.H., and Bunge, H.-P., 2006, Are splash plumes the origin of minor hotspots?: Geology, v. 34, p. 349–352, doi: 10.1130/G22193.1.

Du, Z., Vinnik, L.P., and Foulger, G.R., 2006, Evidence from P-to-S mantle converted waves for a flat "660-km" discontinuity beneath Iceland: Earth and Planetary Science Letters, v. 241, p. 271–280, doi: 10.1016/j.epsl.2005.09.066.

Dueker, K.G., and Sheehan, A.F., 1997, Mantle discontinuity structure from midpoint stacks of converted P to S waves across the Yellowstone hotspot track: Journal of Geophysical Research, v. 102, p. 8313–8327, doi: 10.1029/96JB03857.

Ebinger, C.J., and Sleep, N.H., 1998, Cenozoic magmatism throughout East Africa resulting from impact of a single plume: Nature, v. 395, p. 1788–1791.

Elkins-Tanton, L.T., 2005, Continental magmatism caused by lithospheric delamination, *in* Foulger, G.R., et al., eds., Plates, plumes, and paradigms: Boulder, Colorado, Geological Society of America Special Paper 388, p. 449–462.

Ellsworth, W.L., 1977, Three-dimensional structure of the crust and mantle beneath the island of Hawaii [Ph.D. thesis]: Boston, Massachusetts Institute of Technology, 327 p.

Ernst, R.E., and Buchan, K.L., 2002, Maximum size and distribution in time and space of mantle plumes: Evidence from large igneous provinces, *in* Condie, K.C., et al., eds., Superplume events in Earth's history: Causes and effects: Journal of Geodynamics Special Issue, v. 34, p. 711–714.

Fairhead, M.J., and Wilson, M., 2005, Plate tectonic processes in the south Atlantic Ocean: Do we need deep mantle plumes?, *in* Foulger, G.R., et al., eds., Plates, plumes, and paradigms: Boulder, Colorado, Geological Society of America Special Paper 388, p. 537–554.

Falloon, T.J., Danyushevsky, L.V., Ariskin, A., Green, D.H., and Ford, C.E., 2007a, The application of olivine geothermometry to infer crystallization temperatures of parental liquids: Implications for the temperature of MORB magmas: Chemical Geology, v. 241, p. 153–176.

Falloon, T.J., Green, D.H., and Danyushevsky, L.V., 2007b (this volume), Crystallization temperatures of tholeiite parental liquids: Implications for the existence of thermally driven mantle plumes, *in* Foulger, G.R., and Jurdy, D.M., eds., Plates, plumes, and planetary processes: Boulder, Colorado, Geological Society of America Special Paper 430, doi: 10.1130/2007.2430(12).

Favela, J., and Anderson, D.L., 1999, Extensional tectonics and global volcanism, *in* Boschi, E., et al., eds., Volume problems in geophysics for the new millennium: Bologna, Editrice Compositori, p. 463–498.

Ferrari, L., 2004, Slab detachment control on mafic volcanic pulse and mantle heterogeneity in central Mexico: Geology, v. 32, p. 77–80, doi: 10.1130/G19887.1.

Fitton, J.G., 1987, The Cameroon line, West Africa: A comparison between oceanic and continental alkaline volcanism, *in* Fitton, J.G., and Upton, B.G.J., eds., Alkaline igneous rocks: London, Geological Society of London Special Publication 30, p. 273–291.

Fitton, J.G., 2007 (this volume), The OIB paradox, *in* Foulger, G.R., and Jurdy, D.M., eds., Plates, plumes, and planetary processes: Boulder, Colorado, Geological Society of America Special Paper 430, doi: 10.1130/2007.2430(20).

Fitton, J.G., and Dunlop, H.M., 1985, The Cameroon line, West Africa, and its bearing on the origin of oceanic and continental alkali basalt: Earth and Planetary Science Letters, v.72, p. 23–38, doi: 10.1016/0012-821X(85) 90114-1.

Fitton, J.G., and James, D., 1986, Basic volcanism associated with intraplate linear features: Philosophical Transactions of the Royal Society of London (Series A), v. 317, p. 253–266.

Fitton, J.G., Mahoney, J.J., Wallace, P.J., and Saunders, A.D., 2004, Origin and evolution of the Ontong Java plateau: Introduction, in Fitton, J.G., et al., eds., Origin and evolution of the Ontong Java plateau: London, Geological Society of London Special Publication 229, p. 1–8.

Ford, C.E., Russell, D.G., Craven, J.A., and Fisk, M.R., 1983, Olivine-liquid equilibria: Temperature, pressure and composition dependence of the crystal/liquid cation partition coefficients for Mg, Fe²⁺, Ca, and Mn: Journal of Petrology, v. 24, p. 256–265.

Foulger, G.R., 2002, Plumes, or plate tectonic processes? Astronomy & Geophysics, v. 43, p. 6.19–6.23.

Foulger, G.R., 2003, On the apparent eastward migration of the spreading ridge in Iceland, in Foulger, G.R., et al., eds., Penrose conference plume IV: Beyond the plume hypothesis, 25–29 August, Hveragerdi, Iceland: Boulder, Colorado, Geological Society of America.

Foulger, G.R., Plate tectonic processes, http://www.mantleplumes.org/PTProcesses.html (2004).

Foulger, G.R., 2005, The puzzle of the petrology of oceanic plateaux: EOS, Transactions of the American Geophysical Union, v. Fall Meeting Supplement, p. Abstract, V51B–0569.

Foulger, G.R., 2006a, Do plumes exist? www.mantleplumes.org.

Foulger, G.R., 2006b, Older crust underlies Iceland: Geophysical Journal International, v. 165, p. 672–676, doi: 10.1111/j.1365-246X.2006.02941.x.

Foulger, G.R., and Anderson, D.L., 2005, A cool model for the Iceland hot spot: Journal of Volcanology and Geothermal Research, v. 141, p. 1–22, doi: 10.1016/j.jvolgeores.2004.10.007.

Foulger, G.R., and Natland, J.H., 2003, Is "hotspot" volcanism a consequence of plate tectonics?: Science, v. 300, p. 921–922, doi: 10.1126/science.1083376.

Foulger, G.R., Pritchard, M.J., Julian, B.R., Evans, J.R., Allen, R.M., Nolet, G., Morgan, W.J., Bergsson, B.H., Erlendsson, P., Jakobsdottir, S., Ragnarsson, S., Stefansson, R., and Vogfjord, K., 2001, Seismic tomography shows that upwelling beneath Iceland is confined to the upper mantle: Geophysical Journal International, v. 146, p. 504–530, doi: 10.1046/j.0956-540x.2001.01470.x.

Foulger, G.R., Natland, J.H., and Anderson, D.L., 2005a, Genesis of the Iceland melt anomaly by plate tectonic processes, in Foulger, G.R., et al., eds., Plates, plumes, and paradigms: Boulder, Colorado, Geological Society of America Special Paper 388, p. 595–626.

Foulger, G.R., Natland, J.H., and Anderson, D.L., 2005b, A source for Icelandic magmas in remelted Iapetus crust: Journal of Volcanology and Geothermal Research, v. 141, p. 23–44, doi: 10.1016/j.jvolgeores.2004.10.006.

Foulger, G.R., Natland, J.H., Presnall, D.C., and Anderson, D.L., eds., 2005c, Plates, plumes, and paradigms: Boulder, Colorado, Geological Society of America Special Paper 388, 881 p.

Georgen, J.E., and Lin, J., 2002, Three-dimensional passive flow and temperature structure beneath oceanic ridge-ridge-ridge triple junctions: Earth and Planetary Science Letters, v. 204, p. 115–132, doi: 10.1016/S0012-821X(02)00953-6.

Glen, W., 2005, The origins and early trajectory of mantle plume quasi-paradigm, in Foulger, G.R., et al., eds., Plates, plumes, and paradigms: Boulder, Colorado, Geological Society of America Special Paper 388, p. 91–118.

Gorbatov, A., Widiyantoro, S., Fukao, Y., and Gordeev, E., 2000, Signature of remnant slabs in the North Pacific from P-wave tomography: Geophysical Journal International, v. 142, p. 27–36.

Gordon, R.G., and Royer, J., 2005, Diffuse oceanic plate boundaries, thin viscous sheets of oceanic lithosphere, and Late Miocene changes in plate motion and tectonic regime: Eos (Transactions, American Geophysical Union), Fall meeting supplement, abstract, p. U43B–0836.

Graham, D.W., 2002, Noble gas isotope geochemistry of mid-ocean ridge and ocean island basalts; Characterization of mantle source reservoirs, in Porcelli, D., et al., eds., Noble gases in geochemistry and cosmochemistry: Reviews in mineralogy and geochemistry: Washington, D.C., Mineralogical Society of America, p. 247–318.

Grand, S.P., 1994, Mantle shear structure beneath the Americas and the surrounding oceans: Journal of Geophysical Research, v. 99, p. 11,591–11,621, doi: 10.1029/94JB00042.

Green, D.H., and Falloon, T.J., 2005, Primary magmas at mid-ocean ridges, "hot spots" and other intraplate settings; Constraints on mantle potential temperature, in Foulger, G.R., et al., eds., Plates, plumes, and paradigms: Boulder, Colorado, Geological Society of America Special Paper 388, p. 217–248.

Green, D.H., Falloon, T.J., Eggins, S.M., and Yaxley, G.M., 2001, Primary magmas and mantle temperatures: European Journal of Minerology, v. 13, p. 437–451, doi: 10.1127/0935-1221/2001/0013-0437.

Green, T.H., Green, D.H., and Ringwood, A.E., 1967, The origin of high-alumina basalts and their relationships to quartz tholeiites and alkali basalts: Earth and Planetary Science Letters, v. 2, p. 41–51, doi: 10.1016/0012-821X(67)90171-9.

Griffiths, R.W., and Campbell, I.H., 1991, Interaction of mantle plume heads with the Earth's surface and onset of small-scale convection: Journal of Geophysical Research, v. 96, p. 18,295–18,310.

Grunewald, S., Weber, M., and Kind, R., 2001, The upper mantle under central Europe: Indications for the Eifel plume: Geophysical Journal International, v. 147, p. 590–601, doi: 10.1046/j.1365-246x.2001.01553.x.

Gu, Y.J., and Dziewonski, A.M., 2001, Variations in thickness of the upper mantle transition zone, in long-term observations in the oceans, in Romanowicz, B., Suyehiro, K., and Kawakatsu, H., eds., Long-term observations in the oceans: Current status and perspectives for the future: International Ocean Network, p. 175–180.

Gu, Y.J., Dziewonski, A.M., Weijia, S., and Ekstrom, G., 2001, Models of the mantle shear velocity and discontinuities in the pattern of lateral heterogeneities: Journal of Geophysical Research, v. 106, p. 11,169–11,199, doi: 10.1029/2001JB000340.

Hales, T.C., Abt, D.L., Humphreys, E.D., and Roering, J.J., 2005, Lithospheric instability origin for Columbia River flood basalts and Wallowa Mountains uplift in northeast Oregon: Nature, v. 438, p. 842–845, doi: 10.1038/nature04313.

Harris, R.N., and McNutt, M.K., 2005, Observations of heat flow on hotspot swells, in The great plume debate, 28 August–1 September, Ft. William, Scotland: Washington, D.C., American Geophysical Union.

He, B., Xu, Y.-G., Ching, S.L., Xiao, L., and Wang, Y., 2003, Sedimentary evidence for rapid crustal doming prior to the eruption of the Emeishan floods basalts: Earth and Planetary Science Letters, v. 213, p. 391–405, doi: 10.1016/S0012-821X(03)00323-6.

Hieronymus, C.F., and Bercovici, D., 1999, Discrete alternating hotspot islands formed by interaction of magma transport and lithospheric flexure: Nature, v. 397, p. 604–607, doi: 10.1038/17584.

Hieronymus, C.F., and Bercovici, D., 2000, Non-hotspot formation of volcanic chains: Control of tectonic and flexural stresses on magma transport: Earth and Planetary Science Letters, v. 181, p. 539–554, doi: 10.1016/S0012-821X(00)00227-2.

Hofmann, A.W., and White, W.M., 1982, Mantle plumes from ancient oceanic crust: Earth and Planetary Science Letters, v. 57, p. 421–436, doi: 10.1016/0012-821X(82)90161-3.

Hooper, P.R., Camp, V.E., Reidel, S.P., and Ross, M.E., 2007 (this volume), The origin of the Columbia River flood basalt province: Plume versus non-plume models, in Foulger, G.R., and Jurdy, D.M., eds., Plates, plumes, and planetary processes: Boulder, Colorado, Geological Society of America Special Paper 430, doi: 10.1130/2007.2430(30).

Ingle, S., and Coffin, M.F., 2004, Impact origin for the greater Ontong Java plateau?: Earth and Planetary Science Letters, v. 218, p. 123–134, doi: 10.1016/S0012-821X(03)00629-0.

Ito, G., and Clift, P.D., 1998, Subsidence and growth of Pacific Cretaceous plateaus: Earth and Planetary Science Letters, v. 161, p. 85–100, doi: 10.1016/S0012-821X(98)00139-3.

Ivanov, A.V., 2007 (this volume), Evaluation of different models for the origin of the Siberian Traps, *in* Foulger, G.R., and Jurdy, D.M., eds., Plates, plumes, and planetary processes: Boulder, Colorado, Geological Society of America Special Paper 430, doi: 10.1130/2007.2430(31).

Iyer, H.M., Evans, J.R., Zandt, G., Stewart, R.M., Coakley, J.M., and Roloff, J.N., 1981, A deep low-velocity body under the Yellowstone caldera, Wyoming: Delineation using teleseismic P-wave residuals and tectonic interpretation: Summary: Geological Society of America Bulletin, v. 92, p. 792–798, doi: 10.1130/0016-7606(1981)92<792:ADLBUT>2.0.CO;2.

Jackson, E.D., and Shaw, H.R., 1975, Stress fields in central portions of the Pacific plate: Delineated in time by linear volcanic chains: Journal of Geophysical Research, v. 80, p. 1861–1874.

Jackson, E.D., Shaw, H.R., and Barger, K.E., 1975, Calculated geochronology and stress field orientations along the Hawaiian chain: Earth and Planetary Science Letters, v. 26, p. 145–155, doi: 10.1016/0012-821X(75)90082-5.

Jordan, B.T., 2005, The Oregon high lava plains: A province of counter-tectonic age-progressive volcanism, *in* Foulger, G.R., et al., eds., Plates, plumes, and paradigms: Boulder, Colorado, Geological Society of America Special Paper 388, p. 503–516.

Julian, B.R., 2005, What can seismology say about hot spots?, *in* Foulger, G.R., et al., eds., Plates, plumes, and paradigms: Boulder, Colorado, Geological Society of America Special Paper 388, p. 155–170.

Kamo, S.L., Czamanske, G.K., Amelin, Y., Fedorenko, V.A., Davis, D.W., and Trofimov, V.R., 2003, Rapid eruption of Siberian flood-volcanic rocks and evidence for coincidence with the Permian-Triassic boundary and mass extinction at 251 Ma: Earth and Planetary Science Letters, v. 214, p. 75–91, doi: 10.1016/S0012-821X(03)00347-9.

Karato, S., 1993, Importance of anelasticity in the interpretation of seismic tomography: Geophysical Research Letters, v. 20, p. 1623–1626.

Kay, R.W., and Kay, S.M., 1993, Delamination and delamination magmatism: Tectonophysics, v. 219, p. 177–189, doi: 10.1016/0040-1951(93)90295-U.

Kellogg, L.H., and Wasserburg, G.J., 1990, The role of plumes in mantle helium fluxes: Earth and Planetary Science Letters, v. 99, p. 276–289, doi: 10.1016/0012-821X(90)90116-F.

Keskin, M., 2007 (this volume), Eastern Anatolia: A hot spot in a collision zone without a mantle plume, *in* Foulger, G.R., and Jurdy, D.M., eds., Plates, plumes, and planetary processes: Boulder, Colorado, Geological Society of America Special Paper 430, doi: 10.1130/2007.2430(32).

Korenaga, J., 2005, Why did not the Ontong Java plateau form subaerially?: Earth and Planetary Science Letters, v. 234, p. 385–399, doi: 10.1016/j.epsl.2005.03.011.

Kroenke, L.W., Wessel, P., and Sterling, A., 2004, Motion of the Ontong Java plateau in the hot-spot frame of reference: 122 Ma–present, *in* Fitton, J.G., et al., eds., Origin and evolution of the Ontong Java plateau: London, Geological Society of London Special Publication 299, p. 9–20.

Larson, R.L., 1997, Superplumes and ridge interactions between the Ontong Java and Manihiki plateaus and the Nova-Canton trough: Geology, v. 25, p. 779–782, doi: 10.1130/0091-7613(1997)025<0779:SARIBO>2.3.CO;2.

Laske, G., and Masters, G., 1997, A global digital map of sediment thickness: Eos (Transactions, American Geophysical Union), v. 78, F483.

Leitch, A.M., Cordery, M.J., Davies, G.F., and Campbell, I.H., 1997, Flood basalts from eclogite-bearing mantle plumes: South African Journal of Geology, v. 100, p. 311–318.

Lithgow-Bertelloni, C., and Guynn, J., 2004, Origin of the lithospheric stress field: Journal of Geophysical Research, v. 109, doi: 10.1029/2003JB002467.

Lithgow-Bertelloni, C., and Richards, M.A., 1995, Cenozoic plate driving forces: Geophysical Research Letters, v. 22, p. 1317–1320, doi: 10.1029/95GL01325.

Lourenco, N., Miranda, J.M., Luis, J.F., Ribeiro, A., Mendes Victor, L.A., Madiera, J., and Needham, H.D., 1998, Morpho-tectonic analysis of the Azores volcanic plateau from a new bathymetric compilation of the area:

Marine Geophysical Researches, v. 20, p. 141–156, doi: 10.1023/A:1004505401547.

Lundin, E., and Doré, T., 2005, The fixity of the Iceland "hotspot" on the mid-Atlantic Ridge: Observational evidence, mechanisms and implications for Atlantic volcanic margins, *in* Foulger, G.R., et al., ed., Plates, plumes, and paradigms: Boulder, Colorado, Geological Society of America Special Paper 388, p. 627–652.

Lustrino, M., 2005, How the delamination and detachment of lower crust can influence basaltic magmatism: Earth Science Reviews, v. 72, p. 21–38, doi: 10.1016/j.earscirev.2005.03.004.

Lynch, M.A., 1999, Linear ridge groups: Evidence for tensional cracking in the Pacific Plate: Journal of Geophysical Research, v. 104, p. 29,321–29,333, doi: 10.1029/1999JB900241.

Maclennan, J., and Jones, S.M., 2006, Regional uplift, gas hydrate dissociation and the origins of the Paleocene–Eocene Thermal Maximum: Earth and Planetary Science Letters, v. 245, p. 65–80.

Mao, W., Meng, Y., Shen, G., Prakapenka, V., Campbell, A., Heinz, D., Shu, J., Caracas, R., Cohen, R., Fei, Y., Hemley, R., and Mao, H., 2005, Iron-rich silicates in the Earth's D layer: Proceedings of the National Academy of Sciences, USA, v. 102, p. 9751–9753, doi: 10.1073/pnas.0503737102.

Masaitis, V.L., 1983, Permian and Triassic volcanism of Siberia: Problems of dynamic reconstructions: Zapiski Vserossiiskogo Mineralogicheskogo Obshestva, v. 4, p. 412–425.

McKenzie, D.P., and O'Nions, R.K., 1995, The source regions of ocean island basalts: Journal of Petrology, v. 36, p. 133–159.

McNutt, M.K., Caress, D.W., Reynolds, J., Jordahl, K.A., and Duncan, R.A., 1997, Failure of plume theory to explain midplate volcanism in the southern Austral Islands: Nature, v. 389, p. 479–482, doi: 10.1038/39013.

Meibom, A., and Anderson, D.L., 2004, The statistical upper mantle assemblage: Earth and Planetary Science Letters, v. 217, p. 123–139, doi: 10.1016/S0012-821X(03)00573-9.

Meibom, A., Sleep, N.H., Zahnle, K., and Anderson, D.L., 2005, Models for noble gases in mantle geochemistry: Some observations and alternatives, *in* Foulger, G.R., et al., eds., Plates, plumes, and paradigms: Boulder, Colorado, Geological Society of America Special Paper 388, p. 347–364.

Michael, P.J., 1999, Implications for magmatic processes at Ontong Java plateau from volatile and major element contents of Cretaceous basalt glasses: Geochemistry, Geophysics, Geosystems, v. 1, doi: 10.1029/1999GC000025.

Montelli, R., Nolet, G., Dahlen, F.A., Masters, G., Engdahl, R.E., and Hung, S.-H., 2004, Finite frequency tomography reveals a variety of plumes in the mantle: Science, v. 303, p. 338–343, doi: 10.1126/science.1092485.

Moreau, C., Regnoult, J.-M., Déruelle, B., and Robineau, B., 1987, A new tectonic model for the Cameroon line, central Africa: Tectonophysics, v. 141, p. 317–334, doi: 10.1016/0040-1951(87)90206-X.

Moreira, M., Doucelance, R., Kurz, M.D., Dupre, B., and Allegre, C.J., 1999, Helium and lead isotope geochemistry of the Azores archipelago: Earth and Planetary Science Letters, v. 169, p. 189–205, doi: 10.1016/S0012-821X(99)00071-0.

Morgan, W.J., 1971, Convection plumes in the lower mantle: Nature, v. 230, p. 42–43, doi: 10.1038/230042a0.

Mukhopadhyay, S., Lassiter, J.C., Farley, K.A., and Bogue, S.W., 2003, Geochemistry of Kauai shield-stage lavas: Implications for the chemical evolution of the Hawaiian plume: Geochemistry, Geophysics, Geosystems, v. 4, doi: 10.1029/2002GC000342.

Müller, R.D., Roest, W.R., Royer, J.-Y., Gahagan, L.M., and Sclater, J.G., 1997, Digital isochrons of the world's ocean floor: Journal of Geophysical Research, v. 102, p. 3211–3214. Data: http://www-sdt.univ-brest.fr/~jyroyer/Agegrid/utig_report.html.

Natland, J.H., 2003, Capture of mantle helium by growing olivine phenocrysts in picritic basalts from the Juan Fernandez Islands: SE Pacific: Journal of Petrology, v. 44, p. 421–456.

Natland, J.H., and Winterer, E.L., 2005, Fissure control on volcanic action in the Pacific, *in* Foulger, G.R., et al., eds., Plates, plumes, and paradigms: Boulder, Colorado, Geological Society of America Special Paper 388, p. 687–710.

Neal, C.R., Mahoney, J.J., Kroenke, L.W., Duncan, R.A., and Petterson, M.G., 1997, The Ontong-Java plateau, *in* Mahoney, J.J., and Coffin, M.F., eds., Large igneous provinces. Continental, oceanic, and planetary flood volcanism: Washington, D.C., American Geophysical Union Geophysical Monograph 100, p. 183–216.

Niu, Y., and O'Hara, M.J., 2003, Origin of ocean island basalts: A new perspective from petrology, geochemistry, and mineral physics considerations: Journal of Geophysical Research, v. 108, p. 2209, doi: 10.1029/2002JB002048.

Niu, Y., Regelous, M., Wendt, I.J., Batiza, R., and O'Hara, M.J., 2002, Geochemistry of near-EPR seamounts: Importance of source vs. process and the origin of enriched mantle component: Earth and Planetary Science Letters, v. 199, p. 327–345, doi: 10.1016/S0012-821X(02)00591-5.

Norton, I.O., 1995, Plate motions in the North Pacific; The 43 Ma nonevent: Tectonics, v. 14, p. 1080–1094, doi: 10.1029/95TC01256.

Norton, I.O., 2007 (this volume), Speculations on Cretaceous tectonic history of the northwest Pacific and a tectonic origin for the Hawaii hotspot, *in* Foulger, G.R., and Jurdy, D.M., eds., Plates, plumes, and planetary processes: Boulder, Colorado, Geological Society of America Special Paper 430, doi: 10.1130/2007.2430(22).

Nunns, A.G., 1983, Plate tectonic evolution of the Greenland-Scotland ridge and surrounding regions, in Bott, M.H.P., et al., eds., Structure and development of the Greenland-Scotland ridge: New York and London, Plenum Press, p. 1–30.

O'Connor, J.M., Stoffers, P., van den Bogaard, P., and McWilliams, M., 1999, First seamount age evidence for significantly slower African plate motion since 19 to 30 Ma: Earth and Planetary Science Letters, v. 171, p. 575–589, doi: 10.1016/S0012-821X(99)00183-1.

O'Hara, M.J., 1975, Is there an Icelandic mantle plume?: Nature, v. 253, p. 708–710, doi: 10.1038/253708a0.

Oreiro, S.G., Cupertino, J.A., Szatmari, P., and Thomaz Filho, A., 2006, Influence of pre-salt alignments in the Cabo Frio High and its surroundings, Santos and Campos basins, SE Brazil: An example of non-plume–related magmatism: http://mantleplumes.org/Brazil.html.

Parman, S.W., Kurz, M.D., Hart, S.R., and Grove, T.L., 2005, Helium solubility in olivine and implications for high $^3He/^4He$ in ocean island basalts: Nature, v. 437, p. 1140–1143, doi: 10.1038/nature04215.

Phipps Morgan, J., Morgan, W.J., and Price, E., 1995, Hotspot melting generates both hotspot volcanism and a hotspot swell?: Journal of Geophysical Research, v. 100, p. 8045–8062, doi: 10.1029/94JB02887.

Pilet, S., Hernandez, J., and Villemant, B., 2002, Evidence for high silicic melt circulation and metasomatic events in the mantle beneath alkaline provinces: The Na-Fe-augitic green-core pyroxenes in the Tertiary alkali basalts of the Cantal massif (French Massif Central): Mineralogy and Petrology, v. 76, p. 39–62, doi: 10.1007/s007100200031.

Pilet, S., Hernandez, J., Bussy, F., and Sylvester, P.J., 2004, Short-term metasomatic control of Nb/Th ratios in the mantle sources of intra-plate basalts: Geology, v. 32, p. 113–116, doi: 10.1130/G19953.1.

Pilot, J., Werner, C.D., Haubrich, F., and Baumann, N., 1998, Palaeozoic and Proterozoic zircons from the mid-Atlantic ridge: Nature, v. 393, p. 676–679, doi: 10.1038/31452.

Presnall, D.C., 1995, Phase diagrams of Earth-forming minerals, in mineral physics and crystallography: A handbook of physical constants, *in* Ahrens, T.J., ed., AGU Reference Shelf, Volume 2: Washington, D.C., American Geophysical Union, p. 248–268.

Presnall, D.C., and Gudfinnsson, G.H., 2005, Carbonatitic melts in the oceanic low-velocity zone and deep upper mantle, *in* Foulger, G.R., et al., eds., Plates, plumes, and paradigms: Boulder, Colorado, Geological Society of America Special Paper 388, p. 207–216.

Putirka, K.D., 2005, Mantle potential temperatures at Hawaii, Iceland, and the mid-ocean ridge system, as inferred from olivine phenocrysts: Evidence for thermally driven mantle plumes: Geochemistry, Geophysics, Geosystems, v. 6, doi: 10.1029/2005GC000915.

Rainbird, R.H., and Ernst, R.E., 2001, The sedimentary record of mantle plume uplift, *in* Ernst, R.E., and Buchan, K.L., eds., Mantle plumes: Their identi-

fication through time: Boulder, Colorado, Geological Society of America Special Paper 352, p. 227–245.

Ritsema, J., 2005, Global seismic maps, *in* Foulger, G.R., et al., eds., Plates, plumes, and paradigms: Boulder, Colorado, Geological Society of America Special Paper 388, p. 11–18.

Ritsema, J., and Allen, R.M., 2003, The elusive mantle plume: Earth and Planetary Science Letters, v. 207, p. 1–12, doi: 10.1016/S0012-821X(02)01093-2.

Ritsema, J., van Heijst, H.J., and Woodhouse, J.H., 1999, Complex shear wave velocity structure imaged beneath Africa and Iceland: Science, v. 286, p. 1925–1928, doi: 10.1126/science.286.5446.1925.

Ritter, J.R.R., Achauer, U., Christensen, U.R., and the Eifel Plume Team, 2000, The teleseismic tomography experiment in the Eifel region, central Europe: Design and first results: Seismological Research Letters, v. 71, p. 437–443.

Roberge, J., Wallace, P.J., White, R.V., and Coffin, M.F., 2005, Anomalous uplift and subsidence of the Ontong Java plateau inferred from CO_2 contents of submarine basaltic glasses: Geology, v. 33, p. 501–504, doi: 10.1130/G21142.1.

Robinson, J.E., and Eakins, B.W., 2006, Calculated volumes of individual shield volcanoes at the young end of the Hawaiian ridge: Journal of Volcanology and Geothermal Research, v. 151, p. 309–317, doi: 10.1016/j.jvolgeores.2005.07.033.

Sager, W.W., 2005, What built Shatsky rise, a mantle plume or ridge tectonics?, *in* Foulger, G.R., et al. eds., Plates, plumes, and paradigms: Boulder, Colorado, Geological Society of America Special Paper 388, p. 721–734.

Sager, W.W., Jinho, K., Klaus, A., Nakanishi, M., and Khankishiyeva, L.M., 1999, Bathymetry of Shatsky rise, northwest Pacific Ocean; Implications for ocean plateau development at a triple junction: Journal of Geophysical Research, v. 104, p. 7557–7576, doi: 10.1029/1998JB900009.

Sandwell, D.T., and Fialko, Y., 2004, Warping and cracking of the Pacific plate by thermal contraction: Journal of Geophysical Research, v. 109, p. doi: 10.1029/2004JB003091.

Sandwell, D.T., Winterer, E.L., Mammerickx, J., Duncan, R.A., Lynch, M.A., Levitt, D.A., and Johnson, C.L., 1995, Evidence for diffuse extension of the Pacific plate from Pukapuka ridges and cross-grain gravity lineations: Journal of Geophysical Research, v. 100, p. 15,087–15,089, doi: 10.1029/95JB00156.

Sandwell, D.T., Anderson, D.L., and Wessel, P., 2005, Global tectonic maps, *in* Foulger, G.R., et al., eds., Plates, plumes, and paradigms: Boulder, Colorado, Geological Society of America Special Paper 388, p. 1–10.

Schilling, J.-G., 1973, Iceland mantle plume: Nature, v. 246, p. 141–143, doi: 10.1038/246141a0.

Schilling, J.-G., Sigurdsson, H., Davis, A.N., and Hey, R.N., 1985a, Easter microplate evolution: Nature, v. 317, p. 325–331, doi: 10.1038/317325a0.

Schilling, J.-G., Thompson, G., Kingsley, R., and Humphris, S., 1985b, Hotspot-migrating ridge interaction in the south Atlantic: Nature, v. 313, p. 187–191, doi: 10.1038/313187a0.

Sensarma, S., Hoernes, S., and Mukhopadhyay, D., 2004, Relative contributions of crust and mantle to the origin of Bijli Rhyolite in a palaeoproterozoic bimodal volcanics sequence (Dongargarh group), central India: Proceedings of the Indian Academy of Sciences (Earth and Planetary Science), v. 113, p. 619–648.

Shapiro, M.N., Soloviev, A.V., and Ledneva, G.V., Is there any relation between the Hawaiian-Emperor seamount chain bend at 43 Ma and the evolution of the Kamchatka continental margin?: http://www.mantleplumes.org/Kamchatka2.html.

Sharp, W.D., and Clague, D.A., 2006, 50-Ma initiation of Hawaiian-Emperor bend records major change in Pacific plate motion: Science, v. 313, p. 1281–1284, doi: 10.1126/science.1128489.

Shaw, H.R., 1973, Mantle convection and volcanic periodicity in the Pacific; Evidence from Hawaii: Geological Society of America Bulletin, v. 84, p. 1505–1526, doi: 10.1130/0016-7606(1973)84<1505:MCAVPI>2.0.CO;2.

Shaw, H.R., and Jackson, E.D., 1973, Linear island chains in the Pacific: Result of thermal plumes or gravitational anchors?: Journal of Geophysical Research, v. 78, p. 8634–8652.

Sheth, H.C., 2005, From Deccan to Réunion: No trace of a mantle plume, *in* Foulger, G.R., et al., eds., Plates, plumes, and paradigms: Boulder, Colorado, Geological Society of America Special Paper 388, p. 477–502.

Sheth, H.C., 2006a, Plume-related regional pre-volcanic uplift in the Deccan Traps: Absence of evidence, evidence of absence: http://www.mantleplumes.org/DeccanUplift.html.

Sheth, H.C., 2006b, "Large Igneous Provinces (LIPS)": Definition, recommended terminology, and a hierarchical classification: http://www.mantleplumes.org/LIPclass.html.

Sheth, H.C., 2007a, "Large igneous provinces (LIPs)": Definition, recommended terminology, and a hierarchical classification: Earth Science Reviews, in press.

Sheth, H.C., 2007b (this volume), Plume-related regional prevolcanic uplift in the Deccan Traps: Absence of evidence, evidence of absence, *in* Foulger, G.R., and Jurdy, D.M., eds., Plates, plumes, and planetary processes: Boulder, Colorado, Geological Society of America Special Paper 430, doi: 10.1130/2007.2430(36).

Silver, P.G., Behn, M.D., Kelley, K., Schmitz, M., and Savage, B., 2006, Understanding cratonic flood basalts: Earth and Planetary Science Letters, v. 245, p. 190–201, doi: 10.1016/j.epsl.2006.01.050.

Simkin, T., and Siebert, L., 1994, Volcanoes of the world: Tucson, Arizona, Geoscience Press, 349 p.

Skogseid, J., Planke, S., Faleide, J.I., Pedersen, T., Eldholm, O., and Neverdal, F., 2000, NE Atlantic continental rifting and volcanic margin formation, *in* Nottvedt, A., ed., Dynamics of the Norwegian Margin: London, Geological Society of London Special Publication 167, p. 295–326.

Smith, A.D., 2003, A re-appraisal of stress field and convective roll models for the origin and distribution of Cretaceous to Recent intraplate volcanism in the Pacific basin: International Geology Review, v. 45, p. 287–302.

Smith, A.G., 2007 (this volume), A plate model for Jurassic to Recent intraplate volcanism in the Pacific Ocean basin, *in* Foulger, G.R., and Jurdy, D.M., eds., Plates, plumes, and planetary processes: Boulder, Colorado, Geological Society of America Special Paper 430, doi: 10.1130/2007.2430(23).

Smith, W.H.F., and Sandwell, D.T., 1997, Global sea floor topography from satellite altimetry and ship depth soundings: Science, v. 277, p. 1956–1962, doi: 10.1126/science.277.5334.1956.

Sobolev, A.V., Hofmann, A.W., Sobolev, S.V., and Kikogosian, I.K., 2005, An olivine-free mantle source of Hawaiian shield basalts: Nature, v. 434, p. 590–597, doi: 10.1038/nature03411.

Steinberger, B., Sutherland, R., and O'Connell, R.J., 2004, Prediction of Emperor-Hawaii seamount locations from a revised model of global plate motion and mantle flow: Nature, v. 430, p. 167–173, doi: 10.1038/nature02660.

Stuart, F.M., Lass-Evans, S., Fitton, J.G., and Ellam, R.M., 2003, Extreme ^{3}He/^{4}He in picritic basalts from Baffin Island: The role of a mixed reservoir in mantle plumes: Nature, v. 424, p. 57–59, doi: 10.1038/nature01711.

Stuart, W.D., Foulger, G.R., and Barall, M., 2007 (this volume), Propagation of the Hawaiian-Emperor volcano chain by Pacific plate cooling stress, *in* Foulger, G.R., and Jurdy, D.M., eds., Plates, plumes, and planetary processes: Boulder, Colorado, Geological Society of America Special Paper 430, doi: 10.1130/2007.2430(24).

Tanton, L.T.E., and Hager, B.H., 2000, Melt intrusion as a trigger for lithospheric foundering and the eruption of the Siberian flood basalts: Geophysical Research Letters, v. 27, p. 3937–3940, doi: 10.1029/2000GL011751.

Tarduno, J.A., and Cottrell, R.D., 1997, Paleomagnetic evidence for motion of the Hawaiian hotspot during formation of the Emperor seamounts: Earth and Planetary Science Letters, v. 153, p. 171–180, doi: 10.1016/S0012-821X(97)00169-6.

Taylor, B., 2006, The single largest oceanic plateau: Ontong Java–Manihiki–Hikurangi: Earth and Planetary Science Letters, v. 241, p. 372–380, doi: 10.1016/j.epsl.2005.11.049.

Trampert, J., Deschamps, F., Resovsky, J., and Yuen, D., 2004, Probabilistic tomography maps chemical heterogeneities throughout the lower mantle: Science, v. 306, p. 853–856, doi: 10.1126/science.1101996.

Turcotte, D.L., 1974, Membrane tectonics: Geophysical Journal of the Royal Astronomical Society, v. 36, p. 33–42.

Turcotte, D.L., and Oxburgh, E.R., 1973, Mid-plate tectonics: Nature, v. 244, p. 337–339, doi: 10.1038/244337a0.

van der Hilst, R.D., and de Hoop, M.V., 2005, Banana-doughnut kernels and mantle tomography: Geophysical Journal International, v. 163, p. 956–961.

van Wijk, J.W., Huismans, R.S., Ter Voorde, M., and Cloetingh, S.A.P.L., 2001, Melt generation at volcanic continental margins: No need for a mantle plume?: Geophysical Research Letters, v. 28, p. 3995–3998, doi: 10.1029/2000GL012848.

van Wijk, J.W., van der Meer, R., and Cloetingh, S.A.P.L., 2004, Crustal thickening in an extensional regime: Application to the mid-Norwegian Vøring margin: Tectonophysics, v. 387, p. 217–228, doi: 10.1016/j.tecto.2004.07.049.

Vogt, P.R., and Jung, W.-Y., 2005, Paired (conjugate) basement ridges: Spreading axis migration across mantle heterogeneities?, *in* Foulger, G.R., et al., eds., Plates, plumes, and paradigms: Boulder, Colorado, Geological Society of America Special Paper 388, p. 555–580.

Vogt, P.R., and Jung, W.-Y., 2007 (this volume), Origin of the Bermuda volcanoes and Bermuda Rise: History, observations, models, and puzzles, *in* Foulger, G.R., and Jurdy, D.M., eds., Plates, plumes, and planetary processes: Boulder, Colorado, Geological Society of America Special Paper 430, doi: 10.1130/2007.2430(27).

von Herzen, R.P., Cordery, M.J., Detrick, R.S., and Fang, C., 1989, Heat-flow and the thermal origin of hot spot swells—The Hawaiian swell revisited: Journal of Geophysical Research, v. 94, p. 13,783–13,799.

Watts, A.B., Weissel, J.K., Duncan, R.A., and Larson, R.L., 1988, Origin of the Louisville ridge and its relationship to the Eltanin fracture zone system: Journal of Geophysical Research, v. 93, p. 3051–3077.

Wilson, J.T., 1963, A possible origin of the Hawaiian Islands: Canadian Journal of Physics, v. 41, p. 863–870.

Wolfe, C.J., Solomon, S.C., Silver, P.G., VanDecar, J.C., and Russo, R.M., 2002, Inversion of body-wave delay times for mantle structure beneath the Hawaiian islands: Results from the PELENET experiment: Earth and Planetary Science Letters, v. 198, p. 129–145, doi: 10.1016/S0012-821X(02)00493-4.

Xu, Y.-G., and He, B., 2007 (this volume), Thick, high-velocity crust in the Emeishan large igneous province, southwestern China: Evidence for crustal growth by magmatic underplating or intraplating, *in* Foulger, G.R., and Jurdy, D.M., eds., Plates, plumes, and planetary processes: Boulder, Colorado, Geological Society of America Special Paper 430, doi: 10.1130/2007.2430(39).

Yaxley, G.M., and Green, D.H., 1998, Reactions between eclogite and peridotite: Mantle refertilisation by subduction of oceanic crust: Schweizerische Mineralogische und Petrographische Mitteilungen, v. 78, p. 243–255.

MANUSCRIPT ACCEPTED BY THE SOCIETY 31 JANUARY 2007

DISCUSSION

26 December 2006, Rex H. Pilger

Foulger (this volume) writes that the Easter melting anomaly is located close to a ridge triple junction and is associated with a microplate. Pilger and Handschumacher (1981) showed that the Easter "hotspot" could have produced both the Tuamotu and Nazca ridges as well as the Sala y Gomez trace, especially if the Hawaiian-Emperor bend were older than 43 Ma, significantly predating the triple junction and the Easter plate. The oblique trends of the Tuamotu and Nazca ridges to magnetic chrons

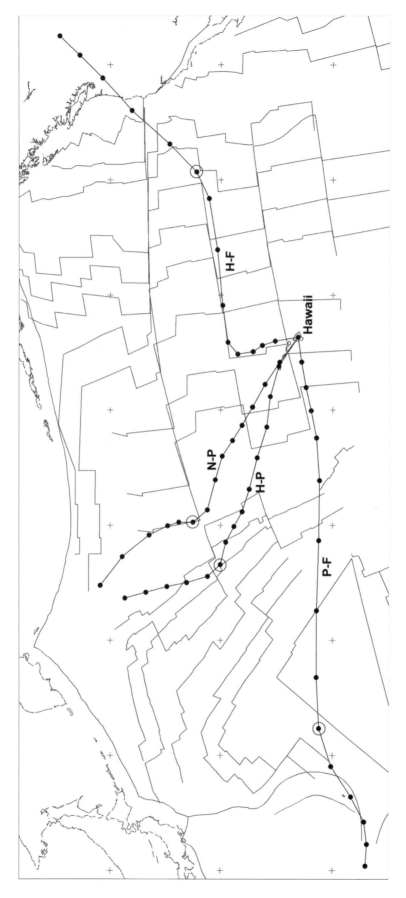

Figure D-1. Calculated loci at 5-m.y. intervals, from Hawaii: H-F—Hawaii to Farallon; H-P—Hawaiian "hotspot" relative to Pacific plate; N-P—North American to Pacific plate; P-F—Pacific to Farallon plate. Circled nodes are at 45 Ma. Gray crosses are latitude-longitude graticules at fifteen-degree intervals; the cross immediately southeast of Hawaii is at 15°S, 150°W. Gray line segments are magnetic isochrons from Müller et al. (1977).

implies that the first stress mechanism proposed by Beutel and Anderson (this volume) is not applicable, but the second, melting anomaly, mechanism (also part of the plate model of Foulger) could still apply.

Foulger also states, "The direction of propagation [of the Hawaiian-Emperor chain] changed by ~60° at ca. 47 Ma, at the time of the 'bend.' . . . No corresponding change in the direction of Pacific plate motion occurred at this time." There is indeed no evidence of a change in Pacific plate motion relative to adjacent plates at 43 Ma. However, there are ca. 47-Ma kinks in motion loci of all adjacent continental plates relative to Pacific plate (Norton, 2000, Fig. 10). North and South American–Pacific loci show comparable deflections to the Hawaiian-Emperor bend.

Loci of Pacific and Farallon (-Juan de Fuca) plate motion and North American and Hawaiian "hotspot" motion are shown in Figure D-1, interpolated at 5-m.y. intervals, using parameters from Pilger (2007). The circled nodes indicate the 45-Ma reconstructions and bends (finer interpolation would show bends from 45 to 48 Ma). A possible explanation for these correspondences is presented in Pilger (2007).

The oft-stated lack of correspondence of the bend with plate motions is apparently erroneous, but this does not invalidate the plate hypothesis for the origin of anomalous volcanic traces.

8 February 2007, Gillian R. Foulger and William W. Sager

In his comment of 26 December, Pilger makes two points. First, he writes that a fixed Easter "hotspot" could have produced the Tuamotu, Nazca, and Sala y Gomez ridges. In a region containing widespread volcanism it is to be expected that some can be found that conform with fixed hotspot models. However, the Easter and Sala y Gomez chains emanate from the southern edge of the Easter microplate, which is unlikely to be coincidence.

Pilger's second comment is to take issue with a statement in Foulger (this volume) that there was no change in plate motion corresponding to the Hawaiian-Emperor bend. Models of plate motion based on the Hawaiian-Emperor and Louisville chains only, and assuming the fixed hotspot hypothesis predict a large change. However, other reconstructions do not, e.g., that of Raymond et al. (2000), which is based on hotspot propagation models for the Indian and Atlantic plates, reconstructing them to the Pacific plate using relative spreading models from magnetic lineations.

Pilger argues for a significant shift in plate motion based on kinks in his plate motion trends for the South and North American plates relative to the Pacific. However, the contemporaneous changes in his Figure D-1 for plates adjacent to the Pacific are small or nonexistent. Furthermore, the paths of the North and South American plates relative to the Pacific plate are highly uncertain, as a result of errors in the long chains of relative plate motions that must be used to make such reconstructions and uncertainty concerning the long-term rigidity of Antarctica as a single plate. These uncertainties are large and are not shown in Figure D-1.

The fact that the bend in the Hawaiian-Emperor chain is accompanied by very small changes, if any, in the major fracture zones of the Pacific is a first-order observation that cannot be readily dismissed. The 60° change in trend in the Hawaiian-Emperor chain, now known to have occurred at ca. 50 Ma (Sharp and Clague, 2006), can correspond to only a small change in plate motion, if any. Atwater (1989) attributes a small change in plate motion at about this time to the breakup of the Farallon plate into north (Vancouver plate) and south parts. The small change in direction she postulates is accompanied by only very small changes in the fracture zones. If the plate changed direction, new plate rotation poles would be set up, and large changes in the transform faults would be expected. The orientations of fracture zones are, after all, prima facie testimony of the direction of plate motion, in contrast to other data that usually require careful processing to extract information on plate motion direction.

The time of the bend correlates with a radical change of behavior of the melting anomaly, when it ceased rapid southerly migration with respect to the geomagnetic pole (Tarduno et al., 2003; Sager, this volume). The evidence is consistent with most of the north-south motion implied by the Emperor seamounts being due to migration of the locus of melt extraction, and not by plate motion, whereas the reverse is the case for the Hawaiian chain. In the deep-rooted, fixed hotspot hypothesis, the occurrence of this change in behavior simultaneously with a major plate motion change would have to be coincidence. We find explanation of these observations as coincidence unlikely. The conclusion is inevitable: the bend in the Hawaiian-Emperor volcano chain is probably largely due to a change in migration behavior of the melt extraction anomaly, and not to a change in plate motion direction.

REFERENCES CITED

Atwater, T., 1989. Plate tectonic history of the northeast Pacific and western North America, *in* Winterer, E.L., et al., eds., The eastern Pacific Ocean and Hawaii: Boulder, Colorado, Geological Society of America Geology of North America N, p. 21–72.

Beutel, E., and Anderson, D.L., 2007 (this volume), Ridge-crossing seamount chains: A nonthermal approach, *in* Foulger, G.R., and Jurdy, D.M., eds., Plates, plumes, and planetary processes: Boulder, Colorado, Geological Society of America Special Paper 430, doi: 11.1130/2007.2430(19).

Foulger, G.R., 2007 (this volume), The plate model for the genesis of melting anomalies, *in* Foulger, G.R., and Jurdy, D.M., eds., Plates, plumes, and planetary processes: Boulder, Colorado, Geological Society of America Special Paper 430, doi: 11.1130/2007.2430(01).

Müller, R.D., Roest, W.R., Royer, J.-Y., Gahagan, L.M., and Sclater, J.G., 1997, Digital isochrons of the world's ocean floor: Journal of Geophysical Research, v. 102, p. 3211–3214. Data: http://www-sdt.univ-brest.fr/~jyroyer/Agegrid/utig_report.html.

Norton, I.O., 2000, Global hotspot reference frames and plate motion, *in* Richards, M.A., et al., eds., The history and dynamics of global plate motions: Washington, D.C., American Geophysical Union Geophysical Monograph 121, p. 339–357.

Pilger, R.H., 2007, The Bend: Origin and significance: Bulletin of the Geological Society of America, v. 119, p. 302–313.

Pilger, R.H., and Handschumacher, D.W., 1981, The fixed hotspot hypothesis and origin of the Easter–Sala y Gomez trace: Geological Society of America Bulletin, v. 92, p. 437–446.

Raymond, C.A., Stock, J.M., and Cande, S.C., 2000, Fast Paleogene motion of the Pacific hotspots from revised global plate circuit constraints, *in* Richards, M.A., et al., eds., History and dynamics of plate motions: Washington, D.C., American Geophysical Union Geophysical Monograph 121, p. 359–375.

Sager, W.W., 2007 (this volume), Divergence between paleomagnetic and hotspot-model–predicted polar wander for the Pacific plate with implications for hotspot fixity, *in* Foulger, G.R., and Jurdy, D.M., eds., Plates, plumes, and planetary processes: Boulder, Colorado, Geological Society of America Special Paper 430, doi: 10.1130/2007.2430(17).

Sharp, W.D., and Clague, D.A., 2006, 50-Ma initiation of Hawaiian-Emperor bend records major change in Pacific plate motion: Science, v. 313, p. 1281–1284.

Tarduno, J.A., Duncan, R.A., Scholl, D.W., Cottrell, R.D., Steinberger, B., Thordarson, T., Kerr, B.C., Neal, C.R., Frey, F.A., Torii, M., and Claire Carvallo, C., 2003, The Emperor seamounts: Southward motion of the Hawaiian hotspot plume in Earth's mantle: Science, v. 301, p. 1064–1069.

The Geological Society of America
Special Paper 430
2007

Origins of the plume hypothesis and some of its implications

Norman H. Sleep*

Department of Geophysics, Stanford University, Stanford, California 94305, USA

ABSTRACT

Hotspots are regions of voluminous volcanism, such as Hawaii and Iceland. However, direct tests by mid-mantle tomography and petrology of lavas do not conclusively resolve that mantle plumes and mantle high temperature cause these features. Indirect and model-dependent methods thus provide constraints on the properties of plumes, if they exist. Classical estimates of the excess (above mid-ocean ridge basalt values) potential temperature of the upwelling mantle (before melting commences) are 200–300 K. The estimate for Iceland utilizes the thickness of the oceanic crust. Estimates for Hawaii depend on the melting depth beneath old lithosphere. Flux estimates for Iceland depend on the kinematics of the ridge and those of Hawaii or the kinematics and dynamics of the Hawaiian swell. These techniques let one compute the global volume and heat flux of plumes. The flux is significant—approximately one-third of the mantle has cycled through plumes, extrapolating the current rate over geological time. However, obvious vertical tectonics may not be evident for many hotspots. For example, the relatively weak Icelandic hotspot (one-sixth of the volume flux of Hawaii) would produce only modest volcanism and swell uplift if it impinged on the fast-moving Pacific plate or the fast-spreading East Pacific Rise.

Keywords: plumes, hotspots, Hawaii, Iceland, swells

INTRODUCTION

The Hawaii-Emperor seamount chain is a prominent volcanic feature in the middle of the Pacific plate. Iceland is a region of voluminous volcanism on the Mid-Atlantic Ridge. I spent considerable time early in my career attempting to find alternative causes to plumes for these features. By 1987, I had found my alternative hypotheses contrived and wanting. I acknowledge that there are still multiple viable alternatives to thermal plumes (e.g., Foulger et al., 2005a,b), and that data sets that relate to one hypothesis may not bear on another. I do not attribute all off-axis volcanism to plumes. Lithospheric cracks may well tap low-volume source regions at normal mantle temperatures. Cracks are also necessary for plume magma to ascend in midplate regions.

Plumes meet the minimum requirement for a scientific hypothesis in that they putatively explain a class of observations on hotspots. They lead to testable predictions. Sleep (2006) reviewed deep aspects of the plume hypothesis starting with convection at the base of the lithosphere. I concentrate on geologically observable surface features in this article. I first discuss a well-posed form of the plume hypothesis to obtain testable implications. Plumes, as I envision, are mainly thermal features from convection heated from below. Hot material rises buoyantly through the mantle through cylindrical low-viscosity conduits.

*E-mail: norm@pangea.stanford.edu.

Sleep, N.H., 2007, Origins of the plume hypothesis and some of its implications, *in* Foulger, G.R., and Jurdy, D.M., eds., Plates, plumes, and planetary processes: Geological Society of America Special Paper 430, p. 29–45, doi: 10.1130/2007.2430(02). For permission to copy, contact editing@geosociety.org. ©2007 The Geological Society of America. All rights reserved.

The deep part of the issue would be settled if one could re-solve the structure at mid-mantle depths. Seismic data do provide evidence that such conduits exist in the expected places (Montelli et al., 2004). However, the features are near the limit of resolution (e.g., van der Hilst and de Hoop, 2005). These data are far from resolving whether the plumes are thermal or chemical features.

Petrological studies of lavas bear on the shallow part of the hypothesis. They have the potential to determine whether the mantle beneath hotspots is in fact hotter than the mantle that supplies most mid-oceanic ridge axes and whether the source regions are chemically different. Green and Falloon (2005), for example, state that there is no systematic difference in source temperature between mid-ocean ridge basalt (MORB) and hotspot lavas. Putirka (2005) contends that hotspot sources are in fact hotter than MORB source. I make no attempt in this article to derive source temperatures from magma compositions.

In the absence of firm seismological and petrological evidence, earth scientists use data sets that indirectly constrain the properties of plumes, if they exist. In this article, I apply heat and mass balance constraints to quantify the implications of the plume hypothesis. I center my arguments on the prominent hotspots, Iceland and Hawaii. I use a reliable data set, topography, as much as possible. Conversely, I point out lines of evidence that currently do not have the resolution to provide strong constraints.

THERMAL PLUME HYPOTHESIS

I discuss plumes as thermal features to put the hypothesis in context. I start with melting temperature and melt volume as the hypothesis relates to voluminous volcanism. I then discuss the flow of plume material along the base of the lithosphere with regard to Iceland and Hawaii. I discuss starting plume heads in the last part of this section.

Adiabatic Melting and Mantle Source Temperature

Plumes are supposedly hotter than the "ordinary" mantle that ascends at mid-ocean ridges. The temperature difference is considered to be greater than the ambient temperature variations in the rest of the asthenosphere. That is, the hypothesis implies one can, to the first order, partition temperature variation into "normal" mantle near the MORB adiabat, a plume adiabat, and sinking slabs.

I present a linearized version of pressure-release melting to obtain semi-schematic diagrams and to consistently define terms. The fraction of melting

$$T_p \equiv T - \gamma Z, \qquad (1)$$

where γ is the solid adiabatic gradient. I use a linear expression for the fraction of melting,

$$f = \frac{T_p - T_0 - \frac{\partial T_m}{\partial Z} Z}{T_1}, \qquad (2)$$

where T_0 is the solidus at the surface, T_1 the temperature difference between the solidus and the liquidus, and $\partial T_m/\partial Z$ is the gradient of melting temperature with depth.

Ascending mantle material is initially solid with a potential temperature T_s. It starts to melt at a depth

$$Z_s \equiv (T_s - T_0) \left(\frac{\partial T_m}{\partial Z} \right)^{-1}. \qquad (3)$$

A balance between specific heat and latent heat determines the temperature within an abiabatic column above this depth:

$$\rho C(T_s - T_p) = f \rho L, \qquad (4)$$

where ρ is density, C is specific heat per unit mass, and L is latent heat per unit mass. One solves (2) and (4) for the temperature and the melt fraction (Fig. 2):

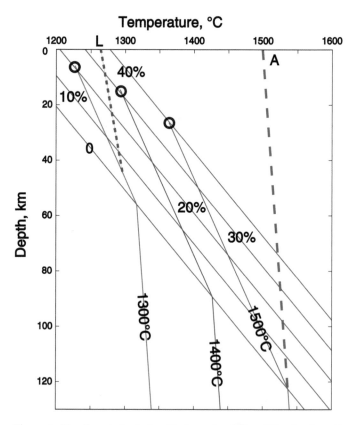

Figure 1. Simple petrological grid shows isopleths of the fraction of partial melting and the paths of abiabatic ascent in temperature-depth space. The paths at a ridge axis terminate at the base of the crust (circles). The potential temperature of the source before melting is the solid adiabat extrapolated to the surface (dashed line A). Magma ascending along the liquid adiabat (path L) becomes superheated.

$$f = \frac{T_s - T_0 - \dfrac{\partial T_m}{\partial Z}}{T_1 + \dfrac{L}{C}} . \qquad (5)$$

The ratio of latent heat to specific heat L/C has dimensions of temperature.

Finally, the total thickness of segregated melt from the column is

$$Z_c = \int_{Z_{top}}^{Z_f} \frac{f}{1-f} \, df, \qquad (6)$$

where Z_{top} is the depth to the top of the column and the factor $1/(1-f)$ accounts (to the first order) for the geometrical effect that the melt that segregates and ascends is replaced by more ascending material (Fig. 2). At sufficiently fast-spreading ridge axes, the base of the oceanic crust defines the top of the column. That is, $Z_c = Z_{top}$. Elsewhere, the thermal (or perhaps chemical) base of the lithosphere defines the top.

Figures 1 and 2 use parameters calibrated to give 6-km-thick crust at "normal" ridge axes where the source potential temperature is 1300 °C. The surface solidus is 1132 °C. The melting point gradient is 3 K km⁻¹. The temperature L/C and the temperature T_1 are both 365.5 °C. The solid adiabatic gradient is 0.3 K km⁻¹. Melting begins at 56 km depth at normal ridge axes.

I obtain example classical estimates for the source temperature differences for Iceland and Hawaii from the petrological grids for Iceland and Hawaii. They do not involve detailed petrology. They effectively give the average temperature anomaly of the region in the core of the plume that actually melts. (Plank et al., 1995, discuss the semantics and systematics of the average fraction of melting in a column.) I use a rounded estimate of 26 km for Iceland. (This value is 20 km thicker than normal crust. It is necessary to average over the edifice because plume material melts within the rising plume and then spreads laterally.) This excess thickness implies that the Icelandic source region is 200 K hotter than normal mantle. Voluminous melting beneath Hawaii begins at a rounded depth of 116 km (60 km greater than normal mantle), which implies an excess temperature of 180 K. Analysis using more sophisticated petrological grids yields somewhat higher excess temperatures. For example, Watson and McKenzie (1991) obtain ~280 K from numerical models of melting beneath Hawaii. Sobolev et al. (2005) consider the detailed sequence of melting beneath Hawaii. Recycled oceanic crust melts first and reacts with surrounding

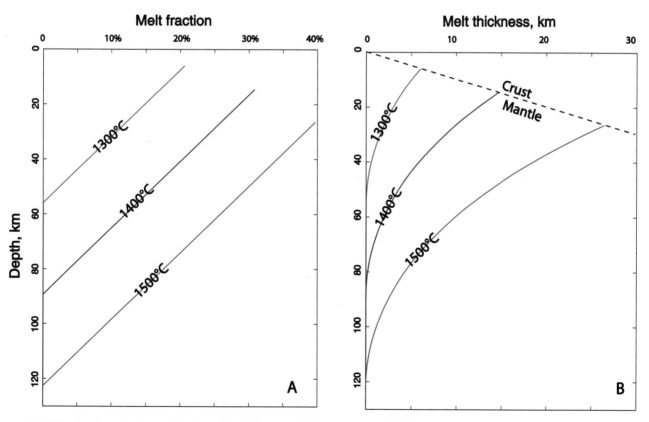

Figure 2. (A) The fraction of melt as a function of depth for potential temperatures of 1300, 1400, and 1500 °C. The lines end at the base of the oceanic crust (dashed line in panel B). (B) The total thickness of melt produced during adiabatic ascent. The curves end where the thickness equals the depth to the base of the oceanic crust (dashed line).

peridotite to form pyroxenite. The pyroxenite then melts to form voluminous Ni-rich magmas. They obtain an excess temperature of 250–300 K.

I reiterate the assumptions for the petrological grid to introduce alternatives to thermal plumes. (1) The ascending material is geometrically that needed to make the oceanic lithosphere. The velocity is upward and no material recirculates. (2) Much of the melt segregates and ascends. (3) The material is homogeneous and the composition is the same beneath hotspots and normal ridges. The validity of these assumptions and alternatives are potentially detectable from petrological studies.

Edge-driven convection could conceivably recirculate the material beneath the ridge axis, leading to much more melting than implied by the model (e.g., King and Anderson, 1998). For example, 26-km-thick ocean crust beneath Iceland could form a hot adiabatic column as assumed in my model. It could conceivably form by recirculating a column of normal-temperature mantle (that produces 6 km of crust) 26/6 = 4.3 times. The composition of the melt in the two cases should differ (e.g., Plank et al., 1995).

The second assumption is valid for mass-balance calculations as long as most of the melt segregates and ascends. One cannot appeal to inefficient melt segregation to explain the difference between Iceland and normal ridges. Using the example thickness of 6 and 26 km, 20 km of melt would have to be retained in the mantle beneath normal ridges. If it were distributed over the upper 100 km of the lithosphere, there would be 20% partial melt. This feature would be obvious from seismology.

The third assumption is obviously incorrect in detail: radiogenic isotopes indicate that the mantle is heterogeneous and has been so for billions of years. Still one can construct a petrological grid for each heterogeneous domain. The heat balance in (4) still applies in an integral sense over the domains. Small heterogeneities melt over different depth intervals as they ascend. Latent heat cools the first domains that melt, and heat flows in from the surrounding unmelted regions (Sleep, 1984). This mechanism enhances the fraction of melting where it is already occurring and suppresses melting elsewhere.

Major element (including water and maybe CO_2) variations from normal source composition are needed to produce voluminous volcanism at normal mantle temperatures. I discuss three useful end-members. First as already noted, a small, easily melted domain does not affect the heat balance. The surrounding material maintains its temperature at the solid adiabat. This effect explains low-volume hydrous magmas but not voluminous volcanism.

Second, I consider that the ascending mantle is a eutectic composed of the minimum melting material on the grid. The parameter T_1 is then 0 rather than approximately L/C, which is ~1, and T_0 is unchanged. This modification has the effect of doubling the melt production in (5). That is, 26-km-thick crust cannot be obtained from normal temperature material with the properties of the near-solidus melts at normal ridges.

Third, the solidus of the material may be less than the ridge solidus. In terms of the grid, one may reduce the surface solidus T_0 by ~200 K rather than increasing the source potential temperature T_s by that amount and leave T_1 unchanged.

Finally to appraise the thermal plume hypothesis, one would like to determine the source potential temperature T_s of the material before significant melting begins. However, complicated magma processes at depth make this estimate far from straightforward. MORB is the partly pooled series of melts from the adiabatic column. Melt ascending on the liquid adiabat is superheated and may assimilate mantle wall rock. The melts pool within the axial magma chamber and fractionally crystallize. Midplate source regions may cool by conduction into the overlying lithosphere and melts may have complex histories within deep and high-level magma chambers.

Careful sampling does show some enriched MORB samples away from hotspots on the slow spreading Mid-Atlantic Ridge, as expected if local heterogeneities are ubiquitous (Donnelly et al., 2004). Many low-volume magmas do not reach the surface. They freeze within the mantle, producing new heterogeneities (Donnelly et al., 2004). These domains are sources of enriched MORB when they remelt beneath ridge axes. Radiogenic isotopes indicate they have ~300 m.y. survival times. Off-axis, these domains are likely sources for low-volume magmas that are not associated with plumes.

Iceland

I apply the petrological grid formalism to obtain the flux of the on-ridge hotspot Iceland and then discuss some geometrical complications. The plume supplies material that melts and eventually forms lithosphere along a length L of the axis (Fig. 3). That is, the volume flux is

$$V = U_R LD, \qquad (7)$$

where U_R is the full spreading rate of the ridge, L is the length of ridge axis supplied by the plume, and D the thickness of lithosphere formed from plume material. The thickness D is ~100 km, the depth where voluminous melting begins. Petrology and the thermal thickness of the lithosphere away from the axis provide further constraints (Schilling, 1991).

About a 1000-km length of ridge is involved (Fig. 3). The aseismic Iceland-Greenland and Iceland-Faeroe ridges extend to the adjacent continental margins. The full spreading rate is ~20 mm yr^{-1}. These values give a volume flux of 63 m^3 s^{-1}.

The geometry of mantle flow beneath Iceland presents complications in determining the source temperature from rocks. Plume material ascends beneath the hotspot and flows laterally along the ridge axis (e.g., Albers and Christensen, 2001). The entire hotspot is, to the first order, the product of an adiabatic column, but the local melts pooled in a crustal magma chamber are not. For example, the mantle material at the distal ends of the hotspot (the Reykjanes Ridge on the south) has already partially melted in the plume conduit.

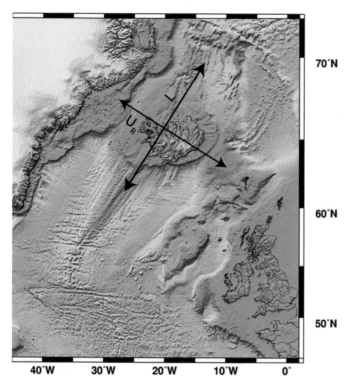

Figure 3. Topographic map of the Iceland area after Smith and Sandwell (1997). The hotspot strongly affects the region between the Faeroe Islands and Greenland for over 1000 km of strike distance. The starting plume head affected the Irish Sea and the west side of Greenland (not shown). Stretched continental margins occupy a much smaller surface area.

Hawaii

The concept of buoyancy flux is relevant to off-ridge hotspots. In terms of the plume, the buoyancy flux is

$$B = \Delta \rho V, \tag{8}$$

where $\Delta \rho$ is the density contrast of the plume material with normal mantle and V is the volume flux through the plume.

One obtains this quantity indirectly from "swells" along hotspot tracks. Hawaii is the best example. The swell is more than 1000 km wide and over a kilometer high along the axis of the chain (Fig. 4). The measured buoyancy flux from the swell is

$$B = (\rho_m - \rho_w)U_P W H, \tag{9}$$

where ρ_m is the density of the mantle, ρ_w is the density of ocean water, and U_P is the velocity of the plate relative to the hotspot. I express the cross-sectional area of the swell compactly as the product of the width W with the height H. Estimation of these quantities requires correction of relief associated with the volcanic edifice and its flexural moats. In addition, the plume material is fluid, not a rigid layer attached to the plate. The estimate

in (9) thus does not precisely give the desired plume quantity in (8) (Ribe and Christensen, 1994).

The Hawaii-Emperor chain is the clearest hotspot track (Fig. 4). Any hypothesis to explain this feature must account for the Hawaii swell as well as volcanism along the axis of the chain. The swell is a major submarine uplift. Its full width is 1500 km and its maximum height is 1.4 km. Its cross-sectional area is 1400 km^2 (Sleep, 1992). The plate velocity is 83 mm yr^{-1}, giving a swell buoyancy flux of 8.9 Mg s^{-1}.

To compare volume flux and buoyancy flux, I represent the density change in terms of the temperature change:

$$\Delta \rho = \rho \alpha \Delta T, \tag{10}$$

where ρ is the density of average mantle (3400 kg m^{-3}), α is the volume thermal expansion coefficient (3 × 10^{-5} K^{-1}, and ΔT is the temperature contrast (here ~250 K) between the plume material and the normal mantle. These values give the volume flux for Hawaii of 350 m^{-3} s^{-1}. It is also useful to have the excess heat per volume of material:

$$E = \Delta C \Delta T, \tag{11}$$

where C is specific heat per unit mass. The ratio of density change to excess heat,

$$\frac{\Delta \rho}{E} = \frac{\alpha}{C}, \tag{12}$$

involves only measurable physical parameters. The uncertainty in these parameters is much less than the uncertainty in volume or buoyancy flux estimates.

The "nose" of the swell extends upstream from the hotspot by ~400 km (Fig. 4). I initially investigated its kinematics to dis-

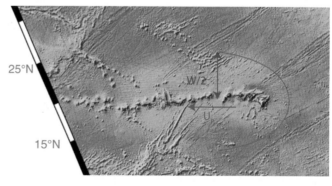

Figure 4. The light blue region of the Hawaiian swell is evident on the topographic map after Smith and Sandwell (1997). The red "parabola" from the kinematic model outlines the nose. The swell becomes more symmetrical north-south when one plots the difference between actual seafloor depth and the expected depth of the age of the seafloor (Phipps Morgan et al., 1995). Note that there is no increase in the rate of volcanism associated with the youngest edifice. The "Big Island" of Hawaii appears larger because the crust is loaded on only its west side (Watts and ten Brink, 1989).

prove the involvement of a plume. However, a plume readily explains the existence and shape of the swell. I discuss both a kinematic model and a more sophisticated dynamic model.

The kinematic model illustrates the geometry (Sleep, 1990a, 1992). Close to the hotspot, plume material flows radially into the asthenosphere from the hotspot. Plate motion sweeps the plume material and the asthenosphere downstream. The depth-average velocity of the ponded plume material is

$$\vec{U} = \frac{V}{2\pi Ar}\hat{r} + \frac{U_P}{2}\hat{x}, \tag{13}$$

where A is the thickness of the layer of spreading plume material, r is horizontal distance from the center of the plume, $\hat{r}$ is the radial unit vector, and $\hat{x}$ is the unit vector in the downstream direction. The factor of 2 in the second right-hand term arises because the velocity from drag of the plate goes from plate velocity at the top of the plume material to near the hotspot frame at its base. A stagnation point where the radial velocity from the plume balances the velocity from the drag of the plate occurs upstream where

$$s = \frac{V}{\pi A U_P}. \tag{14}$$

The parameters in the equation are grossly known. The volume flux is 350 m^3 s^{-1}, the thickness of flowing material is 100 km, and the plate velocity is 83 mm yr^{-1}. These values yield a distance from the nose to the hotspot of 425 km, which is acceptable. The streamline through the stagnation point separates material supplied by the plume and normal asthenosphere. Its approximately parabolic shape fits the swell, particularly to the north (Fig. 4).

Ribe and Christensen (1994) considered buoyant fluid ponded at the base of the moving plate. This dynamic modeling better represents the physics of the swell. I retain the thickness A of the plume material and the distance to the nose of the swell s as dimensional parameters. I let the base of the lithosphere be flat (see Sleep [1996] for a discussion of the effect of relief at the base of the lithosphere). The flux (volume per horizontal length) of material away from the plume is then dimensionally,

$$F \approx \frac{\partial A}{\partial x}\frac{\Delta\rho g A^3}{\eta_P} \approx \frac{\Delta\rho g A^4}{s\eta_P}, \tag{15}$$

where g is the acceleration of gravity, η_P is the viscosity of the plume material, and the slope of the bottom of the plume material $\partial A/\partial x$ in the upstream direction drives flow. The volume flux is the product of F and the circumference of a circle of radius s around the plume, which is dimensionally,

$$V = 2\pi s = \frac{\Delta\rho g A^4}{\eta_P}. \tag{16}$$

Solving, the thickness of plume material,

$$A = \left[\frac{V\eta_P}{\Delta\rho g}\right]^{1/4}, \tag{17}$$

depends weakly on the parameters and is independent of plate velocity. The distance s to the nose of the swell is obtained by noting that the upstream velocity is dimensionally the plate velocity U_P,

$$U_P = \frac{F}{A} = \frac{\Delta\rho g A^3}{s\eta_P}. \tag{18}$$

Solving (17) and (18) yields the nose distance

$$s = \frac{1}{U_P}\left[\frac{\Delta\rho g V^3}{\eta P}\right]^{1/4}. \tag{19}$$

The kinematic and dynamic models differ in that the thickness of plume material in (17) is not constant in the latter. The distance to the nose of the swell is inversely proportional to plate velocity in both models. Note that the models do not explicitly assume that thermal buoyancy drives the flow. Part of the uplift of the swell may be due to the buoyant residuum from the partial melting that produced the volcanic chain (Phipps Morgan et al., 1995).

Finally, I treat Hawaii at 90 Ma as an on-ridge hotspot. In Figure 8 of the reconstruction of Norton (this volume), the hotspot occupies ~300 km of ridge axis. The full spreading rate is unknown as this time is during the long Cretaceous interval of normal magnetic polarity. I estimate it to be 180 mm yr^{-1}. I let the thickness of affected lithosphere be 100 km. These estimates yield a volume flux of 170 m^3 s^{-1}, which is somewhat less than my estimate for the current flux.

Volume of Starting Plume Heads

Starting plumes heads impinge on the base of the lithosphere. One would like to include their long-term average in estimates of global plume flux. In the standard plume hypothesis, starting heads produce radial dike swarms and flood basalts. The total volume of material in the plume head, once ponded, is

$$Q = \pi R^2_H D_H, \tag{20}$$

where R_H is the radius of the region underplated by the plume head and D_H is the thickness of the underplated material. Estimation of either quantity is not straightforward. Plume material spreads laterally beneath regions of thin lithosphere. Dike swarms are not truly radial (McHone et al., 2005), which is expected, as the ambient stress in the plate adds tensorially to the stress associated with the plume. This effect is likely to deflect the distal parts of the swarm, where plume-related stresses are weak. Hill et al. (1992) obtain an average heatflow of 1.2 TW from starting heads as an average for the past 100–200 m.y.

GLOBAL FLUX ESTIMATE

Continuing with kinematic implications of the plume hypothesis, I discuss the global flux of hotspots. Davies (1988) and Sleep (1990a) attempted to quantify the vigor of individual hotspots and the global amount of heat and mass transfer from plumes. Their compilations are still cited, not because of high accuracy, but because of the dearth of significant subsequent improvement.

Both Davies (1988) and Sleep (1990a) summed the flux of their individual hotspots. The more vigorous measurable examples, like Hawaii, contributed a significant fraction to their fluxes. They did not attempt to estimate the flux from starting plume heads. I use a more recent global estimate of Anderson (2002), who includes the flux from starting plume heads from Hill et al. (1992). Plume tails currently supply a heat flux of 2.3 TW and starting plume heads average 1.2 TW, giving a total of 3.5 TW. From my experience in estimating the fluxes of individual hotspots, I give a qualitative estimate of the uncertainty as ±30%, provided the gross concept is correct. This uncertainty is small enough that it does not affect the gist of the issues that I raise below.

The global volume flux is convenient for discussion of topics involving the lithosphere. The volume heat capacity ρC is ~4 $\times 10^6$ J m^{-3} K^{-1}, which yields a volume flux of 2900 m^3 s^{-1} from (12). The mass flux is convenient for the whole mantle, as density increases with depth. For a shallow mantle density of 3400 kg m^{-3}, the mass flux is 10^7 kg s^{-1}.

The flux is significant in terms of available reservoirs; it recycles the mass of the mantle, 4×10^{24} kg, in 12.7 billion years. Equivalently, the flux cycles 1/13 of the mass of the mantle in a billion years, or 35% of it since the Earth formed at 4.5 Ga, extrapolating the present rate.

The estimated heat flux, 3.5 TW, is a significant fraction of the global mantle heat flux of 36 TW. The actual heat budget of plumes is greater (Anderson, 2002; Labrosse, 2002, 2003; Bunge, 2005; Mittelstaedt and Tackley, 2006). The bottom hot thermal boundary layer of the mantle warms cool subducted material to the MORB adiabat before it imparts the excess temperature of plumes. Both heat fluxes are comparable, so the total heat flux to plume from the base of the mantle is ~7 TW.

This flux allows modeling of the thermal history of the core (Anderson, 2002; Labrosse, 2002, 2003; Nimmo et al., 2004). The specific heat per unit mass of the core is about half that of the mantle, 0.625×10^3 J kg^{-1} K^{-1}, and the mass of the core is 2 $\times 10^{24}$ kg. A heat flux of 7 TW cools the core at 176 K b.y.$^{-1}$ in the absence of contributions from radioactive decay and latent heat. This rate requires that the core cools faster than the mantle. Radioactive decay of potassium in the core is a possible heat source. A chemically dense layer at the base of the mantle may contain radioactive elements that augment the heat from the core (Davaille, 1999; Kellogg et al., 1999). There is no hard evidence demonstrating a hidden chemical reservoir, but U and Th antineutrino detectors will soon have the capability of detecting a heat source from a chemically dense "dregs" layer at the base of the mantle (Araki et al., 2005; Fiorentini et al., 2005). This technique is especially applicable if the layer is locally thicker (passive cusps entrained by the plume or buoyant hot "lava-lamp" upwellings) beneath hotspots, as seafloor detectors could resolve lateral variations in antineutrino flux within the Pacific basin. Precise seismic tomography could conceivably independently resolve these features and the deepest conduit regions of the plumes. Potassium antineutrino detectors are not yet practical.

It is also illustrative to compare the rate at which plumes recirculate mantle with the rate at which plate processes segregate it. Modern oceanic crust is ~6 km thick; the underlying depleted residuum is ~50 km thick (Klein and Langmuir, 1987; Plank and Langmuir, 1992; Klein, 2003). White et al. (1992) give a precise estimate of 3.3 km^2 yr^{-1} for the global rate of seafloor production. I use the rounded value of ~3 km^2 yr^{-1} as a recent long-term estimate. The rates of crustal production and residuum production are therefore 570 and 5300 m^3 s^{-1}, respectively. This total volume is only twice the estimated volume flux of plume material. At the current rate and depth of melting, a mass equivalent to 70% of the mantle has passed through the melting zone of the lithosphere at ridges since 4.5 Ga. The actual fraction is higher, as the melting depth was higher in the past when the mantle was hotter. Rates of plate tectonics on the early Earth conceivably were faster (slower or nonexistent) than at present.

The lack of an obvious water source is relevant to "wet spot" hypotheses (e.g., Bonath, 1990). Estimates for the amount of water 1700×10^{10} mol yr^{-1} carried down with subducted crust (Staudigel, 2003; cf. Rüpke et al., 2004, who estimate 1.8 times this value) are similar to the flux through arc volcanoes (Oppenheimer, 2003; Wallace, 2005). Even the sign of the net long-term flux is unknown. I agree that moderately water-rich regions (700 ppm H$_2$O, compared with ~150 ppm in depleted mantle; Saal et al., 2002) may exist and contribute to medium-volume hotspots like the Azores (Asimow et al., 2004).

ISSUES WITH THE PLUME HYPOTHESIS

I cannot in a reasonably short article attempt to refute or even summarize all the objections and alternatives to the plume hypothesis. As I noted in the introduction, the direct lines of evidence, including mid-mantle tomography and lava petrology, do not have sufficient resolution to obtain the potential temperature of the mantle source. I concentrated on reliable data sets, including the thickness of crust beneath Iceland, the depth of extensive melting beneath Hawaii, and the existence of the Hawaiian swell. I continue with other lines of evidence that bear on the plume hypothesis, but are currently inadequate to provide definitive tests.

Plumes as Dynamic Features

Some confusion arises from cartoon models of plumes where the hotspot is a point source of heat rather than a source of buoyant material. The actual dynamics of plumes are more complicated, but lead to observable features. (Here semantics clouds thinking. "Spot" in English can mean "point" or "region" like a spot on a dog or a sunspot. Note that the French term is "*le point chaud*" not "*la tache chaude.*")

First, the strong dependence of viscosity on temperature explains both the excess volcanism and its spatial localization. The bottom thermal boundary layer in the mantle acts as a planar conduit. Plume material flows radially inward toward the tail conduit. The tail conduit is a low-viscosity channel. Plume material impinges on the base of the lithosphere and it spreads laterally away from the plume conduit as a buoyant fluid. I considered this effect with regard to the Hawaiian swell and briefly with the Reykjanes Ridge.

Plumes, like hurricanes in meteorology and sunspots, focus attention on part of the larger global flow pattern. Like rigid plate tectonics, earlier fixed hotspot theory was an approximation. It is inevitable that the source of plumes at depth and their mid-mantle conduits advect with the rest of the flow. This attribute is predictable from fluid dynamics (Steinberger, 2000; Steinberger et al., 2004). The main difficulties are obtaining accurate relative plate velocities and track velocities (Gripp and Gordon, 2002). Steinberger et al. (2004) jointly analyzed hotspot tracks, relative velocities, and plume advection and found the results consistent with the existence of plumes.

Delamination Alternative

I discuss a well-posed nonplume hypothesis for Hawaii that has simple implications. A crack propagates through the plate, producing volcanism at its tip and causing the lower lithospheric mantle to delaminate on its flanks over a region perpendicular to the track forming the swell. Delamination to produce the swell and the volcanism have testable implications that distinguish it from plumes. I summarize them here. Some data sets are consistent with either delamination or a plume. Others need to be better resolved to be definitive.

Plate cracking is evident along the Hawaiian chain. Solomon and Sleep (1974) pointed out the strike of the chain aligns perpendicular to the expected axis of intraplate tension (Stuart et al., this volume). En echelon lineaments of volcanic edifices align along the trends of the numerous chains, including Hawaii (e.g., Jackson and Shaw, 1975). These features are compatible with a propagating crack, but also with plumes. Plume-derived melts need cracks to ascend to produce surface hotspots. These cracks preferentially align with the local intraplate stress.

I use the 1400-m uplift at the axis of the Hawaiian swell as an example. The amount of thinning of the lithosphere by delamination is essentially that of rejuvenation models of hotspots (Crough, 1983). For an example, I choose integers that are perfect squares: changing 81-m.y. lithosphere into 25-m.y. lithosphere produces an uplift of 1400 m. (Depth in meters varies as 350 times the square root of age in millions of years in my examples. The lithosphere thickness in kilometers is ~12 times the square root of age in millions of years.) The base of the lithosphere of 95-Ma lithosphere would thin from ~117 to ~69 km if the uplift is 1400 m.

The delaminated lithosphere needs to sink. Ambient mantle upwells to replace the lithosphere that delaminates. Its source temperature cannot be greater than that of the underlying asthenosphere. The material immediately underlying the lithosphere is just ambient asthenosphere. With good tomographic resolution, shallow lateral variation of seismic velocity would be confined to about the depth to the pre-delamination base of the lithosphere.

The temperature anomaly of the delaminated lithosphere is equivalent to that of subducted of 16-Ma lithosphere, $\sqrt{16} = \sqrt{81} - \sqrt{25}$. There is in fact a broad positive geoid anomaly around Hawaii. This feature might indicate dense delaminated material, as slabs produce positive long-wavelength geoid anomalies. However, the low viscosity and buoyancy of the plume tail conduit produces a similar geoid anomaly (Richards et al., 1988).

Tomography at mid-mantle depths would resolve the issue. One needs only the sign of the anomaly. A plume produces a narrow negative seismic velocity anomaly whereas delaminated material produces a broad positive one. Montelli et al. (2004) resolve a negative anomaly beneath Hawaii, as do Lei and Zhao (2006).

Uplift and Subsidence from Hotspots

Returning to Hawaii, the excess elevation of the swell is evident even before one accounts for the dependence of seafloor depth on age (Fig. 4). The correction is straightforward and modest in the nose region of the swell (plate 1 *in* Phipps Morgan et al., 1995). One estimates the uplift rate of a point on the crust on the nose of the swell by assuming that the nose propagates across the seafloor with the hotspot velocity. This predicted rate of uplift in the past few million years could be appraised by careful studies of benthic fossils obtained from cores. Such a study could eliminate the possibility that the swell is an older feature unrelated to the current hotspot.

The behavior of the seafloor after the passage of a hotspot provides information on the underlying process. The long-term subsidence of oceanic islands has been known since the time of Darwin. However, quantifying this process is not simple. We may have a good constraint on seafloor and edifice ages. We may have paleodepth indications back in geological time. I show that it is difficult to resolve the effects of plumes from alternatives using seafloor subsidence data. The melting sequence beneath Hawaii discussed by Sobolev et al. (2005) indicates that the buoyancy is mainly thermal.

Ponded plume material produces modest uplifts. For example, a ponded layer 100-km thick with a temperature contrast of

250 K produces 1060 m of uplift. The material spreads laterally near the hotspot and transfers its heat into the overlying lithosphere. I use the 1400-m uplift at the axis of the Hawaiian swell in examples. The buoyancy causing uplift at the active hotspot is partially from ponded plume material and partially from lithospheric thinning. The excess heat of the plume thins the lithosphere downstream, and the heat anomaly thereafter moves with the lithosphere.

Studies of ancient edifices need to account for the general subsidence of seafloor with age. That is, one must resolve the additional cooling from the loss of heat added by plumes. Mathematically, the O'Connell relationship governs the net increase rate of seafloor depth with time:

$$\frac{DS}{Dt} = \left(\frac{\rho_m}{\rho_m - \rho_w} \right) \left(\frac{\alpha}{\rho C} \right) (q - q_b), \qquad (21)$$

where the substantive derivative indicates subsidence of a place on the seafloor as would be sampled by drilling, the first term represents the effect of isostasy for subsidence occurring beneath water, and the third term is the difference between the heatflow out of the top of plate and heatflow into the bottom. A rejuvenation model treats cooling as a half-space, so there is no heatflow into the bottom except during the rejuvenation event. The seafloor depth is then

$$S = S_{\text{ridge}} + \left(\frac{2\alpha\Delta T_L \rho_m}{\rho_m - \rho_w} \right) \left(\frac{\kappa t}{\pi} \right)^{1/2}, \qquad (22)$$

where S_{ridge} is the depth at the ridge axis, ΔT_L is the temperature contrast between the surface and the base of the lithosphere, t is apparent plate age, and κ is thermal diffusivity.

A plume model is more complicated. Uplift occurs when the plume material ponds beneath the lithosphere. Subsidence occurs later if the plume material spreads laterally and thins. Finally, subsidence occurs from the heatflow escaping from the top of the plate. The plume material is buoyant relative to the underlying mantle. This suppresses small-scale "stagnant-lid" convection into the base of the plate. The long-term subsidence thus differs from ordinary lithosphere, where stagnant-lid convection inputs heat to the bottom.

I use two end-member models to illustrate the difficulties in resolving the effect of plumes. At one end, the lithosphere is not thermally affected by the hotspot and subsides with the square root of its crustal age. At the other end, it behaves as rejuvenated lithosphere. The predicted difference in subsidence for a midplate plume is small. I use these models, as a best case. If one cannot resolve the predicted difference for Hawaii, one is unlikely to resolve the difference for a weaker midplate hotspot.

For example, consider lithosphere now near the Hawaii-Emperor bend that was rejuvenated 50 m.y. ago from a crustal age of 81 m.y. to a rejuvenated age of 25 m.y. (I use perfect squares for ages and retain extra digits on depths to aid in following the calculations.) Over the past 50 m.y., the rejuvenated

area subsided 1281 m and the reference region subsided 856 m, using the square-root subsidence model. The difference, 425 m, would be readily apparent if we knew the age of a surface eroded to sea level on a seamount exactly at the time of the passage of the hotpsot. Resolving this extra subsidence with real data is problematic. In an ideal case, we would have the current depths at each site of sea-level erosion surfaces formed at the same time, so eustatic sea-level variation does not cause uncertainty. An erosion surface is likely for a seamount chain, but less likely for a reference site away from the chain.

More realistically, erosion beveled the edifice downstream of the hotspot. The time taken to erode accrues significant source error unless there are tight age constraints. For example, if the beveled surface formed 10 m.y. downstream on rejuvenated lithosphere, it would have already subsided 320 m. This crust would have subsided $1281 - 320 = 961$ m in the subsequent 40 m.y. If one incorrectly interpreted that the surface formed at the hotspot at 50 Ma, one would conclude that the crust subsided about the amount predicted for ordinary lithosphere, 856 m.

On-ridge hotspots are also difficult to analyze precisely. I again use the half-space model for illustration. It may apply near the axis where lithosphere is underlain by plume material. It does not apply far from the axis where normal lithosphere lies beneath the plate. The analytical formula for subsidence (22) implies that the depth change from the axis to aged seafloor depends linearly on the half-space temperature. Making the plume source mantle as hot as reasonable from the petrological grid, 1600 °C versus 1300 °C for normal mantle, implies a subsidence of 430 m at 1 m.y. versus 350 m. For example, the two regions subside 3010 and 2450 m, respectively, in 49 m.y. The difference of 560 m is readily resolved if we find a subsided sea-level surface formed right at the axis. The time to erode to sea level, however, is an even more serious problem than with midplate hotspots. For example, a surface beveled 4 m.y. from the ridge axis would have already subsided 860 and 700 m, respectively, more that the expected difference between plume and ordinary lithosphere.

Clift (2005) analyzed data obtained from paleontological studies of cores obtained by deep-sea drilling. He correctly concluded that he did not see large anomalous elevation changes in deep-marine data from ancient hotspot edifices. The resolution of his data is inadequate to resolve small differences predicted between rejuvenated and unaffected lithosphere. These compiled data partition depths to 0–150 m, 150–500 m, 500–2000 m, 2000–4000 m, and deeper. Only near-sea-level erosion surfaces and the current depth provide reliable information.

The sizes of the deeper bins are larger than the expected amounts of subsidence. In addition, the frustrating compilation method produces spurious precision (exactly 2000 m) when the paleo-depth in a borehole changes, say, from the 500–2000-m bin to the 2000–4000-m bin. It also presumes that the depths of water masses inhabited by index benthic organisms did not vary with time. A careful reexamination of available core is certainly warranted.

Hotspot Track Continuity

Physically the plume source is a thermal boundary layer beneath a much thicker region, the rest of the mantle. In this case, the flux into the tail conduit depends on the properties of the deep part of the mantle, including flow stirred by plates and slabs and thermal anomalies from sunken slabs. It does not depend otherwise on the condition of the shallow mantle. Starting plume heads and plume tails thus impinge on a variety of surface tectonic environments (Fig. 5).

One would thus expect plumes to track across a variety of environments. Equivalently, geologists call both Hawaii and Iceland "hotspots." This nomenclature implies that both long-lived features are due to a common mechanism. Sleep (1990a,b, 1992, 2002a) argued that some hotspots have evolved from off-ridge to on-ridge whereas others have moved from the axis into plate interiors.

The trend through the Monteregian hills, the New England seamounts, the Corner seamounts, and Great Meteor seamount is the best example (Fig. 5). If this interpretation is correct, it is strong evidence for a deep cause. Tristan-Gough (ridge leaving), Réunion (ridge crossing), Louisville (ridge approaching), Ninetyeast (ridge leaving), and Foundation (ridge approaching) are additional examples.

Hawaii should be added to the list as a ridge-leaving hotspot. Norton (this volume) shows it as an on-ridge hotspot at ca. 90 Ma, when the present northernmost part of the track formed. The other "half" of the track (formed before the hotspot crossed the ridge axis) was thus on another plate and is not now adjacent to the northern Emperor chain.

McHone and Butler (1996) doubts that a plume produced the New England seamount track. He gives well-posed points that apply in general to hotspot tracks. They are based on valid data. I summarize them, give related objections by other authors, and briefly reply.

1. There is no age progression evident on the land part of the track in New England. Furthermore, Bear seamount on the continental rise is older: 120 Ma (Swift et al., 1986), or ~20 m.y. older than the age expected from a simple track model. These observations are correct, but are expected from a dynamic model of a plume that is a source of hot buoyant material. From (19), the time for a place on the crust to track from the nose of the swell to the plume is, dimensionally,

$$t_{note} \equiv \frac{s}{U_P} = \frac{1}{U_P^2} \left[\frac{\Delta \rho g V^3}{\eta_P} \right]^{1/4}, \qquad (23)$$

which becomes large when the plate velocity U_P is small. This time is ~5 m.y. for Hawaii, where the track velocity is 8.3 mm yr^{-1}. This scale time indicates how long hot plume material remains beneath a patch of lithosphere.

I obtain the nose time, from parameters estimated by Sleep (1990b). The buoyancy flux for the western seamounts is about one-quarter that of Hawaii, ~2 Mg s^{-1}. The track velocity about half that of Hawaii, 4.7 mm yr^{-1}. The nose time for the western seamounts is a factor of ~$2^{1.25} = 2.8$, or 12 m.y.

Sublithospheric relief may have caused plume material to spread upstream (east) when the plume was beneath New England. The base of the lithosphere shallows to the east from the craton north of Montréal, across old Appalachian continental boundary, and to the younger passive margin. This slope acted as an upside-down drainage pattern to the east. Bear seamount formed ahead of the plume from buoyant material that cascaded up the lithospheric escarpment at the passive margin.

A general age progression does exist when the track is viewed as a whole. Baksi (2005) pointed out that the radiometric ages for the track by Duncan (1984) are not up to modern standards. Stratigraphic (Swift et al., 1986) and pa-

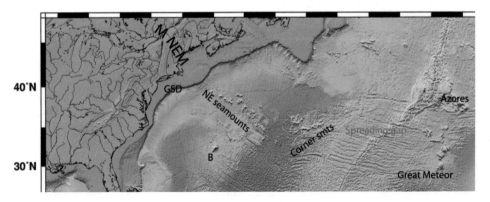

40°N

30°N

Figure 5. Topographic map after Smith and Sandwell (1997) shows the Monteregian hotspot track. The older end of the track becomes evident in the Monteregian hills (M) at the edge of the craton. It produced igneous events in New England (NEM). It crossed the passive margin. The Great Stone Dome (GSD) is an intrusion into passive margin sediments on the flanks of the track. Flow of plume material over the lithospheric escarpment beneath the passive margin lead to pressure-release melting. The New England (NE) seamounts formed in the middle of the plate. The track then jumped to the on-ridge Corner seamounts. It crossed the ridge axis and is now forming the midplate Great Meteor seamount. The gap between the New England and Corner seamounts is dynamic in origin (Sleep, 2002a). The plume became close enough to the ridge that plume material flowed along the base of the lithosphere toward the axis. The gap between Corner and Great Meteor seamounts is kinematic in origin, arising from seafloor spreading that has occurred after the track crossed the axis. This interpretation illustrates that hotspot tracks are somewhat oblivious of shallow tectonic features, and that on-ridge and off-ridge hotspots share a common cause.

leontological ages (Uchupi et al., 1970), however, support the progression. Collection of more samples, particularly by drilling, and their dating with modern radiometric methods and paleontology is warranted.

2. The track is not straight. True. Given its duration, one would not expect the velocity of the Atlantic plate to stay constant.

3. The land volcanism follows local features. This statement is true when the track is viewed on a local scale. It is also true of the marine track (Beutel, 2005), and is to be expected, in that dikes are perpendicular to the most tensile stress and tend to follow local weaknesses. There is a tendency for flexural stresses to cause regions of horizontal tension and horizontal compression at the passive margin and hence domains of enhanced and sparse volcanism.

4. The Early Jurassic magmas in New England are similar to the lavas associated with the track. There is no obvious chemical signature of the plume. There are other volcanic provinces, including those in Newfoundland and the southeastern United States. True. I would associate the Early Jurassic rocks with the Bahama–Cape Verde starting plume head centered in southern Georgia and the break-up of the Atlantic. Newfoundland may be part of the Azores track and the southeastern United States part of the Bermuda track. I reiterate that I consider plumes to be mainly thermal features.

5. The uplift associated with the track is small. True, and I have already noted that plume material produces only modest uplifts.

6. There is Eocene to Miocene coralline algae on Mytilus seamount (Swift et al., 1986), which indicates a near sealevel surface at that time. It also indicates an episode of renewed volcanism in Tertiary, well after the track passed. The algal fossils are mixed with Cretaceous fossils (Uchupi et al., 1970, who quote an internal Woods Hole Oceanographic Institution report by J.M. Zeigler). An organic reef grew on this seamount, and the dredged fossils are likely to be in place. They indicate the death age of the surface reef, not an age of volcanic activity. Similar fossils from Bear seamount are less likely to be in place. (The report by J.M. Zeigler is now declassified. The Cretaceous and Tertiary tentative dates refer to different taxa of fossils from the same samples. The older non-algal dates may be more reliable. Too little information is given for further evaluation.) The preplate-theory (Zeigler) papers correlated New England seamounts with Bermuda. In general, feeble volcanism at a later date does not preclude earlier presence of a plume.

Why It Is Hard to See Evidence of Other Pacific Plumes

In the plume hypothesis, Hawaii is a vigorous plume in the middle of a fast-moving plate. It leaves a clear track. Iceland is a much less vigorous plume, by a factor of ~6. It impinges at a slow-spreading ridge axis. This scenario allows the maximum amount of pressure-release melting. The thickened old lithosphere on each side of the ridge confines the buoyant plume material, channeling it along the axis. These favorable circumstances do not exist for many hotspots, and the ages of many seafloor edifices are unknown.

I place the Iceland plume, with its one-sixth of the Hawaiian (volume and buoyancy) flux, beneath the Pacific plate as a thought experiment. Its buoyancy flux produces a swell cross-section of $1400/6 = 230$ km^2. The swell nose from (19) is a factor of $(1/6)^{3/4} = 0.26$ closer to the plume, or ~100 km. The swell thus is confined to the area of the volcanic edifices and their flexural moat, which obscures its existence. The thickness of the plume material beneath the hotspot A is a factor of $(1/6)^{1/4} = 0.63$ of that beneath Hawaii.

When I place the Iceland plume beneath the East Pacific Rise, where it is spreading ~8 times faster than at Iceland, it can supply only $1000/8 = 125$ km of ridge length with a strong flux. More likely, the buoyant plume material spreads along the axis, producing a broader region of mildly enhanced volcanism. Kinematically, a plume cannot stay near a fast ridge axis like one does near Iceland. The "absolute" velocities of ridge axes are much greater in the Pacific than in the Atlantic.

CONCLUSIONS

The plume hypothesis is a straightforward explanation of the salient features of hotspots. Hot buoyant material ascends from deep within the mantle. It ponds at the base of the lithosphere and spreads laterally. Interaction with the drag of the plate provides a ready explanation for the Hawaiian swell. It also explains numerous flow-line features between a near-ridge hotspot and the ridge axis (Morgan, 1978; Sleep, 2002a,b; Kopp et al., 2003). Buoyant material at the base of the lithosphere flows in that direction. Numerical modeling of this flow is not easy because of the strong dependence of buoyancy and viscosity on temperature.

The excess temperature of plume material explains voluminous magma production. Starting plume heads explain the sudden outbreak of magmatism of broad, previously quiescent regions. Plume tails heated at a basal boundary layer explain hotspot tracks. The deep parts of this process are at the limit of tomographic resolution. I do not have a good constraint on the source depth of plumes. With regard to the mid-mantle, plumes have been putatively resolved where expected beneath many hotspots (Montelli et al., 2004), but this claim is controversial (e.g., van der Hilst and de Hoop, 2005). Numerical codes cannot yet handle the viscosity contrasts and resolution necessary for plumes to self-organize in three-dimensional flow with plates.

It is thus helpful to go over the shallow implications of plume theory. We know the depth of the seafloor and its age over most of the ocean basins. I contend that one should concentrate on the most reliable data sets first. These are the current bathymetry and gravity data sets. The latter is useful for inferring bathymetry where data are sparse and for locating the swells of weak hotspots. Seismic data need improved resolution at lithospheric and asthenospheric depths, that is, where hot plume material may occur in the asthenosphere versus thinned lithosphere. Limited S-wave splitting data are compatible with the flow pattern expected for the Hawaiian swell (Walker et al., 2005) but are not available globally. Paleobathymetric data provide only gross constraints. The scatter in heatflow data precludes their use in finding temperature anomalies beneath old seafloor.

The gross bathymetry of hotspots provides estimates of the volume of material involved with plumes and with any alternative mechanism. The global volume flux, 2900 m^3 s^{-1}, and the heat flux, 3.5 TW, are large in many respects. They constitute a major heat sink on the deep mantle. If we extrapolate over time, a significant fraction (one-third) of the mantle has gone through plumes in the Earth's history. Mass requirements alone preclude nonrenewable low-volume sources for chemical melting anomalies. Conversely, most of the mantle has melted at least once at a ridge axis and been subducted. Sobolev et al. (2005) contend that this recycled crust along with elevated temperature control the chemistry of Hawaiian magmas.

Conversely, some effects of plumes are small and local, as they constitute only 10% of the global heat flux, which is particularly true on the fast-moving Pacific plate. The Iceland plume would have only subtle features if it impinged there or at the East Pacific Rise. The vertical tectonics associated with plumes is also modest. It is difficult to resolve subsidence along hotspot tracks from the ordinary subsidence of seafloor. It is also difficult to resolve the timing and amount of ancient uplift of the deep seafloor. The poor resolution of much of relevant data is cause for improving analytic techniques and collecting more data, particularly seismic, paleobathymetry, and the ages of seafloor edifices. In particular, we still do not know the age of many volcanic edifices. Unreliable dates make it difficult to appraise the age progression of the New England track.

ACKNOWLEDGMENTS

This work is supported by National Science Foundation grant EAR-0406658. I thank Gillian Foulger, Don Anderson, and two anonymous reviewers for helpful comments.

REFERENCES CITED

Albers, M., and Christensen, U.L., 2001, Channeling of plume flow beneath mid-ocean ridges: Earth and Planetary Science Letters, v. 187, p. 207–220, doi: 10.1016/S0012-821X(01)00276-X.

Anderson, O.L., 2002, The power balance at the core-mantle boundary: Physics of the Earth and Planetary Interiors, v. 131, p. 1–17, doi: 10.1016/S0031-9201(02)00009-2.

Araki, T., and many others, 2005, Experimental investigation of geologically produced antineutrinos with KamLAND: Nature, v. 436, p. 499–503, doi: 10.1038/nature03980.

Asimow, P.D., Dixon, J.E., and Langmuir, C.H., 2004, A hydrous melting and fractionation model for mid-ocean ridge basalts: Application to the mid-Atlantic ridge near the Azores: Geochemistry, Geophysics, Geosystems, v. 5, art. no. Q01E16.

Baksi, A.K., 2005, Evaluation of radiometric ages pertaining to rocks hypothesized to have been derived by hotspot activity, in and around the Atlantic, Indian, and Pacific oceans, *in* Foulger, G.R., et al., eds., Plates, plumes, and paradigms: Boulder, Colorado, Geological Society of America Special Paper 388, p. 55–70.

Beutel, E.K., 2005, Stress-induced seamount formation at ridge-transform intersections, *in* Foulger, G.R., et al., eds., Plates, plumes, and paradigms: Boulder, Colorado, Geological Society of America Special Paper 388, p. 581–593.

Bonath, E., 1990, Not so hot "hot spots" in the oceanic mantle: Science, v. 250, p. 107–111, doi: 10.1126/science.250.4977.107.

Bunge, H.P., 2005, Low plume excess temperature and high core heat flux inferred from non-adiabatic geotherms in internally heated mantle circulation models: Physics of the Earth and Planetary Interiors, v. 153, p. 3–10, doi: 10.1016/j.pepi.2005.03.017.

Clift, P.D., 2005, Sedimentary evidence for moderate mantle temperature anomalies associated with hotspot volcanism, *in* Foulger, G.R., et al., eds., Plates, plumes, and paradigms: Boulder, Colorado, Geological Society of America Special Paper 388, p. 279–287.

Crough, S.T., 1983, Hotspot swells: Annual Review of Earth and Planetary Sciences, v. 11, p. 165–193, doi: 10.1146/annurev.ea.11.050183.001121.

Davaille, A., 1999, Two-layer thermal convection in miscible viscous fluids: Journal of Fluid Mechanics, v. 379, p. 223–253, doi: 10.1017/S0022112099008003322.

Davies, G.F., 1988, Ocean bathymetry and mantle convection. 1. Large-scale flow and hotspots: Journal of Geophysical Research, v. 93, p. 10,467–10,480.

Donnelly, K.E., Goldstein, S.L., Langmuir, C.H., and Spiegelman, M., 2004, Origin of enriched ocean ridge basalts and implications for mantle dynamics: Earth and Planetary Science Letters, v. 226, p. 347–366, doi: 10.1016/j.epsl.2004.07.019.

Duncan, R.A., 1984, Age progressive volcanism in the New England seamounts and the opening of the central Atlantic Ocean: Journal of Geophysical Research, v. 89, p. 9980–9990.

Fiorentini, G., Lissia, M., Mantovani, F., and Vannucci, R., 2005, Geo-neutrinos: A new probe of Earth's interior: Earth and Planetary Science Letters, v. 238, p. 235–247, doi: 10.1016/j.epsl.2005.06.061.

Foulger, G.R., Natland, J.H., and Anderson, D.L., 2005a, A source for Icelandic magmas in remelted Iapetus crust: Journal of Volcanology and Geothermal Research, v. 141, p. 23–44, doi: 10.1016/j.jvolgeores.2004.10.006.

Foulger, G.R., Natland, J.H., and Anderson, D.L., 2005b, Genesis of the Iceland anomaly by plate tectonic processes, *in* Foulger, G.R., et al., eds., Plates, plumes, and paradigms: Boulder, Colorado, Geological Society of America Special Paper 388, p. 595–625.

Green, D.H., and Falloon, T.J., 2005, Primary magmas at mid-ocean ridges, "hotspots," and other intraplate settings: Constraints on mantle potential temperature, *in* Foulger, G.R., et al., eds., Plates, plumes, and paradigms: Boulder, Colorado, Geological Society of America Special Paper 388, p. 595–625.

Gripp, A.E., and Gordon, R.G., 2002, Young tracks of hotspots and current plate velocities: Geophysical Journal International, v. 150, p. 321–361, doi: 10.1046/j.1365-246X.2002.01627.x.

Hill, R.I., Campell, I.H., Davies, G.F., and Griffiths, R.W., 1992, Mantle plumes and continental tectonics: Science, v. 256, p. 186–193, doi: 10.1126/science.256.5054.186.

Jackson, E.D., and Shaw, H.R., 1975, Stress fields in central portions of the Pacific plate: Delineated in time by linear volcanic chains: Journal of Geophysical Research, v. 80, p. 1861–1874.

Kellogg, L.H., Hager, B.H., and van der Hilst, R.D., 1999, Compositional stratification in the deep mantle: Science, v. 283, p. 1881–1884, doi: 10.1126/science.283.5409.1881.

King, S.D., and Anderson, D.L., 1998, Edge-driven convection: Earth and Planetary Science Letters, v. 160, p. 289–296.

Klein, E.M., 2003, Geochemistry of the igneous oceanic crust, *in* Holland, H.D., Turekian, K.K., and Rudnick, R.L., eds., Treatise on geochemistry, Volume 3: Oxford, Elsevier Pergamon, p. 433–463.

Klein, E.M., and Langmuir, C.H., 1987, Global correlations of ocean ridge basalt chemistry with axial depth and crustal thickness: Journal of Geophysical Research, v. 92, p. 8089–8115.

Kopp, H., Kopp, C., Phipps Morgan, J., Flueh, E.R., Weinrebe, W., and Morgan, W.J., 2003. Fossil hot spot–ridge interaction in the Musicians seamount province: Geophysical investigations of hotspot volcanism at volcanic elongated ridges: Journal of Geophysical Research, v. 108, doi: 10.1029/2002JB002015.

Labrosse, S., 2002, Hotspots, mantle plumes and core heat loss: Earth and Planetary Science Letters, v. 199, p. 147–156, doi: 10.1016/S0012-821X(02)00537-X.

Labrosse, S., 2003, Thermal and magnetic evolution of the Earth's core: Physics of the Earth and Planetary Interiors, v. 140, p. 127–143, doi: 10.1016/j.pepi.2003.07.006.

Lei, J., and Zhao, D., 2006, A new insight into the Hawaiian plume: Earth and Planetary Science Letters, v. 241, p. 438–453, doi: 10.1016/j.epsl.2005.11.038.

McHone, J.G., and Butler, J.R., 1996, Constraints on the mantle plume model for Mesozoic alkaline intrusions in northeastern North America: Canadian Mineralogist, v. 34, p. 325–334.

McHone, J.G., Anderson, D.L., Beutel, E.K., and Fialko, Y.A., 2005, Giant dikes, rifts, flood basalts, and plate tectonics: A contention of mantle models, *in* Foulger, G.R., et al., eds., Plates, plumes, and paradigms: Boulder, Colorado, Geological Society of America Special Paper 388, p. 401–420.

McKenzie, B., and Bickle, M.J., 1988, The volume and composition of melt generated by extension of the lithosphere: Journal of Petrology, v. 29, p. 625–679.

Mittelstaedt, E., and Tackley, P.J., 2006, Plume heat flow is much lower than CMB heat flow: Earth and Planetary Science Letters, v. 241, p. 202–210, doi: 10.1016/j.epsl.2005.10.012.

Montelli, R., Nolet, G., Dahlen, F.A., Masters, G., Engdahl, E.R., and Hung, S.-H., 2004, Finite-frequency tomography reveals a variety of plumes in the mantle: Science, v. 303, p. 338–343, doi: 10.1126/science.1092485.

Morgan, W.J., 1978, Rodriguez, Darwin, Amsterdam, . . .: A second type of hotspot island: Journal of Geophysical Research, v. 83, p. 5355–5360.

Nimmo, F., Price, G.D., Brodholt, J., and Gubbins, D., 2004, The influence of potassium on core and geodynamo evolution: Geophysical Journal International, v. 156, p. 363–376, doi: 10.1111/j.1365-246X.2003.02157.x.

Norton, I.O., 2007 (this volume), Speculations on Cretaceous tectonic history of the northwest Pacific and a tectonic origin for the Hawaii hotspot, *in* Foulger, G.R., and Jurdy, D.M., eds., Plates, plumes, and planetary processes: Boulder, Colorado, Geological Society of America Special Paper 430, doi: 10.1130/2007.2430(22).

Oppenheimer, C., 2003, Volcanic degassing, *in* Holland, H.D., Turekian, K.K., Rudnick, R.L., eds., Treatise on geochemistry, Volume 3: Oxford, Elsevier Pergamon, pp. 1511–1535.

Phipps Morgan, J., Morgan, W.J., and Price, E., 1995, Hotspot melting generates both hotspot volcanism and a hotspot swell?: Journal of Geophysical Research, v. 100, p. 8045–8062, doi: 10.1029/94JB02887.

Plank, T., and Langmuir, C.H., 1992, Effects of the melting regime on the composition of the oceanic crust: Journal of Geophysical Research, v. 97, p. 19,749–19,770.

Plank, T., Spiegelman, M., Langmmuir, C.H., and Forsyth, D.W., 1995, The meaning of "mean F": Clarifying the mean extent of melting at ocean ridges: Journal of Geophysical Research, v. 100, p. 15,045–15,052.

Putirka, K.D., 2005, Mantle potential temperatures at Hawaii, Iceland, and the mid-ocean ridge system, as inferred from olivine phenocrysts: Evidence for thermally driven mantle plumes, Geochemistry, Geophysics, Geosystems, v. 6, art. no. Q05L08.

Ribe, N., and Christensen, U.R., 1994, Three-dimensional modelling of plume-lithosphere interaction: Journal of Geophysical Research, v. 99, p. 669–682, doi: 10.1029/93JB02386.

Richards, M.A., Hager, B.H., and Sleep, N.H., 1988, Dynamically supported geoid highs over hotspots: Observation and theory: Journal of Geophysical Research, v. 93, p. 7690–7708.

Rüpke, L.H., Phipps Morgan, J., Hort, M., and Connolly, J.A.D., 2004, Serpentine and the subduction zone water cycle: Earth and Planetary Science Letters, v. 223, p. 17–34, doi: 10.1016/j.epsl.2004.04.018.

Saal, A.E., Hauri, E.H., Langmuir, C.H., and Perfit, M.R., 2002, Vapour undersaturation in primitive mid-ocean-ridge basalt and the volatile content of Earth's upper mantle: Nature, v. 419, p. 451–455, doi: 10.1038/nature01073.

Schilling, J.G., 1991, Fluxes and excess temperatures of mantle plumes inferred from their interaction with migrating midocean ridges: Nature, v. 352, p. 397–403, doi: 10.1038/352397a0.

Sleep, N.H., 1984, Tapping of magmas from ubiquitous mantle heterogeneities—An alternative to mantle plumes: Journal of Geophysical Research, v. 89, p. 29–41.

Sleep, N.H., 1990a, Hotspots and mantle plumes: Some phenomenology: Journal of Geophysical Research, v. 95, p. 6715–6736.

Sleep, N.H., 1990b, Monteregian hotspot track: A long-lived mantle plume: Journal of Geophysical Research, v. 95, p. 21,983–21,990.

Sleep, N.H., 1992, Hotspot volcanism and mantle plumes: Annual Review of Earth and Planetary Sciences, v. 20, p. 19–43, doi: 10.1146/annurev.ea.20.050192.000315.

Sleep, N.H., 1996, Lateral flow of plume material ponded at sublithospheric depths: Journal of Geophysical Research, v. 101, p. 28,065–28,083.

Sleep, N.H., 2002a, Ridge-crossing mantle plumes and gaps in tracks: Geochemistry, Geophysics, Geosystems, v. 3, doi:10.1029/2001GC000290.

Sleep, N.H., 2002b, Local lithospheric relief associated with fracture zones and ponded plume material: Geochemistry, Geophysics, Geosystems, v. 3, doi:10.1029/2001GC000376.

Sleep, N.H., 2006, Mantle plumes from top to bottom: Earth-Science Reviews, v. 77, p. 231–271, doi: 10.1016/j.earscirev.2006.03.007.

Smith, W.H.F., and Sandwell, D.T., 1997, Global seafloor topography from satellite altimetry and ship depth soundings: Science, v. 277, p. 1957–1962.

Sobolev, A.V., Hofmann, A.W., Sobolev, S.V., and Nikogosian, I.K., 2005, An olivine-free mantle source of Hawaiian shield basalts: Nature, v. 434, p. 590–597, doi: 10.1038/nature03411.

Solomon, S.C., and Sleep, N.H., 1974, Some simple physical models for absolute plate motions: Journal of Geophysical Research, v. 79, p. 2557–2567.

Staudigel, H., 2003, Hydrothermal alteration processes in the oceanic crust, *in* Holland, H.D., Turekian, K.K., and Rudnick, R.L., eds., Treatise on geochemistry, Volume 3: Oxford, Elsevier Pergamon, p. 1511–1535.

Steinberger, B., 2000, Plumes in a convecting mantle: Models and observations for individual hotspots: Journal of Geophysical Research, v. 105, p. 11,127–11,152, doi: 10.1029/1999JB900398.

Steinberger, B., Sutherland, R., and O'Connell, R.J., 2004, Prediction of Emperor-Hawaii seamount locations from a revised model of global plate motion and mantle flow: Nature, v. 430, p. 167–173, doi: 10.1038/nature02660.

Stuart, W.D., Foulger, G.R., and Barall, M., 2007 (this volume), Propagation of the Hawaiian-Emperor volcano chain by Pacific plate cooling stress, *in* Foulger, G.R., and Jurdy, D.M., eds., Plates, plumes, and planetary processes: Boulder, Colorado, Geological Society of America Special Paper 430, doi: 10.1130/2007.2430(24).

Swift, S.A., Ebinger, C.J., and Tucholke, B.E., 1986, Seismic stratigraphy correlation across the New England seamounts, western North Atlantic Ocean: Geology, v. 14, p. 346–349, doi: 10.1130/0091-7613(1986)14<346:SSCATN>2.0.CO;2.

Uchupi, R., Philleps, J.D., and Prada, K.E., 1970, Origin and structure of New-England–seamount-chain: Deep Sea Research, v. 17, p. 483–494.

van der Hilst, R.D., and de Hoop, M.V., 2005, Banana-doughnut kernels and mantle tomography: Geophysical Journal International, v. 163, p. 956–961.

Walker, K.T., Bokelmannn, G.H.R., Klemperer, S.I., and Nyblade, A., 2005, Shear wave splitting around hotspots: Evidence for upwelling-related mantle flow?, in Foulger, G.R., et al., eds., Plates, plumes, and paradigms: Boulder, Colorado, Geological Society of America Special Paper 388, p. 171–192.

Wallace, P.J., 2005, Volatiles in subduction zone magmas: Concentrations and fluxes based on melt inclusion and volcanic gas data: Journal of Volcanology and Geothermal Research, v. 140, p. 217–240, doi: 10.1016/j.jvolgeores.2004.07.023.

Watson, S., and Mckenzie, D., 1991, Melt generation by plumes—A study of Hawaiian volcanism: Journal of Petrology, v. 32, p. 501–537.

Watts, A.B., and ten Brink, U.S., 1989, Crustal structure, flexure, and subsidence history of the Hawaiian Islands: Journal of Geophysical Research, v. 94, p. 10,473–10,500.

White, R.S., McKenzie, D., and O'Nions, R.K., 1992, Oceanic crustal thickness from seismic measurements and rare earth element inversions: Journal of Geophysical Research, v. 97, p. 19,683–19,715.

MANUSCRIPT ACCEPTED BY THE SOCIETY 31 JANUARY 2007

DISCUSSION

12 November 2006, Don L. Anderson

In contrast to Sleep I spent considerable time early in my career attempting to *prove* the existence of plumes and to rationalize the numerous paradoxes and failures of predictions. By 1987, I had found the plume hypothesis to be contrived and wanting (Anderson, 1981a,b, 1984, 1985, 1987) and had started to develop alternative models involving mantle heterogeneity, fertile diapirs (chemical plumes), and an asthenosphere near or above the solidus. Early plume advocates dismissed alternatives, but these were mainly propagating crack models, involving homogeneous cold mantle. The mantle was assumed to be isothermal and to require importation of core heat to melt, except at ridges and arcs. A complete theory of mantle magmatism requires treatment of both the lithosphere and the underlying mantle. If the mantle is assumed to be homogeneous and isothermal, and to be well below the melting point (standard assumptions in the plume hypothesis) then one's attention is naturally drawn to the lower mantle and core. But when the effects of pressure are considered (which Sleep ignores), then there are also problems with the deep thermal and bottom-up explanations for surface magmatism. Narrow thermal plumes do not spontaneously form in realistic non-Boussinesq mantles, and one is forced to either just assume that they exist, as Sleep does, or to force them to exist by injecting hot fluids into a tank of fluid (in the absence of realistic pressure gradients).

Plumes originally met the minimum requirement for a scientific hypothesis in that they lead to testable predictions (Morgan, 1972). None of the predictions in this classic paper have been confirmed, and most of the assumptions have been shown to be wrong or implausible. The predictions include the fixity of hotspots, the parallelism of volcanic chains, the dimensions of plumes, the number of predicted Hawaii-sized features, the global heat and magma fluxes, the temperatures of magmas, independence from surface features, driving forces of plate tectonics, origin of continental break-up forces, and the predicted uplift and local heatflow. The flux estimates in Sleep (2006), in fact, are below Morgan's acceptable range: "the ridges would close up." Morgan also predicted the style of mantle convection; narrow hot upwellings were compensated by broad diffuse downwellings. This scenario is the exact opposite of our current understanding of mantle convection: narrow downgoing slabs drive the plates, and internal heating gives rise to broad diffuse upwellings. We now also know that many midplate volcanic features were formed at plate boundaries and were stranded by later plate reorganizations. The hypothesis has been repeatedly modified to satisfy each new observation, and it is no longer possible to test the hypothesis or even to define plumes. Furthermore, linear island chains and age progressions are not a unique prediction of the plume hypothesis. More seriously, Sleep does not even attempt to address the physics, which appears to rule out plumes or to justify the strong heating from below that underlies the hypothesis. In Sleep's article, the term "shallow fertile anomaly" could replace every use of "plume," with no contradiction with any observation. Where Sleep uses "hot" he could equally well have used "low melting temperature" or "fertile." Low seismic velocities are taken as unambiguous evidence for "plume" and evidence against delamination. In fact, delaminated eclogite, or cold peridotite with H_2O or CO_2 also have low seismic velocities. The fertile blob hypothesis also satisfies the apparent ability of melting anomalies to cross ridges.

It is instructive that recent attempts to identify "real plumes" (Courtillot et al., 2003) do not use any thermal criteria (magma temperature, heatflow, precursory uplift, lithospheric erosion) or any of the predicted properties of Morgan (1972), but use only proxy criteria that have been assigned to plumes assuming that they are plumes. The criteria used, in fact, also apply to fertile blobs, and some apply to other or all shallow tectonic processes.

Although the original plume hypothesis was testable and falsifiable, the new versions are not. Sleep agrees that many of his selected observations have alternate explanations, that many hotspots are not plumes, and that even his own calculations could apply to low melting point material rather than to thermal plumes. Because many observations, such as volume flux, seismic velocity, uplift, and composition (including helium isotopes), are satisfied by features that all agree are not plumes (e.g., Courtillot et al., 2003; Anderson, 2005a), there is the danger that one can pick and choose among the many proposed hotspots to find those that satisfy any particular set of criteria, and then to use these to prove the existence of plumes. Sleep focuses on Iceland and Hawaii, but these also have alternate explanations that involve tectonics, compositional heterogeneity, delamination, ponding, and non-peridotite or upper-mantle sources. There is no tomographic evidence that they extend into the lower mantle.

13 November 2006, Norman H. Sleep

I thank Don Anderson for his comments. The intent of my article is to lay out some features of modern plume theory and to point out the dearth and poor resolution of available data. That is, I made an argument from partial ignorance that it is premature to jettison the plume concept.

I agree that plume theory has evolved since the paper of Morgan (1972). I also agree that the deep processes that allegedly feed plumes are not well understood. I did not address this issue in my article; it is partly disjoint from the existence of plumes at mid-mantle and sublithospheric depths. Seismic tomography and petrology provide direct evidence. The plume hypothesis is testable, and tests will be possible when such data become more reliable and better understood.

Current mid-mantle tomography is fuzzy, but does resolve low-velocity features in expected places. The possibility of small amounts of partial melt from minor components makes it difficult to relate seismic velocity changes to temperature changes. Current work certainly cannot resolve chemical from thermal features on the scale of plumes.

In general, petrology is a potent way to examine the chemistry and temperature of the rock that actually melts. However, petrological work has yet to produce agreement on the composition and source temperature beneath hotspots. Herzberg et al. (2007) provide an update. Lacking this agreement, I concentrated on the depth of melting beneath Hawaii and the volume of melting beneath Iceland. Higher temperatures from plumes have the correct sign and magnitude. My approach delineates the minimal requirements for a chemical alternative.

My main modification to the original plume theory is to concentrate on the implications of plumes as regions of hot buoyant upwelling fluid. Plume material is indeed a modest item in the global heat and mass budget—flow driven by plates and slabs dominates. Plumes advect in the rest of the flow. Hotspot fixity is not a prediction of my form of the theory. Non-fixity is potentially predictable from fluid dynamics, as is the deviation of plume conduits from vertical cylinders.

I concentrated on processes that affect bathymetry in my article. Plume material ponds beneath the lithosphere and flows laterally. To be sure, a continuous supply of chemically buoyant material would behave similarly, but not identically. The sign of the slope of the base of the lithosphere is known near ridge axes and along passive margins. Dynamic calculations in three dimensions quantify lateral flow. I chose the Hawaiian swell as a

resolved feature. Tomography might resolve the structure near the base of the lithosphere and thus indicate chemical versus thermal differences.

Yes, an advocate of plumes must pick and choose. That is, one needs to use data to tentatively sort out primary hotspots underlain by plumes, secondary hotspots from lateral flow away from plumes, and low-volume volcanoes that tap sources at ambient mantle temperature. Right now, bathymetric tracks and lineations meagerly constrain speculation. Many submarine edifices are still unsampled and undated.

24 December 2006, Geoffrey F. Davies

A complementary presentation of arguments relating to the existence of mantle plumes is given by Davies (2005). I quote the abstract:

The existence of at least several plumes in the Earth's mantle can be inferred with few assumptions from well-established observations. As well, thermal mantle plumes can be predicted from well-established and quantified fluid dynamics and a plausible assumption about the Earth's early thermal state. Some additional important observations, especially of flood basalts and rift-related magmatism, have been shown to be plausibly consistent with the physical theory. Recent claims to have detected plumes using seismic tomography may comprise the most direct evidence for plumes, but plume tails are likely to be difficult to resolve definitively and the claims need to be well tested. Although significant questions remain about its viability, the plume hypothesis thus seems to be well worth continued investigation. Nevertheless there are many non-plate–related magmatic phenomena whose association with plumes is unclear or unlikely. Compositional buoyancy has recently been shown potentially to substantially complicate the dynamics of plumes, and this may lead to explanations for a wider range of phenomena, including "headless" hotspot tracks, than purely thermal plumes. (p. 1541)

I have no significant disagreement with Sleep's article. I just think the inference of some plumes from observations is more direct and straightforward than is perhaps implied by him. Also, the basic physics of thermal plumes is by now well understood and well quantified, with a significant number of predictions being quantitatively confirmed by observations. A summary of important predictions and relevant observations is given by Campbell and Davies (2006).

1 January 2006, Don L. Anderson

Sleep develops a passive version of the plume hypothesis that parallels in many respects the fertile blob model. Hotspot fixity is not a prediction of his form of the plume theory. Plumes advect in the rest of the flow, pond beneath the lithosphere, and flow laterally. These are the same as predictions of the fertile blob model. However, blobs are basically upper-mantle features; if they are entrained in a broad asthenospheric counterflow channel, then they will drift counter to plate motions and will appear to define a fixed reference system for each plate. Plumes have more complex trajectories.

Compositional and melt-induced buoyancy not only substantially complicate the plume hypothesis, but can potentially remove the perceived need for plumes. The melting point of fertile mafic blobs is about 200° C lower than for peridotite, about the same magnitude as excess temperatures required for plumes. Realistic convection simulations with large plates, internal heating, and temperature-dependent properties raise the background temperature of the mantle, again lessening the need to import high temperatures from deep in the mantle. The delaminated-lithosphere hypothesis is different from the delaminated-crust hypothesis. Ambient mantle upwells to replace the material that delaminates. Its source temperature is that of the underlying asthenosphere but need not be the same as mid-ocean ridge basalt (MORB) temperatures at spreading ridges. Mantle under thick, long-lived plates can be hotter than the shallow mantle under mature ridges.

One source of fertile blobs is crustal delamination. Over-thickened basaltic crust does not just form in mountain belts. It can form in island arcs and at ridge-ridge-ridge triple junctions, where 3-D focusing of magma occurs. Oceanic plateaus, with thick crust, may form at such places. Underplating and ponding may also be involved if extrusion is prohibited by the stress state of the lithosphere. Subduction of seamount chains and aseismic ridges also introduce thick mafic sections into the mantle. By contrast, the plume hypothesis assumes that the upper mantle is homogenous, cold, and roughly isothermal so that excessive melting requires importing high absolute temperature. In the delamination and fertile blob hypotheses, melting anomalies result from increased fertility or homologous temperature rather than excess absolute temperature. The fertile blob hypothesis makes different predictions about mantle temperature, heatflow, and uplift timing compared to the thermal plume hypothesis, but many of the effects of a buoyant blob are the same as those of a hot plume head. Mafic blobs may be tens of kilometers in dimension and therefore differ from pyroxenite-vein or marble-cake models.

Tomography is unlikely to resolve the hot plume versus fertile blob issue. Both give negative shear-velocity anomalies, and it is agreed that the tails, if any, are unresolvable. Poisson's ratio and attenuation may be able to distinguish between the two, and there are already some claims in this direction. One needs more than the sign of the velocity anomaly. Delaminated crustal material or subducted aseismic ridges at near-ambient temperatures will also have low seismic velocities.

The excess fertility of blob material can explain voluminous magma production. Changes in lithospheric stress, and the presence of a blob, can explain the sudden outbreak of magmatism in previously quiescent regions. Pre-existing plate boundaries and fracture zones, underlain by subducted seamount chains or delaminated arc crust, explain most hotspot tracks. Others can be attributed to the stress state of the plate or to incipient plate boundaries. The absence of hotspot tracks leading away from most large igneous provinces is then easy to understand.

2 January 2007, Dean C. Presnall

Sleep mentions that temperature is a critically important variable in the plume debate, and petrology should be able to help. As several major plume candidates (Iceland, Azores, Tristan, Galápagos, Afar, Easter, Bouvet) lie on or very close to ocean ridges, tight constraints on temperatures of MORB generation along ridges would settle at least a major part of the controversy. Unfortunately, petrologists have been debating the temperature and depth ranges of MORB generation for about 40 years, with no significant convergence of conclusions.

The MORB model that is probably the most widely accepted (Klein and Langmuir, 1987) assumes a very wide range of magma-generation temperatures, and this has nourished the concept of hot plumes. However, in a global reexamination of MORB glass compositions, Presnall and Gudfinnsson (2007) found that the model of Klein and Langmuir is not supported by data from any ridge segment. Instead, the systematics of the Na_8 and Fe_8 parameters used by Klein and Langmuir (1987) are beautifully consistent with solidus phase relations in the CaO-MgO-Al_2O_3-SiO_2-Na_2O-FeO-Fe_2O_3 system at a globally uniform potential temperature of ~1240–1260° C and a pressure range of ~0.9–1.5 GPa (Presnall et al., 2002). As this seven-component system contains all the major minerals in the mantle and about 99% of the composition of both the source (lherzolite or some mixture of lherzolite and basalt) and the extracted basaltic melts, it provides constraints that are particularly robust.

For all ridges, including Iceland, the modeling of Presnall and Gudfinnsson (2007) replaces large potential temperature variations and hot plumes with mantle heterogeneity and uniformly low temperatures of magma generation.

REFERENCES CITED

Anderson, D.L., 1981a, Hotspots, basalts and the evolution of the mantle: Science, v. 213, p. 82–89.

Anderson, D.L., 1981b, Rise of deep diapirs: Geology, v. 9, p. 7–9 and v. 10, p. 561–562.

Anderson, D.L., 1984, The Earth as a planet: Paradigms and paradoxes: Science, v. 223, p. 347–355.

Anderson, D.L., 1985, Hotspot magmas can form by fractionation and contamination of MORB: Nature, v. 318, p. 145–149.

Anderson, D.L., 1987, The depths of mantle reservoirs, *in* Mysen, B.O., ed., Magmatic processes: University Park, Pennsylvania, Geochemical Society Special Publication 1, p. 3–12.

Anderson, D.L., 2005b, Self-gravity, self-consistency, and self-organization in geodynamics and geochemistry, *in* van der Hilst, R.D., et al., eds., Earth's deep mantle: Structure, composition, and evolution: Washington, D.C., American Geophysical Union Geophysical Monograph 160, p. 165–186.

Campbell, I.H., and Davies, G.F., 2006, Do mantle plumes exist?, Episodes, v. 29, p. 162–168.

Courtillot, V., Davaille, A., Besse, J., and Stock, J.M., 2003, Three distinct types of hotspots in the Earth's mantle: Earth and Planetary Science Letters, v. 205, p. 295–308.

Davies, G.F., 2005, A case for mantle plumes: Chinese Science Bulletin, v. 50, p. 1541–1554.

Herzberg, C., Asimow, P.D., Arndt, N., Niu, Y., Lesher, C.M., Fitton, J.G., Cheadle, M.J., and Saunders, A.D., 2007, Temperatures in ambient mantle and plumes: Constraints from basalts, picrites, and komatiites: Geochemistry, Geophysics, Geosystems, v. 8, art. no. Q02006, doi: 10.1029/2006GC001390.

Klein, E.M., and Langmuir, C.H., 1987, Global correlations of ocean ridge basalt chemistry with axial depth and crustal thickness: Journal of Geophysical Research, v. 92, p. 8089–8115.

Morgan, W.J., 1972, Deep mantle convection plumes and plate motions: Bulletin of the American Association of Petroleum Geology, v. 56, p. 203–213.

Presnall, D.C., Gudfinnnsson, G.H., and Walter, M. J., 2002, Generation of mid-ocean ridge basalts at pressures from 1 to 7 Gpa: Geochimica et Cosmochimica Acta, v. 66, p. 2073–2090.

Presnall, D.C., and Gudfinsson, G., 2007, Global Na8-Fe8 Systematics of MORBs: Implications for mantle heterogeneity, temperature, and plumes, Abstract EGU2007-A-00436, European Geophysical Union General Assembly, 15–20 April, Vienna.

Sleep, N.H., 2006, Mantle plumes from top to bottom: Earth Science Reviews, v. 77, p. 231–271.

Sleep, N.H., 2007 (this volume), Origins of the plume hypothesis and some of its implications, *in* Foulger, G.R., and Jurdy, D.M., eds., Plates, plumes, and planetary processes: Boulder, Colorado, Geological Society of America Special Paper 430, doi: 10.1130/2007.2430(02).

The Geological Society of America
Special Paper 430
2007

The eclogite engine: Chemical
geodynamics as a Galileo thermometer

Don L. Anderson*

California Institute of Technology, Seismological Laboratory 252-21,
Pasadena, California 91125, USA

ABSTRACT

Migrating and incipient ridges and triple junctions sample the heterogeneous mantle created by plate tectonics and crustal stoping. The result is a yo-yo vertical convection mode that fertilizes, cools, and removes heat from the mantle. This mode of mantle convection is similar to the operation of a Galileo thermometer (GT).[1] The GT mode of small-scale convection, as applied to the mantle, differs from the Rayleigh-Taylor (RT) instability of a homogeneous fluid in a thermal boundary layer. It involves stoping of over-thickened continental crust and the differences in density and melting behavior of eclogites and peridotites in the mantle. The fates of subducted and delaminated crust, underplated basalt, and peridotite differ because of differences in scale, age, temperature, melting temperature, chemistry, thermal properties, and density. Cold subducted oceanic crust—as eclogite—although denser than ambient mantle at shallow depths, may become less dense or neutrally buoyant somewhere in the upper mantle and transition zone, and may be gravitationally trapped to form mafic eclogite-rich blobs or layers. Detached lower continental crust starts out warmer; it thermally and gravitationally equilibrates at shallower depths than do slabs of cold mature lithosphere. The density jumps at the depths of 400 and 650 km act as barriers. Trapped eclogite is heated by conduction from the surrounding mantle and its own radioactivity. It is displaced, entrained, and melted as it warms up to ambient mantle temperature. Both the foundering and the re-emergence of mafic and ultramafic blobs create midplate magmatism and uplift. Mantle upwellings and partially molten blobs need not be hotter than ambient mantle or from a deep thermal boundary layer. The fertile blobs drift slowly in the opposite direction to plate motions—the counterflow model—thereby maintaining age progressions and small relative motions between hotspots. Large-scale midplate volcanism is due to mantle fertility anomalies, such as large chunks of delaminated crust or subducted seamount chains, or to the release of accumulated underplate when the plate experiences flexure or pre-breakup extension.

---

*E-mail: dla@gps.caltech.edu.

[1] A Galileo thermometer is a liquid-filled container with suspended objects in it. As the liquid changes temperature, the suspended objects rise and fall to stay at the position where their density is equal to that of the surrounding liquid, or until they hit a barrier. The controlling parameter is buoyancy; an object's mass relative to the mass of the liquid displaced. If the object's mass is greater than the mass of liquid displaced, the object will sink. If the object's mass is less than that of liquid displaced, it will rise. Different blobs settle at different depths, for a while.

Anderson, D.L., 2007, The eclogite engine: Chemical geodynamics as a Galileo thermometer, *in* Foulger, G.R., and Jurdy, D.M., eds., Plates, plumes, and planetary processes: Geological Society of America Special Paper 430, p. 47–64, doi: 10.1130/2007.2430(03). For permission to copy, contact editing@geosociety.org.

Eclogite can have lower shear velocities than volatile-free peridotite and will show up
in seismic tomograms as low-velocity, or red, regions, even when cold and dense. This
model removes the paradoxes associated with deep thermal RT instabilities, propagating cracks and small-scale thermal convection. It explains such observations as relative fixity of melting spots, even though the fertile blobs are shallow.

Keywords: delamination, small-scale convection, plumes, blobs, LIPs, eclogite

INTRODUCTION

The subduction of oceanic lithosphere drives convection
and creates chemical heterogeneity in the mantle. Subduction
and conduction of heat through the surface boundary layer are
the main mechanisms for cooling the mantle. The delamination
or stoping of lower continental crust also affects the dynamics,
temperature, and composition of the mantle. It is a completely
different mechanism of cooling and convection than is traditionally considered (see the Appendix). Most of the volcanism
of the world is associated with plate boundaries; ~24 km^3 of
magma is extracted from the mantle each year at plate margins,
mostly at mid-ocean ridges. Only minor amounts of magma
(1–2 km^3/yr) are removed by intraplate (also called mid-plate,
hotspot, or anomalous) volcanism. That magmatism is geographically widespread implies that partial melting of the mantle is a common phenomenon, even though the melting
temperature of dry pyrolite is very high. The volume of anomalous volcanism is much less than the amount of oceanic crust
that is recycled back into the mantle. Melting in the asthenosphere is widespread, but magma may not always be able to rise
to the surface. Minor fluctuations in temperature and fertility
(eclogite content) are expected to give large variations in melting for a mantle that is close to the average melting temperature.

Most, if not all, volcanic regions lie in areas of extensional
lithospheric stress. Melt is extracted from the mantle through
dikes that form in such regions, often by magma fracture and
shattering of the country rock (stoping). A necessary but not sufficient condition for magmatism is extensional stress, modulated
by lithospheric fabric and architecture. A more controversial issue is the required condition for large amounts of melting in so-
called hotspot areas. It may be mantle fertility rather than high
absolute temperature or water content and may involve magma
ponding and episodic release of melt. It may also involve the stability of over-thickened continental crust. Extending regions
and rift zones are not all expected to be equally productive of
magma.

DEFINITIONS

Eclogite, as used here, is a garnet- and clinopyroxene-rich
rock of unspecified origin. It includes jadeite-poor materials
usually referred to as *garnet clinopyroxenites.* Mantle eclogites
and pyroxenites have a variety of origins and can be metamor-

phic, igneous, or reaction products (Lee et al., 2006). Eclogites
may be high-pressure crystallized melts, remnants of subducted
oceanic crust, or the residue of partial melting of such crust.
Some mantle eclogites may be products of metamorphism of
mafic lower continental crust, that is, gabbroic to anorthositic
protoliths, and some may be high-pressure garnet-pyroxene cumulates. Secondary eclogites can form by the interaction of
eclogite partial melts with peridotites, or the intrusion of basalt
into lower crustal cumulates.

Piclogite is an assemblage that includes eclogites and peridotites, and low- and high-melting temperature components.
The components may be kilometers in dimension and separation. Piclogite does not necessarily come in hand-specimen-
sized samples. The term "piclogite" was coined because eclogites
have a specific composition and origin in the metamorphic literature. Piclogites may be thought of as assemblages that have
less olivine and orthopyroxene than peridotites, and are of variable fertility and density. It is unlikely that fertile blobs in the
mantle are pure eclogite; they are likely to be attached to or surrounded by peridotite and therefore form piclogite packages that
can become buoyant as they warm up.

Stoping is a process by which magma intrudes; blocks of
wall rock break off and sink. In *large-scale magmatic stoping*
(Daly, 1933), magma rises and shatters but does not melt the surrounding rocks, which sink, making room for more magma to
rise. This theory was instrumental in explaining the structure of
many igneous rock formations.

THE MANTLE AS AN ENGINE

The type of convection proposed in this article differs so
dramatically from the type presently being pursued by convection modelers (see the Appendix) that a different metaphor, or
conceptual model, is needed. The pot-on-the-stove metaphor is
not appropriate. Large changes in properties due to compression, phase changes, heat exchangers, condensers, pistons, and
repetitious cycles are essential to the operation of an engine.
Thermodynamic consistency, material properties that depend on
both temperature and pressure, low efficiency, a way to expel
waste heat, and obedience to the second law of thermodynamics distinguish a real engine from a perpetual motion machine
and some models of mantle convection. Mantle convection is
often simulated with fluids that have constant properties or properties that depend only on temperature, no large volume or ma-

terial property changes, no phase changes, inconsistent thermo-dynamic approximations, and, sometimes, violations of the second law. It is useful to think of mantle convection as a self-consistent thermodynamic cycle or an engine, rather than as an ad hoc cycle that is driven by thermal expansion alone at constant pressure (see the Appendix).

The mantle is a machine that converts energy into mechanical forces and motions. It is therefore an engine. In the eclogite engine, the working fluids are not water and steam but basalt, eclogite, and magma, and packages of the same. At the top of the system, magma and basalt are converted to dense eclogite by cooling and compression. This "fluid" is further compressed—adiabatically—by sinking into the mantle, which is, in effect, an infinite heat source. The eclogite package is heated by conduction from the heat bath, and heated by "internal combustion" (radioactivity). It expands and changes phase, causing it to further decompress and rise. At the end of the stroke, it discharges some heat and magma through the surface heat sink and the cycle repeats. This process cools the mantle and brings magma to the surface.

As the surface layer thickens by cooling, underplating, intrusion, extrusion, and compression, it can become unstable. If the upper part is buoyant, or stiff, the lower portion can peel off or founder (stope) and be replaced by asthenosphere or magma. This stage is accompanied by magmatism. Delaminated material will melt as it warms up by conduction from ambient mantle. This process reverses the density contrast and the material rises back to the surface; ascent increases the melting and further lowers the density. A second stage of magmatism is the result. This process is also a form of convection; it can be called *yo-yo tectonics,* an up-down motion controlled from the top and by gravity. Relative buoyancy is mainly controlled by phase changes, gabbro to eclogite, and melting. Density variations associated with these phase changes are an order of magnitude greater than can be achieved by thermal expansion, and they can be more abrupt than normal thermal expansion. Industrial engines do not work well without the large volume changes associated with pressure and phase changes.

Normal thermal convection is driven by thermal expansion, which gives a gradual decrease of density with a rise in temperature. The eclogite engine is driven by more abrupt changes in density at the basalt-eclogite-melt phase boundaries and the rapid decrease in density of the melt phase with decreasing pressure. The basalt-eclogite transition occurs over a finite interval (Ringwood and Green, 1966) of temperature and pressure but it is still abrupt, and large, compared to thermal expansion–induced density changes, and it is added to other thermal effects. Eclogite trapped at a chemical or phase boundary will be denser than the overlying mantle. It therefore takes some time to heat up to neutral buoyancy and more time before the accumulated buoyancy causes it to rise. Parts of a deep layer can also be displaced upward by subsequent subduction and delamination, rather than by local buoyancy and a Rayleigh-Taylor (RT) in-stability. As a displaced blob rises, it may melt. This kind of convection does not require a superadiabatic gradient or a deep thermal boundary layer. The ascent is terminated by density or strength barriers. The further step of intrusion and eruption requires the cooperation of the outer shell. Underplating, or ponding, may lead to crustal thickening and future instabilities and melt release.

UNDERLYING ASSUMPTIONS

A basic premise of this article is that the large volume changes associated with partial melting, and other phase changes, and the large density differences between eclogites and peridotites may be as important as thermal expansion in mantle dynamics. Another premise is that melt volumes and melt compositions are more a function of mantle composition and lithology than of absolute temperature, and that the mantle is lithologically heterogeneous. This assumption contrasts with the homogeneous mantle and melting models of McKenzie and Bickle (1988) and Langmuir et al. (1992) that dominate the current petrological literature (see the Appendix and other articles in this volume; for a complete guide to the literature of mantle geodynamics and geochemistry, see http://www.mantleplumes.org/). An important but often overlooked result of petrology and phase relations is that melt compositions can be similar for a wide variety of starting materials (Presnall, 1969; Jacques and Green, 1980). In contrast to standard models, absolute temperature variations may have almost nothing to do with the so-called global arrays of geochemical parameters, the volumes of basalts erupted at melting anomalies, or the driving forces of mantle convection. The assumptions underlying conventional models of mantle convection and mantle geochemistry are given in Anderson (2005a,b), Tackley et al. (2005), and in the Appendix.

Large magma fluxes can be produced from eclogite-rich regions of the mantle (piclogite) without high absolute temperatures (Takahashi et al., 1998; Yasuda and Fujii, 1998). Takahashi and Nakajima (2002) suggested that some Hawaiian basalts were produced from large eclogite blocks with an inferred excess potential temperature of ~100 °C. However, high potential temperatures for Hawaii are implied with the standard homogeneous peridotitic source assumption and the assumption that a given magma is a unique product of a given composition and temperature.

It is mantle heterogeneity that mainly explains the diversity of mid-ocean ridge basalts (MORBs), and melt volume and elevation anomalies. There are numerous ways to introduce heterogeneity into the mantle, but I concentrate on the basalt and eclogite blobs that get into the mantle by processes other than subduction. Convective homogenization, another theme of standard models, is of secondary importance (e.g., Anderson, 2002d; Meibom and Anderson, 2004). Plate tectonics, recycling, crustal stoping, and differentiation serve to dehomogenize the mantle, and these are ongoing processes.

RT INSTABILITIES

In a cooling fluid, the surface thermal boundary layer becomes unstable at relatively short times and thicknesses. Part of the outer boundary layer of the mantle is compositionally buoyant, which extends the conduction gradient to greater depths. When the temperature dependence of viscosity is taken into account, the boundary layer becomes thicker still. The buoyancy and extra stiffness require more force, and more cooling time, to remove it. The average thickness of the boundary layer of the mantle has been estimated to be 280 km (Kaula, 1983), whereas the maximum thickness for a purely thermal boundary layer in an homogenous fluid having mantle properties is more like 100 km. Kaula (1983) also estimated that the potential temperature at the fully convective depth was ~1410 ± 180 °C, which is higher than petrological estimates derived from magmas—which may come from shallower depths—but is consistent with the thicker conduction layer. Under these conditions, the lower part of the boundary layer may peel off, delaminate, or founder, leaving the colder, lower density and stiffer part behind.

With modern estimates of mantle and melting temperatures (see http://www.mantleplumes.org/Temperature.html) and their variations, it is difficult to avoid localized melting and the partial melting explanation of the asthenosphere (e.g., Anderson and Sammis, 1970; Lambert and Wyllie, 1970). The mantle is hotter, and solidi can be lower and more variable, than generally assumed. The shallow mantle is replenished from above, in part by the insertion of low-melting-temperature mafic material lost from the base of continents. A mass balance calculation can be done for this mechanism. A homogeneous, partially molten peridotitic asthenosphere is expected to drain rapidly, but a mantle with variable melting temperatures and unconnected fertile regions can develop melt retention buoyancy. The partial melt explanation for regions of the asthenosphere with the lowest seismic velocity does not violate global tomographic studies, as these studies average out the extremes of velocity variations.

MASS BALANCE

Oceanic crust is created and destroyed at about the same rates. It is not clear if this process represents a cycle or if oceanic crust is lost from the system over periods of billions of years or longer, to be replaced by melts from the displaced upper mantle. One billion years of oceanic crust formation or subduction is equivalent to a 70-km-thick layer of eclogite. This material may be in the transition region, and there is no mass balance calculation or tomographic image that prohibits the process, or that requires oceanic crustal recycling. Material is also removed from and added to the continental crust.

The total subduction rate of oceanic crust is ~20 km³/yr. Midplate magmatism is roughly 1–2 km³/yr, comparable to the rates of crustal delamination and seamount subduction. Scholl (2006) estimates 2.5 km³/yr for the underside erosion of continents at marine margins and 2–3 km³/yr of erosion plus delamination at continental collisions. Some of this material may be dragged beneath the continent and underplated, but there is a much closer match between rates of midplate volcanism and continental crustal recycling than there is with rates of oceanic crustal recycling.

Larger and hotter chunks may be involved in the stoping part of the process than would be the case with subduction of normal oceanic crust. This material may be responsible for fertile melting anomalies, in addition to contributing trace element and isotopic inhomogeneities to the mantle. This eclogite, or subterranean, cycle serves to fertilize and cool the mantle.

LAMINATING THE MANTLE

If the upper mantle is compositionally variable, convection and mantle temperatures will be quite different from standard pictures involving a single surface thermal boundary layer and a vigorously stirred mantle with a high Rayleigh number. Cooling of the mantle, top-down convection, and plate tectonics can still occur. Small-scale features in the mantle are below the resolution of global tomographic and geoid data but are resolvable with high-resolution geophysical data (Anderson, 2005a,b, 2006). For example, many robust mantle reflectors occur above a depth of 1300 km (Deuss and Woodhouse, 2002), implying some sort of stratification. Scatterers and less robust (less reproducible) reflectors imply, in addition, a blobby mantle. All of this detail is averaged out in global tomographic images.

The density and shear-wave seismic velocity of crustal and mantle minerals and rocks at standard temperature and pressure (STP) are arranged according to increasing density in Figure 1. This approximates the situation in an ideally chemically stratified mantle. P and T effects may change the ordering and the velocity and density jumps but do not change the overall picture. Eclogite can settle to various levels, depending on composition; the eclogite bodies that can sink to greater depths because of their density and size have low seismic velocities compared to other rocks with similar density. MORB-eclogite contains stishovite at high pressure and may sink deeper than other eclogites. Subducted and delaminated material eventually heats up and may rise, even if it does not lie in a thermal boundary layer, creating a gigantic Galileo thermometer (Fig. 2).

The Earth—and the mantle—is stratified by density, composition, and phase (Anderson, 2002d). It is also stratified by viscosity. A low-viscosity zone in the upper mantle—a natural result of the depth variation of P and T—was prominent in early discussions of the structure of mantle flow, particularly plate-driven shallow counterflow (Elsasser, 1969; Jacoby, 1970). It is important in the present discussion because it explains why hotspots show such small relative motions and mimic the behavior of fixed thermal anomalies (see the section on counterflow in the Appendix).

DENSITY VERSUS DEPTH

The possible density crossover, at depth, between eclogite and peridotite (e.g., Anderson, 1979a,b, 1989a,b) is well known

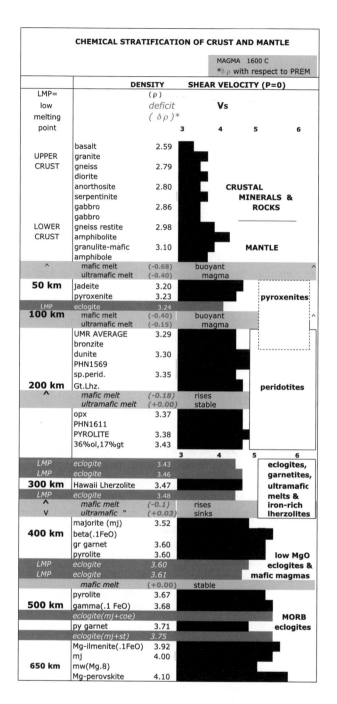

Figure 1. Crustal and mantle minerals arranged in order of increasing density. The densities and shear velocities for standard temperature and pressure are given. The depth scale is approximate. Along the mantle geotherm, the order will change because eclogite will be above the solidus. The densities of mafic and ultramafic melts are given in terms of the density deficit at 1600 °C and high pressure with respect to the Preliminary Reference Earth Model (PREM; Dziewonski and Anderson, 1981). Because of the high compressibility of magma, the density deficit increases rapidly with decreasing depth. Partially molten eclogite diapirs become increasingly buoyant as they rise. Some magmas are neutrally buoyant at depth in the mantle. Multiple discontinuities and low-velocity zones (eclogite or magma) are predicted in a chemically stratified mantle. These features may not be obvious in global tomography, but they shown up in high-resolution reflection, receiver function, and scattering experiments. The layering shown is a snapshot. Eclogite becomes buoyant as it collects heat from the adjacent mantle. Delaminated eclogite will be 400–700 °C colder than ambient mantle, whereas oceanic crust can be 1300 °C colder. Deeper layers can be displaced upward by later generations of subduction and delamination and by ridge spreading. In a layered mantle, the temperature gradient is superadiabatic (a conductive geotherm), unless there is radioactive heating or insertion of cold dense material. Subadiabatic gradients may occur at depth in this situation. In general, one cannot infer temperature simply by assuming an adiabatic gradient below a depth of ~200 km. Convection in this kind of system is similar to a lava lamp and is quite different from a pot-on-a-stove or thermal plume convection, or convection in a homogeneous fluid. The system is driven by cooling from above. Abbreviations: gr—grossular; gt—garnet; Gt. Lhz.—garnet lherzolite; mj—majorite; mw—magnesiowüstite; ol—olivine; opx—orthopyroxene; PREM—reference mantle; sp. perid—spinel peridotite; UMR—ultramafic rock. Magma densities are from Sakamaki et al. (2006). Rock and mineral densities are from standard sources and databases and Lee et al. (2006).

dering in an ideally density-stratified mantle. If the density contrast between layers is greater than the thermal expansivity times the temperature fluctuations, then the stratification is stable (see the Appendix). Figure 1 shows that eclogites can have lower shear velocities than some peridotites of the same density. In a chemically stratified mantle, an eclogite blob can have similar or lower shear velocities than the peridotite that it is in density equilibrium with. A profound chemical or density boundary can have a small, or even a negative, velocity jump. The perceived absence of large seismic velocity jumps in global tomographic models of the mantle, except at the depths of 400 and 650 km, has led many to suppose that the mantle is homogeneous (e.g., Albarède and van der Hilst, 2002a,b; Bercovici and Karato, 2003). The detection of low shear velocity regions in the mantle might also lead one to speculate that these are high-temperature or partially molten regions, but they could also be compositional. Shear velocity does not correlate well with density. Visual inspection of selected tomographic cross-sections can be misleading.

PROBLEMS WITH ECLOGITE

Can eclogite or eclogite-rich material be an important magma source in the mantle? Two assumptions about eclogite sources are responsible for the perception that it cannot. The first is that eclogite is a well-defined rock type that is denser than the

but continues to be controversial (Ringwood, 1975; Irifune and Ringwood, 1993; Hirose et al., 1999; Litasov et al., 2004; Cammarano et al., 2005). The variation with depth of in situ density of eclogite and peridotite is discussed or plotted in some of these papers, which come to contradictory conclusions about the fates of eclogite and slabs in the mantle. A density crossover may lead to the formation of garnetite-rich layers or blobs in the transition zone (TZ).

What will a chemically stratified mantle look like to a seismologist? I have tabulated the measured or inferred zero-pressure densities and shear velocities of mantle minerals and mineral assemblages in Figure 1. This tabulation gives the approximate or-

mantle at all depths (e.g., Ringwood, 1975); it therefore sinks readily and rapidly into the lower mantle and is removed from the upper mantle. The second problem with eclogite is the belief that almost complete melting is required to yield a basalt. A corollary of these concerns is the belief that partial melts of eclogite will rapidly drain away, as they are predicted to do in a homogeneous peridotite mantle; extensive melting is therefore impossible. A contradictory assumption is that the fertile eclogitic portion of the mantle is distributed as small-scale lamellae or veins, the result of vigorous mantle convection and stirring (Allègre and Turcotte, 1986). Meibom and Anderson (2004) and Anderson (2006) argued that eclogite exists as discrete large blobs in the mantle, and that the homogeneity of MORB is a result of magma blending, and sampling of melt aggregations (SUMA), not chaotic stirring in the solid state. Fertile veins in a shallow peridotite can form a permeable network but isolated fertile blobs need not (see also http://www.mantleplumes.org/LowerCrust.html). A partially molten diapir may also rise faster than melt can escape (Anderson, 1981; Marsh et al., 1981).

Eclogites and basalts—which probably constitute no more than ~7% of the mantle (Anderson, 1989a)—enter the mantle as large chunks of oceanic crust and lower continental crustal cumulates and metamorphics. The oceanic crust is cold, thin, and initially buoyant; the lower crust is warm and can be thick and dense. In both cases, the eclogite, once in the mantle, will be surrounded by warmer refractory peridotite that, in general, has higher melting temperatures and low subsolidus permeabilities. Piclogites are assemblages of eclogites and peridotites, but the eclogite may occur in dispersed blobs rather than as grains or veins; it can be a chunky stew or gumbo[2] instead of a plum pudding or marble cake, or a rock type. If the chunks are small, they are not likely to rapidly sink through the mantle. Under these conditions, a fertile diapir can melt extensively as it warms up at depth and as it rises into the shallow mantle (Marsh et al., 1981; Ghods and Arkani-Hamed, 2002). This property is called "melt-retention buoyancy." The main difference between this model and melting of peridotite with fertile streaks or grains is the scale and the starting temperature of the mafic components. Extensive melting of eclogite at depth is possible because melts are much less mobile and more dense at high pressure and because the dispersed eclogite is surrounded by the more abundant peridotite. A garnet peridotite or pyrolite source region has centimeter-sized fertile streaks, whereas a piclogite source region may have fertile patches on the scale of kilometers to tens of kilometers, and these need not be interconnected. Melting reduces the density of eclogite, and this density reduction increases rapidly as the blob starts to rise, partly because magma is very compressible. At low pressure, magmas are 10–20% less dense than their source rock, but at depths on the order of 400

km, the density differential can be a few percent or less (Sakamaki et al., 2006).

A third problem with the eclogite source hypothesis is that high magnesium melts, such as noncumulate picrites and komatiites, may require an ultramafic and high-temperature source. Highly magnesian melts (high-Mg basalts, picrites, and komatiites) are rare, but they are found in several large igneous provinces and in Hawaii. Their rarity implies that the conditions for high-temperature melting and extraction of dense melts are also rare. Mantle temperatures are higher and more variable than usually assumed (Kaula, 1983; Anderson, 2000); melting temperatures are also variable, and can be lower than for dry peridotite. Melts derived from eclogite may react with the surrounding peridotite to form secondary pyroxenitic sources (Yaxley and Green, 1998; Sobolev et al., 2005), olivine being consumed by this reaction. The low-melting-temperature "eclogite" component of the mantle may actually be refertilized peridotite. The "eclogite" imprint is transferred by melts from eclogite partial melts reacting with and refertilizing peridotite, which is then the source of the picrites. Clearly, the diversity and volumes of mantle magmas depend on more than absolute temperature.

GALILEO THERMOMETER

Delamination and buoyant decompression melting mimic the behavior of a Galileo thermometer. A Galileo thermometer (GT) is a sealed liquid-filled container, with a number of suspended objects in it. It is heated or cooled from the sides by ambient air. If the blobs absorb sunlight, then they will behave as if internally heated. The relative buoyancies of the blobs change with time because the fluid temperature changes with time; the opposite is the case for mantle blobs.

The masses in the mantle are recycled, initially cold blobs, inserted from the top, that have different properties from the surrounding mantle. They change density rapidly as they warm up because of thermal expansion, partial melting, and other phase changes (Fig. 2). The mantle is cooled from the top and heated from within. The blobs are cooled from the top, sink, and are then heated by the mantle. A blob sinks because it is intrinsically denser than the top of the mantle and because it is cold. It rises as it warms if it becomes buoyant at temperatures lower than ambient mantle temperatures. This condition will be the case if the blob is intrinsically less dense than ambient mantle, or if it melts or undergoes phase changes at temperatures lower than ambient mantle. At any given time, there are blobs at different levels, just as in a GT. Basalt and eclogite blobs may contain more radioactive elements than do peridotite blobs and may heat up faster by internal heating than cold harzburgite blobs.

There are several differences between the mode of convection being discussed here and GT and RT instabilities. The heating and cooling cycles of eclogite near its solidus are not symmetrical, and the launching of eclogite into the mantle may

---

[2]A New Orleans version of a celebrated seafood stew from Provence, made with an assortment of fish and shellfish, onions, tomatoes, white wine, olive oil, garlic, saffron, and herbs. The stew is ladled over thick slices of French bread.

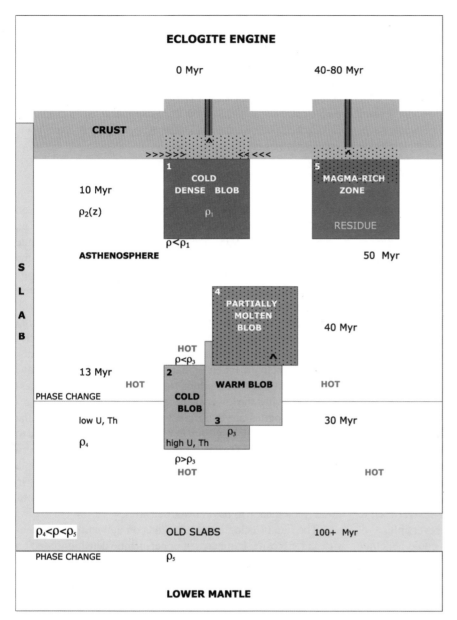

Figure 2. Schematic illustration of the operation of the eclogite cycle. Cold slabs (yellow) sink to the base of the transition region, displacing warmer or less-dense material upward. They are not treated in this chapter. (1) Over-thickened continental crust or underplated basalt cools and converts to dense eclogite, which is unstable in the shallow mantle; it may take 10 m.y. for a sufficiently thick eclogite-rich (piclogite) root to form and detach (all times are approximate). (2) The cold piclogite assemblage sinks rapidly to a depth of neutral buoyancy or is trapped by a phase-change density jump in ambient mantle. The blob is surrounded by hot ambient mantle and also has higher radioactive content than does the ambient mantle. The blob warms up gradually. (3) Liftoff occurs as a result of heating, melting, or displacement by subsequent downwellings. (4) Upwelling may be accompanied by partial melting and a further density reduction; melts drain to the top of the blob and may react with the surrounding peridotite. (5) The blob delivers melt to the shallow mantle and surface. The whole cycle may take 40–80 m.y., which is much faster than the plume cycle, which relies on the small amount of heat leaving the core. The relative densities (ρ) are shown. The approximate time scales of the various processes (cooling, delamination, heating, melt percolation) are given in Jull and Kelemen (2001), Ghods and Arkani-Hamed (2002), Raddick et al. (2002), and Anderson (2005a), and in standard texts. Also see http://www.mantleplumes.org/TopPages/LithThinTop.html and http://www.mantleplumes.org/LithGravInstab.html.

be caused by tectonic rather than buoyancy forces. Because of adiabatic decompression melting and viscous heating, the upwelling may be a runaway process.

The viability of the GT mode of convection and cooling of the mantle depends on the relative densities of the mantle and the blobs. The various parts of the cycle are classical physics problems; growth of instabilities at a cooling surface, sinking and rising of spherical blobs in a viscous fluid, heating of a sphere by conduction and internal radioactivity, and melt segregation in a rising diapir. I concentrate instead on the conditions that will allow this mode of small-scale convection rather than that of rapid removal of eclogite to the lower mantle, or rapid homogenization by stretching and folding, and the thermal implications.

THERMAL IMPLICATIONS

In a chemically stratified, internally heated mantle that undergoes periodic injections of cold slabs and warm delaminated lower continental crust, the temperature gradient will be complex; it will not be a simple adiabat or a conduction geotherm. Deep-mantle temperatures can be higher than in a homogenous convecting mantle, and there will be subadiabatic temperature gradients at depth (Jeanloz and Morris, 1987). Mattern et al. (2005) inferred a subadiatic temperature gradient and a decreasing iron content from 660 to 1300 km in depth and significant heterogeneity from 800 to 1300 km in depth. These gradients can be understood if this region, not D″, is the graveyard—

temporary or permanent—of cold dense MORB-eclogite sinkers (Wen and Anderson, 1997). Cold slabs will displace the surrounding mantle, which may melt (Fig. 2).

FATE OF ECLOGITE IN THE MANTLE

A series of papers (see below), starting with Ringwood (1975) argue that eclogite will sink deep into the mantle, rather than being trapped above depths of 650 km or at the 1000-km discontinuity, or other shallow levels, as I have argued. These papers assume that the top of the oceanic crust, MORB, is representative of all eclogites, and that if eclogite is denser than pyrolite it will sink, and pyrolite provides a good density model for the mantle. In other words, local density differences alone control the fate of eclogite. The fate of eclogite in the mantle depends, in part, on the relative densities between eclogite and the surrounding mantle, and also on the relative sizes of eclogite blobs. The mantle, particularly the transition region, may be eclogite-rich (Anderson, 1975) and denser than pyrolite (Cammarano et al., 2005). The upper mantle contains fertile peridotite, residual peridotite, dunite, and various kinds of mafic materials and melts (Fig. 1); the dominant lithology probably changes with depth. Plausible end-member lithologies range in density from 3.3 to 3.65 g/cm^3. The upper mantle may involve mixtures or layers of these; the denser lithologies will tend to lie deeper. Convection can still occur in the presence of a stabilizing density gradient, with an accompanying nonadiabatic temperature gradient, but it is unlike any kind of mantle convection that has been treated to date. It is top-down convection, similar to a GT, composed of sinking and rising blobs. It is not the whole-mantle vigorous convection that homogenizes the mantle, as is often envisaged, but it can cool the mantle and deliver melt to the surface. The stratification is temporary and reversible.

Hirose et al. (1999) suggest that if MORB could penetrate the buoyancy barrier between the depths of 650 and 720 km (in a pyrolite mantle), it could sink farther into the lower mantle. This scenario assumes that eclogite is unmodified as it sinks into the mantle (i.e., it stays unaltered and does not lose a silica-rich melt), and that some mechanism exists for dragging it through higher-density mantle. It also assumes that pyrolite is a good model for the TZ and lower mantle. Lee et al. (2004) found that the lower mantle is 2–4% denser than pyrolite, which would reverse the conclusion of Hirose et al. (1999) regarding the ability of eclogite to sink into the lower mantle.

Nishihara et al. (2005) conclude that MORB-eclogite is denser than pyrolite in the TZ; other eclogites are less MgO- and SiO$_2$-rich and should be less dense. Pyrolite may be a good model for the shallow mantle, but it is not a good match to the seismic velocities in the TZ and lower mantle (Lee et al., 2004; Cammarano et al., 2005). A pyrolite lithology has higher seismic velocities in the TZ and below than are obtained from seismology. This observation suggests more garnet, and perhaps more FeO, in the TZ than there is in pyrolite. The seismic data

also suggest a gradient with depth in the TZ, with, possibly, more high-pressure olivine and FeO-rich phases at depth (Mattern et al., 2005). The TZ and the phase change discontinuities vary laterally (Ishii and Tromp, 2004; Deuss et al., 2006), as expected if the TZ is the dumping ground, even temporarily, of eclogite. Other evidence for chemical stratification and barriers near depths of 650 and 1000 km is summarized in Anderson (2002c,d, 2005a,b) (see also the Appendix and http://www.mantleplumes.org/Eclogite.html).

RELATIVE DENSITIES

The viability of the GT and fertile blob models depends on density contrasts between eclogite and ambient mantle (e.g., Yasuda and Fujii, 1998), the size and viscosity of the eclogite components (e.g., Meibom and Anderson, 2004), and the temperature at which a density reduction sets in because of partial melting. The above papers focus on the densities and fates of subducted oceanic upper crust. Other eclogites in the mantle have deep crustal protoliths, and some are secondary, forming from reactions of eclogite melts with peridotites. Some eclogites may be cumulates that have been invaded by basaltic melts. Subsolidus eclogite is denser than most peridotites but may have a significantly lower solidus temperature; some eclogites-restites have had the lower-melting-temperature components removed. Eclogites are therefore variable in density and in the temperature of initial melting. Eclogite is commonly assumed to have a density of 3.5 g/cm^3 (Leitch and Davies, 2001) or even greater (Sleep, this volume), but this value is too high to be a good average. Eclogites have measured or inferred densities in the range from 3.30 to 3.65 g/cm^3. Packages of eclogite and peridotite-piclogites will have lower densities.

The fate of eclogite also depends on ambient mantle density and viscosity. Estimates of upper-mantle densities based on abyssal and obducted peridotites and xenoliths from the continental lithosphere range from 3.30 to 3.40 g/cm^3. Dunites have densities of 3.35–3.50 g/cm^3. These values may not be representative of the density of the deeper mantle. An intrinsic density of 3.3 g/cm^3 may be appropriate for the depleted residue that accumulates at the top of the mantle but is probably too low to represent the bulk of the mantle below some 200 km. This range of intrinsic densities corresponds to a temperature variation of 2000 °C and strongly implies that the mantle must be chemically stratified. Available samples of mantle rocks may be biased toward the more buoyant ones and the shallow mantle.

Kogiso et al. (2003, 2004) and Pertermann and Hirschmann (2003) calculate an STP solidus density of eclogite that is only 2–4% denser than many garnet peridotites and pyrolite and substantially less than the 4.5–6% contrast that is commonly assumed. Additionally, it is not just the local density contrast that controls the fate of eclogite in the mantle. It is the integrated density, the volume of the sinking or rising mass, and the surrounding viscosity. There is also a spread of densities of ~3% among various types of lherzolites and dunites. It is therefore not obvious that eclogites are substantially denser than the material of the upper mantle or the TZ. In fact, the TZ may be eclogite-rich (Anderson, 1979a,b, 1989a). Eclogites may contain more radioactive elements than do upper-mantle peridotites.

An upper-mantle blob with 10–20% subsolidus eclogite of density 3.44 g/cm^3 must be 50–300 °C hotter than an ambient pyrolite mantle to have neutral buoyancy, which would imply partial, or even complete, melting of the eclogite. Smaller excess temperatures are required if ambient mantle is more fertile and iron-rich, or if the eclogite is garnet-poor. No excess temperature is required if eclogite is sufficiently above its solidus. Eclogite can be less dense than commonly assumed, and the excess temperature required for a buoyant blob to support a given eclogite load is markedly less than usually assumed. An eclogite-bearing blob may conceivably become buoyant with no excess temperature, that is, having potential temperatures in the range inferred from MORBs and from geophysical data. But it needs to be stressed that fertile streaks or blobs in the mantle need not be intrinsically buoyant to be involved in mantle melting.

CONSISTENCY WITH HIGH-RESOLUTION SEISMOLOGY

Global tomographic images have little resolving power, but higher resolution and spectral domain and correlation techniques are available. The mantle contains many seismic discontinuities above 1300 km in depth (Deuss and Woodhouse, 2002). Deep reflection seismology shows a complicated and variable structure with single and double reflections ranging from 640 to 720 km in depth near the base of the TZ (Deuss et al., 2006). The spectral amplitude of the lateral density variations of the mantle exhibits a maximum at ~600 km in depth. The density of the mantle TZ has negative or zero correlation with shear and compressional wave velocities; TZ velocity-velocity and velocity-density correlations are completely different from the overlying and underlying regions (Ishii and Tromp, 2004). Gu et al. (2001) showed that there is a distinct change in seismic velocity variations, near 650 km in depth, with large variations above and small variations below. These observations are inconsistent with thermal variations alone or with whole-mantle convection. Subduction of cold eclogite into the TZ will lower the shear wave seismic velocity there (Fig. 1) but will have little effect on the density until the material warms up (the eclogite displaces similar-density material, as in a GT). An eclogite-rich TZ may also explain the velocity jumps at 410 and 650 km, which are too small and too large, respectively, to be entirely due to phase changes in pyrolite. Garnet does not undergo a phase change near 400 km, so it dilutes the jumps created by phase transitions in olivine and orthopyroxene at depths of 400 and 500 km. The low shear wave seismic velocity of eclogite and its low melting temperature will create large tomographic variations in the TZ, particularly as the distribution of eclogite is not expected to be uniform.

Seismic profiles calculated for a peridotite mantle, compared to the Preliminary Reference Earth Model (PREM) and other seismic models, have lower velocities above 400 km, larger jumps near 410 km, lower TZ gradients, lower jumps around 650 km, and strong velocity gradients directly below (Cammarano et al., 2005). The lower mantle is denser than pyrolite (Lee et al., 2004). The upper mantle is full of seismic scatterers (Anderson, 2006), and sometimes there is a low-velocity zone atop the 410-km discontinuity (Song et al., 2004). The nature of the upper-mantle discontinuities is not constant from place to place (Deuss et al., 2006), but there is little evidence that these variations are due to large variations in temperature. These observations are all consistent with an upper mantle having variable amounts of eclogite (Anderson, 1979a, 1989a; Anderson and Bass, 1986). The seismic observations are inconsistent with a homogeneous pyrolite mantle with recycled material sinking readily to the core-mantle boundary.

The evidence summarized above and in my earlier papers supports the idea that the TZ is a density filter and is a barrier to subduction. This evidence is much stronger than the hints of through-going convection based on visual inspection of a few selected low-resolution tomographic cross-sections. The proposal that eclogite, volatiles, and radioactive elements are trapped in and above the TZ is the opposite of the standard model and that of Bercovici and Karato (2003), which assumes that water and large-ion lithophile elements (including U, Th, and K) are primarily in the lower mantle and that ascending—rather than descending—plumes are filtered by the TZ. In those models, the upper mantle is heated strongly from below, whereas in the model proposed here, convection is driven by secular cooling of the mantle and slabs, and upper-mantle radioactivity. The mantle is fertilized from above, not from below. The fertile parts of the TZ are predicted to start out cool and then warm up. High-resolution seismic studies and those that measure impedance rather than just shear velocity are consistent with a stratified and blobby mantle and with the predictions of the GT mode of convection.

DISCUSSION

A mantle with phase changes at various depths and with variably melting, large-sized constituents exhibits behavior that is completely different from a homogenous or layered fluid or a well-mixed solid. It is not well approximated by any of the reservoir and box models or convection calculations that have been done to date (see the Appendix). Fertile blobs rise or fall, depending on their size and the time since they were inserted into the mantle. Upwelling that involves melting is a runaway process because of adiabatic decompression melting and the high compressibility of the melt phase. Upwellings do not have to have high absolute temperatures, and they do not have to start in a deep thermal boundary layer heated from below. Their isotopic signatures do not evolve from a depleted-MORB or primordial starting signature. The predicted lithologic variability of this kind of mantle is in agreement with various kinds of petrological and geochemical data (Natland, 1989, this volume; Jull and Kelemen, 2001; Agranier et al., 2005). The GT model replaces the isolated reservoir models of mantle geochemistry and the one- and two-layer models of geodynamics, which are incompatible (Albarède and van der Hilst, 2002a,b).

The upper mantle may be composed of transient blobs as well as being density stratified (Brueckner and van Roermund, 2004). I have concentrated on the density differences between various eclogites and between eclogite and ambient mantle (which is not necessarily pyrolite). There are equally important density and fertility differences among natural peridotites; the density differences correspond to a temperature range of at least 300 °C in a lherzolite mantle (David Green, personal commun., 2005). The shallow mantle may be the refractory residue after melting out of basaltic crust and therefore intrinsically buoyant (the perisphere). Low-density peridotites (depleted or iron-poor) may never get cold enough to sink very far, or for very long, in a mantle composed of denser peridotites or piclogites. At depth, it would be difficult to make subsolidus eclogites buoyant within depleted lherzolite or pyrolite, but there is no reason to suspect that these lithologies represent ambient mantle at all depths. If the mantle is stratified by intrinsic density, then denser material than pyrolite may occupy the deeper upper mantle and the TZ as well as the lower mantle. In the extreme case, the mantle is continuously stratified by density, melting point, and volatility throughout the accretional, differention, and cooling processes. Some of this stratification may be irreversible (Anderson, 2002d).

The shallow fertile blob source for melting anomalies has many interesting consequences and satisfies data regarding relative motions of hotspots (e.g., Cuffaro and Doglioni, this volume). Shallow sources and shallow return flow imply faster plate motions and migration of ridges over fertile mantle. This mode of mantle convection solves several long-standing problems, including how ridges can be continuously productive and how the asthenosphere survives if it is constantly leaking. This problem has been used as evidence against the partially molten asthenosphere hypothesis and in favor of the plume hypothesis (see http://www.bioedonline.org/news/news.cfm?art = 2687). However, ridge migration and a supply of material from internally cycled eclogite prevents the asthenosphere from slowly wasting away. The fact that all rifts and margins are not equally magmatic has also been used as an argument against nonplume models. A blobby mantle solves this problem. It also addresses issues recently raised against all current models of mantle dynamics (Tackley, 2006). An unresolved issue is the melting behavior, and therefore density, of a plausible range of eclogite compositions as a function of pressure. The types of convection and heterogeneity discussed here provide challenges to geodynamic modelers and seismologists. It may require a fundamental change in the assumptions and ground rules behind current geodynamic modeling (Anderson, 2002c; see the Appendix).

ACKNOWLEDGMENTS

This article is one of a series on implications of the top-down hypothesis for mantle dynamics and geochemistry. The top-down, or plate, hypothesis is being developed in collaboration with Jim Natland, Anders Meibom, and Gillian Foulger. Interactions with Dean Presnall, Jerry Winterer, Warren Hamilton, Seth Stein, Alan Smith, Cin-Ty Lee, and contributors to http://www.mantle-plumes.org/ have been, and continue to be, invaluable.

Jim Natland reviewed the preliminary manuscript and stimulated a complete rewrite, as did the reviewers of the present article. My interest in delamination was started by a series of discussion seminars at Rice University, Houston, and by conversations with Alan Levander, Adrian Lenardic, Richard Gordon, Gene Humphreys, and Cin-Ty Lee. The manuscript benefited enormously from comments and reprints by Dave Green, John Hernlund, and Jim Natland. The information on density borrows heavily from papers by Cin-Ty Lee and Dave Green.

APPENDIX: COMPARISON WITH OTHER PROPOSED MECHANISMS

The reviewers of this article suggested that I compare the GT mode of mantle convection with previous models, stressing that which is new and different. Since this discussion did not fit in well with the main text, I collect this information, plus other background material, in this appendix. I also review the plate model.

Until recently, most convection simulations have ignored melting, gravitational differentiation, delamination, heterogeneous fluids, and density stratification; these form the essence of the present chapter. GT convection has some similarities with previous studies in the areas of planetology and petrology, where differentiation and gravitational stratification are well-known concepts. The present mantle is often considered to be homogeneous, or, at most, to consist of only two layers, either separated at 650 km or at the D″ boundary. The GT model assumes that material inserted into the mantle as large blocks will tend to settle to a depth of neutral buoyancy, usually at a phase boundary of ambient mantle but also at chemical boundaries. On average, the intrinsic density of the mantle increases with depth. This model contrasts with models that assume vigorous stirring and rapid homogenization.

Tackley and Stevenson (1993) discuss a mechanism for spontaneously creating buoyant upwellings by perturbations in the melt content of a homogeneous layer of the upper mantle. This process has been termed a "buoyant decompression melting instability" (Raddick et al., 2002), a potentially important source of intraplate volcanism. A similar mechanism, but at lower temperature, will operate if the melting temperature of the mantle varies, as in GT convection. John Hernlund discusses a "drippy mode" of convection that is related to the RT mode of delamination (http://geodyn.ess.ucla.edu/~hernlund/smscconv.html).

The melt instability discussed here differs from those treated above, in the following ways:

1. The instability is triggered and/or nucleated by cold low-melting-temperature blobs heating up toward ambient temperature, not by a statistical variation in temperature or melt content;
2. Much more melt is created by a given rise in temperature than in a uniform peridotite layer;
3. The fertile blob is surrounded by subsolidus impermeable low-homologous-temperature peridotite;
4. There is a natural way to replenish the upwelled materials: the melting zone is replenished by downwelling instabilities starting within and below the overlying plate;
5. The instability can operate in an isothermal mantle;
6. The instability can initiate at temperatures that are lower than ambient; and
7. The scale is set by the scale of the heterogeneity.

Many authors have suggested that material from a deep, dense layer can be displaced upward by sinking slabs, usually thinking of the D″ region above the core. If the upper mantle and TZ are chemically stratified, then the upward displacement may result in melting—and more buoyancy—which is unlikely for deeper dense layers.

A geological analog to the GT and "yo-yo" mechanism is dunk tectonics (Brueckner and van Roermund, 2004), used to explain the cycle that takes crustal material to depths of ~150 km and then returns it to the surface as ultra-high pressure rocks containing coesite and eclogite. Dunk tectonics, in fact, may be a minor side effect of GT. In both cases, shearing, thrusting, and entrainment may be more important than strictly fluid dynamic instabilities, although both have been analyzed in terms of RT and density instabilities.

The reheated blob and yo-yo mechanisms (Figure 2) superficially resemble the thermochemical oscillatory dome mechanism of Davaille (1999) that occurs in chemically stratified fluids. The system she studied experimentally consists of superposed layers of homogeneous fluids with different viscosities and densities that convect in response to heating from below. The effects of pressure, melting, and phase changes on material properties and internal heating are not simulated; these are serious shortcomings. Upwellings from the deeper denser layer are caused by very high temperatures and are a form of thermochemical convection. This mode differs from the current model in that temperatures must be even higher than in models of thermal convection because the intrinsic density contrast must be overcome. The most important parameter controlling the dynamics of layered convection and accretional differentiation is the "buoyancy number" **B,** defined as the ratio of the chemical buoyancy (the intrinsic density difference between layers or blobs) to the total thermal buoyancy caused by the temperature contrast across the tank. This geometry can produce oscillatory domes, large thermochemical plumes, or families of small secondary plumes, depending on the density contrast. When the lower layer is thin and potentially buoyant, thermochemical plumes rise from the interface and entrain filaments of chemically dense material from the lower layer. A second mode consists of large domes of fluid that oscillate over the whole height of the tank. A third mode consists of small plumes that are generated at the top of a rising dome.

In the eclogite engine, intrinsic density contrasts are overcome by phase changes, and **B** varies with time and depth. For certain parameter ranges and compositions of the differentiation products, the layering can be irreversible (Anderson, 2002a). In GT mode, the fertile blobs start out cold, but they are much warmer than subducted oceanic crust. Heating is by a combination of heat conducted into the blob from the surrounding fluid and, over a longer time scale, by internally generated heat in the blob (Fig. 2). Mantle heterogeneities are introduced from the top, not the bottom. The GT mode owes its existence to smaller-scale chemical heterogeneity and variations in intrinsic density, thermal expansion, viscosity, and melting intervals. Because of the effects of pressure on the coefficient of thermal expansion and melt density, the buoyancy parameter decreases with depth, making irreversible chemical stratification more likely than reversible stratification, except in the upper parts of the mantle. Secular cooling can also reverse the direction of net mass flux; for example, from upward to downward transport of mafic material. **B** is the most important parameter in GT mode and in convection calculations that include differentiation and the possibility of mantle layering—reversible and irreversible. The homologous temperature replaces the absolute temperature as the fundamental temperature scale. In most simulations, the Rayleigh number and the absolute temperature are the dominant parameters.

A layered mantle can convect and remove mantle and core heat, but this convection does not resemble the broad cellular convection, with steady 2D convection cells, that is depicted in textbooks and perceived to be the source of conflict between geochemistry and geophysics (e.g., Albarède and van der Hilst, 2002a,b; Bercovici and Karato, 2003; Anderson, 2005a,b). Superimposed on large-scale convection and subduction-driven flow are various smaller scales of convection. In the present model, this scale is set by petrology and tectonics, not by fluid dynamic considerations, as in most treatments of small-scale thermal or melt-induced convection.

Bercovici and Karato (2003), and others, argue that the mantle is chemically homogeneous because they see no evidence for chemical boundaries and thermal boundary layers. They rely on selected low-resolution global seismic tomographic cross-sections and qualitative visual tomographic interpretations, and ignore the high-resolution seismic evidence for a complex mantle that has multiple reflections and scatterers (e.g., Anderson, 2005b). They do not define how they would recognize layered convection if it existed. They argue that layered mantle models suffer from dynamic inconsistencies, because any enriched lower layer has most of the heat-producing elements, the depleted overlying layer is heated almost entirely along its base, and this kind of convection can be ruled out. But there is no evidence for an enriched lower mantle; it is purely an assumption and violates mass-balance calculations (Anderson, 1989a). A radioactive-rich lower mantle will overheat and overturn, resulting in most of the radioactivity being transferred to the upper layers. But it is more likely that the large-ion radioactive elements were zone-refined into the shallow mantle during the planetary accretion stage. If most of the radioactivity is in the crust and upper mantle, then layered models cannot be ruled out so casually. Bercovic and Karato (2003) attack artificial and unphysical models and do not discuss thermodynamically self-consistent models, plausible heating modes, and quantitative tomographic interpretations.

Yo-Yo Tectonics

Eclogite that has settled into the upper mantle will be trapped at various depths above the 650-km discontinuity (Anderson, 1989b; 2005b; Hirose et al., 1999); molten eclogite will density-equilibrate near 400 km (see Sakamaki et al., 2006; Fig. 1). Only the coldest, thickest, driest, and most SiO_2-rich eclogites are candidates for sinking deeper than 650 km. To do so, they must first decouple from any attached depleted harzburgite.

The large density jump 400 km deep in the ambient mantle is mainly due to phase changes in olivine and orthopyroxene. Eclogite does not undergo phase changes at the same depths. Some eclogites and magmas are therefore likely to be trapped at the 400-km depth, causing a low shear velocity anomaly. Large eclogite blobs are more likely to control their own fate than are the fertile veins in peridotites. Any chemical stratification involving cold eclogite and other low-melting-temperature blobs is temporary, because the melting associated with heating involves large density changes. Hot or partially melted eclogite will equilibrate at shallower depths and may rise and spread out beneath the lithosphere (Fig. 1).

Dry peridotitic melts and komatiites may be denser than the mantle between 250 and 410 km. Eclogite-rich blobs warming up in a peridotite subsolidus mantle probably can retain their melts, although they will react with the peridotite. There is a density reduction as garnet in eclogite is converted to magma, but initially the melts have little buoyancy (Fig. 1). The complete cycle is shown schematically in Figure 2.

The Plate Process

A generalization of plate tectonics that emphasizes the role of near-surface phenomena and features such as plates, lithosphere, asthenosphere, tectonics, and eclogite blobs, is called the PLATE (plate, lithosphere, and asthenosphere tectonics and eglogite), or plate, paradigm; cooling of the surface, extensional stress, recycling, homologous temperature, lithospheric architecture, counterflow, delamination, and the buoyancy parameter are the important parameters and processes. Plate motions drive a shallow mantle counterflow—counter to plate motions; the thickness and viscosity of the counterflow channel are important parameters. In the thermal—or standard —models, the main parameter—essentially the only parameter— is absolute temperature; plate motions and mantle convection are driven from below. The present article elaborates on one part of the top-down (Anderson, 2001, 2005a), or plate, paradigm. A key element is that the volume of melt produced from a given volume of mantle is controlled by the fraction of low-melting-temperature material, not by the absolute temperature. The temperatures in D″ and the conditions at the core-mantle boundary are interesting, but they have little to do with surface magmatism, including large igneous provinces (LIPs; Anderson, 2005a). Other elements of the plate process have been developed in Anderson (2001, 2002a,b, 2005a,b, 2006), Foulger et al. (2005a,b), and Foulger (this volume). Counterflow calculations are given in Harper (1978) and Chase (1979); these calculations predict small relative motions of entrained blobs and therefore of surface hotspots. Earlier work on the stability and composition of near-surface layers (stoping or delamination potential) is given in Daly (1933), Ringwood and Green (1966), Green and Ringwood (1967), and Ringwood (1975), which give detailed discussions of these themes. What emerges is a novel form of convection that is not captured by laboratory and computer simulations of convection in a homogeneous or layered fluid, mainly heated from below or uniformly heated from within. It is more akin to diapirism (Presnall and Helsley, 1982). Key aspects of the top-down process, some extracted from the above papers, are described below. (See also http://www.mantleplumes.org/Top Pages/SelfOrganisedTop.html; http://www.mantleplumes.org/ Convection.html.)

High pressure stiffens the deep parts of a silicate planet, and low temperature stiffens the outer shell. If there are continents and stiff and buoyant regions at the surface, they control the pattern of convection, the cooling rate, the aspect ratio of convection, and the lag between heat generation and surface heatflow. The cooling surface layer and the surface boundary condition is the active and self-organizing part of the system, and it drives and organizes the motions in the passive interior, including the shallow counterflow, which is opposite to plate motions. The average thermal gradient in the interior will be subadiabatic under these conditions.

If the "fluid" is a mixture of several components with different intrinsic densities, melting behavior, phase changes, and other properties, it can evolve to a chemically layered system. The deeper part of the stratification will be irreversible if the intrinsic density contrasts or pressure are great enough. If there are a large number of layers, cooling is primarily by conduction, unless part of the cool surface layer can detach (subduct, delaminate, founder, or stope) and cool off the interior; these latter methods for cooling the interior and providing melt to the surface are seldom addressed.

The processes of planetary accretion, melt extraction, and

accretional differentiation placed most of the radioactive elements into the materials that became the crust and upper mantle. The processes that heat and cool the mantle, and create and dissipate buoyancy, are most effective at the top.

Counterflow

Arguments against alternatives to plume mechanisms are usually based on the propagating crack idea, or on the perceived fixity of hotspots. Tackley (2006), in reviewing mechanisms for hotspots, recently posed the following questions: Why do the Pacific hotspots exhibit little relative motion as the plate moves over them? If they are caused by propagating cracks, why do all the cracks propagate at the same rate? Plate boundaries tend to be linear; why, then, are flood basalts, and hotspots that occur at spreading centers, not linear, following the plate boundary?

Shallow counterflow mechanisms explain the small relative motions of the Pacific hotspots and the measurable relative motions between groups of hotspots on different plates. Heterogeneities in the upper-mantle flow in the opposite direction to the overlying plate and at a small fraction of the plate velocity. This mechanism gives tracks, on a given plate, that are parallel. Statistical compatibility tests show that the ages and trends of volcanic chains are incompatible with fixed hotspots (Wang and Liu, 2006). Furthermore, hotspot rates are consistently lower than those predicted by best-fitting absolute plate motion models. This deficit may be explained by fertile spots in the mantle moving systematically opposite to the plate motion, as predicted by counterflow models of mantle convection (Harper, 1978; Chase, 1979). The locations, trends, and relative motions of melting anomalies are then due to asthenospheric flow that is almost exactly opposite to plate motion, as in the lubrication models of plate tectonics (Harper, 1978) and in recent models of relative motions (Wang and Liu, 2006). Dense sinkers that bottom out at upper-mantle phase changes may also exhibit little relative motion, particularly if they return to the surface rather quickly. Cuffaro and Doglioni (this volume) and Norton (this volume) have shown that shallow sources satisfy plate reconstruction data.

Plate boundaries and fracture zones tend to be linear, as do so-called hotspot tracks. Many flood basalts and hotspots are not linear, because they occur at triple junctions and are transient in nature. Fertile streaks in the upper mantle do not need to follow plate boundaries, nor do they need to be linear. The intersections of fertile blobs with ridges, triple junctions, reactivated sutures, and new plate boundaries, and the configurations of tensile stress domains are what control the location and shape of melting anomalies. Nevertheless, much of the fabric of the seafloor is parallel to plate motions, and linear volcanic chains are expected for incipient—and dying—plate boundaries. It is not obvious why the linearity of long-lived volcanic constructs and the nonlinearity of short-lived constructs and those at triple junctions are arguments against plate tectonic control and shallow origins of "mid-plate" phenomena (e.g., Tackley, 2006). Like-

wise, it is not clear why the absence of LIPs along some extending margins is an argument for plumes.

Ground Rules

Many concepts and scalings of mantle dynamics are based on small-scale laboratory experiments and simplified theory motivated by experiments, which have become so engrained in the collective consciousness of geodynamicists that they are taken for granted. Tackley et al. (2005) summarize these concepts as fourteen rules that dictate the current directions of fluid dynamic research. These rules do not include the issues raised in this article and in other recent papers that stress the top-down nature of mantle convection, self-consistent thermodynamics, and the onset of phase-change–induced instabilities (Anderson, 2005a,b). These new concepts can be formalized into fourteen alternative rules:

1. The basic structure of the mantle does not involve steady flow, or convection cells with narrow boundary layers and an adiabatic interior;
2. The volume dependence and pressure dependence of thermal properties are as important as temperature dependence;
3. The planform of three-dimensional convection may be controlled by lithospheric architecture and mantle chemical heterogeneity rather than by fluid dynamic instabilities of a homogeneous fluid: plates and slabs organize or even control mantle convection;
4. The volume dependence of such properties as viscosity, conductivity, and thermal expansivity makes deep-mantle upwellings broader, increases the horizontal wavelength and time constants of deep flow, and contributes to irreversible chemical stratification; pressure suppresses the importance of lower-mantle convection in determining surface processes;
5. The radioactive elements are not uniformly distributed throughout the mantle nor in the various components;
6. Plate tectonic forces, including magma fracture, can break the lithosphere: plates are not a rigid lid nor a uniform or permanent boundary condition;
7. Phase transitions, including eclogite-basalt-magma, in a multicomponent mantle can induce instabilities and permanent or episodic chemical layering and blobby-type heterogeneity;
8. Continents, including their motions, and the inherent instability of thick continental crust cannot be ignored;
9. Lithological variations and variations in heat productivity and viscous dissipation can cause local heating;
10. Convection models must be thermodynamically self-consistent;
11. Lithologic variations affect seismic velocities: current scaling parameters between temperature, density, and seismic velocity are inadequate and misleading;
12. The locations and scales of mantle instabilities are set by

the locations and scales of lithologic heterogeneity rather than by fluid dynamics of a homogeneous fluid;

13. The scales of chemical heterogeneities in the mantle control the fate of such heterogeneities and their geochemical signatures; and

14. The upper mantle is close to or above the melting temperature of some of its components.

Mantle convection is a nonlinear process with outcomes that depend strongly on initial and boundary conditions and other assumptions; actual mantle dynamics may violate the collective consciousness and conventional wisdom of geodynamicists operating by the rules of the current paradigm.

Geochemical Models

Geochemistry cannot constrain the depth or locations of the sources of basalts. These sources are usually cast as boxes or reservoirs connected by arrows, and other information is used to suggest a location. In the standard models of mantle geochemistry, the long-wavelength, or dispersed, parts of mantle heterogeneity (melting anomalies) are best accounted for by the presence of mantle hot spots (Agranier et al., 2005). These hotspots in turn are attributed to the return of deeply subducted ancient oceanic crust from the lower mantle back into the upper mantle (Hofmann and White, 1982). The injection of lithospheric plates continuously introduces differentiated material into the mantle that either sinks to the core-mantle boundary (Hofmann and White, 1982) or is mixed into the shallow mantle (Christensen and Hofmann, 1994). A common assumption is that, upon subduction, 5- to 10-km-thick sequences typical of oceanic crust will be mixed into ambient mantle (Christensen and Hofmann, 1994), thereby creating a homogeneous source. It is considered inescapable that recycled material is multiply stretched and refolded by mantle convection (Agranier et al., 2005), and that old plates find their way back to the asthenospheric upper mantle, where they become an intrinsic part of the well-mixed MORB source (the so-called convecting upper mantle). Delamination of subcontinental lithosphere and metasomatism at subduction zones are also assumed to create heterogeneities that mantle convection repeatedly folds and stretches (Allègre and Turcotte, 1986).

Isotopic mantle heterogeneities along mid-ocean ridges that cannot be attributed to stretching and folding of heterogeneities by convection are attributed to overprinting of the mid-ocean ridges by injections of deep mantle (Hamelin et al., 1984). Melting anomalies represent hot material from the deep mantle.

In contrast to this type of model is the statistical upper-mantle assemblage model (SUMA), in which the mantle is intrinsically heterogeneous at all scales, but small-scale heterogeneity is averaged out by the melting and sampling process at ridges and is mainly evident only in smaller-scale samples—ocean island volcanoes, seamounts, and xenoliths—and dense sampling programs (Meibom and Anderson, 2004). Larger-scale hetero-

geneities, such as chunks of delaminated crust, are not averaged out or mixed back into the convecting mantle. They can be resampled along the global spreading ridge system and as mid-plate melting anomalies that are not particularly hot (Anderson, 2005a,b). Chemical heterogeneity and magma volume anomalies are due to the heterogeneous nature of the mantle, not primarily to temperature variations. Relatively fixed hotspots are due to the slow counterflow of the asthenosphere (Harper, 1978; Chase, 1979).

Usual treatments of terrestrial magmatism also consider that there are only three possible causes for melting in the mantle: decompression, heating, and insertion of volatiles. If parts of the upper mantle are already above the solidus, then adiabatic ascent or an increase in temperature can increase the amount of melting. The presence of water may significantly reduce the melting temperature, but the degree of melting is small between the wet and dry solidi due to the high solubility of water in melt. A partially molten asthenosphere may develop melting instabilites, driven by thermal, melt, and depletion buoyancies. Asthenosphere can also passively upwell in response to spreading and to the delamination of continental crust or lithosphere. Volcanism associated with melting instabilities and shallow passive upwelling of fertile mantle are two mechanisms for creating melting anomalies. It is almost universally assumed, however, that large amounts of melting require mantle temperatures that are much higher than average. Attempts are then made to rationalize the major-element chemistry of "hotspot" magmas in terms of larger degrees and depths of melting caused by increased temperatures (Langmuir et al., 1992).

There are other mechanisms for increasing the melt content of a given region of the mantle and for initiating a melt or buoyancy instability (Tackley and Stevenson, 1993). Variations in melting temperature—or homologous temperature—and lithology can cause instabilities. These instabilities are triggered by cold fertile blobs that are warming up to ambient temperature rather than by regions that are hotter than ambient mantle. They are buoyant because they contain a large melt fraction, a large depleted peridotite component, or both.

Fertile Blobs versus Hot Plumes

LIPs are produced at high eruption rates, but there is no evidence that they are produced from particularly hot mantle (Clift, 2005). Hot upwellings would be expected to produce highly magnesian lavas, but these are rather rare in LIPs. The locations of LIPs along mobile belts, sutures, and triple junctions suggest that shallow-mantle fertility, release of ponded melts, edge effects, and focusing may be responsible for the large volumes and short durations (King and Anderson, 1998; Foulger and Anderson, 2005; Foulger et al., 2005a,b). These locations also suggest that crustal stoping may be involved (Anderson, 2005a; see also Daly, 1933). Delamination produces asthenospheric upwelling and decompression melting as the root is detached; it produces magmatism again as the material returns to the shallow mantle.

Several workers have proposed that large magma fluxes can be produced from eclogite-bearing mantle without assuming high potential temperatures (Takahashi et al., 1998; Yasuda and Fujii, 1998; Leitch and Davies, 2001). Takahashi et al. (1998) argued that LIPs could result from partial melting of eclogite with a potential temperature not greatly in excess of the MORB adiabat. Takahashi and Nakajima (2002) suggested that the Koolau component in Hawaii can be produced from large eclogite blocks, but the inferred excess potential temperature is only ~100° C. This possibility means that high absolute temperatures can be replaced by high homologous temperatures. Partial melts of eclogite can explain the silica-enriched compositions of some LIPs (Takahashi et al., 1998), and extensive melting of eclogite can explain magmas with no garnet residue signature. The cited papers all consider that the eclogite is delivered by deep-seated thermal plumes. The present article argues that such plumes are unnecessary.

REFERENCES CITED

Agranier, A., Blichert-Toft, J., Graham, D., Debaille, V., Schiano, P., and Albarède, F., 2005, The spectra of isotopic heterogeneities along the mid-Atlantic ridge: Earth and Planetary Science Letters, v. 238, p. 96–109, doi: 10.1016/j.epsl.2005.07.011.

Albarède, F., and van der Hilst, R.D., 2002a, Zoned mantle convection: Geochimica et Cosmochimica Acta, v. 66, p. A12, supplement.

Albarède, F., and van der Hilst, R.D., 2002b, Zoned mantle convection: Philosophical Transactions of the Royal Society of London, v. A360, p. 2569–2592.

Allègre, C.J., and Turcotte, D.L., 1986, Implications of a two-component marble-cake mantle: Nature, v. 323, p. 123–127, doi: 10.1038/323123a0.

Anderson, D.L., 1975, Chemical plumes in the mantle: Geological Society of America Bulletin, v. 86, p. 1593–1600, doi: 10.1130/0016-7606(1975)86<1593:CPITM>2.0.CO;2.

Anderson, D.L., 1979a, The upper mantle TZ: Eclogite?: Geophysical Research Letters, v. 6, p. 433–436.

Anderson, D.L., 1979b, Chemical stratification of the mantle: Journal of Geophysical Research, v. 84, p. 6297–6298.

Anderson, D.L., 1981, Rise of deep diapers: Geology, v. 9, p. 7–9, doi: 10.1130/0091-7613(1981)9<7:RODD>2.0.CO;2.

Anderson, D.L., 1989a, Theory of the Earth: Boston, Blackwell Scientific, 366 p.

Anderson, D.L., 1989b, Where on Earth is the crust?: Physics Today, v. 42, p. 38–46.

Anderson, D.L., 2000, Thermal state of the upper mantle; No role for mantle plumes: Geophysical Research Letters, v. 27, p. 3623–3626, doi: 10.1029/2000GL011533.

Anderson, D.L., 2001, Top-down tectonics?: Science, v. 293, p. 2016–2018, doi: 10.1126/science.1065448.

Anderson, D.L., 2002a, How many plates?: Geology, v. 30, p. 411–414, doi: 10.1130/0091-7613(2002)030<0411:HMP>2.0.CO;2.

Anderson, D.L., 2002b, Occam's razor: Simplicity, complexity, and global geodynamics: Proceedings of the American Philosophical Society, v. 146, p. 56–76.

Anderson, D.L., 2002c, Plate tectonics as a far-from-equilibrium self-organized system, in Stein, S., and Freymueller, J.T., eds., Plate boundary zones: Washington, D.C., American Geophysical Union Geodynamics Monograph 30, p. 411–425.

Anderson, D.L., 2002d, The case for the irreversible chemical stratification of the mantle: International Geology Review, v. 44, p. 97–116.

Anderson, D.L., 2005a, Large igneous provinces, delamination, and fertile mantle: Elements, v. 1, p. 271–275.

Anderson, D.L., 2005b, Self-gravity, self-consistency, and self-organization in geodynamics and geochemistry, in van der Hilst, R.D., et al., eds., Earth's deep mantle: Structure, composition, and evolution: Washington, D.C., American Geophysical Union Geophysical Monograph 160, p. 165–186.

Anderson, D.L., 2006, Speculations on the nature and cause of mantle heterogeneity: Tectonophysics, v. 416, p. 7–22, doi: 10.1016/j.tecto.2005.07.011.

Anderson, D.L., and Bass, J.D., 1986, Transition region of the Earth's upper mantle: Nature, v. 320, p. 321–328, doi: 10.1038/320321a0.

Anderson, D.L., and Sammis, C.G., 1970, Partial melting in the upper mantle: Physics of the Earth and Planetary Interiors, v. 3, p. 41–50, doi: 10.1016/0031-9201(70)90042-7.

Bercovici, D., and Karato, S., 2003, Whole mantle convection and transition-zone water filter: Nature, v. 425, p. 39–44, doi: 10.1038/nature01918.

Brueckner, H.K., and van Roermund, H.L.M., 2004, Dunk tectonics: A multiple subduction/eduction model for the evolution of the Scandinavian Caledonides: Tectonics, v. 23, p. 1–20, doi: 10.1029/2003TC001502.

Cammarano, F., Deuss, A., Goes, S., and Giardini, D., 2005, One-dimensional physical reference models for the upper mantle and TZ: Combining seismic and mineral physics constraints: Journal of Geophysical Research, v. 110, p. B01306, doi: 10.1029/2004JB003272.

Chase, C.G., 1979, Asthenospheric counterflow: A kinematic model: Geophysical Journal of the Royal Astronomical Society, v. 56, p. 1–18.

Christensen, U.R., and Hofmann, A.W., 1994, Segregation of subducted oceanic crust in the convecting mantle: Journal of Geophysical Research, v. 99, p. 19,867–19,884, doi: 10.1029/93JB03403.

Clift, P.D., 2005, Sedimentary evidence for moderate mantle temperature anomalies associated with hotspot volcanism, in Foulger, G.R., et al., eds., Plates, plumes and paradigms: Boulder, Colorado, Geological Society of America Special Paper 388, p. 279–288. Also see http://www.mantleplumes.org/SedTemp.html.

Cuffaro, M., and Doglioni, C., 2007 (this volume), Global kinematics in deep versus shallow hotspot reference frames, in Foulger, G.R., and Jurdy, D.M., eds., Plates, plumes, and planetary processes: Boulder, Colorado, Geological Society of America Special Paper 430, doi: 10.1130/2007.2430(18).

Daly, R.A., 1933, Igneous rocks and the depths of the Earth: New York, Mc-Graw Hill, 598 p.

Davaille, A., 1999, Simultaneous generation of hotspots and superswells by convection in a heterogeneous planetary mantle: Nature, v. 402, p. 756–760, doi: 10.1038/45461.

Deuss, A., and Woodhouse, J.H., 2002, A systematic search for mantle discontinuities using SS-precursors: Geophysical Research Letters, v. 29, doi: 10.1029/2002GL014768.

Deuss, A., Redfern, S., Chambers, K., and Woodhouse, J.H., 2006, The nature of the 660-km discontinuity in Earth's mantle from global seismic observations of PP precursors: Science, v. 311, p. 198–201, doi: 10.1126/science.1120020.

Dziewonski, A.M., and Anderson, D.L., 1981, Preliminary reference Earth model: Physics of the Earth and Planetary Interiors, v. 25, p. 297–356, doi: 10.1016/0031-9201(81)90046-7.

Elsasser, W.M., 1969, Convection and stress propagation in the upper mantle, in Runcorn, S.K., ed., The application of modern physics to the Earth and planetary interiors: New York, Wiley, p. 223–249.

Foulger, G.R., 2007 (this volume), The "plate" model for the genesis of melting anomalies, in Foulger, G.R., and Jurdy, D.M., eds., Plates, plumes, and planetary processes: Boulder, Colorado, Geological Society of America Special Paper 430, doi: 10.1130/2007.2430(01).

Foulger, G.R., and Anderson, D.L., 2005, A cool model for the Iceland hotspot: Journal of Volcanology and Geothermal Research, v. 141, p. 1–22, doi: 10.1016/j.jvolgeores.2004.10.007.

Foulger, G.R., Natland, J.H., and Anderson, D.L., 2005a, A source for Icelandic magmas in remelted Iapetus crust: Journal of Volcanology and Geothermal Research, v. 141, p. 23–44, doi: 10.1016/j.jvolgeores.2004.10.006.

Foulger, G.R., Natland, J.H., Presnall, D.C., and Anderson, D.L., eds., 2005b. Plates, plumes and paradigms: Boulder, Colorado, Geological Society of America Special Paper 388, 881 p.

Ghods, A., and Arkani-Hamed, J., 2002, Effect of melt migration on the dynamics and melt generation of diapirs ascending through asthenosphere: Journal of Geophysical Research, v. 107, p. 2026, doi: 10.1029/2000JB000070.

Green, D.H., and Ringwood, A., 1967, The genesis of basaltic magmas: Contributions to Mineralogy and Petrology, v. 15, p. 103–190, doi: 10.1007/BF00372052.

Gu, Y.J., Dziewonski, A.M., Su, W.-J., and Ekström, G., 2001. Models of the mantle shear velocity and discontinuities in the pattern of lateral heterogeneities: Journal of Geophysical Research, v. 106, p. 11,169–11,199.

Hamelin, B., Dupre, B., and Allègre, C.J., 1984, Lead-strontium isotopic variations along the East Pacific rise and the mid-Atlantic ridge: Earth and Planetary Science Letters, v. 67, p. 340–350, doi: 10.1016/0012-821X(84)90173-0.

Harper, J.F., 1978, Asthenosphere flow and plate motions: Geophysical Journal of the Royal Astronomical Society, v. 55, p. 87–110.

Hirose, K., Fei, Y., Ma, Y., and Mao, H.K., 1999, The fate of subducted basaltic crust in the Earth's lower mantle: Nature, v. 397, p. 53–56, doi: 10.1038/16225.

Hofmann, A.W., and White, W.M., 1982, Mantle plumes from ancient oceanic crust: Earth and Planetary Science Letters, v. 57, p. 421–436, doi: 10.1016/0012-821X(82)90161-3.

Irifune, T., and Ringwood, A.E., 1993, Phase transformations in subducted oceanic crust and buoyancy relationships at depths of 600–800 km in the mantle: Earth and Planetary Science Letters, v. 117, p. 101–110, doi: 10.1016/0012-821X(93)90120-X.

Ishii, M., and Tromp, J., 2004, Constraining large-scale mantle heterogeneity using mantle and inner-core sensitive normal modes: Physics of the Earth and Planetary Interiors, v. 146, p. 113–124, doi: 10.1016/j.pepi.2003.06.012.

Jacoby, W.R., 1970, Instability in the upper mantle and global plate movements: Journal of Geophysical Research, v. 75, p. 5671–5680.

Jacques, A., and Green, D., 1980, Anhydrous melting of peridotite at 0–15 Kb pressure and the genesis of tholeitic basalts: Contributions to Mineralogy and Petrology, v. 73, p. 287–310, doi: 10.1007/BF00381447.

Jeanloz, R., and Morris, S., 1987, Is the mantle geotherm subadiabatic?: Geophysical Research Letters, v. 14, p. 335–338.

Jull, M., and Kelemen, P.B., 2001, On the conditions for lower crustal convective instability: Journal of Geophysical Research, v. 106, p. 6423–6446, doi: 10.1029/2000JB900357.

Kaula, W.M., 1983, Minimum upper mantle temperature variations consistent with observed heat flow and plate velocities: Journal of Geophysical Research, v. 88, p. 10,323–10,332.

King, S.D., and Anderson, D.L., 1998, Edge-driven convection: Earth and Planetary Science Letters, v. 160, p. 289–296, doi: 10.1016/S0012-821X(98)00089-2.

Kogiso, T., Hirschmann, M.M., and Frost, D.J., 2003, High-pressure partial melting of garnet pyroxenite: Possible mafic lithologies in the source of ocean island basalts: Earth and Planetary Science Letters, v. 216, p. 603–617, doi: 10.1016/S0012-821X(03)00538-7.

Kogiso, T., Hirschmann, M.M., and Pertermann, M., 2004, High-pressure partial melting of mafic lithologies in the mantle: Journal of Petrology, v. 45, p. 2407–2422, doi: 10.1093/petrology/egh057.

Lambert, I.B., and Wyllie, P.J., 1970, Low-velocity zone of the Earth's mantle: Incipient melting caused by water: Science, v. 169, p. 764–766, doi: 10.1126/science.169.3947.764.

Langmuir, C.H., Klein, E.M., and Plank, T., 1992, Petrological systematics of mid-ocean ridge basalts: Constraints on melt generation beneath mid-oceanic ridges, *in* Phipps Morgan, J., et al., eds., Mantle flow and melt generation at mid-oceanic ridges: Washington, D.C., American Geophysical Union Geophysical Monograph 71, p. 183–280.

Lee, C.-T., Cheng, X., and Ulyana Horodyskyj, U., 2006, The development and refinement of continental arcs by primary basaltic magmatism, garnet py-roxenite accumulation, basaltic recharge and delamination: Insights from the Sierra Nevada, California: Contributions to Mineralogy and Petrology, v. 151, p. 222–242, doi: 10.1007/s00410-005-0056-1.

Lee, K.K.M., O'Neill, B., Panero, W., Shim, S.-H., Benedetti, L., and Jeanloz, R., 2004, Equations of state of the high-pressure phases of a natural peridotite and implications for the Earth's lower mantle: Earth and Planetary Science Letters, v. 223, p. 381–393, doi: 10.1016/j.epsl.2004.04.033.

Leitch, A.M., and Davies, G.F., 2001, Mantle plumes and flood basalts: Enhanced melting from plume ascent and an eclogite component: Journal of Geophysical Research, v. 106, p. 2047–2059, doi: 10.1029/2000JB900307.

Litasov, K., Ohtani, E., Suzuki, A., Kawazoe, T., and Funakoshi, K., 2004, Absence of density crossover between basalt and peridotite in the cold slabs passing through 660 km discontinuity: Geophysical Research Letters, v. 31, doi: 10.1029/2004GL021306.

Marsh, B.D., Morris, S., and Anderson, D.L., 1981, Comment and reply on "Rise of deep diapirs": Geology, v. 9, p. 7–9 and 559–561.

Mattern, E., Matas, J., Ricard, Y., and Bass, J., 2005, Lower mantle composition and temperature from mineral physics and thermodynamic modeling: Geophysical Journal International, v. 160, p. 973–990, doi: 10.1111/j.1365-246X.2004.02549.x.

McKenzie, D., and Bickle, M.J., 1988, The volume and composition of melt generated by extension of the lithosphere: Journal of Petrology, v. 29, p. 625–679.

Meibom, A., and Anderson, D.L., 2004, The statistical upper mantle assemblage: Earth and Planetary Science Letters, v. 217, p. 123–139, doi: 10.1016/S0012-821X(03)00573-9.

Natland, J.H., 1989, Partial melting of a lithologically heterogeneous mantle, *in* Saunders, A.D., and Norry, M., eds., Magmatism in the ocean basins: London, Geological Society of London Special Publication 42, p. 41–77.

Natland, J.H., 2007 (this volume), ΔNb and the role of magma mixing at the East Pacific Rise and Iceland, *in* Foulger, G.R., and Jurdy, D.M., eds., Plates, plumes, and planetary processes: Boulder, Colorado, Geological Society of America Special Paper 430, doi: 10.1130/2007.2430(21).

Nishihara, Y., Aoki, I., Takahashi, E., Matsukage, K.N., and Funakoshi, K., 2005, Thermal equation of state of majorite with MORB composition: Physics of the Earth and Planetary Interiors, v. 148, p. 73–84, doi: 10.1016/j.pepi.2004.08.003.

Norton, I.O., 2007 (this volume), Speculations on Cretaceous tectonic history of the northwest Pacific and a tectonic origin for the Hawaii hotspot, *in* Foulger, G.R., and Jurdy, D.M., eds., Plates, plumes, and planetary processes: Boulder, Colorado, Geological Society of America Special Paper 430, doi: 10.1130/2007.2430(22).

Pertermann, M., and Hirschmann, M., 2003, Anhydrous partial melting experiments on MORB-like eclogite: Phase relations, phase compositions and mineral-melt partitioning of major elements at 2–3 GPa: Journal of Petrology, v. 44, p. 2173–2201, doi: 10.1093/petrology/egg074.

Presnall, D.C., 1969, The geometrical analysis of partial fusion: American Journal of Science, v. 267, p. 1178–1194.

Presnall, D.C., and Helsley, C.E., 1982, Diapirism of depleted peridotite—A model for the origin of hot spots: Physics of the Earth and Planetary Interiors, v. 29, p. 148–160, doi: 10.1016/0031-9201(82)90069-3.

Raddick, M.J., Parmentier, E.M., and Scheirer, D.S., 2002, Buoyant decompression melting: A possible mechanism for intraplate volcanism: Journal of Geophysical Research B, v. 107, doi: 10.1029/2001JB000617.

Ringwood, A.E., 1975, Composition and petrology of the Earth's mantle: New York, McGraw Hill, 618 p.

Ringwood, A.E., and Green, D.H., 1966, An experimental investigation of the gabbro-eclogite transformation and some geophysical implications: Tectonophysics, v. 3, p. 383–427, doi: 10.1016/0040-1951(66)90009-6.

Sakamaki, T., Suzuki, A., and Ohtani, E., 2006, Stability of hydrous melt at the base of the Earth's upper mantle: Nature, v. 439, p. 192–194, doi: 10.1038/nature04352.

Scholl, D., 2006, http://gsa.confex.com/gsa/2006AM/finalprogram/abstract_110054.htm.

Sleep, N.H., 2007 (this volume), Origins of the plume hypothesis and some of its implications, *in* Foulger, G.R., and Jurdy, D.M., eds., Plates, plumes, and planetary processes: Boulder, Colorado, Geological Society of America Special Paper 430, doi: 10.1130/2007.2430(02).

Song, T.R.A., Helmberger, D.V., and Grand, S., 2004, Low velocity zone atop the 410 seismic discontinuity in the northwestern US: Nature, v. 427, p. 530–533, doi: 10.1038/nature02231.

Tackley, P.J., 2006, Heating up the hotspot debates: Science, v. 313, p. 1240–1241.

Tackley, P.J., and Stevenson, D.J., 1993, A mechanism for spontaneous self-perpetuating volcanism on the terrestrial planets, *in* Stone, D.B., and Runcorn, S.K., eds., Flow and creep in the solar system: Observations, modeling and theory: Dordrecht, Kluwer, p. 307–321.

Tackley, P.J., Xie, S., Nakagawa, T., and Hernlund, J.W., 2005, Numerical and laboratory studies of mantle convection: Philosophy, accomplishments and thermo-chemical structure and evolution, *in* van der Hilst, R.D., et al., eds., Earth's deep mantle: Structure, composition, and evolution: Washington, D.C., American Geophysical Union Geophysical Monograph 160, p. 83–99, doi:10.1029/160GM07.

Takahashi, E., Nakajima, K., and Wright, T.L., 1998, Origin of the Columbia River basalts: Melting model of heterogeneous plume head: Earth and Planetary Science Letters, v. 162, p. 63–80, doi: 10.1016/S0012-821X(98) 00157-5.

Takahashi, E., and Nakajima, K., 2002, Melting process in the Hawaiian plume, *in* Takahashi, E., et al., eds., Hawaiian volcanoes: Deep underwater perspectives: Washington, D.C., American Geophysical Union Geophysical Monograph 128, p. 403–418.

Wang, S., and Liu, M., 2006, Moving hotspots or reorganized plates?: Geology, v. 34, p. 465–468, doi: 10.1130/G22236.1

Wen, L., and Anderson, D.L., 1997, Slabs, hotspots, cratons and mantle convection revealed from residual seismic tomography in the upper mantle: Physics of the Earth and Planetary Interiors, v. 99, p. 131–143, doi: 10.1016/S0031-9201(96)03162-7.

Yasuda, A., and Fujii, T., 1998, Ascending subducted oceanic crust entrained within mantle plumes: Geophysical Research Letters, v. 25, p. 1561–1564, doi: 10.1029/98GL01230.

Yaxley, G.M., and Green, D.H., 1998, Reactions between eclogite and peridotite: Mantle refertilisation by subduction of oceanic crust: Schweizerische Mineralogische Petrographische Mitteilung, v. 78, p. 243–255.

Manuscript Accepted by the Society 31 JANUARY 2007

DISCUSSION

2 February 2007, Don L. Anderson

Phipps Morgan in his comments of 31 January regarding Yamamoto et al. (this volume, Chapter 10) raises "major challenges" to plate tectonic models. He favors a bottom-up, or plume, explanation for refertilizing the asthenosphere. He and his colleagues adopt a boundary-layer form of whole-mantle convection, with fertile streaks introduced from below. This comment addresses these challenges, and calls attention to new publications relevant to the top-down and yo-yo mechanisms.

"Why does the MORB [mid-ocean ridge basalt] source have less thermal variability than that introduced into the mantle by subducting slabs . . . given the temperature difference between a subducting slab and the MORB source region?" The simple answer is that slabs mainly cool other parts—deeper parts—of the mantle, whereas delamination from drifting continents, the yo-yo or water bucket effect, cools and removes heat from shallower areas. Neither process affects the temperatures in the surface boundary layer except at the delamination site and where the fertile blobs re-emerge to deliver their heat and magma.

Most models assume that MORB sources lie on a mantle adiabat and that variations in magma temperatures reflect lateral variations in mantle temperature. The potential temperature of a magma is calculated by cooling it by adiabatic expansion to $P = 0$. This approach does not imply that the mantle has an adiabatic gradient, nor does it yield the deep mantle geotherm. In the perisphere and Galileo thermometer concepts, the mantle is chemically stratified, with intrinsic density increasing with depth. It therefore has a conduction geotherm, modified by the passage of dense and buoyant sinkers. This geotherm starts out at about 10° C/km and decreases to about 4°C/km at 200 km depth. The adiabatic gradient is about 0.3°C/km in a solid mantle heated entirely from below. In an internally heated mantle the gradient starts out superadiabatic and becomes subadiabatic at depth. A mantle cooled by subduction and delamination can also exhibit temperature maxima. Under these conditions the geotherm will intersect the solidus but will continue to be superadiabatic at greater depths. If melting occurs at 200 km, then the temperature at 250 km, say, may be 200°C higher. The asthenosphere can be hotter than the MORB potential temperature. Slabs mainly settle at greater depths and serve to cool the overlying mantle from below rather than internally. They displace deeper material upward, just as delaminated material is replaced by hotter asthenosphere. In steady-state boundary layer models, heated from below, one does not think in these terms. The sites of delamination and subduction cooling also migrate around the surface. They are not constantly cooling one part of the mantle.

Much of the upper mantle is cooled from below by bottomed-out slabs and cooled from within by delaminating blobs. These blobs are colder than ambient mantle but hotter than slabs; they may contain most of the radioactivity of the upper mantle. They are constantly sinking and rising, delivering heat to the surface. The temperature of the surrounding mantle is buffered by the melting temperatures of the low-melting constituents, as well as by the near-constant temperature of the lower continental crust. A partially molten asthenosphere, or one with partially molten blobs, cannot be treated as a homogeneous subsolidus ideal fluid; melting anomalies have other causes than high absolute temperature.

"Why are there sharp transitions between volcanic and non-volcanic margins . . . [and] a rapid transition from OIB-like . . . basalts to . . . MORB during the initial formation of an ocean basin?" (Phipps Morgan, 31 January discussion of Yamamoto et al., this volume, chapter 10). This question is addressed by Foulger et al. (2005), Foulger and Anderson (2005), and Natland (this volume). The shallow mantle along reopened sutures and collisional mountain belts is not uniform and is not the same as the mantle at mature ridges because of trapped buoyant oceanic crust and delaminated continental crust. The mantle is not the well-stirred homogeneous fluid or marble-cake that is implied by the phrase "the convecting upper mantle."

A classic argument for deep-mantle plumes involves the noble gases. I have argued that high ^{3}He/^{4}He ratios can be generated and preserved in shallow melt-depleted mantle (Anderson, 1998). The implication is that high primordial/(radiogenic, nucleogenic) noble gas ratios (e.g., ^{3}He/^{4}He, ^{22}Ne/^{21}Ne), characteristic of sources that have been attributed to plumes, can be achieved by involving a previously melted (depleted) mantle source rather than isolated, non-degassed primordial mantle. This refractory source ("focal zone," or FOZO) can be a cumulate, or the continental or oceanic lithosphere. There is now experimental evidence in support of this trapping mechanism, for gases in general and for the noble gases in particular (Schiano et al., 2006; Heber et al., 2007). There are also new mass balance calculations in support of the delamination model from several groups (Plank, 2005; Hawkesworth and Kemp, 2006; David Scholl, personal commun., 2006). The mass flux exceeds the "hotspot" flux; the lower crust is recycled at much greater rates than the upper crust and must be an important component of the mafic blobs in the mantle.

Addendum, 8 February 2007, Don L. Anderson

In their comment of 5 February 2007, concerning the article by Sheth (this volume), Hooper and Widdowson write that I assert that if "any one suggested result of the plume model is not obvious . . . the plume model is therefore void and [has] to be abandoned." I have nowhere said or implied anything like this. I *have* shown, repeatedly, that none of the predictions and assumptions of the plume hypothesis have been confirmed. Regarding convergence, the Deccan Traps erupted while India was converging with Asia; a plume was not splitting India or driving India away from Asia.

In his comment of 7 Feburary 2007, on Sheth (this volume), Widdowson defines a mantle plume as "a static, spatially restricted, mantle melting anomaly." He is confusing melting anomaly or hotspot with plume; they are not the same (Anderson and Schramm, 2005). Except for the word "static" instead of "upwelling" or "buoyant," this definition applies to a fertile or low-melting-point blob, or delamination or rift-induced asthenospheric melting and upwelling. These are the mechanisms that best explain surface melting anomalies. A mantle plume, by contrast, is a thermal anomaly from a deep thermal boundary layer, for which there is no evidence.

REFERENCES CITED

Anderson, D.L., 1998, A model to explain the various paradoxes associated with mantle noble gas geochemistry: Proceedings of the National Academy of Sciences, USA, v. 95, p. 9087–9092.

Anderson, D.L., and Schramm, K.A., 2005, Global hotspot maps, *in* Foulger, G.R., et al., eds. Plates, plumes, and paradigms: Boulder, Colorado, Geological Society of America Special Paper 388, p. 19–29, doi: 10.1130/2005.2388(03).

Foulger, G.R., and Anderson, D.L., 2005, A cool model for the Iceland hotspot: Journal of Volcanology and Geothermal Research, v. 141, p. 1–22.

Foulger, G.R., Natland, J.H., and Anderson, D.L., 2005, A source for Icelandic magmas in remelted Iapetus crust: Journal of Volcanology and Geothermal Research, v. 141, p. 23–44.

Hawkesworth, C.J., and Kemp, A.I.S., 2006, The differentiation and rates of generation of the continental crust: Chemical Geology, v. 226, p. 134–143.

Heber, V.S., Brooker, R.A., Kelley, S.P., and Wood, B.J., 2007, Crystal-melt partitioning of noble gases (helium, neon, argon, krypton, and xenon) for olivine and clinopyroxene: Geochimica et Cosmochimica Acta, v. 71, p. 1041–1071.

Natland, J.H., 2007, ΔNb and the role of magma mixing at the East Pacific Rise and Iceland, *in* Foulger, G.R., and Jurdy, D.M., eds., Plates, plumes, and planetary processes: Boulder, Colorado, Geological Society of America Special Paper 430, doi: 10.1130/2007.2430(21).

Plank, T., 2005, Constraints from thorium/lanthanum on sediment recycling at subduction zones and the evolution of the continents: Journal of Petrology, v. 46, p. 921–944, doi:10.1093/petrology/egi005.

Schiano, P., Provost, A., Clocchiatti, R., and Faure, F., 2006, Transcrystalline melt migration and Earth's mantle: Science, v. 314, p. 970–974.

Sheth, H.C., 2007 (this volume), Plume-related regional prevolcanic uplift in the Deccan Traps: Absence of evidence, evidence of absence, *in* Foulger, G.R., and Jurdy, D.M., eds., Plates, plumes, and planetary processes: Boulder, Colorado, Geological Society of America Special Paper 430, doi: 10.1130/2007.2430(36).

Yamamoto, M., Phipps Morgan, J., and Morgan, W.J., 2007 (this volume, Chapter 10), Global plume-fed asthenosphere flow—II: Application to the geochemical segmentation of mid-ocean ridges, *in* Foulger, G.R., and Jurdy, D.M., eds., Plates, plumes, and planetary processes: Boulder, Colorado, Geological Society of America Special Paper 430, doi: 10.1130/2007.2430(10).

The Geological Society of America
Special Paper 430
2007

Plate velocities in the hotspot reference frame

W. Jason Morgan*
Department of Earth and Planetary Sciences, 20 Oxford Street,
Harvard University, Cambridge, Massachusetts 02138, USA
Jason Phipps Morgan
Department of Earth and Atmospheric Sciences, Snee Hall,
Cornell University, Ithaca, New York 14853-1504, USA

ABSTRACT

We present a table giving the "present-day" (average over most recent ~5 m.y.) azimuths of tracks for fifty-seven hotspots, distributed on all major plates. Estimates of the azimuth errors and the present-day rates for those tracks with age control are also given. An electronic supplement contains a discussion of each track and references to the data sources. Using this table, the best global solution for plates moving in a fixed hotspot reference frame has the Pacific plate rotating about a pole at 59.33°N, 85.10°W with a rate that gives a velocity at this pole's equator of 89.20 mm/yr (−0.8029 °/m.y.). Errors in this pole location and rate are on the order of ±2°N, ±4°W, and ±3 mm/yr, respectively. The motions of other plates are related to this through the NUVEL-1A model.

The large number of close, very short tracks in the Pacific superswell region precludes all hotspots being rooted near the core-mantle boundary. In general, we think the asthenosphere is hotter than the mantle just below it (in the sense of potential temperature). Asthenosphere is very hot—it is brought up from the core-mantle boundary by plumes. The mantle is cooled by downgoing slabs, and a convective stability is established whereby mantle rises only at plumes and sinks only at trenches. We propose that this normal mantle geotherm is overwhelmed by much-larger-than-average mantle upwelling in superswell areas, making many short-lived instabilities in the upper mantle. Because soft asthenosphere so decouples plates from the mantle below, instabilities in the upper mantle (even above the 660-km discontinuity) are relatively fixed in comparison to plate motions. With the mantle velocity contribution being minor, tracks are parallel to and have rates set by plate velocities.

Keywords: hotspots, mantle plume, plate motion, superswell

---

*E-mail: wjmorgan@princeton.edu.

Morgan, W.J., and Phipps Morgan, J., 2007, Plate velocities in the hotspot reference frame, *in* Foulger, G.R., and Jurdy, D.M., eds., Plates, plumes, and planetary processes: Geological Society of America Special Paper 430, p. 65–78, doi: 10.1130/2007.2430(04). For permission to copy, contact editing@geosociety.org. ©2007 The Geological Society of America. All rights reserved.

INTRODUCTION

The bulk of this article is an electronic supplement that gives the supporting data for each entry in Table 1.[1] This table is the main difference between this work and previous papers solving for present-day plate motions in a hotspot reference frame. In a future paper, we plan to describe in detail the algorithm we applied to the data to solve for the best velocity model; here we briefly describe the algorithm and go directly to our main findings. One of the novel features of our method is the use of "azimuth-only" data when solving for the optimum plate velocity solution—the relative velocities between the plates supplies the velocity information needed, and the greater errors in track rates do not contaminate the more accurately determined track-azimuth data. An azimuth-only technique was recently incorporated in the latest paper by Gripp and Gordon (2002). What distinguishes our work from theirs is our much larger, more globally distributed data set. Note, in their final ("best") solution, Gripp and Gordon (2002) include with their azimuth-only data the rate data from two tracks: Hawaii and Society. We think both of these tracks have rates biased toward high values—these rates were determined with (early, some of the very first) K-Ar measurements, not the more accurate $^{40}Ar/^{39}Ar$ measurements. Their final numbers thus give a velocity for the Pacific plate that is too high, and consequently their fit of tracks in European, African, and Indian regions is worse than in their earlier models. (If they kept their pole the same and slowed down the Pacific, their fit to tracks in these distant regions would become very good.)

Table 1 summarizes all the discussion presented in the supplement. It gives the latitude and longitude of each hotspot and the plate it is on; a weight (which will be defined in the next paragraph); the observed azimuth of its (present-day) track and an estimated error of this azimuth; where possible, the measured velocity (with estimated error); and in the final columns the azimuth and rate predicted by our model. We tried to determine the azimuth over as short an interval as possible, usually over a 5-m.y. interval on fast-moving plates (Pacific, Nazca) and over 10 m.y. on slower plates. If the interval is longer, this is noted in the discussion of the track in the electronic supplement. We give a model azimuth and rate even for places where we cannot measure an azimuth (e.g., at the questionable Kilimanjaro): These places are interesting, and knowing a direction predicted by consistency on the same plate could aid in recognizing a pattern (e.g., if Etna had a track, where would you most likely find it?). We have not considered how deep the source of a hotspot is located (we think there are too many in the list and several that are too close to one another for all of them to have deep origins). A finite-motion reconstruction of long-lived tracks is needed to test the permanence of a source—this list was only used to test the instantaneous present-day motions. However, the results we obtain with our model show that the hotspots in this list provide a very useful reference frame for present-day motions of plates over something "fixed" to the mantle. We think the asthenosphere almost completely decouples the plates from a much more fixed mantle below the asthenosphere, and even minor hotspots that may not originate at the core-mantle boundary appear to have negligible motion with respect to this general frame in our instantaneous inversion.

The weight **w** is a number between 1.0 and 0.2; it is our estimate of the accuracy of the azimuth of the track of a hotspot. The weight is based on the estimated error of the azimuth of the track (σ_{azim}), with downward adjustment of the weight at some tracks based on qualitative criteria, as discussed in the corresponding section of the electronic supplement. The general rule for assigning weights is: $(\sigma_{azim} \leq 8°) \Rightarrow \mathbf{w} = 1.0$, $(8° < \sigma_{azim} \leq 10°) \Rightarrow \mathbf{w} = 0.8$, $(10° < \sigma_{azim} \leq 12°) \Rightarrow \mathbf{w} = 0.5$, $(12° < \sigma_{azim} \leq 15°) \Rightarrow \mathbf{w} = 0.3$, and $(15° < \sigma_{azim}) \Rightarrow \mathbf{w} = 0.2$. If no direction of a track can be determined, the weight is zero and instead a quality letter is given in Table 1. "A" means almost certainly a hotspot but no track (e.g., Etna, Tristan da Cunha); "B" means perhaps a hotspot but not too certain (e.g., Massif Central); and "C" means most likely not a hotspot, even though some characteristics may suggest one (e.g., Jan Mayen). Those with a C ranking are not listed in Table 1. The weight **w** has no meaning in regards to the rate; only rate error bars give indication of the reliability of a rate.

SUPERSWELL REGION

The determination of azimuths and rates as described in the electronic supplement was generally quite straightforward, except in the central Pacific, with its very large number of tracks, many very close together, and many of apparently short duration (~5–10 m.y.). In this section we summarize our findings in the Pacific superswell region, roughly defined as the region enclosed by the Easter, Marquesas, Samoa, and Foundation features. This region is marked by shallower-than-normal seafloor (by ~500 m), numerous volcanic chains, and a warmer-than-typical mantle (e.g., McNutt and Fischer, 1978; Sichoix et al., 1998). The anomalous nature of this area has lent much support to nonplume models of mantle convection (the Web site <httm://www.mantle-plumes.org/> describes many of these models and has discussions about them). We think the essence of these nonplume models is most clearly presented by Anderson (2005, p. 44): "The plate hypothesis assumes that the upper mantle is near the melting point and is variable in fertility, temperature, and solidus temperature. A small change in temperature, volatile content, and composition can have a large effect on melt volumes for a near-solidus mantle. Plate and plate tectonic-induced perturbations can generate 'melting anomalies'." (Anderson uses "plate tectonic-induced perturbations" to mean overall plate tension, with volcano alignments showing stress orientation.)

---

[1]GSA Data Repository Item 2007090, Discussion of each entry in Table 1 with supporting data, is available on request from Documents Secretary, GSA, P.O. Box 9140, Boulder, CO 80301-9140, USA, or editing@geosociety.org, at www.geosociety.org/pubs/ft2007.htm.

TABLE 1. AZIMUTH AND RATE OF EACH HOTSPOT TRACK

Hotspot	Plate	Latitude (°N)	Longitude (°E)	Weight	Az$_{obs}$ (°)	±Error (°)	V$_{obs}$ (mm/yr)	±Error (mm/yr)	Az$_{mdl}$ (°)	V$_{mdl}$ (mm/yr)
Eifel	eu	50.2	6.7	1.0	082	8	12	2	080	5
Iceland	eu	64.4	−17.3	0.8	075	10	5	3	072	3
Azores	eu	37.9	−26.0	0.5	110	12	N.D.*	N.D.	079	6
Massif Central	eu	45.1	2.7	B	N.D.	N.D.	N.D.	N.D.	081	5
Etna	eu	37.8	15.0	A	N.D.	N.D.	N.D.	N.D.	083	6
Baikal	eu	51.0	101.0	0.2	080	15	N.D.	N.D.	100	5
Hainan	ch	20.0	110.0	A	000	15	N.D.	N.D.	N.D.	N.D.
Hoggar	af	23.3	5.6	0.3	046	12	N.D.	N.D.	045	9
Tibesti	af	20.8	17.5	0.2	030	15	N.D.	N.D.	042	11
Jebel Marra	af	13.0	24.2	0.5	045	8	N.D.	N.D.	045	13
Afar	af	7.0	39.5	0.2	030	15	16	8	044	16
Cameroon	af	−2.0	5.1	0.3	032	3	15	5	062	14
Madeira	af	32.6	−17.3	0.3	055	15	8	3	061	4
Canary	af	28.2	−18.0	1.0	094	8	20	4	070	5
Great Meteor	af	29.4	−29.2	0.8	040	10	N.D.	N.D.	101	4
Cape Verde	af	16.0	−24.0	0.2	060	30	N.D.	N.D.	087	8
St. Helena	af	−16.5	−9.5	1.0	078	5	20	3	075	15
Tristan da Cunha	af	−37.2	−12.3	A	N.D.	N.D.	N.D.	N.D.	080	17
Gough	af	−40.3	−10.0	0.8	079	5	18	3	079	17
Vema	af	−32.1	6.3	B	N.D.	N.D.	N.D.	N.D.	067	17
Discovery	af	−43.0	−2.7	1.0	068	3	N.D.	N.D.	073	17
Shona	af	−51.4	1.0	0.3	074	6	N.D.	N.D.	071	17
Réunion	af	−21.2	55.7	0.8	047	10	40	10	043	17
Comores	af	−11.5	43.3	0.5	118	10	35	10	047	17
Kilimanjiro	af	−3.0	37.5	B	N.D.	N.D.	N.D.	N.D.	047	16
Karisimbi	af	−1.5	29.4	B	N.D.	N.D.	N.D.	N.D.	049	16
Mt. Rungwe	af	−8.3	33.9	B+	N.D.	N.D.	N.D.	N.D.	049	16
Marion	an	−46.9	37.6	0.5	080	12	N.D.	N.D.	106	9
Crozet	an	−46.1	50.2	0.8	109	10	25	13	102	9
Ob-Lena	an	−52.2	40.0	1.0	108	6	N.D.	N.D.	107	9
Kerguelen	an	−49.6	69.0	0.2	050	30	3	1	096	9
Heard	an	−53.1	73.5	0.2	030	20	N.D.	N.D.	094	9
Balleny	an	−67.6	164.8	0.2	325	7	N.D.	N.D.	052	6
Scott	an	−68.8	−178.8	0.2	346	5	N.D.	N.D.	044	5
Erebus	an	−77.5	167.2	A	N.D.	N.D.	N.D.	N.D.	037	4
Peter I	an	−68.8	−90.6	B	N.D.	N.D.	N.D.	N.D.	124	1
Martin Vaz	sa	−20.5	−28.8	1.0	264	5	N.D.	N.D.	259	19
Fernando do Noronha	sa	−3.8	−32.4	1.0	266	7	N.D.	N.D.	260	19
Ascension	sa	−7.9	−14.3	B	N.D.	N.D.	N.D.	N.D.	257	19
Guyana	sa	5.0	−61.0	B	N.D.	N.D.	N.D.	N.D.	266	18
Iceland	na	64.4	−17.3	0.8	287	10	15	5	292	16
Bermuda	na	32.6	−64.3	0.3	260	15	N.D.	N.D.	261	18
Yellowstone	na	44.5	−110.4	0.8	235	5	26	5	235	17
Raton	na	36.8	−104.1	1.0	240	4	30	20	239	18
Azores	na	37.9	−26.0	0.3	280	15	N.D.	N.D.	284	18
Anyuy	na	67.0	166.0	B−	N.D.	N.D.	N.D.	N.D.	157	7
Lord Howe	au	−34.7	159.8	0.8	351	10	N.D.	N.D.	001	63
Tasmantid	au	−40.4	155.5	0.8	007	5	63	5	000	66
Eastern Australia	au	−40.8	146.0	0.3	000	15	65	3	006	70
Cocos-Keeling	au	−17.0	94.5	0.2	028	6	N.D.	N.D.	033	70
Juan Fernandez	nz	−33.9	−81.8	1.0	084	3	80	20	081	62
San Felix	nz	−26.4	−80.1	0.3	083	8	N.D.	N.D.	080	61
Easter	nz	−26.4	−106.5	1.0	087	3	95	5	097	61
Galapagos	nz	−0.4	−91.6	1.0	096	5	55	8	086	48
Galapagos	co	−0.4	−91.6	0.5	045	6	N.D.	N.D.	045	81
Louisville	pa	−53.6	−140.6	1.0	316	5	67	5	300	76
Foundation	pa	−37.7	−111.1	1.0	292	3	80	6	283	88
Macdonald	pa	−29.0	−140.3	1.0	289	6	105	10	295	88
Arago	pa	−23.4	−150.7	1.0	296	4	120	20	298	88

(continued)

TABLE 1. *Continued*

Hotspot	Plate	Latitude (°N)	Longitude (°E)	Weight	Az$_{obs}$ (°)	±Error (°)	V$_{obs}$ (mm/yr)	±Error (mm/yr)	Az$_{mdl}$ (°)	V$_{mdl}$ (mm/yr)
North Austral	pa	−25.6	−143.3	**B**	293	3	75	15	296	88
Maria/Southern Cook	pa	−22.2	−154.0	0.8	300	4	N.D.	N.D.	299	88
Samoa	pa	−14.5	−169.1	0.8	285	5	95	20	301	88
Crough	pa	−26.9	−114.6	0.8	284	2	N.D.	N.D.	285	89
Pitcairn	pa	−25.4	−129.3	1.0	293	3	90	15	291	89
Society	pa	−18.2	−148.4	0.8	295	5	109	10	297	89
Marquesas	pa	−10.5	−139.0	0.5	319	8	93	7	295	88
Caroline	pa	4.8	164.4	1.0	289	4	135	20	299	89
Hawaii	pa	19.0	−155.2	1.0	304	3	92	3	302	80
Guadalupe	pa	27.7	−114.5	0.8	292	5	80	10	294	54
Cobb	pa	46.0	−130.1	1.0	321	5	43	3	317	44
Bowie	pa	53.0	−134.8	0.8	306	4	40	20	327	42

Notes: Columns correspond to the hotspot name and the plate its track is on, its location (latitude, longitude), its weight **w** (see the introduction in the text for an explanation of **w**), the observed azimuth Az$_{obs}$ (usually from the most recent ~5–10 m.y.) with estimated error, the observed rate of motion V$_{obs}$ with estimated error, and the azimuth Az$_{mdl}$ and rate V$_{mdl}$ predicted by our model. N.D.—not determined. Plate names: af—African plate; an—Antarctic plate; au—Australian plate; ch—China plate; co—Cocos plate; eu—Eurasian plate; na—North American plate; nz—Nazca plate; pa—Pacific plate; sa—South American plate.

We think the geodynamic nature of this region is perhaps best interpreted as shown in Figure 4 of Courtillot et al. (2003) or Figure 14 of Davaille et al. (2003). An important variation of this interpretation is shown in Figure 2 of Jellinek et al. (2003) or Figure 4 of Gonnermann et al. (2004). In these figures, we see from shadowgraphs of narrow plumes rising from a hot bottom boundary layer that the bases and rising columns of plumes are swept toward the regions with the greatest amount of uprising—there are thus many, many more plumes in a superswell region. This general pattern is portrayed even more clearly, but more schematically, in Figure 17 of Jellinek and Manga (2004). (Their Fig. 17 is reproduced here as our Fig. 3.) Our interpretation of the superswell region is that a very large amount of upward transport from the core/mantle boundary region overwhelms the normal geothermal profile of the Earth and leads to conditions not present above narrow rising plumes. We think for almost all the mantle, asthenosphere (being brought up by plumes from the core/mantle boundary region) is warmer than the mantle below the asthenosphere; this excess of (potential) temperature makes the asthenosphere stable against convection from just below. (That is, a mid-ocean rise has no 2-D mantle "roll" rising from deep beneath it; instead the needed material flows in horizontally from asthenosphere supplied at the nearest plume.) But perhaps an extreme excess of a broad rising plume in the superswell region creates a mantle with the same (potential) temperature as the asthenosphere above, and instabilities that do not exist elsewhere can exist here. In any case, this is a very anomalous, very interesting region; understanding it will likely lead to a big advance in geodynamics. We shall return to these issues in the last section of this article.

FINDINGS

What can we conclude from the information presented above and in the electronic supplement? First, we show two figures from our analysis, which uses the observed directions of hotspot tracks in Table 1 to solve for motions of plates over a fixed mantle reference frame. From these figures, we show how well a fixed reference frame fits the tracks on all plates. We then return to the superswell problem.

The plate motion in a fixed reference frame was found in the following manner. The motions of the major plates are tied together using NUVEL-1A rotations (DeMets et al., 1994). Solving for the three parameters of one plate, the motions of all plates are determined. We took a starting trial of three parameters (ω_x, ω_y, ω_z) for the Pacific, used NUVEL-1A to compute the angular velocities of other plates and thus predict the azimuths of tracks at each hotspot in our data set, and then searched for the three parameters that minimized a misfit measure. Our measure was the sum of the weight of a given track times the absolute value of the difference between the observed azimuth and the computed azimuth (i.e., minimized $\Sigma\, \mathbf{w}_i \times |\,AZ_{obs,i} - AZ_{pred,i}\,|$). (Almost any minimizing function would give essentially the same result. This scheme downplays outliers, as extreme differences are not squared in the sum.) Azimuth-only data are all that are needed because the velocities are determined by the NUVEL-1A model. If the choice of pole position (of the Pacific) is correct but the angular velocity is wrong, tying to other plates would seriously mismatch the azimuths of motion there. (An azimuth-only scheme was used by Gripp and Gordon, 2002, in one version of their solution, but then they included rate data for their final HS3 model.)

The advantage of azimuth-only data is that azimuths are more accurately observed, and there are no systematic errors, which may be present in rate measurements. Many of the tracks are poorly dated, and K-Ar measurements may have unknown systematic errors. Also, how does one devise a weighting function to include a track with an accurate azimuth but poorly determined rate in with the other tracks? We have used azimuths only, and shall test the solution by comparing it with the measured rates. (A systematic comparison of the measured rates, and their error estimates, with model-predicted-rates is a major component of our paper-in-progress. Here we have only a visual comparison, as shown later in the article in Fig. 2B.)

An important concern is that azimuths on the Pacific and Nazca plates are determined from recent (0 to ~5 m.y.) parts of the tracks, whereas on the slower plates, the azimuth is generally defined over a longer period of time (0 to ~10 m.y.), and on some of the poorest defined tracks (Hoggar, Tibesti, Afar, Cape Verde, and Kerguelen) over up to ~30 m.y. (The fast-moving Australian tracks are averaged over a longer period for a different reasons: (1) there are no seamounts in the Tasmantid chain younger than 5 Ma, and (2) the very straight-line alignment of the inland volcanics used to find the azimuth of the East Australian track range in age from 8 to 17 Ma.) However, several of the slow tracks are quite narrowly defined in azimuth (Eifel, Canary, St. Helena, Martin Vaz, Fernando do Noronha, Yellowstone, and Raton), and there is no obvious kink in their tracks to suggest a significant long-term and/or short-term difference.

Figure 1 shows how the bootstrap method was used to estimate the uncertainty in the latitude, longitude, and rate of the Pacific plate motion over the mantle. From the data table of fifty-seven hotspots, weights, and azimuths (Table 1), we generated many synthetic data sets, each with fifty-seven values randomly selected from the actual data set. In these synthetic sets, some hotspots appear multiple times and some are omitted —the bootstrap method uses a random, non-emptying selection process. For each synthetic set, there is a search for the ω_x, ω_y, and ω_z that minimizes the misfit measure. To visualize this process, the latitude and longitude of the best fit for each synthetic set is plotted as a dot in Figure 1A. The average of all these points (each point being the best-fit ω for a given synthetic data set) is the best pole for the original data set; the dispersion of points measures the error of this best-fit. (No information on rates is displayed in Fig. 1A.) The concentric ellipses are of sizes that contain 68, 90, and 95% of the points; these show the confidence intervals. The figure was drawn with only 330 synthetic samples (many more were used to fix the orientation and sizes of the ellipses). With this smaller number, individual points are discernable; for example, fourteen points are outside the 95% confidence ellipse, very close to 5% of the total 330 points. The streaks and lines are not random variations; they occur because a best fit for a given synthetic set is strongly influenced by which plates end up with many repeats of a datum and which end up with very few (or none).

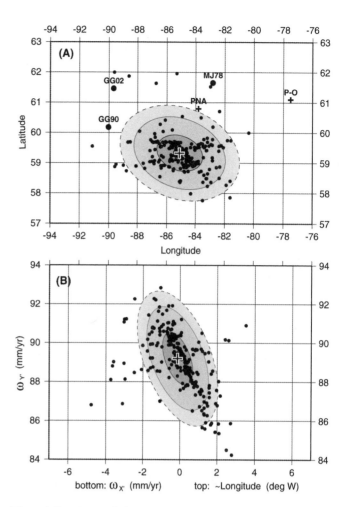

Figure 1. Bootstrap solution pole positions for present-day Pacific plate motion in the fixed mantle reference frame, based on the data set in Table 1. The scatter of the bootstrap solutions shows the error in determining the average; 68, 90, and 95% confidence ellipses are shown. The best solution is 59.33°N, 85.10°W with an angular velocity that produces 89.20 mm/yr at the equator of this pole. (A) The 95% confidence ellipse of the pole position has the following parameters: semimajor axis, 2.11 arc-deg; semi-minor axis, 1.50 arc-deg; azimuth, 109.2°. (B) A slice in angular velocity space which shows the magnitude of rotation rate versus longitude. The component ω_y' is parallel to the magnitude; ω_x' changes the direction of the vector in the east-west direction. The scale is set so that the horizontal distance here is approximately the same as the longitude scale in panel A. Angular velocity is less if the pole position is farther from the Pacific (and vice versa), minimizing the change in linear velocity of points in the Pacific. Other solutions are marked by black circles and crosses: GG90—pole determined by Gripp and Gordan (1990); GG02—pole determined by Gripp and Gordan (2002); P-O—Pacific-only pole; MJ78—pole determined by Minster and Jordan (1978); PNA—Pacific, North American, and Australian pole. See the text for a discussion of these solutions.

Figure 1B is a slice in angular velocity space through the center of the ellipsoid of bootstrap solutions of $\omega_x\omega_y\omega_z$. This $(\omega_x\omega_y\omega_z)$ was rotated to get $(\omega_{x'}\omega_{y'}\omega_{z'})$, where the component $\omega_{y'}$ is parallel to the best-fit vector and $\omega_{x'}$ and $\omega_{z'}$ are tangent to the map projection shown in Figure 1A. The component $\omega_{y'}$ is the magnitude of the rotation rate (within 1%, there is a small cosine-effect); $\omega_{x'}$ is perpendicular to this and acts to change the orientation of the angular velocity in the east-west direction. The scale of Figure 1B was chosen such that a change in $\omega_{x'}$ will measure the same horizontal distance as the corresponding change in longitude in Figure 1A. Again, ellipses are drawn that would contain 68, 90, and 95% of the points if the points fit a normal distribution. If the pole position is a little west, closer to the Pacific plate, its angular velocity is greater. The combination of radius and angular velocity interact with each other in a manner that keeps the linear velocities of points in the Pacific about the same for small changes in the pole position.

Figure 2 compares our model with the observations listed in Table 1. The top panel shows the predicted azimuths (black lines) of our model and observed azimuths (gray lines). The agreement with the azimuth data is quite remarkable. There are fifty-seven tracks with an observable azimuth listed in Table 1. If the nine poorly defined tracks assigned a weight of 0.2 and the six additional tracks where the modeled plate velocity is less than 7 mm/yr are excluded, we are left with forty-three "quality" tracks. Of these forty-three, only seven (16%) have a difference between observed and modeled azimuth greater than 15° (the uncertainty threshold we used for setting $\mathbf{w} = 0.2$). These are the Cameroon, Comores, Marion, Louisville, Samoa, Marquesas, and Bowie tracks. In the corresponding sections of the electronic supplement (see footnote 1), possible special circumstances that might explain why the observed azimuths differ from the model azimuths are discussed for these tracks. Figure 2B shows arrows whose lengths are proportional to velocity. It shows the model velocities (black arrows) and the observed velocities (gray arrows). If a velocity has not been observed, a gray line without an arrowhead is drawn in the direction of the observed azimuth. All fast velocities are in the Pacific or Australian regions; all other plates have very slow velocities over the mantle. The fit to the velocity data is not as good as that for the azimuths. On several tracks (Caroline, Guadalupe, Macdonald, Arago, Society, Easter, Canary, Réunion, and Crozet), the observed rate is significantly higher than the model rate. (Of these tracks, only Canary and Easter have rates based on $^{40}Ar/^{39}Ar$ dating; rates for the other seven are based on K-Ar.) We have come to the conclusion that the azimuth data set is a much cleaner set and the velocity set less reliable. Velocities are harder to determine—late-stage volcanism, delayed volcanism penetrating the lithosphere, and argon loss in samples caused by small amounts of weathering all systematically shift the ages toward too young (and the rates toward too high). The most striking feature of Figure 2B is the contrast between the very high velocities in the Pacific area and very small, even near-zero, velocities on the other plates. Our model not only duplicates the

very slow velocities on these plates but even correctly predicts the azimuths of tracks on these plates.

Our pole position at 59.33°N, 85.10°W is very similar to that of earlier papers (e.g., Minster and Jordan, 1978; Gripp and Gordon, 1990, 2002). Our confidence ellipse is smaller, but that is a minor difference resulting from the use of a larger data set. What is really different is the rate of angular velocity—ours (89.20 mm/yr or −0.8028°/yr) is significantly less than, say, that of Gripp and Gordon (2002). With this slower rotation about essentially the same pole, the predicted directions of motion of all other plates fit the observed azimuths of their hotspot tracks. If the angular velocity of the Pacific is too fast, it adds a westward component to the motion of all plates, which can result in up to ~180° difference between predicted and observed tracks on very slow moving plates. We tried several variations based on the data in Table 1. In one case, we used only the best nineteen azimuths determined (those with weight $\mathbf{w} = 1$), in another case we used thirty-four very good azimuths (those with $\mathbf{w} = 1$ or 0.8), and in a third case we used all fifty-seven azimuths but gave them all a weight of 1 irrespective of the $\mathbf{w}$ values listed in the table. These cases gave solutions 59.38°N, 83.20°W, 87.40 mm/yr; 59.01°N, 83.45°W, 87.96 mm/yr; and 59.66°N, 87.47°W, 89.77 mm/yr, respectively. All cases fell within the 90% confidence ellipse shown in Figure 1A. (The error ellipses of these cases were all larger than our best solution shown in Figure 1A, generally by a factor between 1.5 and 2.)

We then chose a data set consisting of the thirty-seven tracks that were not in the Pacific; that is, omitting those tracks on the Pacific, Nazca, or Cocos plates. This data set gave a solution of 59.32°N, 85.00°W, 89.18 mm/yr—almost identical to our solution using all the data. Thus we find no difference from the pole of non-Pacific hotspots and the pole using all tracks. Next we chose a data set for which only the twenty tracks on the Pacific, Nazca, or Cocos plates were used. These gave a solution of 61.1°N, 77.5°W, 83.2 mm/yr but with very large uncertainties: the 95% ellipse had a semi-major axis of 24 arc-deg and a semi-minor axis of 8 arc-deg, with the long-axis along an azimuth of 069°. (This Pacific-only pole is marked with a cross labeled "P-O" in Figure 1A.) This large uncertainty was expected from the azimuth-only technique; with the data essentially all on the same plate, the lines perpendicular to the tracks are all nearly parallel, so that there are no cross-cutting constraints to refine the position in the long direction. We would need velocity versus distance data to pin down the best pole on this long streak, just as both fracture-zone trends and spreading rates are needed to accurately fix a pole for the relative motion between two plates. We chose a slightly larger data set, the twenty Pacific tracks above plus Yellowstone and Raton on the North American plate and the East Australian, Tasmantid, and Lord Howe tracks on the Australian plate (reasoning that these tracks are nearby and might be part of a Pacific set). This choice gave a solution of 60.8°N, 83.8°W, 89.2 mm/yr with smaller but still large uncertainties, for 95%, the semi-major axis is 12 arc-deg, semi-minor axis 7 arc-deg, and azimuth 040°. (This Pacific plus North

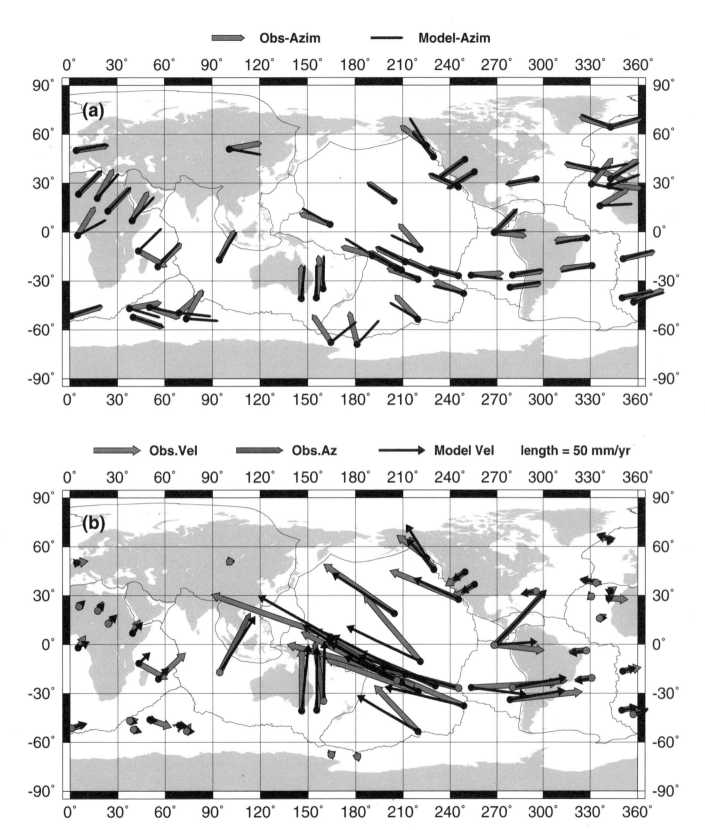

Figure 2. (A) Thick gray lines with constant length show observed azimuths of hotspot tracks, and thin black lines show predicted azimuths for each entry in Table 1. (B) The length of each arrow is proportional to velocity. Black arrows show predicted velocities and gray arrows (with arrowheads) show observed velocities. If the velocity of a track could not be determined, a gray line with no arrowhead is drawn in the direction of the observed azimuth (with a length the same as that for the model velocity).

America plus Australia pole is marked with a cross labeled "PNA" in Fig. 1A.) However, there was no need to find the pole for Pacific-only motion ourselves; papers by Minster and Jordan (1978) and Gripp and Gordon (1990, 2002) have already done so. The Minster and Jordan (1978) and Gripp and Gordon (1990) papers use only data from the Pacific, Cocos, and Nazca plates plus one other data point—the azimuth of the Yellowstone track in North America. The Gripp and Gordon (2002) paper uses data from the Pacific area and Yellowstone plus one other data point —the azimuth of the Martin Vaz/Trindade trend on the South American plate. These three poles are shown in Figure 1A, marked MJ78, GG90, and GG02. They are all within ~3 arc-deg of our pole that best fits the non-Pacific set: the thirty-seven tracks not on the Pacific, Cocos, or Nazca plates. (Recall that the non-Pacific pole is practically on top of our pole that best fits all fifty-seven tracks.)

We have shown and discussed only the pole and rate of the Pacific plate, but motions of other plates are directly tied to this Pacific motion. Global velocities are given in Table 2, obtained by adding each plate's difference to Pacific motion as given in NUVEL-1A (DeMets et al., 1994).

We conclude, for present-day motion, that a single reference frame fits the entire world—the hotspots as a group in the Pacific do not drift relative to a group centered on Africa. The various poles for the Pacific tracks are ~3° from our pole for the non-Pacific tracks, just outside our error ellipse, but our pole is well within the larger error ellipses of the three earlier studies. The rates about these poles are different; our rate is ~89 mm/yr, the rates of Minster and Jordan (1978) and Gripp and Gordon (1990, 2002) average about 20% faster (~110 mm/yr). If the hotspots on opposite sides were to drift relative each other, we find it very strange they would both have the same pole position but have different rates about this pole—why not a totally different pole if one side moves relative to the other? Note also that our slower rate is close to the average rate of Pacific motion for

the past 47 Ma—the higher rate of other models needs to have been much slower earlier or the Pacific tracks would get to the Emperor bend too soon. We conclude that earlier studies have chosen too high a velocity, biased by systematics in the rate measurements of the hotspot tracks.

Could the rates of track migration on the Pacific plate be high because the mantle and hotspot pipes themselves are being pushed away from the Japan-Marianas centers of subduction? That is, the high velocity measurements in the Pacific are correct but there is an added component, as explained in several papers (e.g., Steinberger and O'Connell, 1998, 2000; Steinberger, 2000). We think this pipe-drift effect is real, but that the rates of hotspot drift are not ~20 mm/yr, as is needed to explain the difference between our model and the earlier models. In a simple mass balance, a 200-km-thick slab descending into the lower mantle at 100 mm/yr creates a volume flux that can be matched by a 2000-km-thick slab of lower mantle moving horizontally at 10 mm/yr, or, with half the flow going in each direction, at ~5 mm/yr. This effect might have a strong influence at some sites (e.g., Caroline), but could not cause all the tracks in the Pacific to drift west-to-east at the ~20 mm/yr needed to reconcile the earlier models to ours. Furthermore, the ~5 mm/yr is probably on the high side—even a 5-mm/yr drift of hotspots would play havoc with the strikingly good model-to-data fit of azimuths on the slower plates.

SOME INFERENCES

Given that we find strong support for a model of rigid plates over a fixed mantle reference frame, we discuss two issues that came up during our compilation of the data for each track. The first is the closeness of pairs of hotspots, the second is the super-swell problem.

There are a number of hotspots that are very near other hotspots—so close that it suggests they cannot be independently anchored deep in the mantle at the core-mantle boundary. In particular, the following pairs of hotspots are close together: Tristan/Gough, Madeira/Canary, Eifel/Massif Central, Kerguelen/Heard, Eastern Australia/Tasmantid, Easter/Crough, and Cobb/Pratt-Welker (Bowie). A possible explanation for this close pairing that allows for the fixity of hotspots is as follows. Suppose in the initial, flood-basalt stage of a plume, a large pipe 300 km across develops to rapidly carry a large volume to the surface. The excess temperature of the ascending plume causes zone refinement in the mantle surrounding the pipe. The higher temperature and increased diffusion rate extracts volatiles from the surrounding mantle, leaving a wall a few tens of kilometers thick. Without the volatiles, this wall is tougher than normal mantle. Then, as the large plume flux fades away, the 300-km-diameter tube is too large for the flux it carries. The tube starts to collapse, flattening from a round to an elongate elliptical cross-section. As the upward flux continues to decrease, opposite sides of the tube might touch each other, sealing off the center. But the two distal portions would remain open because the 10+-km-

TABLE 2. PLATE MOTIONS IN HOTSPOT-FIXED FRAME

Plate	Latitude (°N)	Longitude (°E)	Angular rate (°/m.y.)	Maximum speed (mm/yr)
Africa	43.07	−24.56	0.1544	17.16
Antarctica	70.18	82.26	0.0989	10.99
Arabia	31.00	13.06	0.5205	57.83
Australia	20.14	40.28	0.6990	77.67
Caribbean	−20.80	−56.84	0.0808	8.98
Cocos	22.32	−116.24	1.3343	148.26
Eurasia	82.54	−125.53	0.0619	6.88
Juan de Fuca	−31.32	60.04	0.8540	94.89
Nazca	49.95	−95.74	0.5634	62.60
North America	−51.87	−48.70	0.1628	18.08
Pacific	−59.33	94.90	0.8028	89.20
Philippine	−45.71	−23.83	0.9928	110.31
South America	−75.22	100.24	0.1749	19.43

Note: Plates tied to one another using relative motions of NUVEL-1A.

thick tough rind flexes with a minimum radius, not deforming into a tight crease that would seal the tips. What observations would test such a scenario? Three observations might be: (1) the pairing would occur long after a flood basalt stage; (2) at the very end of a plume's life, the flux might become too low to supply two conduits and one might disappear by starvation; and (3) the plane between the two tips would be oriented perpendicular to the maximum horizontal compressive stress in the lower mantle.

How can the superswell region, with its (1) shallower-than-normal seafloor, (2) numerous short volcanic chains that appear to turn on and off with time scales of ~10 m.y. rather than the ~100 m.y. typical of many hotspots, and (3) a warmer-than-typical mantle (see the section on the superswell region above for details) be reconciled with our model of a global, mantle-fixed reference frame? We assume a very weak asthenosphere that almost completely decouples the plates from the mantle below. (This decoupling is less complete beneath old parts of continents; consequently, plates with large areas of shield and/or platform move slower over the mantle than do largely oceanic plates.) In a recent paper on mantle viscosity, Mitrovica and Forte (2004, Fig. 1c,d) find a low-viscosity channel below the lithosphere, a higher viscosity just beneath the asthenosphere, and a much higher viscosity in the lower mantle (until very near the core-mantle boundary). In the averages we calculated when we re-chose the depth of boundaries between regions, we get from their Figures 1c and d the following values for the viscosity: $\eta = 5 \times 10^{19}$ Pa s between 100 km and 300 km, $\eta = 5 \times 10^{20}$ Pa s between 300 km and 660 km, and $\eta \approx 2 \times 10^{22}$ Pa s in the lower mantle. This general pattern, with a low-viscosity asthenosphere channel and higher-viscosity lower mantle and with approximately the same ratios between high and low viscosities, was also found by Panasyuk and Hager (2000, Fig. 5a and b). In our analysis of horizontal flow in an asthenospheric channel, we found an average viscosity of the asthenosphere of 7×10^{18} Pa s (caption to Fig. 16 *in* Yamamoto et al., this volume, chapter 9). (In this analysis, we assumed the base of the asthenosphere was at a depth of 300 km, and the thickness of the asthenosphere was ~200 km, varying slightly as the lithosphere thickened with age. If the asthenosphere viscosity were greater than 7×10^{18} Pa s, there would be too much pile up and corresponding lift of the seafloor because of the pressure gradient created by plate drag as the Pacific moves westward toward the Mariana and Japan trenches.)

Using these viscosities and thicknesses (~1×10^{19} Pa s in a 200-km-thick asthenosphere, ~5×10^{20} Pa s between ~400 km and 660 km, and ~2×10^{22} Pa s in a ~2000-km-thick lower mantle), we get roughly the following results. If a hotspot source is well anchored near 660 km at the base of the transition zone, its drift relative to the fixed frame would be only ~1/20 of the horizontal motion of the plate above (assuming uniform shear stress in the layers). If it is anchored at the base of the mantle, its drift would be even less, as determined by horizontal flow in the lower mantle needed for mass balance of what is being injected at subduction zones. (However, much of the horizontal transport

in the lower mantle could be very high velocities in the low-viscosity D″ layer and not slow horizontal velocities distributed throughout the lower mantle.) The fixity of hotspots has been one of the main arguments favoring their sources being near the core-mantle boundary. We have shown in this article that this fixity is even more prominent than previously thought, but if it is due to a near-inviscid asthenosphere decoupling the plates from all mantle below, hotspot fixity is no longer a reason to place the sources at the base of the mantle. However, we think there are other arguments favoring the deep origin of plumes.

In our model, all asthenosphere is created by plumes bringing material up from D″. This process requires a very large flux: the total upward flow in all plume pipes is ~300 km³/yr—the same as the downward flux of all slabs. The asthenosphere is very hot—it is the temperature of D″ minus only the adiabatic effect as the plume rises to asthenospheric depth. The subducting slabs carry coldness into the mantle. These are gradually warmed up by conduction from the surrounding mantle (extracting heat from a larger volume), by internal radioactivity over their trip time (on the order of billion years), plus the added input for the short time material is in D″ before coming back to the asthenosphere. This pattern of convection would produce an asthenosphere hotter than the mantle below it. Downward conduction of heat from the asthenosphere would make an isothermal temperature gradient in the transition-zone region between the base of the asthenosphere and the 660-km discontinuity (and somewhat below the 660-km discontinuity), or at least it would reduce the temperature increase with depth to below that for an adiabatic gradient. Thus the mode of convection we are suggesting would act as an on/off switch: if the plumes bring up enough hot material from D″ depths to make a hot asthenosphere, then this process turns off all other convection from the mid-mantle or 660-km discontinuity into the upper mantle. That is, adiabatic cooling of material starting from mid-mantle or 660-km depths would result in material that is cooler than that above it, so such modes of convection (such as rising from below the East Pacific Ridge to supply the spreading ridge with material) would be turned off. This model requires that a sufficient quantity of material be brought up by plumes and that this material spreads out easily from where there are sources of asthenosphere (above plumes) to where there are sinks of asthenosphere (mid-ocean ridges). We do not know how much material is "sufficient," but once a limit is reached, plumes-make-asthenosphere becomes the only mode of convection.

We think the plate tectonic cycle adjusts its rate to equal what is brought up by plumes. If plume rates were to decrease, the asthenosphere would thin, plate drag would increase, and subduction rates slow down. Contrariwise, if plumes were more vigorous, plate drag would decrease and movement toward subduction zones would be faster. There may be cycles of fluctuations tens of millions of years long. For example, at the time of a flood basalt, the asthenosphere may temporarily increase in volume and the asthenosphere may then thin as this excess is used up (the Cretaceous super-cycle may be such a case)—but the long-term rate is set by the plumes. Many consequences of this pattern of flow are explored in Yamamoto et al., (this volume, chapter 9).

The main effect of plumes is to remove heat from the earth's interior, not to make volcanism at the surface (although without this volcanism plumes would be near-invisible). This pattern of flow can efficiently extract heat and set the rate for plate tectonics, but how can it be applied to the superswell region? Why would a small area of the Pacific (only 5% of the Earth's surface) have a quarter of the world's hotspot tracks? A slightly larger region (~10% of the Earth's surface) on the African side has another quarter of all hotspots. If the long-term convection scheme remains more or less the same for some long period of time, the rising plumes could be swept into a central region. The pattern shown schematically in Figure 3 (adapted from Fig. 17 *in* Jellinek and Manga, 2004) is a highly schematic interpretation of the plume/tank experiments shown in Figure 4 of Gonnermann et al. (2004). This same idea is nicely illustrated in Figure 11 of Steinberger and O'Connell (1998), and is clearly stated in Steinberger's thesis (Steinberger, 1996, p. 76): "This raises another possibility—plumes might be stationary where they are, not because flow in the mantle is so slow, but because they already have been advected into large-scale stationary upwellings in the lower mantle."

Convective overturn removes interior heat by bringing mantle close to the top surface, where its heat is lost through the surface by conduction. We think the asthenosphere is hotter than mantle beneath it, cutting off any upwelling from the transition zone into the asthenosphere except in rising plumes and possibly the superswell region, as discussed below. This restriction does not mean that there are no up-and-down motions in the mantle except sinking slabs and rising plumes. Rather, any secondary interior flow induced by denser slabs and buoyant plumes will not lead to heat loss through the surface. In this sense, slow interior flow is only of secondary importance to Earth's convective heat transport.

For most of the Earth, a hotspot at the surface is an indication of a plume rising from the deep. The exception is in the superswell region. If a plume, or group of plumes, remains in one place in the mantle for hundreds of millions of years, the outward conduction of heat from the rising pipe would spread out into the surrounding mantle. If it conducts outward ~100 km in 100 m.y., it would conduct outward ~300 km in 1 b.y. If rheology depends on temperature and volatile content, the high temperature alone would make the mantle surrounding a pipe less viscous, but a zone-refining effect as discussed above regarding hotspot pairs would increase the viscosity of the region around a pipe. The net result may be the mantle around a pipe is hotter but not significantly less viscous than other parts of the mantle. Thus in the superswell region, the usual stable geotherm pattern (where asthenosphere is hotter than mantle below) is overwhelmed. Here relatively shallow convective instabilities could occur as temperatures at the 660-km discontinuity and the mantle above might be hotter than asthenosphere (because of the long-term outward conduction of heat into the mantle from the stable pipes). Short-term instabilities could then rise into the asthenosphere, creating short-lived tracks. In our model of viscosity structure of the upper mantle (from Panasyuk and Hager, 2000; Mitrovica and Forte, 2004; Yamamoto et al., this volume, chapter 9), the mantle between ~400 and 660 km is considerably more viscous than asthenosphere and would have low velocities compared to those for fast plate motion. Thus all observed tracks at the surface would be nearly parallel to plate motion, and all would have migration rates very close to the velocity of the plate, because to be rooted in the lower transition zone would be nearly as fixed as being rooted at the core-mantle boundary. The major difference would be that the shallow-rooted sources would have shorter lifespans than the deeper-rooted sources, and they would not begin with a flood basalt stage. Such a scheme (the main features of which could be validated with numerical modeling) reconciles the model of plumes rising from the core-mantle boundary as the primary pattern of mantle convection with the conundrum of so many hotspots, many short-lived, and many close together, as was forcefully argued in the book *Plates, plumes, and paradigms* (Foulger et al., 2005).

ACKNOWLEDGMENTS

WJM thanks the Department of Earth and Planetary Sciences at Harvard University, Cambridge, Massachusetts, for welcoming him as a visiting researcher. For the tracks on land, the map collection at the Princeton University library, Princeton, New Jersey, was invaluable. For tracks in the oceans, the synthetic bathymetry of W.H.F. Smith and D.T. Sandwell (version 7.2) was our primary source. Blowup plots of the bathymetry defining the ocean tracks were easy to make because of the GMT software of P. Wessel and W.H.F. Smith (1998), and we espe-

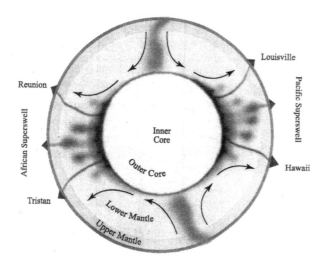

Figure 3. Schematic diagram illustrating how long-term subduction in fixed places (here the Andes and western Pacific) could push and concentrate the plumes into the superswell regions. Adapted from Jellinek and Manga (2004).

cially thank W.H.F. Smith for showing us a work-around to a plotting difficulty we encountered. Ajoy Baksi showed us how to use the metadata to evaluate the accuracy of an age date. We thank Gill Foulger, Richard Gordon, Dietmar Müller, and Bernhard Steinberger for their thoughtful reviews, which added significantly to our manuscript. WJM thanks the Alexander von Humboldt Foundation and Princeton University and JPM thanks the National Science Foundation and Deutsche Forschungsgemeinschaft (DFG) for support during the course of this work.

REFERENCES CITED

An additional 289 references referring to specific tracks are in the electronic supplement.

Anderson, D.L., 2005, Scoring hotspots: The plume and plate paradigms, *in* Foulger, G.R., et al., eds., Plates, plumes, and paradigms: Boulder, Colorado, Geological Society of America Special Paper 388, p. 31–54.

Courtillot, V., Davaille, A., Besse, J., and Stock, J., 2003, Three distinct types of hotspots in the Earth's mantle: Earth and Planetary Science Letters, v. 205, p. 295–308, doi: 10.1016/S0012-821X(02)01048-8.

Davaille, A., Le Bars, M., and Carbonne, C., 2003, Thermal convection in a heterogeneous mantle: Comptes rendus Geoscience, v. 335, p. 141–156.

DeMets, C., Gordon, R.G., Argus, D.F., and Stein, S., 1994, Effect of recent revisions to the geomagnetic reversal time scale on estimates of current plate motions: Geophysical Research Letters, v. 21, p. 2191–2194.

Foulger, G.R., Natland, J.H., Presnall, D.C., and Anderson, D.L., eds., 2005, Plates, plumes, and paradigms: Boulder, Colorado, Geological Society of America Special Paper 388, 881 p.

Gonnermann, H.M., Jellinek, A.M., Richards, M.A., and Manga, M., 2004, Modulation of mantle plumes and heat flow at the core mantle boundary by plate-scale flow: Results from laboratory experiments: Earth and Planetary Science Letters, v. 226, p. 53–67, doi: 10.1016/j.epsl.2004.07.021.

Gripp, A.E., and Gordon, R.G., 1990, Current plate velocities relative to the hotspots incorporating the NUVEL-1 global plate motion model: Geophysical Research Letters, v. 17, p. 1109–1112.

Gripp, A.E., and Gordon, R.G., 2002, Young tracks of hotspots and current plate velocities: Geophysical Journal International, v. 150, p. 321–361, doi: 10.1046/j.1365-246X.2002.01627.x.

Jellinek, A.M., and Manga, M., 2004, Links between long-lived hotspots, mantle plumes, D″, and plate tectonics: Reviews of Geophysics, v. 42, art. no. RG3002, doi: 10.1029/2003RG000144.

Jellinek, A.M., Gonnermann, H.M., and Richards, M.A., 2003, Plume capture by divergent plate motions: Implications for the distribution of hotspots,

geochemistry of mid-ocean ridge basalts, and estimates of the heat flux at the core-mantle boundary: Earth and Planetary Science Letters, v. 205, p. 361–378, doi: 10.1016/S0012-821X(02)01070-1.

McNutt, M.K., and Fischer, K.M., 1978, The South Pacific superswell, *in* Keating, B.H., et al., eds., Seamounts, islands, and atolls: Washington, D.C., American Geophysical Union Geophysical Monograph 43, p. 25–34.

Minster, J.B., and Jordan, T.H., 1978, Present-day plate motions: Journal of Geophysical Research, v. 83, p. 5331–5354.

Mitrovica, J.X., and Forte, A.M., 2004, A new inference of mantle viscosity based upon joint inversion of convection and glacial isostatic adjustment data: Earth and Planetary Science Letters, v. 225, p. 177–189, doi: 10.1016/j.epsl.2004.06.005.

Panasyuk, S.V., and Hager, B.H., 2000, Inversion for mantle viscosity profiles constrained by dynamic topography and the geoid, and their estimated errors: Geophysical Journal International, v. 143, p. 821–836.

Sichoix, L., Bonneville, A., and McNutt, M.K., 1998, The seafloor swells and superswell in French Polynesia: Journal of Geophysical Research, v. 103, p. 27,123–27,133, doi: 10.1029/98JB02411.

Steinberger, B.M., 1996, Motion of hotspots and changes of the Earth's rotation axis caused by a convecting mantle [Ph.D. thesis]: Cambridge, Massachusetts, Harvard University, 203 p.

Steinberger, B.M., 2000, Plumes in a convecting mantle: Models and observations for individual hotspots: Journal of Geophysical Research, v. 105, p. 11,127–11,152, doi: 10.1029/1999JB900398.

Steinberger, B., and O'Connell, R.J., 1998, Advection of plumes in mantle flow: Implications for hotspot motion, mantle viscosity and plume distribution: Geophysical Journal International, v. 132, p. 412–434, doi: 10.1046/j.1365-246x.1998.00447.x.

Steinberger, B., and O'Connell, R.J., 2000, Effects of mantle flow on hotspot motion, *in* Richards, M.A., et al., eds., The history and dynamics of global plate motions: Washington, D.C., American Geophysical Union Geophysical Monograph 121, p. 377–398.

Wessel, P., and Smith, W.H.F., 1998, New improved version of Generic Mapping Tools released: Eos (Transactions, American Geophysical Union), v. 79, 579. See also http://gmt.soest.hawaii.edu/ (accessed 15 May 2006).

Yamamoto, M., Phipps Morgan, J., and Morgan, W.J., 2007 (this volume, chapter 9), Global plume-fed asthenosphere flow—1: Motivation and model development, *in* Foulger, G.R., and Jurdy, D.M., eds., Plates, plumes, and planetary processes: Boulder, Colorado, Geological Society of America Special Paper 430, doi: 10.1130/2007.2430(09).

MANUSCRIPT ACCEPTED BY THE SOCIETY 31 JANUARY 2007

DISCUSSION

27 January 2007, Alan D. Smith

A difficulty arises in the plum-pudding plume model (Morgan and Phipps Morgan, this volume; Yamamoto et al., this volume, chapters 9 and 10) with regard to Pt-Os isotope systematics. The $^{186}Os/^{188}Os$ ratios in the mid-ocean ridge basalt (MORB)-source mantle, as indicated by the isotopic compositions of abyssal peridotites ($^{186}Os/^{188}Os = 0.119830$—$0.119838$), are generally lower than in intraplate volcanic rocks (Hawaiian picrites and Gorgona komatiites; $^{186}Os/^{188}Os = 0.119831$—$0.119850$). In standard plume models, in which MORB and in-

traplate volcanic rocks are derived from distinct isolated reservoirs, the isotopic differences are explained by the addition of approximately 0.5 wt% outer-core material ($^{186}Os/^{188}Os = 0.119870$) to a plume (Brandon and Walker, 2005, and references therein).

In the plum-pudding plume model, the asthenosphere is fed by plumes generated from the D″ layer at the core-mantle boundary. MORB and intraplate volcanic rocks should therefore have similar $^{186}Os/^{188}Os$, unless a core-derived Os component could be selectively removed on melting of a plume. Such a scenario would seem unlikely, as MORB and intraplate volcanic rocks have lower Os content than do mantle peridotites, suggesting Os partitions into the mantle on melting. A hypothetical plume comprising 30% recycled basaltic oceanic crust, 69.5% oceanic lithosphere, and 0.5% outer-core material, would have an Os content of ~3.8 ppb, and would be unlikely to be depleted of Os by generation of ocean island basalt (OIB)/picrites, which have 0.03–1.0 ppb Os. The similar Os contents estimated for MORB-source mantle (~3.3 ppb Os) and plumes would fit the plum-pudding plume model, but the variation in $^{186}Os/^{188}Os$ is not consistent with such a model.

31 January 2007, Warren B. Hamilton

The fixed-plume concept—narrow buoyant jets of hot material rise from the deep mantle and burn through overpassing lithosphere plates to produce tracks of volcanoes that define the motions of the plates over the bulk Earth—was developed in large part by W.J. Morgan and was anchored to the younging of the Hawaiian chain of volcanoes and seamounts east-southeastward from the Emperor elbow. The speculations by Morgan and Phipps Morgan (this volume) take this notion to extreme form. The material and excess heat [none is needed] of the globe-girdling asthenosphere are carried from the basal to the shallow mantle by scores of plumes. Almost all mentioned intraplate volcanism, much intraplate deformation, and much plate-margin volcanism is a direct or indirect product of plume-borne heat and material. The tens of thousands of small Pacific seamounts are unmentioned, and it is unclear why only a sizeable fraction of spreading-ridge magmatism is a byproduct of plumes.

The concept of globally fixed plumes has become progressively harder to defend against voluminous contrary evidence, including multidisciplinary disproof of the critical notion that the Emperor chain of seamounts tracked pre-Hawaiian-chain motion of the Pacific plate, but Morgan and Phipps Morgan (this volume) update the rationalization that plumes are approximately fixed from the deep mantle to the surface, and thus that plume tracks, as inferred from whatever surface features can be attributed to hypothetical plumes, define an absolute-motion framework for lithosphere plates. They show that the azimuths of their selection of hotspot tracks approximately accord with a fixed global frame. A temporal progression along a track is not required: many of their tracks are undated, or have ages incompatible with predicted progressions, or have reliable dates demonstrating simultaneous activity throughout their lengths. Hypothetical track velocities are assigned from a plate model that assumes the postulated fixed frame. The descriptions and rationalizations of "hotspots" in the electronic supplement (for this GSA Data Repository item no. 2007090 description, see footnote 1 in Morgan and Phipps Morgan, this volume) summarize the substance behind the inferences.

Most of the hotspots chosen are in the oceans, where severe selectivity is applied. "Tracks" are identified by the fit of their azimuths to the model. Some azimuths of possible tracks that do not fit are explained by diversion of plume materials along fracture zones, spreading ridges, or lithosphere changes. Jan Mayen "is not a hotspot but rather [is] due to channeled asthenosphere flow from Iceland" (electronic supplement, p. 6). The Azores define no track but must record a plume, so a track close to the wanted direction is rationalized by assuming that some of the islands are "formed by channeled flow from the hotspot to the mid-Atlantic, and that Santa Maria is due to flow from the hotspot to a minor spreading center along the East Azores Fracture Zone" (electronic supplement, p. 5). The purported hotspot track defined by Réunion and Mauritius fits the model poorly, so its azimuth is adjusted because "Mauritius [magmatism] is 'pulled toward' the fracture zone at the eastern edge" (electronic supplement, p. 22). Pukapuka trends in the right direction but is too fast, so it "marks a narrow 'river' of asthenosphere" (electronic supplement, p. 77). These adjusted and hypothetical azimuths are plotted as "observed" on the figures.

The North American plate model is constrained only by selective use of sparse data that fit. Absolute motion of the plate is defined by Morgan and Phipps Morgan (this volume) with the supposed tracks of four hotspots, Iceland, Bermuda, Yellowstone, and Raton. Iceland tomography has failed to identify a deep-seated plume, but in their electronic supplement (p. 5), Morgan and Phipps Morgan assume the Mid-Atlantic Ridge to be moving over a plume such that the model-required three-quarters of the hypothetical motion can be assigned to the American side, and this division is termed "observed." The purported Bermuda track is defined by several vaguely dated and widely separated on-land uplifts and minor igneous occurrences, selected because they fit the desired rough trend to end at Oligocene Bermuda (electronic supplement, p. 36). Detailed seismic tomography, by both proponents and opponents of plume hypotheses, shows the thermal effect of the purported Yellowstone hotspot to be limited to the upper mantle, but the temporally erratic east-northeastward late Neogene magmatic progression is used as a track by Morgan and Phipps Morgan (electronic supplement, p. 36), because it fits, whereas the temporally more uniform Brothers progression, in a different direction from the same origin over the same time span, is unmentioned. Raton, which consists of a line drawn to connect widely separated late Neogene volcanic fields, was proposed as a track only because the line is in the desired direction. Morgan and Phipps Morgan (electronic supplement, p. 39) term it "the best track on North America" with which to define azimuth —but the many dates they report are mostly late Miocene through Quaternary in all fields along the 600-km length. The only common denominator of these four very different purported manifestations of plumes is that each was selected, from arrays of otherwise possible candidates, because its azimuth fit fixed-plume predictions. The fact that these hoped-for tracks do fit thus is irrelevant as evidence.

Onland Africa is no better. Hoggar, Tibesti, and Jebel Marra are designated, and plotted, as tracks, even though there is no relevant age progression along any of them, because they have appropriate orientations (electronic supplement, p. 10–11). A track is assumed for Afar. The nonfit of Cameroon magmatism is given plumespeak rationalization. The large East African volcanoes are assumed to cap trackless plumes.

The whole-mantle-convection and superplume model favored by Morgan and Phipps Morgan (this volume), with its long-fixed subduction systems bounding a constant-width Pacific Ocean on both sides, is disproved by, among other features, broad global spreading patterns that are independent of inferred frameworks. The Pacific must be shrinking, by rollback of its bounding hinges, at some large fraction of the areal rate at which the subduction-free Atlantic is expanding.

31 January 2007, Jason Phipps Morgan and W. Jason Morgan

In response to Smith's comment of 27 January, we point out that differing Re-Os isotope systematics of OIB and MORB were used by us as observational support for progressive melt-extraction from a plume-fed asthenosphere (Phipps Morgan, 1999; Phipps Morgan and Morgan, 1999). A key aspect of the Os-isotopic systems is that they imply, because Os is compatible and abundant in olivine (or sulphide micro-inclusions in olivine—a topic of much recent exploration), that the melts of low-abundance mantle components that are rich in radiogenic Os-isotopes do not re-equilibrate with the bulk Os-isotope ratios of the mantle they ascend through. This failure to re-equilibrate implies that both OIB and MORB are composed of a pooling of melts produced by selective melting of a heterogeneous source, and that these melts do not equilibrate, on average, with the average trace-element composition of the mantle during their ascent to the surface. If a small mass fraction of a core-derived Pt-radiogenic Os component does indeed exist in OIB, but not MORB, then we would argue that this is a low-solidus component that is being stripped by plume-melt extraction. (Note that any residue of partial melting of this trace component may become so refractory that it does not melt again during later ascent beneath a mid-ocean ridge [Phipps Morgan, 2001], and also that any olivine crystallization during ascent of an OIB/picrite would reduce the Os concentration in the ascending magma, so that Smith's mass balance conclusion is suspect.) However, we think that the detection of differences between Pt-radiogenic Os between OIB and MORB is still being assessed.

In response to Hamilton's 31 January 2007 comment about poorly defined hotspot tracks on several continents, Africa in particular, we agree these tracks are poor. We think the reason hotspot "tracks" are so poorly expressed on continents is that first, pressure-release plume melting is greatly reduced beneath thick continental lithosphere in comparison to plume melting beneath typically thinner oceanic lithosphere. Second, drainage upward and laterally of plume-fed material beneath the base of a continent may induce secondary decompression melting that is spatially removed from the plume-stem source, as discussed by Ebinger and Sleep (1998).

Our model for recent absolute plate motions asks a simple question—is the observed pattern of "geologically recent" volcanic lineaments on the oceanic portions of plates consistent with known NUVEL1A (DeMets et al., 1994) relative plate motions and an assumed "fixed" source of deep plume upwelling beneath each of these oceanic lineaments? We find that the hypothesis works surprisingly well, and therefore we essentially used the well-defined volcanic tracks on the oceanic portions of the plates, plus known relative plate motions, to constrain plate motions above continental hotspots, such as Hoggar. This method means that the inferred motion of North America is not being dominated by the few potential hotspot traces within North America, but rather by the global fit to the many better-defined azimuths on the oceanic portions of all plates. Although we describe each potential continental hotspot "track" in the supplement, they have extremely little weight on the inferred pattern for recent absolute plate motions. Similarly, even though we describe in the supplement a few places where we adjust azimuths by a bit for the individual geologic reasons that we describe for each of these adjustments, these corrections have very little effect on the absolute plate motions we infer, as discussed at some length in the text. (They do not change the predicted motions, just add more "noise" to the comparison of absolute motions with our estimates for the tracks.)

We hope it is clear that we do not think that intraplate volcanism only occurs above the stems of plumes upwelling from much deeper mantle. Lateral flow and drainage is an essential aspect of a plume-fed asthenosphere, and there is the potential for much melting of plume-fed asthenosphere far from the stems of the plumes in which this material first upwelled toward the surface, and even from the spreading of the buoyant hotspot-melting–created roots to hotspot swells (Phipps Morgan et al., 1995). This observation does not mean that the concept of plumes should be discarded, simply that it needs to be extended to the more complete conceptual model of slabs, plumes, and a plume-fed asthenosphere.

REFERENCES CITED

Brandon, A.D., and Walker, R.J., 2005, The debate over core-mantle interaction: Earth and Planetary Science Letters, v. 232, p. 211–225.

DeMets, C., Gordon, R.G., Argus, D.F., and Stein, S., 1994, Effect of recent revisions to the geomagnetic reversal time scale on estimate of current plate motions: Geophysical Research Letters, v. 21, p. 2191–2194.

Ebinger, C.J., and Sleep, N.H., 1998, Cenozoic magmatism throughout east Africa resulting from impact of a single plume: Nature, v. 395, p. 788–791.

Morgan, W.J., and Phipps Morgan, J., 2007 (this volume), Plate velocities in the hotspot reference frame, in Foulger, G.R., and Jurdy, D.M., eds., Plates, plumes, and planetary processes: Boulder, Colorado, Geological Society of America Special Paper 430, doi: 10.1130/2007.2430(04).

Phipps Morgan, J., 1999, The isotope topology of individual hotspot arrays: Mixing curves or melt extraction trajectories?: Geochemistry, Geophysics, Geosystems, v. 1, art. no. 1999GC000004.

Phipps Morgan, J., 2001, The thermodynamics of the pressure-release melting of a veined plum-pudding mantle: Geochemistry, Geophysics, Geosystems, v. 2, art. no. 2000GC000049.

Phipps Morgan, J., and Morgan, W.J., 1999, Two-stage melting and the geochemical evolution of the mantle: A recipe for mantle plum-pudding: Earth and Planetary Science Letters, v. 170, p. 215–239.

Phipps Morgan, J., Morgan, W.J., and Price, E., 1995, Hotspot melting generates both hotspot volcanism and a hotspot swell?: Journal of Geophysical Research, v. 100, p. 8045–8062.

Yamamoto, M., Phipps Morgan, J., and Morgan, W.J., 2007 (this volume, chapter 9), Global plume-fed asthenosphere flow—I: Motivation and model development, *in* Foulger, G.R., and Jurdy, D.M., eds., Plates, plumes, and planetary processes: Boulder, Colorado, Geological Society of America Special Paper 430, doi: 10.1130/2007.2430(09).

Yamamoto, M., Phipps Morgan, J., and Morgan, W.J., 2007 (this volume, chapter 10), Global plume-fed asthenosphere flow—II: Application to the geochemical segmentation of mid-ocean ridges, *in* Foulger, G.R., and Jurdy, D.M., eds., Plates, plumes, and planetary processes: Boulder, Colorado, Geological Society of America Special Paper 430, doi: 10.1130/2007.2430(10).

The Geological Society of America
Special Paper 430
2007

Implications of lower-mantle structural heterogeneity for the existence and nature of whole-mantle plumes

Edward J. Garnero*

*School of Earth and Space Exploration, Bateman Physical Sciences Building,
F-Wing, Arizona State University, Tempe, Arizona 85287-1404, USA*

Thorne Lay

*Earth and Planetary Sciences Department, 1156 High Street,
University of California, Santa Cruz, California 95064, USA*

Allen McNamara

*School of Earth and Space Exploration, Bateman Physical Sciences Building,
F-Wing, Arizona State University, Tempe, Arizona 85287-1404, USA*

ABSTRACT

Recent seismological studies demonstrate the presence of strong deep-mantle elastic heterogeneity and anisotropy, consistent with a dynamic environment having chemical anomalies, phase changes, and partially molten material. The implications for deep-mantle plume genesis are discussed in the light of the seismological findings. Nearly antipodal large low–shear velocity provinces (LLSVPs) in the lowermost mantle beneath the Pacific Ocean and Africa are circumscribed by high-velocity regions that tend to underlie upper-mantle downwellings. The LLSVPs have sharp boundaries, low V_S/V_P ratios, and high densities; thus, they appear to be chemically distinct structures. Elevated temperature in LLSVPs may result in partial melting, possibly accounting for the presence of ultra-low-velocity zones detected at the base of some regions of LLSVPs. Patterns in deep-mantle fast shear wave polarization directions within the LLSVP beneath the Pacific are consistent with strong lateral gradients in the flow direction. The thermal boundary layer at the base of the mantle is a likely location for thermal instabilities that form plumes, but geodynamical studies show that the distribution of upwellings is affected when piles of dense chemical heterogeneities are present. The location of lowermost mantle plume upwellings is predicted to be near the boundaries of the large thermochemical complexes comprising LLSVPs. These observations suggest that any large mantle plumes rising from the deep mantle that reach the surface are likely to be preferentially generated in regions of distinct mantle chemistry, with nonuniform spatial distribution. This hypothesis plausibly accounts for some attributes of major hotspot volcanism.

Keywords: core-mantle boundary, D″, plumes, thermochemical piles, hotspots

*E-mail: garnero@asu.edu.

Garnero, E.J., Lay, T., and McNamara, A., 2007, Implications of lower-mantle structural heterogeneity for the existence and nature of whole-mantle plumes, *in* Foulger, G.R., and Jurdy, D.M., eds., Plates, plumes, and planetary processes: Geological Society of America Special Paper 430, p. 79–101, doi: 10.1130/2007.2430(05).

INTRODUCTION

The depth of origin of the source of long-lived hotspot volcanism has been of great interest to geological scientists for decades (e.g., Morgan, 1971). This question intersects nearly all Earth science disciplines, and hence continues to attract active debate. The most common interpretation is that thermal plumes rise from an internal mantle thermal boundary layer and sustain hotspot activity. As long as heat is flowing into the base of the mantle from the core—an apparent requirement for long-term maintenance of the geodynamo (e.g., Buffett, 2002)—a thermal boundary layer should be present at the base of the lower mantle. This boundary layer is commonly invoked as the source of deep mantle plumes (see Lay, 2005), consistent with the early notions advanced by Morgan. Certainly, cylindrical plumes commonly initiate from the basal boundary layers in numerical and experimental convection experiments with basal heating or basal injection of fluid (e.g., Davies, 1990; Olson and Kincaid, 1991; Farnetani and Richards, 1994; van Keken, 1997; Farnetani and Samuel, 2005; Lin and van Keken, 2005), carrying heat and any unique isotopic signatures from the boundary layer to the surface. However, demonstrating that such plumes rise ~2900 km from the core-mantle boundary (CMB), traversing the Earth's entire mantle, has proven challenging. Many discussions of this problem invoke very simple notions of the lower-mantle boundary layer, at odds with recent seismic findings. Our goal is to place the question of deep-mantle plume genesis in the context of current seismological and geodynamical ideas about lower-mantle structure and processes. We will avoid the issue of connecting specific hotspot observations to plumes or alternate explanations, as that is addressed in detail elsewhere in this volume (e.g., Sleep, this volume). Our focus is on the implications of seismically defined lower-mantle structures for the occurrence and characteristics of any plumes that do rise from the lowermost mantle.

A number of fields (e.g., seismology, geodynamics, and geochemistry) have presented arguments and some evidence for deep mantle plumes (e.g., Ji and Nataf, 1998; Lin and van Keken, 2005; Montelli et al., 2006; Wen, 2006), but the issue is still under debate, and an increasing number of studies find that some hotspots do not require origins in the lower mantle (e.g., Cserepes and Yuen, 2000; Foulger and Pearson, 2001; Foulger et al., 2001; Courtillot et al., 2003). In this article, we focus primarily on the elastic structure of the deep mantle derived by seismic methods and the dynamics of plume initiation, stability, fixity, and longevity in the presence of the large-scale chemical heterogeneity suggested by the seismic results, including the fact that hotspots are typically only found away from regions of subduction. Over the past several years, a variety of deep-mantle structural characteristics have emerged from high-resolution imaging with broadband seismic data. These structures include chemically distinct provinces beneath the Pacific Ocean and Africa, thin ultra-low-velocity zones (ULVZs) at the CMB, deep-mantle seismic wave anisotropy, and variable occurrence and topography of the D″ seismic velocity discontinuity. This article considers these deep-mantle findings, exploring their implications for the possibility of lowermost mantle plume origination. Key seismological observations are summarized in the next section, followed by a section that considers the chemistry and dynamics of these features. This description provides a framework for considering the implications for any deep mantle plumes that may rise to the surface.

LOWERMOST MANTLE SEISMIC VELOCITY STRUCTURE

It has been known for decades that relatively high seismic velocities in the deep mantle tend to underlie past or present subduction zones, whereas lower-than-average wavespeeds are commonly found beneath the Pacific Ocean, the southern Atlantic Ocean, and Africa (e.g., Dziewonski, 1984; Hager et al., 1985). The distribution of lower-mantle velocities is quite consistent among recent tomographic S-wave velocity (V_S) models, but there is less consistency among P-wave velocity (V_P) models (Fig. 1). The differences between large-scale lower-mantle V_S and V_P heterogeneity have led to the inference that the origin of the velocity perturbations is not solely thermal (e.g., Masters et al., 2000). Unfortunately, until the V_P maps are better resolved (as indicated either by agreement between results from different studies or by demonstration that a particular study has produced the most robust results), it is difficult to confidently separate chemical and thermal effects based on the patterns of heterogeneity at the present time.

It has recently been demonstrated that the expected primary lower-mantle mineral—$(Mg,Fe)SiO_3$, magnesium silicate in perovskite structure (Pv)—should undergo a phase transformation at pressure-temperature conditions within a few hundred kilometers above the CMB (Murakami et al., 2004; Oganov and Ono, 2004; Tsuchiya et al., 2004; Lay et al., 2005). Pv transforms into a post-perovskite structure (pPv) that is predicted to be accompanied by a V_S increase of several percent, but little change in V_P. This difference in response could be one cause of decoupling of variations in V_S and V_P in the lowermost mantle. The V_S/V_P ratio should be highest in high–shear velocity regions because the phase transition will occur at shallower depths in cool regions that have higher seismic velocities to begin with. One challenge in seeking this behavior is uncertainty in the reference level for measuring velocity anomalies; for example, the increase in temperature in the thermal boundary layer above the CMB will tend to lower seismic velocities, with more pronounced effects on S-wave velocity than for P-wave velocities. As seen in Figure 1, most seismic models tend to have means of zero at a given depth, which affects inferences about relative velocity behavior significantly.

The circum-Pacific band of high shear velocities apparent in Figure 1 is plausibly linked to occurrence of pPv in regions with relatively low temperatures below present-day and historic subduction zones. If this link is the case, the large low–shear ve-

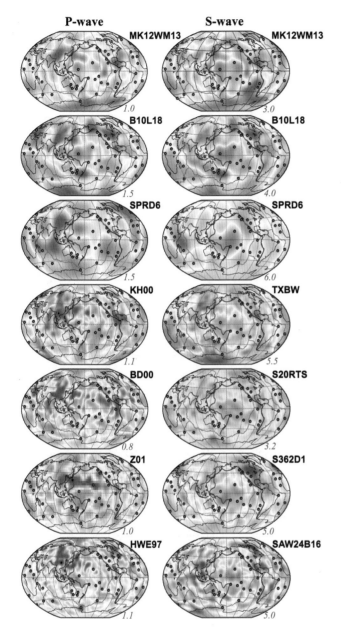

Figure 1. Tomographically derived P-wave (left column) and S-wave (right column) velocity perturbations at the base of the mantle. Red and blue colors indicate lower and higher velocities than global averages, respectively. Color scales are not uniform for the different models; the peak-to-peak value is indicated in the lower right of each map (blue number). Model names are given in the upper right and correspond to the following studies: MK12WM13 (Su and Dziewonski, 1997), B10L18 (Masters et al., 2000), SPRD6 (Ishii and Tromp, 2004), KH00 (Kárason and van der Hilst, 2001), TXBW (Grand, 2002), BD00 (Becker and Boschi, 2002), S20RTS (Ritsema and van Heijst, 2000), Z01 (Zhao, 2001), S362D1 (Gu et al., 2001), HWE97 (van der Hilst et al., 1997), and SAW24B16 (Mégnin and Romanowicz, 2000). More information comparing many of these models is found in Becker and Boschi (2002). Hotspot locations (from Steinberger, 2000) are shown as red-filled circles. Plate boundaries are magenta lines, but convergent boundaries are shown in blue.

locity provinces (LLSVPs) beneath Africa and the Pacific might have no pPv or only a very thin layer of it, and may be relatively warm. The Pv-pPv phase boundary has a large positive Clapeyron slope that would allow large lateral variations of thickness of a layer of pPv within the boundary layer to be caused by large-scale thermal variations. The LLSVPs are basically isolated from locations where subduction has occurred over the past 200 m.y., which is commonly invoked as an indication of control on the deep seismic heterogeneity by large-scale mid-mantle convection coupled to the shallow subduction history. The relatively low shear velocities in LLSVPs can thus be attributed to a combination of relatively high temperature and lack of pPv, but there are indications that there is also a chemical anomaly present in the LLSVPs.

Several free-oscillation studies have found evidence for lateral variation in large-scale lowermost-mantle density distribution (Ishii and Tromp, 1999, 2004; Kuo and Romanowicz, 2002; Trampert et al., 2004). Although debate continues on this topic (Romanowicz, 2001; Masters and Gubbins, 2003), indications are that a density increase is associated with the strongest V_S reductions located in LLSVPs (e.g., Ishii and Tromp, 1999; Trampert et al., 2004). Simultaneous analysis of V_S and V_P behavior further suggests that LLSVPs have bulk sound velocity anomalies (increases) that are anticorrelated with the low–shear velocity anomalies (e.g., Masters et al., 2000). These observations suggest that LLSVPs are chemically distinct from the surrounding mantle. This possibility immediately complicates the interpretation of these regions, because chemical differences can also affect the occurrence of the pPv phase change, and some compositional effects, such as iron enrichment, tend to reduce shear velocity as much or more than high temperature does at high pressures. The presence of iron or aluminum can also affect the phase transition pressure and sharpness (see Lay et al., 2005), although the magnitude of such effects is being debated. Sorting out these tradeoffs requires more detailed structural information than provided by tomography alone.

Portions of LLSVPs have been characterized at relatively short scale lengths (e.g., study regions spanning 500–1000 km laterally) using forward modeling of body wave travel times and waveforms. For example, a significant number of LLSVP margins (Fig. 2) show strong evidence for an abrupt lateral transition over a few hundred kilometers or less between the LLSVP and surrounding mantle. The sharpness of the LLSVP margins supports the notion that there is a chemical contribution because thermal gradients should be more gradual. Additionally, weak reflections from a velocity decrease in the upper portion of the LLSVP in the Pacific may indicate a chemical boundary (Lay et al., 2006) or a phase boundary within chemically distinct LLSVP material (K. Ohta, personal communication). Additional internal layering within the LLSVP beneath the Pacific has been inferred from seismic wavefield reflectivity resolved by stacking a large number of seismic data (Lay et al., 2006). In Lay et al. (2006), the northern portion of the LLSVP beneath the Pacific Ocean is found to have a sharp velocity increase overlying a

Garnero et al.

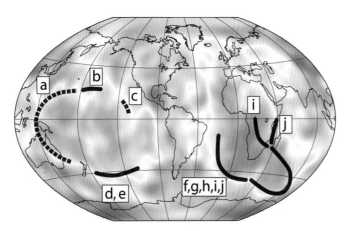

Figure 2. Map showing lowermost-mantle V_S perturbations from model TXBW (Grand, 2002; color scale is the same as in Fig. 1), along with locations where seismic studies have inferred distinct edges to the large low–shear velocity provinces (LLSVPs; thick black lines) using waveform and/or travel time analyses. Lower-cased letters in boxes indicate specific studies for different regions: a (He et al., 2006), b (Luo et al., 2001), c (Bréger and Romanowicz, 1998), d (To et al., 2005), e (Ford et al., 2006), f (Ni and Helmberger, 2001), g (Wen et al., 2001), h (Ni et al., 2002), i (Ni and Helmberger, 2003a,b), j (Wang and Wen, 2004). Solid lines indicate regions with sharp lateral boundaries, and dashed lines indicate regions where the LLSVP edges are only loosely resolved by travel time analysis.

sharp decrease that was attributed to forward and reverse transformations of Pv to pPv (as predicted by Hernlund et al., 2005). This region is also underlain by a mild ULVZ (Avants et al., 2006). Lateral depth variations of the mapped perovskite phase boundaries within the LLSVP are consistent with a lateral increase in temperature toward the LLSVP margin in the central Pacific region.

In addition to evidence for sharp boundaries of the two LLSVPs, several forward modeling studies have advocated a chemical origin for isolated seismic heterogeneities at small to intermediate scales (500–1000 km; e.g., Wysession et al., 1994; Bréger and Romanowicz, 1998), as well as at small scales (one to tens of kilometers) that scatter seismic waves and contribute to high-frequency coda energy (e.g., Hedlin and Shearer, 2000; Earle and Shearer, 2001). Some of the small-scale scattering features at the CMB have been attributed to partial melt of deep-mantle material (e.g., Vidale and Hedlin, 1998; Wen and Helmberger, 1998, Rost et al., 2005), owing primarily to magnitude of the requisite velocity reductions and a 3:1 $V_S:V_P$ velocity reduction ratio. It is difficult to attribute large deep-mantle velocity reductions (e.g., >10%) to any expected deep-mantle materials in the absence of some level of melt; in fact, the possibility of partial melt in the deep mantle has been advocated as an explanation for the occurrence of thin ULVZs right above the CMB (Williams and Garnero, 1996; Revenaugh and Meyer, 1997). ULVZs appear to exist preferentially beneath low-V_S regions, including the two LLSVPs (e.g., Garnero et al., 1998; Williams et al., 1998). This preference is hard to quantify be-

cause the global distribution of ULVZ structure is not presently well constrained; less than half of the surface area of the CMB has been even qualitatively characterized (Thorne and Garnero, 2004). However, the CMB in some isolated spots has been analyzed in great detail, suggesting partial melt in small domes, small dense zones with a flat top, and even small pockets right beneath the CMB with anomalous properties (Helmberger et al., 1998, 2000; Wen and Helmberger, 1998; Wen, 2000; Rost and Revenaugh, 2001; Rost et al., 2005). These structures all point to complex processes occurring down to small scales, undoubtedly related to high temperatures at the base of the mantle thermal boundary layer.

Lowermost-mantle seismic wave anisotropy may also offer clues to deep-mantle chemistry and dynamics; as suggested by seismic studies (e.g., see Kendall and Silver, 1998; Lay et al., 1998; Kendall, 2000), mineral physics calculations (e.g., Stixrude, 1998; Karki et al., 1999; Mainprice et al., 2000; Wentzcovitch et al., 2004; Hirose, 2006; Hirose et al., 2006), as well as by geodynamics experiments (e.g., McNamara et al., 2001, 2002, 2003). Several seismological studies have mapped geographical changes in the fast propagation direction of deep-mantle shear waves (Russell et al., 1998; Garnero et al., 2004; Wookey et al., 2005; Rokosky et al., 2006). In one case, geometrical patterns in fast propagation directions have been interpreted as being related to lowermost-mantle boundary layer convective currents that may involve flow into a boundary layer upwelling, possibly related to a plume that rises to the Hawaiian hotspot (Russell et al., 1998). Given that there is a first-order correlation between the distribution of surface hotspots and the locations of the LLSVPs (e.g., Thorne et al., 2004; any correlation is less apparent for P-wave velocity heterogeneity) at the base of the mantle (Fig. 1), it is reasonable to seek any evidence for plumes extending through the mantle above these regions.

Direct seismic imaging of any deep mantle plumes is very difficult, primarily owing to the expected small dimension of the plume conduit (e.g., <500 km) compared to the long seismic wave propagation paths (typically >5000 km; see, e.g., Nataf, 2000; Dahlen, 2004). Most tomographic efforts have not directly imaged vertically continuous deep mantle plumes or their relationship to LLSVPs, as the minimum lateral wavelength of resolvability is commonly >1000 km. One notable exception is the study by Montelli et al. (2004), in which several surface hotspots are inferred to be underlain by low V_P values extending down to the CMB (this observation is currently under active debate; see Dahlen and Nolet, 2005; de Hoop and van der Hilst, 2005). The seismological community may eventually converge on models that either support or refute the existence of whole-mantle low-velocity plume conduits. However, deep-mantle seismic plume detection may be almost impossible if plume temperature does not significantly exceed the surrounding mantle (e.g., Farnetani, 1997; Farnetani and Samuel, 2005), giving too small an elastic velocity signature. If the velocity perturbations are, in fact, strong enough, there is some hope to image plume features if wavepath coverage is dense enough (e.g.,

Tilmann et al., 1998), but at present, this imaging does not appear to have been done convincingly.

Less direct approaches, such as correlation studies of surface hotspot distributions and deep-mantle velocity patterns have been pursued over several decades (e.g., Morgan, 1971; Hager et al., 1985; Thorne et al., 2004), but such analyses do not unequivocally constrain or require the existence of whole-mantle plumes.

LOWER-MANTLE CHEMISTRY AND DYNAMICS

Several conceptual models have been developed in recent years to explain the observed LLSVPs. It proves dynamically difficult to account for the huge low-velocity anomalies beneath Africa and the Pacific in an isochemical mantle, even in models that impose a large-scale pattern of downwelling by employing geologically recent plate velocities as surface boundary conditions (e.g., Bunge et al., 1998; McNamara and Zhong, 2005). In dynamical models that lack a thermochemical component, plumes tend to organize into clustered networks of thin upwellings (plume clusters) that form away from downwelling regions (e.g., Schubert et al., 2004; McNamara and Zhong, 2005). Although current research is assessing whether regions of plume clusters may resemble the large, low-velocity anomalies in the lowermost mantle when viewed through the blurred "eyes" of seismic tomography (Ritsema et al., 2007), it appears that thermochemical models of mantle convection provide the best explanation for the existence of the LLSVPs.

Thermochemical conceptual models that strive to explain the dynamics related to LLSVPs beneath Africa and the Pacific typically fall into two categories. Both invoke a large volume of anomalously dense mantle material; however, they differ in terms of the relative buoyancy and geologic longevity of the chemical anomaly.

The superplume hypothesis typically describes the large, low-velocity anomalies as being due to the presence of large, upward-doming plumes of the more-dense material (e.g., Davaille, 1999; Forte and Mitrovica, 2001; Davaille et al., 2002, 2005). These models are characterized by the denser component having a net positive buoyancy (the thermal buoyancy exceeds the negative buoyancy associated with the intrinsic density anomaly) such that it becomes unstable and forms large superplumes that are currently rising in the mantle. It has been shown (e.g., Davaille et al., 2002) that these structures may rise and sink many times before ultimately being well mixed into the background mantle. Smaller-scale thermal plumes can originate from the tops of these large domes, entraining some of the chemically distinct material in the dense dome. This entrainment may, in turn, explain the anomalous chemistry observed at ocean island basalts (OIBs; e.g., Hofmann, 1997, Fitton, this volume). One aspect of this model is that the present instance in Earth's history has the more-dense material actively rising, as opposed to other times in the past in which this material was either a stratified layer or sinking after a previous ascent.

Another, similar superplume phenomenon observed in geodynamical modeling involves the presence of long-lived, stable superplumes (McNamara and Zhong, 2004a). If the anomalously dense mantle component has a higher intrinsic viscosity (~100×) than the less-dense material, it may form large dome structures that migrate laterally across the lower mantle. These structures can maintain a vertical height, and they do not experience the rising and sinking observed in laboratory studies (e.g., Davaille et al., 2002). However, it is difficult to put forth a mineral physics explanation that would provide the necessary intrinsic viscosity increase to the dense material, so such a model is not favored here.

Although the basic superplume model remains a viable hypothesis, here we focus on the second category of thermochemical mantle hypotheses, which involve the presence of long-lived, stable, dense piles of chemically distinct material (e.g., Christensen and Hofmann, 1994; Tackley, 1998, 2002; Kellogg et al., 1999; Jellinek and Manga, 2002, 2004; Ni et al., 2002; McNamara and Zhong, 2004a,b, 2005; Nakagawa and Tackley, 2005; Tan and Gurnis, 2005). In these models, piles of dense material maintain a near-neutral (slightly negative) buoyancy, and as a result, they are passively swept aside by downwelling flow and are focused beneath upwelling regions. Piles tend to form large, ridge-like structures that have thermal plumes originating from their peaks that entrain a small fraction of the more-dense material.

McNamara and Zhong (2005) performed numerical thermochemical calculations in a 3-D spherical geometry with Earth's recent plate history imposed as surface boundary conditions (120 m.y. over eleven stages of plate motions, as provided by Lithgow-Bertelloni and Richards, 1998). The calculation employed the Boussinesq approximation with constant thermodynamic properties; however, depth-dependent thermal conductivity was explored and found to have only a minimal effect on the resulting thermal and chemical structures. A depth- and temperature-dependent rheology that included a thirty-fold increase across the transition zone was used. The initial condition included a flat, 255-km-thick more-dense layer and a steady-state temperature field derived from an axisymmetric thermochemical calculation. The imposed plate history acted to guide the formation of downwellings in historical subduction regions, which resulted in the focusing of the lower-mantle dense material into piles beneath Africa and the Pacific. These are the same regions characterized by the observed LLSVPs (see Fig. 3A and D). This modeling demonstrated that it is dynamically feasible that global flow patterns derived from the history of subduction can focus a dense component into thermochemical structures that, to first order, resemble the present-day LLSVP configuration.

Our preferred interpretation of LLSVPs is that they are large, dense thermochemical piles stabilized by upwelling currents that are downwelling-induced return flow. The temperature within and around the pile depends on several uncertain factors, like thermal conductivity and the degree of viscous heating. If we assume that deep-mantle piles have some temporal stability, then it is reasonable to assume they are denser than surrounding mantle (e.g., at a minimum, 2–5% denser). This assumption is consistent with the suggestion from some seismic studies of increased density in these locations (Ishii and Tromp, 2004; Trampert et al., 2004). The possibility of Fe-enrichment in D″ would decrease shear velocity and result in density elevation, so the chemical heterogeneity could be residual material from the core-formation process or accumulation of core-mantle reaction products. Subducted mid-ocean ridge basalt (MORB) is also ex-

pected to be denser than surrounding material at lower-mantle conditions and thus may account for the dense LLSVPs, assuming that MORB has accumulated progressively in the lowermost mantle (Hirose et al., 1999; K. Ohta, personal communication).

The detailed structure (e.g., topography) of the sides and top of chemically distinct LLSVP material can play a significant role in the style and morphology of local upwelling currents and plume initiation (e.g., Jellinek and Manga, 2002, 2004; McNamara and Zhong, 2005). Numerical calculations show upward convective return flow guided by the LLSVP margins. Basal heating and internal flow of the LLSVP cause the boundaries between the surrounding mantle and LLSVP to be particularly hot (see Fig. 3B). If partial melt is indeed the origin of ULVZ structure, we would expect the edges of LLSVP structure to have the highest occurrence of ULVZ structure. As previously mentioned, the geographical distribution of ULVZ structure at present is not known in great enough detail to document such a spatial correlation. It is noteworthy, however, that two recent high-resolution studies detailing ULVZ structure are both near (and within) LLSVP margins: a double-array stacking study of ScS (a core-reflected S-wave) beneath the northern margin of the LLSVP beneath the Pacific Ocean (Avants et al., 2006; Lay et al., 2006), and a multiple vespagram analysis of ScP (an S-wave that converts to a P-wave upon reflection at the CMB) beneath the southwest margin of the same LLSVP (Rost et al., 2005, 2006). The strongest lateral gradients in tomographically derived V_S structures are near the margins of the LLSVPs (consistent with pile "edges"), and these regions of strong gradients are found to statistically correlate with surface hotspot locations (Thorne et al., 2004).

These findings are consistent with the conceptual model put forth in Figures 3C and 4. Large thermochemical piles are deflected away from downwellings by subduction currents and are swept to concentrate beneath upwelling return flow. LLSVP piles may thus be the key to the long-term history of subduction. LLSVP topographical features near their margins guide upwelling and serve as sites of thermal boundary layer instabili-

(A) Compositionally distinct, dense piles

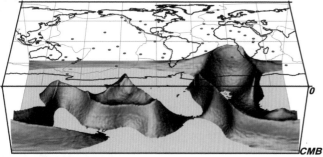

(B) Highest temperatures

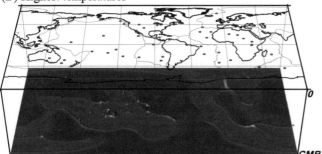

(C) Dense piles and temperatures

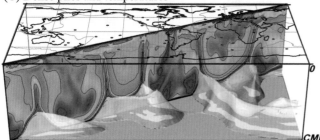

(D) Shear velocity heterogeneity

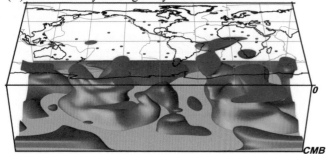

Figure 3. Continents, plate boundaries, and hotspots are shown on maps at the top of four boxes that represent the area of the whole globe, and the volume of the mantle from the surface to the core-mantle boundary (CMB). (A) Compositionally distinct, dense piles from the geodynamic calculation of McNamara and Zhong (2005). (B) The locations of the hottest temperatures in the mantle for the calculation of panel A are shown. Isotemperature contour is 0.98 for the calculation that spans temperatures from 0 to 1. The hottest temperatures are within the piles and are typically near the edges. (C) A temperature cross-section is shown, along with the piles from panel A in a transparent gray, and the hottest CMB temperatures of panel B are included (faint red stripes within the piles). Pile topography guides plume upwellings. (D) Shear velocity heterogeneity from Ritsema and van Heijst (2000) filtered to maximum spherical harmonic degree l = 8, with iso-velocity contours at −0.3% (red) and 0.5% (blue). Dense piles in the geodynamic calculation of panel A are geographically distributed similarly to the low velocities (red) in panel D.

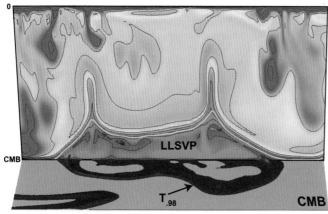

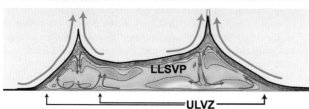

Figure 4. A close-up of the thermochemically dense pile beneath the Pacific Ocean in Figure 3A. (A) A cross-section from the surface to the core-mantle boundary (CMB) displays temperature variations, with the yellow line denoting the boundary of the chemically distinct material in the pile. We identify the thermochemical anomaly as the large low–shear velocity province (LLSVP). Part of the CMB surface is shown in front of the cross-section, along with an isotemperature contour (at 0.98, as in Fig. 3). (B) The same cross-section, but only the pile, with a more expanded color scale (colors span $T = 0.7$ to 1.0). Convective motions are indicated by the arrows. The hottest zones may invoke partial melt of LLSVP material, either at the CMB (denoted as an ultra-low velocity zone [ULVZ] in the figure), or in some isolated locations farther up within the LLSVP.

ties. Ascending plumes from the LLSVP margins may carry distinct chemical tracers from the deep mantle and CMB. OIB geochemistry for major hotspots favors recycled slab material as a significant source (e.g., Hofmann, 1997), which is consistent with past and/or ongoing subduction of slabs to the base of mantle and concentration of slab materials into the piles (e.g., Christensen and Yuen, 1984; Hager et al., 1985; Hutko et al., 2006). Thus, LLSVPs can be viewed as a by-product of whole-mantle convection, with physical segregation of dense material in the boundary layer. This process could occur today even if slabs temporarily go stagnant in the transition zone because of the difficulty of penetrating the 670-km phase boundary (Mitrovica and Forte, 1997) before they avalanche into the lower mantle. Of course, not all slab material has to penetrate into the deep mantle, and the LLSVPs may be comprised of slab material subducted long ago.

Seismological evidence for reflections down to 1000 km beneath southwest Pacific subduction zones is consistent with the penetration of MORB-bearing material into the lower mantle (S. Rost, personal communication). Sequestration of dense

MORB material (e.g., Hirose et al., 1999; Tan and Gurnis, 2005; K. Ohta, personal communication) may account for the chemically distinct nature of the LLSVP material. This concept certainly requires geochemical assessment, as it is the only approach to establishing the temporal isolation of the LLSVP reservoir.

CONCLUSIONS

Recent deep-mantle seismological findings give rise to the hypothesis that any deep mantle plumes will originate near the margins of LLSVPs at the base of the mantle. Chemically distinct and dense LLSVP piles may be organized underneath large-scale upwellings associated with return flow from subduction-induced downwellings. The margins of the LLSVP at the CMB are the hottest locations in the mantle and may contain partial melt at the CMB that is imaged as ULVZ structure. The LLSVP and ULVZ structures may contain important isotopic signatures that become entrained in plumes that rise from boundary layer instabilities on the LLSVP margins. The recent data and models do not demonstrate that whole-mantle plumes exist. However, the emerging understanding of lower-mantle structure and processes does provide guidance as to where and why any plume rising from the deep mantle will originate, and how they may sample thermally and chemically distinct source regions other than right at the CMB.

ACKNOWLEDGMENTS

This work was supported in part by the U.S. National Science Foundation under grants EAR-0125595, EAR-0453884, EAR-0453944, EAR-0510383, and EAR-0456356. Editors Gillian Foulger and Donna Jurdy and reviewers Anne Hofmeister and Dion Heinz made numerous helpful suggestions that improved the article. The authors thank Thorsten Becker for posting the tomography models from Becker and Boschi (2002) on his Web page and for software to expand the spherical harmonics, and Jeroen Ritsema for plotting software. Figures 1–3 were made with the aid of Generic Mapping Tools (Wessel and Smith, 1998).

REFERENCES CITED

Avants, M., Lay, T., and Garnero, E.J., 2006, A new probe of ULVZ S-wave velocity structure: Array stacking of ScS waveforms: Geophysical Research Letters, v. 33, p. L07314, doi: 10.1029/2005GL024989.

Becker, T.W., and Boschi, L., 2002, A comparison of tomographic and geodynamic mantle models: Geochemistry, Geophysics, and Geosystems, v. 3, doi: 2001GC000168.

Bréger, L., and Romanowicz, B., 1998, Thermal and chemical 3D heterogeneity in D″: Science, v. 282, p. 718–720, doi: 10.1126/science.282.5389.718.

Buffett, B.A., 2002, Estimates of heat flux in the deep mantle based on the power requirements for the geodynamo: Geophysical Research Letters, v. 29, doi: 10.1029/2001GL014649.

Bunge, H.-P., Richards, M.A., Lithgow-Bertelloni, C., Baumgardner, J.R., Grand, S.P., and Romanowicz, B.A., 1998, Time scales and heterogeneous structure in geodynamic Earth models: Science, v. 280, p. 91–95.

Christensen, U.R., and Hofmann, A.W., 1994, Segregation of subducted oceanic crust in the convecting mantle: Journal of Geophysical Research, v. 99, p. 19,867–19,884, doi: 10.1029/93JB03403.

Christensen, U.R., and Yuen, D.A., 1984, The interaction of a subducting lithospheric slab with a chemical or phase boundary: Journal of Geophysical Research, v. 89, p. 4389–4402.

Courtillot, V., Davaille, A., Baesse, J., and Stock, J., 2003, Three distinct types of hotspots in the Earth's mantle: Earth and Planetary Science Letters, v. 205, p. 295–308, doi: 10.1016/S0012-821X(02)01048-8.

Cserepes, L., and Yuen, D.A., 2000, On the possibility of a second kind of mantle plume: Earth and Planetary Science Letters, v. 183, p. 61–71, doi: 10.1016/S0012-821X(00)00265-X.

Dahlen, F.A., 2004, Resolution limit of traveltime tomography: Geophysical Journal International, v. 157, p. 315–331, doi: 10.1111/j.1365-246X.2004.02214.x.

Dahlen, F.A., and Nolet, G., 2005, Comment on "On sensitivity kernels for wave equation transmission tomography": Geophysical Journal International, v. 163, p. 949–951.

Davaille, A., 1999, Simultaneous generation of hotspots and superswells by convection in a heterogeneous planetary mantle: Nature, v. 402, p. 756–760, doi: 10.1038/45461.

Davaille, A., Girard, F., and Le Bars, M., 2002, How to anchor hotspots in a convecting mantle?: Earth and Planetary Science Letters, v. 203, p. 621–634, doi: 10.1016/S0012-821X(02)00897-X.

Davaille, A., Stutzmann, E., Silveira, G., Besse, J., and Courtillot, V., 2005, Convective patterns under the Indo-Atlantic box: Earth and Planetary Science Letters, v. 239, p. 233–252, doi: 10.1016/j.epsl.2005.07.024.

Davies, G.F., 1990, Mantle plumes, mantle stirring, and hotspot chemistry: Earth and Planetary Science Letters, v. 99, p. 94–109, doi: 10.1016/0012-821X(90)90073-7.

de Hoop, M.V., and van der Hilst, R.D., 2005, Reply to comment by F.A. Dahlen and G. Nolet: Geophysical Journal International, v. 163, p. 952–955.

Dziewonski, A.M., 1984, Mapping the lower mantle—Determination of lateral heterogeneity in P-velocity up to degree and order 6: Journal of Geophysical Research, v. 89, p. 5929–5942.

Earle, P.S., and Shearer, P.M., 2001, Distribution of fine-scale mantle heterogeneity from observations of Pdiff coda: Bulletin of the Seismological Society of America, v. 91, p. 1875–1881, doi: 10.1785/0120000285.

Farnetani, C.G., 1997, Excess temperature of mantle plumes: The role of chemical stratification across D″: Geophysical Research Letters, v. 24, p. 1583–1586, doi: 10.1029/97GL01548.

Farnetani, C.G., and Richards, M.A., 1994, Numerical investigations of the mantle plume initiation model for flood basalt events: Journal of Geophysical Research, v. 99, p. 13,813–13,833, doi: 10.1029/94JB00649.

Farnetani, C.G., and Samuel, H., 2005, Beyond the thermal plume paradigm: Geophysical Research Letters, v. 32, p. L07311, doi: 1029/2005GL022360.

Fitton, J.G., 2007 (this volume), The OIB paradox, *in* Foulger, G.R., and Jurdy, D.M., eds., Plates, plumes, and planetary processes: Boulder, Colorado, Geological Society of America Special Paper 430, doi: 10.1130/2007.2430(20).

Ford, S.R., Garnero, E.J., and McNamara, A.K., 2006, A strong lateral shear velocity gradient and anisotropy heterogeneity in the lowermost mantle beneath the southern Pacific: Journal of Geophysical Research, v. 111, p. B03306, doi: 10.1029/2004JB003574.

Forte, A.M., and Mitrovica, J.X., 2001, Deep-mantle high viscosity flow and thermochemical structure inferred from seismic and geodynamic data: Nature, v. 410, p. 1049–1056, doi: 10.1038/35074000.

Foulger, G.R., and Pearson, D.G., 2001, Is Iceland underlain by a plume in the lower mantle? Seismology and helium isotopes: Geophysical Journal International, v. 145, p. F1–F5, doi: 10.1046/j.0956-540x.2001.01457.x.

Foulger, G.R., Pritchard, M.J., Julian, B.R., Evans, J.R., Allen, R.M., Nolet, G., Morgan, W.J., Bergsson, B.H., Rlendsson, P., Jakobsdottir, S., Ragnarsson, S., Stefansson, R., and Vogfjörd, K., 2001, Seismic tomography shows that upwelling beneath Iceland is confined to the upper mantle: Geophysical Journal International, v. 146, p. 504–530, doi: 10.1046/j.0956-540x.2001.01470.x.

Garnero, E.J., Revenaugh, J., Williams, Q., Lay, T., and Kellogg, L.H., 1998, Ultralow velocity zone at the core-mantle boundary, *in* Gurnis, M., et al., eds., The core-mantle boundary region: Washington, D.C., American Geophysical Union, p. 319–334.

Garnero, E.J., Maupin, V., Lay, T., and Fouch, M.J., 2004, Variable azimuthal anisotropy in Earth's lowermost mantle: Science, v. 306, p. 259–261, doi: 10.1126/science.1103411.

Grand, S.P., 2002, Mantle shear-wave tomography and the fate of subducted slabs: Philosophical Transactions of the Royal Society of London A, v. 360, p. 2475–2491.

Gu, Y.J., Dziewonski, A.M., Su, W.J., and Ekström, G., 2001, Models of the mantle shear velocity and discontinuities in the pattern of lateral heterogeneities: Journal of Geophysical Research, v. 106, p. 11,169–11,199, doi: 10.1029/2001JB000340.

Hager, B.H., Clayton, R.W., Richards, M.A., Comer, R.P., and Dziewonski, A.M., 1985, Lower mantle heterogeneity, dynamic topography and the geoid: Nature, v. 313, p. 541–546, doi: 10.1038/313541a0.

He, Y., Wen, L., and Zheng, T., 2006, Geographic boundary and shear wave velocity structure of the "Pacific anomaly" near the core-mantle boundary beneath western Pacific: Earth and Planetary Science Letters, v. 244, p. 302–314, doi: 10.1016/j.epsl.2006.02.007.

Hedlin, M.A.H., and Shearer, P.M., 2000, An analysis of large-scale variations in small-scale mantle heterogeneity using Global Seismic Network recordings of precursors to PKP: Journal of Geophysical Research, v. 105, p. 13,655–13,673, doi: 10.1029/2000JB900019.

Helmberger, D.V., Wen, L., and Ding, X., 1998, Seismic evidence that the source of the Iceland hotspot lies at the core-mantle boundary: Nature, v. 396, p. 251–255, doi: 10.1038/24357.

Helmberger, D.V., Ni, S., Wen, L., and Ritsema, J., 2000, Seismic evidence for ultra low velocity zones beneath Africa and the Atlantic Ocean: Journal of Geophysical Research, v. 105, p. 23,865–23,878, doi: 10.1029/2000JB900143.

Hernlund, J.W., Thomas, C., and Tackley, P.J., 2005, A doubling of the post-perovskite phase boundary and the structure of the lowermost mantle: Nature, v. 434, p. 882–886, doi: 10.1038/nature03472.

Hirose, K., 2006, Postperovskite phase transition and its geophysical implications: Reviews of Geophysics, v. 44, p. RG3001, doi: 10.1029/2005RG000186.

Hirose, K., Fei, Y., Ma, Y., and Mao, H.-K., 1999, The fate of subducted basaltic crust in the Earth's lower mantle: Nature, v. 397, p. 53–56, doi: 10.1038/16225.

Hirose, K., Karato, S.-I., Cormier, V.F., Brodholt, J.P., and Yuen, D.A., 2006, Unsolved problems in the lowermost mantle: Geophysical Research Letters, v. 33, p. L12S01, doi:10.1029/2006GL025691.

Hofmann, A.W., 1997, Mantle geochemistry: The message from oceanic volcanism: Nature, v. 385, p. 219–229, doi: 10.1038/385219a0.

Hutko, A., Lay, T., Garnero, E.J., and Revenaugh, J.S., 2006, Seismic detection of folded, subducted lithosphere at the core-mantle boundary: Nature, v. 441, p. 333–336, doi: 10.1038/nature04757.

Ishii, M., and Tromp, J., 1999, Normal-mode and free-air gravity constraints on lateral variations in velocity and density of Earth's mantle: Science, v. 285, p. 1231–1236, doi: 10.1126/science.285.5431.1231.

Ishii, M., and Tromp, J., 2004, Constraining large-scale mantle heterogeneity using mantle and inner-core sensitive normal modes: Physics of the Earth and Planetary Interiors, v. 146, p. 113–124, doi: 10.1016/j.pepi.2003.06.012.

Jellinek, A.M., and Manga, M., 2002, The influence of a chemical boundary layer on the fixity, spacing, and lifetime of mantle plumes: Nature, v. 418, p. 760–763, doi: 10.1038/nature00979.

Jellinek, A.M., and Manga, M., 2004, Links between long-lived hot spots, mantle plumes, D″, and plate tectonics: Reviews of Geophysics, v. 42, doi: 10.1029/2003RG000144.

Ji, Y., and Nataf, H.-C., 1998, Detection of mantle plumes in the lower mantle by diffraction tomography: Hawaii: Earth and Planetary Science Letters, v. 159, p. 99–115, doi: 10.1016/S0012-821X(98)00060-0.

Kárason, H., and van der Hilst, R.D., 2001, Tomographic imaging of the lowermost mantle with differential times of refracted and diffracted core phases (PKP, Pdiff): Journal of Geophysical Research, v. 106, p. 6569–6588, doi: 10.1029/2000JB900380.

Karki, B.B., Wentzcovitch, R.M., de Gironcoli, S., and Baroni, S., 1999, First-principles determination of elastic anisotropy and wave velocities of MgO at lower mantle conditions: Science, v. 286, p. 1705–1709, doi: 10.1126/science.286.5445.1705.

Kellogg, L.H., Hager, B.H., and van der Hilst, R.D., 1999, Compositional stratification in the deep mantle: Science, v. 283, p. 1881–1884, doi: 10.1126/science.283.5409.1881.

Kendall, J.-M., 2000, Seismic anisotropy in the boundary layers of the mantle, *in* Karato, S.-I., et al., eds., Earth's deep interior: Mineral physics and tomography from the atomic to the global scale: Washington, D.C., American Geophysical Union, p. 133–159.

Kendall, J.-M., and Silver, P.G., 1998, Investigating causes of D″ anisotropy, *in* Gurnis, M., et al., eds., The core-mantle boundary region: Washington, D.C., American Geophysical Union, p. 97–118.

Kuo, C., and Romanowicz, B., 2002, On the resolution of density anomalies in the Earth's mantle using spectral fitting of normal mode data: Geophysical Journal International, v. 150, p. 162–179, doi: 10.1046/j.1365-246X.2002.01698.x.

Lay, T., 2005, The deep mantle thermo-chemical boundary layer: The putative mantle plume source, *in* Foulger, G.R., et al., eds., Plates, plumes and paradigms: Boulder, Colorado, Geological Society of America Special Paper 388, p. 193–205.

Lay, T., Garnero, E.J., Williams, Q., Kellogg, L., and Wysession, M.E., 1998, Seismic wave anisotropy in the D″ region and its implications, *in* Gurnis, M., et al., eds., The core-mantle boundary region: Washington, D.C., American Geophysical Union, p. 299–318.

Lay, T., Heinz, E., Ishii, M., Shim, S.H., Tsuchiya, T., Tsuchiya, J., Wentzcovich, R., and Yuen, D., 2005, Multidisciplinary impact of the lower mantle perovskite phase transition: Transactions of the American Geophysical Union, v. 86, p. 1–5.

Lay, T., Hernlund, J., Garnero, E.J., and Thorne, M.S., 2006, A post-perovskite lens and D″ heat flux beneath the central Pacific: Science, v. 314, p. 1272–1276, doi: 10.1126/science.1133280.

Lin, S., and van Keken, P.E., 2005, Multiple volcanic episodes of flood basalts caused by thermochemical mantle plumes: Nature, v. 436, p. 250–252, doi: 10.1038/nature03697.

Lithgow-Bertelloni, C., and Richards, M.A., 1998, The dynamics of Cenozoic and Mesozoic plate motions: Reviews of Geophysics, v. 36, p. 27–78, doi: 10.1029/97RG02282.

Luo, S.-N., Ni, S., and Helmberger, D.V., 2001, Evidence for a sharp lateral variation of velocity at the core-mantle boundary from multipathed PKPab: Earth and Planetary Science Letters, v. 189, p. 155–164, doi: 10.1016/S0012-821X(01)00364-8.

Mainprice, D., Barruol, G., and Ben Ismail, W., 2000, The seismic anisotropy of the Earth's mantle: From single crystal to polycrystal, *in* Karato, S.-I., et al., eds., Earth's deep interior: Mineral physics and tomography from the atomic to the global scale: Washington, D.C., American Geophysical Union, p. 237–264.

Masters, G., and Gubbins, D., 2003, On the resolution of density within the Earth: Physics of the Earth and Planetary Interiors, v. 140, p. 159–167, doi: 10.1016/j.pepi.2003.07.008.

Masters, G., Laske, G., Bolton, H., and Dziewonski, A.M., 2000, The relative behavior of shear velocity, bulk sound speed, and compressional velocity in the mantle: Implications for chemical and thermal structure, *in* Karato, S.-I., et al., eds., Earth's deep interior: Mineral physics and tomography from the atomic to the global scale: Washington, D.C., American Geophysical Union, p. 63–87.

McNamara, A.K., and Zhong, S., 2004a, Thermochemical structures within a spherical mantle: Superplumes or piles?: Journal of Geophysical Research, v. 109, p. B07402, doi: 10.1029/2003JB002847.

McNamara, A.K., and Zhong, S., 2004b, The influence of thermochemical convection on the fixity of mantle plumes: Earth and Planetary Science Letters, v. 222, p. 485–500, doi: 10.1016/j.epsl.2004.03.008.

McNamara, A.K., and Zhong, S., 2005, Thermochemical piles under Africa and the Pacific: Nature, v. 437, p. 1136–1139, doi: 10.1038/nature04066.

McNamara, A.K., Karato, S.-I., and van Keken, P.E., 2001, Localization of dislocation creep in the lower mantle: Implications for the origin of seismic anisotropy: Earth and Planetary Science Letters, v. 191, p. 85–99, doi: 10.1016/S0012-821X(01)00405-8.

McNamara, A.K., van Keken, P.E., and Karato, S.-I., 2002, Development of anisotropic structure in the Earth's lower mantle by solid-state convection: Nature, v. 416, p. 310–314, doi: 10.1038/416310a.

McNamara, A.K., van Keken, P.E., and Karato, S.-I., 2003, Development of finite strain in the convecting lower mantle and its implications for seismic anisotropy: Journal of Geophysical Research, v. 108, p. 2230, doi: 10.1029/2002JB001970.

Mégnin, C., and Romanowicz, B., 2000, The shear velocity structure of the mantle from the inversion of body, surface, and higher modes waveforms: Geophysical Journal International, v. 143, p. 709–728, doi: 10.1046/j.1365-246X.2000.00298.x.

Mitrovica, J.X., and Forte, A.M., 1997, Radial profile of mantle viscosity: Results from joint inversion of convection and post-glacial rebound observables: Journal of Geophysical Research, v. 102, p. 2751–2769, doi: 10.1029/96JB03175.

Montelli, R., Nolet, G., Dahlen, F.A., Masters, G., Engdahl, E., and Hung, S., 2004, Finite-frequency tomography reveals a variety of plumes in the mantle: Science, v. 303, p. 338–343, doi: 10.1126/science.1092485.

Montelli, R., Nolet, G., Dahlen, F.A., and Masters, G., 2006, A catalogue of deep mantle plumes: New results from finite-frequency tomography: Geochemistry, Geophysics, and Geosystems, v. 7, p. Q11007, doi: 10.1029/2006GC001248.

Morgan, W.J., 1971, Convection plumes in the lower mantle: Nature, v. 230, p. 42–43, doi: 10.1038/230042a0.

Murakami, M., Hirose, K., Kawamura, K., Sata, N., and Ohishi, Y., 2004, Post-perovskite phase transition in $MgSiO_3$: Science, v. 304, p. 855–858, doi: 10.1126/science.1095932.

Nakagawa, T., and Tackley, P.J., 2005, The interaction between the post-perovskite phase change and a thermo-chemical boundary layer near the core-mantle boundary: Earth and Planetary Science Letters, v. 238, p. 204–216, doi: 10.1016/j.epsl.2005.06.048.

Nataf, H.-C., 2000, Seismic imaging of mantle plumes: Annual Review of Earth and Planetary Sciences, v. 28, p. 391–417, doi: 10.1146/annurev.earth.28.1.391.

Ni, S., and Helmberger, D.V., 2001, Horizontal transition from fast (slab) to slow (plume) structures at the core-mantle boundary: Earth and Planetary Science Letters, v. 187, p. 301–310, doi: 10.1016/S0012-821X(01)00273-4.

Ni, S., and Helmberger, D.V., 2003a, Seismological constraints on the South African superplume; Could be the oldest distinct structure on Earth: Earth and Planetary Science Letters, v. 206, p. 119–131, doi: 10.1016/S0012-821X(02)01072-5.

Ni, S., and Helmberger, D.V., 2003b, Ridge-like lower mantle structure beneath South Africa: Journal of Geophysical Research, v. 108, p. 2094, doi: 10.1029/2001JB001545.

Ni, S., Tan, E., Gurnis, M., and Helmberger, D.V., 2002, Sharp sides to the African superplume: Science, v. 296, p. 1850–1852, doi: 10.1126/science.1070698.

Oganov, A.R., and Ono, S., 2004, Theoretical and experimental evidence for a post-perovskite phase of $MgSiO_3$ in Earth's D″ layer: Nature, v. 430, p. 445–448, doi: 10.1038/nature02701.

Olson, P., and Kincaid, C., 1991, Experiments on the interaction of thermal convection and compositional layering at the base of the mantle: Journal of Geophysical Research, v. 96, p. 4347–4354.

Revenaugh, J.S., and Meyer, R., 1997, Seismic evidence of partial melt within a possibly ubiquitous low-velocity layer at the base of the mantle: Science, v. 277, p. 670–673, doi: 10.1126/science.277.5326.670.

Ritsema, J., and van Heijst, H.J., 2000, Seismic imaging of structural heterogeneity in Earth's mantle: Evidence for large-scale mantle flow: Science Progress, v. 83, p. 243–259.

Ritsema, J., McNamara, A.K., and Bull, A.L., 2007, Tomographic filtering of geodynamic models: Implications for model interpretation and large-scale mantle structure: Journal of Geophysical Research, v. 112, p. B01303, doi: 10.1029/2006JB004566.

Rokosky, J., Lay, T., and Garnero, E.J., 2006, Small-scale lateral variations in azimuthally anisotropic D″ structure beneath the Cocos plate: Earth and Planetary Science Letters, v. 248, p. 411–425, doi: 10.1016/j.epsl.2006.06.005.

Romanowicz, B., 2001, Can we resolve 3D density heterogeneity in the lower mantle?: Geophysical Research Letters, v. 28, p. 1107–1110.

Rost, S., and Revenaugh, J.S., 2001, Seismic detection of rigid zones at the top of the core: Science, v. 294, p. 1911–1914, doi: 10.1126/science.1065617.

Rost, S., Garnero, E.J., Williams, Q., and Manga, M., 2005, Seismic constraints on a possible plume root at the core-mantle boundary: Nature, v. 435, p. 666–669, doi: 10.1038/nature03620.

Rost, S., Garnero, E.J., and Williams, Q., 2006, Fine scale ultra-low velocity zone structure from high-frequency seismic array data: Journal of Geophysical Research, v. 111, p. B09310, doi: 10.1029/2005JB004088.

Russell, S.A., Lay, T., and Garnero, E.J., 1998, Seismic evidence for small-scale dynamics in the lowermost mantle at the root of the Hawaiian hotspot: Nature, v. 396, p. 255–258, doi: 10.1038/24364.

Schubert, G., Masters, G., Olson, P., and Tackley, P.J., 2004, Superplumes or plume clusters?: Physics of the Earth and Planetary Interiors, v. 146, p. 147–162, doi: 10.1016/j.pepi.2003.09.025.

Sleep, N.H., 2007 (this volume), Origins of the plume hypothesis and some of its implications, *in* Foulger, G.R., and Jurdy, D.M., eds., Plates, plumes, and planetary processes: Boulder, Colorado, Geological Society of America Special Paper 430, doi: 10.1130/2007.2430(02).

Steinberger, B., 2000, Plumes in a convecting mantle: Models and observations for individual hotspots: Journal of Geophysical Research, v. 105, p. 11,127–11,152, doi: 10.1029/1999JB900398.

Stixrude, L., 1998, Elastic constants and anisotropy of $MgSiO_3$ perovskite, periclase, and SiO_2 at high pressure, *in* Gurnis, M., et al., eds., The core-mantle boundary region: Washington, D.C., American Geophysical Union, p. 83–96.

Su, W.J., and Dziewonski, A.M., 1997, Simultaneous inversion for 3-D variations in shear and bulk velocity in the mantle: Physics of the Earth and Planetary Interiors, v. 100, p. 135–156, doi: 10.1016/S0031-9201(96)03236-0.

Tackley, P.J., 1998, Three-dimensional simulations of mantle convection with a thermochemical CMB boundary layer: D″?, *in* Gurnis, M., et al., eds., The core-mantle boundary region: Washington, D.C., American Geophysical Union, p. 231–253.

Tackley, P.J., 2002, Strong heterogeneity caused by deep mantle layering: Geochemistry, Geophysics, and Geosystems, v. 3, p. 1024, doi: 10.1029/2001GC000167.

Tan, E., and Gurnis, M., 2005, Metastable superplumes and mantle compressibility: Geophysical Research Letters, v. 32, p. L20307, doi: 10.1029/2005GL024190.

Thorne, M., and Garnero, E.J., 2004, Inferences on ultralow-velocity zone structure from a global analysis of SPdKS waves: Journal of Geophysical Research, v. 109, p. B08301, doi: 10.1029/2004JB003010.

Thorne, M., Garnero, E.J., and Grand, S., 2004, Geographic correlation between hot spots and deep mantle lateral shear-wave velocity gradients: Physics of the Earth and Planetary Interiors, v. 146, p. 47–63, doi: 10.1016/j.pepi.2003.09.026.

Tilmann, F.J., McKenzie, D., and Priestley, K.F., 1998, P and S wave scattering from mantle plumes: Journal of Geophysical Research, v. 103, p. 21,145–21,163, doi: 10.1029/98JB01070.

To, A., Romanowicz, B., Capdeville, Y., and Takeuchi, N., 2005, 3D effects of sharp boundaries at the borders of the African and Pacific superplumes: Observation and modeling: Earth and Planetary Science Letters, v. 233, p. 137–153, doi: 10.1016/j.epsl.2005.01.037.

Trampert, J., Deschamps, F., Resovsky, J., and Yuen, D.A., 2004, Probabilistic tomography maps chemical heterogeneities throughout the mantle: Science, v. 306, p. 853–856, doi: 10.1126/science.1101996.

Tsuchiya, T., Tsuchiya, J., Umemoto, K., and Wentzcovitch, R.M., 2004, Phase transition in $MgSiO_3$ perovskite in the Earth's lower mantle: Earth and Planetary Science Letters, v. 224, p. 241–248, doi: 10.1016/j.epsl.2004.05.017.

van der Hilst, R.D., Widyantoro, S., and Engdahl, E.R., 1997, Evidence for deep mantle circulation from global tomography: Nature, v. 386, p. 578–584, doi: 10.1038/386578a0.

van Keken, P.E., 1997, Evolution of starting mantle plumes: A comparison between numerical and laboratory models: Earth and Planetary Science Letters, v. 148, p. 1–11, doi: 10.1016/S0012-821X(97)00042-3.

Vidale, J.E., and Hedlin, M.A.H., 1998, Evidence for partial melt at the core-mantle boundary north of Tonga from the strong scattering of seismic waves: Nature, v. 391, p. 682–685, doi: 10.1038/35601.

Wang, Y., and Wen, L., 2004, Mapping the geometry and geographic distribution of a very-low velocity province at the base of the Earth's mantle: Journal of Geophysical Research, v. 109, p. B10305, doi: 10.1029/2003JB002674.

Wen, L., 2000, Intense seismic scattering near the Earth's core-mantle boundary beneath the Comoros hotspot: Geophysical Research Letters, v. 27, p. 3627–3630, doi: 10.1029/2000GL011831.

Wen, L., 2006, A compositional anomaly at the Earth's core-mantle boundary as an anchor to the relatively slowly moving surface hotspots and as source to the DUPAL anomaly: Earth and Planetary Science Letters, v. 246, p. 138–148, doi: 10.1016/j.epsl.2006.04.024.

Wen, L., and Helmberger, D.V., 1998, Ultra-low velocity zones near the core-mantle boundary from broadband PKP precursors: Science, v. 279, p. 1701–1703, doi: 10.1126/science.279.5357.1701.

Wen, L., Silver, P., James, D., and Kuehnel, R., 2001, Seismic evidence for a thermo-chemical boundary layer at the base of the Earth's mantle: Earth and Planetary Science Letters, v. 189, p. 141–153, doi: 10.1016/S0012-821X(01)00365-X.

Wentzcovitch, R.M., Karki, B.B., Cococcioni, M., and de Gironcoli, S., 2004, Thermoelastic properties of $MgSiO_3$-perovskite: Insights on the nature of the Earth's lower mantle: Physical Review Letters, v. 92, p. 018501-1–018501-4, doi: 10.1103/PhysRevLett.92.018501.

Wessel, P., and Smith, W.H.F., 1998, New, improved version of Generic Mapping Tools released: Eos (Transactions, American Geophysical Union), v. 79, p. 579, doi: 10.1029/98EO00426.

Williams, Q., and Garnero, E.J., 1996, Seismic evidence for partial melt at the base of Earth's mantle: Science, v. 273, p. 1528–1530, doi: 10.1126/science.273.5281.1528.

Williams, Q., Revenaugh, J.S., and Garnero, E.J., 1998, A correlation between ultra-low basal velocities in the mantle and hot spots: Science, v. 281, p. 546–549, doi: 10.1126/science.281.5376.546.

Wookey, J., Kendall, J.-M., and Rumpker, G., 2005, D″ anisotropy from differential S-ScS splitting: Geophysical Journal International, v. 161, p. 829–838, doi: 10.1111/j.1365-246X.2005.02623.x.

Wysession, M.E., Bartkó, L., and Wilson, J.B., 1994, Mapping the lowermost mantle using core-reflected shear waves: Journal of Geophysical Research, v. 99, p. 13,667–13,684, doi: 10.1029/94JB00691.

Zhao, D., 2001, Seismic structure and origin of hotspots and mantle plumes: Earth and Planetary Science Letters, v. 192, p. 251–265, doi: 10.1016/S0012-821X(01)00465-4.

MANUSCRIPT ACCEPTED BY THE SOCIETY 31 JANUARY 2007

DISCUSSION

25 December 2006, Don L. Anderson

If someone points out to you that your pet theory of the universe is in disagreement with Maxwell's equations—then so much the worse for Maxwell's equations. If it is found to be contradicted by observation—well, these experimentalists do bungle things sometimes. But if your theory is found to be against the second law of thermodynamics I can give you no hope; there is nothing for it but to collapse in deepest humiliation.

Sir Arthur Stanley Eddington,
The nature of the physical world (1915)

Most scientific paradigms survive until a better paradigm comes along. Rarely, a paradigm is abandoned because some overlooked physics shows that it is impossible, or improbable. The concepts of Ptolemy, aether, phlogiston, and caloric, the most famous of the classical physics paradigm shifts, occurred when the concepts became so contrived and convoluted that people lost interest. But an idea that violates thermodynamics should be abandoned immediately. Seismological studies paired with fluid injection experiments and Boussinesq simulations cannot answer the question of whether the lowermost mantle is a plausible source for surface hotspots. Low velocities, high velocities, ultraslow velocities, and anisotropy have all been used as "evidence" for plumes (see the references in Garnero et al., this volume). Plumes have been proposed to emanate from the tops of superplumes, rather than from D″. We have also been told that lack of clear tomographic evidence for plumes in the lower mantle is due to lack of resolution or coverage. Now we are told that the most likely locations of mantle plumes are over the boundaries between slow and fast regions. There is no way that the plume hypothesis can be falsified with these many and contradictory options. So much for the observational aspect. The commonly used Boussinesq approximation for mantle convection is not as soundly based as Maxwell's equations, and it certainly ignores thermodynamics. A thermal boundary layer (TBL) at the base of the mantle certainly exists, but this is not a sufficient condition to form deep narrow plumes, as currently envisaged. A TBL serves to conduct heat out of the core, but whether it can form sufficiently buoyant plume heads to break out of the lower mantle, or develop 100- to 200-km-wide plume tails in a reasonable amount of time depends on the material parameters, which depend on composition, pressure, core heat, mantle radioactivity, and convective vigor. If the local Rayleigh number is less than about 1000, the TBL will not go unstable. If the coefficient of thermal expansion is low, the TBL may never develop enough buoyancy to escape. When pressure is taken into account, the dimensions of buoyant instabilities are on the order of thousands, not hundreds, of kilometers. If the intrinsic density of D″ is as little as 1% higher than the rest of the mantle, it will be permanently trapped, but this does not mean that it will have a simple structure. It is surprising that none of the discussions in this volume that argue for deep-mantle plumes men-

tion the effects of pressure on material properties and the ability of internal heating and background mantle convention to prevent or destroy the instabilities.

As long as heat is flowing into the base of the mantle, a TBL should be present, but the ratio of core heat to mantle heat and the local (pressure-dependent) Rayleigh number are the key parameters, not the mere existence of a TBL. Cylindrical plumes in numerical and laboratory experiments usually involve localized basal heating, the instantaneous creation of a hot sphere, or basal injection of fluid; pressure and background convection effects are ignored and thermal effects exaggerated (see, e.g., the articles by Sleep, this volume, and King and Redmond, this volume). The "expected small plume conduit dimension (e.g., <500 km)" (Garnero et al., this volume) is based on experiments that ignore pressure effects or that impose the dimension ("experimentalists do bungle things sometimes").

The smoking gun against deep-mantle plumes is thermodynamics. All the critical thermodynamic parameters depend on volume, but these are ignored in all calculations that yield narrow whole-mantle plumes (the Boussinesq approximation). None of the fluid dynamic calculations used by the authors to support their view take into account the effect of pressure on thermal expansivity and its role in chemically stratifying the mantle and stabilizing deep thermal structures. The few non-Boussinesq calculations that have been done do not use self-consistent thermodynamic relations, but even so, they do not predict plume-like dimensions and timescales at D″ depths. At upper-mantle pressures, thermal expansivity is high, and chemical stratification is reversible. However, simple scaling relations show that pressure increases conductivity and viscosity—and spatial and temporal scales—and decreases the coefficient of thermal expansion and the local Rayleigh number. Even slight compositional effects can make deep dense layers permanent and complex.

27 December 2006, Geoffrey F. Davies

The factor that invalidates Anderson's claims is the temperature dependence of viscosity. It is the reason plumes form head-and-tail structures, as illustrated in Davies (1999). The effect is actually considerably stronger at higher pressure, because the activation enthalpy may be two or three times larger at the base of the mantle than near the surface. Plume heads are large mainly because they have to displace high-viscosity surroundings to rise, but plume tails can be narrow, because the lower-viscosity plume fluid can flow up a pre-existing path. The other factors Anderson mentions have some effect but are not dominant.

The problems with Anderson's arguments were detailed in Davies (2005). Either Davies's criticisms should be refuted or Anderson's arguments should not be repeated, because they appear to be quite invalid. By the way, plenty of plume models have been done with uniform basal heating; see, for example, Davies (1999) and Leitch and Davies (2001).

29 December 2006, Don L. Anderson

I thank Geoff Davies for providing these references and allowing me to clear up possibly confusing points. He has put his finger on the essence of the issue separating those who think mantle plumes are obvious and inevitable from those who remain skeptical of their physical basis. The issue regarding the temperature dependence of viscosity is not straightforward and is regarded as a paradox (e.g., Nataf, 1991; Lenardic and Kaula, 1994). When this effect is taken into account—in its entirety—the upper TBL becomes thicker, making the upper mantle hotter, and the lower mantle acquires a negative or subadiabatic temperature gradient; the lower TBL becomes *colder*. Melting is more likely to occur in the upper mantle, and to greater depths, than is the case with constant viscosity. Cavity plumes are less likely to form at the base of the system unless the boundary layers interact.

The temperature dependence of viscosity is a two-edged sword. If applied to the lower TBL in isolation, it would seem to make cavity plumes more likely. But it also makes the upper TBL stiffer, longer-lived, and with a larger temperature drop. When this upper layer goes unstable, it cools the lower mantle and the lower TBL, making them stiffer (Lenardic and Kaula, 1994). Much of the lower mantle develops a subadiabatic or negative temperature gradient because of internal heating and slab cooling, increasing the viscosity with depth. Ironically, it has been argued that hot plumes from the deep TBL will heat and thin the plate, but sinking of the cold surface TBL and cooling of the base of the system, the parallel effect, has been ignored and may be more important, because of the temperature dependence of viscosity!

When the possibilities of melting and differentiation are allowed for in a convection calculation, the mantle can become chemically stratified (Tackley and Xie, 2002). The various components (e.g., eclogite and refractory residue in this case) collect or re-collect at levels of neutral buoyancy and survive or regenerate there for billions of years (see also Anderson, this volume). Buoyant material in the shallow mantle also extends the surface TBL, making the upper mantle hotter still. Chemical layering is facilitated by pressure dependence of thermal properties and reduces the Rayleigh number and the vigor of convection, particularly at depth.

The equations in Davies (2005) regarding the effect of temperature and pressure on viscosity are identical to those in Chapter 7 of *Theory of the Earth* (Anderson, 1989). There is no disagreement there. The standard Arrhenius form and the various terms—the pre-exponential, the activation volume, and the activation energy—are derived. It was determined (Anderson, 1989, p. 133) that the viscosity should increase by about a factor of 60 to 80, due to compression across the lower mantle, at constant *T*. The total variation across the mantle involves a large decrease due to the temperature rise in the upper mantle and a possibly smaller decrease at the lower TBL. The mid-mantle effect is uncertain. The viscosity jump across discontinuities may be negative.

As Davies points out, there are many calculations in the plume literature of uniform bottom heating, and also of localized heating and injections of hot fluid. If we have only bottom heating, no pressure effects, and no radioactivity, then plumes are inevitable if the heating is strong enough; the core is the only heat source in this model and its heat is removed by plumes. But plumes, plates and convection, and D″ should not be treated separately (Lenardic and Kaula, 1994). The papers by Tozer (1973), Kaula (1983), Nataf (1991), Lenardic and Kaula (1994), and Tackley and Xie (2002) collectively make the point that one cannot treat one variable (viscosity), one parameter (temperature), one region of the mantle (D″), one mode of heating and/or cooling (core heat), or one boundary condition independently; the whole parameter space and system must be treated together, in a self-consistent way. This is required by thermodynamics and far-from-equilibrium systems, such as convection.

All simulations of narrow plumes involve injection of hot fluid through a circular orifice, uniform or localized heating from below, or the instantaneous creation of a hot sphere as the initial condition; plumes and their dimensions are imposed by the investigator, rather than being natural fluid dynamic instabilities in a realistic setup that resembles the mantle. In other studies, a plume is just assumed to exist and its properties are investigated.

The five remarkable papers cited above, mostly overlooked, plus *Theory of the Earth* (Anderson, 1989), form the basis of the present discussion. Petrology and self-consistent fluid dynamics appear to explain the thick average TBL thickness (280 km) and the mean mantle potential temperature ($1410 \pm 180°$ C) derived by Kaula (1983) from geophysical and plate tectonic data. These values exceed expectations from cooling of a homogeneous fluid or the temperatures of mid-ocean ridge basalt (MORB). The temperatures and depths are consistent with 'hotspot' magmas being derived from within or just below the surface TBL, but do not rule out the existence of fertile blobs (Anderson, this volume; Beutel and Anderson, this volume).

30 December 2006, Geoffrey F. Davies

Anderson's points are either invalid, confused, or make little difference.

Upper TBL. A temperature-dependent viscosity can stiffen the top TBL, make it thicker, and raise the internal temperature. However this is only true if the TBL becomes immobile (stagnant lid regime). The mantle's top TBL is mobile, because it is broken into moving plates that can subduct. In this case Anderson's arguments do not apply, though they would in any case only change the details, not the general principles described below. Besides, how could you reconcile a 280-km-thick TBL with seismological constraints on the thickness of the oceanic lithosphere, and with seafloor subsidence and heat flow? I think you cannot.

Subadiabatic Gradient. It is well known that the vertical temperature gradient between TBLs is subadiabatic, regardless of heating mode or the nature of the top TBL. Recent estimates are that the temperature is 100–300° C lower than adiabatic near the bottom of the mantle. This temperature reduction will raise the viscosity significantly relative to an adiabatic profile, as Anderson says, but it is only part of the uncertainty in deep-mantle viscosity, as I discuss below.

Bottom TBL and Plume Formation. The viscosity and temperature above the bottom TBL are not the only determinants of whether plumes will occur. The other major determinant is the temperature at the core-mantle boundary, which changes only on billion-year timescales. The minimum viscosity in the TBL occurs at this boundary, and is likely to be two to five orders of magnitude lower than the overlying ambient mantle, depending on the temperature increase through the boundary layer, which is commonly estimated to be 1000° C or more. Even an increase of 500° C is ample to generate plumes (Leitch and Davies, 2001).

The role of the viscosity above the TBL is to control the timing and size of an instability that can begin to rise—the larger the viscosity, the larger the blob must be before it can detach (Griffiths and Campbell, 1990; Davies, 1999). Once a blob is detached and rising, the behavior of material following from the TBL depends on its viscosity, which, as noted above, depends on the core-mantle boundary temperature, not the temperature above the TBL. Because it is likely to be two or more orders of magnitude lower in viscosity, it will form a narrow conduit, as noted in previous discussion and demonstrated by Davies (1999).

The viscosity above the TBL could be changed by one to two orders of magnitude without changing this general behavior. Only quantitative details would change, such as the exact dimensions of the initial head and the following tail.

If the lower mantle were cooler and more viscous than we have thought, as Anderson advocates, then the temperature difference between the mantle and core would be greater, which would cause stronger plumes. The higher viscosity would mean the plume heads took longer to develop and would be larger.

Viscosity Increase through the Lower Mantle. Anderson's discussion of the depth dependence of viscosity seems to be confused. Mid-mantle viscosity is constrained by post-glacial rebound and subduction zone geoids to be around 10^{22} Pa s (e.g., Mitrovica, 1996). Deep-mantle viscosity is unlikely to be much more than an order of magnitude greater than this value, or it would affect the rebound or Earth's rotation noticeably.

I do not know how Anderson gets an isothermal increase of viscosity by only a factor of 60–80 over the lower mantle from his formulas. His $(\partial \ln \eta / \partial \ln \rho) = 40$–$48$ (η is viscosity, and ρ is density) (Anderson, 1989, p. 133) and a density increase from 4400 to 5500 kg/m^3 through the lower mantle yields an increase by a factor of 7500–44,000. This increase is larger than the observational constraints seem to permit, so evidently these formulas are not good guides (e.g., Davies, 2005).

Anderson's isothermal increase by a factor of 60–80 (Anderson, 1989) would seem to imply little increase or even a decrease along the actual (subadiabatic) mantle temperature profile, which would not serve his cause of suppressing mantle plumes.

Anderson's formula for viscosity on p. 133 of his book (Anderson, 1989), to which he refers, is not correct: viscosity $\sim (G/\sigma)^n$, $n = 1$–3 (σ is deviatoric stress). If G is the rigidity (shear modulus) as used on the previous page, then this ratio is dimensionless. Presumably Anderson meant strain rate $\sim (\sigma/G)^n$, which yields viscosity $\sim (G^n/\sigma^{(n-1)})$. Even so, this term is only one factor in the full expression for strain rate (see equation 6.10.3 in Davies, 1999). G is a useful scaling factor, not a rigorous predictor. Similarly, Anderson's equation relating activation volume to the depth dependence of G on the previous page is no more than a rough, possibly very rough guide, given the sensitivity of viscosity to activation volume.

Uniform Bottom Heating. Anderson lumps uniform bottom heating in with other types of boundary condition and repeats his claim that modelers' boundary conditions have predetermined plumes and their dimensions, rather than leaving the fluid dynamics to determine the outcome. Let it be clear: uniform bottom heating does not determine plume dimensions or other characteristics—it leaves that to the fluid dynamics.

The amount of bottom heating does predetermine the occurrence of plumes, it is true, and that is because the physics requires them under the conditions prescribed. Those conditions are, in the better experiments (laboratory or numerical), tailored to the conditions near the core-mantle boundary, as best we understand them. Anderson has yet to make a persuasive case that those conditions might be very different.

Anderson does a disservice to modelers by claiming or implying that other factors have not been considered, such factors as changes of properties with depth, radioactivity, and the existence of another TBL. They have. If one understands the physics, one can sensibly understand the usefulness and relevance of the models. Furthermore, it is not true that modelers have not included a vertical viscosity gradient in their lower mantles—they have.

If the Hawaiian hotspot chain did not exist, and if several hotspot chains did not emerge from flood basalt provinces, there might be some point to the debate, but models of thermal plumes give a good quantitative account of those phenomena. I do not, however, claim that thermal plumes can explain everything (see Davies, 2005).

General Comment. The physics of thermal convection, and of plumes in particular, is well understood and not too hard to follow. Readers are referred to Davies (1999). The points made here have been made before. The productive debates have moved on to other things, such as the role of compositional variations and the influence of subduction.

31 December 2006, Don L. Anderson

What the general reader of these pages wants to know is: Does fluid dynamics prove that mantle plumes must exist, and does it rule out alternative explanations for melting anomalies? Davies and I both agree that it does neither. Are self-consistent, self-organized simulations better than ad hoc parameters and tightly constrained experiments? Yes. Does the newly found complexity in D″ imply that the source of plumes has been found? Is the mantle almost entirely heated from below? Of course not. When we look at volcanoes can we ignore the lowermost mantle? Are the early arguments, assumptions, predictions, and experiments for plumes still valid? Is it possible that the shallow mantle is hotter and more variable in fertility and melting point than generally assumed? Here, Davies and I diverge.

The best way for the nonspecialist to understand these issues is to look at the series of calculations published, for example, by Tackley (1998, 2002), Lowman et al. (2001), Davies (2005), and Phillips and Bunge (2005), who demonstrate the extreme sensitivity to initial and boundary conditions that is characteristic of nonlinear chaotic systems. These investigators usually start with the simple case discussed by Leitch and Davies (2001), Davies (2005), and Campbell and Davies (2006) (hereafter DLC)—bottom heating with constant properties. Nice little mushroom forests appear, as expected. Then radioactive heating is introduced, and temperature- and depth-dependent properties, and then, continents. The mushrooms disappear and the mantle heats up and becomes unsteady. Then melting and differentiation are introduced. Layers and blobs appear. Then large plates and large-aspect-ratio convection cells are allowed. The mantle gets hotter still and fusible blobs melt. Then secular cooling is thrown in. The whole system changes as each new element is introduced, which is the nature of self-organized, far-from-equilibrium, or chaotic systems. The system also flips spontaneously from one state to another.

The plates may control mantle temperature. In some cases, the system jams and heats up, and one has to crack the plates to allow subduction, and the whole mantle reorganizes again. The lower TBL is cooled if the plates sink to the bottom. Modelers who have got this far do not much discuss narrow, hot, stationary, long-lived, or radial plumes. The mantle is plenty hot, fertile, and variable, so that many alternative explanations of melting anomalies are now on the table. The claims that narrow plumes are an inevitable result of a TBL, that plumes are the way the core gets rid of its heat, and that hotspots are independent of other forms of mantle convection and plate tectonics have not received numerical validation, although calculations can easily be designed that satisfy these claims.

Delamination of the lower part of the plate, not yet included in any global simulation, can also fertilize and cool the mantle. When delamination occurs, warm asthenosphere rushes up to fill the gap and we get a volcano. When the delaminated blob heats up, it melts, and we can get another volcano or volcanic chain. The more realistic the fluid dynamics gets, the more plausible are layering, delamination, shallow return flow, and fertile blob mechanisms. In the extreme case, it is the outer shell that is the regulator, not mantle viscosity or even temperature.

These points are well illustrated in Parmentier et al. (1994), Tackley and Xie (2002), Grigné et al. (2005), Phillips and Bunge (2005), and Nakagawa and Tackley (2006). These are realistic mantle simulations (hereafter RMS)[1] performed in wide boxes or spherical shells that can self-organize. These papers show how the various effects (internal heating, plates, continents, melting, layering, secular cooling, 3-D) affect the background temperatures and mantle motions that plumes must endure. Not surprisingly, these studies do not validate the results, assumptions, and boundary conditions in DLC. Plumes were initially proposed because of perceived shortcomings in the simplified convection and tectonic models existing at the time. It was then thought that the lower mantle was rigid or had very high viscosity, and that plates were rigid.

Bottom-heated, axisymmetric, and injection experiments (DLC) have been useful for understanding idealized thermal plumes. Indeed, it was these studies that caused the wider community to embrace the plume idea. Plumes were an elegant and easy-to-understand solution to midplate volcanism. However, it is a disservice to the wider community to imply that plumes must always exist, and that they provide the only explanation for Hawaii and the like, particularly as more realistic calculations (laboratory experiments cannot cover the appropriate conditions) plus lithologic heterogeneities provide alternative ways to form hot mantle and melting anomalies. For example, if the mantle is only 100° C hotter than assumed in the plume literature, and the melting point of blobs is 200° C lower, then plumes are not needed.

Realistic convection simulations confirm that the mantle runs hotter and more episodically than homogeneous or bottom-heated models (Phillips and Bunge, 2005). Fe-enrichment in D″ means that excess plume temperatures must be much higher than DLC assumes; otherwise, material cannot rise out of the region. Pressure-dependent properties do the same thing. Decreased thermal expansion coefficient and increased conductivity do not automatically preclude plumes, but the intrinsic density of D″, its thickness, and its buoyancy parameter can preclude their escape. The required temperature contrasts may become larger than available. Even a 1–2% density excess may stabilize D″, but large structures are required to generate the buoyancy needed to make plumes rise.

Plate tectonics involves recycling and the introduction of low-melting-point constituents into the mantle, forming layers and blobs. A stiff outer shell prevents magmatism except where permitted by extension or delamination, and it also causes the temperature to rise. In other words, melting anomalies, the very reasons that plumes were introduced in the first place, are potentially explained by models with realistic properties and without plumes. An internally heated mantle, cooled from the top, will have large rising regions caused by slowly developing buoyancy and displacement by sinkers, but these are not plumes as conventionally defined; they are the normal convection that plumes were invented to augment.

---

[1]The RMS papers and associated figures can be found conveniently by inputting the search string author+convection into Google or Google Images.

There are still surprises out there, as one expects in self-organized far-from-equilibrium systems that are allowed to do their own thing. A small change (one crack, one continent) can change everything, as elegantly shown by Gurnis, Bercovici, Phillips, Bunge, Tackley, Lenardic, Lowman, King, Hansen, Conrad, Hager, and their collaborators (see references on www.MantlePlumes.org). These authors have repeatedly reminded us that mantle convection is not only a branch of chaos theory but also a branch of thermodynamics; thus, all the parameters and boundary conditions are interconnected, and self-consistency is essential. RMS studies show that self-consistent models run hot and unsteadily, and have large plates and large aspect ratio convection. A single injection or axisymmetric experiment simply cannot be so generalized; too many degrees of freedom have been removed. Until the RMS calculations were done, we did not even ask whether the mantle organized the plates or the plates organized the mantle, or whether things oscillated. In bottom-heated cylinders, these issues are not even raised; one knows what will happen, and where.

It is interesting to note the progression in geodynamic thinking, from lower-boundary control (bottom heating, plumes) to mantle self-regulation (the Tozer effect) to control by the plates (plate bending, top-down tectonics). The top boundary condition, ignored or simplified until recently, may act as a template but may also organize and drive mantle convection, and localize magmatism. Fretting about details of D″ may be beside the point for volcanoes if the upper mantle is as hot and variable as realistic simulations suggest.

31 December 2006, Edward J. Garnero, Thorne Lay, and Allen McNamara

We thank Don Anderson for his comments on our paper and Geoff Davies for his responses. First, let us reiterate that the purpose of our paper was to consider (not to prove or disprove) the possible connections between recent deep-mantle seismic findings, state-of-the-art numerical geodynamical calculations, and geographical systematics for upwellings and plume initiation. We drew attention to the remarkable seismic evidence for "sharp edges" to the large low-shear-velocity provinces (LLSVPs) in the deepest mantle (note our explicitly dynamically neutral terminology), which cannot be attributed solely to temperature gradients. Thus, a distinct chemical component to the LLSVPs is highly likely. This possibility motivates consideration of implications of thermochemical boundary layers for plume initiation in contrast to the standard isochemical TBL behavior discussed in most plume scenarios.

Although we discussed studies that advance the interpretation that LLSVPs are "superplumes" that rise in the mantle because of intrinsic thermal or chemical buoyancy, we do not advocate such models. The fact that LLSVPs are not beneath past or present subduction locations is consistent with subduction-related currents sweeping the LLSVP material into "piles" and maintaining their strong lateral margins (between LLSVP material and adjacent non-LLSVP mantle). In this scenario, the deep mantle must have significant convection currents. We appeal to a density increase of the LLSVPs, as suggested by a few seismic studies, recognizing that this increase is an issue of debate. Thus, our article takes the perspective that LLSVPs are relatively stable deep-mantle features with configurations sustained by the past few hundred million years of mantle circulation. Thus, some connection between deep mantle and surface structures is worth examining.

Geographically correlating phenomena at the top and bottom of the mantle certainly leaves one wanting better constraints on structure between the boundaries of the mantle. The reduction of seismic resolution in the mid-mantle (compared to the upper and lower boundary layers) cannot be used for or against any favored hypothesis. Nonetheless, it is significant that commonly designated hotspot populations are twice as likely to overlie the regions of strongest D″ lateral shear-velocity gradients than regions with the lowest velocities. They are increasingly unlikely to overlie regions with high velocities (Thorne et al., 2004). The strongest shear-velocity gradients in tomographic studies are coincident with LLSVP edges; thus, hotspots are observationally more likely to be situated above LLSVP edges. We recognize the difficult issue of what one calls a "hotspot" (e.g., Courtillot et al., 2003), but that problem is beyond the scope of our current article. The thermochemical geodynamic calculations driven by historical subduction patterns result in configurations of large chemical piles in close accord with seismic observations and predict concentrations of TBL instabilities on the margins of the piles, as hot boundary layer material is swept up onto the pile edges. Thus, the seismic observations, geodynamic models, and crude correlations with surface phenomena give a provocative new perspective on how deep-mantle heterogeneity may plausibly influence some surface volcanism.

Anderson points out that the geodynamic calculations we consider involve approximations; in particular, pressure dependence of the coefficient of thermal expansion is neglected in the Boussinesq approximation. Numerical thermochemical models with compressibility and pressure-dependent thermal expansion have been conducted (e.g., Tackley, 1998; Hansen and Yuen, 2000), which show qualitatively similar results to our calculations; specifically, thermal plumes rising from thermochemical piles. As Anderson notes, the length scales of thermal instabilities tend to be larger in such models than in calculations with constant thermal expansion coefficients. But, for systems at least partially heated from below, decreased thermal expansion coefficient does not intrinsically preclude the development of plumes; the calculations yield fewer, larger plumes (Davies's commentary of 30 December on this problem is also quite relevant). The calculations that we report suggest that deep-mantle thermal instabilities will be geographically rooted near LLSVP margins. When it becomes viable to compute fully thermochemical spherical convection models with depth-dependent parameters, our expectation is that the configuration of upwellings will remain the same, only with larger-scale plume initiation at LLSVP margins. This prediction will need to be explored in the future. The destabilizing effects of post-perovskite phase transitions must also be considered, as such changes increase the potential for development of boundary layer instabilities, possibly countering any inhibiting effects of pressure dependence. However, at this time, the calculations have not been done, so speculation is appropriate for commentary, not for inclusion in a paper. We do note that some of the speculation in Anderson's comments of 25, 29, and 31 December appear at odds with the experience of actual numerical calculations, so his assertions about the dynamics are also premature.

Finally, realistic simulation of mantle circulation is a long-term research objective in geodynamics: All current models make some simplifying assumptions based on either computational considerations or lack of constraint on various thermodynamic parameters. There is often a resulting disconnect between practitioners who compute numerical models and geophysicists who speculate on plausible complexities not incorporated into those models. Similarly, there are large uncertainties in relating findings from different disciplines, as in our case of exploring interpretations of seismic observations by geodynamic models constrained by plate tectonic histories. Any claims of uniqueness would be laughable, but arguments by assertion without computational validation are of little merit as well.

31 December 2006, Norman H. Sleep

I comment on two of Anderson's points: (1) The thermal expansion coefficient in the deep mantle is so low that convection heated from below would be sluggish, and (2) any chemical-density stratification would overwhelm thermal expansion.

With regard to the thermal expansion coefficient, it is useful to review the parameterized convection equation. The convective heatflow q from a thin boundary layer is

$$q = Ak\Delta T \left[\frac{\rho g \alpha \Delta T}{\kappa \eta} \right]^{1/3}, \qquad (1)$$

where k is thermal conductivity, ρ is density, g is the acceleration of gravity, α is the thermal expansion coefficient, and κ is the thermal diffusivity. The dimensionless multiplicative constant A is on the order of 1; it depends on the boundary condition at the core-mantle boundary or the interface between "dregs" (chemically dense regions at the base of the mantle with no geometry or origin implied) and normal mantle. ΔT is the temperature contrast that actually drives convection and η is the viscosity, which is a weighted average between the low-viscosity hot boundary layer and the conducting interior (Davaille and Jaupart, 1993a,b, 1994; Solomatov, 1995; Solomatov and Moresi, 2000). The product $\rho \alpha$ rather than just α oc-

curs in the equation, which partly offsets the effect of the decrease of thermal expansion with pressure. In addition, this product is raised to the 1/3 power, making the heatflow insensitive to its value, provided it does not become zero or negative. To obtain the long-term thermal expansion coefficient, one needs to include the effect of multiphase systems, in which density depends on the partition of components among phases with temperature rather than on just the expansion of isochemical phases.

Anderson is correct that compositional variations can overwhelm thermal expansion. The Earth's core-mantle boundary and the rock-air interface are obvious extreme cases. The evolution of the lowermost mantle over time is relevant to plume convection heated from below. The core likely formed hot in the wake of the moon-forming impact. Conduction from the core thus has heated the lowermost mantle over geological time. In the absence of convection, the hot region at the base of the mantle would be quite thick, scaling to $\sqrt{\kappa t_E}$, where t_E is the age of the Earth. The heatflow from the core would scale crudely with the square root of the age of the Earth. One needs to account for the cooling of the core over time to obtain a better conductive model.

Overall, chemical stratification in the lower mantle leads to predictable but complex behavior. Anderson's "dregs" layer is gradually stratified, so it does not convect internally. It is quite thick, so that a vigorous thermal boundary layer does not form above it. However, seismologists and petrologists have yet to constrain strongly the properties of conceivable chemically dense regions at the base of the mantle. We do not even know whether the dregs layer is thin (i.e., corresponding to the D″ layer), or thick, as in a lava lamp. We do not know how dregs formed to begin with and whether current mantle processes enhance or disrupt stratification. Given this state of ignorance and physical complexity, observable manifestations of deep processes remain relevant. These manifestations include hotspot tracks and the possible detection of plumes by tomography.

2 January 2007, Don L. Anderson

The opposite of a vigorously convecting, well-stirred, high-Rayleigh-number homogeneous mantle—as usually modeled— is a mantle stratified by intrinsic density. The possibility that the mantle is chemically stratified is usually dismissed outright, particularly by modelers. Compositional stratification is plausible and merits more attention (Anderson, this volume).

In his comment of 31 December, Sleep has added insight into this issue and the formation of "dregs" layers at chemical and viscosity boundaries. Much of the chemical layering was probably contemporaneous with accretion and Moon formation, but subsequent cooling can also create stratification, by the dregs mechanism (e.g., by light material leaving the core and newly dense basalt-eclogite in the proto-crust returning to the transition region). At some point in Earth's evolution, presumably as a result of cooling, the deep-subduction mode of plate tectonics kicked in; this mode returns surface material back to the mantle and displaces deeper mantle upward. But there are ways to form a chemically distinct D'' layer without importing upper-crustal material from the surface.

Although the core-mantle boundary is the most obvious place (apart from the surface) to collect dregs from the mantle and dross from the core, it is not the only candidate boundary for collecting debris from Earth accretion and differentiation. As the mantle cools, the thick basaltic crust converts to eclogite, which sinks and collects at the 410- or 650-km discontinuity, depending on its temperature and major-element chemistry. Very cold SiO_2-rich MORB eclogite may sink into the deeper mantle, until it reaches a density or viscosity barrier. Mantle viscosity may jump by two or three orders of magnitude at depths of ~1000 km and ~2000 km (Forte and Mitrovica, 2001). Big chunks of delaminate are not easily stirred back down to meter- and centimeter-sized pieces. Olivine-rich cumulates and restites are buoyant and collect under the crust as a perisphere. The crust and proto-crust are the dross of terrestrial differentiation and the core is the dregs. These are just a few examples of possible chemical layers.

Plausible chemical differences among mantle lithologies give huge density differences compared to thermal expansion. Overall, chemical stratification in the mantle leads to complex behavior that is not necessarily predictable. The delamination scenario is particularly interesting. Delaminated blobs are fertile, fusible, and initially dense. They will form dregs in the mantle, where they become neutrally buoyant. They do not necessarily form a continuous dense layer. They then heat up and approach ambient mantle temperature, melt, and become buoyant. Their fates depend on their sizes. This scenario has not yet received numerical validation.

Sleep is correct in pointing out the need to generalize the expansion coefficient in a multiphase rock. One example is the gradual heating of garnet-majorite that may reside in the transition region as a result of the delamination of the lower part of the mafic crust of an overlying continent. The conversions of majorite back to garnet + pyroxene and then of garnet to magma give large density reductions. These effects, and compositional variations, can overwhelm thermal expansion.

The interpretation of tomography is more ambiguous than usually appreciated. Cold peridotite with CO_2 (Presnall and Gudfinnsson, 2005, 2007) can be dense and yet have low seismic velocities. Eclogite has low shear-wave velocities compared to dry peridotite of the same density; dense sinkers of eclogite can have low shear-wave velocities and may be mistaken for hot rising plumes. Refractory peridotite has high seismic velocity but low density; in tomographic images it appears blue but is not sinking. Tomography is not a thermometer.

Sleep, however, is optimistic about the possible detection of plumes and the relationship of volcanic chains to D''. In bottom-heated, but otherwise realistic, convection calculations, plumes, when they exist, are more like wandering strands of cooked spaghetti than rigid, upright rods, as often illustrated in cartoons, or axisymmetric cylinders, as they appear when modeled in isolation from mantle flow. They would be invisible to tomography. However, plumes should spread out below the 650-km discontinuity and the lithosphere and be detectable. But there is no seismic evidence for this (Deuss, this volume). If plumes exceed ~1000 km in dimension they would overlap the normal scale of mantle convection and the broad upwellings that are intrinsic to an internally heated mantle in the absence of plumes.

2 January 2007, Scott D. King

Unwary readers should take warning that ordinary language undergoes modification to a high-pressure form when applied to the interior of the Earth. A few examples of equivalents follow:

High Pressure Form	Ordinary Meaning
Certain	Dubious
Undoubtedly	Perhaps
Positive proof	Vague suggestion
Unanswerable argument	Trivial objection
Pure iron	Uncertain mixture of all the elements

The discussion points by Anderson and Davies on the paper by Garnero et al. (this volume) remind me of the quote above from Birch (1952, p. 234). I refer readers to my own article in this volume, in which I present calculations that illustrate many of the effects mentioned in this discussion thread.

Let me begin with the rather obvious observation that the Earth is round. The surface area of the core-mantle boundary is approximately one-quarter of the surface, favoring a larger TBL at the base of the mantle than at the surface, all other things being equal. As Anderson reminds us, all things are not equal.

Anderson and Davies both appeal to temperature-dependent mantle rheology to support their views. Kellogg and King (1997), van Keken (1997), and Davies (1999) show that the large plume head, narrow tail structure is a natural consequence of convection heated from below with an Arrhenius form of rheology and otherwise uniform properties (e.g., Boussinesq convection). Few if any papers in the past decade have used a constant viscosity, so this objection begs the question. I demonstrate the effects of internal heating (uniform), pressure-dependent coefficient of thermal expansion (the major effect of compressible convection; c.f. Ita and King, 1994), upper-mantle phase transformations, and a deep stabilized layer at the base of the mantle (phase change or compositional) with a strong temperature-dependent viscosity, including an increase in viscosity with depth. These factors reduce the peak geoid, topographic, and heatflow anomalies, bringing the calculations closer to the observations, but produce deep plumes. It is worth noting that many other calculations have used temperature-dependent rheology with basal heating and in many cases internal heating and phase changes (e.g., Kiefer and Hager, 1992; Farnetani and Richards, 1994, 1995; Davies, 1995; Farnetani et al., 1996; Farnetani, 1997; Kellogg and King, 1997; King, 1997; van Keken, 1997; Leitch et al.,

1998; Leitch and Davies, 2001; Goes et al., 2004; Davies, 2005; Lin and van Keken, 2006a,b; Zhong, 2006), and it seems past time to put that objection to rest. A significant amount of work has been done.

Anderson reminds us that in a complex system, such as the mantle, it may be dangerous to make simplifying approximations that do not allow for a self-consistent formulation. However, his arguments beg the question as he proceeds to base his own arguments on at best the same inconsistent calculations or at worst simple theory that does not account for nonlinear feedback. As an example, any effect that decreases the lower-mantle Rayleigh number will cause the lower mantle to heat up, lowering the viscosity and increasing the flow. It is exactly because the mantle is a complex, self-organizing system, that such thought experiments, which do not consider feedback mechanisms, are every bit as dangerous (if not more so) as inconsistent calculations. A model as complex as reality is likely to yield very little understanding, because it will be as unwieldy as reality (e.g., Oreskes et al., 1994).

Although Anderson points out the effect of the increasing adiabatic temperature on viscosity, he does not mention the pressure-dependent effect (activation volume) that trades off with the adiabatic temperature and is not well constrained experimentally. Viscosity models of the Arrhenius form can be compared with other models of mantle viscosity (e.g., King and Masters, 1992). The argument presented by Anderson on the temperature effect of rheology has been used by many of us to explain the geoid and topography inversions for the past two decades and is consistent with what he often calls the standard model. This model has been included in many plume calculations, including most of the ones cited above, so it also begs the question. The lower mantle is made up of perovskite and ferropericlase, and most laboratory rheology measurements are made on olivine or olivine analogues. So self-consistency is in the eye of the beholder.

Ironically, the discussion has little to do with the observations in the original article of Garnero et al. Seismic observations at the base of the mantle are complex and not obviously consistent or inconsistent with any of the models in their original form. They are leading to interesting new ideas (e.g., Ishii and Tromp, 1999; Gurnis et al., 2000; Le Bars and Davaille, 2004; Schubert et al., 2004; Farnetani and Samuel, 2005) that one could envision being reconciled with Anderson's, Davies's, or a hybrid view of the mantle. As for Anderson's arguments regarding the modification of the plume theory, remember Kuhn's (1970) work—scientific theories are always modified when presented with new observations. It seems to me that the proliferation of observations and ideas, including the chapter by Garnero et al. in this volume, show that the scientific process is working well.

3 January 2007, Geoffrey F. Davies

Anderson, in his 31 December comment, shifts from debating specifics to a general discussion of mantle dynamics. His representation of my views is inaccurate. I have never made such unqualified claims as "plumes must always exist, and they provide the only explanation for Hawaii, and the like." I therefore summarize here my actual views, for the record.

In my book (Davies, 1999) I argue that mantle convection can be usefully viewed as driven by two TBLs, with rather different dynamical styles, that interact to a substantial degree. Clearly plates from the top TBL are a major driving force, and clearly they affect plumes. Nevertheless some phenomena, Hawaii being the outstanding example, are quite well explained quantitatively by the thermal plume model, apparently with only secondary effects from the rest of the mantle system. That is why models of isolated plumes are a useful approximation to some aspects of the mantle system.

Sleep and I were among the first to conclude, from inferred plume fluxes, that the mantle is only secondarily heated from below, and therefore mainly heated by internal radioactivity (with some secular cooling). Indeed, for many years I have modeled the role of plates in the mantle system using only internal radioactive heating and excluding bottom heating and plumes. The point, of course, is to isolate those parts of the system that can be usefully isolated. It is therefore incorrect to claim that I do not account for radioactivity in the mantle on the basis of plume models tailored to isolate the plume phenomenon.

The mantle system clearly has complications, but that does not mean that parts of it cannot be usefully approximated by simpler models. Our understanding of the system has advanced considerably using this standard scientific approach.

Of course ultimately we would like to include all the main phenomena in one model. This is not yet possible, because models including both plates and plumes must be three dimensional, but 3-D models do not yet have the high resolution necessary for accurate modeling of plumes, although they are steadily approaching this goal.

Although Anderson refers to some recent models as being more realistic, it is important to appreciate that there are still important aspects that are not well understood or well constrained. This problem applies particularly to the top TBL, mainly because the rheology of the lithosphere is complicated and still not well characterized. Also, no model can accurately predict absolute mantle temperatures because of uncertainties in mantle rheology. Just because there is a relative progression of temperatures among some models does not provide a strong criterion for accepting or rejecting particular models.

Regarding alternatives to plumes, any model that involves a passive upper mantle responding to lithospheric changes would need to account for uplift preceding some flood basalt eruptions (e.g., Hooper et al., this volume, and references therein). Some other aspects of the mantle system, such as the role of plates in organizing mantle flow and the role of chemical heterogeneities (both as passive tracers and as active influences on buoyancy and melting), have long been discussed and modeled by practitioners. It has also long been appreciated that the mantle system is interconnected, self-organizing, self-regulating, far from equilibrium, and unsteady, and that thermodynamics and the role of the plates are important. I refer readers to my book (Davies, 1999) for a broad summary of the long history of work conducted by many people on these subjects.

The art of modeling the mantle system, as with any complicated system, is to construct models that are instructive, because they include important physics, while not depending strongly on parts that are poorly understood. This construction involves judgment and is therefore legitimately a matter of debate. There has always been active debate about plumes and alternatives to, or elaborations of, them. The useful role of this volume is to continue that traditional debate.

Scott King (his 2 January comment) points out in more detail than I did (30 December) that many models of plumes take account of things Anderson (31 December) claims have been disregarded. Anderson further claims (2 January) that "The possibility that the mantle is chemically stratified is usually dismissed outright, particularly by modelers." This assertion is incorrect and potentially misleading and damaging to forward progress. Possible compositional stratification, either in D″ (Christensen and Hofmann, 1994) or in the lower third of the mantle (Kellogg et al., 1999), has been a major theme of debate and modeling in recent years. Possible present or past compositional differences between the upper and lower mantle have also emerged from recent modeling (e.g., Ogawa, 2003; Xie and Tackley, 2004; Davies, 2006). The rest of Anderson's comment is unquantified speculation.

5 January 2007, Don L. Anderson

There has been a tendency to regard plumes as a distinct, secondary mode of convection . . . such a mode of flow has never been observed in any self-consistent numerical or laboratory experiment. (Larsen and Yuen, 1997, p. 1995)

King's comments are valid, but he and King and Redmond (this volume) are discussing and modeling normal mantle convection, albeit with enforced axisymmetry. Mantle plumes were invented as an alternate, or addition, to this broad-scale convection, which is driven by internal radioactivity, surface cooling, and plate tectonics. Calculations of mantle convection are indeed getting more realistic, but do they confirm the mantle plume hypothesis (i.e., an independent, narrow, plume mode of convection that is responsible for hotspots)?

The question of whether deep mantle upwellings can be independent, narrow, hot, and fast—as required in the mantle plume hypothesis—is at the core of more fundamental questions: Are the locations and dimensions of volcanic chains controlled by lithosphere and mantle heterogeneity, **or** by localized high absolute temperature and rapid upwelling? Do hotspots require a deep source of heat and material?

Larsen and Yuen (1997, p. 1995) addressed this problem:

The enigma of . . . nearly stationary plumes . . . in mantle convection arises in the hotspot hypothesis . . . separation of time scales between

the fast plume and adjacent mantle is necessary and, in fact, was invoked by Morgan in his original concept of plumes. . . . plume studies have usually modeled a plume in isolation from the rest of the mantle. . . . Upwelling plumes always occur as part of the main convecting system (rather than independently). In particular, there is a problem of obtaining hotspot-like plumes, which must satisfy the requirements of being fast as compared to the ambient mantle circulation and fairly thin. . . . Mantle plumes have peak ascending velocities of 20 meters/yr.

What distinguishes mantle plumes, as conventionally defined —and as widely perceived outside of the geodynamics/convection community—from normal convection upwellings is higher temperatures, higher ascent velocities, and much smaller dimensions. Initially, mantle plumes were also thought to be stationary. Mantle plumes differ from alternative mechanisms in being entirely thermal in nature and in having a source deep in the mantle (although this idea is continually being modified in the face of observation, e.g., Courtillot et al., 2003). Starting plumes differ in uplift history from alternative mechanisms. It is these characteristics that stimulated the question "Do mantle plumes exist?" In this debate, no one is challenging the existence of convection, broad upwellings, or small-scale features that can be attributed to extension and fertile blob scales. Kuhn (1970) noted this tendency to talk past one another when a paradigm is challenged.

My comments were specifically addressed only to those studies that isolate the lower TBL from the rest of the system, or that argue for narrow plumes or neglect such effects as feedback: "these [effects] are ignored in all calculations *that yield narrow whole-mantle plumes*. None of the fluid dynamic calculations used . . . *to support their view* take [these] into account . . . [more realistic calculations] do not predict plume-like dimensions and timescales at D″ depths" (my comment of 25 December in this discussion thread; emphasis is new).

I do not suggest that all studies have ignored pressure, layering, or internal heating. I referred to studies that allowed self-organization and did not support the widely held narrow-plume assumption. These studies, however, do not address the small-scale and other characteristics of hotspots and volcanic chains that motivated the plume hypothesis.

Recent fluid dynamic simulations (Zhong, 2006; King and Redmond, this volume) and most of the above comments and references refer to normal mantle convection or superplumes, not to the original mantle plume hypothesis. Broad plumes are not what Sleep, Larsen, Yuen, and Olsen, for example, are modeling. Serious attempts have been made to rescue the original narrow-plume hypothesis (Larsen and Yuen, 1997), but these models violate other constraints. Although many workers have abandoned the small-scale stationary plume idea, many others attribute the difficulty in observing plumes to their very small size.

Although there are exceptions, most models assume whole-mantle convection. The more interesting calculations allow for chemical stratification (e.g., Tackley and Xie, 2002). Quite often, scaling between shear velocity and density is assumed, instead of an appropriate thermodynamic scaling via volume

(Birch, 1952). In seismology, low shear velocity is still usually attributed to high temperature and low density.

There are various mechanisms for causing uplift before and concurrent with volcanism (e.g., delamination). The plume hypothesis is unique in requiring major uplift many millions of years before the eruptions. Models that involve fertile blobs do not require such precursory uplift.

Birch (1952) developed the machinery for self-consistent treatments of mantle dynamics. Kuhn (1970) did argue that scientific theories can be modified, but as a prelude to his discussion of why the concepts of Ptolemy, aether, and phlogiston failed; those concepts became so contrived and amended, and had so many versions, that people lost interest. The existence of a chemically layered mantle with large, dense thermochemical features (piles) at the base is not being disputed; in fact it was predicted. It is the association of these with surface volcanism that might be regarded as unquantified speculation. The shallow mantle and the transition region also have suggestive correlations with tectonics. It is interesting that the plume hypothesis—motivated by the idea of a buoyant hot D''—now requires importation of cold dense downwellings from the surface. The latter may occur, but it does make the plume hypothesis immune to new observations and theory.

6 January 2007, Alexei Ivanov

Davies (his comment of 3 January) writes "Regarding alternatives to plumes, any model that involves a passive upper mantle responding to lithospheric changes would need to account for uplift preceding some flood basalt eruptions (e.g., Hooper et al., this volume, and references therein)." The Siberian Traps were not preceded by uplift (e.g., Ivanov, this volume, and references therein).

Davies refers to uplift preceding some flood basalts but neglects the absence of uplift preceding others. The Siberian Traps and Columbia River flood basalts are 4×10^6 km^3 and 0.2×10^6 km^3 in volume, respectively. Which example would seem more important to the debate regarding plumes and alternatives?

There is a mathematical rule that in a complex system, evidence can always be found in support of a hypothesis. Probably we all take advantage of this.

6 January 2007, Geoffrey F. Davies

Anderson's latest comment (5 January) reiterates his opinion but adds little to the continuing debate. However it does illustrate the current level of debate and disagreement among modelers, which is sufficiently diverse to allow Anderson to choose studies that support his contentions. This diversity is just the current manifestation of a debate that has always been vigorous, despite charges to the contrary.

The debates about plumes and the accumulation of new observations have led many to consider variations on Morgan's initial proposals. For example, in my opinion there was never a good rationale, or need, for plumes to be rigorously fixed, rather than simply slow-moving (see Davies, 2005). However this process of learning and modifying has been portrayed by some as rendering plumes arbitrarily adaptable and therefore untestable and unscientific. There have certainly been many poorly motivated proposals invoking plumes, which I would join in criticizing, but there has been a core of quantitative work that makes quantitative predictions. A summary of some significant predictions and relevant observations has recently been given by Campbell and Davies (2006).

An irony here is that two charges that have been made—lack of consideration of alternatives and arbitrarily malleable hypotheses—are mutually contradictory. They cannot both be true. Science never ties up every last loose end of observation, especially in studies of very complicated subjects, such as Earth. Thus there is always some level of uncertainty to nourish dissenters. Ultimately it is a matter of judgment as to when to consider the issue decided and move on, though all conclusions are conditional and subject to later modification or replacement.

22 January 2007, Edward J. Garnero, Thorne Lay, and Allen McNamara

Debate about the nature of hotspot volcanism intrinsically raises the question of what the large-scale configuration of mantle convection is. As perhaps the foremost problem in global geophysics, it comes as no surprise that strongly held and conflicting perspectives of this issue persist despite extensive recent advances of our understanding of Earth's internal structure and processes. Enthusiasm ebbs and flows for end-member scenarios of whole-mantle or layered-mantle convection, and there is, as yet, no consensus other than agreement that the most likely scenario involves a more complex thermochemical system than either end-member. We think it is fair to state that many deep-Earth geophysicists find the evidence favoring large-scale mixing of the mantle more compelling than evidence for strongly layered convection, but probably the strongest statement we would defend is that no line of evidence yet precludes significant flow between the upper and lower mantles.

Given that perspective, our chapter highlights some of the exciting deep-mantle high-resolution seismic findings in the context of state-of-the-art numerical geodynamical calculations that do allow upper-mantle flow to influence the deep mantle. Simply put, we addressed this question: If plumes originate from the deep mantle, what are their possible geographic systematics, given the recent seismic and geodynamic analyses? The extensive debate spawned by our article raised several important factors, most of which were beyond the article's intended scope, but all of which bear on the fundamental question of what the configuration of the mantle dynamic system is.

Our article highlighted 3-D spherical numerical calculations of McNamara and Zhong (2005), which explore thermochemical dynamics. These calculations assumed an initially dense, chemically distinct layer in the lowermost several hundred kilometers of the mantle, adopted the Boussinesq approximation, incorporated temperature-dependent viscosity, and used a roughly equal ratio of basal to internal heating. The past 119 Ma of plate motions from Lithgow-Bertelloni and Richards (1998) were imposed as a surface flow boundary condition such that upper-mantle downwellings spatially control the deep-mantle flow that interacts with the chemically stratified layer. The hot, dense material of that layer is swept into large piles under upwelling return flow, yielding a configuration of large piles of chemically distinct material that have strong spatial affinity to LLSVPs in the deep mantle observed by seismology.

Our intention was not to predict plumes or deep-mantle plume-hotspot connections, but the computations yield boundary layer instabilities on the edges of the chemical piles that rise as plumes in the 3-D flow (admittedly, the detailed character of the instabilities is not fully resolved and does depend on the Boussinesq approximation, although the physics of compressibility will likely not change the results in general). The simulations show that the dense-pile material influences the plume distribution. This configuration is in general agreement with the empirically observed tendency for hotspot volcanism to overlay lateral margins of deep-mantle, low-velocity provinces (Thorne et al., 2004). Thus, plumes are a consequence of our calculations, not an input design. There is no injection of material at the base of the model, both internal and bottom heating are present, and unlike most earlier models, we explicitly incorporated initial chemical stratification in the system. The present discussion thread should not confuse the readers about what is actually in our article. Even if the descent of slab material is more inhibited or limited than in the calculations, the general flow pattern and implications for where plume instabilities might arise is unlikely to change.

The take-home message is that the presence of dense thermochemical piles in the deep mantle can influence the location of boundary layer upwellings, providing a geometric distribution of boundary layer instabilities that is absent in mantle flow models without piles. Thus, although vertical continuity of flow to the surface is not directly constrained, it is attractive to consider the possible connection between upwellings on pile margins (whether these involve continuous plume conduits or fragmented plumes or blobs) and hotspots at Earth's surface. Passive rifting at ridges may cause separate, relatively shallow upwellings that have no direct connection to lower boundary layer instabilities, so we focus on the possible linkage to hotspots. The hotspot research community may find it valuable to keep an eye on developments in deep-mantle research as the plume debate progresses.

REFERENCES CITED

Anderson, D.L., 1989, Theory of the Earth: Boston, Blackwell Scientific, 366 p. http://caltechbook.library.caltech.edu/14/.

Anderson, D.L., 2007 (this volume), The Eclogite engine: Chemical geodynamics as a Galileo thermometer, *in* Foulger, G.R., and Jurdy, D.M., Plates, plumes, and planetary processes: Boulder, Colorado, Geological Society of America Special Paper 430, doi: 10.1130/2007.2430(03).

Beutel, E., and Anderson, D.L., 2007 (this volume), Ridge-crossing seamount chains: A nonthermal approach, *in* Foulger, G.R., and Jurdy, D.M., Plates, plumes, and planetary processes: Boulder, Colorado, Geological Society of America Special Paper 430, doi: 10.1130/2007.2430(19).

Birch, A.F., 1952, Elasticity and constitution of the Earth's interior: Journal of Geophysical Research, v. 57, p. 227–286.

Campbell, I.H., and Davies, G.F., 2006, Do mantle plumes exist?: Episodes, v. 29, p. 162–168.

Christensen, U.R., and Hofmann, A.W., 1994, Segregation of subducted oceanic crust in the convecting mantle: Journal of Geophysical Research, v. 99, p. 19,867–19,884.

Courtillot, V., Davaille, A., Baesse, J., and Stock, J., 2003, Three distinct types of hotspots in the Earth's mantle: Earth and Planetary Science Letters, v. 205, p. 295–308.

Davaille, A., and Jaupart, C., 1993a, Thermal convection in lava lakes: Geophysical Research Letters, v. 20, p. 1827–1830.

Davaille, A., and Jaupart, C., 1993b, Transient high-Rayleigh-number thermal convection with large viscosity variations: Journal of Fluid Mechanics, v. 253, p. 141–166.

Davaille, A., and Jaupart, C., 1994, The onset of thermal convection in fluids with temperature-dependent viscosity: Application to the oceanic mantle: Journal of Geophysical Research, v. 99, p. 19,853–19,866.

Davies, G.F., 1995, Penetration of plates and plumes though the mantle transition zone: Earth and Planetary Science Letters, v. 133, p. 507–516.

Davies, G.F., 1999, Dynamic Earth: Plates, plumes and mantle convection: Cambridge, Cambridge University Press, 460 p.

Davies, G.F., 2005, A case for mantle plumes: Chinese Science Bulletin, v. 50, p. 1541–1554. http://www.mantleplumes.org/WebDocuments/ChineseSciBull3papers2005.pdf.

Davies, G.F., 2006, Gravitational depletion of the early Earth's upper mantle and the viability of early plate tectonics: Earth and Planetary Science Letters, v. 243, p. 376–382.

Deuss, A., 2007 (this volume), Seismic observations of transition-zone discontinuities beneath hotspot locations, *in* Foulger, G.R., and Jurdy, D.M., Plates, plumes, and planetary processes: Boulder, Colorado, Geological Society of America Special Paper 430, doi: 10.1130/2007.2430(07).

Farnetani, C.G., 1997, Excess temperature of mantle plumes; The role of chemical stratification across D″: Geophysical Research Letters, v. 24, p. 1583–1586.

Farnetani, C.G., and Richards, M.A., 1994, Numerical investigations of the mantle plume initiation model for flood basalt events: Journal of Geophysical Research, v. 99, p. 13,813–13,833.

Farnetani, C.G., and Richards, M.A., 1995, Thermal entrainment and melting in mantle plumes: Earth and Planetary Science Letters, v. 136, p. 251–267.

Farnetani, C.G., and Samuel, H., 2005, Beyond the thermal plume paradigm: Geophysical Research Letters, v. 32, art. no. L07311, doi: 10.1029/2005 GL022360.

Farnetani, C.G., Richards, M.A., and Ghiorso, M.S., 1996, Petrological models of magma evolution and deep crustal structure beneath hotspots and flood basalt provinces: Earth and Planetary Science Letters, v. 143, p. 81–94.

Forte, A.M., and Mitrovica, J.X., 2001, Deep-mantle high-viscosity flow and thermochemical structure inferred from seismic and geodynamic data: Nature, v. 410, p. 1049–1056, doi: 10.1038/35074000.

Garnero, E.J., Lay, T., and McNamara, A., 2007 (this volume), Implications of lower-mantle structural heterogeneity for existence and nature of whole-mantle plumes, *in* Foulger, G.R., and Jurdy, D.M., Plates, plumes, and planetary processes: Boulder, Colorado, Geological Society of America Special Paper 430, doi: 10.1130/2007.2430(05).

Goes, S., Cammarano, F., and Hansen, U., 2004, Synthetic seismic signature of thermal mantle plumes: Earth and Planetary Science Letters, v. 218, p. 403–419.

Griffiths, R.W., and Campbell, I.H., 1990, Stirring and structure in mantle plumes, Earth and Planetary Science Letters, v. 99, p. 66–78.

Grigné, C., Labrosse, S., and Tackley, P.J., 2005, Convective heat transfer as a function of wavelength: Implications for the cooling of the Earth: Journal of Geophysical Research, v. 110, art. no. B03409, doi: 10.1029/2004 JB003376.

Gurnis, M., Mitrovica, J.X., Ritsema, J., van Heijst, H.J., 2000, Constraining mantle density structure using geological evidence of surface uplift rates; The case of the African superplume: Geochemistry, Geophysics, Geosystems, v. 1, art. no. 1999G000035.

Hansen, U., and Yuen, D.A., 2000, Extended-Boussinesq thermal-chemical convection with moving heat sources and variable viscosity: Earth and Planetary Science Letters, v. 176, p. 401–411.

Hooper, P.R., Camp, V., Reidel, S., and Ross, M., 2007 (this volume), The origin of the Columbia River flood basalt province: Plume versus nonplume models, *in* Foulger, G.R., and Jurdy, D.M., Plates, plumes, and planetary processes: Boulder, Colorado, Geological Society of America Special Paper 430, doi: 10.1130/2007.2430(30).

Ishii, M., and Tromp, J., 1999, Normal-model and free-air gravity constraints on lateral variations in velocity and density of the earth's mantle: Science, v. 285, p. 1231–1236.

Ita, J.J., and King, S.D., 1994, The sensitivity of convection with an endothermic phase change to the form of governing equations, initial conditions, aspect ratio, and equation of state: Journal of Geophysical Research, v. 99, p. 15,919–15,938.

Ivanov, A., 2007 (this volume), Evaluation of different models for the origin of the Siberian Traps, *in* Foulger, G.R., and Jurdy, D.M., Plates, plumes, and planetary processes: Boulder, Colorado, Geological Society of America Special Paper 430, doi: 10.1130/2007.2430(31).

Kaula, W.M., 1983, Minimal upper mantle temperature variations consistent with observed heat flow and plate velocities: Journal of Geophysical Research, v. 88, p. 10,323–10,332.

Kellogg, L.H., and King, S.D., 1997, The effect of temperature dependent viscosity on the structure of new plumes in the mantle: Results of a finite element model in a spherical, axisymmetric shell: Earth and Planetary Science Letters, v. 148, p. 13–26.

Kellogg, L.H., Hager, B.H., and van der Hilst, R.D., 1999, Compositional stratification in the deep mantle: Science, v. 283, p. 1881–1884.

Kiefer, W.H., and Hager, B.H., 1992, Geoid anomalies and dynamic topography from convection in cylindrical geometry: Applications to mantle plumes on Earth and Venus: Geophysical Journal International, v. 108, p. 198–214.

King, S.D., 1997, Geoid and topographic swells over temperature-dependent thermal plumes in spherical-axisymmetric geometry: Geophysical Research Letters, v. 24, p. 3093–3096.

King, S.D., and Masters, G., 1992, An inversion for radial viscosity structure using seismic tomography: Geophysical Research Letters, v. 19, p. 1551–1554.

King, S.D., and Redmond, H.L., 2007 (this volume), The structure of thermal plumes and geophysical observations, *in* Foulger, G.R., and Jurdy, D.M., Plates, plumes, and planetary processes: Boulder, Colorado, Geological Society of America Special Paper 430, doi: 10.1130/2007.2430(06).

Kuhn, T.S., 1970, The structure of scientific revolutions (2nd edition): Chicago, University of Chicago Press, 206 p.

Larsen, T.B., and Yuen, D.A., 1997, Ultrafast upwelling bursting through the upper mantle: Earth and Planetary Science Letters, v. 146, p. 393–400.

Le Bars, M., and Davaille, A., 2004, Whole layer convection in a heterogeneous planetary mantle: Journal of Geophysical Research, v.109, art. no. B03403.

Leitch, A.M., and Davies, G.F., 2001, Mantle plumes and flood basalts: Enhanced melting from plume ascent and an eclogite component: Journal of Geophysical Research, v. 106, p. 2047–2059.

Leitch, A.M., Davies, G.F., and Wells, M., 1998, A plume head melting under a rifted margin: Earth and Planetary Science Letters, v. 161, p. 161–177.

Lenardic, A., and Kaula, W.M., 1994, Tectonic plates, D″ thermal structure, and the nature of mantle plumes: Journal of Geophysical Research, v. 99, art. no. 94JB00466, p. 15,697–15,708.

Lin, S.-C., and van Keken, P.E., 2006a, Dynamics of thermochemical plumes: 1. Plume formation and entrainment of a dense layer: Geochemistry, Geophysics, Geosystems, v. 7, art. no. Q02006.

Lin, S.-C., and van Keken, P.E., 2006b, Dynamics of thermochemical plumes: 2. Complexity of plume structures and its implications for mapping mantle plumes: Geochemistry, Geophysics, Geosystems, v. 7, Q02006.

Lithgow-Bertelloni, C., and Richards, M.A., 1998, The dynamics of Cenozoic and Mesozoic plate motions: Reviews of Geophysics, v. 36, p. 27–78.

Lowman, J., King, S., and Gable, C., 2001, The influence of tectonic plates on mantle convection patterns, temperature and heat flow: Geophysical Journal International, v. 146, p. 619, doi:10.1046/j.1365-246X.2001 .00471.

McNamara, A.K., and Zhong, S., 2005, Thermochemical piles under Africa and the Pacific: Nature, v. 437, p. 1136–1139.

Mitrovica, J.X., 1996, Haskell (1935) revisited: Journal of Geophysical Research, v. 101, p. 555–569.

Nakagawa, T., and Tackley, P.J., 2006, Three-dimensional structures and dynamics in the deep mantle: Effects of post-perovskite phase change and deep mantle layering: Geophysical Research Letters, v. 33, art. no. L12S11, doi: 10.1029/2006GL025719.

Nataf, H-C., 1991, Mantle convection, plates, and hotspots; Tectonophysics, v. 187, p. 361–371.

Ogawa, M., 2003, Chemical stratification in a two-dimensional convecting mantle with magmatism and moving plates: Journal of Geophysical Research, v. 108, p. 2561.

Oreskes, N., Shrader-Frechette, K., and Belitz, K., 1994, Verification, validation, and confirmation of numerical models in the Earth sciences: Science, v. 263, p. 641–646.

Parmentier, E.M., Sotin, C., and Travis, B.J., 1994, Turbulent 3-D thermal convection in an infinite Prandtl number, volumetrically heated fluid— Implications for mantle dynamics: Geophysical Journal International, v. 116, p. 241–251.

Phillips, B.R., and Bunge, H.-P., 2005, Heterogeneity and time dependence in 3D spherical mantle convection models with continental drift: Earth and Planetary Science Letters, v. 233, p. 121–135.

Presnall, D.C., and Gudfinnsson, G.H., 2005, Carbonate-rich melts in the oceanic low-velocity zone and deep mantle, *in* Foulger, G.R., et al., eds., Plates, plumes, and paradigms: Boulder, Colorado, Geological Society of America Special Paper 388, p. 207–216, doi: 10.1130/2005.2388(13).

Presnall, D.C., and Gudfinnsson, G., 2007, Global Na8-Fe8 Systematics of MORBs: Implications for mantle heterogeneity, temperature, and plumes, abstract EGU2007-A-00436, European Geophysical Union General Assembly, 15–20 April, Vienna.

Samuel, H., and Farnetani, C.G., 2003, Thermochemical convection and helium concentrations in mantle plumes: Earth and Planetary Science Letters, v. 207, p. 39–56.

Schubert, G., Masters, G., Olson, P., and Tackley, P., 2004, Superplumes or plume clusters?: Physics of Earth and Planetary Interiors, v. 146, p. 147–162.

Sleep, N.H., 2007 (this volume), Origins of the plume hypothesis and some of its implications, *in* Foulger, G.R., and Jurdy, D.M., Plates, plumes, and planetary processes: Boulder, Colorado, Geological Society of America Special Paper 430, doi: 10.1130/2007.2430(02).

Solomatov, V.S., 1995, Scaling of temperature- and stress-dependent viscosity convection: Physics of Fluids, v. 7, p. 266–274.

Solomatov, V.S., and Moresi, L.-N., 2000, Scaling of time-dependent stagnant lid convection: Application to small-scale convection on Earth and other terrestrial planets: Journal of Geophysical Research, v. 105, p. 21,795–21,817.

Tackley, P.J., 1998, Three-dimensional simulations of mantle convection with a thermal-chemical boundary layer D", *in* Gurnis, M., et al., eds., The core-mantle boundary region: Washington, D.C., American Geophysical Union Geodynamics Monograph 28, p. 231–253.

Tackley, P.J., 2002, Strong heterogeneity caused by deep mantle layering: Geochemistry, Geophysics, Geosystems, v. 3, doi: 10.1029/2001GC000167.

Tackley, P.J., and Xie, S., 2002, The thermo-chemical structure and evolution of Earth's mantle: constraints and numerical models: Philosophical Transactions of the Royal Society of London A, v. 360, p. 2593–2609.

Thorne, M., Garnero, E.J., and Grand, S., 2004, Geographic correlation between hot spots and deep mantle lateral shear- wave velocity gradients: Physics of the Earth and Planetary Interiors, v. 146, p. 47–63.

Tozer, D.C., 1973, Thermal plumes in the Earth's mantle: Nature, v. 244, p. 398–400.

Van Keken, P.E., 1997, On entrainment in starting mantle plumes: Earth and Planetary Science Letters, v. 148, p. 1–12.

Xie, S., and Tackley, P.J., 2004, Evolution of U-Pb and Sm-Nd systems in numerical models of mantle convection and plate tectonics: Journal of Geophysical Research, v. 109, doi: 10.1029/2004JB003176.

Zhong, S., 2006, Constraints on thermochemical convection of the mantle from plume heat flux, plume excess temperature, and upper mantle temperature: Journal of Geophysical Research, v. 111, art. no. B04409.

The Geological Society of America
Special Paper 430
2007

The structure of thermal plumes and geophysical observations

Scott D. King*
Hannah L. Redmond
Department of Earth and Atmospheric Sciences, 550 Stadium Mall Drive,
Purdue University, West Lafayette, Indiana 47907, USA

ABSTRACT

We present a sequence of numerical experiments studying thermal plumes that arise from the basal thermal boundary layer in a spherical axisymmetric convecting fluid, reporting the geoid, topographic (swell), and heatflow anomalies. We consider the influence of spherical geometry, temperature and pressure dependence of rheology, internal heat generation, pressure-dependent coefficient of thermal expansion, the Clapeyron slope of the endothermic phase transformation, and compositional layering at the base of the mantle on the structure and dynamics of thermal plumes and resulting geophysical anomalies. This fluid system simplifies a number of complexities within the Earth (e.g., it is not compressible, does not include plate motions, enforces axisymmetric symmetry, and has a simple equation of state); however, this system is computationally tractable and provides an opportunity to evaluate each effect in a systematic fashion. With the assumptions listed above, we are formally unable to reproduce the observed hotspot swell, heatflow, and geoid anomalies from the resulting thermal plumes. The calculations that most closely approach the observed geoid, topographic, and heatflow anomalies contain all the effects mentioned above. One particularly interesting aspect of these calculations is that when we include a depth-dependent coefficient of thermal expansion, the plume conduits in the lower 500 km of the mantle are more than 200 km in diameter wider than with an otherwise identical, uniform coefficient of thermal expansion calculation. We speculate that the lack of mobile plates in our calculations, which would both advect upwelling material away from the plume conduit and change the global cooling mechanism, is a major shortcoming and that it may be possible to reconcile thermal plume calculations with the geoid, topographic, and heatflow observations if we included mobile plates.

Keywords: hotspot, geoid, topographic swell, plume

INTRODUCTION

Thermal plumes originating at the core-mantle boundary have become the standard explanation for hotspot volcanism (e.g., Morgan, 1971; Burke and Wilson, 1976; Davies, 1988; Griffiths and Campbell, 1990, 1991; Sleep, 1990; Davies and Richards, 1992; Hill et al., 1992; Schubert et al., 2001). The plume hypothesis has benefited from elegant fluid dynamical theory and tank models (e.g., Whitehead and Luther, 1975; Olson and Singer, 1985; Griffiths, 1986; Griffiths and Campbell, 1990,

*E-mail: sking@purdue.edu.

King, S.D., and Redmond, H.L., 2007, The structure of thermal plumes and geophysical observations, *in* Foulger, G.R., and Jurdy, D.M., eds., Plates, plumes, and planetary processes: Geological Society of America Special Paper 430, p. 103–120, doi: 10.1130/2007.2430(06). For permission to copy, contact editing @geosociety.org. ©2007 The Geological Society of America. All rights reserved.

1991; Davaille, 1999; Davaille and Vatterville, 2005). However, the primary geophysical observations of heatflow, geoid, and topographic swell anomalies are difficult to reproduce in laboratory experiments, as are the effects of pressure-dependent material properties. Calculations of thermal plumes have focused on temperature-dependent rheology, phase transformations, and equation of state (e.g., Kiefer and Hager, 1992; Sleep, 1992, 1994, 1996; Kellogg and King, 1993, 1997; Farnetani and Richards, 1994, 1995; Davies, 1995; Schubert et al., 1995; Farnetani et al., 1996; Ratcliff et al., 1996; Farnetani, 1997; King, 1997; van Keken, 1997; Goes et al., 2004; Farnetani and Samuel, 2005; Lin and van Keken, 2006a,b; Zhong, 2006). Given improvements in our understanding of the properties of the mantle in the past decade, it is useful to reassess how phase transformations, rheology, equation of state, and composition influence the structure, vigor, and other observable properties of plumes in the mantle. The hypothesis tested here is that by including these effects, we can produce geoid, topographic, and heatflow anomalies that are consistent with the observed hotspot anomalies. While the influence of these parameters has been studied in general convection calculations, less attention has been paid to mantle plumes and the resulting geophysical anomalies. Here we build systematically, varying one parameter at a time, and focus specifically on comparing the anomalies produced by thermal plumes with observations at hotspots. This work builds on previous work by a variety of investigators, including Kiefer and Hager (1992), Davies (1995), Schubert et al. (1995), King (1997), and Goes et al. (2004). We do not discuss the effect of chemical layering in the mid-mantle or a fully self-consistent equation of state (c.f. Ita and King, 1994), both of which should be investigated in future studies.

The first question is why instabilities from the core-mantle boundary should be expected. This problem can be addressed in a number of different ways, and although the following remarks are not meant to be exhaustive, they provide the reader with a flavor of the arguments and further references. First, thermal boundary layer instabilities, usually referred to as "plumes" in the fluid dynamical literature, are a natural consequence of the cooling of the Earth (c.f. Schubert, 1979; Davies, 1980). It is generally accepted that Earth was strongly heated during formation (c.f. Safronov, 1978; Wetherill, 1990). This initially hot mantle cools, largely by plate-scale mantle flow and cooling of the surface boundary layer (c.f. Davies, 1988). As the mantle cools, the temperature difference between the mantle and core grows, and a thermal boundary layer between the mantle and core forms (e.g., Stacey and Loper, 1983). As this thermal boundary layer grows, it reaches a point where it becomes gravitationally unstable, and instabilities form at this thermal boundary layer (c.f. Stacey and Loper, 1983; Olson et al., 1987; Sleep et al., 1988; further details can be found in Stacey and Loper, 1983, and Schubert et al., 2001). These instabilities take the form of cylindrical plume-like upwellings, where the morphology depends on the rheology (e.g., Kellogg and King, 1997). All of these studies rely on fairly simple fluid approximations of the

mantle (i.e., incompressible, chemically homogeneous, and constant coefficient of thermal expansion). Estimates of the temperature contrast across the core-mantle boundary can be made by extrapolating the melting temperature of iron from the inner core boundary through the outer core and extrapolating the geotherm through the mantle to the core-mantle boundary (e.g., Anderson, 2002), leading to an estimate of a 1000–1800 °C temperature change between the mantle and the core. This temperature difference is presumably concentrated at the base of the mantle (e.g., Anderson, 2002; Buffett, 2003). This is a large temperature difference (with a fairly large uncertainty), and it is not only a problem for the thermal evolution of the core and Earth's magnetic field (Buffett, 2003), but one for the mantle plume hypothesis, because the geochemistry and petrology of hotspot lavas require a fairly small increase in potential temperature of at most several hundred degrees at the melting source (e.g., Farnetani, 1997) and a small heatflow anomaly (e.g., DeLaughter et al., 2005). Recent calculations show that the lower mantle may be subadiabatic with a superadiabatic region above the coremantle boundary (e.g., Bunge et al., 2001; Monnereau and Yuen, 2002; Zhong, 2006); this possibility reduces the problem to some extent. This problem will be examined further in the discussion.

There is an alternative to the thermal plume hypothesis. Nakagawa and Tackley (2004) show that a thermochemical model of the lower mantle could develop broad superplumes consistent with the long wavelength seismic structure, and the thermal evolution develops isolated reservoirs that would be consistent with the geochemical observations. Monnereau and Yuen (2002) show similar broad, sluggish features in compressible spherical convection calculations with depth-dependent rheology and a depth-dependent coefficient of thermal expansion. Schubert et al. (2004) suggest that the seismic data are consistent with a spatial cluster of smaller plumes.

The second line of reasoning for the inevitability of plumes comes from scaling analysis. The Rayleigh number of a fluid governs the convective vigor of the fluid (c.f. Chandrasekhar, 1961). One can think of the Rayleigh number as the ratio of the two forces that are balanced in viscous flow: buoyancy and viscous drag. The Rayleigh number takes the form:

$$\mathrm{Ra} = \frac{\rho g \alpha T D^3}{\kappa \eta}, \qquad (1)$$

where ρ is the average mantle density, g is the acceleration due to gravity, α is the coefficient of thermal expansion, ΔT is the temperature drop across the fluid, D is the depth of the fluid, κ is the thermal diffusivity, and η is the viscosity. Typical ranges of the values of these parameters (and references) are given in Table 1. The Rayleigh number does not reflect the changes of the properties of the fluid with pressure or temperature; hence as one studies more complex fluid systems, the Rayleigh number becomes a limited diagnostic and one must be careful to define how the variable properties are used in the definition of the Rayleigh number. For example, in thermochemical systems, a second "chemical" Rayleigh number can be introduced (e.g., Farnetani and Richards, 1995), and the dynamics of the fluid can be mapped out with these two parameters. A limitation of Rayleigh number scaling is that it cannot predict the interaction or feedback between parameters. For example, consider a fluid with a substantial source of heat below that is cooling only by conduction. Conduction is a less efficient means of heat transport than convection, and if the rate of heat being supplied from below exceeds the rate at which heat can be conducted through the fluid, the fluid temperature will rise. If, like the mantle, the fluid has a rheology that is a strong function of temperature, as the temperature rises, the viscosity decreases, and the fluid will eventually become convectively unstable. It is increasingly apparent that there are nonlinear interactions between the depth-dependent and temperature-dependent properties that are not well understood at this time (c.f. Dubuffett and Yuen, 2000; Matyska and Yuen, 2005).

Third, both stability analysis and numerical models of mantle convection show that 3-D quasi-cylindrical upwellings (i.e., plumes) are the dominant planform of convection in spherical geometry (c.f. Busse, 1975; Zebib et al., 1980, 1985; Busse and Riahi, 1982; Bercovici et al., 1989; Tackley et al., 1993, 1994; Ratcliff et al., 1995; Zhang and Yuen, 1995; Bunge et al., 1996, 1997; Farnetani and Samuel, 2005; Zhong, 2006). In all calculations presented, plume-like upwellings are observed, although most of these models use a Boussinesq approximation and no chemical layering.

It is worth noting that there is a potentially viable alternative model in which the properties of the lower mantle lead to thermal instabilities at the base of the mantle that are long-wavelength, large-scale superplumes (c.f. Wen and Anderson, 1997a,b; Davaille, 1999; Nakagawa and Tackley, 2004; Anderson, 2005a,b) that may form over a timescale of hundreds of millions of years (or longer). Within the framework of this hypothesis, hotspots would be caused by local compositional anomalies (c.f. Anderson, 2005a), instabilities within the upper mantle from deep superplumes (Courtillot et al., 2003), or lithospheric effects (King and Ritsema, 2000). Davaille (1999) and Courtillot et al. (2003) suggest that this mode of convection could operate in

conjunction with a deep, narrow plume mode. We do not investigate this hypothesis in this study.

Geophysical observations related to hotspots include the geoid, topographic, and heatflow anomalies in addition to the hotspot track itself (e.g., Crough and Jurdy, 1980; Crough, 1983; Davies, 1988; Sleep, 1990; Goes et al., 2004; Anderson, 2005a; Anderson and Schramm, 2005; DeLaughter et al., 2005; Sandwell et al., 2005; Zhong, 2006). There are also important geochemical observations from hotspot lavas that provide insight into composition, temperature, and melting history of hotspot lavas, and these have been investigated in several numerical studies (e.g., Ribe and Christensen, 1994; Farnetani and Richards, 1995; Farnetani et al., 1996; Samuel and Farnetani, 2003). We will not consider melting here, although clearly one of the most critical aspects of hotspots is the associated volcanic activity. Calculations with an endothermic phase transformation and internal heating show that the majority of the lower mantle may be subadiabatic with a superadiabatic temperature gradient near the core-mantle boundary (e.g., Bunge et al., 2001; Monnereau and Yuen, 2002; Zhong, 2006). As much as 300–500 °C of the temperature difference across the core may be explained by subadiabaticity of the lower mantle.

Seismologists have imaged velocity anomalies beneath hotspot regions (Nataf and Van Decar, 1993; Wolfe et al., 1997; Wolfe, 1998; Bijwaard and Spakman, 1999; Foulger et al., 2000; Humphreys et al., 2000; Allen et al., 2002; Ritsema and Allen, 2003; Montelli et al., 2004). There is a tradeoff between the magnitude of a seismic anomaly and the width of the anomaly in seismic imaging (e.g., Allen and Tromp, 2005), and there is limited resolution for depths greater than 400 km in studies that specifically attempt to image anomalies below hotspots, because of the restricted geometry of the stations and receivers (Wolfe et al., 1997; Wolfe, 1998; Foulger et al., 2000; Humphreys et al., 2000; Allen et al., 2002).

The specific observations that we use to constrain the plume calculations are the heatflow, geoid, and topographic anomalies and, to a lesser extent, the width of the seismic anomaly beneath hotspots, which we take from seismic tomography (e.g., Wolfe, 1998; Foulger et al., 2000; Humphreys et al., 2000; Allen et al.,

TABLE 1. PARAMETERS USED IN THIS STUDY

Parameter	Symbol	Value	Reference
Depth (km)	D	2891	
Gravitational acceleration (m/s²)	g	9.8	
Average mantle viscosity (Pa s)	η	10^{20}–10^{22}	Mitrovica (1996)
Density (kg/m³)	ρ	3700–5500	Dziewonski and Anderson (1981)
Thermal conductivity (W/m/K)	k	3.0	
Thermal diffusivity (m²/s)	$\kappa = \dfrac{k}{\rho c_p}$	0.9–1.2×10^{-6}	
Temperature change (nonadiabatic) (K)	ΔT	2000	Schubert et al. (2001)
Coefficient of thermal expansion (1/K)	α	0.5–4.0×10^{-5}	Chopelas and Boehler (1992); Chopelas (2000)

Notes: In most cases, these parameters are functions of temperature, pressure, and composition. The values shown are global average estimates. Functional forms are discussed in the text.

2002; Allen and Tromp, 2005). As will become apparent, exact values of the geophysical observations are not critical for this exercise, because many of our plume calculations create anomalies that are an order of magnitude larger than the hotspot observations. Given the uncertainties currently associated with many of the parameters, we do not see much value in attempting to vary the model parameters until we achieve specific values of the hotspot observations. The goal here is to evaluate and illustrate the importance of the effects.

Heatflow anomalies at hotspot swells are small. Davies (1988) estimates that plume flow makes up less than 10% of the total heatflow of the surface of the Earth, and DeLaughter et al. (2005) suggest that it is significantly less than this value. Sleep (1990) estimates that the total contribution to the global surface heatflow of the Earth from ~40 plumes is 4 mW/m^2. Heatflow measurements are sparse, even for some of the best-studied hotspots; however, heatflow observations from Hawaii suggest that the peak anomaly is less than 10–20 mW/m^2 (von Hertzen et al., 1989) and probably is significantly less (DeLaughter et al., 2005). The possibility that heat is redistributed by hydrothermal circulation in the crust (Harris et al., 2000) cannot be ruled out and further complicates the interpretation of the heatflow observations. We consider a peak heatflow anomaly less than 10 mW/m^2 as consistent with the observations, although we note that for many hotspots, this may be a generous estimate.

The geoid anomaly associated with hotspot swells is on the order of a few meters to at most 10 m (e.g., Sleep, 1990). Although there are significantly better geoid models available now with the GRACE data, the results presented by Sleep do not change significantly. Therefore, we consider peak geoid anomalies less than 10 m as consistent with the observations. With both the geoid and topographic anomalies, it is important to remember that the plume calculations here do not include the volcanic construction or underplating of magma beneath the lithosphere, which could be a significant part of the shorter-wavelength anomaly. When looking at the geoid and topographic swell data, we are looking at longer-wavelength (i.e., 1,000–2,000 km) anomalies that most likely reflect the dynamic contribution to the geoid and topography rather than from the crustal or lithospheric structure (c.f. Richards et al., 1988; Sleep, 1990).

Topographic swell anomalies estimated for twenty-six hotspots range from 500 to 2500 m, with most swells in the 500- to 1200-m range and the largest swell over Ethiopia (Crough, 1983). Hotspots occur on ridges, young and old oceanic lithosphere, and beneath continents. There is no clear correlation between the magnitude of the hotspot swell and the lithosphere type or thickness, which again suggests that the swell is reflecting deeper mantle dynamics and not crustal or lithospheric structure. We consider peak topographic swells of less than 2000 m as consistent with the hotspot observations, although we note that there are many hotspot swells that are significantly smaller.

Although there is some ambiguity in the imaged seismic structures in the upper mantle beneath hotspots, and there is a tradeoff between the width and the magnitude of the anomalies,

a 200-km-wide anomaly in the upper mantle appears to be consistent with the seismic observations (Goes et al., 2004). Seismic studies image a low-velocity anomaly in the upper mantle that may be as narrow as 100 km (e.g., Foulger et al., 2000; Allen et al., 2002), but our calculations have plume material ponding beneath the lithosphere, because there is no plate flow advecting this material away from the plume tail in our calculations. Thus we are cautious about interpreting the detailed upper-mantle structure in these calculations. We also point out that Goes et al. (2004) show that when converting temperature structures from thermal plume calculations to seismic velocity anomalies, the resulting seismic velocity anomalies are significantly broader than the thermal anomalies.

Although there are a number of calculations that have investigated the effect of rheology (e.g., Kellogg and King, 1997; van Keken, 1997; Goes et al., 2004; Schott and Yuen, 2004) and phase transformations (Liu, 1994; Bina and Liu, 1995; Davies, 1995; Schubert et al., 1995; Goes et al., 2004) on mantle plumes, many of these investigations focused on one parameter in isolation and few report geoid, topographic, and heatflow anomalies. In this work, we examine the structure of thermal plumes in a set of conditions consistent with our current knowledge of the mantle, looking at the geoid, topographic, and heatflow anomalies predicted from the calculations. Given the range of values spanned by many of the parameters relevant to mantle convection (e.g., Table 1), we avoid reporting only those cases where the resulting plumes are consistent with the hotspot observations. Instead, we consider a range of values for each parameter that spans the uncertainties associated with the parameter, noting values where the resulting plume anomalies satisfy hotspot observations as well as those that do not. It makes sense to present this investigation in a series of steps, adding one additional parameter at a time.

METHOD

The approach that we take is to use 2-D spherical axisymmetric calculations using the finite element code SCAM (Kellogg and King, 1997). The axis of symmetry occurs about the pole at $\theta = 0$, where θ is the co-latitude. This geometry enforces a symmetry and stability on the resulting thermal plumes. The temperature and velocity are functions of the radius, r, and the co-latitude, θ, and are independent of the longitude, φ. For problems with variable viscosity, the equations for momentum and mass conservation are given by:

$$-\nabla p + \nabla \cdot (\eta \dot{\varepsilon}) = RaT\hat{r} \qquad (2)$$

and

$$\nabla \cdot u = 0, \qquad (3)$$

where p is the dynamic pressure; η is the viscosity, which can be a function of temperature, pressure, composition, and/or strain-rate; Ra is the Rayleigh number (defined in equation 1); T is the temperature, u is the velocity, and

$$\dot{\varepsilon} = \frac{1}{2}\left(u_{i,j} + u_{j,i}\right) \tag{4}$$

is the deviatoric deformation rate.

The conservation of energy equation in a spherical axisymmetric geometry is given by:

$$\frac{\partial T}{\partial t} + u_r \frac{\partial T}{\partial r} + \frac{u_\theta}{r}\frac{\partial T}{\partial \theta} = \frac{1}{r^2}\frac{\partial}{\partial r}\left(\kappa r^2 \frac{\partial T}{\partial r}\right) + \frac{1}{r^2 \sin\theta}\frac{\partial}{\partial \theta}\left(\kappa \sin\theta \frac{\partial T}{\partial \theta}\right) + H, \tag{5}$$

where T is the temperature, t is time, κ is the thermal diffusivity, u_r and u_θ are the radial and azimuthal components of velocity, and H is the rate of internal heating. Equation 5 is solved using streamline-upwind Petrov-Galerkin shape functions (Hughes, 1987) with bilinear elements. This method is second-order accurate in space.

We use an Arrhenius form for the temperature-dependent part of the viscosity law:

$$\eta(\hat{T}) = \eta_o exp\left(\frac{E^*}{\hat{T}+T_o}\right) exp\left(-\frac{E^*}{1+T_o}\right), \tag{6}$$

where η_o is the normalized pre-exponential viscosity; $\eta(\hat{T})$ is the effective viscosity; $\hat{T}$ is the dimensionless temperature; and E^* is the activation energy divided by $R\Delta T$, where ΔT is the temperature scaling factor and T_o is the temperature offset. The second exponential scales $\eta(\hat{T})$ so that $\eta(1.0) = \eta_o$. In these calculations, we use an activation energy of 350 kJ/mole unless otherwise noted (c.f. Karato and Wu, 1993).

The method for calculating topographic and geoid anomalies is described in Kiefer and Hager (1992) and King and Hager (1994). The geoid anomaly, δN, is scaled by:

$$\delta N = \frac{G\rho_m \alpha \Delta T D^2}{g} \delta N', \tag{7}$$

where G is Newton's gravitational constant, ρ_m is the average mantle density, g is the acceleration due to gravity, α is the coefficient of thermal expansion, ΔT is the temperature drop across

the fluid, D is the depth of the fluid, and the values of the constants are given in Table 1. The topographic anomaly, δh, is given by:

$$\delta h = \left(\frac{\rho_m}{\rho_S}\right)\alpha\Delta T D \delta h', \tag{8}$$

where ρ_s is the density of the surface layer.

We use a 2-D rather than a 3-D geometry because of the size constraints of the computational grid needed to ensure that our solutions are well resolved. The resolution needed before the peak geoid, topographic, and heatflow anomalies show signs of converging is a factor of three greater than the resolution used in most 3-D calculations (Table 2). With temperature-dependent rheology, the issue of resolution is even more significant than it is with a constant viscosity fluid, because the viscosity changes by several orders of magnitude over less than 100 km within the plume geometries in this study. Two limitations of the 2-D spherical axisymmetric geometry are that we cannot consider a moving plate and the plume is stationary. Davaille and Vatterville (2005) show the effects of time-dependence on plumes in a series of tank experiments, and we caution that steady-state plumes may produce larger anomalies than do transient plumes. In the calculations presented here, the upwelling material carried by the plumes spreads out symmetrically beneath the top thermal boundary layer. This scenario undoubtedly overestimates the topographic and geoid swell for the Pacific hotspots, where some of the upwelling material should be advected in the direction of plate motion (c.f. Ribe and Christensen, 1994). The calculations here then are more directly comparable to hotspots over slow moving or stationary plates; however, we do not see any reason that the general trends we find would be invalidated in three dimensions. The moving plate will advect some of the plume material in the direction of plate motion (e.g., Ribe and Christensen, 1994), and the peak anomalies should be smaller than those calculated here. There is no correlation between the velocity of the plate and the hotspot topographic anomaly (see Table 3 *in* Sleep, 1990).

We also present our results at steady-state conditions. Although it is unlikely that the mantle is in a state of thermal equilibrium, or steady state, presenting our results at steady state

TABLE 2. MAXIMUM GEOID, TOPOGRAPHIC, AND HEATFLOW ANOMALIES FROM VARIOUS GRIDS FOR CONVECTION CALCULATIONS USING A TEMPERATURE-DEPENDENT SPHERICAL SHELL

Grid name	Grid size $N_\theta \times N_r$	Heatflow anomaly (mW/m²)	Topographic anomaly (m)	Geoid anomaly (m)
A	50 × 64	21.1	5272.5	151.3
B	75 × 96	18.1	4941.9	145.7
C	100 × 128	17.2	4835.7	144.3
D	125 × 160	16.5	4733.3	142.5
E	150 × 192	16.0	4700.5	142.2
F	175 × 224	15.7	4649.3	141.3

gives us a well-defined condition to compare results from one set of calculations to another. If, for example, we chose a time early in the thermal evolution of one calculation and later in the thermal evolution of another calculation, we risk mapping changes in thermal evolution into our comparison of physical properties. We avoid this problem by using Picard iteration (e.g., Cuvelier et al., 1986), which solves the equations of motion for the steady-state solution (i.e., $\partial T/\partial t = 0$ in Equation 5). When calculations are time dependent, we use an explicit time-stepping scheme and allow the solution to reach a quasi-steady or statistically steady state.

Table 2 presents the calculated maximum geoid, topographic, and heatflow anomalies from a typical 2-D axisymmetric calculation as a function of computational grid size. The calculation has temperature-dependent rheology (e.g., equation 6), with a Rayleigh number (based on the interior viscosity) of 10^6 and uniform thermodynamic properties. Although the grid resolution does not affect the overall planform of the temperature field, it has a significant impact on the values of the heatflow, geoid, and dynamic topography. Grid A (50×64 elements) is typical of the resolution used in many 3-D spherical calculations. We present all of the results in this study using grid E (150×192 elements).

RESULTS

We begin by presenting a comparison of constant-viscosity convection calculations in 2-D Cartesian and spherical axisymmetric geometries (Fig. 1). In both cases, the Rayleigh number is 10^6; there is no internal heating; the fluid is heated from below and cooled from above; and the top, bottom, and sides are uniform free-slip boundaries. Although the assumptions used in this calculation are far from describing the properties of the Earth, they provide a mathematically well-defined starting point and correspond to the assumptions used in the original scaling arguments for plume theory. The most striking difference between the calculations is the internal temperature of the fluid and the temperature contrast between the upwellings and downwellings. In the Cartesian case, the volume average temperature is 0.5 with symmetric upwellings and downwellings, whereas in the spherical case, the volume average temperature is 0.14 and the upwelling is much stronger than the downwelling. In spherical geometry, most of the heat from the base of the fluid is advected up the plume tail. The difference between these calculations is largely the result of the difference in surface area between the top and bottom of the spherical shell. In the Cartesian case, the surface area of the top and bottom boundaries is the same. In the spherical case, the surface area of the bottom is approximately one-quarter of the surface area of the top ($[R_{CMB}/R_{Surf}]^2$). To be in equilibrium, the total heat into the bottom shell is equal to the total heat out the top. Because of the difference in surface areas, the heatflux (heat per unit area) into the bottom of the shell is much larger than the heatflux out of the top of the shell. As a result, because the thickness of the boundary layer is proportional to the heatflux, the bottom boundary layer is thicker than the top boundary layer and the upwelling carries more heat than the downwelling.

Although the effects of spherical geometry are well understood (c.f. Jarvis et al., 1995; King, 1997), it is worth beginning the discussion of the plume calculations by pointing out that the spherical nature of planets favors the development of a large thermal boundary layer at the bottom of the mantle because of the difference in surface areas. It is also noteworthy that the constant-viscosity convection calculation in spherical axisymmetric geometry (Fig. 1) does not resemble any cartoon or model of convection in a planetary body. Furthermore, note that it has been well established (e.g., Nataf and Richter, 1982) that temperature-dependent rheology will increase the average interior temperature with asymmetric top and bottom boundary layers. This rise will have the effect of creating an asymmetry in

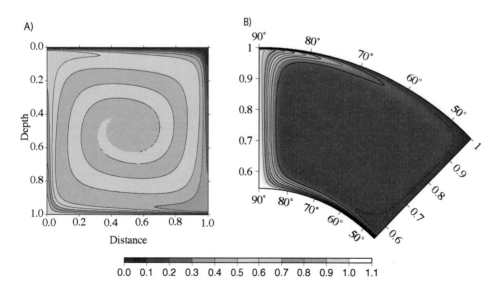

Figure 1. Temperature fields from steady-state constant-viscosity (Rayleigh number 10^6) convection calculations in (A) a 1×1 Cartesian box and (B) a spherical axisymmetric shell with an inner radius of 0.55 and 45° co-latitude.

the boundary layers in Cartesian geometry, increasing the interior temperature and decreasing the effect of the bottom thermal boundary layer in the spherical shell geometry. Thus, with temperature-dependent viscosity or a viscous lid to simulate the effect of temperature dependence, the difference between Cartesian and spherical shell upwellings is less dramatic than shown here. For the constant viscosity, spherical axisymmetric shell with a Rayleigh number of 10^6, the heatflow anomaly over the plume is more than 196 mW/m^2, the peak geoid anomaly is 280 m, and the peak topographic swell anomaly is 11,000 m.

The next step is to include temperature-dependent rheology, the subject of prior investigations, particularly with respect to starting plumes (e.g., Olson et al., 1987, 1993; Sleep et al., 1988; Kellogg and King, 1997; van Keken, 1997); however, these studies did not examine the geoid, topographic, and heatflow anomalies. We use equation 6 for the temperature dependence of the rheology, and now use the viscosity based on the volume-averaged temperature as the viscosity in the definition of the Rayleigh number. We also include a high-viscosity shell with a thickness of 90 km and a viscosity that is 10^3 times the background viscosity to create a strong lithosphere. Kiefer and Hager (1992) show that the geoid and topography are insensitive to the viscosity of the lithosphere, as long as it is greater than 10^3 times the background viscosity. In this case, we see a dramatic difference in the steady-state temperature field for the temperature-dependent plume calculation (Fig. 2) compared with constant viscosity calculation. The structure of the plume is largely unaffected but now a significant thermal boundary layer forms at the surface and the average internal temperature (0.43) is significantly greater than the constant viscosity case (0.14). The heatflow anomaly over this plume is just over 16 mW/m^2, the peak geoid anomaly is ~142 m, and the peak topographic swell anomaly exceeds 4700 m. These values are still considerably larger than observations at hotspots. As the Rayleigh number increases (Table 3, rows 1–5), the geoid and topographic anomalies decrease, as has previously been shown by Kiefer and Hager (1992); however, the heatflow anomaly increases, so by increasing the Rayleigh number, we cannot bring all of the anomalies associated with the plumes in these calculations into the range of the geophysical observations at hotspots.

We recognize that the mantle of the Earth is not strictly heated from below but is also heated from within by the decay of radioactive elements. Although the distribution of radiogenic elements may not be uniform throughout the mantle (c.f. Anderson, 2005a,b), we restrict this investigation to the simplest case of uniform heat generation throughout the mantle. The next series of calculations (Fig. 3) uses temperature-dependent viscosity and a volume-averaged Rayleigh number of 10^6 in spherical shell that is identical to the calculation from Figure 2, except that a volumetric heat source term (i.e., H in equation 5) is included. We define the Rayleigh number based on the δT across the shell and vary the rate of internal heating (Table 3, rows 6–10). Convection with strong internal heating does not have steady-state solutions, and so here we run the calculations with

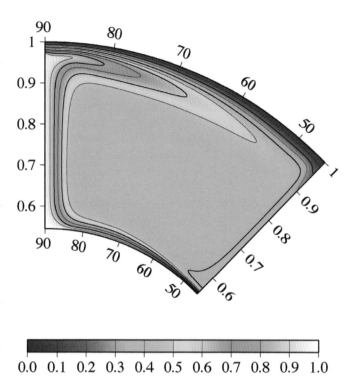

Figure 2. Temperature field from a steady-state temperature-dependent viscosity calculation, with a volume-averaged Rayleigh number 10^6 in a spherical shell with an inner radius of 0.55 and 45° co-latitude. The activation energy used in the viscosity law is 350 kJ/m^3. The viscosity used in the definition of the Rayleigh number is calculated from the volume-averaged temperature.

an explicit time-stepping scheme until the shell achieves a quasi-steady state. With the highest rate of internal heating (1.5 mW/m^3), the heatflow anomaly over the plume is ~13 mW/m^2, the maximum geoid anomaly is 16.4 m, and the maximum topographic swell anomaly is 3745 m. This rate of internal heating is still nearly a factor of ten smaller than that estimated for the present-day Earth (Schubert et al., 2001); however, solutions with higher rates of internal heating are strongly time dependent, making quantitative comparisons difficult. We plot the temperature excess in the plume as a function of depth (i.e., temperature along the axis divided by the adiabatic temperature) in Figure 3D. Two effects are clear: first, the temperature excess is reduced as the amount of internal heating increases (comparing the different curves); and second, the temperature excess as a function of depth decreases throughout the lower part of the domain. This decrease is a well-documented effect of internal heating (e.g., Bunge et al., 2001; Goes et al., 2004; Zhong, 2006). If we extrapolate the trend of the results in Table 3 to a realistic rate of internal heat generation for the Earth, the results of the plume calculations approach the range of the hotspot observations. It is important to remember that these calculations have a stagnant lid and do not have moving plates. Both of these conditions probably contribute to the large thermal anomaly below the lithosphere. A stagnant lid is less efficient at removing

TABLE 3. RESULTS FROM AXISYMMETRIC PLUME CALCULATIONS

Rayleigh number*	Internal heating rate (W/kg)[†]	Lower-mantle viscosity ratio	Coefficient of thermal expansion (functional form)[§]	Clapeyron slope (MPa/K)	Average temperature	Heatflow anomaly (mW/m²)	Topographic anomaly (m)	Geoid anomaly (m)
10^6	0.0	1	Constant	0.0	0.43	16.0	4700.5	142.1
3×10^6	0.0	1	Constant	0.0	0.47	20.7	3770.1	91.9
10^7	0.0	1	Constant	0.0	0.51	28.2	3040.7	55.7
3×10^7	0.0	1	Constant	0.0	0.55	31.8	2287.9	19.4
10^8	0.0	1	Constant	0.0	0.60	43.1	2187.8	11.4
10^6	0.2×10^{-12}	1	Constant	0.0	0.46	16.1	4438.5	123.4
10^6	0.4×10^{-12}	1	Constant	0.0	0.50	15.9	4185.4	104.9
10^6	0.6×10^{-12}	1	Constant	0.0	0.54	15.3	3939.8	85.6
10^6	0.8×10^{-12}	1	Constant	0.0	0.66	15.0	4039.0	46.8
10^6	1.0×10^{-12}	1	Constant	0.0	0.66	12.9	3745.1	16.4
10^6	0.8×10^{12}	1	Constant	0.0	0.66	15.0	4039.0	46.8
10^6	0.8×10^{12}	10	Constant	0.0	0.55	20.3	4154.9	36.2
10^6	0.8×10^{-12}	30	Constant	0.0	0.45	27.9	3380.4	19.4
10^6	0.8×10^{-12}	100	Constant	0.0	0.40	33.3	2766.0	6.0
10^6	0.8×10^{-12}	30	Constant	0.0	0.45	27.9	3380.4	19.4
10^6	0.8×10^{-12}	30	$(r/r_0)^3$	0.0	0.42	27.7	3161.3	48.9
10^6	0.8×10^{-12}	30	$(r/r_0)^{3**}$	0.0	0.51	30.1	2633.9	14.3
10^6	0.8×10^{-12}	30	$(r/r_0)^3$	0.0	0.42	27.7	3161.3	48.9
10^6	0.8×10^{-12}	30	$(r/r_0)^3$	−1.0	0.42	27.6	3173.3	44.4
10^6	0.8×10^{-12}	30	$(r/r_0)^3$	−2.0	0.42	27.4	3187.3	41.7
10^6	0.8×10^{-12}	30	$(r/r_0)^3$	−2.8	0.43	27.1	3168.4	28.1
10^6	0.8×10^{-12}	30	$(r/r_0)^3$	−2.8	0.33	15.0	2358.1	25.6
10^6	2.4×10^{-12}	30	$(r/r_0)^3$	−2.8	0.54	12.9	2130.3	13.4
10^6	4.8×10^{-12}	30	$(r/r_0)^3$	−2.8	0.55	12.9	2141.9	12.2

*Volume averaged.
[†]Current estimates of mantle heat generation rate are on the order of 7.0×10^{-12} W/kg (Schubert et al., 2001).
[§]Volume-averaged value of coefficient of thermal expansion is used in Rayleigh number.
**Surface value of coefficient of thermal expansion is used in Rayleigh number.

heat from the interior (e.g., Gurnis, 1989); hence, we do not expect to be able to use an earth-like rate of internal heat generation with stagnant lid calculations. Plate-like surface boundary conditions will cool the interior more efficiently (e.g., Gurnis, 1989) and could lead to the formation of larger instabilities at the core-mantle boundary. It is interesting to note that the width of the thermal anomaly of the plume tail does not change as the rate of internal heating increases; however, the magnitude of the thermal anomaly decreases relative to the increasing average internal temperature of the shell.

We next consider the viscosity of the lower mantle. Theory shows that the geoid and dynamic topography are sensitive to the radial mantle viscosity profile and observations from subduction environments (e.g., Hager, 1984; Chen and King, 1998), plumes (e.g., Richards et al., 1988; Kiefer and Hager, 1992), global geoid inversions using seismic tomography models (e.g., King and Masters, 1992), and glacial isostatic adjustment stud-ies (e.g., Mitrovica, 1996) are consistent with a factor of 10–100 viscosity contrast between the upper and lower mantles. A result of the temperature-dependent rheology used in these calculations is that the viscosity of the lower mantle is actually lower than that of the upper mantle, because of the radial temperature profile. Hence, a modest radial viscosity increase with pressure would be needed to maintain a uniform viscosity with depth. Therefore, our next set of calculations (Table 3, rows 11–14) examine the effect of the lower-mantle viscosity on the geoid, topography, and heatflow, using the same temperature-dependent rheology and a volume-averaged Rayleigh number of 10^6, where the viscosity in the Rayleigh number is the volume-averaged viscosity, and the value of the internal heat generation term is 0.8×10^{-12} W/kg (e.g., Table 3, row 9). To this set of parameters, we now add a lower-mantle viscosity increase of 10, 30, and 100, a range of lower mantle viscosities consistent with subduction studies (e.g., Hager, 1984), global inversions of the flow

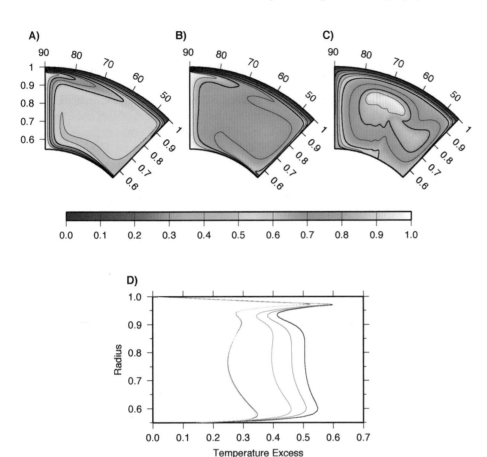

Figure 3. Temperature fields from steady-state temperature-dependent viscosity calculations as in Figure 2 with (A) 22% internal heating (0.4×10^{-12} W/kg internal heating rate), (B) 33% internal heating (0.8×10^{-12} W/kg internal heating rate), and (C) 100% internal heating (no bottom heating). (D) Plot of temperature excess along the center of the plume (i.e., $T_{\theta = 0} - T_{ave}$) as a function of radius for the calculations with 0.2×10^{-12} W/kg internal heating rate (black), 0.4×10^{-12} W/kg internal heating rate (red), 0.6×10^{-12} W/kg internal heating rate (green), 0.8×10^{-12} W/kg internal heating rate (blue), and 1.0×10^{-12} W/kg internal heating rate (yellow).

driven by seismic velocity (e.g., King and Masters, 1992), and glacial isostatic adjustment (e.g., Mitrovica, 1996). Increasing the lower-mantle viscosity decreases the peak geoid and topographic anomalies, as has been shown before (c.f. Richards et al., 1988; Kiefer and Hager, 1992); however, it also decreases the average internal temperature of the shell and increases the heatflow anomaly associated with the plume. We could lower the heatflow anomaly by increasing the rate of internal heating, which in this calculation is only ~10% of the estimated rate of internal heat generation for the mantle. This modification would increase the average internal temperature of the shell, reducing the temperature drop across the bottom thermal boundary layer and the heatflow anomaly associated with the plume (Table 3, rows 11–14). However, to keep our experiments simple and to clearly present the effect of each parameter, we did not attempt to change the viscosity and the rate of internal heating simultaneously. Because these calculations have a stagnant lid and lack plate tectonics, it is impossible to generate completely realistic thermal conditions from these calculations.

An important aspect of the equation of state of mantle minerals is that the coefficient of thermal expansion decreases with depth (c.f. Chopelas and Boehler, 1992; Chopelas, 2000). Thus, a temperature anomaly of 200 °C in the lower mantle will not have the same buoyancy force as the same temperature anomaly in the upper mantle. We next consider this effect by taking the calculation with a factor of 30 viscosity contrast from Table 3 (row 13) and adding a coefficient of thermal expansion of the form:

$$\alpha(r) = \alpha_0 (r/r_0)^3 \qquad (9)$$

where $\alpha(r)$ is the coefficient of thermal expansion as a function of depth, α_0 is the surface coefficient of thermal expansion, r_0 is the surface radius, and r is the radius at depth. This functional form reduces the coefficient of thermal expansion by a factor of 6.25 from the surface to the core-mantle boundary. When changing the coefficient of thermal expansion, it is important to carefully consider what value of the coefficient of thermal expansion to use in the definition of the Rayleigh number. We note that we do not consider a fully self-consistent equation of state (Ita and King, 1994). Although more consistent, such calculations are intensive and rely on parameters that are as yet poorly constrained. In our previous work (Ita and King, 1994), we found that the primary effect of the fully self-consistent equation of state came from the variable coefficient of thermal expansion, which is not surprising, because the coefficient of thermal expansion is contained in the buoyancy force term and buoyancy is the driving force of convection. We present results both for those cases in which we use the surface value of the coefficient of thermal expansion in the Rayleigh number as well as for cases using a volume-averaged coefficient of thermal expansion (Table 3, rows 15–17). These calculations become time dependent, so an explicit time-stepping algorithm is used; however, the planform of the plume and anomalies remain stable in spite of the time dependence, which takes the form of small instabilities in the boundary layers away from the central upwelling. Using the surface or the volume-averaged coefficient of thermal expansion in the Rayleigh number gives two different values of the Rayleigh number for the same calculation. Conversely, if we hold the Rayleigh number fixed, as we have been with these calculations, the two solutions with different definitions of coefficient of thermal expansion in the Rayleigh number will follow the trend of changing the Rayleigh number. The Rayleigh number based on the surface value of the coefficient of thermal expansion is larger than that based on the volume-averaged coefficient of thermal expansion, and the results in Table 3, rows 15–17, do indeed follow the trends for increasing the Rayleigh number that were presented in rows 1–5.

To better illustrate the difference between the solutions, we subtract the reference field (i.e., the constant coefficient of thermal expansion calculation with temperature-dependent viscos-

ity, a rate of internal heating of 0.8×10^{-12} W/kg, and an increase in viscosity of a factor of 30 at 660 km; i.e., the case in Table 3 with a viscosity ratio of 30, row 13) from the temperature fields from the variable coefficient of thermal expansion calculations. Figure 4A is the reference calculation temperature field. Figure 4B is the case with a variable coefficient of thermal expansion and a volume-averaged coefficient of thermal expansion used in the definition of the Rayleigh number, and Figure 4C is the temperature field from the variable coefficient of thermal expansion with the surface value used in the definition of the Rayleigh number. While the contour plots of the fields look similar, differences can be more readily seen by looking at a temperature profile at a constant radius taken 300 km above the core-mantle boundary (Fig. 4D) or by looking at the difference between the reference temperature field (Fig. 4A) and the variable coefficient of thermal expansion temperature fields (Fig. 4E and F, volume-averaged and surface coefficient of thermal expansion, respectively). A decrease in the coefficient of thermal expansion with increasing pressure, coupled with the increase in lower-mantle viscosity has a profound influence on the thermal structure of the plume. First, using the surface value of the co-

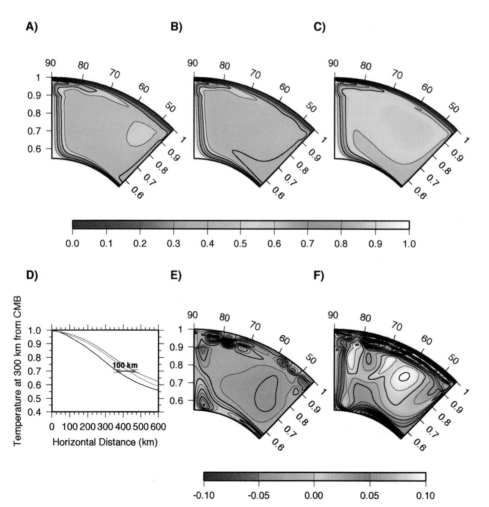

Figure 4. Temperature fields from steady-state temperature-dependent viscosity calculations with 0.8×10^{-12} W/kg internal heating rate, and a thirty-fold increase in viscosity in the lower mantle with (A) constant coefficient of thermal expansion with depth (the reference case); (B) coefficient of thermal expansion decreasing as r^3, Rayleigh number based on the surface value of the coefficient of thermal expansion; and (C) coefficient of thermal expansion decreasing as r^3, volume-averaged Rayleigh number 10^6. (D) Plot of temperature at 300 km above the core-mantle boundary (CMB) as a function of distance from the plume axis for the three cases above: black—constant coefficient of thermal expansion; red—volume-averaged, variable coefficient of thermal expansion; blue—variable coefficient of thermal expansion with surface value used for scaling. (E) Difference between temperature fields from the reference model in panels A and B. (F) Difference between temperature fields from the reference model in panels A and C.

efficient of thermal expansion (Fig. 4C and F) leads to an interior that is hotter than the reference case, although the structure of the plume remains almost unchanged. The difference is dominated by the increase in the Rayleigh number that results from using the high surface value of the coefficient of thermal expansion, and it is difficult to see any changes in the plume. In the calculation for which the volume-averaged coefficient of thermal expansion is used (Fig. 4B and E), the effect on the plume is still pronounced; the bottom half of the plume conduit (and the basal thermal boundary layer) is hotter (by ~200 °C maximum) and broader (by ~200 km in diameter) than the constant coefficient of thermal expansion calculation, whereas the top half of the plume conduit is cooler and more narrow than in the calculation with a constant coefficient of thermal expansion. A larger temperature contrast in the lower mantle is required to achieve a similar magnitude buoyancy force because of the reduction in the coefficient of thermal expansion with depth, and we observe that the plume is hotter in the lower mantle and cooler in the upper mantle, which compensates, at least partially, for the change in the coefficient of thermal expansion with depth. The temperatures in the plume tail for the variable coefficient of thermal expansion cases are significantly lower than in the reference case. Once again, the geoid, topographic, and heatflow anomalies are significantly larger (in most cases by a factor of two) than the values for the upper limit of the hotspot observations (Table 3, rows 15–17).

Although the effect of phase transformations has been included in a number of previous investigations (e.g., Farnetani and Richards, 1994; Christensen, 1995; Davies, 1995; Schubert et al., 1995), and because the planform of the plumes has changed significantly with the inclusion of a variable coefficient of thermal expansion, for completeness we consider here the effect of an endothermic phase transformation. The effect of the exothermic transformation of olivine to wadslayite (which occurs at ~410 km depth) will enhance the vertical motion of the plume in the upper mantle (c.f. Christensen, 1995), and we do not include it here. We consider Clapeyron slopes of –1.0, –2.0, and –2.8 MPa/K for the phase transformation from ringwoodite to perovskite + ferropericlase (which occurs at ~660 km). We implement the phase transformation using the phase function formulation described by Christensen (1995). This formulation assumes that we can treat the 660-km discontinuity (or more broadly, the bottom of the transition zone) as a single univariant phase change. Deuss et al. (2006) show that *PP* and *SS* precursors show different behavior in the same location, which cannot be explained by temperature variations in the region and requires additional phase transformations and/or compositional differences. Several studies have considered the effect of temperature on the endothermic ringwoodite to spinel + ferropericlase boundary, showing that the effect in slabs and plumes could be different (Liu, 1994; Bina and Liu, 1995; Davies, 1995). In spite of potential problems, the formulation we use is a common formulation in numerical calculations of mantle convection, and recognizing the potential limitations, we forge ahead. The cur-

rent best estimate of the Clapeyron slope for this phase transformation is –2.8 MPa/K (Hirose, 2002). Because the garnet component of the mantle undergoes a broad, nearly continuous set of phase transformations through the transition zone, the effect of the garnet component of the mantle is approximated by reducing the Clapeyron slope of the olivine system by a mass-weighted average of the two components. Thus, the most appropriate value of the Clapeyron slope is probably between –1.0 and –2.0 MPa/K. The heatflow, geoid, and topographic anomalies are presented as a function of the Clapeyron slope in Table 3, rows 18–21. The influence of the Clapeyron slope is most pronounced on the geoid, because the thermal structure of the plumes is most affected by the phase change in the transition zone and the geoid includes a term that is the integral of the density anomalies with depth. As with the previous calculations, the magnitudes of the swell observations from these plume calculations are larger (generally by a factor of two) than the upper limit of the hotspot observations.

We finally examine a stably-stratified D″ layer. There have been speculations that the D″ layer is both a chemical layer (e.g., Christensen and Hofmann, 1994; Farnetani, 1997) and a stably stratified layer caused by the post-perovskite phase transformation (c.f. Matyska and Yuen, 2005). To model a stably stratified D″ layer, we take a simple approach and use a high-viscosity layer for the bottom 90 km of the shell. Although we do not intend to suggest that the lowermost mantle has such a high viscosity, the high-viscosity layer acts like a stable chemical or phase-change boundary near the core-mantle boundary. The inclusion of a stably stratified layer has a profound effect on both the thermal structure of the plume and the average internal temperature of the fluid. This effect is best illustrated by comparing calculations with and without the stably stratified layer that have variable coefficient of thermal expansion, an increase in viscosity of a factor of 30 at 660 km, a rate of internal heating of 0.8×10^{-12} W/kg, and a Clapeyron slope of –2.8 MPa/K (e.g., Table 3, rows 18–21). The difference in internal temperature of 0.42 for the no-layer calculation (Fig. 5A) versus 0.32 for the calculation with a layer is immediately apparent from the color shading (Fig. 5A); however, the difference between the magnitudes of the thermal anomalies of the plumes is also striking and potentially more important. This difference is reflected in the reduction in the heat flow anomaly from 27 to 15 mW/m^2 for the no-layer and layer calculations, respectively. The geoid decreases from 28 to 25 m while the topography decreases from 3168 to 2358.1 m for the no-layer and layer calculations, respectively. This result is consistent with the findings of Farnetani (1997) and Goes et al. (2004). Note that the effect of a thin layer at the base of the mantle is different than the effect of a thicker layer in the mid-lower mantle or transition zone (e.g., Wen and Anderson, 1997a,b; Kellogg et al., 1999; Nakagawa and Tackley, 2004). This thin layer could have a significantly different effect on the planform of lower thermal boundary layer instabilities, leading to broad superplumes that may not be directly related to hotspots.

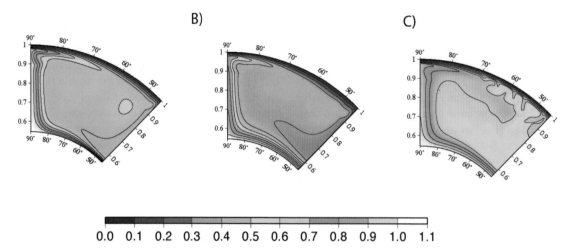

Figure 5. Temperature fields from steady-state temperature-dependent viscosity calculations with 0.8×10^{-12} W/kg rate internal heating, and a thirty-fold increase in viscosity in the lower mantle, depth-dependent coefficient of thermal expansion, and endothermic phase transformation with (A) no basal layer, (B) ~90 km stably stratified layer, and (C) ~90 km stably stratified layer and increased rate of internal heating of 4.8×10^{-12} W/kg.

As this variation represents the final parameter in our modeling exercise, we wish to explore whether we can broaden the scope of these plume calculations by further varying the rate of internal heat generation to bring the geophysical anomalies more in line with the hotspot observations. It is important to recall that the value of internal heat generation we have been using is significantly smaller than realistic estimates for the Earth, so by increasing this value, we are still well within the estimated parameter range of Earth-like internal heating. We also consider calculations with internal heat generation rates of 0.8×10^{-12} W/kg and 2.4×10^{-12} W/kg, which represents ~30% of the estimated mantle heat production rate, and 4.8×10^{-12} W/kg, which represents ~70% of the estimated mantle heat production rate. The resulting peak geoid, topographic, and heatflow anomalies for all three deep-layer calculations are shown in Table 3, rows 22–24. The temperature field from the calculation with an internal heat generation rate of 4.8×10^{-12} W/kg is shown in Figure 5C. Although there is clear evidence in Figure 5C that small-scale convection in the upper mantle is inhibited by the phase transformation away from the axis where the upwelling plume forms, the central upwelling plume still penetrates the phase boundary. The resulting geoid, topographic, and heatflow anomalies illustrate that it is very challenging (at best) to produce thermal plumes that are consistent with the geophysical observations at hotspots, a result consistent with the recent work of Goes et al. (2004).

DISCUSSION

Although the geometry of the axisymmetric calculations shown here ensures that the resulting thermal structures will take the form of plumes, both theory (e.g., Busse, 1975; Busse

and Riahi, 1982) and 3-D spherical calculations (e.g., Bercovici et al., 1989; Tackley et al., 1993, 1994; Bunge et al., 1996, 1997; Zhong, 2006) show that plumes are the upwelling planform of spherical shell convection. It is possible that the lower mantle has deep instabilities that more closely resemble broad superplumes (e.g., Nakagawa and Tackley, 2004), but the observed broad structures could also be consistent with clusters of plumes (e.g., Schubert et al., 2004), or there could be a mixed mode of superplumes (or domes) and more conventional plumes (e.g., Davaille, 1999; Courtillot et al., 2003). The question addressed here is: Under what conditions are the thermal plumes in these calculations consistent with the observations at hotspot swells? The straightforward answer is that the geoid, topographic, and heatflow anomalies at hotspots are challenging to match with thermal plume calculations. Although we have avoided "tuning" these models by continuing to vary the parameters to get a calculation that is consistent with the observations, the point is that if hotspots are the result of thermal plumes, these plumes must be weaker than the plumes in our calculations. It is important to point out that computational studies are showing that the structure of instabilities from the core-mantle boundary in more complex fluid systems are diverse and do not always fit the traditional plume model (e.g., Farnetani and Samuel, 2005; Lin and van Keken, 2006b).

Perhaps the most interesting result from this sequence of calculations is the change in the plume conduit shape with the inclusion of pressure-dependent coefficient of thermal expansion, as shown in Figure 4. The plume conduit becomes broader at the base of the mantle and smaller in the upper mantle compared to that found in the calculation using a constant coefficient of thermal expansion. The deep mantle has less thermal buoyancy for a given thermal anomaly; hence, instabilities at the base of the mantle must be larger than they would be with a constant coefficient of thermal expansion to achieve the same buoyancy force. The resulting calculations are time dependent and the structure of the plume, particularly the plume conduit, is more variable with depth. The broadening of the plume conduit with depth has interesting implications for seismic imaging. Although this effect is modest, in the overall picture, it has several important implications for mantle plumes. One of the difficulties encountered when applying mantle plumes to hotspot volcanism is that the estimated temperature contrast at the core-mantle boundary leads to plume upwellings that are too hot when compared to the excess temperatures estimated from hotspot-related volcanism. This effect is reduced to some extent by the subadiabaticity of the lower mantle (Bunge et al., 2001; Zhong, 2006). The variable coefficient of thermal expansion reduces the temperature in the uppermost part of the plume, compared with estimates of plume excess temperature that come from uniform coefficient of thermal expansion calculations. Second, previous work has shown that the impact of the endothermic phase transformation becomes more pronounced when coupled with a decrease in the coefficient of thermal expansion (Ita and King, 1994). Previous investigations of the interactions of plumes and an endothermic phase transformation have found little or no evidence that the phase change from ringwoodite to perovskite + ferropericlase will inhibit a rising plume (Schubert et al., 1995), and our results confirm that this is still the case, even when we include the pressure effect of the coefficient of thermal expansion.

When we include a stable layer at the base of the mantle, the geoid, topographic, and heatflow anomalies associated with the plumes approach the upper limit of the hotspot observations. There have been speculations that the D″ layer is both a chemical layer (e.g., Christensen and Hofmann, 1994; Farnetani, 1997) and a stably stratified layer caused by a phase change (c.f. Matyska and Yuen, 2005). Although our approach does not include all of the relevant physics of these hypotheses, which have been studied in more detail by others, it illustrates the important effect that a stably stratified layer can have on the thermal structure of plumes. This stably stratified layer at the base of the mantle leads to a thick, stable conductive thermal layer with a smaller, active thermal boundary layer above it. The upwelling plume that forms has a smaller temperature contrast compared with the surrounding mantle than is found in an identical calculation without a stably stratified layer at the base of the mantle. It is possible that if we were able to perform calculations with even larger rates of internal heating, we could achieve a similar plume thermal structure without a stable layer at the base of the mantle, so we are cautious that this phenomenon may be the result of a limitation of our calculations.

The next result is rather obvious: there must be a thermal boundary layer at the core-mantle boundary for deep-mantle plumes to form there. For example, in the calculation with 100% internal heating (Fig. 3C), no upwelling plume forms. The best estimate of the temperature mismatch between extrapolating from the inner core boundary through the core to the base of the

mantle and extrapolating the geotherm through the mantle gives a 1000–1800 °C temperature drop across the core-mantle boundary (Anderson, 2002). This range depends on a number of assumptions; however, given the large value, it seems challenging to completely eliminate it. In fact, our results suggest that this difference is currently too large to be contained in a thermal boundary layer without somehow stabilizing the layer (either chemically or by a phase transformation) and isolating part of the temperature contrast, as in our calculations with a stable D″ layer (Fig. 5C). This result is also consistent with the findings of Farnetani (1997), Goes et al. (2004), and Lin and van Keken (2006a). The large temperature contrast between the core and mantle forces the consideration of some kind of thermal boundary layer instability, and the alternative to the kinds of plume structures shown here would be broad superplume structures (e.g., Nakagawa and Tackley, 2004; Anderson, 2005a,b). We do not investigate this mode of convection here.

Additionally, without a stiff, viscous lid, the resulting geoid, topographic, and heatflow anomalies are impossible to reconcile with observations (e.g., Fig. 1). Although this is a well-known result (e.g., Kiefer and Hager, 1992), the calculations here reinforce the importance of a viscous lid. Yet because hotspots occur on all types of lithosphere—from cratons to ridges—and we do not see a significant correlation between the hotspot anomalies (e.g., geoid, topography, heatflow) and the type of lithosphere, we do not consider the strength or thickness of the lithosphere to be a primary controlling factor in hotspot swell properties.

We have not reported the width of the plumes from these calculations, in large part because to compare these results with seismic images, we need to convert the temperature anomalies to seismic velocity anomalies. Our results are comparable with those of Goes et al. (2004), and we estimate that our calculations will produce seismic anomalies in the range of 500–900 km in diameter in the lower mantle. These diameters are wider than those in the seismic images presented by Montelli et al. (2004) and also the plume tails from typical cartoons and calculations of mantle plumes. Increasing the Rayleigh number of the calculations would decrease the width of the boundary layers (c.f. Schubert et al., 2001); however, increasing the Rayleigh number will also increase the heatflow anomaly (c.f. Table 3, rows 1–5).

One parameter that we have not considered is variable thermal conductivity and the possibility of radiative thermal conductivity (Hofmeister, 1999). Convection calculations show that variable thermal conductivity leads to more sluggish convection in the lower mantle and to broader, stronger plumes (e.g., Dubuffet and Yuen, 2000; van den Berg et al., 2001). The nonlinear interaction between variable viscosity and thermal conductivity with phase transformations is being investigated by several researchers, and it is not possible to properly add variable thermal conductivity and produce meaningful results here because of the nonlinear feedback. At present, it appears that variable thermal conductivity will produce larger anomalies over plumes, making it more difficult to match the hotspot observations with plume calculations; however this problem requires further investigation.

CONCLUSIONS

There are several scenarios under which deep-mantle plumes would not form: (1) a core heatflux that is so small that no significant boundary layer forms at the base of the mantle or (2) a compositional change, phase change, and/or combination of pressure-dependent effects that give rise to broad, deep superplumes (e.g., Davaille, 1999; Kellogg et al., 1999; Monnereau and Yuen, 2002; Courtillot et al., 2003; Nakagawa and Tackley, 2004). We do not evaluate superplume-like models here, although superplume models require either coupling with the upper mantle to explain hotspots (e.g., Courtillot et al., 2003) or a mechanism independent of superplumes (e.g., Anderson, 2005a,b). We find some cases of thermal plumes that produce unacceptably large geophysical anomalies, including cases with constant mantle viscosity or temperature-dependent rheology and no internal heat generation, which are easy to rule out as geophysically unreasonable. However, most of the calculations presented here have geophysical anomalies that are a factor of two to four larger than hotspot observations, and they only approach the upper limit of the range of observations at hotspots when we include depth-dependent mantle viscosity, a significant rate of internal heat generation, a depth-dependent coefficient of thermal expansion, and a stable D″ layer. Strictly speaking, we have not been able to produce hotspot geophysical anomalies with thermal plumes. We speculate that the inability to include mobile plates in the spherical axisymmetric domain is a significant limitation because (1) upwelling material ponds beneath the boundary layer and is not advected away from the plume by the plate-scale flow (c.f. Ribe and Christensen, 1994), and (2) the stagnant-lid mode of convection does not allow for realistic cooling of the lithosphere (c.f. Gurnis, 1989). Recent 3-D spherical results with a compressible mantle are able to match plume heatflux and plume excess temperature (Zhong, 2006). The 2-D axisymmetric geometry also enforces a symmetry on the fluid system that could influence our results, and the steady-state assumption, necessary to compare one result with another without confusing time-dependent effects with material property-dependent effects, could be important. Several recent computational studies have shown that with more Earth-like parameters, instabilities from the core-mantle boundary are time dependent, and some of the features of the plume model (e.g., large heads and long-lived tails) are not always present in every instability (Farnetani and Samuel, 2005; Lin and van Keken, 2006a,b). We also point out that the calculations we present are incompressible. In a compressible fluid with a variable coefficient of thermal expansion, the temperature contrast at the base of the fluid may be significantly larger than the temperature contrast as the plume nears the surface (e.g., Monnereau and Yuen, 2002), which could reduce the thermal anomaly associated with plumes;

however, these calculations also often lead to partially layered flows with broad superplumes.

Perhaps the most interesting result is the change in the plume conduit (i.e., tail) shape with the inclusion of a pressure-dependent coefficient of thermal expansion. The plume tail is broader at the base of the mantle, and the overall thermal anomaly in the plume tail at any given depth is smaller than in the calculations using a constant coefficient of thermal expansion. We speculate that the width of the thermal structures (~950 km in diameter 300–500 km above the core-mantle boundary) in these calculations should be resolvable by seismic imaging, especially bearing in mind that the seismic anomaly will be broader than the thermal anomaly (c.f. Goes et al., 2004). The shape of the plumes in these calculations differs from the lollipop-like shape seen in many tank experiments.

Finally, an increase in mantle viscosity with depth, decrease in the coefficient of thermal expansion with depth, a significant fraction of internal heating, and a stably stratified D″ are all critical to producing plumes with heatflow, geoid, and topographic anomalies that approach the range of the hotspot observations. Plume calculations using spatially uniform material properties are impossible to reconcile with the hotspot observations. This observation is important, because many of the simple scaling arguments used in plume theory are based on such assumptions of uniform properties. Given the uncertainties in the important parameters, it is not possible to say that these calculations either prove or rule out mantle plumes as the source of hotspot anomalies. What is clear from these calculations is that the complexities of the mantle rheology and equation of state reduce what would otherwise be enormous anomalies associated with mantle plumes. The results here are best interpreted in terms of understanding how individual parameters affect observable geophysical anomalies rather than taking any specific calculation as a "best model" for mantle plumes.

ACKNOWLEDGMENTS

We thank Don L. Anderson, Hans-Peter Bunge, Gillian Foulger, Donna Jurdy, and Walter Kiefer for helpful and thoughtful reviews. The ideas presented here benefited from discussions at the Great Plume Debate (Fort William, Scotland, September 2005). The Generic Mapping Tools (GMT) software package (Wessel and Smith, 1991) was used in this work and is gratefully acknowledged. Support from National Science Foundation grant EAR-0207222 and National Aeronautics and Space Administration grant NAG5-13371 is acknowledged.

REFERENCES CITED

Allen, R.M., and Tromp, J., 2005, Resolution of regional seismic models: Squeezing the Iceland anomaly: Geophysical Journal International, v. 161, p. 373–386, doi: 10.1111/j.1365-246X.2005.02600.x.

Allen, R.M., Nolet, G., Morgan, W.J., Vogfjord, K., Bergsson, B.H., Erlendsson, P., Foulger, G.R., Jakobsdottir, S., Julian, B.R., Pritchard, M., Ragnarsson,

S., and Stefansson, R., 2002, Imaging the mantle beneath Iceland using integrated seismological techniques: Journal of Geophysical Research, v. 107, p. 2325, doi: 10.1029/2001JB000595.

Anderson, D.L., 2005a, Scoring hotspots: The plume and plate paradigms, *in* Foulger, G.R., et al., eds., Plates, plumes, and paradigms: Boulder, Colorado, Geological Society of America Special Paper 388, p. 31–54.

Anderson, D.L., 2005b, Self-gravity, self-consistency, and self-organization in geodynamics and geochemistry, *in* van der Hilst, R.D., et al., eds., Earth's deep mantle: Structure, composition, and evolution: Washington, D.C., American Geophysical Union Geophysical Monograph 160, p. 165–186.

Anderson, D.L., and Schramm, K.A., 2005, Global hotspot maps, *in* Foulger, G.R., et al., eds., Plates, plumes and paradigms: Boulder, Colorado, Geological Society of America Special Paper 388, p. 19–29.

Anderson, O.L., 2002, The power balance at the core-mantle boundary: Physics of the Earth and Planetary Interiors, v. 131, p. 1–17, doi: 10.1016/S0031-9201(02)00009-2.

Bercovici, D., Schubert, G., Glatzmaier, G.A., and Zebib, A., 1989, Three-dimensional thermal convection in a spherical shell: Journal of Fluid Mechanics, v. 206, p. 75–104, doi: 10.1017/S0022112089002235.

Bijwaard, H., and Spakman, W., 1999, Tomographic evidence for a narrow whole-mantle plume below Iceland: Earth and Planetary Science Letters, v. 166, p. 121–126, doi: 10.1016/S0012-821X(99)00004-7.

Bina, C.R., and Liu, M., 1995, A note on the sensitivity of mantle convection models to composition-dependent phase relations: Geophysical Research Letters, v. 22, p. 2565–2568, doi: 10.1029/95GL02546.

Buffett, B.A., 2003, Thermal state of the Earth's core: Science, v. 299, p. 1675–1677, doi: 10.1126/science.1081518.

Bunge, H.-P., Richards, M.A., and Baumgardner, J.R., 1996, Effect of depth-dependent viscosity on the planform of mantle convection: Nature, v. 379, p. 436–438, doi: 10.1038/379436a0.

Bunge, H.-P., Richards, M.A., and Baumgardner, J.R., 1997, A sensitivity study of three-dimensional spherical mantle convection at 10^8 Rayleigh number: Effects of depth-dependent viscosity, heating mode, and an endothermic phase change: Journal of Geophysical Research, v. 102, p. 11,991–12,008.

Bunge, H.-P., Ricard, Y., and Matas, J., 2001, Non-adiabaticity in mantle convection: Geophysical Research Letters, v. 28, p. 879–882, doi: 10.1029/2000GL011864.

Burke, K., and Wilson, J.T., 1976, Hotspots on the Earth's surface: Scientific American, v. 235, p. 46–57.

Busse, F.H., 1975, Patterns of convection in spherical shells: Journal of Fluid Mechanics, v. 72, p. 67–85, doi: 10.1017/S0022112075002947.

Busse, F.H., and Riahi, N., 1982, Patterns of convection in spherical shells, part 2: Journal of Fluid Mechanics, v. 123, p. 283–301, doi: 10.1017/S0022112082003061.

Chandrasekhar, S., 1961, Hydrodynamic and hydromagnetic stability: Oxford, Clarendon Press, 652 p.

Chen, J., and King, S.D., 1998, The influence of temperature- and depth-dependent viscosity on geoid and topography profiles from models of mantle convection: Physics of the Earth and Planetary Interiors, v. 106, p. 75–91, doi: 10.1016/S0031-9201(97)00110-6.

Chopelas, A., 2000, Thermal expansivity of mantle-relevant magnesium silicates derived from vibrational spectroscopy at high pressure: American Mineralogist, v. 85, p. 270–278.

Chopelas, A., and Boehler, R., 1992, Thermal expansivity in the lower mantle: Geophysical Research Letters, v. 19, p. 1983–1986.

Christensen, U.R., 1995, Effects of phase transitions on mantle convection: Annual Review of Earth and Planetary Sciences, v. 23, p. 65–88, doi: 10.1146/annurev.ea.23.050195.000433.

Christensen, U.R., and Hofmann, A.W., 1994, Segregation of subducted oceanic crust in the convecting mantle: Journal of Geophysical Research, v. 99, p. 19,867–19,884, doi: 10.1029/93JB03403.

Courtillot, V., Davaille, A., Besse, J., and Stock, J., 2003, Three distinct types of hotspots in the Earth's mantle: Earth and Planetary Science Letters, v. 205, p. 295–308, doi: 10.1016/S0012-821X(02)01048-8.

Crough, S.T., 1983, Hotspot swells: Annual Review of Earth and Planetary Sciences, v. 11, p. 165–193, doi: 10.1146/annurev.ea.11.050183.001121.

Crough, S.T., and Jurdy, D.M., 1980, Subducted lithosphere, hotspots, and the geoid: Earth and Planetary Science Letters, v. 48, p. 15–22, doi: 10.1016/0012-821X(80)90165-X.

Cuvelier, C., Segal, A., and van Steenhoven, A.A., 1986, Finite element methods and the Navier-Stokes equations: Norwell, Massachusetts, D. Reidel, 483 p.

Davaille, A., 1999, Simultaneous generation of hotspots and superswells by convection in a heterogeneous planetary mantle: Nature, v. 402, p. 756–760, doi: 10.1038/45461.

Davaille, A., and Vatterville, J., 2005, On the transient nature of mantle plumes: Geophysical Research Letters, v. 32, art. no. L14309, doi: 10.1029/2005GL023029.

Davies, G.F., 1980, Thermal histories of convective Earth models and constraints on radiogenic heat production in the Earth: Journal of Geophysical Research, v. 85, p. 2517–2530.

Davies, G.F., 1988, Ocean bathymetry and mantle convection; 1, Large-scale flow and hotspots: Journal of Geophysical Research, v. 93, p. 10,467–10,480.

Davies, G.F., 1995, Penetration of plates and plumes though the mantle transition zone: Earth and Planetary Science Letters, v. 133, p. 507–516, doi: 10.1016/0012-821X(95)00039-F.

Davies, G.F., and Richards, M.A., 1992, Mantle convection: Journal of Geology, v. 100, p. 151–206.

DeLaughter, J., Stein, C.A., and Stein, S., 2005, Hotspots: A view from the swells, *in* Foulger, G.R., et al., eds., Plates, plumes, and paradigms: Boulder, Colorado, Geological Society of American Special Paper 388, p. 257–278.

Deuss, A., Redfern, S.A., Chambers, K., and Woodhouse, J.H., 2006, The nature of the 660-kilometer discontinuity in Earth's mantle from global seismic observations of PP precursors: Science, v. 311, p. 198–201, doi: 10.1126/science.1120020.

Dubuffet, F., and Yuen, D.A., 2000, A thick pipe-like heat-transfer mechanism in the mantle: Nonlinear coupling between 3-D convection and variable thermal conductivity: Geophysical Research Letters, v. 27, p. 17–20, doi: 10.1029/1999GL008338.

Dziewonski, A.M., and Anderson, D.L., 1981, Preliminary reference earth model (PREM): Physics of the Earth and Planetary Interiors, v. 25, p. 297–356, doi: 10.1016/0031-9201(81)90046-7.

Farnetani, C.G., 1997, Excess temperature of mantle plumes; The role of chemical stratification across D″: Geophysical Research Letters, v. 24, p. 1583–1586, doi: 10.1029/97GL01548.

Farnetani, C.G., and Richards, M.A., 1994, Numerical investigations of the mantle plume initiation model for flood basalt events: Journal of Geophysical Research, v. 99, p. 13,813–13,833, doi: 10.1029/94JB00649.

Farnetani, C.G., and Richards, M.A., 1995, Thermal entrainment and melting in mantle plumes: Earth and Planetary Science Letters, v. 136, p. 251–267, doi: 10.1016/0012-821X(95)00158-9.

Farnetani, C.G., and Samuel, H., 2005, Beyond the thermal plume paradigm: Geophysical Research Letters, v. 32, art. no. L07311, doi: 10.1029/2005GL022360.

Farnetani, C.G., Richards, M.A., and Ghiorso, M.S., 1996, Petrological models of magma evolution and deep crustal structure beneath hotspots and flood basalt provinces: Earth and Planetary Science Letters, v. 143, p. 81–94, doi: 10.1016/0012-821X(96)00138-0.

Foulger, G.R., Pritchard, M.J., Julian, B.R., Evans, J.R., Allen, R.M., Nolet, G., Morgan, W.J., Bergsson, B.H., Erlendsson, P., Jakobsdottir, S., Ragnarsson, S., Stefansson, R., and Vogfjord, K., 2000, The seismic anomaly beneath Iceland extends down to the mantle transition zone and no deeper: Geophysical Journal International, v. 142, p. F1–F5, doi: 10.1046/j.1365-246x.2000.00245.x.

Goes, S., Cammarano, F., and Hansen, U., 2004, Synthetic seismic signature of thermal mantle plumes: Earth and Planetary Science Letters, v. 218, p. 403–419, doi: 10.1016/S0012-821X(03)00680-0.

Griffiths, R.W., 1986, Thermals in extremely viscous fluids: Journal of Fluid Mechanics, v. 166, p. 115–138, doi: 10.1017/S002211208600006X.

Griffiths, R.W., and Campbell, I.H., 1990, Stirring and structure in mantle starting plumes: Earth and Planetary Science Letters, v. 99, p. 66–78, doi: 10.1016/0012-821X(90)90071-5.

Griffiths, R.W., and Campbell, I.H., 1991, On the dynamics of long-lived plume conduits in the convecting mantle: Earth and Planetary Science Letters, v. 103, p. 214–227, doi: 10.1016/0012-821X(91)90162-B.

Gurnis, M., 1989, A reassessment of the heat transport by variable viscosity convection with plates and lids: Geophysical Research Letters, v. 16, p. 179–182.

Hager, B.H., 1984, Subducted slabs and the geoid: Constraints on mantle rheology and flow: Journal of Geophysical Research, v. 89, p. 6003–6016.

Harris, R.N., von Herzen, R.P., McNutt, M.K., Garven, G., and Jordahl, K., 2000, Submarine hydrogeology of the Hawaiian archipelagic apron; 1, Heat flow patterns north of Oahu and Maro Reef: Journal of Geophysical Research, v. 105, p. 21,353–21,369, doi: 10.1029/2000JB900165.

Hill, R.I., Campbell, I.H., Davies, G.F., and Griffiths, R.W., 1992, Mantle plumes and continental tectonics: Science, v. 256, p. 186–193, doi: 10.1126/science.256.5054.186.

Hirose, K., 2002, Phase transitions in pyrolitic mantle around 670-km depth: Implications for upwelling of plumes from the lower mantle: Journal of Geophysical Research, v. 107, p. 2078, doi: 10.1029/2001JB000597.

Hofmeister, A.M., 1999, Mantle values of thermal conductivity and the geotherm from phonon lifetimes: Science, v. 283, p. 1699–1706, doi: 10.1126/science.283.5408.1699.

Hughes, T.J.R., 1987; The finite element method: Linear static and dynamic finite element analysis: Englewood Cliffs, New Jersey, Prentice-Hall, 803 p.

Humphreys, E.D., Dueker, K.G., Schutt, D.L., and Smith, R.B., 2000, Beneath Yellowstone: Evaluating plume and nonplume models using teleseismic images of the upper mantle: GSA Today, v. 10, p. 1–7.

Ita, J.J., and King, S.D., 1994, The sensitivity of convection with an endothermic phase change to the form of governing equations, initial conditions, aspect ratio, and equation of state: Journal of Geophysical Research, v. 99, p. 15,919–15,938, doi: 10.1029/94JB00852.

Jarvis, G.T., Glatzmaier, G.A., and Vangelov, V.I., 1995, Effects of curvature, aspect ratio and plan form in two- and three-dimensional spherical models of thermal convection: Geophysical and Astrophysical Fluid Dynamics, v. 79, p. 147–171.

Karato, S., and Wu, P., 1993, Rheology of the upper mantle: a synthesis: Science, v. 260, p. 771–778, doi: 10.1126/science.260.5109.771.

Kellogg, L.H., and King, S.D., 1993, Effect of mantle plumes on the growth of D″ by reaction between the core and the mantle: Geophysical Research Letters, v. 20, p. 379–382.

Kellogg, L.H., and King, S.D., 1997, The effect of temperature-dependent viscosity on the structure of new plumes in the mantle: Results of a finite element model in a spherical, axisymmetric shell: Earth and Planetary Science Letters, v. 148, p. 13–26, doi: 10.1016/S0012-821X(97)00025-3.

Kellogg, L.H., Hager, B.H., and van der Hilst, R.D., 1999, Compositional stratification in the deep mantle: Science, v. 283, p. 1881–1884, doi: 10.1126/science.283.5409.1881.

Kiefer, W.H., and Hager, B.H., 1992, Geoid anomalies and dynamic topography from convection in cylindrical geometry: Applications to mantle plumes on Earth and Venus: Geophysical Journal International, v. 108, p. 198–214.

King, S.D., 1997, Geoid and topographic swells over temperature-dependent thermal plumes in spherical axisymmetric geometry: Geophysical Research Letters, v. 24, p. 3093–3096, doi: 10.1029/97GL53154.

King, S.D., and Hager, B.H., 1994, Subducted slabs and the geoid: 1) Numerical calculations with temperature-dependent viscosity: Journal of Geophysical Research, v. 99, p. 19,843–19,852, doi: 10.1029/94JB01552.

King, S.D., and Masters, G., 1992, An inversion for radial viscosity structure using seismic tomography: Geophysical Research Letters, v. 19, p. 1551–1554.

King, S.D., and Ritsema, J., 2000, African hotspot volcanism: Small-scale convection in the upper mantle beneath cratons: Science, v. 290, p. 1137–1140, doi: 10.1126/science.290.5494.1137.

Lin, S.-C., and van Keken, P.E., 2006a, Dynamics of thermochemical plumes: 1. Plume formation and entrainment of a dense layer: Geochemistry, Geophysics, Geosystems, v. 7, art. no. Q02006, doi: 10.1029/2005GC001071.

Lin, S.-C., and van Keken, P.E., 2006b, Dynamics of thermochemical plumes: 2. Complexity of plume structures and its implications for mapping mantle plumes: Geochemistry, Geophysics, Geosystems, v. 7, art. no. Q03003, doi: 10.1029/2005GC001072.

Liu, M., 1994, Asymmetric phase effects and mantle convection patterns: Science, v. 264, p. 1904–1907, doi: 10.1126/science.264.5167.1904.

Matyska, C., and Yuen, D.A., 2005, The importance of radiative heat transfer on superplumes in the lower mantle with the new post-perovskite phase change: Earth and Planetary Science Letters, v. 234, p. 71–81, doi: 10.1016/j.epsl.2004.10.040.

Mitrovica, J.X., 1996, Haskell (1935) revisited: Journal of Geophysical Research, v. 101, p. 555–569, doi: 10.1029/95JB03208.

Monnereau, M., and Yuen, D.A., 2002, How flat is the lower-mantle temperature gradient?: Earth and Planetary Science Letters, v. 202, p. 171–183, doi: 10.1016/S0012-821X(02)00756-2.

Montelli, R., Nolet, G., Dahlen, F.A., Masters, G., Engdahl, E.R., and Hung, S.H., 2004, Finite-frequency tomography reveals a variety of plumes in the mantle: Science, v. 303, p. 338–343, doi: 10.1126/science.1092485.

Morgan, W.J., 1971, Convection plumes in the lower mantle: Nature, v. 230, p. 42–43, doi: 10.1038/230042a0.

Nakagawa, T., and Tackley, P.J., 2004, Thermo-chemical structure in the mantle arising from a three-component convective system and implications for geochemistry: Physics of the Earth and Planetary Interiors, v. 146, p. 125–138, doi: 10.1016/j.pepi.2003.05.006.

Nataf, H.C., and Richter, F.M., 1982, Convection experiments in fluids with highly temperature-dependent viscosity and the thermal evolution of planets: Physics of the Earth and Planetary Interiors, v. 29, p. 320–329, doi: 10.1016/0031-9201(82)90020-6.

Nataf, H.C., and van Decar, J., 1993, Seismological detection of a mantle plume?: Nature, v. 364, p. 115–120, doi: 10.1038/364115a0.

Olson, P.L., and Singer, H., 1985, Creeping plumes: Journal of Fluid Mechanics, v. 158, p. 511–531, doi: 10.1017/S0022112085002749.

Olson, P.L., Schubert, G., and Anderson, C., 1987, Plume formation in the D″-layer and the roughness of the core-mantle boundary: Nature, v. 327, p. 409–415, doi: 10.1038/327409a0.

Olson, P.L., Schubert, G., and Anderson, C., 1993, Structure of axisymmetric plumes: Journal of Geophysical Research, v. 98, p. 6829–6844.

Ratcliff, J.T., Schubert, G., and Zebib, A., 1995, Three-dimensional variable viscosity convection of an infinite Prandtl number Boussinesq fluid in a spherical shell: Geophysical Research Letters, v. 22, p. 2227–2230, doi: 10.1029/95GL00784.

Ratcliff, J.T., Schubert, G., and Zebib, A., 1996, Effects of temperature-dependent viscosity on thermal convection in a spherical shell: Physica D Nonlinear Phenomena, v. 97, p. 242–252, doi: 10.1016/0167-2789(96)00150-9.

Ribe, N.M., and Christensen, U.R., 1994, 3-Dimensional modeling of plume-lithosphere interaction: Journal of Geophysical Research, v. 99, p. 669–682, doi: 10.1029/93JB02386.

Richards, M.A., Hager, B.H., and Sleep, N.H., 1988, Dynamically supported geoid highs over hotspots: Observation and theory: Journal of Geophysical Research, v. 93, p. 7690–7708.

Ritsema, J., and Allen, R.M., 2003, The elusive mantle plume: Earth and Planetary Science Letters, v. 207, p. 1–12, doi: 10.1016/S0012-821X(02)01093-2.

Safronov, V.S., 1978, The heating of the Earth during its formation: Icarus, v. 33, p. 3–12, doi: 10.1016/0019-1035(78)90019-2.

Samuel, H., and Farnetani, C.G., 2003, Thermochemical convection and helium concentrations in mantle plumes: Earth and Planetary Science Letters, v. 207, p. 39–56, doi: 10.1016/S0012-821X(02)01125-1.

Sandwell, D., Anderson, D.L., and Wessel, P., 2005, Global tectonic maps, *in* Foulger, G.R., et al., eds., Plates, plumes, and paradigms: Boulder, Colorado, Geological Society of America Special Paper 388, p. 1–10.

Schott, B., and Yuen, D.A., 2004, Influences of dissipation and rheology on mantle plumes coming from the D″-layer: Physics of the Earth and Planetary Interiors, v. 146, p. 139–145, doi: 10.1016/j.pepi.2003.07.026.

Schubert, G., 1979, Subsolidus convection in the mantles of terrestrial planets: Annual Review of Earth and Planetary Sciences, v. 7, p. 289–342, doi: 10.1146/annurev.ea.07.050179.001445.

Schubert, G., Anderson, C., and Goldman, P., 1995, Mantle-plume interaction with an endothermic phase change: Journal of Geophysical Research, v. 100, p. 8245–8256, doi: 10.1029/95JB00032.

Schubert, G., Turcotte, D.L., and Olson, P., 2001, Mantle convection in the Earth and planets: Cambridge, Cambridge University Press, 940 p.

Schubert, G., Masters, G., Olson, P., and Tackely, P., 2004, Superplumes or plume clusters?: Physics of the Earth and Planetary Interiors, v. 146, p. 147–162, doi: 10.1016/j.pepi.2003.09.025.

Sleep, N.H., 1990, Hotspots and mantle plumes: Some phenomenology: Journal of Geophysical Research, v. 95, p. 6715–6736.

Sleep, N.H., 1992, Time dependence of mantle plumes: Some simple theory: Journal of Geophysical Research, v. 97, p. 20,007–20,019.

Sleep, N.H., 1994, Lithospheric thinning by midplate mantle plumes and thermal history of hot plume material ponded at sublithospheric depths: Journal of Geophysical Research, v. 99, p. 9327–9343, doi: 10.1029/94JB00240.

Sleep, N.H., 1996, Lateral flow of hot plume material ponded at sublithospheric depths: Journal of Geophysical Research, v. 101, p. 28,065–28,083, doi: 10.1029/96JB02463.

Sleep, N.H., Richards, M.A., and Hager, B.H., 1988, Onset of mantle plumes in the presence of preexisting convection: Journal of Geophysical Research, v. 93, p. 7672–7689.

Stacey, F.D., and Loper, D.E., 1983, The thermal boundary layer interpretation of D″ and its role as a plume source: Physics of the Earth and Planetary Interiors, v. 33, p. 45–55, doi: 10.1016/0031-9201(83)90006-7.

Tackley, P.J., Stevenson, D.J., Glatzmaier, G.A., and Schubert, G., 1993, Effects of an endothermic phase transition at 670 km depth in a spherical model of convection in the Earth's mantle: Nature, v. 361, p. 699–704, doi: 10.1038/361699a0.

Tackley, P.J., Stevenson, D.J., Glatzmaier, G.A., and Schubert, G., 1994, Effects of multiple phase transitions in a 3-D spherical model of convection in the Earth's mantle: Journal of Geophysical Research, v. 99, p. 15,877–15,901, doi: 10.1029/94JB00853.

van den Berg, A.P., Yuen, D.A., and Steinbach, V., 2001, The effects of variable thermal conductivity on mantle heat transfer: Geophysical Research Letters, v. 28, p. 875–878, doi: 10.1029/2000GL011903.

van Keken, P.E., 1997, On entrainment in starting mantle plumes: Earth and Planetary Science Letters, v. 148, p. 1–12, doi: 10.1016/S0012-821X(97)00042-3.

von Hertzen, R.P., Cordery, M.J., Dietrick, R.S., and Fang, C., 1989, Heatflow and thermal origin of hotspot swells: The Hawaiian swell revised: Journal of Geophysical Research, v. 94, p. 13,783–13,799.

Wen, L., and Anderson, D.L., 1997a, Layered mantle convection: A model for geoid and topography and seismology: Earth and Planetary Science Letters, v. 146, p. 367–377, doi: 10.1016/S0012-821X(96)00238-5.

Wen, L., and Anderson, D.L., 1997b, Slabs, hotspots, cratons and mantle convection revealed from residual seismic tomography in the upper mantle: Physics of the Earth and Planetary Interiors, v. 99, p. 131–143, doi: 10.1016/S0031-9201(96)03162-7.

Wessel, P., and Smith, W.H.F., 1991, Free software helps map and display data: Eos (Transactions, American Geophysical Union), v. 72, p. 441–446, doi: 10.1029/90EO00319.

Wetherill, G.W., 1990, Formation of the Earth: Annual Review of Earth and Planetary Sciences, v. 18, p. 205–256, doi: 10.1146/annurev.ea.18.050190.001225.

Whitehead, J.A., and Luther, D.S., 1975, Dynamics of laboratory diapir and plume models: Journal of Geophysical Research, v. 80, p. 705–717.

Wolfe, C.J., 1998, Prospecting for hotspot roots: Nature, v. 396, p. 212–213, doi: 10.1038/24258.

Wolfe, C.J., Bjarnason, I.Th., Van Decar, J.C., and Solomon, S.C., 1997, Seismic structure of the Iceland mantle plume: Nature, v. 385, p. 245–247, doi: 10.1038/385245a0.

Zebib, A., Schubert, G., and Straus, J.M., 1980, Infinite Prandtl number thermal convection in a sherical shell: Journal of Fluid Mechanics, v. 97, p. 257, doi: 10.1017/S0022112080002558.

Zebib, A., Goyal, A.K., and Schubert, G., 1985, Convective motions in a spherical shell: Journal of Fluid Mechanics, v. 152, p. 39–48, doi: 10.1017/S0022112085000556.

Zhang, S., and Yuen, D.A., 1995, The influence of lower mantle viscosity stratification of 3D spherical-shell mantle convection: Earth and Planetary Science Letters, v. 132, p. 157–166, doi: 10.1016/0012-821X(95)00038-E.

Zhong, S., 2006, Constraints on thermochemical convection of the mantle from plume heat flux, plume excess temperature, and upper mantle temperature: Journal of Geophysical Research, v. 111, art. no. B04409, doi: 10.1029/2005JB003972.

MANUSCRIPT ACCEPTED BY THE SOCIETY 31 JANUARY 2007

DISCUSSION

6 January 2007, Don L. Anderson

One way or another, the mantle convects. There is still discussion about whether the mode is whole mantle or layered and if it is driven primarily from the top. Radioactivity, secular cooling, and strong plates influence convection and cause upwellings to be broad and migratory. Pressure and compositional effects cause the deep mantle to have very broad features. In any case, broad upwellings exist in the mantle, if only to replace the sinking slabs or to get rid of internal radioactive heat. These upwellings would exist even if the mantle were not heated from below and even if there were no lower thermal boundary layer (TBL). In the fluid dynamic literature buoyant upwellings are called plumes.

It was suggested that plate tectonics and normal mantle convection could not explain hotspots, melting anomalies, and volcanic chains, because normal mantle was too cold. A different small-scale mode of convection was proposed (Morgan, 1972). The defining characteristics of mantle plumes have changed with time, but great depth, a TBL origin, high temperatures, rapid ascent rates, and low viscosity are the invariants. Initially, mantle plumes were treated as fixed, whole-mantle, chemically primitive or enriched, and responsible for lifting and breaking continents and for driving plate tectonics. Plumes have been associated with anomalies in bathymetry, heatflow, gravity, chemistry, magma volume, and location. An anomaly is with regard to some reference model. The reference model logically is one that has plate tectonics and normal mantle convection; anomalies are features that fall outside the normal range. Unfortunately, the reference model in discussions of plumes involves a homogeneous isothermal mantle that has uniform depth, geoid, and magma supply at ridges. All positive anomalies in elevation, geoid, and magma supply are then attributed to mantle plumes. The article by King and Redmond (this volume) provides a more realistic reference model.

There has been a tendency to regard mantle plumes as a distinct, secondary mode of convection, but this mode has never been observed in any self-consistent numerical or laboratory experiment (Larsen and Yuen, 1997a). Plume studies have usually modeled a plume in isolation from the rest of the mantle. King and Redmond (this volume) do not do this, but they use the term "plume" in the fluid dynamic sense, not as a distinct secondary mode of convection.

There has long been a problem of obtaining hotspot-like plumes (Tozer, 1973; McKenzie and Weiss, 1975; Larsen and Yuen, 1997a,b), which must satisfy the requirements of being fast (up to 20 m/year) and hot (>200° C), compared to ambient mantle circulation, and narrow (mantle plumes are usually modeled as 200- to 300-km-diameter vertical cylinders). There is also the issue of whether the lower boundary condition can be treated in isolation in a self-organizing system and whether plumes can even form or survive in a convecting mantle (Nataf, 1991; Lenardic and Kaula, 1994; Anderson, 2002, 2005).

In alternative models, hotspot and mid-plate volcanism are not due to a separate small-scale mode of convection; the dimensions of volcanoes and island chains are controlled by the lithosphere (cracks, rifts, dikes) and the dimensions of fertile inhomogeneities in the mantle; a partially molten asthenosphere is the source of magma.

The article by King and Redmond (this volume) is an important contribution to normal mantle convection, not to plume theory as ordinarily understood. A TBL at the base of the mantle is inevitable but narrow; rapid upwellings under mid-plate volcanoes are not. If upwellings are 1000 km in radius, they are not the localized jets that J. Tuzo Wilson and W. Jason Morgan proposed, and they are not confined to the mantle just under the volcanoes. Normal mantle upwellings from internal heating are expected to be of this dimension, and it is no surprise that volcanoes of all types, including mid-ocean ridges, tend to be in these regions.

If King and Redmond (this volume) endorse the idea of plate-scale convection, as their calculations indicate, as an explanation for features that others have attributed to narrow thermal plumes, then they are implying that mantle plumes as conventionally defined are unnecessary. Normal mantle con-

vection and a partially molten asthenosphere have always been alternatives to plumes; it was the perceived narrowness and fixity of hotspots and the assumption of a cold isothermal mantle that motivated the idea that something other than plate tectonics and plate-scale flow was going on.

It is now known that magmatism can occur anywhere, not just in extending, young, thin lithosphere or plume locales (e.g., Hofmann and Hart, 2007). These locations include old oceanic lithosphere, deep ridges, slow spreading ridges, cold mantle, and near cold downwellings. If the coldest mantle can melt, then normal mantle, which can be 100° C hotter, can melt a lot. Fertile blobs, then, with melting points 200° C lower than that for peridotite, can be the source of melting anomalies. An assumption in the plume hypothesis is that the mantle everywhere is subsolidus except where there are ridges or plumes.

REFERENCES CITED

Anderson, D.L., 2002, Plate tectonics as a far-from-equilibrium self-organized system, *in* Stein, S., and Freymueller, J.T., eds., Plate boundary zones: Washington, D.C., American Geophysical Union Geodynamics Monograph 30, p. 411–425.

Anderson, D.L. 2005. Self-gravity, self-consistency, and self-organization in geodynamics and geochemistry, *in* van der Hilst, R.D., et al., eds., Earth's deep mantle: Structure, composition, and evolution: Washington, D.C., American Geophysical Union Geophysical Monograph 160, p. 165–186.

Hofmann, A., and Hart, S.R. 2007, Another nail in which coffin?: Science, v. 315, p. 39–40.

King, S.D., and Redmond, H.L., 2007 (this volume), The structure of thermal plumes and geophysical observations, *in* Foulger, G.R., and Jurdy, D.M., eds., Plates, plumes, and planetary processes: Boulder, Colorado, Geological Society of America Special Paper 430, doi: 10.1130/2007.2430(06).

Larsen, T.B., and Yuen, D.A., 1997a, Ultrafast upwelling bursting through the upper mantle: Earth and Planetary Science Letters, v. 146, p. 393–400.

Larsen, T.B., and Yuen, D.A., 1997b, Fast plumeheads: Temperature-dependent versus non-Newtonian rheology: Geophysical Research Letters, v. 24, p. 1995–1998.

Lenardic, A., and Kaula, W.M., 1994, Tectonic plates, D″ thermal structure, and the nature of mantle plumes: Journal of Geophysical Research, v. 99, no. 94JB00466, p. 15,697–15,708.

McKenzie, D.P., and Weiss, N., 1975, Speculations on the thermal and tectonic history of the Earth: Geophysical Journal of the Royal Astronomical Society, v. 42, p. 131–174. Excerpt from: http://www.mantleplumes.org/McKenzie%2BWeiss1974.html.

Morgan, W.J., 1972, Deep mantle convection plumes and plate motions: Bulletin of the American Association of Petroleum Geologists, v. 56, p. 203–213.

Nataf, H-C., 1991, Mantle convection, plates, and hotspots: Tectonophysics, v. 187, p. 361–371.

Tozer, D.C., 1973, Thermal plumes in the Earth's mantle: Nature, v. 244, p. 398–400. http://www.mantleplumes.org/WebDocuments/Tozer_Nature.pdf.

The Geological Society of America
Special Paper 430
2007

Seismic observations of transition-zone discontinuities beneath hotspot locations

Arwen Deuss*

*Bullard Labs and Institute of Theoretical Geophysics, University of Cambridge,
Madingley Road, Cambridge CB3 0EZ, UK*

ABSTRACT

The seismic structure of the transition-zone discontinuities was studied beneath the forty-nine hotspot locations of the catalog of Courtillot et al. (2003), using a global data set of *SS* precursors. Some of these hotspots are proposed to originate from plumes rising in the upper mantle or from the core-mantle boundary region. I found thin transition zones in approximately two-thirds of the twenty-six hotspot locations for which precursor observations could be made. This observation agrees with the expectation for the olivine phase transition of a systematically thin transition zone in high-temperature regions. Other hotspot locations showed a clear deepening of both the 410- and 660-km discontinuities, which is consistent with a phase transition from majorite garnet to perovskite at a depth of 660 km. Predictions from mineral physics suggest that this transition is more important than the olivine phase transition in regions with high mantle temperatures. So, a hotspot location with a deep 410-km discontinuity in combination with either a shallow or deep 660-km discontinuity might be consistent with hot upwellings rising from the lower into the upper mantle. Hotspot locations with a shallow 410-km discontinuity are not in agreement with a positive thermal anomaly from the surface down to the mantle transition zone. This new interpretation of seismic discontinuities in the transition zone has important implications for our understanding of geodynamics in potential mantle plume locations.

Keywords: mantle plumes, mantle discontinuities, transition zone, seismology, phase transitions

INTRODUCTION

Hotspots (melting anomalies) at the Earth's surface are often attributed to hot thermal plumes in the mantle underneath (Morgan, 1971). Although hot upwellings are essential features of mantle convection, it is not necessarily the case that all hotspots are caused by mantle plumes. Alternative explanations have been proposed, such as propagating cracks, abandoned ridges, or other examples of thinned lithosphere above a partially molten mantle (e.g., Anderson, 2000, and see http://www.mantleplumes.org). It is important to determine which hotspots are due to deep-mantle plumes, and which might be the result of shallow processes.

The depth of origin of the hotspots is also widely debated. Geochemical data suggest that some hotspots, such as Hawaii, arise from the lower mantle (e.g., Hofmann, 1997). The issue is

---

*E-mail: deuss@esc.cam.ac.uk.

Deuss, A., 2007, Seismic observations of transition-zone discontinuities beneath hotspot locations, *in* Foulger, G.R., and Jurdy, D.M., eds., Plates, plumes, and planetary processes: Geological Society of America Special Paper 430, p. 121–136, doi: 10.1130/2007.2430(07). For permission to copy, contact editing @geosociety.org.

further complicated by the difficulty of imaging narrow plumes using seismic tomography (Nataf, 2000). Although subduction zones are close to seismically active plate boundaries and show up as high-wavespeed anomalies, narrow low-wavespeed anomalies are not major features in global tomographic models. Nevertheless, some recently developed global models show evidence for vertically continuous negative wavespeed anomalies beneath a few hotspots, extending from the Earth's surface down to either the transition zone or the core-mantle boundary (CMB) (Ritsema et al., 1999; Montelli et al., 2004, 2006). There have also been a number of regional studies, for example, in Iceland (Foulger et al., 2000, 2001). In locations where the tomographic models suggest that the hotspot is due to a mantle plume, it is also important to obtain constraints on the depth extent of the mantle plume (i.e., upper or lower mantle) using the seismic structure of the transition-zone discontinuities.

Courtillot et al. (2003) suggested five possible criteria to determine whether a hotspot has a very deep origin. These include: (1) long-lived tracks, (2) traps or flood basalts at their initiation, (3) high bouyancy-flux values, (4) high ^{3}He/^{4}He isotope ratios, and (5) anomalously low shear wavespeeds. Here, I propose the detailed structure of the transition zone as an additional criterion. This criterion has been used before (e.g., Shen et al., 1998; Li et al., 2000, 2003a,b), but I show that its interpretation is more complicated than was previously thought.

The transition zone separates the Earth's upper mantle from the lower mantle and is bounded by discontinuities at ~410- and 660-km depths. The characteristics of these discontinuities, such as their depths and wavespeed jumps, are determined by the local temperature and composition. Thus, detailed studies of the transition-zone discontinuities can be used to study lateral variations in temperature and composition of the Earth's mantle. Mantle plumes, with higher temperatures than those in the surrounding mantle, should affect the detailed structure of the

transition-zone discontinuities. Most seismic studies of transition-zone structure base their interpretations on phase transitions of olivine (Helffrich, 2000; Shearer, 2000), for which the transition zone is thinner in hotter regions and thicker in colder regions. Previous searches for evidence of high-temperature anomalies underlying hotspots looked for evidence of a thinner transition zone in these locations (see, e.g., Li et al., 2003a,b).

A recent seismic study of the 660-km discontinuity suggests that mineral phase transitions in garnet must be taken into account in its interpretation (Deuss et al., 2006). The effects of the multiple phase transitions in olivine and garnet are most pronounced at high temperature (Weidner and Wang, 1998; Hirose, 2002). In particular, the 660-km discontinuity becomes deeper in hotter regions. In combination with a deepening of the 410-km discontinuity at high temperature, the effect of garnet is to leave the transition-zone thickness in hot regions unaffected (Fig. 1). Ponding of plume material beneath the 660-km discontinuity could also take place, which would enhance the thermal effects of the positive temperature anomaly on the depth of the 660-km discontinuity.

Here, I use precursors to the seismic phase *SS* to study the transition-zone thickness in potential high-temperature regions in detail, by focusing on hotspot locations. To do this, I measure transition-zone thicknesses for all these locations and compare the results with predictions based on recent information about the phase diagrams of garnet-olivine mixtures. I also compare my results with locations where seismic tomography models find vertically continuous negative wavespeed anomalies in the upper mantle.

SS PRECURSORS

A global data set of *SS* precursors was used. This type of data has been used extensively in previous studies to determine

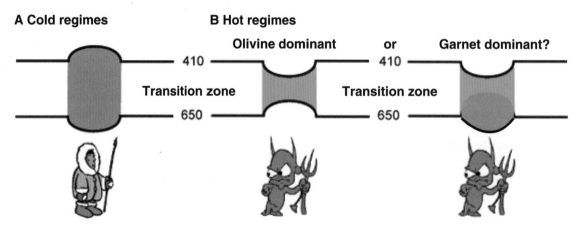

Figure 1. Cartoon showing the different possible behaviors of seismic discontinuities in the transition zone in different thermal regimes. (A) Olivine phase transitions produce thick transition zones in cold regions (such as subduction zones) and (B) thin transition zones in hot regions (such as mantle plumes). However, garnet might modify the behavior of the 660-km discontinuity in hot regions, leading to lack of anomalies in transition-zone thickness. Adapted from http://www.mantleplumes.org.

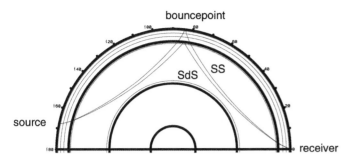

Figure 2. Ray paths of *SS* (reflected at the surface) and the precursors *SdS* (here reflected at *d* = 660-km discontinuity).

the global characteristics of the transition-zone discontinuities (e.g., Shearer, 2000). The precursors arrive before the major *SS* shear wave because they are reflected not from the Earth's surface but from a discontinuity below the bouncepoint of the *SS* wave (Fig. 2). Precursors are named *SdS,* where *d* is the depth of the reflector. *SS* precursors are useful for probing the mantle discontinuities globally, beneath both oceans and continents.

SS precursors have a complex Fresnel zone of ~1000 km (Neele et al., 1997). The large size of the Fresnel zone can lead to underestimates in discontinuity topography, in particular if the topography has a smaller wavelength than the Fresnel zone. Mantle plumes are predicted to be quite narrow features, potentially much smaller than the Fresnel zones of the *SS* precursors. Receiver functions would be more suitable than *SS* precursors, as they have smaller Fresnel zones, and they have been used before to investigate transition-zone discontinuities around hotspot locations (Shen et al., 1998; Li et al., 2000, 2003a).

Unfortunately, many oceanic receiver functions are not located directly above hotspot locations, and only a few hotspots have been studied using this technique. *SS* precursors are the only data currently available with a sufficient global coverage in the oceans to enable a study of all hotspot locations. Model S20RTS (Ritsema et al., 1999) shows continuous low-wavespeed features, which might be associated with mantle plumes, of sizes comparable to or larger than the Fresnel zones of the *SS* precursors. Also, the recent shear wave velocity model of Montelli et al. (2006) shows slow features with radii of more than 500 km in certain regions. Thus, mantle plumes are potentially much wider features than previously thought, justifying the use of a global data set of *SS* precursors to search for seismic signatures of potential mantle plumes beneath hotspot locations.

The data were collected from earthquakes in the Incorporated Research Institutions for Seismology (IRIS) catalog and included all events with depths of less than 75 km and magnitude $6.0 \leq Mw \leq 7.0$. This search yielded 1625 events with an epicentral distance range of $100° \leq \Delta \leq 160°$. Arrival times were measured using cross correlation with a reference pulse to ensure that only high-quality data were included and to determine the polarities of the *SS* phases. The selected data set contained 8054 seismograms. These seismograms were band-pass filtered to pass

periods of 15–75 s, resulting in a Fresnel-zone radius of ~1000 km. This scale is similar to the resolution of global tomographic model S20RTS (Ritsema et al., 1999), which was one of the models used for comparison with my *SS* precursor observations.

The *SS* precursors have small amplitudes, only a few percent of the main *SS*-phase amplitudes, and are often not visible on individual seismograms. Therefore, I stacked large numbers of seismograms to suppress incoherent noise and make the precursors visible. First, the traces were aligned on the arrival time of *SS,* the polarity was reversed if necessary, and each trace was normalized to its maximum *SS* amplitude. Then the traces were stacked in the slowness-time domain. The stacks were converted into a trace in which the stacking slowness is time-dependent, the time dependence being chosen to maximize the amplitudes of reflections from a continuous range of depths.

I also determined the robustness of the precursors by computing 95% confidence levels for the stacks using the bootstrap resampling algorithm of Efron and Tibshirani (1991). The 95% confidence levels are the 2σ error boundaries around the stacked traces. I only interpreted precursors for which the lower confidence levels are larger than zero. The confidence levels are omitted from the plotted results (e.g., see Fig. 3) for clarity, but the parts of the stacked trace for which the lower confidence level is larger than zero are colored to highlight the robust reflectors. The depth of the discontinuity was determined by measuring the travel-time difference between *SS* (t_{SS}) and the precursor *SdS* (t_{SdS}). I used a deconvolution technique to determine the travel-time differences, by iteratively deconvolving the precursor time window by the main *SS* pulse (see Chambers et al., 2005). My measurements were corrected for crustal and mantle structure using the crustal model CRUST5.1 (Mooney et al., 1995) and the shear-wavespeed model S20RTS (Ritsema et al., 1999).

TRANSITION-ZONE OBSERVATIONS

I computed *SS* precursor stacks for all forty-nine globally distributed hotspots from the catalog of Courtillot et al. (2003), which incorporates the catalog of Sleep (1990) (Table 1). As discussed above, *SS* precursors have a complex Fresnel zone (Neele et al., 1997), and the plumes could be quite narrow features. To get the most favorable situation for detecting mantle plumes, it is necessary to compute the precursor stacks centered around each hotspot location. These stacks should lead to better results than comparing hotspot locations with a global map of transition-zone discontinuity observations, as was done by Li et al. (2003b).

The stacks were initially computed in the time domain, and the 410- and 660-km discontinuity arrival times are measured as travel-time differences with the main *SS* phase. To obtain discontinuity depths, a mantle wavespeed model is needed to convert the travel-time differences to depths. In addition, a tomographic velocity model and crustal model are used to correct for local 3-D structure. However, this method causes the exact discontinuity depths to be dependent on the chosen tomo-

TABLE 1. TRANSITION-ZONE OBSERVATIONS FROM *SS* PRECURSORS

Location of hotspot	Transition zone measured time difference (s)	410 measured time (s)	660 measured time (s)	Transition zone inferred thickness (km)	410 inferred depth (km)	660 inferred depth (km)	Tomographic model S20RTS	Tomograpic model FF	Bouyancy flux
Azores	73	155	228	243.7	401.8	645.6	Ridge	d	1.1
Baja*	70	165	235	235.6	419.0	654.7	Ridge		0.3
Balleny		159	D		412.5	D			
Bermuda	72	155	227	241.0	413.0	654.0			1.1
Bowie*	71	162	233	237.1	414.7	651.8	Transition zone	mm	0.3
Canary	69	154	223	228.2	405.2	633.4		d	1
Cape Verde	76	151	227	255.0	389.7	644.7		d	1.6
Comores*	71	158	229	240.2	411.0	651.2			
Crozet	70	155	225	234.2	404.0	638.2	Ridge	d	0.5
Darfur	67	155	222	224.2	401.2	625.4			
Discovery*	65	160	225	219.8	429.0	648.8			0.5
Easter*	75	159	234	250.8	412.1	662.9	Transition zone	d	3
Fernando	72	155	227	243.4	410.9	654.3			0.5
Hawaii	74	154	228	245.7	405.2	650.9	Transition zone	pd	8.7
Iceland		156	D		395.8	D	Transition zone	d	1.4
Jan Mayen		155	D		399.3	D			
Juan de Fuca*	70	162	232	232.5	415.6	648.1	Ridge	mm	0.3
Kerguelen	68	157	225	228.0	404.9	632.9	Ridge	d	0.5
Louisville		158	D		415.4	D	Transition zone	pd	0.9
Lord Howe		D	225		D	631.3			0.9
MacDonald*	71	157	228	235.6	412.5	648.1	Transition zone		3.3
Marquesas*	69	158	227	231.7	415.6	647.3	Ridge		3.3
New England	68	156	224	228.0	416.3	644.3			0.5
Pitcairn*	69	160	229	231.9	420.1	652.0	Ridge		3.3
Raton*	71	162	233	236.9	419.1	656.0			
Réunion	74	155	229	248.0	408.6	656.6		pd	1.9
Samoa*	70	157	227	232.5	414.4	646.9	Transition zone	d	1.6
San Felix	75	155	230	254.5	405.7	660.2	Ridge		1.6
Tahiti*	72	158	230	242.2	414.0	656.2		d	3.3
Tristan		160			418.4		Ridge		1.7
Vema*	63	161	224	212.9	429.0	641.9			
Yellowstone	75	158	233	253.9	408.3	662.2			1.5
Plume average	71	157	228	237.2	411.0	648.2			
Global average	72	156	228	242.5	410.1	652.6			

Notes: Observations were made using our data set, for hotspot locations from the Courtillot et al. (2003) catalog. The locations labeled with * have a deeper inferred 410-km discontinuity in both the measured times and the corrected depths, and have average or thin transition zones. Tomographic information from shear wave velocity model S20RTS is also included, labeled to the depth of the slow velocity anomalies, i.e., ridge or transition zone (Ritsema and Allen, 2003). Interpretations from finite frequency tomography (Montelli et al., 2006) are also included. Bouyancy flux is taken from Sleep (1990). Abbreviations: d—deep plume; D—discontinuities with a double peak; FF—finite frequency tomography; mm—mid-mantle plume; pd—potentially deep plume.

graphic model. Therefore, I regarded the time measurements as the "raw" measurements, and the computed and corrected depths as interpretations.

The stacks are presented here in the time domain because these are the most direct, raw measurements (Fig. 3). Of the forty-nine locations, only thirty-two showed robust reflectors

(within the 95% confidence levels) from the transition-zone discontinuities. The other seventeen locations did not have enough observations to give robust results and are omitted here. The remaining thirty-two hotspot locations show clear reflections from both the 410- and 660-km discontinuities. In some places, for example, Iceland, either the 410- or 660-km reflection is

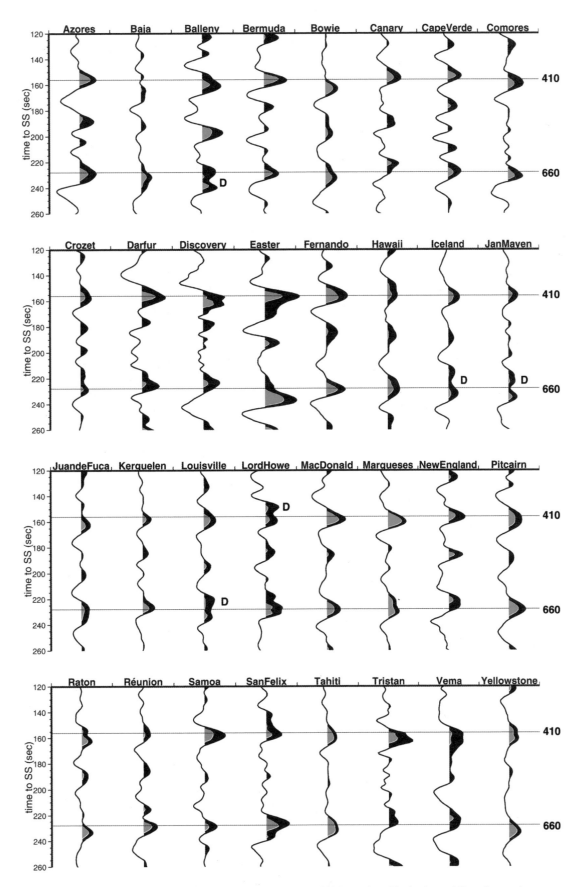

Figure 3. *SS* precursor stacks for locations of the Courtillot et al. (2003) catalog. The horizontal lines denote the average arrival times for reflections from the 410- and 660-km discontinuities for the global data set, of which the hotspot locations are a subset. The red coloration shows the area between the lower 95% confidence level and the zero line, denoting robust reflectors. "D" marks the locations with a double peak from the 410-or 660-km discontinuity.

characterized by a double peak. Double or broad peaks could be due to the effect of multiple phase transitions (Simmons and Gurrola, 2000; Deuss and Woodhouse, 2001; Deuss et al., 2006), caused by the focusing and defocusing by narrow topographic variations (Chaljub and Tarantola, 1997; Neele et al., 1997), or caused by ponding of plume material below the 660-km discontinuity (Vinnik et al., 1997). Detailed seismic modeling will be needed to investigate what is causing the double peaks here, which is beyond the scope of this chapter. The locations with double peaks are shown for completeness in Figure 3, but have not been used to measure discontinuity depths or transition-zone thicknesses. Transition-zone thickness has been measured for twenty-six locations. Table 1 shows both the measured time differences and the inferred depths for those locations with single reflections from the 410- and 660-km discontinuities.

The hotspot precursor stacks are a subset of a global data set of *SS* precursor stacks, which has good data coverage in both continental and oceanic regions. The average values of $t_{SS} - t_{S410S}$ and $t_{SS} - t_{S660S}$ are 156 s and 228 s, respectively, for the global data set. These times are indicated with horizontal lines in Figure 3, and lead to an average transition zone travel-time difference ($t_{S410S} - t_{S660S}$) of 72 s, which corresponds to a thickness of 242.5 km and is in agreement with other *SS* precursor studies (Flanagan and Shearer, 1998; Gu et al., 1998). The average depths corresponding to the 410- and 660-km arrival times are 410.1 and 652.6 km. The average times and depths are used in this study to test whether observations in hotspot locations differ from those for the average Earth. It is difficult to determine when a depth is significantly greater than the average. I have chosen not to specify a minimum difference from the average, but to list all my measurements to enable readers to make their own decisions. The details of my observations are discussed in the next two sections, and possible interpretations are offered.

MINERAL PHYSICAL INTERPRETATION

The transition-zone discontinuities have usually been interpreted in terms of phase transitions in olivine (Anderson, 1967; Ringwood, 1975; Ito and Takahashi, 1989). A transition from olivine (α-phase) to a spinel crystal structure (β-phase) occurs at 410-km depth, from spinel to ringwoodite (γ-phase) at 520-km depth, and from ringwoodite to perovskite and magnesiowustite at 660-km depth. The olivine phase transition at 410 km has a positive Clapeyron slope (pressure increases with temperature), which leads to a deeper transition in hotter regions. The post-spinel phase transition at 660 km has a negative Clapeyron slope and will lead to a shallower transition depth in hotter regions. Defining the transition-zone thickness as the distance between the 410- and 660-km discontinuity depths predicts thinner transition zones in hotter regions. This thickness is the seismic signature that most previous studies of mantle plumes have sought.

However, olivine phase transitions are not the only ones important in the Earth's mantle; other minerals must be taken into account as well. The commonly used pyrolite mantle model contains 60% olivine and 40% garnet at transition-zone pressures (Ringwood, 1975; Irifune and Ringwood, 1987). Most seismic studies have ignored the influence of garnet when interpreting seismic observations of the transition-zone discontinuities. However, recent mineral physics studies suggest that garnet phase transitions play a major role in the detailed structure of the discontinuities at 520 and 660 km (Weidner and Wang, 2000). At 520-km depth, garnet transforms to Ca-perovskite, and at ~660-km depth, majoritic garnet changes into perovskite. The majorite phase transition at 660 km could be more important than the olivine phase transition at high temperatures (Weidner and Wang, 1998; Hirose, 2002). The majorite phase transition has a positive Clapeyron slope, which would lead to a deeper 660-km discontinuity in hotter areas and thus a relatively temperature-independent transition-zone thickness.

Figure 4 shows the different phase transitions near 660-km depth. The 410-km discontinuity is not affected by garnet phase transitions. Water might influence the depth and sharpness of the 410-km discontinuity. Bercovici and Karato (2003) suggested the existence of a potentially global water-rich layer near 400 km. However, recent experimental work shows that seismically observable effects of water are only expected at mantle temperatures lower than 1200 °C and at water concentrations close to

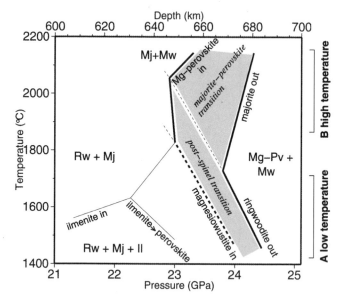

Figure 4. Phase diagram for a pyrolitic mantle at a depth of 600–700 km, after Hirose (2002). The shaded field indicates the perovskite-forming phase transitions that determine the characteristics of the 660-km discontinuity. (A) At low temperatures, perovskite forms from the post-spinel phase in olivine, with a negative Clapeyron slope. (B) At high temperatures, perovskite forms from the breakdown of majorite garnet, with a positive Clapeyron slope. Thin lines show additional phase transitions forming ilmenite and perovskite at shallower depth. Ca-perovskite is also present in all phase fields. Il—ilmenite; Mj—majorite garnet; Mw—magnesiowustite; Pv—perovskite; Rw—ringwoodite.

saturation level (Frost and Dolejs, 2006). Mantle plumes are expected to have a much higher temperature than 1200 °C, so the effect of water can be ignored. The presence of ferric iron may also play a role in broadening the discontinuity, but is as yet unquantified. I will assume that the olivine phase transition alone describes the effects of temperature on this boundary.

Thus, hot thermal regimes are characterized by a deep 410-km discontinuity, but could have either a shallower (olivine dominant) or deeper (majorite-garnet dominant) 660-km dis-

continuity. The dominance of either olivine or garnet is highly dependent on temperature and the exact aluminum (Al) content of the mantle (Weidner and Wang, 1998), and a detailed interpretation can only be made after more mineral physical data become available. Here, I compare my seismic observations with both interpretations.

The arrival times, depths of the discontinuities, and transition-zone thicknesses are plotted in Figure 5 for the twenty-six locations with a single reflector from both 410- and 660-km

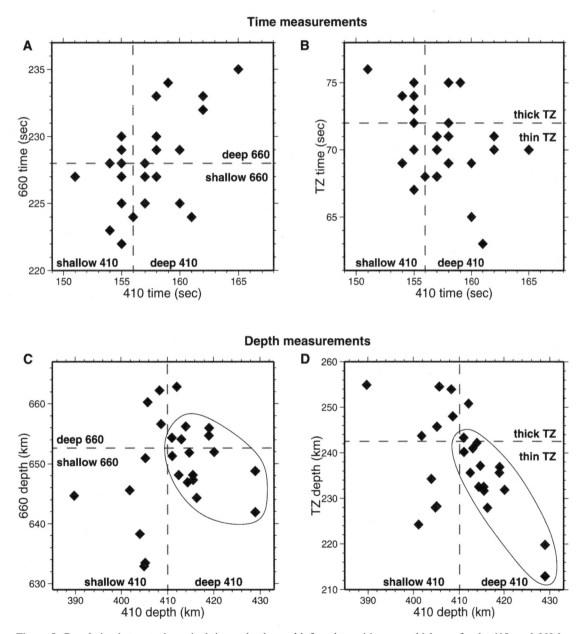

Figure 5. Correlation between the arrival times, depths, and inferred transition-zone thickness for the 410- and 660-km discontinuities. Dashed horizontal and vertical lines indicate the average values for our global data set. (A, B) Raw arrival time measurements; (C,D) computed depths, which are corrected for local mantle and crustal structure. The circled clusters in the depth plots are in agreement with the mineral physical signature that would be predicted in potential mantle plume locations; the remaining point in the right top quadrants outside the cluster is Easter hotspot.

discontinuities. Also shown are horizontal and vertical lines denoting the average values for the discontinuity depths and the transition-zone thickness for the global data set; these lines divide each panel in four quadrants. Previous studies have interpreted the combination of a deep 410-km discontinuity and a shallow 660-km discontinuity (and thus a thin transition zone) as evidence for a high temperature, which corresponds to the measurements in the lower right quadrants only of each panel in Figure 5. However, the combination of a deep 410-km discontinuity with a deep 660-km discontinuity and an average transition-zone thickness can also be due to high temperatures. This scenario enhances the field of possible mantle plume observations to both the lower and top right quadrants of each panel in Figure 5. It is important to note that variations in the 410- and 660-km discontinuity depths could also be due to local temperature variations in the transition zone only, and a deep 410-km discontinuity does not necessarily prove the existence of vertically continuous high-temperature anomalies. The results in this study should only be interpreted as part of a multidisciplinary effort to determine whether mantle plumes exist beneath hotspot locations.

The average value of $t_{SS} - t_{S410S}$ in the global data set is 156 s. Nineteen locations (from a total of thirty-one locations with simple reflections) have larger time differences than 156 s (Fig. 5A and B). When these differential times are corrected for local mantle heterogeneity and the corresponding depths are computed, eighteen hotspot locations still show a deeper 410-km discontinuity than the average depth of 410.1 km, and the depths seem to cluster more (Fig. 5C and D). Note that some of these locations have double peaks at 660 km, so no measurement of the transition-zone thickness is possible. There are twenty-six locations with both 410- and 660-km discontinuity observations, which are shown in Figure 5.

The locations with a deep 410-km discontinuity have both shallow and deep 660-km discontinuity observations (Fig. 5C). The combination of the 410- and 660-km depths leads to a thin or average transition-zone thickness (Fig. 5D) in all but one of the locations with a deep 410-km discontinuity. All these estimated locations could be due to a positive thermal anomaly, using the mineral physical interpretation described above; they are labeled with an asterisk in Table 1, and the cluster of these locations is circled in Figure 5C and D. The Easter hotspot is the only exception, having a significantly thicker than average transition zone, which could still be due to a high-temperature anomaly if the Clapeyron slope of the majorite garnet transition at 660-km depth is larger than the Clapeyron slope of the olivine transition at 410-km depth. If this difference in slopes is indeed the case, then the Easter hotspot will have the largest positive thermal anomaly of all thirteen potential mantle plume locations. The remaining thirteen hotspot locations with shallower 410-km discontinuities either are not due to positive thermal anomalies in the transition zone or have thermal anomalies too narrow to be imaged using *SS* precursors.

The depth of the 660-km discontinuity offers a new method for measuring temperature. Assuming that depth differences are caused only by temperature differences (ignoring compositional heterogeneity), then the locations with the deepest 660-km discontinuities should have the highest temperatures (Fig. 4). These locations are in the part of the phase diagram where the majorite-perovskite transition, which has a positive Clapeyron slope, is dominant. Locations with a shallow 660-km discontinuity might still be in the olivine-dominated part of the phase diagram and therefore have lower temperatures. Thus, locations in the top right quadrants of Figure 5A and C could be due to higher temperatures than those in the bottom right quadrants. Easter hotspot, which has the deepest 660-km discontinuity, would then have the largest positive temperature anomaly.

It is important to realize that only about two-thirds of the hotspot locations show a deep 410-km discontinuity, and the other one-third shows a shallow 410-km discontinuity. These measurement do not provide statistically significant evidence for wide high-temperature anomalies in the transition zone beneath hotspot locations; from these data alone, it is not possible to draw conclusions either in favor of or against the existence of mantle plumes. The discontinuity measurements presented here should only be interpreted in combination with other geophysical and geochemical data sets, and in particular, with other seismological results, such as tomographic models (see below).

COMPARISON WITH SEISMIC TOMOGRAPHY

Vertically continuous negative shear-wavespeed anomalies in the upper mantle, expected for plumes, have been found in some recent global tomographic models (Ritsema and Allen, 2003; Montelli et al., 2004, 2006). A vertically continuous negative shear-wavespeed anomaly in the upper mantle below a hotspot will lead to larger-than-average travel-time measurements ($t_{SS} - t_{SdS}$) for both the 410- and 660 km reflections in *SS* precursors, even if there is no deepening of the discontinuities. A large $t_{SS} - t_{S410S}$ could be due to a deep 410-km discontinuity, a low shear wavespeed in the mantle above, or a combination of both. When using a tomographic model to correct *SS* precursor times for local 3-D structure, the large $t_{SS} - t_{S410S}$ times will be reduced by the correction procedure if the 3-D tomographic model has a negative wavespeed anomaly. It is difficult to separate these two competing possibilities, especially as the corrections are sensitive to the specific tomographic model. Therefore, the measurement of absolute discontinuity depths in supposedly hot areas, which will have both negative wavespeed anomalies and deep 410-km discontinuities, is complicated. Here I compare the *SS* precursor observations with two recent shear wave tomography models. Bouyancy flux (Sleep, 1990) is also included in Table 1 for comparison with the *SS* precursor observations.

Global Shear Wave Model S20RTS

Model S20RTS (Ritsema et al., 1999) is obtained using shear waves, surface waves, and normal-mode splitting functions. It has a horizontal resolution of ~1000 km, which is similar to that of my *SS* precursor data set, making comparison with my shear-wave observations straightforward. Vertically continuous negative shear-wavespeed anomalies in the upper mantle down to the transition zone are present only beneath Afar, Bowie, Easter, Hawaii, Iceland, Louisville, McDonald, and Samoa (Ritsema and Allen, 2003). A number of other hotspot locations, on or nearby mid-oceanic ridges, show negative shear-wavespeed anomalies in the upper 200 km only (see Table 1). The remaining hotspot locations are without a negative wavespeed anomaly in the underlying upper mantle.

For the hotspot locations with negative wavespeed anomalies in the transition zone, I find large $t_{SS} - t_{S410S}$ values from the 410-km discontinuity in most cases (e.g., Bowie, Easter, Louisville, MacDonald, Samoa). After correcting these arrival times using model S20RTS and computing the corresponding depths, these locations still have deeper-than-average 410-km discontinuities. Of the remaining three locations, Afar does not show robust *SS* precursor reflections, Iceland has an average 410-km arrival time, and Hawaii has an early arrival time. Bowie, McDonald, and Samoa also have thin transition zones. Easter has a deep 660-km discontinuity, which might also be consistent with a high temperature if the garnet phase transition is dominant in this location. Louisville shows a double reflection around 660-km depth, and no interpretation can be made. It is promising that two different methods with similar resolution, that is, precursor observations of transition-zone discontinuities (this study) and global tomography (Ritsema and Allen, 2003), find evidence for mantle plumes in a number of the same locations.

Finite-Frequency Model PRI-S05

Montelli et al. (2004, 2006) derived a compressional and a shear-wavespeed model using finite-frequency kernels, from which they claimed that a number of hotspots have deep (i.e., lower mantle) or CMB origins. Although the significance of finite-frequency kernels as compared to ray theory has been disputed (van der Hilst and de Hoop, 2005), comparison with these tomographic results is included for completeness. Table 1 includes the interpretation of Montelli et al. (2006) of the depth origin of mantle plumes below hotspot locations for which robust *SS* precursor observations were made in this study. The horizontal resolution of model PRI-S05 is on the order of a few hundreds of kilometers, which is much smaller than the resolution of model S20RTS and the *SS* precursor observations. It is likely that small-scale features of model PRI-S05 cannot be seen in the *SS* precursor data because of averaging over large areas.

Azores, Cape Verde, and Canary all have a shallow 410-km discontinuity and thus no evidence for a positive thermal anomaly. These hotspot locations show quite narrow plumes in the images of Montelli et al. (2006), so either the potential mantle plumes are too narrow to be seen with the *SS* precursors or there is no positive thermal anomaly present beneath these locations. Crozet and Kerguelen show a large low-wavespeed anomaly near the CMB and are interpreted as deep plumes in PRI-S05. These hotspot locations show a thin transition zone but have a shallow 410-km discontinuity and are therefore not interpreted as deep-mantle plumes in the *SS* precursor study.

Bowie, Juan de Fuca, and Yellowstone are characterized by shallow, broad low-wavespeed anomalies down to the mid-mantle, which are quite wide in the transition zone. The *SS* precursors for Bowie and Juan de Fuca show deepening of the 410-km discontinuity and a thin transition, and are identified as potential mantle plume locations. Yellowstone has a much lower wavespeed anomaly in PRI-S05 than do Bowie and Juan de Fuca. Dueker and Sheehan (1997) found a deepening of the 410-km discontinuity below the Yellowstone hotspot using receiver functions and determined that the 660-km discontinuity topography was unrelated to the 410-km discontinuity. There is an ambiguity in the *SS* precursor observations for Yellowstone, as it has a large value for $t_{SS} - t_{S410S}$, but after correcting for 3-D structure, the discontinuity depth is shallow. Model PRI-S05 also shows an ambiguity, and this hotspot has not been included in the table of Montelli et al. (2006).

Samoa, Cook Island (close to MacDonald in Table 1), and Tahiti are closely spaced and have strong low-wavespeed anomalies that seem to originate from the Pacific superplume. All three locations are also identified as potential mantle plume locations in the *SS* precursor study. Tahiti is the strongest plume in model PRI-S05 and has deep 410- and 660-km discontinuities in the *SS* precursor observations, in agreement with a large positive thermal anomaly in the mantle transition zone. Samoa has a deep 410-km discontinuity and a thin transition zone. Mac-Donald is not mentioned in the table of Montelli et al. (2006), but its location is very close to Cook Island, and the *SS* precursors show a deep 410-km discontinuity and thin transition zone, in agreement with a positive thermal anomaly.

Easter hotspot has low-wavespeed anomaly in model PRI-S05, which is quite wide in the transition zone, extending down to the CMB. The *SS* precursor data for this hotspot show a deep 410-km discontinuity and the deepest 660-km discontinuity, suggesting the presence of a large positive thermal anomaly in the transition zone. Easter hotspot is one of the few hotspots that is common on the lists of "major" or "CMB" hotspots by Courtillot et al. (2003), Ritsema and Allen (2003), and Montelli et al. (2004, 2006), so it is encouraging that the *SS* precursor data agree with tomographic models and other geophysical and geochemical data sets.

Hawaii is one of the few hotspot locations that has been investigated in regional seismic studies, which found evidence for a low-wavespeed anomaly (Priestley and Tilmann, 1999) and also a thin transition zone (Li et al., 2000). Model PRI-S05 shows a large low-wavespeed anomaly in the upper mantle, but in the lower mantle, it is not the giant plume that it appears to be from observations, such as bouyancy flux. The *SS* precursor data show no evidence for a positive thermal anomaly (average transition-zone thickness and shallow 410- and 660-km discontinuities). The transition-zone thickness measurements made using receiver functions by Li et al. (2000) show a plume that is narrower than the *SS* precursor Fresnel zone, so it is perhaps not surprising that this signature cannot be seen in the *SS* precursors.

Iceland has been the subject of a long debate regarding the existence of mantle plumes and has had the advantage of a regional network that has been used in a number of studies. Shen et al. (1998) found a thin transition zone using receiver functions and interpreted this zone as evidence for a lower-mantle origin of the Iceland plume. However, a more recent interpretation of virtually the same data set concluded that there was evidence for a thin transition zone caused by a deep 410-km discontinuity, but that the 660-km discontinuity was "flat" beneath Iceland (Du et al., 2006). Model PRI-S05 shows a strong low-wavespeed anomaly in the upper mantle, in agreement with regional tomographic models (Foulger et al., 2000, 2001), but this signature disappears below 1000 km. The *SS* precursor data show no evidence for a deep 410-km discontinuity, and the 660-km discontinuity is characterized by a double peak, making it impossible to measure one discontinuity depth. The double peaks could be due to the averaging over a large area in the *SS* precursor stack, but it is interesting to note that Du et al. (2006) show a broadening of the 660-km discontinuity in some stacks. This effect could be due to the combination of the olivine and garnet phase transitions, but more detailed seismic modeling will be required to fully understand the complexities of the transition-zone discontinuity observations beneath Iceland.

CONCLUSIONS

Measurements of the transition-zone discontinuities in the hotspot locations of the catalog of Courtillot et al. (2003) show evidence for late arrivals and deeper-than-average depths for the 410-km discontinuity in about two-thirds of the cases. A deep 410-km discontinuity agrees with the predicted deepening of the olivine phase transition in high-temperature regimes. Most locations with deep 410-km discontinuities have thinner or average transition-zone thicknesses and either shallow or deep 660-km discontinuities. Mineral physical information (Weidner and Wang, 1998; Hirose, 2002) suggests that either scenario could be consistent with high temperatures. Using these criteria, about half of the hotspots were identified as possible mantle plume locations. A number of these potential plume locations are consistent with negative wavespeed anomalies in the overlying mantle, according to the tomographic models by Ritsema and Allen (2003) and Montelli et al. (2006).

One-third of the hotspot locations shows a shallow or average 410-km discontinuity depth and no evidence for a thin transition zone. These locations are either not consistent with a positive thermal anomaly in the transition zone, or the anomaly is too narrow to be imaged using *SS* precursors. It is important to note that *SS* precursors have large Fresnel zones (~1000 km radius), making it difficult to image narrow plumes. Receiver functions would be sensitive to narrow features of ~100-km radius, but unfortunately, there are only a few receivers close

enough to hotspot locations to be used for such an analysis. Future deployment of ocean-bottom seismometers will be needed to obtain more detailed constraints on the seismic observability of thermal signatures in the transition-zone discontinuities. It is not possible to use the current data set alone to draw general conclusions in favor of or against the existence of mantle plumes. Nevertheless, the results presented here add another "column" to the hotspot tables and give further evidence in deciding which of the hotspot locations might be due to mantle plumes.

ACKNOWLEDGMENTS

This project was supported by a research grant from the Royal Society. Jeff Gu and Bruce Julian are thanked for constructive reviews, and Gillian Foulger for inviting me to attend the "Great Plume Debate" in 2005, which encouraged me to work on this topic.

REFERENCES CITED

Anderson, D.L., 1967, Phase changes in the upper mantle: Science, v. 157, p. 1165–1173, doi: 10.1126/science.157.3793.1165.

Anderson, D.L., 2000, The thermal state of the upper mantle; No role for mantle plumes: Geophysical Research Letters, v. 27, p. 3623–3626, doi: 10.1029/2000GL011533.

Bercovici, D., and Karato, S., 2003, Whole-mantle convection and the transition-zone water filter: Nature, v. 425, p. 39–44, doi: 10.1038/nature01918.

Chaljub, E., and Tarantola, A., 1997, Sensitivity of *SS* precursors to topography on the upper-mantle 660-km discontinuity: Geophysical Research Letters, v. 24, p. 2613–2616, doi: 10.1029/97GL52693.

Chambers, K., Deuss, A., and Woodhouse, J.H., 2005, Reflectivity of the 410-km discontinuity from *PP* and *SS* precursors: Journal of Geophysical Research, v. 110, B02301, doi: org/10/1029/2004JB003345.

Courtillot, V., Davaille, A., Besse, J., and Stock, J., 2003, Three distinct types of hotspots in the Earth's mantle: Earth and Planetary Science Letters, v. 205, p. 295–308, doi: 10.1016/S0012-821X(02)01048-8.

Deuss, A., and Woodhouse, J.H., 2001, Seismic observations of splitting of the mid-transition zone discontinuity in Earth's mantle: Science, v. 294, p. 354–357, doi: 10.1126/science.1063524.

Deuss, A., Redfern, S.A.T., Chambers, K., and Woodhouse, J.H., 2006, The nature of the 660-km discontinuity in earth's mantle from global seismic observations of *PP* precursors: Science, v. 311, p. 198–201, doi: 10.1126/science.1120020.

Du, Z., Vinnik, L., and Foulger, G., 2006, Evidence from P-to-S mantle converted waves for a flat "660-km" discontinuity beneath Iceland: Earth and Planetary Science Letters, v. 241, p. 271–280, doi: 10.1016/j.epsl.2005.09.066.

Dueker, K.G., and Sheehan, A.F., 1997, Mantle discontinuity structure from midpoint stacks of converted P to S waves across the Yellowstone hotspot track: Journal of Geophysical Research, v. 102, p. 8313–8327, doi: 10.1029/96JB03857.

Efron, B., and Tibshirani, R., 1991, Statistical data analysis in the computer age: Science, v. 253, p. 390–395, doi: 10.1126/science.253.5018.390.

Flanagan, M.P., and Shearer, P.M., 1998, Global mapping of topography on transition zone velocity discontinuities by stacking of *SS* precursors: Journal of Geophysical Research, v. 103, p. 2673–2692, doi: 10.1029/97JB03212.

Foulger, G.R., Pritchard, M., Julian, B., Evans, J., Allen, R., Nolet, G., Morgan, W., Bergsson, G., Erlendsson, P., Jakobsdttir, S., Ragnarsson, S., Stefansson, R., and Vogfjird, K., 2000, The seismic anomaly beneath Iceland extends down to the mantle transition zone and no deeper: Geophysical Journal International, v. 142, p. F1, doi: 10.1046/j.1365-246x.2000.00245.x.

Foulger, G.R., Pritchard, M., Julian, B., Evans, J., Allen, R., Nolet, G., Morgan, W., Bergsson, G., Erlendsson, P., Jakobsdttir, S., Ragnarsson, S., Stefansson, R., and Vogfjird, K., 2001, Seismic tomography shows that upwelling beneath Iceland is confined to the upper mantle: Geophysical Journal International, v. 146, p. 504, doi: 10.1046/j.0956-540x.2001.01470.x.

Frost, D.J., and Dolejs, D., 2006, Experimental determination of the effect of H_2O on the 410-km seismic discontinuity: Earth and Planetary Science Letters, v. 256, p. 182–195, doi: 10.1016/j.epsl.2007.01.023.

Gu, Y., Dziewonski, A.M., and Agee, C.B., 1998, Global de-correlation of the topography of transition zone discontinuities: Earth and Planetary Science Letters, v. 157, p. 57–67, doi: 10.1016/S0012-821X(98)00027-2.

Helffrich, G., 2000, Topography of the transition zone seismic discontinuities: Reviews of Geophysics, v. 38, p. 141–158, doi: 10.1029/1999RG 000060.

Hirose, K., 2002, Phase transitions in pyrolitic mantle around 670-km depth: Implications for upwelling of plumes from the lower mantle: Journal of Geophysical Research, v. 107, doi: 10.1029/2001JB000597.

Hofmann, A.W., 1997, Mantle geochemistry: The message from oceanic volcanism: Nature, v. 385, p. 219–229, doi: 10.1038/385219a0.

Irifune, T., and Ringwood, A.E., 1987, Phase transformations in a harzburgite composition to 26 GPa: Implications for dynamical behavior of the subducting slab: Earth and Planetary Science Letters, v. 86, p. 365–376, doi: 10.1016/0012-821X(87)90233-0.

Ito, E., and Takahashi, E., 1989, Post-spinel transformations in the system Mg_2SiO_4-Fe_2SiO_4 and some geophysical implications: Journal of Geophysical Research, v. 94, p. 10,637–10,646.

Li, X., Kind, R., Priestly, K., Sobolev, S.V., Tilmann, F., Yuan, X., and Weber, M., 2000, Mapping the Hawaiian plume conduit with converted seismic waves: Nature, v. 405, p. 938–941, doi: 10.1038/35016054.

Li, X., Kind, R., and Yuan, X., 2003a, Seismic study of upper mantle and transition zone beneath hotspots: Physics of the Earth and Planetary Interiors, v. 136, p. 79–92, doi: 10.1016/S0031-9201(03)00021-9.

Li, X., Kind, R., Yuan, X., Sobolev, S.V., Hanka, W., Ramesh, D., Gu, Y., and Dziewonski, A., 2003b, Seismic observations of narrow plumes in the oceanic upper mantle: Geophysical Research Letters, v. 30, p. 67-1–67-4, doi: 10.1029/2002GL015411.

Montelli, R., Nolet, G., Masters, G., Dahlen, F., and Hung, S.-H., 2004, Finite-frequency tomography reveals a variety of plumes in the mantle: Science, v. 303, p. 338–343, doi: 10.1126/science.1092485.

Montelli, R., Nolet, G., Dahlen, F., and Masters, G., 2006, A catalogue of deep mantle plumes: New results from finite-frequency tomography: Geochemistry, Geophysics, Geosystems, v. 7, art. no. Q11007, doi: 10.1029/2006GC001248.

Mooney, W.D., Laske, G., and Masters, G., 1995, A new global crustal model at 5 × 5 degrees: CRUST5.0: Eos (Transactions, American Geophysical Union), v. 76, p. F421.

Morgan, W., 1971, Convection plumes in the lower mantle: Nature, v. 230, p. 42–43, doi: 10.1038/230042a0.

Nataf, H.-C., 2000, Seismic imaging of mantle plumes: Annual Review of Earth and Planetary Sciences, v. 28, p. 391–417, doi: 10.1146/annurev.earth .28.1.391.

Neele, F., de Regt, H., and VanDecar, J., 1997, Gross errors in upper-mantle discontinuity topography from underside reflection data: Geophysical Journal International, v. 129, p. 194–204.

Priestley, K., and Tilmann, F., 1999, Shear-wave structure of the lithosphere above the Hawaiian hotspot from two-station Rayleigh wave phase velocity measurements: Geophysical Research Letters, v. 26, p. 1493–1496, doi: 10.1029/1999GL900299.

Ringwood, A.E., 1975, Composition and petrology of the Earth's mantle: New York, McGraw-Hill, 618 p.

Ritsema, J., and Allen, R., 2003, The elusive mantle plume: Earth and Planetary Science Letters, v. 207, p. 1–12, doi: 10.1016/S0012-821X(02)01093-2.

Ritsema, J., van Heijst, H., and Woodhouse, J., 1999, Complex shear wave velocity structure imaged beneath Africa and Iceland: Science, v. 286, p. 1925–1928, doi: 10.1126/science.286.5446.1925.

Shearer, P.M., 2000, Upper mantle seismic discontinuities, in Karato, S., et al., eds., Earth's deep interior: Mineral physics and tomography from the atomic to the global scale: Washington, D.C., American Geophysical Union Geophysical Monograph 117, p. 115–131.

Shen, Y., Solomon, S., Bjarnason, I., and Wolfe, C., 1998, Seismic evidence for a lower-mantle origin of the Iceland plume: Nature, v. 395, p. 62–65, doi: 10.1038/25714.

Simmons, N.A., and Gurrola, H., 2000, Multiple seismic discontinuities near the base of the transition zone in the Earth's mantle: Nature, v. 405, p. 559–562, doi: 10.1038/35014589.

Sleep, N., 1990, Hotspots and mantle plumes: Some phenomenology: Journal of Geophysical Research, v. 95, p. 6715–6736.

van der Hilst, R., and de Hoop, M., 2005, Banana-doughnut kernels and mantle tomography: Geophysical Journal International, v. 163, p. 956–961.

Vinnik, L., Chevrot, S., and Montagner, J.P., 1997, Evidence for a stagnant plume in the transition zone?: Geophysical Research Letters, v. 24, p. 1007–1010, doi: 10.1029/97GL00786.

Weidner, D.J., and Wang, Y., 1998, Chemical- and Clapeyron-induced bouyancy at the 660 km discontinuity: Journal of Geophysical Research, v. 103, p. 7431–7441, doi: 10.1029/97JB03511.

Weidner, D.J., and Wang, Y., 2000, Phase transformations: Implications for mantle structure, in Karato, S., et al., eds., Earth's deep interior: Mineral physics and tomography from the atomic to the global scale: Washington, D.C., American Geohysical Union Monograph 117, p. 215–235.

MANUSCRIPT ACCEPTED BY THE SOCIETY 31 JANUARY 2007

DISCUSSION

21 December 2006, Don L. Anderson

The data in the article by Deuss (this volume) show that there is no significant difference in transition zone (TZ) thicknesses between hotspot and non-hotspot regions. The average TZ thickness under hotspots is 237 ± 10 km, which is within 0.5σ of the global average. Yellowstone, Hawaii, Réunion, and Easter have thicker TZs than the global mean, suggesting colder than average temperatures, at least for an olivine-dominated mantle. In general, however, the depths of the discontinuities and the TZ thicknesses show no correlation with hotspots. This elusiveness of plumes to seismic detection is usually attributed to the small size of plume tails and poor resolution (rather than to the absence of plumes). But other larger-scale plume-related predictions can be tested. In particular, plumes and plume heads are predicted to spread out under the lithosphere and under the 650-km discontinuity—if the barrier in question has a negative Clapeyron slope—and these features are well within the ability of surface waves and *SS* precursors, respectively, to detect (e.g., Anderson et al., 1992a,b). Many authors have suggested that plumes originate from the so-called lower-mantle superplumes. If so, much of the mantle under Africa and the Pacific should have shallow 410-km discontinuities and warm transition regions—which do not seem to be the case.

It is useful to compare the means and standard deviations for the various data sets now available. The average thickness of the TZ for hotspots is 237 ± 10 km; the global average is 242 ± 20 km. From a statistical point of view, the two populations are indistinguishable. Thirteen hotspots were separated out by Deuss (this volume) as possible plumes. The TZ thickness for this subset is 234 ± 13 km, essentially identical to the other two populations. Most of these hotspots have small swells and buoyancy fluxes (Courtillot et al., 2003) and can be related to tectonic features. If plumes originate in the lower mantle, the TZ should be particularly thin under the "primary plumes" (Hawaii, Iceland, Easter, Afar, and Réunion; see Courtillot et al., 2003; Anderson, 2005). The average TZ thickness for these is 243 ± 9 km (Iceland data are from Du et al., 2006; Afar data from Nyblade et al., 2000). A TZ thickness of 244 ± 19 km is obtained for the Afar region, consistent with the global average and with other hotspots. This thickness suggests that the pronounced thermal anomaly beneath Afar that exists in the shallow mantle probably does not extend as far down as the TZ (Nyblade et al., 2000). This possible failure to reach the TZ is significant, because Afar is one of the highest-scoring hotspots, as far as having strong plume credentials, and has some of the lowest upper-mantle velocities in the world.

There is doubt regarding whether TZ thicknesses can be used as a thermometer, but the depth of the 410-km discontinuity has fewer complications and uncertainties. The average depth of this discontinuity for the thirteen plume candidates is 417 ± 6 km; the total hotspot list (Deuss, this volume, Table 1) gives 411 ± 9 km, statistically the same as the global average. The inferred temperature standard deviation is small, about ±75° C (the total range is −100 to +250° C).

Deuss states that it is not possible to use *SS* data to draw general conclusions in favor or against the existence of plumes, but she may be too pessimistic. A long-lived plume will heat up the surrounding mantle over distances much greater than the nominal 100- to 150-km radius of a plume. In fact, some seismologists claim to see seismic evidence for many plumes (what they mean is, evidence for low velocity in the mantle, usually having lateral dimensions of >10°). Some studies (see references in Deuss, this volume) attribute all very large low-velocity regions to plumes. The features that are labeled plumes in these papers have lateral dimensions of 10° to >20° at TZ depths. The fact that the *SS* data do not see variations in TZ thickness or "410" depths suggests that temperature variations are slight, certainly not plume-like. The present results can be used to rule out lateral spreading of plumes beneath the 650-km discontinuity and can rule out the high temperatures for the larger features that have been attributed to plumes because of their low seismic velocity. It is common to attribute lack of evidence for plumes to lack of resolution, but it may be due to the absence of plumes.

Other studies (see references in Deuss, this volume, and www.mantleplumes.org) have shown that the 650-km discontinuity is remarkably immune to variations in 410-km discontinuity depths, surface tectonics and magmatism, and upper-mantle seismic velocities and inferred temperatures. The lack of a strong anticorrelation of discontinuity depths suggests that olivine is not the dominant TZ phase or that the high-temperature thermal anomalies inferred at shallow depths do not extend through the TZ. A plausible interpretation of all of the data is that there are no large temperature excursions at TZ depths under surface hotspots or associated with low-seismic-velocity regions in the mantle. A developing consensus is that many seismic features are due to composition, not temperature. A minority view is that features that have been attributed to plumes may actually have a shallow origin and may be, in large part, due to lithology or high homologous temperature, not high absolute temperature. Deuss (this volume) supports that view.

26 January 2007, Benoît Tauzin

Most seismic observations of the TZ discontinuities have been interpreted in terms of phase transitions in olivine. In the light of recent developments in high-pressure mineral physics, Deuss (this volume) proposes to include the effect of garnet in the interpretation. This effect is null on the 410-km discontinuity but can be important for the discontinuities at 520- and 660-km depths. Taking into account the garnet phase transition, almost any reasonable mantle transition zone (MTZ) thickness variation can be reconciled with a hot thermal plume, if the 410-km discontinuity is depressed. The author fairly states that *SS* wave data alone cannot be used to draw general conclusions in favor of, or against the existence of, plumes.

Deuss studied the forty-nine hotspot locations of the catalogue of Courtillot et al. (2003) using *SS* precursors. She observed robust precursors generated at the 410- and 660-km discontinuities beneath twenty-six stations. Among these twenty-six stations, two-thirds showed evidence for a deepening of the 410-km discontinuity. The Clapeyron slope of the olivine phase transition at 410 km suggests that this deepening is compatible with a hot thermal anomaly. Such a hot thermal anomaly could be produced by a thermal plume, although a small anomaly that is not vertically continuous through the MTZ would produce the same effect.

SS precursor data have small amplitudes and are very sensitive to the approach used in their analysis. Deuss (this volume) uses a bootstrap resampling technique to estimate uncertainties in the amplitude of the stacks. This technique can also be used to estimate errors in traveltimes. Gu and Dziewonski (2002) showed that, even in well-sampled regions, the traveltime uncertainties may occasionally reach 3 s (~10 km). Taking into account traveltime uncertainties might reduce the number of stations for which a deepening of the 410-km discontinuity is statistically significant.

The traveltime observations beneath hotspots are compared with a global average. Some studies report a thinner MTZ under oceans than beneath continents (–5 km compared with +8 km relative to the global average; Gu et al., 2002). As most of the studied hostposts are located in oceanic regions, one can wonder whether the observed traveltime anomalies would remain if compared with an oceanic average. Global *SS* precursor maps do not show evidence of MTZ thickness variations that might be associated with thermal plumes within Atlantic, Indian, and Pacific oceans (Li et al., 2003).

One-third of the observations show no evidence for a deepening of the 410-km discontinuity and remain to be explained. Some of these observations have traveltime differences (t_{ss} – t_{s410s}) close to the global average and may be attributed to a failure of *SS* precursors to resolve narrow-plume conduits. However, Cape Verde and Canary have $t_{ss} - t_{s410s}$ traveltimes smaller than the global average (up to –5 s for Cape Verde). If correct, this observation indicates that either the mantle above the MTZ is much faster than average, or there is a strong, cold anomaly at 410-km depth that is picked up, despite the broad X-shaped Fresnel zone of *SS* precursors. In either case, the observation seems difficult to reconcile with that of a deep mantle plume by Montelli et al. (2006).

Currently, *SS* precursors and *Pds/Sdp* converted waves put better constraints on the thickness of the MTZ than on the absolute depths of discontinuities. If the effect of the garnet phase transition is significant, MTZ thickness measurements alone are unable to detect thermal plumes. Further experimental work is needed to estimate the relative contribution of olivine and garnet to the topography of the 660-km discontinuity. With new experimental results, and if future progress in seismic tomography allows us to estimate better the absolute depths of seismic discontinuities, there is a hope that *SS* precursors and *Pds/Sdp* data sets will become more accurate thermometers of the TZ. In the meantime, many features of these data sets that are not yet explained could give additional constraints on upper-mantle structure. For example, anomalous phases are sometimes observed just after the 410-km phase (see the negative polarities at the Azores, Comores, and Easter in Fig. 3 of Deuss, this volume). Similar features are observed on *Sdp* records (Vinnik and Farra, 2006) and suggest the existence of low-velocity anomalies within the TZ.

27 January 2007, Lev P. Vinnik

In my view, the seismic literature on plumes is heavily contaminated by controversial data. The main problem of *SS* precursors, a method that is very popular, is that lateral resolution of these data is low, perhaps a couple of thousand kilometers at best, and application of it to objects ten times smaller cannot bring anything but confusion. I doubt whether estimates of the thickness of the TZ made by this method are correct for several reasons, but mostly because *SS* is not the minimum time phase. Comparing the results of this method with another method (*Ps* converted phases in *P* receiver functions) revealed almost no correlation (Chevrot et al., 1999).

P receiver functions have better resolution, but this simple instrument can, in some hands, be harmful. Among results for plumes I would single out those obtained for Iceland, for at least two reasons. First, the data were analyzed by at least two independent groups, and the tilted conduit and topography on the 660-km discontinuity reported by one group were not confirmed by the other group (Du et al., 2006). Without independent checking, the tilted conduit and deformed 660-km discontinuity beneath Iceland claimed by the first group would have become favorite subjects for citation and imitation. Second, the network in Iceland is quite large, and there are several independent tomographic studies of the region, which means that topography on the 410-km discontinuity could be separated from the effects of volumetric velocity variations. In most other regions this opportunity is lacking.

Of course, tomography of Iceland was conducted by assuming that the mantle beneath 400 km is laterally homogeneous. This assumption is certainly untrue, but the related effects have not been evaluated by anyone. Another approach to the problem of discontinuity topography is to measure the differential time between the *Ps* phases related to the 410- and 660-km discontinuities. The thickness of a hot TZ and the related differential time should be small relative to the standard. This popular wisdom neglects the fact that the *P*- and *S*-wave velocities in the TZ of hotspots could be anomalous also. The study of Du et al. (2006) demonstrates that this possibility is of extreme importance for the issue of topography at the 660-km discontinuity. Unfortunately, sufficiently accurate data on velocities in the TZ beneath most hotspots are unavailable.

Low-velocity layers in the TZ and immediately above the 410-km discontinuity are very important for plumes, but I do not know whether these topics have been presented adequately. The author of a review paper should separate signals from noise in the literature, but this is difficult, both politically and scientifically.

27 January 2007, Don L. Anderson

The comments by Vinnik (27 January) are valid, but they do not change the conclusion of Deuss (this volume) and many previous studies that there is no correlation between TZ thickness and hotspots. This conclusion is based on many different phases and many different authors and methods. The only robust conclusion is that oceanic regions have a thinner TZ than for continents, and many subduction zones have thicker TZ (see http://www.mantleplumes.org/TransitionZone.html for references). Thus, TZ thickness joins heatflow and magma temperature as having little correlation with melting anomalies.

There are numerous reports of low-velocity zones (LVZs) atop the 410- and 650-km discontinuities; these are quite consistent with eclogite sinkers being trapped at these depths (Anderson, this volume) but not with hot upwelling plumes. Hot plumes will spread out beneath endothermic phase changes, but not above phase boundaries. Low seismic velocities are consistent with eclogite or magma, but also with CO_2-bearing peridotite (Presnall and Gudfinnsson, 2007). Low seismic velocity and high magma volume are not proxies for high absolute temperature or low density. Even if plumes are only 100–200 km across, their effect on the surrounding mantle and their spreading out in the shallow mantle will affect the long-wavelength temperature and structure of the mantle (Morgan and Phipps Morgan, this volume; King and Redmond, this volume). However, the low seismic velocities of solid or partially molten eclogite blobs are not the result of excess absolute temperature, and their influence is relatively local. Eclogite blobs and peridotite infiltrated with eclogite melt can have low seismic velocities even if colder than ambient mantle. All-in-all, the seismic data are consistent with a top-down fertile blob model but not with a bottom-up thermal plume model.

1 February 2007, Yu Jeffrey Gu

Global decorrelation of TZ discontinuities has long been a mystery based on earlier studies of *SS* precursors (Shearer, 1993; Flanagan and Shearer, 1998; Gu et al., 1998, 2003). Explanation of this enigmatic result ranged from a poorly behaved 410-km discontinuity to inaccuracies in traveltime measurements and/or mantle shear-velocity corrections. By systematically examining forty-nine hotspot locations (twenty-six of them relatively well resolved), Deuss (this volume) underlines the critical effect of majorite-garnet phase transformation on the 660-km discontinuity. This hypothesis highlights the complexities within the TZs and offers a viable explanation for the lack of a better anti-correlation between its two major phase boundaries.

The author introduces an important new criterion to determine the depth of origin of thermal plumes, emphasizing the topography of the 410-km discontinuity—the simpler of the two TZ phase boundaries. This hypothesis combines pre-existing knowledge about the exothermic phase change of olivine (Anderson, 1967) with statistically significant observations of a locally depressed 410-km discontinuity (as well as a normal-to-thin transition zone) at a substantial fraction (eighteen out of twenty-six) of potential plume locations. The Deuss article represents a key contribution to the identification and understanding of mantle plumes.

From a data perspective, this article uses *SS* precursors as the main observables on the existence and/or vertical extent of thermal plumes. This choice is appropriate, as the differential traveltimes of underside reflections can be highly sensitive to TZ thickness, as well as to the vertical shear velocity at comparable depths beneath the midpoints of the source-receiver pairs. However, questions have surfaced in recent years regarding the accuracy of the structure and/or topography inferred from *SS* precursors because of the mini-max nature of reflected waves and their wide Fresnel zones at long periods (Chaljub and Tarantola, 1997; Neele et al., 1997). Shearer et al. (1999) addressed some of the potential biases through a multi-scale resolution analysis. By inverting for synthetic differential travel times, they showed

that a topographic inversion using long-period *SS* precursor observations is virtually immune to smaller-scale artifacts at a major subduction zone. That being said, a hot thermal mantle plume is more difficult to detect than a cold subducted oceanic lithosphere for at least two reasons: (1) a potentially smaller lateral dimension and (2) the effect of wavefront healing.

The Fresnel zone of the *SS* precursors used in this study has a lateral dimension of 1000–1500 km, which is generally considered larger than the nominal resolution of a primary plume (Courtillot et al., 2003; Deuss, this volume). However, the footprints of the *SS* precursors are comparable to or smaller than those for secondary plumes beneath many hotspot locations. Even for some primary plumes or hotspots—for example, Iceland—high-resolution receiver function studies have shown reasonably coherent measurements of *Pds* (*d* for a discontinuity) across multiple bins that span ~1000 km laterally (Shen et al., 2003; Du et al., 2006). An anomaly of such dimensions can be effectively imaged by a targeted approach (Deuss, this volume) using *SS* precursors. The combination of heat dissipation from the ascending plume and the potential ponding of plume material at the base of the 660-km discontinuity (because of the endothermic phase transition of olivine) also aid long-period observations, such as *SS* precursors. The effect of wavefront healing is difficult to gauge, though it is reasonable to assume that models generally underestimate the absolute values of velocity and discontinuity topography in the presence of hot thermal plumes.

The study of Deuss (this volume) finds that approximately two-thirds of the twenty-six reasonably resolved hotspot locations have a depressed 410-km discontinuity and a normal or thinner TZ. This general criterion for detecting mantle plumes could potentially benefit from high-resolution methods using receiver functions at, for example, the Hawaii hotspot. These different methods could be complementary, as a recent global study by Lawrence and Shearer (2006) showed remarkable agreements between *Pds*- and *SdS*-derived TZ thickness variations (Fig. D-1). Each method has its caveats. Low resolution remains the Achilles' heel for *SS* precursors, though recent studies by Schmerr and Garnero (2006) and An et al. (2007) have shown some promise by utilizing small averaging caps (both studies), by including higher frequency signals (former study), and by introducing high-resolution least-squares radon transforms (latter study). In the study of An et al. (2007), the preliminary topography measurements at four out of five primary hotspot locations satisfy the TZ structure criterion proposed by Deuss (this volume). In comparison, *Sdp* or *Pds* waves can achieve a significantly greater resolution (100–500 km), though the interpretation of their arrival times is strongly influenced by both *P* and *S* velocities across a given discontinuity. Unfortunately, the *P* velocity resolution at TZ depths leaves much to be desired. A bigger issue is data coverage, as *Pds* and *Sdp* waves rely solely on seismic station locations that are, at present, largely continental.

The moral is that there is no perfect method for detecting and interpreting thermal plumes that consist of distinct types, chemical signatures, and surface expressions. Until more ocean-bottom

A)

Gu et al., 1998

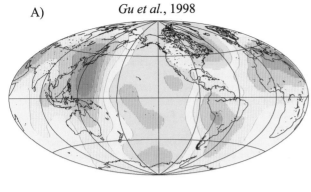

B)

This Study

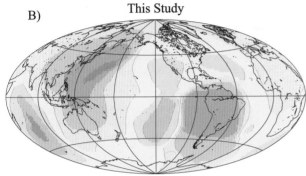

C)

Flannagan & Shearer, 1998

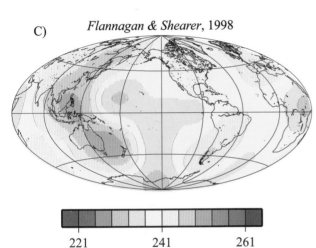

221   241   261

MTZ Thickness (km)

Figure D-1. Smoothed maps of transition zone thickness for (A) Gu et al. (1998), (B) Lawrence and Shearer (2006), and (C) Flanagan and Shearer (1998). MTZ—mantle transition zone. Reproduced from Lawrence and Shearer (2006) with the authors' permission.

seismographs are deployed around the world, *SS* precursors will continue to be one of the most effective global imaging tools to determine the existence and/or vertical extent of mantle plumes.

9 February 2007, Arwen Deuss

The purpose of my article is twofold: (1) to add another seismological data set to the mantle plume debate and (2) to show

that positive thermal anomalies would be consistent with both thin and normal TZs, complicating our search for mantle plumes. *SS* precursors have been used in this study, as they can be focused on proposed hotspot locations. It will be important to add *Pds* receiver functions, and other data types, but none of these data types currently has the global coverage to study a large proportion of oceanic hotspot locations.

It is interesting to note, however, that the comments on my article by Anderson, Tauzin, Vinnik, and Gu show that the same data set can be interpreted as either showing significant evidence in favor of mantle plumes (Deuss, this volume, and comment by Gu), or against mantle plumes (comments by Anderson, Tauzin, and Vinnik). Personal opinion apparently strongly determines the preferred interpretation of a complex data set. Hopefully, at some point enough data will become available, with unequivocal statistical significance, so that we can all agree on the same interpretation.

REFERENCES CITED

An, Y., Gu, Y.J., and Sacchi, M., 2007, Imaging mantle discontinuities using least-squares Radon transform: Journal of Geophysical Research, in press.

Anderson, D.L., 1967, Phase changes in the upper mantle: Science, v. 157, p. 1165–1173.

Anderson, D.L., 2005, Scoring hot spots: The plume and plate paradigms, *in* Foulger, G.R., et al., eds., Plates, plumes, and paradigms: Boulder, Colorado, Geological Society of America Special Paper 388, p. 31–54, doi: 10.1130/2005.2388(04).

Anderson, D.L., 2007 (this volume), The Eclogite engine: Chemical geodynamics as a Galileo thermometer, *in* Foulger, G.R., and Jurdy, D.M., eds., Plates, plumes, and planetary processes: Boulder, Colorado, Geological Society of America Special Paper 430, doi: 10.1130/2007.2430(03).

Anderson, D.L., Tanimoto, T., and Zhang, Y.-S., 1992a, Plate tectonics and hotspots: The third dimension: Science, v. 256, p. 1645–1650.

Anderson, D.L., Zhang, Y.-S., and Tanimoto, T., 1992b, Plume heads, continental lithosphere, flood basalts and tomography, *in* Storey, B.C., et al., Magmatism and the causes of continental break-up: Geological Society [London] Special Publication 68, p. 99–124.

Chaljub, E., and Tarantola, A., 1997, Sensitivity of SS precursors to topography on the upper-mantle 660-km discontinuity: Geophysical Research Letters, v. 24, p. 2613–2616.

Chevrot, S.I., Vinnik, L.P., and Monttagner, J.-P., 1999, Global scale analysis of the mantle *Pds* phases: Journal of Geophysical Research, v. 104, p. 20,203–20,219.

Courtillot, V., Davaille, A., Besse, J., and Stock, J.M., 2003, Three distinct types of hotspots in the Earth's mantle: Earth and Planetary Science Letters, v. 205, p. 295–308.

Deuss, A., 2007, Seismic observations of transition-zone discontinuities beneath hotspot locations, *in* Foulger, G.R., and Jurdy, D.M., eds., Plates, plumes, and planetary processes: Boulder, Colorado, Geological Society of America Special Paper 430, doi: 10.1130/2007.2430(07).

Du, Z., Vinnik, L.P., and Foulger, G.R., 2006, Evidence from P-to-S mantle converted waves for a flat "660-km" discontinuity beneath Iceland: Earth and Planetary Science Letters, v. 241, p. 271–280.

Flanagan, M.F., and Shearer, P.M., 1998, Global mapping of topography on transition zone velocity discontinuities by stacking SS precursors: Journal of Geophysical Research, v. 103, p. 2673–2692.

Gu, Y. J., and Dziewonski, A.M., 2002, Global variability of transition zone thickness: Journal of Geophysical Research B, v. 107, p. 2135, doi: 0.1029/2001JB000489.

Gu, Y.J., Dziewonski, A.M., and Agee, C.B., 1998, Global de-correlation of the topography of transition zone discontinuities: Earth and Planetary Science Letters, v. 157, p. 57–67.

Gu, Y.J., Dziewonski, A.M., and Ekström, G., 2003, Simultaneous inversion for mantle shear velocity and topography of transition zone discontinuities: Geophysical Journal International, v. 154, p. 559–583.

King, S.D., and Redmond, H.L., 2007 (this volume), The structure of thermal plumes and geophysical observations, *in* Foulger, G.R., and Jurdy, D.M., eds., Plates, plumes, and planetary processes: Boulder, Colorado, Geological Society of America Special Paper 430, doi: 10.1130/2007.2430(06).

Lawrence, J.F., and Shearer, P.M., 2006, A global study of transition zone thickness using receiver functions: Journal of Geophysical Research, v. 111, art. no. B06307, doi: 10.1029/2005JB003973.

Li, X., Kind, R., Yuan, X., Sobolev, S.V., Hanka, W., Ramesh, D.S., Gu, Y.J., and Dziewonski, A.M., 2003, Seismic observation of narrow plumes in the oceanic upper mantle: Geophysical Research Letters, v. 30, p. 1334, doi: 10.1029/2002GL015411.

Montelli, R., Nolet, G., Dahlen, F.A., and Masters, G., 2006, A catalogue of deep mantle plumes: New results from finite-frequency tomography: Geochemistry, Geophysics, Geosystems, v. 7, art. no. Q11007, doi: 10.1029/2006 GC001248.

Morgan, W.J., and Phipps Morgan, J., 2007 (this volume), Plate velocities in the hotspot reference frame, *in* Foulger, G.R., and Jurdy, D.M., eds., Plates, plumes, and planetary processes: Boulder, Colorado, Geological Society of America Special Paper 430, doi: 10.1130/2007.2430(04).

Neele, F., de Regt, H., and Van Decar, J., 1997, Gross errors in upper-mantle discontinuity topography from underside reflection data: Geophysical Journal International, v. 129, p. 194–204.

Nyblade, A., Knox, R., and Gurrola, H., 2000, Mantle transition zone thickness beneath Afar: Implications for the origin of the Afar hotspot: Geophysical Journal International, v. 142, p. 615.

Presnall, D.C., and Gudfinnsson, G., 2007, Global Na8-Fe8 systematics of MORBs: Implications for mantle heterogeneity, temperature, and plumes, abstract EGU2007-A-00436, European Geophysical Union General Assembly, 15–20 April, Vienna.

Schmerr, N., and Garnero, E.J., 2006, Investigation of upper mantle discontinuity structure beneath the central Pacific using SS precursors: Journal of Geophysical Research, v. 111, art. no. B08305, doi: 10.1029/2005 JB004197.

Shearer, P.M., 1993, Global mapping of upper mantle reflectors from long-period SS precursors: Geophysical Journal International, v. 115, p. 578–604.

Shearer, P.M., Flanagan, M.F., and Hedlin, A.H., 1999, Experiments in migration processing of SS precursor data to image upper mantle discontinuity structure: Journal of Geophysical Research, v. 104, p. 7229–7242.

Shen, Y., Wolfe, C.J., and Solomon, S.C., 2003, Seismological evidence for a mid-mantle discontinuity beneath Hawaii and Iceland: Earth and Planetary Science Letters, v. 214, p. 143–151.

Vinnik, L.P., and Farra, V., 2006, S velocity reversal in the mantle transition zone: Geophysical Research Letters, v. 33, art. no. L18316, doi: 10.1029/2006GL027120.

The Geological Society of America
Special Paper 430
2007

Lower-mantle material properties and convection models of multiscale plumes

Ctirad Matyska*

Department of Geophysics, Faculty of Mathematics and Physics, Charles University,
V Holešovičkách 2, 180 00 Prague 8, Czech Republic

David A. Yuen

Department of Geology and Geophysics and Minnesota Supercomputing Institute,
University of Minnesota, 117 Pleasant Street, SE, 599 Walter Library,
Minneapolis, Minnesota 55455-0219, USA

ABSTRACT

We present the results of numerical mantle convection models demonstrating that dynamical effects induced by variable mantle viscosity, depth-dependent thermal expansivity, radiative thermal conductivity at the base of the mantle, the spinel to perovskite phase change and the perovskite to post-perovskite phase transition in the deep mantle can result in multiscale mantle plumes: stable lower-mantle superplumes are followed by groups of small upper-mantle plumes. Both radiative thermal conductivity at the base of the lower mantle and a strongly decreasing thermal expansivity of perovskite in the lower mantle can help induce partially layered convection with intense shear heating under the transition zone, which creates a low-viscosity zone and allows for the production of secondary mantle plumes emanating from this zone. Large-scale upwellings in the lower mantle, which are induced mainly by both the style of lower-mantle viscosity stratification and decrease of thermal expansivity, control position of central upper-mantle plumes of each group as well as the upper-mantle plume-plume interactions.

Keywords: partially layered convection, shear heating, temperature- and pressure-dependent viscosity, depth-dependent thermal expansivity, post-perovskite phase transition, radiative heat transfer in D″

INTRODUCTION

During the past decades, geophysicists have pondered over the causes and consequences of surface hotspots within the framework of plate tectonics. Morgan (1971) developed the seminal idea of thin plumes emanating from the core-mantle boundary and their rapid passage to the surface. Such a mecha-

nism corresponds to some burgeoning fluid-dynamical ideas of boundary-layer instabilities and their finite-amplitude growth to diapirs and thin vertical upwellings (Howard, 1964; Goldstein et al., 1990). The catalogs of surface hotspots (e.g., Burke and Wilson, 1976) were compiled in the 1970s, and since then it has been recognized that some of the hotspots might have origins in the transition zone (Allègre and Turcotte, 1985) or in continen-

*E-mail: ctirad.matyska@mff.cuni.cz.

Matyska, C., and Yuen, D.A., 2007, Lower mantle material properties and convection models of multiscale plumes, *in* Foulger, G.R., and Jurdy, D.M., eds., Plates, plumes, and planetary processes: Geological Society of America Special Paper 430, p. 137–163, doi: 10.1130/2007.2430(08). For permission to copy, contact editing@geosociety.org. ©2007 The Geological Society of America. All rights reserved.

tal edges via small-scale convection in the upper mantle (King and Anderson, 1995). The secondary instabilities from the transition zone were modeled successfully in the 1990s by Honda et al. (1993), Steinbach and Yuen (1994), and Cserepes and Yuen (1997), who showed the importance of sufficiently high resolution in the transition zone in capturing these thinner plumes with a width of about 50 km. Courtillot et al. (2003) also consider that the population of hotspots over the Earth's surface should have their origins at different depths. This idea is also supported by recent seismic imaging by Montelli et al. (2004) and Zhao (2004). Implicit in this type of reasoning is the assumption of a relatively stationary state of mantle convection occurring today.

In the lower mantle, large structures with low seismic velocities were found under Africa and the central Pacific by Dziewonski's pioneering work (Dziewonski, 1984; Su and Dziewonski, 1991) exhibiting axial symmetry (Matyska, 1995) and corresponding to the long-wavelengths geoid anomalies and two geographical groups of surface hotspots (e.g., Crough and Jurdy, 1980; Stefanick and Jurdy, 1984; Richards and Hager, 1988; Matyska et al., 1998). These lower-mantle structures were called "superplumes" and recognized by Maruyama (1994) as a late-stage development of mantle evolution. They should be joined with the secondary smaller plumes generated at the boundary between the lower- and the upper-mantle at a depth of 670 km.

The basic physics of heat transfer in the Earth's mantle can be critically dependent on various combinations of mantle properties. We emphasize that this physics is nonlinear and thus one cannot estimate at all the effects of some material changes a priori without carrying out numerical experiments. The main purpose of our paper is to point out that the multiscale nature of mantle plumes in numerical models is a consequence of the richness and complexity of mantle physical properties and processes, such as phase transitions in both the upper and deep mantle. We demonstrate this point by showing some illustrative examples, drawn from both simple and complex models in the long-time regime of mantle convection. One cannot obtain a scenario of multiscale plumes using a simple physical model. Simple steady-state models without phase transitions may work to some degree in the upper mantle, but still they would have problems explaining volcanoes associated with subducting slabs, which require thermal-chemical Rayleigh-Taylor instabilities (Gerya and Yuen, 2003) or secondary convective instabilities (Honda et al., 2002).

Although thermal-chemical convection has long been recognized as being an integral component of mantle convection (Hansen and Yuen, 1989, 1994; Tackley, 1998; Nakagawa and Tackley, 2005; Tan and Gurnis, 2005), we restrict ourselves to the use of thermal convection. Our purely thermal models have the following geophysical attributes:

1. Phase transitions both at the depth of 670 km and in the lower mantle (Iitaka et al., 2004; Murakami et al., 2004; Oganov and Ono, 2004; Tsuchiya et al., 2004);
2. Radiative thermal conductivity in the lower mantle (Lubi-

mova, 1958; Matyska et al., 1994; van den Berg et al., 2002; Matyska and Yuen, 2005, 2006);
3. Temperature- and depth-dependent viscosity, where the depth-dependent viscosity has a peak in the mid-lower-mantle (Mitrovica and Forte, 2004);
4. Depth-dependent thermal coefficient of expansion (Chopelas and Boehler, 1992; Katsura et al., 2005); and
5. Nonlinear feedback effects, such as viscous dissipation and its coupling to temperature-dependent viscosity.

Other effects, not considered here, which are nonetheless very important, are non-Newtonian rheology (e.g., Larsen and Yuen, 1997), grain-size–dependent rheology (Solomatov, 1996, 2001; Korenaga, 2005) and grain-size–dependent thermal conductivity (Hofmeister, 2005).

MODEL DESCRIPTION AND THEORETICAL BACKGROUND

The fundamental laws of conservation describing transfer of heat in a dynamic Earth under the fluid approximation together with rheological behavior and an equation of state are described in this section (for details, see the electronic lecture notes displayed on the Web page http://geo.mff.cuni.cz/~cm/geoterm.pdf ; see also the monographs, e.g., by Ranalli, 1995, and Schubert et al., 2001).

Equation of Continuity (Conservation of Mass)

The conservation of mass is written as:

$$\frac{\partial \rho}{\partial t} + \nabla \cdot (\rho v) = 0, \tag{1}$$

where ρ is the density, v is the velocity of motion, and t is the time. The symbol ∇ denotes the nabla operator and the dot $\cdot$ is the scalar product.

Momentum Equation

The momentum equation in a nonrotating earth model is:

$$\nabla \cdot \tau + \rho g = \rho \frac{\partial v}{\partial t} + \rho v \cdot \nabla v, \tag{2}$$

where τ is the Cauchy stress tensor and g is the gravity acceleration.

Conservation of Moment of Momentum

Conservation of the moment of momentum is given by:

$$\tau = (\tau)^T, \tag{3}$$

where ^{T} denotes transpose of a matrix.

Rheological Relationship

The stress tensor is considerd in the form:

$$\tau = -p\boldsymbol{I} + \sigma(\boldsymbol{v}), \quad \lim_{\boldsymbol{v} \to 0} \sigma(\boldsymbol{v}) = 0, \qquad (4)$$

where p is the pressure, $\boldsymbol{I}$ is the identity tensor, and σ is a non-pressure part of the stress tensor. We will apply this system to the Newtonian fluid, that is:

$$\sigma = \eta(\nabla\boldsymbol{v} + (\nabla\boldsymbol{v})^T - \frac{2}{3}\nabla \cdot \boldsymbol{v}\boldsymbol{I}) \equiv 2\eta(e - \frac{1}{3}\nabla \cdot \boldsymbol{v}\boldsymbol{I}), \qquad (5)$$

where η is the dynamic viscosity and e the strain-rate tensor.

Heat Equation

For the heat equation, we have:

$$\rho T\left(\frac{\partial s}{\partial t} + \boldsymbol{v} \cdot \nabla s\right) = \nabla \cdot (k\nabla T) + \sigma : \nabla\boldsymbol{v} + Q, \qquad (6)$$

where T is the absolute temperature, s is the entropy per unit mass, k is the thermal conductivity, and Q are the volumetric heat sources; the symbol : denotes the total scalar product of the second-order tensors. The first term on the right-hand side of equation 6 describes conduction of heat and the second term the dissipation of heat.

If we assume that there is a reference hydrostatic state characterized by $\boldsymbol{v} = 0$ in which the hydrostatic pressure p_0, hydrostatic density ρ_0, and hydrostatic gravity acceleration g_0 are related by

$$\nabla p_0 = \rho_0 \boldsymbol{g}_0, \qquad (7)$$

and, moreover, that pressure deviations $\Pi = p - p_0$ are negligible in the heat equation, the transfer of heat in a homogeneous material (i.e., entropy may be considered as a function of only p and T) is then described by the well-known equation:

$$\rho c_p \frac{\partial T}{\partial t} = \nabla \cdot (k\nabla T) - \rho c_p \boldsymbol{v} \cdot \nabla T - \rho v_r \alpha Tg + \sigma : \nabla\boldsymbol{v} + Q, \qquad (8)$$

where c_p is the isobaric specific heat, α is the thermal expansion coefficient, and v_r denotes the radial component of velocity. The left-hand side of equation 8 represents local changes of heat balance; the second (third) term on the right-hand side describes advection of heat (adiabatic heating and/or cooling).

Equation of State

The equation of state gives the density as a function of the pressure and temperature:

$$\rho = \rho(p, T). \qquad (9)$$

Today, with the advances made in computational quantum mechanics (Tsuchiya et al., 2005) $\rho(p,T)$ can be constructed readily as a look-up table and be part of the physical setup for modeling (Jacobs et al., 2006).

Approximations

The widely used Boussinesq approximation linearizes these basic laws near the reference hydrostatic state (e.g., Spiegel and Veronis, 1960; Ogura and Phillips, 1962; Schubert et al., 2001). If we neglect density changes caused by the pressure deviations $\Pi = p - p_0$, we may linearize the state equation with respect to the temperature deviations $T - T_0$, where T_0 is a reference temperature, and write:

$$\rho = \rho_0(1 - \alpha(T - T_0)). \qquad (10)$$

This approximation thus means that the influence of hydrostatic pressure (as well as temperature T_0) on density is hidden in a spatial dependence of the reference density ρ_0. For example, Monnereau and Yuen (2002) used in the role of depth-variable density model ρ_0 the PREM model (Dziewonski and Anderson, 1981).

The reference density ρ_0 is assumed to be a time-independent function. Considering only the largest term in the equation of continuity, that is, neglecting thermal expansion, we arrive at the simplified equation:

$$\nabla \cdot (\rho_0 \boldsymbol{v}) = 0 \qquad (11)$$

(see also Jarvis and McKenzie, 1980). After putting equations 7 and 10 into the momentum equation 2, we get:

$$-\nabla\Pi + \nabla \cdot \sigma - \rho_0\alpha(T - T_0)\boldsymbol{g}_0 + \rho_0(\boldsymbol{g} - \boldsymbol{g}_0) = \rho_0\left(\frac{\partial \boldsymbol{v}}{\partial t} + \boldsymbol{v} \cdot \nabla\boldsymbol{v}\right), \qquad (12)$$

where we have neglected the quadratic term $-\rho_0\alpha(T - T_0)(\boldsymbol{g} - \boldsymbol{g}_0)$ on the left-hand side and the thermal expansion on the right-hand side, that is, the changes of the inertial force caused by the thermal expansion. Note that the deviation of the gravity acceleration $\boldsymbol{g} - \boldsymbol{g}_0$ is due to the self-gravitation of the Earth. The magnitude of the term $\rho_0(\boldsymbol{g} - \boldsymbol{g}_0)$ in the mantle is much lower than that of the buoyancy term $-\rho_0\alpha(T - T_0)\boldsymbol{g}_0$ except for the longest wavelengths (Ricard et al., 1984); therefore, it does not influence substantially the basic physics of mantle thermal convection. For this reason, we omit the self-gravitation term throughout the rest of this study.

Simplification of the heat equation consists of replacing ρ by ρ_0:

$$\rho_0 c_p \frac{\partial T}{\partial t} = \nabla \cdot (k\nabla T) - \rho_0 c_p \boldsymbol{v} \cdot \nabla T - \rho_0 v_r \alpha Tg_0 + \sigma : \nabla\boldsymbol{v} + Q. \qquad (13)$$

The system of equations 11–13 is referred to as the *anelastic liquid approximation* of the basic laws of conservation. However, it is common to neglect compressibility in the equation of continuity (11) and to replace it simply by:

$$\nabla \cdot v = 0. \tag{14}$$

The obtained system of equations 12–14 is then usually called the *extended Boussinesq approximation* (e.g., Christensen and Yuen, 1985), which is suitable for general mantle convection studies because shear heating and adiabatic heating and/or cooling are the substantial physical mechanisms influencing temperature and velocity patterns. However, the role played by compressibility in equation 11 is minor except for regions of phase transitions.

The *classical Boussinesq approximation*, although an oversimplification for mantle convection studies, is suitable for many fluids in laboratory conditions and represents a further substantial simplification of the studied system of equations. The reference density ρ_0, the reference gravity acceleration g_0, the thermal expansion coefficient α, the isobaric specific heat c_p, and the thermal conductivity k are constant and the abovementioned system is applied again to the Newtonian fluid with a constant dynamic viscosity η. Moreover, both dissipation $\sigma : \nabla v$ and adiabatic heating and/or cooling $-\rho_0 v_r \alpha T g_0$ are not taken into account. We thus get the system:

$$\nabla \cdot v = 0, \tag{15}$$

$$-\nabla \Pi + \eta \nabla^2 v - \rho_0 \alpha (T - T_0) g_0 = \rho_0 \left(\frac{\partial v}{\partial t} + v \cdot \nabla v \right), \tag{16}$$

$$\frac{\partial T}{\partial t} = \kappa \nabla^2 T - v \cdot \nabla T + \frac{Q}{\rho_0 c_p}, \tag{17}$$

where $\kappa = k/\rho_0 c_p$ is the thermal diffusivity. Two reasons why the extended Boussinesq approximation has gained popularity is that it is easy to convert a Boussinesq code to extended-Boussinesq code, and computationally, this model is much faster than the anelastic approximation (Jarvis and McKenzie, 1980).

We now introduce new dimensionless variables (denoted by *) by means of the relations:

$$\mathbf{r} = dr^*, t = \frac{d^2}{\kappa_s} t^*, v = \frac{\kappa_s}{d} v^*, \Pi = \frac{\eta_s \kappa_s}{d^2} \Pi^*, T = T_s + (T_b - T_s) T^*, \tag{18}$$

where $\mathbf{r}$ is the position vector and d is the characteristic dimension of the system—for example, the thickness of the mantle in mantle convection problems or the vertical dimension of the fluid layer in problems in Cartesian geometry—and the subscript s denotes surface values of corresponding quantities, whereas b denotes their bottom values. The system of equations 12–14 in dimensionless variables thus reads:

$$\nabla^* \cdot v^* = 0, \tag{19}$$

$$-\nabla^* \Pi^* + \nabla^* \cdot \left(\frac{\eta}{\eta_s} (\nabla^* v^* + (\nabla^* v^*)^T) \right)$$
$$+ Ra_s \frac{\alpha}{\alpha_s} (T^* - T^*_0) e_r = Pr_s^{-1} \left(\frac{\partial v^*}{\partial t^*} + v^* \cdot \nabla^* v^* \right), \tag{20}$$

with e_r being the radial unit vector,

$$\frac{\partial T^*}{\partial t^*} = \nabla^* \cdot \left(\frac{k}{\kappa_s} \nabla^* T^* \right) - v^* \cdot \nabla^* T^* + \frac{Raq_s}{Ra_s}$$
$$- Di_s \frac{\alpha}{\alpha_s} \left(T^* + \frac{T_s}{T_b - T_s} \right) v^*_r + \frac{Di_s}{Ra_s} \frac{\eta}{\eta_s} (\nabla^* v^* + (\nabla^* v^*)^T) : \nabla^* v^*, \tag{21}$$

where we introduced the (surface) Prandtl number $Pr_s = \eta_s / \rho_0 \kappa_s$, the (surface) Rayleigh number $Ra_s = \rho_0 \alpha_s (T_b - T_s) g_0 d^3 / \eta_s \kappa_s$, the (surface) Rayleigh number for heat sources $Raq_s = \rho_0 \alpha_s g_0 Q d^5 / \eta_s \kappa_s k_s$, and the (surface) dissipation number $Di_s = \alpha_s g_0 d / c_p$. Note that the dimensionless heating term $R = Raq_s / Ra_s = \dot{Q} d^2 / k_s (T_b - T_s)$.

As the Prandtl number is extremely high (more than 10^{20}, something like 10^{22} at least) for mantle convection applications, we may use the infinite Prandtl number approximation; that is, we may replace equation 20 by:

$$-\nabla^* \Pi + \nabla^* \cdot \left(\frac{\eta}{\eta_s} (\nabla^* v^* + (\nabla^* v^*)^T) \right) + Ra_s \frac{\alpha}{a_s} (T^* - T^*_0) e_r = 0. \tag{22}$$

From the physical point of view, this approximation means that the inertial force is negligible. As for the other numbers, in most of the models of this study we have used $Ra_s = 10^7$, $Di_s = 0.5$, and $R = 3$, which are standard values for mantle material properties (e.g., Turcotte and Schubert, 2002). Enhanced Joule heating (Braginskii and Meitlis, 1987) caused by an increase of electrical conductivity at the base of the mantle from electronic transitions (Badro et al., 2004; Li et al., 2005) in the D″ layer can also influence the thermal-electrical coupling between the core and the lower mantle. We, however, neglect Joule heating in this study, as it should play a remarkable role only if it reaches very high magnitudes (Matyska and Moser, 1994).

Cartesian Geometry

In this section, we describe the equations that were used to obtain numerical models presented in this study. They are written in Cartesian coordinates (x, z), where x is a dimensionless horizontal coordinate and z denotes the dimensionless depth, that is, $z = 0$ at the surface and $z = 1$ at the bottom of a convecting layer. We can now obtain velocity field satisfying the equation of continuity 19 by expressing velocity in the form:

$$v^* \equiv (v_x^*, v_z^*) = \left(\frac{\partial \psi}{\partial z}, -\frac{\partial \psi}{\partial x}\right). \qquad (23)$$

where $\psi = \psi(x,z)$ is the *stream function*. As

$$v^* \cdot \nabla^* \psi = 0, \qquad (24)$$

it is clear that the isolines of ψ are the streamlines of velocity.

The momentum equation 22 can now be rewritten as:

$$-\frac{\partial \Pi^*}{\partial x} + \frac{\partial}{\partial x}\left[2\frac{\eta}{\eta_s}\frac{\partial^2\psi}{\partial x \partial z}\right] + \frac{\partial}{\partial z}\left[\frac{\eta}{\eta_s}\left(\frac{\partial^2\psi}{\partial z^2} - \frac{\partial\psi}{\partial x^2}\right)\right] = 0, \qquad (25)$$

$$-\frac{\partial \Pi^*}{\partial z} + \frac{\partial}{\partial z}\left[2\frac{\eta}{\eta_s}\frac{\partial^2\psi}{\partial x \partial z}\right] + \frac{\partial}{\partial x}\left[\frac{\eta}{\eta_s}\left(\frac{\partial^2\psi}{\partial z^2} - \frac{\partial^2\psi}{\partial x^2}\right)\right] =$$
$$Ra_s\frac{\alpha}{\alpha_s}(T^* - T_0^*). \qquad (26)$$

After applying the operator $\partial/\partial z$ to equation 25, the operator $\partial/\partial x$ to equation 26, and subtracting both equations, we obtain the final form of the momentum equation:

$$\left(\frac{\partial^2}{\partial z^2} - \frac{\partial^2}{\partial x^2}\right)\left[\frac{\eta}{\eta_s}\left(\frac{\partial^2\psi}{\partial z^2} - \frac{\partial^2\psi}{\partial x^2}\right)\right] + 4\frac{\partial^2}{\partial x \partial z}\left[\frac{\eta}{\eta_s}\frac{\partial^2\psi}{\partial x \partial z}\right] =$$
$$-Ra_s\frac{\partial}{\partial x}\left[\frac{\alpha}{\alpha_s}(T^* - T_0^*)\right]. \qquad (27)$$

The heat equation 21 now reads:

$$\frac{\partial T^*}{\partial t^*} = \frac{\partial}{\partial x}\left(\frac{k}{k_s}\frac{\partial T^*}{\partial x}\right) + \frac{\partial}{\partial z}\left(\frac{k}{k_s}\frac{\partial T^*}{\partial z}\right) - \frac{\partial\psi}{\partial z}\frac{\partial T^*}{\partial x} + \frac{\partial\psi}{\partial x}\frac{\partial T^*}{\partial z} + R$$

$$-Di_s\frac{\alpha}{\alpha_s}\left(T^* + \frac{T_s}{T_b - T_s}\right)\frac{\partial\psi}{\partial x} + \frac{Di_s}{Ra_s}\frac{\eta}{\eta_s}\left[\left(\frac{\partial^2\psi}{\partial z^2} - \frac{\partial^2\psi}{\partial z^2}\right)^2 + 4\left(\frac{\partial^2\psi}{\partial x \partial z}\right)^2\right] =$$

$$= \frac{k}{k_s}\nabla^{*2}T^* + \frac{1}{k_s}\frac{\partial k}{\partial z}\frac{\partial T^*}{\partial z} + \frac{1}{k_s}\frac{\partial k}{\partial T^*}(\nabla^*T^* \cdot \nabla^*T^*) - \frac{\partial\psi}{\partial z}\frac{\partial T^*}{\partial x} +$$

$$\frac{\partial\psi}{\partial x}\frac{\partial T^*}{\partial z} + R$$
$$-Di_s\frac{\alpha}{\alpha_s}\left(T^* + \frac{T_s}{T_b - T_s}\right)\frac{\partial\psi}{\partial x} + \frac{Di_s}{Ra_s}\frac{\eta}{\eta_s}\left[\left(\frac{\partial^2\psi}{\partial z^2} - \frac{\partial^2\psi}{\partial x^2}\right)^2 + 4\left(\frac{\partial^2\psi}{\partial x \partial z}\right)^2\right], \qquad (28)$$

as we consider $k = k(z, T^*)$ (see equation 42 below). We can see that this heat equation is strongly nonlinear, as it contains the nonlinear terms describing the following effects: nonlinear diffusion of heat caused by the temperature dependence of thermal conductivity, horizontal and vertical advection of heat, adiabatic heating and/or cooling, and dissipation of heat. Additional nonlinearity, which appears in equation 27, is caused by the temperature-dependence of viscosity (see equations 40 and 41). In principle, nonlinear creep mechanisms, such as dislocation creep characterized by a dependence of viscosity on the strain-rate tensor, can also be taken into account, but dislocation creep dominates near cold slabs and not in hot lower-mantle regions (McNamara et al., 2001); nevertheless it can still be important in olivine in the upper mantle.

The two scalar equations 27 and 28 for the two scalar unknowns ψ and T^* thus describe thermal convection in a 2-D Cartesian box $<0,a> \times <0,1>$, where a is the aspect ratio of the box. To complete the equations, we need to add boundary conditions. We consider all boundaries to be impermeable with a free-slip:

$$\psi = \frac{\partial^2\psi}{\partial x^2} = 0 \quad \text{for} \quad x = 0, a \quad \text{and} \quad \psi = \frac{\partial^2\psi}{\partial z^2} = 0 \quad \text{for} \quad z = 0, 1. \qquad (29)$$

We assume no horizontal heat flow at the sidewalls

$$\frac{\partial T^*}{\partial x} = 0 \quad \text{for} \quad x = 0, a \quad \text{and} \quad T^* = 0 \quad \text{for} \quad z = 0, T^* = 1 \quad \text{for} \quad z = 1. \qquad (30)$$

In other words, the boundary conditions at the sidewalls are reflecting.

The nonlinearities of the dynamical system described by equations 27–30 are the reason why both the time evolution of the system and its spatial properties are crucially dependent on values of the physical properties. For example, it is well known that the Rayleigh number is the key parameter controlling the chaos in the system (e.g., Turcotte, 1992; Schubert et al., 2001). In this study, we also demonstrate that depth- and temperature-dependence of physical properties, such as viscosity, thermal expansivity, and/or conductivity, are important factors for the length scales of convecting fluid and its time-behavior as well.

Phase Changes

Phase changes in multicomponent systems, such as mantle minerals, are characterized by the existence of zones in which two or more phases coexist. For simplicity, we assume that the depth span of these zones is negligible and that we may describe them as the phase interfaces with jumps of density and entropy. Lateral variations of temperature generate undulations of phase-interface topography, which represent substantial additional buoyancy force caused by the density jump $\Delta\rho$. Moreover, the jump in the entropy results in the release or consumption of latent heat when material flows through the phase-change interfaces. We neglect the density jumps in the continuity equation, as it can only change the velocity by a few percent.

If Γ is the Clapeyron slope of such an interface, its undulation h (measured downward) caused by the temperature difference $T - T_0$ is approximately

$$h = \Gamma(T - T_0)/\rho_o g_0. \qquad (31)$$

The buoyancy of the undulation can thus be described by means of the additional pressure $-\Delta\rho g_0 h$, which has the character of an external force in the momentum equation. In the first-order approximation, we may thus add the force term

$$f = \frac{\Delta\rho}{\rho_0} \Gamma(T - T_0)\delta(r - r_p)e_r \qquad (32)$$

into the left-hand side of equation 12 or the force term

$$f^* = \frac{\Delta\rho}{\alpha_s\rho^2_0 g d} \Gamma Ra_s(T^* - T^*_0)\delta(r^* - r^*_p)e_r \qquad (33)$$

into the left-hand side of equation 20. Here δ is the Dirac δ-function and r_p is the radial distance of the phase interface with temperature T_0.

From a formal point of view, we only replace the thermal expansivity α by

$$\alpha' = \alpha + \frac{\Delta\rho}{\rho^2_0 g_0} \Gamma\delta(r - r_p) \qquad (34)$$

or α/α_s by

$$\frac{\alpha'}{\alpha_s} = \frac{\alpha}{\alpha_s} + \frac{\Delta\rho}{\alpha_s\rho^2_0 g_0 d} \Gamma\delta(r^* - r^*_p) \equiv \frac{\alpha}{\alpha_s} + P\delta(r^* - r^*_p), \qquad (35)$$

where $P = \Delta\rho\Gamma/\alpha_s\rho^2_0 g_0 d$ is called the *phase buoyancy parameter* (see also Christensen and Yuen, 1985). It is clear that the same replacement of the thermal expansivity in the adiabatic heating and/or cooling term of the heat equation then includes the latent heat release or consumption.

Thermal Expansivity, Viscosity, and Radiative Heat Transfer

We have used the two profiles of the depth-dependent thermal expansivity. The first model has been parameterized to

$$\frac{\alpha}{\alpha_s} = \frac{8}{(2 + z)^3}, \qquad (36)$$

where the thermal expansivity decreases by 8/27 across the whole mantle (see, e.g., Zhao and Yuen, 1987). Quite recently, Katsura et al. (2005) announced that Anderson-Grüneisen parameter of perovskite is close to 10, and thus the thermal expansivity decrease over the lower mantle should be higher. By approximating a linear depth dependence of density in both the upper and the lower mantles, and assuming that Anderson-Grüneisen parameter of the upper-mantle minerals is close to 5, which corresponds to the estimates for olivine (Chopelas and Boehler, 1992), we arrive at the following relations (see Fig. 1):

$$\frac{\alpha}{\alpha_s} = (1 + 0.78z)^{-5} \text{ if } 0 \leq z \leq 0.23, \qquad (37)$$

$$\frac{\alpha}{\alpha_s} = 0.44(1 + 0.35(z - 0.23))^{-10} \text{ if } 0.23 < z \leq 1. \qquad (38)$$

Note that the magnitude of thermal expansivity at the top of the lower mantle should be $\sim 3.5 \times 10^{-5}$ K^{-1} according to Katsura et al. (2005).

We have considered the depth dependence of viscosity in the form used by Hanyk et al. (1995; see also Figure 1):

$$\frac{\eta_0(z)}{\eta_s} = 1 + 214.3z \exp(-16.7(0.7 - z)^2), \qquad (39)$$

which gives rise to the lower-mantle viscosity maximum at around 1800 km. Such a maximum is consistent with dynamic geoid and postglacial rebound modeling (Ricard and Wuming, 1991; Forte and Mitrovica, 2001; Mitrovica and Forte, 2004).

To take into account the temperature dependence of viscosity as well, we include the temperature-dependent part:

$$f(T^*) = \exp\left(10.0\left(\frac{0.6}{0.2 + T^*} - 1\right)\right) \qquad (40)$$

in the Arrhenius form of a thermally activated process (e.g., Davies, 1999; see also Matyska and Yuen, 2006). The composite temperature- and depth-dependent viscosity, which has been considered in complex models, is given by

$$\frac{\eta(z, T^*)}{\eta_s} = \frac{\eta_0(z)}{\eta_s} \min\{100, \max\{0.01, f(T^*)\}\}, \qquad (41)$$

that is, the temperature dependence of dimensionless viscosity is confined by the limits 0.01 and 100, so that the momentum equation may be solved by a conjugate-gradient iterative solver.

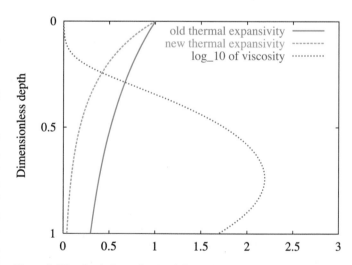

Figure 1. The depth dependence of dimensionless thermal expansivity (see equation 36) for the old thermal expansivity profile and for the new thermal expansivity profile (equations 37 and 38). The remaining curve denotes the decadic logarithm of the depth-dependent part of the dimensionless viscosity (see equation 39).

As one of the main aims of this study is to demonstrate potential creation of the lower-mantle superplumes by means of the radiative transfer of heat, we have neglected the depth and temperature dependence of phonon thermal conductivity (Hofmeister, 1999) and considered the total thermal conductivity in the form:

$$\frac{k(z, T^*)}{k_s} = 1 + g(z)\left(\frac{T_s}{T_b - T_s} + T^*\right)^3, \qquad (42)$$

where the prefactor $g(z)$ enables inclusion of the depth dependence of radiative heat transfer caused by the changes of composition, opacity, and the like.

The set of equations 27–30, which describes our numerical models, thus consists of the fourth-order linear elliptic equation 27 for the streamfunction ψ with the boundary conditions of equation 29 (with both laterally and vertically varying coefficients because of the presence of $\eta(z,T^*)/\eta_s$) and the nonlinear time-evolutionary advection-diffusion equation 28 for the temperature T^* with the boundary conditions described by equation 30. The depth changes of thermal expansivity result in the changes of the buoyancy forcing term in the elliptic equation as well as in the changes of adiabatic heating and/or cooling in the time-evolutionary equation. The inclusion of radiative transfer of heat by means of the temperature-dependent thermal conductivity then changes the ratio between diffusion and advection of heat.

Computations have been carried out in a wide box with an aspect ratio of 10 to avoid the influence of side boundaries. We have used 1281×129 equally distributed nodal points for a second-order finite difference scheme in space. The elliptic equation has been solved by the conjugate-gradient iterative scheme, and a second-order Runge-Kutta scheme has been applied for the time-stepping in the time-evolutionary equation.

RESULTS

Basic Physics of Simple Models: Influence of the Depth Dependence of Mantle Properties

We begin in Figure 2 by illustrating the style of mantle convection without the presence of major phase transitions but with the extended Boussinesq approximation. We vary a whole gamut of depth-dependent properties in the thermal expansion coefficient and viscosity. The top panel is a typical snapshot of temperature field in the long-time regime for constant material properties and $Ra_s = 10^6$. The dissipation number is 0.5. Convection is rather chaotic with many cold downwellings and hot upwellings. We can see that the role played by adiabatic heating and/or cooling is substantial in such a model; plume heads disappear before reaching the top boundary layer of convection because of the rapid cooling due to constant thermal expansivity (Zhao and Yuen, 1987). This extended Boussinesq model thus does not correspond to the classical Boussinesq idea of plumes

generating locally hot material, which then rises and interacts with the cold upper boundary layer (lithosphere). This main feature of convection is not substantially changed when the Rayleigh number is increased to 10^7 in the second panel. The only remarkable difference from the previous case is that influence of the internal heating is minor, and thus we get more symmetry between the cold and hot anomalies.

The middle panel shows the case when the depth-dependent viscosity (see equation 39) is taken into account. Because an increase of viscosity corresponds to a relative decrease of buoyancy in the momentum equation, convection is then characterized by fewer big stable plumes, emerging from the lower thermal boundary layer (see also Hansen et al., 1993, for 2-D models and Cserepes, 1993, for 3-D modeling). It is of great geophysical interest that there is a high lateral temperature contrast between the plume and a very cold ambient mantle in the lower part of the convecting layer, but this contrast becomes small at the top of the model. Such a large thermal contrast in the lower mantle may induce seismic velocity contrasts of a couple of percentage points (e.g., Ni and Helmberger, 2003). The cold thermal boundary layer at the top of the model is unstable and acts as a source of cold "lumps," which slowly fall to the "viscosity hill" (e.g., Ricard and Wuming, 1991; Mitrovica and Forte, 2004) in the mid-lower mantle.

The bottom two panels in Figure 2 represent cases with decreasing thermal expansivity in the lower mantle. They reveal that a similar stabilization of convection cells, creation of big plumes, and lowering of the average temperature also can be produced by a decrease of thermal expansivity in the lower mantle, as it causes a decrease of buoyancy. There is, however, one substantial difference: low thermal expansivity at the bottom of the model also means a low adiabatic cooling of the bottom part of (super)plumes. It results in higher plume temperatures, and thus very hot plume heads (see also Zhao and Yuen, 1987) are able to reach and interact with the cold upper boundary layer.

To deal with more realistic models, where the critical dependence of mantle heat transfer on combinations of mantle properties is demonstrated, we incorporate the effects of variable viscosity, radiative thermal conductivity, and the two major phase transitions in the upper and lower mantles, which are illustrated in Figure 3. These more realistic models allow for a greater variety of dynamical possibilities.

Complex Models

First, we show the results for the old thermal expansivity profile, decreasing by 8/27 across the mantle (see equation 36), depth- and temperature-dependent viscosity, and the two phase transitions characterized by the buoyancy parameters $P_{670} = -0.15$, $P_{D''} = 0.10$. Typical snapshots of temperature are shown in Figure 4, thermal anomalies obtained by subtracting out the horizontally averaged temperature are shown in Figure 5, and corresponding stream functions are shown in Figure 6. The results in the top panel were obtained for the constant thermal con-

No phase changes, k=1, Di=0.5, R=3

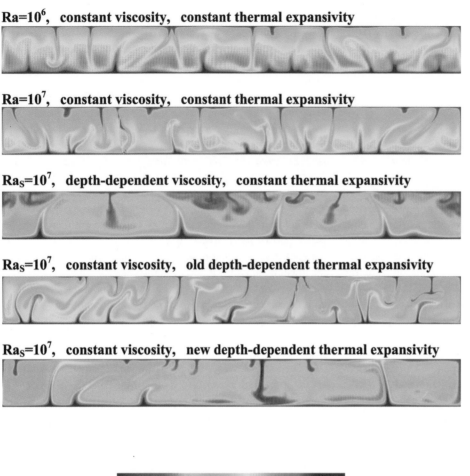

Ra=10⁶, constant viscosity, constant thermal expansivity

Ra=10⁷, constant viscosity, constant thermal expansivity

Ra$_S$=10⁷, depth-dependent viscosity, constant thermal expansivity

Ra$_S$=10⁷, constant viscosity, old depth-dependent thermal expansivity

Ra$_S$=10⁷, constant viscosity, new depth-dependent thermal expansivity

Figure 2. Typical snapshots of the temperature field for a long time scale (on the order of 10^9 years). No phase changes were included. The red color represents the maximum temperature, whereas the dark blue color denotes the minimum temperature. The medium temperature is given by the cold-to-warm transition from green to yellow. The panels show the effect of a change of the surface Rayleigh number from 10^6 to 10^7 in the two top panels and inclusion of the depth-dependence of viscosity or thermal expansivity in the remaining panels. See Figure 1 and equation 36 for the old thermal expansivity, equations 37 and 38 for the new thermal expansivity, and equation 39 for the depth-dependent viscosity.

$T^*=0$ $T^*=1$

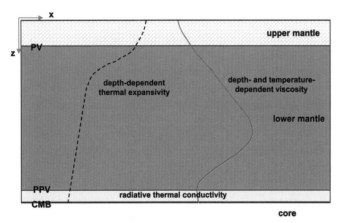

Figure 3. The major geophysical features of the complex models used for generating multiscale plumes. CMB—core-mantle boundary; PPV—postperovskite; PV—perovskite.

ductivity $k = 1$ considered for the entire mantle. We can see that convection is partially layered and the consequence of the temperature dependence of viscosity is the formation of large cold anomalies that are able to penetrate the phase boundary at 670 km. The highest velocity of downwellings is reached in the lower mantle. This downward flow is balanced at 670 km by small plumes originating below this interface, from which hot material is ejected into the upper mantle. The counterpart of the downwellings in the lower mantle are the plumes, which are bigger than those in the upper mantle and are mainly visible in the bottom half of the lower mantle.

The next three panels show the influence of increasing radiative transfer of heat through the D″ layer, which is parameterized by the coefficient g adjacent to the cube of absolute temperature in the radiative thermal conductivity (see equation 42). An important consequence of this phenomenon is an in-

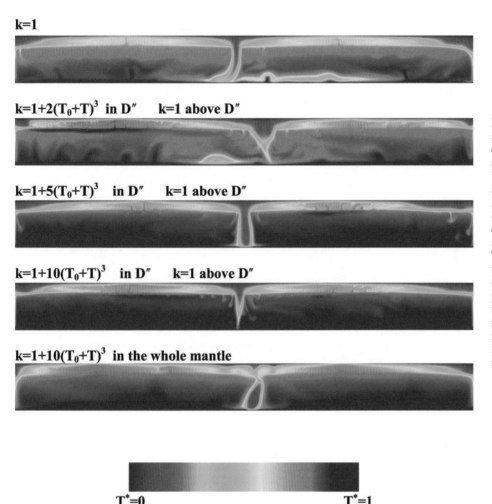

Old thermal expansivity $P_{670} = -0.15$ $P_{D''} = 0.10$

k=1

$k=1+2(T_0+T)^3$ in D″   k=1 above D″

$k=1+5(T_0+T)^3$ in D″   k=1 above D″

$k=1+10(T_0+T)^3$ in D″   k=1 above D″

$k=1+10(T_0+T)^3$ in the whole mantle

$T^* = 0$ $T^* = 1$

Figure 4. Typical snapshots of the temperature field for a long time scale. The old depth-dependent thermal expansivity according to equation 36 was considered. Viscosity was both depth and temperature dependent (see equations 39–41). An endothermic phase change with $P = -0.15$ at the depth of 670 km and an exothermic phase change with $P = 0.10$ at the depth of 2650 km were included. Thermal conductivity was considered in the form of equation 42, where $g(z) = 0$ (no radiative heat transfer) in the top panel and $g(z) = 10$ (strong global radiative heat transfer) in the last panel. In the remaining panels, radiative heat transfer is considered only below the depth 2650 km, with g equal to 2, 5, and 10, respectively.

crease of average temperature in the lower mantle and, subsequently, a much lower temperature contrast between the lower-mantle (super)plumes and the ambient mantle. The vigor of convection is increased together with an enhancement of partial layering.

An outstanding result of these models is the creation of a very hot, thin layer just below 670 km, which is again the source of small upper-mantle plumes. Jetting of hot material into the upper mantle is periodically repeated (inverse flushing events) in several places, thus resulting in the appearance of many upper-mantle plumes, which are then drifted horizontally by an upper-mantle mean flow toward the central location of the main upper-mantle plume. Positions of the central upper-mantle plumes are, however, controlled by the lower-mantle super-plumes. Stability of the lower-mantle superplumes, the attraction of the upper-mantle plumes to one place, and subsequent plume-plume interactions may then explain why the upper-mantle plumes are able to act as a "fixed" sublithospheric heat

source for a long time. Figure 7 illustrates that this hot layer below the 670-km discontinuity and the upper-mantle plumes together form the regions of the lowest viscosities in our models, which correspond to the second asthenosphere, as revealed by Kido and Čadek (1997) under oceanic regions (see also Mitrovica and Forte, 2004).

We also computed the model in which the radiative heating term is taken into account in the whole mantle. Such an overall increase of thermal conductivity further boosted the convective velocities, where cold downwellings in the lower mantle are in the form of huge blobs and the average temperature of the lower mantle decreases. However, lateral temperature contrast between the upper-mantle plumes and the ambient upper mantle becomes smaller.

Figure 8 illustrates that the shear heating generated inside the convecting layers is maximal in downwellings because of the combination of high viscosities together with high rates of deformation; the secondary maxima are usually reached

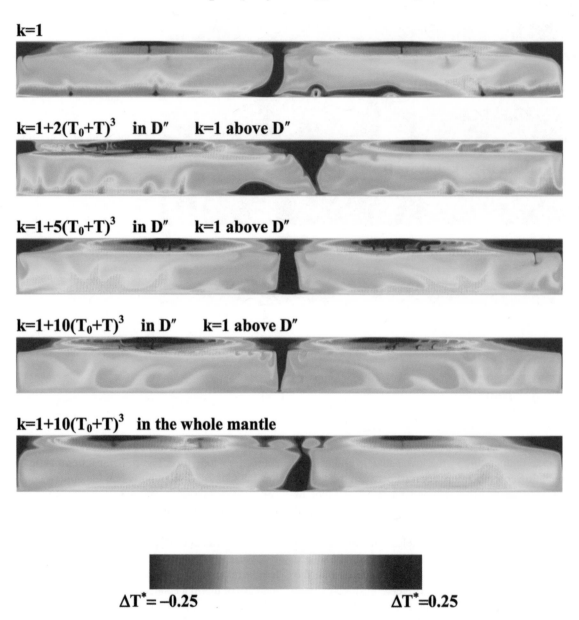

Old thermal expansivity $P_{670} = -0.15$ $P_{D''} = 0.10$

k=1

k=1+2(T₀+T)³ in D″ k=1 above D″

$k=1+2(T_0+T)^3$ in D″ k=1 above D″

$k=1+5(T_0+T)^3$ in D″ k=1 above D″

$k=1+10(T_0+T)^3$ in D″ k=1 above D″

$k=1+10(T_0+T)^3$ in the whole mantle

$\Delta T^* = -0.25$ $\Delta T^* = 0.25$

Figure 5. Residual temperature (the difference between the local temperature and horizontally averaged temperature) for the cases shown in Figure 4.

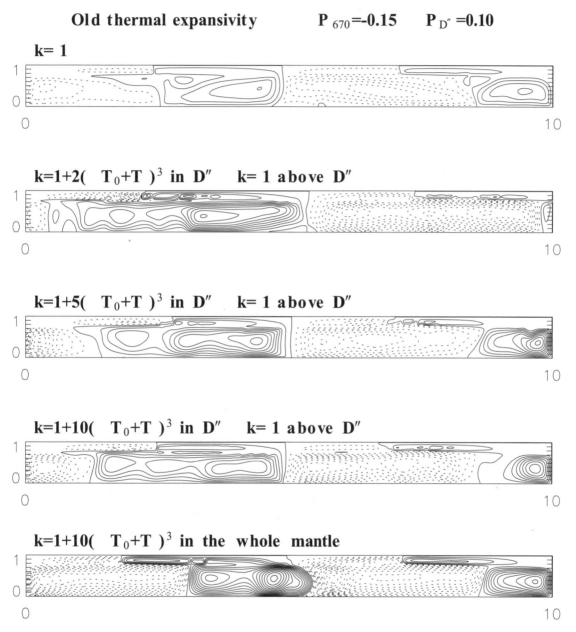

Figure 6. Streamlines of the model flow for the cases shown in Figure 4. The streamfunction is equal to zero on all sides of the box, and the contour interval for streamlines is 50 in dimensionless units. The contour interval is the same in all panels. The solid lines display the zero and negative values and the dashed lines show the positive values of the stream-function.

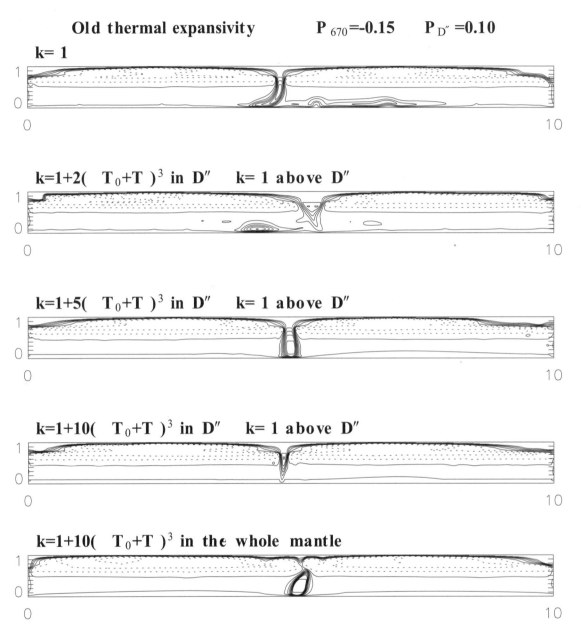

Figure 7. Decadic logarithm of dimensionless viscosity for the cases shown in Figure 4. Dashed lines show negative values and solid lines mark zero and positive values. The contour interval is 0.5 in all panels.

Old thermal expansivity $P_{670} = -0.15$ $P_{D''} = 0.10$

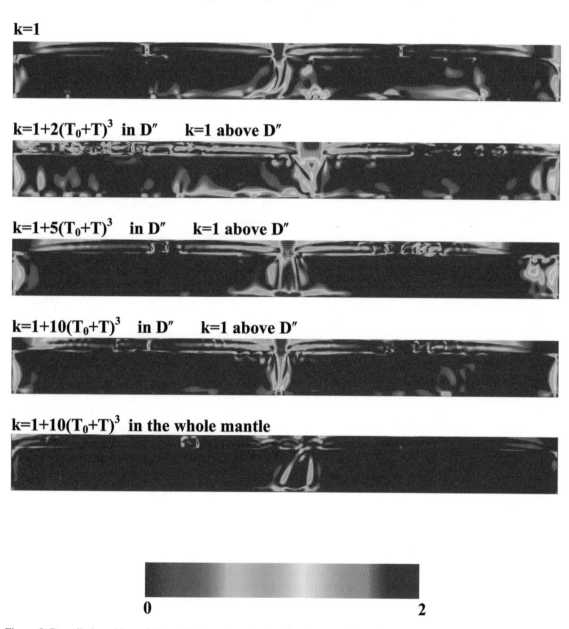

k=1

k=1+2(T$_0$+T)^3 in D''   k=1 above D''

k=1+5(T$_0$+T)^3 in D''   k=1 above D''

k=1+10(T$_0$+T)^3 in D''   k=1 above D''

k=1+10(T$_0$+T)^3 in the whole mantle

0   2

Figure 8. Decadic logarithm of dimensionless shear heating for the cases shown in Figure 4.

inside the upper-mantle plume regions, thus generating the hot layer under the transition zone, as mentioned above. Thus we can see clearly in Figure 8 that shear heating can be locally much higher than the averaged bulk heating, and so it should not be neglected in physical descriptions of complex models where there exist many regions with stagnation points at which flows undergo severe deformation.

Recent measurements on perovskite (Katsura et al., 2005) indicate that decrease of thermal expansivity with increasing pressure is much higher than previously estimated for olivine (Chopelas and Boehler, 1992). For this new thermal expansivity profile in the lower mantle—with a decrease of thermal expansivity in the lower mantle by a factor of eleven—we have considered various models with different types of mantle thermal conductivities, ranging from constant thermal conductivity to prevailing radiative thermal conductivity. As for the old thermal expansivity profile, we have also studied a suite of intermediate cases, in which the radiative conductivity is present only in the D'' layer and is varied from twice to ten times the constant thermal conductivity value. Figures 9 and 10 show the temperature fields and the residual temperature fields. Because the decrease of thermal coefficient of expansion results in a decrease in the average temperature, the lower-mantle plumes can be discerned much more easily than in the previous cases shown in Figures 7 and 8. It is clear that the influence of radiative thermal conductivity and depth-dependent thermal expansivity can be quite profound on the development of the lower-mantle upwellings, because the reference model with constant thermal conductivity results in clusters of smaller, unstable lower-mantle plumes. In contrast, a substantial increase of thermal conductivity in the D'' layer gives rise to the production of lower-mantle superplumes. Broad upwellings are much more prominent, with a thin radiative thermal conductive layer in the D'' layer than with the radiative thermal conductivity distributed throughout the mantle.

In these models, we also obtained flow reversals (see the streamlines shown in Fig. 11) at 670 km boundary and partial layered convection. Flow reversals can generate a considerable amount of mechanical heating. There is a low-viscosity region in the vicinity of the transition zone (see Fig. 12), due to the injection of hot lower-mantle material. Shear heating (Fig. 13) is higher than in the previous models, with a smaller decrease in the thermal expansivity. There are even places located below the 670-km boundary that are overheated by viscous heating to very high temperatures, exceeding the dimensionless temperature of unity, which is the temperature at the core-mantle boundary (we have used a periodic color scale and thus these sites are portrayed by dark blue inside dark red regions). The upper-mantle plumes are again small but very hot.

Note that the convection pattern at 670 km is sensitive to the magnitude of the phase buoyancy parameter P_{670}, buoyancy of the lower-mantle superplumes, and viscosity stratification, because a low-viscosity zone below the interface between the mantles facilitates horizontal flow. We used lower "new" thermal expansivity (see equations 37 and 38) in the second set of models and obtained partial layering for $P_{670} = -0.08$, which is in good agreement with the values estimated for the spinel to perovskite phase change. However, in the first set of models with the old thermal expansivity profile (see equation 36), buoyancy of the lower-mantle superplumes is higher, and we demonstrated that similar partial layering of convection can be obtained for $P_{670} = -0.15$, that is, with slightly overestimated magnitude of the buoyancy parameter.

The change of the model behavior in the case in which the radiative thermal conductivity is considered in the whole mantle (see the bottom panels of Figures 9–13) is now much more remarkable. The combined effect of the small value of thermal expansivity (i.e., small adiabatic gradient) with higher velocities of convection results in a well-developed lower thermal boundary layer. Subsequently, the average lower-mantle temperature is lower, temperature contrast between the superplume and the ambient mantle is higher, and the superplumes are thinner. Moreover, the amount of material passing through the 670-km interface is higher because of higher buoyancy. This combination finally results in thicker upper-mantle plumes. The consequence is that the surface Nusselt number (i.e., the surface heat flow) is rather high, as it varies between 20 and 45. However, typical surface Nusselt numbers of the models with high thermal conductivity confined to D'' layer are lower and usually oscillate between 10 and 15. It is interesting that the model with radiative transfer of heat in the whole mantle yields similar bimodal plumes to those obtained by van Keken et al. (1992) for an ad hoc stratification of rheology and constant thermal conductivity.

To understand better the small-scale features of upper-mantle plumes, we zoomed into the plumes in the upper mantle in Figures 14 and 15 for both thermal expansivity models. It is obvious that many upper-mantle plumes come from the transition zone. The overall morphology of the upper-mantle upwellings is similar in both models, with the upper mantle being slightly hotter for the more steeply decreasing thermal expansivity. Only in the case for which the radiative thermal conductivity prevails throughout the mantle is the upper-mantle plume directly fed from the deep lower mantle (see bottom of Fig. 15). There is also evidence of overheating (see the dark blue color in Fig. 9) in some local areas, corresponding to the roots of upper-mantle plumes. Although such overheating need not be fully realistic for the Earth, it points to the importance of shear heating in convection dynamics.

To demonstrate the time scale of upper-mantle plume dynamics, we present several snapshots of temperature showing the temporal evolution of the upper-mantle plumes in the case for which the thermal conductivity in the D'' layer is dominated by the radiative term, whereas radiative heat transfer is negligible above the D'' layer (see Fig. 16). Time steps between two subsequent panels correspond to ~15 m.y. in dimensional time. We note that non-Newtonian rheology can result in shorter time scales of about a few million years in the upper mantle (Larsen

New thermal expansivity   P$_{670}$ = -0.08   P$_{D''}$ = 0.05

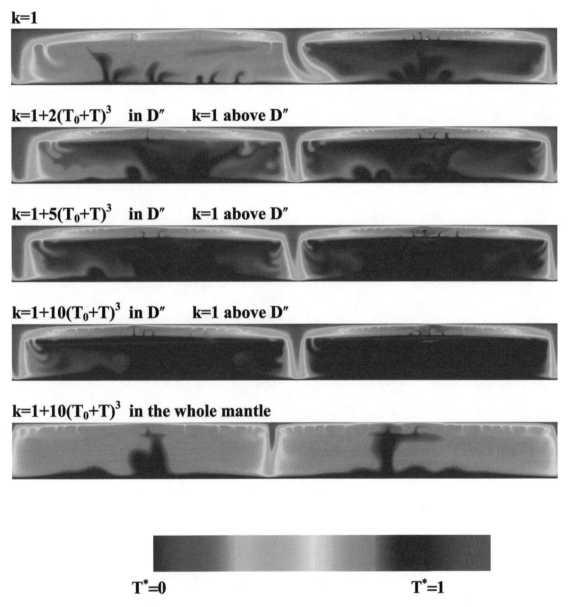

Figure 9. Typical snapshots of the temperature field for a long time scale. The new depth-dependent thermal expansivity according to equations 37 and 38 was considered. Viscosity was both depth and temperature dependent (see equations 39–41). An endothermic phase change with $P = -0.08$ at the depth of 670 km and an exothermic phase change with $P = 0.05$ at the depth of 2650 km were included. Thermal conductivity was considered in the form of equation 42, where $g(z) = 0$ (no radiative heat transfer) in the top panel and $g(z) = 10$ (strong global radiative heat transfer) in the last panel. In the remaining panels, radiative heat transfer is considered only below the depth 2650 km with g equal to 2, 5, and 10, respectively. We used a periodic color scale, that is, dark blue below the transition zone shows where dimensionless temperatures are slightly higher than 1, which corresponds to the temperature at the core-mantle boundary.

New thermal expansivity $P_{670} = -0.08$ $P_{D''} = 0.05$

k=1

k=$1+2(T_0+T)^3$ in D″   k=1 above D″

k=$1+5(T_0+T)^3$ in D″   k=1 above D″

k=$1+10(T_0+T)^3$ in D″   k=1 above D″

k=$1+10(T_0+T)^3$ in the whole mantle

$\Delta T^* = -0.25$ $\Delta T^* = 0.25$

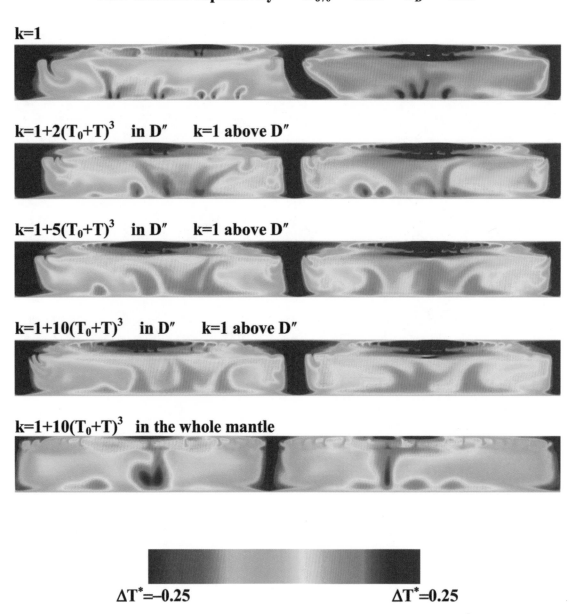

Figure 10. Residual temperature (the difference between the local temperature and horizontally averaged temperature) for the cases shown in Figure 9.

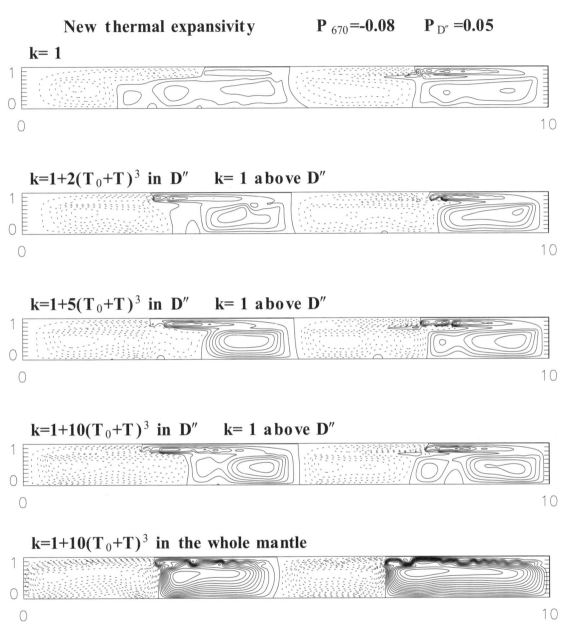

Figure 11. Streamlines of the model flow for the cases shown in Figure 9. The streamfunction is equal to zero on all sides of the box and the contour interval of streamlines is 50 in dimensionless units. The contour interval is the same in all panels. The solid lines display the zero and negative values and the dashed lines show the positive values of the stream-function.

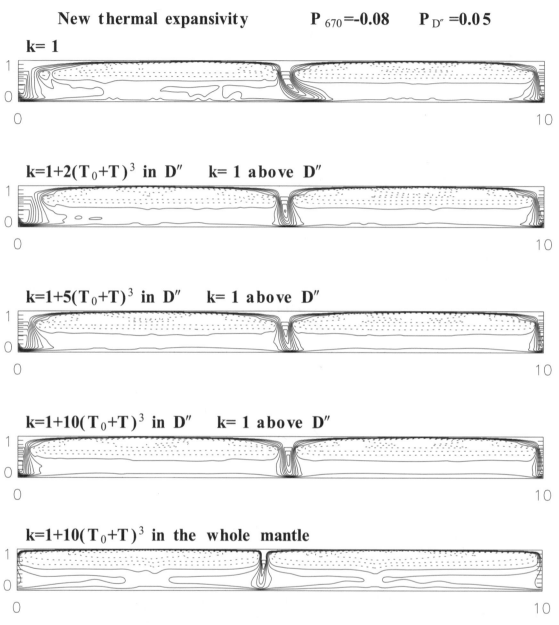

Figure 12. Decadic logarithm of dimensionless viscosity for the cases shown in Figure 9. Dashed lines show negative values and solid lines mark zero and positive values. The contour interval is 0.5 in all panels.

New thermal expansivity $P_{670} = -0.08$ $P_{D''} = 0.05$

k=1

k=1+2$(T_0+T)^3$ **in D″**   **k=1 above D″**

k=1+5$(T_0+T)^3$ **in D″**   **k=1 above D″**

k=1+10$(T_0+T)^3$ **in D″**   **k=1 above D″**

k=1+10$(T_0+T)^3$ in the whole mantle

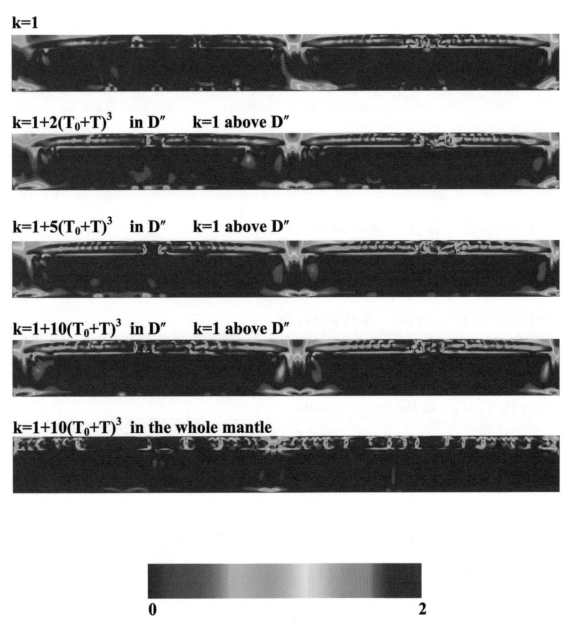

0   2

Figure 13. Decadic logarithm of dimensionless shear heating for the cases shown in Figure 9. Note that the amount of radiogenic heating due to chondritic abundance is ~10.

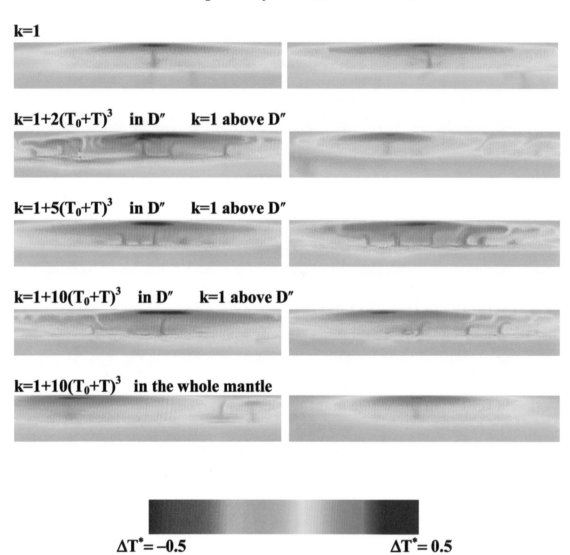

Old thermal expansivity $P_{670} = -0.15$ $P_{D''} = 0.10$

k=1

k=1+2(T₀+T)³   in D″   k=1 above D″

k=1+5(T₀+T)³   in D″   k=1 above D″

k=1+10(T₀+T)³   in D″   k=1 above D″

k=1+10(T₀+T)³   in the whole mantle

ΔT* = −0.5   ΔT* = 0.5

Figure 14. Zoom of the upper-mantle plumes. Left (right) column is for the cases shown in the left (right) part of Figure 4. Vertical range is ~1200 km, and the horizontal length corresponds to ~7200 km. The difference between the local temperature and horizontally averaged temperature is displayed.

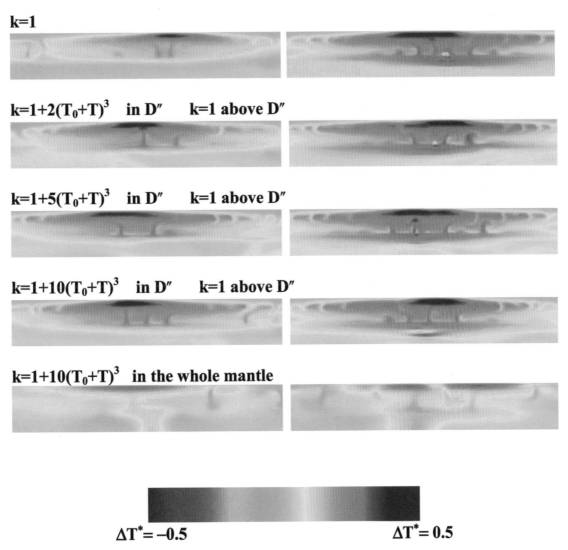

New thermal expansivity $P_{670} = -0.08$ $P_{D''} = 0.05$

k=1

k=1+2$(T_0+T)^3$ in D''   k=1 above D''

k=1+5$(T_0+T)^3$ in D''   k=1 above D''

k=1+10$(T_0+T)^3$ in D''   k=1 above D''

k=1+10$(T_0+T)^3$ in the whole mantle

$\Delta T^* = -0.5$ $\Delta T^* = 0.5$

Figure 15. Zoom of the upper-mantle plumes. Left (right) column is for the cases shown in the left (right) part of Figure 9. Vertical range is ~1200 km, and the horizontal length corresponds to ~7200 km. The difference between the local temperature and horizontally averaged temperature is displayed.

New thermal expansivity P_{670} = -0.08 $P_{D''}$ = 0.05

$k=1+10(T_0+T)^3$ in D'', k=1 above D'';   time steps = 0.00005

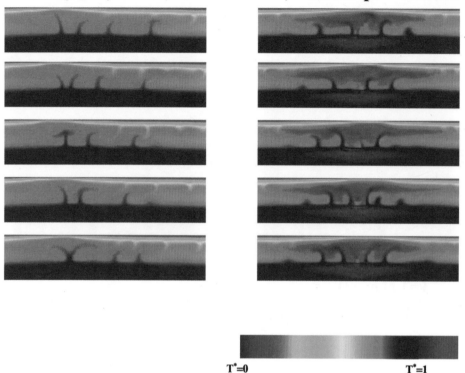

Figure 16. Time series of snapshots of the upper-mantle plumes, in which the depth-dependent thermal expansivity according to equations 37 and 38 was considered. An endothermic phase change with $P = -0.08$ at the depth of 670 km and an exothermic phase change with $P = 0.05$ at the depth of 2650 km were included. Thermal conductivity was considered in the form of equation 42, where radiative heat transfer is considered only below the depth 2650 km with $g = 10$. Vertical range is ~1100 km, and the horizontal length corresponds to ~5300 km. Dimensionless time differences between two subsequent snapshots is 0.00005, corresponding to ~15 m.y. We used a periodic color scale, that is, dark blue below the transition zone shows where the dimensionless temperatures are slightly higher than 1.

$T^*=0$ $T^*=1$

and Yuen, 1997). In our model, the temperature of the plume roots below the 670-km interface is very high—it can even be locally slightly higher than the temperature of the core-mantle boundary, which results in faster plume-plume interactions and more turbulent plume evolution in comparison to places where the upper-mantle plume roots are colder below the transition zone.

CONCLUSIONS

In the past decade, seismic imaging of the mantle has improved immensely because of the great advances in data acquisition (Grand et al., 1997; Ritsema et al., 1999), theoretical developments in finite-frequency effects (Montelli et al., 2004; Zhou et al., 2005), and computational hardware and numerical techniques (Komatitsch and Tromp, 2002a,b). Both superplumes in the lower mantle and smaller upper-mantle plumes have now been unveiled by body waves (Zhao, 2001, 2004; Lei and Zhao, 2005; Pilidou et al., 2005) and surface waves (Zhou et al., 2004). These images undoubtedly have provoked a revision in the traditional concept of how mantle upwellings should appear, because we have been so ingrained by ideas inculcated

in the 1980s from steady-state laboratory experiments, using simple fluids and steady point sources of heating (e.g., Whitehead and Luther, 1975; Olson and Singer, 1985). These recent tomographic images have revealed clearly the multiscale nature of mantle plumes, which we have portrayed schematically in Figure 17. Implicit in this drawing is the multiscale nature in both time and space, as mantle flow is intrinsically very time-dependent because of the nonlinear physics in the transport properties and the phase transitions, and the feedback loops between the various processes. In Figure 17, we see the different origins of the different spatial scales of mantle plumes. Such a change of spatial scales between the upper and lower mantles is also consistent with corresponding change of spatial tomographic spectra (Dziewonski, 2000).

In numerical models of mantle convection, the superplumes in the lower mantle can be created and stabilized in both time and space by the viscosity stratification (e.g., Hansen et al., 1993) and increased thermal conductivity (e.g., Matyska et al., 1994; van den Berg et al., 2001, 2002; Dubuffet et al., 2002; Matyska and Yuen, 2006; Naliboff and Kellogg, 2006), which should be present due to radiative heat transfer. However, radiative trans-

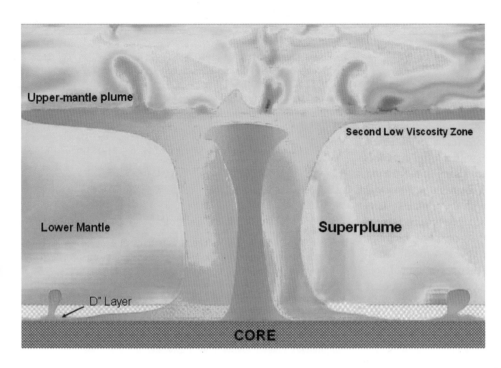

Figure 17. Sketch showing the multi-scale nature of mantle plumes, which involves the creation of the lower-mantle superplume from the D″ layer and generation of the upper-mantle plumes from the low-viscosity layer below 670 km.

fer of heat in the deep mantle is still a matter of controversy. It should be important or even dominant in olivine and perovskite (Hofmeister 1999, 2005; Badro et al., 2004; Gibert et al., 2005) as well as in post-perovskite (W.L. Mao, written commun.), but a decrease of radiative heat transfer with increasing pressure was observed for magnesiowüstite (Goncharov et al., 2006). Stabilization of superplumes enables forward studies of their properties, such as the adiabaticity of the superplume mode of heat transfer (Matyska and Yuen, 2001). Here we showed that a decrease of thermal expansivity with depth is another factor facilitating the existence and stability of the lower-mantle superplumes. Moreover, increased thermal conductivity at the base of the mantle together with intensive shear heating at the 670-km depth is able to create the low-viscosity zone (Kido and Čadek, 1997; Mitrovica and Forte, 2004) acting as the source of the upper-mantle smaller plumes, which are stabilized by interaction with the lower-mantle superplumes. Note that there also can be chemical heterogeneities associated with the seismically observed superplumes (Ishii and Tromp, 1999; Trampert et al., 2004), which can be modeled in the framework of thermal-chemical convection (e.g., Tan and Gurnis, 2005). Together with iron-rich patches in post-perovskite (Mao et al., 2004, 2006), these chemical heterogeneities could also stabilize lower-mantle convection. It is still a problem to distinguish between thermal and chemical heterogeneities; for example, the magnitude of density heterogeneities inferred directly from tomographic models is similar to that obtained from thermal convection models (Matyska and Yuen, 2002).

Smaller plumes with a greater propensity for time dependence because of the lower viscosity and thermal conductivity caused by water content have shorter lifetimes (e.g., Davaille and Vatteville, 2005) and emerge rapidly from the transition zone, which is further aided by non-Newtonian rheology in the upper mantle. In other words, the mantle plumes not only have a multiscale spatial nature, but also richness in the temporal spectrum, which can span over several orders of magnitude. From the Taylor hypothesis (Zaman and Hussain, 1981; Meneveau and Sreenivasan, 1991) in fluid mechanics, we would expect that in a strongly nonlinear regime, larger-scale coherent structures would have longer lifetimes than would the smaller-scale features. In the interpretation of tomographic images, one must then keep in mind that the classical picture of a mantle plume with a long conduit connecting to its source at a thermal boundary layer is only valid for a short time before the detachment process occurs. This criterion is especially pertinent in the upper mantle, where the lifetime of plumes may be only of the order of 10 m.y., from boundary-layer stability estimates based on the physical properties of the transition zone (e.g., Howard, 1964). In this connection, Maruyama (1994) has offered compelling geological arguments concerning the long lifetime of the superplume under the central Pacific.

Our aim in this article is to demonstrate that by using realistic physics, such as the depth-dependent thermal expansivity, variable viscosity, radiative thermal conductivity, and phase transitions, one can also produce a rich spectrum in both the spatial and temporal scales of mantle upwellings for realistic mantle conditions without going into the hard turbulent thermal convection regime (Yuen et al., 1993). Although our modeling was in two dimensions, the recent 3-D findings in a Cartesian box with an aspect-ratio of 6 × 6 × 1 by Kameyama and Yuen

(2006) show that convection behavior in the presence of the post-perovskite phase transition in three dimensions is similar to the results obtained in two dimensions. Therefore, we maintain that our results may be representative of convection in the Earth's mantle. Indeed, one is hard pressed to do the same in laboratory experiments with simple fluids and in a heated-from-below configuration, even at very high Rayleigh numbers. We hope that this article will stimulate more realistic numerical modeling of mantle plumes in the future, with a resurgent focus on the transient nature of mantle plumes (King and Ritsema, 2000; Davaille and Vatteville, 2005) and their interactions with the lithosphere (Thoraval et al., 2006). With the relentless drive toward petascale computing (Cohen, 2005), geodynamicists can overcome the current numerical limitations in speed, memory, and data storage and can address in due course some of the issues raised above for both thermal and thermal-chemical convection, which is a much more difficult computational problem that has the potential to broaden the complexity of plume modeling (Farnetani and Samuel, 2005).

ACKNOWLEDGMENTS

We highly appreciate constructive reviews and comments by Gillian Foulger, Donna Jurdy, Arie van den Berg, Garrett Ito, and Scott King. We thank Cesar da Silva, Renata Wentzcovitch, Charley Kameyama, Shige Maruyama, and Marc Monnereau for their discussions with us. We also thank Donna Jurdy for her encouragement in writing this article and Ying-Chun Liu for her drawings. This research has been supported by National Science Foundation Cooperative Studies of the Earth's Deep Interior and Information Technology Research grants, by the Czech Science Foundation grant 205/06/0580, and by the research project MSM 0021620800 of the Czech Ministry of Education.

REFERENCES CITED

Allègre, C.J., and Turcotte, D.L., 1985, Geodynamic mixing in the mesosphere boundary layer and the origin of oceanic islands: Geophysical Research Letters, v. 12, p. 207–210.

Badro, J., Rueff, J.P., Vankó, G., Monaco, G., Fiquet, G., and Guyot, F., 2004, Electronic transitions in perovskite: Possible non-convecting layers in the lower mantle: Science, v. 305, p. 383–386, doi: 10.1126/science.1098840.

Braginskii, S.I., and Meitlis, V.P., 1987, Overheating instability in the lower mantle near the boundary with the core: Izvestiya: Earth Physics, v. 23, p. 646–649.

Burke, K., and Wilson, J.T., 1976, Hotspots on the Earth's surface: Scientific American, v. 235, p. 46–57.

Chopelas, A., and Boehler, R., 1992, Thermal expansivity in the lower mantle: Geophysical Research Letters, v. 19, p. 1983–1986.

Christensen, U.R., and Yuen, D.A., 1985, Layered convection induced by phase transitions: Journal of Geophysical Research, v. 90, p. 10,291–10,300.

Cohen, R.E., ed., 2005, High-performance computing requirements for the computational solid Earth sciences: http://www.geo-prose.com/computational _SES.html.

Courtillot, V., Davaille, A., Besse, J., and Stock, J., 2003, Three distinct types of hotspots in the Earth's mantle: Earth and Planetary Science Letters, v. 205, p. 295–308, doi: 10.1016/S0012-821X(02)01048-8.

Crough, S.T., and Jurdy, D.M., 1980, Subducted lithosphere, hotspots, and the geoid: Earth and Planetary Science Letters, v. 48, p. 15–22, doi: 10.1016/ 0012-821X(80)90165-X.

Cserepes, L., 1993, Effect of depth-dependent viscosity on the pattern of mantle convection: Geophysical Research Letters, v. 20, p. 2091–2094.

Cserepes, L., and Yuen, D.A., 1997, Dynamical consequences of mid-mantle viscosity stratification on mantle flows with an endothermic phase transition: Geophysical Research Letters, v. 24, p. 181–184, doi: 10.1029/ 96GL03917.

Davaille, A., and Vatteville, J., 2005, On the transient nature of mantle plumes: Geophysical Research Letters, v. 32, art. no. L14309, doi: 10.1029/2005 GL023029.

Davies, G.F., 1999, Dynamic Earth: Cambridge, Cambridge University Press, 458 p.

Dubuffet, F., Yuen, D.A., and Rainey, E.S.G., 2002, Controlling thermal chaos in the mantle by positive feedback from radiative thermal conductivity: Nonlinear Processes in Geophysics, v. 9, p. 311–323.

Dziewonski, A.M., 1984, Mapping the lower mantle: Determination of lateral heterogeneities in P velocity up to degree and order 6: Journal of Geophysical Research, v. 89, p. 5929–5952.

Dziewonski, A.M., 2000, Global seismic tomography: Past, present and future, *in* Boschi, E., et al., eds., Problems in geophysics for the new millennium: Bologna, Editrice Compositori, p. 289–350.

Dziewonski, A.M., and Anderson, D.L., 1981, Preliminary reference Earth model: Physics of the Earth and Planetary Interiors v. 25, p. 297–356, doi: 10.1016/0031-9201(81)90046-7.

Farnetani, C.G., and Samuel, H., 2005, Beyond the thermal plume paradigm: Geophysical Research Letters, v. 32, art. no. L07311, doi: 10.1029/2005 GL022360.

Forte, A.M., and Mitrovica, J.X., 2001, Deep-mantle high-viscosity flow and thermochemical structure inferred from seismic and geodynamic data: Nature, v. 410, p. 1049–1056, doi: 10.1038/35074000.

Gerya, T.V., and Yuen, D.A., 2003, Rayleigh-Taylor instabilities from hydration and melting propel "cold plumes" at subduction zones: Earth and Planetary Science Letters, v. 212, p. 47–62, doi: 10.1016/S0012-821X(03)00265-6.

Gibert, B., Schilling, F.R., Gratz, K., and Tommasi, A., 2005, Thermal diffusivity of olivine single crystals and a dunite at high temperature: Evidence for heat transfer by radiation in the upper mantle: Physics of the Earth and Planetary Interiors, v. 151, p. 129–141.

Goldstein, R.J., Chiang, H.D., and See, D.L., 1990, High–Rayleigh number convection in a horizontal enclosure: Journal of Fluid Mechanics, v. 213, p. 111–126, doi: 10.1017/S0022112090002245.

Goncharov, A.F., Struzhkin, V.V., and Jacobsen, S.D., 2006, Reduced radiative conductivity of low-spin (Mg,Fe)O in the lower mantle: Science, v. 312, p. 1205–1208, doi: 10.1126/science.1125622.

Grand, S.P., van der Hilst, R., and Widiyantoro, S., 1997, Global seismic tomography: A snapshot of convection in the Earth: GSA Today, v. 7, p. 1–7.

Hansen, U., and Yuen, D.A., 1989, Dynamical influences from thermal-chemical instabilities at the core-mantle boundary: Geophysical Research Letters, v. 16, p. 629–632.

Hansen, U., and Yuen, D.A., 1994, Effects of depth-dependent thermal expansivity on the interaction of thermal-chemical plumes with a compositional boundary: Physics of the Earth and Planetary Interiors, v. 86, p. 205–221, doi: 10.1016/0031-9201(94)05069-4.

Hansen, U., Yuen, D.A., Kroening, S.E., and Larsen, T.B., 1993, Dynamical consequences of depth-dependent thermal expansivity and viscosity on mantle circulations and thermal structure: Physics of the Earth and Planetary Interiors, v. 77, p. 205–223, doi: 10.1016/0031-9201(93)90099-U.

Hanyk, L., Moser, J., Yuen, D.A., and Matyska, C., 1995, Time-domain approach for the transient responses in stratified viscoelastic Earth: Geophysical Research Letters, v. 22, p. 1285–1288, doi: 10.1029/95GL01087.

Hofmeister, A.M., 1999, Mantle values of thermal conductivity and the geotherm from phonon lifetimes: Science, v. 283, p. 1699–1706, doi: 10.1126/science.283.5408.1699.

Hofmeister, A.M., 2005, Dependence of diffusive radiative transfer on grain-size, temperature and Fe-content: Implications for mantle processes: Journal of Geodynamics, v. 40, p. 51–72, doi: 10.1016/j.jog.2005.06.001.

Honda, S., Yuen, D.A., Balachandar, S., and Reuteler, D., 1993, Three-dimensional instabilities of mantle convection with multiple phase transitions: Science, v. 259, p. 1308–1311, doi: 10.1126/science.259.5099.1308.

Honda, S., Saito, M., and Nakakuki, T., 2002, Possible existence of small-scale convection under the back arc: Geophysical Research Letters, v. 29, art. no. 2043, doi: 10.1029/2002GL015853.

Howard, L.N., 1964, Convection at high Rayleigh numbers, *in* Görtler, H., ed., Proceedings of the 11th International Congress of Applied Mechanics: Berlin, Heidelberg, and New York, Springer, p. 1109–1115.

Iitaka, T., Hirose, K., Kawamura, K., and Murakami, M., 2004, The elasticity of the $MgSiO_3$ post-perovskite phase in the Earth's lowermost mantle: Nature, v. 430, p. 442–444, doi: 10.1038/nature02702.

Ishii, M., and Tromp, J., 1999, Normal-mode and free-air gravity constraints on lateral variations in velocity and density of Earth's mantle: Science, v. 285, p. 1231–1236, doi: 10.1126/science.285.5431.1231.

Jacobs, M.H.G., van den Berg, A.P., and de Jong, B.H.W.S., 2006, The derivation of thermo-physical properties and phase equilibria of silicate materials from lattice vibrations: Application to convection in the Earth's mantle: Computer Coupling of Phase Diagrams and Thermochemistry, v. 30, p. 131–146.

Jarvis, G.T., and McKenzie, D.P., 1980, Convection in a compressible fluid with infinite Prandtl number: Journal of Fluid Mechanics, v. 96, p. 515–583, doi: 10.1017/S002211208000225X.

Kameyama, M., and Yuen, D.A., 2006, 3-D convection studies on the thermal state in the lower mantle with post-perovskite phase transition: Geophysical Research Letters, v. 33. art. no. L12S10, doi: 10.1029/2006GL025744.

Katsura, T., Yokoshi, S., Shastkiy, A., Okube, M., Fukui, H., Ito, E., Tomioka, N., Sugita, M., Hagiya, K., Kuwata, O., Ohtsuka, K., Nozawa, A., and Funakoshi, K., 2005, Precise determination of thermal expansion coefficient of $MgSiO_3$ perovskite at the top of the lower mantle conditions, *in* 3rd workshop on Earth's mantle composition, structure, and phase transitions, 30 August–3 September, Saint Malo, France, abstract.

Kido, M., and Čadek, O., 1997, Inferences of viscosity from the oceanic geoid: Indication of a low viscosity zone below the 660-km discontinuity: Earth and Planetary Science Letters, v. 151, p. 125–137, doi: 10.1016/S0012-821X(97)81843-2.

King, S.D., and Anderson, D.L., 1995, An alternative mechanism of flood basalt formation: Earth and Planetary Science Letters, v. 136, p. 269–279, doi: 10.1016/0012-821X(95)00205-Q.

King, S.D., and Ritsema, J., 2000, African hot spot volcanism: Small-scale convection in the upper mantle beneath cratons: Science, v. 290, p. 1137–1140, doi: 10.1126/science.290.5494.1137.

Komatitsch, D., and Tromp, J., 2002a, Spectral element simulations of global seismic wave propagation—I: Validation: Geophysical Journal International, v. 149, p. 390–412, doi: 10.1046/j.1365-246X.2002.01653.x.

Komatitsch, D., and Tromp, J., 2002b, Spectral element simulations of global seismic wave propagation—II. Three-dimensional models, oceans, rotation and self-gravitation: Geophysical Journal International, v. 150, p. 303–318, doi: 10.1046/j.1365-246X.2002.01716.x.

Korenaga, J., 2005, Firm mantle plumes and the nature of the core-mantle boundary region: Earth and Planetary Science Letters, v. 232, p. 29–37, doi: 10.1016/j.epsl.2005.01.016.

Larsen, T.B., and Yuen, D.A., 1997, Fast plumeheads: Temperature-dependent versus non-Newtonian rheology: Geophysical Research Letters, v. 24, p. 1995–1998, doi: 10.1029/97GL01886.

Lei, J., and Zhao, D., 2005, Global P-wave tomography: On the effect of various mantle and core phases: Physics of the Earth and Planetary Interiors, v. 154, p. 44–69, doi: 10.1016/j.pepi.2005.09.001.

Li, L., Brodholt, J.P., Stackhouse, S., Weidner, D.J., Alfredsson, M., and Price, G.D., 2005, Electronic spin state of ferric iron in Al-bearing perovskite in the lower mantle: Geophysical Research Letters, v. 32, art. no. L17307, doi: 10.1029/2005GL023045.

Lubimova, A.H., 1958, Thermal history of the earth with consideration of the variable thermal conductivity of the mantle: Geophysical Journal of the Royal Astronomical Society v. 1, p. 115–134.

Mao, W.L., Shen, G., Prakapenka, V.B., Meng, Y., Cambell, A.J., Heinz, D.L., Shu, J., Hemley, R.J., and Mao, H., 2004, Ferromagnesian postperovskite silicates in the D″ layer of the Earth: Proceedings of the National Academy of Sciences, USA, v. 101, p. 15,867–15,869, doi: 10.1073/pnas.0407135101.

Mao, W.L., Mao, H., Sturhahn, W., Zhao, J., Prakapenka, V.B., Meng, Y., Shu, J., Fei, Y., and Hemley, R.J., 2006, Iron-rich post-perovskite and the origin of ultralow-velocity zones: Science, v. 312, p. 564–565, doi: 10.1126/science.1123442.

Maruyama, S., 1994, Plume tectonics: Journal of the Geological Society of Japan, v. 100, p. 24–49.

Matyska, C., 1995, Axisymmetry of mantle aspherical structures: Geophysical Research Letters, v. 22, p. 521–524, doi: 10.1029/94GL03218.

Matyska, C., and Moser, J., 1994, Heating in the D″-layer and the style of mantle convection: Studia Geophysica et Geodaetica, v. 38, p. 286–292.

Matyska, C., and Yuen, D.A., 2001, Are mantle plumes adiabatic?: Earth and Planetary Science Letters, v. 189, p. 165–176, doi: 10.1016/S0012-821X(01)00361-2.

Matyska, C., and Yuen, D.A., 2002, Bullen's parameter η: A link between seismology and geodynamical modelling: Earth and Planetary Science Letters, v. 198, p. 471–483, doi: 10.1016/S0012-821X(01)00607-0.

Matyska, C., and Yuen, D.A., 2005, The importance of radiative heat transfer on superplumes in the lower mantle with the new post-perovskite phase change: Earth and Planetary Science Letters, v. 234, p. 71–81, doi: 10.1016/j.epsl.2004.10.040.

Matyska, C., and Yuen, D.A., 2006, Lower mantle dynamics with the post-perovskite phase change, radiative thermal conductivity, temperature- and depth-dependent viscosity: Physics of the Earth and Planetary Interiors, v. 154, p. 196–207, doi: 10.1016/j.pepi.2005.10.001.

Matyska, C., Moser, J., and Yuen, D.A., 1994, The potential influence of radiative heat transfer on the formation of megaplumes in the lower mantle: Earth and Planetary Science Letters, v. 125, p. 255–266, doi: 10.1016/0012-821X(94)90219-4.

Matyska, C., Yuen, D.A., Breuer, D., and Spohn, T., 1998, Symmetries of volcanic distributions on Mars and Earth and their mantle plume dynamics: Journal of Geophysical Research, v. 103, p. 28,587–28,597, doi: 10.1029/1998JE900002.

McNamara, A.K., Karato, S.-I., and van Keken, P.E., 2001, Localization of dislocation creep in the lower mantle: Implications for the origin of seismic anisotropy: Earth and Planetary Science Letters, v. 191, p. 85–99, doi: 10.1016/S0012-821X(01)00405-8.

Meneveau, C., and Sreenivasan, K.R., 1991, The multifractal nature of turbulent energy-dissipation: Journal of Fluid Mechanics, v. 224, p. 429–484, doi: 10.1017/S0022112091001830.

Mitrovica, J.X., and Forte, A.M., 2004, A new inference of mantle viscosity based upon joint inversion of convection and glacial isostatic adjustment data: Earth and Planetary Science Letters, v. 225, p. 177–189, doi: 10.1016/j.epsl.2004.06.005.

Monnereau, M., and Yuen, D.A., 2002, How flat is the lower-mantle temperature gradient?: Earth and Planetary Science Letters, v. 202, p. 171–183, doi: 10.1016/S0012-821X(02)00756-2.

Montelli, R., Nolet, G., Dahlen, F.A., Masters, G., Engdahl, E.R., and Hung, S.-H., 2004, Finite-frequency tomography reveals a variety of plumes in the mantle: Science, v. 303, p. 338–343, doi: 10.1126/science.1092485.

Morgan, W.J., 1971, Convection plumes in the lower mantle: Nature, v. 230, p. 42–43, doi: 10.1038/230042a0.

Murakami, M., Hirose, K., Kawamura, K., Sata, N., and Ohishi, Y., 2004, Post-perovskite phase transition in MgSiO₃: Science, v. 304, p. 855–858, doi: 10.1126/science.1095932.

Nakagawa, T., and Tackley, P.J., 2005, The interaction between the post-perovskite phase change and a thermo-chemical boundary layer near the core-mantle boundary: Earth and Planetary Science Letters, v. 238, p. 204–216, doi: 10.1016/j.epsl.2005.06.048.

Naliboff, J.B., and Kellogg, L.H., 2006, Dynamic effects of a step-wise increase in thermal conductivity and viscosity in the lowermost mantle: Geophysical Research Letters, v. 33, art. no. L12S09, doi:10.1029/2006GL025717.

Ni, S.D., and Helmberger, D.V., 2003, Seismological constraints on the South African superplume; Could be the oldest distinct structure on Earth: Earth and Planetary Science Letters, v. 206, p. 119–131, doi: 10.1016/S0012-821X(02)01072-5.

Oganov, A.R., and Ono, S., 2004, Theoretical and experimental evidence for a post-perovskite phase of MgSiO₃ in Earth's D″ layer: Nature, v. 430, p. 445–448, doi: 10.1038/nature02701.

Ogura, Y., and Phillips, N.A., 1962, Scale analysis of deep and shallow convection in the atmosphere: Journal of Atmospheric Science, v. 19, p. 173–179, doi: 10.1175/1520-0469(1962)019<0173:SAODAS>2.0.CO;2.

Olson, P.L., and Singer, H.A., 1985, Creeping plumes: Journal of Fluid Mechanics, v. 158, p. 511–538, doi: 10.1017/S0022112085002749.

Pilidou, S., Priestley, K., Debayle, E., and Gudmundsson, O., 2005, Rayleigh wave tomography in the North Atlantic: High resolution images of the Iceland, Azores and Eifel mantle plumes: Lithos, v. 79, p. 453–474, doi: 10.1016/j.lithos.2004.09.012.

Ranalli, G., 1995, Rheology of the Earth: London, Chapman and Hall, 413 p.

Ricard, Y., and Wuming, B., 1991, Inferring viscosity and the 3-D density structure of the mantle from geoid, topography and plate velocities: Geophysical Journal International, v. 105, p. 561–572.

Ricard, Y., Fleitout, L., and Froidevaux, C., 1984, Geoid heights and lithospheric stresses for a dynamic Earth: Annales de Géophysique, v. 2, p. 267–286.

Richards, M.A., and Hager, B.H., 1988, Dynamically supported geoid highs over hotspots: Observations and theory: Journal of Geophysical Research, v. 93, p. 7690–7708.

Ritsema, J., van Heijst, H.J., and Woodhouse, J.H., 1999, Complex shear wave velocity structure imaged beneath Africa and Island: Science, v. 286, p. 1925–1928, doi: 10.1126/science.286.5446.1925.

Schubert, G., Turcotte, D.L., and Olson, P., 2001, Mantle convection in the Earth and planets: Cambridge, Cambridge University Press, 940 p.

Solomatov, V.S., 1996, Can hotter mantle have a larger viscosity?: Geophysical Research Letters, v. 23, p. 937–940, doi: 10.1029/96GL00724.

Solomatov, V.S., 2001, Grain size–dependent viscosity convection and the thermal evolution of the Earth: Earth and Planetary Science Letters, v. 191, p. 203–212, doi: 10.1016/S0012-821X(01)00426-5.

Spiegel, E.A., and Veronis, G., 1960, On the Boussinesq approximation for a compressible fluid: Astrophysical Journal, v. 131, p. 442–447, doi: 10.1086/146849.

Stefanick, M., and Jurdy, D.M., 1984, The distribution of hot spots: Journal of Geophysical Research, v. 89, p. 9919–9925.

Steinbach, V., and Yuen, D.A., 1994, Melting instabilities in the transition zone: Earth and Planetary Science Letters, v. 127, p. 67–75, doi: 10.1016/0012-821X(94)90198-8.

Su, W.-J., and Dziewonski, A.M., 1991, Predominance of long-wavelength heterogeneity in the mantle: Nature, v. 352, p. 121–126, doi: 10.1038/352121a0.

Tackley, P.J., 1998, Three-dimensional simulations of mantle convection with a thermochemical CMB boundary layer: D″?, *in* Gurnis, M. et al., eds., The core-mantle boundary region: Washington, D.C., American Geophysical Union, p. 231–253.

Tan, E., and Gurnis, M., 2005, Metastable superplumes and mantle compressibility: Geophysical Research Letters, v. 32, art. no. L20307, doi: 10.1029/2005GL024190.

Thoraval, C., Tommasi, A., and Doin, M.-P., 2006, Plume-lithosphere interaction beneath a fast moving plate: Geophysical Research Letters, v. 33, art. no. L01301.

Trampert, J., Deschamps, F., Resovsky, J., and Yuen, D.A., 2004, Probabilistic tomography maps chemical heterogeneities throughout the lower mantle: Science, v. 306, p. 853–856, doi: 10.1126/science.1101996.

Tsuchiya, T., Tsuchiya, J., Umemoto, K., and Wentzcovitch, R.M., 2004, Phase transition in MgSiO₃ perovskite in the Earth's lower mantle: Earth and Planetary Science Letters, v. 224, p. 241–248, doi: 10.1016/j.epsl.2004.05.017.

Tsuchiya, J., Tsuchiya, T., and Wentzcovitch, R.M., 2005, Vibrational and thermodynamic properties of MgSiO₃ post-perovskite: Journal of Geophysical Research, v. 110, art. no. B02204, doi: 10.1029/2004JB003409.

Turcotte, D.L., 1992, Fractals and chaos in geology and geophysics: Cambridge, Cambridge University Press, 221 p.

Turcotte, D.L., and Schubert, G., 2002, Geodynamics (2nd edition): Cambridge, Cambridge University Press, 456 p.

van den Berg, A.P., Yuen, D.A., and Steinbach, V., 2001, The effects of variable thermal conductivity on mantle heat-transfer: Geophysical Research Letters, v. 28, p. 875–878, doi: 10.1029/2000GL011903.

van den Berg, A.P., Yuen, D.A., and Allwardt, J.R., 2002, Non-linear effects from variable thermal conductivity and mantle internal heating: Implications for massive melting and secular cooling of the mantle: Physics of the Earth and Planetary Interiors, v. 129, p. 359–375, doi: 10.1016/S0031-9201(01)00304-1.

van Keken, P.E., Yuen, D.A., and van den Berg, A.P., 1992, Pulsating diapiric flows: Consequences of vertical variations of mantle creep laws: Earth and Planetary Science Letters, v. 112, p. 179–194, doi: 10.1016/0012-821X(92)90015-N.

Whitehead, J.A., and Luther, D.S., 1975, Dynamics of laboratory diapir and plume models: Journal of Geophysical Research, v. 80, p. 705–717.

Yuen, D.A., Hansen, U., Zhao, W., Vincent, A.P., and Malevsky, A.V., 1993, Hard turbulent thermal convection and thermal evolution of the mantle: Journal of Geophysical Research, v. 98, p. 5355–5373.

Zaman, K.B., and Hussain, A.K., 1981, Taylor hypothesis and large-scale coherent structures: Journal of Fluid Mechanics, v. 112, p. 379–396, doi: 10.1017/S0022112081000463.

Zhao, D., 2001, Seismic structure and origin of hotspots and mantle plumes: Earth and Planetary Science Letters, v. 192, p. 251–265, doi: 10.1016/S0012-821X(01)00465-4.

Zhao, D., 2004, Global tomographic images of mantle plumes and subducting slabs: Insight into deep Earth dynamics: Physics of the Earth and Planetary Interiors, v. 146, p. 3–34, doi: 10.1016/j.pepi.2003.07.032.

Zhao, W., and Yuen, D.A., 1987, The effects of adiabatic and viscous heatings on plumes: Geophysical Research Letters, v. 14, p. 1223–1227.

Zhou, Y., Dahlen, F.A., and Nolet, G., 2004, Three-dimensional sensitivity kernels for surface wave observables: Geophysical Journal International, v. 158, p. 142–160, doi: 10.1111/j.1365-246X.2004.02324.x.

Zhou, Y., Dahlen, F.A., Nolet, G., and Laske, G., 2005, Finite-frequency effects in global surface-wave tomography: Geophysical Journal International, v. 163, p. 1087–1111.

Manuscript Accepted by the Society 31 January 2007

DISCUSSION

25 December 2006, Gillian R. Foulger

It has been remarked that the heating mode in the article by Matyska and Yuen (this volume) is not clear, the figures give the impression that the mantle is assumed to have no radioactivity, and the 2-D models are driven entirely by heating from below. Would the authors care to clarify the thermal boundary conditions, and the bearing any simplifying assumptions have on the contribution of core heat and the resulting Rayleigh number?

6 January 2007, Ctirad Matyska and David A. Yuen

Our convection models are partly heated from within, as we have used the dimensionless heat source term $R = 3$, which is equivalent to about 25% of the chondritic value. Moreover, with radiative thermal conductivity, there is a tendency for overheating in the lower mantle for a chondritic mantle. The bottom boundary is isothermal, and thus a substantial amount of heat comes from below. This heating mode could be caused by a decrease of the core gravitational potential energy due to inner-core formation, ohmic dissipation, and perhaps some radioactivity from potassium. Moreover, the core was heated during Earth's formation, and its cooling represents another source of heat, which flows from the core to the mantle. However, we did not compute cooling of the core, and thus the temperature of the bottom boundary does not change with time in our computations.

The surface Rayleigh number was set to 10^7; we note, however, the concept of Rayleigh number becomes vague in the presence of depth-dependent thermal expansivity, depth- and temperature-dependent viscosity, and radiative thermal conductivity. An "effective" Rayleigh number is much lower than the surface value, especially at the bottom of the lower mantle—something like 10^5. The combined effect of the depth-dependent mantle material properties resulted in the creation of the lower-mantle superplumes with very low lateral temperature variations, followed by classical thin plumes emerging from the 670-km boundary.

REFERENCE CITED

Matyska, C., and Yuen, D.A., 2007 (this volume), Lower mantle material properties and convection models of multiscale plumes, *in* Foulger, G.R., and Jurdy, D.M., eds., Plates, plumes, and planetary processes: Boulder, Colorado, Geological Society of America Special Paper 430, doi: 10.1130/2007.2430(08).

The Geological Society of America
Special Paper 430
2007

Global plume-fed asthenosphere flow—I:
Motivation and model development

Michiko Yamamoto*
Jason Phipps Morgan*
*Institute for the Study of the Continents, Department of Earth and Atmospheric Sciences,
Cornell University, Snee Hall, Ithaca, New York 14853-1504, USA*
W. Jason Morgan
*Department of Earth and Planetary Sciences, 20 Oxford Street,
Harvard University, Cambridge, Massachusetts 02138, USA*

ABSTRACT

This study explores a conceptual model for mantle convection in which buoyant and low-viscosity asthenosphere is present beneath the relatively thin lithosphere of ocean basins and regions of active continental deformation, but is less well developed beneath thicker-keeled continental cratons. We start by summarizing the concept of a buoyant plume-fed asthenosphere and the alternative implications this framework has for the roles of compositional and thermal lithosphere. We then describe the sinks of asthenosphere made by forming compositional lithosphere at ridges, by plate cooling wherever the thermal boundary layer extends beneath the compositional lithosphere, and by drag-down of buoyant asthenosphere along the sides of subducting slabs. We also review the implied origin of hotspot swell roots by melt-extraction from the hottest portions of upwelling plumes, analogous to the generation of compositional lithosphere by melt-extraction beneath a spreading center. The plume-fed asthenosphere hypothesis requires an alternative to "distinct source reservoirs" to explain the differing trace element and isotopic characteristics of ocean island basalt (OIB) and mid-ocean ridge basalt (MORB) sources; it does so by having the MORB source be the plum-depleted and buoyant asthenospheric leftovers from progressive melt-extraction within upwelling plumes, while the preferential melting and melt-extraction of more-enriched plum components is what makes OIB of a given hotspot typically fall within a tubelike geometric isotope topology characteristic of that hotspot. (The distinct plum components result from the subduction of chemically differing sediments, basalts, and residues to hotspot and mid-ocean ridge melt extraction.) Using this conceptual framework, we construct a thin-spherical-shell finite element model with a ~100-km-scale mesh to explore the possible structure of global asthenosphere flow. Lubrication theory approximations are used to solve for the flow profile in the vertical direction. We assess the correlations between predicted flow and geophysical observations, and conclude by noting current limitations in the model and the reason why we currently neglect the influence of subcontinental plume upwelling for global asthenosphere flow.

Keywords: mantle convection, asthenosphere, mantle geochemistry

*E-mails: Yamamoto, my83@cornell.edu; Phipps Morgan, jp369@cornell.edu.

Yamamoto, M., Phipps Morgan, J., and Morgan, W.J., 2007, Global plume-fed asthenosphere flow—I: Motivation and model development, *in* Foulger, G.R., and Jurdy, D.M., eds., Plates, plumes, and planetary processes: Geological Society of America Special Paper 430, p. 165–188, doi: 10.1130/2007.2430(09). For permission to copy, contact editing@geosociety.org. ©2007 The Geological Society of America. All rights reserved.

INTRODUCTION

For the past dozen years, we have been actively exploring the consequences of a conceptual model of mantle convection in which discrete upwelling plumes feed a weak asthenospheric boundary layer underlying the relatively thin oceanic lithosphere (Fig. 1). In this scenario of mantle convection (Phipps Morgan and Smith, 1992; Phipps Morgan et al., 1995a,b; Phipps Morgan, 1997, 1998, 1999; Phipps Morgan and Morgan, 1999), a plume-fed asthenosphere is viewed as an almost inevitable consequence of the fact that the hottest and most buoyant mantle will rise as plumes (Morgan, 1971) until its ascent is impeded at the base of the lithosphere. At this point, plume material will pond and flow laterally beneath the base of the lithosphere until it: (1) increases in viscosity either by cooling or by drying out through decompression melting to be transformed into part of the overlying lithosphere; or (2) is entrained and dragged down by a subducting slab.

This conceptual model is not new: the key idea was first presented in 1972 by Deffeyes (Deffeyes, 1972) as an implication of Morgan's mantle plume hypothesis (Morgan, 1971). The conceptual model shares many aspects of conventional mantle flow models. In it, the cooling, growth, and subduction of oceanic lithosphere is the heat engine that ultimately drives mantle convection (Turcotte and Oxburgh, 1967; Elsasser, 1971). We also accept and build from the conventional interpretation of seismological evidence that slabs often subduct deep into the man-

tle, e.g., this scenario is a variant of "whole mantle flow." However, this scenario differs from the conventional model in its important roles for plume upwelling; the rheological and density effects of melt-extraction; and the importance of the mantle having a fine-scale marble-cake or plum-pudding lithologic variation as a byproduct of the geochemical heterogeneity introduced by the persistent subduction of compositionally distinct sediments, ocean crust, and lithosphere, and subsequent stirring and stretching of these lithologic heterogeneities during mantle flow.

This scenario for asthenosphere flow also has some aspects in common with previous suggestions for a shallow upper-mantle counterflow from regions of trench supply of slab inputs to the upper mantle to ridge consumption of upper mantle. Schubert and Turcotte (1972) presented the basic hypothesis, while Harper (1978), Schubert et al. (1978), Chase (1979), and Parmentier and Oliver (1979) developed models based upon this hypothesis to predict the global pattern of asthenosphere counterflow. There is a major similarity between our preferred model and these previous works—typically there is counterflow of asthenosphere away from a subduction zone, although in our scenario it is much less. (Tonga is one place where our models differ in a testable way from these counterflow models; in our models, the predicted asthenosphere flow is nearly trench parallel—as opposed to the trench-perpendicular prediction of counterflow models—as it heads southward toward the southern ridge system.) In our scenario, the counterflow is entirely within the asthenosphere. The material that counterflows is the fraction of

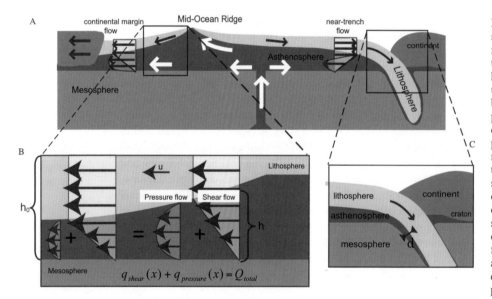

Figure 1. (A) Cartoon of the proposed scenario for asthenosphere flow within mantle convection. A low-viscosity asthenosphere layer is present between higher-strength overlying lithosphere and underlying mesosphere beneath ocean basins, where the lithosphere is thin in comparison to the lithospheric roots beneath continental cratons. Upwelling mantle plumes replenish the asthenosphere with material that is hotter and more buoyant than underlying mesosphere. Asthenosphere is consumed through accretion onto overlying lithosphere and by drag-down of entrained asthenosphere sheets at the sides of subducting slabs. Boundary conditions for asthenosphere flow are shown for an ocean-continent margin (left) and a subduction zone (right). At an ocean-continent margin, the net asthenosphere plus lithosphere flow is equal to the absolute plate velocity times the total thickness of lithosphere and asthenosphere. At a subduction zone, assuming that only the slab subducts, the downgoing flux is equal to the absolute plate velocity times the lithosphere thickness. (B) Cartoon showing a process leading to dynamic asthenosphere flow. The lithosphere grows as it cools, consuming asthenosphere. Yet in the plate-spreading direction in this cartoon, the net flux of lithosphere plus asthenosphere is constant. To conserve mass, a dynamic pressure gradient must develop within the asthenosphere, which induces an additional component of pressure-driven flow. (C) The asthenosphere sink caused by slab drag-down. Lithosphere subducts, dragging a thin asthenosphere layer into mesosphere. In this study, the thickness d of the dragged layer is set to 15 km, following the boundary layer analysis in Phipps Morgan and Morgan (1999) and the supporting lab and numerical experiments of Hasenclever (2004) and Phipps Morgan et al. (2007).

asthenosphere that, because of its buoyancy, resists drag-down by the slab descending into the deeper mantle (i.e., the subducted slab does not counterflow). Contrariwise, in pure counterflow models, the source of new upper-mantle material is the injection of the entire slab into this upper mantle—an assumption that now seems disproven by tomographic images of subducting slabs that are conventionally interpreted to subduct through the upper mantle and often deep into the lower mantle.

There are several big differences between plume-fed asthenosphere flow and counterflow models; in the plume-fed flow scenario, slabs remove subducted lithosphere from the shallow mantle (instead of reinjecting "asthenosphere" into the upper mantle, as in pure counterflow models), and in the plume-fed flow scenario, it is upwelling mantle plumes that provide the supply of new asthenosphere to the shallow mantle. For further discussion of the similarities and differences between these scenarios, see Phipps Morgan and Smith (1992) and Phipps Morgan et al. (1995b).

We realize that readers are likely to be unfamiliar with how this conceptual model presents a coherent alternative. Thus we begin this article with an overview of this scenario and how it involves the reinterpretation of several basic observations on hotspots and mid-ocean ridge volcanism. In general, we highlight observations that postdate a previous paper (Phipps Morgan et al., 1995b), discussing observational evidence that favors the existence of a plume-fed asthenosphere beneath the ocean basins.

The basic element of the conceptual model is that the mantle's upward mass balance for the downward subduction of cold slabs is by upwelling in mantle plumes followed by lateral flow in a hot, weak, and buoyant asthenospheric layer. The picture of focused convective upwelling in mantle plumes is essentially the same as in Morgan's original plume hypothesis (Morgan, 1971, 1972). Flow in the asthenosphere is necessary to redistribute the plume-fed asthenosphere from the localized regions of plume upwelling to the regions where asthenosphere is consumed by lithosphere formation at mid-ocean ridges (a large concentrated sink), subsequent off-axis lithosphere growth by surface cooling (a large diffuse sink), and asthenospheric dragdown or entrainment along the sides of dense subducting slabs (a smaller concentrated sink).

The three main issues of contention for this conceptual model are: (1) Is there a shallow asthenospheric layer that is hotter than underlying mantle in terms of its potential temperature, and hence weaker and more buoyant? (2) Is the upwelling plume flux large enough to counterbalance asthenosphere consumption by plate subduction? (3) If plumes feed ridges, then how are the distinct geochemical differences between hotspot (plume) basalts and mid-ocean ridge basalts (MORBs) created and maintained? Each of these issues has been discussed and resolved to our provisional satisfaction in previous publications. However, because these successful resolutions have been published separately, the potential of the conceptual model to provide a cohesive internal framework for understanding the flow and chemical evolution of the mantle is likely to have been underappreciated.

Asthenosphere

The existence of a shallow asthenospheric region beneath oceanic plates is central to our model. Much of the evidence for such a hot, weak, and buoyant asthenosphere is so familiar that it is perhaps too easily taken for granted. The oldest and strongest observation is that the seismic low-velocity zone (LVZ) has been used to imply that lower-viscosity mantle is to be found beneath oceanic lithosphere and the thinner regions of continental lithosphere. (The observation has two parts. The first is the existence of a seismic LVZ between ~80 and 300 km depths [Gutenberg, 1959; Dziewonski and Anderson, 1981] in average 1-D global models that, being a global average, are biased toward the average velocity structure beneath the 60% of the world lying beneath oceanic lithosphere. The second part is that the lithosphere beneath Proterozoic and Archean regions of the continents is seismically fast in the depth interval of the global LVZ. Anderson, 1989, has a good summary of these well-established seismic observations.)

Recent observations that highlight these well-known seismic characteristics are shown in Figures 2 and 3. Figure 2 (Gaherty et al., 1999) shows the average 1-D seismic velocity structure along transects that cross largely ocean seafloor (central Pacific in PA5, Philippine Sea in PHB3) and an Archean Shield province in Australia (AU3). A region with lower seismic shear wavespeeds begins ~90–110 km beneath oceanic lithosphere, at depths explicable as the thickness that an oceanic plate would cool in ~60–100 m.y. at Earth's surface. Beneath the Archean part of Australia, the seismically fast region persists to depths of ~250–300 km, where a much smaller LVZ is found. Figure 3 (Ekstrom and Dziewonski, 1998; Nettles, 2005) shows a complementary map of global seismic shear wavespeed vari-

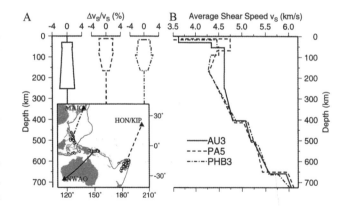

Figure 2. Mean 1-D shear-wave velocity structures over paths crossing the subcontinental Australian craton (path AU3 to station NWAO), suboceanic Philippine Sea back-arc basin (path PHB3 to station MAIO), and suboceanic Pacific (PA5 to station HON/KIP) upper mantle. (A) Anisotropic deviations about the mean shear speeds (in all three cases, the vertical shear wavespeed V_{SV} is slower than the horizontal shear wavespeed V_{SH}). Inset shows the map locations of the AU3, PHB3, and PA5 travel paths resulting in each 1-D velocity model. (B) Mean shear speeds for each region. Modified from Gaherty et al. (1999, Fig. 4).

Yamamoto et al.

S-wavespeed variations, 070 km **S-wavespeed variations, 100 km**

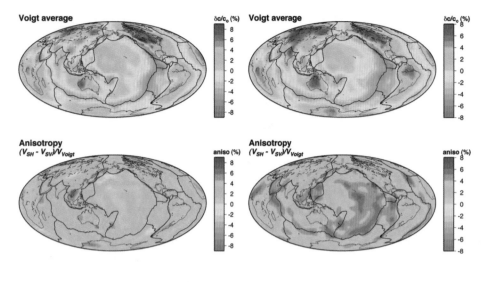

S-wavespeed variations, 150 km

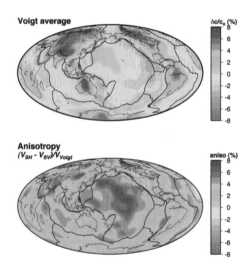

Figure 3. Global shear wavespeed variations determined by Nettles (2005) using the approach of Ekstrom and Dziewonski (1998) for depths of 70, 100, and150 km. Variations in isotropic shear wavespeed are shown in the top map of each pair of maps, with anisotropic variations from the isotropic average shown in the bottom map. The isotropic shear wavespeeds are shown with respect to the global average at each depth; no average has been removed from the anisotropy maps. Modified from Nettles (2005, Fig. 5.4).

ations as imaged by surface waves in the ~50- to 200-km-depth interval. Here the roots of Archean and Proterozoic continental regions are typically 5% faster (blue in the figure) in comparison to the slower (red) regions beneath oceanic seafloor and areas of active continental deformation. Even though the lateral averaging of this map in suboceanic regions is 1000 km, so that only the large-scale oceanic plate cooling with age is evident within the ocean basins, this figure clearly shows this first-order difference between subcontinental and suboceanic regions. Note that even the Australian and Indian continental shields on the rapidly moving Indo-Australian plate have fast seismic wavespeeds, which means this is an effect associated with roots and/or keels that move with continents, not with possible long-lived and deeper mantle structures associated with the memory effect of a Pangaea supercontinent. The cartoon structure that we

infer (Fig. 1) for the differences between continental and oceanic lithosphere is similar to that shown by Gung et al. (2003).

Plume-Fed Asthenosphere

We think the likeliest way to form the asthenosphere is for it to be plume-fed, i.e., fed by naturally occurring buoyant upwellings of hotter-than-average mantle. In this conceptual model, we imagine the asthenospheric layer to be hotter than underlying mantle because it is formed from the most recent upwellings of hotter-than-average mantle; buoyant upwellings that have displaced downward any preexisting cooler and denser asthenosphere as they spread out beneath the base of the lithosphere. Most likely, the asthenosphere is hotter in an absolute sense than its lithospheric lid or mantle base, so that it conductively loses heat through both its top and base. If not hotter in an absolute sense than underlying mantle, then it remains hotter in terms of its potential temperature (its temperature relative to the upper mantle adiabatic temperature gradient of ~0.3–0.4 K/km), because downward heat conduction can only cool the asthenosphere to the same temperature as its underlying mantle and no cooler. Thus the base of the asthenosphere will be intrinsically buoyant with respect to underlying mantle, because its potential density is lower. Concepts of potential temperature and potential density (density corrected for variations along an adiabat) are well appreciated in oceanography, as oceans typically exhibit horizontal isopycnal flow that occurs in layers stratified by increasing potential density with increasing depth. However, these concepts are uncommon in mantle convection.

Melt-Extraction–Induced Strengthening of Mantle: Effects on the Generation of a Compositional Lithosphere and on Forming Restite Roots to Hotspot Swells

Here our thinking has been highly influenced by Shun Karato's suggestion in 1986 (Karato, 1986) that the formation and extraction of a partial melt, because it dehydrates the residue by preferential concentration of water into the melt, increases the viscosity of the residue by more than an order of magnitude. Karato's logic is that because intracrystalline water has a significant weakening effect on the rheology of mantle olivine, melt-extraction from a peridotite may induce, by preferential extraction of incompatible intracrystalline water into the melt phase, an increase in the viscosity of the left-over dehydrated restite. Phipps Morgan (1987, p. 1240) suggested that this effect might lead to the formation of a more viscous (~10^{21} Pa-s) mantle region from ~10–70 km depths beneath mid-ocean ridges in which plate-spreading–induced viscous pressure gradients would be strong enough to focus melt to the spreading axis. Later, Phipps Morgan (1994, 1997) used the term "compositional lithosphere" to distinguish between thermal lithosphere that grows by heat loss through the seafloor and the stronger-than-asthenosphere ~70-km-thick restite layer created by decompression melting and melt extraction at a spreading center.

At that time, we realized that hotspot melting might also lead to the creation of a similar root underlying a hotspot swell, as the peridotites in the hottest central regions of an upwelling plume melt beneath normal oceanic lithosphere (Phipps Morgan et al., 1995a). The rate of subsequent lateral spreading of the Hawaiian swell was used to infer the viscosity of the restitic swell root to be on the order of 3×10^{20}–10^{21} Pa-s (Phipps Morgan et al., 1995a). Soon after, Greg Hirth and Dave Kohlstedt wrote an influential paper that documents and reaffirms the basic plausibility of melt-extraction–linked strengthening of restitic mantle and then discusses several applications of this idea for the growth and evolution of oceanic lithosphere (Hirth and Kohlstedt, 1996).

The two most important implications of these ideas for asthenosphere flow are:

1. Mid-ocean ridges are the sites where enough asthenosphere is consumed by decompression-melting–induced desiccation during plate extension to create a ~60-km-thick compositional lithosphere underlying the ocean crust. In comparison to lithospheric growth by conductive heat loss through the seafloor, this effect leads to much more focused asthenospheric consumption at mid-ocean ridges. It also leads to the existence of a large region between mid-ocean ridges and ca. 40-Ma seafloor where asthenosphere does not further accrete to the lithosphere, as asthenosphere does not accrete to the cooling plate until the thermal boundary layer extends beneath the approximately 60-km thickness of the compositional lithosphere (Phipps Morgan, 1994, 1997; Hirth and Kohlstedt, 1996; Yale and Phipps Morgan, 1998; Lee et al., 2005).

2. The plume-flux estimated by the sizes of hotspot swells (Davies, 1988; Sleep, 1990) is a mis-estimate. In their scenario (Fig. 4A), the regional uplift of a hotspot swell reflects the entire upwelling flux of buoyancy plume material—all of the upwelling plume material is refracted (i.e., bent) and dragged beneath the moving oceanic lithosphere. (This scenario also supposes that the ambient sublithospheric "asthenosphere" is stronger than upwelling plume material, so that the resistance of the surrounding ~10^{20} Pa-s mantle to intrusion by the spreading plume material [Olson, 1990] ultimately limits the lateral spreading of the low-viscosity swell root.)

However, in our preferred conceptual model (Fig. 4B), the swell root is not the total plume flux. Rather, it reflects only a small fraction of the upwelling plume flux—the hottest central flux of the upwelling plume. If the hotspot swell root consists of higher-viscosity restite resulting from sufficient melt-extraction to dry the most heavily melted peridotite within the upwelling and melting plume, then the size of a hotspot swell reflects just the fraction of upwelling plume material that melts enough to create a strong swell root. (In this conceptual model, shown in Fig. 4B, the swell root is more viscous than the surrounding less-melted or unmelted asthenosphere, so that it is the viscosity of the swell root itself [~3×10^{20} Pa-s beneath Hawaii] that ultimately limits its spreading. See further discussion in Phipps Morgan et al., 1995a.)

Note that in this hypothesis for making a swell root, the buoyancy of the swell root is only due in part to its depleted composition, because the swell root is also hotter than surrounding asthenosphere. Because the swell root is more dehydrated and more viscous than surrounding asthenosphere, both the thermal and the compositional buoyancy of the swell root support the relief of the overlying hotspot swell. (Below we assess in more detail the question of how buoyancy changes in response to melt extraction.) This aspect of the restite swell-root model has been recently misstated in critiques of the model in Davies (1999, p. 316) and Schubert et al. (2001, p. 513). Both critiques incorrectly presuppose that all of the buoyancy sup-

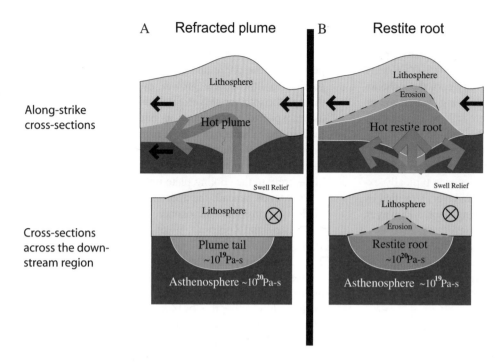

Figure 4. Cartoons showing the rheological differences between (A) the refracted plume and (B) restite-root scenarios for the uplift of a hotspot swell. For each scenario, a schematic cross-section is shown along the island-chain (top) and perpendicular to the island-chain (bottom). The key conceptual difference between the refracted plume and restite-root hypotheses is that the refracted plume is the entire flux of hot low-viscosity plume material that is bent and dragged between the overriding lithosphere and underlying higher-viscosity asthenosphere, whereas the restite-root is the hot, most viscous restite residue to hotspot melt-extraction that is created when partial melt-extraction dries and thereby strengthens the restite so that it is hotter, yet more viscous than underlying cooler-but-wetter asthenosphere. In the restite-root scenario, only the hottest core of an upwelling plume is likely to undergo enough partial melt-extraction to become more-viscous restite; the cooler rim of upwelling plume material becomes asthenosphere (Phipps Morgan et al., 1995b).

porting the hotspot swell in the restite swell-root model must be due to its depleted composition. In fact, in the more detailed assessment in Phipps Morgan et al. (1995a) of the amounts of buoyancy generated by hotspot melting, the conclusion was that the thermal and chemical buoyancy in the swell root and the chemical buoyancy of the thickened crust should all contribute roughly equal amounts to the relief of the swell.

For example, the differences between the estimate by Schubert et al. (2001) of depletion swell-support and ours stems from their use of a much smaller estimate for Hawaiian volcanic production, and, to a lesser degree, from their assumption of a 33% larger typical Hawaiian swell. Phipps Morgan et al. (1995a, p. 8060) present the reasoning behind an estimate of the basalt production rate at Hawaii of 0.25 km^3/year. Note that this estimate is very similar to the more recent estimate of Robinson and Eakin (2006) of 0.21 km^3/year, while Schubert et al. (2001) use the production rate of 0.1 km^3/year that is apparently based on a 1970s estimate (BVSP, 1981) of Hawaiian crustal thickening that was inferred prior to the collection of a good seismic cross-section across the Hawaiian chain. Similarly, Schubert et al. (2001) use an estimate by Sleep (1990) of 8700 kg/s for the buoyancy flux associated with the Hawaiian swell that corresponds to the maximum values estimated by us (cross-sections B and C in Table 1 of Phipps Morgan et al., 1995a), while the average of profiles A–K in Table 1 of Phipps Morgan et al. (1995a) implies a swell buoyancy flux of only 6070 kg/s. With our estimate of 0.25 km^3/year of basalt production linked to the creation of the Hawaiian swell, the reasoning in Phipps Morgan et al. (1995b) would imply that the swell root's depletion buoy-

ancy supports 2285 kg/s and the its thermal buoyancy supports 2665 kg/s of the swell buoyancy flux, and off-island basaltic intrusions support the remaining 900 kg/s of the time-averaged Hawaiian swell buoyancy flux. Davies's (1999) conclusions ultimately depend upon his use of an estimate for basalt production at Hawaii of 0.03 km^3/year, a factor of three lower than that used by Schubert et al. (2001) and almost a factor of eight lower than our preferred estimate of 0.25 km^3/year for the recent rate of Hawaiian volcanism or Robinson and Eakins's (2006) similar recent estimate of 0.21 km^3/year. (Somewhat curiously, Davies, 1999, refers to Phipps Morgan et al., 1995a, for his 0.03 km^3/year estimate, apparently basing his estimate upon the seamount-chain cross-section area shown in Table 1 and Fig. 2b of the latter paper multiplied by the ~0.1 m/year motion of the Pacific Plate over the Hawaiian hotspot. However, Fig. 2a of Phipps Morgan et al., 1995a, shows seismic measurements by ten Brink and Brocher, 1987, that imply that profile D, with a topographic cross-sectional area of 343 km^2, is actually associated with crustal thickening of ~12 km over a width of ~200 km, i.e., it is associated with a magmatic cross-sectional area of 2400 km^2 and an implied volcanic production rate of 0.24 km^3/year instead of 0.03 km^3/year.) If Davies had used Robinson and Eakins' (2006) or our preferred rates for Hawaiian volcanism and for the buoyancy flux associated with the Hawaiian swell, then his approach would also have led to our preferred conclusions.

Finally, a highly viscous swell root should be able to more efficiently excavate the base of the overlying oceanic lithosphere, which will augment swell buoyancy as it replaces cooler lithosphere instead of warm asthenosphere by the hot swell root.

If 20 km of lithosphere is being eroded beneath the Hawaiian swell, this effect would augment the relief above the Hawaiian swell by a further ~20–30%. Lithospheric thinning below a hotspot swell has been largely discounted on the basis of studies of potential thinning above a hot low-viscosity plume (Monnereau et al., 1993; Davies, 1994), with the possible exception of more poorly resolved 3-D experiments reported by Moore et al. (1999). Interestingly, these experiments suggest the potential for greater lithospheric thinning than seen in 2-D experiments. It is still unclear whether the thinning seen in these experiments is an inadvertent byproduct of insufficient grid resolution to properly model the erosion of a strongly temperature-dependent lithosphere. (The experiments used a ~5-km mesh spacing; Bill Moore, personal commun., 2005). However, a recent seismic study reports evidence for noticeable ~20- to 40-km lithospheric erosion beneath the center of the Hawaiian swell (Li et al., 2004; see also Laske et al., this volume), suggesting that the issue of thermomechanical lithospheric erosion should be revisited even though this seismic study also implies that lithospheric erosion cannot be the underlying support of the initial and distal uplift of the swell.

Compositional lithosphere also provides a plausible rationale for why hotspot swells are not seen on seafloor younger than ca. 40 Ma—if the thermal lithosphere is significantly thinner than the compositional lithosphere, then the restitic swell-root material will spread along with the base of the layer of similar-viscosity restitic compositional lithosphere, resulting in much broader-scale lateral flow. In this case, the same amount of swell-root buoyancy will spread beneath a much larger area of seafloor, inducing a much broader but smaller-amplitude topographic signal. For example, the superswell region of the South Pacific (McNutt and Fischer, 1987; Adam and Bonneville, 2005) could be a region where several such thin swell-root puddles have coalesced beneath 60-Ma off-axis seafloor to form the regional topographic anomaly of the superswell. This is also a potential explanation for why the Hawaiian swell largely disappears after crossing the Mendocino fracture zone, as the Hawaiian plume always upwelled beneath young (60-Ma) seafloor during the time it was forming the Emperor seamount chain (Phipps Morgan et al., 1995b).

Because the total plume flux is not constrained by the size of a hotspot swell in our preferred model for the formation of hotspot swells, we prefer to estimate plume fluxes in two different ways: (1) from the relative amounts of hotspot volcanism that they produce; and (2) by the assumption that the net upward plume flux of new asthenosphere is of the same order of magnitude as the net subduction rate of oceanic lithosphere, e.g., the rate at which sinking slabs are returning former asthenosphere to the deep mantle. The second estimate assumes that the thickness of the asthenosphere is quasi-steady state. If instead there have been periods of higher-than-average plume upwelling, such as Larson has suggested to occur in the Cretaceous (Larson 1991a,b; Larson and Olson, 1991) or periods of faster plate subduction, then the thickness of the asthenosphere would

be expected to vary through time. (These implications are more extensively discussed in Phipps Morgan et al., 1995b, which explores the possibility that feedback can occur if the suboceanic asthenospheric layer becomes sufficiently thin so that asthenospheric drag becomes a significant resisting force to plate motions.) If plate motions and subduction rates slow because of thinning of the lubricating asthenospheric layer, then, for a constant rate of plume resupply, the volume/thickness of the asthenosphere would grow, leading to less asthenospheric drag and faster plate motions and subduction. (Phipps Morgan et al., 1995b, also discuss the implication that the relatively small area of subcratonic lithosphere in India and Australia relative to the area of the subducting Indo-Australian slab in contact with subasthenospheric mantle implies that viscous mantle resistance to slab subduction, not subcontinental viscous drag, is the dominant factor limiting plate speeds for plates attached to a subducting slab.)

Recent Geophysical Observations of Oceanic Compositional Lithosphere and Swell Roots

Three recent geophysical studies appear to be imaging the melt-extraction–induced transition from asthenosphere to more viscous restite. At 17°S along the southern East Pacific Rise, joint electromagnetic and seismic observations in the Mantle Electromagnetic and Tomography (MELT) experiments appear to be imaging the dehydration front at ~60 km that is predicted to be induced by mid-ocean ridge melting and melt-extraction (Evans et al., 2005). A recent global survey searching for Ps conversion depths beneath the ocean basins has imaged two similar sharp steps of ~5–10% in the shear seismic velocity structure, one at ~60 km beneath the Central Indian Ridge, which has similar crustal thickness to the East Pacific Rise (Rainer Kind, personal commun., 2006), and the other at ~80 km beneath the thicker crust and presumably hotter and more deeply melting spreading center beneath the Reykjanes Peninsula (Kumar et al., 2005). Intriguingly, the same seismic imaging technique has recently found a similar sharp Ps conversion front at 140 km directly beneath Kileauea, the center of the Hawaiian plume and swell uplift (Rainer Kind, personal commun., 2006). We think this reflection may be showing the depth where melt-extraction from the hottest central region of rising Hawaiian plume material is starting to create a viscous swell root.

Evolution of Thermal and Depletion Buoyancy during Progressive Melt-Extraction

The effect that melt-extraction has on the buoyancy of the residue also plays a crucial role in the development of a buoyant and weak plume-fed asthenosphere. Mass is conserved during mantle melting, but density is not—both a less-dense melt phase and a less-dense residue to melt-extraction will be produced during typical mid-ocean ridge and plume melting. In the mid-1970s, it was recognized that mantle peridotite becomes more buoyant during decompression melting; progressive melt-extraction transforms it from denser peridotite to less-dense harzburgite (Boyd and McCallister, 1976; Oxburgh and Parmentier, 1977; Jordan, 1979).

There are two processes leading to the reduction in peridotite density. The denser minerals garnet and clinopyroxene are preferentially consumed during partial melting of peridotite, decreasing their relative abundance in the restite residue of melt-extraction, hence reducing its density. In addition, in the Fe-Mg solid solutions of typical mantle minerals, the heavier element Fe preferentially partitions into the melt phase while the lighter element Mg preferentially partitions into the solid residue. Oxburgh and Parmentier (1977) noted the importance of these effects for the buoyancy of oceanic lithosphere. The effect is smallest for shallow mid-ocean ridge melting that mostly happens in the ~25- to 60-km depth interval where spinel peridotite is the stable peridotite lithology, and the effect is about twice as large for deeper hotspot melting that occurs mostly at depths greater than ~60 km where garnet peridotite is the stable metamorphic phase assemblage. Oxburgh and Parmentier (1977) parameterized the effect of progressive melt-extraction on the density of the mantle residue as reducing the density of a spinel peridotite residue by $0.03\rho_m f$ and the density of a garnet peridotite residue by $0.06\rho_m f$, where f is the fraction of the melt extracted from the original mantle (f is also known as the net depletion of the mantle residue). These effects are shown in Figure 5, which shows that 10% melt extraction from a garnet peridotite reduces the density of the residue by an amount equivalent to heating the residue by 200 °C. If, instead, the mantle consists of a plum-pudding mixture of ~80% peridotite and ~20% eclogite (Phipps Morgan and Morgan, 1999) created by recycling the products of mid-ocean ridge and hotspot melting back into the mantle by slab subduction (the geochemical implications of this hypothesis are discussed below), then progressive preferential melting of lower-solidus temperature eclogite plums is thought to reduce the density of the aggregate eclogite + peridotite mixture by $0.07\rho_m f$ to $0.10\rho_m f$ (see Fig. 5). The reduction in density is likely to be even larger than that for melting a garnet peridotite, because the eclogite fraction contains a much larger proportion of dense mineral phases (e.g., garnet) that are consumed during partial melting.

Melting also consumes latent heat, which will reduce the temperature of the residue that has melted. However, the cooling effect of melting and melt-extraction will increase the density of the residue by a lesser amount than the compositional depletion effect of melt-extraction reduces the density of the residue. The latent heat needed to melt mantle silicates to a liquid phase is equivalent to the heat associated with heating a solid silicate by ~600 °C (for details, see discussions in Hess, 1989; Phipps Morgan, 2001). This means that the reduction in thermal buoyancy due to the latent heat consumed by melting a fraction f is equal to $f\rho_m\alpha(600°C)$, where $\alpha = 3 \times 10^{-5}$ °C^{-1} is the coefficient of thermal expansion. Thus the combined effect of the temperature and depletion effects on melting of a spinel peridotite is to reduce its density by $\rho_m(0.03 - \alpha600°)f = \rho_m(0.003 - 0.0018)f = 0.012\rho_m f$. Progressive melt-extraction from a garnet peridotite will reduce its density by $0.042\rho_m f$, and progressive melt-extraction that only melts the eclogite plums of a plum-pudding mantle will reduce its density by $\sim 0.052\rho_m f - 0.082\rho_m f$. In other words, an upwelling mantle plume that is 200 °C hotter than average mantle will, after having its garnet peridotite melt by 5%, have a buoyancy equivalent to it being 270 °C instead of 200 °C hotter than average mantle. Preferential melt-extraction from the eclogite plums of an eclogite-peridotite plum pudding will have a 25–100% larger effect than melt-extraction from a garnet periditite.

This compositional buoyancy effect will encourage the stabilization of a low-density depleted asthenospheric layer just below the base of the lithosphere. Note that once a cold subducting slab has returned to the ~80–100 km depths where basaltic ocean crust transforms to denser (garnet-rich) eclogite, then the compositional density of the eclogite + peridotite slab will tend to be higher than that of ambient asthenosphere, because the ocean-island basalt (OIB)-eclogite fraction of the subducting slab is larger than the volume fraction of eclogite veins in the depleted asthenosphere. Below the asthenospheric layer, the compositional density of an average subducting slab will be roughly that of average mantle—but the slab will stay denser than surrounding mantle as long as it remains cooler.

Finally, the preferential extraction of eclogite-pyroxenite veins during deep plume melting is not likely to affect the viscosity of the residue of melt-extraction by nearly as much as would shallower partial melting that also includes melt-extraction from the peridotite fraction of the upwelling mantle. The viscosity of the marblecake assemblage will be dominated by the viscosity of the easiest-to-creep major mineral, olivine, which forms half to two-thirds of the peridotite matrix. As long as the higher-solidus "damp" (but nominally anhydrous!) peridotite fraction does not melt enough to dry out during deep plume melting, the marblecake's viscosity will remain weak and asthenospheric.

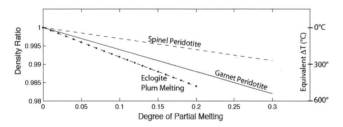

Figure 5. A comparison of the buoyancy associated with progressive melt-extraction from a spinel peridotite, garnet peridotite, and from melt-extraction solely from an eclogite component that makes up 20% of a peridotite-eclogite plum-pudding mantle. The fractional density change is shown at left, and the equivalent temperature increase needed to create the same density change is shown at right. The *x*-axis shows the total amount of progressive melt-extraction (depletion).

Asthenosphere Entrainment by Subducting Slabs: Implications for Asthenosphere Return Flow from Subduction Zones, Viscous Plate-Mantle Coupling, and Lower Mantle Flow

In our preferred conceptual model, the asthenosphere forms a hot and weak puddle that underlies suboceanic lithosphere (which is always thin in this context) and thin regions of subcontinental lithosphere, such as much of North America from the Cordillera westward, as has been proposed on the basis of heat-flow observations (Lewis et al., 2003; Hyndman et al., 2005).

We propose that only a thin (~15–20 km) sheet of asthenosphere is entrained and pulled downward at each side of the subducting slab (Phipps Morgan et al., 2007). This type of flow-structure has not yet been typically seen in numerical models of global mantle flow; instead, in global flow calculations, plume material is efficiently dragged down by plate subduction. We think it has not yet been seen not because it should not happen, but rather because current global models have insufficient numerical resolution of variable viscosity flow to properly capture the dynamics of asthenosphere entrainment by subducting slabs.

Figure 6A shows an example of a numerical experiment (Hasenclever, 2004; Phipps Morgan et al., 2007) with a temperature-dependent asthenosphere and mantle viscosity in which the subduction zone has ~4-km grid resolution in the region where a finger of entrained asthenosphere should develop if the asthenosphere is hotter (hotter potential temperature) than underlying mantle. In this relatively well-resolved case, a roughly ~20-km-wide finger of ~10^{19} Pa-s asthenosphere is entrained beneath the subducting slab in an entrainment pattern that can be quantitatively modeled by a simple boundary layer theory. Because relatively little asthenosphere is entrained by slab subduction, the bulk of the asthenosphere is not subducted and instead has a simple counterflow away from the subduction zone in the basic structure predicted by the 1972 channel flow model of Schubert and Turcotte (1972).

In contrast, Figure 6B shows an example of a poorly resolved numerical experiment (Hasenclever, 2004; Phipps Morgan et al., 2007) with a uniform 30-km grid resolution typical of some of the best-resolved current global studies. In this case, as a byproduct of resolution that is too poor to properly model entrainment, much more asthenosphere is dragged down by slab subduction than should be. We think this particular computational artifact plays a big role in why current global numerical models do not exhibit strong hints of our preferred flow scenario (the other reason being that current computational models also poorly resolve the strength of focused upwelling in hot lower-viscosity plumes.) Our preferred view of asthenosphere entrainment has been quantified and reproduced in both laboratory tank experiments with a moving and subducting plate and in well-resolved numerical experiments. These models and the boundary layer theory described in Phipps Morgan et al. (2007; see also Phipps Morgan and Morgan, 1999) imply that a roughly ~15- to

25-km-thick finger of ~10^{19} Pa-s asthenosphere should be entrained and subducted along with a typical subducting slab, with the remainder of the asthenosphere flowing laterally away from the subduction zone. In the numerical model we develop below to model asthenosphere flow, the entrainment of asthenosphere by subducting slabs will be treated as one of the several exter-

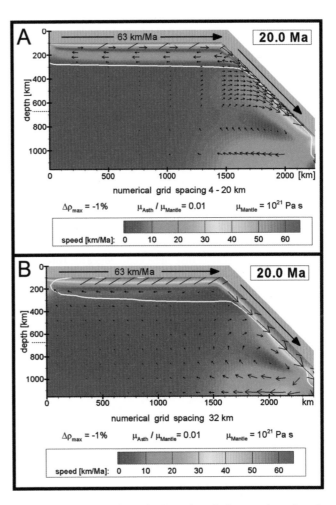

Figure 6. Comparison of (A) the dynamics of a buoyant low-viscosity asthenosphere layer beneath moving and subducting plate and (B) the different calculated flow for the same conditions but with insufficient numerical resolution to properly resolve the dynamics of the drag-down of a thin sheet of asthenosphere by the subducting slab, an insufficient resolution characteristic of many recent global flow calculations. Both runs show the evolution after 20 m.y. of plate subduction, starting with the same initial conditions. An asthenosphere that is simultaneously less dense and less viscous (e.g., hotter) than underlying mantle exhibits limited asthenosphere entrainment by slab subduction and the development of classic Schubert and Turcotte–like (1972) asthenosphere return flow away from the trench. However, calculations using an excessively coarse numerical grid (panel B) cannot resolve the thin entrainment layer and thus tend to drag down too much asthenosphere, resulting in a completely different flow pattern that implies the absence of asthenospheric return flow. Modified from Phipps Morgan et al. (2007, Fig. 4, panels *a* and *d*).

nal boundary conditions on asthenosphere flow. (The numerical model for global asthenosphere flow and its boundary conditions will be presented after the next section.)

Two-Stage Melting: Why Hotspot and MORBs Can Share the Same Mantle Plume Source and Yet Preserve Distinct Geochemical Signatures

At first sight, the idea that mantle plumes have brought up and subsequent asthenosphere flow has laterally redistributed almost all of the material currently melting beneath mid-ocean ridges may seem difficult to reconcile with the well-known geochemical differences (e.g., Hofmann, 1997, 2002) between MORBs and their OIB cousins. In general, the typical MORB melts from source material that is less rich in incompatible elements (elements that easily partition into a melt phase during partial melting) than is the source of OIB. The MORB source is also, on average, more isotopically depleted than the sources of OIB, meaning that this relative depletion in incompatible elements has been relatively ubiquitous and long-lived.

Although the difference between average MORB and OIB chemistry has been recognized since solid radioisotopic measurements first started to be systematically collected in the 1960s, it is also important to note that mid-ocean ridges also erupt a significant amount of enriched-MORB (E-MORB) similar in incompatible element and isotopic composition to many OIBs (e.g., Donnelly et al., 2004). Likewise, in multidimensional isotope plots, arrays of basalts from the same hotspot are typically distributed within a tubelike pattern characteristic of that hotspot (Hart et al., 1992; Phipps Morgan, 1999)—this hotspot array tube (HART) structure has one end that is usually more enriched than any MORB or E-MORB, whereas the other, more-depleted end often verges on a depleted and incompatible-element–poor composition common to many MORBs (e.g., Hart et al., 1992; Phipps Morgan, 1999).

The conventional way to explain these systematics has been to imagine they are the result of additive "pollution" of the MORB source by mixing in small amounts of underlying OIB reservoir(s). In the conventional conceptual model, MORB comes from a well-mixed, depleted, and shallow reservoir into which a smaller fraction of different, enriched OIB reservoirs is mixed to form OIB and E-MORB. (If each individual OIB source component comes from its own well-mixed reservoir, then at least four different OIB reservoirs are now needed to explain the EM-1, EM-2, high U/Pb (HIMU), and ^{3}He-rich flavors found to differing degrees in different HARTs.) In this conceptual model, the MORB source is normal depleted mantle remaining after extraction of the continental crust, while the OIB source is made from MORB source material that has been polluted by the addition of smaller amounts of material from several additional enriched OIB-source reservoirs, depending upon the particular hotspot.

There is an obvious alternative to the idea that the OIB source comes from adding enriched "flavors" to MORB source

material. What if the converse happens instead? What if the MORB source is made by somehow subtracting enriched flavors from the OIB source? When the subtraction of enriched flavors is imperfect, it results in an E-MORB rather than a MORB being formed during mid-ocean ridge melting.

A subtractive method for creating the depleted MORB source fits extremely well into the worldview in which the asthenospheric source for mid-ocean ridge melting is supplied by upwelling mantle plumes. This subtractive process works when the mantle is lumpy or lithologically variable, made up of a veined mixture of components with differing bulk compositions and melting solidii, with solidii typically lowest for the most incompatible-element–enriched and volitile-enriched components. Each lithologic lump will also have differing concentrations of trace elements and can evolve isotopically independently of its neighbors in a particular parcel of mantle. In our preferred conceptual model, subtraction occurs when the easiest-to-melt components of the upwelling plume material partially melt during ascent to create OIBs associated with hotspot volcanism. The enriched melts subtracted from upwelling plume mantle are visible as the enriched OIBs observed to be the type-example products of hotspot volcanism. These OIB partial melts are, however, only a few percent of the mass of the original upwelling plume material; the leftovers after OIB melting and extraction from upwelling plumes still make up more than ~95% by volume of the upwelling plume material. It is this new asthenosphere, "cleansed" by OIB melt-extraction of most of its easiest-to-melt and most-enriched plums, that makes up the ubiquitous depleted MORB source tapped to decompression melt during upwelling beneath a mid-ocean ridge. Note that the idea of a subtractive rather than additive origin for MORB has also been recently championed by Bercovici and Karato (2003), who, however, argue that a water + melt-linked filter on material ascending from the transition zone is the cleansing agent for MORB depletion. We prefer the interpretation that OIB (and E-MORB) melting is the subtractive filter, with observed OIB and E-MORBs having the correct trace-element and isotopic compositions for them to reflect the earlier stages of progressively melting and stripping rising plume material of the components that give OIB (and E-MORB) its distinctive geochemical signature. Note that the frequent occurrence of OIB-like E-MORBs is a natural byproduct of incomplete plume-melting–related removal of lower-solidus plums containing the OIB flavors (a mechanism for this process would be low or nonexistent plume melting of material upwelling at the cooler plume rim). Thus Fitton's "OIB Paradox" (Fitton, this volume) is actually supporting evidence for a plume-fed plum-pudding asthenosphere, as discussed by Phipps Morgan and Morgan (1999). Further supporting evidence is the existence of depleted components in both the OIB and MORB sources (e.g., Fitton, this volume).

We have explored this framework for interpreting the origin of oceanic basalts in several studies during the past decade, and other workers are also beginning to explore these ideas (Ito and Mahoney, 2005a,b). Here is first an example summarizing findings supporting the idea that progressive melt-extraction occurs in the sources of hotspot basalts, and then an example showing how the same process, on a global scale, can create the observed average differences between MORB and OIB/E-MORB source compositions.

Figure 7 shows an example from Phipps Morgan (1999; data courtesy of Al Hofmann) in which the observed isotopic variability in Hawaiian basalts is explained as a byproduct of progressive melt-extraction from a plum-pudding source. Note how the most-depleted portion of the predicted isotopic array of plume basalts is becoming MORB-like in isotopic composition—although it does not project toward the HIMU end of the suite of Pacific MORB compositions. This conceptual model explains the basic structure of the array of basalt compositions as reflecting a trajectory of progressive melt-extraction from the plum-pudding Hawaiian source instead of mixing between enriched and depleted source materials.

If progressive melt-extraction is the cause of isotopic heterogeneity of Hawaiian basalts, then the degree of source enrichment of each basalt should roughly correlate with its depth of melting, with the deepest melts being the most enriched, while melts produced at shallower depths in the melting column are more depleted because their source was depleted by prior, deeper melt-extraction. However, if the plume's temperature is hotter in its center than at its edges, as seems likely, then enriched components will begin to melt at shallower depths in the cooler rim of the plume than they do in the plume's hot central core.

Of course, mixing can occur between melts generated at different points along the melt-extraction trajectory. Furthermore, deeper enriched melts are out of equilibrium with their surrounding peridotitic wallrock during their ascent to the surface, which will induce (by their fluxlike behavior) the surrounding wallrock to also partially melt. The addition of wallrock melts will also produce a mixing-like overprint to the isotopic structure of the Hawaiian melt-extraction trajectory (Phipps Morgan and Connolly, 2004). Local mixing between pairs of source points along a melt-extraction trajectory provides a simple explanation for the otherwise confusing finding of many different apparent Pb-Pb mixing pairs in the sources of Kea basalts (Abouchami et al., 2000, 2005).

Figure 8 shows an example from Phipps Morgan and Morgan (1999) that illustrates that progressive melt-extraction can create isotopically distinct enriched OIB/E-MORB and depleted MORB sources as a natural byproduct of flow and melting processes within a plume-fed asthenosphere. Here the average observed OIB and MORB trace-element and isotopic compositions are reproduced by a simple recipe for mantle evolution that assumes that the MORB source has been a plume-fed asthenosphere during the entire evolution of the Earth, and that present-day mantle heterogeneities were generated, to a first approximation, by the continual recycling through plate subduction of the same basalts and continental erosional byproducts that are being recycled by modern subduction processes. This simple recipe, with the additional assumption that plate tectonics has al-

ways been Earth's mode of heat loss, is enough to reproduce the observed present-day differences between average MORB and OIB. (The corollary to the assumption that plate tectonics has always been the way the mantle has lost its heat is that rates of plate creation and recycling should scale with the square of Earth's heat loss; e.g., Phipps Morgan, 1997.)

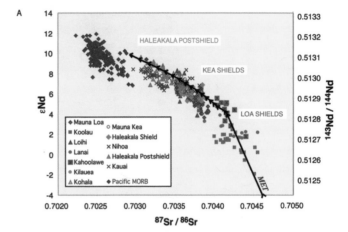

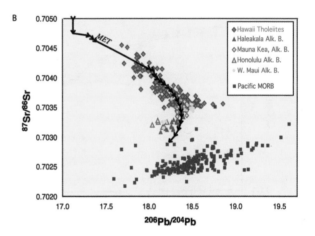

Figure 7. The observed $^{87}Sr/^{86}Sr$-$^{143}Nd/^{144}Nd$-$^{206}Pb/^{204}Pb$ pattern of the array of Hawaiian basalts (A.F. Hofmann, personal commun., 1998) is compared to a simple melt extraction trajectory (MET) model for the progressive melt extraction from a five-component Hawaiian mantle source. For the MET model, the five melting components have isotopic compositions EM (enriched mantle; $^{87}Sr/^{86}Sr$:$^{143}Nd/^{144}Nd$:$^{206}Pb/^{204}Pb$ = 0.7050:0.51161:17.1), ZMORB2 (recycled MORB; $^{87}Sr/^{86}Sr$: $^{143}Nd/^{144}Nd$:$^{206}Pb/^{204}Pb$ = 0.7019:0.51341:17.2), HIMU (high U/Pb; $^{87}Sr/^{86}Sr$:$^{143}Nd/^{144}Nd$:$^{206}Pb/^{204}Pb$ = 0.7030:0.51243:23.), ORES (residues to earlier plume melt-extraction; $^{87}Sr/^{86}Sr$:$^{143}Nd/^{144}Nd$: $^{206}Pb/^{204}Pb$ = 0.7016:0.51351:15.), and PRIM ("primitive" mantle; $^{87}Sr/^{86}Sr$:$^{143}Nd/^{144}Nd$:$^{206}Pb/^{204}Pb$ = 0.7035:0.51264:18.1). The two panels show the $^{87}Sr/^{86}Sr$:$^{143}Nd/^{144}Nd$:$^{206}Pb/^{204}Pb$ observations for Hawaiian basalts and the depleted portion of the model MET. Each melt increment composition is marked by an arrowhead. This figure demonstrates that the MET generated by progressively melting a five-component mantle is a 1-D trajectory in $^{87}Sr/^{86}Sr$:$^{143}Nd/^{144}Nd$:$^{206}Pb/$$^{204}Pb$ isotope space, and that it is possible to reproduce the general isotopic characteristic of the array of Hawaiian basalts by progressive melt-extraction from a multicomponent mantle. Alk. B.—alkaline basalt; MORB—mid-ocean ridge basalt.

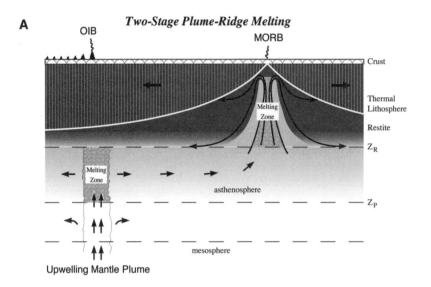

A

OIB　　　***Two-Stage Plume-Ridge Melting***　　　MORB

Crust

Thermal Lithosphere

Melting Zone

Restite

Z_R

Melting Zone

asthenosphere

Z_P

mesosphere

Upwelling Mantle Plume

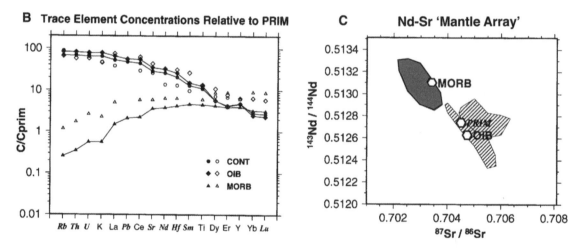

B Trace Element Concentrations Relative to PRIM

C/Cprim

● ○ CONT
◆ ◇ OIB
▲ △ MORB

Rb Th U K La Pb Ce Sr Nd Hf Sm Ti Dy Er Y Yb Lu

C Nd-Sr 'Mantle Array'

$^{143}Nd / ^{144}Nd$

MORB

PRIM

OIB

$^{87}Sr / ^{86}Sr$

Figure 8. (A) Melting scenario implied by our preferred conceptual model, in which material in an upwelling mantle plume first melts beneath a hotspot to create, from its leftovers, asthenosphere, which is the depleted mid-ocean ridge basalt (MORB) source that upwells passively and melts a second time beneath a spreading center. The melting regions of the plume and spreading center are shown by wiggly black lines. Heavy black lines and arrows show schematic flow patterns from the plume through the spreading center. Z_R is the initial depth of ridge melting, and Z_P is the initial depth of plume melting. Melting at a ridge starts at roughly the minimum depth reached during plume melting beneath off-axis lithosphere. (B) Trace-element concentrations of average ocean island basalt (OIB), continental crust (CONT), and MORB. Trace-element abundances are normalized by corresponding primitive mantle values. MORB is characteristically depleted in its highly incompatible elements with respect to OIB and CONT. Two-stage melting produces this MORB depletion. Open symbols show observed values. Solid symbols joined by solid lines show the present-day OIB, CONT, and MORB compositions (normalized to "primitive" mantle [PRIM] concentration = 1) that are generated by the sample recipe discussed in Phipps Morgan and Morgan (1999). (C) The Nd-Sr isotopic mantle array. A first-order relation between Sr and Nd isotope data for OIB (hatched region), MORB (shaded area), and continental crust (not shown) is that their Sr and Nd isotopic evolution is correlated, consistent with the MORB coming from an isotopically depleted source. (Bulk-earth evolution is shown by the PRIM hexagon.) Two-stage melting of a plum-pudding mantle also produces isotopically enriched OIB and isotopically depleted MORB (hexagons show present-day average OIB and MORB compositions that are generated by the recipe of Phipps Morgan and Morgan, 1999). From Phipps Morgan and Morgan (1999), where an expansive description of this model is presented.

The recipe predicts that the present-day mantle should consist of ~10–20% eclogitic recycled basalt and sediment-lithology plums dispersed in a matrix of variably depleted peridotite restites, as shown in Figure 9, of which the dominant fraction (about half the mantle) are highly-depleted harzburgites that have melted at least once beneath a mid-ocean ridge, and maybe 5–15% are relatively primitive peridotites that have experienced only minor amounts of melt-extraction during their 4.5-Ga residence in the mantle. The eclogite plums are also predicted to be highly heterogeneous, with only a few percent of the mantle containing more than 80% of its most-incompatible elements in recycled OIB and continental sediment lithologies. Note that this mixed plum-pudding lithology is also consistent with observations of seismic scattering and inferences of bulk mantle composition (Helffrich and Wood, 2001). In essence, this model is just a marblecake or plum-pudding variant of Ringwood's pyrolite compositional model of the mantle, but with the recycled basaltic plums never becoming compositionally re-homogenized with surrounding harzburgite, so they instead become a heterogeneous gneisslike mantle assemblage dominated in volume by its peridotite fraction. If these heterogeneities are well-folded into a mantle marblecake (Allègre and Turcotte, 1986) at a 1-km scale length, then significant heat can diffuse between lithologies to shape the evolution of pressure-release melting (Sleep, 1984; Phipps Morgan, 2001).

If preferential melt-extraction or convective stirring leads to the development of an uneven distribution of plum components in the mantle, then there may be density differences linked to this large-scale chemical heterogeneity, with buoyant plume upwellings preferentially sampling compositionally buoyant regions of the mantle. However, if only fine-scale (e.g., ~1 km or less) heterogeneity persists in the mantle, then regional temperature variations will dominate the large-scale buoyancy distribution within the mantle, and plumes will preferentially sample from the hotter regions of an isochemical, yet still compositionally heterogeneous, mantle.

Hydrogen (e.g., water) and perhaps helium can also diffuse to a geochemically significant degree between neighboring lithologies during gigayears of mantle convection, yet, unlike heat, not tend to diffuse between neighbors during ten-thousand-times briefer episodes of pressure-release melting. Some potential effects of heat and volatile diffusion between components are discussed in more depth in Phipps Morgan (2001), and potential implications for systematic rare-gas differences between MORB and OIB/E-MORB (Phipps Morgan and Morgan, 2001) are currently being prepared for publication. Finally, this conceptual model offers a simple explanation for why the Earth's geoid varies by ±100 m, yet the ocean basins do not show the ±1–2 km of dynamic topography needed to produce this geoid if stress-support topography at the top surface of the mantle were to compensate the seismically inferred mass anomalies via stresses associated with internal viscous deformation (see Thoraval et al., 1990). If, instead, dynamic deflections at the base of a buoyant asthenosphere compensate the stresses associated with deeper flow, then large geoid variations may be associated with small amounts of geoid-linked dynamic topography (Ravine and Phipps Morgan, 1996; Ravine, 1997).

We hope this extended review has given the reader a better feel for the conceptual model that we wish to further explore in this study and has also shown how this alternative conceptual framework may be able to reconcile diverse observations and conundrums on the structure and evolution of the mantle. Further discussion of the differences between this conceptual model and other scenarios is found in Phipps Morgan et al. (2007).

Note that the conceptual model of a buoyant and weak plume-fed asthenosphere may be right or wrong, but it cannot be half correct. If the asthenosphere spreads out everywhere and is more buoyant than underlying mantle, which it will tend to displace downward, then buoyant plumes must be the only source of the most-buoyant, hence shallowest, suboceanic asthenosphere. If the most-buoyant asthenosphere does not spread out more or less evenly, then upwelling at places other than plumes

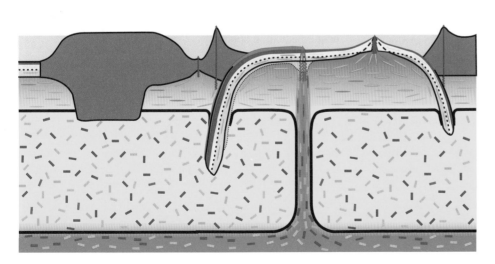

Figure 9. Cartoon showing our preferred view of how a plume-fed asthenosphere layer may shape the large-scale structure of slab-subduction–driven mantle convection, and how this flow pattern can be compatible with the observed geochemical differences between hotspot (ocean island basalts) and mid-ocean ridge basalts.

will also be necessary—with no logical reason for why this passive upwelling would be more buoyant than typical mantle. Next we discuss the additional steps to quantitatively determine the asthenosphere flow predicted for the present-day configurations of ridges, trenchs, and continental cratons.

Physical Model for Global Asthenosphere Flow

To model global asthenosphere flow, we have developed a thin-spherical-shell finite element model based on the lubrication theory paradigm used by Yale and Phipps Morgan (1998) to explore the effects of regional ridge-hotspot interactions. In the physical model, suboceanic asthenosphere fills a low-viscosity channel bounded above by very-high-viscosity lithosphere and below by higher-viscosity mesosphere. This asthenosphere will flow to transport material from its plume sources to where it is consumed by lithosphere growth and by drag-down at subduction zones. It will also flow in response to shear drag from above by moving plates. Asthenosphere is assumed to be brought up by mantle plumes to replenish asthenosphere consumption (sinks) by plate growth and subduction.

Asthenosphere consumption is due to three tectonic activities. One is plate accretion at mid-ocean ridge; another is drag-down next to subducting plates at trenches where asthenosphere is entrained downward by subducting lithosphere; and the third is attachment to the base of the aging, cooling, and thickening lithosphere. Present-day plate velocities over a hotspot reference frame determine the lithosphere-motion–induced shear flux within the asthenosphere. (In this study, we make the additional simplifying assumption that the higher viscosity and more slowly moving base of the asthenosphere has no horizontal motion.) Asthenosphere is continually consumed by being converted to lithosphere or dragged down at subduction zones—at mid-ocean ridges by melt-extraction that makes a ~60-km-thick compositional lithosphere layer more viscous than underlying asthenosphere, by underplating oceanic lithosphere when it cools beyond the 60-km thickness of the compositional lithosphere, and by mechanical drag-down around subducting plates.

To construct the simplest possible physical model that includes the necessary complexity that we choose to explore, additional simplifying assumptions are still needed. Thus we will further assume that the rate of introduction of new asthenosphere from mantle plumes is exactly equal to the rate that asthenosphere is being consumed by lithosphere growth and subduction. (Note that this extra simplification of assuming a system in steady state is not inherent in the conceptual model. There could be times when more asthenosphere is supplied by upwelling plumes than is destroyed—e.g., mid-Cretaceous—and times when asthenosphere consumption by lithosphere creation and subduction exceeds plume resupply, in which case, the thickness of the asthenosphere would wax and wane; Phipps Morgan et al., 1995a). This assumption provides a strong constraint on the boundary conditions for the physical model, and also strongly shapes the pattern of global flow. For example, one implication of this scenario is that there is no broad upwelling deep beneath a mid-ocean ridge; instead, all ridge material comes up at hotspots and flows laterally within the asthenosphere to supply the demand for material at the spreading ridges.

While hotspots replenish asthenosphere consumed by lithosphere growth and subduction, plate velocities over the hotspot reference frame shape, by flow-induced by drag of overlying lithosphere, the shear flux within the asthenosphere. When a plate is stationary with respect to the hotspot reference frame, there is no plate-motion–induced asthenosphere shear, and lithosphere grows through passive accretion by cooling asthenosphere directly below the lithosphere. In contrast, when a plate moves over the hotspot reference frame, its motion induces asthenosphere shear that varies within the lithosphere-asthenosphere system geographically, requiring dynamic (pressure-driven) flow to conserve mass (Fig. 9). In the plate-spreading direction, asthenosphere flow is due to both pressure-driven flow and shear flow. In the ridge-parallel direction, dynamic flow is required to supply asthenosphere material to growing lithosphere. The changing thickness of athenosphere/lithosphere at continental margins creates a different type of asthenosphere sink, as the net horizontal flux within the asthenosphere adjacent to the moving plate margin must match the horizontal flux of material within the migrating asthenosphere-lithosphere margin.

The net lithosphere plus asthenosphere flow through each horizontal column is:

$$q_{\text{shear}}(x) + q_{\text{pressure}}(x) = Q_{\text{total}}, \qquad (1)$$

where the vertically integrated shear flow is:

$$q_{\text{shear}}(x) = U_a(h_0 - h(x)) + \frac{1}{2} U_a h(x). \qquad (2)$$

Here, U_a is the absolute plate spreading rate, h_0 is the depth of the base of the asthenosphere, $h(x)$ is the asthenosphere channel thickness, and $(h_0 - h(x))$ is the lithosphere thickness. Within the asthenosphere, the lubrication theory approximation to the momentum equation applies:

$$\nabla P = \mu \frac{\partial^2 u}{\partial z^2}, \qquad (3)$$

where the pressure gradient is expressed in terms of u, the vector velocity field; z is the distance from the asthenosphere-mesosphere boundary; and μ is the asthenosphere viscosity. This equation is integrated twice over the thickness of asthenosphere for the pressure flux,

$$q_{\text{pressure}} = -\frac{h^3(x)}{12\mu} \nabla P. \qquad (4)$$

Using equations 1 and 4,

$$\nabla P = -\frac{12\mu}{h^3(x)}\left(Q_{\text{total}} - q_{\text{shear}}(x)\right) \qquad (5)$$

defines the horizontal pressure variation, from which the pressure-driven flux can be determined using equation 4.

Solution Method

The finite element method is used to solve the above equations. The mesh used in this study has nodes spaced ~100 km apart. (The reason why this grid resolution was chosen is that the lubrication theory approximation is only valid at scale lengths greater than the thickness of the asthenosphere channel; thus, use of a higher-resolution grid would not provide a higher-resolution picture of the true flow.) The calculation needs five input parameters: plate boundary shapes and plate velocities (Fig. 10), the asthenosphere viscosity, thicknesses of lithosphere and asthenosphere (Fig. 11), sinks of asthenosphere (Fig. 12), and relative hotspot strength (Fig. 13).

Plate velocities over fixed hotspots are based on our currently preferred hotspot track inversion (see Morgan and Phipps Morgan, this volume), which for completeness is given in Table 1. The asthenosphere viscosity is set to 1.59×10^{19} Pa-s. (This nominal value could of course be changed to result in more or less dynamic topography associated with a given amount of pressure-driven asthenosphere flow. In fact, 7×10^{18} Pa-s is a value that better matches the observed dynamic topography. The value was chosen so that 1 dimensionless pressure unit in the program would correspond to 1 m of seafloor uplift.) Oceanic lithosphere younger than 36 Ma is assumed to be 60-km-thick compositional lithosphere created by melt-extraction at a spreading center (Phipps Morgan, 1997), while older lithosphere thickens thermally with thickness proportional to $10\text{km} \times \sqrt{\text{age}}$ (Ma). Lithosphere in continental areas is assumed to be 190, 220, and 250 km thick for Paleozoic/Cenozoic, Proterozoic, and Archean, respectively. The depth h_0 of the base of the asthenosphere has different values for each ocean, based on observed SS-S bounce-point delays (Woodward and Masters, 1991), which implies that the Pacific basin has thicker asthenosphere and the Atlantic basin has thinner asthenospere than other ocean basins. We assumed 330 km for the Pacific, 250 km for the Atlantic, and 300 km for the other areas. However, these changes in the basal depth of the asthenosphere h_0 result in only small changes in the pressure distribution compared to the assumption of a uniform depth for the base of the asthenosphere and do not change the overall flow pattern. In contrast, different amounts of asthenosphere drag-down by subducting slabs produce big differences in pressure and flow patterns, as seen by comparing Figure 14A with 14B. Here we choose the uniform value of 15 km for the thickness of the downdragged (entrained) asthenosphere sheet at the slab surfaces that is suggested by the analysis and experiments in Phipps Morgan et al. (2007).

Determination of Relative and Absolute Plume Strengths

Once one abandons the convenient, but in our opinion, misguided assumption that the relief of a hotspot swell reflects the entire upwelling flux of a mantle plume, then the least-well-constrained inputs to a model for global asthenosphere flow are the upwelling fluxes of mantle plumes. As long as an oceanic plate moves relatively rapidly over a mantle plume, then one can hope that the volume of hotspot volcanism may (roughly) correlate to the upwelling flux of the plume, although variations in lithosphere thickness (age) and variations in plume temperature will obviously also affect pressure-release melting of a mantle plume (White, 1993; Phipps Morgan, 1997).

However, in regions like the Atlantic and Southern ocean basins, where plates move slowly, it is hard to distinguish recent from older volcanism, and even this rough proxy for plume flux becomes extremely difficult to estimate. Thus we use a two-part

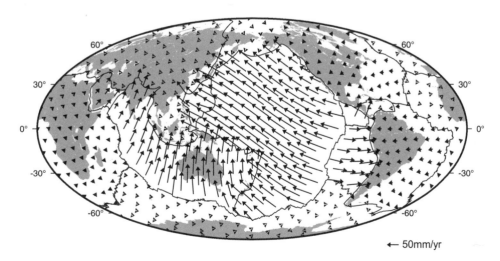

Figure 10. The absolute plate velocity model used as the lithosphere motion boundary condition on asthenosphere flow. Table 1 gives Euler poles describing the absolute velocity of each plate. (The determination of these absolute plate motions is presented in Morgan and Phipps Morgan, this volume.)

← 50mm/yr

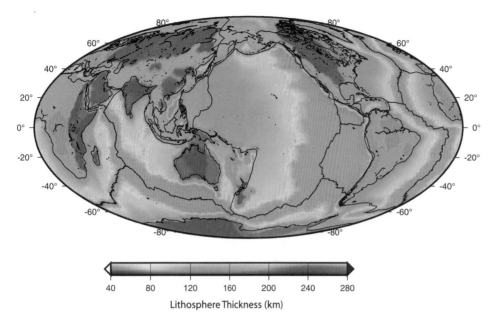

Figure 11. Map of lithosphere thicknesses used in this study. For oceanic lithosphere younger than 36 Ma, a lithosphere thickness of 60 km is adopted, which is appropriate for compositional lithosphere created by melt extraction at mid-ocean ridges (Phipps Morgan, 1997). Once the thermal lithosphere boundary layer penetrates deeper than the compositional lithosphere at ages older than ca. 36 Ma, the thicker thermal lithosphere boundary layer thickness of $10\sqrt{age}\ [Ma]$ [km] is used (see Fig. 1B).

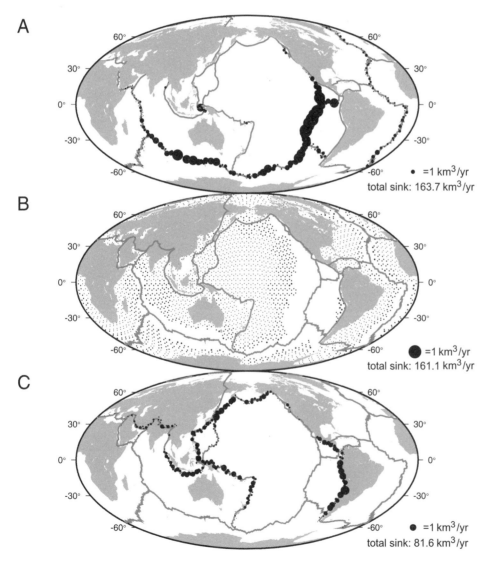

Figure 12. Present-day asthenosphere sinks that lead to a present-day net rate of asthenosphere consumption of 406.4 km³/year. (A) Asthenosphere sink caused by the creation of 60-km-thick compositional lithosphere creation at mid-ocean ridges. (B) Asthenosphere sink due to lithosphere growth by plate cooling. (C) Sink caused by drag-down of a ~15-km-thick asthenosphere sheet at both sides of subducting slabs (see Fig. 1C).

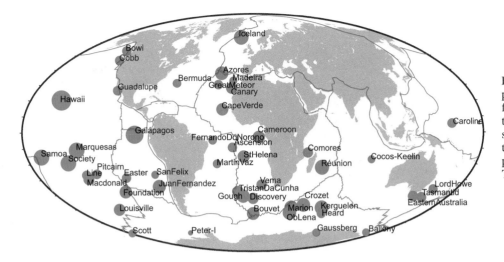

Figure 13. Relative strengths of mantle plumes in our preferred asthenosphere flow model. Red circles show the positions of mantle plumes used in this study. The area of each circle is proportional to the relative strength of each plume. Plume strengths are listed in Table 2.

strategy to estimate the upwelling plume-fluxes shaping present-day asthenosphere flow. In the Atlantic Ocean basin, we tune relative strengths of plumes in the basin to fit the resultant pressure distribution to the geoid maps and to try to match the well-recognized geochemical province boundary between Iceland-influenced and Azores-influenced ridge segments along the Mid-Atlantic Ridge (cf. Schilling, 1986), and the geochemical province boundary in the region of the equatorial Romanche fracture zone (Dupre and Allègre, 1983; Zindler and Hart, 1986). Figure 15 illustrates this tuning process. With no plume beneath Iceland, there would be no geoid high there, and there would also be significant northward asthenosphere flow across the Romanche fracture zone, with or without plume upwelling in the South Atlantic basin (bottom panels of Fig. 15). If there were only an Iceland plume in the Atlantic basin with a flux large enough to fill the entire North Atlantic part of the basin, then sub-Atlantic asthenosphere flow would be linked to a geoid high around Iceland with strong north-south astheno-

sphere flow in the north and south-north flow in the South Atlantic (Fig. 15, top left). With relatively strong Iceland and Azores plumes, and relatively strong Madeira/Canaries and Cape Verde plumes, then one can reproduce the observed geoid high surrounding Iceland, a transition between Iceland-influenced and Azores-influenced asthenosphere along the northern mid-Atlantic Ridge, and an equatorial boundary between North Atlantic and South Atlantic plume-influenced sections of the ridge (Fig. 15, top right).

Note that in this preferred configuration, Iceland retains a strong geoid anomaly because of the combination of a strong plume-source upwelling in a relatively confined basin between Greenland and Norway, whereas other portions of the ridge have relatively muted predicted geoid anomalies—as observed. However, because we use these constraints to tune the relative plume fluxes, we cannot use these constraints as independent data to compare against model predictions. Thus, where possible, we prefer to use independent plume-flux estimates to determine rel-

TABLE 1. PLATE MOTIONS IN HOTSPOT-FIXED FRAME

Plate	Latitude (°N)	Longitude (°E)	Angular rate (°/m.y.)	Maximum speed (mm/year)
Africa	43.07	−24.56	0.1544	17.16
Antarctica	70.18	82.26	0.0989	10.99
Arabia	31.00	13.06	0.5205	57.83
Australia	20.14	40.28	0.6990	77.67
Caribbean	−20.80	−56.84	0.0808	8.98
Cocos	22.32	−116.24	1.3343	148.26
Eurasia	82.54	−125.53	0.0619	6.88
Juan de Fuca	−31.32	60.04	0.8540	94.89
Nazca	49.95	−95.74	0.5634	62.60
North America	−51.87	−48.70	0.1628	18.08
Pacific	−59.33	94.90	0.8028	89.20
Philippine	−45.71	−23.83	0.9928	110.31
South America	−75.22	100.24	0.1749	19.43

Source: Values from Morgan and Phipps Morgan (2007).

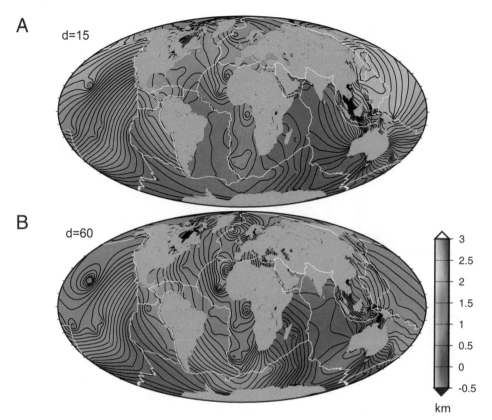

Figure 14. Example showing two calculations in which we vary the thickness of the sheets of asthenosphere that are downdragged by subducting slabs, all other parameters being those of our preferred model. (A) Our preferred thickness for the sheets of downdragged asthenosphere is equivalent to 15 km of asthenosphere subducting on each side of the slab (top panel)—this thickness is based upon numerical experiments like those shown in Fig. 6 and a simple boundary layer theory (Phipps Morgan and Morgan, 1999; Phipps Morgan et al., 2007). (B) Increasing the drag-down thickness by subducting slabs to an equivalent of 60 km of drag-down on each side results in a flow pattern with much larger predicted contrasts between the bathymetry of the Indian Ocean and other oceans, contrasts that are not observed along the global spreading system.

ative plume strengths, and for plumes rising beneath more rapidly moving plates, we have used Phipps Morgan's (1997) and White's (1993) estimates of hotspot magma production in the Pacific and Indian ocean basins as a proxy for the relative strengths of plumes upwelling beneath these basins. We are able to combine the Atlantic and rest of the world estimates of relative plume strengths with the additional assumption that the total upwelling plume flux is equal to the present-day rate of asthenosphere consumption, which we estimated (see Fig. 12) to be 406 km^3/year. Figure 14 and Table 2 show the plume strengths of our preferred prediction of global asthenosphere flow.

Predicted Asthenosphere Flowfield and Pressure Distribution

Figure 16 shows the resulting global map of predicted asthenosphere flow. Strong flow occurs beneath the Pacific basin because of high plate speeds and strong ridgeward fluxes from the equatorial Pacific superswell region (McNutt and Fischer, 1987) and Hawaii. In contrast, the Atlantic shows rather weaker asthenosphere flow because of its much slower plate motions, smaller ridge-consumption of asthenosphere, and the intermediate-to-weak strengths of its plumes. The Indian Ocean basin exhibits strong predicted asthenosphere flow in the central part of the basin, with lesser flow rates in the north and south. Note that the predicted pattern of asthenosphere counterflow near subduc-

tion zones ranges from parallel (e.g., Tonga, Chile) to perpendicular (e.g., northwest Pacific trenches) to the nearest trench.

Tests of the Predicted Asthenosphere Flowfield and Pressure Distribution

How can this model be tested? One direct approach is by looking at the patterns of flow directions in the asthenosphere, which should, in principal, be relatable to the direction of seismic wavespeed anisotropy. (The main caveat here is that the azimuthal seismic wavespeed anisotropy, as measured by Rayleigh and Love surface waves, is also highly sensitive to the strong azimuthal anisotropy in the ocean lithosphere that must be appropriately removed.) We compare a few regional examples of this approach in part 2 of this study (Yamamoto et al., this volume, chapter 10), but we defer a systematic global comparison to a later study.

A second approach is to compare the predicted pressure gradients in the asthenosphere with seafloor bathymetry, as deviatoric asthenosphere pressures should relate to the portion of the topography in the ocean basins not due to plate cooling with age or the perturbations associated with hotspots and hotspot swells. The predicted global pressure distribution does show a relative pressure high in the western Pacific, compatible with seafloor flattening at old ages and also has pressure lows in the wakes of the rapidly moving Australian and Indian cratons, a pattern con-

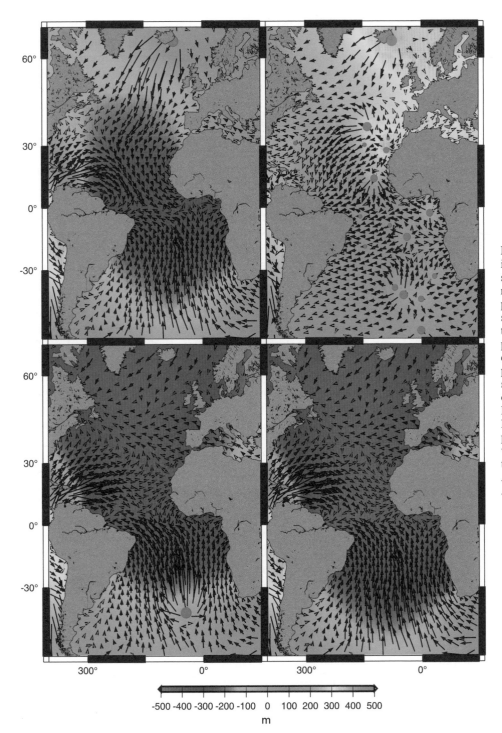

Figure 15. Example illustrating the tuning process used to infer the relative strengths of Atlantic plumes. The bottom right panel shows flow if there were no plumes feeding sub-Atlantic asthenosphere, whereas the bottom left shows the situation with only a south Atlantic plume source, resulting in a similar predicted flow pattern with lower associated pressure gradients. The top left panel shows the flow predicted if there were only the Iceland plume, which would lead to this plume filling the North Atlantic. The top-right model shows our preferred relative plume strengths, which were tuned to make along-ridge isotope boundaries between the Iceland and Azores plumes, and in the equatorial Atlantic.

sistent with predictions based on simpler analytical idealizations of this process (Phipps Morgan and Smith, 1992; Phipps Morgan et al., 1995b).

However, if the asthenosphere is buoyant with respect to underlying mantle, then an asthenospheric pressure high should not only lead to dynamic stress-supported uplift of the seafloor (known as "dynamic topography") but also to a much larger am-

plitude stress-supported depression of the base of the buoyant asthenosphere layer. We propose to call this effect "dynamic isostasy," because the thicker root of low-density asthenosphere associated with the downward deflection of the base of the asthenosphere beneath a region of high dynamic pressure is exactly the magnitude of the deflection that would be predicted if one assumed that the dynamic seafloor uplift were isostatically

TABLE 2. HOTSPOT AND PLUME LOCATIONS AND STRENGTHS

Hotspot	Latitude (°N)	Longitude (°E)	Strength*
Iceland	64.4	−17.3	0.60
Azores	37.9	−26.0	0.40
Bermuda	32.4	−64.0	0.08
Madeira	32.7	−17.0	0.00
Canary	28.2	−18.0	0.15
Great Meteor	29.4	−29.2	0.05
Cape Verde	14.9	−24.3	0.40
Cameroon	−2.0	5.1	0.30
Fernando De Noronha	−4.0	−32.3	0.08
Ascension	−7.9	−14.3	0.10
St. Helena	−15.6	−7.0	0.30
Martin Vaz	−20.5	−28.9	0.08
Vema	−32.1	6.3	0.15
Tristan Da Cunha	−37.2	−12.3	0.25
Gough	−40.3	−10.0	0.36
Discovery	−43.0	−2.7	0.20
Bouvet	−51.4	1.1	0.35
Ob-Lena	−52.2	40.0	0.40
Marion	−46.9	37.6	0.40
Crozet	−26.1	50.2	0.40
Kerguelen	−49.6	69.0	0.50
Heard	−53.1	73.5	0.10
Réunion	−21.2	55.7	0.60
Comores	−11.5	43.3	0.15
Lord Howe	−34.7	159.8	0.10
Tasmantid	−40.4	155.5	0.15
Eastern Australia	−40.8	146.0	0.15
Cocos-Keeling	−17.0	94.5	0.05
Caroline	4.8	156.4	0.10
Hawaii	19.1	−155.1	2.00
Samoa	−14.6	−168.2	0.80
Marquesas	−10.5	−139.0	0.80
Society	−17.9	−148.2	0.80
Macdonald	−29.0	−140.3	0.20
Line	−27.0	−139.0	0.10
Pitcairn	−25.2	−129.1	0.10
Louisville	−50.9	−138.1	0.25
Foundation	−37.0	−113.0	0.35
Easter	−26.8	−107.6	0.10
Galapagos	−0.4	−91.6	1.30
Juan Fernandez	−33.9	−81.8	0.40
San Felix	−26.4	−80.1	0.05
Guadalupe	26.8	−112.4	0.10
Cobb	46.2	−130.0	0.10
Bowie	53.3	−135.7	0.10
Scott	−68.8	−178.0	0.05
Peter I	−68.8	−90.7	0.01
Gaussberg	−66.8	89.3	0.05
Balleny	−67.6	164.8	0.05

*A plume strength of 1 corresponds to an upwelling plume flux of 29.4 km³/year, implying a total plume upwelling flux of 406 km³/year.

compensated by a thickened root of lower-density asthenosphere. This mode of compensation of an asthenospheric pressure gradient is the mode that maximizes deformation in the low-viscosity asthenosphere layer while minimizing flow within the underlying higher-viscosity mantle. The complexity it induces in a determination of asthenospheric flow is that the depth of the asthenosphere channel becomes a function of pressure. Because these initial calculations assume a fixed depth for the base of the asthenosphere, their pressure predictions provide only a qualitative assessment of the lateral pressure variations associated with asthenospheric flow. As the asthenosphere's resistance to lateral flow depends on the third power of the thickness of the asthenospheric channel (equation 4), lateral pressure gradients will be reduced at pressure highs where the asthenosphere is thickened by the effects of dynamic isostasy and will be increased at pressure lows where the asthenosphere thins because of uplift of its base. Furthermore, if the base of the buoyant asthenosphere is also compensating relief due to normal stresses arising from the deeper mantle flow that creates a "deep-mantle" geoid, then the thickness of the asthenosphere layer thickness should vary from this effect, too. The calculated flow field is less sensitive to this second-order effect than is pressure. Because of this insensitivity, we focus on testing the predicted flow-field in this initial determination of global asthenosphere flow, and defer a detailed exploration of the predicted pressure field to a later study in which we will treat the effects of dynamic isostasy.

A third method of testing the model is that regions of asthenosphere fed by each hotspot can be determined, and the boundaries between these regions should be evident at mid-ocean ridges as geochemically distinct provinces of MORBs. This third test is the simplest to do now, as it takes advantage of the wealth of geochemical data that have already been collected along the global ridge system. Thus this approach will be the main focus of the companion article to this study (Yamamoto et al., this volume, chapter 10). In the rest of this article, we discuss other nongeochemical implications of the asthenosphere flowfield predicted by this model and then finish by noting current limitations in our treatment of subcontinental plumes and the assumption of a steady-state pattern of asthenosphere flow.

The Easter Island Vortex

One of the most curious features of the predicted pattern of asthenosphere flow shown in Figure 16 occurs near the fastest-spreading part of the ridge spreading system near Easter Island. A clear but small vortex in the predicted flow pattern forms here as a byproduct of the pattern of plate spreading and is strengthened by the presence of the nearby Foundation plume. (However, a weaker vortex pattern with the same spatial pattern forms even when we remove the Foundation plume from the model, which shows that the effect is primarily due to the pattern of plate spreading in this region.)

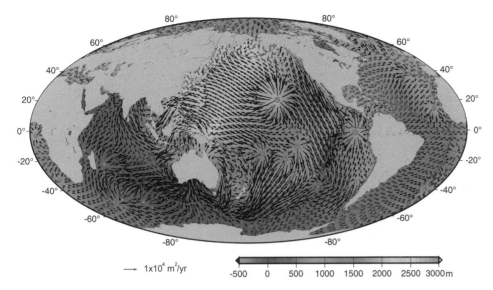

Figure 16. Our preferred global asthenosphere flow model. Arrows represent total flux (pressure-induced flow plus shear-induced flow), and colors show the lateral pressure field associated with this flow expressed as dynamic topography. The magnitude of the pressure field scales with the assumed asthenosphere viscosity. Here we chose 1.6×10^{19} Pa-s (e.g., too high a value), so that 1 pressure unit would correspond to 1 m of dynamic topography. Our preferred value for the average viscosity of the asthenosphere would be one-third to half as large, which would produce ~1 km of dynamic topography near the subduction zones in the western Pacific. Note that this dynamic topography would be compensated by opposing dynamic relief on the base of the asthenosphere, so that only a small geoid anomaly would be associated with this topography. (An example of this type of dynamic isostasy is seen in the numerical experiment shown in Fig. 6A.)

Two microplates, Easter and Juan Fernandez, are found in this fastest-spreading area. Schouten et al. (1993) proposed a microplate driving hypothesis they called the "roller bearing" model. They suggested that a microplate rotates as a rigid block between plates moving in opposite directions and presented analysis to dispute the hypothesis that basal asthenospheric flow–induced shear stresses could be a significant driving force for the observed internal rotation of microplates. Neves et al. (2003) tested the roller bearing model with numerical experiments using the lithospheric stress pattern inferred from observations at Easter microplate, and also concluded that asthenosphere flow could not force microplate rotation, but instead should act as a resisting torque. They raised the possibility of an asthenosphere vortex as a possible driving force, but discarded this possibility on the grounds of its seemingly contrived geometry. However, given that our predicted asthenosphere flow pattern shows a convenient vortex in this region of microplate activity, we speculate that the vortex may help force microplate rotation to some degree, or at the very least make asthenosphere drag be much less of a rotation-resisting force. This pattern also makes us wonder whether the persistent creation of microplates may be an indirect consequence of a persistant asthenosphere flow vortex beneath this region of the spreading center.

Subcontinental Plumes

Subcontinental plumes have been essentially ignored in this study, because we consider that plume-fed asthenosphere will behave quite differently there than in the suboceanic case. For example, seismic tomography shows very steep velocity gradients around continental hotspots and LVZs form narrow corridors to points on the boundary of continent and ocean (Ritsema and Allen, 2003). This observation suggests to us that a pipe-like pattern of lateral plume-fed flow exists beneath continental lithosphere instead of the flow pattern in a sheet of low-viscosity asthenosphere layer that exists beneath ocean basins. We think the difference in flow pattern is related to the greater thickness of continental lithosphere and the stronger lateral variations in the depth of the base of continental tectosphere. As long as plume material is surrounded by more-viscous mantle, then its ability to form and flow within a low-viscosity conduit will dominate its flow pattern. Only when a pond or puddle of low-viscosity asthenosphere has formed can lateral asthenosphere flow be described by the equations discussed in this study. Thus we imagine the drainage of asthenosphere from beneath continents to occur within drainage channels of thinner tectosphere (i.e., beneath the relatively thinner lithosphere of ancient sutures or failed rifts between cratons, such as the Cameroon Line; Ebinger and Sleep, 1998). For this initial study, we simply neglect the contribution of subcontinental hotspots, and we most regret the loss of a contribution from an Afar plume. In future work, we hope to merge this approach for suboceanic flow with a better characterization of subcontinental plume-fed asthenosphere drainage through incorporation of the "upside-down drainage" approach used by Sleep and coworkers.

Current Limitations and Future Improvements

The predicted flow pattern shown in Figure 16 is based on the assumption of steady state. It assumes that the asthenosphere has had a steady-state thickness, the plate geometry has not changed over time, hotspot fluxes have remained constant through time, and the suboceanic viscosity is uniform throughout the asthenosphere layer. We hope in future work to correct the oversimplification of a steady-state plate geometry to include plate motion evolution, the opening of the continents and the shrinkage of ocean basins, and to explore what impact secular changes in the thickness of the asthenosphere have on predicted flow. With an iterative solution technique, one can also improve the current neglect of the effects that dynamic isostasy will have in shaping the depth of the base of the asthenosphere, as discussed in the previous section. We anticipate that a more realistic parameterization of these boundary conditions will improve the model's accuracy, at a cost of even greater model complexity. A further possibility with this model is to estimate the plume flux distribution using additional constraints from the observed directions of seismic anisotropy. We remain hopeful that the asthenosphere flow modeling technique developed in the present study is a useful tool to further test whether the paradigm of plume hotspots and lateral flow within a plume-fed asthenosphere can be reconciled with and provide a framework for understanding the flow and melting of the upper mantle.

ACKNOWLEDGMENTS

We acknowledge Morgan A.L. Crooks, Jörg Hasenclever, Walter Smith, and Mara Yale for their help during the 12 years of gestation of this work. We also thank Don Anderson, Gill Foulger, Donna Jurdy, and Lars Rüpke for helpful reviews.

REFERENCES CITED

Abouchami, W., Galer, S.J.G., and Hofmann, A.W., 2000, High precision lead isotope systematics of lavas from the Hawaiian Scientific Drilling Project: Chemical Geology, v. 169, p. 187–209, doi: 10.1016/S0009-2541 (00)00328-4.

Abouchami, W., Hofmann, A.W., Galer, S.J.G., Frey, F.A., Eisele, J., and Feigenson, M., 2005, Lead isotopes reveal bilateral asymmetry and vertical continuity in the Hawaiian mantle plume: Nature, v. 434, p. 851–856, doi: 10.1038/nature03402.

Adam, C., and Bonneville, A., 2005, Extent of the South Pacific superswell: Journal of Geophysical Research, v. 110, doi:10.1029/2004JB3465.

Allègre, C.J., and Turcotte, D.L., 1986, Implications of a two-component marble-cake mantle: Nature, v. 323, p. 123–127, doi: 10.1038/323123a0.

Anderson, D.L., 1989, Theory of the Earth: London, Blackwell Scientific, 366 p.

Bercovici, D., and Karato, S., 2003, Whole-mantle convection and the transition-zone water filter: Nature, v. 425, p. 39–44, doi: 10.1038/nature01918.

Boyd, F.R., and McCallister, R.H., 1976, Densities of fertile and sterile garnet peridotites: Geophysical Research Letters, v. 3, p. 509–512.

BVSP (members of the Basaltic Volcanism Study Project), 1981, Basaltic volcanism on the terrestrial planets: New York, Pergamon Press, 1286 p.

Chase, C.G., 1979, Asthenospheric counterflow: A kinematic model: Geophysical Journal of the Royal Astronomical Society, v. 56, p. 1–18.

Davies, G.F., 1988, Ocean bathymetry and mantle convection, 1, large-scale flow and hotspots: Journal of Geophysical Research, v. 93, p. 10,467–10,480.

Davies, G.F., 1994, Thermomechanical erosion of the lithosphere by mantle plumes: Journal of Geophysical Research, v. 99, p. 15,709–15,722, doi: 10.1029/94JB00119.

Davies, G.F., 1999, Dynamic Earth: Cambridge, Cambridge University Press, 458 p.

Deffeyes, K.S., 1972, Plume convection with an upper mantle temperature inversion: Nature, v. 240, p. 539–544, doi: 10.1038/240539a0.

Donnelly, K.E., Goldstein, S.L., Langmuir, C.H., and Spiegelman, M., 2004, Origin of enriched ocean ridge basalts and implications for mantle dynamics: Earth and Planetary Science Letters, v. 226, p. 347–366.

Dupre, B., and Allègre, C.J., 1983, Pb-Sr isotope variation in Indian Ocean basalts and mixing phenomena: Nature, v. 303, p. 142–146, doi: 10.1038/303142a0.

Dziewonski, A.M., and Anderson, D.L., 1981, Preliminary Earth reference model: Physics of the Earth and Planetary Interiors, v. 25, p. 297–356, doi: 10.1016/0031-9201(81)90046-7.

Ebinger, C.J., and Sleep, N.H., 1998, Cenozoic magmatism throughout east Africa resulting from impact of a single plume: Nature, v. 395, p. 788–791, doi: 10.1038/27417.

Ekstrom, G., and Dziewonski, A., 1998, The unique anisotropy of the Pacific upper mantle: Nature, v. 394, p. 168–172, doi: 10.1038/28148.

Elsasser, W.M., 1971, Sea-floor spreading as thermal convection: Journal of Geophysical Research, v. 76, p. 1101–1112.

Evans, R.L., Hirth, G., Baba, K., Forsyth, D.W., Chave, A., and Mackie, R., 2005, Geophysical evidence from the MELT area for compositional controls on oceanic plates: Nature, v. 437, p. 249–252, doi: 10.1038/nature04014.

Fitton, J.G., 2007 (this volume), The OIB paradox, *in* Foulger, G.R., and Jurdy, D.M., eds., Plates, plumes, and planetary processes: Boulder, Colorado, Geological Society of America Special Paper 430, doi: 10.1130/2007.2430(20).

Gaherty, J.B., Kato, M., and Jordan, T.H., 1999, Seismological structure of the upper mantle: A regional comparison of seismic layering: Physics of the Earth and Planetary Interiors, v. 110, p. 21–41, doi: 10.1016/S0031-9201(98)00132-0.

Gung, Y., Panning, M., and Romanowicz, B., 2003, Global anisotropy and the thickness of continents: Nature, v. 422, p. 707–711.

Gutenberg, B., 1959, Physics of the Earth's interior: New York, Academic Press, 240 p.

Harper, J.F., 1978, Asthenosphere flow and plate motions: Geophysical Journal of the Royal Astronomical Society, v. 55, p. 87–110.

Hart, S.R., Hauri, E.H., Oschmann, L.A., and Whitehead, J.A., 1992, Mantle plumes and entrainment: Isotopic evidence: Science, v. 256, p. 517–520, doi: 10.1126/science.256.5056.517.

Hasenclever, J., 2004, Implications of a weak and buoyant asthenosphere for entrainment and mantle flow at a subduction zone [Ph.D. thesis]: Kiel, Germany, Leibnitz Institut für Meerskunde, Kiel University, 93 p.

Helffrich, G.R., and Wood, B.J., 2001, The Earth's mantle: Nature, v. 412, p. 501–507, doi: 10.1038/35087500.

Hess, P.C., 1989, Origins of igneous rocks: Cambridge, Massachusetts, Harvard University Press, 336 p.

Hirth, G., and Kohlstedt, D.L., 1996, Water in the oceanic upper mantle: Implications for rheology, melt extraction, and the evolution of the lithosphere: Earth and Planetary Science Letters, v. 144, p. 93–108, doi: 10.1016/0012-821X(96)00154-9.

Hofmann, A.W., 1997, Mantle geochemistry: The message from oceanic volcanism: Nature, v. 385, p. 219–229.

Hofmann, A.W., 2003, Sampling mantle heterogeneity through oceanic basalts: Isotopes and trace elements, *in* Carlson, R.W., ed., The mantle and core, volume 2: Treatise on geochemistry: Oxford, Elsevier-Pergamon, p. 61–101.

Hyndman, R.D., Currie, C.A., and Mazzotti, S.P., 2005, Subduction zone backarcs, mobile belts, and orogenic heat: GSA Today, v. 15, p. 4–10.

Ito, G., and Mahoney, J.J., 2005a, Flow and melting of a heterogeneous mantle: 2. Implications for a chemically nonlayered mantle: Earth and Planetary Science Letters, v. 230, p. 47–63, doi: 10.1016/j.epsl.2004.10.034.

Ito, G., and Mahoney, J.J., 2005b, Flow and melting of a heterogeneous mantle: 1. Method and importance to the geochemistry of ocean island and mid-ocean ridge basalts: Earth and Planetary Science Letters, v. 230, p. 29–46, doi: 10.1016/j.epsl.2004.10.035.

Jordan, T.H., 1979, Mineralogies, densities, and seismic velocities of garnet lherzolites and their geophysical implications, *in* Boyd, F.R., and Meyer, H.O.A., eds., Proceedings of the 2nd International Kimberlite Conference, vol. 2: Washington, D.C., American Geophysical Union, p. 1–14.

Karato, S., 1986, Does partial melting reduce the creep strength of the upper mantle?: Nature, v. 319, p. 309–310, doi: 10.1038/319309a0.

Kumar, P., Kind, R., Hanka, W., Wylegalla, K., Reigber, C., Yuan, X., Woelnern, I., Schwintzer, P., Fleming, K., Dahl-Jensen, T., Larsen, T.B., Schweitzer, J., Priestley, K., Gudmundsson, O., and Wolf, D., 2005, The lithosphere-asthenosphere boundary in the north-west Atlantic region: Earth and Planetary Science Letters, v. 236, p. 249–257, doi: 10.1016/j.epsl.2005 .05.029.

Larson, R.L., 1991a, Geological consequences of super plumes: Geology, v. 19, p. 963–966, doi: 10.1130/0091-7613(1991)019<0963:GCOS>2.3.CO;2.

Larson, R.L., 1991b, Latest pulse of the Earth: Evidence for a mid-Cretaceous super plume: Geology, v. 19, p. 547–550, doi: 10.1130/0091-7613(1991) 019<0547:LPOEEF>2.3.CO;2.

Larson, R.L., and Olson, P., 1991, Mantle plumes control magnetic reversal frequency: Earth and Planetary Science Letters, v. 107, p. 437–447, doi: 10.1016/0012-821X(91)90091-U.

Laske, G., Phipps Morgan, J., and Orcutt, J.A., 2007 (this volume), The Hawaiian SWELL pilot experiment—Evidence for lithosphere rejuvenation from ocean bottom surface wave data, *in* Foulger, G.R., and Jurdy, D.M., eds., Plates, plumes, and planetary processes: Boulder, Colorado, Geological Society of America Special Paper 430, doi: 10.1130/2007.2430(11).

Lee, C.-T.A., Lenardic, A., Cooper, C.M., Niu, F., and Levander, A., 2005, The role of chemical boundary layers in regulating the thickness of continental and oceanic thermal boundary layers: Earth and Planetary Science Letters, v. 230, p. 379–395, doi: 10.1016/j.epsl.2004.11.019.

Lewis, T.J., Hyndman, R.D., and Flueck, P., 2003, Heat flow, heat generation, and crustal temperatures in the northern Canadian Cordillera: Thermal control of tectonics: Journal of Geophysical Research, v. 108, doi:10.1029/ 2002JB002090.

Li, X., Kind, R., Yuan, X., Woelbern, I., and Hanka, W., 2004, Rejuvenation of the lithosphere by the Hawaiian plume: Nature, v. 427, p. 827–829, doi: 10.1038/nature02349.

McNutt, M.K., and Fischer, K.M., 1987, The South Pacific superswell, *in* Keating, B.H., et al., eds., Seamounts, islands, and atolls: Washington, D.C., American Geophysical Union Geophysical Monograph 43, p. 25–34.

Monnereau, M., Rabinowicz, M., and Arquis, E., 1993, Mechanical erosion and reheating of the lithosphere: A numerical model for hotspot swells: Journal of Geophysical Research, v. 98, p. 809–823.

Moore, W.B., Schubert, G., and Tackley, P., 1999, The role of rheology in lithospheric thinning by mantle plumes: Geophysical Research Letters, v. 26, p. 1073–1076, doi: 10.1029/1999GL900137.

Morgan, W.J., 1971, Convection plumes in the lower mantle: Nature, v. 230, p. 42.

Morgan, W.J., 1972, Plate motions and deep mantle convection, in Shagam, R., et al., eds., Studies in Earth and space sciences (The Harry H. Hess volume): Boulder, Colorado, Geological Society of America Memoir 132, p. 7–22.

Morgan, W.J., 1981, Hotspot tracks and the opening of the Atlantic and Indian Oceans, *in* Emiliani, C., ed., The oceanic lithosphere: New York, Wiley, p. 443–487.

Morgan, W.J., and Phipps Morgan, J., 2007 (this volume), Plate velocities in the hotspot reference frame, *in* Foulger, G.R., and Jurdy, D., eds., Plates, plumes, and planetary processes: Boulder, Colorado, Geological Society of America Special Paper 430, doi: 10.1130/2007.2430(04).

Nettles, M., 2005, Anisotropic velocity structure of the mantle beneath North America [Ph.D. thesis]: Cambridge, Massachusetts, Harvard University, 253 p.

Neves, M.C., Searle, R.C., and Bott, M.H.P., 2003, Easter microplate dynamics: Journal of Geophysical Research, v. 108, doi: 10.1029/2001JB000908.

Olson, P., 1990, Hot spots, swells, and mantle plumes, *in* Ryan, M.P., ed., Magma transport and storage: New York, John Wiley, p. 33–51.

Oxburgh, E.R., and Parmentier, E.M., 1977, Compositional and density stratification in oceanic lithosphere—Causes and consequences: Journal of the Geological Society of London, v. 133, p. 343–355.

Parmentier, E.M., and Oliver, J.E., 1979, A study of shallow mantle flow due to the accretion and subduction of lithospheric plates: Geophysical Journal of the Royal Astronomical Society, v. 57, p. 1–22.

Phipps Morgan, J., 1987, Melt migration beneath mid-ocean spreading centers: Geophysical Research Letters, v. 14, p. 1238–1241.

Phipps Morgan, J., 1994, The effect of mid-ocean ridge melting on subsequent off-axis hotspot upwelling and melting: Eos (Transactions of the American Geophysical Union), Spring meeting supplement, v. 75, p. 336.

Phipps Morgan, J., 1997, The generation of a compositional lithosphere by mid-ocean ridge melting and its effect on subsequent off-axis hotspot upwelling and melting: Earth and Planetary Science Letters, v. 146, p. 213–232, doi: 10.1016/S0012-821X(96)00207-5.

Phipps Morgan, J., 1998, Thermal and rare gas evolution of the mantle: Chemical Geology, v. 145, p. 431–445, doi: 10.1016/S0009-2541(97)00153-8.

Phipps Morgan, J., 1999, The isotope topology of individual hotspot basalt arrays: Mixing curves or melt extraction trajectories?: Geochemistry, Geophysics, Geosystems, v. 1, doi: 1029/1999GC000004.

Phipps Morgan, J., 2001, Thermodynamics of pressure release melting of a veined plum pudding mantle: Geochemistry, Geophysics, Geosystems, v. 2, doi:10.1029/2000GC000049.

Phipps Morgan, J., and Connolly, J., 2004, Are MORB and OIB produced by a hybrid flux-melting process instead of "pure" pressure-release melting?: Eos (Transactions of the American Geophysical Union), Fall meeting supplement, v. 85, abstract V22A-08.

Phipps Morgan, J., and Morgan, W.J., 1999, Two-stage melting and the geochemical evolution of the mantle: A recipe for mantle plum-pudding: Earth and Planetary Science Letters, v. 170, p. 215–239, doi: 10.1016/S0012-821X(99)00114-4.

Phipps Morgan, J., and Morgan, W.J., 2001, The two-stage melting hypothesis for the relationship between the OIB and MORB sources also provides a nice resoluiton to the "He-Paradox" (+Ne +Ar), *in* European Union of Geosciences XI Abstract Programme: Cambridge, Cambridge Publications, p. 450.

Phipps Morgan, J., and Smith, W.H.F., 1992, Flattening of the seafloor depth-age curve as a response to asthenospheric flow: Nature, v. 359, p. 524–527, doi: 10.1038/359524a0.

Phipps Morgan, J., Morgan, W.J., and Price, E., 1995a, Hotspot melting generates both hotspot volcanism and a hotspot swell?: Journal of Geophysical Research, v. 100, p. 8045–8062, doi: 10.1029/94JB02887.

Phipps Morgan, J., Morgan, W.J., Zhang, Y.-S., and Smith, W.H.F., 1995b, Observational hints for a plume-fed sub-oceanic asthenosphere and its role in mantle convection: Journal of Geophysical Research, v. 100, p. 12753–12768, doi: 10.1029/95JB00041.

Phipps Morgan, J., Hasenclever, J., Hort, M., Rüpke, L., and Parmentier, E.M., 2007, On subducting slab entrainment of buoyant asthenosphere: Terra Nova, v. 19, p. 167–173.

Ravine, M.A., 1997, Investigations into aspects of mantle viscosity and dynamics [Ph.D. thesis]: San Diego, University of California at San Diego–SIO, 181 p.

Ravine, M.A., and Phipps Morgan, J., 1996, Inversion for radial mantle viscosity with a layered constraint: A better fit to dynamic topography?: Eos (Transactions, American Geophysical Union), v. 77, p. F721.

Ritsema, J., and Allen, R., 2003, The elusive mantle plume: Earth and Planetary Science Letters, v. 207, p. 1–12, doi: 10.1016/S0012-821X(02)01093-2.

Robinson, J.E., and Eakins, B.W., 2006, Calculated volumes of individual shield volcanoes at the young end of the Hawaiian Ridge: Journal of Volcanology and Geothermal Research, v. 151, p. 309–317.

Schilling, J.-G., 1986, Geochemical and isotopic variation along the Mid-Atlantic Ridge axis from 79°N to 0°N, *in* Vogt, P.R., and Tucholke, B.E., eds., The western Atlantic region: Boulder, Colorado, Geological Society of America Geology of North America M, p. 137–153.

Schouten, H., Klitgord, K.D., and Gallo, D.G., 1993, Edge-driven microplate kinematics: Journal of Geophysical Research, v. 98, p. 6689–6702.

Schubert, G., and Turcotte, D.L., 1972, One-dimensional model of shallow mantle convection: Journal of Geophysical Research, v. 77, p. 945–951.

Schubert, G., Yuen, D., Froidevaux, C., Fleitout, L., and Souriau, M., 1978, Mantle circulation with partial shallow return flow: Effects on stresses in oceanic plates and topography of the sea floor: Journal of Geophysical Research, v. 83, p. 745–758.

Schubert, G., Turcotte, D.L., and Olson, P., 2001, Mantle convection in the Earth and planets: Cambridge, Cambridge University Press, 940 p.

Sleep, N.H., 1984, Tapping of magmas from ubiquitous mantle heterogeneities: An alternative to mantle plumes?: Journal of Geophysical Research, v. 89, p. 10,029–10,042.

Sleep, N.H., 1990, Hotspots and mantle plumes: Some phenomenology: Journal of Geophysical Research, v. 95, p. 6715–6736.

ten Brink, U.S., and Brocher, T.M., 1987, Multichannel seismic evidence for a subcrustal intrusive complex under Oahu and a model for Hawaiian volcanism: Journal of Geophysical Research, v. 92, p. 13,687–13,707.

Turcotte, D.L., and Oxburgh, E.R., 1967, Finite amplitude convection cells and continental drift: Journal of Fluid Mechanics, v. 28, p. 29–42, doi: 10.1017/S0022112067001880.

White, R.S., 1993, Melt production rates in mantle plumes: Philosophical Transactions of the Royal Society of London, Series A: Mathematical and Physical Sciences, v. 342, p. 137–153.

Woodward, R.L., and Masters, G., 1991, Global upper mantle structure from long-period differential travel times: Journal of Geophysical Research, v. 96, p. 6351–6377.

Yale, M.M., and Phipps Morgan, J., 1998, Asthenosphere flow model of hotspot-ridge interactions: A comparison of Iceland and Kerguelen: Earth and Planetary Science Letters, v. 161, p. 45–56, doi: 10.1016/S0012-821X(98)00136-8.

Yamamoto, M., Phipps Morgan, J., and Morgan, W.J., 2007 (this volume, chapter 10), Global plume-fed asthenosphere flow—II: Application to the geochemical segmentation of mid-ocean ridges, *in* Foulger, G.R., and Jurdy, D., eds., Plates, plumes, and planetary processes: Boulder, Colorado, Geological Society of America Special Paper 430, doi: 10.1130/2007.2430(10).

Zindler, A., and Hart, S., 1986, Chemical geodynamics: Annual Review of Earth and Planetary Sciences, v. 14, p. 493–571, doi: 10.1146/annurev.ea.14.050186.002425.

MANUSCRIPT ACCEPTED BY THE SOCIETY 31 JANUARY 2007

The Geological Society of America
Special Paper 430
2007

Global plume-fed asthenosphere flow—II: Application to the geochemical segmentation of mid-ocean ridges

Michiko Yamamoto*
Jason Phipps Morgan*
*Institute for the Study of the Continents, Department of Earth and Atmospheric Sciences,
Cornell University, Snee Hall, Ithaca, New York 14853-1504, USA*
W. Jason Morgan
*Department of Earth and Planetary Sciences, 20 Oxford Street,
Harvard University, Cambridge, Massachusetts 02138, USA*

ABSTRACT

Asthenosphere plume-to-ridge flow has often been proposed to explain both the existence of geochemical anomalies at the mid-ocean ridge segments nearest an off-axis hotspot and the existence of apparent geochemical provinces within the global mid-ocean spreading system. We have constructed a thin-spherical-shell finite element model to explore the possible structure of global asthenosphere flow and to determine whether plume-fed asthenosphere flow is compatible with present-day geochemical and geophysical observations. The assumptions behind the physical flow model are described in the companion paper to this study. Despite its oversimplifications (especially the steady-state assumption), Atlantic, Indian, and Pacific mid-ocean ridge isotope geochemistry can be fit well at medium and long wavelengths by the predicted global asthenosphere flow pattern from distinct plume sources. The model suggests that the rapidly northward-moving southern margin of Australia, not the Australia-Antarctic discordance, is the convergence zone for much plume material in the southern hemisphere. It also suggests a possible link between the strike of asthenosphere flow with respect to a ridge axis and along-axis isotopic peaks.

Keywords: asthenosphere flow, plume-ridge interaction, geochemistry

INTRODUCTION

It is generally believed that mid-ocean ridges passively sample the shallow mantle beneath the ridge axes. This shallow layer, the asthenosphere, is where the geochemical anomalies found in mid-ocean ridge basalts (MORBs) are also assumed to originate. We think the asthenosphere also plays an important role in mantle flow because of its buoyancy and low viscosity. Asthenosphere flow can transport chemical features from a source hotspot to a mid-ocean ridge, thereby linking geophysical flow to geochemical observations. Along-ridge migration of asthenosphere was first suggested by Vogt (1971), Vogt and Johnson (1972, 1975), and Schilling (1973), who discussed asthenosphere flow along the Reykjanes Ridge. Morgan (1978)

*E-mails: Yamamoto, my83@cornell.edu; Phipps Morgan, jp369@cornell.edu.

Yamamoto, M., Phipps Morgan, J., and Morgan, W.J., 2007, Global plume-fed asthenosphere flow—II: Application to the geochemical segmentation of mid-ocean ridges, *in* Foulger, G.R., and Jurdy, D.M., eds., Plates, plumes, and planetary processes: Geological Society of America Special Paper 430, p. 189–208, doi: 10.1130/2007.2430(10). For permission to copy, contact editing@geosociety.org. ©2007 The Geological Society of America. All rights reserved.

proposed the possibility of asthenospheric channels between off-axis hotspots and ridges, and asthenosphere migration in any direction to mid-ocean ridges has also been considered. Now asthenosphere flow is a frequently used mechanism to link observed geophysical and geochemical anomalies at mid-ocean ridges to hotspot-ridge connections.

Many sites for hotspot-ridge asthenospheric connections have been proposed. Studies of the Iceland-Reykjanes Ridge connection initiated the concept of plume-fed asthenosphere flow. Now a key Iceland-related question is how far away from Iceland the plume material is spreading (Poreda et al., 1986; Fitton et al., 1997; Taylor et al., 1997; Peate et al., 2001). Schilling et al. (1985) and Hanan et al. (1986) studied isotopic variation along the southern Mid-Atlantic Ridge and proposed that isotopic peaks at ridge sites nearest to off-ridge hotspots are observational evidence of flow channels linking these deep mantle plumes to the ridge axis. Niu et al. (1999) proposed that lateral asthenosphere flow from Hawaii to the East Pacific Rise (as suggested by Phipps Morgan et al., 1995) left East Pacific Rise basalts with both enriched and depleted lithologies. Using gravity and topographic data, Small (1995) proposed hotspot-ridge asthenospheric connections at Louisville, Discovery, Shona, and Kerguelen. Marks and Stock (1994) suggested the need for along-axis asthenosphere flow from hotspots to produce morphologic variation in the Pacific-Antarctic Ridge. Sempere et al. (1997) and Small et al. (1999) explained variations in geophysical data on the South East Indian Ridge by asthenosphere flow from the Amsterdam–St. Paul hotspot. These previous geophysical and geochemical studies of plume-ridge interaction discussed local hotspot-ridge connections. From these few examples, it is difficult to visualize a global pattern of asthenosphere flow.

This study assumes that plume-fed asthenosphere flow is the potential cause of much of the geochemical segmentation observed along the global mid-ocean ridge system. In the companion paper to this study (Yamamoto et al., this volume, chapter 9), we construct a physical and numerical model for the global flow of buoyant plume-fed material within a thin and laterally heterogeneous asthenosphere layer based on the lubrication theory idealization discussed by Phipps Morgan et al. (1995). In this model, asthenosphere is assumed to be brought up by mantle plumes to replenish asthenosphere consumption (sinks) by plate growth and subduction. Asthenosphere consumption is due to three tectonic activities. The first and largest is plate accretion at mid-ocean ridges, the second is drag-down next to subducting plates at trenches where asthenosphere is entrained by subducting lithosphere, and the third is attachment to the base of the aging and thickening lithosphere. Within this conceptual framework, asthenosphere flow is also shaped by the drag of overriding moving lithosphere; a mechanism explored in earlier models of global asthenosphere trench-to-ridge counterflow that are reviewed in Yamamoto et al. (this volume, chapter 9). Next, we explore the model's predicted global asthenosphere flow pattern to test how consistent it is with the spatial patterns of geochemical variations observed at mid-ocean ridges. We find that this conceptually simple model suggests a possible mechanism for the observed boundaries between geochemical provinces of MORB volcanism.

RESULTS

Figure 1 is a map of the pattern of global asthenosphere flow predicted by this model (see Yamamoto et al., this volume, chapter 9, for a detailed description of the assumptions that underlie this prediction). Strong flow occurs beneath the Pacific basin, because of high plate speeds and the strong ridgeward fluxes from the equatorial Pacific superplume region and Hawaii. In contrast, the Atlantic shows rather weak flows because of its slow spreading rate and the intermediate-to-weak strengths of its plumes. The Indian Ocean basin has strong as-

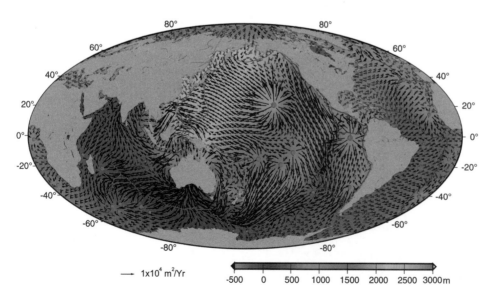

Figure 1. Global asthenosphere flow model. Arrows represent total flux (pressure-induced flow plus shear-induced flow), and color shows the lateral pressure field associated with this flow expressed as dynamic topography. This preferred solution for the predicted large-scale pattern of global asthenosphere flow is based on the absolute plate motions shown in Table 1 and relative plume strengths shown in Table 2 of Yamamoto et al. (this volume, chapter 9), which contains a detailed description of the modeling approach and model assumptions.

thenosphere flow in the center and reduced flow to north and south.

The predicted flow is quite sensitive to the total plume supply within each basin, as this determines whether the basin will push or pull asthenosphere from its neighboring basins. At a smaller scale, the relative plume fluxes between neighboring plumes will affect details of their exact regions of influence along the global spreading system. Yamamoto et al. (this volume, chapter 9) contains some examples illustrating these sensitivities.

How can this model be tested? A promising geochemical test is that regions of asthenosphere fed by each hotspot can be determined, and the boundaries between these regions should be evident at mid-ocean ridges as geochemically distinct provinces of MORBs. This test is the focus of this article (see Figs. 2–6 below). As described in the companion article (Yamamoto et al., this volume, chapter 9), velocities of present-day asthenosphere flow are calculated using present-day plate motions, plate geometries, and hotspot strengths—the steady-state assumption. From these asthenosphere velocities, a flow-line is then tracked between each point and its plume source, and an age of the asthenosphere at this point is also determined. We used this method to trace flow in all of the asthenosphere; on subsequent figures, the resulting plume sources for only those places where asthenosphere is predicted to be supplied by plume upwelling more recently than 100 Ma will be shown as separate hotspot provinces.

Of course a plate evolution model with varying ocean basin shapes and changing plate motions, plate geometries, and plume strengths should be used for flow tracking, but it is too daunting a task for this first cut at global modeling. There has not been a major plate motion reorganization during the most recent ca. 50 Ma (except perhaps in the westernmost Pacific). Thus we assume that present-day asthenosphere flow velocities are crudely representative of asthenosphere velocities during the past ca. 50 Ma to make these maps. Where this assumption is obviously bad (e.g., Gulf of Aden; the oldest asthenosphere regions in the Arctic Ocean, western Atlantic, east of Australia, and south of Indonesia), the predictions will also be bad. One further simplification was used to generate these figures. The plume strengths we found for continental hotspots (e.g., Hoggar, Tibesti) were all very weak; thus they were all set to zero because they contributed little to the asthenosphere flow within the ocean basins. (See Yamamoto et al., this volume, chapter 9, for further discussion of the reasons we currently neglect the effects of subcontinental plume upwelling.) Again, where a continental hotspot could contribute significantly as a source of suboceanic asthenosphere (e.g., Gulf of Aden), inaccuracies will result.

IMPLICATIONS FOR THE STRUCTURE OF GLOBAL ASTHENOSPHERE FLOW

The flow from Iceland covers the northernmost part of the Atlantic (Fig. 2). The flow from Iceland going to the south finally meets flow fed from other plumes at 47°N, where Iceland's flow is deflected to the west, merging with other flows from the

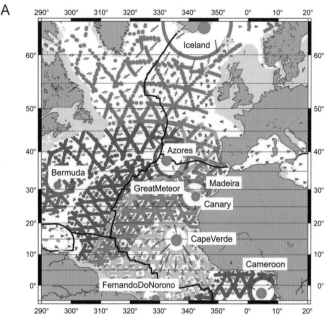

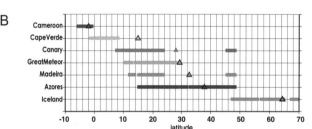

Figure 2. Plume provinces in the North Atlantic and plume influences along the northern Mid-Atlantic Ridge. (A) Plume provinces in the North Atlantic. Different colors show different plume sources calculated by forward-tracking from the source plumes, under the assumption that the calculated present-day flow pattern represents steady-state asthenosphere flow. Only one color may be visible in an area where in fact several plume materials run though, because colors may overprint. (B) Plume influence along the ridge. This plot shows the plume sources for the regions along the ridge. The same color coding is used for plumes as in panel A. Triangles show the source plume's latitude (or longitude, depending on the axis scale in the figure panel).

Azores, Madeira, and Canary hotspots. The Iceland material also reaches eastward of the Mid-Atlantic Ridge to reach England and Scandinavia, but here asthenosphere flow is nearly stagnant. The Iceland material spreads northward to 75°N within 60 m.y. However, the influence would reach farther north if given a longer time in which to spread.

The Azores plume feeds a sectorlike region to its southwest. Along the ridge, it meets Iceland plume material at 47°N. Azores-fed asthenosphere stretches between 15°N and 48°N along the Mid-Atlantic Ridge (Fig. 2B). Eastward of the Azores, the Madeira and Canary plumes block eastward flow from the Azores. Madeira and Canary-fed asthenosphere follows similar paths that avoid the center of the Azores plume's influence and pressure. Their asthenosphere flows both northward to the 47°N region of westward flow and also southwestward to link up with material from the Great Meteor plume. Flow from Canary meets asthenosphere from Cape Verde at ~9°N along the Mid-Atlantic Ridge. Cape Verde–sourced asthenosphere fans out from Cape Verde's latitude of 15°N toward the equator. This flow results in westward-directed flow at the equator, so that north Atlantic asthenosphere does not enter beneath the south Atlantic basin.

Asthenosphere flow beneath the south Atlantic basin is dominated by east-west–directed flow except for small radial patterns near hotspots (Fig. 3). The flow direction is constantly east to west on the southernmost Mid-Atlantic Ridge, because almost all of the south Atlantic plumes are on the eastern side of the ocean basin. Cameroon, St. Helena, and Fernando de Noronha generate the equatorial flow that blocks asthenosphere of north Atlantic origin from entering the south Atlantic.

Figure 3 suggests that a plume province boundary also exists in the south Atlantic near 25°S. Asthenosphere north of the boundary is supplied with material mixed from Cameroon, Fernando, Ascension, St. Helena, and Martin Vaz; regions south of the boundary are supplied by Vema, Tristan da Cunha, Gough,

Discovery, and Bouvet. In the north, St. Helena appears to be the main asthenosphere source for the ridge, whereas Cameroon's flow area tapers off toward the ridge to become too narrow at the ridge to be a major source. Other plumes in the north province are located west of the ridge, and their material flows to the west —thus they do not influence the ridge's geochemistry. South of the 25°S boundary, source hotspots are located to the east of the ridge, and all of them except Vema supply material that flows westward to cross the ridge. Tristan da Cunha and Gough have similar patterns. Their main stream fans westward to supply the ridge between 25° and 50°S. One side branch flows eastward to the western coast of Africa, one to the Antarctic, and a third flows south of Africa to the Indian Ocean. Flow from Discovery has three similar branches. Bouvet-fed asthenosphere mainly flows to the Antarctic, but a smaller branch goes into the Indian Ocean basin. The plume material of Tristan da Cunha, Gough, Discovery, and Bouvet together flows into the Indian Ocean, where it is pulled eastward by the pressure gradient in the southern Indian Ocean through the Rodrigues triple junction and finally converging to the south of Australia.

The flow tracking shows that the Rodrigues triple junction may be a distant floodgate for many hotspots (Fig. 4). Asthenosphere fed from the Tristan-group hotspots (Tristan da Cunha, Gough, Bouvet, and Discovery in the Atlantic) and also Crozet, Marion, Ob-Lena, and Réunion in Indian Ocean all pass beneath the region of this triple junction. The Tristan group's flow track divides Indian Ocean into two provinces. The northern province is mainly filled with material from the Réunion and Comores (and Afar) plumes. This flow tracking suggests that Réunion material alone supplies the Central Indian Ridge, whereas the Comores (and Afar) plumes are the source region for the Carlsberg Ridge and the Gulf of Aden. Flow from Réunion does not overlap with flow from these others.

In the south Indian Ocean province, five hotspots, Ob-Lena, Marion, Crozet, Kerguelen, and Heard, are located close in latitude near 50°S and grouped into two longitudinal regions of 30°E for Ob-Lena, Maria, and Crozet, and 60°E for Kerguelen and Heard. All asthenosphere fed from these hotspots heads eastward (pushed by the excess of South Atlantic plume supply of asthenosphere) to finally converge to the south of Australia at 110°E (Fig. 5A). Kerguelen and Heard, closer to this convergence point than are the other hotspots, feed flow that goes straight through the Southeast Indian Ridge. In contrast, Marion, Crozet, and Ob-Lena feed winding and splitting flow that avoids the region fed by Kerguelen and Heard (Figs. 4 and 5). Their branches have two contact points with the Southeast Indian Ridge, one west of 80°E, the other east of 98°E (Fig. 5B). The direction of asthenosphere flow beneath the Southeast Indian Ridge is consistently perpendicular to the ridge, whereas along-ridge flow dominates beneath the Southwest Indian Ridge.

In the Pacific basin, Hawaiian plume material feeds a large area beneath the northern Pacific, whereas other plumes produce bands of material that fill up the rest of the northern area (Fig. 6A). The hotspots along the western coast of the North Ameri-

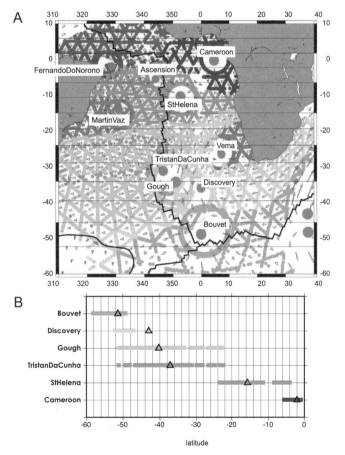

Figure 3. (A) Plume provinces in the south Atlantic and (B) plume influences along the southern Mid-Atlantic Ridge. See caption to Figure 2 for description of plotting conventions.

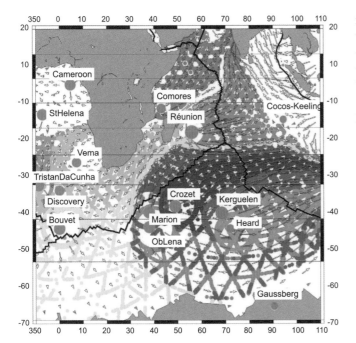

Figure 4. Plume provinces in the southern Indian Ocean and along the Southwest Indian Ridge. See caption to Figure 2A for description of plotting conventions. Note that asthenosphere in this region is fed from many different plume sources, including those in the south Atlantic that cross the spreading center near the Rodrigues triple junction. For plume influences along the Southwest Indian Ridge, see Figures 3B and 5B.

rection of flow beneath the East Pacific Rise and Pacific-Antarctic Ridge changes with latitude; there is ridge-parallel flow north of 15°S, across-ridge flow between 15°S and the southern end of the East Pacific Rise, and ridge-parallel flow beneath most of the Pacific-Antarctic Ridge.

An intriguing respect of this predicted asthenosphere flow pattern is the vortex beneath the East Pacific Rise near 30°S—the fastest-spreading region in the world, where two microplates currently exist. This area influences flow from many plumes beneath the Pacific plate (Fig. 6B), namely, Hawaii, Bowie, Cobb, Marquesas, Pitcairn, Easter, and Foundation. All are located north of or close to this region. The flow vortex is a very persistent feature in our model, occurring even when the relative balance between Pacific plume strengths is changed or when there are other plume sources in the Pacific. We think it is due to the large ridge consumption of asthenosphere in this region and the large pull of material beneath the Indian and Pacific basins to the great sink south of Australia.

DISCUSSION

The calculated asthenosphere flow pattern can be compared to geochemical province boundaries along the mid-ocean ridge

can continent, Bowie, Cobb, and Guadalupe, produce strong and narrow southeastward along-coast flow to the East Pacific Rise. Bowie and Cobb's flow tracks overlap, and both cross the East Pacific Rise at 15°N. Flow from Guadalupe goes along the ridge with material from the Galapagos plume until it meets Bowie and Cobb's across-ridge flow. The plumes beneath the superswell region (McNutt and Fischer, 1987), Marquesas, Society, Pitcairn, Line, and Macdonald, recharge most of the southern part of the Pacific basin. Foundation and Louisville plume material flows southwestward along the edge of flow from the superswell.

Plume material originating beneath the Pacific plate appears to have two main destinations. One is toward the western Pacific trenches, which is the main destination for material from the Hawaiian, Caroline, Samoa, Marquesas, and Society hotspots. The other destination is the southern continental margin of Australia, which is also the destination for asthenosphere from plumes in the south Indian basin. This rapidly moving continental margin is an asthenospheric sink that pulls asthenosphere fed from plumes beneath the Pacific plate and from the Scott and Balleny plumes beneath the Antarctic plate, in addition to several Indian plumes. Other Pacific-basin hotspots beneath the Nazca plate, namely Galapagos, San Felix, and Juan Fernandez, feed flow that does not terminate south of Australia. Instead, they flow to the south, then turn westward after passing beneath the Chile Rise, just reach the southern part of East Pacific Rise, then turn southward once more (Fig. 6A). The di-

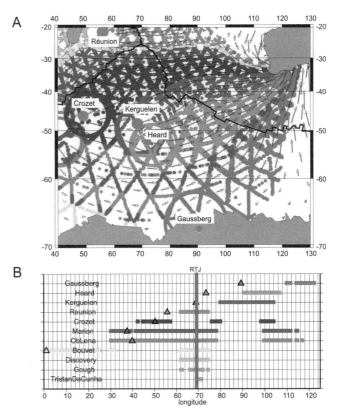

Figure 5. (A) Plume provinces in the southern Indian Ocean and (B) plume influences along the Southeast Indian Ridge. See caption to Figure 2 for description of plotting conventions. RTJ—Rodrigues triple junction.

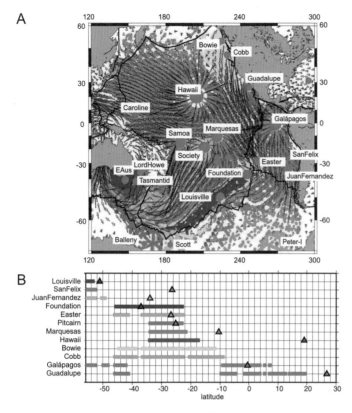

Figure 6. (A) Plume provinces in the Pacific Ocean and (B) plume influences along the East Pacific Rise. See caption to Figure 2 for description of plotting conventions. Note that the fastest spreading ridge on East Pacific Rise (near 30°S) pulls a large amount of plume material from beneath the Pacific plate. Note, too, the vortex pattern of asthenosphere flow beneath this region.

mid-ocean ridge system (typically at least one sample per 100 km of along-ridge length) do exhibit smaller-scale variability that is interesting to compare to the province structure predicted by the flow model.

Next we discuss specific regional aspects of the flow pattern, comparing isotopic and/or chemical patterns along the mid-ocean ridge to model inferences. We assume that the connections between hotspots and mid-ocean ridges (see Figs. 8–11 below) have existed long enough so that the ridge material has had time to flow from its source plume. In general this assumption seems plausible, because most plume sources are calculated to reach the ridge they cross within 30 m.y., and during this time, plate motions have remained relatively stable.

Atlantic Ocean Basin

Our results imply that Atlantic athenosphere has four distinct domains separated by three internal boundaries. All our predicted asthenosphere province boundaries are east-west oriented. They cross the Mid-Atlantic Ridge at 48°N, 4°S, and 25°S respectively. Dosso et al. (1993) compiled $^{87}Sr/^{86}Sr$, La/Sm, and Nb/Zr data along the northern Mid-Atlantic Ridge between 10° and 74°N (Fig. 7). Along-ridge variations in these trace-element ratios show four peaks at 64°N, 45°N, 39°N, and 14°N. Because these four positive peaks appear at the latitude of Iceland, Azores, and Cape Verde, their hotspot-ridge connections were suggested by Phipps Morgan et al. (1995).

Our model supports the idea that the peak at 64°N results from the injection of the Iceland material, and the peaks at 45°N and 39°N are from the Azores. However, the model does not support the proposed relationship between the peak at 14°N and the Cape Verde hotspot. Instead it suggests that the changes at 14°N may be related to the injection of Canary, Madeira, or Great Meteor plume material within a narrow region at this latitude. Several geochemical observations do not favor a possible link between the Cape Verde plume and the 14°N section of the Mid-Atlantic Ridge. For example, the material at 14°N differs from the Cape Verde hotspot basalts in its $^4He/^3He$ signature. The $^4He/^3He$ ratio suddenly rises at 14°N (Staudacher et al., 1989; Bonatti et al., 1992), whereas low $^4He/^3He$ values in Cape Verde have been reported (Christensen et al., 2001; Doucelance et al., 2003; however, this evidence is not conclusive, see later discussion regarding St. Paul and Kerguelen). Furthermore, Schilling et al. (1994) studied Pb-Nd-Sr systematics along the ridge and found that the 14°N anomaly is more similar to the 1.7°N anomaly and is distinct from the signature of hotspots in the North Atlantic. They attributed the 1.7°N anomaly to the Sierra Leone hotspot at 5°N, 10° south of Cape Verde. However, if the 14°N anomaly is also due to flow of material from the Sierra Leone plume, it would need the reverse sense of flow than that predicted from our model. The plume source list for our flow calculation does not include a possible Sierra Leone plume, but the southward flux north of the equator is quite strong, so that adding a Sierra Leone plume is not likely to change the

system and to maps of seismic wavespeeds and anisotropy in the shallow upper mantle. These comparisons show promising agreement between the calculated flow pattern and these geochemical and seismic observations. For example, maps of seismic azimuthal anisotropy suggest a south-north flow direction in the area between Australia and Antarctica, and potential west-east flow between Hawaii and the East Pacific Rise (Montagner and Tanimoto, 1990; Leveque et al., 1998; Montagner and Guillot, 2002; Gaboret et al., 2003), both consistent with our results. In the Atlantic, Ritsema and Allen (2003) suggest that the asthenosphere beneath the Mid-Atlantic Ridge at 14°N lies in the fastest velocity zone along this ridge, while our flow model suggests that the 14°N is the farthest point from the source hotspot. In other words, both are consistent with the possibility that 14°N is the coldest region along the Mid-Atlantic Ridge.

Although comparisons between seismic wavespeed maps and flow may indicate the most extreme asthenosphere province boundaries (e.g., the Australian-Antarctic discordance), in general, seismic velocities are slow beneath the spreading center system and thus cannot be easily used to locate more subtle asthenosphere province boundaries. However, geochemical isotopic province boundaries along the reasonably well-sampled

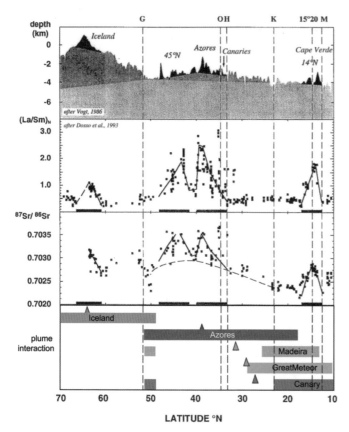

Figure 7. Along-axis bathymetric (Vogt, 1986) and geochemical (Dosso et al., 1993) variations along the northern Mid-Atlantic Ridge. All geochemical "spikes" along the ridge seem to be associated with ridge segments that are closest to a neighboring hotspot. Phipps Morgan et al. (1995) suggested the interaction of each spike and the hotspot at the same latitude. This study supports their interpretation for regions surrounding Iceland and the Azores; however, it does not support their suggestion for a flow-connection between the Cape Verde plume and the 14°N spike. G—Gibbs fracture zone; H—Hayes fracture zone; K—Kane fracture zone; M—Marathon fracture zone; O—Oceanographer fracture zone; 15°20′–15°20′ fracture zone; from Phipps Morgan et al. (1995).

pattern. We think it more likely that either Canary or Madeira plume material may be intruding into the Mid-Atlantic Ridge here, or that this anomaly represents the northern limit of the Cape Verde plume's influence on the ridge axis—but there remains the question of why its influence should be more pronounced here than to the south.

Westward asthenosphere flow dominates the south Atlantic—the effects of hotspots east of the ridge are predicted to be seen at roughly the same ridge latitude as that of the source hotspots. Observed anomalies in $^4He/^3He$, La/Sm, and Nb/Zr are evident at the latitudes corresponding to Ascension, St. Helena, Tristan da Cunha, and Gough; however, the $^{206}Pb/^{204}Pb$ peak is not seen in the Tristan da Cunha and Gough area (Fig. 8), possibly because they have an originally low ratio (Dickin, 1995). All data, except He, show no signal between 20°S and 30°S, a region with no nearby hotspots (Schilling et al., 1985; Hanan et al., 1986; Graham et al., 1992). Discovery plume material injected at 43°S is predicted to flow southward to reach the ridge south of 47°S, in agreement with the observations of Douglass et al. (1999). A flow boundary in the south Atlantic runs perpendicular to the ridge at 25°S, where the St. Helena flux meets the Tristan da Cunha and Gough fluxes. In this region, the asthenosphere flux becomes more stagnant. This area was studied by Graham et al. (1996), who focused on He anomalies. They discovered an He signal at 26°S and speculated that more than two mantle components may mix or meet in that region.

Indian Ocean Basin

Our model suggests two asthenosphere domains in the Indian Ocean. The boundary separating the Indian Ocean into northern and southern parts goes though the Rodriguez triple junction. The Rodriguez triple junction region receives flow from the Discovery, Tristan da Cunha, and Gough (Tristan group) plumes in the south Atlantic, although the long distance and predicted long asthenospheric travel time between these plumes and this section of the ridge make this inference more speculative than the other inferences in this article. However, we cannot dismiss this possibility, because the triple junction has existed since the upper Cretaceous, which is significantly longer than the time for material to travel from the Tristan group and the triple junction, and because this inference is consistent with geochemical evidence. The triple junction is reported to be a peculiar region in isotopic composition; basalts from here have relatively high $^{87}Sr/^{86}Sr$ and low $^{206}Pb/^{204}Pb$ ratios, and these ratios differ strongly from the isotopic chemistry of adjacent ridges (Price et al., 1986; Mahoney et al., 1989, 1992). Michard et al. (1986) suggested that the Rodriguez triple junction basalt source was produced by mixing with a mantle of extremely low $^{206}Pb/^{204}Pb$ and low $^{143}Nd/^{144}Nd$ ratios but intermediate-to-high $^{207}Pb/^{206}Pb$ and $^{87}Sr/^{86}Sr$ ratios. This characterization is like that of Tristan-group basalts (Graham et al., 1992; Dickin, 1995). Even if doubts remain about the connection between the south Atlantic and Rodriguez triple junction, our results and isotopic studies agree that the Rodriguez triple junction lies at the geophysical and geochemical boundary between distinct Indian Ocean asthenosphere provinces.

The connection between the south Atlantic and Indian oceans seems to be associated with the Sr-Pb Dupal anomaly (Dupre and Allègre, 1983; Hart, 1984). The Dupal anomaly shows an isotopic similarity between the Tristan group, Kerguelen, and Ninetyeast Ridge. Grouping Ninetyeast with Kerguelen-plume volcanism is acceptable, because the old and dead Ninetyeast Ridge was formed prior to ca. 46 Ma, when the Ninetyeast Ridge was near the plume and mid-ocean ridge (Royer and Sandwell, 1989). Therefore it is natural that Ninetyeast Ridge chemistry shows Kerguelen's signature, but the signature is from old rather than recent ridgeward flow.

The three mid-ocean ridges within the Indian Ocean basin,

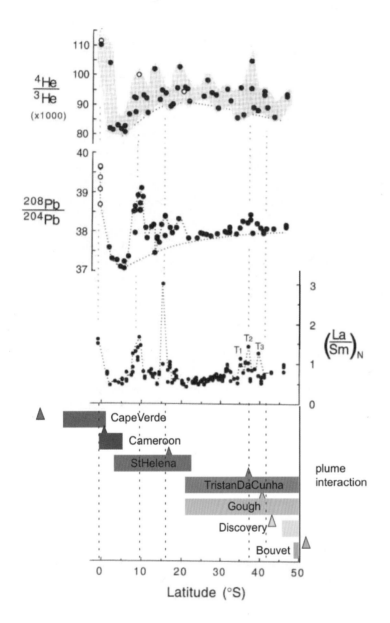

Figure 8. Axial variation of ^{4}He/^{3}He, ^{208}Pb/^{204}Pb, and (La/Sm)$_{N}$ along the southern Mid-Atlantic Ridge (Graham et al., 1992). Geochemical spikes occur at the same latitude along the ridge as that of nearby hotspots and show no variation—except in ^{4}He/^{3}He—along the ridge in the region between 20°S and 30°S, where there are no off-axis hotspots.

the Central Indian Ridge, Southwest Indian Ridge, and Southeast Indian Ridge, each have their own characteristic geometric and geochemical relationship with their source hotspots. The Central Indian Ridge shows a simple signature, reflecting a single Réunion plume source. It has $^{87}Sr/^{86}Sr$, $^{206}Pb/^{204}Pb$, and $^{143}Nd/^{144}Nd$ signals similar to those of Réunion basalts that appear on the ridge at the same latitude as the hotspot (Mahoney et al., 1989). This phenomenon requires eastward flow from the Réunion, which is a feature of our model.

The northern limit of the Réunion flow on the ridge is located at ~8°S, which corresponds to the axial fissuring area that is the morphological transition zone between the Central Indian Ridge and the Carlsberg Ridge, although Mahoney et al. (1989) positioned the limit on the Marie Celeste fracture zone at 16°S, and assigned the 70 km south of the Marie Celeste fracture zone to Réunion's influence, based on the detectable Réunion-like isotope signal. This 70-km-long ridge section is the closest to the hotspot in the sampling points along the ridge; therefore, it is reasonable that the area shows such a strong peak, as the area could well have a strong asthenosphere flow channel connecting it to the Réunion hotspot (Morgan, 1978).

The Southwest Indian Ridge has three nearby hotspots: one is the Marion hotspot, only 180 km away from the ridge axis; one is the Ob-Lena, located very close to the Marion; and the third is the Crozet, ~500 km away from the ridge. Southwest Indian Ridge isotope data do not provide direct evidence of the connection between the near-ridge hotspots and the isotopic signals, in spite of their proximity. There are big spikes in Sr, Pb, and Nd near 40°E (Mahoney et al., 1992), and these isotopic compositions are not like those of either Marion or Crozet. (Mahoney et al., 1992, considered these signals to arise from old Madagascaran lithosphere stranded in that area.) Actually, small peaks in Sr and Nd resembling Marion's composition do exist, but not at the closest point on the ridge to the Marion; instead they are ~550 km east of it. Other places along the ridge show no chemical traces similar to nearby hotspots.

The asthenosphere flow model suggests a possible correlation between local asthenosphere flow directions and the existence (or absence) of hotspot signal in this Southwest Indian Ridge area. The region having Marion's signal is the only place with across-ridge flow from the Marion source hotspot. All other sites along the Southwest Indian Ridge are regions with a strong along-ridge flow, where plume material from Marion, Ob-Lena, and Crozet flows together along-axis. This flow perhaps mixes the material and reduces the geochemical variation along the ridge.

Interaction between Kerguelen-Heard hotspots and Southeast Indian Ridge has been proposed by many earlier studies (Dosso et al., 1988; Small, 1995; Johnson et al., 2000; Mahoney et al., 2002; Weis and Frey, 2002), and our model also indicates that Kerguelen-Heard material flows to the Southeast Indian Ridge and extends over a rather large area. Asthenosphere flow at the Southeast Indian Ridge is consistently ridge-perpendicular in our model (Fig. 5). The isotope variability here (Fig. 9) is bigger than along the Southwest Indian Ridge but less than seen

at the southern Mid-Atlantic Ridge near St. Helena (Fig. 8), another region of persistent across-axis asthenosphere flow. We think its smaller amplitude in comparison to that on the Mid-Atlantic Ridge may reflect the greater distances between the ridge and its source hotspots. Even the closest, Kerguelen, is 1400 km away from the ridge, and Marion, the most remote source, is 6600 km away.

Although across-ridge flow is predicted by our model, many previous studies suggested the existence of eastward along-ridge flow, inferred from the variations in ridge depth, morphology, and segment geometry east of Amsterdam–St. Paul (Sempere et al., 1997; Graham et al., 1999; Small et al., 1999). Specifically, these studies suggested that Amsterdam–St. Paul is the source plume for along-ridge flow, resulting in a mantle temperature gradient from "hot" west of Amsterdam–St. Paul to "cold" at the Australian-Antarctic discordance. (In contrast to these authors, we think Amsterdam–St. Paul is not a separate plume but a manifestation of plume-to-ridge flow from the Kerguelen plume.) This change in sub-ridge temperature structure was considered to be the cause of the transformation from an axial high with shallow depth and longer segmentation along the western Southeast Indian Ridge to a median valley with short segmentation at the Australian-Antarctic discordance (Ying and Cochran, 1996; Sempere et al., 1997). Spreading-rate variations cannot be the cause, because the spreading rate remains constant along this ridge. However, along-ridge flow is not a unique way to explain this variation in thermal structure. With the same logic (that the distance from a plume source to the ridge will shape the temperature of the asthenosphere feeding a given section of the ridge), our model also suggests a similar gradient from hotter west to colder east. Furthermore, the dynamic topography calculated from pressure distribution (Fig. 1) deepens to the east along the ridge, in agreement with observed axial depth variations.

Some previous studies appear to be in conflict with the along-ridge flow hypothesis. Mahoney et al. (2002) studied isotopic variations along the Southeast Indian Ridge and found that the regional isotope pattern cannot be explained by the eastward flow. If eastward flow exists, then variations in isotopic ratios should show a monotonic change to the east, but in fact the variations show several peaks along the ridge (Fig. 9). Ying and Cochran (1996) studied the detailed morphology of the Southeast Indian Ridge and revealed that the ridge continues to become shallower away from Amsterdam–St. Paul to the east, and that an axial valley is present just east of Amsterdam–St. Paul. Therefore, they concluded that the Amsterdam–St. Paul plume was not the source of along-ridge flow to east. They also located the axial high between 82°E and 104°E. This region is exactly where Kerguelen influences the ridge in our model. Kerguelen is the strongest and the nearest hotspot to the Southeast Indian Ridge; therefore, the thermal effects causing the axial-high morphology are most plausibly related to Kerguelen.

Our across-ridge flow hypothesis has at least one weakness. Across-ridge flow does not intuitively explain the "V" shapes on

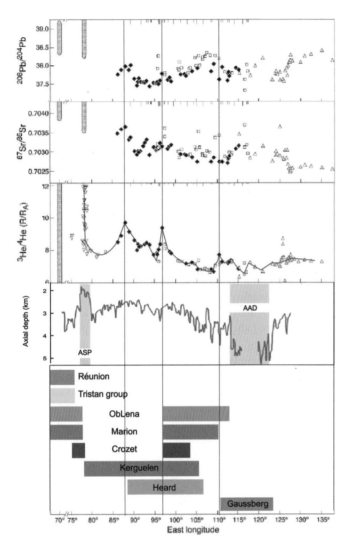

Figure 9. Longitudinal patterns of ^{206}Pb/^{204}Pb, ^{87}Sr/^{86}Sr, (^{3}He/^{4}He)/ (^{3}He/^{4}He)$_{atm}$, and axial depth (Mahoney et al., 2002) along the Southeast Indian Ridge. The along-ridge isotopic peak appears to the east of St. Paul hotspot, suggesting that St. Paul is not the source for eastward-flowing asthenosphere flow along the Southeast Indian Ridge. Note that the western edge of a plume's influence lies at each ^{3}He/^{4}He peak. The large influence area of the Kerguelen plume is consistent with the idea that this is the strongest plume in this region and agrees with the ridge's shallow topography and axial high morphology (Ying and Cochran, 1996) in the region of Kerguelen plume influence. AAD—Australia-Antarctic discordance; ASP—Amsterdam–St. Paul plateau.

the seafloor along the ridge, which have been proposed to be traces of eastward mantle propagation (Sempere et al., 1997). However these shapes have migrated for small distances along-axis (~150 km) compared with "V"-shaped features on Pacific-Antarctic Ridge or northern Mid-Atlantic Ridge (>1000 km). Perhaps these small shapes are not the surface expression of asthenosphere flow but instead a more superficial phenomenon caused by to persistent eastward ridge propagation away from the central shallow region.

Pacific Ocean Basin

The East Pacific Rise geochemistry is characterized by much less variability compared to other ridges. Only rather broad and low peaks in ^{206}Pb/^{204}Pb and ^{87}Sr/^{86}Sr between 15°S and 21°S are resolvable (Macdougall and Lugmair, 1986; Bach et al., 1994; Niu et al., 1999; Lehnert et al., 2000). This isotopic variation has not been explained as being due to plume-ridge connections because of the absence of the nearby hotspots, and it cannot be explained by relationships with second-order segmentation, because there are no such systematic relationships. Bach et al. (1994) suggested these Pb and Sr peaks relate to either an incipient mantle plume near 17°S or large-scale mantle heterogeneity. Mahoney et al. (1994) interpreted these peaks together with a narrow He peak at 17°S as an expression of a mantle flow with a discrete heterogeneity that enters the ridge system at 15.8°S and then migrates to 20.7°S. However, our model suggests the 15–20°S peak is associated with a region of across-ridge flow. Model flow directions change along the ridge, with along-ridge flow north of 15°S and across-ridge flow between 15°S and 25°S. This configuration is consistent with the relationship between flow direction and isotopic peaks seen in the Indian Ocean basin, although opposite to the inference of Mahoney et al. (1994) of along-ridge flow south of 15°S.

The results of Vlastelic et al. (1999) match our predicted flow models well. They compiled Nd, Sr, and Pb data along the East Pacific Rise and Pacific-Antarctic Ridge, and created a δNd-Sr and δSr-Pb diagnostic to discriminate between ridge source provinces (Fig. 10). According to this discriminant, the ridge can be divided into four distinct geochemical regions with a data gap between the Juan Fernandez microplate and the Vacquier fracture zone. The four divisions are, starting from the north: (1) the area from 15°N to north of Easter(δ(Nd-Sr) > 0, δ(Sr-Pb) > 0); (2) the area from south of Easter to the southern boundary of the Juan Fernandez microplate (δ(Nd-Sr) < 0, δ(Sr-Pb) < 0); (3) the area from south of Eltanin fracture zone to Udintsev fracture zone (δ(Nd-Sr) > 0, δ(Sr-Pb) > 0); and (4) the ridge south of Udintsev fracture zone (δ(Nd-Sr) < 0, δ(Sr-Pb) < 0). In other words, the ridge south of Easter generally has the isotopic signature of (δ(Nd-Sr) < 0, δ(Sr-Pb) < 0), but only around the Eltanin fracture zone do ridge basalts show δ(Nd-Sr) > 0 and δ(Sr-Pb) > 0. Castillo et al. (1998) also reported a possible change in the ridge's mantle source at the Heezen transform of the Eltanin fracture zone system.

The other two ridges connecting to the East Pacific Rise—the Galapagos Rise and the Chile Rise—were sorted by Vlastelic into the group with δ(Nd-Sr) > 0 and δ(Sr-Pb) > 0. These isotopic characteristics are also explicable in terms of our flow model if asthenosphere from South Pacific plume sources has δ(Nd-Sr) > 0 and δ(Sr-Pb) > 0 while asthenosphere from North Pacific or Nazca plumes has δ(Nd-Sr) < 0 and δ(Sr-Pb) < 0. This is a possible explanation for why the regions around the Eltanin fracture zone and the southern Chile Rise have the same geo-

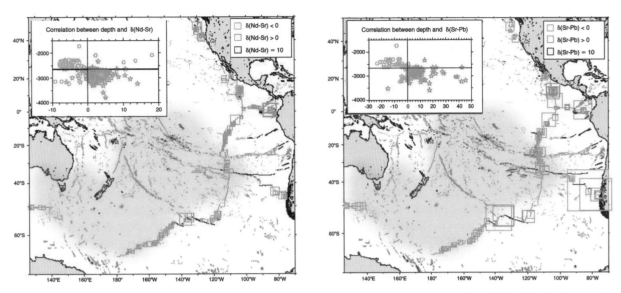

Figure 10. δ(Sr-Pb) along the Pacific spreading system. The color of the square is related to the sign of the deviation from a reference line: green indicates a positive deviation and red shows a negative deviation. The size of each square is proportional to the absolute deviation value. The inset at top left shows δ(Sr-Pb) versus ridge depth. From Vlastelic et al. (1999).

chemical signature—because both regions are predicted to have the same Juan Fernandez plume source material.

Geophysical implications of an asthenosphere flow vortex underlying the East Pacific Rise between 25°S and 35°S (Fig. 6A) are discussed in the companion study in this volume (Yamamoto et al., this volume, chapter 9). The region above this vortex is characterized by the fastest present-day spreading rates (160 mm/year). A large flux of asthenosphere is being pulled into this area to replenish the asthenosphere-consuming creation of compositional lithosphere at the world's fastest-spreading mid-ocean ridge. The Easter hotspot is the nearest of the involved hotspots. However Easter plume signals are not observed; instead, high $^{87}Sr/^{86}Sr$ ratios were reported on the ridge at about the latitude of the Easter hotspot, although Easter itself has $^{87}Sr/^{86}Sr$ values presumed to be too low to be the source of that anomaly (Macdougall and Lugmair, 1986). This isotopic signal could result from contamination with asthenosphere from the south Pacific isotopic and thermal anomaly (Staudigel et al., 1991) region to the west that is recognized as one of the highest-value $^{87}Sr/^{86}Sr$ plume sources in the world. This suggestion agrees with the predicted asthenosphere flowpaths, which show both Marquesas and Pitcairn plume materials being mixed into this area.

Australian-Antarctic Discordance

Many previous studies have suggested along-ridge asthenosphere flow toward the Australian-Antarctic discordance and downwelling beneath this discordance in a "cold spot." The Australian-Antarctic discordance's rough topography, closely spaced transform faults, regional gravity low, and low magnetic amplitudes are considered to reflect regionally cooler mantle temperature and lower than normal melt production beneath it, which led many scientists to the concept of mantle downwelling beneath this region. The west-pointing and east-pointing "V" shapes on both sides of the discordance have also been cited as evidence of flow converging toward the Australian-Antarctic discordance. Geochemically, a sharp compositional boundary between Pacific and Indian MORBs is also found beneath the eastern Australian-Antarctic discordance.

Although the many observations seem to be consistent with asthenosphere flow into the Australian-Antarctic discordance, our model suggests a slight variation of this scenario: perhaps asthenosphere flow is converging to the south of the rapidly moving Australian continental lithosphere, not along the ridge to the Australian-Antarctic discordance ridge segments themselves (Figs. 5A and 6A). In our model, the "suction" forcing this asthenosphere flow is due to the very fast northward motion of Australia—the large change in lithosphere thickness between the craton and oceanic lithosphere at Australia's southern continental margin creates a northward moving void that must be replenished by northward flow of asthenosphere. Currently this region is the location where thick continental cratonic lithosphere is moving most rapidly away from an ocean basin. The geophysical argument for along-ridge flow convergence beneath the Australian-Antarctic discordance is that the discordance contains the maximum or minimum value in topography, gravity, and seismic anomalies along the Pacific-Indian Ridge system. However, if the entire basin between Australia and Antarctica is considered as a whole, then the maximum or minimum geophysical anomalies lie south of Australia (i.e., north of the mid-ocean ridge) rather than at the ridge axis segments of the Australian-Antarctic discordance. For example, residual topography and gravity anomalies south of Australia are lower than these anomalies within the Australian-Antarctic discordance (Veevers, 1982; Kido and Seno, 1994; Gurnis et al., 2000). Furthermore, the results of Christie et al. (1998) indicate that lateral mantle migration to the Australian-Antarctic discordance is constrained to a narrow, relatively shallow region, directly beneath the easternmost segment of the spreading axis. Deeper south-to-north asthenosphere flow beneath the region does not conflict with these observations, and yet still predicts the existence of a strong geochemical boundary along the ridge segments within the Australian-Antarctic discordance.

Flow Direction and He Isotopic Anomalies

Comparing the predicted flow directions and MORB isotopic anomalies along the ridge axis shows the following potential empirical relationships between them:

1. Across-ridge flow projects the asthenosphere source plume compositions onto the ridge axis;
2. Across-ridge flow generates He isotope peaks at the leading edge of each plume province, with the amplitude of the peak apparently unrelated to the distance between the source plume and ridge axis;
3. Along-ridge asthenosphere flow results in little to no variability in MORB isotopic signals; and
4. The above three processes do not occur when the hotspot is located just beneath or close to the ridge: in this case, there is strong evidence for along-ridge flow with the isotopic peaks reflecting the plume's mantle source composition.

Along-ridge asthenosphere flow is prominent at the Southwest Indian Ridge, Pacific-Antarctic Ridge, and the northern parts of the East Pacific Rise. All of these regions show very flat patterns in their Pb-Sr-Nd isotope ratios (Mahoney et al., 1992; Lehnert et al., 2000; Vlastelic et al., 2000). This correlation occurs even when the hotspots are located rather near the ridge (e.g., Southwest Indian Ridge, northern East Pacific Rise), whereas the across-ridge asthenosphere flow produces isotopic peaks on the ridge even after there has been long-distance travel from the source hotspots to the ridge (e.g., Central Indian Ridge, southern Mid-Atlantic Ridge, Southeast Indian Ridge). Vlastelic et al. (2000) show a good example. They noted that the Udintsev fracture zone is a boundary between a pattern of isotopic variability north of the fracture zone and one of little isotopic variability south of the fracture zone. Our model predicts that the Udintsev fracture zone is a boundary between across-ridge asthenosphere flow direction to its north and along-ridge flow to its south. Perhaps the asthenosphere flow direction at a ridge (across-axis versus along-axis) may be a key factor underlying these observed MORB isotope systematics. Along-ridge

flow may better mix material from multiple plume sources, thus subduing the distinct influence of individual plumes.

It is well known that He isotopes exhibit peculiar behavior that is decoupled from those of other isotopes (e.g., Rubin and Mahoney, 1993). The reasons for this distinction are not yet explained in terms of any global empirical rules. We suggest that the He signal may be strongly related to across-ridge asthenosphere flow. For example, along the southern Mid-Atlantic Ridge, He peaks appear not only at hotspot-related areas but also at areas without hotspots, although Pb, Nb/Zr, and La/Sm peaks appear at only the latitude of the hotpots and have no peaks elsewhere (Graham et al., 1992; Fig. 8). The East Pacific Rise (Fig. 11) has a narrow He peak at 17°S (Mahoney et al., 1994), which is the only place having across-ridge flow along the East Pacific Rise. The Southeast Indian Ridge shows periodic peaks in He variation along the ridge (Fig. 9 *in* Graham et al., 2001). Figures 9 and 11 indicate that the locations with these He peaks seem to be well correlated with province boundaries between different plume sources. However, when the plume lies directly beneath the spreading center (e.g., Iceland), then the He-peak is much broader and slowly decays away with increasing distance from the plume.

The good agreement between the locations of He peaks along the Southeast Indian Ridge and the predicted boundaries between regions of plume influence on Southeast Indian Ridge (Fig. 9) suggests that there is a common mechanism causing these He peaks. From the east, one He peak would be linked to flow from Gaussberg, one to the Crozet group, one to Heard, and the central one to Kerguelen instead of the proposed Amsterdam–St. Paul hotspot located near the western edge of the Kerguelen plume influence. (In the simplest interpretation consistent with these results, Amsterdam–St. Paul hotspot volcanism would not be fed from a deep underlying plume of mantle upwelling, but instead would lie at the distal end of a region of lateral plume-to-ridge asthenosphere flow between Kergulen and the Southeast Indian Ridge; see Yale and Phipps Morgan, 1998. This scenario has the conceptual strength that it makes the influence of Kerguelen on the ridge like that of its neighboring plumes, just with a larger influence along the ridge because of Kerguelen's larger plume flux.) Therefore, as the flow model suggests their physical connection, this high ^{3}He content would originate in Kergulen plume, not in a separate Amsterdam–St. Paul plume.

Kerguelen's He has been observed to have a lower ^{3}He/^{4}He ratio than does MORB. Some studies have suggested that Kerguelen's basaltic volcanism is contaminated by old materials in the thick lithosphere just before eruption; thus they estimate the He ratio in the Kerguelen source composition to be higher than the observed ratio (Barling et al., 1994; Hilton et al., 1995; Nicolaysen et al., 2001; Coffin et al., 2002; Frey et al., 2002; Mahoney et al., 2002). However, no study has mentioned the effect of the thick lithosphere overlying the Kerguelen hotspot on the He content in lateral plume-to-ridge flow from the Kerguelen plume. We suggest that the limited melting at Kerguelen

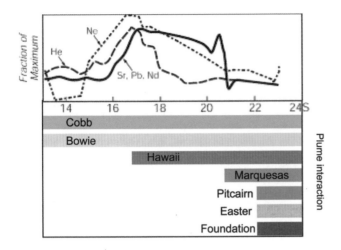

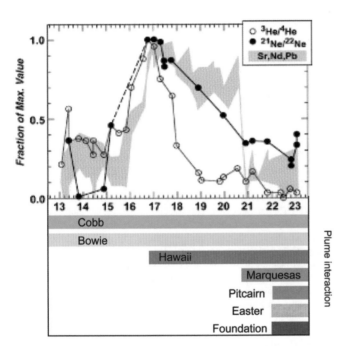

Figure 11. ^{3}He/^{4}He, ^{21}Ne/^{22}Ne, and Sr, Nd, and Pb isotopes in mid-ocean ridge basalts as a function of their latitude along the East Pacific Rise (Mahoney et al., 1994). Isotope values are displayed as a fraction of their individual minimum-to-maximum isotopic values along this section of the global ridge system. Color bars show plume influences along the ridge.

keeps much of the He from the Kerguelen plume source in the (plume-fed) asthenosphere, to be more completely extracted when it ascends again to melt beneath the Southeast Indian Ridge.

Recent seismic tomography and geochemical studies have discussed whether ^{3}He/^{4}He truly is a fingerprint of major (or deep-rooted) plumes (Foulger and Pearson, 2001; Meibom et al., 2003). We are highly influenced by the fact that the Iceland plume influences Sr, Pr, and He isotopes along the Reykjanes Ridge (e.g., Breddam et al., 2000; Hilton et al., 2000), and that

the Iceland plume source can be richer in He when it first upwells and melts than other asthenosphere that melts beneath a spreading center. We favor the interpretation that high ^{3}He/^{4}He ratios, when seen, are diagnostic of the dominance of an He contribution from melt-extraction from a component of more primitive mantle that makes up a fraction of the upwelling mantle plum-pudding, but that plumes also contain low ^{3}He/^{4}He components (e.g., enriched radiogenic plum components), so that the absence of high ^{3}He/^{4}He hotspot basalts does not necessarily imply the absence of a deep plume. This interpretation is discussed in more detail by Phipps Morgan and Morgan (1999).

Transport Distance and the Survival of a Plume's Isotope Signal

The Crozet group has two output sites along the Southeast Indian Ridge. One is north of Amsterdam–St. Paul, the other far to the east, near 116°E. The predicted eastern output region has a gentle Pb peak at the middle of the flow range (Mahoney et al., 2002), whereas the northern contact region shows no visible peaks (Mahoney et al., 1989; Lehnert et al., 2000), although both routes involve about the same distance of lateral asthenosphere flow. A possible difference between them may be caused by the difference in lithospheric thickness along their route to the output point. The route to the northern output point is consistently capped with relatively young and thin lithosphere. However, the asthenosphere flowpath to the eastern output region underlies old, thick lithosphere, whose thickness is constant or increasing until ~1000 km before the ridge. Niu et al. (1999) explained how an enriched component could possibly survive over the long distance between Hawaii and East Pacific Rise.

If thick lithosphere suppresses decompression melting during the course of lateral asthenospheric transport, then the plume lithology can be preserved in spite of a long horizontal travel path—as long as it is progressive melt-extraction, and melt-extraction alone that removes the distinctive plume-component signals from the plume-fed asthenosphere that forms the MORB source (Phipps Morgan, 1997, 1999; Phipps Morgan and Morgan, 1999). If this conceptual framework is true, then the Crozet group would be in a similar or even more advantageous situation than long-distance transport from Hawaii to the East Pacific Rise, because the predicted flow from the Crozet group consistently occurs beneath thick, old lithosphere, whereas lithosphere along the route from Hawaii gradually thins toward the East Pacific Rise. Therefore, we think it feasible that the Pb peak at 116°E could originate from the Crozet group.

SUMMARY AND ASSESSMENT OF THIS EXERCISE IN GEOCHEMICAL PATTERN MATCHING

This initial comparison between our predicted flow-field for suboceanic asthenosphere and observed geochemical province boundaries along the global spreading system has yielded intriguing but not definitive results. In the Atlantic Ocean basin, a one-to-one match is often possible—but in this basin, several poorly determined parameters in the flow model have been tuned to match the regional geochemical province boundaries. Nevertheless, it is encouraging that a match was possible. In the other ocean basins, no such tuning was necessary, because relative hotspot upwelling fluxes beneath moving plates could be crudely constrained by rates of the past 5–10 m.y. of magma production at their associated hotspots. (See White, 1993, and Phipps Morgan, 1997, for two similar estimates that were obtained using somewhat different geophysical approaches). In these basins, the model asthenosphere flow predictions raise several intriguing possibilities to explain observed geochemical variations along the spreading center. They suggest that the Australian-Antarctic discordance is not a region of along-ridge downwelling, but rather simply a region of convergence between Indian and Pacific plume material that is being driven by the rapid northward migration of the Australian craton. It suggests that cross-ridge asthenosphere flow and along-ridge flow may be associated with different geochemical fingerprints. The model also suggests that individual plumes, as their upwelled plume material flows across a ridge axis, may generate He anomalies along the leading edge of each flow domain. We remain hopeful that the asthenosphere flow modeling technique developed in this work and the companion study (Yamamoto et al., this volume, chapter 9) will be a useful tool to further test whether the paradigm of a plume-fed asthenosphere will ultimately provide a better conceptual framework with which to understand the flow and melting of Earth's upper mantle.

ACKNOWLEDGMENTS

We acknowledge Morgan A.L. Crooks, Jörg Hasenclever, Walter Smith, and Mara Yale for their help during the 12 years of gestation of this work. We also thank Don Anderson, Gill Foulger, Donna Jurdy, and Lars Rüpke for helpful reviews.

REFERENCES CITED

Bach, W., Hegner, E., Erzinger, J., and Satir, M., 1994, Chemical and isotopic variations along the superfast spreading East Pacific Rise from 6° to 30°S: Contributions to Mineralogy and Petrology, v. 116, p. 365–380, doi: 10.1007/BF00310905.

Barling, J., Goldstein, S.L., and Nicholls, I.A., 1994, Geochemistry of Heard Island (southern Indian Ocean): Characterization of an enriched mantle component and implications for enrichment of the sub-Indian Ocean mantle: Journal of Petrology, v. 35, p. 1017–1053.

Bonatti, E., Peyve, A., Kepezhinskas, P., Kurentsova, N., Seyler, M., Skolotnev, S., and Udintsev, G., 1992, Upper mantle heterogeneity below the mid-Atlantic Ridge, 0–15°N: Journal of Geophysical Research, v. 97, p. 4461–4476.

Breddam, K., Kurz, M.D., and Storey, M., 2000, Mapping out the conduit of the Iceland mantle plume with helium isotopes: Earth and Planetary Science Letters, v. 176, p. 45–55, doi: 10.1016/S0012-821X(99)00313-1.

Castillo, P.R., Natland, J.H., Yaoling, N., and Lonsdale, P.F., 1998, Sr, Nd, and Pb isotopic variation along the Pacific-Antarctic risecrest, 53–57°S: Implications for the composition and dynamics of the South Pacific upper man-

tle: Earth and Planetary Science Letters, v. 154, p. 109–125, doi: 10.1016/S0012-821X(97)00172-6.

Christensen, B.P., Holm, P.M., Jambon, A., and Wilson, J.R., 2001, Helium, argon and lead isotopic composition of volcanics from Santo Antao and Fogo: Cape Verde Islands: Chemical Geology, v. 178, p. 127–142, doi: 10.1016/S0009-2541(01)00261-3.

Christie, D.M., West, B.P., Douglas, G.P., and Hanan, B.B., 1998, Chaotic topography, mantle flow and mantle migration in the Australian-Antarctic discordance: Nature, v. 394, p. 637–643, doi: 10.1038/29226.

Coffin, M.F., Pringle, M.S., Duncan, R.A., Gladczenko, T.P., Storey, M., Mueller, R.D., and Gahagan, L.A., 2002, Kerguelen hotspot magma output since 130 Ma: Journal of Petrology, v. 43, p. 1121–1139, doi: 10.1093/petrology/43.7.1121.

Dickin, A.P., 1995, Radiogenic isotope geology (2nd edition): Cambridge, Cambridge University Press, 490 p.

Dosso, L., Bougault, H., Beuzart, P., Calvez, J.Y., and Joron, J.L., 1988, The geochemical structure of the South-East Indian Ridge: Earth and Planetary Science Letters, v. 88, p. 47–59, doi: 10.1016/0012-821X(88)90045-3.

Dosso, L., Bougault, H., and Joron, J.L., 1993, Geochemical morphology of the North mid-Atlantic Ridge, 10–24°N: Trace element-isotope complementarity: Earth and Planetary Science Letters, v. 120, p. 443–462, doi: 10.1016/0012-821X(93)90256-9.

Doucelance, R., Escrig, S., Moreira, M., Gariepy, C., and Kurz, M.D., 2003, Pb-Sr-He isotope and trace element geochemistry of the Cape Verde Archipelago: Geochimica et Cosmochimica Acta, v. 67, p. 3717–3733, doi: 10.1016/S0016-7037(03)00161-3.

Douglass, J., Schilling, J.G., and Fontignie, D., 1999, Plume-ridge interactions of the Discovery and Shona mantle plumes with the southern mid-Atlantic Ridge (40–55°S): Journal of Geophysical Research, v. 104, p. 2941–2962, doi: 10.1029/98JB02642.

Dupre, B., and Allègre, C.J., 1983, Pb-Sr isotope variation in Indian Ocean basalts and mixing phenomena: Nature, v. 303, p. 142–146, doi: 10.1038/303142a0.

Fitton, J.G., Saunders, A.D., Norry, M.J., Hardarson, B.S., and Tayler, R.N., 1997, Thermal and chemical structure of the Iceland plume: Earth and Planetary Science Letters, v. 153, p. 197–208, doi: 10.1016/S0012-821X(97)00170-2.

Foulger, G.R., and Pearson, D.G., 2001, Is Iceland underlain by a plume in the lower mantle? Seismology and helium isotopes: Geophysical Journal International, v. 145, p. F1–F5, doi: 10.1046/j.0956-540x.2001.01457.x.

Frey, F.A., Nicolaysen, K., Kubit, B.K., Weis, D., and Gret, A., 2002, Flood basalt from Mont Tourmente in the central Kerguelen Archipelago: The change from transitional to alkalic basalt at ~25 Ma: Journal of Petrology, v. 43, p. 1367–1387, doi: 10.1093/petrology/43.7.1367.

Gaboret, C., Forte, A.M., and Montagner, J.P., 2003, The unique dynamics of the Pacific Hemisphere mantle and its signature on seismic anisotropy: Earth and Planetary Science Letters, v. 208, p. 219–233, doi: 10.1016/S0012-821X(03)00037-2.

Graham, D.W., Jenkins, W.J., Schilling, J.G., Thompson, G., Kurz, M.D., and Humphris, S.E., 1992, Helium isotope geochemistry of mid-ocean ridge basalts from the South Atlantic: Earth and Planetary Science Letters, v. 110, p. 133–147, doi: 10.1016/0012-821X(92)90044-V.

Graham, D.W., Castillo, P.R., Lupton, J.E., and Batiza, R., 1996, Correlated He and Sr isotope ratios in South Atlantic near-ridge seamounts and implications for mantle dynamics: Earth and Planetary Science Letters, v. 144, p. 491–503, doi: 10.1016/S0012-821X(96)00172-0.

Graham, D.W., Johnson, K.T.M., Priebe, L.D., and Lupton, J.E., 1999, Hotspot-ridge interaction along the Southeast Indian Ridge near Amsterdam and St. Paul islands: Helium isotope evidence: Earth and Planetary Science Letters, v. 167, p. 297–310, doi: 10.1016/S0012-821X(99)00030-8.

Graham, D.W., Lupton, J.E., Spera, F.J., and Christie, D.M., 2001, Upper-mantle dynamics revealed by helium isotope variations along the Southeast Indian Ridge: Nature, v. 409, p. 701–703, doi: 10.1038/35055529.

Gurnis, M., Louis, M., and Mueller, R.D., 2000, Models of mantle convection incorporating plate tectonics: The Australian region since the Cretaceous, *in* Richards, M.A., et al., eds., The history and dynamics of global plate motions: Washington, D.C., American Geophysical Union Geophysical Monograph 121, p. 211–238.

Hanan, B.B., Kingsley, R.H., and Schilling, J.G., 1986, Pb isotope evidence in the South Atlantic for migrating ridge-hotspot interactions: Nature, v. 322, p. 137–144, doi: 10.1038/322137a0.

Hart, S.R., 1984, A large-scale isotope anomaly in the Southern Hemisphere mantle: Nature, v. 309, p. 753–757, doi: 10.1038/309753a0.

Hilton, D.R., Barling, J., and Wheller, G.E., 1995, Effect of shallow-level contamination on the helium isotope systematics of ocean-island lavas: Nature, v. 373, p. 330–333, doi: 10.1038/373330a0.

Hilton, D.R., Thirlwall, M.F., Tayler, R.N., Murton, B.J., and Nichols, A., 2000, Controls on magmatic degassing along the Reykjanes Ridge with implications for the helium paradox: Earth and Planetary Science Letters, v. 183, p. 43–50, doi: 10.1016/S0012-821X(00)00253-3.

Johnson, K., Graham, D.W., Rubin, K., Nicolaysen, K., Scheirer, D., Forsyth, D., Baker, E., and Douglas-Priebe, L., 2000, Boomerang Seamount: The active expression of the Amsterdam–St. Paul hotspot: Southeast Indian Ridge: Earth and Planetary Science Letters v. 183, p. 245–259, doi: 10.1016/S0012-821X(00)00279-X.

Kido, M., and Seno, T., 1994, Dynamic topography compared with residual depth anomalies in oceans and implications for age-depth curves: Geophysical Research Letters, v. 21, p. 717–720, doi: 10.1029/94GL00305.

Larson, R.L., 1991, Geological consequences of super plumes: Geology, v. 19, p. 963–966, doi: 10.1130/0091-7613(1991)019<0963:GCOS>2.3.CO;2.

Lehnert, K., Su, Y., Langmuir, C.H., Sarbas, B., and Nohl, U., 2000, A global geochemical database structure for rocks: Geochemistry, Geophysics, Geosystems, v. 1, doi: 1999GC000026.

Leveque, J.J., Debayle, E., and Maupin, V., 1998, Anisotropy in the Indian Ocean upper mantle from Rayleigh- and Love-waveform inversion: Geophysical Journal International, v. 133, p. 529–540, doi: 10.1046/j.1365-246X.1998.00504.x.

Macdougall, J.D., and Lugmair, G.W., 1986, Sr and Nd isotopes in basalts from the East Pacific Rise: Significance for mantle heterogeneity: Earth and Planetary Science Letters, v. 77, p. 273–284, doi: 10.1016/0012-821X(86)90139-1.

Mahoney, J.J., Natland, J.H., White, W.M., Poreda, R., Bloomer, S.H., Fisher, R.L., and Baxter, A.N., 1989, Isotopic and geochemical provinces of the western Indian Ocean spreading centers: Journal of Geophysical Research, v. 94, p. 4033–4052.

Mahoney, J.J., LeRoex, A.P., Peng, A., Fisher, R.L., and Natland, J.H., 1992, Southwestern limits of Indian Ocean Ridge mantle and the origin of low $^{206}Pb/^{204}Pb$ mid-ocean ridge basalt: Isotope systematics of the central Southwest Indian Ridge (17–50°E): Journal of Geophysical Research, v. 97, p. 19,771–19,790.

Mahoney, J.J., Sinton, J.M., Kurz, M.D., Macdougall, J.D., Spencer, K.J., and Lugmair, G.W., 1994, Isotope and trace element characteristics of a superfast spreading ridge: East Pacific rise, 13–23°S: Earth and Planetary Science Letters, v. 121, p. 173–193, doi: 10.1016/0012-821X(94)90039-6.

Mahoney, J.J., Graham, D.W., Christie, D.M., Johnson, K.T.M., Hall, L.S., and Vonderhaar, D.L., 2002, Between a hotspot and a cold spot: Isotopic variation in the Southeast Indian Ridge athenosphere, 86°E–118°E: Journal of Petrology, v. 43, p. 1155–1176, doi: 10.1093/petrology/43.7.1155.

Marks, K.M., and Stock, J.M., 1994, Variation in ridge morphology and depth-age relationships on the Pacific-Antarctic ridge: Journal of Geophysical Research, v. 99, p. 531–541, doi: 10.1029/93JB02760.

McNutt, M.K., and Fischer, K.M., 1987, The South Pacific superswell, *in* Keating, B.H., et al., eds., Seamounts, islands, and atolls: Washington, D.C., American Geophysical Union Geophysical Monograph 43, p. 25–34.

Meibom, A., Anderson, D.L., Sleep, N.H., Frei, R., Chamberlain, C.P., Hren, M.T., and Wooden, J.L., 2003, Are high $^3He/^4He$ rations in oceanic basalts an indicator of deep-mantle plume components?: Earth and Planetary Science Letters, v. 208, p. 197–204, doi: 10.1016/S0012-821X(03)00038-4.

Michard, A., Montigny, R., and Schlich, R., 1986, Geochemistry of the mantle beneath the Rodriguez triple junction and the South-East Indian Ridge: Earth and Planetary Science Letters, v. 78, p. 104–114, doi: 10.1016/0012-821X(86)90176-7.

Montagner, J.P., and Guillot, L., 2002, Seismic anisotropy and global geodynamics: Plastic deformation of minerals and rocks: Washington, D.C., Mineralogical Society of America Reviews of Mineralogy and Geochemistry 51, p. 353–385.

Montagner, J.P., and Tanimoto, T., 1990, Global anisotropy in the upper mantle inferred from the regionalization of phase velocities: Journal of Geophysical Research, v. 95, p. 4797–4819.

Morgan, W.J., 1978, Rodriguez, Darwin, Amsterdam . . .: A second type of hotspot island: Journal of Geophysical Research, v. 83, p. 5355–5360.

Nicolaysen, K., Bowring, S., Frey, F., Weis, D., Ingle, S., Pringle, M.S., and Coffin, M.F., 2001, Provenance of Proterozoic garnet-biotite gneiss recovered from Elan Bank, Kerguelen Plateau, southern Indian Ocean: Geology, v. 29, p. 235–238.

Niu, Y., Collerson, K.D., Batiza, R., Wendt, J.I., and Regelous, M., 1999, Origin of enriched type mid-ocean ridge basalt at ridges far from mantle plumes: The East Pacific Rise at 11°20′N: Journal of Geophysical Research, v. 104, p. 7067–7087, doi: 10.1029/1998JB900037.

Peate, D.W., Hawkesworth, C.J., van Calsteren, P.W., Taylor, R.N., and Murton, B.J., 2001, ^{238}U-^{230}Th constraints on mantle upwelling and plume-ridge interaction along the Reykjanes Ridge: Earth and Planetary Science Letters, v. 187, p. 259–272, doi: 10.1016/S0012-821X(01)00266-7.

Phipps Morgan, J., 1997, The generation of a compositional lithosphere by mid-ocean ridge melting and its effect on subsequent off-axis hotspot upwelling and melting: Earth and Planetary Science Letters, v. 146, p. 213–232, doi: 10.1016/S0012-821X(96)00207-5.

Phipps Morgan, J., 1999, The isotope topology of individual hotspot basalt arrays: Mixing curves or melt extraction trajectories?: Geochemistry, Geophysics, Geosystems, v. 1, doi: 1029/1999GC000004.

Phipps Morgan, J., and Morgan, W.J., 1999, Two-stage melting and the geochemical evolution of the mantle: A recipe for mantle plum-pudding: Earth and Planetary Science Letters, v. 170, p. 215–239, doi: 10.1016/S0012-821X(99)00114-4.

Phipps Morgan, J., Morgan, W.J., Zhang, Y.-S., and Smith, W.H.F., 1995, Observational hints for a plume-fed sub-oceanic asthenosphere and its role in mantle convection: Journal of Geophysical Research, v. 100, p. 12,753–12,768, doi: 10.1029/95JB00041.

Poreda, R., Schilling, J.-G., and Craig, H., 1986, Helium and hydrogen isotopes in ocean-ridge basalts north and south of Iceland: Earth and Planetary Science Letters, v. 78, p. 1–17, doi: 10.1016/0012-821X(86)90168-8.

Price, R.C., Kennedy, A.K., Sneeringer, M.R., and Frey, F.A., 1986, Geochemistry of basalts from the Indian Ocean triple junction: Implications for the generation and evolution of Indian Ocean ridge basalts: Earth and Planetary Science Letters, v. 78, p. 379–396, doi: 10.1016/0012-821X(86)90005-1.

Ritsema, J., and Allen, R., 2003, The elusive mantle plume: Earth and Planetary Science Letters, v. 207, p. 1–12, doi: 10.1016/S0012-821X(02)01093-2.

Royer, J.-Y., and Sandwell, D.T., 1989, Evolution of the Eastern Indian Ocean since the late Cretaceous: Constraints from GEOSAT altimetry: Journal of Geophysical Research, v. 94, p. 13,755–13,782.

Rubin, K., and Mahoney, J., 1993, What's on the plume channel?: Nature, v. 362, p. 109–110, doi: 10.1038/362109a0.

Schilling, J.-G., 1973, Iceland mantle plume: Geochemical study of Reykjanes Ridge: Nature, v. 242, p. 565–571, doi: 10.1038/242565a0.

Schilling, J.-G., Thompson, G., Kingsley, R.H., and Humphris, S.E., 1985, Hotspot-migrating ridge interaction in the South Atlantic: Nature, v. 313, p. 187–191, doi: 10.1038/313187a0.

Schilling, J.-G., Hanan, B.B., McCully, B., and Kingsley, R.H., 1994, Influence of the Sierra Leone mantle plume on the equatorial mid-Atlantic Ridge: A Nd-Sr-Pb isotopic study: Journal of Geophysical Research, v. 99, p. 12,005–12,028, doi: 10.1029/94JB00337.

Sempere, J.C., Cochran, J.R., and SEIR Scientific Team, 1997, The southeast Indian Ridge between 88°E and 118°E: Variations in crustal accretion at constant spreading rate: Journal of Geophysical Research, v. 102, p. 15,489–15,505, doi: 10.1029/97JB00171.

Small, C., 1995, Observation of ridge-hotspot interactions in the Southern Ocean: Journal of Geophysical Research, v. 100, p. 17,931–17,946, doi: 10.1029/95JB01377.

Small, C., Cochran, J.R., Sempere, J.C., and Christie, D.M., 1999, The structure and segmentation of the South Indian Ridge: Marine Geology, v. 161, p. 1–12, doi: 10.1016/S0025-3227(99)00051-1.

Staudacher, T., Sarda, P., Richardson, S.H., Allègre, C.J., Sagna, I., and Dmitriev, L.V., 1989, Noble gases in basalt glasses from a mid-Atlantic Ridge topographic high at 14°N: Geodynamic consequences: Earth and Planetary Science Letters, v. 96, p. 119–133, doi: 10.1016/0012-821X(89)90127-1.

Staudigel, H., Park, K.-H., Pringle, M.S., Rubenstone, J.L., Smith, W.H.F., and Zindler, A., 1991, The longevity of the South Pacific isotopic and thermal anomaly: Earth and Planetary Science Letters, v. 102, p. 24–44, doi: 10.1016/0012-821X(91)90015-A.

Taylor, R.N., Thirlwall, M.F., Murton, B.J., Hilton, D.R., and Gee, M.A.M., 1997, Isotopic constraints on the influence of the Icelandic plume: Earth and Planetary Science Letters, v. 148, p. E1–E8, doi: 10.1016/S0012-821X(97)00038-1.

Veevers, J.J., 1982, Australian-Antarctic depression from the mid-ocean ridge to adjacent continents: Nature, v. 295, p. 315–317, doi: 10.1038/295315a0.

Vlastelic, I., Aslanian, D., Dosso, L., Bougault, H., Olivet, J.L., and Geli, L., 1999, Large-scale chemical and thermal division of the Pacific mantle: Nature, v. 399, p. 345–350, doi: 10.1038/20664.

Vlastelic, I., Dosso, L., Bougault, H., Aslanian, D., Geli, L., Etoubleau, J., Bohn, M., Joron, J.L., and Bollinger, C., 2000, Chemical systematics of an intermediate spreading ridge: The Pacific-Antarctic Ridge between 56°S and 66°S: Journal of Geophysical Research, v. 105, p. 2915–2936, doi: 10.1029/1999JB900234.

Vogt, P.R., 1971, Asthenosphere motion recorded by the ocean floor south of

Iceland: Earth and Planetary Science Letters, v. 13, p. 153–160, doi: 10.1016/0012-821X(71)90118-X.

Vogt, P.R., 1986, Plate 8A. MAR: Plate boundary synthesis, in Vogt, P.R., and Tucholke, B.E., eds., The western Atlantic region: Boulder, Colorado, Geological Society of America Geology of North America M, plate 8 (supplemental maps).

Vogt, P.R., and Johnson, L., 1972, Seismic reflection survey of an oblique aseismic basement trend on the Reykjanes Ridge: Earth and Planetary Science Letters, v. 15, p. 248–254, doi: 10.1016/0012-821X(72)90170-7.

Vogt, P.R., and Johnson, L., 1975, Transform faults and longitudinal flow below the midoceanic ridge: Journal of Geophysical Research, v. 80, p. 1399–1428.

Weis, D., and Frey, F.A., 2002, Submarine basalts of the northern Kerguelen Plateau: Interaction between the Kerguelen plume and the Southeast Indian Ridge revealed at ODP SITE 1140: Journal of Petrology, v. 43, p. 1287–1309, doi: 10.1093/petrology/43.7.1287.

White, R.S., 1993, Melt production rates in mantle plumes: Philosophical Transactions of the Royal Society of London, Physical Sciences and Engineering, v. 342, p. 137–153.

Yale, M.M., and Phipps Morgan, J., 1998, Asthenosphere flow model of hotspot-ridge interactions: A comparison of Iceland and Kerguelen: Earth and Planetary Science Letters, v. 161, p. 45–56, doi: 10.1016/S0012-821X(98)00136-8.

Yamamoto, M., Phipps Morgan, J., and Morgan, W.J., 2007 (this volume, chapter 9), Global plume-fed asthenosphere flow—I: Motivation and model development, in Foulger, G.R., and Jurdy, D.M., eds., Plates, plumes, and planetary processes: Boulder, Colorado, Geological Society of America Special Paper 430, doi: 10.1130/2007.2430(09).

Ying, M., and Cochran, J.R., 1996, Transitions in axial morphology along the Southeast Indian Ridge: Journal of Geophysical Research, v. 101, p. 15,849–15,866, doi: 10.1029/95JB03038.

MANUSCRIPT ACCEPTED BY THE SOCIETY 31 JANUARY 2007

DISCUSSION

17 January 2007, Don L. Anderson

Increasing temperature decreases viscosity, seismic velocity, thermal conductivity, and density, and increases thermal expansivity and seismic attenuation. Pressure has the opposite effect and, in addition, increases the melting point. The net effect is that there will be a low-viscosity, low-velocity zone (LVZ) in the upper mantle that has a high homologous temperature. The presence of an asthenosphere is not a mystery, even in a homogeneous mantle with normal pressure and temperature gradients. A LVZ with a lid may also collect low-density, low-melting-point debris from mantle differentiation and crustal delamination and will therefore be laterally heterogeneous with respect to lithology, chemistry, and melting point.

A heterogeneous **partially molten asthenosphere**[1] is an alternative to the deep-mantle plume model. But what is the mechanism for replacing the material that leaks out of it at hotspots?

[1]**Text in boldface** can retrieve the appropriate references when entered into Google.

Yamamoto et al. (this volume, chapters 9 and 10) argue that it is replaced by upwelling plumes from the deepest thermal boundary layer (TBL). I have argued that the delamination of lower continental and arc crust, and the subduction of buoyant aseismic ridges accounts for more than the amount of magma involved in hotspot magmatism (Anderson, this volume). The subduction of oceanic crust involves an order of magnitude more material, but this material may sink deeper into the mantle and may be inappropriate for recycling as ocean island basalt (OIB) because of subduction-zone processing and chemistry. It, however, does displace deeper material upward that may fuel the ridges. If irreversible deep subduction started late in Earth's history, at current rates, 70 km of basalt and/or eclogite will accumulate in or below the transition region every 1 b.y., and this material will displace deep material upwards.

A thermal maximum and a **subadiabatic mantle** are expected in a mantle heated by radioactive decay and cooled by sinking slabs; plumes and bottom heating are not required. Lateral transport of material in the asthenosphere, including entrained fertile blobs, is intrinsic to the counterflow associated with plate motions, but not with whole-mantle convection. Recycling of foundered lower crust and subducted aseismic ridges is probably the primary source of fertile **olivine-free mantle and melting anomalies;** these mechanisms place long, linear mafic heterogeneities into the mantle.

In contrast to plume models, plates are driven by surface and slab forces that induce mantle convection, including counterflow (Harper, 1978; Chase, 1979). This process is consistent with anisotropy measurements (see also http://caltechbook .library.caltech.edu/14/). In places where the mantle contains a melt phase or fertile streaks, volcanic chains may be the result of lithospheric stress and magma fracture rather than thermal erosion. Water and CO_2 affect the strength of the lithosphere and the melting temperature and viscosity of the mantle. The homologous temperature therefore controls the dynamics of the mantle and the amount of melting.

In the plume hypothesis (Yamamoto et al., this volume, chapters 9 and 10), hot material from the core-mantle boundary rises through narrow plumes, creating a hot LVZ. The plumes are compensated by sinking of the whole mantle—in the original plume hypothesis—or by slabs. Asthenospheric flow is away from hotspots. The upper mantle is homogenized by chaotic stirring.

The alternative **top-down tectonics** model is essentially the inverse of this process. Slabs and plates, not plumes, drive the system. Radioactive heating, crustal stoping, spreading, and material displaced by slabs cause broad passive upwellings. LVZs are the result of the dominance of temperature over pressure and partial melting. Ubiquitous but not universal magmatism in extending regions implies a heterogeneous shallow mantle with high but variable homologous temperature; the upper mantle has not been homogenized. Very low seismic velocities and viscosities are due to mineralogy (eclogite), volatiles, fertility, and partial melting. The mantle is refertilized from the top by subduction of young or thick oceanic crust and delaminated continental crust, and from the bottom by the abovementioned broad upwelling. Fertile blobs can be tens of kilometers in extent. The closure of back-arc basins and the trapping of mantle wedge material are also involved. In fact, the similarity of back-arc basin and continental flood basalts and OIB has suggested to some (Hooper et al., this volume) that the first two are caused by plumes.

In the **shallow counterflow** model, flow is anti-parallel to plate motions and toward "hotspots" and upwells at thin spots. Ridges, trenches, and continents migrate over the upper mantle, fertilizing it, and sampling it at extending- and thin-plate regions. Melt potential of the mantle is a function of eclogite and volatile contents, not absolute temperature. Eclogite sinkers and recycled material with volatiles have low melting points and low seismic velocities. Focused upwelling is controlled by the lid, including fracture zones. Delaminated crust, even when cold, melts because of conductive heating; mantle displaced upward melts because of pressure release. The similarity of "hotspot" magmas and the lower continental crust is evidence that crustal delamination is one source of "plume components."

Alternatives to the plume conjecture are now well developed and testable. They involve much more than propagating cracks, a usual strawman (e.g., Hooper et al., this volume). It is no longer acceptable to just assume that narrow deep-mantle plumes must exist "because there are no alternatives" and that all melting anomalies and low-velocity regions of the mantle must have high absolute temperatures.

19 January 2007, Alan D. Smith

The plum-pudding plume model of Yamamoto et al. (this volume, chapters 9 and 10) derives mid-ocean ridge basalt (MORB) and OIB from a common reservoir, and is thus different from standard plume models (e.g., Hofmann and White, 1982), which are based on the concept of isotopic differences between MORB and OIB requiring distinct sources for these categories of volcanism. In this regard, the model of Yamamoto et al. is similar to the plate model (e.g., Foulger, this volume), in which subducted oceanic crust is mixed directly with the convecting mantle, and differences between MORB and OIB result from sampling of recycled materials and their host mantle (Meibom and Anderson, 2004; Smith, 2005). Why, then, invoke the plume stage at all?

The geochemical argument for plumes rests essentially on the interpretation of $^3He/^4He$ ratios. Hofmann and White (1982) suggested isolation of subducted oceanic crust in plume sources shortly after remixing of such material with the convecting mantle had been indicated from banding in orogenic lherzolites (Polvé and Allègre, 1979). However, interpretation of high $^3He/^4He$ ratios in OIB as signifying elevated ^3He abundances from a primitive mantle reservoir (e.g., Allègre et al., 1983), led to the plume model being added to subsequent considerations of recycling crust into the convecting mantle (e.g., Allègre and Turcotte, 1986). Thereafter, crustal recycling into the convecting mantle was relegated to an explanation for heterogeneity in the MORB source (e.g., Saunders et al., 1988), and in the plum-pudding plume model the concept becomes redundant, as the entire geochemical heterogeneity in the mantle results from plumes.

The flaw in the plume logic is, however, that high $^3He/^4He$ ratios can arise from low abundances of ^4He rather than high abundances of ^3He (e.g., Anderson, 1998, 2000), which removes any requirement for an ultra-deep origin for intraplate volcanism. Which model, then, provides the more comprehensive explanation for intraplate volcanism? The plum-pudding plume model has an advantage over the standard plume model in accounting for the distribution of geochemical heterogeneities throughout the asthenosphere, without having to invoke a multitude of plumes or ad hoc plumbing arrangements to explain such features as nonlinear age progressions in ocean island chains. However, like all plume models, it lacks a mechanism for tapping such sources away from hotspots. The plate model, however, provides a mechanism for the distribution of recycled material throughout the mantle, as well as a means of tapping such material via plate interactions.

20 January 2007, Don L. Anderson

The plum-pudding plume models and the marble cake models derive MORB and OIB from a common reservoir and are thus different from the standard reservoir models, which are based on the concept of isotopic differences between MORB and OIB requiring distinct isolated reservoirs (Smith comment, 19 January). In the marble cake model subducted oceanic crust is mixed thoroughly into the so-called convecting mantle to maintain the apparent homogeneity of the MORB reservoir. Meibom and Anderson (2004) and Smith (2005) argue that a heterogeneous mantle can provide homogeneous basalts by the sampling process. If the mantle is chemically stratified (Anderson, this volume), thorough chaotic mixing is unlikely.

Recycled NMORB that is processed through subduction zones may not be a good candidate for the enriched mantle (EM) components of the mantle (Niu et al., 2002; Lustrino and Dallai, 2003). If deep subduction started only in Neoproterozoic time (Stern, 2005), then oceanic crust may not be involved at all in modern magmatism. But slabs will displace deeper mantle upward, which is a possible source for mid-ocean ridges. There are other plausible and voluminous sources of mafic material for OIB and other "hotspot" magmas that cannot be ignored but are overlooked by Yamamoto et al. (this volume, chapters 9 and 10). These include (1) delaminated lower continental and arc crust, (2) mantle wedge material, (3) the slow upwelling of mantle displaced by sinking slabs, and (4) subduction of young buoyant oceanic lithosphere or aseismic ridges (Fig. D-1).

The fertile blob model (Anderson, this volume) is similar to the plum-pudding model but the "plums" may be tens of kilo-meters in extent. These fertile blobs may collect in and below the asthenosphere. Smaller blobs can be entrained in the counterflow that flows toward hotspots and ridges. Mass balance and geochemistry do not require that oceanic crust be recycled or reused at hotspots. Thus, recycled oceanic crust need not be involved in either OIB or MORB magmatism. There is no need for deep-mantle plumes to explain either the physical or chemical properties of the shallow mantle.

23 January 2007, Alan D. Smith

Models for the isotopic evolution of recycled oceanic crust need to take into consideration both dehydration and melting of the subducting slab (Smith, 2005). Slab dehydration will result in recycled basaltic crust evolving to high μ (HIMU) isotopic compositions, dehydrated sediment will evolve to EM2 (terminology of Zindler and Hart, 1986). Higher geothermal gradients and subduction of young oceanic crust favor a greater role for slab melting in the Archean (Martin, 1986). Oceanic crust that has undergone melting in subduction zones evolves toward the depleted mantle (DM) isotopic component of Zindler and Hart (1986). The convecting mantle may thus contain two types of recycled oceanic crust, with the common component between MORB and OIB being recycled dehydrated oceanic crust (Smith, 2005).

The isotopic component EM1 is also found in MORB and OIB, and I agree with Anderson (comment of 20 January) that this component may be equated with continental lower crust (see also Lustrino and Dallai, 2003), although continental mantle is a further option. Such materials can be delaminated and/or eroded directly into the shallow mantle during continental col-

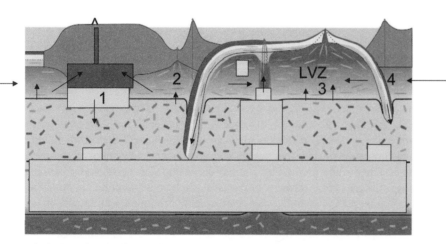

1 delaminated crust, 2 wedge, 3 broad upwelling,

4 young oceanic crust

Figure D-1. The delamination-fed, shallow counterflow model of Anderson (this volume). Fertile low-melting-point blobs warm up with time and melt. These replace the hot thermal plumes of the Yamamoto et al. model (Chapters 9 and 10, this volume). The thermal maximum in the shallow mantle is a natural result of internal radioactive heating and deep cooling by slabs. Melting anomalies are due to locally high homologous temperature, not high absolute temperature. Modified from the plume-fed asthenosphere model of Yamamoto et al. (Chapter 9, this volume).

lision or rifting events, without any need for cycling through plume sources, as depicted by McKenzie and O'Nions (1983). Entrainment of delaminated continental lower crust into ocean ridge upwelling is a possible mechanism for the generation of oceanic plateaus, but EM1 is not a ubiquitous mantle component, and I would regard the difference between MORB and OIB to result from tapping of the recycled melted oceanic crustal component beneath ocean ridges.

31 January 2007, Jason Phipps Morgan

In response to Smith's comment (23 January) that we neglect subduction-related transformations in the scenario of a plume-fed asthenosphere, I agree that formal modeling of slab-dehydration and slab-melting processes are missing from the quantitative geochemical evolution model described in Phipps Morgan and Morgan (1999). There we noted our surprise that even with the neglect of these geologic processes the models could still reproduce many of the geochemical patterns seen in OIB, enriched MORB (E-MORB), and MORB, with obvious exceptions for the elements that have been proposed to be fractionated by slab dehydration (e.g., Ce versus Pb). Obviously, the processes that create the different components in a plum-pudding model need to be identified—here I only wish to note that it is likely that many enriched components exist in much more extreme forms in the mantle than is seen by their "flavoring" of erupted basalts. For example, melt inclusions within end-member HIMU basalts can be much more extreme in isotopic composition than their host basalt (Saal et al., 2005).

If we focus on MORB rather than OIB or E-MORB, we are left with several questions to be explained by a passive (non-plume-fed) upper mantle. Why does the MORB source have far less thermal variability than that introduced into the mantle by subducting slabs? To me a major challenge for both Anderson's perisphere concept and conventional upper-mantle MORB source scenarios is how to so evenly homogenize mantle temperature beneath the global ridge system given the ~600- to 700°-C average temperature difference between a subducting slab and the MORB source region and the fact that the MORB-source mantle, given its predicted concentrations of radioactive elements, would only warm from internal radioactivity by ~20–100 °C/b.y. Likewise, a similar challenge for nonplume-fed scenarios for the MORB source is to explain why the source temperatures and crustal thicknesses are so similar beneath mid-ocean ridges that are near long-time subducting slabs—e.g., Juan de Fuca, Chile Rise—and those far from slabs. If plume-fed, then the MORB source would be relatively uniform in temperature, as more-buoyant plume material would tend to displace any denser mantle downward so it could not upwell beneath ridges. Why are there sharp lateral transitions between volcanic and nonvolcanic rifted margins, and why is there a sharp temporal transition from OIB-like "rift" basalts to the eruption of MORB during the initial formation of an ocean

basin? These considerations suggest that the MORB source is relatively warm and uniform in temperature, and that MORB-source asthenosphere is absent beneath nonvolcanic rifted margins until it can enter by lateral flow of asthenosphere from beneath the thinner lithosphere of an adjacent ocean basin.

Regarding Anderson's comments, please note that we do not just view plume upwelling as replacing the mass "that leaks out at hotspots," but rather as the much larger upward counterflow to subducting slabs. However, we still view the subduction of a surface-cooled slab as both the main driving force for deep-mantle flow and the means of reinjecting the lithologic and chemical differentiation created by near-surface melting and hydrous interactions back into the mantle, where subsolidus mantle flow may stretch and stir, but not homogenize, this geochemical plum pudding. In many other aspects we basically agree with the concepts of Anderson and Smith of a "streaky" plum-pudding mantle as the source of oceanic volcanism, and we agree with the emerging geochemical consensus that the selective melting of lower-solidus streaks within the mantle seems to be the right explanation for the observed differences between different "flavors" of OIB and MORB.

REFERENCES CITED

Allègre, C.J., and Turcotte, D.L., 1986, Implications of a two component marble cake mantle: Nature, v. 323, p. 123–127.

Allègre, C.J., Staudacher, T., Sarda, P., and Kunz, M., 1983, Constraints on evolution of the Earth's mantle from rare gas systematics: Nature, v. 303, p. 762–766, doi: 10.1038/303762a0.

Anderson, D.L., 1998, The helium paradoxes: Proceedings of the National Academy of Sciences, USA, v. 95, p. 4822–4827.

Anderson, D.L., 2000, The statistics and distribution of helium in the mantle: International Geology Review, v. 42, p. 289–311.

Anderson, D.L., 2007 (this volume), The Eclogite engine: Chemical geodynamics as a Galileo thermometer, *in* Foulger, G.R., and Jurdy, D.M., eds., Plates, plumes, and planetary processes: Boulder, Colorado, Geological Society of America Special Paper 430, doi: 10.1130/2007.2430(03).

Chase, C.G., 1979, Asthenospheric counterflow: A kinematic model: Geophysical Journal of the Royal Astronomical Society, v. 56, p. 1–18.

Foulger, G.R., 2007 (this volume), The "plate" model for the genesis of melting anomalies, *in* Foulger, G.R., and Jurdy, D.M., eds., Plates, plumes, and planetary processes: Boulder, Colorado, Geological Society of America Special Paper 430, doi: 10.1130/2007.2430(01).

Harper, J.F., 1978, Asthenosphere flow and plate motions: Geophysical Journal of the Royal Astronomical Society, v. 55, p. 87–110.

Hofmann, A.W., and White, W.M., 1982, Mantle plumes from ancient oceanic crust: Earth and Planetary Science Letters, v. 57, p. 421–436.

Hooper, P.R., Camp, V.E., Reidel, S.P., and Ross, M.E., 2007 (this volume), The origin of the Columbia River flood basalt province: Plume versus nonplume models, *in* Foulger, G.R., and Jurdy, D.M., eds., Plates, plumes, and planetary processes: Boulder, Colorado, Geological Society of America Special Paper 430, doi: 10.1130/2007.2430(30).

Lustrino, M., and Dallai, L., 2003, On the origin of EM–I end-member: Neues Jahrbuch für Mineralogie Abhandlung, v. 179, p. 85–100.

Martin, H., 1986, Effect of steeper Archean geothermal gradient on geochemistry of subduction zone magmas: Geology, v. 14, p. 753–756.

McKenzie, D., and O'Nions, R.K., 1983, Mantle reservoirs and ocean island basalts: Nature, v. 301, p. 229–231.

Meibom, A., and Anderson, D.L., 2004, The statistical upper mantle assemblage: Earth and Planetary Science Letters, v. 217, p. 123–139.

Niu, Y.-L., Regelous, M., Wendt, I., Batiza, R., and O'Hara, M.J. (2002). Geochemistry of near-EPR seamounts: Importance of source vs. process and the origin of enriched mantle component: Earth and Planetary Science Letters, v. 199, p. 327–345.

Phipps Morgan, J., and Morgan, W.J., 1999, Two-stage melting and the geochemical evolution of the mantle: A recipe for mantle plum-pudding: Earth and Planetary Science Letters, v. 170, p. 215–239.

Polvé, M., and Allègre, C.J., 1979, Orogenic lherzolite complexes studied by ^{87}Rb-^{87}Sr: A clue to understanding mantle convection processes?: Earth and Planetary Science Letters, v. 51, p. 71–93.

Saal, A.E., Hart, S.R., Shimizu, N., Hauri, E.H., Layne, G.D., and Eiler, J.M., 2005, Pb isotopic variability in melt inclusions from the EMI-EMII-HIMU mantle end-members and the role of the oceanic lithosphere: Earth and Planetary Science Letters, v. 240, p. 605–620.

Saunders, A.D., Norry, M.J., and Tarney, J., 1988, Origin of MORB and chemically-depleted mantle reservoirs: Trace element constraints: Journal of Petrology, special volume 1988, p. 415–445.

Smith, A.D., 2005, The streaky-mantle alternative to mantle plumes and its bearing on bulk-Earth geochemical evolution, *in* Foulger, G.R., et al., eds., Plates, plumes, and paradigms: Boulder, Colorado, Geological Society of America Special Paper 388, p. 303–325, doi:10.1130/2005.2388(19).

Stern, R.J., 2005, Evidence from ophiolites, blueschists, and ultrahigh-pressure metamorphic terranes that the modern episode of subduction tectonics began in Neoproterozoic time: Geology, v. 33, p. 557–560.

Yamamoto, M., Phipps Morgan, J., and Morgan, W.J., 2007 (this volume, chapter 9), Global plume-fed asthenosphere flow—I: Motivation and model development, *in* Foulger, G.R., and Jurdy, D.M., eds., Plates, plumes, and planetary processes: Boulder, Colorado, Geological Society of America Special Paper 430, doi: 10.1130/2007.2430(09).

Yamamoto, M., Phipps Morgan, J., and Morgan, W.J., 2007 (this volume, chapter 10), Global plume-fed asthenosphere flow—II: Application to the geochemical segmentation of mid-ocean ridges, *in* Foulger, G.R., and Jurdy, D.M., eds., Plates, plumes, and planetary processes: Boulder, Colorado, Geological Society of America Special Paper 430, doi: 10.1130/2007.2430(10).

Zindler, A., and Hart, S.R., 1986, Chemical geodynamics: Annual Reviews of Earth and Planetary Sciences, v. 14, p. 493–571, doi: 0.1146/annurev.ea.14.050186.002425.

The Geological Society of America
Special Paper 430
2007

The Hawaiian SWELL pilot experiment—Evidence for lithosphere rejuvenation from ocean bottom surface wave data

Gabi Laske*

*Cecil H. and Ida M. Green Institute of Geophysics and Planetary Physics,
Scripps Institution of Oceanography, University of California San Diego,
9500 Gilman Drive, La Jolla, California 92093-0225, USA*

Jason Phipps Morgan

*Department of Earth & Atmospheric Sciences, Snee Hall,
Cornell University, Ithaca, New York, 14853-1504, USA*

John A. Orcutt

*Cecil H. and Ida M. Green Institute of Geophysics and Planetary Physics,
Scripps Institution of Oceanography, University of California San Diego,
9500 Gilman Drive, La Jolla, California 92093-0225, USA*

ABSTRACT

During the roughly year-long Seismic Wave Exploration in the Lower Lithosphere (SWELL) pilot experiment in 1997/1998, eight ocean bottom instruments deployed to the southwest of the Hawaiian Islands recorded teleseismic Rayleigh waves with periods between 15 and 70 s. Such data are capable of resolving structural variations in the oceanic lithosphere and upper asthenosphere and therefore help understand the mechanism that supports the Hawaiian Swell relief. The pilot experiment was a technical as well as a scientific feasibility study and consisted of a hexagonal array of Scripps Low-Cost Hardware for Earth Applications and Physical Oceanography (L-CHEAPO) instruments using differential pressure sensors. The analysis of eighty-four earthquakes provided numerous high-precision phase velocity curves over an unprecedentedly wide period range. We find a rather uniform (unaltered) lid at the top of the lithosphere that is underlain by a strongly heterogeneous lower lithosphere and upper asthenosphere. Strong slow anomalies appear within ~300 km of the island chain and indicate that the lithosphere has most likely been altered by the same process that causes the Hawaiian volcanism. The anomalies increase with depth and reach well into the asthenosphere, suggesting a sublithospheric dynamic source for the swell relief. The imaged velocity variations are consistent with thermal rejuvenation, but our array does not appear to have covered the melt-generating region of the Hawaiian hotspot.

Keywords: lithosphere, surface waves, dispersion, seismic tomography, Hawaii hotspot

---

*E-mail: glaske@ucsd.edu.

Laske, G., Phipps Morgan, J., and Orcutt, J.A., 2007, The Hawaiian SWELL pilot experiment—Evidence for lithosphere rejuvenation from ocean bottom surface wave data, *in* Foulger, G.R., and Jurdy, D.M., eds., Plates, plumes, and planetary processes: Geological Society of America Special Paper 430, p. 209–233, doi: 10.1130/2007.2430(11). For permission to copy, contact editing@geosociety.org. ©2007 The Geological Society of America. All rights reserved.

INTRODUCTION

The Hawaiian hotspot and its island chain are thought to be the textbook example of a hotspot located over a deep-rooted mantle plume (Wilson, 1963; Morgan, 1971). Because plume material is expected to ascend in a much more viscous surrounding mantle, it is expected to stagnate near the top and exhibit a sizable plume head that eventually leads to the uplift of the overlying seafloor (e.g., Olson, 1990). A hotspot on a stationary plate may then develop a dome-shaped swell (e.g., Cape Verde), whereas a plate moving above a plume would shear it and drag some of its material downstream, creating an elongated swell (Olson, 1990; Sleep, 1990). Hawaii's isolated location within a plate, away from plate boundaries, should give scientists the opportunity to test most basic hypotheses on plume-plate interaction and related volcanism. Yet the lack of many crucial geophysical data has recently revived the discussions on whether even the Hawaiian hotspot volcanism is related to a deep-seated mantle plume or is rather an expression of propagating cracks in the lithosphere (Natland and Winterer, 2005). Similarly, the dominant cause of the Hawaiian Swell relief has not yet been conclusively determined. At least four mechanisms have been proposed (see, e.g., Phipps Morgan et al., 1995; Fig. 1)—thermal rejuvenation, dynamic support, compositional buoyancy, and propagating crack—but none of them is universally accepted as the single dominant mechanism. All these mechanisms create a buoyant lithosphere and so can explain the bathymetric anomalies, but they have distinct geophysical responses, and each model currently appears to be inconsistent with at least one observable.

Possible Causes for Swell Relief

In the *thermal rejuvenation model,* the lithosphere reheats and thins when a plate moves over a hotspot (Fig. 1A). It explains the uplift of the seafloor and the age-dependent subsidence of seamounts along the Hawaiian Island chain (Crough, 1978; Detrick and Crough, 1978). This model was reported to be consistent with gravity and geoid anomalies, and observations suggest a compensation depth of only 40–90 km (instead of the 120 km for 90-Ma-old lithosphere). Initially, rapid heating of the lower lithosphere (40–50 km) within 5 m.y. and subsequent cooling appeared broadly consistent with heatflow data along the swell (von Herzen et al., 1982), though Detrick and Crough (1978) had recognized that the reheating model does not offer a mechanism for the rapid heating. The heatflow argument was later revised when no significant anomaly was found across the swell southeast of Midway (von Herzen et al., 1989), though the interpretation of those data is still the subject of debate (M. McNutt, personal commun., 2002). The thermal rejuvenation model has received extensive criticism from geodynamicists, as it is unable to explain the rapid initial heat loss by conduction alone, and modeling attempts fail to erode the lithosphere significantly if heating were the only mechanism involved (e.g., Ribe and Christensen, 1994; Moore et al., 1998). The *dynamic support model* is a result of early efforts to reconcile gravity and bathymetry observations of the Hawaiian Swell (Watts, 1976). Ponding, or pancaking, of ascending hot asthenosphere causes an unaltered lithosphere to rise. A moving Pacific plate shears the ponding mantle material and drags it along the island chain, thereby causing the elongated Hawaiian Swell (Olson, 1990;

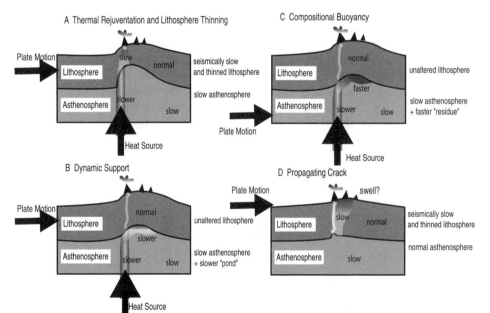

Figure 1. Concept figure for four possible mechanisms for hotspot swells. Thermal rejuvenation and the propagating crack model predict a significant impact on the lithosphere but the latter is associated with normal asthenosphere. The dynamic support and the compositional buoyancy model have an unaltered lithosphere but an anomalous asthenosphere.

Sleep, 1990). The compensation depth for this model remains at 120 km depth. An unaltered lithosphere is, however, inconsistent with the heatflow data along the swell (von Herzen et al., 1989) and the geoid. A recent hybrid model—dynamic thinning—in which secondary convection in the ponding asthenosphere erodes the lithosphere downstream (Ribe, 2004), appears to find support in a recent seismic study (Li et al., 2004). The third model, *compositional buoyancy,* was suggested by Jordan (1979) and is based on the idea that the extraction of melt by basaltic volcanism leaves behind a buoyant, low-density mantle residue (see also Robinson, 1988). Of the models described here, this is the only one that predicts high seismic velocities in the lithosphere. At this point, it is unclear whether a fourth model of *propagating cracks* in the lithosphere could produce enough buoyant material for a swell (J. Winterer, personal commun., 2007). A cracking lithosphere would most likely have the seismic signature of some degree of rejuvenation, but the asthenosphere below should be normal (Fig. 1D).

The Hawaiian Hotspot and Seismic Tomography

Seismology provides useful tools to identify and image the seismic imprint of a mantle plume or other source for hotspot volcanism. Assuming thermal derivatives, $\partial v/\partial T$, near $1 \times 10^{-4} K^{-1}$ (Karato, 1993), thermal plumes with excess temperatures of a few 100 K give rise to changes of upper mantle seismic velocities by a few percent, which should be resolvable by modern seismic tomography. Nevertheless, progress has been slow, especially in the imaging of a Hawaiian plume. Global body wave tomographic models often display a low-velocity anomaly near Hawaii in the upper mantle (e.g., Grand et al., 1997), and a recent study cataloged the seismic signature of plumes (Montelli et al., 2006) to reassess heat and mass fluxes through plumes (Nolet et al., 2006). However, such models typically have poor depth resolution in the upper few 100 km unless the data set contains shallow-turning phases or surface waves (which both cited studies do not have). Further complicating imaging capabilities with global data is the fact that the width of the plume conduit is expected to be on the order of only a few 100 km. Such a small structure is near the limits of data coverage, the model parameterization, and the wavelength of the probing seismic waves, and proper imaging may require the use of a finite-frequency approach (Montelli et al., 2006). Surface waves should be capable of sensing a shallow wide plume head, but global dispersion maps at 60 s, with signal wavelengths of 250 km, largely disagree on even the approximate location of a possible low-velocity anomaly near Hawaii (e.g., Laske and Masters, 1996; Trampert and Woodhouse, 1996; Ekström et al., 1997; Ritzwoller et al., 2004; Maggi et al., 2006). The reason for this is that the lateral resolution of structure around Hawaii is rather poor, due to the lack of permanent broadband seismic stations.

Regional body wave tomography using temporary deployments of broadband arrays has come a long way in imaging plumelike features on land (e.g., Wolfe et al., 1997; Keyser et al.,

2002; Schutt and Humphreys, 2004), but similar studies at Hawaii are extremely limited because of the nearly linear alignment of the islands (e.g., Wolfe et al., 2002). Such studies usually also do not have the resolution within the lithosphere and shallow asthenosphere to distinguish between the four models proposed for the swell uplift, but surface waves studies do. The reheating and the propagating crack models cause low seismic velocities in the lower lithosphere, whereas normal velocities would be found for the dynamical support model (Fig. 1). The compositional buoyancy model predicts high velocities, which are claimed to have been found by Katzman et al. (1998) near the end of a corridor between Fiji/Tonga and Hawaii. We would be able to distinguish between the reheating and the propagating crack models, as the latter leaves seismic velocities in the asthenosphere unchanged, whereas a plume would lower the velocities in the reheating model. Surface wave studies along the Hawaiian Islands have found no evidence for lithospheric thinning (Woods et al., 1991; Woods and Okal, 1996; Priestley and Tilmann, 1999), though shear velocities in the lithosphere appear to be at least 2.5% lower between Oahu and Hawaii than downstream between Oahu and Midway. These studies used the two-station dispersion measurement technique between only one station pair. It has been argued that the resulting dispersion curves in this case may be biased high, because laterally trapped waves along the swell may not have been accounted for properly (Maupin, 1992). What is obviously needed are constraints from crossing ray paths that can only be obtained from broadband observations on ocean bottom instruments deployed around the Hawaiian Swell.

Prior to the Mantle Electromagnetic and Tomography (MELT) experiment (Forsyth et al., 1998) across the relatively shallow East Pacific Rise, extensive long-term deployments were not possible because of the prohibitively high power demand of broadband seismic equipment. In 1997, we received National Science Foundation funding to conduct a year-long proof-of-concept deployment for our proposed Seismic Wave Exploration in the Lower Lithosphere (SWELL) experiment near Hawaii (Fig. 2). Eight of our Low-Cost Hardware for Earth Applications and Physical Oceanography (L-CHEAPO) instruments (Willoughby et al., 1993) were placed in a hexagonal array across the southwestern margin of the Hawaiian Swell to record Rayleigh waves at periods beyond the microseism band (15 s and longer). Unlike in the MELT experiment that used a combination of three-component seismometers and pressure sensors, the sole sensor used in our deployment was a broadband Cox-Webb pressure variometer that is commonly known as a differential pressure gauge (DPG; Cox et al., 1984). The use of such sensors was met with some skepticism, and the interested reader is referred to GSA Data Repository, Appendix B.[1] The proximity to the Ocean Seismic Network (OSN) borehole seismometer test site at borehole 843B of the Ocean Drilling Pro-

---

[1]GSA Data Repository item 2007091, Appendixes A, B, and C, is available at www.geosociety.org/pubs/ft2007.htm, or on request from editing@geosociety.org, Documents Secretary, GSA, P.O. Box 9140, Boulder, CO 80301, USA.

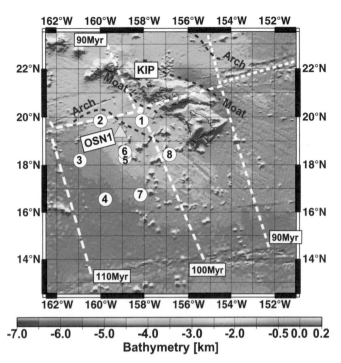

Figure 2. Site location map of the SWELL pilot experiment, which collected data continuously from April 1997 through May 1998. The array covered the southwestern margin of the Hawaiian Swell, which is characterized by its shallow bathymetry. Also marked are the ocean seismic network pilot borehole OSN1 (February through June 1998) and permanent broadband station KIP (Kipapa) of the global seismic network (GSN) and GEOSCOPE. Dashed lines mark the age of the ocean floor (Müller et al., 1997).

gram (ODP) south of Oahu allowed us to compare our data with observatory quality broadband seismometer data collected by much more expensive seafloor equipment (Vernon et al., 1998). To support or refute the dynamic support model for the Hawaiian Swell, structure has to be recovered reliably down to at least

130 km. It is therefore essential to measure dispersion successfully down to at least 20 mHz (see Figs. 20 and 21 in Appendix B for details). GSA Data Repository, Appendix A, describes the field program. It turns out that the collected data set is of an unprecedented bandwidth, quality, and richness in signal that has gone beyond our expectations to retrieve the average structure beneath the pilot array (Laske et al., 1999). In the following, we present data examples, dispersion curves along two-station legs, and a 3D-model across the margin of the Hawaiian Swell. The model is nonunique, and we discuss possible aspects that can influence the retrieval of a model. Finally, we discuss the consistency of our model with several other geophysical observables.

DATA EXAMPLES

Spectra to Assess Signal-to-Noise Characteristics

During the deployment from April 1997 through May 1998, we recorded eighty-four shallow teleseismic events at excellent signal-to-noise levels. The azimuthal data coverage is as good as any 1-year-long deployment can achieve (Laske et al., 1999). For many of these events, we are able to measure the dispersion at periods between 17 and 60 s, sometimes even beyond 70 s. Figure 3 shows an example of ambient noise and earthquake spectra. On the high-frequency end, the SWELL stations exhibit pronounced microseism peaks centered at ~0.2 Hz. Equally large is the noise at infragravity frequencies below 0.015 Hz (see also Webb, 1998), which limits our ability to measure dispersion at very long periods. Nevertheless, the earthquake signal stands out clearly above the noise floor at frequencies below 0.15 Hz. Signal can be observed down to at least 0.015 Hz (at site #3), which may not have been achieved on previous ocean bottom seismometer (OBS) deployments. Comparing the spectra with those at station KIP (Kipapa, on Oahu), it is quite clear that the earthquake generated observable signal at frequencies below

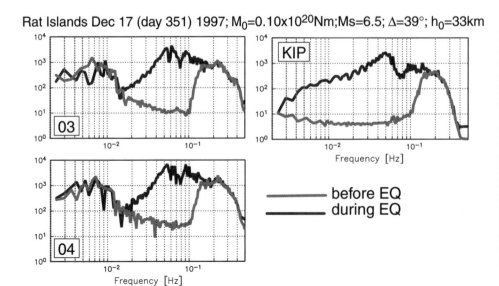

Rat Islands Dec 17 (day 351) 1997; $M_0=0.10\times10^{20}$Nm;Ms=6.5; Δ=39°; h_0=33km

Figure 3. Ambient noise and earthquake amplitude spectra for the Rat Island event shown in Laske et al. (1999), at sites #3 and #4. Also shown are the spectra for the very-broadband Wielandt-Streckeisen STS-1 vault seismometer at the permanent station KIP, the global seismic installation with possibly the lowest long-period vertical-component noise levels. Spectra are calculated using 28-min-long boxcar windows before and during the event. The instrument response is not removed to avoid possible numerical contamination near the rolloff ends of the responses. Δ—epicentral distance; EQ—earthquake; h_0—source depth; M_0—scalar seismic moment; M_s—surface wave magnitude.

0.01 Hz, but the noisy environment on the ocean floor did not allow us to observe this. It is somewhat curious but not well understood that the long-period noise floor at KIP is one of the lowest, if not the lowest, of all global seismic network (GSN) stations.

Figure 4 compares our spectra with others collected during the OSN1 pilot deployment. As for the Rat Island event, the spectra at KIP show that the event generated observable signal far below 0.01 Hz. The signal-to-noise ratio is not as good as that of the Rat Island event, which was closer to the stations and whose surface wave magnitude was larger. Nevertheless, we are able to observe signal on the SWELL instruments to frequencies below 0.02 Hz. Also shown are the spectra at the very-broadband Teledyne-Geotech KS54000 borehole seismometer at OSN1. The KS54000 is often used at GSN stations as an alternative to the STS-1. At this instrument, the noise floor grows above the signal level at ~0.006 Hz, and one could be misled to believe that this is infragravity noise. A broadband Güralp CMG-3T seismometer that was buried just below the seafloor

(Collins et al., 1991) appears to be much quieter. The KS54000 was deployed 242 m below the seafloor in a borehole that reached through 243 m of sediments and 70 m into the crystalline basement (Collins et al., 1991; Dziewonski et al., 1991). During a test-deployment of this sensor at our test facility at Piñon Flat Observatory, near Palm Springs, California, the seismometer had problems with long-period noise, and it was conjectured that water circulating in the borehole caused the noise (F. Vernon, personal commun., 2000). It is obviously possible to achieve an impressive signal-to-noise ratio with buried OBS equipment, but such deployment methods are probably prohibitively costly for large-scale experiments. A CMG-3T deployed on the seafloor exhibits high noise levels in the infragravity band and probably does not allow us to analyze long-period signal beyond what is achieved on the SWELL DPG. Note that the pressure signal from the earthquake is quite different from the ground motion signal, but the crossover of noise and earthquake signals occurs at similar frequencies, though the overall

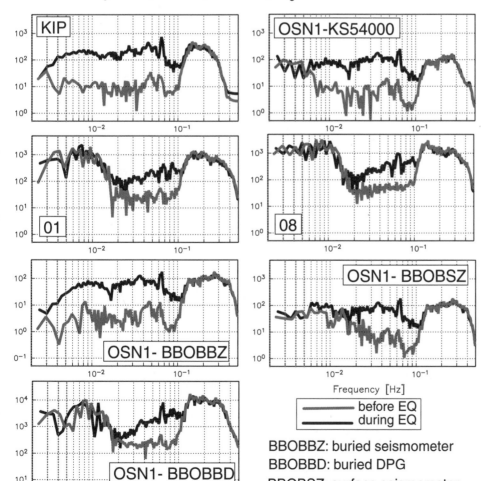

Off Southern Chile, Apr 01, 98; 22:43:00 UTC;
h_0=9km; Δ=97°; M_s=6.0; M_0=0.12x10^{20}Nm

Figure 4. Noise and signal amplitude spectra calculated for an earthquake off the coast of southern Chile, at sites #1 and #8. Also shown are spectra at land-station KIP, from the very-broadband bolehole sensor (KS54000) at OSN1, and from OSN1 broadband buried and surface instruments. BBOBS stands for broadband ocean bottom seismometer. For details, see Figure 3.

signal-to-noise ratio appears to be slightly better in ground motion. Also shown are the spectra of the buried DPG, which are virtually identical to the unburied ones. Burying a pressure sensor therefore does not appear to have any benefits. Regarding the seismic bandwidth, our data are favorably compatible with that of the MELT experiment (Forsyth et al., 1998).

Time Series to Assess Signal Coherence

Figure 5 shows the record sections for two earthquakes off the coast of Chile that were ~1000 km apart. Except for the record at site #5 for the April 1998 event of Figure 4, the SWELL records compare well with those at stations KIP and OSN1. We notice that some of the energy at periods shorter than 25 s appears to be diminished at stations KIP, #1, and #8, implying a local increase in attenuation or diffraction, though some of this reduction may also be explained by source radiation. Figure 6 shows examples for three events in Guatemala. Great waveform coherency is apparent, even for smaller events. The overall good

signal-to-noise conditions in our deployment allows us to analyze events with surface wave magnitudes down to $M_S = 5.5$.

We notice some noise contamination, e.g., at station #5 for the December 1997 Guatemala and April 1998 Chile events, and #3 for the March 1998 Guatemala event. The noise is extremely intermittent, typically lasting for a few hours, is confined to a narrow band at ~30 s (though this band varies with time), and has one or two higher harmonics. The noise does not compromise data collection severely, but some individual phase measurements have to be discarded, as we do not attempt to correct for the noise. This problem has not been noticed before, as we were the first group to use this equipment for observing long-period signals. After carefully analyzing the nature of the noise, we conclude that its origin is most likely not environmental but instrumental and is due to two beating clocks on the datalogger and the sensor driver boards.

Figure 6 suggests that subtle relative waveform delays are repeatable. The traces of stations #1, #2, and #8 are delayed, though the delay at #2 is small, and those of #4 and #7 are clearly

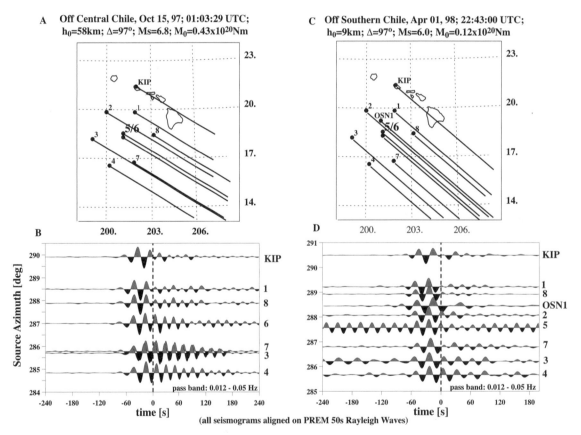

Figure 5. Record sections of two earthquakes off the coast of Chile. Records are shown for our SWELL sites as well as for the observatory-quality stations KIP and OSN1. The records are aligned relative to the preliminary reference Earth model (PREM) 50-s Rayleigh wave arrival times (Dziewonski and Anderson, 1981). They are band-pass filtered using a zero-phase shift five-step Butterworth filter in the frequency band indicated in the section. Records are not corrected for instrumental effects, i.e., phase shifts between KIP and differential pressure gauges (DPGs) may not be due to structure. Differences in the waveforms at sites #1, #8, and KIP are most likely due to structural variations near Hawaii. The record of OSN1 for the April 1998 event is shifted upward for better comparison. For notation of source parameters, see Figure 3.

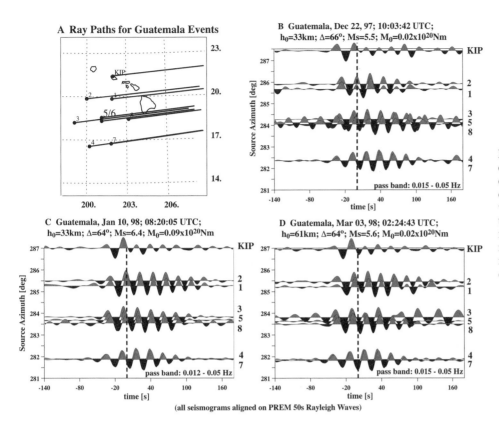

Figure 6. Record sections of three earthquakes in Guatemala. The epicentral distance was ~65° for all events. The December 1997 and the March 1998 events were more than four times smaller than the January 1998 event. Noise observed for these events is transient, nearly harmonic, and affects individual instruments only and not the whole array. For details, see Figure 5.

advanced. The delay between #1/#8 and #4/#7 amounts to 5.7 s. In principle, the delay can have been accumulated anywhere between Guatemala and the array, but if the slow structure was far from Hawaii, the record at #3 should also be delayed. A similar delay can be found for events from Venezuela, Colombia, and other events in the northern quadrant. We do not observe this delay for earthquakes whose rays do not cross the islands before arriving at the array (i.e., the events in Chile, Tonga, Fiji, and along the western Pacific Ocean). Taking into account the reduced amplitudes at #1 and #8 for the Chile events, we infer a strong anomaly near the islands, with a maximum extent possibly beyond sites #1 and #8, but likely diminished. Because #4 and #7 are not affected, the delay may obviously be associated with a thickened crust beneath the Hawaiian Ridge (see Fig. 22 in GSA Data Repository, Appendix C, referenced in footnote 1). The dominant period in the seismograms is ~22 s. At a phase velocity of ~4 km/s, the observed delay amounts to a phase velocity anomaly of at least 6.5%. A thickened crust can explain only ~2% of the anomaly but not much more. Rayleigh waves at these periods are sensitive to upper-mantle structure down to at least 60 km, and we gather first evidence that a low-velocity body in the mantle causes our observations.

PHASE MEASUREMENTS ACROSS THE PILOT ARRAY

Our phase velocity analysis involves three steps: (1) measure frequency-dependent phase, (2) determine phase velocity curves, and (3) invert phase velocity curves for structure at depth. For each event, we measure the frequency-dependent phase at one station with respect to those of all the others, using the transfer function technique of Laske and Masters (1996). A multitaper approach improves bias conditions in the presence of noise and provides statistical measurement errors. From the phase data, we then determine phase velocities. We seek to apply methods that do not require the knowledge of structure between earthquake sources and our array. For example, incoming wavefronts can be fit to all phases measured in a station subarray to determine average velocities within this array (e.g., Stange and Friederich, 1993; Laske et al., 1999). A multiparameter fit allows the wavefronts to have simple or complex shapes and oblique arrival angles (Alsina and Snieder, 1993). The latter accounts for the fact that lateral heterogeneity between source and the array refracts waves away from the source-receiver great circles. Fitting spherical instead of plane waves significantly improves the fit to our data and provides more consistent off-great-circle arrival angles, but more complicated wavefronts are not necessary for circum-Pacific events. Events occurring in the north Atlantic, Indian Ocean, or Eurasia exhibit highly complex waveforms that are sometimes not coherent across the array. Such events are associated with waves traveling across large continental areas and most likely require the fitting of complex wavefronts, a process that is highly nonunique (e.g., Friederich et al., 1994). We therefore discard such events. We are left with fifty-eight mainly circum-Pacific events for which stable phase velocity estimates are possible. We use the triangle technique in a later section to validate the 2-D phase velocity variations resulting from a comprehensive two-station approach.

The two-station approach lets us best assess lateral variations across the array without having to resort to modeling structure outside the pilot array. This approach requires earthquakes that share the same great circle as a chosen two-station leg. Because this condition is almost never achieved, we have to choose a maximum off-great-circle tolerance, which is done individually for each station leg. Station #2 was operating only during the second deployment, so the maximum allowed angle of 20° is relatively high. The tolerance for other legs can be as low as 8° and still provide as many as eight earthquakes. An off-great-circle approach of 20° effectively shortens the actual travel path by 6%. We correct for this contraction to avoid phase velocity estimates to be biased high. We also have to take into account off-great-circle propagation caused by lateral refraction. With the spherical wave fitting technique, we rarely find approaches away from the great-circle direction by more than 5°. The average is 2.6°, which accounts for a 0.1% bias. This is within our measurement uncertainties, and we therefore do not apply additional corrections. Events with larger arrival angles, such as the great March 25, 1998, Balleny Island event, are typically associated with complicated waveforms due either to the source process, relative position of the array with respect to the radiation pattern, or propagation effects. We therefore exclude such events (a total of eight) from the analysis.

LATERAL VARIATIONS ACROSS THE SWELL PILOT ARRAY

Figure 7 shows path-averaged dispersion curves for two nearly parallel two-station legs. Both legs are roughly aligned with the Hawaiian Ridge, but while leg 1–8 is on the swell, leg 3–4 is in the deep ocean and is thought to traverse unaltered ca. 110-Ma-old lithosphere. The dispersion curve for leg 1–8 is based on data from eight events (Aleutian Islands, Kamchatka, Kuril Islands, and Chile), whereas that for 3–4 is based on six events. The two curves are significantly different, with the leg 1–8 curve being nearly aligned with the Nishimura and Forsyth (1989) (N&F) prediction for extremely young lithosphere, whereas the leg 3–4 curve is slightly above the N&F curve for lithosphere older than 110 Ma. Also shown is the dispersion curve obtained by Priestley and Tilmann (1999) (P&T) between the islands of Oahu and Hawaii along the Hawaiian Ridge. Their curve is slightly lower than our 1–8 curve and lies just outside our measurement errors. The fact that the P&T curve is lower than the 1–8 curve is expected, because the largest mantle anomalies associated with plume-lithosphere interaction should be found along the Hawaiian Ridge. With ~5% at 40 s, the difference in dispersion between legs 3–4 and 1–8 is remarkable, considering that the associated structural changes occur over only 350 km, but it is not unrealistic. We are somewhat cautious to interpret isolated two-station dispersion curves, because lateral heterogeneity away from the two-station path

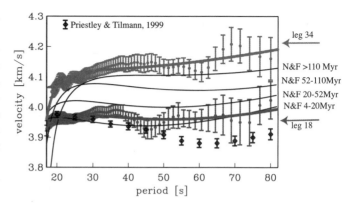

Figure 7. Path-averaged phase velocity along the two parallel station legs 1–8 and 3–4, together with the curves calculated for the best-fitting models obtained in our inversions (Figs. 10 and 11). The error bars reflect 1σ variations of several dispersion curves obtained for the same two-station leg. Also shown are the age-dependent phase velocities by Nishimura and Forsyth (1989) (N&F) and observed phase velocities by Priestley and Tilmann (1999) between the islands of Oahu and Hawaii.

and azimuthal anisotropy along the path have an impact on path-averaged two-station dispersion. The analysis of crossing paths in Figure 8 helps diminish this deficiency. Perhaps an indication that the bias cannot be severe is the fact that other parallel two-station legs that have entirely different azimuths exhibit similar heterogeneity (e.g., legs 2–1 and 4–7). Results from crossing two-station legs scatter somewhat but are marginally consistent. The most obvious and dominant feature is a pronounced velocity gradient from the deep ocean toward the islands. This gradient can be observed at all periods but is strongest at longer periods.

In principle, the observation of lower velocities near the islands would be consistent with changes in crustal structure, but a thickened oceanic crust could account for no more than 1.5%. There is no evidence that the crust changes dramatically across the array (see GSA Data Repository, Appendix C, referenced in footnote 1). A change in water depth across the array has some impact, but only at periods shorter than 30 s. The influence of water depth can be ruled out here, because the effect has the opposite sign, i.e., a decreasing water depth increases velocities. Because long periods are affected more than short periods, anomalies at depth must be distributed either throughout the lithosphere or a pronounced anomaly is located in the lower lithosphere or deeper. Rayleigh waves at 50 s are most sensitive to shear velocity near a depth of 80 km, but the anomaly could reach as deep as 150 km, or deeper (Fig. 20 in GSA Data Repository, Appendix B). A marked increase in measurement errors beyond ~67 s/15 mHz is associated with the fact that dispersion measurements become uncertain when the signal wavelength approaches the station spacing. We therefore expect a degradation of resolution at depths below 150 km.

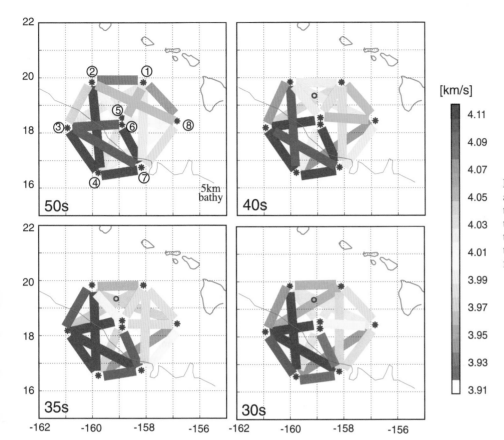

Figure 8. Path-averaged phase velocities across the SWELL pilot array, as functions of period. The most prominent feature is a strong velocity gradient across the SWELL margin, with lower velocities found near the islands. Modified from Laske et al. (1999).

INVERSION FOR STRUCTURE AT DEPTH

To retrieve structure at depth, we perform two-step inversions. First we determine path-averaged depth profiles along each two-station leg. All profiles are then combined in an inversion for 3-D structure.

Surface waves are sensitive to shear and compressional velocities, V_S, V_P, and density, ρ, but the most dominant and best-resolved parameter is V_S (see Fig. 20 in GSA Data Repository, Appendix B). To limit the number of model parameters for a well-conditioned inverse problem, tomographers often ignore sensitivity to V_P and ρ. Such a strategy could lead to biased models in which shallow V_P structure can be mapped into deeper V_S structure. We prefer to scale the kernels for V_P and ρ and include them in a single kernel for V_S, using the following scaling:

$$\tilde{A} \cdot \delta\alpha = (1/1.7)\tilde{B} \cdot \delta\beta$$
$$\tilde{R} \cdot \delta\rho = (1/2.5)\tilde{B} \cdot \delta\beta \qquad (1)$$

The scaling factors have been determined in both theoretical and experimental studies (e.g., Anderson et al., 1968; Anderson and Isaak, 1995) for high temperatures and low pressures such as we find in the upper mantle. They are applicable as long as strong compositional changes or large amounts of melt (i.e.,

>10%) do not play a significant role. We use a modified N&F model for 52- to 110-Ma-old lithosphere as starting model. It is parameterized in seventeen constant layers whose thickness is 7 km near the top but then increases with depth to account for the degrading resolution. Because the 90-s data are sensitive to structure beyond 200 km, our bottom layer is 50 km thick and ends at 245 km. Velocities retrieved at these depths are extremely uncertain and are excluded from later interpretation, but including such a layer in the inversion avoids artificial mapping of deep structure into shallower layers. The crust is adjusted using the model described in Table 1 and Figure 22 in GSA Data Repository, Appendix C. We also adjust for two-station path-averaged water depths.

We seek smooth variations to the starting model that fit our data to within an acceptable misfit, χ^2/N, where $\chi = x_d - x_t$, x_d is the datum, x_t the prediction, and N the number of data. Formally, we seek to minimize the weighted sum of data prediction error, χ^2, and model smoothness, $\delta\mathbf{m}$:

$$\chi^2 + \mu \,|\mathbf{m}^T\partial^T\partial\mathbf{m}|, \qquad (2)$$

where $\mathbf{m}$ is the model vector, and μ the smoothing or regularization parameter. Superscript T denotes transpose. The tradeoff between the two terms is shown in Figure 9. The shape of the tradeoff curve depends on the data errors as well as on the com-

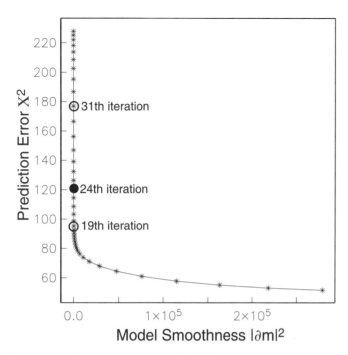

Figure 9. Tradeoff curve for station leg 1–8. Displayed are the data prediction error and model smoothness as functions of the regularization parameter, μ. The location of the final model (twenty-fourth iteration) is marked as well as the range of acceptable models that lie within the model error range of Figure 10. The chosen models have misfits, χ^2/N, between 1.0 and 1.9.

position of the data set, but the resulting optimal model is actually similar to the one shown here. In practice, models that are very close to the minimum of equation 2 are highly oscillatory, and we choose smoother models. Model errors can be obtained from the data errors through a formal singular-value decomposition or by Monte Carlo forward modeling. Here we show the range of acceptable models along the trade-off curve. The final model has a misfit, χ^2/N, of 1.3, and so is slightly inconsistent with the data.

The final model in Figure 10 is significantly slower than the N&F model for 52- to 110-Ma-old lithosphere below ~30 km. Our model follows that of the N&F model for 20- to 52-Ma-old lithosphere down to ~120 km, below which it remains somewhat slower. Although the velocities are relatively poorly constrained at depths below 170 km, the difference from the N&F model at shallower depths is significant and indicates that the cooling lithosphere has been altered at its base through secondary processes. Models derived from surface waves are nonunique. If we had chosen fewer layers, such as the two-layer parameterization of Priestley and Tilmann (1999), the resulting velocity above 80 km might be similar to their velocity, which is close to the velocity of the preliminary reference Earth model (PREM; Dziewonski and Anderson, 1981). Below 80 km, our model is significantly faster than the P&T model, which is in agreement with the fact that our dispersion curve is systematically faster

than theirs. Inversions can get caught in a local minimum, and the model presented here may not be the actual solution to minimizing equation 2. In Figure 10B, we show the final model for a different starting model that is rather unrealistic but helps illuminate how the final model depends on the starting model. This model (model B) is virtually identical to our preferred model (model A) down to 70 km but then oscillates more significantly around the N&F model for 20–52 Ma. Higher velocities are found down to ~150 km, whereas much lower velocities are found below that, though they remain above the Priestley and Tilmann (1999) velocities. The misfit of this model is slightly less than that of model A ($\chi^2 = 1.19$), but we nevertheless discard it as an improbable solution. In a hypothesis test, we remove one deep layer after another and test the misfit. We would expect that the misfit does not decrease dramatically initially, because of the decreased sensitivity at great depth. This is the case for model A, for which the misfit increases by 1.6% when omitting the bottom layer. For model B this increase is 40%. This large increase means that the bottom slow layer is required to counteract the effects of high, shallower velocities to fit the data. Including structure of only the upper thirteen layers (down to 125 km) of model A gives a misfit of 1.7, whereas that of model B gives 12.9 and is clearly inconsistent with our data.

Figure 11 shows the model obtained along the two-station leg 3–4. Shear velocities are significantly higher than along station leg 1–8, by ~4.5% in the lithosphere and 6% in the asthenosphere at 150 km depth. Below ~70 km, velocities roughly follow those of PREM, where the velocity increase at ~200 km is uncertain in our model. At nearly 4.8 km/s, the velocities found in the upper lithosphere are unusually high but are required to fit the dispersion curve in Figure 7. They are not unphysical and have been observed beneath the Canadian Shield (Grand and Helmberger, 1984) and in laboratory experiments (Jordan, 1979; Liebermann, 2000). The azimuth of the station leg is roughly aligned with past and present-day plate motion directions between 60 and 95°. Strong azimuthal anisotropy has been found in the eastern Pacific Ocean (e.g., Montagner and Tanimoto, 1990; Larson et al., 1998; Laske and Masters, 1998; Ekström, 2000), and we find evidence that azimuthal anisotropy is ~3% in the southwestern part of our array, away from the Hawaiian Swell. The velocities shown here may therefore be those associated with the fast direction of azimuthal anisotropy, though this interpretation would also include velocities in the asthenosphere, where mantle flow is assumed to align anisotropic olivine.

The combined interpretation of all dispersion data shown in Figure 8 provides the final 3-D model for isotropic velocity variations (Figs. 12 and 13). Although small-scale variations are most likely imaging artifacts caused by sparse path coverage, the most striking feature is a strong velocity gradient across the swell margin, starting at a depth of ~60 km, while the upper lithosphere is nearly uniform. The gradient amounts to ~1% across the array at a depth of 60 km but increases with depth to nearly 8% at 140 km. Along a profile across the swell margin,

A: Realistic Starting Model

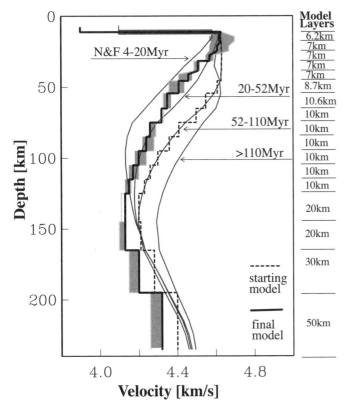

B: Unrealistic Starting Model

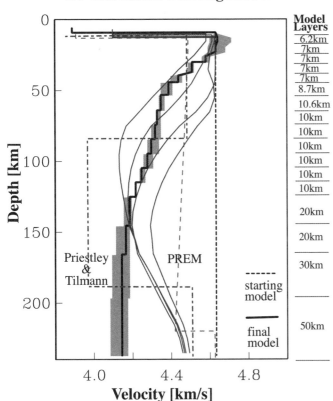

Figure 10. Shear velocity models for the two-station leg 1–8. (A) Model obtained using the modified Nishimura and Forsyth (1989) 52- to 110-m.y. starting model. The predictions for this model are shown in Figure 7. The shaded area marks the range of models along the tradeoff curve that still fit the data to a given misfit (see Fig. 9). (B) Model obtained using a constant velocity as starting model. In the upper ~75 km, the final model is very similar to the model in panel A but is faster down to ~150 km and then is significantly slower. Also shown are the preliminary reference Earth model (PREM), the age-dependent models by Nishimura and Forsyth (1989), and the model by Priestley and Tilmann (1999) between the islands of Oahu and Hawaii.

←

we find clear evidence that the on-swell lower lithosphere has either been eroded from 90 to 60 km or has lower seismic velocities, which is consistent with its rejuvenation by lithosphere-plume interaction. Our results appear to be in conflict with those of Priestley and Tilmann (1999), who find no evidence for lithospheric thinning along the Hawaiian Ridge. On the other hand, their model includes only two layers in the depth range shown here, the upper one being 75 km thick and representing the entire lithosphere. The velocity in their upper layer is 4.48 km/s, which is lower than what we find in the upper 40 km but larger below that. Whether our model is consistent with an eroded lithosphere is addressed in a later section, but we clearly find some type of rejuvenation.

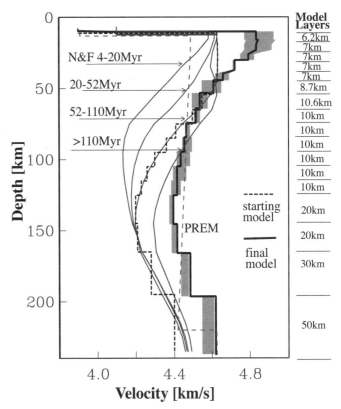

Figure 11. Shear velocity model for the two-station leg 3–4. For details, see Figure 10. PREM—preliminary reference Earth model.

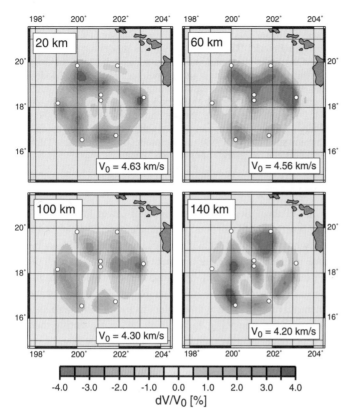

Figure 12. Final 3-D model of shear velocity variation across the SWELL pilot array from the inversion of all two-station dispersion curves. Variations are shown at four depths and are given in percentages with respect to the velocities of the Nishimura and Forsyth (1989) model for 52- to 110-Ma-old lithosphere (given in the right bottom corner of each plot as V_0).

The base of the lithosphere is not defined in our modeling, which does not explicitly include discontinuity kernels. But our suggestion of a doming lithosphere-asthenosphere boundary (LAB) is consistent with the results from a recent receiver function study that reaches into our array (Li et al., 2004). Their earlier study (Li et al., 2000), which samples the mantle beneath the island of Hawaii, places the LAB ~120 km deep. Li et al. (2004) argue that the lithosphere thins away from the island of Hawaii and is only 50 km thick beneath Kauai, lending support for the hybrid dynamic support–lithosphere erosion model. Beneath a rejuvenated lithosphere, we find a pronounced on-swell anomaly centered at a depth of 140 km in the asthenosphere. The anomaly could reach deeper than 200 km, where our data lose resolution. This slow anomaly is consistent with the asthenosphere identified by Priestley and Tilmann (1999), though they give a somewhat lower velocity of 4.03 km/s. The anomaly found in the low-velocity body is ~4.5% slower than the off-swell, probably unaltered asthenosphere (our off-swell velocities are consistent with the velocities of PREM). Though not well resolved, our image suggests that we sense the bottom of the asthenosphere in the southwestern half of our array. Priestley and Tilmann (1999) placed the bottom of the asthenosphere at ~190 km beneath the Hawaiian Islands, though this is somewhat uncertain.

VALIDATION OF THE MODEL WITH OTHER APPROACHES

The two-station approach is appealing for several reasons. It readily provides path-averaged dispersion estimates along two-station legs without having to know details in earthquake

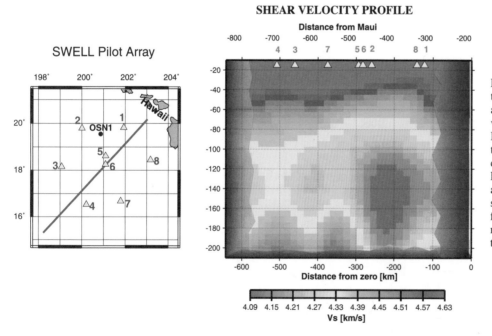

Figure 13. Shear velocity profile across the 3-D model of Figure 12. Velocities along the profile represent averages over velocities within 50 km of the profile. Imaging capabilities are reduced toward the end of the profile because of lack of data (e.g., the apparent thickening of the lithosphere east of sites #1 and #8). Variations in the lithosphere and asthenosphere are clearly imaged. "Distance from zero" refers to the distance from the northeastern end of the line marked in the map.

source mechanisms. Having crossing paths available, it may provide detailed insight into lateral structural variations. Problems arise, however, in cases where unmodeled effects become significant. These include off-great-circle approach caused by lateral refraction between earthquakes and the array. We can validate our model by testing it against results when using the tripartite approach in which we fit incoming spherical waves to the phase within station triangles. This is a low-resolution approach laterally, but the advantage is that off-great-circle propagation is included in the modeling and so may not bias the resulting model. The velocity maps in Figure 14A are significantly smoothed versions of the ones from the two-station method in Figure 8, but the basic features of velocity variations are consistent: there is a significant gradient across the swell margin, and the gradient appears most pronounced at long periods. The fact that the velocity difference at 50 s between triangles 3–4–6 and 1–8–6 is only 1.5% indicates that the extreme velocity differences must be confined to the edges of our array and likely extend beyond. The maps in Figure 14B indicate that errors are largest at long periods, but the errors are small compared to observed variations. Because station #2 was operating only during the second deployment but all three stations have to provide a clean seismogram for a given earthquake, the number of earthquakes for triangles involving station #2 is reduced.

In the presence of azimuthal anisotropy, the velocities shown in Figure 14 represent true average isotropic velocities only in cases of good data coverage. We therefore check our results against inversions when azimuthal anisotropy is included in the modeling. The azimuthally varying phase velocity is parameterized as a truncated trigonometric power series:

$$c(\Psi) = c_i + a_1 \cos(2\Psi) + a_2 \sin(2\Psi) \\ + a_3 \cos(4\Psi) + a_4 \sin(4\Psi), \qquad (3)$$

where Ψ is the azimuth, the a_i are known local linear functionals of the elastic parameters of the medium (Smith and Dahlen, 1973; Montagner and Nataf, 1986), and c_i is the azimuth-independent average (or isotropic) phase velocity.

Solving equation 3 is straightforward, and in cases of adequate data coverage, the results for c_i should be consistent with those of Figure 14. Figure 15 shows that this is indeed the case for most of the periods considered, except at long periods, for which the number of reliable data decreases. When solving equation 3, we search for five times as many unknowns as in the isotropic case. In cases of sparse data coverage, an inversion can yield anisotropic models that fit the data extremely well but are unnecessarily complicated or physically unrealistic. Most realistic petrological models have one dominant symmetry axis that may be oriented arbitrarily in 3-D space. For all such models, the contribution of the 4Ψ terms is relatively small for Rayleigh waves. We see from Figure 15 that ignoring the 4Ψ terms yields consistent results for c_i as well as the strength of anisotropy. The only time when results from anisotropic modeling including or excluding the 4Ψ terms diverge is at long periods beyond 65 s, where results are also different regardless of whether anisotropy is considered at all. In these cases of sparse data, ignoring strong azimuthal anisotropy yields biased values for c_i. On the other hand, with few data available, the fits become uncertain, yielding phase velocity distributions that strongly oscillate with azimuth, which is especially so for the 4Ψ fits. Such strong variations have to be discarded as numerically unstable as well as unphysical. Overall, the test here demonstrates that we obtain reasonably unbiased velocities when we ignore anisotropy.

The general good agreement of results regardless of whether azimuthal anisotropy is included in the modeling gives us confidence that the frequency-dependent phase velocities in this study and their implications for structure at depth are very well constrained. The modeling of the azimuth dependence of phase velocity in terms of 3-D anisotropic structure is beyond the scope of this article, but preliminary modeling suggests that mantle flow in the asthenosphere follows the plate motion direction off the swell but is disturbed on the swell (manuscript in preparation).

Both the two-station and the triangle approaches use only subsets of data. Because of the presence of noise or transient problems with individual stations, our database rarely contains earthquakes for which we can measure phase at all eight sta-

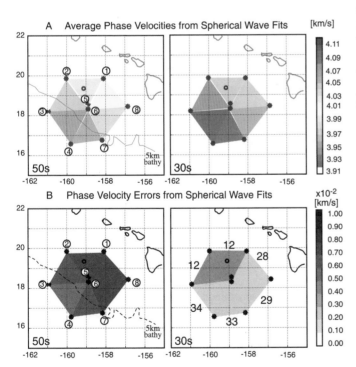

Figure 14. (A) Lateral phase velocity variations obtained with the station triangle method, at two periods. The maps are obviously smoothed versions of those in Figure 8, but the velocity gradient across the swell margin is still observed. (B) Error maps. The errors are largest at long periods but remain below 0.007 km/s. The velocity gradient across the swell margin is therefore significant. The number of earthquakes used for each station triangle is given in the map for 30 s.

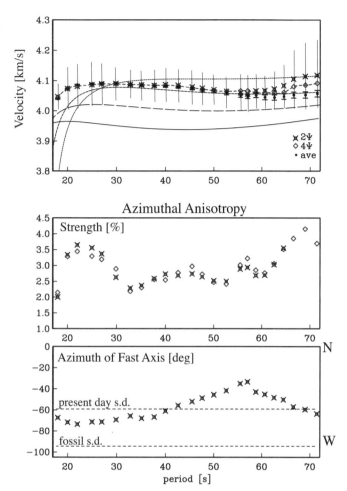

Figure 15. Average phase velocities for station triangle 3–4–6. Shown are the results for the isotropic station triangle fit for Figure 14 as well as the c_i terms (in equation 3) when fitting order 2 and 2/4 azimuthal anisotropy. Vertical bars mark the minimum and maximum variations of phase velocities in the 2Ψ fit. The Nishimura and Forsyth (1989) dispersion curves are shown for reference. Also shown are the strength of anisotropy obtained for the order 2 and 2/4 fits as well as the direction of fast phase velocity for the order 2 fit. Results agree overall, except at long periods, where the number of constraining data decreases. s.d.— spreading direction.

tions. Both methods also strictly provide images within the array but give no information on structure outside of it, though we have already discussed evidence that anomalies reach to the outside of our array. In a last consistency test, we embed our entire data set of nearly two thousand phase measurements in our global database (Bassin et al., 2000). The global data set includes nearly 20,000 high-quality hand-picked minor and major arc and great circle data and arrival angle data that enhance small-scale resolution (Laske and Masters, 1996). In a global inversion, contributions to our SWELL data from lateral heterogeneity between seismic sources and the array are implicitly included in the modeling. The highest frequency in our global data set is currently 17 mHz, which is near the long-period limit

of the SWELL data set. We choose 16 mHz (62.5 s) for our test. All phase and arrival angle data are used in an inversion for a global phase velocity map that is parameterized in half-degree equal-area cells. We use nearest neighbor smoothing in a least-squares iterative QR scheme (e.g., van der Sluis and van de Vorst, 1987). The resulting maps in Figure 16 clearly show that the SWELL data help image a low-velocity region that is not resolved by the current global network of permanent seismic stations. With station KIP being, until recently, the only site in the area that has delivered high-quality data, not enough crossing rays are available to resolve structure at wavelengths much below 1000 km. The imaged velocity contrast between the deep ocean and the swell reaches 8%, which is consistent with what we found with the two-station method. Being able to image structure outside the array, we also notice that the low-velocity anomaly extends well to the northeast of our array, most likely beyond the Hawaiian Islands. This observation is roughly consistent with Wolfe et al. (2002), who find a pronounced low-velocity anomaly extending from OSN1 to the Hawaiian Islands and from Oahu south to the northern end of the island of Hawaii. We are therefore confident that the results in our two-station approach are robust features and trace a profoundly altered lithosphere and asthenosphere beneath the Hawaiian Swell. A possible asymmetry of the low-velocity anomaly, which is more pronounced to the southwest of the island of Hawaii than to the northeast, is intriguing but is consistent with a similar asymmetry in bathymetry.

DISCUSSION

Resolution Limits and Significance of Results

The skeptical reader may wonder whether our data are precise enough to constrain the deep structure reliably. Our measurement errors increase at periods longer than 50 s. The sensitivity kernel for 50-s Rayleigh waves to shear velocity at depth peaks at ~80 km (Fig. 20 in GSA Data Repository, Appendix B, referenced in footnote 1). However, this does not imply that our data cannot resolve deeper structure. Rather, the combination of all kernels at periods 50 s and shorter provides sensitivity beyond 100 km (see the Backus-Gilbert test in Fig. 20 in GSA Data Repository, Appendix B).

The rejuvenation of the lithosphere in Figure 13 is therefore extremely well constrained by our data, because high-precision data are required only at periods shorter than 30 s. Resolution below 120 km deteriorates somewhat, for three reasons: (1) the sensitivity kernels spread out over greater depths for longer periods, so that deep structure is smeared out over a depth range greater than a few tens of km; (2) at periods longer than 50 s, the station spacing of 220 km is about a signal wavelength, and measurement accuracy deteriorates; and (3) at periods significantly beyond 60 s, ambient noise conditions for some earthquakes increases measurement uncertainties. We should stress, however, that our errors are most likely conservative compared

Global Dataset

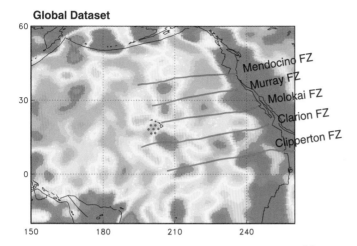

Global Dataset + SWELL

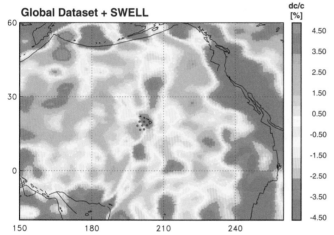

Figure 16. North Pacific section of the global phase velocity map at 16 mHz obtained when inverting the global data set only (top) and when including the SWELL data (bottom). Because of inadequate station distribution, the global data set lacks resolution near Hawaii. The SWELL data dramatically improve resolution and help image a low-velocity region that extends from the SWELL array east beyond the islands. c—phase velocity; FZ—fracture zone.

to those of other studies (see Fig. 7). We do not apply any smoothing or other conditioning along the dispersion curves, but our errors are still less than 2%, which is a third of the anomaly found in the asthenosphere. The question arises whether this strong, possibly plume-related anomaly is required to fit our data. We had discussed in Figures 10 and 11 that the leg 1–8 dispersion curve would be marginally consistent with 1.7% higher velocities around 100 km, with an associated 0.5% velocity reduction at a depth of 50 km. However, such a model would require unrealistically low velocities below 150 km. We have not found a model for leg 1–8 that exhibits velocities at 120 km as high as those along leg 3–4 and infer that this anomaly is indeed real. On the other hand, asthenosphere velocities along leg 3–4 are PREM-like, i.e., near normal, and velocities cannot be lowered significantly. Flow-induced anisotropy along leg 3–4

could account for some of the high off-swell velocity. This would lower the isotropic velocity contrast across the swell margin, as there is no evidence that this flow extends to leg 1–8. The difference in anisotropy would lend support to a swell-scale mantle dynamical process.

Comparison with SWELL Magnetotelluric Data

During the first 7.5 months of the deployment, Constable and Heinson (2004) collected seafloor magnetotelluric (MT) data with a seven-station array that roughly overlapped with ours. The major features in their model include a resistive lithosphere underlain by a conductive lower mantle, and a narrow, conductive "plume" connecting the surface of the islands to the lower mantle. They argue that their data require this plume, which is located just to the northwest of our array but outside of it. It has a radius of less than 100 km and contains 5–10% of melt. Unfortunately, our model does not cover this area. Constable and Heinson did not find any evidence for a lowering of shallow (60 km) resistivity across the swell and therefore argue against lithosphere reheating and thinning as proposed by Detrick and Crough (1978). In fact, resistivity appears to slightly increase in the upper 50 km. Because of the high resistivities found in the lithosphere (100–1000 Ωm), they place an upper bound of 1% melt at a depth of 60 km, where our lithosphere is thinnest, and argue for a hot dry lithosphere (1450–1500 °C) compared to a cooler (1300 °C) off-swell lithosphere. They estimate that a melt fraction of 3–4% could explain a 5% reduction in seismic velocities (Sato et al., 1989), but it would also reduce the resistivity to 10 Ωm, which is not observed. Using temperature derivatives given by Sato et al. (1989), Constable and Heinson estimate that an increase of mantle temperature from 0.9 to 1.0 of the melting temperature (150–200 K in our case) can also cause a 5% velocity increase in our model but would not cause electrical resistivity to drop to 10 Ωm. The authors therefore propose a thermally rejuvenated but not eroded lithosphere that would be consistent with both seismic and MT observations. On the other hand, the estimates of Sato et al. (1989) were obtained in high-frequency laboratory experiments, and Karato (1993) argues that taking into account anelastic effects can increase the temperature derivatives for seismic velocities by a factor of two. In this case, much smaller temperature variations are required to fit the seismic model. Constable and Heinson (2004) do not attempt to reconcile the seismic and MT model below 150 km, but it is worth mentioning that their model exhibits a gradient to lower resistivity near the low-velocity body in the asthenosphere. Anelastic effects become most relevant at greater depths, below 120 km, when attenuation increases in the asthenosphere. As dramatic as our seismic model appears, it is nevertheless physically plausible. Modeling attempts that include thermal, melt, and compositional effects reveal that no melt is required to explain our model below 120 km, and depletion through melt extraction could explain the lower velocities above it (S. Sobolev, personal commun., 2004).

Comparison with Bathymetry and Geoid

Both model parameterization and regularization used in the inversion influence the resulting velocity model, especially the amplitude of velocity anomalies. We can test the physical consistency of our model with other geophysical observables, such as the bathymetry in the region. Our test is based on the assumption that the regional lithosphere and asthenosphere are isostatically compensated, i.e., there is neither uplift nor subsidence. We also assume that the causes for our observed velocity anomalies are predominantly of thermal origin, in which case we can apply the velocity-density scaling of equation 1 to convert δV_S to density variations. We assume Pratt isostacy and search for the optimum depth of compensation that is most consistent with observed lateral variations in bathymetry along the profile in Figure 13. We find that a compensation depth of ~130 km is most consistent with the observed bathymetry (Fig. 17). Taking into account deeper structure grossly overpredicts variations in bathymetry, whereas shallower compensation depths are unable to trace slopes in bathymetry. With a compensation depth of 130 km, the low-velocity anomaly in the asthenosphere would then give rise to uplift unless it is compensated by dense material farther down. Katzman et al. (1998) argued that Hawaii is underlain by dense residue material that may be capable of sinking. On the other hand, the exact V_S-to-ρ scaling is relatively poorly known. Karato (1993) argues that anelastic and anharmonic effects significantly alter the temperature derivatives for velocity. In regions of high attenuation (low-Q), such as the asthenosphere, the correction due to anelasticity roughly doubles. In this case, temperature anomalies as well as density anomalies have to be corrected downward, for a given shear velocity anomaly, or dln V_S/dln ρ needs to be increased. In principle, we would need to reiterate our inversions using different scaling factors,

but here we only discuss the effects. Karato (1993) indicates that when taking anelastic and anharmonic effects into account, dln V_S/dln ρ decreases from roughly 4.4 at 100 km to 4.0 at 200 km. If we then assume an average scaling of 4.0 over the whole depth range of our model, the predicted compensation depth deepens to 170 km, because shear velocity variations now have a reduced effect on bathymetry. This would include the anomaly in the asthenosphere without requiring compensating material at greater depth. We find no justifiable strategy to raise the compensation depth to 90 km or above that would be consistent with lithospheric thinning as proposed by Detrick and Crough (1978). Rather, the results here are roughly in agreement with the dynamic support model of Watts (1976) that places the compensation depth at 120 km.

We also test our model against the geoid. For Pratt compensation, the geoid anomaly, ΔN, is:

$$\Delta N = \frac{-2\pi G}{g} \left\{ \int_0^h (\rho_w - \rho_0)zdz + \int_h^W (\rho(z) - \rho_0)zdz \right\}, \quad (4)$$

where G is the gravitational constant, g acceleration of gravity, ρ_0 a reference density, ρ_w is the density of water, h the water depth, and W the compensation depth. Equation 4 only holds if the area is isostatically compensated. We are somewhat cautious about this test, because deeper structure in our model now has a graver impact than shallow structure, but at the same time, model errors are also greater. Figure 18 shows the observed geoid anomalies from model OSU91A1F (Rapp et al., 1991) and the anomalies predicted from our velocity model. The exact base level caused by our model is somewhat uncertain, because our data do not constrain structure of extremely long wavelength (e.g., harmonic degrees $l = 3$). As can be seen, taking into account structure above 110 km is most consistent with the geoid, east of the –400 km mark. A compensation depth of 120 km therefore appears to be roughly in agreement with both bathymetry and geoid, which validates the approach assumed here. To the west of the –400 km, our model grossly overpredicts the geoid, and we have no immediate explanation for this discrepancy. Changing the velocity-density scaling relationship has only little impact overall and no impact at all on the optimal compensation depth. Our model implies an excess mass above 110 km, because lower compensation depths cause no changes. Velocity anomalies at great depth are somewhat uncertain, but it is hard to find a compelling reason to conclude that velocities at shallower depths are wrong. Even if we assume that the model resulting from our two-station dispersion is biased toward fast velocities off the swell, the model from the tripartite method still implies the same overall inconsistency (low above the swell, high off the swell). As mentioned above, Katzman et al. (1998) find high velocities near Hawaii that correlate with a bathymetric and geoid high to the east of our profile mark of –200 km. To the west of the –300 km mark, they find a strong negative anomaly in the mid-upper mantle that our technique is unable to im-

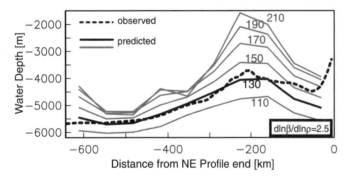

Figure 17. Observed bathymetry along the profile marked in Figure 13. Also shown is the predicted bathymetry derived from the shear velocity model. We assume that the lithosphere is isostatically compensated above the compensation depth given by the labels on each curve. Assuming a deep compensation depth, we overpredict the bathymetry, whereas a depth of ~130 km matches it quite well. A shallower compensation depth is also inconsistent with the bathymetry. β—shear velocity; ρ—density.

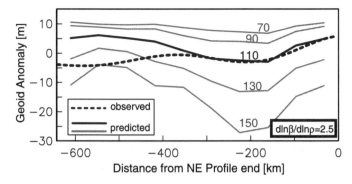

Figure 18. Observed geoid anomalies from model OSU91A1F (Rapp et al., 1991). Only harmonic degrees *l* = 3 and above are considered. Also shown are geoid anomalies predicted from our model. The compensation depth for each curve is given by the label. Pratt isostatic compensation is calculated with respect to the preliminary reference Earth model (PREM). A baseline of 7 m was added to the predictions to best match the geoid undulations between –400 and 0 km along the profile, as our data are insensitive to very-long-wavelength structure. β—shear velocity; ρ—density.

age because of its depth. Such an anomaly would most likely compensate our shallow "excess mass."

SUMMARY

During the 1997/1998 SWELL pilot experiment, we recorded Rayleigh waves on differential pressure sensors on the seafloor at an unprecedented signal level that allows us to image the lithosphere and asthenosphere beneath the Hawaiian Swell to depths below 150 km. The relatively inexpensive equipment is reliable in one-year deployments without significant maintenance.

We find pronounced lateral variations across the margin of the swell. In the deep ocean, velocities in the asthenosphere closely follow those of reference Earth model PREM, and are significantly higher than what is found along the island chain (Priestley and Tilmann, 1999). Velocities in the lid are higher than in PREM and also higher than in the Nishimura and Forsyth (1989) model for mature, 100-Ma-old lithosphere. Velocity variations along a profile across the swell margin suggest that the lithosphere on the swell has undergone a rejuvenation process.

Comparison of the velocities with those found in laboratory experiments and the results of a concurrent magnetotelluric study suggest that the anomalies are caused by thermal effects, and that the amount of melt cannot exceed 1% in the altered lithosphere at a depth of 60 km. Our model is consistent with thermal rejuvenation and is in some disagreement with Priestley and Tilmann (1999), who find no significant rejuvenation beneath the Hawaiian Islands. The seismic images bear the signature of a thermally rejuvenating lithosphere, but our model is inconsistent with significant amounts of melt beneath the onswell lithosphere, speaking against a mechanically eroded lithosphere that is proposed in the lithosphere thinning model (Fig.

1), unless the thinning is restricted to within 100 km of the islands. The comparison with local bathymetry and the geoid shows that our model is inconsistent with a shallow compensation depth as implied by this model, at least in the area covered by our array. We find a deeper compensation depth as suggested by the dynamic support model, but the latter does not account for the velocity variations we find in the lithosphere. If the area around Hawaii is isostatically compensated, we propose a hybrid thermal rejuvenation–dynamic thinning model in which the lithosphere near a possible plume head may be mechanically unaltered but thermally rejuvenated (Fig. 19). This model could also explain seismic evidence found by Li et al. (2004) for thinning downstream, in an area that is not covered by our data.

Our data are inconsistent with the other models proposed for the Hawaiian Swell uplift and volcanism. The data lend no support for the compositional buoyancy model, which requires high seismic velocities, unless plume-lithosphere interaction involves a very large area that extends well beyond the Hawaiian Swell. Off the swell, we find evidence for seismically fast material that is in conflict with the geoid, for compensation depths of 120 km or shallower. Katzman et al. (1998) find deeper low-velocity anomalies in the upper mantle, and it has been suggested that these are the signature of secondary shallow mantle convection. Lastly, our data are also inconsistent with a cracked lithosphere as the source of the Hawaiian volcanism, as this model has no suggestion for the low-velocity anomaly found in the asthenosphere. The SWELL pilot study covered only a small area of the Hawaiian Swell and cannot address some of the fundamental questions related to the possibly plume-related Hawaiian volcanism.

Rayleigh waves are extremely useful tools to investigate the shallow (less than 200 km) lithosphere-asthenosphere system, which remains elusive to standard teleseismic body wave tomography. However, fundamental-mode Rayleigh waves in the

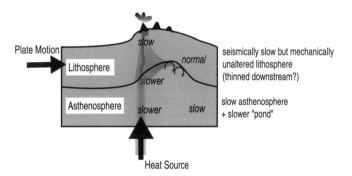

Figure 19. Concept figure for a possible mechanism for the Hawaiian swells that is most consistent with our data. The model is a hybrid thermal rejuvenation–dynamic support/thinning model. The lithosphere is rejuvenated by reheating but not mechanically eroded. The associated compensation depth would be 120 km. Mechanical thinning of the lithosphere may occur downstream, as proposed by Li et al. (2004), in an area that is not covered by our data.

period range shown here do not constrain structure in the transition zone. Unlike the analysis of receiver functions, our surface wave analysis cannot support or disprove the lower-mantle origin of a proposed mantle plume. SWELL is now part of the Plume-Lithosphere-Undersea-Mantle experiment (PLUME; Laske et al., 2006). This experiment involves the occupation of ten land and seventy ocean bottom sites that are well distributed over a 1000-km-wide area around Hawaii. The combination of all techniques mentioned above will give us the unprecedented opportunity to collect excellent seismic constraints that will help us resolve one of the most tantalizing questions in plate tectonics: Is the Hawaiian hotspot volcanism fed by a deep-seated mantle plume or not?

ACKNOWLEDGMENTS

We thank the L-CHEAPO team at IGPP (Cris Hollinshead, Paul Zimmer, and Dave Willoughby) for making this experiment possible. The team deployed and recovered the instruments on the University of Hawaii's R/V *Moana Wave*. We thank the officers and crew of this vessel for three excellent cruises. The operators of OSN1, GEOSCOPE, GSN, IRIS and its data management center provided data and instrument responses used in this study. Many thanks go to Stephan Sobolev for his thermomechanical calculations for our model. Laske wishes to thank Dave Sandwell and Jerry Winterer for numerous fruitful discussions. Reviews by Walter Zürn, Zhijun Du, and editors Gillian Foulger and Donna Jurdy are greatly appreciated and helped to improve this chapter. This research was financed by National Science Foundation grants OCE-95-29707 and OCE-99-06836.

REFERENCES CITED

Alsina, D., and Snieder, R., 1993, A test of the great circle approximation in the analysis of surface waves: Geophysical Research Letters, v. 20, p. 915–918.

Anderson, O.L., and Isaak, M., 1995, Elastic constants of mantle minerals at high temperature, *in* Ahrens, T.J., ed., Mineral physics and crystallography: Washington, D.C., American Geophysical Union Reference Shelf 2, p. 64–97.

Anderson, O.L., Schreiber, E., Liebermann, R.C., and Soga, M., 1968, Some elastic constant data on minerals relevant to geophysics: Reviews of Geophysics, v. 6, p. 491–524.

Backus, G., and Gilbert, F., 1968, The resolving power of gross Earth data: Journal of the Royal Astronomical Society, v. 16, p. 169–205.

Bassin, C., Laske, G., and Masters, G., 2000, The current limits of resolution for surface wave tomography in North America: Eos (Transactions, American Geophysical Union), v. 81, p. F897.

Brocher, T.M., and ten Brink, U.S., 1987, Variations in oceanic layer 2 elastic velocities near Hawaii and their correlation to lithosphere flexure: Journal of Geophysical Research, v. 92, p. 2647–2661.

Collins, J.A., Duennebier, F., and Shipboard Science Party, 1991, Site survey and underway geophysics, in Winkler, W., ed., Proceedings of the Ocean Drilling Program, initial reports, v. 136/137: College Station, Texas A&M University, p. 27–36.

Collins, J.A., Vernon, F.L., Orcutt, J.A., Stephen, R.A., Peal, K.R., Wooding, F.B., Spiess, F.N., and Hildebrand, J.A., 2001, Broadband seismology on the oceans: Lessons from the Ocean Seismic Network Pilot Experiment:

Geophysical Research Letters, v. 28, p. 49–52, doi: 10.1029/2000 GL011638.

Constable, S., and Heinson, G., 2004, Hawaiian hot-spot swell structure from seafloor MT sounding: Tectonophysics, v. 389, p. 111–124, doi: 10.1016/ j.tecto.2004.07.060.

Cox, C., Deaton, T., and Webb, S., 1984, A deep-sea differential pressure gauge: Journal of Atmospheric and Oceanic Technology, v. 1, p. 237–246, doi: 10.1175/1520-0426(1984)001<0237:ADSDPG>2.0.CO;2.

Crough, S.T., 1978, Thermal origin of midplate hot-spot swells: Journal of the Royal Astronomical Society, v. 55, p. 451–469.

Davies, G.F., 1988, Ocean bathymetry and mantle convection, 1: Large-scale flow and hotspots: Journal of Geophysical Research, v. 93, p. 10,467–10,480.

Detrick, R.S., and Crough, S.T., 1978, Island subsidence, hot spots, and lithospheric thinning: Journal of Geophysical Research, v. 83, p. 1236–1244.

Dziewonski, A.M., and Anderson, D.L., 1981, Preliminary reference Earth model: Physics of the Earth and Planetary Interiors, v. 25, p. 297–356, doi: 10.1016/0031-9201(81)90046-7.

Dziewonski, A., Wilkens, R.H., Firth, J.V., and Shipboard Science Party, 1991, Background and objectives of the ocean seismographic network, and leg 136 drilling results, *in* Winkler, W., ed., Proceedings of the Ocean Drilling Program, initial reports, v. 136/137: College Station, Texas A&M University, p. 3–8.

Ekström, G., Tromp, J., and Larson, E., 1997, Measurements and global models of surface wave propagation: Journal of Geophysical Research, v. 102, p. 8137–8157, doi: 10.1029/96JB03729.

Ekström, G., 2000, Mapping the lithosphere and asthenosphere with surface waves: Lateral structure and anisotropy, *in* Richards, M.A., et al., eds., The history and dynamics of global plate motions: Washington, D.C., American Geophysical Union Geophysical Monograph 121, p. 239–255.

Ellsworth, W.L., and Koyanagi, R.Y., 1977, Three-dimensional crust and mantle structure of Kilauea Volcano, Hawaii: Journal of Geophysical Research, v. 82, p. 5379–5394.

Forsyth, D.W., Weeb, S.C., Dorman, L.M., and Shen, Y., 1998, Phase velocities of Rayleigh waves in the MELT experiment on the East Pacific Rise: Science, v. 280, p. 1235–1238, doi: 10.1126/science.280.5367.1235.

Friederich, W., Wielandt, E., and Stange, S., 1994, Non-plane geometries of seismic surface wavefields and their implications for regional surface-wave tomography: Geophysical Journal International, v. 119, p. 931–948.

Grand, S.P., and Helmberger, D.V., 1984, Upper mantle shear structure of North America: Journal of the Royal Astronomical Society, v. 76, p. 399–438.

Grand, S., van der Hilst, R.D., and Widiyantoro, S., 1997, Global seismic tomography: A snapshot of convection in the Earth: GSA Today, v. 7, p. 1–7.

Hill, D.P., 1969, Crustal structure of the island of Hawaii from seismic-refraction measurements: Bulletin of the Seismological Society of America, v. 59, p. 101–130.

Hill, D.P., and Zucca, J.J., 1987, Geophysical constraints on the structure of Kilauea and Mauna Loa volcanoes and some implications for seismomagmatic processes, *in* Decker, R.W., et al., eds., Volcanism in Hawaii: Washington, D.C., U.S. Geological Survey Professional Paper 1350, p. 903–917.

Jordan, T.H., 1979, Mineralogies, densities, and seismic velocities of garnet lherzolites and their geophysical implications, *in* Boyd, F.R., and Meyer, H.O.A., eds., The mantle sample: Inclusions in kimberlites and other volcanics (vol. 2): Washington, D.C., American Geophysical Union, p. 1–14.

Karato, S., 1993, Importance of anelasticity in the interpretation of seismic tomography: Geophysical Research Letters, v. 20, p. 1623–1626.

Katzman, R., Zhao, L., and Jordan, T.H., 1998, High-resolution, two-dimensional vertical tomography of the central Pacific mantle using *ScS* reverberations and frequency-dependent travel times: Journal of Geophysical Research, v. 103, p. 17,933–17,971, doi: 10.1029/98JB00504.

Keyser, M., Ritter, J.R.R., and Jordan, M., 2002, 3D shear-wave velocity structure of the Eifel plume, Germany: Earth and Planetary Sciences Letters, v. 203, p. 59–82, doi: 10.1016/S0012-821X(02)00861-0.

Larson, E.W.F., Tromp, J., and Ekström, G., 1998, Surface-wave polarization data and global anisotropic structure: Geophysical Journal International, v. 132, p. 654–666, doi: 10.1046/j.1365-246X.1998.00452.x.

Laske, G., and Masters, G., 1996, Constraints on global phase velocity maps by long-period polarization data: Journal of Geophysical Research, v. 101, p. 16,059–16,075, doi: 10.1029/96JB00526.

Laske, G., and Masters, G., 1998, Surface-wave polarization data and global anisotropic structure: Geophysical Journal International, v. 132, p. 508–520, doi: 10.1046/j.1365-246X.1998.00450.x.

Laske, G., Phipps Morgan, J., and Orcutt, J.A., 1999, First results from the Hawaiian SWELL pilot experiment: Geophysical Research Letters, v. 26, p. 3397–3400, doi: 10.1029/1999GL005401.

Laske, G., Collins, J.A., Wolfe, C.J., Weeraratne, D., Solomon, S., Detrick, R.S., Orcutt, J.A., Bercovici, D.A., and Hauri, E.H., 2006, The Hawaiian PLUME project successfully completes its first deployment: Eos (Transactions of the American Geophysical Union), Fall supplement, v. 87, art. no. V13B-0657.

Li, X., Kind, R., Priestley, K., Sobolev, S.V., Tilmann, F., Yuan, X., and Weber, M., 2000, Mapping the Hawaiian plume with converted seismic waves: Nature, v. 405, p. 938–941, doi: 10.1038/35016054.

Li, X., Kind, R., Yuan, X., Wölbern, I., and Hanka, W., 2004, Rejuvenation of the lithosphere by the Hawaiian plume: Nature, v. 427, p. 827–829, doi: 10.1038/nature02349.

Liebermann, R.C., 2000, Elasticity of mantle minerals (Experimental studies), *in* Karato, S.-I., et al., eds., Earth's deep interior; Mineral physics and tomography from the atomic to the global scale: Washington, D.C., American Geophysical Union Geophysical Monograph 117, p. 181–199.

Lindwall, D.A., 1988, A two-dimensional seismic investigation of crustal structure under the Hawaiian Islands near Oahu and Kauai: Journal of Geophysical Research, v. 93, p. 12,107–12,122.

Lindwall, D.A., 1991, Old Pacific crust near Hawaii: A seismic view: Journal of Geophysical Research, v. 96, p. 8191–8203.

Maggi, A., Debayle, E., Priestley, K., and Barruol, G., 2006, Multimode surface waveform tomography of the Pacific Ocean: A closer look at the lithospheric cooling signature: Geophysical Journal International, v. 166, p. 1384–1397.

Maupin, V., 1992, Modelling of laterally trapped surface waves with application to Rayleigh waves in the Hawaiian Swell: Geophysical Journal International, v. 110, p. 553–570.

Montagner, J.-P., and Nataf, H.-C., 1986, A simple method for inverting the azimuthal anisotropy of surface waves: Journal of Geophysical Research, v. 91, p. 511–520.

Montagner, J.-P., and Tanimoto, T., 1990, Global anisotropy in the upper mantle inferred from the regionalization of phase velocities: Journal of Geophysical Research, v. 95, p. 4797–4819.

Montelli, R., Nolet, G., Dahlen, F.A., and Masters, G., 2006, A catalog of deep mantle plumes: New results from finite-frequency tomography: Geochemistry, Geophysics, Geosystems, v. 7, art. no. Q11007, doi: 10.1029/2006GC001248.

Moore, W.B., Schubert, G., and Tackley, P., 1998, Three-dimensional simulations of plume-lithosphere interaction at the Hawaiian Swell: Science, v. 279, p. 1008–1011, doi: 10.1126/science.279.5353.1008.

Morgan, W.J., 1971, Convection plumes in the lower mantle: Nature, v. 230, p. 42–43, doi: 10.1038/230042a0.

Müller, R.D., Roest, W.R., Royer, J.-Y., Gahagan, L.M., and Sclater, J.G., 1997, Digital isochrons of the world's ocean floor: Journal of Geophysical Research, v. 102, p. 3211–3214, doi: 10.1029/96JB01781.

Natland, J.H., and Winterer, E.L., 2005, Fissure control on volcanic action in the Pacific, *in* Foulger, G.R., et al., eds., Plates, plumes, and paradigms: Boulder, Colorado, Geological Society of America Special Paper 388, p. 687–710.

Nishimura, C.E., and Forsyth, D.W., 1989, The anisotropic structure of the upper mantle in the Pacific: Geophysical Journal, v. 96, p. 203–229.

Nolet, G., Karato, S.-I., and Montelli, R., 2006, Plume fluxes from seismic to-

mography: Earth and Planetary Sciences Letters, v. 248, p. 685–699, doi: 10.1016/j.epsl.2006.06.011.

Okubo, P.G., Benz, H.M., and Chouet, B.A., 1997, Imaging the crustal magma sources beneath Mauna Loa and Kilauea volcanoes, Hawaii: Geology, v. 25, p. 867–870, doi: 10.1130/0091-7613(1997)025<0867:ITCMSB>2.3.CO;2.

Olson, P., 1990, Hot spots, swells, and mantle plumes, *in* Ryan, M.P., ed., Magma transport and storage: New York, John Wiley, p. 33–51.

Phipps Morgan, J., Morgan, W.J., and Price, E., 1995, Hotspot melting generates both hotspot volcanism and a hotspot swell?: Journal of Geophysical Research, v. 100, p. 8045–8062, doi: 10.1029/94JB02887.

Priestley, K., and Tilmann, F., 1999, Shear-wave structure of the lithosphere above the Hawaiian hot spot from two-station Rayleigh wave phase velocity measurements: Geophysical Research Letters, v. 26, p. 1493–1496, doi: 10.1029/1999GL900299.

Raitt, R.W., 1956, Seismic-refraction studies of the Pacific Ocean basin. Part 1: Crustal thickness of the central equatorial Pacific: Geological Society of America Bulletin, v. 67, p. 1623–1640, doi: 10.1130/0016-7606(1956)67[1623:SSOTPO]2.0.CO;2.

Rapp, R.H., Wang, Y.M., and Pavlis, N.K., 1991, The Ohio State 1991 geopotential and sea surface topography harmonic coefficient models: Columbus, Department of Geodetic Science and Surveying, Ohio State University Report 410, 108 p.

Renkin, M., and Sclater, J.G., 1988, Depth and age in the North Pacific: Journal of Geophysical Research, v. 93, p. 2919–2935.

Ribe, N.M., 2004, Through thick and thin: Nature, v. 427, p. 793–795, doi: 10.1038/427793a.

Ribe, N.M., and Christensen, U.R., 1994, Three-dimensional modelling of plume-lithosphere interaction: Journal of Geophysical Research, v. 99, p. 669–682, doi: 10.1029/93JB02386.

Ritzwoller, M.H., Shapiro, N.M., and Zhong, S.-J., 2004, Cooling history of the Pacific lithosphere: Earth and Planetary Science Letters, v. 226, p. 69–84, doi: 10.1016/j.epsl.2004.07.032.

Robinson, E.M., 1988, The topographic and gravitational expression of density anomalies due to melt extraction in the uppermost oceanic mantle: Earth and Planetary Sciences Letters, v. 90, p. 221–228, doi: 10.1016/0012-821X(88)90102-1.

Ryall, A., and Bennett, D.L., 1968, Crustal structure of southern Hawaii related to volcanic processes in the upper mantle: Journal of Geophysical Research, v. 73, p. 4561–4582.

Sato, H., Sacks, S., and Murase, T., 1989, The use of laboratory velocity data for estimating temperature and partial melt fraction in the low-velocity zone: Comparison with heat flow and electrical conductivity studies: Journal of Geophysical Research, v. 94, p. 5689–5704.

Schutt, D.L., and Humphreys, E.D., 2004, P and S wave velocity and V_P/V_S in the wake of the Yellowstone hot spot: Journal of Geophysical Research, v. 109, art. no. B01305, doi: 10.1029/2003JB002442.

Shor, G.G., 1960, Crustal structure of the Hawaiian Ridge near Gardner Pinnacles: Bulletin of the Seismological Society of America, v. 50, p. 563–573.

Sleep, N.H., 1990, Hotspots and mantle plumes: Some phenomenology: Journal of Geophysical Research, v. 95, p. 6715–6736.

Smith, M.L., and Dahlen, F.A., 1973, The azimuthal dependence of Love and Rayleigh wave propagation in a slightly anisotropic medium: Journal of Geophysical Research, v. 78, p. 3321–3333.

Stange, S., and Friederich, W., 1993, Surface wave dispersion and upper mantle structure beneath southern Germany from joint inversion of network recorded teleseismic events: Geophysical Research Letters, v. 20, p. 2375–2378.

Trampert, J., and Woodhouse, J.H., 1996, High-resolution global phase velocity distributions: Geophysical Research Letters, v. 23, p. 21–24, doi: 10.1029/95GL03391.

van der Sluis, A., and van de Vorst, H.A., 1987, Numerical solution of large sparse linear systems arising from tomographic problems, *in* Nolet, G., ed., Seismic tomography: Dordrecht, D. Reidel Publishing, p. 49–83.

Vernon, F.L., Collins, J.A., Orcutt, J.A., Stephen, R.A., Peal, K., Wolfe, C.J., Hildebrand, J.A., and Spiess, F.N., 1998, Evaluation of teleseismic waveforms and detection thresholds from the OSN pilot experiment: Eos (Transactions, American Geophysical Union), v. 79, p. F650.

von Herzen, R.P., Detrick, R.S., Crough, S.T., Epp, D., and Fehn, U., 1982, Thermal origin of the Hawaiian Swell: Heat flow evidence and thermal models: Journal of Geophysical Research, v. 87, p. 6711–6723.

von Herzen, R.P., Cordery, M.J., Detrick, R.S., and Fang, C., 1989, Heat flow and the thermal origin of hot spot swell: The Hawaiian Swell revisited: Journal of Geophysical Research, v. 94, p. 13,783–13,799.

Watts, A.B., 1976, Gravity and bathymetry in the central Pacific Ocean: Journal of Geophysical Research, v. 81, p. 1533–1553.

Watts, A.B., ten Brink, U.S., Buhl, P., and Brocher, T.M., 1985, A multichannel seismic study of lithospheric flexure across the Hawaiian-Emperor seamount chain: Nature, v. 315, p. 105–111, doi: 10.1038/315105a0.

Webb, S.C., 1998, Broadband seismology and noise under the ocean: Reviews of Geophysics, v. 36, p. 105–142, doi: 10.1029/97RG02287.

Willoughby, D.F., Orcutt, J.A., and Horwitt, D., 1993, A microprocessor-based ocean-bottom seismometer: Bulletin of the Seismological Society of America, v. 83, p. 190–217.

Wilson, J.T., 1963, A possible origin of the Hawaiian Islands: Canadian Journal of Physics, v. 41, p. 863–868.

Wolfe, C.J., Bjarnason, I.T., vanDecar, J.C., and Solomon, S.C., 1997, Seismic structure of the Iceland mantle plume: Nature, v. 385, p. 245–247, doi: 10.1038/385245a0.

Wolfe, C.J., Solomon, S.C., Silver, P.G., vanDecar, J.C., and Russo, R.M., 2002, Inversion of body-wave delay times for mantle structure beneath the Hawaiian Islands; Results from the PELENET experiment: Earth and Planetary Science Letters, v. 198, p. 129–145, doi: 10.1016/S0012-821X(02)00493-4.

Woods, M.T., and Okal, E.A., 1996, Rayleigh-wave dispersion along the Hawaiian Swell: A test of lithospheric thinning by thermal rejuvenation at a hotspot: Geophysical Journal International, v. 125, p. 325–339.

Woods, M.T., Leveque, J.-J., Okal, E.A., and Cara, M., 1991, Two-station measurements of Rayleigh wave group velocity along the Hawaiian Swell: Geophysical Research Letters, v. 18, p. 105–108.

MANUSCRIPT ACCEPTED BY THE SOCIETY 31 JANUARY 2007

DISCUSSION

5 January 2007, James H. Natland and Edward L. Winterer

Laske et al. (this volume) provide evidence for lithospheric rejuvenation of a portion of the Hawaiian Swell southwest of Mauna Loa. They evaluate four models and settle on a "hybrid thermal rejuvenation–dynamic thinning" model to explain their data and those of Li et al. (2004) elsewhere along the chain. They reject models of compositional buoyancy and lithospheric fracture (Natland and Winterer, 2005) in favor of one in which the lithosphere and upper asthenosphere are modified by heat arriving from below in a narrow conduit—in short, a plume, or vertical conveyor belt. They prefer such a conduit even though admitting that their own field area did not encompass the region of melt generation beneath Hawaii and would not include the vertical conduit they drew in the diagram for their model. Following convention, they prefer a deep source for the heat, even though the tomographic model does not extend below 200 km. Inclusion of a thermal conduit in their model may be permissible but is not justified from their data set.

Laske et al. (this volume) reject lithospheric fracture, which they say has "no suggestion for the low-velocity anomaly found in the asthenosphere." This is a minimalist interpretation of the nexus of hypotheses offered by Natland and Winterer (2005) to explain Hawaii. These hypotheses combined not just fracture, but also focusing of asthenospheric counterflow in response to the geometry of the Hawaiian Ridge and redistribution of mass within the lithosphere as a consequence of large-scale melt production and eruption of lava. Natland and Winterer (2005) suggested a modification of the gravitational anchor hypothesis of Shaw and Jackson (1973); namely, that the Hawaiian Ridge is keeled by dense dunite-wehrlite cumulates that crystallize during differentiation of Hawaiian tholeiite, and that these cumu-

lates tend to sink more or less by the mechanism of Jull and Kelemen (2001; but originally Daly, 1914, 1933) into warm but refractory mantle that was produced by extraction of abyssal tholeiite near the East Pacific Rise. These cumulates displace and deflect asthenospheric counterflow, triggering melting in the lee of the advancing Hawaiian Ridge.

This hypothesis is not a simple fracture hypothesis, and it does offer a suggestion that Laske et al. (this volume) seem to have overlooked. In essence, it turns the plume conveyor belt on its side; fertile material is fed in laterally. The experiment described by Laske et al. (this volume) was merely a partial investigation of this downtrend lee, and is insufficient to resolve how warm mantle and/or partial melt became concentrated there. A plume is but one possibility for the convective arrangement southwest of Hawaii. The critical question is whether a vertical conduit is essential to explain the geophysical data.

6 January 2007, James H. Natland

I question whether a model that assumes that the convective process is entirely thermal is appropriate. Petrology says otherwise. At least two types of compositional heterogeneity will influence density relationships: (1) density contrasts among solid rock (the mantle consists of different lithologies) and (2) the distribution of melt.

These types are closely related. Thus the mantle beneath Hawaii does not have merely an identical composition of residual abyssal peridotite left over from the partial melting of mid-ocean ridge basalt at the East Pacific Rise ($\rho = 3.35$ in Table A-1 of Laske et al., this volume). Ultramafic cumulates produced by differentiation of Hawaiian tholeiite, mainly dunite, are also present in great volume beneath the islands (Clague and Denlinger, 1994) and to considerable depth, as borne out by studies of ultramafic xenoliths (e.g., Jackson, 1968; Jackson and Wright, 1970; Sen and Presnall, 1986; Chen et al., 1991; Sen et al., 2005). That such rocks will founder when warm (e.g., >600° C; Jull and Kelemen, 2001) can hardly be doubted, because their normative densities (Niu and Batiza, 1991) are dominated by iron-rich olivine (Fo$_{65-85}$), making them as much as 10% denser than abyssal peridotite (with Fo$_{91-92}$), which is also present in the xenolith suites. Mafic gabbro associated with ultramafic cumulates will accelerate the sinking when it converts to eclogite.

No one has ever seen, let alone examined, a physical specimen of any mantle source rock for Hawaiian tholeiite. All presumptions about the typical extent and temperature of partial melting of a uniform Hawaiian source are model dependent (e.g., the original pyrolite model of Green and Ringwood, 1967). However, the source is likely a composite of diverse lithologies with different densities, the proportions of which determine the relative fertility of the mantle. Thus primitive Hawaiian tholeiite may derive in large measure from eclogite (recycled ocean crust) in the mantle source (Lassiter and Hauri, 1998; Sobolev et al., 2000, 2005) or from refractory peridotite infused with basaltic melt (Falloon et al., this volume). This nonuniqueness adds great complexity to the consideration of any geophysical model for Hawaiian volcanism and its thermal effects on the lithosphere.

Eclogite is often considered to be a component in a plume source, but in fact there are no constraints whatsoever on its distribution in the asthenosphere. It could be widespread; in a layer; or unevenly distributed in strips, dipping slabs, small blobs, big blobs, blob clusters, or columns. Experimental studies indicate that eclogite has a lower range of melting temperatures than does peridotite; that pods or schlieren of it will contain some partial melt even when adjacent peridotite contains none; and that when any adjacent peridotite finally begins to melt, those pods or schlieren will contain a considerable fraction of melt. Fluctuations in melt volume along the Hawaiian chain thus could as well result from fluctuations in the proportion of eclogite in a laterally convecting source in the upper asthenosphere as from thermal, compositional, or volumetric perturbations of an ascending plume.

This possibility militates against thermal convective interpretations of lithospheric structure at Hawaii, because volcanism proves that melt, whether it is derived from eclogite or peridotite, is always present and it cannot be ignored. As long as the porosity structure allows, it will rise and then break out to the surface whenever and wherever the stress field on the plate allows. If partial melt is widespread in a low-velocity layer beneath the lithosphere (Anderson and Spetzler, 1970; Presnall and Gudfinnsson, 2005), differential ponding will likely result from patterns of flow in the convecting upper mantle and the action of the lithospheric plate itself as an impermeable barrier. Values of 5% or more of partial melt distributed in the mantle near Hawaii, as revealed by magnetotelluric experiments (Constable and Heinson, 2004), are not surprising but do not prove the existence of a plume.

Long ago, Daly (1914) described basalt as "the bringer of

heat." The presence of eclogite in the source brings the mechanism of transfer of heat into the domain of temperatures of common basaltic liquids; high potential temperatures acting on a homogeneous peridotite source are not necessary to explain Hawaii (e.g., Anderson and Natland, 2005; Falloon et al., this volume) or the shape of the lithosphere. Thus the main mechanism of rejuvenation of the lithosphere, and of underplating (e.g., McNutt and Bonneville, 2000), is injection of basaltic dikes into the base of the lithosphere.

Xenolith diversity indicates that such basalt is only rarely primitive magma with a high temperature (Sen et al., 2005). The process clearly starts at the zero-age end of the chain at Loihi (Clague, 1988) and the South Arch volcanic field. Therefore, lithospheric erosion along a portion of the chain is not a matter of convective overturn of homogeneous peridotite in the solid state. Instead, the deep Hawaiian lithosphere transforms into something rheologically different, because it contains either basaltic melt or, where it is cooler, at least the cumulus products of such basaltic melt. These products on the average will remain more plastic than abyssal peridotite, still and always being nearer their melting and/or crystallization temperatures and therefore having lower shear velocities than for abyssal peridotite. Perhaps the lithosphere near Kauai is thin (Li et al., 2004) because of sinking of a mixed mass of abyssal peridotite and dense cumulates from the lower lithosphere into the convecting upper mantle.

9 January 2007, Don L. Anderson

Laske and her colleagues are to be commended for mounting this remarkable and successful experiment. The seismological conclusions are well founded and appropriately conservative. The data itself cannot address a lower-mantle origin of a proposed mantle plume, as the authors state, but they can test the hypothesis that a plume, as conventionally defined, exists. It can also test alternate hypotheses—including excess fertility—as proposed in the discussion contributions by Winterer and Natland, which are quite different than the ones criticized by Laske et al. (this volume).

Although the experiment is off-axis from the conjectured plume track, it is close enough to see a plume head or lateral flow of a hot plume, if these in fact exist, as the authors apparently believe (http://mahi.ucsd.edu/Gabi/plume.html). Plume theory has been refined for more than 30 years and offers very testable predictions, even if plume tails are too small to resolve. For example, the region around Hawaii should look like the diagrams in Campbell and Davies (2006), which include the refinements and modifications that have been made to the hypothesis up to this time (http://www.mantleplumes.org/WebDocuments/Episodes06-plumes.pdf). Other authors envision even longer-distance effects away from the plume axis. In the plume model, the traveltimes of vertically traveling S waves should be long, and the attenuation should be high. The transition region should be thin, and the plume should also spread out beneath the 650-km discontinuity.

A thermal plume differs from other explanations for Hawaii in being a very strong and hot active upwelling. An active upwelling, in contrast to a passive upwelling, spreads out beneath the plate; an upwelling at a dike, at a ridge, or in response to delamination (or "lithospheric erosion"), is focused toward the eruption site, or the region of thin or extending lithosphere. A plume will have a broad pancake or mushroom-shaped low-velocity region, concentric about the center of the upwelling or the region of active magmatism. There is no evidence of this shape in the current data for Hawaii. A passive upwelling will be cone- or wedge-shaped, focusing and narrowing as it rises, rather than spreading out, much as is observed at ridges, and ridge-centered Iceland (Wolfe et al., 1997; Foulger et al., 2001). It will also have little impact on downstream heatflow. In the fertile blob-counterflow model, the melting anomaly will come in from the northwest and will not be a large, hot, circular feature centered on Hawaii-Loihi. In the delamination and self-propagating-volcanic-chain models the upwelling will be local, linear, progressive, and will have asthenosphere-like temperatures, slightly higher than average because of the insulating effect of the large and long-lived Pacific plate. It will have an eclogite imprint on the chemistry; upwelling rates and chemistry will be affected by pre-existing features, such as fracture zones (FZ). Normal crust and mantle may exist very close to the eruption site.

Other seismic observations are consistent with shallow and lateral flow mechanisms, and with the absence or smallness of the effects seen by the SWELL experiment. Multiple *ScS* waves bouncing between the surface and the core have normal travel times (Best et al., 1975; Sipkin and Jordan, 1975; Julian, 2005) and attenuation, and the transition zone (410–650 km) thickness shows no thermal thinning (Deuss, this volume). Thus the breadth, depth, and magnitude of the Hawaiian anomaly are, to some extent, already constrained.

It is useful to recall that a purely thermal explanation for the high magma production rate at Hawaii in a small area requires an upwelling velocity of ~50 cm/year, temperature excesses of up to 300° C, and lateral flow of plume material out to more than 500 km. The upwelling is very narrow in the deep mantle but very broad near the surface.

Removal of the lower part of the plate may trigger upwelling, as in alternative models, rather than the reverse. Fertile material may be brought in laterally (horizontal conveyor belt), rather than in a narrow, vertical cylinder that spreads out laterally. Low shear velocities are often attributed to hot buoyant plumes, but eclogite-bearing blobs, or regions with CO_2, can also have low velocities and cause melting anomalies, even if not particularly hot or buoyant.

There is therefore sufficient motivation to consider alternate mechanisms and not just restrict attention to the plume and crack models, both of which, in their pure form, assume a homogeneous isothermal reference mantle, which is assuredly not the case. The implication in Laske et al. (this volume) is that a plume origin of some sort is not in dispute, just the depth. The follow-up experiment is named "PLUME," but one would hope that serious nonplume and nonthermal explanations will be assessed, such as fertile blobs, self-perpetuating volcanism, and delamination—not just cartoonish or strawman versions of these ideas.

11 January 2005, Edward L. Winterer and James H. Natland

FZs and lithosphere structure must be considered in any model of Hawaiian volcanism. The magnetic anomaly offset at the Molokai fracture zone, one of the longest in the north Pacific, spans a 16-Ma difference in ages; the lithosphere should be thinner by at least 10 km beneath Kauai. The most prominent part of the Hawaiian Swell and the largest volcanoes are between the Molokai fracture zone and the Murray fracture zone. The volume of basalt along the entire Emperor-Hawaiian chain reaches its peak at the Molokai fracture zone (Van Ark and Lin, 2004). These cannot be coincidences; for one thing, thin lithosphere allows the asthenosphere at its solidus to well up further and to melt more extensively by adiabatic decompression in this region.

A complex of northeast-trending Cretaceous seamounts lies west of the island of Hawaii (Eakins and Robinson, undated map), suggesting that this part of the plate has long been vulnerable to intrusion and the lithosphere petrologically modified. This region is where the SWELL pilot experiment was carried out. The data of Li et al. (2004) show steps at places that correspond to the Molokai fracture zone and the small-offset Maui fracture zone between Maui and Hawaii. The evidence for thin lithosphere to the northwest of Hawaii is therefore not necessarily evidence for lithospheric thinning. The lithosphere there is still not very thin, and the thinnest parts are far from Hawaii and the chain axis. Besides thermal rejuvenation, thin lithosphere may be inherited (e.g., an effect of lithospheric age or fabric) or result from athermal thinning, stretching, and/or delamination. Nor is thin lithosphere evidence for rejuvenation unless the prior thickness is known. In this example, a good case can be made that the lithosphere was thin to begin with, and that prior seamount formation made it more vulnerable to current melting. Thus North Arch volcanism (Clague et al., 1990; Frey et al., 2000) indicates very young and widespread melt productivity near Kauai. Perhaps lithospheric enrichment or refertilization occurred when the Musician seamounts were produced.

The Molokai fracture zone is a transtensional band some 300 km wide (Searle et al., 1993), narrowing toward the islands. Changes in lithosphere thickness across it likely ramp up in several smaller steps rather than thickening in one abrupt step. This suggestion agrees with Fig. 2 of Li et al. (2004), which shows a long ramp at the base of the lithosphere, shallowing by some 50 km northward, with marked steps of length 10–20 km at the main Molokai fracture zone (between Molokai and Oahu) and the smaller Maui fracture zone. This known blocky architecture of the plate is not treated in Li et al. (2004), and results instead in their depicting a smooth asthenospheric bulge and proposing gradual heating of lithosphere by a plume passing beneath. Control of lithosphere thickness by pre-Hawaii plate architecture could also simplify the history and not require a plume.

The Koolau-Lanai-Kahoolawe isotopic anomaly (Basu and Faggart, 1996) and the peak volume of Hawaiian magmatism coincide at the intersection of the Molokai fracture zone with the Hawaiian ridge. The basalts of these islands exhibit greater scatter in isotope signatures than elsewhere along the Hawaiian chain. Some of this scatter might result from introduction of seawater into the crust of the fracture zone or the mantle underneath, but it may also indicate susceptibility of fissured and irregularly shaped lithosphere to prior modification by off-axis seamount magmatism. Furthermore, the FZ today may act as a dam or a conduit for magma, which may facilitate removal of material at the base of the plate by partial melting. The volcanoes in the chain to the northwest have smaller volumes and more limited isotopic variability (Basu and Faggart, 1996).

These conjectures are consistent with low temperatures for melting and differentiation beneath the islands, as revealed by ultramafic xenoliths. The deepest (highest-pressure) xenoliths from Hawaii suggest that the in situ temperature near the base of the lithosphere reached a maximum of 1350° C, or 1260° C, if the effects of volatiles are considered (Sen et al., 2005); this value is 50–300° C lower than predicted by plume models. Estimated temperatures at the lithosphere-asthenosphere transition beneath Oahu are not significantly different from those of normal 90-Ma lithosphere that has not been affected by a hot plume. The Hawaiian lithosphere therefore is not unusually hot (e.g., Green et al., 2001; Green and Falloon, 2005; Presnall and Gudfinnsson, 2005). In addition, several studies suggest that low-melting-point eclogite may be involved in the Hawaiian source (Hauri, 1995; Sobolev et al., 2000, 2005; Ren et al., 2005), particularly the Koolau volcano (Hauri, 1995), which is also the most enriched of the Koolau-Lanai-Kahoolawe anomaly. Eclogite is even necessary for plumes, if they exist, to work (Cordery et al., 1997). However, the distribution of eclogite in the mantle is unknown, and its connection to plumes is not demonstrated; it may simply be distributed in the shallow asthenosphere to begin with (see the 6 January comment by Natland).

Laske and coworkers need to consider these factors in developing models to explain their data; at this stage it is premature to claim that their results are inconsistent with the hypothesis of a propagating fracture at Hawaii and instead are evidence for a mantle plume.

8 February 2005, Gabriele Laske and John A. Orcutt

Natland and Winterer, in their comments of 5 and 11 January, feel that we took a minimalist approach to reconcile our seismic model with their model of a propagating crack in the lithosphere. They point out that we may have overlooked the role of counterflows in the asthenosphere in their model. Natland also questions whether it is appropriate to assume a purely thermal model for our seismic anomalies.

Addressing the second point first, Anderson (9 January) comments that our interpretation is appropriately conservative. Recall that we find low seismic anomalies in the lower lithosphere as well as in the asthenosphere. We believe that these anomalies are sufficiently well constrained to search for possible causes. Perhaps the best-understood cause for seismic velocity anomalies are thermal effects, and anelastic effects are the next perturbation to this most simplistic idea (Karato, 1993). Partial melt indeed changes seismic velocity dramatically, perhaps more so than temperature anomalies do (e.g., Sato et al., 1989). Compositional changes, such as the abundance of eclogite suggested by Natland and Winterer, also influence seismic velocity. As they emphasize, eclogite may or may not be very abundant in the asthenosphere. Of these causes, the change in composition is probably the most speculative, which leaves temperature and melt fraction. Lacking enough constraints, seismologists usually try to reconcile their data with temperature variations alone, and it turns out that T does not have to change unrealistically to fit our data. Our results are supported by the electromagnetic study of Constable and Heinson (2004). At this point, it is difficult to reconcile melt fractions of more than 1% with electromagnetic and seismic data farther than 300 km from the islands.

This difficulty alone may or may not speak against a mechanical erosion or an injection of the base of the lithosphere with basaltic dikes, but our tests of the seismic model against bathymetry and the geoid support our hypothesis that the lithosphere is not mechanically eroded. We agree that a thin lithosphere is not necessarily a result of rejuvenation, as Winterer and Natland point out. However, a mechanically thin lithosphere is inconsistent with our model, at least for the part of the swell covered by the pilot deployment, which includes the South Arch volcanoes. This argument holds only if the greater area is isostatically compensated. As we pointed out, we find some inconsistency with the geoid in the deep ocean that remains to be explained. Winterer and Natland's definition of "thin" may actually agree with ours, if we allow the "thin" to be altered at the base but not asthenosphere-like.

We find a pronounced anomaly in the asthenosphere. Any model to explain the Hawaiian Swell has to involve at least this region (i.e., models with sources confined to the lithosphere do not work). As Anderson (January 9) points out, the presence of an anomaly in the asthenosphere does not necessarily refute the specific model of Natland and Winterer (2005), which also predicts some changes in the asthenosphere through the horizontal supply of fertile material and which we have omitted in Figure 1 in our chapter. Perhaps we have used Natland and Winterer's reference in the wrong context, but the figure caption does not say that panel D describes their model. If a horizontal conveyor supplies fertile material, then the accumulated material in the asthenosphere has to cause a seismic anomaly of 8%, 300 km away from the islands. Sobolev et al. (2005) argue that the enrichment from recycled crust is found only near the proposed plume center, which has undergone melting, but is insignificant near the plume edge, where our array is located. Recall that the electromagnetic study supports significant melt fractions near the islands.

As Anderson points out, our pilot study is not appropriate to search for a deep-mantle plume and we never say it is. Our pilot experiment is appropriate to search for causes of the Hawaiian Swell, as we have discussed in our article.

REFERENCES CITED

Anderson, D.L., and Natland, J. H., 2005. A brief history of the plume hypothesis and its competitors: Concept and controversy, *in* Foulger, G.R., et al., eds., Plates, plumes, and paradigms: Boulder, Colorado, Geological Society of America Special Paper 388, p. 119–145, doi: 10.1130/2005.2388 (08).

Anderson, D.L., and Spetzler, H., 1970, Partial melting and the low-velocity zone: Physics of the Earth and Planetary Interiors, v. 4, p. 62–64.

Basu, A.R., and Faggart, B.E., 1996. Temporal variation in the Hawaiian mantle plume: The Lanai anomaly, the Molokai fracture zone and a seawater-altered lithospheric component, in Hawaiian volcanism, *in* Basu, A., and Hart., S., Earth processes: Reading the isotopic code: Washington, D.C., American Geophysical Union Geophysical Monograph 95, p. 149–159.

Best, W.J., Johnson, L.R., and McEvilly, T.V., 1975, ScS and the mantle beneath Hawaii: Eos (Transactions, American Geophysical Union), v. 1147, p. 1147.

Campbell, I., and Davies, G., 2006, Do mantle plumes exist?: Episodes, v. 29, p. 162–168.

Chen, C.-H., Presnall, D.C., and Stern, R.J., 1991, Petrogenesis of ultramafic xenoliths from the 1800 Kaupulehu flow, Hualalai volcano, Hawaii: Journal of Petrology, v. 33, p. 163–202.

Clague, D.A., 1988. Petrology of ultramafic xenoliths from Loihi Seamount, Hawaii: Journal of Petrology, v. 29, p. 1161–1186.

Clague, D.A., and Denlinger, R.P., 1994, Role of olivine cumulates in destabilizing the flanks of Hawaiian volcanoes: Bulletin of Volcanology, v. 56, p. 425–434.

Clague, D.A., Holcomb, R.T., Sinton, J.M., Detrick, R.S., and Torresan, M.R., 1990, Pliocene and Pleistocene alkalic flood basalts on the seafloor north of the Hawaiian Islands: Earth and Planetary Science Letters, v. 98, p. 175–191.

Constable, S.C., and Heinson, G.S., 2004, Haiwaiian hot-spot swell structure from seafloor MT sounding: Tectonophysics, v. 389, p. 111–124.

Cordery, M.J., Davies, G.F., and Campbell, I.H., 1997, Genesis of flood basalts from eclogite-bearing mantle plumes: Journal of Geophysical Research, v. 102, p. 20,179–20,197.

Daly, R.A., 1914. Igneous rocks and their origin: New York, McGraw-Hill, 563 p.

Daly, R.A., 1926. Our mobile Earth: New York, Charles Scribner's Sons, 342 p.

Daly, R.A., 1933. Igneous rocks and the depths of the Earth: New York, McGraw-Hill (Hafner Reprint, 1962), 598 p.

Deuss, A., 2007, Seismic observations of transition-zone discontinuities beneath hotspot locations, *in* Foulger, G.R., and Jurdy, D.M., eds., Plates, plumes, and planetary processes: Boulder, Colorado, Geological Society of America Special Paper 430, doi: 10.1130/2007.2430(07).

Eakins, B., and Robinson, J.E., undated map, Bathymetry and topography of the Hawaiian Islands: Menlo Park, California, U.S. Geological Survey.

Falloon, T.J., Green, D.H., and Danyushevsky, L.V., 2007, Crystallization temperatures of tholeiite parental liquids: Implications for the existence of thermally driven mantle plumes, *in* Foulger, G.R., and Jurdy, D.M., eds., Plates, plumes, and planetary processes: Boulder, Colorado, Geological Society of America Special Paper 430, doi: 10.1130/2007.2430(12).

Foulger, G.R., Pritchard, M.J., Julian, B.R., Evans, J.R., Allen, R.M., Nolet, G., Morgan, W.J., Bergsson, B.H., Erlendsson, P., Jakobsdottir, S., Ragnarsson, S., Stefansson, R., and Vogfjord, K., 2001, Seismic tomography shows

that upwelling beneath Iceland is confined to the upper mantle: Geophysical Journal International, v. 146, p. 504–530.

Frey, F.A., Clague, D.A., Mahoney, J.J., and Sinton, J.M., 2000, Volcanism at the edge of the Hawaiian plume: Petrogenesis of submarine alkalic lavas from the North Arch volcanic field: Journal of Petrology, v. 41, p. 667–691.

Green, D.H., and Ringwood, A.E., 1967, The genesis of basaltic magmas: Contributions to Mineralogy and Petrology, v. 15, p. 103–190.

Green, D.H., Falloon, T.J., Eggins, S.E., and Yaxley, G.M., 2001, Primary magmas and mantle temperatures: European Journal of Mineralogy, v. 13, p. 437–451.

Green, D.H., and Falloon, T.J., 2005, Primary magmas at mid-ocean ridges, "hot spots" and other intraplate settings: Constraints on mantle potential temperature, *in* Foulger, G.R., et al., eds., Plates, plumes and paradigms: Boulder, Colorado, Geological Society of America Special Paper 388, p. 217–247, doi: 10.1130/2005.2388(14).

Hauri, E., 1995, Major-element variability in the Hawaiian mantle plume: Nature, v. 382, p. 415–419.

Jackson, E.D., 1968, The character of the lower crust and upper mantle beneath the Hawaiian Islands, *in* Beneš, K., ed., 23rd International Geological Congress, Prague, Proceedings: Prague, Academia, v. 1, p. 135–160.

Jackson, E.D., and Wright, T.L., 1970, Xenoliths in the Honolulu volcanic series, Hawaii: Journal of Petrology, v. 11, p. 405–430.

Julian, B.R., 2005, What can seismology say about hot spots?, *in* Foulger, G.R., et al., eds., Plates, plumes, and paradigms: Boulder, Colorado, Geological Society of America Special Paper 388, p. 155–170, doi: 10.1130/2005 .2388(10).

Jull, M., and Kelemen, P.B., 2001, On the conditions for lower crustal convective instability: Journal of Geophysical Research, v. 106, p. 6423–6446.

Karato, S., 1993. Importance of anelasticity in the interpretation of seismic tomography: Geophysical Research Letters, v. 20, p. 1623–1626.

Laske, G., Phipps Morgan, J., and Orcutt, J.A., 2007, The Hawaiian SWELL pilot experiment—Evidence for lithosphere rejuvenation from ocean bottom surface wave data, *in* Foulger, G.R., and Jurdy, D.M., eds., Plates, plumes, and planetary processes: Boulder, Colorado, Geological Society of America Special Paper 430, doi: 10.1130/2007.2430(11).

Lassiter, J.C., and Hauri, E.H., 1998, Osmium-isotope variations in Hawaiian lavas: Evidence for recycled oceanic lithosphere in the Hawaiian plume: Earth and Planetary Science Letters, v. 164, p. 483–496.

Li, X.-Q., Kind, R., Yuan, X., Wölbern, I., and Hanka, W., 2004, Rejuvenation of the lithosphere by the Hawaiian plume: Nature, v. 427, p. 827–829.

McNutt, M.K, and Bonneville, A., 2000, A shallow, chemical origin for the Marquesas swell: Geochemistry, Geophysics, Geosystems, v. 1, art. no. 1999 GC00028.

Natland, J.H., and Winterer, E.L., 2005, Fissure control on volcanic action in the Pacific, *in* Foulger, G.R., et al., eds., Plates, plumes, and paradigms: Boulder, Colorado, Geological Society of America Special Paper 388, p. 687–710, doi: 10.1130/2005.2388(39).

Niu, Y.-L., and Batiza, R., 1991, In situ densities of MORB melts and residual mantle: Implications for buoyancy forces beneath mid-ocean ridges: Journal of Geology, v. 99, p. 767–775.

Presnall, D.C., and Gudfinnsson, G., 2005. Carbonatitic melts in the oceanic low-velocity zone and deep mantle, *in* Foulger, G.R., et al., eds., Plates, plumes, and paradigms: Boulder, Colorado, Geological Society of America Special Paper 388, p. 207–216, doi: 10.1130/2005.2388(13).

Presnall, D.C., and Gudfinnsson, G., 2007, Global Na8-Fe8 Systematics of MORBs: Implications for mantle heterogeneity, temperature, and plumes, abstract EGU2007-A-00436, European Geophysical Union General Assembly, 15–20 April, Vienna.

Presnall, D.C., Gudfinnsson, G.H., and Walter, M.J., 2002, Generation of mid-ocean ridge basalts at pressures from 1 to 7 GPa: Geochimica et Cosmochimica Acta, v. 66, p. 2073–2090.

Ren, Z.-Y., Ingle, S., Takahashi, E., Hirano, N., and Hirata, T., 2005, The chemical structure of the Hawaiian mantle plume: Nature, v. 436, p. 837–840.

Sato, H., Sacks, S., and Murase, T., 1989. The use of laboratory velocity data for estimating temperature and partial melt fraction in the low-velocity zone: Comparison with heat flow and electrical conductivity studies: Journal of Geophysical Research, v. 94, p. 5689–5704.

Searle, R.C., Holcomb, R.T., Wilson, J.B., Holmes, M.L., Whittington, R.J., Kappel, E.S., McGregor, B.A., and Shor, A.N., 1993, The Molokai fracture zone near Hawaii, and the Late Cretaceous change in Pacific/Farallon spreading direction, *in* Pringle, M.S., et al., eds., The Mesozoic Pacific: Geology, tectonics, and volcanism: Washington, D.C., American Geophysical Union Geophysical Monograph 77, p. 155–169.

Sen, G., and Presnall, D.C., 1986, Petrogenesis of dunite xenoliths from Koolau volcano, Oahu, Hawaii: Implications for Hawaiian volcanism: Journal of Petrology, v. 27, p. 197–217.

Sen, G., Keshav, S., and Bizimis, M., 2005, Hawaiian mantle xenoliths and magmas: Composition and thermal character of the lithosphere: American Mineralogist, v. 90, p. 871–887.

Shaw, H.R., and Jackson, E.D., 1973, Linear island chains in the Pacific: Result of thermal plumes or gravitational anchors?: Journal of Geophysical Research, v. 78, p. 8634–8652.

Sipkin, S.A., and T.H. Jordan, 1975, Lateral heterogeneity of the upper mantle determined from the travel times of ScS: Journal of Geophysical Research, v. 80, p. 1474–1484.

Sobolev, A.V., Hofmann, A.W., and Nikogosian, I.K., 2000, Recycled oceanic crust observed in "ghost plagioclase" within the source of Mauna Loa lavas: Nature, v. 404, p. 986–990.

Sobolev, A.V., Hofmann, A.W., Sobolev, S.V., and Nikogosian, I.K., 2005, An olivine-free mantle source of Hawaiian shield basalts: Nature, v. 434, p. 590–597.

Van Ark, E., and Lin, J., 2004, Time variation in igneous volume flux of the Hawaii-Emperor hotspot seamount chain: Journal of Geophysical Research, v. 109, doi: 10.1029/2003JB002949.

Wolfe, C.J., Bjarnason, I.Th., VanDecar, J.C., and Solomon, S.C., 1997, Seismic structure of the Iceland mantle plume: Nature, v. 385, p. 245–247.

The Geological Society of America
Special Paper 430
2007

Crystallization temperatures of tholeiite parental liquids: Implications for the existence of thermally driven mantle plumes

Trevor J. Falloon*
School of Earth Sciences and Centre for Marine Science,
University of Tasmania, Private Bag 79, Hobart, Tasmania 7001, Australia
David H. Green*
Research School of Earth Sciences, Australian National University,
Mills Road, Acton, Canberra 0200, ACT, Australia
Leonid V. Danyushevsky*
School of Earth Sciences and Australian Research Council Centre for Excellence
in Ore Deposits, University of Tasmania, Private Bag 79, Hobart, Tasmania 7001, Australia

ABSTRACT

To compare magmatic crystallization temperatures between ocean island basalt (OIB) proposed to be plume-related and normal mid-ocean ridge basalt (MORB) parental liquids, we have examined and compared in detail three representative magmatic suites from both ocean island (Hawaii, Iceland, and Réunion) and mid-ocean ridge settings (Cocos-Nazca, East Pacific Rise, and Mid-Atlantic Ridge). For each suite we have good data on both glass and olivine phenocryst compositions, including volatile (H_2O) contents. For each suite we have calculated parental liquid compositions at 0.2 GPa by incrementally adding olivine back into the glass compositions until a liquid in equilibrium with the most-magnesian olivine phenocryst composition is obtained. The results of these calculations demonstrate that there is very little difference (a maximum of ~20 °C) between the crystallization temperatures of the parental liquids (MORB 1243–1351 °C versus OIB 1286–1372 °C) when volatile contents are taken into account.

To constrain the depths of origin in the mantle for the parental liquid compositions, we have performed experimental peridotite-reaction experiments at 1.8 and 2.0 GPa, using the most magnesian of the calculated parental MORB liquids (Cocos-Nazca), and compared the others with relevant experimental data utilizing projections within the normative basalt tetrahedron. The mantle depths of origin determined for both the MORB and OIB suites are similar (MORB 1–2 GPa; OIB 1–2.5 GPa) using this approach.

Calculations of mantle potential temperatures (T_P) are sensitive to assumed source compositions and the consequent degree of partial melting. For fertile lherzolite sources, T_P for MORB sources ranges from 1318 to 1488 °C, whereas T_P for ocean island tholeiite sources (Hawaii, Iceland, and Réunion) ranges from 1502 °C (Réunion) to 1565 °C (Hawaii). The differences in T_P values between the hottest MORB and ocean island tholeiite sources are ~80 °C, significantly less than predicted by the ther-

*E-mails: Falloon, trevor.falloon@utas.edu.au; Green, David.H.Green@anu.edu.au; Danyushevsky, l.dan@utas.edu.au.

Falloon, T.J., Green, D.H., and Danyushevsky, L.V., 2007, Crystallization temperatures of tholeiite parental liquids: Implications for the existence of thermally driven mantle plumes, *in* Foulger, G.R., and Jurdy, D.M., eds., Plates, plumes, and planetary processes: Geological Society of America Special Paper 430, p. 235–260, doi: 10.1130/2007.2430(12). For permission to copy, contact editing@geosociety.org. ©2007 The Geological Society of America. All rights reserved.

mally driven mantle plume hypothesis. These differences disappear if the hotspot magmas are derived by smaller degrees of partial melting of a refertilized refractory source. Consequently the results of this study do not support the existence of thermally driven mantle plumes originating from the core-mantle boundary as the cause of ocean island magmatism.

Keywords: MORB, Hawaii, Réunion, Iceland, Kilauea, olivine, peridotite melting experiments, mantle potential temperatures, primary magmas, mantle plumes

INTRODUCTION

The composition and temperature at which olivine crystallizes from a mantle-derived parental liquid at a low pressure is one of the key constraints on any model of magma genesis (e.g., Green and Ringwood, 1967; Sobolev and Danyushevsky, 1994), as these parameters are necessary for estimating potential temperatures of the source mantle. Olivine is the first phase to crystallize at low pressure from any mantle-derived melt that is in chemical equilibrium with peridotite at source depths. This is due to the rapid expansion of the olivine phase volume at lower pressures, demonstrated by numerous experimental studies on model melt compositions. As both olivine crystallization temperature and the composition of liquidus olivine are sensitive to the composition of the crystallizing melt, it is possible to calculate both, given the composition of melt alone. Conversely, given an olivine composition, it is possible to test whether an observed glass or rock composition is in equilibrium with the olivine, by using empirically calibrated functions for equilibria between melt and olivine end-members. Such a calibration is referred to as an olivine geothermometer (e.g., Roeder and Emslie, 1970; Ford et al., 1983; Herzberg and O'Hara, 2002).

The application of olivine geothermometry to determine the composition of parental liquids is of particular significance to our understanding of the causes of ocean island volcanism, including inferred hot mantle plumes. The hypothesis of thermally driven mantle plumes derived from the core-mantle boundary predicts a significant temperature contrast between upwelling plume material and ambient upper mantle. Consequently decompression melts derived from the mantle plume materials should be ~200–300 °C hotter than melts derived from ambient upper mantle (McKenzie and Bickle, 1988). If the thermally driven mantle plume hypothesis is correct, then we should expect to find evidence from olivine crystallization temperatures that parental liquids to Hawaii olivine tholeiites, a typical ocean island magma, are significantly hotter than parental liquids to mid-ocean ridge basalt (MORB) olivine tholeiites. We should also find evidence for very different pressures and degrees of partial melting. The determination of olivine crystallization temperatures of parental melts is also the first step necessary for estimating mantle potential temperatures (T_p).

In this article, we calculate the parental liquids for a representative range of tholeiite compositions from both mid-ocean ridge (Cocos-Nazca, East Pacific Rise, and Mid-Atlantic Ridge) and ocean island settings (Hawaii, Réunion, and Iceland). We find that there are no significant differences (a maximum of ~20 °C) between crystallization temperatures of MORB and ocean island basalt (OIB) parental tholeiite liquids, when the hottest parental liquids from each setting are compared. To constrain the depth of origins in the mantle for the parental liquid compositions, we have performed experimental peridotite-reaction experiments at 1.8 and 2.0 GPa (see Appendix 1), using the most magnesian of the calculated parental MORB liquids (Cocos-Nazca), and compared the others with relevant experimental data utilizing projections within the normative basalt tetrahedron. The mantle depths of origin determined for both the MORB and OIB suites are similar (MORB 1–2 GPa; OIB 1–2.5 GPa) using this approach. Finally we present model calculations for the source mantle T_P of the calculated parental liquids. These calculations are significantly dependent on the chosen models of source compositions and thus on the inferred degree of partial melting and the magnitude of latent heat of melting in the models. Below we first outline the rationale of our approach before discussing in detail our case studies from Hawaii, Réunion, Iceland, and MORB.

RATIONALE OF APPROACH

Our approach to estimating the temperatures of tholeiite magmas is based on the use of olivine geothermometers to calculate the compositions of parental liquids. Unmodified parental liquids rarely erupt, and evidence for their existence is only preserved in the mineralogy of magnesian olivine phenocrysts enclosed in phyric magma (rock) compositions. An olivine geothermometer is therefore used to reconstruct the composition of the parental liquid, and its temperature of crystallization, by adding back olivine in incremental equilibrium steps (e.g., 0.01 wt%; see appendix in Danyushevsky et al., 2000, for a detailed explanation) into an evolved liquid (glass) composition. This use of an olivine geothermometer assumes that olivine crystallizes fractionally, i.e., olivine is chemically isolated from the melt when it is formed. Along with the assumption of fractional crystallization, the parental liquid calculation also requires the following:

1. A composition of an evolved melt within the olivine-only field. This melt should be either a natural glass composition

or an aphyric whole-rock composition that represents a liquid.

2. Establishment of the composition of the most-magnesian olivine phenocryst or microphenocryst for the suite, as a target for the olivine addition calculations. In any magma suite, there will be found a range in olivine phenocryst compositions, if present (Danyushevsky et al., 2002). Most large olivine phenocrysts show normal zoning from core Mg# values (Mg# = $100 \times X_{Mg}/(X_{Mg} + X_{Fe})$], where X_{Fe} and X_{Mg} are cation fractions of Fe^{2+} and Mg, respectively) higher than those in equilibrium with the erupted evolved liquid composition. Olivine in equilibrium with the erupted melt is usually present as both discrete microphenocrysts and rims on the more-magnesian phenocryst cores. In some suites, the magnesian olivine phenocrysts are xenocrystic and could either represent (i) disaggregated cumulate material from previously erupted magmas, which may or may not have similar magma compositions to the composition of interest or (ii) lithospheric wallrock samples detached and disaggregated into magmas or melts that have moved through the lithosphere toward crustal magma chambers. Minor element (Ca, Al, Cr, and Ni) or trace element contents of olivine may be used to evaluate phenocryst versus xenocryst relationships with respect to the host magma (Norman and Garcia, 1999). Also in most suites, it is possible to identify and analyze melt inclusions in magnesian olivine phenocrysts that demonstrate that these olivine phenocrysts are related to the evolved, erupted liquid composition, via the process of crystal fractionation. Where phenocryst and microphenocryst relationships are established, it is possible to use an olivine geothermometer to incrementally add back olivine in small equilibrium steps to obtain a parental composition that is in equilibrium with the most-magnesian phenocryst composition observed (Irvine, 1977; Albarede, 1992; Danyushevsky et al., 2000).

3. An estimate of volatile content of the evolving melt (especially H_2O) and its effect on the crystallization temperature and composition. Anhydrous calculations for magmas that have small amounts of H_2O can lead to large differences in crystallization temperatures between parental liquid compositions (e.g., Falloon and Danyushevsky, 2000). In this article, we use the model of Falloon and Danyushevsky (2000) to estimate the effect of H_2O, on olivine liquidus temperatures. The effect of H_2O on the value of the equilibrium constant for iron-magnesium exchange between olivine and liquid ($K_D = [(X_{Fe})^{Ol}/(X_{Fe})^L]/[(X_{Mg})^{Ol}/(X_{Mg})^L]$, where Ol is olivine, L is melt, and X_{Fe} and X_{Mg} are cation fractions of Fe^{2+} and Mg, respectively) is less well known but experimental and theoretical studies (Ulmer, 1989; Putirka, 2005; Toplis, 2005) suggest that the effect is very small for the likely H_2O contents of tholeiite parental liquids (<1 wt%) and will not introduce a significant error into calculations using olivine geothermometers under anhydrous conditions.

4. An estimate of melt oxidation state. Two components are

necessary: (i) an oxygen fugacity and (ii) a model to calculate the Fe^{2+}/Fe^{3+} ratio of the melt for the given oxygen fugacity (see Danyushevsky and Sobolev, 1996, for a detailed discussion). Differences in oxygen fugacity will not have a serious effect on conclusions reached about temperature differences between parental compositions (~40 °C for four orders of magnitude difference in the oxygen fugacity).

5. An estimate of olivine crystallization pressure. This estimate can be determined from studies of primary fluid inclusions in olivine phenocrysts (e.g., Anderson and Brown, 1993; Sobolev and Nikogosian, 1994).

6. And finally, the choice of an appropriate olivine geothermometer. In this article, we use the geothermometer of Ford et al. (1983), as it is the most appropriate for modeling olivine crystallization.

A more detailed review of the use of olivine geothermometers in calculating parental liquid compositions is provided by Falloon et al. (2007).

CASE STUDIES OF HOTSPOT OLIVINE THOLEIITE MAGMAS

Hawaii

Kilauea Volcano. Kilauea is an active shield-building volcano on the island of Hawaii. The magmatic rocks of Kilauea volcano have been the subject of a large number of detailed studies. These studies provide us with a relatively comprehensive set of data on whole rocks, glass, and olivine phenocryst compositions, including volatile (H_2O) contents. These data make Kilauea an ideal subject for our approach of calculating parental liquids for an ocean island tholeiite series. There are two aspects of the Kilauean data set that are of particular significance for our purposes. The first is the 1959 summit eruption of Kilauea, as this event was a very intensely studied eruption both in terms of volcanology and petrology. The volcanological observations in particular recorded the height and temperatures of lava fountaining, and these data, combined with associated petrological data on glasses and volatiles, allow us to model the temperatures of eruptive magmas (see below). The second is the recovery of quenched glasses from the submarine Puna rift (Clague et al., 1991; Clague et al., 1995). These glasses are the most magnesian (~15 wt% MgO) so far recovered for an ocean island tholeiite series. In addition, the Puna ridge glasses contain as microphenocrysts the most-magnesian olivine compositions (Mg# 90.7) reported from Kilauea. Below we examine the 1959 eruption in detail to gain insights on likely volatile contents and pressures of olivine crystallization for Kilauea volcano. We then use the constraints derived from the 1959 eruption to examine the range of liquid compositions parental to the Puna ridge glass compositions.

Kilauea Summit Eruption of 1959. The 1959 summit eruption lasted from 14 November to 20 December 1959, forming a cinder cone (Puu Puai) and a large lava lake in Kilauea Iki pit crater. The eruption consisted of seventeen phases of high fountaining and lava output. Of particular importance is that observations were made of lava fountaining heights (maximum observed 580 m) and temperatures (using optical pyrometers; maximum observed fountain temperature 1190 °C; Ault et al., 1961). In addition, the eruptive products of individual fountaining events were sampled. The most-magnesian glass (S-5g; Table 1) for a historical Kilauea eruption was sampled on 18 November 1959 from the hottest lava fountaining and the most-olivine-rich magma of the 1959 eruption (Murata and Richter, 1966). This eruption on 18 November provides us with a unique sample for which we have a liquid composition and its eruption temperature (1190 °C). Using the olivine geothermometer of Ford et al. (1983) in combination with the model of Falloon and Danyushevsky (2000), we calculate a olivine liquidus temperature of 1190 °C, if the glass composition S-5g (also known as Iki-22 in some publications) was a liquid at 0.2 GPa containing 0.93 wt% H_2O. This calculation was performed at 0.2 Gpa, as this is the maximum pressure of olivine crystallization for the 1959 eruption, based on the study of primary CO_2 fluid inclusions by Anderson and Brown (1993). The volatile content is calculated so as to cause the appropriate liquidus depression to

match the observed lava fountaining temperature, using the model of Falloon and Danyushevsky (2000)—in the absence of 0.93 wt% H_2O, the calculated liquidus temperature for S-5g is 1261 °C, i.e., 71 °C above the observed eruption temperature. This model volatile content results in an H_2O/K_2O value of 1.94, close to the maximum observed values in olivine hosted melt inclusions from the eruption (Anderson and Brown, 1993; see Fig. 1) and values in the submarine Puna ridge glasses (Clague et al., 1991; Clague et al., 1995). During crystal fractionation, both H_2O and K_2O should be highly incompatible and thus should increase together in an evolving magma. The lack of correlation between H_2O and K_2O (Fig. 1) indicates that significant degassing of H_2O has occurred (Wallace and Anderson, 1998). It is therefore appropriate to take the higher H_2O/K_2O values as representative of the undegassed magma. Using the calculated H_2O/K_2O value for sample S-5g and applying it to the lowest MgO glass from this eruptive episode (sample F-3g, Table 1), we calculate a temperature of 1064 °C at 0.2 GPa, which closely matches the recorded temperature of the coolest fountains (1060 °C; Ault et al., 1961). In summary, we believe that our modeling using the Ford et al. (1983) geothermometer in combination with the model of Falloon and Danyushevsky (2000) can accurately reproduce the eruption temperatures observed during the

TABLE 1. GLASS COMPOSITIONS CHOSEN FOR STUDY

Number	1	2	3	4	5	6	7	8	9	10	11	12
Sample name	F-3g	S-5g	57-11c	57-13g	57-15D	57-9F	182-7	REg	ODP896A	D20-3	Vema	NO42
(wt%)												
SiO_2	50.12	48.69	48.83	48.7	49.59	51.65	51.24	47.96	49.07	49.15	49.66	48.32
TiO_2	3.52	2.47	1.92	1.9	2.09	2.27	2.17	2.98	0.64	0.96	0.86	0.78
Al_2O_3	13.36	12.39	11.38	10.99	12.03	13.43	13.4	14.2	16.08	16.99	17.77	15.81
Fe_2O_3	2.49	1.52	1.6	1.52	1.38	1.04	1.36	1.78	0.67	0.60	0.49	1.06
FeO	10.0	10.02	11.25	10.53	9.98	7.93	9.30	9.3	8.53	7.61	6.61	8.41
MnO	0.18	0.18	N.D.	N.D.	N.D.	N.D.	0.16	0.2	0.12	0.13	0.13	0.18
MgO	5.68	10.12	13.29	14.92	12.53	9.73	8.42	7.09	9.44	9.88	9.51	9.80
CaO	9.61	10.95	9.01	8.67	9.46	10.63	10.7	11.3	13.67	12.13	12.45	13.79
Na_2O	2.57	2.03	1.65	1.63	1.77	2.03	2.16	2.62	1.62	2.41	2.32	1.63
K_2O	0.71	0.48	0.3	0.32	0.33	0.36	0.31	0.76	0.02	0.02	0.06	0.04
P_2O_5	0.39	0.24	0.20	0.19	0.21	0.23	0.19	0.29	0.03	0.05	0.05	0.09
Cr_2O_3	N.D.	N.D.	N.D.	N.D.	N.D.	N.D.	N.D.	N.D.	0.07	N.D.	N.D.	N.D.
H_2O	1.38	0.93	0.58	0.63	0.64	0.7	0.6	1.52	0.05	0.07	0.10	0.10
Mg#	50.3	64.3	67.8	71.6	69.1	68.6	61.7	57.6	66.4	69.8	71.9	67.5
Temp*	1064	1190	1269	1301	1253	1189	1157	1100	1200	1223	1207	1202
Oliv Eq[†]	74.8	85.0	86.6	88.7	87.4	87.1	82.9	81.1	86.2	88.3	89.3	86.9

Data sources: Glasses 1 and 2, Murata and Richter (1966); glasses 3–6, Clague et al. (1995); glass 7, Garcia et al. (1985); glass 8, Tilley et al. (1971); glass 9, glass ODP Leg 896A-27R-1, 124-130, pc.15, McNeill and Danyushevsky, 1996; glass 10, Danyushevsky et al. (2003); glass 11, Sobolev et al. (1989); glass 12, Sigurdsson et al. (2000).

Notes: Fe_2O_3 and FeO contents of all glasses have been calculated at QFM + 0.5 log units (glasses 1–8, 12) or QFM – 0.5 log units (glasses 9–11) at 0.2 GPa using the model of Borisov and Shapkin (1990). H_2O contents are either analyzed in the glass (glasses 9–11) or calculated from K_2O contents (glasses 1–8, 12; see text for discussion). All calculations performed with software PETROLOG (Danyushevsky, 2001). N.D.—no data.

*Calculated olivine liquidus temperature (°C) at 0.2 GPa using the models of Ford et al. (1983) and Falloon and Danyushevsky (2000).

[†]Calculated equilibrium olivine composition at the 0.2-GPa liquidus temperature.

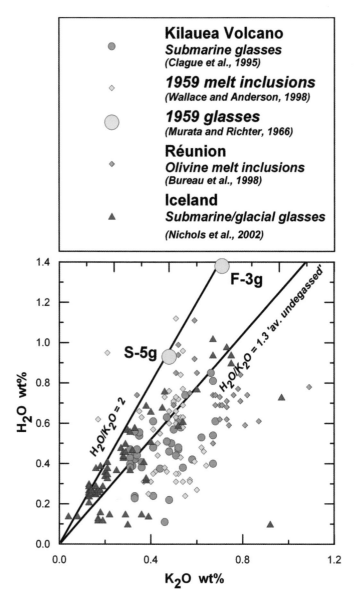

Figure 1. H_2O versus K_2O contents of glasses and melt inclusions from ocean island basalt "hotspot" tholeiite series magmas. The H_2O contents of the 1959 glasses are calculated using the observed eruption temperatures as a constraint (see text for discussion). The line "av. undegassed" is from the study of Wallace and Anderson (1998).

1959–1960 eruptions of Kilauea volcano. We therefore use a pressure of 0.2 GPa and a H_2O/K_2O value of 1.94 for the purposes of calculating parental liquid compositions to Kilauea volcano (see below).

Puna Ridge Glasses. In Figure 2, the FeO^T and MgO contents of glasses recovered from the Puna ridge are compared with glasses and whole-rock compositions with >8.5 wt% MgO from Kilauea volcano. The blue arrow in Figure 2 delineates an olivine control line defined by the whole-rock compositions of the 1959 summit eruption. This control line defines mixtures of

variably fractionated liquids and olivine crystals with an average composition of ~Mg# 87 (Murata and Richter, 1966). The whole-rock composition (S-5) for the glass S-5g was the most-olivine-rich magma of the 1959 eruption (Fig. 2). As can be seen from Figure 2, the Puna ridge glasses display a significant range in FeO^T and MgO contents compared to the whole-rock data analyzed from Kilauea volcano. Four of the Puna ridge glasses are of particular interest (Table 1). Sample 57–11c is the most-mag-

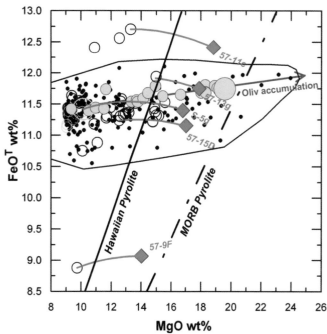

Figure 2. FeO^T versus MgO contents of whole rocks and glasses >8.5 wt% MgO from Kilauea volcano, Hawaii. Symbols and lines: small yellow circles—whole-rock compositions from the 1959 Kilauea summit eruption; small open yellow circles—glass compositions from the 1959 Kilauea summit eruption; large open yellow circle—glass S-5g; large yellow circle—whole-rock composition S-5; blue line—an olivine control line representing olivine accumulation of ~Mg# 87; small black circles—other Kilauean whole-rock compositions; open black circles—other Kilauean glass compositions; red diamonds—calculated 0.2-GPa parental liquid compositions (see Table 2); red lines—olivine addition paths (reverse of fractional crystallization); solid black line—locus of solidus melts for Hawaiian pyrolite composition (Green and Falloon, 1998); dash-dot black line—locus of solidus melts for mid-ocean ridge basalt (MORB) pyrolite composition (Green and Falloon, 1998). Both lines are calculated using the equations of Herzberg and O'Hara (2002). Data sources: 1959 glasses, Murata and Richter (1966), Helz (1987); 1959 whole rocks, Murata and Richter (1966), Basaltic Volcanism Study Project (1981); other Kilauea glasses, Moore (1965), Clague et al. (1991, 1995), Garcia et al. (1996, 2003); other Kilauea whole rocks, MacDonald and Eaton (1955), Tilley (1960), Muir and Tilley (1963), Moore (1965), Murata and Richter (1966), Richter and Moore (1966), Moore and Evans (1967), Aramaki and Moore (1969), Moore and Koyanagi (1969), Gunn (1971), Wright (1971), Wright and Fiske (1971), Helz (1980), Basaltic Volcanism Study Project (1981), Wilkinson and Hensel (1988), Garcia et al. (1992, 1998, 2000, 2003); Clague et al. (1995), Mangan et al. (1995), Chen et al. (1996), Norman and Garcia (1999), Quane et al. (2000).

nesian glass of a high-FeO^T subset of glasses (Fig. 2), whereas sample 57–9F is the only example of a relatively low-FeO^T glass. As can be seen from Figure 2, both the high- and low-FeO^T glasses have no matching whole-rock compositions. That is, so far, no magma composition has been analyzed from Kilauea volcano that is consistent with olivine accumulation into a high or low-FeO^T liquid composition. Glass samples 57–13g and 57–15D are important, as they both contain olivine microphenocrysts (<20 μm) with core compositions ranging to Mg# 90.7, the most magnesian observed from Kilauea volcano. As the olivines are small microphenocrysts, we can be reasonably confident that these microphenocryst have crystallized from a parental liquid related to the evolved glass composition. As well, both samples 57–13g and 57–15D have FeO^T contents within the array defined by whole-rock compositions from Kilauea volcano (Fig. 2).

Calculation of Parental Melts for Kilauea Volcano. In Table 2 and Figure 2, we present the results of parental liquid compositions at 0.2 GPa using a target olivine composition of Mg# 90.7. We have performed calculations for S-5g and the four Puna ridge glasses. Note that in Figure 2, the liquid lines of descent produced by adding olivine incrementally back into the glasses produce curved as opposed to straight lines. This observation emphasizes the point made by Irvine (1977) and Albarede

(1992) that extrapolation of olivine control lines (due to olivine accumulation) will result in incorrect calculated parental liquid compositions. Calculated parental compositions range in MgO contents from 14 to 18.8 wt% with crystallization temperatures ranging from 1286 to 1372 °C. As we do not know whether the target olivine is appropriate for both the high-FeO^T and low-FeO^T glasses, the best-constrained parental compositions are for glasses 57–13g and 57–15D. The parental liquids for these two compositions have 17–18 wt% MgO and crystallization temperatures of 1340–1355 °C.

Mauna Loa Volcano. The parental composition for Mauna Loa volcano was calculated in a similar manner to the Kilauea parental compositions. We used the most-MgO-rich glass recovered from the submarine southwest rift of Mauna Loa volcano (glass 182-7, Table 1). Sample 182-7 is particularly significant, as it contains the most-magnesian olivine phenocryst composition (Mg# 91.3) obtained from Mauna Loa volcano. Using this target olivine, we calculate a parental liquid composition containing 17.53 wt% MgO at a temperature of 1354 °C (Table 2). As can be seen from Table 2, the parental liquid and temperature is essentially identical to those for samples 57–13g and 57–15D from Kilauea volcano. We believe that this result further supports the conclusion that samples 57–13g and 57–15D provide the most reliable estimates for Kilauea parental liquid compositions.

TABLE 2. CALCULATED MODEL PARENTAL LIQUIDS AT 0.2 GPA

Number	Kilauea					Mauna Loa	Réunion		MORB		Iceland
	1	2	3	4	5	6	7	8	9	10	11
Sample name	S-5g	57-11c	57-13g	57-15D	57-9F	182-7	REg	ODP896A	D20-3	Vema	NO42
(wt%)											
SiO$_2$	47.24	47.53	48.01	48.48	50.44	48.74	46.23	47.70	48.29	49.36	47.00
TiO$_2$	2.03	1.62	1.73	1.83	2.02	1.67	2.31	0.54	0.86	0.83	0.64
Al$_2$O$_3$	10.21	9.59	10.04	10.54	11.95	10.30	10.98	13.46	15.25	17.17	13.03
Fe$_2$O$_3$	1.58	1.80	1.64	1.51	1.12	1.54	1.51	0.85	0.69	0.51	1.40
FeO	9.96	10.77	10.26	9.78	8.05	9.38	9.91	8.56	7.69	6.68	8.15
MnO	0.15	N.D	N.D	N.D	N.D	0.12	0.15	0.10	0.12	0.12	0.15
MgO	16.79	18.80	17.86	17.00	14.00	17.53	16.17	15.86	13.93	10.86	16.75
CaO	9.03	7.59	7.92	8.28	9.46	8.22	8.73	11.44	10.88	12.03	11.37
Na$_2$O	1.67	1.39	1.49	1.55	1.81	1.66	2.02	1.36	2.16	2.24	1.34
K$_2$O	0.39	0.25	0.29	0.29	0.32	0.24	0.59	0.02	0.02	0.06	0.03
P$_2$O$_5$	0.20	0.17	0.18	0.18	0.20	0.15	0.22	0.03	0.04	0.05	0.07
Cr$_2$O$_3$	N.D.	N.D.	N.D.	N.D.	N.D.	N.D.	N.D.	0.06	N.D.	N.D.	N.D.
H$_2$O	0.76	0.49	0.57	0.56	0.62	0.46	1.18	0.04	0.06	0.10	0.08
Mg#	75.0	75.7	75.6	75.6	75.6	76.9	74.4	76.8	76.3	74.3	78.6
Temp*	1335	1372	1355	1341	1286	1354	1323	1351	1320	1243	1361
Oliv Eq[†]	90.70	90.70	90.70	90.70	90.70	91.30	90.65	91.60	91.50	90.50	92.40
% Oliv[§]	17.6	15.7	8.6	12.4	11.0	23.1	22.7	16.3	10.3	3.4	17.5

Notes: Data sources as for Table 1. All calculations performed with software PETROLOG (Danyushevsky, 2001). N.D.—no data; MORB—mid-ocean ridge basalt.
*Calculated olivine liquidus temperature at 0.2 GPa using the models of Ford et al. (1983) and Falloon and Danyushevsky (2000).
[†]Calculated equilibrium olivine composition at the 0.2-GPa liquidus temperature.
[§]Amount of olivine added in incremental steps to obtain equilibrium with the target olivine compositions (see text for discussion).

Réunion: The 1939 Eruption of Piton de la Fournaise

The island of Réunion consists of two shield volcanoes, Piton des Neiges (3069 m) and Piton de la Fournaise (2631 m). The former is extinct and deeply eroded, whereas the latter is an active volcano. Picrites and olivine basalts from both volcanoes are referred to as the oceanite series. To model the composition of a parental magma for the oceanite series lavas of Réunion, we have chosen the melt composition from the oceanite magma eruption of 6 January 1939, from Piton de la Fournaise volcano. The whole-rock composition of the 1939 magma had 15.21 wt% MgO, and the melt component of the magma, represented by glass REg, has 7.18 wt% MgO (Table 1). The target olivine for our calculations is the most-magnesian olivine phenocryst composition reported from the oceanite lavas of Réunion of Mg# 90.65 (Fretzdorff and Haase, 2002). The H_2O contents of olivine hosted melt inclusions in oceanite series lavas from Réunion suggest parental liquids have H_2O/K_2O values of ~2 (Bureau et al., 1998; Fig. 1), essentially identical to Kilauea liquids. The calculated parental liquid to the REg glass composition at 0.2 GPa, has ~16 wt% MgO and a crystallization temperature of 1323 °C (Table 2), within the range calculated for parental liquids to Kilauea.

Iceland: Theistareykir Volcanic System in Northeastern Iceland

The most-magnesian olivine phenocryst composition that has been reported from Iceland has Mg# of 92.4 (Sigurdsson et al., 2000). This olivine phenocryst comes from the Borgarhraun lava flow, which is part of the Theistareykir volcanic system and has an average whole-rock MgO content of 12 wt% (Maclennan et al., 2003). The liquid composition is represented by glass NO42, which has 9.80 wt% MgO (Sigurdsson et al., 2000). The water content in this glass has not been analyzed but can be estimated from its K_2O content. Nichols et al. (2002) analyzed H_2O contents of submarine and subglacial pillow basalts and concluded the Icelandic mantle source is relatively enriched in water. As can be seen from Figure 1, water contents in Icelandic lavas are very similar to Kilauea and Réunion. For the purposes of our calculation of a liquid composition parental to the glass NO42, we have used a H_2O/K_2O value of 2.5, which is close to the undegassed value for Icelandic magmas. The calculated parental liquid (Table 2) has 16.75 wt% MgO and a crystallization temperature of 1361 °C at 0.2 GPa.

COMPARISON BETWEEN "HOTSPOT" AND SPREADING-RIDGE THOLEIITES

The thermally driven plume hypothesis predicts that there should exist a significant thermal anomaly between upwelling plume material and the temperature of ambient upper mantle. There is a general consensus that MORBs are the result of melting from upwelling ambient upper mantle at oceanic spreading ridges. It would therefore seem a straightforward matter to compare parental liquid olivine crystallization temperatures between OIB and MORB. It is common in the literature to find the temperature of 1280 °C quoted and used as a sort of benchmark for T_P of the upwelling ambient upper mantle after the study of McKenzie and Bickle (1988). However, it is important to recognize that this mantle T_P of 1280 °C represents an average, not the hottest, MORB mantle T_P. Using this average temperature, it would appear obvious that a thermal anomaly exists. Indeed our own calculations for Kilauea volcano (see Table 5 below) would indicate a thermal excess of ~285 °C, compared to an ambient upper mantle T_P of 1280 °C, a result consistent with the thermal plume hypothesis. We, however, strongly disagree with the use of an average temperature for MORB when there is clearly a range in olivine crystallization temperatures for MORB glasses. This spread in values is demonstrated by the range in FeO^T versus MgO contents of MORB glasses (Fig. 3). In Figure 3 we have plotted the FeO^T and MgO contents of 682 MORB glass analyses from the study of Danyushevsky (2001). These glass compositions were all determined by electron microprobe analysis using the glass standard VG-2 (U.S. National Museum no. 111240/2; Jarosewich et al., 1980). Also plotted in Figure 3 are FeO^T and MgO contents of 189 glass compositions from the Petdb database (petdb.ldeo.columbia.edu). These 189 glasses were all determined by electron microprobe analysis (we note that many reported glasses in the Petdb database are in fact glass-rich rocks and whole-rock analyses) and have MgO contents >9.5 wt%. Consequently all these glasses (liquids) are in the olivine-only phase field at low pressure. As can be seen from Figure 3, the MORB glasses from the two independent data sets display a significant range in FeO^T contents, and consequently there must exist a range in temperatures of MORB parental liquids. This temperature range must occur, because if all calculation parameters are held constant, higher FeO^T contents (of liquids of MORB-like chemical composition) require higher temperatures to be in equilibrium with the same olivine composition.

In this article, we wish to compare temperatures between the hottest of MORB parental liquids and the hottest of the OIB parental liquids, not averages. We believe it is appropriate to take the hottest of MORB parental liquids as representative of melts derived from adiabatic upwelling of ambient upper mantle. This decision is reasonable because there are several processes —e.g., nonadiabatic upwelling, cooling, mantle-melt reaction— that could conceivably lead to lower temperatures in MORB parental liquids. If the lowest temperatures are taken as representative of ambient upper mantle, this choice provides no explanation for why hotter MORBs exist, apart from the circular argument that they are plume related and represent a depleted, MORB-source-like component within the plume. The discussion illustrates the difficulty of testing the deep-mantle thermal plume hypothesis. In its most clearly stated form, the hypothesis asserts that spreading centers sample modern well-mixed upper mantle with normal MORB (N-MORB) trace element signatures. Mantle plumes sample compositionally distinct (OIB trace element and isotopic characteristics), heterogeneous mantle sources residing in the deep mantle. To the extent that these distinctions between magma and source characteristics are abandoned, the hypothesis requires modification or rejection. Just as circularity in argument is introduced if MORB-type (geochemically) and picritic magmas indicating high temperature are said to be influenced by (or be part of) a nearby plume, so circularity of argument is introduced by blurring of geochemical distinctions. The enriched, OIB-like geochemical signatures at many spreading centers with no obvious nearby plume are often attributed to a plume-component but without any justification. For these reasons, we have sought to establish

magma temperatures and model mantle potential temperatures for the classical, geodynamically defined "fixed" hotspots and on geodynamically defined spreading centers, and the latter samples chosen for detailed scrutiny have N-MORB or depleted MORB (D-MORB) geochemistry. Anticipating our possible conclusion, if we can find no significant temperature difference between parental MORB and hotspot magmas, we seriously undermine the deep-mantle thermal plume model but are left with evidence for heterogeneous mantle, geochemical differences between MORB and OIB sources, and the hotspot source and/or cause below and detached from the moving lithospheric plate.

To compare crystallization temperatures with MORB, we have calculated parental liquid compositions at 0.2 GPa for three representative MORB glasses with differing FeO^T contents (Table 1, sample numbers 9–11). These glasses are:

1. The most-magnesian glass (Table 1, number 9) from Ocean Drilling Program (ODP) site 896A, drilled into ca. 5-Ma crust formed by the Cocos-Nazca spreading ridge (herein referred to as ODP896A; McNeill and Danyushevsky, 1996). This glass contains microphenocrysts with core compositions ranging to Mg# 91.6.
2. Glass D20–3 (Table 1, number 10) from the Siqueiros fracture zone, East Pacific Rise (Danyushevsky et al., 2003). The most-magnesian olivine phenocryst observed in the Siqueiros fracture zone has Mg# of 91.5.
3. The composition (Table 1, glass 11) of a magnesian olivine-hosted glass inclusion from the Vema fracture zone on the

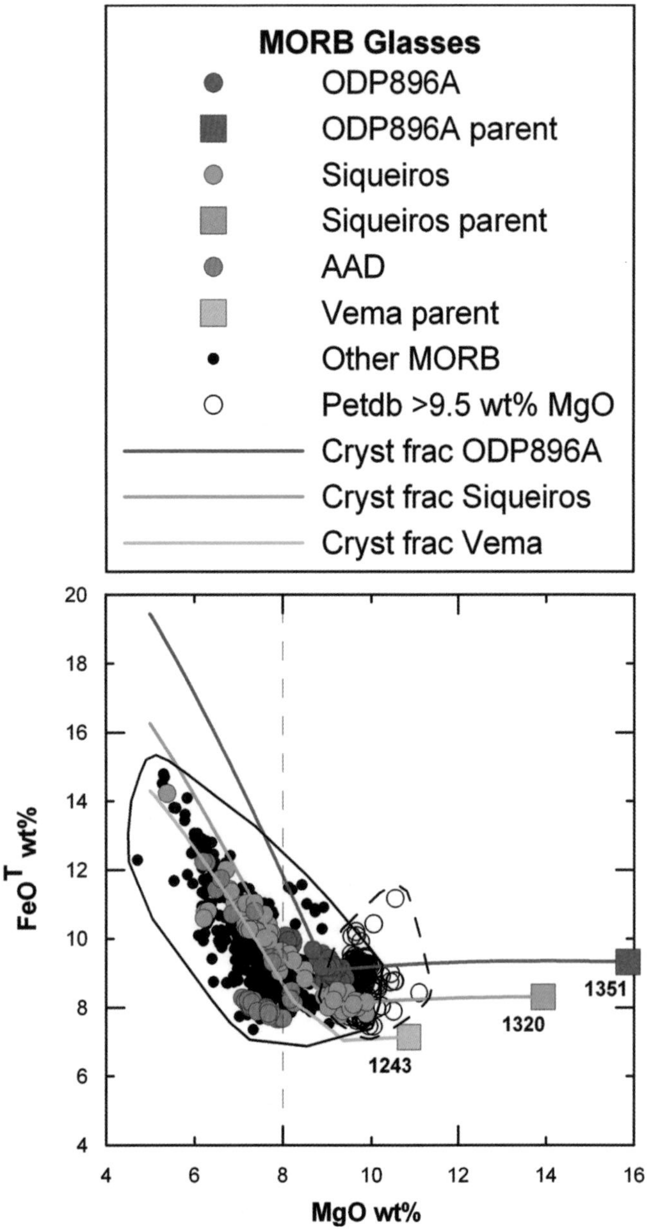

Figure 3. FeO^T versus MgO wt% of mid-ocean ridge basalt (MORB) glass compositions. Symbols, lines, and fields: Blue square—parental liquid to MORB glass ODP896A (Table 2, number 8); green square—parental liquid to MORB glass D20–3 (Table 2, number 9); cyan square—parental liquid to MORB glass Vema, Mid-Atlantic Ridge (Table 2, number 10); blue circles—glass compositions from hole ODP896A, Cocos-Nazca (Danyushevsky, 2001); green circles—glass compositions from Siqueiros fracture zone, East Pacific Rise (Danyushevsky et al., 2003); magenta circles—glass compositions from the Australian-Antarctic discordance (AAD; Danyushevsky, 2001); small black circles—MORB glasses from the study of Danyushevsky (2001); small open black circles—glasses from the Petdb database with >9.5 wt% MgO, as determined by electron microprobe analysis (petdb .ldeo.columbia.edu); solid blue, green, and cyan lines—calculated liquid lines of descent for each respective parental liquid compositions undergoing crystal fractionation of olivine, clinopyroxene, and plagioclase at 0.2 GPa. Calculations were performed using the software PETROLOG and the methods of Danyushevsky (2001). MORB glass field delineated by solid black line encompasses the range of FeO^T and MgO contents for 682 glasses from Danyushevsky (2001). MORB glass field delineated by dashed black line encompasses the range of FeO^T and MgO contents of 189 glasses with >9.5 wt% MgO, as determined by electron microprobe analysis from the Petdb database (petdb.ldeo.columbia.edu). Temperatures of MORB parental liquid compositions are calculated using the olivine geothermometer of Ford et al. (1983) at 0.2 GPa (see text and Table 2 for details). Dashed line at 8 wt% MgO is the MORB reference MgO content after Klein and Langmuir (1987).

Mid-Atlantic Ridge from the study of Sobolev et al. (1989). This glass inclusion composition is important, as it is representative of a relatively low-FeO^T liquid compared with the other parental MORB liquid compositions. The maximum olivine phenocryst composition observed in the Vema fracture zone has Mg# of 90.5.

The H_2O contents listed in Table 1 have all been determined on the glasses themselves. In Table 2 and Figure 3 (see also Fig. 5 below), we present the results of our parental liquid calculations for all three compositions. The range in MgO contents in parental liquids varies from ~11 to 16 wt% MgO, and crystallization temperatures at 0.2 GPa range from 1243 to 1351 °C.

A significant result of our calculations (Table 2) is that the olivine crystallization temperature of the representative ODP896A parental liquid is significantly hotter than the commonly quoted ambient upper-mantle T_P value of 1280 °C and is higher than the values inferred for MORB by Presnall et al. (2002), based on experimental studies on simple systems (average mantle potential temperature required ~1260 °C). We have chosen this particular high-FeO^T MORB glass from ODP896A simply because we have detailed information on both its mineralogy and geochemistry. Such detailed information, required for our calculations, is currently not available from the vast majority of studies that report MORB glass geochemistry. The ODP896A glass has normal N-MORB geochemistry (L.V. Danyushevsky, unpubl. data, 2006), and we believe it is representative of other high-FeO^T MORB glasses for which we do not have detailed information on mineralogy. Over half the glasses in the Petdb database ($n = 80$) with MgO contents >9.5 wt% have $FeO^T > 9$ wt%, similar to the ODP896A glass. The average FeO^T of these high FeO^T (>9 wt%) glasses is 9.3 ± 0.3, thus the ODP896A glass ($FeO^T = 9.1$) has slightly below-average FeO^T compared to these glasses. These high-FeO^T glasses are sampled from a range of "normal" crustal thickness spreading ridges (Carslberg Ridge, Chile Ridge, Easter Microplate rifts, East Pacific Rise, Galapagos Spreading Center, Juan De Fuca Ridge, Mid-Atlantic Ridge, and Red Sea) and are associated with low-FeO^T glasses both in time and space. Although primitive high-MgO and high-FeO^T glasses, such as ODP896A, are rarely sampled at spreading centers, glass compositions derived from such parents are relatively common. Dmitriev et al. (1985), using an eight-component discriminant function, determined that at least 37% of MORB glasses from the Mid-Atlantic Ridge were derived from such high-FeO^T parental compositions.

If we use our approach and calculate parental compositions for all the high-MgO glasses (>9.5 wt%; $n = 189$) from the Petdb database, using K_2O abundances as an estimate for H_2O contents and an olivine target composition of Mg# 91.5, then calculated parental liquids at 0.2 GPa display a significant range in both MgO contents (13–18 wt%) and temperatures (1300–1430 °C); however, the averages (15.1 ± 0.9 wt% MgO; 1343 ± 19 °C) are very close to the calculated values for the ODP896A

parental composition for which we have detailed petrological information. To determine the true range in composition and crystallization temperatures of MORB parental liquids requires more detailed petrological and mineralogical studies from a range of different spreading ridges. The uncertainty in calculations for which only electron probe analyses of glass compositions are available is that we do not know the actual H_2O contents, nor do we have any idea of the maximum Mg# of olivine phenocrysts. It may well be that we are underestimating the temperatures of some of the high-FeO^T glasses, as some picrite suites with similar MORB geochemistry have olivines up to Mg# 93–94 (e.g., Baffin and Gorgona Island picrites). It is a circular argument that assigns these high temperature picrite suites to the influence of mantle plumes: they may well be derived from normal ambient upper mantle.

In summary, therefore, we believe that when comparisons are made between olivine crystallization temperatures of parental liquids for OIB and MORB magmas, it is the temperatures inferred for parental liquids for high-FeO^T MORB glass suites that should be used to test the thermal plume hypothesis. The result of our calculations suggest there are no significant differences in the olivine crystallization temperatures when OIB and MORB parental liquids are compared (the differences between MORB and the best-constrained calculations for Kilauea volcano are essentially zero, and the Icelandic parent is only 10 °C hotter). If other picrite suites with MORB geochemistry are used in this comparison, then OIB parental liquids would be significantly cooler than MORB parental liquids.

MANTLE POTENTIAL TEMPERATURES

As mentioned previously, the calculation of parental liquid compositions is only the first step in the process of calculating a mantle T_P. To calculate mantle T_P values, it is necessary to make a number of assumptions concerning source compositions, mantle melting, and the melt segregation process. It is beyond the scope of this article to present any comprehensive argument or defense of any particular mantle T_P. What we do in this section is present some simple mantle T_P calculations based on the following assumptions:

1. For comparative purposes, we first assume that the mantle sources for both OIB and MORB primary melts are of pyrolite (lherzolite) composition, with bulk CaO and Al_2O_3 contents within the range of ~3–4 wt%. We believe the assumption of a peridotitic (pyrolite) source is justified by the magnesian contents of both the calculated parental liquids and co-existing olivines (Table 3). The production of primary melts of MgO contents of between 10 and 19 wt% MgO, and olivines of Mg# 90.5–92.4 requires an olivine-rich peridotite source. Many studies of Hawaiian volcanism have argued the need for recycled crustal components in the mantle source in the form of eclogite and/or pyroxenite (Takahashi and Nakajima, 2002; Sobolev et al., 2005).

However, numerous experimental melting studies have demonstrated that eclogite and pyroxenite source components are incapable of producing any of the calculated parental liquids unless there has either been extensive reaction with peridotite before melt segregation (Green and Falloon, 2005) or, alternatively, mixing with a dominant peridotite melt component before eruption (Sobolev et al., 2005). In Figure 2, we have plotted the locus of FeO^T and MgO contents of initial melts (i.e., at the solidus) for both Hawaiian and MORB pyrolite compositions (Green and Falloon, 1998), using the equations of Herzberg and O'Hara (2002). Melts formed above the solidus by higher degrees of melting lie to the right of these lines. As can be seen from Figure 2, it is impossible for a MORB pyrolite composition source to produce melt compositions with appropriate FeO^T and MgO contents that match the calculated parental liquid compositions to Kilauea volcano. This observation was used as an argument by Herzberg and O'Hara (2002) for more-magnesian parental liquids for Hawaii. For a mantle composition, such as MORB pyrolite (7.55 wt% FeO^T), to be a source for Kilauea parental liquids, MgO contents, in the case of the 57-13 g parental liquid, would need to be >20 wt% MgO with olivine of >Mg# 91.9 at 0.2 GPa. So far there is no evidence for such magnesian olivine compositions from Hawaiian volcanoes. However, if the mantle source has FeO^T and MgO contents similar to Hawaiian pyrolite composition (9.5 wt% FeO^T; Green and Falloon, 1998), then the calculated parental liquid compositions are possible primary melts of this mantle composition.

The Hawaiian pyrolite composition is believed to be a result of a complex history of previous melt extraction followed by mantle refertilization by small-degree melts from recycled subducted oceanic crust interacting with refractory lithosphere (harzburgite) and normal mantle (Yaxley and Green, 1998; Green and Falloon, 2005). We have previously suggested that, rather than assuming that the source composition (Hawaiian pyrolite) should resemble the most-fertile mantle-derived lherzolites with 3–4 wt% CaO and Al_2O_3 (Green and Falloon, 1998), and thus yielding parental tholeiitic picrites by 30–40% melting (see Table 5), a more appropriate source is a refertilized harzburgite with ~1.5–2.0 wt% CaO and Al_2O_3 (Green et al., 2001). Such a harzburgitic source would be compositionally buoyant if it is at a temperature similar to the enclosing MORB pyrolite, and on upwelling would yield ~10% tholeiitic picrite with harzburgite residue. We therefore make alternative assumptions in our mantle T_P calculations that either a very fertile lherzolite (Hawaiian pyrolite) or a refertilized harzburgite is a suitable mantle source composition for OIB primary melts. The two assumptions demonstrate the importance of the melt fraction and latent heat of melting estimates to calculation of mantle potential temperature. For MORB, we assume a composition similar to peridotite MM-3 (Baker and Stolper, 1994; see Fig. 5 below) and

MORB pyrolite (Green et al., 1979) as suitable mantle source compositions—both of these are lherzolite compositions with 3–4 wt% CaO and Al_2O_3.

2. We assume that the melting process can be closely modeled by simple batch melting, and that the compositions of unmodified primary melts segregate from their mantle sources as fully integrated and equilibrated melt compositions. Thus, based on this assumption, we should be able to find at some P and T a close match in composition between our calculated parental liquids and experimental melts from our assumed mantle source if the parental liquids are indeed primary mantle derived melts. Justification for this assumption is presented in Green and Falloon (1998), and in the case of Kilauea, is supported by the petrogenetic model presented by Eggins (1992a,b). Our assumption of batch melting means that we can simply calculate the degree of partial melting of our mantle source composition by using the abundances of the relatively incompatible minor elements TiO_2, Na_2O, and K_2O in our calculated parental liquids.

3. We ignore the possible effect of the relatively minor amounts of H_2O and CO_2 that are contained in our calculated parental compositions, on the positions of mantle melting cotectics within the normative tetrahedron. We are not focusing on the C-H-O–controlled solidus and incipient melting regime but on the onset of the major melting regime (Green and Falloon, 1998, 2005). The main effect of H_2O is to produce a higher degree of partial melt at a given temperature compared to volatile-free conditions. However, when melt compositions are compared on an anhydrous basis, no significant differences are apparent either in the compositions of experimental melts or in their normative positions (for H_2O approximately <1% in the melt).

4. We assume that the mantle source upwells adiabatically and undergoes partial melting when it crosses its solidus. This assumption is very important, as mantle T_P calculations presented will be in error if upwelling is not adiabatic (c.f., Ganguly, 2005). For these models of tholeiitic picrite genesis, we assume that the "solidus" is actually the entry into the major melting regime of the peridotite-C,H,O system (see assumption 3 above; Green and Falloon, 1998, 2005; Fig. 3).

5. To calculate the amount of heat loss caused by melting of the mantle source, we have used the values for the latent heating of melting (100 cal/g) and heat capacity (0.3 cal/g/deg) recommended by Langmuir et al. (1992). For the two alternative source compositions (lherzolite and harzburgite) for Hawaiian or other hotspot magmas, the heat loss calculated is very different, such that ΔT_P for different melt fractions is ~3.3 °C per 1% melt (see Table 5).

6. We assume a mantle solid adiabat of 10 °C/GPa (Birch, 1952) and liquid adiabat of 30 °C/GPa (McKenzie and Bickle, 1988).

Depths of Mantle Equilibration

To determine mantle depths of mantle equilibration of our calculated parental liquids, we have projected their compositions from both olivine and diopside end-members into the normative tetrahedron and compared them with experimentally determined mantle melting cotectics (Figs. 4 and 5; Tables 3 and 4). In Table 5, we also present a pressure of mantle equilibration using the experimentally calibrated empirical model of Albarede (1992). Both methods give reasonably consistent results.

Kilauea Volcano. In Figure 4, the glasses and parental compositions from Kilauea are shown projected from olivine (Fig. 4A) or diopside (Fig. 4B) and compared with experimentally determined cotectics on the Hawaiian pyrolite composition. Green and Falloon (2005) have previously shown that compositions from a range of Hawaiian volcanoes define arrays of compositions in the olivine projection consistent with equilibrium with harzburgite residues but with differing Ca/Al ratios. In Figure 4A, it can be seen that the mantle source for the 1959 Kilauea summit eruption must have a higher Ca/Al value compared to the source for the Puna ridge glasses and other historical eruptions of Kilauea (e.g., data of Norman and Garcia, 1999; Norman et al., 2002). The Hawaiian pyrolite was based on the 1959 eruption compositions, and thus the Ca/Al ratio of Hawaiian pyrolite matches the S-5g glass in Figure 4A but is too high for the Puna ridge glasses.

In Figure 4B, the calculated parental compositions show a range of mantle equilibration pressures from ~1 GPa (57-9F) to 2.5 GPa (57-11c). The two best-constrained parental composi-

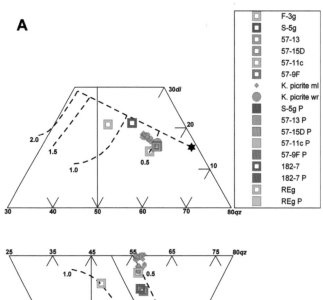

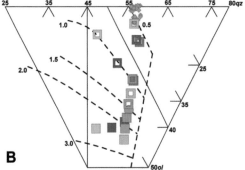

Figure 4. Projection of partial melting trends for Hawaiian pyrolite composition (dashed curves, 0.5–3.0 GPa), i.e., liquid compositions lying on lherzolite (olivine [ol] + orthopyroxene [opx] + clinopyroxene [cpx] + spinel) and harzburgite (ol + opx + chromian-spinel) within the normative basalt tetrahedron (Falloon and Green, 1988; Falloon et al., 1988; T.J. Falloon, unpubl. data, 2006). Projections are from (A) olivine and (B) diopside. Glasses and calculated parental liquid compositions (P) are from Tables 2 and 3. Kilauea picrite whole-rock (K. picrite wr) and melt inclusions (K. picrite ml) are from Norman and Garcia (1999) and Norman et al. (2002). Sample numbers in the key are from Tables 1 and 2.

TABLE 3. COMPARISON OF MODEL KILAUEA PARENTAL LIQUIDS WITH EXPERIMENTAL MELT COMPOSITIONS FROM HW-40 AT 2 GPA UNDER ANHYDROUS CONDITIONS

Glass name	T-2018	57-15D	57-13g	T-650
(wt%)				
SiO_2	48.15	48.48	48.01	50.19
TiO_2	2.35	1.83	1.73	1.74
Al_2O_3	10.76	10.54	10.04	8.61
Fe_2O_3	N.A.	1.51	1.64	N.A.
FeO	10.16	9.78	10.26	9.34
MgO	16.27	17.00	17.86	19.96
CaO	9.38	8.28	7.92	7.40
Na_2O	1.87	1.55	1.49	1.40
K_2O	0.39	0.29	0.29	0.32
P_2O_5	N.D.	0.18	0.18	N.D.
H_2O	N.D.	0.56	0.57	N.D.
Temp*	1500	1506	1523	1550
Mg#	74.1	75.6	75.6	79.2

Data sources: experimental glasses T-2018 and T-650, Falloon et al. (1988).
Note: N.D.—no data; N.A.—not applicable.
*The anhydrous olivine liquidus temperatures calculated at 2 GPa using the model of Ford et al. (1983) and assuming all Fe as Fe^{2+}.

tions (57-13g and 57-15D) are both consistent with a pressure of ~2 GPa and closely match experimental compositions from Hawaiian pyrolite at this pressure (Table 3). This pressure of mantle equilibration matches the expected pressure at the base of the oceanic lithosphere (~60 km) beneath Hawaii, based on geophysical studies (Eaton and Murata, 1960). Thus the parental liquid compositions calculated for 57-13g and 57-15D are consistent with the model in which a compositionally buoyant diapir (i.e., plume) is emplaced into the base of the oceanic lithosphere, and melt fractions are integrated and equilibrated with a harzburgite residue before segregation via melt channeling through the oceanic lithosphere to crustal magma reservoirs at ~0.2 GPa (~6 km) depth beneath Kilauea volcano. In such a model, the garnet signature observed in Hawaiian magmas is either the result of the integration of very small melt fractions derived from equilibrium with garnet over a range of pressures

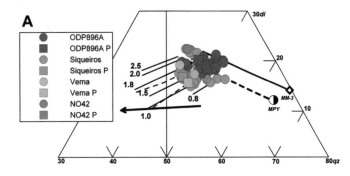

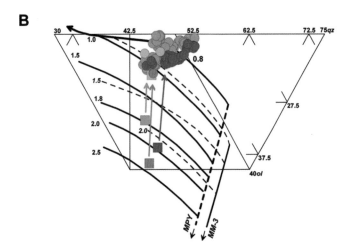

TABLE 4. COMPARISON OF MODEL ODP896A PARENTAL LIQUID WITH EXPERIMENTAL MELTS FROM MM-3 AND MPY-87 AT 2 GPA UNDER ANHYDROUS CONDITIONS

Melt name	T-4107	ODP896A	T-4190
(wt%)			
SiO_2	47.75	47.7	48.30
TiO_2	0.61	0.54	0.74
Al_2O_3	13.96	13.46	13.40
Fe_2O_3	N.A.	0.85	N.A.
FeO	8.49	8.56	8.01
MnO	N.D.	0.10	0.15
MgO	15.37	15.86	15.95
CaO	11.83	11.44	11.48
Na_2O	1.40	1.36	1.57
K_2O	N.A.	0.02	N.A.
P_2O_5	N.A.	0.03	N.A.
Cr_2O_3	0.34	0.06	0.36
H_2O	N.D.	0.04	N.D.
Temp*	1460	1474	1475
Mg#	76.3	76.8	78.0

Data sources: experimental glasses T-4107 and T-4190 (see appendix Table A3).
Note: N.D.—no data; N.A.—not applicable.
*The anhydrous olivine liquidus temperatures calculated at 2 GPa using the model of Ford et al. (1983) and assuming all Fe as Fe^{2+}.

Figure 5. Projection of partial melting trends for MORB-pyrolite (MPY; thick solid lines, 0.8–2.5 GPa) and MM-3 lherzolite composition (thin dashed lines, 1.0–2.0 GPa) within the normative basalt tetrahedron (Falloon and Green, 1988; Falloon et al., 1999; Falloon et al., 2001; T.J. Falloon, unpubl. data, 2006; Appendix 1). Solid line with arrow is the locus of liquids in equilibrium with plagioclase lherzolite (olivine + orthopyroxene + clinopyroxene + plagioclase ± chromianspinel) at 1 GPa—such liquids are low in normative diopside and olivine. Projections are from (A) olivine and (B) diopside. Glasses and calculated parental liquid compositions (P) are from Tables 2 and 3. Data for ODP896A, Siqueiros, and Vema glasses are from L.V. Danyushevsky (unpubl. data, 2006). Colored arrows represent olivine control lines from the respective parental liquid compositions.

(Eggins, 1992a,b) or is a source feature derived by reaction between melts caused by the incipient melting of eclogite and refractory harzburgite (Yaxley and Green, 1998).

However, the low-Fe (57-9F) and high-Fe (57-11c) glasses are inconsistent with this model. The calculated parental compositions to these glasses indicate pressures of ~1 GPa for 57-9F and ~2.5 GPa for 57-11c. It is very difficult to relate the differences in inferred pressures of mantle equilibration to any known mantle melting or reaction process, as both these glasses are colinear with glasses 57-13g and 57-15D in the projection from olivine in Figure 4A. That is, all the Puna ridge parental glass compositions define a single olivine control line in both Figure 4A and B, whereas Figure 2 indicates that it is impossi-

ble for them to be related by olivine fractionation because of their strong differences in FeO^T content. Some process other than mantle melting or reaction must be invoked to explain the compositions of glasses 57-11c and 57-9F. As both these glass compositions have no known matching whole-rock compositions, we consider them to be anomalous and therefore the mantle T_P calculated for their parental liquids should be treated with caution.

Mauna Loa. The parental composition to the glass 182-7 is also consistent with a mantle equilibration pressure of 2 GPa. This observation further supports the view that 2 GPa represents the pressure at the base of the oceanic lithosphere and that major element compositions of parental liquids reflect equilibration with a harzburgite mantle source at this pressure.

Réunion and Iceland. Both parental compositions give depths of mantle equilibration of ~2.5 GPa (Figs. 4B and 5B).

MORB. In Figure 5, we compare the parental compositions of our three representative MORB glass compositions with experimentally determined cotectics from melting experiments on peridotite composition, MM-3 and MORB pyrolite. The parental compositions define a range of pressures (~1–2 Gpa; Table 5) correlated with their FeO^T contents. In particular, the calculated parental composition to the glass ODP896A shows a very close match to experimental reaction compositions using the parental composition for the same glass calculated by McNeill and Danyushevsky (1996) and peridotite MM-3 (Table 4, Appendix 1). This result gives support to the assumption that

TABLE 5. PETROGENETIC SUMMARY AND MODEL CALCULATIONS OF MANTLE POTENTIAL TEMPERATURES

Parent	T (0.2 GPa)*	T (final)†	ΔT (adiabat)§	P (final)¶	P (final)‡	MgO (wt%)**	H_2O (wt%)††	Oliv Mg# (0.2 GPa)§§	Oliv Mg# (final)¶¶	Mantle residue‡‡	F (%)***	T_P†††
MORB												
ODP896A	1351	1441	36	2.00	2.01	15.86	0.04	91.6	90.5	Sp LHz	0.20	1488
Siqueiros	1320	1400	32	1.80	1.65	13.93	0.06	91.5	90.4	Sp LHz	0.13	1425
Vema	1243	1286	16	1.10	1.16	10.86	0.10	90.5	89.8	Sp LHz	0.13	1318
Iceland												
NO42	1361	1468	41	2.40	2.34	16.75	0.08	92.4	91.3	Hz	0.17	1501
Hawaii												
S-5g	1335	1436	41	2.20	2.27	16.79	0.76	90.7	89.4	Hz	0.33	1524
											0.12	*1454*
57-13g	1355	1451	39	2.10	2.16	17.86	0.57	90.7	89.5	Hz	0.40	1563
											0.14	*1477*
57-15D	1341	1422	33	1.80	1.93	17.00	0.56	90.7	89.7	Hz	0.38	1531
											0.14	*1451*
57-11c	1372	1488	47	2.50	2.44	18.8	0.49	90.7	89.3	Hz	0.44	1610
											0.16	*1516*
57-9F	1286	1326	16	1.00	1.23	14.00	0.62	90.7	90.1	Hz	0.34	1429
											0.12	*1356*
Mauna Loa												
182-7	1354	1445	37	2.00	1.92	17.53	0.46	91.3	90.3	Hz	0.42	1565
											0.15	*1475*
Réunion												
REg	1323	1440	48	2.50	2.52	16.17	1.18	90.65	89.1	Hz	0.26	1502
											0.09	*1445*

Notes: Values in italics for F and for T_P refer to calculations based on a refractory Hawaiian Pyrolite composition (harzburgite + 10% picrite). Hz—harzburgite ± Cr-spinel; MORB—mid-ocean ridge basalt; Sp LHz—spinel lherzolite.

*Calculated olivine crystallization temperature of the parental liquids at 0.2 GPa (see Table 2).

†Calculated olivine crystallization at the pressure of mantle equilibration; see the column P (final)‡.

§Difference between the temperature of the parental liquid if it rises to the surface along a liquid adiabat and the olivine crystallization temperature at 0.2 GPa, i.e., a measure of possible superheat of the magma.

¶Pressure of mantle equilibration estimated from the position of the parental liquid compositions within the normative tetrahedron (see text for discussion).

‡Pressure of mantle equilibration estimated by the empirical calibration of Alabarede (1992).

**MgO wt% of the calculated parental liquids (Table 2).

††H_2O wt% of the calculated parental liquids (Table 2).

§§Calculated equilibrium liquidus olivine at 0.2 GPa (Table 2).

¶¶Calculated equilibrium liquidus olivine at the pressure of mantle equilibration; see column P (final)¶.

‡‡Residue mineralogy at the pressure and temperature of mantle equilibration.

***Degree of partial melting estimated for the parental compositions.

†††Calculated mantle potential temperature derived from $T_{final} + \Delta T_{fusion} - \Delta T_{solid\ adiabat}$, where ΔT_{fusion} is the latent heat of melting based on the model source and derived F%, and $\Delta T_{solid\ adiabat}$ is the adiabatic cooling from P_{final} to surface along the olivine adiabat.

mantle melting can be modeled by simple batch melting, and that major elements of primary melts reflect the pressure of last equilibration with the mantle and do not represent a mixture of independent melt fractions finally assembled in a crustal magma chamber.

Summary. The results of this study, based on reasonable assumptions concerning source compositions and melting processes, indicate very little differences in pressures of mantle equilibration between MORB and OIB parental liquids (1–2 GPa versus 1–2.5 GPa). We have used the mantle equilibration

pressures and the models of Ford et al. (1983) and Falloon and Danyushevsky (2000) to calculate temperatures of equilibration with olivine at the pressures indicated from the normative tetrahedron analysis. These temperatures are listed in Table 5 ($T_{(final)}$). As was the case with temperatures at 0.2 GPa, there is very little difference in temperatures between these parental compositions derived from similar pressures of mantle equilibration. The best-constrained compositions from Kilauea and Mauna Loa give temperatures and pressures of mantle equilibration essentially identical to the hottest MORB composition

ODP896A (1422–1451 °C versus 1441 °C). These results strongly suggest that there is unlikely to be any significant differences in source mantle T_P between OIB and MORB.

Calculation of Mantle T_P

Based on the assumptions listed previously (see above), we present in Table 5 our results of mantle T_P calculations for the calculated parental compositions. The highest calculated MORB source T_P is 1488 °C, for Réunion 1502 °C, for Iceland 1501 °C, and the best-constrained estimate for Hawaii is 1565 °C. These calculations assume a source pyrolite (3–4 wt% CaO and Al_2O_3) and require a high degree of melting of such a fertile source to yield harzburgite residue for OIB parental magmas. These simple but nonunique results demonstrate that there does not exist a significantly large difference in mantle T_P between OIB and MORB mantle. If the degree of melting for Hawaii is the lower of the values presented in Table 5, caused by a more refractory source than Hawaiian pyrolite, then the T_P differences between OIB and MORB source mantle would be essentially zero.

CONCLUSIONS

In this study, we have modeled the reverse of olivine fractionation to calculate parental liquids for a range of OIB and MORB glasses. Our results suggest that there is very little difference in either the temperature of crystallization or the pressure and temperature of mantle equilibration between parental liquids in OIB and MORB settings. This result strongly suggests that it is unlikely that there is a significant difference in values of source mantle T_P. This conclusion is supported by simple nonunique calculations of mantle T_P based on a number of reasonable assumptions.

ACKNOWLEDGMENTS

We thank Gillian Foulger, Kaj Hoernle, Donna Jurdy, and Mike Walter for their constructive reviews of this article.

APPENDIX 1. EXPERIMENTAL PERIDOTITE-REACTION EXPERIMENTS ON A MODEL PARENTAL LIQUID COMPOSITION AT 1.8 AND 2.0 GPA

Introduction

In this appendix, we present experimental results (see Tables A1–A3) of peridotite-reaction experiments between peridotite MM-3 (Baker and Stolper, 1994) and the parental liquid to glass ODP896A (glass ODP Leg 896A-27R-1, 124–130, pc.15, calculated by McNeill and Danyushevsky, 1996, at 0.01 MPa) at 1.8 and 2.0 GPa. At 2.0 GPa, we also present the results

of reaction experiments using the magnesian MORB composition ARP74 10-16 (Bougault et al., 1979).

Experimental and Analytical Techniques

The starting compositions (ODP896A, ARP74, and MM-3; Table A1) were prepared from a mixture of analytical grade oxides and carbonates (Ca, Na) ground under acetone in an agate mortar. This mixture was pelletized and sintered overnight (~16–20 h) at 950 °C. An appropriate amount of synthetic fayalite was then added to the sintered mix, and the mixture was again ground under acetone, before storage in glass vials in an oven at 110 °C. Experiments were performed using standard piston-cylinder techniques in the High Pressure Laboratory, formerly housed in the School of Earth Sciences, University of Tasmania. All experiments used NaCl/Pyrex assemblies with graphite heaters, fired pyrophyllite and alumina spacers, mullite and alumina surrounds, graphite capsules, and a $W_{97}Re_3/W_{75}Re_{25}$ thermocouple. The thermocouple enters the assembly through a composite two- and four-bore alumina sheath. The thermocouple was separated and protected from the graphite capsule by a 1-mm alumina disc. No pressure correction was applied to the thermocouple calibration. The thermocouple junction is formed by crossing the thermocouple wires utilizing the four-bore alumina sheath, which forms the top 5 mm of the alumina thermocouple sheath. All experimental components and starting materials were stored in an oven at 110 °C. Experiments were performed using the hot piston-out technique, and pressures are accurate to within ± 0.1 GPa. Temperature was controlled to within ± 1 °C of the set point using a Eurotherm™ type 818 controller.

At the end of each experiment, the entire experimental charge was mounted and sectioned longitudinally before polishing. Experimental run products were analyzed either by wavelength dispersive microanalysis using Cameca SX-50™ microprobes housed in the Central Science Laboratory, University of Tasmania (operating conditions: 15 KV, 20 ηA) or energy-dispersive microanalysis using a Cameca MICROBEAM™ microprobe housed in the Research School of Earth Sciences, Australian National University (operating conditions: 15 KV, 5 ηA). All glass analyses have been normalized to the composition of international glass standard VG-2 (Jarosewich et al., 1980), which was analyzed together with the glasses under the same analytical conditions.

Our experimental results are presented in Tables A2 and A3. In Table A3, electron microprobe analyses of selected run products and mass balance calculations are presented. Table A3 demonstrates that our run products are essentially homogeneous and produce good mass balance with low sums of residuals squared. Our run times of 24 h are sufficient to have produced experimental run products closely matching equilibrium assemblages for the respective bulk compositions used.

TABLE A1. STARTING COMPOSITIONS USED (WT%)

Composition	SiO$_2$	TiO$_2$	Al$_2$O$_3$	FeO	MnO	MgO	CaO	Na$_2$O	Cr$_2$O$_3$
MM-3	45.50	0.11	3.98	7.18	0.13	38.30	3.57	0.31	0.68
ODP896A	47.90	0.57	13.87	9.29	0.10	15.15	11.60	1.40	0.07
ARP74	50.82	0.83	15.20	8.12	N.A.	10.25	12.21	2.09	N.A.

Note: N.A.—not applicable.

TABLE A2. EXPERIMENTAL RUN DATA ON PERIDOTITE REACTION EXPERIMENTS

Number	Run number	Run temperature* (°C)	Time (hours, minutes)	Basalt (wt%)[†]	ML$_i$/ML$_f$[§]	Basalt composition	Phase assemblage
1.8 GPa							
1	T-4154	1425	24	0.26	0.68	ODP896A	Ol+Opx+Cpx+L
2	T-4153	1450	24	0.31	0.61	ODP896A	Ol+Opx+L
2.0 GPa							
3	T-4107	1450	24	0.44	0.81	ODP896A	Ol+Opx+Cpx+L
4	T-4190	1450	24	0.37	0.68	ARP74	Ol+Opx+L
5	T-4191	1475	24	0.30	0.56	ARP74	Ol+Opx+L

Notes: Cpx—clinopyroxene; L—liquid; Ol—olivine; Opx—orthopyroxene.
*Recorded temperature of the experiment.
[†]The wt. fraction of basalt composition added as a layer in the peridotite reaction experiments.
[§]The wt. fraction ratio between the initial amount of added basalt component to the final amount of melt component in the experiment obtained by mass balance (Table A3). Values <1 indicate that the peridotite component has melted and contributed to the melt phase during reaction. Thus the final equilibrium assemblage is the result of a melting and reaction process not a crystallization and reaction process (values >1).

TABLE A3. COMPOSITIONS OF EXPERIMENTAL RUN PRODUCTS

Run number	Phase	MB BC*	MB MM-3[†]	Type	SiO$_2$	TiO$_2$	Al$_2$O$_3$	Cr$_2$O$_3$	FeO	MnO	MgO	CaO	Na$_2$O
1.8 GPa													
T-4154	Olivine	0.424(4)	0.55(1)	N.A.	40.81	N.D.	N.D.	N.D.	8.52	0.21	50.11	0.34	N.D.
	Orthopyroxene	0.17(1)	0.22(3)	N.A.	54.51	0.07	4.86	1.25	5.07	0.1	31.29	2.76	0.09
	Clinopyroxene	0.02(1)	0.06(3)	N.A.	52.37	0.09	5.66	1.47	4.55	0.06	23.37	12.08	0.35
	Glass	0.38(1)	0.17(2)	N.A.	48.09	0.65	14.16	0.38	8.12	0.12	14.8	12.09	1.49
	Residual	0.0217	0.1699										
T-4153	Olivine	0.395(6)	0.56(1)	N.A.	41.43	N.D.	N.D.	N.D.	7.94	0.12	50.16	0.33	N.D.
	Orthopyroxene	0.09(1)	0.18(2)	N.A.	54.51	0.07	4.86	1.25	5.07	0.1	31.29	2.76	0.09
	Glass	0.51(1)	0.25(1)	N.A.	48.6	0.56	12.83	0.54	8.21	0.11	16.31	11.51	1.23
	Residual	0.0552	0.1662										
2.0 GPa													
T-4107	Olivine	0.306(5)	0.56(2)	N.A.	40.9	N.D.	N.D.	N.D.	9.27	N.D.	49.68	0.15	N.D.
	Orthopyroxene	0.11(2)	0.20(5)	N.A.	54.93	N.D.	4.63	0.69	5.51	N.D.	31.78	2.45	N.D.
	Clinopyroxene	0.04(2)	0.05(5)	N.A.	51.92	N.D.	5.47	1.04	4.97	N.D.	24.88	10.02	0.17
	Glass	0.54(1)	0.20(3)	N.A.	47.75	0.61	13.96	0.34	8.49	N.D.	15.37	11.83	1.4
	Residual	0.0341	0.3622										
T-4190	Olivine	0.263(4)	0.55(1)	A(3)	41.3(3)	N.D.	N.D.	0.25(2)	8.4(1)	0.10(4)	50.2(3)	0.31(0)	N.D.
	Orthopyroxene	0.185(7)	0.21(2)	A(3)	54.87(9)	0.09(2)	4.3(1)	1.1(1)	4.8(1)	0.13(4)	32.0(1)	2.5(1)	0.12(1)
	Glass	0.548(5)	0.23(2)	A(5)	48.3(1)	0.74(4)	13.4(1)	0.36(2)	8.0(1)	0.15(4)	15.95(7)	11.48(9)	1.57(3)
	Residual	0.0256	0.2603										
T-4191	Olivine	0.324(6)	0.56(1)	A(3)	41.3(3)	N.D.	N.D.	0.36(2)	8.0(1)	0.10(2)	50.3(7)	0.5(3)	N.D.
	Orthopyroxene	0.14(10	0.17(2)	A(3)	55.6(2)	0.07(4)	3.4(2)	0.88(7)	4.83(7)	0.09(4)	33.0(1)	2.02(2)	0.09(2)
	Glass	0.535(8)	0.26(1)	A(5)	48.8(2)	0.66(3)	12.66(7)	0.43(3)	7.8(2)	0.15(2)	17.20(5)	10.8(1)	1.37(2)
	Residual	0.0636	0.1447										

Notes: Mass balance in weight fraction was performed using least squares linear regression and the software PETMIX. "Residual" refers to the square of the sum of the residuals. Numbers in parentheses next to each analysis or mass balance are 1σ in terms of the last units cited; e.g., 0.540(9) refers to 0.540 ± 0.009. A(*n*)—analysis is an average with the number of analyses used to calculate the average given in the parentheses. N.D.—not determined; N.A.—not applicable.
*Mass balance using the bulk composition of the experiment.
[†]Mass balance using the peridotite composition MM-3 only.

REFERENCES CITED

Albarede, F., 1992, How deep do common basaltic magmas form and differentiate?: Journal of Geophysical Research, v. 97, p. 10,997–11,009.

Anderson, A.T., and Brown, G.G., 1993, CO_2 and formation pressures of some Kilauean melt inclusions: American Mineralogist, v. 78, p. 794–803.

Aramaki, S., and Moore, J.G., 1969, Chemical composition of prehistoric lavas at Makaopuhi Crater, Kilauea volcano, and periodic change in alkali content of Hawaiian tholeiitic lavas: Bulletin of the Earthquake Research Institute, v. 47, p. 257–270.

Ault, W.U., Eaton, J.P., and Richter, D.H., 1961, Lava temperatures in the 1959 Kilauea eruption and cooling lake: Geological Society of America Bulletin, v. 72, p. 791–794, doi: 10.1130/0016-7606(1961)72[791:LTITKE]2.0.CO;2.

Baker, M.B., and Stolper, E.M., 1994, Determining the compositions of high-pressure melts using diamond aggregates: Geochimica et Cosmochimica Acta, v. 58, p. 2811–2827, doi: 10.1016/0016-7037(94)90116-3.

Basaltic Volcanism Study Project, 1981, Basaltic volcanism on the terrestrial planets: New York, Pergamon Press, 1286 p.

Birch, F., 1952, Elasticity and constitution of the earth's interior: Journal of Geophysical Research, v. 57, p. 227–286.

Borisov, A.A., and Shapkin, A.I., 1990, A new empirical equation rating Fe^{3+}/Fe^{2+} in magmas to their composition, oxygen fugacity, and temperature: Geochemistry International, v. 27, p. 111–116.

Bougault, H., Cambon, P., Corre, O., Joron, J.L., and Treuil, M., 1979, Evidence for variability of magmatic processes and upper mantle heterogeneity in the axial region of the mid-Atlantic Ridge near 22°N and 36°N: Tectonophysics, v. 55, p. 11–34, doi: 10.1016/0040-1951(79)90333-0.

Bureau, H., Pineau, F., Métrich, N., Semet, M.P., and Javoy, M., 1998, A melt and fluid inclusion study of the gas phase at Piton de la Fournaise volcano (Réunion Island): Chemical Geology, v. 147, p. 115–130, doi: 10.1016/S0009-2541(97)00176-9.

Chen, C.-Y., Frey, F.A., Rhodes, J.M., and Easton, R.M., 1996, Temporal geochemical evolution of Kilauea volcano: Comparison of Hilina and Puna basalt: Earth processes, in Basu, A., and Hart, S.R., eds., Reading the isotopic code: Washington, D.C., American Geophysical Union Geophysical Monograph 95, p.161–181.

Clague, D.A., Weber, W.S., and Dixon, J.E., 1991, Picritic glasses from Hawaii: Nature, v. 353, p. 553–556, doi: 10.1038/353553a0.

Clague, D.A., Moore, J.G., Dixon, J.E., and Friesin, W.B., 1995, Petrology of submarine lavas from Kilauea's Puna Ridge, Hawaii: Journal of Petrology, v. 36, p. 299–349.

Danyushevsky, L.V., 2001, The effect of small amounts of H_2O on crystallization of mid-ocean ridge and backarc basin magmas: Journal of Volcanology and Geothermal Research, v. 110, p. 265–280, doi: 10.1016/S0377-0273(01)00213-X.

Danyushevsky, L.V., and Sobolev, A.V., 1996, Ferric-ferrous ratio and oxygen fugacity calculations for primitive mantle-derived melts: Calibration of an empirical technique: Mineralogy and Petrology, v. 57, p. 229–241, doi: 10.1007/BF01162360.

Danyushevsky, L.V., Della-Pasqua, F.N., and Sokolov, S., 2000, Re-equilibration of melt inclusions trapped by magnesian olivine phenocrysts from subduction-related magmas: Petrological implications: Contributions to Mineralogy and Petrology, v. 138, p. 68–83, doi: 10.1007/PL00007664.

Danyushevsky, L.V., Sokolov, S., and Falloon, T.J., 2002, Melt inclusions in olivine phenocrysts: Using diffusive re-equilibration to determine the cooling history of a crystal, with implications for the origin of olivine-phyric volcanic rocks: Journal of Petrology, v. 43, p. 1651–1671, doi: 10.1093/petrology/43.9.1651.

Danyushevsky, L.V., Perfit, M.R., Eggins, S.M., and Falloon, T.J., 2003, Crustal origin for coupled "ultra-depleted" and "plagioclase" signatures in MORB olivine-hosted melt inclusions: Evidence from the Siqueiros Transform Fault, East Pacific Rise: Contributions to Mineralogy and Petrology, v. 144, p. 619–637.

Dmitriev, L.V., Sobolev, A.V., Sushevskaya, N.M., and Zpunny, S.A., 1985,

Abyssal glasses, petrologic mapping of the oceanic floor and "geochemical Leg" 82, in Bougault, H., et al., eds., Initial reports of the Deep Sea Drilling Project (vol. LXXXII): Washington, D.C., U.S. Government Printing Office, p. 509–518.

Eaton, J.P., and Murata, K.J., 1960, How volcanoes grow: Science, v. 132, p. 925–938, doi: 10.1126/science.132.3432.925.

Eggins, S.M., 1992a, Petrogenesis of Hawaiian tholeiites: 1, Phase equilibria constraints: Contributions to Mineralogy and Petrology, v. 110, p. 387–397, doi: 10.1007/BF00310752.

Eggins, S.M., 1992b, Petrogenesis of Hawaiian tholeiites: 2, Aspects of dynamic melt segregation: Contributions to Mineralogy and Petrology, v. 110, p. 398–410, doi: 10.1007/BF00310753.

Falloon, T.J., and Danyushevsky, L.V., 2000, Melting of refractory mantle at 1.5, 2 and 2.5 GPa under anhydrous and H_2O-undersaturated conditions: Implications for high-Ca boninites and the influence of subduction components on mantle melting: Journal of Petrology, v. 41, p. 257–283, doi: 10.1093/petrology/41.2.257.

Falloon, T.J., and Green, D.H., 1988, Anhydrous melting of peridotite from 8 to 35 kb and the petrogenesis of MORB: Journal of Petrology, special lithosphere issue, p. 379–414.

Falloon, T.J., Green, D.H., Hatton, C.J., and Harris, K.L., 1988, Anhydrous partial melting of fertile and depleted peridotite from 2 to 30 kbar and application to basalt petrogenesis: Journal of Petrology, v. 29, p. 257–282.

Falloon, T.J., Green, D.H., Danyushevsky, L.V., and Faul, U.H., 1999, Peridotite melting at 1.0 and 1.5 GPa: An experimental evaluation of techniques using diamond aggregates and mineral mixes for determination of near-solidus melts: Journal of Petrology, v. 40, p. 1343–1375, doi: 10.1093/petrology/40.9.1343.

Falloon, T.J., Danyushevsky, L.V., and Green, D.H., 2001, Reversal experiments on partial melt compositions produced by peridotite-basalt sandwich experiments: Journal of Petrology, v. 42, p. 2363–2390, doi: 10.1093/petrology/42.12.2363.

Falloon, T.J., Danyushevsky, L.V., Ariskin, A., Green, D.H., and Ford, C.E., 2007, The application of olivine geothermometry to infer crystallization temperatures of parental liquids: Implications for the temperature of MORB magmas, in Campbell, I.H., and Kerr, A.C., eds., The great plume debate: Testing the plume theory: Chemical Geology, special issue, v. 241, p. 207–233.

Ford, C.E., Russell, D.G., Craven, J.A., and Fisk, M.R., 1983, Olivine-liquid equilibria: Temperature, pressure and composition dependence of the crystal/liquid cation partition coefficients for Mg, Fe^{2+}, Ca and Mn: Journal of Petrology, v. 24, p. 256–265.

Fretzdorff, S., and Haase, K.M., 2002, Geochemistry and petrology of lavas from the submarine flanks of Réunion Island (western Indian Ocean): Implications for magma genesis and the mantle source: Mineralogy and Petrology, v. 75, p. 153–184, doi: 10.1007/s007100200022.

Ganguly, J., 2005, Adiabatic decompression and melting of mantle rocks: An irreversible thermodynamic analysis: Geophysical Research Letters, v. 32, no. 6, LO6312, doi: 10.1029/2005GL022363.

Garcia, M.O., Rhodes, J.M., Wolfe, E.W., Ulrich, G.E., and Ho, R.A., 1992, Petrology of lavas from episodes 2–47 of the Pu'u O'o eruption of Kilauea volcano, Hawaii: Evaluation of magmatic processes: Bulletin of Volcanology, v. 55, p. 1–16, doi: 10.1007/BF00301115.

Garcia, M.O., Hulsebosch, T.P., and Rhodes, J.M., 1995, Olivine-rich submarine basalts from the southwest rift zone of Mauna Loa volcano: Implications for magmatic processes and geochemical evolution, in, Rhodes, J. M., and Lockwood, J.P., eds., Mauna Loa revealed: Structure, composition, history, and hazards: Washington, D.C., American Geophysical Union Geophysical Monograph 92, p. 219–239.

Garcia, M.O., Rhodes, J.M., Trusdell, F.A., and Pietruszka, A.J., 1996, Petrology of lavas from the Pu'u O'o eruption of Kilauea volcano: III. The Kupaianaha episode (1986–1992): Bulletin of Volcanology, v. 58, p. 359–379, doi: 10.1007/s004450050145.

Garcia, M.O., Ito, E., Eiler, J.M., and Pietruszka, A.J., 1998, Crustal contamination of Kilauea volcano magma revealed by oxygen isotope analysis of glass and olivine from Pu'u O'o eruption lavas: Journal of Petrology, v. 39, p. 803–817, doi: 10.1093/petrology/39.5.803.

Garcia, M.O., Pietruszka, A.J., Rhodes, J.M., and Swanson, K., 2000, Magmatic processes during the prolonged Pu'u O'o eruption of Kilauea volcano, Hawaii: Journal of Petrology, v. 41, p. 967–990, doi: 10.1093/petrology/41.7.967.

Garcia, M.O., Pietruszka, A.J., and Rhodes, J.M., 2003, A petrologic perspective of Kilauea volcano's summit magma reservoir: Journal of Petrology, v. 44, p. 2313–2339, doi: 10.1093/petrology/egg079.

Green, D.H., and Falloon, T.J., 1998, Pyrolite: A Ringwood concept and its current expression, in Jackson, I.N.S., ed., The Earth's mantle: Composition, structure, and evolution: Cambridge, Cambridge University Press, p. 311–380.

Green, D.H., and Falloon, T.J., 2005, Primary magmas at mid-ocean ridges, "hot spots" and other intraplate settings: Constraints on mantle potential temperature, in Foulger, G.R., et al., eds., Plates, plumes and paradigms: Boulder, Colorado, Geological Society of America Special Paper 388, p. 217–247, doi: 10.1130/2005.2388(14).

Green, D.H., and Ringwood, A.E., 1967, The genesis of basaltic magmas: Contributions to Mineralogy and Petrology, v. 15, p. 103–190, doi: 10.1007/BF00372052.

Green, D.H., Hibberson, W.O.H., and Jaques, A.L., 1979, Petrogenesis of mid-ocean ridge basalts, in McElhinny, M.W., ed., The Earth: Its origin, structure, and evolution: London, Academic Press, p. 265–290.

Green, D.H., Falloon, T.J., Eggins, S.E., and Yaxley, G.M., 2001, Primary magmas and mantle temperatures: European Journal of Mineralogy, v. 13, p. 437–451, doi: 10.1127/0935-1221/2001/0013-0437.

Gunn, B.M., 1971, Trace element partitioning during olivine fractionation of Hawaiian basalts: Chemical Geology, v. 8, p. 1–13, doi: 10.1016/0009-2541(71)90043-X.

Helz, R.T., 1980, Crystallization of Kilauea Iki lava lake as seen in drill core recovered in 1967–1979: Bulletin of Volcanology, v. 43, p. 675–701.

Helz, R.T., 1987, Diverse olivine types in lava of the 1959 eruption of Kilauea volcano and their bearing on eruption dynamics, in Dekker, R.W., et al., eds., Volcanism in Hawaii: Boulder, Colorado, Geological Survey of America Professional Paper 1350, p. 691–722.

Herzberg, C., and O'Hara, M.J., 2002, Plume-associated ultramafic magmas of Phanerozoic age: Journal of Petrology, v. 43, p. 1857–1883, doi: 10.1093/petrology/43.10.1857.

Irvine, T.N., 1977, Definition of primitive liquid compositions for basic magmas: Washington, D.C., Carnegie Institution Year Book, v. 76, p. 454–461.

Jarosewich, E.J., Nelen, J.A., and Norberg, J.A., 1980, Reference samples for electron microprobe analyses: Geostandards Newsletter, v. 4, p. 257–258.

Klein, E.M., and Langmuir, C.H., 1987, Global correlations of ocean ridge basalt chemistry with axial depth and crustal thickness: Journal of Geophysical Research, v. 92, p. 8089–8115.

Langmuir, C.H., Klein, E.M., and Plank, T., 1992, Petrological systematics of mid-ocean ridge basalts: Constraints on melt generation beneath ocean ridges, in Morgan, J.P., et al., eds., Mantle flow and melt generation at mid-ocean ridges: Washington, D.C., American Geophysical Union Geophysical Monograph 71, p. 183–280.

MacDonald, G.A., and Eaton, J.P., 1955, Hawaiian volcanoes during 1953: U.S. Geological Survey Bulletin, v. 1021-D, p. 127–166.

Maclennan, J., McKenzie, D., Hilton, F., Gronvöld, K., and Shimizu, N., 2003, Geochemical variability in a single flow from northern Iceland: Journal of Geophysical Research, v. 108, no. B1, p. 2007, doi: 10.1029/2000JB000142.

Mangan, M.T., Heliker, C.C., Hofmann, A.W., Mattox, T.N., Kauahikaua, J.P., and Helz, R.T., 1995, Episode 49 of the Pu'u O'o-Kupaianaha eruption of Kilauea volcano: Breakdown of a steady-state eruptive area: Bulletin of Volcanology, v. 57, p. 127–135.

McKenzie, D., and Bickle, M.J., 1988, The volume and composition of melt generated by extension of the lithosphere: Journal of Petrology, v. 29, p. 625–679.

McNeill, A.W., and Danyushevsky, L.V., 1996, Composition and crystallization temperatures of primary melts from Hole 896A basalts: Evidence from melt inclusion studies, in Alt, J.C., et al., eds., Proceedings of the Ocean Drilling Program, Scientific Results (vol. 148): Washington, D.C., U.S. Government Printing Office, p. 21–35.

Moore, J.G., 1965, Petrology of deep sea basalt near Hawaii: American Journal of Science, v. 263, p. 40–52.

Moore, J.G., and Evans, B.W., 1967, The role of olivine in the crystallization of the prehistoric Makaopuhi tholeiitic lava lake, Hawaii: Contributions to Mineralogy and Petrology, v. 15, p. 202–223, doi: 10.1007/BF01185342.

Moore, J.G., and Koyanagi, R.Y., 1969, The October 1963 eruption of Kilauea volcano, Hawaii: Washington, D.C., U.S. Geological Survey Professional Paper, 614-C, p. 1–13.

Muir, I.D., and Tilley, C.E., 1963, Contributions to the petrology of Hawaiian basalts, part 2. The tholeiitic basalts of Mauna Loa and Kilauea, with chemical analyses by J.H. Scoon: American Journal of Science, v. 261, p. 111–128.

Murata, K.J., and Richter, D.H., 1966, Chemistry of the lavas of the 1959–60 eruption of Kilauea volcano, Hawaii: Washington, D.C., U.S. Geological Survey Professional Paper 537-A, p. 1–26.

Nichols, A.R.L., Carroll, M.R., and Höskuldsson, Á., 2002, Is Iceland hot spot also wet? Evidence from the water contents of undegassed submarine and subglacial pillow basalts: Earth and Planetary Science Letters, v. 202, p. 77–87, doi: 10.1016/S0012-821X(02)00758-6.

Norman, M.D., and Garcia, M.O., 1999, Primitive magmas and source characteristics of the Hawaiian plume: Petrology and geochemistry of shield picrites: Earth and Planetary Science Letters, v. 168, p. 27–44, doi: 10.1016/S0012-821X(99)00043-6.

Norman, M.D., Garcia, M., Kamenetsky, V.S., and Nielson, R.L., 2002, Olivine-hosted olivine melt inclusions in Hawaiian picrites: Equilibration, melting and plume source characteristics: Chemical Geology, v. 183, p. 143–168, doi: 10.1016/S0009-2541(01)00376-X.

Presnall, D.C., Gudfinnsson, G.H., and Walter, M.J., 2002, Generation of mid-ocean ridge basalts at pressures from 1 to 7 GPa: Geochimica et Cosmochimica Acta, v. 66, p. 2073–2090, doi: 10.1016/S0016-7037(02)00890-6.

Putirka, K.D., 2005, Mantle potential temperatures at Hawaii, Iceland, and the mid-ocean ridge system, as inferred from olivine phenocrysts: Evidence for thermally driven mantle plumes: Geochemistry, Geophysics, Geosystems, v. 6, art. no. Q05L08, doi:10.1029/2005GC000915.

Quane, S.L., Garcia, M.O., Guillou, H., and Hulsebosch, T.P., 2000, Magmatic history of the east rift zone of Kilauea volcano, Hawaii, based on drill core SOH1: Journal of Volcanology and Geothermal Research, v. 102, p. 319–338, doi: 10.1016/S0377-0273(00)00194-3.

Richter, D.H., and Moore, J.G., 1966, Petrology of the Kilauea Iki lava lake, Hawaii: Washington, D.C., U.S. Geological Survey Professional Paper, v. 537-B, p. 1–26.

Roeder, P.L., and Emslie, R.F., 1970, Olivine-liquid equilibrium: Contributions to Mineralogy and Petrology, v. 29, p. 275–289, doi: 10.1007/BF00371276.

Sigurdsson, I.A., Steinthorsson, S., and Grönvold, K., 2000, Calcium-rich melt inclusions in Cr-spinels from Borgarhraun, northern Iceland: Earth and Planetary Science Letters, v. 183, p. 15–26, doi: 10.1016/S0012-821X(00)00269-7.

Sobolev, A.V., and Danyushevsky, L.V., 1994, Petrology and geochemistry of boninites from the north termination of the Tonga Trench: Constraints on the generation conditions of primary high-Ca boninite magmas: Journal of Petrology, v. 35, p. 1183–1211.

Sobolev, A.V., and Nikogosian, I.K., 1994, Petrology of long-lived mantle plume magmatism: Hawaii, Pacific, and Réunion Island: Indian Ocean: Petrologiya, v. 2, p. 131–168.

Sobolev, A.V., Danyushevsky, L.V., Dmitriev, L.V., and Sushchevskaya, N.M., 1989, High-alumina magnesian tholeiite as the primary basalt magma at midocean ridge: Geochemistry International, v. 26, p. 128–133.

Sobolev, A.V., Hofmann, A.W., Sobolev, S.V., and Nikogosian, I.K., 2005, An olivine-free mantle source of Hawaiian shield basalts: Nature, v. 434, p. 590–597, doi: 10.1038/nature03411.

Takahashi, E., and Nakajima, K., 2002, Melting process in the Hawaiian plume: An experimental study, *in* Takahashi, E., et al., eds., Hawaiian volcanoes: Deep underwater perspectives: Washington, D.C., American Geophysical Union Geophysical Monograph 128, p. 403–417.

Tilley, C.E., 1960, Kilauea magma: Geological Magazine, v. 97, p. 494–497.

Tilley, C.E., Thompson, R.N., Wadsworth, W.J., and Upton, B.G.J., 1971, Melting relations of some lavas of Réunion Island, Indian Ocean: Mineralogical Magazine, v. 38, p. 344–352, doi: 10.1180/minmag.1971.038.295.09.

Toplis, M.J., 2005, The thermodynamics of iron and magnesian partitioning between olivine and liquid: Criteria for assessing and predicting equilibrium in natural and experimental systems: Contributions to Mineralogy and Petrology, v. 149, p. 22–39, doi: 10.1007/s00410-004-0629-4.

Ulmer, P., 1989, The dependence of the Fe^{2+}-Mg cation partitioning between olivine and basaltic liquid on pressure, temperature and composition: Contributions to Mineralogy and Petrology, v. 101, p. 261–273, doi: 10.1007/BF00375311.

Wallace, P.J., and Anderson, A.T., 1998, Effects of eruption and lava drainback on the H_2O contents of basaltic magmas at Kilauea volcano: Bulletin of Volcanology, v. 59, p. 327–344, doi: 10.1007/s004450050195.

Wilkinson, J.F.G., and Hensel, H.D., 1988, The petrology of some picrites from Mauna Loa and Kilauea volcanoes, Hawaii: Contributions to Mineralogy and Petrology, v. 98, p. 326–345, doi: 10.1007/BF00375183.

Wright, T.L., 1971, Chemistry of Kilauea and Mauna Loa lava in space and time: Washington, D.C., U.S. Geological Survey Professional Paper, v. 735, p. 1–40.

Wright, T.L., and Fiske, R.S., 1971, Origin of the differentiated and hybrid lavas of Kilauea volcano, Hawaii: Journal of Petrology, v. 12, p. 1–65.

Yaxley, G.M., and Green, D.H., 1998, Reactions between eclogite and peridotite: Mantle refertilization by subduction of oceanic crust: Schweizerische Mineralogische und Petrographische Mitteilungen, v. 78, p. 243–255.

MANUSCRIPT ACCEPTED BY THE SOCIETY 31 JANUARY 2007

DISCUSSION

1 January 2007, Dean C. Presnall

The only unequivocal magma composition is one that is produced from a direct analysis of a glass. The validity of calculated compositions, such as those produced by Falloon et al. (this volume) is always dependent on a series of assumptions. Despite thousands of microprobe analyses of mid-ocean ridge basalt (MORB) glasses, no compositions that show a trend of olivine-controlled crystallization have ever been reported, either from "normal" ridge segments or Iceland (over 400 glass analyses just from Iceland). In contrast, at Kilauea, Hawaii, direct analyses have shown the existence of a clear trend of olivine-controlled crystallization (Clague et al., 1991, 1995). MORB generation (including Iceland) at low and relatively uniform pressures (0.9–1.5 GPa) and potential temperatures (~1240–1260° C) (Presnall et al., 1979, 2002; Presnall and Gudfinnsson, 2007) explains both the inverse and positive Na8-Fe8 correlations as direct melts from a heterogeneous mantle. In addition, it avoids the conundrum of the complete absence of olivine-controlled fractionation.

12 January 2007, James H. Natland

The fatal assumption behind estimation of temperatures of crystallization using olivine-liquid FeO^T-MgO relationships is that the most forsteritic olivine in a rock, or surmised to be representative of a liquid in equilibrium with the mantle, is probably not related to the host liquid composition along a single closed-system liquid line of descent. That assumption is the entire basis for the procedure of adding incrementally more forsteritic olivine into liquid compositions until an "equilibrium" liquid is reached, but it is wrong. This is because most picrites are hybrid rocks, the results of magma mixing, which can be established by careful studies of crystallization histories. The mixing is of two types: (1) mixing between primitive magma strains near or somewhere above their melt sources in the mantle and (2) mixing between primitive and differentiated magma strains.

An example of mixing of primitive magma strains is sample D20-3 in Table 1 of Falloon et al. (this volume), from Siqueiros fracture zone on the East Pacific Rise. The glass analysis is from Danyushevsky et al. (2003), but an interesting aspect of the history of this sample is that it was obtained from almost exactly the location of an earlier dredged picrite (Batiza et al., 1977) with a glass composition (Natland, 1989, Table 1, analysis SD7-C) that is identical to within parameters considered by Melson et al. (1976) to represent material from the same eruption. The glass contains olivine dendrites and plagioclase spherulites (Natland, 1980), and thus it is on a two-phase cotectic. Even though the rock is a picrite, the glass is not an olivine-controlled liquid. Nor indeed is even the most-magnesian MORB glass (Presnall et al., 2002). Olivine in SD7-C, at least, commonly occludes Cr-spinel within skeletal embayments that also contain glass. Some of the spinel is zoned. But most importantly it has a substantial range in Cr# but little in Mg# (Fig. 3 *in* Natland, 1989), indicating crystallization from similarly magnesian parental liquids but with significant differences in Al_2O_3 content (Poustovetov and Roeder, 2000). Such differences cannot be the result of olivine crystallization and must reflect mixing at about the same temperature of melt strains derived from a heterogeneous source.

Pertinent here is that two other chemically distinctive picrites were obtained in the same dredge haul (one of these during the later expedition). The one studied by Natland (1989) in detail (SD-7A) has spinel with a similar range in Cr# as SD7-C but at systematically lower Mg#; associated olivine is also a bit more iron rich, and there are rare plagioclase phenocrysts. The glass has higher FeO^T, and if one were to add olivine incrementally back into it to, say, a nominal primitive olivine composition of Fo_{91}, then the estimated parental crystallization temperature (and glass MgO content) would be higher than that of SD-7C. But from the minerals actually in the rock, there is no indication that it even makes sense to do this addition. The cooler picrite is the more iron-rich. The two samples did not derive from the same mantle composition, even though they erupted side by side and maybe at almost the same time. They simply sampled the array of primitive liquid compositions across the melting domain in two different ways.

A second example is provided by Icelandic picrites. Many, including samples from the Borgarhraun flow studied by Falloon et al. (this volume), contain phenocrysts of plagioclase and clinopyroxene (e.g., Slater et al., 2001; Maclennan et al., 2003). Primitive basalt from site 896 on the Costa Rica rift also contains plagioclase phenocrysts (Shipboard Scientific Party, 1993). These basalt samples are proof positive that the liquids did not follow olivine-controlled liquid lines of descent. Besides the phenocrysts, melt inclusions in Icelandic picrites provide substantial evidence for mixing of primitive melt strains as well as evidence for pyroxenite in the melt source (Sigurdsson et al., 2000; Slater et al., 2001; Foulger et al., 2005).

Next, Hawaii, particularly Kilauea and Puna ridge, provides examples of mixing between primitive and differentiated liquids. Such mixing has been amply demonstrated by petrological studies of the Kilauea rift system (e.g., Wright and Fiske, 1971). It has the effect of elevating FeO^T contents of the hybrid beyond that which the primitive mixing component could have reached by differentiation of olivine alone. Such mixing results in artificially high temperature estimates. Clague et al. (1995) discussed this mixing in the paper that also presented the composition of the most primitive Hawaiian glass (a grain with no phenocrysts in a cored, thinly bedded sand). This type of mixing is also evident in the mineralogy of similar but more iron-rich, tholeiitic picrites from the Juan Fernandez Islands (Natland, 2003). Another possibility is that some Hawaiian and Icelandic melt strains derive from eclogitic components in mantle sources (e.g., Sobolev et al., 2000, 2005; Natland, this volume) that may be either more or less iron-rich than commonly construed mantle peridotite. This variation in source petrology will contribute to heterogeneity of melt strains contributing to erupted picrites and add further complexity to estimation of crystallization and potential temperatures. Therefore, no one should try to estimate these temperatures until the full crystallization histories of the rocks are understood.

27 January 2007, Keith D. Putirka and J. Michael Rhodes

Falloon et al. (this volume) attempt to show that magmas from mid-ocean ridges (MORB) and ocean islands (ocean island basalt [OIB]), or "plumes," have similar mantle potential temperatures (T_p). Their strategy contains five errors and an inconsistency. Errors 1 and 5 derive from a misunderstanding of the thermal implications of the plume model, errors 2 and 3 lead to overestimates of ambient mantle T_p, error 4 leads to an underestimate of hot spot T_p, and all errors minimize mantle excess temperatures ($T_{ex} = T_p^{OIB} - T_p^{MORB}$):

1. Their physical model is wrong. Plumes represent point sources, so only the highest temperatures at any OIB are relevant to T_p^{OIB}, which repeats an error in Green et al. (2001). Falloon et al. (this volume) also presume that only the highest MORB temperatures represent ambient mantle, contradicting our understanding of the linkages between mantle temperatures, bathymetry, and geochemistry (Langmuir et al., 1992)—and no alternative explanation is provided.

2. In Falloon et al. (this volume) MORB with low-moderate FeO^T are ignored. However, it is impossible to generate low–moderate FeO^T MORB from high FeO^T MORB by fractionation of olivine ± plagioclase (their 1243° C and 1320° C trends cannot be derived by fractionation from their 1351° C trend). These low–moderate FeO^T MORBs must reflect differences in ambient T_p^{MORB} and cannot be ignored when estimating ambient T_p; source heterogeneity does not alleviate the problem, because olivine thermometry is independent of source composition (Putirka, 2005).

3. Falloon et al. (this volume) overestimate mean MORB FeO^T (at 9.3% FeO^T, for MORB glasses with >9.5% MgO [$n = 80$]). But there are at least 137 glasses in PetDB with MgO > 9.5%, and they average 8.5% FeO^T. And because olivine fractionation does not affect FeO^T, there is no reason to exclude whole rocks with >9.5% MgO, which combined with glasses yield an average of 8.2% FeO^T for MORB ($n = 192$).

4. Falloon et al. (this volume) assume a very low $Fe^{2+}O$ for Hawaii. At Mauna Loa, sample 182-7 has 9.38% FeO—much lower than Hawaiian picrites, which average 10.4–10.6% $Fe^{2+}O$ (Herzberg and O'Hara, 2002; Putirka et al., 2007). This underestimate of $Fe^{2+}O$ at Hawaii artificially reduces T_p^{Hawaii} by >70° C.

5. The authors state that plume T_{ex} must be 200–300° C. Some models of excess bathymetry suggest that $T_{ex} > 160°$ C (Sleep, 1990; Schilling, 1991), but the thermal plume model only requires that over a given depth range, the Rayleigh number is above critical. At Iceland, where acoustic anomalies extend to 670 km (Foulger et al., 2005), T_{ex} of a few tens of degrees Centigrade would be more than sufficient to support thermally driven active upwelling.

6. Falloon et al. (this volume) are inconsistent in that they use the highest FeO (and T) at MORB but not at OIB.

7. Concluding statements aside, Falloon et al. (this volume) present a convincing case that some ocean islands have very high T_{ex}. If we eliminate only error 6, they derive a minimum T_{ex} of 122° C at Hawaii; taking the average of their T estimates at Hawaii and MORB, their minimum T_{ex} is 126° C (Falloon et al., this volume, Table 5). Had Falloon et al. used observed fO_2 and FeO for Mauna Loa (Rhodes and Vollinger, 2004), their primitive magma would have 20.6% MgO (assuming their implied $K_D(Fe-Mg)^{ol-liq} = 0.318$; Table 2 in their article), and their minimum T_{ex} would be ~70° C hotter.

Their article is the third attempt by this group to argue that at Hawaii $T_{ex} = 0°$ C. Green et al. (2001) suggested that because Siqueiros (along the East Pacific Rise) and Hawaii yield olivine phenocrysts with similarly high forsterite (Fo) contents, that T_p must be similar. Putirka (2005), however, showed that because Hawaiian lavas contain more FeO, Hawaiian T_{ex} is at least 220° C. More recently, Falloon et al. (2007) suggested that the T_p estimates of Putirka (2005) at Hawaii were too high due to model error. But Putirka et al. (2007) show that Hawaii, Iceland, and Samoa have high T_{ex} regardless of which thermometer is used (and that the Ford et al., 1983, thermometer has systematic error not present in Putirka et al., 2007, or Beattie, 1993). Finally, Putirka et al. (2007) demonstrate that:

1. Excluding Iceland, MORBs exhibit a T_p range of 140° C with a standard deviation of ±34° C;
2. The MORB T_p range is 210° C when Iceland is included—consistent with Langmuir et al. (1992); and
3. Hawaii, Iceland, and Samoa have T_p values that do not overlap with MORB within 2σ.

It is thus safe to conclude not just that the mantle thermally convects, but that convection currents drive intraplate volcanism.

Finally, although we disagree with the conclusions of Falloon et al. (this volume), unlike Natland (see his comment of 12 January), we do not find fault with their general approach. We agree with Natland that the strong linear Hawaiian trends are mixing trends and not liquid lines of descent (Rhodes and Vollinger, 2004). Nonetheless, Natland's suggestion that parental magmas are lower in FeO, and bear little relationship to the mixing trends, is a red herring. Perusal of Figure 11 in Rhodes and Vollinger (2004) will make this point clear.

28 January 2007, Dean C. Presnall

In their comment of 27 January, Putirka and Rhodes accept the arguments of Langmuir et al. (1992) that mantle temperature, bathymetry, and geochemistry are linked. However, a global examination of MORB glass analyses in the Smithsonian database shows no such linkages (Presnall and Gudfinnsson, 2007). Instead, the observed Na8-Fe8 systematics of MORBs match the systematics of melts at the lherzolite ± basalt solidus of the CaO-MgO-Al_2O_3-SiO_2-Na_2O-FeO system in the narrow pressure-temperature range of 0.9–1.5 GPa and 1240–1260° C (Presnall et al., 2002; Presnall and Gudfinnsson, 2007). This low and globally uniform potential temperature along all ridges (including Iceland) is not consistent with the existence of hot plumes (Galápagos, Iceland, Azores, Tristan, Bouvet, Afar, Easter) on or close to ridges.

28 January 2007, Don L. Anderson

The statements and conclusions in Putirka and Rhodes (comment of 27 January) require linkages between mantle temperatures, bathymetry, and composition (Langmuir et al., 1992) and seismic velocity that may not exist (Presnall and Gudfinnsson, 2007; Anderson, this volume). It is traditional to attribute melting, bathymetric, petrologic, and tomographic anomalies to variations in absolute temperature of a common parent rock. This simplistic approach underlies all plume speculations and much of the current discussion.

In their comment, Putirka and Rhodes attempt to show that magmas from mid-ocean ridges and ocean islands have mantle sources with distinctly different potential temperatures (T_p) and similar compositions. They do not define "plume," "normal mantle," or "potential temperature," but definitions can be extracted from their discussion. The inferred temperature differences are within plate tectonic and normal convection expectations. They may appear large in the context of the isothermal-mantle assumption that underlies the plume hypothesis.

Putirka and Rhodes make a series of critical but hidden assumptions;

1. Adiabatically corrected magma temperatures give the potential temperature at the source and also define the potential temperature and temperature gradient of the underlying mantle;
2. Variations in the temperatures of magmas are due to lateral variations in mantle temperature (and are not due to different depths or lithologies in and below the surface boundary layer); and
3. MORB from mature spreading ridges bracket the allowable temperature range of "normal mantle."

It is evident that they consider temperature excesses of greater than a few tens of degrees to be sufficient to define a plume. Low acoustic velocities also uniquely define, to them, hot buoyant upwellings. The MORB range is considered to be about 200° C and is usually qualified as being from "ridges that are unaffected by plumes." This qualification excludes new ridges, shallow ridges, and ridge segments near arbitrarily defined "hotspots." It also ignores the variations in older MORB. They argue that only the hottest temperatures are diagnostic of plume temperatures, because plumes are from point sources. Although it is true that many of the experiments upon which the plume model is based use "point sources"—injection of a hot fluid through a narrow tube—it is not evident that there are such point sources in the mantle. "Normal mantle" in these studies is defined as isothermal and homogeneous, and the thermal effects of plates and continents are ignored.

The actual range in mantle temperatures is likely to be much greater than is observed along mature and "normal"-depth ridges. Regions of the mantle that have been covered by large plates for a long time are expected to be hotter than average (this effect has been called "continental insulation," but it also applies to large oceanic plates). Thus, the mantle under Hawaii, Iceland, Samoa, and Afar is expected to be hotter than along the East Pacific Rise.

But lateral potential-temperature variations are not the only cause of magma temperature variations. There is no reason why the geotherm cannot cross the solidus within a conduction layer, particularly if the outer layers of the Earth are chemically buoyant. With a conduction gradient of 10° C/km, a typical MORB can start to form at a depth of 120 km while magma with an excess temperature of 100° C can start to form some 10 km deeper (the lithosphere is about half the thickness of the thermal boundary layer [TBL]). If both of these sources are then brought adiabatically to the surface, the latter will melt more and will appear to have a higher potential temperature. In neither case is it necessary that the source or underlying mantle be on an adiabat or be homogeneous.

The expected excess temperatures of mantle plumes are on the order of 1200° C, the temperature rise across a deep TBL. The usual explanation of the much smaller observed temperature excesses is that only the top of the deep TBL is involved. But the usual type of plume discussed in the geodynamics literature is a cavity plume, which involves the lowest-viscosity, lowest-density, and highest-temperature part.

The authors assume that excess bathymetry and "acoustic anomalies" are due entirely to excess temperature. They argue that some islands have "very high T_{ex}." (~120° C at Hawaii); this value is only 10% of the temperature excess expected from a deep mantle plume. However, a modest temperature excess is quite consistent with normal variations in the shallow mantle from plate tectonic processes and with variations in the surface boundary layer.

Putirka et al. (2007) argue that current spreading ridges exhibit a petrological T_p range of 140–210° C (±34° C)—depending on how the data are filtered. This range is only half the global long-wavelength T_p range inferred from geophysics (Kaula, 1983; Anderson, 2000), but it still allows temperature excesses of ~270° C (2σ), sufficient to take the mantle from a state of incipient melting to extensive melting, even if the large range of lithologically plausible melting temperatures is ignored. It is probable, however, that melting and tomographic anomalies mainly depend on lithology, and only secondarily on absolute temperature (Anderson, this volume; Foulger, this volume; Natland, this volume). But even the thermal and heatflow arguments do not require a deep mantle plume.

31 January 2007, James H. Natland

An illustration of Hawaiian picrite (Fig. D-1) may reduce the redness of the herring surmised by Putirka and Rhodes in their comment of 27 January, from my earlier comment. Two sorts of data may be obtained from picrites: bulk compositions and mineral compositions. Without the latter, inferences about incremental addition of olivine to liquids or olivine accumulation are subject to ambiguity. Bulk compositions do not reveal any of the potential combination of differentiated or primitive magma strains that may contribute to the whole. Ideally, one would like to have compositions of all liquids that contributed to the bulk rock.

The Nuuanu landslide was a major sector collapse of Oahu's Koolau volcano, which was accompanied by one or more enormous submarine vitric pyroclastic eruptions. The eruptives cascaded out and around large downslumped blocks of the volcano (Clague et al., 2002) and over the seafloor for a distance of about 200 km, where beds of the glass-rich material cap the crest of the Hawaiian arch (Stephen et al., 2003). There, at Ocean Drilling Program (ODP) site 1223, amid several thin turbidites and mudstones (Garcia et al., 2006), two beds of glassy vitric tuff, each well indurated and several meters thick, were cored. The two beds consist mainly of angular shards of basaltic glass variably altered at their rims and, given the overall state of alteration of the rock, unusually fresh olivine. In bulk composition they are olivine-rich tholeiites and picrites (10.9–15.8% MgO; Shipboard Scientific Party, 2003). With ~20% fresh glass, much of it enclosing olivine and Cr-spinel, they reveal in detail how Hawaiian picrite forms.

Most olivine-rich tholeiites and picrites from Kilauea and other Hawaiian volcanoes (Clague et al., 1995; Norman and Garcia, 1999) also have FeO contents (10–11%) within the range of Mauna Kea and Mauna Loa picrites. But some Kilauea glass has FeO contents as low as 8%, matching low Nuuanu values, and some glass inclusions in olivine (Clague et al., 1995) have even lower FeO contents (5.6–7.2%). Basalt glass with low FeO content (5.6–9.%) clearly exists at Kilauea volcano and within individual thin sections of Nuuanu picritic vitric tuff. It is not a red herring. These liquids probably had more-magnesian parental precursors that may even have crystallized olivine as forsteritic as Fo_{91} (e.g., lowest dashed line in Fig. D-1), but for Nuuanu, this suggestion is totally conjectural. Based on the equilibrium relationship with Fo_{91} in Figure D-1 (Rhodes and Vollinger, 2004), such liquid may only have had MgO of ~13%, and a correspondingly lower eruptive temperature than any calculated at equilibrium with Fo_{91} for Hawaii by Putirka (2005) and Putirka et al. (2007). The Mg# of this liquid (~72) is about that of the most primitive MORB liquids.

Thus without well-documented crystallization histories, you cannot pin down which melt strain in a picrite produced Fo_{91}, which is almost never present in the rock anyway. Many olivine phenocrysts from Kilauea and other Hawaiian picrites (Wilkinson and Hensel, 1988; Clague et al., 1995; Norman and Garcia, 1999) are much more iron-rich (Fo_{85-78}); accumulation of such olivine undoubtedly occurs there and elsewhere. However, from Figure D-1, if you follow the procedure advocated by Putirka and Rhodes, then tiny Juan Fernandez has to be significantly hotter than huge Hawaii. Instead, accumulation of iron-rich olivine (Natland, 2003; represented by the upper dashed lines in Fig. D-1) is clearly the reason why the bulk rocks there have higher FeO contents than found in Kilauea tholeiites. It is not a consequence of higher potential temperature. (Based on my own mineral data, the same consideration applies to Samoa.)

Mixing between primitive and differentiated compositions is the most serious bugbear in the calculation of primitive parental liquid compositions and potential temperatures. Ironically,

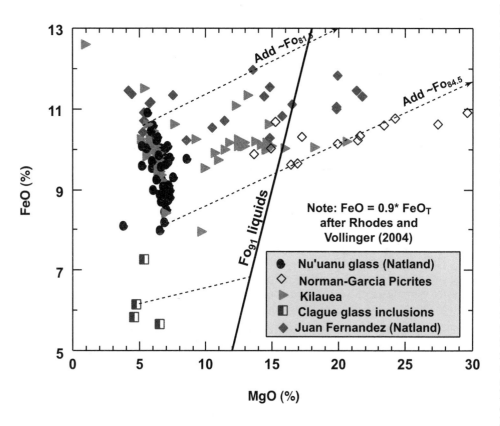

Figure D-1. Plot of FeO versus MgO for several suites of glass samples. None of the glass compositions from site 1223 and from dive samples from collapsed Nuuanu blocks nearer Oahu (Clague et al., 2002) are picritic. The picritic bulk compositions thus result from olivine accumulation. All glasses are differentiated, with MgO tightly clustered from 5 to 7%. FeO content increases from 8 to 11% as MgO decreases. This trend indicates a cotectic, not olivine-controlled, liquid line of descent. Olivine compositions are $Fo_{85.4-75.5}$, with Fo correlating generally with Mg# of enclosing glass. In Figure 11 of Rhodes and Vollinger (2004), the FeO content of samples from the Hawaii Scientific Drilling Project (HSDP)-2 drill site (Mauna Kea and Mauna Loa) varies from 10–11%, and the MgO value extends to 30%. The FeO content is at the high end of the range of the Nuuanu glasses. Inasmuch as original FeO content of liquids that produced the olivine and the olivine compositions themselves are unknown, all HSDP picrites thus could be produced by olivine addition to substantially differentiated cotectic liquids. Subtrends with higher FeO could simply indicate accumulation of more iron-rich olivine, such as that at Nuuanu.

Rhodes and Vollinger (2004, p. 24) made exactly the same argument, observing "if our interpretation is correct, and of general application, the common practice of back-calculating to putative primary magma compositions through addition of incremental amounts of equilibrium olivine, may well lead to incorrect results and spurious correlations with MgO." Precisely.

20 February 2007, T.J. Falloon,
L. Danyushevsky, and D.H. Green

In response to Presnall's comment of 1 January, we reiterate that our purpose is to identify the highest temperature liquids among MORB or OIB. We used glass compositions, analysed by electron microprobe, and calculated parental compositions by incremental addition of equilibrium olivine where the glass contains both its liquidus olivine and more magnesian olivine, using the latter as the target or limit justified by the petrography. We make no other assumptions.

Presnall comments "Despite thousands of microprobe analyses of MORB glasses, no compositions that show a trend of olivine-controlled crystallization have ever been reported." This statement is not true. It arises from inadequate coverage of published work and from the way Presnall and Gudfinnsson (2007)

select their glass data for petrogenetic interpretation. In our chapter we plot 190 glass analyses from mid-ocean ridge settings with >9.5% MgO, from the PetDB database. Among these, there are clearly glasses that lie in the ol+liq field at eruption pressures and temperatures (e.g., high-magnesian glasses from Siqueiros; Fig. D-2). In addition we have clear mineral-composition evidence from selected sites (Siqueiros, ODP hole 896A), which demonstrate the sequence of appearance of phases as ol (+rare spinel), ol+plag, ol+plag+cpx. Experimental demonstration of this sequence goes back at least to the paper of Green et al. (1979). We are not the first to identify this range of MORB glasses extending into the ol+liq field. In the petrogenetic analysis of global MORB glass chemical compositions of Presnall and Gudfinnsson (2007), the authors state that "the LKP (Langmuir et al., 1992) procedure of retaining only analyses with MgO values between 5 and 8.5% has been followed." The authors further restrict their dataset to the Smithsonian database. Langmuir et al. (1992) wished to select glasses lying on the ol+plag+cpx fractionation surface at low pressure and which approximate to a linear trend in several oxide versus oxide variation diagrams, particularly MgO versus FeO and MgO versus Na_2O. They wished to avoid more-magnesian glasses (>8.5% MgO), which depart from this multiply-saturated surface. Presnall and Gudfinnsson (2007) exclude higher temperature, MgO-rich glasses from their consideration and then state that there is no evidence for them, in spite of other authors' publications of such glasses.

In advocating a relatively low-temperature and low-pressure origin for MORB, Presnall and Gudfinnsson (2007) argue that parental MORB lie on the ol, opx, cpx, plag, and/or sp saturation surface at 0.9–1.5 GPa, based on consideration of compositions interpolated at MgO = 8.0%. Neither these nor higher-temperature glasses with >9.5% MgO lie precisely on the 1-GPa ol, opx, cpx, plag, sp mutiply-saturated surface when projected into multicomponent normative projections as in our article, Green et al. (2001), or Green and Falloon (2005)—they lie at higher normative diopside and lower normative olivine than the multiphase cotectic. In addition the glasses (Presnall and Gudfinnsson, 2007) at MgO = 8% are not sufficiently magnesian to be in equilibrium with mantle olivine (>Fo$_{89}$). The more magnesian parental glasses derived in our article by olivine addition to Fo$_{91-91.5}$ have *eruption* temperatures greater than the restricted T_p = 1240–1260° C of Presnall and Gudfinnsson (2007), further demonstrating the error in the low-temperature MORB model.

In his comment of 28 January, Presnall does not address our article directly but disputes the argument of Purtika and Rhodes (their comment of 27 January) for variable (including >1400° C) T_p in mid-ocean ridge settings. We infer variable eruption temperatures, *P, T,* and percentage melting along ridges, but we consider that attribution of this variability to differences in T_p, differences in the departure from adiabatic upwelling, or source compositions, is premature. We expect that all these factors have roles, and choice among them requires a much greater database of primitive melt compositions as assessed by methods used in our article.

In his comment of 12 January, Natland expresses reservations regarding our methodology of adding olivine to olivine-saturated glasses to infer more-magnesian and olivine-rich parental magmas. In our article we use only glass (liquid) compositions that contain olivines matching liquidus olivine compositions (derived from experimentally calibrated Fe/Mg partitioning; Ford et al., 1983) and contain, in addition, more-magnesian olivine phenocrysts and microphenocrysts. Our interpretation is consistent with the petrographic observations and is the simplest interpretation. We agree with Natland that most picrites are mixtures of crystals, particularly olivine, and the interpretation of liquid composition requires petrographic information on mineral compositions and their relation to crystal fractionation and possibly magma mixing (Danyushevsky et al., 2002). However, we are not using compositions of picrite rocks (i.e., rocks with >10–15% modal olivine). We are using glasses and only the highest-temperature and most-magnesian glasses lying in the ol+liq phase field at their liquidus temperatures. We demonstrate this in Figure D-2, and note that Natland's glass in the picrite D20-3 lies on the ol+liq trend of Figure D-2, and some other glasses from Natland (1980, 1989) fall on the ol+plag trend of Figure D-1 between 8 and 9% MgO.

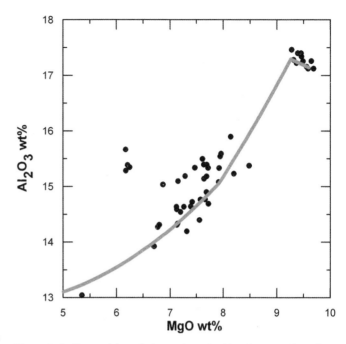

Figure D-2. Composition of glasses from the Siqueiros transform from Danyushevsky et al. (2003), and a fractionation trend (gray line) calculated from the most magnesian glass following the model (PETROLOG) of Danyushevsky (2001). This plot clearly demonstrates that the most magnesian glasses from Siqueiros are saturated in olivine only.

We also agree with Natland that most MORB glasses are evolved and the processes of evolution include crystal fractionation, "AFC" (assimilation, fractionation, crystallization) processes, reactive porous flow, magma mixing, and the like. This complexity is documented by Natland (1980, 1989) and clouds the interpretations of Presnall and Gudfinnsson (2007) and of Langmuir et al. (1992) using arrays of evolved glasses with 5.–8.5% MgO. However, by choosing the glasses as discussed, we identify the highest temperature melts (i.e., the most primitive observed liquids). There is no conflict between the recognition of the complexity of processes that a high-temperature picritic melt may undergo as it cools, reacts, crystallizes, and quenches to crystals + glass, and the identification of the highest temperature liquid (i.e., the liquid approaching or reaching an end-member "primitive magma").

We respond to Natland's detailed comments on the Siqueiros locality in a data repository item supplement to this comment.[1] Natland (1989) documents evidence for mixing and reaction of high-temperature melts with crystalline inclusions, notably An-rich plagioclase. For a liquid already crystallizing olivine (+spinel), the energy required to dissolve phases with which the

[1]GSA Data Repository Item 2007193, Expanded discussion and reply on Falloon et al. (2007) article, is available on request from Documents Secretary, GSA, P.O. Box 9140, Boulder, CO 80301-9140, USA, or editing@geosociety.org, at www.geosociety.org/pubs/ft2007.htm.

liquid is undersaturated is provided by exothermic olivine ± spinel crystallization at the site of dissolution. This process may also lead to entrapment of ephemeral and local melt inclusions and zoned spinels, reflecting the competition between diffusive homogenization of melt and growth of olivine and spinel (Danyushevsky et al., 2003). Natland summarizes evidence for more than one parental magma type in the Siqueiros sampling area. We also argue for significant source heterogeneity and complex evolution of primitive magmas at crustal and subcrustal depths.

In responding to Natland's comments on Iceland and Hawaii, we recognize complexity in the evolution of primitive melts, but our methodology, using both petrography and PETROLOG (Danyushevsky, 2001; Danyushevsky et al., 2003) modeling, identifies the highest-temperature melts in each sampled setting and explores the implications of such melts for constraining the actual melting and melt segregation process (*P, T,* and residual phases) from a lherzolite or harzburgite source. Natland comments that "it is possible that Hawaiian and Icelandic melt strains derive from eclogitic components in mantle sources." Pyroxenitic or eclogite heterogeneity in mantle sources of MORB, and particularly OIB, may permit melting at temperatures below the peridotite solidus, but such melts can rarely reach the surface. Normally they react out of existence in the enclosing peridotite, "fertilizing" the latter and possibly acting as a flux by lowering the solidus temperature and causing melting of the new refertilized mantle. Such liquids will be picritic, controlled in their major-element composition by the residual phases of lherzolite and/or harzburgite but reflecting in their trace-element (incompatible elements) and isotopic composition their precursor (eclogite+residual peridotite) histories (Yaxley and Green, 1998; Yaxley, 2000).

Concerning the comments of Putirka and Rhodes (27 January), in the GSA data repository item (see footnote 1 in this discussion), we have inserted material appropriately. The comments by Anderson and Natland (31 January) address the comment of Putirka and Rhodes but do not directly address our article. We are in general agreement with them. Most importantly, Natland provides a figure and discussion illustrating the differences between olivine fractional crystallization and olivine phenocryst accumulation, refuting the approach of Putirka (2005) and Putirka et al. (2007).

Contrary to the statements of Putirka and Rhodes, we do not use a particular model with respect to hotspots and we are careful to document the three sequential steps followed to infer:

1. Liquidus temperature for the most magnesian glasses (Table 1 *in* Falloon et al., this volume);
2. Parental magmas to these glasses and *P* and *T* conditions of compatibility with peridotite residue (Tables 2, 3, and 4); and
3. Mantle potential temperatures, with emphasis on the assumptions with respect to melt fraction/source composition and latent heat of melting and their effect on T_p estimation (Table 5).

We do not use mean temperatures or compositions, as these are meaningless when each magma batch, upwelling column or diapir, ridge segment, seamount, or volcano may have a different compositional, source *P* and *T,* or nonadiabatic ascent path to eruption. We reject the numerical and computational approach, which compiles and manipulates thousands of analyses of rocks and glasses to seek arithmetic algorithms to define geodynamic models, ignoring petrological information and physical reality. A more detailed discussion of Putirka (2005) and Putirka et al. (2007) is presented by Falloon et al. (2007).

Putirka and Rhodes incorrectly state that we ignore MORB with low to moderate FeO^T contents. We illustrate three MORBs with different FeO, different eruption and source *T* and *P,* and thus different inferred T_p values. We do not advocate a unique ambient mantle T_p or plume T_p, but demonstrate the lack of any evidence for differences between the range of eruption temperatures or source *T* and *P* for MORB and hotspot primitive magmas.

In their point 5, Putirka and Rhodes (27 January) suggest that thermally buoyant plumes require minimal temperature differences from normal mantle. Other presentations of the deep-mantle plume hypothesis have weakened the original constraints of a fixed point source (plumes are deflected, flow along ridge axes, are intermittent, etc.) and unique source compositions (depleted mantle in plumes; enriched MORB on ridges; OIB chemistry in rifts, seamounts, and ridges) but the prediction and assertion of high magma eruption temperatures at hotspots has remained a cornerstone of the hypothesis (e.g., Campbell, 2006). Although in their comment Putirka and Rhodes appear to relax this difference to within the spread of MORB sources (see also Anderson's comment of 28 January), elsewhere (Putirka, 2005; Putirka et al., 2007) they support T_p for MORB of >1400° C and excess T_p for "plumes" of ~200° C. Such values for T_p should produce peridotitic komatiite magmas at hotspots with ~60–70% melting of mantle lherzolite and residual dunite of Fo_{93-94}. These magmas are not seen.

Contrary to their statement in their point 6 (Putirka and Rhodes, 27 January), we use the observed range of FeO and eruption temperature for glasses and derive a similar range of eruption and segregation temperatures and pressures for both mid-ocean ridge and hotspot settings. Their point 7 and following text distorts our article and previous papers, and we refer readers to them and to Falloon et al. (2007) for detailed discussions of the methods of Putirka (2005) and Putirka et al. (2007) for inferring potential temperature and depths of origin of magmas. Both Putirka papers calculate increasingly large errors in estimating liquidus temperatures with increasing MgO content.

We agree with the comments of Anderson (28 January) concerning the Putirka and Rhodes discussion, and more importantly, we agree with his emphasis on compositional rather than thermal variability as causes for relative density, buoyancy, and seismic properties within the upper mantle. In Green et al. (2001) and Green and Falloon (2005) we specifically advocate a key role for refractory, buoyant, old subducted slabs as causes for topographic and magmatic aspects of "hotspots." We agree that the fuller exploration of the geodynamic consequences of mantle compositional heterogeneity is timely.

We have to our satisfaction tested and denied a key prediction (oft-quoted as a confirmation) of the deep-mantle thermal plume hypothesis—that their magmatic expression at hotspots is characterized by much-higher-temperature magmas and by greater depths and extents of melting than apparent in the "normal" mantle upwelling and melting at mid-ocean ridges. We find on the contrary that the ranges of magmatic temperatures in "hotspot" and mid-ocean ridge settings are similar, and observed differences in magmatic products are a consequence of compositional heterogeneity in the upper mantle.

REFERENCES CITED

Anderson, D.L., 2000, Thermal state of the upper mantle; No role for mantle plumes: Geophysical Research Letters, v. 27, p. 3623–3626.

Anderson, D.L., 2007 (this volume), The eclogite engine: Chemical geodynamics as a Galileo thermometer, *in* Foulger, G.R., and Jurdy, D.M., eds., Plates, plumes, and planetary processes: Boulder, Colorado, Geological Society of America Special Paper 430, doi: 10.1130/2007.2430(03).

Batiza, R., Rosendahl, B., and Fisher, R.L., 1977, Evolution of oceanic crust, 3, Petrology and chemistry of basalts from the East Pacific Rise and Siqueiros transform fault: Journal of Geophysical Research, v. 92, p. 265–276.

Beattie, P., 1993, Olivine-melt and orthopyroxene-melt equilibria: Contributions to Minerology and Petrology, v. 115, p. 103–111.

Campbell, I.H, 2006, Large igneous provinces and the mantle plume hypothesis: Elements, v. 1, p. 265–269.

Clague, D.A., Weber, W., and Dixon, J. E., 1991, Picritic glasses from Hawaii: Nature, v. 353, p. 553–556.

Clague, D.A., Moore, J.G., Dixon, J.E., and Friesen, W.B., 1995, Petrology of submarine lavas from Kilauea's Puna Ridge, Hawaii: Journal of Petrology, v. 36, p. 299–349.

Clague, D.A., Moore, J.G., and Davis, A.S., 2002, Volcanic breccia and hyaloclastite in blocks from the Nu'uanu and Wailau landslides, Hawaii, *in* Takahashi, E., et al., eds., Hawaiian volcanoes: Deep underwater perspectives: Washington, D.C., American Geophysical Union Geophysical Monograph 128, p. 279–296.

Danyushevsky, L.V. 2001, The effect of small amounts of H_2O on crystallization of mid-ocean ridge and backarc basin magmas: Journal of Volcanology and Geothermal Research, v. 110, p. 265–280.

Danyushevsky, L.V., Sokolov, S., and Falloon, T.J., 2002, Melt inclusions in olivine phenocrysts: Using diffusive re-equilibration to determine the cooling history of a crystal, with implications for the origin of olivine-phyric volcanic rocks: Journal of Petrology, v. 43, p. 1651–1671.

Danyushevsky, L.V., Perfit, M.R., Eggins, S.M., and Falloon, T.J., 2003, Crustal origin for coupled "ultra-depleted" and "plagioclase" signatures in MORB olivine-hosted melt inclusions: Evidence from the Siqueiros Transform Fault, East Pacific Rise: Contributions to Mineralogy and Petrology, v. 144, p. 619–637.

Falloon, T.J., Danyushevsky, L.V., Ariskin, A., Green, D.H., and Ford, C.E., 2007, The application of olivine geothermometry to infer crystallization temperatures of parental liquids: Implications for the temperatures of MORB magmas: Chemical Geology, v. 241, no. 3–4, p. 207–233.

Falloon, T.J., Green, D.H., and Danyushevsky, L.V., 2007 (this volume), Crystallization temperatures of tholeiite parental liquids: Implications for the existence of thermally driven mantle plumes, *in* Foulger, G.R., and Jurdy, D.M., eds., Plates, plumes, and planetary processes: Boulder, Colorado,

Geological Society of America Special Paper 430, doi: 10.1130/2007.2430(12).

Ford, C.E., Russell, D.G., Craven, J.A., and Fisk, M.R., 1983, Olivine-liquid equilibria: Temperature, pressure and composition dependence of the crystal/liquid cation partition coefficients for Mg, Fe^{2+}, Ca and Mn: Journal of Petrology, v. 24, p. 256–265.

Foulger, G.R., 2007 (this volume), The plate model for the genesis of melting anomalies, *in* Foulger, G.R., and Jurdy, D.M., eds., Plates, plumes, and planetary processes: Boulder, Colorado, Geological Society of America Special Paper 430, doi: 10.1130/2007.2430(01).

Foulger, G.R., Natland, J.H., and Anderson, D.L., 2005, Genesis of the Iceland melt anomaly by plate tectonic processes, *in* Foulger, G.R., et al., eds., Plates, plumes and paradigms: Boulder, Colorado, Geological Society of America Special Paper 388, p. 595–626, doi: 10.1130/2005.2388(35).

Garcia, M.O., Sherman, S.B., Moore, G.F., Goll, R., Popova-Goll, I., Natland J.H., and Acton, G., 2006, Frequent landslides from Ko'olau Volcano: Results from ODP Hole 1223A: Journal of Volcanology and Geothermal Research, v. 151, p. 251–268.

Green, D.H., and Falloon, T.J., 2005, Primary magmas at mid-ocean ridges, "hotspots," and other intraplate settings: Constraints on mantle potential temperature, *in* Foulger, G.R., et al., eds., Plates, plumes and paradigms: Boulder, Colorado, Geological Society of America Special Paper 388, p. 217–247, doi: 10.1130/2005.2388(14).

Green, D.H., Hibberson, W.O., and Jaques, A.L., 1979, Petrogenesis of mid-ocean ridge basalts, *in* McElhinny, M.W., ed., The Earth: Its origin, structure and evolution: London, Academic Press, p. 265–290.

Green, D.H., Falloon, T.J., Eggins, S.M., and Yaxley, G.M., 2001, Primary magmas and mantle temperatures: European Journal of Mineralogy, v. 13, p. 437–451.

Herzberg, C., and O'Hara, M.J., 2002, Plume-associated magmas of Phanerozoic age: Journal of Petrology, v. 43, p. 1857–1883.

Kaula, W.M., 1983, Minimum upper mantle temperature variations consistent with observed heat flow and plate velocities: Journal of Geophysical Research, v. 88, p. 10,323–10,332.

Langmuir, C.H., Klein, E.M., and Plank, T., 1992, Petrological systematics of mid-ocean ridge basalts: Constraints on melt generation beneath ocean ridges, *in* Morgan, J.P., et al., eds., Mantle flow and melt generation at mid-ocean ridges: Washington, D.C., American Geophysical Union Geophysical Monograph 71, p. 183–280.

Maclennan, J., McKenzie, D., Grönvold, K., Shimizu, N., Eiler, J.M., and Kitchen, N., 2003, Melt mixing and crystallization under Theistareykir, northeast Iceland: Geochemistry, Geophysics, Geosystems, v. 4, p. 1–40, doi: 10.1029/2003GC005588.

Melson, W.G., Vallier, T.L., Wright, T.L., Byerly, G.R., and Nelen, J.A., 1976, Chemical diversity of abyssal volcanic glass erupted along Pacific, Atlantic, and Indian Ocean sea-floor spreading centers: Washington, D.C., American Geophysical Union Geophysical Monograph 19, p. 351–368.

Natland, J.H., 1980, Crystal morphologies in basalts dredged and drilled from the East Pacific Rise near 9°N and the Siqueiros Fracture Zone, *in* Melson, W.G., et al., eds., Initial reports of the Deep Sea Drilling Project, volume 54: Washington, D.C., U.S. Government Printing Office, p. 605–633.

Natland, J.H., 1989, Partial melting of a lithologically heterogeneous mantle: Inferences from crystallization histories of magnesian abyssal tholeiites from the Siqueiros Fracture Zone, *in* Saunders, A.D., and Norry, M.J., eds., Magmatism in the ocean basins: London, Geological Society of London Special Publication 42, p. 41–70.

Natland, J.H., 2003, Capture of mantle helium by growing olivine phenocrysts in picritic basalts from the Juan Fernandez Islands, SE Pacific: Journal of Petrology, v. 44, p. 421–456.

Natland, J.H., 2007 (this volume), ΔNb and the role of magma mixing at the East Pacific Rise and Iceland, *in* Foulger, G.R., and Jurdy, D.M., eds., Plates, plumes, and planetary processes: Boulder, Colorado, Geological Society of America Special Paper 430, doi: 10.1130/2007.2430(21).

Norman, M.D., and Garcia, M.O., 1999. Primitive magmas and source characteristics of the Hawaiian plume: Petrology and geochemistry of shield picrites: Earth and Planetary Science Letters, v., 168, p. 27–44.

Poustovetov, A., and Roeder, P.L., 2000, The distribution of Cr between basaltic melt and chromian spinel as an oxygen geobarometer: Canadian Mineralogist, v. 39, p. 309–317.

Presnall, D.C., and Gudfinnsson, G., 2007, Origin of the oceanic lithosphere: Journal of Petrology, in press.

Presnall, D.C., Dixon, J.R., O'Donnell, T.H., and Dixon, S.A., 1979, Generation of mid-ocean ridge tholeiites: Journal of Petrology, v. 20, p. 3–35.

Presnall, D.C., Gudfinnsson, G.H., and Walter, M.J., 2002, Generation of mid-ocean ridge basalts at pressures from 1 to 7 GPa: Geochimica et Cosmochimica Acta, v. 66, p. 2037–2090.

Putirka, K., 2005, Mantle potential temperatures at Hawaii, Iceland, and the mid-ocean ridge system, as inferred from olivine phenocrysts: Evidence for thermally-driven mantle plumes: Geochemistry, Geophysics, Geosystems, v. 6, doi: 10.1029/2005GCGC000915.

Putirka, K.D., Perfit, M., Ryerson, F.J., and Jackson, M.G., 2007, Ambient and excess mantle temperatures, olivine thermometry, and active vs. passive upwelling: Chemical Geology, v. 241, p. 177–206.

Rhodes, J.M., and Vollinger, M.J., 2004, Composition of basaltic lavas sampled by phase-2 of the Hawaii Scientific Drilling Project: Geochemical stratigraphy and magma types, Geochemistry, Geophysics, Geosystems, v. 5, doi: 10.1029/2002GC000434.

Schilling, J.G., 1991, Fluxes and excess temperatures of mantle plumes inferred from their interaction with migrating mid-ocean ridges: Nature, v. 352, p. 397–403.

Shipboard Scientific Party, 1993, Site 896, *in* Alt, J.C., et al., eds., Proceedings of the Ocean Drilling Program, initial reports, volume 148: College Station, Texas, Ocean Drilling Program, p. 123–192.

Shipboard Scientific Party, 2003, Site 1223, *in* Stephen, R.A., et al., eds., Proceedings of the Ocean Drilling Program, initial reports, volume 200: College Station, Texas, Ocean Drilling Program, p. 1–159.

Sigurdsson, I.A., Steinthorsson, S., and Gronvold, K., 2000, Calcium-rich melt inclusions in Cr-spinels from Goprgarhraun, northern Iceland: Earth and Planetary Science Letters, v. 183, p. 15–26.

Slater, L., McKenzie, D., Grönvold, K., and Shimizu, N., 2001. Melt generation and movement beneath Theistareykir, NE Iceland: Journal of Petrology, v. 42, p. 321–354.

Sleep, N.H., 1990, Hotspots and mantle plumes: Some phenomenology: Journal of Geophysical Research, v. 95, p. 6715–6736.

Sobolev, A.V., Hofmann, A.W., and Nikogosian, I.K., 2000. Recycled oceanic crust observed in "ghost plagioclase" within the source of Mauna Loa lavas: Nature, v. 404, p. 986–990.

Sobolev, A.V., Hofmann, A.W., Sobolev, S.V., and Nikogosian, I.K., 2005. An olivine-free mantle source of Hawaiian shield basalts: Nature, v. 434, p. 590–597.

Stephen, R.A., Kasahara, J., Acton, G., et al., eds., 2003, Proceedings of the Ocean Drilling Program, initial reports, volume 200: College Station, Texas, Ocean Drilling Program, 178 p., doi: 10.2973/odp.proc.ir.200.2003.

Wilkinson, J.G.F., and Hensel, H.D., 1988, The petrology of some picrites from Mauna Loa and Kilauea volcanoes, Hawaii: Contributions to Mineralogy and Petrology, v., 98, p. 326–345.

Wright, T.L., and Fiske, R.S., 1971, Origin of the differentiated and hybrid lavas of Kilauea volcano, Hawaii: Journal of Petrology, v. 12, p. 1–65.

Yaxley, G.M., 2000, Experimental study of the phase and melting relations of homogeneous basalt + peridotite mixtures and implications for the petrogenesis of flood basalts: Contributions to Mineralogy and Petrology, v. 139, p. 326–338.

Yaxley, G.M., and Green, D.H., 1998, Reactions between eclogite and peridotite: Mantle refertilisation by subduction of oceanic crust: Schweizerische Mineralogische und Petrographische Mitteilungen, v. 78, p. 243–255.

The Geological Society of America
Special Paper 430
2007

Potential effects of hydrothermal circulation and magmatism on heatflow at hotspot swells

Carol A. Stein*

*Department of Earth and Environmental Sciences, University of Illinois at Chicago,
845 West Taylor Street, Chicago, Illinois 60607-7059, USA*

Richard P. Von Herzen*

*Department of Geology and Geophysics, Woods Hole Oceanographic Institution,
Clark 244B, MS#22, Woods Hole, Massachusetts 02543, USA,
and Department of Earth Sciences, University of California at Santa Cruz,
Earth and Marine Science Building, Santa Cruz, California 95064-1077, USA*

ABSTRACT

The lack of high heatflow values at hotspots has been interpreted as showing that the mechanism forming the associated swells is not reheating of the lower half of oceanic lithosphere. Alternatively, it has recently been proposed that the hotspot surface heatflow signature is obscured by fluid circulation. We re-examine closely spaced heatflow measurements near the Hawaii, Réunion, Crozet, Cape Verde, and Bermuda hotspots. We conclude that hydrothermal circulation may redistribute heat near the swell axes, but it does not mask a large and spatially broad heatflow anomaly. There may, however, be heatflow perturbations associated with the cooling of igneous intrusions emplaced during hotspot formation. Although such effects may raise heatflow at a few sites, the small heatflow anomalies indicate that the mechanisms producing hotspots do not significantly perturb the thermal state of the lithosphere.

Keywords: heatflow, swell, hotspot, mantle plume

INTRODUCTION

After the recognition that hotspots, regions of excess volcanism and higher elevation compared to surrounding areas, result from the rise of material through the mantle to the Earth's surface beneath both oceans and continents, marine geothermal measurements were among the first geophysical techniques used to assess their nature. Initially it was proposed that significant reheating of the bottom half of the lithosphere over a distance of ~300 km in diameter was required to match the depth anomalies at hotspots (e.g., Crough, 1978). A consequence of this mechanism is unusually high surface heatflow, which should be measurable by established methodologies. The first detailed study of the Hawaiian hotspot found high mean heatflow (Von Herzen et al., 1982), but subsequent studies showed that the anomaly was overestimated because the reference model predicted significantly lower heatflow than observed in unperturbed older lithosphere (Von Herzen et al., 1989; Stein and Stein, 1993). The conclusion was that lithospheric reheating is not a major factor in hotspot formation. Thus, possible

*E-mails: Stein, cstein@uic.edu; Von Herzen, rvonh@whoi.edu.

Stein, C.A., and Von Herzen, R.P., 2007, Potential effects of hydrothermal circulation and magmatism on heatflow at hotspot swells, *in* Foulger, G.R., and Jurdy, D.M., eds., Plates, plumes, and planetary processes: Geological Society of America Special Paper 430, p. 261–274, doi: 10.1130/2007.2430(13). For permission to copy, contact editing@geosociety.org. ©2007 The Geological Society of America. All rights reserved.

mechanisms producing the swells must be consistent with small or essentially zero heatflow anomalies, such as those expected for dynamic models (e.g., Ribe and Christensen, 1994).

However, it has been recently proposed (Harris et al., 2000a,b) that hydrothermal circulation may obscure the signal from the processes forming hotspots. Harris et al. (2000a, p. 21,368) suggested that "fluid flow has the potential to obscure basal heatflow patterns associated with the Hawaiian hot spot," which would "not bode well for capturing the form and magnitude of this signature." Most hydrothermal circulation occurs in young oceanic crust (0 to ca. 65 Ma), resulting in lower measured heatflow compared to conductive cooling models of the lithosphere. Heatflow transferred by hydrothermal circulation in older oceanic lithosphere is thought to be much less (e.g., Stein and Stein, 1994; Von Herzen, 2004). However, as suggested above, if the overall rough basement and islands and seamounts associated with hotspots cause fluid flow to occur both in the igneous basement and perhaps in the relatively permeable volcaniclastic sediments, then "capturing the maximum basal heat flux may be confounded by fluid flow" (Harris et al., 2000a, p. 21,368). Evidence on the magnitude and extent of hydrothermal heat transfer may be provided by the detailed heatflow at the five hotspots on crust older than ~65 Ma.

In this article, we first examine heatflow anomalies near these hotspots. The anomalies are computed relative to the Global Depth and Heatflow reference model (Stein and Stein, 1992) and with crustal ages from Mueller et al. (1997). Second, we examine which sites may be affected by hydrothermal circulation and thus mask higher heatflow. Last, we consider heatflow perturbations associated with the cooling of igneous structures that result from the hotspot processes, including the igneous material emplaced on the surface of pre-existing oceanic crust, and intruded sills at subcrustal depths.

ESTIMATED HEATFLOW ANOMALIES FROM HOTSPOTS

Although the depth in the Earth from which hotspots originate is still debated (e.g., Sleep, 2003; Anderson, 2005; Foulger, this volume; Garnero et al., this volume), a geophysical challenge is to separate the effects of mechanisms forming hotspots from shallower lithospheric processes, including the "normal" lithospheric cooling as the plate ages. Lithospheric interactions with the upward flow of magma, the absolute velocity of the plate, and near-surface physical processes (including excess volcanism and hydrothermal circulation) complicate the task of understanding the deeper mantle processes leading to the formation of hotspots. Also, because heat is largely transferred by conduction through the lithosphere, excess heat loss associated with hotspot formation at a given location may occur long after the hotspot upwelling has ceased. Although hotspots are associated with anomalously shallow bathymetry, it is unclear whether their surface heatflow is anomalously high. Depending on the answer to this question, different mechanisms have been proposed for hotspot formation.

Many mechanisms have been suggested to explain hotspot formation. An important observation to model is the anomalously shallow bathymetry and its subsequent subsidence (e.g., DeLaughter et al., 2005). For example, the reheating model requires significant upwelling of mantle material that rapidly thins the bottom half of the overlying plate (Crough, 1978, 1983). For the dynamic plume model that incorporates a large viscosity contrast between the plume and overlying lithosphere, most reheating is at the base of the lithosphere (e.g., Liu and Chase, 1989; Ribe and Christensen, 1994). The compositional buoyancy model predicts shallower depths from relatively low-density magmas emplaced either within the lithosphere or near its base (e.g., Robinson, 1988; Sleep, 1994; Phipps Morgan et al., 1995). Different amounts of reheating and excess heatflow with time since hotspot formation are expected from these different mechanisms. Hence, heatflow data may help to discriminate among them.

For the Hawaiian chain, on the relatively fast-moving Pacific plate, matching the observed uplift and subsequent subsidence by the reheating mechanism requires that the bottom third to half of the oceanic lithosphere is rapidly reheated to asthenospheric temperatures over a zone 300 km wide centered on the axis (Von Herzen et al., 1982). At the time reheating begins, the surface heatflow anomaly should be zero, but as the plate moves away from the upwelling, the heating of the lower lithosphere propagates upward by conduction. The predicted anomaly gradually increases to a maximum of ~25 mW m^{-2} ~15–20 m.y. after reheating and then gradually decreases. The heatflow anomaly should be a maximum at the axis of the chain and decrease to zero over ~600 km normal to the axis (Fig. 1A).

Although anomalously high heatflow was initially reported for the Hawaiian chain (Von Herzen et al., 1982), consistent with significant lithospheric reheating, subsequent data and analysis (Von Herzen et al., 1989; Stein and Stein, 1993) showed that the expected Gaussian distribution of anomalous heat flux orthogonal to the axis did not exist, and many of the apparent anomalies along axis result from comparing the data to reference thermal models that underestimate heatflow elsewhere. Thus, the implication is that heatflow data do not support swell formation by extensive lithospheric heating.

The small heatflow anomaly and absence of the pattern expected for lithospheric reheating at Hawaii and other marine hotspots (Réunion [Bonneville et al., 1997], Bermuda [Detrick et al., 1986], Cape Verde [Courtney and White, 1986], and Crozet [Courtney and Recq, 1986]) imply that uplift may result from the dynamic effects of rising plumes. Their thermal effects are concentrated at the base of the lithosphere and hence would raise surface heatflow at most slightly, given that tens of millions of years are required for the conduction of heat to the surface. For example, Ribe and Christensen (1994) predicted heatflow anomalies for Hawaii of less than 1 mW m^{-2} (Fig. 1B). The heat-

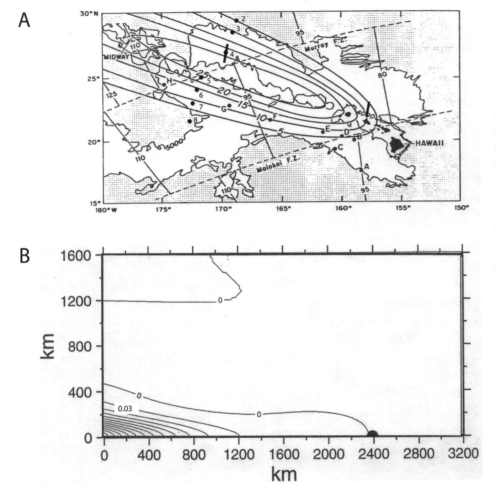

Figure 1. Predicted heatflow anomalies from reheating and dynamic plume models. (A) The reheating model predicts heatflow anomalies up to ~25 mW m^{-2} (Von Herzen et al., 1982) in contrast to (B) a dynamic plume model of less than 1 mW m^{-2} (Ribe and Christensen, 1994). The location of heatflow measurements for Hawaii from Von Herzen et al. (1982, 1989) and Harris et al. (2000a) are shown in panel A. In panel B, the predicted surface heatflow from an upwelling plume (location indicated by the solid semicircle) beneath a moving plate is shown for up to 2400 km downstream of where the plume has passed (horizontal axis) and for up to 1600 km perpendicular (vertical axis) from the path of the plume. F.Z. —fracture zone.

flow anomaly associated with compositional buoyancy depends on whether the magma is near the base of the lithosphere (e.g., Sleep, 1994) or at relatively shallow depth in the lithosphere (e.g., Phipps Morgan et al., 1995).

DATA

Of the five oceanic hotspots (Hawaii, Réunion, Cape Verde, Bermuda, and Crozet) with closely spaced "pogo"[1] heatflow measurements (Fig. 2), Hawaii is the best studied (Von Herzen et al., 1982, 1989; Harris et al., 2000a). The ca. 81-Ma hotspot track (e.g., Keller et al., 1995; Clouard and Bonneville, 2005; Norton, this volume) presently has an absolute velocity in the

hotspot frame of ~104 mm/year (Gripp and Gordon, 2002). Heatflow measurements on ca. 84- to ca. 118-Ma crust show neither the expected anomalously high heatflow near the axis parallel to the relative motion of the hotspot track nor a decrease from the center of the axis of the swell toward the edges (Figs. 3A,B and 4A). The anomalies are ~5–10 mW m^{-2} higher than expected for similar aged lithosphere (Von Herzen et al., 1989; Stein and Stein, 1992). These anomalies may reflect lithosphere, before perturbation by the hotspot processes, having a somewhat higher heatflow than typical, but still within the observed range of variability for its age (DeLaughter et al., 2005).

Réunion is the location of most recent volcanism for a ca. 65-Ma hotspot track (e.g., Bonneville et al., 1997) with a present-day ~13 mm/year absolute velocity in the hotspot reference frame (Gripp and Gordon, 2002). The heatflow measurements were made in the form of long (~200 and ~400 km) parallel profiles ~100 km apart extending westward from the Mascarene Ridge (perpendicular to the hotspot track) on crust affected by hotspot volcanism ~15–20 m.y. ago (Fig. 2B). The crustal age is

---

[1]For "pogo" measurements, the equipment is initially lowered to the seafloor, and then after heatflow is determined at a location it is raised slightly above the seafloor and towed to other nearby sites, allowing for closely spaced measurements typically for ~24 hours, before the equipment is returned to the ship. Hence, it is similar to jumping with a pogo stick.

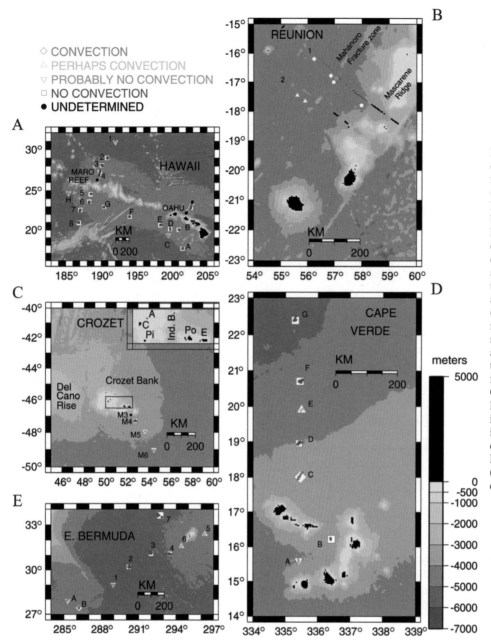

Figure 2. Bathymetry and sites with closely spaced heatflow measurements for (A) Hawaii, (B) Réunion, (C) Crozet, (D) Cape Verde, and (E) Bermuda. Von Herzen (2004) examined heatflow and bathymetry to infer whether the sites have hydrothermal circulation (red; diamond), probably have hydrothermal circulation (orange; upward-pointing triangle), probably have conductive heat transfer (green; downward-pointing triangle), or have conductive heat transfer (blue; square). Sites without a determination are shown as black circles. Most sites have conductive heat transfer. The associated numbers and letters next to the heatflow locations are the site names from the published heatflow papers. For Crozet, the upper-right inset is an enlargement of the region with the islands indicated by the small box. The eastern islands (Po—Île de la Possession; E—Île de l'Est) are separated by the Indivat Basin (Ind. B.) from the western islands (Pi—Île des Pingouins; C—Île aux Cochons; and A—Îles des Apotres).

ca. 65 Ma west of the Mahanoro fracture zone. To the east, the crustal ages from Mueller et al. (1997) are ~10–15 m.y. younger than those used by Bonneville et al. (1997; derived from Dyment, 1991) to analyze their heatflow measurements. With the revised younger crustal ages, many measured heatflow values to the east of the Mahanoro fracture zone, which were initially interpreted as greater than the expected average by Bonneville et al. (1997), are now about that expected, thus suggesting no heatflow anomaly. The heatflow anomaly is largest west of the Mahanoro fracture zone (~6 mW m^{-2} for the northern profile, although it may be affected by hydrothermal circulation) and decreases toward the Mascarene Ridge. Hence, the ex-

pected increase in the heatflow anomaly toward the axis of the hotspot from a lithospheric reheating mechanism is not observed (Fig. 3C,D).

Unlike the Hawaiian and Réunion hotspots with clear and long volcanic signatures, volcanism for Crozet on the Antarctic plate started at ca. 8 Ma and lacks a clear track (Fig. 2C). The shallow Crozet Bank appears to have been emplaced on old, mechanically thick crust different from the adjacent Del Cano Rise, formed on or near the Southwest Indian ridge before anomaly 24 time (Goslin and Diament, 1987; Recq et al., 1998). The Crozet Bank's eastern islands (Île de la Possession and Île de l'Est) are separated by the Indivat basin from the western islands

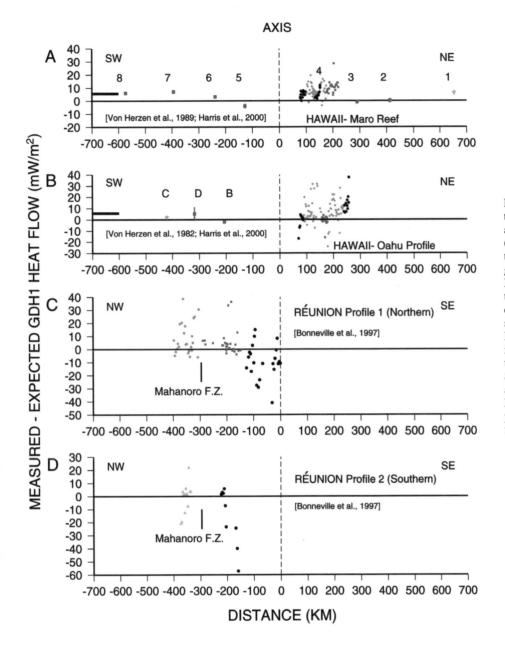

Figure 3. Heatflow anomaly with distance for profiles taken perpendicular to a hotspot track for Hawaii and Réunion (see locations from Figure 2A,B). Although the heatflow may be ~5 mW m^{-2} greater than predicted by the Global Depth and Heatflow reference model, the heatflow anomaly does not show the expected Gaussian-type pattern from significant lithospheric reheating with a large maximum at the axis (0 km). Similarly, few sites show evidence of hydrothermal circulation (Von Herzen, 2004). Symbols as for Figure 2. Horizontal bar at left in panels A and B represents mean heatflow anomaly on incoming Pacific plate before reaching hotspot. F.Z.—fracture zone.

(Île des Pingouins, Île aux Cochons, and Îles des Apotres). The eastern islands are usually thought to be the oldest, with the oldest dated volcanism at ca. 8 Ma, but detailed paleomagnetic work indicates that volcanism also occurred from 5 to 0.5 Ma on the eastern island of Possession (Camps et al., 2001). Also, the western island of Apotres has volcanism dated as old as 5.5 Ma (Giret et al., 2003). Both western and eastern islands have some Holocene volcanism. The relatively few heatflow measurements for the Crozet hotspot are all on ca. 68-Ma crust extending southeast from the eastern islands (Courtney and Recq, 1986) and presumably into crust unperturbed by the hotspot activity. Of the three reliable multipenetration sites, M4, with the shallowest seafloor, has a heatflow anomaly of ~14 mW m^{-2} (Fig. 4B). The other two sites, M5 and M6, farther away from

the volcanism, have heatflow approximately equal to that expected for the crustal age. It is difficult to interpret one isolated measurement, M3, at the base of the shallowing volcanic construct of the archipelago, where the heatflow is ~35 mW m^{-2} greater than expected. This heatflow value is discussed later in the article.

Two additional heatflow surveys are on swells perturbing old crust. Cape Verde is on the African plate (Courtney and White, 1986) and Bermuda (Detrick et al., 1986) is on the North American plate (Fig. 2D,E). Some characteristics of Cape Verde are similar to those of Crozet. Volcanism producing the islands is relatively recent on ca. 125-Ma crust. The easternmost islands formed ca. 10–20 Ma, with the western ones younger than 8 Ma (Courtney and White, 1986). The absolute velocity in the hotspot

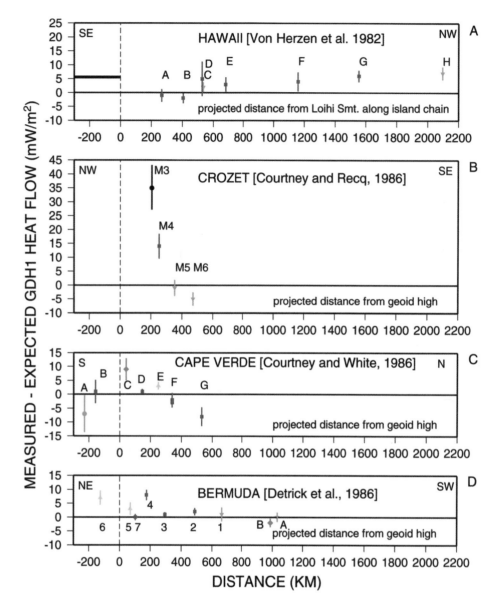

Figure 4. Heatflow anomaly with distance for profiles taken parallel to a hotspot track for (A) Hawaii. (B) Crozet, (C) Cape Verde, and (D) Bermuda have no clear hotspot track, so distances for the sites are measured from the local geoid maximum in the approximate direction of the heatflow sites. Symbols as for Figure 2. Horizontal bar at left in panel A represents mean heatflow anomaly on incoming Pacific plate before reaching hotspot. GDH1—Global Depth and Heatflow reference model.

reference frame for Cape Verde is somewhat higher (~21 mm/year) than for Crozet (~6 mm/year) (Gripp and Gordon, 2002). As with Crozet, the heatflow sites are located on a profile of approximately similar age crust extending from near the volcanic islands into crust unperturbed by the hotspot activity. Of the seven sites with closely spaced measurements (Fig. 2D), site A was considered unreliable because of the difficulty penetrating the seafloor with the heatflow probe, the lack of measured thermal conductivity, and likelihood of bottom erosion to the seafloor (Courtney and White, 1986). We have included the values here, using an estimated thermal conductivity of 1 W/(m K). The heatflow anomaly is relatively high (~9 mW m^{-2}) at site C,

within a few hundred kilometers of the maximum geoid high (Fig. 4C).

Bermuda (Fig. 2E), on the North American plate (with an absolute velocity of ~38 mm/year in the hotspot reference frame; Gripp and Gordon, 2002), is a shallow region without volcanism for ~33 m.y. Bermuda may be either a "one-shot" hotspot or perhaps part of a track with a very large spatial gap in volcanism (e.g., Vogt and Jung, this volume). Compared to the other four hotspot regions, the "pogo" measurements at Bermuda (Detrick et al., 1986) have a similarly small maximum anomaly (~8 mW m^{-2}). Also, a heatflow value determined from a drill hole on the island of Bermuda (Hyndman et al., 1974) has a small (~5 mW m^{-2}) anomaly. The heatflow data do not display the broad anomaly pattern expected for significant thermal rejuvenation of the lithosphere (Fig. 4D), although the mean heatflow values are systematically distributed with distance from Bermuda.

HYDROTHERMAL CIRCULATION

"Fluid flow has the potential to obscure basal heatflow patterns associated with the Hawaiian hot spot" (Harris et al., 2000a, p. 21,368). Hydrothermal circulation is assumed to transfer relatively little heat in oceanic lithosphere older than ca. 65 Ma (e.g., Stein and Stein, 1994; Von Herzen, 2004), about the minimum crustal age in these surveys. However, even in old crust, a few areas have been interpreted as having hydrothermal circulation (e.g., Noel and Hounslow, 1988; Fisher and Von Herzen, 2005). Also, areas where isolated igneous basement peaks penetrate or are only shallowly buried by the sediment cover are thought to be the main flow path for fluids between the overlying ocean and the crystalline crust (e.g., Harris et al., 2004). This effect has been observed on 3.5-Ma crust on the Juan de Fuca plate (Mottl et al., 1998) and inferred from heatflow measured on 83-Ma crust of the northwestern Atlantic (Embley et al., 1983). Studies suggest that pore water circulation removes heat from the surrounding area, extending a few to tens of kilometers distance from (near-) outcrops of basement rock (e.g., Stein and Stein, 1997; Fisher et al., 2003).

An additional factor for hotspots is that if hydrothermal circulation removes large amounts of heat near seamounts, significant pore water flow may also occur within the relatively highly permeable volcaniclastic sediments. The permeabilities of sediment obtained for the Canary Islands seafloor by Urgeles et al. (1999) are as high as ~10^{-16} m^2. However, the mostly thick sediments at the hotspots we discuss should prevent any significant vertical flow of water (e.g., Spinelli et al., 2004) and thus any significant advective heat transfer. Slow horizontal pore water velocities may occur, given this sediment permeability, but modeling of flow in oceanic crystalline basement suggests that flow sufficient to transport significant amounts of heat requires higher permeabilities, at a minimum roughly three to five orders of magnitude larger (Fisher et al., 2003; Stein and Fisher, 2003; Hutnak et al., 2006; A. Fisher, personal commun., 2006). For the sediments, for some models, Harris et al. (2000b, p. 21,382) "found that horizontal permeabilities as large as 10^{-11} m^2 were needed to produce significant perturbations to the surface heatflow."

To test this hypothesis of significant hydrothermal circulation, we use the analysis of fifty-eight high-quality closely spaced heatflow measurements ("pogo" surveys; Von Herzen, 2004). This approach used the correlation of heatflow variations with the shape of the basaltic basement and sediment cover to infer whether areas have hydrothermal circulation. Von Herzen (2004) classified the sites in four categories: (1) having hydrothermal circulation, (2) perhaps having hydrothermal circulation, (3) uncertain if all conductive, and (4) conductive heat

transfer. Forty-four of the fifty-eight pogo surveys are on or near the swells discussed in this article (Fig. 2). For the surveys associated with hotspots, Von Herzen (2004) suggested that twenty-four (~55%) have conductive heatflow, eight (~18%) probably have conductive heatflow, eight (~18%) have convection, and four (~9%) probably have convection.

About 27% of the surveys near the swells were interpreted as having or probably having convection. Most are within ~200 km of the swell axes (Figs. 3 and 4). These sites include five that are close to the Hawaiian axis (site #4, two sections of the Maro Reef profile, and two sections of the Oahu profile; Fig. 2A), the two closest sites to Bermuda (sites #5 and #6), and site C from Cape Verde, ~40 km from the geoid high and closest to the youngest region of volcanism. Conversely, most heatflow sites off axis are apparently not affected by hydrothermal circulation. Thus, it is doubtful that hydrothermal circulation masks a high and wide heatflow anomaly associated with lithospheric reheating mechanism for the formation of the swells, but it may occur near the swell axes.

VOLCANISM

Although magmatic activity is associated with the formation of hotspot swells, its thermal effects are typically not included in heat flux estimates. Crustal thickness at hotspots is significantly greater than for normal oceanic crust (e.g., White et al., 1992), although the conditions producing the melt have large uncertainties (e.g., Sallarès and Calahorrano, this volume). The emplacement of the additional hot material at relatively shallow depths should have a thermal signature. Although some excess crust is associated with the near-surface volcanism constructing the islands, some hotspots have a deeper sill structure intruded near the base of oceanic crust. Such sills are deduced from seismic data for Hawaii (ten Brink and Brocher, 1987) (Fig. 5A,B) and the Marquesas (Caress et al., 1995; McNutt and Bonneville, 2000), but others do not appear to have obvious underplating (e.g., Crozet; Recq et al., 1998). Next, we examine the heat flux associated with the cooling of the island/seamount chain and the underplated sill for Hawaii, and the implications for the other heatflow surveys on hotspots.

Total Heat Contribution

The contribution of the cooling magma to the geothermal budget in part depends on its production with time. We estimate the production rate at the Hawaiian hotspot from the magma cross-section geometry (Fig. 5B) and the absolute plate velocity of ~100 mm/year. Assuming that erosion of Oahu has reduced its initial volume by a factor ranging between 0.3 and 0.5, the mean magma volume generation/unit time (V) is ~3–6 m^3/s, comparable to other estimates of the rate of Hawaiian magmatism (Swanson, 1972; Shaw, 1973; Wadge, 1980).

The total thermal power (P) from the magma is its latent heat from solidification (PL) plus the cooling (PT). For the

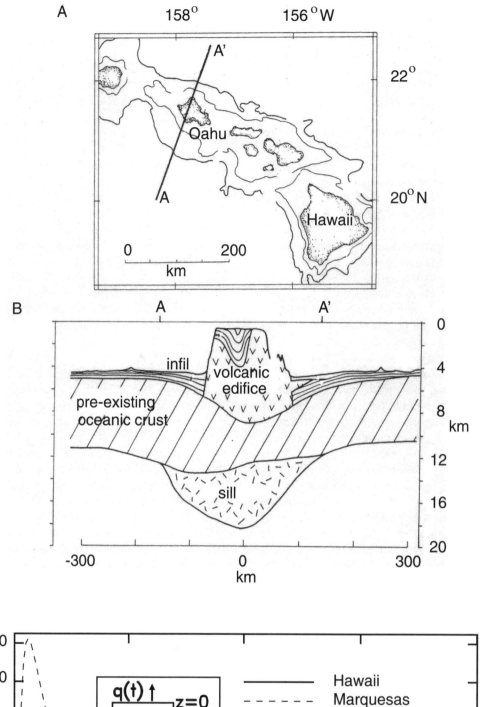

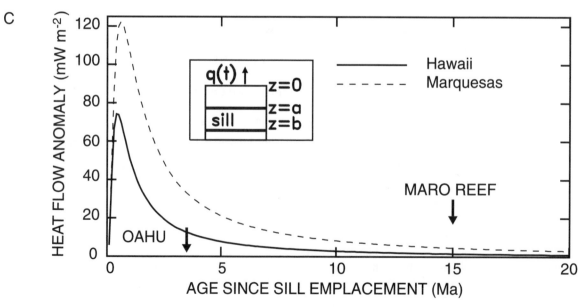

"island" region, the total thermal power is given in equation 1 and values for the parameters from Table 1:

$$P = PL + PT = V\rho \, (L + c \, \Delta T) = 1\text{--}2 \times 10^4 \text{ MW.} \quad (1)$$

Additional heat comes from the deeper sill that has an average thickness of 2.5 km. We estimate that the volume of the sill is about half that of the edifice of the islands before erosion. Although the emplacement of the underplated material may consist of thinner sills that average to a lower effective initial temperature, for our calculations, we assume a single sill initially at the solidus temperature (~1000 °C above the intruded rock), to maximize its thermal effects. Thus, including the contribution of both the "island" and the sill, the total power is 1.5 – 3 × 10⁴ MW (50% greater than that estimated in equation 1). Another way of examining this issue is to calculate the power released from the cooling for a volume of magma, 1-m² area with a thickness of 2.5 km. The total power, 8.25 × 10¹² J, for the square-meter area divided by ~10 m.y. of cooling produces an average heatflow of ~26 mW m⁻² or three times that if the "island" and eroded material are included.

The Sill Model

The cooling of a magmatic underplated region should have a longer-term effect compared to the near-surface volcanism. At mid-ocean ridges, heatflow rapidly decreases from cooling lava (Johnson and Hutnak, 1997), and heat is quickly dissipated by seawater. Similarly, on land, atmospheric cooling dissipates heat after an eruption. The surface heatflow from the underplated area with time is calculated assuming 1-D cooling of an infinitely extending horizontal sill, appropriate because of the large ratio of width to thickness, such as observed near Oahu, Hawaii. The analytical solution for the surface heat flux (q) with time (t) is given in Von Herzen et al. (1989, equation 5; initially obtained from Carslaw and Jaeger, 1959) as:

$$q(t) = [k\Delta T/(\pi\kappa t)^{1/2}]\{\exp[-a^2/(4\kappa t)] - \exp[-b^2/(4\kappa t)]\}, \quad (2)$$

where a and b are the depths of the upper and lower sill surfaces, respectively and the other parameters are from Table 1.

The deep sill structure found by ten Brink and Brocher (1987) is ~200 km wide with an average thickness of 2.5 km, and we presume it was formed by intruded magma associated with the hotspot. The heat flux over time is illustrated in Figure 5C, for values of the top (a) and the bottom (b) of the sill equal

---

Figure 5. Sill intrusions and heatflow anomalies. (A) Location of seismic study across Oahu shown in panel B. (B) Cross-section with volcanic sill and edifice above (from ten Brink and Brocher, 1987). (C) Heatflow anomaly expected for the observed sills (with an average thickness of 2.5 km) under Oahu and (with an average thickness of 5 km) under the Marquesas. Heatflow for Hawaii is a maximum at ca. 0.75 Ma and decreases with increasing age.

TABLE 1. PHYSICAL PROPERTIES ASSUMED FOR THE COOLING MAGMA

Symbol	Term	Value
ρ	Density	3000 kg/m³
c	Heat capacity	800 J/kg-K
k	Thermal conductivity	2.2 W/m-K
κ	Thermal diffusivity	8×10^{-7} m²/s
L	Latent heat of melting	3×10^5 J/kg
ΔT	Temperature drop after solidification	1000 K

to 7 and 9.5 km, respectively, relative to the seafloor depths away from the edifice. The surface heatflow for the deep sill is zero at emplacement and then rapidly increases to a maximum of ~75 mW m⁻² at ~0.5 m.y. after emplacement, but subsequently rapidly decreases with increasing age since emplacement. This model predicts excess heatflow of ~10 mW m⁻² at ca. 3.5 Ma (near Oahu) but only ~2 mW m⁻² at ca. 15 Ma (near Maro Reef). However, the observed heatflow anomalies (measured less than expected from GDH1; Stein and Stein, 1992) along the Oahu and Maro Reef profiles (Fig. 6) to ~150 km from the islands are about the same value and similar to other measurements for the Hawaiian chain (~5–10 mW m⁻²).

After correcting the heatflow values at Oahu and Maro Reef for thermal blanketing caused by deposition of cold sediment (e.g., Von Herzen and Uyeda, 1963; Hutchison, 1985), the data are consistent with the suggestion that additional heat is derived from an intrusive body. The "corrected" heatflow (Harris et al., 2000a) at Oahu is anomalous by ~24 mW m⁻² and at Maro by ~12 mW m⁻², approximately the expected difference in heatflow at these locations modeled by conductive cooling of the sill intrusion. However, the amount of the correction depends greatly on such factors as the sedimentation rate, thermal conductivity of the sediments, and assumed boundary conditions at the sediment/igneous crust interface. Most oceanic sites have sedimentation rates that are too low to require corrections, and for those with somewhat thicker sediment, the corrections are typically small (a few percent). In contrast, for the measurement sites closest to Oahu and Maro Reef, ~2 km of sediment have been deposited in the moat formed by the "loading" of the islands on the crust. The calculated correction is large if the heatflow at the sediment-basement interface varies with normal conductive cooling of oceanic lithosphere, as assumed by Harris et al. (2000a). However, the correction is somewhat smaller if vigorous hydrothermal circulation occurs within the top of the igneous oceanic crust (Wang and Davis, 1992). Additional uncertainties for the calculation are introduced if significant hydrothermal circulation occurs within permeable near-surface volcaniclastic sediments. Thus, for Oahu, a case can be made for or against additional heat from igneous intrusions, depending on assumptions.

Magmatic activity may explain the higher heatflow at two sites near Crozet and perhaps one at Cape Verde. For Crozet, at station M3 the heatflow anomaly is ~35 mW m⁻², ~50 km from

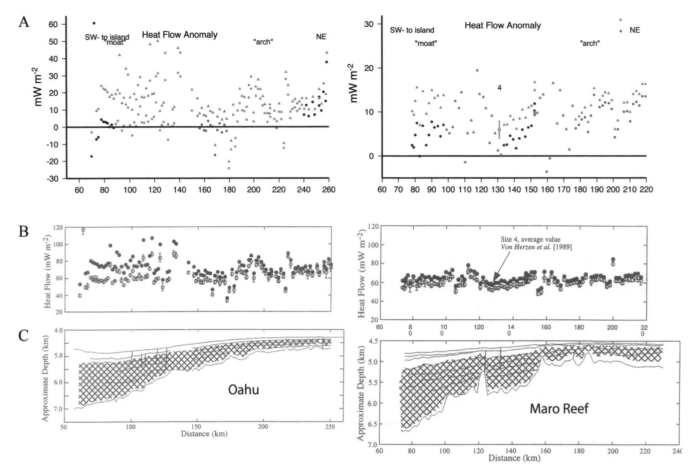

Figure 6. (A) Heatflow anomalies, (B) measured and sediment-corrected heatflow, and (C) depth-converted seismic reflection profiles for (left) Oahu and (right) Maro Reef. The top panels show the heatflow anomalies for the measured (solid circles) and sediment-corrected (purple triangles) heatflow. Blue circles—no hydrothermal circulation inferred; red circles—hydrothermal circulation inferred; black circles—no determination of circulation. The middle panel shows the measured (open circles) and sediment-corrected (solid circles) heatflow values. The location of these Hawaiian measurements are shown in Figure 2A. The middle and lower panels are from Harris et al. (2000a, Figs. 10 and 11).

the Holocene volcanism on Île de l'Est; and at M4, the heatflow anomaly is ~15 mW m^{-2}, ~100 km distant (Fig. 4B). The interpretation at M3 is problematic, because only one reliable measurement was made, but the site is clearly on the slope of the volcanic structure. At Cape Verde, station C, the ~10-mW m^{-2} heatflow anomaly is within ~100 km of Sao Vincent, an island with Holocene volcanism (Fig. 2D).

DISCUSSION AND CONCLUSIONS

Reanalysis of closely spaced heatflow surveys near five hotspots confirms that the heatflow anomalies are quite small, generally less than ~10 mW m^{-2}, and do not have the Gaussian-shaped distribution of values extending ~600 km away from the swell axis expected from the lithospheric reheating models. This finding accords with previous studies (Von Herzen et al., 1989; Stein and Stein, 1993; DeLaughter et al., 2005) showing that sig-

nificant lithospheric reheating does not occur, and the upwelling plume of material must have a substantially smaller thermal imprint than the suggested 300-km diameter (e.g., Crough, 1978).

Issues with Hydrothermal Circulation

It seems possible that hydrothermal circulation may redistribute some heat within ~200 km of the swell axis. However, it does not mask a large and spatially broad heatflow anomaly expected from reheating of a significant portion of lower lithosphere (e.g., Crough, 1978).

Hydrothermal circulation occurs at the two sites nearest to Bermuda's volcanic edifice. Unlike the four other hotspots with present-day volcanism, the most recent volcanic activity here is dated at ca. 33 Ma (Vogt and Jung, this volume). Bermuda does not appear to be associated with low mantle seismic velocities, as inferred at the other four active hotspots (e.g., Zhao, 2004;

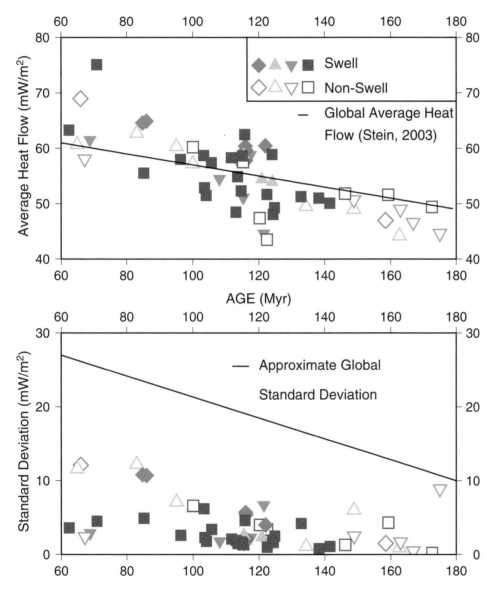

Figure 7. Heatflow data with crustal age for the swell (solid symbols) and non-swell (open symbols) "pogo" surveys analyzed by Von Herzen (2004). (A) The average heatflow from each survey with age is similar to the observed global average heatflow (Stein, 2003). (B) The standard deviations of the pogo surveys are significantly less than the global average. Colors as for Figure 2. Two sites from Von Herzen (2004) are not included in this figure. One located near a seamount has extremely high and variable heatflow (presumably due to fluid flow) and the other, in the Gulf of Mexico, has a large uncertainty in crustal age.

Dziewonski, 2005; Ritsema, 2005). The three sites with the highest heatflow anomalies are located near the base of the volcanic edifice (site #4), and the others (sites #5 and #6) are higher on the structure (Fig. 2C). The first is interpreted as having conductive heat transfer but the other two as perhaps having convection (Von Herzen, 2004). Hence, some higher than expected heatflow near the volcanic edifices may result from convection, rather than from volcanic processes.

Heatflow and Absolute Plate Velocity

The maximum heatflow anomalies for all five hotspots are about the same, yet the duration of the hotspot tracks and the absolute plate velocities and age are quite different. Maximum heatflow anomalies for the two long hotspot tracks of Hawaii and Réunion are approximately the same as observed at the volcanically active, more concentrated, large volcanic highs of Crozet and Cape Verde (Fig. 2). Despite the large range of absolute plate velocities for these five hotspots (6–104 mm/year) there is no clear trend of larger heatflow anomalies with decreasing absolute plate velocity, as might be expected if volcanic activity occurred over a longer period of time in one region.

Pogo Surveys and the Global Heatflow Data

We have examined how the heatflow on and off swells varies with lithospheric age and compares to the global data set (Fig. 7A). Swell sites are those sites on the regions of shallow depths associated with the hotspots discussed in this article. Off-swell sites include those away from the shallow depth regions and others discussed in Von Herzen (2004). On average, pogo surveys have heatflow values similar to global averages for oceanic lithosphere.

Overall, the standard deviations for the pogo sites (Fig. 7B) are much lower than for widely spaced measurements on similarly aged crust (Stein and Stein, 1994). There is a slight decrease in the standard deviations for the pogo surveys with increasing age. These observations suggest that although hydrothermal circulation transfers relatively little heat in old oceans, its effect is still slightly decreasing with age, even in Mesozoic crust. Von Herzen (2004) reached a similar conclusion based on the decreasing percentage with increasing crustal age of pogo survey sites on old oceanic lithosphere that were interpreted to have hydrothermal circulation.

Future Work

The obvious difficulty in interpreting the data for swells is the low density of heatflow measurements. For example, although the Hawaiian Swell has the most extensive geothermal, geological, and geophysical data compared to other hotspots, it is still difficult to determine the extent and thermal effects of hotspot magmatism. It is interesting to note that a high measured heatflow anomaly for Hawaii occurs at the northeastern ends of the Oahu and Maro Reef profiles (Fig. 6). These sites are on the relatively thinly sedimented flexed arches that have been associated with relatively recent and extensive seafloor volcanism (Clague et al., 2002). Similarly, the highest heatflow for Crozet (M3; Fig. 2C) is on the volcanic edifice near recent volcanism.

Additional work on hotspots would be useful. As far as we know, no pogo heatflow measurements have been carried out on the Marquesas. Seismic data from the Marquesas hotspot (Caress et al., 1995, Fig. 4) indicate a similarly shaped, but substantially larger, subcrustal intrusion compared to Hawaii, with ~10 km maximum thickness at the center, thinning to ~3–4 km at 150 km distance to either side. Given this intrusion and the recent age of the chain (less than 6 Ma; Desonie et al., 1993), the Marquesas may be an excellent location for future study (Fig. 5C).

ACKNOWLEDGMENTS

We thank Andy Fisher and Seth Stein for important discussions. Constructive reviews by Gillian Foulger, Rob Harris, Donna Jurdy, and Will Sager significantly improved the article. Also, we thank Gillian Foulger for organizing the stimulating Chapman Conference, *The Great Plume Debate,* and inviting this contribution. The GMT software package (Wessel and Smith, 1998) was used to make some of the figures in this article.

REFERENCES CITED

Anderson, D.L., 2005. Scoring hotspots: The plume and plate paradigms, *in* Plates, plumes, and paradigms, Foulger, G.R., et al., eds., Plates, plumes, and paradigms: Boulder, Colorado, Geological Society of America Special Paper 388, p. 31–54, doi: 10.1130/2005.2388(04).

Bonneville, A., Von Herzen, R.P., and Lucazeau, F., 1997, Heatflow over Réunion hot spot track; Additional evidence for thermal rejuvenation of oceanic lithosphere: Journal of Geophysical Research, v. 102, p. 22,731–22,747, doi: 10.1029/97JB00952.

Camps, P., Henry, B., Prevot, M., and Faynot, L., 2001, Geomagnetic paleosecular variation recorded in Plio-Pleistocene volcanic rocks from Possession Island (Crozet Archipelago, southern Indian Ocean): Journal of Geophysical Research, v. 106, p. 1961–1971, doi: 10.1029/2000JB900370.

Caress, D.W., McNutt, M.K., Detrick, R.S., and Mutter, J.C., 1995, Seismic imaging of hotspot-related crustal underplating beneath the Marquesas Islands: Nature, v. 373, p. 600–603, doi: 10.1038/373600a0.

Carslaw, H.S., and Jaeger, J.C., 1959. Conduction of heat in solids: Oxford, Oxford University Press, 510 p.

Clague, D.A., Uto, K., Satsake, K., and Davis, A.S., 2002. Eruption style and flow emplacement in the submarine North Arch volcanic field, Hawaii, *in* Takahashi, E., et al., eds., Hawaiian volcanoes: Deep underwater perspectives: Washington, D.C., American Geophysical Union Geophysical Monograph 128, p. 65–84.

Clouard, V., and Bonneville, A., 2005. Age of seamounts, islands, and plateaus on the Pacific plate, *in* Foulger, G.R., et al., eds., Plates, plumes, and paradigms: Boulder, Colorado, Geological Society of America Special Paper 388, p. 71–90, doi: 10.1130/2005.2388(06).

Courtney, R.C., and Recq, M., 1986, Anomalous heatflow near the Crozet Plateau and mantle convection: Earth and Planetary Science Letters, v. 79, p. 373–384, doi: 10.1016/0012-821X(86)90193-7.

Courtney, R.C., and White, R.S., 1986, Anomalous heatflow and geoid across the Cape Verde Rise: Evidence for dynamic support from a thermal plume in the mantle: Geophysical Journal of the Royal Astronomical Society, v. 87, p. 815–867.

Crough, S.T., 1978, Thermal origin of mid-plate hot-spot swells: Geophysical Journal of the Royal Astronomical Society, v. 355, p. 451–469.

Crough, S.T., 1983, Hotspot swells: Annual Review of Earth and Planetary Sciences, v. 11, p. 165–193, doi: 10.1146/annurev.ea.11.050183.001121.

DeLaughter, J.E., Stein, C.A., and Stein, S., 2005. Hotspots: A view from the swells, *in* Foulger, G.R., et al., eds., Plates, plumes, and paradigms: Boulder, Colorado, Geological Society of America Special Paper 388, p. 257–278, doi: 10.1130/2005.2388(16).

Desonie, D.L., Duncan, R.A., and Natland, J.H., 1993, Temporal and geochemical variability of volcanic products of the Marquesas hotspot: Journal of Geophysical Research, v. 98, p. 17,649–17,665.

Detrick, R.S., Von Herzen, R.P., Parsons, B., Sandwell, D., and Dougherty, M., 1986, Heatflow observations on the Bermuda Rise and thermal models of midplate swells: Journal of Geophysical Research, v. 91, p. 3701–3723.

Dyment, J., 1991. Structure et évolution de la lithosphère océanique dans l'océan Indien: Apport des anomalies magnétiques [Ph.D. thesis]: Strasbourg, France, University of Strasbourg, 374 p.

Dziewonski, A.M., 2005. The robust aspects of global seismic tomography, *in* Foulger, G.R., et al., eds., Plates, plumes, and paradigms: Boulder, Colorado, Geological Society of America Special Paper 388, p. 147–154, doi: 10.1130/2005.2388(09).

Embley, R.W., Hobart, M.A., Anderson, R.N., and Abbott, D., 1983, Anomalous heatflow in the northwest Atlantic: A case for continued hydrothermal circulation in 80-m.y. crust: Journal of Geophysical Research, v. 88, p. 1067–1074.

Fisher, A.T., Davis, E.E., Hutnak, M., Spiess, V., Zühlsdorff, L., Cherkaoui, A., Christiansen, L., Edwards, K.M., Macdonald, R., Villinger, H., Mottl, M., Wheat, C.G., and Becker, K., 2003, Hydrothermal circulation across 50 km on a young ridge flank: The role of seamounts in guiding recharge and discharge at a crustal scale: Nature, v. 421, p. 618–621, doi: 10.1038/nature 01352.

Fisher, A.T., and Von Herzen, R.P., 2005, Models of hydrothermal circulation within 106 Ma seafloor: Constraints on the vigor of fluid circulation and crustal properties below the Madeira Abyssal Plain: Geochemistry, Geophysics, Geosystems, v. 6, doi: 10.1029/2005GC001013.

Foulger, G.R., 2007 (this volume), The "plate" model for the genesis of melting anomalies, *in* Foulger, G.R., and Jurdy, D.M., eds., Plates, plumes, and planetary processes: Boulder, Colorado, Geological Society of America Special Paper 430, doi: 10.1130/2007.2430(01).

Garnero, E.J., Lay, T., and McNamara, A., 2007 (this volume), Implications of lower-mantle structural heterogeneity for existence and nature of whole-mantle plumes, *in* Foulger, G.R., and Jurdy, D.M., eds., Plates, plumes, and planetary processes: Boulder, Colorado, Geological Society of America Special Paper 430, doi: 10.1130/2007.2430(05).

Giret, A., Weis, D., Zhou, X., Cottin, J.-Y., and Tourpin, S., 2003. Géologie des îles Crozet (Geology of the Crozet Islands): Géologues, v. 137, Union Française des Géologues, Paris, 15–23.

Goslin, J., and Diament, M., 1987, Mechanical and thermal isostatic response of the Del Cano Rise and Crozet Bank (southern Indian Ocean) from altimetry data: Earth and Planetary Science Letters, v. 84, p. 285–294, doi: 10.1016/0012-821X(87)90093-8.

Gripp, A.E., and Gordon, R.G., 2002, Young tracks of hotspots and current plate velocities: Geophysical Journal International, v. 150, p. 321–361, doi: 10.1046/j.1365-246X.2002.01627.x.

Harris, R.N., Von Herzen, R.P., McNutt, M.K., Garven, G., and Jordahl, K., 2000a, Submarine hydrogeology of the Hawaiian archipelagic apron 1. Heatflow patterns north of Oahu and Maro Reef: Journal of Geophysical Research, v. 105, p. 21,353–21,369, doi: 10.1029/2000JB900165.

Harris, R.N., Garven, G., Georgen, J., McNutt, M.K., Christiansen, L., and Von Herzen, R.P., 2000b, Submarine hydrogeology of the Hawaiian archipelagic apron 2. Numerical simulations of coupled heat transport and fluid flow: Journal of Geophysical Research, v. 105, p. 21,371–21,385, doi: 10.1029/2000JB900164.

Harris, R.N., Fisher, A.T., and Chapman, D., 2004, Seamounts induce large fluid fluxes: Geology, v. 32, p. 725–728, doi: 10.1130/G20387.1.

Hutchison, I., 1985, The effects of sedimentation and compaction on oceanic heatflow: Geophysical Journal of the Royal Astronomical Society, v. 82, p. 439–459.

Hutnak, M., Fisher, A.T., Zuhlsdorff, L., Spiess, V., Stauffer, P., and Gable, C.W., 2006, Hydrothermal recharge and discharge guided by basement outcrops on 0.7–3.6 Ma seafloor east of the Juan de Fuca Ridge: Observations and numerical models: Geochemistry, Geophysics, Geosystems, v. 7, doi: 10.1029/2006GC001242.

Hyndman, R.D., Muecke, G.K., and Aumento, F., 1974, Deep Drill 1972. Heat-flow and heat production in Bermuda: Canadian Journal of Earth Sciences, v. 11, p. 809–818.

Johnson, P., and Hutnak, M., 1997, Conductive heat loss in recent eruptions at mid-ocean ridges: Geophysical Research Letters, v. 24, p. 3089–3092, doi: 10.1029/97GL02998.

Keller, R.A., Duncan, R.A., and Fisk, M.R., 1995. Geochemistry and ^{40}Ar/^{39}Ar geochronology of basalts from ODP leg 145 (North Pacific transect), *in* Rea, D.K., et al., eds., Proceedings of the Ocean Drilling Program, Scientific Results (vol. 145): College Station, Texas, Ocean Drilling Program, p. 333–344.

Liu, M., and Chase, C.G., 1989, Evolution of midplate hotspot swells: Numerical solutions: Journal of Geophysical Research, v. 94, p. 5571–5584.

Mottl, M.J., Wheat, G., Baker, E., Becker, N., Davis, E., Feely, A.R., Grehan, A., Kadko, D., Lilley, M., Massoth, G., Moyer, C., and Sansone, F., 1998, Warm springs discovered on 3.5 Ma oceanic crust, eastern flank of the Juan de Fuca Ridge: Geology, v. 26, p. 51–54, doi: 10.1130/0091-7613(1998) 026<0051:WSDOMO>2.3.CO;2.

McNutt, M., and Bonneville, A., 2000. A shallow, chemical origin for the Marquesas Swell: Geochemistry, Geophysics, Geosystems, v. 1, doi: 10.1029/1999GC00028.

Mueller, R.D., Roest, W.R., Royer, J.-Y., Gahagan, L.M., and Sclater, J.G., 1997. Digital isochrons of the world's ocean floor: Journal of Geophysical Research, v. 102, p. 3211–3214 (ftp://ftp.es.usyd.edu.au/pub/agegrid/, version 1.6).

Noel, M., and Hounslow, M.W., 1988, Heatflow evidence for hydrothermal convection in Cretaceous crust of the Madiera Abyssal Plain: Earth and Planetary Science Letters, v. 90, p. 77–86, doi: 10.1016/0012-821X(88) 90113-6.

Norton, I.O., 2007 (this volume), Speculations on Cretaceous tectonic history of the northwest Pacific and a tectonic origin for the Hawaii hotspot, *in* Foulger, G.R., and Jurdy, D.M., eds., Plates, plumes, and planetary processes: Boulder, Colorado, Geological Society of America Special Paper 430, doi: 10.1130/2007.2430(22).

Phipps Morgan, J., Morgan, W.J., and Price, E., 1995. Hotspot melting generates both hotspot volcanism and a hotspot swell?: Journal of Geophysical Research, v. 100, p. 8045–8062.

Recq, M., Gosslin, J., Charvis, P., and Operto, S., 1998, Small-scale crustal variability within an intraplate structure: The Crozet Bank (southern Indian Ocean): Geophysical Journal International, v. 134, p. 145–156, doi: 10.1046/j.1365-246x.1998.00530.x.

Ribe, N.M., and Christensen, U.R., 1994, Three-dimensional modeling of plume-lithosphere interaction: Journal of Geophysical Research, v. 99, p. 669–682, doi: 10.1029/93JB02386.

Ritsema, J., 2005. Global seismic structure maps, *in* Foulger, G.R., et al., eds., Plates, plumes, and paradigms: Boulder, Colorado, Geological Society of America Special Paper 388, p. 11–18, doi: 10.1130/2005.2388(02).

Robinson, E.M., 1988, The topographic and gravitational expression of density anomalies due to melt extraction in the uppermost oceanic mantle: Earth and Planetary Science Letters, v. 90, p. 221–228, doi: 10.1016/0012-821X (88)90102-1.

Sallarès, V., and Calahorrano, A., 2007 (this volume), Geophysical characterization of mantle melting anomalies: A crustal view, *in* Foulger, G.R., and Jurdy, D.M., eds., Plates, plumes, and planetary processes: Boulder, Colorado, Geological Society of America Special Paper 430, doi: 10.1130/2007.2430(25).

Shaw, H.R., 1973, Mantle convection and volcanic periodicity in the Pacific: Evidence from Hawaii: Geological Society of America Bulletin, v. 84, p. 1505–1526, doi: , doi: 10.1130/0016-7606(1973)84<1505:MCAVPI> 2.0.CO;2.

Sleep, N.H., 1994, Lithospheric thinning by midplate mantle plumes and the thermal history of hot plume material ponded at sublithospheric depths: Journal of Geophysical Research, v. 99, p. 9327–9343, doi: 10.1029/94JB00240.

Sleep, N.H., 2003, Mantle plumes?: Astronomy & Geophysics, v. 44, p. 1.11–1.13, doi: 10.1046/j.1468-4004.2003.44111.x.

Spinelli, G.A., Giambalvo, E.R., and Fisher, A.T., 2004. Sediment permeability, distribution, and influences on fluxes in oceanic basement, *in* Davis, E.E., and Elderfield, E.H., eds., Hydrogeology of the oceanic lithosphere: Cambridge, Cambridge University Press, p. 151–188.

Stein, C.A., 2003, Heatflow and flexure at subduction zones: Geophysical Research Letters, v. 30, doi: 10.1029/2003GL018478.

Stein, C.A., and Stein, S., 1992, A model for the global variation in oceanic depth and heatflow with lithospheric age: Nature, v. 359, p. 123–129, doi: 10.1038/359123a0.

Stein, C.A., and Stein, S., 1993, Constraints on Pacific midplate swells from global depth-age and heatflow-age models, *in* Pringle, M., et al., eds., The Mesozoic Pacific: Washington, D.C., American Geophysical Union Geophysical Monograph 77, p. 53–76.

Stein, C.A., and Stein, S., 1994, Constraints on hydrothermal heat flux through the oceanic lithosphere from global heatflow: Journal of Geophysical Research, v. 99, p. 3081–3095, doi: 10.1029/93JB02222.

Stein, C.A., and Stein, S., 1997, Estimation of lateral hydrothermal flow distances from spatial variations in heatflow: Geophysical Research Letters, v. 24, p. 2323–2326, doi: 10.1029/97GL02319.

Stein, J.S., and Fisher, A.T., 2003. Observations and models of lateral hydrothermal circulation on a young ridge flank: Numerical evaluation of thermal and chemical constraints: Geochemistry, Geophysics, Geosystems, v. 4, doi: 10.1029/2002GC000415.

Swanson, D.A., 1972, Magma supply rate of Kilauea volcano, 1952–1971: Science, v. 175, p. 169–170, doi: 10.1126/science.175.4018.169.

ten Brink, U.S., and Brocher, T.M., 1987, Multichannel seismic evidence for a subcrustal intrusive complex under Oahu and a model for Hawaiian volcanism: Journal of Geophysical Research, v. 92, p. 13,687–13,707.

Urgeles, R., Canals, M., Roberts, J., and SNV "Las Palmas" Shipboard Party, 1999, Fluid flow from pore pressure measurements off La Palma, Canary Islands: Journal of Volcanology and Geothermal Research, v. 94, p. 305–321, doi: 10.1016/S0377-0273(99)00109-2.

Vogt, P.R., and Jung, W.-Y., 2007 (this volume), Origin of the Bermuda volcanoes and Bermuda Rise: History, observations, models, and puzzles, *in* Foulger, G.R., and Jurdy, D.M., eds., Plates, plumes, and planetary processes: Boulder, Colorado, Geological Society of America Special Paper 430, doi: 10.1130/2007.2430(27).

Von Herzen, R.P., 2004. Geothermal evidence for continuing hydrothermal circulation in older (60 m.y.) ocean crust, *in* Davis, E.E., and Elderfield, E.H., eds., Hydrogeology of the oceanic lithosphere: Cambridge, Cambridge University Press, p. 414–447.

Von Herzen, R.P., and Uyeda, S., 1963, Heatflow through the eastern Pacific Ocean floor: Journal of Geophysical Research, v. 68, p. 4219–4250.

Von Herzen, R.P., Detrick, R.S., Crough, S.T., Epp, D., and Fehn, U., 1982, Thermal origin of the Hawaiian swell: Heatflow evidence and thermal models: Journal of Geophysical Research, v. 87, p. 6711–6723.

Von Herzen, R.P., Cordery, M.J., Detrick, R.S., and Fang, C., 1989, Heatflow and the thermal origin of hotspot swells: The Hawaiian Swell revisited: Journal of Geophysical Research, v. 94, p. 13,783–13,799.

Wadge, G., 1980, Output rate of magma from active central volcanoes: Nature, v. 288, p. 253–255, doi: 10.1038/288253a0.

Wang, K., and Davis, E.E., 1992, Thermal effects of marine sedimentation in hydrothermally active areas: Geophysical Journal International, v. 110, p. 70–78.

Wessel, P., and Smith, W.H.F., 1998, New, improved version of generic mapping tools released: Eos (Transactions, American Geophysical Union), v. 79, p. 579, doi: 10.1029/98EO00426.

White, R.S., McKenzie, D., and O'Nions, R.K., 1992, Oceanic crustal thickness from seismic measurements and rare earth element inversions: Journal of Geophysical Research, v. 97, p. 19,683–19,715.

Zhao, D., 2004, Global tomographic images of mantle plumes and subducting slabs: Insight into deep Earth dynamics: Physics of the Earth and Planetary Interiors, v. 146, p. 3–34, doi: 10.1016/j.pepi.2003.07.032.

MANUSCRIPT ACCEPTED BY THE SOCIETY 31 JANUARY 2007

The Geological Society of America
Special Paper 430
2007

Crustal geotherm in southern Deccan basalt province, India: The Moho is as cold as adjoining cratons

P. Senthil Kumar*
Rajeev Menon
G. Koti Reddy

National Geophysical Research Institute, Uppal Road, Hyderabad 500007, India

ABSTRACT

Deccan basalts have long been considered to be a product of Réunion plume–Indian lithosphere interaction, ca. 65 Ma. However, recent studies call into question their plume origin. In this study, we investigate whether there is a thermal signature of the plume in the present-day thermal regime of the Deccan crust. Our study is limited to the southeastern Deccan province, where surface heatflow values are abundant and the radiogenic heat contribution of the underlying crust (ca. 2.7- to 2.5-Ga eastern Dharwar craton) can be well constrained from its exposed crustal cross-section. Surface heatflow varies from 33 to 73 mW m^{-2}, with a mean of 45 mW m^{-2}. Heat production of the Deccan basalts (tholeiite) is 0.39 μW m^{-3}, and heat production of the basement rocks is assigned on the basis of the data from the Dharwar craton. The basement Dharwar crust is composed of middle-to-late Archaean greenstone belts, gneisses, granites, and granulites, whose heat production has been determined from >1500 sites, which belong to various crustal depth layers (greenschist, amphibolite, and granulite facies) of the exposed crustal section. The data suggest a radiogenic heat contribution of Deccan crust of ~38 mW m^{-2}, implying a Moho heatflow of ~7 mW m^{-2} and Moho temperature of ~280 °C, which are similar to those in the adjoining Archaean Dharwar craton. Therefore, it appears that there is no thermal trace of the supposed plume in the present-day crustal thermal regime of the southeastern Deccan basalt province.

Keywords: heatflow, heat production, crustal geotherm, Deccan basalt, Réunion plume, Dharwar craton, India

INTRODUCTION

Continental crust affected by a mantle plume would have an elevated surface heatflow and Moho heatflow, leading to a perturbed lithospheric geotherm relative to the adjoining provinces unaffected by the plume. However, such a thermal anomaly would decay with time. The Deccan basalt province (Fig. 1) is a possible example, as several researchers have attributed its origin to ca. 65-Ma Réunion plume (or hotspot), although this inference has been severely debated in recent years (see Sheth, 2005). In this study, we aim to understand the nature of the present-day crustal geotherm beneath the southern Deccan basalt

*E-mail: senthilngri@yahoo.com.

Kumar, P.S., Menon, R., and Reddy, G.K., 2007, Crustal geotherm in southern Deccan basalt province, India: The Moho is as cold as adjoining cratons, *in* Foulger, G.R., and Jurdy, D.M., eds., Plates, plumes, and planetary processes: Geological Society of America Special Paper 430, p. 275–284, doi: 10.1130/2007.2430(14). For permission to copy, contact editing@geosociety.org. ©2007 The Geological Society of America. All rights reserved.

province, in an effort to assess the possibility of a thermal trace of Réunion plume that was thought to be responsible for the formation of 68- to 63-Ma Deccan basalts and their derivatives. Bonneville et al. (1997) determined heatflow across the Réunion hotspot track (Mascarene Ridge, located northeast of Réunion Island) and found a significantly higher heatflow of 6–8 mW m^{-2} than the reference heatflow predicted for the seafloor age. Therefore, in this study, we examine whether such an anomalous heatflow (although this value is low) can be seen in the present-day crustal thermal regime of the Deccan province. Modeling the present-day crustal thermal structure of the Deccan province is possible, as an adequate number of heatflow determinations are available for its southeastern part (Fig. 1), and crustal radiogenic heat production can be estimated from an exposed crustal cross-section of the Dharwar craton (Figs. 2 and 3) that forms the basement of the southern Deccan basalt province.

In the Deccan province, Roy and Rao (2000) presented new heatflow values, in addition to compiling the existing heatflow data (Figs. 1 and 4). Also, Roy and Rao (1999) modeled the present-day crustal geotherm of southeastern Deccan province, on the basis of heatflow measurements from the deep boreholes in the 1993 Killari (or Latur) earthquake area (Fig. 1) and a few heat production measurements from its basement rocks; these authors also concluded that the surface heatflow of the Deccan province is low, similar to the adjoining areas, and that the thermal transient related to the Deccan volcanic episode had decayed. Subsequent to this study, there have been several improvements to the understanding of the crustal structure, composition, and evolution of the basement Dharwar craton (see Kumar and Reddy, 2004). A better understanding of the heatflow, heat production, and crustal thermal structure of the Dharwar craton has been possible owing to further studies (e.g., Ray et al., 2003; Kumar and Reddy, 2004). In particular, Kumar and Reddy (2004) have established a new well-constrained crustal radiogenic heat production model for the entire Dharwar craton from the ~30-km-thick exposed crustal cross-section, measuring

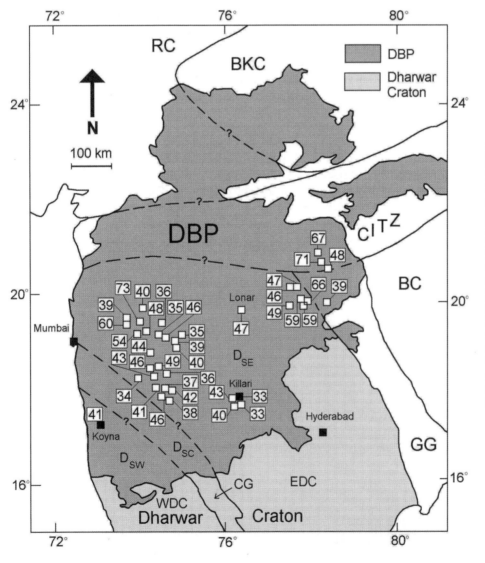

Figure 1. Map showing the Deccan basalt province (DBP) and surface heatflow values. The basalt cover is underlain by the Dharwar (western Dharwar craton [WDC], Closepet Granite batholith [CG], and eastern Dharwar craton [EDC]), Bastar (BC), Bundelkhand (BKC), Rajasthan (RC) cratons, and the intervening Proterozoic fault systems (Godavai graben [GG] and central Indian tectonic zone [CITZ]). Surface heatflow values (mW m^{-2}) shown inside the boxes have been determined by Gupta and Gaur (1984) and Roy and Rao (2000). The various crustal provinces of the Dharwar craton, such as WDC, CG, and EDC form the basements of southwestern (D_{SW}), south-central (D_{SC}), and southeastern (D_{SE}) Deccan basalt province, respectively, from west to east.

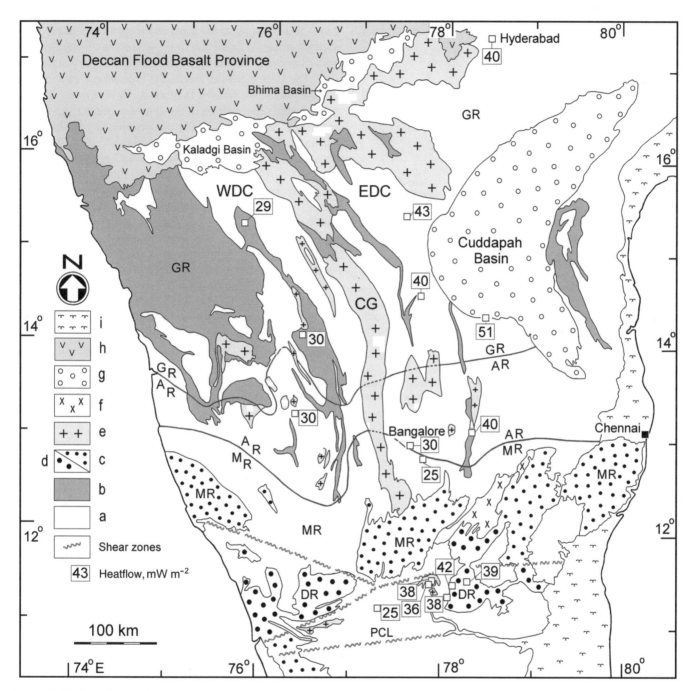

Figure 2. Geological map of the Dharwar craton, showing ~30-km-thick oblique crustal cross-section (see Kumar and Reddy, 2004). The four metamorphic facies regions (GR, AR, MR, and DR) expose the rock types belonging to four metamorphic facies—greenschist, amphibolite, metasomatized granulite (composed of medium-P granulites), and depleted granulite facies (composed of high-P granulites), respectively, from north to south. Rock types: a—gneisses; b—greenstone belts or supracrustal belts; c—medium-P granulites composed of high-grade gneisses and enderbites with subordinate amounts of mafic granulites and high-grade metasedimentary rocks; d—high-P granulites composed predominantly of enderbites with little mafic granulites; e—granites; f—alkaline rocks, mainly syenites and carbonatites; g—Proterozoic sedimentary rocks; h— ca. 65-Ma Deccan basalts; i—alluvial cover. The Dharwar craton is divided into western Dharwar craton (WDC) and eastern Dharwar craton (EDC), with the Closepet Granite batholith (CG) marking their boundary. The Palghat-Cauvery lineament (PCL) defines the southern boundary of the Dharwar craton. Also, the Dharwar craton forms the basement of the southern Deccan flood basalt province (see also Fig. 3). Surface heat-flow values (in boxes) were determined by Gupta et al. (1991), Roy and Rao (2000), and Ray et al. (2003).

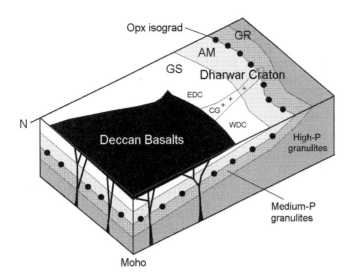

Figure 3. Schematic crustal structure of the Deccan basalt province and Dharwar craton. Note that the various metamorphic facies layers of the Dharwar craton underlie the Deccan basalts. WDC—western Dharwar craton (WDC), Closepet Granite batholith (CG), and eastern Dharwar craton (EDC) form the basements to the southwestern, south-central, and southeastern Deccan provinces, respectively (see also Fig. 1). Metamorphic facies layers, greenschist (GS), amphibolite (AM), and granulite (GR) layers dip northward. The GR is composed of two layers, the medium-*P* granulites (also known as the metasomatized granulites) and the high-*P* granulites (the depleted granulites). The orthopyroxene (Opx) isograd marks the diffuse boundary between amphibolite-facies and granulite-facies rocks.

m.y., between 63 and 68 Ma, with a maximum eruption rate ca. 66–65 Ma (see Pande, 2002). The Deccan basalt sequence is thin (~500 m) along its margins, and it apparently thickens toward its center. In particular, it is ~4 km thick along the Narmada-Son region that forms a part of the central Indian tectonic zone. The Deccan basalts overlie the Precambrian basement of cratons and mobile belts of the Indian shield (Fig. 1). The exposed crustal cross-section of the Dharwar craton (e.g., Kumar and Reddy, 2004; Figs. 2 and 3), composition of granulite xenoliths in the Deccan basalts (e.g., Dessai et al., 2004), borehole studies (e.g., Gupta et al., 1999), and other geophysical studies (e.g., NGRI, 1978; Mohan and Kumar, 2004) clearly indicate that the Dharwar craton forms the southern basement of the Deccan province and probably terminates along the central Indian tectonic zone.

The southern Deccan basalt province is further divided into three subprovinces, the southwestern, south-central, and southeastern provinces (Fig. 1), where the rocks of western Dharwar craton, Clospet Granite batholith, and eastern Dharwar craton form the basements, from west to east, respectively (Figs. 1–3). The crustal structure of the western Dharwar craton, Clospet Granite batholith, and eastern Dharwar craton have been well constrained in the Dharwar craton (see Kumar and Reddy, 2004). These crustal provinces expose ~30-km-thick crust over a lateral distance of >500 km and likely extend beneath the Deccan basalts up to the central Indian tectonic zone (Figs. 1–3).

several hundreds of sites for heat production by all the crustal rocks. These new studies open vistas to refine our understanding of the present-day crustal thermal structure of southern Deccan basalt province. The thermal signatures of hotspot anomalies are usually found to be very low (e.g., DeLaughter et al., 2005), and therefore, the need to have an improved crustal thermal model that can resolve low heatflow anomalies is paramount. Unfortunately, the thermal structure of other parts of Deccan province cannot be modeled because of too few heatflow and heat production measurements from its basement rocks (e.g., Menon et al., 2003) belonging to the Rajasthan, Bundelkhand, and Bastar cratons (Fig. 1).

GEOLOGY

The Deccan basalt province (Fig. 1) is spread over an area of ~500,000 km^2, which consists predominantly of tholeiitic basalts, with subordinate amounts of alkaline and picritic basalts. A detailed account on the geology of Deccan basalt province can be found in Subbarao (1999). The total thickness of the entire Deccan basalt sequence is ~3.4 km (see Mahoney et al., 2000), containing three subgroups and eleven formations, of which the basalts of Ambenali Formation are largely exposed on the southern part of the Deccan province (e.g., Mitchell and Widdowson, 1991). Eruption of the Deccan basalts lasted for ~5

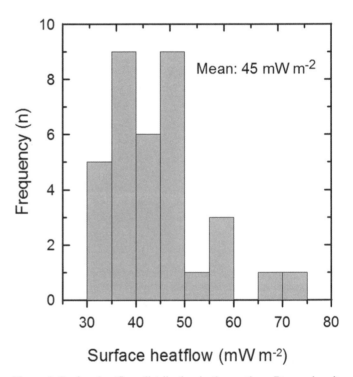

Figure 4. Surface heatflow distribution in the southern Deccan basalt province (Gupta and Gaur, 1984; Roy and Rao, 2000). See Figure 1 for the locations of surface heatflow determinations. All measurements except one belong to the southeastern Deccan province.

The exposed section consists of greenschist, amphibolite, meta-somatized granulite (medium-*P* granulites), and depleted granulite (high-*P* granulites) facies layers, which are successively exposed from north to south (Figs. 2 and 3), implying a northward tilting of ~1–5°. The thicknesses of these layers have also been well constrained by geothermo-barometric data obtained from metamorphic mineral assemblages. Surface geology, structure, metamorphic *P-T* data, and seismic geophysical data provide constraints to model the crustal lithology of the Dharwar craton (Kumar and Reddy, 2004) as well as the southern Deccan basalt province (Fig. 3).

CRUSTAL THERMAL STRUCTURE

The main objective of this study is to estimate the crustal radiogenic heat production, Moho heatflow, and crustal geotherm of the southern Deccan province to assess the thermal trace of the hypothesized mantle plume. To achieve this, we utilize (1) available surface heatflow measurements, (2) heat production estimates of the Deccan basalt cover and the basement and the rocks of exposed crustal section of the Dharwar craton, (3) the crustal radiogenic heat contribution to surface heatflow, and (4) the pressure- and temperature-dependent thermal conductivity model of Ketcham (1996). These provide a reliable temperature-depth profile of the crust (e.g., Ketcham, 1996).

Surface Heatflow

Surface heatflow measurements have been carried out by Gupta and Gaur (1984) and Roy and Rao (2000) for the southern parts of the Deccan basalt province (Figs. 1 and 4). These authors have determined heatflow values using the geothermal gradient measurements in deep boreholes and the thermal conductivity of associated rocks. We do not consider heatflow values determined through geochemical methods (e.g., Shankar, 1988), because of their large uncertainties. In the southeastern Deccan province, surface heatflow varies from 33 mW m^{-2} to 73 mW m^{-2}, with an average of 45 mW m^{-2} (Fig. 4). These values, in general, are restricted to the part (southeastern Deccan province), where the rocks of the Eastern Dharwar craton form the basement (Fig. 1). Only one heatflow value (41 mW m^{-2}) is available from the Koyna region (southwestern province), where the rocks of the western Dharwar craton are the likely basement. The heatflow values of Deccan province are broadly similar to those observed in the Dharwar craton (e.g., Roy and Rao, 2000; Fig. 2).

Heat Production

Heat production of the Deccan basalts and the basement rocks are given in Table 1. Heat production of Deccan basalts (tholeiites) is calculated from the published geochemical data that contain K, U, and Th abundances (Chandrasekharam et al., 1999; Mahoney et al., 2000; Sheth et al., 2004). Mean heat pro-

duction of the tholeiites is 0.39 µW m^{-3}, which is similar to heat production of basalts in general. For the basement, we consider that the rocks of the western Dharwar craton, Closepet Granite, and eastern Dharwar craton of the Dharwar craton underlie the Deccan basalts in the southwestern, south-central, and southeastern parts of the southern Deccan province (Figs. 1 and 3), respectively. Heat production of rocks of the various depth levels (e.g., greenschist-, amphibolite-, and granulite-facies layers) of the western Dharwar craton, Clospet Granite batholith, and eastern Dharwar craton has been determined by Kumar and Reddy (2004) and are listed in Table 1. It is evident from Table 1 that (1) rocks of the western and eastern Dharwar cratons are characterized by distinct heat production characteristics; (2) the greenschist- and amphobolite-facies level gneisses and granites of the western Dharwar craton produce less heat than do their eastern Dharwar craton counterparts; (3) granites produce more heat than do the gneisses; (4) granulite-facies rocks produce the least amount of heat; (4) heat production decreases with increasing metamorphic grade, specifically, across the amphibolite-to-granulite facies transition; and (5) heat production in Closepet Granite batholith increases systematically from the deeper (granulite-facies) to shallower (greenschist-facies) levels, conforming to the evolution of I-type granite magmas, and is represented in Table 1 and Figure 5.

Crustal Contribution to Surface Heatflow

When the heat production of major crustal rocks and their proportions are known, it is possible to calculate the crustal contribution to surface heatflow. The crustal contribution model envisaged by Roy and Rao (1999) for southeastern Deccan province consists of four layers: 8-km-thick migmatite-granite, 9-km-thick gneiss-granite, 3-km-thick transition zone rocks (amphibolite to granulite facies), and 17-km-thick granulite layers, from the top to bottom. Considering the limited number of heat production data available at that time, Roy and Rao (1999) suggested a crustal radiogenic heat contribution of 31 mW m^{-2}. However, recent studies in the Dharwar craton (e.g., Kumar and Reddy, 2004) mandate improvement to the above crustal model. The exposed crustal section of the Dharwar craton (Figs. 2 and 3) provides the proportion of rocks in the crust beneath the Deccan basalts and their heat production (Tables 1 and 2). We consider an average thickness of the southern Deccan crust of ~36 km (e.g., Rai et al., 2003) containing 1-km-thick basalt cover (although it varies from ~2 km in the west coast to ~0.5 km or less along its boundary) on the top, and followed by a 35-km-thick Dharwar crust. The exposed crustal section suggests that the Deccan basement crust is composed of four metamorphic facies layers—greenschist, amphibolite, metasomatized granulite, and depleted granulite (Fig. 3), whose thicknesses are likely to be ~4 km, 8 km, 8 km, and 15 km, respectively. The greenschist-facies layer beneath the Deccan basalt is likely to be thicker than in the northernmost Dharwar craton, because of the decrease in the amount of uplift and erosion, from south to north. An aver-

TABLE 1. HEAT PRODUCTION DATA USED IN THIS STUDY

Depth level	Rock type	Number of analyses	Heat production (μW m^{-3})
	Deccan basalt province		
Surface layer	Tholeiitic basalts[1,2,3]	48	0.39
	Western Dharwar craton		
Greenschist facies			
	Greywackes[4]	35	1.13
	Gneisses	47	1.01
	Granites	72	2.59
Amphibolite facies			
	Mafic volcanic rocks[5,6]	21	0.19
	Felsic volcanic rocks	5	1.51
	Gneisses	89	1.13
	Granites	78	1.65
Granulite facies			
	Enderbites (low *P*)*	68	0.66
	Enderbites (high *P*)†	16	0.34
	Closepet Granite batholith		
Greenschist facies			
	Porphyritic monzogranite	215	3.53
	Anatectic granites	120	3.41
Amphibolite facies			
	Porphy. monzogranite	123	1.78
	Anatectic granites	8	2.67
Granulite facies*			
	Porphy. monzogranite	31	1.26
	Anatectic granites	5	5.16
	Eastern Dharwar craton		
Greenschist facies			
	Mafic volcanic rocks[7,8]	22	0.09
	Felsic volcanic rocks[8]	6	1.34
	Gneisses	25	2.40
	Granites[9]	24	2.60
Amphibolite facies			
	Mafic volcanic rocks	5	0.09
	Felsic volcanic rocks	15	2.24
	Gneisses	228	2.36
	Granites	43	4.19
Granulite facies			
	Enderbites (medium *P*)*	33	0.35
	Enderbites (high *P*)†	32	0.16

Sources: Authors' published data: Kumar (2002); Kumar and Srinivasan (2002); Ray et al. (2003); Kumar and Reddy (2004). Other published geochemical data: 1. Chandrasekharam et al. (1999); 2. Mahoney et al. (2000); 3. Sheth et al. (2004); 4. Naqvi et al. (1988); 5. Naqvi et al. (1983); 6. Atal et al. (1978); 7. Hanuma Prasad et al. (1997); 8. Naqvi et al. (2002); 9. Roy and Rao (1999).

Notes: Heat production was calculated from K, U, and Th data using the conversion factors of Rybach (1988).

*Metasomatized granulites.

†Depleted granulites.

age thickness of 4 km has been considered for the greenschist-facies layer, and the rest are the same as for the Dharwar craton. The major rock types constituting each layer are given in Table 2, and their heat production in Table 1. Heat production of various rock types, weighted by their abundances, has been used in arriving at gross heat production for the five layers (including the basalt layer on the top) and is envisaged in Table 2. Gross

heat production of these layers and their layer thicknesses have been used to calculate the crustal contribution to heatflow, which is 24 mW m^{-2} for the southwestern part of the Deccan basalt province and 38 mW m^{-2} for the southeastern Deccan basalt province, indicating a significant variation across the province. Also, the estimate for the southeastern part is significantly higher than that of Roy and Rao (1999), who modeled only the

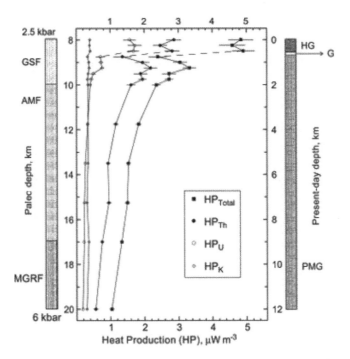

Figure 5. Heat-production variation with depth in the Closepet Granite batholith, which may be applicable to the heat production of the upper to middle crust of the south-central Deccan province. Note that the heat production decreases gradually with increasing depth, which is characteristic of I-type granite batholiths. Heat production of the anatectic granites shown in Table 1 are not considered here, as these are small plutons occurring along the boundary of the batholith. The ~12-km-thick granite batholith is exposed over strike length of ~300 km, cutting across the greenschist (GSF)-, amphibolite (AMF)-, and metasomatized granulite (MGRF)-facies host gneisses and greenstone belts (see also Fig. 1). The batholith is composed predominantly of porphyritic monzogranite (PMG) and lesser amounts of homogeneous granite (HG) that occurs in the upper level of the batholith. Field observations suggest that the PMG gives rise to HG by filter-pressing petrologic process through the fracture zones in the gap (G), where host gneisses are dominant. HP—heat production. For more details, see Kumar et al. (2003) and Kumar and Reddy (2004).

southeastern crust. Crustal contribution of the south-central part has not been determined, where the Closepet Granite batholith may be the basement. The omission is due to the large uncertainty in the nature of depth distribution of heat production as well as the thickness of the batholith beneath the Deccan cover, although heat production has been well determined for the batholith in the exposed craton (Table 1 and Fig. 5). Therefore, our crustal contribution estimates are limited to the southwestern and southeastern parts of the province (Table 2).

Crustal Geotherms

Surface heatflow, crustal heat production and *P*- and *T*-dependent thermal conductivity constrain temperature variation with depth in the Deccan crust (e.g., Ketcham, 1996), assuming a steady-state thermal condition. The average surface heatflow

of the southeastern Deccan province is 45 mW m^{-2} (Fig. 4), and it represents the region where rocks of the eastern Dharwar craton form the basement (Fig. 1). Therefore, crustal thermal modeling has been restricted to the southeastern part only (Fig. 6). The average surface heatflow (45 mW m^{-2}) and the crustal contribution to heatflow (38 mW m^{-2}), would imply a Moho heatflow of 7 mW m^{-2} beneath the southeastern Deccan crust (Fig. 6), which is similar to the lower limits of Moho heatflow inferred in the Dharwar craton (Kumar and Reddy, 2004), as well as in Archaean cratons elsewhere (e.g., Jaupart et al., 1998). Also, the modeled steady-state crustal geotherm of Deccan crust is broadly similar to the Dharwar crust (Fig. 6). The Moho temperature is ~280 °C, which is as low as that for the Dharwar crust (Fig. 6), and is lower than that estimated (~350 °C) by Roy and Rao (1999). More surface heatflow measurements in the southwestern and south-central parts would be required to model the crustal geotherm using the crustal heat production models presented in this work.

DISCUSSION

It is clear from this study that the surface heatflow, Moho heatflow, crustal radiogenic heat contribution, and crustal temperatures of the southeastern Deccan crust are similar to the thermal structure of the adjoining Dharwar craton. The Moho of the Deccan crust is as cold as that of the Dharwar crust. Therefore, it appears that there is no apparent thermal trace of a Réunion plume or any other thermal anomaly responsible for the genesis of the Deccan basalts, in the present-day thermal regime. It is possible that the thermal transient might have decayed, considering its ca. 65-Ma age, as suggested by previous authors (e.g., Roy and Rao, 2000). Alternatively, the thermal anomaly, if any, would be so low that it cannot be detected in the present-day crustal thermal regime. For this case, it is necessary to address the magnitude of thermal anomaly one would expect today, had there been a thermal plume beneath the Deccan province ca. 65 Ma. A close look at thermal scenarios of hotspot swells (which are mostly younger than the Deccan) occurring on the ocean floor may provide better insights on whether to expect a thermal anomaly over hotspots affected by volcanism through time and is the best way to assess the magnitude of such an anomaly.

DeLaughter et al. (2005) lucidly summarizes the heatflow observations over hotspot swells (e.g., Hawaii, Réunion, Iceland). Heatflow observations across the Mascarene Ridge (near Réunion Island), which was probably affected by the Réunion hotspot ca. 15 Ma, point to a very low heatflow anomaly of 6–8 mW m^{-2}, above that expected for the seafloor cooling plate model (Bonneville et al., 1997). Similar studies across the Hawaiian Swell also do not indicate any significant heatflow anomaly; the inferred heatflow anomaly being on the order of 5–10 mW m^{-2}, which is indeed very difficult to resolve, considering the uncertainties in the cooling plate models or the off-swell reference heatflow data (Von Herzen et al., 1989). The Iceland hotspot also does not exhibit any significant heatflow

TABLE 2. HEAT PRODUCTION DISTRIBUTION AND CRUSTAL CONTRIBUTION FOR HEATFLOW OF THE SOUTHWESTERN AND SOUTHEASTERN DECCAN BASALT PROVINCES

	Southwestern Deccan basalt province			Southeastern Deccan basalt province		
Rock type	Heat production (μW m^{-3})	Volume (%)	Weighted heat production (μW m^{-3})	Heat production (μW m^{-3})	Volume (%)	Weighted heat production (μW m^{-3})
Surface layer (0–1 km; 1-km thick)						
Deccan basalts	0.39	100	0.39	0.39	100	0.39
Gross heat production			0.39			0.39
Greenschist-facies layer (1–5 km; 4-km thick)						
Supracrustal belts						
(a) Graywackes	1.13	40	0.45	N.A.	N.A.	N.A.
(b) Mafic volcanic rocks	0.19	7.5	0.01	0.09	9	0.01
(c) Felsic volcanic rocks	1.51	7.5	0.11	1.34	1	0.01
Granites	2.59	5	0.13	2.60	10	0.26
TTG gneisses	1.01	40	0.40	2.40	90	2.16
Gross heat production			1.10			2.44
Amphibolite-facies layer (5–13 km; 8-km thick)						
Supracrustal belts						
(a) Mafic volcanic rocks	0.19	7.5	0.01	0.09	4.5	0.004
(b) Felsic volcanic rocks	1.51	7.5	0.11	2.24	0.5	0.01
Granites	1.65	5	0.08	4.19	30	1.26
TTG gneisses	1.13	80	0.90	2.36	65	1.53
Gross heat production			1.10			2.80
Metasomatized granulite-facies layer (13–21 km; medium-P granulites; 8-km thick)						
Enderbites	0.65	100	0.65	0.35	100	0.35
Gross heat production			0.65			0.35
Depleted granulite-facies layer (21–36 km; high-P granulites; 15-km thick)						
Enderbites	0.34	100	0.34	0.16	100	0.16
Gross heat production			0.34			0.16
Total crustal contribution (mW m^{-2})			23.9			37.8

Notes: The crustal rocks of the western and eastern parts of the Dharwar craton form the basement of the southwestern and southeastern Deccan basalt provinces, respectively. N.A.—not applicable; TTG—tonalite-trondhjemite-granodiorite.

anomaly (Stein and Stein, 2003). The same holds true for other hotspots, such as Bermuda Rise and Cape Verde (see DeLaughter et al., 2005, and the references cited there). The heatflow anomaly of 5–10 mW m^{-2} can be considered characteristic of hotspot swells, based on the above studies. However, these values are very sensitive to the type of reference seafloor heatflow data considered (see DeLaughter et al., 2005, for further discussion). Also, these small anomalies are related to younger hotspot activity than the ca. 65-Ma Deccan event. Thus, if we do assume that the Deccan province was affected by the Réunion hotspot, arguably, the expected present-day heatflow anomaly should be much less than that inferred (6–8 mW m^{-2}) across the Mascarene Ridge. Furthermore, if we assume 2 m.y. as the period of stay of the Deccan basalt province over a Réunion plume (considering the high basalt eruption rates) before the Indian plate drifted northward (e.g., stationary plume–moving plate situation), the duration of heating would also have been short. In all probability, this brevity would not result in high heatflow anomalies, unlike the stationary plume–stationary plate situations, where the duration of heating is expected to be long.

The Moho heatflow inferred for the southeastern Deccan province (7 mW m^{-2}) is very low and is similar to the lower bounds of the Moho heatflow inferred for the Dharwar craton (Kumar and Reddy, 2004). Also, this Moho heatflow value is lower than that inferred for the exposed eastern Dharwar craton (~17 mW m^{-2}) located to the south of southeastern Deccan province. This apparent difference appears to be due to the thickening of the high-heat–producing greenschist-facies layer in the Deccan province. Even if one contends that the Moho heatflow of the Deccan province bears the thermal trace of the plume (7 mW m^{-2} plus 6 or 8 mW m^{-2}), one cannot distinguish this trace from the inferred Moho heatflow values of the adjoining cratons. Furthermore, the underlying uncertainties in the crustal heat production and thermal conductivity models and the

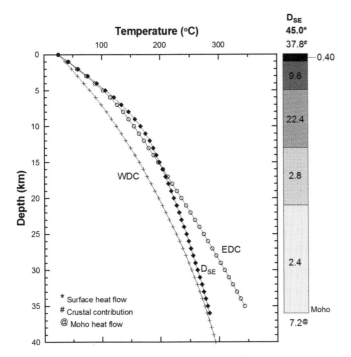

Figure 6. Crustal contribution to surface heatflow and the crustal geotherm of the southeastern Deccan crust (D_{SE}). The crustal geotherms of the western Dharwar craton (WDC) and eastern Dharwar craton (EDC) are modeled using the data of Kumar and Reddy (2004) and are shown here for comparison. Note that the Deccan geotherm is no different from the Dharwar craton geotherms. The crustal contribution to surface flow (shown on the right) is modeled using Table 2. The average surface heatflow (45 mW m^{-2}) and the crustal contribution (38 mW m^{-2}) predict the Moho heatflow of 7 mW m^{-2}.

errors in the surface heatflow determinations reduce our ability to resolve the low heatflow anomaly of 6–8 mW m^{-2} or less, in the crustal thermal regime of the Deccan basalt province.

ACKNOWLEDGMENTS

The authors sincerely thank the reviewers, Robert N. Harris, Arthur H. Lachenbruch, and Richard A. Ketcham, for their helpful comments. Ketcham's suggestions have improved the text significantly. Thanks are due to Gillian R. Foulger (one of the editors) for excellent editorial processing and cheerful encouragement; Hetu C. Sheth for useful discussions; T. Seshunarayana, C. Shankar, and G. Ramacharyulu for their support; and V.P. Dimri, director of the National Geophysical Research Institute, Hyderabad, India, for permission to publish this article.

REFERENCES CITED

Atal, B.S., Bhalla, N.S., Lall, Y., Mahadevan, T.M., and Udas, G.R., 1978, Radioactive element distribution in the granulite terrains and Dharwar schist belts of Peninsular India, *in* Windley, B.F., and Naqvi, S.M., eds., Archaean geochemistry: Amsterdam, Elsevier, p. 205–220.

Bonneville, A., Von Herzen, R.P., and Lucazeau, F., 1997, Heat flow over hotspot track: Additional evidence for thermal rejuvenation of oceanic lithosphere: Journal of Geophysical Research, v. 102, B10, p. 22,731–22,747, doi: 10.1029/97JB00952.

Chandrasekharam, D., Mahoney, J.J., Sheth, H.C., and Duncan, R.A., 1999, Elemental and Nd–Sr–Pb isotope geochemistry of flows and dikes from the Tapi rift, Deccan flood basalt province, India: Journal of Volcanology and Geothermal Research, v. 93, p. 111–123, doi: 10.1016/S0377-0273(99)00081-5.

DeLaughter, J.E., Stein, C.A., and Stein, S., 2005, Hotspots: A view from the swells, *in* Foulger, G.R., et al., eds., Plates, plumes, and paradigms: Boulder, Colorado, Geological Society of America Special Paper 388, p. 257–278, doi: 10.1130/2005.2388(16).

Dessai, A.G., Markwick, A., Vaselli, O., and Downes, H., 2004, Granulite and pyroxenite xenoliths from the Deccan Trap: Insight into the nature and composition of the lower lithosphere beneath cratonic India: Lithos, v. 78, p. 263–290, doi: 10.1016/j.lithos.2004.04.038.

Gupta, H.K., Rao, R.U.M., Srinivasan, R., Rao, G.V., Reddy, G.K., Dwivedy, K.K., Banerjee, D.C., Mohanty, R., and Satyasaradhi, Y.R., 1999, Anatomy of surface rupture zones of two stable continental region earthquakes, 1967 Koyna and 1993 Latur, India: Geophysical Research Letters, v. 26, p. 1985–1988, doi: 10.1029/1999GL900399.

Gupta, M.L., and Gaur, V.K., 1984, Surface heat flow and probable evolution of Deccan volcanism: Tectonophysics, v. 105, p. 309–318, doi: 10.1016/0040-1951(84)90210-5.

Gupta, M.L., Sharma, S.R., and Sundar, A., 1991, Heat flow and heat generation in the Archaean Dharwar cratons and implications for the southern Indian Shield geotherm and lithospheric thickness: Tectonophysics, v. 194, p. 107–122, doi: 10.1016/0040-1951(91)90275-W.

Hanuma Prasad, M., Krishna Rao, B., Vasudev, V.N., Balaram, R., and Balaram, V., 1997, Geochemistry of Archaean bimodal volcanic rocks of the Sandur supracrustal belt, southern India: Journal of the Geological Society of India, v. 49, p. 307–322.

Jaupart, C., Mareschal, J.-C., Frottier, L.G., and Davaille, A., 1998, Heat flow and thickness of the lithosphere in the Canadian Shield: Journal of Geophysical Research, v. 103, p. 15269–15286, doi: 10.1029/98JB01395.

Ketcham, R.A., 1996, Distribution of heat-producing elements in the upper and middle crust of southern and west central Arizona: Evidence from core complexes: Journal of Geophysical Research, v. 101, p. 13,611–13,632, doi: 10.1029/96JB00664.

Kumar, P.S., 2002, Radioelemental distribution in the Dharwar craton, South India: Implications for the evolution of upper continental crust and heat generation in the craton [Ph.D. thesis]: Hyderabad, India, Osmania University, 217 p.

Kumar, P.S., and Reddy, G.K., 2004, Radioelements and heat production of an exposed Archaean crustal cross-section, Dharwar craton, south India: Earth and Planetary Science Letters, v. 224, p. 309–324, doi: 10.1016/j.epsl.2004.05.032.

Kumar, P.S., and Srinivasan, R., 2002, Fertility of Late Archaean basement granite in the vicinity of U-mineralized Neoproterozoic Bhima basin, peninsular India: Current Science, v. 82, p. 571–575.

Kumar, P.S., Basavalingu, B., and Srinivasan, R., 2003, Origin of the Closepet Granite in the light of fabric, mineralogy and radioelemental composition, *in* Mohan, A., ed., Milestones in petrology: Bangalore, India, Geological Society of India Memoir 52, p. 229–254.

Mahoney, J.J., Sheth, H.C., Chandrasekharam, D., and Peng, Z.X., 2000, Geochemistry of flood basalts of the Toranmal section, northern Deccan Traps, India: Implications for regional Deccan stratigraphy: Journal of Petrology, v. 41, p. 1099–1120, doi: 10.1093/petrology/41.7.1099.

Menon, R., Kumar, P.S., Reddy, G.K., and Srinivasan, R., 2003, Radiogenic heat production of late Archaean Bundelkhand granite and some Proterozoic gneisses and granitoids of central India: Current Science, v. 85, p. 634–638.

Mitchell, C., and Widdowson, M., 1991, A geological map of the southern Deccan Traps, India, and its structural implications: Journal of Geological Society of London, v. 148, p. 495–505.

Mohan, G., and Kumar, M.R., 2004, Seismological constraints on the structure and composition of western Deccan volcanic province from converted phases: Geophysical Research Letters, v. 31, art. no. L02601, doi: 10.1029/2003GL018920.

Naqvi, S.M., Condie, K.C., and Allen, P., 1983, Geochemistry of some unusual early Archaean sediments from the Dharwar craton: Precambrian Research, v. 22, p. 125–147, doi: 10.1016/0301-9268(83)90061-X.

Naqvi, S.M., Sawkar, R.H., Subba Rao, D.V., Govil, P.K., and Gnaneshwar Rao, T., 1988, Geology and geochemistry and tectonic setting of Archaean greywackes from Karnataka nucleus, India: Precambrian Research, v. 39, p. 193–216, doi: 10.1016/0301-9268(88)90042-3.

Naqvi, S.M., Manikyamba, C., Gnaneshwar Rao, T., Subba Rao, D.V., Ram Mohan, M., and Srinivasa Sarma, D., 2002, Geochemical and isotopic constraints of Neoarchaean fossil plume for evolution of volcanic rocks of Sandur greenstone belt, India: Journal of the Geological Society of India, v. 60, p. 27–56.

NGRI (National Geophysical Research Institute), 1978, Gravity map series of India, scale 1:1,500,000: Hyderabad, India, NGRI.

Pande, K., 2002, Age and duration of the Deccan Traps, India: A review of radiometric and palaeomagnetic constraints: Proceedings of the Indian Academy of Sciences, v. 111, p. 115–123.

Rai, S.S., Priestley, K., Suryaprakasam, K., Srinagesh, D., Gaur, V.K., and Du, Z., 2003, Crustal shear velocity structure of the south Indian Shield: Journal of Geophysical Research, v. 108, 2088, doi: 10.1029/2002JB001776.

Ray, L., Kumar, P.S., Reddy, G.K., Roy, S., Rao, G.V., Srinivasan, R., and Rao, R.U.M., 2003, Higher mantle heat flow in a Precambrian granulite province: Evidence from southern India: Journal of Geophysical Research, v. 108, art. no. B02084, doi: 10.1029/2001JB000688.

Roy, S., and Rao, R.U.M., 1999, Geothermal investigations in the 1993 Latur earthquake area, Deccan volcanic province, India: Tectonophysics, v. 306, p. 237–252, doi: 10.1016/S0040-1951(99)00051-7.

Roy, S., and Rao, R.U.M., 2000, Heat flow in the Indian Shield: Journal of Geophysical Research, v. 105, p. 25,587–25,604, doi: 10.1029/2000JB900257.

Rybach, L., 1988, Determination of heat production rate, *in* Haenel, R., et al., eds., Handbook of terrestrial heat flow determination: Dordrecht, Kluwer, p. 125–142.

Shankar, R., 1988, Heat flow map of India and discussions on its geological and economic significance: Indian Minerals, v. 42, p. 89–110.

Sheth, H.C., 2005, From Deccan to Réunion: No trace of a mantle plume, *in* Foulger, G.R., et al., eds., Plates, plumes, and paradigms: Boulder, Colorado, Geological Society of America Special Paper 388, p. 477–501.

Sheth, H.C., Mahoney, J.J., and Chandrasekharam, D., 2004, Geochemical stratigraphy of Deccan flood basalts of the Bijasan Ghat section, Satpura Range, India: Journal of Asian Earth Sciences, v. 23, p. 127–139, doi: 10.1016/S1367-9120(03)00116-0.

Stein, C., and Stein, S., 2003, Mantle plumes: Heat flow near Iceland: Astronomy & Geophysics, v. 44, p. 8–10, doi: 10.1046/j.1468-4004.2003.44108.x.

Subbarao, K.V., ed., (1999), Deccan volcanic province: Bangalore, India, Geological Society of India Memoir 43, 947 p.

Von Herzen, R.P., Cordery, M.J., Detrick, R.S., and Fang, C., 1989, Heat flow and thermal origin of hot spot swells: The Hawaiian Swell revisited: Journal of Geophysical Research, v. 94, B10, p. 13,783–13,799.

MANUSCRIPT ACCEPTED BY THE SOCIETY 31 JANUARY 2007

The Geological Society of America
Special Paper 430
2007

A quantitative tool for detecting alteration in undisturbed rocks and minerals—I: Water, chemical weathering, and atmospheric argon

Ajoy K. Baksi*

*Department of Geology and Geophysics, Louisiana State University,
Baton Rouge, Louisiana 70803, USA*

ABSTRACT

Alteration of undisturbed igneous material used for argon dating work, is the most common cause of incorrect (low) estimates of the time of crystallization. Identification of alteration has relied on qualitative and subjective (optical) methods. For $^{40}Ar/^{39}Ar$ dating, I introduce a new parameter—the alteration index (A.I.)—to *quantitatively* assess alteration. This looks to the quantity of ^{36}Ar (atmospheric argon) released in such studies. A non-dimensional parameter is used, relating the ^{36}Ar levels to that of ^{39}Ar for K-rich phases (K-feldspar, biotite, whole-rock basalt), and to ^{37}Ar for Ca-rich phases (plagioclase feldspar and hornblende). Water contains large amounts of dissolved argon derived from the atmosphere. During chemical weathering, ^{36}Ar carried by water is introduced into the silicate phases of rocks. All common alteration minerals contain water; their ^{36}Ar contents are ~100–1000 times higher than in anhydrous silicate phases. Incipient alteration, undetected by current tests, is unequivocally recognized by the A.I. method. In $^{40}Ar/^{39}Ar$ stepheating studies, the plateau steps (if any) release argon from the least altered sites. The A.I. of plateau steps for fresh, subaerial, material yields the cut-off value for detecting alteration. *Partial loss of $^{40}Ar^*$ from altered samples may result in statistically acceptable plateaus that underestimate the true crystallization age by ~2–10%.* Many ages are invalid as accurate estimates of the age of crystallization (a) based on statistical analysis of the apparent ages on plateau/isochron plots and/or (b) ages derived from altered phases within the sample. At subduction zones, the hydrated slab cycles substantial quantities of (atmospheric) argon into the mantle. Monitoring ^{36}Ar levels in fresh (mafic and intermediate) rocks should serve as a sensitive tool in elucidating the role of water driven off the subducted slab in triggering magmatism in convergent zone settings.

Keywords: chemical weathering, diffusional loss, argon ages, water, atmospheric argon

It is my intention to cite experience, then to show by reasoning why this experience is constrained to act in this manner. And this is the rule according to which speculators as to natural effects have to proceed.
Leonardo da Vinci (manuscript E, folio 55r, ca. 1513)

*E-mail: akbaksi@yahoo.com.

Baksi, A.K., 2007, A quantitative tool for detecting alteration in undisturbed rocks and minerals—I: Water, chemical weathering, and atmospheric argon, *in* Foulger, G.R., and Jurdy, D.M., eds., Plates, plumes, and planetary processes: Geological Society of America Special Paper 430, p. 285–303, doi: 10.1130/2007.2430(15). For permission to copy, contact editing@geosociety.org. ©2007 The Geological Society of America. All rights reserved.

INTRODUCTION

Obtaining precise and accurate radiometric data plays a critical role in earth sciences in unravelling the timing, duration, and rates of geological processes/phenomena. The most widely used techniques are the argon dating methods, utilizing the presence of measurable quantities of potassium in most rocks and minerals, and application to a wide temporal range (~2 ka to 4.5 Ga; Renne et al., 1997; Renne, 2000) to which the method has been utilized. The ^{40}Ar/^{39}Ar stepheating technique (Merrihue and Turner, 1966) permits evaluation of the possible (partial) loss of ^{40}Ar*, and/or the presence of excess argon in terrestrial material (Lanphere and Dalrymple, 1971, 1976). The reproducibility of the resulting step ages determines the accuracy and precision of ages as estimates of the time of crystallization of igneous rocks and minerals. Numerous ^{40}Ar/^{39}Ar stepheating ages in the literature do not meet the basic "mathematical" requirements in this regard. All such ages should be statistically acceptable at the 95% confidence level. Many "ages" related to hotspot-generated rocks have been shown to be statistically invalid (Baksi, 1999, 2005a).

For the argon dating methods, the daughter product is a gas and can escape from rocks/minerals under various physico-chemical conditions. The effect of raising the ambient temperature, leading to diffusion loss of ^{40}Ar* and lowered K-Ar dates, has been known for many decades (e.g., Hart, 1964). Effort has been directed to recover accurate time and (cooling) rate information (see McDougall and Harrison, 1999, and references therein). Alteration (chemical weathering), ubiquitous in geological materials, can lead to lowered argon ages by the formation of "leaky" (altered) minerals that do not retain 100% of argon over geological time, or by episodic loss of ^{40}Ar* at the time of the weathering processes (see reviews in Dalrymple and Lanphere, 1969; McDougall and Harrison, 1999).

Arguably, the outstanding achievement of the argon dating methods was the establishment of the geomagnetic polarity time scale (GPTS), which led to the plate tectonic revolution (Glen, 1982). The K-Ar based GPTS (Fig. 1) was based on dating specimens (primarily whole-rock basalts and sanidine separates) carefully selected for freshness on the basic of petrographic examination (e.g., McDougall and Chamalaun, 1966). Recent astrochronological and ^{40}Ar/^{39}Ar dating studies have shown that the K-Ar derived GPTS underestimates the ages of most geomagnetic reversals over the past ~5 m.y. by 5–7% (Fig. 1). The low (incorrect) K-Ar dates result, in part, from dating altered whole-rock material which had suffered minor (~5–7%) loss of ^{40}Ar* (Baksi, 1995). *The most careful selection of fresh samples (based on optical methods) for very young subaerial rocks is inadequate.* Dating older submarine material that has been in an environment suited to alteration/weathering for tens of millions of years is fraught with problems. Thus, the age of the bend in the Hawaiian-Emperor chain, based primarily on whole-rock material (Dalrymple and Clague, 1976), indicated an age of ca. 43 Ma; recent work on mineral separates suggests it is ~15% older (Sharp and Clague, 2006).

The ^{36}Ar content of analyses has been used as a guiding factor to detecting alteration (Baksi, 1974a; Roddick, 1978). The method remains underutilized, perhaps due to the lack of a firm theoretical basis for such usage. Herein, this technique is placed on a quantitative basis; the causes of a marked increase in ^{36}Ar contents during alteration are investigated. "The effects of weathering and alteration usually can be recognized in hand specimens or thin sections" (Dalrymple and Lanphere, 1969, p. 146). Advances in this respect have been considerable (see Devigne, 1998), but the method remains semiquantitative (at best), and its applicability varies widely. The partial loss of an inert gas (^{40}Ar*) from rocks/minerals (often at the sub-parts-per-trillion level) must be traced by sensitive tools. Chemical methods seem

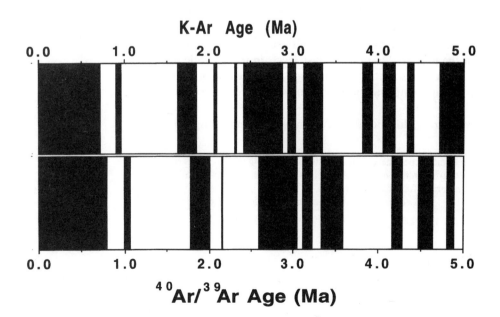

Figure 1. The geomagnetic polarity time scale for 5–0 Ma; black—normal polarity; white—reversed polarity. The K-Ar based version yields ages for chron boundaries that are 5–7% younger than those based on ^{40}Ar/^{39}Ar dating. This discrepancy results, in part, from K-Ar dates on altered subaerial whole-rock basalts (see text). K-Ar scale from Harland et al. (1990); ^{40}Ar/^{39}Ar scale from Baksi (1995).

well suited; the water content of mafic whole-rocks has been utilized with limited success (e.g., Kaneoka, 1972). Diffusion loss of ^{40}Ar* at the ppt level is being sought by tools that operate, at best, at the ppm level. The argon data itself, the quantity of one isotope (^{36}Ar) routinely measured in gas samples, are best suited for detecting alteration. Other causes may cause lowered argon ages, including devitrification and exsolution processes. Occasionally, weathering associated with hydrothermal solutions leads to "ages" that are too old; these are not investigated here.

I first review the statistical rules to be observed, then proceed to examine the role of ^{36}Ar in the understanding of alteration/weathering effects. Ages must be listed with estimated errors, both of which are statistically robust. Only such radiometric "numbers" should be used to derive the timing, and rates of various geological phenomena/processes. I shall denote (in figures) and list errors at the one-sigma level. Ages are reported relative to the calibrations of Renne et al. (1998) and where necessary, converted to the decay constants, isotopic abundances, listed by Steiger and Jager (1977). All plateau/isochron ages with probability values <0.05 (2σ or 95% confidence interval) are rejected as accurate estimates of crystallization age. In cases where the goodness of fit parameter (hereafter F) is >1, but the corresponding probability value is >0.05, the error in the plateau isochron age is multiplied by $F^{1/2}$ (cf. Roddick, 1978).

METHODOLOGY

Statistical Tests

The ^{40}Ar/^{39}Ar stepheating method gives a series of apparent ages, which are then examined to recover possible crystallization ages for igneous rocks. The data sets are often examined on an isochron plot. The goodness of fit parameter (York, 1969) is calculated for the resulting straight line, bearing in mind the number of data points (N) involved. This (F) parameter is used with χ^2 tables to arrive at the probability that the scatter of the points around the isochron results (entirely) from the experimental errors in measurement of the isotopic ratios (peaks). If this probability value is <0.05, excess scatter of the data points is demonstrated—geological error is present. For most undisturbed igneous phases, violation of the "closed system" assumption has occurred, indicating gain or loss of the parent/daughter after formation (time $t = 0$). For argon dating work, the most common cause is loss of the gaseous daughter product.

Many igneous rocks and minerals show little or no excess ^{40}Ar, as evidenced by isochron plots, yielding initial ^{40}Ar/^{36}Ar ratios (on isochron plots) that are statistically indistinguishable from the atmospheric argon value (295.5; Nier, 1950). For such material, the use of the age spectrum is justified (see Dalrymple, 1991). Statistics for data evaluation are similar to those applied to isochrons (see Baksi, 2005a) and are easier to understand and evaluate for the average reader. Age spectra with plateau sections that show statistically defined variations at the 95% confidence level can be identified, if the age spectrum is drawn to a proper scale (see www.mantleplumes.org/ArAr.html). Readers can satisfy themselves as to the probable validity of the plateau age and, if necessary, turn to the detailed examination of the isotopic data sets. *It is critical that authors/journals make all such data available within the manuscript itself or at a supplementary site that can be accessed by readers.* The basics of testing the robustness of isochron data were placed on a firm footing over three decades ago. Berger and York (1970) examined the factors underlying the scatter of points in their replicate analyses of a whole-rock basalt. Important cases in which these principles were disregarded are outlined below.

^{36}Ar—The Key to Detecting Alteration in Rocks and Minerals

In K-Ar work, the ^{36}Ar (atmospheric argon) seen in analyses is derived from two sources: (1) the "initial" argon within the sample itself and (2) "blank" drawn from the fact that the atmosphere contains ~0.003% of ^{36}Ar. Careful control of blanks is a prerequisite to using the level of ^{36}Ar in analyses for quantitative purposes. For much of the early (pre-1980) work, this is not the case, though it was demonstrated that such data sets could be achieved with diligent vacuum techniques (Hayatsu and Carmichael, 1970, 1977; Baksi, 1973, 1974a,b; Hayatsu and Palmer, 1975; Hall and York, 1978). With the advent of microdating techniques and the utilization of ultrasensitive mass spectrometers with low and controlled levels of blank (York et al., 1981, 1984), ^{36}Ar levels in analyses reflect that within the sample itself.

Over three decades ago, it was noted that altered wholerock basalts showed considerably higher amounts of ^{36}Ar than does fresh material (Baksi, 1974a,b). Fresh whole-rock basaltic material subjected to K-Ar dating show ~2×10^{-10} cm^3 STP g^{-1} of ^{36}Ar (Baksi, 1974a,b; Hall and York, 1978). More felsic material is expected to contain ~3 times as much ^{36}Ar (see Lux, 1987). Both fresh and marine water contain large amounts of dissolved argon drawn from the atmosphere that currently contains ~0.93% of argon (Kellas, 1895). During chemical weathering, silicate phases interact with water; the latter contains ~1×10^{-6} cm^3 STP g^{-1} of dissolved ^{36}Ar (Bieri et al., 1968), about 5000 times higher than in fresh anhydrous silicates. Large amounts of all three isotopes of argon are transferred over to the (altered) mineral, and its isotopic composition should reflect that of the atmosphere at the time of alteration. Because the ^{40}Ar$_{At}$ level cannot be detected uniquely, i.e., distinguished from ^{40}Ar*, ^{36}Ar (which is ~5 times more plentiful than ^{38}Ar) is used to detect alteration. The ^{36}Ar content of a rock or mineral is sharply raised during the alteration/weathering process. Incipient alteration, unobservable by other (chemical or optical) techniques, can be pinpointed by a marked increase in the ^{36}Ar content of the rock or mineral. This feature forms the basis of detecting alteration in rocks and minerals. The role of water in raising the ^{36}Ar level in rocks was recognized over 20 years ago (Baksi, 1982), in seeking to explain K-Ar isochron results (Hayatsu and Carmichael, 1970) on interbedded sediments and basalts from Cape Breton Island, Canada.

The level of atmospheric argon observed in rocks and minerals used for K-Ar dating varies widely. I shall refer to this component by looking at ^{36}Ar in the sample (see Fig. 2). McDougall and Harrison (1999) observe that high levels of atmospheric argon are seen in platy minerals (e.g., micas) as compared to those with prismatic habits. I note a marked correlation between the level of structural water in the material and its ^{36}Ar content. The lowest levels are seen in anhydrous material (K-feldspar, plagioclase feldspar, whole-rock basalt), somewhat higher levels in minerals containing low levels of water (hornblende), and the highest levels in those containing high levels of water (micas). Alteration products contain higher amounts of structural water than does the parent fresh material, and carry significantly higher levels of ^{36}Ar. Figure 2 shows a wide variation—an order of magnitude or more—in the ^{36}Ar contents of all minerals. These variations result in (large) part from dating of samples that have suffered alteration; values for fresh anhydrous minerals should fall close to (or lower than) the minimum values seen in Figure 2. In particular, plagioclase feldspar and whole-rock basalts, commonly used for argon dating work, are prone to alteration. Leaching the crushed samples in acid in an ultrasonic bath is successful in removing alteration products and improves the quality of the results. The acid reacts with (and removes) surface alteration and penetrates into microfractures where incipient alteration has taken place.

In ^{40}Ar/^{39}Ar dating work, the weight of the sample utilized and the quantity of gas released in each step are often not enumerated; further, the K content of the material is generally not measured. It is then not possible to calculate the absolute amount of ^{36}Ar released in either stepheating or total fusion studies. The neutron irradiation of the specimen produces ^{39}Ar from ^{39}K, as well as ^{37}Ar from ^{40}Ca. The quantity of these isotopes of argon, routinely measured during mass spectrometry, permits calculation of the K and Ca content of the specimen. The efficiency of production of ^{39}Ar$_K$ and ^{37}Ar$_{Ca}$ are dependent on the total neutron flux to which the sample was exposed, measured by J, the irradiation parameter. I shall utilize the quantity of these isotopes released in each step (normalized for the level of neutron irradiation) as a proxy for the weight of the sample under study. The ^{36}Ar measured must be corrected for blanks and contributions from side (interfering) reactions for argon produced from Ca and Cl within the sample. Ideally, one should measure the quantity of ^{36}Ar released (in cm^3 STP g^{-1} or moles g^{-1}) in all argon dating experiments.

Setting Up the Alteration Index

A single parameter for detecting alteration in all rocks and minerals may not be feasible. I look to establish the modalities for silicates commonly used for argon dating. It involves quantitatively assessing the amount of ^{36}Ar present in the material under study. Typical calculations are shown in the Appendix.

Whole-Rock Basalts. Fresh whole-rock basalts in K-Ar work, wherein careful vacuum techniques were utilized, contain ~2×10^{-10} cm^3 STP g^{-1} of ^{36}Ar (Baksi, 1974a,b; Hall and York, 1978). In K-Ar dating, a wide range of values is noted (Fig. 2), resulting from analysis of partially altered rocks. In basalts, the

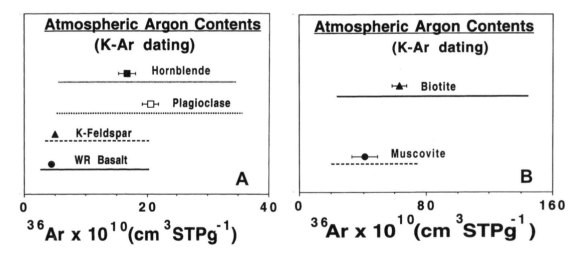

Figure 2. The atmospheric Ar contents of various silicate phases as obtained from K-Ar dating studies. Results shown in units of ^{36}Ar, with the mean and standard error on the mean, along with a line showing the range of values listed in McDougall and Harrison (1999); single outliers omitted for clarity. (A) Listing for anhydrous silicate phases and hornblende—the large range of values in each case is thought to result from dating of altered material. (B) Results for hydrated phases—the micas. The latter contain about ten times more ^{36}Ar than found in the anhydrous phases, resulting from the high solubility of (atmospheric) Ar in water. WR—whole rock. Modified from McDougall and Harrison (1999).

fine-grained mesostasis carries most of the K (Mankinen and Dalrymple, 1972). On alteration, this substance forms clay minerals that contain substantially greater amounts of ^{36}Ar. I shall relax the upper cut-off value to $6 \times 10^{-10} cm^3$ STP g^{-1} of ^{36}Ar in K-Ar analyses to denote use of fresh material for dating purposes. This method has been used with success to narrow down the range of good K-Ar dates for tholeiitic rocks (Baksi, 1987, 1989; Baksi and Hoffman, 2000).

Plateau sections in $^{40}Ar/^{39}Ar$ stepheating studies of undisturbed rocks and minerals are taken to reflect crystallization ages (see McDougall and Harrison, 1999). However, numerous cases in the literature show otherwise. Recovery of statistically acceptable $^{40}Ar/^{39}Ar$ plateau ages, shown to be considerably younger than crystallization values, prompted a search for a technique to detect alteration in rocks and minerals. A significant question is how to recognize correct $^{40}Ar/^{39}Ar$ ages. This is best attempted by comparison with ages obtained by more robust techniques, such as U-Pb dating. Ages may differ by ~1%, due, in part, to errors in the currently used values of the decay constants in argon work and/or the ages of the monitors (standards) that are utilized (Min et al., 2000; Renne, 2000). For young rocks, comparison may be possible with the results obtained by the astrochronological approach (Baksi et al., 1992; Baksi, 1994; Renne et al., 1994) or with historical notes (Renne et al., 1997).

In incremental heating studies on whole rocks, nonplateau steps with low (incorrect) ages invariably show higher amounts of ^{36}Ar than the plateau steps. $^{40}Ar/^{39}Ar$ dating results for the Gettysburg Sill, northeastern United States, illustrate important features. The age spectra for three specimens yield different plateau ages, all valid from the statistical viewpoint (Fig. 3A).

The age of 200 Ma (Baksi, 2003) is close to the U-Pb age (201 Ma; Dunning and Hodych, 1990). The other "ages" (Sutter and Smith, 1979) are significantly younger, resulting from partial loss of $^{40}Ar^*$ from altered sites in the material dated. To verify this assertion, a suitable alteration index (A.I.) needs to be set up—I utilize the $^{36}Ar/^{39}Ar$ ratio, corrected for $^{36}Ar_{Ca}$. All values are normalized to J = 0.0100, measuring the production of ^{39}Ar from ^{39}K. Figure 3B shows the results for the Gettysburg Sill samples. The plateau steps in the work of Sutter and Smith (1979) show ~20–50 times more ^{36}Ar than unaltered GS-14 (Baksi, 2003) and are clearly altered. The K content of these rocks is taken to be subequal. In wider application, the K content of basalts used for argon dating may vary by a factor of ~10 (0.2–2.0% K). This will lead to a similar variation in A.I., resulting from differing K contents, and not from alteration effects. Whenever possible, the A.I. must be adjusted for K contents, and I utilize an intermediate value of K = 0.65%. In the absence of this correction, a variation of ~3 in the A.I. would result, reflecting K contents, compared to the 10- to 1000-fold increase caused by the presence of altered material. The latter may be removed by acid leaching. Use of HF on material with ferromagnesian phases can cause loss of $^{40}Ar^*$ without a concomitant loss of K, and the use of HNO_3 (Baksi and Archibald, 1997) is recommended.

To derive a cut-off value for the A.I. on fresh basalts, I look at $^{40}Ar/^{39}Ar$ stepheating work on a variety of subaerial whole rocks, ranging in age from <1 Ma to ca. 120 Ma, and with K contents varying from ~0.1 to 3.0%. All samples were washed with HNO_3 and yielded plateau ages for a substantial fraction of the gas released. The A.I. (Fig. 4) shows a relatively small range, il-

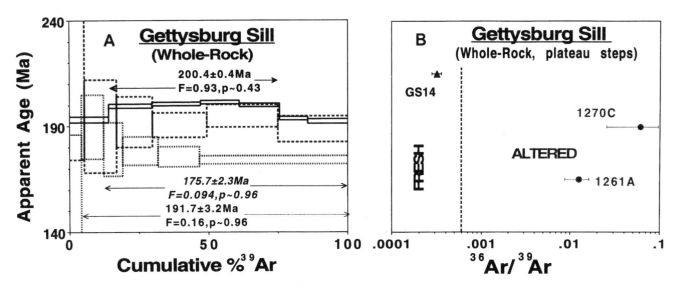

Figure 3. $^{40}Ar/^{39}Ar$ dating results for whole-rock samples from the Gettysburg Sill. (A) Age spectra for three splits; plateau sections delineated by arrows, F = goodness of fit parameter, p = probability of occurrence. One specimen (solid lines; Baksi, 2003) yields the correct age of ca. 200 Ma. Two samples (dotted lines; Sutter and Smith, 1979) yield statistically acceptable plateaus but are measurably younger. (B) Assessment of the alteration state for the plateau steps of all three samples. Average alteration index (A.I.) values and the associated standard error on the mean are shown on a log scale. Sample GS14 of Baksi (2003) shows about twenty to fifty times less ^{36}Ar than in samples 1261A and 1270C of Sutter and Smith (1979). The latter are clearly altered, even for the plateau (best) sites.

lustrating the efficacy of our approach. The average value of ~0.00035 is in agreement with that noted in the work of Hall and York (1978) on ca. 50-ka (fresh) rocks. Whole-rock basalts exhibiting A.I. >0.0006 are altered. The Gettysburg Sill samples of Sutter and Smith (1979) exhibit A.I. values twenty to one hundred times higher and are highly altered. The ^{36}Ar content of the (fresh) rocks used in Figure 4 leads to an average value of ~4 × 10^{-10} cm^3 STP g^{-1}; a tholeiite (RM82–5) from the Rajmahal Traps, India, contains 90% less ^{36}Ar. These values are within the guidelines developed over the years for fresh samples. It is critical that the stepheating process, at intermediate and high laboratory extraction temperatures, is able to derive argon from (relatively) unaltered sites, verifiable by noting their A.I. values. Whole-rock diabase intrusions from the Newark Trend Basins, United States (Sutter and Smith, 1979), show A.I. >0.02, thirty times the cut-off value. The plateau steps in these experiments typically contain >100 × 10^{-10} cm^3 STP g^{-1} of ^{36}Ar. These rocks are grossly altered, and the conclusions of Sutter and Smith (1979) regarding the timing of their intrusion and the opening of the central Atlantic Ocean must be discounted.

Plagioclase Feldspar. In K-Ar work, this mineral should show low ^{36}Ar content because it is anhydrous (see Fig. 2A). It alters readily to form clay minerals. The latter are hydrous and contain considerably higher amounts of ^{36}Ar. The large range of ^{36}Ar values noted in K-Ar dating of this mineral (Fig. 2A) results from analysis of altered material. Thus, analyses on plagioclase separates from the Deccan Traps, India (Courtillot et al., 1986), show a correlation between increase in ^{36}Ar contents (alteration) and lowered K-Ar dates (Fig. 5).

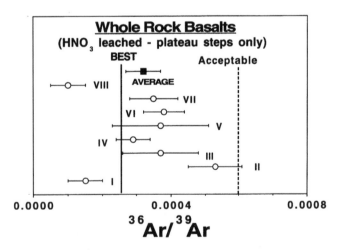

Figure 4. Assessing the alteration index (A.I.) for fresh whole-rock basalts, normalized to J = 0.0100 and K = 0.65%, (average values and standard error on the mean shown on a *linear* scale). The average A.I. value is ~0.00035, and the cut-off for fresh samples is relaxed to <0.0006. Source of data: I, alkali basalts ca. 0.8 Ma—Baksi et al. (1992); II, tholeiite ca. 1.8 Ma—Baksi (1994); III, tholeiites ca. 2.15 Ma—Baksi and Hoffman (2000); IV, alkali basalts ca. 9.7 Ma—Baksi et al. (1993); V, tholeiites ca. 65 Ma—Baksi (2001); VI–VIII, andesitic basalt, alkali basalt, tholeiite (ca. 116–119 Ma—Rajamhal Traps, India) —A.K. Baksi (unpubl. data, 2001).

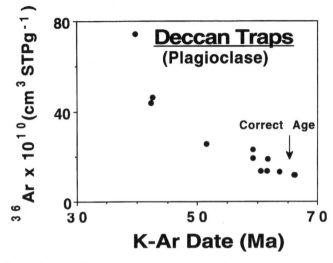

Figure 5. K-Ar dates on plagioclase separates from the Deccan Traps (Courtillot et al., 1986) assessed for alteration. Samples yielding "ages" <65 Ma exhibit large amounts of ^{36}Ar, are altered, and have suffered partial loss of ^{40}Ar*.

The ^{40}Ar/^{39}Ar study of Knight et al. (2003) on the Rajahmundry Traps, India, is our starting point; plagioclase separates were leached with HNO$_3$ prior to analysis and gave plateau ages of ca. 65 Ma. Utilizing the A.I. developed above, it is noted the nonplateau steps show high levels of ^{36}Ar (see Fig. 6A). The age spectrum technique helps obtain a plateau age from the best (least altered) sites. Two specimens (RA99.02 and 99.06/2) are altered throughout (see Baksi, 2005b) and are not considered further. There is considerable variation (factor of about ten) in the A.I. for the best phases (plateau steps) for the different samples (Fig. 6A), resulting from differing K contents. To overcome this problem, a new A.I. for low-K, high-Ca samples is defined: ^{36}Ar/^{37}Ar. The Ca content of the plagioclase serves as a proxy for the weight of the sample. For most plagioclase from mafic rocks, the Ca content varies by about a factor of two (~An$_{40}$ to An$_{80}$). Figure 6B uses this new A.I. (normalized to J = 0.0100) for the Rajahmundry samples. The data sets are tightened up considerably; the total variation is a factor of about three. Changes caused by alteration raise the A.I. by a factor of about ten to one thousand.

Alteration products in plagioclase separates can be removed by acid leaching of crushed material (see Dalrymple and Lanphere, 1969). Plagioclase separates from the central Atlantic magmatic province rocks in Africa and North America have been analyzed in various forms. HF-treated samples (Hames et al., 2000) show the lowest ^{36}Ar concentrations (Fig. 7A). Five samples yield plateau ages of 199–201 Ma and low A.I. A sixth sample, SCD-9, shows higher A.I. and yields a barely acceptable plateau age that is measurably younger at ca. 196 Ma (Fig. 7B). This sample is altered, shows disturbance even in the best plateau steps, and helps to form a sharp A.I. cutoff (<0.00006) for fresh plagioclase. The central Atlantic magmatic province plagioclase samples of Deckart et al. (1997), which are un-

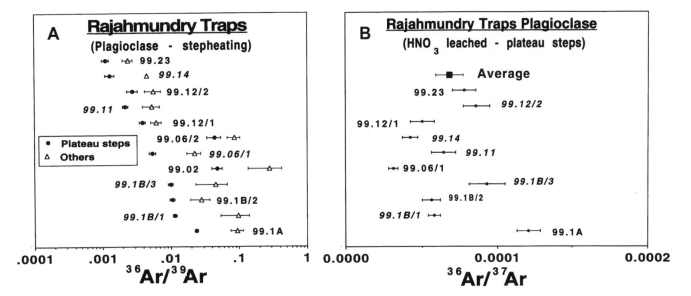

Figure 6. $^{40}Ar/^{39}Ar$ stepheating studies on HNO_3-leached plagioclase separates from the Rajahmundry Traps, India (data from Knight et al., 2003). Average alteration index (A.I.) values shown with associated standard error on the mean (SEM). (A) Results assessed for their ^{36}Ar contents (normalized to ^{39}Ar) on a log scale. The plateau steps show significantly lower amounts of ^{36}Ar than for other steps. The A.I. for the plateau steps differ considerably. (B) ^{36}Ar levels normalized to ^{37}Ar (see text); plateau step averages, with associated SEM, shown on a *linear* scale. The range of A.I. is reduced considerably. Sample numbers as in Knight et al. (2003).

treated by acid, show much higher A.I. (Fig. 8A): about ten to one hundred times that of fresh samples. Their "ages" are not proper plateaus, as based on statistical observations, and are thought to be ~5–10 m.y. too young (Baksi, 2003). The hot HNO_3 leached central Atlantic magmatic province plagioclase samples of Verati et al. (2005) show lower values of A.I. (Fig.

8B), but are three to five times that of the freshest samples; plateau ages are older than those of Deckart et al. (1997). The quantity of (low-temperature) gas in each set of experiments that do not constitute the plateau and are derived from altered (high A.I.) sites varies substantially. For untreated samples, this gas constitutes ~20% (Deckart et al., 1997), falling to 3–8% for the

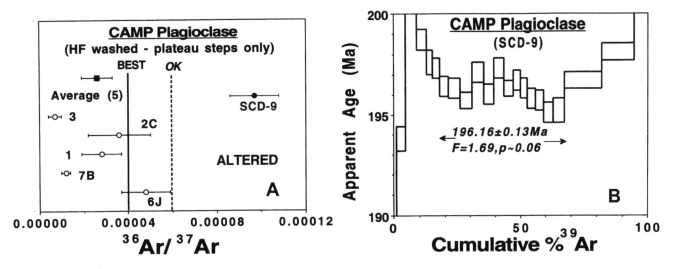

Figure 7. (a) Assessment of the alteration index (A.I.) for HF-leached plagioclase from central Atlantic magmatic province (CAMP) rocks (normalized to J = 0.0100; data from Hames et al., 2000). Averages and associated standard error on the mean for plateau steps shown on a *linear* scale. Five samples yield a low A.I. cut-off (~0.00004) for fresh samples; SCD-9 shows a higher A.I. value. (B) The age spectrum for SCD-9 (symbols and legends as in Fig. 3A) yields a barely acceptable plateau and an "age" that is ~4 m.y. too young (see text). This sample is altered; a hard cut-off value for A.I. <0.00006 excludes altered plagioclase.

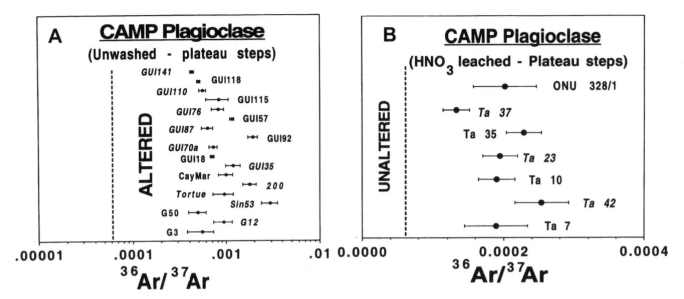

Figure 8. Alteration index (A.I.) values for central Atlantic magmatic province (CAMP) plagioclase samples. (A) Average and associated standard error on the mean (SEM) for plateau steps shown on a log scale for acid-unwashed samples (Deckart et al., 1997). These show about ten to fifty times higher amounts of ^{36}Ar than the cut-off value (~0.00006) and are altered. (B) Average and associated SEM are shown on a *linear* scale for the plateau steps of HNO$_3$-leached samples (Verati et al., 2005).). A.I. values are about three to five times higher than the recommended cut-off value; these samples yield plateau ages ~3–5 m.y. older than those of Deckart et al. (1997).

HNO$_3$-leached samples (Verati et al., 2005), and <2% for the HF-leached samples (Hames et al., 2000). The A.I. values of Hames et al. (2000) are somewhat lower than those of the HNO$_3$-washed samples of Knight et al. (2003) from the Rajahmundry Traps. The latter chose not to leach their samples with HF, as this process often produces "fluffy" material on the outside of the plagioclase crystals that is difficult to date properly (P.R. Renne, pers. comm., 2006). The Rajahmundry Traps rocks may contain higher amounts of ^{36}Ar, as they were formed from intracanyon flows in estuarine or shallow marine conditions (Baksi et al., 1994; Baksi, 2001, 2005b).

I recommend an A.I. cut-off value of 0.00006 for screening plagioclase samples. Values <0.00004 (based on the work of Hames et al., 2000; Fig. 7A) denote the freshest material. Samples with A.I. values >0.00006 show alteration and yield lower ages (see specimen SCD-9 above). The A.I. method shows different degrees of alteration for plagioclase samples used for defining the central Atlantic magmatic province event. Samples washed in HNO$_3$ yield statistically valid plateau ages but may be marginally too young. Until these ages can be verified by some other technique, it is premature to attempt fine resolution of the duration of the central Atlantic magmatic province event in North America, Africa, South America, and Europe. Alternately, only the ages of samples showing A.I. <0.00006 (washed in HF?) serve as accurate estimates of the time of crystallization. The results of Hames et al. (2000) on ca. 200-Ma samples yield ^{36}Ar values of ~1 × 10^{-10} cm^3 STP g^{-1}, lower than the minimum values observed in K-Ar dating (Fig. 2A). The ca. 1.2-Ma Alder Creek (rhyolite) plagioclase (Nomade et al., 2005) shows ^{36}Ar

~0.5 × 10^{-10} cm^3 STP g^{-1}. This value is <20% of the lowest value noted in K-Ar dating (see Fig. 2A).

K-feldspar. This mineral is resistant to chemical weathering and in its high-temperature forms—sanidine and anorthoclase—serves as good material for dating volcanic rocks. It is anhydrous and shows low amounts of atmospheric argon (Fig. 2A). Acid washing of mineral grains serves to remove surface alteration, with dilute HF being the reagent of choice (Dalrymple and Lanphere, 1969).

I begin with the study of Min et al. (2000), attempting cross-calibration of the K-Ar and U-Pb dating schemes and utilizing minerals from the ca. 1.1-Ga Palisade rhyolite. A number of single, clear grains of K-feldspar yield good ^{40}Ar/^{39}Ar plateaus, averaging 1088 Ma (Fig. 9A). Other multigrain (cloudy) samples yield a statistically acceptable plateau but are measurably younger (Fig. 9B). There is no reason for rejecting the latter as crystallization ages, beyond the qualitative (optical) observation they may be altered. (It is noted that the Ca/K ratio of the cloudy grains is about five times those of the clear grains). Figure 10A looks at the A.I. by observing the ^{36}Ar/^{39}Ar ratios for all steps, normalized to J = 0.0100. For the (cloudy) grains, yielding younger ages of ca. 1068–1077 Ma, the nonplateau steps show significantly higher quantities of ^{36}Ar. For the plateau steps, the argon was drawn from the freshest sites within the cloudy grains. The average value of the A.I. is shown in Figure 10B, for the plateau steps, of both the clear (1088 Ma) and cloudy (1068–1077 Ma) grains. Alteration of the latter, shown by about a twenty-fold increase in the A.I., led to an associated ~2% loss of ^{40}Ar*. In evaluating high-temperature K-feldspar ages for pos-

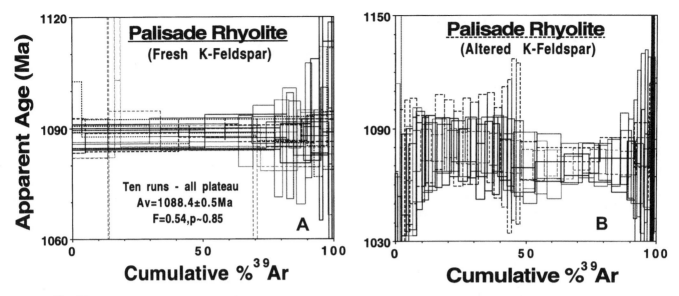

Figure 9. ^{40}Ar/^{39}Ar age spectra for K-feldspar from the Palisade rhyolite (after Min et al., 2000). (A) Results for ten runs on clear grains. All yield excellent plateaus, with an average age of 1088 Ma. (B) Results for five sets of cloudy grains. These yield plateaus that are statistically acceptable (probability >0.05) but are measurably younger at 1068–1077 Ma, each with errors of ±2–3 m.y.

sible alteration, the A.I. must be normalized to a single value—chosen to be 0.0100. K-feldspars used for argon dating may have K values varying from ~3 to 15%. Whenever possible, the A.I. should be adjusted to a single value (chosen to be 10.0%). For K-feldspars from different geological environments, K contents may not be reported and could vary by a factor of about two, having an equivalent effect on the A.I. This amount is small compared to the changes we seek in detecting alteration, where

the ^{36}Ar content (and hence A.I.) will be raised ten- to a hundred-fold.

The ^{36}Ar content of the fresh, 1.2-Ma (Alder Creek rhyolite) sanidine (Renne et al., 1998; Nomade et al., 2005), is ~1 × 10^{-10} cm^3 STP g^{-1}, lower than all the values noted in K-Ar dating (see Fig. 2A). The clear K-feldspar grains (1088 Ma) from the Palisade rhyolite (Min et al., 2000) yield values of ~20 × 10^{-10} cm^3 STP g^{-1}, at the high end of that observed in K-Ar dat-

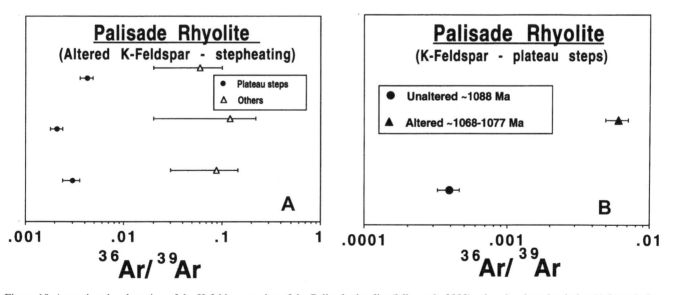

Figure 10. Assessing the alteration of the K-feldspar grains of the Palisade rhyolite (Min et al., 2000) using the alteration index (A.I.) technique. Average values are shown with associated standard error on the mean on log scales. (A) For the cloudy grains, nonplateau steps show significantly higher amounts of ^{36}Ar than do the plateau steps. (B) Values for the plateau steps only on the clear (1088 Ma) and cloudy (1068–1077 Ma) grains. The latter show about twenty times more ^{36}Ar than found in the clear grains, reflecting alteration.

ing (Fig. 2A). This variation indicates that fixing a single sharp (A.I.) cut-off for freshness may not be possible, as ^{36}Ar contents appear to be dependent on the age of the sample. This possibility is investigated further below.

Biotite. This mineral finds wide use for argon dating purposes. Alteration (to chlorite) lowers the K and also increases the Ca content of the mineral. ^{36}Ar contents of this mineral are high in K-Ar dating work (see Fig. 2B); the average value (~60 × 10^{-10} cm^3 STP g^{-1}) is biased to the high side by many samples showing very high values (see McDougall and Harrison, 1999, Figure 2-6). The latter apparently resulted from the analyses of samples with measurable chloritization. This alteration product contains significantly more water (~10 wt%) than fresh biotite (~4 wt%).

Biotites used as standards (with low Ca/K values, hence unaltered) have been carefully analyzed. I utilize blank corrected values from work on tens-of-milligram splits (Baksi et al., 1996) and laser work on a few flakes (Renne et al., 1998; Daze et al., 2003). These show ^{36}Ar values (in units of 10^{-10} cm^3 STP g^{-1}), of ~35–60 (see Fig. 11), in the range expected for fresh biotite. For unaltered material from Ordovician K-bentonites (Min et al., 2001), ^{36}Ar levels are 10–20 × 10^{-10} cm^3 STP g^{-1}, comparable to the lowest 5% listed for K-Ar dating by McDougall and Harrison (1999). Two tools may be used to detect alteration in biotite. The first is to look at ^{37}Ar/^{39}Ar ratios in analyses. Elevated values (>0.02—i.e., Ca >0.3 wt%) indicate the presence of alteration products, chlorite and/or epidote. The second is to use the A.I. method. However, pinpointing a cut-off value for freshness is more difficult for this hydrous mineral (as well as for hornblende—see below), because of much higher levels of ^{36}Ar than in anhydrous minerals. For biotite, the parameter would be

the same as for K-feldspar, namely ^{36}Ar/^{39}Ar; the normalization should be to J = 0.0100 and K = 8.0%. A.I. values for standards fall in the range 0.0006–0.0013 and somewhat lower for the Ordovician K-bentonite (see Fig. 11). This is a relatively large range of values; I shall use 0.00130 as the upper cut-off value for fresh biotite specimens for total fusion studies. For step-heating work, a lower cut-off of ~0.0005 seems appropriate for argon derived from the best (plateau) steps.

Hornblende. This mineral is commonly used for argon dating. I evaluate the homogeneity of samples, looking at total fusion age and chemical composition (Ca/K ratios). Comprehensive data sets are available for laser ^{40}Ar/^{39}Ar work on one to three grain splits of standards MMHb-1, PP-20, and NL-25 (Renne et al., 1998; Renne, 2000). For MMHb-1, ages differ by >10 m.y. (precision of ages ± 1 m.y.), and grains differ in Ca/K content by ~30%. Splits of PP-20 (a repreparation of standard Hb3Gr) differ by 20 m.y. (precision of ages ± 3 m.y.), and Ca/K values differ in value by a factor of about two. NL-25 shows Ca/K values differing by a factor of about three; individual ages differ by ~100 m.y. (precision of ages ± 15 m.y.). These observations on the best specimens (standards) raise questions about the homogeneity of hornblende samples in general.

^{36}Ar contents, based on K-Ar dating (Fig. 2A), are intermediate between those of anhydrous minerals (feldspars) and hydrous minerals (micas), in line with its structural water content. The standards listed above have ^{36}Ar contents (units of 10^{-10} cm^3 STP g^{-1}) of ~10 for MMHb-1 and ~15–20 for PP-20 and NL-25. The A.I. for this mineral could be tied to its K or Ca content. The latter is the better choice, for K contents may vary by a factor of about five. Thus, ^{36}Ar/^{39}Ar values are ~0.0010 (MMHb-1), ~0.0020 (PP-20), and ~0.0095 (NL-25). ^{36}Ar/^{37}Ar values (Fig. 12) are less scattered; MMHb-1 show slightly lower values (~0.00030), than the older PP-20 and NL-25 (~0.00045–0.00060). Stepheating work on MMHb-1 (Harrison, 1981), reveals ^{36}Ar/^{37}Ar (ratio) variation (Fig. 12). Nonplateau steps carrying a very small percentage of the total gas show low ages and high A.I. values (>0.001); the plateau (least-altered) steps show <0.0002. My tentative cut-off A.I. (^{36}Ar/^{37}Ar) value in looking for unaltered samples is 0.0002, for the plateau steps.

DISCUSSION

The techniques outlined above have been utilized to critically evaluate published results for rocks linked to hotspot activity (Baksi, this volume, Chapter 16). The focus therein is to (1) verify plateau sections meet proper statistical requirements and (2) detect alteration by the A.I. technique. For material showing substantial alteration, plateau ages are liable to be underestimates of the time of crystallization. Here, after relating the A.I. method to previous work, I look at (1) important failures to evaluate isotopic results by statistical methods, (2) the mode of applying the atmospheric argon correction and the relationship of the argon isotopes added to silicate phases, (3) the variation of chemical weathering through space and time, and

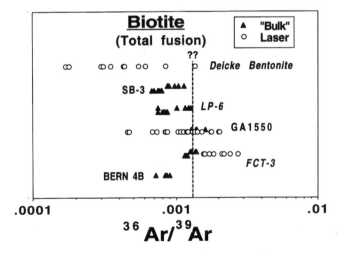

Figure 11. Alteration index (A.I.) values for biotite, ^{36}Ar values normalized to ^{39}Ar, J = 0.0100, K = 8.0%, shown on a log scale. Results shown for total fusion on ~10-mg splits (bulk—blank corrected) and few flakes (laser) on standards. Tentative cut-off for total fusion analyses on fresh material is set at 0.00130 (see text). Data from Baksi et al. (1996); Renne et al. (1998); Min et al. (2001); Daze et al. (2003).

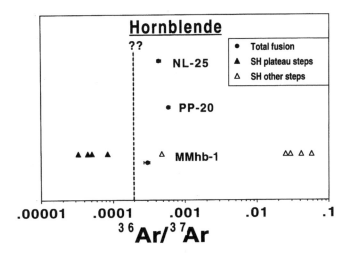

Figure 12. Alteration index (A.I.) for hornblende samples; ^{36}Ar is normalized to ^{37}Ar and J = 0.0100. Results for total fusion (average and associated standard error on the mean) and stepheating on standards, shown on a log scale. For MMhb-1, values >0.0005 represent nonplateau steps; plateau steps show values <0.0002. A tentative cut-off for fresh material is set at 0.0002 for plateau steps (see text). SH—stepheating. Data from Harrison (1981); Renne et al. (1998); and Renne (2000).

(4) the importance of subduction in cycling atmospheric argon back into the mantle.

Extending Previous Work

Over thirty years ago, it was observed that altered wholerock basalts from the ca. 16-Ma Columbia River basalts, United States, contain significantly more atmospheric argon than fresh

material (Baksi, 1974a). Even for very young (ca. 70-ka) material, alteration raised the ^{36}Ar content of the rock significantly (Baksi, 1974b). Roddick (1978) looked at the sources of atmospheric argon in argon dating studies. He emphasized the role of variable blanks in smearing isochron plots (see also Baksi, 1973; Hayatsu and Carmichael, 1977). Roddick (1978) examined the results on whole-rocks from the Ferrar magmatic province, Antarctica (Fleck et al., 1977). It was noted that specimen 64.01 contained much less ^{36}Ar than 27.17; the former apparently gave the proper crystallization age, whereas the latter did not. There was no attempt to quantify the ^{36}Ar content of these samples; instead, the researchers looked at the atmospheric argon content of the various steps; the latter is determined, in part, by the K content and age of the sample. The A.I. measures the ^{36}Ar content quantitatively, or makes use of a nondimensional parameter, for ^{40}Ar/^{39}Ar dating results. Thus, it is noted that though 64.01 and 27.17 give acceptable plateaus (see Fig. 13A), neither correctly reflects the age of crystallization (ca. 180 Ma). The A.I. of these samples explains the different ages—*both* samples are altered (Fig. 13B). The plateau ages of 64.01 and 27.17 underestimate the crystallization age by ~3% and ~15%, respectively. ^{36}Ar contents (in units of 10^{-10} cm^3 STP g^{-1}) are ~29 (overall) and ~23 (plateau steps) for 64.01. For 27.17, the corresponding values are 220 and 160, respectively. These values should be compared to the cut-off for fresh samples of <6. The stepheating procedure shows that both samples are altered pervasively. A third specimen (27.46) gave a statistically acceptable plateau age of ca. 132 Ma, which underestimates the crystallization value by ~30%. The A.I. value of its plateau steps is 0.062, slightly higher than that of the very altered 27.17 (0.02), but not sufficiently high to explain the further lowering

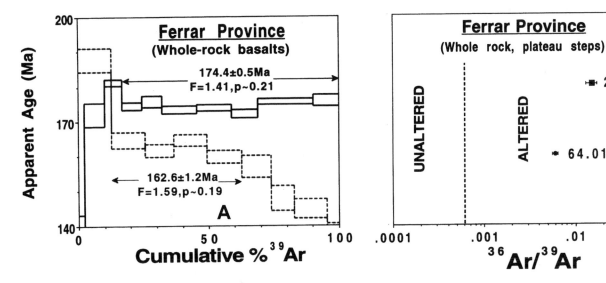

Figure 13. Assessment of the ^{40}Ar/^{39}Ar stepheating work of Fleck et al. (1977) on whole-rock basalts from the Ferrar magmatic province. (A) Age spectra for two samples yielding statistically acceptable plateau ages. Solid lines—64.01; dashed lines—27.17. Symbols and notations are as in Figure 3A. (B) alteration index (A.I.) for the plateau steps, average and associated standard error on the mean shown on a log scale. Both samples are altered and yield incorrect (low) plateau ages.

of the age by ~30 m.y. Not all cases of lowered ages for undisturbed igneous rocks result from alteration. Thin section examination (Fleck et al., 1977) shows that the matrix of 27.46 is made up largely of devitrified glass. Careful petrographic examination of samples prior to dating must remain standard practice.

Further Statistical Observations

Proper utilization of the goodness of fit parameter to plateau/isochrons is often overlooked. The fundamental assumptions underlying the use of the method (e.g., Faure, 1986) are that all samples (1) must be of the same age, (2) should be "comagmatic" —show the same initial isotopic ratio, and (3) must have remained closed to gain or loss of parent/daughter since the time of the event being dated. Three cases are examined.

The first is the dating of a "Permo-Triassic boundary crater" (Becker et al., 2004), which has been questioned (Renne et al., 2004). The age spectrum (Fig. 14A) shows excess statistical scatter; the system is "disturbed" (gain or loss of parent/daughter isotopes). The A.I. plot (Fig. 14B) shows the plagioclase is badly altered, containing about one hundred times too much ^{36}Ar, and cannot yield a valid age. The purported age of 250.1 ± 4.5 Ma (Becker et al., 2004) is arithmetically and statistically invalid, does not take into account the disturbed nature of the sample, and does not accurately elucidate the age of Bedout Crater.

Incorrect use of statistical parameters in other types of isotopic dating is not uncommon. In a Sm-Nd study, Basu et al. (1981) presented a ca. 3800-Ma isochron for rocks from the Singhbum craton, eastern India. Moorbath et al. (1986) questioned this age. $^{40}Ar/^{39}Ar$ dating (Baksi et al., 1987) indicated the batholithic complex was ca. 3300 Ma in age. Subsequently, Sharma et al. (1994) revised the Sm-Nd age to 3292 ± 26 Ma.

The isochron analysis of the Basu et al. (1981; Fig. 15A) shows excess scatter ($p < 0.05$). Considerable geological error is present; rocks collected ~50 km distant from one another are unlikely to be coeval and/or comagmatic—i.e., show the same age and identical initial Nd isotopic ratios at time $t = 0$. Allègre et al. (1999) presented a Re-Os isochron age of 65.60 ± 0.15 Ma for rocks from the Deccan Traps. Samples were collected from locations separated by ~700 km (see Allègre et al., 1999, Fig. 1); such distant rocks are unlikely to be coeval and/or comagmatic. Figure 15B shows the resulting isochron; excess scatter of data is revealed and the age must be rejected as an accurate estimate of the time of extrusion of the Deccan Traps. Gross scatter results from breakdown of all three conditions for isochron plots. Two sets of replicate analysis on a sample (see Figure 15B, inset) do not define a single point on the isochron, and the tie-line between the replicate analysis of AP1030 runs oblique to the isochron. The sample in question has suffered gain or loss of Re/Os since ca. 65 Ma.

The mathematics underlying probability theory (and hence statistics) was placed on a firm footing by the work of Pierre Fermat and Blaise Pascal ~350 years ago. In a letter dated July 29, 1654, Pascal (1952, p. 475) wrote to Fermat, "I can no longer doubt the truth of my results after finding myself in such wonderful agreement with you." Also, "I see indeed that truth is the same at Toulouse and at Paris." Statistical truths are universal, and we disregard them at our peril.

Weathering Profiles—Cryptomelane

The search for datable minerals in weathering profiles has drawn considerable attention, with cryptomelane being the material of choice. I examine some aspects of one such effort (Feng

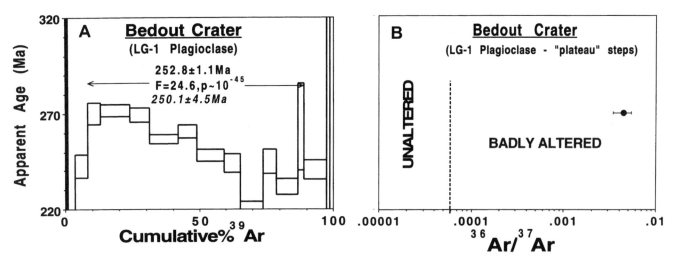

Figure 14. Critical examination of $^{40}Ar/^{39}Ar$ dating results on the Bedout Crater (Becker et al., 2004). (A) Age spectrum plot for a plagioclase separate, with the authors' plateau section delineated by arrows and listed in italics. Symbols and legends as in Figure 3A. Statistical evaluation yields values listed in bold letters. Excess scatter is present, and no accurate age was recovered. (B) The average alteration index (A.I.) and standard error on the mean of plateau steps, shown on a log scale. The specimen is grossly altered, showing about one hundred times more ^{36}Ar than found in fresh material, and cannot be expected to give an accurate crystallization age.

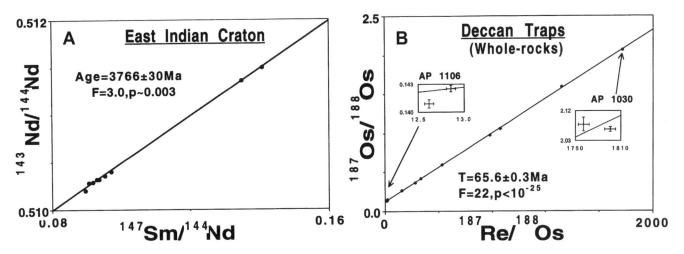

Figure 15. (A) Sm-Nd whole-rock isochron for the Singbhum craton, eastern India (Basu et al., 1981). Symbols as in Figure 3A. The goodness of fit parameter (F) and associated probability value indicate geological error is present. The ca. 3800-Ma age was subsequently corrected to ca. 3300 Ma. (B) Re-Os whole-rock isochron for the Deccan Traps, India (after Allègre et al., 1999). Symbols as in Figure 3A. The goodness of fit parameter shows gross geological error and negates the high-precision "age" as an accurate value for the time of formation of the Deccan Traps. Replicate analyses of individual samples (insets) show excess scatter and suggest postcrystallization movement of Re and/or Os.

and Vasconcelos, 2001); $^{40}Ar/^{39}Ar$ stepheating experiments were carried out on <1-Ma samples from Queensland, Australia. The method is, at best, a semiquantitative tool for arriving at soil formation rates. Eighteen age spectra were used to arrive at plateau ages by Feng and Vasconcelos (2001)—see their Figure 2. Six of these involve two steps each and, by definition, cannot yield plateau ages. Another ten fail the statistical (F value) test to be used as accurate estimates of the time of formation. Cryptomelane's chemical composition, $KMn_8O_{16}.nH_2O$, suggests it should display high A.I. values. $^{36}Ar/^{39}Ar$ ratios for the plateau steps fall in the range 0.0004–0.0010, about the same as unaltered whole-rock material (<0.0006). Cryptomelane does not contain structural water; during overnight bakeout at ~200 °C under high vacuum conditions, the water is lost, and ^{36}Ar levels (and hence A.I. values) fall to low levels.

Weathering and the Paleoatmosphere

One of the most interesting problems addressed by argon workers is to recover points on the von Weiszacker trajectory. This curve is the $^{40}Ar/^{36}Ar$ ratio in the Earth's atmosphere from ca. 4.5 Ga to today, as it has increased from <1 (cf. Renne, 2000) to 295.5 (Nier, 1950). In 1937, Carl von Weiszacker noted the presence of "excess" ^{40}Ar in the Earth's atmosphere, which he suggested resulted from the escape of this isotope from rocks and minerals, wherein it was produced by the (then-unknown) decay of ^{40}K (Dalrymple and Lanphere, 1969). The search for ancient atmospheric argon in rocks and minerals has been directed at material that has interacted substantially with the atmosphere at the Earth's surface. I look at the dating of pyroxene separates from the cumulate zone of a layered tholeiitic sill in the Canadian Shield (Hanes et al., 1985). A section of their

stepheating results gave an isochron age of 2.7 Ga and an initial argon isotopic ratio of 258 ± 3 (Fig. 16A). Hanes et al. (1985, p. 956) argued for deuteric amphibole in these mantle-derived rocks formed by "interaction of circulating seawater readily available in the submarine volcanic environment." Further, "this seawater would carry argon equilibrated with the atmosphere, and thus the trapped initial argon in the amphibole . . . provide(s) information on the atmospheric $^{40}Ar/^{36}Ar$ ratio at . . . 2.7 Ga." I examine the A.I. of their deuteric amphibole normalized to ^{39}Ar (Fig. 16B). It shows the degree of alteration (hydration) of anhydrous phases, with about a twenty-fold increase in ^{36}Ar levels, resulting from interaction of the pre-existing silicate phases with seawater. The work of Hanes et al. (1985) yields an important point on the von Weiszacker trajectory.

Chemical Weathering (Alteration) through Space and Time

The intensity of chemical weathering is dependant on climate (geographic location). Foland et al. (1993) presented results on the Kirkpatrick Basalt, Antarctica. A plagioclase separate and a glass concentrate yield plateau ages of 179.4 ± 0.4 and 179.2 ± 0.2 Ma, respectively (Fig. 17). Both show substantial loss of $^{40}Ar*$. Their results are evaluated for freshness of the material dated using the A.I. method (Fig. 17). The plagioclase sample shows A.I. values falling by three orders of magnitude as the laboratory extraction temperature is raised. The high-temperature steps, drawing gas from unaltered section, yield a plateau age. The glass concentrate shows a similar drop in A.I. during the stepheating. Stepheating work on acid-untreated samples can recover crystallization ages for samples from the interior (chemically unaltered) sections of the crystals on high-latitude samples. Further $^{40}Ar/^{39}Ar$ work on Antarctic samples

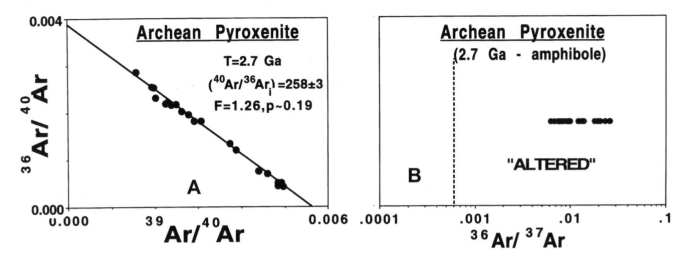

Figure 16. Sampling the paleoatmosphere at 2.7 Ga. Analysis of a pyroxenite (whole-rock) by Hanes et al. (1985). (A) Twenty-two steps from stepheating experiments lie on an isochron (acceptable statistics) with an age of 2.7 Ga, initial Ar ratio of 258. Symbols and legends as in Figure 3A. Ar was derived from deuteric amphibole formed by the rock's interaction with seawater at 2.7 Ga (Hanes et al., 1985). (B) Alteration index (A.I.) normalized to ^{39}Ar for the isochron steps, shown on a log scale. The pyroxenite has been grossly altered.

should confirm low levels of alteration. Knight and Renne (2005) report ^{40}Ar/^{39}Ar plateau ages spanning ca. 180–174 Ma for the Ferrar Dolerite; these could represent emplacement ages and/or cooling ages for plagioclase. Alternately, magmatism may have been narrowly focused at ca. 180 Ma; if so, the A.I. of ca. 180-Ma samples should be markedly lower than those yielding the younger ages.

High-temperature K-feldspars are resistant to chemical weathering. The A.I. of material of differing ages, from ca. 2 ka to 1.1 Ga, are normalized for (1) the level of neutron irradiation (J = 0.0100) and (2) K = 10.0% (see Fig. 18). A clear pattern emerges—alteration appears to increase with the time of exposure. An interesting test would be to carry out ^{40}Ar/^{39}Ar step-

heating work on > 200 Ma K-feldspar, varying the intensity of the HF leaching procedure.

The reasons underlying the roughly fifty-fold increase in ^{36}Ar content of ca. 1.1-Ga K-feldspar, compared to the ca. 2-ka one (Fig. 18) will be considered in detail elsewhere. It may result from chemical weathering (alteration), either in an episodic manner, or slowly over the life of the mineral. In either case, the ^{40}Ar/^{36}Ar ratio used to apply the atmospheric argon correction should be <295.5. This reduced value would lower the amount of atmospheric ^{40}Ar in such material, raising its age correspondingly. The latter amounts to ca. 1–2 Ma for the ca. 1.1-Ga K-feldspar, not sufficient to explain the difference between the U-Pb and ^{40}Ar/^{39}Ar ages noted by Min et al. (2000). A critical,

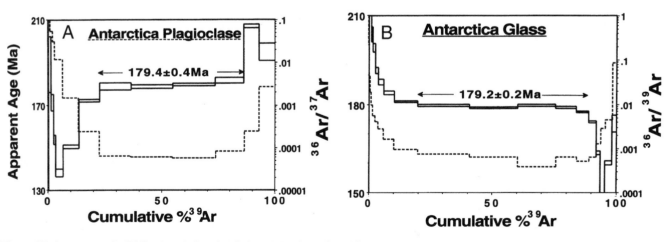

Figure 17. Age spectra (solid lines) and alteration index (A.I.) plots (dotted lines on a log scale) for (A) a plagioclase separate and (B) a glass concentrate sample from Antarctica (data from Foland et al., 1993). Each yields a good plateau age of ca. 180 Ma. Altered low-temperature sites show high A.I. values. The plateau steps, drawn from fresh sites, show about one thousand times less ^{36}Ar.

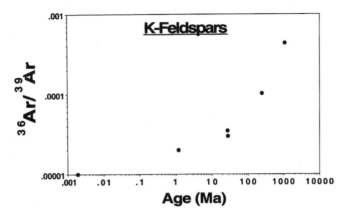

Figure 18. Alteration index (A.I.) of K-feldspars giving plateau ages from different localities, shown on a log-log scale. An increase in A.I. with age is noted. Data points used: 1.9 ka—79 A.D. Vesuvius eruption (Renne et al., 1997); 1.2 Ma—Alder Creek Rhyolite (Nomade et al., 2005); 28.0 and 28.3 Ma—Fish Canyon Tuff and Taylor Creek Rhyolite (Renne et al., 1998); 250 Ma—Permo-Triassic boundary ash (Renne et al., 1995); 1088 Ma—Palisade Rhyolite (clear grains; Min et al., 2000).

though unresolved, problem is whether chemical weathering of samples over tens of millions of years causes partial loss of $^{40}Ar^*$. Thus, the ca. 1.1-Ga and ca. 250-Ma K-feldspars shown in Figure 18 are ~1% younger than U-Pb ages on these same rocks/ash bed (Renne et al., 1995; Min et al., 2000). These $^{40}Ar/^{39}Ar$ ages are reported relative to the calibrations of Renne et al. (1998), which yield ages that are older than other intercalibration studies (Roddick, 1983; Lanphere et al., 1990; Baksi et al., 1996). Further, the age of GA1550 biotite, the primary standard used by Renne et al. (1998), has a K determination made by isotope dilution technique, but the absolute argon calibration has been questioned (Lanphere and Baadsgaard, 2004). Efforts to revise the decay constants used for argon dating (e.g., Min et al., 2000; Kwon et al., 2002) need to be augmented by other considerations. First is the possible lowering of argon ages by alteration for samples >200 Ma in age. Second, the absolute age(s) of standards used for K-Ar dating need to be worked out more fully (cf. Lanphere and Baadsgaard, 2004). Finally, the possibility of ^{39}Ar recoil loss from standards—thereby making the ages used for K-Ar work and $^{40}Ar/^{39}Ar$ dating unequal (Baksi et al., 1996)—needs to be examined.

The Role of Argon in Subduction Zone–Related Studies

Large amounts of argon dissolved in ocean water are recycled into the Earth's interior at subduction zones (e.g., Sumino et al., 2005). Order of magnitude calculations suggest that plate tectonics of the current style over ~4 b.y. would recycle a substantial fraction of the Earth's atmosphere. Large amounts of atmospheric argon are added to silicates in the mantle wedge by the dehydration process. This process will raise the level of ^{36}Ar seen in magmas so generated, and in the mafic and intermediate rocks formed.

Over three decades ago, attempts were made to look for rocks showing $(^{40}Ar/^{36}Ar)_I$ differing from the atmospheric value (Dalrymple, 1969; Krummenacher, 1970; Baksi, 1974a,b). The Mt. Sakurajima, Japan, lavas of 1951 (subduction zone area) showed $(^{40}Ar/^{36}Ar)_I$ ~292 (Dalrymple, 1969); this value is measurably lower than the atmospheric value of 295.5 (Nier, 1950).

Cherdyntsev and Shitov (1967) noted that gases released from volcanoes in the Kamchatka area show ^{36}Ar in excess of that in the atmosphere by 1–6%. They suggested that argon from Earth's ancient atmosphere was stored in the upper mantle in rocks and released when the rocks are melted. Both rocks and volcanic gases from areas close to subduction zones may then show "excess" ^{36}Ar. Water recycled into the mantle in such areas carries ~1 × 10^{-6} cm^3 STP g^{-1} of ^{36}Ar (~3–4 km depth, 4 °C; Bieri et al., 1968). As the subducted slab heats up to ~50 °C, loss of ~50% of the dissolved argon may be expected, based on the temperature alone (Bieri et al., 1968; Hamme and Emerson, 2004). Increased pore pressure may hold the argon loss down to lower levels. As the subducted slab dehydrates, substantial quantities of ^{36}Ar (>10^{-7} cm^3 STP g^{-1}?) are carried into the overlying mantle wedge. Some of this gas will escape as volcanic emanations, possibly showing $^{40}Ar/^{36}Ar$ ratios <295.5, as noted in the Kamchatka area (Cherdyntsev and Shitov, 1967). As magmatism is induced in these areas, kinetic effects in the rock-water interaction (see Behrens and Zhang, 2001, and references therein) could result in isotopic fractionation favoring the lighter isotopes. Lavas (mafic and, more notably, those of intermediate composition) should show high levels of ^{36}Ar and possibly $(^{40}Ar/^{36}Ar)_I$ <295.5.

Argon studies should be carried out on fresh lavas from some key areas (Japan? Central America?) where a wealth of trace-element geochemical analyses are available. Low levels of ^{36}Ar (<10^{-9} cm^3 STP g^{-1}) should be seen for rocks (basalts?) showing no Nb-Ta anomalies on spidergram plots. By contrast (intermediate?) rocks with spidergrams displaying troughs at Nb-Ta should show higher amounts of ^{36}Ar. A useful parameter to monitor is the $^{36}Ar/^{38}Ar$ ratio of these rocks. Andesites incorporating material from the subducted slab may show argon isotopic fractionation effects (i.e., $^{36}Ar/^{38}Ar$ ratios >5.35, the atmospheric argon value; Nier, 1950).

CONCLUSIONS

Incorrect low argon ages for undisturbed igneous rocks commonly result from alteration (chemical weathering) of the phases used for dating, alteration not evident using the methods routinely used for such purposes. Optical methods are qualitative and subjective in nature. A quantitative method for assessing the alteration of material utilized for argon dating has been developed, making use of the data routinely gathered during mass spectrometry. This method looks at the critical role water plays in chemical weathering processes. Large amounts of dissolved atmospheric argon in the water interact with the very low levels seen in rocks, resulting in the ^{36}Ar levels in the silicate phases being raised by factors of ten to one thousand; all altered phases are hydrous and consequently exhibit high levels of ^{36}Ar. The method has been applied with success to K-Ar dating results (Baksi, 1987, 1989; Baksi and Hoffman, 2000). Here it is extended to ^{40}Ar/^{39}Ar dating studies, using nondimensional parameters. The A.I. involves looking at the ^{36}Ar/^{39}Ar ratio for relatively K-rich material (K-feldspar, whole-rock basalt, and micas) and at the ^{36}Ar/^{37}Ar ratio for Ca-rich material (plagioclase feldspar and hornblende). These parameters must be normalized for the total production of ^{39}Ar and ^{37}Ar, using a single value for the irradiation parameter (J = 0.0100 here) and for a single K value for a given mineral or whole-rock work. The use of this technique may be problematical where substantial amounts of atmospheric argon are added to the samples during the neutron irradiation procedure (e.g., Roddick, 1983).

Chemical weathering (alteration) is strongly dependent on both location and the age of the sample. Water is known to be pervasive through much of the crust, and most material, regardless of age, will show alteration. Selected ca. 200-Ma rocks from Antarctica show low degrees of alteration. ^{36}Ar contents for K-feldspar appears to be time dependent. It is critical that all argon data be reviewed for alteration using the A.I. test. As shown here, it is possible to obtain statistically valid plateau ages that underestimate crystallization ages by ~2–15% ("[not] all ^{40}Ar/^{39}Ar ages are equal[ly valid]; some are more equal than others"). Results on seafloor rocks are considered in the companion article (Baksi, this volume, Chapter 16). For whole-rock basalts and plagioclase feldspar, raising the A.I. to more than five times the cut-off value for freshness lowers the argon age by a few percent; greater amounts of alteration can lower argon ages by ~10%, though still yielding statistically acceptable plateau sections in ^{40}Ar/^{39}Ar stepheating studies.

Large amounts of argon dissolved in seawater are recycled back into the Earth's interior at subduction zones. These areas serve to transfer ^{36}Ar from the atmosphere back into the mantle. Dehydration of the subducted slab and subsequent melting of rocks in the mantle wedge play a critical role in magmatism in such areas. Careful measurement of the level of ^{36}Ar in fresh island arc magmatic material should provide a sensitive tool for elucidating the nature and role of the fluids driven off during the subduction process. Rocks formed by incorporation of material from the subducted slab should show relatively high ^{36}Ar levels and perhaps ^{36}Ar/^{36}Ar ratios >5.35. Andesites formed by Bowen type fractional crystallization of mafic magmas should show lower amounts of ^{36}Ar and ^{36}Ar/^{38}Ar ratios of ~5.35.

ACKNOWLEDGMENTS

I thank the editors of this volume for this opportunity to express some thoughts on improving the precision and accuracy of argon ages. This article is dedicated to family members, for their encouragement and support. Critical comments and suggestions made by Gillian Foulger and the official reviewers shaped this manuscript. I express my most sincere gratitude to Brent Dalrymple; his work and unselfish willingness to extend help at all times serve as an inspiration.

APPENDIX

I illustrate the mode of calculating the alteration index for anhydrous silicate phases in ^{40}Ar/^{39}Ar dating studies.

For whole-rock basalt and K-feldspar:

$$A.I. = [(^{36}Ar/^{39}Ar)_M - (^{36}Ar/^{37}Ar)_{Ca} \times (^{37}Ar/^{39}Ar)_C] \times (J/0.01) \times (B/D),$$

where the term in square brackets corrects for ^{36}Ar derived from Ca isotopes during neutron irradiation; $(^{36}Ar/^{37}Ar)_{Ca}$ is a reactor constant, ~0.00025–0.00028; $(^{36}Ar/^{39}Ar)_M$ is the measured isotope ratio; $(^{37}Ar/^{39}Ar)_C$ is the measured isotope ratio, corrected for decay of ^{37}Ar; B is the K content (%) of specimen; and D is the K content for normalization purposes, 0.65% for whole-rock material and 10.0% for K-feldspar.

Whole-Rock Basalt

The cut-off value for fresh material is A.I. <0.0006 (see Fig. 4):

1. *Unaltered.* Andesitic basalt RM82–11 (Mahoney et al., 1983). Specimen VI, see Figure 4. Dated at Queen's University, Kingston, Ontario, Canada (A.K. Baksi, unpubl. data, 2001). $(^{36}Ar/^{37}Ar)_{Ca} = 0.000254$, J = 0.00686, K = 0.75%. (Plateau) step 6, age ca. 116 Ma. $(^{36}Ar/^{39}Ar)_M = 0.00077$, $(^{37}Ar/^{39}Ar)_C = 1.515$. A.I. = [0.00077 – 0.000254 × 1.515] × (0.00686/0.01) × (0.75/0.65) = 0.00030.

2. *Altered.* Whole-rock basalt 64.01 of Fleck et al. (1977). See Figure 13. $(^{36}Ar/^{37}Ar)_{Ca} = 0.000265$, J = 0.00775, K = 1.45%. 780 °C (plateau) step, age ca. 174 Ma. $(^{36}Ar/^{39}Ar)_M = 0.0025$, $(^{37}Ar/^{39}Ar)_C = 1.372$. A.I. = [0.0025 – 0.000265 × 1.372] × (0.00775/0.01) × (1.45/0.65) = 0.0037.

Note the ten-fold increase in A.I. for the altered sample, compared to unaltered RM82–11.

K-Feldspar

A single cut-off value for this mineral is not proposed, as ^{36}Ar contents appear to be age dependant (see text).

3. *Unaltered (clear grain).* The Palisade Rhyolite (Min et al., 2000); see Figure 10. $(^{36}Ar/^{37}Ar)_{Ca} \sim 0.00027$, J = 0.01518, K ~7%. Age 1088 Ma. (Plateau) step 30692–201C. $(^{36}Ar/^{39}Ar)_M = 0.00050$, $(^{37}Ar/^{39}Ar)_C = 0.0000$. A.I. = [(0.00050 – 0.00027 × 0.0)] × (0.01518/0.01) × (7/10) = 0.00053.

4. *Altered (cloudy grains).* The Palisade Rhyolite (Min et al., 2000); see Figure 10. $(^{36}Ar/^{37}Ar)_{Ca} \sim 0.00027$, J = 0.01518,

K ~7%. Age ca. 1068 Ma. (Plateau) step 30392–23J. $({}^{36}Ar/{}^{39}Ar)_M = 0.00399$, $({}^{37}Ar/{}^{39}Ar)_C = 0.023$. A.I. = $[0.00399 - 0.00027 \times 0.023] \times (0.01518/0.01) \times (7/10) = 0.00424$.

Note the eight-fold increase in A.I. and rise in the Ca/K $({}^{37}Ar/{}^{39}Ar)_C$ ratio for the altered material.

Plagioclase Feldspar

A cut-off value for fresh material is set at A.I. <0.00006 (see Fig. 7A):

$$A.I. = [({}^{36}Ar/{}^{39}Ar)_M - ({}^{36}Ar/{}^{37}Ar)_{Ca} \times ({}^{37}Ar/{}^{39}Ar)_C] \times (J/0.01) / ({}^{37}Ar/{}^{39}Ar)_C.$$

Note that Ca/K ~2 × $({}^{37}Ar/{}^{39}Ar)_C$ for all reactor facilities:

5. *Unaltered.* Central Atlantic magmatic province plagioclase (Hames et al., 2000), specimen SCD-2C (see Fig. 7). $({}^{36}Ar/{}^{37}Ar)_{Ca} = 0.00027$, J = 0.001734. 680 °C (plateau) step, age ca. 200 Ma. $({}^{36}Ar/{}^{39}Ar)_M = 0.0102$, $({}^{37}Ar/{}^{39}Ar)_C = 22.43$. A.I. = $[0.0102 - 0.00027 \times 22.43] \times (0.001734/0.01)/(22.43) = 0.000032$.

6. *Altered.* Central Atlantic magmatic province plagioclase (Deckart et al., 1997), GUI35 (see Fig. 9). $({}^{36}Ar/{}^{37}Ar)_{Ca} = 0.00027$, J = 0.0307. 830 °C (plateau) step, age ca. 194 Ma. $({}^{36}Ar/{}^{39}Ar)_M = 0.00756$, $({}^{37}Ar/{}^{39}Ar)_C = 10.65$. A.I. = $[0.00756 - 0.00027 \times 10.65] \times (0.0307/0.01)/(10.65) = 0.00135$.

Note the roughly forty-fold increase in A.I. for the altered sample. SCD-2C was leached with HF prior to dating, whereas GUI35 was not treated with acid.

REFERENCES CITED

Allègre, C.J., Birck, J.L., Capmas, F., and Courtillot, V., 1999, Age of the Deccan Traps using $^{187}Re/^{187}Os$ systematics: Earth and Planetary Science Letters, v. 170, p. 197–204, doi: 10.1016/S0012-821X(99)00110-7.

Baksi, A.K., 1973, K-Ar dating: Loading techniques in argon extraction work and sources of air contamination: Canadian Journal of Earth Sciences, v. 10, p. 1678–1684.

Baksi, A.K., 1974a, Isotopic fractionation of a loosely held atmospheric argon component in the Picture Gorge basalts: Earth and Planetary Science Letters, v. 21, p. 431–438, doi: 10.1016/0012-821X(74)90183-6.

Baksi, A.K., 1974b, K-Ar study of the SP Flow: Canadian Journal of Earth Sciences, v. 11, p. 1350–1356.

Baksi, A.K., 1982, ^{40}Ar-^{39}Ar incremental heating studies on a suite of "disturbed" rocks from Cape Breton Island, Canada: Detection of a deformational episode: Journal of the Geological Society of India, v. 23, p. 267–276.

Baksi, A.K., 1987, Critical evaluation of the age of the Deccan Traps, India: Implications for flood-basalt volcanism and faunal extinction: Geology, v. 15, p. 147–150, doi: 10.1130/0091-7613(1987)15<147:CEOTAO>2.0.CO;2.

Baksi, A.K., 1989, Reevaluation of the timing and duration of extrusion of the Imnaha, Picture Gorge and Grande Ronde basalts, Columbia River basalt group, *in* Reidel, S.P., and Hooper, P.R., eds., Volcanism and tectonism in the Columbia River flood-basalt province: Boulder, Colorado, Geological Society of America Special Paper 239, p. 105–111.

Baksi, A.K., 1994, Concordant seafloor spreading rates from geochronology, astrochronology and space geodesy: Geophysical Research Letters, v. 21, p. 133–136, doi: 10.1029/93GL03534.

Baksi, A.K., 1995, Fine-tuning the radiometrically derived geomagnetic polarity time scale for 0–10 Ma: Geophysical Research Letters, v. 22, p. 457–460, doi: 10.1029/94GL03214.

Baksi, A.K., 1999, Revaluation of plate motion models based on hotspot tracks in the Atlantic and Indian oceans: Journal of Geology, v. 107, p. 13–26, doi: 10.1086/314329.

Baksi, A.K., 2001, The Rajahmundry Traps, Andhra Pradesh; Evaluation of their petrogenesis relative to the Deccan Traps: Proceedings of the Indian Academy of Sciences, v. 110, p. 397–407.

Baksi, A.K., 2003, Critical evaluation of $^{40}Ar/^{39}Ar$ ages for the central Atlantic magmatic province: Timing, duration and possible migration of magmatic centers, *in* Hames, W.E., et al., eds., The central Atlantic magmatic province: Insights from fragments of Pangea: Washington, D.C., American Geophysical Union Geophysical Monograph 136, p. 77–90.

Baksi, A.K., 2005a, Evaluation of radiometric ages pertaining to rocks hypothesized to have been derived by hotspot activity, in and around the Atlantic, Indian and Pacific oceans, *in* Foulger, G.R., et al., eds., Plates, plumes and paradigms: Boulder, Colorado, Geological Society of America Special Paper 388, p. 55–70.

Baksi, A.K., 2005b, $^{40}Ar/^{39}Ar$ dating of the Rajahmundry Traps, Eastern India, and their relationship to the Deccan Traps: Discussion: Earth and Planetary Science Letters, v. 239, p. 368–373.

Baksi, A.K., 2007 (this volume, Chapter 16), A quantitative tool for detecting alteration in undisturbed rocks and minerals—II: Application to argon ages related to hotspots, *in* Foulger, G.R., and Jurdy, D.M., eds., Plates, plumes, and planetary processes: Boulder, Colorado, Geological Society of America Special Publication 430, doi: 10.1130/2007.2430(16).

Baksi, A.K., and Archibald, D.A., 1997, Mesozoic igneous activity in the Maranhao province, northern Brazil; $^{40}Ar/^{39}Ar$ evidence for separate episodes of magmatism: Earth and Planetary Science Letters, v. 151, p. 139–153, doi: 10.1016/S0012-821X(97)81844-4.

Baksi, A.K., and Hoffman, K.A., 2000, On the age and morphology of the Réunion event: Geophysical Research Letters, v. 27, p. 2997–3000, doi: 10.1029/2000GL008536.

Baksi, A.K., Archibald, D.A., Sarkar, S.N., and Saha, A.K., 1987, ^{40}Ar-^{39}Ar incremental heating study of mineral separates from the Early Archaen east Indian craton: Implications for the thermal history of a section of the Singbhum Granite batholithic complex: Canadian Journal of Earth Sciences, v. 24, p. 1985–1993.

Baksi, A.K., Hsu, V., McWilliams, M.O., and Farrar, E., 1992, $^{40}Ar/^{39}Ar$ dating of the Brunhes-Matuyama geomagnetic field reversal: Science, v. 256, p. 356–357, doi: 10.1126/science.256.5055.356.

Baksi, A.K., Byerly, G.R., Chan, L.H., and Farrar, E., 1994, Intracanyon flows in the Deccan Province, India—Case-history of the Rajahmundry Traps: Geology, v. 22, p. 605–608.

Baksi, A.K., Hoffman, K.A., and Farrar, E., 1993, A new calibration point for the late Miocene section of the geomagnetic polarity time scale: $^{40}Ar/^{39}Ar$ dating of lava flows from Akaroa volcano, New Zealand: Geophysical Research Letters, v. 20, p. 667–670.

Baksi, A.K., Archibald, D.A., and Farrar, E., 1996, Intercalibration of $^{40}Ar/^{39}Ar$ dating standards: Chemical Geology, v. 129, p. 307–324, doi: 10.1016/0009-2541(95)00154-9.

Basu, A.R., Ray, S.L., Saha, A.K., and Sarkar, S.N., 1981, Eastern Indian 3800-million-year-old crust and early mantle differentiation: Science, v. 212, p. 1502–1506, doi: 10.1126/science.212.4502.1502.

Becker, L., Poreda, R.J., Basu, A.R., Pope, K.O., Harrison, T.M., Nicholson, C., and Isaky, R., 2004, Bedout: A possible end Permian impact crater offshore northwestern Australia: Science, v. 304, p. 1469–1476, doi: 10.1126/science.1093925.

Behrens, H., and Zhang, Y., 2001, Ar diffusion in hydrous silicate melts: Implications for volatile diffusion mechanisms and fractionation: Earth and Planetary Science Letters, v. 192, p. 363–376, doi: 10.1016/S0012-821X (01)00458-7.

Berger, G.W., and York, D., 1970, Precision of the $^{40}Ar/^{39}Ar$ dating technique: Earth and Planetary Science Letters, v. 9, p. 39–44, doi: 10.1016/0012-821X(70)90021-X.

Bieri, R.H., Koide, M., and Goldberg, E.D., 1968, Noble gas contents of marine waters: Earth and Planetary Science Letters, v. 4, p. 329–340, doi: 10.1016/0012-821X(68)90060-5.

Cherdyntsev, V.V., and Shitov, Y.U., 1967, ^{36}Ar excess in volcanic and postvolcanic gases of the USSR: Geokhimiya, v. 5, p. 618–620.

Courtillot, V., Besse, J., Vandamme, D., Montigny, R., Jaeger, J.-J., and Capetta, H., 1986, Deccan flood basalts at the Cretaceous-Tertiary boundary?: Earth and Planetary Science Letters, v. 80, p. 361–374, doi: 10.1016/0012-821X (86)90118-4.

Dalrymple, G.B., 1969, $^{40}Ar/^{36}Ar$ analyses of historic lava flows: Earth and Planetary Science Letters, v. 6, p. 47–55, doi: 10.1016/0012-821X(69) 90160-5.

Dalrymple, G.B., 1991, The age of the Earth: Stanford, California, Stanford University Press, 474 p.

Dalrymple, G.B., and Clague, D.A., 1976, Age of the Hawaiian-Emperor bend: Earth and Planetary Science Letters, v. 31, p. 313–329, doi: 10.1016/0012-821X(76)90113-8.

Dalrymple, G.B., and Lanphere, M.A., 1969, Potassium-Argon dating: San Francisco, Freeman, 258 p.

Daze, A., Lee, J.K.W., and Villeneuve, M., 2003, An intercalibration study of the Fish Canyon sanidine and biotite $^{40}Ar-^{39}Ar$ standards and some comments on the age of the Fish Canyon Tuff: Chemical Geology, v. 199, p. 111–127, doi: 10.1016/S0009-2541(03)00079-2.

Deckart, K., Feraud, G., and Bertrand, H., 1997, Age of Jurassic continental tholeiites of French Guyana/Suriname and Guinea: Implications to the opening of the central Atlantic Ocean: Earth and Planetary Science Letters, v. 150, p. 205–220, doi: 10.1016/S0012-821X(97)00102-7.

Devigne, J.E., 1998, Atlas of micromorphology of mineral alteration and weathering: Canadian Mineralogist Special Publication 3, 494 p.

Dunning, G.R., and Hodych, J.P., 1990, U/Pb zircon and baddeleyite ages for the Palisades and Gettysburg sills of the northeastern United States: Implications for the age of the Triassic/Jurassic boundary: Geology, v. 18, p. 795–798, doi: 10.1130/0091-7613(1990)018<0795:UPZABA>2.3.CO;2.

Faure, G., 1986, Principles of isotope geology: New York, John Wiley and Sons, 589 p.

Feng, Y., and Vasconcelos, P., 2001, Quaternary continental weathering geochronology by laser-heating $^{40}Ar/^{39}Ar$ analysis of supergene cryptomelane: Geology, v. 29, p. 635–638, doi: 10.1130/0091-7613(2001)029 <0635:QCWGBL>2.0.CO;2.

Fleck, R.J., Sutter, J.F., and Elliott, D.H., 1977, Interpretation of discordant $^{40}Ar/^{39}Ar$ age-spectra of Mesozoic tholeiites from Antarctica: Geochimica et Cosmochimica Acta, v. 41, p. 15–32, doi: 10.1016/0016-7037(77)90184-3.

Foland, K.A., Fleming, T.H., Heimann, A., and Elliott, D.H., 1993, Potassium-argon dating of fine-grained basalts with massive Ar loss: Application of the $^{40}Ar/^{39}Ar$ technique to plagioclase and glass from the Kirkpatrick Basalt, Antarctica: Chemical Geology, v. 107, p. 173–190, doi: 10.1016/0009-2541(93)90109-V.

Glen, W., 1982, The road to Jaramillo: Stanford, California, Stanford University Press, 459 p.

Hall, C.M., and York, D., 1978, K-Ar and $^{40}Ar/^{39}Ar$ age of the Laschamp geomagnetic polarity reversal: Nature, v. 274, p. 462–464, doi: 10.1038/274462a0.

Hames, W.E., Renne, P.R., and Ruppel, C., 2000, New evidence for geologically instantaneous emplacement of the earliest Jurassic central Atlantic magmatic province basalts of the North American margin: Geology, v. 28, p. 859–862, doi: 10.1130/0091-7613(2000)28<859:NEFGIE>2.0.CO;2.

Hamme, R.C., and Emerson, S.R., 2004, The solubility of neon, nitrogen and argon in distilled water and seawater: Deep-Sea Research, v. 51, p. 1517–1528, doi: 10.1016/j.dsr.2004.06.009.

Hanes, J.A., York, D., and Hall, C.M., 1985, An $^{40}Ar/^{39}Ar$ geochronological and electron microprobe investigation of an Archean pyroxenite and its bearing on ancient atmospheric compositions: Canadian Journal of Earth Sciences, v. 22, p. 947–958.

Harland, W.B., Armstrong, R.L., Cox, A.V., Craig, L.E., Smith, A.G., and Smith, D.G., 1990, A geologic time scale 1989: Cambridge, Cambridge University Press, 263 p.

Harrison, T.M., 1981, Diffusion of ^{40}Ar in hornblende: Contributions to Mineralogy and Petrology, v. 78, p. 324–331, doi: 10.1007/BF00398927.

Hart, S.R., 1964, The petrology and isotopic-mineral age relations of a contact zone in the Front Range, Colorado: Journal of Geology, v. 72, p. 493–525.

Hayatsu, A., and Carmichael, C.M., 1970, K-Ar isochron method and initial argon ratios: Earth and Planetary Science Letters, v. 8, p. 71–76, doi: 10.1016/0012-821X(70)90102-0.

Hayatsu, A., and Carmichael, C.M., 1977, Removal of atmospheric argon contamination and use and misuse of the K-Ar isochron method: Canadian Journal of Earth Sciences, v. 14, p. 337–345.

Hayatsu, A., and Palmer, H.C., 1975, K-Ar isochron study of the Tudor Gabbro, Grenville Province, Ontario: Earth and Planetary Science Letters, v. 25, p. 208–212, doi: 10.1016/0012-821X(75)90197-1.

Kaneoka, I., 1972, The effect of hydration on the K/Ar ages of volcanic rocks: Earth and Planetary Science Letters, v. 14, p. 216–220, doi: 10.1016/0012-821X(72)90009-X.

Kellas, A., 1895, On the percentage of argon in atmospheric and respired air: Proceedings of the Royal Society of London, v. 59, p. 66–68.

Knight, K.B., and Renne, P.R., 2005, Evidence for extended (5–10 Ma) emplacement of Ferrar Dolerite from $^{40}Ar-^{39}Ar$ geochronology: EOS (Transactions, American Geophysical Union), v. 85, abstract v23A-0684.

Knight, K.B., Renne, P.R., Halkett, A., and White, N., 2003, $^{40}Ar/^{39}Ar$ dating of the Rajahmundry Traps, Eastern India, and their relationship to the Deccan Traps: Earth and Planetary Science Letters, v. 208, p. 85–99, doi: 10.1016/S0012-821X(02)01154-8.

Krummenacher, D., 1970, Isotopic composition of argon in modern surface volcanic rocks: Earth and Planetary Science Letters, v. 8, p. 109–117, doi: 10.1016/0012-821X(70)90159-7.

Kwon, J., Min, K., Bickel, P.J., and Renne, P.R., 2002, Statistical methods for jointly estimating the decay constant of ^{40}K and the age of a dating standard: Mathematical Geology, v. 34, p. 457–474, doi: 10.1023/A:1015035228810.

Lanphere, M.A., and Baadsgaard, H., 2004, Reply to comment on "Precise K–Ar, $^{40}Ar/^{39}Ar$, Rb– Sr and U–Pb mineral ages from the 27.5 Ma Fish Canyon Tuff reference standard": Chemical Geology, v. 211, p. 389–390, doi: 10.1016/j.chemgeo.2004.03.003.

Lanphere, M.A., and Dalrymple, G.B., 1971, A test of the $^{40}Ar/^{39}Ar$ age spectrum technique on some terrestrial materials: Earth and Planetary Science Letters, v. 12, p. 359–372.

Lanphere, M.A., and Dalrymple, G.B., 1976, Identification of excess ^{40}Ar by the $^{40}Ar/^{39}Ar$ age spectrum technique: Earth and Planetary Science Letters, v. 32, p. 141–148, doi: 10.1016/0012-821X(76)90052-2.

Lanphere, M.A., Dalrymple, G.B., Fleck, R.J., and Pringle, M.S., 1990, Intercalibration of mineral standards for K-Ar and $^{40}Ar/^{39}Ar$ age measurements: EOS (Transactions, American Geophysical Union), v. 71, p. 1658.

Lux, G., 1987, The behavior of noble gases in silicate liquids: Solution, diffusion, bubbles and surface effects, with applications to natural samples: Geochimica et Cosmochimica Acta, v. 51, p. 1549–1560, doi: 10.1016/0016-7037(87)90336-X.

Mahoney, J.J., Macdougall, J.D., Lugmair, G.W., and Gopalan, K., 1983, Kerguelen hotspot source for Rajmahal Traps and Ninetyeast Ridge: Nature, v. 303, p. 385–389.

Mankinen, E.A., and Dalrymple, G.B., 1972, Electron microprobe evaluation of terrestrial basalts for whole-rock K-Ar dating: Earth and Planetary Science Letters, v. 17, p. 89–94, doi: 10.1016/0012-821X(72)90262-2.

McDougall, I., and Chamalaun, F.H., 1966, Geomagnetic polarity scale of time: Nature, v. 212, p. 1415–1418, doi: 10.1038/2121415a0.

McDougall, I., and Harrison, T.M., 1999, Geochronology and thermochronology by the $^{40}Ar/^{39}Ar$ method: New York, Oxford University Press, 269 p.

Merrihue, C., and Turner, G., 1966, Potassium-argon dating by activation with fast neutrons: Journal of Geophysical Research, v. 71, p. 2852–2857.

Min, K.W., Mundil, R., Renne, P.R., and Ludwig, K.R., 2000, A test for systematic errors in $^{40}Ar/^{39}Ar$ geochronology, through comparison with U/Pb analysis of a 1.1 Ga rhyolite: Geochimica et Cosmochimica Acta, v. 64, p. 73–98, doi: 10.1016/S0016-7037(99)00204-5.

Min, K.W., Renne, P.R., and Huff, W.D., 2001, $^{40}Ar/^{39}Ar$ dating of Ordovician K-bentonites in Laurentia and Baltoscandia: Earth and Planetary Science Letters, v. 185, p. 121–134, doi: 10.1016/S0012-821X(00)00365-4.

Moorbath, S., Taylor, P.N., and Jones, N.W., 1986, Dating the oldest terrestrial rocks: Facts and fiction: Chemical Geology, v. 57, p. 63–86, doi: 10.1016/0009-2541(86)90094-X.

Nier, A.O., 1950, A redetermination of the relative abundances of the isotopes of carbon, nitrogen, oxygen, argon and potassium: Physical Review, v. 77, p. 789–793, doi: 10.1103/PhysRev.77.789.

Nomade, S., Renne, P.R., Vogel, N., Deino, A.L., Sharp, W.D., Becker, T.A., Jaouni, A.R., and Mundil, R., 2005, Alder Creek sanidine (ACs-2): A Quaternary $^{40}Ar/^{39}Ar$ dating standard tied to the Cobb Mountain geomagnetic event: Chemical Geology, v. 218, p. 315–338, doi: 10.1016/j.chemgeo.2005.01.005.

Pasacal, B., 1952, The provinical letters, Pensées, scientific treatises: London, Encyclopedia Brittanica Great Books of the Western World 33, 487 p.

Renne, P.R., 2000, $^{40}Ar/^{39}Ar$ age of plagioclase from Acapulco meteorite and problem of systematic errors in cosmochronology: Earth and Planetary Science Letters, v. 175, p. 13–26, doi: 10.1016/S0012-821X(99)00287-3.

Renne, P.R., Deino, A.L., Walter, R.C., Turrin, B.D., Swisher, C.C., Becker, T.A., Curtis, G.H., Sharp, W.D., and Jaouni, A.R., 1994, Intercalibration of astronomical and radioisotope time: Geology, v. 22, p. 783–786, doi: 10.1130/0091-7613(1994)022<0783:IOAART>2.3.CO;2.

Renne, P.R., Zichao, Z., Richards, M.A., Black, M.T., and Basu, A.R., 1995, Synchrony and causal relations between Permian-Triassic boundary crises and Siberian Trap volcanism: Science, v. 269, p. 1413–1416, doi: 10.1126/science.269.5229.1413.

Renne, P.R., Sharp, W.D., Deino, A.L., Orsi, G., and Civetta, L., 1997, $^{40}Ar-^{39}Ar$ dating into the historical realm; Calibration against Pliny the Younger: Science, v. 277, p. 1279–1280, doi: 10.1126/science.277.5330.1279.

Renne, P.R., Swisher, C.C., Deino, A.L., Karner, D.B., Owens, T.L., and DePaolo, D.J., 1998, Intercalibration of standards, absolute ages and uncertainties in $^{40}Ar-^{39}Ar$ dating: Chemical Geology, v. 145, p. 117–152, doi: 10.1016/S0009-2541(97)00159-9.

Renne, P.R., Melosh, H.J., Farley, K.A., Reimold, W.U., Koeberl, C., Rampino, M.R., Kelley, S.P., and Ivanov, B.A., 2004, Is Bedout an impact structure? Take 2: Science, v. 306, p. 610–611, doi: 10.1126/science.306.5696.610.

Roddick, J.C., 1978, The application of isochron diagrams in $^{40}Ar-^{39}Ar$ dating: A discussion: Earth and Planetary Science Letters, v. 41, p. 233–244, doi: 10.1016/0012-821X(78)90014-6.

Roddick, J.C., 1983, High precision intercalibration of $^{40}Ar-^{39}Ar$ standards: Geochimica et Cosmochimica Acta, v. 47, p. 887–898, doi: 10.1016/0016-7037(83)90154-0.

Sharma, M., Basu, A.R., and Ray, S.L., 1994, Sm-Nd isotopic and geochemical study of the Archaen tonalite-amphibolite association from the eastern Indian Craton: Contributions to Mineralogy and Petrology, v. 117, p. 45–55, doi: 10.1007/BF00307728.

Sharp, W.D., and Clague, D.A., 2006, 50-Ma initiation of Hawaiian-Emperor bend records major change in plate motion: Science, v. 313, p. 1281–1284, doi: 10.1126/science.1128489.

Steiger, R.H., and Jager, E., 1977, Subcommission on geochronology: Convention on the use of decay constants in geo- and cosmochronology: Earth and Planetary Science Letters, v. 36, p. 359–362, doi: 10.1016/0012-821X(77)90060-7.

Sumino, H., Mizukami, T., and Wallis, S.R., 2005, Slab-derived noble gases preserved in wedge mantle peridotite from the Sanbagawa belt, Shikoku, Japan: EOS (Transactions, American Geophysical Union), v. 85, no. 47, abstract v41A-1428.

Sutter, J.F., and Smith, T.E., 1979, $^{40}Ar/^{39}Ar$ ages of diabase intrusions from Newark trend basins in Connecticut and Maryland: Initiation of central Atlantic rifting: American Journal of Science, v. 279, p. 808–831.

Verati, C., Bertrand, H., and Feraud, G., 2005, The farthest record of the central Atlantic magmatic province into West African craton: Precise $^{40}Ar/^{39}Ar$ dating and geochemistry of Taoudenni basin intrusives (northern Mali): Earth and Planetary Science Letters, v. 235, p. 391–407.

York, D., 1969, Least squares fitting of a straight line with correlated errors: Earth and Planetary Science Letters, v. 5, p. 320–324, doi: 10.1016/S0012-821X(68)80059-7.

York, D., Hall, C.M., Yanase, Y., Hanes, J.A., and Kenyon, W.J., 1981, $^{40}Ar/^{39}Ar$ dating of terrestrial minerals with a continuous laser: Geophysical Research Letters, v. 8, p. 1136–1138.

York, D., Hall, C.M., Gaspar, M.J., and Lynch, M., 1984, Laser-probe $^{40}Ar/^{39}Ar$ dating with ultra-sensitive mass spectrometer: EOS (Transactions, American Geophysical Union), v. 65, p. 303.

MANUSCRIPT ACCEPTED BY THE SOCIETY 31 JANUARY 2007

The Geological Society of America
Special Paper 430
2007

A quantitative tool for detecting alteration in undisturbed rocks and minerals—II: Application to argon ages related to hotspots

Ajoy K. Baksi*

*Department of Geology and Geophysics, Louisiana State University,
Baton Rouge, Louisiana 70803, USA*

ABSTRACT

Alteration of undisturbed igneous material used for argon dating work often results in inaccurate estimates of the crystallization age. A new quantitative technique to detect alteration has been developed (see Baksi, this volume, Chapter 15), utilizing the ^{36}Ar levels observed in rocks and minerals. The method is applied to data in the literature for rocks linked to hotspot activity.

For subaerial rocks, argon dating results are critically examined for the Deccan Traps, India. The duration of volcanic activity and its coincidence in time with the K-T boundary are shown to be uncertain. The bulk of dated seafloor material (recovered from the Atlantic, Indian, and Pacific oceans) proves to be altered. Ages determined using large (hundreds of milligram) samples are generally unreliable, due to inclusion of altered phases. Such analyses include studies suggesting an age of ca. 43 Ma for the bend in the Hawaiian-Emperor chain. More recent attempts, using much smaller subsamples (~10 mg) that have been acid leached to remove alteration products, are generally of higher reliability. Plagioclase separates sometimes yield reliable results. However, many whole-rock basalts from the ocean floor yield ages that are, at best, minimum estimates of the time of crystallization. Most "rates of motions," calculated from hotspot track ages, are shown to be invalid.

Seafloor rocks are recovered at considerable expense but often are not suitable for dating by the argon methods. Most are severely altered by prolonged contact with seawater. A method is recommended for testing silicate phases prior to attempts at argon dating. This involves a quantitative determination of the ^{36}Ar content of the material at hand; dating phases without pretreatment—leaching with HNO_3 for material containing ferromagnesian phases, and HF for feldspars—is strongly discouraged.

Keywords: hotspots, argon ages, alteration, Deccan Traps, oceanic tracks

Truth alone was the daughter of time.

Leonardo da Vinci (manuscript M, ca. 1500)

---

*E-mail: akbaksi@yahoo.com.

Baksi, A.K., 2007, A quantitative tool for detecting alteration in undisturbed rocks and minerals—II: Application to argon ages related to hotspots, *in* Foulger, G.R., and Jurdy, D.M., eds., Plates, plumes, and planetary processes: Geological Society of America Special Paper 430, p. 305–333, doi: 10.1130/2007.2430(16).

INTRODUCTION

The $^{40}Ar/^{39}Ar$ dating stepheating technique (Merrihue and Turner, 1966) is the most widely applied radiometric tool in earth sciences. For mafic rocks, it may be the only tool to yield precise and accurate age information. Reproducibility of the step ages permits evaluation of the accuracy and precision of resulting estimates of the crystallization age of igneous rocks and minerals. Such ages must be listed with associated errors. For $^{40}Ar/^{39}Ar$ dating, the formula of Dalrymple et al. (1981) is used to calculate the random errors in ages. Systematic and random errors must not be associated with one another. Alteration causes partial loss of $^{40}Ar^*$ and leads to age estimates that are systematically too young.

Numerous $^{40}Ar/^{39}Ar$ stepheating ages in the literature do not meet the basic mathematical requirements as precise and accurate estimates of time of crystallization. Many ages related to hotspot-generated rocks have been shown to be statistically invalid (see Baksi, 1999, 2005). The availability of such unreliable numbers in the literature has led to problems. The field geologists' adage "I wouldn't have believed it, if I hadn't seen it," has been turned into "I wouldn't have seen it, if I hadn't believed it." About fifty years ago, Irving Langmuir foresaw the extreme situation in this regard and termed it "pathological science." As elaborated therein (see Langmuir, 1989, p. 43), "people are tricked into false results . . . by wishful thinking."

I shall investigate the effects of alteration, with applications to rocks recovered from submarine environments linked to hotspot activity. McDougall and Harrison (1999, p. 33) observe that "Very few igneous rocks from the deep-sea environment meet the criteria of freshness and crystallinity for K-Ar dating, as hydrothermal alteration and submarine weathering result in development of chlorite, smectite and other clay minerals, as well as calcite, at the expense of primary high-temperature minerals."

METHODOLOGY

Radiometric ages play a critical role in earth sciences in elucidating the timing, duration, and rates of various processes. For such calculations, the ages should be accurate and precise. For post-Paleozoic rocks, I suggest that a 1σ precision better than $\pm1.0\%$ of the age is a prerequisite. In cases where the random error is estimated to be >2%, "ages" should not be used for calculation of absolute plate velocities.

Argon ages available in the literature on undisturbed rocks are evaluated as valid estimates of the time of crystallization. First, they are tested for statistical robustness, as outlined in earlier efforts (Baksi, 1999, 2005). Next, the state of chemical alteration of the material dated will be evaluated by the alteration index (A.I.) technique; *silicate phases that have suffered detectable alteration can yield statistically valid plateau/isochron sections that underestimate the time of crystallization by ~2–10%* (Baksi, this volume, Chapter 15). Statistical tests are often disregarded (see Baksi, this volume, Chapter 15). Sometimes, the statistics appear to be "too good," i.e., probability values are well in excess of >0.95. Isolated cases may be due to chance, but generally critical examination is called for. The case of the Gettysburg sill, northeastern United States, has been examined elsewhere (Baksi, 1991). The study of Turrin et al. (1994) on the Alder Creek rhyolite sanidine is examined. Both the age spectrum and isochron plots (Fig. 1) yield probability values that are very high. (The original statistics of Turrin et al.,

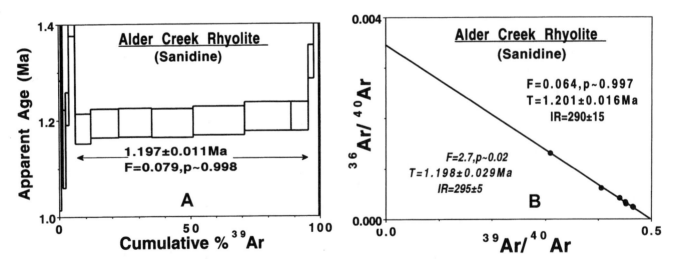

Figure 1. $^{40}Ar/^{39}Ar$ dating results on the Alder Creek (rhyolite) sanidine (after Turrin et al., 1994). (A) Age spectrum with the plateau section delineated; age listed with 1σ error, the goodness of fit parameter (F), and the probability of occurrence (p). The plateau is "too good," with p ~0.998. (B) The isochron plot for the same data set. IR—initial ($^{40}Ar/^{36}Ar$) ratio; T—age; F and p as above. This age overlaps the plateau value. The points fit the line too well, with p ~0.997. The isochron listings of Turrin et al. (1994), shown in italics, are in error.

1994, for the isochron plot are in error). Reducing the step age errors to ~30% lowers the probability values to ~0.5. The error associated with the final age is reduced proportionally.

The bulk of this effort is directed toward ^{40}Ar/^{39}Ar stepheating data. Guidelines for statistical evaluation of plateau sections and isochron figures have been outlined elsewhere (Baksi, 1999, 2005). These techniques are utilized here to test for statistical robustness of the data. In particular, the goodness of fit parameter (F) is used, along with the number of steps involved, to arrive at a probability figure (p) making use of χ^2 tables. Where $p < 0.05$ (95% confidence level test), excess scatter beyond estimated analytical errors is indicated. Geological error is present, and the calculated age is not an accurate estimate of the time of crystallization. The second set of tests examines the A.I. of the plateau (least-altered) steps and looks at the ^{36}Ar content of the sample. A parameter, normalized to the K content of certain phases (whole rocks, K-feldspar, and mica) and to the Ca content of others (plagioclase feldspar and hornblende), is used. The methodology, rationale, and efficacy of the method has been outlined elsewhere (Baksi, this volume, Chapter 15).

^{40}Ar/^{39}Ar ages are reported relative to those of standards preferred by Renne et al. (1998). Errors are quoted and shown in figures at the 1σ level. Where necessary, ages have been recalculated to the decay constants and isotopic abundances suggested by Steiger and Jager (1977). The A.I. cut-off for fresh whole-rock basalt is ^{36}Ar/^{39}Ar < 0.0006, and that for plagioclase is ^{36}Ar/^{37}Ar < 0.00006; ^{36}Ar content for fresh whole-rock basalt is $< 6 \times 10^{-10}$ cm^3 STP g^{-1} (Baksi, this volume, Chapter 15).

RESULTS

For proper statistical evaluation of ages, as well as A.I. calculation, full data sets (isotopic ratios) are required. These are often not available in the literature; in many instances, such data sets were not made available, even after repeated requests made to the editors of journals and to the concerned authors.

Subaerial Material—Continental Flood Basalts

These provinces have been genetically linked to hotspot activity. Further, they are said to be responsible for global faunal extinction events. Critical examination of the timing and duration of flood basalt volcanism in numerous cases is hampered by lack of access to the detailed data sets. A large body of data is available for the Deccan Traps, India, and these are considered in detail. Results generated at different laboratories are reduced to a single base, the calibrations preferred by Renne et al. (1998). The Deccan province has been linked to the K-T boundary extinction event (Courtillot et al., 1988; Duncan and Pyle, 1988; Hofmann et al., 2000). The age of the K-T boundary, relative to the monitor ages used here, is placed at 65.58 Ma (see Knight et al., 2003). ^{40}Ar/^{39}Ar plateau ages that pass the relevant statistical test for validity must also be unaltered, as based on A.I. tests (Baksi, this volume, Chapter 15).

Deccan Traps, Whole-Rock Basalts. Isotopic data for the results of Duncan and Pyle (1988), were obtained (R.A. Duncan, personal commun., 1989). Numerous step ages and plateau values (Fig. 2) differ from those listed in Duncan and Pyle (1988). Most of the plateau ages fail the relevant statistical tests. Only three pass this test: 65.4 ± 0.3 (TEM-004), 67.5 ± 0.6 (CAT-021), and 67.0 ± 0.4 Ma (MAP-056). The latter two are significantly older than the K-T boundary. TEM-004 lies ~1.2 km below CAT-021 (Duncan and Pyle, 1988; Fig. 2) but yields a measurably younger age (by 2.1 m.y., whereas the 95% confidence interval is ±1.3 m.y). Also, TEM-004 lies ~1.6 km below MAP-056 (Duncan and Pyle, 1988; Fig. 2) but yields a measurably younger age (by 1.6 m.y., whereas the 95% confidence interval is ±1.0 m.y). These spurious ages violate the principle of superposition and must be rejected. Based on the A.I. of their plateau (least-altered) steps (Fig. 3A), all rocks are significantly altered, and no accurate ages were recovered.

The results of Venkatesan et al. (1993) are examined in agespectrum form (Fig. 4), using the relevant data sets (K. Pande, personal commun., 1996). Only two (of eight) ages pass the statistical test, i.e., for IG82-39 and MB81-24. The A.I. test (Fig. 3B) shows that the rocks are somewhat fresher than those of Duncan and Pyle (1988). The two rocks that pass the statistical test for plateau ages are quite altered; the resulting ages can only be used as minimum values. No proper crystallization ages were recovered from the study of Venkatesan et al. (1993).

Deccan Traps, Plagioclase Feldspar. Courtillot et al. (1988) presented ages from the eastern sections of the Deccan. Detailed isotopic results were obtained (G. Feraud, personal commun., 1998). All five specimens pass the relevant statistical tests (Fig. 5). The whole rock shows loss of ^{40}Ar* and fails the A.I. test for freshness (0.0034 ± 0.0004, compared to <0.0006 for fresh material); its age is significantly younger than the K-T boundary, although it is from one of the lowermost flows in the area. Plagioclase ages are: DK0103, 66.6 ± 0.4 Ma; and for the Narmada section (moving up stratigraphic sequence): NA16, 67.6 ± 0.9; NA17, 64.9 ± 0.3; and NA18, 63.6 ± 0.3 Ma. The last (a dike) appears to be altered (Fig. 6), and its age is rejected. Specimens NA16 and NA17 are slightly altered and suggest that volcanism in this area continued for ~2–3 m.y. The DK specimen is least altered; its age, and that of NA16, suggest that the oldest parts of the Deccan Traps (overlying the Lameta and Bagh beds) are >1 m.y. older than the K-T boundary.

Hofmann et al. (2000) presented ^{40}Ar/^{39}Ar ages on rocks from the Western Ghats section. Plagioclase separates gave statistically acceptable plateaus, but are "too good," with probability values >0.99 in many cases (see Fig. 7). Errors in step ages have been overestimated by factors of about three. If so, JW5 (64.86 ± 0.24 Ma) is measurably younger than the overlying JW7 (65.77 ± 0.29 Ma). The A.I. plot (Fig. 6B) indicates that most samples have suffered minor alteration, and ages should approximate crystallization values. Further discussion is not attempted here. It is unclear how their ages (obtained relative to standard

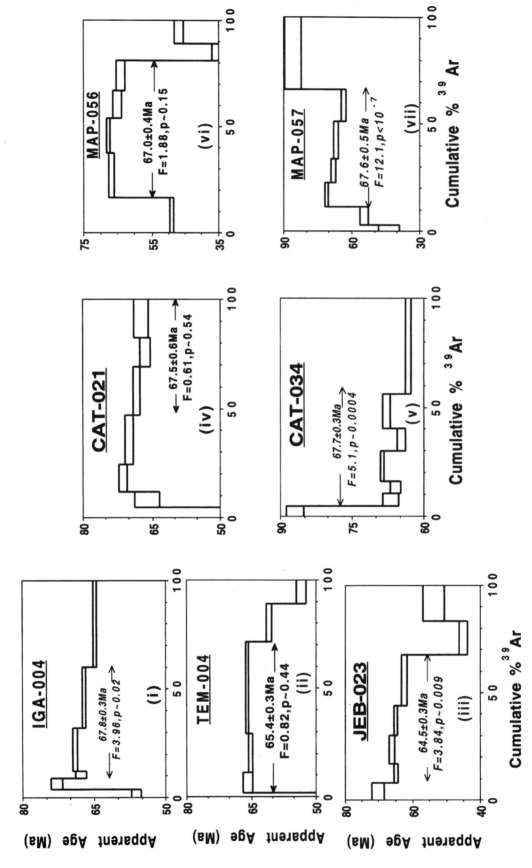

Figure 2. $^{40}Ar/^{39}Ar$ age spectra on whole-rock basalts from the Deccan Traps, India (Duncan and Pyle, 1988). Symbols as in Figure 1A. Specimens i–vii are in stratigraphic succession. Acceptable ages shown in italics are rejected on statistical grounds. Ages shown in italics are rejected on statistical grounds. Acceptable ages for three specimens (TEM-004, CAT-021, and MAP-056) are out of stratigraphic order.

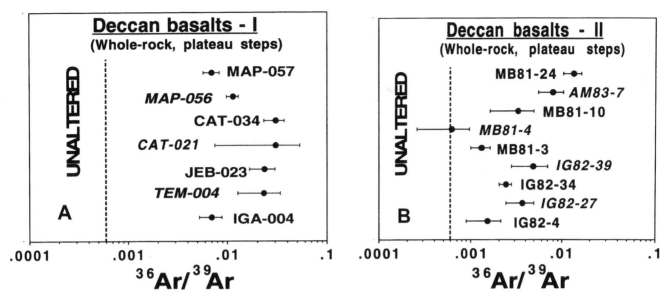

Figure 3. Assessing the alteration state of whole-rock basalts from the Western Ghats section. Average alteration index (A.I.) values and the standard error on the mean are shown for plateau steps (best sites) on log scales. (A) Rocks of Duncan and Pyle (1988)—see Figure 2. All rocks are altered, and plateau ages are rejected. (B) Rocks of Venkatesan et al. (1993)—see Figure 4. With a few exceptions, samples are altered, and plateau ages are rejected.

Hb3Gr at 1072 Ma) should be converted to those preferred by Renne et al. (1998). The work of Roddick (1983) suggests ages should be increased by ~1.0%, whereas that of Renne (2000) indicates ages need to be decreased by ~0.6%. (The analyses of Renne, 2000, on PP-20, yield an age of 1067 Ma and not 1073–1074 Ma, relative to Fish Canyon sanidine at 28.02 Ma).

Allègre et al. (1999) presented a whole-rock Re-Os isochron age of 65.60 ± 0.15 Ma. This estimate is not acceptable as a crystallization value, as the results appear to violate all three assumptions necessary for isochron ages (Baksi, this volume, Chapter 15). The thickest sections of the Deccan Traps (e.g., the Western Ghats composite section) primarily show reversed magnetic polarity and are hypothesized to have formed during chron 29r (see Baksi, 1994). Critical examination of all ages leaves only a few accurate measures of the time of crystallization. Some are measurably older than chron 29r (66.17–65.33 Ma), and sections of the eastern Deccan appear to be >1 m.y. older than the K-T boundary (65.58 Ma). There is no unequivocal evidence that the most voluminous sections were formed during chron 29r, or that the extrusion of the Deccan Traps is coincident in time with, and was responsible for, the K-T boundary faunal extinctions.

The Oceanic Realm

The main emphasis is on rocks from hotspot tracks and purported connections to plume activity. I look to results generated with old-fashioned set-ups first, in which large samples (hundreds of milligrams), generally not acid leached, were used and yielded poor results. With the advent of modern instrumentation

(York et al., 1984), sample sizes were reduced to tens of milligrams; many were handpicked under binocular microscope and/or subjected to acid leaching. These contain fresh(er) material, display lower A.I. values, and yield better ages than the larger samples.

$^{40}Ar/^{39}Ar$ stepheating work on terrestrial whole-rock basalts (or purified groundmass) should be carried out on powdered material. Use of chunks (large pieces) permits altered material within the sample to remain undegassed, leads to very high amounts of atmospheric argon in the analyses, and to incorrect ages (see Baksi, 1974a). The material should be dated in fairly coarse fractions (~10–60 mesh size), because fine grinding introduces ^{36}Ar into the sample (Baksi, 1974a). For fresh whole-rock material, single pieces >20 g have been dated successfully by the K-Ar method (Baksi, 1974b).

Evaluation of Some Early Work. K-Ar dates are almost invariably minimum estimates of crystallization ages, based on high ^{36}Ar contents. Attention is directed to $^{40}Ar/^{39}Ar$ studies. Seidemann (1978) analyzed near-ridge pillow basalts containing excess ^{40}Ar. Figure 8 looks at the A.I. on two splits of T3–71-D-148, identified as being more and less altered (Table 2 *in* Seidemann, 1978). Both are clearly altered, with the fresher sample showing an order of magnitude less ^{36}Ar than the more altered one. Walker and McDougall (1982) carried out K-Ar and $^{40}Ar/^{39}Ar$ studies of pillow basalt material from the Dabi volcanics of Papua New Guinea. (^{36}Ar values are listed in units of 10^{-10} cm^3 STP g^{-1}.) The boninites showing ^{36}Ar ~2–6 are less altered than the tholeiites (^{36}Ar ~10–90). Boninite 38.2B (Fig. 9A) shows partial loss of $^{40}Ar*$, and a marginal (<50% of the total gas) plateau age of ca. 50 Ma (not ca. 54 Ma, as listed by

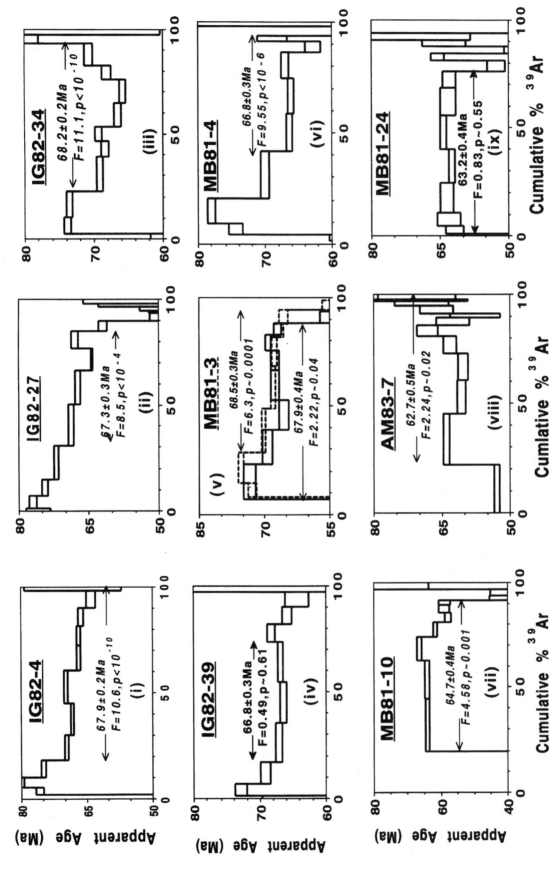

Figure 4. ^{40}Ar/^{39}Ar age spectra on whole-rock basalts from the Deccan Traps, India (Venkatesan et al., 1993). Symbols as in Figure 1A. Samples i–ix are in stratigraphic succession. Ages shown in italics are rejected on statistical grounds. Only two specimens (IG82-39 and MB81-24) give acceptable plateau ages.

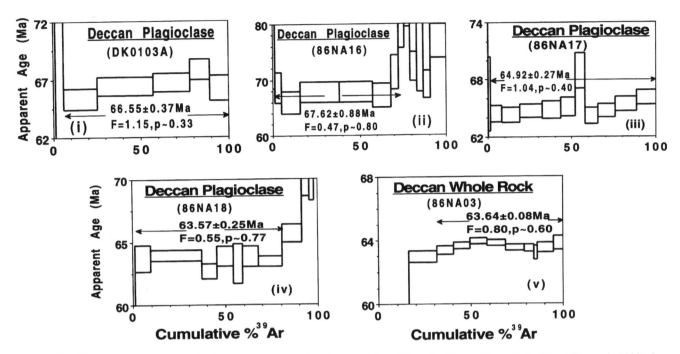

Figure 5. ^{40}Ar/^{39}Ar age spectra on four plagioclase separates and a whole-rock basalt from the Deccan Traps, India (Courtillot et al., 1988). Symbols as in Figure 1A. All give statistically acceptable plateau ages.

Walker and McDougall, 1982). These steps exhibit high A.I. values (Fig. 9B). Tholeiite 28.3 does not yield a "meaningful plateau age of 58.9 ± 1.1 Ma" (Walker and McDougall, 1982, p. 2185) for the steps delineated by arrows in Figure 9A. Initial steps show very high ^{36}Ar levels, introduced during the neutron irradiation procedure. The intermediate-temperature steps from the relatively unaltered phases of the rock show no plateau age

and are altered (see Fig. 9). The conclusion that "the ages of the tholeiitic and boninitic volcanics are almost identical at 58.9 ± 1.1 Ma" (Walker and McDougall, 1982, p. 2188) is unfounded.

Oceanic Plateaus. These formations are linked to hotspot activity, and it is critical to obtain accurate radiometric information. Sinton et al. (1998) attempted ^{40}Ar/^{39}Ar dating of samples from the Caribbean area, and presented plateau ages of

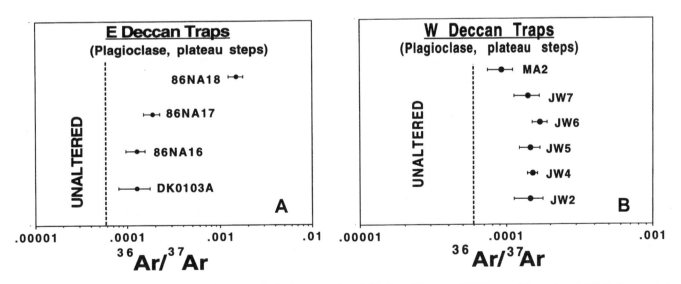

Figure 6. Assessing the state of alteration of Deccan plagioclase samples of (A) Courtillot et al. (1988)—see Figure 5, and (B) Hofmann et al. (2000)—see Figure 7. Average alteration index (A.I.) and standard error on the mean of plateau steps shown on log scales. Most samples show minor alteration, and plateau ages should be close to crystallization values.

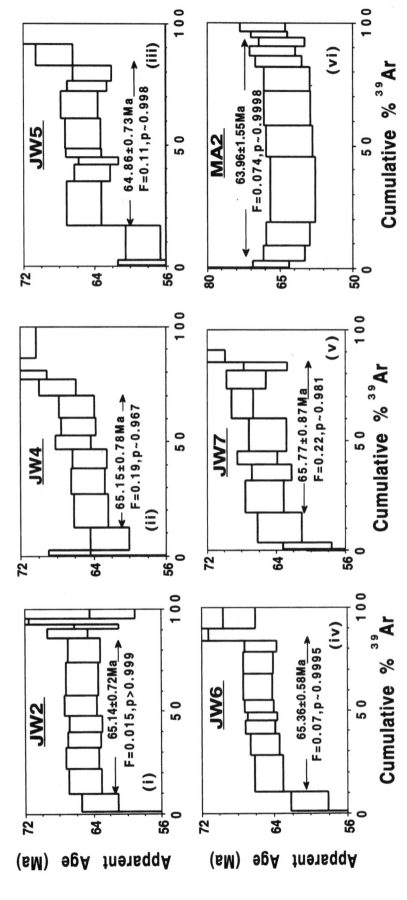

Figure 7. $^{40}\text{Ar}/^{39}\text{Ar}$ age spectra on six plagioclase separates from the Deccan Traps, India (after Hofmann et al., 2000). Ages reported relative to 1072 Ma for Hb3Gr (see text). Symbols as in Figure 1A. Specimens i–vi are in stratigraphic succession. The statistics are too good ($p > 0.96$); step age errors have been overestimated.

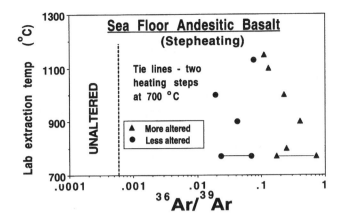

Figure 8. Assessment of the freshness of two altered seafloor andesitic basalts used for $^{40}Ar/^{39}Ar$ study (Seidemann, 1978). Alteration index (A.I. on a log scale) plotted versus laboratory extraction temperature. The results confirm the modality of using the $^{36}Ar/^{39}Ar$ ratio as an A.I.

ca. 75–95 Ma. The Gorgona and Curaçao samples (92–97, 79BE-73, and 79KV-9) gave isochrons with initial $^{40}Ar/^{36}Ar$ ratios <295.5, caused by disturbed K-Ar systems (see Lanphere and Dalrymple, 1978) and their ages are rejected. The isochron plots of Sinton et al. (1998) permit assessment of the freshness of the rocks (Fig. 10). All rocks show alteration, and their "ages" can, at best, serve as minimum estimates. Rocks from the Deep Sea Drilling Project (DSDP) Leg 150 and Gorgona Island (150-11-2 and 92-27) are severely altered. There is no accurate radiometric evidence of two stages of magmatism at ca. 90 and ca. 76 Ma (Sinton et al., 1998), nor of volcanics in the Dominican Republic and Costa Rica formed over the Galapagos hotspot.

Mahoney et al. (1993) listed ages for the Ontong Java plateau, suggesting it was formed at 122–123 Ma in a short interval of time. The isotopic data for this very important work are not available for inspection. The isochron plot for 130-807C-84R-6 (0–3) (Mahoney et al., 1993, Fig. 2) permits A.I. calculation. The results are compared to a similar rock (RM82-5; Mahoney et al., 1983) from the Rajmahal Traps, India (Fig. 11). The latter yields step ages of higher precision (Fig. 11A) and is substantially fresher than the Ontong Java plateau rock (Fig. 11B). The Mahoney et al. (1993) age is a minimum estimate; the Ontong Java plateau was formed at >122 Ma; the duration of this event remains unknown.

Sinton et al. (1997) looked at a suite of rocks on Nicoya peninsula, Costa Rica. Using geochemical, isotopic, and radiometric data, they linked these rocks to the Galapagos plume and argued for their formation mostly in the time frame ca. 90–84 Ma. A full table of isotopic data is not available; their Figure 2 (isochron plots) permits evaluation of their ages and freshness of the material dated (Fig. 12). None of these rocks should yield true estimates of the crystallization age, because even the (best) plateau steps are derived from altered sites. Both plagioclase samples show alteration and cannot be expected to give accurate crystallization ages. There are no accurate ages of ca. 88, ca. 83, and ca. 64 Ma (Sinton et al., 1997), and genetic connections to the Galapagos plume are questioned.

All three cases examined above involved dating small quantities of material using ultra sensitive mass spectrometers and low-blank extraction apparatus. Modern instrumentation, per se, does not guarantee better (correct) age results. The samples in these studies were dated without acid leaching to remove alteration products. It is unsurprising that such submarine rocks did not yield accurate estimates of crystallization age.

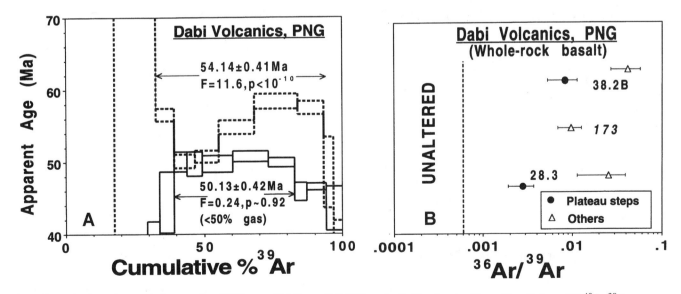

Figure 9. Evaluation of argon dating results of Walker and McDougall (1982) on the Dabi volcanics, Papua New Guinea. (A) $^{40}Ar/^{39}Ar$ age spectra for two whole-rock samples. Solid lines—38.2B; dashed lines—28.3; symbols as in Figure 1A. Neither sample yields a plateau age. (B) Alteration index (A.I.) values shown with associated standard error on the mean on a log scale. High values for plateau steps indicate alteration.

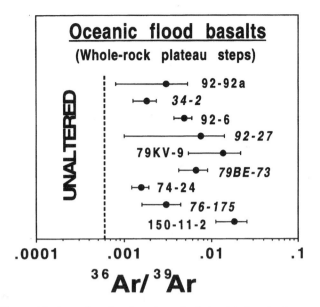

Figure 10. Assessing the alteration of whole-rock basalts from the Caribbean area, dated by the $^{40}Ar/^{39}Ar$ method (Sinton et al., 1998). Alteration index (A.I.) and standard error on the mean for plateau steps are shown on a log scale. All rocks are altered, and no reliable crystallization ages were obtained.

Rocks from the Indian Ocean. Basalts from the Ninetyeast Ridge were subjected to argon dating (Duncan, 1978, 1991). The purported plateau ages have been demonstrated to show gross excess scatter (Baksi, 1999, 2005). K-Ar dates (Duncan, 1978) show high amounts of ^{36}Ar, which is unsurprising, as Duncan

(1978) reports observation of smectite and chlorite in thin sections. Such rocks cannot yield accurate crystallization ages. The A.I. of the best $^{40}Ar/^{39}Ar$ (plateau) steps (Fig. 13A) confirms that all rocks are altered. A later study (Duncan, 1991) used even more heavily altered rocks (see Fig. 13A). All of the site 756–758 rocks are severely altered, with ^{36}Ar contents thirty to one hundred times higher than that for fresh basalts. Further, the ages of Duncan (1991) have been shown to be invalid on statistical grounds (Baksi, 1999, 2005). The oft-cited age progression for the Ninetyeast Ridge and its connection to the Kerguelen hotspot, as based on the work of Duncan (1978, 1991), is rejected, because no reliable crystallization ages were recovered.

Duncan and Hargraves (1990) carried out whole-rock dating of basalts from the Chagos-Laccadive Ridge and the Mascarene plateau. Their ages show excess scatter in isochron plots and age spectra (Baksi, 1999, 2005). The A.I. plot (Fig. 13B) shows most of them are severely altered, and their ages must be rejected. NB1-1 appears to be mildly altered, and its age serves as an estimate of its time of crystallization (see Baksi, 2005). The conclusions of Duncan and Hargraves (1990) regarding the temporal tracking of the Réunion hotspot are discounted.

Rocks from the Atlantic Ocean. The ages of Duncan (1984) for the New England seamounts have been shown to be statistically invalid (Baksi, 1999, 2005). The A.I. shows that the amphibole specimens are altered, except for the last two steps of the Nashville seamount sample. Two steps cannot define a plateau, but it is likely that the crystallization age is ca. 83 Ma. For the whole rocks, the A.I. of the plateau steps are ~0.005 (Atlantis II) and ~0.0025 (Bear); the cut-off for fresh samples is <0.0006 (Baksi, this volume, Chapter 15). The conclusions of

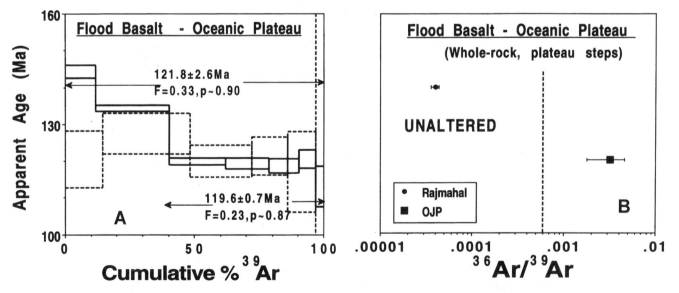

Figure 11. Evaluation of $^{40}Ar/^{39}Ar$ dating of the Ontong Java plateau (OJP; Mahoney et al., 1993); results compared to a similar basalt from India (Baksi, unpubl. data, 2001). (A) Age spectra for OJP (dotted lines) and RM82-5 (solid lines). Symbols as in Figure 1A. RM82-5 yields step ages with smaller errors and a more precise plateau age. (B) The alteration index (A.I.) for plateau steps, average and standard error on mean shown on a log scale. The OJP rock is altered and yields a minimum age for the time of formation.

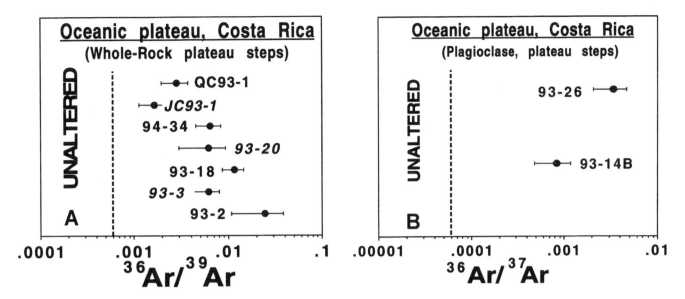

Figure 12. Evaluating the ^{40}Ar/^{39}Ar work of Sinton et al. (1997) on rocks from the Nicoya peninsula, Costa Rica. Average alteraton index (A.I.) and standard error on the mean for plateau steps are shown on a log scale. (A) Whole-rock samples are altered, showing three to fifty times more ^{36}Ar than found in fresh material. Plateau ages for these samples are rejected. (B) Two plagioclase separates; both are altered, and the plateau ages of Sinton et al. (1997) are rejected as accurate estimates of the time of crystallization.

Duncan (1984) regarding age progression in the New England seamounts, and its associated hotspot, are unfounded.

Statistical analysis of age spectra and isochrons (Baksi, 1999, 2005) shows the ages of O'Connor and Duncan (1990) for the track of the Tristan da Cunha hotspot in the south Atlantic Ocean are unfounded. The A.I. of the plateau steps is evaluated in Figure 14A. Only two rocks (93-11-8 and 93-14-1) are fresh, but these did not yield valid plateau/isochron ages (see Baksi, 2005). Other samples are badly altered (Fig. 14A) and cannot yield valid crystallization ages. The conclusions of O'Connor

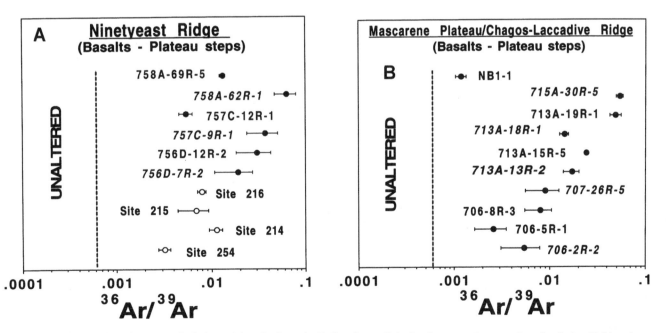

Figure 13. Testing the alteration state of whole-rock basalts from the Indian Ocean linked to hotspots. Average alteration index (A.I.) and standard error on the mean for plateau steps are shown on log scales. (A) Samples from the Ninetyeast Ridge, analyzed by Duncan (1978, 1991). Both sets of rocks are altered. All plateau ages listed by the author are rejected. (B) Rocks from the Chagos-Laccadive Ridge, analyzed by Duncan and Hargraves (1990). Samples are badly altered, and the plateau ages are rejected. Only NB1-1 appears to be relatively fresh.

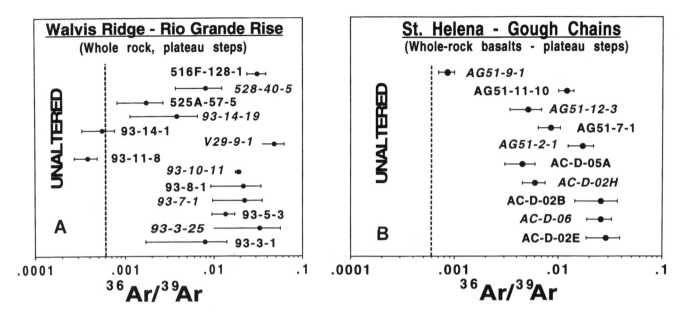

Figure 14. Assessing the alteration state of the volcanic rocks from hotspot tracks in the Atlantic Ocean. Average alteration index (A.I.) and standard error on mean of the plateau steps are shown on log scales. (A) Rocks from the Walvis Ridge–Rio Grande Rise (O'Connor and Duncan, 1990). Most samples are badly altered and cannot yield valid crystallization ages. (B) Rocks from the St. Helena–Gough chains (O'Connor and Le Roex, 1992). Most samples do not yield accurate estimates of the crystallization age (see Fig. 15).

and Duncan (1990), tracing the Tristan da Cunha hotspot on the African and South American plates, are negated. O'Connor and Le Roex (1992) presented $^{40}Ar/^{39}Ar$ stepheating ages for the St. Helena and Gough volcanic chains. Statistical examination of their age spectra (Fig. 15) shows that six out of nine results do not yield proper plateaus, and the ages are rejected. The A.I. of most of the rocks are high (Fig. 14B), indicating alteration; no accurate crystallization ages were obtained for these rocks. AG51-9-1 is minimally altered, and its age should be ca. 19.1 Ma. For the other rocks, ages are not known accurately. Conclusions regarding the reconstruction and motion of the African plate over the hotspot-plume systems and quantitative calculation of plate motions (O'Connor and Le Roex, 1992) are not justified.

Pacific Ocean and Surroundings

There are a large number of purported hotspot tracks in the Pacific Ocean. Many of these have been dated, and the results are widely quoted and used in the literature. The list of Clouard and Bonneville (2005) of hotspot-related papers proved useful for locating the relevant data sets. A preliminary investigation based on statistical appraisal of plateaus (Baksi, 2004) showed that many of the ages reported are invalid as proper estimates of time of crystallization.

Pacific Northwest, United States. Duncan (1982) suggested that the Yellowstone hotspot track could be traced to an island chain that collided with North America, forming the Coast Range of Oregon and Washington. Statistical examination of the relevant $^{40}Ar/^{39}Ar$ stepheating data (Baksi, 2005) indi-

cated that none of the ages were proper estimates of the time of formation of these rocks. For K-Ar dates (Table 1 *in* Duncan, 1982), most rocks contain $>20 \times 10^{-10}$ cm^3 STP g^{-1} of ^{36}Ar and are altered. For $^{40}Ar/^{39}Ar$ work, A.I. values on D80-RB-31 and D80-CV-25 are ~0.015, compared to <0.0006 for fresh material. At low extraction temperatures, D78-SR-10 shows severe alteration; the two highest-temperature sites tap into less-altered sites (A.I. ~0.002). Sample D78-SR-1 is altered, based on its very high ^{36}Ar content. The stepheating experiment shows high levels of ^{36}Ar in the low–intermediate temperature steps. The two highest-temperature steps are drawn from sites that are somewhat fresher (A.I. ~0.001). Two steps cannot define a plateau age; the minimum age of these rocks is ca. 55 Ma. Accurate ages for these rocks in the Pacific Northwest are unknown; there are no age data for a hotspot track for this island chain or of a link to the Yellowstone hotspot.

Seamounts in the North-Central Pacific. Pringle (1993) presented K-Ar and $^{40}Ar/^{39}Ar$ data on rocks from the Musicians seamounts. K-Ar data are assessed for freshness; with a few exceptions (plagioclase separates from Haydn and West Mendelssohn), all show ^{36}Ar values greater than ten times that for fresh material. The A.I. values for $^{40}Ar/^{39}Ar$ total fusion on the whole-rock samples are twenty to one hundred times higher than the cutoff value of 0.0006. All whole-rock samples are badly altered. The total fusion work on plagioclase separates shows better results; A.I. values are three to ten times higher than for fresh material (cut-off value <0.00006). These samples were leached in warm 3–6 N HCl (Pringle, 1993); it has been shown that this procedure is not effective in *totally* removing all traces of alteration

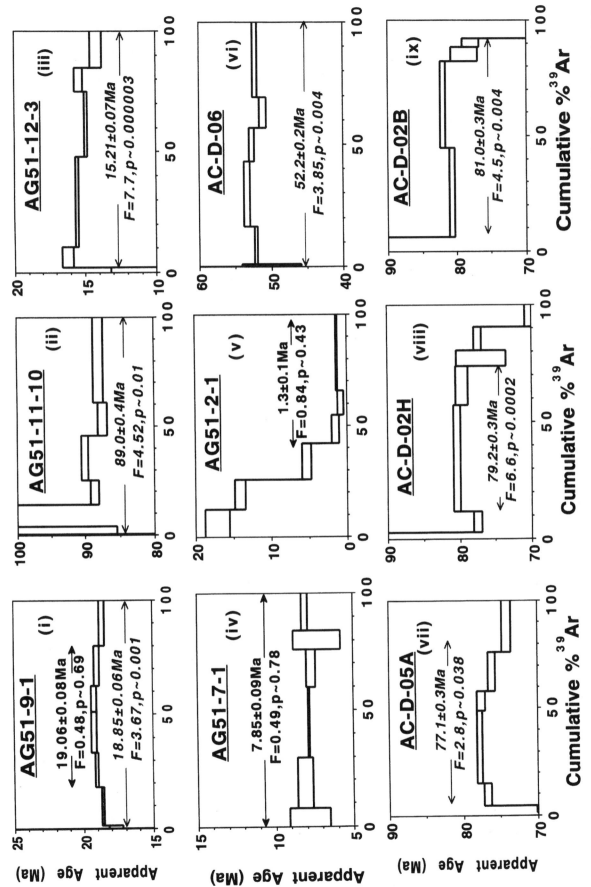

Figure 15. $^{40}Ar/^{39}Ar$ age spectra for nine whole-rock samples from the St. Helena and Gough volcanic lines dated by O'Connor and Le Roex (1992). Symbols as in Figure 1A. Only three samples (51-9-1, 51-7-1, 51-2-1) pass the relevant statistical tests for validity—ages in italics are rejected on statistical grounds.

(Baksi, this volume, Chapter 15). The stepheating work of Pringle (1993) reveals further details on A.I. plots (Fig. 16). The plagioclase separates are altered (Fig. 16A). The plateau ages are, at best, minimum values for the time of crystallization. The whole-rock material (Fig. 16B) shows very high amounts of ^{36}Ar, as these altered rocks were dated in core (chunk) form. All ages are rejected as accurate estimates of the time of crystallization.

Western Pacific Seamounts. Ozima et al. (1977) presented ^{40}Ar/^{39}Ar dating results for six whole-rock samples dredged from Guyots. Numerous step ages are incorrectly reported (see Baksi, 2004). Further, F was put equal to $[SUMS/(N-2)]^{1/2}$, where N is the number of steps or data points used, and SUMS is the sum of the residuals on the isochron fit (see York, 1969; Roddick, 1978), whereas F should be equal to $SUMS/(N-2)$. This erroneous definition reduced many straight lines with excess scatter (errorchrons) to acceptable isochrons. A summary of their age spectra (calculated with the $(^{40}$Ar/^{36}Ar)$_i$ values used by the authors) is listed in Table 1.

Only two samples yield statistically acceptable ($p > 0.05$) plateau ages. The Seiko Guyot sample shows an initial argon ratio below the atmospheric value. This number is unacceptable and results from dating of disturbed (altered) rocks (see Lanphere and Dalrymple, 1978). The results of Ozima et al. (1977) are evaluated for alteration by the A.I. technique (Fig. 17A). All six rocks are severely altered, containing about fifty to two hundred times more ^{36}Ar than found in fresh samples and cannot yield proper crystallization ages. Ozima et al. (1978) analyzed three DSDP rocks; these show ^{36}Ar values more than two hundred times higher than for fresh basalts. The authors noted that "Microscopic examination did not give any positive evidence for (such) K-bearing alteration products" and "preliminary electron micro-

TABLE 1. RESULTS FOR ^{40}AR/^{39}AR STEPHEATING STUDIES ON WHOLE-ROCK SAMPLES FROM THE WESTERN PACIFIC

Sample (Guyot)	$(^{40}$Ar/^{36}Ar)$_i$	Plateau age (Ma)	F (p)
Wilde	296 ± 7	96.1 ± 0.7	0.68 (~0.61)
Lamont	334 ± 9	88.3 ± 3.0	0.26 (~0.77)
Scripps	312 ± 4	100.1 ± 0.4	6.0 (~0.0004)
Renard	327 ± 6	98.0 ± 0.8	10.4 (<10^{-5})
Makarov	302 ± 7	98.5 ± 1.3	3.31 (0.02)
Seiko	256 ± 19	No plateau	N.A.

Notes: Age spectra calculated using the $(^{40}$Ar/^{36}Ar)$_i$ values of Ozima et al. (1977). F is the goodness of fit parameter; p is the corresponding probability value (see text). Errors listed at the 1σ level. N.A.—Not applicable.

probe analysis showed that K residues [*sic*]—essentially along the grain boundaries" (Ozima et al., 1978, p. 702). The latter indicates alteration. The A.I. test (Fig. 17B) shows that the rocks are very badly altered (about two hundred times the cut-off value); such rocks should not be used for argon dating work.

Saito and Ozima (1977) analyzed samples from the Line volcanic chain, the Suiko seamount, and the Necker Rise. A summary of their results is listed in Table 2. Many of their ages are unacceptable for the same reasons as outlined above for the work of Ozima et al. (1977). Isochron ages for 128D, 133D, 137D-9, and 144D were calculated with subatmospheric (<295.5) initial argon ratios. Only three samples yield statistically acceptable ages (i.e., $p > 0.05$). The A.I. for all samples (plateau steps only) is shown in Figure 18. All rocks are severely altered, containing about fifty to five hundred times more ^{36}Ar than found in fresh basaltic material. No crystallization ages were recovered.

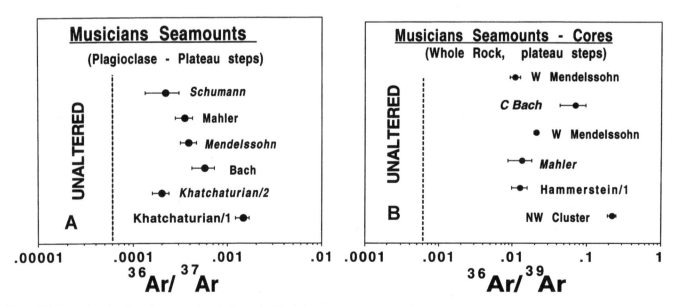

Figure 16. Assessing the alteration state of rocks from the Musicians seamounts dated by Pringle (1993). The average alteration index (A.I.) and standard error on the mean for plateau steps are shown on log scales. (A) Results for plagioclase separates; all samples are altered. (B) Results for whole rocks; very high A.I. values show that samples are badly altered. No accurate ages were recovered.

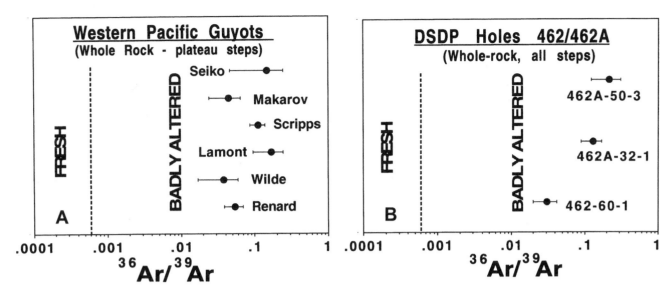

Figure 17. Assessing the alteration of whole-rock western Pacific samples. The average alteration index (A.I.) and standard error on the mean are shown on log scales. (A) Plateau steps for rocks from Guyots (Ozima et al., 1977); all rocks are badly altered, and no proper ages were recovered. (B) Evaluation of basalts and an altered dolerite from Deep Sea Drilling Project (DSDP) hole 462/462A, analyzed by Ozima et al. (1978). The rocks are badly altered, containing about two hundred times the amount of ^{36}Ar of fresh basaltic material and display the efficacy of the A.I. technique in detecting alteration, which is not readily discernible by optical techniques.

Galapagos. Whole-rock material from this area was analyzed using modern instrumentation (Sinton et al., 1996). Most samples are young (<10 Ma), and the resulting plateau ages appear to have acceptable statistics but show high error estimates. Their A.I. (listed in parentheses) shows that all rocks have suffered alteration. Specimen 17-4 (~0.0012) appears to be less altered, whereas samples 1-46, 4-19, 5-1, 10-5, and 20-2 (~0.005–

0.008) are more altered, containing about ten times more ^{36}Ar than found in fresh material (<0.0006). Their ages are minimum estimates of the time of crystallization. The conclusions of Sinton et al. (1996) regarding the velocity of the Nazca plate relative to the Galapagos hotspot and possible changes in Pacific hotspot motion are rejected, as they are based on inaccurate radiometric ages.

TABLE 2. RESULTS FOR ^{40}AR/^{39}AR STEPHEATING STUDIES ON WHOLE-ROCK SAMPLES FROM THE LINE ISLANDS, NECKER RISE, AND SUIKO SEAMOUNT

Sample number	(^{40}Ar/^{36}Ar)$_i$	Plateau age (Ma)	F (p)
119D	293 ± 13	63.9 ± 0.3	3.9 (~0.004)
128D	275 ± 9	No plateau	N.A.
130D	299 ± 3	73.9 ± 1.8	1.15 (~0.33)
133D	269 ± 3	No plateau	N.A.
134N	308 ± 10	46.7 ± 0.4	6.62 (<10^{-5})
137D-10	277 ± 11	50.4 ± 0.8	4.67 (~0.003)
137D-12	275 ± 21	49.9 ± 0.4	10.7 (<10^{-6})
137D-9	246 ± 8	No plateau	N.A.
144D	278 ± 9	80.1 ± 2.4	0.91 (~0.44)
142D	281 ± 6	No plateau	N.A.
KH68-3	314 ± 3	60.1 ± 0.1	0.31 (~0.91)

Notes: Age spectra calculated using the (^{40}Ar/^{36}Ar)$_i$ values of Saito and Ozima (1977). F is the goodness of fit parameter; *p* is the corresponding probability value (see text). Errors listed at the 1σ level. N.A.—Not applicable.

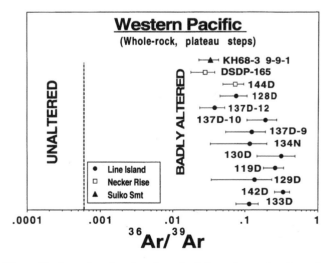

Figure 18. Assessing the alteration of whole-rock samples from the Line volcanic chain, Suiko seamount, and Necker Rise (see Table 2). The average alteration index (A.I.) for the best sites within the rocks (plateau steps) and standard error on the mean are shown on a log scale. All rocks are badly altered, and the ages are rejected as accurate values for the time of crystallization.

Hawaiian-Emperor Chain. The age progression for this island-seamount is included in almost every introductory textbook. Subaerial rocks are altered based on their ^{36}Ar contents. I look at the older ages (primarily ^{40}Ar/^{39}Ar stepheating ages on whole-rock material) that were used to suggest an age of ca. 43 Ma and its subsequent use for plate tectonic velocities and other calculations (over the fixed Hawaiian plume). The plateaus are not questioned on a statistical basis. I evaluate the alteration state of the material used in earlier studies to estimate the age of the bend in the chain.

K-Ar dating and ^{40}Ar/^{39}Ar total fusion work on samples from the Koko seamount led to an age of ca. 47 Ma (Clague and Dalrymple, 1973). A single basalt sample (44-5) shows about three to five times more atmospheric argon than in fresh material. In ^{40}Ar/^{39}Ar total fusion studies, two basaltic samples (43-71 and 43-80) yield A.I. values three to five times higher than for unaltered material. The samples are altered, and the minimum age of the Koko seamount is 47 Ma.

Clague et al. (1975) carried out K-Ar and ^{40}Ar/^{39}Ar studies on alkalic basalts from the northern end of the Hawaiian Ridge and the southernmost part of the Emperor seamounts. The Yuryaku seamount lies close to the bend in the chain. The basalts from this location are partially altered, and Clague et al. (1975) suggested that ^{40}Ar/^{39}Ar methods yielded better results than K-Ar dating. Two samples gave plateau ages of ca. 42 Ma, and their total fusion ages fell in the range 42–49 Ma with 1σ errors of ±2–9 m.y. The A.I. of the plateau steps on these rocks (Fig. 19) indicates that the best sites within the rocks are altered. The average age of ca. 42–44 Ma for the Yuryaku seamount is a minimum value. Dalrymple and Clague (1976) studied alkalic and tholeiitic basalts from the Diakakuji and Kinmei seamounts. The former lies almost exactly at the bend in the chain, and the latter lies ~200 km to the NNW. K-Ar analysis yielded low ages, as the rocks are altered. ^{40}Ar/^{39}Ar total fusion ages on basalts and plagioclase separates fell in the range ca. 42–46 Ma. The A.I. for these samples (see Fig. 19) show ^{36}Ar/^{39}Ar > 0.01 (for basalts; the cut-off value is <0.0006 for fresh rocks) and ^{36}Ar/^{37}Ar ~0.0007–0.0010 (for plagioclase; the cut-off value is <0.00006). These samples are altered, and total fusion ages can, at best, serve as minimum values for the time of crystallization.

Four rocks were subjected to ^{40}Ar/^{39}Ar stepheating. For sample 52-50 from the Kinmei seamount, the A.I. of the plateau steps (Fig. 19) is about ten times higher than acceptable for fresh rocks; the age of 38 Ma is an underestimate of the time of crystallization. In summary, an age of 43 Ma is, at best, a minimum value for the age of the bend in the Hawaiian-Emperor chain. More recent work on mineral separates (Sharp and Clague, 2006) gives an age of 50 Ma for the initiation of the bend. These samples were acid-washed prior to analyses. Plagioclase show A.I. values of ~0.00005; the material is fresh, and their plateau ages are good estimates of crystallization values. However, the suggestion of Sharp and Clague (2006)—that the earlier ages

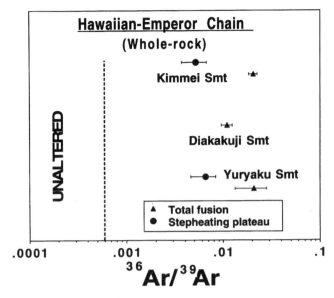

Figure 19. Assessing the accuracy of ^{40}Ar/^{39}Ar ages for rocks used to obtain an age of 43 Ma for the bend in the Hawaiian-Emperor chain. Data from Clague and Dalrymple (1973), Clague et al. (1975), and Dalrymple and Clague (1976). Alteration index (A.I.) for ^{40}Ar/^{39}Ar analyses on rocks from the Yuryaku, Diakakuji, and Kinmei seamounts are shown, with standard error on the mean, on a log scale. Total fusion values show that rocks are altered. Stepheating analyses on the Kinmei and Yuryaku seamounts show lower A.I. values, but plateau steps are derived from altered sites or phases. The age of the bend must be >43 Ma. Smt—seamount.

leading to an age of ca. 43 Ma for the bend was due to dating postshield material—appears to be incorrect. Rather, as outlined above, it resulted from the dating of altered material. Further, the large and rapid changes in plate velocity implied by Sharp and Clague (2006, Fig. 2) cannot be correct. This "model" is predicated on their use of an age of ca. 77 Ma for the Detroit seamount, based on the work of Duncan and Keller (2004). The specimens dated by Duncan and Keller (2004) are clearly altered; A.I. values for whole-rock basalt and plagioclase are more than ten times higher than cut-off values for fresh material.

Gilbert Ridge and Tokelau Seamounts. Koppers and Staudigel (2005) determined ages for these features, which have sharp (60°) bends, similar to that in the Hawaiian-Emperor chain. All three bends should be synchronous at ca. 43–47 Ma, if they were formed by Pacific plate motion over stationary hotspots. The ^{40}Ar/^{39}Ar data suggest that the bends in the Gilbert Ridge and Tokelau seamounts were formed at ca. 67 and ca. 57 Ma, respectively. Koppers and Staudigel (2005) conclude that *their findings are not compatible with the stationary hotspot paradigm.* The mineral separates from rocks in both areas give plateaus that generally meet the statistical requirements for validity and appear to be relatively unaltered. Koppers and Staudigel (2005) also reported plateau ages for eleven groundmass separates that had been acid leached. Scrutiny of these data show

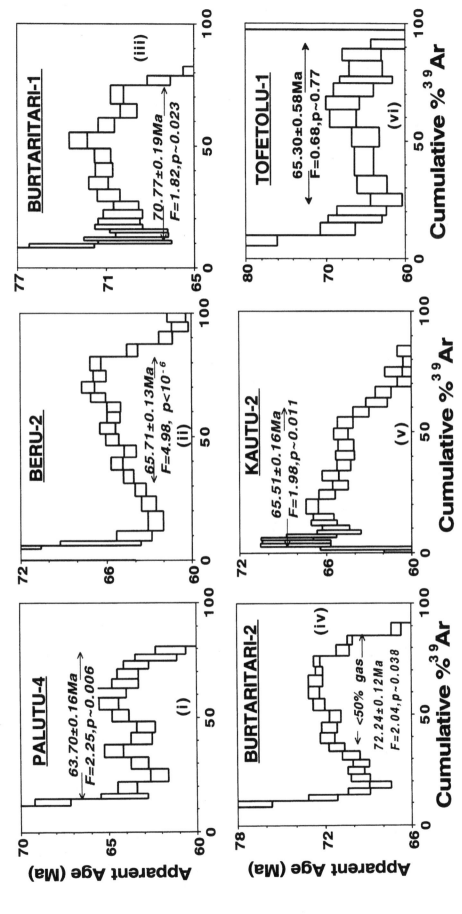

Figure 20. $^{40}Ar/^{39}Ar$ age spectra for acid-leached groundmass samples from the Gilbert Ridge, analyzed by Koppers and Staudigel (2005). Symbols as in Figure 1A. All except the Tofetolu-1 sample show excess scatter ($p < 0.05$) and are disturbed specimens. They do not yield valid plateau ages as stated by Koppers and Staudigel (2005). The plagioclase separate ages on the Gilbert Ridge (Koppers and Staudigel, 2005) appear to be statistically valid.

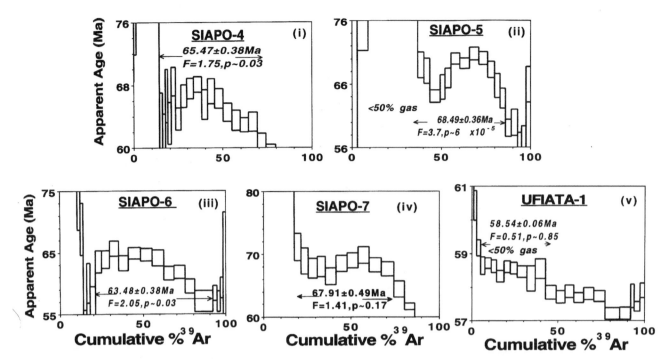

Figure 21. ^{40}Ar/^{39}Ar age spectra for acid-leached groundmass samples from the Tokelau seamounts, analyzed by Koppers and Staudigel (2005). Symbols as in Figure 1A. All except Siapo-7 show excess scatter ($p < 0.05$), and are disturbed specimens; Ufiata-1 displays at best a marginal (<50% gas) plateau. They do not yield valid plateau ages as stated by Koppers and Staudigel (2005). The plagioclase separate ages on the Tokelau seamounts (Koppers and Staudigel, 2005) appear to be statistically valid.

that only one sample each from Tofetolu and Siapo meet the statistical requirements for validity (see Figures 20 and 21). Their A.I. suggests that most of the samples are quite fresh. The Tofetolu-1 and Siapo-7 samples gave statistically valid plateau ages, but these may not be correct estimates of time of crystal-

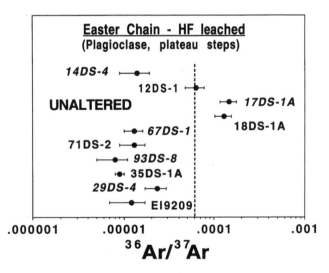

Figure 22. Assessing the alteration state of plagioclase samples from the Easter chain, dated by O'Connor et al. (1995). Average alteration index (A.I.) and standard error on the mean of plateau steps shown on a log scale. All samples are quite fresh, proving the efficacy of HF in removing alteration products.

lization. The Tofetolu-1 groundmass gave a plateau age of 65.30 ± 0.58 Ma, whereas the plagioclase separate gave a measurably older age of 67.11 ± 0.24 Ma. The plateau age of Siapo-7 groundmass cannot be confirmed, because no plateau ages for plagioclase from this location are available (A.A.P. Koppers, personal commun., 2005). Plagioclase separates yield better ages than whole-rock samples. It is therefore critical to examine all aspects of a recent work on dating crystalline groundmass separates (Koppers et al., 2000), from the statistical and A.I. points of view. These detailed data sets are currently unavailable.

Many workers, including Koppers and Staudigel (2005), use FCT-3 Biotite as the monitor, with an age of 28.03 Ma, referencing the work of Renne et al. (1998). The latter did not analyze FCT-3 Biotite. The correct age of FCT-3 Biotite relative to the calibrations of Renne et al. (1998) is ca. 28.23 Ma (see Baksi et al., 1996; Baksi, 2003).

Easter Chain Volcanism. O'Connor et al. (1995) presented ages for volcanism along this chain. They utilized plagioclase separates leached in 6% HF prior to ^{40}Ar/^{39}Ar dating. Many samples contained excess argon, but careful stepheating led to (statistically speaking) good plateau ages. Most of the A.I. values fall in the range 0.00002–0.00006; i.e., plagioclase samples were unaltered (see Fig. 22). Careful selection of samples, their pretreatment with acid, and good laboratory techniques can lead to accurate ages. An intriguing facet of their work is the recovery of some plateau ages that are "too good" (see Fig. 23). The simplest solution is to reduce the estimated errors in each step

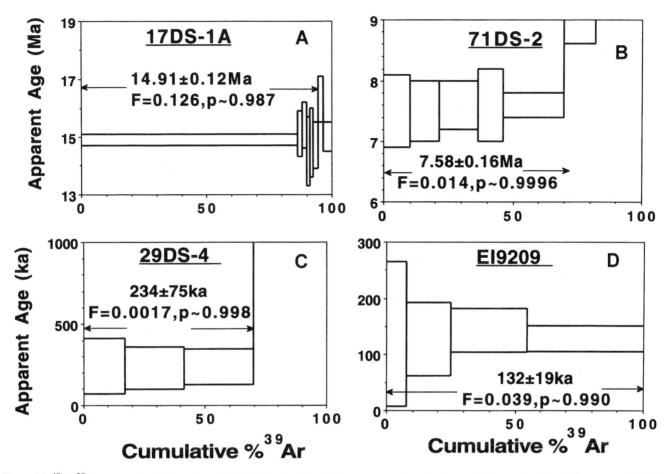

Figure 23. $^{40}Ar/^{39}Ar$ age spectra for four (out of ten) HF-leached plagioclase separates from the Easter chain, analyzed by O'Connor et al. (1995). Symbols as in Figure 1A. All plateau ages are statistically valid, but the statistics are "too good," suggesting that the errors reported by these authors should be reduced to ~20%.

by a factor of about five. The standard error on the mean for the plateau ages would be reduced correspondingly.

CONCLUSIONS

Summary on the Validity of Hotspot-Relatd Ages

The ages of most of the major continental flood basalt provinces are being narrowed by the use of $^{40}Ar/^{39}Ar$ dating techniques. Care must be used to select fresh, acid-leached specimens, and the resulting plateau/isochrons must be rigorously tested for statistical validity, as well as by the A.I. technique for freshness of the phases dated. In numerous cases, confusion or difficulty has resulted from the lack of full tables of analytical data. Journals must insist on the authors making such data sets available in print or on request. For elucidating the duration of such volcanic and/or magmatic events, as well as possible temporal overlap with other geological phenomena (e.g., faunal extinction events), all material must be dated by a single technique. If the $^{40}Ar/^{39}Ar$ method is chosen, the samples should be neu-

tron irradiated in a single batch (cf. McWilliams et al., 1992) to eliminate errors associated with the uncertainty in the determination of the irradiation (J) parameter. Lack of such data does not permit unequivocal confirmation of the oft-postulated link between flood basalt volcanism and global faunal extinction events.

Very few seafloor rocks yield proper crystallization ages based on argon dating studies (see Table 3). This observation is unsurprising in light of their alteration state, resulting from prolonged contact with a medium conducive to alteration—seawater. Analyses prior to ca. 1990 generally yielded unreliable ages (e.g., Duncan, 1978, 1991). Whole-rock material was dated without acid leaching, and grains were not selected for freshness utilizing binocular examination. Use of material in chunk or mini-core form (e.g., Pringle, 1993) caused further problems. During this early period, work on mineral separates gave somewhat better ages. The A.I. method reveals that samples were almost invariably altered and cannot be expected to give accurate estimates of the time of crystallization. Work carried out in the past 10–15 years used modern instrumentation, permitting use of smaller subsamples (often <10 mg). Even in these cases, alteration (particularly in whole-rock material) can escape detection, as confirmed by evaluation of the alteration index of each specimen. The studies of Sinton et al. (1996, 1998) used altered material, and the resulting ages are not proper estimates of the time of crystallization. The necessity for acid leaching all specimens prior to dating is emphasized. The work of Koppers and Staudigel (2005) indicates that HF treatment of plagioclase and HNO_3 leaching of whole rocks can generally remove altered material. Attention must be directed to calculation of A.I. for all steps, and it is critical that the resulting ages be statistically evaluated for validity on age-spectrum and/or isochron plots. The whole-rock ages of Koppers and Staudigel (2005) are not valid estimates of crystallization age; their plagioclase ages are proper estimates, and hence their conclusions regarding the timing of bends in the Gilbert Ridge and Tokelau seamounts (asynchronous with that of the Hawaiian-Emperor chain) appear valid.

Recommendations for Argon Dating of SeaFloor Rocks

Seafloor rocks are recovered at considerable expense, and it is tempting to attempt radiometric dating of all material. The argon dating methods remain the best tool. In many cases, the material recovered is severely altered, and dating should not be attempted. "If the reliability of a sample is not certain beforehand, then the results will be ambiguous" (Dalrymple and Lanphere, 1969, p. 184). This guideline has often not been followed; the literature has numerous ages for seafloor rocks that are not accurate measures of their time of crystallization.

The following procedure is recommended in approaching such work in the future:

1. Carefully screen all samples petrographically and by microscopic examination.
2. Using only the freshest material, work with coarse crushed samples (see above) and not with chunks of whole rocks.
3. Carry out acid leaching of all specimens prior to dating, using HF for feldspar samples and HNO_3 for material containing ferromagnesian phases (see Baksi and Archibald, 1997).
4. Following total fusion on a weighed amount of material that has not been neutron irradiated, calculate the amount of ^{36}Ar present (by the manometric mode of mass spectrometery; see Baksi, 1973).
5. $^{40}Ar/^{39}Ar$ stepheating work should only be attempted on the freshest material, selected following the guidelines outlined here and in Baksi (this volume, Chapter 15).
6. Plagioclase separates leached with HF give better results than whole-rock material.

TABLE 3. SUMMARY OF THE VALIDITY OF PUBLISHED ARGON AGES ON SEAFLOOR ROCKS LINKED TO HOTSPOT (TRACKS)

Sample geographic location	Statistical evaluation of plateau/isochron	Alteration state of material dated	Validity of ages
Caribbean flood basalts	Poor; not all data sets available	Poor (see Fig. 10)	All ages are minimum values
Ontong Java plateau	OK(?); but few or no data sets available	Poor (see Fig. 11)	Age appears to be a minimum value
Costa Rica	Poor; not all data sets available	Poor (see Fig. 12)	All ages are minimum values
Ninetyeast Ridge	Very poor (see Baksi, 2005)	Very poor (see Fig. 13A)	No proper ages recovered
Mascarene plateau–Chagos Ridge	Very poor (see Baksi, 2005)	Very poor (see Fig. 13B)	Only one specimen gives a proper age
New England seamounts	Poor (see Baksi, 2005)	Poor (see text)	Ages are mimimum values
Walvis Ridge–Rio Grande Rise	Poor (see Baksi, 2005)	Poor (see Fig. 14A)	Ages are mimimum values
St. Helena–Gough chains	Poor (see Fig. 15)	Poor (see Fig. 14B)	Only one specimen gives a proper age
Coast Range, Oregon	Poor (see Baksi, 2005)	Poor (see text)	Ages are mimimum values
Musicians seamounts	Generally good	Poor (see Fig. 16)	Ages are mimimum values
Western Pacific Guyots	Very poor (see Table 1)	Very poor (see Fig. 17A)	No proper ages recovered
Line Island/Suiko seamounts	Very poor (see Table 2)	Very poor (see Fig. 18)	No proper ages recovered
Hawaiian-Emperor chain (1)	Generally good	Poor (see Fig. 19)	Incorrect ages
Hawaiian-Emperor chain (2)	Good	Generally good	Ages appear valid
Tokelau seamounts (whole rock)	Poor (see Fig. 20)	Generally good (see text)	No proper ages recovered; plagioclase ages are valid
Gilbert Ridge (whole rock)	Poor (see Fig. 21)	Generally good (see text)	No proper ages recovered; plagioclase ages are valid
Easter chain	Good (see Fig. 23)	Very good (see Fig. 22)	Good ages; errors overestimated?

Notes: For the Hawaiian-Emperor chain, (1) refers to the work of Clague and Dalrymple (1973), Clague et al. (1975), Dalrymple and Clague (1976), and Duncan and Keller (2004); (2) refers to the work of Sharp and Clague (2006).

7. All results must be critically examined for statistical validity and for freshness of the phases dated, using the A.I. technique.

With careful work along these lines a (small) body of reliable crystallization ages will emerge. These data can then be examined for trails of hotspot tracks, possible coincidence with other geological phenomena, and calculation of the duration and/or rates of magmatic events.

ACKNOWLEDGMENTS

Comments and suggestions made by the editors of this volume and the official reviewers helped in shaping this manuscript. I thank those who made available tables of analytical data from their published papers. Carlo Pedretti kindly identified the exact sources of da Vinci's quotations. I express my gratitude to the late John Reynolds; his work helped establish modern inert gas work, and his imprint on scientists utilizing the argon dating techniques is indelible.

REFERENCES CITED

Allègre, C.J., Birck, J.L., Capmas, F., and Courtillot, V., 1999, Age of the Deccan Traps using ^{187}Re-^{187}Os systematics: Earth and Planetary Science Letters, v. 170, p. 197–204, doi: 10.1016/S0012-821X(99)00110-7.

Baksi, A.K., 1973, Quantitative unspiked argon runs in K-Ar dating: Canadian Journal of Earth Sciences, v. 10, p. 1415–1419.

Baksi, A.K., 1974a, Isotopic fractionation of a loosely held atmospheric argon component in the Picture Gorge Basalts: Earth and Planetary Science Letters, v. 21, p. 431–438, doi: 10.1016/0012-821X(74)90183-6.

Baksi, A.K., 1974b, K-Ar study of the SP flow: Canadian Journal of Earth Sciences, v. 11, p. 1350–1356.

Baksi, A.K., 1991, U-Pb zircon and baddeleyite ages for the Palisade and Gettysburg Sills of the northeastern United States—Implications for the age of the Triassic-Jurassic boundary—Comment: Geology, v. 19, p. 860–861.

Baksi, A.K., 1994, Geochronological studies on whole-rock basalts, Deccan Traps, India: Evaluation of the timing of volcanism relative to the K-T boundary: Earth and Planetary Science Letters, v. 121, p. 43–56, doi: 10.1016/0012-821X(94)90030-2.

Baksi, A.K., 1999, Revaluation of plate motion models based on hotspot tracks in the Atlantic and Indian oceans: Journal of Geology, v. 107, p. 13–26, doi: 10.1086/314329.

Baksi, A.K., 2003, Critical evaluation of ^{40}Ar/^{39}Ar ages for the central Atlantic mamatic province: Timing, duration and possible migration of magmatic centers, *in* Hames, W.E., et al., eds., The central Atlantic magmatic province: Insights from fragments of Pangea: Washington, D.C., American Geophysical Union Geophysical Monograph 136, p. 77–90.

Baksi, A.K., 2004, Critical evaluation of radiometric ages used for tracking hotspots in the Pacific Ocean: EOS (Transactions, American Geophysical Union), v. 85, fall meeting supplement, no. 47, abstract no. 51B-0544.

Baksi, A.K., 2005, Evaluation of radiometric ages pertaining to rocks hypothesized to have been derived by hotspot activity, in and around the Atlantic, Indian and Pacific oceans, *in* Foulger, G.R., et al., eds., Plates, plumes, and paradigms: Boulder, Colorado, Geological Society of America Special Paper 388, p. 55–70.

Baksi, A.K., 2007 (this volume, chapter 15), A quantitative tool for detecting alteration in undisturbed rocks and minerals—I: Water, chemical weathering, and atmospheric argon, *in* Foulger, G.R., and Jurdy, D.M., eds., Plates,

plumes, and planetary processes: Boulder, Colorado, Geological Society of America Special Publication 430, doi: 10.1130/2007.2430(15).

Baksi, A.K., and Archibald, D.A., 1997, Mesozoic igneous activity in the Maranhao province, northern Brazil: ^{40}Ar/^{39}Ar evidence for separate episodes of magmatism: Earth and Planetary Science Letters, v. 151, p. 139–153, doi: 10.1016/S0012-821X(97)81844-4.

Baksi, A.K., Archibald, D.A., and Farrar, E., 1996, Intercalibration of ^{40}Ar/^{39}Ar dating standards: Chemical Geology, v. 129, p. 307–324, doi: 10.1016/0009-2541(95)00154-9.

Clague, D.A., and Dalrymple, G.B., 1973, Age of Koko seamount, Emperor seamount chain: Earth and Planetary Science Letters, v. 17, p. 411–415, doi: 10.1016/0012-821X(73)90209-4.

Clague, D.A., Dalrymple, G.B., and Moberly, R., 1975, Petrography and K-Ar ages of dredged volcanic rocks from the western Hawaiian ridge and the southern Emperor seamount chain: Geological Society of America Bulletin, v. 86, p. 991–998, doi: 10.1130/0016-7606(1975)86<991:PAKAOD>2.0.CO;2.

Clouard, V., and Bonneville, A., 2005, Ages of seamounts, islands, and plateaus on the Pacific plate, *in* Foulger, G.R., et al., eds., Plates, plumes, and paradigms: Boulder, Colorado, Geological Society of America Special Paper 388, p. 71–90.

Courtillot, V., Feraud, G., Maluski, H., Vandamme, D., Moreau, M.G., and Besse, J., 1988, Deccan flood basalts and the Cretaceous-Tertiary boundary: Nature, v. 333, p. 843–846, doi: 10.1038/333843a0.

Dalrymple, G.B., and Clague, D.A., 1976, Age of the Hawaiian-Emperor bend: Earth and Planetary Science Letters, v. 31, p. 313–329, doi: 10.1016/0012-821X(76)90113-8.

Dalrymple, G.B., and Lanphere, M.A., 1969, Potassium-argon dating: San Francisco, Freeman, 258 p.

Dalrymple, G.B., Alexander, E.C., Lanphere, M.A., and Kraker, G.P., 1981, Irradiation of samples for ^{40}Ar/^{39}Ar dating using the Geological Survey TRIGA reactor: Reston, Virginia, U.S. Geological Survey Professional Paper 1176, 55 p.

Duncan, R.A., 1978, Geochronology of basalts from the Ninetyeast ridge and continental dispersion in the eastern Indian Ocean: Journal of Volcanology and Geothermal Research, v. 4, p. 283–305, doi: 10.1016/0377-0273(78)90018-5.

Duncan, R.A., 1982, A captured island chain in the Coast Range of Oregon and Washington: Journal of Geophysical Research, v. 87, p. 10,827–10,837.

Duncan, R.A., 1984, Age progressive volcanism in the New England seamounts and the opening of the central Atlantic Ocean: Journal of Geophysical Research, v. 89, p. 9980–9990.

Duncan, R.A., 1991, Age distribution and volcanism along aseismic ridges in the eastern Indian Ocean: Proceedings of the Ocean Drilling Program, v. 121, p. 507–517.

Duncan, R.A., and Hargraves, R.B., 1990, ^{40}Ar-^{39}Ar geochronology of basement rocks from the Mascarene plateau, Chagos bank and Maldive ridge: Proceedings of the Ocean Drilling Program, v. 115, p. 43–51.

Duncan, R.A., and Keller, R.A., 2004, Radiometric ages for basement rocks from the Emperor seamounts: Geochemistry, Geophysics, Geosystems, v. 5, Q08L03, doi: 10.1029/2004GC000704.

Duncan, R.A., and Pyle, D.G., 1988, Rapid extrusion of the Deccan flood basalts at the Cretaceous-Tertiary boundary: Nature, v. 333, p. 841–843, doi: 10.1038/333841a0.

Hofmann, C., Feraud, G., and Courtillot, V., 2000, ^{40}Ar/^{39}Ar dating of mineral separates and whole rocks from the Western Ghats lava pile: Further constraints on duration and age of the Deccan Traps: Earth and Planetary Science Letters, v. 180, p. 13–27, doi: 10.1016/S0012-821X(00)00159-X.

Knight, K.B., Renne, P.R., Halkett, A., and White, N., 2003, ^{40}Ar/^{39}Ar dating of the Rajahmundry Traps, eastern India, and their relationship to the Deccan Traps: Earth and Planetary Science Letters, v. 209, p. 257, doi: 10.1016/S0012-821X(03)00078-5.

Koppers, A.A.P., and Staudigel, H., 2005, Asynchronous bends in Pacific seamount trails: A case for extensional volcanism?: Science, v. 307, p. 904–907, doi: 10.1126/science.1107260.

Koppers, A.A.P., Staudigel, H., and Wijbrans, J.R., 2000, Dating crystalline groundmass separates of altered Cretaceous seamount basalts by the ^{40}Ar/^{39}Ar incremental heating technique: Chemical Geology, v. 166, p. 139–158, doi: 10.1016/S0009-2541(99)00188-6.

Langmuir, I., 1989, Pathological science (transcribed and edited by Hall, R.N.): Physics Today, v. 42, p. 36–48.

Lanphere, M.A., and Dalrymple, G.B., 1978, The use of ^{40}Ar/^{39}Ar data in evaluation of disturbed K-Ar systems: Reston, Virginia, U.S. Geological Survey Open-File Report 78-701, p. 241–243.

Mahoney, J.J., Macdougall, J.D., Lugmair, G.W., and Gopalan, K., 1983, Kerguelen hotspot source for Rajmahal Traps and Ninetyeast Ridge?: Nature, v. 303, p. 385–389, doi: 10.1038/303385a0.

Mahoney, J.J., Storey, M., Duncan, R.A., Spencer, K.J., and Pringle, M., 1993, Geochemistry and age of the Ontong Java plateau, *in* Pringle, M.S., et al., eds., The Mesozoic Pacific: Geology, tectonics, and volcanism: Washington, D.C., American Geophysical Union Geophysical Monograph 77, p. 233–261.

McDougall, I., and Harrison, T.M., 1999, Geochronology and thermochronology by the ^{40}Ar/^{39}Ar method: New York, Oxford University Press, 269 p.

McWilliams, M.O., Baksi, A.K., Bohor, B.F., Izett, G.A., and Murali, A.V., 1992, High-precision relative ages of K/T boundary events in North America and Deccan Trap volcanism in India: EOS (Transactions, American Geophysical Union), v. 73, p. 363.

Merrihue, C., and Turner, G., 1966, Potassium-argon dating by activation with fast neutrons: Journal of Geophysical Research, v. 71, p. 2852–2857.

O'Connor, J.M., and Duncan, R.A., 1990, Evolution of the Walvis ridge–Rio Grande rise hot spot systems: Implications for African and South American plate motions over plumes: Journal of Geophysical Research, v. 95, p. 17,475–17,502.

O'Connor, J.M., and Le Roex, A.P., 1992, South Atlantic hot-spot plume systems: 1. Distribution of volcanism in time and space: Earth and Planetary Science Letters, v. 113, p. 343–364, doi: 10.1016/0012-821X(92)90138-L.

O'Connor, J.M., Stoffers, P., and McWilliams, M.O., 1995, Time-space mapping of Easter Chain volcanism: Earth and Planetary Science Letters, v. 136, p. 197–212, doi: 10.1016/0012-821X(95)00176-D.

Ozima, M., Honda, M., and Saito, K., 1977, ^{40}Ar-^{39}Ar ages of guyots in the western Pacific and discussion of their evolution: Geophysical Journal of the Royal Astronomical Society, v. 51, p. 475–485.

Ozima, M., Saito, K., and Takagami, Y., 1978, ^{40}Ar-^{39}Ar geochronological studies on rocks drilled at holes 462 and 462A, Deep Sea Drilling Project Leg 61: Initial Reports of the Deep Sea Drilling Project (Volume 61): Washington, D.C., U.S. Government Printing Office, p. 701–703.

Pringle, M.S., 1993, Age progressive volcanism in the Musicians seamounts; A test of the hotspot hypothesis for the late Cretaceous Pacific, *in* Pringle, M.S., et al., eds., The Mesozoic Pacific: Geology, tectonics and volcanism: Washington, D.C., American Geophysical Union Geophysical Monograph 77, p. 187–215.

Renne, P.R., 2000, ^{40}Ar/^{39}Ar age of plagioclase from Acapulco meteorite and problem of systematic errors in cosmochronology: Earth and Planetary Science Letters, v. 175, p. 13–26.

Renne, P.R., Swisher, C.C., Deino, A.L., Karner, D.B., Owens, T.L., and DePaolo, D.J., 1998, Intercalibration of standards, absolute ages and uncertainties in ^{40}Ar-^{39}Ar dating: Chemical Geology, v. 145, p. 117–152, doi: 10.1016/S0009-2541(97)00159-9.

Roddick, J.C., 1978, The application of isochron diagrams in ^{40}Ar-^{39}Ar dating: A discussion: Earth and Planetary Science Letters, v. 41, p. 233–244.

Roddick, J.C., 1983, High precision intercalibration of ^{40}Ar-^{39}Ar standards: Geochimica et Cosmochimica Acta, v. 47, p. 887–898, doi: 10.1016/0016-7037(83)90154-0.

Saito, K., and Ozima, M., 1977, ^{40}Ar-^{39}Ar geochronological studies on submarine rocks from the western Pacific area: Earth and Planetary Science Letters, v. 33, p. 353–369, doi: 10.1016/0012-821X(77)90087-5.

Seidemann, D.E., 1978, ^{40}Ar/^{39}Ar studies of deep-sea igneous rocks: Geochimica et Cosmochimica Acta, v. 42, p. 1721–1734, doi: 10.1016/0016-7037(78)90258-2.

Sharp, W.D., and Clague, D.A., 2006, 50-Ma initiation of the Hawaiian-Emperor bend records major change in Pacific plate motion: Science, v. 313, p. 1281–1284, doi: 10.1126/science.1128489.

Sinton, C.W., Christie, D.M., and Duncan, R.A., 1996, Geochronology of Galapagos seamounts: Journal of Geophysical Research, v. 101, p. 13,689–13,700, doi: 10.1029/96JB00642.

Sinton, C.W., Duncan, R.A., and Denyer, P., 1997, Nicoya peninsula, Costa Rica: A single suite of Caribbean oceanic plateau magmas: Journal of Geophysical Research, v. 102, p. 15,507–15,520, doi: 10.1029/97JB00681.

Sinton, C.W., Duncan, R.A., Storey, M., Lewis, J., and Estrada, J.J., 1998, An oceanic flood basalt province within the Caribbean plate: Earth and Planetary Science Letters, v. 155, p. 221–235, doi: 10.1016/S0012-821X(97)00214-8.

Steiger, R.H., and Jager, E., 1977, Subcommission on geochronology: Convention of the use of decay constants in geo- and cosmochronology: Earth and Planetary Science Letters, v. 36, p. 359–362, doi: 10.1016/0012-821X(77)90060-7.

Turrin, B.D., Donnelly-Nolan, J.M., and Hearn, B.C., Jr., 1994, ^{40}Ar-^{39}Ar ages from the rhyolite of Alder Creek, California: Age of the Cobb Mountain normal polarity subchron revisited: Geology, v. 22, p. 251–254, doi: 10.1130/0091-7613(1994)022<0251:AAAFTR>2.3.CO;2.

Venkatesan, T.R., Pande, K., and Gopalan, K., 1993, Did Deccan volcanism predate the Cretaceous/Tertiary boundary?: Earth and Planetary Science Letters, v. 119, p. 181–189, doi: 10.1016/0012-821X(93)90015-2.

Walker, D.A., and McDougall, I., 1982, ^{40}Ar/^{39}Ar and K-Ar dating of altered glassy volcanic rocks: The Dabi volcanics, P.N.G: Geochimica et Cosmochimica Acta, v. 46, p. 2181–2190, doi: 10.1016/0016-7037(82)90193-4.

York, D., 1969, Least squares fitting of a straight line with correlated errors: Earth and Planetary Science Letters, v. 5, p. 320–324, doi: 10.1016/S0012-821X(68)80059-7.

York, D., Hall, C.M., Gaspar, M.J., and Lynch, M., 1984, Laser-probe ^{40}Ar/^{39}Ar dating with ultra-sensitive mass spectrometer: EOS (Transactions, American Geophysical Union), v. 65, p. 303.

MANUSCRIPT ACCEPTED BY THE SOCIETY 31 JANUARY 2007

DISCUSSION

17 November 2006, Romain Meyer, Michael Abratis, and Henry Rauche

In addition to Baksi's exceptionally detailed description of alteration problems resulting in geological Ar-Ar age errors, a recent study of igneous rocks from the Central European volcanic province (Abratis et al., 2007) clearly points to an additional process responsible for wrong radiometric Ar ages. In this study, all data except for one outlier range between 20 and 14 Ma. The outlier, with an age of 29 Ma, is from a volcanic dike in the northeastern Rhön, close to the Thuringian Forest.

The sample location is in the Werra district, a region in Germany famous for its former potash mining industry. These Permian (Zechstein) potash salt deposits were penetrated in the Cenozoic by basaltic magma dikes. This Central European volcanic province magmatism has been linked in the literature by many authors to a postulated mantle plume arriving in the subcontinental lithosphere, on the basis of mantle tomography, geochemistry, and radiogenic age data (see references in Lustrino and Wilson, 2007).

Despite the tremendous thermal and tectonic stresses that must have accompanied injection of the magmas, the salt layers remained nearly unchanged. However, geochemical salt–magma interactions occurred (e.g., diffusion of K and Ar from the salt into the magma is apparent). The K and Ar geochemistry of these basalts illustrates such assimilation and/or contamination via (1) potash salt in the basaltic magma and (2) fluid circulation in the proximal zone of the basaltic dikes (Steinmann et al., 1999). Basalts having undergone these interactions are enriched in highly radiogenic Ar from the salt deposit and cannot indicate a real crystallization age.

Thus, it is not only processes that occur after crystallization (e.g., alteration) that can affect apparent radiogenic ages, but as is proposed for geochemical mantle plume "fingerprints," melt–crust rock interactions can mask the initial signal. The lithology of the rocks through which the magma rises to the Earth's surface can cause inaccurate estimates of the crystallization ages. Even if basalts are petrographically fresh, statistically acceptable ages of Central European volcanic province rocks from central Germany are difficult to obtain due to (1) the widespread distribution of Zechstein salt deposits and (2) the likelihood that only the central parts of large dikes that penetrated into salt deposits are not affected by highly radiogenic Ar from the salt. A good knowledge of the continental crust below continental volcanic provinces will permit better constraint of the intrinsic geochemical heterogeneity—in space and time—of mantle melts.

15 January 2007, Ajoy K. Baksi

Meyer and colleagues raise an important point that was inadvertently left out in the final revision of my manuscript. Namely, "alteration" can, on occasion, lead to ^{40}Ar/^{39}Ar ages that are too old (see also Dalrymple and Lanphere, 1969). This problem appears to beset intrusive rocks and is illustrated by study of results on a Deccan Trap sample from India.

A dike from the Panvel area (D-921; see Baksi, 1994) was analyzed using a system with very low argon blanks (Baksi et al., 1992). The results are presented in Figure D-1. The whole-rock specimen contains excess argon, and the isochron yields an age of ca. 62 Ma (i.e., ca. 3 Ma after the main pulse of Deccan volcanism). The alteration index (A.I.) values for the isochron steps in Figure D-1B fall in the range 0.04–0.0015. The cutoff for freshness is <0.0006; the rock is "altered" and should not yield the correct crystallization age. However, it is suggested that the correct age is ca. 62 Ma.

Note that the A.I. technique is successful in pinpointing some cases of alteration in sills—see the section dealing with the Gettysburg Sill in Baksi (this volume, chapter 15). Further investigation is required to fully understand the possible application of the A.I. technique to sills and dikes in general.

24 January 2007, Alexei Ivanov

Baksi (this volume, chapters 15 and 16) suggests a very interesting approach for determining secondary alteration of samples

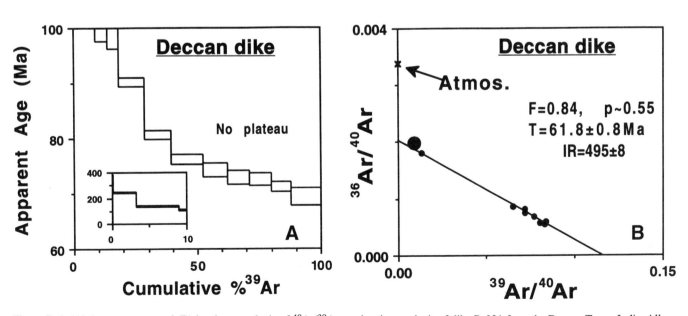

Figure D-1. (A) Age spectrum and (B) isochron analysis of ^{40}Ar/^{39}Ar stepheating analysis of dike D-921 from the Deccan Traps, India. All errors shown at the 1σ level. The rock contains "excess argon" and has an age of ca. 62 Ma. F—goodness of fit parameter; p—probability of occurrence; T—age; IR—initial (^{40}Ar/^{36}Ar) ratio.

through measurements of the amount of [36]Ar. Briefly, in the present-day atmosphere [36]Ar comprises up to 0.338% of the total argon or as much as ~0.003% of the total atmosphere. The mantle and mantle-derived volcanic rocks are thought to contain significantly less [36]Ar. Argon is soluble in water. Thus, chemically weathered minerals contain an excess of atmospheric [36]Ar. On the basis of this line of reasoning, and a number of representative examples from K-Ar and [40]Ar/[39]Ar studies, Baksi (this volume, chapter 15) develops a set of equations and cutoff values for different minerals and basaltic rock compositions to estimate their degree of secondary alteration. Below is rather an addendum to Baksi than a critique of his approach.

Baksi (this volume, chapters 15 and 16; 20 January comment to Hooper et al., this volume; 15 January comment to Ivanov, this volume) may be too strict in accepting or rejecting published ages. I think that there cannot be a single cutoff value, even for the same type of dated material, as is evident from analyses of plagioclases from the central Atlantic magmatic province and the Rajahmundry Traps (Figs. 6B and 7A in Baksi, this volume, chapter 15). This lack of a unique cutoff value may be especially relevant for whole-rock samples, because the whole rock is a mixture of different minerals, each of which may be characterized by its own range of cutoff values. For instance, the occurrence of biotite in the whole rock will shift A.I. toward higher values. Small crystals of primary magmatic mica (biotite?) may occur in some dolerites (see Fig. 12 in Ivanov, this volume).

Baksi (this volume, chapters 15 and 16) acknowledges that recycling of atmospheric water through subduction may create high [36]Ar in mantle-wedge-derived volcanic rocks. The water recycling may also be responsible, in origin, of at least some flood basalt provinces (see Ivanov, this volume; 17 January comment to Hooper et al., this volume). Thus, we may expect high [36]Ar in the flood basalts, too. This possibility necessitates performing special methodical [40]Ar/[39]Ar studies, which may include dating of three matrix aliquots, two leached with HNO_3 and HF, and one unleached, from visually altered and fresh samples from island-arcs, flood basalt provinces, and continental alkaline volcanic rock occurrences.

Another important question is the style and timing of alteration. Was it continuous or episodic, and if episodic, when did the episodic alteration happen? The concept of timing of episodic alteration is illustrated in Figure D-2 with three models for a hypothetical 250-Ma sample:

1. Alteration event at 200 Ma with 100% loss of radiogenic argon,
2. Alteration event at 240 Ma with 50% loss of radiogenic argon, and
3. Alteration event at 249 Ma with 100% loss of radiogenic argon.

Model 1 shows a 20% decrease of apparent age relative to the true crystallization age, whereas models 2 and 3 yield less promi-

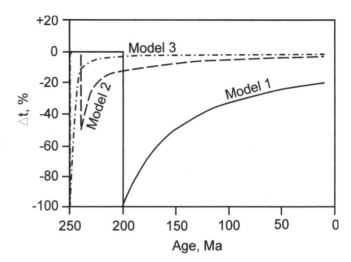

Figure D-2. Dependence of the difference between true and apparent ages on timing of episodic alteration and degree of associated radiogenic argon loss. $\Delta t = (t_{apparent}/t_{true} - 1) \gg \times 100$. See text for description of the models.

nent apparent age decreases (3.5% and 0.4%, respectively). Of course, Figure D-2 gives an oversimplified view of the problem. Alteration (chemical weathering) redistributes potassium from primary minerals to tiny secondary clay minerals, which are less retentive of radiogenic argon and suffer from the nuclear-irradiation–related [39]Ar recoil problem. This is probably the case for the plagioclase sample SCD-9, which shows a complex argon release pattern (Fig. 7A in Baksi, this volume, chapter 15).

In summary:

1. Samples of magmatic rocks and minerals with statistically acceptable plateaus and isochron [40]Ar/[39]Ar ages and A.I. values below the cutoff values represent true crystallization ages.
2. Samples of magmatic rocks and minerals with statistically acceptable plateaus and isochron [40]Ar/[39]Ar ages and A.I. values above the cutoff values are suspected of having apparent ages that are too young. They may yield true crystallization ages if they (a) are derived from a magmatic source with recycled atmospheric water through subduction, or (b) were altered soon after crystallization.
3. The A.I. approach is both simple and powerful. It should be used routinely in [40]Ar/[39]Ar dating for additional assessment of the age reliability.

29 January 2007, Ajoy K. Baksi

Ivanov raises pertinent questions regarding the utility of the A.I. method. Clarification is offered on a number of issues. First, for plagioclase feldspars from mafic material, the cutoff value of A.I. < 0.00006 for freshness appears to be valid. The results on HF-washed samples from the central Atlantic magmatic province

(Hames et al., 2000) show this validity best (Baksi, this volume, chapter 15), and those on the hot HNO₃-leached samples (Verati et al., 2005) may be valid. The latter should be reanalyzed following HF leaching. The acid-unleached samples (Sebai et al. 1991; Deckart et al., 1997) do not give proper plateaus (Baksi, 2003) and fail the A.I. test for freshness (Baksi, this volume, chapter 15). The Rajahmundry Traps data sets referred to by Ivanov (comment of 24 January) are considerably tightened up by rejection of samples showing high A.I. values (see Baksi, 2005a). On extrusion into a shallow-marine or estuarine environment (Baksi, 2001, 2005b), these rocks incorporated higher quantities of ³⁶Ar than most whole-rock basalts.

Ivanov (19 January) suggests magmatism in the Siberian Traps area spanned ~25 m.y. Some (U-Pb) radiometric data appears to support this hypothesis. All ⁴⁰Ar/³⁹Ar (plateau/isochron) ages must be critically evaluated for statistical validity, as well as freshness of the material dated by the A.I. technique (Baksi, 2005a, this volume, chapters 15 and 16). Dalrymple et al. (1995) noted that some of their samples were altered. Most of the whole rocks dated by Venkatesan et al. (1997) show alteration. All the "ages" of Walderhaug et al. (2005) must be rejected on statistical grounds; their isochrons, in particular, show gross excess scatter, and in some instances nonconsecutive steps were used for straight-line fitting. In rare instances, altered samples may yield argon ages that are accurate; this could result from (episodic) alteration very soon after crystallization. It is currently not possible to unequivocally identify the time of alteration in rocks and minerals. Argon ages on samples that have clearly suffered alteration (A.I. values more than five times the cutoff value for freshness—see Baksi, this volume, chapter 15), should not be utilized as accurate estimates of the time of crystallization.

Earlier, the A.I. method was not recommended for subduction zone rocks (Baksi, this volume, chapter 15). Figure D-3 looks to its application to tholeiites and boninites from the Izu-Bonin-Marianas arc in the Pacific Ocean. The plateau steps of Cosca et al. (1998) show elevated A.I. values, but less than initially envisaged (Baksi, this volume, chapter 15). During the (high-temperature) melt generation process, much of the ³⁶Ar in the water driven off the subducted slab escapes and is released from volcanoes. Ivanov suggests some flood basalts may contain contributions from (earlier) subducted material. Melting of such rocks with A.I. values similar to those in Figure D-3 would release much of the ³⁶Ar; furthermore, the solubility of argon in such melts is low (Lux, 1987) and would produce (flood) basalts with low (<0.0006?) A.I. values.

The A.I. method should initially be applied only to mafic, extrusive material. Ivanov expresses concern regarding the cutoff recommended for alkali basalts that may contain minor quantities of hornblende and/or biotite. The Bengal Trap alkali basalt "Debagram" contains primary biotite (Baksi, 1995), was used to construct Figure 4 in Baksi (this volume, chapter 15), and passes the A.I. test for freshness. Ivanov discusses alteration as being an episodic process. I suggest chemical weathering for *extrusive* rocks is (quasi)-continuous (see Fig. 18 in Baksi, this

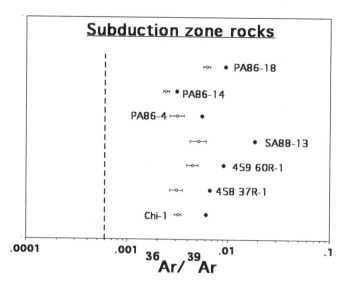

Figure D-3. Alteration index (A.I.) values for rocks from the Izu-Bonin-Marianas Trench (data from Cosca et al., 1998). Average values for plateau steps (open circles) and total gas values (filled circles), shown with associated standard error on the mean values, on a log scale. Dotted line shows cutoff for unaltered basaltic material. High values (i.e., high ³⁶Ar contents) result not from alteration but from interaction with water driven off the subducted slab (see Baksi, this volume, chapter 15).

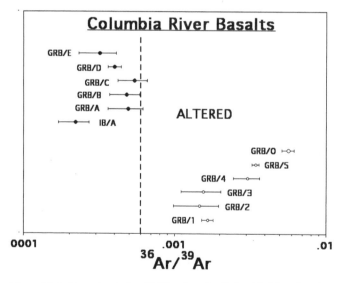

Figure D-4. Alteration index (A.I.) values for Columbia River basalts; average values for plateau steps shown with associated standard error on the mean shown on a log scale. GRB—Grande Ronde basalt; IB—Imnaha basalt. Results of Long and Duncan (1983) shown as open circles; all samples are altered and resulting ages are rejected. Results on HNO₃-washed samples shown as filled circles (A.K. Baksi, unpubl. data, 1993). All samples show A.I. < 0.0006 (cutoff for freshness). The resulting age of ca. 16.4 Ma (IB) is preferred over the earlier result of Baksi and Farrar (1990) on an acid-untreated split of rock. Ages of ca. 16.4–16.0 Ma (GRB) are in agreement with the results of Baksi and Farrar (1990).

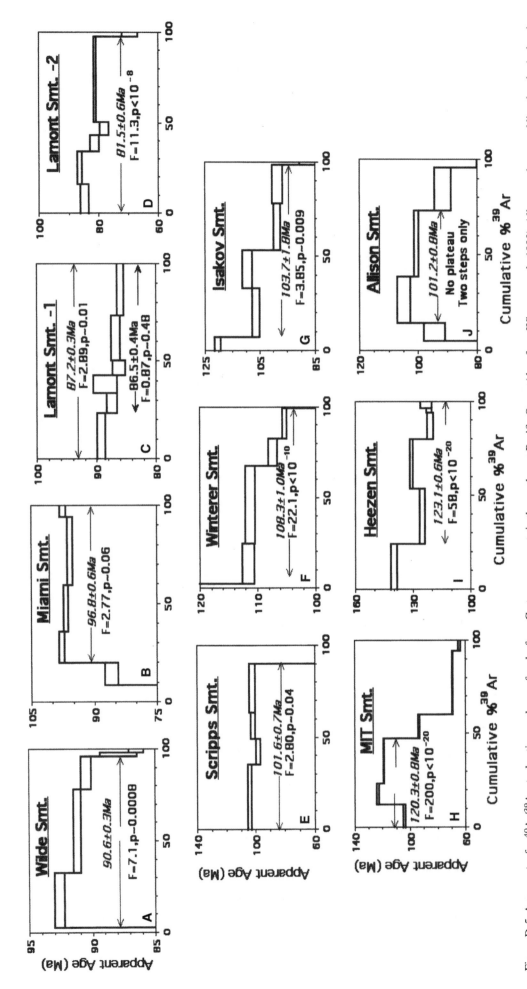

Figure D-5. Age spectra for $^{40}Ar/^{39}Ar$ stepheating analyses of rocks from Cretaceous guyots in the northwest Pacific Ocean (data from Winterer et al., 1993). All errors shown and listed at the 1σ level. Plateau sections and ages postulated by the authors are shown in italics. The corresponding statistical parameters are shown in bold. F—goodness of fit parameter; p—probability of occurrence. All of these ages are rejected, as p < 0.15—the cutoff value suggested by Sharp and Clague (2006) for statistical validity. This value is more stringent than the initial value of p < 0.05 suggested earlier (Baksi, 1999, 2005) and is preferred here. Only steps 2–6 of Lamont Seamount-1 yields a statistically acceptable age.

volume, chapter 15). Many altered continental and oceanic basalts show A.I. values higher than the Izu-Bonin-Marianas arc rocks (see Figs. 3 and 13 in Baksi, this volume, chapter 15, and Figs. 10–19 in Baksi, this volume, chapter 16). Continuous alteration at low temperatures leads to higher A.I. values than seen in higher-temperature–generated subduction-zone rocks.

The high. A.I. values in flood basalts result from alteration and lead to lowered ages, which is illustrated using results from the Columbia River basalt. Hooper et al. (2002) and Hooper (2004) argue for >90% of the volcanism occurring between ca. 16.1 and 15.0 Ma, utilizing in part the ages of Long and Duncan (1983). Figure D-4 examines the A.I. values for the Long and Duncan work. The rocks are altered and yield (incorrect) low ages. Material recovered from the deepest borehole in the Grande Ronde basalt (GRB/O) gave step ages of ca. 10 Ma; ~30% loss of $^{40}Ar^*$ resulted from gross alteration of the material dated. An alternative viewpoint is that most sections of the Columbia River basalt are ca. 0.5 to 1.0 Ma older than envisaged by Hooper (see Baksi, 1993; discussion by Baksi of Hooper et al., this volume). This estimation was based, in large part, on a ca. 17.5-Ma age for the Imnaha basalt (Fig. 1a *in* Baksi and Farrar, 1990). The F value (see caption for Fig. D-1 for definition) for the plateau section is 2.3, the probability of occurrence ~0.02, and the age must be rejected (Baksi, this volume, chapter 15). The other ages of Baksi and Farrar (1990), adjusted to the calibrations of Renne et al. (1998), are 16.3–16.0 Ma for the R_1 through N_2 magnetostratigraphic units. Figure D-4 shows the A.I. for Columbia River basalt rocks analyzed following HNO_3 leaching of crushed whole-rock material (A.K. Baksi, unpubl. data, 1993). Ages are ca. 16.3 Ma for the top of the Imnaha basalt, and for the Grande Ronde basalt they agree with the earlier ages of ca. 16.4–16.0 Ma. All samples show acceptable A.I. values, are unaltered, and thus the ages are more reliable than the acid-untreated rocks of Baksi and Farrar (1990).

In summary, I agree with Hooper et al. (see discussion of Hooper et al., this volume) that most of the Columbia River basalt volcanism occurred in less than ~0.75 m.y., but at ca. 16.5–16.0 Ma (cf. Jarboe et al., 2006) not 16.1–15.0 Ma. The age of 16.6–16.5 Ma for the Steens basalt is not in dispute.

Scrupulous attention to $^{40}Ar/^{39}Ar$ ages (1) using only plateau/isochron ages that are statistically acceptable and (2) on rocks that are unaltered—using the A.I. test—can help resolve (magneto)stratigraphic problems. Samples (feldspars, whole-rocks, in particular) should be leached with acid (see Baksi, this volume, chapter 16) prior to dating.

4 February 2007, Ajoy K. Baksi

I make a final comment on my second, application, chapter. In evaluating the argon ages for hotspot tracks in the Pacific Ocean (Baksi, this volume, chapter 16), I overlooked the data for seamounts in the Northwest Pacific (Winterer et al., 1993). These ages are of importance, as the authors hypothesize ~40 m.y. of

volcanism on the Darwin Rise—a superswell. Earlier (Baksi, 2004), some of these ages were shown to be untenable on the basis of statistical evaluation of the plateaus postulated by these authors. Herein, I evaluate all ten sets of ages put forth by Winterer et al. (1993), looking to both the statistical validity of plateau sections of the age spectra, as well as the alteration state (i.e., A.I.) of the whole-rock samples utilized by these authors.

Step ages were calculated following Dalrymple et al. (1981), using the isotopic data in Table 1 of Winterer et al. (1993). Some step ages differ from those listed by these authors. No effort was made to adjust the ages from those reported relative to MMhb-1 at 520.4 to those preferred by Renne et al. (1998). The statistical analysis is straightforward—see Figure D-5. Only a single plateau was recovered, for Lamont seamount-1, for a different set of steps than those used by Winterer et al. (1993). Most cases show very low probability values, notably Wilde/Lamont-2/Winterer/Isakov/MIT/Heezen seamounts.

The two-step plateau for Allison seamount is rejected, as a minimum of three steps must form a plateau. The A.I. for each sample (Fig. D-6) was calculated as $^{36}Ar/^{39}Ar$, normalized for the production of ^{39}Ar from ^{39}K (see Baksi, this volume, chapter 15). All samples are altered, most showing greater than ten times the amount of ^{36}Ar expected for fresh whole-rock material. The statisically acceptable plateau age of Lamont-1 must be rejected, as the sample is badly altered. The finding that all samples are altered is in line with the observations that "extensive (residual) alteration in some of (the rocks)" and "many lavas . . .

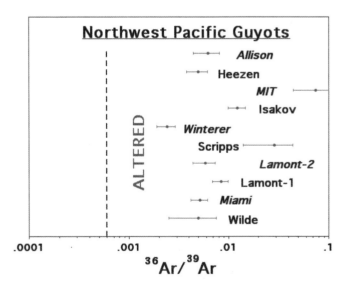

Figure D-6. Assessing the alteration state of rocks from the Cretaceous guyots of the northwest Pacific Ocean dated by Winterer et al. (1993). The alteration index (A.I.) and standard error on the mean for plateau steps are shown on a log scale. The cutoff value for fresh samples is A.I. < 0.0006 (dotted line)—see Baksi (this volume, chapter 15). All samples are altered, and no accurate estimates of the time of crystallization can be recovered from these rocks.

are largely transformed to smectite, zeolite and authigenic K-feldspar" (Winterer et al., 1993, p. 310). Such specimens should not be dated without recourse to acid leaching to remove alteration products (Baksi, this volume, chapter 16). In light of the ~100-m.y. exposure of these rocks to seawater, pervasive alteration is expected. The radiometric work of Winterer et al. (1993) yielded no valid ages and cannot be used to test predictions using seamount magnetic data, hotspot models, or the Pacific apparent polar wander path for the time period ca. 120–80 Ma.

REFERENCES CITED

Abratis, M., Mädler, J., Hautmann, S., Leyk, H.-J., Meyer, R., Lippolt, H.J., and Viereck-Götte, L., 2007, Two distinct Miocene age ranges of basaltic rocks from the Rhön and Heldburg areas (Germany) based on $^{40}Ar/^{39}Ar$ step heating data: Chemie der Erde—Geochemistry, doi: 10.1016/j.chemer .2006.03.003. Available online: www.sciencedirect.com/science/journal/ 00092819.

Baksi, A.K., 1993, A new geomagnetic polarity time scale for the period 0–17 Ma, based on $^{40}Ar/^{39}Ar$ plateau ages for selected field reversals: Geophysical Research Letters, v. 20, p. 1607–1610.

Baksi, A.K., 1994, Geochronological studies on whole-rock samples from the Deccan Traps, India: Evaluation of the timing of volcanism relative to the K-T boundary: Earth and Planetary Science Letters, v. 121, p. 43–56.

Baksi, A.K., 1995, Petrogenesis and timing of volcanism in the Rajmahal flood basalt province, northeastern India: Chemical Geology, v. 121, p. 73–90.

Baksi, A.K., 1999, Reevaluation of plate motion models based on hotspot tracks in the Atlantic and Indian Oceans: Journal of Geology, v. 107, p. 13–26.

Baksi, A.K., 2001, The Rajahmundry Traps, Andhra Pradesh; Evaluation of their petrogenesis relative to the Deccan Traps: Proceedings of the Indian Academy of Sciences, v. 110, p. 397–407.

Baksi, A.K., 2003, Critical evaluation of $^{40}Ar/^{39}Ar$ ages for the Central Atlantic Magmatic Province: Timing, duration and possible migration of eruptive centers, in Hanes, W.E., McHare, J.G., Renne, P.R., and Ruppel, C., eds., The Central Atlantic Magmatic Province: Insights from fragments of Pangea: Washington, D.C., American Geophysical Union Geophysical Monograph 136, p. 77–90.

Baksi, A.K., 2004, Critical evaluation of radiometric ages used for tracking hotspots in the Pacific Ocean: Eos (Transactions, American Geophysical Union), v. 85, no. 47, fall meeting supplement, abstract V51B-0544.

Baksi, A.K., 2005a, Evaluation of radiometric ages pertaining to rocks hypothesized to have been derived by hotspot activity, in and around the Atlantic, Indian, and Pacific Oceans, in Foulger, G.R., et al., eds., Plates, plumes, and paradigms: Boulder, Colorado, Geological Society of America Special Paper 388, p. 55–70, doi: 10.1130/2005.2388(05).

Baksi, A.K., 2005b, $^{40}Ar/^{39}Ar$ dating of the Rajahmundry Traps, eastern India and their relationship to the Deccan Traps: Discussion: Earth and Planetary Science Letters, v. 239, p. 368–373.

Baksi, A.K., 2007 (this volume, chapter 15), A quantitative tool for evaluating alteration in undisturbed rocks and minerals—I: Water, chemical weathering and atmospheric argon, in Foulger, G.R., and Jurdy, D.M., eds., Plates, plumes, and planetray processes: Boulder, Colorado, Geological Society of America Special Paper 430, doi: 10.1130/2007.2430(15).

Baksi, A.K., 2007 (this volume, chapter 16), A quantitative tool for evaluating alteration in undisturbed rocks and minerals—II: Application to argon ages related to hotspots, in Foulger, G.R., and Jurdy, D.M., eds., Plates, plumes, and planetray processes: Boulder, Colorado, Geological Society of America Special Paper 430, doi: 10.1130/2007.2430(16).

Baksi, A.K., and Farrar, E., 1990, Evidence for errors in the geomagnetic polarity time scale at 17–15 Ma: $^{40}Ar/^{39}Ar$ dating of basalts from the Pacific Northwest, USA: Geophysical Research Letters, v. 17, p. 1117–1120.

Baksi, A.K., Hsu, V., McWilliams, M.O., and Farrar, E., 1992, $^{40}Ar/^{39}Ar$ dating of the Brunhes-Matuyama geomagnetic field reversal: Science, v. 256, p. 356–357.

Cosca, M.A., Arculus, R.J., Pearce, J.A., and Mitchell, J.G., 1998, $^{40}Ar/^{39}Ar$ and K-Ar geochronological constraints for the inception and early evolution of the Izu-Bonin-Mariana arc system: Island Arc, v. 7, p. 579–595.

Dalrymple, G.B., and Lanphere, M.A., 1969, Potassium-argon dating: San Francisco, W.H. Freeman and Co., 258 p.

Dalrymple, G.B., Alexander, E.C., Lanphere, M.A., and Kraker, G.P., 1981, Irradiation of samples for $^{40}Ar/^{39}Ar$ dating using the Geological Survey TRIGA Reactor: Reston, Virginia, United States Geological Survey Professional Paper 1176, 55 p.

Dalrymple, G.B., Czamanske, G.K., Fedorenko, V.A., Simonov, O.N., Lanphere, M.A., and Likachev, A.P., 1995, A reconaissance $^{40}Ar/^{39}Ar$ geochronologic study of ore-bearing and related rocks, Siberian Russia: Geochimica et Cosmochimica Acta, v. 59, p. 2071–2083.

Deckart, K., Feraud, G., and Bertrand, H., 1997, Age of Jurassic continental tholeiites of French Guyana/Suriname and Guinea: Implications to the opening of the Central Atlantic Ocean: Earth and Planetary Science Letters, v. 150, p. 205–220.

Hames, W.E., Renne, P.R., and Ruppel, C., 2000, New evidence for geologically instantaneous emplacement of the earliest Jurassic Central Atlantic magmatic province basalts of the North American margin: Geology, v. 28, p. 859–862.

Hooper, P.R., 2004, Ages of the Steens and Columbia River flood basalts and their relationship to extension-related calc-alkalic volcanism in eastern Oregon: Reply: Geological Society of America Bulletin, v. 116, p. 249–250.

Hooper, P.R., Binger, G.B., and Lees, K.R., 2002, Ages of the Steens and Columbia River flood basalts and their relationship to extension-related calc-alkalic volcanism in eastern Oregon: Geological Society of America Bulletin, v. 114, p. 43–50.

Hooper, P.R., Camp, V.E., Reidel, S.P., and Ross, M.E., 2007 (this volume), The origin of the Columbia River flood basalt province: Plume versus non-plume models, in Foulger, G.R., and Jurdy, D.M., eds., Plates, plumes, and planetary processes: Boulder, Colorado, Geological Society of America Special Paper 430, doi: 10.1130/2007.2430(30).

Ivanov, A.V., 2007, Evaluation of different models for the origin of the Siberian Traps, in Foulger, G.R., and Jurdy, D.M., eds., Plates, plumes, and planetary processes: Boulder, Colorado, Geological Society of America Special Paper 430, doi: 10.1130/2007.2430(31).

Jarboe, N.A., Coe, R.S., Renne, P.R., and Glen, J.M., 2006, $^{40}Ar/^{39}Ar$ ages of the Early Columbia River basalt group: Determining the Steens Mountain geomagnetic polarity reversal (R0-N0) as the top of the C5Cr chron and the Imnaha Normal (N0) as the C5Cn.3n chron: Eos (Transactions, American Geophysical Union), v. 87, no. 52, fall meeting supplement, abstract V51D-1702.

Long, P.E., and Duncan, R.A., 1983, $^{40}Ar/^{39}Ar$ ages of the Columbia River Basalt from deep boreholes in south-central Washington: Eos (Transactions, American Geophysical Union), v. 64, abstract, p. 90.

Lustrino, M., and Wilson, M., 2007, The circum-Mediterranean anorogenic Cenozoic igneous province: Earth-Science Reviews, v. 81, p. 1–65.

Lux, G., 1987, The behavior of noble gases in silicate liquids: Solution, diffusion, bubbles and surface effects, with applications to natural samples: Geochimica et Cosmochimica Acta, v. 51, p. 1549–1560.

Renne, P.R., Swisher, C.C., Deino, A.L., Karner, D.B., Owens, T.L., and DePaolo, D.J., 1998, Intercalibration of standards, absolute ages and uncertainties in $^{40}Ar/^{39}Ar$ dating: Chemical Geology, v. 145, p. 117–152.

Sebai, A., Feraud, G., Bertrand, H., and Hanes, J.A., 1991, $^{40}Ar/^{39}Ar$ dating and geochemistry of tholeiitic magmatism related to the early opening of the central Atlantic rift: Earth and Planetary Science Letters v. 104, p. 455–472.

Sharp, W.D., and Clague, D.A., 2006, 50-Ma initation of the Hawiian-Emperor

Bend records major change in Pacific plate motion: Science, v. 313, p. 1281–1284.

Steinmann, M., Stille, P., Bernotat, W., and Knipping, B., 1999, The corrosion of basaltic dykes in evaporites: Ar–Sr–Nd isotope and rare earth elements evidence: Chemical Geology, v. 153, p. 259–279.

Venkatesan, T.R., Kumar, A., Gopalan, K., and A'Mukhamedov, A.I., 1997, ^{40}Ar-^{39}Ar ages of Siberian basaltic volcanism: Chemical Geology, v. 138, p. 303–310.

Verati, C., Bertrand, H., and Feraud. G., 2005, The farthest record of the Central Atlantic Magmatic Province into West African craton: Precise ^{40}Ar/^{39}Ar dating and geochemistry of Taoudenni basin intrusives (northern Mali): Earth and Planetary Science Letters, v. 235, p. 391–407.

Walderhaug, H.J., Eide, E.A., Scott, R.A., Inger, S., and Golionko, E.G., 2005, Paleomagnetism and ^{40}Ar/^{39}Ar geochronology from the South Taimyr igneous complex, Arctic Russia: A middle-late Triassic magmatic pulse after Siberian flood-basalt volcanism: Geophysical Journal International, v. 163, p. 501–517.

Winterer, E.L., Natland, J.H., van Waasbergen, R.J., Duncan, R.A., McNutt, M.K., Wolfe, C.J., Premoli Silva, I., Sager, W.W., and Sliter, W.V., 1993, Cretaceous guyots in the Northwest Pacific: An overview of their geology and geophysics, *in* Pringle, M.S., et al., eds., The Mesozoic Pacific: Geology, tectonics, and volcanism: Washington, D.C., American Geophysical Union Geophysical Monograph 77, p. 307–334.

The Geological Society of America
Special Paper 430
2007

Divergence between paleomagnetic and hotspot-model–predicted polar wander for the Pacific plate with implications for hotspot fixity

William W. Sager*

Department of Oceanography, Texas A&M University,
317 Eller Building, College Station, Texas 77843, USA

ABSTRACT

If mantle plumes (hotspots) are fixed in the mantle and the mantle reference frame does not move relative to the spin axis (i.e., true polar wander), a model of plate motion relative to the hotspots should predict the positions of past paleomagnetic poles. Discrepancies between modeled and observed poles thus may indicate problems with these assumptions, for example, that the hotspots or spin axis have shifted. In this study, I compare paleomagnetic and hotspot-model–predicted apparent polar wander paths (APWP) for the Pacific plate. Overall, the two types of APWP have similar shapes, indicating general agreement. Both suggest ~40° total northward drift of the Pacific plate since ca. 123 Ma. Offset between paleomagnetic and hotspot-predicted poles is small for the past ca. 49 Ma, consistent with fixed hotspots during that time, but the offsets are large (6–15°) for earlier times. These differences appear significant for the Late Cretaceous and early Cenozoic. During the period 94–49 Ma, the hotspot model implies the paleomagnetic pole should have drifted ~20° north without great changes in rate. Measured paleomagnetic poles, however, indicate rapid polar motion between 94 and 80 Ma and a stillstand from 80 to 49 Ma. Comparison with global synthetic APWP suggests that the 94- to 80-Ma polar motion may be related to true polar wander. The stillstand indicates negligible northward motion of the Pacific plate during the formation of the Emperor seamounts. This observation is drastically different from most accepted Pacific plate motion models and requires rethinking of western Pacific tectonics. If the Emperor seamounts show relative motion of the plate relative to the Hawaiian hotspot, the implied southward hotspot motion is ~19°. Lack of a diagnostic coeval phase of polar wandering in global APWP and consideration of the significance of the Hawaiian-Emperor bend imply that true polar wander is probably not the cause. Likewise, mantle-flow models do not readily explain the large southward drift of the hotspot or its inferred large westward velocity component. Thus, current models for the formation of the Emperor seamounts appear inadequate, and new ideas and further study are needed. Comparison of the Pacific APWP with a global APWP, both rotated into the Antarctic reference frame, shows an offset of ~10°,

*E-mail: wsager@ocean.tamu.edu.

Sager, W.W., 2007, Divergence between paleomagnetic and hotspot-model–predicted polar wander for the Pacific plate with implications for hotspot fixity, *in* Foulger, G.R., and Jurdy, D.M., eds., Plates, plumes, and planetary processes: Geological Society of America Special Paper 430, p. 335–357, doi: 10.1130/2007.2430(17).

implying problems with plate circuits connecting Antarctica with surrounding plates. This result suggests that caution is required when predicting trends of hotspot seamount chains using plate circuits through Antarctica.

Keywords: paleomagnetism, polar wander, Pacific, hotspots, Hawaiian seamounts, Emperor seamounts

INTRODUCTION

How did the Hawaiian-Emperor seamount chain (Fig. 1) come into being and what is the significance of the bend where the two chains meet? For many years, the answer seemed obvious. Wilson (1963) explained the Hawaiian chain and its linear age progression as volcanism that occurred on the Pacific plate as it drifted over a mantle plume (often given the generic name "hotspot") that was either fixed or moving slowly relative to the mantle. Morgan (1971, 1972) noted similarities in trend between the Hawaiian-Emperor and three other Pacific seamount chains (Cobb-Bowie, Austral-Cook-Gilbert-Ellice, and Tuamotu-Line; Fig. 1) and suggested that these and other linear seamount chains were all formed by plumes that were nearly fixed in a stable lower mantle. Studies of linear seamount chains in other oceans gave similar results, supporting the fixed hotspot hypothesis (e.g., Duncan, 1981; Morgan, 1983; Müller et al., 1993). Owing to its simplicity and the fact that it made tectonic predictions that seemed consistent with observed geology, the fixed hotspot hypothesis became widely accepted. Indeed, most introductory textbooks in geology and oceanography written within the past several decades contain a figure showing the Hawaiian-Emperor chain, explaining its formation to be the result of plate motion over a nearly fixed plume, with the bend having resulted from a change in plate motion.

Today the picture is not so clear. Questions have arisen about the number and even existence of deep-mantle plumes (e.g., Anderson, 2000, 2005; Courtillot et al., 2003; Foulger and Natland, 2003), and several lines of evidence suggest that hotspots are anything but fixed. Because it is the archetype of hotspot seamount chains, the Hawaiian-Emperor chain is the nexus of many such observations and arguments. Drilling of the Emperor chain by the Deep Sea Drilling Project (DSDP) produced paleomagnetic data showing that Suiko seamount formed ~7° north of the current latitude of the hotspot (Kono, 1980). Subsequent paleomagnetic measurements from other Emperor seamounts cored by the Ocean Drilling Program (ODP) confirmed and augmented this observation, showing a progressive offset of paleolatitude along the chain from nearly zero near the bend to ~13° at the north end (Tarduno et al., 2003). These findings dovetail with mantle modeling studies that imply the hotspots should move with mantle flow (e.g., Steinberger, 2000; Steinberger et al., 2004) and reconstructions of Pacific plate motion derived from hotspot tracks in other oceans that fail to reproduce the Hawaiian-Emperor bend and show significantly

less northward motion during the Emperor seamounts period (e.g., Cande et al., 1995; DiVenere and Kent, 1999; Raymond et al., 2000). As a result there is an ongoing re-examination of the fixed hotspot hypothesis in general and the meaning of the Hawaiian-Emperor bend in particular (e.g., Norton, 1995, this volume; Sharp and Clague, 2006). The outcome of this debate is of wide interest because of its implications for mantle properties, behavior, and the flux of deep volcanism to the Earth's surface.

Paleomagnetism is often used to examine plate motion because it gives an axially symmetric, absolute reference frame tied to the Earth's spin axis. If fixed or slowly moving, mantle plumes can also be used as an absolute reference frame, so a comparison of predictions from the two is a useful way to examine hotspot motion and related phenomena (e.g., Gordon, 1987; Besse and Courtillot, 2002). The Pacific plate is ideal for such comparisons because plate motion has been rapid and it contains well-defined linear seamount chains, including the Hawaiian-Emperor chain, which has been used as the basis for numerous hotspot plate motion models. Moreover, updates of both reference frames are available. For paleomagnetic data, I call upon recent compilations of the Pacific paleomagnetic apparent polar wander path (APWP) (Sager, 2006; Beaman et al., 2007). For models of Pacific plate motion relative to the hotspots, I examine several reports, but focus on a recent, well-documented update (Wessel et al., 2006). In this article, I explore the similarities and differences between paleomagnetic observations and hotspot model predictions. The results suggest a general agreement overall and close agreement for the past ~49 m.y., but imply significant relative motion between the two reference frames for earlier times.

BACKGROUND

Apparent and True Polar Wander

An APWP is a time series of paleomagnetic poles showing pole movement relative to a particular plate or collection of plates. It is usually calculated by averaging many paleomagnetic poles from individual geologic formations grouped by age (e.g., Irving and Irving, 1982; Besse and Courtillot, 2002; Schettino and Scotese, 2005). APWP construction makes two fundamental assumptions: the time-averaged geomagnetic field is dipolar and all sites for which data are averaged are on the same rigid plate. The first assumption allows the paleomagnetic inclination

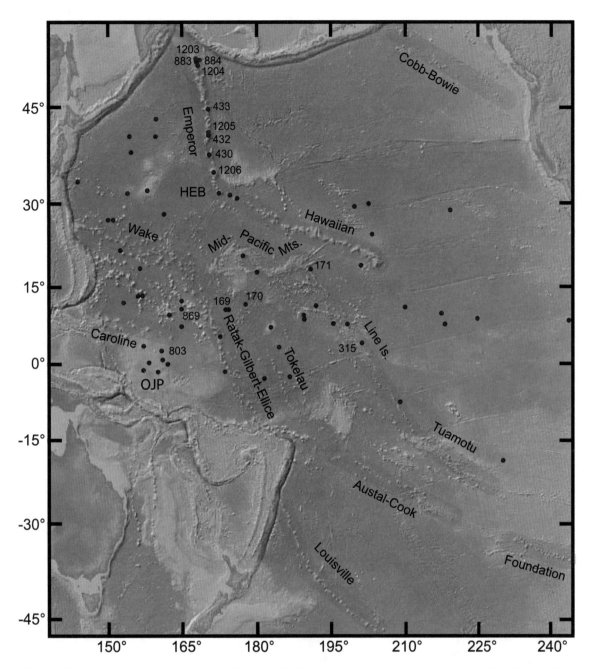

Figure 1. Hotspot seamount chains on the Pacific plate. Highlighted areas show seamount chains mentioned in text. Dots show sampling locations for paleomagnetic data used in apparent polar wander path calculations (Sager, 2006; Beaman et al., 2007). Numbers denote Deep Sea Drilling Project and Ocean Drilling Program sites on the Emperor chain and other data used for revision of the 94-Ma pole (Table 1). Note that sample locations are largely restricted to the northern hemisphere. HEB—Hawaiian-Emperor bend; OJP—Ontong Java plateau.

and declination at a given site to be translated simply into paleo-latitude and the azimuth of the spin axis (e.g., Butler, 1992). Various studies have found this so-called geocentric axial dipole assumption to be a good first-order approximation through time, with nondipole fields having simple zonal form (i.e., axially symmetric) and usually averaging less than 5% (e.g., McElhinny et al., 1996; Merrill and McFadden, 2003; Courtillot and

Besse, 2004) to 10% (e.g., Torsvik and Van der Voo, 2002) during the past 200 m.y. A 5% contribution from a low-order zonal nondipole field (for example g^0_2 or g^0_3) would cause a perturbation of only ~4° in paleomagnetic inclination or ~2° in paleolatitude compared to that calculated using the dipole field assumption (Merrill et al., 1996). The second assumption is important because plates with different drift histories give divergent APWP, with possibly confusing results if poles from more than one independent plate are combined without proper reconstruction of relative motions.

Polar wandering, the apparent movement of paleomagnetic poles with time, happens for several reasons. Most polar wander occurs as a result of plate motion. If a plate's motion is described by a pole of rotation, the same rotation will affect the paleomagnetic pole, and the resulting polar path follows a small circle concentric with the rotation axis (Fig. 2A). Naturally, if a plate's history is described by more than one rotation pole, the APWP will assume more complex shapes. In general, an APWP consists of a series of small-circle segments congruent with rotation poles that describe the motion of the plate relative to the spin axis (Gordon et al., 1984).

Another cause of polar wandering is actual motion of the spin axis relative to the Earth, a phenomenon termed true polar wander (TPW). TPW can occur owing to changes in the density structure of the mantle, which cause the maximum principle axis of inertia to shift and the spin axis to follow (e.g., Goldreich and Toomre, 1969). Whereas APWP for different plates are usually disparate owing to different plate drift histories, TPW is a globally coherent phenomenon. If TPW occurred in the absence of plate motion, all plates would have matching APWP.

A third cause of apparent polar wander is large changes to the average nondipole components of the geomagnetic field, which would change paleomagnetic directions without concomitant changes in site location. As stated previously, there is little evidence that this phenomenon has caused significant po-

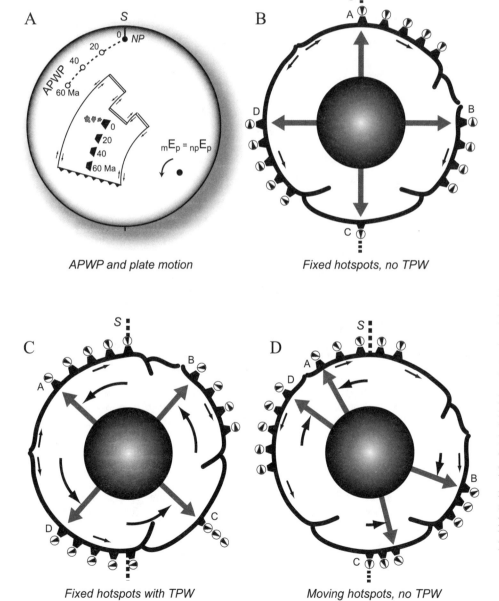

Figure 2. Cartoon showing relationships among polar wander, hotspot drift, and true polar wander (TPW). (A) Movement of a plate relative to the spin axis is described by a rotation pole $_{np}E_p$ (here "np" stands for north pole). The apparent polar wander path (APWP) traces a small-circle segment concentric on the rotation pole for the time period for which the rotation pole applies. If a hotspot forms a seamount chain and there is no drift of the hotspot relative to the spin axis, the seamounts formed by the hotspot also trace a small circle concentric on the same rotation pole. If there is no TPW, the spin axis is fixed relative to the mantle and the rotation pole describing the motion of the plate relative to the mantle, $_mE_p$, is the same as $_{np}E_p$. (B) If hotspots (points A–D) are fixed relative to the mantle and there is no TPW, all hotspot paleolatitudes will remain the same through time. The arrows indicate the paleomagnetic inclination, which is zero (horizontal magnetization) at the equator (hotspots B and D) and vertical at the poles (hotspots A and C). (C) If the hotspots A–D are fixed relative to the mantle, but the mantle shifts relative to the spin axis (i.e., TPW), seamount paleolatitudes (and paleomagnetic poles) will change through time, but in a globally coherent manner. (D) If there is no TPW, but the hotspots move independently relative to the spin axis with time, the hotspot paleolatitudes (and paleomagnetic poles) shift with time, but in a manner that is not consistent across the Earth. Redrawn from Carlson et al. (1988).

APWP and plate motion

Fixed hotspots, no TPW

Fixed hotspots with TPW

Moving hotspots, no TPW

lar wander, because most studies have concluded that long-term nondipole fields have been small for the late Mesozoic and Cenozoic (McElhinny et al., 1996; Courtillot and Besse, 2004).

If a hotspot is fixed relative to the spin axis and there is no TPW, the APWP and the hotspot-track volcanoes both follow small circles concentric with the pole (Fig. 2A). Under these conditions, a model of plate motion based on hotspot tracks can be used to predict the APWP. If either assumption is invalid, this will not be true, so differences between hotspot-track models and APWP can be interpreted as TPW or departures from hotspot fixity. For example, on an Earth with fixed hotspots but no TPW (Fig. 2B), hotspot paleolatitudes will not change. If hotspots are fixed but TPW occurs, hotspot track seamount paleolatitudes (and paleopoles) will change in a globally consistent manner (Fig. 2C). If there is no TPW, but the hotspots move relative to one another, the paleolatitudes (and paleopoles) will not show a globally consistent pattern (Fig. 2D).

Pacific Apparent Polar Wander Path

Most major continental plates have relatively dense areal and time coverage with paleomagnetic data (e.g., Irving and Irving, 1982; Besse and Courtillot, 2002; Schettino and Scotese, 2005), but development of the Pacific APWP has lagged, owing to the inaccessibility of the plate. Furthermore, whereas APWP from continental plates bordering the Atlantic Ocean can be augmented by assimilating rotated paleomagnetic data from adjacent plates (e.g., Besse and Courtillot, 2002; Schettino and Scotese, 2005), such improvement is not possible for the Pacific plate, because continental plates cannot be linked to the Pacific by direct seafloor spreading, except for Antarctica, which has few data and is characterized by uncertainty about its long-term rigidity (e.g., Acton and Gordon, 1994).

Pacific plate inaccessibility has also affected the type of data used for APWP calculations. Because most of the plate is covered by water, Pacific paleomagnetic data are mostly from modeling of magnetic anomalies and paleomagnetic studies of azimuthally unoriented ocean drilling cores. Both types of data are only rarely used for continental plate APWP, because fully oriented data from rock outcrops are considered more reliable. As a result, the Pacific APWP still has significant uncertainties.

Magnetic anomaly studies include models of seamount magnetic anomalies (e.g., Francheteau et al., 1970), which give both inclination and declination data. They also include determinations of the skewness (asymmetry) of marine magnetic lineations, a quantity that is related to paleomagnetic inclination (Schouten and Cande, 1976). Core data, which are rarely oriented in azimuth and therefore give only paleomagnetic inclination (and paleolatitude), are derived mainly from sedimentary or basalt core samples (Cox and Gordon, 1984; Gordon, 1990). Almost all such data have issues with systematic errors, complicating interpretation of the APWP. A detailed discussion of these errors is beyond the scope of this article, but can be found elsewhere (Sager and Pringle, 1988; Sager, 2006; Beaman et al.,

2007). In defining the Pacific APWP (Fig. 3), we have looked for consistency among data, combining different types when possible (Sager, 2006; Beaman et al., 2007).

Early studies of the Pacific APWP relied mainly on data derived from the modeling of seamount magnetic anomalies (Francheteau et al., 1970; Harrison et al., 1975; Gordon, 1983; Sager and Pringle, 1988). These data are suitable for showing the gross features of the APWP and were used extensively when few other data were available, but today they are considered the most problematic data type owing to potential systematic errors. Seamount data showed that the Pacific plate has drifted ~30° northward since Cretaceous time (Francheteau et al., 1970) and that the APWP has a north-south trend from Late Cretaceous to present, a significant bend, and an east-west trend for earlier times (Gordon, 1983; Sager and Pringle, 1988). The ~30° of northward drift interpreted from seamount models by Francheteau et al. (1970) was used by Morgan (1971) as confirmation for his model of Pacific plate motion relative to the hotspots, which implied a similar amount of northward motion.

A compilation of basalt core paleomagnetic data confirmed the ~30° of northward drift of the Pacific plate, as indicated by seamount studies, and suggested southward motion during the Jurassic and Early Cretaceous, giving the APWP an overall "fishhook" shape (Cox and Gordon, 1984). This finding was bolstered by similar results from magnetic lineation skewness and sedimentary core data (Larson and Sager, 1992; Larson et al., 1992). The formation of the fishhook shape by southward-followed-by-northward plate motion is explained in Figure 4.

The fishhook shape is seen in the APWP shown in Figure 3 (poles given in Table 1). This path is based mainly on three studies: Larson and Sager (1992) for the 139- and 145-Ma poles, Sager (2006) for the 123- to 80-Ma poles, and Beaman et al. (2007) for the post-80-Ma portion. The 139- and 145-Ma skewness poles from Larson and Sager (1992) are used to define the old end of the APWP, which is poorly known, because data of this age are scarce and somewhat contradictory. Early Cretaceous and Jurassic basalt core data give a large range of paleolatitudes that are consistent with the 145-Ma skewness pole, the 123-Ma basalt core pole, or are somewhere in between (Sager, 2006). Because sediment core data also imply southward motion of the Pacific plate for this time (e.g., Larson et al., 1992), I believe that the 145-Ma pole is a reasonable starting point for the Late Jurassic–Early Cretaceous APWP. The amount of southward motion is poorly constrained because of uncertainties about the accuracy of the skewness poles. The exact location of the Late Jurassic–age pole is also uncertain, because skewness data give two significantly different pole positions, depending on whether one assumes a contribution from "anomalous skewness" (a cause of mismatch within coeval skewness data sets—the cause of which is poorly understood; Larson and Sager, 1992). If the skewness poles calculated with a contribution from anomalous skewness are considered, the Pacific APWP fishhook is wide, and Late Jurassic–Early Cretaceous poles are located in North America. If the solutions without

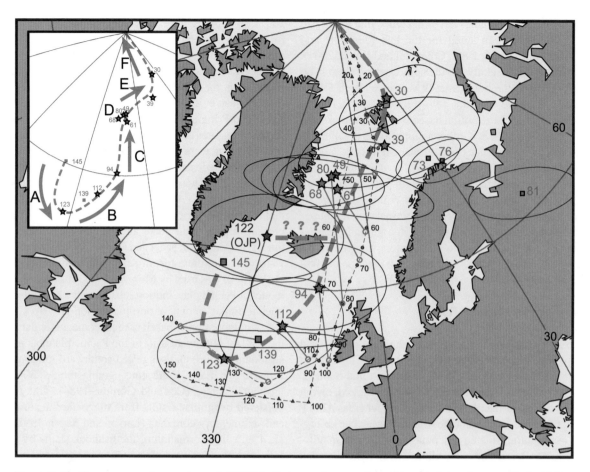

Figure 3. Pacific apparent polar wander path (APWP). Red stars denote pole positions defining the most likely APWP (Sager, 2006; Beaman et al., 2007), shown by the heavy dashed blue line. Poles are surrounded by 95% confidence ellipses and labeled by age in Ma. Blue star denotes Ontong Java plateau (OJP) pole, which is considered anomalous (Sager, 2006). Green squares show poles determined from magnetic lineation skewness (73-, 76-, and 81-Ma poles from Petronotis and Gordon, 1999, and Vasas et al., 1994; 139- and 142-Ma poles from Larson and Sager, 1992). The Late Cretaceous skewness poles appear anomalous compared to other data (Beaman et al., 2007). Thin dashed lines show predicted polar wander path from plate/hotspot motion models of Duncan and Clague (1985) (purple with triangles) and Wessel et al. (2006) (blue with dots). Triangle and dot symbols show predicted pole positions at 5-m.y. intervals, labeled every 10 m.y. Red lines show offset between paleomagnetic and hotspot-model–predicted poles. Inset sketch map shows interpreted phases of polar wander. Plot is an equal-area map. Numbers are pole ages in Ma.

anomalous skewness are used (as in Fig. 3), the fishhook is narrower and the APWP may double back almost upon itself (Larson and Sager, 1992).

Later paleomagnetic poles are all hybrids, calculated by combining core and magnetic anomaly–derived data in varying amounts, although the recent studies have minimized the use of noncore data (Sager, 2006; Beaman et al., 2007). For example, the 123- and 112-Ma pole positions were based mainly on basalt core data, which constrain the pole latitude well but give poor constraints on the pole longitude. Consequently, declinations from seamount magnetic anomaly models of appropriate age were used to help constrain the pole longitudes (Sager, 2006). The 94- and 80-Ma poles are also based on basalt core and seamount model declination data, but with the addition of

some sediment core data (Sager, 2006). The 94-Ma pole (Fig. 5; Table 1) is revised here from the 92-Ma pole published in Sager (2006) by including oriented sediment core data from ODP site 869 (Sager et al., 1995). In the previous calculation, pole error bounds were large because basalt core data near this age are few (Sager, 2006). The addition of the sediment data reduced the error ellipse greatly without changing the pole position significantly.

Paleomagnetic poles for 68, 61, 49, 39, and 30 Ma are based mainly on azimuthally unoriented sediment core paleomagnetic data but include information from basalt cores and seamount anomaly model declinations (Beaman et al., 2007). Data from anomaly skewness studies were not included in these pole calculations because Late Cretaceous skewness poles diverge from

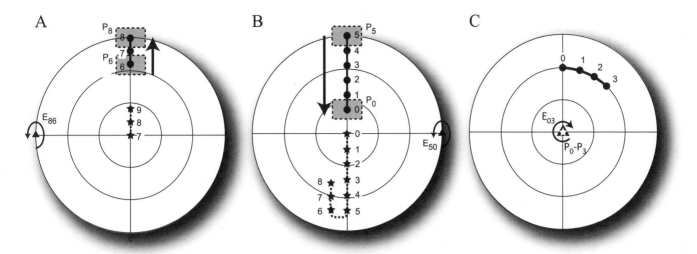

Figure 4. Cartoon explaining apparent polar wander path (APWP) fishhook shape. Plots show geographic pole viewed from above. (A) Southward drift of plate from ages 8–6, described by rotation around Euler pole E_{86}, moves plate from locations P_6 to P_8. Points 6–8 (filled circles) show motion of marker point on plate. This motion causes APWP (stars) with ages increasing toward the plate. (B) Northward drift of plate from ages 5–0, described by rotation around Euler pole E_{50}, moves plate from locations P_5 to P_0. Points 5–0 show motion of marker point on plate. This motion causes APWP with ages increasing away from plate. Overall shape of APWP is a fishhook that indicates southward motion followed by northward motion of plate. (C) If the Euler pole (E_{03}) coincides with the spin axis, no polar wander occurs, and the time series of paleomagnetic poles does not appear to move (a stillstand). If later plate tectonic motion occurs such that the recorded Euler pole is moved away from the spin axis, the APWP will have a cluster of poles ($P_3 = P_2 = P_1 = P_0$) at some point that is not at the spin axis.

the APWP determined from other data (Fig. 3). The reasons for this discrepancy are not understood, so it was deemed prudent to use only the most consistent data (Beaman et al., 2007).

Another discrepancy is the separation of the 122-Ma Ontong Java pole from the north-south trend of the APWP (Fig. 3). In analyses of Early Cretaceous basalt core paleomagnetic data,

it was found that data from Ontong Java plateau appear anomalous and show ~10° less northward motion than data from elsewhere in the north Pacific (Sager, 2006). This difference was attributed to tectonic displacement of Ontong Java plateau from the rest of the Pacific, but there is no accepted tectonic model to explain this discrepancy. Furthermore, younger Late Cretaceous

TABLE 1. PACIFIC PALEOMAGNETIC POLES

	Pole location		95% Confidence				Data weights (%)		
Age (Ma) ± std. dev.	Latitude (°N)	Longitude (°E)	Maj.	Min.	Azim.	N	S	B	D
[†]29.5 ± 2.5	80.1	24.4	6.1	2.6	91	15	87	13	0
[†]39.2 ± 2.3	75.8	14.6	10.8	4.3	96	9	69	0	31
[†]48.6 ± 3.8	73.4	350.0	7.7	3.4	77	13	77	10	13
[†]61.2 ± 3.2	71.8	350.9	11.4	2.9	101	14	38	17	45
[†]68.3 ± 1.7	72.4	344.5	7.3	3.1	91	10	71	3	26
[*]79.9 ± 2.8	73.2	349.2	7.9	4.8	106	15	17	41	42
94.2 ± 2.6	60.5	345.9	8.2	5.1	83	11	36	21	43
[*]112.2 ± 3.6	55.6	334.9	7.7	5.5	67	11	0	53	47
[*]120.5 ± 1.8 (Ontong Java plateau)	65.3	331.0	9.0	4.9	75	10	0	50	50
[*]122.7 ± 4.4	50.0	329.1	8.6	4.6	77	40	0	55	45
[§]136–141	53.0	334.0	11.1	1.0	78				
[§]142–149	60.4	321.5	10.8	1.7	74				

Notes: Std. dev.—standard deviation; Maj.—major semi-axis of confidence ellipse; Min.—minor semi-axis of ellipse; Azim.—azimuth of major semi-axis, clockwise from north; Data weights are percentage of weight for a particular data type in the determination of the pole location; N—total number of independent data; S—sediment core data weight; B—basalt core data weight; D—seamount model declination data weight.
*Pole from Sager (2006), with 94.2-Ma pole revised as explained in text.
[†]Pole from Beaman et al. (2007).
[§]Pole from Larson and Sager (1992).

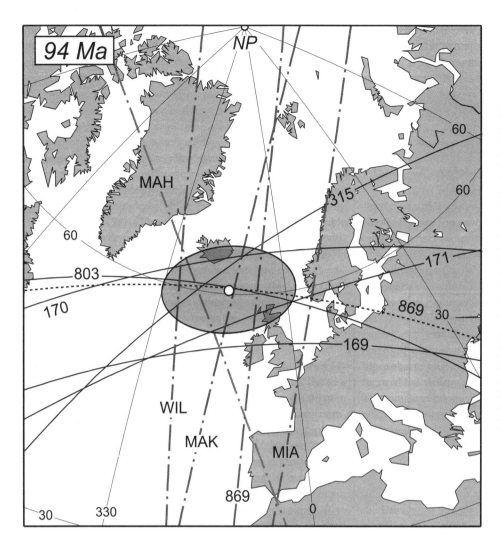

Figure 5. Revised Late Cretaceous mean paleomagnetic pole for 94 Ma. Data used for the pole calculation are the same as those for the 92-Ma pole of Sager (2006) but with the addition of sedimentary inclination and declination data from Ocean Drilling Program (ODP) site 869 (Sager et al., 1995). Solid arcs show the locus of the paleomagnetic pole inferred from the mean paleomagnetic inclination from basalt or igneous core data. Dashed arc shows the same from site 869 sediment data. Nearly vertical dash-dot lines show pole location inferred from seamount anomaly model declination data. Open circle is mean pole position, and surrounding ellipse is 95% confidence region (Table 1). Numbers give Deep Sea Drilling Project and ODP site numbers. MAH, WIL, MAK, and MIA are abbreviations for seamount names (see Sager, 2006, for data and sources). Equal-area projection. MAH —Mahler; MAK—Makarov; MIA— Miami; *NP*—north pole; WIL—Wilde.

and early Cenozoic sediment data from Ontong Java plateau also appear to show less northward motion (Hall and Riisager, 2006). It is unclear what has caused this dichotomy, but one implication is that some portion of the Pacific plate may have experienced ~10° less northward motion. Unfortunately, this hypothesis is difficult to test, because there are few reliable, well-dated paleomagnetic data from south of the equator. Indeed, as Figure 1 shows, data used to derive the Pacific APWP are located almost entirely in the north Pacific. If the Pacific APWP curves to the west as shown in Figure 3 (dashed line to the Ontong Java plateau pole) and the total northward motion is thus no more than ~20°, interpretations in this article that rely on poles >80 Ma in age (those farther south) are unreliable. Interpretations for 80-Ma and younger poles, however, will not change significantly.

Pacific Hotspot Models

In the years after Wilson's (1963) hotspot hypothesis was published, alternative models were proposed for the formation of the Hawaiian Islands and Hawaiian-Emperor chain. One considered the melting anomaly to be caused by a propagating crack that caused asthenosphere melting and magma ascension to the surface (Jackson and Shaw, 1975; Jackson et al., 1975). Another called upon a crack and gravitational anchor (i.e., a negatively buoyant melt residuum whose sinking caused local convection and the rise and melting of asthenospheric material; Shaw, 1973; Shaw and Jackson, 1973). However, once Morgan (1971) published his fixed hotspot model, it was rapidly accepted as the correct explanation, based on the simplicity of the model, respect for Morgan's status as one of the formulators of the plate tectonics paradigm, and the fact that it seemed to follow logically from simple plate tectonics (Glen, 2005).

In the three decades plus since Morgan (1971) published his model of Pacific plate motion relative to the hotspots, many other authors have refined the Pacific model or created similar models for other oceans. On the whole, most have been derived in a similar manner: a series of stage rotation poles were determined to fit segments of seamount chains thought to be coeval and stage pole start/stop ages and rotation rates and/or angles were calculated from often sparse and sometimes inaccurate age data. For the Pacific, the main differences stem from different choices of which segments of seamount chains to be fit and the number and accuracy of age data used to determine pole ages and rotation angles. For example, Morgan (1971) used two rotation poles for his model, using primary trend and age control from the Hawaiian-Emperor chain. At the time, age data for the Hawaiian-Emperor chain were almost nonexistent. As new and revised radiometric dates for these and other Pacific seamounts became available, the model has been revised numerous times (Jarrard and Clague, 1977; Duncan and Clague, 1985; Fleitout and Moriceau, 1992; Wessel and Kroenke, 1997; Harada and Hamano, 2000; Raymond et al., 2000; Wessel et al., 2006).

Many of these models were constructed with a ca. 43-Ma date for the Hawaiian-Emperor Bend, based on the once widely accepted date for that feature from Clague and Dalrymple (1975). Recent geochronology studies indicate that this date is too young (Sharp and Clague, 2006), and many workers now accept an age of 47–50 Ma (e.g., Sharp and Clague, 2006; Wessel et al., 2006). The modeled age of the northern Emperor chain has also seen a significant change. Models published prior to the mid-1990s mostly assumed an age of 70–75 Ma for the northern terminus of the Emperor chain, based on the age of oldest sediments recovered by DSDP drilling from Meiji seamount, the northernmost Emperor seamount. Newer models use an age of ca. 81 Ma, based on radiometric dates from basalts cored from ODP site 883 on Detroit seamount (Keller et al., 1995).

It has been and continues to be problematic to make a model for Pacific seamount chains older than the northern Emperor seamounts (>81 Ma), because the connection between older and younger chains is tenuous. Moreover, older western Pacific seamount chains tend to be short and have overlaps and inconsistencies in trends and age progressions. As a result, it has been difficult to construct a consistent plate motion model (e.g., Koppers et al., 2003). Models that go farther back in time than 81 Ma assume that the Line Islands chain is copolar and coeval to the Emperor chain and that the Mid-Pacific Mountains (connected to the Line chain) and/or Wake seamounts show older plate motion over the hotspots (e.g., Duncan and Clague, 1985; McNutt and Fischer, 1987; Wessel and Kroenke, 1997; Kroenke et al., 2004). The Mid-Pacific Mountains–Line Islands Bend is assigned an age of ca. 90 Ma, and the Mid-Pacific Mountains and Wake seamounts trends take the model back to ca. 140 Ma. Although plausible, this model for earlier Pacific plate motion has significant uncertainties because of the complexity of the Mid-Pacific Mountains and Line Islands (e.g., Winterer and Sager, 1995) as well as uncertainties in dates and alignments of other pre-81-Ma seamount chains (Koppers et al., 2003).

Recently, several investigators have used a different approach to modeling Pacific plate motion. Rather than determine stage poles based on seamount trends, Harada and Hamano (2000) and Wessel et al. (2006) solved for total reconstruction rotation poles that fit seamount positions within a variable number of linear Pacific chains. Having determined a series of rotation poles, an age model was calculated by fitting an age-distance function to all available seamount dates (Wessel et al., 2006). This approach has distinct advantages: it is less dependent on choices for copolar segments and break points between stage poles, and it allows the model to make an optimal fit for all age data.

Two ramifications of the model are important to note. In its construction, all hotspots were considered as equal mantle plumes, despite conclusions by others that the constellation of Pacific plumes may contain both primary (deep)- and secondary (shallow)-sourced plumes (Courtillot et al., 2003). If models of upper-mantle flow are correct, differences in source depth could result in different implied volcanic propagation rates for different seamount chains (Doglioni et al., 2005; Cuffaro and Doglioni, this volume). The Wessel et al. (2006) model assumes the same propagation rate for all Pacific hotspots, but perhaps the discrepancies resulting from different source depths are hidden within the relatively large uncertainties in the average pole rotation rates. Another important implication is that the assumption of fixity for Pacific plate hotspots is remarkably good. Deviations of individual seamount chains from small circles concentric around the model rotation poles are very small (typically <1°). Whatever it is that these seamount chains describe, it is consistent. If the model does indeed represent motion of the plate over a series of hotspots, they show very little relative motion.

COMPARISON OF PALEOMAGNETIC AND HOTSPOT MODEL APWP

Pacific paleomagnetic apparent polar wander can be divided into six different phases, shown in Figure 3 by changes in direction or wander velocity. Moving forward in time, segment A consists of southward polar motion that implies southward motion of the Pacific plate during the Late Jurassic and Early Cretaceous. The 139-Ma skewness pole lies between the 123- and 112-Ma poles, seemingly out of order, but this pole has large east-west uncertainty, so it plausibly lies somewhere near the dashed blue line in Figure 3 that represents the short end of the fishhook. Segments B and C show northward motion of the pole and plate during the Mid- and Late Cretaceous. From existing data, it appears the plate turned around at ca. 123 Ma. Segment C is distinguished from B by a doubling of the implied rate of polar drift, with the pole shifting at ~1°/m.y. between the 94- and 80-Ma poles. Phase D is a stillstand, with the paleomagnetic pole showing negligible motion between 80 and 49 Ma. After 49 Ma, the pole began moving northward again (segment E) and shifted direction at ca. 30 Ma (segment F).

Most segments of an APWP are thought to result from periods of stable plate motion, with changes in APWP direction or speed caused by shifting boundary forces on the plate edges (e.g., Gordon et al., 1984). It appears that a plausible connection can be made between segments of the Pacific APWP and its tectonic history. During the Early Cretaceous, it is likely that the Pacific plate was small (Fig. 6) and surrounded by spreading ridges (Hilde et al., 1976). Without connections to Pacific-rim subduction zones, the plate should have moved relatively slowly and could have changed directions. This scenario appears consistent with the relatively slow polar motion (~0.5°/m.y.) of APWP segments A and B. Sometime during the Mid- or Late Cretaceous, the Pacific plate became engaged with western Pacific subduction zones. The timing of this event is highly uncertain, because the record of the western Pacific plate has been subducted, the western extent of the Pacific plate at that time is unknown, and the event probably occurred during the Cretaceous Quiet Period, so there are no seafloor spreading magnetic lineations to show changes in plate motion. It is plausible that the connection occurred during the Early Cretaceous and explains the turnaround in motion ca. 123 Ma.

The stillstand from 80 to 49 Ma implies that the plate moved nearly east-west (i.e., the pole of rotation that describes plate motion relative to the spin axis was located near the spin axis; Fig. 4). This motion may have occurred because the western edge of the Pacific plate began to subduct while the northern edge was still separated from northern subduction zones by an intervening plate (Fig. 6), which seems to have been the case until well into the Cenozoic (e.g., Lonsdale, 1988). Northward motion would have resumed (segment E) when the Pacific plate northern edge began to subduct into the Aleutian Trench. Dating of the Aleutian Arc implies that this subduction occurred during the Eocene (Scholl et al., 1986), which coincides with the recommencement of polar wander between 49 and 39 Ma (Fig. 3). The shift in APWP direction ca. 30 Ma may have resulted from changing plate-boundary forces once the eastern Pacific plate came into contact with North America and the San Andreas transform system began to form (Atwater, 1989).

Despite differences in data and methods used to derive models of Pacific plate motion relative to the hotspots, APWP predicted by these models are similar. The models all show predicted pole positions trending southward in the north Atlantic, more-or-less along the prime meridian (Fig. 7). Scatter in predicted pole positions increases with age; for 20 Ma, the 95% confidence circle of the mean of predicted poles is only ~2°, but for 80 Ma, this circle has increased to ~6°. All models show rapid polar motion throughout the Late Cretaceous and Cenozoic, many with a kink in the path corresponding to the Hawaiian-Emperor Bend. The total displacement from the spin axis is ~12° for 40 Ma and ~30° at 80 Ma, the difference reflecting polar motion during the time of the Emperor chain formation.

Predicted polar wander prior to the Emperor chain is poorly known, and most plate motion models do not treat this period. Two models are plotted in Figure 3 (Duncan and Clague, 1985; Wessel et al., 2006), and both show the APWP bending to the west ca. 100 Ma. The earlier model is simpler and uses a single rotation pole to describe Early Cretaceous plate motion, whereas the latter contains several kinks where different sets of seamount trends define segments of the rotation model (Wessel et al., 2006).

Comparison of the paleomagnetic APWP with the predicted APWP from hotspot models shows excellent correlation for the

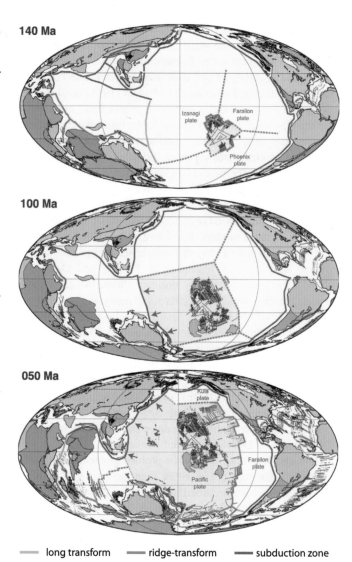

Figure 6. Sketch maps showing plate boundaries surrounding the Pacific plate at three times in the past. Dashed lines show the most uncertain boundaries. Blue arrows show subduction of Pacific plate. Reconstructed continents and magnetic lineations from Lawver et al. (2003). Pacific lineations and features were backtracked relative to the Indo-Atlantic plates using a plate circuit. Plate boundaries were taken from many sources.

past ~50 m.y. but significant discrepancies for earlier times (Fig. 3). The predicted APWP of Wessel et al. (2006) has a 30-Ma pole well within the 95% confidence ellipse of the 30-Ma paleomagnetic pole, the 40-Ma predicted pole is very close to the 39-Ma paleomagnetic pole, and the 50-Ma predicted pole is within the confidence ellipse of the 49-Ma paleomagnetic pole. In other words, the two paths are statistically indistinguishable for much of the Cenozoic.

A major offset between predicted and measured polar paths occurs from 80 to 49 Ma, during which time the paleomagnetic APWP shows a stillstand whereas the hotspot-derived APWP

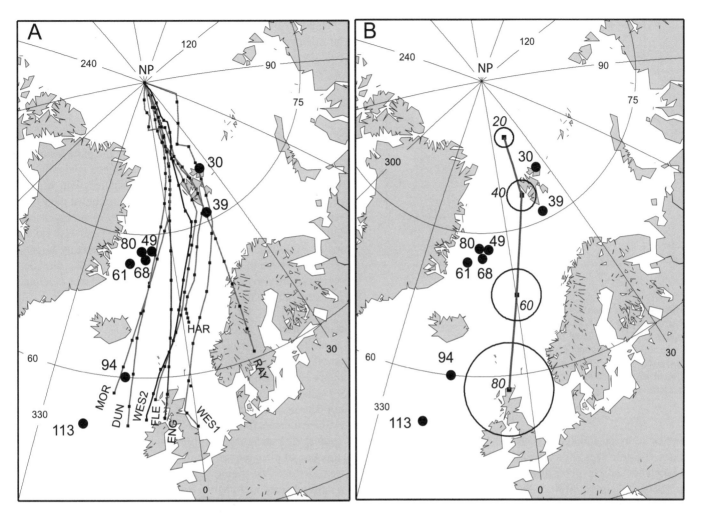

Figure 7. Predicted Pacific apparent polar wander path (APWP) from various models for the past 80 Ma, assuming hotspots fixed relative to the spin axis. Filled circles are paleomagnetic mean poles (Fig. 3), labeled by age in Ma. (A) APWP predicted by each model, with dots at 5-m.y. intervals. (B) Average positions (and 95% confidence region around the mean) for the 20-, 40-, 60-, and 80-Ma predicted poles. Abbreviations for models: MOR—Morgan (1971); DUN—Duncan and Clague (1985); WES1—Wessel and Kroenke (1997); WES2—Wessel et al. (2006); FLE—Fleitout and Moriceau (1992); ENG—Engebretson et al. (1984); HAR—Harada and Hamano (2000); RAY—Raymond et al. (2000). Numbers are pole ages in Ma.

indicates significant polar motion. Even with the large paleomagnetic pole confidence ellipses and allowing for several degrees of uncertainty in the predicted pole position (Wessel et al., 2006), the two paths appear distinct by 61 Ma (Fig. 8). In all, the hotspot-model–predicted APWP shows a maximum difference of 15° compared to the paleomagnetic APWP. This difference implies ~19° of southward motion for the Hawaiian hotspot from 80 to 49 Ma, because the paleomagnetic data imply that the plate did not move northward during the time that the Emperor seamounts formed. (Note that because of spherical geometry, offset values are different depending on the location of the points being compared. Thus, the offset between paleomagnetic poles in the north Atlantic is different from that in seamount locations in the central Pacific.)

Farther back in time, the details of the hotspot-model–predicted APWP diverge from one another and the paleomagnetic APWP because of the poorly known connections to earlier seamounts. Neither predicted APWP implies a spurt of rapid apparent polar wander to match that observed between the 94- and 80-Ma paleomagnetic poles. Prior to that time, both predicted that APWP trend westward, broadly consistent with the hook in the paleomagnetic APWP. The Wessel et al. (2006) model comes closest to matching the paleomagnetic data, with the predicted APWP implying slow polar motion and slow northward motion of the plate from ca. 125 to 90 Ma. The oldest part of the predicted APWP implies mostly east-west polar motion with a slight southward component, and thus the predicted model implies a turnaround in plate motion at about the same time as

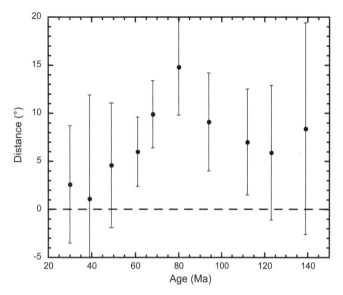

Figure 8. Arc distance between a paleomagnetic pole (Table 1) and its corresponding position on the hotspot model apparent polar wander path (APWP; calculated from Wessel et al., 2006). Error bars are an estimate of the amount of paleomagnetic pole 95% confidence ellipse traversed along a great-circle path connecting the two points. Dashed line at zero represents agreement of the two poles. Uncertainty in the hotspot model is not represented.

indicated by paleomagnetic data. Although the 145-Ma paleomagnetic pole suggests a large difference between the predicted and paleomagnetic polar paths, the accuracy of paleomagnetic poles of this age is uncertain, and the discrepancy is much less for the 139-Ma pole.

DISCUSSION

On the whole, the trend of the paleomagnetic APWP and that predicted from hotspot-derived plate motion models are broadly similar. The hotspot-derived plate motion models predict a large amount (~40°) of northward motion of the Pacific plate since Mid-Cretaceous time, similar to the paleomagnetic APWP. Furthermore, hotspot-derived models that describe motion prior to the Emperor chain predict a significant change in apparent polar wander direction, with earlier polar motion nearly east-west (Duncan and Clague, 1985), perhaps with a slight component in the north-south direction, indicating Late Jurassic–Early Cretaceous plate motion to the south (Wessel et al., 2006). Although there appears to be general agreement, there are large offsets (>5°) between the paleomagnetic and hotspot-model–predicted APWP (Fig. 8). Both APWP show a Cretaceous turnabout in plate motion, but paleomagnetic data may indicate greater southward motion of the plate during the Late Jurassic–Early Cretaceous and that the APWP bend is sharper. Furthermore, the hotspot-model–predicted APWP indicates relatively uniform northward motion, with over 30° of polar motion since 80 Ma, whereas the paleomagnetic APWP shows significant

changes in the rate of polar motion, with rapid polar wander between 94 and 80 Ma and a stillstand from 80 to 49 Ma. The paleomagnetic data imply only ~17° of northward motion since 80 Ma.

Which APWP Differences Are Significant?

Although there is >5° offset between hotspot-predicted and paleomagnetic APWP poles for much of Pacific plate history, it is not immediately clear how large a difference is significant. The paleomagnetic poles have been calculated with usually more than ten different data from different locations; thus, they represent a large area average. Uncertainty (95% confidence) ellipses are mostly <5° on the minor semi-axis and <10° on the major semi-axis. Moreover, the short axis is usually aligned nearly north-south, so the uncertainty is least in the direction corresponding to paleolatitude differences. Hotspot-predicted poles are outside the paleomagnetic pole confidence ellipses for the 61-, 68-, 80-, 94-, and 145-Ma poles, but within the uncertainties of the 30-, 39-, 49-, 113-, 123-, and 139-Ma poles (Figs. 3 and 8). Those hotspot-predicted poles that are within the paleomagnetic pole confidence ellipses of the same age are not statistically distinct.

Uncertainties for the hotspot-predicted pole positions are not determined for most models; however, Wessel et al. (2006) estimated errors for their model for the present back to 67 Ma. Typical 95% uncertainty ellipses are highly elongated along the direction of plate motion but narrow in the perpendicular direction, because the minor semi-axis is controlled by the easily measured seamount chain geometry, whereas the major semi-axis is constrained by sparse age data that define the volcanic migration rate along track. Uncertainty ellipses for the older, Emperor chain poles are ~8° in length versus 2–3° in width and slightly smaller for the late Cenozoic. Taking into account these estimated errors, the hotspot-predicted and paleomagnetic poles for 61 Ma may not be distinct, because there is probably a significant overlap of uncertainty ellipses. Uncertainties were not estimated for older hotspot model poles (Wessel et al., 2006), but if similar to those for 67 Ma, the differences between 68- and 94-Ma poles may not be as significant as suggested by Figure 8, because the hotspot-model–predicted poles have large north-south uncertainties (and confidence ellipses would overlap at least a little). However, given the large distance between the hotspot-model–predicted and paleomagnetic poles for 80 Ma, this offset is large enough to be distinct.

Given the large uncertainty for the paleomagnetic APWP prior to ca. 123 Ma and the significant uncertainties in connecting pre-Emperor age (ca. 81-Ma) seamounts to Late Cretaceous–Cenozoic plate motion models, the cause and significance of the differences between observed and predicted APWP are unclear. There may well be significant differences in the amount of northward motion and timing implied by the two different polar paths, but conclusions based on those differences are premature, given the data and model uncertainties. More paleomagnetic data are needed for the Late Jurassic and Early Cretaceous, as are better models of hotspot motion for the same period.

Although the preceding discussion seems to suggest that the two polar paths are almost indistinguishable, the APWP differences for the Late Cretaceous are both large and systematic. Furthermore, implications for the motion of the Pacific plate imply markedly different tectonics for this time period.

Implications of Late Cretaceous–Early Cenozoic Polar Wander

As noted previously, the paleomagnetic and hotspot-model–predicted APWP are indistinguishable for times 49 Ma and after. In contrast, the most notable and significant differences occur from 94 to 49 Ma. The distance between the 94- and 80-Ma paleomagnetic poles (~13°) and the age difference imply rapid northward polar motion at ~1°/m.y. (Sager, 2006). Although there is relatively large uncertainty in the distance of polar motion (±7°) and the ages of poles (2–3 m.y.), the finding of rapid polar motion seems robust, having been noted in other analyses of similar data sets (e.g., Cottrell and Tarduno, 2003) and in a largely independent paleomagnetic data set derived from seamount magnetic anomalies (Sager and Koppers, 2000). The reason for the rapid polar motion is not known. It implies rapid northward motion of the Pacific plate (Cottrell and Tarduno, 2003), but there is no indication that during the Mid-Cretaceous, the Pacific plate was engaged on its northern boundary in a subduction zone that would have given it a large northward component of motion (Fig. 6). Sager and Koppers (2000) contended that the rapid polar motion resulted from TPW. Although some investigators found paleomagnetic evidence from global composite APWP to support this idea (Prévot et al., 2000), others disagree (e.g., Torsvik et al., 2002; Cottrell and Tarduno, 2003). The result is that there is no consensus about the source of this rapid polar shift.

If the cause is TPW, the shift should be found in paleomagnetic data across the globe. Comparing Pacific paleomagnetic poles with others from the rest of the globe can be difficult because of the differing plate motions. A number of authors have made interplate comparisons by removing plate motions defined relative to the hotspots. Having done so, the remaining polar wander is sometimes interpreted as TPW (i.e., wander of the spin axis relative to the hotspots and/or mantle). Figure 9 shows that two global average mantle APWP indicate motion of the spin axis toward the Pacific Ocean during the Late Cretaceous and early Cenozoic. Figure 10 explains how a shift of the spin axis toward the Pacific causes apparent northward motion of the plate, which is what the 94- to 80-Ma pole shift suggests. The Besse and Courtillot (2002) APWP implies ~8° of motion from 90 to 67 Ma, and the Prévot al. (2000) APWP indicates ~16° from 80 to 65 Ma. Both curves were constructed with 20-m.y. averages, which tend to smooth abrupt changes in polar wander. Nevertheless, if the hotspot-based plate motion models used in those studies for backtracking the paleomagnetic data are cor-

rect, both APWP support the idea that the rapid 94- to 80-Ma shift in Pacific paleomagnetic poles may have been caused by a shift in the spin axis (i.e., TPW).

The finding of a polar stillstand from 80 to 49 Ma is startling, because it implies a very different tectonic motion than previously thought during the period corresponding to formation of the Emperor seamounts. The nearly north-south trend of these seamounts (and other similar chains; e.g., Wessel et al., 2006) has been used to infer that Pacific plate motion was largely north-south during this period. Indeed, this implied northward drift is so ingrained that dozens of western Pacific tectonic models have incorporated this ~20° of northward motion (e.g., Hilde et al., 1976; Engebretson et al., 1984). In contrast, the paleomagnetic APWP implies negligible (~4° or less) north-south motion during the Emperor seamounts period. This is a drastic difference with revolutionary implications. If the paleomagnetic APWP is correct, many tectonic models will have to be rethought.

Paleolatitudes for the Hawaiian-Emperor chain (estimated from the distance between seamounts and the APWP poles; Fig. 11) agree with the paleolatitude trend in basalt core data from the Emperor chain (Tarduno et al., 2003). Agreement is not surprising, because the Emperor data were used in the APWP calculations, but the point is that a larger set of paleomagnetic data from widespread locations gives the same results as data from drill cores recovered from four Emperor seamounts. The paleolatitude of the plate and Hawaiian hotspot apparently changed in the same linear manner during the time corresponding to the 80- to 49-Ma poles. After this time, the implied hotspot paleolatitude was the same as its present location, in agreement with published findings (e.g., Sager, 1984; Sager et al., 2005).

In a graph of total northward motion (Fig. 11), paleomagnetic data agree with the hotspot-predicted latitude of the Hawaiian-Emperor seamounts (Wessel et al., 2006) from 30 to 49 Ma. In contrast, the 61- to 80-Ma poles predict little northward drift while the hotspot-predicted northward drift increases rapidly, leading to an offset of up to ~18° at 80 Ma. Equatorial-band sediments from Cretaceous DSDP cores from the western Pacific (see Sager, 1984) agree with the amount of northward drift implied by the paleomagnetic poles (Fig. 11). Recent estimates of paleolatitude from equatorial-band sediments for the Cenozoic (Parés and Moore, 2005) show a similar trend as the paleomagnetic poles and seamount chains but with an offset of ~3° to the south. The agreement of two independent types of paleolatitude data (paleomagnetic and paleoequator) imply that the paleomagnetic data are accurate, although the 3°-Cenozoic offset suggests a potential for a small, systematic bias in one or the other. One possibility is that the equatorial sediment band, which is related to equatorial currents, may not have been precisely at the equator. Another is that the geocentric axial dipole assumption for paleomagnetic data may be slightly inaccurate owing to long-term nondipole components in the time-averaged geomagnetic field (McElhinny et al., 1996; Courtillot and Besse, 2004).

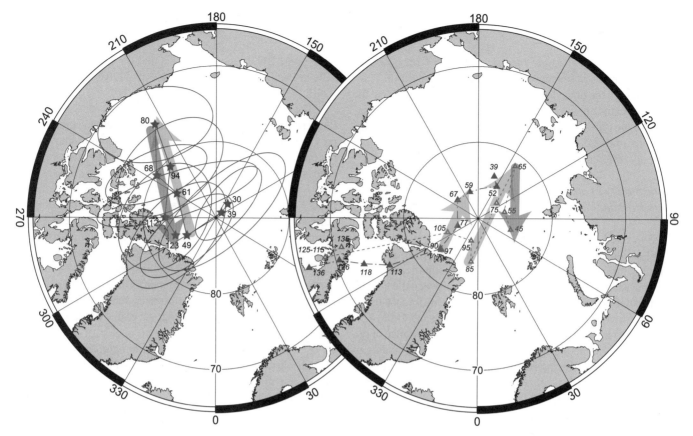

Figure 9. Polar wander in the hotspot reference frame. (Left) Pacific paleomagnetic poles (stars) and 95% confidence ellipses (Table 1) were reconstructed using a hotspot-based plate motion model (Wessel et al., 2006). (Right) Global true polar wander (TPW) curves. Filled triangles and dotted line show a composite apparent polar wander path constructed from continental plates, reconstructed into the African plate reference frame and backtracked using a model of drift of the African plate relative to the hotspots (Besse and Courtillot, 2002). Open triangles and dash-dot line show another, similar polar wander path constructed using only volcanic rock paleomagnetic data (Prévot et al., 2000). If the hotspots form a mantle reference frame (i.e., have small relative motions), there has been no TPW, and long-term nondipole geomagnetic field components are small, then the paleomagnetic poles should reconstruct to the spin axis. Polar wander in the hotspot reference frame is frequently interpreted as TPW (e.g., Andrews, 1985; Gordon, 1987; Besse and Courtillot, 2002). Pink arrows show implied spin axis motion toward the Pacific hemisphere, whereas light blue arrows show motion in the opposite sense. Numbers are pole ages in Ma.

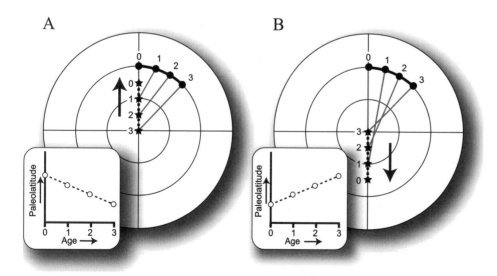

Figure 10. Cartoon explaining the effect of true polar wander (TPW) on paleolatitude. Plots show the geographic pole (looking down from above) and polar wander (movement of the spin axis) in the hotspot reference frame, calculated by backtracking paleomagnetic poles using models of plate motion relative to the hotspots. Filled circles represent a seamount chain that has been created by a hotspot at location 0. Plate motion has no northward component, so the Euler pole describing the plate motion relative to the hotspot is located at the spin axis, and older seamounts have the same latitude as the hotspot. (A) If TPW moves the spin axis (paleomagnetic poles) toward the hotspot, paleolatitudes appear to increase for younger seamounts (inset). (B) If TPW moves the spin axis away from the hotspot, paleolatitudes appear to decrease in younger seamounts. The example in panel B mimics observations of paleolatitudes in the Emperor seamounts (Tarduno et al., 2003), implying the paleolatitude shift can be explained by motion of the spin axis toward the Pacific.

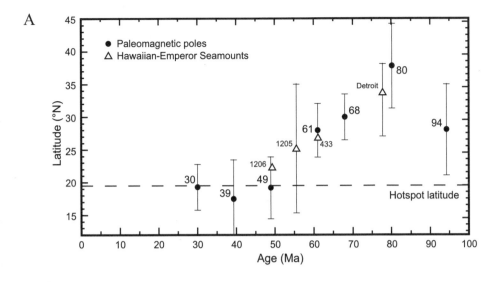

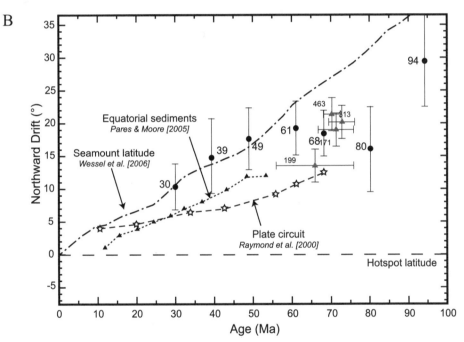

Figure 11. (A) Paleolatitude implied by Pacific paleomagnetic data versus age. Filled circles show estimated paleolatitudes of Hawaiian-Emperor seamounts, determined by distance from paleomagnetic poles to seamount sites (estimated from Wessel et al., 2006, model). Open triangles show paleolatitudes from Deep Sea Drilling Project/Ocean Drilling Program basalt drill cores (Tarduno et al., 2003). (B) Northward drift implied by Pacific paleomagnetic data versus age. Filled circles show northward drift implied by paleomagnetic poles. Dash-dot line and dots show northward drift of Hawaiian-Emperor seamounts with time (from Wessel et al., 2006). Small dashed line and triangles denote northward drift shown by equatorial sediments (Parés and Moore, 2005). Gray triangles show additional estimates of northward drift from equatorial sediments (Sager and Bleil, 1987). Heavy dashed line and stars are estimates of seamount latitude from plate circuit model of Raymond et al. (2000). Numbers are pole ages in Ma.

The simplest explanation for the discrepancy between paleomagnetic and hotspot-predicted APWP for this period is that the hotspot moved rapidly south while the plate moved mostly east-west. Two mechanisms for such a shift have been proposed. Some investigators have argued that the hotspot itself shifted ~13° in latitude as a result of flow in the mantle (e.g., Steinberger, 2000; Steinberger and O'Connell, 2000; Tarduno et al., 2003; Steinberger et al., 2004). Others have posited TPW, a shift of the entire mantle (and embedded hotspots) relative to the spin axis (e.g., Gordon and Cape, 1981; Duncan and Storey, 1992). With paleomagnetic and paleolatitude data from only one plate, it is difficult to distinguish between these two hypotheses. TPW should give a coherent shift of hotspots over the entire globe,

whereas mantle flow should yield globally inconsistent motions (Fig. 2).

For TPW to cause the observed apparent southward motion of the hotspot, the spin axis must have moved away from the Pacific (Figs. 9 and 10). Of the two global synthetic APWP shown in Figure 9, one implies a small shift (~8°) of the spin axis away from the Pacific from 65 to 45 Ma (Prévot et al., 2000), but the other does not (Besse and Courtillot, 2002). Given the magnitude of the shift implied by Pacific paleomagnetic data, it seems probable that such a large shift would appear prominently in global pole compilations, even with severe averaging. Furthermore, TPW as an explanation for the southward motion of the Hawaiian hotspot is unsatisfying, because it does not explain the Hawaiian-Emperor Bend. With TPW, the Pacific plate and mantle would move together, and there would be no cause for an apparent change of plate motion relative to the Hawaiian hotspot. Thus, if TPW is posited to be the reason that the paleolatitudes change, then the Hawaiian-Emperor Bend implies the coincident action of some other phenomenon.

Hotspot motion within the mantle as an explanation for the apparent rapid hotspot drift also has difficulties. The paleomagnetic data imply ~19° of paleolatitude change for the Hawaiian hotspot, whereas favored mantle-flow models explain only ~12–13° of change (Steinberger, 2000; Steinberger and O'Connell, 2000; Steinberger et al., 2004; Steinberger and Antretter, 2006). It is unclear whether such models can readily explain the ~50% greater amount of paleolatitude shift with reasonable adjustments to model parameters. Even more problematic than the drift rate is the direction of hotspot motion. Existing mantle-flow models have the Hawaiian hotspot moving south or southeast, responding to mantle flow toward eastern Pacific upwelling. The north-south trend of apparent hotspot motion during the formation of the Emperor seamounts, however, implies a significant westward component of hotspot velocity.

The motion of the hotspot relative to the mantle, $_m\mathbf{V}_h$ (bold indicates a vector), is the difference of the vector representing plate motion relative to the hotspot, $_h\mathbf{V}_p$, and the vector representing plate motion relative to the mantle, $_m\mathbf{V}_p$ (Fig. 12). The trend of the Emperor seamounts gives an estimate of $_h\mathbf{V}_p$ for the Late Cretaceous and early Cenozoic. Although the velocity of the Pacific plate relative to the mantle is unknown, it can be assumed that it was either similar to the late Cenozoic motion (NNW) or nearly east-west because of the paleomagnetic paleolatitude constraints (i.e., negligible northward motion). Given paleomagnetic pole uncertainties, these two assumptions may not be significantly different. Either is reasonable, because the plate was probably engaged in western Pacific subduction zones that would have pulled it westward, but was probably not yet subducting into the Aleutian Trench to provide a northward component of motion (Fig. 6). The vector diagram in Figure 12 indicates that if the Pacific plate had a significant westward motion, the hotspot must have had a similar component of westward motion, so that the two nearly canceled to produce seamounts with a nearly north-south trend. The amount of westward hotspot motion inferred therefore depends on the amount of westward motion assumed for the plate. The length and age

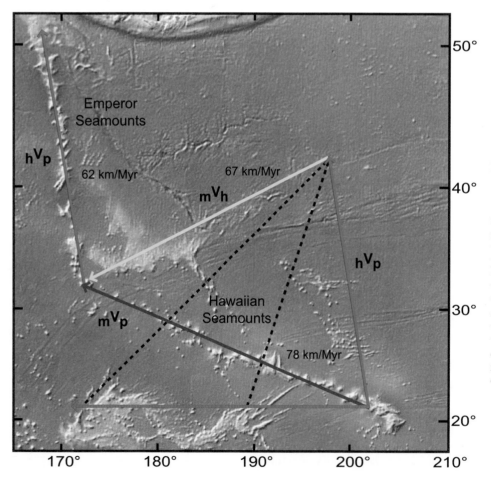

Figure 12. Sketch of motion vectors indicating Hawaiian hotspot drift during the formation of the Emperor seamounts. Motion of plate relative to hotspot, $_h\mathbf{V}_p$ (red vector), given by trend of Emperor seamounts. Motion of plate relative to mantle (assumed fixed relative to spin axis), $_m\mathbf{V}_p$ (purple vector), is assumed to be same as at present (Hawaiian chain). Sum is motion of hotspot relative to the mantle, $_m\mathbf{V}_h$ (yellow vector), which has a large westward component. Horizontal vector at bottom (magenta) shows Pacific plate motion if the plate had no northward component of velocity. Dashed-line vectors show predicted motion of hotspot relative to mantle if Pacific plate motion had no northward component. Different dashed lines correspond to different westward velocities. Background is a shaded relief plot of Hawaiian-Emperor chain bathymetry.

span of the Hawaiian chain gives approximate average progression rate of 78 km/m.y., whereas the Emperor chain suggests an average progression rate of 62 km/m.y. Adding these estimated vectors gives a velocity of ~67 km/m.y. at an azimuth of ~224° (Fig. 12), nearly at right angles to the predicted direction of Hawaiian hotspot motion (Steinberger, 2000). Although this estimate is crude because of the gross averages for velocity vectors and planar approximation, it clearly shows that the hotspot probably moved in a southwesterly direction.

In sum, large differences between measured and hotspot-predicted APWP for the Late Cretaceous to early Cenozoic are a challenge for existing explanations of hotspot drift during the formation of the Emperor chain. Both TPW and hotspot motion caused by mantle flow have drawbacks. Neither is disproved, certainly, but perhaps these difficulties indicate that other hypotheses should be considered more carefully. Several authors in this volume (Foulger, this volume; Norton, this volume; Smith, this volume; Stuart et al., this volume) and elsewhere (Natland and Winterer, 2005) suggest that the Hawaiian-Emperor chain (and other Pacific seamount chains) could have formed from a propagating crack that was initiated by and whose path has been determined by changing stresses applied by shifting plate

boundary forces. A crack, for example, might not suffer from the problem that the paleolatitude shift of the Emperor seamounts was rapid and that the Hawaiian-Emperor Bend implies an abrupt change in hotspot motion.

Comparison of Pacific and Antarctic APWP

Several studies have used motion models of Indian and Atlantic ocean plates relative to the hotspots to predict the motion of the Pacific plate relative to the hotspots (Stock and Molnar, 1987; Cande et al., 1995; DiVenere and Kent, 1999; Raymond et al., 2000). This calculation is done by connecting the plates with models of relative motion derived from spreading boundaries. For the Pacific plate, the most direct connection to plates outside the Pacific rim is through Antarctica. The idea is that if the hotspots constitute a fixed constellation of mantle position markers, motion of the Pacific plate reconstructed from Indian and Atlantic plate motion will match that derived from hotspot tracks on the Pacific plate itself. Such analyses are usually able to match the younger Hawaiian seamounts reasonably well, but fail to show the amount of northward motion suggested by the Emperor seamounts (Stock and Molnar, 1987; Cande et al.,

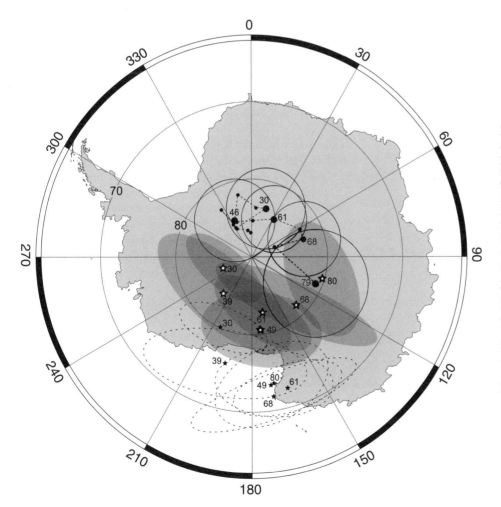

Figure 13. Comparison of the Antarctic and Pacific plate apparent polar wander paths. Antipoles of Pacific paleomagnetic poles are shown as solid stars, with 95% confidence ellipses traced by dashed lines. Open stars and gray confidence ellipses show the same poles rotated into the Antarctic reference frame using a model of the relative motion of the Pacific and Antarctic plates derived from seafloor spreading on the Pacific-Antarctic Ridge (Mayes et al., 1990; Cande et al., 1995; Tebbens and Cande, 1997). Antarctic polar wander path (small and large filled circles connected by dotted line) is synthetic polar wander path for continents rotated into Antarctic reference frame (Besse and Courtillot, 2002). Large circles are 95% confidence regions for Antarctic poles. Numbers are pole ages in Ma.

1995; DiVenere and Kent, 1999; Raymond et al., 2000). One interpretation is that this discrepancy shows interhotspot motion, but another is that the plate circuit is flawed because relative motion has occurred on an undefined plate boundary hidden within Antarctica or the southern Pacific plate (Acton and Gordon, 1994).

With a refined Pacific paleomagnetic APWP, it is possible to test the Antarctic plate circuit by rotating the APWP into the Antarctic reference frame (closing up the opening of the Pacific-Antarctic Ridge) and comparing with the Antarctic APWP. Because the Antarctic APWP is poorly defined owing to sparse data, the comparison must be made with a composite APWP constructed for the continents and rotated to Antarctica (Besse and Courtillot, 2002). This synthetic APWP is located mostly in east Antarctica, on the Atlantic side of the 120°- and 300°-meridians (Fig. 13). In contrast, Pacific APWP south poles are rotated 8–14° toward Antarctica by closure of Late Cretaceous and Cenozoic spreading on the Pacific-Antarctic Ridge (using the spreading model of Cande et al., 1995, updated by Tebbens and Cande, 1997, with the Late Cretaceous pole of Mayes et al., 1990). Except for the 80-Ma pole, the rotated Pacific poles are all on the Pacific side of the 120°- and 300°-meridians, separated by ~10° from the Antarctic APWP. The uncertainty ellipses for the 39- to 68-Ma Pacific poles do not overlap the uncertainty circles of coeval poles for the Antarctic APWP and probably would not overlap significantly even if uncertainties for the plate rotations were explicitly included. The systematic offset has been noted by other authors (Acton and Gordon, 1994; Andrews et al., 2004) and probably indicates problems with the plate circuit through Antarctica. It is unclear whether the problem occurs with the spreading models for the Pacific-Antarctic or Southwest Indian Ridge (which brings the continental APWP to Antarctica from Africa) or whether the problem is poorly documented plate boundaries in the south Pacific plate or Antarctica (Acton and Gordon, 1994). Whatever the cause, the systematic mismatch of most Pacific and Antarctic paleomagnetic poles suggests that reconstructions of Pacific hotspot tracks using Indo-Atlantic plate motions and a plate circuit through Antarctica (or vice versa) and interpretations of interhotspot motion made from them may not be reliable.

SUMMARY AND CONCLUSIONS

Comparison of the Pacific plate paleomagnetic APWP with that derived from models of plate motion relative to the hotspots show differences of up to 15°. Cenozoic poles from 30 to 49 Ma agree within uncertainties, indicating no significant difference between the paleomagnetic and hotspot reference frames for that period. Uncertainties in Late Jurassic and Early Cretaceous paleomagnetic poles and plate motion models relative to the hotspots make the significance of offsets between the two reference frames prior to ca. 123 Ma uncertain, even though those differences range from 6° to 9°. The offset during 112–61 Ma appears significant and shows a maximum for the 80-Ma pole.

Whereas the hotspot plate motion model shows nearly constant northward motion from ca. 95 Ma into the Cenozoic, the paleomagnetic APWP has a period of rapid polar wander from 94 to 80 Ma, followed by a stillstand from 80 to 49 Ma. The cause of the rapid polar motion from 94 to 80 Ma is uncertain, although comparison with global paleomagnetic polar wander curves suggests that it may have resulted at least partly from TPW. It is less clear whether TPW is an adequate explanation for the offset between reference frames for 80–49 Ma. Although one global TPW curve shows similar spin axis motion, TPW does not readily explain the large difference in plate motion relative to the hotspots implied by the Hawaiian-Emperor Bend. One must call upon both TPW and another mechanism to explain the bend. A change in hotspot motion is a simpler explanation. Paleomagnetic data indicate that the 80-Ma offset is 15° between APWP, implying ~19° of southward motion of the Hawaiian hotspot. This amount is considerably greater than the southward drift implied by previous studies. Furthermore, consideration of average velocity vectors for the Pacific plate and Hawaiian hotspot indicate that the hotspot had a westward velocity that nearly equaled the westward motion of the Pacific plate. The rapid implied drift of the hotspot relative to the mantle may prove challenging to explain by-mantle flow, because the implied drift velocity is ~50% greater than published models, and it is almost at right angles to modeled flow directions. If neither TPW nor mantle flow are an adequate explanation for the apparent hotspot drift, it is appropriate to consider other explanations.

ACKNOWLEDGMENTS

This study would not have been possible without core samples collected from the Pacific plate by the Deep Sea Drilling Project and the Ocean Drilling Program over 35 years. The author is indebted to the many scientists whose paleomagnetic results were accumulated in the Pacific apparent polar wander path. Funding for the Pacific paleomagnetic analyses were provided in part through U.S. Science Support Program grants to the author to participate in individual Ocean Drilling Program cruises and in part by the Jane and R. Ken Williams '45 Chair in Ocean Drilling Science, Education, and Technology, which is currently held by the author. I thank Carlo Doglioni, Gill Foulger, Stuart Hall, Donna Jurdy, Jim Ogg, Jerry Winterer, and an anonymous reviewer for helpful suggestions for improving the chapter.

REFERENCES CITED

Acton, G.D., and Gordon, R.G., 1994, Paleomagnetic tests of Pacific plate reconstructions and implications for motion between hotspots: Science, v. 263, p. 1246–1254, doi: 10.1126/science.263.5151.1246.

Anderson, D.L., 2000, The thermal state of the upper mantle; No role for mantle plumes: Geophysical Research Letters, v. 27, p. 3623–3626, doi: 10.1029/2000GL011533.

Anderson, D.L., 2005, Scoring hotspots: The plume and plate paradigms, in Foulger, G.R., et al., eds., Plates, plumes, and paradigms: Boulder, Colorado, Geological Society of America Special Paper 388, p. 31–54, doi: 10.1130/2005.2388(04).

Andrews, J.A., 1985, True polar wander: An analysis of Cenozoic and Mesozoic paleomagnetic poles: Journal of Geophysical Research, v. 90, p. 7737–7750.

Andrews, D.L., Gordon, R.G., and Horner-Johnson, B.C., 2004, Paleomagnetic tests of global plate reconstructions with fixed and moving hotspots: Eos (Transactions, American Geophysical Union), v. 85, fall meeting supplement, abstract no. V51B–0540.

Atwater, T., 1989, Plate tectonic history of the northeast Pacific and western North America, *in* Winterer, E.L., et al., eds., The eastern Pacific Ocean and Hawaii: Boulder, Colorado, Geological Society of America Decade of North American Geology, v. N, p. 21–72.

Beaman, M., Sager, W., Acton, G., Lanci, L., and Pares, J., 2007, Improved Late Cretaceous and early Cenozoic paleomagnetic apparent polar wander path for the Pacific Plate: Earth and Planetary Science Letters, in press.

Besse, J., and Courtillot, V., 2002, Apparent and true polar wander and the geometry of the geomagnetic field over the last 200 Myr: Journal of Geophysical Research, v. 107, art. no. 2300, doi: 10.1029/2000JB000050.

Butler, R.F., 1992, Paleomagnetism: Boston, Blackwell Scientific, 319 p.

Cande, S.C., Raymond, C.A., Stock, J., and Haxby, W.F., 1995, Geophysics of the Pitman fracture zone and Pacific-Antarctic plate motions during the Cenozoic: Science, v. 270, p. 947–953.

Carlson, R.L., Sager, W.W., and Jurdy, D.M., 1988, Plate tectonics reference frames workshop report, December 1988: College Station, Texas, Texas A&M University.

Clague, D.A., and Dalrymple, G.B., 1975, Cretaceous K-Ar ages of volcanic rocks from the Muscians seamounts and the Hawaiian ridge: Geophysical Research Letters, v. 2, p. 305–308.

Cottrell, R.D., and Tarduno, J.A., 2003, A Late Cretaceous pole for the Pacific plate: Implications for apparent and true polar wander and the drift of hotspots: Tectonophysics, v. 362, p. 321–333, doi: 10.1016/S0040-1951(02)00643-1.

Courtillot, V., and Besse, J., 2004, A long-term octupolar component in the geomagnetic field? (0–200 million years B.P.), *in* Channel, J.E.T., et al., eds., Timescales of the geomagnetic field: Washington, D.C., American Geophysical Union Geophysical Monograph 145, p. 59–74.

Courtillot, V., Davaille, A., Besse, J., and Stock, J., 2003, Three distinct types of hotspots in the Earth's mantle: Earth and Planetary Science Letters, v. 205, p. 295–308, doi: 10.1016/S0012–821X(02)01048–8.

Cox, A., and Gordon, R.G., 1984, Paleolatitudes determined from paleomagnetic data from vertical cores: Reviews of Geophysics and Space Physics, v. 22, p. 47–72.

Cuffaro, M., and Doglioni, C., 2007 (this volume), Global kinematics in deep versus shallow hotspot reference frames, *in* Foulger, G.R., and Jurdy, D.M., eds., Plates, plumes, and planetary processes: Boulder, Colorado, Geological Society of America Special Paper 430, doi: 10.1130/2007.2430(18).

DiVenere, V., and Kent, D.V., 1999, Are the Pacific and Indo-Atlantic hotspots fixed? Testing the plate circuit through Antarctica: Earth and Planetary Science Letters, v. 170, p. 105–117, doi: 10.1016/S0012-821X(99)00096-5.

Doglioni, C., Green, D.H., and Mongelli, F., 2005, On the shallow origin of hotspots and the westward drift of the lithosphere, *in* Foulger, G.R., et al., eds., Plates, plumes, and paradigms: Boulder, Colorado, Geological Society of America Special Paper 388, p. 735–749, doi: 10.1130/2005.2388(42).

Duncan, R.A., 1981, Hotspots in the southern oceans—An absolute frame of reference for motion of the Gondwana continents: Tectonophysics, v. 74, p. 29–42, doi: 10.1016/0040-1951(81)90126-8.

Duncan, R.A., and Clague, D.A., 1985, Pacific plate motion recorded by linear volcanic chains, *in* Stehli, F.G., and Uyeda, S., eds., The ocean basins and margins: New York, Plenum, p. 89–121.

Duncan, R.A., and Storey, M., 1992, The life cycles of Indian Ocean hotspots, *in* Kidd, R.B., et al., eds., Synthesis of results from scientific drilling in the Indian Ocean: Washington, D.C., American Geophysical Union Geophysical Monograph 70, p. 91–103.

Engebretson, D.C., Cox, A., and Gordon, R.G., 1984, Relative motions between oceanic plates of the Pacific basin: Journal of Geophysical Research, v. 89, p. 10,291–10,310.

Fleitout, L., and Moriceau, C., 1992, Short-wavelength geoid, bathymetry, and the convective pattern beneath the Pacific Ocean: Geophysical Journal International, v. 110, p. 6–28.

Foulger, G.R., 2007 (this volume), The "plate" model for the genesis of melting anomalies, *in* Foulger, G.R., and Jurdy, D.M., eds., Plates, plumes, and planetary processes: Boulder, Colorado, Geological Society of America Special Paper 430, doi: 10.1130/2007.2430(01).

Foulger, G.R., and Natland, J.H., 2003, Is "hotspot" volcanism a consequence of plate tectonics?: Science, v. 300, p. 921–922, doi: 10.1126/science.1083376.

Francheteau, J., Harrison, C.G.A., Sclater, J.G., and Richards, M.L., 1970, Magnetization of Pacific seamounts: A preliminary polar curve for the northeastern Pacific: Journal of Geophysical Research, v. 75, p. 2035–2061.

Glen, W., 2005, The origins and early trajectory of the mantle plume quasi-paradigm, *in* Foulger, G.R., et al., eds., Plates, plumes, and paradigms: Boulder, Colorado, Geological Society of America Special Paper 388, p. 91–117, doi: 10.1130/2005.2388(07).

Goldreich, P., and Toomre, A., 1969, Some remarks on polar wandering: Journal of Geophysical Research, v. 74, p. 2555–2567.

Gordon, R.G., 1983, Late Cretaceous apparent polar wander path of the Pacific plate: Evidence for rapid shift of the Pacific hotspots with respect to the spin axis: Geophysical Research Letters, v. 10, p. 709–712.

Gordon, R.G., 1987, Polar wandering and paleomagnetism: Annual Review of Earth and Planetary Sciences, v. 15, p. 567–593, doi: 10.1146/annurev.ea.15.050187.003031.

Gordon, R.G., 1990, Test for bias in paleomagnetically determined paleolatitudes from Pacific plate Deep Sea Drilling Project sediments: Journal of Geophysical Research, v. 95, p. 8397–8404.

Gordon, R.G., and Cape, C., 1981, Cenozoic latitudinal shift of the Hawaiian hotspot and its implications for true polar wander: Earth and Planetary Science Letters, v. 55, p. 37–47, doi: 10.1016/0012-821X(81)90084-4.

Gordon, R.G., Cox, A., and O'Hare, S., 1984, Paleomagnetic Euler poles and the apparent polar wander and absolute motion of North America since the Carboniferous: Tectonics, v. 3, p. 499–537.

Hall, S., and Riisager, P., 2006, Palaeomagnetic palaeolatitudes of the Ontong Java plateau from 120 to 55 Ma: Implications for the apparent polar wander path of the Pacific plate: Geophysical Journal International, v. 169, p. 455–470, doi: 10.1111/j.1365-246X.2007.03338.X.

Harada, Y., and Hamano, Y., 2000, Recent progress on the plate motion relative to the hotspots, *in* Richards, M., et al., eds., The history and dynamics of global plate motions: Washington, D.C., American Geophysical Union Geophysical Monograph 121, p. 327–338.

Harrison, C.G.A., Jarrard, R.D., Vacquier, V., and Larson, R.L., 1975, Palaeomagnetism of Cretaceous Pacific seamounts: Geophysical Journal of the Royal Astronomical Society, v. 42, p. 859–882.

Hilde, T.W.C., Isezaki, N., and Wageman, J.M., 1976, Mesozoic seafloor spreading in the north central Pacific, *in* Sutton, G.H., et al., eds., The geophysics of the Pacific Ocean basin and its margin: Washington, D.C., American Geophysical Union Geophysical Monograph 19, p. 205–226.

Irving, E., and Irving, G.A., 1982, Apparent polar wander paths Carboniferous through Cenozoic and the assembly of Gondwana: Geophysical Surveys, v. 5, p. 141–188, doi: 10.1007/BF01453983.

Jackson, E.D., and Shaw, H.R., 1975, Stress fields in the central portions of the Pacific plate: Delineated in time by linear volcanic chains: Journal of Geophysical Research, v. 80, p. 1861–1874.

Jackson, E.D., Shaw, H.R., and Bargar, K.E., 1975, Calculated geochronology and stress field orientations along the Hawaiian chain: Earth and Planetary Science Letters, v. 26, p. 145–155, doi: 10.1016/0012-821X(75)90082-5.

Jarrard, R.D., and Clague, D.A., 1977, Implications of Pacific island and seamount ages for the origin of volcanic chains: Reviews of Geophysics and Space Physics, v. 15, p. 57–76.

Keller, R.A., Duncan, R.A., and Fisk, M.R., 1995, Geochemistry and $^{40}Ar/^{39}Ar$

geochronology of basalts from ODP Leg 145, *in* Rea, D.K., et al., eds., Proceedings of the Ocean Drilling Program, scientific results, volume 145: College Station, Texas, Ocean Drilling Program, p. 333–344.

Kono, M., 1980, Paleomagnetism of DSDP Leg 55 basalts and implications for the tectonics of the Pacific plate, *in* Initial Reports of the Deep Sea Drilling Project, volume 55: Washington, D.C., U.S. Government Printing Office, p. 737–752.

Koppers, A.A.P., Staudigel, H., Pringle, M.S., and Wijbrans, J.R., 2003, Short-lived and discontinuous intraplate volcanism in the South Pacific: Hot spots or extensional volcanism?: Geochemistry, Geophysics, Geosystems, v. 4, art. no. 1089, doi: 10.1029/2003GC000533.

Kroenke, L.W., Wessel, P., and Sterling, A., 2004, Motion of the Ontong Java plateau in the hot spot frame of reference: 122 Ma–present, *in* Fitton, J.G., et al., eds., Origin and evolution of the Ontong Java plateau: London, Geological Society of London Special Publication 229, p. 9–20.

Larson, R.L., and Sager, W.W., 1992, Skewness of magnetic anomalies M0 to M29 in the northwestern Pacific, *in* Larson, R.L., et al., eds., Proceedings of the Ocean Drilling Program, scientific results, volume 129: College Station, Texas, Ocean Drilling Program, p. 471–481.

Larson, R.L., Steiner, M.B., Erba, E., and Lancelot, Y., 1992, Paleolatitudes and tectonic reconstructions of the oldest Pacific portion of the Pacific plate: A comparative study, *in* Larson, R.L., et al., eds., Proceedings of the Ocean Drilling Program, scientific results, volume 129: College Station, Texas, Ocean Drilling Program, p. 615–631.

Lawver, L.A., Dalziel, I.W.D., Gahagan, L.M., Martin, K.M., and Campbell, D.A., 2003, The Plates 2003 atlas of plate reconstructions (750 Ma to present day), Plates Progress Report 280–0703: Austin, Texas, University of Texas Institute for Geophysics Technical Report 190, 97 p.

Lonsdale, P., 1988, Paleogene history of the Kula plate: Offshore evidence and onshore implications: Geological Society of America Bulletin, v. 100, p. 733–754, doi: 10.1130/0016-7606(1988)100<0733:PHOTKP>2.3.CO;2.

Mayes, C.L., Lawver, L.A., and Sandwell, D.T., 1990, Tectonic history and new isochron chart of the South Pacific: Journal of Geophysical Research, v. 95, p. 8543–8567.

McElhinny, M.W., McFadden, P.L., and Merrill, R.T., 1996, The time-averaged paleomagnetic field 0–5 Ma: Journal of Geophysical Research, v. 101, p. 25,007–25,027, doi: 10.1029/96JB01911.

McNutt, M.K., and Fischer, K.M., 1987, The South Pacific superswell, *in* Keating, B.H., et al., eds., Seamounts, islands, and atolls: Washington, D.C., American Geophysical Union Geophysical Monograph 43, p. 25–34.

Merrill, R.T., and McFadden, P.L., 2003, The geomagnetic axial dipole field assumption: Physics of the Earth and Planetary Interiors, v. 139, p. 171–185, doi: 10.1016/j.pepi.2003.07.016.

Merrill, R.T., McElhinny, M.W., and McFadden, P.L., 1996, The magnetic field of the Earth: New York, Academic Press, 531 p.

Morgan, W.J., 1971, Deep mantle convection plumes in the lower mantle: Nature, v. 230, p. 42–43, doi: 10.1038/230042a0.

Morgan, W.J., 1972, Deep mantle convection plumes and plate motions: American Association of Petroleum Geologists Bulletin, v. 56, p. 203–213.

Morgan, W.J., 1983, Hotspot tracks and the early rifting of the Atlantic: Tectonophysics, v. 94, p. 123–139, doi: 10.1016/0040-1951(83)90013-6.

Müller, R.D., Royer, J.-Y., and Lawver, L.A., 1993, Revised plate motions relative to the hotspots from combined Atlantic and Indian ocean hotspot tracks: Geology, v. 21, p. 275–278, doi: 10.1130/0091-7613(1993)021<0275:RPMRTT>2.3.CO;2.

Natland, J.H., and Winterer, E.L., 2005. Fissure control on volcanic action in the Pacific, *in* Foulger, G.R., et al., eds., Plates, plumes, and paradigms: Boulder, Colorado, Geological Society of America Special Paper 388, p. 687–710, doi: 10.1130/2005.2388(39).

Norton, I.O., 1995, Plate motions in the North Pacific: The 43 Ma nonevent: Tectonics, v. 14, p. 1080–1094, doi: 10.1029/95TC01256.

Norton, I.O., 2007 (this volume), Speculations on Cretaceous tectonic history of the northwest Pacific and a tectonic origin for the Hawaii hotspot, *in* Foul-

ger, G.R., and Jurdy, D.M., eds., Plates, plumes, and planetary processes: Boulder, Colorado, Geological Society of America Special Paper 430, doi: 10.1130/2007.2430(22).

Parés, J.M., and Moore, T.C., 2005, New evidence for the Hawaiian hotspot plume motion since the Eocene: Earth and Planetary Science Letters, v. 237, p. 951–959, doi: 10.1016/j.epsl.2005.06.012.

Petronotis, K.E., and Gordon, R.G., 1999, A Maastrichtian paleomagnetic pole for the Pacific plate from a skewness analysis of marine magnetic anomaly 32: Geophysical Journal International, v. 139, p. 227–247, doi: 10.1046/j.1365-246X.1999.00901.x.

Prévot, M., Mattern, E., Camps, P., and Dagnieres, M., 2000, Evidence for a 20° tilting of the Earth's rotation axis 110 million years ago: Earth and Planetary Science Letters, v. 179, p. 517–528, doi: 10.1016/S0012-821X(00)00129-1.

Raymond, C.A., Stock, J.M., and Cande, S.C., 2000, Fast Paleogene motion of the Pacific hotspots, *in* Richards, M.A., et al., eds., The history and dynamics of global plate motions: Washington, D.C., American Geophysical Union Geophysical Monograph 121, p. 359–375.

Sager, W.W., 1984, Paleomagnetism of Abbott seamount and implications for the latitudinal shift of the Hawaiian hotspot: Journal of Geophysical Research, v. 89, p. 6271–6284.

Sager, W.W., 2006, Cretaceous paleomagnetic apparent polar wander path for the Pacific plate calculated from Deep Sea Drilling Project and Ocean Drilling Program basalt cores: Physics of the Earth and Planetary Interiors, v. 156, p. 329–349, doi: 10.1016/j.pepi.2005.09.014.

Sager, W.W., and Bleil, U., 1987, Latitudinal shift of Pacific hotspots during the Late Cretaceous and early Tertiary: Nature, v. 326, p. 488–490, doi: 10.1038/326488a0.

Sager, W.W., and Koppers, A.A.P., 2000, Late Cretaceous polar wander of the Pacific plate: Evidence of a true polar wander event: Science, v. 287, p. 455–459, doi: 10.1126/science.287.5452.455.

Sager, W.W., and Pringle, M.S., 1988, Mid-Cretaceous to early Tertiary apparent polar wander path of the Pacific plate: Journal of Geophysical Research, v. 93, p. 11,753–11,771.

Sager, W.W., Tarduno, J.A., and MacLeod, C.J., 1995, Paleomagnetism of Cretaceous sediments, ODP hole 869B, western Pacific Ocean: Magnetic polarity, paleolatitude, and a paleomagnetic pole, *in* Sager, W.W., et al., eds., Proceedings of the Ocean Drilling Program, scientific results, volume 143: College Station, Texas, Ocean Drilling Program, p. 405–418.

Sager, W.W., Lamarche, A.J., and Kopp, C., 2005, Paleomagnetic modeling of seamounts near the Hawaiian-Emperor bend: Tectonophysics, v. 405, p. 121–140, doi: 10.1016/j.tecto.2005.05.018.

Schettino, A., and Scotese, C.R., 2005, Apparent polar wander paths for the major continents (200 Ma to the present day): A palaeomagnetic reference frame for global plate tectonic reconstructions: Geophysical Journal International, v. 163, p. 727–759, doi: 10.1111/j.1365-246X.2005.02638.x.

Scholl, D.W., Vallier, T.L., and Stevenson, A.J., 1986, Terrane accretion, production and continental growth: A perspective based on the origin and tectonic fate of the Aleutian–Bering Sea region: Geology, v. 14, p. 43–47, doi: 10.1130/0091-7613(1986)14<43:TAPACG>2.0.CO;2.

Schouten, H., and Cande, S.C., 1976, Palaeomagnetic poles from marine magnetic anomalies: Geophysical Journal of the Royal Astronomical Society, v. 44, p. 567–575.

Sharp, W.D., and Clague, D.A., 2006, 50-Ma initiation of Hawaiian-Emperor bend records major change in Pacific plate motion: Science, v. 313, p. 1281–1284, doi: 10.1126/science.1128489.

Shaw, H.R., 1973, Mantle convection and volcanic periodicity in the Pacific: Evidence from Hawaii: Geological Society of America Bulletin, v. 84, p. 1505–1526, doi: 10.1130/0016-7606(1973)84<1505:MCAVPI>2.0.CO;2.

Shaw, H.R., and Jackson, E.D., 1973, Linear island chains in the Pacific: Result of thermal plumes or gravitational anchors?: Journal of Geophysical Research, v. 78, p. 8634–8652.

Smith, A.D., 2007 (this volume), A plate model for Jurassic to Recent intraplate volcanism in the Pacific Ocean basin, *in* Foulger, G.R., and Jurdy, D.M., eds., Plates, plumes, and planetary processes: Boulder, Colorado, Geological Society of America Special Paper 430, doi: 10.1130/2007.2430(23).

Steinberger, B., 2000, Plumes in a convecting mantle: Models and observations for individual hotspots: Journal of Geophysical Research, v. 105, p. 11,127–11,152, doi: 10.1029/1999JB900398.

Steinberger, B., and Antretter, M., 2006, Conduit diameter and buoyant rising speed of mantle plumes: Implications for the motion of hot spots and shape of mantle plumes: Geochemistry, Geophysics, Geosystems, v. 7, p. 1–25, doi: 10.1029/2006GC001409.

Steinberger, B., and O'Connell, R.J., 2000, Effects of mantle flow on hotspot motion, *in* Richards, M.A., et al., eds., The history and dynamics of global plate motions: Washington, D.C., American Geophysical Union Geophysical Monograph 121, p. 377–398.

Steinberger, B., Sutherland, R., and O'Connell, R.J., 2004, Prediction of Emperor-Hawaii seamount locations from a revised model of global plate motion and mantle flow: Nature, v. 430, p. 167–173, doi: 10.1038/nature02660.

Stock, J.M., and Molnar, P., 1987, Revised history of early Tertiary plate motion in the southwest Pacific: Nature, v. 325, p. 495–499, doi: 10.1038/325495a0.

Stuart, W.D., Foulger, G.R., and Barall, M., 2007 (this volume), Propagation of the Hawaiian-Emperor volcano chain by Pacific plate cooling stress, *in* Foulger, G.R., and Jurdy, D.M., eds., Plates, plumes, and planetary processes: Boulder, Colorado, Geological Society of America Special Paper 430, doi: 10.1130/2007.2430(24).

Tarduno, J.A., Duncan, R.A., Scholl, D.W., Cottrell, R.D., Steinberger, B., Thordarson, T., Kerr, B.C., Neal, C.R., Frey, F.A., Torii, M., and Carvallo, C., 2003, The Emperor seamounts: Southward motion of the Hawaiian hotspot plume in the Earth's mantle: Science, v. 301, p. 1064–1069, doi: 10.1126/science.1086442.

Tebbens, S.F., and Cande, S.C., 1997, Southeast Pacific tectonic evolution from early Oligocene to present: Journal of Geophysical Research, v. 102, p. 12,061–12,084, doi: 10.1029/96JB02582.

Torsvik, T.H., and Van der Voo, R., 2002, Refining Gondwana and Pangea palaeogeography: Estimates of Phanerozoic non-dipole (octupole) fields: Geophysical Journal International, v. 151, p. 771–794, doi: 10.1046/j.1365-246X.2002.01799.x.

Torsvik, T.H., Van der Voo, R., and Redfield, T.F., 2002, Relative hotspot motions versus true polar wander: Earth and Planetary Science Letters, v. 202, p. 185–200, doi: 10.1016/S0012-821X(02)00807-5.

Vasas, S.M., Gordon, R.G., and Petronotis, K.E., 1994, New paleomagnetic poles for the Pacific plate from analysis of the shapes of anomalies 33n and 33r: Eos (Transactions, American Geophysical Union), v. 75, fall meeting supplement, abstracts, p. 203.

Wessel, P., and Kroenke, L.W., 1997, A geometric technique for relocating hotspots and refining absolute plate motions: Nature, v. 387, p. 365–369, doi: 10.1038/387365a0.

Wessel, P., Harada, Y., and Kroenke, L.W., 2006, Toward a self-consistent, high-resolution absolute plate motion model for the Pacific: Geochemistry, Geophysics, Geosystems, v. 7, art. no. Q03L12, doi: 10.1029/2005GC00100.

Wilson, J.T., 1963, A possible origin of the Hawaiian Islands: Canadian Journal of Physics, v. 41, p. 863–870.

Winterer, E.L., and Sager, W.W., 1995, Synthesis of drilling results from the Mid-Pacific Mountains: Regional context and implications, *in* Sager, W.W., et al., eds., Proceedings of the Ocean Drilling Program, scientific results, volume 143: College Station, Texas, Ocean Drilling Program, p. 497–535.

MANUSCRIPT ACCEPTED BY THE SOCIETY 31 JANUARY 2007

DISCUSSION

4 January 2007, Edward (Jerry) L. Winterer

Sager has rendered us a service by carefully segregating the three types of data used in making apparent polar wander paths (APWP) for the Pacific plate:

1. Measurements of magnetic inclination on drill cores (mainly basalt cores, not subject to compaction, but including sediment cores);

2. Seamount magnetism, which can yield inclination and declination for seamounts assumed to be uniformly magnetized (mainly those emplaced during the Cretaceous Normal Superchron); and

3. Skewness of magnetic anomalies. If true polar wander is not a chimera and actually exists, it cannot be detected using data from the Pacific plate.

Sager then compares this APWP with the results of the other approach to constructing a polar wandering path, via the orientation and age sequences along age-progressive seamount chains —the so-called hotspot method. The difficulties here arise from assumptions about long-term fixity of hotspots (loci of melting in the upper mantle and eruption of seamount-constructing lavas). Many chains (e.g., Samoa, Ellice, Gilbert, Marshall, Puka Puka) are not age-progressive, and the rates of progression are inconsistent from one parallel chain to another. In at least one chain, the Emperors, the core data show that the volcanoes in the chain were each erupted at a different latitude—another violation of the fixity assumption. Sager supplies a plausible explanation for this difference in latitudes by vectoring different motions of plate versus mantle, plate versus hotspot, and hotspot versus mantle.

The Emperor data free hotspots from any notion of fixity during 80–49 Ma, and published data from other south-trending chains (e.g., Ellice, Marshall, Line; Davis et al., 2002; Koppers et al., 2003) extend that age span back to at least 86 Ma. None of these chains is age progressive, and the Line Islands show two short bursts of volcanism (86–81 Ma and 73–68 Ma) on two parallel parts of the chain 1200 and >4000 km long. Plumes are totally inadequate sources for this style of volcanism, whereas tensional cracks, continuous and intermittent, opening through the lithosphere along south-trending lines and filled from fusible sources (e.g., eclogite) in the partially molten asthenosphere easily answer the requirements. The timing of this era of north-south cracking is most plausibly the inception of subduction of the western edges of the Pacific plate as the adjacent plates to the west were consumed. The direction of Pacific plate motion during this epoch is hard to discover, but the Hawaii-parallel Line Islands cross trend is an obvious candidate, being consistent with the magnetic data for little latitude change during this time interval.

The end of this epoch (ca. 50 Ma) is the time of India-Eurasia collision, whose "escape" effects spread eastward to the Pacific. It also marks the establishment of the Aleutian arc in the North Pacific, and it initiated a new dominant direction of Pacific plate motion (Natland and Winterer, 2005) and a new orientation of tensional stresses in the plate normal to the direction of plate motion. The switch between the two regimes is marked by the Hawaii-Emperor Bend.

5 February 2007, William W. Sager

I thank Winterer for highlighting the existing uncertainty about the interpretation of Pacific seamount chains and plate motions as hotspots. He reminds us that many of the Pacific seamount chains do not have simple age progressions (if they have age progressions at all), and that published paleomagnetic data from the Emperor chain imply that the Hawaiian hotspot (and others by implication) was not fixed in latitude. Further, Winterer suggests that stress cracking of the plate is an alternative that is not constrained to have simple age progressions or to be fixed in any particular spot (Natland and Winterer, 2005).

Being keenly aware of the changing views of the hotspot hypothesis—and being uncertain myself of what mechanism is the best explanation—I chose to remain neutral about the mechanisms of the Hawaiian-Emperor chain formation. I say that we should consider alternative hypotheses, but I went no further, in part because I saw such an extrapolation to be beyond the scope of my study. It is interesting to consider what the Wessel et al. (2006) plate motion model entails and implies. That study is clearly an outgrowth of the fixed hotspot hypothesis, although the authors do not strictly make that assertion. Their model only requires that seamount chains are a fixed distance apart, but not that the melting anomalies have strictly linear propagation rates or that the anomalies are fixed relative to anything except one another. Thus, one interpretation could be a constellation of plumes that maintain nearly fixed distances from one another, yet shift relative to other plumes (e.g., Tarduno and Gee, 1995). Another might be a set of tensional cracks that form with a regular spacing (Natland and Winterer, 2005).

It is also interesting to note that the Wessel et al. (2006) model is mostly based on two seamount chains, Hawaiian-Emperor and Louisville, and not on many of the other seamount chains mentioned by Winterer in his comment. Some of those other seamount chains may be copolar (fit the same small-circle trends) with the Hawaiian-Emperor and Louisville chains but do not show the same simple age progressions. It is only in the Cenozoic that their model is able to use more than those two chains. Although this model is at times representative of broader

motion, mostly it is a Louisville and Hawaiian-Emperor model. Those two seamount chains, at least, seem to behave most coherently in the manner ascribed to plate motion over hotspots.

Two of the most important implications of my study are (1) the Wessel et al. (2006) hotspot model does a decent job of predicting the overall plate motion recorded by paleomagnetism, and (2) the Hawaiian-Emperor Bend is not what we thought for over three decades. Overall, the shapes of paleomagnetic and hotspot-predicted APWP are similar. Thus, the plate motion model seems to be recording plate motion in some manner; although I leave it to the reader to infer what that might be (especially if not fond of the hotspot hypothesis).

The second point is very important in that it represents a wholesale overturn of accepted Pacific plate motions, and because it must say something important about the behavior of the Hawaiian melting anomaly. A few years ago, like almost everybody else, I thought that Pacific plate motion was nearly north-south during the period of the Emperor Chain formation. But new paleomagnetic data (Tarduno et al., 2003; Sager, this volume) suggest otherwise. The paleomagnetic data are consistent with little or no northward motion during that time, so the bend does not represent much of a change in plate motion. There may have been a small change in plate motion, as marked by changes in fracture zone trends corresponding to ca. 45–50 Ma (Atwater, 1989), but nothing like the ~60° change implied by the Hawaiian-Emperor Bend. This apparent, large change in plate motion is woven into many western Pacific tectonic interpretations and now must be culled out. Furthermore, I find it interesting that the oldest (northern) end of the Emperor chain, where it changes trend from southeast to south-trending, is approximately the same age as a change recorded in seafloor magnetic lineations during Chron 33 (Atwater, 1989), similar to the Hawaiian-Emperor Bend. The simplest explanation for this set of coincidences is that the melting anomaly responded to changes in plate motion, which is essentially the thesis of the article by Norton (this volume) and the views expressed by Winterer. I must admit that the two contentions above seem to be in conflict. The one suggests the plate motion model is mostly correct, and the other suggests that at least one part is highly flawed. I can only hope that future data will help solve this apparent paradox.

REFERENCES CITED

Atwater, T., 1989, Plate tectonic history of the northeast Pacific and western North America, *in* Winterer, E.L., et al., eds., The eastern Pacific Ocean and Hawaii: Boulder, Colorado, Geological Society of America Decade of North American Geology N, p. 21–72.

Davis, A.S., Gray, L.B., and Clague, D.A., 2002, The Line Islands revisited: New 40-Ar/39-Ar geochronologic evidences for episodes of volcanism due to lithospheric extension: Geochemistry, Geophysics, Geosystems, v. 3, doi: 10.1029/2001GC000190.

Koppers, A.A.P., Staudigel, H., Pringle, M.S., and Wijbrans, J.R., 2003, Short-lived and discontinuous intraplate volcanism in the South Pacific: Hot spots or extensional volcanism?: Geochemistry, Geophysics, Geosystems, v. 4, p. 1089, doi: 10.1029/2003GC000533.

Natland, J.H., and Winterer, E.L., 2005, Fissure control on volcanic action in the Pacific, *in* Foulger, G.R., et al., eds., Plates, plumes, and paradigms: Boulder, Colorado, Geological Society of America Special Paper 388, p. 687–710, doi: 10.1130/2005.2388(39).

Norton, I.O., 2007 (this volume), Speculations on Cretaceous tectonic history of the northwest Pacific and a tectonic origin for the Hawaii hotspot, *in* Foulger, G.R., and Jurdy, D.M., eds., Plates, plumes, and planetary processes: Boulder, Colorado, Geological Society of America Special Paper 430, doi: 10.1130/2007.2430(22).

Sager, W.W., 2007 (this volume), Divergence between paleomagnetic and hotspot-model–predicted polar wander for the Pacific plate with implications for hotspot fixity, *in* Foulger, G.R., and Jurdy, D.M., eds., Plates, plumes, and planetary processes: Boulder, Colorado, Geological Society of America Special Paper 430, doi: 10.1130/2007.2430(17).

Tarduno, J.A., and Gee, J., 1995, Large-scale motion between Pacific and Atlantic hotspots: Nature, v. 78, p. 477–480.

Tarduno, J.A., Duncan, R.A., Scholl, D.W., Cottrell, R.D., Steinberger, B., Thordarson, T., Kerr, B.C., Neal, C.R., Frey, F.A., Torii, M., and Carvallo, C., 2003, The Emperor seamounts: Southward motion of the Hawaiian hotspot plume in the Earth's mantle: Science, v. 301, p. 1064–1069, doi: 10.1126/science.1086442.

Wessel, P., Harada, Y., and Kroenke, L.W., 2006, Toward a self-consistent, high-resolution absolute plate motion model for the Pacific: Geochemistry, Geophysics, Geosystems, v. 7, art. no. Q03L12, doi: 10.1029/2005GC00100.

The Geological Society of America
Special Paper 430
2007

Global kinematics in deep versus shallow hotspot reference frames

Marco Cuffaro*
Carlo Doglioni*
Dipartimento di Scienze della Terra, Sapienza Università di Roma,
Piazzale Aldo Moro 5, 00185 Rome, Italy

ABSTRACT

Plume tracks at the Earth's surface probably have various origins, such as wet spots, simple rifts, and shear heating. Because plate boundaries move relative to one another and relative to the mantle, plumes located on or close to them cannot be considered as reliable for establishing a reference frame. Using only relatively fixed intraplate Pacific hotspots, plate motions with respect to the mantle in two different reference frames, one fed from below the asthenosphere, and one fed by the asthenosphere itself, provide different kinematic results, stimulating opposite dynamic speculations. Plates move faster relative to the mantle if the source of hotspots is taken to be the middle-upper asthenosphere, because hotspot tracks would then not record the entire decoupling occurring in the low-velocity zone. A shallow intra-asthenospheric origin for hotspots would raise the Pacific deep-fed velocity from a value of 10 cm/year to a faster hypothetical velocity of ~20 cm/year. In this setting, the net rotation of the lithosphere relative to the mesosphere would increase from a value of 0.4359°/m.y. (deep-fed hotspots) to 1.4901°/m.y. (shallow-fed hotspots). In this framework, all plates move westward along an undulated sinusoidal stream, and plate rotation poles are largely located in a restricted area at a mean latitude of 58°S. This reference frame seems more consistent with the persistent geological asymmetry that suggests a global tuning of plate motions related to Earth's rotation. Another significant result is that along east- or northeast-directed subduction zones, slabs move relative to the mantle in the direction opposed to the subduction, casting doubts on slab pull as the first-order driving mechanism of plate dynamics.

Keywords: plate motions, reference frames, shallow hotspots, westward drift of the lithosphere

*E-mails: marco.cuffaro@uniroma1.it; carlo.doglioni@uniroma1.it.

Cuffaro, M., and Doglioni, C., 2007, Global kinematics in deep versus shallow hotspot reference frames, *in* Foulger, G.R., and Jurdy, D.M., eds., Plates, plumes, and planetary processes: Geological Society of America Special Paper 430, p. 359–374, doi: 10.1130/2007.2430(18). For permission to copy, contact editing @geosociety.org. ©2007 The Geological Society of America. All rights reserved.

INTRODUCTION

Absolute plate motions represent movements of plates relative to the mesosphere. To describe displacements of the lithosphere, two different absolute frameworks are used, the hotspots and the mean lithosphere. The first is based on the assumption that hotspots are fixed relative to the mesosphere and to one another (Morgan, 1972; Wilson, 1973). The second is defined by the no-net-rotation condition (NNR; Solomon and Sleep, 1974), and it is assumed that there is uniform coupling between the lithosphere and the asthenosphere. Both absolute reference frames are referred to the mesosphere, and any difference between the mean-lithosphere and the hotspot frames is interpreted as a net rotation of the lithosphere with respect to the mesosphere (Forsyth and Uyeda, 1975). When plate motions are measured in the classic hotspot reference frame, the lithosphere shows a net westward rotation (Bostrom, 1971; O'Connell et al., 1991; Ricard et al., 1991; Gripp and Gordon, 2002; Crespi et al., 2007).

This so-called westward drift has been so far considered only as an average motion of the lithosphere, due to the larger weight of the Pacific plate in the global plate-motion computation. But the westward drift also persists when plate motions are computed relative to Antarctica (Le Pichon, 1968; Knopoff and Leeds, 1972). Moreover, and more importantly, it is supported by independent geological and geophysical asymmetries along subduction zones and rifts, showing a global tuning and not just an average asymmetry (Doglioni et al., 1999, 2003). To check whether the westward drift is only an average casual component or a globally persistent signature, we analyze the different kinematics resulting from different hotspot reference frames.

Hotspot tracks have been used for computing the motion of plates relative to the mantle. For this purpose, it is fundamental to know (1) whether hotspots are fixed relative to the mantle, (2) whether they are fixed relative to one another, and (3) from what depth hotspots are fed. Hotspots have been used often uncritically, regardless of their real nature. Looking at maps of hotspots (e.g., Anderson and Schramm, 2005), plumes occur both in intraplate settings or close to or along plate boundaries. Hotspot reference frames have been used and misused, possibly because their volcanic tracks have been considered monogenic and with similar source depths. A number of models have been produced to quantify the relative motion among hotspots and their reliability for generating a reference frame. Rejuvenating volcanic tracks at the Earth's surface may be a result of intraplate plumes (e.g., Hawaii), retrogradation of subducting slabs, migration of back-arc spreading, along-strike propagation of rifts (e.g., east Africa), or propagation of transform faults with a transtensive component (Chagos?). All those volcanic trails may have different depths of their mantle sources, and they should be differentiated (Fig. 1).

Plate boundaries are by definition moving relative to one another and relative to the mantle (e.g., Garfunkel et al., 1986; Doglioni et al., 2003). Therefore, any hotspot located along a plate boundary cannot be used for the reference frame. For example, Norton (2000) grouped hotspots into three main families that have very little internal relative motion (Pacific, Indo-Atlantic, and Iceland). In fact, he concluded that a global hotspot reference frame is inadequate, because Pacific hotspots move relative to Indo-Atlantic hotspots and to Iceland. Because Indo-Atlantic hotspots and the Iceland hotspot are located along ridges, they do not satisfy the required fixity. In his analysis, Pacific plate hotspots are reasonably fixed relative to one another during the past 80 m.y., and they are located in intraplate settings. Therefore, they are unrelated to plate-margin processes and do not move with any margin. Screening of volcanic tracks to be used for the hotspot reference frame provides a very limited number of hot-lines, and only the Pacific ones satisfy the requirements.

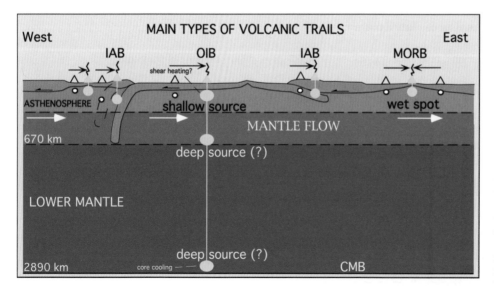

Figure 1. The main volcanic chains at the Earth's surface may have different origins and depths. The thin arrows indicate the direction of migration of volcanism with time. Filled triangles represent the youngest volcanic products. Volcanic trails originating on ridges may be wetspots (sensu Bonatti, 1990) and fed from a fluid-rich asthenosphere. The hotspots located on plate boundaries are not fixed by definition, because both ridges and trenches move relative to one another and with respect to the mantle. Pacific hotspots, regardless of their source depth, are located within the plate and are virtually the only ones that can be considered reliable for a hotspot reference frame. CMB—core-mantle boundary; IAB—island arc basalt; MORB—mid-ocean ridge basalt; OIB—ocean island basalt; open triangles—extinct volcanoes.

Hotspots may have short (<15 m.y.) or long (>50 m.y.) time gaps between their emplacement and the age of the oceanic crust on which they reside. A shorter time frame suggests a closer relation with the formation of the oceanic crust, particularly when (1) the location is persistently close to the ridge and (2) ridges form on both sides of the rifts (Doglioni et al., 2005). Therefore, ridge-related plumes should move with a speed close to the velocity of the plate boundary relative to the mantle. Although moving relative to one another, hotspots always have a speed slower than plate motions and have been considered useful for a reference frame (e.g., Wang and Wang, 2001). However, the velocity of plate boundaries tends to be slower than the velocity of the relative plate motion among pairs of plates. For example, the Mid-Atlantic Ridge moves westward at rates comparable to the relative motion between the Pacific and Atlantic hotspots, but this intra-hotspot motion could be related to the motion of the Mid-Atlantic Ridge.

Moreover, assuming a deep source for the hotspots, several models have been computed to infer deep-mantle circulation (e.g., Steinberger and O'Connell, 1998; Steinberger, 2000). These models argue that volcanic tracks move opposite to plate motions. However, this conclusion may be regarded again as a problem of reference. For example, in the NNR reference frame, Africa moves "east," opposite relative to Ascencion and Tristan da Cunha, but in HS3-NUVEL1A (Gripp and Gordon, 2002), Africa moves in the same direction due "west," although at different velocity. Therefore, the assumption that seamounts in hotspot tracks always move opposite to plate motions is misleading if not wrong.

In most of the models so far published on mantle circulation and hotspot reference frames, two main issues are disregarded: (1) plumes have different origins and different kinematic weights for the reference frames; and (2) in the cases of plumes that are shallow asthenospheric features, this second condition determines a different kinematic scenario with respect to the deep-mantle circulation pattern.

Accumulating evidence suggests that hotspots are mostly shallow features (Bonatti, 1990; Smith and Lewis, 1999; Anderson, 2000; Foulger, 2002; Foulger et al., 2005). For example, Atlantic hotspots might be interpreted more as wetspots rather than hot-lines, as suggested by Bonatti (1990). An asthenospheric source richer in fluids that lower the melting point can account for the overproduction of magma. Propagating rifts (hot-lines, etc.) are shallow phenomena, which are not fixed to any deep mantle layer. The only hotspots that should be relevant to the reference frame are those located within plate. For a compelling petrological, geophysical, and kinematic analysis on the shallow origin of plumes, see Foulger et al. (2005). In this book, some data are presented that support a shallow source depth for hotspots (e.g., upper mantle, asthenosphere, base lithosphere). Several theoretical models have been proposed to explain the different settings, such as rift zones, fluids in the asthenosphere, shear heating at the lithosphere-asthenosphere decoupling zone, and lateral mantle compositional variations. All these models

could be valid, but applied to different cases. Therefore, we disagree with the practice of using uncritically all so-called hotspots, because their different origin can corrupt the calculation of lithosphere-mantle relative motion.

In this article, we present current plate motions relative to a shallow hotspot framework, similar to Crespi et al. (2007). Moreover, because two fixed points are geometrically sufficient to construct a kinematic reference frame, we use only Pacific intraplate hotspots, which are significantly fixed relative to one another (Gripp and Gordon, 2002). We obtain angular velocities that imply a different plate kinematics than the one obtained with the HS3-NUVEL1A plate kinematic model (Gripp and Gordon, 2002). Unlike Wang and Wang (2001), we find a much faster net rotation of the mean lithosphere with respect to HS3-NUVEL1A.

DECOUPLING IN THE ASTHENOSPHERE

The asthenosphere is anisotropic, having the main orientation of crystals along the sense of shear (e.g., Barruol and Granet, 2002; Bokelmann and Silver, 2002). The asthenosphere is present all over the Earth (Gung et al., 2003) and shows an upper low-velocity zone that is more or less pronounced (Calcagnile and Panza, 1978; Thybo, 2006). This layer may have a viscosity far lower (Scoppola et al., 2006) than estimates for the whole asthenosphere (e.g., Anderson, 1989), and it should engineer the main decoupling between lithosphere and the underlying mesosphere.

The origin of intraplate Pacific magmatism is rather obscure, and its source depth and the mechanism of melting is still under discussion (Foulger et al., 2005). Beause the Pacific is the fastest plate, shear heating along the basal décollement has been interpreted as a potential mechanism for generating localized hotspot tracks (Fig. 2B).

Kennedy et al. (2002) have shown how mantle xenoliths record a shear possibly located at the lithosphere-asthenosphere interface. This observation supports the notion of flow in the upper mantle and some decoupling at the base of the lithosphere, as indicated by seismic anisotropy (Russo and Silver, 1996; Doglioni et al., 1999; Bokelmann and Silver, 2000). The fastest plate on Earth in the hotspot reference frame (i.e., the Pacific) is the one affected by the most widespread intraplate magmatism.

It is noteworthy that the fastest plate, the Pacific, overlies the asthenosphere with the mean lowest viscosity (5×10^{17} Pa s; Pollitz et al., 1998), and possibly the most undepleted mantle, and therefore prone to melt. Because of the melting characteristics of peridotite with minor amounts of carbon and hydrogen (lherzolite-[C + H^{+}O] system), the asthenosphere is already partly molten (e.g., Schubert et al., 2001), with $T \sim 1430°$ C (e.g., Green and Falloon, 1998; Green et al., 2001). The rise of T of only few tens of degrees will increase the extent of melting which, in a deforming material, will migrate toward the surface. We postulate that locally, the viscosity of the asthenosphere can also increase (e.g., 10^{19} Pa s) because of refractory geochemi-

A

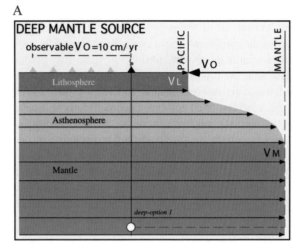

B

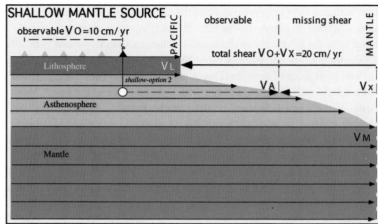

Figure 2. The Hawaiian volcanic track indicates that there is decoupling between the magma source and the lithosphere, which is moving relatively toward the WNW. (A) If the source is below the asthenosphere (e.g., in the subasthenospheric mantle), the track records the entire shear between lithosphere and mantle. (B) In the case of an asthenospheric source for the Hawaiian hotspot, the volcanic track does not record the entire shear between the lithosphere and subasthenospheric mantle, because part of it operates below the source (deep, missing shear). Moreover, the larger decoupling implies larger shear heating, which could be responsible for scattering the punctiform Pacific intraplate magmatism. After Doglioni et al. (2005). See text for definitions of the velocities VA, VL, VM, VO, and VX.

cal anisotropy, or decrease because of locally higher water activity. Shear stress could be irregularly distributed in such inhomogeneous materials, and consequently, higher shear heating (Shaw, 1973) may be locally developed to generate punctiform magmatism. However, other models for the asthenospheric temperature can be devised (Foulger and Anderson, 2006).

Doglioni et al. (2005) modeled the shear heating between the lithosphere and asthenosphere as a possible source for Hawaiian-type magmatism. In that model, it was assumed the asthenosphere behaves as a Couette flow (Turcotte and Schubert, 2002). Following that model (i.e., the shear heating localized in the middle of the flow) we started from the assumption that the source of this type of hotspot could be positioned close to the half thickness of the asthenosphere. The asthenosphere has been shown to be a heterogeneous layer by a large number of geophysical and petrological models (e.g., Panza, 1980; Anderson, 2006; Thybo, 2006) in which composition and viscosity may change laterally. Areas with viscosity higher than normal in the asthenospheric décollement should generate greater shear heating.

In such a model, punctuated and stiffer mantle sections would be able to generate sufficient extra T for asthenospheric melting. These mantle anisotropies, whenever shearing started, remained quite fixed relative to one another. According to Norton (2000) and Gripp and Gordon (2002), these intraplate Pacific plate hotspots satisfy the requirement of relative fixity, at least for the past few million years.

PLATE MOTIONS RELATIVE TO THE DEEP AND SHALLOW HOTSPOTS

Most of the hotspots used are not fixed; nor do they represent a fixed reference frame, because they are located on plate margins, such as moving ridges (e.g., Galapagos, Easter Island, Iceland, Ascension), transform faults (Réunion), above subduction zones, or continental rifts (Afar), all features that are moving relative to one another and relative to the mantle.

In contrast, Pacific hotspots are reasonably fixed relative to one another, and their volcanic tracks can be used for the hotspot reference frame. WNW motion of the Pacific plate relative to the underlying mantle is inferred from the Hawaiian and other major intraplate hotspot tracks (Marquesas, Society, Pitcairn, Samoan, and Macdonald), which suggest an average velocity of ~103–118 mm/year, and also move along the same trend (290°–300°WNW).

Following the hypothesis of deep-fed hotspots, after assuming that shear is distributed throughout the asthenospheric channel (Fig. 2A), and providing the velocity V_L of the Pacific lithosphere toward the ESE (110°–120°) is slower than that of the underlying subasthenospheric mantle V_M ($V_M > V_L$), the relative velocity V_O corresponding to the WNW delay of the lithosphere is:

$$V_O = V_L - V_M. \tag{1}$$

For the case of Hawaii, the observed linear velocity is $V_O = 103$ mm/year, corresponding to the propagation rate of the Hawaiian volcanic track (Fig. 2A).

The HS3-NUVEL1A (Gripp and Gordon, 2002) plate-motion model with respect to the mantle is based on the deep-fed hotspot hypothesis. Gripp and Gordon (2002) compute absolute plate motions, estimating eleven segment trends and two propagation rates for volcanic tracks and presenting a set of absolute angular velocities consistent with the relative plate-motion model NUVEL-1A (DeMets et al., 1990, 1994). Volcanic propagation rates used by Gripp and Gordon (2002) are those of Hawaii and Society, both on the Pacific plate, and they found a Pacific angular velocity of 1.0613°/m.y. about a pole located at 61.467°S, 90.326°E (Table 1 and Fig. 3). Another simple way to reproduce the HS3-NUVEL1A angular velocities consists of adding the Pacific plate Euler vector, estimated by Gripp and Gordon (2002), to the relative plate-motion model NUVEL-1A (DeMets et al., 1990, 1994), as Cuffaro and Jurdy (2006) also did to incorporate motions of microplates in the deep-fed hotspot framework.

If the location of the Hawaiian melting spot is in the middle of the asthenosphere (Fig. 2B) instead of the lower mantle (Fig. 2A), it would imply that the shear recorded by the volcanic track at the surface is only that occurring between the asthenospheric source and the top of the asthenosphere, i.e., only half of the total displacement, if the source is located in the middle of the asthenosphere.

Under this condition, the velocity recorded at the surface is:

$$V_O = V_L - V_A, \tag{2}$$

with

$$V_A = V_X + V_M, \tag{3}$$

where $V_O = 103$ mm/year is still the observed propagation rate of the volcanic track (e.g., Hawaii), V_A is the velocity recorded at the shallow source of the hotspot, and V_X is that part of the velocity that is not recorded, due to the missing shear measurement.

Substituting equation 3 in equation 2, we have:

$$V_O = V_L - V_M - V_X \tag{4}$$

and

$$V_O + V_X = V_L - V_M. \tag{5}$$

The observed velocity $V_O = 103$ mm/year of Hawaii is the velocity of total displacement if the magmatic source is located in the deep mantle, whereas it represents only half of the total shear if the source is located in the middle of the asthenosphere. In that case, to refer plate motions again with respect to the mesosphere, the velocity V_X has to be added to the observed velocity V_O (Fig. 2B), as in equation 5. If the source of Pacific hotspots is in the middle of the asthenosphere, half of the lithosphere–subasthenospheric mantle relative motion is unrecorded, which means that the total relative displacement of the Hawaiian hotspot would amount to about $V_O + V_X = 200$ mm/year (Fig. 2B).

Under the hypothesis of a shallow source for Pacific hotspots, located in the middle of the asthenosphere, and referring to the HS3–NUVEL1A methods (Gripp and Gordon, 2002), Pacific plate rotation would occur about a pole located at 61.467°S, 90.326°E, but with a rate of 2.1226°/m.y. Adding this Pacific Euler vector to the NUVEL-1A relative plate-motion model (DeMets et al., 1990, 1994) results absolute plate motions with respect to the shallow hotspot reference frame (Table 1 and Fig. 4). Moreover, referring to geometrical factors proposed by Argus and Gordon (1991), and using methods described by Gordon and Jurdy (1986) and Jurdy (1990), we computed net rotation of the lithosphere relative to the mesosphere, which, under the shallow hotspot hypothesis, amounts to ~1.4901°/m.y. (Table 1), and is higher than that computed by Gripp and Gordon (2002) (0.4359°/m.y., deep hotspot condition, Table 1).

This faster velocity for the Pacific plate has these basic consequences: (1) it extends westward drift of the lithosphere to all

TABLE 1. GLOBAL PLATE MOTIONS WITH RESPECT TO THE DEEP AND SHALLOW HOTSPOT REFERENCE FRAMES

| | Deep source | | | Shallow source | | |
| | Euler pole | | ω | Euler pole | | ω |
Plate	°N	°E	(°/m.y.)	°N	°E	(°/m.y.)
Africa	−43.386	21.136	0.1987	−61.750	76.734	1.2134
Antarctica	−47.339	74.514	0.2024	−59.378	86.979	1.2564
Arabia	2.951	23.175	0.5083	−46.993	56.726	1.2393
Australia	−0.091	44.482	0.7467	−38.865	62.780	1.4878
Caribbean	−73.212	25.925	0.2827	−65.541	82.593	1.3216
Cocos	13.171	−116.997	1.1621	−42.844	−135.856	0.9818
Eurasia	−61.901	73.474	0.2047	−62.352	87.511	1.2647
India	3.069	26.467	0.5211	−46.051	57.930	1.2563
Juan de Fuca	−39.211	61.633	1.0122	−51.452	72.836	2.0104
North America	−74.705	13.400	0.3835	−67.520	79.790	1.4094
Nazca	35.879	−90.913	0.3231	−71.733	91.649	0.7824
Pacific	−61.467	90.326	1.0613	−61.467	90.326	2.1226
Philippine	−53.880	−16.668	1.1543	−68.889	25.661	1.9989
South America	−70.583	80.401	0.4358	−64.176	88.125	1.4925
Scotia	−76.912	52.228	0.4451	−66.654	84.271	1.4877
Lithosphere	−55.908	69.930	0.4359	−60.244	83.662	1.4901

Note: Angular velocities of the deep-fed hypothesis come from the HS3-NUVEL1A absolute plate kinematic model (Gripp and Gordon, 2002).

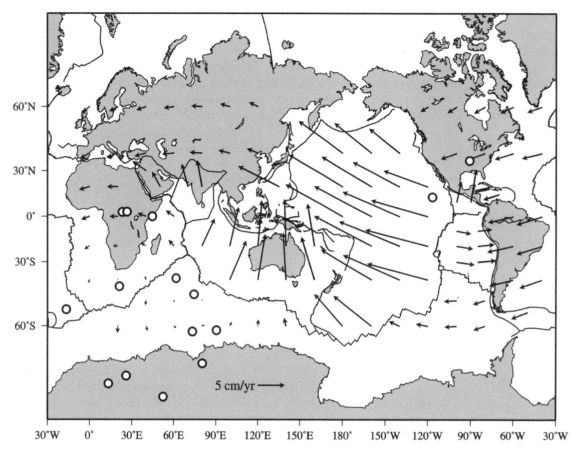

Figure 3. Current velocities with respect to the deep hotspot reference frame. Open circles are the rotation poles. Data from HS3-NUVEL1A (Gripp and Gordon, 2002).

plates (Fig. 4), (2) it more than doubles the westward drift compared to that of the deep hotspot reference frame, and (3) it increases the shear heating within the asthenosphere.

SHALLOW HAWAIIAN PLUME

There is evidence that the propagation rate of Pacific hotspots or seamount tracks has varied with time, even with jumps back and forth and oblique propagation relative to plate motions with respect to the mantle. This variability casts doubt on both the notion of absolute plate motions computed in the hotspot reference frame and the nature of the magmatism itself, e.g., deep plume, or rather shallow plumes generated by cracks or boudins of the lithosphere (Winterer and Sandwell, 1987; Sandwell et al., 1995; Lynch, 1999; Natland and Winterer, 2003) filled by a mantle with compositional heterogeneity and no demonstrable thermal anomaly in hotspot magmatism relative to normal mid-oceanic ridges.

Janney et al. (2000) described a velocity of the Pukapuka volcanic ridge (interpreted as either a hotspot track or a leaky fracture zone), located in the eastern central Pacific, between 5 and 12 m.y., of ~200–300 mm/year. They also inferred a shal-

low mantle source for Pacific hotspots based on their geochemical characteristics.

Relative plate motions can presently be estimated with great accuracy using space geodesy data (e.g., Robbins et al., 1993; Heflin et al., 2004), refining the earlier NUVEL-1A plate-motion model (DeMets et al., 1990, 1994).

The East Pacific Rise, separating the Pacific and Nazca plates, opens at a rate of 128 mm/year just south of the equator (e.g., Heflin et al., 2004). At the same latitude, shortening along the Andean subduction zone, where the Nazca plate subducts underneath South America, has been computed to be ~68 mm/year. When inserted in a reference frame in which the Hawaiian hotspot is considered fixed and positioned in the subasthenospheric mantle, these relative motions imply that the Nazca plate is moving eastward relative to the subasthenospheric mantle at ~25 mm/year (Fig. 7, option 1 *in* Doglioni et al., 2005). If we assume that the source of Pacific intraplate hotspots is instead in the middle asthenosphere and half of the lithosphere–subasthenospheric mantle relative motion is missing in the Hawaiian track (Fig. 2B), the movement could rise to 200 mm/year, as also suggested by some segments of the Pukapuka volcanic ridge (Janney et al., 2000). Note that in this configuration,

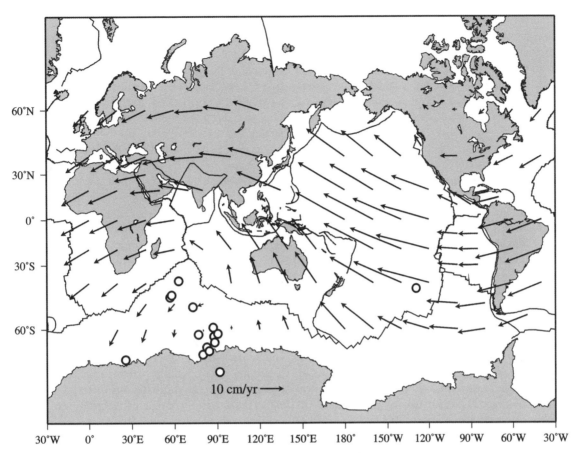

Figure 4. Present-day plate velocities relative to the shallow hotspot reference frame, incorporating the NUVEL-1A relative plate-motion model (DeMets et al., 1990, 1994). Note that in this frame, all plates have a westward velocity component. Open circles are the rotation poles.

Nazca would instead move west relative to the mantle at 72 mm/year (Fig. 7, option 2 *in* Doglioni et al., 2005), and therefore all three plates would move westward relative to the sub-asthenospheric mantle.

This last case agrees with the east-west–trending shearwave splitting anisotropies beneath the Nazca plate, turning north-south when encroaching on the Andean Slab, which suggests eastward mantle flow relative to the overlying plate (Russo and Silver, 1994). This flow could also explain the low dip of the Andean Slab. Both factors suggest relative eastward mantle flow. Similar eastward mantle flow was proposed for the North American plate (Silver and Holt, 2002). The low dip of the Andean Slab has alternatively been attributed to the young age of the subducting lithosphere. However, the oceanic age has been proved not to be sufficient to explain the asymmetry between westerly-directed (steep and deep) versus easterly-directed (low dip and shallow) subduction zones (Cruciani et al., 2005). In fact, the geographically related asymmetry persists even where the same lithosphere (regardless oceanic or continental) subducts in both sides, such as in the Mediterranean orogens (Doglioni et al., 1999).

Another consequence of having a shallower source for Hawaiian magmatism is that the westward motion of the Pacific

plate increases to a velocity faster than the spreading rate of the East Pacific Rise (Fig. 7, option 2 *in* Doglioni et al., 2005). A shallow, intra-asthenospheric origin of Pacific hotspots provides a kinematic frame in which all mid-ocean ridges move westward. As a consequence, the ridge migrates continuously over a fertile mantle, which presents a possible explanation for the endless source of mid-ocean ridge basalts (MORB), which have a relatively constant composition. Moreover, the rift generates melting and consequently increases the viscosity of the residual mantle moving beneath the eastern side of the ridge, providing a mechanism for maintaining higher coupling at the lithosphere base, and retarding the plate to the east (Doglioni et al., 2003, 2005).

DISCUSSION AND CONCLUSIONS

We have computed plate motions with respect to a shallow hotspot reference frame, making a comparison with the HS3-NUVEL1A results (Gripp and Gordon, 2002) and showing that shallow sources for hotspots produces different plate kinematics, i.e., new, faster plate motions with respect to the mesosphere than those previously calculated. Moreover in the deep hotspot frame, rotation poles are largely scattered, and most of the plates move toward the west, except for Nazca, Cocos, and Juan de Fuca plates. On the contrary, relative to the shallow hotspot framework, all plates move westerly, and rotation poles are mostly located in a restricted area at a mean latitude of 58°S. Furthermore, we computed a faster net rotation of the lithosphere for the case of a shallow-fed hotspot, which is useful to compute plate motions in the mean-lithosphere reference frame (NNR; Jurdy, 1990).

The mean lithosphere is also the framework for space geodesy applications to plate tectonics (Heflin et al., 2004). Most of the geodesy plate-motion models are referred to the NNR frame (Sella et al., 2002; Drewes and Meisel, 2003). The International Terrestrial Reference Frame (ITRF2000; Altamimi et al., 2002) is the framework in which site velocities are estimated. The ITRF2000 angular velocity is defined using the mean lithosphere. As suggested by Argus and Gross (2004), it would be better to estimate site positions and velocities relative to hotspots, continuing first to estimate velocity in the ITRF2000 and then adding the net-rotation angular velocity.

The deep and shallow hotspot interpretations generate two hotspot reference frames. In the case of deep-mantle sources for the hotspots, there still are few plates moving eastward relative to the mantle (Fig. 3), whereas in the case of shallow mantle sources, all plates move "westward," although at different velocities (Fig. 4). The kinematic and dynamic consequences of the shallow reference frame are so unexpected that it could be argued that they suggest that plumes are instead fed from the deep mantle. However, the shallow reference frame fits better observed geological and geophysical asymmetries which indi-

cates a global tuning (i.e., a complete "westward" rotation of the lithosphere relative to the mantle) rather than a simple average of plate motions (i.e., where the westward drift is only a residual of plates moving both westward and eastward relative to the mantle). In fact, geological and geophysical signatures of subduction and rift zones independently show a global signature, suggesting a complete net westward rotation of the lithosphere and a relative eastward motion of the mantle that can kinematically be inferred only from the shallow hotspot reference frame.

Plates move along a sort of mainstream depicting a sinusoid (Doglioni, 1990, 1993; Crespi et al., 2007; Fig. 5), which is largely confirmed by present space geodesy plate kinematics (e.g., Heflin et al., 2004). Global shearwave splitting directions (Debayle et al., 2005) are quite consistent with such undulate flow, deviating from it at subduction zones, which should represent obstacles to relative mantle motion. In fact, along this flow, west-directed subduction zones are steeper than those that are east- or northeast-directed, and associated orogens are characterized by lower structural and topographic elevations, back-arc basins, or by higher structural and morphological elevation and no back-arc basins (Doglioni et al., 1999). The asymmetry is striking when comparing western and eastern Pacific subduction zones, and it has usually been interpreted as related to the age of the downgoing oceanic lithosphere, i.e., older, cooler, and denser on the western side. However these differences persist elsewhere, regardless of the age and composition of the downgoing lithosphere, e.g., in the Mediterranean Apennines and Carpathians versus the Alps and Dinarides, or in the Banda and

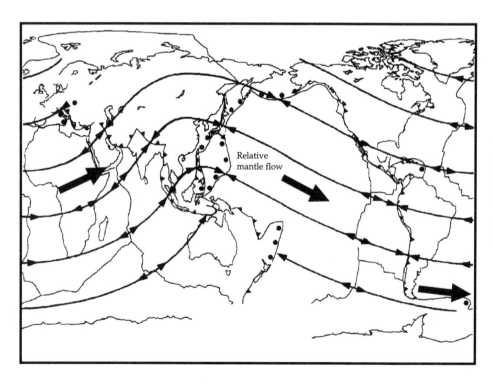

Figure 5. Connecting the directions of absolute plate motions that we can infer from large-scale rift zones or convergent belts from the past 40 Ma, we observe a coherent sinusoidal global flow field, along which plates appear to move at different relative velocities in the geographic coordinate system. After Doglioni (1993).

Sandwich arcs, where even continental or zero-age oceanic lithosphere is almost vertical along west-directed subduction zones. Rift zones are also asymmetric, with the eastern side more elevated by ~100–300 m worldwide (Doglioni et al., 2003).

The westward drift of the lithosphere implies that plates have a general sense of motion and that they are not moving randomly. If we accept this postulate, plates move along this trend at different velocities toward the west relative to the mantle along the flow lines of Figure 5, which undulate and are not exactly oriented east-west. In this view, plates would be more or less detached with respect to the mantle, as a function of the decoupling at their base. The degree of decoupling would be mainly controlled by the thickness and viscosity of the asthenosphere. Lateral variations in decoupling could control the variable velocity of the overlying lithosphere (Fig. 6). When a plate moves faster westward with respect to an adjacent plate to the east, the resulting plate margin is extensional; when it moves

faster westward with respect to the adjacent plate to the west, their common margin will be convergent (Fig. 6).

The kinematic frame of shallow Pacific hotspots (Fig. 4) constrains plate motions as entirely polarized toward the west relative to the deep mantle. This framework provides a fundamental observation along east- or northeast-directed subduction zones. In fact, with this reference frame, the slab tends to move out relative to the mantle, but subduction occurs because the upper plate overrides the lower plate faster. This scenario argues against slab pull as the main mechanism for driving plate motions, because the slab does not move into the mantle. In this view, slabs are rather passive features (Fig. 7). This kinematic reconstruction is coherent with the frequent intraslab down-dip extension earthquake focal mechanisms that characterize east- or northeast-directed subduction zones (e.g., Isacks and Molnar, 1971). It is generally assumed that oceanic plates travel faster than plates with large fractions of continental lithosphere. However,

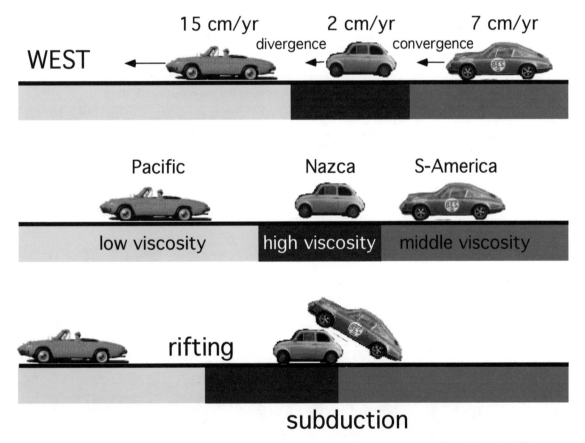

Figure 6. Cartoon illustrating that plates (cars) move along a common trail (e.g., the lines of Fig. 5) but with different velocities toward the west, as indicated by the westward drift of the lithosphere relative to the mantle. The differential velocities control the tectonic environment and result from different viscosities in the decoupling surface, i.e., the asthenosphere. There is extension when the western plate moves westward faster with respect to the plate to the east, whereas convergence occurs when the plate to the east moves westward faster with respect to the plate to the west. When the car in the middle is "subducted," the tectonic regime switches to extension, because the car to the west moves faster, e.g., the Basin and Range.

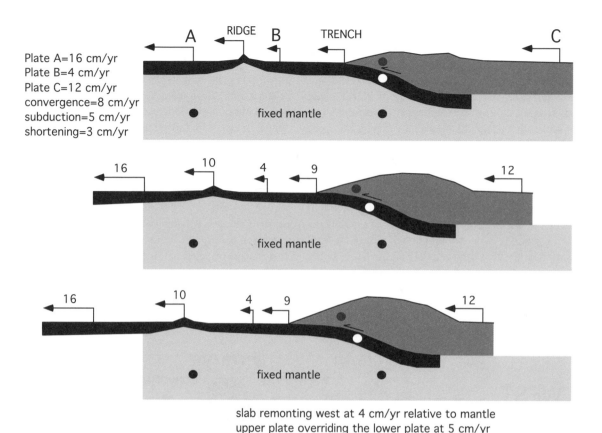

Plate A=16 cm/yr
Plate B=4 cm/yr
Plate C=12 cm/yr
convergence=8 cm/yr
subduction=5 cm/yr
shortening=3 cm/yr

slab remonting west at 4 cm/yr relative to mantle
upper plate overriding the lower plate at 5 cm/yr

Figure 7. Cartoon assuming a Pacific plate (plate A) moving at 16 cm/year. When plate motions are considered relative to the shallow hotspot reference frame, the slabs of east- or northeast-directed subduction zones may move out of the mantle. This scenario is clearly the case for Hellenic subduction and, in the shallow hotspot reference frame, also for Andean subduction. This kinematic evidence for slabs moving out of the mantle casts doubt on slab pull as the driving mechanism of plate motions.

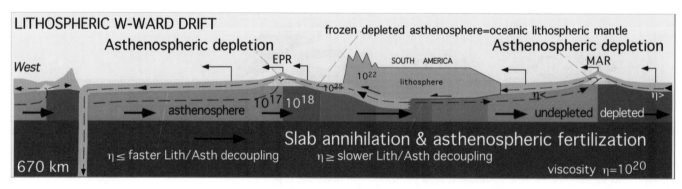

Figure 8. Model for the upper-mantle cycle in the case of the shallow Pacific hotspot reference frame. The lower the asthenospheric viscosity is, the faster the westward displacement of the overlying plate. The asthenospheric depletion at oceanic ridges makes the layer more viscous and decreases the lithosphere-asthenospheric decoupling, and the plate to the east is then slower. The oceanic lithosphere subducting eastward enters the asthenosphere, where it could partly melt again to refertilize the asthenosphere. West-directed subduction provides deeper circulation. After Doglioni et al. (2006a).

Gripp and Gordon (2002), even in the deep hotspot reference frame, have shown that the South American plate is moving faster than the purely oceanic Nazca plate. Another common assumption is that plates move away from ridges, but again, in the deep reference frame, Africa is moving toward the Mid-Atlantic Ridge, although slower than is South America. Moreover, Africa is moving away from the Hellenic subduction zone. In the shallow reference frame, these observations are accentuated and become unequivocal. Another typical assumption is that plates with attached slabs move faster, but the Pacific plate moves at ~1.06°/m.y., much faster in terms of absolute velocity than the Nazca plate (~0.32°/m.y.). The Pacific and Nazca plates have roughly the same percentage of attached slab (37% and 34%, respectively).

Therefore, in the case of a shallow origin for Pacific hotspots, westward drift implies a generalized counterflow of the underlying mantle (Fig. 8). With such an asymmetric flow, upper-mantle circulation would be constrained in this frame but disturbed by subduction and rift zones (Doglioni et al., 2006a,b). The fertile asthenosphere coming from the west melts and depletes along the ridge. Continuing its travel to the east, the depleted asthenosphere is more viscous and lighter (Doglioni et al., 2005). Subduction zones directed to the east or NNE, along the mantle counterflow, might refertilize the upper mantle, whereas west-directed subduction zones would instead penetrate deeper into the mantle.

The global-scale asymmetry of tectonic features and the westward drift of the lithosphere support a rotational component for the origin of plate tectonics (Scoppola et al., 2006). The westward drift could be the combined effect of three processes: (1) tidal torques acting on the lithosphere and generating a westerly-directed torque that decelerates Earth's spin; (2) downwelling of denser material toward the bottom of the mantle and in the core, slightly decreasing the moment of inertia and speeding up Earth's rotation and only partly counterbalancing tidal drag; and (3) thin (3- to 30-km) layers of very-low-viscosity hydrate channels in the asthenosphere. It is suggested that shear heating and mechanical fatigue self-perpetuate one or more channels of this kind, providing the necessary decoupling zone of the lithosphere (Scoppola et al., 2006) in the upper asthenosphere.

ACKNOWLEDGMENTS

We extend many thanks for suggestions and discussions to F. Antonucci, M. Crespi, G. Foulger, D. Jurdy, F. Innocenti, and F. Riguzzi. Critical readings by D. Argus, J. Stock, and B. Steinberger were much appreciated. Many of the figures were made with the Generic Mapping Tools of Wessel and Smith (1995).

REFERENCES CITED

Altamimi, Z., Sillard, P., and Boucher, C., 2002, ITRF2000: A new release of the International Terrestrial Reference Frame for earth sciences applications: Journal of Geophysical Research, v. 107, art. no. 2214, doi: 10.1029/2001JB000561.

Anderson, D.L., 1989, Theory of the Earth: Boston, Blackwell Scientific, 366 p.

Anderson, D.L., 2000, Thermal state of the upper mantle; No role for mantle plumes: Geophysical Research Letters, v. 27, p. 3623–3626, doi: 10.1029/2000GL011533.

Anderson, D.L., 2006, Speculations on the nature and cause of mantle heterogeneity: Tectonophysics, v. 416, p. 7–22, doi: 10.1016/j.tecto.2005.07.011.

Anderson, D.L., and Schramm, K.A., 2005, Global hotspot maps, *in* Foulger, G.R., et al., eds., Plates, plumes, and paradigms: Boulder, Colorado, Geological Society of America Special Paper 388, p. 19–29, doi: 10.1130/2005.2388(3).

Argus, D.F., and Gordon, R.G., 1991, No-net-rotation model of current plate velocities incorporating plate motion model NUVEL-1: Geophysical Research Letters, v. 18, p. 2039–2042.

Argus, D.F., and Gross, R.S., 2004, An estimate of motion between the spin axis and the hotspots over the past century: Geophysical Research Letters, v. 31, art. no. L006614, doi: 10.1029/2004GL019657.

Barruol, G., and Granet, M., 2002, A Tertiary asthenospheric flow beneath the southern French Massif Central indicated by upper mantle seismic anisotropy and related to the west Mediterranean extension: Earth and Planetary Science Letters, v. 202, p. 31–47, doi: 10.1016/S0012-821X(02)00752-5.

Bokelmann, G.H.R., and Silver, P.G., 2000, Mantle variation within the Canadian Shield: Travel times from the portable broadband Archean–Proterozoic Transect 1989: Journal of Geophysical Research, v. 105, p. 579–606, doi:10.1029/1999JB900387.

Bokelmann, G.H.R., and Silver, P.G., 2002, Shear stress at the base of shield lithosphere: Geophysical Research Letters, v. 29, p. 2091, doi: 10.1029/2002GL015925.

Bonatti, E., 1990, Not so hot "hot spots" in the oceanic mantle: Science, v. 250, p. 107–110, doi: 10.1126/science.250.4977.107.

Bostrom, R.C., 1971, Westward displacement of the lithosphere: Nature, v. 234, p. 536–538, doi: 10.1038/234536a0.

Calcagnile, G., and Panza, G., 1978, Crust and upper mantle structure under the Baltic Shield and Barents Sea from the dispersion of Rayleigh waves: Tectonophysics, v. 47, p. 59–71, doi: 10.1016/0040-1951(78)90151-8.

Crespi, M., Cuffaro, M., Doglioni, C., Giannone, F., and Riguzzi, F., 2007, Space geodesy validation of the global lithospheric flow: Geophysical Journal International, v. 169, p. 498–506.

Cruciani, C., Carminati, E., and Doglioni, C., 2005, Slab dip vs. lithosphere age: No direct function: Earth and Planetary Science Letters, v. 238, p. 298–310, doi: 10.1016/j.epsl.2005.07.025.

Cuffaro, M., and Jurdy, D.M., 2006, Microplate motions in the hotspot reference frame: Terra Nova, v. 18, p. 276–281.

Debayle, E., Kennett, B., and Priestley, K., 2005, Global azimuthal seismic anisotropy and the unique plate-motion deformation of Australia: Nature, v. 433, p. 509–512, doi: 10.1038/nature03247.

DeMets, C., Gordon, R.G., Argus, D.F., and Stein, S., 1990, Current plate motions: Geophysical Journal International, v. 101, p. 425–478.

DeMets, C., Gordon, R.G., Argus, D.F., and Stein, S., 1994, Effect of recent revisions to the geomagnetic reversal time scale on estimates of current plate motions: Geophysical Research Letters, v. 21, p. 2191–2194, doi: 10.1029/94GL02118.

Doglioni, C., 1990, The global tectonic pattern: Journal of Geodynamics, v. 12, p. 21–38, doi: 10.1016/0264-3707(90)90022-M.

Doglioni, C., 1993, Geological evidence for a global tectonic polarity: Journal of the Geological Society of London, v. 150, p. 991–1002.

Doglioni, C., Harabaglia, P., Merlini, S., Mongelli, F., Peccerillo, A., and Piromallo, C., 1999, Orogens and slabs vs their direction of subduction: Earth Science Reviews, v. 45, p. 167–208, doi: 10.1016/S0012-8252(98)00045-2.

Doglioni, C., Carminati, E., and Bonatti, E., 2003, Rift asymmetry and continental uplift: Tectonics, v. 22, doi: 10.1029/2002TC001459.

Doglioni, C., Green, D., and Mongelli, F., 2005, On the shallow origin of hotspots and the westward drift of the lithosphere, *in* Foulger, G.R., et al., eds., Plates, plumes, and paradigms: Boulder, Colorado, Geological Society of America Special Paper 388, p. 735–749, doi: 10.1130/2005.2388(42).

Doglioni, C., Carminati, E., and Cuffaro, M., 2006a, Simple kinematics of subduction zones: International Geological Review, v. 48, p. 479–493.

Doglioni, C., Cuffaro, M., and Carminati, E., 2006b, What moves slabs?: Bollettino Geofisica Teorica e Applicata, v. 47, p. 224–247.

Drewes, H., and Meisel, B., 2003, An actual plate motion and deformation model as a kinematic terrestrial reference system: Geotechnologien Science Report, v. 3, p. 40–43.

Forsyth, D.W., and Uyeda, S., 1975, On the relative importance of the driving forces of plate motion: Geophysical Journal of the Royal Astronomical Society, v. 43, p. 163–200.

Foulger, G.R., 2002, Plumes or plate tectonic processes?: Astronomy and Geophysics, v. 43, v. 19–23.

Foulger, G.R., and Anderson, D.L., 2006, The Emperor and Hawaiian volcanic chains: How well do they fit the plume hypothesis?: http://www.mantle plumes.org/Hawaii.html.

Foulger, G.R., Natland, J.H., Presnall, D.C., and Anderson, D.L., eds., 2005, Plates, plumes, and paradigms: Boulder, Colorado, Geological Society of America Special Paper 388, 881 p.

Garfunkel, Z., Anderson, C.A., and Schubert, G., 1986, Mantle circulation and the lateral migration of subducted slabs: Journal of Geophysical Research, v. 91, p. 7205–7223.

Gordon, R.G., and Jurdy, D.M., 1986, Cenozoic global plate motions: Journal of Geophysical Research, v. 91, p. 12,384–12,406.

Green, D.H., and Falloon, T.J., 1998, Pyrolite: A ringwood concept and its current expression, *in* Jackson, I.N.S., ed., The Earth's mantle; Composition, structure, and evolution: Cambridge, Cambridge University Press, p. 311–380.

Green, D.H., Falloon, T.J., Eggins, S.M., and Yaxley, G.M., 2001, Primary magmas and mantle temperatures: European Journal of Mineralogy, v. 13, p. 437–451, doi: 10.1127/0935-1221/2001/0013-0437.

Gripp, A.E., and Gordon, R.G., 2002, Young tracks of hotspots and current plate velocities: Geophysical Journal International, v. 150, p. 321–364, doi: 10.1046/j.1365-246X.2002.01627.x.

Gung, Y., Panning, M., and Romanowicz, B., 2003, Global anisotropy and the thickness of continents: Nature, v. 422, p. 707–711, doi: 10.1038/nature 01559.

Heflin, M.B., et al., 2004, GPS time series: http://sideshow.jpl.nasa.gov/mbh/series.html.

Isacks, B., and Molnar, P., 1971, Distribution of stresses in the descending lithosphere from a global survey of focal-mechanism solutions of mantle earthquakes: Review of Geophysics, v. 9, p. 103–174.

Janney, P.E., Macdougall, J.D., Natland, J.H., and Lynch, M.A., 2000, Geochemical evidence from the Pukapuka volcanic ridge system for a shallow enriched mantle domain beneath the South Pacific Superswell: Earth and Planetary Science Letters, v. 181, p. 47–60, doi: 10.1016/S0012-821X (00)00181-3.

Jurdy, D.M., 1990, Reference frames for plate tectonics and uncertainties: Tectonophysics, v. 182, p. 373–382, doi: 10.1016/0040-1951(90) 90173-6.

Kennedy, L.A., Russell, J.K., and Kopylova, M.G., 2002, Mantle shear zones revisited: The connection between the cratons and mantle dynamics: Geology, v. 30, p. 419–422, doi: 10.1130/0091-7613(2002)030<0419: MSZRTC>2.0.CO;2.

Knopoff, L., and Leeds, A., 1972, Lithospheric momenta and the deceleration of the Earth: Nature, v. 237, p. 93–95, doi: 10.1038/237093a0.

Le Pichon, X., 1968, Sea-floor spreading and continental drift: Journal of Geophysical Research, v. 73, p. 3661–3697.

Lynch, M.A., 1999, Linear ridge groups: Evidence for tensional cracking in the Pacific plate: Journal of Geophysical Research, v. 104, p. 29,321–29,334, doi:10.1029/1999JB900241.

Morgan, W.J., 1972, Plate motions and deep mantle convection, *in* Studies in earth and space sciences: Boulder, Colorado, Geological Society of America Memoir 132, p. 7–22.

Natland, J.H., and Winterer, E.L., 2003, What really happened in the Pacific?: Penrose Conference Plume IV: Beyond the plume hypothesis, 25–29 August, Hveragerdi, Iceland, abstracts.

Norton, I.O., 2000, Global hotspot reference frames and plate motion, *in* Richards, M.A., et al., eds., The history and dynamics of global plate motions: Washington, D.C., American Geophysical Union Geophysical Monograph 121, p. 339–357.

O'Connell, R., Gable, C.G., and Hager, B., 1991, Toroidal-poloidal partitioning of lithospheric plate motions, *in* R. Sabadini, et al., eds., Glacial isostasy, sea-level and mantle rheology: Dordrecht, Kluwer Academic, p. 535–551.

Panza, G.F., 1980, Evolution of the Earth's lithosphere, *in* Davies, P.A., and Runcorn, S.K., eds., Mechanisms of continental drift and plate tectonics: New York, Academic Press, p. 75–87.

Pollitz, F.F., Burgmann, R., and Romanowicz, B., 1998, Viscosity of oceanic asthenosphere inferred from remote triggering of earthquakes: Science, v. 280, p. 1245–1249, doi: 10.1126/science.280.5367.1245.

Ricard, Y., Doglioni, C., and Sabadini, R., 1991, Differential rotation between lithosphere and mantle: A consequence of lateral viscosity variations: Journal of Geophysical Research, v. 96, p. 8407–8415.

Robbins, J.W., Smith, D.E., and Ma, C., 1993, Horizontal crustal deformation and large scale plate motions inferred from space geodetic techniques, *in* Smith, D.E., and Turcotte, D.L., eds., Contributions of space geodesy to geodynamics: Crustal dynamics: Washington, D.C., American Geophysical Union Geodynamics Series 23, p. 21–36.

Russo, R.M., and Silver, P.G., 1994, Trench-parallel flow beneath the Nazca plate from seismic anisotropy: Science, v. 263, p. 1105–1111, doi: 10.1126/science.263.5150.1105.

Russo, R.M., and Silver, P., 1996, Cordillera formation, mantle dynamics, and the Wilson cycle: Geology, v. 24, p. 511–514, doi: 10.1130/0091-7613 (1996)024<0511:CFMDAT>2.3.CO;2.

Sandwell, D.T., Winterer, E.L., Mammerickx, J., Duncan, R.A., Lynch, M.A., Levitt, D.A., and Johnson, C.L., 1995, Evidence for diffuse extension of the Pacific plate from Pukapuka ridges and cross-grain gravity anomalies: Journal of Geophysical Research, v. 100, p. 15,087–15,099, doi: 10.1029/95JB00156.

Schubert, G., Turcotte, D.L., and Olson, P., 2001, Mantle convection in the Earth and planets: Cambridge, Cambridge University Press, 940 p.

Scoppola, B., Boccaletti, D., Bevis, M., Carminati, E., and Doglioni, C., 2006, The westward drift of the lithosphere: A rotational drag?: Geological Society of America Bulletin, v. 118, p. 199–209, doi: 10.1130/B25734.1.

Sella, G.F., Dixon, T.H., and Mao, A., 2002, REVEL: A model for recent plate velocity from space geodesy: Journal of Geophysical Research, v. 107, art. no. 2081, doi: 10.1029/2000JB000033.

Shaw, H.R., 1973, Mantle convection and volcanic periodicity in the Pacific: Evidence from Hawaii: Geological Society of America Bulletin, v. 84, p. 1505–1526, doi: 10.1130/0016-7606(1973)84<1505:MCAVPI>2.0.CO;2.

Silver, P.G., and Holt, W.E., 2002, The mantle flow field beneath western North America: Science, v. 295, p. 1054–1057, doi: 10.1126/science.1066878.

Smith, A.D., and Lewis, C., 1999, The planet beyond the plume hypothesis: Earth Science Reviews, v. 48, p. 135–182, doi: 10.1016/S0012-8252(99) 00049-5.

Solomon, S., and Sleep, N.H., 1974, Some simple physical models for absolute plate motions: Journal of Geophysical Research, v. 79, p. 2557–2567.

Steinberger, B., 2000, Plumes in a convecting mantle: Models and observations for individual hotspots: Journal of Geophysical Research, v. 105, B5, p. 11,127–11,152, doi: 10.1029/1999JB900398.

Steinberger, B., and O'Connell, R.J., 1998, Advection of plumes in mantle flow; Implications for hotspot motion, mantle viscosity and plume distribution: Geophysical Journal International, v. 132, p. 412–434, doi: 10.1046/j.1365-246x.1998.00447.x.

Thybo, H., 2006, The heterogeneous upper mantle low velocity zone: Tectonophysics, v. 416, p. 53–79, doi: 10.1016/j.tecto.2005.11.021.

Turcotte, D.L., and Schubert, G., 2002, Geodynamics: Cambridge, Cambridge University Press, 456 p.

Wang, S., and Wang, R., 2001, Current plate velocities relative to hotspots: Implications for hotspot motion, mantle viscosity and global reference frame: Earth and Planetary Science Letters, v. 189, p. 133–140, doi: 10.1016/S0012-821X(01)00351-X.

Wessel, P., and Smith, W.H.F., 1995, New version of the Generic Mapping Tools (GMT) version 3.0 released: Eos (Transactions of the American Geophysical Union), v. 76, p. 329, doi: 10.1029/95EO00198.

Wilson, J.T., 1973, Mantle plumes and plate motions: Tectonophysics, v. 19, p. 149–164, doi: 10.1016/0040–1951(73)90037-1.

Winterer, E.L., and Sandwell, D.T., 1987, Evidence from en-echelon cross-grain ridges for tensional cracks in the Pacific plate: Nature, v. 329, p. 534–537, doi: 10.1038/329534a0, doi: 10.1038/329534a0.

MANUSCRIPT ACCEPTED BY THE SOCIETY 31 JANUARY 2007

DISCUSSION

4 January 2007, Federica Riguzzi

The basic idea of this article is to analyze the impact of two different reference frames in plate kinematics and some consequent geodynamic implications. The definition of alternative absolute reference frames, including variable hotspot source depths, implicitly assumes variable net lithospheric westward rotations, and vice versa. Though not crucial in geodesy, in fact geodesists are concerned to define more rigorous lithospheric (terrestrial) reference frames (Dermanis, 2001). The question is significant in geodynamics, because it can reconcile some independent geological and/or geophysical evidence and open new and interesting questions.

From a geodetic point of view, the establishment of global geodetic networks aims to provide a unified way to describe the positions of points on the Earth's surface. Terrestrial Reference Frames (TRFs) are essentially conventional kinematic reference frames, because there is the need to overcome the variability due to Earth's rotation and to take into account plate motions. TRFs provided by the International Earth Rotation and Reference System Service consist of coordinates and velocities of the observing sites anchored to the NNR-NUVEL1A geodynamic model (Altamimi et al., 2002). They are no-net-rotation (NNR) or, in other words, strictly linked to the lithosphere, thus allowing accurate estimations only of surface relative motions.

When we want to represent absolute plate motions, the motion of the plates relative to the deep mantle, we assume the latter deforms slowly enough to constitute a reference independent from the plates themselves and the hotspot tracks recording the relative motion between lithosphere and mantle. Hotspot reference frame (HSRF) recent plate motion models (Gripp and Gordon, 2002) find a global westward rotation of the lithospheric NNR frame with respect to the absolute (or deep mantle) frame of up to 0.44° m.y.$^{-1}$.

Even if the transition from pure NNR to HSRF systems may be regarded as a simple linear transformation involving velocities $\vec{V}_{HSRF} = \vec{V}_{HNNR} + \vec{V}_{netrot}$, the estimation of net rotation depends somewhat on the assigned hotspot source depths. The article by Cuffaro and Doglioni (this volume) shows that the shallower the hotspot sources, the more polarized plate motion

is expected to be, with respect to the mantle; assuming asthenospheric hotspot sources, all the plates have a westward component of motion reaching 1.49° m.y.$^{-1}$ and reconciling well with independent geological evidence (Doglioni, 1990, 1993).

In support of this view, it has been recently shown that a fast net rotation estimate (corresponding to shallow hotspot sources) matches, in a statistical sense, remarkably well with some large-scale geological constraints (Crespi et al., 2007).

30 January 2007, Warren B. Hamilton

Most Euler poles of current relative rotation between large lithosphere plates are at high latitudes, so a substantial part of present plate motion can be expressed as differential spin velocity. But do the motions sum to zero in a whole-Earth frame, or is there a net drift; and if the latter, is it a transient phenomenon (as, due to the evolving self-organization of plate motions; cf. Anderson, 2007), or is there a unidirectional spin term (tidal drag?) in plate motions? A net westward drift of lithosphere, with some retrograde motions, relative to the bulk Earth, is required by the popular hotspot reference frame for plate motions, but, as many papers and discussions in this volume and its predecessor (Foulger et al., 2005) show, the weak evidence cited in favor of fixed hotspots is contradicted by much else.

Doglioni and his colleagues (e.g., Cuffaro and Doglioni, this volume; Crespi et al., 2007) have speculated for many years that plate tectonics is a product of differential westward motion of lithosphere plates, decoupled across a very weak distributed-shear asthenosphere from the main mass of the mantle, in response to tidal drag. This mechanism is viewed as a substitute for, not a modification of, other proposed modes of plate propulsion. A gravitational drive by subduction is specifically rejected, and some apparently subducting plates are postulated to be rising from the mantle, not sinking into it. The present article by Cuffaro and Doglioni (this volume) seeks a reference frame wherein all plates move westward, and finds it by assuming that only Pacific hotspots are fixed, and in the low upper mantle rather than the deep mantle, and that the Hawaiian-hotspot-track velocity on the Pacific plate is only half the velocity of the plate over the source, the other half of the motion being smeared out by shearing in the asthenosphere. (The embedding within un-moving low upper mantle of local sources of heat and melt to feed plumes for 50 m.y. is not addressed.) This assumption is termed the "shallow hotspot reference frame," and that space-geodesy vectors of relative motion can be transposed into this frame (as they can be into any frame) is wrongly claimed by Crespi et al. (2007) to validate the concept.

These westward-drift assumptions are derived from other assumptions. Doglioni and colleagues, including Cuffaro and Doglioni (this volume), have long claimed that, because of differential shear, west-dipping subducting slabs are steeper than east-dipping ones. Lallemand et al. (2005), not cited by Cuffaro and Doglioni (this volume), addressed this claim and disproved it in detail.

Were the model of Cuffaro and Doglioni (this volume) valid, westward-subducting slabs that penetrate below the asthenosphere (which most do, although this seems contrary to the model) should be overpassed by westward-moving lithosphere, and should appear geometrically as though dipping eastward at depth, and as plated down eastward onto the 660-km discontinuity. The opposite is the case; for example, lithosphere is plated down westward for as much as 2000 km under China from the lower limit of west-dipping western Pacific subduction systems (Huang and Zhao, 2006).

Other objections to the model can be raised on the basis of its incompatibility with geologically and geophysically observed features of subduction systems.

1 February 2007, Marco Cuffaro and Carlo Doglioni

We thank Warren Hamilton for his comment, which allows us to clarify a few issues in our article. First, we affirm that plumes

should be differentiated by whether they are intraplate or steadily located close to plate margins, which are, by definition, moving relative to one another, and relative to the mantle (e.g., Garfunkel et al., 1986). Because practically all intraplate plumes are on the Pacific plate, we used those hotspot tracks (e.g., Norton, 2000; Gripp and Gordon, 2002) as a coherent reference frame. Starting from the idea that Pacific plumes are sourced from the asthenosphere (e.g., Smith and Lewis, 1999; Doglioni et al., 2005; Foulger et al., 2005), the consequence of this interpretation would be that westward drift of the lithosphere is not just an average rotation dictated by the Pacific plate, but is rather a global rotation relative to the mantle. This conclusion is more consistent with the geometric (Doglioni, 1994; Doglioni et al., 1999; Mariotti and Doglioni, 2000; Garzanti et al., 2007; Lenci and Doglioni, 2007), kinematic (Doglioni et al., 2006a; Crespi et al., 2007), and dynamic (Marotta and Mongelli, 1998; Doglioni et al., 2006b, 2007; Scoppola et al., 2006) observations of plate tectonics and subduction zones in particular. Subduction dip is just one parameter of subduction zones. There are a number of other observable features that have to be taken into account, such as morphological and structural elevation, metamorphism, magmatism, dip of the foreland monocline (Fig. D-1), the gravimetric and heatflow signatures, and the type of rocks involved in the prism or orogen. All these signatures support global systematics of the sort we describe.

However, because the aim of our article is not to discuss the differences between orogens and subduction zones as a function of their polarity, we did not quote the paper by Lallemand et al. (2005), and, contrary to Hamilton's statement of 30 January, Lallemand et al. (2005) accept the existence of global westward drift of the lithosphere. They only argue that slab dip is not sig-

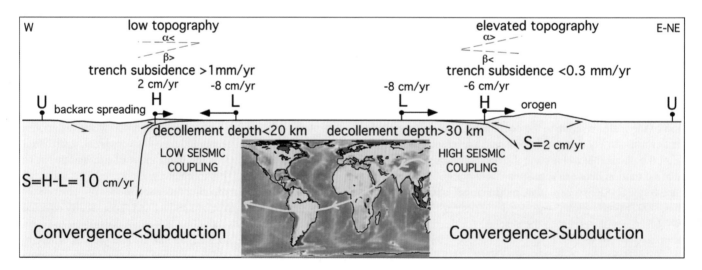

Figure D-1. Assuming point U is fixed the upper plate, along west-directed subduction zones, the subduction hinge H mostly diverges relative to U, whereas it converges along east-directed subduction zones. L—lower plate. Note that the subduction S is larger than the convergence along west-directed slabs, providing larger volumes for mantle recycling, whereas S is smaller for the east-directed case. The two end-members of hinge behavior are respectively accompanied on average by low and high topography, steep and shallow foreland monoclines, faster and slower subsidence rates in the trench or foreland basin, single and double verging orogens, and the like, highlighting a first-order worldwide subduction asymmetry along the flow lines of plate motions, as indicated in the inset (Doglioni et al., 2007).

nificantly influenced by the polarity of subduction. But their analysis is misleading and different from what is suggested in a number of alternative articles in which slab dip is measured not simply comparing east- versus west-directed subduction zones, but is measured along the undulated flow of absolute plate motions (e.g., Doglioni et al., 1999), and the definition of west- versus east- or northeast-directed is rather related to whether subduction accords with this flow. Moreover, their analysis subdivides the slab into a shallow (<125 km) and a deeper part (>125 km). This subdivision is ambiguous for a number of reasons. The east- or northeast-directed subduction zones have mostly continental lithosphere in the upper plate, and the dip of the shallowest 125 km is mostly constrained by the thickness and shape of the upper plate. Moreover, oblique or lateral subduction zones, such as the Cocos plate underneath Central America, are, from geometrical constraints, steeper (>50°) than frontal subduction zones (e.g., Chile), like the lateral ramp of a thrust.

In Cruciani et al. (2005) we reached similar conclusions to Lallemand et al. (2005) and find no correlation between slab age and dip of the slab. Our analysis stopped at about 250 km depth, because east- or northeast-directed subduction zones do not have systematic seismicity at deeper depth, apart from a few areas where seismicity notoriously appears to be concentrated between 630 and 670 km, close to the lower boundary of the upper mantle. The origin of these deep isolated earthquakes remains obscure (e.g., mineral phase change, blob of detached slab, higher shear stress), and therefore they cannot represent a simple geometric prolongation of the shallow part of the slab. Therefore, the deep dip of the slab based on seismicity cannot be compared between west- versus east- or northeast-directed subduction zones, simply because most of the east- or northeast-directed slabs do not show continuous seismicity deeper than 250 km. High-velocity bodies suggesting the presence of slabs in tomographic images often do not match slab seismicity.

Moreover, Lallemand et al. (2005) note that steeper slabs occur where the upper plate is oceanic, whereas shallower slabs occur where the upper plate is continental. However, the majority of east- or northeast-directed subduction zones worldwide have continental lithosphere in the upper plate, confirming the asymmetry we proposed. Apart from these issues, west-directed subduction zones, compared to east- or northeast-directed slabs, still maintain a number of fundamental differences, e.g., they are steeper, deeper (or at least they present more coherent slab-related seismicity from the surface down to the 670-km discontinuity), and they show opposite down-dip seismicity. Northern Japan is an exception, having a shallow dip; however, the subduction hinge there has started to invert, and it migrates toward the upper plate (Mazzotti et al., 2001). The back-arc basin is shrinking, and the system is losing the typical character of west-directed subduction zones in which the subduction hinge retreats relative to the upper plate.

In our article we do not address the problem of whether slabs penetrate into the lower mantle, because it is not relevant to our work. Slab pull is also not treated for the same reason. Moreover, we do not reject the negative buoyancy of the oceanic lithosphere as a fundamental component of plate tectonics, but we argue against considering slab pull to be the main driving force of plate tectonics. Apart from the kinematic counterarguments presented in our article, the inferred slab pull described in the literature is larger than the yield strength of the lithosphere under extension (i.e., the Pacific plate should have been broken by the pull) and is not sufficiently high to generate the observed slab rollback (Doglioni et al., 2006b).

REFERENCES CITED

Altamimi, Z., Sillard, P., and Boucher, C., 2002, The impact of no-net-rotation condition on ITRF2000: Geophysical Research Letters, v. 30, p. 1064, doi: 10.1029/2002GL016279.

Anderson, D.L., 2007, The new theory of the Earth: Cambridge, Cambridge University Press, 384 p.

Crespi, M., Cuffaro, M., Doglioni, C., Giannone, F., and Riguzzi, F., 2007, Space geodesy validation of the global lithospheric flow: Geophysical Journal International, v. 168, p. 491–506.

Cruciani, C., Carminati, E., Doglioni, C., 2005, Slab dip vs. lithosphere age: No direct function: Earth and Planetary Science Letters, v. 238, p. 298–310.

Cuffaro, M., and Doglioni, C., 2007 (this volume), Global kinematics in deep versus shallow hotspot reference frames, *in* Foulger, G.R., and Jurdy, D.M., eds., Plates, plumes, and planetary processes: Boulder, Colorado, Geological Society of America Special Paper 430, doi: 10.1130/2007.2430(18).

Dermanis, A., 2001, Global reference frames: Connecting observation to theory and geodesy to geophysics, IAG 2001 scientific assembly "Vistas for Geodesy in the New Millenium," 2–8 September, Budapest. http://der .topo.auth.gr/DERMANIS/ENGLISH/Index_ENGLISH.htm.

Doglioni, C., 1990, The global tectonic pattern: Journal of Geodynamics, v. 12, p. 21–38.

Doglioni, C., 1993, Geological evidence for a global tectonic polarity: Journal of the Geological Society of London, v. 150, p. 991–1002.

Doglioni, C., 1994, Foredeeps versus subduction zones: Geology, v. 22, p. 271–274.

Doglioni, C., Harabaglia, P., Merlini, S., Mongelli, F., Peccerillo, A., and Piromallo, C., 1999, Orogens and slabs vs their direction of subduction: Earth Sciences Review, v. 45, p. 167–208.

Doglioni, C., Green, D., and Mongelli, F., 2005, On the shallow origin of hotspots and the westward drift of the lithosphere, *in* Foulger, G.R., et al., eds., Plates, plumes, and paradigms: Boulder, Colorado, Geological Society of America Special Paper 388, p. 735–749, doi: 10.1130/2005.2388(42).

Doglioni, C., Carminati, E., and Cuffaro, M., 2006a, Simple kinematics of subduction zones: International Geological Review, v. 48, p. 479–493.

Doglioni, C., Cuffaro, M., and Carminati, E., 2006b, What moves plates? Bollettino di Geofisica Teorica e Applicata, v. 47, no. 3, p. 227–247.

Doglioni, C., Carminati, E., Cuffaro, M., and Scrocca, D., 2007, Subduction kinematics and dynamic constraints: Earth-Science Reviews, doi: 10.1016/j.earscirev.2007.04.001.

Foulger, G.R., Natland, J.H., Presnall, D.C., and Anderson, D.L., eds., 2005, Plates, plumes, and paradigms: Boulder, Colorado, Geological Society of America Special Paper 388, 881 p.

Garfunkel, Z., Anderson, C.A., and Schubert, G., 1986, Mantle circulation and the lateral migration of subducted slabs: Journal of Geophysical Research, v. 91, p. 7205–7223.

Garzanti, E., Doglioni, C., Vezzoli, G., and Andò, S., 2007, Orogenic belts and orogenic sediment provenances: Journal of Geology, v. 115, p. 315–334.

Gripp, A.E., and Gordon, R.G., 2002, Young tracks of hotspots and current plate velocities: Geophysical Journal International, v. 150, p. 321–361.

Huang, J., and Zhao, D., 2006, High-resolution mantle tomography of China and surrounding regions: Journal of Geophysical Research, v. 111, no. B9, art. no. 305, 21 p.

Lallemand, S., Heuret, A., and Boutelier, D., 2005, On the relationships between slab dip, back-arc stress, upper plate absolute motion, and crustal nature in subduction zones: Geochemistry, Geophysics, Geosystems, v. 6, art. no. 17.

Lenci, F., and Doglioni, C., 2007, On some geometric prism asymmetries, *in* Lacombe, O., et al., eds., Thrust belts and foreland basins: From fold kinematics to hydrocarbon systems: New York, Springer Frontiers in Earth Sciences 24, p. 41–60.

Mariotti, G., and Doglioni, C., 2000, The dip of the foreland monocline in the Alps and Apennines: Earth and Planetary Science Letters, v. 181, p. 191–202.

Marotta, A., and Mongelli, F., 1998, Flexure of subducted slabs: Geophysical Journal International, v. 132, p. 701–711.

Mazzotti, S., Henry, P., and Le Pichon, X., 2001, Transient and permanent deformation of central Japan estimated by GPS-2. Strain partitioning and arc-arc collision: Earth and Planetary Science Letters, v. 184, p. 455–469.

Norton, I.O., 2000. Global hotspot reference frames and plate motion, *in* Richards, M.A., et al., eds., The history and dynamics of global plate motions: Washington, D.C., American Geophysical Union Geophysical Monograph 121, p. 339–357.

Scoppola, B., Boccaletti, D., Bevis, M., Carminati, E., and Doglioni, C., 2006, The westward drift of the lithosphere: A rotational drag?: Geological Society of America Bulletin, v. 118, no. 1/2, p. 199–209, doi: 10.1130/B25734.1.

Smith, A.D., and Lewis, C., 1999, The planet beyond the plume hypothesis: Earth Sciences Review, v. 48, p. 135–182.

The Geological Society of America
Special Paper 430
2007

Ridge-crossing seamount chains: A nonthermal approach

Erin K. Beutel*

Department of Geology, College of Charleston, 66 George Street,
Charleston, South Carolina 29424, USA

Don L. Anderson

Geological and Planetary Sciences, California Institute of Technology,
MC 170-25, 1200 East California Boulevard, Pasadena, California 91125, USA

ABSTRACT

In this article we examine whether it is viable to form an age-progressive ridge-crossing seamount chain using a nonplume mechanism. Nonthermal melt sources considered include fertile mantle blobs and subsolidus mantle while lithospheric stresses generated at the ridge and at ridge-transform intersections (RTIs) are tapped to bring the mantle to the surface. Finite element models, analog models, and an analysis of the Tristan de Cunha chain all show that ridge-crossing seamount chains may be created using these mechanisms. Essentially, as a ridge migrates or reorganizes, excess magmatism may appear to switch sides of the ridge as areas of extensional stress at the RTI migrate with the ridge.

Keywords: plume, ridge, hotspot, transform, Tristan de Cunha, Atlantic, finite element

INTRODUCTION

Hypotheses for ridge-crossing seamount chains have focused on the interaction of mantle plumes with migrating ridge segments (Sleep 2002b; DePaolo and Manga, 2003). These hypotheses call for a stationary plume of hot mantle material that rises to the base of the lithosphere and creates a chain of age-progressive seamounts in the middle of the plate. As a migrating ridge approaches the plume, buoyant mantle material rises along the base of the lithosphere like a helium balloon along a cathedral ceiling to the ridge, and age-progressive chains of seamounts are created on both sides of the ridge from a single hotspot source. Eventually the ridge moves away from the plume and seamount production switches across the ridge to the other plate, ceases to generate seamounts on the ridge, and continues to generate seamounts on the new plate only (e.g., Small,

1995; Kincaid et al., 1996; Ribe, 1996; Sleep 2002b; DePaolo and Manga, 2003). This scenario results in the formation of an asymmetrical V-shaped array of seamounts, such as the Tristan de Cunha chain in the South Atlantic (Fig. 1).

Because current nonthermal models of seamount formation are unable to explain this geometry (e.g., propagating cracks are presumed to be unable to cross ridges, and asthenospheric heterogeneities are presumed to be moving rapidly), ridge-crossing seamount chains have been cited as evidence for plumes of material that remain relatively stationary in comparison to the plates. However, if the mantle is lithologically heterogeneous and contains areas with different melting points on a scale from kilometers to hundreds of kilometers and/or if ridge-transform intersections (RTIs) are involved, then nonthermal ridge-crossing seamount chains are not only viable, but also likely. In this article we propose a mechanism for the development of ridge-

*E-mail: beutele@cofc.edu.

Beutel, E.K., and Anderson, D.L., 2007, Ridge-crossing seamount chains: A nonthermal approach, *in* Foulger, G.R., and Jurdy, D.M., eds., Plates, plumes, and planetary processes: Geological Society of America Special Paper 430, p. 375–386, doi: 10.1130/2007.2430(19). For permission to copy, contact editing@geosociety.org.

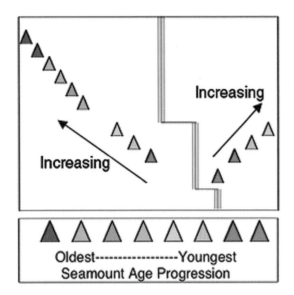

Figure 1. Cartoon of a ridge-crossing seamount chain geometry in the Atlantic. Triple lines indicate ridge position, and arrows indicate increasing age of seamounts. After Sleep (2002b).

crossing seamount chains that does not invoke deep-mantle plumes yet still accounts for the four primary elements necessary for a viable hypothesis for the origin of seamount chains ranging in size from two or three seamounts to seamount chains spanning a 100 m.y. The primary elements included in this mechanism are a source of melt, a means of bringing melt to the surface, an asymmetrical ridge-crossing seamount chain, and an explanation for the age-progressive nature of the chain. Vogt and Jung (2005) proposed a similar mechanism using fertile mantle or anomalous mantle patches and migrating ridges, but did not account for the asymmetrical nature of some of the chains. We address that issue here using seamount emplacement off-ridge, ahead of RTIs.

MELT SOURCE

We focus on two alternative sources for the melt: (1) the mantle has fertile or low-melting-point regions, or (2) the mantle is in a nascent melting condition. In the first case, the mantle is heterogeneous and contains fertile (e.g., compositionally different) streaks or blobs, some of which are large and produce major outpourings of magma when tapped (i.e., when the stress condition in the overlying plate permits dikes and volcanoes to form). Fertile mantle streaks and blobs require lower temperatures or lower ascent rates to produce melt volumes similar to those modeled for infertile, deep-mantle sources.

In the second case, the mantle may be relatively homogeneous and is at, or near, its melting point. Under these conditions melt is created any time pressure drops or when the lithosphere extends and causes passive upwelling to occur. In both cases the stress conditions of the lithosphere will determine the form of

the melt intrusion—ponds, sills, or dikes and volcanoes. Elements of these scenarios have been discussed previously by various authors (e.g., Sleep, 1997; Anderson, 1998; McNutt and Bonneville, 1999; Favela and Anderson, 2000; Yaxley, 2000; Natland and Dick, 2001; Niu et al., 2001; Sleep, 2002b; Meibom and Anderson, 2004). Variants of these scenarios can be imagined. The asthenosphere is not expected to be precisely isothermal, and small variations in temperature for a mantle near its solidus can produce large variations in melt content to be tapped by localized stresses and cracks. Because ridges themselves are moving, stresses and cracks associated with the ridges may tap mantle melt in a V-shaped pattern that appears to cross the ridge without invoking a deep-mantle source.

MELT TRANSPORT

Melt is transported through the oceanic lithosphere to the surface to form seamounts in the form of dikes (e.g., Gudmundsson, 1990). Dikes require a horizontal least-compressive stress regime plus a source of melt. The overall state of stress in oceanic lithosphere is usually compressional, except near ridges; it is necessary for a change in the local stress field to occur for dike intrusion to commence (e.g., Richardson and Solomon, 1979; Wiens and Stein, 1984; Zoback, 1992). Otherwise, we have ponding, underplating, and sill intrusion. A change in the stress field of an oceanic plate from compressional to tensional may be due to localized upwelling of low-density material beneath the plate (Sleep, 1997, 2002a,b; Tentler, 2003); large-scale tectonic forces, such as changes in subduction and plate reorganizations; plate flexure (Hieronymus and Bercovici, 2000); and regional stress patterns associated with ridges, transforms, microplates, and seamounts (e.g., Bergman and Solomon, 1992; Shah and Buck, 2003; Craddock et al., 2004; Neves et al., 2004). Thermal contraction of a cooling plate can explain some magmatism (Sandwell and Fialko, 2004). To account for a series of injection events that result in a chain of seamounts, these changes in the stress regime must be ongoing or self-sustaining for the lifetime of the chain, usually less than 15 m.y.

Previous hypotheses for seamount chain emplacement invoke a rising hot, buoyant deep-mantle plume, to thin, weaken, and uplift the lithosphere. The buoyancy of the ponded magma creates a localized horizontal tensional stress field that overcomes the normal compressive state of oceanic lithosphere (Sleep, 1997, 2002b). The ongoing supply of hot material from the plume tail sustains the buoyancy and maintains the tensional stress field. The episodic nature of the emplacement of the seamount chain is a result of the influence of the extruded and underplated material on the stress state combined with possible flux variations in the plume tail (e.g., Richards et al., 1989; Campbell, 2001; Courtillot et al., 2003). The melt source is the hot buoyant plume.

Other models propose that age-progressive seamount chains may be the result of tensional stresses, tectonic cracking of the oceanic plate, or horizontal dike propagation extending preexisting features (e.g., Favela and Anderson, 2000; Hieronymus and Bercovici, 2000; Fairhead and Wilson, 2005; Natland and Winterer, 2005). The development of a chain results from the propagation of the crack or dike in response to the stress field across the plate. Melt is either already available from underplated magma or from fertile blobs, and/or is created through decompression melting of the mantle in response to the extension of the crust. In the fertile blob model, melting creates buoyancy, and the melting is sustained by adiabatic decompression as the material rises (Raddick et al., 2002). In these models it is the relative motion between the elements that is important, not the absolute motion, as in the fixed plume model.

Horizontal tensional stress in a plate can also be created at and ahead of RTIs when slip along the transform fault is impeded (Lachenbruch and Thompson, 1972; Fujita and Sleep, 1978; Phipps Morgan and Parmentier, 1984; West et al., 1999; Beutel, 2005; van Wijk and Blackman, 2005). Large-magnitude tensional stress fields may cause decompression melting of fertile mantle regions and of mantle in a nascent melt condition. The resulting melt is then either transported to the surface through the cracks formed by the tensional stress, or it is ponded and increases the horizontal tensional stress until dikes develop and seamount formation occurs (Beutel, 2005). The applicability of this model for creating lithospheric extension is suggested by the observation that many seamount chains intersect, or originate, near RTIs (e.g., Graham et al., 1999; Hekinian et al., 1999; Johnson et al., 2000; Klingelhöfer et al., 2001).

Large areas of extension develop in the lithosphere when transform slip is impeded (Beutel, 2005; Fig. 2). If the mantle is already partially molten, or if magma is ponded beneath the plate, then the condition for dike development is simply that the least-compressive axis is horizontal. If the mantle is entirely subsolidus, then decompression is required to generate melt. Decompression and melt generation are expected to occur in response to lithospheric extension, such as occurs in the neighborhood of RTIs. Development of seamount chains rather than linear ridges is postulated to be a result of episodic strengthening and weakening of the transform fault due to changes in the local stress state, which occurs partially as a result of the injection of magma into the crust (Sleep, 2002b). Sustainability results from the ongoing presence of mantle at, or near, its solidus or the presence of a fertile mantle blob near the ever-changing stress fields at RTIs. Large amounts of melt probably require a fertile blob that is already above its melting point or pre-eruptive ponding.

MODELS

Based on our understanding of nonthermal melt sources and the transport of melt to the surface, we have developed two models by which nonplume ridge-crossing seamount chains may be generated. These models are based on the following assumptions: because of welding or tectonic forces, most transforms experience periods of decreased slip; when transform slip is impeded, extension is concentrated at, and ahead of, the RTI (Fig. 2); if the mantle is fertile or near its solidus, extension at an RTI will result in the formation of a seamount; and ridges migrate and reorganize. We also examine these four processes for the Tristan de Cunha seamount chain in the South Atlantic.

Model 1—Ridge Reorganization

Model 1 is illustrated in Figure 2. We examine the potential for a ridge-crossing geometry for seamount chains formed during a ridge reorganization over a near-solidus mantle. Because the mantle is near solidus it is assumed that concentrated horizontal tensional stress in the lithosphere will result in decompression melting, dike formation, and seamount emplacement. Finite element models of an evolving ridge were constructed to determine the location of these tensional stresses and the resultant seamount pattern that would emerge during a ridge reorganization. More than twelve initial time frames were constructed, and the six most relevant are shown in a forward model of a reorganizing ridge. Figure 2 presents the finite element model results as maximum stress type by color (red is tensional and blue is compressional), maximum and minimum stress vectors, and conceptual models of seamount emplacement and oceanic crustal ages—and therefore motion—are shown adjacent to the stress maps. The exact model parameters are given in the appendix. Stresses applied to the model consist chiefly of ridge-perpendicular gravity forces applied to the whole of the plate; thus plate motion is initially perpendicular to the ridge and becomes more oblique to the overall ridge over time (Fig. 2):

0 m.y.: Three north-south–trending ridge segments are connected by weak transforms. The oceanic plate is under relatively little east-west stress.

10 m.y.: A change in plate motion direction results in northeast-southwest ridge-push forces and the strengthening of both transform faults. Large areas of tensional stress are concentrated ahead of, and at, the RTIs. Seamounts are emplaced in older crust ahead of the RTIs.

15 m.y.: In response to changing plate motions a new northwest-southeast–oriented ridge propagates between the southern two ridge segments, eradicating the transform. The northern transform is modeled as weak. Extension is concentrated where the new northwest-southeast–trending ridge intersects the middle north-south–trending ridge segment. Seamounts are emplaced on the western plate only.

25 m.y.: Restrengthening of the northern transform results in tensional stresses at and ahead of the RTIs there. New seamounts are emplaced.

35 m.y.: A new northwest-southeast–trending ridge segment propagates over the northern transform. Large tensional stresses are concentrated where the north-south ridge segments intersect the northwest-southeast–trending segments, and more seamounts are emplaced on the western plate.

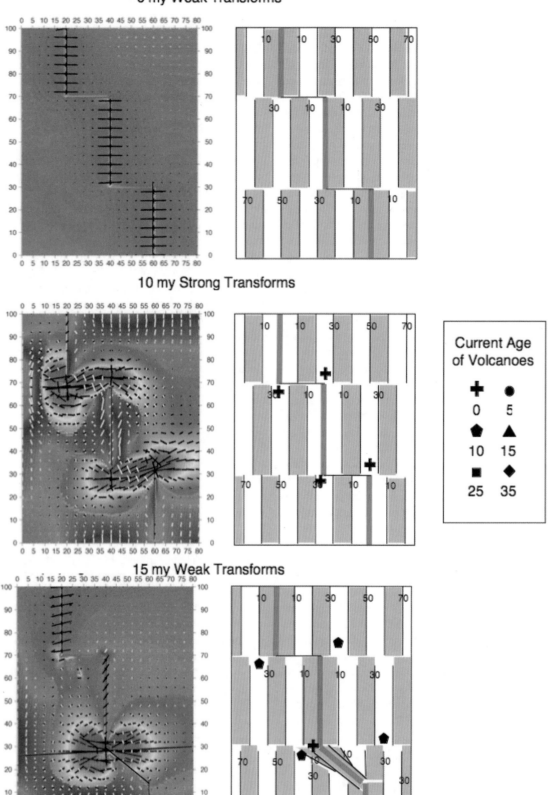

Figure 2. Results of finite element modeling of the nascent-melting case. Panels show a series of 2-D models of a reorganizing ridge. Background colors indicate maximum stress intensity, warmer colors indicate larger stresses, white and black bars indicate maximum and minimum stress vectors (white: compressional; black: tensional). Panels to the right of the stress maps/vector plots show modeled ocean ages and approximate location of seamounts caused by tensional stress ahead of ridge-transform intersections. Ages are Ma.

25 my Strong Transforms

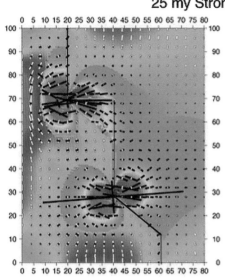
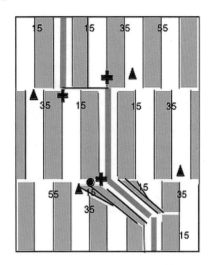

35 my Weak Transforms

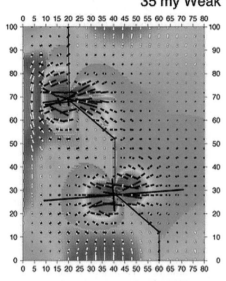
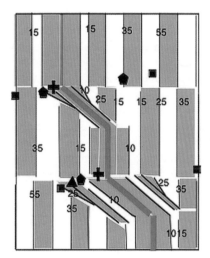

Current Age
of Volcanoes

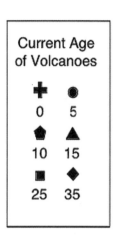

✚	⬤
0	5
⬠	▲
10	15
◼	◆
25	35

45 my Weak Transforms

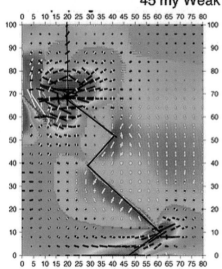
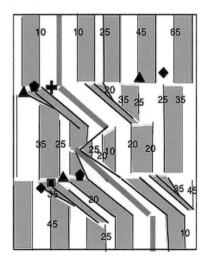

45 m.y.: A new, weak transform forms between the north-west-southeast–trending ridge segments. Extension is still concentrated at the intersection of the north-south– and northwest-southeast–trending ridge segments. Seamounts are emplaced on the western plate at the intersection.

The resulting seamount pattern is a pair of chains that increase in age away from the ridge and appear to change from emplacement on both sides of the ridge to emplacement only on the west side. This process results in the same seamount pattern that has been attributed to a ridge overrunning a hotspot (Sleep, 2002b). Recall that we only have information about the relative positions of ridges and fertile mantle blobs. The only difference between the generation of seamounts due to ridge reorganization over a subsolidus mantle and those generated by a plume is that a plume "burns" through the overlying lithosphere and thereby can emplace seamounts at will, whereas this model shows that the same results can be achieved by stress in the lithosphere releasing subsolidus mantle from below to form the seamounts. However, if a plume did exist under the modeled ridge, melt would likely still be guided by the stress fields modeled, making differentiating a plume source from an upper-mantle source even more difficult.

Although no specific seamount chain is modeled in Figure 2, the ridge reorganization that is modeled is similar to ongoing and past ridge reorganizations in the Pacific. The high density of midplate seamounts in the Pacific compared with the Atlantic suggests that the mantle beneath the plates in the Pacific is closer to its solidus than is mantle beneath the Atlantic plates, or that the plates in the Atlantic are under a greater degree of horizontal compression.

Model 2—A "Stationary" Fertile Blob

In this section we present both a conceptual model and a simple analysis of the Tristan de Cunha seamount chain. The conceptual model (Fig. 3) takes the ridge geometry and motion proposed in Sleep (2002b), applies the finite element–modeled stresses to the ridge, and moves the ridge over a fertile blob. A fertile blob that is stationary relative to the overlying lithosphere will have the same seamount pattern as a plume, as long as a stress field in the lithosphere exists to tap into it. In this model we show that the extensional stress fields ahead of RTIs combined with the ridge itself may produce a pattern of ridge-crossing seamounts similar to that of a plume.

In the second fertile blob model (Fig. 4) we reconstructed the South Atlantic basin ridges for the past 80 m.y. using the oceanic ages of Müller et al. (1997), the ocean floor gravity signature of Smith and Sandwell (1997), and some seamount ages (O'Connor et al., 1999). A stationary fertile blob relative to a fixed reference frame (the border of the model) was also added. Once again, by assuming that seamounts can be formed at the ridge and in older oceanic crust ahead of RTIs, we were able to

recreate the ridge-crossing chain. We acknowledge that at this time the active island of Tristan is not well understood, but submit that the location of Tristan is difficult for any model to place; its location near a strongly defined fracture zone suggests that lithospheric forces similar to those described at the RTIs may be responsible.

Model 2a. In this conceptual model (Fig. 3) we apply the finite element results of extension at, or ahead of, RTIs to a ridge migrating (in relative terms—both may be moving) over a fertile mantle region. This model assumes that seamounts only form when there is strong tensional stress above a fertile mantle region, and the tensional stress results from impeded slip on a transform fault. For ease of illustration, the fertile mantle blob in Model 2 is shown as stationary, but no actual stability relative to the Earth's core is implied, simply a point fixed in the mantle relative to a moving ridge.

Figure 3 is a schematic diagram that illustrates the modeling results separated into discrete time intervals. During the first four time periods (T1–T4) a fertile mantle region is intersected by a series of RTIs as the ridge moves to the northwest. This process results in a single chain of seamounts on the western plate, the spacing of which is determined by the production rate at, and the spacing of, the RTIs. The seamounts move to the northwest relative to one another, but to the west relative to the ridge. In time interval T5 the ridge overrides the fertile mantle region, and chains of seamounts are created on both sides of the ridge. Finally, the ridge moves off the fertile mantle region, and growth of the seamount chain on the western plate is terminated. In time interval T8 the fertile mantle region is intersected by an RTI on the eastern side of the ridge, and a new series of seamounts is initiated. The resultant seamount distribution gives the impression that the seamount chain has crossed the ridge.

Model 2b. The time series shown in Figure 4 is a reconstruction of the south Atlantic showing all areas of increased volcanism (seamounts and ridges), the ages of the ocean floor as determined by Müller et al. (1997), and the ages of the main ridge-crossing Tristan seamount chain (O'Connor et al., 1999). Also shown is the approximate location and geometry of the ridge, as determined from the Müller et al. (1997) data, and an area of fertile mantle that may have contributed to the formation of seamounts both on the ridge and at, and ahead of, seamounts. Note that the change from seamount production on both sides of the ridge to just the eastern side of the ridge is associated with changes in ridge geometry. All seamounts shown were either created at the ridge or could have been created ahead of a RTI in older crust (Fig. 4):

Figure 3. Time series for a conceptual model of a ridge and a fertile mantle region migrating with respect to a fixed point on the eastern (African) plate. Ridge geometry and motion are taken from Sleep (2002b). A—Africa; ANT—Antarctica; MA—motion absolute; SA—South America.

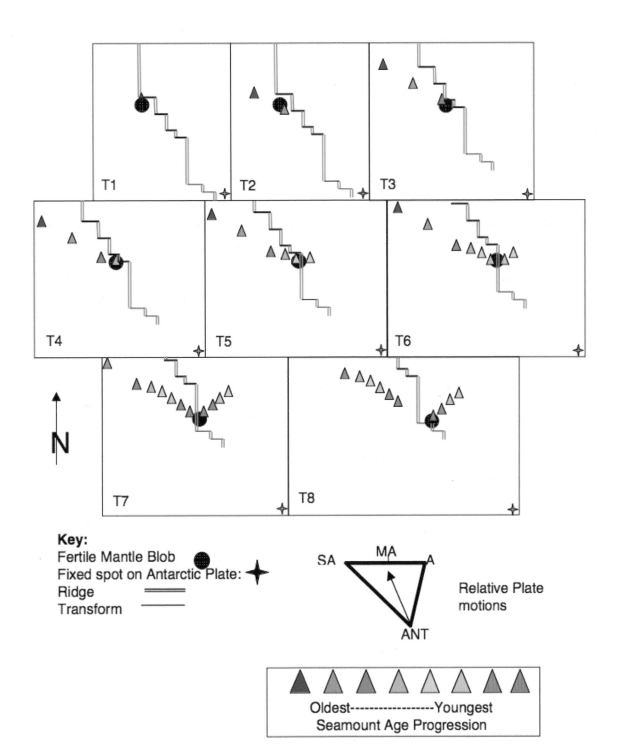

Key:

Fertile Mantle Blob ●

Fixed spot on Antarctic Plate: ✦

Ridge ══

Transform ──

Relative Plate motions

SA — MA — A
ANT

Oldest------------------Youngest
Seamount Age Progression

N

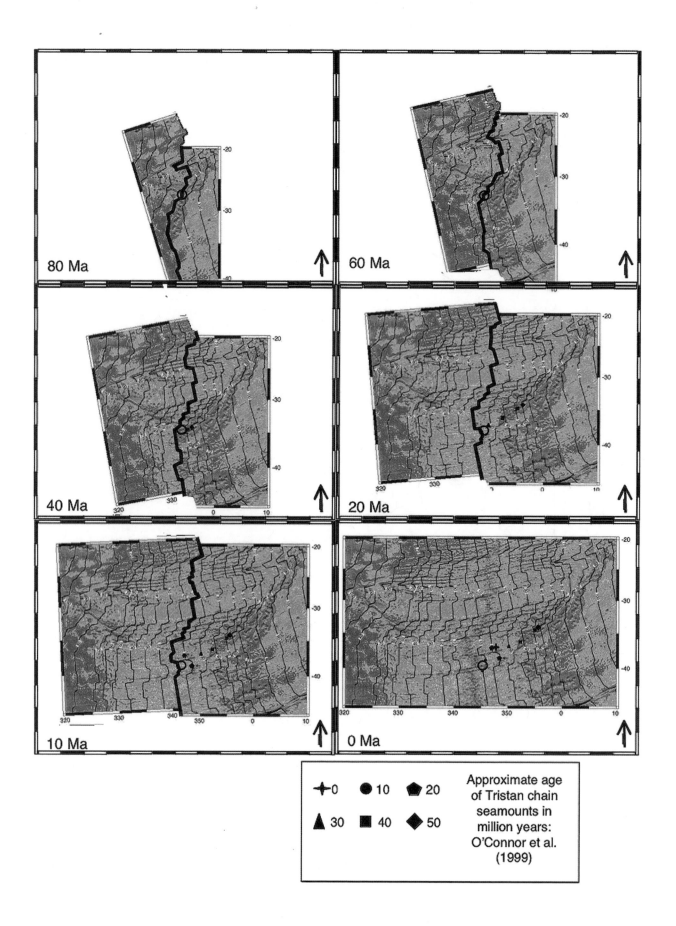

✛ 0	● 10	⬠ 20
▲ 30	■ 40	◆ 50

Approximate age of Tristan chain seamounts in million years: O'Connor et al. (1999)

80–60 Ma: It appears that a large area of fertile mantle could have created portions of the Walvis Ridge at the ridge between 80 and 70 Ma, and then between 70 and 60 Ma that portion of the oceanic crust slid past the series of east-stepping transforms. This process would have resulted in repeated injection of magma.

60–40 Ma: During this time period the transforms become more distinct, no ridge propagation patterns are detected, but some degree of reorganization must have taken place if the Müller et al. (1997) age lines are correct. The majority of seamount activity also switches from the South American plate to the African plate at this time. A reinjection of previously injected crust would account for some of these changes, as would the location of the fertile blob under and ahead of RTIs on the African plate during ridge reorganizations and impeded transform slip.

40–20 Ma: A comparatively quiet time, the South American plate has few distinct seamounts, but rather boasts an overall increased heat and/or plate thickness, as indicated by the gravity signature. The African plate has some distinct seamounts, which appear to be have formed near the RTIs. Ridge migration over the fertile blob would account for the seamounts at the RTI. The decrease in Africa's motion at this time (O'Connor et al., 1999) would have affected the stress field at the ridge and may account for the difference in the style of magma injection.

20–10 Ma: Increased westward drift of the ridge results in the continued production of seamounts on the African plate at RTIs.

10–0 Ma: Westward drift of the ridge puts the fertile mantle blob on the very fringes of the ridge and generates seamounts at the RTIs on the long offset transforms. Tristan is created along an old transform.

Note that we only tracked one ridge-crossing seamount chain. However, given that the area was surrounded by inward-dipping subduction zones only 230 Ma, it is likely that the area is rife with subduction-derived fertile mantle blobs.

OTHER SEAMOUNT GEOMETRIES

Our first model, of lithospheric extension above a mantle near its solidus, accounts for the generation of numerous seamount chains, some of which appear to cross the ridge. This seamount geometry is similar to that seen in the Pacific basin.

Figure 4. Reconstruction of Atlantic plate motions and ridge location for the past 80 m.y. using the age grid from Müller et al. (1997), gravity data from Smith and Sandwell (1997), and hotspot ages from O'Connor et al. (1999). Plate location relative to a fixed frame is modeled based on seamount production at the ridge and a fixed fertile blob (designated by an open circle), the circle is held fixed relative to the black and white box around each time period. Note that most of the seamounts are created on the ridge.

The second model, with its single ridge-crossing seamount chain, has a geometry that more closely resembles that of seamount chains in the Atlantic basin. The principles of RTI-generated seamounts can also be applied to other observed geometries. The combination of on- and off-ridge localized extension with mantle compositional anomalies may result in a great number of seamount geometries. For example, short seamount chains may result from small regions of mantle fertility and/or the migration of a ridge away from the fertile region, whereas large areas of melt represented by aseismic ridges and large igneous provinces may represent large regions of fertility or mantle in a nascent melting state. The more numerous seamounts and faster spreading rates in the south Pacific indicate that different mantle and lithosphere conditions in the region may affect seamount formation, including strong thermal contraction of the lithosphere, a plate close to the tensile state, and/or numerous fertile patches or proximity to the nascent melting state.

LARGE IGNEOUS PROVINCES

Plume-based models for ridge-crossing hotspots involve the separation of postulated plume heads (large igneous provinces [LIPs]) from the ends of their tails (volcanic chains) by actively spreading ridges. Most volcanic chains, however, do not start at a LIP, and most LIPs are not associated with a volcanic chain. Perhaps the best-documented case is the separation of the Kerguelen plateau from the Ninetyeast Ridge and the Rajmahal basalts (Coffin et al., 2002; Kent et al., 2002; Weis et al., 2002). To link the basaltic outpourings on mainland India with the Kerguelen plateau and Broken Ridge a series of ridge jumps are postulated to have occurred as the Kerguelan hotspot drifted slowly to the south (Antretter et al., 2002; Kent et al., 2002). Other models involve multiple plumes (e.g., Coffin et al., 2002). In this article we draw attention to the links between LIPs, ridges, and transform faults, whereby many of the proposed splits between "plume heads" and "plume tails" may be the result of ridge-reorganizations and transform-fault interactions rather than fixed thermal anomalies crossing ridges. There is a prevalence of "coincidental" relationships between supposed hotspot features and tectonic features: (1) the Ninety-east Ridge lies along an extensive offset of seafloor magnetic anomalies—a fossil transform fault; (2) Réunion Island is located on the intersection of an abandoned ridge and a fracture zone; (3) Mauritius developed on Paleocene fossil spreading centers, and Réunion and Mauritius were transported away from each other by a fracture zone that lies between them; and (4) the Ontong Java and Shatsky plateaus are thought to have been created at triple junctions (e.g., Sager, 2005).

DISCUSSION

Seamount chains that appear to cross mid-ocean ridges do not require a plume-based model; they can be explained by a combination of ridge dynamics and mantle heterogeneities. Our models demonstrate that volcanism can migrate from one side of a ridge to the other when tensional stress, at and ahead of RTIs, is considered. We further suggest that the nature of the melt sources may also affect seamount-chain geometry. Fertile mantle regions and those at or near their liquidi may produce larger volumes of magma than does the surrounding depleted mantle or "hot" mantle (McKenzie and Bickle, 1988).

The recognition that hotspots move relative to one another, and relative to the geomagnetic reference frame, led to the concept of drifting plumes, the predicted consequences of which are no different from those of passive fertile heterogeneity (Silver et al., 2006). In contrast to mantle-plume models, a stress-controlled mechanism for magma release is better able to explain the rapid volcanism that builds LIPs and the rapid switching on and off of magmatism along volcanic chains. Furthermore, the size of a fertile heterogeneity is not important, because changes in lithospheric stress and the extension of the lithosphere control and localize the volcanism. However, fertile anomalies are not required to be small and may exist as large 3-D compositional blobs in the mantle, such that lithospheric stress concentrations may result in long-term seamount chains.

The melting of fertile patches of mantle requires no additional heat input or heat from the Earth's core to explain even the largest volumes of melt produced. The largest LIP on Earth is the Ontong-Java plateau. If the 20-km-thick crust there resulted from draining an area three times larger than the plateau itself (as a result of focusing at the apex of a triple junction), and if 20% melting was involved, then only a 30-km-thick section of the mantle may have been involved. The normal mantle geotherm is usually close to or above the solidus in the depth range of ~30 to ~50 km. Clift (2005) has shown that the subsidence patterns of many oceanic plateaus attributed to plumes are consistent with normal mantle temperatures and not the elevated temperatures expected for plumes. This observation is consistent with the model we propose here, which attributes ridge-crossing seamount chains and LIPs to the extraction of melt from regions of fertility and does not invoke greatly elevated temperature.

CONCLUSIONS

Many so-called hotspot tracks lie along pre-existing fracture zones and transform faults or emerge from RTIs. This circumstance, along with the inability of propagating cracks to cross ductile zones associated with active spreading ridges, led us to explore mechanisms for the migration of stress conditions as an explanation for the anomalous age progressive volcanism. A fertile region in the mantle can have effects similar to a hotspot (plume). Finite element models demonstrate the viability of off-ridge tensional-stress migration associated with RTIs. Combined with fertile mantle and/or mantle in the nascent melting condition, such areas of tensional stress may create ridge-crossing seamount chains without the need to invoke a mantle plume. This possibility has been shown using both analog mod-

els and the reconstruction of the Tristan de Cunha seamount chain.

Many of the geochemical arguments for "plume" compositional components could apply equally well to passive compositional hetereogeneities in the mantle (fertile blobs). Our geometric models do not rule out a fixed thermal plume source for ridge-crossing seamount chains, but they demonstrate that such a model is not required. Differentiating between deep-seated mantle plumes and those generated by tensional stresses in the lithosphere is complicated and often debated. In addition to ongoing debates about the viability of various geochemical parameters for determining the depth of the mantle melt origin, it has been pointed out that the same tensional forces in the lithosphere that could generate seamounts from shallow melts could also affect where deep-seated melts penetrate the lithosphere. Perhaps the most telling aspect of this debate is that differentiating between the deep-source models and the shallow-source models is difficult. For specific chains, making more actual measurements of stress in the ocean crust and dating more seamounts may tell us which seamount chains have been affected by the lithospheric stresses, but the question of the depth of the magma source is ultimately left to those who study mineral physics and the ability of material to penetrate the phase boundaries of the Earth.

ACKNOWLEDGMENTS

Many thanks to Jeff Karson and David Naar for their extremely helpful reviews, much appreciated. Also thanks to Norm Sleep for pointing out some pertinent literature. The editors, Gillian Foulger and Donna Jurdy, were also very helpful in the preparation of the manuscript.

APPENDIX: MODEL PARAMETERS AND GEOMETRY

Model Type

Two-dimensional plane-strain elastic finite element model, program by Gobat and Atkinson (1996).

Model Parameters

Applied Forces: Applied forces are ridge-perpendicular and applied to the whole of the plate, based on the modeled age of the crust. As new ridges are propagated to the southwest from the north-south–trending ridges, new forces are applied perpendicular to the ridges that are now trending northwest-southeast. The basic applied forces are:

	Age (Ma)				
	0	10	20	30	40
Force applied (N/m)	2×10^{12}	1.23×10^{11}	9.17×10^{10}	7.1×10^{9}	5.1×10^{8}

Strength: The model consists of material of two end-member strengths, weak and strong. The strong material is three orders of magnitude stronger than the weak material. This relationship was based on the strong decoupling expected between ridge material (weak) and oceanic crustal material (strong). Other researchers (Richardson and Solomon, 1979) have demonstrated that the exact ratio is not important, as long as the ratio is greater than one. Because this model was testing end-member conditions, the transform was alternately modeled as both weak and strong. If the model of Richardson and Solomon (1979) holds true for transforms, the exact ratio will not matter, and our test models indicate this insensitivity to be the case.

Boundary Conditions

The model is fixed in space along its eastern edge. Numerous configurations were tested to determine the configuration with the least edge effects. By fixing the model on its eastern edge we allow the model as a whole to move while still providing a baseline to move against, similar to the apparent movement of the entire south Atlantic basin away from the pinned African plate.

REFERENCES CITED

Anderson, D.L., 1998, The scales of mantle convection: Tectonophysics, v. 284, p. 1–17, doi: 10.1016/S0040-1951(97)00169-8.

Antretter, M., Steinberger, B., Heider, F., and Soffel, H., 2002, Paleolatitudes of the Kerguelen hotspot: New paleomagnetic results and dynamic modeling: Earth and Planetary Science Letters, v. 203, p. 635–650, doi: 10.1016/S0012-821X(02)00841-5.

Bergman, E.A., and Solomon, S.C., 1992, On the strength of oceanic fracture zones and their influence on the intraplate stress field: Journal of Geophysical Research B, v. 97, p. 15,365–15,377.

Beutel, E.K., 2005, Stress induced seamount formation at ridge-transform-intersections, *in* Foulger, G.R., et al., eds., 2005, Plates, plumes, and paradigms: Boulder, Colorado, Geological Society of America Special Paper 388, p. 581–594, doi: 10.1130/2005.2388(34).

Campbell, I.H., 2001, The identification of ancient mantle plumes, *in* Ernst, R.E., and Buchan, K.L., eds., Mantle plumes: Their identification through time: Boulder, Colorado, Geological Society of America Special Paper 352, p. 5–21, doi: 10.1130/2005.2352(1).

Clift, P.D., 2005, Sedimentary evidence for moderate mantel temperature anomalies associated with hotspot volcanism, *in* Foulger, G.R., et al., eds., Plates, plumes, and paradigms: Boulder, Colorado, Geological Society of America Special Paper 388, p. 279–288, doi: 10.1130/2005.2388(17).

Coffin, M.F., Pringle, M.S., Duncan, R.A., Gladczenko, T.P., Storey, M., Mueller, R.D., and Gahagan, L.A., 2002, Kerguelen hotspot magma output since 130 Ma: Journal of Petrology, v. 43, p. 1121–1139, doi: 10.1093/petrology/43.7.1121.

Courtillot, V., Davaille, A., Besse, J., and Stock, J., 2003, Three distinct types of hotspots in the Earth's mantle: Earth and Planetary Science Letters, vol. 205, p. 295–308.

Craddock, J.P., Farris, D.W., and Roberson, A., 2004, Calcite-twinning constraints on stress-strain fields along the Mid-Atlantic ridge, Iceland: Geology, v. 32, p. 49–52, doi: 10.1130/G19905.1.

DePaolo, D.J., and Manga, M., 2003, Deep origin of hotspots: The mantle plume model: Science, v. 300, p. 920–921, doi: 10.1126/science.1083623.

Fairhead, J.D., and Wilson, M., 2005, Plate tectonic processes in the central, equatorial and southern Atlanic Ocean: Do we need deep mantle plumes,

in Foulger, G.R., et al., eds., Plates, plumes, and paradigms: Boulder, Colorado, Geological Society of America Special Paper 388, p. 537–554, doi: 10.1130/2005.2388(32).

Favela, J., and Anderson, D.L., 2000, Extensional tectonics and global volcanism, *in* Boschi, E., et al., eds., Problems in geophysics for the new millennium: Bologna, Italy, Editrice Compositori, p. 463–498.

Fujita, K., and Sleep, N.H., 1978, Membrane stresses near mid-ocean ridge-transform intersections: Tectonophysics, v. 50, p. 207–221, doi: 10.1016/0040-1951(78)90136-1.

Gobat, J., and Atkinson, D., 1996, FElt: San Diego, Univeristy of California, 12 p.

Graham, D.W., Johnson, K.T.M., Priebe, L.D., and Lupton, J.E., 1999, Hotspot-ridge interaction along the Southeast Indian ridge near Amsterdam and St. Paul islands: Helium isotope evidence: Earth and Planetary Science Letters, v. 167, p. 297–310, doi: 10.1016/S0012-821X(99)00030-8.

Gudmundsson, A., 1990, Emplacement of dikes, sills and crustal magma chambers at divergent plate boundaries: Tectonophysics, v. 176, p. 257–275, doi: 10.1016/0040-1951(90)90073-H.

Hekinian, R., Stoffers, P., Ackermand, D., Revillon, S., Maia, M., and Bohn, M., 1999, Ridge-hotspot interaction: The Pacific-Antarctic ridge and the Foundation seamounts: Marine Geology, v. 160, p. 199–233, doi: 10.1016/S0025-3227(99)00027-4.

Hieronymus, C.F., and Bercovici, D., 2000, Non-hotspot formation of volcanic chains: Control of tectonic and flexural stresses on magma transport: Earth and Planetary Science Letters, v. 181, p. 539–554, doi: 10.1016/S0012-821X(00)00227-2.

Johnson, K.T.M., Graham, D.W., Rubin, K.H., Nicolaysen, K., Scheirer, D.S., Forsyth, D.W., Baker, E.T., and Douglas-Priebe, L.M., 2000, Boomerang seamount; The active expression of the Amsterdam–St. Paul hotspot Southeast Indian ridge: Earth and Planetary Science Letters, v. 183, p. 245–259.

Kent, R.W., Pringle, M.S., Mueller, R.D., Saunders, A.D., and Ghose, N.C., 2002, $^{40}Ar/^{39}Ar$ geochronology of the Rajmahal basalts, India, and their relationship to the Kerguelen plateau: Journal of Petrology, v. 43, p. 1141–1153, doi: 10.1093/petrology/43.7.1141.

Kincaid, C., Schilling, J.G., and Gable, C., 1996, The dynamics of off-axis plume-ridge interaction in the uppermost mantle: Earth and Planetary Science Letters, v. 137, p. 29–43, doi: 10.1016/0012-821X(95)00201-M.

Klingelhöfer, F., Minshull, T.A., Blackman, D.K., Harben, P., and Childers, V., 2001, Crustal structure of Ascension Island from wide-angle seismic data: Implications for the formation of near-ridge volcanic islands: Earth and Planetary Science Letters, v. 190, p. 41–56, doi: 10.1016/S0012-821X(01)00362-4.

Lachenbruch, A.H., and Thompson, G.A., 1972, Oceanic ridges and transform faults: Their intersection angles and resistance to plate motion, Earth and Planetary Science Letters, vol. 15, p. 116–122.

McKenzie, D., and Bickle, M.J., 1988, The volume and composition of melt generated by extension of the lithosphere: Journal of Petrology, v. 29, p. 625–679.

McNutt, M., and Bonneville, A., 1999, A shallow, chemical origin for the Marquesas Swell: Geochemistry, Geophysics, Geosystems, v. 1, 17 p.

Meibom, A., and Anderson, D.L., 2004, The statistical upper mantle assemblage: Earth and Planetary Science Letters, v. 217, p. 123–139, doi: 10.1016/S0012-821X(03)00573-9.

Müller, R.D., Roest, W.R., Royer, J.V., Gahagan, L.M., and Sclater, J.G., 1997, Digital isochrons of the world's ocean floor: Journal of Geophysical Research, v. 102, p. 3211–3214, doi: 10.1029/96JB01781.

Natland, J.H., and Dick, H.J.B., 2001, Formation of the lower ocean crust and the crystallization of gabbroic cumulates at a very slowly spreading ridge: Journal of Volcanology and Geothermal Research, v. 110, p. 191–233, doi: 10.1016/S0377-0273(01)00211-6.

Natland, J.H., and Winterer, E.L., 2005, Fissure control on volcanic action in the Pacific, *in* Foulger, G.R., et al., eds., Plates, plumes, and paradigms: Boulder, Colorado, Geological Society of America Special Paper 388, p. 687–710, doi: 10.1130/2005.2388(39).

Neves, M.C., Bott, M.H.P., and Searle, R.C., 2004, Patterns of stress at mid-ocean ridges and their offsets due to sea floor subsidence: Tectonophysics, v. 386, p. 223–242, doi: 10.1016/j.tecto.2004.06.010.

Niu, Y., Bideau, D., Hekinian, R., and Batiza, R., 2001, Mantle compositional control on the extent of mantle melting, crust production, gravity anomaly, ridge morphology, and ridge segmentation: A case study at the Mid-Atlantic ridge 33–35 degrees N: Earth and Planetary Science Letters, v. 186, p. 383–399, doi: 10.1016/S0012-821X(01)00255-2.

O'Connor, J.M., Stoffers, P., van den Bogaard, P., and McWilliams, M., 1999, First seamount age evidence for a significantly slower African plate motion since 19 to 30 Ma: Earth and Planetary Science Letters, v. 171, p. 575–589, doi: 10.1016/S0012-821X(99)00183-1.

Phipps Morgan, J., and Parmentier, E.M., 1984, Lithospheric stress near a ridge-transform intersection: Geophysical Research Letters, v. 11, p. 113–116.

Raddick, M.J., Parmentier, E.M., and Scheirer, D.S., 2002, Buoyant decompression melting; A possible mechanism for intraplate volcanism: Journal of Geophysical Research B, v. 107, 14 p., doi: 10.1029/2001JB000617.

Ribe, N.M., 1996, The dynamics of plume-ridge interaction 2. Off-ridge plumes: Journal of Geophysical Research B, v. 101, p. 16,195–16,204, doi: 10.1029/96JB01187.

Richards, M.A., Duncan, R.A., and Courtillot, V.E., 1989, Flood basalts and hot-spot tracks: Plume heads and tails: Science, v. 246, no. 4926, p. 103–107.

Richardson, R.M., and Solomon, S.C., 1979, Tectonic stress in the plates: Reviews of Geophysics and Space Physics, v. 17, p. 981–1019.

Sager, W.W., 2005, What built Shatsky rise, a mantle plume or ridge tectonics?, *in* Foulger, G.R., et al., eds., Plates, plumes, and paradigms: Boulder, Colorado, Geological Society of America Special Paper 388, p. 721–734, doi: 10.1130/2005.2388(41).

Sandwell, D., and Fialko, Y., 2004, Warping and cracking of the Pacific plate by thermal contraction: Journal of Geophysical Research, v. 109, 12 p., doi: 10.1029/2004JB003091.

Shah, A.K., and Buck, W.R., 2003, Plate bending stresses at axial highs, and implications for faulting behavior: Earth and Planetary Science Letters, v. 211, p. 343–356, doi: 10.1016/S0012-821X(03)00187-0.

Silver, P.G., Behn, M.D., Kelley, K., Schmitz, M., and Savage, B., 2006, Understanding cratonic flood basalts: Earth and Planetary Science Letters, v. 245, p. 190–201, doi: 10.1016/j.epsl.2006.01.050.

Sleep, N.H., 1997, Lateral flow and ponding of starting plume material: Journal of Geophysical Research B, v. 102, p. 10,001–10,012, doi: 10.1029/97JB00551.

Sleep, N.H., 2002a, Local lithospheric relief associated with fracture zones and ponded plume material: Geochemistry, Geophysics, Geosystems, v. 3, 17 p.

Sleep, N.H., 2002b, Ridge-crossing mantle plumes and gaps in tracks: Geochemistry, Geophysics, Geosystems, v. 3, 33 p.

Small, C., 1995, Observations of ridge-hotspot interactions in the Southern Ocean: Journal of Geophysical Research, v. 100, p. 17,931–17,946, doi: 10.1029/95JB01377.

Smith, W.H.F., and Sandwell, D.T., 1997, Global sea floor topography from satellite altimetry and ship depth soundings: Science, v. 277, p. 1956–1962, doi: 10.1126/science.277.5334.1956.

Tentler, T., 2003, Analogue modeling of overlapping spreading centers; Insights into their propagation and coalescence: Tectonophysics, v. 376, p. 99–115, doi: 10.1016/j.tecto.2003.08.011.

van Wijk, J.W., and Blackman, D.K., 2005, Deformation of oceanic lithosphere near slow-spreading ridge discontinuities: Tectonophysics, v. 407, p. 211–225, doi: 10.1016/j.tecto.2005.08.009.

Vogt, P.R., and Jung, W.L., 2005, Paired basement ridges: Spreading axis migration across mantle heterogeneities, *in* Foulger, G.R., et al., eds., Plates, plumes, and paradigms: Boulder, Colorado, Geological Society of America Special Paper 388, p. 555–579, doi: 10.1130/2005.2388(33).

Weis, D., Frey, F.A., Schlich, R., Schaming, M., Montigny, R., Damasceno, D., Mattielli, N., Nicolaysen, K.E., and Scoates, J.S., 2002, Trace of the Kerguelen mantle plume; Evidence from seamounts between the Kerguelen Archipelago and Heard Island, Indian Ocean: Geochemistry, Geophysics, Geosystems, v. 3, 20 p.

West, B.P., Lin, J., and Christie, D.M., 1999, Forces driving ridge propagation: Journal of Geophysical Research B, v. 104, p. 22,845–22,858, doi: 10.1029/1999JB900154.

Wiens, D.A., and Stein, S.A., 1984, Intraplate seismicity and stresses in young oceanic lithosphere: Journal of Geophysical Research B, v. 89, p. 11,442–11,464.

Yaxley, G.M., 2000, Experimental study of the phase and melting relations of homogeneous basalt + peridotite mixtures and implication for the petrogenesis of flood basalts: Contributions to Minerology and Petrology, v. 139, p. 326–338, doi: 10.1007/s004100000134.

Zoback, M.L., 1992, First- and second-order patterns of stress in the lithosphere: The world stress map project: Journal of Geophysical Research B, v. 97, p. 11,703–11,728.

MANUSCRIPT ACCEPTED BY THE SOCIETY 31 JANUARY 2007

The Geological Society of America
Special Paper 430
2007

The OIB paradox

J. Godfrey Fitton*

School of GeoSciences, University of Edinburgh, Grant Institute,
West Mains Road, Edinburgh EH9 3JW, UK

ABSTRACT

Ocean island basalt (OIB) and OIB-like basalt are widespread in oceanic and continental settings and, contrary to popular belief, most occur in situations where mantle plumes cannot provide a plausible explanation. They are readily distinguished from normal mid-ocean ridge basalt (N-MORB) through ΔNb, a parameter that expresses the deviation from a reference line (ΔNb = 0) separating parallel Icelandic and N-MORB arrays on a logarithmic plot of Nb/Y versus Zr/Y. Icelandic basalts provide a useful reference set because (1) they are by definition both enriched mid-ocean ridge basalt (E-MORB) and OIB, and (2) they represent a larger range of mantle melt fractions than do intraplate OIBs. Virtually all N-MORB has ΔNb < 0, whereas all Icelandic basalts have ΔNb > 0. E-MORB with ΔNb > 0 is abundant on other sections of ridge, notably in the south Atlantic and south Indian oceans. E-MORB and N-MORB from this region form strongly bimodal populations in ΔNb, separated at ΔNb = 0, suggesting that mixing between their respective mantle sources is very limited. Most OIBs and basalts from many small seamounts, especially those formed on old lithosphere, also have ΔNb > 0. HIMU OIB (OIB with high $^{206}Pb/^{204}Pb$ values and therefore a high-μ [U/Pb] source) has higher ΔNb on average than does EM (enriched mantle) OIB, consistent with the presence of recycled continental crust (which has ΔNb < 0) in the EM source. Although EM OIBs tend to have the lowest values, most still have ΔNb > 0, suggesting that a relatively Nb-rich component (probably subducted ocean crust) is present in all OIB sources. The OIB source components seem to be present on all scales, from small streaks or blobs of enriched material (with positive ΔNb) carried in the upper-mantle convective flow and responsible for small ocean islands, some seamounts, and most E-MORB, to large mantle upwellings (plumes), inferred to be present beneath Hawaii, Iceland, Réunion, and Galápagos. It is not possible to identify a point on this continuum at which mantle plumes (if they exist) become involved, and it follows that OIB cannot be a diagnostic feature of plumes. The geochemical similarity of allegedly plume-related OIB and manifestly nonplume OIB is the first part of the OIB paradox. Continental intraplate transitional and alkali basalt in both rift and nonrift (e.g., Cameroon line) settings usually has positive ΔNb and is geochemically indistinguishable from OIB. Continental volcanic rift systems erupt OIB-like basalt, irrespective of whether they are apparently plume-driven (e.g., East Africa, Basin and Range), passive (e.g., Scottish Midland Valley) or somewhere between (e.g., North Sea basin). Magma erupted in passive rifts must have its source in the upper

*E-mail: Godfey.Fitton@ed.ac.uk.

Fitton, J.G., 2007, The OIB paradox, *in* Foulger, G.R., and Jurdy, D.M., eds., Plates, plumes, and planetary processes: Geological Society of America Special Paper 430, p. 387–412, doi: 10.1130/2007.2430(20). For permission to copy, contact editing@geosociety.org. ©2007 The Geological Society of America. All rights reserved.

mantle, and yet it is always OIB-like. N-MORB–like magma is only erupted when rifting progresses to continental break-up and the onset of seafloor spreading. Continental OIB-like magma is frequently erupted almost continuously in the same place on a moving lithospheric plate for tens of millions of years, suggesting that its source is coupled in some way to the plate, and yet the Cameroon line (where continental and oceanic basalts are geochemically indistinguishable) suggests that the source is sublithospheric. The causes and sources of continental OIB-like magma remain enigmatic and form the second part of the OIB paradox.

Keywords: intraplate magmatism, basalt, niobium, MORB, Iceland

INTRODUCTION

Plate tectonics can account for the volume and composition of magmas erupted at plate boundaries (mid-ocean ridges and above subduction zones) but not those erupted within plates. Intraplate magmatism is responsible for ocean islands and some continental volcanoes, and also frequently accompanies continental rifting. The observation that some ocean islands and seamounts form time-progressive chains led to the hypothesis that they are the products of convective plumes originating in the lower mantle (Morgan, 1971). Ocean island basalt (OIB) is chemically and isotopically distinct from basalt erupted at normal segments of mid-ocean ridges (normal mid-ocean ridge basalt; N-MORB), and this distinction is frequently cited as evidence that the two basalt types originate in different parts of the mantle. The N-MORB source has been depleted (compared to primitive mantle) in those elements that are incompatible in mantle phases, largely through the extraction of the continental crust. By contrast, the OIB source is consistently less depleted in incompatible elements, and the corresponding isotopic differences require ~1–2 b.y. to develop (e.g., Hofmann, 1997).

Ocean crust forms at mid-ocean ridges through the passive upwelling of the asthenosphere, and so the composition of N-MORB is taken to represent that of the upper mantle. OIB, on the other hand, is the characteristic basalt type erupted at hotspots and is often assumed to originate in lower mantle brought to the surface in mantle plumes. A hotspot, in this context, is any localized occurrence of anomalous (usually, but not always, intra-plate) magmatism not easily explained by plate tectonic processes. The common assumption that hotspots are synonymous with mantle plumes and that OIB is therefore diagnostic of plumes is clearly questionable (e.g., Anderson and Schramm, 2005). It is not easy, for example, to reconcile the large number of isolated islands and seamounts (all of them hotspots) with a mantle-plume origin. Wessel and Lyons (1997) estimate that there are around 70,000 seamounts with heights >1 km in the Pacific Ocean alone, and most of these are not aligned in chains. A significant proportion of these seamounts are likely to be made of OIB (e.g., Niu and Batiza, 1997). OIB-like basalt is frequently erupted in continental rift systems, and this fact is also difficult to explain through mantle plumes, especially when rifting appears to be a passive response to plate stresses.

Hypotheses for OIB formation fall into three categories: (1) mantle plumes; (2) dispersed blobs or streaks of incompatible-element–enriched material in the depleted upper mantle; and (3) a layer of shallow mantle (the perisphere; Anderson, 1995) that is enriched in incompatible elements compared to the deeper parts of the upper mantle. In the plume hypothesis, subducted and dehydrated ocean crust is stored in the deep mantle and eventually returned to the surface in mantle plumes (Hofmann and White, 1982). The core-mantle boundary, the 660-km discontinuity, and discontinuities within the lower mantle have all been suggested as possible storage sites. The resulting mixture of peridotite and eclogite or pyroxenite melts on decompression, because it is both hot and fertile. In earlier versions of the plume hypothesis, OIBs were thought to be generated from the melting of primitive (undepleted) mantle, but the wide range of OIB compositions makes this untenable (see the review of this subject by Hofmann, 1997).

The second (blobs or streaks) hypothesis also appeals to mantle enrichment through the addition of subducted ocean crust, but here the enriched components are dispersed in a depleted matrix to form the convecting upper mantle. Small degrees of melting will tend to sample the enriched and more easily fusible parts preferentially, leading to the formation of OIB, whereas higher-degree melting under mid-ocean ridges will homogenize the mixture and produce more depleted (N-MORB) magma (e.g., Sleep, 1984; Fitton and James, 1986; Meibom and Anderson, 2004; Ito and Mahoney, 2005).

The Anderson (1995) perisphere hypothesis explains OIB as the product of melting of a global, weak, enriched layer that lies immediately beneath the lithosphere. Seafloor spreading drags this layer aside, allowing deeper and more-depleted upper mantle to rise to the surface under mid-ocean ridges.

The objective of this article is to address the question of the nature and whereabouts of the OIB source(s) through a survey of the scale and distribution of OIB and OIB-like basalt. To do this, it is necessary to establish reliable quantitative criteria for discriminating between OIB and N-MORB and then to apply these criteria to basalt erupted in oceanic and continental settings. The approach used is that developed by Fitton et al. (1997) to distinguish Icelandic basalt from N-MORB and uses Nb, Zr, and Y data.

DATA SOURCES

The usefulness of the discrimination technique applied here is limited by analytical consistency in the data used. Except where otherwise stated, the data used in this study were produced by X-ray fluorescence (XRF) spectrometry in the XRF laboratory, Edinburgh University, using analytical conditions described in Fitton et al. (1998). The concentration of Nb is particularly critical and was determined with long count times on samples with Nb contents <5 ppm. Analytical precision in these samples is ± 0.1 ppm. Older Edinburgh data were corrected for subsequent changes in the recommended values for international reference standards. Most of the mid-ocean ridge data were taken from the literature but, because of the low concentrations of Nb in N-MORB, only recent (mostly post-1990) data were used, and then only where analytical results for international reference standards were given. Where necessary, the analyses were adjusted to bring them into line with standard values used in Edinburgh. Typical values (in ppm) obtained on four basalt reference standards in the Edinburgh XRF laboratory are given in Table 1. Hitherto unpublished data used in this study are available from the GSA Data Repository.[1]

OCEANIC VOLCANISM

Iceland

Iceland sits on the Mid-Atlantic Ridge and, because it is both an ocean island and an anomalous segment of mid-ocean ridge, it provides a link between OIB and MORB. It is widely regarded as the best example of plume-ridge interaction, although Foulger et al. (2005a,b) have proposed a nonplume explanation in which the excess magmatism results from the melting of upper mantle that has been fertilized through the addition of ancient subducted oceanic crust trapped in the Caledonian suture. Variation in Icelandic basalt composition has long been ascribed to mixing between depleted N-MORB and enriched plume-derived OIB magma (e.g., Hanan and Schilling, 1997). Correlated chemical and isotopic variation in Icelandic basalt requires a heterogeneous mantle source with depleted and enriched components (Hémond et al., 1993; Thirlwall et al., 2004), but there is strong Pb-isotope and trace-element evidence that the depleted component is not derived from the ambient upper mantle but instead forms an intrinsic part of the Iceland plume (Thirl-

[1]GSA Data Repository item 2007092, an Excel file containing data used in this article, is available at www.geosociety.org/pubs/ft2007.htm, or on request from editing@geosociety.org, Documents Secretary, GSA, P.O. Box 9140, Boulder, CO 80301, USA.

TABLE 1. TYPICAL ANALYSES OF REFERENCE STANDARDS OBTAINED IN THE EDINBURGH UNIVERSITY XRF LABORATORY

Standard	Nb	Zr	Y
BIR-1	0.6	16.2	16.1
BCR-1	13.0	192	38.4
BHVO-1	19.8	175	27.4
BE-N	116	268	30.4

Note: Data are in ppm.

wall, 1995; Fitton et al., 1997, 2003). Thus, Iceland provides a unique area in which to study the nature of the OIB source. Iceland's location allows the mantle source to be sampled through large-degree melting beneath the rift axes and smaller-degree melting off-axis, in contrast to most other ocean islands, where only small-degree melts are produced.

Figure 1 shows the compositional range of Icelandic basalt on a logarithmic plot of Nb/Y versus Zr/Y. The Icelandic data define a linear array, with incompatible-element–depleted basalt at the low-Zr/Y end and enriched transitional to alkaline basalt at the other end. The most-enriched basalts are from off-axis locations and from the propagating tip of the eastern rift zone; the rest of the samples are from the actively spreading rift zones, with the most-depleted rocks being picrites from the Reykjanes peninsula and the northern rift zone. Fitton et al. (1997) showed that the Iceland array can be modeled by variable degrees of fractional melting of a heterogeneous mantle source. MORB sampled at locations away from the influence of hotspots (N-MORB) plots on a parallel array at lower Nb/Y and clearly cannot be the depleted end-member in the Iceland array. As with the Iceland array, the N-MORB array also reflects degree of melting. The smallest-degree melts (highest Zr/Y) are from the slow-spreading southwest Indian (Robinson et al., 2001) and Gakkel (Muhe et al., 1997) ridges. The highest-degree melts (lowest Zr/Y) are from 57.5°–61°N on the Reykjanes Ridge, south of the Iceland geochemical anomaly, where hotter-than-normal mantle has resulted in oceanic crust 8–10 km thick (Smallwood et al., 1995).

It is clear from Figure 1 that both the depleted and enriched source components of Icelandic basalt must be enriched in Nb compared to the N-MORB source. This relative enrichment is a characteristic feature of OIB, which tends to have positive Nb anomalies on primitive-mantle–normalized diagrams (Hofmann, 1997) and is the reason why relative Nb (and Ta) abundances in basalt have been used to discriminate between ancient tectonic environments (e.g., Meschede, 1986). The abundance of most elements in the depleted upper mantle (the source of N-MORB) can be modeled adequately by mass-balance calculations in which average continental crust is subtracted from primitive mantle (Hofmann, 1988). This procedure, however, fails to account for the low abundance of Nb in the upper mantle, because

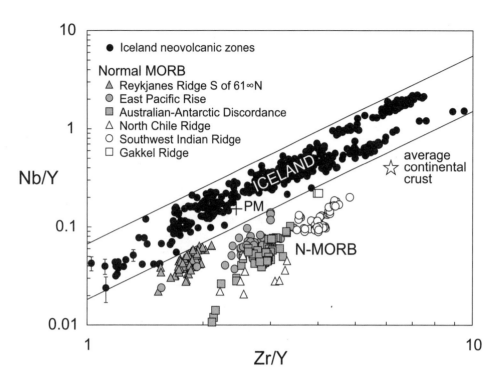

Figure 1. Nb/Y and Zr/Y variation for Icelandic basalt (MgO >5 wt.%) and normal mid-ocean ridge basalt (N-MORB). Data sources for N-MORB are given in Fitton et al. (1997) with additional data from the Australian-Antarctic discordance (Kempton et al., 2002), north Chile Ridge (Bach et al., 1996), and Gakkel Ridge (Muhe et al., 1997). The + symbol represents an estimate of primitive mantle (PM) composition (McDonough and Sun, 1995). The average crust composition is from Rudnick and Fountain (1995) and Barth et al. (2000). The southwest Indian Ridge data can be found in GSA Data Repository (see footnote 1). After Fitton et al. (1997, 2003).

both continental crust and N-MORB are depleted in Nb compared to similarly incompatible elements. For example, both the continental crust and N-MORB have lower Nb/La than does primitive mantle (Rudnick et al., 2000; Hofmann, 2004).

Figure 1 shows clearly the relative deficiency in Nb in N-MORB and continental crust; the composition of primitive mantle (PM) lies within the Iceland array, whereas both N-MORB and average continental crust plot below it. The missing Nb is probably stored in subducted and dehydrated ocean crust, possibly in rutile, and ultimately recycled as the OIB source (Hofmann, 1997; Rudnick et al., 2000). The observation that both N-MORB and average continental crust are deficient in Nb appears to be at odds with the observation that N-MORB is not depleted in Nb compared with U, which has a similar bulk partition coefficient during mantle melting (Hofmann et al., 1986; Hofmann, 2004). Both N-MORB and OIB have Nb/U ≈50 (Hofmann, 2004), a value that is significantly higher than the chondritic ratio of 30, whereas the average continental crust has subchondritic Nb/U (5.6; Barth et al., 2000). The conclusions drawn in the present article, however, are based on the empirical observation that Icelandic basalt and N-MORB form distinct arrays on a logarithmic plot of Nb/Y versus Zr/Y and are therefore unaffected by the Nb/U problem.

The parallel lines on Figure 1 mark the upper and lower limits of the Iceland data array. The lower of these separates the Iceland and N-MORB data and, as will be shown, is a useful reference line when discussing the global distribution of N-MORB and OIB. Fitton et al. (1997) defined a parameter, ΔNb, that expresses the excess or deficiency in Nb relative to this line:

$$\Delta Nb = 1.74 + \log(Nb/Y) - 1.92 \log(Zr/Y).$$

ΔNb is unaffected by degree of melting, which controls the position of a sample along the array but not across it (Fitton et al., 1997). Otherwise, arrays of data representing variable degrees of melting (such as the Iceland array) would not be linear. Vertical position within the array is a function of the relative Nb abundance in the source. ΔNb is also unaffected by fractional crystallization of olivine and plagioclase, because Nb, Zr, and Y are highly incompatible in these phases. This is not entirely true of augite, in which Y is moderately, and Zr slightly, compatible. Consequently, ΔNb falls slightly during prolonged fractional crystallization of augite and so, to eliminate this effect, rock samples with <5 wt% MgO have been excluded from this study. In the following sections, data from a variety of mid-ocean ridge and intraplate settings are plotted on the Nb/Y-Zr/Y diagram and compared with the Iceland array. Data from Iceland are a useful reference set because Icelandic basalt is generated over a much wider melting range than are OIBs elsewhere.

N-MORB and E-MORB

Iceland, being part of the Mid-Atlantic Ridge, is composed of MORB, but Icelandic basalt is more enriched in highly incompatible elements than is N-MORB and is also distinct in its radiogenic isotope ratios. Such basalts are described as enriched MORB (E-MORB). Iceland represents by far the world's most voluminous occurrence of E-MORB, and the excess magmatism there is generally ascribed to plume-ridge interaction. Other examples of anomalous ridge segments composed of E-MORB, though much smaller in volume, tend to be topographically higher than normal segments and are often located close to ocean islands or large seamounts. For this reason they are also often assumed to result from plume-ridge interaction (e.g., Douglass et al., 1999), even though the case for plumes is far less compelling than in the example of Iceland.

There are no universally accepted geochemical criteria in the literature for defining N- and E-MORB. Samples tend to be assigned to one type or the other on an ad hoc basis, and those that are transitional between the two are sometimes described as T-MORB (e.g., Mahoney et al., 1994). The distinction between N- and E-MORB is generally based on enrichment in K and other highly incompatible elements relative to moderately incompatible elements, such as Ti or P. Thus, Mahoney et al. (1994) define T-MORB as MORBs with Rb/Nd $\geq$ 0.15, whereas Hall et al. (2006) refer to samples with $K_2O/TiO_2 \geq 0.1$ as E-MORB and comment that this is roughly equivalent to the Mahoney et al. (1994) definition of T-MORB. Niu et al. (2002) use K/Ti > 0.11 to define E-MORB. Other authors use relative depletion in light rare-earth elements (LREE) to discriminate between N- and E-MORB. Mahoney et al. (2002), for example, use chondrite-normalized La/Sm, (La/Sm)n > 0.8 to define E-MORB but emphasize that a continuum of compositions exists between N- and E-MORB. All of the criteria that have been used suffer from the same inability to distinguish between the effects of source enrichment and degree of melting. A small melt fraction from a depleted (low-K/Ti) source could have the same K/Ti as a large melt fraction from an enriched (high-K/Ti) source. La/Sm is likewise sensitive to degree of melting, though to a lesser extent than is K/Ti. ΔNb is insensitive to degree of melting and, in the following discussion, I show that it is a more useful discriminant between N- and E-MORB because it reflects source composition alone.

To assess the similarity between Icelandic basalt and much less voluminous E-MORB, I use data from the southern Mid-Atlantic and southwest and central Indian ridges. These ridges form a continuous spreading center around the south of Africa and are noted for the abundance of E-MORB. Figure 2 shows the distribution and composition of dredged MORB samples, color-coded for ΔNb (positive in red, negative in yellow). Samples represented by red points plot with Icelandic basalt (E-MORB) and the yellow points with N-MORB. Note that the samples with positive ΔNb tend to cluster in regions of anomalously shallow ridge, close to ocean islands and large seamounts.

The data points on the Nb/Y-Zr/Y plots in Figure 2 appear to form clusters separated by the reference line (ΔNb = 0), which is confirmed by the histogram of ΔNb for the whole data set. The basalt population is strongly bimodal, with positive and negative peaks separated by a trough centered on ΔNb = 0. This is an important observation, because it suggests that the Iceland reference line is of global significance. Furthermore, E-MORB and N-MORB form discrete populations rather than a continuum, and there appears to be limited mixing between the E-MORB

and N-MORB mantle sources. That the E-MORB data plot in the Iceland array (ΔNb > 0) suggests that the E-MORB source has an excess of Nb similar to that in the source of Icelandic basalt. Recycled subducted ocean crust seems to be involved in both, but this need not imply the involvement of mantle plumes in the generation of E-MORB, as is often assumed (e.g., Douglass et al., 1999). Passive blobs or streaks of recycled crust carried around in the upper-mantle flow could equally well explain the distribution and composition of E-MORB in Figure 2.

Figure 3 shows ΔNb for the samples used in Figure 2, plotted against (La/Sm)n (Sm/Nd normalized to the ratio in chondrite meteorites or primitive mantle). (La/Sm)n is a measure of relative enrichment or depletion of LREE on chondrite-normalized REE patterns and is often used to discriminate between N- and E-MORB. The data form two distinct clusters, separated at ΔNb $\approx$ 0 and (La/Sm)n $\approx$ 0.8. Mahoney et al. (2002) use (La/Sm)n = 0.8 to distinguish between N- and E-MORB on the Southeast Indian Ridge and note that the two types form a continuum. The histogram in Figure 3 shows no bimodality in (La/Sm)n, in contrast to ΔNb, which is strongly bimodal (Fig. 2). Thus ΔNb is a much better discriminant between N- and E-MORB than any parameter used previously, probably because it reflects fundamental differences in their mantle sources and is insensitive to degree or depth of melting.

E-MORB tends to occur on elevated ridge segments close to ocean islands or large seamounts, as in the southern Mid-Atlantic and southwest and central Indian ridges (Fig. 2), but this association does not always hold. The Southeast Indian Ridge, for example, is remote from ocean islands (Fig. 4) and composed mostly of N-MORB, but E-MORB is locally abundant, especially in the eastern part, where the water depth is greater (Mahoney et al., 2002; Fig. 4). Basalt samples dredged from the Southeast Indian Ridge appear to form a continuum on a logarithmic plot of Nb/Y versus Zr/Y, and the distribution of ΔNb is only weakly bimodal (Fig. 4), in contrast to the strongly bimodal distribution seen in data from the southern Mid-Atlantic and southwest and central Indian ridges (Fig. 2). This weak bimodality suggests that mantle is heterogeneous on a smaller scale beneath the Southeast Indian Ridge and that the enriched domains are less able to dominate the melt zone. This was also noted by Mahoney et al. (2002), who commented that N- and E-MORB samples were often retrieved in the same dredge haul.

Ocean Islands

There is a tendency in the literature to equate ocean islands with mantle plumes. This seems not unreasonable in the case of those islands (e.g., Hawaii, Réunion) that lie on the ends of long, time-progressive chains of islands and seamounts, but is debatable for many if not most islands. The composition of OIB is shown in Figure 5. The OIB data set includes samples from most major islands and island groups and therefore reflects the range of OIB composition. Data from the Galápagos Islands are not

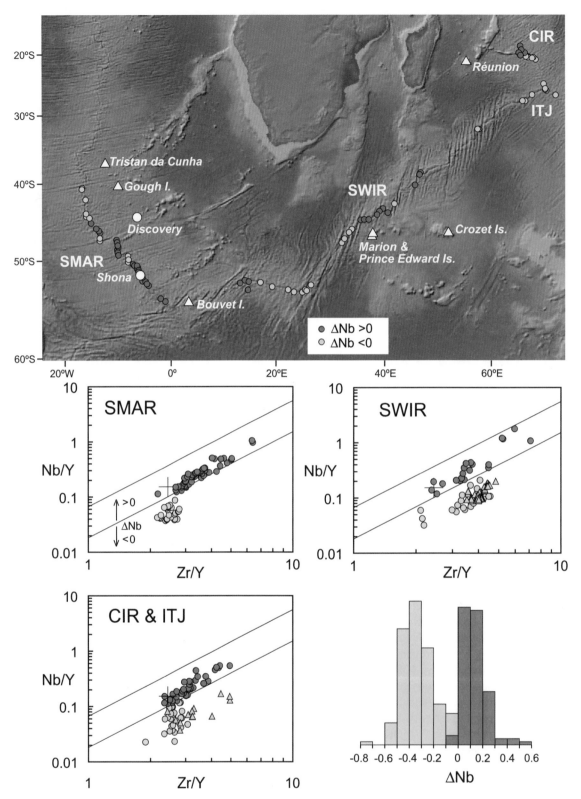

Figure 2. Map of the south Atlantic and south Indian oceans showing the location of mid-ocean ridge basalt samples for which reliable Nb, Zr, and Y data are available. These data are shown on the three plots below the map. SMAR—south Mid-Atlantic Ridge (data from le Roux et al., 2002). SWIR—Southwest Indian Ridge (circles: data from Janney et al., 2005; triangles: unpublished data from 57°E and 66°E, GSA Data Repository [see footnote 1]). CIR—Central Indian Ridge (circles: data from Murton et al., 2005, and Nauret et al., 2006). ITJ—Indian Ocean triple junction (triangles: data from Price et al., 1986, and Chauvel and Blichert-Toft, 2001). Parallel lines on the plots mark the limits of the Iceland array shown in Figure 1; the + symbol represents primitive mantle. ΔNb is the deviation, in log units, from the reference line separating the Iceland array from N-MORB (the lower of the two parallel lines). The points on the map and the data points are color coded for positive (red) and negative (yellow) values of ΔNb. Three SMAR samples with slightly negative ΔNb (–0.02 or greater) are included with the positive group, because they are clearly different from the negative samples. The histogram shows the distribution of ΔNb in the whole data set (254 analyses). Note that the samples with ΔNb > 0 are located on anomalously shallow segments of ridge close to ocean islands or large seamounts. The location of the Discovery and Shona "plumes" is from Douglass et al. (1999).

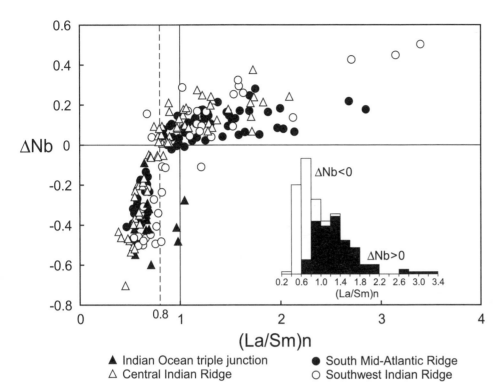

Figure 3. ΔNb plotted against (La/Sm)n for the same basalt samples used in Figure 2 (data sources as in Fig. 2). (La/Sm)n is primitive-mantle–normalized La/Sm (normalizing values from McDonough and Sun, 1995). The vertical line at (La/Sm)n = 1 separates light rare earth element (LREE)–depleted samples (<1) from LREE-enriched samples (>1); (La/Sm)n = 0.8 (broken vertical line) is the value used by Mahoney et al. (2002) to separate normal and enriched mid-ocean ridge basalts on the Southeast Indian Ridge. Note that the data form two distinct clusters separated by ΔNb ≈ 0 and (La/Sm)n ≈ 0.8. The inset histogram shows the distribution of (La/Sm)n in the data set (202 analyses); samples with ΔNb < 0 (white) and ΔNb > 0 (black). (La/Sm)n does not separate the two clusters as effectively as does ΔNb (Fig. 2).

included in the data set because, as will be shown later, these islands are unique and not typical OIB. Most of the data plot within and at the high-Zr/Y end of the Iceland array, implying that OIB and Icelandic basalt have similar mantle sources but that the OIB source has generally been melted to a smaller degree than has the mantle under Iceland. This observation is consistent with Iceland being located on a mid-ocean ridge, whereas most ocean islands form over thick lithosphere.

The similarity between OIB and Icelandic basalt is shown in the histograms of ΔNb values in Figure 5. The two populations have very similar distributions, but the OIB data extend outside the Iceland array to higher and lower values. Of the fifty-three samples (out of a total of 768) with negative ΔNb, twenty-four are from the Hawaiian Islands, seven are from Tahiti, five from Kerguelen, four from the Caroline Islands, and four from the Cameroon line. There is no obvious relationship between ΔNb and size of island, nor with whether or not it forms part of a time-progressive chain. The Hawaiian Islands, for example, span the whole range of ΔNb (Fig. 6). All of the samples from the Hawaiian Islands that plot around the lower (reference) line (ΔNb = 0) are from the shield-forming stage of their respective islands, whereas those plotting around the upper line are from the rejuvenated stage, as has been noted by Frey et al. (2005). Figure 6 also shows data from the Emperor seamounts (from Regelous et al., 2003). These data, representing the composition of the Hawaiian mantle source from 85 to 42 Ma, plot around the lower line, along with the shield-forming lavas from the islands, suggesting that these values reflect the long-term composition of the Hawaiian plume (Frey et al., 2005).

Zindler and Hart (1986) used radiogenic isotope ratios (Sr, Nd, and Pb) to identify three extreme OIB types (HIMU, which is OIB with high $^{206}Pb/^{204}Pb$ values—and therefore a high-μ [U/Pb] source; enriched mantle [EM]-1; and EM-2) representing mantle reservoirs thought to originate, respectively, through the recycling of subducted ocean crust, lower continental crust, and upper continental crust (or marine sediment). Mixing between these end-members and the depleted MORB-source mantle (DMM) can, in principle, account for the isotopic variation seen in all oceanic basalts. When plotted in $^{87}Sr/^{86}Sr$–$^{143}Nd/^{144}Nd$–$^{206}Pb/^{204}Pb$ space, data from individual islands or island groups form arrays that suggest mixing between the various mantle reservoirs. DMM is rarely at the end of these arrays, however, and they fan out instead from a common or focal zone, which Hart et al. (1992) proposed as a fifth mantle reservoir (FOZO) residing in the lower mantle or at the core-mantle boundary. Stracke et al. (2005) noted that OIBs with the most radiogenic lead isotope ratios fall into two groups: HIMU types and those close to FOZO in isotopic composition. They argued that both the FOZO and HIMU reservoirs represent recycled subducted ocean crust, but that the rare HIMU type is not a common mixing end-member in other OIB arrays and MORB. Their redefined FOZO, however, is a ubiquitous component common to both OIB and MORB.

Variation of ΔNb in OIB correlates significantly with the isotopic character of the respective island or island group. Willbold and Stracke (2006) have compiled a set of trace-element data representing the three extreme OIB types and have used these, in conjunction with isotope data, to argue that all OIB

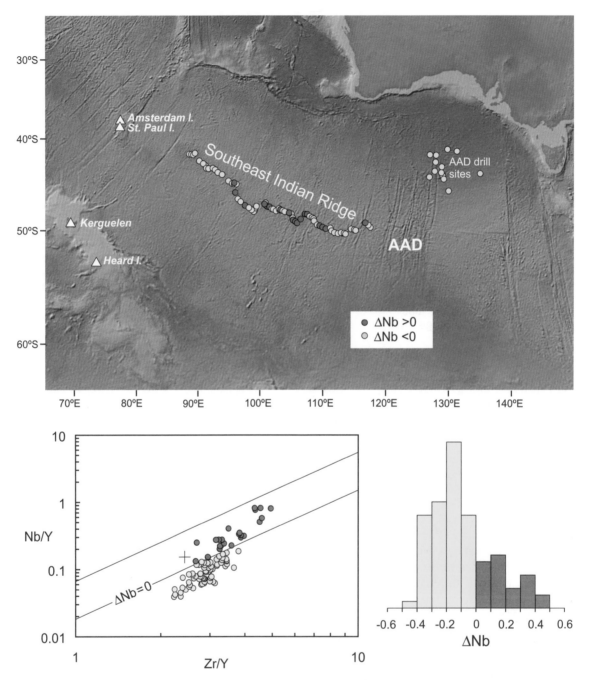

Figure 4. Map of the Southeast Indian Ridge showing the distribution and composition of dredged samples. Nb data from Mahoney et al. (2002); Zr and Y data from D.M. Christie (unpubl. data). All the data are available in the GSA Data Repository (see footnote 1). The location and composition (Kempton et al., 2002) of samples from the Australian-Antarctic discordance (AAD) drill sites (Fig. 1) are also shown, but these data are not included in the plots in this figure. Parallel lines on the plots mark the limits of the Iceland array shown in Figure 1; the + symbol represents primitive mantle. ΔNb is the deviation, in log units, from the reference line separating the Iceland array from N-MORB (the lower of the two parallel lines). The points on the map and the data points are color coded for positive (red) and negative (yellow) values of ΔNb. The histogram shows the distribution of ΔNb in the data set (ninety-five samples).

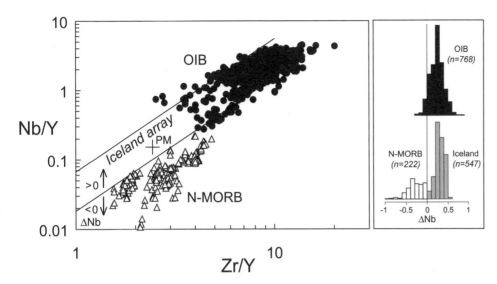

Figure 5. Comparison of ocean island basalt (OIB) data (J.G. Fitton and D. James, GSA Data Repository, see footnote 1) with the Iceland array and normal mid-ocean ridge basalt (N-MORB) composition (Fig. 1). The OIB data set includes samples from most major ocean islands and island groups but excludes data from the Galápagos Islands. PM is the composition of primitive mantle (McDonough and Sun, 1995). The histogram compares the distribution of ΔNb in OIB, Icelandic basalt, and N-MORB.

sources contain subducted ocean crust (the redefined FOZO of Stracke et al., 2005), and that the two EM types contain additional components derived from the continental crust. Figure 7, based on the Willbold and Stracke (2006) data set, shows histograms of ΔNb in OIB representing the three extreme types. HIMU OIB has generally higher ΔNb and, apart from two samples with extreme values, a more restricted range than EM-1 and EM-2. This characteristic is consistent with the HIMU component being subducted and dehydrated ocean crust, because such crust is likely to have excess Nb retained in rutile-bearing eclogite (Rudnick et al., 2000). Recycled continental crust will have

negative ΔNb (Fig. 1), and addition of this component will therefore lower the ΔNb in OIB and account for the slightly negative values in Figure 5. The positive values of ΔNb in most EM-type OIBs is, however, consistent with the dominance of subducted oceanic crust in their source.

From the histograms in Figure 5 it is clear that ΔNb is an excellent discriminant between N-MORB and OIB; at least as good any other geochemical parameter, including isotope ratios. The near-normal distribution of ΔNb in OIB suggests that little, if any, mixing takes place between the OIB source and the ambient upper mantle, as was also noted in the case of E-MORB

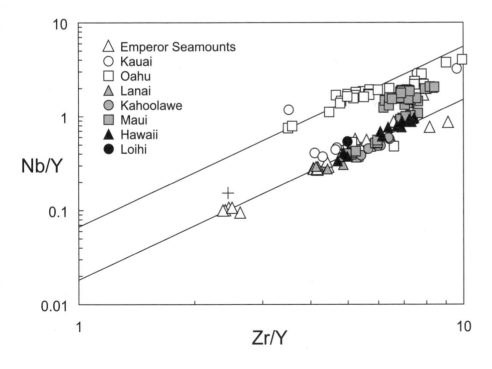

Figure 6. Composition of basalt samples from the Hawaiian Islands (J.G. Fitton, GSA Data Repository, see footnote 1) and Emperor seamounts (data from Regelous et al., 2003). The two parallel lines mark the limits of the Iceland array; the + represents the composition of primitive mantle (see Fig. 1). Hawaiian basalts plotting on the lower line (ΔNb = 0) are from the shield-forming stage of their respective islands; those plotting on the upper line are from the rejuvenated stage (Frey et al., 2005). The Emperor seamount data, representing the composition of the Hawaiian mantle source from 85 to 42 Ma, plot around the lower line with the shield-forming lavas from the islands.

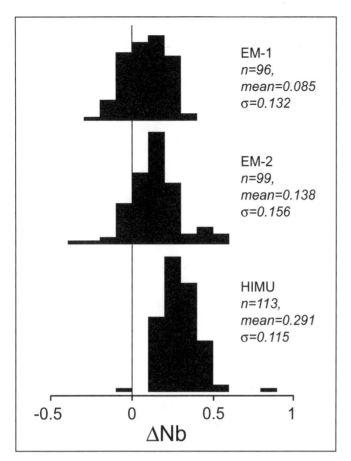

EM-1
n=96,
mean=0.085
σ=0.132

EM-2
n=99,
mean=0.138
σ=0.156

HIMU
n=113,
mean=0.291
σ=0.115

-0.5   0   0.5   1

ΔNb

Figure 7. Histograms showing the distribution of ΔNb in extreme examples of high-μ (HIMU), enriched mantle (EM)-1, and EM-2 ocean island basalt. Data from Willbold and Stracke (2006), filtered to exclude samples with MgO < 5 wt%.

and N-MORB. A similar conclusion was reached by Hart et al. (1992) on the basis of radiogenic isotope ratios in OIB. As was noted earlier, these authors showed that many OIBs plot on sublinear arrays in Sr-Nd-Pb isotopic space, but the depleted upper mantle is rarely a mixing end-member of these arrays. Only basalts from Iceland and the Galápagos Islands seemed to show mixing with the upper mantle, and both of these sit on or close to mid-ocean ridges. The data in Figure 1 show little, if any, mixing between Icelandic basalt and N-MORB, but what about the Galápagos Islands?

The Galápagos Islands formed through the interaction of a hotspot with the Galápagos spreading center, currently to the north of the archipelago. Relative motion of the hotspot and spreading center over the past 20 Ma has resulted in hotspot magmas being erupted preferentially on the Cocos plate to the north and the Nazca plate to the south at different times (Werner et al., 2003). Proximity of hotspot and spreading center has led to the formation of the Cocos and Carnegie ridges, respectively north and south of the spreading center. Because of its long history of magmatism and association with time-progressive aseis-

mic ridges, the Galápagos hotspot is generally interpreted as due to a mantle plume.

Geochemical studies on basalt from the Galápagos Islands (Geist et al., 1988; White et al., 1993) suggest that the plume has a depleted, N-MORB–like, core partly surrounded on the north, west, and south by a horseshoe-shaped OIB-like outer zone. Mixing between these two end-member components can account for the chemical and isotopic variation seen in Galápagos basalts. Geist et al. (1988) propose that the depleted core of the plume is ambient upper mantle entrained during ascent. If the depleted basalts are the products of N-MORB–source upper mantle, then they should have negative ΔNb. Apparently similar depleted basalts in Iceland have also been interpreted as the result of mixing with ambient upper mantle (Schilling et al., 1982), but these basalts have positive ΔNb (Fig. 1) and therefore more likely represent melts from an intrinsic depleted component of the Iceland plume (Fitton et al., 2003).

The suite of Galápagos samples used by White et al. (1993) was analyzed for Nb, Zr, and Y at the XRF laboratory, Edinburgh University, and the results plotted in Figure 8A, along with data from the Cocos and Carnegie ridges (from Harpp et al., 2005) and East Pacific Rise N-MORB (from Mahoney et al., 1994). The Galápagos samples show a wide scatter extending to negative ΔNb, unlike basalts from Iceland (Fig. 1). Figure 8B shows that ΔNb correlates negatively with $^{143}Nd/^{144}Nd$, as would be expected from mixing between the Galápagos plume and ambient upper (N-MORB–source) mantle represented by the East Pacific Rise data. Interestingly, the Cocos and Carnegie ridge data plot almost entirely within the Iceland array in Figure 8A, suggesting that mixing between plume and upper mantle is a recent phenomenon. Figure 9A shows the location of samples on the Cocos and Carnegie ridges, color coded for ΔNb. All the samples dredged from the ridges have ΔNb > 0 (as in Iceland); the few samples with ΔNb < 0 were dredged from the seafloor adjacent to the ridges. Figure 9B is an enlargement of part of Figure 9A showing ΔNb for each of the Galápagos Islands. The zonation noted by Geist et al. (1988) is clearly reflected in ΔNb, with an N-MORB–like central area surrounded to the north, west, and south by a zone with ΔNb > 0.

The Galápagos Islands seem to be unique among ocean islands in the involvement of depleted upper mantle in their formation. No other island or island group contains basalt with such strongly negative ΔNb (Fig. 5; Galápagos data were not included in the OIB compilation). Basalt from the shield-forming stage on the Hawaiian Islands has ΔNb ≈ 0 ± 0.1 (Fig. 6), but isotopic data show that these low values cannot be due to the involvement of the N-MORB source. The rejuvenated-stage basalts (with strongly positive ΔNb) have radiogenic-isotope ratios closer to those in Pacific N-MORB than do the shield-forming basalts (Frey et al., 2005). Blichert-Toft et al. (1999) have shown that Hf-isotope ratios in Hawaiian shield-forming basalts identify a recycled pelagic sediment component in their mantle source, which is consistent with low ΔNb in these basalts (Frey et al., 2005).

Seamounts

The largest ocean islands, such as Iceland, Hawaii, Réunion, and the Galápagos Islands, are associated with time-progressive aseismic ridges or seamount chains that often extend back in time to link with large igneous provinces. A mantle-plume origin provides the most plausible explanation for such islands. At the other end of the scale are small islands and seamounts with no obvious connection to mantle plumes. Figure 5 includes many small islands, and these share with the larger islands the OIB characteristic of positive ΔNb, but no seamount data are included in the compilation.

A geochemical study carried out by Niu and Batiza (1997) on samples dredged from small seamounts on the flanks of the East Pacific Rise provides ideal data for the present survey of OIB distribution. Niu and Batiza (1997) showed that the seamounts have an extraordinary range in composition, from highly depleted N-MORB to enriched alkali basalt, and conclude that their mantle source must be heterogeneous on a small scale. Niu and Batiza (1997) and Niu et al. (2002) argue that the region must be underlain by a two-component mantle with an enriched and easily melted component dispersed as physically distinct domains in a more depleted and refractory matrix. The enriched component, they suggest, is recycled subducted ocean lithosphere (crust and metasomatically enriched lower lithospheric mantle).

Figure 10 shows that seamount data from Niu and Batiza (1997) plot in distinct N-MORB and OIB (or E-MORB) groups, separated almost perfectly by the $\Delta Nb = 0$ reference line and with a weakly bimodal distribution of ΔNb. About 20% of the data plot in the OIB group. The accompanying map shows that OIB-type seamounts are interspersed with N-MORB–type seamounts and show no systematic geographical distribution. Niu and Batiza (1997) note that some individual seamounts contain both OIB- and N-MORB–type basalt (Fig. 10) and therefore require a mantle source that is heterogeneous on a scale of several hundred meters.

The northern East Pacific Rise seamounts studied by Niu and Batiza (1997) were erupted on very young ocean crust, close to the ridge axis, and are therefore part of the ridge itself. Many (possibly most) seamounts, however, formed farther from ridge axes, and data from examples of these are provided by the Hall et al. (2006) study of the Rano Rahi seamount field off the southern East Pacific Rise. The distribution and composition of these seamounts and of samples dredged from the spreading axes in the region are shown in Figure 11. It is clear that N-MORB dominates both the axes and the seamounts. Mahoney et al. (1994) identify all but six of the southern East Pacific Rise samples as N-MORB on the basis of their definition of T-MORB (Rb/Nd > 0.15), but only one sample has $\Delta Nb > 0$ (Fig. 11). Of the eight samples with $\Delta Nb > 0$ from the spreading axes around the Easter microplate, seven are from segments adjacent to Easter Island. Four of the thirty samples from the Rano Rahi seamounts have $\Delta Nb > 0$, and ^{40}Ar-^{39}Ar dating of two of these (one from the cluster of three in the north and the sample in the south) shows that these two seamounts were erupted onto older oceanic crust (3 and 2.5 Ma, respectively) than were all the other dated seamounts (Hall et al., 2006). The other two samples with $\Delta Nb > 0$ have not been dated. This relationship between ΔNb and lithosphere age is not apparent in the Pukapuka Ridges; the one sample with $\Delta Nb > 0$ (Fig. 11) was erupted onto younger crust than were the other seamounts.

It is clear from the histograms in Figure 11 that the spread of ΔNb is much larger in the seamount data than in samples from the spreading axes. This spread is probably because individual seamounts only sample their underlying mantle, whereas axial magma reservoirs accumulate magma from a larger volume of mantle and then homogenize it, as was suggested by Niu et al. (2002) for the northern East Pacific Rise seamounts and adjacent ridge axis. The high values and strongly bimodal distribution of ΔNb seen in the southern Mid-Atlantic and southwest and central Indian ridges (Fig. 2) are not seen in the data in Figure 11. This difference suggests that the enriched domains in the mantle beneath the southern East Pacific Rise and adjacent areas are smaller and less abundant than they are in the mantle beneath the south Atlantic and Indian oceans, and are therefore unable to dominate the melt zones beneath ridge axes and seamounts.

The Rano Rahi seamounts were erupted onto crust as old as 3 Ma (Hall et al., 2006), but seamounts elsewhere in the Pacific Ocean were erupted onto much older crust. Seamounts off the California coast, for example, have been shown to be 7–11 Ma younger than the underlying ocean crust (Davis et al., 2002), and volcanoes dated 1–8 Ma have been found on 135-Ma ocean crust in the northwest Pacific Ocean off Japan (Hirano et al., 2001, 2006). Alkali basalt is the dominant rock type in both of these examples, and in both cases the magmatism has been ascribed to extension or fracturing of the lithosphere. They have been described respectively as a "different" and a "new" kind of intraplate magmatism, and used to argue that oceanic intraplate magmatism need not be related to mantle plumes. McNutt (2006) has used the Japanese example to argue for a re-examination of the plume hypothesis. Data from these two occurrences plot within the Iceland array on a Nb/Y versus Zr/Y diagram (Fig. 12) and are compositionally indistinguishable from OIB. This is an important observation, because it shows that very small-degree melting of the upper mantle, as would be expected beneath thick lithosphere, produces OIB-like magma ($\Delta Nb > 0$) and is, therefore, sampling only the enriched mantle domains and not the depleted matrix ($\Delta Nb < 0$).

CONTINENTAL VOLCANISM

Comparing continental and oceanic basalt is complicated by the effects of contamination with the continental crust, which could have the effect of blurring the geochemical distinction between N-MORB–like and OIB-like magmas. However, the use of the Nb/Y-Zr/Y plot avoids this problem, because both N-MORB and average continental crust have negative ΔNb

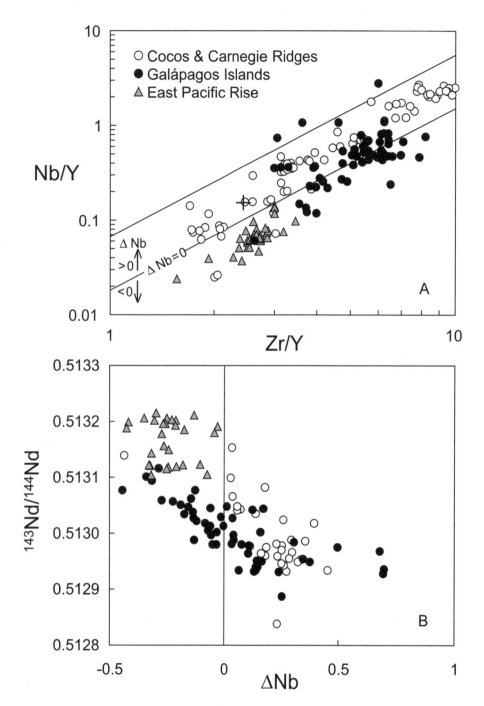

Figure 8. (A) Comparison of Nb/Y and Zr/Y from the Galápagos Islands (Thompson et al., 2004, and J.G. Fitton, GSA Data Repository [see footnote 1]) with data from the Cocos and Carnegie ridges (Harpp et al., 2005) and East Pacific Rise (Mahoney et al., 1994). The two parallel lines mark the limits of the Iceland array; the + represents the composition of primitive mantle (see Fig. 1). Some of the Galápagos basalt samples have significantly negative ΔNb, unlike other OIB (see Fig. 5, which does not include the Galápagos data). These samples probably represent small-degree melts of a normal mid-ocean ridge basalt source. Note that the data from the Cocos and Carnegie ridges plot within the Iceland array; the samples with ΔNb < 0 were dredged from the seafloor adjacent to the ridges (Fig. 9). This observation suggests that mixing with ambient upper mantle is a recent phenomenon in the Galápagos Islands. (B) Variation of ^{43}Nd/^{144}Nd with ΔNb in the same suite of samples. Isotope data are from White et al. (1993), Galápagos Islands; Werner et al. (2003), Cocos and Carnegie ridges; and Mahoney et al. (1994), East Pacific Rise. The good negative correlation probably results from mixing between Galápagos plume mantle and ambient upper mantle (Geist et al., 1988).

Figure 9. Maps of the Galápagos Islands and adjacent seafloor showing the location of samples in Figure 8. (A) The location and character (ΔNb) of samples dredged on and around the Cocos and Carnegie ridges. Note that the samples with ΔNb < 0 were dredged from the seafloor adjacent to the ridges; samples from the ridges have ΔNb > 0 (similar to Iceland). (B) An enlargement of the area in the white box in panel A summarizing the character of the basalts on each of the Galápagos Islands. Islands in the center of the archipelago have ΔNb ≈ 0 or <0. These values are consistent with the observation of Geist et al. (1988) and White et al. (1993) that the Galápagos hotspot has a normal mid-ocean ridge basalt–like core partly surrounded by a horseshoe-shaped ocean island basalt–like outer zone.

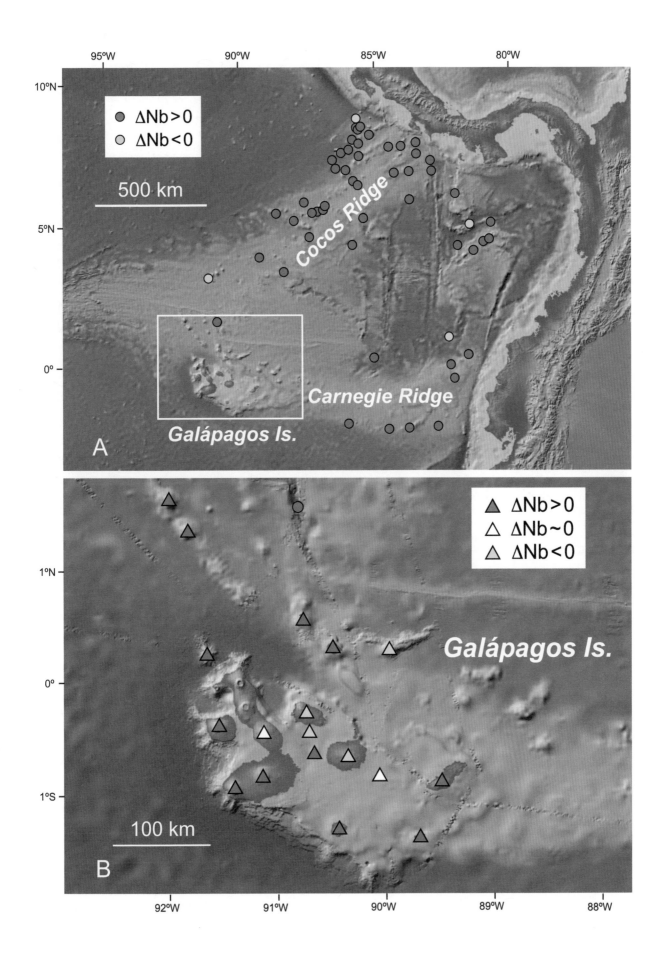

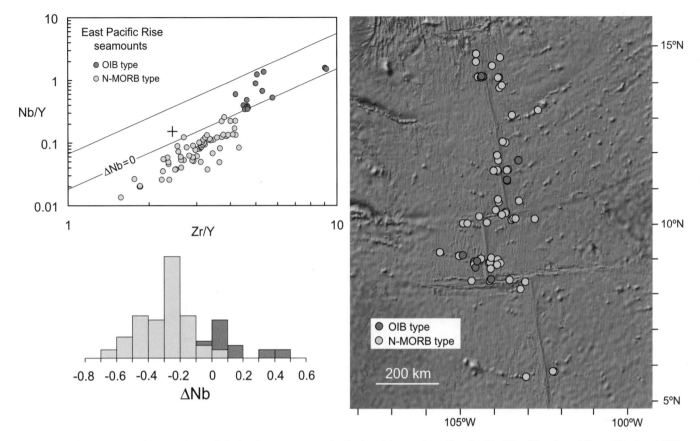

Figure 10. The location and composition of dredged seamounts on the flanks of the northern East Pacific Rise (data from Niu and Batiza, 1997). The two parallel lines mark the limits of the Iceland array; the + represents the composition of primitive mantle (see Fig. 1). The seamounts vary in composition from highly depleted normal mid-ocean ridge basalt (N-MORB) to enriched alkali basalt (Niu and Batiza, 1997) and fall into two distinct N-MORB (yellow) and ocean island basalt (OIB; red) compositional types separated almost perfectly by the ΔNb = 0 reference line. The histogram (color coded as for the data points) shows the weakly bimodal distribution of ΔNb in the data set (eighty samples); the distribution of N-MORB- and OIB-type seamounts is shown on the map.

(Fig. 1). Contamination of N-MORB–like magma with continental crust can never produce hybrid magmas with positive values of ΔNb, although OIB-like magma could, if it were contaminated with enough continental crust, have ΔNb < 0. The contaminated basalt would, however, be readily distinguished from N-MORB through its high concentrations of most highly incompatible elements. Another useful feature of the Nb/Y-Zr/Y plot is that Nb, Zr, and Y are high-field-strength elements and therefore relatively immobile during weathering, alteration, and low-grade metamorphism.

The Cameroon Line

The Cameroon line in West Africa (Fitton, 1987; Déruelle et al., 1991) provides a valuable and unique link between ocean island and continental alkaline magmatism. It consists of a Y-shaped chain of volcanoes extending for 1600 km, from the Atlantic island of Annobon to the interior of the African continent, and has been essentially continuously active since the late

Cretaceous. Magmatic activity from 66 to ca. 30 Ma is represented by plutonic complexes, and the more recent activity by volcanic edifices. Cameroon line magmatism has been essentially alkaline throughout its history. The plutons are composed of gabbro, syenite, and alkali granite, and the volcanic rocks range in composition from alkali basalt to trachyte and alkali rhyolite. There is no consistent age progression among the volcanic and plutonic centers, and the most recent volcanic activity (Mt. Cameroon in 2000) was in the middle of the line. Mt. Cameroon is composed mostly of alkali basalt and basanite and, at 4070 m, is one of the largest volcanoes in Africa.

The origin of the Cameroon line remains unexplained (see Déruelle et al., 1991, for a review of proposed origins). There is no evidence for extensional faulting on either the oceanic or continental sectors. Its longevity and large volume of magma suggest a plume origin, but the plume would need to have a sheet-like form and have remained stationary with respect to the African plate for the past 66 m.y. The importance of the Cameroon line is that it allows a lithospheric source for the continental alkaline

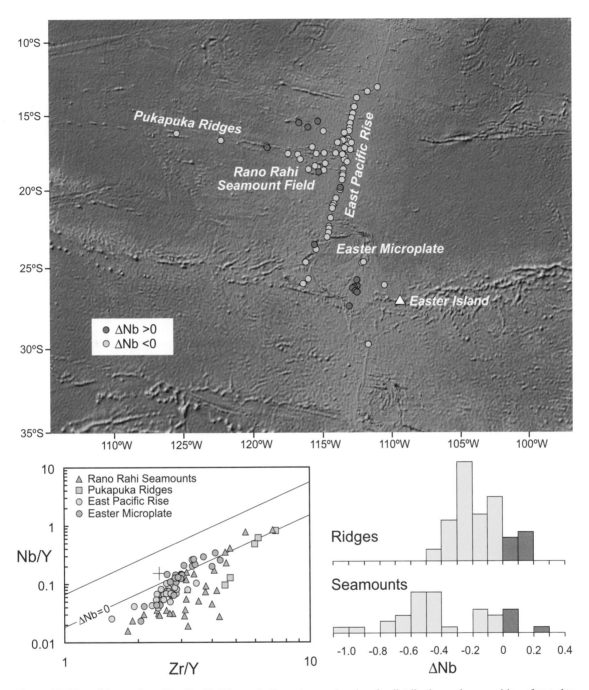

Figure 11. Map of the southern East Pacific Rise and adjacent areas showing the distribution and composition of samples dredged from the Rano Rahi seamount field (Hall et al., 2006), the Pukapuka Ridges (Janney et al., 2000), the axis of the southern East Pacific Rise (Mahoney et al., 1994), and the spreading centers around the Easter microplate (Haase, 2002). Parallel lines on the Nb/Y versus Zr/Y plot mark the limits of the Iceland array shown in Figure 1; the + symbol represents primitive mantle composition. ΔNb is the deviation, in log units, from the reference line separating the Iceland array from N-MORB (the lower of the two parallel lines). Points on the map and the data points are color coded for positive (red) and negative (yellow) values of ΔNb. The histograms compare the distribution of ΔNb in seamounts (thirty-four samples) with that in the spreading axes (fifty-four samples); note the greater spread of ΔNb in the seamounts. Most of the axial samples with ΔNb > 0 are from parts of the ridge closest to Easter Island.

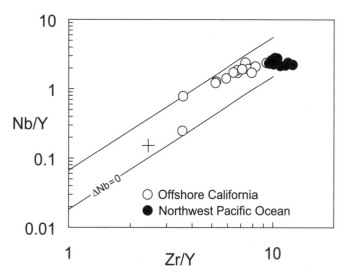

Figure 12. Composition of alkali basalt from two groups of seamounts in the Pacific Ocean thought to originate through lithospheric extension and fracture. The seamounts off the California coast (Davis et al., 2002) are 7–11 m.y. older than the underlying ocean crust; those from the northwest Pacific, off Japan (Hirano et al., 2006), formed at 8–1 Ma on 135-Ma ocean crust. Parallel lines on the plot mark the limits of the Iceland array shown in Figure 1; the + symbol represents primitive mantle. Basalt samples from each group of seamounts plot within the Iceland array and are compositionally indistinguishable from ocean island basalt.

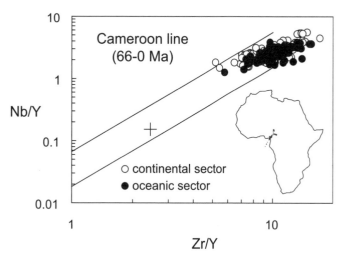

Figure 13. Composition of basalt from the oceanic and continental sectors of the Cameroon line (J.G. Fitton, GSA Data Repository, see footnote 1). The two parallel lines mark the limits of the Iceland array shown in Figure 1; the + represents the composition of primitive mantle. Inset map of Africa shows the outcrop of Cameroon line volcanic rocks. The data plotted are from basalt samples ranging in age from ca. 30 to 0 Ma; the older rocks (not plotted) are plutonic. The basalts from the continental sector are indistinguishable in composition from those from the oceanic sector ocean island basalt and therefore must have a sublithospheric source (Fitton and Dunlop, 1985).

magma to be ruled out. Fitton and Dunlop (1985) showed that basalt from the continental sector is identical to that in the oceanic sector, despite the two sectors being formed on lithosphere that differs considerably in age and geological history. Whatever the cause of Cameroon line magmatism, its source must be sublithospheric. Data from Cameroon line basalts are plotted in Figure 13, which shows that the oceanic and continental populations are indistinguishable and that the continental basalts are therefore OIB-like (positive ΔNb).

RIFT SYSTEMS

Continental extension and rifting is often accompanied by the eruption of alkali basalt that is generally OIB-like in composition. This scenario is illustrated in Figure 14, which shows data from four rift systems. In two of these (East Africa and western United States) rifting has been accompanied by ~1 km of uplift, but the other two rifts remained close to sea level throughout their active phase. The East African rifts and the western United States both need some means of dynamic support to account for their elevation, and mantle plumes provide a plausible mechanism. The volume of primary basaltic magma needed to account for the observed and inferred volume of basic and evolved volcanic rocks in the Kenya rift is far too high to be accounted for solely by decompression melting accompanying extension (Latin et al., 1993). Some means of actively feeding mantle into the melt zone is needed, and this observation is consistent with a plume origin.

Despite their tectonic differences, basalts from the four rift systems are remarkably similar and OIB-like (Fig. 14). The only basalts with significantly negative ΔNb are from the early (16- to 5-Ma) phase of magmatism in the Basin and Range province of the western United States. Fitton et al. (1991) and Kempton et al. (1991) have shown that these basalts probably inherited their chemical and isotopic characteristics from a subcontinental lithospheric mantle source enriched by fluids expelled from a subducted slab. The enriched lithospheric mantle component is best represented by the lamproites emplaced in several parts of the western United States shortly before the onset of Basin and Range magmatism. The composition of some of these lamproites is shown in Figure 14. The sudden onset of OIB-like Basin and Range magmatism at ca. 5 Ma coincided with the beginning of the recent phase of uplift on the Colorado plateau (Lucchitta, 1979).

Africa has many examples of long-lived alkaline magmatism (e.g., Cameroon line, East African rifts), which may be due in part to the African plate being virtually stationary in the hotspot reference frame since ca. 35 Ma (Burke, 1996). Ebinger and Sleep (1998) have suggested that a single mantle plume centered under Afar could be responsible for prolonged magmatism in East Africa and the Cameroon line. A mantle plume, however, cannot explain Permo-Carboniferous rift magmatism in the Scottish Midland Valley, which lasted for ~70 m.y. but stayed close to sea level throughout (Read et al., 2002). Basalt erupted

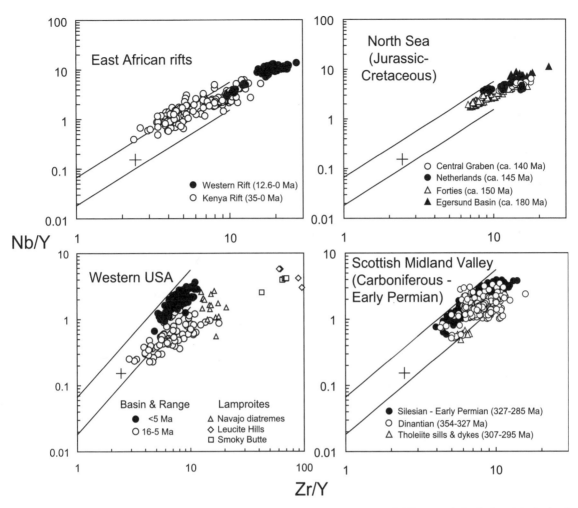

Figure 14. Composition of basalt from four continental rift systems. The two parallel lines mark the limits of the Iceland array shown in Figure 1; the + represents the composition of primitive mantle. Kenya rift data are from Macdonald et al. (2001); Western rift data from James (1995); North Sea data from Latin (1990) and Latin et al. (1990); Western United States data from Fitton et al. (1991); Smoky Butte data from Fraser et al. (1985); Scottish Midland Valley data from Smedley (1986) and Wallis (1989). Data from Ph.D. theses and other unpublished data and from Fitton et al. (1991) are available in the GSA Data Repository (see footnote 1). Ages of volcanism from Ebinger (1989; Western rift); Latin et al. (1990; North Sea basin); Fitton et al. (1991; Basin and Range); Macdonald (2003; Eastern rift); Upton et al. (2004; Scottish Midland Valley). With the exception of the early (16- to 5-Ma) Basin and Range basalts, most of the basalts are ocean island basalt–like. The composition of the early Basin and Range basalts is best explained by mixing with an enriched component, represented by lamproites, in the subcontinental lithospheric mantle.

over this time interval is OIB-like in its incompatible-element abundances and Sr-Nd-Pb-isotope ratios (Smedley, 1986, 1988; Wallis, 1989; Upton et al., 2004) and also has generally positive ΔNb (Fig. 14). Unlike the African rifts, which formed on a plate that was stationary with respect to hotspots, the Scottish Midland Valley moved ~15° northward across the Equator during the 70 m.y. when it was volcanically active (Smith et al., 1981; Lawver et al., 2002; Fig. 15). The ~15° northward motion of Scotland is inferred mostly from paleomagnetic evidence (e.g., Irving, 1977) and must be a minimum estimate, because it takes no account of east-west motion.

In contrast to the Scottish Midland Valley, magmatism in the North Sea rifts may not have been an entirely passive response to extension. Late Jurassic–Early Cretaceous extension and volcanism was immediately preceded by a phase of domal uplift, which interrupted an earlier (Permian–Early Jurassic) phase of basin subsidence (Underhill and Partington, 1993). These authors propose that this ~1000-km-diameter dome was caused by a warm, diffuse, and transient plume head. If this uplift really was caused by a mantle plume, then it must have had the form of a buoyant blob of mantle, quite unlike the long-lived features thought to be responsible for Hawaii, Réunion, and Ice-

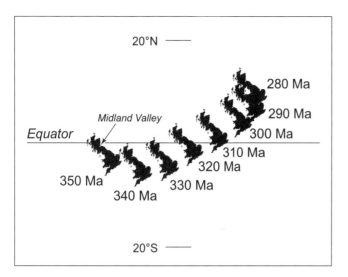

Figure 15. Location of the Scottish Midland Valley relative to the equator during the period of rifting and magmatism (Early Carboniferous–Early Permian), based on Lawver et al. (2002). The Midland Valley moved ~15° northward across the equator during the 70 m.y. when it was volcanically active.

land. This raises the question of why the transient "plume" should have been located so precisely under the center of a developing rift basin. Whatever its cause, subsidence of the dome was accompanied by Late Jurassic–earliest Cretaceous extension, leading to the formation of three rift valleys radiating from the dome center (Underhill and Partington, 1993).

The Rift-to-Drift Transition

Rifting is a precursor to continent break-up and the formation of new ocean basins, and at some point in this process OIB-like magmatism, if present, must give way to the eruption of N-MORB. This transition is difficult to investigate because the volcanic rocks involved are under water and buried beneath thick piles of continental-margin sediments. Several Deep Sea Drilling Project (DSDP) and Ocean Drilling Program (ODP) legs have been devoted to investigating the continent–ocean transition. Two of these (DSDP Leg 80 on Goban Spur in the northeast Atlantic Ocean, and ODP Leg 210 on the Newfoundland margin) have recovered samples of the earliest ocean-floor basalts. The location of the drill sites and the composition of the basalt samples recovered are shown in Figure 16. At both locations, basalt erupted at the point of continental separation is clearly N-MORB.

DISCUSSION

The data presented in the previous sections show that Icelandic basalt invariably has more Nb, relative to Zr and Y, than does N-MORB. A reference line separating Icelandic basalt from

N-MORB on a plot of Nb/Y versus Zr/Y (ΔNb = 0) is also an effective discriminant between OIB in general and N-MORB (Fig. 5). Basalt with positive ΔNb (OIB) occurs in the ocean basins on all scales, from large islands with time-progressive hotspot trails (e.g., Iceland, Hawaii, Réunion) to tiny seamounts (Figs. 10–12). The former may be the product of deep-mantle plumes, but the latter clearly are not, and yet they seem to have very similar mantle sources. A clear bimodality in ΔNb in basalt from sections of mid-ocean ridge affected by hotspots (Fig. 2) suggests only limited mixing between the OIB and N-MORB sources. This distinction of source appears to be a general feature of OIB (Fig. 5) and is supported by isotopic studies (e.g., Hart et al., 1992; Blichert-Toft et al., 1999). Only on the Galápagos Islands (Fig. 8) do melts from the OIB and N-MORB sources appear to mix to a significant extent.

The OIB source contains enriched components (HIMU, EM-1, and EM-2) mixed in variable proportions. Additionally, the source of Icelandic basalt contains at least one depleted component. Variable degrees of melting beneath Iceland effectively sample both the enriched and depleted components, and therefore the mantle must be heterogeneous on a scale that is smaller than the melt zone. It is clear from Figure 1 that the depleted component in the source of Icelandic basalt, with positive ΔNb, cannot be the ambient upper mantle, which has negative ΔNb. Frey et al. (2005) showed that the source of Hawaiian basalt contains a similar depleted component. Storage of magma in large reservoirs tends to homogenize magma, so that large basaltic flows in Iceland have a fairly uniform composition, representing a blend of the depleted and enriched parts of the source (Hardarson and Fitton, 1997). The magma composition is biased toward the more enriched and fusible components.

The upper-mantle source of N-MORB must likewise be heterogeneous, as can be seen from the composition and distribution of near-axis seamounts on the northern East Pacific Rise (Fig. 10). Most of the seamounts have N-MORB composition, but a few are made of OIB-like basalt, suggesting that the upper mantle contains blobs of OIB-source mantle (positive ΔNb) in a depleted matrix with negative ΔNb. Small-degree melting of the upper mantle beneath old, thick oceanic lithosphere only samples the enriched blobs and not the depleted matrix (Fig. 12), consistent with the former being more easily fusible than the latter. The OIB and N-MORB sources, therefore, appear to share a common set of enriched components but, in the case of Iceland and Hawaii, differ in their depleted component.

Enriched components (recycled subducted ocean crust with or without material derived from the continental crust) must be present in the convecting upper mantle and also in the source of deeper-mantle upwellings (plumes) responsible for the larger islands. Ocean islands and seamounts thought to result from mantle plumes and those formed by the melting of fusible blobs carried passively in the upper-mantle flow are indistinguishable on trace-element (including ΔNb) or isotopic criteria. E-MORB from segments of the south Atlantic and Indian ocean ridges (Fig. 2), for example, is at least as likely to result from enriched

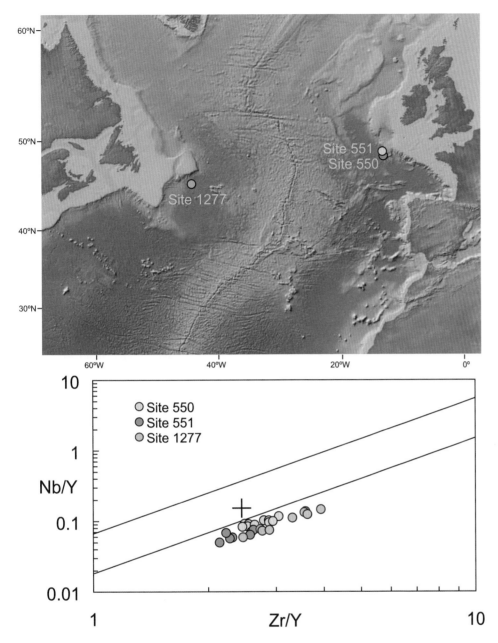

Figure 16. Map showing the location of Deep Sea Drilling Project and Ocean Drilling Program drill sites that have penetrated the earliest ocean crust on nonvolcanic ocean margins. The composition of basalt samples recovered at these three sites is shown on the diagram below the map. Data for Sites 550 and 551 from Kempton et al. (2000) and J.G. Fitton, GSA Data Repository (see footnote 1); data for Site 1277 from Robertson (2007). The two parallel lines mark the limits of the Iceland array shown in Figure 1; the + represents the composition of primitive mantle. All the samples have negative ΔNb and appear to be normal mid-ocean ridge basalt.

blobs in the upper mantle as from mantle plumes. Moreover, E-MORB in many other areas (e.g., the eastern Southeast Indian Ridge, Fig. 4), thousands of kilometers from the nearest hotspots, is clearly not formed by mantle plumes.

The depleted component in the OIB source is less accessible than the enriched components, because it is likely to be more refractory and therefore only sampled at larger degrees of melting than are represented by most OIBs. It is not surprising, therefore, that the clearest evidence for its existence is provided by basalt from Iceland (Fitton et al., 1997, 2003) and Hawaii (Frey et al., 2005). Subducted oceanic lithospheric mantle provides a plausible source for the depleted component in the Iceland plume (Skovgaard et al., 2001). The upper-mantle depleted component

is the principal source of N-MORB and evolved through the formation of the continental crust and, crucially for the development of negative ΔNb, the removal of dehydrated subducted ocean crust.

Continental intraplate magmatism is much more difficult to explain. Large, long-lived continental alkaline volcanic provinces can be explained by mantle plumes only in cases where they form on stationary continents, such as Africa. A mantle-plume origin is arguably the only plausible explanation for magmatism in the East African rift system, given the volumes of magma produced and the scale of regional uplift (e.g., Latin et al., 1993; Ebinger and Sleep, 1998; Macdonald et al., 2001). Whether the same is true for other African alkaline provinces, however, is debatable. The Cameroon line, for example, has been active for 66 m.y. without any systematic shift in the locus of activity. The magma source seems to have been fixed very precisely to the African plate over this period.

Passive extension of continental lithosphere leading to rift formation and decompression melting of the upper mantle might be expected to produce at least some magma with N-MORB composition, but it never does. Contamination of N-MORB–like parental magma with continental crust cannot produce OIB-like hybrids, because the contaminated magma would have negative ΔNb (Fig. 1), whereas rift basalts have positive ΔNb (Fig. 14). The geochemical similarity of continental rift basalts and OIB, most strikingly demonstrated by the Cameroon line (Fig. 13), also rules out a source in the continental lithospheric mantle (Fitton and Dunlop, 1985). The source of most rift-zone magmatism must, therefore, be sublithospheric and, in cases of passive rifting, must originate in the upper mantle.

Basaltic magmatism in the Mesozoic North Sea basin was accompanied by doming, which Underhill and Partington (1993) ascribe to a blob-like, transient mantle plume head. A mantle plume could account for the OIB-like composition of the basalts (Fig. 14), but it seems an unlikely coincidence for a plume originating deep in the mantle to arrive at the surface under a developing rift basin. Late Jurassic–earliest Cretaceous extension in the North Sea basin resulted in lithospheric thinning by a factor of 1.6 at most (Barton and Wood, 1984), which is insufficient to cause melting in peridotite with an upper-mantle potential temperature of 1300° C (Latin et al., 1990). Unless the mantle was significantly hotter, passive upwelling would allow only the more fusible (i.e., enriched) parts of the upper mantle to melt, and these, by analogy with seamounts, would melt to produce OIB-like magma. It would take much larger degrees of extension, decompression, and melting for the depleted matrix to begin to melt and thereby produce N-MORB–like magmas, as clearly happens at the initiation of seafloor spreading (Fig. 16). This requirement may explain the complete absence of N-MORB–like magma from passive continental rifts. Thus basalt erupted in passive rifts, and on most ocean islands and some seamounts, may share a common source in small-scale blobs or streaks of enriched material in the depleted upper mantle.

Magmatism in the North Sea basin might fit a simple model of lithospheric extension leading to melting of enriched components in the upper mantle, because magma volumes are small and localized, and melting was associated with extension (Latin et al., 1990). Other rift basins are not so simple. The Permo-Carboniferous Midland Valley of Scotland (Fig. 14) is part of a much larger rift system that extends across the North Sea into north Germany and northward into the Oslo rift. OIB-like magmatism in the Midland Valley was more voluminous and widespread than in the Mesozoic North Sea basin and persisted almost continuously for ~70 m.y. in a belt across central Scotland (Upton et al., 2004). The volume of igneous rocks (~6000 km³; Tomkieff, 1937) and duration of magmatism cannot be reconciled with simple stretching, and yet there is no evidence for

mantle-plume activity. The Midland Valley area remained close to sea level throughout its magmatic history (Read et al., 2002) and, over this period, drifted at least 1700 km northward (Fig. 15). The magma source appears to be fixed to the lithosphere, as in the case of the Cameroon line, and its origin is equally enigmatic.

The selective sampling of enriched components in the upper mantle during extension of the lithosphere is clearly not a viable explanation for magmatism in all passive continental rifts. Nor can it explain magmatism in the Cameroon line. Nonplume explanations for OIB-like magmatism in the Cameroon line and in some passive rifts require a fortuitous concentration of enriched blobs or streaks in the underlying upper mantle and also some mechanism to keep them supplied and melting for very long periods. This problem is circumvented in the Anderson (1995) perisphere hypothesis, which explains OIB as the melt product of a global, weak, enriched layer that lies immediately beneath the lithosphere. Such a layer provides a potentially inexhaustible supply of OIB-like magma wherever the lithosphere is ruptured. The hypothesis predicts that OIB-like magmatism should persist through the continental rifting stage and into the early stages of seafloor spreading, until the perisphere is locally exhausted under the new ocean basin. Continental rifting alone would not be able to exhaust the perisphere. A test of the hypothesis is provided by the composition of basalt from the continent–ocean transition on passive margins. That this basalt is clearly N-MORB (Fig. 16) seriously undermines the perisphere hypothesis.

CONCLUSIONS

OIB (sensu stricto) occurrences can be explained satisfactorily by a combination of mantle plumes and passive enriched streaks and blobs in the convecting upper mantle. Plumes currently provide the best explanation for large islands with time-progressive ridges and seamount chains (e.g., Hawaii, Réunion, Iceland, Galápagos) but not smaller islands and isolated seamounts. E-MORB is formed from OIB-type mantle when a spreading center encounters a mantle plume (as in Iceland) or a passive enriched blob in the upper mantle. The so-called Shona plume (Fig. 2) is probably an example of the latter. E-MORB and N-MORB form bimodal populations (Fig. 2), suggesting only limited mixing between enriched and depleted components in the upper mantle. A continuum exists between small seamounts and large islands, and geochemistry seems unable to distinguish between OIB formed from passive blobs and that formed from mantle plumes. The enriched components appear to be the same in both. It is not possible to identify the point on this continuum at which mantle plumes (if they exist) become involved, and it follows that OIB cannot be a diagnostic feature of plumes. The geochemical similarity of allegedly plume-related OIB and manifestly nonplume OIB is the first part of the OIB paradox.

Continental intraplate transitional and alkali basalts are identical in composition to OIB, but their origin is even more enigmatic than is the origin of true OIB. They are frequently erupted on the same small area of crust over tens of millions of years on moving plates. Their source appears to move with the plates but is sublithospheric. Passive rift systems, if volcanic, always tap the OIB source and only erupt N-MORB following continental break-up and the onset of seafloor spreading. Some OIB and some continental OIB-like basalts appear to share a common plume origin, but this commonality cannot be invoked in the majority of continental rifts, any more than it can be invoked in the majority of ocean islands and seamounts. Why do passive continental rifts never erupt N-MORB? The coupled problems of location of the subcontinental OIB source and the mechanisms by which continental OIB-like magmas are produced remain unresolved. None of the proposed models for OIB generation (plumes, enriched blobs in the upper mantle, or an enriched perisphere) is able, either singly or in combination, to explain the global distribution of OIB and OIB-like basalts. This failure is the second part of the OIB paradox. OIB and OIB-like basalts are widespread in oceanic and continental settings and, contrary to popular belief, most occurrences are in situations where mantle plumes cannot provide a plausible explanation.

ACKNOWLEDGMENTS

Many of the data used in this study are from unpublished Ph.D. theses, and I thank their authors (Dodie James, Dave Latin, Cynthia Robinson, Pauline Smedley, and Sue Wallis) for permission to include their data in the GSA Data Repository. I am also very grateful to Dodie James for her help in producing the OIB data set; to Alexander McBirney and Bill Leeman for supplying basalt samples from the Galápagos and Hawaiian islands, respectively; and to Dave Christie for permission to use unpublished Zr and Y data from the Southeast Indian Ridge. Thoughtful and constructive reviews from John Dixon, Gill Foulger, Al Hofmann, and John Mahoney considerably improved the manuscript, and I thank John Dixon, John Mahoney, Jim Natland, and John Underhill for fruitful discussions. Fieldwork in Iceland and subsequent analysis of rock samples was supported by research grants (GST/02/673, GR9/01897, and GR2/12769) from the UK Natural Environment Research Council.

REFERENCES CITED

Anderson, D.L., 1995, Lithosphere, asthenosphere, and perisphere: Reviews of Geophysics, v. 33, p. 125–149, doi: 10.1029/94RG02785.

Anderson, D.L., and Schramm, K.A., 2005, Global hotspot maps, *in* Foulger, G.R., et al., eds., Plates, plumes, and paradigms: Boulder, Colorado, Geological Society of America Special Paper 388, p. 19–29, doi: 10.130/2005.2388(03).

Bach, W., Erzinger, J., Dosso, L., Bollinger, C., Bougault, H., Etoubleau, J., and Sauerwein, J., 1996, Unusually large Nb-Ta depletions in North Chile ridge basalts at 36°50′ to 38°56′S: Major element, trace element, and isotopic data: Earth and Planetary Science Letters, v. 142, p. 223–240, doi: 10.1016/0012-821X(96)00095-7.

Barth, M.G., McDonough, W.F., and Rudnick, R.L., 2000, Tracking the budget of Nb and Ta in the continental crust: Chemical Geology, v. 165, p. 197–213, doi: 10.1016/S0009- 2541(99)00173-4.

Barton, P., and Wood, R., 1984, Tectonic evolution of the North Sea Basin: Crustal stretching and subsidence: Geophysical Journal of the Royal Astronomical Society, v. 79, p. 987–1022.

Blichert-Toft, J., Frey, F.A., and Albarède, F., 1999, Hf isotope evidence for pelagic sediments in the source of Hawaiian basalts: Science, v. 285, p. 879–882, doi: 10.1126/science.285.5429.879.

Burke, K., 1996, The African plate: South African Journal of Geology, v. 99, p. 339–410.

Chauvel, C., and Blichert-Toft, J., 2001, A hafnium isotope and trace element perspective on melting of the depleted mantle: Earth and Planetary Science Letters, v. 190, p. 137–151, doi: 10.1016/S0012-821X(01)00379-X.

Davis, A.S., Clague, D.A., Bohrson, W.A., Dalrymple, G.B., and Greene, H.G., 2002, Seamounts at the continental margin of California: A different kind of oceanic intraplate volcanism: Geological Society of America Bulletin, v. 114, p. 316–333, doi: 10.1130/0016-7606(2002)114<0316:SATCMO>2.0.CO;2.

Déruelle, B., Moreau, C., Nkoumbou, C., Kambou, R., Lissom, J., Njonfang, E., Ghogumu, R.T., and Nono, A., 1991, The Cameroon line: A review, *in* Kampunzu, A.B, and Lubala, R.T., eds., Magmatism in extensional structural settings. The Phanerozoic African plate: Heidelberg, Springer-Verlag, p. 274–327.

Douglass, J., Schilling, J.-G., and Fontignie, D., 1999, Plume- ridge interactions of the Discovery and Shona mantle plumes with the southern Mid-Atlantic Ridge (40°–55°S): Journal of Geophysical Research, v. 104, p. 2941–2962, doi: 10.1029/98JB02642.

Ebinger, C.J., 1989, Tectonic development of the western branch of the East African rift system: Geological Society of America Bulletin, v. 101, p. 885–903, doi: 10.1130/0016-7606(1989)101<0885:TDOTWB>2.3.CO;2.

Ebinger, C.J., and Sleep, N.H., 1998, Cenozoic magmatism throughout east Africa resulting from impact of a single plume: Nature, v. 395, p. 788–791, doi: 10.1038/27417.

Fitton, J.G., 1987, The Cameroon line, West Africa: A comparison between oceanic and continental alkaline volcanism, *in* Fitton, J.G., and Upton, B.G.J., eds., Alkaline igneous rocks: London, Special Publications of the Geological Society of London, 30, p. 273–291.

Fitton, J.G., and Dunlop, H.M., 1985, The Cameroon line, West Africa, and its bearing on the origin of oceanic and continental alkali basalt: Earth and Planetary Science Letters, v. 72, p. 23–38, doi: 10.1016/0012-821X(85)90114-1.

Fitton, J.G., and James, D., 1986, Basic volcanism associated with intraplate linear features: Philosophical Transactions of the Royal Society of London A, v. 317, p. 253–266.

Fitton, J.G., James, D., and Leeman, W.P., 1991, Basic magmatism associated with late Cenozoic extension in the western United States: Compositional variation in space and time: Journal of Geophysical Research, v. 96, p. 13,693–13,711.

Fitton, J.G., Saunders, A.D., Norry, M.J., Hardarson, B.S., and Taylor, R.N., 1997, Thermal and chemical structure of the Iceland plume: Earth and Planetary Science Letters, v. 153, p. 197–208, doi: 10.1016/S0012-821X(97)00170-2.

Fitton, J.G., Saunders, A.D., Larsen, L.M., Hardarson, B.S., and Norry, M.J., 1998, Volcanic rocks from the southeast Greenland margin at 63°N: Composition, petrogenesis and mantle sources, *in* Saunders, A.D., et al., eds., Proceedings of the Ocean Drilling Program, scientific results, v. 152, p. 331–350.

Fitton, J.G., Saunders, A.D., Kempton, P.D., and Hardarson, B.S., 2003, Does depleted mantle form an intrinsic part of the Iceland plume?: Geochemistry, Geophysics, Geosystems, v. 4, art. no. Q01032, doi: 10.1029/2002GC000424.

Foulger, G.R., Natland, J.H., and Anderson, D.L., 2005a, A source for Icelandic magmas in remelted Iapetus crust: Journal of Volcanology and Geothermal Research, v. 141, p. 23–44, doi: 10.1016/j.jvolgeores.2004.10.006.

Foulger, G.R., Natland, J.H., and Anderson, D.L., 2005b, Genesis of the Icelandic melt anomaly by plate tectonic processes, *in* Foulger, G.R., et al., eds., Plates, plumes, and paradigms: Boulder, Colorado, Geological Society of America Special Paper 388, p. 595–625, doi: 10.1130/2005.2388(35).

Fraser, K.J., Hawkesworth, C.J., Erlank, A.J., Mitchell, R.H., and Scott-Smith, B.H., 1985, Sr, Nd and Pb isotope and minor element geochemistry of lamproites and kimberlites: Earth and Planetary Science Letters, v. 76, p. 57–70, doi: 10.1016/0012-821X(85)90148-7.

Frey, F.A., Huang, S., Blichert-Toft, J., Regelous, M., and Boyet, M., 2005, Origin of depleted components in basalt related to the Hawaiian hot spot: Evidence from isotopic and incompatible element ratios: Geochemistry, Geophysics, Geosystems, v. 6, art. no. Q02L07, doi: 10.1029/2004GC 000757.

Geist, D.J., White, W.M., and McBirney, A.R., 1988, Plume-asthenosphere mixing beneath the Galapagos archipelago: Nature, v. 333, p. 657–660, doi: 10.1038/333657a0.

Haase, K., 2002, Geochemical constraints on magma sources and mixing processes in Easter microplate MORB (SE Pacific): A case study of plume-ridge interaction: Chemical Geology, v. 182, p. 335–355, doi: 10.1016/S0009-2541(01)00327-8.

Hall, L.S., Mahoney, J.J., Sinton, J.M., and Duncan, R.A., 2006, Spatial and temporal distribution of a *C*-like asthenospheric component in the Rano Rahi seamount field, East Pacific Rise, 15°–19°S: Geochemistry, Geophysics, Geosystems, v. 7, art. no. Q03009, doi: 10.1029/2005GC000994.

Hanan, B.B., and Schilling, J.-G., 1997, The dynamic evolution of the Iceland mantle plume: The lead isotope perspective: Earth and Planetary Science Letters, v. 151, p. 43–60, doi: 10.1016/S0012-821X(97)00105-2.

Hardarson, B.S., and Fitton, J.G., 1997, Mechanisms of crustal accretion in Iceland: Geology, v. 25, p. 1043–1046, doi: 10.1130/0091-7613(1997)025 <1043:MOCAII>2.3.CO;2.

Harpp, K.S., Wanless, V.D., Otto, R.H., Hoernle, K., and Werner, R., 2005, The Cocos and Carnegie aseismic ridges: A trace element record of long-term plume–spreading center interaction: Journal of Petrology, v. 46, p. 109–133, doi: 10.1093/petrology/egh064.

Hart, S.R., Hauri, E.H., Oschmann, L.A., and Whitehead, J.A., 1992, Mantle plumes and entrainment: Isotopic evidence: Science, v. 256, p. 517–520, doi: 10.1126/science.256.5056.517.

Hémond, C., Arndt, N., Lichtenstein, U., Hofmann, A.W., Oskarsson, N., and Steinthorsson, S., 1993, The heterogeneous Iceland plume: Nd–Sr–O isotopes and trace element constraints: Journal of Geophysical Research, v. 98, p. 15,833–15,850.

Hirano, N., Kawamura, K., Hattori, M., Saito, K., and Ogawa, Y., 2001, A new type of intra-plate volcanism; Young alkali-basalts discovered from the subducting Pacific plate, northern Japan Trench: Geophysical Research Letters, v. 28, p. 2719–2722, doi: 10.1029/2000GL012426.

Hirano, N., Takahashi, E., Yamamoto, J., Abe, N., Ingle, S.P., Kaneoka, I., Hirata, T., Kimura, J.-I., Ishii, T., Ogawa, Y., Machido, S., and Suyehiro, K., 2006, Volcanism in response to plate flexure: Science, v. 313, p. 1426–1428, doi: 10.1126/science.1128235.

Hofmann, A.W., 1988, Chemical differentiation of the Earth: The relationship between mantle, continental crust, and oceanic crust: Earth and Planetary Science Letters, v. 90, p. 297–314, doi: 10.1016/0012-821X(88)90132-X.

Hofmann, A.W., 1997, Mantle geochemistry: The message from oceanic volcanism: Nature, v. 385, p. 219–229, doi: 10.1038/385219a0.

Hofmann, A.W., 2004, Sampling mantle heterogeneity through oceanic basalts: Isotopes and trace elements, *in* Carlson, R.W, ed., The mantle and core, volume 2 of Treatise on geochemistry (eds. Holland, H.D, and Turekian, K.K.): Oxford, Elsevier-Pergamon, p. 61–101.

Hofmann, A.W., and White, W.M., 1982, Mantle plumes from ancient oceanic crust: Earth and Planetary Science Letters, v. 57, p. 421–436, doi: 10.1016/0012-821X(82)90161-3.

Hofmann, A.W., Jochum, K.P., Seufert, M., and White, W.M., 1986, Nb and Pb in oceanic basalts: New constraints on mantle evolution: Earth and Planetary Science Letters, v. 79, p. 33–45, doi: 10.1016/0012-821X(86) 90038-5.

Irving, E., 1977, Drift of the major continental blocks since the Devonian: Nature, v. 270, p. 304–309, doi: 10.1038/270304a0.

Ito, G., and Mahoney, J.J., 2005, Flow and melting of a heterogeneous mantle. 1: Method and importance to the geochemistry of ocean island and mid-

ocean ridge basalts: Earth and Planetary Science Letters, v. 230, p. 29–46, doi: 10.1016/j.epsl.2004.10.035.

James, D.E., 1995, The geochemistry of feldspar-free volcanic rocks [Ph.D. thesis]: Milton Keynes, UK, Open University, 317 p.

Janney, P.E., Macdougall, J.D., Natland, J.H., and Lynch, M.A., 2000, Geochemical evidence from the Pukapuka volcanic ridge system for a shallow enriched mantle domain beneath the South Pacific Superswell: Earth and Planetary Science Letters, v. 181, p. 47–60, doi: 10.1016/S0012-821X (00)00181-3.

Janney, P.E., le Roex, A.P., and Carlson, R.W., 2005, Hafnium isotope and trace element constraints on the nature of mantle heterogeneity beneath the central southwest Indian ridge (13°E to 47°E): Journal of Petrology, v. 46, p. 2427–2464, doi: 10.1093/petrology/egi060.

Kempton, P.D., Fitton, J.G., Hawkesworth, C.J., and Ormerod, D.S., 1991, Isotopic and trace element constraints on the composition and evolution of the lithosphere beneath the southwestern United States: Journal of Geophysical Research, v. 96, p. 13,713–13,735.

Kempton, P.D., Fitton, J.G., Saunders, A.D., Nowell, G.M., Taylor, R.N., Hardarson, B.S., and Pearson, G., 2000, The Iceland plume in space and time: A Sr-Nd-Hf-Pb study of the North Atlantic rifted margin: Earth and Planetary Science Letters, v. 177, p. 255–271, doi: 10.1016/S0012-821X (00)00047-9.

Kempton, P.D., Pearce, J.A., Barry, T.L., Fitton, J.G., Langmuir, C., and Christie, D.M., 2002, εNd vs. εHf as a geochemical discriminant between Indian and Pacific MORB-source mantle domains: Results from ODP Leg 187 to the Australian-Antarctic discordance: Geochemistry, Geophysics, Geosystems, v. 3, art. no. Q01074, doi: 10.1029/2002GC000320.

Latin, D.M., 1990, The relationship between extension and magmatism in the North Sea basin [Ph.D. thesis]: Edinburgh, University of Edinburgh, 404 p.

Latin, D.M., Dixon, J.E., and Fitton, J.G., 1990, Rift-related magmatism in the North Sea basin, *in* Blundell, D.J., and Gibbs, A.D., eds., Tectonic evolution of the North Sea rifts: Oxford, Oxford University Press, p. 101–144.

Latin, D.M., Norry, M.J., and Tarzey, R.J.E., 1993, Magmatism in the Gregory rift, East Africa—Evidence for melt generation by a plume: Journal of Petrology, v. 34, p. 1007–1027.

Lawver, L.A., Dalziel, I.W.D., Gahagan, L.M., Martin, K.M., and Campbell, D., 2002, PLATES 2002 atlas of plate reconstructions (750 Ma to present day): University of Texas Institute for Geophysics. Powerpoint animation: http://www.ig.utexas.edu/research/projects/plates/.

le Roux, P.J., le Roex, A.P., Schilling, J.G., Shimizu, N., Perkins, W.W., and Pearce, N.J.G., 2002, Mantle heterogeneity beneath the southern Mid-Atlantic ridge: Trace element evidence for contamination of ambient asthenospheric mantle: Earth and Planetary Science Letters, v. 203, p. 479–498, doi: 10.1016/S0012-821X(02)00832-4.

Lucchitta, I., 1979, Late Cenozoic uplift of the southwestern Colorado Plateau and adjacent lower Colorado River region: Tectonophysics, v. 61, p. 63–95, doi: 10.1016/0040-1951(79)90292-0.

Macdonald, R., 2003, Magmatism of the Kenya Rift Valley: A review: Transactions of the Royal Society of Edinburgh: Earth Sciences, v. 93, p. 239–253.

Macdonald, R., Rogers, N.W., Fitton, J.G., Black, S., and Smith, M., 2001, Plume-lithosphere interactions in the generation of the basalts of the Kenya Rift, East Africa: Journal of Petrology, v. 42, p. 877–900, doi: 10.1093/petrology/42.5.877.

Mahoney, J.J., Sinton, J.M., Macdougall, J.D., Spencer, K.J., and Lugmair, G.W., 1994, Isotope and trace element characteristics of a super-fast spreading ridge: East Pacific Rise, 13–23°S: Earth and Planetary Science Letters, v. 121, p. 173–193, doi: 10.1016/0012-821X(94)90039-6.

Mahoney, J.J., Graham, D.W., Christie, D.M., Johnson, K.T.M., Hall, L.S., and Vonderhaar, D.L., 2002, Between a hotspot and a cold spot: Isotopic variation in the southeast Indian ridge asthenosphere, 86°E–118°E: Journal of Petrology, v. 43, p. 1155–1176, doi: 10.1093/petrology/43.7.1155.

McDonough, W.F., and Sun, S.-s., 1995, The composition of the Earth: Chemical Geology, v. 120, p. 223–253, doi: 10.1016/0009-2541(94)00140-4.

McNutt, K.K., 2006, Another nail in the plume coffin?: Science, v. 313, p. 1394–1395, doi: 10.1126/science.1131298.

Meibom, A., and Anderson, D.L., 2004, The statistical upper mantle assemblage: Earth and Planetary Science Letters, v. 217, p. 123–139, doi: 10.1016/S0012-821X(03)00573-9.

Meschede, M., 1986, A method of discriminating between different types of mid-ocean ridge basalts and continental tholeiites with the Nb-Zr-Y diagram: Chemical Geology, v. 56, p. 207–218, doi: 10.1016/0009-2541(86)90004-5.

Morgan, W.J., 1971, Convection plumes in the lower mantle: Nature, v. 230, p. 42–43, doi: 10.1038/230042a0.

Muhe, R., Bohrmann, H., Garbe-Schonberg, D., and Kassens, H., 1997, E-MORB glasses from the Gakkel Ridge (Arctic Ocean) at 87° N: Evidence for the Earth's most northerly volcanic activity: Earth and Planetary Science Letters, v. 152, p. 1–9, doi: 10.1016/S0012-821X(97)00152-0.

Murton, B.J., Tindle, A.G., Milton, J.A., and Sauter, D., 2005, Heterogeneity in southern central Indian ridge MORB: Implications for ridge–hot spot interaction: Geochemistry, Geophysics, Geosystems, v. 6, art. no. Q03E20, doi:10.1029/2004GC000798.

Nauret, F., Abouchami, W., Galer, S.J.G., Hofmann, A.W., Hémond, C., Chauvel, C., and Dyment, J., 2006, Correlated trace element-Pb isotope enrichments in Indian MORB along 18–20°S, Central Indian Ridge: Earth and Planetary Science Letters, v. 245, p. 137–152, doi: 10.1016/j.epsl.2006.03.015.

Niu, Y.L., and Batiza, R., 1997, Trace element evidence from seamounts for recycled oceanic crust in the eastern Pacific mantle: Earth and Planetary Science Letters, v. 148, p. 471–483, doi: 10.1016/S0012-821X(97)00048-4.

Niu, Y.L., Regelous, M., Wendt, I.J., Batiza, R., and O'Hara, M.J., 2002, Geochemistry of near-EPR seamounts: Importance of source vs. process and the origin of enriched mantle component: Earth and Planetary Science Letters, v. 199, p. 327–345, doi: 10.1016/S0012-821X(02)00591-5.

Price, R.C., Kennedy, A.K., Riggs-Sneeringer, M., and Frey, F.A., 1986, Geochemistry of basalts from the Indian-Ocean triple junction—Implications for the generation and evolution of Indian-Ocean ridge basalts: Earth and Planetary Science Letters, v. 78, p. 379–396, doi: 10.1016/0012-821X(86)90005-1.

Read, W.A., Browne, M.A.E., Stephenson, D., and Upton, B.G.J., 2002, Carboniferous, in Trewin, N.H., ed., The geology of Scotland: London, Geological Society of London, p. 251–299.

Regelous, M., Hofmann, A.W., Abouchami, W., and Galer, S.J.G., 2003, Geochemistry of lavas from the Emperor Seamounts, and the geochemical evolution of Hawaiian magmatism from 85 to 42 Ma: Journal of Petrology, v. 44, p. 113–140, doi: 10.1093/petrology/44.1.113.

Robertson, A.H.F., 2007, Evidence of continental breakup from the Newfoundland rifted margin (Ocean Drilling Program Leg 210): Lower Cretaceous seafloor formed by exhumation of subcontinental mantle lithosphere and the transition to seafloor spreading, in Tucholke, B.E., et al., eds., Proceedings of the Ocean Drilling Program, scientific results, volume 210: College Station, Texas, Ocean Drilling Program, p. 1–69, doi: 10.2973/odp.proc.sr.210.104.2007.

Robinson, C.J., Bickle, M.J., Minshull, T.A., White, R.S., and Nichols, A.R.L., 2001, Low degree melting under the Southwest Indian Ridge: The roles of mantle temperature, conductive cooling and wet melting: Earth and Planetary Science Letters, v. 188, p. 383–398, doi: 10.1016/S0012-821X(01)00329-6.

Rudnick, R.L., and Fountain, D.M., 1995, Nature and composition of the continental-crust—A lower crustal perspective: Reviews of Geophysics, v. 33, p. 267–309, doi: 10.1029/95RG01302.

Rudnick, R.L., Barth, M., Horn, I., and McDonough, W.F., 2000, Rutile-bearing eclogites: Missing link between continents and depleted mantle: Science, v. 287, p. 278–281, doi: 10.1126/science.287.5451.278.

Schilling, J.-G., Meyer, P.S., and Kingsley, R.H., 1982, Evolution of the Iceland hotspot: Nature, v. 296, p. 313–320, doi: 10.1038/296313a0.

Skovgaard, A.C., Storey, M., Baker, J., Blusztajn, J., and Hart, S.R., 2001, Osmium-oxygen isotopic evidence for a recycled and strongly depleted component in the Iceland mantle plume: Earth and Planetary Science Letters, v. 194, p. 259–275, doi: 10.1016/S0012-821X(01)00549-0.

Sleep, N.H., 1984, Tapping of magmas from ubiquitous mantle heterogeneities—An alternative to mantle plumes: Journal of Geophysical Research, v. 89, p. 29–41.

Smallwood, J.R., White, R.S., and Minshull, T.A., 1995, Sea-floor spreading in the presence of the Iceland plume: The structure of the Reykjanes Ridge at 61°40′N: Journal of the Geological Society of London, v. 152, p. 1023–1029.

Smedley, P.L., 1986, Petrochemistry of Dinantian volcanism in northern Britain [Ph.D. thesis]: Edinburgh, University of Edinburgh, 341 p.

Smedley, P.L., 1988, Trace element and isotopic variations in Scottish and Irish Dinantian volcanism: Evidence for an OIB-like mantle source: Journal of Petrology, v. 29, p. 413–443.

Smith, A.G., Hurley, A.M., and Briden, J.C., 1981, Phanerozoic paleocontinental world maps: Cambridge, Cambridge University Press, 102 p.

Stracke, A., Hofmann, A.W., and Hart, S.R., 2005, FOZO, HIMU, and the rest of the mantle zoo: Geochemistry, Geophysics, Geosystems, v. 6, art. no. Q05007, doi: 10.1029/2004GC000824.

Thirlwall, M.F., 1995, Generation of the Pb isotopic characteristics of the Iceland plume: Journal of the Geological Society of London, v. 152, p. 991–996.

Thirlwall, M.F., Gee, M.A.M., Taylor, R.N., and Murton, B.J., 2004, Mantle components in Iceland and adjacent ridges investigated using double-spike Pb isotope ratios: Geochimica et Cosmochimica Acta, v. 68, p. 361–386, doi: 10.1016/S0016-7037(03)00424-1.

Thompson, P.M.E., Kempton, P.D., White, R.V., Kerr, A.C., Tarney, J., Saunders, A.D., Fitton, J.G., and McBirney, A., 2004, Hf-Nd isotope constraints on the origin of the Cretaceous Caribbean plateau and its relationship to the Galápagos plume: Earth and Planetary Science Letters, v. 217, p. 59–75, doi: 10.1016/S0012-821X(03)00542-9.

Tomkieff, S.I., 1937, Petrochemistry of the Scottish Carboniferous–Permian igneous rocks: Bulletin Volcanologique, Série 2, v. 1, p. 59–87.

Underhill, J.R., and Partington, M.A., 1993, Jurassic thermal doming and deflation in the North Sea: Implications of the sequence stratigraphic evidence, in Parker, R.J., ed., Petroleum geology of northwest Europe: Proceedings of the 4th conference: London, Geological Society of London, p. 337–345.

Upton, B.G.J., Stephenson, D., Smedley, P.M., Wallis, S.M., and Fitton, J.G., 2004, Carboniferous and Permian magmatism in Scotland, in Wilson, M., et al., eds., Permo-Carboniferous magmatism and rifting in Europe: London, Geological Society of London Special Publication 223, p. 195–217.

Wallis, S.M., 1989, Petrology and geochemistry of Upper Carboniferous–Lower Permian volcanic rocks in Scotland [Ph.D. thesis]: Edinburgh, University of Edinburgh, 374 p.

Werner, R., Hoernle, K., Barckhausen, U., and Hauff, F., 2003, Geodynamic evolution of the Galápagos hot spot system (Central East Pacific) over the past 20 m.y.: Constraints from morphology, geochemistry, and magnetic anomalies: Geochemistry, Geophysics, Geosystems, v. 4, art. no. Q01108, doi: 10.1029/2003GC000576.

Wessel, P., and Lyons, S., 1997, Distribution of large Pacific seamounts from Geosat/ERS-1: Implications for the history of intraplate volcanism: Journal of Geophysical Research, v. 102, p. 22,459–22,475, doi: 10.1029/97JB01588.

White, W.M., McBirney, A.R., and Duncan R.A., 1993, Petrology and geochemistry of the Galapagos Islands—Portrait of a pathological mantle plume: Journal of Geophysical Research, v. 98, p. 19,533–19,563.

Willbold, M., and Stracke, A., 2006, Trace element composition of mantle end-members: Implications for recycling of oceanic and upper and lower continental crust: Geochemistry, Geophysics, Geosystems, v. 7, art. no. Q04004, doi: 10.1029/2005GC001005.

Zindler, A., and Hart, S.R., 1986, Chemical geodynamics: Annual Review of Earth and Planetary Sciences, v. 14, p. 493–571, doi: 10.1146/annurev.ea.14.050186.002425.

MANUSCRIPT ACCEPTED BY THE SOCIETY 31 JANUARY 2007

DISCUSSION

10 January 2007, James H. Natland

Since OIB-like lavas occur on so many features that are not age-progressive linear island chains, such geochemistry does not by itself require a deep mantle plume anywhere else. (Natland and Winterer, 2005, p. 688)

We suspect that geochemistry will not deliver the silver bullet for proving or disproving plumes. (Hofmann and Hart, 2005, p. 40)

Do these two epigraphs and discovery of the ocean island basalt (OIB) paradox by Fitton (this volume) mean that we all agree? Those of us who are old enough to have been trained as petrologists rather than geochemists will certainly recall the alkaline olivine-basalt volcanic association of Turner and Verhoogen (1960). They divided the association into two types, one on the continents, the other in the ocean basins, treated the petrogenesis of lavas of both under one general discussion, and noted certain features in common, such as development of feldspathoidal differentiates and the presence of ultramafic xenoliths in the basalts. Turner and Verhoogen (1960) also recognized a separate association of tholeiitic and alkalic olivine-basalts in both the ocean basins (Hawaiian province) and on continents (Hebridean province), another association of tholeiitic flood basalts (Deccan, Karoo, Columbia River), and yet another of potash-rich volcanic rocks (western African rift). Today, we recognize both continental and oceanic occurrences of all of these, even the last (e.g., Davis et al., 2002; Hirano et al., 2006), and many geochemists see the plume hypothesis as one way to link them, even to the extent that hypothetical plume heads can be flattened and distended at the base of the lithosphere for thousands of kilometers away from linear island chains and can feed seamounts near distant ridges (Niu et al., 1999).

The OIB paradox, as Fitton (this volume) describes it, is simply a renewed recognition that the alkalic olivine-basalt association of continents and ocean basins is a valid way to frame the problem of the petrogenesis of those lavas. However, instead of peering at it from the continents outward, use of the terms "OIB" and "OIB-like" makes the ocean basins the frame of reference. The term "OIB" itself is a misnomer (Natland and Winterer, 2005), because it refers to rocks that can only be sampled on foot, whereas it clearly occurs on thousands of both tall and short seamounts, most of them not parts of linear island chains. The difficulty is compounded by application of OIB to many tholeiites, and discovery that so-called enriched mid-ocean ridge basalt (E-MORB) at ridges resembles certain OIBs. Because some E-MORBs are indeed nepheline normative alkalic olivine basalt, we could as well say that the association of tholeiitic and alkalic olivine basalt extends to spreading ridges themselves, and not just islands like Hawaii and rift provinces like the Hebridean. Obviously, if we are faced with using E-MORB for alkalic olivine basalt on a spreading ridge, and OIB for an identical rock on a seamount or island, and OIB-like for the same rock in a continental rift, we have a curious and confusing problem with terminology. The old rock name, which does not reference locality even symbolically with an acronym, is clearly better, and so, probably, is the Turner and Verhoogen (1960) approach to discussing their petrogenesis.

Fitton (this volume) tackles the E-MORB problem by proposing use of his ΔNb criterion as a basis for definition. I am surprised that a procedure developed primarily to distinguish Icelandic tholeiite from depleted normal MORB (N-MORB) should work so well for so many locales where the contrasting rock types are, or at least include, alkalic olivine basalt (positive ΔNb) on the one hand and depleted abyssal tholeiite (now negative ΔNb, by definition) on the other. But E-MORB has never been systematically or consistently defined before, so this effort is a step in the right direction. However, I believe that classification in terms of a single parameter based on trace elements oversimplifies the problem, and that the general petrological (e.g., normative) character of the rocks should always be established and borne in mind. I am also impressed that outside of Iceland, trends on a ΔNb diagram (log Zr/Y versus log Nb/Y) shown by Fitton (this volume) for the ocean basins are almost systematically oblique to the line for ΔNb = 0. This trend would show up even more dramatically, for example, along the southern superfast East Pacific Rise, had Fitton included existing data for Garret fracture zone in the evaluation. Such oblique trends and other geochemical attributes are evidence that magma mixing between disparate alkalic and tholeiitic end-members has occurred in all of these regions. Exclusion of differentiated compositions (e.g., andesites, trachytes, rhyolites) also makes it impossible to assess whether such magmas have mixed in small proportions with any of the basalts. This mixing would tend to flatten data arrays, making them parallel to ΔNb = 0, usually at positive ΔNb (Natland, this volume).

31 January 2007, Jason Phipps Morgan and W.J. Morgan

This article notes that E-MORB, OIB-like basalts, are often erupted at mid-ocean ridges remote from known sites of the plume upwelling that provides the proposed source material for OIB volcanism. Fitton (this volume) interprets this "OIB paradox" to show that "plume" material can somehow make it into the source of some mid-ocean ridge (and continental) basalt sources where "mantle plumes cannot provide a plausible explanation." This situation is not a paradox in our preferred scenario of a plume-fed asthenosphere. If most of the mantle upwelling at plumes has its most-incompatible-element-rich components tapped at least to a small degree by plume melting, then OIB will be more incompatible-element-rich and more isotopically enriched than is the typical MORB created by melting the asthenospheric leftovers to plume melt-extraction (Phipps Morgan, 1999; Phipps Morgan and Morgan, 1999). If, however, the cooler rim-material of mantle plumes does not pressure-release melt during plume upwelling but becomes part of the asthenosphere, then it will melt when it starts to ascend beneath a mid-ocean ridge, making E-MORB (Phipps Morgan and Morgan [1999] discuss this observation and its resolution by a plume-fed plum-pudding asthenosphere).

Lateral flow within a plume-fed asthenosphere also offers an explanation for why both OIB and MORB have similar "flavors" within a given geographic area (e.g., the Dupal anomaly shared by both hotspot OIB and MORB within the Indian Ocean province), which is explored in more detail by Yamamoto et al. (this volume, chapter 10). Likewise, the preferential melting of easiest-to-melt plum components during lateral flow and upward-drainage beneath a lithospheric thin-belt provides a conceptually simple explanation for the Cameroon Line (Ebinger and Sleep, 1998). A plume-fed asthenosphere is also consistent with—although not required by—OIB-like basalts erupting during the initial phase of continental rifting. Lateral flow of plume-fed asthenosphere does, however, provide a simple explanation for the often-abrupt transition from rift OIB to drift N-MORB volcanism during continental rifting (Reston and Phipps Morgan, 2004). Note too, that in the plume-fed asthenosphere scenario plumes are not much hotter on average than the plume-fed MORB source, but both plume material and the plume-fed asthenosphere are hotter than the average underlying mantle. Thus these "OIB paradoxes" are only paradoxes in scenarios in which plumes are a small-mass-flux addition to the asthenosphere; they are to be expected in the scenario of a plume-fed asthenosphere.

31 January 2007, Don L. Anderson

Phipps Morgan and Morgan (comment of 31 January) argue that a plume-fed asthenosphere can explain one of the paradoxes associated with the standard version of the plume paradigm—the fact that E-MORB occurs at ridges unaffiliated with hotspots. This explanation also removes the rationale for "correcting" or filtering ridge data to avoid the influence of plumes. It opens up the narrow chemical spectrum that has long been attributed to depleted upper mantle so that it now includes chemical attributes that have been assigned to hotspots, or OIB. The concept of the convecting mantle has been associated with mid-ocean ridges and MORB; this concept is shorthand for "well-stirred homogeneous mantle." The idea behind this nomenclature is that inhomogeneities introduced into the upper mantle rapidly get stirred into it. The rationale is that homogeneous MORB requires a homogeneous source, which itself is a fallacy (Meibom and Anderson, 2003).

The arguments given by Phipps Morgan and Morgan in their comment (31 January) also apply to the mafic blob models in which the low-melting-point constituents are introduced from above (delamination, off-scaping of the lower crust, tectonic erosion, abandoned arcs and mantle wedges, and so on), as in the Galileo thermometer model (Anderson, this volume). Melting anomalies are then due to low-melting eclogite, rather than

to lower-mantle plumes. The amount of melt depends on the fertility of the mantle, not the absolute temperature. It is the homologous temperature (temperature divided by melting temperature) that is important. Eclogite blobs are dense at their starting (crustal) temperatures but become buoyant and partially melted at ambient mantle temperatures. They then deliver heat and magma to the surface. They can also get entrained in the asthenosphere counterflow, which sweeps them toward ridges and hotspots. In this respect, the fertile blob model differs from the plume-fed asthenosphere model; flow is toward, not away from, hotspots.

The concerns regarding the isothermal nature of MORB are addressed in my comment of 28 January 2007, on the chapter by Falloon et al. (this volume). MORB does seem to be extracted from a common depth and temperature region, but this source need not be the "top of the adiabatic convecting mantle." It is likely to be the depth of the intersection of the solidus with the conduction geotherm; the thermal boundary layer can extend deeper than this intersection. Variations in magma temperatures can be due to variations in depth, or lithology as well as to lateral variations in the geotherm.

7 February 2007, J. Godfrey Fitton

Natland's neat summary of Turner and Verhoogen's (1960) view of basalt associations, together with the other discussion comments and those of the article by Yamamoto et al. (this volume, chapter 10), nicely encapsulate the OIB paradox. No single model, nor even a plausible combination of models, can yet explain all the occurrences of OIB and OIB-like basalt. Heterogeneous asthenosphere, whether fed from below via plumes or enriched from above via delamination and other processes, might be able to provide an inexhaustible source of both OIB and MORB, but only with the application of some special pleading in critical examples. Two of these examples, cited by Phipps Morgan and Morgan serve to illustrate this point.

Phipps Morgan and Morgan claim that lateral flow of plume-fed asthenosphere can provide a simple explanation for the abrupt transition from OIB-like rift magmatism to oceanic N-MORB. This flow, however, requires that the onset of seafloor spreading and N-MORB magmatism be triggered by the arrival of plume-fed asthenosphere when the developing continental rift connects to the global mid-ocean ridge network, right at the point of continental separation (Reston and Phipps Morgan, 2004). The other example is the Cameroon line, a volcanic lineament that has been randomly and virtually continuously active for the past 66 Ma, composed largely of OIB on its oceanic sector and of compositionally indistinguishable basalt on the African mainland. Morgan and Phipps Morgan (this volume) appeal again to lateral flow of plume-fed asthenosphere, this time beneath a lithospheric thin-belt. They offer no explanation for why the lithosphere should be thin along the Cameroon line, or why it has remained thin for 66 Ma. There is no evidence for rifting; its continental volcanic edifices are built on uplifted basement rocks. Solving the mystery of the origin of this enigmatic feature may one day help to resolve the OIB paradox.

REFERENCES CITED

Anderson, D.L., 2007 (this volume), The eclogite engine: Chemical geodynamics as a Galileo thermometer, *in* Foulger, G.R., and Jurdy, D.M., eds., Plates, plumes, and planetary processes: Boulder, Colorado, Geological Society of America Special Paper 430, doi: 10.1130/2007.2430(03).

Davis, A.S., Gray, L.B., Clague, D.A., and Hein, J.R., 2002, The Line Islands revisited: New $^{40}Ar/^{39}Ar$ geochronologic evidence for episodes of volcanism due to lithospheric extension: Geochemistry, Geophysics, Geosystems, v. 3, p. 1–28, doi:10.1029/2001GC000190.

Ebinger, C.J., and Sleep, N.H., 1998, Cenozoic magmatism throughout east Africa resulting from impact of a single plume: Nature, v. 395, p. 788–791.

Falloon, T.J., Green, D.H., and Danyushevsky, L.V., 2007 (this volume), Crystallization temperatures of tholeiite parental liquids: Implications for the existence of thermally driven mantle plumes, *in* Foulger, G.R., and Jurdy, D.M., eds., Plates, plumes, and planetary processes: Boulder, Colorado, Geological Society of America Special Paper 430, doi: 10.1130/2007.2430(12).

Fitton, J.G., 2007 (this volume), The OIB paradox, *in* Foulger, G.R., and Jurdy, D.M., eds., Plates, plumes, and planetary processes: Boulder, Colorado, Geological Society of America Special Paper 430, doi: 10.1130/2007.2430(20).

Hirano, N., Takahashi, E., Yamamoto, J., Abe, N., Ingle, S.P., Kaneoka, I., Hirata, T., Kimura, J.-I., Ishii, T., Ogawa, Y., Machnida, S., and Suyehiro, S., 2006, Volcanism in response to plate flexure: Science, v. 313, p. 1426–1428.

Hofmann, A.W., and Hart, S.R., 2005, Another nail in which coffin?: Science, v. 351, p. 39–40. http://www.mantleplumes.org/WebDocuments/Which-Coffin.pdf.

Meibom, A., and Anderson, D.L., 2003, The statistical upper mantle assemblage: Earth and Planetary Science Letters, v. 217, p. 123–139.

Morgan, W.J., and Phipps Morgan, J., 2007 (this volume), Plate velocities in the hotspot reference frame, *in* Foulger, G.R., and Jurdy, D.M., eds., Plates, plumes, and planetary processes: Boulder, Colorado, Geological Society of America Special Paper 430, doi: 10.1130/2007.2430(04).

Natland, J.H., 2007 (this volume), ΔNb and the role of magma mixing at the East Pacific Rise and Iceland, *in* Foulger, G.R., and Jurdy, D.M., eds., Plates, plumes, and planetary processes: Boulder, Colorado, Geological Society of America Special Paper 430, doi: 10.1130/2007.2430(21).

Natland, J.H., and Winterer, E.L., 2005, Fissure control on volcanic action in the Pacific, *in* Foulger, G.R., et al., eds., Plates, plumes, and paradigms: Boulder, Colorado, Geological Society of America Special Paper 388, p. 687–710, doi: 10.1130/2005.2388(39).

Niu, Y., Collerson, K.D., Batiza, R., Wendt, J.I., and Regelous, M., 1999, The origin of E-type MORB at ridges far from mantle plumes: The East Pacific Rise at 11°20′N: Journal of Geophysical Research, v. 104, p. 7067–7087.

Phipps Morgan, J., 1999, Isotope topology of individual hotspot basalt arrays: Mixing curves or melt extraction trajectories?: Geochemistry, Geophysics, Geosystems, v. 1, 1999GC000004.

Phipps Morgan, J., and Morgan, W.J., 1999, Two-stage melting and the geochemical evolution of the mantle: A recipe for mantle plum-pudding: Earth and Planetary Science Letters, v. 170, p. 215–239.

Reston, T.J., and Phipps Morgan, J., 2004, Continental geotherm and the evolution of rifted margins: Geology, v. 32, p. 133–136.

Turner, F.J., and Verhoogen, J., 1960, Igneous and metamorphic petrology: New York, McGraw-Hill, 694 p.

Yamamoto, M., Phipps Morgan, J., and Morgan, W.J., 2007 (this volume, chapter 10), Global plume-fed asthenosphere flow—II: Application to the geochemical segmentation of mid-ocean ridges, *in* Foulger, G.R., and Jurdy, D., eds., Plates, plumes, and planetary processes: Boulder, Colorado, Geological Society of America Special Paper 430, doi: 10.1130/2007.2430(10).

The Geological Society of America
Special Paper 430
2007

ΔNb and the role of magma mixing
at the East Pacific Rise and Iceland

James H. Natland*

Rosenstiel School of Marine and Atmospheric Science,
University of Miami, Miami, Florida 33149, USA

ABSTRACT

ΔNb is a geochemical construct of Fitton et al. (1997) based on two trace-element ratios, Nb/Y and Zr/Y, plotted against one another on a log-log diagram. ΔNb is defined in relation to a diagonal line on the diagram separating Nb-enriched Icelandic tholeiite from Nb-depleted mid-ocean ridge basalt (MORB). Neither crystallization differentiation nor partial melting of a peridotite mantle source can drive compositions below the line, with negative ΔNb, to locations above the line, with positive ΔNb, or vice versa. The parameter ΔNb thus is taken to indicate distinct mantle sources.

Crystallization differentiation along the East Pacific Rise has almost no effect on ΔNb. The low and similar partition coefficients for Y, Zr, and Nb during shallow crystallization differentiation ensure that data points for primitive and evolved basalt as well as rare iron-enriched andesite and dacite plot virtually atop one another on the diagram, with all differentiates retaining negative ΔNb. Only enriched MORB (E-MORB) obtained from both ridge segments and the summits of some near-ridge seamounts, which in the extreme includes alkalic olivine basalt, has positive ΔNb. Normal MORB and E-MORB together produce a curving array that crosses the line on the diagram, and this crossing trend indicates heterogeneity in the source.

At Iceland, on the other hand, tholeiitic ferrobasalt and abundant andesite, dacite, and rhyolite are widely separated from primitive basalt on such a diagram, plotting at successively higher Nb/Y and Zr/Y at positive ΔNb. The most enriched rocks at Iceland are rhyolites with approximately the geochemistry of the inferred common mantle component, FOZO (focal zone). These rocks cannot be related to each other by either crystallization differentiation or partial melting of a peridotite source. Calculations show that mixing of only 1–2% rhyolite or ~5% evolved basalt with primitive basalt can shift Nb/Y, Zr/Y, and in certain circumstances ΔNb to higher values; it can also shift chondrite-normalized patterns of rare earth elements in primitive basalt from negative to positive slopes, a general attribute of Icelandic lava suites and the progression along Reykjanes Ridge approaching Iceland from the south. Both isotopic ratios and other trace-element concentrations and ratios are consistent with such mixing.

At Iceland, the usual differentiated and enriched mixing component is ferrobasalt with TiO_2 contents so high that it cannot be derived from commonly construed mantle peridotite; a Ti phase such as ilmenite, titanomagnetite, or—at high pressure—

*E-mail: jnatland@rsmas.miami.edu.

Natland, J.H., 2007, ΔNb and the role of magma mixing at the East Pacific Rise and Iceland, *in* Foulger, G.R., and Jurdy, D.M., eds., Plates, plumes, and planetary processes: Geological Society of America Special Paper 430, p. 413–449, doi: 10.1130/2007.2430(21). For permission to copy, contact editing@geosociety
.org.

rutile, must be in the source. Most picrites, on the other hand, have lower amounts of Nb, Y, Zr, and TiO_2 than primitive MORB. These attributes have led previous workers to propose a source in recycled ocean crust for both types of basalt, brought to the vicinity of Iceland in a mantle plume.

However, given the generally enriched character of all Icelandic basalt and the association with rhyolite, I suggest instead that they all are linked to a common source and that they represent successive fractional melts of gabbroic cumulates originally produced during differentiation of ancient granitic continental crust. Much later, during the early stages of the current rifting episode in the North Atlantic, the dense gabbro sank into the warming upper mantle by means of convective instability, converting to eclogite. As Greenland separated from Europe, the eclogite experienced decompression partial melting that contributed to the eruption of flood basalt, and some of it is retained in the Iceland melt source to this day. The high ΔNb of the basalt at Iceland is thus less an indication of mantle source heterogeneity than it is of mixing between depleted basalt and petrogenetically unrelated mafic to silicic differentiates with an ultimate origin in ancient continental crust. It is not evidence for the existence of a mantle plume.

Keywords: Iceland, silicic magmas, ferrobasalt, mid-ocean ridge basalt, delta niobium

It is an old idea that the more pointedly and logically we formulate a thesis, the more irresistibly it cries out for its antithesis.

Hermann Hesse, 1943, *Das Glasperlenspiel,*
translated in 1969 as *The Glass Bead Game*
by Richard and Clara Winston

INTRODUCTION

ΔNb was invented by Fitton et al. (1997) to provide a clear-cut discriminant between mid-ocean ridge basalt (MORB) and Icelandic tholeiite, with potential application to other places (Baksi, 2001; Fitton, this volume). It is based on two trace-element ratios, Nb/Y and Zr/Y, when plotted against one another on a log-log diagram, which can be called the ΔNb diagram. My attempts to use it to interpret several flood basalt provinces raise problems of both definition and interpretation. Briefly, these stem from (1) restricting the perspective just to certain types of basalt and (2) assuming that their geochemical signature comes strictly from peridotite plus variable amounts of crystallization differentiation. An important difficulty turns out to be identifying the effects of magma mixing between primitive and strongly differentiated (e.g., ferrobasaltic to rhyolitic) components, or alternatively assimilation of silicic materials in the crust by basaltic magma, something that occurs commonly in flood basalt provinces, and seeing through this to an original mantle signal. Such mixing can shift ΔNb. The boundary, $\Delta Nb = 0$, itself appears to be somewhat arbitrary, because its formulation depends on distinguishing between an array of just tholeiitic basalts (MgO > 5%) from Iceland and depleted normal MORB (N-MORB) (Fitton et al., 1997). Add less magnesian (including silicic) rocks to Iceland, and the full panoply of transitional and enriched MORB (T- and E-MORB) plus some silicic rocks to

depleted N-MORB from, say, the East Pacific Rise and its near-ridge seamounts or the Indian Ocean, and the picture changes.

High ionic potential (high ratios of charge, or valence, to ionic radius) means that Nb, Zr, and Y are almost insoluble in water, and thus not readily removed from crystal lattices during alteration or metamorphism; their high valence (+3 to +4) also means that they require coupled substitutions with other elements to maintain charge balance during crystallization. Consequently they have very low partition coefficients for silicate minerals at the melt stage (they partition strongly into melts). Thus these elements are ideal for the study of magmatic processes, even if the rocks are substantially altered. Because they can now be measured to a high degree of precision by at least a couple of techniques, they serve better for this purpose than elements that really do move around during alteration (e.g., the monovalent large-ion lithophile elements K, Rb, and Cs and, to a lesser extent, divalent Ba). This is especially important where volcanic glass is lacking, as in most subaerial exposures, and in older provinces where alteration is likely to be significant. Potential correlations with rare earth elements, which are also nearly immobile during alteration, and the isotopic ratios of Sr, Nd, and Pb are important. Because of their high charge relative to radii, the elements Zr and Nb, as well as Hf and Ta, have been termed high field strength elements (HFSE), and Y is usually grouped with them.

As I said, Nb, Zr, and Y strongly partition into melts during crystal-liquid separation processes, but not quite equally, and herein is their main utility. Partitioning theory can be used to establish how they should behave under particular conditions of partial melting and crystallization differentiation. If one or another of these elements, but usually Nb, does not conform to theory, the anomaly, which is to say the extent of the difference between theory and fact, can be attributed to source heterogeneity. Thus ΔNb, derived from a ΔNb diagram, is a measure of one such anomaly. Fitton et al. (1997) used positive values, especially high ones, to reveal what they termed "the thermal and compositional structure of the head of the ancestral Iceland plume."

Of the three elements, Nb is most strongly partitioned into melt during shallow crystallization differentiation, Zr is next, and Y is least partitioned, principally because Y has a slightly higher partition coefficient for clinopyroxene (D = 0.467) than for the other low-pressure silicate minerals (D = 0.17 to much less; e.g., Bédard, 1994). In essence, however, all three elements are strongly preferentially concentrated into residual liquids during shallow crystallization differentiation; thus the ratios Y/Zr and Nb/Zr do not change significantly during formation of gabbroic cumulates. That is, bulk distribution coefficients involving minerals that precipitate from basaltic magma into gabbro are extremely low for Nb and Zr, and only slightly higher for Y. The picture is somewhat different at higher pressure, inasmuch as Y partitions strongly into garnet (D = 2.8; Johnson, 1998).

Sometimes it is useful to think of the behavior of Y, Zr, and Nb in comparison to the particular rare earth elements (REE) that behave similarly (they have similar partition coefficients), but until recently have been far more difficult to analyze. Respective REE with similar behavior are, for Nb, the light REE, La and Ce; for Zr, Nd and Sm; and for Y, the heavy REE, Er and Yb. Thus during shallow crystallization differentiation of a single parental MORB, all REE except Eu (which is both divalent and trivalent, and thus has a high average partition coefficient for plagioclase depending on oxidation state) should increase in step on chondrite-normalized or mantle-normalized REE diagrams. Consequently, at many places in the ocean crust we can select a group of moderately to strongly differentiated depleted MORB with nearly parallel REE patterns (with a narrow range of, say, Ce/Yb), even to compositions as evolved as andesite (Fig. 1A), and we should expect the HFSE ratio Nb/Y not to change among these samples either (Fig. 1B). On a ΔNb diagram, data points should cluster in a small region regardless of their extent of differentiation (Fig. 1C). On the other hand, if the REE patterns of the basalts vary in steepness or there are crossing REE patterns, we should expect significant spread of data arrays on ΔNb diagrams. This is indeed what occurs at Iceland (light blue data field in Fig. 1C) and along different portions of the global system of spreading ridges (orange fields).

Thus the two ratios Nb/Y and Zr/Y respectively provide comparisons to the steepness of light/heavy and intermediate/heavy portions of REE diagrams. If we equate flat REE patterns to relative depletion and strongly sloping REE patterns (a positive linear correlation between Nb/Y and Zr/Y on the ΔNb dia-

gram) to relative enrichment, Nb/Y of basalt in particular is an index of the state of enrichment of mantle sources, *assuming that nothing has modified the signal imparted by partial melting in the mantle, and also assuming that the mantle itself is a single composition of peridotite.*

Formally, ΔNb is the distance above or below the sloping dotted line drawn in Figure 1C, which separates data fields for Icelandic tholeiites (light blue field) and MORB (orange fields), the latter including basalt from Reykjanes Ridge adjacent to Iceland (Fitton et al., 1997). The formula of the line is $1.92 \times \log(Zr/Y) = \log(Nb/Y) + 1.74$, so that for given samples $\Delta Nb = 1.74 + \log(Nb/Y) - 1.92 \times \log(Zr/Y)$. Positive values are Icelandlike; negative values are MORB-like.

However, bear in mind that ΔNb diagrams compare *ratios* on a log-log diagram. This gets away from dealing with concentrations altogether when presenting the results of models. Thus Fitton et al. (1997) came up with very tight linear trends and established a compositional dichotomy between N-MORB and Icelandic tholeiite with literally no overlap. Their partial melting models (blue lines in Fig. 1C), constructed assuming a homogeneous mantle and no interactions with crustal rocks, produced nearly linear relationships for liquids and solid residues for both spinel and garnet peridotite that closely parallel the line given by their equation. Note the more pronounced effect among fractional melts in the garnet stability field (G), an indication of the high partition coefficient of Y for garnet. However, also note that this does not drive the model trend across the line for ΔNb = 0. The "goodness" of the fit was enhanced by selecting among the array of possibilities between batch melting and fractional melting the proper amount of melt (2%) retained in the melt source following each step of incremental removal of partial melt. A similar set of models could be constructed assuming lower ΔNb, and this would explain the MORB array. By partial melting, at least, the Icelandic field with higher ΔNb could not be derived from depleted MORB mantle, and the source heterogeneity of trace elements is proven.

This sort of evidence for REE provided the initial geochemical rationale for the model of a plume beneath Iceland (Schilling, 1973a). Other workers then endorsed the concept, and Fitton et al. (1997) assumed its existence in order to use ΔNb to modify the model for an Iceland plume. Bear in mind that the assumption of mantle homogeneity on which the Fitton et al. (1997) melting models are based seemingly contradicts the conclusion that the mantle is heterogeneous in trace elements. One must thus further assume that the trace elements are present in such low concentrations that they behave as ideal dilute solutions and do not influence melting relationships among principal silicate phases. For the purposes of modeling, this is certainly convenient, but it leaves unexamined the question of whether the contrast in trace elements actually does indicate a difference in compositions of phases, or even the presence of other important melting phases, either of which could affect melting relationships, and it provides no formal comparison to melting models based on major oxides.

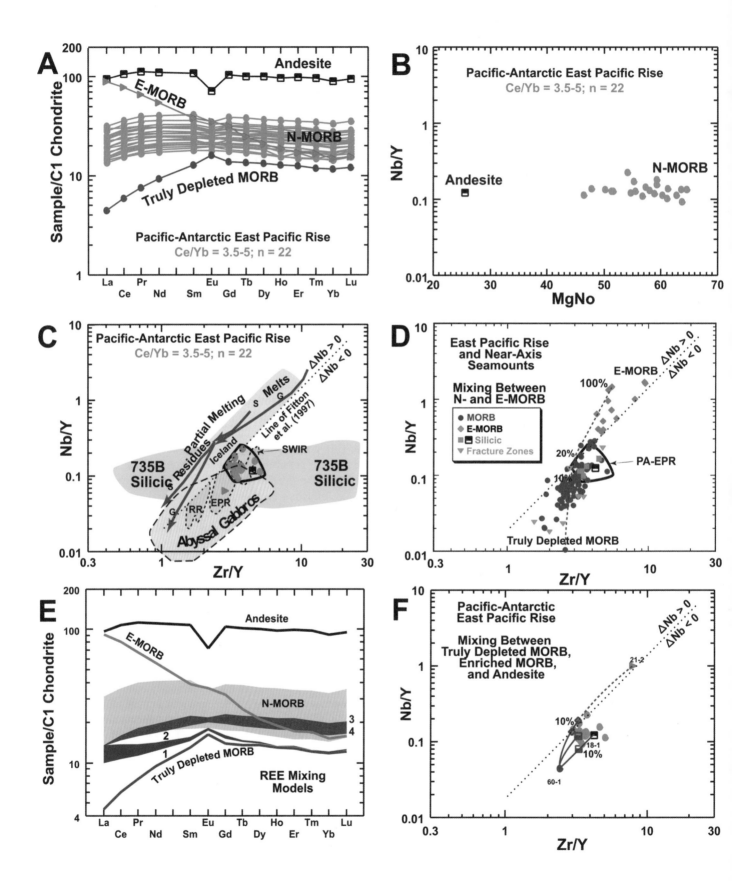

Now, a decade later, larger and more inclusive data sets reveal that trends for the East Pacific Rise, Reykjanes and Kolbeinsey ridges, Iceland, and other flood basalt provinces are oblique to the trends and line of Fitton et al. (1997), and that those of the flood basalt provinces flatten at higher Zr/Y. Evidently something other than partial melting is going on. By examining the inclusive data sets, I have discovered that, at all places I have investigated in detail (Iceland, the Karoo of South Africa, and the Deccan of India), what can be termed "end-member" mixing, between primitive and strongly differentiated magma types (viz., basalt and rhyolite), or its equivalent cast in terms of assimilation (contamination) processes, is very important, although it produces effects of different degree on ΔNb at the different places. The curvature of the trends is mimicked by simple model mixing calculations. Although such end-member mixing was already known from isotopes for the Deccan and the Karoo, the more general prevalence of it even among depleted compositions along spreading ridges has not previously been shown, nor have its potential consequences for interpretation of mantle geochemistry been considered. Quite simply, the likelihood of basalt-rhyolite magma mixing at these places, and even mixing between certain types of basalt, seriously compromises any ability to close in on the primary mantle geochemical attributes of parental basaltic magma. The effect is least along the East Pacific Rise, larger at Iceland, and most important at the continental flood basalt provinces where basalts erupt through ancient granitic crust. In short, so-called enriched mantle signatures at these places, or similarities to E-MORB, which are often interpreted to support the mantle plume hypothesis, are instead a signature of rhyolite or other strongly differentiated melt mixed into or contaminating depleted basalt. At all these places, the petrological and geochemical problem now should properly shift from concerns about the mantle to the origin and differentiation of rhyolite and associated strongly differentiated basalt and the mechanism of their interaction with primitive basaltic magma ascending through the crust.

In this article, I provide the crux of the mixing argument by comparing the East Pacific Rise and Iceland. In a paper in preparation, I consider the Karoo and Deccan provinces and the Deccan in light of subsequent volcanism along Chagos-Laccadive Ridge and Mascarene plateau, down to the active volcanic island Réunion. I develop an alternative to the modern hypothesis of plume heads and tails espoused initially by Morgan (1981), Campbell and Griffiths (1990, 1993), and Griffiths and Campbell (1991) to explain these and other places. The place to begin, however, is the East Pacific Rise.

Throughout, I have used data files available in the Petrological Database of the Ocean Floor (PetDB; available at http://www.petdb.org) of Lamont-Doherty Earth Observatory, and Geochemistry of Rocks of the Oceans and Continents (GeoRoc; available at http://georoc.mpch-mainz.gwdg.de) of the Max Plank Institut für Chemie.

THE EAST PACIFIC RISE

About equal proportions, roughly 6% each, of silicic lava (MgO < 5% and SiO_2 = 53–67%) and E-MORB (MgO > 5%; $[K/Ti]_N$ > 0.18, including alkalic olivine basalt) have been sampled along the axis of the East Pacific Rise between the Siqueiros and Clipperton fracture zones. Most eruptions occur along a shallow and very narrow fissured depression <1 km wide, sometimes called a neovolcanic zone, that is situated almost directly and ~1 km above a barely wider seismically detectable axial melt lens that is interpreted to be a persistent, indeed steady-state, feature that lies just beneath the base of sheeted dikes (Sinton and Detrick, 1992; Harding et al., 1993; Kent et al., 1993). From a single exposure through a partial section of fast-spreading crust at Hess Deep in the eastern equatorial Pacific including high-level gabbros, the melt lens is inferred to contain, *most of the time* (between eruptions), highly differentiated iron-rich basaltic magma capable of precipitating gabbronorite and oxide gabbro cumulates (Natland and Dick, 1996).

Figure 1. (A) Chondrite-normalized rare earth element (REE) concentrations in basalt from ridge segments on the Pacific-Antarctic East Pacific Rise near the Udintsev and Eltanin fracture zones (data from PetDB, the Lamont Petrology Database, obtained by Yao-Ling Niu). Basalt from major ridge segments (red dots) has slightly concave-downward and quite parallel REE patterns. One basalt from a short intratransform eruptive center (blue) is more strongly depleted in light REE, and one seamount basalt (green), is relatively enriched. The REE pattern of an andesite (half-filled black squares) parallels the group in red, but has a pronounced Eu anomaly. MORB—mid-ocean ridge basalt; N-MORB—normal MORB. (B) MgNo versus Nb/Y for the samples in panel A. Symbols are as in panel A. Nb/Y does not change during shallow crystallization differentiation that causes MgNo to decrease. (C) ΔNb diagram (log-log Zr/Y versus Nb/Y) for the samples of panels A and B (symbols as in panel A) compared with fields for other spreading ridges (RR—Reykjanes Ridge; EPR—East Pacific Rise; SWIR—Southwest Indian Ridge) as shown in Fitton et al. (1997). ΔNb is defined in the text. The line for ΔNb = 0, the Iceland data field, and model trends for melts and residues are also from Fitton et al. (1997). The model trends assume a homogeneous starting material in the spinel peridotite (S) and garnet peridotite (G) stability fields. The data fields for abyssal gabbros and 735B silicic rocks (diorites, tonalities, trondhjemites) are from Natland and Dick (2002). (D) ΔNb diagram for basaltic glasses from the East Pacific Rise, for near-axis seamounts between the Siqueiros and Clipperton fracture zones (symbol code in diagram), and for the Pacific-Antarctic East Pacific Rise (PA-EPR), from PetDB (the Lamont Petrology Database). Symbols as in panel A. See text for the explanation of the dashed mixing trend. Note the location of enriched MORB (E-MORB), generally above the line for ΔNb = 0, and other MORB, including truly depleted MORB, below the line. (E) Chondrite-normalized rare earth diagram based on panel A, showing field-bounded model mixing patterns 1–4 (see text) superimposed. (F) ΔNb diagram showing the effects of mixing truly depleted MORB with 10% E-MORB, 10% andesite, and a combination of the two using just the data for the Pacific-Antarctic East Pacific Rise depicted in panels A and B.

The gabbros are crosscut in places by tonalite or trondhjemite veins and dikelets. These rocks are present at the top of the gabbros because the melt lens is nearest hydrothermal reaction zones in fractured basaltic crust near the base of the dikes, and thus it tends to contain, *on average,* the coolest, therefore most strongly differentiated, magmas in the gabbroic layer.

At the eruptive stage, the dynamics of spreading at a fast-spreading ridge constrain most magma migrating from the mantle and the consolidating gabbroic layer of the crust to flow from a broad base more than 100 km wide (MELT Seismic Team, 1998) toward and through the narrow melt lens. Thus primitive to moderately differentiated magma encounters the melt lens and necessarily entrains and mixes with the cool and strongly differentiated liquids there contained (Natland and Dick, 1996). The economy of basalt along the East Pacific Rise therefore consists of three extremes: (1) depleted primitive MORB derived from the mantle; (2) E-MORB, also derived from the mantle; and (3) strongly differentiated mafic and silicic magma, the melt composition of the latter being sodic rhyolite. In every geochemical respect, the depleted basalt that we call N-MORB lies between these extremes, and its average, even after screening out E-MORB, as many geochemists do, is not itself an end-component.

Because almost all melt migration paths lead to narrow melt lenses, depleted primitive MORB derived from the mantle almost never erupts above the melt lens without first mixing with some fraction of E-MORB and, at the melt lens, strongly differentiated mafic and silicic magma. Sometimes magma entry into the melt lens from the sides and below purges the strongly evolved magma from the lens, and it erupts as composite basalt-andesite lava flows (Natland, 1991a). Where primitive MORB does erupt without such mixing, or at least with less such mixing, is along the floors of transform faults, on very near-ridge seamounts, and at the tips of rifts propagating within micro-plates (e.g., Natland, 1980; Allan et al., 1989, 1996; Sinton et al., 1991; Perfit et al., 1996; Niu et al., 2002a). These are all places where melt lenses have not been detected seismically. Basalt at such locations tends to have MgO > 8% (some are picrites), to be strongly depleted in light REE (Fig. 1A), and to have low Nb/Y and Zr/Y and the most negative ΔNb (Fig. 1D). The great majority of depleted MORB that erupts along the neovolcanic zone has flatter REE patterns, some with negative Eu anomalies, higher Nb/Y and Zr/Y, and higher, albeit still negative, ΔNb, than much basalt from transform faults. E-MORB is enriched in light REE, but has steep REE patterns overall (Fig. 1A), and they lack Eu anomalies. Most silicic lavas have approximately the composition of iron-rich andesite; they have high concentrations of REE, but still flat patterns and pronounced Eu anomalies. As already noted, these evolved rocks differ but little from typical N-MORB of the neovolcanic zones in Nb/Y, Zr/Y, and ΔNb.

To distinguish the primitive and more strongly depleted basalt of transform faults and seamounts from those that erupt above melt lenses, I shall term the former "truly depleted MORB" and postulate that it is representative of depleted mantle regionally along the East Pacific Rise. *Truly depleted* means, for example, exhibiting strongly negative slopes to the left-hand portion of chondrite-normalized REE (the light REE; see example in Fig. 1A), but also extremely low concentrations in glasses of, e.g., K_2O (0.01–0.03%), Ba (1–2 ppm), and Rb (0.1–0.3 ppm). N-MORB erupted above melt lenses consequently begins its existence as truly depleted MORB and becomes slightly more enriched because of almost inescapable mixing with small amounts of E-MORB and evolved melt-lens magma. It is also more differentiated (lower MgO and MgNo) because of residence within the cooling lower ocean crust. Because E-MORB exists, however, the extent of magma mixing between enriched and truly depleted magmas must be variable, even allowing fairly strongly enriched alkalic olivine basalt to erupt some of the time along or very close to the ridge axis (Batiza et al., 1977; Natland and Melson, 1980). Based on the statistics of rock compositions, the mantle consequently may itself be compartmentalized into a majority (~90%) of truly depleted material and a minority (~10%) of enriched material (the plum-pudding mantle).

The alternative might be to consider that the statistically most prominent group of basalts (neovolcanic N-MORB) represents most of the mantle, and that transform faults and seamounts tend to sample the minority extremes of truly depleted and enriched mantle. But why should this be? The nearly universal occurrence of melt lenses along the East Pacific Rise and their arrangement directly beneath fissured neovolcanic zones along the ridge axes seemingly compels mixing among magmas derived over broad regions, whereas seamounts and small volcanic structures in transform faults provide spot samples of the underlying mantle without magmas having to pass through 4 km of partly molten gabbro and a melt lens. The extent and scale of mantle diversity thus is blurred along ridge axes, and revealed at seamount clusters and small intratransform volcanoes. Basalt erupted along ridge axes is still depleted, but not nearly as depleted as basalt that bypasses the axial mixing regime. The mixing ensures higher Nb/Y, Zr/Y, and ΔNb among axial N-MORB even before significant differentiation occurs to produce parallel REE patterns at nearly constant Nb/Y and Zr/Y as MgNo decreases. Mixing is the reason that the array of hybrids at the East Pacific Rise in Figure 1D is oblique to the line of Fitton et al. (1997). Simple calculations will show this.

First, does the oblique array result from mixing between truly depleted MORB and E-MORB only? The answer is no. With increasing proportion of E-MORB, a mixing calculation for the extreme compositions shown in Figure 1A predicts at first very little change in Zr/Y as Nb/Y increases—a nearly vertical relationship. At ~20% addition of E-MORB, the calculated trend flattens a bit. This is because on a log-log diagram of ratios, mixing trends curve and the proportional influence of small amounts of E-MORB on low concentrations of truly depleted MORB is very high. Thus initially not too much E-MORB is required in mixes to increase ΔNb significantly. Perhaps 3–5% of E-MORB in the mix with truly depleted MORB will place the hybrid within or close to the main data cluster. Nevertheless, no matter which truly depleted MORB is chosen, the resultant mixing trend misses the field for axial N-MORB at the Pacific-Antarctic East Pacific Rise. Because the mixing trend shown is calculated using the truly depleted MORB with highest Zr/Y, the misfit would be worse using any other truly depleted composition with lower Zr/Y. Furthermore, whereas mixing of 6–11% of E-MORB into truly depleted MORB will flatten REE patterns (models 1 and 2, respectively, bounding the brown field in Fig. 1E), the light REE portions of the calculated patterns are concave upward rather than downward.

To obtain concave downward patterns, mixing between some variety of silicic magma and truly depleted MORB is required. In Figure 1E, model 3 (bounding the top of the blue field), is a calculated mix of 90% truly depleted MORB and 10% andesite from Figure 1. The andesite itself, which has ~57% SiO_2, is approximately a 2:1 mixture between a very iron-rich basaltic liquid with ~50% SiO_2 and rhyodacite with ~70% SiO_2 (cf. Natland, 1991a). To emphasize this hybrid aspect of its composition in this discussion, I shall henceforth term it "ferroandesite." That is, a possible mix with ferroandesite works out to ~92% truly depleted MORB, 5% ferrobasalt, and 3% rhyodacite (model 4). This is almost equivalent to 93% truly depleted MORB, 5% E-MORB, and 2% rhyodacite, which is to say that finding a unique mixing solution is probably futile. The resulting shifts in Nb/Y, Zr/Y, and ΔNb are shown in Figure 1F.

If the concave downward REE patterns of most N-MORB are produced by small degrees of mixing with strongly differentiated liquids in axial melt lenses, the concave downward shape of those patterns is clearly tied to the origin of those siliceous melts. The problem is complex and poorly understood. Crystallization differentiation, partial melting of hydrous mafic ocean crust, and late-stage liquid immiscibility have all been proposed to explain the origin of rhyodacite and affiliated ferrobasalt produced experimentally from MORB (Dixon and Rutherford, 1979; Spulber and Rutherford, 1983; Juster et al., 1989). The REE distributions are probably dependent on the influence of accessory phases that crystallize during the later stages of differentiation of basalt and gabbro (DeLong and Chatelain, 1990), namely magmatic amphibole (pargasite), alkali feldspar, ilmenite, titanomagnetite, apatite, zircon, and perhaps others. Ilmenite in particular has a high partition coefficient for Y, and at some level its fractionation must be important in causing reduction in TiO_2 contents and increases in SiO_2 contents among MORB silicic residua. This could be the cause for the small but distinct divergence of silicic compositions to the right in the data array for the East Pacific Rise shown in Figure 1D. Such divergence coupled with mixing pulls the overall MORB array without E-MORB into an orientation parallel to the line of Fitton et al. (1997).

Mixing between primitive MORB and its own differentiated residua has rarely been considered, let alone understood to be an impediment to understanding MORB mantle geochemistry. However differentiated depleted MORB may be, it is almost uni-versally interpreted as a probe of the mantle and related to peridotite by simple combinations of partial melting and crystallization differentiation. On a ΔNb diagram, the potential effects of mixing may seem subtle. However, consider one general result from the drilling of abyssal gabbro, namely that most of them are *ad*cumulates, with little or no retained interstitial melt (Natland et al., 1991; Natland and Dick, 1996, 2002). Because HFSE are strongly excluded from crystals in melts, and almost all residual melt is then expelled from the lower ocean crust by a combination of crystal compaction and adcumulus growth, the effective HFSE budget in the ocean crust is transferred almost entirely to the upper crust of basalts and dikes. Almost all Nb, for example, is there and not in the gabbros. But how did it get there? It was first strongly concentrated in residual liquids at the melt lens, and there entrained by more primitive basalts en route to eruption, which removed it from the gabbros. Given that gabbros represent perhaps 75% of the ocean crust, this implies a roughly fourfold enrichment of Nb in basalts and dikes compared with whatever its average concentration was in all magmas supplied to the crust from the mantle (both N- and E-MORB and their hybrids produced in the mantle). However, because of the modest retention of Zr in clinopyroxene, and the stronger retention of Y in both clinopyroxene and ilmenite, transfer of these two elements from the bulk of gabbroic cumulates into basalts and dikes is not as efficient as that of Nb. On the average, then, the Nb/Y and Zr/Y of basalts and dikes along spreading ridges will be higher than in truly depleted MORB supplied from the mantle; thus average MORB will not precisely indicate mantle ratios.

Even so, whatever the mechanism of the origin of strongly differentiated melt in the ocean crust, and its potential for mixing with primitive basalt, the impact on Nb/Y, Zr/Y, and ΔNb is not really very great along the East Pacific Rise. Wholesale shift of depleted MORB from negative to strongly positive (Icelandic) values of ΔNb is not evident in the N-MORB spectrum of compositions.

The ΔNb diagram is not ideal for displaying the type of mixing described so far or for illustrating its prevalence. Any consideration of actual concentrations is more informative. Let at least one axis show a concentration. In Figure 2, log(Nb/Y) versus Zr, two data sets for the East Pacific Rise are considered, one combining three provinces for which high-precision ICP-MS data are available, the other all data, however measured, for samples from the well-studied segment of the East Pacific Rise between the Siqueiros and Clipperton transform faults (~8.5°–10.5°N), including off-axis seamounts. Both have the same features, two general trends, between truly depleted MORB and E-MORB, the other branching less steeply toward the higher Zr concentrations of strongly differentiated compositions, mainly ferroandesite. Seamount data conform chiefly to the former; axial N-MORB is more evenly divided, with prongs of data heading in both directions. The branch pointing toward higher Zr indicates the prevalence of mixing between primitive basalt and silicic compositions along the axial neovolcanic zone of the East Pacific Rise.

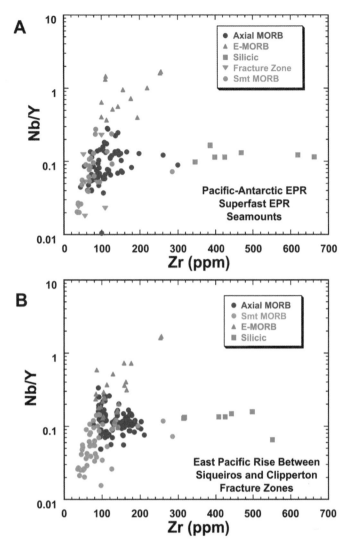

Figure 2. Zr (ppm) versus Nb/Y diagrams for (A) samples from the Pacific-Antarctic East Pacific Rise (EPR) and seamounts near the superfast portion of the East Pacific Rise and (B) the East Pacific Rise between the Siqueiros and Clipperton fracture zones. All data are from PetDb (the Lamont Petrology Database). Symbol keys are in each diagram. MORB—mid-ocean ridge basalt; E-MORB—enriched MORB; Smt—seamount.

Note also that the Nb/Y values of the ferroandesite compositions are ~0.1 (ΔNb ~ 0). This appears to represent the maximum effectiveness of all aspects of differentiation in the ocean crust, including mixing to produce Nb enrichments in an environment dominated by truly depleted MORB. However, much higher Nb/Y and ΔNb values occur at other places.

Could arrays between truly depleted MORB and E-MORB represent differences in extent of partial melting? Not according to the models of Fitton et al. (1997); not, that is, if the mantle consists of a single homogeneous peridotite. Partial melting of a plum-pudding mantle, however, to produce trends parallel to the model trends of E-MORB and N-MORB, but within the

regional or local arrays of E-MORB and N-MORB separately, and subsequent blending of depleted and enriched magma strains, might well obscure their separate Fittonian compositional vectors in Figure 1, C and D, and steepen the overall trend. Mixing with ferroandesite distorts the picture even further.

Fitton et al. (1997) based their MORB array in Figure 1C on depleted compositions (N-MORB; MgO > 5%) from three widely separate portions of the ocean floor, one in the Atlantic, one in the Pacific, and one in the Indian Ocean. Their modeling suggests that the difference between the Southwest Indian Ridge and Reykjanes Ridge, with the East Pacific Rise in between, is a matter of increasing the extent of partial melting, i.e., in response to cooler mantle supplying ridges in the Indian Ocean, warmer mantle beneath the East Pacific Rise, and even warmer mantle near Iceland. This is also the general conclusion reached by Klein and Langmuir (1987) and by Langmuir et al. (1992) based on their melt column modeling and consideration of regional differences in parental values of Na_2O and total iron as FeO (Na_8 and Fe_8, both corrected for fractionation to 8% MgO).

The crystallization histories of porphyritic basalt from the Siqueiros fracture zone, however, suggest that the full range of parental compositions (the "global array" of Klein and Langmuir, 1987) is actually available within the reach of a single melting domain beneath the East Pacific Rise (Natland, 1989), and capable of supplying melt strains of quite variable composition to a single basalt. Like most xenolith suites, Alpine peridotites, and fracture zone assemblages of abyssal peridotite, the mantle source of MORB cannot literally be homogeneous at all scales. It must be lithologically heterogeneous on the scale of a single melting domain containing an array of fertile (lherzolitic) to refractory (nearly harzburgitic) lithologies, all of which can produce some type of basaltic liquid at the same temperature in a melting domain—a large range altogether—at a single location along the ridge axis. Magma mixing beneath the ridge axis produces a preponderance of slightly less depleted axial N-MORB, compared with truly depleted MORB, that gives a false impression of mantle homogeneity despite the likely heterogeneity.

In Figure 1D, the array of truly depleted MORB spans both the original array for the East Pacific Rise of Fitton et al. (1997) in Figure 1C and that of Reykjanes Ridge. In accord with Natland (1989), Figure 1D therefore indicates the presence of relatively fertile lherzolite and refractory harzburgite together in the same mantle beneath the East Pacific Rise. The Fitton et al. (1997) partial melting trend exists, but more than that, it *pre-exists* in peridotites on a local scale in the mantle. Thus some prior melting event, presumably an ancient one like formation of continental crust, explains the current local variability. I take the scattering of truly depleted MORB to be representative of the lithologic variability of depleted mantle beneath the East Pacific Rise. The depleted portion of the mantle beneath the East Pacific Rise is more like that of Reykjanes Ridge than previously realized, but it is leavened both with more fertile, yet still depleted, peridotite, and with the sources of E-MORB. Discovery of very refractory shallow harzburgite in the mantle transition at Hess Deep (Arai and Matsukage, 1996; Dick and Natland, 1996) that is still associated with fairly typical N-MORB accords with this view. The variability indicated by the separate MORB data fields in Figure 1C and the Na_8 evaluations of Klein and Langmuir (1987) and Langmuir et al. (1992) suggest that the lithological variability of the mantle may also exist on a regional scale that has nothing to do with either current average extents of melting or mantle temperatures (Presnall et al., 2002).

Overall, ΔNb diagrams provide far more compelling evidence for mantle source heterogeneity than for homogeneity beneath the East Pacific Rise and, together with these other techniques, provide convincing evidence that three varieties of magma mixing are important beneath spreading ridges: (1) mixing among magma strains derived from varieties of depleted peridotite, (2) mixing between these magmas and strains of E-MORB, and (3) mixing of primitive products of 1 and 2 with silicic melt in the vicinity of axial melt lenses. The latter two are evident in Figure 2. Models of partial melting based on the assumption of a homogeneous mantle do not encompass all the phenomena revealed by careful evaluation of ΔNb diagrams. The quantity ΔNb nevertheless is based on a circumstantial distinction between Icelandic tholeiite and certain depleted MORB that does not properly display its typical variation, in part because of properties of the diagram that tend to obscure or minimize the appearance of mixing relationships with enriched MORB and silicic lavas that until now have been left out of the diagrams.

ICELAND AND ADJACENT RIDGES

Magma Mixing at Iceland

Besides being an emergent portion of the Mid-Atlantic Ridge and having crust up to 40 km thick, Iceland has a significant proportion, perhaps 10%, of silicic differentiates, much of it rhyolite, distributed among its basalts (Walker, 1963). The silicic rocks are mainly concentrated at Iceland's several dozen extinct and active central volcanoes, and are widely considered to be differentiates of the tholeiitic basalt that abounds at Iceland (e.g., Carmichael, 1964). I shall challenge this conclusion. The differentiation is thought to take place in large magma chambers beneath the central volcanoes, and the remains of some of these are exposed as plutonic complexes where older volcanoes are cut by glaciers. Lateral drainage of one such magma reservoir was responsible for the large Laki fissure eruption in 1783. Crystallization differentiation may also produce intermediate silicic rocks—andesite (also called icelandite) and dacite—but they probably result more commonly from mixing between basalt and rhyolite. Evidence for such mixing includes composite lava flows and dikes, quenched lava with differently colored blebs of basaltic and silicic material (so-called emulsion rocks), occurrences of pillows injected into coarse-grained silicic plutons, and complex interfingering of "acid" and "basic" constituents at interfaces between silicic and basaltic magmas that crystallized

to coarse grain size in zoned plutons (e.g., Gibson and Walker, 1964; Walker and Skelhorn, 1966; Yoder, 1973; Sigurdsson and Sparks, 1981; Marshall and Sparks, 1984; MacDonald et al., 1987). However, such intimacy is no proof of a relationship between the two by crystallization differentiation in the crust.

Among Iceland's basalts, picrites are least removed from primitive parental compositions derived by partial melting in the mantle. They typically occur as small flows on fissure systems in between the central volcanoes (e.g., Breddam, 2002), where their feeders bypass most contact with the central vents. They are picrites in the sense that they have accumulated olivine phenocrysts, and thus have MgO contents up to 23%, but plagioclase and clinopyroxene phenocrysts occur in many of the same rocks (e.g., Hansteen, 1991; Sigurdsson et al., 2000; Slater et al., 2001; Maclennan et al., 2003). The most magnesian Icelandic glass so far found, from Kistufell in central Iceland, has only 10.56% MgO (Breddam, 2002). Everything more magnesian than this likely has accumulated olivine. In further discussion, I shall term all analyses with >10% MgO "picritic," meaning "likely to have accumulated olivine phenocrysts," even though the International Union of Geological Sciences definition of picrite based on chemistry has MgO ≥ 12%.

At Iceland as well as at several other flood basalt provinces, including Deccan and Karoo, basalt compositions vary because of both their extent of differentiation and their parental attributes. One parental attribute that varies strongly is TiO_2 content. Petrologists thus distinguish low-Ti from high-Ti basalts at these places, but in fact there is a full continuum between them. Thus there are lower-Ti and higher-Ti basalts, and some rather arbitrary boundary between the two has to be chosen. At Iceland, both types include picrites (Fig. 3A); those with highest TiO_2 are similar to ferropicrites of other flood basalt provinces (e.g., Gibson et al., 2000). Whether all picrites crystallized from similarly primitive liquids might be questioned, but because TiO_2 also increases in residual compositions with extent of differentiation, an appropriate boundary should take into account both the effects of olivine accumulation in samples with high MgO content and those of subsequent multiphase cotectic differentiation in samples with lower MgO content.

At Iceland, a significant proportion of picritic basalt has TiO_2 contents < 1.4% (Fig. 3A) and appears to form a group. If we take this group to have genetic significance, 1.4% would be a reasonable maximum near-parental value for TiO_2 contents at, say, MgO = 8%. Then for samples with <8% MgO we can assume a MORB-like multiphase cotectic relationship and calculate the parental parameters, termed Ti_8, Fe_8, and Na_8, correcting all differentiated basalts by means of average slopes to MgO = 8% using the linear relationships of Klein and Langmuir (1987) and Langmuir et al. (1992). Actual suite-specific calculations could be made using somewhat different slopes, but for the overall data, a very large data set that includes data from dozens of localities for which the slopes are uncertain, this is neither practical nor feasible, and it would not result in significantly different estimated parental values.

The highest TiO_2 contents are in basalt with ~3.5% MgO content. Thus for all basalt samples with >3.5% MgO, I define low-Ti basalt as having <1.4% Ti_8 and higher-Ti basalt as everything with greater Ti_8, and I distinguish these by symbol in Figure 3A. Ti_8 can be calculated for picrites by assuming a predominant olivine-addition control (nominally ~Fo_{86} based on a linear regression of Al_2O_3, not present in olivine, to 0% MgO for all picrites) and correcting TiO_2 contents by olivine subtraction. For this calculation, after considering other binary variation diagrams, I include samples with MgO contents between 8 and 10% with picritic compositions that are likely olivine controlled; thus the break between the two methods of calculation is at 8% MgO contents. I define rocks with analyses showing <3.5% MgO and/or >55% SiO_2 as silicic (andesite, dacite, and rhyolite) and exclude them from the calculation. Figure 3A thus shows the differences between all basalts related to low-Ti and higher-Ti parents regardless of extent of differentiation or olivine accumulation. Among more than 2400 analyses of basalt in the GeoRoc Iceland data file, analyses of higher-Ti basalt exceed

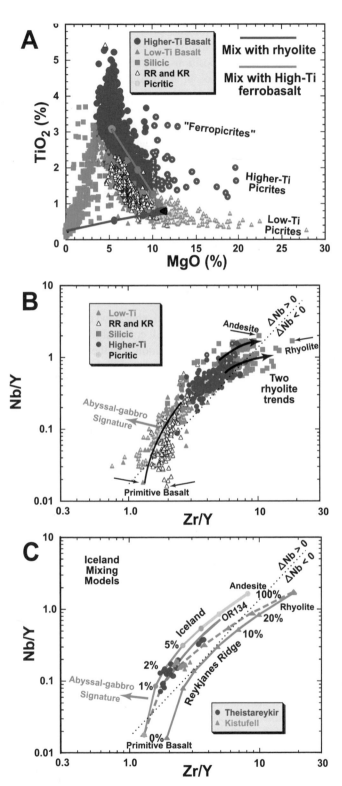

Figure 3. General geochemical relationships of basalt from Iceland and the adjacent Reykjanes and Kolbeinsey ridges (RR and KR). All data are from GeoRoc (Geochemistry of Rocks of the Oceans and Continents of the Max Plank Institut für Chemie) and PetDb (the Lamont Petrology Database). Symbol keys are in each diagram. The Iceland samples are divided into low-Ti and higher-Ti groups, as discussed in the text. The picritic samples have MgO ≥ 10% and are separately indicated by a smaller yellow dot superimposed on the symbols for low-Ti and higher-Ti compositions. Silicic rocks (>55% SiO_2 or <3.5% MgO) are mainly differentiates of higher-Ti basalt, although some are hybrids of basalt and rhyolite. (A) MgO versus TiO_2 diagram. The boundary for discerning low-Ti from higher-Ti basalt is based on picritic compositions, but is extrapolated to analyses with <10% MgO based on computation of Ti_8 (low-Ti basalt has Ti_8 < 1.4%). On this basis, most glass analyses from Reykjanes and Kolbeinsey ridges are low-Ti basalt. The diagram also shows two hypothetical mixing trends, one of basalt with rhyolite and one of basalt with a representative differentiated higher-Ti basalt. Only a few analyses appear to represent mixing of significant proportions of rhyolite with low-Ti basalt, although mixing with small amounts of rhyolite may be commonplace. Regardless, strict mixing with rhyolite cannot increase the TiO_2 content of low-Ti basalt. (B) ΔNb diagram for the samples depicted in panel A, with the same symbols. The line for ΔNb = 0 is indicated by a dotted line. The solid curving line separates most data for low-Ti picrite from Iceland from data for low-Ti basalt from Reykjanes and Kolbeinsey ridges. The displacement of low-Ti basalt from Iceland to lower Zr/Y is one indication of the involvement of abyssal gabbro from recycled ocean crust in the mantle source (see text for further explanation and references). The short arrows indicate the samples used for mixing calculations in panel C. The curving arrows among the silicic lavas highlight two trends among extended differentiates with somewhat different Nb/Y. (C) ΔNb diagram showing calculated mixing trends (blue and green curves) between the samples indicated by arrows in panel B. Tick marks along the curves indicate 1%, 2%, 5%, 10%, and 20% proportions of the silicic mixing component to each of two primitive basalts, one a low-Ti basalt from Iceland, the other a low-Ti basalt from Reykjanes Ridge. Data points for high-MgO basalt from Theistareykir (northern Icelandic rift) and Kistufell (central Iceland) volcanoes are plotted for comparison with the model mixing lines.

those of low-Ti basalt by a factor of ~1.5, but fewer than 5% of them are picritic, whereas nearly 25% of analyses of low-Ti basalt are picritic.

By these criteria, almost all basalt from Reykjanes and Kolbeinsey ridges, adjacent to Iceland to the south and north, respectively, is low-Ti basalt. Only a few rocks analyzed from Reykjanes Ridge have >1.4% Ti_8, and these are all north of the southern edge of the Icelandic shelf; there are none from Kolbeinsey Ridge. Many of the analyses are of glasses, none of which is picritic. Only four Reykjanes and Kolbeinsey whole-rock analyses are picritic. As far as is known, therefore, the entire region lacks picrite glass (>12% MgO). Nevertheless, Reykjanes and Kolbeinsey basalt can be distinguished from low-Ti Icelandic basalt. It is more MORB-like in general attributes, with somewhat higher SiO_2 and lower Na_2O at given MgO contents, lower Zr/Y, and somewhat higher REE concentrations.

On a ΔNb plot (Fig. 3B), all data from both Iceland and adjacent ridges form broad, curving arrays, with most primitive basalt including picrite lying on the steeper left-hand portions of the arrays. Basalt from Reykjanes and Kolbeinsey ridges has a sharper curvature than low-Ti Icelandic basalt; thus the Reykjanes and Kolbeinsey samples fall toward higher Zr/Y for given Nb/Y, with many having negative ΔNb. Samples from Reykjanes Ridge north of 61°N, close to the Iceland shelf, systematically have positive ΔNb and were screened from inclusion with depleted basalts of Reykjanes Ridge south of 61°N by Fitton et al. (1997), because only the latter are considered a variety of depleted MORB. However, whether sampled from south or north of 61°N, all basalt from Reykjanes Ridge is nonpicritic, has low Ti_8, and stands apart from low-Ti Icelandic basalt in terms of ΔNb. Therefore, I group them together to illustrate that they actually form a curving array on a ΔNb diagram, and, very like basalt from the East Pacific Rise (when E-MORB are included), they cross the line for ΔNb = 0. Basalts north of Iceland along Kolbeinsey Ridge do the same, but without a gradient to positive ΔNb as the Iceland shore is approached; samples with both positive and negative ΔNb occur in close proximity.

Higher-Ti basalt falls along the flatter portion of the Iceland array at positive ΔNb, plotting at about the location of E-MORB for the East Pacific Rise. Silicic compositions are concentrated toward the upper right at high Nb/Y, with rhyolite per se having the highest Zr/Y, but they have both positive and negative ΔNb. They partly overlap higher-Ti basalts, including some picrites. The silicic compositions appear to break into two trends at highest Zr/Y and Nb/Y; the trend with higher Nb/Y consists mainly of samples from Snæfellsnes, an off-axis locality in western Iceland, and Örafæjökull, a volcano of the eastern rift.

Based on the previous evaluation of the East Pacific Rise, the great separation between most of the primitive basalt at one end of the diagram and rhyolite at the other cannot be the result of closed-system crystallization differentiation. Nor can these placements be the result of differences in extent of partial melting of a peridotitic source; rhyolite is not a melt of peridotitic mantle. By either simple process, rhyolite at Iceland is therefore shown to be unrelated to primitive (including picritic) low-Ti basalt.

Figure 3C shows estimated effects of mixing between two silicic compositions, one an andesite and the other a rhyolite, and two primitive basalt compositions (small arrows). The compositions were selected because they lie at the extremes of the data array. Because this is a log-log, ratio-ratio diagram, the mixing trends curve, but it is fair to say that their shapes and the locations of particular intermediate mixing proportions depend strongly on actual end-member concentrations. The calculated curves thus provide examples only. Nevertheless, the effects of even a small amount of mixing, say 1–2% of either the andesite or the rhyolite, with primitive depleted basalt is clearly sufficient to effect a strong change in ΔNb. After that, a higher proportion of silicic material in the mixes accomplishes less change in ΔNb; the mixing curves flatten until they roughly parallel the line of Fitton et al. (1997) for ΔNb = 0, with one curve above and the other just about on the line. The calculated mixing trends match the curving data array quite well (Fig. 3B). Data for primitive basalt from two locations on Iceland, Theistareykir (northern rift) and Kistufell (central Iceland, over the thickest crust), nearly parallel the line of Fitton et al. (1997) in Figure 3C, and thus could conform to their partial-melting model. But their trend is also well reproduced by a calculated mixing trend (dashed green line) between more primitive basalt and 1–3% rhyolite.

A question raised in review is whether several of the Icelandic low-Ti basalts shown in Figure 3B, including the one I have used for this calculation, actually have negative ΔNb. The concern is whether the Nb measurements are of sufficient quality at low concentrations, because the data are, for example, neither recently obtained nor normalized, as is the practice of Fitton et al. (1997) and Fitton (this volume), to a well-known standard (G. Fitton, 2006, personal commun.). The particular sample also fails a test of internal consistency with other trace-element determinations, notably the ratio Nb/La, an indication that Nb has not been properly measured. For this sample, I agree. Curving trends notwithstanding, there is indeed no evidence that any Icelandic basalt properly analyzed as a whole rock has negative ΔNb. Some olivine-hosted melt inclusions from Reykjanes peninsula have negative ΔNb (Gurenko and Chaussidon, 1995), but again, Nb measurements in inclusions at low concentrations are probably suspect. Other olivine-hosted inclusions from Theistareykir volcano in northern Iceland (Slater et al., 2001) may also have slightly negative ΔNb (Fig. 3B), based on more precisely measured ion microprobe determinations of La and estimates of Nb made by assuming Nb/La from Theistareykir whole-rock compositions (Stracke et al., 2003; ppm Nb = 1.13 × La – 012; correlation coefficient r = 0.97).

Whether some few data fall slightly below the line for ΔNb = 0 is beside the point. The more important question is: Why is the Iceland trend so flat? Apart from rhyolite, this comes down to asking why Icelandic picrites with very low concentrations of Nb, Y, and Zr nevertheless have lower Zr/Y than basalt from

Reykjanes and Kolbeinsey ridges. More generally, what is it about Iceland that makes its most primitive basalt different in this way from that of its adjacent ridges?

My calculations also show that large shifts in Nb/Y and Zr/Y, and in some cases shifts in ΔNb from negative to positive values, can be accomplished by mixing small amounts of rhyolite into primitive, depleted basalt at Iceland and along its adjacent ridges. However, this is only because, contrary to widespread and long-standing opinion, Icelandic rhyolite is not related to the basalt either by partial melting of a homogeneous peridotite mantle or by subsequent closed-system crystallization differentiation. It has a very different origin, as discussed later. Thus the situation is quite unlike that entailed by silicic compositions at the East Pacific Rise, which are too similar to basalts in Nb/Y and Zr/Y to affect ΔNb significantly by mixing and are plausibly derived from parental depleted MORB by crystallization differentiation.

We need to consider next whether the most important mixing is not necessarily between low-Ti basalt and rhyolite but between low-Ti and higher-Ti basalt. This has to be appraised if only because rhyolite has almost no TiO_2, and simple mixing between it and low-Ti basalt cannot produce higher-Ti basalt. Rhyolite also has very low Sr concentrations; thus mixing with it could not change $^{87}Sr/^{86}Sr$.

A number of workers postulate that many nonpicritic Icelandic basalts, most of them tholeiitic in the norm, are actually E-MORB, "alkalic," or "enriched," yet that rhyolites are derived from them by crystallization differentiation processes (e.g., Gunn, 2005). Favoring this is the juxtaposition on the ΔNb diagram of many higher-Ti basalts, including picrites, with silicic compositions. We have already seen how little effect crystallization differentiation has on shifting locations of basaltic and silicic lavas from the East Pacific Rise on a ΔNb diagram. If higher-Ti Icelandic basalt plots at the same positions as andesite, dacite, and rhyolite on such a diagram, mixing of primitive low-Ti basalt with any of those magma types could potentially serve to shift their Nb/Y and Zr/Y. This possibility is similar to the inference we have already made about multicomponent mixing along the East Pacific Rise, but at Iceland there are greater contrasts in the trace-element ratios between potential *basaltic* mixing end-members.

One distinction between rocks described as E-MORB at Iceland and the East Pacific Rise needs emphasis. Designation of any Icelandic rocks from higher-Ti basalt to rhyolite that plot atop each other on a ΔNb diagram as "alkalic" or comparable to E-MORB is somewhat misleading, because the basalts among them are all hypersthene- or even quartz-normative tholeiites, many of them ferrobasalts. This is a hallmark of the entire north Atlantic igneous province and is consistent with the abundance of silicic differentiates among the basalts. The entire concept of a "tholeiitic differentiation series" stems from the early, classic studies of these rocks.

Along the East Pacific Rise, E-MORB has more classically alkalic tendencies or attributes, including lower SiO_2 than the vastly more abundant N-MORB and nepheline in the norms of those with highest K_2O. The extreme compositions, at least, are undersaturated with respect to silica, and thus are not tholeiitic. From our prior discussion, remember that so-called transitional basalt, or T-MORB, is likely a mixture between truly depleted basalt and Ne-normative alkalic basalt. The transitional hybrid may yet be hypersthene-normative, but the mixing end-members can still clearly be identified (Fig. 1D). On near-ridge seamounts, the alkalic-basalt end-member typically differentiates toward trachytic residua rather than rhyolite (e.g., Engel et al., 1965; Gee et al., 1991).

At Iceland, mixing with rhyolite also produces some geochemical attributes among higher-Ti basalts similar to those of Pacific E-MORB, and this has produced confusion. The difficulty is in discerning whether basalt-rhyolite mixing has happened, whether alternatively such properties are intrinsic attributes of the basaltic mixing end-member, or, over the population of higher-Ti basalt, both. Functionally, in a place where mixing between basalt and silicic magma is so obviously an important process, it has to be excluded as an explanation for the geochemical attributes of any particular basalt in order to assign those properties to a source in the mantle.

To frame this difficulty in a more general way, the E-MORB conundrum at Iceland is but one expression of a general attribute of Icelandic lavas, namely that their extent of differentiation based on virtually any standard of comparison—e.g., MgO content, MgNo, SiO_2 content, etc.—correlates with their degree of trace-element and isotopic enrichment (cf. Stracke et al., 2003). No matter what geochemical parameter is chosen, the average moderately differentiated basalt at Iceland is more enriched than primitive basalt, andesite is more enriched than moderately differentiated basalt, and rhyolite is more enriched than andesite. This should not be the case if shallow processes dominated by crystallization differentiation link all of these rock types, and it is also clearly unrelated to normative attributes of the rocks. O'Hara (1973) first noted this property of Icelandic basalt in connection with variability approaching Iceland along Reykjanes Ridge (Hart et al., 1973; Schilling, 1973a), and he therefore proposed that shallow differentiation processes alone shift the ratios of trace elements and isotopes. Schilling (1973b), however, rebutted this view in favor of mantle source heterogeneity based on partitioning theory applied to mantle silicate minerals. Nevertheless, Schilling's argument applies to parental liquid compositions, and thus fails to explain the correlation, which according to his argument can then be only circumstantial.

Today, many more data exist. Thus, whereas trends along Reykjanes Ridge toward Iceland are gradational, and thus potentially explainable by along-axis mixing of two distinctive types of material at either end of the ridge, at Iceland proper the problem is compounded by the evident simultaneous availability of depleted low-Ti and enriched high-Ti basalt for short distances along the main Icelandic rifts, with high-Ti basalt associated with rhyolite in central volcanoes. Thus, rather than restricting evaluation of heterogeneity simply to basalt so that we consider only mantle sources as explaining it, the data insist that the correlation of enrichment with differentiation extends to rhyolite and that *local* mixing and assimilation involving strongly differentiated compositions in the crust must be important in producing geochemical diversity among Iceland's basaltic rocks. Recently, attempts have been made to select basalt samples to avoid such processes, but the effects of even tiny amounts of rhyolite mixed into primitive basalt are so striking that one may question whether assertions of a successful screening (e.g., Slater et al., 2001; Maclennan et al., 2003; Stracke et al., 2003) are too optimistic.

That rhyolite represents the extreme of so-called "enrichment" at Iceland is revealed simply by comparison of isotopes of Sr, Nd, and Pb (Fig. 4). The diagrams compare basalts with $^{87}Sr/^{86}Sr$ more and less radiogenic than 0.7034, picritic basalts (chiefly low-Ti with MgO > 10%), and dacite-rhyolite (>62% SiO_2). The silicic compositions partly overlap the more radiogenic (higher-Ti) basalts, and everything with $^{87}Sr/^{86}Sr$ > 0.7034, including rhyolite, falls in the range of the Hart et al. (1992) focal zone (FOZO) mantle samples (the boundaries are from Bell and Tilton, 2002). But rhyolite claims the most enriched extreme on all diagrams. Both most-enriched basalt and rhyolite plot well away from Reykjanes-Kolbeinsey low-Ti basalts as well as Icelandic low-Ti basalt and picrite.

FOZO has been interpreted to indicate a common component in the mantle (Hart et al., 1992; Stracke et al., 2005; cf. Hanan and Graham, 1996). The surprise here is that *rhyolite is the FOZO-like component at Iceland*, with isotopic values matching the extremes of enrichments in basalt. As in Figure 3B, the data for rhyolite split into two trends, and the basalts with more radiogenic isotopes replicate these trends. If one ignores the silicic lavas and refers only to basalts, a mantle origin for the isotopic data arrays would be the only possibility to consider. But including the silicic lavas opens the possibility that the isotopic arrays were produced by shallow magma mixing among several magma types, with rhyolite and other silicic differentiates the most enriched. The inference that high-Ti basalt owes its enrichment to something in mantle peridotite must be questioned, and the possibilities of mixing with silicic magma or crustal anatexis eliminated before an origin for the heterogeneity in peridotite can be accepted.

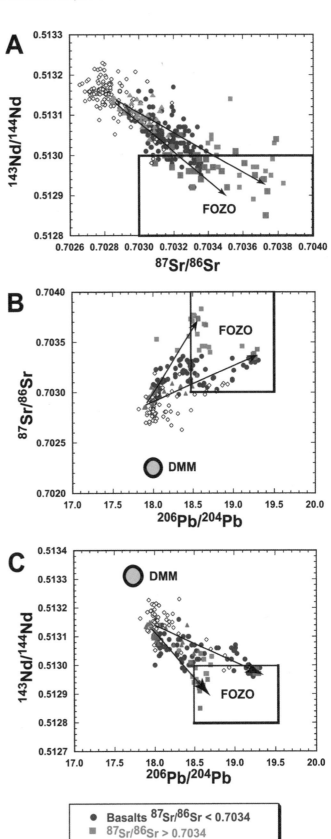

Figure 4. Some isotopic systematics of basalts from Iceland and adjacent ridges. (A) $^{87}Sr/^{86}Sr$ versus $^{143}Nd/^{144}Nd$. (B) $^{206}Pb/^{204}Pb$ versus $^{87}Sr/^{86}Sr$. (C) $^{206}Pb/^{204}Pb$ versus $^{143}Nd/^{144}Nd$. Data are from GeoRoc (Geochemistry of Rocks of the Oceans and Continents of the Max Plank Institut für Chemie) and PetDB (the Lamont Petrology Database) plus some more recently published data sets. Locations of depleted mid-ocean ridge basalt mantle (DMM) and focal zone (FOZO) are from Bell and Tilton (2002). See key for symbol explanations. In all panels Icelandic dacites and rhyolites (SiO_2 > 62%) plus other lithologies with $^{87}Sr/^{86}Sr$ > 0.7034 occupy the region of FOZO. The arrows show two possible mixing trends with ends in rhyolite of different isotopic composition that correspond to the two trends among silicic compositions in Figure 3B.

Drawing this more sharply into focus, we might concede that the lineage from some (not all) higher-Ti basalt to some (not all) rhyolite could well be a shallow differentiation sequence (data points for many of them plot atop each other on the ΔNb diagram); further, that *any* representative of this lineage could mix with low-Ti basalt to shift ΔNb from negative to positive; and consequently, that the ΔNb diagram provisionally reveals three fundamental basalt types in this region with different sources: (1) depleted low-Ti tholeiite that erupts along rifts between central volcanoes, (2) depleted Reykjanes-Kolbeinsey low-Ti tholeiite that dominates spreading ridges beyond the Icelandic shelf, and (3) higher-Ti tholeiite that is abundantly associated with its own differentiated rhyolite and other silicic materials in central volcanoes. These correspond precisely to the three components based on isotopes inferred by Kempton et al. (2000) for Iceland. Quoting these authors, but rearranging the components, we have "1) a depleted component within the Iceland plume; 2) a depleted sheath surrounding the plume; and 3) one (or more likely a small range of) enriched component(s) within the Iceland plume" (p. 255). We could then say that mixing between components 1 and 3 occurs at Iceland and that mixing between components 2 and 3 occurs along the Reykjanes Ridge. Only the third component has distinctly enriched characteristics, but it is probably misleading to equate a ferrobasalt-rhyolite association with the E-MORB of alkalic characteristics that erupts elsewhere along spreading ridges. If higher-Ti ferrobasalt and rhyolite ultimately have a common enriched source, we must eventually come to understand how this geochemically consistent range of materials originated with the following attributes: (1) strong differentiation in bulk, (2) hypersthene- to quartz-normativity, (3) high Ti content, (4) rhyolitic residua, and (4) enriched geochemistry (e.g., high ΔNb, high $^{87}Sr/^{86}Sr$, high light and heavy REE content).

Influence of Magma Mixing on REE

Basalt-rhyolite mixing at Iceland produces precisely the same effects on REE as on ΔNb. The original argument of Schilling (1973a) that Reykjanes Ridge records progressive enrichment in light REE approaching Iceland was based on a diagram of chondrite-normalized Yb_N, which records the extent of crystallization differentiation versus $(La/Sm)_N$, which records source heterogeneity. (Rather than the subscript $_N$, Schilling used the subscript $_{E.F.}$, which stands for *enrichment factor*). A very similar trend shows up when ΔNb or Nb/Y is plotted versus latitude. The full GeoRoc data file for Iceland, including silicic compositions, is depicted in Figure 5A, as well as PetDB listings for basalt from Reykjanes and Kolbeinsey ridges. The diagram shows modest spread in Yb_N for basalts (they are little to moderately differentiated), yet much variability in both Icelandic and Reykjanes-Kolbeinsey $(La/Sm)_N$ (which should not change at all during crystallization differentiation). Schilling's (1973a) argument is illustrated for the field in blue, expanded in Figure 5B, with plotted points restricted to the Reykjanes-

Kolbeinsey data and one comparative data set for Theistareykir and Krafla volcanoes in central and northern Iceland (Maclennan et al., 2003). The vertical spread of data accountable by crystallization differentiation (variation in Yb_N) is given for one primitive, depleted starting composition by the arrow labeled "gabbro fractionation."

The lateral spread in $(La/Sm)_N$, which Schilling (1973a) argued indicates source heterogeneity, is also reproduced by potential basalt-rhyolite mixing trends using two different starting compositions and two rhyolites (OR135 and SNS17, respectively from Örafæjökull volcano and the Snæfellsnes peninsula of western Iceland). These compositions represent the two trends for rhyolites in Figures 3B and 4. About 3% admixture of either rhyolite is sufficient to replicate both the Reykjanes-Kolbeinsey and the Theistareykir-Krafla data arrays. Among the basalts, mixing will also produce increases in $^{87}Sr/^{86}Sr$ that correlate generally with $(La/Sm)_N$ (Fig. 5C).

Standard REE diagrams place some restrictions on potential magma mixing at Iceland. Crossing patterns occur in the Theistareykir-Krafla data set of Maclennan et al. (2003; Fig. 6A); among these I have chosen light REE–depleted basalt TH8372, with 11.7% MgO contents, to represent a near-liquid primitive composition. I have calculated REE patterns for different proportions of mixing with differentiated high-Ti ferrobasalt OR134, again from Örafæjökull, with its steep REE pattern (Fig. 6A), and rhyolite R-06, which has an equally steep pattern (compare the flatter pattern of the andesite from the East Pacific Rise in Fig. 1A), but with a pronounced Eu anomaly. Note the similarity in REE patterns of OR134 and E-MORB W03-21-2 from the East Pacific Rise. Note also the similarly fanning array of the entire data field for Reykjanes and Kolbeinsey ridges, but that the heavy REE concentrations are higher than in TH9372. This difference could not be produced by mixing with any variety of differentiated lava, basaltic or rhyolitic, and thus reflects different sources of primitive and depleted compositions between Iceland and its adjacent ridges.

Small admixtures of either to TH9372 will very readily flatten the estimated REE patterns (Fig. 6B and C), but with the rhyolite admixture the patterns develop a steadily increasing Eu anomaly. Not all Icelandic rhyolite has so pronounced an Eu anomaly, but it is fair to say that individual Theistareykir-Krafla basalts lack such an anomaly or, if they do not, it is very small. Put another way, the maximum extent of possible rhyolite admixture into primitive basalt is probably ~5%, with relative REE concentrations (including a small Eu anomaly) and Nb/Y (Fig. 7) thereafter staying constant during shallow crystallization differentiation.

One thing rhyolite mixing cannot do, however, is increase the TiO_2 contents of any basaltic magma. The contrasting effects of possible mixing of TH8372 with both rhyolite R-O6 and high-Ti basalt OR-134 are shown in Figure 3A. Whereas high proportions of mixing between rhyolite and some basalt may well explain silicic compositions on the diagram with 4–6% MgO contents, no high-Ti basalt or picrite is likely to result from

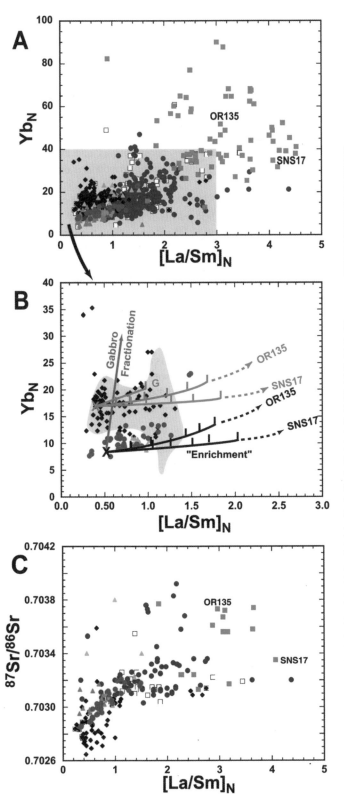

Figure 5. Some rare earth element systematics for Iceland and adjacent ridges. See key for symbols. RR—Reykjanes Ridge; KR—Kolbeinsey Ridge. Data are from GeoRoc (Geochemistry of Rocks of the Oceans and Continents of the Max Plank Institut für Chemie) and PetDB (the Lamont Petrology Database) plus some more recently published data sets. Low-Ti and high-Ti basalts are given the same symbol, as are silicic rocks (>55% SiO_2 or <3.5% MgO); otherwise symbols are based on locations. (A) Chondrite-normalized $(La/Sm)_N$ versus Yb_N. The area with the blue background is blown up in panel B. Silicic samples OR135 (Örafæjökull) and SNS17 (Snæfellsnes) fall in the separate trends for silicic lavas indicated in Figures 3B and 4. (B) Inset of panel A showing calculated mixing trends from two starting basalt compositions with two different rhyolites, samples OR135 (Örafæjökull) and SNS17 (Snæfellsnes), in the proportions indicated by the successive tick marks, 1%, 2%, 5%, 10%, 20%, and 50%. The yellow field is that of Schilling (1973a) for Reykjanes Ridge. His trend for "gabbro fractionation" (shallow crystallization differentiation of clinopyroxene, plagioclase, and olivine) is also indicated. The trend for, e.g., Theistareykir is reproduced by 1–5% mixing with rhyolite OR135. (C) $(La/Sm)_N$ versus $^{87}Sr/^{86}Sr$ for Iceland and adjacent ridges. Symbols are as in panel A. Data for primitive basalt from Theistareykir show a strong correlation between the two ratios and point directly toward rhyolite OR135, although the changes in ratios are not proportionally related on this ratio-ratio diagram. At other locations, e.g., Snaefellsnes, samples with $(La/Sm)_N$ >1 have nearly constant $^{87}Sr/^{86}Sr$ because mixing trends usually curve on ratio-ratio diagrams, depending on the concentrations of the elements or isotopes and the proportions of mixing.

differentiation of a mixture between low-Ti basalt and rhyolite. And more than 20% admixture of titanian ferrobasalt OR134 is required to drive the composition of TH8372 into the realm of higher-Ti basalt. On balance, mixing with high-Ti basalt is more likely than mixing with rhyolite to explain shifts in ΔNb and $(La/Sm)_N$ among primitive Icelandic basalts, but some basalt-rhyolite mixing undoubtedly occurs, and some, probably most, higher-Ti basalt should be viewed as having an independent and more enriched source than low-Ti basalt. However, the geochemistry of that source is likely diluted among the many Ice-landic basaltic rocks that are hybrids between higher-Ti and low-Ti compositions.

The independence of the two basalt types is revealed well by considering the actual concentrations of magma types potentially involved in mixing, using—as I did earlier for the East Pacific Rise—diagrams of Zr versus Nb/Y. In Figure 8A and B, I plot low-Ti and higher-Ti basalts (plus differentiates), and in Figure 8C I provide the mixing models for comparison. There is no higher-Ti basalt with low Nb/Y, but there is some low-Ti basalt with high Nb/Y. The curvature of the low-Ti basalt data field in Figure 8A is well and almost identically matched by mixing between depleted compositions and either titanian ferrobasalt or rhyolite, but different proportions of the two are required.

In a diagram of $^{87}Sr/^{86}Sr$ versus Zr (Figure 8D), samples with Zr > 200 ppm fall into two groups that have some geographic significance, viz., $^{87}Sr/^{86}Sr$ is about the same for all differentiated samples from the Vestmann Islands, but lower than among differentiated samples from Örafæjökull. Thus again, silicic compositions fall into two groups, separated by the upper dashed line. Comparable locations for the line in terms of Nb/Y are shown in Figure 8A and B. Basalt with less radiogenic $^{87}Sr/^{86}Sr$ than any higher-Ti basalt is delimited by another dashed line in Figure 8D and labeled Group 1. Again, this is an indication that higher-Ti basalt is distinctive from the outset, being somewhat more enriched isotopically and in other aspects of geochemistry than uncontaminated low-Ti basalt. In Figure 8A, note the distinctly lower Zr contents of Icelandic low-Ti basalt

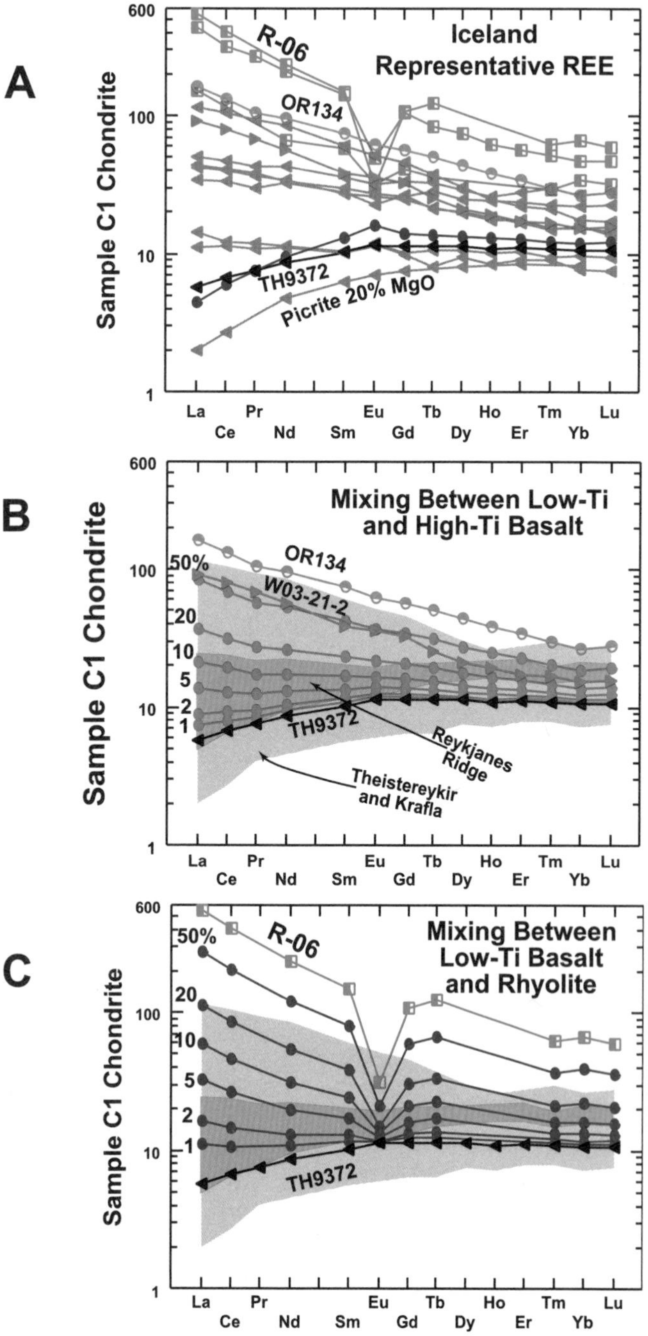

Figure 6. Diagrams showing the results of calculations performed on rare earth elements (REE) at Iceland for potential mixing between primitive depleted tholeiites and both higher-Ti basalt and rhyolite. (A) Representative chondrite-normalized REE patterns for Iceland used for mixing calculations and comparisons. Symbols: gray triangles—basalts with fanning REE patterns from Theistareykir volcano that span the range of $(La/Sm)_N$ for Theistareykir in Figure 5C; black triangles—most depleted nonpicritic Theistareykir sample TH9372 used in calculations; half-filled red squares—icelandite and rhyolites with negative Eu anomalies, including sample R-06, which is used in the calculations; half-filled light blue circles—high-Ti basalt OR134 (Örafæjökull) also used in the calculations; green triangles—sample Westward 03-21-2, enriched mid-ocean ridge basalt (E-MORB), from the Pacific-Antarctic East Pacific Rise, plotted for comparison; and blue dots—sample Westward 60-02, truly depleted MORB, from the Pacific-Antarctic East Pacific Rise (also plotted in Fig. 1A). (B) Calculated mixing proportions (patterns with red dots) between TH9372 and 1%, 2%, 5%, 10%, 20%, and 50% high-Ti basalt OR134. Note the similarity between the 50:50 mixture and E-MORB Westward 03-21-2 from the Pacific-Antarctic East Pacific Rise. The mixing does not result in Eu anomalies. The gray background spans the fanning range of Theistareykir basalts in panel A. The blue background spans the range of REE in basalt samples from Reykjanes Ridge. These have similar but more restricted fanning patterns and a limited range at somewhat higher concentrations of heavy REE. (C) Calculated mixing proportions (patterns with blue dots) between TH9372 and rhyolite R-06 in the same proportions as in panel B. The background fields are as in panel B. Notice that even small proportions of mixing readily produce negative Eu anomalies.

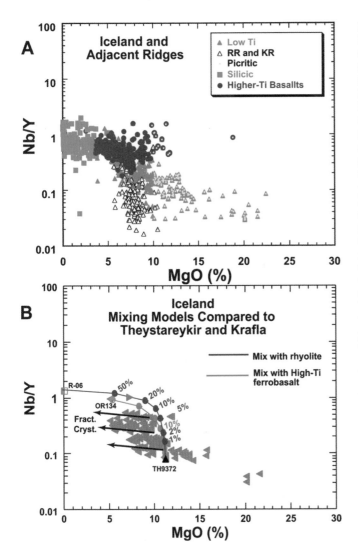

Figure 7. MgO versus Nb/Y for Iceland and the adjacent Reykjanes and Kolbeinsey ridges (RR and KR). (A) All data from GeoRoc (Geochemistry of Rocks of the Oceans and Continents of the Max Plank Institut für Chemie) and PetDb (the Lamont Petrology Database). Symbols are given in the key and are as in Figure 3A. Since crystallization differentiation does not change Nb/Y, the difference between low-Ti and higher-Ti basalt at Iceland is produced in another way. (B) Curving trends that show the effects of mixing different proportions of rhyolite R-06 and high-Ti basalt OR134 with primitive basalt TH9372, as depicted for REE in Figure 6. All data for Theistareykir (gray triangles) are shown for comparison. Most Theistareykir picritic compositions (MgO ≥ 10%) have low Nb/Y. Much Theistareykir basalt with <10% MgO has higher Nb/Y. This indicates that most olivine accumulation among picrites occurs before a change in Nb/Y is superimposed during shallow differentiation, i.e., it is not a signal of mantle heterogeneity. In shallow differentiating magma chambers, then, Nb/Y is increased by mixing with either rhyolite or high-Ti basalt along trends given by the red and blue dots. The spread in data at Theistareykir occurs because of a combination of mixing and shallow fractional crystallization (black arrows). A maximum of ~15% admixture of material similar to high-Ti basalt OR134, or of 5% rhyolite R-06, or some combination of the two with primitive basalt is suggested by the mixing trends. This would steepen the rare earth element (REE) patterns but produce at most only slight Eu anomalies in Figure 6C (none in Fig. 6B). The full spread in Theistareykir light REE without significant Eu anomalies follows from subsequent shallow crystallization differentiation that drops MgO contents from ~11% in liquids to ~6%.

than in similarly low-Ti basalt of Reykjanes and Kolbeinsey ridges. This adds to the picture of source heterogeneity among depleted compositions in the North Atlantic already evident from consideration of ΔNb and REE diagrams.

To summarize, heterogeneity among the sources of basalt in the Iceland region does exist, but ΔNb is not a geographical discriminant. Northward along Reykjanes Ridge approaching Iceland, one type of depleted basalt with negative ΔNb appears increasingly to admix with higher-Ti basalt (or differentiates of it) with positive ΔNb. Kolbeinsey Ridge north of Iceland has the same type of low-Ti basalt, but with no indication of increasing ΔNb southward approaching Iceland. This abrupt discontinuity is probably the result of an offset of the Mid-Atlantic Ridge at the Tjörnes fracture zone. Quite another type of depleted basalt but also with near-zero or positive ΔNb is present at Iceland. This commonly mixes with the same type of higher-Ti basalt and probably also with rhyolite. The Reykjanes-Kolbeinsey basalt type with negative ΔNb is not present between the northern and southern shelves of Iceland.

Other Trace Elements

Chauvel and Hémond (2000) used pairs of trace-element ratios to illustrate their thesis that recycled abyssal gabbro is present as eclogite in the Iceland melt source. The pairs they chose combine Y, REE, and HFSE in various combinations, including a ΔNb diagram, and all embody small contrasts in bulk distribution coefficients during shallow crystallization differentiation and partial melting of peridotite. Their general point is that the mantle ratios for most of their pairs are very restricted, that this consistency should be imparted to mantle melts, and thus that spreads in the ratios entail some phase control on partitioning beyond that possible with only a peridotite source. Mixing between melts derived from peridotite and recycled abyssal gabbro cumulates is one way to explain the diversity in trace-element ratios. Their argument was restricted to variations among basalts.

Figure 9A–F compares several of these ratios plotted against Zr/Y for comparison to ΔNb, but with silicic compositions included. As in the ΔNb diagram, the data arrays slope, and trends for the primitive basalt suites of Theistareykir and Kistufell fall along the arrays. The principal groupings of low-Ti basalt, higher-Ti basalt, Reykjanes-Kolbeinsey basalt, and silicic compositions occupy the same relative positions that they do on a ΔNb diagram, so the same arguments made earlier about potential mixing relationships can be made for each of them; the spreads could be caused by mixing between depleted low-Ti basalt and either evolved higher-Ti basalt or rhyolite. The ratios Rb/Cs and La/Ta do not vary between primitive Icelandic basalt

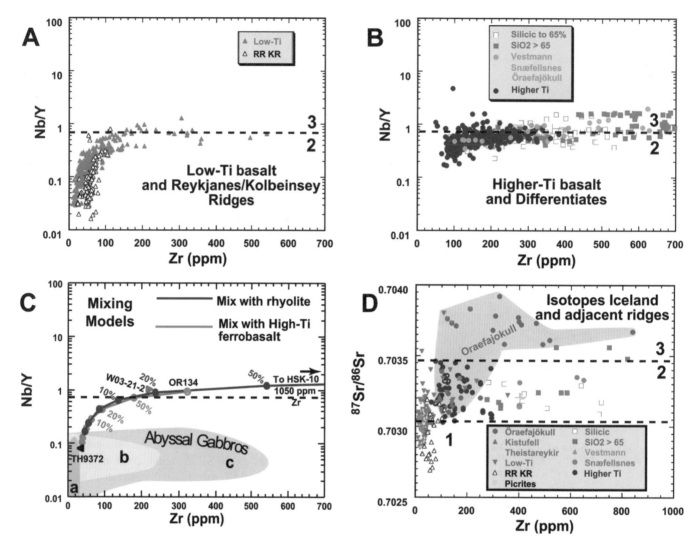

Figure 8. (A–C) Zr versus Nb/Y and (D) versus $^{87}Sr/^{86}Sr$ for Iceland and the adjacent Reykjanes and Kolbeinsey ridges (RR and KR). The symbols are given in the key to each diagram. The dashed line divides samples corresponding to the two groups (Group 3 above and Group 2 below the line) of differentiated lavas with differing Nb/Y, shown in Figure 3B; differing $^{87}Sr/^{86}Sr$, shown in Figure 4; and differing $(La/Sm)_N$, shown in Figure 5, and is repeated on all diagrams here. (A) Low-Ti basalt, including samples from Kistufell and Theistareykir compared with low-Ti samples from Reykjanes and Kolbeinsey ridges. Most samples follow curving trends toward higher Nb/Y and Zr concentrations, but level off in Nb/Y at Zr concentrations of ~150 ppm. The curving trends cannot be produced by crystallization differentiation at low or high pressure. Note the offset of the Reykjanes and Kolbeinsey samples toward higher Zr than in low-Ti basalt from Iceland. (B) Higher-Ti basalt and differentiates distinguished by composition as given in the key. None of these has low Nb/Y. The differentiates comprise two groups (2 and 3), mostly below and above the dashed line, respectively. (C) Mixing calculations between primitive basalt TH9372 and high-Ti basalt OR134 (red curve with % high-Ti basalt indicated at dots) and high-Zr rhyolite HSK-10, with 1050 ppm Zr (blue curve with % rhyolite dots). The samples with 20–50% mixing resemble silicic compositions with 55–65% SiO_2 in B, suggesting that many of those could be hybrid rocks. A data field for abyssal gabbros from ODP Hole 735B (Natland and Dick, 2002) is broken into (a) troctolite and olivine gabbro (orange); (b) gabbro and disseminated oxide gabbro (yellow); and (c) oxide gabbro (light green). Most primitive Icelandic low-Ti basalt (in A, with Nb/Y = 0.03–0.10) matches olivine gabbro and troctolite in having very low Zr concentrations. (D) Data broken down partly by location and divided by lines separating Group 1 (depleted Icelandic and Reykjanes-Kolbeinsey tholeiites), Group 2 (with values of $^{87}Sr/^{86}Sr$ similar to and including the values of higher-Zr silicic differentiates from Snæfellsnes), and Group 3 (with values of $^{87}Sr/^{86}Sr$ similar to those of silicic compositions from Öræfæjökull). The complete Öræfæjökull field, which spans Groups 2 and 3, is shown with a yellow background.

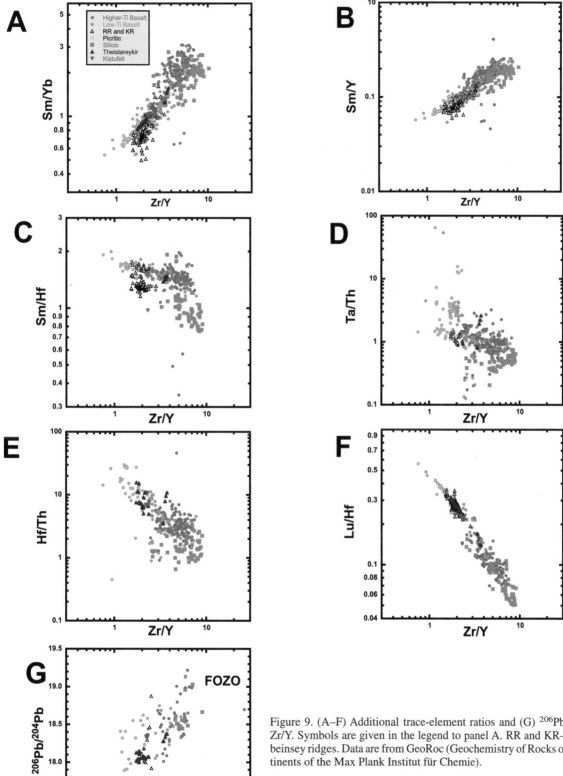

Figure 9. (A–F) Additional trace-element ratios and (G) $^{206}Pb/^{204}Pb$ plotted versus Zr/Y. Symbols are given in the legend to panel A. RR and KR—Reykjanes and Kolbeinsey ridges. Data are from GeoRoc (Geochemistry of Rocks of the Oceans and Continents of the Max Plank Institut für Chemie).

and rhyolite, but that means they are still consistent with the possibility of such mixing. Particular suites such as that of Theistareykir (Stracke et al., 2003) have Ba/Th diminishing with increasing Zr/Y, thus pointing toward rhyolite, but the GeoRoc data set as a whole does not show a consistent relationship between the two ratios, perhaps indicating interlaboratory analytical differences.

Slater et al. (2001) observed far greater variability among light REE (and elements with similar partition coefficients) than among heavy REE (and elements with similar partition coefficients) in phenocryst-hosted melt inclusions in basalt from Theistareykir, and they also observed that such variability cannot be caused by shallow crystallization differentiation; again, absolute concentrations of these elements can be changed appreciably by fractionation, but not their ratios. The observed variability of the melt inclusions, however, is exactly that shown by the spreads in any of the trace-element ratios in Figure 9. Slater et al. (2001) concluded that the "obvious explanation of this behavior is that it results from fractional melting" (p. 345). I say instead that the explanation is magma mixing involving primitive depleted basalt and strongly differentiated basalt or rhyolite. This will also shift ratios of strongly incompatible elements to the more compatible elements and is the more obvious choice for Iceland, where such mixing is abundantly described in the literature. Slater et al. (2001) used principal component analysis to establish that most variance in trace elements of Theistareykir basalt depends on a component dominated by light REE, with the second component dominated by the heavy REE. This is because the light REE are more affected by a silicic admixture or contaminant than the heavy REE (Fig. 6). Beyond that, the case for mixing implied by principal component analysis is quite eloquent, predicting, among other things, greater variance of melt inclusions than host basalts; it is just that the mixing is not among instantaneous fractional melts of a peridotite mantle source.

Stracke et al. (2003) noticed strong correlations among incompatible trace-element concentrations and ratios, isotopic ratios, and aspects of major-element composition in basalt from Theistareykir. They explained these correlations using a melt column model in which enriched basalt is derived from low partial melt fractions at higher pressure and depleted basalt is derived from greater partial melt fractions at lower pressure. This is a melt column model similar to that of Klein and Langmuir (1987) for MORB. At this juncture, needless to say, I believe instead that the observed correlations of trace-element ratios that are indicative in general of "enrichment," namely increases in major oxides such as TiO_2 and K_2O, and the decreasing CaO/Al_2O_3 and SiO_2/FeO noted by Stracke et al. (2003), can all alternatively be explained by mixing between primitive basalt and either strongly differentiated Icelandic ferrobasalt or silicic magma.

Overall, this means that melt fraction, pressure, and potential temperature cannot be adduced for these rocks by inversion especially of REE patterns to specific mantle compositions. I judge that most such patterns are too strongly influenced by

shallow mixing processes to allow us to infer the intrinsic properties of whatever aliquot of them represents a mantle source. Besides, at Iceland if anywhere, the isotopically multicomponent source is not likely to be lithologically homogeneous (Anderson and Natland, 2005), as inversion models require (McKenzie and Bickle, 1988), and it may not even be peridotite.

A potential shortcoming of principal component analysis using mainly REE is that by itself it does not seem able to discriminate between the effects of mixing involving a generalized silicic end-component and mixing with an E-MORB or more generalized alkalic end-component of the type I have discussed for the East Pacific Rise. In both cases, the light REE have greater variance than the heavy REE. Thus Slater et al. (2001) obtained fairly similar first and second principal components for places as disparate as Iceland, the FAMOUS and MARK areas in the central North Atlantic, and the Gorda and Juan de Fuca Ridges in the Pacific.

The mixing story has to be compatible with data on isotopes. Briefly, there are strong correlations (or anticorrelations) between $^{87}Sr/^{86}Sr$, $^{143}Nd/^{144}Nd$, $^{176}Hf/^{177}Hf$, and $^{206}Pb/^{204}Pb$ with Zr/Y, with rhyolite occupying the more radiogenic end of each data field at high Zr/Y. Considering just $^{87}Sr/^{86}Sr$, tholeiitic ferrobasalt, andesite (icelandite), and dacite have much higher concentrations of Sr (400–600 ppm) than rhyolite (<100 ppm), and thus seem likelier to increase $^{87}Sr/^{86}Sr$ by mixing with undifferentiated low-Ti tholeiite (~50–70 ppm Sr). A rough calculation indicates that $^{87}Sr/^{86}Sr$ can be increased from 0.7028 to 0.7031, which is fairly representative of Icelandic tholeiite, by mixing between basalt with ~50 ppm Sr and ~10% of anything at Iceland with ~500 ppm Sr.

Central to this story, two mixing components that differ by 450 ppm Sr cannot be related by shallow crystallization differentiation involving plagioclase. That is, in general among Icelandic tholeiites, Sr behaves like K, Rb, Ba, or any other incompatible element; it actually tracks Ti_8 fairly closely. However, once whatever primitive magma is emplaced at crustal levels, crystallization differentiation involving plagioclase takes effect, and thus Sr concentrations remain nearly constant *locally* through most subsequent stages of differentiation, buffered to nearly constant levels principally by the contrasting partitioning controls of cotectic plagioclase and clinopyroxene. The local constancy represents a *preset* but regionally quite variable parental Sr concentration. Among parental liquids, however, Sr is not buffered by plagioclase, which indicates a distinctive evolution below the depth of plagioclase stability, that is, beneath the seismically measured crust in the upper mantle. Most high-Ti basalt, or primitive (ferropicritic) precursors (but very likely differentiated high-Ti ferrobasalt itself), passes untrammeled through the mantle-crust transition at Iceland.

Finally, Stracke et al. (2003) indicated that the enriched component involved in magma mixing at Theistareykir has HIMU isotopic attributes (high $^{206}Pb/^{204}Pb$, indicative of high U and Th concentrations in melt sources). Later Stracke et al. (2005) modified this view, construing the HIMU localities in the ocean basins to be few (St. Helena, Tubuai) and not to include Iceland. Elsewhere, including Iceland, FOZO should be considered the more usual mixing end-component with somewhat high ^{206}Pb/^{204}Pb. Silicic magma at Iceland has this attribute (Figs. 3 and 9G).

Gabbro-Sourced Basalt at Iceland?

Chauvel and Hémond (2000) and Breddam (2002) found similarities between the trace-element and isotope geochemistry of primitive Icelandic tholeiite, mainly picrite, and depleted abyssal gabbro. They proffered the model that enriched Icelandic tholeiite is derived by partial melting of upper ocean crust (basalts and dikes) in the melt source, whereas low-Ti basalt derives from partial melting of lower ocean crust (gabbro cumulates) and mantle harzburgite together. Thus a "complete section of ocean crust" seems to be present in the melt source, probably as eclogite, an interpretation adopted by Foulger et al. (2005a,b) in order to speculate on how such material might have been emplaced beneath Iceland in consequence of plate tectonic processes. Both Chauvel and Hémond (2000) and Breddam (2002), however, favored recycling of ocean crust from the lower mantle after its emplacement there by deep subduction, and subsequent entrainment in an ascending mantle plume.

I wished to discover whether primitive Iceland basalt with lower Zr/Y than Reykjanes-Kolbeinsey basalt (Fig. 3B) matches the general description of gabbro-sourced basalt provided by Chauvel and Hémond (2000) and Breddam (2002). The comparison depends on a general similarity between the trace-element patterns of primitive Icelandic tholeiite and abyssal gabbro cumulates. A key indicator is that the basalts should have high Sr/REE, or Sr/Nd, in consequence of the high partition coefficient for Sr for plagioclase that originally crystallized in the ocean crust now presumed to have been recycled down and up through the mantle beneath Iceland.

GeoRoc contains fifty-eight analyses of samples from Iceland with Zr/Y < 1.8, that is, less than in Reykjanes-Kolbeinsey basalt, and almost all of them are picritic (>10% MgO). These have 0.40% < Ti_8 < 0.75%, with an average Ti_8 = 0.63%. These are values so low that they are matched by only a tiny fraction of MORB. They are distributed mainly in the Reykjanes and northern volcanic zones, with only two analyzed from the eastern volcanic zone. Of these samples, thirty-nine have also been analyzed for Sr and REE. Marked positive Sr anomalies show up well in trace-element patterns of all these picritic lavas when normalized to primitive mantle (Fig. 10A), as they do in modern abyssal olivine-gabbro and troctolite adcumulates (Fig. 10B). Indeed, they are more pronounced than in any of the several samples in which Sr anomalies were discovered by Chauvel and Hémond (2000) and Breddam (2002), all of which have higher Zr/Y. No such anomalies are present in typical MORB or in higher-Ti Icelandic tholeiite. Figure 10A also shows that the Sr anomalies in Icelandic picrites are as strongly pronounced as ones in mineral-hosted melt inclusions from northern Iceland (Sigurdsson et al., 2000; Slater et al., 2001) and olivine-hosted

melt inclusions from Mauna Loa, Hawaii (Sobolev et al., 2000). Sigurdsson et al. (2000) inferred that some of the Icelandic inclusions, those that have $CaO/Al_2O_3 > 1$, require a pyroxenite rather than a peridotite melt source, and Sobolev et al. (2000) inferred an origin for the Hawaiian inclusions in eclogite, with the Sr anomaly indicating the effect of "ghost" plagioclase occult in a formerly gabbroic assemblage now too deep beneath the Pacific lithospheric plate for plagioclase to be stable. Even though the Sr anomalies of Icelandic picrites are positive, the Sr concentrations themselves are still low (average 66 ± 20 ppm), about half that in the MORB suite from the Pacific-Antarctic East Pacific Rise considered in Figure 1. Another point of comparison is the low Zr/Y itself (Foulger et al., 2005a,b), which in abyssal olivine gabbro is a consequence of the concentration of these two elements in cumulus clinopyroxene and the absence of a significant residual melt fraction in rocks that are adcumulates.

In general, potential gabbro-sourced basalt, mainly picritic, is the predominant truly primitive and ostensibly highest-temperature, lowest-Ti tholeiite at Iceland. It occurs along all the principal rifts, although it is less common on the eastern rift, and it is distinct from the low-Ti basalt of Reykjanes and Kolbeinsey ridges. It has geochemical attributes similar to the low-Ti mineral-hosted melt inclusions also occurring in Icelandic low-Ti picrites.

Precise comparisons between the trace-element patterns of picrites and abyssal gabbros is difficult because the former are mainly basaltic liquids with some percentage of accumulated olivine, whereas the most nearly similar gabbros are mainly assemblages of accumulated minerals with almost no retained liquid. If the latter are construed as a melt source for the former, then some partitioning model needs to be developed to infer the trace-element concentrations in partial melts of those gabbros; the gabbro and the partial melt will not be equivalent. We should be comparing model melts derived from gabbro with picrites. Thus, in Figure 10A, the aggregate of trace-element patterns, apart from Sr, slopes consistently downward to the left, toward lower Ba, Th, and Nb, whereas among the gabbros the distribution through the heavy and intermediate REE is flatter, and it abruptly drops downward toward Ba, Th, and Nb. Among the picrites, Nb and La have about the same normalized concentrations, whereas among the gabbros normalized Nb is less than La. Does this mean that the picrite melt source is relatively enriched in Nb (thus providing even highly primitive melts with positive ΔNb)? Should we even expect that these details of the gabbro patterns could be retained in partial melts of recycled gabbro? Should they be retained if melting occurs in the eclogite facies? The gabbros depicted in Figure 10B are from a particular location and crystallized from only a small spectrum of the global diversity of MORB. Other abyssal gabbros may be different.

Let us suppose, however, that the patterns for picrites fairly mimic those of their source. Then the Sr anomaly is best explained by derivation of the picrites from a gabbroic or eclogite source. But then the downward trend toward the left among picrite REE is also an indication that the principal phase in

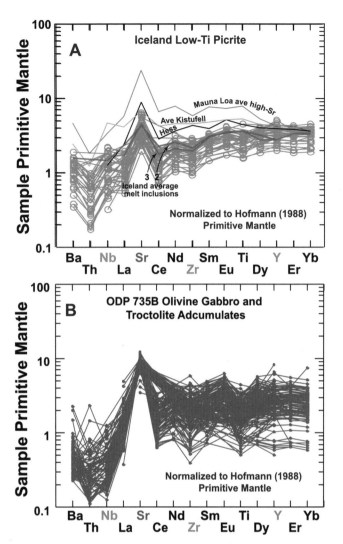

Figure 10. Spider diagrams for trace elements comparing compositions of (A) Icelandic picrites with Zr/Y < 1.2 (from GeoRoc, or Geochemistry of Rocks of the Oceans and Continents of the Max Plank Institut für Chemie) with (B) abyssal olivine gabbro and troctolite adcumulates with Zr/Y < 1.2 from ODP Hole 735B, Southwest Indian Ridge (Natland and Dick, 2002, Appendix B). Data are normalized to Hofmann's (1988) primitive mantle composition. Also plotted in A are average Mauna Loa high-Sr glass inclusions (Sobolev et al., 2000); average primitive olivine tholeiite from Kistufell volcano, central Iceland (Breddam, 2002); averages of Groups 2 and 3 glass inclusions in Icelandic picrites (Gurenko and Chaussidon, 1995; Sigurdsson et al., 2000; Slater et al., 2001); and an average of gabbro cumulates from Hess Deep (Pederson et al., 1996, J. Natland, unpubl. data).

the source containing REE, namely clinopyroxene, had an even steeper downward slope in the same direction (e.g., Johnson et al., 1990). This is not an attribute of the gabbros shown in Figure 10B. The source gabbros of the picrites crystallized from more refractory liquids than typical MORB, with stronger depletions in light REE.

At the same time, Ba, Th, and Nb are more enriched with respect to REE in the Icelandic picrites than in the abyssal gab-

bros. Thus positive ΔNb of the picrites is an attribute indicating some, albeit very slight, relative enrichment in all of these elements with respect to depletion in REE when compared with abyssal gabbro. Chauvel and Hémond (2000) argued that positive ΔNb in primitive Icelandic tholeiite actually indicates depletion in Zr, but the full trace-element patterns suggest that two processes acted simultaneously: depletion in Zr along with light and intermediate REE and a bit of enrichment in Ba, Th, and Nb. Neither is an attribute of abyssal gabbro, at least those shown in Figure 10B.

How was the slight enrichment in Ba, Th, and Nb accomplished? Here the very low concentrations of the particular elements and the difficulties of analyzing them precisely at such low concentrations come into play. A very slight amount of mixing with or assimilation of rhyolite or other silicic material cannot be excluded. Neither can slight incorporation of an alkalic E-MORB component. For that matter, neither can slight incorporation of incremental fractional melts derived from peridotite, even if the principal trace-element attributes are established by partial melting of a thoroughly depleted eclogite. On balance, I favor the option that some slight amount of rhyolite incorporation has occurred, because it is a demonstrated process at Iceland, it is consistent with the mixing trends expected on the basis of the trace elements and isotopes elaborated earlier, and because of the lithological heterogeneity that I believe can be expected of eclogite remnants recycled from either oceanic or continental crust, from which partial melts must aggregate. I shall return to this point shortly.

Origin of Three Basalt Types at Iceland and Adjacent Ridges

Figure 11 shows values of Ti_8—that is, TiO_2 contents—corrected for fractionation to nominal parental compositions with 8% MgO, plotted against similarly corrected values of Fe_8 and Na_8. This includes picrites, as explained earlier. The arbitrary distinction between low-Ti and higher-Ti basalts with a boundary at 1.4% Ti_8 usefully discriminates parental picritic compositions (Fig. 3A) and links all subsequent differentiates to the range of parental basalt types at Iceland. Based on prior discussion, the complete gradation between extreme low-Ti and high-Ti basalt types may be a consequence of mixing, but this diagram shows that there is an underlying and considerable parental variability. However, a true enriched high-Ti parental basalt type with, say, >4% Ti_8 and picritic variants does indeed exist at Iceland, as does depleted low-Ti parental basalt with <0.5% Ti_8. Bearing in mind that the diagrams are based on whole-rock compositions, some of the scatter in Fe_8 and Na_8 may result from alteration (e.g., to sodic spilite or basalt partly transformed to iron-rich clay minerals, this proportionately reducing MgO contents and enhancing the fractionation correction), basalt-rhyolite mixing (lower Fe_8), or addition of plagioclase phenocrysts (again, lower Fe_8). The principal observation to make, however, is that Ti_8 varies by about a factor of ten at essentially constant Fe_8 and Na_8, with the latter centered on low values of 1.5–2.5%.

The constancy of Na_8 and Fe_8 over the wide range in Ti_8 is especially difficult to explain by melting models that use peridotite as a starting composition. Langmuir et al. (1992) produced a strong correlation in their models between Na_8 (from 3.8 to 1.6% as the aggregate melt fraction increased) and Ti_8 (from 1.8 to 1.0%) using a starting composition with 0.15% TiO_2, about the same as in the starting composition (McDonough and Sun, 1995) assumed by Fitton et al. (1997). However, compare the data fields in Figure 11 with the fields of the global array for MORB, shown here for Indian Ocean MORB, and the general trends, shown as red arrows, for the melt column models of Klein and Langmuir (1987) and Langmuir et al. (1992). The Indian Ocean MORB field in Figure 11B is broken into three portions corresponding to three parental basalt types with respectively higher Na_8 (Mahoney et al., 1989; Natland, 1991b). The MORB global array is provided by an increasing extent of

partial melting of a homogeneous peridotite source under the influence of temperature, with the highest temperature represented by samples with least Ti_8 and Na_8 of Type 1 Indian MORB. Nevertheless, all MORB has < 2% Ti_8, and most has < 1.4% Ti_8, consistent with the models of Langmuir et al. (1992). Klein and Langmuir (1987) attributed basalt from Reykjanes and Kolbeinsey ridges to the low-Ti_8 end of the global array (corresponding to Type 1 Indian Ocean MORB), and in their model this is related to other more sodic and titanian parental MORB by a higher average extent of partial melting of peridotite near warmer mantle plumes. Later, I shall propose an alternative.

The melt column model, however, does not reproduce the very high Ti_8 at nearly constant Na_8 of much Icelandic tholeiite. Indeed, by itself it contravenes the modeling of peridotite partial melting of Fitton et al. (1997) that was designed to show mantle heterogeneity using $ΔNb$. Partitioning relationships during melting between basaltic melt and major mantle silicate phases, including garnet during fractional melting (Fig. 1C), cannot produce such extreme variability in Ti_8 either in the liquid or in residual peridotite. The melting of commonly construed peridotite in the upper mantle is obviously inadequate to account for the Ti_8 variability of Icelandic basalt. Stracke et al. (2003) also developed a melt column model for Theistareykir, but did not consider basalt in which $Ti_8 > 2\%$. Even so, neither their model nor the data from Theistareykir resemble either the global or local trends for MORB shown in Figure 11.

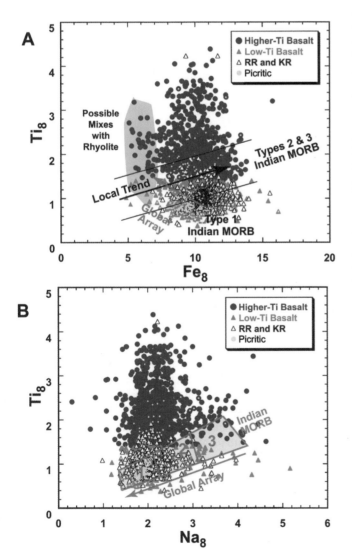

Figure 11. Parental characteristics of basalt from Iceland and the adjacent Reykjanes and Kolbeinsey ridges (RR and KR). Symbols are given in the keys. Silicic compositions are excluded. The parameters Ti_8, Na_8, and Fe_8 are corrected to the nominal comparative values of 8% MgO both for samples with <10% MgO using formulas of Klein and Langmuir (1987) and for samples with >10% MgO by correction to accumulation of composition Fo_{86}, as described in the text. (A) Fe_8 versus Ti_8. An approximate data field for low-Ti Indian mid-ocean ridge basalt (MORB) is given by the dashed red circle and coincides with both low-Ti Icelandic basalt and low-Ti basalt from Reykjanes and Kolbeinsey ridges. The parallel lines bound the remaining Indian MORB. The red arrow gives a general trend for the global MORB array in the Indian Ocean, with decreasing integrated extent of melting in the melt column models of Klein and Langmuir (1987) and Langmuir et al. (1992) trending toward higher Ti_8 and lower Fe_8. The longer black arrow shows the extent of the "local trend" in the Indian Ocean. Neither the global array nor the MORB local trends match the greater contrast between low-Ti and higher-Ti Icelandic basalt. Discounting the effects of the MORB local trend, many Icelandic basalts have somewhat higher Fe_8 and much higher Ti_8 than primitive MORB. Some Iceland basalts with very low Fe_8 could be either hybrids with silicic lava or rich in plagioclase phenocrysts. (B) Na_8 versus Ti_8. The skewed red box with the blue background provides a comparison to Indian MORB, with the sequence Type 3 to Type 2 to Type 1 representing subdivisions of the global array, which correspond to the increasing extent and average depth of partial melting in the melt column models of Klein and Langmuir (1987) and Langmuir et al. (1992). Many Icelandic basalts have much higher Ti_8 than any MORB, but in the majority of them this is at low Na_8, representing high, not low, extents of partial melting in MORB melting models.

I propose that the Ti_8 variability reflects the influence of a readily melted Ti phase in the source. Add this phase to olivine, pyroxenes, and an aluminous silicate (plagioclase, spinel, garnet). For the highest values of Ti_8, say above 3%, only the oxide minerals ilmenite (~45% TiO_2) and titanomagnetite (20–30% TiO_2) at low pressure, and rutile (pure TiO_2) at high pressure, are likely to have a strong influence on this aspect of melt compositions. Magmatic pargasitic amphibole with up to 3.5% TiO_2 in a somewhat hydrous source might also contribute. However, none of these minerals are usually construed to occur even in mildly enriched (or metasomatized) peridotite sources of tholeiitic basalt. They were not, for example, in the models of Fitton et al. (1997).

If not in peridotite, where could the Ti phase or phases reside? In the ocean crust, gabbro cumulates contain as much as 20% oxide gabbro. These have as much as 30% ilmenite and titanomagnetite combined, plus perhaps 1–2% pargasite (e.g., Natland et al., 1991; Natland and Dick, 1996, 2001). These are at the low-temperature end of gabbroic differentiation in slowly spreading ocean crust, and typically occur as deformed seams crosscutting nearly oxide-free olivine gabbro. They also occur in close proximity to tonalite/trondhemite veins. This is because the two represent similar and perhaps even identical stages of differentiation, especially if they are complements produced by liquid immiscibility (Natland et al., 1991; Natland and Dick, 2001, 2002). In any case, they are loci of highest concentrations of incompatible elements (Natland and Dick, 2002), especially HFSE locked by coupled substitution into clinopyroxene and large-ion lithophile elements in sodic to potassic feldspars. The silicic veins, which contain zircons, have the highest HFSE concentrations of all (Niu et al., 2002b). If this association, or any similar association produced in a continental setting, were to wind up in the mantle by means of subduction or any other process, the HFSE-enriched zones of oxide gabbro and silicic veins would simultaneously provide the earliest melts during fractional melting, and also the most enriched basaltic compositions, even if the assemblage were transformed to the eclogite facies.

Both Chauvel and Hémond (2000) and Breddam (2002) have proposed aspects of this hypothesis. They postulate that melting of "complete sections of ocean crust" is required to explain the basalt geochemical variability at Iceland. They proposed that low-Ti basalts carry the trace-element and isotopic signatures of abyssal gabbros, whereas higher-Ti basalts carry those of altered abyssal tholeiites occurring as flows and dikes in the ocean crust. They assumed that all of this material is present in the melt source via recycling, first through subduction down through the entire mantle and then up again in an ascending deep mantle plume (cf. Leitch and Davies, 2001).

Foulger et al. (2005a,b) supported one aspect of this geochemical model by noting the strong similarity of many low-Ti basalts in concentrations and ratios of Y, Nb, Zr, and Ti to abyssal olivine gabbro and troctolite, and the difficulty of extracting these concentrations and proportions from peridotite. I have already discussed the problem of positive ΔNb in Icelandic pic-

rites. A greater difficulty, however, lies in explaining higher-than-expected concentrations of Nb in higher-Ti basalts. For this, Foulger et al. (2005a) proposed that E-MORB like that of the East Pacific Rise is sufficiently present in bulk or average ocean crust to provide a preferential enhanced increment of Nb (and similar elements) to initial fractional melts of ocean crust subducted and abandoned in a proto-Icelandic zone of collision of continental lithosphere.

Now, after considering that melts with both high Ti_8 and high ΔNb are required to explain higher Ti basalt, and the possibility that low-Ti Icelandic picrite derives from olivine gabbro cumulates transformed to eclogite, I suggest another possibility. First, a more enriched material than ocean crust is indeed present among the sources of Icelandic basalt. But I suggest that its lithology is more like that of oxide-rich titanian ferrogabbro veined with silicic materials than like E-MORB of alkalic affinity. Partial to complete melting of this material produces the essence of high-Ti basalt, and this blends in varying proportions with low-Ti basaltic magma derived from one or more depleted sources beneath Iceland. I suggest that oxide-rich gabbro was originally part of deep continental crust, and formed during ancient differentiation of sialic components of that crust. I propose that some of this material is now present as delaminated lower mafic continental crust in the central north Atlantic beneath Iceland. Also, some silicic material is still intimately associated with this delaminated continental crust in the Icelandic melt source, perhaps as veins, pods, or Skaergaard-like "sandwich horizons" (Wager and Deer, 1939) in gabbro or granulite of gabbroic composition, or at least as partially assimilated country rock; these contribute to the abundance of Iceland's rhyolites, set its isotopic signature, and give it Nb/Y higher by an order of magnitude than that of silicic material in the ocean crust. Thus the origin of the enriched characteristics, including high ΔNb, is tied to the origin both of titanian basaltic magma and of silicic differentiates in ancient continental crust and is a consequence of modern blending of diverse strains of magma extracted either fractionally or by batch melting from this heterogeneous protolith. Therefore, Icelandic picrites are derived by more extended fractional melting from residual olivine gabbro, or its eclogitic equivalent, after extraction of the components of oxide gabbro and associated felsic veins.

The heterogeneous protolith is the key to the mixing relations described herein. It effectively provides a pre-differentiated and premixed body of rock, with all stages of differentiation present even in short sections of the gabbroic or eclogitic assemblages, from which the character of mixing end-members is provided. An olivine gabbro that is intimately veined around grain boundaries with tonalite or trondhjemite cannot fail, even as eclogite, to produce a partial melt without at least a slight signal of a silicic component in its geochemistry. Fractional melting models have always been based on an assumption of the lithologic heterogeneity of the melt source. As I have argued elsewhere (Natland, 1989), however, this is unlikely even for the depleted peridotite sources of MORB, and for gabbro cumulates it is contravened by almost every description of such rocks in the field that has ever been written. Therefore begin with the assumption that the source is lithologically heterogeneous at the outset, and the models will at least to that extent correspond better to nature. Adding such complexity to the source simplifies our picture of the melting process by placing all source lithologies in a single body of rock within the melting domain. It also means that the sequence from rhyolite through higher-Ti ferrobasalt to primitive olivine tholeiite at Iceland is fundamentally a partial melting sequence and, as a consequence, that both rhyolite and higher-Ti ferrobasalt do indeed cross from the mantle into the crust at depths of 10–40 km beneath Iceland. This is implicit in the model of Fitton et al. (1997), except that it cannot occur if the source is merely peridotite.

Reykjanes and Kolbeinsey low-Ti basalts still appear to have a peridotite source, but it may actually be that of detached or delaminated subcontinental lithosphere rather than the high-temperature end of the MORB global melting array. The low-Ti basalt of Iceland approaches abyssal gabbro in its HFS trace-element attributes, and this could still indicate the presence of some sort of trapped or detached ocean crust in melt sources beneath Iceland. Altogether, however, this source appears to be more refractory than typical abyssal gabbro, and slightly more enriched; thus I favor, albeit provisionally, an interpretation that locates its provenance also within the lower continental crust, and thus in immediate proximity to oxide-gabbro and silicic components inferred from high-Ti basalts and enriched icelandite, dacite, and rhyolite. ΔNb relationships show no significant present-day influence of depleted MORB mantle, or even of peridotite sources of Reykjanes and Kolbeinsey basalt, in melt sources directly beneath Iceland, in accord with conclusions based on isotopes of Kempton et al. (2000). I thus offer the tentative conclusion that Iceland presents an array of basaltic rocks derived from sources originally in lower continental crust, but probably not peridotite, within a "sheath" of basaltic rocks derived from delaminated subcontinental peridotite.

ON DELAMINATION

In addition to subduction, I suggest that fundamental heterogeneity of the upper mantle is caused by separation and disaggregation of continental crust and subcontinental lithosphere, and that this was the case in the north Atlantic. I offer this as an alternative to a mantle plume for Iceland and the north Atlantic igneous province. I take this opportunity to explain more fully how I envision this process, and how it relates to the petrogenesis of Icelandic tholeiites and silicic volcanic rocks discussed here.

As scientists, we prefer simple models, or perhaps have biases favoring simple ideas. One of these is that rupture of continents produces knife-edge discontinuities—granitic continents on two sides, depleted MORB ocean crust in the middle. But this is hardly likely to be the case, given the existence of microcontinents like the Seychelles (see Holmes, 1928); submerged

banks of continental crust coated with flood basalt and peraluminous rhyolite like Rockall Bank (Roberts, 1975; Hitchen et al., 1997) or with alkalic basalt and trachyte like Jan Mayen just north of Iceland in the north Atlantic (Talwani and Eldholm, 1977; Maaløe et al., 1986; Kodaira et al., 1998); and the direct petrochemical evidence for incorporation of components of continental crust in the basalts of Kerguelen in the Indian Ocean (e.g., Frey et al., 2002). Beneath each granitic outlier, some thickness of lower continental crust must reside, and beneath that, a portion of subcontinental lithospheric mantle. No great stretch of the imagination is required to postulate that smaller bits and pieces of continental material become entrained in convecting upper mantle that moves in around all of these abandoned fragments and microcontinents to supplant drifting continents.

Delamination is an odd word, implying, I would think, the stripping off of a layer along a curved surface concentric with the surface of the Earth, like peeling the plastic from a driver's license. Yet it is the word that has been in use now for nearly two decades to explain the wide distribution of certain types of chemical heterogeneity in the crust of the Indian Ocean in the lee of drifting greater India (e.g., Mahoney et al., 1989). The term has also been applied to orogens to mean the sinking of pieces of the lower crust into the mantle (Kay and Kay, 1991, 1993; Rudnick, 1995). Synonyms for this process, which nevertheless have nothing to do with layers, are *foundering* (Daly, 1926), *major stoping* (Daly, 1933), *peeling off* (Schott and Schmeling, 1998), *dripping into mantle downwellings* (Zandt et al., 2004), *arc-root foundering* (Zandt et al., 2004), and *convective instability* (Jull and Kelemen, 2001; Saleeby et al., 2003). Anderson (2005 and this volume) invokes eclogite foundering from the base of drifting continents as a major mechanism imparting heterogeneity to the upper mantle.

Most workers have emphasized that the density of mafic lower crustal igneous assemblages and some granulites can be higher than that of mantle rock, although Daly construed this in terms of the contrast between solid mafic rock and his presumed vitreous basaltic substratum. Nowadays, the contrast is between rocks in the solid state; viz., eclogite and garnet pyroxenite in the lower crust are denser than peridotite. Jull and Kelemen (2001) emphasize that sinking of dense lower crustal materials will happen only at elevated temperatures (at least 500 °C but more usually 700 °C or higher), but that this could occur under conditions at arcs, rifted continental margins, and regions of continents undergoing extension. This mainly means that all granulite associations, which certainly occur pervasively in continental crust at depth, could have been subject to such stresses at some point in their histories, and that therefore a significant fraction of continental crust may have recycled back into the mantle by this process and subduction combined (Hamilton, 2003).

Daly (1926, 1933), whose views on tectonic processes were closely tied to an important question of his day, namely isostasy, was certainly the first to indicate that uplift would accompany the detachment and sinking of blocks of dense lower crust into the substratum. This is nothing other than the casting off of ballast, a process known to anyone familiar with boats and ships, and, it seems to me, a plausible alternative to the hypothesis of thermal uplift thought to accompany plume impact (e.g., Campbell, 2005). For example, recently some workers have used it to explain uplift of the Colorado plateau (Bird, 1979) and the Sierra Nevada (Ducea and Saleeby, 1996), and high-temperature metamorphism in the Himalayas (Bird, 1978). Daly (1933) also emphasized the link between major stoping and assimilation or partial melting of continental crust in the broad picture of differentiation of the Earth, a view that hardly differs from modern treatments (e.g., Rudnick, 1995; Jull and Kelemen, 2001; Saleeby et al., 2003; Bédard, 2006).

How should we picture these sinking blocks? If they are eclogite, do we mean simple bimineralic eclogite (transformed basalt) or the full eclogite facies (transformed anything as long as it has dense garnet)? Do we limit ourselves by thinking, for example, that subducted ocean crust can be treated as a single composition, average MORB? Average MORB, after all, is a differentiated rock. How did it get that way?

We have now drilled long sections of lower ocean crust where the full variability of gabbroic differentiation, from olivine-rich troctolite to tonalite or trondhjemite, is juxtaposed by complex high-temperature crystal plastic deformation in every 50 m of the section, and in many 10 m portions of the section (Dick et al., 2000). Based on comparisons to dredge collections, and an understanding of tectonic processes of rift valleys at ridge-transform intersections, this must be typical (e.g., Natland and Dick, 2002). Subduction returns most of this material to the interior of the Earth, and it represents two-thirds or more of the ocean crust. Yet our models of the melting relations of subducted ocean crust are based on experiments on single primitive compositions of MORB lava. Now we have an additional way of imagining how crustal material enters the convecting mantle, namely delamination, and we are entitled to wonder how far such material might sink, and the manner in which it might reappear at the surface of the Earth.

The petrological complexity of the lower ocean crust on so small a scale is important because, even though our knowledge of lower continental crust of Archean age is limited, that crust surely is no less complicated (e.g., Bédard, 2006). Thus the lower ocean crust universally follows a course of high-iron (Fenner-type) differentiation that leads to the formation of oxide gabbro cumulates, which comprise roughly 20% of the average lower crustal section (Dick et al., 2000; Natland and Dick, 2002). Some of these rocks contain up to 30% magmatic oxides (Bloomer et al., 1991)—ilmenite and titanomagnetite in a ratio of ~5:1 (Natland et al., 1991). These minerals will transform to rutile in the eclogite facies; indeed, Alpine eclogite transformed from abyssal oxide gabbro contains up to 20% rutile. The parental molten material for these rocks is as close to a universal Earth magma as any material that exists today. By analogy, we can infer that if voluminous tholeiitic basalt was important to the production of continental crust during the Archean—and

especially if, as Bédard (2006) argues, it was emplaced in thick plateaus and was ultimately parental to the widespread tonalite-trondhjemite-granodiorite (TTG) suite of the uppermost crust—oxide gabbro and/or rutile eclogite was produced along the way. If, for example, the standard depletions in TiO_2, Ta, Y, and Nb observed in the TTG suite require separation of one or more titanian phases, either as cumulus minerals during crystallization differentiation or as restite minerals during partial melting, those minerals are not merely present but abundant in the lower continental crust. The high densities of titanian minerals will also contribute to their tendency to detach and sink into the convecting mantle under appropriate conditions of lower-crustal convective instability (i.e., temperatures at or above those of the granulite facies).

Composite intrusive events in an environment of shear characterizes the lower ocean crust at slowly spreading ridges (Dick et al., 2000), juxtaposing all lithologies produced at every stage of differentiation at scales of meters to centimeters. Thus much of the lower ocean crust bears no resemblance to the zoned open-system layered intrusion conceptually based on Skaergaard that is depicted in the standard Penrose ophiolite model (Natland and Dick, 2001). Instead, at slowly spreading ridges, troctolite and olivine gabbro obtained from transform faults are laced with hundreds of seams of oxide gabbro and riddled with TTG dikes. Crystal-plastic deformation to form porphyroclastic, gneissic, and mylonitic textures was extensive and occurred at the magmatic stage, probably at temperatures ranging from 700 to 1000 °C (e.g., Natland et al., 1991); many contacts between the different lithologies are along sharp planar shear surfaces. Differentiation was concurrent with deformation. Bédard's (2006) picture of the development of the Archean lower continental crust by repetitive high-temperature and partly magmatic superposition of one foundered block on top of another differs from this chiefly in scale and occurrence at high pressure, across a high-grade metamorphic isograd.

Foundering of large blocks of Archean lower continental crust during a later orogenic or rifting event thus will carry all stages of magmatic differentiation recorded by those rocks, and probably also some metasedimentary material, into the convecting mantle. No block of dense gabbro granulite or eclogite will descend without carrying something granitic with it. The geochemical avatars of "enrichment" will be concentrated in the most strongly differentiated rocks of the subsiding masses—oxide gabbros and proximal tonalite-trondhjemite veins—and by and large they will be continental in character. These will contribute to the earliest melt fractions in those blocks as they descend to hotter mantle. Those liquids will likely depart their original hosts because of their buoyancy, and will join any streams of basaltic magma that may be ascending from primitive mantle below or around them, modifying their geochemistry, intruding, reintruding, and intruding again all overlying rocks. If the descending blocks become hot enough, a low-temperature basaltic melt fraction could be produced that is rich in iron and titanium extracted from the oxide gabbros (or rutile eclogite), and this will add to primitive magmas ascending from other sources in the vicinity, producing enrichment in iron and titanium in otherwise primitive compositions. This type of mixing, between primitive and strongly differentiated basaltic magmas, has long been documented along slowly spreading ridges (e.g., Rhodes and Dungan, 1979; Rhodes et al., 1979), and the mechanism for it—repetitive injection of primitive basalt into deforming rock riddled with seams of incompletely crystallized oxide gabbro—has been discerned in the long sections of gabbro cores obtained by drilling (Natland and Dick, 2002). I propose that this occurs in active flood basalt provinces, but on a glorified scale compared with mid-ocean ridges.

Into what sort of mantle do the sinking blocks of lower continental crust descend? First, they will descend into less dense refractory subcontinental lithosphere, the residual mantle left from extraction of continental crust, which commonly makes up the roots of Archean cratons (Jordan, 1979; Gaul et al., 2000; O'Reilly and Griffin, 2006). The refractory character is evidenced by highly magnesian olivine (Fo_{92-94}) and a low proportion of clinopyroxene in xenoliths, which are commonly harzburgite. The thicknesses of subcontinental lithosphere beneath Archean cratons (archons) based on xenoliths are 160–250 km (O'Reilly and Griffin, 2006). I suggest that low-Ti basalt from Reykjanes and Kolbeinsey ridges derives from partial melting of peridotite like this, left behind in the North Atlantic following continental rifting. Physically associated eclogite masses derived from diverse gabbro foundered from lower continental crust, but concentrated beneath Iceland, either add to the volume of melts derived from subcontinental peridotite or are the only sources of Icelandic basalt.

In the view of Klein and Langmuir (1987), basalt from Reykjanes and Kolbeinsey ridges represents the low-Na_8, low-Ti_8, high-Fe_8, and high-Si_8 end of the MORB global array (Type 1 MORB in Fig. 11). In their melt column model, such basalt represents partial melting of a homogeneous peridotite source at the highest temperatures and greatest average depths followed by varying degrees of crystallization differentiation, and this view has been used to support the plume model for Iceland. The alternative interpretation is that occurrence of such basalt depends on bulk lithologic heterogeneity in the melt source, namely a predominance of relatively refractory and infertile, nearly harzburgitic, peridotite rather than more fertile lherzolite (Jaques and Green, 1980; Natland, 1989; Sweeney et al., 1991; Green et al., 2001; Presnall et al., 2002; Anderson and Natland, 2005; Green and Falloon, 2005). This heterogeneity is, I believe, provided by continental rifting, which is then the fundamental mechanism for the association of primitive tholeiitic picrite, higher-Ti ferrobasalt, and rhyolite at Iceland, all the geochemical evidence for mixing that I outlined earlier, and the contrast between Icelandic tholeiite and the slightly different but nevertheless very distinctive basalt from the adjacent Reykjanes and Kolbeinsey ridges.

For East Greenland, where the volcanism of the north Atlantic igneous province began in the Paleogene, this type of

magma mixing did not require foundered blocks of lower crustal material. Those rocks were already there. Intact lower continental crust heated by intensive dike injection was subject to in situ partial melting of low-temperature gabbroic constituents, which inevitably mixed with ascending magma. However, for basalt erupted later across the north Atlantic and at Iceland, foundered, stoped, peeled off, or otherwise delaminated blocks of lower crust were essential.

To investigate this hypothesis, greater attention to continental xenolith suites in and around flood basalt provinces to confirm the presence and petrochemical character of oxide gabbros or equivalent material in the granulite and eclogite facies becomes important. On the other hand, even if such material is rarely found, perhaps because of the efficiency of its incorporation into flood basalt, high-Ti basalt such as that at Iceland is very difficult to explain without something like the mechanism proposed here. It does not derive from crystallization differentiation of primitive, low-Ti basalt, nor certainly does affiliated rhyolite. On its own merits, then, the widespread occurrence of high-Ti basalt in continental flood basalt provinces indicates that the lower reaches of ancient continental crust commonly contain oxide-rich gabbro or its equivalent in the eclogite facies, and in this way supports the plateau model of formation of Archean continental crust proposed by Bédard (2006) and numerous prior workers whom he cites.

Compared with the study of plumes, theoretical consideration of the delamination of continental lithosphere and the convective instability of lower continental crust is in its infancy. Perhaps the simplest way to consider these phenomena is to view the lithosphere as a thermal boundary layer cooled from the top. In appropriate plate tectonic circumstances this boundary layer produces instabilities of varying dimensions that allow dense crustal rocks to sink into deeper mantle. In effect, this is the plume hypothesis turned upside down. However, the lithosphere is by no means uniform in its lithologic composition, the distribution of radioactive elements within it (Andreoli et al., 2006), or its responsiveness to the thermal or chemical heterogeneity of the convecting asthenosphere over which the plates move. Proterozoic and younger formerly mobile belts separate many Archean cratons, and much continental crust consists of sutured terranes of the remnants of arcs, back-arc basins, and accretionary sediment prisms sometimes pinched between collided continents. How and where large concentrations of lower continental crust might subside into the underlying mantle and later reappear to influence ridge and midplate volcanism are open questions that require the insights of geology, geophysics, igneous and metamorphic petrology, and geochemistry to resolve.

CONCLUSIONS

I recapitulate briefly as follows.

1. Closed-system crystallization differentiation of depleted MORB from primitive tholeiite to silicic compositions dif-

ferentiated as dacite produces no spread of data points on ΔNb diagrams.

2. Such spread of data points on ΔNb diagrams as exists along the East Pacific Rise results from magma mixing among three end-members: truly depleted N-MORB, E-MORB of alkalic character, and—very slightly—high-SiO_2 dacite-rhyolite produced by shallow differentiation; only mixing with E-MORB causes shifts from negative to positive ΔNb.

3. Consideration of likely mixing end-members points to the existence of an extreme truly depleted MORB, found principally in transform faults and near-axis seamounts, as the principal depleted magma type along the East Pacific Rise; commonly construed N-MORB is a hybrid combining truly depleted MORB (~90%), E-MORB (~5%), and some silicic differentiates (~5%) in axial magma chambers and conduit systems.

4. For Iceland, on the other hand, ΔNb diagrams show a large spread in data points corresponding to the sequence low-Ti tholeiite, high-Ti tholeiite, and silicic differentiates including rhyolite, none of which can therefore be related to each other by closed-system crystallization differentiation in the crust.

5. In all trace-element and isotopic respects, at Iceland rhyolite occupies the place in mixing models formerly attributed to a generalized enriched component in the mantle; in detail, compositions ranging from strongly differentiated higher-Ti ferrobasalt to dacite comprise an aggregate of enriched compositions involved in the mixing.

6. At Iceland, silicic lavas isotopically resemble the formerly construed pervasive mantle component, FOZO; mixing of most depleted Icelandic tholeiite with silicic melts at levels of 5–10% can account for their isotopic variability.

7. Parental TiO_2 (Ti_8) at Iceland varies by a factor of ten, from ~0.4% to 4%; melt column models involving peridotite cannot explain this.

8. Instead a titanian phase is required in melt sources to explain the most titanian parental compositions, and this, in turn, entails a gabbroic rather than peridotitic assemblage in melt sources, one containing ilmenite, titanomagnetite, or (in the eclogite facies) rutile.

9. Primitive low-Ti Icelandic picrite, which in many respects is more depleted than primitive MORB, has trace-element attributes indicating a source in olivine gabbro or troctolite, or their eclogitic equivalents, in which concentrations of trace elements and their ratios are specified by compositions of cumulus plagioclase (e.g., high Sr/REE) and clinopyroxene (e.g., very low Zr/Y), in adcumulates with very low proportions of trapped residual melt.

10. Nearly as depleted but still geochemically distinct tholeiite erupts along Reykjanes and Kolbeinsey ridges adjacent to Iceland, but the source of these is a refractory peridotite previously inferred to represent the result of extended melt column (high-T, high-P) melting of a uniform peridotite source beneath spreading ridges.

11. Instead, I prefer to think that their mantle source was comparatively refractory and depleted to begin with, and that this entails no particularly higher temperature or degree of melting than that of more typical MORB derived elsewhere from more fertile mantle.

12. The thickened Iceland crust results in part from the addition of gabbroic or eclogitic components to the melt source that give rise to both low-Ti and higher-Ti Icelandic tholeiite.

13. I propose that convective instability and foundering of lower continental crust into refractory subcontinental lithospheric mantle provides the necessary combination of melt sources to explain the array of volcanic products at Iceland and along Reykjanes and Kolbeinsey ridges, and that Paleogene detachment of rigid upper crust and lithosphere from these rocks during initial continental rifting in the north Atlantic allowed retention of the necessary assemblages in the modern convecting mantle to explain the geochemistry of the current volcanism at Iceland.

ACKNOWLEDGMENTS

First, I offer heartfelt thanks to originators and compilers of both GeoRoc and PetDB. Your databases are providing the community a tremendous service, and it is a whole lot easier referencing your files than dozens of individual studies.

A long time ago, alas, Dave Clague and I developed the kernel of the argument about mixing between basalt and rhyolite at Iceland, which ran aground in review. I thank Dave for his early inspiration and effort on behalf of this idea. I would also like to acknowledge the ideas of Mike Rhodes about mixing in the realm of MORB, which have continued to influence my thinking long after we sailed together. The ideas about abyssal gabbros were developed in conjunction with Henry Dick, Bob Fisher, Sherm Bloomer, and many shipboard colleagues. I am grateful to them all. As always in the past few years, Gillian Foulger, Don Anderson, Dean Presnall, and Warren Hamilton helped, advised, and encouraged in countless ways. Angelo Peccerillo, Hetu Sheth, and Godfrey Fitton heroically reviewed the much longer and more complicated version of this paper. Regrettably, for the sake of length, the parts about Deccan and Karoo were left on the cutting-room floor. However, like James Bond, they will be back. Gill Foulger gracefully arbitrated the reviews. I thank the National Science Foundation and various minions of the Ocean Drilling Program for their support of work on the East Pacific Rise and the Indian Ocean.

REFERENCES CITED

Allan, J.F., Batiza, R., Perfit, M.R., Fornari, D.J., and Sack, R.O., 1989, Petrology of lavas from the Lamont seamount chain and adjacent East Pacific Rise, 10°N: Journal of Petrology, v. 30, p. 1245–1298.

Allan, J., Falloon, T., Pedersen, R., Lakkapragada, B.S., Natland, J.H., and Malpas, J., 1996, Petrology of selected Leg 147 basaltic lavas and dikes, *in* Mével, C., et al., eds., Proceedings of the Ocean Drilling Program, Scientific Results, v. 147: College Station, Texas, Ocean Drilling Program, p. 173–186.

Anderson, D.L., 2005, Large igneous provinces, delamination, and fertile mantle: Elements, v. 1, p. 271–275.

Anderson, D.L., 2007 (this volume), The eclogite engine: Chemical geodynamics as a Galileo thermometer, in Foulger, G.R., and Jurdy, D.M., eds., Plates, plumes, and planetary processes: Boulder, Colorado, Geological Society of America Special Paper 430, doi: 10.1130/2007.2430(03).

Anderson, D.L., and Natland, J.H., 2005, A brief history of the plume hypothesis and its competitors: Concept and controversy, in Foulger, G.R., et al., eds., Plates, plumes, and paradigms: Boulder, Colorado, Geological Society of America Special Paper 388: p. 119–145.

Andreoli, M.A.G., Hart, R.J., Ashwal, L.D., and Coetzee, H., 2006, Correlations between U, Th content and metamorphic grade in the western Namaqualand Belt, South Africa, with implications for radioactive heating of the crust: Journal of Petrology, v. 47, p. 1095–1118, doi: 10.1093/petrology/egl004.

Arai, S., and Matsukage, K., 1996, Petrology of gabbro-troctolite-peridotite complex from Hess Deep, equatorial Pacific: Implications for mantle-melt interaction within the oceanic lithosphere, in Mével, C., et al., eds., Proceedings of the Ocean Drilling Program, Scientific Results, v. 147: College Station, Texas, Ocean Drilling Program, p.135–155.

Baksi, A.K., 2001, Search for a deep mantle component in mafic lavas using a Nb-Y-Zr plot: Canadian Journal of Earth Sciences, v. 38, p. 813–824, doi: 10.1139/cjes-38-5-813.

Batiza, R., Rosendahl, B.R., and Fisher, R.L., 1977, Evolution of oceanic crust 3: Petrology and chemistry of basalts from the East Pacific Rise and Siqueiros transform fault: Journal of Geophysical Research, v. 82, p. 266–276.

Bédard, J.H., 1994, A procedure for calculating the equilibrium distribution of trace elements among the minerals of cumulate rocks, and the concentration of trace elements in the coexisting liquids: Chemical Geology, v. 118, p. 143–153, doi: 10.1016/0009-2541(94)90173-2.

Bédard, J.H., 2006, A catalytic delamination-driven model for coupled genesis of Archaean crust and sub-continental lithospheric mantle: Geochimica et Cosmochimica Acta, v. 70, p. 1188–1214, doi: 10.1016/j.gca.2005.11.008.

Bell, K., and Tilton, G.R., 2002, Probing the mantle: The story from carbonatites: Eos (Transactions, American Geophysical Union), v. 83, p. 273, 276–277.

Bird, P., 1978, Initiation of intracontinental subduction in the Himalaya: Journal of Geophysical Research, v. 83, p. 4975–4987.

Bird, P., 1979, Continental delamination and the Colorado Plateau: Journal of Geophysical Research, v. 84, p. 7561–7571.

Bloomer, S.H., Meyer, P.S., Dick, H.J.B., Ozawa, K., and Natland, J.H., 1991, Textural and mineralogic variations in gabbroic rocks from Hole 735B, in Von Herzen, R.P., and Robinson, P.T., et al., eds., Proceedings of the Ocean Drilling Program, Scientific Results, v. 118: College Station, Texas, Ocean Drilling Programm, p. 21–39.

Breddam, K., 2002, Kistufell: Primitive melt from the Iceland mantle plume: Journal of Petrology, v. 43, p. 345–373, doi: 10.1093/petrology/43.2.345.

Campbell, I.H., 2005, Large igneous provinces and the mantle plume hypothesis: Elements, v. 1, p. 265–269.

Campbell, I.H., and Griffiths, R.W., 1990, Implications of mantle plume structure for the evolution of flood basalts: Earth and Planetary Science Letters, v. 99, p. 79–93, doi: 10.1016/0012-821X(90)90072-6.

Campbell, I.H., and Griffiths, R.W., 1993, The evolution of the mantle's chemical structure: Lithos, v. 30, p. 389–399, doi: 10.1016/0024-4937(93)90047-G.

Carmichael, I.S.E., 1964, The petrology of Thingmuli, a Tertiary volcano in eastern Iceland: Journal of Petrology, v. 5, p. 435–460.

Chauvel, C., and C. Hémond, 2000, Melting of a complete section of recycled oceanic crust: Trace element and Pb isotopic evidence from Iceland: Geochemistry, Geophysics, Geosystems, v. 1, 1999GC000002.

Daly, R.A., 1926, Our mobile Earth: New York, Charles Scribner's Sons, 342 p.

Daly, R.A., 1933, Igneous Rocks and the Depths of the Earth: New York, McGraw-Hill (Hafner Reprint, 1962), 598 p.

DeLong, S.E., and Chatelain, C., 1990, Trace-element constraints on accessory-phase saturation in evolved MORB magma: Earth and Planetary Science Letters, v. 101, p. 206–215, doi: 10.1016/0012-821X(90)90154-P.

Dick, H.J.B., and Natland, J.H., 1996, Late-stage melt evolution and transport in the shallow mantle beneath the East Pacific Rise, in Mével, C., et al., Proceedings of the Ocean Drilling Program, Scientific Results, v. 147: College Station, Texas, Ocean Drilling Program, p. 103–134.

Dick, H.J.B., Natland, J.H., Alt, J.C., Bach, W., Bideau, D., Gee, J.S., Haggas, S., Hertogen, J.G.H., Hirth, G., Holm, P.M., Ildefonse, B., Iturrino, G.J., John, B.E., Kelley, D.S., Kikawa, E., Kingdon, A., LeRoux, P.J., Maeda, J., Meyer, P.S., Miller, D.J., Naslund, H.R., Niu, Y.-L., Robinson, P.T., Snow, J., Stephen, R.A., Trimby, P.W., Worm, H.-U., and Yoshinobu, A., 2000, A long in situ section of the lower ocean crust: Results of ODP Leg 176 drilling at the Southwest Indian Ridge: Earth and Planetary Science Letters, v. 179, p. 31–51, doi: 10.1016/S0012-821X(00)00102-3.

Dixon, S., and Rutherford, M.J., 1979, Plagiogranites and late-stage immiscible liquids in ophiolites and mid-ocean ridge suites: An experimental study: Earth and Planetary Science Letters, v. 45, p. 45–60, doi: 10.1016/0012-821X(79)90106-7.

Ducea, M., and Saleeby, J., 1996, Buoyancy sources for a large, unrooted mountain range, the Sierra Nevada, California: Evidence from xenolith thermobarometry: Journal of Geophysical Research, v. 101, p. 8229–8244, doi: 10.1029/95JB03452.

Engel, A.E.J., Engel, C.G., and Havens, R.G., 1965, Chemical characteristics of oceanic basalts and the upper mantle: Bulletin of the Geological Society of America, v. 76, p. 719–734, doi: 10.1130/0016-7606(1965)76[719:CCOOBA]2.0.CO;2.

Fitton, J.G., 2007 (this volume), The OIB paradox, in Foulger, G.R., and Jurdy, D.M., eds., Plates, plumes, and planetary processes: Boulder, Colorado, Geological Society of America Special Paper 430, doi: 10.1130/2007.2430(20).

Fitton, J.G., Saunders, A.D., Norry, M.J., Hardarson, B.S., and Taylor, R.N., 1997, Thermal and chemical structure of the Iceland plume: Earth and Planetary Science Letters, v.153, v. 197–208.

Foulger, G.R., Natland, J.H., and Anderson, D.L., 2005a, Genesis of the Iceland melt anomaly by plate tectonic processes, in Foulger, G.R., et al., eds., Plates, plumes, and paradigms, Boulder, Colorado, Geological Society of America Special Paper 388, p. 595–625.

Foulger, G.R., Natland, J.H., and Anderson, D.L., 2005b, A source for Icelandic magmas in remelted Icelandic crust: Journal of Volcanology and Geothermal Research, v. 141, p. 23–44, doi: 10.1016/j.jvolgeores.2004.10.006.

Frey, F.A., Weis, D., Borisova, A., and Xu, G., 2002, Involvement of continental crust in the formation of the Cretaceous Kerguelen Plateau: New perspectives from ODP Leg 120 Sites: Journal of Petrology, v. 43, p. 1207–1239, doi: 10.1093/petrology/43.7.1207.

Gaul, O.F., Griffin, W.L., O'Reilly, S.Y., and Pearson, N.J., 2000, Mapping olivine composition in the lithospheric mantle: Earth and Planetary Science Letters, v. 182, p. 223–235, doi: 10.1016/S0012-821X(00)00243-0.

Gee, J., Staudigel, H., and Natland, J.H., 1991, Geology and petrology of Jasper Seamount: Journal of Geophysical Research, v. 96, p. 4083–4106.

Gibson, I.L., and Walker, G.P.L., 1964, Some composite rhyolite/basalt lavas and related composite dykes in eastern Iceland: Proceedings of the Geological Associaton, v. 3, p. 301–308.

Gibson, S.A., Thompson, R.N., and Dickin, A.P., 2000, Ferropicrites: Geochemical evidence for Fe-rich streaks in upwelling mantle plumes: Earth and Planetary Science Letters, v. 174, p. 355–374, doi: 10.1016/S0012-821X(99)00274-5.

Green, D.H., and Falloon, T.J., 2005, Primary magmas at mid-ocean ridges, "hotspots," and other intraplate settings: Constraints on mantle potential temperature, in Foulger, G.R., et al., eds., Plates, plumes, and paradigms, Boulder, Colorado, Geological Society of America Special Paper 388, p. 217–247.

Green, D.H., Falloon, T.J., Eggins, S.M., and Yaxley, G.M., 2001, Primary magmas and mantle temperatures: European Journal of Mineralogy, v. 13, p. 437–451, doi: 10.1127/0935-1221/2001/0013-0437.

Griffiths, R.W., and Campbell, I.H., 1991, Interaction of mantle plume heads with the Earth's surface, and onset of small-scale convection: Journal of Geophysical Research, v. 96, p. 18,295–18,310.

Gunn, B.M., 2005, Atlantic Ocean volcanism, OIBs, tholeiites and basanites, *in* Gunn, B.M., Geochemistry of igneous rocks, GeoKem: An electronic reference text of igneous geochemistry, http://www.geokem.com/OIB-volcanic-iceland.html.

Gurenko, A.A., and Chaussidon, M., 1995, Enriched and depleted primitive melts included in olivine from Icelandic tholeiites: Origin by continuous melting of a single mantle column: Geochimica et Cosmochimica Acta, v. 59, p. 2905–2917.

Hamilton, W.B., 2003, An alternative Earth: GSA Today, v. 13, p. 4–12, doi: 10.1130/1052-5173(2003)013<0004:AAE>2.0.CO;2.

Hanan, B.B., and Graham, D.W., 1996, Lead and helium isotopic evidence for a common deep source of mantle plumes: Science, v. 272, p. 991–995, doi: 10.1126/science.272.5264.991.

Hansteen, T.H., 1991, Multi-stage evolution of the picritic Maelifell rocks, SW Iceland: Constraints from mineralogy and inclusions of glass and fluid in olivine: Contributions to Mineralogy and Petrology, v. 109, p. 225–239, doi: 10.1007/BF00306481.

Harding, A., Kent, G.C., and Orcutt, J.A., 1993, A multichannel seismic investigation of upper crustal structure at 9 N on the East Pacific Rise: Implications for crustal accretion: Journal of Geophysical Research, v. 98, p. 13,925–13,944.

Hart, S.R., Schilling, J.-G., and Powell, J.L., 1973, Basalts from Iceland and along the Reykjanes Ridge: Sr Isotope Geochemistry: Nature: Physical Science, v. 246, p. 104–107.

Hart, S.R., Hauri, E.H., Oschmann, L.A., and Whitehead, J.A., 1992, Mantle plumes and entrainment: Isotopic evidence: Science, v. 256, p. 517–520, doi: 10.1126/science.256.5056.517.

Hitchen, K., Morton, A.C., Mearns, E.W., Whitehouse, M., and Stoker, M.S., 1997, Geological implications from geochemical and isotopic studies of Upper Cretaceous and Lower Tertiary igneous rocks around the northern Rockall Trough: Journal of the Geological Society of London, v. 154, p. 517–521.

Hofmann, A.W., 1988, Chemical differentiation of the Earth: The relationship between mantle, continental crust, and oceanic crust: Earth and Planetery Science Letters, v. 90, p. 297–314, doi: 10.1016/0012-821X(88)90132-X.

Holmes, A., 1928, Radioactivity and earth movements: Transactions of the Geological Society of Glasgow, v. 18, p. 559–606.

Jaques, A.L., and Green, D.H., 1980, Anhydrous melting of periodotite at 0–15 Kb pressure and the genesis of tholeiitic basalts: Contributions to Mineralogy and Petrology, v. 73, p. 287–310, doi: 10.1007/BF00381447.

Johnson, K.T.M., 1998, Experimental determination of partition coefficients for rare earth and high-field-strength elements between clinopyroxene, garnet, and basaltic melt at high pressure: Contributions to Mineralogy and Petrology, v. 133, p. 60–68, doi: 10.1007/s004100050437.

Johnson, K.T.M., Dick, H.J.B., and Shimizu, N., 1990, Melting in the oceanic upper mantle: An ion microprobe study of diopsides in abyssal peridotites: Journal of Geophysical Research, v. 95, p. 2661–2678.

Jordan, T.L., 1979, Mineralogies, densities, and seismic velocities of garnet lherzolites and their geophysical implications, *in* Boyd, F.R., and Meyer, H.O.A., eds., The mantle sample: Inclusions in kimberlites and other volcanics: Proceedings of the 2nd International Kimberlite Conference, v. 2: Washington, D.C., American Geophysical Union, p. 1–14.

Jull, M., and Kelemen, P.B., 2001, On the conditions for lower crustal convective instability: Journal of Geophysical Research, v. 106, p. 6423–6446, doi: 10.1029/2000JB900357.

Juster, T., Grove, T.L., and Perfit, M.R., 1989, Experimental constraints on the generation of FeTi basalts, andesites and rhyodacites at the Galapagos Spreading Center, 85°W and 95°W: Journal of Geophysical Research, v. 94, p. 9251–9247.

Kay, R.W., and Kay, S.M., 1991, Creation and destruction of the lower continental crust: International Journal of Earth Sciences, v. 80, p. 259–278.

Kay, R.W., and Kay, S.M., 1993, Delamination and delamination magmatism: Tectonophysics, v. 219, p. 177–189, doi: 10.1016/0040-1951(93)90295-U.

Kempton, P.D., Fitton, J.G., Saunders, A.D., Nowell, G.M., Taylor, R.N., Hardarson, B.S., and Pearson, G., 2000, The Iceland plume in space and time: A Sr-Nd-Pb-Hf study of the north Atlantic rifted margin: Earth and Planetary Science Letters, v. 177, p. 255–271, doi: 10.1016/S0012-821X(00)00047-9.

Kent, G.M., Harding, A.J., and Orcutt, J.A., 1993, Distribution of magma beneath the East Pacific Rise between the Clipperton Transform and the 9° 17′N Deval from forward modeling of common-depth-point data: Journal of Geophysical Research, v. 98, p. 13,945–13,969.

Klein, E.M., and Langmuir, C.H., 1987, Global correlations of ocean ridge basalt chemistry with axial depth and crustal thickness: Journal of Geophysical Research, v. 92, p. 8089–8115.

Kodaira, S., Mjelde, R., Gunnarslon, K., Shiobara, H., and Shimamura, H., 1998, Structure of the Jan Mayen microcontinent and implications for its evolution: Geophysical Journal International, v. 132, p. 383–400, doi: 10.1046/j.1365-246x.1998.00444.x.

Langmuir, C.M., Klein, E.M., and Plank, T., 1992. Petrological systematics of mid-ocean ridge basalts: Constraints on melt generation beneath ocean ridges, *in* Phipps Morgan, J., Blackman, D.K., and Sinton, J.M., eds., Mantle flow and melt generation at mid-ocean ridges: Washington, D.C., American Geophysical Union, Geophysical Monograph 71, p. 183–280.

Leitch, A.M., and Davies, G.F., 2001, Mantle plumes and flood basalts: Enhanced melting from plume ascent and an eclogite component: Journal of Geophysical Research, v. 106, p. 2047–2059, doi: 10.1029/2000JB900307.

Maaløe, S., Sorensen, I.B., and Hertogen, J., 1986, The trachybasaltic suite of Jan Mayen: Journal of Petrology, v. 27, p. 439–466.

MacDonald, R., Sparks, R.S.J., Sigurdsson, H., Mattey, D.P., McGarvie, D.W., and Smith, R.L., 1987, The 1875 eruption of Askja Volcano, Iceland: Combined fractional crystallization and selective contamination in the generation of rhyolitic magma: Mineralogical Magazine, v. 51, p. 183–202, doi: 10.1180/minmag.1987.051.360.01.

Maclennan, J., McKenzie, D., Grönvold, K., Shjimizu, N., Eiler, J.M., and Kitchen, N., 2003, Melt mixing and crystallization under Theistareykir, northeast Iceland: Geochemistry, Geophysics, Geosystems, v. 4, n. 11, 8624, 40 p., doi:10.1029/2003GC000558.

Mahoney, J.J., Natland, J.H., White, W.M., Poreda, R., Bloomer, S.H., Fisher, R.L., and Baxter, A.N., 1989, Isotopic and geochemical provinces of the western Indian Ocean spreading centers: Journal of Geophysical Research, v. 94, p. 4033–4052.

Marshall, L.A., and Sparks, R.S.J., 1984, Origins of some mixed-magma and net-veined ring intrusions: Journal of the Geological Society of London, v. 141, p. 171–182.

McDonough, W.F., and Sun, S.-S., 1995, The composition of the Earth: Chemical Geology, v. 120, p. 2223–2253.

McKenzie, D., and Bickle, J., 1988, The volume and composition of melt generated by extension of the lithosphere: Journal of Petrology, v. 29, p. 625–679.

MELT Seismic Team, 1998, Imaging the deep seismic structure beneath a mid-ocean ridge: The MELT experiment: Science, v. 280, p. 1215–1218, doi: 10.1126/science.280.5367.1215.

Morgan, W.J., 1981, Hotspot tracks and the opening of the Atlantic and Indian oceans, *in* Emiliani, C., ed., The sea, v. 7: New York, Wiley, p. 443–487.

Natland, J.H., 1980, Effect of axial magma chambers beneath spreading centers on the compositions of basaltic rocks, *in* Rosendahl, B.R., et al., eds. Initial Reports of the Deep Sea Drilling Project, v. 54: Washington, D.C., U.S. Government Printing Office, p. 833–850.

Natland, J.H., 1989, Partial melting of a lithologically heterogeneous mantle: Inferences from crystallization histories of magnesian abyssal tholeiites from the Siqueiros Fracture Zone, *in* Saunders, A.D., and Norry, M., eds., Magmatism in the ocean basins: Geological Society of London Special Publication 42, p. 41–77.

Natland, J.H., 1991a. Crystallization and mineralogy of ocean basalts, *in* Floyd, P.A., ed., Oceanic basalts: London, Blackie and Sons, p. 62–93.

Natland, J.H., 1991b, Indian Ocean crust, *in* Floyd, P.A., ed., Oceanic basalts: London, Blackie and Sons, p. 289–310.

Natland, J.H., and Dick, H.J.B., 1996, Melt migration through high-level gabbroic cumulates of the East Pacific Rise at Hess Deep: The origin of magma lenses and the deep crustal structure of fast-spreading ridges, *in* Mével, C., et al., eds., Proceedings of the Ocean Drilling Program, Scientific Results, v. 147: College Station, Texas, Ocean Drilling Program, p. 21–58.

Natland, J.H., and Dick, H.J.B., 2001, Formation of the lower ocean crust and the crystallization of gabbroic cumulates at a very slowly spreading ridge: Journal of Volcanology and Geothermal Research, v. 110, p. 191–233, doi: 10.1016/S0377-0273(01)00211-6.

Natland, J.H., and Dick, H.J.B., 2002, Stratigraphy and composition of gabbros drilled at ODP Hole 735B, Southwest Indian Ridge: A synthesis, *in* Natland, J.H., et al., eds., Proceedings of the Ocean Drilling Program, Scientific Results, v. 176: College Station, Texas, Ocean Drilling Program, p. 1–69, http://www.odp.tamu.edu/publications/176_SR/VOLUME/SYNTH/SYNTH.PDF.

Natland, J.H., and Melson, W.G., 1980, Compositions of basaltic glasses from the East Pacific Rise and Siqueiros fracture zone near 9 N, *in* Rosendahl, B.R., et al., eds., Initial Reports of the Deep Sea Drilling Project, v. 54: Washington, D.C., U.S. Government Printing Office, p. 705–723.

Natland, J.H., Meyer, P.S., Dick, H.J.B., and Bloomer, S.H., 1991, Magmatic oxides and sulfides in gabbroic rocks from Hole 735B and the later development of the liquid line of descent, *in* Von Herzen, R.P., et al., eds., Proceedings of the Ocean Drilling Program, Scientific Results, v. 118: College Station, Texas, Ocean Drilling Program, p. 75–111.

Niu, Y.-L., Regelous, M., Wendt, I., Batiza, R., and O'Hara, M.J., 2002a, Geochemistry of near-EPR seamounts: Importance of source vs. process and the origin of enriched mantle component: Earth and Planetary Science Letters, v. 199, p. 327–345, doi: 10.1016/S0012-821X(02)00591-5.

Niu, Y.-N., Gilmore, T., Mackie, S., Greig, A., and Bach, W., 2002b, Mineral chemistry, whole-rock compositions and petrogenesis of Leg 176 gabbros: Data and discussion, *in* Natland, J.H., et al., eds., Proceedings of the Ocean Drilling Program, Scientific Results, v. 176, p. 1–60, http://www.odp.tamu.edu/publications/176_SR/chap_08/chap_08.htm.

O'Hara, M.J., 1973, Non-primary magmas and dubious mantle plume beneath Iceland: Nature, v. 243, p. 507–508, doi: 10.1038/243507a0.

O'Reilly, S.Y., and Griffin, W.L., 2006, Imaging global chemical and thermal heterogeneity in the subcontinental lithospheric mantle with garnets and xenoliths: Geophysical implications: Tectonophysics, v. 416, p. 289–309, doi: 10.1016/j.tecto.2005.11.014.

Pedersen, R.B., Malpas, J., and Falloon, T., 1996, Petrology and geochemistry of gabbroic and related rocks from Site 894, Hess Deep, *in* Mével, C., Gillis, K., Allan, J., and Meyer, P.S., eds., Proceedings of the Ocean Drilling Program, Scientific Results, v. 147: College Station, Texas, Ocean Drilling Program, p. 3–20.

Perfit, M.R., Fornari, D.J., Casey, J., Kastens, K.A., Kirsk, P.A., Edwards, M., Ridley, W.I., Shuster, R., Paradis, S., and Xia, C., 1996, Recent volcanism in the western Siqueiros transform fault: Eruption of picritic and MgO-rich basalts and implications for MORB magmagenesis: Earth and Planetary Science Letters, v. 141, p. 91–108, doi: 10.1016/0012-821X(96)00052-0.

Presnall, D.C., Gudfinnsson, G.H., and Walter, M.J., 2002, Generation of mid-ocean ridge basalts at pressures from 1 to 7 GPa: Geochimica et Cosmochimica Acta, v. 66, p. 2073–2090, doi: 10.1016/S0016-7037(02)00890-6.

Rhodes, J.M., and Dungan, M.A., 1979, The evolution of ocean floor basaltic magmas, *in* Talwani, M., Harrison, C.G.A., and Hayes, D.E., eds., Deep drilling results in the Atlantic Ocean: Ocean crust, Maurice Ewing Series 2: Washington, D.C., American Geophysical Union, p. 239–244.

Rhodes, J.M., Dungan, M.A., Blanchard, D.P., and Long, P.E., 1979, Magma mixing at mid-ocean ridges: Evidence from basalts drilled near 22°N on the Mid-Atlantic Ridge: Tectonophysics, v. 55, p. 35–61, doi: 10.1016/0040-1951(79)90334-2.

Roberts, D.G., 1975, Marine geology of the Rockall Plateau and Trough: Philosophical Transactions of the Royal Society of London, Series A, v. 278, p. 447–509.

Rudnick, R.L., 1995, Making continental crust: Nature, v. 378, p. 571–578, doi: 10.1038/378571a0.

Saleeby, J., Ducea, M., and Clemens-Knott, D., 2003, Production and loss of a high-density batholithic root, southern Sierra Nevada, California: Tectonics, v. 22, doi:10.1029/2002TC001374, p. 3-1–3-24.

Schilling, J.-G., 1973a, Iceland mantle plume: Geochemical study of Reykjanes ridge: Nature, v. 242, p. 565–571, doi: 10.1038/242565a0.

Schilling, J.-G., 1973b, Iceland mantle plume: Nature, v. 246, p. 141–143, doi: 10.1038/246141a0.

Schott, B., and Schmeling, H., 1998, Delamination and detachment of a lithospheric root: Tectonophysics, v. 296, p. 225–247, doi: 10.1016/S0040-1951(98)00154-1.

Sigurdsson, H., and Sparks, R.S.J., 1981, Petrology of rhyolitic and mixed magma ejecta from the 1875 eruption of Askja, Iceland: Journal of Petrology, v. 22, p. 41–84.

Sigurdsson, I.A., Steinthorsson, S., and Grønvold, K., 2000, Calcium-rich melt inclusions in Cr-spinels from Borgarhraun, northern Iceland: Earth and Planetary Science Letters, v. 183, p. 15–26, doi: 10.1016/S0012-821X(00)00269-7.

Sinton, J.M., and Detrick, R.S., 1992, Mid-ocean ridge magma chambers: Journal of Geophysical Research, v. 97, p. 197–216.

Sinton, J.M., Smaglik, S.M., Mahoney, J.J., and MacDonald, K.C., 1991, Magmatic processes at superfast spreading mid-ocean ridges: Glass compositional variations along the East Pacific Rise, 113°W, 23°S: Journal of Geophysical Research, v. 96, p. 6133–6155.

Slater, L., McKenzie, D., Grønvold, K., and Shimizu, N., 2001, Melt generation and movement beneath Theistareykir, NE Iceland: Journal of Petrology, v. 42, p. 321–354, doi: 10.1093/petrology/42.2.321.

Sobolev, A.V., Hofmann, A.W., and Nikogosian, I.K., 2000, Recycled oceanic crust observed in "ghost plagioclase" within the source of Mauna Loa lavas: Nature, v. 404, p. 986–990, doi: 10.1038/35010098.

Spulber, S.D., and Rutherford, M.J., 1983, The origin of rhyolite and plagiogranite in oceanic crust: An experimental study: Journal of Petrology, v. 24, p. 1–25.

Stracke, A., Zindler, A., Salters, V.J.M., McKenzie, D., Blichert-Toft, J., Albarède, F., and Grønvold, K., 2003, Theistareykir revisited: Geochemistry, Geophysics, Geosystems, v. 4, 49 p., doi:10.1029/2002GC002001.

Stracke, A., Hofmann, A.W., and Hart, S.R., 2005, FOZO, HIMU, and the rest of the mantle zoo: Geochemistry, Geophysics, Geosystems, v. 6, 20 p., doi:10.1029/2004GC000824.

Sweeney, R.J., Falloon, T.J., Green, D.H., and Tatsumi, Y., 1991, The mantle origins of Karoo picrites: Earth and Planetary Science Letters, v. 107, p. 256–271, doi: 10.1016/0012-821X(91)90075-S.

Talwani, M., and Eldholm, O., 1977, Evolution of the Norwegian-Greenland Sea: Bulletin of the Geological Society of America, v. 88, p. 969–999, doi: 10.1130/0016-7606(1977)88<969:EOTNS>2.0.CO;2.

Wager, L.R., and Deer, W.A., 1939, Geological investigations in East Greenland, part 3: The petrology of the Skaergaard intrusion, Kangerdlugssuaq, East Greenland: Meddelelser om Grønland, v. 105, p. 1–352.

Walker, G.P.L., 1963, The Breiddalur central volcano eastern Iceland: Quarterly Journal of the Geological Society of London, v. 119, p. 29–63.

Walker, G.P.L., and Skelhorn, R.R., 1966, Some associations of acid and basic igneous rocks: Earth-Science Reviews, v. 2, p. 93–109, doi: 10.1016/0012-8252(66)90024-9.

Yoder, H.S., 1973, Contemporaneous basaltic and rhyolitic magmas: American Mineralogist, v. 58, p. 153–171.

Zandt, G., Gilbert, H., Owens, T.J., Ducea, M., Saleeby, J., and Ge Jones, C.H., 2004, Active foundering of a continental arc root beneath the southern Sierra Nevada in California: Nature, v. 431, p. 41–46, doi: 10.1038/nature02847.

MANUSCRIPT ACCEPTED BY THE SOCIETY JANUARY 31, 2007

DISCUSSION

25 January 2007, J. Godfrey Fitton

The Iceland melting anomaly is widely held to be the product of hot and fertile mantle supplied by a mantle plume. Recycled subducted ocean crust, for which the ΔNb parameter (Fitton et al., 1997; Fitton, this volume) is intended as a proxy, provides a plausible fertile component. In a radical alternative hypothesis, Natland (this volume) proposes that melting of lithologies originating in detached or delaminated continental lithospheric mantle and lower crust can provide virtually the whole spectrum of Icelandic magmas. Specifically, he recognizes four distinct magma types in and around Iceland:

1. Depleted basalt erupted on the Reykjanes and Kolbeinsey ridges (south and north of Iceland, respectively) is produced by melting subcontinental lithospheric mantle.
2. Depleted Icelandic picrite results from the melting of depleted olivine gabbro.
3. Evolved Fe-Ti–rich ferrobasalt and icelandite result from the melting of oxide-rich gabbro.
4. Rhyolite magma is produced by melting silicic veins in the gabbro protolith of magma types 2 and 3.

In Natland's model, all Icelandic magmas are the product of mixing between these four end-members, and fractional crystallization plays only a minor role in their evolution. The model is based principally on four questionable assertions:

1. The array of Icelandic data used by Fitton et al. (1997) to define ΔNb only appears to be linear because evolved rocks have been excluded. Adding intermediate and silicic compositions emphasizes the curvature in a mixing array between depleted basalt and the Fe-Ti–rich or silicic end-members. Figure D-1 shows a second-order regression line fitted to the data used by Fitton et al. (1997). The line is almost straight and parallel to the $\Delta Nb = 0$ reference line. Linear regression gives a virtually identical line with an equally good fit ($r^2 = 0.929$ for both). There is, therefore, no statistically significant curvature in the data array.
2. Mixing primitive depleted basalt ($\Delta Nb < 0$) with as little as 1–2% of evolved andesite or rhyolite is sufficient to change ΔNb from negative to positive. A judicious choice of end-members is required to make this work. Figure D-1 shows Natland's mixing line between primitive "Icelandic" and silicic end-members. The primitive end-member has 0.3 ppm Nb and 21.9 ppm Zr, implying a source that is much more depleted than the source of normal mid-ocean ridge basalt and yet is capable of melting at normal temperatures to supply >90% of the mass of Iceland (Fig. D-1). The silicic end-member is a trachyte (not andesite as stated by Natland) from the off-axis mildly alkaline Snæfellsnes volcano. It contains 122 ppm Nb and 623 ppm Zr and is a very unlikely rock type to

expect in significant amounts in the lower continental crust. Mixing between the two end-members gives a very poor fit to the data of Fitton et al. (1997). Natland acknowledges this by assigning a different origin (melting of olivine gabbro) to depleted Icelandic picrite (those rocks with Zr/Y < 1.8 on Fig. D-1). Mixing between the primitive end-member and average continental crust is incapable of reproducing the composition of Icelandic basalt (Fig. D-1).

3. ΔNb is insensitive to fractional crystallization, and therefore, basaltic magma with $\Delta Nb > 0$ cannot be parental to Icelandic rhyolite with $\Delta Nb < 0$. This is simply not so, as can easily be shown from the Rayleigh fractionation equation,

$$C_L/C_0 = F^{D-1},$$

where C_L is the concentration of a trace element in an evolving liquid, C_0 its initial concentration, F the fraction of liquid remaining, and D the bulk distribution coefficient. Applying this equation to Nb, Zr, and Y gives three equations that can be combined and rearranged to show that

$$[\log (Nb/Y)_L - \log (Nb/Y)_0]/[\log (Zr/Y)_L - \log (Zr/Y)_0] = (D_{Nb} - D_Y)/(D_{Zr} - D_Y).$$

The form of this equation shows that any fractional crystallization path on a plot of log (Nb/Y) versus log (Zr/Y) will be a straight line with a slope of $(D_{Nb} - D_Y)/(D_{Zr} - D_Y)$. Because $D_{Nb} \sim 0$ for most low-pressure mineral assemblages, the fractional crystallization path will have a shallower slope than the $\Delta Nb = 0$ reference line (1.92) when $D_{Zr}/D_Y < {\sim}0.5$ and will have a slope of unity (constant Nb/Zr) when $D_{Zr}/D_Y = 0$. Thus, fractional crystallization of any assemblage in which Y is more than twice as compatible as Zr will result in a reduction of ΔNb in the evolved magma. The large arrow in Figure D-1 represents 80% fractional crystallization ($F = 0.2$) of an assemblage in which $D_{Nb} = 0$, $D_{Zr} = 0.13$, and $D_Y = 0.96$, equivalent to a mixture of 40% clinopyroxene (D_{Zr} and D_Y from Ewart and Griffin, 1994) and 60% of phases (e.g., feldspars) in which all three elements have $D = 0$. Fractional crystallization of clinopyroxene-bearing assemblages can readily account for the negative values of ΔNb found in some Icelandic rhyolites. Clinopyroxene joins the low-pressure crystallizing assemblage at ~5 wt% MgO in Icelandic tholeiites, which is why Fitton et al. (1997) excluded basalt samples with <5 wt% MgO. Partial melting of hydrated basaltic crust (amphibolite) is also likely to produce rhyolite with $\Delta Nb < 0$ because amphiboles in equilibrium with silicic magmas also have $D_Y \gg D_{Zr}$ (Ewart and Griffin, 1994).
4. A tenfold range in TiO_2 content in parental Icelandic basalts cannot be explained by partial melting of peridotite. This is

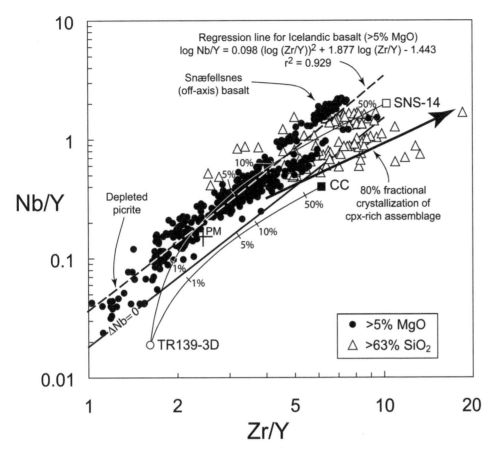

Figure D-1. The composition of Icelandic basic (Fitton et al., 1997) and silicic (GeoRoc) volcanic rocks compared with average continental crust (CC, from Rudnick and Fountain, 1995; Barth et al., 2000), and primitive mantle (PM, from McDonough and Sun, 1995). Basic volcanic rocks (>5 wt% MgO) form a linear array parallel to the ΔNb = 0 reference line (from Fitton et al., 1997). Mixing lines between the depleted (TR139-3D) and enriched (SNS-14; wrongly identified as SNS-17 in GeoRoc) end-members proposed by Natland and between the depleted end-member and average continental crust are also shown. The large arrow shows that the silicic rocks could be derived by fractional crystallization from average Icelandic basalt.

valid only if magmas are produced by equilibrium melting of a homogeneous mantle source. Titanium is only moderately incompatible in mantle phases, and so a tenfold abundance variation in primary melt would require more than a tenfold variation in degree of melting. But melting is not an equilibrium process, and radiogenic-isotope ratios show that the source cannot be homogeneous. Fractional melting of a heterogeneous mantle source will strip out the more fusible and Ti-rich material first, leaving a progressively more Ti-depleted peridotite residue to contribute during advanced stages of melting. These two extremes are represented in Iceland by, respectively, small volumes of enriched off-axis basalt (e.g., Snæfellsnes) and even smaller volumes of depleted picrite erupted in the rift axes (Hardarson and Fitton, 1997; Fitton et al., 2003). Most on-axis melting, however, will produce melt that reflects the bulk composition of the mantle, and this melt will evolve by fractional crystallization to produce the ubiquitous Icelandic tholeiite (Fig. D-1). Natland's model implicitly appeals to fractional melting, but his source is a mixture of depleted olivine gabbro and enriched ferrogabbro, both derived from the lower continental crust.

The foundations of Natland's model are unsound. He invokes an ad hoc assemblage of lithologies inferred to exist as major components of the lower continental crust, and for which there is little evidence. These lithologies were transferred to the convecting upper mantle during continental breakup and provide a constant source of tholeiitic magmatism that has lasted for >55 m.y. and shows no sign of exhaustion. By contrast, the plume model for Icelandic magmatism requires only two lithologies: variably depleted peridotite and an enriched component that could be eclogite. This assemblage is known to be fed back into the upper mantle through subduction, and its recycling through a mantle plume provides a plausible and virtually inexhaustible supply of mantle that is both fertile and hot. It can explain all the first-order geochemical features of Iceland with far fewer and more plausible assumptions and postulated components than Natland requires. Natland's hypothesis would not survive the application of Occam's razor.

I thank the editors for special permission to post a slightly longer comment than normally allowed.

2 February 2007, James H. Natland

Fitton's sense is that I have advocated a random set of end-members just to suit my model (see his comment of 25 January). No. They all go together, as they do in nature. We must stop thinking that each source lithology exists in physical isolation, in reservoirs. They never do. In this case, there is good reason for them to be together, and this allows a new, unifying, simplifying, and far from ad hoc hypothesis for Icelandic petrogenesis. The key is that the *one* source (call it the lower continental-crust assemblage) is polylithologic. This is far from arbitrary, just common sense, and simple.

I stated from the start that the Iceland trend was flatter than for mid-ocean ridge basalt (MORB); I never tried to explain negative ΔNb in rhyolite by crystallization differentiation. Fitton offered enriched eclogite as the source of high-Ti basalt, which was one of the main points of my paper. In addition, his attempt to explain rhyolite does not make rhyolite.

1. Fitton's linear regression only makes the flat trend for Iceland even more explicit. The flat trend is largely a result of the low-Ti olivine tholeiites and picrites derived, as I argue, from olivine gabbro cumulates or equivalent eclogite. Curvature on a ΔNb diagram within data for basalt is evident, especially for the northern part of the eastern rift. The curve partially overlaps the data from Kolbeinsey Ridge just to the north, which includes samples with negative ΔNb. The depleted sample I used for my calculation is from a submerged portion of the Iceland platform abutting the Tjörnes fracture zone, which offsets Iceland from the Kolbeinsey Ridge, suggesting that the depleted peridotite source for the Kolbeinsey Ridge extends some distance beneath northern Iceland.

2. Fitton notes that the "andesite" from Snæfellsnes is actually a trachyte; this was an error in GeoRoc. It hardly matters. Other andesites, dacites, or rhyolites could be used to make the same point. Fitton worries about the temperature of melting of the source of the primitive mixing component. Extent of depletion has nothing to do with melting temperature. Low-Ti basaltic melt can be produced by partial melting of refractory, near-harzburgitic peridotite at the same temperature and nearly to the same extent as basalt from fertile lherzolite (Jaques and Green, 1980; Natland, 1989; Presnall et al., 2002). I did not say and do not know what percentage of Iceland is derived from such a melt type or its differentiated residua. I suggested that partial melting of eclogite transformed from depleted olivine-gabbro and oxide-gabbro cumulates is responsible for most of Iceland. Eclogite has a lower melting temperature than peridotite (Anderson, 2005; Foulger et al., 2005).

3. I did not discuss derivation of rhyolite with negative ΔNb. I am concerned mainly with trends among basalts and how they might be influenced by silicic contamination or mixing. Fitton's scheme would not make rhyolite. It cannot increase SiO_2 or reduce FeOT and TiO_2 from amounts in the starting basalt. It has to involve oxide minerals, which have higher partition coefficients for Nb and Zr than for Y (compilation of Bédard, 1994). The question is not whether Icelandic rhyolite is derived from some basalt by crystallization differentiation but whether it is derived in this way from Icelandic tholeiite today. Any of it with higher $^{87}Sr/^{86}Sr$,

which is most of it, cannot be; on this basis, no rhyolite in Iceland is related by crystallization differentiation to primitive olivine tholeiite and picrite. Much of the rhyolite in the Deccan and Karoo flood basalt provinces has negative ΔNb, but it has even more obviously continental isotopic signatures. It is derived by partial melting of granitoids, granulites, or amphibolites that were produced by crystallization differentiation of basalt, metamorphism, and partial melting of metabasite through long geological time (Betton, 1979; Hawkesworth et al., 1984; Mahoney, 1988; Bédard, 2006). These plot at the same location as Icelandic rhyolite on a ΔNb diagram, but they are clearly not derived from Mesozoic flood basalt by crystallization differentiation. Iceland is a geochemically feebler version of this and of associations in Greenland and Scotland. Along the East Pacific Rise, crystallization differentiation of primitive basalt does not significantly change Nb/Y or Zr/Y and thus ΔNb. The same should be true even if the parental basalt has a positive ΔNb. At Iceland, therefore, why is there such a spread in these parameters just among basalts with >5% MgO? It cannot be done by crystallization differentiation, but it can by mixing with rhyolite, andesite, or rhyolite-contaminated ferrobasalt, even if it is produced along the lines of Fitton's model. Remember also that $^{87}Sr/^{86}Sr$ increases generally to the right on a ΔNb diagram among Icelandic basalts.

4. Partial melting of commonly construed mantle peridotite does not produce high-Ti basalt. This fails utterly to explain the absence of any kind of trends and the high Ti_8 of many Icelandic tholeiites. Fitton can argue in terms of partition coefficients and fractional melting, but a far simpler explanation is to allow the presence of a titanian phase in the source, which likely means that it is (was) gabbroic. Because in the end Fitton does suggest eclogite for the "enriched component," why not regard it as a facies (rather than high-pressure average MORB) and include among its precursors oxide gabbro with a titanian phase? Oxide gabbro will almost certainly accompany any body of olivine gabbro adcumulates in the protolith.

Fitton's initial and final comments concern the fourfold diversity of lithologies I invoke as sources for Iceland and adjacent ridges, which he terms an assemblage so ad hoc that it "would not survive the application of Occam's Razor." I emphasize that in the lower continental crust, where I believe three of the four originate, they occur together and will stay together should portions of lower continental crust sink into subcontinental mantle (the fourth). Accepting that tholeiitic high-iron differentiation characterizes many layered intrusions, olivine gabbro cumulates and oxide gabbros will be present together in plutonic assemblages in the continental crust wherever such differentiation has taken place. A lot of it did take place to make, e.g., ancient granitoids (Bédard, 2006), Bushveld felsites, Phanerozoic island arcs, etc. Later remelting of such composite assemblages would produce parallel strains of quite different partial melts. This is the simplest explanation for the duality of Icelandic basalt and much of the mixing I infer. Also, Icelandic rhyolite need not come simply from partial melting of gabbro; other lithologies might be involved, and their melting or assimilation would result in high Nb/Y and Zr/Y with positive ΔNb. Thus, during foundering of deep gabbroic constituents of continental crust, something granitic (granitoidal, gneissic, migmatitic, granulitic) may descend as well, all the rocks sinking into refractory, often harzburgitic, subcontinental lithosphere. Basalts from Reykjanes and Kolbeinsey ridges have the least Ti_8 and Na_8 of all MORB, and recent interpretation is that this is not so much a consequence of high temperature but of a refractory, more nearly harzburgitic, source (Natland, 1989; Presnall et al., 2002). Where do we *know* such refractory lithosphere occurs in abundance? Why, beneath the continents! Following rifting, all source lithologies are unavoidably involved in rift-related partial melting. This is why similar low-Ti basalt occurs near the Seychelles microcontinent in the Indian Ocean (Natland, 1991).

The gentle dispensation of the editors allowed me to match the length of Fitton's discussion. Many thanks.

4 February 2007, J. Godfrey Fitton

Placing the four postulated end-member protoliths for Icelandic basalt in close proximity in the lower continental crust and lithospheric mantle does not make them any less ad hoc. They still need to be conveniently similar in composition to Icelandic basalt. Lower continental crust is mafic in composition (Rudnick and Fountain, 1995), but there the similarity to Icelandic basalt ends. Icelandic basalt and lower continental crust differ in two crucial respects. First, crustal rocks tend to be deficient in Nb (e.g., low Nb/La and negative ΔNb), whereas Icelandic basalts have higher Nb/La (Hémond et al., 1993) and positive ΔNb. Natland requires at least one component with a high concentration of Nb and strongly positive ΔNb. Second, Icelandic basalt and ancient lower crustal rocks have very different Sr–Nd–Pb isotope ratios. This is evident from the extreme isotopic composition of the earliest (61 Ma) basalts erupted on the southeast Greenland continental margin before the opening of the north Atlantic Ocean (Fitton et al., 2000). The lower crustal granulites with which these are inferred to have been contaminated cannot be a significant component in the source of modern Icelandic basalt.

Natland proposes that his assemblage of lower continental crust and lithospheric mantle lithologies became detached and incorporated into the convecting upper mantle during the rifting episode that led to the formation of the north Atlantic Ocean. How much time is required for this assemblage to warm up enough to produce basaltic melt on decompression? Excess magmatism in the region would be expected to postdate continental separation at 55–54 Ma, but this is not so. Large-volume tholeiitic magmatism started synchronously across a 2000-km-wide area stretching from Baffin Island through Greenland to Scotland at around 61 Ma (Saunders et al., 1997). The sudden, widespread onset of magmatism before continental separation is predicted by the plume hypothesis but is difficult to reconcile with Natland's alternative model.

6 February 2007, James H. Natland

I invoked gabbro/eclogite, not granulite, as a source for Icelandic basalt. Granulite and felsic xenoliths (Rudnick and Fountain, 1995) are similar to Icelandic rhyolite with negative ΔNb on a ΔNb diagram (round symbols, Fig. D-2); the felsic rocks have >70% SiO_2. Other symbols are Canadian Archaean and Proterozoic gabbros (Owens and Dymek, 1992; Li et al., 2000; Kerr, 2003; Leatherdale et al., 2003). Eight of these are oxide-apatite gabbronorites with high TiO_2 (4.6–10.6%), total iron as Fe_2O_3 (19.7–41.6%), and P_2O_5 (2.8–5.4%) contents; they are ilmenite-magnetite-apatite cumulates. Others are mainly leucogabbro with low oxide contents. Most of the gabbros have positive ΔNb; the others overlap felsic xenoliths. The entire suite resembles enriched Icelandic high-Ti basalt and rhyolite in ΔNb. The gabbros have other suitable attributes (high Sr and Ba and steep REE patterns). On this basis, I see no reason to abandon my hypothesis that gabbroic material, particularly oxide gabbro, and something silicic (granitoidal, gneissic, migmatitic, granulitic) from the lower continental crust, left by foundering in the mantle during rifting, comprise the source of high-Ti Icelandic basalts.

I accept the more general point that it is early in the game to accept without reservation continental or subcontinental sources for Iceland; we do not yet have a full assessment of lithologies in those sources, their geochemical variability, or the mechanisms by which they attained that variability (crystallization, anatexis, melting of amphibolite, etc.). We do not fully understand partitioning of trace elements such as Nb and Ta during metamorphism to eclogite or into liquids during partial melting of eclogite (e.g., Barth et al., 2000; Rudnick et al., 2000). I am not particularly concerned with the somewhat high La/Nb of granulite xenoliths and most of the gabbros compared with Icelandic basalt (they actually partially overlap). This is a slight dissimilarity among strong similarities. Misfit isotopic ratios in Eocene basalt from a locality in Greenland only means that the Iceland source is now different; granulites and continental gabbros can be depleted or enriched. How a feebly enriched but originally continental source came to remain under Iceland is still conjectural. Obviously, a granitic crust had a stronger influence on basalt compositions in Greenland and Scotland than it does at Iceland now; most of the granite has been removed. The province is smaller now, as it should be.

Jull and Kelemen (2001) summarize evidence that gabbro and pyroxenite produced in the differentiation of island arcs are not abundant in deep sections of the crust; this is why, they argue, such rocks founder into the deeper mantle. Gabbros sensu stricto are also not among the xenolith types in Rudnick and Fountain (1995), although some may be included in their average mafic granulites.

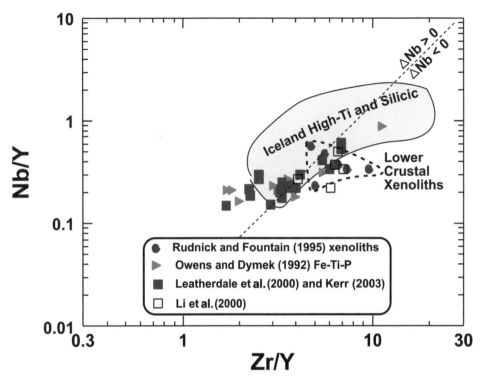

Figure D-2. ΔNb diagram (Zr/Y versus Nb/Y) for Canadian Archaean and Proterozoic felsites, granulites, and oxide gabbros with high total iron as Fe_2O_3, TiO_2, and P_2O_5 contents (Fe-Ti-P gabbros) compared to the data field for Iceland high-Ti tholeiites and silicic lavas (andesites, dacites, and rhyolites), from Natland (this volume). Symbols and data sources are given in the key to the figure.

Daly (1926, 1933) invoked foundering of the roots of mountains to explain their elevation and pointed out that the earliest protocontinental crust no longer exists; it foundered back into the Earth's interior. This foundering of lower crust continues. How can we ignore these potential sources for modern volcanic suites? Why suppose that all unusual compositions are tied up in plumes? We have become used to viewing subduction as the only mechanism for introducing heterogeneity into the mantle and plumes as the only mechanism for retrieving it. Neither mountain building nor continental drift has ever been a clean process. Drifting continents produce plenty of flotsam and jetsam in their wake. This is the minimal conclusion that geochemists are starting to draw in all the discussion about delamination leaving its geochemical signature on ridge basalts in the Indian Ocean. The serious question is how large the scale of this might be. The distributions of radioactive heat, fusible material, and structural weakness in continents are not uniform. The process of manufacture of continental crust is very uneven, which is why rifting is favored where places are still warm, weak, and refulgent with easy-to-melt materials. It is also why, after continents drift away, some magmatic provinces in ocean basins, such as Iceland, are bigger than others.

I thank Fitton for his efforts first at reviewing my article and then offering comments that compelled me to expand and clarify my arguments. We may not agree, but our discourse through these and many private communications has always been friendly, courteous, informative, and very valuable.

REFERENCES CITED

Anderson, D.L., 2005, Large igneous provinces, delamination, and fertile mantle: Elements, v. 1, p. 271–275.

Barth, M.G., McDonough, W.F., and Rudnick, R.L., 2000, Tracking the budget of Nb and Ta in the continental crust: Chemical Geology, v. 165, p. 197–213.

Bédard, J.H., 1994, A procedure for calculating the equilibrium distribution of trace elements amongst the minerals of cumulate rocks, and the concentration of the trace elements in co-existing liquids: Chemical Geology, v. 118, p. 143–153.

Bédard, J.H., 2006, A catalytic delamination-driven model for coupled genesis of Archaean crust and sub-continental lithospheric mantle: Geochemica et Cosmochimica Acta, v. 70, p. 1188–1214.

Betton, P.J., 1979, Isotopic evidence for crustal contamination in the Karroo rhyolits of Swaziland: Earth and Planetary Science Letters, v. 45, p. 263–274.

Daly, R.A., 1926, Our mobile Earth: New York, Charles Scribner's Sons, 342 p.

Daly, R.A., 1933, Igneous rocks and the depths of the Earth: New York, Hafner Reprint, 1968, 598 p.

Ewart, A., and Griffin, W.L., 1994, Application of proton-microprobe data to trace-element partitioning in volcanic rocks: Chemical Geology, v. 117, p. 51–284.

Fitton, J.G., 2007 (this volume), The OIB paradox, *in* Foulger, G.R., and Jurdy, D.M., eds., Plates, plumes, and planetary processes: Boulder, Colorado, Geological Society of America Special Paper 430, doi: 10.1130/2007.2430(03).

Fitton, J.G., Saunders, A.D., Norry, M.J., Hardarson, B.S., and Taylor, R.N., 1997, Thermal and chemical structure of the Iceland plume: Earth and Planetary Science Letters, v. 153, p. 197–208.

Fitton, J.G., Larsen, L.M., Saunders, A.D., Hardarson, B.S., and Kempton, P.D., 2000, Palaeogene continental to oceanic magmatism on the SE Greenland continental margin at 63°N: A review of the results of Ocean Drilling Program Legs 152 and 163: Journal of Petrology, v. 41, p. 951–966.

Fitton, J.G., Saunders, A.D., Kempton, P.D., and Hardarson, B.S., 2003, Does depleted mantle form an intrinsic part of the Iceland plume?: Geochemistry, Geophysics, Geosystems, v. 4, art no. Q01032, doi:10.1029/2002GC000424.

Foulger, G.R., Natland, J.H., and Anderson, D.L., 2005, Genesis of the Iceland melt anomaly by plate tectonic processes, *in* Foulger, G.R., et al., eds., Plates, plumes, and paradigms: Boulder, Colorado, Geological Society of America Special Paper 388, p. 595–625, doi: 10.1130/2005.2388(35).

Hardarson, B.S., and Fitton, J.G., 1997, Mechanisms of crustal accretion in Iceland: Geology, v. 25, p. 1043–1046.

Hawkesworth, C.J., Marsh, J.S., Duncan, A.R., Erlank, A.J., and Norry, M.J., 1984, The role of continental lithosphere in the generation of the Karoo volcanic rocks: Evidence from combined Nd and Sr-isotope studies: Special Publication of the Geological Society of South Africa, v. 13, p. 341–354.

Hémond, C., Arndt, N., Lichtenstein, U., Hofmann, A.W., Oskarsson, N., and Steinthorsson, S., 1993, The heterogeneous Iceland plume: Nd-Sr-O isotopes and trace element constraints: Journal of Geophysical Research, v. 98, p. 15,833–15,850.

Jaques, A.L., and Green, D.H., 1980. Anhydrous melting of periodotite at 0–15 kb pressure and the genesis of tholeiitic basalts: Contributions to Mineralogy and Petrology, v. 73, p. 287–310.

Jull, M., and Kelemen, P.B., 2001, On the conditions for lower crustal convective instability: Journal of Geophysical Research, v. 106, p. 6423–6446.

Kerr, A., 2003, Nickeliferous gabbroic intrusions of the Pants Lake area, Labrador, Canada: Implications for the development of magmatic sulfides in mafic systems: American Journal of Science, v. 303, p. 221–258.

Leatherdale, S.M., Maxeiner, R.O., and Amsdell, K.M., 2003, Petrography and geochemistry of the Love Lake Leucogabbro, Swan River complex, Peter Lake Domain, northern Saskatchewan: Saskatchewan Geological Survey Summary of Investigations, v. 2, p. 1–17.

Li, C., Lightfoot, P.C., Amelin, Y., and Naldrett, A.J., 2000, Contrasting petrological and geochemical relationships in the Voisey's Bay and Mushuau intrusions, Labrador, Canada: Implications for ore genesis: Economic Geology, v. 95, p. 771–799.

Mahoney, J.J., 1988. Deccan Traps, *in* Macdougall, J.D., ed., Flood basalts: Dordrecht, Kluwer, p. 151–194.

McDonough, W.F., and Sun, S.-s., 1995, The composition of the Earth: Chemical Geology, v. 120, p. 223–253.

Natland, J.H., 1989, Partial melting of a lithologically heterogeneous mantle: Inferences from crystallization histories of magnesian abyssal tholeiites from the Siqueiros Fracture Zone, *in* Saunders, A.D., and Norry, M., eds., Magmatism in the ocean basins: London, Geological Society of London Special Publication 42, p. 41–77.

Natland, J.H., 1991, Indian Ocean crust, *in* Floyd, P.A., ed., Oceanic basalts: Glasgow, Blackie, p. 63–93.

Natland, J.H., 2007 (this volume), ΔNb and the role of magma mixing at the East Pacific Rise and Iceland, *in* Foulger, G.R., and Jurdy, D.M., eds., Plates, plumes, and planetary processes: Boulder, Colorado, Geological Society of America Special Paper 430, doi: 10.1130/2007.2430(21).

Owens, B.E., and Dymek, R.F., 1992. Fe-Ti-P-rich rocks and massif anorthosite: Problems of interpretation illustrated from the Labrieville and St. Urbain Plutons, Quebec: Canadian Mineralogist, v. 30, p. 163–190.

Presnall, D.C., Gudfinnsson, G.H., and Walter, M.J., 2002, Generation of midocean ridge basalts at pressures from 1 to 7 GPa: Geochimica et Cosmochimica Acta, v. 66, p. 2073–2090.

Rudnick, R.L., and Fountain, D.M., 1995, Nature and composition of the continental crust—A lower crustal perspective: Reviews of Geophysics, v. 33, p. 267–309.

Rudnick, R.L., Barth, M., Horn, I., and McDonough, W.F., 2000, Rutile-bearing refractory eclogites: Missing link between continents and depleted mantle: Science, v. 287, p. 278–281.

Saunders, A.D., Fitton, J.G., Kerr, A.C., Norry, M.J., and Kent, R.W., 1997, The North Atlantic Igneous Province, *in* Mahoney, J.J., and Coffin, M.F., Large igneous provinces: Washington, D.C., American Geophysical Union Geophysical Monograph 100, p. 45–93.

The Geological Society of America
Special Paper 430
2007

Speculations on Cretaceous tectonic history of the northwest Pacific and a tectonic origin for the Hawaii hotspot

Ian O. Norton*

ExxonMobil Upstream Research, P.O. Box 2189, Houston, Texas 77252, USA

ABSTRACT

Current interpretations of Cretaceous tectonic evolution of the northwest Pacific trace interactions between the Pacific plate and three other plates, the Farallon, Izanagi, and Kula plates. The Farallon plate moved generally eastward relative to the Pacific plate. The Izanagi and Kula plates moved generally northward relative to the Pacific plate, with Izanagi the name given to the northward-moving plate prior to the Cretaceous normal polarity superchron and the name Kula applied to the postsuperchron plate. In this article I suggest that these names apply to the same plate and that there was only one plate moving northward throughout the Cretaceous. I suggest that the tectonic reorganization that has previously been interpreted as formation of a new plate, the Kula plate, at the end of the superchron was actually a plate boundary reorganization that involved a 2000 km jump of the Pacific–Farallon–Kula/Izanagi triple junction. Because this jump occurred during a time of no magnetic reversals, it is not possible to map or date it precisely, but evidence suggests mid-Cretaceous timing. The Emperor Trough formed as a transform fault linking the locations of the triple junction before and after the jump. The triple junction jump can be compared with an earlier jump of the triple junction of 800 km that has been accurately mapped because it occurred during the Late Jurassic formation of the Mesozoic-sequence magnetic lineations. The northwest Pacific also contains several volcanic features, such as Hawaii, that display every characteristic of a hotspot, although whether deep mantle plumes are a necessary component of hotspot volcanism is debatable. Hawaiian volcanism today is apparently independent of plate tectonics, i.e., Hawaii is a center of anomalous volcanism not tied to any plate boundary processes. The oldest seamounts preserved in the Hawaii-Emperor chain are located on Obruchev Rise at the north end of the Emperor chain, close to the junction of the Aleutian and Kamchatka trenches. These seamounts formed in the mid-Cretaceous close to the spreading ridge abandoned by the 2000 km triple junction jump. Assuming that Obruchev Rise is the oldest volcanic edifice of the Hawaiian hotspot and thus the site of its initiation, the spatial and temporal coincidence between these events suggests that the Hawaii hotspot initiated at the spreading ridge that was abandoned by the 2000 km jump of the triple junction. This implies a tectonic origin for the hotspot. Other volcanic features in the northwest Pacific also appear to have tectonic origins. Shatsky Rise is known to have formed on the migrating Pacific-Farallon-Izanagi triple junction during the Late Jurassic–Early

*E-mail: ian.o.norton@exxonmobil.com.

Norton, I.O., 2007, Speculations on Cretaceous tectonic history of the northwest Pacific and a tectonic origin for the Hawaii hotspot, *in* Foulger, G.R., and Jurdy, D.M., eds., Plates, plumes, and planetary processes: Geological Society of America Special Paper 430, p. 451–470, doi: 10.1130/2007.2430(22). For permission to copy, contact editing@geosociety.org. ©2007 The Geological Society of America. All rights reserved.

Cretaceous, not necessarily involving a plume-derived hotspot. Models for the formation of Hess Rise have included hotspot track and anomalous spreading ridge volcanism. The latter model is favored in this article, with Hess Rise forming on a ridge axis possibly abandoned as a result of a ridge jump during the superchron. Thus, although a hotspot like Hawaii could be associated with a deep mantle plume today, it would appear that it and other northwest Pacific volcanic features originally formed as consequences of shallow plate tectonic processes.

Keywords: hotspots, Hawaii, north Pacific, plate tectonics

INTRODUCTION

The Hawaii Hotspot

Hawaii, with its long age-progressive chain of volcanic islands and seamounts, has been regarded as the type example of a hotspot since Wilson (1963) first proposed the concept. Hotspots have been invoked as the source for many other sites of excess volcanism, such as other island or seamount chains and oceanic plateaus. Since Morgan (1971, 1972) suggested that hotspots like Hawaii are fed by mantle plumes originating possibly as deep as the core-mantle boundary, hotspots have been associated with plumes and have typically been regarded as features originating from deep enough within the Earth that they develop essentially independent of lithospheric influences such as plate motions. This separation of plumes or hotspots from plate tectonics has persisted in spite of indicators to the contrary, such as the observation by Aslanian et al. (1994) that many, if not most, of the global ocean's large volcanic plateaus are associated with triple junctions. Subsequent work has shown that many hotspots in fact have a plate tectonic origin (see the summary by Anderson, 2005). Tomographic interpretations suggest that the Hawaii hotspot may presently be located on a mantle plume (Lei and Zhao, 2006). Even if a plume is associated with the hotspot today, it is not clear whether the plume existed from the time of inception of the hotspot. Geochemical evidence, for instance, shows that seamounts at the old end of the Hawaii-Emperor chain have an oceanic signature that is typical of mid-ocean ridge basalts (MORB) but not typical of mantle plumes (Keller et al., 2000). However, the tectonic setting of the northwest Pacific at the time of initiation of the Hawaii hotspot is poorly known, so the relationships between spreading ridges and the early hotspot are not obvious. In this article I reexamine the plate tectonic history of the area, especially the far northwest corner of the Pacific plate, and suggest that the hotspot initiated at a ridge axis abandoned during plate boundary reorganizations. This implies that Hawaii can join the growing list of hotspots whose origin can be associated with plate tectonics.

OVERVIEW OF NORTHWEST PACIFIC TECTONICS

Tectonic information from the northwest Pacific comes from the mapping of oceanic spreading magnetic lineations,

fracture zones, and other topographic features (Figs. 1 and 2). Figure 1 is a view of the satellite gravity using the current version of the data (Sandwell, 2005) described by Sandwell and Smith (1997). Although bathymetry maps are available for the area (e.g., Mammerickx and Sharman, 1988; Mammerickx, 1989), there are large areas with only scattered data. Satellite-derived gravity provides a regionally consistent and detailed (one-minute grid interval) view of the seafloor structure. In this area of mostly thin sediment cover (Mammerickx and Sharman, 1988), a good correlation between gravity and basement topography is expected (Smith and Sandwell, 1997). The image processing used in Figure 1, with artificial illumination from the northwest, is designed to enhance subtle features in the data; fracture zones are especially clear. In Figure 2 the magnetic lineations are identified, with a more subdued gravity background. The most prominent feature is the Hawaii-Emperor chain, accentuated in the figure by the negative gravity anomaly due to the loading-induced flexural moat (Watts, 2001), but for now we will concentrate on other tectonic information. Magnetic lineations (Fig. 2) are in two groups. Jurassic to Early Cretaceous lineations of the Mesozoic (M) sequence lie to the west. Lineations to the east start with chron 34, of Campanian age, and young to the north toward the Aleutian Trench and also off the figure to the east toward North America. These two zones of magnetically dated oceanic crust are separated by large areas of seafloor created during the Cretaceous normal polarity superchron (abbreviated as simply superchron in this article).

Campanian and younger magnetic lineations were the first to be mapped in the northwest Pacific, by Pitman and Hayes (1968) and Grow and Atwater (1970). These authors showed that these lineations display two trends. Lineations that strike approximately north-south young to the east and record spreading along the Pacific-Farallon Ridge, with the Farallon plate moving east relative to the Pacific. The direction of relative motion between the Pacific and Farallon plates is recorded by the Mendocino, Pioneer, and other large fracture zones (Fig. 2). Atwater (1989), Atwater et al. (1993), and Searle et al. (1993) showed that these fracture zones indicate a change in spreading direction of ~30° during the superchron, with a further change of 10°–15° during chron 33–75 Ma according to the timescale used in this article, that of Gradstein et al. (2004). Lineations directly south of the Aleutian Trench strike east-west. They young to the north and record spreading at a ridge that separated the

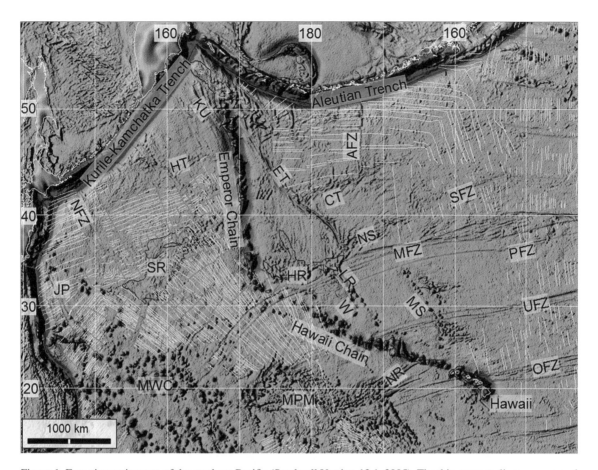

Figure 1. Free air gravity map of the northern Pacific (Sandwell Version 13.1, 2005). The thin magenta lines are mapped fracture zones, the thin yellow lines identified magnetic lineations. AFZ—Amlia fracture zone; CT—Chinook Trough; ET—Emperor Trough; HR—Hess Rise; HT—Hokkaido Trough; JP—Japanese Group seamounts; KU—Kruzenstern fracture zone; LR—Liliuokalani Ridge; MFZ—Mendocino fracture zone; MS—Musicians seamounts; MPM—Mid Pacific Mountains; MWC—Marcus Wake chain; NFZ—Nosappu fracture zone; NR—Necker Ridge; NS—Non Surveyor feature; OFZ—Molokai fracture zone; PFZ—Pioneer fracture zone; SFZ—Surveyor fracture zone; SR—Shatsky Ridge; UFZ—Murray fracture zone. Magnetic lineations (identified in Fig. 2) are from compilations maintained by Larry Lawver and Lisa Gahagan at the Plates Project, University of Texas–Austin, and my own updates digitized from Nakanishi et al. (1989) and Atwater (1989). Mercator projection; scale bar is for approximately the latitude of Hess Rise.

Pacific from a plate that has since been subducted. This plate was named the Kula plate by Grow and Atwater (1970). The Kula and Farallon lineations meet at a magnetic bight that traces the progress of the Pacific-Farallon-Kula triple junction (the heavy solid black line in Fig. 2). Fracture zone orientations and magnetic lineation ages have been used to derive the details of plate motions between these three plates by Engebretson (1984), Engebretson et al. (1984a,b, 1987), Atwater and Severinghaus (1989), and Atwater et al. (1993).

M-sequence lineations, termed the Japanese and Hawaiian lineations by Larson and Chase (1972), are seen mostly west of the Hawaii-Emperor chain in Figure 2. These lineations track evolution of spreading centers from formation of the Pacific plate at 170 Ma (Bartolini and Larson, 2001) through chron M0 at 125 Ma. The Hawaiian lineations strike northwest-southeast and young to the east. They record spreading between the Pacific

and Farallon plates. The Japanese lineations strike northeast-southwest and young to the northwest toward the Japan-Kamchatka subduction zone. Early workers who reported on these lineations (Hayes and Pitman, 1970; Larson and Chase, 1972; and Larson and Pitman, 1972) pointed out that, like the Kula plate, a plate that has since been subducted must have been moving away from the Pacific plate to form the Japanese lineations. These workers assumed that this plate was the Kula plate. Woods and Davies (1982), however, suggested that it was a different plate that they named the Izanagi, and that the Kula plate had come into existence only at chron 32b (72 Ma). Woods and Davies (1982) used this age because chron 32b was the oldest postsuperchron magnetic anomaly mapped in this area at the time; Mammerickx and Sharman (1988) and Atwater and Severinghaus (1989) extended the mapping of the Kula lineations to chron 34 (83 Ma), as shown in Figure 3. Later in this article

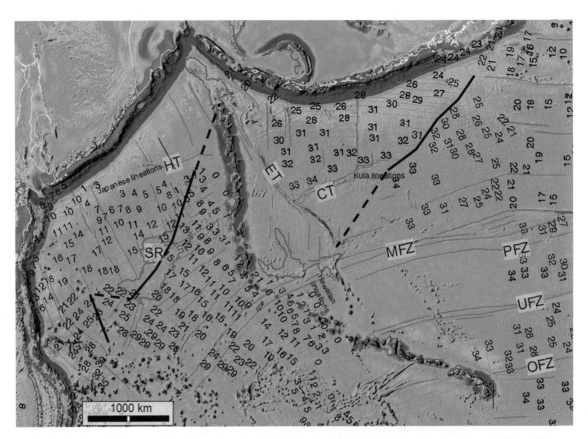

Figure 2. Identified magnetic lineations in the north Pacific, labeled with chron numbers. The heavy black lines track triple junctions, dashed where inferred. The Japanese and Hawaiian lineations of the M sequence are mostly west of the Hawaii-Emperor chain; they are numbered without the M prefix. Abbreviations as in Figure 1.

I will discuss whether the Kula and Izanagi are in fact the same plate. For now, though, I will use the name Kula for the chron 34 and younger plate and Izanagi for the M-sequence plate that moved generally north to northwest away from the Pacific.

Hilde et al. (1976 and 1977), Handschumacher et al. (1988), Sager et al. (1988), Sharman and Risch (1988), and Nakanishi et al. (1989, 1992a,b) showed that the Hawaii and Japanese lineations meet at a magnetic bight that traces the motion of the Pacific-Farallon-Izanagi triple junction (the heavy solid black line in Fig. 2) and that for ~1000 km the path of the triple junction follows the axis of Shatsky Rise, a prominent bathymetric feature in this part of the Pacific. More recent papers (Nakanishi et al., 1999; Sager et al., 1999) emphasize magnetic lineations on Shatsky Rise itself, showing that the evolution of the rise was directly tied to spreading ridge axis intersections at the Pacific-Farallon-Izanagi triple junction. Detailed interpretations of the plate kinematics implied by the Hawaii-Japanese lineations were offered by Engebretson (1984), Engebretson et al. (1984b, 1987), and Nakanishi et al. (1989). These authors showed that the spreading rates varied from 37 to 79 mm/yr on the ridge systems involved and that the Izanagi plate may have been separated into two plates moving in slightly different directions for

some of the spreading history. Sager et al. (1988) and Nakanishi et al. (1989) showed that there was at least one significant jump of the triple junction to a new location, shown by the dashed line at the southwest end of Shatsky Rise in Figure 2. This 800 km jump occurred at M21 time (147 Ma) and was approximately coincident with a change in direction of relative motion of 25°–30° between the Pacific and Izanagi plates, although there was no simultaneous Pacific-Farallon motion vector change (Sager et al., 1988; Nakanishi et al., 1999). The excess volcanism that created Shatsky Rise began at the new triple junction location (Sager et al., 1988). A mantle plume as the source for the anomalous volcanism that produced Shatsky Rise has been a common model for the origin of the rise since the hotspot paradigm was first proposed, as summarized by Sager (2005). Sager (2005) points out, however, that a plume origin for a hotspot has several inconsistencies. One is the coincidence between the triple junction jump at M21, the change in Pacific-Izanagi motion direction, and the initiation of Shatsky Rise. If a plume was associated with the initiation of Shatsky Rise, it is not easy to explain why the motion direction of the Izanagi plate changed at the same time (Sager, 2005). Limited geochemical data (Mahoney et al., 2005) and interpretations of heatflow (Kotelkin et al., 2004; Verzhbitskii and

Kononov, 2004) on Shatsky Rise suggest a spreading ridge rather than a plume origin for the rise, although, as Sager (2005) concludes, there is not yet enough information to allow us to decide on a plume, hotspot, or ridge origin.

There are several other topographic features in this part of the Pacific that bear on a discussion of the tectonic evolution of the area. The Emperor Trough (Fig. 1) is a dramatic linear topographic feature more than 1500 km long and up to 100 km wide with more than 2000 m of relief in places (Mammerickx and Sharman, 1988). Larson and Chase (1972) and Hilde et al. (1977) postulated that it formed as a transform fault during the superchron. In the tectonic scenarios of Rea and Dixon (1983) and Mammerickx and Sharman (1988), the trough is shown as a spreading center, but no estimates of how much seafloor was created while it was a spreading center are given. Woods and Davies (1982) and Atwater (1989) treat it as a rift, but also do not give quantitative estimates of rift duration or amount of extension. East of the Emperor Trough is the east-west-trending Chinook Trough. This discontinuous structure lies close to the southern edge of the Kula spreading lineations (Figs. 1–3). Woods and Davies (1982), noting that the trough is approxi-

mately parallel to the Mendocino fracture zone, suggested that the Chinook Trough was originally a Pacific-Farallon transform fault that served as the locus for spreading that initiated when the Kula plate formed and commenced its northward motion. Mammerickx and Sharman (1988) and Atwater (1989) agreed with this interpretation. To the west of the Emperor chain on Cretaceous quiet zone seafloor there is another topographic feature, Hokkaido Trough (Fig. 1), that morphologically resembles Chinook Trough in this figure. It trends approximately parallel to the Japanese M-sequence lineations and perpendicular to fracture zones, suggesting an origin related to post-M-sequence seafloor spreading. In the rather complicated tectonic model of Mammerickx and Sharman (1988), Hokkaido Trough is interpreted as a failed oceanic rift in which rifting propagated westward along the trough. North of Hokkaido Trough and still on quiet zone seafloor are several northwest-trending fracture zones, including the Kruzenstern. Because these fracture zones are approximately parallel to the spreading direction during the time represented by the Japanese lineations, they are assumed here to have formed by Pacific-Izanagi spreading during the superchron.

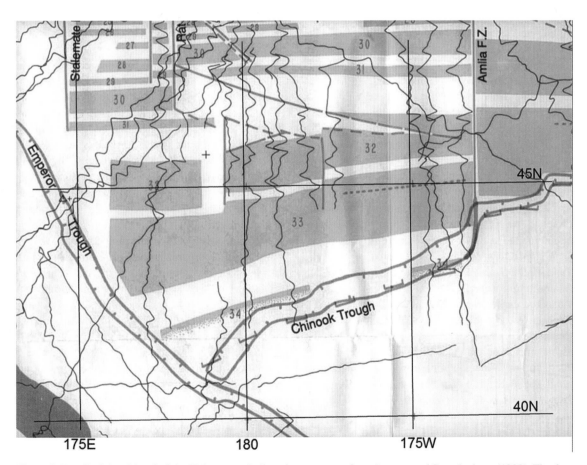

Figure 3. Detail of the old end of the Kula magnetic lineation sequence from Atwater and Severinghaus (1989). The data show magnetic anomalies plotted perpendicular to ship tracks, with red bands showing interpreted and identified lineations. The Stalemate fracture zone shown in this map is named the Buldir by Mammerickx and Sharman (1988).

Hess Rise is another large topographic feature. It is located within Cretaceous quiet zone seafloor, so details of the age of surrounding seafloor are unknown except for the Aptian to Campanian bounding ages of the superchron. Age data from the rise itself are sparse. Vallier et al. (1983) suggest an uppermost Aptian to Albian age (112–100 Ma), based on the stratigraphic ages of sediments cored in four Deep Sea Drilling Program (DSDP) wells on the rise. Radiometric ages for the rise summarized by Clouard and Bonneville (2005) show an age of 87.1 Ma for a seamount near the southern margin of the rise. Taking this age plus the uppermost Aptian age for the northern rise from Vallier et al. (1983) and other younger ages from the Wentworth seamounts, which lie between Hess Rise and the Hawaiian chain, Pringle and Dalrymple (1993) suggested that the rise formed over a fixed hotspot, with the rise moving north over the hotspot. Sager (2005) points out that the age progression along Shatsky Rise and Hess Rise is consistent with the concept that both of these features formed over the same hotspot, although with a tortuous path. Verzhbitskii and Kononov (2004) suggest that Hess Rise, like Shatsky Rise, formed along the track of the Pacific-Farallon-Izanagi triple junction during the superchron. Vallier et al. (1983), noting that northern Hess Rise parallels the Hawaii lineations to the west and that in their interpretation all of Hess Rise formed simultaneously (i.e., there was no age progression), prefer a model whereby Hess Rise was created at the Pacific-Farallon spreading center. This is the model adopted in this article. The possible formation of Hess Rise at a ridge axis during a spreading direction change is similar to models proposed for south Pacific island chains by Hieronymus and Bercovici (2000), who relate these features to plate stress changes with a ready supply of melt.

The Hawaii-Emperor chain, stretching from Meiji seamount to the Hawaiian Islands (Fig. 1), is the most dominant morphologic feature in the northwest Pacific. The hotspot model for origin of the chain, first proposed by Wilson (1963), has remained the leading explanation for its creation (Clouard and Bonneville, 2005; Duncan et al., 2006). Meiji is thought to be the oldest preserved seamount in the Hawaii-Emperor seamount chain (Duncan and Keller, 2004). It is located at the junction between the Kamchatka and Aleutian trenches. We will examine the age data for this area later in this article.

As summarized earlier, the data available for constraining the tectonic scenarios for the development of the northwest Pacific are concentrated on the areas and times represented by the Hawaii-Japanese lineations and the Kula lineations. As Atwater (1989) pointed out, it is not obvious how plate boundaries evolved through the superchron. Variations in published models center on different interpretations of Emperor Trough and on the origin of the seafloor between Chinook Trough and the Mendocino fracture zone. Larson and Chase (1972), Hilde et al. (1977), and Smith (2003) interpret Emperor Trough as a transform fault that linked spreading on the Kula lineations with spreading on since-subducted seafloor near the north end of the Emperor seamounts. Woods and Davies (1982) suggested that Emperor Trough formed as a rift during early stages of Kula plate initiation. Because in their model the Kula plate initiated along the Chinook Trough, the portion of the Emperor Trough that lies to the south of Chinook Trough was explained as a rift, although they did not give an estimate of the total amount of extension. In a variation of this model, Rea and Dixon (1983) introduced an extra plate, the Chinook plate, as a way of creating the seafloor between Chinook Trough and the Mendocino fracture zone. In their model the Emperor Trough is shown as a rift with motion across the feature lasting from 83 to 72 Ma. Rea and Dixon (1983) suggest that the Emperor Trough accommodated 132 km of extension (12 mm/yr for 11 m.y.) south of the Chinook Trough and more than 800 km (78 mm/yr for 11 m.y.) north of the Chinook Trough. There is no morphologic evidence for such very different tectonic regimes along the Emperor Trough on either side of Chinook Trough (Fig. 1), casting some doubt on the interpretation of different plates on either side of the Chinook Trough. Mammerickx and Sharman (1988) present a tectonic scenario for the area that also includes the separate Chinook plate and substantial extension across the Emperor Trough (64 mm/yr for 13 m.y. or a total of 700 km), which is unlikely based on preserved morphology. In addition, Atwater et al. (1993) discount the existence of the Chinook plate, because it would require a large section of the Mendocino fracture zone to be a strike-slip boundary between the Chinook and Pacific plates, for which there is no evidence.

In this article I suggest that the tectonic evolution of this area of the Pacific can be fairly simply understood in terms of the evolution of the triple junction that moved through the area. The motion of this triple junction is known from preserved magnetic lineations during both M-sequence time and Kula lineation time. It moved to the northeast during both these times (actually NNW prior to M21), as shown by the heavy black lines in Figure 2. The heavy dashed lines in this figure show the triple junction track across quiet zone seafloor as postulated in this article. I suggest that during the early part of the superchron the triple junction continued moving northeast away from the M-sequence seafloor. It then jumped nearly 2000 km to the southeast in a reorganization of plate boundaries before continuing its northeast motion along the post–chron 34 magnetic bight that records its motion. One implication of this model is that there was always just one plate moving generally northward away from the Pacific and that the Kula and Izanagi plates are in fact the same plate, as originally suggested by Larson and Chase (1972) and Hilde et al. (1977). Another implication comes from the tracking of spreading centers along the northern boundary of the Pacific plate. When the triple junction jumped to the southeast, it left an abandoned spreading center that I suggest was the site of excess volcanism that initiated the Hawaii hotspot and started development of the Emperor seamount chain. Tectonic evolution is illustrated with a series of maps depicting the evolution of the spreading plate boundaries, and then this evolution is used to draw inferences about the tectonic origin of the Hawaii hotspot.

Tectonic Evolution of the Northwest Pacific

The fracture zone and magnetic isochron data summarized earlier provide the primary constraint on deciphering the tectonic history of the northwest Pacific. Because there are no isochron data from the large area of quiet zone seafloor, it is worth taking a closer look at data from either side of this area. M-sequence lineations are well mapped with fairly dense data (Nakanishi et al., 1989; Sager et al., 1999), but data for the young side of the superchron are not as good. Figure 3 is a detail of the data used to map these younger lineations, taken directly from Plate 1 of Atwater and Severinghaus (1989). Chron 31 shows the east-west strike that is displayed by all the younger Kula lineations. Older lineations, though, are more broken up, with several pseudo-faults mapped by Atwater and Severinghaus, especially in chron 32. There is a difference of ~15° in strike between chrons 34 and 31 as mapped, although there are few data constraining the mapping of chron 34. The structuring seen in mapped lineations could be caused by readjustments of the spreading centers to this 15° change in spreading direction. It could also be the cause of the rough-smooth transition mapped by Mammerickx and Sharman (1988). These authors mapped a transition between rough and smooth seafloor in chron 32b and

associated it with a change in spreading rate, but it is possible that it represents ridge reorientations responding to changes in spreading direction. As can be seen in Figure 3, data for precise mapping of chron 34 are rather sparse, but it appears that Kula-Pacific relative motion underwent a 15° change in direction at about the same time as Pacific-Farallon motion, mentioned earlier. Chron 34 is not perpendicular to the Emperor Trough, as would be expected if the Emperor Trough were a transform active at chron 34 time, but the west end of Chinook Trough is perpendicular to the Emperor Trough where they meet. This is consistent with the Chinook Trough's having been formed at the location of a spreading ridge axis perpendicular to the Emperor Trough, if the Emperor Trough is assumed to be a transform.

The model for tectonic evolution of the north Pacific presented in this article is illustrated by a series of figures showing progressive changes in the Pacific, Izanagi, Kula, and Farallon plate boundaries. I use the gravity and tectonic data map in Figure 2 as a base. The figures step though time, showing areas of the seafloor appearing as they were created. Figure 4 is a copy of Figure 2 with the Hawaii-Emperor chain shaded as a visual reminder that this feature is mostly younger than the seafloor on which it was built. Also shaded is the area north and west of the trench. This is done to emphasize that the seafloor we are deal-

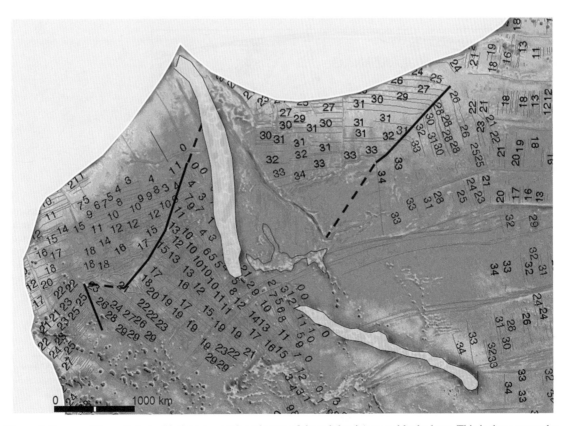

Figure 4. Same as Figure 2, but with the area north and west of the subduction zone blanked out. This is done to emphasize that the area of the Pacific that we are dealing with was tectonically active though it was a long way from the margin. Also shaded is the Hawaii-Emperor chain as a visual reminder that it was created after the times of concern in the following figures.

ing with was a long way from any continental margin when it formed, as shown in Figure 5. This is a Pacific-wide reconstruction for 80 Ma using the plate circuit approach for calculating relative plate positions (rotation poles from Norton, 1995). It is not possible to calculate the position of the Pacific plate relative to the surrounding continents earlier than ca. 90 Ma, because the Pacific was totally surrounded by subduction zones prior to that time. The plate circuit can be used only for times since 90 Ma, when the Pacific became attached to New Zealand and rifted from West Antarctica (Cande et al., 1995). It would be useful if the fixed hotspot reference frame assumption could be used for earlier plate reconstructions, but this assumption can no longer be regarded as valid, as shown for the Hawaii hotspot by Tarduno and Cottrell (1997), Tarduno et al. (2003), and Doubrovine and Tarduno (2004).

The record of the motion of the Pacific-Farallon-Izanagi triple junction and the formation of Shatsky Rise during M-sequence time is well understood (Sager 2005). Following is a summary of pertinent tectonic information from that time span:

1. The triple junction moved to the NNW from M29, the oldest magnetic lineation in the M sequence (157 Ma), to M21 (147 Ma), then moved northeast to M0, the youngest lineation in the M sequence (125 Ma).

2. Shatsky Rise formed on the migrating triple junction over a time span of ~20 m.y., from M21 (147 Ma) to M1 (127 Ma) (Nakanishi et al., 1999; Sager et al., 1999; Sager, 2005).

3. There was at least one significant jump of the triple junction to a new location, shown by the dashed line at the southwest end of Shatsky Rise in Figure 2. This 800 km jump occurred at M21 time and was approximately coincident with a change of relative motion of 25°–30° between the Pacific and Izanagi plates (Sager et al., 1988; Nakanishi et al., 1989). Significantly, there was no coincident change in Pacific-Farallon relative motion.

4. Geochemical and heatflow data suggest a MORB-like rather than a mantle plume origin, for basalts on the rise (Kotelkin et al., 2004; Verzhbitskii and Kononov, 2004; Mahoney et al., 2005), although a plume origin cannot be unequivocally ruled out (Sager, 2005).

Figure 6 shows the plate configuration at M0 time (125 Ma, the beginning of the Cretaceous normal polarity superchron) for the area. The Pacific plate boundaries are constrained by magnetic

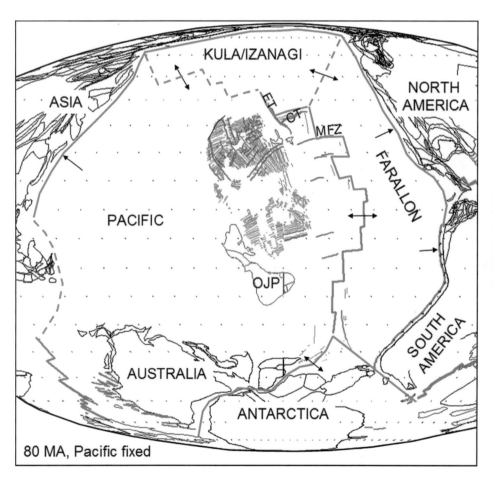

Figure 5. Plate reconstruction for 80 Ma. The arrows indicate relative motions across plate boundaries (red lines), which are dashed where inferred. The green lines are magnetic lineations, magenta lines fracture zones. OJP—Ontong Java plateau.

lineations; the Pacific-Izanagi-Farallon boundaries are drawn assuming a ridge-ridge-ridge triple junction. Figure 7 shows the plate boundaries for ~15 m.y. later, with ridge axis positions extrapolated using spreading rates at the end of the M sequence given by Nakanishi et al. (1989). This extrapolation ignores possible complications posed by Hokkaido Trough if the trough is associated with a ridge jump or change in spreading rate. The ridge-transform geometry is copied from that at M0 time. Note that the Pacific-Farallon spreading center straddles Hess Rise, with the ridge axis north of the Mendocino fracture zone located along northern Hess Rise and the ridge axis south of the Mendocino fracture zone located along Liliuokalani Ridge, which is a feature apparently in structural continuity with the rise. This coincidence of inferred spreading ridges and a greater Hess Rise suggests that the rise was formed by excess volcanism associated with the spreading process. Whether a ridge jump was also involved is not known, although it does seem to be associated with a spreading direction change of ~15° early in the superchron. As discussed earlier and as shown by Atwater et al. (1993), there must have been spreading ridge reorganizations, including ridge jumps, during the superchron. As part of these reorganizations, the total offset across the Mendocino fracture zone increased from 300 km at the end of M-sequence time to nearly 1400 km at chron 34 time.

Figure 8 shows the plate boundaries later in the superchron, at ca. 90 Ma. The scenario presented is similar to that proposed by Larson and Chase (1972), Hilde et al. (1977), and Smith (2003). The Pacific-Farallon spreading center is extrapolated backward from the chron 34 configuration, i.e., this assumes that the ridge-transform configuration seen at chron 34 was formed by ridge jumps prior to (or at) 90 Ma. The Pacific–Kula/Izanagi plate boundary is drawn assuming that spreading on this boundary proceeded from 110 Ma (Fig. 7) in a NNW direction relative to the Pacific, with motion parallel to the Kruzenstern and nearby parallel fracture zones. The triple junction would have migrated to the northeast, following the dashed black line. The area of seafloor shown by the crosshatch pattern is postulated to be Pacific seafloor that was captured by the Farallon plate at the time of the next figure.

Figure 9 shows interpreted spreading patterns for a short time, probably a few million years, after the time of Figure 8. It is proposed that the triple junction jumped 2000 km from point A, where it was in Figure 8, to point B in Figure 9. In this scenario, the Emperor Trough was created as a sinistral transform fault separating Pacific crust from crust that is now part of the Kula/Izanagi plate. This crust would have been captured from the Farallon plate or, for the crosshatched area in Figure 8, from the Pacific. This triple junction jump is a larger version of the

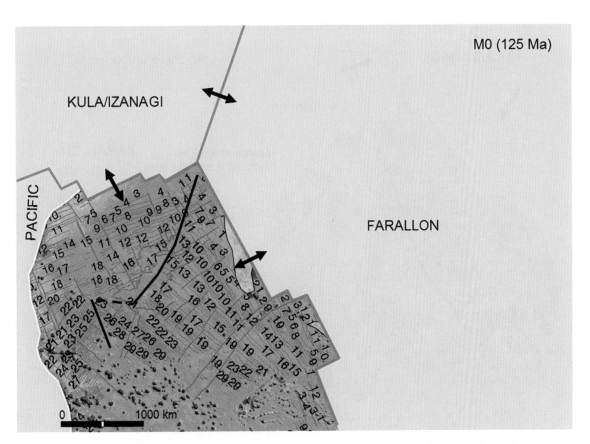

Figure 6. Tectonic setting at M0 time, 125 Ma. The heavy red lines are plate boundaries; the arrows show relative motion directions.

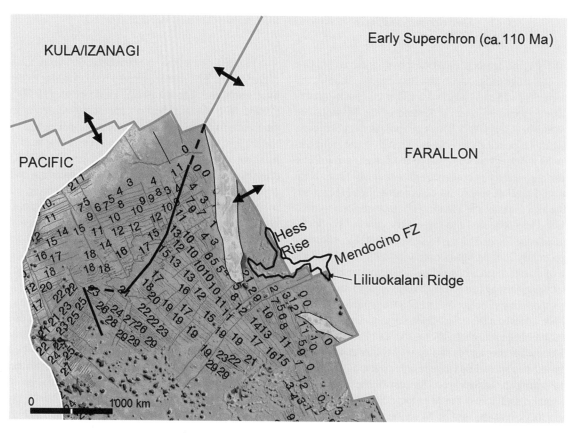

Figure 7. Tectonic setting at ca. 110 Ma.

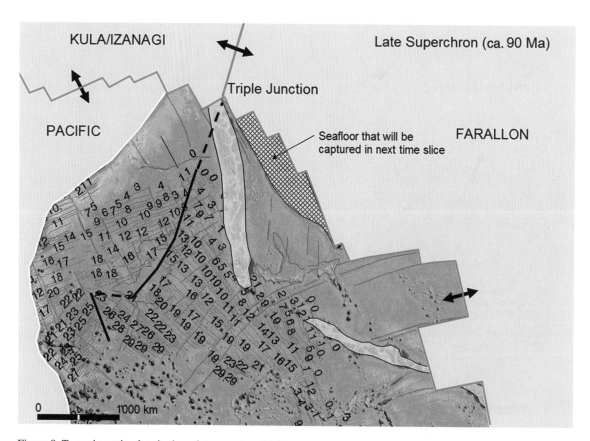

Figure 8. Tectonic setting late in the quiet zone at ca. 90 Ma.

jump that happened at M21 time (see Fig. 2). There would have been significant reorganizations on the boundary between the Farallon and Kula/Izanagi plates, but the crust involved has all been subducted so we have no data to show what happened.

Figures 10 and 11 show, respectively, the tectonic setting at chron 34, 84 Ma, and chron 32, 71 Ma. At 84 Ma (Fig. 10), the Farallon plate was moving almost directly east relative to the Pacific plate. Farallon motion changed by 15° to N75°E by 71 Ma (Engebretson et al., 1984a,b; Fig. 11). At 84 Ma, the Kula/Izanagi plate was moving NNW, but was starting to change direction to more nearly directly north, the motion direction it achieved by chron 32 (Fig. 11). As discussed earlier and shown in Figure 3, the change of 15° in Pacific–Kula/Izanagi motion after chron 34 is not well constrained. In the tectonic scenario presented here, with the Emperor Trough as a transform during the superchron, there would have been a total change of ~35° in Pacific–Kula/Izanagi motion direction from the time of the triple junction jump (motion direction N35°W; Fig. 9) to chron 32 time (Fig. 11), when motion was directly north. For comparison, the total change in Pacific-Farallon motion during the superchron was ~25°, from N65°E at M0 to east-west at chron 34 (directions measured along the Mendocino fracture zone, Fig. 2). In the tectonic scenario presented here, the Chinook Trough formed as the spreading ridge reorganized in response to chang-

ing spreading directions close to chron 34 time. This means that it is not as significant a tectonic feature as suggested by Woods and Davies (1982), who argued that it formed at the site of first formation of the Kula plate. The ridge-transform configuration on the Pacific-Kula boundary shown in Figure 11 at the northern end of the Stalemate fracture zone is speculative. It must have been similar to what is shown, however, because the spreading fabric around Meiji seamount is all northwest and probably formed in the earlier spreading phase (Fig. 10). An implication of this model is that Meiji seamount could have formed on an abandoned spreading center. Because Meiji is the oldest preserved seamount of the Hawaii-Emperor chain, a further implication is that the Hawaii hotspot originated on this abandoned spreading center, i.e., has a tectonic origin.

DISCUSSION

In the simple model proposed here, the Kula and Izanagi plates are actually the same plate. Consideration of the general tectonic setting of the area lends support to this hypothesis. Magnetic lineations and fracture zones tell us that from the beginning of M-sequence time in the Jurassic until well into the Cenozoic, there was a plate moving in a generally northward direction (in present-day coordinates) away from the Pacific

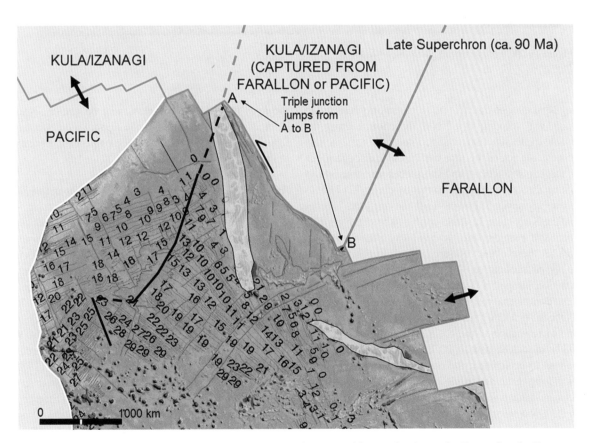

Figure 9. Tectonic setting at ca. 90 Ma after the triple junction jumped from point A to point B, creating the Emperor Trough as a sinistral transform fault.

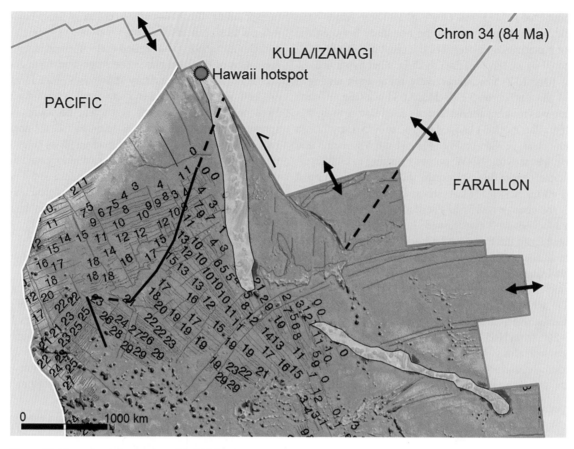

Figure 10. Tectonic setting at chron 34, 84 Ma.

plate. During M-sequence time this plate, commonly referred to as the Izanagi, was moving in a NNW direction. During the superchron the triple junction changed location (all models for the tectonic evolution of this area must include a scenario for initiation of this triple junction), and shortly before chron 34 (84 Ma), the motion direction changed, so that by chron 32 (71 Ma) motion was directly north. This northward-moving plate is commonly called the Kula plate. It would appear to be much simpler to think in terms of a single plate moving away from the Pacific plate through all these events. Referring back to Figure 5, the Pacificwide reconstruction for 80 Ma, it can be seen that when the triple junction jumped it was more than 4000 km (40° latitude) from the continental margin flanking the northern edge of the Pacific Ocean. If a new plate was formed at 84 Ma (chron 34), the time of formation of the Kula plate in the Woods and Davies (1982) model, a new 4000-km-long plate boundary would have had to be formed; the new boundary would have been 6000 km long if the new triple junction location had been included. I suggest that a reorganization of existing boundaries, even with a 2000 km relocation of the triple junction, is a more likely scenario.

The hypothesis presented here, that the triple junction jump was coincident with initiation of the Hawaii hotspot, implies that

the ages for these two events must be similar. Knowledge of seafloor age at either end of the jump would constrain the timing of the jump. The available age control for these events is examined in the next section.

Age of Seafloor at Either End of Triple Junction Jump

The timing of the triple junction jump proposed here, occurring as it did during the superchron, is poorly constrained. At best the data allow us to speculate that the triple junction was located near the north end of the Emperor chain before relocating to the southern end of the Emperor Trough. If the age of the seafloor at either end of the Emperor Trough was known, this would provide an age for the jump. This location is on seafloor created by Pacific-Farallon spreading north of the Mendocino fracture zone and at the west end of a structured area named the "Non Surveyor" by Mammerickx and Sharman (1988). The unmapped ridge jumps (including the one proposed here) between M0 and chron 34 along this plate boundary (Engebretson et al., 1984b; Atwater et al., 1993) introduce additional uncertainty in estimating the seafloor age. Nevertheless, some idea of this age can be obtained by examining available data. Engebretson et al. (1984b) calculated parameters of motion between the Pacific,

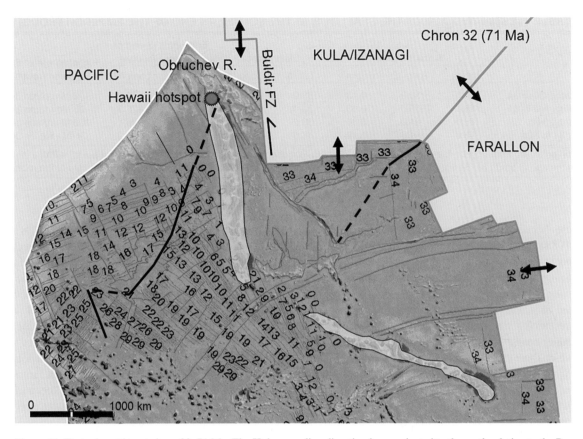

Figure 11. Tectonic setting at chron 32, 71 Ma. The Kula spreading direction has reoriented to the north relative to the Pacific plate.

Farallon, and Kula plates that can be used for interpolation or extrapolation of an age estimate. Choices are interpolation between chrons M0 and 34 generated by Pacific-Farallon spreading and extrapolation of chron 34 and younger spreading rates backward to the Emperor Trough for both Pacific-Farallon and Pacific-Kula spreading. Engebretson et al. (1984b) calculated two possible rates for chron M0–34 Pacific-Farallon motion, depending on whether ridge jumps occurred north or south of the Mendocino fracture zone. Table 1 summarizes the results of the interpolation and extrapolation calculations. Age estimates for the southern end of the Emperor Trough from the four different calculations range from 96 to 106 Ma.

Another estimate of the time of the jump can be obtained from extrapolation of plate motion rates calculated from the M-sequence lineations to the position of the abandoned triple junction at the north end of the Emperor chain. A simple method that incorporates data from both the Japanese and Hawaiian lineations is to use the calculated motion of the triple junction. Nakanishi et al. (1989) show velocity triangles for the Pacific-Izanagi-Farallon triple junction for several times, with chron M3 (129 Ma) the youngest. This diagram shows the triple junction moving at 4 cm/yr to the NNE. The youngest lineation with a well-defined magnetic bight, M1 (127 Ma), is ~1100 km from

Meiji seamount, which at 4 cm/yr yields a traveltime of the triple junction to Meiji of 27.5 m.y., or a time of arrival of 99.5 Ma. The uncertainties in this estimate are large; it assumes constant and symmetric spreading within the superchron and assumes that no complications were introduced by possible spreading ridge reorganizations at Hokkaido Trough that are not possible to map with existing data.

The structure of the northern end of the Emperor chain is shown in Figure 12. Meiji is the northernmost seamount; it is located on Obruchev Rise, an elongated plateau that trends northwest-southeast and includes Detroit seamount. The 4 km isobath around Obruchev Rise is highlighted in Figure 12. This covers an area of ~57,000 km^2. Although Obruchev Rise is not separated from the Hawaii-Emperor chain in the global catalog of large igneous provinces by Coffin and Eldholm (1994), this area is comparable in size to some oceanic plateaus listed, such as the Cape Verdes or Conrad rise. There are several structural grains to consider in Figure 4. On the east side of the figure, the Rat and Buldir fracture zones strike north-south; these features track Pacific-Kula relative motion. The northern Emperor seamounts strike N15°W, as shown by the pale blue dashed line. The northwest grain of the seafloor structure is apparent in the area west of the Buldir fracture zone. The Kruzenstern fracture zone, which

TABLE 1. ESTIMATED AGE OF THE SEAFLOOR AT THE SOUTHERN END OF THE EMPEROR TROUGH

	Rotation pole			Start point		End point (S end of Emperor Trough)		Implied age of S end of Emperor Trough
	Lat °N	Long °E	deg/m.y.	Lat °N	Long °E	Lat °N	Long °E	
Pacific-Farallon, M0 to 34	65.0	56.0	0.43	38.0	−167.5	35.6	−176.2	102 Ma (chron 34 + 18 m.y.)
Pacific-Farallon, M0 to 34	65.0	56.0	0.63	38.0	−167.5	35.5	−175.7	96 Ma (chron 34 +12 m.y.)
Pacific-Farallon, 34 to 25	66.0	64.0	0.36	38.0	−167.5	35.3	−176.1	106 Ma (chron 34 + 22 m.y.)
Pacific-Kula, 34 to 32b	18.0	111.0	0.49	44.0	−177.0	37.2	−178.3	96 Ma (chron 34 + 12 m.y.)

Note: Rotation poles are from Engebretson et al. (1984a). These poles are used to rotate the "start point" to the "end point," with the implied age estimated from the age of chron 34 (84 Ma) plus the number of million years required for the rotation at the degrees/m.y. rate of the rotation pole.

traces Pacific–Izanagi/Kula relative motion (Mammerickx and Sharman, 1988) for at least part of superchron time, strikes N30°W. This is the same strike as the northern end of Emperor Trough. Joining these two features is Tenji seamount, which shows a strong linear grain perpendicular to both the Kruzenstern fracture zone and the Emperor Trough; this structural relationship resembles what would be expected if Tenji was a fossil spreading center. This is not the postulate here, because Tenji (for which there are no direct age data) should be ~20 million years younger than its surrounding seafloor, but the relationship could bear further investigation. The Stalemate fracture zone strikes N60°W. Shown in white numbers on Figure 12 are magnetic lineations 23 (51 Ma) to 20 (43 Ma) and a postulated extinct ridge (RR) as mapped by Lonsdale (1988). Lonsdale (1988) interpreted these lineations to have formed during large

plate motion direction changes perhaps associated with the bend in the Hawaii-Emperor chain. Mapping of these lineations, however, is tenuous, because they are based on data from just a few ship tracks and incorporate extremely asymmetric spreading (Atwater, 1989). The southwest flank of Obruchev Rise also strikes N60°W. This trend could be a reflection of relative motion between the early Hawaii hotspot and the Pacific plate. This trend is 30° different from the trend of the Emperor seamounts. This change in the apparent motion of the hotspot relative to the Pacific, which is half the angle of the more famous Hawaii-Emperor bend, has not, as far as I am aware, been incorporated into any published models of Pacific-hotspot relative motion, but it probably should be. Finally, visible in Figure 12 is a weak N25°W structural grain running obliquely across Obruchev Rise toward the intersection of the Aleutian and Kamchatka trenches

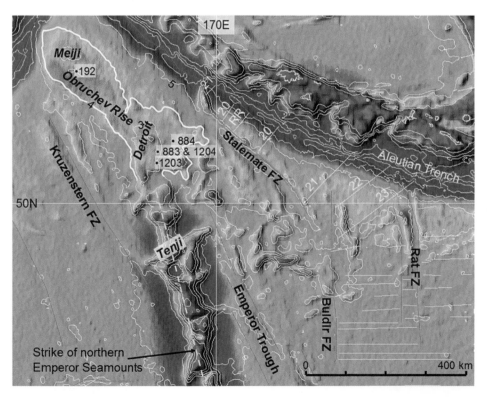

Figure 12. Detailed view of the northern end of the Emperor chain. The colored background is the satellite gravity (Sandwell, 2005). The white contours are from the Smith and Sandwell (1997) global gridded bathymetry, contour interval 1 km. The numbered points on Obruchev Rise refer to DSDP and ODP drilling sites. The white labeled lines in the Aleutian Trench are magnetic lineations identified by Lonsdale (1988). RR—ridge axis.

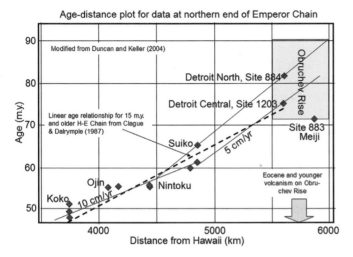

Figure 13. Age of seamounts in the Emperor chain plotted versus distance from Hawaii along the Hawaii-Emperor (H-E) chain. Modified from Duncan and Keller (2004).

(one of these features cuts through the *R* of the label *Rise* in Fig. 13). The origin of this grain is not known, although if it is fault related, it could be a fruitful area for future studies of relationships between hotspot tracks and stress-induced fracturing, as has been proposed for some hotspot tracks (Smith, 2003, 2005; Natland and Winterer, 2005).

Age of the Northern Emperor Seamounts

There is no reliable basement age data for Meiji seamount itself. Dalrymple et al. (1980) reported a minimum K-Ar age of 61.9 ± 5 Ma from basalts cored at DSDP site 192 on the seamount. Fossil assemblages in overlying sediments yield a Maastrichtian age (68–70 Ma; Worsley, 1973). These ages are younger than those for the next seamount down the chain, Detroit seamount (Fig. 12), for which there are reliable age data, so unless the general assumption about age progression along the chain is incorrect, these ages do not reflect the age of formation of Meiji seamount (Duncan and Keller, 2004). Basalts from Ocean Drilling Program (ODP) site 884 on the northeast flank of Detroit seamount yield a ^{40}Ar/ ^{39}Ar plateau age of 81.2 ± 1.3 Ma for a plagioclase-free component and an isochron age of 80.0 ± 0.9 Ma (Keller et al., 1995). This age is consistent with the reversed magnetic polarity measured at the site (Keller et al., 1995). Later sampling by Tarduno et al. (2003) and Duncan and Keller (2004) from ODP sites 1203 and 1204, located on the seamount summit (Fig. 4), yielded younger ages. These authors report a mean age of 75.82 ± 0.62 Ma from plateaus in ^{40}Ar/ ^{39}Ar incremental heating spectra from three whole-rock basalt samples and two feldspar separates at site 1203. This age agrees with a Campanian age (75–76 Ma) obtained from nannofossils found in sedimentary beds within the basement sequence at the site (Tarduno et al., 2002). Basalt samples from site 1204 (Fig.

4) did not yield reliable radiometric ages, but a 71–76 Ma (late Campanian) biostratigraphic age from overlying sediments is similar to ages measured from site 1203. Site 884, which yielded the older age of 80–81 Ma, is situated ~48 km northeast of site 1204 on the flank of Detroit seamount. Duncan and Keller (2004) refer to this location as "Detroit North." These authors suggest that drilling on seamount summits will of necessity sample the younger shield-building stage of seamount formation. They suggest that the older age is a result of sampling of an older volcanic portion of the seamount, implying that the minimum age for initial formation of Detroit seamount is 81 Ma.

Age data for the sample locations shown in Figure 12 and other area data are summarized in Figure 13, a diagram modified from Duncan and Keller (2004). This figure shows reported ages versus distance from Hawaii. The symbols show the age data summarized earlier, with trend lines (solid lines) of 5 cm/yr and 10 cm/yr motion relative to the Pacific. Also shown with a dashed line is the trend line for all data older than 15 Ma for the chain, from Clague and Dalrymple (1987). The rectangle represents Obruchev Rise. The width of the rectangle represents the 500 km length (NW-SE axis) of the rise. Rectangle height represents age, ranging from the Maastrichtian sediment age at site 883 to an age of close to 90 Ma found by linear extrapolation from Suiko through Detroit North. As shown by the arrow symbol at lower right in Figure 13, the Obruchev Rise area has experienced significant posthotspot volcanism, as evidenced by ash layers found within the sedimentary column (summary in Duncan and Keller, 2004), Eocene dates on some basalt samples (Duncan and Keller, 2004), and lava flows seen in seismic data (Kerr et al., 2005). The northeast flank of Obruchev Rise is also the site of an unusual sedimentary package, the Meiji Drift, which is an Oligocene and younger sequence of fans and contourites sourced from the northwest (Scholl et al., 1977, 2003; Kerr et al., 2005).

Other evidence that Obruchev Rise was created close to a spreading center comes from gravity and geochemical data. As can be seen in Figure 1 and in more detail in Figure 12, the free-air gravity signature of Obruchev Rise is subdued when compared to the signatures of younger seamounts in the Emperor chain. Younger seamounts are rimmed by a negative gravity anomaly (blue in the figure) that is the gravity effect of a flexural moat, caused by the loading of a seamount onto preexisting crust (Watts, 2001). The presence of a moat can be used as an indicator of relative age. The lack of a flexural moat around a seamount or oceanic plateau is an indication that it may have formed close to, or on, a spreading ridge (Watts, 2001). Shatsky Rise (Fig. 1) is a good example. The lack of a flexural moat around Obruchev Rise suggests that it formed close to a recently active spreading center. One problem with this simple interpretation, however, is the location of Obruchev Rise close to the Aleutian and Kamchatka trenches. Obruchev Rise is located within the topographic high associated with flexural downwarping of the Pacific plate into the subduction zone, so it is possible that any flexural moat around Obruchev Rise is obscured by this extra flexural effect (Watts, 2006, personal commun.). There is other evidence that Obruchev Rise formed close to a spreading center. Geochemical evidence presented by Keller et al. (2000), particularly strontium, lead, and neodymium ratios, are consistent with the formation of Meiji and Detroit seamounts close to a ridge axis. As inferred in Figure 11, Obruchev Rise may actually have formed at an abandoned spreading center left behind as north-directed spreading became established.

In summary, in this section several approaches to understanding age control for the suggested tectonic events have been attempted. Using seafloor spreading information, the age of ocean crust at either end of the Emperor Trough is estimated in the range 96–106 Ma. Radiometric and biostratigraphic data for the northern Emperor seamounts suggest a minimum age for Meiji seamount of ca. 90 Ma. Thus oceanic crust ages tend to be slightly older than the estimated oldest seamount age, although, as already emphasized, uncertainties in these estimates are large and not easily quantified. More data are needed from the area to corroborate the suggested age of 90 Ma for the triple junction jump and initiation of the Hawaii hotspot. Gravity and geochemical data are consistent with the initial formation of Obruchev Rise close to a spreading center. In the scenario presented here, this spreading center would have been the one abandoned by the triple junction jump.

Tectonics of the Kamchatka–Aleutian Trench Intersection

Obruchev Rise is inferred to be the oldest preserved volcanic feature in the Hawaii-Emperor seamount chain. Whether there were older seamounts in the chain that have been subducted is, of course, not known. There is no evidence that any seamount material has been accreted to Kamchatka, because the accretionary material preserved there is all arc-sourced (Soloviev et al., 2002, 2006). However, if the inference from gravity data, isotopic data, and the plate boundary scenario presented earlier that Obruchev Rise formed at a ridge axis is correct, it is highly likely that this is in fact the oldest volcanic feature in the chain. It is feasible that a preexisting hotspot crossed the ridge axis, but then we must say that it is coincidence that none of the older seamounts are preserved.

The sharp corner between the Kamchatka and Aleutian subduction zones north of Obruchev Rise (Fig. 14) could be a result of collision of Obruchev Rise with the subduction zone. Indirect evidence for this is in Komandorsky basin, the basin across the subduction zone north of Meiji seamount. This basin is very young, Miocene to Pliocene, in age (Baranov et al., 1991). It has a northwest-southeast structural grain (Baranov et al., 1991; the grain is visible in Fig. 14) that is parallel to Pacific plate motion (the magenta line in Fig. 14). It is possible that the basin was rifted open by the collision of Obruchev Rise with Kamchatka when Kamchatka was farther east. If Kamchatka was originally part of what is now the Shirshov Ridge, when the Pacific plate moved along the path annotated in Figure 14, Obruchev Rise would have collided with it ca. 7–8 Ma. Because

Obruchev Rise is an oceanic plateau, it may have been difficult to subduct, and it may have pushed Kamchatka to the west, opening Komandorsky basin. Early collision of this portion of the Pacific plate containing Obruchev Rise with the Aleutian-Kamchatka system may also explain the existence of the Meiji Tongue. Mentioned earlier, this Neogene sediment wedge was deposited by bottom currents sourced from continental terranes to the north (Scholl et al., 1977; Kerr et al., 2005). It is possible that these source terranes were the ancestral Shirshov Ridge or even Bowers Ridge, an enigmatic structure (Marlow et al., 1990) preserved today on the north side of the Aleutian Trench. Sediment sourced from either or both of these structures could have been deposited on the Pacific plate as it moved by. Full explanation of this tectonic scenario, however, is beyond the scope of this article; it will be developed elsewhere.

CONCLUSIONS

This article presents a model for the tectonic evolution of the northwest Pacific and initiation of the Hawaii hotspot. Until better age control for the tectonic events discussed is available, however, this model is still, as the article title implies, more speculation than observation-based fact. Data that could help to constrain the model are as follows:

1. Data for the age of oceanic crust at either end of the Emperor Trough. If this feature is a transform fault, oceanic crust at either end of the trough should be the same age.
2. A difference between this model and models in which the Chinook Trough is the initiation rift for the Kula plate is found in the predicted age of crust on either side of the trough. In the model of Woods and Davies (1982), the Chinook Trough separated new crust created when the Kula plate started moving north from older Pacific crust. Their model predicts an age difference of ~20 m.y. for crust north or south of the western end of the Chinook Trough, while the model in this paper predicts a very similar age for this crust. Age data from this region would help to differentiate between the models.
3. The Stalemate fracture zone is located adjacent to the Aleutian Trench. It trends approximately parallel to the Kruzenstern and other fracture zones that, in the model presented here, were created by seafloor spreading during middle to late superchron time. This suggests that the Stalemate fracture zone should also be of mid-Cretaceous age. Magnetic lineations on the northeast side of the Stalemate fracture zone mapped by Lonsdale (1988) yield a Paleocene to Eocene age for this crust, implying a similar age for the Stalemate fracture zone. Better age constraints from this region would help to show whether the Stalemate is, in fact, an older feature. This would help to show that the crust in the area around the northern Emperor seamounts is all of superchron age and was created by Pacific-Izanagi spreading.
4. Determination of a similar age for the onset of the Hawaii

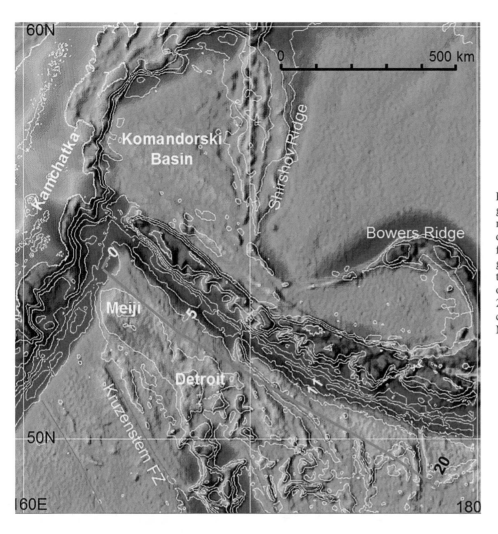

Figure 14. Detailed view of the satellite gravity (Sandwell, 2005), including the northern Emperor chain and Komandorsky basin. The white contours are from the Smith and Sandwell (1997) global gridded bathymetry, contour interval 1 km. The magenta line is the path of the Pacific relative to Asia for the last 20 m.y., with times in m.y. annotated, calculated from the plate circuit poles in Norton (1995).

hotspot and cessation of spreading when the triple junction jumped would validate the model.

5. Between Hess Rise and the Emperor Trough (Fig. 1) is a triangular zone of seafloor with a subtle north-south grain; some of the lineaments are highlighted. If this grain was caused by north-south seafloor spreading, it does not fit any existing tectonic models, including the one presented here. More age and structural data are needed from this area to understand whether its tectonic evolution will require further updates to northwest Pacific tectonic history.

As presented here, the Jurassic through Cretaceous tectonic evolution of the north Pacific involved just three plates. These were the Pacific, Farallon, and Kula/Izanagi plates. There is no reason to invoke extra plates like the Chinook (Rea and Dixon, 1983) or to have one plate, the Izanagi, bordering the northern Pacific plate during the Jurassic and the Early Cretaceous and another, the Kula, from Late Cretaceous time onward. Changes in spreading direction implied by changes in fracture zone and magnetic lineation strike can all be accounted for with a simple three-plate system. Anomalous topographic features in the area were created in several ways: Shatsky Rise by excess volcanism at a triple junction, Hess Rise by excess volcanism at a ridge-transform intersection perhaps associated with a spreading direction change, Chinook Trough by ridge reorganization also associated with a spreading direction change, Emperor Trough as a fracture zone, and Meiji seamount as a spreading ridge associated both with a spreading direction change and perhaps, at the ridge left behind by a ridge jump. This tectonic reorganization, which took place near the end of the superchron, is probably linked to changes in motion of the entire Pacific plate. This was the time when subduction along the southern boundary of the Pacific plate ceased and the Pacific rifted from West Antarctica (Cande et al., 1995), which implies major plate motion direction changes at this time.

The tectonic scenario for the northwest Pacific leads to the suggestion that the spreading center abandoned as a result of the triple junction jump became the site of anomalous volcanism that evolved into the Hawaii hotspot. This implies that the hotspot originated through shallow, i.e., asthenospheric or lithospheric, processes. There is no need to invoke a mantle plume for the early history of the hotspot, although it is quite possible that a mantle plume later intersected this shallow structure and the combined center of anomalous volcanism evolved into the Hawaii hotspot as it exists today. One intriguing possibility is that a mantle plume combined with the shallow hotspot close to the time of the Hawaii-Emperor bend, ultimately causing the hotspot to change motion direction relative to the Pacific plate.

ACKNOWLEDGMENTS

I thank Gillian Foulger for the encouragement to write this paper. It arose from conversations with Will Sager and John Tarduno at the American Geophysical Union's Chapman Conference, The Great Plume Debate: The Origin and Impact of LIPs and Hot Spots, and their insight is appreciated. I also thank Brian Bell, Ian Campbell, Gillian Foulger, Warren Hamilton, Dean Presnall, Dave Sandwell, and Tony Watts for stimulating discussions. Insightful reviews from Will Sager, David Scholl, and John Tarduno were extremely helpful in improving the manuscript. I thank Lisa Gahagan and Larry Lawver of the Plates Project, University of Texas Institute for Geophysics, for permission to use their magnetic lineation and fracture zone compilations and Dave Sandwell for permission to use his gravity data in the figures. I also thank Doug Robertson and Bob Brovey for their help in generating the figures and ExxonMobil for permission to publish.

REFERENCES CITED

Anderson, D.L., 2005, Scoring hotspots: The plume and plate paradigms, in Foulger, G., et al., eds., Plates, plumes, and paradigms: Boulder, Colorado, Geological Society of America Special Paper 388, p. 31–54.

Aslanian, D., Sahabi, M., Olivet, J.L., Goslin, J., Gilg Capar, L., Maia, M., Recq, M., Beuzart, P., and Geli, L., 1994, Genesis of topographic anomalies in the oceans, Part 1: The role of triple junctions: Eos (Transactions, American Geophysical Union), v. 75, no. 16, Supplement, p. 321.

Atwater, T., 1989, Plate tectonic history of the northeast Pacific and western North America, in Winterer, E.L., Hussong, D.M., and Decker, R.W., eds., The Geology of North America, v. N: The Eastern Pacific Ocean and Hawaii. Boulder, Colorado, Geological Society of America, p. 21–70.

Atwater, T., and Severinghaus, J., 1989, Tectonic maps of the northeast Pacific, in Winterer, E.L., Hussong, D.M., and Decker, R.W., eds., The Geology of North America, v. N: The Eastern Pacific Ocean and Hawaii: Boulder, Colorado, Geological Society of America, p. 15–20.

Atwater, T., Sclater, J., Sandwell, D., Severinghaus, J., and Marlow, M.S., 1993, Fracture zone traces across the North Pacific Cretaceous Quiet Zone and their tectonic implications, in Pringle, M.S., et al., eds., The Mesozoic Pacific: Geology, tectonics and volcanism: Washington, D.C., American Geophysical Union, Geophysics Monograph 77, p. 137–154.

Baranov, B.V., Seliverstov, N.J., Murav'ev, A.V., and Muzarov, E.L., 1991, The Komandorsky Basin as a product of spreading behind a transform plate boundary: Tectonophysics, v. 199, p. 237–269, doi: 10.1016/0040-1951 (91)90174-Q.

Bartolini, A., and Larson, R.L., 2001, Pacific microplate and the Pangea supercontinent in the Early to Middle Jurassic: Geology, v. 29, p. 735–738, doi: 10.1130/0091-7613(2001)029<0735:PMATPS>2.0.CO;2.

Cande, S.C., Raymond, C.A., Stock, J., and Haxby, W.F., 1995, Geophysics of the Pitman Fracture Zone and Pacific-Antarctic plate motions during the Cenozoic: Science, v. 270, p. 947–953, doi: 10.1126/science.270.5238.947.

Clague, D.A., and Dalrymple, G.B., 1987, The Hawaiian-Emperor volcanic chain, Part 1: Geologic evolution: Reston, Virginia, U.S. Geological Survey, Professional Paper 1350, p. 5–54.

Clouard, V., and Bonneville, A., 2005, Ages of seamounts, islands and plateaus on the Pacific plate, in Foulger, G., et al., eds., Plates, plumes, and paradigms: Boulder, Colorado, Geological Society of America Special Paper 388, p. 71–90.

Coffin, M.F., and Eldholm, O., 1994, Large igneous provinces: Crustal structure, dimensions, and external consequences: Reviews of Geophysics, v. 32, p. 1–36, doi: 10.1029/93RG02508.

Dalrymple, G.B., Lanphere, M.A., and Natland, J.H., 1980, K-Ar minimum age for Meiji Guyot, Emperor seamount chain, in Jackson, E.D., et al., eds., Initial reports of the Deep Sea Drilling Project, v. 55, p. 677–683.

Doubrovine, P.V., and Tarduno, J.A., 2004, Late Cretaceous paleolatitude of the Hawaiian Hot Spot: New paleomagnetic data from Detroit Seamount (ODP Site 883): Geochemistry, Geophysics, Geosystems, v. 5, p. Q11L04, doi: 10.1029/ 2004GC000745.

Duncan, R.A., and Keller, R.A., 2004, Radiometric ages for basement rocks from the Emperor Seamounts, ODP Leg 197: Geochemistry, Geophysics, Geosystems, v. 5, p. Q08L03, doi: 10.1029/2004GC000704.

Duncan, R.A., Tarduno, J.A., and Scholl, D.W., 2006, Leg 197 synthesis: Southward motion and geochemical variability of the Hawaiian hotspot, in Duncan, R.A., et al., eds., ODP, Scientific results, v. 197, p. 1–39.

Engebretson, D.C., 1984, Relative motion between oceanic and continental plates in the Pacific Basin: Boulder, Colorado, Geological Society of America Supplement Paper 206, 59 p.

Engebretson, D.C., Cox, A., and Thompson, G.A., 1984a, Correlation of plate motions with continental tectonics; Laramide to Basin-Range: Tectonics, v. 3, p. 115–119.

Engebretson, D.C., Cox, A., and Gordon, R.G., 1984b, Relative motions between oceanic plates of the Pacific Basin. S.: Thomas Crough memorial: Journal of Geophysical Research, v. 89B, p. 10,291–10,310.

Engebretson, D.C., Cox, A., and Debiche, M., 1987, Reconstructions, plate interactions, and trajectories of oceanic and continental plates in the Pacific Basin, in Monger, J.W.H., and Francheteau, J., eds., Circum-Pacific orogenic belts and evolution of the Pacific Ocean basin: Washington, D.C., American Geophysical Union, Geodynamics Series, v. 18, p. 19–27.

Gradstein, F.M., Ogg, J.G., and Smith, A.G., 2004, A geologic time scale 2004: Cambridge, England, Cambridge University Press. Also available at http://www.stratigraphy.org/.

Grow, J.A., and Atwater, T., 1970, Mid-Tertiary tectonic transition in the Aleutian Arc: Geological Society of America Bulletin, v. 81, p. 3715–3722, doi: 10.1130/0016-7606(1970)81[3715:MTTITA]2.0.CO;2.

Handschumacher, D.W., Sager, W.W., Hilde, T.W.C., and Bracey, D.R., 1988, Pre-Cretaceous tectonic evolution of the Pacific plate and extension of the geomagnetic polarity reversal time scale with implications for the origin of the Jurassic "Quiet Zone": Tectonophysics, v. 155, p. 365–380, doi: 10.1016/0040-1951(88)90275-2.

Hayes, D.E, and Pitman, W.C., III, 1970, Magnetic lineations in the North Pacific, in Hays, J.D., ed., Geological investigations of the north Pacific: Boulder, Colorado, Geological Society of America Memoir 126, p. 291–314.

Hieronymus, C.F., and Bercovici, D., 2000, Non-hotspot formation of volcanic chains: Control of tectonic and flexural stresses on magma transport: Earth and Planetary Science Letters, v. 181, p. 539–554, doi: 10.1016/S0012-821X(00)00227-2.

Hilde, T.W.C., Isezaki, N., and Wageman, J.M., 1976, Mesozoic sea-floor spreading in the North Pacific, in Sutton., G.H., Manghnani, M.H., and Moberly, R., eds., The geophysics of the Pacific Ocean basin and its margin: Washington, D.C., American Geophysical Union, Geophysical Monograph 19, p. 205–226.

Hilde, T.W.C., Uyeda, S., and Kroenke, L., 1977, Evolution of the western Pacific and its margin: Tectonophysics, v. 38, p. 145–165, doi: 10.1016/0040-1951(77)90205-0.

Keller, R.A., Fisk, M.R., and White, W.M., 2000, Isotopic evidence for Late Cretaceous plume-ridge interaction at the Hawaiian hotspot: Nature, v. 405, p. 673–676, doi: 10.1038/35015057.

Keller, R.A., Duncan, R.A., and Fisk, M.R., 1995, Geochemistry and $^{40}Ar/^{39}Ar$ geochronology of basalts from ODP Leg 145: Scientific Results of Ocean Drilling Program, v. 145, p. 333–344.

Kerr, B.C., Scholl, D.W., and Klemperer, S.L., 2005, Seismic stratigraphy of Detroit Seamount, Hawaiian-Emperor seamount chain: Post-hot-spot shield-building volcanism and deposition of the Meiji drift: Geochemistry, Geophysics, Geosystems, v. 6., p. Q07L10, doi: 10.1029/2004 GC000705.

Kotelkin, V.D., Lobkovskii, L.I., Verzhbitskii, E.V., and Kononov, M.V., 2004, A geodynamical model for the formation of the Shatsky Rise, Pacific Ocean: Oceanology (Moscow), v. 44, p. 257–260.

Larson, R.L., and Chase, C.G., 1972, Late Mesozoic evolution of the western Pacific Ocean: Geological Society of America Bulletin, v. 83, p. 3627–3644, doi: 10.1130/0016-7606(1972)83[3627:LMEOTW]2.0.CO;2.

Larson, R.L., and Pitman, W.C., III, 1972, World-wide correlation of Mesozoic magnetic anomalies and its implications: Geological Society of America Bulletin, v. 83, p. 3645–3662, doi: 10.1130/0016-7606(1972)83[3645:WCOMMA]2.0.CO;2.

Lei, D., and Zhao, D., 2006, A new insight into the Hawaiian plume: Earth and Planetary Science Letters, v. 241, p. 438–453, doi: 10.1016/j.epsl.2005.11.038.

Lonsdale, P., 1988, Paleogene history of the Kula plate: Offshore evidence and onshore implications: Geological Society of America Bulletin, v. 100, p. 733–754, doi: 10.1130/0016-7606(1988)100<0733:PHOTKP>2.3.CO;2.

Mahoney, J.J., Duncan, R.A., Tejada, M.L.G., Sager, W.W., and Bralower, T.J., 2005, Jurassic-Cretaceous boundary age and mid-ocean-ridge–type mantle source for Shatsky Rise: Geology, v. 33, p. 185–188, doi: 10.1130/G21378.1.

Mammerickx, J., 1989, Large-scale undersea features of the northeast Pacific, *in* Winterer, E.L., Hussong, D.M., and Decker, R.W., eds., The geology of North America, v. N: The eastern Pacific Ocean and Hawaii: Boulder, Colorado, Geological Society of America, p. 5–13.

Mammerickx, J., and Sharman, G.F., 1988, Tectonic evolution of the North Pacific during the Cretaceous quiet period: Journal of Geophysical Research, v. 93B, p. 3009–3024.

Marlow, M.S., Cooper, A.K., Dadisman, S.V., Geist, E.L., and Carlson, P.R., 1990, Bowers Swell: Evidence for a zone of compressive deformation concentric with Bowers Ridge, Bering Sea: Marine and Petroleum Geology, v. 7, p. 398–409, doi: 10.1016/0264-8172(90)90017-B.

Morgan, W.J., 1971, Convective plumes in the lower mantle: Nature, v. 230, p. 42–43, doi: 10.1038/230042a0.

Morgan, W.J., 1972, Plate motions and deep mantle convection, *in* Shagam, R., et al., eds., Studies in Earth and space sciences: A volume in honor of Harry Hammond Hess: Boulder, Colorado, Geological Society of America Memoir 132, p. 7–22.

Nakanishi, M., Tamaki, K., and Kobayashi, K., 1989, Mesozoic magnetic anomaly lineations and seafloor spreading history of the Northeastern Pacific: Journal of Geophysical Research, v. 94, p. 15,437–15,462.

Nakanishi, M., Tamaki, K., and Kobayashi, K., 1992b, A new Mesozoic isochron chart of the northwestern Pacific Ocean: Paleomagnetic and tectonic implications: Geophysical Research Letters, v. 19, p. 693–696.

Nakanishi, M., Tamaki, K., and Kobayashi, K., 1992a, Magnetic anomaly lineations from Late Jurassic to Early Cretaceous in the west-central Pacific Ocean: Geophysical Journal International, v. 109, p. 701–719.

Nakanishi, M., Sager, W.W., and Klaus, A., 1999, Magnetic lineations within Shatsky Rise, Northwest Pacific Ocean: Implications for hot spot–triple junction interaction and oceanic plateau formation: Journal of Geophysical Research, v. 104B, p. 7539–7556, doi: 10.1029/1999JB900002.

Natland, J.H., and Winterer, E.L., 2005, Fissure control on volcanic action in the Pacific, *in* Foulger, G., et al., eds., Plates, plumes, and paradigms: Boulder,

Colorado, Geological Society of America Special Paper 388, p. 687–710, doi: 10.1130/2005.2388(39).

Norton, I.O., 1995, Plate motions in the North Pacific: The 43 Ma non-event: Tectonics, v. 14, p. 1080–1094, doi: 10.1029/95TC01256.

Pitman, W.C., and Hayes, D.E., 1968, Seafloor spreading in the Gulf of Alaska: Journal of Geophysical Research, v. 73, p. 6571–6580.

Pringle, M.S., and Dalrymple, G.B., 1993, Geochronological constraints on a possible hot spot origin for Hess Rise and the Wentworth seamount chain, *in* The Mesozoic Pacific: Geology, tectonics and volcanism: Washington, D.C.: American Geophysical Union, Geophysics Monograph, v. 77, p. 263–277.

Rea, D.K., and Dixon, J.M., 1983, Late Cretaceous and Paleogene tectonic evolution of the North Pacific Ocean: Earth and Planetary Science Letters, v. 65, p. 145–166, doi: 10.1016/0012-821X(83)90196-6.

Sager, W.W., 2005, What built Shatsky Rise, a mantle plume or ridge tectonics? *in* Foulger, G., et al., eds., Plates, plumes, and paradigms: Boulder, Colorado, Geological Society of America Special Paper 388, p. 721–733.

Sager, W.W., Handschumacher, D.W., Hilde, T.W.C., and Bracey, D.R., 1988, Tectonic evolution of the northern Pacific Plate and Pacific-Farallon-Izanagi triple junction in the Late Jurassic and Early Cretaceous (M21–M10): Tectonophysics, v. 155, p. 345–364, doi: 10.1016/0040-1951(88)90274-0.

Sager, W.W., Jinho, K., Klaus, A., Nakanishi, M., and Khankishiyeva, L.M., 1999, Bathymetry of Shatsky Rise, Northwest Pacific Ocean: Implications for ocean plateau development at a triple junction: Journal of Geophysical Research, v. 104B, p. 7557–7576, doi: 10.1029/1998JB900009.

Sandwell, D.T., and Smith, W.H.F., 1997, Marine gravity anomaly from Geosat and ERS 1 satellite altimetry: Journal of Geophysical Research, v. 102B, p. 10,039–10,054, doi: 10.1029/96JB03223.

Sandwell, D.T., 2005, Global gravity. Available at ftp://topex.ucsd.edu/pub/global_grav_1min/.

Scholl, D.W., Hein, J.R., Marlow, M., and Buffington, E.C., 1977, Meiji sediment tongue: North Pacific evidence for limited movement between the Pacific and North American plates: Geological Society of America Bulletin, v. 88, p. 1567–1576, doi: 10.1130/0016-7606(1977)88<1567:MSTNPE>2.0.CO;2.

Scholl, D.W., Stevenson, A.J., Noble, M.A., and Rea, D.K., 2003, The Meiji Drift body and Late Paleogene–Neogene paleoceanography of the North Pacific–Bering Sea region, *in* Prothero, D.R., Ivany, L.C., and Nesbitt, E.A., eds., From greenhouse to icehouse: The marine Eocene–Oligocene transition: New York, Columbia University Press, p. 119–153.

Searle, R.C., Holcomb, R.T., Wilson, J.B., Holmes, M.L., Whittington, R.J., Kappel, E.S., McGregor, B.A., and Shor, A.N., 1993, The Molokai fracture zone near Hawaii, and the Late Cretaceous change in Pacific/ Farallon spreading direction, *in* Pringle, M.S., et al., eds., The Mesozoic Pacific: Geology, tectonics and volcanism: Washington, D.C., American Geophysical Union, Geophysics Monograph 77, p. 155–169.

Sharman, G.F., and Risch, D.L., 1988, Northwest Pacific tectonic evolution in the Middle Mesozoic: Tectonophysics, v. 155, p. 331–344, doi: 10.1016/0040-1951(88)90273-9.

Smith, A.D., 2003, A re-appraisal of stress field and convective role models for the origin and distribution of Cretaceous to Recent intraplate volcanism in the Pacific basin: International Geological Review, v. 45, p. 287–302.

Smith, A.D., 2005, Stress fields and the distribution of intraplate volcanism in the Pacific Basin. Available at http://www.mantleplumes.org/Pacific-Cracks.html.

Smith, W.H.F., and Sandwell, D.T., 1997, Global sea floor topography from satellite altimetry and ship depth soundings: Science, v. 277, p. 1956–1962, doi: 10.1126/science.277.5334.1956.

Soloviev, A.V., Shapiro, M.N., Garver, J.I., Shcherbinina, E.A., and Kravchenko-Berezhnoy, I.R., 2002, New age data from the Lesnaya Group: A key to understanding the timing of arc-continent collision, Kamchatka, Russia: The Island Arc, v. 11, p. 79–90, doi: 10.1046/j.1440-1738.2002.00353.x.

Soloviev, A., Garver, J.I., and Ledneva, G., 2006, Cretaceous accretionary complex related to Okhotsk–Chukotka subduction, Omgon Range, Western Kamchatka, Russian Far East: Journal of Asian Earth Sciences, v. 27, p. 437–453, doi: 10.1016/j.jseaes.2005.04.009.

Tarduno, J.A., and Cottrell, R.D., 1997, Paleomagnetic evidence for motion of the Hawaiian hotspot during formation of the Emperor seamounts: Earth and Planetary Science Letters, v. 153, p. 171–180, doi: 10.1016/S0012-821X(97)00169-6.

Tarduno, J.A., Duncan, R.A., and Scholl, D.W., 2002, Proceedings of the Ocean Drilling Program, v. 197: Initial Report: College Station, Texas, Ocean Drilling Program, 92 p.

Tarduno, J.A., Duncan, R.A., Scholl, D.W., Cottrell, R.D., Steinberger, B., Thordarson, T., Kerr, B.C., Neal, C.R., Frey, F.A., Torii, M., and Carvallo, C., 2003, The Emperor Seamounts: Southward motion of the Hawaiian hotspot plume in Earth's mantle: Science, v. 301, p. 1064–1069, doi: 10.1126/science.1086442.

Vallier, T.L., Dean, W.E., Rea, D.K., and Thiede, J., 1983, Geologic evolution of Hess Rise, central North Pacific Ocean: Geological Society of America Bulletin, v. 94, p. 1289–1307, doi: 10.1130/0016-7606(1983)94<1289: GEOHRC>2.0.CO;2.

Verzhbitskii, E.V., and Kononov, M.V., 2004, Heat flow and geodynamics of the Shatsky and Hess Rises, Pacific Ocean: Oceanology (Moscow), v. 44, p. 404–411.

Watts, A.B., 2001, Isostasy and flexure of the lithosphere: Cambridge, England, Cambridge University Press, 458 p.

Wilson, J.T., 1963, A possible origin of the Hawaiian Islands: Canadian Journal of Physics, v. 41, p. 863–870.

Woods, M.T., and Davies, G.F., 1982, Late Cretaceous genesis of the Kula plate: Earth and Planetary Science Letters, v. 58, p. 161–166, doi: 10.1016/0012-821X(82)90191-1.

Worsley, J.R., 1973, Calcareous nannofossils: Leg 19 of the Deep Sea Drilling Project: Initial Reports of the DSDP, v. 19, p. 741–650.

MANUSCRIPT ACCEPTED BY THE SOCIETY JANUARY 31, 2007

The Geological Society of America
Special Paper 430
2007

A plate model for Jurassic to Recent intraplate volcanism in the Pacific Ocean basin

Alan D. Smith*

Department of Geological Sciences, University of Durham, Durham DH1 3LE, UK

ABSTRACT

Reconstruction of the tectonic evolution of the Pacific basin indicates a direct relationship between intraplate volcanism and plate reorganizations, which suggests that volcanism was controlled by fracturing and extension of the lithosphere. Middle Jurassic to Early Cretaceous intraplate volcanism included oceanic plateau formation at triple junctions (Shatsky Rise, the western Mid-Pacific Mountains) and a diffuse pattern of ocean island volcanism (Marcus Wake, Magellan seamounts) reflecting an absence of any well-defined stress field within the Pacific plate. The stress field changed in the Early Cretaceous when accretion of the Insular terrane to the North American Cordillera and the Median Tectonic arc to New Zealand stalled migration of the Pacific-Farallon and Pacific-Phoenix ocean ridges, leading to the generation of the Ontong Java, Manahiki, Hikurangi, and Hess Rise oceanic plateaus. Plate reorganizations in the Late Cretaceous resulted from the breakup of the Phoenix and Izanagi plates through collision of the Pacific-Phoenix ocean ridge with the southwest margin of the basin and development of island arc–marginal basin systems in the northwestern part of the basin. The Pacific plate nonetheless remained largely bounded by spreading centers, and intraplate volcanism followed preexisting lines of weakness in the plate fabric (Line Islands) or resulted from fractures generated by ocean ridge subduction beneath island arc systems (Emperor chain). The Pacific plate began to subduct under Asia in the Early Eocene as inferred from the record of accreted material along the Japanese margin. Further changes to the stress field at this time resulted from abandonment of the Kula-Pacific and the North New Guinea (Phoenix)–Pacific ridges and from development of the Kamchatkan and Izu-Bonin-Mariana arcs, leading to the generation of the Hawaiian chain as a propagating fracture. The final major change in the stress field occurred in the Late Oligocene as a result of breakup of the Farallon into the Cocos and Nazca plates, which caused a hiatus in Hawaiian volcanism; initiated the Sala y Gomez, Foundation, and Samoan chains; and terminated the Louisville chain. The correlations with tectonic events are compatible with shallow-source models for the origin of intraplate volcanism and suggest that the three principal categories of volcanism, intraplate, arc, and ocean ridge, all arise from plate tectonic processes, unlike in plume models, where intraplate volcanism is superimposed on plate tectonics.

Keywords: Pacific basin, Jurassic to Recent history, intraplate volcanism, accreted terranes, plate tectonics

*E-mail: muic2000@yahoo.com.

Smith, A.D., 2007, A plate model for Jurassic to Recent intraplate volcanism in the Pacific Ocean basin, *in* Foulger, G.R., and Jurdy, D.M., eds., Plates, plumes, and planetary processes: Geological Society of America Special Paper 430, p. 471–495, doi: 10.1130/2007.2430(23). For permission to copy, contact editing@geosociety.org.

INTRODUCTION

Before the mantle plume model was widely invoked, intraplate volcanism in the Pacific basin (Fig. 1) was attributed to shallow mantle convection or propagating fractures induced by stresses acting on the Pacific plate (e.g., Jackson et al., 1972; Jackson and Shaw, 1975; Bonatti and Harrison, 1976; Turcotte and Oxburgh, 1976; Walcott, 1976). One of the most advanced explanations was the model of Jackson and Shaw (1975), who suggested that volcanism along the Hawaiian chain reflected injection of magma into the lithosphere as the plate underwent rotation due to either changing forces along its boundaries or variations in coupling between the lithosphere and the asthenosphere. Such models sought to understand intraplate volcanism within the framework of plate tectonics; however, with the increasing popularity of the hotspot model, the emphasis shifted to interpreting the intraplate volcanic record as the result of mantle plumes. Intraplate volcanism thus was imposed on plate tectonics by processes in the deep mantle or at the core-mantle boundary. However, after three decades of promotion of the plume model, there is little consensus with regard to the number of plumes, few examples of intraplate volcanism have been shown to conform to the predictions of the plume head–tail model, and many examples of intraplate volcanism can be fit only by invoking ad hoc variations to the plume model (e.g., Clouard and Bonneville, 2001; Anderson, 2005). Even then the plume model constitutes only a partial explanation for intraplate volcanism, because it cannot account for the vast majority of seamounts across the ocean basin (Natland and Winterer, 2005) or for volcanism in areas of plate fracturing remote from postulated hotspots (Hirano et al., 2006). Instead of seeking alternative explanations only when a plume model cannot be fit, we should ask whether plate tectonics or a "plate model" (Anderson, 1998; Foulger and Natland, 2003; Foulger, this volume) could provide a comprehensive explanation for intraplate volcanism rather than having two mechanisms for the generation of one category of volcanism.

In the plate model, subducted oceanic crust is remixed with the convecting mantle, as in the statistical upper mantle assemblage model of Meibom and Anderson (2004) and the streaky mantle model of Smith (2005), rather than being isolated in plume sources. The same recycled geochemical components are present as in plume models, but their distribution and sampling are different. Large regions of the shallow mantle are close to or at the peridotite solidus, and ocean island basalt–like melts are pervasive in the asthenosphere. The location of intraplate volcanism is controlled by the extension or fracturing of plates, which allows melt to be released (Anderson, 1998; Favela and Anderson, 2000; Foulger and Natland, 2003; Foulger, this volume; Stuart et al., this volume). The volume of melt is controlled by the size, percentage, and orientation of streaks of recycled crust in the region of mantle being tapped (Maclennan, 2004; Meibom and Anderson, 2004; Smith, 2005). Fracturing is related to the stresses acting on the plate, which include thermal stress from cooling, loading stresses from the presence of topographic features, and stresses imposed from plate interactions along the margins of the plate (e.g., Hieronymus and Bercovici, 2000; Natland and Winterer, 2005). Because boundary stresses may be transmitted great distances (e.g., Scotese et al., 1988), the plate model predicts a relationship between intraplate volcanism and the tectonic record. Correlations between the volcanic record and events around the basin margin have been suggested for the Hawaiian-Emperor chain (Jackson et al., 1972; Jackson and Shaw, 1975; Smith, 2003) and for oceanic plateaus in the mid-Cretaceous (Filatova, 1998; Tagami and Hasebe, 1999), although the mid-Cretaceous event has been considered only with the prior assumption of a plume origin. The aim of this article is to examine how the entire record of intraplate volcanism in the Pacific basin could be interpreted in a plate model. Following Natland and Winterer (2005), I have divided the volcanic record into three stages—Middle Jurassic to Early Cretaceous, Late Cretaceous to Eocene, and Eocene to Recent—and examined the relationship with plate configuration to determine the role of plate tectonic processes at each stage.

RECONSTRUCTION OF INTRAPLATE VOLCANISM IN THE PACIFIC BASIN

Previous plate reconstructions of the Pacific basin have been based mainly on the hotspot reference frame. The latter is not valid in a plate model because plumes are not required in the model, and volcanism can occur along preexisting lines of weakness in the plate fabric, independent of the direction of plate movement. The tectonic history and record of intraplate volcanism in the basin have thus been reconstructed (Fig. 2) from the present-day ocean floor topography (Smith and Sandwell, 1997), plate reconstructions based on seafloor magnetic anomalies (Scotese et al., 1988), paleomagnetic data for the Emperor chain and Ontong Java plateau, and the record of intraplate volcanism in accreted terranes around the basin margin.

MIDDLE JURASSIC TO EARLY CRETACEOUS

Plate Configuration at the Time of Formation of the Pacific Plate

The Pacific plate originated in the central paleo-Pacific basin between 175 and 170 Ma in the Middle Jurassic, soon after the initial stages of opening of the Atlantic Ocean (Bartolini and Larson, 2001) and coincident with the closure of the Mongol-Okhotsk basin in the western part of the basin (Nokleberg et al., 2000). The Pacific plate was bordered by the Izanagi, Farallon, and Phoenix plates to the west, east, and south, respectively, although the orientations of ocean ridge systems at this stage are speculative (Fig. 2A). In the Middle Jurassic, the Izanagi, Farallon, and Phoenix plates were of approximately equal size, but the occurrence of Tethyan faunas in older accreted oceanic assemblages such as the Cache Creek terrane of western

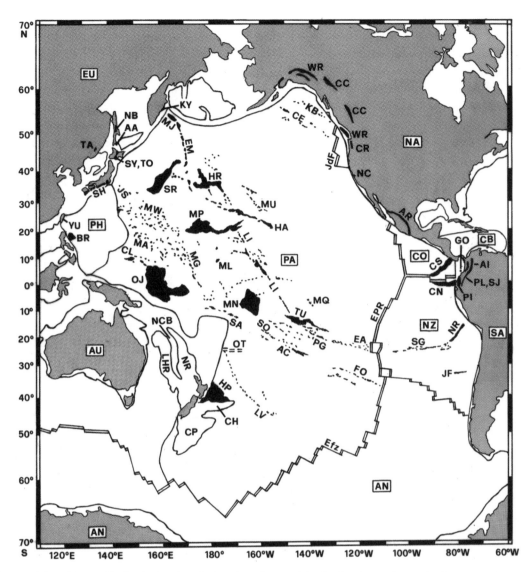

Figure 1. Map of the Pacific basin showing the location of ocean island volcanism and oceanic plateaus on the present-day seafloor and in accreted terranes around the basin margin. Plates: AN—Antarctica; AU—Australia-India; CB—Caribbean; CO—Cocos; EU—Eurasia; NA—North America; NZ—Nazca; PA—Pacific; PH—Philippine; SA—South America. Submerged plateaus: CH—Chatham Rise; CP—Campbell plateau; LHR—Lord Howe Rise; NR—Norfolk Ridge. Basins: NCB—New Caledonia basin. Ocean ridges: EPR—East Pacific Rise; JdF—Juan de Fuca; OT—Osbourn Trough. Fracture zones: Efz—Eltanin fracture zone. Seamounts, island chains, plateaus, and rises, with ages in Ma, from Clouard and Bonneville (2005) except where indicated: AC—Austral-Cook (29–0); BR—Benham Rise (ca. 49); CL—Caroline (14–1); CE—Cobb-Eickelberg (26–2); CN—Carnegie Ridge (15–0); CS—Cocos Ridge (20–0); EM—Emperor (81–46); FO—Foundation (21–2); HA—Hawaiian (46–0); HP—Hikurangi plateau (>85; Mortimer and Parkinson, 1996); HR—Hess Rise (111–87); JF—Juan Fernandez (0–4; Baker et al., 1987); JS—Japanese seamounts (108–71); KB—Kodiak-Bowie (25–0); LI—Line Islands (128–35); LV—Louisville (77–25); MA—Magellan seamounts (129–74); MG—Marshall Gilbert (138–56); MJ—Meiji (>81); ML—Magellan Rise (135–100); MN—Manihiki plateau (123–110); MP—Mid-Pacific Mountains (>128–73); MQ—Marquesas (6–1); MU—Musicians (98–64); MW—Marcus Wake (120–78); NR—Nazca Ridge (>25; O'Connor et al., 1995); OJ—Ontong Java (122–105; Fitton et al., 2004); PG—Pitcairn-Gambier (11–0); SA—Samoa (14–5); SG—Sala y Gomez (22–0; O'Connor et al., 1995); SO—Society (5–0); SR—Shatsky Rise (143–129); TU—Tuamotu (>50–20). Accretionary terranes or assemblages containing ocean island or plateau material, with ages of active intraplate volcanism (compiled from Monger and Berg, 1987; Silberling et al., 1987; Taira et al., 1989; Sun et al., 1998; Reynaud et al., 1999; Nokleberg et al., 2000): AA—Anvina (185–110); AI—Amaime (160–130); AR—Arperos (150–110); CC—Cache Creek (340–200); CR—Crescent (62–49); GO—Gorgona-Serranía de Baudó (88–73); KY—Kronotskiy (83–25?); NB—Nabilsky (185–110); NC—Nicasio (140–135); PI—Piñón (>95–80); PL, SJ—Pallatanga-San Juan (ca. 123); SH—Shimanto (132–65); SY—Sorachi-Yezo (163–147); TA—Taukha (170–150); TO—Tokoro (142–110); WR—Wrangell (232–228); YU—Yuli (>110–97).

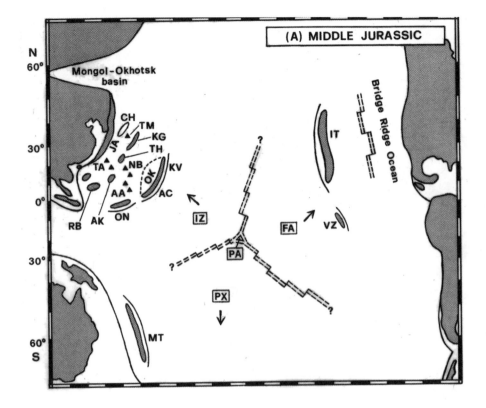

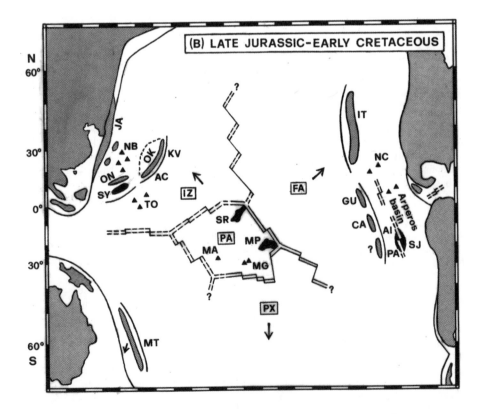

Figure 2. Plate reconstruction of the Pacific basin from the Middle Jurassic (panel A) to the Middle Eocene (panel F). Active intraplate volcanism is shown in black. Ocean ridges, shown by solid lines, are defined by magnetic anomalies; ocean ridges, indicated by dashed lines, are speculative. Reconstructions are based on Scotese et al. (1988), with modifications as indicated for each time frame. Plates: AN—Antarctic; CB—Caribbean; FA—Farallon (n = north, s = south); HK—Hikurangi; IV—Iruney-Vatuna; IZ—Izanagi; KU—Kula; MB—Marie Byrd; MO—Moa; NG—North New Guinea; PA—Pacific; PH—Philippine; PX—Phoenix. Ocean ridges: OT—Osbourn Trough; PA-MN—Pacific-Manahiki; TR—Tasman; VO—Vetlovka. Ocean islands or seamounts (triangles): AA—Anvina; CR—Crescent; EM—Emperor; HA—Hawaiian; JS—Japanese seamounts; LI—Line Islands; LV—Louisville; KY—Kronotskiy; MA—Magellan; MG—Marshall-Gilbert; MU—Musicians; MW—Marcus-Wake; NB—Nabilsky; NC—Nicasio; SH—Shimanto; TA—Taukha; TM—Tamba; TO—Tokoro; TU—Tuamotu; YU—Yuli. Oceanic plateaus (shaded = active, unshaded = inactive): CH—Chichibu; CP—Caribbean; GO—Gorgona-Serranía de Baudó; HP—Hikurangi; HR—Hess Rise; OJ—Ontong Java; MJ—Meiji; ML—Magellan; MN—Manahiki; MP—Mid-Pacific Mountains; PA—Pallatanga; PI—Piñón; SJ—San Juan; SR—Shatsky Rise; SY—Sorachi. Crustal blocks: AK—Abukuma–South Kitakami; CH—Chatham Rise; IT—Insular terrane; JA—Japan and Sikhote-Alin; G—Kurosegawa; LHR—Lord Howe Rise; OK—Okhotia (ocean floor and arc composite; dashed line indicates the extent of crust considered part of the Okhotsk block); NR—Norfolk Ridge; RB—Reed Bank; TH—Tahin. Arcs: AC—Academy of Sciences; AV—Achivayam-Valaginskaya; CA—Central America; CGA—Caribbean Great arc; EP—East Philippines–Daito Ridge; GU—Guerrero;

continued

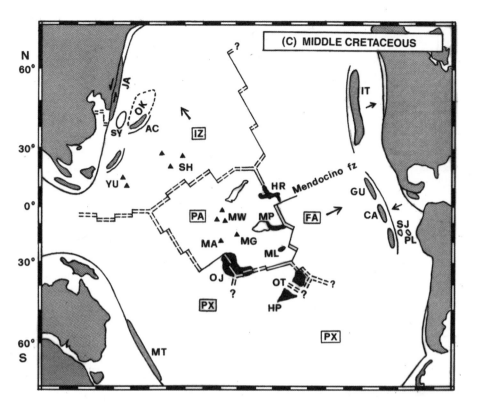

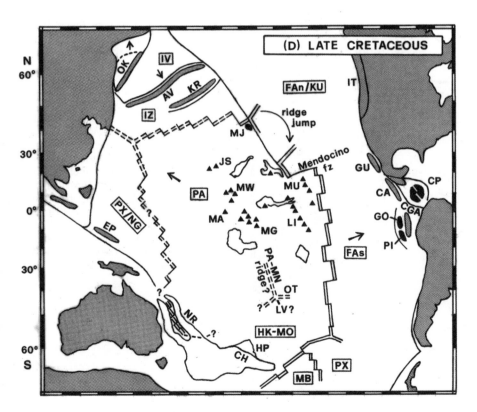

Figure 2 *continued*

IBM—Izu-Bonin-Mariana; KR—Kronot-skaya; KV—Kvakhon; MC—Macuchi; MT—Median Tectonic; ON—Oku-Niikappu; VZ—Vizcaino.

(A) Middle Jurassic (ca. 170 Ma). The Pacific plate formed in the center of the paleo-Pacific basin. Island arcs fringed much of the basin (after Maruyama et al., 1989; Muir et al., 1995; Nokleberg et al., 2000). Intraplate volcanism was limited to ocean island or seamount volcanism in marginal basins in the northwestern Pacific basin. The sequence of terrane accretion that led to the formation of the continental crust of Japan and Sikhote-Alin follows the reconstruction of Osozawa (1998) for the continental margin of Asia before strike-slip displacement of terranes in the mid-Cretaceous. (B) Late Jurassic–Early Cretaceous (150–130 Ma). The Pacific plate attained a rhombic outline. The Shatsky Rise and western Mid-Pacific Mountains formed in conjunction with triple junctions along the Pacific-Izanagi and Pacific-Farallon ocean ridges. Island arc systems in the northwestern part of the basin underwent rifting, leading to the generation of the Sorachi oceanic plateau. Development of island arc systems to the west of Central America led to the formation of the Arperos basin (Freydier et al., 1996). (C) Mid-Cretaceous (120–100 Ma). Oceanic plateau formation along the northeast and SSW margins of the Pacific plate coincided with plate reorganizations along the adjacent basin margins as a result of accretion of the Insular terrane to North America and the Median Tectonic arc to New Zealand. The position of the Manahiki plateau is based on Joseph et al. (1993). The latitude for the Ontong Java plateau is based on Riisager et al. (2003). (D) Late Cretaceous (84–82 Ma). Major plate reorganizations took place in the northern and southern Pacific basin, with fragmentation of the Izanagi plate and the extension of the Pacific

continued

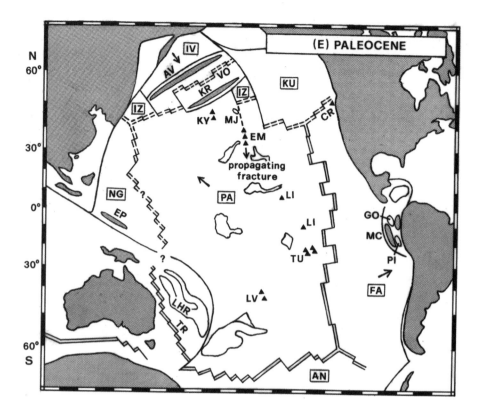

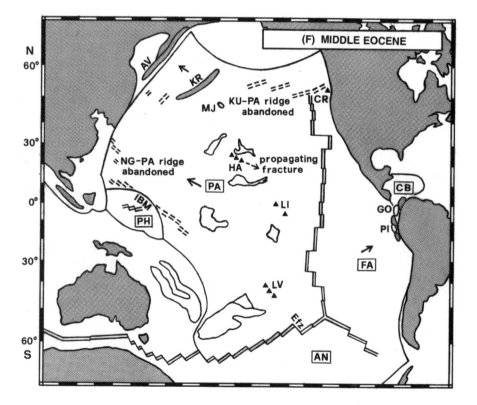

Figure 2 *continued*

plate southward on the abandonment of spreading on the Osbourn Trough. The plate configuration along the Antarctic margin is from Larter et al. (2002). Remnants of the Phoenix plate were trapped in the western and southern Pacific basin, where they became the North New Guinea plate and part of the Antarctic plate, respectively. The North New Guinea–Pacific Ridge is suggested to have continued into the New Caledonia basin, with the Pacific plate boundary extending across the Norfolk Ridge to the Osbourn Trough in the early Late Cretaceous. The boundary between the Australian and North New Guinea plates follows Honza and Fujioka (2004). The plate configuration in the northwestern part of the basin has been drawn to fit the ocean ridge migration sequence along the Japanese margin described by Osozawa (1997). (E) Paleocene (ca. 59 Ma). The North New Guinea–Pacific Ridge approached the continental margin of Asia and possibly linked with the Tasman Ridge between Australia and the Lord Howe Rise. The Kula-Pacific Ridge began to subduct under island arc systems in the northwestern part of the basin (configuration after Konstantinovskaia, 2000, 2001), leading to the generation of the Emperor chain as a propagating fracture. (F) Middle Eocene (ca. 46 Ma). This panel illustrates the plate configuration immediately after the Early Eocene reorganization. The Hawaiian chain evolved as a propagating fracture, whereas volcanism along the Louisville chain was controlled by the development of the Eltanin fracture zone (Efz). The outlines of the Philippine and North New Guinea plates are based on Hall (2002).

Canada (e.g., Orchard et al., 2001) (Fig. 1) implies an earlier plate configuration dominated by a large eastward-subducting Farallon plate to allow for transport of terranes across the paleo-Pacific basin.

At the time of formation of the Pacific plate, the Farallon plate subducted beneath the Insular (super)terrane to form the Gravina-Nutzotin-Gambier arc in the northeastern part of the basin (Nokleberg et al., 2000). The Insular terrane is a composite of the Alexander terrane, which comprises crust rifted from eastern Gondwana in the middle Paleozoic (Gehrels and Saleeby, 1987) and late Paleozoic–Mesozoic arc strata overlain by intraplate basalts (the Karmutsen-Nikolai Formation) of the Wrangell terrane (Fig. 1). In plate models the Karmutsen-Nikolai Formation can be interpreted as an arc-rift assemblage, as suggested by Barker et al. (1989). The Insular terrane was separated from North America by the Bridge River Ocean, which began to close in the Middle Jurassic as a result of westward movement of North America on the opening of the Atlantic. The existence of an island arc–marginal basin system to the south of the Insular terrane in the Middle Jurassic (Fig. 2A) is suggested by strata in the Vizcaino peninsula of western Mexico (Moore, 1985). Strata in the Guerrero terrane of Mexico and the Western Cordillera of Colombia indicate that the arc system off Central America was well established by the Late Jurassic–Early Cretaceous (Tardy et al., 1994). However, the Vizcaino-Guerrero arc system is unlikely to have extended south of the latitude of the Arica elbow in South America, where subduction of the Farallon plate beneath the continental margin began to generate the Early Andean magmatic province in the Middle Jurassic (e.g., Oliveros et al., 2006).The Guerrero-Vizcaino arc was separated from North and Central America by oceanic crust of the Arperos basin (Freydier et al., 1996) (Fig. 2A and B). The relationship of the Vizcaino-Guerrero arc to the Insular terranes is unknown on account of uncertainties in the width of the Arperos basin and the paleolatitude of the Insular terrane. The debate regarding the latter has been summarized by Cowan et al. (1997): paleomagnetic evidence suggests accretion of the Insular terrane at a latitude of 30° to 40°N, but a lack of evidence for suitable fault systems along which subsequent northward displacement could have taken place has led to suggestions of accretion at present-day latitudes. The position for the Insular terrane ~10° north of the Guerrero arc, illustrated in Figure 2, follows the model of Butler et al. (2001), which advocates moderate postaccretion latitudinal displacements of 1000 km. In the Arperos basin, the intraplate record begins in the Early Cretaceous, with intraplate basaltic rocks in central Mexico interpreted as fragments of accreted ocean islands or an aseismic ridge (Ortiz and Martinez, 1993; Freydier et al., 1996). Early Cretaceous basaltic rocks of the Amaime Formation of Colombia, the Pallatanga and San Juan groups of western Ecuador (Lapierre et al., 2000; Kerr et al., 2002), and the Nicasio terrane of the western United States (Silberling et al., 1987) may also have originated in the Arperos basin (Fig. 2B).

Plate reconstructions for the northwest Pacific basin in the Middle Jurassic–Early Cretaceous have portrayed the Farallon-Izanagi Ridge as migrating northward along the continental margin of Asia, followed by the accretion of oceanic plateaus, seamounts, and microcontinent blocks from the Izanagi plate (e.g., Maruyama and Seno, 1986; Isozaki et al., 1990; Kinoshita, 1995). The microcontinents include the crustal blocks bordering the South China Sea (Reed Bank, Macclesfield Bank, Palawan), the Kurosegawa and Abukuma–South Kitakami belts of Japan, and the Tahin terrane of Sikhote-Alin, which were derived from the Tethyan region in the late Paleozoic (Maruyama et al., 1989; Metcalfe, 1996). However, evidence for accreted intraoceanic arc remnants in Hokkaido (the Oku-Niikappu Complex) suggests that the northwest Pacific basin was more like the present-day Philippine Sea in containing a series of arc and marginal basin systems (Ueda and Miyashita, 2005). The latter model is supported by the occurrence of blocks of volcanic rocks of arc origin in the mid-Cretaceous Yuli belt accretionary complex of Taiwan (Sun et al., 1998) and by Jurassic–Early Cretaceous arc volcanic rocks on the Academy of Sciences Rise and in the Kvakhon terrane along the eastern margin of the Okhotsk block (Watson and Fujita, 1985). The Academy of Sciences Rise–Kvakhon arc must have lain at low latitudes on account of Tethyan faunas in the Late Jurassic–Early Cretaceous limestones that cap guyots in the Nabilsky terrane (Nokleberg et al., 2000) along the suture between the Eurasian continent and the Okhotsk block. The southwest continuation of the Academy of Sciences Rise–Kvakhon arc is suggested to be the Oku-Niikappu arc of Ueda and Miyashita (2005), remnants of which were accreted to Hokkaido in the mid-Cretaceous. Intraplate volcanism within the Middle–Late Jurassic marginal basins is represented by basalts in the Taukha terrane of Sikhote-Alin, the Tamba terrane of Japan, and the Anvina terrane of Sakhalin (Nokleberg et al., 2000; Koizumi and Ishiwatari, 2006) (Fig. 2A).

The terrane record on New Zealand also suggests the existence of an island arc–marginal basin system along the southwest margin of the Pacific basin in the Middle Jurassic–Early Cretaceous (Bradshaw, 1989). Remnants of the Mesozoic arc, which is thought to have been constructed along the western margin of an earlier Permian arc (the Brook Street arc; Bishop et al., 1985), are found along the Median tectonic zone of New Zealand (Kimbrough et al., 1993; Muir et al., 1995).

Oceanic Plateaus and Diffuse Island Chains

The oldest intraplate volcanism on the Pacific plate is represented by the Early Cretaceous Shatsky Rise and the western Mid-Pacific Mountains oceanic plateaus and by seamounts in the Magellan and Northern Marshall groups (Fig. 2B). The Shatsky Rise formed in conjunction with a triple junction between the Pacific, Izanagi, and Farallon plates over a period of 16 m.y. between 143 and 127 Ma (Nakanishi et al., 1999; Sager, 2005), and the chemistry of the volcanism is like that of mid-ocean ridge basalt (Mahoney et al., 2005). The Mid-Pacific Mountains have a complex morphology, suggesting that they are comprised of a series of ridges and seamounts (Winterer and

Sager, 1995; Natland and Winterer, 2005). The radiometric ages for the Mid-Pacific Mountains range from 128 to 73 Ma (Clouard and Bonneville, 2005), indicating a protracted evolution. Because only the central and eastern parts have been dated, volcanism may have overlapped formation of the Shatsky Rise and have formed in a triple junction setting with respect to the Pacific-Farallon and Pacific-Phoenix ridges (Fig. 2B). Splayed ridges on the western (Early Cretaceous or older) and eastern (Late Cretaceous) section of the Mid-Pacific Mountains have no counterpart in current intraplate volcanism in the Pacific basin, and their origin is unknown (Natland and Winterer, 2005). Linear east-west- and northwest-southeast-trending age progressions have been suggested for the Magellan and Marshall seamounts in plume models (e.g., Koppers et al., 2001). However, other studies have noted that there is no paleolatitudinal pattern consistent with a plume origin and have argued that the volcanism has no systematic age distribution as a result of the plate's lack of a well-defined stress field (Sager et al., 1993; Natland and Winterer, 2005).

Basaltic sequences of the Kamuikotan Complex in the Late Jurassic–Early Cretaceous Sorachi terrane of Japan have been suggested to represent a conjugate oceanic plateau to the Shatsky Rise (Kimura et al., 1994; Kimura, 1997). However, the Sorachi terrane had accreted to the continental margin by 119 Ma (Kimura et al., 1994), which would require a very high rate of plate movement of 25 cm/yr^{-1} for transport from a central Pacific location. An origin for basalts in the Sorachi terrane as a result of forearc rifting has also been proposed from the presence of arc volcanic rocks within the terrane (Takashima et al., 2002). The position of the Sorachi terrane adjacent to the Oku-Niikappu Complex raises the possibility of a relationship to the rifting of the Oku-Niikappu arc described by Ueda and Miyashita (2005) (Fig. 2B). An oceanic plateau (the Okhotsk arch) has also been suggested to underlie the north central Sea of Okhotsk (Watson and Fujita, 1985) and could potentially be linked with volcanism in the Sorachi terrane as part of the rifting of the arc system separating the Jurassic marginal basins from the Izanagi plate.

Formation of the Ontong Java plateau also likely occurred in conjunction with an ocean ridge system (Hussong et al., 1979). Lineations on the plateau have been suggested to correspond to both southwest-northeast- (Winterer, 1976; Winterer and Nakanishi, 1995; Gladczenko et al., 1997) and northwest-southeast-trending (Neal et al., 1997) ridge axes. When plume models were developed for the plateau, age data were available for only three drill sites and appeared to indicate formation of the main plateau from 124 to 121 Ma, followed by volcanism on the eastern lobe at 90 Ma (Mahoney et al., 1993; Neal et al., 1997). The narrow range of ages, along with correlations with sites in nearby basins and with sections on Malatia and Santa Isobel islands were used to infer rapid eruption (e.g., Tarduno et al., 1991; Coffin and Gahagan, 1995); however, if the ridge system had been oriented northwest-southeast, the sampling localities would have lain close to the ridge axis, and formation of the plateau could have taken place over a longer time span (Smith, 2003). Recent Ar-Ar dating of new drill sites from ODP leg 192, discussed in Fitton et al. (2004), has yielded ages of 122–105 Ma. Although the younger ages have been suggested to be an artifact of analysis, which would also negate the 90 Ma episode of volcanism (Fitton et al., 2004), paleontological evidence suggests that eruption continued to 112 Ma. The model suggested in Figure 2C is thus of eruption of both the main plateau and the eastern lobe over northwest-southeast- and east-west-trending segments of the Pacific-Phoenix Ridge over a time span of ~10 m.y.

The Hess Rise (110–100 Ma) also formed on an ocean ridge (Pacific-Farallon), and it has been suggested to result from confinement of ridge upwelling by transform faulting along the Shatsky and Mendocino fracture zones (Windom et al., 1981; Rea and Dixon, 1983). The tectonic setting of the Manahiki plateau (123–110 Ma) is uncertain, although the structure of the plateau has been related to a jump of the Pacific-Farallon-Phoenix (Tongevara) triple junction that intersected the plateau during its early stages of formation (Joseph et al., 1993; Viso et al., 2005). Implied in such models is a plume origin whereby the plateau was already forming as a result of hotspot activity at the time of the triple junction jump. An eastern conjugate to the Manahiki plateau (the Mollendo Ridge) was postulated to have subducted under South America from a Late Cretaceous magmatic gap in the Peruvian Andes (Soler et al., 1989). Similarly, the Hikurangi plateau, which lies off the New Zealand margin (Mortimer and Parkinson, 1996), has been suggested as a southern conjugate of the Manahiki plateau (Sutherland and Hollis, 2001). Rifting of the Manahiki and Hikurangi plateaus is considered to have occurred along the Osbourn Trough, a slowly spreading ridge that now lies nearly equidistant from the plateaus in the southwest Pacific (Billen and Stock, 2000). Recent models have suggested that the Ontong Java and Manahiki-Hikurangi plateaus are part of a single large igneous province that was rifted at ca. 112 Ma by a triple junction west of the Osbourn Trough (Taylor, 2006; Worthington et al., 2006).

Mid-Cretaceous Plate Reorganization

The generation of oceanic plateaus in the Early Cretaceous coincided with a protracted phase of plate reorganization along the eastern and southwestern margins of the Pacific basin. Accretion of the Insular terrane to western North America began in the Late Jurassic (McClelland et al., 1992) and involved closure of basins now represented by the Hozameen, Bridge River, and Methow-Tyaughton terranes (Monger and Berg, 1987). Final closure of marginal basins did not occur until around 100 Ma, however, from the timing of arc activity in the Spences Bridge Group along the western margin of North America (Thorkelson and Smith, 1989). Closure of the Arperos basin also took place in the Early Cretaceous, with the island arc system along the west of the basin accreting to form western Mexico and Central America (Fig. 2D). The oldest volcanism in the Gravina-Nutzotin-Gambier arc on the Insular terrane arc occurred at 110 Ma (Nokleberg et al., 2000), with subduction reestablishing under the continental margin to generate the Coast Plutonic Complex, Sierra Nevada, and Salinia and Peninsular ranges batholiths in the Late Cretaceous (e.g., Butler et al., 2001). Subduction of the Farallon plate was therefore disrupted between 110 and 100 Ma, corresponding to the timing of formation of the Hess Rise. Similarly, accretion of the island arc system in the southwestern Pacific basin to the New Zealand margin (Rangitata orogeny) began in the Late Jurassic and terminated during the Early Cretaceous (Muir et al., 1995). The interruption of subduction of the Phoenix plate from 130 to 110 Ma as a result of the accretion event (Spörli and Ballance, 1989) corresponds to the timing of formation of the Ontong Java, Manahiki, and Hikurangi plateaus.

In contrast to the reorganizations along the northeast and southwest margins, there was a greater continuity of tectonic regime in the northwest and southeast Pacific basin. The South American and Asian margins both experienced strike-slip faulting in the mid-Cretaceous, but subduction continued beneath arc systems in the northwest Pacific basin and along the Peru-Chile margin (Watson and Fujita, 1985; Taira et al., 1989; Osozawa, 1998; Scheuber and Gonzalez, 1999). The evidence from the North American and New Zealand margins thus suggests that formation of the mid-Cretaceous oceanic plateaus may have been linked to the interruption of subduction, causing migration of ocean ridge systems bounding the subducting plate to slow or stall. Large volumes of melt could have been generated from entrainment of large streaks of recycled crust in the convecting mantle into the ocean ridge upwelling (Meibom and Anderson, 2004; Korenaga, 2005; Smith, 2005). The slowing of ridge migration would have caused the excess melt to no longer be accommodated by the formation of normal oceanic crust, thereby leading to the generation of an oceanic plateau. The generation of the Ontong Java, Manahiki, Hikurangi, and Hess Rise plateaus may therefore have marked the first major change in the stress field of the Pacific plate. The jamming of subduction zones along the northeast and southwest margins of the basin would also explain the lack of movement of the Pacific plate suggested by paleomagnetic data (Tarduno and Sager, 1995) without having to invoke pinning of the plate by plumes.

LATE CRETACEOUS TO EARLY EOCENE

The Fate of the Pacific-Phoenix Ridge and Onset of Subduction of the Pacific Plate

In the mid-Cretaceous, the Farallon plate separated into north and south parts along the Mendocino fracture zone, and the Pacific-Farallon South Ridge became oriented north-south. The Pacific plate extended to the southeast by spreading along the Osbourn Trough and/or a series of ocean ridge jumps (Joseph et al., 1993; Worthington et al., 2006). Expansion of the Pacific plate westward from the Early Cretaceous reconstruction would also have resulted in the collision of the Izanagi-Pacific Ridge with the margin of Asia, as well as the onset of subduction of the Pacific plate. The latter event was suggested to have

occurred around 83 Ma by Uyeda and Miyashiro (1974), from extrapolation of the Kula-Farallon Ridge configuration of Larson and Chase (1972). The fate of the Pacific-Phoenix Ridge was not considered by Uyeda and Miyashiro (1974), but it presumably migrated southward along the western Pacific margin in models like theirs. Subduction of the Pacific-Phoenix Ridge was suggested to occur at 105 Ma in the southwestern part of the basin, followed by rifting of Zealandia (New Zealand, Campbell plateau, Chatham Rise, Lord Howe Rise, and Norfolk Ridge) from eastern Gondwana in the early Late Cretaceous (Bradshaw, 1989). Other models suggest that the Pacific-Phoenix Ridge was abandoned close to the continental margin, with rifting of Zealandia resulting from capture of the subducted slab of the Phoenix plate and the overlying crustal blocks by the Pacific plate and their movement with it (Luyendyk, 1995). The western margin of the basin may thus have been characterized by transform faulting or extension, possibly with ocean ridge systems separating the Pacific, Australian, and Indian Ocean plates (Wells, 1989; Coney, 1990).

Models for subduction of the Pacific plate in the Late Cretaceous are consistent with evidence from accreted fragments in subduction zone assemblages in Japan and Sakhalin for the passage of the Kula (Izanagi)–Pacific ocean ridge system northward along the continental margin from 95 to 40 Ma (e.g., Kimura et al., 1992; Kinoshita, 1995). However, a second spreading system was proposed by Osozawa (1992, 1997) to have intersected the Kyushu margin at 50 Ma and migrated southward. This second ridge was suggested to be the North New Guinea–Pacific Ridge along the northern margin of the North New Guinea plate. The latter plate was hypothesized from the Cenozoic tectonics of the Philippine Sea by Seno and Maruyama (1984) because the inferred northerly motion of the Pacific plate in the hotspot reference frame is incompatible with subduction beneath the northern margin of the Philippine plate in the Early Eocene. The argument regarding the motion of the Pacific plate does not apply in the plate model, but the record of convergence in Papua New Guinea and the Philippines supports the existence of the North New Guinea plate. Northern New Guinea represents an island arc beneath which seafloor north of southern New Guinea was subducted in the Late Cretaceous–Early Miocene (Seno and Maruyama, 1984; Cullen and Pigott, 1989). Similarly, the eastern Philippines and Daito Ridge of the Philippine plate represent a Late Cretaceous to Paleocene arc to which the western Philippines were accreted from the Australian plate in the Early Eocene (Honza and Fujioka, 2004). Locating such arcs along the southern margin of the Pacific plate would have resulted in accretion to southern Asia. The occurrence of widespread boninitic volcanism along the Izu-Bonin-Mariana arc in the Early Eocene is also consistent with the presence of an ocean ridge system in the western Pacific supplying young, thermally anomalous oceanic crust parallel to the arc system during its early stages of evolution. Without the North New Guinea plate, a mantle plume has to be invoked to provide the thermal anomaly, as in the model of Macpherson and Hall (2001).

The name of the North New Guinea plate reflects its geographical location, but implies that it represents a new plate formed after the Pacific plate had begun to subduct along the western margin of the basin. Rather than invoking a new plate, it is suggested that the North New Guinea plate was a remnant of the Phoenix plate (Fig. 2D). The ocean ridge migration sequence along the Japanese margin may thus reflect the collision of the Izanagi-Pacific-Phoenix triple junction with the continental margin, with the implication that subduction of the Pacific plate under Asia did not begin until the Early Eocene (Osozawa, 1992) (Fig. 2E and F). The model also raises the possibility of a link between the tectonic evolution of the North New Guinea plate and the rifting of Zealandia. Opening of the New Caledonia basin occurred around 90 Ma, with separation of the Norfolk Ridge from the Lord Howe Rise (Sdrolias et al., 2003). If the spreading system in the New Caledonia basin was an extension of the North New Guinea–Pacific Ridge and the New Caledonia basin was linked with the Osbourn Trough by faulting across the southern Norfolk Ridge, the Pacific plate may have been bounded almost entirely by ocean ridge systems in the early Late Cretaceous (Fig. 2D). The Pacific plate would thus have continued to lack a well-defined stress direction as in the Middle Jurassic–Early Cretaceous; hence the continuation of diffuse ocean island volcanism in the Marcus Wake, Magellan, and Marshall islands through the early Late Cretaceous.

Late Cretaceous ocean island volcanism in the Line Islands chain was more linear than in the Marcus Wake, Magellan, and Marshall islands, but lacked any linear age progression and was cut by several crosstrends (Natland and Winterer, 2005). Similarities in the orientation of the Line Islands chain to earlier spreading centers on the Pacific-Farallon Ridge were noted by Winterer (1976), and the volcanism can be explained by reactivation of earlier lines of weakness in the plate fabric (Davis et al., 2002). A prominent lineation oriented N32°W in the central, mid-Cretaceous section of the Mid-Pacific Mountains parallels the Line Islands trend and is also interpreted to result from reactivation of an earlier abandoned spreading center. Limited dating of volcanism in the eastern Mid-Pacific Mountains suggests that the volcanism may in part have been orthogonal to the Pacific–Farallon South Ridge, and thus related to thermal contraction of the oceanic lithosphere (Natland and Winterer, 2005).

Early Campanian Plate Reorganization and Breakup of the Izanagi Plate

A major plate reorganization occurred in the Pacific basin at 84–82 Ma (early Campanian), which included splitting of the Kula plate from the Farallon North plate (Woods and Davies, 1982). Volcanism along the Emperor chain was initiated at this time, as inferred from dating of Detroit seamount (Keller et al., 1995), which suggests an age of greater than 81 Ma for Meiji seamount at the northern tip of the chain (Fig. 3). In hotspot models, the orientation of the Emperor chain is used to infer northward motion of the Pacific plate between 81 and 50 Ma,

which is explained by subduction of the Kula plate being faster than the spreading on the Kula-Pacific Ridge, causing the motion of the Kula plate to be imparted to the Pacific plate (Lonsdale, 1988a). The Kula plate, as depicted by Lonsdale (1988a), includes the Izanagi plate in the northwestern Pacific basin, and therein lies a difficulty, because reconstructions of the northwest Pacific basin from the terrane accretion record suggest that the Izanagi plate began to fragment with the development of the Kronotskaya and Achaivayam-Valaginskaya island arcs between 89 and 83 Ma (Konstantinovskaia, 2000, 2001) (Fig. 2D). The timing corresponds to the final docking of accreted terranes now under the Sea of Okhotsk (Soloviev et al., 2006), such that breakup of the Izanagi plate may have been precipitated by the plate reorganization along the basin margin (Fig. 2E). However, regardless of the cause of breakup, the reduction in size of the Izanagi plate in conjunction with subduction of young oceanic crust beneath the Kronotskaya arc would likely have slowed spreading on the Pacific-Izanagi Ridge, making it unlikely that motion of the Pacific plate was controlled by plates in the northern part of the basin.

A further difficulty with hotspot models is that paleomagnetic evidence indicates formation of the Emperor chain at least 15° north of the present-day location of the supposed Hawaiian plume (Tarduno and Cottrell, 1997; Sager, 2002; Doubrovine and Tarduno, 2004). The paleomagnetic position of 43°N for Detroit seamount (Sager, 2002) lies close to the position of the Pacific-Farallon-Izanagi triple junction in the plate reconstruction of Rea and Dixon (1983). The proximity of initial Emperor magmatism to the Kula-Pacific Ridge has been suggested by other plate reconstructions (Mammerickx and Sharman, 1988), guyot morphology (Caplan-Auerbach et al., 2000), and the geochemical signatures of the oldest lavas in the Emperor chain (Keller et al., 2000; Regelous et al., 2003). Meiji seamount may thus have been generated in a triple junction setting between the Pacific, Izanagi, and Farallon North plates, with volcanism interrupted by the plate reorganization at 84–82 Ma such that the edifice did not reach the size attained by other oceanic plateaus (Smith, 2003).

The Kula (Izanagi)–Pacific Ridge configuration pertaining to the initiation of the Emperor chain has been refined by Norton (this volume), who suggests that the triple junction associated with the formation of Meiji seamount jumped to a position northeast of the Hess Rise, with the Emperor Trough acting as a sinistral transform fault between the old and new triple junctions (Fig. 2D). The Kula (Izanagi)–Pacific spreading center then reorientated east-west and began to collide with the Kronotskaya arc. Following the reconstructions of Konstantinovskaia (2001) and Norton (this volume), the Emperor chain is envisaged to result from tearing of the Pacific plate caused by a section of the Kula (Izanagi)–Pacific Ridge becoming trapped outboard of the Kronotskaya arc. The Pacific plate would therefore have subducted under the western Kronotskaya arc, whereas the Kula (Izanagi) plate subducted under the eastern section of the arc (Fig. 2E). Weakening of the Pacific plate by lithospheric load-

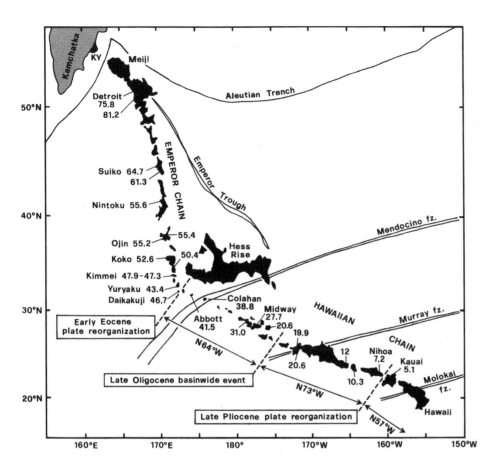

Figure 3. Morphological relationship of the Hawaiian-Emperor chain (bathymetric outline according to the 4 km isobath) to events around the Pacific basin margin. Ages (in Ma) are from Keller et al. (1995), Clouard and Bonneville (2005), and Sharp and Clague (2006). KY—Kronotskiy terrane.

ing from Meiji seamount may also have facilitated initiation of the Emperor chain. Ocean island fragments in the subduction assemblage of the Kronotskaya arc (Watson and Fujita, 1985) may be related to similar plate fracturing by convergent margin geometry. The timing of volcanism in the Kronotskiy terrane is constrained only from the Late Cretaceous to the Oligocene. However, because the island fragments now lie at the same latitude as Meiji seamount and accreted around 10 Ma (Fig. 3), they may have lain ~900 km to the west of Meiji as inferred from current plate velocities and convergence trends, and are speculated to have been related to the subduction of a more westerly segment of the Kula-Pacific Ridge segment beneath the Kronotskaya arc (Fig. 2E).

Major plate reorganizations also took place in the southern Pacific basin in the middle Late Cretaceous, including the abandonment of the Osbourn Trough following accretion of the Hikurangi plateau to the Chatham Rise (Worthington et al., 2006). As a result of this event, the Pacific plate extended southward by the capture of the Hikurangi and Moa plates between the Osbourn Trough and Antarctica at ca. 82 Ma (Sutherland and Hollis, 2001) (Fig. 2D). The Tasman Sea also began to open at 85–80 Ma by spreading along the Tasman Ridge, which was linked to the Pacific-Antarctic Ridge by transform faulting (Kamp, 1986; Schellart et al., 2006). The Pacific plate may have

been bounded by arc systems along the New Guinea–New Zealand margin at this time; however, if the Tasman Ridge continued into the North New Guinea–Pacific Ridge, Zealandia may have belonged to the Pacific plate until the return of the crustal blocks to the Australian plate on abandonment of the Tasman Ridge in the Early Eocene (Fig. 2E and F).

The Louisville chain has been related to the Eltanin fracture zone in several models (Hayes and Ewing, 1971; Larson and Chase, 1972; Watts et al., 1988). Seafloor magnetic anomalies suggest formation of the fracture zone as a result of the early Campanian plate reorganizations (Fig. 4). Only Eocene to Oligocene volcanism along the chain extrapolates directly into the fracture zone, but the proximity of the Louisville chain to the Osbourn Trough (Figs. 1 and 4) raises the possibility that early volcanism along the chain was related to the triple junction responsible for rifting of the Ontong Java and Manahiki plateaus in the model of Worthington et al. (2006) (Fig. 2D). As for Meiji seamount, the ocean ridge system may have been abandoned during the early stages of Louisville volcanism, which may then have propagated to the southeast toward fracture zones along the Pacific-Antarctic Ridge close to the Chatham Rise prior to the reorganizations along the spreading center in the Early Eocene, which resulted in control by fracturing along the Eltanin fracture zone.

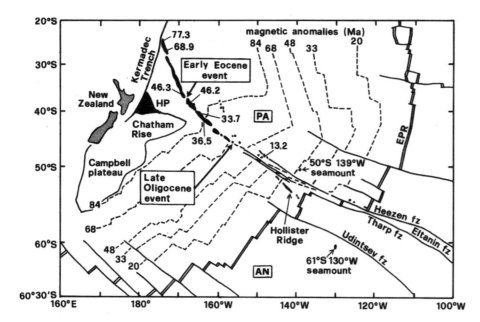

Figure 4. Map showing the relationship of volcanism along the Louisville chain to the Eltanin fracture zone (after Smith and Sandwell, 1997; Vlastelic et al., 1998). Magnetic lineations are from Müller et al. (1997), ages (in Ma) from Clouard and Bonneville (2005). Plates: PA—Pacific; AN—Antarctica.

Farallon and Caribbean Plates

Late Cretaceous–Early Eocene intraplate volcanism on the Farallon and Caribbean plates includes Paleocene seamounts of the Crescent terrane of the northwestern United States (Duncan, 1982), the Caribbean plateau, and oceanic assemblages in the Western Cordillera and coastal provinces of Colombia and Ecuador (e.g., Kerr et al., 1997; Reynaud et al., 1999). Seamounts of the Crescent terrane were generated when the Kula-Farallon and Pacific-Farallon ridges were in close proximity to North America and have been suggested to result from rifting related to migration of the Kula-Farallon Ridge along the continental margin (Babcock et al., 1992). The origin of the Caribbean plateau has generally been considered to have involved generation over the Galápagos hotspot at ca. 90 Ma, followed by collision of the plateau with the east-dipping Caribbean Great Arc some 10 m.y. later (e.g., Kerr et al., 1997; Sinton et al., 1998). Subduction then had to switch to the eastern margin of the plateau as it moved between North and South America. However, difficulties with the accretionary model for the Caribbean plateau were pointed out by James (2006), who argues for an in situ origin. The latter is adopted in this study, with the Caribbean plateau depicted as forming in a back-arc setting behind the Caribbean Great Arc (Fig. 2D).

Accreted intraplate volcanic rocks in coastal Ecuador (the Piñón Formation), on Gorgona Island, and in the Serranía de Baudó province of Colombia were originally considered part of the Caribbean province, but were later attributed to the Sala y Gomez hotspot (e.g., Kerr and Tarney, 2005). The occurrence of komatiites on Gorgona Island has been considered evidence for a plume origin (e.g., Arndt et al., 1997). However, generation of komatiite melts under hydrous melting regimes in supra–subduction zone environments similar to those invoked for boninites

has been proposed by Parman and Grove (2005). Volcanism in the Gorgona province, as in the Kamuikotan Complex of the Sorachi terrane and the Karmutsen-Nikolai Formation of the Wrangell terrane, may thus have a common origin related to arc rifting rather than core-mantle boundary processes.

EARLY EOCENE TO RECENT

Early Eocene Plate Reorganization

The bend between the Hawaiian and Emperor chains has been considered to mark a change in motion of the Pacific plate from northerly to west-northwest in plume models, in response to either the collision of India with Asia (e.g., Clague and Dalrymple, 1989) or the onset of subduction of the Pacific plate beneath the Philippine plate (Seno and Maruyama, 1984). However, it has been argued that Tethyan events would have had little effect on the motion of the Pacific plate (Richards and Lithgow-Bertelloni, 1996), and ages of 49–47 Ma for the earliest boninitic magmatism in the Izu-Bonin-Mariana arc (e.g., Cosca et al., 1998) indicate the onset of subduction beneath the Philippine plate before the earlier accepted age of 43 Ma for the Hawaiian-Emperor bend. Other studies have pointed out a lack of corresponding tectonic events around the basin margin that would have been expected had plate motion changed at 43 Ma (Herron, 1972; Norton, 1995). Interactions with the Philippine plate have been reconsidered in plume models from reevaluations of the age of the Hawaiian-Emperor Bend. Sharp and Clague (1999) suggested an age of ca. 47 Ma for the bend, which corresponds to the age of Daikakuji seamount, which may be considered the westernmost expression of the Hawaiian chain, but this was later increased to 50 Ma (Sharp and Clague, 2006) from extrapolation of the age of 47.9 Ma for postshield alkaline lavas on Kimmei seamount. The uncertainties in the timing reflect a degree of nonlinearity in the age progressions, along with sampling of only the later stages of volcanism on the edifices. However, even if the initiation of the Hawaiian-Emperor Bend and the Izu-Bonin-Mariana arc were contemporaneous, it is unlikely that slab pull and ridge push forces would have changed rapidly enough to cause a change in plate direction (Favela and Anderson, 2000). Rather, the continuity of magnetic lineations along the Pacific-Farallon Ridge (Foulger and Anderson, 2005) suggests that there was no change in Pacific plate motion at either 43 or 50 Ma.

The stress field of a plate, however, would have responded rapidly to a change in plate interactions. In a plate model, the development of the Izu-Bonin-Mariana arc, which was likely associated with abandonment of the North New Guinea-Pacific Ridge (Fig. 2F), represents only one of three major tectonic changes in the Middle Eocene. In addition to the potential Early Eocene onset of subduction of the Pacific plate under the continental margin of Asia (Fig. 2E and F), the stress field of the Pacific plate may also have changed as a result of plate interactions in the northwest Pacific basin (Smith, 2003; Shapiro et al., 2006). At ca. 46 Ma, subduction of the the Iruney-Vatuna basin ceased and the polarity of subduction under the Achaivayam-Valaginskaya arc flipped as a result of collision of the arc with the continental margin (Konstantinovskaia, 2000; Shapiro et al., 2006) (Fig. 2E). The change in subduction polarity also coincided with or followed soon after the cessation of activity on the Kula (Izanagi)–Pacific Ridge (Byrne, 1979). Thereafter the Izanagi and Kula remnants became part of the Pacific plate, which subducted to form the central Kamchatkan volcanic belt. The effect of the Early Eocene events along the margins of Kamchatka and the Philippine plate would have been to change the stress field in the west of the Pacific basin from compressional to extensional perpendicular to plate motion, which would have allowed formation of the Hawaiian chain as a propagating fracture (Stuart et al., this volume).

Late Oligocene and Miocene–Pliocene Plate Reorganizations

After the Early Eocene plate reorganization, volcanism along the Hawaiian chain was sparse until Midway Island was formed at 28 Ma, and then there was a hiatus between 27 and 21 Ma west of the Murray fracture zone (Norton, 1995) (Fig. 3). Volcanism then shifted to a slightly different trend with a different pole of rotation (Epp, 1984; Koppers et al., 2001) until ca. 5 Ma. The increase in volcanic output at the Murray fracture zone may indicate structural control, and the hiatus that preceded it coincided with Late Oligocene basinwide tectonic events including the division of the Farallon plate into the Cocos and Nazca plates at 25–23 Ma (Tebbens and Cande, 1997; Lonsdale, 2005), reorganizations along the Antarctic-Pacific Ridge (Kamp, 1991), and the initial opening of the Japan Sea (28 Ma; Jolivet et al., 1995). The bathymetry of the Hawaiian

chain shows a further break between the islands of Nihoa (7 Ma) and Kauai (5 Ma), which corresponds to the timing of plate reorganizations along the Fiji margin (Cox and Engebretson, 1985). The Late Oligocene and Miocene–Pliocene events have similar significance in other island chains on the Pacific plate. Volcanism along the Samoan chain (23–0 Ma) lies in a region of deformation of the Pacific plate near where the margin changes from subduction along the Tonga Trench to a transform boundary along the Fijian margin (Natland and Winterer, 2005). The Cook and Foundation chains were initiated during the Late Oligocene event, and volcanism in the Austral and Marquesas Islands began during the Miocene–Pliocene event (Natland and Winterer, 2005).

Volcanism along the Louisville chain also shows a marked decrease in output around 25 Ma (Géli et al., 1998). In plume models, the present-day Louisville hotspot is considered to lie beneath a seamount at 50°S, 139°W (Lonsdale, 1988b), such that the chain appears to curve away from the Eltanin fracture zone. However, a single seamount should not be considered evidence for a plume, and although the seafloor topography map of Smith and Sandwell (1997) shows few seamounts in the region, there are others at comparable distances from the Pacific-Antarctica Ridge that have no relationship to any island chain in the region, such as the seamount at 61°S, 130°W south of the Udintsev fracture zone. Volcanism along the Louisville chain is thus suggested to have been terminated by the Late Oligocene plate reorganizations, with the sporadic formation of small seamounts since then related to axial ridge processes or reactivation of other fracture zones in the region.

The fission of the Farallon plate into Cocos and Nazca plates was accompanied by reorientation of the East Pacific Rise orthogonal to the orientation of subduction. This reorganization of spreading centers provides strong evidence for the transmission of stress from the convergent margin across the plate (Natland and Winterer, 2005). Relationships between convergent margin geometry and intraplate features have also been suggested for the Nazca plate (Isacks and Barazangi, 1977; Anderson, 1998; Favela and Anderson, 2000; Smith, 2003). Between 10° and 40°S, the convergent margin is characterized by two regions of flat (15° dip) subduction, marked by an absence of Quaternary volcanism in southern Peru and central Chile (Isacks and Barazangi, 1977) (Fig. 5). The Juan Fernandez Ridge has been noted to coincide with a change in slab dip at 33°S, and the Nazca Ridge and Sala y Gomez chains to extrapolate to changes in slab geometry at 15° and 28°S. In conventional models, the assumption of a plume origin for the volcanism has caused the changes in slab geometry to be attributed to the subduction of topographic features associated with the intraplate volcanism (e.g., Yáñez et al., 2002). Such a relationship may apply to the Nazca Ridge, because the volcanism was terminated by the Late Oligocene plate reorganization. However, the slab flexure at 33°S cannot have resulted from subduction of the Juan Fernandez Ridge, because the older section of the ridge curves northeast and intersects the convergent margin at 31°S (Fig. 5).

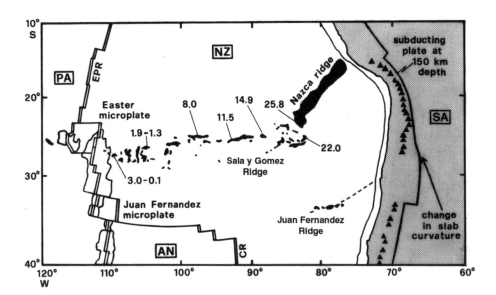

Figure 5. Relationship of intraplate volcanic features on the Nazca plate to slab geometry between 10° and 40°S along the South American margin. Quaternary continental arc volcanism is marked by black triangles. The position of the subducted slab at 150 km depth follows Isacks and Barazangi (1977), modified following Cahill and Isacks (1992). The dashed line indicates the trace of the Juan Fernandez Ridge as depicted by Yáñez et al. (2002). Ages are shown for the Sala y Gomez and Nazca ridges in Ma (O'Connor et al., 1995). Plates: AN—Antarctica; NZ—Nazca; PA—Pacific; SA—South America. Ocean ridges: CR—Chile Rise; EPR—East Pacific Rise.

Similarly, if the Sala y Gomez chain was initiated as a result of the Late Oligocene plate reorganization, the age progression along the chain suggests that it has yet to undergo subduction, so the chain cannot be causing the change in slab geometry. The region between 25° and 28°S to which the Sala y Gomez chain extrapolates is characterized by a change in curvature of the slab from concave upward to convex upward (Cahill and Isacks, 1992). The zone of slab deformation would thus be more extensive than the abrupt flexuring of the slab at 15° and 33°S. Transmission of stress from warping of the slab may thus explain the diffuse distribution of volcanism along the Sala y Gomez chain compared to the more linear Juan Fernandez chain.

CONCLUSIONS

Intraplate volcanism shows a relationship to plate tectonic events and lithospheric architecture throughout the history of the Pacific Ocean basin, which suggests that plate tectonics controls the volcanism rather than that the volcanism controls plate interactions, as has been proposed for some stages of basin evolution in plume models. In the Middle Jurassic–Early Cretaceous, the accreted terrane record indicates that extensive island arc–marginal basin systems existed in the northwest, northeast, and southwest margins of the basin. Major plate tectonic events such as the opening of the Atlantic caused tectonic reorganizations within the marginal basin systems, such that stresses were not transmitted directly to the Pacific plate. The Pacific plate thus grew symmetrically with no well-defined stress direction until the middle Early Cretaceous, such that apart from generation of oceanic plateaus by focusing melts at triple junctions, intraplate volcanism on the plate lacked any linear trend or age progression. Basins along the northeastern and southwestern parts of the Pacific basin closed in the Early Cretaceous, halting the subduction of the Farallon and Phoenix plates. Stresses were trans-

mitted across these plates, slowing the migration of spreading on the Pacific-Farallon and Pacific-Phoenix ridges, leading to the generation of the Hess Rise, Ontong Java, and Manahiki-Hikurangi oceanic plateaus.

Modeling of plate configurations for the Late Cretaceous–Early Eocene is critically dependent on the timing of the onset of subduction of the Pacific plate beneath Asia. Conventional interpretations that the latter event occurred around 83 Ma should be reevaluated, because the fate of the Pacific-Phoenix Ridge has not been considered. If the Pacific-Phoenix Ridge subducted beneath the western Pacific margin in the Late Cretaceous, the North New Guinea plate has to be invoked to explain the Cenozoic tectonic evolution of the western Pacific. Alternatively, following the interpretation of Osozawa (1992, 1997) that two ocean ridges migrated along the Japanese margin in the Late Cretaceous–Paleogene, an alternative hypothesis can be constructed whereby a fragment of the Phoenix plate was trapped in the western Pacific, where it subsequently became the North New Guinea plate. In this second, preferred model, subduction along the western margin of the Pacific plate did not commence until the Early Eocene. Other complexities arise from the uncertainties in the south Pacific plate configuration and from the possible relationship of volcanism along the Louisville chain to ocean ridge systems west of the Osbourn Trough. The Late Cretaceous–Early Eocene nonetheless appears to have been characterized by three contrasting styles of ocean island volcanism: diffuse volcanism in the Marcus-Wake, Magellan, and Marshall islands suggests that the plate continued to lack any well-defined stress field until at least the early Campanian plate reorganization, linear non-age-progressive volcanism in the Line Islands followed preexisting lines of weakness in the plate fabric, and the Emperor chain may have been related to tearing of the Pacific plate as a result of subduction geometry under island arc systems in the northwestern part of the basin.

The Pacific plate attained its modern configuration in the Early Eocene as a result of the onset of subduction under Asia, the accretion of arc systems along the Kamchatkan margin and the abandonment of the Kula-Pacific Ridge, and the initiation of the Izu-Bonin-Mariana arc and the demise of the Pacific–North New Guinea Ridge. The latter two events changed the stress field in the western Pacific from compressional to extensional, giving rise to the Hawaiian chain as a propagating fracture. Subsequent ocean island volcanism was linear as a result of the development of a principal stress direction from subduction of the Pacific plate, but shows a strong relationship to Late Oligocene and Miocene–Pliocene plate reorganizations. The key to understanding the volcanic record is thus in reconstruction of the tectonic history of the basin rather than in modeling of hypothetical plume events.

ACKNOWLEDGMENTS

I thank D.L. Anderson, P. Castillo, G. Foulger, J. Natland, W. Stuart, and two anonymous reviewers for their comments.

REFERENCES CITED

Anderson, D.L., 1998, The EDGES of the mantle, *in* Gurnis et al., eds., The core-mantle boundary region: Washington, D.C., American Geophysical Union, Geodynamics Series 28, p. 255–271.

Anderson, D.L., 2005, Scoring hotspots: The plume and plate paradigms, *in* Foulger, G., et al., eds., Plates, plumes, and paradigms: Boulder, Colorado, Geological Society of America Special Paper 388, p. 31–54, doi: 10.1130/2005.2388(04).

Arndt, N.T., Kerr, A.C., and Tarney, J., 1997, Dynamic melting in plume heads: The formation of Gorgona komatiites and basalts: Earth and Planetary Science Letters, v. 146, p. 289–301, doi: 10.1016/S0012-821X(96)00219-1.

Babcock, R.S., Burmester, R.F., Engebretson, D.C., Warnock, A., and Clark, K.P., 1992, A rifted margin origin for the Crescent basalts and related rocks in the Northern Coast Range volcanic province, Washington and British Columbia: Journal of Geophysical Research, v. 97, p. 6799–6821.

Baker, P.E., Gledhill, A., Harvey, P.K., and Hawkesworth, C.J., 1987, Geochemical evolution of the Juan Fernandez Islands, SE Pacific: Journal of the Geological Society, London, v. 144, p. 933–944.

Barker, F., Sutherland-Brown, A., Budahn, J.R., and Plafker, G., 1989, Back-arc with frontal-arc component origin of Triassic Karmutsen basalt, British Columbia, Canada: Chemical Geology, v. 75, p. 81–102, doi: 10.1016/0009-2541(89)90022-3.

Bartolini, A., and Larson, R.L., 2001, Pacific microplate and the Pangea supercontinent in the Early to Middle Jurassic: Geology, v. 29, p. 735–738, doi: 10.1130/0091-7613(2001)029<0735:PMATPS>2.0.CO;2.

Billen, M.I., and Stock, J., 2000, Morphology and origin of the Osbourn Trough: Journal of Geophysical Research, v. 105, p. 13,481–13,489, doi: 10.1029/2000JB900035.

Bishop, D.G., Bradshaw, J.D., and Landis, C.A., 1985, Provisional terrane map of South Island, New Zealand, *in* Howell, D.G., ed., Tectonostratigraphic terranes of the Circum Pacific region: Houston, Circum Pacific Council for Energy and Mineral Resources, Earth Science Series v. 1, p. 515–521.

Bonatti, E., and Harrison, C.G.A., 1976, Hot lines in the Earth's mantle: Nature, v. 263, p. 402–404, doi: 10.1038/263402a0.

Bradshaw, J.D., 1989, Cretaceous geotectonic patterns in the New Zealand region: Tectonics, v. 8, p. 803–820.

Butler, R.F., Gehrels, G.E., and Kodama, K.P., 2001, A moderate translation alternative to the Baja British Columbia hypothesis: GSA Today, June, p. 1–10.

Byrne, T., 1979, Late Paleocene demise of the Kula-Pacific spreading centre: Geology, v. 7, p. 341–344, doi: 10.1130/0091-7613(1979)7<341:LPDOTK>2.0.CO;2.

Cahill, T., and Isacks, B.L., 1992, Seismicity and shape of the subducted Nazca plate: Journal of Geophysical Research, v. 97, p. 17,503–17,529.

Caplan-Auerbach, J., Duennebier, F., and Ito, G., 2000, Origin of intraplate volcanoes from guyot heights and oceanic paleodepth: Journal of Geophysical Research, v. 105, p. 2679–2697, doi: 10.1029/1999JB900386.

Clague, D.A., and Dalrymple, G.B., 1989, Tectonics, geochronology, and origin of the Hawaiian-Emperor volcanic chain, *in* Winterer, E.L., Hussong, D.M., and Decker, R.W., eds., The geology of North America: The Eastern Pacific Ocean and Hawaii: Boulder, Colorado, Geological Society of America DNAG v. N, p. 187–217.

Clouard, V., and Bonneville, A., 2001, How many Pacific hotspots are fed by deep-mantle plumes?: Geology, v. 29, p. 695–698, doi: 10.1130/0091-7613(2001)029<0695:HMPHAF>2.0.CO;2.

Clouard, V., and Bonneville, A., 2005, Ages of seamounts, islands, and plateaus on the Pacific plate, *in* Foulger, G., et al., eds., Plates, plumes, and paradigms, Geological Society of America Special Paper 388, p. 71–90, doi:10.1130/2005.2388(06).

Coffin, M.F., and Gahagan, L.M., 1995, Ontong Java and Kerguelen plateaux: Cretaceous Icelands?: Journal of the Geological Society of London, v. 152, p. 1047–1052.

Coney, P.J., 1990, Terranes, tectonics and the Pacific rim, *in* Wiley, T.J., Howell, D.G., and Wong, F.L., eds., Terrane analysis of China and the Pacific rim: Circum-Pacific Council for Energy and Mineral Resources, Earth Science Series v. 13, p. 49–69.

Cosca, M.A., Arculus, R.J., Pearce, J.A., and Mitchell, J.G., 1998, [40]Ar/[39]Ar and K-Ar geochronological age constraints for the inception and early evolution of the Izu-Bonin-Mariana arc system: The Island Arc, v. 7, p. 579–595.

Cowan, D.S., Brandon, M.T., and Garver, J.I., 1997, Geologic tests of hypotheses for large coastwise displacements—A critique illustrated by the Baja British Columbia controversy: American Journal of Science, v. 297, p. 117–173.

Cox, A., and Engebretson, D., 1985, Change in motion of Pacific plate at 5 Myr BP: Nature, v. 313, p. 472–474, doi: 10.1038/313472a0.

Cullen, A.B., and Pigott, J.D., 1989, Post-Jurassic tectonic evolution of Papua New Guinea: Tectonophysics, v. 162, p. 291–302, doi: 10.1016/0040-1951(89)90250-3.

Davis, A.S., Gray, L.B., Clague, D.A., and Hein, J.R., 2002, The Line Islands revisited: New [40]Ar/[39]Ar geochronologic evidence for episodes of volcanism due to lithospheric extension: Geochemistry, Geophysics, Geosystems, v. 3, doi: 1029/2001GC000190.

Doubrovine, P.V., and Tarduno, J.A., 2004, Late Cretaceous paleolatitude of the Hawaiian hot spot: New paleomagnetic data from Detroit seamount (ODP site 883): Geochemistry, Geophysics, Geosystems, v. 5, Q11L04, doi:10.1029/2004GC000745.

Duncan, R.A., 1982, A captured island chain in the coast ranges of Oregon and Washington: Journal of Geophysical Research, v. 87, p. 10,827–10,837.

Epp, D., 1984, Possible perturbations to hotspot traces and implications for the origin and structure of the Line Islands: Journal of Geophysical Research, v. 89, p. 11,273–11,286.

Favela, J., and Anderson, D.L., 2000, Extensional tectonics and global volcanism, *in* Boschi, E., Ekstrom, G., and Morelli, A., eds., Problems in geophysics for the new millennium: Bologna, Edrice Compositori, p. 463–498.

Filatova, N.I., 1998, Evolution of Cretaceous active continental margins and their correlation with other global events: The Island Arc, v. 7, p. 253–270, doi: 10.1046/j.1440-1738.1998.00175.x.

Fitton, J.G., Mahoney, J.J., Wallace, P.J., and Saunders, A.D., 2004, Origin and evolution of the Ontong Java plateau: Introduction, *in* Fitton, J.G., et al., eds., Origin and evolution of the Ontong Java plateau: Geological Society, London, Special Publication 229, p. 1–8.

Foulger, G.R., 2007 (this volume), The "plate" model for the genesis of melting anomalies, *in* Foulger, G.R., and Jurdy, D.M., eds., Plates, plumes, and planetary processes: Boulder, Colorado, Geological Society of America Special Paper 430, doi: 10.1130/2007.2430(01).

Foulger, G.R., and Anderson, D.L., 2005, The Emperor and Hawaiian volcanic chains: How well do they fit the plume hypothesis?: http://www.mantle-plumes.org.

Foulger, G.R., and Natland, J.H., 2003, Is "hotspot" volcanism a consequence of plate tectonics?: Science, v. 300, p. 921–922, doi: 10.1126/science.1083376.

Freydier, C., Martinez R.J., Lapierre, H., Tardy, M., and Coulon, C., 1996, The Early Cretaceous Arperos basin (western Mexico): Geochemical evidence for an aseismic ridge formed near a spreading centre: Tectonophysics, v. 259, p. 343–367.

Gehrels, G.E., and Saleeby, J.B., 1987, Geologic framework, tectonic evolution, and displacement history of the Alexander terrane: Tectonics, v. 6, p. 151–173.

Géli, L., Aslanian, D., Olivet, J.-L., Vlastelic, I., Dosso, L., Guillou, H., and Bougault, H., 1998, Location of Louisville hotspot and origin of Hollister ridge: Geophysical constraints: Earth and Planetary Science Letters, v. 164, p. 31–40, doi: 10.1016/S0012-821X(98)00217-9.

Gladczenko, T.P., Coffin, M.F., and Eldholm, O., 1997, Crustal structure of the Ontong Java plateau: Modelling of new gravity and existing seismic data: Journal of Geophysical Research, v. 102, p. 22,711–22,729, doi: 10.1029/97JB01636.

Hall, R., 2002, Cenozoic geological and plate tectonic evolution of SE Asia and the SW Pacific: Computer-based reconstructions, model and animations: Journal of Asian Earth Sciences, v. 20, p. 353–431, doi: 10.1016/S1367-9120(01)00069-4.

Hayes, D.E., and Ewing, M., 1971, The Louisville ridge—A possible extension of the Eltanin fracture zone, *in* Reid, J.L., ed., Antarctic oceanology I: Washington, D.C., American Geophysical Union, Antarctic Research Series v. 15, p. 223–228.

Herron, E.M., 1972, Sea-floor spreading and the Cenozoic history of the east-central Pacific: Bulletin of the Geological Society of America, v. 83, p. 1671–1692, doi: 10.1130/0016-7606(1972)83[1671:SSATCH]2.0.CO;2.

Hieronymus, C.F., and Bercovici, D., 2000, Non-hotspot formation of volcanic chains: Control of tectonic and flexural stresses on magma transport: Earth and Planetary Science Letters, v. 181, p. 539–554, doi: 10.1016/S0012-821X(00)00227-2.

Hirano, N., Takahashi, E., Yamamoto, J., Abe, N., Ingle, S.P., Kaneoka, I., Hirata, T., Kimura, J.-I., Ishii, T., Ogawa, Y., Machida, S., and Suyehiro, K., 2006, Volcanism in response to plate flexure: Science, v. 313, p. 1426–1428, doi: 10.1126/science.1128235.

Honza, E., and Fujioka, K., 2004, Formation of arcs and backarc basins inferred from the tectonic evolution of Southeast Asia since the Late Cretaceous: Tectonophysics, v. 384, p. 23–53, doi: 10.1016/j.tecto.2004.02.006.

Hussong, D.M., Wipperman, L.K., and Kroenke, L.M., 1979, The crustal structure of the Ontong Java and Manahiki plateaus: Journal of Geophysical Research, v. 84, p. 6003–6010.

Isacks, B.L., and Barazangi, M., 1977, Geometry of Benioff zones: Lateral segmentation and downbending of the subducted lithosphere, *in* Talwani, M., and Pitman, W.C., III, eds., Island arcs, deep sea trenches and back-arc basins: Washington, D.C., American Geophysical Union, Maurice Ewing Series, v. 1, p. 99–114.

Isozaki, Y., Maruyama, S., and Furuoka, F., 1990, Accreted oceanic materials in Japan: Tectonophysics, v. 181, p. 179–205, doi: 10.1016/0040-1951(90)90016-2.

Jackson, E.D., and Shaw, H.R., 1975, Stress fields in central portions of the Pacific plate: Delineated in time by linear volcanic chains: Journal of Geophysical Research, v. 80, p. 1861–1874.

Jackson, E.D., Silver, E.A., and Dalrymple, G.B., 1972, Hawaiian-Emperor Chain and its relation to Cenozoic circumpacific tectonics: Bulletin of the Geological Society of America, v. 83, p. 601–618, doi: 10.1130/0016-7606(1972)83[601:HCAIRT]2.0.CO;2.

James, K.H., 2006, Arguments for and against the Pacific origin of the Caribbean plate: Discussion, finding for an inter-American origin: Geologica Acta, v. 4, p. 279–302.

Jolivet, L., Shibuya, H., and Fournier, M., 1995, Paleomagnetic rotations and the Japan Sea opening, *in* Taylor, B., and Natland, J., eds., Active margins and marginal basins of the Western Pacific: Washington, D.C., American Geophysical Union, Monograph 88, p. 355–369.

Joseph, D., Taylor, B., Shor, A.N., and Yamazaki, T., 1993, The Nova-Canton Trough and the Late Cretaceous evolution of the central Pacific, *in* Pringle, M.S., et al., eds., The Mesozoic Pacific: Geology, tectonics and volcanism: Washington, D.C., American Geophysical Union, Geophysical Monograph 77, p. 171–185.

Kamp, P.J.J., 1986, Late Cretaceous–Cenozoic tectonic development of the southwest Pacific region: Tectonophysics, v. 121, p. 225–251, doi: 10.1016/0040-1951(86)90045-4.

Kamp, P.J.J., 1991, Late Oligocene Pacific-wide tectonic event: Terra Nova, v. 3, p. 65–69.

Keller, R.A., Duncan, R.A., and Fisk, M.R., 1995, Geochemistry and ^{40}Ar/^{39}Ar geochronology of basalts from ODP leg 145. Proceedings of the Ocean Drilling Program, Scientific Results, v. 145, p. 333–344.

Keller, R.A., Fisk, M.A., and White, W.M., 2000, Isotopic evidence for Late Cretaceous plume-ridge interaction at the Hawaiian hotspot: Nature, v. 405, p. 673–676, doi: 10.1038/35015057.

Kerr, A.C., and Tarney, J., 2005, Tectonic evolution of the Caribbean and northwestern South America: The case for accretion of two Late Cretaceous oceanic plateaus: Geology, v. 33, p. 269–272, doi: 10.1130/G21109.1.

Kerr, A.C., Tarney, J., Marriner, G.F., Nivia, A., and Saunders, A.D., 1997, The Caribbean-Colombian Cretaceous igneous province: The internal anatomy of an oceanic plateau, *in* Mahoney, J.J., and Coffin, M.F., eds., Large igneous provinces: Continental, oceanic, and planetary flood volcanism: Washington, D.C., American Geophysical Union, Geophysical Monograph 100, p. 123–144.

Kerr, A.C., Aspden, J.A., Tarney, J., and Pilatasig, L.F., 2002, The nature and provenance of accreted oceanic terranes in western Ecuador: Geochemical and tectonic constraints: Journal of the Geological Society of London, v. 159, p. 577–594.

Kimbrough, D.L., Tulloch, A.J., Geary, E., Coombs, D.S., and Landis, C.A., 1993, Isotopic ages from the Nelson region of South Island New Zealand: Crustal structure and definition of the Median Tectonic Zone: Tectonophysics, v. 225, p. 433–448, doi: 10.1016/0040-1951(93)90308-7.

Kimura, G., 1997, Cretaceous episodic growth of the Japanese islands: The Island Arc, v. 6, p. 52–68, doi: 10.1111/j.1440-1738.1997.tb00040.x.

Kimura, G., Rodzdestvenskiy, V.S., Okumura, K., Melinikov, O., and Okamura, M., 1992, Mode of mixture of oceanic fragments and terrigenous trench fill in an accretionary complex: Example from southern Sakhalin: Tectonophysics, v. 202, p. 361–374, doi: 10.1016/0040-1951(92)90120-U.

Kimura, G., Sakakibara, M., and Okamura, M., 1994, Plumes in central Panthalassa?: Deductions from accreted oceanic fragments in Japan: Tectonics, v. 13, p. 905–916, doi: 10.1029/94TC00351.

Kinoshita, O., 1995, Migration of igneous activities related to ridge subduction in Southwest Japan and the East Asian continental margin from the Mesozoic to the Paleogene: Tectonophysics, v. 245, p. 25–35, doi: 10.1016/0040-1951(94)00211-Q.

Koizumi, K., and Ishiwatari, A., 2006, Ocean plateau accretion inferred from Late Paleozoic greenstones in the Jurassic Tamba accretionary complex, southwest Japan: Island Arc, v. 15, p. 58–83, doi: 10.1111/j.1440-1738.2006.00518.x.

Konstantinovskaia, E.A., 2000, Geodynamics of an Early Eocene arc-continent collision reconstructed from the Kamchatka orogenic belt, NE Russia: Tectonophysics, v. 325, p. 87–105, doi: 10.1016/S0040-1951(00)00132-3.

Konstantinovskaia, E.A., 2001, Arc-continent collision and subduction reversal in the Cenozoic evolution of the Northwest Pacific: An example from Kamchatka (NE Russia): Tectonophysics, v. 333, p. 75–94, doi: 10.1016/S0040-1951(00)00268-7.

Koppers, A.A.P., Morgan, J.P., Morgan, J.W., and Staudigel, H., 2001, Testing the fixed hotspot hypothesis using $^{40}Ar/^{39}Ar$ age progressions along seamount trails: Earth and Planetary Science Letters, v. 185, p. 237–252, doi: 10.1016/S0012-821X(00)00387-3.

Korenaga, J., 2005, Why did not the Ontong Java plateau form subaerially?: Earth and Planetary Science Letters, v. 234, p. 385–399, doi: 10.1016/j.epsl.2005.03.011.

Lapierre, H., Bosch, D., Dupuis, V., Polvé, M., Maury, R.C., Hernandez, J., Monié, P., Yeghicheyan, D., Jaillard, E., Tardy, M., Mercier de Lépinay, B., Mamberti, M., Desmet, A., Keller, F., and Sénebier, F., 2000, Multiple plume events in the genesis of the peri-Carbbean Cretaceous oceanic plateau province: Journal of Geophysical Research, v. 105, p. 8403–8421, doi: 10.1029/1998JB900091.

Larson, R.L., and Chase, C.G., 1972, Late Mesozoic evolution of the western Pacific Ocean: Bulletin of the Geological Society of America, v. 83, p. 3627–3644, doi: 10.1130/0016-7606(1972)83[3627:LMEOTW]2.0.CO;2.

Larter, R.D., Cunningham, A.P., Barker, P.F., Gohl, K., and Nitsche, F.O., 2002, Tectonic evolution of the Pacific margin of Antarctica 1: Late Cretaceous tectonic reconstructions: Journal of Geophysical Research, v. 107, doi: 10.1029/2000JB000052.

Lonsdale, P., 1988a, Paleogene history of the Kula plate: Offshore evidence and onshore implications: Bulletin of the Geological Society of America, v. 100, p. 733–754, doi: 10.1130/0016-7606(1988)100<0733:PHOTKP>2.3.CO;2.

Lonsdale, P., 1988b, Geography and history of the Louisville hotspot chain in the southwest Pacific: Journal of Geophysical Research, v. 93, p. 3078–3104.

Lonsdale, P., 2005, Creation of the Cocos and Nazca plates by fission of the Farallon plate: Tectonophysics, v. 404, p. 237–264, doi: 10.1016/j.tecto.2005.05.011.

Luyendyk, B.P., 1995, Hypothesis for Cretaceous rifting of east Gondwana caused by subducted slab capture: Geology, v. 23, p. 373–376, doi: 10.1130/0091-7613(1995)023<0373:HFCROE>2.3.CO;2.

Maclennan, J., 2004, Melting anomalies: T_p, X_c, or v?: UKNPS meeting, http://www.mantleplumes.org.

Macpherson, C.G., and Hall, R., 2001, Tectonic setting of Eocene boninite magmatism in the Izu-Bonin-Mariana forearc: Earth and Planetary Science Letters, v. 186, p. 215–230, doi: 10.1016/S0012-821X(01)00248-5.

Mahoney, J.J., Storey, M., Duncan, R.A., Spencer, K.J., and Pringle, M., 1993, Geochemistry and age of the Ontong Java plateau, in Pringle, M.S., et al., eds., The Mesozoic Pacific: Geology, tectonics and volcanism: Washington, D.C., American Geophysical Union, Geophysical Monograph 77, p. 233–261.

Mahoney, J.J., Duncan, R.A., Tejada, M.L.G., Sager, W.W., and Bralower, T.J., 2005, Jurassic–Cretaceous boundary age and mid-ocean-ridge–type mantle source for Shatsky Rise: Geology, v. 33, p. 185–188, doi: 10.1130/G21378.1.

Mammerickx, J., and Sharman, G.F., 1988, Tectonic evolution of the North Pacific during the Cretaceous quiet period: Journal of Geophysical Research, v. 93, p. 3009–3024.

Maruyama, S., and Seno, T., 1986, Orogeny and relative plate motions: Example of the Japanese islands: Tectonophysics, v. 127, p. 305–329, doi: 10.1016/0040-1951(86)90067-3.

Maruyama, S., Liou, J.G., and Seno, T., 1989, Mesozoic and Cenozoic evolution of Asia, in Ben-Avraham, Z., ed., The evolution of the Pacific Ocean margins: New York, Oxford University Press, p. 75–99.

McClelland, W.C., Gehrels, G.E., and Saleeby, J.B., 1992, Upper Jurassic–Lower Cretaceous basinal strata along the Cordilleran margin: Implications for the accretionary history of the Alexander-Wrangellia-Peninsular terrane: Tectonics, v. 11, p. 823–835.

Meibom, A., and Anderson, D.L., 2004, The statistical upper mantle assemblage: Earth and Planetary Science Letters, v. 217, p. 123–139, doi: 10.1016/S0012-821X(03)00573-9.

Metcalfe, I., 1996, Pre-Cretaceous evolution of SE Asian terranes, in Hall, R., and Blundell, D., eds., Tectonic evolution of Southeast Asia: Geological Society of London Special Publication 106, p. 97–122.

Monger, J.W.H., and Berg, H.C., 1987, Lithotectonic terrane map of western Canada and southeastern Alaska, in Silberling, N.J., and Jones, D.L., eds., Lithotectonic terrane map of the North American Cordillera: Reston, Virginia, U.S. Geological Survey, Map 1874B, scale 1:2,500,000, 31 p.

Moore, T.E., 1985, Stratigraphy and tectonic significance of the Mesozoic tectonostratigraphic terranes of the Vizcaino peninsula, Baja California Sur, Mexico, in Howell, D.G., ed., Tectonostratigraphic terranes of the Circum Pacific region: Houston, Circum Pacific Council for Energy and Mineral Resources, Earth Science Series, v. 1, p. 315–329.

Mortimer, N., and Parkinson, D., 1996, Hikurangi plateau: A Cretaceous large igneous province in the southwest Pacific Ocean: Journal of Geophysical Research, v. 101, p. 687–696, doi: 10.1029/95JB03037.

Muir, R.J., Weaver, S.D., Bradshaw, J.D., Eby, G.N., and Evans, J.A., 1995, The Cretaceous Separation Point batholith, New Zealand: Granitoid magmas formed by melting of mafic lithosphere: Journal of the Geological Society, London, v. 152, p. 689–701.

Müller, R.D., Roest, W.R., Royer, J.-Y., Gahagan, L.M., and Sclater, J.G., 1997, Digital isochrons of the world's ocean floor: Journal of Geophysical Research, v. 102, p. 3211–3214, doi: 10.1029/96JB01781.

Nakanishi, M., Sager, W.W., and Klaus, A., 1999, Magnetic lineations within Shatsky Rise, northwest Pacific Ocean: Implications for hotspot–triple junction interaction and oceanic plateau formation: Journal of Geophysical Research, v. 104, p. 7539–7556, doi: 10.1029/1999JB900002.

Natland, J.H., and Winterer, E.L., 2005, Fissure control on volcanic action in the Pacific, in Foulger, G., et al., eds., Plates, plumes, and paradigms: Boulder, Colorado, Geological Society of America Special Paper 388, p. 687–710, doi: 10.1130/2005.2388(39).

Neal, C.R., Mahoney, J.J., Kroenke, L.W., Duncan, R.A., and Petterson, M.G., 1997, The Ontong Java plateau, in Mahoney, J.J., and Coffin, M.F., eds., Large igneous provinces: Continental, oceanic, and planetary flood volcanism: Washington, D.C., American Geophysical Union, Geophysical Monograph 100, p. 183–216.

Nokleberg, W.J., Parfenov, L.M., Monger, J.W.H., Norton, I.O., Khanchuk, A.I., Stone, D.B., Scotese, C.R., Scholl, D.M., and Fujita, K., 2000, Phanerozoic tectonic evolution of the circum–North Pacific: Reston, Virginia, U.S. Geological Survey Professional Paper 1626, 133 p.

Norton, I.O., 1995, Plate motions in the North Pacific: The 43 Ma nonevent: Tectonics, v. 14, p. 1080–1094, doi: 10.1029/95TC01256.

Norton, I.O., 2007 (this volume), Speculations on Cretaceous tectonic history of the northwest Pacific and a tectonic origin for the Hawaii hotspot, in Foulger, G.R., and Jurdy, D.M., eds., Plates, plumes, and planetary processes: Boulder, Colorado, Geological Society of America Special Paper 430, doi: 10.1130/2007.2430(22).

O'Connor, J.M., Stoffers, P., and McWilliams, M.O., 1995, Time-space mapping of Easter Chain volcanism: Earth and Planetary Science Letters, v. 136, p. 197–212, doi: 10.1016/0012-821X(95)00176-D.

Oliveros, V., Féraud, G., Aguirre, L., Fornari, M., and Morata, D., 2006, The Early Andean magmatic province (EAMP): $^{40}Ar/^{39}Ar$ dating on Mesozoic volcanic and plutonic rocks from the coastal Cordillera, northern Chile: Journal of Volcanology and Geothermal Research, v. 157, p. 311–330, doi: 10.1016/j.jvolgeores.2006.04.007.

Orchard, M.J., Cordey, F., Rui, L., Bamber, E.W., Mamet, B., Struik, L.C., Sano, H., and Taylor, H.J., 2001, Biostratigraphic and biogeographic constraints on the Carboniferous to Jurassic Cache Creek terrane in central British Columbia: Canadian Journal of Earth Sciences, v. 38, p. 551–578, doi: 10.1139/cjes-38-4-551.

Ortiz, L.E., and Martinez, J., 1993, Evidence of Cretaceous hot-spot intra-plate magmatism in the central segment of the Guerrero terrane, in Ortega, F., Coney, P.J., Centeno, E., and Gómez, A., eds., Proceedings of the First Circum-Pacific and Circum Atlantic Terrane Conference, Mexico City, UNAM, p. 110–112.

Osozawa, S., 1992, Double ridge subduction recorded in the Shimanto accretionary complex, Japan, and plate reconstruction: Geology, v. 20, p. 939–942, doi: 10.1130/0091-7613(1992)020<0939:DRSRIT>2.3.CO;2.

Osozawa, S., 1997, The cessation of igneous activity and uplift when an actively spreading ridge is subducted beneath an island arc: The Island Arc, v. 6, p. 361–371, doi: 10.1111/j.1440-1738.1997.tb00046.x.

Osozawa, S., 1998, Major transform duplexing along the eastern margin of Cretaceous Eurasia, *in* Flower, M.F.J., et al., eds., Mantle dynamics and plate interactions in East Asia: Washington, D.C., American Geophysical Union, Geodynamics Series 27, p. 245–257.

Parman, S.W., and Grove, T.L., 2005, Komatiites in the plume debate, *in* Foulger, G., et al., eds., Plates, plumes, and paradigms: Boulder, Colorado, Geological Society of America Special Paper 388, p. 249–256, doi:10.1130/2005.2388(15).

Rea, D.K., and Dixon, J.M., 1983, Late Cretaceous and Paleogene tectonic evolution of the North Pacific Ocean: Earth and Planetary Science Letters, v. 65, p. 145–166, doi: 10.1016/0012-821X(83)90196-6.

Regelous, M., Hofmann, A.W., Abouchami, W., and Galer, S.J., 2003, Geochemistry of lavas from the Emperor seamounts, and the geochemical evolution of Hawaiian magmatism from 85 to 42 Ma: Journal of Petrology, v. 44, p. 113–140, doi: 10.1093/petrology/44.1.113.

Reynaud, C., Jaillard, E., Lapierre, H., Mamberti, M., and Mascle, G.H., 1999, Oceanic plateau and island arcs of southwestern Ecuador: Their place in the geodynamic evolution of northwestern South America: Tectonophysics, v. 307, p. 235–254, doi: 10.1016/S0040-1951(99)00099-2.

Richards, M.A., and Lithgow-Bertelloni, C., 1996, Plate motion changes, the Hawaiian-Emperor bend, and the apparent success and failure of geodynamic models: Earth and Planetary Science Letters, v. 137, p. 19–27, doi: 10.1016/0012-821X(95)00209-U.

Riisager, P., Hall, S., Antretter, M., and Zhao, X., 2003, Paleomagnetic paleolatitude of Early Cretaceous Ontong Java Plateau basalts: Implications for Pacific apparent and true polar wander: Earth and Planetary Science Letters, v. 208, p. 235–252, doi: 10.1016/S0012-821X(03)00046-3.

Sager, W.W., 2002, Basalt core paleomagnetic data from Ocean Drilling Program site 883 on Detroit Seamount, northern Emperor seamount chain, and implications for the paleolatitude of the Hawaiian hotspot: Earth and Planetary Science Letters, v. 199, p. 347–358, doi: 10.1016/S0012-821X(02)00590-3.

Sager, W.W., 2005, What built Shatsky Rise, a mantle plume or ridge tectonics? *in* Foulger, G., et al., eds., Plates, plumes, and paradigms: Boulder, Colorado, Geological Society of America Special Paper 388, p. 721–733, doi:10.1130/2005.2388(41).

Sager, W.W., Duncan, R.A., and Handschumacher, D.W., 1993, Paleomagnetism of the Japanese and Marcus-Wake seamounts, western Pacific Ocean, *in* Pringle, M.S., et al., eds., The Mesozoic Pacific: Geology, tectonics and volcanism: Washington, D.C., American Geophysical Union, Geophysical Monograph 77, p. 401–435.

Schellart, W.P., Lister, G.S., and Toy, V.G., 2006, A Late Cretaceous and Cenozoic reconstruction of the Southwest Pacific region: Tectonics controlled by subduction and slab rollback processes: Earth-Science Reviews, v. 76, p. 191–233, doi: 10.1016/j.earscirev.2006.01.002.

Scheuber, E., and Gonzalez, G., 1999, Tectonics of the Jurassic–Early Cretaceous magmatic arc of the north Chilean coastal Cordillera (22°–26°S): A story of crustal deformation along a convergent plate boundary: Tectonics, v. 18, p. 895–910, doi: 10.1029/1999TC900024.

Scotese, C.R., Gahagan, L.M., and Larson, R.L., 1988, Plate reconstruction of the Cretaceous and Cenozoic ocean basins: Tectonophysics, v. 155, p. 27–48, doi: 10.1016/0040-1951(88)90259-4.

Sdrolias, M., Müller, R.D., and Gaina, C., 2003, Tectonic evolution of the southwest Pacific using constraints from backarc basins: Sydney, Geological Society of Australia, Special Paper 372, p. 343–359.

Seno, T., and Maruyama, S., 1984, Paleogeographic reconstruction and origin of the Philippine Sea: Tectonophysics, v. 102, p. 53–84, doi: 10.1016/0040-1951(84)90008-8.

Shapiro, M.N., Soloviev, A.V., and Ledneva, G.V., 2006, Is there any relation between the Hawaiian-Emperor seamount chain bend at 43 Ma and the evolution of the Kamchatka continental margin?: http://www.mantleplumes.org.

Sharp, W.D., and Clague, D.A., 1999, A new older age of ~47 Ma for the Hawaiian-Emperor bend: Eos (Transactions, American Geophysical Union), v. 90, p. F1196.

Sharp, W.D., and Clague, D.A., 2006, 50-Ma initiation of Hawaiian-Emperor bend records major change in Pacific plate motion: Science, v. 313, p. 1281–1284, doi: 10.1126/science.1128489.

Silberling, N.J., Jones, D.L., Blake, M.C., and Howell, D.G., 1987, Lithotectonic terrane map of the western conterminous United States, *in* Silberling, N.J., and Jones, D.L., eds., Lithotectonic terrane map of the North American Cordillera: Reston, Virginia, U.S. Geological Survey, Map 1874C, scale 1:2,500,000, 43 p.

Sinton, C.W., Duncan, R.A., Storey, M., Lewis, J., and Estrada, J.J., 1998, An oceanic flood basalt province within the Caribbean plate: Earth and Planetary Science Letters, v. 155, p. 221–235, doi: 10.1016/S0012-821X(97)00214-8.

Smith, A.D., 2003, A reappraisal of stress field and convective roll models for the origin and distribution of Cretaceous to Recent intraplate volcanism in the Pacific basin: International Geology Review, v. 45, p. 287–302.

Smith, A.D., 2005, The streaky-mantle alternative to mantle plumes and its bearing on bulk-Earth geochemical evolution, *in* Foulger, G., et al., eds., Plates, plumes, and paradigms: Boulder, Colorado, Geological Society of America Special Paper 388, p. 303–325, doi:10.1130/2005.2388(19).

Smith, W.H.F., and Sandwell, D., 1997, Global sea floor topography from satellite altimetry and sparse shipboard bathymetry: Science, v. 277, p. 1956–1961, doi: 10.1126/science.277.5334.1956.

Soler, P., Carlier, G., and Marocco, R., 1989, Evidence for the subduction and underplating of an oceanic plateau beneath the south Peruvian margin during the Late Cretaceous: Structural implications: Tectonophysics, v. 163, p. 13–24, doi: 10.1016/0040-1951(89)90114-5.

Soloviev, A., Garver, J.I., and Ledneva, G., 2006, Cretaceous accretionary complex related to Okhotsk-Chukotka suduction, Omgon Range, western Kamchatka, Russian Far East: Journal of Asian Earth Sciences, v. 27, p. 437–453, doi: 10.1016/j.jseaes.2005.04.009.

Spörli, K.B., and Ballance, P.F., 1989, Mesozoic ocean floor /continent interaction and terrane configuration, southwest Pacific area around New Zealand, *in* Ben-Avraham, Z., ed., The evolution of the Pacific Ocean margins: New York, Oxford University Press, p. 176–190.

Stuart, W.D., Foulger, G.R., and Barall, M., 2007 (this volume), Propagation of the Hawaiian-Emperor volcano chain by Pacific plate cooling stress, *in* Foulger, G.R., and Jurdy, D.M., eds., Plates, plumes, and planetary processes: Boulder, Colorado, Geological Society of America Special Paper 430, doi: 10.1130/2007.2430(24).

Sun, C.-H., Smith, A.D., and Chen, C.-H., 1998, Nd-Sr isotopic and geochemical evidence on the protoliths of exotic blocks in the Juisui area, Yuli Belt, Taiwan: International Geology Review, v. 40, p. 1076–1087.

Sutherland, R., and Hollis, C., 2001, Cretaceous demise of the Moa plate and strike-slip motion at the Gondwana margin: Geology, v. 29, p. 279–282, doi: 10.1130/0091-7613(2001)029<0279:CDOTMP>2.0.CO;2.

Tagami, T., and Hasebe, N., 1999, Cordilleran-type orogeny and episodic growth of continents: Insights from the circum-Pacific continental margins: The Island Arc, v. 8, p. 206–217, doi: 10.1046/j.1440-1738.1999.00232.x.

Taira, A., Tokuyama, H., and Soh, W., 1989, Accretion tectonics and evolution of Japan, *in* Ben-Avraham, Z., ed., The evolution of the Pacific Ocean margins: New York, Oxford University Press, p. 100–123.

Takashima, R., Nishi, H., and Yoshida, T., 2002, Geology, petrology and tectonic setting of the Late Jurassic ophiolite in Hokkaido, Japan: Journal of Asian Earth Sciences, v. 21, p. 197–215, doi: 10.1016/S1367-9120(02)00028-7.

Tarduno, J.A., and Cottrell, R.D., 1997, Paleomagnetic evidence for motion of the Hawaiian hotspot during formation of the Emperor seamounts: Earth and Planetary Science Letters, v. 153, p. 171–180, doi: 10.1016/S0012-821X(97)00169-6.

Tarduno, J.A., and Sager, W.W., 1995, Polar standstill of the mid-Cretaceous Pacific plate and its geodynamic implications: Science, v. 269, p. 956–959.

Tarduno, J.A., Sliter, W.V., Kroenke, L., Leckie, M., Mayer, H., Mahoney, J.J., Musgrave, R., Storey, M., and Winterer, E.L., 1991, Rapid formation of Ontong Java plateau by Aptian mantle plume volcanism: Science, v. 254, p. 399–403, doi: 10.1126/science.254.5030.399.

Tardy, M., Lapierre, H., Freydier, C., Coulon, C., Gill, J.-B., Mercier de Lepinay, B., Beck, C., Martinez R.J., Talavera, M.O., Ortiz, H.E., Stein, G., Bourdier, J.-L., and Yta, M., 1994, The Guerrero suspect terrane (western Mexico) and coeval arc terranes (the Greater Antilles and the Western Cordillera of Colombia): A late Mesozoic intra-oceanic arc accreted to cratonal America during the Cretaceous: Tectonophysics, v. 230, p. 49–73, doi: 10.1016/0040-1951(94)90146-5.

Taylor, B., 2006, The single largest oceanic plateau: Ontong Java–Manahiki–Hikurangi: Earth and Planetary Science Letters, v. 241, p. 372–380, doi: 10.1016/j.epsl.2005.11.049.

Tebbens, S.F., and Cande, S.C., 1997, Southeast Pacific tectonic evolution from early Oligocene to Present: Journal of Geophysical Research, v. 102, p. 12,061–12,084, doi: 10.1029/96JB02582.

Thorkelson, D., and Smith, A.D., 1989, Arc and intraplate volcanism in the Spences Bridge Group: Implications for Cretaceous tectonics in the Canadian Cordillera: Geology, v. 17, p. 1093–1096, doi: 10.1130/0091-7613(1989)017<1093:AAIVIT>2.3.CO;2.

Turcotte, D.L., and Oxburgh, E.R., 1976, Stress accumulation in the lithosphere: Tectonophysics, v. 35, p. 183–199, doi: 10.1016/0040-1951(76)90037-8.

Ueda, H., and Miyashita, S., 2005, Tectonic accretion of a subducted intraoceanic remnant arc in Cretaceous Hokkaido, Japan, and implications for evolution of the Pacific northwest: The Island Arc, v. 14, p. 582–598, doi: 10.1111/j.1440-1738.2005.00486.x.

Uyeda, S., and Miyashiro, A., 1974, Plate tectonics and the Japanese islands: A synthesis: Geological Society of America Bulletin, v. 85, p. 1159–1170, doi: 10.1130/0016-7606(1974)85<1159:PTATJI>2.0.CO;2.

Viso, R.F., Larson, R.L., and Pockalny, R.A., 2005, Tectonic evolution of the Pacific-Phoenix-Farallon triple junction in the South Pacific Ocean: Earth and Planetary Science Letters, v. 233, p. 179–194, doi: 10.1016/j.epsl.2005.02.004.

Vlastelic, I., Dosso, L., Guillou, H., Bougault, H., Geli, L., Etoubleau, J., and Joron, J.L., 1998, Geochemistry of the Hollister Ridge: Relation with the Louisville hotspot and the Pacific-Antarctic ridge: Earth and Planetary Science Letters, v. 160, p. 777–793, doi: 10.1016/S0012-821X(98)00127-7.

Walcott, R.I., 1976, Lithospheric flexure, analysis of gravity anomalies, and the propagation of seamount chains, *in* Manghnani, M.H., and Moberly, R., eds., The geophysics of the Pacific Ocean Basin and its margin: Washington, D.C., American Geophysical Union, Geophysical Monograph 19, p. 431–438.

Watson, B.F., and Fujita, K., 1985, Tectonic evolution of Kamchatka and the Sea of Okhotsk and implications for the Pacific basin, *in* Howell, D.G., ed., Tectonostratigraphic terranes of the Circum Pacific region: Houston, Circum Pacific Council for Energy and Mineral Resources, Earth Science Series v. 1, p. 333–348.

Watts, A.B., Weissel, J.K., Duncan, R.A., and Larson, R.L., 1988, Origin of the Louisville Ridge and its relationship to the Eltanin Fracture Zone system: Journal of Geophysical Research, v. 93, p. 3051–3077.

Wells, R.E., 1989, Origin of the oceanic basement of the Solomon Islands arc and its relationship to the Ontong Java plateau—Insights from Cenozoic plate motion models: Tectonophysics, v. 165, p. 219–235, doi: 10.1016/0040-1951(89)90048-6.

Windom, K.E., Seifert, K.E., and Vallier, T.L., 1981, Igneous evolution of Hess Rise: Petrologic evidence from DSDP leg 62: Journal of Geophysical Research, v. 86, p. 6311–6322.

Winterer, E.L., 1976, Anomalies in the tectonic evolution of the Pacific, *in* Sutton, G.H., Manghnani, M.H., and Moberly, R., eds., The geophysics of the Pacific Ocean Basin and its margins: Washington, D.C., American Geophysical Union, Geophysical Monograph 19, p. 269–278.

Winterer, E.L., and Nakanishi, M., 1995, Evidence for a plume-augmented, abandoned spreading centre on Ontong Java plateau: Eos (Transactions, American Geophysical Union), v. 76, p. F617.

Winterer, E.L., and Sager, W.W., 1995, Synthesis of drilling results from the Mid-Pacific Mountains: Regional context and implications, *in* Winterer, E.L., et al., eds., Proceedings of the Ocean Drilling Program, Scientific Results, v. 143, p. 497–535.

Woods, M.T., and Davies, G.F., 1982, Late Cretaceous genesis of the Kula plate: Earth and Planetary Science Letters, v. 58, p. 161–166, doi: 10.1016/0012-821X(82)90191-1.

Worthington, T.J., Hekinian, R., Stoffers, P., Kuhn, T., and Hauff, F., 2006, Osbourn Trough: Structure, geochemistry and implications of a mid-Cretaceous paleospreading ridge in the South Pacific: Earth and Planetary Science Letters, v. 245, p. 685–701, doi: 10.1016/j.epsl.2006.03.018.

Yáñez, G., Cembrano, J., Pardo, M., Ranero, C., and Selles, D., 2002, The Challenger–Juan Fernández–Maipo major tectonic transition of the Nazca-Andean subduction system at 33–34°S: Geodynamic evidence and implications: Journal of South American Earth Sciences, v. 15, p. 23–38, doi: 10.1016/S0895-9811(02)00004-4.

Manuscript Accepted by the Society January 31, 2007

DISCUSSION

27 December 2006, Keith H. James

Figure 2B of Smith (this volume) shows the Caribbean Great Arc above west-dipping subduction. If one believes the Caribbean plate came from the Pacific, then the literature shows this arc above east-dipping subduction. Arrival of the Caribbean plateau from the west choked the zone in the Albian, causing subduction reversal.

Figure 2D (Smith, this volume) indicates formation of the Caribbean plate in the late Cretaceous. Obducted Jurassic crust around the Caribbean is taken to indicate Late Jurassic–Early Cretaceous spreading. This would accord with Atlantic and Gulf of Mexico opening. The Caribbean plateau seems to have formed during several pulses: 130–120 Ma, 90–88 Ma, and ca. 78–76 Ma. It is centered on the Beata Ridge. Analogy with the Scotia plate (West Scotia Ridge) suggests that this was the locus of early Atlantic–Caribbean spreading between the Americas (James, 2005a). In this model the Caribbean plate and its plateau formed in place.

Figure 2D (Smith, this volume) and the text indicate backarc spreading behind the (western) "Caribbean Great Arc." The Scotia analogy (East Scotia Ridge) would suggest back-arc spreading centered on the Aves Ridge, behind the Lesser Antilles (James, 2005a).

I have used analogy with Iceland and Manihiki to explain Caribbean plateau thickening above the Beata Ridge. Association of Ontong Java with a ridge, noted by Smith (this volume), corroborates this. The Iceland model also suggests a mechanism of Caribbean plate definition.

The Great Caribbean Arc concept holds that the Aruba–Blanquilla island chain, the Lesser Antilles, the Aves Ridge, and the Greater Antilles from Cuba to the Virgin Islands formed in the Pacific as a north-south–trending intraoceanic arc ca. 130–120 Ma (e.g., Bouysse et al., 1990). As this entered the Caribbean area, the northern and southern parts became distributed along the boundaries of the Caribbean and on Cuba. There are major problems with this hypothesis.

First, in order for the Caribbean to migrate into place, the continental block of Chortis needs to be elsewhere. It is supposed to have rotated from southwestern Mexico and, in a process never explained, to have jumped onto the rear end of the migrating plate. The geologies of Chortis and southwest Mexico are not compatible (Keppie and Morán-Zenteno, 2005). Jurassic rift lineaments on Chortis (Guayape-Patuca faults) remain parallel to regional (inter-American) coeval structures, showing that no rotation has occurred (James, 2006a,b). Chortis has always been on the western end of the Caribbean plate.

Crustal thicknesses, gravity data, high-silica ignimbrites, xenoliths, and quartz sands also indicate continental blocks below southern Central America. They are covered by thrusted Mesozoic oceanic and volcanic rocks and Cenozoic volcanic/sedimentary rocks. The blocks are among continental fragments dispersed around the Caribbean during Jurassic–Early Cretaceous rift/drift. The Scotia analogy illustrates this, but more importantly, I have compiled a variety of data that indicate that the Caribbean is at least one-third continental (http://www.mantleplumes.org/WebDocuments/CaribbeanPlateau.pdf). The presence of continent below the whole of Central America, on the western end of the Caribbean plate, denies any plate migration through the area.

The idea that southern Central America is intraoceanic has provoked inverted reasoning. Vogel et al. (2004) attribute high-silica rocks in Costa Rica to "continentalization" while at the same time noting crustal thickness of 40 km, low seismic velocities, and geochemical similarity to continental rocks. The same explanation is given for a Santonian tonalite on Aruba, "which was in the Pacific at that time" (White et al., 1999). Andesites are used to signify continental crustal input (original definition and my textbooks). Their presence in southern Central America is used to call this into doubt in recent texts.

Second, supposedly allochthonous (Pacific-derived) elements of the arc found along northern South America show gradational geological continuity with continental equivalents to the south. They bear chemical evidence of continental input as far back as the Albian, long before the "Great Arc" is supposed to have approached the Caribbean area. Elements of the "Great Arc" along northern South America and on Cuba carry Paleozoic zircons.

My understanding is that there never was a "Caribbean Great Arc" in the Pacific. Instead, volcanic activity occurred around and within the Caribbean plate. It probably began along with spreading (Jurassic volcaniclastic rocks in Costa Rica, upper Jurassic volcanic rocks on Cuba, Hispaniola, Puerto Rico, and La Désirade and volcaniclastic rocks at least as old as Albian in the northeastern Lesser Antilles).

Data I have compiled show that Caribbean volcanism paused in the Albian and Cenomanian and ceased (except for the free faces in the east and west) in the Middle Eocene. Each pause was accompanied by uplift to wavebase and development of an erosional unconformity covered by shallow marine limestones.

The approximate coincidence of the Albian and Cenomanian Caribbean episodes with plate reorganizations highlighted by Smith (this volume) is significant. The Middle Eocene Pacific/Caribbean episodes are exactly coeval. Around the Caribbean, mixtures of oceanic, volcanic, and continental-margin rocks were emplaced in continent-verging thrust systems (see James, 2005b, for a discussion). Ophiolite/arc nappes up to 5 km thick and hundreds of kilometers long occur in Cuba and Venezuela. This event appears to have been short and violent. It was a coeval, regional affair recorded in the Caribbean, Mexico, Colombia, Ecuador, and Peru. This was recognized as long ago as 1938 (Hess, 1938), but it is ignored by models that show the Caribbean plate diachronously interacting with north and south boundaries as it migrated eastward.

How much uplift occurred? Ocean floor and deep faunal assemblages ("radiolarian" cherts, pelagic forams) ended up on continental margins.

Some questions remain:

1. "Large regions of shallow mantle are close to or at the peridotite solidus. Intraplate volcanism is related to plate reorganizations, suggesting that it is controlled by fracturing and extension that allows melt to be released (Favela and Anderson, 2000; Foulger and Natland, 2003; Foulger, this volume)." The Caribbean is in an extensional setting: fracture zones in the Atlantic and Pacific diverge toward the Caribbean. The same is true east and west of the Scotia plate. I see decompression melting occurring in this extensional habitat. What caused extension in the Pacific? If extension results in decompression melting and plateau formation, what causes coeval convergence along plate margins? I wonder if change of density, once begun, results in runaway melting ("meltdown").

2. How confident does the author feel about modern geochemical distinctions of mid-ocean ridge basalt (MORB), intraplate, and margin products? I suspect that esoteric data (often without statistical analysis) are selected to fit preferred ideas. The sampling bias of dated rocks from Ontong Java, noted by Smith (this volume), makes me wonder if a similar bias enters into distinctions of oceanic igneous/volcanic rocks.

29 December 2006, Rex H. Pilger

There are a few points Smith (this volume) has made that are deserving of clarification:

1. Smith writes, "The final major change in the stress field occurred in the Late Oligocene as a result of breakup of the Farallon into the Cocos and Nazca plates, and caused a hiatus in Hawaiian volcanism, initiated the Sala y Gomez, Foundation, and Samoan chains, and terminated the Louisville chain." Pilger and Handschumacher (1981) showed that the Sala y Gomez chain may have originated from the same melting anomaly responsible for the Tuamotu and Nazca ridges. This would have occurred when southward motion of the anomaly relative to the overlying Pacific and Farallon plates resulted in the anomaly's displacement from beneath the spreading center, crossing under a transform fault, to be entirely under the Farallon plate (before its fragmentation).

2. Further, the Foundation chain may significantly predate fragmentation of the Farallon plate; its calculated older extension (to ca. 40 Ma) corresponds with isotopic ages from seamounts proximate to the Austral Islands (Pilger, 2003). Nevertheless, the paucity in seamounts between the Austral region and the Foundation chain could represent the stress mechanism history Smith proposes.

3. Smith writes,

The Juan Fernandez ridge was noted to coincide with a change in slab dip at 33°S. . . . In conventional models, the assumption of a plume origin for the volcanism has resulted in the changes in slab geometry being attributed to subduction of topographic features associated with the intraplate volcanism . . . the slab flexure at 33°S cannot result from subduction of the Juan Fernandez Ridge as the older section of the ridge curves northeast and intersects the convergent margin at 31°S.

Projection of the Juan Fernandez Ridge to the low-angle subduction segment beneath central Chile includes an elongate seamount at ~32.8°S, 73.7°W and intersection of the further projected ridge with the Chilean coast at ~32.5°S, supported by the calculated locus of Nazca plate motion relative to the "hotspot" (e.g., Pilger, 1984). In Figure D-1, bathymetry from Google Earth Plus (2006) shows the same feature originally interpreted as part of the Juan Fernandez "hotspot trace" and inferred to be genetically related to the low-angle subduction segment (also supported by the volcanic gap documented by published isotopic age dates; Pilger, 1984). The northeast-trending feature at ~31.5°S, 73.7°W could well be a paleofracture zone, as its orientation parallels other mapped fractures of the Easter plate, instead of a continuation of the Juan Fernandez trace, as Smith has apparently interpreted it.

4. Smith also noted that eastern extrapolation of the Sala y Gomez chain coincides with a change in slab geometry at

28°S. Because in his model the chain formed as a result of Farallon plate fragmentation, this extrapolation would be irrelevant. The melting anomaly model for the Sala y Gomez-Nazca trace has the same implications.

These observations remove some of the supporting evidence for Smith's model but do not invalidate it, especially if the stress mechanism is accompanied by sublithospheric melting anomalies as Beutel and Anderson (this volume) propose in their second model and has been argued for previously (Pilger and Handschumacher, 1981; see also Pilger, 2007).

15 January 2007, Alan Smith

The reconstruction of an arc fringing the Arperos basin in Figure 2 (Smith, this volume) was based on Tardy et al. (1994), who correlated the Guerrero terrane of Mexico with the Western Cordillera of Colombia on the basis of similar ages and magmatic evolution sequences. Tardy et al. (1994) also included correlations with the Greater Antilles (not used in the reconstruction), which suggests influence by the plume model. I therefore agree with James that the reconstruction contains elements reminiscent of models in which the Caribbean plateau originates over the Galapagos hotspot, although it was not intended to imply such an origin for the plateau. I have no objection to the model that James outlines in his comment. If Central America formed in situ, the Arperos basin would not extend south of the Guerrero terrane in Figure 2 (Smith, this volume).

The growth of the Pacific plate was controlled by the motion of the Izanagi, Farallon, and Phoenix plates until the Late Cretaceous. The latter were large, long-lived oceanic plates that were likely subject to strong slab-pull forces. Oceanic plateau formation occurred along ocean ridge systems as a result of melt focusing at triple junctions (Georgen and Lin, 2002) or entrainment of fertile mantle into the ocean ridge upwelling (Korenaga, 2005). The geochemistry of several plateaus indicates an EM1 source component, such that the fertile component may be equated with continental mantle or lower continental crust entrained in the shallow mantle (Smith and Lewis, 1999; Anderson, 2005). Formation of ocean island chains is envisaged to result from lithospheric fracturing allowing the escape of melts generated from shearing of volatile-rich sources in the asthenosphere (Smith and Lewis, 1999; Doglioni et al., 2005).

I consider there has been considerable bias in the focus of the plume model on island chains such as Hawaii and large igneous provinces such as the Ontong Java plateau. As noted by Okal and Batiza (1987) regarding the difficulty of applying hotspot models to the south-central Pacific: "one starts wondering whether hotspot theory would have emerged as it did, had more cases of Austral-type chains been documented (especially if located in more accessible regions) early in the game." Similarly, Natland and Winterer (2005) note the thousands of seamounts of the Pacific Ocean floor that do not show any regular

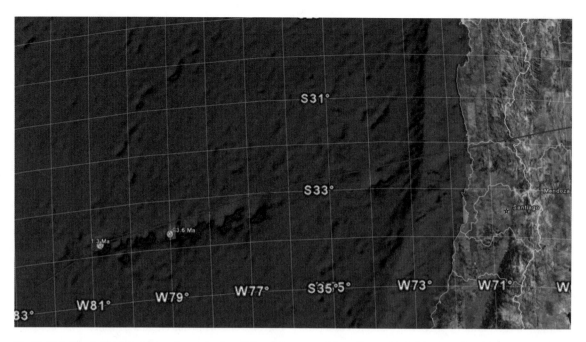

Figure D-1. Shaded bathymetry and topography of the region around Juan Fernandez Ridge, together with the calculated hotspot motion locus using the parameters of Pilger (2007) corrected to the timescale of Gradstein et al. (2004). Projection of the ridge through the seamount at ~32.8°S, 73.7°W would produce a coastal intersection near 32.5°S. Age date citations are in Pilger (2003).

distribution but are capped by OIB-type basalts. Such features suggest the sources of OIB are widely distributed throughout the shallow mantle, and the simplest way to accomplish this would be to mix subducted oceanic crust directly into the convecting mantle (Meibom and Anderson, 2004; Smith, 2005).

The Pilger and Handschumacher (1981) model attributes the Tuamotu plateau and Nazca Ridge to a Morgan-type hotspot that subsequently formed the Sala y Gomez chain. However, such a model does not explain differences in morphology between the Nazca Ridge and the Sala y Gomez chain (Woods and Okal, 1994). The Nazca Ridge may also extend to the Roggeveen Rise at 31°S, 91°W, 1000 km further southwest than commonly depicted, which would further preclude any relationship with the Sala y Gomez chain (Woods and Okal, 1994). A more complex scenario than a single melting anomaly may also be required from the Tuamotu plateau, where the identification of three sets of intersecting ridges may indicate protracted growth over several stages (Natland and Winterer, 2005).

Correlating the Austral and Foundation chains would produce a more linear age progression favorable to a plume model, but there is no volcanic record from 131°W to 140°W to support such a model. Instead, I suggest the Austral chain may be comprised of ridges aligned with the Line Islands and Puka-puka trends proposed for the Tuamotu plateau by Natland and Winterer (2005).

I agree with the coordinates given by Pilger for intersection of the Juan Fernandez chain with the continental margin. The track depicted in Figure 5 (Smith, this volume) should intersect the continental margin approximately 1° further south of the position shown, but this does not affect the interpretation of Pliocene–Recent volcanism along the chain being controlled by slab geometry.

The reorganization of the Pacific-Farallon Ridge at ca. 25 Ma has been attributed to the transmission of stresses from the convergent margin (Natland and Winterer, 2005). After the reorganization, the Nazca plate would have been stressed by the slab flexure into which I extrapolate the Sala y Gomez chain. The origin of the chain is thus related to both the plate reorganization and convergent margin geometry.

16 January 2007, Rex H. Pilger

In the context of the plate model, the differences in morphology between the Nazca Ridge and the Sala y Gomez chain are a result of the former having formed from a "hotspot" (or melting anomaly) located beneath a spreading center and the latter having formed within a single plate of the anomaly and as a result of intraplate stresses focused by the "hotspot" (as explained by Pilger and Handschumacher, 1981).

The Tuamotu Ridge, which has a similar morphology to its conjugate, the Nazca Ridge, is the only part of the Tuamotu "plateau" to have formed at a spreading center. The remainder of the complex consists of multiple isolated islands and seamounts indicative of formation within a plate. In the context of the plate model, a large fertile zone is required to explain the Tuamotu complex. However, the concentration of the islands, seamounts, and ridge on the south side of the Austral fracture zone is understandable given the significant difference in age and plate thickness (older and greater on the north side): Pressure release melting occurs progressively within the fertile zone as the fracture zone passes over it (Raddick et al., 2002; Pilger, 2003), and focused extensional stresses produce the Tuamotu and Nazca ridges (with part of the fertile zone extending beneath the spreading center).

The older seamounts near the Austral chain could represent the same fertile zone responsible for the Foundation chain. The gap between the Austral and Foundation chains would then be indicative of a period of compressive stress that did not tap the melting anomaly. The older seamounts near the Austral chain are closer to the oldest portions of the Foundation chain than they are to the Line Islands chain(s), and few older seamounts occur between them.

Implicit in this interpretation is recognition that the fertile zones responsible for the principal chains of the Pacific have not moved significantly with respect to one another (although they may be moving collectively relative to the spin axis and the melting anomalies of the Atlantic and Indian oceans; Pilger, 2003). The origin of the fertile zones remains uncertain; however, the kinematic arguments emphasized in my comments do not rule out either shallow or deep ("plume") origins; neither do they prefer either mechanism. Nevertheless, the importance of intraplate stresses in controlling magmatism still appears to be significant.

21 January 2007, Alan Smith

The term "hotspot" is misleading. Petrological evidence suggests that the sources of OIB are volatile-rich and melt within the temperature range of normal mantle (e.g., Green and Falloon, 2005; Falloon et al., this volume). Nor is there evidence for significant thermal anomalies from oceanic swells (e.g., De-Laughter et al., 2005).

An alternative model for the Tuamotu plateau is that initial volcanism occurred along a trend contemporaneous with the Line Islands. A second set of seamount ridges was then generated orthogonal to the NNW-oriented Pacific-Farallon spreading center (then along the Roggeveen and Mendoza rises; Mammerickx et al., 1980) akin to Puka-puka Ridges. The Pacific-Farallon Ridge then jumped westward and realigned NNE, with a third set of seamount ridges forming orthogonal to the new spreading center. The Tuamotu plateau is where the three sets of ridges intersect (Natland and Winterer, 2005). Such an origin is consistent with the observations of Ito et al. (1995) that the northwestern part of the plateau formed 600 km away from the Pacific-Farallon spreading center, although the southeastern part of the plateau may have formed closer to the spreading center (Okal and Cazenave, 1985).

Melting anomalies may appear quasifixed between different ocean basins without having a deep mantle origin. In a plate model, the sources of intraplate volcanism lie within the asthenosphere (e.g., Cuffaro and Doglioni, this volume). Motion between melting anomalies in the Pacific and Atlantic/Indian oceans is the result of differences in flow regime in the asthenosphere beneath the plates in these basins (Smith and Lewis, 1999). Changes in plate configuration will alter the drift of the melting anomalies. Rapid motion of melting anomalies may result from propagation of fractures.

22 January 2007, Don L. Anderson

Rex Pilger's comment of 16 January 2007 rightly notes that in the context of the plate model, a large fertile zone is required to explain some melting anomalies and that concentrations of islands and seamounts along fracture zones are understandable given a significant difference in age and plate thickness. This also applies to Hawaii (Winterer and Natland, comment of 11 January on chapter by Laske et al., this volume), particularly the peak in activity at the Molokai fracture zone. Pressure release melting occurs within the fertile zone as the fracture zone passes over it; fertile blobs also rise and melt beneath thin plates and spreading centers. I also agree that intraplate stress ultimately controls the location of magmatism and the access to fertile zones. In contrast to the plume model, the amount of melting depends on the fertility, or eclogite content, of the underlying mantle, and the duration or size of the melting anomaly depends on the volume of the fertile blob or streak, not the absolute temperature.

Pilger also points out that implicit in this interpretation is that the fertile zones responsible for the volcanic chains of the Pacific have not moved significantly with respect to one another (although they may move collectively relative to the spin axis and the melting anomalies of the Atlantic and Indian oceans). This is exactly the prediction of the asthenospheric counterflow model; embedded fertile blobs or streaks (from subducted aseismic ridges) under a given plate will move parallel to one another and antiparallel to the overriding plate. This is not the case for whole-mantle convection or for weak plumes. The origin of the fertile material is discussed in Foulger et al. (2005) and Anderson (this volume; comment of 20 January on Yamamoto et al., this volume). The sources include delaminated crust, trapped and abandoned mantle wedges, and subducted seamount chains; these account for the total volume of so-called hotspots (which Smith in his comment rightly points out are not hot; magma volume is not a proxy for high absolute temperature).

In the shallow counterflow model, the flow is directed away from lithospheric sinks and toward hotspots, not radially away from them as in the model of Yamamoto et al. (this volume); anisotropy should mimic plate motions and not be radial to hotspots (Fig. D-2). The flow lines in Figure D-2 are consistent with the anisotropy and can be compared with those in Yamamoto et al. (this volume).

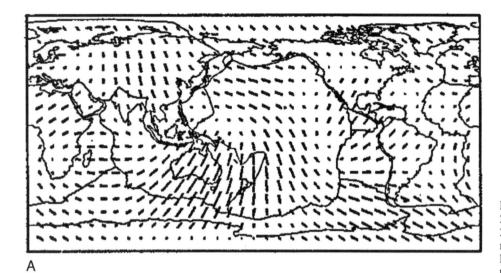

A

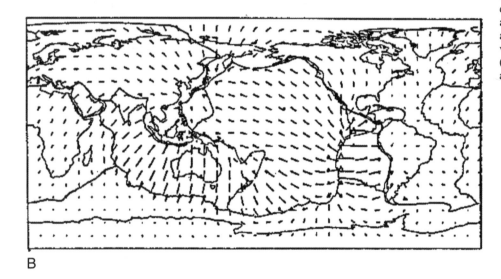

B

Figure D-2. (A) Azimuthal anisotropy of 200-s Rayleigh waves. (B) Flow lines at 260 km depth for the upper-mantle kinematic flow model. This model includes a low-viscosity channel in the upper mantle (see Chapter 15 http://caltechbook .library.caltech.edu/14/). Fertile blobs embedded in the counterflow will define a fixed reference system for each plate and will move slowly compared to plate velocities, consistent with observations (Wang and Liu, 2006). The vectors in B are about one-fifth of the plate velocity.

REFERENCES CITED

Anderson, D.L., 2005, Large igneous provinces, delamination, and fertile mantle: Elements, v. 1, p. 271–275.

Anderson, D.L., 2007 (this volume), The eclogite engine: Chemical geodynamics as a Galileo thermometer, *in* Foulger, G.R., and Jurdy, D.M., eds., Plates, plumes, and planetary processes: Boulder, Colorado, Geological Society of America Special Paper 430, doi: 10.1130/2007.2430(03).

Beutel, E., and Anderson, D.L., 2007 (this volume), Ridge-crossing seamount chains: A nonthermal approach, *in* Foulger, G.R., and Jurdy, D.M., eds., Plates, plumes, and planetary processes: Boulder, Colorado, Geological Society of America Special Paper 430, doi: 10.1130/2007.2430(19).

Bouysse, P., Westercamp, D., and Andreieff, P. 1990, The Lesser Antilles Island Arc: Proceedings of the Ocean Drilling Program, Scientific Results, v. 110, p. 29–44.

Cuffaro, M., and Doglioni, C., 2007 (this volume), Global kinematics in deep versus shallow hotspot reference frames, *in* Foulger, G.R., and Jurdy, D.M., eds., Plates, plumes, and planetary processes: Boulder, Colorado, Geological Society of America Special Paper 430, doi: 10.1130/2007.2430(18).

DeLaughter, J., Stein, C.A., and Stein, S., 2005, Hotspots: A view from the swells, in Foulger, G.R., et al., eds., Plates, plumes, and paradigms: Boulder, Colorado, Geological Society of America Special Paper 388, p. 257–278.

Doglioni, C., Green, D.H., and Mongelli, F., 2005, On the shallow origin of hotspots and the westward drift of the lithosphere, *in* Foulger, G.R., et al., eds., Plates, plumes, and paradigms: Boulder, Colorado, Geological Society of America Special Paper 388, p. 735–749, doi:10.1130/2005.2388(42).

Falloon, T., Green, T.H., and Danushevsky, L.V., 2007 (this volume), Crystallization temperatures of tholeiite parental liquids: Implications for the existence of thermally driven mantle plumes, *in* Foulger, G.R., and Jurdy, D.M., eds., Plates, plumes, and planetary processes: Boulder, Colorado, Geological Society of America Special Paper 430, doi: 10.1130/2007 .2430(12).

Favela, J., and Anderson, D.L., 1999, Extensional tectonics and global volcanism, in Boschi, E., Ekstrom, G., and Morelli, A., eds., Editrice Compositori, Volume Problems in Geophysics for the New Millenium: Bologna, Italy, p. 463–498.

Foulger, G.R., 2007 (this volume), The plate model for the genesis of melting anomalies, *in* Foulger, G.R., and Jurdy, D.M., eds., Plates, plumes, and planetary processes: Boulder, Colorado, Geological Society of America Special Paper 430, doi: 10.1130/2007.2430(01).

Foulger, G.R. and Natland, J.H., 2003, Is "hotspot" volcanism a consequence of plate tectonics?: Science v. 300, p. 921–922.

Foulger, G.R., Natland, J.H., and Anderson, D.L., 2005, A source for Icelandic magmas in remelted Icelandic crust: Journal of Volcanology and Geothermal Research, v. 141, p. 23–44.

Georgen, J.E., and Lin, J., 2002, Three-dimensional passive flow and temperature structure beneath oceanic ridge-ridge-ridge triple junctions: Earth and Planetary Science Letters, v. 204, p. 115–132.

Google Earth Plus, 2006, Version 4.0 1657 (beta): http://earth.google.com/.

Gradstein, F.M., Ogg, J.G., and Smith, A.G., eds., 2004, Geologic time scale 2004: Cambridge, Cambridge University Press, 610 p.

Green, D.H., and Falloon, T.J., 2005, Primary magmas at mid-ocean ridges, "hotspots," and other intraplate settings: Constraints on mantle potential temperature, *in* Foulger, G.R., et al., eds., Plates, plumes, and paradigms: Boulder, Colorado, Geological Society of America Special Paper 388, p. 217–247, doi: 10.1130/2005.2388(14).

Hess, H.H., 1938, Gravity anomalies and island arc structure with particular reference to the West Indies: American Philosophical Society Proceedings, v. 79, p. 71–96.

Ito, G., McNutt, M., and Gibson, R.L., 1995, Crustal structure of the Tuamotu Plateau, 15oS, and implications for its origin: Journal of Geophysical Research, v. 100, p. 8097–8114.

James, K.H., 2005a, A simple synthesis of Caribbean geology: Transactions, 16th Caribbean Geological Conference, Barbados: Caribbean Journal of Earth Sciences, v. 39, p. 71–84.

James, K.H., 2005b, Palaeocene to middle Eocene flysch-wildflysch deposits of the Caribbean area: A chronological compilation of literature reports, implications for tectonic history and recommendations for further investigation: Transactions, 16th Caribbean Geological Conference, Barbados: Caribbean Journal of Earth Sciences, v. 39, p. 29–46.

James, K.H., 2006a, Arguments for and against the Pacific origin of the Caribbean Plate: Discussion, finding for an inter-American origin, *in* Iturralde-Vinent, M.A., and Lidiak, E.G., eds., Caribbean plate tectonics: Geologica Acta, v. 4, p. 279–302.

James, K.H., 2006b, Structural Geology, *in* Bundschuh, J., and Alvarado, G.E., eds., Central America: Geology, Resources and Hazards, Leiden, The Netherlands, Taylor & Francis, p. 277–321.

Keppie, J.D., and Morán-Zenteno, D.J., 2005, Tectonic implications of alternative Cenozoic reconstructions for Southern Mexico and the Chortis block: International Geology Review, v. 47, p. 473–491.

Korenaga, J., 2005, Why did not the Ontong Java plateau form subaerially?: Earth and Planetary Science Letters, v. 234, p. 385–399.

Laske, G., Phipps Morgan, J., and Orcutt, J.A., 2007 (this volume), The Hawaiian SWELL pilot experiment—Evidence for lithosphere rejuvenation from ocean bottom surface wave data, *in* Foulger, G.R., and Jurdy, D.M., eds., Plates, plumes, and planetary processes: Boulder, Colorado, Geological Society of America Special Paper 430, doi: 10.1130/2007.2430(11).

Mammerickx, J., Herron, E., and Dorman, L., 1980, Evidence for two fossil spreading ridges in the southeast Pacific: Geological Society of America Bulletin, v. 91, p. 263–271.

Meibom, A., and Anderson, D.L., 2004, The statistical upper mantle assemblage: Earth and Planetary Science Letters, v. 217, p. 123–139.

Natland, J.H., and Winterer, E.L., 2005, Fissure control on volcanic action in the Pacific, *in* Foulger, G.R., et al., eds., Plates, plumes, and paradigms: Boul-

der, Colorado, Geological Society of America Special Paper 388, p. 687–710, doi: 10.1130/2005.2388(39).

Okal, E.A., and Batiza, R., 1987, Hotspots: The first 25 years, *in* Keating, B.H., et al., eds., Seamounts, islands, and atolls: Washington, D.C., American Geophysical Union Geophysical Monograph 43, p. 1–11.

Okal, E.A., and Cazenave, A., 1985, A model for the plate tectonic evolution of the east-central Pacific based on Seasat investigations: Earth and Planetary Science Letters, v. 72, p. 99–116.

Pilger, R.H., 1984, Cenozoic plate kinematics, subduction and magmatism: South American Andes: Geological Society of London Journal, v. 141, p. 794–802.

Pilger, R.H., 2003, Geokinematics: Prelude to geodynamics: Berlin, Springer-Verlag, 338 p.

Pilger, R.H., 2007, The Bend: Origin and significance, Geological Society of America Bulletin, v. 119, p. 302–313.

Pilger, R.H., and Handschumacher, D.W., 1981, The fixed hotspot hypothesis and origin of the Easter–Sala y Gomez trace: Geological Society of America Bulletin, v. 92, p. 437–446.

Raddick, M.J., Parmentier, E.M., and Scheirer, D.S., 2002, Buoyant decompression melting: A possible mechanism for intraplate volcanism: Journal of Geophysical Research, v. 107, no. B10, art no. 2228, doi: 10.1029/2001JB000617.

Smith, A.D., 2005, The streaky-mantle alternative to mantle plumes and its bearing on bulk-Earth geochemical evolution, *in* Foulger, G.R., et al., eds., Plates, plumes, and paradigms: Boulder, Colorado, Geological Society of America Special Paper 388, p. 303–325, doi:10.1130/2005.2388(19).

Smith, A.D., 2007 (this volume), A plate model for Jurassic to Recent intraplate volcanism in the Pacific Ocean basin, *in* Foulger, G.R., and Jurdy, D.M., eds., Plates, plumes, and planetary processes: Boulder, Colorado, Geological Society of America Special Paper 430, doi: 10.1130/2007.2430(23).

Smith, A.D., and Lewis, C., 1999, The planet beyond the plume hypothesis: Earth Science Reviews, v. 48, p. 135–182.

Tardy, M., Lapierre, H., Freydier, C., Coulon, C., Gill, J.-B., Mercier De Lepinay, B., Beck, C., Martinez, R.J., Talavera, M.O., Ortiz, H.E., Stein, G., Bourdier, J.-L., Yta, M. 1994, The Guerrero suspect terrane (western Mexico) and coeval arc terranes (the Greater Antilles and the Western Cordillera of Colombia): A late Mesozoic intra-oceanic arc accreted to cratonal America during the Cretaceous: Tectonophysics, v. 230, p. 49–73.

Vogel, T.A., Patino, L.C., Guillermo, G.E., and Gans, P.B, 2004, Silicic ignimbrites within the Costa Rican volcanic front: evidence for the formation of continental crust: Earth and Planetary Science Letters, v. 226, p. 149–159.

Wang, S., and Liu, M., 2006. Moving hotspots or reorganized plates?: Geology, v. 34, p. 465–468.

White, R.V., Tarney, J., Kerr, A.C., Saunders, A.D., Kempton, P.D., Pringle, M.S., and Klaver, G.T., 1999, Modification of an oceanic plateau, Aruba, Dutch Caribbean: Implications for the generation of continental crust: Lithos, v. 46, p. 43–68.

Woods, M.T., and Okal, E., 1994, The structure of the Nazca ridge and the Sala y Gomez seamount chain from dispersion of Rayleigh waves: Geophysical Journal International, v. 117, p. 205–222.

Yamamoto, M., Morgan, W.J., Phipps Morgan, J., 2007 (this volume), Global plume-fed asthenosphere flow—II: Application to the geochemical segmentation of mid-ocean ridges, *in* Foulger, G.R., and Jurdy, D.M., eds., Plates, plumes, and planetary processes: Boulder, Colorado, Geological Society of America Special Paper 430, doi: 10.1130/2007.2430(10).

The Geological Society of America
Special Paper 430
2007

Propagation of the Hawaiian-Emperor volcano chain by Pacific plate cooling stress

William D. Stuart*
U.S. Geological Survey, Menlo Park, California 94025, USA
Gillian R. Foulger
Department of Earth Sciences, Durham University, Durham DH1 3LE, UK
Michael Barall
U.S. Geological Survey, Menlo Park, California 94025, USA

ABSTRACT

The lithosphere crack model, the main alternative to the mantle plume model for age-progressive magma emplacement along the Hawaiian-Emperor volcano chain, requires the maximum horizontal tensile stress to be normal to the volcano chain. However, published stress fields calculated from Pacific lithosphere tractions and body forces (e.g., subduction pull, basal drag, lithosphere density) are not optimal for southeast propagation of a stress-free, vertical tensile crack coincident with the Hawaiian segment of the Hawaiian-Emperor chain. Here we calculate the thermoelastic stress rate for present-day cooling of the Pacific plate using a spherical shell finite element representation of the plate geometry. We use observed seafloor isochrons and a standard model for lithosphere cooling to specify the time dependence of vertical temperature profiles. The calculated stress rate multiplied by a time increment (e.g., 1 m.y.) then gives a thermoelastic stress increment for the evolving Pacific plate. Near the Hawaiian chain position, the calculated stress increment in the lower part of the shell is tensional, with maximum tension normal to the chain direction. Near the projection of the chain trend to the southeast beyond Hawaii, the stress increment is compressive. This incremental stress field has the form necessary to maintain and propagate a tensile crack or similar lithosphere flaw and is thus consistent with the crack model for the Hawaiian volcano chain.

Keywords: Pacific plate, Hawaiian-Emperor, thermoelastic

INTRODUCTION

The two main explanations for the Hawaiian-Emperor volcano chain are mantle plume models (Morgan, 1971; Sleep, 1992) and lithosphere propagating crack models (Dana, 1849; Jackson and Shaw, 1975; Clague and Dalrymple, 1989). Both require northwest motion of the Pacific plate with respect to Hawaii (Fig. 1). In plume models, the plume top is the magma source, and the observed age-progressive sequence of volcanoes is produced by a buoyant plume column that is fixed with respect to the deep mantle. In crack models, a vertical tensile crack through the lithosphere coincides with the Hawaiian volcano

*E-mail: stuart@usgs.gov.

Stuart, W.D., Foulger, G.R., and Barall, M., 2007, Propagation of the Hawaiian-Emperor volcano chain by Pacific plate cooling stress, *in* Foulger, G.R., and Jurdy, D.M., eds., Plates, plumes, and planetary processes: Geological Society of America Special Paper 430, p. 497–506, doi: 10.1130/2007.2430(24). For permission to copy, contact editing@geosociety.org. ©2007 The Geological Society of America. All rights reserved.

chain, and magma is produced by decompression melting at depth near the crack-tip stress concentration, currently at Hawaii. The location of this stress concentration depends on the plate-wide stress field, which in turn is controlled by tractions and body forces acting on the plate as a whole. Because such a stress field is approximately stationary with respect to the boundary around the entire circumference of the plate, it follows that the chain of progressively older volcanoes to the northwest results from motion of Pacific plate material relative to the plate's evolving boundary. In other words, the preferred reference frame for Pacific plate motion is not the mantle, but instead is the slowly changing boundary of the Pacific plate.

Propagating crack models for the Hawaiian chain have not, to date, been developed beyond qualitative form, but one can infer the symmetry that the model's driving stress field must have. If the presently propagating crack lacks shear strength, cuts the entire lithosphere, and coincides with part or all of the Hawaiian chain, the chain straightness implies the existence of a plate stress field symmetric to the crack plane in its neighborhood. This is because such a tensile crack propagates in its own plane under this condition. (In fact, neither the Hawaiian nor the Emperor chain is exactly straight on a small scale, and the Hawaiian chain has changed propagation direction in the last few m.y. [Wessel and Kroenke, 2000].) Chain straightness also implies that the crack is nearly parallel to the Pacific plate velocity vector with respect to its boundary; otherwise older portions of the active crack would be carried into unfavorable areas of the stress field. Finally, the southeast crack tip would be near the location where crack-normal stress vanishes or becomes compressive.

Solomon and Sleep (1974) and Sandwell et al. (1995) suggested, using a symmetry argument, that the opposed pulls of subducted slabs under the Aleutian Islands and in the southwest Pacific could induce maximum horizontal tension perpendicular to the Hawaiian chain near the chain. However, global or Pacific plate models in which lithosphere stress is calculated from tractions and gravity body forces (Richardson et al., 1979; Wortel et al., 1991; Bai et al., 1992; Steinberger et al., 2001; Yoshida et al., 2001; Lithgow-Bertelloni and Guynn, 2004) generally have horizontal stress unfavorable to the crack model. The stress models either lack maximum tension along and normal to the Hawaiian chain or they have tensile stress along the southeast projection of the chain, ahead of the present position of Hawaii. In some cases the model stress fields seem to be too spatially variable to support a simple crack model. Perhaps the computed stress fields most consistent with a crack model are Lithgow-Bertelloni and Guynn's (2004) cases TD0 and TD5, which are based on lateral heterogeneity of the lithosphere.

In this article we calculate the present stress rate field due to cooling of the Pacific plate as it moves away from the East Pacific Rise. In contrast to those who performed the earlier studies of slab pull, we find that the horizontal stress rate field near the Hawaiian segment from this thermal source has symmetry and sign favorable to lithosphere crack models.

THERMOELASTIC STRESS FIELD

The present-day total thermoelastic stress field cannot be calculated directly from only current Pacific plate properties, because the plate has unknown initial conditions and an unknown history of cooling, accretion at its base and ridge boundaries, and ablation at subduction zones. Instead, we calculate the present instantaneous time derivative of stress at particle positions. However, our subsequent discussion is perhaps more intuitive when expressed in terms of stress, so we multiply our computed stress rate by 1 m.y. to obtain a stress increment whose numerical value is the same as that of the stress rate. This stress increment is an approximation to the true thermoelastic stress increment that would be obtained by allowing the plate geometry to change slightly during the passage of 1 m.y.

First we represent the Pacific plate as a finite-element spherical shell (in spherical coordinates) where elements are elastic and the shell lateral boundary approximates the actual plate boundary as shown in Figure 1. Then we assign to each element the temperature derivative with respect to time determined from a standard analytic equation for vertical conductive heat transfer in a cooling half-space (Turcotte and Schubert, 2002) according to seafloor age interpolated from seafloor isochron lines (Müller et al., 1997) (Fig. 1). The bottom boundary of the elastic plate is defined to be at the 800 °C isotherm as calculated from the cooling equation. When the 800 °C isotherm corresponds to depths > 50 km (i.e., the plate is older than 50 m.y.), we assign a shell depth of 50 km. The 50 km maximum depth is a compromise between depths estimated two ways. With respect to the passage of seismic waves, the Pacific plate lithosphere attains a steady-state thickness of 80–100 km for lithosphere older than ~100 m.y. (Watts, 2001; Turcotte and Schubert, 2002). Also, the depth of low-density compensation for the Hawaiian swell (discussed later) is ~100 km (Turcotte and Schubert, 2002). On the other hand, ~the elastic flexural thickness of the lithosphere near Hawaiian-Emperor volcanoes is 20–30 km (Watts, 2001). For global oceanic lithosphere older than ~70 m.y., cooling of a 95-km-thick plate with basal temperature 1450 °C is more consistent with heatflow and seafloor depth data than a half-space model (Stein and Stein,

---

Figure 1. Bathymetry and topography of the Pacific seafloor and the surrounding area (Simkin et al., 2006). Seafloor age (Müller et al., 1997) is shown by isochron lines labeled in Ma. The continuous black line is the boundary of elastic shell approximation to the Pacific plate as shown in Figures 2 and 3. The continuous red lines near the model boundary are the present-day spreading East Pacific Rise and Juan de Fuca Ridge (JDF). The Hawaiian swell is light blue and has ~1 km maximum elevation above the surrounding seafloor. The volcano chains shown are Hawaiian-Emperor; Louisville (L); Marquesas (M); Pitcairn (P); Society (S); and Samoa (Sa). Me is Meiji seamount.

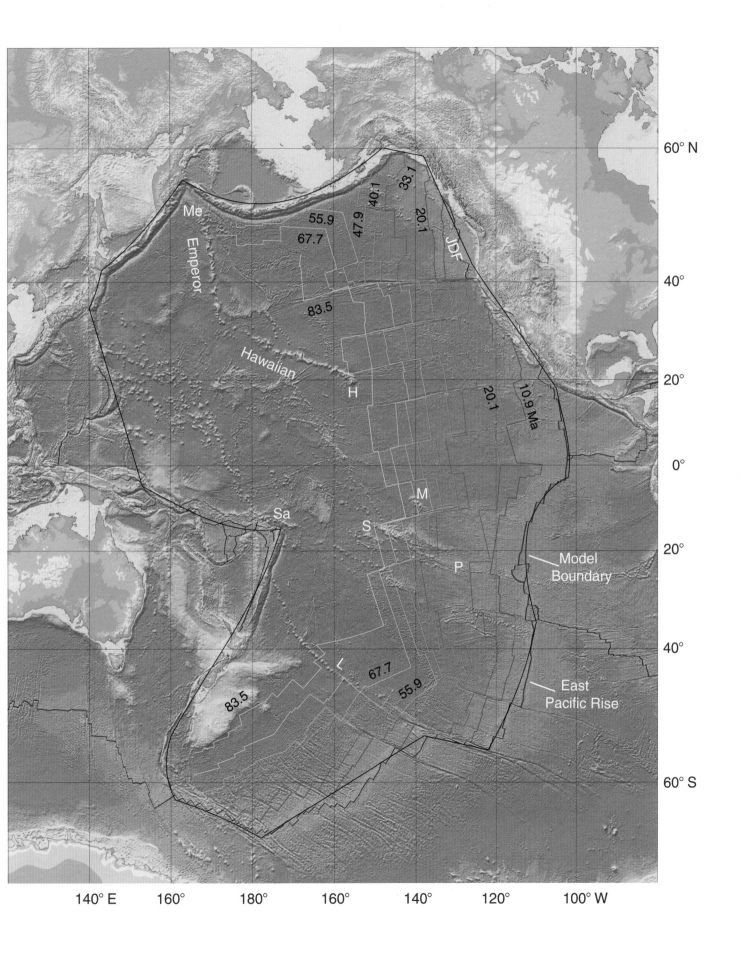

60° N

40°

20°

0°

20°

40°

60° S

140° E   160°   180°   160°   140°   120°   100° W

Me

Emperor

Hawaiian

Sa

55.9
67.7

47.9
40.1

33.1
20.1

JDF

83.5

H

20.1

10.9 Ma

M

S

P

Model
Boundary

East
Pacific Rise

83.5

L
67.7

55.9

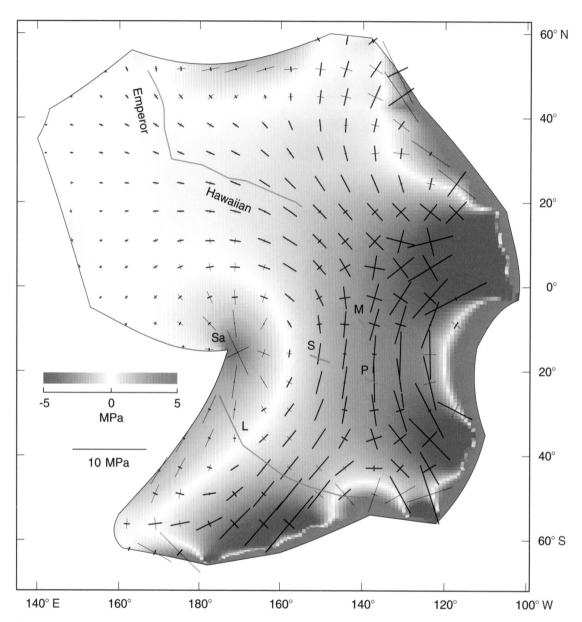

Figure 2. Calculated increment of horizontal stress field for the Pacific plate for the second finite element layer from the top for 1 m.y. Horizontal principal stress vectors are shown by orthogonal line pairs, thin lines for tension, thick lines for compression. Principal stress vectors greater than 10 MPa are not shown. The magnitude of horizontal mean normal stress is shown by color, with maximum tension bright red and maximum compression bright blue; magnitudes greater than 5 MPa are plotted as 5 MPa. The green lines are the volcano chains marked in Figure 1. L—Louisville; M—Marquesas; P—Pitcairn; S—Society; Sa—Samoa.

1992). Our result is an approximation to the solution of a more realistic cooling problem in which the upper part of the lithosphere would be elastic and the lower part would become more anelastic with depth.

The shell has 62,192 3-D quadratic elements (not membrane) and is divided into four layers of equal thickness vertically but variable thickness horizontally. Thermal diffusivity is 8×10^{-7} m^2 s^{-1}, volume coefficient of thermal expansion 3.1×10^{-5} °C^{-1}, initial temperature 1300 °C, Poisson's ratio 0.25, and shear modulus 30 GPa. Boundary conditions are zero traction on all surfaces, zero radial displacement at the shell bottom, and a minimal displacement constraint to prevent rigid motion. More pervasive displacement boundary conditions are not appropriate, because over geologic time all parts of the Pacific plate have moved with respect to surrounding plates. We did not study the consequences of replacing the zero vertical displacement condition at the shell bottom with a force condition to simulate isostatic equilibrium.

Figures 2 and 3 show principal stress vectors and mean normal stress for the calculated 1 m.y. increment of horizontal stress for shell layers 2 and 4 (bottom). In layer 2 (Fig. 2) near the East Pacific Rise and the Juan de Fuca Ridge (Fig. 1), rapid thermal contraction produces strong tensile stress. This strong contraction, in turn, compresses the adjacent shell material on the left to about the middle of the plate. From there to the subduction zones, the stress field becomes more tensional. The west half of the Hawaiian chain has slight tension normal to the chain. The Louisville chain has tension near its two ends, and compression along its center. Stresses are compressive near the Marquesas, Pitcairn, and Society chains of the south Pacific superswell area. Layer 4 (Fig. 3) is similar except that the stress field is overall more tensional. The prominent tensional region near Samoa, caused by a notch in the Pacific plate boundary, occurs in all layers. Principal stresses in the interior of layer 1 (not shown) are orientated similar to those in layer 2 but are compressive except near Samoa, near the plate boundary notch at the east end of the Louisville chain, in a small area adjacent to the southernmost East Pacific Rise, and in a small area adjacent to the Juan de Fuca Ridge. In a depth-averaged sense, one may think of thermal contraction along part of the shell rim as stretching the rest of the shell rim. At a particular horizontal location away from a ridge, however, thermal contraction at the plate bottom produces tensile stress near the plate bottom and compressive stress near the top.

DISCUSSION

Thermoelastic Stress Field

The Hawaiian chain is qualitatively consistent in direction and length with the crack model if the plate stress is predominantly thermoelastic and the incremental thermoelastic stress and the unknown total thermoelastic stress have similar form. Near the Hawaiian volcano chain, the incremental stress in layers 2–4 (layer 3 stress is intermediate between the stresses in Figs. 2 and 3) is nearly symmetric with respect to the chain, with principal tension normal to the chain and principal compression parallel to it. Of all straight trial cracks that can be drawn on Figures 2 and 3, cracks close to the Hawaiian line are the longest that have principal tension normal to them over their full length. The Hawaiian line is also parallel to Pacific plate motion (relative to its ridge boundary), so straight cracks close to the Hawaiian line stay in a favorable stress field for the longest time. On a southeast extrapolation of the Hawaiian line, the principal stresses become compressive and rotate clockwise and thus may explain the location of the Hawaiian hotspot as the position beyond which an equilibrium straight crack cannot propagate under the applied plate stress. The Louisville chain, like the Hawaiian-Emperor chain, is age progressive from east to west (Clouard and Bonneville, 2005). On Figures 2 and 3, the maximum tensile stress at the east end of the Louisville chain, near the East Pacific Rise, is oriented approximately normal to the volcano chain, consistent with a tensile crack theory for the chain end. On the other hand, the principal stress directions near the east-to-west age-progressive Marquesas, Pitcairn, and Society chains (Clouard and Bonneville, 2005) are inconsistent with a simple crack theory.

If calculated thermoelastic stress rates are applied for a 10 m.y. time interval, the magnitudes of thermoelastic tensile stresses in layer 4 normal to and near the Hawaiian chain are ~10 MPa. Their magnitude is then comparable to the tensional stresses for the two favorable cases of Lithgow-Bertelloni and Guynn (2004) for the region near the Hawaiian chain. The stress magnitudes near the Hawaiian chain for other plate models are quite variable, but are usually less than ~100 MPa.

Because the incremental stress field by itself makes the crack model consistent with the observed Hawaiian chain, the superposition of actual stresses from all other sources must resemble the incremental form for a crack model to apply. The possibilities are (1) nonthermoelastic stresses are negligible, and total and incremental thermoelastic stresses have the same form, or (2) nonthermoelastic stresses are not negligible, but combine in a fortuitous way with the total thermoelastic stress to have the necessary form. A special case of the second possibility is that nonthermoelastic stresses, total thermoelastic stress, and incremental thermoelastic stress all have the same form. Conversely, a crack model independently known to be correct and driven by a known total thermoelastic stress places bounds on plate tractions and body forces.

For a long-term stable plate boundary configuration (and this has probably persisted since as long ago as the Late Oligocene), the current total thermoelastic stress would be approximately equal to the sum of incremental stresses, allowing for motion of plate material with respect to the boundary, calculated over time. Thus the total and incremental stress fields would have nearly the same symmetry with respect to the Hawaiian line, but the position of stress sign reversal along the Hawaiian line is uncertain.

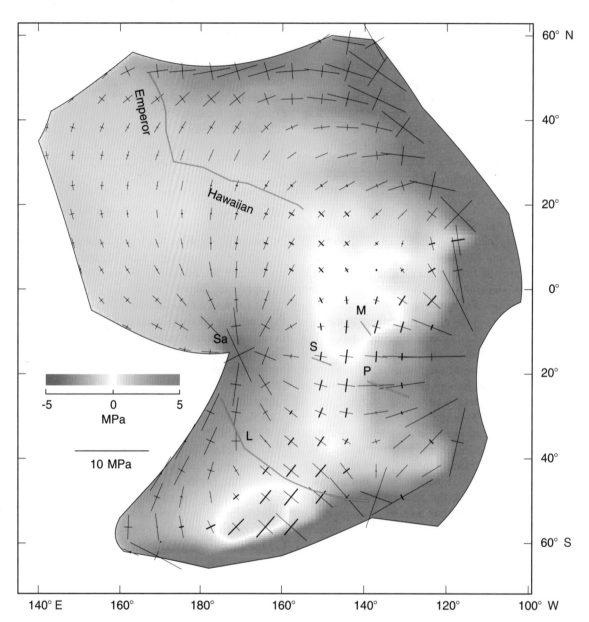

Figure 3. Calculated increment of horizontal stress field for the Pacific plate for the bottom layer for 1 m.y. Horizontal principal stress vectors are shown by orthogonal line pairs, thin lines for tension, thick lines for compression. Principal stress vectors greater than 10 MPa are not shown. The magnitude of horizontal mean normal stress is shown by color, with maximum tension bright red and maximum compression bright blue; magnitudes greater than 5 MPa are plotted as 5 MPa. The green lines are the volcano chains marked in Figure 1. L—Louisville; M—Marquesas; P—Pitcairn; S—Society; Sa—Samoa.

Emperor Chain, Hawaiian-Emperor Bend, and Hawaiian Swell

If the stress rate results we have described are a good approximation to the actual total stress field, Figures 2 and 3 imply that the Emperor chain is not a propagating tensile crack in a thermoelastic stress field at present because the direction of maximum tension is inclined to the crack plane. From ca. 80 Ma to 47 Ma, the date of the Hawaiian-Emperor Bend (Tarduno et al., 2003), the Pacific plate was much smaller than it is today and had ridge boundaries on the north (Pacific-Kula), east (Pacific-Farallon), and southwest (Pacific–North New Guinea) (Hall, 1997; Smith, 2003, this volume). The Emperor chain is thought to have originated at the ridge axis between the Pacific plate and the Kula (Izanagi) plate as a result of excess volcanism associated with a 30° change in spreading direction from 84 to 71 Ma (Norton, this volume). This volcanism built the Meiji seamount, the oldest in the Emperor chain. One explanation for the Emperor chain is that it was a crack near the Kula Ridge that propagated southeastward until the ridge became inactive around 47 Ma. A simple mechanical history for the Hawaiian-Emperor crack would then be that the Emperor chain propagated first inside the Pacific-Kula Ridge thermal boundary layer and then, near 47 Ma, found itself in the outer field of the East Pacific Rise thermal boundary layer. In other words, the cause of the Hawaiian-Emperor Bend may have been a rapid change in the thermoelastic stress field associated with the disappearance of ridge segments. In a thermoelastic stress field, no fixed latitude is necessarily implied for Emperor chain volcanoes, but the observed latitude decrease with time (Tarduno et al., 2003; Sager, this volume) has yet to be simulated in an evolving thermoelastic crack model. On the other hand, the observed bend of the Hawaiian-Emperor chain appears not to correspond to an appropriate change in Pacific plate motion (Norton, 1995).

The Hawaiian swell, a region of elevated seafloor ~1200 km wide, ~2500 km long, and coincident with the younger half of the Hawaiian chain (Fig. 1), has been attributed to (1) a thermal source (Crough, 1983), and (2) buoyant residuum left behind from melt extraction and underplated onto the base of the lithosphere, spreading viscously with time (Phipps Morgan et al., 1995). A thermal anomaly along the swell has not been observed (von Herzen et al., 1989). It has been suggested that its absence may be due to hydrothermal circulation (McNutt, 2002), but DeLaughter et al. (2005) and Stein and von Herzen (this volume) suggest that hydrothermal circulation is unlikely to be able to camouflage the heatflow anomalies expected from thermal plumes. Crack models attribute the swell to model 2, i.e., buoyant residuum, a theory that is supported by the observation that the volume of the swell is roughly proportional to the volume of melt extracted along the chain (Phipps Morgan et al., 1995).

CONCLUDING REMARKS

The present work represents a first step toward quantifying the crack model that was suggested for the Hawaiian volcanic chain as far back as Dana (1849). The crack model is appealing because several first-order features of the Hawaiian and Emperor chains that are inconsistent with the plume model or require surprising coincidences may be consistent with the crack model. These include the inception of the Emperor chain on a ridge, the lack of a "plume head" large igneous province, the ~60° change in propagation direction that occurred at ca. 47 Ma, the rapid southward migration of the Emperor hotspot prior to this, and the lack of the heatflow anomaly expected for a plume.

Nevertheless, the current lithosphere crack models are difficult to test. There are no solved boundary value problems available to enable us to study the conditions for crack propagation, melt production, and melt transport, and in particular, the spatial extent of the crack is unknown. Neither is it clear, given the current technology, how the very small extension rates expected at the surface above a crack could be measured geodetically in the Hawaii region. The crack model is a member of a general class of lithosphere flaw models that could be constructed based on an elongate weak zone under the Hawaiian chain. Other flaw models of this class would have the self-preservation, propagation, and stress concentration properties attributed to the crack model and may be more consistent with first-order observations.

ACKNOWLEDGMENTS

We thank D. Anderson, J. Moore, J. Savage, N. Sleep, S. Stein, and D. Jurdy (co-editor) for helpful suggestions that improved the manuscript.

REFERENCES CITED

Bai, W., Vigny, C., Ricard, Y., and Froidevaux, C., 1992, On the origin of deviatoric stresses in the lithosphere: Journal of Geophysical Research, v. 97, p. 11,729–11,737.

Clague, D.D., and Dalrymple, G.B., 1989, Tectonics, geochronology, and origin of the Hawaiian-Emperor volcanic chain, *in* Decker, R.W., ed., Geology of North America, v. N: Boulder, Colorado, Geological Society of America, p. 188–217.

Clouard, V., and Bonneville, A., 2005, Ages of seamounts, islands, and plateaus on the Pacific plate, *in* Foulger, G.R., et al., eds., Plates, plumes, and paradigms: Boulder, Colorado, Geological Society of America Special Paper 388, p. 71–90, doi: 10.1130/2005.2388(06).

Crough, S.T., 1983, Hotspot swells: Annual Review of Earth and Planetary Sciences, v. 11, p. 165–193, doi: 10.1146/annurev.ea.11.050183.001121.

Dana, J.D., 1849, U.S. Exploring Expedition during the years 1838–1842 under the command of Charles Wilkes, U.S.N.: Geology, v. 10, p. 307–336.

DeLaughter, J.E., Stein, C.A., and Stein, S., 2005, Hotspots: A view from the swells, *in* Foulger, G.R., et al., eds., Plates, plumes, and paradigms: Boulder, Colorado, Geological Society of America Special Paper 388, p. 257–278, doi: 10.1130/2005.2388(16).

Hall, R., 1997, Cenozoic plate tectonic reconstructions of SE Asia, *in* Fraser, A.J., Matthews, S.J., and Murphy, R.W., eds., Petroleum geology of southeast Asia: Geological Society of London Special Publication 126, p. 11–23.

Jackson, E.D., and Shaw, H.R., 1975, Stress fields in central portions of the Pacific plate: Delineated in time by linear volcanic chains: Journal of Geophysical Research, v. 80, p. 1861–1874.

Lithgow-Bertelloni, C., and Guynn, J.H., 2004, Origin of the lithospheric stress field: Journal of Geophysical Research, v. 109, B01408, doi:10,1029/2003JB002467.

McNutt, M., 2002, Heat flow variations over Hawaiian swell controlled by near-surface processes, not plume properties, *in* Takahashi, E., et al., eds., Hawaiian volcanoes: Deep underwater perspectives: Washington, D.C., American Geophysical Union, Geophysical Monograph Series, v. 128, p. 365–372.

Morgan, W.J., 1971, Convection plumes in the lower mantle: Nature, v. 230, p. 42–43, doi: 10.1038/230042a0.

Müller, R.D., Roest, W.R., Royer, J.-Y., Gahagan, L.M., and Sclater, J.G., 1997, Digital isochrons of the world's ocean floor: Journal of Geophysical Research, v. 102, p. 3211–3214, doi: 10.1029/96JB01781.

Norton, I.O., 1995, Plate motions in the North Pacific: The 43 Ma nonevent: Tectonics, v. 14, p. 1080–1094, doi: 10.1029/95TC01256.

Norton, I.O., 2007 (this volume), Speculations on the Cretaceous tectonic history of the northwest Pacific and a tectonic origin for the Hawaii hotspot, *in* Foulger, G.R., and Jurdy, D.M., eds., Plates, plumes, and planetary processes: Boulder, Colorado, Geological Society of America Special Paper 430, doi: 10.1130/2007.2430(22).

Phipps Morgan, J., Morgan, W.J., and Price, E., 1995, Hotspot melting generates both hotspot volcanism and a hotspot swell?: Journal of Geophysical Research, v. 100, p. 8045–8062, doi: 10.1029/94JB02887.

Richardson, R.M., Solomon, S.C., and Sleep, N.H., 1979, Tectonic stress in the plates: Reviews of Geophysics, v. 17, p. 981–1019.

Sager, W.W., 2007 (this volume), Divergence between paleomagnetic and hotspot model predicted polar wander for the Pacific plate with implications for hotspot fixity, *in* Foulger, G.R., and Jurdy, D.M., eds., Plates, plumes, and planetary processes: Boulder, Colorado, Geological Society of America Special Paper 430, doi: 10.1130/2007.2430(17).

Sandwell, D.T., Winterer, E.L., Mammerickx, J., Duncan, R.A., Lynch, M.A., Levitt, D.A., and Johnson, C.L., 1995, Evidence for diffuse extension of the Pacific plate from Pukapuka ridges and cross-grain gravity lineations: Journal of Geophysical Research, v. 100, p. 15,087–15,099, doi: 10.1029/95JB00156.

Simkin, T., Tilling, R.I., Vogt, P.R., Kirby, S., Kimberly, P., and Stewart, D.B., 2006, This dynamic planet: World map of volcanoes, earthquakes, impact craters, and plate tectonics: Reston, Virginia, U.S. Geological Survey, Geologic Investigations Series Map I-2800, scale 1:30,000,000 at the equator.

Sleep, N.H., 1992, Hotspot volcanism and mantle plumes: Annual Review of Earth and Planetary Sciences, v. 20, p. 19–43, doi: 10.1146/annurev.ea .20.050192.000315.

Smith, A.D., 2003, A re-appraisal of stress field and convective roll models for the origin and distribution of Cretaceous to Recent intraplate volcanism in the Pacific basin: International Geology Review, v. 45, p. 287–302.

Smith, A.D., 2007 (this volume), A shallow-source plate model for intraplate

volcanism in the Panthalassan and Pacific Ocean basins, *in* Foulger, G.R., and Jurdy, D.M., eds., Plates, plumes, and planetary processes: Boulder, Colorado, Geological Society of America Special Paper 430, doi: 10.1130/2007.2430(23).

Solomon, S.C., and Sleep, N.H., 1974, Some simple physical models for absolute plate motions: Journal of Geophysical Research, v. 79, p. 2557–2567.

Stein, C.A., and Stein, S.A., 1992, A model for the global variation in oceanic depth and heat-flow with lithospheric age: Nature, v. 359, p. 123–129, doi: 10.1038/359123a0.

Stein, C.A., and Von Herzen, R.P., 2007 (this volume), Potential effects of hydrothermal circulation and magmatism on heat flow at hotspot swells, *in* Foulger, G.R., and Jurdy, D.M., eds., Plates, plumes, and planetary processes: Boulder, Colorado, Geological Society of America Special Paper 430, doi: 10.1130/2007.2430(13).

Steinberger, B., Schmeling, H., and Marquart, G., 2001, Large-scale lithospheric stress field and topography induced by global mantle circulation: Earth and Planetary Science Letters, v. 186, p. 75–91, doi: 10.1016/S0012-821X (01)00229-1.

Tarduno, J.A., Duncan, R.A., Scholl, D.W., Cottrell, R.D., Steinberger, B., Thordarson, T., Kerr, B.C., Neal, C.R., Frey, F.A., Torii, M., and Carvallo, C., 2003, The Emperor seamounts: Southward motion of the Hawaiian hotspot plume in Earth's mantle: Science, v. 301, p. 1064–1069, doi: 10.1126/science.1086442.

Turcotte, D.L., and Schubert, G., 2002, Geodynamics (2nd edition): New York, Cambridge University Press, 456 p.

von Herzen, R.P., Cordery, M.J., Detrick, R.S., and Fang, C., 1989, Heat-flow and the thermal origin of hot spot swells—The Hawaiian swell revisited: Journal of Geophysical Research, v. 94, p. 13,783–13,799.

Watts, A.B., 2001, Isostasy and flexure of the lithosphere: New York, Cambridge University Press, 458 p.

Wessel, P., and Kroenke, L.W., 2000, Ontong Java Plateau and late Neogene changes in Pacific plate motion: Journal of Geophysical Research, v. 105, p. 28,255–28,277, doi: 10.1029/2000JB900290.

Wortel, M.J.R., Remkes, M.J.N., Govers, R., Cloetingh, S.A.P.L., and Meijer, P.Th., 1991, Dynamics of the lithosphere and the intraplate stress field: Royal Society of London Philosophical Transactions, ser. A, v. 337, p. 111–126.

Yoshida, M., Honda, S., Kido, M., and Iwase, Y., 2001, Numerical simulation for the prediction of the plate motions: Effects of lateral viscosity variations in the lithosphere: Earth: Planets and Space, v. 53, p. 709–721.

MANUSCRIPT ACCEPTED BY THE SOCIETY JANUARY 31, 2007

DISCUSSION

12 January 2007, James H. Natland and Edward L. Winterer

We agree with the model of Stuart and others (this volume). Here, rather fancifully, is why.

Figure D-1 is an oblique Mercator projection of part of the Pacific plate about the Hawaiian pole of rotation (68°N, 75°W). Red lines highlight major linear volcanic chains. The chart is on end with lines of latitude vertical so that aspects of bilateral symmetry, which we are used to identifying in the human figure, stand out. Phenomena in the mantle that influence the plate surface and are fixed will produce trends parallel to lines of latitude

on the moving plate. Several linear chains are parallel. White lines are portions of three lines of latitude, one of which is the plate-rotation equator (PRE on the horizontal reference line at the top). There, the East Pacific Rise is spreading most rapidly, 149 km/m.y. (Hey et al., 2004). The two longest linear volcanic chains on the plate, the Hawaiian (H) and Louisville (L) ridges, are surprisingly, to within about 1° of latitude (vertical arrow), symmetrical about the PRE. Note that transform seismicity of the Eltanin fracture zone system (focal spheres) aligns with the Louisville seamounts.

A serrated or W-shaped arrangement of contour lines showing depth to the seafloor about the PRE also shows symmetry,

Figure D-1. Oblique Mercator projection of bathymetry of the central Pacific covering much of the Pacific plate, about the Hawaiian pole of rotation at 68°N, 75°W, but rotated 90° counterclockwise to reveal components of symmetry in the distribution of linear volcanic chains and bathymetric elevations near the East Pacific Rise. See text for discussion and explanation of annotations.

but not the expected U-shaped square-root-of-time age–depth relationship, especially with respect to older arrangements of spreading ridges and fracture zones (fine white lines). The Pacific plate thus rides over two fixed topographic anomalies beyond the Marquesas (M) and Austral (A) chains, which are equidistant from and nearly bilaterally symmetrical about the PRE. The Tuamotu and Society (SO) chains straddle the PRE (and further west, the Caroline chain, C) and so continue the symmetry. This region has been termed "SOPITA" (South Pacific isotopic and thermal anomaly; Staudigel et al., 1991), but it is clearly divided into subregions. The isotopic anomaly extends through near-ridge seamount provinces (Janney et al., 2000; Hall et al., 2006) to a fairly wide segment of the East Pacific Rise (Mahoney et al., 1994); enriched mantle sources of the seamounts are shallow and are not affiliated with a plume.

Finally, the horizontal belt of seismicity at the Tonga-Kermadec Trench is along a line of longitude in this projection. The active ends of both the Samoan chain (S, near the trench) and the Hawaiian chain are both almost precisely along this line of longitude. The Samoan chain and the curving corner of the Tonga Trench are also close to the PRE.

We believe that these aspects of Pacific volcanism are best explained by fracture propagation (Natland and Winterer, 2005) and the thermoelastic cooling model of Stuart et al. (this volume; cf., Sandwell and Fialko, 2004). If one rotates Figure 2 of Stuart et al. (this volume) to the orientation of Figure D-1, stress couplets near Hawaii parallel the chain (also the PRE) and are determined by a combination of cooling of the lithosphere (compressive regime, blue) and regions of tension (red) on the rise and at trenches. The maximum tensional stress occurs where the Tonga Trench changes direction by 90° and becomes a transform fault.

In this configuration, the stress field "bows" across the Pacific lithosphere between the Tonga Trench and Juan de Fuca Ridge. The Hawaiian Ridge sits precisely on the crest of the bow. A similar bow exists between the Tonga Trench and the Pacific–Antarctic East Pacific Rise near the Eltanin fracture zones. However, here, the stress field does not match the condition for tensional crack propagation except very near the ridge axis. Thus, only seamounts of the Louisville Ridge have ever formed near the ridge at the western end of the Eltanin fracture zones. The Samoan chain sits precisely in the maximum tensional regime that is produced by bending and disruption of the Pacific plate at the curving corner of the Tonga Trench (Natland, 1980).

Plume advocates now must explain why deep-mantle plumes that are premised to start in the lower mantle, and which are supposedly independent of plate motion, happen to "know" (1) where the Pacific plate is spreading most rapidly so that many aspects of midplate volcanism, including the Hawaiian and Louisville ridges, are symmetrical about the PRE; (2) the precise location of the curving corner of the Tonga Trench, to make Samoa; and (3) why stress regimes most favorable to fracture propagation coincidentally occur at the ends of the two longest linear chains on the plate.

REFERENCES CITED

Hall, L.S., Mahoney, J.J., Sinton, J.M., and Duncan, R.A., 2006, Spatial and temporal distribution of a C-like asthenospheric component in the Rano Rahi seamount field, East Pacific Rise, 15°–19°S: Geochemistry, Geophysics, Geosystems, v. 7, p. 1–27, doi:10.1029/2005GC000994.

Hey, R.N., Baker, E.T., Bohnenstiehl, D., Massoth, G.J., Kleinrock, M., Martinez, F., Naar, D., Pardee, D., Lupton, J., Feely, R.A., Gharib, J., Resing, J., Rodrigo, C., Sansone, F., and Walker, S.L., 2004, Tectonic/volcanic segmentation and controls on hydrothermal venting along the Earth's fast seafloor spreading system, EPR 27 degrees–32 degrees S: Geochemistry, Geophysics, Geosystems, v. 5, doi:10.1029/2004GC000764.

Janney, P.E., Macdougall, J.D., Natland, J.H., and Lynch, M.A., 2000, Geochemical evidence from the Pukapuka volcanic ridge system for a shallow
enriched mantle domain beneath the South Pacific Superswell: Earth and
Planetary Science Letters, v. 181, p. 47–60.

Mahoney, J.J., Sinton, J.M., Kurz, M.D., Macdougall, J.D., Spencer, K.J., and
Lugmair, G.W., 1994, Isotope and trace element characteristics of a superfast spreading ridge, East Pacific Rise, 13-23°S: Earth and Planetary Science Letters, v. 121, p. 173–193.

Natland, J.H., 1980, The progression of volcanism in the Samoan linear volcanic chain: American Journal of Science, v. 280A, Jackson Volume, p. 709–
735.

Natland, J.H., and Winterer, E.L., 2005, Fissure control on volcanic action in the
Pacific, *in* Foulger, G.R., et al., eds., Plates, plumes, and paradigms: Boulder, Colorado, Geological Society of America Special Paper 388, p. 687–
710, doi: 10.1130/2005.2388(39).

Sandwell, D., and Fialko, Y., 2004, Warping and cracking of the Pacific plate
by thermal contraction: Journal of Geophysical Research, v. 109, art. no.
B10411, doi:10.1029/2004JB003091.

Staudigel, H., Park, K.-H., Pringle, M., Rubenstone, J.L., Smith, W.H.F., and
Zindler, A., 1991, The longevity of the South Pacific isotopic and thermal
anomaly: Earth and Planetary Science Letters, v. 102, p. 24–44.

Stuart, W.D., Foulger, G.R., and Barall, M., 2007 (this volume), Propagation of
the Hawaiian-Emperor volcano chain by Pacific plate cooling stress, *in* Foulger, G.R., and Jurdy, D.M., eds., Plates, plumes, and planetary processes:
Boulder, Colorado, Geological Society of America Special Paper 430, doi:
10.1130/2007.2430(24).

The Geological Society of America
Special Paper 430
2007

Geophysical characterization of mantle melting anomalies: A crustal view

Valentí Sallarès*
*Unidad de Tecnología Marina–CMIMA, Consejo Superior de
Investigaciones Científicas (CSIC), Barcelona, Spain*
Alcinoe Calahorrano
*Instituto de Ciencias del Mar–CMIMA, Consejo Superior de
Investigaciones Científicas (CSIC), Barcelona, Spain*

ABSTRACT

At present there is no single "unified theory" capable of explaining the variety of geological, geophysical, and geochemical observations that characterize what is generically known as hotspot magmatism. An increasing number of geophysical and geochemical observations disagree with the predictions of the conventional thermal plume model, in which excess melting is due mainly to high mantle temperatures. Other parameters such as the presence of water or the composition of the mantle source have been shown to be as important as temperature in controlling the structure and physical properties of the igneous crust. In this article we first emphasize the importance of doing proper velocity and density modeling, including comprehensive uncertainty analysis, to determine how well resolved the geophysical parameters actually are. We show that in some cases the contribution of velocity-derived lateral crustal density variations can be sufficiently significant to account for the observed gravity and topography anomalies without calling for noticeable mantle density contrasts. Next we show that the comparison of crustal geometry obtained along age-progressive volcanic tracks enables temporal variations of the hotspot-ridge distance to be estimated. Finally we use a 2-D mantle melting model to illustrate the effect of different mantle melting parameters on the resulting crustal structure. The tests made indicate that it is difficult to find a plausible combination of mantle temperature, upwelling rate, melt productivity, and thickness of the melting zone to explain either the high-velocity, underplated bodies frequently described at midplate settings or the lack of a positive crustal thickness–velocity (H-V_p) correlation found at igneous provinces originated on-ridge. We suggest that the main parameter controlling the generation of volcanic underplating is the presence of a lithospheric lid limiting the extent of the mantle melting zone, whereas the H-V_p anticorrelation can be related to the presence of a major element heterogeneity, such as eclogite derived from recycled oceanic lithosphere, in the mantle source.

Keywords: crustal structure, seismic velocity, density, mantle melting parameters

*Address: Unidad de Tecnología Marina–CMIMA, Consejo Superior de Investigaciones Científicas, Passeig Maritim de la Barceloneta 37–49, 08003 Barcelona, Spain; e-mail: vsallares@cmima.csic.es.

Sallarès, V., and Calahorrano, A., 2007, Geophysical characterization of mantle melting anomalies: A crustal view, *in* Foulger, G.R., and Jurdy, D.M., eds., Plates, plumes, and planetary processes: Geological Society of America Special Paper 430, p. 507–524, doi: 10.1130/2007.2430(25). For permission to copy, contact editing@geosociety.org. ©2007 The Geological Society of America. All rights reserved.

INTRODUCTION

Large igneous provinces, oceanic plateaus, flood basalts, and aseismic ridges constitute some of the largest manifestations of terrestrial magmatism, but their origin and the melting processes and underlying mantle dynamics necessary to generate them are still poorly understood. From here on, we refer to this type of magmatism as "hotspot magmatism" to denote a locality where excess magma is produced, not as a sign of high temperature. Similarly, we refer to its diverse surface expressions using the generic term "igneous provinces" and to the source of both as a "melting anomaly." One of the common characteristics of hotspot magmatism is that it does not necessarily occur near the boundaries between tectonic plates, and therefore it does not seem to be directly related to the main plate tectonic cycle. Whereas normal oceanic crust is presumed to originate by decompression melting of normal oceanic mantle (~1300 °C temperature, pyrolitic in composition) upwelling passively beneath a spreading center, and arc volcanism is widely accepted to be the remnant of melts originating within the mantle wedge by water expelled from the dehydrating subducting slab at con-

vergent margins, hotspot magmatism can occur nearly anywhere on Earth (Fig. 1).

The most widely accepted model to explain the occurrence of hotspot magmatism is the thermal plume model (Wilson, 1963; Morgan, 1971), in which hot, buoyant mantle plumes resulting from thermal instabilities at the core-mantle boundary rise to the surface and produce large amounts of melt. The presence of deep mantle plumes is consistent with continuous low-velocity anomalies underlying several active hotspots that extend from the surface to the lower mantle, as shown by global tomography models (e.g., Zhao, 2001; Montelli et al., 2004). Given that they are anchored in the deep mantle, the kinematics of mantle plumes must be decoupled from the main plate tectonic cycle so that they can qualitatively explain the apparent fixity of the hotspot framework, the widespread occurrence of intraplate magmatism, and the formation of age-progressive volcanic tracks. In the thermal plume model, elevated temperatures cause mantle melting, and an excess of MgO-rich (olivine-rich) melt is mostly emplaced to form thick, high-velocity igneous crust (e.g., White and McKenzie, 1989). Igneous provinces are generally characterized by prominent topography and gravity

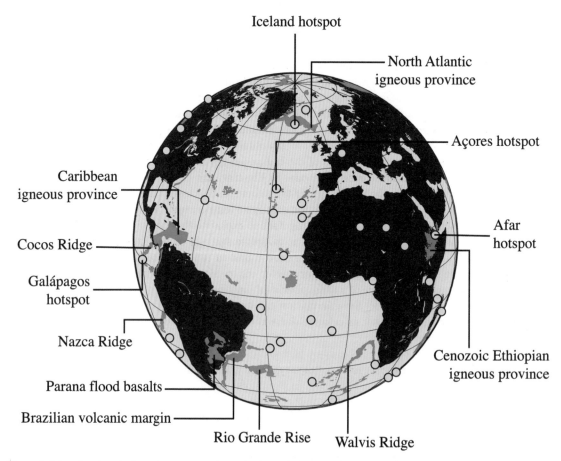

Figure 1. Map showing the large igneous provinces (red areas) that have been recognized around the Atlantic Ocean. The yellow circles indicate the locations of currently active hotspots (modified from Korenaga, 2004).

anomalies (e.g., Anderson et al., 1973). Seismic investigations show, in turn, that potential field anomalies are partly or totally compensated by the presence of thick crustal roots (e.g., Darbyshire et al., 2000; McNutt and Bonneville, 2000; Korenaga et al., 2001; Sallarès et al., 2003).

Although there is general agreement that long-lived, deep-seated thermal mantle plumes constitute an attractive explanation for the occurrence of hotspot magmatism, an increasing number of observations do not agree with the predictions of this classical model. It is interesting to note, for instance, that first-order hotspots such as Iceland or Galápagos show low-velocity tomography anomalies that are confined to the upper mantle only (Montelli et al., 2004), and the estimated lower-crustal velocity for a number of igneous provinces (e.g., Kerguelen, the Galápagos, the north Atlantic igneous province) is lower than expected for melting of hotter than normal mantle (e.g., Charvis and Operto, 1999; Korenaga et al., 2000; Sallarès et al. 2003).

What seems evident at present is that excess mantle temperature is not the only factor that may explain enhanced hotspot magmatism; in some special cases, it has been questioned if temperature is significant at all (e.g., Green et al., 2001; Green and Falloon, 2005). Mantle composition, water content, upwelling rates, and the melting process itself have been shown to be as important as temperature in controlling the amount of melt produced as well as the geophysical and geochemical properties of igneous rocks (e.g., Ito et al., 1999; Maclennan et al., 2001; Cushman et al., 2004). Enhanced melting has been attributed to the presence of components with lower melting points, such as eclogite derived from recycled oceanic crust, or to the influence of deep damp melting for a number of hotspots, including Hawaii (e.g., Sobolev et al., 2000), Açores (e.g., Schilling et al., 1983; Bonatti, 1990), Iceland (e.g., Ito et al., 1999; Foulger and Anderson, 2005), and the Galápagos (e.g., Cushman et al., 2004; Sallarès et al., 2005). It is therefore likely that excess melting (and thicker crust) is not produced by a single, well-defined mechanism (i.e., a thermal mantle plume) but it can be the product of several different processes that must be examined and characterized case by case.

In this article we emphasize the importance of using an integrated approach combining geophysical methods with geological and geochemical observations to systematically characterize the structure, physical properties, melt volume, and composition of the crustal igneous rocks that are the final product of any melting process. Systematic crustal characterization will help constrain the parameters that govern the mantle melting process in each particular case, and can be used as input into geodynamic models to gain quantitative understanding of the underlying mantle dynamics.

The rest of this article is organized into two sections. In the first section, we begin by summarizing geophysical observations at different igneous provinces, their differences, and their relationships to the mode of crustal emplacement. Then we describe a procedure for obtaining crustal velocity and density models and discuss the importance of performing accurate modeling together with a comprehensive analysis of model parameter uncertainties in order to reliably estimate the structure and physical properties of the igneous crust. In the following section, we present recent advances in the characterization of mantle melting parameters based on crustal structure information. Then we explain how to estimate the intensity of a melting anomaly based on crustal thickness observations. Finally we illustrate the effects of the different parameters that characterize mantle melting on the resulting seismic structure of the igneous crust and compare the predictions of the mantle melting model with observations made at different igneous provinces.

STRUCTURE AND PHYSICAL PROPERTIES OF IGNEOUS PROVINCES

From a geophysical point of view, igneous provinces are segments of anomalous oceanic crust characterized by (1) bathymetric swells showing shallow and rough seafloor over areas ranging from 10^4 to 10^6 km^2; (2) associated prominent, wide potential field anomalies encompassing the volcanic edifices; and (3) estimated melt volumes that can exceed 10^7 km^3, emplaced to form thicker than normal igneous crust (e.g., Coffin and Eldholm, 1994). Whereas normal oceanic crust generated at spreading centers is around 6–7 km thick (White et al., 1992), the crust of igneous provinces is generally thicker than 15 km (Charvis et al., 1999; Ye et al., 1999; Grevemeyer et al., 2001; Sallarès and Charvis, 2003), and in extreme cases like Iceland or the Ontong Java plateau, it can exceed 30 km (Darbyshire et al., 1998; Miura et al., 2004). In terms of seismic velocity (V_p) (and likewise density), the igneous crust comprises two layers. The upper layer (oceanic layer 2, including sediments and upper crust) is characterized by large vertical V_p (density) gradients, with V_p ranging from ~2.0 km/s to ~6.5 km/s, whereas the lower layer (oceanic layer 3) shows much more uniform V_p (6.5–7.5 km/s).

Depending on whether the mantle melting anomaly is located near or far from a spreading center, the internal architecture of the crust as well as the relative thickness of the two layers can vary considerably. To illustrate these effects, in Table 1 we show a compilation of seismic observations made at several igneous provinces, some of which are believed to have originated off-ridge and others on-ridge. Thus, when a melting anomaly is located beneath or in the vicinity of a spreading center, the crust is thought to be generated in a single step. The mantle wells up in response to seafloor spreading and melts from the base of the mantle melting zone to the surface. The mode of crustal emplacement and the resulting crustal structure are in this case similar to those of normal oceanic crust. Some examples of this are the Ontong-Java, north Atlantic, and Galápagos volcanic provinces (see Table 1). As shown in this table, the main difference is the thicker crust with respect to that of normal oceanic crust. Regardless of crustal thickness variations, the ratio of layer 2 thickness versus total crustal thickness is quite uniform (around 25%), indicating that layer 3 accommodates most of the crustal overthickening (Mutter and Mutter, 1993), and the mean layer 3 velocity is similar to that of normal oceanic crust (White et al. 1992).

When the melting anomaly is midplate, far from spreading centers, the total melt production is limited by the presence of a mechanical boundary (i.e., a cold lithospheric lid) that restricts the minimum depth at which melt can be generated and thus the total amount of melting. This is the reason that midplate igneous provinces (e.g., Hawaii, Réunion) tend to have thinner crust than those generated beneath spreading centers (e.g., Iceland, Ontong-Java), even if the melting anomaly is potentially stronger. In this case, part of the melt is believed to accumulate at the base of the crust, thickening it by crustal underplating, and part is extruded and piled on top of the preexisting plate. This is the case, for example, for the Hawaii, Canary, Réunion, and Marquesas islands (see Table 1). The ratio of layer 2 thickness to total crustal thickness is higher than in the case of igneous provinces generated at a spreading center (around 50%), and the underplated bodies are typically characterized by moderate thickness (~5 km) and high V_p (in excess of 7.3 km/s).

In general, the V_p/density gradient of layer 2 largely reflects lithological variations in the uppermost crust and diminishing alteration and porosity or fracturing of igneous rocks with depth (Bratt and Purdy, 1984; Detrick et al., 1994). The physical properties of layer 3 and those of the underplated bodies are thus considered good proxies for those of the mostly unaltered, crack-free gabbroic rocks that probably predominate at lower crustal levels (e.g., Kelemen and Holbrook, 1995). It is therefore important to confidently determine layer 3 V_p and density as well as their uncertainties in order to link these "measurable values" with the parameters that govern mantle melting. In the next sections we describe a possible procedure for doing this.

Crustal Seismology

Seismic Modeling and Uncertainty Analysis. Wide-angle reflection and refraction seismics (WAS) constitute the geophysical method most commonly used to study the structure of the crust. Depending on the acquisition layout and the modeling technique, the method has the potential to yield velocity models with a spatial resolution better than 1 km. In a classical marine survey, a number of ocean bottom seismometers (OBSs) deployed on the seafloor record successive airgun shots generated at the sea surface. WAS data are represented by offset-traveltime diagrams called record sections that show seismic phases refracted within the crust and upper mantle and reflected at the main discontinuities (Fig. 2B). In contrast to multichannel seismics (MCS), WAS record sections do not provide direct images of the sub-seafloor structure, so it is necessary to construct velocity models to account for the data. At present, the most widely used modeling technique is traveltime tomography (e.g., Zelt and Barton, 1998; Korenaga et al., 2000), in which the velocity field and the reflectors' geometry are simultaneously calculated from the joint inversion of traveltimes of reflected and refracted phases (Fig. 2C).

Owing to the nature of the joint refraction and reflection inverse problem, there are two main issues that have to be addressed before interpreting the results. The first is related to the nonuniqueness of the inversion solution, which means that the final solution depends on the starting velocity model. The second concerns estimating the model uncertainties, which is essential to confidently relate V_p to mantle melting parameters. It has been shown that the only practical way to address both issues is to perform Monte Carlo–type analysis (e.g., Tarantola, 1987). The degree of dependence of the solution obtained on the

TABLE 1. CHARACTERISTICS OF A NUMBER OF IGNEOUS PROVINCES

Igneous province	Location	Total crustal thickness (TCT)	Mean layer 3 velocity	Layer 2/ TCT	Underplated body thickness	Underplated body velocity	Emplacement
Hawaii-Emperor	Hawaii (1)	15 km	6.9 km/s	45%	3 km	7.7 km/s	Off-ridge
Canary	Gran Canaria (2)	20 km	6.9 km/s	50%	5 km	7.5 km/s	Off-ridge
	Tenerife (3, 4)	13 km	6.9 km/s	60%	?	>7.3 km/s	Off-ridge
Marquesas	Marquesas Islands (5)	11 km	7.0 km/s	50%	6 km	7.7 km/s	Off-ridge
Réunion	Réunion Island (6)	13 km	6.8 km/s	50%	3 km	7.5 km/s	Off-ridge
Kerguelen	Ninetyeast Ridge (7)	18 km	7.0 km/s	23%	5 km	7.5 km/s	Off-ridge
	N. Kerguelen (8)	24 km	7.0 km/s	26%	—	—	On-ridge
Ontong-Java plateau	Solomon Islands (9)	34 km	7.1 km/s	27%	—	—	On-ridge
North Atlantic volcanic province	Central Iceland (10)	35 km	7.0 km/s	21%	—	—	On-ridge
	SE Greenland (11)	27 km	6.9 km/s	26%	—	—	On-ridge
Galápagos volcanic province	Carnegie Ridge (12)	19 km	6.85 km/s	26%	—	—	On-ridge
	Cocos Ridge (13)	17 km	6.95 km/s	25%	—	—	On-ridge
	Malpelo Ridge (13)	19 km	6.85 km/s	25%	—	—	On-ridge
Nazca Ridge	East Nazca Ridge (14)	17 km	7.0 km/s	27%	—	—	On-ridge

Sources: (1) Watts and Ten Brink, 1989; (2) Ye et al., 1999; (3) Watts et al., 1997; (4) Canales et al., 2000; (5) Caress et al., 1995; (6) Charvis et al., 1999; (7) Grevemeyer et al., 2001; (8) Charvis and Operto, 1999; (9) Miura et al., 2004; (10) Darbyshire et al., 1998; (11) Korenaga et al., 2000; (12) Sallarès et al., 2003; (13) Sallarès et al., 2005; (14) Hampel et al., 2004.

initial model can be assessed by conducting a number of inversions with a variety of randomly generated initial models and noisy data sets (e.g., Sallarès et al., 2005). If all models have the same probability and the initial models cover the full region of non-null probability within the parameter space, the a posteriori covariance of the solutions obtained can be interpreted as a statistical measure of the solution uncertainty (Tarantola, 1987). One alternative to Monte Carlo analysis is to calculate the errors from the model Hessian (e.g., Hobro et al., 2003). However, the diagonal elements of the a posteriori covariance calculated from the Hessian can be interpreted as a posteriori errors (uncertainty) only when the problem is linearizable in all the regions of significant posteriori probability density, which is not easy to estimate a priori (Tarantola, 1987).

A third issue is related to the inherent tradeoff between depth and velocity parameters in reflection tomography (Bickel, 1990). This is especially important if the purpose is to determine the velocity of oceanic layer 3. As stated earlier, the velocity gradient within layer 3 is small, so refracted waves do not penetrate deep into that layer. Therefore, both layer 3 velocity and the geometry of the crust-mantle boundary (i.e., the Moho) must be determined using Moho reflections only, and the tradeoff must be estimated by comparing the results of the inversion using different values of the depth-kernel weighting parameter (Korenaga et al., 2000).

Velocity Structure of Igneous Provinces. Most existing melting models assume that once mantle melts, the magma rises through the lithosphere and is eventually emplaced to form igneous crust. The crustal volume, as well as the physical properties of the igneous rocks produced, depends on the temperature, composition, and upwelling rate of the mantle source. Infinite lithospheric stretching over normal-temperature pyrolitic mantle produces melt with a composition like that of mid-ocean ridge basalts (MORB) that crystallizes to form igneous crust with a mean V_p of ~7.0 km/s (McKenzie and Bickle, 1988; White et al., 1992). As potential temperature (T_p) rises above normal, the MgO content of the melt also increases, basically due to its larger olivine component. The V_p of igneous crust is mainly controlled by this MgO content: Higher MgO content implies higher V_p. White and McKenzie (1989) showed that if T_p is increased by ~200 °C above normal, the mean velocity of igneous crust increases by as much as 0.2–0.3 km/s. Therefore, if thick igneous crust is produced by the melting of anomalously hot mantle, there should be a positive correlation between V_p and crustal thickness. This is apparently the case for igneous provinces emplaced at midplate, far from spreading centers, such as Hawaii (Watts and ten Brink, 1989), Canary (Ye et al., 1999; Canales et al., 2000), the Marquesas (Caress et al., 1995), Réunion (Charvis et al., 1999), and the Ninetyeast Ridge (Grevemeyer et al., 2001) (Table 1), where the presence of thick, high-velocity crustal roots or underplated bodies has been attributed to the influence of postulated thermal anomalies.

In contrast, velocity models for igneous provinces near spreading centers, such as Kerguelen (Charvis and Operto,

1999), the north Atlantic volcanic province (Darbyshire et al., 1998; Korenaga et al., 2000), and the Cocos, Carnegie, and Malpelo aseismic ridges of the Galápagos volcanic province (Sallarès et al., 2003, 2005), do not show higher than normal lower crustal velocity, but rather show velocity that progressively increases between the top and the base of layer 3, similar to that observed in normal oceanic crust (e.g., White et al., 1992). The lack of a positive correlation between V_p and crustal thickness does not agree, therefore, with the predictions of the conventional thermal plume model.

However, it is important to perform comprehensive uncertainty analysis in order to relate the seismic structure to the parameters that characterize mantle melting. Historically, the classical way of obtaining velocity models from seismic data is by means of forward modeling. That requires varying the lower crustal velocity and Moho geometry by trial and error until an "acceptable" fit to the data is attained. Under those circumstances, it is not possible to formally quantify model parameter uncertainty. This makes it difficult to know to what extent lower crustal velocity is well resolved by the data or, in other words, what the tradeoff is between crustal velocity and crustal thickness. Without this information, it is not possible to confidently relate the seismic parameters to those controlling the mantle melting process. An error of 3% for a typical mean lower crustal velocity of 7.0 km/s, for example, means an error of 0.2 km/s, which almost completely precludes any possible petrological inference based on crustal velocity.

In contrast, joint refraction and reflection traveltime tomography enable Monte Carlo–type uncertainty analysis to be performed, including the effect of both traveltime choice errors and the dependence of the solution obtained on the initial model. This is the case for the models showing low-velocity lower crust in the Greenland margin (Korenaga et al., 2000) and the Galápagos volcanic province (Sallarès et al., 2005). In these cases, alternatives or modifications to the thermal anomaly model need to be considered in explaining the geophysical observations. This is discussed in the section headed "Constraining the Parameters That Govern Mantle Melting."

Gravity Modeling and Compensation of Topography

Constructing Density Models. Gravity analysis is commonly performed to estimate the density structure of the crust and upper mantle in hotspot-influenced areas (e.g., Ito and Lin, 1995; Darbyshire et al., 2000; Korenaga et al., 2001). It is important to characterize mantle density variations that may be associated with the presence of melting anomalies, which is critical to improve constraint on their nature. However, constructing a density model based on gravity data is prone to nonuniqueness, so it can become arbitrary unless a priori information is adequately incorporated into the modeling (e.g., see Barton, 1986). A common way of doing this consists of subtracting the effects of the seafloor and an estimated crust-mantle boundary interface from the free-air gravity anomaly and attributing the remaining

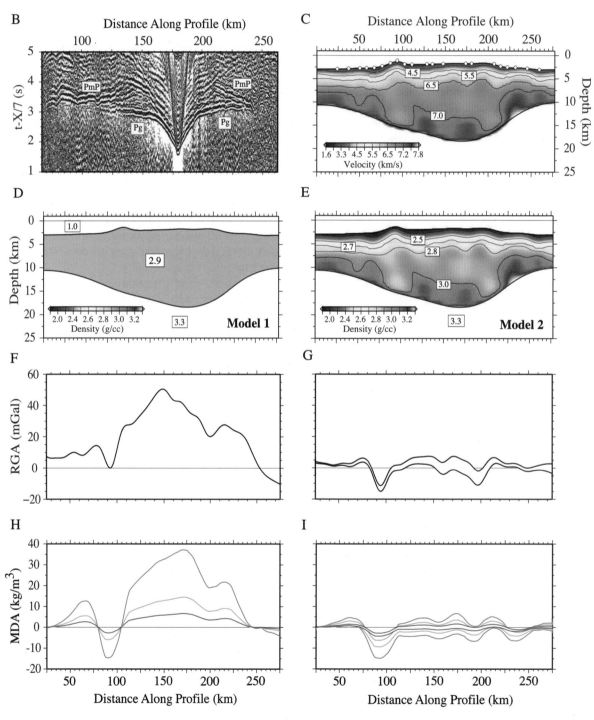
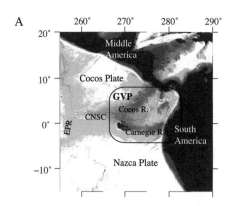

anomaly to density variations in the upper mantle (e.g., Ito and Lin, 1995). This type of approach assumes, however, that crustal density is laterally uniform along the transect, which in some cases might be an oversimplification.

Clear examples in which lateral density variations can be significant are profiles crossing aseismic ridges from side to side. In such cases the crustal density distribution of the adjacent oceanic basins is different from that of the thickened ridge segments, as has been shown in the case of the Galápagos volcanic province (Sallarès et al., 2005). When possible, it is therefore preferable to include reliable information on the lateral crustal density variations by jointly inverting or combining seismic and gravity data (e.g., Nielsen and Jacobsen, 2000; Korenaga et al., 2001). Seismic modeling is probably the best way of constraining the crustal structure, and then velocity-derived density models can be constructed based on existing velocity-density empirical relationships (e.g., Birch, 1961; Carlson and Herrick, 1990; Christensen and Mooney, 1995). However, using the velocity model as a benchmark to infer the crustal and upper mantle density distribution requires that (1) both the velocity field and the crust-mantle boundary geometry be well constrained by seismic data and (2) the velocity-density conversion be sufficiently accurate.

As explained earlier, model parameter uncertainty can be quantified by performing a Monte Carlo–type analysis. Velocity uncertainty can then be propagated to density and to the calculated gravity anomaly. This way it is possible to accurately correct the crustal contribution to the gravity field, including the effects of topography, geometry of the crust-mantle boundary, and vertical or lateral crustal density variations. The remaining gravity anomaly can be confidently attributed to mantle density contrasts. Concerning the velocity-density conversion, for oceanic crust it is reasonable to assume one-step crustal production from mantle melting, so the velocity-density conversion is quite well constrained and relatively simple compared with that of continental crust, where a large variety of rock types exist (e.g., Christensen and Mooney, 1995). This is especially true for the crack-free, almost unaltered oceanic layer 3, where velocity-density relationships for oceanic crust (e.g., Birch, 1961;

Carlson and Herrick, 1990) can be confidently used to derive density from velocity.

Joint Gravity and Topography Compensation Analysis. Topography compensation analysis is frequently carried out in combination with gravity modeling. Successful completion of this exercise, similar to gravity modeling, requires incorporating a priori information on crustal structure, preferably a velocity-derived density model and crust-mantle boundary geometry based on WAS data modeling. As in the case of many gravity studies, the effect of lateral crustal density variations in the observed topography is often neglected. Isostatic equilibrium of topography is generally assumed to be a product of crustal thickness variations (i.e., Airy isostasy), lateral variations in the mantle density above a given compensation depth (i.e., Pratt isostasy), or, more likely, a combination of both (e.g., Ito and Lin, 1995).

As discussed in the previous section, however, neglecting lateral crustal density variations is not a valid assumption in the case of igneous provinces. As an example to illustrate the potential effect of lateral crustal density variations in the observed topography and gravity field, in Figure 2 we compare the predicted gravity anomaly and the calculated mantle density across the Cocos Ridge (Galápagos volcanic province) for two models with the crustal geometry shown in Figure 2C but different crustal density distributions. The calculation was repeated for three different compensation depths: 50, 100, and 200 km. Model 1 (Fig. 2D) has a uniform crustal density of 2900 kg/m³, whereas the crustal density of Model 2 (Fig. 2E) was derived from the velocity distribution shown in Figure 2C, using the velocity-density conversion law of Carlson and Herrick (1990), $\rho = 3.61 - 6.0/V_p$, for the upper crust, and that of Birch (1961), $\rho = (V_p + 1.0)/2.67$, for the lower crust. The reference mantle density is 3300 kg/m³ in both models.

For Model 1, the maximum residual gravity anomaly is ~40–50 mGal (Fig. 2F) and the predicted mantle density anomaly beneath the thickened part of the crust is as high as 30–40 kg/m³ for $Z = 50$ km (Fig. 2H). In contrast, the uncertainty in the predicted residual gravity anomaly for Model 2 (Fig. 2G), which was obtained using the uncertainty of the velocity model of Figure 2C, is of the order of the remaining gravity anomaly. Simi-

Figure 2. (A) Bathymetric map of the Galápagos volcanic province (GVP) in the easternmost Pacific Ocean. Shaded areas indicate shallower seafloor than predicted based on plate cooling models. The location of the seismic profile across the Cocos Ridge is shown in red. CNSC—Cocos-Nazca spreading center; EPR—East Pacific Rise. (B) Example of a seismic record section registered at an Ocean Bottom Seismometer (OBS) during the SALIERI-2001 cruise in the Galápagos volcanic province. Pg and PmP denote seismic phases refracted within the crust and reflected at the crust-mantle boundary, respectively. (C) Seismic tomography model obtained across the Cocos Ridge in the Galápagos volcanic province. The velocity model and Moho geometry correspond to the average of 100 Monte Carlo inversions (see text for details). The open circles indicate OBS locations. (D) and (E) Density models along the profile shown in Figure 2C. Model 1 has the Moho geometry obtained from seismic tomography (Fig. 2C) and a uniform crustal density of 2900 kg/m³. Model 2 has the same Moho geometry as Model 1, but density has been derived from seismic velocity (Fig. 2C) using the relationships of Carlson and Herrick (1990) and Birch (1961) for oceanic layers 2 and 3. (F) and (G) Residual gravity anomaly (RGA), which is the difference between the observed free-air gravity anomaly and the calculated gravity anomaly for Models 1 and 2. The white band in panel G shows the gravity anomaly uncertainty inferred from the Monte Carlo analysis. (H) and (I) Mantle density anomaly (MDA) corresponding to Models 1 and 2, inferred from topography compensation analysis. The calculation has been repeated for different values of the compensation depth, $Z = 50$ km (red line), 100 km (orange line), and 200 km (blue line). In panel I, the white bands between the colored lines denote mantle density uncertainty inferred from the Monte Carlo analysis for each compensation depth.

larly, estimated mantle density anomalies necessary to account for the observed topography are statistically insignificant, ranging from ±5 kg/m³ for $Z = 50$ km to ±1 kg/m³ for $Z = 200$ km (Fig. 2I). This means that the contribution of lateral crustal density variations is sufficiently significant to account for the observed gravity and topography anomalies. If we do not consider the effect of lateral crustal density variations, we might be tempted to assign an upper mantle origin to the remaining anomaly, which is unsupported by the data. In summary, we reiterate that very different density models can satisfy a given gravity or topography constraint, so it is crucial to include a priori information provided by WAS models and to assess the uncertainty in model parameters before interpreting the results.

In this situation it is not possible to reliably infer mantle density information based on gravity and topography data because if there is a mantle density anomaly, it is smaller than the uncertainty in the methods. Considering lateral crustal density (velocity) variations, especially in regions with laterally varying structure such as oceanic plateaus or aseismic ridges, is thus essential to confidently correct the crustal contributions to the gravity field and to lithospheric buoyancy.

Density Structure of Igneous Provinces. Gravity anomalies and topographic swells typically encompassing igneous provinces are believed to be sustained by a combination of crustal thickening and sublithospheric mantle density anomalies (e.g., Oxburgh and Parmentier, 1977; Phipps Morgan et al., 1995), but the relative importance of each factor in accounting for the observed anomalies remains unclear. Sublithospheric compensation models consider the primary source of buoyancy to be located in the mantle. Possible sources of buoyancy may be mantle density variations associated with thermal anomalies (White and McKenzie, 1989), those caused by melt depletion (Phipps Morgan et al., 1995), or a combination of both.

One example of the latter is the mantle plum-pudding model of Phipps Morgan and Morgan (1999), in which a heterogeneous, hot mantle composed of incompatible element–rich veins within a more depleted matrix wells up and melts in a two-stage process. The first stage corresponds to hotspot melting, in which deep, low-degree melts are extracted from the mantle mixture. Because incompatible element–rich veins are easier to melt, proportionally more of the incompatible elements are extracted from these components, forming ocean island basalts (OIB) as well as a residual column composed of a mixture of leftovers that is hot, buoyant, and depleted in composition compared with normal mantle. In the second stage, depleted MORB are extracted beneath spreading centers from remelting of the mixture of leftovers. In that case, the residual mantle column beneath hotspot swells would be expected to show detectable velocity and density anomalies. Similar models based on flow and melting of a heterogeneous mantle have been proposed to explain the geochemistry of OIB and MORB as well (Ito and Mahoney, 2005). A recent study performed along the Galápagos hotspot–affected segment of the Cocos-Nazca spreading center

(Fig. 2A) indicated that the shoaling of the topographic swell and the decreasing gravity anomaly are sustained by (1) crustal thickening (which accounts for ~50% of the anomaly), (2) thermal buoyancy (~30%), and (3) chemical buoyancy arising from melt depletion (~20%) (Canales et al., 2002). The contribution of mantle anomalies is likely to be significant for oceanic swells located above active hotspots, as indicated by the presence of striking low-velocity, deep-seated anomalies shown by global tomography models (Zhao, 2001; Montelli et al., 2004).

Alternatives to sublithospheric compensation are crustal compensation models, in which the swell is mainly supported by lateral variations of crustal density and thickness, with a minor contribution from mantle density anomalies. This may be the case for igneous provinces located away from the zone of direct hotspot influence. One example is the Marquesas swell, where the buoyancy of the material underplating the island chain has been shown to be able to support the swell almost completely (McNutt and Bonneville, 2000). Another example is the case of the Galápagos volcanic province (Fig. 2), where the upper mantle density anomaly, if any, must be small, practically undetectable, based on gravity and topography modeling. This means that the significance of compositional mantle buoyancy due to melt extraction and depletion may be, at least in some special cases, less than suggested by theoretical studies.

The importance of a proper crustal correction has also been pointed out in connection with the gravity highs observed at ocean-continent transition zones. These gravity highs have typically been associated with high-density crustal roots, in good agreement with the high-velocity lower crust frequently described beneath the transition zones of rifted continental margins (e.g., Kelemen and Holbrook, 1995). The presence of high-velocity, high-density lower crust has usually been explained by the melting of hotter than normal mantle. However, it has recently been demonstrated that the velocity-derived lower crustal density of the eastern Greenland margin is too low to account for the observed gravity high (Korenaga et al., 2001). The most likely explanations are (1) the presence of a dense Fe-rich shallow mantle beneath the margin (i.e., a fertile, mild mantle source) and (2) denser upper crust compared with normal oceanic crust in the ocean-continent transition zone.

GEOPHYSICAL CONSTRAINTS THAT CHARACTERIZE THE MANTLE MELTING PROCESS

A key connection between the physical properties that we can "measure" in volcanic rocks and those of the mantle source material is the mantle melting process. It is therefore essential to establish experimental relationships between the parameters governing the mantle melting process and the physical properties of the resulting crust and include them in mantle melting models to enable direct comparisons between the geophysical observations and the predictions made for a given set of parameters.

In this section we first explain how we estimate variations in the amount of magma supplied (i.e., the volumetric melt flux) and the relative distance between a mid-oceanic ridge and the center (i.e., the point of maximum intensity) of a melting anomaly. Then we use a steady-state, 2-D mantle melting model to illustrate the effects of the different parameters that characterize mantle melting (upwelling rate, melt productivity, water content, presence of a lithospheric lid, composition) on the resulting seismic structure of the igneous crust, and we compare the predictions of the mantle melting model with observations made at different igneous provinces (Table 1).

Constraining the Volumetric Melt Flux Provided by Melting Anomalies and Its Temporal Variations

If we assume that all melts generated within the mantle melting zone are emplaced as seismically observable igneous crust (McKenzie and Bickle, 1988), the excess of crustal thickness compared to normal oceanic crust can be used to estimate the additional melt flux provided by a melting anomaly (e.g., Ito et al., 1997; Sallarès and Charvis, 2003). This value is called the volumetric melt flux for extra crustal production (Q_v) and can be taken as a measure of the intensity of the melting anomaly. Q_v can be calculated from crustal thickness measurements by integrating the excess crustal production along a given profile, as follows:

$$Q_v = U \int_W (h_c(x) - h)dx, \qquad (1)$$

where U is the spreading rate, W is the width of the overthickened crustal segment, χ is along-axis distance, $h_c(\chi)$ is the crustal thickness measured at χ, and h represents normal crustal thickness.

It has to be noted, however, that Q_v is a measure of the volume of melt "instantaneously" provided by a melting anomaly, so there are two conditions that must be met to correctly estimate this value: (1) the χ-coordinate must be parallel to an isochron, and (2) the limits of integration, $\pm W/2$, must encompass the full overthickened crustal section between two contiguous oceanic basins. As we explained in the first section of this article, WAS is probably the best method for precisely estimating the crustal geometry along a profile. In a case in which the seismic profile is parallel to an isochron, we can directly integrate the overthickening to calculate the volumetric melt flux provided by the melting anomaly in a given period of time. Unfortunately, this type of case is rare. In addition, the width of the overthickened crustal section is difficult to determine in the absence of seismic data.

A good opportunity to estimate both values arises when melting anomalies interact with nearby oceanic spreading centers, which seems to be the case for at least 21 of the 30–50 identified present-day hotspots (Ito et al., 2003). Bathymetric and gravity data show that the along-axis width of the hotspot-affected region can be very large, especially for hotspots located just beneath or near the ridge axis, such as Iceland (~1500 km; Ito et al., 1996) or Galápagos (~1300 km; Ito et al., 1997). Long-lasting hotspot-ridge interactions generate volcanic edifices that form conjugate age-progressive tracks at both sides of the spreading center, for instance, the V-shaped Cocos and Carnegie ridges in the Galápagos volcanic province. By comparing the crustal geometry across different segments of the volcanic tracks it is possible to estimate the temporal variations of the melting anomaly intensity, which can in turn be interpreted in terms of the relative distance between the hotspot center and the spreading axis (Sallarès and Charvis, 2003).

To a first order, the along-isochron Moho geometry can be estimated with a function decaying linearly, from H at $\chi = 0$ to h at $\chi = \pm W/2$. Under these circumstances, equation 1 can be then resolved analytically and is reduced to

$$Q_v = \frac{(H - h)UW}{2}, \qquad (2)$$

where H is the maximum crustal thickness along the ridge axis in a given period of time.

If we assume that temporal variations of the maximum potential intensity of the melting anomaly are primarily the result of variations in the distance between the melting anomaly center and the spreading center, Q_v must be maximum when the melting anomaly is ridge-centered and asymptotic to zero for increasing hotspot-ridge distance. This tendency can be represented by the following function:

$$Q_v = Q_M exp(-\beta y), \qquad (3)$$

where Q_M represents the maximum potential volumetric melt flux (i.e., for a ridge-centered anomaly), y is the relative hotspot-ridge distance, and β is a factor that determines the shape of the function. This function does not represent a rigorously based physical model but is only an assumed idealization. Then, defining $\chi = Q_v/Q_M$, and comparing this value for two different profiles, we obtain

$$y\log\chi_0 = y_0\log\chi, \qquad (4)$$

where χ (χ_0) represents the relative intensity of the along-axis melting anomaly when it is located at y (y_0) distance from the spreading center.

It has been shown that W depends primarily on U at the time of crustal accretion (e.g., Ito and Lin, 1995). In the case that U does not vary significantly over a given period of time, we can approximate $\chi \sim (H - h)/(H_M - h)$, and the relative distance between the hotspot center and the spreading axis can be estimated based on equation 4.

In conclusion, if we know the maximum potential crustal thickness that can be generated by a melting anomaly, H_M, the hotspot-ridge distance, y_0, and the maximum along-axis crustal thickness, H_0, in a given period of time, equation 4 provides a practical way to place first-order constraints on the relative intensity of the along-axis melting anomaly, χ_1, and thus on the hotspot-ridge distance, y_1, at any time for which we know the maximum along-axis crustal thickness, H_1.

One example of this application is with regard to the Galápagos volcanic province, where the comparison of crustal geometry along the present-day axis of the Cocos-Nazca spreading center and across the Cocos, Carnegie, and Malpelo volcanic tracks (Fig. 2A) has allowed constraint of (1) the temporal variations of the volumetric melt flux, (2) the relative intensity of the Galápagos melting anomaly compared with the Hawaiian and Icelandic ones, (3) the relative motion of the Cocos-Nazca spreading center with respect to the Galápagos hotspot, and (4) the tectonic evolution of the Galápagos volcanic province during the last ~20 m.y. (Sallarès and Charvis, 2003). Although these estimates are only approximate and depend heavily on several assumptions, the remarkable agreement between the results obtained and those of a number of previous studies using independent magnetic and Global Positioning System data (e.g., Barkhausen et al., 2001; Trenkamp et al., 2002) suggests that the approach works and is suitable for placing first-order constraints on the geodynamic evolution of the Galápagos volcanic province. Although not all igneous provinces are well suited to this type of study, the procedure can probably be extrapolated to other provinces characterized by the presence of age-progressive volcanic tracks, such as the Hawaiian-Emperor seamounts and island chain.

Constraining the Parameters That Govern Mantle Melting

Characterizing mantle melting anomalies based on crustal structure information is an approach that has received increasing attention during the past fifteen years. The first attempt to quantitatively relate seismic crustal structure to mantle melting parameters was made by White and McKenzie (1989). Their approach consisted of relating variations in crustal thickness (H) and V_p with changes in mantle potential temperature for different factors of lithospheric stretching. They showed that if the temperature of the asthenosphere increases up to 1480 °C, some 200 °C above normal, the percentage of MgO increases systematically from ~10% to 18%. Thus, a systematic change in the MgO content of the melt with mantle temperature is predicted to cause systematic changes in V_p and density of the igneous rocks formed on rifted margins. White et al. (1992) combined seismic observations with a rare earth element inversion technique to infer the amount of mantle melting generated as well as the depth interval over which it occurred. They concluded that the crustal structure and geochemical composition of oceanic and continental igneous provinces were results of "decompres-

sion melting of abnormally hot mantle brought to the base of the lithosphere by plumes."

A second step in relating the seismic structure of the igneous crust to mantle melting parameters was made by Kelemen and Holbrook (1995). Their key contribution was to establish a function relating V_p in igneous rocks to the mean pressure (P) of melting and melt fraction (F) of the parental melt V_p(F, P) by applying multiple linear regression to the melt data of Kinzler and Grove (1992). The relationship obtained was tested by comparing its predictions with experimental results not included in the regression. Kelemen and Holbrook (1995) concluded that the high-velocity rocks of the East Coast margin were the result of a thermal anomaly. Korenaga et al. (2002) followed the same approach to derive a more accurate relationship compiling an extensive set of high-quality experiments of mantle peridotites (e.g., Kinzler and Grove, 1992; Hirose and Kushiro, 1993; Kinzler, 1997). They developed a 1-D steady-state mantle melting model including the effects of a preexisting lithospheric lid and active mantle upwelling to compare predicted crustal thickness and lower crustal seismic velocity with seismic observations made at the southeastern Greenland margin. Their main conclusion was that the thick igneous crust with 6.9–7.0 km/s average lower crustal V_p results from active upwelling of normal-temperature mantle. Alternatively, they showed that the effect of major element heterogeneity in the mantle source can also be a key factor accounting for the excess of magmatism.

An option that is not considered in the model of Korenaga et al. (2002) is the possible influence of deep damp melting between the dry and wet solidus for a volatile-bearing mantle. Damp melting has been suggested to account for a significant part of the total volume of melt, even if the melting rate is an order of magnitude lower than that of dry melting (Hirth and Kohlstedt, 1996; Braun et al., 2000), if it is coupled with vigorous upwelling at the base of the mantle melting zone (>70 km deep). Numerical models indicate that the viscosity increase associated with dehydration prevents buoyancy forces from contributing significantly to mantle upwelling above the dry solidus, indicating that active upwelling must be restricted only to the damp melting zone (Ito et al., 1999). Deep damp melting has been proposed as the main source of magmatism, for instance, in the Galápagos volcanic province (Cushman et al., 2004) and Iceland (Maclennan et al., 2001).

Sallarès et al. (2005) modified the quantitative approach of Korenaga et al. (2002) to simulate mantle melting for a 2-D, steady-state, triangular melting regime resulting from mantle corner flow (e.g., Plank and Langmuir, 1992). They incorporated the effect of deep damp melting (Hirth and Kohlstedt, 1996; Braun et al., 2000). This model assumes perfect mixing and focusing of melts to generate oceanic crust, and it restricts active upwelling to beneath the dry solidus (Ito et al., 1999). The average degree (pressure) of melting is calculated as the average degree (pressure) of melting of all the individual parcels of mantle pooled in the crust (Forsyth, 1993; Plank et al., 1995), assum-

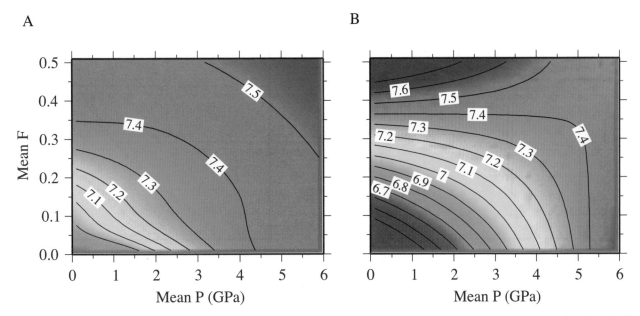

Figure 3. (A) Contours of predicted seismic velocity (in km/s) for mantle melts as a function of their mean pressure (P) and fraction (F) of melting using the multiple linear regression of data from melting experiments of mantle peridotites. (B) Same as panel A, but for a hypothetical fertile mantle melt composed of 70% depleted pyrolitic mantle and 30% mid-ocean ridge basalt (modified from Korenaga et al., 2002).

ing a linear melting function as mantle wells up. The total volume of melts allows crustal thickness to be estimated, and the mean pressure and degree of melting are used to predict V_p based on the multilinear regression of Korenaga et al. (2002) for mantle peridotites (Fig. 3A).

In Figure 4 we illustrate the potential effect of the different mantle melting model parameters with regard to both H and mean lower crustal V_p in order to provide a quantitative framework for comparison with seismic observations made at different igneous provinces (Table 1). Key parameters used are the mantle upwelling ratio, χ (i.e., the ratio of mantle upwelling to seafloor spreading or plate velocity); the melt productivity within the dry and damp melting regions, $\Gamma_d = 10\%–20\%/GPa$ (McKenzie, 1984; Langmuir et al., 1992) and $\alpha v \delta \, \Gamma_w \sim 1\%/GPa$ (Braun et al., 2000), respectively; the thickness of the damp mantle melting zone, $\Delta z = 50–75$ km (Hirth and Kohlstedt, 1996; Braun et al., 2000); the upwelling decay between the base and the top of the damp melting zone, $\alpha \sim 0.2$ (Ito et al., 1999); and the thickness of the preexisting oceanic lithosphere, $b = 0–70$ km.

For reference, in Figure 4A and B we show the predicted H and V_p versus mantle T_p for a melting model with $\Gamma_d = 15\%/$GPa, $\Gamma_w = 1\%/GPa$, $\Delta z = 50$ km, $\alpha = 0.2$, and $b = 0$ km. The different lines plotted correspond to different mantle upwelling ratios at the base of the mantle melting zone. The solid blue line corresponds to $\chi = 0$, i.e., to passive dry mantle melting with no contribution from deeper damp melting. For a "normal mantle temperature" of $T_p \sim 1300$ °C the model predicts $H \sim 7$ km with $V_p \sim 7.1$ km/s, in agreement with global compilations for normal

oceanic crust (e.g., White et al., 1992). A 200 °C increase in T_p would produce a ~30-km-thick crust with $V_p \sim 7.3$ km/s, consistent with the calculations of White and McKenzie (1989). Incorporating progressively higher mantle upwelling ratios with identical melting parameters results in consistently thicker crust and slightly higher V_p for a given T_p. The latter is due to the fact that damp melting is extracted at high melting pressures and low melting fractions, resulting in lower SiO_2 and higher MgO contents in mantle melts (Fig. 3A). Thus, active upwelling ($\chi = 20$) of damp mantle source with a normal T_p of only 1300 °C could also account for a 23-km-thick crust with $V_p \sim 7.15$ km/s.

Figure 4C–F shows the potential effect of Γ_d and Γ_w on the resulting crustal structure for $\chi = 5$. Given that active upwelling is restricted to beneath the dry melting zone, the effect of Γ_d variations in both H and V_p is minor compared with those of Γ_w. Thus, for a $T_p = 1300$ °C, a 6% increase in Γ_d would result in a crust ~3 km thicker, whereas an increase of only 1.5% in Γ_w would produce a crust ~7 km thicker. The model also predicts an increase of V_p with increasing melt productivity, which is more significant in the case of damp melting. However, the increase would be ~0.1 km/s at the most.

Another parameter with a notable effect on crustal structure is the thickness of the damp melting zone (Δz). Figure 4G and H shows that a 90-km-thick damp melting zone (with $\chi = 5$ and $T_p \sim 1300$ °C) could generate a crustal thickening, ΔH, of ~ 6 km, with a V_p increase of 0.1 km/s.

In summary, it is interesting to note that it does not seem possible to find a plausible combination of T_p, χ, Γ_d, Γ_w, and Δz to explain either (1) crustal roots (underplated bodies) with $V_p >$

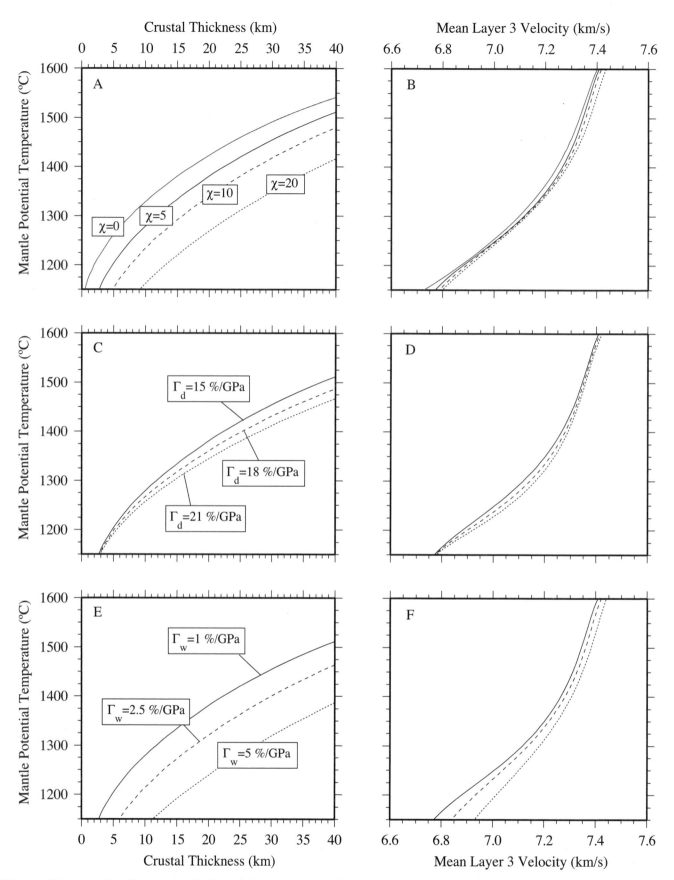

Figure 4. Diagrams of predicted crustal thickness (left) and mean layer 3 velocity (right) versus T_p of the mantle source using the 2-D, steady state, mantle melting model of Sallarès et al. (2005) and the $V_p(F, P)$ relationship of Korenaga et al. (2002) for mantle peridotite (Fig. 3A). Reference model parameters are $\Gamma_d = 15\%/GPa$, $\Gamma_w = 1\%/GPa$, $\chi = 10$, $\Delta z = 50$ km, $\alpha = 0.2$, and $b = 0$ km (see text for definitions). (A) and (B) Effect of increasing mantle upwelling ratio at the base of the mantle melting zone, $\chi = 0$ (blue line, dry melting only), $\chi = 5$ (solid black line), $\chi = 10$ (dashed line), and $\chi = 20$ (dotted line). (C) and (D) Effect of increasing melt productivity in the dry melting zone, $\Gamma_d = 15\%/GPa$ (solid line), $\Gamma_d = 18\%/GPa$ (dashed line), and $\Gamma_d = 21\%/GPa$ (dotted line). (E) and (F) Effect of increasing melt productivity in the damp melting zone, $\Gamma_w = $

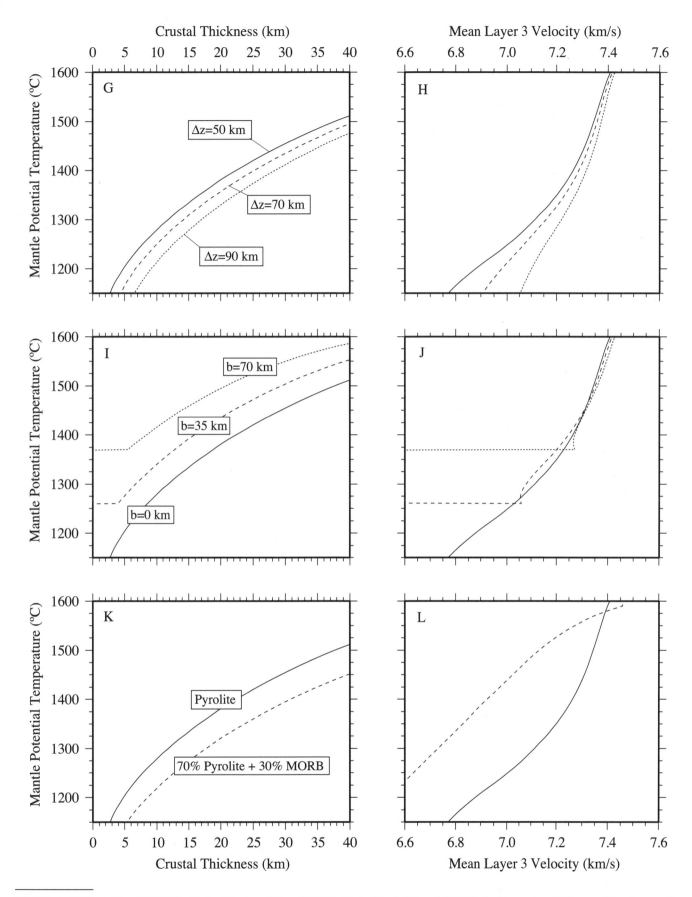

1%/GPa (solid line), Γ_w = 2.5%/GPa (dashed line), and Γ_w = 5%/GPa (dotted line). (G) and (H) Effect of increasing thickness of the damp melting zone, Δz = 50 km (solid line), Δz = 70 km (dashed line), and Δz = 90 km (dotted line). (I) and (J) Effect of increasing thickness of the preexisting lithospheric lid, b = 0 km (solid), b = 35 km (dashed), and b = 70 km (dotted). (K) and (L) Effect of mantle source heterogeneity. The solid lines have been calculated with the reference model parameters and the V_p(F, P) relationship of Figure 3A, and the dashed line for a hypothetical fertile mantle source composed of 70% depleted pyrolitic mantle and 30% mid-ocean ridge basalt (MORB) (Fig. 3B).

7.5 km/s that have been described from numerous igneous provinces (Table 1) or (2) the *H*-V_p anticorrelation found at other igneous provinces such as the Galápagos, Kerguelen, or north Atlantic volcanic provinces (Table 1). It is therefore necessary to consider the effect of other mantle melting parameters when trying to explain these observations.

A common characteristic of igneous provinces showing high-velocity, underplated bodies is that most are thought to have originated in midplate settings (Table 1). Thus, the presence of a lithospheric lid may play a role in determining the amount and properties of the melts generated. The lid will prevent shallow melting from occurring, resulting in limited amounts of higher-pressure and lower-fraction melting (higher MgO) within the damp melting zone, and one would therefore expect a relative crustal overthickening with high-velocity cumulates (e.g., Kelemen and Holbrook, 1995). The effect of a preexisting lithospheric lid of 35 km (with $\chi = 5$ to enable direct comparison with the previous tests) is shown in Figure 4I and J. Interestingly, the model predicts that in this case the occurrence of damp melting with a normal $T_p \sim 1300$ °C would result in a ΔH of ~5 km, in remarkable agreement with seismic observations made at intraplate oceanic provinces like Hawaii, Réunion, the Marquesas, or the Canary Islands (Table 1). However, no significant difference is visible in the predicted V_p (~7.1 km/s). The reason for this lack of sensitivity regarding the predicted V_p is that a thicker lid also increases Na_2O and FeO contents. The result is to produce components of lower seismic velocity, counterbalancing the effect of the MgO content increase. The V_p(F, P) relationship of Korenaga et al. (2002) takes into account this balance between the different compositional effects. For the same melting parameters, a $T_p > 1375$ °C would be necessary to generate melting beneath a thicker lithospheric lid of $b = 70$ km. In this case, a moderate temperature increase of less than 100 °C above normal ($T_p \leq 1400$ °C) would generate a ΔH of 7–8 km, with a significantly higher V_p of 7.3 km/s. As stated earlier, still higher V_p could be obtained with a thicker damp melting zone in combination with a slightly higher Γ_w.

However, none of the melting parameters checked up to this point can explain the lack of a positive *H*-V_p correlation found in several igneous provinces, such as the Galápagos, Kerguelen, and the north Atlantic (Table 1). It has previously been suggested that an alternative explanation for this observation could be the presence of a compositional heterogeneity in the mantle source. As explained earlier, the V_p(F, P) relationship of Figure 3A is valid only for mantle compositions similar to that of pyrolite, but too few melting experiments with source compositions different from pyrolite exist to allow us to develop a quantitative model including the effect of source heterogeneities. The only attempt to illustrate the potential effect of major element heterogeneity in the mantle source was made by Korenaga et al. (2002). They developed a relationship between V_p and the mean fraction and pressure of melting for a hypothetical source composed of 70% depleted pyrolitic mantle (Kinzler, 1997) and 30%

MORB (Hofmann, 1988) (Fig. 3B). The presence of a fertile component in the mantle source (i.e., the MORB) results in higher amounts of melting with lower MgO and considerably higher FeO content for a given T_p, compared with a pyrolitic source. This effect is clearly observed in Figures 4K and L. For $\chi = 5$ and a $T_p = 1300$ °C, an 18-km-thick crust with 6.7–6.8 km/s would be generated, in contrast with the 12 km and 7.1 km/s that would result from pyrolite melting.

In summary, the only plausible way of explaining the *H*-V_p anticorrelation seems to be the presence of a striking major-element heterogeneity, which could be related to the presence of a fertile components in the mantle source, such as eclogite derived from recycled oceanic lithosphere. The possible influence of such a fertile anomaly in the mantle source to explain the geochemical and geophysical observations has been suggested for a number of hotspots, including Iceland (Korenaga and Kelemen, 2000; Foulger and Anderson, 2005; Foulger et al., 2005), Hawaii (Hauri, 1996; Sobolev et al., 2000), and the Galápagos (Sallarès et al., 2005). What currently seems evident is that the causes of hotspot magmatism are probably manifold. It appears clear to us that excess mantle temperature, even if coupled with deep damp melting, is not the only way of explaining the crustal structure observed at many igneous provinces. In some cases, as in Galápagos, observations are even contrary to what would be expected from the melting of hotter than normal mantle.

However, it has to be noted that it is not easy to find a mechanism capable of explaining the upwelling of an Fe-rich mantle source. Subducted oceanic crust is denser than normal pyrolitic mantle above the bottom of the mantle transition zone (Ringwood and Irifune, 1988), so it will tend to lie at this level of neutral buoyancy unless it is sent back to the surface by an elevating stream. Korenaga (2004) proposed sublithospheric convection driven by surface cooling as a plausible mechanism in the framework of continental breakup settings such as the north Atlantic volcanic province. We do not think that the same mechanism can be applied to intraplate hotspots, so we prefer to restrict our discussion to what can be deduced from the crustal-based approach.

Another process that could play a role in determining the seismic velocity of the igneous crust is the occurrence of subcrustal fractionation. When magmas fractionate, part of the high-velocity minerals (e.g., olivine, clinopyroxene) are segregated from the melts, and therefore the seismic velocities of the resulting igneous rocks could be lower than those expected for the primary, mantle-derived melt. However, experiments on fractional crystallization modeling at high pressures (Korenaga et al., 2002) showed that the effect of subcrustal fractionation is significant only for $V_p > 7.5$ km/s. In the case of normal $V_p < 7.2$ km/s, this effect is negligible.

A promising interdisciplinary approach to inference of the parameters controlling mantle melting consists of comparing the observed geochemical signatures, physical properties, and Moho geometry of the igneous crust, with predictions based on experimental data with fluid-dynamically consistent melting models for different source compositions. A first attempt was recently performed by Ito and Mahoney (2005). These authors developed a method that simulates fractional melting with thermodynamically consistent melting functions for a heterogeneous mantle composed of enriched and depleted mantle peridotite and pyroxenite. They demonstrated that variations in both mantle flow and lithospheric thickness between mid-ocean ridges and hotspots may lead to significant differences between the isotopic and trace element characteristics of MORB and OIB, independent of any compositional differences between their respective mantle sources.

SUMMARY AND CONCLUSIONS

In this work we advocate an interdisciplinary approach to inference of the parameters that govern mantle melting processes and the characteristics of the melting anomalies that comprise hotspot magmatism, based on crustal structure information obtained by means of seismic and gravity methods. Owing to the nonuniqueness of velocity and density modeling, comprehensive uncertainty analysis must be performed to quantitatively interpret the geophysical models as well as to use them to assess mantle melting processes. This linkage is crucial to relate the properties that can be measured in rocks to the underlying mantle properties and dynamics.

Combining crustal seismology and gravity modeling enables the velocity and density structure and the Moho geometry of the igneous crust to be constructed. We show with the example of the Cocos Ridge in the Galápagos volcanic province that the crustal contribution to the gravity field can be more significant than is usually thought. In this particular case, lateral variations of the velocity-derived crustal density model account for most of the observed gravity and topography anomalies without the need to call for anomalous mantle densities. The remaining gravity and topography anomalies, if any, are under the uncertainty threshold of the method.

The example of the Galápagos volcanic province is also used to show that the comparison of crustal thickness variations along age-progressive volcanic tracks can be used to approximate the temporal variations in the volumetric melt flux provided by a melting anomaly, enabling us, in turn, to estimate variations on the relative distance between the spreading axis and the hotspot center.

The connection between the crustal structure of the igneous crust and the parameters governing mantle melting can be established based on models that simulate mantle melting and incorporate empirical relationships between the seismic velocity of igneous rocks and the pressure and fraction of melting. We did several tests using a 2-D, steady-state mantle melting model and different combinations of mantle melting parameters to illustrate that the presence of water in the mantle source, a lithospheric lid limiting the extent of the mantle melting zone, and the composition of the source itself can be as important as man-

tle temperature in controlling the amount of melt produced as well as the physical properties of the igneous crust. On one hand, we showed that the origin of ~5-km-thick, high-velocity underplated bodies frequently described at midplate settings (e.g., Hawaii, Réunion, Marquesas, Canary Islands) could be explained by the presence of a preexisting lithospheric lid on top of a damp (or fertile) melting anomaly with no significant thermal effect. On the other hand, we illustrated that the lack of the positive H-V_p correlation described in a number of igneous provinces originated in the vicinity of spreading centers (e.g., the Galápagos, Kerguelen, Nazca, and the north Atlantic) is difficult to reconcile with high temperature of a pyrolite mantle source. A likely explanation may be the heterogeneity of a major element, such as eclogite derived from recycled oceanic lithosphere, in the mantle source.

Additional melting experiments with source compositions different from pyrolite are required to improve quantitative assessment of the melting process, including the effect of source heterogeneities. Considering predictions of the geochemical signatures of igneous rocks based on experimental data with fluid-dynamically consistent melting models constitutes a promising approach to improve our understanding of mantle melting and dynamics.

As a summary of the work performed, we now list the main steps of a suitable procedure to characterize mantle melting anomalies based on this crustal approach.

1. Tomography-based modeling of WAS data to estimate the seismic structure (Moho geometry, velocity field) of the crust.
2. Monte Carlo–type analysis of model parameter uncertainties.
3. Velocity-derived density modeling of gravity data to determine the crustal contribution to topography and the gravity field.
4. Estimation of the mantle density and isostatic compensation depth corrected for the different crustal contributions (topography, Moho geometry, crustal density variations).
5. Propagation of seismic parameter uncertainties to predicted gravity and inferred mantle density.
6. Calculation of the volumetric flux of melting anomalies and its temporal variation by comparing the seismic structure along age-progressive volcanic tracks.
7. Inference of the relative significance of mantle melting parameters (temperature, upwelling ratio, damp melting, source composition) in the anomalous crustal production process on the basis of existing empirical relationships between bulk crustal velocity and the mean pressure and degree of melting.
8. If possible, comparison of the observed geochemical signatures and seismic structure of the igneous crust with predictions based on experimental data with fluid-dynamically consistent melting models for different source compositions.

ACKNOWLEDGMENTS

The work presented here benefited from fruitful discussions with Ph. Charvis and E.R. Flueh. Constructive reviews by editors Gillian R. Foulger and Donna Purdy, as well as Tim Minshull and an anonymous reviewer, helped to significantly improve the initial version of the manuscript. A significant part of the data processing and modeling that constitute the basis of the discussion was completed during part of the four-year visit of the authors to the Géosciences Azur Laboratory, Villefranche-sur-Mer, France. A.C. was supported by a grant from the Institute de Recherche pour le Développement (IRD) during that time. The WAS data used to construct the models were acquired during the German-French PAGANINI-1999 SALIERI-2001 surveys. Special thanks to Unai, both authors' second half.

REFERENCES CITED

Anderson, R.N., McKenzie, D., and Sclater, J.G., 1973, Gravity, bathymetry and convection in the Earth: Earth and Planetary Science: Letters, v. 18, p. 391–407.

Barkhausen, U., Ranero, C.R., von Huene, R., Cande, S.C., and Roeser, H.A., 2001, Revised tectonic boundaries in the Cocos plate off Costa Rica: Implications for the segmentation of the convergent margin and for plate tectonic models: Journal of Geophysical Research, v. 106, p. 19, 207–19,220.

Barton, P.J., 1986, The relationship between seismic velocity and density in the continental crust—A useful constraint?: Geophysical Journal of the Royal Astronomical Society of London, v. 87, p. 195–208.

Bickel, S.H., 1990, Velocity-depth ambiguity of reflection traveltimes: Geophysics, v. 55, p. 266–276, doi: 10.1190/1.1442834.

Birch, F., 1961, The velocity of compressional waves in rocks to 10 kilobars, part 2: Journal of Geophysical Research, v. 66, p. 2199–2224.

Bonatti, E., 1990, Not so hot "hot spots" in the oceanic mantle: Science, v. 250, no. 4977, p. 107–111, doi: 10.1126/science.250.4977.107.

Bratt, S.R., and Purdy, G.M., 1984, Structure and variability of oceanic crust in the flanks of the East Pacific Rise between 11° and 13°N: Journal of Geophysical Research, v. 89, p. 6111–6125.

Braun, M.G., Hirth, G., and Parmentier, E.M., 2000, The effects of deep damp melting on mantle flow and melt generation beneath mid-oceanic ridges: Earth and Planetary Science Letters, v. 176, p. 339–356, doi: 10.1016/S0012-821X(00)00015-7.

Canales, J.P., Dañobeitia, J.J., and Watts, A.B., 2000, Wide-angle seismic constraints on the internal structure of Tenerife, Canary Islands: Journal of Volcanology and Geothermal Research, v. 103, p. 65–81, doi: 10.1016/S0377-0273(00)00216-X.

Canales, J.P., Ito, G., Detrick, R.S., and Sinton, J., 2002, Crustal thickness along the western Galápagos Spreading Center and compensation of the Galápagos Swell: Earth and Planetary Science Letters, v. 203, p. 311–327, doi: 10.1016/S0012-821X(02)00843-9.

Caress, D.W., Nutt, M.K., Detrick, R.S., and Mutter, J.C., 1995, Seismic imaging of hotspot-related crustal underplating beneath the Marquesas islands: Nature, v. 373, p. 600–603, doi: 10.1038/373600a0.

Carlson, R.L., and Herrick, C.N., 1990, Densities and porosities in the oceanic crust and their variations with depth and age: Journal of Geophysical Research, v. 95, p. 9153–9170.

Charvis, Ph., and Operto, S., 1999, Structure of the Cretaceous Kerguelen Volcanic Province (southern Indian Ocean) from wide-angle seismic data: Journal of Geodynamics, v. 28, p. 51–71.

Charvis, Ph., Laesanpura, A., Gallart, J., Hirn, A., Lépine, J.-C., de Voogd, B., Hello, Y., Pontoise, B., and Minshull, T.A., 1999, Spatial distribution of hotspot material added to the lithosphere under La Réunion, from wide-angle seismic data: Journal of Geophysical Research, v. 104, p. 2875–2893, doi: 10.1029/98JB02841.

Christensen, N.I., and Mooney, W.D., 1995, Seismic velocity structure and composition of the continental crust: A global view: Journal of Geophysical Research, v. 100, p. 9761–9788, doi: 10.1029/95JB00259.

Coffin, M.F., and Eldholm, O., 1994, Large igneous provinces: Crustal structure, dimensions and external consequences: Reviews of Geophysics, v. 32, p. 1–36, doi: 10.1029/93RG02508.

Cushman, B.J., Sinton, J.M., Ito, G., and Dixon, J.E., 2004, Glass compositions, plume-ridge interaction, and hydrous melting along the Galápagos Spreading Center, 90°30′W to 98°W: Geochemistry, Geophysics, Geosystems, v. 5, Q089E17, doi:10.1029/2004GC000709.

Darbyshire, F.A., Bjarnason, I.Th., White, R.S., and Flovenz, O.G., 1998, Crustal structure above the Iceland mantle plume imaged by the ICEMELT refraction profile: Geophysical Journal International, v. 135, p. 1131–1149, doi: 10.1046/j.1365-246X.1998.00701.x.

Darbyshire, F.A., White, R.S., and Priestley, K.F., 2000, Structure of the crust and uppermost mantle of Iceland from a combined seismic and gravity study: Earth and Planetary Science Letters, v. 181, p. 409–428, doi: 10.1016/S0012-821X(00)00206-5.

Detrick, R.S., Collins, J., Stephen, R., and Swift, S., 1994, In situ evidence for the nature of the layer 2/3 boundary in oceanic crust: Nature, v. 370, p. 288–290, doi: 10.1038/370288a0.

Forsyth, D.W., 1993, Crustal thickness and the average depth and degree of melting in fractional melting models of passive flow beneath mid-ocean ridges: Journal of Geophysical Research, v. 98, p. 16,073–16,079.

Foulger, G.R., and Anderson, D.L., 2005, A cool model for the Iceland hotspot: Journal of Volcanology and Geothermal Research, v. 141, p. 1–22, doi: 10.1016/j.jvolgeores.2004.10.007.

Foulger, G.R., Natland, J.H., and Anderson, D.L., 2005, A source for Icelandic magmas in remelted Iapetus crust: Journal of Volcanology and Geothermal Research, v. 141, p. 23–44, doi: 10.1016/j.jvolgeores.2004.10.006.

Green, D.H., and Falloon, T.J., 2005, Primary magmas at mid-ocean ridges, "hot spots" and other intraplate settings: Constraints on mantle potential temperature, in Foulger, G.R., et al., eds., Plates, plumes, and paradigms: Boulder, Colorado, Geological Society of America Special Paper 385, p. 217–248.

Green, D.H., Falloon, T.J., Eggins, S.M., and Yaxley, G.M., 2001, Primary magmas and mantle temperatures: European Journal of Mineralogy, v. 13, p. 437–451, doi: 10.1127/0935-1221/2001/0013-0437.

Grevemeyer, I., Flueh, E.R., Reichert, C., Bialas, J., Kläschen, D., and Kopp, C., 2001, Crustal architecture and deep structure of the Ninetyeast Ridge hotspot trail from active-source ocean bottom seismology: Geophysical Journal International, v. 144, p. 414–431, doi: 10.1046/j.0956-540X.2000.01334.x.

Hampel, A., Kukowski, N., Bialas, J., Huebscher, C., and Heinbockel, R., 2004, Ridge subduction at an erosive margin—The collision zone of the Nazca Ridge in southern Peru: Journal of Geophysical Research, v. 109, p. B02101, doi: 10.1029/2003JB002593.

Hauri, E.H., 1996, Major element variability in the Hawaiian mantle plume: Nature, v. 382, p. 415–419, doi: 10.1038/382415a0.

Hirose, K., and Kushiro, I., 1993, Partial melting of dry peridotites at high pressures: Determination of composition of melts segregated from peridotite using aggregates of diamond: Earth and Planetary Science Letters, v. 114, p. 477–489, doi: 10.1016/0012-821X(93)90077-M.

Hirth, G., and Kohlstedt, D.L., 1996, Water in the oceanic upper mantle: Implications for rheology, melt extraction and the evolution of the lithosphere: Earth and Planetary Science Letters, v. 144, p. 93–108, doi: 10.1016/0012-821X(96)00154-9.

Hobro, J., Singh, S.C., and Minshull, T.A., 2003, Three-dimensional tomographic inversion of combined reflection and refraction seismic traveltime data: Geophysical Journal International, v. 152, p. 79–93, doi: 10.1046/j.1365-246X.2003.01822.x.

Hofmann, A.W., 1988, Chemical differentiation of the Earth: The relationship between mantle, continental crust and oceanic crust: Earth and Planetary Science Letters, v. 90, p. 297–314, doi: 10.1016/0012-821X(88)90132-X.

Ito, G., and Lin, J., 1995, Mantle temperature anomalies along the present and paleoaxes of the Galápagos Spreading Center as inferred from gravity analyses: Journal of Geophysical Research, v. 100, p. 3733–3745, doi: 10.1029/94JB02594.

Ito, G., and Mahoney, J.J., 2005, Flow and melting of a heterogeneous mantle, 1, Method and importance to the geochemistry of ocean island and mid-ocean ridge basalts: Earth and Planetary Science Letters, v. 230, p. 29–46, doi: 10.1016/j.epsl.2004.10.035.

Ito, G., Lin, J., and Gable, C.W., 1996, Dynamics of mantle flow and melting at a ridge-centered hotspot: Iceland and the Mid-Atlantic Ridge: Earth and Planetary Science Letters, v. 144, p. 53–74, doi: 10.1016/0012-821X(96)00151-3.

Ito, G., Lin, J., and Gable, C.W., 1997, Interaction of mantle plumes and migrating mid-ocean ridges: Implications for the Galápagos plume-ridge system: Journal of Geophysical Research, v. 102, p. 15,403–15,417, doi: 10.1029/97JB01049.

Ito, G., Shen, Y., Hirth, G., and Wolfe, C.J., 1999, Mantle flow, melting and dehydration of the Iceland mantle plume: Earth and Planetary Science Letters, v. 165, p. 81–96, doi: 10.1016/S0012-821X(98)00216-7.

Ito, G., Lin, J., and Graham, D., 2003, Observational and theoretical studies of the dynamics of mantle plume–mid-ocean ridge interaction: Reviews of Geophysics, v. 41, no. 4, p. 1017, doi: 10.1029/2002RG000117.

Kelemen, P.B., and Holbrook, W.S., 1995, Origin of thick, high-velocity igneous crust along the U.S. East Coast margin: Journal of Geophysical Research, v. 100, p. 10,077–10,094, doi: 10.1029/95JB00924.

Kinzler, R.J., 1997, Melting of mantle peridotite at pressures approaching the spinel to garnet transition: Application to mid-ocean ridge basalt petrogenesis: Journal of Geophysical Research, v. 102, p. 852–874, doi: 10.1029/96JB00988.

Kinzler, R.J., and Grove, T.L., 1992, Primary magmas of mid-ocean ridge basalts, 2, Applications: Journal of Geophysical Research, v. 97, p. 6907–6926.

Korenaga, J., 2004, Origin of the Iceland hotspot and the North Atlantic Igneous Province: http://www.mantleplumes.org.

Korenaga, J., and Kelemen, P.B., 2000, Major element heterogeneity in the mantle source of the North Atlantic igneous province: Earth and Planetary Science Letters, v. 184, p. 251–268.

Korenaga, J., Holbrook, W.S., Kent, G.M., Kelemen, P.B., Detrick, R.S., Larsen, H.-C., Hopper, J.R., and Dahl-Jensen, T., 2000, Crustal structure of the southeast Greenland margin from joint refraction and reflection seismic tomography: Journal of Geophysical Research, v. 105, p. 21,591–21,614, doi: 10.1029/2000JB900188.

Korenaga, J., Holbrook, W.S., Detrick, R.S., and Kelemen, P.B., 2001, Gravity anomalies and crustal structure at the southeast Greenland margin: Journal of Geophysical Research, v. 106, p. 8853–8870, doi: 10.1029/2000JB900416.

Korenaga, J., Kelemen, P.B., and Holbrook, W.S., 2002, Methods for resolving the origin of large igneous provinces from crustal seismology: Journal of Geophysical Research, v. 107, no. B9, 2178, doi:10.1029/2001JB001030.

Langmuir, C.H., Klein, E.M., and Plank, T., 1992, Petrological systematics of mid-ocean ridge basalts: Constraints on melt generation beneath mid-oceanic ridges, in Phipps-Morgan, J., Blackman, D.K., and Sinton, J.M., eds., Mantle flow and melt generation at mid-oceanic ridges: Washington, D.C., American Geophysical Union, Geophysical Monograph Series, v. 71, p. 183–280.

Maclennan, J., McKenzie, D., and Gronvold, K., 2001, Plume-driven upwelling under central Iceland: Earth and Planetary Science Letters, v. 194, p. 67–82, doi: 10.1016/S0012-821X(01)00553-2.

McKenzie, D., 1984, The generation and compaction of partially molten rock: Journal of Petrology, v. 25, p. 713–765.

McKenzie, D., and Bickle, M.J., 1988, The volume and composition of melt generated by extension of the lithosphere: Journal of Petrology, v. 29, p. 625–679.

McNutt, M.K., and Bonneville, A., 2000, A shallow, chemical origin, for the Marquesas swell: Geochemistry, Geophysics, Geosystems, v. 1, no. 6, doi: 10.1029/1999GC000028.

Miura, S., Suyehiro, K., Shinohara, M., Takahashi, N., Araki, E., and Taira, A., 2004, Seismological structure and implications of collision between the Ontong Java Plateau and Solomon Island Arc from ocean bottom seismometer-airgun data: Tectonophysics, v. 389, p. 191–220, doi: 10.1016/j.tecto.2003.09.029.

Montelli, R., Nolet, G., Dahlen, F.A., Masters, G., Engdahl, E.R., and Hung, S.-H., 2004, Finite-frequency tomography reveals a variety of plumes in the mantle: Science, v. 303, no. 5656, p. 338–343, doi: 10.1126/science.1092485.

Morgan, W.J., 1971, Convection plumes in the lower mantle: Nature, v. 230, p. 42–43, doi: 10.1038/230042a0.

Mutter, C.Z., and Mutter, J.C., 1993, Variations in thickness of Layer 3 dominate oceanic crustal structure: Earth and Planetary Science Letters, v. 117, p. 295–317, doi: 10.1016/0012-821X(93)90134-U.

Nielsen, L., and Jacobsen, B.H., 2000, Integrated gravity and wide-angle seismic inversion for 2-D crustal modelling: Geophysical Journal International, v. 140, p. 222–232, doi: 10.1046/j.1365-246x.2000.00012.x.

Oxburgh, E.R., and Parmentier, E.M., 1977, Compositional and density stratification in oceanic lithosphere—Causes and consequences: Journal of the Geological Society of London, v. 133, p. 343–355.

Phipps Morgan, J., and Morgan, W.J., 1999, Two-stage melting and the geochemical evolution of the mantle: A recipe for mantle plum-pudding: Earth and Planetary Science Letters, v. 170, p. 215–239, doi: 10.1016/S0012-821X(99)00114-4.

Phipps Morgan, J., Morgan, W.J., and Price, E., 1995, Hotspot melting generates both hotspot volcanism and a hotspot swell?: Journal of Geophysical Research, v. 100, p. 8045–8062, doi: 10.1029/94JB02887.

Plank, T., and Langmuir, C.H., 1992, Effects of melting regime on the composition of the oceanic crust: Journal of Geophysical Research, v. 97, p. 19,749–19,770.

Plank, T., Spiegelman, M., Langmuir, C.H., and Forsyth, D.H., 1995, The meaning of "mean F": Clarifying the mean extent of melting at ocean ridges: Journal of Geophysical Research, v. 100, p. 15,045–15,052, doi: 10.1029/95JB01148.

Ringwood, A.E., and Irifune, T., 1988, Nature of the 650-km seismic discontinuity—Implications for mantle dynamics and differentiation: Nature, v. 331, p. 131–136, doi: 10.1038/331131a0.

Sallarès, V., and Charvis, Ph., 2003, Crustal thickness constraints on the geodynamic evolution of the Galápagos volcanic province: Earth and Planetary Science Letters, v. 214, no. 3–4, p. 545–559, doi: 10.1016/S0012-821X(03)00373-X.

Sallarès, V., Charvis, Ph., Flueh, E.R., and Bialas, J., 2003, Seismic structure of Cocos and Malpelo ridges and implications for hotspot-ridge interaction: Journal of Geophysical Research, v. 108, no. 2564, doi: 10.1029/2003JB002431.

Sallarès, V., Charvis, Ph., Flueh, E.R., Bialas, J., and the SALIERI Scientific Party, 2005, Seismic structure of the Carnegie ridge and the nature of the Galápagos hotspot: Geophysical Journal International, v. 161, no. 3, p. 763–788, doi: 10.1111/j.1365-246 X.2005.02592.x.

Schilling, J.-G., Zajac, M., Evans, R., Johnson, T., White, W., Devine, J.C., and Kingsley, R., 1983, Petrological and geochemical variations along the Mid-Atlantic ridge from 29°N to 73°N: American Journal of Science, v. 283, p. 510–586.

Sobolev, A., Hofmann, A.W., and Nikogosian, I.K., 2000, Recycled oceanic crust observed in "ghost plagioclase" within the source of the Mauna Loa lavas: Nature, v. 404, p. 986–990, doi: 10.1038/35010098.

Tarantola, A., 1987, Inverse problem theory: Methods for data fitting and model parameter estimation: Amsterdam, Elsevier Science, 613 p.

Trenkamp, R., Kellogg, J.N., Freymueller, J.T., and Mora, H.P., 2002, Wide plate margin deformation, southern Central America and northwestern South America, CASA GPS observations: Journal of South American Sciences, v. 15, p. 157–171, doi: 10.1016/S0895-9811(02)00018-4.

Watts, A.B., and ten Brink, U.S., 1989, Crustal structure, flexure, and subsidence history of the Hawaiian Islands: Journal of Geophysical Research, v. 94, p. 10,473–10,500.

Watts, A.B., Pierce, C., Collier, J., Dalwood, R., Canales, J.P., and Henstock, T.J., 1997, A seismic study of lithospheric flexure in the vicinity of Tenerife, Canary Islands: Earth and Planetary Science Letters, v. 146, p. 431–447, doi: 10.1016/S0012-821X(96)00249-X.

White, R.S., and McKenzie, D., 1989, Magmatism at rift zones: The generation of volcanic continental margins and flood basalts: Journal of Geophysical Research, v. 94, p. 7685–7794.

White, R.S., McKenzie, D., and O'Nions, R.K., 1992, Oceanic crustal thickness from seismic measurements and rare earth element inversions: Journal of Geophysical Research, v. 97, p. 19,683–19,715.

Wilson, J.T., 1963, A possible origin of the Hawaiian Islands: Canadian Journal of Physics, v. 41, p. 863–870.

Ye, S., Canales, J.P., Rihm, R., Danobeitia, J.J., and Gallart, J., 1999, A crustal transect through the northern and northeastern part of the volcanic edifice of Gran Canaria, Canary Islands: Journal of Geodynamics, v. 28, p. 3–26, doi: 10.1016/S0264-3707(98)00028-3.

Zelt, C.A., and Barton, P.J., 1998, 3D seismic refraction tomography: A comparison of two methods applied to data from the Faeroe Basin: Journal of Geophysical Research, v. 103, p. 7187–7210, doi: 10.1029/97JB03536.

Zhao, D., 2001, Seismic structure and origin of hotspots and mantle plumes: Earth and Planetary Science Letters, v. 192, p. 251–265, doi: 10.1016/S0012-821X(01)00465-4.

MANUSCRIPT ACCEPTED BY THE SOCIETY JANUARY 31, 2007

The Geological Society of America
Special Paper 430
2007

The North Atlantic Igneous Province:
A review of models for its formation

Romain Meyer*
Katholieke Universiteit Leuven, Afd. Geologie,
Celestijnenlaan 200E, 3001 Leuven-Heverlee, Belgium
Jolante van Wijk*
Los Alamos National Laboratory, Earth and Environmental Sciences Division,
Los Alamos, New Mexico 87545, USA
Laurent Gernigon*
Norges Geoligoske Undersøkelse, Geological Survey of Norway,
Leiv Eirikssons Vei 39, 7491 Trondheim, Norway

ABSTRACT

The mantle plume concept is currently being challenged as an explanation for North Atlantic Igneous Province formation. Alternative models have been suggested, including delamination, meteorite impact, small-scale rift-related convection, and chemical mantle heterogeneities. We review available data sets on uplift, strain localization, age and chemistry of igneous material, and tomography for the North Atlantic Igneous Province and compare them with predictions from the mantle plume and alternative models. The mantle plume concept is quite successful in explaining the formation of the North Atlantic Igneous Province, but unexplained aspects remain. Delamination and impact models are currently not supported. Rift-related small-scale convection models appear to be able to explain volcanic rifted margin volcanism well. However, the most important problem that nonplume models need to overcome is the continuing, long-lived melt anomaly extending via the Greenland and Faeroe ridges to Iceland. Mantle heterogeneities resulting from an ancient subducted slab are included in plate tectonic models to explain the continuing melt production as an alternative to the mantle plume model, but there are still uncertainties related to this idea that need to be solved.

Keywords: North Atlantic Igneous Province, volcanic rifted margins, LIP formation, mantle plume versus alternatives, continent breakup

*E-mails: mail@romain-meyer.eu; jolante@lanl.gov; Laurent.Gernigon@ngu.no.

Meyer, R, van Wijk, J., and Gernigon, L., 2007, The North Atlantic Igneous Province: A review of models for its formation, *in* Foulger, G.R., and Jurdy, D.M., eds., Plates, plumes, and planetary processes: Geological Society of America Special Paper 430, p. 525–552, doi: 10.1130/2007.2430(26). For permission to copy, contact editing@geosociety.org. ©2007 The Geological Society of America. All rights reserved.

INTRODUCTION: CHALLENGING
THE MANTLE PLUME CONCEPT

Continental breakup at the Paleocene–Eocene transition marked the culmination of an ~350 m.y. period of predominately extensional deformation in the northern North Atlantic subsequent to the Caledonian orogeny (e.g., Ziegler, 1988; Doré et al., 1999). During this period, numerous sedimentary basins developed that can now be found on the North Atlantic continental margins. Basin formation was not accompanied by significant magmatism, except around Eurasia-Greenland breakup time. The magmatic events prior to and during continental separation, and the postbreakup continuous activity of the Iceland melting anomaly, have resulted in one of the largest large igneous provinces (LIPs) in the world: the North Atlantic Igneous Province. The North Atlantic Igneous Province was long ago recognized by the significant regional on-shore distribution of volcanic features, including flood basalt traps and mafic and ultramafic complexes (Hutton, 1788). For the past forty years, numerous seismic studies and commercial and scientific drilling have shown that this magmatic province also extends off-shore. It is well developed along the Kolbeinsey and Reykjanes spreading ridges, which are still anomalously productive, and the breakup axis of the Northeast Atlantic, which is highly volcanic. Over the last decade or so, tomographic studies have provided information on the mantle seismic velocity structure beneath the area, and a 3-D picture of the mantle below the North Atlantic Igneous Province has emerged. The low-seismic-velocity anomaly beneath parts of the province is now generally thought to be linked to its formation.

As a result of intense scientific and economic interest, the North Atlantic Igneous Province is the most thoroughly documented LIPs on Earth. From current observations, a regional picture emerges of widespread early Paleogene magmatism extending from the Charlie Gibbs fracture zone in the south to the Senja Fracture Zone in the north on the eastern side of the North Atlantic. On the western side of the North Atlantic, the North Atlantic Igneous Province extends from south of East Greenland northward to the Greenland Fracture Zone and includes parts of West Greenland and Baffin Island, bordering Baffin Bay. It includes the Greenland, Iceland, and Faeroes ridges; the Kolbeinsey and Reykjanes spreading systems; and Iceland itself (e.g., Upton, 1988; White and McKenzie, 1989; Coffin and Eldholm, 1994; Saunders et al., 1997) (Fig. 1). The area covered by flood basalts, both on-shore and off-shore, may represent ~1.3×10^6 km^2, with a volume of ~1.8×10^6 km^3. Including magmatic underplating and other intrusions, it could reach 5×10^6 to 1×10^7 km^3 (Eldholm and Grue, 1994).

The scientific progress made in the last few decades has led to an exciting development in the context of the present volume: The plume concept as an explanation for North Atlantic Igneous Province formation is now being challenged. The reason is that not all plume model predictions have yet been supported by observations. The lack of a clear plume trail and the absence of a distinct lower-mantle continuation of the upper-mantle seismic wavespeed anomaly are arguments against the plume model as an explanation for North Atlantic Igneous Province formation (e.g., Foulger, 2005). Not only is a mantle plume source for LIP formation challenged; the "plume concept" itself is a topic of debate as well. For discussion of this aspect, we refer the reader to Anderson and Natland (2005) and Foulger (2005).

Following Morgan (1971), most workers have explained the Early Paleogene volcanism of the North Atlantic Igneous Province in terms of lithospheric impingement of the proto-Iceland mantle plume, and a wide variety of ideas on plume size, plume origin and path, and the possibility of multiple plume heads or pulsating plumes have been discussed in the literature in this context (e.g., Lawver and Müller, 1994; White et al., 1995; Nadin et al., 1997; Saunders et al., 1997; Torsvik et al., 2001). This evolution of the mantle plume concept has been addressed by several authors (e.g., Anderson and Natland, 2005; Foulger, 2005), and alternative hypotheses on North Atlantic Igneous Province formation have been formulated over the past ten years or so. These include moderately elevated mantle temperatures, fertile patches in the upper mantle, and small-scale convection (e.g., King and Anderson, 1995, 1998; Boutilier and Keen, 1999; van Wijk et al., 2001; Korenaga, 2004; Anderson and Natland, 2005; Foulger and Anderson, 2005; Foulger et al., 2005; Lundin and Doré, 2005; http://www.mantleplumes.org).

The objective of this review article is to test existing geodynamic models for North Atlantic Igneous Province and LIP formation against available geochemical, structural, and geophysical data. Predicted features of the various geodynamic models include uplift history, timing, duration, location and chemical signature of magma that is produced, and strain localization (rift formation). There is a wealth of data available that allow us to test these model predictions. We summarize the data sets in the next sections. We then discuss the characteristic predictions of the main models for North Atlantic Igneous Province formation and compare model predictions with observations. We conclude our study by exploring an alternative, non-mantle-plume model to explain the formation of this province.

GEOLOGICAL, GEOCHEMICAL,
AND GEOPHYSICAL OBSERVATIONS

Rifting Episodes

The Northeast Atlantic domain has undergone multiphase evolution, from rifting to continental rupture (e.g., Ziegler, 1988; Doré et al., 1999). Extensional deformation began with postorogenic collapse in East Greenland, the northern United Kingdom, and Norway (Séranne and Séguret, 1987; Strachan, 1994) as early as Devonian times. It continued with widespread Permo-Triassic rifting in Greenland, Norway, off-shore United Kingdom, off-shore Ireland, the North Sea, and East Greenland (Ziegler, 1988; Coward, 1995; Doré et al., 1999) (Fig. 2). Magmatism has been associated with this rift phase. There is evidence from East

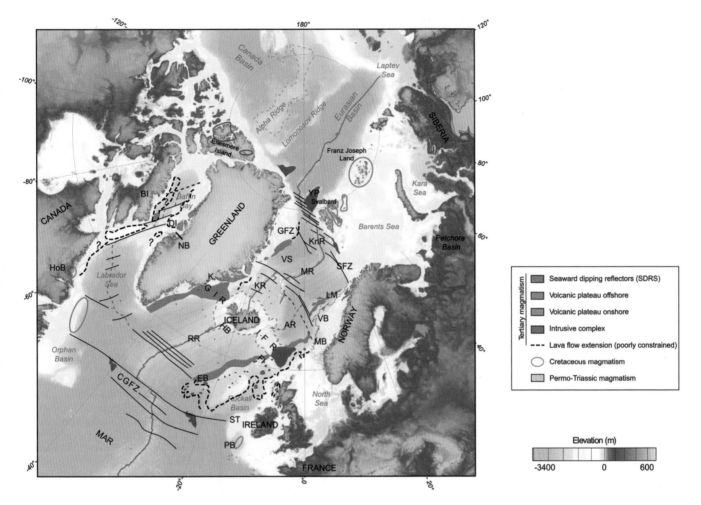

Figure 1. The North Atlantic Igneous Province. Tertiary on-shore and off-shore magmatism is indicated, as well as older Cretaceous and Permo-Triassic magmatism, interpreted magnetic anomalies, (extinct) spreading ridges, and fracture zones. AR—Aegir Ridge; BI—Baffin Island; CGFZ—Charlie Gibbs Fracture Zone; DI—Disko Island; EB—Edoras Bank; FI—Faeroe Islands; GFZ—Greenland Fracture Zone; GIR—Greenland-Iceland Ridge; HB—Hatton Bank; HoB—Hopedale Basin; IFR—Iceland-Faeroe Ridge; K—Kangerlussuaq; KnR—Knipovich Ridge; KR—Kolbeinsey Ridge; LM—Lofoten margin; MAR—Mid-Atlantic Ridge; MB—Møre Basin; MR—Mohns Ridge; NB—Nuussuaq Basin; PB—Porcupine Basin; RR—Reykjanes Ridge; SFZ—Senja Fracture Zone; ST—Slyne Trough and Erris Basin; VB—Vøring Basin; VS—Vesteris Seamount; YP—Yermak Plateau.

Greenland, the southwestern Olso Rift and the North Sea area (Fig. 2; Surlyk, 1990; Heeremans and Faleide, 2004). Following the Permian rift phase, several more rifting episodes occurred in the Jurassic. Limited early Middle Jurassic rifting is documented north to off-shore mid-Norway (Blystad et al., 1995; Koch and Heum, 1995; Dancer et al., 1999; Erratt et al., 1999). Rifting along the Porcupine and Slyne basins first occurred in the Permo-Triassic and then again in the Late Jurassic (e.g., Tate, 1993; Chapman et al., 1999; Dancer et al., 1999).

A Middle Jurassic uplift event occurred in the central part of the North Atlantic (Doré et al., 1999) (Fig. 2). This was a non-breakup-related magmatic event in a period of broad uplift (Fig. 2). The Late Jurassic period is usually defined as a major rifting episode that affected the entire North Atlantic (Doré et al., 1999). It is recognized in the Central Graben, Viking Graben,

and Moray Firth Basin (Erratt et al., 1999) of the North Sea rift system. The effects of this event possibly extended north to the Møre Basin and the Haltenbanken, as well as off-shore Norway (Koch and Heum, 1995), and it may also have influenced the East Greenland basins (Surlyk, 1990). The northernmost expression of this rift system is documented in the Norwegian part of the Barents Sea (Faleide et al., 1993).

Early Cretaceous extension (Fig. 2) is often interpreted as a separate extensional event and represents a significant geodynamic change in the evolution of the North Atlantic rifted basins (Doré et al., 1999). This event marks the end of the rift activity in the North Sea and the development of a new northeast-southwest-oriented rift axis stretching from South Rockall to Lofoten in Norway and to the Barents Sea. At the same time, a northwest-southeast-oriented rift developed from the Iberia-

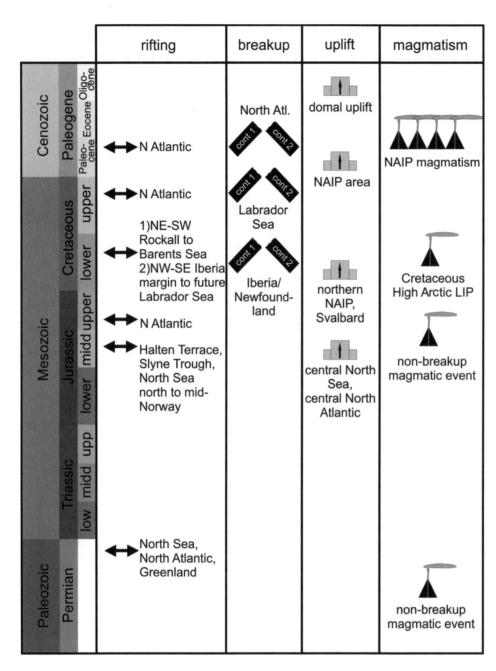

Figure 2. Schematic overview of the tectonic history of the North Atlantic Igneous Province (NAIP) region since Permian times. Several separate rifting episodes have affected the area since the Devonian extensional deformation that followed orogenic collapse: a widespread Permian-Triassic rift event is documented in the North Sea, North Atlantic, and Greenland. Limited early Middle Jurassic rifting is documented north to off-shore of mid-Norway. A major rifting episode affected the North Atlantic region (including the North Sea area) in the Late Jurassic. Prior to the breakup-related uplift event, a Jurassic period of uplift affected the central North Atlantic. This uplift was related to a period of magmatism; several magmatic episodes (relatively minor in volume) preceded the formation of the North Atlantic Igneous Province. Continental breakup in the Northeast Atlantic followed the Paleocene rift event. See text for references.

Newfoundland margins to the future Labrador Sea. West Iberia broke away from the Grand Banks of Newfoundland in the Early Cretaceous (e.g., Driscoll et al., 1995). A nonbreakup magmatic event of Lower to Upper Cretaceous age related to a period of uplift is documented in the Hopedale Basin and Svalbard areas. This magmatic event is defined as having formed an independent LIP (the so-called Cretaceous High Arctic Large Igneous Province) (Maher, 2001) (Fig. 2).

The Late Cretaceous phase of extension leading to seafloor spreading in Late Paleocene– Early Eocene time is well recognized in the North Atlantic (Doré et al., 1999). In Norway, this episode started around early Campanian time but was not nec-

essarily synchronous all along the margin segments (Gernigon et al., 2003). The onset of formation of oceanic crust in the Labrador Sea following the trend of the South Rockall-Hatton margin is still controversial. While older models propose that seafloor spreading in the Labrador Sea had already started at chron 33 in the Late Cretaceous (Roest and Srivastava, 1989), recent studies suggest opening in the Paleocene (Chalmers and Laursen, 1995).

Continental separation between Scandinavia and Greenland (Fig. 2) commenced in Early Eocene time (C24, ca. 53 Ma). After breakup, early seafloor spreading immediately northeast of Iceland first occurred along the now extinct Aegir Ridge until anomaly C13 (Mosar et al., 2002), or maybe longer, till C10 (Breivik et al., 2006). Subsequently, seafloor spreading along the incipient Kolbeinsey Ridge began (Vogt et al., 1980; Nunns, 1983).

Extensional deformation in the northern Atlantic domain thus occurred long before commencement of the final Late Cretaceous–Paleocene rift phase, which resulted in continental separation.

Magmatism: Pre-Breakup- and Breakup-Related

Norwegian Margins. Ocean Drilling Program (ODP) efforts on the mid-Norway margin at the Vøring plateau (leg 104) recovered volcanic rock successions that record the magmatic activity from the early continental breakup phase. The volcanic sequence is divided into an upper and lower series (US and LS). The US includes the seaward-dipping reflector sequences (SDRSs), interpreted as tilted subaerial and transitional-type tholeiitic mid-ocean ridge basalt (MORB) lava flows with interbedded volcaniclastic sediments (Eldholm et al., 1987). The LS comprises a sequence of intermediate extrusive flow units and volcaniclastic sediments erupted prior to the onset of MORB-type magmatism and derived, at least partially, from continental source material. A similar lithology, of a US basaltic SDRS and an underlying LS of strongly crustally contaminated units has been sampled at the southeast Greenland margin (ODP leg 152, 1993, and leg 163, 1995) (Larsen et al., 1999; Saunders et al., 1999).

Available geochemical data for southeast Greenland and the Vøring Plateau show that the plateau US and the south Greenland US are chemically and isotopically rather similar. The LS from both areas are fundamentally different from each other in many respects (e.g., LS southeast Greenland $^{87}Sr/^{86}Sr$ < 0.702 and LS Vøring Plateau $^{87}Sr/^{86}Sr$ > 0.710) (Meyer et al., 2005). The marked difference in the geochemical signatures of the southeast Greenland and Vøring plateau samples points to a substantial difference in either the prebreakup crustal composition at the two localities or to different styles of mantle-crust interaction.

West Greenland and Baffin Island. Igneous rocks are recorded from Disko to Svartenhuk on-shore West Greenland (Larsen and Pedersen, 1988), as well as on Baffin Island and in the Labrador and Baffin Sea areas (e.g., Skaarup et al., 2006). The Nuussuaq Basin extends from Disko to Svartenhuk. Volcanism started here in Middle Paleocene time in a subsiding marine environment, so the earliest volcanic rocks are hyaloclastite breccias. After a significant Early Paleocene uplift, later volcanism was almost entirely subaerial (Dam et al., 1998). The volcanic rocks are divided into three formations (Gill et al., 1992). The lower formation consists mainly of picritic and other olivine-rich tholeiitic basalts, with local evidence of crustal contamination. The other two formations are composed of olivine and plagioclase phyric olivine tholeiitic basalts, basaltic tuffs, transitional basalts, and acidic ignembrites.

An extraordinary characteristic of the West Greenland and Baffin Island lavas in the North Atlantic Igneous Province is the high ratio of picrites to basalts of the erupted volumes (30–50 vol% compared to, e.g., 15 vol% in East Greenland) (Gill et al., 1992; Larsen et al., 1992; Holm et al., 1993). Definitions of picrites are not always unambiguous (Le Bas, 2000); Gill et al. (1992) defined all basaltic rocks with MgO concentrations higher than 10 wt% as picrites. The MgO content of the rocks analyzed by Gill et al. (1992) ranges between 15 and 30 wt%. Petrologists debate the nature of these rocks: Do the melts represent unmodified high-MgO mantle melts (e.g., Clarke, 1970; Clarke and O'Hara, 1979), or are they derived from a primary mantle magma modified by one or more subsequent processes, e.g., olivine accumulation (Hart and Davis, 1979)? There is no doubt that the high-MgO (up to 30 wt%) picrites have been influenced by olivine accumulation (Gill et al., 1992). The picrites on Baffin Island can be classified according to the Kent and Fitton (2000) classification scheme for the Paleogene British igneous province (BIP) into M2 magmas with slight enrichment in light rare earth elements ($[La/Sm]_N$ of 1–1.2; Robillard et al., 1992) and M3 magmas ($[La/Sm]_N$ of 0.6–0.7; Robillard et al., 1992) (Fig. 3B). Most picrites have a radiogenic isotopic geochemistry ($\varepsilon Nd_{t=60}$ > +3.4 and $^{86}Sr/^{87}Sr_{(Pd)}$ > 0.7031; Holm et al., 1993), similar to the present-day Atlantic MORB and/or the Iceland basalts (Fig. 3A).

East Greenland Margin. Prior to breakup and opening of the Iceland basin, the southeast Greenland margin formed one system with the West Hatton volcanic margin. The southeast Greenland margin is characterized by a well-developed package of SDRSs in a wide zone across the shelf and in the adjacent off-shore region. ODP legs 152 and 163 drilling transects show that SDRSs were formed by subaerial or shallow marine lavas. Drilling during ODP legs 152 and 163 sampled the feather edge and the central part of the SDRS, recovering mainly basaltic lavas (Larsen et al., 1994; Duncan et al., 1996). Site 917 penetrated basalts and dacites of Late Paleocene age (ca. 61 Ma). These LS magmas represent the prebreakup lava flows. The parent magmas evolved by assimilation- and fractional crystallization–type processes in continental crustal reservoirs. Trace element and radiogenic isotope compositions indicate that the contaminant changed through time, from lower-crustal granulite to a mixture of granulite and amphibolite (Fitton et al., 2000) (Fig. 3A). The degree of contamination decreased rapidly after continental separation. Dominantly normal MORB (N-MORB) basalts with a few "Icelandic" basalt flows erupted during the prebreakup phase, but the source of the postbreakup magmas was clearly an "Icelandic" mantle.

Off-Shore U.K. and the British Igneous Province. The geochemistry and the mantle source of the British Igneous Province have been discussed in several publications (e.g., Thompson et al., 1982; Kent and Fitton, 2000). The composition of the on-shore and, to a lesser extent, off-shore Scottish basalts is relatively well known. Geochemical studies defined three successive Paleogene magma types: M1–M3 (Kerr, 1995;

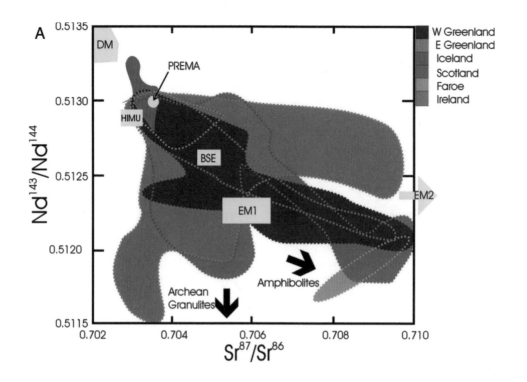

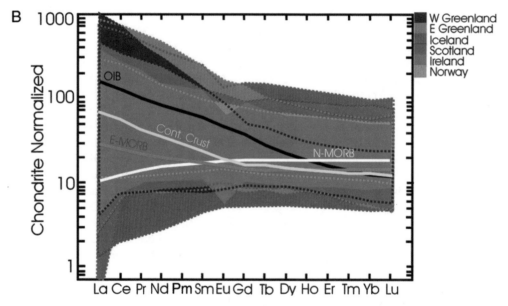

Figure 3. Geochemical overview of North Atlantic Igneous Province magmas. (A) ^{143}Nd/^{144}Nd versus ^{87}Sr/^{86}Sr diagram for igneous products of the province. Isotopic differences are due to the contrast in Nd and Sr isotopic composition between old continental crust (e.g., gneiss, amphibolites) and mantle magma sources. Variations of ^{143}Nd/^{144}Nd and ^{87}Sr/^{86}Sr in crustal rocks are related to age and the contrasting geochemical behavior of Sm and Nd relative to Rb and Sr. The mantle plume model suggests that the correlation of initial Nd and Sr isotopes represents mixing lines between "mantle zoo" reservoirs of distinct chemistry and age. BSE—bulk silicate earth; DM—depleted MORB (mid-ocean ridge basalt) mantle; HIMU—high μ (i.e., high U/Pb subducted oceanic crust); EM1 and EM2—enriched mantle 1 (recycled pelagic ocean floor sediments or subcontinental lithosphere) and 2 (subducted continental material); PREMA—prevalent mantle. Fields from Hart (1984) and Hart and Zindler (1989). Alternatively, the trend may result from mixing by assimilation or contamination of high-Sm/Nd, low-Rb/Sr mantle melts with old low-Sm/Nd, high-Rb/Sr crustal material, as has been shown in some areas below the North Atlantic Igneous Province (Scotland). See the section headed "Exploring Alternatives" for a discussion. Data are from the Max Planck Institut für Chemie, Geochemistry Division, GeoRoc database. (B) Masuda-Coryell diagram comparing the rare earth element (REE) compositions of the areas below the North Atlantic Igneous Province. The REE patterns are controlled by the source geochemistry and crystal-melt equilibria. Melting processes (e.g., partial melting) affect the content of REE but not the shape of the C1 normalized pattern, reflecting the source composition. All areas below the North Atlantic Igneous Province include both melts from an enriched mantle source and magmas generated from a depleted mantle. Data are from the Max Planck Institut für Chemie, Geochemistry Division, GeoRoc database, and the lines for normalized mid-ocean ridge basalt (N-MORB), enriched MORB (E-MORB), and ocean island basalt (OIB) are representative patterns from Sun and McDonough (1989). The continental crust pattern is from Taylor and McLennan (1985). Data are normalized to the C1 chondrite composition of Sun and McDonough (1989).

Kent and Fitton, 2000). A volumetrically insignificant fourth magma type (M4) is preserved as dikes and plugs. M1 and M2 are characterized by major element geochemistry similar to that of transitional alkali basalts, whereas the M3 magmas have tholeiitic basalt major element chemistry (Kent and Fitton, 2000). The rare earth element (REE) geochemistry of the M1 (high Ce/Y) magmas is defined by a pattern of slight enrichment in light REE (LREE), while the M2 (moderate Ce/Y) magmas have a rather flat REE pattern (around 20 × chondritic), and the M3 (low Ce/Y) magmas have a pattern similar to that of N-MORB (Fig. 3B). Chambers and Fitton (2000) showed that the earliest tapped mantle source on Mull was N-MORB, followed by an "Icelandic" mantle, before returning to an N-MORB source. The isotopic geochemistry of most British Igneous Province rocks confirms strong interaction between mantle melt and crust during the formation of these magmas (Fig. 3A). Sr, Nd, Hf, and Pb isotopic compositions not only shed light on crust-mantle interactions but are also able to reveal the crustal contaminant (lower crustal gneisses, upper crustal amphibolites, upper crustal sediments) by means of their distinctive isotopic signatures (e.g., Geldmacher et al., 2002; Troll et al., 2004). Dickin (1981) even suggested that in the British Igneous Province the Pb isotopic system provides no information on the mantle source but mainly provides information on the nature of the contaminating continental crust. Such crustal geochemical continental contamination or assimilation trends can be seen in all the subareas of the North Atlantic Igneous Province. Only after the deduction of these mantle-crust interaction processes, some of which are still unknown, can precise information on possible mantle plume influences be deduced.

The northeast Irish basalts of the British Igneous Province have also been divided into three formations. The volcanic activity there was episodic, and the majority of the mantle magmas were contaminated by assimilation of crustal rocks (Fig. 3A) (Kerr, 1995; Barrat and Nesbitt, 1996). With the exception of the Causeway Member, most have convex-up REE patterns. These are interpreted as being due to residual garnet in the mantle source (Fig. 3B). This suggests that at the start of Paleogene volcanism, the melt regime was controlled by a thick lithosphere that thinned with time such that the Causeway tholeiitic MORB-like basalts were produced at shallower levels. Barrat and Nesbitt (1996) suggest that the return to the convex-up patterns of the Upper Formation shows that simple models of lithospheric stretching and rifting are not able to explain the Antrim situation.

Faeroe Islands. The Faeroe Islands rest on top of the subsided, continental Faeroe-Rockall Plateau, which is separated from the British Isles by the deep rift basins of the Faeroe-Shetland Channel and the Rockall Trough (Fig. 1). The islands expose up to 3 km of flood basalts—the Faeroe Plateau Lava Group—which is also divided into three basalt formations (Waagstein, 1988). All basaltic lavas have been erupted subaerially and are tholeiitic (Fig. 3A). The lowermost formation differs from the olivine tholeiites in the upper and middle formations in be-

ing silica-oversaturated and more crust-contaminated (Fig. 3A). The lower formation is capped by a coal-bearing sequence, dated as Paleocene (Lund, 1983), which is evidence for a break or decrease in volcanic activity. So far, no hiatus has been documented between the middle and upper formations. A mantle source change occurs at this transition; while all of the lower and middle formation tholeiites are slightly LREE enriched, the upper formation is a mixture of LREE-depleted and LREE-enriched tholeiites (Gariépy et al., 1983).

Irish Volcanic Margin. The Irish margins are characterized by a nonvolcanic margin to the southeast and by a Late Paleocene–Early Eocene volcanic margin to the northwest, which forms the southern edge of the North Atlantic Igneous Province (see Fig. 1) (Barton and White, 1995, 1997). The West Hatton volcanic margin exhibits a typical volcanostratigraphic sequence similar to that of the Norwegian volcanic margin.

DSDP leg 81 sampled the Irish SDRS tholeiitic basalt lava flows, extruded under subaerial or very shallow marine conditions. The lavas are all consistently MORB-like in composition (Harrison and Merriman, 1984).

The Hatton Basin and the West Hatton margin are part of the Rockall-Hatton Plateau, a continental fragment between the Rockall Basin and the oceanic crust, east of the C24 magnetic anomaly. From Hitchen (2004) we know that Albian sediments are present beneath the thin Cenozoic section in the western part of the Hatton Basin. These sediments are on-lapped by Paleocene lava flows that issued from igneous centers, clearly characterized by a circular-shaped gravity anomaly (e.g., Hitchen, 2004; Geoffroy et al., this volume). They were recently drilled by the British Geological Survey and comprise basalt, gabbro, and andesite (Hitchen, 2004).

Arctic, Svalbard, and Barents Sea. Svalbard contains a record of magmatic events dating back to the Mesozoic. The only direct evidence for Paleogene volcanic activity at Svalbard is minor tuff input in the Central Basin.

Off-shore volcanism at the Yermak Plateau (Fig. 1) of northern Spitsbergen may be Paleogene (Harland et al., 1997). Neogene to Recent volcanism can be divided, after Harland and Stephens (1997), into two distinct groups: Miocene plateau lavas and Pleistocene volcanics with hydrothermal activity. Ages of 10–12 Ma for the plateau basalts are based on K-Ar dating (Prestvik, 1978).

In this area, an older volcanic episode has been recognized and defined as an independent LIP (the so-called Cretaceous High Arctic Large Igneous Province) (Maher, 2001) (Fig. 2). Flood basalts, dikes, and sills with Early to Late Cretaceous ages are found in the Canadian Arctic Island (Estrada, 1998). Recent Ar-Ar dating from northernmost Greenland suggests an age of ca. 80–85 Ma for the magmatism, which is older than previous whole-rock K-Ar measurements indicated (Kontak et al., 2001; Estrada and Henjes-Kunst, 2004).

The aseismic Alpha Ridge extends northward beneath the Arctic Ocean. This ridge is volcanic and partly of Cretaceous age, with an alkaline rock sample dated using the Ar-Ar technique at 82 ± 1 Ma (C33r) (Jokat, 2003). Lower Cretaceous basaltic rocks are known in Svalbard, Kongs Karl Land, and Franz Josef Land (Bailey and Rasmussen, 1997). Tholeiitic basalts in Franz Josef Land have been dated at 117 ± 2.5 Ma (C34n) by Pumhösl (1998) using Ar-Ar techniques. Also at Svalbard, abundant doleritic Jurassic and Cretaceous intrusions are known with a few volcanic lava flows. Burov et al. (1975) suggest that quartz dolerites typically have ages around 144 ± 5 Ma, and younger olivine dolerites have ages around 105 ± 5 Ma. However, like the Prestvik (1978) ages, they need to be confirmed with state-of-the-art dating methods. From seismic data it is known that sills are present in the central and northern part of the Russian Barents Sea. Preliminary age data indicate that these widely distributed outcrops were formed as part of the same event, but the data are not sufficiently precise to allow us to determine whether the occurrences were truly contemporaneous or whether there is any systematic age progression from the Arctic region to the North Atlantic Igneous Province.

Postbreakup Magmatism: Iceland and the Oceanic Part of the North Atlantic Igneous Province

The seismic crust beneath Iceland is estimated to reach a thickness of 38–46 km in some places. Part of the Icelandic crust is possibly continental (Foulger et al., 2005). Two thick-crust ridges flank Iceland: the Greenland-Iceland and Iceland-Faeroe ridges (Fig. 1), with a Moho depth of ~25–30 km (Richardson et al., 1998; Smallwood et al., 1999). Iceland is composed mainly of tholeitic basalts with smaller amounts of alkali basalts, rhyolites, and obsidians. It is widely known that the $^{87}Sr/^{86}Sr$ isotope data for Iceland (and the Reykjanes Ridge) show significantly higher concentrations than those of N-MORB (e.g., O'Nions and Pankhurst, 1974). The geochemical signature of Icelandic igneous rocks is characterized by high $^3He/^4He$ ratios relative to the atmospheric ratio, Ra. The highest noncosmogenic $^3He/^4He$ isotope ratios of ~42 Ra are found in Iceland (Hilton et al., 1999; Breddam, 2002). The major geochemistry of basalts from the nearby ridge segments, Vesteris seamount, the Jan Mayen area, and the previously described Paleogene successions (Fig. 3A and B) indicate that the upper mantle in most of the Northeast Atlantic has the same chemical characteristics as the current Iceland source (Saunders et al., 1997). The isotopic data of the whole North Atlantic Igneous Province show that most of the breakup igneous rocks and the Iceland magmas derived from an enriched mantle source and not from a depleted mantle (e.g., North Atlantic MORB). The fact that such enriched magmas are limited, with some exceptions, such as Iceland, to the breakup time strongly supports the idea that the mantle source below the North Atlantic Igneous Province was heterogeneous. After the breakup magmatism, this source was depleted, and it is now producing the typical North Atlantic MORB.

The oceanic crust formed at Kolbeinsey Ridge (Fig. 1) north of Iceland is 1.0–2.5 km thicker than normal oceanic crust (Kodaira et al., 1998). The crust is up to 10 km thick at the Reyk-

janes Ridge axis south of Iceland (Smallwood et al., 1995), which is also anomalously thick for this slow-spreading ridge. The Reykjanes Ridge shows further remarkable features. While it is slow spreading, it lacks a rift valley in its northern part. Furthermore, basement ridges are observed that cut obliquely across lines of equal age. They are arranged symmetrically about the spreading boundary and form a V-shaped pattern (Vogt, 1971; Ito, 2001). These features have also been documented at the Kolbeinsey Ridge, north of Iceland (S.M. Jones et al., 2002; Vogt and Jung, 2005). Lateral transport of plume-derived material from Iceland southward along the Reykjanes Ridge has been suggested to explain the ridge bathymetry (which slopes up toward Iceland) and the V-shaped ridges (e.g., White et al., 1995; Ito, 2001). Alternatively, passive upper-mantle heterogeneities could have played a role in the formation of the V-shaped structures (S.M. Jones et al., 2002; Vogt and Jung, 2005).

Magmatism: Age and Overview

The North Atlantic Igneous Province shows a wide range of magmatic ages of Paleogene volcanic rocks (Fig. 4). For this overview we used the most reliable available data, the compilation of geochronological data on the province by Torsvik et al. (2001), and included data from Iceland (Moorbath et al., 1968; Ross and Mussett, 1976; Hardarson et al., 1997; Foulger, 2006). Note, for example, the single data point of the mid-Norwegian margin, where several publications suggest a range of ages. The Vøring plateau magmas have been inconsistently dated by Rb-Sr isochrones to 57.8 ± 1.0 Ma (LeHuray and Johnson, 1989) and 63 ± 19 Ma (Taylor and Morton, 1989). Since recent Ar-Ar geochronology investigations, these LS magmas are believed to be much younger, 56–55 Ma (Sinton et al., 1998). However, we note that there is currently debate on the reliability even of part of this data set (Baksi, 2005 and this volume, both chapters). The uncertainty in age determination, in combination with logistical limitations, has left a large part of the North Atlantic Igneous Province region unsampled or undersampled.

The present data set (Fig. 4) suggests several robust patterns: (1) there is evidence of prolonged magmatism after continental separation on Iceland and in the northern parts of the West and East Greenland margins, while the southern part of the Northeast Atlantic margins seem devoid of postbreakup magmatism, and (2) Paleogene, prebreakup magmatism (ca. 60 Ma) occurs in all regions (except for Iceland) and is middle to late synrift (Fig. 2). The absence of postbreakup magma samples along most of the northwest Atlantic margins away from the Greenland, Iceland, and Faeroe ridges is in concert with the rapidly decreasing oceanic crustal thickness after breakup. It indicates that the melting source was depleted at this time.

Volume of Magmatism

Igneous rocks emplaced during the breakup event comprise three main units: (1) voluminous extrusive complexes, includ-

Magmatic ages (My)

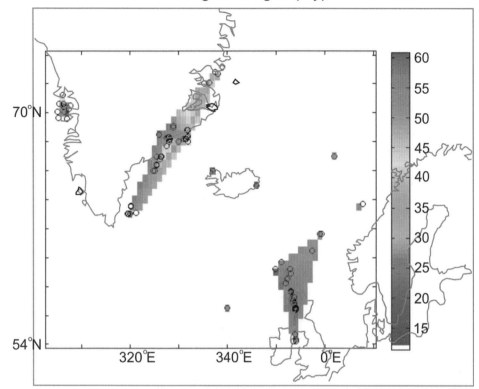

Figure 4. Cenozoic ages of North Atlantic Igneous Province igneous rocks. See text for references and explanation. This is a linear interpolation of the data points onto a regular grid with a 1° by 0.25° resolution. Data set: Moorbath et al. (1968), Ross and Mussett (1976), Brooks and Gleadow (1977), Gleadow and Brooks (1979), Bugge et al. (1980), Dickin (1981), Dickin and Jones (1983), Mussett (1986), Gibson et al. (1987), Thompson et al. (1987), Noble et al. (1988), Hitchen and Ritchie (1993), Nevle et al. (1994), Upton et al. (1995), Pearson et al. (1996), Hardarson et al. (1997), Hirschmann et al. (1997), Price et al. (1997), Sinton and Duncan (1998), Sinton et al. (1998), Storey et al. (1998), Tegner et al. (1998), Gamble et al. (1999), Tegner and Duncan (1999), and Foulger (2006).

ing the SDRS; (2) thick initial oceanic or transitional crust, often with a high-velocity lower-crustal body (LCB); and (3) intrusives. Between individual margin segments, there is variability in the extent and volume of these magmatic features. For example, the thickness and surface area covered by SDRSs, the thickness of the LCB, and the thickness of the initially formed oceanic crust vary considerably.

At the northern Norwegian Lofoten-Vesterålen margin, for example (Fig. 1), the initial oceanic crust is ~12–15 km thick (Tsikalas et al., 2005). A LCB is documented seaward of the shelf edge. To the south, the mid-Norwegian Vøring margin is characterized by an increased but along-strike-variable initial ocean crust thickness (e.g., Skogseid and Eldholm, 1995). Packages of SDRSs are commonly present at this margin, and the thickness of the LCB varies along strike (Mjelde et al., 2005; Ebbing et al., 2006). At the Møre margin (Fig. 1), there is evidence of a LCB (e.g., Raum, 2000) and SDRS (e.g., Planke et al., 2000).

The volcanic central region of Baffin Island is characterized by SDRSs. To the north the margin is nonvolcanic, and SDRSs are absent (Skaarup et al., 2006). In the south, Baffin magmatism is limited and the margin is interpreted as nonvolcanic (Skaarup et al., 2006). The conjugate margins of the Labrador Sea, southern Labrador, and southern West Greenland are nonvolcanic (Chian et al., 1995). The existence of SDRSs at Disko Island remains controversial (Geoffroy et al., 1998; Chalmers et al., 1999).

Along the East Greenland margin, transects image LCBs (Korenaga et al., 2000; Holbrook et al., 2001). The crustal thickness varies (Holbrook et al., 2001) from 30 to 33 km at the postulated hotspot-proximal zone to 10–20 km away from the region. Maximum crustal thickness (~30 km) decreases over the period 10–12 Ma just after the C24 magnetic anomalies to only 8–9 km. On a crustal scale, massive and scattered LCBs are recognized in several parts of the Hatton Basin and along the West Hatton margin, but are not present everywhere (Fowler et al., 1989; Barton and White, 1997; Vogt et al., 1998).

The thickness of the LCBs not only varies laterally (Tsikalas et al., 2005; Ebbing et al., 2006), but there is also discussion regarding its magmatic nature (Ebbing et al., 2006; Gernigon et al., 2006). A widely accepted interpretation of these bodies is magmatic underplating, representing both ponded magmatic material trapped beneath the Moho and magmatic sills injected into the lower crust (Furlong and Fountain, 1986; White and McKenzie, 1989). Some recent studies, however, suggest an alternative explanation. Gernigon et al. (2006) propose that the LCB characteristics may be partly explained by the presence of preexisting high-velocity rocks such as eclogites or migmatites. Ebbing et al. (2006) propose that the LCBs could be remnants of the Caledonian root.

The nonmagmatic hypothesis has significant implications. A nonmagmatic interpretation of the LCBs would lower the estimated North Atlantic Igneous Province melt volumes (Eldholm and Grue, 1994) significantly. This, in turn, would affect most conventional models for LIP formation, which generally link melt volumes to potential mantle temperature (White and McKenzie, 1989).

From available seismic, gravity, and magnetic data, a picture emerges of a melt volume that varies significantly along strike of the North Atlantic margins, but that is clearly largest near the Iceland anomaly, as shown by the great oceanic crustal thickness there (Greenland–Iceland–Faeroe ridges). The postbreakup oceanic crustal thickness shows a similar pattern of excessive magma production at the Greenland–Iceland–Faeroe ridges and present-day Iceland and increased crustal production at the Reykjanes and Kolbeinsey spreading ridges (see the earlier section on postbreakup magmatism).

Mantle Seismic Velocity Structure

Tomographic studies agree on an upper-mantle low-seismic-velocity anomaly beneath Iceland (e.g., Bijwaard and Spakman, 1999; Ritsema et al., 1999; Foulger et al., 2001; Montelli et al., 2004). This anomaly is centered beneath east central Iceland and is approximately cylindrical in the top 250 km but tabular in shape at greater depth (Foulger et al., 2001). A narrow anomaly seems to be embedded in a wider (1000 km or more), weaker anomaly beneath the Iceland region. In the lower mantle the strong anomaly is no longer present; anomalies there are only a fraction of the strength of the upper-mantle anomaly and in most studies are discontinuous with the upper-mantle anomaly (Ritsema et al., 1999). Although the temperature dependence of wavespeeds in the mantle decreases with depth, making plumes more difficult to image deeper down, a sharp change is not expected at the base of the upper mantle. It has been claimed that several plumes have been observed extending into the deep mantle elsewhere, suggesting that the lack of continuity at Iceland is real (Montelli et al., 2004). The tomographic images thus provide no evidence for an upwelling, deep-mantle plume (Montelli et al., 2004).

Uplift History of the Margins

It is generally accepted that the North Atlantic Igneous Province experienced uplift and exhumation during Early Paleogene time (cf. Saunders et al., 2007), but a regional picture of these vertical movements and their amplitudes is far from well constrained. The currently available information is summarized later, and a regional overview of well-constrained Paleocene–Eocene uplifted areas in the North Atlantic domain is shown in Figure 5. We note that this compilation is far from being complete; we omitted poorly constrained data as well as subsidence patterns along the incipient margin. The breakup-related phase of uplift was preceded by several other periods of uplift (Fig. 2) in the North Atlantic domain since collapse of the Caledonian orogeny. After continental separation, domal uplift affected the North Atlantic (Fig. 2). The locations of domes in-

clude southern and northern Norway, the northern United Kingdom, Svalbard, and Finnmark (e.g., Rohrman and van der Beek, 1996).

Regional Early Paleogene (prevolcanic) uplift has been demonstrated by field and/or seismic data in West Greenland (Dam et al., 1998), the Faeroes-Shetland Basin (Nadin et al., 1997; Sørensen, 2003), the North Sea Basin (Mudge and Jones, 2004), the Porcupine Basin (Jones et al., 2001), the Celtic and Central Irish Sea (Murdoch et al., 1995; Corcoran and Doré, 2005), the Hatton Basin (Hitchen, 2004), and the Norwegian margin (Skogseid and Eldholm, 1989).

Along the North Atlantic Igneous Province, an epeirogenic event is well documented in the latest Cretaceous–earliest Paleocene before the breakup and the main volcanic phase. This prebreakup event coincides with a regional hiatus and erosion features along the outer Vøring Basin (Skogseid et al., 1992). During this period, the Møre Basin experienced the rapid and episodic influx of coarse sediment in response to regional exhumation and denudation of the mainland (Martinsen et al., 1999). In East Greenland, isolated outcrops show Upper Cretaceous and Lower Paleocene off-shore shales and submarine channel turbidites that are unconformably overlain by fluvial conglomerate and sandstones, implying a similar dramatic shallowing there as well (Larsen et al., 1999).

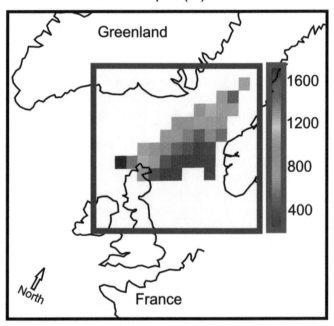

Figure 5. Transient Paleocene-Eocene boundary uplift in the North Atlantic Igneous Province. The East Greenland uplift is not shown here, nor are several locations that were insufficiently well constrained. Linear interpolation of the selected data points onto a regular grid with a 2° by 2° resolution. Compilation of data shown here from Nadin et al. (1997), Knox (1998), Jones et al. (2001), Deptuck et al. (2003), Jones and White (2003), Ren et al. (2003), Ceramicola et al. (2005), and Mackay et al. (2005).

West Greenland is not included in Figure 5, but there uplift has been documented 5–10 Ma before the onset of volcanism. The uplift (up to 1.3 km) was associated with fluvial peneplenation, exhumation of deep marine sediments, valley incision, and catastrophic deposition (Dam et al., 1998). Around the British Isles, apatite fission track studies have identified two exhumation events: an early Cenozoic and a late Cenozoic episode (Green et al., 2001). Subsidence analysis in the Moray Firth Basin (Mackay et al., 2005) suggests transient uplift initiated in Early Paleocene time. The basin records contemporaneous changes from hemipelagic to clastic sedimentation in the northern North Sea and coincides with a marine regression (Ahmadi et al., 2003). In the Faeroe-Shetland Basin, the Early Paleocene uplift is recognized and coincides with the Danian Maureen formation deposition in the North Sea. A Middle Paleocene phase associated with a major change in sediment provenance is recognized as well (Naylor et al., 1999).

At the onset of Northeast Atlantic breakup in the Early Eocene, widespread volcanism and local erosion and uplift occurred along the breakup axis of the Mid-Norwegian margin. Tectonic subsidence curves show uplift around the time of rift-drift transition on the western side of the Vøring Basin and the Vøring Marginal High (Skogseid et al., 1992). Around Ireland and in the North Sea, a maximum transient uplift occurred in Late Paleocene–Early Eocene time, followed by a rapid post–Early Eocene subsidence (Nadin et al., 1997; Green et al., 2001; Jones et al., 2001; Mackay et al., 2005). In the Faeroe-Shetland Basin, the Late Paleocene–Early Eocene phase coincided with a strong unconformity and prograding sands and mudstone sequences off-shore Faeroes (Sørensen, 2003). Along the Northeast Greenland shelf, five sequences of eastward-prograding wedges document a strong uplift of the Greenlandic mainland from Paleocene to Early Eocene time (Tsikalas et al., 2005). Figure 5 suggests a pattern of increased transient uplift toward the future breakup axis. The data are too sparse to definitely point to a concentration around the inferred paleoposition of the Iceland anomaly.

GEODYNAMIC MODELS FOR NORTH ATLANTIC IGNEOUS PROVINCE FORMATION

The formation of the North Atlantic Igneous Province can be related only to geodynamic models explaining and predicting all the previously described observations. Because the plume concept is challenged today, alternative models have been proposed as sources for LIPs. In this section we discuss the most prominent models, including delamination, impact, fertile mantle, small-scale convection, and mantle plume.

Lithosphere Delamination

One mechanism that can explain the production of huge amounts of basalt involves a link between melt production and delamination of the lower lithosphere (Fig. 6A). The mechanism

of delamination that is considered here as a possible explanation for North Atlantic Igneous Province formation involves the rapid foundering of the mantle part of the lithosphere into the convective mantle. Under specific conditions, the lithospheric mantle can become insufficiently buoyant or viscous to survive as lithosphere, and detach (Bird, 1979; Bird and Baumgardner, 1981). There are two conditions that must be met before delamination can occur: the lithospheric mantle has to be gravitationally unstable, and its viscosity must be low enough to allow flow (Elkins-Tanton and Hager, 2000; Schott et al., 2000). Several conditions may result in a sufficient increase in lithosphere-asthenosphere density contrast for delamination to occur (e.g., see Kay and Kay, 1993). These include melt intrusions that subsequently freeze, cooling of mantle lithosphere by shortening and mountain building, or a compositional difference. Viscosity decrease can be caused by melt intrusions (Carlson et al., 2005).

Gentle removal of the base of the lithospheric mantle would probably have no expression in the overlying crust or vertical movements, but rapid delamination is expected to influence the remaining overlying part of the lithosphere. Potential consequences are increased heatflow and generation of delamination-related magmas; crustal subsidence during delamination preceding short-lived, postdelamination rapid uplift of the surface; and changes in tectonic style to an extensional stress regime (e.g., England and Houseman, 1989; Elkins-Tanton and Hager, 2000; Morency and Doin, 2004; Carlson et al., 2005). The geochemical content of igneous rocks related to delamination is characterized by high-potassium magmas such as lamprophyres, leucitites, and absarokites (Kay et al., 1994; Ducea and Saleeby, 1998; Farmer et al., 2002; Elkins-Tanton and Grove, 2003). Delamination will not result in steady-state melting for tens of m.y.; instead, a short pulse is expected (Elkins-Tanton, 2005). Delamination can occur over a range of scales, and modeling studies suggest that a wide range of melt volumes may result (Elkins-Tanton, 2005).

As far as we know, the amount of crustal subsidence and subsequent uplift that can be expected from delamination has not yet been quantified. We assume that the size and area of delamination will influence the amount and distribution of vertical movements and melt production (Elkins-Tanton, 2005). The sequence of events during delamination are thus different (opposite) from that typically proposed to accompany the arrival of a mantle plume: subsidence is followed by uplift and magmatism, while a mantle plume is expected to cause uplift prior to volcanism.

The long extensional phases that ultimately resulted in continental separation between Norway and East Greenland were preceded by a mountain-building phase, the Caledonian orogeny. So the circumstances that could lead to a gravitational unstable lithospheric root and possibly delamination were present. However, there is no documentation of substantial magmatism following the orogeny, and the time lapse between the mountain-building phase and the Paleocene–Eocene volcanic event is too large (several 100 m.y.) to expect any relationship between

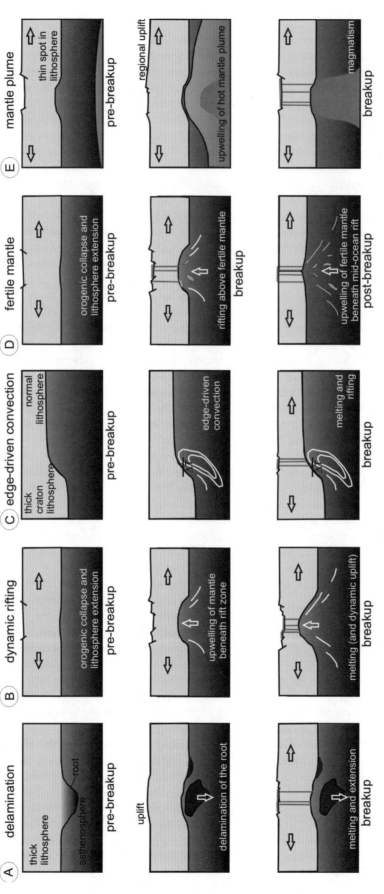

Figure 6. Schematic overview of different models suggested for large igneous province or North Atlantic Igneous Province formation. (A) Delamination of thick lithosphere root. Uplift is a response to the delamination, followed by magmatism and extension. (B) Dynamic rift model. Passive rifting follows Caledonian collapse. Warm mantle material wells up beneath the rift zone to fill the space created by extension, resulting in decompression melting. (C) Edge-driven convection model. Convection occurs due to lateral temperature differences between craton and noncraton lithosphere. This brings large volumes of mantle material into the melt window. (D) Fertile mantle concept. (E) Fertile mantle model. A remnant of subducted oceanic lithosphere), resulting in volcanic margins and later in continued magmatism at Iceland. (E) Mantle plume model. Mantle plume material fills an existing thin spot in the lithosphere, causing uplift and excessive melting.

them. Moreover, the chemical contents of North Atlantic Igneous Province igneous rocks are not in agreement with the expected chemical signature of lavas following from the mechanism of delamination, and the tectonic evolution of the North Atlantic Igneous Province (mountain building, extension, uplift, magmatic phase) is not in concert with delamination (mountain building, magmatism and uplift, extension). We therefore reject the delamination mechanism as an explanation for formation of the North Atlantic Igneous Province.

Extraterrestrial Body Impact

The idea that meteorite impacts may initiate volcanic eruptions and form LIPs on Earth was first proposed by Rogers (1982). He suggested that the collision of a large extraterrestrial body with Earth could cause the formation of an impact crater with a diameter on the order of 100 km and penetrate the lithosphere into the asthenosphere. Massive volcanism would result and extend to outside the crater, so the crater would be obliterated. Another consequence of an impact is that it could cause the formation of a long-lived mantle plume, with an extended, secondary period of additional melting (see also Abbott and Isley, 2002). This concept has been applied to continental flood basalt provinces (for example, the Siberian Traps; A.P. Jones et al., 2002) as well as oceanic plateau formation (e.g., Ingle and Coffin, 2004).

Igneous rocks are very common near large impacts craters. However, numerical simulations of a large asteroid impact in the last decades have produced contradicting predictions regarding the amount of melt that could result from an impact (Ivanov and Melosh, 2003; A.P. Jones et al., 2005a). Ivanov and Melosh (2003) find that impact-induced melt volume is limited and cannot explain LIPs unless a large impact occurs on very warm lithosphere, such as a mid-ocean ridge, which is unlikely. A.P. Jones et al. (2005a), on the other hand, using a different set of geotherms and melt calculation methods, find impact melt volumes that are large enough to explain the origin of the largest LIPs on Earth.

Apart from the impact crater (which can be obliterated) and possible subsequent top-down-induced plume formation, there are no tectonic processes related to the impact scenario that we can compare with observations of the North Atlantic Igneous Province. Deposits that are being related to impact events, such as quartz spherules, have been found near Disko Island in central West Greenland (A.P. Jones et al., 2005b). In the North Atlantic Igneous Province, melt eruptions took place over an elongated zone of ~3200 km length following a long period of extension. After breakup, there was continuous melt production at Iceland. Even if a large projectile impacted this extended area with steep geotherms in the Late Paleocene just prior to breakup, the elongated shape of the eruption zone does not support an impact scenario for the rifted margins of the North Atlantic Igneous Province. The Iceland hotspot would, in this scenario, represent the top-down-induced mantle plume. Lack of more data and

model development to support an impact scenario for formation of the North Atlantic Igneous Province inhibits a definite rejection or acceptance of this scenario.

Fertile Mantle: Chemical Anomalies

A heterogeneous mantle source has been suggested by several authors to explain the variation in rift-related melt production and geochemical contents of igneous rocks along rifted margins. Korenaga (2004) suggests a fertile mantle heterogeneity as the source of the North Atlantic Igneous Province. Recycled oceanic crust subducted during the closure of the Iapetus Ocean and subsequently segregated at the 660 km discontinuity was eventually brought toward the surface from the lower mantle in a quite complex mixing process. In this model, both the upper and lower mantle were involved in mantle mixing. Fertile material may have accumulated near the upper-lower mantle boundary during subduction. During continental rifting, plate-driven flow assists the entrainment process, and upwelling of the fertile mantle will result in a high degree of melting and emplacement of flood basalts. This process of counterupwelling fertile material could continue for tens of millions of years according to Korenaga (2004). Because recycled oceanic crust with possibly ultramafic cumulates is involved in the convection process, high $^3He/^4He$ ratios such as those found in the North Atlantic Igneous Province are expected.

Subducted slab material is suggested to have played a role in the scenario developed by Foulger and Anderson (2005) and Foulger et al. (2005) (Fig. 6D). Their model suggests that the Iapetus oceanic lithosphere was probably very young (~50 m.y.) at the time of subduction, and therefore buoyant, and that subduction was flat. Subducted oceanic crust might have been trapped within the continental lithosphere. Subsequently, a metamorphic transformation could have transformed the subducted material to eclogite (Foulger and Anderson, 2005). Upon extension of the lithosphere, the fertile material would have been tapped, and excessive melt production would be expected. The geochemical contents of igneous samples of the North Atlantic Igneous Province, including Iceland, are well predicted by this scenario. The melt volume produced at the latitude of Iceland, however, is such that more than one "normal thickness" of oceanic crust is required, and Foulger et al. (2005) suggest that imbrication would have provided the necessary amount of oceanic crust. In this model, the longevity of anomalous volcanism at the Mid-Atlantic Ridge at the latitude of Iceland is attributed to its location on a Caledonian suture, which could run transversely across the North Atlantic. The Mid-Atlantic Ridge near Iceland could have been migrating transversely over this structure since breakup time, which could explain the continuous production of large volumes of melt below Iceland and the shorter period of high melt generation at the volcanic margins of the North Atlantic Igneous Province. The model by Korenaga (2004) does not include a specific mechanism that explains the single short pulse of volcanism at the rifted margins of the North

Atlantic Igneous Province; it implicitly assumes that the crustal inventory was drained quickly, while it continued beneath Iceland. It does predict time-varying melt productivity, because the amount of oceanic crustal material embedded in the mantle matrix is likely to have varied.

A subducted slab is by nature inhomogeneous, because it includes altered N-MORB to ocean island basalts (OIB) and gabbros, depleted oceanic lithospheric mantle, and continental sediments. The model of Foulger et al. (2005) finds support in the inference of McKenzie et al. (2004) that correlations between isotope ratios and elemental concentrations point to the presence of subducted OIB of Caledonian age beneath Theistareykir, North Iceland. Similar to the model of Phipps Morgan and Morgan (1999), the enriched OIB-like components could have been heterogeneously distributed in the mantle by veins etc. in the underlying lherzolite. A separate geochemical mantle reservoir in the deeper mantle is no longer needed to explain OIB-like magmas from Iceland (McKenzie et al., 2004; Foulger et al., 2005).

These eclogite models offer alternative explanations for the genesis of ferrobasalts. These are difficult to explain using classical mantle models, in which they are considered to be the products of remelting of the thick Icelandic crust (Oskarsson et al., 1982). Yasuda et al. (1994) show that initial melts produced by small degrees of partial melting of an eclogite will have an andesitic composition. Basaltic andesites are reported from several initial North Atlantic Igneous Province magma series, e.g., the LS at the Vøring margin. Higher degrees of partial melting (10%–30%) will produce ferrobasaltic melts (Ito and Kennedy, 1974), and 60%–80% melting is required to produce Icelandic melts (Foulger et al., 2005). In the present case, such high degrees of melting might have resulted after a long time, when the Caledonian slab was thermally relaxed and so was closer to its solidus (and perhaps even partially molten) than the surrounding lherzolite (Foulger et al., 2005).

Fertile mantle models could explain the chemistry of Iceland magmas and the longevity of the Iceland melt anomaly, although there is still much uncertainty about the ancient suture and subducted slab. Rifting-induced upwelling of fertile mantle material would explain the coexistence of rifting and magmatism in the North Atlantic. The models do not present a mechanism for prebreakup uplift (Fig. 5). The low seismic wavespeed anomaly beneath the Iceland region, in this scenario, would not have been caused by a mantle plume but instead would have resulted from plate tectonic–related processes such as compositional heterogeneity or a trace of partial melt due to the presence of excess volatiles (Foulger et al., 2001).

Small-Scale and Edge-Driven Convection

Many LIPs, including the North Atlantic Igneous Province, can be related to Archean craton boundaries and lithospheric discontinuities, according to King and Anderson (1995) and Anderson (2005). The edge-driven convection scenario (King and Anderson, 1995, 1998; King, 2005) suggests that thick cratonic lithosphere adjacent to thinner or normal lithosphere may control edge-driven convection cells, which produce flood basalt province formations. This step in lithosphere thickness and corresponding variations in thermal structure would induce small-scale convection of mantle material and flow of previously unmelted mantle into the melting zone (Fig. 6C). Several numerical modeling studies suggest that such a step in lithosphere thickness indeed results in small-scale convection below the edge of the craton (King and Anderson, 1995, 1998) and in increased and focused upwelling (Pascal et al., 2002), and thus in decompression melting. When the melts are formed, a mechanism is required in order for the melts to reach the surface. A favored mechanism for this process is plate separation (King and Anderson, 1995); extension and faulting of the lithosphere will facilitate melt transport toward the surface.

King and Anderson (1995, 1998) find that the amount of upwelling and melting depends on the thickness contrast between the craton and normal or thinner lithosphere, the rate of plate separation, and the thermal structure of the mantle. Material may flow into the melt zone from large depths (>400 km; King, 2005). Detailed melt calculations are not incorporated in the models, and it is therefore not known to what extent the chemical signature of edge-induced melts would differ from or agree with mantle plume related melts. The models by King (2005) and King and Anderson (1995, 1998) do not consider vertical movements of the crust, so we do not know the pattern of uplift/subsidence that can be expected from this scenario. Other studies, however, suggest that uplift of the margins of the extended zone can be expected, which can be attributed to the effects of a thick cratonic lithosphere positioned next to hot asthenosphere below a rift (Vågnes and Amundsen, 1993; Kelemen and Holbrook, 1995).

The edge-driven convection scenario explains several North Atlantic Igneous Province characteristics well. It explains its location (close to the Baltica and Laurentian cratons), its elongated shape, and relation to the rift structure. It also provides a mechanism for generating more melt than is expected from passive rifting alone (White and McKenzie, 1989). It does not directly explain the short period of eruptions; this should be a lithospheric extension–controlled aspect. Also, the timing of volcanism (middle to late synrift) is extension-related and not a component of the edge-driven convection process alone. The scenario has problems explaining the continuous excessive melt production on Iceland.

Modeling experiments (van Wijk et al., 2001) suggest that excess igneous crustal thickness at rifted margins may result from active upwelling rates that are larger than the plate separation rates during breakup, so that more mantle material flows into the melting zone than in passive rift models such as that described, e.g., by Bown and White (1995). This may lead to the melting of large amounts of normal-temperature mantle around breakup time. Buoyancy-driven upwelling is inherent to the rifting process in this model, but is strongly influenced by how the lithosphere deforms during rifting, which, in turn, is influenced by factors such as prerift lithosphere structure, rheology, thermal structure of the lithosphere and mantle, and extension rates (van Wijk et al., 2001, 2004; Corti et al., 2002). This rifting model predicts high eruption rates starting a short time before continental separation (middle to late synrift), with most of the melting taking place in a short period of time around breakup (Fig. 6B). The model does not incorporate the entire upper mantle, but is restricted to the lithosphere and a thin layer of asthenosphere material, so it is not known from what depth material might flow into the rift zone. Dependent on the inherited lithosphere architecture, dynamic uplift of the margin induced by small-scale convection may occur prior to breakup in this model (van Wijk et al., 2004). This rift-related, small-scale convection model has been designed to study the prebreakup history of Atlantic-type passive rifted margins and can thus not provide an explanation for North Atlantic Igneous Province postbreakup events or the continuous excessive melt production at Iceland. Application of this model to the North Atlantic Igneous Province, especially the mid-Norwegian Vøring margin, suggests that characteristic preseparation features (extension, vertical movements, timing and duration of magmatism) are well explained by it (van Wijk et al., 2004), provided that (pre-)rift conditions were favorable.

Boutilier and Keen (1999) and Keen and Boutilier (2000) have found that upwelling of upper-mantle material by divergent plate motions is insufficient to explain the observed melt volumes and postbreakup oceanic crustal thickness of the North Atlantic Igneous Province. A thin (~50 km) Iceland plume sheet layer with a thermal anomaly of ~100 °C is needed in these models to generate large thicknesses of postbreakup igneous crust on the North Atlantic volcanic margins. In these models, small-scale convection continued below the margin edge for some time after breakup. This suggests that vertical motions of the margins could have been expected around breakup time. Because the plume sheet layer would soon have been depleted below the margins, Keen and Boutilier (2000) conclude that formation of the Greenland-Iceland and Iceland-Faeroe ridges requires a more deeply sourced thermal anomaly.

Nielsen and Hopper (2002) also use an upper-mantle convection model to study the interaction of a sublithospheric hot plume sheet with rift-driven flow. They specifically focus on how weak the mantle must be in order to induce small-scale convection and enhance melt production during breakup, while after breakup a stable system with realistic oceanic crustal thickness is established. Their models confirm that small-scale convection can occur during rifting and breakup. To produce thick oceanic crust, the model requires a thin layer of hot material underlying the lithosphere. The temperature anomalies associated with this layer are ~100–200 °C, and after breakup the layer must be quickly exhausted so that a steady-state oceanic spreading ridge system is established. The model by Nielsen and Hopper (2002) thus requires the passive influence of a mantle plume and spread of a plume head for rapid emplacement of warm mantle material in lithospheric thin spots. However, the estimated melt volumes predicted in this study and in the studies of breakup-related magmatism by Boutilier and Keen (1999) assume that the LCBs represent magmatic material, an idea that is presently challenged.

Convection models agree on a possible increased melt volume around breakup in comparison with prior kinematic modeling studies. They disagree on whether fertile mantle, elevated (sublithospheric sheetlike, or ambient) mantle temperatures, large extension rates, or cratonic boundaries are necessary conditions for explaining the vertical movements and melt characteristics of the North Atlantic Igneous Province. They have focused on different aspects of the rift system and rift evolution, but are still being developed. We find that all models provide an explanation for North Atlantic Igneous Province volcanic margin evolution, but the continuous large melt production on Iceland is difficult to explain.

Mantle Plume

The mantle plume concept was first defined by Morgan (1971). In this concept, ~20 mantle plumes originating in the deep mantle bring warm material from the core-mantle boundary to the base of the lithosphere. Once arrived, mantle material flows radially away from the plume. Stresses thereby imposed on the base of the lithosphere will drive plate tectonics. Morgan (1971) related the British Igneous Province to the Iceland plume and suggested that breakup of the Atlantic could have been caused by the line-up of hotspots in this basin (Tristan, Azores, Iceland). He also recognized the regional high topography around each hotspot and suggested a relation with the upwelling material. In the decades following this proposal, the mantle plume concept has been refined and adapted frequently (for an overview see Courtillot et al., 2003; Anderson and Natland, 2005). It is beyond the scope of this article to discuss all adapted scenarios; we will limit ourselves to important adaptations for the North Atlantic Igneous Province (Fig. 6E).

Campbell and Griffiths (1990) and Griffiths and Campbell (1990, 1991), for example, determined the thermal anomaly associated with upwelling plumes, as well as the area affected by the arrival of a plume below the lithosphere. Their starting plume model requires thermal anomalies of ~300 ± 100 °C at the source of the plume, the core-mantle boundary.

Plume head temperatures at the top of the mantle are then up to 250 °C. The plume head may have a diameter of up to 1200 km in the upper mantle, but when the plume head collapses on reaching the base of the lithosphere, it may spread out and affect an area 2000–2500 km across. The starting plume model also predicts vertical movements of the surface (Campbell, 2005). A maximum elevation of ~500–1000 m is predicted for a plume head with a temperature anomaly of ~100 °C. Uplift thus precedes volcanism. This uplift is expected to be followed by subsidence as the plume head spreads beneath the lithosphere. At the margins of the spreading plume head, uplift continues at this time, and the pattern of uplift and subsidence might thus be quite complex. This is also predicted by numerical modeling experiments described by Burov and Guillou-Frottier (2005). They show that upon interaction with the lithosphere, the plume head may cause both small- and large-scale vertical movements of the surface, and uplift as well as subsidence. The resulting topography may be asymmetric and is influenced by lithosphere structure and rheology: a very thick and cold lithosphere will result in different topographic wavelengths from a warm, thin lithosphere.

An important aspect of the mantle plume concept is the mechanism by which the plume penetrates the lithosphere, or decompression melting actually reaches the surface. White and McKenzie (1989) proposed that this can be accomplished by the coincidence between lithosphere extension and thinning and plumes. In the North Atlantic, a period of lithosphere extension predated the arrival of the postulated Iceland plume. Griffiths and Campbell (1991) and Thoraval et al. (2006) suggest that small-scale instabilities that develop in plume heads might ease penetration. Turner et al. (1996) suggest that plume penetration is not a necessary mechanism, but that melting may occur by heating the lithosphere above a mantle plume instead of in the mantle plume itself. Other studies have shown that it is difficult to erode the lower lithosphere with a hot mantle plume (e.g., Ribe and Christensen, 1994; Moore et al., 1998, 1999; Jurine et al., 2005).

The plume concept also predicts a trail of volcanism that reflects the relative motions of the plume and the lithospheric plates. If a plume originates at the core-mantle boundary, it is assumed to have a fixed position with respect to the moving plates and to cause a volcanic trail at the surface of the Earth. Several studies have estimated the position of the Iceland proto-plume through time (e.g., Lawver and Müller, 1994; Torsvik et al., 2001). Details of the timing, location, structure, and composition of the postulated plume in the Early Paleogene, however, remain controversial. Lundin and Doré (2005) questioned the validity of the idea of pinpointing the suggested plume locality.

Another aspect of the mantle plume concept, of importance to the study of the North Atlantic Igneous Province, is stretching of the lithosphere above the plume. Vertical movements associated with plume impingement would place the lithosphere under stress, and zones of both compression and extension may result above the plume head (Burov and Guillou-Frottier, 2005). Whether this rift zone would be elongated, as in the North Atlantic, is not known.

Some descriptions of the plume concept (e.g., Morgan, 1971; Campbell and Griffiths, 1990; Griffiths and Campbell, 1990, 1991; Farnetani and Samuel, 2005) assume that the plume originates at the core-mantle boundary. Because of a clear head-tail structure, tomographic images are expected to show this structure, and some plumes can actually be traced to the deep mantle (Montelli et al., 2004), although there is debate on this finding (van der Hilst and de Hoop, 2005). The Iceland anomaly is not one of this group of deep plumes, which detracts from the plume concept as an explanation for North Atlantic Igneous

Province formation according to Foulger et al. (2005). Farnetani and Samuel (2005) recently calculated the wavespeed anomaly predicted from geodynamic models of thermochemical plumes. They found that a wide range of plume shapes and seismic wavespeed anomalies is possible: a plume head is not always found in their models, and the head-tail structure can be absent. The great variety of plume shapes and sizes translates into a large variety of predicted styles of tomographic image according to this study. The wavespeed anomaly below Iceland would be consistent with a deep-seated plume if such variations in plume shape and style are indeed possible. Some authors have suggested that upper-mantle plumes can be fed by lower-mantle upwellings (e.g., Goes et al., 1999). This would be an alternative explanation for the discontinuous anomaly beneath Iceland.

Material may flow away from the postulated mantle plume laterally over large distances according to the starting plume model of Sleep (1996, 1997) (Fig. 6E). This model proposes that plume material flows underneath the lithosphere and prefers to follow thin spots. Thin spots are formed, for example, by rifting of the lithosphere that might leave relief on its underside (Thompson and Gibson, 1991). This model could provide a mechanism for geological effects of plumes far beyond the location of impingement, and it is proposed as a possible explanation for the huge area affected by North Atlantic Igneous Province volcanism following the rift zones in the North Atlantic. This scenario is poorly supported, though, by the data reported by Doré and Lundin (2005). These authors find, for example, that the LCB has not filled preexisting thin spots in the crust, but terminates abruptly against margin-perpendicular lineaments.

The geochemistry of the North Atlantic Igneous Province is consistent with the plume concept, and to a large extent with the two-stage model presented by Saunders et al. (1997). We refer to their study for an overview of present views of the genesis of the North Atlantic Igneous Province from the plume point of view. The plume scenario further explains uplift history, the upper mantle tomographic results and the distribution and timing of melting. Lack of a time-progressive volcanic track in the Iceland region is a scenario that is not supported by observations, and there is also debate about the continuation of the upper-mantle low seismic wavespeed anomaly into the lower mantle. Furthermore, it is important to stress that recent studies suggest that the mantle plume scenario for North Atlantic Igneous Province formation cannot be tested (e.g., Foulger et al., 2005). This is because many of the adaptations that mantle plume models have undergone in the past decades seem designed specifically to achieve a better fit with observations of the North Atlantic Igneous Province (Foulger, 2005). As a result, the concept cannot be disproved.

EXPLORING ALTERNATIVES

Are the alternative (i.e., non-mantle-plume) models that are at hand today able to explain the different aspects of North Atlantic Igneous Province formation? Delamination and impact models do not appear to be good candidates to explain this LIP, and must at this stage be rejected. The small-scale convection and dynamic rift models do not address the ongoing melt production on Iceland, and therefore need to be combined with other hypotheses. This role could be filled by a mantle plume that has had limited influence in creating the volcanic margins but has facilitated continuing excessive melting on Iceland. Alternatively, a mantle heterogeneity resulting from plate tectonic processes could have acted as a source of continuing igneous activity.

Some studies argue that alternative, nonplume explanations for the chemical signal of the North Atlantic Igneous Province melts may be valid. The geochemical model for the Earth's mantle that has prevailed over the past thirty years (e.g., Morgan, 1971; Hofmann, 1988) involves an enriched deep mantle reservoir (the OIB source) overlain by an undepleted lower primordial mantle on top of which lies the depleted upper mantle reservoir (the MORB source). These chemically isolated reservoirs are genetically connected through the early evolution of the Earth. During melt extraction to form the incompatible element–rich continental crust, the upper mantle was depleted in these elements compared to the primordial mantle (e.g., Hofmann, 1988). In earlier models, the mantle was of only lherzolite composition (harzburgite as a restite of MORB), but studies of mantle peridotites on surface outcrops have pointed to a basalt-pyroxenite-veined harzburgite-lherzolite rock as the major mantle component (Polvé and Allègre, 1980; Suen and Frey, 1987). For the last two decades, several varieties of the alternative "plum-pudding" or "marble-cake" mantle model have been suggested (e.g., Davies, 1981; Zindler et al., 1984; Allègre and Turcotte, 1986; Allègre and Lewin, 1995; Phipps Morgan and Morgan, 1999; Kellogg et al., 2002; Meibom and Anderson, 2004; Albarède, 2005). These models suggest that a high degree of chemical and isotopic heterogeneity characterizes many regions of the mantle. The mantle is, in these models, a heterogeneous assemblage of depleted residues and enriched, recycled subducted oceanic crust, lithosphere, and sediments. Primordial material may still be present in the mantle in small, strongly sheared, and refolded domains (Albarède, 2005). The models differ in various aspects, including whether separate mantle reservoirs exist (the upper and lower mantle).

Phipps Morgan and Morgan (1999) proposed that the mantle contains pyroxenite veins on scales of 1 m to 10 km that span the entire global isotopic range of Nd, Pb, and Os data in erupted OIB and MORB. They concluded that it is possible to produce enriched OIB-like melts in the upper mantle provided the source has enough embedded veins. However, in such a veined (enriched) mantle, the melting of a depleted MORB is much more problematic. MORB-like magmas can be generated from such a mantle only after a prior melting process that produced OIB melts. This primary melting phase has to subtract vein components from the source and deplete the mantle (Phipps Morgan and Morgan, 1999). The North Atlantic Igneous Province offers the opportunity to test competing chemical mantle models, because it is an area where the mantle source contributed not only to flood magmatism (often considered to be related to mantle plumes) but also to magmatism associated with continental breakup.

Most of the Paleogene magmas erupted in the North Atlantic Igneous Province have been interpreted on the basis of the classical layered mantle or the isolated reservoirs mantle model. The chemical signatures result either from a plume mantle source or from the N-MORB mantle reservoir at the outer envelope of the postulated plume (Fig. 3A) (e.g., Saunders et al., 1997). A diffuse chemical "bipolar" trend can be seen in most of the subareas of the North Atlantic Igneous Province (Fig. 3A and B). In the British Igneous Province, the proportion of melts with an "Icelandic" chemical fingerprint increases with time at the base, then later reverts to lavas with N-MORB chemistry (e.g., Hamilton et al., 1998; Kent and Fitton, 2000).

Other petrological and geochemical data commonly attributed to a mantle plume during the formation of the North Atlantic Igneous Province are the picritic, high-Mg magmas (Gill et al., 1992; Larsen and Pedersen, 2000; Upton et al., 2002). Most of the picritic magmas can be associated with continental breakup and with the opening of the Labrador Sea and the North Atlantic at Greenland, Scotland, and Baffin Island. The eruption of large volumes of picrite in West Greenland has generally been related to an abnormally hot local mantle. A problem with this interpretation is that such a huge volume (30–50 vol%) of picrites erupted nowhere else in the North Atlantic Igneous Province. The "doughnut" shape of the Griffiths and Campbell (1990) model well explains the peripheral eruption of the North Atlantic Igneous Province. However, the model predicts that the high-temperature magmas can be produced only by the 150-km-wide hot central plume jet, which is not in agreement with the fixed plume position (Gill et al., 1992). Different explanations have been proposed for the huge amount of picrites, such as a separate, short-lived "Davis Strait plume" (McKenzie and Bickle, 1988). However, why such a hotter plume should be short-lived has not been explained to date. The actual Iceland mantle anomaly provides little support for such plumelike temperatures (Foulger and Anderson, 2005).

A plausible alternative explanation for the formation of the picrites in Greenland and in other North Atlantic Igneous Province regions was suggested by Fram et al. (1998). They proposed that the high Mg content of the magmas reflected their generation at moderately high average pressure as a result of the thick lithosphere at that time. This, combined with the clear evidence for olivine accumulation in some West Greenland high-MgO magmas, weakens the case for the two-stage plume pulse model of Saunders et al. (1997). However, the Fram et al. (1998) interpretation cannot rule out the Saunders et al. (1997) model. The accumulation of olivine in the source points to a possible scenario recently published by Foulger et al. (2005) to explain the geochemistry and large melt volumes in the Iceland region. These authors point out that the expected radial symmetry of geochemical signatures predicted for a plume (e.g., Schilling et al., 1983) is not observed in Iceland. Geochemical discontinuities occur in Iceland across relatively minor structures, and the melting temperatures of the most primitive Icelandic melts are calculated to be similar to MORB formation temperatures (Breddam, 2002).

Major geochemical problems in relating LIPs to postulated mantle plumes are the identification and quantitative estimation of the contribution of crustal and lithospheric sources to the magmatism. However, such studies are essential for us to ultimately understand which of the geochemical variations observed in basalts reflect the expected intrinsic variations of deep-mantle sources. Cautious isotopic studies (see Fig. 3A) showed that for most classical isotopic systems (Rb, Sr, Pb) the differences between mixing lines of depleted MORB source with enriched mantle reservoirs and mixing lines between mantle sources with continental crust signatures are not visible. For North Atlantic Igneous Province magmas, this is even more important, because many lavas erupted through continental crust and show contamination by the crust (Fig. 3A). The OIB paradox (Fitton, this volume) illustrates the problem of distinguishing subcontinental and/or continental sources for OIB magmas using standard geochemical systems and models.

Recently the geochemistry of helium has been assumed to reflect evidence for mantle plumes. Helium has two isotopes: ^{3}He, a primordial isotope incorporated in the Earth during planetary accretion, and ^{4}He, an isotope largely produced by radioactive decay of U and Th isotopes. Basalts related to postulated mantle plumes have high ^{3}He/^{4}He ratios compared to MORB. This observation is assumed to be an explicit indicator of a primitive, undegassed lower-mantle source component with a high ^{3}He content. However, deep-mantle sources are not the only possible explanation for high ^{3}He/^{4}He ratios (Meibom et al., 2003). He concentrations in rocks believed to reflect deep mantle sources should be higher compared to melts from the depleted MORB mantle. However, the He content in OIB is often two to three orders of magnitude lower than in MORB (Foulger et al., 2005). A key process in He geochemistry is oceanic crust formation, because the incompatible mother isotopes of ^{4}He (U and Th) are extracted from the source mantle (e.g., Albarède, 1998; Anderson, 1998). Thus, ultramafic cumulates are possibly low–U + Th sources. Ultramafic cumulates could be part of the inhomogeneous mantle source proposed by Phipps Morgan and Morgan (1999) and might provide high He ratios to North Atlantic Igneous Province erupted basalts. The extreme He data of Stuart et al. (2003) from East Greenland picrites confirms this scenario, because rocks with the highest MgO content have the highest He ratios.

Can the volume and timing of magmatic activity be explained by alternative models? Both the volume and timing (middle to late synrift) observed along the North Atlantic Igneous Province volcanic margins can be explained by dynamic rift-related processes (King and Anderson, 1995, 1998; van Wijk et al., 2001). Mantle heterogeneities resulting from plate tectonic processes (Korenaga, 2004; Foulger and Anderson, 2005; Foulger et al., 2005) have been proposed to account for the ongoing excessive melt production at the Mid-Atlantic, Reykjanes, and Kolbeinsey ridges and Iceland, but there are still uncertainties concerning the location of the proposed ancient suture zone and old slab distribution. Uplift of the surface might be explainable by dynamic rift processes as well, but more work is needed to investigate that issue. The low seismic velocity anomaly in the upper mantle beneath the Iceland region has been suggested to result from plate tectonic processes (Foulger et al., 2001), but further investigation of this issue, too, is required to build a convincing case.

CONCLUSION

The mantle plume concept is currently most successful in explaining the formation of the North Atlantic Igneous Province. However, it has been suggested that this is largely due to the adaptations that have been made to the site-specific model over the years to better explain observations of the province. There are unexplained aspects of the concept that remain to be addressed, while alternative models also need to be further matured.

Alternative models (e.g., rift-related small-scale convection) are being developed that appear to be able to explain volcanic rifted margin volcanism well. However, the most important problem that nonplume models need to overcome is the continuing, long-lived melt anomaly extending from the Greenland and Faeroe ridges to Iceland. Mantle heterogeneities resulting from ancient subducted slabs are invoked in plate tectonic models to explain the continuing melt production, but further proof of this is required.

Existing data sets and geodynamic concepts are incomplete, which hinders a more conclusive statement on whether the mantle plume or alternative models can be accepted or rejected. For example, (1) the lateral distribution of magmatic material is poorly known, and there is still debate regarding the composition of the LCB; (2) there is uncertainty regarding several characteristics of the melts, such as the source temperatures for the picritic melts; and (3) the degree of homogeneity of the mantle is controversial. Ongoing development of the different models and the availability of increasingly more reliable data should help to resolve the problem of the challenged mantle plume concept in the near future.

ACKNOWLEDGMENTS

We would like to thank G. Foulger, D. Jurdy, P. Vogt, and an anonymous reviewer for their detailed and constructive comments and Jan Hertogen for discussion and pre-reading. Comments by A. Breivik, L. Geoffroy, and S. Jones during formal e-mail discussion improved the clarity of this article. R.M. is supported by Fonds du Wetenschappelijk Onderzoek-Vlaanderen (Project G.0008.04) and Bourse de formation-recherche 05/133 Luxembourg in the framework of the European Science Foundation EUROMARGINS program. J.v.W. was supported by National Science Foundation grant 0527215 and LDRD.

REFERENCES CITED

Abbott, D.H., and Isley, A.E., 2002, Extraterrestrial influences on mantle plume activity: Earth and Planetary Science Letters, v. 205, p. 53–62, doi: 10.1016/S0012-821X(02)01013-0.

Ahmadi, Z., Sawyers, M., Kenyon-Roberts, S., Stanworth, C.W., Kugler, K.A., Kristensen, J., and Fugelli, E.M.G, 2003, Paleocene, *in* Evans, D., et al., eds., The millennium atlas: Petroleum geology of the central and northern North Sea, p. 235–259.

Albarède, F., 1998, Time-dependent models of U-Th-He and K-Ar evolution and the layering of mantle convection: Chemical Geology, v. 145, p. 413–429, doi: 10.1016/S0009-2541(97)00152-6.

Albarède, F., 2005, The survival of mantle geochemical heterogeneities, *in* Earth's deep mantle: Structure, composition and evolution: Washington, D.C., American Geophysical Union, Geophysical Monograph Series, v. 160, p. 27–46.

Allègre, C.J., and Lewin, E., 1995, Isotopic systems and stirring times of the earth's mantle: Earth and Planetary Science Letters, v. 136, p. 629–646, doi: 10.1016/0012-821X(95)00184-E.

Allègre, C.J., and Turcotte, D.L., 1986, Implications of a two-component marble-cake mantle: Nature, v. 323, p. 123–127, doi: 10.1038/323123a0.

Anderson, D.L., 1998, The helium paradoxes: Proceedings of the National Academy of Sciences of the United States of America, v. 95, p. 4822–4827, doi: 10.1073/pnas.95.9.4822.

Anderson, D.L., 2005, Scoring hotspots: The plume and plate paradigms, *in* Foulger, G.R., et al., eds., Plates, plumes, and paradigms: Boulder, Colorado, Geological Society of America Special Paper 388, p. 31–54.

Anderson, D.L., and Natland, J.H., 2005, A brief history of the plume hypothesis and its competitors: Concept and controversy, *in* Foulger, G.R., et al., eds., Plates, plumes, and paradigms: Boulder, Colorado, Geological Society of America Special Paper 388, p. 119–145.

Bailey, J.C., and Rasmussen, M.H., 1997, Petrochemistry of Jurassic and Cretaceous tholeiites from Kongs Karls Land, Svalbard, and their relation to Mesozoic magmatism in the Arctic: Polar Research, v. 16, p. 37–62.

Baksi, A.K., 2005, Evaluation of radiometric ages pertaining to rocks hypothesized to have been derived by hotspot activity, in and around the Atlantic, Indian, and Pacific Oceans, *in* Foulger, G.R., et al., eds., Plates, plumes, and paradigms: Boulder, Colorado, Geological Society of America Special Paper 388, p. 55–70.

Baksi, A.K., 2007a (this volume), A quantitative tool for detecting alteration in undisturbed rocks and minerals—I: Water, chemical weathering, and atmospheric argon, *in* Foulger, G.R., and Jurdy, D.M., eds., Plates, plumes, and planetary processes: Boulder, Colorado, Geological Society of America Special Paper 430, doi: 10.1130/2007.2430(15).

Baksi, A.K., 2007b (this volume), A quantitative tool for detecting alteration in undisturbed rocks and minerals—II: Application to argon ages related to hotspots, *in* Foulger, G.R., and Jurdy, D.M., eds., Plates, plumes, and planetary processes: Boulder, Colorado, Geological Society of America Special Paper 430, doi: 10.1130/2007.2430(16).

Barrat, J.A., and Nesbitt, R.W., 1996, Geochemistry of the Tertiary volcanism of Northern Ireland: Chemical Geology, v. 129, p. 15–38, doi: 10.1016/0009-2541(95)00137-9.

Barton, A.J., and White, R.S., 1995, The Edoras Bank margin: Continental breakup in the presence of a mantle plume: Journal of the Geological Society of London, v. 152, p. 971–974.

Barton, A.J., and White, R.S., 1997, Crustal structure of Edoras Bank continental margin and mantle thermal anomalies beneath the North Atlantic: Journal of Geophysical Research, v. 102, p. 3109–3129, doi: 10.1029/96JB03387.

Bijwaard, H., and Spakman, W., 1999, Tomographic evidence for a narrow whole mantle plume below Iceland: Earth and Planetary Science Letters, v. 166, p. 121–126, doi: 10.1016/S0012-821X(99)00004-7.

Bird, P., 1979, Continental delamination and the Colorado Plateau: Journal of Geophysical Research, v. 84, p. 7561–7571.

Bird, P., and Baumgardner, J., 1981, Steady propagation of delamination events: Journal of Geophysical Research, v. 86, p. 4891–4903.

Blystad, P., Brekke, H., Færseth, R.B., Larsen, B.T., Skogseid, J., and Tørudbakken, B., 1995, Structural elements of the Norwegian continental shelf, part II, The Norwegian Sea region: Norwegian Petroleum Directorate Bulletin, v. 8, p. 1–45.

Boutilier, R.R., and Keen, C.C., 1999, Small-scale convection and divergent plate boundaries: Journal of Geophysical Research, v. 104, p. 7389–7403, doi: 10.1029/1998JB900076.

Bown, J.W., and White, R.S., 1995, Effect of finite extension rate on melt generation at rifted continental margins: Journal of Geophysical Research, v. 100, p. 18,011–18,029, doi: 10.1029/94JB01478.

Breddam, K., 2002, Kistufell: Primitive melt from the Iceland mantle plume: Journal of Petrology, v. 43, p. 345–373, doi: 10.1093/petrology/43.2.345.

Breivik, A.J., Mjelde, R., Faleide, J.I., and Murai, Y., 2006, Rates of continental breakup magmatism and seafloor spreading in the Norway Basin–Iceland plume interaction: Journal of Geophysical Research, v. 111, doi: 10.1029/2005JB004004.

Brooks, C.K., and Gleadow, A.J.W., 1977, A fission-track age for the Skeargaard intrusion and the age of the East Greenland basalts: Geology, v. 5, p. 539–540, doi: 10.1130/0091-7613(1977)5<539:AFAFTS>2.0.CO;2.

Bugge, T., Prestvik, T., and Rokoengen, K., 1980, Lower Tertiary volcanic rocks off Kristiansund–Mid Norway: Marine Geology, v. 35, p. 277–286, doi: 10.1016/0025-3227(80)90121-8.

Burov, E., and Guillou-Frottier, L., 2005, The plume head–continental lithosphere interaction using a tectonically realistic formulation for the lithosphere: Geophysical Journal International, v. 161, p. 469–490, doi: 10.1111/j.1365-246X.2005.02588.x.

Burov, Y.P., Krasil'scikov, A.A., Firsov, D.V., and Klubov, B.A., 1975, The age of Spitsbergen dolerites (from isotopic dating): Arbok-Norsk Polarinstitutt, v. 1975, p. 101–108.

Campbell, I.H., 2005, Large igneous provinces and the mantle plume hypothesis: Elements, v. 1, p. 265–269.

Campbell, I.H., and Griffiths, R.W., 1990, Implications of mantle plume structure for the evolution of flood basalts: Earth and Planetary Science Letters, v. 99, p. 79–93, doi: 10.1016/0012-821X(90)90072-6.

Carlson, R.W., Pearson, D.G., and James, D.E., 2005, Physical, chemical, and chronological characteristics of continental mantle: Reviews of Geophysics, v. 43, p. RG1001, doi: 10.1029/2004RG000156.

Ceramicola, S., Stoker, M., Praeg, D., Shannon, P.M., De Santis, L., Hoult, R., Hjelstven, B.O., Laberg, S., and Mathiesen, A., 2005, Anomalous Cenozoic subsidence along the "passive" continental margin from Ireland to mid-Norway: Marine and Petroleum Geology, v. 22, p. 1045–1067, doi: 10.1016/j.marpetgeo.2005.04.005.

Chalmers, J.A., and Laursen, K.H., 1995, Labrador Sea: The extent of continental crust and the timing of the start of sea-floor spreading: Marine and Petroleum Geology, v. 12, p. 205–217, doi: 10.1016/0264-8172(95)92840-S.

Chalmers, J.A., Whittaker, R.C., Skaarup, N., Pulvertaft, T.C.R., Geoffroy, L., Skuce, A.G., Angelier, J., Gelard, J.P., Lepvrier, C., and Olivier, P., 1999, The coastal flexure of Disko (West Greenland): Onshore expression of the "oblique reflectors": Discussion and reply: Journal of the Geological Society, London, v. 156, p. 1051–1055.

Chambers, L.M., and Fitton, J.G. 2000, Geochemical transitions in the ancestral Iceland plume: Evidence from the Isle of Mull Tertiary volcano, Scotland: Journal of the Geological Society, London, v. 157, p. 261–263.

Chapman, T.J., Broks, T.M., Corcoran, D.V., Duncan, L.A., and Dancer, P.N., 1999, The structural evolution of the Erris Trough, offshore Northwest Ireland, and implications for hydrocarbon generation: Petroleum Geology of Northwest Europe: Proceedings of the 5th conference, v. 5, p. 455–469.

Chian, D., Keen, C., Reid, I., and Louden, K.E., 1995, Evolution of the nonvolcanic rifted margins: New results from the conjugate margins of the Labrador Sea: Geology, v. 23, p. 589–592, doi: 10.1130/0091-7613(1995) 023<0589:EONRMN>2.3.CO;2.

Clarke, D.B., 1970, Tertiary basalts of Baffin Bay: Possible primary magma from the mantle: Contributions to Mineralogy and Petrology, v. 25, p. 203–224, doi: 10.1007/BF00371131.

Clarke, D.B., and O'Hara, M.J., 1979, Nickel, and the existence of high-MgO liquids in nature: Discussion: Earth and Planetary Science Letters, v. 44, p. 153–158, doi: 10.1016/0012-821X(79)90016-5.

Coffin, M.F., and Eldholm, O., 1994, Large igneous provinces: Crustal structure, dimensions, and external consequences: Reviews of Geophysics, v. 32, p. 1–36, doi: 10.1029/93RG02508.

Corcoran, D.V., and Doré, A.G., 2005, A review of techniques for the estimation of magnitude and timing of exhumation in offshore basins: Earth-Science Reviews, v. 72, p. 129–168, doi: 10.1016/j.earscirev.2005.05.003.

Corti, G., van Wijk, J., Bonini, M., Sokoutis, D., Cloetingh, S., Innocenti, F., and Manetti, P., 2002, Transition from continental breakup to seafloor spreading: How fast, symmetric and magmatic: Geophysical Research Letters, v. 30, doi: 10.1029/2003GL017374.

Courtillot, V., Davaille, A., Besse, J., and Stock, J., 2003, Three distinct types of hotspots in the Earth's mantle: Earth and Planetary Science Letters, v. 205, p. 295–308, doi: 10.1016/S0012-821X(02)01048-8.

Coward, M., 1995, Structural and tectonic setting of the Permo-Triassic basins of Northwest Europe, *in* Boldy, S.A.R., ed., Permian and Triassic rifting in Northwest Europe: Geological Society, London, Special Publications 91, p. 7–39.

Dam, G., Larsen, M., and Sonderholm, M., 1998, Sedimentary response to mantle plumes: Implications for Paleocene onshore successions: West and East Greenland: Geology, v. 26, p. 207–210.

Dancer, P.N., Algar, S.T., and Wilson, I.R., 1999, Structural evolution of the Slyne Trough: Petroleum Geology of Northwest Europe: Proceedings of the 5th conference, v. 5, p. 445–453.

Davies, G.F., 1981, Earth's neodymium budget and structure and evolution of the mantle: Nature, v. 290, p. 208–213, doi: 10.1038/290208a0.

Deptuck, M.E., MacRae, R.A., Shimeld, J.W., Williams, G.L., and Fensome, R.A., 2003, Revised upper Cretaceous and lower Paleogene lithostratigraphy and depositional history of the Jeanne d'Arc Basin, offshore Newfoundland, Canada: AAPG Bulletin, v. 87, p. 1459–1483, doi: 10.1306/ 050203200178.

Dickin, A.P., 1981, Isotope geochemistry of Tertiary igneous rocks from the Isle of Skye, NW Scotland: Journal of Petrology, v. 22, p. 155–189.

Dickin, A.P., and Jones, N.W., 1983, Isotopic evidence for the age and origin of pitchstones and felsites, Isle of Eigg, NW Scotland: Journal of the Geological Society, London, v. 140, p. 691–700.

Doré, A.G., and Lundin, E.R., 2005, Challenges and controversies in exploration of the NE Atlantic margin: 25th Annual Gulf Coast Section SEPM Foundation, Program and Abstracts, p. 18–21.

Doré, A.G., Lundin, E.R., Jensen, L.N., Birkeland, Ø., Eliassen, P.E., and Fichler, C., 1999, Principal tectonic events in the evolution of the Northwest European Atlantic margin: Petroleum Geology of Northwest Europe: Proceedings of the 5th conference, v. 5, p.41–61.

Driscoll, N.W., Hogg, J.R., Christie-Blick, N., and Karner, G.D., 1995, Extensional tectonics in the Jeanne d'Arc Basin, offshore Newfoundland: Implications for the timing of breakup between Grand Banks and Iberia, *in* Scrutton, R.A., et al., eds., The tectonics, sedimentation and palaeoceanography of the north Atlantic region: London, Geological Society of London, p. 1–28.

Ducea, M., and Saleeby, J., 1998, A case for delamination of the deep batholithic crust beneath the Sierra Nevada, California: International Geology Review, v. 40, p. 78–93.

Duncan, R.A., Larsen, H.C., Allan, J.F., et al., 1996, Proceedings of the Ocean Drilling Program, Initial Reports, Southeast Greenland margin, covering Leg 163 of the cruises of the drilling vessel JOIDES Resolution: Proceedings of the Ocean Drilling Program, Initial Reports, v. 163, 623 pp.

Ebbing, J., Lundin, E., Olesen, O., and Hansen, E.K., 2006, The mid-Norwegian margin: A discussion of crustal lineaments, mafic intrusions, and remnants of the Caledonian root by 3D density modeling and structural interpretation: Journal of the Geological Society, London, v. 163, p. 47–59, doi: 10.1144/0016-764905-029.

Eldholm, O., and Grue, K., 1994, North Atlantic volcanic margins: Dimensions and production rates: Journal of Geophysical Research, v. 99, p. 2955–2968, doi: 10.1029/93JB02879.

Eldholm, O., Thiede, J., Taylor, E., et al., 1987, Summary and preliminary conclusions, ODP Leg 104: Proceedings of the Ocean Drilling Program, Initial Reports, v. 104, p. 751–771.

Elkins-Tanton, L.T., 2005, Continental magmatism caused by lithospheric delamination, *in* Foulger, G.R., et al., eds., Plates, plumes, and paradigms: Boulder, Colorado, Geological Society of America Special Paper 388, p. 449–461.

Elkins-Tanton, L.T., and Grove, T.L., 2003, Evidence for deep melting of hydrous, metasomatized mantle: Pliocene high potassium magmas from the Sierra Nevadas: Journal of Geophysical Research, v. 108, doi: 10.1029/ 2002JB002168.

Elkins-Tanton, L.T., and Hager, B.H., 2000, Melt intrusion as a trigger for lithospheric foundering and the eruption of the Siberian flood basalts: Geophysical Research Letters, v. 27, p. 3937–3940, doi: 10.1029/2000GL011751.

England, P.C., and Houseman, G.A., 1989, Extension during continental convergence, with application to the Tibetan Plateau: Journal of Geophysical Research, v. 94, p. 17,561–17,579.

Erratt, D., Thomas, G.M., and Wall, G.R.T., 1999, The evolution of the central North Sea Rift: Petroleum Geology of Northwest Europe: Proceedings of the 5th Conference, v. 5, p. 63–82.

Estrada, S., 1998, Basaltic dykes in the Kap Washington and Frigg Fjord areas (North Greenland): Polarforschung, v. 68, p. 19–23.

Estrada, S., and Henjes-Kunst, F., 2004, Volcanism in the Canadian High Arctic related to the opening of the Arctic Ocean: Zeitschrift der Deutschen Geologischen Gesellschaft, v. 154, p. 579–603.

Faleide, J.I., Vågnes, E., and Gudlaugsson, S.T., 1993, Late Mesozoic–Cenozoic evolution of the South-western Barents Sea in a regional rift–shear tectonic setting: Marine and Petroleum Geology, v. 10, p. 186–214, doi: 10.1016/0264-8172(93)90104-Z.

Farmer, G.L., Glazner, A.F., and Manley, C.R., 2002, Did lithospheric delamination trigger late Cenozoic potassic volcanism in the southern Sierra Nevada, California?: Geological Society of America Bulletin, v. 114, p. 754–768, doi: 10.1130/0016-7606(2002)114<0754:DLDTLC>2.0.CO;2.

Farnetani, C.G., and Samuel, H., 2005, Beyond the thermal plume paradigm: Geophysical Research Letters, v. 32, doi: 10.1029/2005GL022360.

Fitton, J.G., 2007 (this volume), The OIB paradox, *in* Foulger, G.R., and Jurdy, D.M., eds., Plates, plumes, and planetary processes: Boulder, Colorado, Geological Society of America Special Paper 430, doi: 10.1130/2007 .2430(20).

Fitton, J.G., Larsen, L.M., Saunders, A.D., Hardarson, B.S., and Kempton, P.D., 2000, Palaeogene continental to oceanic magmatism on the SE Greenland continental margin at 63°N: A review of the results of Ocean Drilling Program Legs 152 and 163: Journal of Petrology, v. 41, p. 951–966, doi: 10.1093/petrology/41.7.951.

Foulger, G.R., 2005, Mantle plumes: Why the current scepticism?: Chinese Science Bulletin, v. 50, p. 1555–1560, doi: 10.1360/982005-919.

Foulger, G.R., 2006, Older crust underlies Iceland: Geophysical Journal International, v. 165, p. 672–676, doi: 10.1111/j.1365-246X.2006.02941.x.

Foulger, G.R., and Anderson, D.L., 2005, A cool model for the Iceland hot spot: Journal of Volcanology and Geothermal Research, v. 141, p. 1–22, doi: 10.1016/j.jvolgeores.2004.10.007.

Foulger, G.R., Pritchard, M.J., Julian, B.R., Evans, J.R., Allen, R.M., Nolet, G., Morgan, W.J., Bergsson, B.H., Evlendsson, P., Jakobsolóttir, S., Ragharsson, S., and Vogfjørd, K., 2001, Seismic tomography shows that upwelling beneath Iceland is confined to the upper mantle: Geophysical Journal International, v. 146, p. 504–530, doi: 10.1046/j.0956-540x.2001.01470.x.

Foulger, G.R., Natland, J.H., and Anderson, D.L., 2005, Genesis of the Iceland melt anomaly by plate tectonic processes, *in* Foulger, G.R., et al., eds., Plates, plumes, and paradigms: Boulder, Colorado, Geological Society of America Special Paper 388, p. 595–625.

Fowler, S.R., White, R.S., Spence, G.D., and Westbrook, G.K., 1989, The Hatton Bank continental margin, II, Deep structure from two-ship expanding spread seismic profiles: Geophysical Journal of the Royal Astronomical Society, v. 96, p. 295–309.

Fram, M.S., Lesher, C.E., and Volpe, A.M., 1998, Mantle melting systematics: Transition from continental to oceanic volcanism on the Southeast Greenland margin: Proceedings of the Ocean Drilling Program, Scientific Results, v. 152, p. 373–386.

Furlong, K.P., and Fountain, D.M., 1986, Continental crustal underplating: Thermal considerations and seismic-petrologic consequences: Journal of Geophysical Research, v. 91, p. 8285–8294.

Gamble, J.A., Wysoczanski, R.J., and Meighan, I.G., 1999, Constraints on the age of the British Tertiary Volcanic Province from ion microprobe U-Pb (SHRIMP) ages for acid igneous rocks from NE Ireland: Journal of the Geological Society, London, v. 156, p. 291–299.

Gariépy, C., Ludden, J., and Brooks, C., 1983, Isotopic and trace element constraints on the genesis of the Faeroe lava pile: Earth and Planetary Science Letters, v. 63, p. 257–272, doi: 10.1016/0012-821X(83)90041-9.

Geldmacher, J., Troll, V.R., Emeleus, C.H., and Donaldson, C.H., 2002, Pb-isotope evidence for contrasting crustal contamination of primitive to evolved magmas from Ardnamurchan and Rum: Implications for the structure of the underlying crust: Scottish Journal of Geology, v. 38, p. 55–61.

Geoffroy, L., Gelard, J.P., Lepvrier, C., and Olivier, P., 1998, The coastal flexure of Disko (West Greenland), onshore expression of the "oblique reflectors": Journal of the Geological Society, London, v. 155, p. 463–473.

Geoffroy, L., Aubourg, C., Callot, J.P., and Barrat, A., 2007 (this volume), Mechanisms of crustal growth in large igneous provinces: The north Atlantic province as a case study, *in* Foulger, G.R., and Jurdy, D.M., eds., Plates, plumes, and planetary processes: Boulder, Colorado, Geological Society of America Special Paper 430, doi: 10.1130/2007.2430(34).

Gernigon, L., Ringenbach, J.C., Planke, S., Le Gall, B., and Jonquet-Kolsto, H., 2003, Extension, crustal structure and magmatism at the outer Vøring Basin, Norwegian margin: Journal of the Geological Society, London, v. 160, p. 197–208.

Gernigon, L., Lucazeau, F., Brigaud, F., Ringenbach, J.-C., Le Gall, B., and Planke, S., 2006, Interpretation and integrated modelling along a volcanic rifted margin: Deep structures and breakup processes along the Vøring margin (Norway): Tectonophysics, v. 412, p. 255–278, doi: 10.1016/j.tecto.2005.10.038.

Gibson, D., McCormick, A.G., Meighan, I.G., and Halliday, A.N., 1987, The British Tertiary Igneous Province: Young Rb-Sr ages for the Mourne Mountains granites: Scottish Journal of Geology, v. 23, p. 221–225.

Gill, R.C.O., Pedersen, A.K., and Larsen, J.G., 1992, Tertiary picrites in west Greenland: Melting at the periphery of a plume? *in* Storey, B.C., Alabaster, T., and Pankhurst, R.J., eds., Magmatism and the causes of continental break-up: New evidence from West Greenland: Geological Society of London Special Publication 68, p. 335–348.

Gleadow, A.J.W., and Brooks, C.K., 1979, Fission track dating, thermal histories and tectonics of igneous intrusions in East Greenland: Contributions to Mineralogy and Petrology, v. 71, p. 45–60, doi: 10.1007/BF00371880.

Goes, S., Spakman, W., and Bijwaard, H., 1999, A lower mantle source for central European volcanism: Science, v. 286, p. 1928–1931, doi: 10.1126/science.286.5446.1928.

Green, P.F., Thomson, K., and Hudson, J.D., 2001, Recognition of tectonic events in undeformed regions: Contrasting results from the Midland Platform and East Midland Shelf, Central England: Journal of the Geological Society, London, v. 158, p. 59–73.

Griffiths, R.W., and Campbell, I.H., 1990, Stirring and structure in mantle starting plumes: Earth and Planetary Science Letters, v. 99, p. 66–78, doi: 10.1016/0012-821X(90)90071-5.

Griffiths, R.W., and Campbell, I.H., 1991, Interaction of mantle plume heads with the Earth's surface and onset of small-scale convection: Journal of Geophysical Research, v. 96, p. 18,295–18,310.

Hamilton, M.A., Pearson, D.G., Thompson, R.N., Kelley, S.P., and Emeleus, C.H., 1998, Rapid eruption of Skye lavas inferred from precise U-Pb and Ar-Ar dating of the Rum and Cuillin plutonic complexes: Nature, v. 394, p. 260–263, doi: 10.1038/28361.

Hardarson, B.S., Fitton, J.G., Ellam, R.M., and Pringle, M.S., 1997, Rift relocation—A geochemical and geochronological investigation of a paleo-rift in northwest Iceland: Earth and Planetary Science Letters, v. 153, p. 181–196, doi: 10.1016/S0012-821X(97)00145-3.

Harland, W.B., and Stephens, C.F., 1997, Neogene–Quaternary history: Memoirs of the Geological Society of London, v. 17, p. 418–435.

Harland, W.B., Anderson, L.M., and Manasrah, D., 1997, The geology of Svalbard: Memoirs of the Geological Society of London, v. 17, 521 p.

Harrison, R.K., and Merriman, R.J., 1984, Petrology, mineralogy, and chemistry of basaltic rocks: Leg 81: Initial Reports of the Deep Sea Drilling Project, v. 81, p. 743–774.

Hart, S.R., 1984, A large-scale isotope anomaly in the southern hemisphere mantle: Nature, v. 309, p. 753–757, doi: 10.1038/309753a0.

Hart, S.R., and Davis, K.E., 1979, Nickel, and the existence of high-MgO liquids in nature: Reply: Earth and Planetary Science Letters, v. 44, p. 159–161, doi: 10.1016/0012-821X(79)90017-7.

Hart, S.R., and Zindler, A., 1989, Constraints on the nature and development of chemical heterogeneities in the mantle: Fluid Mechanics of Astrophysics and Geophysics, v. 4, p. 261–387.

Heeremans, M., and Faleide, J.I., 2004, Late Carboniferous–Permian tectonics and magmatic activity in the Skagerrak, Kattegat and the North Sea, *in* Wilson, M., et al., eds., Permo-Carboniferous magmatism and rifting in Europe: Geological Society of London Special Publication 223, p. 157–176.

Hilton, D.R., Gronvold, K., Macpherson, C.G., and Castillo, P.R., 1999, Extreme ^{3}He/^{4}He ratios in Northwest Iceland: Constraining the common component in mantle plumes: Earth and Planetary Science Letters, v. 173, p. 53–60, doi: 10.1016/S0012-821X(99)00215-0.

Hirschmann, M.M., Renne, P.R., and McBirney, A.R., 1997, ^{40}Ar-^{39}Ar dating of the Skaergaard intrusion: Earth and Planetary Science Letters, v. 146, p. 645–658, doi: 10.1016/S0012-821X(96)00250-6.

Hitchen, K., 2004, The geology of the UK Hatton-Rockall margin: Marine and Petroleum Geology, v. 21, p. 993–1012, doi: 10.1016/j.marpetgeo.2004.05.004.

Hitchen, K., and Ritchie, J.D., 1993, New K-Ar ages, and a provisional chronology, for the offshore part of the British Tertiary Igneous Province: Scottish Journal of Geology, v. 29, p. 73–85.

Hofmann, A.W., 1988, Chemical differentiation of the Earth: The relationship between mantle, continental crust and oceanic crust: Earth and Planetary Science Letters, v. 90, p. 297–314, doi: 10.1016/0012-821X(88)90132-X.

Holbrook, W.S., Larsen, H.C., Korenaga, J., Dahl-Jensen, T., Reid, I.D., Kelemen, P.B., Hopper, J.R., Kent, G.M., Lizarralde, D., Bernstein, S., and Detrick, S., 2001, Mantle thermal structure and active upwelling during continental breakup in the North Atlantic: Earth and Planetary Science Letters, v. 190, p. 251–266, doi: 10.1016/S0012-821X(01)00392-2.

Holm, P.M., Gill, R.C.O., Pedersen, A.K., Larsen, J.G., Hald, N., Nielsen, T.F.D., and Thirlwall, M.F., 1993, The Tertiary picrites of West Greenland: Contributions from "Icelandic" and other sources: Earth and Planetary Science Letters, v. 115, p. 227–244, doi: 10.1016/0012-821X(93)90224-W.

Hutton, J., 1788. Theory of the Earth; or, An investigation of the laws observable in the composition, dissolution, and restoration of land upon the globe: Transactions of the Royal Society of Edinburgh.

Ingle, S., and Coffin, M.F., 2004, Impact origin of the greater Ontong Java Plateau?: Earth and Planetary Science Letters, v. 218, p. 123–134, doi: 10.1016/S0012-821X(03)00629-0.

Ito, G., 2001, Reykjanes "V"-shaped ridges originating from a pulsing and dehydrating mantle plume: Nature, v. 411, p. 681–684, doi: 10.1038/35079561.

Ito, K., and Kennedy, G.C., 1974, The composition of liquids formed by partial melting of eclogites at high temperatures and pressures: Journal of Geology, v. 82, p. 383–392.

Ivanov, B.A., and Melosh, H.J., 2003, Impacts do not initiate volcanic eruptions: Eruptions close to the crater: Geology, v. 31, p. 869–872, doi: 10.1130/G19669.1.

Jokat, W., 2003, Seismic investigations along the western sector of Alpha Ridge, central Arctic Ocean: Geophysical Journal International, v. 152, p. 185–201, doi: 10.1046/j.1365-246X.2003.01839.x.

Jones, A.P., Price, G.D., Price, N.J., DeCarli, P.S., and Clegg, R.A., 2002, Impact induced melting and the development of large igneous provinces: Earth and Planetary Science Letters, v. 202, p. 551–561, doi: 10.1016/S0012-821X(02)00824-5.

Jones, A.P., Wünemann, K., and Price, G.D., 2005a, Modeling impact volcanism as a possible origin for the Ontong Java Plateau, in Foulger, G.R., et al., eds., Plates, plumes, and paradigms: Boulder, Colorado, Geological Society of America Special Paper 388, p. 711–720.

Jones, A.P., Kearsley, A.T., Friend, C.R.L., Robin, E., Beard, A., et al., 2005b, Are there signs of a large Palaeocene impact, preserved around Disko bay, W. Greenland?: Nuussuaq spherule beds origin by impact instead of volcanic eruption?: Geological Society of America Special Paper 384, p. 281–298.

Jones, S.M., and White, N., 2003, Shape and size of the starting Iceland plume swell: Earth and Planetary Science Letters, v. 216, p. 271–282, doi: 10.1016/S0012-821X(03)00507-7.

Jones, S.M., White, N., and Lovell, B., 2001, Cenozoic and Cretaceous transient uplift in the Porcupine basin and its relationship to a mantle plume: Geological Society of London Special Publication 188, p. 345–360.

Jones, S.M., White, N., and Maclennan, J., 2002, V-shaped ridges around Iceland: Implications for spatial and temporal patterns of mantle convection: Geochemistry, Geophysics, Geosystems, v. 3, no. 10, p. 1059, doi: 10.1029/2002GC000361.

Jurine, D., Jaupart, C., Brandeis, G., and Tackley, P.J., 2005, Penetration of mantle plumes through depleted lithosphere: Journal of Geophysical Research, v. 110, doi: 10.1029/2005JB003751.

Kay, R.W., and Kay, S.M., 1993, Delamination and delamination magmatism: Tectonophysics, v. 219, p. 177–189, doi: 10.1016/0040-1951(93)90295-U.

Kay, S.M., Coira, B., and Viramonte, J., 1994, Young mafic back arc volcanic rocks as indicators of continental lithospheric delamination beneath the Argentine Puna plateau, central Andes: Journal of Geophysical Research, v. 99, p. 24,323–24,339, doi: 10.1029/94JB00896.

Keen, C.E., and Boutilier, R.R., 2000, Interaction of rifting and hot horizontal plume sheets at volcanic margins: Journal of Geophysical Research, v. 105, p. 13,375–13,387, doi: 10.1029/2000JB900027.

Kelemen, P.B., and Holbrook, W.S., 1995, Origin of thick, high-velocity igneous crust along the U.S. East Coast margin: Journal of Geophysical Research, v. 100, p. 10,077–10,094, doi: 10.1029/95JB00924.

Kellogg, J.B., Jacobsen, S.B., and O'Connell, R.J., 2002, Modeling the distribution of isotopic ratios in geochemical reservoirs: Earth and Planetary Science Letters, v. 204, p. 183–202, doi: 10.1016/S0012-821X(02)00981-0.

Kent, R.W., and Fitton, J.G., 2000, Mantle sources and melting dynamics in the British Paleogene igneous province: Journal of Petrology, v. 41, p. 1023–1040, doi: 10.1093/petrology/41.7.1023.

Kerr, A.C., 1995, The geochemistry of the Mull-Morvern lava succession, NW Scotland: An assessment of mantle sources during plume-related volcanism: Chemical Geology, v. 122, p. 43–58, doi: 10.1016/0009-2541(95)00009-B.

King, S.D., 2005, North Atlantic topographic and geoid anomalies: The result of a narrow ocean basin and cratonic roots? in Foulger, G.R., et al., eds., Plates, plumes, and paradigms: Boulder, Colorado, Geological Society of America Special Paper 388, p. 653–664.

King, S.D., and Anderson, D.L., 1995, An alternative mechanism for flood basalt formation: Earth and Planetary Science Letters, v. 136, p. 269–279, doi: 10.1016/0012-821X(95)00205-Q.

King, S.D., and Anderson, D.L., 1998, Edge-driven convection: Earth and Planetary Science Letters, v. 160, p. 289–296, doi: 10.1016/S0012-821X(98)00089-2.

Knox, R.W.O'B., 1998, The tectonic and volcanic history of the North Atlantic region during the Paleocene–Eocene transition: Implications for NW European and global biotic events, in Aubrey, M.P., Lucas, S.G., and Berggren, W.A., eds., Late Paleocene–Early Eocene climatic and biotic events in the marine and terrestrial records: New York, Columbia University Press, p. 91–102.

Koch, J.O., and Heum, O.R., 1995, Exploration trends of the Halten Terrace: 25 years of petroleum exploration in Norway: Norwegian Petroleum Society Special Publication 4, p. 235–251.

Kodaira, S., Mjelde, R., Gunnarsson, K., Shiobara, H., and Shimamura, H., 1998, Evolution of oceanic crust on the Kolbeinsey Ridge, north of Iceland, over the past 22 Myr: Terra Nova, v. 10, p. 27–31, doi: 10.1046/j.1365-3121.1998.00166.x.

Kontak, D.J., Jensen, S.M., Dostal, J., Archibald, D.A., and Kyser, T.K., 2001, Cretaceous mafic dyke swarm, Peary Land, northernmost Greenland: Geochronology and petrology: Canadian Mineralogist, v. 39, p. 997–1020.

Korenaga, J., 2004, Mantle mixing and continental breakup magmatism: Earth and Planetary Science Letters, v. 218, p. 463–473, doi: 10.1016/S0012-821X(03)00674-5.

Korenaga, J., Holbrook, W.S., Kent, G.M., Detrick, R.S., Larsen, H.C., Hopper, F.R., and Dahl-Jensen, T., 2000, Crustal structure of the Southeast Greenland margin from joint refraction and reflection seismic tomography: Journal of Geophysical Research, v. 105, p. 21,591–21,614, doi: 10.1029/2000JB900188.

Larsen, L.M., and Pedersen, A.K., 1988, Investigations of Tertiary volcanic rocks along the south coast of Nuussuaq and in eastern Disko, 1987: Report: Geological Survey of Greenland, v. 140, p. 28–32.

Larsen, L.M., and Pedersen, A.K., 2000, Processes in high-Mg, high-T magmas: Evidence from olivine, chromite and glass in Paleogene picrites from West Greenland: Journal of Petrology, v. 41, p. 1071–1098, doi: 10.1093/petrology/41.7.1071.

Larsen, L.M., Pedersen, A.K., Pedersen, G.K., and Piasecki, S., 1992, Timing and duration of early Tertiary volcanism in the north Atlantic, in Storey, B.C., Alabaster, T., and Pankhurst, R.J., eds., Magmatism and the causes of continental break-up: New evidence from West Greenland: Geological Society of London Special Publication 68, p. 321–333.

Larsen, H.C., Saunders, A.D., Clift, P.D., and Leg 152 Scientific Party, 1994, Proceedings of the Ocean Drilling Program: Initial reports: East Greenland margin: covering Leg 152 of the cruises of the drilling vessel JOIDES Resolution: Proceedings of the Ocean Drilling Program, Initial Reports, v. 152, 977 p.

Larsen, L.M., Fitton, J.G., and Saunders, A.M., 1999, Composition of volcanic rocks from the Southeast Greenland margin, Leg 163: Major and trace element geochemistry: Proceedings of the Ocean Drilling Program, Scientific Results, v. 163, p. 63–75.

Lawver, L.A., and Müller, R.D., 1994, Iceland hotspot track: Geology, v. 22, p. 311–314, doi: 10.1130/0091-7613(1994)022<0311:IHT>2.3.CO;2.

Le Bas, M.J., 2000, IUGS reclassification of the high-Mg and picritic volcanic rocks: Journal of Petrology, v. 41, p. 1467–1470.

LeHuray, A.P., and Johnson, E.S., 1989, Rb-Sr systematics of Site 642 volcanic rocks and alteration minerals: Proceedings of the Ocean Drilling Program: Scientific Results, v. 104, p. 437–448.

Lund, J., 1983, Biostratigraphy of interbasaltic coals, in Bott, M.H.P., et al., eds., Structure and development of the Greenland-Scotland Ridge: New methods and concepts: New York, Plenum Press, p. 417–423.

Lundin, E.R., and Doré, A.G., 2005, Fixity of the Iceland "hotspot" on the Mid-Atlantic Ridge: Observational evidence, mechanisms, and implications for Atlantic volcanic margins, in Foulger, G.R., et al., eds., Plates, plumes, and paradigms: Boulder, Colorado, Geological Society of America Special Paper 388, p. 627–651.

Mackay, L.M., Turner, J., Jones, S.M., and White, N.J., 2005, Cenozoic vertical motions in the Moray Firth Basin associated with initiation of the Iceland Plume: Tectonics, v. 24, no. 5, T C5004, doi: 10.1029/2004TC001683.

Maher, H.D., 2001, Manifestations of the Cretaceous High Arctic Large Igneous Province in Svalbard: Journal of Geology, v. 109, p. 91–104, doi: 10.1086/317960.

Martinsen, O.J., Boen, F., Charnock, M.A., Mangerud, G., and Nottvedt, A., 1999, Cenozoic development of the Norwegian margin 60–64°N: Sequences and sedimentary response to variable basin physiography and tectonic setting: Petroleum Geology of Northwest Europe: Proceedings of the 5th Conference, v. 5, p. 293–304.

McKenzie, D., and Bickle, M.J., 1988, The volume and composition of melt generated by extension of the lithosphere: Journal of Petrology, v. 29, p. 625–679.

McKenzie, D., Stracke, A., Blichert-Toft, J., Albarède, F., Gronvold, K., and O'Nions, R.K., 2004, Source enrichment processes responsible for isotopic anomalies in oceanic island basalts: Geochimica et Cosmochimica Acta, v. 68, p. 2699–2724, doi: 10.1016/j.gca.2003.10.029.

Meibom, A., and Anderson, D.L., 2004, The statistical upper mantle assemblage: Earth and Planetary Science Letters, v. 217, p. 123–139, doi: 10.1016/S0012-821X(03)00573-9.

Meibom, A., Anderson, D.L., Sleep, N.H., Frei, R., Chamberlain, C.P., Hren, M.T., and Wooden, J.L., 2003, Are high ^{3}He/^{4}He ratios in oceanic basalts an indicator of deep mantle-plume components?: Earth and Planetary Science Letters, v. 208, p. 197–204, doi: 10.1016/S0012-821X(03)00038-4.

Meyer, R., Hertogen, J., Pedersen, R.B., Viereck-Götte, L., and Abratis, M., 2005, Geochemical signatures of rift-related volcanism in the Vøring Plateau and the S.E. Greenland margin: Geophysical Research Abstracts, v. 7, p. 05734.

Mjelde, R., Raum, T., Myhren, B., et al., 2005, Continent-ocean transition on the Vøring Plateau, NE Atlantic, derived from densely sampled ocean bottom seismometer data: Journal of Geophysical Research, v. 110, no. B5, Bo5101, doi: 10.1029/2004JB003026.

Montelli, R., Nolet, G., Dahlen, F.A., Masters, G., Engdahl, R.E., and Hung, S.-H., 2004, Finite-frequency tomography reveals a variety of plumes in the mantle: Science, v. 303, p. 338–343, doi: 10.1126/science.1092485.

Moorbath, S., Sigurdsson, H., and Goodwin, R., 1968, K-Ar ages of the oldest exposed rocks in Iceland: Earth and Planetary Science Letters, v. 4, p. 197–205.

Moore, W.B., Schubert, G., and Tackley, P.J., 1998, Three-dimensional simulations of plume-lithosphere interaction at the Hawaii swell: Science, v. 279, p. 1008–1011, doi: 10.1126/science.279.5353.1008.

Moore, W.B., Schubert, G., and Tackley, P.J., 1999, The role of rheology in lithospheric thinning by mantle plumes: Geophysical Research Letters, v. 26, p. 1071–1076.

Morency, C., and Doin, M.P., 2004, Numerical simulations of the mantle lithosphere delamination: Journal of Geophysical Research, v. 109, doi: 10.1029/2003JB002414.

Morgan, W.J., 1971, Convection plumes in the lower mantle: Nature, v. 230, p. 42–43, doi: 10.1038/230042a0.

Mosar, J., Lewis, G., and Torsvik, T.H., 2002, North Atlantic sea-floor spreading rates: Implications for the Tertiary development of inversion structures of the Norwegian-Greenland Sea: Journal of the Geological Society, London, v. 159, p. 503–515.

Mudge, D.C., and Jones, S.M., 2004, Palaeocene uplift and subsidence events in the Scotland-Shetland and North Sea region and their relationship to the Iceland plume: Journal of the Geological Society, London, v. 161, p. 381–386.

Murdoch, L.M., Musgrove, F.W., and Perry, J.S., 1995, Tertiary uplift and inversion history in the North Celtic Sea Basin and its influence on source rock maturity, *in* Croker, P.F., and Shannon, P.M., eds., The petroleum geology of Ireland's offshore basins: Geological Society of London Special Publication 93, p. 297–319.

Mussett, A.E., 1986, ^{40}Ar-^{39}Ar step-heating ages of the Tertiary igneous rocks of Mull, Scotland: Journal of the Geological Society, London, v. 143, p. 887–896.

Nadin, P.A., Kusznir, N.J., and Cheadle, M.J., 1997, Early Tertiary plume uplift of the North Sea and Faeroe-Shetland basins: Earth and Planetary Science Letters, v. 148, p. 109–127, doi: 10.1016/S0012-821X(97)00035-6.

Naylor, P.H., Bell, B.R., Jolley, D.W., Durnall, P., and Fredsted, R., 1999, Palaeogene magmatism in the Faeroe-Shetland Basin: Influences on uplift history and sedimentation, *in* Fleet, A.J., and Boldy, S.A.R., eds., Petroleum geology of Northwest Europe: Proceedings of the 5th conference, Geological Society, London, p. 545–559.

Nevle, R.J., Brandriss, M.E., Bird, D.K., McWilliams, M.O., and O'Neil, J.R., 1994, Tertiary plutons monitor climate change in East Greenland: Geology, v. 22, p. 775–778, doi: 10.1130/0091-7613(1994)022<0775:TPMCCI>2.3.CO;2.

Nielsen, T.K., and Hopper, J.R., 2002, Formation of volcanic rifted margins: Are temperature anomalies required?: Geophysical Research Letters, v. 29, doi: 10.1029/2002GL015681.

Noble, R.H., Macintyre, R.M., and Brown, P.E., 1988, Age constraints on Atlantic evolution: Timing of magmatic activity along the E Greenland continental margin, *in* Morton, A.C., and Parson, L.M., eds., Early Tertiary volcanism and the opening of the NE Atlantic: Geological Society of London Special Publication 39, p. 201–214.

Nunns, A.G., 1983, Plate tectonic evolution of the Greenland-Scotland Ridge and surrounding regions: NATO Conference Series IV: Marine Sciences, v. 8, p. 11–30.

O'Nions, R.K., and Pankhurst, R.J., 1974, Rare-earth element distribution in Archaean gneisses and anorthosites, Godthab area, West Greenland: Earth and Planetary Science Letters, v. 22, p. 328–338, doi: 10.1016/0012-821X(74)90142-3.

Oskarsson, N., Sigvaldason, G.E., and Steinthorsson, S., 1982, A dynamic model of rift zone petrogenesis and the regional petrology of Iceland: Journal of Petrology, v. 23, p. 28–74.

Pascal, C., van Wijk, J.W., Cloetingh, S.A.P.L., and Davies, G.R., 2002, Effect of lithosphere thickness heterogeneities in controlling rift localization: Numerical modeling of the Oslo Graben: Geophysical Research Letters, v. 29, doi: 10.1029/2001GL014354.

Pearson, D.G., Emeleus, C.H., and Kelley, S.P., 1996, Precise ^{40}Ar/^{39}Ar age for the initiation of Palaeogene volcanism in the Inner Hebrides and its regional significance: Journal of the Geological Society of London, v. 153, p. 815–818.

Phipps Morgan, J., and Morgan, W.J., 1999, Two-stage melting and the geochemical evolution of the mantle: A recipe for mantle plum-pudding: Earth and Planetary Science Letters, v. 170, p. 215–239, doi: 10.1016/S0012-821X(99)00114-4.

Planke, S., Symonds, P.A., Alvestad, E., and Skogseid, J., 2000, Seismic volcanostratigraphy of large-volume basaltic extrusive complexes on rifted margins: Journal of Geophysical Research, v. 105, p. 19,335–19,351, doi: 10.1029/1999JB900005.

Polvé, M., and Allègre, C.J., 1980, Orogenic lherzolite complexes studied by ^{87}Rb-^{87}Sr: A clue to understand the mantle convection process?: Earth and Planetary Science Letters, v. 51, p. 71–93.

Prestvik, T., 1978, Cenozoic plateau lavas of Spitsbergen: A geochemical study: Arbok-Norsk Polarinstitutt, v. 1977, p. 129–142.

Price, S., Brodie, J., Whitham, A., and Kent, R., 1997, Mid-Tertiary rifting and magmatism in the Traill Ø region, East Greenland: Journal of the Geological Society, London, v. 154, p. 419–434.

Pumhösl, H., 1998, Petrographische und geochemische Untersuchungen an den Basalten der Insel Salisbury, Franz Josef-Land, Russische Arktis: Mitteilungen der Oesterreichischen Mineralogischen Gesellschaft, v. 143, p. 218–219.

Raum, T., 2000, Crustal structure and evolution of the Faeroe, Møre and Vøring margins from wide-angle seismic and gravity data [thesis]; University of Bergen, Norway.

Ren, S., Faleide, J.I., Eldholm, O., Skogseis, J., and Gradstein, F., 2003, Late Cretaceous–Paleocene tectonic development of the NW Vøring Basin: Marine and Petroleum Geology, v. 20, p. 177–206, doi: 10.1016/S0264-8172(03)00005-9.

Ribe, N.M., and Christensen, U.R., 1994, Three-dimensional modeling of plume-lithosphere interaction: Journal of Geophysical Research, v. 99, p. 669–682, doi: 10.1029/93JB02386.

Richardson, K.R., Smallwood, J.R., White, R.S., Snyder, D., and Maguire, P.K.H., 1998, Crustal structure beneath the Faeroe Islands and the Faeroe-Iceland Ridge: Tectonophysics, v. 300, p. 159–180, doi: 10.1016/S0040-1951(98)00239-X.

Ritsema, J., van Heijst, H.J., and Woodhouse, J.H., 1999, Complex shear wave velocity structure imaged beneath Africa and Iceland: Science, v. 286, p. 1925–1928, doi: 10.1126/science.286.5446.1925.

Robillard, I., Francis, D., and Ludden, J.N., 1992, The relationship between E- and N-type magmas in the Baffin Bay lavas: Contributions to Mineralogy and Petrology, v. 112, p. 230–241, doi: 10.1007/BF00310457.

Roest, W.R., and Srivastava, S.P., 1989, Sea-floor spreading in the Labrador Sea: A new reconstruction: Geology, v. 17, p. 1000–1003, doi: 10.1130/0091-7613(1989)017<1000:SFSITL>2.3.CO;2.

Rogers, G.C., 1982, Oceanic plateaus as meteorite impact signatures: Nature, v. 299, p. 341–342, doi: 10.1038/299341a0.

Rohrman, M., and van der Beek, P., 1996, Cenozoic postrift domal uplift of North Atlantic margins: An asthenospheric diapirism model: Geology, v. 24, p. 901–904, doi: 10.1130/0091-7613(1996)024<0901:CPDUON>2.3.CO;2.

Ross, J.G., and Mussett, A.E., 1976, ^{40}Ar/^{39}Ar dates for spreading rates in eastern Iceland: Nature, v. 259, p. 36–38, doi: 10.1038/259036a0.

Saunders, A.D., Fitton, J.G., Kerr, A.C., Norry, M.J., and Kent, R.W., 1997, The North Atlantic Igneous Province, in Mahoney, T.J., and Coffin, M.F., eds., Large igneous provinces: Continental, oceanic and planetary flood volcanism: Washington, D.C., American Geophysical Union, Geophysical Monograph 100, p. 45–93.

Saunders, A.D., Kempton, O.D., Fitton, J.G., and Larsen, L.M., 1999, Sr, Nd, and Pb isotopes and trace element geochemistry of basalts from the Southeast Greenland margin: Proceedings of the Ocean Drilling Program, Scientific Results, v. 163, p. 77–93.

Saunders, A.D., Jones, S.M., Morgan, L.A., Pierce, K.L., Widdowson, M., and Xu, Y., 2007, Regional uplift associated with continental large igneous provinces: The roles of mantle plumes and the lithosphere: Chemical Geology, v. 241, p. 282–318.

Schilling, J.G., Meyer, P.S., and Kingsley, R.H., 1983, Rare earth geochemistry of Iceland basalts: Spatial and temporal variations: NATO Conference Series IV, Marine Sciences, v. 8, p. 319–342.

Schott, B., Yuen, D., and Schmeling, H., 2000, The significance of shear heating in continental delamination: Physics of the Earth and Planetary Interiors, v. 118, p. 273–290, doi: 10.1016/S0031-9201(99)00159-4.

Séranne, M., and Séguret, M., 1987, The Devonian basins of western Norway: Tectonics and kinematics of an extending crust, in Coward, M.P., Dewey, J.F., and Hancock, P.L., eds., Continental extensional tectonics: Geological Society of London Special Publication 28, p. 537–548.

Sinton, C.W., and Duncan, R.A., 1998, ^{40}Ar-^{39}Ar ages of lavas from the Southeast Greenland margin, ODP Leg 152, and the Rockall Plateau, DSDP Leg 81: Proceedings of the Ocean Drilling Program, Scientific Results, v. 152, p. 387–402.

Sinton, C.W., Hitchen, K., and Duncan, R.A., 1998, ^{40}Ar-^{39}Ar geochronology of silicic and basic volcanic rocks: Geological Magazine, v. 135, p. 161–170, doi: 10.1017/S0016756898008401.

Skaarup, N., Jackson, H.R., and Oakey, G., 2006, Margin segmentation of Baffin Bay / Davis Strait, eastern Canada, based on seismic reflection and potential field data: Marine and Petroleum Geology, v. 23, p. 127–144, doi: 10.1016/j.marpetgeo.2005.06.002.

Skogseid, J., and Eldholm, O., 1989, Vøring Plateau continental margin: Seismic interpretation, stratigraphy and vertical movements: Proceedings of the Ocean Drilling Program, Scientific Results, v. 104, p. 993–1030.

Skogseid, J., and Eldholm, O., 1995, Rifted continental margin off mid-Norway, in Banda, E., Torné, M., and Talwani, M., eds., Rifted ocean-continent boundaries: Dordrecht, Kluwer, p. 147–153.

Skogseid, J., Pedersen, T., and Larsen, V.B., 1992, Vøring Basin: Subsidence and tectonic evolution: Norwegian Petroleum Society Special Publication 1, p. 55–82.

Sleep, N.H., 1996, Lateral flow of hot plume material ponded at sublithospheric depths: Journal of Geophysical Research, v. 101, p. 28,065–28,083, doi: 10.1029/96JB02463.

Sleep, N.H., 1997, Lateral flow and ponding of starting plume material: Journal of Geophysical Research, v. 102, p. 10,001–10,012, doi: 10.1029/97JB00551.

Smallwood, J.R., White, R.S., and Minshull, T.A., 1995, Sea-floor spreading in the presence of the Iceland mantle plume: The anomalous heatflow and structure of the Reykjanes Ridge at 61°40′ N: Journal of the Geological Society, London, v. 152, p. 1023–1029.

Smallwood, J.R., Staples, R.K., Richardson, K.R., and White, R.S., 1999, Crust generated above the Iceland mantle plume: From continental rift to oceanic spreading center: Journal of Geophysical Research, v. 104, p. 22,885–22,902, doi: 10.1029/1999JB900176.

Sørensen, A.B., 2003, Cenozoic basin development and stratigraphy of the Faeroes area: Petroleum Geoscience, v. 9, p. 189–207.

Storey, M., Duncan, R.A., Pedersen, A.K., Larsen, L.M., and Larsen, H.C., 1998, ^{40}Ar/^{39}Ar geochronology of the West Greenland Tertiary volcanic province: Earth and Planetary Science Letters, v. 160, p. 569–586, doi: 10.1016/S0012-821X(98)00112-5.

Strachan, R.A., 1994, Evidence in north-east Greenland for Late Silurian–Early Devonian regional extension during the Caledonian Orogeny: Geology, v. 22, p. 913–916, doi: 10.1130/0091-7613(1994)022<0913:EINEGF>2.3.CO;2.

Stuart, F.M., Lass-Evans, S., Fitton, J.G., and Ellam, R.M., 2003, Extreme 3He/4He in picritic basalts from Baffin Island: The role of a mixed reservoir in mantle plumes: Nature, v. 424, p. 57–59, doi: 10.1038/nature01711.

Suen, C.J., and Frey, F.A., 1987, Origins of the mafic and ultramafic rocks in the Ronda Peridotite Massif: Earth and Planetary Science Letters, v. 85, p. 183–202.

Sun, S.S., and McDonough, W.F., 1989, Chemical and isotopic systematics of oceanic basalts: Implications for mantle composition and processes, in Saunders, A.D., and Norry, M.J., eds., Magmatism in the ocean basins: Geological Society of London Special Publication 42, p. 313–345.

Surlyk, F., 1990, Timing, style and sedimentary evolution of late Palaeozoic–Mesozoic extensional basins of East Greenland, in Hardman, R.F.P., and Brooks, J., eds., Tectonic events responsible for Britain's oil and gas reserves: Geological Society of London Special Publication 55, p. 107–125.

Tate, M.P., 1993, Structural framework and tectono-stratigraphic evolution of the Porcupine Seabight Basin, offshore western Ireland: Marine and Petroleum Geology, v. 10, p. 95–123, doi: 10.1016/0264-8172(93)90016-L.

Taylor, P.N., and Morton, A.C., 1989, Sr, Nd and Pb isotope geochemistry of the upper and lower volcanic series at Site 642: Proceedings of the Ocean Drilling Program, Scientific Results, v. 104, p. 429–435.

Taylor, S.R., and McLennan, S.M., 1985, The continental crust: Its composition and evolution: Oxford, Blackwell Scientific, 312 p.

Tegner, C., and Duncan, R.A., 1999, ^{40}Ar-^{39}Ar chronology for the volcanic history of the Southeast Greenland rifted margin: Proceedings of the Ocean Drilling Program, Scientific Results, v. 163, p. 53–62.

Tegner, C., Duncan, R.A., Bernstein, S., Bird, D.K., Brooks, C.K., and Storey, M., 1998, ^{40}Ar-^{39}Ar geochronology of Tertiary mafic intrusions along the east Greenland rifted margin: Relation to flood basalt and the Iceland hotspot track: Earth and Planetary Science Letters, v. 156, p. 75–88, doi: 10.1016/S0012-821X(97)00206-9.

Thompson, P., Mussett, A.E., and Dagley, P., 1987, Revised ^{40}Ar-^{39}Ar age for granites of the Mourne Mountains, Ireland: Scottish Journal of Geology, v. 23, p. 215–220.

Thompson, R., and Gibson, S., 1991, Sub-continental mantle plumes, hot-spots and pre-existing thinspots: Journal of the Geological Society, London, v. 148, p. 973–977.

Thompson, R.N., Dickin, A.P., Gibson, I.L., and Morrison, M.A., 1982, Elemental fingerprints of isotopic contamination of Hebridean Palaeocene mantle-derived magmas by Archaen sial: Contributions to Mineralogy and Petrology, v. 79, p. 159–168, doi: 10.1007/BF01132885.

Thoraval, C., Tommasi, A., and Doin, M.P., 2006, Plume-lithosphere interaction beneath a fast moving plate: Geophysical Research Letters, v. 33, doi: 10.1029/2005GL024047.

Torsvik, T.H., Mosar, J., and Eide, E.A., 2001, Cretaceous–Tertiary geodynamics: A North Atlantic exercise: Geophysical Journal International, v. 146, p. 850–866, doi: 10.1046/j.0956-540x.2001.01511.x.

Troll, V.R., Donaldson, C.H., and Emeleus, C.H., 2004, Pre-eruptive magma mixing in ash-flow deposits of the Tertiary Rum Igneous Centre, Scotland: Contributions to Mineralogy and Petrology, v. 147, p. 722–739, doi: 10.1007/s00410-004-0584-0.

Tsikalas, F., Eldholm, O., and Faleide, J.I., 2005, Crustal structure of the Lofoten-Vesterålen continental margin, off Norway: Tectonophysics, v. 404, p. 151–174, doi: 10.1016/j.tecto.2005.04.002.

Turner, S., Hawkesworth, C., Gallagher, K., Stewart, K., Peate, D., and Mantovani, M., 1996, Mantle plumes, flood basalts, and thermal models for melt generation beneath continents: Assessment of a conductive heating model and application to the Paraná: Journal of Geophysical Research, v. 101, p. 11,503–11,518, doi: 10.1029/96JB00430.

Upton, B.G.J., 1988, History of Tertiary igneous activity in the N Atlantic borderlands, *in* Morton, A.C., and Parson, L.M., eds., Early Tertiary volcanism and the opening of the NE Atlantic: Geological Society of London Special Publication 39, p. 429–453.

Upton, B.G.J., Emeleus, C.H., Rex, D.C., and Thirlwall, M.F., 1995, Early Tertiary magmatism in NE Greenland: Journal of the Geological Society of London, v. 152, p. 959–964.

Upton, B.G.J., Skovgaard, A.C., McClurg, J., Kirstein, L., Cheadle, M., Emeleus, C.H., Wadsworth, W.J., and Fallick, A.E., 2002, Picritic magmas and the Rum ultramafic complex, Scotland: Geological Magazine, v. 139, p. 437–452, doi: 10.1017/S0016756802006684.

Vågnes, E., and Amundsen, H., 1993, Late Cenozoic uplift and volcanism on Spitsbergen—Caused by mantle convection: Geology, v. 21, p. 251–254, doi: 10.1130/0091-7613(1993)021<0251:LCUAVO>2.3.CO;2.

van der Hilst, R.D., and de Hoop, M.V., 2005, Banana-doughnut kernels and mantle tomography: Geophysical Journal International, v. 163, p. 956–961.

van Wijk, J.W., Huismans, R.S., Ter Voorde, M., and Cloetingh, S.A.P.L., 2001,

Melt generation at volcanic continental margins: No need for a mantle plume?: Geophysical Research Letters, v. 28, p. 3995–3998, doi: 10.1029/2000GL012848.

van Wijk, J.W., van der Meer, R., and Cloetingh, S.A.P.L., 2004, Crustal thickening in an extensional regime: Application to the mid-Norwegian Vøring margin: Tectonophysics, v. 387, p. 217–228, doi: 10.1016/j.tecto.2004.07.049.

Vogt, P.R., 1971, Asthenosphere motion recorded by the ocean floor south of Iceland: Earth and Planetary Science Letters, v. 13, p. 153–160, doi: 10.1016/0012-821X(71)90118-X.

Vogt, P.R., and Jung, W.-Y., 2005, Paired basement ridges: Spreading axis migration across mantle heterogeneities? *in* Foulger, G.R., et al., eds., Plates, plumes, and paradigms: Boulder, Colorado, Geological Society of America Special Paper 388, p. 555–579.

Vogt, P.R., Johnson, G.L., and Kristjansson, L., 1980, Morphology and magnetic anomalies north of Iceland: Journal of Geophysics, v. 47, p. 67–80.

Vogt, U., Makris, J., O'Reilly, B.M., et al., 1998, The Hatton bank and continental margin: Crustal structure from wide-angle seismic and gravity data: Journal of Geophysical Research, v. 103, p. 12,545–12,566, doi: 10.1029/98JB00604.

Waagstein, R., 1988, Structure, composition and age of the Faeroe basalt plateau, *in* Morton, A.C., and Parson, L.M., eds., Early Tertiary volcanism and the opening of the NE Atlantic: Geological Society of London Special Publication 39, p. 225–238.

White, R.S., and McKenzie, D.P., 1989, Magmatism at rift zones: The generation of volcanic continental margins and flood basalts: Journal of Geophysical Research, v. 94, p. 7685–7729.

White, R.S., Bown, J.W., and Smallwood, J.R., 1995, The temperature of the Iceland plume and origin of outward-propagating V-shaped ridges: Journal of the Geological Society of London, v. 152, p. 1039–1045.

Yasuda, A., Fujii, T., and Kurita, K., 1994, Melting phase relations of an anhydrous mid-ocean ridge basalt from 3 to 20 GPa: Implications for the behaviour of subducted oceanic crust in the mantle: Journal of Geophysical Research, v. 99, p. 9401–9414, doi: 10.1029/93JB03205.

Ziegler, P.A., 1988, Evolution of the Arctic–North Atlantic and the western Tethys: Tulsa, Oklahoma, American Association of Petroleum Geologists Memoir 43, 198 p.

Zindler, A., Staudigel, H., and Batiza, R., 1984, Isotope and trace element geochemistry of young Pacific seamounts: Implications for the scale of upper mantle heterogeneity: Earth and Planetary Science Letters, v. 70, p. 175–195, doi: 10.1016/0012-821X(84)90004-9.

MANUSCRIPT ACCEPTED BY THE SOCIETY JANUARY 31, 2007

DISCUSSION

31 January 2007, Laurent Geoffroy

Any effort to summarize the available data from the north Atlantic igneous province is welcome, and we should thank Meyer et al. (this volume) for their valuable contribution. However, I would like to make the following comments, the first on a minor aspect and the second concerning a much more general point.

The South-East Baffin Controversy. Meyer et al. (this volume) refer to the argument concerning the existence of an onshore exposed seaward dipping reflector (SDR) along the west Green-land margin north of ~70°N (Chalmers et al., 1999). I do not think that this is a matter of debate any longer. Geoffroy et al. (1998, 2001) were correct in recognizing that the Disko-Svartenhuk west Greenland coastal area exposes a thick fan-shaped inner prism of syntectonic seaward-dipping basalts and tuffs that essentially forms an upper crustal structure typical of volcanic passive margins worldwide. The recent confirmation of an SDR located on the conjugate margin offshore from Baffin Island fully supports the view that the south Baffin Bay margins are of volcanic type (Skaarup et al., 2006). However, Chalmers et al. (1999) rightly point out the geodynamic importance of the Ungava transform splay faults in the tectonic structure of the Disko and Nuussuaq areas. The west Greenland controversy was compounded by a general misunderstanding about the tectonic setting of SDR wedges across volcanic margins. According to Chalmers et al. (1999), if we consider the west Greenland coastal flexure to be an SDR, this implies a transition to the oceanic crustal domain across the coastline (conflicting with hopes for the petroleum potential of this area). This ignores the fact that, at volcanic margins, the innermost SDR prisms are often continental, as suggested by Roberts et al. (1979) and confirmed in later studies (e.g., Tard et al., 1991). One can truly consider the west Greenland volcanic margin at the latitude of Ubekendt-Ejland and Svartenhuk (Geoffroy et al., 2001, Plate 3) as representing the most suitable place to study in detail the development of a partly eroded continental inner-SDR and the relationships linking the pre-breakup and breakup-related volcanic formations with pre-existing sedimentary basins. More generally, in discussing this aspect, we might add that there is little point in interpreting offshore geology at margins without a detailed examination of adjacent onshore data. By the same token, offshore observations (when available) are just as crucial for interpreting onland geology.

Distribution and Timing of the North Atlantic Large Igneous Province Volcanism. The authors proposed that the distribution of large igneous province (LIP)-related magmatism in the North Atlantic (their Fig. 1) is somewhat surprising compared to former views, suggesting a much wider geographical extent (see, for example, Fig. 8 in White and McKenzie, 1989). Could they specify the range of data used for this otherwise quite useful compilation? Did they take into account the extensively sill-intruded basins that are fully part of the LIP? This is not merely a point of detail because the authors' diagram appears to confine most of the magmatism to the breakup area, in good agreement with the mantle-melting model developed by Van Wijk et al. (2001). This distribution is also discussed in the text, where the authors state that the "Paleocene magmatism is middle or late syn-rift." From this, we can understand that they are referring to previous Cretaceous rifting in the North Atlantic (their Fig. 2). According to such a presentation of the data, magmatic activity would be a direct and natural consequence of a late-stage stretching of the lithosphere. This would be surprising because, apart from the breakup zones where Eocene SDR are developed, very little extension took place during the emplacement of most of the North Atlantic Paleocene-related extrusives. These formations are mainly associated with horizontal dilatation caused by dike injection and thus have no relation to tectonic stretching and thinning. In many cases, traps at LIPs can be seen to seal former sedimentary rifts and basins. For example, this is exactly the pattern observed in the Afar area, where the Ethiopian and Yemen traps form thick horizontal piles over wide areas, lying directly on top of the basement or capping the older Mesozoic rifts. I do not think that we can so easily discount such a robust and generalized observation.

Whatever the mantle-melting mechanism (and mantle plumes are certainly not the only possible explanation, as the authors rightly state), we still need to explain why a major and sudden pulse of magmatism occurred simultaneously over such a wide area. At the same time, this magmatism is coeval with a stress pattern that is unrelated to the regime acting during premagmatic Mesozoic tectonic events (Geoffroy et al., this volume). This type of scenario strongly resembles a major dynamic plate-scale perturbation of a metastable system, i.e., the buoyant mantle located beneath the rigid and heterogeneous lithosphere.

1 February 2007, Stephen Jones

Postbreakup Magmatism: Iceland and the Oceanic Part of the North Atlantic Igneous Province. Several authors have pointed out that the geometry of the V-shaped ridges might be explained by passive upwelling of a thermal halo surrounding Iceland and containing outward-dipping thermal/compositional anomalies instead of by lateral advection of thermal/compositional anomalies within a plume head (Jones et al., 2002a; Vogt and Jung, 2005). Previously, the main difficulty in deciding between the lateral and vertical transport models was a lack of seismic images with sufficient vertical resolution to distinguish a slow sublithospheric plume head layer about 100 km thick from a slow thermal halo filling the entire upper mantle (Jones et al., 2002a). The situation is now improving. Pilidou et al. (2005) presented a Rayleigh wave tomography model that resolves a regional sublithospheric slow layer whose base lies at a depth of less than 300 km. The balance of the seismic evidence now appears to favor the lateral outflow model.

Lithospheric Delamination. Constraints on the degree of lithospheric delamination and regional uplift are available for the British igneous province. The top of the melting region that supplied the British igneous province has been estimated at about 70 km from modeling of rare earth element distributions (Brodie and White, 1995; White and McKenzie, 1995). The British igneous province was emplaced within the Sea of the Hebrides Trough, which formed by modest rifting (total strain less than 2) in the Jurssasic, and no significant rifting occurred coeval with British igneous province emplacement (Emeleus and Bell, 2005). Lithospheric thinning to 70 km is not easily attributed to mild Jurassic rifting, making it more likely that some lithospheric erosion was associated with the mantle circulation that led to British igneous province emplacement during the Middle Paleocene. Field observations show that uplift of the British igneous province itself occurred before and during emplacement, and some of this uplift could be related to lithospheric erosion. However, uplift histories interpreted from sedimentary basins surrounding the British igneous province show relatively little regional uplift coeval with British igneous province emplacement. Instead, most of the regional uplift (up to 1 km) occurred after British igneous province activity, peaking close to the Paleocene-Eocene boundary and Europe-Greenland breakup time (a summary and references can be found in Maclennan and Jones, 2006; Saunders et al., 2007). Thus, in the case of the British igneous province, it is likely that some delamination did occur, but it was not directly associated with major uplift outside the province itself.

Fertile Mantle. The authors point out that models involving fertile mantle and no thermal anomaly cannot explain pre-breakup uplift. An equally important problem is that such models cannot explain the present-day north Atlantic bathymetric swell, which has a diameter of around 2000 km and a central amplitude of 2 km (excluding the contribution of unusually thick crust; Jones et al., 2002b; Conrad et al., 2004). The topographic swell also correlates with a positive free-air gravity anomaly. It has long been recognized that such broad swells with correlated gravity anomalies must be supported by low-density material within the mantle (e.g., Anderson et al., 1973). Even the more skeptical workers agree that it is difficult to explain the 2000-km topographic swell without a thermal anomaly of up to about 100 °C (Foulger and Anderson, 2005).

Small-Scale and Edge-Driven Convection. The conclusion that small-scale convection by itself cannot explain the north Atlantic igneous province is supported by the regional uplift history. Although the uplifted swell that developed close to the Paleocene-Eocene boundary is not well constrained over its entire area, data from the well-explored basins around the British Isles clearly show that uplift of 100 m to 1 km occurred at a distance of 1000 km from the Europe-Greenland breakup zone (Maclennan and Jones, 2006). Such regional uplift is not easily explained by small-scale convection centered on the breakup zone.

Mantle Plume. It is true that mantle plume models applied to the north Atlantic igneous province differ in many details. However, one common proposal is that a large amount of hot mantle was convectively emplaced beneath the entire north Atlantic igneous province during the Middle Paleocene. The hot asthenosphere first generated the British and West Greenland igneous provinces and then remained beneath the area to generate the Europe-Greenland volcanic margins (Larsen and Saunders, 1998). This model implies major regional uplift at or before the Middle Paleocene and persisting until the Early Eocene. The well-resolved uplift history surrounding shows that the model cannot be correct because major regional uplift (up to 1–2 km over a diameter comparable to the north Atlantic igneous province) is not observed until the Late Paleocene, after emplacement of both the British and Greenland provinces (Maclennan and Jones, 2006; Saunders et al., 2007). It is very difficult to explain both the igneous and the sedimentary record of uplift by a single plume head impact. Whatever model for the north Atlantic igneous province is accepted in future, it must explain both the igneous record and the record of uplift. My view is that it will be very difficult to do so without advection of hot mantle beneath the lithosphere.

6 February 2007, Romain Meyer, Jolante van Wijk, and Laurent Gernigon

We thank Geoffroy and Jones for their constructive comments on our chapter. The comments confirm the conclusion of the re-view, i.e., "existing datasets and geodynamic concepts are incomplete, which hinders a more conclusive statement on whether or not the mantle plume or alternative models can be accepted or rejected." The second comment from Geoffroy concerns the distribution and timing of north Atlantic igneous province magmatism, (Fig. 1 in Meyer et al., this volume). We initially chose a conservative approach while constructing this figure (map of the north Atlantic igneous province with distribution of volcanism) and used only published, strongly founded data for its compilation. Following an e-mail discussion with Geoffroy, we decided to modify the map; in some areas, the volcanic rocks are likely more extensive than our initial map suggested. Note that the boundaries for the localities of the Rockall, Faeroes, and Baffin magmatism are still uncertain.

In our chapter we have related north Atlantic igneous province magmatism to the timing of the pre-breakup rift phase (Fig. 2 in Meyer et al., this volume). By doing so we did, however, not intend to relate all magmatism to the process of rifting, and we agree with Geoffroy that such a relation is far from well constrained. The origin of such magmatic formations away from the breakup zone (as in the UK–Rockall area) is not discussed in detail in our chapter; we refer to Geoffroy et al. (this volume) for a discussion on and possible explanation for this phenomenon.

REFERENCES CITED

Anderson, R.N., McKenzie, D.P., and Sclater, J.G., 1973. Gravity, bathymetry and convection within the Earth: Earth and Planetary Science Letters, v. 18, p. 391–407.

Brodie, J., and White, N., 1995. The link between sedimentary basin inversion and igneous underplating, *in* Buchanan, J., and Buchanan, P., eds., Basin inversion: London, Geological Society of London Special Publication 88, p. 21–38.

Chalmers, J.A., Whittaker, C., Skaarup, N., and Pulvertaft, T.C.R., 1999. Discussion on the coastal flexure of Disko (West Greenland), onshore expression of the "oblique reflectors": Journal of the Geological Society, London, v. 156, p. 1051–1055.

Conrad, C.P., Lithgow-Bertelloni, C., and Louden, K.E., 2004, Iceland, the Farallon slab, and dynamic topography of the North Atlantic: Geology, v. 32, p. 177–180.

Emeleus, C.H., and Bell, B.R., 2005, The Palaeogene volcanic districts of Scotland: British Geological Survey, Keyworth, 214 p.

Foulger, G.R., and Anderson, D.L., 2005, A cool model for the Iceland hot spot: Journal of Volcanology and Geothermal Research, v. 141, p. 1–22.

Geoffroy, L., Gélard, J.P., Lepvrier, C., and Olivier, P., 1998, The coastal flexure of Disko (West Greenland), onshore expression of the "oblique reflectors": Journal of the Geological Society of London, v. 155, p. 463–473.

Geoffroy, L., Callot, J.-P., Scaillet, S., Skuce, A.G., Gélard, J.-P., Ravilly, M., Angelier, J., Bonin, B., Cayet, C., Perrot-Galmiche, K., and Lepvrier, C., 2001, Southeast Baffin volcanic margin and the North American–Greenland plate separation: Tectonics, v. 20, p. 566–584.

Geoffroy, L., Aubourg, C., Callot, J.-P., and Barrat, J.A., 2007 (this volume), Mechanisms of crustal growth in large igneous provinces: The north Atlantic province as a case study, *in* Foulger, G.R., and Jurdy, D.M., eds., Plates, plumes, and planetary processes: Boulder, Colorado, Geological Society of America Special Paper 430, doi: 10.1130/2007.2430(34).

Jones, S.M., White, N., and Maclennan, J., 2002a, V-shaped ridges around Iceland: Implications for spatial and temporal patterns of mantle convection: Geochemistry, Geophysics, Geosystems, v. 3, doi:10.1029/2002GC000361.

Jones, S.M., White, N., Clarke, B.J., Rowley, E., and Gallagher, K., 2002b, Present and past influence of the Iceland Plume on sedimentation: London, Geological Society of London Special Publication 188, p. 345–360.

Larsen, H.C., and Saunders, A.D., 1998, Tectonism and volcanism at the southeast Greenland rifted margin: A record of plume impact and later continental rupture, *in* Saunders, A.D., et al., eds., Proceedings of the Ocean Drilling Program, scientific results: College Station, Texas, Ocean Drilling Program, v. 152, p. 503–533.

Maclennan, J.C., and Jones, S.M., 2006. Regional uplift, gas hydrate dissociation and the origins of the Paleocene/Eocene thermal maximum: Earth and Planetary Science Letters, v. 245, p. 65–80.

Meyer, R., van Wijk, J., and Gernigon, L., 2007 (this volume), The north Atlantic igneous province: A review of models for its formation, *in* Foulger, G.R., and Jurdy, D.M., eds., Plates, plumes, and planetary processes: Boulder, Colorado, Geological Society of America Special Paper 430, doi: 10.1130/2007.2430(26).

Pilidou, S., Priestley, K., Debayle, E., and Gudmundsson, O., 2005. Rayleigh wave tomography in the North Atlantic: High resolution images of the Iceland, Azores and Eifel mantle plumes: Lithos, v. 79, p. 453–474.

Roberts, D.G., Morton, A.C., Murray, J.W., and Keene, J.B., 1979, Evolution of volcanic rifted margins: Synthesis of Leg 81 results on the West margin of Rockall plateau: Initial Reports Deep Sea Drilling Project, v. 81, p. 883–911.

Saunders, A.D., Jones, S.M., Morgan, L.A., Pierce, K.L., Widdowson, M., and Xu Y., 2007, Regional uplift associated with continental large igneous provinces: The roles of mantle plumes and the lithosphere: Chemical Geology, v. 241, p. 282–318.

Skaarup, N., Jackson, H.R., and Oakey, G., 2006, Margin segmentation of Baffin Bay/Davis Strait, eastern Canada based on seismic reflection and potential field data: Marine and Petroleum Geology, v. 23, p. 127–144.

Tard, F., Masse, P., Walgenwitz, F., and Gruneisen, P., 1991, The volcanic passive margin in the vicinity of Aden, Yemen: Bulletin des Centres de Recherche et d'Exploration-Production Elf-Aquitaine, v. 15, p. 1–9.

Van Wijk, J.W., Huismans, R.S., ter Voorde, M., and Cloetingh, S.A.P.L., 2001, Melt generation at volcanic continental margins: No need for a mantle plume?: Geophysical Research Letters, v. 28, p. 3995–3998.

Vogt, P.R., and Jung, W.-Y., 2005, Paired basement ridges: Spreading axis migration across mantle heterogeneities?, *in* Foulger, G.R., et al., eds., Plates, plumes, and paradigms: Boulder, Colorado, Geological Society of America Special Paper 388, p. 555–579, doi: 10.1130/2005.2388(33).

White, R.S., and McKenzie, D.P., 1989, Magmatism at rift zones: The generation of volcanic continental margins and flood basalts: Journal of Geophysical Research, v. 94, p. 7685–7729.

White, R.S., and McKenzie, D., 1995. Mantle plumes and flood basalts: Journal of Geophysical Research, v. 100, p. 17,543–17,585.

The Geological Society of America
Special Paper 430
2007

Origin of the Bermuda volcanoes and the Bermuda Rise: History, observations, models, and puzzles

Peter R. Vogt*
Marine Science Institute, University of California, Santa Barbara, California 93106-6150, USA
Woo-Yeol Jung
Code 7420, Naval Research Laboratory, Washington, D.C. 20375-5320, USA

ABSTRACT

Cores recovered on Deep Sea Drilling Program leg 43 and on Bermuda itself, together with geophysical data (anomalies in basement depth, geoid, and heatflow) and modeling have long suggested that the uplift forming the Bermuda Rise, as well as the initial igneous activity that produced the Bermuda volcanoes, began ca. 47–40 Ma, during the early to middle part of the Middle Eocene. Some authors attribute 65 Ma igneous activity in Mississippi and 115 Ma activity in Kansas to a putative "Bermuda hotspot" or plume fixed in the mantle below a moving North America plate. While this is more or less consistent with hotspot traces computed from "absolute motion" models, the hotspot or plume must resemble a blob in a lava lamp that is turned off for up to 25 million years at a time, and/or be heavily influenced by lithosphere structure. Moreover, Cretaceous igneous activity in Texas and Eocene intrusions in Virginia then require separate mantle "blobs."

The pillow lavas forming the original Bermuda shield volcano have not been reliably dated, and the three associated smaller edifices have not been drilled or dated. A well-dated (ca. 33–34 Ma) episode of unusually titaniferous sheet intrusion in the Bermuda edifice was either triggered by platewide stress changes or reflects local volcanogenic events deep in the mantle source region. The high Ti and Fe of the Bermuda intrusive sheets probably relate to the very high-amplitude magnetic anomalies discovered on the islands. Numerical models constrained by available geophysical data attribute the Bermuda Rise to some combination of lithospheric reheating and dynamic uplift. While the relative contributions of these two processes cannot yet be wholly separated, three features of the rise clearly distinguish it from the Hawaiian swell: (1) the Bermuda Rise is elongated at right angles to the direction of plate motion; (2) there has been little or no subsidence of the rise and the volcanic edifice since its formation—in fact, rise uplift continued at the same site from the late Middle Eocene into the Miocene; and (3) the Bermuda Rise lacks a clear, age-progressive chain. We infer that the Bermuda Rise and other Atlantic midplate rises are supported by anomalous asthenosphere, upwelling or not, that penetrates the thermal boundary layer and travels with the overlying plate.

The elongation along crustal isochrons of both the Bermuda volcanoes and the Bermuda Rise and rise development mostly within a belt of rougher, thinner crust and

---

*E-mail: ptr_vogt@yahoo.com.

Vogt, P.R., and Jung, W.-Y., 2007, Origin of the Bermuda volcanoes and the Bermuda Rise: History, observations, models, and puzzles, *in* Foulger, G.R., and Jurdy, D.M., eds., Plates, plumes, and planetary processes: Geological Society of America Special Paper 430, p. 553–591, doi: 10.1130/2007.2430(27). For permission to copy, contact editing@geosociety.org. ©2007 The Geological Society of America. All rights reserved.

seismically "slower" upper mantle—implying retention of gabbroic melts at the ancient Mid-Atlantic Ridge axis—suggest that the mantle lithosphere may have helped localize rise development, in contradiction to plume models. The Bermuda Rise area is seismically more active than its oceanic surroundings, preferentially along old transform traces, possibly reflecting a weaker upper mantle lithosphere.

We attribute the "Bermuda event" to a global plate kinematic reorganization triggered by the closing of the Tethys and/or the associated gravitational collapse into the lower mantle of subducted slabs that had been temporarily stagnant near the 660 km mantle discontinuity. The widespread onset of sinking slabs required simultaneous upwelling for mass balance. In addition, the global plate kinematic reorganization was accompanied by increased stress in some plate interiors, favoring magma ascent along fractures at structurally weak sites. We suggest that the Bermuda event and concomitant igneous activity in Virginia, West Antarctica, Africa, and other regions were among such upwellings, but structurally influenced by the lithosphere, and probably originated in the upper mantle.

Drilling a transect of boreholes across and along the Bermuda Rise to elucidate turbidite offlap during rise formation might discriminate between a widely distributed mantle source (such as a previously subducted slab) and a narrow plume whose head (or melt root) spreads out quasi-radially over time, generating an upward and outward expanding swell.

Keywords: Bermuda, Bermuda Rise, midplate swells, midplate volcanism, plate reorganization

INTRODUCTION

The Bermuda Rise (Fig. 1), capped by a cluster of extinct volcanoes (Fig. 2, bottom) dominated by the Bermuda Pedestal, dominates the western North American basin (the western central north Atlantic). The rise forms an oval basement swell that apparently rose, with concomitant volcanism, in Middle Eocene through Oligocene, perhaps into Miocene, times. Geoscientists from the nineteenth century to the middle of the twentieth were already interested in the geology of Bermuda and its pedestal (see Foreman, 1951, for a review of early interpretations). However, with the advent of plate tectonics, attention was directed to the associated Bermuda Rise, which has become in many ways the archetypical—and to date best studied—regional basement swell developed in preexisting and apparently slow-moving (~15–30 mm/yr) oceanic lithosphere, variously contrasting with

and comparable to the Hawaiian-Emperor island or seamount chain and its associated basement swell, formed and still developing within fast-moving (~90 mm/yr) Pacific lithosphere. While much of the geological history of the igneous activity is reasonably well known, the underlying geophysical processes are subjects of long-standing and ongoing debate: For example, is the apparently "unifying hypothesis" valid—that both Bermuda and Hawaii are caused by mantle plumes or hotspots, differing only insofar as the Bermuda "plume" is more like what we see in a lava lamp, with intermittent blobs of hot mantle (e.g., Morgan, 1972; Crough, 1978)? Of the many plumes that have been postulated, Bermuda, although throughout the past thirty years it has appeared in most "hotspot catalogues" (Steinberger, 2000, is a rare exception), rates practically at the bottom in terms of consistency with the deep mantle plume hypothesis (e.g., Courtillot et al., 2003; Anderson, 2005; Anderson and Schramm,

---

Figure 1. (Bottom) Bathymetry (Smith and Sandwell, 1997) of Bermuda and vicinity, with heatflow stations (W/m²) from Hyndman et al. (1974) and Detrick et al. (1986), location of DSDP site 386 from Shipboard Party (1979), and seafloor spreading magnetic anomaly (M0) from Klitgord and Schouten (1986). The solid line shows the outer limit of seismic reflector A^v, caused by volcanogenic sediments of Bermudan origin (Tucholke and Mountain, 1979). The dashed lines indicate major fracture zones as interpreted from geophysical data. The contour interval is 0.5 km. F.Z.—fracture zone. (Top) The Greater Bermuda Rise region: Bathymetry (Smith and Sandwell, 1997), key magnetic lineations (Müller et al., 1993), DSDP leg 43 drill site locations (stars); heatflow values (Detrick et al., 1986; Louden et al., 1987); residual depth anomaly (white contours at 200 m intervals; Sclater and Wixon, 1986); residual geoid anomaly (in meters, dashed black lines); trace of the Kane fracture zone (F.Z., dashed line; after Jaroslow and Tucholke, 1994); and predicted track (red) of the North America plate over a fixed Bermuda hotspot (Duncan, 1984), with the predicted present hotspot location shown by the large red circle. The heatflow station on southern Bermuda Rise shows the value of Detrick et al. (1986), "(49.9*)" the same original data, but recalculated by Louden et al. (1987). HAP—Hatteras Abyssal Plain; MS—Muir seamount; NAP—Nares Abyssal Plain; NES—New England seamounts; SAP—Sohm Abyssal Plain.

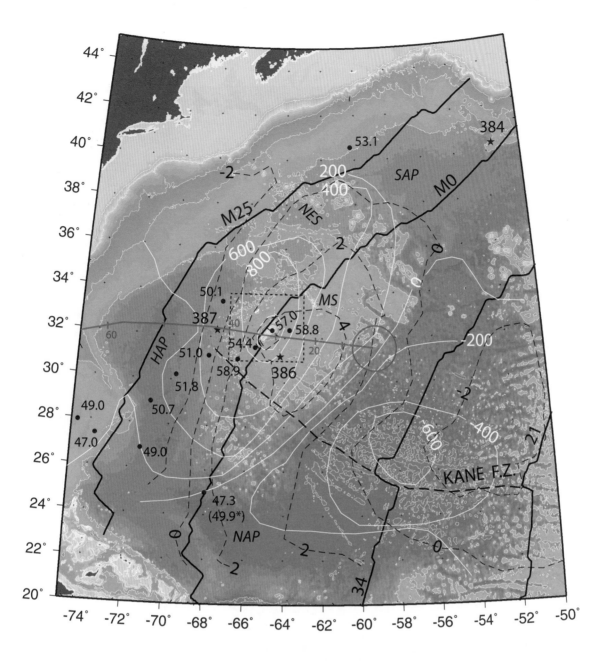
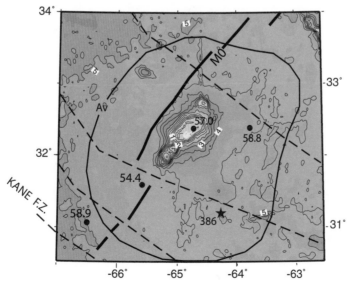

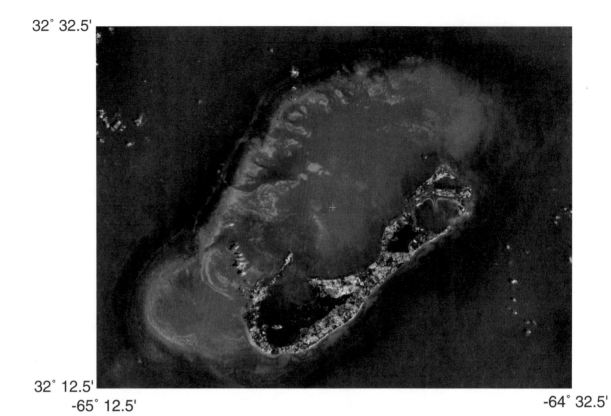

32° 32.5'

32° 12.5'
-65° 12.5'   -64° 32.5'

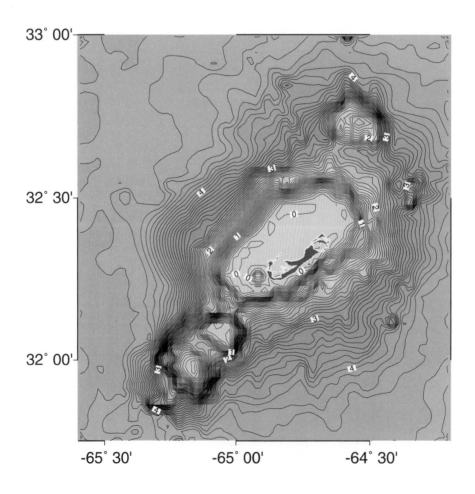

2005). Is the Bermuda Rise an expression of thermal expansion of hot underlying mantle, low-density materials intruded into the plate, a low-density buoyant root of partial melt residue, dynamic uplift maintained by convection, or some combination (e.g., Detrick et al., 1986; Sclater and Wixon, 1986; Louden et al., 1987; Sheehan and McNutt, 1989; Sleep, 1990; Phipps Morgan et al., 1995)? There are troubling inconsistencies between the data and any of these mechanisms, and even problems with the data themselves, particularly data on heatflow measured from shallow-penetration deep-sea probes (e.g., Ruppel, 1996, 2000). Problems extracting a mantle thermal signal from heatflow data on the Hawaiian swell (e.g., Harris et al., 2000; McNutt, 2002) probably complicate the heatflow interpretation issue on the Bermuda Rise as well.

In this article we offer an in-depth review of what is known, first, of the geology and geophysics of Bermudan igneous volcanism and, second, of the associated Bermuda Rise. With this information as a backdrop, we then examine the geophysical models that have been presented for the deep (i.e., mantle) processes. The possibility that the postulated "Bermuda hotspot" has a trace that extends into and accounts for the Cretaceous igneous intrusive activity in the Mississippi Embayment area (Cox and Van Arsdale, 1997, 2002) is treated separately from the geophysical models for the present Bermuda Rise. We propose some new ideas to reconcile the differences between Hawaii and Bermuda and between observations and models. Our organization is partly historical: we review the observational database for each parameter and the modeling efforts more or less in their historical order.

The primary timescale used in this article to convert from stratigraphic to absolute ages is the one of Gradstein et al. (2004). We also used this timescale to recompute the age of the original oceanic crust below the Bermuda edifices, and, where we cite specific magnetic anomalies or chrons, their ages based on the 2004 timescale. However, when referencing older work that directly or indirectly depends on the stratigraphic timescale (e.g., work on spreading rates, heatflow relative to crustal age, etc.), we did not adjust the published values. For the most part, such adjustments are small compared to other uncertainties in the values.

BERMUDA: IGNEOUS BIRTH, EROSION, AND SUBSIDENCE

Setting and Morphology

The Bermuda Islands (often referred to simply as "Bermuda," sometimes as "the Bermudas") rise up to 76 m above sea level along the southeastern margin of a northeast-elongated oval bank (Fig. 2) ~50 km long, which has long been considered the truncated stump of a large, extinct shield volcano generally referred to as the Bermuda Pedestal. A thin (15–100 m), mainly Quaternary-aged carbonate cap covers the truncated top of the volcanic pedestal (Vacher et al., 1989; Rowe, 1998). Bermuda is the largest of four evidently related seamounts arranged in a northeast-trending line 100 km long (Fig. 1, bottom), perched on the summit of the broad Bermuda Rise (discussed separately later). The four edifices, starting in the southwest, are Plantagenet Bank (or "Shelf"), later also known as Argus (or "Argos"; Cox, 1959) Bank, Challenger Bank, Bermuda, and Bowditch seamount. Applying the same terminology used by Vogt and Smoot (1984) in their analysis of the Cretaceous Geisha guyots (Japanese seamounts) in the western Pacific, we would call the short Bermuda chain a "cluster," as further discussed later.

The present Bermuda Islands, which occupy only the southeastern 7% of the ~665 km² Bermuda Platform, are composed almost entirely of limestones—variously indurated calcareous dune deposits (aeolianites) originally eroded from biogenic, primarily coral reef limestones during low sea levels of several Pleistocene glaciations, when the entire Bermuda summit platform was exposed. A variety of corals presently grow on submarine parts of the platform, thus making Bermuda the most northerly coral reef habitat in the modern Atlantic Ocean. The present Bermuda Platform consists of a central lagoon averaging ~18 m in depth (the greatest lagoon depth is 25 m, in Devil's Hole), surrounded except at the islands by a coral reef 2–10 m deep (Fig. 2, top). Just beyond and bordering this narrow (~1–1.5-km-wide) reef is a "terrace reef zone" ~1–3 km wide and ~20 m deep. Scattered coral patches within the lagoon rise to within a few meters of sea level. The lagoon, reef, and outer terrace are known collectively as the reef platform (Logan, 1988). At a smaller scale, the pattern of reefs and islands (Fig. 2, top) includes a number of ringlike or semicircular features with diameters of approximately 500–1500 m. The relation, if any, of this morphology to the underlying volcanic "basement" is unknown. Although Bermuda resembles typical Pacific atolls in morphology, the widespread occurrence of aeolianites, especially in the islands, has prompted some authors to call Bermuda, at 32.3°N, a "pseudoatoll."

Four interglacial soil horizons occur within the Pleistocene carbonate sandstones (Livingston, 1944; Harmon et al., 1978). Surficial deposits are mostly eolianite, with isolated sublittoral marine and beach deposits (Vacher and Hearty, 1989). Sea levels dated to approximately the Late Pleistocene, perhaps 1–2 m above present (~80 k.y.; Vacher and Hearty, 1989; Ludwig et al., 1996) and at +20 m (~420 k.y.; Hearty et al., 1999), constrain

Figure 2. (Top) Satellite image of the Bermuda Platform showing shallow (<25 m) water, central lagoon (blue) reef rims (light blue), and islands (light gray and green) (NASA World Wind v. 1.3 at about 40 km altitude). The isolated white specks distant from Bermuda are clouds. (Bottom) Bathmetry of the Bermuda Pedestal, the three smaller associated edifices, and surrounding seafloor, 100 m contour interval. Based on Smith and Sandwell (1997).

not only eustatic sea level fluctuations and global ice volumes, but also geophysical models for the Bermuda Rise (discussed later), most of which make predictions about elevation and later subsidence of the rise. Data collected to date suggest that reef limestone began to accumulate on the slowly subsiding, wave-eroded summit igneous plateau in the early Miocene, thus indicating that net postvolcanic subsidence has been negligible, averaging only about 0.004 mm/yr. At least under parts of Bermuda, the sequence of limestones and interglacial soils is underlain by a lateritic clay horizon ("primary red clay") derived from subaerial weathering of the volcanic basement in a humid tropical or semitropical environment (Moore and Moore, 1946; Foreman, 1951).

The 100-km-long chain (or "cluster," using the terminology of Vogt and Smoot, 1984) of four edifices trends 035 (Johnson and Vogt, 1971; the oval Bermuda edifice itself trends somewhat more easterly, 055), paralleling and ~30–40 km southeast of reversed chron M0 (Fig. 1; Rice et al., 1980). Based on the most recent age estimate for M0 (125.0 ± 1.0 Ma for the Aptian-Barremian transition; Gradstein et al., 2004) and an assumed but poorly constrained spreading half-rate of 2 mm/yr after M0 time, we estimate the age of the oceanic crust under the Bermuda edifices at ca. 123–124 Ma. Thus the Bermudan volcanoes (ca. 45–35 Ma in age; see later) grew from oceanic crust ca. 78–88 Ma in age at the time of volcanism. The apparent control of the old seafloor structure on the linear arrangement, paralleling crustal isochrons, of the Bermuda volcanoes contrasts with a nearly total absence along the Hawaiian chain of any evidence for such control by preexisting crustal structures (e.g., paralleling or perpendicular to Cretaceous transform trends crossing that chain).

Early Bermuda and Bermuda Platform Geophysics

Charles Darwin first suggested that the many isolated Pacific coral reefs are underlain by stumps of eroded former volcanic islands. A similar origin for Bermuda must have occurred to many scientists and naturalists before the twentieth century. The idea of an ancient volcano under Bermuda goes back at least to Verrill, who also believed it was of Triassic age (as quoted by Pirsson, 1914a). The first hint of a volcanic edifice at depth came indirectly: at least as far back as the 1873 Challenger expedition, local declination (compass variation) anomalies were apparently known at Bermuda. Given the dependence of navigation on compass bearings, there was a good practical reason to investigate these anomalies. The first to do so was J.F. Cole (1908), who mapped the magnetic declination throughout the islands in 1905. His "isogon" map revealed short-wavelength spatial frequency anomalies, which Cole merely attributed to "sources of considerable local disturbance." In retrospect, we can consider these data the first sign of highly magnetized igneous rock at shallow depth below Bermuda.

The Department of Terrestrial Magnetism of the Carnegie Institution then sent two expeditions to Bermuda (in 1907 and 1922) to study the magnetic anomaly previously known to exist

there (Fisk, 1927). Attempts were made to reoccupy some of the 1873 Challenger station sites. At the time it was unknown that local declination anomalies could be caused by remanent magnetism; hence the emphasis on repeat measurement. Although Cole had measured only variation of the compass (declination), Fisk included inclination and horizontal intensity, measuring these at five primary and seventy-eight secondary stations. Two decades prior to the invention of the proton precession magnetometer, measurements typically took half an hour per station! By the time of the 1922 measurements, a 1912 borehole had recovered igneous rock from shallow depth (see later), and Fisk (1923) could declare (p. 118) that "the island is a submerged mountain of volcanic origin." Fisk reported horizontal changes of as much as 700 nT over a horizontal distance of only ~6 m. Determining that the magnetization of the limestone and topsoil were inadequate to account for the anomalies, Fisk (1923, p. 119) concluded, "There are [*sic*], therefore, a major or primary source of disturbance lying deep in the lower structure of the submerged mountain."

The next geophysical advance was that of Woollard and Ewing (1939, p. 898), who ran two seismic refraction profiles in the islands, returning depths of 83 and 74 m for the tops of volcanic rocks. The experiment yielded a P-wave speed (4.88 km/s) for this "basement," with the overlying calcareous sandstone yielding 2.68 km/s. These may have been the first seismic refraction experiments on a buried igneous basement in an oceanic venue.

In addition to the two seismic lines, Woollard and Ewing (1939) analyzed both the isogon chart of Cole (1908) and Fisk's measurements (made on the 1907 and 1922 expeditions, reported in Fisk, 1927) of horizontal field strength on the islands, concluding from depth-to-source calculations that the basement depths are about 1200 ft (366 m) or less. Woollard and Ewing (1939, p. 898) went on to conduct "a series of observations on the vertical component of the earth's field and find very high local disturbances both positive and negative in character." Their 1939 *Nature* article is a very short summary, and to our knowledge the actual seismic and magnetic data were never published, owing perhaps to the onset of World War II.

As part of an extensive seismic reconnaissance of the Atlantic, Officer et al. (1952) reported four marine seismic refraction profiles shot across the north central part of the platform (off Hamilton Island) in the summer of 1950 and the spring of 1951. In water depths of 12–15 m, the experiments returned "basement" depths of 35–96 m, and two volcanic basement P-wave speeds of 5.13 and 5.34 km/s. In a footnote, Officer et al. (1952) cite unpublished 1949 (M. Ewing, G.R. Hamilton, F. Press, and J.L. Worzel) unreversed refraction profiles in the same area west of northern Bermuda, returning a basement depth of 90 m, a sediment P-wave speed of 2.7 km/s, and two basement speeds of 5.43 and 4.21 km/s. Historically, probably the most significant discoveries of Officer et al. (1952) were not the structure just below the Bermuda platform, but the relatively thin, albeit overestimated, sediment layer and oceanic crust (Moho at only 10 km bsl, with no "granitic layer") in the North America basin, on the east flank of the Bermuda Rise. Starting a decade later, these surprising results would be understood in terms of seafloor spreading and then plate tectonics. Even though borehole microfossil analysis had found nothing older than Eocene age (Pirsson, 1914a,b; see later), Officer et al. (1952) preferred a Triassic age for the Bermuda volcanic pedestal—completely ignoring the evidence for continental drift (Wegener, 1966), which implied a younger age. Officer et al. (1952, p. 806) anticipated the still unexplained lack of edifice subsidence by concluding that "the volcanic platform was planed to its present depth during a Pleistocene glacial lowering of sea level" and that "it is not necessary to hypothesize subsidence to explain the known geology of Bermuda."

Bermuda Boreholes

Long before seismic methods first measured the thickness of the Bermuda carbonate cap, the latter had already been directly measured when an exploratory well (the "Gibbs Hill Boring") was drilled in 1912 to 1283 ft (391 m) bsl for the owners of a local hotel in the hopes of discovering potable water supplies. Although the borehole failed in this objective, it did prove the presence of a volcanic "stump" under the island. The shallowest volcanogenic rock appeared at 230 ft (70 m) bsl (Pirsson and Vaughan, 1913; Pirsson, 1914a,b; Livingston, 1944; Foreman, 1951). Pirsson's (1914a,b) borehole section noted the oldest microfossils on the volcanic basement as Eocene. Remarkably, this age has stood the test of almost a century! Moore (1942) and Moore and Moore (1946) presented syntheses of Bermuda geology prior to the discovery of plate tectonics.

The second borehole to reach volcanic basement was drilled in 1958 (Rice et al., 1980). Later seismic reflection surveying showed the volcanic "basement" surface to vary from 200 to 25 m bsl, averaging ~76 m bsl (Gees and Medioli, 1970), close to the two refraction results of Woollard and Ewing (1939). The first attempt at age-dating Bermudan volcanism (Gees, 1969) yielded 34 and 52 Ma for two chips of igneous rock from the 1958 borehole.

A new phase of research on Bermudan volcanism was initiated by the continuously cored, research-driven "Deep Drill 1972" borehole, which penetrated 802 m below the ground surface, reaching volcanics at only 26 m bsl and continuing 767 m into the igneous basement (Aumento and Sullivan, 1974; Hyndman et al., 1974, 1979; Reynolds and Aumento, 1974; Rice et al., 1980). The cores revealed "431 submarine pillows or flows of hydrothermally altered tholeiitic composition" and "493 interfingering intrusive limburgitic sheets" of unusual composition (Rice et al., 1980, p. 212). The flows were thin, averaging only 1 m in thickness, and the sheets were preferentially intruded between flows or into preexisting sheets. The 1972 cores have recently been reexamined by M.-C. Williamson of Dalhousie University (Williamson, 2005, personal commun.; Williamson et al., 2006). A paper initially rejected for publication is now available on the Internet (Aumento and Gunn, 2005). A new geochemical analysis of dike rocks recovered in both the 1972

and the 1980 boreholes (see later) was performed by S. Olsen (2005).

Reynolds and Aumento (1974) reported a minimum (K-Ar whole-rock) age of 91 ± 5 Ma for the flows and a more reliable phlogopite (mica) age of 33.5 ± 2 Ma for the sheets. The latter age has been confirmed by new Ar-Ar dating of samples from the same cores (M.-C. Williamson, 2006, personal commun.; Williamson et al., 2006).

The intrusives have been variously described as "lamprophyre" and "nephelenite" dikes, and alternately as "very titaniferous alkaline rock," "melilite melanephelinite," or "Bermudite" (Aumento and Gunn, 2005, based in part on their 1970s work on core samples from Deep Drill 1972). The radiometric ages were averaged from whole-rock K-Ar ages of 47 and 91 Ma for the flows and 32, 14, and 30 Ma for the lamprophyre intrusives. In addition, mineral dates of 27, 36, and 33 Ma were obtained for "phlogopite" samples from the intrusives. The flows are "albitized and chloritized basalts of the OIB [ocean island basalt] type" (Aumento and Gunn, 2005), resembling mid-oceanic ridge basalts (MORB) of the "E-MORB" variety, i.e., somewhat enriched in large-ion lithophile elements. The intrusive sheets (dikes) are highly evolved rocks, with very high TiO_2 concentrations (5.0–6.2%), FeO* exceeding 16%, high P_2O_5 (to 2%) and K_2O, high La/Sm (up to 7.8%) and Sr (550–1050 ppm), and very low SiO_2 (30–40%) and Zr/Nb (as low as 2%). These compositions were taken from Aumento and Gunn (2005); see also Aumento et al. (1974, 1975) and Olsen (2005). Significantly, no subaerially extruded flows were cored, not even at the top of the igneous section, now at 26 m bsl. This observation strongly suggests that the flows were uplifted after they were erupted.

Two research boreholes (1980) were more recently drilled at Government Quarry, in Bailey's Bay on the south shore of Castle Harbor, where the carbonate cap was known to be less than 50 m thick, and over the gravity anomaly maximum at the Naval Air Station Annex (Peckenham, 1981). The latter borehole was abandoned in undrillable unconsolidated sediment at 42 m depth. However, the first borehole was successfully drilled to 98 m depth, well into the igneous basement. Nearly 50 m of submarine melilitic, variously altered pillow lavas were cored, associated with volcaniclastic rocks cut by narrow melilite-bearing intrusives. The igneous rocks, resembling those recovered by Deep Drill 1972, are silica-undersaturated, titanium- and volatile-rich (Peckenham, 1981; Olsen, 2005). Paleomagnetic studies of the lavas were unsuccessful, because inclinations were either not stable or not reproducible. The foramininfera (Orbulina and Globigerinids) in interpillow sediments are mid-Tertiary in age, consistent with other evidence on the age of Bermuda volcanism (Peckenham, 1981).

Based on a careful new analysis, Olsen (2005) concludes that the intrusive dikes (sheets) can have originated only by partial melting of very large-ion-enriched, pyroxenite-veined depleted upper mantle at high pressures of 5 GPa or more corresponding to ~150 km depth. Depths that great clearly put the magma sources in the asthenosphere—in the putative plume head.

The 1980 drillhole was supplemented in 1981 by submersible (Pisces IV) exploration (Peckenham et al., 1982). Three rock samples, two of them igneous (volcaniclastic breccia and porphyritic limburgite), were collected at 600–700 m bsl depths on the southern flank of the edifice. The sample composition is consistent with core materials and volcaniclastics previously recovered at Deep Sea Drilling Program (DSDP) site 386 (discussed later).

Borehole Results from the Surrounding Ocean Floor

In 1975 R/V *Glomar Challenger* drilled a hole (DSDP site 386; Shipboard Scientific Party, 1979) ~140 km southeast of Bermuda (Fig. 1), adding another dimension to the story of island volcanism (Shipboard Party, 1979; Tucholke and Mountain, 1979; Tucholke and Vogt, 1979). The seismic reflecting horizon A^v was found to correspond to the top of coarse volcaniclastic turbidites originating from subaerial (and subtropical) weathering and erosion of the volcanic islands that existed on the Bermuda Pedestal and its two satellite banks to the southwest. A^v extends up to 200 km away from Bermuda; upon approaching the platform, the reflector merges with the acoustically opaque archipelagic apron around the platform base. As far as is known to date, the Bermuda volcanoes did not erupt large amounts of ash: No discrete ash horizons were recovered at DSDP site 386 or 387, and although typically a few percentage points of "altered ash" content were reported in most of the volcaniclastic samples described at site 386, these were always followed by a question mark (Shipboard Party, 1979).

Volcaniclastic turbidites with intermixed shallow-water detritus began to arrive at site 386 in late Middle or early Late Eocene times (ca. 40–36 Ma), indicating that the Bermuda volcano (or volcanoes) had risen to sea level no later than that. The volcaniclastic turbidite subunit (clay, silt, and sand) is more than 162 m thick, with the thickest and coarsest individual turbidites, suggesting that the most intense erosion and/or increased volcanism and/or improved turbidite dispersal pathways to the site happened during the Middle Oligocene.

Interpreting the volcaniclastic turbidite record at site 386 in terms of volcanic episodes is questionable, however. Independent evidence suggests that the sea level fell by 50–75 m during the Middle Oligocene (e.g., Haq et al., 1987; Van Sickel et al., 2004), so the apparent increase in turbidite frequency and thickness at the same time may reflect increased exposure of the volcanic edifice to erosion, not increased volcanic activity. Furthermore, the lack of subaerially erupted basalts—even at 26 m below sea level—in the 1972 drillhole (Aumento et al., 1975; Hyndman et al., 1979) is consistent with uplift of an initially deeper volcanic pile due to (1) edifice inflation associated with intrusion of the sheets or other intrusions (as has apparently happened in the Cape Verde and Canary islands, resulting in subaerial exposures of original oceanic crustal basalts) and/or (2) continuing (subsequent to the eruption of the tholeiitic lavas) uplift of the Bermuda Rise and/or (3) exposure or reexposure because of lowered sea levels.

In their initial interpretation of drill core results, Aumento et al. (1974, p. 455) envisaged "a massive buildup of tholeiitic lavas near the axis of the Mid-Atlantic Ridge at least 91 m.y. ago," at or soon after the formation of the oceanic crust itself at the Mid-Atlantic Ridge axis. This "huge seamount (or island) remained dormant for almost 60 m.y.," sinking below sea level and experiencing protracted low-temperature hydrothermal alteration. The 33 Ma intrusion of lamprophyre sheets, then, according to Aumento et al. (1974), inflated the edifice by almost 40%, causing it to emerge (perhaps for the second time) above sea level. This early interpretation never appeared in a full paper—the 91 Ma date was generally disregarded as unreliable by most later authors—and an early (pre–91 Ma) formation of a large seamount or even an island near the Mid-Atlantic Ridge axis was scarcely supported by site 386 drilling results and is inconsistent with the isostatic seamount-height limitation developed by Vogt (1974, 1979b).

In any case, the lack of subaerial eruptives at shallow depths below Bermuda is atypical of oceanic volcanic islands, where loading-induced subsidence has depressed subaerially erupted rocks up to 1 km or more below sea level (e.g., Hyndman et al., 1979, p. 98). (The Cape Verde and Canary archipelagos show evidence of significant uplifts, however.) Although subaerially erupted Bermudan rocks may well have once existed at higher levels in the edifice—and were then eroded—better evidence of this must await additional coring and analysis of the volcaniclastics in Horizon A^v. So far the only evidence of subaerial or shallow-water eruptions is provided by the few percentage points of material noted as "altered ash?" in shipboard descriptions of the volcaniclastic sands at site 386 (Shipboard Scientific Party, 1979) and the altered tuff described by Foreman (1951), who inferred "volcanic eruptions of the explosive type, both andesitic and basaltic." However, isolated non-Bermuda igneous rocks could have floated (pumice) or been rafted by tree roots (e.g., Emery, 1955) from distant volcanic islands, particularly in the eastern Caribbean.

None of the publications describing the 1972 and 1980 drill cores reported even increased degrees of fracturing or increased vesicle volumes in the shallowest basalt flows, effects that would be expected had those flows been erupted in water depths of less than a few hundred meters (e.g., Moore, 1970; Moore and Schilling, 1973; Duffield, 1978; Schilling, 1986). Although vesicularity was not reported in the drill core descriptions, unusually high vesicle volume probably would have been noted. Alkalic basalts generally contain more volatiles and would show systematically higher vesicularity at any eruption depth (Moore, 1970; Moore and Schilling, 1973), but even the tholeiites cored at Bermuda should be significantly (say, more than ~5%) vesicular—particularly in pillow interiors—if they had been erupted at water depths of less than ~1000 m (see, e.g., Fig. 2 *in* Duffield, 1978, or Fig. 10 *in* Schilling, 1986). Thus, the shallowest recovered flows can scarcely represent the very top of the submarine part of the lava pile, but rather must have been uplifted, whether or not they were covered by additional submarine lava flows subsequently removed by erosion.

The age of the youngest volcaniclastic sediments at site 386 is Late Oligocene (ca. 25 Ma), so the volcanic foundation of Bermuda must have been emergent or shallow water at least until that time. As is true for Hawaiian volcanism, directly erupted materials such as pumice or ash are not abundant, and no discrete ash horizons were cored either at site 386 or at the more distant site 387. "Muscovite" mica is abundant in the youngest volcaniclastic turbidites, but with the rarity of associated quartz is extremely unlikely to have originated from distant continental sources. Very likely this mica is the same phlogopite recovered and dated (see earlier discussion) in the 1972 drillhole on Bermuda, its stratigraphic age at site 386 (Middle to Late Oligocene, ca. 30–25 Ma) in reasonable agreement with the 33.5 ± 2 Ma radiometric age and, further attesting to a later stage of Bermudan igneous activity, the intrusion of the lamprophyre sheeted dikes. The dike ages would be expected to be somewhat older than the stratigraphic age of their erosion products, as observed. If the radiometric age of the altered flows (91 ± 5 Ma) is ignored as unreliable—i.e., the edifice-building stage is dated via the site 386 stratigraphic occurrence of edifice erosion products—the intrusions postdated the shield-building stage by ~5–10 m.y. A better estimate of hiatus length awaits an accurate age determination of the lavas.

Turbidites continued to arrive at the site to the end of the Early Miocene, however, and although primarily composed of carbonate, still contain appreciable amounts of altered ash, heavy minerals, and zeolites, suggesting continued erosion of local outcrops and possibly minor, late-stage explosive activity.

DSDP site 386 cored one other interval—much deeper and older—with abundant volcanogenic components (Shipboard Scientific Party, 1979). A 29-m-thick, primarily upper Cenomanian (ca. 95–94 Ma) bed of zeolitic claystones contains alteration products of volcanic ash, with glass shards indicating that the volcanism was primarily pyroclastic. While this date agrees with the older of the two whole-rock ages (91 Ma) reported by Reynolds and Aumento (1974) for the Bermudan lavas—leading those authors and Peckenham et al. (1982) to suggest that Bermudan volcanism began ca. 90–110 Ma, near the ancient axis of the Mid-Atlantic Ridge—the agreement is most likely fortuitous: the cored lavas and Bermuda Pedestal morphology suggest passive, flow-dominated shield volcanism, not pyroclastic activity. Furthermore, the present elevation of the wave-eroded igneous rock platform under the islands, above the adjacent oceanic basement (~4.5–5 km) is too high for an origin near the axis of the mid-ocean ridge (Vogt, 1979b), and independently supports an off-axis origin for the volcanoes.

Eroding the Original Bermuda Volcanic Islands

The original maximum elevation of Bermuda prior to erosion was estimated at 3.5 km above sea level, based on extrapolation of submarine slopes, by Pirsson (1914a,b). A rough comparison with modern volcanic islands led Vogt (1979b) to reduce this to 3 ± 1 km for Bermuda and roughly 1 km for Plantagenet and Challenger Bank. A more careful analysis of active

volcanic island maximum elevations versus exposed area is shown in Figure 6 *in* Vogt and Smoot (1984). Using the 116 km^2 Bermuda Bank area and the best-fitting line in their figure yields an original Bermuda Island elevation of ~1000 m. The volume of material eroded can be estimated from Figure 7 *in* Vogt and Smoot (1984) as 30 km^3 and the time to reduce the island to sea level as 3 to 10 m.y. for shoreline erosion rates of 2.0 to 0.5 km/m.y., respectively. The magnitude of the erosion time is consistent with the length of time Bermuda erosion products are stratigraphically important at site 386, as discussed earlier.

The estimate of 30 km^3 for the total eroded rock volume, if distributed evenly over the region where Horizon A^v (Fig. 1; Tucholke and Mountain, 1979) can be identified in seismic records, is equivalent to a layer only ~0.3 m thick. This is comparable in magnitude to the minimum detectable thickness of a reflection (1/30 of the dominant seismic wavelength, i.e., ~0.5 m for 100 kHz sound traveling through a 1500 m/s medium; B.E. Tucholke, 2005, personal commun.). However, the actual thickness of A^v is likely to exceed this minimum, and in any case 0.3 m is vastly less than the cored thickness of volcaniclastic material, 162 m, at site 386; even upon reduction to original rock volume and allowing for dilution by nonvolcanic sediments, this is two orders of magnitude greater than 30 cm). Allowing for the inevitable concentration of volcaniclastics in valleys (as at site 386) and near the islands would make the calculated thickness even less elsewhere in the A^v area mapped by Tucholke and Mountain (1979). One interpretation of this paradox is to infer that the Bermuda volcanoes were active long enough to produce the equivalent of many successive emergent islands. However, this would invalidate the earlier erosion time calculation, which assumed that the island was constructed and then eroded.

The total volume of igneous intrusive and extrusive rock produced by the four Bermuda volcanoes is unknown due to the uncertain volume contained in A^v, the uncertain mass intruded into preexisting lithosphere, and the amount of lava, debris, and intrusives filling the flexural depression below the edifices. (The latter could in principle be estimated by flexural models such as that of Sheehan and McNutt, 1989, or by the seismic reflection and refraction methods described in Purdy and Ewing, 1986.) We calculated a minimum volume of ~760 km^3, the edifice relief above the 5 km depth level. Of this total, the main Bermuda edifice accounts for ~72%, Plantagenet and Challenger banks for 8% each, and Bowditch seamount for 12%.

The Bermuda Volcanic Cluster: Comparisons with the Japanese Seamounts

Compared to volcanoes along the Hawaiian chain, the Bermuda cluster (including the three satellite edifices) is a mere pimple, with a total volume only ~10% of a typical Hawaiian edifice. Guyots and seamounts along the 1100-km-long Geisha (now called the Japanese) seamount chain (Vogt and Smoot, 1984), provide a much better morphological match with Bermuda. In length and degree of linearity, the 100-km-long Bermuda cluster (four edifices) closely resembles the Winterer cluster (six edifices) in the Japanese seamounts. Summit plateau break depths along the these guyots (ranging from ca. 104 to 94 Ma in age) rise on average 4650 m above the basement, indistinguishable from the relief of Bermuda's wave-eroded igneous basement above the regional oceanic basement. The platform area of Bermuda (116 km^2) is identical, within measurement error, to that of Charlie Johnson guyot. The ~600 km^3 volume of the Bermuda edifice above the regional basement is comparable to that of Thomas Washington guyot (800 km^3), and the volumes of the three Bermuda satellite edifices are comparable to the ~100 km^3 volumes of some of the smaller Japanese seamounts, e.g., WC-3 and W-2 of Vogt and Smoot (1984).

There are nevertheless four significant differences between the Japanese seamounts and Bermuda—and at present it is unclear if these differences are related. First, as already implied, the Japanese seamounts comprise a number of guyot or seamount clusters arranged in a linear chain, whereas Bermuda constitutes just one cluster. Second, the Bermuda cluster (Fig. 1) appears structurally related to the underlying oceanic basement, trending parallel to crustal isochrons (as is also true of the Bermuda Rise; see later); by contrast, neither the overall trend of the chain nor the trends of any clusters or subclusters bear any relation to the preexisting plate tectonic fabric as defined by magnetic lineations and fracture zones (Fig. 2 *in* Vogt and Smoot, 1984). Third, almost all the Japanese seamounts or guyots (except Makarov guyot; Figure 3A *in* Vogt and Smoot, 1984) exhibit four to six flank rifts per edifice, giving the guyot physiography a starfish character not seen at Bermuda. Finally, the most obvious difference between the two volcanic provinces is postvolcanic subsidence: the Japanese seamounts evidently erupted on the summit of a swell of similar relief as the Bermuda Rise, but in the 100 m.y. that have passed, this swell has long since subsided into the noise. The Bermuda volcanoes and the Bermuda Rise are only half the age of the Japanese seamounts, so would not be expected to have subsided as much. However, neither Bermuda nor the Bermuda Rise appears to have subsided at all since 40 Ma—a major difference with the Japanese seamounts and many other oceanic volcanic lineaments and associated swells (as discussed further later). However, other Atlantic "hotspot"-type volcanic provinces share this lack of subsidence, or have even been elevated; e.g., Holm et al., 2006. However, this difference must relate to the nature and evolution of the Bermuda Rise (discussed later) and/or to the slow plate motion, not to the Bermuda volcanic edifices.

Other Seamounts in the Bermuda Rise Area

Besides the Bermuda cluster, no volcanoes or other vestiges of igneous activity of Cenozoic age have been documented anywhere in the western north Atlantic, including the Bermuda Rise (Fig. 3). The nearest sizeable seamount on the Bermuda Rise is Muir seamount (Fig. 1), an elongated (more or less transform-parallel, unlike Bermuda) edifice ~250–300 km northeast of Bermuda. Like Bermuda, Muir seamount was constructed on

post-M0 crust of ca. 120 Ma age and is located near the crest of the rise. However, Muir is part of a chain of mainly small and widely separated seamounts paralleling the much more conspicuous New England seamount chain, evidently unrelated to Bermuda and therefore not discussed in this article. At least eight sediment cores (the earliest described in Black, 1964) have been recovered from Muir seamount (unpublished compilation by L. Raymond, Woods Hole Oceanographic Institution; B.E. Tucholke, 2005, personal commun.). Core-bottom sediment ages range from Pliocene to Late Cretaceous, the latter ruling out a Middle to Late Eocene age like that inferred for Bermuda (a Late Cretaceous age is shown on the Geologic Map of North America; Reed et al., 2005; Fig. 3). No radiometric dates for Muir seamount itself are available, so the feature may even be Early Cretaceous (but of course no older than the Aptian-aged basement).

BERMUDA RISE

Residual Depth Anomalies and Crustal Age

The northeast-southwest-trending Bermuda volcanoes rise from the crest of a similarly trending, ~1500-km-long, 500- to 1000-km-wide basement rise (or swell) called the Bermuda Rise (Figs. 1, 3, and 4; Johnson and Vogt, 1971). The axis and western limits of the rise lie, respectively, ~900–1200 and 600–800 km ESE of the continental–oceanic crustal transition zone. As a bathymetric feature, the Bermuda Rise is approximately outlined by the 5500 m depth contour (e.g., Fig. 1 *in* Jaroslow and Tucholke, 1994). The rise (swell) is bordered by connected abyssal plains on three sides (Fig. 1): the Sohm in the northeast, the Hatteras to the west, and the Nares to the south, with depths and source distance increasing in that order (e.g., Pilkey and Cleary, 1986). Besides affecting turbidite deposition patterns, the Bermuda Rise has steered abyssal currents—in geologically recent times, the Antarctic bottom water—and influenced sedimentation patterns, e.g., along the deeper southwest (Driscoll and Laine, 1996) and northeast flanks (Laine et al., 1994) of the rise.

Detrick et al. (1986) give the dimensions of the Bermuda Rise bathymetric expression as 900 × 600 km, similar to Johnson and Vogt's (1971) 500 × 300 nautical miles (~925 × 550 km). Other papers cite different but similar dimensions. Of course, horizontal dimensions of any kind of anomaly—except those with sharp edges—are necessarily hard to pin down, and estimates often depend on the "eye of the beholder" and on what is defined as the background.

Because the Bermuda Rise is clearly a feature of the oceanic crust and upper mantle, its substantial sediment cover (Figs. 3 and 4) is of no interest (except as a record of volcanism and uplift; see later) and needs to be "removed" because it affects estimates of residual heatflow and depth anomaly. The total Bermuda Rise sediment cover averages 500–1000 m in thickness (Fig. 4) over most of the central rise but increases to 1000–

NEAR SEAFLOOR GEOLOGY
Sediment age, basement outcrops, and crustal isochrons

Figure 3. Geologic map of the Bermuda Rise area, part of the Geologic Map of North America of Reed et al. (2005): water depth contours (blue) at 1000 m intervals; Cretaceous-Jurassic and Lower-Upper Cretaceous crustal age boundaries (thin black lines). Note that most of the surface geology is dominated by Quaternary (white) and late Tertiary (yellow) hemipelagic sediments. Igneous outcrops (small "v" patterns) are limited to a few fracture zone escarpments (original oceanic basement) and seamounts like Bermuda.

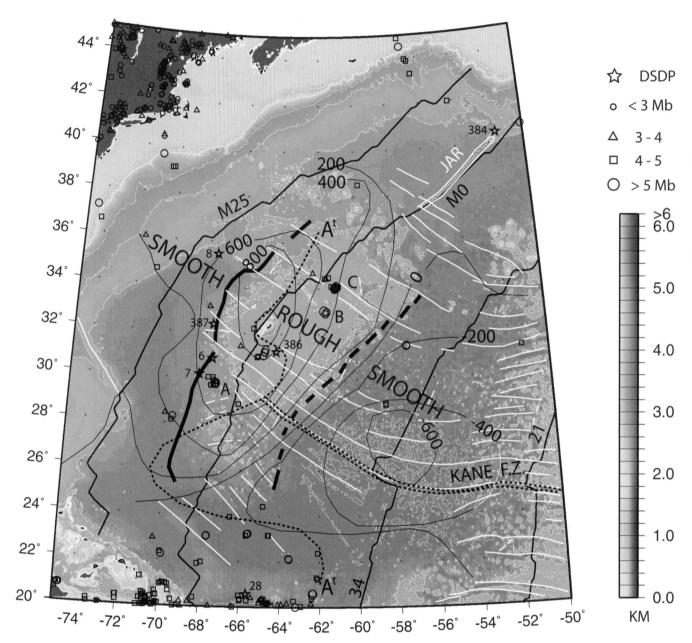

Figure 4. Bathymetry, residual depth contours at 200 m intervals, and boreholes reproduced from Figure 1 (top); western (thick solid line, after Sundvik et al., 1984) and approximate eastern (dashed line) limits of rough oceanic basement topography, created by slow spreading; eastern limits of seismic reflector A^t (dotted lines; Tucholke and Mountain, 1979); turbidites that covered part of the region prior to the uplift of the Bermuda Rise and spread eastward inside the Kane fracture zone (F.Z.) valley (Jaroslow and Tucholke, 1994); J-Anomaly Ridge (JAR) and associated basement escarpment (M0—seafloor spreading magnetic anomaly; Tucholke and Vogt, 1979; Tucholke and Ludwig, 1982); earthquake epicenters (small open circles; NEIC database as of December 2005, with additional older events from Zoback et al., 1986). A, B, and C denote epicenters whose first motion and/or depth could be determined. For events with aftershocks, only the location of the main shock is plotted. DSDP—Deep Sea Drilling Program.

1500 m on the outer northeast, west, and southwest flanks while decreasing to 100–500 m over the middle and outer southeast flanks (Tucholke, 1986). According to Sheehan and McNutt (1989), sediment thickness in the Bermuda region varies from 200 to 1300 m. Rises or swells in oceanic crust far from terrestrial sediment sources, such as many in the Pacific, have accumulated far less sediment than, e.g., the Bermuda Rise, making the shape and extent of the features there much easier to estimate from bathymetry alone.

Bathymetry and basement isochrons alone (Fig. 1) suggest that the oceanic crust underlying the rise (Fig. 1) ranges from ca. 100–105 Ma (roughly midway between chrons M0R and anomaly 34) to ca. 140 Ma (roughly midway between M0R and M25N). Upon allowance for isostatic crustal depression by sediments (e.g., Crough, 1983a), plots of basement depth versus crustal age suggest that the Bermuda Rise depth anomaly (Fig. 1, bottom, and Fig. 5) extends from ca. 80–100 Ma crust west to 150–160 Ma crust and along the axis of the rise a distance of ~1700–2000 km, from ~25°–27°N to 37°–38° N (Figs. 9 and 10 *in* Sclater and Wixon, 1986). The depth anomaly is somewhat asymmetrical in transverse section, with steeper southeastern versus northwestern slopes.

The Bermuda Rise basement between anomalies M0 and 34 was formed at higher average spreading half-rates (~20 mm/yr; recomputed, using the Gradstein et al. [2004]) time scale, from Klitgord and Schouten, 1986) than crust formed since anomaly 34 time and from ca. 135–140 Ma until M0 time. However, any variation of spreading rates during the long normal polarity

chron between magnetic anomalies M0 and 34 is unconstrained by magnetic lineations or other data. It is thus unknown if spreading rates increased at M0 time or somewhat later. Generally, oceanic basement topography formed by faster spreading is smoother. As a dramatic example, the basement on the upper western flanks of the Bermuda Rise becomes sharply smoother west of a slightly time-transgressive boundary aged ca. 135–140 Ma (Fig. 4; Sundvik et al., 1984), a time when half-rates declined to ~7–10 mm/yr (Klitgord and Schouten, 1986). Despite the presumably higher spreading rates, the basement on the eastern flanks of the Bermuda Rise does not become dramatically smoother there (e.g., Vogt and Johnson, 1971; Jaroslow and Tucholke, 1994; Lizarralde et al., 2004). However, existing detailed mapping along the Kane fracture zone corridor (Vogt and Tucholke, 1989) and earlier flow-line profiles to the south (Vogt and Johnson, 1971) suggest that the topography in the western ~30–50% of the M0–34 interval (the "Cretaceous quiet zone") is rougher than in the eastern part and was therefore probably formed at spreading rates less than the average for the interval. This eastern part of the Cretaceous quiet zone also exhibits fewer fracture zones (Fig. 2), based on their gravity expression. In its trend and extent, the Bermuda Rise thus approximately corresponds to a region of rougher basement (Fig. 4), an observation whose possible significance is discussed later.

There is no indication from the character of the basement reflections, from within the sediment column, or from the well-recorded magnetic lineations that intrusive or extrusive activity, if any, penetrated the crust subsequent to its formation by seafloor

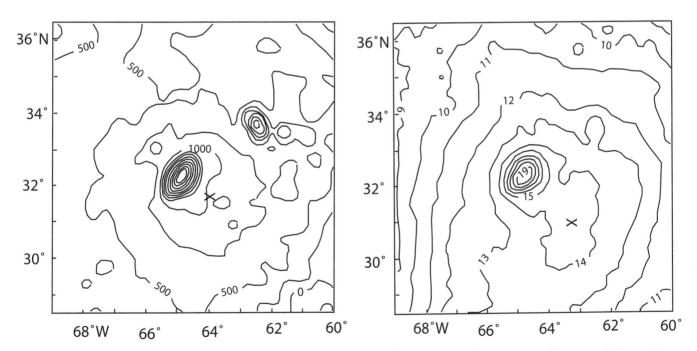

Figure 5. (Left) Residual depth anomalies (in meters) and (right) geoid anomaly (in meters) over the central part of the Bermuda Rise, with our picks (marked by *x*s) of anomaly maxima (slightly southeast of the more dramatic local highs associated with the Bermuda edifice cluster). Modified from Figures 2 and 3 *in* Sheehan and McNutt (1989).

spreading. The only exceptions are a few scattered seamounts and the Bermuda volcanoes (discussed earlier). Furthermore, refraction and wide-angle reflection data from the Bermuda Rise (e.g., Purdy and Ewing, 1986) show no evidence that the oceanic crust under the Bermuda Rise is in any systematic way different from—and specifically thicker than—other oceanic crust of comparable age and spreading rate of formation.

Upon correction for isostatic loading by sediments, the residual depth anomaly, developed in old oceanic crust (from 80–100 to 150–160 Ma; see earlier discussion), reaches 800–1000 m (Fig. 1, top, and Fig. 5) in the region around the volcanoes (Detrick et al., 1986; Sclater and Wixon, 1986; Sheehan and McNutt, 1989; Vogt, 1991). However, the precise magnitude and extent of the anomaly depends on model assumptions and parameters, e.g., sediment thickness and density structure (Crough, 1983a), crustal thickness and density structure, and more generally what is considered "normal" for the crustal ages represented on the Bermuda Rise. For example, if basement depth is compared to the Parsons and Sclater (1977) boundary-layer model, the anomaly on the Bermuda Rise crest is ~1800 m (Sclater and Wixon, 1986; Jaroslow and Tucholke, 1994). An independent estimate for the minimum Bermuda Rise uplift can be obtained from the present elevation of formerly abysssal plain pre–Bermuda Rise turbidites on versus off the rise: The values so obtained (+400 m at DSDP site 387 and +700 m at 386, discussed further later) are consistent with other estimates. The formation of the Bermuda Rise displaced ~0.5–1 million km^3 of water, which should have increased the global sea level ~2 m.

Several studies, starting with Davis et al. (1984), have concluded that the corrected basement even off the Bermuda Rise is significantly too shallow compared to predictions of both the plate and the cooling half-space (boundary layer) models. Sclater and Wixon (1986) found the average depth anomaly in the western Atlantic (including the Bermuda Rise) to be about +200 m relative to the plate model. Detrick et al. (1986) inferred a regional depth anomaly of ~+300–400 m relative to the plate model and +1000 m relative to the boundary layer model.

However, if the two flanks of the north Atlantic south of 52°N are compared, the basement, corrected for sediment loading, averages 557 m (i.e., ~500 m) deeper west of the Mid-Atlantic Ridge versus to the east (Conrad et al., 2004). This result—which is inconsistent with earlier studies, e.g., Fig. 2 *in* Colin and Fleitout (1990)—is independent of lithosphere cooling models and was attributed by Conrad et al. (2004) to continued sinking of the "Farallon slab," which, based on seismic tomography, is thought to underlie the western Atlantic at great depths in the mantle. However, those authors made no effort to remove the effects of midplate swells, which are more numerous and better developed east of the Mid-Atlantic Ridge (e.g., the Cape Verde Rise) than the Bermuda Rise is to the west. At any rate, the interpreted existence of the Farallon slab below the far western north Atlantic is speculative.

Geoid and Gravity Anomalies and Their Relation to Topography

Associated with the Bermuda Rise depth anomaly is an oval geoid high (Fig. 1, top, and Fig. 5) reaching 5–10 m in amplitude, depending on the model (Crough, 1978; Haxby and Turcotte, 1978; Detrick et al., 1986; Sheehan and McNutt, 1989), and a corresponding gravity high, measured relative to the adjacent seafloor, of about +10 mGals (Rabinowitz and Jung, 1986; Sclater and Wixon, 1986). Differences in the calculated geoid high (e.g., 8 ± 2 m, Detrick et al., 1986, versus 5 ± 2 m, Sheehan and McNutt, 1989; Anderson, 2005, cites +5.5 m) reflect different methods of filtering the raw geoid.

More locally, the Bermuda edifice generates the largest free-air gravity anomaly (+350 mGals) in the Atlantic Ocean (Rabinowitz and Jung, 1986), with a local geoid high, above the regional Bermuda Rise high, of +7 m (Fig. 5; Sheehan and McNutt, 1989). Sheehan and McNutt (1989) derived the same elastic thickness (30 ± 5 km) for the mechanical lithosphere downflexed under the "positive load" of the Bermuda volcanoes by two different methods: (1) the admittance (cross-spectral ratio of observed geoid to observed topography in the wave number domain) and (2) linear filters. Black and McAdoo (1988) had previously used the admittance method to compare various seafloor topographic features, including the Bermuda Rise area, and to test whether the admittances at longer wavelengths better fit uplift by convection or by plate temperature.

Both admittance and the linear filtering methods of Sheehan and McNutt (1989) yield similar compensation depths for the "negative load" under the Bermuda Rise (50 ± 5 km and 55 ± 10 km). Prior derivations of Bermuda Rise compensation depth, which ignored wavelength dependence of the anomalous geoid or topography ratio, varied from 40 to 70 km (Crough, 1978) to 80 km (Cazenave et al., 1988) and 100 km (Haxby and Turcotte, 1978; repeated in Turcotte and Schubert, 2002).

Seismicity and Stress

The state of stress in any part of the crust and the underlying mechanical part of the mantle lithosphere can be established by a number of techniques, including borehole elongation and earthquakes of sufficient magnitude to allow determination of first motion. In the western north Atlantic, Zoback et al. (1986) reviewed historical and instrumentally recorded earthquakes and suggested that midplate seismicity is anomalously high in the crust and/or the upper mantle below the Bermuda Rise compared to surrounding oceanic crust. When augmented with epicenters from events postdating their paper (Fig. 4), the anomalously high Bermuda Rise midplate seismicity is reinforced. This result is further supported by historical reports of earthquakes felt at least as far back as the late nineteenth century (Sieberg, 1932; Zoback et al., 1986), with an 1883 event of Modified Mercalli level VI to VII relayed in the latter paper. Of course, such earlier observations would have been biased toward shocks occurring in the vicinity of Bermuda.

Three earthquakes in the Bermuda Rise area (labeled A, B, and C in Fig. 4) were large enough to allow determinations of fault motion and/or focal depth (e.g., Nishenko and Kafka, 1982; see the reviews by Zoback et al., 1986, and Zhu and Wiens, 1991). The largest and most studied event (A, m_b 6.0, March 24, 1978), ~380 km southwest of Bermuda, involved thrust faulting at a focal depth of ~8 km. A m_b 5.1 shock (B) ~300 km east of Bermuda (November 24, 1976) showed strike-slip faulting at a focal depth of 10 km, based on surface wave amplitudes (Nishenko and Kafka, 1982). The December 9, 1987, m_b 5.2 event (C), located ~500 km northeast of Bermuda, involved thrust faulting with a strike-slip component; its focal depth could not be accurately determined. The preferential occurrence of earthquakes in the area of the Bermuda Rise (Fig. 5) shows that the mechanical lithosphere under the rise is under stress, but what is the source of this stress?

The net stress within any part of the mechanical lithosphere, such as under the Bermuda Rise, reflects the combined effects of (1) ridge push resulting from oceanic topography (i.e., the Mid-Atlantic Ridge) and greatest at shallow depths in the plate; (2) thermal stresses remaining from initial cooling; (3) stresses resulting from the loading of volcanic edifices, e.g., the Bermuda volcanoes, most effective near the edifices, or sediments; (4) membrane stresses due to the change of local curvature; (5) basal drag due to motion of the plate with respect to the underlying asthenosphere; and (6) thermoelastic stresses due to hotspot reheating. The latter effect—in the assumed absence of other forces—was modeled by Zhu and Wiens (1991), who concluded that if positive temperature anomalies exist, the resultant thermoelastic stress may be an important stress contributor near hotspots in old oceanic lithosphere. Their modeling predicts a complicated pattern of stresses, with maximum deviatoric stress on the order of 100 MPa, over an extensive part of a swell. Their models further predict extension at depths of less than 10–15 km and compression from there to the base of the mechanical lithosphere. Both thermoelastic and membrane (bending) stresses may have facilitated the rise of magma through tensional fractures in the crust and uppermost mantle to create the Bermuda volcanoes. Zhu and Wiens (1991) list three Bermuda Rise earthquakes (labeled A, B, and C in Fig. 4), whose P axes "suggest horizontal compressional stresses oriented radially with respect to the Bermuda swell, in agreement with our modeling." They also modeled the thermoelastic stress field for the Cape Verde Rise, finding the two first-motion solutions for earthquakes there also in approximate agreement with model predictions.

However, Zhu and Wiens (1991) note that compressive stresses should prevail only deeper than 15 km, in contrast to the 8 km focal depth of the 1978 m_b 6.0 event on the Bermuda Rise (A in Fig. 4), the most energetic shock instrumentally recorded on the rise. The authors suggest that the Bermuda hotspot may have had enough time to heat the upper part of the plate—so as

to account for the discrepancy between model and observations—but this suggestion is not supported by heatflow data discussed at length later.

Alternatively, the Bermuda Rise earthquakes could be occurring largely in response to a northeast- to ENE-oriented maximum horizontal compressive stress that dominates the central and northern Atlantic in all lithosphere older than ca. 20 Ma (Zoback et al., 1986) and is largely due to the ridge-push force. As shown in Figure 2 of Zoback et al. (1986), the stress pattern implied by Bermuda Rise earthquakes is largely consistent with this regional stress field versus a more local thermoelastic field as modeled by Zhu and Wiens (1991). The relative concentration of mid-plate seismicity on the Bermuda Rise may at least in part reflect a thinner, more fractured (hence weaker) crust and more gabbro-rich (hence weaker) upper mantle associated with the crust formed at very slow spreading rates (i.e., in the "rough basement" belt in Fig. 4), as deduced by Lizarralde et al. (2004) and discussed in a previous section. However, whatever process formed and maintains the Bermuda Rise must still play some role in the overall pattern of strain release (Zoback et al., 1986); otherwise enhanced seismicity would be spread northeast and southwest beyond the limits of the rise.

Zoback et al. (1986; see their Fig. 3) further suggested that many of the teleseismic events on the Bermuda Rise are located near fracture zones, implying a simple causal relationship. Given the small data set and the large epicenter location errors, those authors were reserved about this correlation. Remapping fracture zones by satellite radar altimetry-derived free-air gravity anomalies, and adding newer epicenters (courtesy of the National Earthquake Information Center), we can now be somewhat more certain that Zoback et al. (1986) were correct in suggesting that Bermuda Rise earthquakes tend to occur along the traces of ancient fracture zones (Fig. 4). However, this observation does not explain how the processes forming the Bermuda Rise have promoted ongoing fracturing in the oceanic crust or the uppermost mantle below the rise.

Heatflow

If anomalously elevated oceanic crust such as the Bermuda Rise has a thermal origin (e.g., Crough, 1978, 1983b, and later authors), it may, depending on the specific model (discussed further later), exhibit anomalously high degrees of heatflow. Heatflow (Fig. 1) has thus come to be considered an important geophysical constraint on the nature and origin of the mantle processes responsible for Bermuda and the Bermuda Rise, motivating several expeditions to acquire new data from the rise and surrounding parts of the western Atlantic (e.g., Davis et al., 1984; Detrick et al., 1986; Louden et al., 1987), as well as to reanalyze extant databases (e.g., Sclater and Wixon, 1986; Nagahira et al., 1996). Local heatflow variability, typically on horizontal scales of ~5–6 km (e.g., Detrick et al., 1986; Harris et al., 2000), mandates that each heatflow "value" used for modeling a large-scale process like the Bermuda Rise be obtained by averaging a number of closely spaced measurements of both gradient and thermal conductivity, with local basement topography and sediment thickness mapped by seismic reflection data for the purposes of modeling and correcting for the effects of sedimentation rates, spatially varying conductivity structure, and porewater convection within the sediments and/or basement (e.g., McNutt, 2002; von Herzen, 2004). The data discussed here (Fig. 1) all represent averages of many individual penetrations; older, single-penetration values, of dubious value, are not shown.

Starting with Langseth (1969) and as late as the paper by Sclater and Wixon (1986), no significant heatflow high, predicted by some models, was resolvable in archival heatflow databases. However, Detrick et al. (1986) measured heatflow at additional stations, and their measurements, combined with earlier data and arranged in an ENE-trending transect from the western Bahama Islands to Bermuda, did reveal a modest but significant ($8–10$ W/m^2) heatflow high over the rise (Fig. 1). The average heatflow was found to be 57.4 ± 2.6 mW/m^2 on the rise versus 49.5 ± 1.7 mW/m^2 off the rise (Detrick et al., 1986). However, local spatial variability (notably at the "54.4 mW/m^2" site closest to Bermuda, where individual penetrations returned values from ~$37–73$ mW/m^2), raises the possibility that the spatial coverage may be inadequate to define the mean heat flux (Detrick et al., 1986, p. 3709). Noting the correlation of local heatflow with basement topography, Detrick et al. tried, but failed, to account for the heatflow variability by modeling the heatflow refraction effect predicted from pure thermal conductivity. Effects of rough but buried basement topography could, from molecular conduction alone, account for local values ranging up to 25% above the regional mean (von Herzen, 2004).

With the benefit of more recent studies along the Hawaiian moat and swell (e.g., Harris et al., 2000; McNutt, 2002), we speculate that the high degree of local variability observed close to Bermuda reflects porewater flow within the spatially varying thickness of Bermudan volcanogenic debris (discussed earlier; Seismic Reflection Horizon A^v; Figure 1, top; e.g., Shipboard Scientific Party, 1979, and Tucholke and Vogt, 1979; Tucholke and Mountain, 1979). The volcanogenic debris—comprising turbidites—is thicker in valleys. If these sandy sediments serve as recharge zones (downward-percolating seawater), the thermal gradients above them would be reduced, thereby perhaps accounting for the apparent correlation of heatflow with basement topography noted by Detrick et al. (1986).

Even without allowance for the effects of the Bermudan volcanogenic sediments, the generally rougher basement topography of the crust that includes the Bermuda Rise (e.g., Sundvik et al., 1984; Fig. 3 *in* Zoback et al., 1986) would favor higher heatflow via porewater advection (over basement peaks) in the more permeable basement. Von Herzen (2004) estimates that thermally significant hydrothermal circulation has influenced ~$20–30\%$ of heatflow stations on oceanic crust older than 95 Ma. Anomalous heat flux would be expected to extend out to older crust for slower spreading rates (i.e., rougher basement to-

pography) and would involve porewater advection on the order of 10 cm/yr (von Herzen, 2004). Although none of the Bermuda Rise heatflow measurements to date show large deviations from the regional mean, some of the most anomalous (up to an order of magnitude above the regional mean) heatflow values on old crust anywhere lie just south of the Bermuda Rise, on the Nares Abyssal Plain (Embley et al., 1983; Von Herzen, 2004) and are almost certainly the result of porewater advection.

Heatflow measured in the 1972 Bermuda Island borehole ($53–57$ mW/m^2; Hyndman et al., 1974) is comparable to the average value returned from the three ocean floor sites within ~200 km of Bermuda (Fig. 1, bottom; Detrick et al., 1986), even if radioactive heat sources in the Bermuda volcano (specifically the lamprophyre sheets) contribute up to 4 mW/m^2 to the borehole value, as suggested by Hyndman et al. (1974). Borehole temperatures—measured at 8 m intervals to a depth of 360 m— showed a linear increase with depth; the modest gradient and lack of irregularities are consistent with long-term volcanic and hydrothermal quiescence.

In a later study, Louden et al. (1987) found relatively high values of heatflow (53 mW/m^2, 25% higher than predicted by either plate or cooling half-space models), even on old (ca. 163 Ma) crust northwest of the Bermuda Rise (Fig. 1). Comparing this heatflow with other values in the region, they suggest that the entire Mesozoic northwest Atlantic may have a uniformly elevated heatflow. Louden et al. (1987) corrected all heatflow values for this region by using a higher sediment conductivity for the station on anomaly M0 (Fig. 1) and by allowing for lithospheric cooling with age (as predicted by the plate model). This reduced the apparent heatflow high on the Bermuda Rise to a mere 5 mW/m^2 compared to the $8–10$ mW/m^2 derived by Detrick et al. (1986) and so far demonstrable only on the southwest flank of the rise (Louden et al., 1987).

And, while the heatflow southwest of the rise has apparently reached an equilibrium value (~$48–50$ mW/m^2), this is ca. 15% higher than expected for crust of this age, and the basement depths are at least $200–300$ m too shallow. These anomalies were first noted by Davis et al. (1984) and further discussed by Detrick et al. (1986) and Nagahira et al. (1996). While the depth and thermal anomalies over supposedly "normal" ocean crust may cast doubt on the interpretation of data from the rise itself, it may be that the geographic "extent" or influence of the process (discussed later) that formed the Bermuda Rise affected a wider area of crust than suggested by the basement swell dimensions. For example, although Nagahira et al. (1996) deliberately excluded heatflow stations on the Bermuda Rise (as defined by them), they concluded that stations within 600 km of the Bermuda Rise crest suffered the greatest amount of thermal rejuvenation, whereas those least reheated are 900 km distant.

Except for the Bermuda borehole data of Hyndman et al. (1974), all the measurements given earlier were made with shallow-penetration probes. For example, Detrick et al. (1986) used either 3.5 m or 7 m probes, depending on bottom sediment penetrability. Louden et al. (1987) employed a 4-m-long probe.

Even though temperature gradients were shown to be relatively constant along these thermistor strings, there is always nagging doubt about the effects of long-term (many centuries and longer) bottom water temperature changes and other effects that may have biased the results. These doubts have been reinforced by a comparison (Ruppel, 1996, 2000) between deep (320–420 m bsf) ODP leg 164 borehole sediment temperatures in the southwestern north Atlantic (Blake Ridge area) and shallow penetration data from the same region, underlain by oceanic crust estimated to be 175–180 Ma in age. The borehole data suggest heatflow values of ~35 mW/m^2, compared to 48–55 mW/m^2 for the shallow-penetration data (e.g., Fig. 1). This result casts some doubt on the putative anomalously high western Atlantic heatflow—or at least its magnitude—inferred by Davis et al. (1984), Detrick et al. (1986), Louden et al. (1987), and Nagahira et al. (1996). Ruppel (1996) concluded that basement heat flux on the Blake Ridge may be up to 25% lower than the previous estimate of 49 mW/m^2. Given the various error sources, this apparent "excess" heatflow is comparable in sign and magnitude to the result of Detrick et al. (1986), who found heatflow off the Bermuda Rise to be ~15% higher than expected. Similarly, Louden et al. (1987, p. 120) concluded that "heat flow values are elevated by 20–30% throughout the entire region compared to standard one-dimensional conductive cooling models, but are not systematically related to basement depths." Of course, additional borehole thermal data are needed to demonstrate that the discrepancy found by Ruppel (2000) applies to the entire region. However, her borehole results do support a primarily conductive heat transfer, at least in the borehole area.

Deep Structure below the Bermuda Rise: Evidence from Seismic Tomography

High-resolution whole-mantle seismic tomography is still in its infancy (see e.g., Nataf, 2000, for a global synthesis of seismic searches for mantle plumes and also Zhao, 2004; see Montelli et al., 2004, and van der Hilst and de Hoop, 2005, for more recent analyses). However, several studies have inferred the existence of the "Farallon Slab" deep below the region from the eastern United States to the western Bermuda Rise at depths increasing from ~800 km in the west to ~2500 km in the east (e.g., Fig. 19 *in* Zhao, 2004). As noted previously, Conrad et al. (2004) attributed the generally deeper western Atlantic (versus the eastern) to the effects of this deep slab. Subducting slabs are now thought to become stagnant at the 670 km discontinuity—and only much later to detach and sink into the lower mantle (e.g., Fukao et al., 2001). In the upper mantle, modestly slower (by ~0.2–0.5%, perhaps not significant) P-wave speeds extend to ~400 km directly below Bermuda on an east-west vertical cross-section (Fig. 19 *in* Zhao, 2004), suggesting that anomalously warm mantle associated with the Bermuda Rise may extend that deep. Similarly, anomalously "slow" upper mantle extends to ~1000 km depth under a broad region extending from the western Bermuda Rise to the eastern margin of North Amer-

ica. In the shear wavespeed tomography of Ritsema (2005), the mantle below Bermuda is fast at 100 km depth (plausibly the result of the thick, old lithosphere), slow at depths of 300, 600, and 1100 km, and fast at still greater depths, which is broadly consistent with the P-wave model of Zhao (2004). However, Ritsema's model does not resolve a separate anomaly under the Bermuda Rise, which can also be said of most other postulated hotspots or mantle plumes (Fig. 6 *in* Ritsema, 2005). In general, the S wavespeed contrast is highest in the uppermost mantle and lowest below ~600 km depth.

At the very bottom of the mantle (2850 km, layer D″), most of the North America plate, including the western north Atlantic, generally is "fast" in terms of both shear (Ritsema, 2005) and compressional (Dziewonski, 2005) wavespeeds.

Uplift History of the Bermuda Rise: The Sedimentary Record

Deep drilling on the Bermuda Rise at DSDP sites 386 and 387 (Fig. 1) recovered numerous mixed bioclastic and terrigenous turbidites that may well have originated along the North American continental margin, with basement uplift—presumably the formation of the Bermuda Rise—shutting off this sediment source sometime in the Middle Eocene (whose age range is given by Gradstein et al., 2004, as 48.6–37.2 Ma) (Tucholke and Mountain, 1979; Tucholke and Vogt, 1979). The top of this turbidite unit is marked by seismic reflector A^t, whose seaward extent is shown in Figure 4. The offlapping relation for the youngest turbidites—i.e., the cessation of turbiditic sedimentation that occurred earlier on what would become the crest of the Bermuda Rise than farther away on the flanks—is tentatively suggested by the "younging" of the youngest turbidites from east (Middle Eocene at site 386) to west (post–Late Eocene at DSDP site 8). Minimum (i.e., neglecting possible offlap effects) total uplifts of the Bermuda Rise (700 m at site 386 and 400 m at site 387) can be estimated from the present Bermuda Rise elevations of originally presumably nearly horizontal (i.e., abyssal plain) A^t (Tucholke and Vogt, 1979).

At least at DSDP site 386, near the crest of the rise, these turbidites comprise largely biogenic sediments, so whereas an origin along the continental margin—some 1200 km distant—is likely, it is not absolutely certain. At least some turbiditic sediments might also have been derived from adjacent basement highs. If the end of turbidite deposition marked the uplift of the Bermuda Rise, its age immediately preceded or was coeval with Bermuda volcanism, because the oldest certain volcanogenic debris shed from the islands (discussed previously) arrived at site 386 immediately upon cessation of biogenic turbidite deposition.

The Bermuda Rise is crossed by fracture zones, some of which have deep basement valleys that continue for long distances. At least on the central parts of the Bermuda Rise and farther east, these fracture zones show up as negative gravity lineations, indicating basement valleys (Fig. 2). The Kane fracture zone is foremost among those, and due to its depth has accumulated a thicker sediment pile than adjoining oceanic crust. The greater thickness of sediments in fracture zone valleys reflects mass wasting from local basement highs, but also turbidity flows from distant sources. In any case, the sediment record is generally expanded in deeps, allowing better resolution of seismic stratigraphy. Jaroslow and Tucholke (1994) took advantage of these circumstances and analyzed the sediment fill within the Kane fracture zone valley, along which turbidity flows moving southeast (down the flank of the Bermuda Rise) traveled as far as 55.3°W (well beyond the Bermuda Rise), where further travel was stopped by a basement obstruction (dam) within the valley floor. The seismic stratigraphy within the valley fill revealed reflectors, generally associated with former abyssal plain surfaces, postdating the Bermuda volcaniclastic seismic horizon A^v. Because these reflectors must have been nearly horizontal when deposited but are now inclined (down toward the southeast) by amounts that increase with increasing subbottom depth, Jaroslow and Tucholke (1994) were able to demonstrate that uplift of the Bermuda Rise continued into the Miocene, later than had been assumed from earlier observations. Their analysis confirmed that significant uplift of the Bermuda Rise had occurred by Early Oligocene times, but that some 400–500 m of additional uplift must have occurred after ca. 33 Ma, the time of the second (lamprophyre or "Bermudite" sheet) igneous episode discussed earlier. Relative excess uplift of ~200 m was inferred from the middle Early Oligocene to the earliest Miocene, with an additional, albeit perhaps not significant, uplift of 45–50 m during the Early to Late Miocene interval. Notably, the seismic stratigraphy in the Kane fracture zone valley, as interpreted by Jaroslow and Tucholke (1994), reveals no evidence of migratory uplift such as would be expected from relative westward motion of the plate over a mantle hotspot.

Independent evidence of the initiation of Bermuda Rise formation comes from bottom current–controlled sediment deposition, because such currents are steered by bottom topography. In geologically recent times, sediment cores and seismic profiles show a pattern of sediment redeposition caused by the movement of Antarctic bottom water around the Bermuda Rise (Laine et al., 1994; Driscoll and Laine, 1996). Seismic reflection data suggest that bottom currents began to be steered around the Bermuda Rise as early as the Middle Eocene (Ayer and Laine, 1982), which is consistent with other evidence for initiation of Bermuda Rise uplift.

The Southeast Bermuda Deep

A regional bathymetric low, centered ~900 km southeast of the Bermuda Rise crest (Fig. 1), may be genetically related to the origin of the Bermuda Rise (see later), and is therefore included in this review. The feature was discussed both by Sclater and Wixon (1986), who referred to it as a "pronounced counterbalancing low," and by Vogt (1991), who called it the "Anti–Bermuda Rise." We suggest the more appropriate name Southeast Bermuda Deep. The residual depth anomaly of this feature

(Fig. 4 *in* Sclater and Wixon, 1986) is below –600 m but does not reach –800 m, the core of the Southeast Bermuda Deep anomaly roughly defined by an area where average water depths commonly exceed 6000 m. Because sediment thicknesses average less than 100 m (Tucholke, 1986), the basement depth and water depth contours define the shape of the feature almost equally. Chron 34 bisects the central part of the Southeast Bermuda Deep, which is developed in crust ranging from ca. 50 to 90 Ma. Sclater and Wixon (1986) calculated the depth anomaly only east to 40°W, i.e., approximately out to the crest of the Mid-Atlantic Ridge.

Although the amplitude of the Southeast Bermuda Deep low is only two-thirds that of the Bermuda Rise high and is somewhat smaller in platform, both are of comparable extent and amplitude. Furthermore, if the depth anomaly were to be defined on the basis of average corrected basement depths in the western Atlantic instead of from theoretical plate cooling models, the Bermuda Rise and Southeast Bermuda Deep would have about the same amplitudes (Sclater and Wixon, 1986).

A geoid low of –5 to –6 m is associated with the Southeast Bermuda Deep, which is again comparable but opposite in sign to the regional Bermuda Rise geoid high (Sclater and Wixon, 1986). The corresponding free-air gravity low is about –10 mGals relative to "normal" seafloor in the western Atlantic and about –20 mGals relative to that associated with the Bermuda Rise (Rabinowitz and Jung, 1986). The archival heatflow database shows only widely scattered stations in the Southeast Bermuda Deep area. Although these give no indications of a heatflow low, this result may well be inconclusive.

The Southeast Bermuda Deep appears to have no seamounts large enough to be resolved in the Smith and Sandwell (1997) bathymetric grid. The finer-resolution seamount location map (including features less than 1 km in relief) of Epp and Smoot (1989) does not extend far enough southward to cover the Southeast Bermuda Deep. If the deep is related to the Bermuda Rise (e.g., Sclater and Wixon, 1986), it presumably developed, as did the Bermuda Rise, in preexisting oceanic crust during Eocene times. However, it is much more likely that the Southeast Bermuda Deep is unrelated to the Bermuda Rise, reflecting instead the formation of anomalously deep oceanic crust at the ancient Mid-Atlantic Ridge axis. This view is supported by the continuation of relatively deep crust east to the Mid-Atlantic Ridge axis and to the existence of a conjugate region of anomalously deep crust of comparable age east of the ridge. Conjugate deep areas east of the ridge (generally more than 300 m deeper than normal for the ages, but depending on the age depth model) appear in the global depth anomaly maps of, e.g., Colin and Fleitout (1990), Kido and Seno (1994), and DeLaughter et al. (2005). The north Atlantic depth anomaly, calculated from Model GDH2 and kindly shared by DeLaughter et al., is shown in Figure 6. The depth anomaly would probably be even more symmetric about the Mid-Atlantic Ridge had it not been affected by Tertiary swell development that formed and probably continues to maintain (Courtney and White, 1986) the Cape Verde Rise.

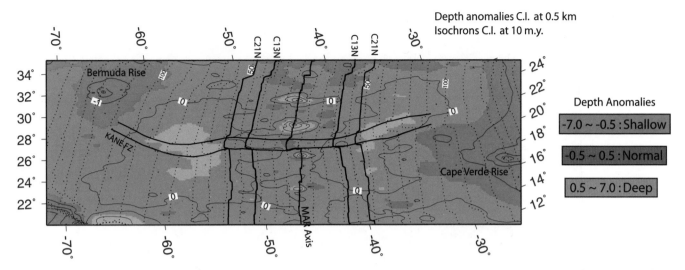

Figure 6. Depth anomaly contours from the western north Atlantic / Bermuda Rise region across the Mid-Atlantic Ridge (MAR) to the conjugate region in the eastern Atlantic. Some fracture zone (FZ) traces are shown (the Kane fracture zone is the southernmost of the two transforms shown) to aid linking conjugate crustal parcels. Note that the southeast Bermuda Deep discussed in the text (Fig. 1, bottom) corresponds to a conjugate deep region east of the Mid-Atlantic Ridge axis, and was therefore probably formed at the ancient Mid-Atlantic Ridge axis. The Cape Verde Rise (swell) probably formed during Tertiary times (volcanism since at least 19 Ma; reviewed in Holm et al., 2006), elevating part of the eastern southeast Bermuda Deep conjugate. Based on the global depth anomaly grid derived by DeLaughter et al. (2005) from their model GDH2. The isochrons are from Müller et al. (1997), with the locations (dotted lines) of anomalies 21 (46.5 Ma) and 13 (33.5 Ma) added to show the transform trends (Africa and North America plate motion directions) at about the time of the formation of the Bermuda Rise and edifice, as well as the time the sheets were intruded into the lava flows. Note the ~30° bend recorded by the western flank of the Kane fracture zone and the adjacent Northern fracture zone at about anomaly 21 time. C.I.—contour interval.

We therefore consider it fairly safe to dismiss the interpretation (e.g., Sclater and Wixon, 1986; Vogt, 1991) of the Southeast Bermuda Deep as the downwelling counterpart of the Bermuda Rise.

IS THERE A BERMUDA HOTSPOT TRACE?

When Morgan (1972) extended J.T. Wilson's hotspot ideas to propose relatively fixed mantle plumes, he attempted to account for elongate traces of progressive age-varying igneous activity (e.g., the Hawaii-Emperor seamount chain) and to use these traces to develop a model for "absolute" motion (i.e., relative to a reference frame in the deeper mantle) of lithospheric plates. The isolated Bermuda volcanoes, perched on the crest of the broad Bermuda Rise, did not fit this pattern and probably for that reason were not discussed in the early hotspot or mantle plume literature. This well-mapped part of the Atlantic shows no evidence of the seamount chain that should exist, younging toward the east and becoming older to the west, according to the "stationary hotspot" or plume model (e.g., Figs. 1 and 3). In fact, the four volcanoes of the short Bermuda cluster are oriented at right angles to the expected trace, as is the Bermuda Rise. Furthermore, the uplift history—most recently clarified by Jaroslow and Tucholke (1994)—shows continuing uplift centered on the Bermuda Rise from the middle Eocene at least into the earliest Miocene. These authors found no evidence of eastward migra-

tion, predicted by the hotspot model, of the uplift center. In fact, there is little or no evidence of any kind for eastward migration.

All three hotspot models (Morgan, 1983; Duncan, 1984; Müller et al., 1993) concur in predicting the present location of a relatively fixed "Bermuda hotspot" at around 32°N, 58°W, ~650 km east of the islands, on the outer edge of the present Bermuda Rise. This predicted hotspot location is ~1450 km from the closest point (~26°N, 45°W) on the present Mid-Atlantic Ridge spreading axis.

The present bathymetry over the region that should have been crossed by a Bermuda hotspot trace from ca. 40 Ma to the present (Fig. 7) shows a small seamount at 31.1°N, 58.2°W, not too far from the predicted present hotspot location. Although lack of data on this feature precludes ruling out the young age expected if this seamount reflected present activity of a Bermuda hotspot, the latter appears extremely unlikely: no earthquake seismicity has been reported there (Fig. 4). Of the three Bermuda Rise–area earthquakes summarized in Zoback et al. (1986), the nearest to the "predicted hotspot" is a $m_b = 5.1$ reverse faulting event ~350 km to the west (33.01°N, 61.66°W), and no seamounts exist between the small seamount and Bermuda (Fig. 7). Most probably the small seamount is Cretaceous in age, forming the southeast end of a sparsely populated northwest-trending chain that includes the relatively large Muir seamount north of Bermuda.

Morgan (1978) suggested that mantle material flows from a near-ridge plume in a subhorizontal conduit toward the spread-

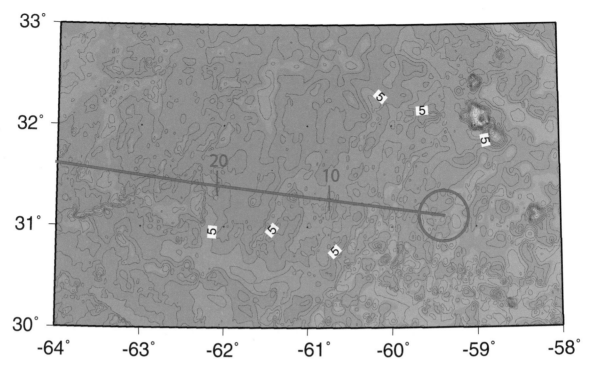

Figure 7. Bathymetry (200 m contour interval, based on Smith and Sandwell, 1997) of the eastern Bermuda Rise region, showing the predicted trace (red; Duncan, 1984, and other models) of igneous activity expected from a continuously active, fixed Bermuda hotspot model. Models predict a present hotspot location of ~32°N, 58°W (red circle), a region lacking seamounts large enough to be detected via radar altimetry. The small seamount at 31.1°N, 58.2°N has not been investigated, but is very likely of Cretaceous age.

ing axis, generating secondary volcanism above the flowing material. By contrast, Sleep (2002) proposed that when plumes are sufficiently close to a spreading axis, the buoyant material flows toward the axis rather than erupting locally, thus forming a gap in the hotspot track. Sleep wrote: "The Bermuda track is a less clear example of a ridge-approaching hot spot with an unclear on-axis hot spot between 20 N and 28 N. . . . [The] activity has ceased near off-axis Bermuda . . . but has not yet clearly commenced at the axis" (p. 2). We are skeptical about this explanation in the absence of any evidence for hotspot influence along this section of the Mid-Atlantic Ridge. If anything, the ridge is deeper than elsewhere in this region (Fig. 6).

The analysis of Bermuda Rise depth and geoid anomalies (Sheehan and McNutt, 1989) shows a modest southeast displacement of the rise culmination relative to Bermuda (Fig. 5). Inspection puts the summit of the depth anomaly of the rise (disregarding Bermuda) at ~31.7°N, 64.0°W, 75–100 km southwest of the Bermuda edifice; the summit of the geoid anomaly is near 31.0°N, 63.3°W, ~200 km from Bermuda (Sheehan and McNutt, 1989, noted this offset, estimating it at 150 km). If these offsets represent the distance the plate has moved since 40 Ma relative to the Bermuda mantle source below the lithosphere, the average rates of migration have been only on the order of 0.2–0.5 mm/yr, with the implied plate motion toward the northwest, ver-

sus the west, predicted by hotspot models (Figs. 1 and 7). However, this type of calculation assumes the edifice was originally built on the summit of the Bermuda Rise, a plausible but probably untestable assumption. The offset might equally be attributed to the outward spread of the swell root, with magmas using a northeast-southwest-trending zone of weakness (based on the model of Phipps Morgan et al. [1995]).

Nevertheless, longer-term "absolute" plate motion models can readily be applied to any spot on any plate so as to predict the trace that point would have made over the mantle below the plates. Morgan and Crough (1979) first applied their plate motion model to a postulated "Bermuda hotspot." They could constrain their model only at one place (Bermuda) and time (40–45 Ma, as discussed earlier), i.e., postulating that this hotspot affected the upper crust only for a geologically short interval of time, during which the Bermuda volcanoes and the Bermuda Rise were formed. This hotspot trace postulated a present hotspot location on the lower east flank of the Bermuda Rise, and a track that increased in age toward the west, passing over the area of the "Cape Fear Arch," a northwest-trending structural high along the Carolina coast, about the time this arch was thought to have originated (Morgan and Crough, 1979). However, Winker and Howard (1977) showed that uplift of the arch continued into the later Neogene, and newer studies (e.g., Stewart and Dennison, 2006) show that the Cape Fear arching occurred too late (Late Oligocene to Middle Miocene) to correspond to the time it is estimated that the area passed over the putative hotspot (ca. 50–60 Ma; Fig. 8). Moreover, subsequent plate motion models have predicted a more southwesterly track for the putative Bermuda hotspot (e.g., Morgan, 1983; Duncan, 1984; Müller et al., 1993). The Cape Fear Arch consequently disappeared from most later treatments of the "Bermuda hotspot" idea. Furthermore, the denudation history of the Appalachians—as revealed by sedimentary deposits along the U.S. middle Atlantic margin (Poag and Sevon, 1989)—does not support any uplift during the time the putative Bermuda hotspot track should have been crossing this area (Paleocene).

However, Morgan and Crough (1979) also correlated the Middle to Late Cretaceous alkalic and ultrabasic igneous activity, particularly the formation of kimberlite pipes, in central Arkansas and around the junction of that state with Louisiana and Mississippi with the predicted position of the North America plate over a fixed "Bermuda hotspot" during that time. This

correlation has been further developed in several subsequent papers, from Crough et al. (1980) to Cox and Van Arsdale (1997, 2002). However, as McHone (1996) has emphasized, postbreakup Jurassic and Cretaceous magmatism along the eastern U.S. margin, along and west of the Appalachians, and in New England scarcely fits the basic model of the North America plate passing over even several plumes. Although the hotspot models of Morgan (1983), Duncan (1984), and Müller et al. (1993) all predict somewhat different traces for the "Bermuda hotspot," particularly prior to 60 Ma, all three of their traces pass through the general area of this igneous province at about the right time, with Duncan's providing the best geometric fit (Fig. 8; Cox and Van Arsdale, 2002). Because the three hotspot models do not include any data from Bermuda or its putative trace, agreement between prediction and observation could be considered either support for the fixed hotspot model or fortuitous, but not circular reasoning of any sort.

More recent compilations by Cox and Van Arsdale (2002) indicate that the igneous activity (including that forming kim-

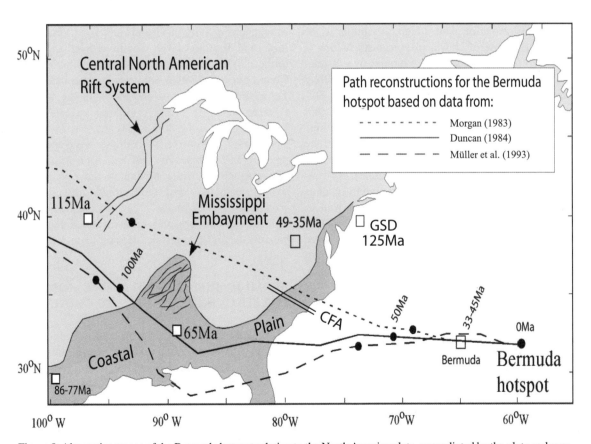

Figure 8. Alternative traces of the Bermuda hotspot relative to the North America plate, as predicted by the plate and mantle motion models of Morgan (1983), Duncan (1984), and Müller et al. (1993), with filled circles indicating the predicted hotspot location at 0, 50, and 100 Ma for each trace. The thin lines within the Mississippi Embayment are the Mississippi Valley graben faults, and the open squares show the average location of the earliest (ca. 115 Ma) and latest (ca. 65 Ma) eruptions along a migratory path of igneous activity. CFA—Cape Fear arch; GSD—Great Stone dome. Modified from Figure 1 *in* Cox and Van Arsdale (2002). See de Boer et al. (1988) and McHone (1996) for more detailed maps of Jurassic and younger igneous activity in the eastern United States and adjacent Canada.

berlites, syenites, peridotites, lamprophyres, and carbonatites) increases in age in the direction and at the approximate rate predicted by the three predicted hotspot traces (Fig. 8). The oldest intrusions date to ca. 115 Ma in eastern Kansas, and the youngest (ca. 65 Ma) occur in central Mississippi. Cox and Van Arsdale (1997, 2002) also attribute the coeval Middle Cretaceous uplift (~1–3 km) and subsequent erosion of the Mississippi Embayment to the "Bermuda hotspot." Nunn (1990) similarly attributed the lesser Sabine uplift, immediately southwest of the Mississippi Embayment uplift, to the Bermuda hotspot.

Noting the lack of evidence for "Bermuda hotspot" activity anywhere between the Bermuda Rise and Mississippi, Cox and Van Arsdale (2002) speculate that the Cretaceous igneous activity and uplift "may have been the result of increased hotspot flux of the typically weak Bermuda hotspot during the Cretaceous superplume mantle event (ca. 120–80 Ma)."

The youngest known igneous activity in the south central United States (central Mississippi) is centered ~32.25°N, 90.3°W (Fig. 8), ~2400 km west of the Bermuda volcanoes (32.3°N, 64.8°W). Even if the putative hotspot trace comprised a great circle, the distance covered in ~25 m.y. (between ca. 65 Ma and ca. 40 Ma) implies an average rate of motion of ~10 cm/yr for this point on the North America plate, a high rate similar to the late Tertiary migration rate of Hawaiian volcanism, which is generally attributed to rapid Pacific plate motion over a relatively fixed Hawaii hotspot. However, there is now compelling evidence that the ca. 81–47 Ma age progression along the Emperor seamounts, formed roughly (considering dating and other errors) over the same time as the 65–40 Ma "fast motion" Bermuda interval, can be explained only by rapid motion of the Hawaii "hotspot" (the locus of melt extraction) (Courtillot et al., 2003; Tarduno et al., 2003; Sharp and Clague, 2006). By contrast, the distance between the ca. 65 Ma Mississippi igneous rocks and the ca. 115 Ma eruptions to the northwest in Kansas (~40°N, 96°W, based on Fig. 1 *in* Cox and Van Arsdale, 2002) is 1000 km, implying an average migration rate of only 2 cm/yr.

The high rate of average North America plate motion from 65 Ma to 40 Ma required by the separation of 40 Ma Bermuda and the 65 Ma Mississippi igneous activity is inconsistent with predicted ages along the Late Cretaceous–Early Tertiary parts of predicted hotspot traces. For example, the Duncan (1984) model trace, which best fits the igneous events of Cox and Van Arsdale (2002), predicts that the Bermuda hotspot should have been located near 32°N, 76°W at 65 Ma. This is 600–700 km east of the central Mississippi igneous centers, a significant misfit even if the hotspot is assumed to be a broad area of upwelling.

A major difficulty with the previously described attempts to link widely separated igneous activity and uplifts to the same fixed hotspot or plume in the mantle is that tens of millions of years (~25 m.y. for Bermuda) have to pass between "pulses" of activity. That is, we are asked to imagine a "plume" resembling a blob in a lava lamp that only rarely produces a hot blob. In the case of the igneous activity and uplift in the Mississippi Embayment area, earlier and later igneous and tectonic events in the

same region must then be explained in other ways; for example, Triassic and Lower Jurassic igneous rocks are attributed to rifting in the Gulf of Mexico, and late Paleozoic intrusions are found around the north end of the Missisissippi Embayment (Cox and Van Arsdale, 2002). Meanwhile, the Kansas igneous rocks were intruded from a region of basement faults of the Proterozoic central North American rift system, while the Late Cretaceous intrusions could have followed reactivated faults of the Late Proterozoic–Early Paleozoic Mississippi graben. To account for such correlation with older underlying fault systems, hotspot proponents like Cox and Van Arsdale have had to postulate that igneous intrusions of hotspot origin can penetrate the crust only when and where a hotspot happens to underlie such structures. It might be simpler to attribute pulses of midplate tectonic and igneous events to episodes of increased stress associated with reorganizations of plate motion. In any case, no one attributes the historic and ongoing seismicity and tectonics in the New Madrid area to an underlying hotspot!

Finally, attempts to correlate the Mississippi-Kansas igneous belt (extending from the 65 Ma and 115 Ma points in Fig. 8) to a Bermuda hotspot trace must necessarily and somewhat arbitrarily disregard or attribute to another hotspot the ca. 49–35 Ma igneous activity in Virginia and West Virginia (Fig. 8; Fullagar and Bottino, 1969; Southworth et al., 1993; Tso and Surber, 2002), as well as the 100-km-wide, 400-km-long Late Cretaceous (87–77 Ma) Balcones igneous belt in southwestern Texas (Fig. 8; Griffin et al., 2005). In the western Balcones (Uvalde County), $^{40}Ar/^{39}Ar$ dating suggests that there were igneous pulses at 82–80 Ma and again at 74–72 Ma (Miggins et al., 2004). The northeast-southwest trend and lack of age progression in the Balcones belt are inconsistent with predicted North America mantle motion in the Late Cretaceous, and scarcely consistent with the Bermuda trace. If interpreted according to "mantle plume"–type models, both the Virginia and Balcones igneous episodes reflect isolated mantle blobs. Alternatively they reflect episodes of intraplate stress (e.g., Anderson, 2002).

BERMUDA RISE: QUANTITATIVE MODELS FOR ITS CURRENT STATE AND ORIGIN

The Bermuda Rise as a bathymetric feature was known for decades before any attempts were made to explain its origin. More or less ignoring or discounting the early twentieth-century works of Alfred Wegener (1966), most U.S. and Soviet marine geoscientists prior to the 1960s presumably considered the rise a pre-Cambrian relic of the early Earth. The first model put forth specifically for the Bermuda Rise was that of Engelen (1964), who proposed that the Wegener-type continental drift that had opened the north Atlantic in Mesozoic times had exposed the sub-Moho level of the American Shield. The pressure reduction in the lee of the drifting continent would have induced, by means of pressure reduction, a phase transition from eclogite to gabbro, thereby generating an "intumescence" and hence a rise.

After the "new global tectonics" had burst onto the scene in the later 1960s, Menard (1969) proposed the first modern explanation for "midplate rises," attributing them to small transient convection cells or other localized disturbances in the mantle, directly below the lithospheric plates. Models creating midplate rises by thermal expansion date from Crough (1978), although these models and the earlier modeling of the mid-oceanic ridge as an effect of thermal expansion (Langseth et al., 1966; Sclater et al., 1971) were prefigured by Alfred Wegener (1966) in the second (1929) edition of his book. Wegener noted that the average depth of ocean basins was less for those basins he had shown had opened more recently. In a rough calculation he showed that the depth differences of the observed magnitude could easily be explained by higher temperatures and greater thermal expansion of the younger mantle between continents that had separated more recently.

Aside from thicker sediment accumulations, which can be dismissed as a cause of the rise, there are basically only a few ways a seafloor swell like the Bermuda Rise (i.e., a basement swell) can have been generated (e.g., Fig. 1 *in* DeLaughter et al., 2005), and some of those can readily be discounted. Seafloor rises or swells can be expressions of (1) isostatically compensated thickened oceanic crust, i.e., crust similar to large igneous plateaus, aseismic ridges, or smaller volcanic masses, such as the Bermuda edifice; (2) flexural upwarping or forebulge construction due to adjacent loads, such as the volcanic edifices imposed nearby on the mechanical lithosphere or ice sheets; (3) reduced density distributed widely within the mantle lithosphere under the bulge (Figs. 9 and 10), by lower-density intrusions (A) and/or by higher temperatures (B, i.e., by thermal expansion) and/or by the mantle melting beneath the hotspot (C); (4) thinning of the lithosphere, e.g., via heating from anomalously hot upwelling asthenosphere or by delamination and sinking of some of the lower lithosphere, and its replacement by hotter material; or (5) dynamic support by upwelling mantle convection under the swell, either attached to the swell or, as in deep mantle plume models, decoupled from it.

Mechanism 1 cannot account for the Bermuda Rise, because seismic refraction data show the rise to be underlain by oceanic crust not significantly thicker than elsewhere (Purdy and Ewing, 1986). In fact, more recent seismic experiments show the crust east of the basement-topographic smooth-rough boundary (Fig. 4) to be ~1.5 km thinner than normal (Lizarralde et al., 2004). Moreover, modeling of the geoid anomaly over the rise (Figs. 1, 5, and 9) clearly indicates a Pratt, not an Airy, type of isostatic compensation (Haxby and Turcotte, 1978). Furthermore, known oceanic rises formed of thickened crust generally have well-marked topographic edges and are not broad and gently sloping.

Mechanism 2 was briefly considered by the senior author ca. 1970 (Johnson and Vogt, 1971), with the idea that the Bermuda Rise and the Appalachians could both be flexural upwarps caused by the thick sediment loads along the North American continental margin. However, subsequent studies of, e.g.,

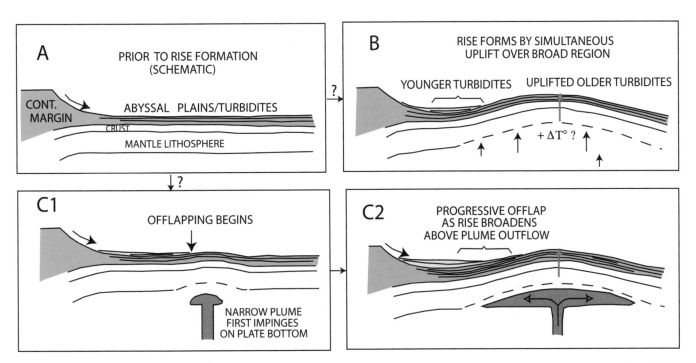

Figure 9. Schematic cross-sections showing the formation of Bermuda Rise (A) by simultaneous uplift over a broad region (B) or, alternatively, by an initially narrow plume (C1) that then spread out below the lithosphere (C2). These two uplift scenarios might be distinguished with a deep-sea drilling transect to recover the progressive offlap pattern of Eocene turbidites.

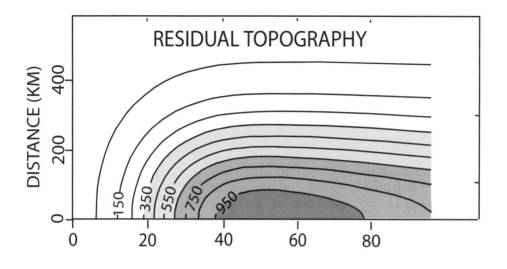

RESIDUAL TOPOGRAPHY

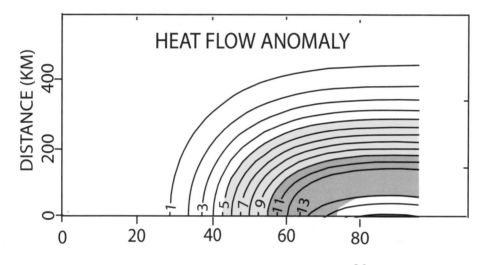

HEAT FLOW ANOMALY

Figure 10. Predicted temporal evolution in anomalous topography (top), heatflow (middle), and geoid (bottom) for a model lithosphere 90 km thick, moving at 15 mm/yr across an underlying Gaussian-shaped heat source, with first contact at t = 0. Dynamic uplift is assumed to be negligible. Adapted from Figure 9 *in* Liu and Chase (1989). Note that this "steady-state" model assumes that the moving plate first encountered and then was modified by a fixed source; actual evidence shows no sign of plate migration over a fixed source and no hotspot trace.

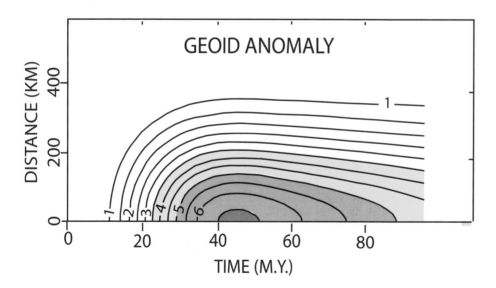

GEOID ANOMALY

seamounts and islands showed the flexural characteristics of the mechanical lithosphere to be incompatible with the large distances from the Bermuda Rise and the Appalachians to the sediment pile. A flexural or forebulge-type origin for the Bermuda Rise can safely be dismissed.

One variant of mechanism 3, widespread intrusions (Withjack, 1979), can perhaps also be excluded as the sole mode of rise formation, because, as also noted by Liu and Chase (1989), volcanic activity has occurred only in one relatively small area of the Bermuda Rise (the Bermuda volcanoes). However, Phipps Morgan et al. (1995) attributed much of swell elevations to lenslike, lower-density regions of plume-generated partial melting under the lithosphere. Inspired by the latter authors, Holm et al. (2006; their Fig. 20) suggested that melts generated at depth below the root of refractory melt under the Cape Verde Rise must have moved around it to reach the surface, accounting for the succession of different extruded compositions and the 200 km displacement of the Cape Verde volcanoes from the rise summit. The corresponding 75–100 km displacement of Bermuda from its rise summit (Fig. 5) might have had a similar origin—whether by this or some other mechanism.

Most of the papers modeling the Bermuda Rise either alone or along with other swells (rises) were published in the period 1978–1990, a "golden age" of rise modeling. The subsequent decline in swell-modeling papers can be ascribed in part to the lack of significantly improved or new data (i.e., diminishing marginal improvement, from the point of view of observables such as depth and geoid anomalies). Marine heatflow research efforts declined, particularly in the United States. Moreover, in more recent years the validity of shallow-penetration probe data for inferring deeper thermal structure has been challenged, as noted earlier (e.g., Ruppel, 2000; McNutt, 2002).

All papers attempting to model the Bermuda Rise in the "plate tectonic era" have adduced, modeled, or tested some combination of mechanisms 3, 4, and 5. The Bermuda Rise, like many other oceanic "midplate swells," represents departures (depths anomalously shallow) from the predicted increase of crustal depth with crustal age predicted by the cooling plate model (e.g., Sclater et al., 1971) and the thickening boundary layer (cooling half-space) model (e.g., Parker and Oldenburg, 1973). These departures from the predicted subsidence (Figs. 1 and 4–6; e.g., DeLaughter et al., 2005), typically begin at crustal ages of ca. 80 Ma, although the plate model fits the data from older seafloor better than does the cooling half-space model. This is well illustrated by the Bermuda Rise (see Fig. 2 *in* Sclater and Wixon, 1986), which is about twice as "anomalous" relative to the latter model. In reconciling the two models, Parsons and McKenzie (1978) distinguished between an upper cooler and therefore rigid "mechanical" lithosphere (boundary layer) and an underlying viscous region constituting a thermal boundary layer. In their model, both "layers" thicken as the square root of crustal age, but under old crust the thermal boundary layer becomes unstable and begins to convect, transporting heat to the base of the mechanical plate and thereby maintaining its thickness. The combined thickness of the two parts of a lithosphere thus defined is ~150–200 km. Numerical models of shallow convection under the plates (starting with McKenzie et al., 1974) also make predictions about the correlation between gravity and depth anomalies, first shown to be approximately satisfied for the Bermuda Rise by Sclater et al. (1975).

Taking a somewhat different tack, Crough (1978) led the efforts to explain this and other swells by thermal expansion within the plate and/or by plate thinning caused by plate motion over hotspots or thermal plumes below the plates (e.g., Crough, 1978, 1983b; Detrick and Crough, 1978). A simple plume model (e.g., Griffiths and Campbell, 1991; Campbell, 2006) explains primarily only the height and horizontal scale of the Bermuda Rise. The relation between geoid anomaly and depth anomaly over, e.g., Hawaii and Bermuda (Crough, 1978; Haxby and Turcotte, 1978) implied that these and other rises were supported by mass deficits at relatively shallow depths (40–70 km), above the thermal boundary layer of Parsons and McKenzie (1978). Crough (1978) and Detrick and Crough (1978) also examined the rates of crustal subsidence for Pacific swells and concluded that hotspots or plumes had "reset" the thermal age to that of 25 Ma crust; i.e., after passing over the hotspot and being reheated at depth, the plate would subsequently share the subsidence of "normal" 25 Ma seafloor. (As reviewed in a previous section, similar swell subsidence has not occurred at Bermuda.)

If plate reheating occurred only by thermal conduction, an unrealistically high temperature anomaly would be required at depth (Detrick and Crough, 1978). This problem was first addressed by Parsons and Daly (1983), whose numerical models of cellular convection suggested that seafloor topography and gravity or geoid anomalies largely reflect the temperature structure in the upper part of the thermal boundary layer, which would participate in convective overturn but then lose heat by conduction into the overlying rigid mechanical plate. The rapid observed subsidence of oceanic crust in Pacific swells could then be explained if, as reasonable, the convecting upper thermal boundary layer does not (or only incompletely) share the motion of the rigid plate overhead. In other words, the observed subsidence would include the effect of thermal contraction of material not moving with the Pacific plate. We note that a Parsons and Daly–type model might explain the difference between the fast subsidence of Pacific swells and the slow subsidence of the Bermuda Rise if we imagine that fast-moving plates shear off the underlying thermal boundary layer convection, while slow-moving plates like the North America plate carry their underlying convection cells with them. Several authors have previously noted that any mantle convection associated with and in some way responsible for the Bermuda Rise must be traveling with the North America plate (Sclater and Wixon, 1986; Vogt, 1991; King and Anderson, 1998).

Although they did no modeling of mantle convection, Sclater and Wixon (1986, p. 269) concluded that the Bermuda Rise could have formed "either by a cooling boundary layer of which the thermal structure is reset by a localized convective jet" (i.e., the type of model proposed by Crough, 1978) or "by a two-layer thermomechanical plate perturbed by upper mantle convection." Sclater and Wixon preferred the plate model because it better explains the residual gravity and depth anomalies, better explains the lack of observed subsidence on older crust, and also explains the lack of a heatflow anomaly. (At the time of their writing, no heatflow anomaly had been resolved on the Bermuda Rise.)

As elaborated earlier, Detrick et al. (1986) collected additional heatflow data that showed a 8–10 mW/m^2 heatflow high over the rise. They modeled these data and the previously known depth (800–1000 m) and geoid (6–8 m) anomalies and compensation depths of ~40–70 km with two-dimensional models containing a low-viscosity zone immediately below the mechanical plate. Cazenave et al. (1988) supported this model by calculating geoid to depth anomaly (admittance) as a function of crustal age. They suggested that swells like the Bermuda Rise, developed in older crust, are underlain by a thinner or more viscous low-viscosity zone. Following Parsons and Daly (1983), Detrick et al. (1986, p. 3701) concluded that the Bermuda Rise anomalies are "consistent with simple convection models in which the lower part of the thermally defined plate acts as the upper part of the thermal boundary layer of convection."

Louden et al. (1987) reduced the possible Bermuda Rise heatflow high to 5 mW/m^2. They attributed the lack of a larger anomaly, despite the large geoid and depth anomalies, to dynamic uplift of the Bermuda Rise by mantle convection and the lack of time for the thermal equilibration. However, they presented no convection models.

Robinson and Parsons (1988 and other papers) concluded from numerical modeling of a low-viscosity zone (asthenosphere) below the conducting lithosphere and a high-viscosity underlying mantle that convective instabilities with Bermuda Rise–scale wavelengths and reasonable heatflow anomalies arise spontaneously below old oceanic lithosphere. The magnitude and onset time of convection were found to be very sensitive to the viscosity contrast (0.40–0.01) between the low-viscosity layer and the overlying lid. In this type of model, the Bermuda Rise is supported by thermally induced buoyancy forces.

In another numerical modeling attempt, Liu and Chase (1989) compared Hawaii with Bermuda, attributing each to Gaussian-shaped thermal perturbations caused by a steady-state plume over which the plate moves at rates of 99 mm/yr (Hawaii; "strong plume") and 15 mm/yr (Bermuda; "weak plume"). Their models (Fig. 10) showed that thermal convection and convective thinning were required for Hawaii, while simple conduction worked for Bermuda. The model parameters assumed by Liu and Chase (1989) were anomalous heat flux at the base of lithosphere, 90–120 mW/m^2; excess plume temperature, 100–150°C; and radial velocity of the plume relative to background convection, 2–4. The predicted evolution of heatflow and topographic and geoid anomalies over the Bermuda Rise (Fig. 10) incorrectly assumes that the North America plate encountered

an existing, steady-state "Bermuda plume" but probably gives a good sense of what simple conductive models can achieve. The models would have to be refined to reproduce the longer than assumed history of rise uplift (Jaroslow and Tucholke, 1994) and the probably smaller than assumed heatflow anomaly (Louden et al., 1987). The same year, Sheehan and McNutt (1989) calculated an effective elastic thickness (30 ± 5 km) of the mechanical lithosphere supporting the Bermuda volcanoes (from the topography and gravity anomaly over the pedestal), as well as the Bermuda Rise compensation depth (55 ± 10 km, from the depth and geoid anomalies). Based in part on these results, they presented a new inversion technique (assuming instantaneous reheating from a point source) to put envelopes on the possible geotherms at the time of reheating (assumed to be 35 Ma) and at present. While their thermal expansion- or conduction-based models fit observations reasonably well without any dynamic uplift component, Sheehan and McNutt concluded (p. 390) that "a thermal convection mechanism is almost certainly required for thinning the lithosphere initially." As in other published models, the minimal or even nonexistent postvolcanic subsidence of Bermuda was only marginally predicted; these difficulties are compounded by the discovery that 400–500 m uplift occurred subsequent to 33 Ma (Jaroslow and Tucholke, 1994).

Assuming Morgan-type narrow plumes, Sleep (1990) analyzed the global hotspot data set—mainly swell heights and dimensions and geoid and heatflow anomalies—to constrain the uplift mechanisms. Using swell cross-sections and shapes, he estimated volume, heat, and buoyancy fluxes. His measured values for Bermuda are as follows: cross-sectional area of swell, 1300 km^2 (1000 km^2 when corrected for asthenospheric compensation); buoyancy flux, assuming 15 mm/yr plate velocity, 1.1 Mg/s (1.5 Mg/s according to Davies, 1988); volume flux (assuming a 225°C excess plume temperature), 48 m^3/s; and stagnation distance, 320 km. Sleep (his Fig. 11) fit (by eye) a "stagnation curve" of 500 km radius to the residual geoid anomaly over the eastern Bermuda Rise. This swell "snout" is interpreted as stagnation streamlines separating Bermuda plume outflow from the ambient asthenosphere below the western Atlantic. In recent years, plume temperature excesses such as 225°C have been challenged, and they are not supported by heatflow data. Furthermore, Sleep's model for the Bermuda swell as produced by a narrow mantle plume under the crest of the Bermuda Rise is inconsistent with the lack of evidence for migration during the last 30–45 m.y., unless the plume head is attached to and has moved with the North America plate.

Phipps Morgan et al. (1995) explored the possibility of a hot, partially molten "swell root" supplying the volcanoes with magma, with the residue spreading out under the plate to elevate the rise. They focused on Hawaii, where the fast-moving plate drags this residue downstream, thus making room for new plume materials to rise and experience decompression melting. Although they did not mention Bermuda, Phipps Morgan et al.

(1995) did examine model predictions for a plume underlying a stationary plate, using the Cape Verde rise and volcanoes as the type example. Given the slow (15 mm/yr) motion of the North America plate at Bermuda, the latter would be closer to the Cape Verde in terms of model predictions. The model of Phipps Morgan et al. (1995) would attribute the relatively meager (compared to Hawaii) Bermuda edifice volume (760 km^3, see earlier discussion), relative to the large swell, to the accumulation of residue below the rise and to the thick, old lithosphere. This accumulation, although spreading laterally to help support the rise, would tend to impede decompression melting by clogging the underside of the plate in the area above the plume (e.g., Fig. 20 *in* Holm et al., 2006).

The lack of a Bermuda hotspot trace, the elongated shape of the rise (discussed further later), and the lack of subsidence at Bermuda led to another class of explanations, which involved convection attached to the North America plate and moving with it (e.g., Sclater and Wixon, 1986; Vogt, 1991). In particular, the deep boundary between the "new" Atlantic oceanic lithosphere and the old North American continental lithosphere would form a major thermal boundary. King and Anderson (1998) analyzed this possibility with numerical convection models, showing that many oceanic basement swells located ~1000 km seaward of continental margins (such as the Bermuda and Cape Verde rises) could be caused by such "edge-driven convection." Such models better account for the northeast orientation of the Bermuda Rise, the development of the swell after the ocean had widened, and the lack of a hotspot trace. However, the apparent connection between Bermuda and the igneous activity in the Mississippi Embayment (Fig. 8; Cox and Van Arsdale, 2002) would then be fortuitous.

NORTHWEST ELONGATION OF BERMUDA RISE AND OTHER GEOGRAPHIC COINCIDENCES

If the northeast-southwest elongation of both the Bermuda volcanoes and the Bermuda Rise cannot be explained as a hotspot trace, how can they be explained? Clearly, the northeast orientation of the four volcanic edifices and even the Bermuda edifice itself suggests structural control—e.g., magma ascent along normal faults formed at the ancient axis of the Mid-Atlantic Ridge. The location of this volcanism at the magnetic lineation bend (from NNE south of Bermuda to northeast to the north; Fig. 1) also hints at structural control. Is there something special about the crustal or upper-mantle structure or composition below Bermuda, or does this igneous activity simply mark the culmination of the underlying mantle blob (source) or thermal anomaly, as might be expected from the near-correspondence of the volcanoes with the Bermuda Rise swell summit (Fig. 5)? Interestingly, the next largest volcano on the Bermuda Rise is the much older Muir seamount (discussed earlier), which is Cretaceous in age but rose from oceanic crust of the Bermuda Rise crest about the same age as that under Bermuda

(Fig. 3). Did this happen by chance, or does it imply some atypical "preconditioning" of the oceanic crust and upper mantle formed around M0 time?

Farther northeast, the same-aged crust that hosts Bermuda and Muir seamounts forms an anomalous, west-dipping escarpment under the Sohm Abyssal Plain (Fig. 4). This escarpment, somewhat diachronous (southwestward) toward younger crust, rises above the abyssal plain in the northeast, forming the J Anomaly Ridge (Fig. 4), a drillhole target at DSDP site 384 (Tucholke and Vogt, 1979). The northeastern, highest part of this basement feature is developed in crust of M2 to M4 age (ca. 125–129 Ma) and is associated with anomalously high magnetization—the J anomaly, a unique feature in the north Atlantic and perhaps globally. However, its timing corresponds to the 122–124 Ma magmatism "concurrent over a large area of the [North American] continent" (McHone, 1996, p. 328). Moreover, within dating errors, the Great Stone dome lamprophyric pluton (Fig. 8) 1100 km northwest of Bermuda was intruded at the same time (Crutcher, 1983) and on the same NNE-NE structural bend replicated by seafloor spreading from the 175 Ma breakup to the 123 Ma crust on which Bermuda is located.

At least 600 km long, the J Anomaly Ridge escarpment can be followed southwest in seismic reflection profiles at least to 38.5°N, 54.5°W, ~1200 km northeast of Bermuda, and involves oceanic crust the same age as that under Bermuda. Is this pure coincidence, or are the crust and upper mantle along this isochron anomalous (e.g., a zone of weakness), favoring igneous activity both at Muir and later at Bermuda? Indeed, a structural or compositional preconditioning would also be consistent with an early (but not demonstrated, given the isotope-dating problems for altered basalt) formation of a "proto"–Bermuda seamount erupting the tholeiite pillows (as originally proposed by Aumento et al., 1974) long before the Eocene uplift of the Bermuda Rise and the intrusion of the lamprophyre sheets. However, the Cretaceous-aged New England seamount chain does not appear to be anomalous where the chain crosses crust of slightly post-M0 age (Figs. 1 and 4).

Also problematic, but similarly hinting at "preconditioning," is the Bermuda Rise elongation more or less along crustal isochrons (Figs. 1 and 4). This correlation may of course be the result of simple chance, in which case the shape of the Bermuda Rise more or less represents the shape of the underlying mantle blob (source) or thermal anomaly. To our knowledge, no other swells developed in preexisting old oceanic crust have been observed to "follow" crustal structures. (As discussed earlier, the Bermuda Rise developed in crust ~78–88 m.y. old at the time of rise formation.) The plate boundary "imprinting" of structure would be expected to influence only the crust and uppermost mantle, with deeper mantle lithosphere, formed gradually off-axis, not so influenced. However, to the extent that mantle lithosphere forms rapidly off-axis, near-axis structural control could be imprinted at greater depth. To a first approximation, the Bermuda Rise happens to involve crust formed at very slow paleospreading rates, and with rough basement topography (Fig. 4). One possible interpretation (besides mere chance) is that "very slow" crust and mantle lithosphere are somehow more "vulnerable" to swell development and magma intrusion. Although little is known about the differences between mantle lithosphere formed at fast versus slow opening rates, Lizarralde et al. (2004) determined the seismic structure along a line across the southern Bermuda Rise. They found the rough (450–1200 m relief), very slow (8 mm/yr half-rate) crust that underlies most of the Bermuda Rise (Fig. 4) along their profile to be ~1.5 km thinner than the crust below the smoother (150–450 m relief), not-so-slow (13–14 mm/yr half-rate) oceanic crust in the western part of the profile. Perhaps significantly, they found the mantle lithosphere below the two regimes to differ as well, at least to the 30–40 km depth limit allowed by their experiment. (To prove beyond a doubt that this anomalous upper-mantle structure was created at the ancient spreading axis, not during Bermuda Rise formation, the seismic experiment would need to be repeated along a conjugate profile in the eastern Atlantic.) They attribute this mantle difference to less complete melt extraction under lower spreading rates at the ancient Mid-Atlantic Ridge axis. In any event, the results of Lizarralde et al. (2004) suggest that the mantle lithosphere is distinctive (retaining more gabbro), at least to 30–40 km depth below the "very rough" belt that more or less corresponds to the Bermuda Rise. How strength varies with depth depends on lithology, and we suggest (following Vink et al., 1984) that mantle lithosphere that retains more gabbroic components is somewhat weaker than a more purely peridotitic mantle and might be more vulnerable to fracturing. With more retained gabbro melts, such mantle would also be more vulnerable to the renewed partial melting associated with swell development. At least reduced strength in the uppermost mantle might help account for the greater seismicity of the Bermuda Rise (Fig. 4; Zoback et al., 1986).

Although the Cape Verde Rise is not elongated in a "structural" direction, it was developed within crust of largely similar age (ca. 155–105 Ma; Williams et al., 1990) as the Bermuda Rise and should, unless modified by rise development processes, have the same crustal and upper-mantle structure as determined by Lizarralde et al. (2004) for the Bermuda Rise. Barring coincidence, this adds credence to a compositional and/or structural "preconditioning" role for the lithosphere, and thereby undermines a simple mantle plume model for these features, although Montelli et al. (2004) adduce tomographic evidence for a Cape Verde plume rising from at least 1900 km depth.

If the apparent correspondence between crustal structures and the location and northeast-southeast elongation of the Bermuda Rise is fortuitous, there are two other possibilities: (1) the rise was formed and is being maintained by edge-driven convection (King and Anderson, 1998), as discussed later, or (2) remobilized, previously subducted lithosphere, including eclogite, provided an elongated mantle source, as proposed for Iceland by Foulger et al. (2005). However, in the latter case the materials (partially molten, less dense mantle?) derived from this source must have become incorporated in the North America plate ca. 40–50 Ma and are now traveling with it.

THE "BERMUDA EVENT": SYMPTOM OF A GLOBAL PLATE TECTONIC REORGANIZATION?

Was the Middle to Late Eocene or Early Oligocene Bermuda Rise formation and volcanic activity just an anomalous event in the upper mantle below the western north Atlantic? We suspect not, suggesting instead that the activity was part of a global plate reorganization episode. This is still speculative and difficult to prove on a planet with ongoing "noise" of igneous episodes and plate reorganizations.

Some degree of global synchronism would be expected if there were episodic reorganizations of mantle dynamics, whether by plumes (e.g., Vogt, 1972, 1975, 1979a) or simply as the effects of episodic and unavoidable plate kinematic or tectonic reorganizations (e.g., Anderson, 2002). For example, Rona and Richardson (1978; see also Patriat and Achache, 1984) listed many Eocene events they attributed to the collision of India with Eurasia, i.e., the closing of the Tethys Ocean.

Plate kinematic reorganizations would generally change intraplate stress fields as well as asthenosphere flow patterns, and perhaps account for the Bermuda event as originating in the lithosphere and the asthenosphere. Although the Bermuda Rise is characterized by somewhat increased seismicity (as discussed earlier) there have been no specific model predictions of how the Eocene plate reorganization could have increased stress or decreased plate strength in areas like the Bermuda Rise or Virginia. Moreover, as also discussed earlier, seismic tomography (Zhao, 2004; Ritsema, 2005) suggests that anomalously "slow" mantle extends to at least 400 km, possibly to 1000 km below the Bermuda Rise area, although an anomaly specifically associated with the mantle below the rise has not been resolved. Alternatively, the postulated relatively abrupt, widespread Eocene onset of gravitational instability as previously subducted lithospheric slabs temporarily stuck (flattened out) near the 660 km discontinuity began sinking into the lower mantle (Lithgow-Bertelloni and Richards, 1998; Fukao et al., 2001) would require upwelling in other areas for mass balance purposes. While this plausibly explains the timing of the Bermuda event, neither the specific location of the Bermuda Rise nor its apparent elongation along crustal isochrons is thereby specifically explained. Whatever the precise mechanism, a global Eocene reorganization of plate motions would be expected to affect plate boundaries and even plate interiors. Compilations of accurately dated tectonic or igneous events are only the first step: temporal correlations do not necessarily translate to causality.

We first ask if Bermuda Rise formation relates to motion of the plate in which the Bermuda Rise and its volcanoes were developed. As exemplified in Figure 6, the direction of spreading between the Africa and North America plate, as recorded by the Kane and other transform fracture traces in the central north Atlantic (see Vogt and Tucholke, 1989, for a more detailed map along the Kane fracture zone corridor), changed by ~30° to a more east-west direction about the time of magnetic anomaly 21, now dated at 46.5 Ma, essentially coeval with the Hawaii-Emperor bend (47 Ma, Tarduno et al., 2003; 50 Ma, Sharpe and Clague, 2006). Within error limits, this was also the time of Bermuda Rise and seamount formation. The coincidence in timing may be explained by a changing stress field acting on the plate, with the changed stress field responsible for opening the lithosphere to intrusions and upwelling (Bermuda Rise formation) and simultaneously forcing a change in plate motion. However, why did other bends not also cause midplate swells and volcanism?

Ages of more distant, possibly coeval "events" (Fig. 11) can be obtained from stratigraphic evidence, from direct radiometric dates, and from magnetic anomalies in the oceans. Comparison among different types of dates depends on a timescale relating stratigraphic time and magnetic polarity reversals to absolute (radiometric) time. In this article, we mostly use the timescale of Gradstein et al. (2004), but Figure 11 is based on the slightly different scale of Palmer and Geissman (1999). Earlier we reviewed the evidence dating the beginning of the Bermuda event (initiation of Bermuda Rise uplift) to some time or time interval within the Middle Eocene, which lasted from 48.6 to 37.2 Ma. The emergence of the Bermuda volcano above sea level (based on DSDP site 386) happened somewhat later, in the early to middle Late Eocene (i.e., ca. 37–35 Ma). However, because the volcanism must have begun before the first erosion products arrived at site 386, Bermuda Rise uplift and volcano birth may have begun at the same time in the Middle Eocene. The sheet intrusion event (33–34 Ma) was plausibly, like the Hawaiian posterosional volcanism, a local phenomenon deep below Bermuda; however, it might also have been triggered by a later global stress reorganization, distinct from the Middle Eocene event.

A limited compilation of more or less coeval events around the world suggests that Bermuda was indeed part of a larger "happening" (Fig. 11). Atlantic events included the extinction of the Mid-Labrador Sea spreading axis (also called Ran Ridge), at least by C13 time (ca. 33–34 Ma) according to Kristoffersen and Talwani (1977) but possibly as early as C18n (ca. 38–39.5 Ma); the very slow spreading prevents reliable identification of the corresponding magnetic lineations. In the Norway basin, a regionally unique linear ridge developed along the Aegir spreading axis at C18n time (Vogt and Jung, 2004), plausibly in response to the "annexation" of the Greenland plate by the North America plate.

On the neighboring continent of North America, felsitic and mafic magmas as well as diatremes—vestiges of the only Cenozoic igneous activity in the eastern United States—were emplaced in one small part of the Virginia and West Virginia Appalachians (Figs. 8 and 11). De Boer et al. (1988) call this the "Shenandoah Igneous Province," pointing out that the same area had been intruded by diverse rock types in the Jurassic and the Early Cretaceous, with upwelling mantle material structurally

localized at the intersection of a modern geothermal high and the major thirty-eighth parallel fracture zone (de Boer et al., 1988). Radiometric dating and paleomagnetic work indicate that the intrusions began ca. 49 Ma, with most of the activity occurring during a few million years, with the latest dates ca. 35 Ma (e.g., Fullagar and Bottino, 1969; Southworth et al., 1993; Tso and Surber, 2002). Nearly all the intrusives are reversely magnetized (Lovlie and Opdyke, 1974), suggesting that most may have been emplaced within less than 1 m.y. (de Boer et al., 1988). Within the various dating uncertainties, this unique igneous episode (Middle Eocene; 45 ± 5 m.y.; de Boer et al., 1988) was Bermuda's precise contemporary, albeit without the formation of a corresponding rise or swell.

Extensive deep coring and "backstripping" calculations in New Jersey and the adjacent continental shelf and margin 600–700 km northwest of Bermuda have revealed a local—probably largely eustatic—sea level history from the Early Cretaceous to the present (van Sickel et al., 2004). The results show that the greatest (100–150 m) pre-Quaternary sea level fluctuations in the last 100 m.y. (from a high ca. 54 Ma to a low at or below modern sea level ca. 46 Ma, and a new high ca. 43 Ma) occurred during the period of interest. Van Sickel et al. (2004) speculate that the Early Eocene high was related to the opening of the Greenland-Norwegian Sea, the succeeding low to the closing of Tethys, which increased ocean basin space by doubling the continental crustal thickness under Tibet.

The only spreading-type plate boundary between the North America and Caribbean plates—and the closest of any to Bermuda—is found in the Cayman Trough. East-west rifting, which continues today, began in the Late Eocene (Rosencrantz et al., 1988).

Across the Mid-Atlantic Ridge from Bermuda, Africa also experienced enhanced midplate igneous activity at many sites ca. 40 Ma (Bailey and Woolley, 2005). Repeated synchronous activity also occurred at many of the same sites ca. 20, 85, and 120 Ma, strongly supporting the concept of global episodes of increased midplate stress effecting tectonic-magmatic reactivation.

Although the more or less persistent volcano-tectonic activity in the North American Cordillera complicates resolution of unique igneous or tectonic events, Oldow et al. (1989, p. 219–220) state that "A dramatic transition in magmatic type occurred throughout the Cordillera about 40 to 42 Ma. This event coincided with a major (global?) change in plate interaction. Timing of this transition corresponds to the age of the bend in the Hawaii-Emperor seamount chain (42 Ma; Dalrymple and Clague, 1976) indicating a change in Pacific plate motion." Bally et al. (1989) identify five Mesozoic–Cenozoic tectonostratigraphic events in the geology of North America and relate them to global plate tectonic reorganizations. The last such event was the "plate reorganization that followed the collision of India with Eurasia in the early Eocene." This event is marked in North America by "the end of compression of the Cordilleran foreland belt, the inception of extensional tectonics in the Basin and Range Province, and of strike-slip tectonics in California,

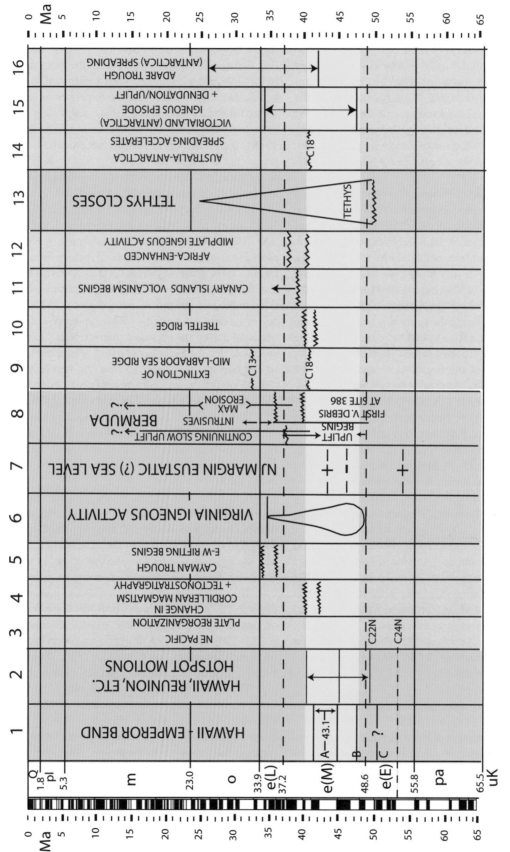

Figure 11. Time correlation diagram showing selected, more or less coeval tectonic or magmatic events at Bermuda and elsewhere in the world. Based on numerous sources, including Dalrymple and Clague (1976), Kristoffersen and Talwani (1977), Rona and Richardson (1978), Oldow et al. (1989), Winterer et al. (1989), Southworth et al. (1993), Palmer and Geissman (1999), Courtillot et al. (2003), Tarduno et al. (2003), Cande and Stock (2004), Gradstein et al. (2004), Van Sickel et al. (2004), Vogt and Jung (2004), Bailey and Woolley (2005), Rocchi et al. (2005), Sharp and Clague (2006), and various data relating to Bermuda, as discussed in this article.

as well as the establishment of major drainage patterns that are responsible for the accumulation of thick Tertiary clastic sequences of the Gulf Coast" (Bally et al., 1989, p. 7).

Winterer et al. (1989) put the Hawaii-Emperor bend age at 43.1 ± 1.4 Ma. Although estimates of bend age had changed little since the early 1970s, Sharp and Clague (2006), using just dates along the Emperor chain, recently suggested that the bend may actually have been initiated as long ago as 50 Ma, while recent drilling (ODP leg 197) indicates that 47 Ma is the most likely bend age (Tarduno et al., 2003). However, given the various uncertainties, the bend age may still correlate with the beginning of uplift of the Bermuda Rise sometime during the Middle Eocene (49–37 Ma). In any case, the change in Pacific plate motion, as shown by the bend, was probably not "instantaneous."

An abrupt acceleration of the motion of Australia away from Antarctica along the young Southeast Indian Ridge can be confidently dated to the beginning of C18n (ca. 38.0–39.5 Ma) (Cande and Stock, 2004). This was preceded according to the same authors by the start of the main episode of Cenozoic east-west Antarctic separation, the initiation of seafloor spreading in the Adare basin, and other plate rearrangements at about chron 20 time (42–45 Ma).

Cenozoic uplift and igneous activity in the west Antarctic rift area (Victoria Land and Ross Sea; Rocchi et al., 2005) bear a striking resemblance to what happened far away in Bermuda: A main denudation-uplift in the Middle Eocene was accompanied by igneous activity dating from 48 to 35 Ma, with a geographic shift in activity ca. 33 Ma. However, unlike at Bermuda, renewed igneous activity began in the Middle Miocene and has continued locally to the present. In contrast, spreading in the Dare Trough occurred from 43 Ma to 26 Ma, closely coinciding with the entire Bermudan episode—rise uplift and igneous activity.

Courtillot et al. (2003; their Fig. 2) compiled "time variations of four significant kinematic, geographic or dynamic indicators of hotspot motion" relating to Hawaii and Réunion hotspot traces and true polar wander. All four parameters show a steplike velocity change ca. 45 Ma (at least sometime between 50 and 40 Ma). Evidently a major reorganization of mantle motion occurred ca. 45 Ma, with the Réunion and Hawaii hotspots moving rapidly toward the equator prior to that time, but slowly if at all subsequently. The relatively sharp Hawaii-Emperor bend implies that this reorganization of mantle motion was a rather abrupt event.

Finally, Vogt (1975) speculated that if plumes originate near the core-mantle boundary, changes in plume convection might influence geomagnetic reversal frequency, and any time lag between frequency change and hotspot expressions at the Earth's surface might yield the rise speed of plume material. He noted an apparent correlation between the time of the Hawaii-Emperor Bend (then dated at 43 Ma) and an increase in average reversal frequency about that time. However, improvements in radiometric dating (Fig. 10) seem to rule out such a correlation: The Hawaii-Emperor Bend is more likely ca. 47 Ma in age (Tarduno, 2005), or even 50 Ma (Sharp and Clague, 2006), whereas reversal frequency appears to have increased ca. 30–31 Ma, with a prior short-lived interval of higher reversal frequency between ca. 39 and –35 Ma, i.e., of "Bermudan age," probably not statistically demonstrable. Moreover, plume modeling suggests a much longer rise time for new plumes (100 m.y.; Campbell, 2006). However, modeling of the geomagnetic dynamo (Glatzmaier et al., 1999) shows that different patterns of geographic variation in heat flux from the core across the core-mantle boundary can affect reversal frequency as well as dipole moment, secular variation, and the duration over which reversals occur. Whereas the correlation of the ca. 40 Ma igneous event to geomagnetic reversal frequency change is unconvincing, it is more compelling for the 80 and 125 Ma limits of the Cretaceous normal superchron (Vogt, 1975; Larson and Olson, 1991; Bailey and Woolley, 2005).

DISCUSSION AND CONCLUSION

What Produced Bermuda and the Bermuda Rise—Why Then and Why There?

As shown by the length of this article and the large number of references, Bermuda and its rise (swell) have been extensively studied, and practically every paper dealing with oceanic swells or rises and hotspots at least mentions Bermuda, and in many cases uses it—in contrast to Hawaii—as the "type example" of a weak hotspot under a slow-moving plate. (The fact that Bermuda has actually been quiescent for ~30 m.y. is generally ignored; better "type examples" would be the Cape Verde and Canary islands, volcanically active and rising from the nearly stationary Africa plate). On balance, almost all models involve some component of heating and thinning of the lower lithosphere, but, as typified by Sheehan and McNutt (1989), some contribution from dynamic uplift (ascending convection) cannot be excluded. The strongest proponents of a dynamic uplift model (Louden et al., 1987) base their case largely on the absence of a significant heatflow anomaly. Most thermal uplift models also have difficulties explaining the lack of subsidence since igneous activity ended, probably prior to 30 Ma, and most such papers predated Jaroslow and Tucholke (1994), who showed evidence for 400–500 m continued uplift after that time, exacerbating the problem. However, Bermuda is by no means unique in terms of lack of subsidence: in the eastern Atlantic volcanic archipelagos and their swells, not only has there been no subsidence, but old oceanic crust was elevated above sea level in the Cape Verde Islands, and submarine lavas associated with the island volcanism are now up to 400–500 m above sea level (e.g., Holm et al., 2006). The continued accumulation of melt residues below these swells (as in the model of Phipps Morgan et al., 1995) best accounts for this pervasive uplift or lack of subsidence where plates are nearly stationary. However, Bermuda and its swell must have been carried some 650 km west, away from any fixed mantle source, since 35 Ma, according to all the major absolute motion models (e.g., Duncan, 1984). A model like that of Phipps Morgan et al. (1995) might attribute this to a melt residue cushion that travels with the plate. To adapt a simple plume model to Bermuda would require a plume source that either traveled with the North America plate or was bent over by the lithosphere or by larger-scale mantle motions (e.g., Steinberger, 2000), perhaps finally breaking its connection in the Early Miocene, when uplift finally ended (according to Jaroslow and Tucholke, 1994).

An alternative, non-plume type explanation for Bermuda is edge-driven convection (King and Anderson, 1998), which accounts for several features not predicted by the various other models: (1) formation of Bermuda when the Atlantic basin was already wide, (2) elongation of the Bermuda Rise parallel to and ~1000 km southeast of the oceanic-continental crustal boundary, and (3) the lack of migration of the Bermuda Rise. Radially symmetric upwelling, no matter how wide the plume head (e.g., Griffiths and Campbell, 1991), would not be expected to produce swells elongated perpendicular to plate motion, as is the case for Bermuda—although we noted that the crust and lithosphere under most of the rise may be structurally or compositionally preconditioned (Fig. 4), having been formed at very slow spreading rates at the ancient ridge axis (Lizarralde et al., 2004). While edge-driven convection models do not explain the apparent ancient continuation (Fig. 8) of a Bermuda hotspot trace via Mississippi (65 Ma) to Kansas (115 Ma), these suggested traces seem highly suspect given the occurrence of igneous activity in Virginia and Texas and the implied bloblike behavior of a Bermuda plume.

Courtillot et al. (2003) place Bermuda in the category of "A-type plumes," i.e., not deep mantle plumes. This accounts for the poor "rating" Bermuda receives (Courtillot et al., 2003, and Anderson, 2005, both give Bermuda a zero) when compared with predictions of deep mantle plume models (some predictions, e.g., the He^3/He^4 ratios for its igneous rocks, have not yet been tested on Bermuda, however). While we concur with this assignment, we find it curious that the otherwise rather similar Cape Verde Rise gets a rather higher rating (a 2 from Courtillot et al., 2003, and a 1 from Anderson, 2005) than Bermuda. It would seem that the main difference between the two is age: the Cape Verde Archipelago and Rise began developing in the Early Miocene and are currently active, while Bermuda's activity ended at that time, and should not be tested against currently active hotspots. Another curious discrepancy involves calculated buoyancy flux (Sleep, 1990), which is only slightly higher for Iceland (1.2 Mg/s) than for Bermuda (1.1 Mg/s), when Iceland receives the highest plume score on the planet: a 4+ or 5 from Courtillot et al. (2003) and an 8? or 12 from Anderson (2005).

Why did the Bermuda swell and volcanism occur when they did? In this article (Fig. 11), we attribute the timing to a ca. 45 Ma mantlewide rearrangement of convection (e.g., Courtillot et al., 2003), perhaps associated with the closing of the Tethys (Rona and Richardson, 1978; Patriat and Achache, 1984) and/or the sinking into the lower mantle of slabs that had accumulated near the 660 km transition zone (Fukao et al., 2001). The loca-

tion of the Bermuda Rise may simply reflect a geochemically more fertile region of the upper mantle. With the asthenosphere close to its pressure melting point, even a modest rise, perhaps triggered by edge-type convection, would trigger substantial partial melting, causing more partial melting, plate thinning, and uplift. Alternatively, the region of partial melting may have been influenced by the preexisting composition and structure of the oceanic mantle lithosphere (Fig. 4).

The debate about the origin of so-called hotspots continues, and professional opinions have moved in recent years toward "plate or shallow origin" and away from "plume or deep origin" causality (e.g., Anderson, 2005). Nevertheless, some tomographic studies have adduced seismic speed anomalies as evidence for at least some deep mantle plumes (e.g., Montelli et al., 2004), results that have been challenged (e.g., van der Hilst and de Hoop, 2005). However, seismic tomography shows that the great majority of hotspots are located above two large regions (the equatorial Pacific and Africa) of slow S-wave (Ritsema, 2005) and slow P-wave (Dziewonski, 2005) in the lowermost mantle. As Dziewonski (2005) notes, there must be some communication between the upper and lower mantle systems. Most of North America and the western Atlantic is underlain by "fast" lowermost mantle, whereas the eastern Atlantic (the Canaries, Madeira, the Cape Verdes) is part of the African "slow" lower-mantle province. While some have attributed the greater abundance of volcanism in Africa and the eastern Atlantic to the more nearly stationary plate, this explanation does not account for a complete lack of midplate volcanism in very slow-moving northern parts of the North America plate, not to speak of the lack of it on many fast-moving plates. Perhaps the movement of the North America plate 650 km farther into the region underlain by high-seismic-wavespeed lower mantle in some way accounted for the shutting off of the anomalous mantle source under Bermuda.

The Two-Stage Igneous History Implied by Available Data

If the 33–34 Ma "Bermudite" (Aumento and Gunn, 2005) intrusive sheets—postdating the shield formation—can be shown to be, as we hypothesize, the Bermudan equivalent of Hawaiian posterosional basalts, Bermuda may affect explanations for the latter. In the Hawaiian case, volcanism continued for up to at least ~5 m.y. (Gurriet, 1987; Clague and Dalrymple, 1989), but with periods of quiescence of 1.0–2.5 m.y. between the early tholeiitic phase and the volumetrically trivial postshield ("posterosional") alkalic phase. The lag for Bermuda is clouded by the still uncertain age of the shield basalt flows; a long quiescent period is possible but not proven, because the continuous arrival of volcaniclastic turbidites at site 386 implies continuous erosion, not necessarily volcanism. If we date the shield formation at 40–45 Ma based on leg 386 volcaniclastic results (Tucholke and Vogt, 1979), the corresponding time from shield to sheets is ~6–12 m.y., within the "ca. 5 m.y." range of Gurriet (1987) and Phipps Morgan et al. (1995). Comparable or even greater (up to

65 m.y.) volcanic durations characterize eastern Atlantic volcanoes, e.g., in the Canary and Cape Verde archipelagos.

Clague and Dalrymple (1986) explained the late eruption of Hawaiian osional volcanics and the 1.0–2.5 m.y. quiescent period in terms of the time it takes for mature Hawaiian shields to be carried by plate motion over the arches associated with the youngest active shields. As the old shields passed over the swell, they would experience flexure and thus faulting, thus causing decompression melting at depth and providing pathways (along faults) for magma ascent. Gurriet (1987) modeled the conductive heating and partial melting of the lowermost lithosphere as it passes over a putative hotspot and attributed the ~5 m.y. spread in eruption ages (Clague and Dalrymple, 1989) to the time required to extract small percentages of partial melt. Gurriet's models could not account for a quiescent period, so he appealed to the swell-overriding idea of Clague and Dalrymple. Phipps Morgan et al. (1995) attributed the ~5 m.y. duration to "post-emplacement spreading and thinning of the swell root," and on the basis of this duration calculated a viscosity for their hypothesized depleted swell root of ~1–3 $\times 10^{20}$ Pa. In their view, melting is stopped below shields because vertical motion is impeded, while progressive, laterally upward migration of the spreading root materials causes decompression melting and volcanism at some distance along the existing chain, thus explaining the observed duration of activity along some chains. Ribe and Christensen (1999) were able to predict a quiescent period of ~4 m.y. between the main shield stage and the arrival of the edifices over a secondary melt zone.

However, we are suspicious of all the models described, because all assume movement across a melting anomaly and therefore do not explain an apparently similar (albeit longer) duration (and possible quiescent period) of volcanism on Bermuda, which is not a chain and sits on a swell that has apparently traveled with the plate.

Why Is There Enhanced Midplate Seismicity on the Bermuda Rise?

As first noticed by Zoback et al. (1986) and substantiated by earthquakes that occurred after they went to press (Fig. 4), the Bermuda Rise is a region of abnormally high (for a midplate oceanic region) seismicity. This is the only known way in which Bermuda and the Bermuda Rise are geophysically active today. The reason for this enhanced seismicity is elusive. In this article we have used the deep seismic reflection results of Lizarralde et al. (2004) to suggest that the upper mantle below the Bermuda Rise is relatively weaker, due to (extrapolating from Vink et al., 1984) greater gabbro retention under slow spreading rates (Klitgord and Schouten, 1986) at the ancient Mid-Atlantic Ridge. In this scenario, there is no need to postulate any swell-related stress enhancement under the Bermuda Rise.

Alternatively, the enhanced seismicity reflects unusual stress conditions associated with swells. Neglecting the various other stress sources likely to be affecting the western north Atlantic lithosphere (e.g., Zoback et al., 1986), Zhu and Wiens (1991) calculated thermoelastic stresses caused by putative hotspot heating. They found some agreement between their predicted deviatoric stress and the few fault-plane solutions and seismicity, both for the Bermuda and the Cape Verde rises. However, it seems unlikely that the Bermuda Rise, whose volcanism and most if not all uplift had ended prior to 30 Ma, would still exhibit the same thermoelastic stresses and midplate seismicity as the higher, younger (and volcanically active) Cape Verde Rise. Perhaps the Cape Verde Rise, also developed within "slow-spreading" crust about the same age as that of the Bermuda Rise, is therefore also weaker and more easily ruptured. However, this cannot be the only factor at work; otherwise much of the crust in the north Atlantic, formed at slow spreading rates, would exhibit higher degrees of midplate seismicity.

Future Research Approaches

How can future research discriminate among the wealth of models so far presented in the literature, and perhaps others still to be developed? First, earthquake-source seismic tomography with even better resolution than, e.g., that in Figure 19 *in* Zhao (2004) may delineate the "slow" region in the upper 400 km of the mantle below the Bermuda Rise. Below some such depth, the mantle must be decoupled from the motion of the North America plate. Explosion-source seismic experiments with ocean bottom seismometer arrays on the rise, expensive and perhaps not feasible due to concerns for marine mammals, would probably be needed to test for a buoyant refractory root (e.g., Phipps Morgan et al., 1995; Holm et al., 2006), which might extend from ~50 km (the depth of swell compensation deduced from geoid data) to 200 km below the rise (the depth of origin of the Bermuda sheets is ≥150 km; Olsen, 2005). Airgun-source multichannel profiling of the type conducted by Lizarralde et al. (2004) across the southern Bermuda Rise might detect cooled intrusions that reached the upper-mantle lithosphere, thus testing the model of Withjack (1979), but any reduced wavespeed anomalies would have to allow for melt retention under the slow spreading rates that formed most of the crust under the Bermuda Rise (Lizarralde et al., 2004). Spreading rate–dependent mantle velocities should change only gradually northward along isochrons, whereas any anomalies associated with the Bermuda Rise should reach *extrema* under the rise summit (e.g., the Xs in Fig. 5).

Additional drilling into the igneous basement of Bermuda and its three smaller satellites (Plantagenet or Argos and Challenger Banks and Bowditch seamount) is essential. Given the geologic complexity we know from volcanic islands (e.g., the Cape Verde Archipelago; Holm et al., 2006), it seems highly unlikely that even the Bermuda edifice was formed by a simple two-stage process of an Eocene tholeiitic shield, followed after ~6–12 m.y. quiescence by 33–34 Ma "Bermudite" (Aumento and Gunn, 2005) sheet intrusions. Hints of other igneous units in the edifice are provided by Foreman (1951), who wondered whether the rather abundant zircon and quartz he found in soils derived from the volcanics could have come from "late intrusions of a more acidic nature." Hints of explosive andesitic or basaltic eruptions are provided by altered tuff pebbles (Foreman, 1951) and small amounts of altered ash at DSDP site 386.

Further, we have assumed—with no direct evidence!—that the three satellite edifices are of the same age as Bermuda. They may have formed as major Pacific seamount rift zones (e.g., Vogt and Smoot, 1984) extending northeast and southwest from the main Bermuda edifice, or they may have erupted from separate conduits extending deep into the upper mantle. All this remains pure speculation until they are drilled, preferably to several km depth, and the core samples analyzed. A detailed magnetic and gravity survey of all four edifice summits and upper flanks should be conducted in advance of any drilling to help us map out the structural layout of intrusive sheets, lava accumulations, central conduits, and flanking volcaniclastic debris aprons.

Several deep holes similar to those drilled at DSDP sites 386 (Fig. 1) and 385 (the New England seamounts; Tucholke and Vogt, 1979) should also be placed around the bases of the four Bermuda edifices, as close as possible to the bases but still practically penetrating the volcaniclastic debris and flows (i.e., the "inner" seismic reflector A^v of Tucholke and Mountain, 1979), to allow us to recover and biostratigraphically date the youngest sediments overlain by the oldest Bermudan rocks. Recovery of larger, less altered rock fragments would also be more likely closer to the base of the volcanic edifices.

A transect of boreholes across and along the Bermuda Rise, just deep enough to sample the oldest hemipelagic sediment just above the Eocene biosiliceous turbidites, should enable us to resolve the detailed spatial-temporal pattern of Bermuda Rise uplift initiation. Such boreholes might also recover the time when bottom currents were first steered by the Bermuda Rise (e.g., Ayer and Laine, 1982). Several boreholes should be placed along the Kane fracture zone to allow us to calibrate the uplift history deduced by Jaroslow and Tucholke (1994) from seismic reflection mapping of more local turbidites. Abyssal plains, with gradients of 1:1000 or less (the present Bermuda Rise is surrounded on three sides by modern or at least Late Pleistocene abyssal plains; Pilkey and Cleary, 1986), should be extremely sensitive to small elevation changes. As depicted in Figure 9, the Middle Eocene turbidite offlap pattern—in time and space—would depend on the uplift mechanism. A plume model predicts rise uplift migrating radially outward from a region above the upwelling plume head, whereas a "distributed source" model with simultaneous partial melting and/or temperature rise would predict simultaneous uplift over the entire area of the present Bermuda Rise. A plume-type model (Griffiths and Campbell, 1991; Campbell, 2006) predicts a possible lag of ~2 m.y. between first uplift above the center of the plume head and uplift on the outer fringes of the expanding asthenosphere (I. Campbell, 2005, personal commun.); such a lag should be recorded by the offlapping turbidites and is probably measurable from bio-

stratigraphic dating of the first hemipelagics deposited on the last turbidites. The model of Campbell (2006) also predicts a lag of ~2–4 m.y. between uplift initiation and onset of volcanism. Current dating (Middle Eocene for onset of uplift; late Middle to early Late Eocene or earlier for the volcanism) make such a lag possible, but not proven. The plume model also predicts a possibly testable early central uplift several hundred meters higher in the beginning, before spreading below the plate flattened the head and reduced swell height in the next ~2 m.y. The "swell root spreading" models of Phipps Morgan et al. (1995; their Fig. 6) make even more specific and testable predictions about the rise uplift and swell radius as a function of time. Uplift resolution might be further refined by correlation of individual turbidites from one borehole to the next. Some of the thicker and compositionally distinctive Quaternary turbidites have been correlated from one core to another in modern abyssal plains surrounding Bermuda (Pilkey and Cleary, 1986). Readers interested in joining the authors in proposing the Bermuda Rise borehole transects we have discussed are welcome to join the authors and others in this effort.

ACKNOWLEDGMENTS

This review article began to unfold after the senior author agreed—on invitation by Gillian Foulger—to create a Bermuda Web page for the Web site www.MantlePlumes.org. We thank Marie-Claude Williamson and B.G. (Bernie) Gunn for useful e-mail exchanges about the age, geochemistry, and lithology of Bermuda borehole samples. We thank John DeLaughter, Carole Stein, and Seth Stein for sharing their GRDC2-based depth anomaly data set. Bob Detrick, Greg McHone, and Marie-Claude Williamson provided valuable reviews. Linda Norton of the Naval Research Laboratory tracked down many of the papers cited; Shaun Hardy of the Carnegie Institution found the H.W. Fisk papers in the Institution's archives.

REFERENCES CITED

Anderson, D.L., 2002, Plate tectonics as a far-from-equilibrium self-organized system, *in* Stein, S., and Freymuller, J., eds., Plate boundary zone: Washington, D.C., American Geophysical Union, Monograph Series 30, p. 411–425.
Anderson, D.L., 2005, Scoring hotspots: The plume and plate paradigms, *in* Foulger, G.R., et al., eds., Plates, plumes, and paradigms: Boulder, Colorado, Geological Society of America Special Paper 388, p. 31–54.
Anderson, D.L., and Schramm, K.A., 2005, Global hotspot maps, *in* Foulger, G.R., et al., eds., Plates, plumes, and paradigms: Boulder, Colorado, Geological Society of America Special Paper 388, p. 19–29.
Aumento, F., and Gunn, B.M., 1975, Geology of the Bermuda Seamount, *in* Geochemistry of igneous rocks. Available at www.geokem.com (see Oceanic Islands Basalts/Atlantic/Bermuda), 10 p., 2005.
Aumento, F., and Sullivan, K.D., 1974, Deep-drill investigations of the oceanic crust in the North Atlantic, *in* Kristjansson, L., ed., Geodynamics of Iceland and the North Atlantic Area: Dordrecht, Holland, D. Reidel, p. 83–103.
Aumento, F., Reynolds, P.H., and Gunn, B.M., 1974, The Bermuda Seamount, a reactivated section of an older oceanic crust [abs.]: Eos (Transactions, American Geophysical Union), v. 55, p. 455.

Aumento, F., Ade-Hall, J.M., and Keen, M.J., 1975, 1974—The year of the Mid-Atlantic Ridge, Reviews of Geophysics and Space Physics, v. 13, p. 53–66.

Ayer, E.A., and Laine, E.P., 1982, Seismic stratigraphy of the northern Bermuda Rise: Marine Geology, v. 49, p. 169–186, doi: 10.1016/0025-3227(82)90035-4.

Bailey, D.K., and Woolley, A.R., 2005, Repeated, synchronous magmatism within Africa: Timing, magnetic reversals, and global tectonics, *in* Foulger, G.R., et al., eds., Plates, plumes, and paradigms: Boulder, Colorado, Geological Society of America Special Paper 388, p. 365–377, doi: 10.1130/2005.2388(22).

Bally, A.T., Scotese, C.R., and Ross, M.I., 1989, North America: Plate-tectonic setting and tectonic elements, *in* Bally, A.W., and Palmer, A.R., eds., The geology of North America, v. A, An overview: Boulder, Colorado, Geological Society of America, p. 1–15.

Black, M., 1964, Cretaceous and Tertiary coccoliths from Atlantic seamounts: Paleontology, v. 7, p. 306–316.

Black, M.T., and McAdoo, D.C., 1988, Spectral analysis of marine geoid heights and ocean depths: Constraints on models of lithosphere and sublithospheric processes: Marine Geophysical Research, v. 10, p. 157–180, doi: 10.1007/BF00310062.

Campbell, I.H., 2006, Large igneous provinces and the mantle plume hypothesis: Elements, v. 1, p. 265–269.

Cande, S.C., and Stock, J.M., 2004, Cenozoic reconstructions of the Australia–New Zealand–South Pacific sector of Antarctica, *in* The Cenozoic southern ocean: Tectonics, sedimentation, and climate change between Australia and Antarctica: Washington, D.C., American Geophysical Union, Geophysical Monograph 151, p. 5–17.

Cazenave, A., Dominh, K., Rabinowicz, M., and Ceuleneer, G., 1988, Geoid and depth anomalies over ocean swells and troughs: Evidence of an increasing trend of the geoid to depth ratio with age of the plate: Journal of Geophysical Research, v. 93, p. 8064–8077.

Clague, D.A., and Dalrymple, G.B., 1986, The geology of the Hawaiian-Emperor volcanic chain, Part I, Geological Evolution: Reston, Virginia, U.S. Geological Survey Professional Paper 1350, p. 5–54.

Clague, D.A., and Dalrymple, G.B., 1989, Tectonics, geochronology, and origin of the Hawaiian-Emperor volcanic chain, *in* Winterer, E., Hussong, D.M., and Decker, R.W., eds., The geology of North America, v. N, The Eastern Pacific Ocean and Hawaii: Boulder, Colorado, Geological Society of America, p. 188–217.

Cole, J.F., 1908, Magnetic declination and latitude observations in the Bermudas: Terrestrial magnetism and atmospheric electricity, v. 13, p. 49–56.

Colin, P., and Fleitout, L., 1990, Topography of the ocean floor: Thermal evolution of the lithosphere and interaction of deep mantle heterogeneities with the lithosphere: Geophysical Research Letters, v. 17, p. 1961–1964.

Conrad, C.P., Lithgow-Bertellini, C., and Louden, K.E., 2004, Iceland, the Farallon slab, and dynamic topography of the North Atlantic: Geology, v. 32, p. 177–180, doi: 10.1130/G20137.1.

Courtillot, V., Davaille, A., Besse, J., and Stock, J., 2003, Three distinct types of hotspots in the Earth's mantle: Earth and Planetary Science Letters, v. 205, p. 295–308, doi: 10.1016/S0012-821X(02)01048-8.

Courtney, R.C., and White, R.S., 1986, Anomalous heat flow and geoid across the Cape Verde Rise: Evidence for dynamic support from a thermal plume in the mantle: Geophysical Journal of the Royal Astronomical Society, v. 87, p. 815–867.

Cox, R.T., and Van Arsdale, R.B., 1997, Hotspot origin of the Mississippi Embayment and its possible impact on contemporary seismicity: Engineering Geology, v. 46, p. 201–216, doi: 10.1016/S0013-7952(97)00003-3.

Cox, R.T., and Van Arsdale, R.B., 2002, The Mississippi Embayment, North America, a first order continental structure generated by the Cretaceous superplume mantle event: Journal of Geodynamics, v. 34, p. 163–176, doi: 10.1016/S0264-3707(02)00019-4.

Cox, W.M., 1959, Bermuda's beginning: Liverpool, England, C. Tisling and Co., 24 p.

Crough, S.T., 1978, Thermal origin of mid-plate hotspot swells: Geophysical Journal of the Royal Astronomical Society, v. 55, p. 451–470.

Crough, S.T., 1983a, The correction for sediment loading on the seafloor: Journal of Geophysical Research, v. 88, p. 6649–6454.

Crough, S.T., 1983b, Hotspot swells: Annual Review of Earth and Planetary Sciences, v. 11, p. 165–193, doi: 10.1146/annurev.ea.11.050183.001121.

Crough, S.T., Morgan, W.J., and Hargraves, B., 1980, Kimberlites: Their relation to mantle hotspots: Earth and Planetary Science Letters, v. 50, p. 260–274, doi: 10.1016/0012-821X(80)90137-5.

Crutcher, T.D., 1983, Baltimore Canyon trough: Seismic expressions of structural styles, *in* Bally, A.W., ed., Studies in geology 15: American Association of Petroleum Geology, p. 20–26.

Dalrymple, G.B., and Clague, D.A., 1976, Age of the Hawaiian-Emperor Bend: Earth and Planetary Science Letters, v. 31, p. 313–329, doi: 10.1016/0012-821X(76)90113-8.

Davies, G.F., 1988, Ocean bathymetry and mantle convection, 1, Large-scale flow and hotspots: Journal of Geophysical Research, v. 93, p. 10,467–10,480.

Davis, E.E., Lister, C.R.B., and Sclater, J.B., 1984, Towards determining the thermal state of old ocean lithosphere: Heat flow measurements from the Blake-Bahama outer ridge; NW Atlantic: Geophysical Journal of the Royal Astronomical Society, v. 78, p. 507–545.

De Boer, J.Z., McHone, J.G., Puffer, J.H., Ragland, P.C., and Whittington, D., 1988, Mesozoic and Cenozoic volcanism, *in* Sheridan, R.E., and Grow, J.A., eds., The geology of North America, v. I-2, The Atlantic continental margin: Boulder, Colorado, Geological Society of America, p. 217–241.

DeLaughter, J.E., Stein, C.A., and Stein, S., 2005, Hotspots: A view from the swells, *in* Foulger, G.R., et al., eds., Plates, plumes, and paradigms: Boulder, Colorado, Geological Society of America Special Paper 388, p. 257–278, doi:10.1130/2005.2388(16).

Detrick, R.S., and Crough, S.T., 1978, Island subsidence, hot spots, and lithospheric thinning: Journal of Geophysical Research, v. 83, p. 1236–1244.

Detrick, R.S., Von Herzen, P.R., Parsons, B., Sandwell, D., and Dougherty, M., 1986, Heat flow observations on the Bermuda Rise: Journal of Geophysical Research, v. 91, p. 3701–3723.

Driscoll, N.W., and Laine, E.P., 1996, Abyssal current influence on the southwest Bermuda Rise and surrounding region: Marine Geology, v. 130, p. 231–263, doi: 10.1016/0025-3227(95)00133-6.

Duffield, W.A., 1978, Vesicularity of basalt erupted at Reykjanes Ridge crest: Nature, v. 274, p. 217–220, doi: 10.1038/274217a0.

Duncan, R.A., 1984, Age progressive volcanism in the New England Seamounts and the opening of the central Atlantic ocean: Journal of Geophysical Research, v. 89, p. 9980–9990.

Dziewonski, A.M., 2005, The robust aspects of global seismic tomography, *in* Foulger, G.R., et al., eds., Plates, plumes, and paradigms: Boulder, Colorado, Geological Society of America Special Paper 388, p. 147–154.

Embley, R.W., Hobart, M.A., Anderson, R.N., and Abbott, D., 1983, Anomalous heat flow in the northwest Atlantic: A case for continued hydrothermal circulation in 80 m.y. crust: Journal of Geophysical Research, v. 88, p. 1067–1074.

Emery, K.O., 1955, Transportation of rocks by driftwood: Journal of Sedimentary Petrology, v. 25, p. 51–57.

Engelen, G.B., 1964, A hypothesis on the origin of the Bermuda Rise: Tectonophysics, v. 1, p. 85–93, doi: 10.1016/0040-1951(64)90030-7.

Epp, D., and Smoot, N.C., 1989, Distribution of seamounts in the North Atlantic: Nature, v. 337, p. 254–257, doi: 10.1038/337254a0.

Fisk, H.W., 1923, The Bermuda anomaly: Eos (Transactions, American Geophysical Union), v. 4, p. 118–120.

Fisk, H.W., 1927, Special field report, magnetic results, 1921–1926, *in* Land Magnetic and Other Observations, 1918–1926: Researches of the Department of Terrestrial Magnetism, v. 6, p. 212–224.

Foreman, F., 1951, Study of some Bermuda rock: Geological Society of America Bulletin, v. 62, p. 1297–1330, doi: 10.1130/0016-7606(1951)62[1297:SOSBR]2.0.CO;2.

Foulger, G.R., Natland, J.H., and Anderson, D.L., 2005, Genesis of the Iceland melt anomaly by plate tectonic processes, *in* Foulger, G.R., et al., eds., Plates, plumes, and paradigms: Boulder, Colorado, Geological Society of America Special Paper 388, p. 595–626.

Fukao, Y., Widiyantoro, S., and Obayashi, M., 2001, Stagnant slabs in the upper and lower mantle transition region: Reviews of Geophysics, v. 39, p. 291–323, doi: 10.1029/1999RG000068.

Fullagar, P.D., and Bottino, M.L., 1969, Tertiary felsite intrusions in the Valley and Ridge province, Virginia: Geological Society of America Bulletin, v. 80, p. 1853–1858, doi: 10.1130/0016-7606(1969)80[1853:TFIITV]2.0.CO;2.

Gees, R.A., 1969, The age of Bermuda Seamount: Maritime Sediments, v. 5, p. 56–57.

Gees, R.A., and Medioli, F., 1970, A continuous seismic survey of the Bermuda Platform: Part I: Castle Harbor: Maritime Sediments, v. 6, p. 21–25.

Glatzmaier, G.A., Coe, R.S., Hongre, L., and Roberts, P.H., 1999, The role of the Earth's mantle in controlling the frequency of geomagnetic reversals: Nature, v. 401, p. 885–890, doi: 10.1038/44776.

Gradstein, F., Ogg, J., and Smith, A., 2004, A geologic time scale: Cambridge, England, Cambridge University Press, 589 p.

Griffin, W.R., Bergman, C., and Leybourne, M.I., Testing magmatic emplacement mechanisms in the Balcones Igneous Province of Texas [Abs.]: Washington, D.C., American Geophysical Union, Chapman Conference Proceedings, p. 78–79.

Griffiths, R.W., and Campbell, I., 1991, Interaction of mantle plume heads with the Earth's surface and onset of small-scale convection: Journal of Geophysical Research, v. 96, p. 18,295–18,310.

Gurriet, P., 1987, A thermal model for the origin of post-erosional alkalic lava, Hawaii: Earth and Planetary Science Letters, v. 82, p. 153–158, doi: 10.1016/0012-821X(87)90115-4.

Haq, B.U., Hardenbol, J., and Vail, P.R., 1987, Chronology of fluctuating sea levels since the Triassic: Science, v. 235, p. 1156–1167, doi: 10.1126/science.235.4793.1156.

Harmon, R.S., Schwartz, H.P., and Ford, D.C., 1978, Late Pleistocene sea level history of Bermuda: Quaternary Research, v. 9, p. 205–218, doi: 10.1016/0033-5894(78)90068-6.

Harris, R.N., Von Herzen, R.P., McNutt, M.K., Garven, G., and Jordahl, K., 2000, Submarine hydrology of the Hawaiian archipelagic apron, Part 1, Heat flow patterns north of Oahu and Maro Reef: Journal of Geophysical Research, v. 105, p. 21,353–21,370, doi: 10.1029/2000JB900165.

Haxby, W.F., and Turcotte, D.L., 1978, On isostatic geoid anomalies: Journal of Geophysical Research, v. 83, p. 5473–5478.

Hearty, P.J., Kindler, P., Cheng, H., and Edwards, R.L., 1999, A +20m middle Pleistocene sea-level highstand (Bermuda and the Bahamas) due to partial collapse of Antarctic ice: Geology, v. 27, p. 375–378, doi: 10.1130/0091-7613(1999)027<0375:AMMPSL>2.3.CO;2.

Holm, P.M., Wilson, J.R., Christensen, B.P., Hansen, L., Hansen, S.L., Hein, K.M., Mortensen, A.K., Pedersen, R., Plesner, S., and Runge, M.K., 2006, Sampling the Cape Verde mantle plume: Evolution of melt compositions on Santa Antao, Cape Verde Islands: Journal of Petrology, v. 47, p. 145–189, doi: 10.1093/petrology/egi071.

Hyndman, R.D., Muecke, G.K., and Aumento, F., 1974, Deep-Drill-1972: Heat flow and heat production in Bermuda: Canadian Journal of Earth Sciences, v. 11, p. 809–818.

Hyndman, R.D., Christensen, N.I., and Drury, M.J., 1979, Seismic velocities, densities, electrical resistivities, porosities and thermal conductivities of core samples from boreholes into the islands of Bermuda and the Azores, *in* Talwani, M., Harrison, C.G., and Hayes, D.E., eds., Deep drilling results in the Atlantic Ocean: Ocean Crust: Washington, D.C., American Geophysical Union, Maurice Ewing Series 2, p. 94–112.

Jaroslow, G.E., and Tucholke, B.E., 1994, Mesozoic–Cenozoic sedimentation in the Kane Fracture Zone, western North Atlantic, and uplift history of the Bermuda Rise: Geological Society of America Bulletin, v. 106, p. 319–337, doi: 10.1130/0016-7606(1994)106<0319:MCSITK>2.3.CO;2.

Johnson, G.L., and Vogt, P.R., 1971, Morphology of the Bermuda Rise: Deep Sea Research, v. 18, p. 605–617.

Kido, M., and Seno, T., 1994, Dynamic topography compared with residual depth anomalies in oceans and implications for age-depth curves: Geophysical Research Letters, v. 21, p. 717–720, doi: 10.1029/94GL00305.

King, S.D., and Anderson, D.L., 1998, Edge-driven convection: Earth and Planetary Science Letters, v. 160, p. 289–296, doi: 10.1016/S0012-821X(98)00089-2.

Klitgord, K.D., and Schouten, H., 1986, Plate kinematics of the central Atlantic, *in* Vogt, P.R., and Tucholke, B.E., eds., The geology of North America, v. M, The Western Atlantic region: Boulder, Colorado, Geological Society of America, p. 351–378.

Kristoffersen, Y., and Talwani, M., 1977, Extinct triple junction south of Greenland and the Tertiary motion of Greenland relative to North America: Geological Society of America Bulletin, v. 88, p. 1037–1049, doi: 10.1130/0016-7606(1977)88<1037:ETJSOG>2.0.CO;2.

Laine, E.P., Gardner, W.D., Richardson, M.J., and Kominz, M., 1994, Abyssal currents and resuspended sediment along the northeastern Bermuda Rise: Marine Geology, v. 119, p. 159–171, doi: 10.1016/0025-3227(94)90146-5.

Langseth, M.G., 1969, The flow of heat from the earth and its global distribution at the surface: American Institute of Aeronautics and Astronautics, 4th Thermophysics Conference, p. 1–10.

Langseth, M.G., LePichon, X., and Ewing, M., 1966, Crustal structure of the Mid-Ocean Ridges, 5, Heat flow through the Atlantic Ocean floor and convection currents: Journal of Geophysical Research, v. 71, p. 5321–5355.

Larson, R.L., and Olson, P., 1991, Mantle plumes control magnetic reversal frequency: Earth and Planetary Science Letters, v. 107, p. 437–447, doi: 10.1016/0012-821X(91)90091-U.

Lithgow-Bertelloni, C., and Richards, M.A., 1998, The dynamics of Cenozoic and Mesozoic plate motions: Reviews of Geophysics, v. 36, p. 27–78, doi: 10.1029/97RG02282.

Liu, M., and Chase, C.G., 1989, Evolution of midplate hotspot swells: Numerical solutions: Journal of Geophysical Research, v. 94, p. 5571–5584.

Livingston, W.L., 1944, Observations on the structure of Bermuda: Geographical Journal, v. 104, p. 40–48, doi: 10.2307/1790028.

Lizarralde, D., Gaherty, J.B., Collins, J.A., Hirth, G., and Kim, S., 2004, Spreading-rate dependence of melt extraction at mid-ocean ridges from mantle seismic refraction data: Nature, v. 432, p. 744–747, doi: 10.1038/nature03140.

Logan, A., 1988, Holocene reefs of Bermuda: Sedimenta, v. 11, p. 1–62.

Louden, K.E., Wallace, D.O., and Courtney, R.C., 1987, Heat flow vs. age for the Mesozoic Northwest Atlantic Ocean: Results from the Sohm abyssal plain and implications for the Bermuda Rise: Earth and Planetary Science Letters, v. 83, p. 109–122, doi: 10.1016/0012-821X(87)90055-0.

Lovlie, R., and Opdyke, N.D., 1974, Rock magnetism and paleomagnetism of some intrusions in Virginia: Journal of Geophysical Research, v. 70, p. 343–349.

Ludwig, K.R., Muhs, D.R., Simmons, K.R., Halley, R.B., and Shinn, E.A., 1996, Sea level records at ca. 80 ka from tectonically stable platforms: Florida and Bermuda: Geology, v. 24, p. 211–214, doi: 10.1130/0091-7613(1996)024<0211:SLRAKF>2.3.CO;2.

McHone, J.G., 1996, Constraints on the mantle plume model for Mesozoic alkaline intrusions in northeastern North America: Canadian Mineralogist, v. 34, p. 325–334.

McKenzie, D.P., Roberts, J.M., and Weiss, N.O., 1974, Towards a numerical simulation: Journal of Fluid Mechanics, v. 62, p. 465–538, doi: 10.1017/S0022112074000784.

McNutt, M.K., 2002, Heat-flow variations over the Hawaiian Swell controlled by near-surface processes, not plume properties, *in* Hawaiian volcanoes: Deep underwater perspectives: Washington, D.C., American Geophysical Union, Geophysical Monograph 128, p. 355–364.

Menard, H.W., 1969, Elevation and subsidence of oceanic crust: Earth Planetary Sciences, v. 6, p. 275–284, doi: 10.1016/0012-821X(69)90168-X.

Miggins, D.P., Blome, C.D., and Smith, D.V., 2004, Preliminary $^{40}Ar/^{39}Ar$

geochronology of igneous intrusions from Uvalde County, Texas: Defining a more precise eruption history for the southern Balcones Volcanic Province: Reston, Virginia, United States Geological Survey Open-File Report 2004–1031, p. 1–31.

Montelli, R., Nolet, G., Dahlen, F.A., Masters, G., Engdahl, E.R., and Hung, S.-H., 2004, Finite-frequency tomography reveals a variety of plumes in the mantle: Science, v. 303, p. 338–343, doi: 10.1126/science.1092485.

Moore, H.B., 1942, The geologic history of the Bermudas: Science, v. 95, p. 551–552, doi: 10.1126/science.95.2474.551.

Moore, H.B., and Moore, D.M., 1946, Preglacial history of Bermuda: Bulletin of the Geological Society of America, v. 57, p. 207–222, doi: 10.1130/0016-7606(1946)57[207:PHOB]2.0.CO;2.

Moore, J.G., 1970, Water content of basalt erupted on the ocean floor: Contributions to Mineralogy and Petrology, v. 28, p. 272–279, doi: 10.1007/BF00388949.

Moore, J.G., and Schilling, J.-G., 1973, Vesicles, water, and sulfur in Reykjanes Ridge basalts: Contributions to Mineralogy and Petrology, v. 41, p. 105–118, doi: 10.1007/BF00375036.

Morgan, W.J., 1972, Plate motions and deep mantle convection: Boulder, Colorado, Geological Society of America Memoir 132, p. 7–22.

Morgan, W.J., 1978, Rodriguez, Darwin, Amsterdam . . . A second type of hotspot island: Journal of Geophysical Research, v. 83, p. 5355–5360.

Morgan, W.J., 1983, Hotspot tracks and the early rifting of the Atlantic: Tectonophysics, v. 94, p. 123–139, doi: 10.1016/0040-1951(83)90013-6.

Morgan, W.J., and Crough, S.T., 1979, Bermuda hotspot and the Cape Fear Arch: Eos (Transactions, American Geophysical Union), v. 60, p. 392–393.

Müller, R.D., Royer, J.-Y., and Lawver, L.A., 1993, Revised plate motions relative to hotspots from combined Atlantic and Indian Ocean hotspot tracks: Geology, v. 21, p. 275–278, doi: 10.1130/0091-7613(1993)021<0275:RPMRTT>2.3.CO;2.

Müller, R.D., Roest, W.R., Royer, Y.-J., Galagher, L.M., and Sclater, J.G., 1997, Digital isochrons of the world's ocean floor, Journal of Geophysical Research B, v. 102, p. 3211–3214, doi: 10.1029/96JB01781.

Nagahira, S., Lister, C.R.B., and Sclater, J.G., 1996, Reheating of old oceanic lithosphere: Deductions from observations: Earth and Planetary Science Letters, v. 139, p. 91–104, doi: 10.1016/0012-821X(96)00010-6.

Nataf, H.-C., 2000, Seismic imaging of mantle plumes: Annual Review of Planetary Science, v. 28, p. 391–417, doi: 10.1146/annurev.earth.28.1.391.

Nishenko, S.P., and Kafka, A.L., 1982, Earthquake focal mechanisms and the intraplate setting of the Bermuda Rise: Journal of Geophysical Research, v. 87, p. 3929–3941.

Nunn, J.A., 1990, Relaxation of continental lithosphere: An explanation for Late Cretaceous reactivation of the Sabine uplift of Louisiana-Texas: Tectonics, v. 9, p. 341–359.

Officer, C.B., Ewing, M., and Wuenschel, P.C., 1952, Seismic refraction measurements in the Atlantic Ocean, Part IV, Bermuda, the Bermuda Rise, and Nares Basin: Bulletin of the Geological Society of America, v. 63, p. 777–808.

Oldow, J.S., Bally, A.W., Ave Lallemant, H.G., and Leeman, W.P., 1989, Phanerozoic evolution of the North American Cordillera, in Bally, A.W., and Palmer, A.R., eds., The geology of North America, v. A, An overview: Boulder, Colorado, Geological Society of America, p. 139–232.

Olsen, S.D., 2005, Petrogenesis of Bermuda's igneous basement [M.S. thesis]: University of Copenhagen, 254 p.

Palmer, A.R., and Geissman, J., compilers, 1999, Geologic time scale: Boulder, Colorado, Geological Society of America.

Parker, R.L., and Oldenburg, D.W., 1973, Thermal model of ocean ridges: Nature, v. 242, p. 137–139.

Parsons, B., and Daly, S., 1983, The relationship between surface topography, gravity anomalies, and temperature structure of convection: Journal of Geophysical Research, v. 88, p. 1129–1144.

Parsons, B., and McKenzie, D.P., 1978, Mantle convection and thermal structure of the plates: Journal of Geophysical Research, v. 83, p. 4485–4496.

Parsons, B., and Sclater, J.G., 1977, An analysis of the variation of ocean floor bathymetry and heat flow with age: Journal of Geophysical Research, v. 82, p. 803–823.

Patriat, P., and Achache, J., 1984, India-Asia collision chronology has implications for crustal shortening and driving mechanism of plates: Nature, v. 311, p. 615–621, doi: 10.1038/311615a0.

Peckenham, J.M., 1981, On the nature and origin of some Paleogene melilitic pillowed lavas, breccias, and intrusives from Bermuda [M.Sc. thesis]: Halifax, Nova Scotia, Dalhousie University, 306 p.

Peckenham, J.M., Ryall, P.J.C., and Schenk, P.E., 1982, The volcanic evolution of Bermuda determined from deep-drilling and submersible observations [abs.]: Eos (Transactions, American Geophysical Union), v. 63, p. 473.

Phipps Morgan, J., Morgan, W.J., and Price, E., 1995, Hotspot melting generates both hotspot volcanism and a hotspot swell?: Journal of Geophysical Research, v. 100, p. 8045–8062, doi: 10.1029/94JB02887.

Pilkey, O.H., and Cleary, W.J., 1986, Turbidite sedimentation in the northwestern Atlantic Ocean basin, in Vogt, P.R., and Tucholke, B.E., eds., The geology of North America, v. M, The Western North Atlantic region: Boulder, Colorado, Geological Society of America, Boulder, 437–450.

Pirsson, L.V., 1914a, Geology of Bermuda Island: The igneous platform: American Journal of Science, v. 38, p. 189–206.

Pirsson, L.V., 1914b, Geology of Bermuda: Petrology of the lavas: American Journal of Science, v. 38, p. 331–344.

Pirsson, L.V., and Vaughan, T.W., 1913, A deep boring in Bermuda Island: American Journal of Science, v. 36, p. 70–71.

Poag, C.W., and Sevon, W.D., 1989, A record of Appalachian denudation in postrift Mesozoic and Cenozoic sedimentary deposits of the U.S. Middle Atlantic continental margin: Geomorphology, v. 2, p. 119–157, doi: 10.1016/0169-555X(89)90009-3.

Purdy, G.M., and Ewing, J., 1986, Seismic structure of the ocean crust, in Vogt, P.R., and Tucholke, B.E., eds., The geology of North America, v. M, The Western North Atlantic region: Boulder, Colorado, Geological Society of America, p. 313–330.

Rabinowitz, P.D., and Jung, W.-Y., Gravity anomalies in the western North Atlantic Ocean, in Vogt, P.R., and Tucholke, B.E., eds., The geology of North America, v. M, The Western North Atlantic region: Boulder, Colorado, Geological Society of America, p. 205–213.

Reed, J.C., Jr., Wheeler, J.O., and Tucholke, B.E., compilers, 2005, Geologic map of North America (chart): Boulder, Colorado, Geological Society of America, scale 1:5,000,000.

Reynolds, P.R., and Aumento, F.A., 1974, Deep Drill 1972: Potassium-argon dating of the Bermuda drill core: Canadian Journal of Earth Sciences, v. 11, p. 1269–1273.

Ribe, N.M., and Christensen, U.R., 1999, The dynamical origin of Hawaiian volcanism: Earth and Planetary Science Letters, v. 171, p. 517–531, doi: 10.1016/S0012-821X(99)00179-X.

Rice, P.D., Hall, J.M., and Opdyke, N.D., 1980, Deep Drill 1972: A paleomagnetic study of the Bermuda Seamount: Canadian Journal of Earth Sciences, v. 17, p. 232–243.

Ritsema, J., 2005, Global seismic structure maps, in Foulger, G.R., et al., eds., Plates, plumes, and paradigms: Boulder, Colorado, Geological Society of America Special Paper 388, p. 11–18.

Robinson, E.M., and Parsons, B., 1988, Effect of a shallow low-viscosity zone on the formation of mid-plate swells: Journal of Geophysical Research, v. 93, p. 3144–3156.

Rocchi, S., Armienti, P., and Di Vincenzo, G., 2005, No plume, no rift magmatism in the West Antarctic Rift, in Foulger, G.R., et al., eds., Plates, plumes, and paradigms: Boulder, Colorado, Geological Society of America Special Paper 388, p. 435–447, doi:10.1130/2005.2338(26).

Rona, P.R., and Richardson, E.S., 1978, Early Cenozoic global plate reorganization: Earth and Planetary Science Letters, v. 40, p. 1–11, doi: 10.1016/0012-821X(78)90069-9.

Rosencrantz, E., Ross, M.I., and Sclater, J.G., 1988, Age and spreading history of the Cayman Trough as determined from depth, heat flow, and magnetic anomalies: Journal of Geophysical Research, v. 93, p. 2141–2157.

Rowe, M.P., 1998, An explanation of the geology of Bermuda: Hamilton, Bermuda, Ministry of the Environment, 30 p.

Ruppel, C., 1996, Hydrates and heat flux on U.S. Atlantic passive margin, leg 164 results: Eos (Transactions, American Geophysical Union, v. 77, no. 46 (Fall Meeting Supplement), p. F726.

Ruppel, C., 2000, Thermal state of the gas hydrate reservoir, *in* Max, M.D., ed., Natural gas hydrate in ocean and permafrost environments: Boston, Kluwer Academic, p. 29–42.

Schilling, J.-G., 1986, Geochemical and isotopic variation along the Mid-Atlantic Ridge axis from 79 deg N to 0 deg N, *in* Vogt, P.R., and Tucholke, B.E., eds., The Geology of North America, v. M, The Western North Atlantic region: Boulder, Colorado, Geological Society of America, p. 137–156.

Sclater, J.G., and Wixon, L., 1986, The relationship between depth and age and heat flow and age in the western North Atlantic, *in* Vogt, P.R., and Tucholke, B.E., eds., The Geology of North America, v. M, The Western North Atlantic region: Boulder, Colorado, Geological Society of America, p. 257–270.

Sclater, J.G., Anderson, R.N., and Bell, M.L., 1971, The elevation of ridges and the evolution of the central eastern Pacific: Journal of Geophysical Research, v. 76, p. 7888–7915.

Sclater, J.G., Lawver, L.A., and Parsons, B., 1975, Comparison of long-wavelength residual elevation and free air gravity anomalies in the North Atlantic and possible implications for the thickness of the lithospheric plate: Journal of Geophysical Research, v. 80, p. 1031–1052.

Sharp, W.D., and Clague, D.A., 2006, 50 Ma initiation of Hawaiian-Emperor Bend records major change in Pacific plate motion: Science, v. 313, p. 1281–1284.

Sheehan, A.F., and McNutt, M.K., 1989, Constraints on thermal and mechanical structure of the oceanic lithosphere at the Bermuda Rise from geoid height and depth anomalies: Earth and Planetary Science Letters, v. 93, p. 377–391, doi: 10.1016/0012-821X(89)90037-X.

Shipboard Scientific Party, Site 386, 1979, Fracture valley sedimentation on the central Bermuda Rise, *in* Tucholke, B.E., and Vogt, P.R., eds., Initial reports of the Deep Sea Drilling Project, v. 43: Washington, D.C., U.S. Government Printing Office, p. 195–321.

Sieberg, A., 1932, Earthquake geography, *in* Gutenberg, B., ed., Handbuch der Geophysik: Berlin, Gebrüder Borntraeger, v. 4, p. 687–1005.

Sleep, N.H., 1990, Hotspots and mantle plumes: Some phenomenology: Journal of Geophysical Research, v. 95, p. 6715–6736.

Sleep, N.H., 2002, Ridge-crossing mantle plumes and gaps in tracks: Geochemistry, Geophysics, Geosystems, v. 8505, doi: 10.1029/2001/GC000290(33).

Smith, H.F., and Sandwell, D.T., 1997, Global seafloor topography from satellite altimetry and ship depth soundings: Science, v. 277, p. 1956–1962, doi: 10.1126/science.277.5334.1956.

Southworth, C.S., Gray, K.J., and Sutter, J.F., 1993, Middle Eocene intrusive rocks of the central Appalachian Valley and Ridge province: Setting, chemistry, and implications for crustal structure: U.S. Geological Survey Bulletin, v. 1839, p. J1–J24.

Steinberger, B., 2000, Plumes in a convecting mantle: Models and observations for individual hotspots: Journal of Geophysical Research, v. 105, p. 11,127–11,152, doi: 10.1029/1999JB900398.

Stewart, K.G., and Dennison, J.M., 2006, Tertiary-to-Recent arching and the age and origin of fracture-controlled lineaments in the Southern Appalachians [abs.]: Geological Society of America Abstracts with Programs, v. 38, no. 3, p. 27.

Sundvik, M., Larson, R.L., and Detrick, R.S., 1984, Rough-smooth basement boundary in the western North Atlantic basin: Evidence for a seafloor-spreading origin: Geology, v. 12, p. 31–34, doi: 10.1130/0091-7613(1984)12<31:RBBITW>2.0.CO;2.

Tarduno, J.A., 2005, On the motion of Hawaii and other mantle plumes [abs.]: American Geophysical Union Chapman Conference Proceedings, v. 15.

Tarduno, J.A., Duncan, R.A., Scholl, D.W., Cottrell, R.D., Steinberger, B., Thor-darson, T., Kerr, B.C., Neal, C.R., Frey, F.A., Torii, M., and Carvallo, C., 2003, The Emperor seamounts: Southward motion of the Hawaiian Hotspot plume in Earth's mantle: Science, v. 301, p. 1064–1069.

Tso, J.L., and Surber, J.D., 2002, Eocene igneous rocks near Monterey, Virginia: A field study: Virginia Minerals, v. 48, p. 25–40.

Tucholke, B.E., 1986, Sediment thickness in the western North Atlantic Ocean, Plate 6 *in* Vogt, P.R., and Tucholke, B.E., eds., The geology of North America, v. M, The Western Atlantic region: Boulder, Colorado, Geological Society of America.

Tucholke, B.E., and Ludwig, W.J., 1982, Structure and origin of the J-Anomaly Ridge, western Atlantic Ocean: Journal of Geophysical Research, v. 87, p. 9389–9407.

Tucholke, B.E., and Mountain, G.S., 1979, Seismic stratigraphy, lithostratigraphy, and paleosedimentation patterns in the North American Basin, *in* Talwani, M., Hay, W., and Ryan, W.B.F., eds., 1979, Deep drilling results in the Atlantic Ocean: Continental margins and paleoenvironment: Washington, D.C., American Geophysical Union, Maurice Ewing Series 3, p. 58–86.

Tucholke, B.E., and Vogt, P.R., 1979, Western North Atlantic: Sedimentary evolution and aspects of tectonic history, *in* Tucholke, B.E., and Vogt, B.E., eds., Initial reports of the Deep Sea Drilling Project, v. 43: Washington, D.C., U.S. Government Printing Office, p. 791–825.

Turcotte, D.L., and Schubert, G., 2002, Geodynamics: Cambridge, England, Cambridge University Press, 456 p.

Vacher, H.L., and Hearty, P., 1989, History of stage 5 sea level in Bermuda: Review with new evidence of a brief rise to present sea level during substage 5a: Quaternary Science Reviews, v. 8, p. 159–168, doi: 10.1016/0277-3791(89)90004-8.

Vacher, H.L., Rowe, M.P., and Garrett, P., 1989, The geological map of Bermuda: London, Oxford Cartographers, and Hamilton, Bermuda, Ministry of Works and Engineering, scale 1:25,000.

Van der Hilst, R.D., and de Hoop, M.V., 2005, Banana-doughnut kernels and mantle tomography: Geophysical Journal International, v. 163, p. 956–961.

Van Sickel, W.A., Kominz, M.A., Miller, K.G., and Browning, J.V., 2004, Late Cretaceous and Cenozoic sea-level estimates: Backstripping analysis of borehole data, onshore New Jersey: Basin Research, v. 16, p. 451–465, doi: 10.1111/j.1365-2117.2004.00242.x.

Vink, G.E., Morgan, W.J., and Zhao, W.-L., 1984, Preferential rifting of continents: A source of displaced terranes: Journal of Geophysical Research, v. 89, p. 10,072–10,076.

Vogt, P.R., 1972, Evidence for global synchronism in mantle plume convection, and possible significance for geology: Nature, v. 240, p. 338–342, doi: 10.1038/240338a0.

Vogt, P.R., 1974, Volcano height and plate thickness: Earth and Planetary Science Letters, v. 23, p. 337–348, doi: 10.1016/0012-821X(74)90123-X.

Vogt, P.R., 1975, Changes in geomagnetic reversal frequency at times of tectonic change: Evidence for coupling between core and upper mantle processes: Earth and Planetary Science Letters, v. 25, p. 313–321, doi: 10.1016/0012-821X(75)90247-2.

Vogt, P.R., 1979a, Global magmatic episodes: New evidence and implications for the steady-state mid-oceanic ridge: Geology, v. 7, p. 93–98, doi: 10.1130/0091-7613(1979)7<93:GMENEA>2.0.CO;2.

Vogt, P.R., 1979b, Volcano height and paleo-plate thickness, *in* Tucholke, B.E., and Vogt, P.R., eds., Initial reports of the Deep Sea Drilling Project, v. 43: Washington, D.C., U.S. Government Printing Office, p. 877–878.

Vogt, P.R., 1991, Bermuda and Appalachian-Labrador rises: Common non-hotspot processes?: Geology, v. 19, p. 41–44, doi: 10.1130/0091-7613(1991)019<0041:BAALRC>2.3.CO;2.

Vogt, P.R., and Johnson, G.L., 1971, Cretaceous sea-floor spreading in the western North Atlantic: Nature, v. 234, p. 22–25, doi: 10.1038/234022a0.

Vogt, P.R., and Jung, W.Y., 2004, Treitel Ridge: A unique inside corner hogback on the west flank of extinct Aegir rift, Norway basin—Tectonic response to abrupt end of North America–Greenland motion?: Geological Society of America Abstracts with Programs, v. 36, p. 504.

Vogt, P.R., and Smoot, N.C., 1984, The Geisha guyots: Multibeam bathymetry and morphometric interpretation: Journal of Geophysical Research, v. 89, p. 11,085–11,107.

Vogt, P.R., and Tucholke, B.E., 1989, North Atlantic basement topography along and across the Mid-Atlantic Ridge, Plate 3 *in* Bally, A.W., and Palmer, A.R., eds., The geology of North America, v. A, An overview: Boulder, Colorado, Geological Society of America.

Von Herzen, R.P., 2004, Geothermal evidence for continuing hydrothermal circulation in older (>60 my) ocean crust, *in* Davis, E.E., and Elderfield, E.H., eds., Hydrogeology of the oceanic lithosphere: Cambridge, England, Cambridge University Press, p. 414–447.

Wegener, A., 1966, The origin of continents and oceans, trans. from the 1929 German edition (Braunschweig, Friedrich Vieweg und Sohn) by J. Biram: New York, Dover, 246 p.

Williams, C.A., Hill, I.A., Young, R., and White, R.S., 1990, Fracture zones across the Cape Verde Rise, NE Atlantic: Journal of the Geological Society of London, v. 147, p. 851–857.

Williamson, M.-C., Villeneuve, M., and Blasco, S., 2006, Deep Drill 1972 revisited: New data on the composition and age of the Bermuda volcanic edifice [abs.]: Eos (Transactions, American Geophysical Union), v. 87, no. 52 (Fall Meeting Supplement), abs. V13A-0646.

Winker, C.D., and Howard, J.D., 1977, Correlation of tectonically deformed shorelines on the southern Atlantic Coastal Plain: Geology, v. 5, p. 123–127, doi: 10.1130/0091-7613(1977)5<123:COTDSO>2.0.CO;2.

Winterer, E.L., Atwater, T.M., and Decker, R.W., 1989, The northeast Pacific and Hawaii, *in* Bally, A.W., and Palmer, A.R., eds., The geology of North America, v. A, An overview: Boulder, Colorado, Geological Society of America, p. 265–298.

Withjack, M., 1979, A convective heat transfer model for lithospheric thinning and crustal uplift: Journal of Geophysical Research, v. 84, p. 3008–3022.

Woollard, G.P., and Ewing, M., 1939, Structural geology of the Bermuda Islands: Nature, v. 143, p. 898.

Zhao, D., 2004, Global tomographic images of mantle plumes and subducting slabs: Insight into deep Earth dynamics: Physics of the Earth and Planetary Interiors, v. 146, p. 3–34, doi: 10.1016/j.pepi.2003.07.032.

Zhu, A., and Wiens, D.A., 1991, Thermoelastic stress in oceanic lithosphere due to hotspot reheating: Journal of Geophysical Research, v. 96, p. 18,323–18,334.

Zoback, M.L., Nishenko, S.P., Richardson, R.M., Hasegawa, H.S., and Zoback, M.D., 1986, Mid-plate stress, deformation, and seismicity, *in* Vogt, P.R., and Tucholke, B.E., eds., The geology of North America, v. M, The Western North Atlantic region: Boulder, Colorado, Geological Society of America, p. 297–312.

Manuscript Accepted by the Society January 31, 2007

Geological Society of America
Special Paper 430
2007

Lithospheric control of Gondwana breakup: Implications of a trans-Gondwana icosahedral fracture system

James W. Sears*

University of Montana, Missoula, Montana 59812, USA

ABSTRACT

Gondwana broke apart along a truncated icosahedral fracture system that minimized total crack length and therefore required the least work to nucleate and propagate new fractures across the supercontinent. The fracture arrangement met conditions imposed by Euler's rule for ordering polyhedrons on a spherical shell. Linear grabens accumulated Permian rift facies along 10,000 km of the fracture system in east Gondwana. Large igneous provinces erupted >100 m.y. later along these fractures. This suggests that widening of existing fractures rather than impingement of deep-mantle plumes triggered outbreaks of flood basalt. The tensile stress field that initiated the fractures was symmetrical with Gondwana and exploited preexisting lithospheric suture zones. The stress field was also symmetrical about the African geoid bulge in the Permian locus of Gondwana. Tensile hoop stress along the Gondwana boundary initiated radial fractures that defined the lateral edges of Australia, India, Arabia, Libya, and northwest Africa. Fractures then evidently propagated inward across Gondwana, spontaneously bending at critical lengths congruent with the tessellation. Fractures later branched outward from the bends to create triple-rift junctions. Plate tectonic processes later exploited the icosahedral fractures to separate the Gondwana daughter continents.

Keywords: Gondwana, supercontinent, icosahedron, hotspot, mantle plume

INTRODUCTION

While it is generally agreed that supercontinents break apart and reassemble in grand tectonic cycles, much controversy surrounds the causes of breakup (Foulger et al., 2005). Following an original idea by J.T. Wilson (1963) that Hawaii was caused by motion of the Pacific lithosphere over a fixed region in the mantle that he termed a "hot spot," the deep-mantle-plume paradigm predicts that superadiabatic plumes rise from the core-mantle boundary to drive continental breakup (Morgan, 1971, 1981; Campbell, 2001). The paradigm proposes the following

sequence of events. Plumes impinge on the base of the lithosphere, forming broad domes (Storey et al., 2001). Plume heads erupt large igneous provinces (LIPs) from three-armed rifts that branch from the domes (Ernst and Buchan, 2001). Two of the rifts propagate outward and link up with older such rifts to break the continent piecemeal; the third rift may form a "failed arm," or "aulocogen" (Burke and Dewey, 1973). Continental fragments then calve away, and seafloor spreading disperses them. Some active volcanic hotspots may represent lingering ascents of thin plume tails at fixed mantle locations (Morgan, 1981). In this paradigm, plume ascents and breakouts are episodic and

*E-mail: james.sears@umontana.edu.

Sears, J.W., 2007, Lithospheric control of Gondwana breakup: Implications of a trans-Gondwana icosahedral fracture system, *in* Foulger, G.R., and Jurdy, D.M., eds., Plates, plumes, and planetary processes: Geological Society of America Special Paper 430, p. 593–601, doi: 10.1130/2007.2430(28). For permission to copy, contact editing@geosociety.org. ©2007 The Geological Society of America. All rights reserved.

depend on deep mantle viscosity and instabilities along the core-mantle boundary (e.g., Steinberger, 2000).

Mantle tomography does not, however, unequivocally demonstrate that plumes cross the mantle transition zone (Foulger et al., 2000). DeWit et al. (1988), Anderson (2001, 2002b), and Hamilton (2003) argue that continental breakup and associated large igneous outbreaks and hotspots are controlled, top-down, by lithospheric processes rather than by rising plumes. Anderson (2005) formalized this opposing view as the "plate paradigm." Continental breakup may be initiated by thermal expansion of ordinary sublithospheric mantle that becomes insulated beneath a sluggish supercontinent. For example, Anderson (1982) showed that the Atlantic-African geoid anomaly coincides with the Permian locus of Pangaea and may represent the residuum of thermally expanded sub-Pangaean mantle. The thermal expansion placed the supercontinent under uniform layer-parallel tension. The supercontinent then rifted apart, with decompression-melt-driven outbreaks of LIPs along rift zones as fragments drifted off the thermal bulge toward retreating trenches. Anomalous hotspot activity that continues at fixed mantle sites within the decaying Atlantic-African geoid anomaly is consistent with this model (Chase, 1979; Crough and Jurdy, 1980; Anderson, 1982; Phillips and Bunge, 2005).

Here I argue that a uniform tensile stress field constrained by Gondwana geometry and boundary conditions initiated the fracturing of Gondwana by Early Permian time in a manner that minimized crack length and therefore minimized the energy required to nucleate and propagate new fractures. The organization of fracture polygons depended on the strength of the Gondwana lithosphere and geometric constraints for tiling a

spherical surface. Plate tectonic processes exploited the initial Early Permian fractures more than 100 m.y. later to widen rifts, release LIPs through secondary decompression melting, and disperse rifted fragments. This model argues against the deep-mantle-plume paradigm and favors Anderson's plate paradigm.

GONDWANA FRACTURE TESSELLATION

Figure 1 presents a standard reconstruction of Gondwana, adapted from DeWit et al. (1988), Golonka et al. (1994), and Lawver et al. (1999). The argument presented in this contribution derives from the recognition that much of the fracture architecture of Gondwana was closely congruent with a precise, energy-minimizing configuration, that of the truncated icosahedron (Sears, 2001; Sears et al., 2005). The truncated icosahedron comprises a polyhedron with twelve pentagonal and twenty hexagonal faces. The pentagonal faces are centered on the vertices of an icosahedron. The buckyball is a familiar example of a truncated icosahedron. Projected onto the Earth's surface, each tile edge of a truncated icosahedral tessellation subtends 23.28° of arc, or ~2600 km.

The Gondwana fracture tessellation included parts of three large pentagons and six large hexagons of the scale and arrangement of a truncated icosahedron at the Earth's surface (Fig. 1). The yellow dots in Figure 1 lie near rift triple junctions separated by 23° of great circle arc. The angles between adjacent arcs are 108° or 120°, the internal angles of pentagonal and hexagonal plates, respectively. The tessellation is intolerant; establishment of a single triple junction defines the distribution of all others. Gondwana fractures with a cumulative length >20,000 km de-

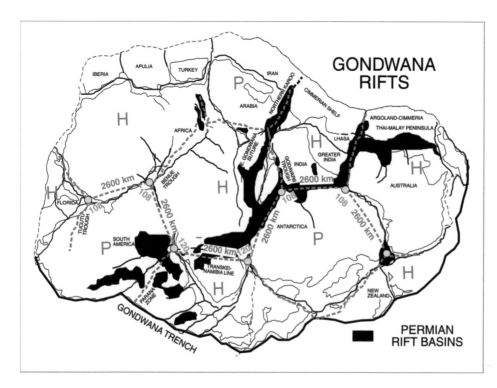

Figure 1. Gondwana reconstruction at 200 Ma, after DeWit et al. (1988), Golonka et al. (1994), and Lawver et al. (1999). The heavy dashed lines define truncated icosahedral tessellation that is congruent with many Gondwana fractures. The black zones are Permian rifts with coal measures, after Bordy and Catuneanu (2002) and Harrowfield et al. (2005). The yellow dots are at the vertices of a precise truncated icosahedron, separated by 23° of great circle arc and at angles of 10° and 120°. P—pentagon; H—hexagon.

fine segments of some sixteen edges of this truncated icosahedral tessellation, highlighted in Figure 1. Older lithospheric sutures were reactivated as rift zones along the Benue Trough, the Transkei-Namibia line, the Parana zone, and east Africa (Vauchez et al., 1997; Vauchez et al., 1998; Tommasi and Vauchez, 2001; Jourdan et al., 2006). These rift zones are approximately congruent with the tessellation, implying that they exerted some control over the orientation of the stress field that initiated the breakup, as discussed in a later section.

The geometric congruence of many Gondwana fractures with a single, rigorously defined tessellation indicates that the fractures formed not piecemeal, but in a uniform, Gondwana-wide stress field prior to dispersal of any daughter continents. Fracture propagation began before Early Permian time. Some 10,000 linear km of grabens that followed the fracture tessellation across east Gondwana accumulated Permian coal measures. Harrowfield et al. (2005) mapped a relict Permian–Triassic rift platform from New Guinea along the western coast of Australia and Antarctica to southern Africa. Bordy and Catuneanu (2002) mapped late Paleozoic Karoo rifts across southern Africa. Hauser et al. (2002) traced the early Permian Karoo rifts north along the Arabia-India rift zone. Şengör and Natal'in (2001) showed that many other Gondwana rifts that are part of the icosahedral pattern were active in Permian and Triassic time.

The Jurassic Karoo and Cretaceous Bunbury, Rajmahal, Godavari, and Parana LIPs all erupted into existing Permian grabens. Clearly, the LIPs did not cause the icosahedral fractures, but rather exploited them >100 m.y. after they had appeared in the geologic record.

HEXAGONAL FRACTURE SYSTEMS

The truncated icosahedral fractures recall the hexagonal tensile fracture patterns of columnar-jointed basalt. Hexagonal joint networks occur in basalt flow interiors due to isotropic layer-parallel thermal stress (Weinberger, 2001). A joint face results from many discrete fracture events as the basalt cools and shrinks and layer-parallel tension accumulates until it exceeds the tensile strength of the crystallized basalt layer (Ryan and Sammis, 1978). A detailed study of columnar joints by DeGraff and Aydin (1987) showed that cracks propagate to a critical length, then commonly bend at 120°. New cracks then propagate either toward or away from the bends to create triple junctions. Each new crack bends when it obtains the critical length and, together with similar cracks, joins a network of hexagonal columns of surprisingly uniform size. A propagating crack will intersect an existing fracture orthogonally because the older fracture forms a free surface for which the principal stresses are parallel and perpendicular (Suppe, 1985).

Hexagonal fracture systems develop in a homogeneous material undergoing uniform layer-parallel tension because they provide the greatest stress relief for the least work to nucleate and propagate cracks (Jagla and Rojo, 2002). A regular hexagonal pattern requires the shortest total crack length to pave a given area and provides the most stable triple junctions, and hexagonal close-packing of fractures best relieves strain between neighboring domains. The energy used for the work of propagating cracks is stored as elastic strain within the volume of the layer.

Stronger layers crack into arrays of larger hexagons with a shorter total crack length. More strain energy is required to initiate the fractures, but because the layer is stronger, it stores more energy before failing. If polygons are sufficiently large to reflect the curvature of a spherical shell, Euler's rule for convex polytopes becomes evident; pentagonal polygons will occupy the twelve vertices of an icosahedron, with intervening hexagonal polygons. As shell strength increases, the sizes of the polygons will increase and the number of hexagons will decrease in a stepwise fashion so as to pave the closed geometry of the sphere. The stepwise nature of the permissible tessellations means that layers with wide ranges of strengths may fracture in similar patterns; threshold strengths must be surpassed before next-sized fracture tessellations are achieved.

The truncated icosahedral fractures evident across much of Gondwana represent the largest hexagons permitted on a spherical tessellation. Gondwana occupies only a portion of a sphere, so has parts of only three pentagons and seven hexagons of the full tessellation. Near its edges, Gondwana fractured into smaller polygons; the discontinuity between the larger and the smaller polygons may represent a threshold strength related to thinning of the lithosphere toward the Gondwana margin.

GONDWANA STRESS TESSELLATION

A tensile stress field that induces a hexagonal array of cracks defines a triangular tessellation, with the vertices of the triangles at the centers of the hexagons (Hills, 1963). The triangular array defines the dual tessellation of the fracture array. (Edges of dual tessellations bisect one another orthogonally, and vertices of dual tessellations occupy the faces of one another.) The vertices of the triangles form null points in the medium; strain increases outward from them to the distance at which the material cracks.

Columnar joints result from shrinkage, whereas Gondwana fractures may have resulted from thermal expansion of the underlying mantle (Anderson, 1982). Both situations induce layer-parallel tension. By analogy with columnar basalt, the stress tessellation for the Gondwana fracture tessellation was its dual, the icosadeltahedron (Fig. 2). This triangular tessellation obeys Euler's rule for convex polytopes; pentamers, with five nearest neighbors, occupy the vertices of an icosahedron, while hexamers, with six nearest neighbors, form the remaining vertices.

Figure 2 shows that the icosadeltahedral stress tessellation followed the Gondwana margin and was surprisingly symmetrical across Gondwana. This configuration provided the most balanced stress distribution and indicates that the intrinsic shape of Gondwana organized the geometry of the tensile stress field. Furthermore, the stress configuration best accommodated existing

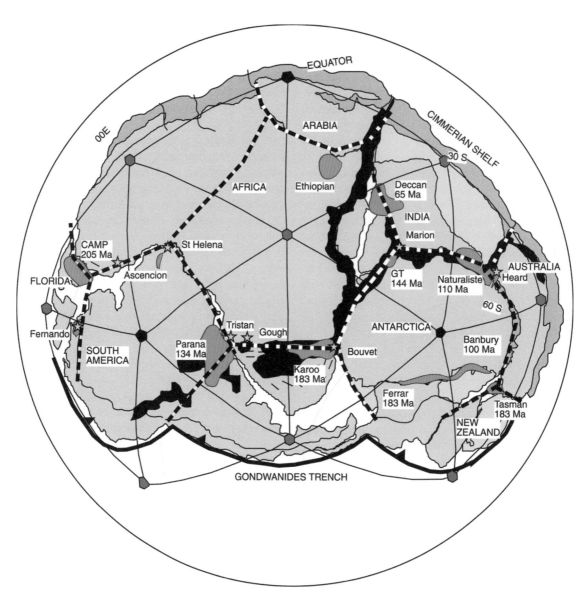

Figure 2. Relationship of Gondwana rift tessellation (heavy dashed lines) and its dual, the icosadeltahedral stress tessellation (thin solid lines), such that vertices of stress tessellation occupy faces of fracture tessellation, and vice versa. The tessellations cross one another orthogonally. The black vertices are pentamers with five nearest neighbors and exactly occupy vertices of the icosahedron at Earth scale, as required by Euler's rule for convex polytopes (see text). The blue vertices are hexamers with six nearest neighbors. Note that stress tessellation follows the northern margin of Gondwana, implying that it was a free surface that guided tensile hoop stress. Thus, Gondwana split on radial fractures along the northern rim. Note the symmetry of stress tessellation across Gondwana. This provided the shortest total fracture length and thus required the least work to break up Gondwana. The yellow stars are major hotspot volcanoes in the modern coordinates indicated by the lines of latitude and longitude. The figure restores Gondwana so that fracture tessellation best fits hotspot tessellation. The red areas are large igneous provinces (LIPs), with eruption dates shown. Note that although the dates range over more than 100 m.y., most LIPs erupted from fractures that restored to single tessellation that is congruent with Permian rifts (black areas). This implies that coherent tessellation dates to before the oldest LIP (205 Ma) and that LIPs erupted from fractures diachronously as later plate tectonics opened fractures.

lithospheric suture zones within Gondwana, opening the east Gondwana, Transkei-Namibia, Parana, and Benue fractures. Those ready-made weak zones could accommodate tensile stresses back to vertices of the stress tessellation without requiring new fractures. New Gondwana fractures that crosscut basement grain at wide angles were nearly perfectly congruent with the ideal truncated icosahedral tessellation.

Tensile hoop stress paralleled the periphery of Gondwana, so most fractures intersected the margins of Gondwana perpendicularly (Fig. 2). These fractures separated New Zealand,

Australia, India, Arabia, northwest Africa, and Central America. Tensile hoop stress forms in response to expansion of an enclosed region, consistent with a uniformly expanding Gondwana. These observations are consistent with the hypothesis of Anderson (1982) that Gondwana insulated the underlying mantle, leading to thermal expansion and shell-parallel extension.

CRACKING SPHERICAL SHELLS

Experiments with drying clay shells provide insight into the formation of polygonal fracture patterns on spherical surfaces (Sears, 2006). Cracks initiate at reentrants in the edge of a drying clay shell. One of these becomes a master crack that zigzags across the shell in segments whose lengths are related to the thickness and strength of the clay (Fig. 3A and B). Branch cracks then propagate from the bends in the master crack to form triple-crack junctions (Fig. 3C). The branch cracks may continue to propagate and bend at the critical length to outline polygons. The cracks in Figure 3 approximate a dodecahedral tessellation.

The cracks result from uniform tension in a drying clay shell. The cracks are Mode I tensile fractures; they must initiate orthogonal to the edge of the shell and also to cracks formed earlier because they constitute free surfaces. Tensile hoop stress follows the margin of the shell and intensifies at reentrants because the sides of the reentrants draw apart as they shrink. Stress concentration at the tip of a crack enables it to propagate. As it propagates, it releases strain energy stored on either side. Once a free fracture surface exists, it forms an expansion crack that can resolve tensile stresses on either side out to a distance that is a function of the strength of the shell.

The tip of a propagating crack bends at ~120° so as to resolve strain in the adjacent region and continues to propagate in the new direction. A bend in a crack forms a reentrant that concentrates stress and initiates a new crack that propagates outward from the bend. The concept of a propagating master crack with secondary cracks branching from bends in a spherical shell fundamentally differs from the paradigm that three-armed cracks form above domes and link together into rifts that eventually separate continents.

In the case of Gondwana, a master crack may have begun at a reentrant along the Cimmerian shelf of north Gondwana between Australia and greater India and zigzagged from west Australia across to southern Africa, with a branch propagating along the Godavari Trough of India. A separate crack may have propagated into Gondwana from a reentrant along the Cimmerian shelf between greater India and Arabia, intersecting the other fracture at Sri Lanka. This second crack followed the east Gondwana suture, which required less work to split than the adjacent lithosphere. These fractures opened grabens in which Early Permian rift facies were deposited, but rifting was not accompanied by mafic igneous activity. This demonstrates that the rifting was not driven by the ascent of mantle plumes or the emplacement of LIPs. The Permian rifts generally paralleled the Permian Gondwanides Trench, suggesting that trench pull may have contributed to their opening. Fractures in west Gondwana opened in Jurassic and Cretaceous time, perhaps in response to trench pull along the Andean margin.

LIPS AND HOT SPOTS

Nine LIPs ranging in age from Early Jurassic to Paleogene erupted as Gondwana rifted apart and its daughter continents dispersed (Ernst and Buchan, 2001). When the continents are gathered into their Gondwana configuration, however, the future sites of the LIPs are congruent with vertices or edges of the fracture tessellation that was already evident in Permian time. This suggests that the fracture tessellation prepared the ascent routes for the eruptive sites, but that LIP outbreaks depended on later effects such as decompression melting as plate tectonic movements widened the fractures and opened conduits for LIP eruptions. Silver et al. (2006) proposed that flood basalts erupt from superheated accumulations of melt beneath continental lithosphere; such conditions may have evolved beneath Gondwana due to insulation, especially if subduction had decreased around Gondwana margins (e.g., Lowman and Gable, 1999; Phillips and Bunge, 2005).

When Gondwana is reconstructed, the fracture tessellation may be superimposed on several major hotspots associated with Late Jurassic or Early Cretaceous rifts and LIPs. Heard, Marion, Bouvet, Gough, Tristan, St. Helena, Ascencion, and Fernando plot within a few degrees of the tessellation, mostly near vertices (Sears et al., 2005; Fig. 2). However, neither older hotspots associated with opening of the central Atlantic Ocean nor younger hotspots associated with the Deccan or Ethiopian LIPs are

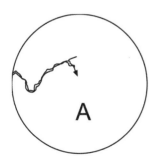
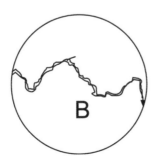

Figure 3. Crack propagation across a drying spherical clay shell, sketched from photographs. (A) and (B) The master crack zigzags across the spherical shell. (C) Branch cracks propagate from bends in the master crack. Some of these also propagate and bend to define polygons. The crack pattern approximates a dodecahedron.

congruent with this position of the tessellation. The congruent hotspots may record a time of drift stagnation of Gondwana that linked the sites in a fixed geographic framework.

Eruption of LIPs may have resulted from decompression melting upon opening of rifts along the fracture tessellation, perhaps augmented by thermal expansion of the upper mantle beneath the insulating supercontinent (e.g., Silver et al., 2006). Lingering hotspots may have been localized by alteration of feeder chimneys in the upper mantle beneath the original sites of the LIP eruptions. Fairhead and Wilson (2006) suggest, alternatively, that some hotspot tracks may be fractures that propagated due to stress instabilities in the widening plates. These

considerations favor lithospheric rather than deep mantle control for Gondwana LIPs and hotspots.

AFRICAN GEOID ANOMALY

Anderson (1982) proposed that the Atlantic-African geoid anomaly marks the Permian footprint of Pangaea, the decaying remnant of thermally expanded mantle that had been insulated beneath the supercontinent. Chase (1979) and Crough and Jurdy (1980) proposed that hotspot activity in the region demonstrates its increased thermal content. Rifting of the fracture tessellation may have coincided with periods of continental drift stagnation.

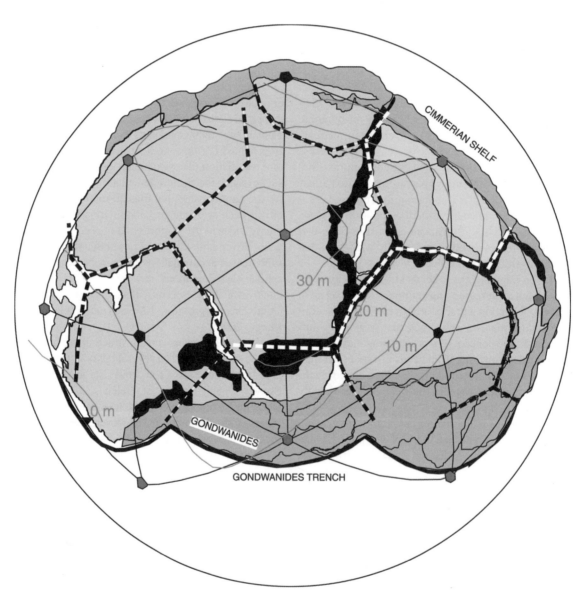

Figure 4. The African geoid anomaly superimposed on Gondwana in Gondwana's Permian position. Note that the contours of the geoid anomaly are generally orthogonal or parallel to stress tessellation and that the anomaly is centered on Gondwana. This is consistent with the hypothesis of Anderson (1982) that the geoid anomaly marks the paleoposition of Gondwana and that Gondwana spread outward from the anomaly as it broke apart.

Paleomagnetism shows that Gondwana moved slowly from 280 to 260 Ma, when it changed direction (Gordon et al., 1979). The Early Permian rift fill in the tessellation fractures correlates with this slow movement. Gondwana increased in velocity from 260 to 210 Ma, then slowed from 210 to 190 Ma (Gordon et al., 1979), when rifting was renewed and Gondwana began to break into daughter continents.

Figure 4 superimposes a contour map of the African part of the geoid anomaly on Gondwana in the mantle position that it may have occupied during Permian time (Golonka et al., 1994). If the geoid bulge was centered on Gondwana as shown, the contours either paralleled or were orthogonal to the fracture and stress tessellations. Anderson (1982) suggested that Gondwana spread radially outward from the geoid high, consistent with the radial components of the stress tessellation. The outward spreading would also have resulted in hoop stress parallel to the geoid contours, consistent with the nonradial components of the stress tessellation. Evidently, when the combination of these stresses was large enough to overcome the strength of the Gondwana lithosphere, it cracked into the pattern that required the shortest total length of new cracks.

THE ICOSAHEDRON IN NATURE

Comparison of the Gondwana tessellation with other natural examples of icosahedral arrangements provides insight into the Gondwana fracture process. In nature, collections of particles or cells commonly surface as a sphere in icosahedral patterns. These include fullerene molecules, blastocysts, colloids, quasicrystals, florets, gumball seedcases, and numerous icosahedral viruses, including HIV and the viruses that cause warts, herpes, and polio (cf. Anderson, 2002a). Because the pentamers form shorter, stronger bonds, viral capsids burst along hexamers (Zandi et al., 2005). The cracks zigzag along polygonal boundaries much like those on fragmenting supercontinents.

Icosahedral configurations solve the classic Thomson problem of minimizing the energy of an array of mutually repulsive coulombic charges on a sphere (Altschuler et al., 1997). The lowest-energy configurations produce truncated icosahedral strain gradients quite similar to the rift patterns seen on Gondwana.

CONCLUSIONS

The self-organized Gondwana fracture tessellation is consistent with the hypothesis of Anderson (1982) that the supercontinent drove its own breakup by insulating the underlying mantle. The thermally expanded mantle lifted Gondwana, placing it under uniform layer-parallel tension. When tension exceeded the strength of the Gondwana lithosphere, it fractured into a symmetrical polygonal pattern commensurate with its strength and conforming to the geometric restrictions of a sphere and to the boundary conditions of the supercontinent. The Permian marine lowstand (Haq, 1995) may record the culmination of Gondwana thermal expansion and uplift (see Anderson, 1982). Likely, the

fractures propagated in zigzag fashion across the supercontinent, bending and branching at critical lengths. The fractures relieved the tension and separated the supercontinent into tiles that could move independently under the influence of plate tectonic processes. Separation of the tiles opened rift valleys and ocean basins and drove decompression melting of the thermally expanded mantle, leading to outbreaks of LIPs and injection of dike swarms. These secondary effects were diachronous and depended on global plate tectonics to exploit the icosahedral fractures.

ACKNOWLEDGMENTS

Discussions with colleagues at the University of Montana and at University College Cork, Leeds University, and Durham University during an academic exchange in 2006 were most helpful in formulating and expressing some of the ideas presented in this manuscript. I especially thank Gillian Foulger, Marge Wilson, Pat Meere, Gray Thompson, and Steve Sheriff for their spirited insight. Comments by reviewers Greg McHone, Sergio Rocchi, and Phil Leat and by editors Donna Jurdy and Gillian Foulger greatly improved the manuscript. The research into continental breakup was partly funded by NSF Grant EAR 0107024.

REFERENCES CITED

Altschuler, L., Williams, T.J., Ratner, E.R., Tipton, R., Stong, R., Dowla, F., and Wooten, F., 1997, Possible global minimum lattice configurations for Thomson's problem of charges on a sphere: Physical Review Letters, v. 78, p. 2681–2685, doi: 10.1103/PhysRevLett.78.2681.

Anderson, D.L., 1982, Hotspots, polar wander, Mesozoic convection and the geoid: Nature, v. 297, p. 391–393, doi: 10.1038/297391a0.

Anderson, D.L., 2001, Top-down tectonics?: Science, v. 293, p. 2016–2018, doi: 10.1126/science.1065448.

Anderson, D.L., 2002a, How many plates?: Geology, v. 30, p. 411–414, doi: 10.1130/0091-7613(2002)030<0411:HMP>2.0.CO;2.

Anderson, D.L., 2002b, Plate tectonics as a far-from-equilibrium self-organized system, *in* Stein, S., and Freymuller, J., eds., Plate boundary zones: Washington, D.C., American Geophysical Union, Geodynamics Series Monograph 30, p. 411–425.

Anderson, D.L., 2005, Scoring hotspots: The plume and plate paradigms, *in* Foulger, G.R., et al., eds., Plates, plumes, and paradigms: Boulder, Colorado, Geological Society of America Special Paper 388, p. 31–54.

Bordy, E.M., and Catuneanu, O., 2002, Sedimentology of the Beaufort-Molteni Karoo fluvial strata in the Tuli basin, South Africa: South African Journal of Geology, v. 105, p. 51–66, doi: 10.2113/1050051.

Burke, K., and Dewey, J.F., 1973, Plume-generated triple junctions: Key indicators in applying plate tectonics to old rocks: Journal of Geology, v. 81, p. 406–433.

Campbell, I.H., 2001, Identification of ancient mantle plumes, *in* Ernst, R.E., and Buchan, K.L., eds., Mantle plumes: Their identification through time: Boulder, Colorado, Geological Society of America Special Paper 352, p. 5–21.

Chase, C.G., 1979, Subduction, the geoid, and lower mantle convection: Nature, v. 282, p. 464–468, doi: 10.1038/282464a0.

Crough, T.J., and Jurdy, D.M., 1980, Subducted lithosphere, hotspots, and the geoid: Earth and Planetary Science Letters, v. 48, p. 15–22, doi: 10.1016/0012-821X(80)90165-X.

DeGraff, J.M., and Aydin, A., 1987, Surface morphology of columnar joints and its significance to mechanics and direction of joint growth: Geological Society of America Bulletin, v. 99, p. 605–617, doi: 10.1130/0016-7606(1987)99<605:SMOCJA>2.0.CO;2.

De Wit, M., Jeffery, M., Bergh, H., and Nicolaysen, L., 1988, Geological map of sectors of Gondwana reconstructed to their disposition ~150 Ma: Tulsa, Oklahoma, American Association of Petroleum Geologists, Tulsa, Oklahoma, scale 1:10,000,000.

Ernst, R.E., and Buchan, K.L., 2001, Large mafic magmatic events through time and links to mantle-plume heads, *in* Ernst, R.E., and Buchan, K.L., eds., Mantle plumes: Their identification through time: Boulder, Colorado, Geological Society of America Special Paper 352, p. 483–566.

Fairhead, J.D., and Wilson, M., 2006, Sea-floor spreading and deformation processes in the South Atlantic Ocean: Are hot spots needed?: www.mantleplumes.org/SAtlantic.html.

Foulger, G.R., Pritchard, M.J., Julian, B.R., Evans, J.R., Allen, R.M., Nolet, G., Morgan, W.J., Bergsson, B.H., Erlendsson, P., Jakobsdóttir, S., Ragnarsson, S., Stefansson, R., and Vogfjörd, K., 2000, The seismic anomaly beneath Iceland extends down to the mantle transition zone and no deeper: Geophysical Journal International, v. 142, p. f1–f5, doi: 10.1046/j.1365-246x.2000.00245.x.

Foulger, G.R., Natland, J.H., Presnall, D.C., and Anderson, D.L., eds., 2005, Plumes, plates, and paradigms: Boulder, Colorado, Geological Society of America Special Paper 388, 861 p.

Golonka, J., Ross, M.I., and Scotese, C.R., 1994, Phanerozoic paleogeographic and paleoclimatic modeling maps, *in* Embry, A.F., et al., eds., Pangaea: Global environments and resources: Calgary, Canadian Society of Petroleum Geologists, p. 1–47.

Gordon, R.G., McWilliams, M.O., and Cox, A., 1979, Pre-Tertiary velocities of the continents: A lower bound from paleomagnetic data: Journal of Geophysical Research, v. 84B, p. 5480–5486.

Hamilton, W.B., 2003, An alternative Earth: GSA Today, v. 13, no. 11, p. 4–12, doi: 10.1130/1052-5173(2003)013<0004:AAE>2.0.CO;2.

Haq, U., 1995, Sea level change: Geotimes, v. 40, p. 45–46.

Harrowfield, M., Holdgate, G.R., Wilson, C.J.L., and McLoughlin, S., 2005, Tectonic significance of the Lambert graben, East Antarctica: Reconstructing the Gondwana rift: Geology, v. 33, p. 197–200, doi: 10.1130/G21081.1.

Hauser, M., Martini, R., Matter, A., Krystyn, L., Peters, T., Stampfi, G., and Zaninetti, L., 2002, The break-up of East Gondwana along the northeast coast of Oman: Evidence from the Batain basin: Geological Magazine, v. 139, p. 145–157, doi: 10.1017/S0016756801006264.

Hills, E.S., 1963, Elements of structural geology: New York, Wiley, 483 p.

Jagla, E.A., and Rojo, A.G., 2002, Sequential fragmentation: The origin of columnar quasihexagonal patterns: Physical Review E, v. 65, 026203, 7 p.

Jourdan, F., Féraud, G., Bertrand, H., Watkeys, M.K., Kampunzu, A.B., and Galle, B.L., 2006, Basement control on dyke distribution in Large Igneous Provinces: Case study of the Karoo triple junction: Earth and Planetary Science Letters, v. 241, p. 307–322, doi: 10.1016/j.epsl.2005.10.003.

Lawver, L.A., Gahagan, L.M., and Dalziel, I.W.D., 1999, A tight fit: Early Mesozoic Gondwana, a plate reconstruction perspective: Memoirs of the National Institute of Polar Research, v. 53, Special Issue, p. 214–229.

Lowman, J.P., and Gable, C.W., 1999, Thermal evolution of the mantle following continental aggregation in 3D convection models: Geophysical Research Letters, v. 26, p. 2649–2652, doi: 10.1029/1999GL008332.

Morgan, W.J., 1971, Convection plumes in the lower mantle: Nature, v. 230, p. 42–43, doi: 10.1038/230042a0.

Morgan, W.J., 1981, Hot spot tracks and the opening of the Atlantic and Indian Oceans, *in* Emiliani, C., ed., The sea, v. 7: New York, Wiley, p. 443–487.

Phillips, B.R., and Bunge, H.-P., 2005, Heterogeneity and time dependence in 3D spherical mantle convection models with continental drift: Earth and Planetary Science Letters, v. 233, no. 1–2, p. 121–135, doi: 10.1016/j.epsl.2005.01.041.

Ryan, M.P., and Sammis, C.G., 1978, Cyclic fracture mechanics in cooling basalt: Geological Society of America Bulletin, v. 89, p. 1295–1308, doi: 10.1130/0016-7606(1978)89<1295:CFMICB>2.0.CO;2.

Sears, J.W., 2001, Icosahedral fracture tessellation of early Mesoproterozoic Laurentia: Geology, v. 29, p. 327–330, doi: 10.1130/0091-7613(2001)029<0327:IFTOEM>2.0.CO;2.

Sears, J.W., 2006, Belt basin: A triskele rift junction, Rocky Mountains, Canada and USA: Northwest Geology, v. 35, p. 77–86.

Sears, J.W., St. George, G.M., and Winne, J.C., 2005, Continental rift systems and anorogenic magmatism: Lithos, v. 80, no. 1–4, p. 147–154, doi: 10.1016/j.lithos.2004.05.009.

Şengör, A.M.C., and Natal'in, B.A., 2001, Rifts of the world, *in* Ernst, R.E., and Buchan, K.L., eds., Mantle plumes: Their identification through time: Boulder, Colorado, Geological Society of America Special Paper 352, p. 389–482.

Silver, P.G., Behn, M.D., Kelley, K., Schmitz, M., and Savage, B., 2006, Understanding cratonic flood basalts: Earth and Planetary Science Letters, v. 245, p. 190–201, doi: 10.1016/j.epsl.2006.01.050.

Steinberger, B., 2000, Plumes in a convecting mantle: Models and observations for individual hotspots: Journal of Geophysical Research, v. 105, no. B5, p. 11,127–11,152, doi: 10.1029/1999JB900398.

Storey, B.C., Leat, P.T., and Ferris, J.K., 2001, The location of mantle-plume centers during the initial stages of Gondwana breakup, *in* Ernst, R.E., and Buchan, K.L., eds., Mantle plumes: Their identification through time: Boulder, Colorado, Geological Society of America Special Paper 352, p. 71–80.

Suppe, J., 1985, Principles of structural geology: Englewood Cliffs, New Jersey, Prentice-Hall, 537 p.

Tommasi, A., and Vauchez, A., 2001, Continental rifting parallel to ancient orogenic belts: An effect of the mechanical anisotropy of the lithospheric mantle: Earth and Planetary Science Letters, v. 185, p. 199–210, doi: 10.1016/S0012-821X(00)00350-2.

Vauchez, A., Barruol, G., and Tommasi, A., 1997, Why do continents break-up parallel to ancient orogenic belts?: Terra Nova, v. 9, p. 62–66, doi: 10.1111/j.1365-3121.1997.tb00003.x.

Vauchez, A., Tommasi, A., and Barruol, G., 1998, Rheological heterogeneity, mechanical anisotropy and deformation of the continental lithosphere: Tectonophysics, v. 296, p. 61–86, doi: 10.1016/S0040-1951(98)00137-1.

Weinberger, R., 2001, Evolution of polygonal patterns in stratified mud during dessication: The role of flaw distribution and layer boundaries: Geological Society of America Bulletin, v. 113, p. 20–31, doi: 10.1130/0016-7606(2001)113<0020:EOPPIS>2.0.CO;2.

Wilson, J.T., 1963, A possible origin of the Hawaiian Islands: Canadian Journal of Physics, v. 41, p. 863–868.

Zandi, R., Reguera, D., Bruinsma, R., Gelbart, W., and Rudnick, J., 2005, Assembly and disassembly of viral capsids: Journal of Theoretical Medicine, v. 6, p. 69–72, doi: 10.1080/10273660500149166.

MANUSCRIPT ACCEPTED BY THE SOCIETY JANUARY 31, 2007

DISCUSSION

17 January 2007, Sergio Rocchi

This article presents a non-plume-paradigmatic, interesting view about the top-down control of plate fragmentation and massive outburst of magma. Two points raised in the chapter are worth emphasizing: (1) the comparison between Gondwana fractures and inherited weak belts within the supercontinent and (2) the age of magmatism later than rifting. A comparison of Gondwana fractures and troughs with former craton borders and continent structural grain clearly supports a prominent role for lithospheric weak linear structures in determining the location of Gondwana rifts and fractures (Vauchez et al., 1997, 1998; Tommasi and Vauchez, 2001). Indeed, both the successful and failed Gondwana fractures are located between two old cratons, e.g., the Benue Trough between West Africa–Sahara–Chad and Congo cratons, the Transkei–Namibia line between Kapvaal (Kalahari) and Zimbabwe/Congo cratons, the Parana line between Amazonian–San Feliciano and Rio de la Plata cratons.

The development of large igneous provinces (LIPs) some 100 Ma after the development of fractures is a significant observation to be coupled with structural-geochronological data from the Karoo triple junction. Here it is demonstrated that magmatism occurred both several tens of millions of years after fracture development and long before it did in the Proterozoic and the Archean (Jourdan et al., 2006). Thus, the triple junction is triple in space but not in time, the arms not being magmatically coeval. Rather, dike emplacement occurred over thousands of millions of years, further supporting a control by lithospheric structures that drove magma ascent over a time magnitude comparable to the lifespan of plate tectonics.

Furthermore, in some cases, the low-volume alkaline rift magmatism was also activated several tens of millions of years after the main rifting process, as for the west Antarctic rift system, where the main rift phase is late Creataceous and the magmatism started in the middle Eocene (Rocchi et al., 2002). This lends further support to the inference that initiation of rifting is not driven by impacting plumes, even when magmatism is low volume, and alternative mechanisms have to be invoked based on the role of inherited lithospheric structures (Salvini et al., 1997; Rocchi et al., 2005).

7 February 2007, James W. Sears

I thank Rocchi for sharing his observations of further evidence that many of the fractures that broke apart Gondwana followed earlier lithospheric weaknesses and that magmatic activity was sporadic along these fractures, in cases following initial rifting by tens of millions of years. His clarifying comment that the Karoo triple junction is triple in space but not in time accentuates the importance of considering the magmatic and structural aspects of continental breakup as separate phenomena. This strengthens the case for lithospheric control of continental breakup and large igneous eruptions. Evidently, the lithosphere breaks up along the least resistant zones that are favorably disposed to the global stress field.

REFERENCES CITED

Jourdan, F., Féraud, G., Bertrand, H., Watkeys, M.K., Kampunzu, A.B., and Galle, B.L., 2006, Basement control on dyke distribution in Large Igneous Provinces: Case study of the Karoo triple junction: Earth and Planetary Science Letters, v. 241, p. 307–322, doi:10.1016/j.epsl.2005.10.003.

Rocchi, S., Armienti, P., D'Orazio, M., Tonarini, S., Wijbrans, J., and Di Vincenzo, G., 2002, Cenozoic magmatism in the western Ross Embayment: Role of mantle plume vs. plate dynamics in the development of the West Antarctic Rift System: Journal of Geophysical Research, v. 107, no. B9, p. 2195, doi: 10.1029/2001JB000515.

Rocchi, S., Di Vincenzo, G., and Armienti, P., 2005, No plume, no rift magmatism in the West Antarctic Rift, *in* Foulger, G.R., et al., eds., Plates, plumes and paradigms: Boulder, Colorado, Geological Society of America Special Paper 388, p. 435–447, doi: 10.1130/2005.2388(26).

Salvini, F., Brancolini, G., Busetti, M., Storti, F., Mazzarini, F., and Coren, F., 1997, Cenozoic geodynamics of the Ross Sea region, Antarctica: Crustal extension, intraplate strike-slip faulting, and tectonic inheritance: Journal of Geophysical Research, v. 102, no. B11, p. 24,669–24,696.

Tommasi, A., and Vauchez, A., 2001, Continental rifting parallel to ancient orogenic belts: An effect of the mechanical anisotropy of the lithospheric mantle: Earth and Planetary Science Letters, v. 185, p. 199–210.

Vauchez, A., Barruol, G., and Tommasi, A., 1997, Why do continents break-up parallel to ancient orogenic belts?: Terra Nova, v. 9, p. 62–66.

Vauchez, A., Tommasi, A., and Barruol, G., 1998, Rheological heterogeneity, mechanical anisotropy and deformation of the continental lithosphere: Tectonophysics, v. 296, p. 61–86.

The Geological Society of America
Special Paper 430
2007

The origin of post-Paleozoic magmatism in eastern Paraguay

Piero Comin-Chiaramonti
Dipartimento di Scienze della Terra dell'Università di Trieste, Via Weiss 8, I-34127 Trieste, Italy
Andrea Marzoli*
Dipartimento di Mineralogia e Petrologia, Padua University, Corso Garibaldi 37, I-35137 Padua, Italy
Celso de Barros Gomes
Anderson Milan
Claudio Riccomini
Instituto de Geociências, Universidade de São Paulo, Rua do Lago 562, CEP 05508-080 São Paulo, Brazil
Victor Fernandez Velázquez
Escola de Artes, Ciências e Humanidades, Universidade de São Paulo, Rua Arlindo Bettio, 03828-000 São Paulo, Brazil
Marta M.S. Mantovani
Instituto de Astronomia, Geofísica e Ciências Atmósferica, Universidade de São Paulo, Rua do Matão 1226,
CEP 05508-900 São Paulo, Brazil
Paul Renne
Department of Earth and Planetary Science, 307 McCone Hall, University of California, Berkeley, California 94720-4767, USA,
and Berkeley Geochronology Center, Berkeley, California 94709, USA
Colombo Celso Gaeta Tassinari
Instituto de Geociências, Universidade de São Paulo, Rua do Lago 562, CEP 05508-080 São Paulo, Brazil
Paulo Marcos Vasconcelos
University of Queensland, Department of Earth Sciences, Brisbane, Queensland 4072, Australia

ABSTRACT

The ages of magmatic rocks are crucial for understanding of the geodynamic relationships among different magmatic events. Between the compressional Andean and the extensional Atlantic systems, Paraguay has been the site of six main taphrogenic events since the end of Paleozoic times. Other than the Paraná flood tholeiites (133–134 Ma; Early Cretaceous, Hauterivian), new high-precision $^{40}Ar/^{39}Ar$ ages show that other alkaline magmatism of various types occurred, namely sodic magmatism at 241.5 ± 1.3 Ma (Middle Triassic, Anisian), 118.3 ± 1.6 Ma (late Early Cretaceous, Aptian), and 58.7 ± 2.4 Ma (Paleocene); and potassic magmatism at 138.9 ± 0.7 (Early Cretaceous, Venginian) and 126.4 ± 0.4 Ma (Early Cretaceous, Barremian). The main geochemical characteristics of the sodic alkaline rock types are systematic Nb-Ta positive anomalies and Sr-Nd isotopes trending to the bulk Earth or the depleted mantle components, contrasting with potassic rocks and tholeiitic basalts that show negative Nb-Ta anomalies and Sr-Nd isotopes trending to the enriched mantle components. The Pb isotope versus Sr-Nd systematics confirm the distinction between potassic rocks enriched in "high-radiogenic" Sr and low in "less radiogenic" Nd-Pb and sodic rocks ranging from depleted components to bulk Earth and transitional to the Paraná flood

*E-mail: andrea.marzoli@unipd.it.

Comin-Chiaramonti, P., Marzoli, A., de Barros Gomes, C., Milan, A., Riccomini, C., Fernandez Velázquez, V., Mantovani, M.M.S., Renne, P., Tassinari, C.C.G., and Marcos Vasconcelos, P., 2007, Origin of post-Paleozoic magmatism in eastern Paraguay, *in* Foulger, G.R., and Jurdy, D.M., eds., Plates, plumes, and planetary processes: Geological Society of America Special Paper 430, p. 603–633, doi: 10.1130/2007.2430(29). For permission to copy, contact editing@geosociety.org. ©2007 The Geological Society of America. All rights reserved.

tholeiites. The occurrence of alkaline, both sodic and potassic (and carbonatititic), and tholeiitic magmatism in the whole Paraná-Angola-Etendeka system, and even in the Andean system, implies appropriate lithospheric sources to generate the various types of magmatic rocks. Therefore, any hypothesis of an asthenospheric plume origin is not compelling; rather, possibly such a plume provided a thermal perturbation and/or a decompressional environment, and possibly mantle sources were driven by Precambrian melts that contaminated and veined the lithosphere. A decompressional environment is inferred as a possible mechanism driven by differential rotation of different subplates in the South America and south Africa plates.

Keywords: Eastern Paraguay, alkaline magmatism, isotopic ages, geochemistry, geodynamic implications

INTRODUCTION

Eastern Paraguay lies in an intercratonic region that includes the westernmost side of the Paraná-Angola-Edendeka system. It is bounded by an anticlinal structure established since the early Paleozoic, the Asunción arch, separating the Paraná basin in the east from the Gran Chaco basin in the west (Almeida, 1983; Comin-Chiaramonti et al., 1997). Notably, Paraguay is located in South America between the compressional Andean and extensional Atlantic systems (Fig. 1; see Gudmundsson and Sambridge, 1998). A Cretaceous rift system, developed in the early Paleozoic mobile belt on the Pacific side of the continent, delimits part of the Brazilian Shield (cf. Lucassen et al., 2002). However, at the latitude of Paraguay, the exact limit between Andean and Atlantic systems is unknown due to the presence of the Chaco-Pantanal Paleogene–Neogene basin. The basement rocks are mainly Precambrian to early Paleozoic granitic intrusions and high- to low-grade metamorphic metasediments. These are considered the northernmost occurrence of the Rio de la Plata craton and the southernmost tip of the Amazon craton (Fulfaro, 1995), in the southern and northern parts of eastern Paraguay, respectively. Post-Paleozoic magmatism, both alkaline and tholeiitic, affected the region, following various pulses of western Gondwana evolution and breakup.

This study offers the first high-precision $^{40}Ar/^{39}Ar$ ages for the ~200 m.y. span of alkaline magmatism that intermittently occurred in eastern Paraguay from Triassic to Paleogene times. The objective is to define the age, duration, and geodynamic significance of the magmatic events from this relatively little-known region. The new data represent new constraints relative to the main magmatic pulses, often in disagreement with older age data (K/Ar and Rb/Sr, mainly). A geochemical review and comparison with the other magmatic occurrences, both alkaline and tholeiitic, in the whole Paraná-Angola-Edendeka system is also presented in order to provide a complete picture of the evolution of western Gondwana before, during, and after its breakup.

SIX MAGMATIC EVENTS IN EASTERN PARAGUAY

Lying on the westernmost side of the Paraná-Angola-Etendeka system, eastern Paraguay represents a magmatic province in and around the Paraná basin where six main magmatic events occurred in a relatively restricted area (i.e., less than 120,000 km^2; Fig. 2) from the end of the Paleozoic to the Cenozoic. This is shown by geological evidence and by previous regional and geochronological studies, from which the following conclusions have been drawn (cf. Comin-Chiaramonti and Gomes, 1995, 2005, and references therein):

1. The Permo-Triassic sodic magmatism of the Alto Paraguay province (255–210 Ma; Gomes et al., 1995, and references therein) is widespread on the southernmost side of the Amazon craton (Fulfaro, 1995; Comin-Chiaramonti et al., 2005a).
2. Potassic alkaline-carbonatitic complexes and dikes from northeastern Paraguay, from the Rio Apa (ca. 142 Ma, as inferred from Gibson et al., 1995a) and Amambay areas (average 141 Ma; Sonoki and Garda, 1988; Eby and Mariano, 1992), predate the tholeiitic flood basalts (Paraná, Serra Geral Formation).
3. The Paraná Serra Geral Formation flood tholeiites and dikes (133 ± 1 Ma according to Renne et al., 1992, 1993, 1996; 137–127 Ma according to Turner et al., 1994, and Stewart et al., 1996) are both represented by high-Ti and low-Ti basalts (cf. Bellieni et al., 1986; Piccirillo and Melfi, 1988).
4. Potassic alkaline complexes and dikes (132–115 Ma; Bitschene, 1987; Comin-Chiaramonti and Gomes, 1995) with subordinate silicocarbonatite flows and dikes are widespread, mainly in the Asunción-Sapucai-Villarrica Graben (central potassic province; Comin-Chiaramonti et al., 1997, 1999).
5. Sodic alkaline complexes, plugs, and dikes (ca. 120 Ma; Comin-Chiaramonti et al., 1992) occur mainly in the Misiones province (San Juan Bautista region), southwestern Paraguay.
6. Paleogene sodic alkaline complexes, plugs, and dikes (66–33 Ma; Bitschene, 1987; Comin-Chiaramonti et al., 1991; Comin-Chiaramonti and Gomes, 1995) crop out on the western side of the Asunción-Sapucai-Villarrica Graben.

The Permo-Triassic rocks form subcircular complexes following a north-south trend and are mainly formed by nepheline syenites and syenites and their effusive equivalents (Comin-

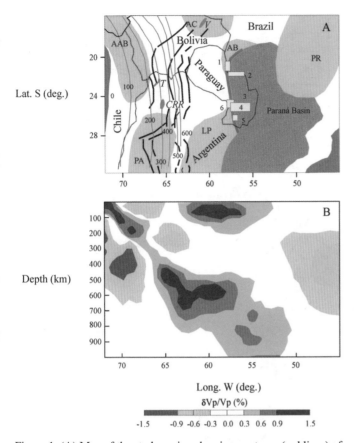

Figure 1. (A) Map of the study region showing contours (red lines) of the depth (km) of the subducting Nazca slab based on seismic data (Gudmundsson and Sambridge, 1998). The heavy black lines outline the Cretaceous rift system, which roughly marks the limit between the Brazilian Shield and the Paleozoic (Pacific) mobile belt (cf. Lucassen et al., 2002). The pink fields delineate the inferred positions of major cratonic fragments below the Phanerozoic cover (after Laux et al., 2005): AAB—Arequipa-Antofalla; AC—Amazon craton; AB—Apa block; PR—Paranapanema; LP—Rio de la Plata; PA—Pampia. Other features: 1—Alto Paraguay sodic alkaline province; 2—Apa and Amabay potassic alkaline province; 3—Paraná Serra Geral flood tholeiites; 4—Asunción-Sapucai-Villarica potassic alkaline central province (ASU); 5—Misiones sodic alkaline province; 6—Asunción sodic alkaline province. For comparison: V—Velasco complexes (Bolivia, 139 Ma; Comin-Chiaramonti et al., 2005b); T—Tusaquillas complexes (Argentina, 144–140 Ma; Cristiani et al., 2005); CRR—alkaline rock types from the Andean Central Rift (Argentina, 90 Ma; Lucassen et al., 2002). (B) Seismic tomography image of Liu et al. (2003) along a profile at ~24°S. The low-velocity feature in the mantle to the east has been interpreted as a fossil mantle plume by VanDecar et al. (1995).

Chiaramonti et al., 2005a). Early Cretaceous alkaline magmatism, both predating and postdating the tholeiitic effusions, is moderately to strongly potassic, represented by rock types from alkali basalt to trachyte and from basanite to phonolite and their intrusive equivalents. They are often associated with carbonatitic rock types (Comin-Chiaramonti and Gomes, 1995; Comin-Chiaramonti et al., 1997, 1999). Early Cretaceous tholeiites are mainly basalts and andesibasalts, both belonging to the high-Ti and the low-Ti suites (Bellieni et al., 1986; Piccirillo and Melfi, 1988; Comin-Chiaramonti and Gomes, 1995; Peate et al., 1999).

The Cretaceous and Paleogene sodic rocks, including ankaratrites, nephelinite, and phonolites, are both characterized by mantle xenoliths (spinel peridotite facies; Comin-Chiaramonti et al., 1992, 2001, and references therein).

GEOLOGICAL-GEOPHYSICAL BACKGROUND

Seismic tomography images show two high-velocity features beneath Paraguay (P-waves; cf. Fig. 1B), up to ~200 km and >450 km in depth, respectively. The latter are probably part of the subducting Nazca slab (Liu et al., 2003). A low-velocity anomaly (Fig. 1B) in the upper mantle and the mantle transition zone was interpreted as a fossil plume by VanDecar et al. (1995). However, thinning of the mantle transition zone has not been observed, and Liu et al. (2003) suggest that either this thermal anomaly does not extend into the mantle transition zone or, alternatively, the observed anomaly is not primarily thermal, but instead dominantly compositional in origin (e.g., "veined" mantle).

In particular, eastern Paraguay, on the westernmost side of the Paraná intercratonic basin (Figs. 1 and 2), is bounded by the north-south Paraguay River Lineament, separating the Paraná basin (east) from the Gran Chaco basin (west) (Almeida, 1983; Comin-Chiaramonti et al., 1997). The basement rocks are mainly Proterozoic to early Paleozoic granitic intrusions, rhyolitic flows, and high- to low-grade metasediments formed during the assembly of Gondwana and the Brasiliano orogenic cycles, corresponding to the Pan African cycle (Wiens, 1986; Kanzler, 1987; Velázquez et al., 1996; Comin-Chiaramonti et al., 2000, and references therein). According to Brito Neves et al. (1999) the basement of the South American Platform displays the lithostructural and tectonic records of three major orogenic collages: the middle Paleoproterozoic or Transamazonic, the late Mesoproterozoic or early Neoproterozoic, and the late Neoproterozoic or Cambrian (the Brasiliano–Pan African collage).

During early to late Mesozoic times, eastern Paraguay was subjected to northeast-southwest-trending crustal extension related to the breakup of western Gondwana and seafloor spreading in the south Atlantic that started at ca. 125–127 Ma (chron M4; Nürnberg and Müller, 1991). Added to these were the effects of relative motions and different angular velocities of various South America subblocks (Unternehr et al., 1988; Turner et al., 1994; Comin-Chiaramonti et al., 1995a,b, 1996; Prezzi and Alonso, 2002). The resulting tectonic pattern controlled the development of graben or half-graben structures in response to the northeast-southwest-directed extension, which lasted until the Oligocene (Comin-Chiaramonti et al., 1992, 1999; Comin-Chiaramonti and Gomes, 1995; Hegarty et al., 1995; Riccomini et al., 2001; Velázquez et al., 2006).

The Alto Paraguay alkaline complexes predate the "central Atlantic magmatic province" event (cf. Iacumin et al., 2003) and were emplaced along the Paraguay belt, a Cambrian suture between the southernmost tip of the Amazon plate and the Paraná block (Ussami et al., 1999). Alternatively, this suture may represent the eastern limit (marked by the Paraguay River) of the Pungoviscana Trough, a large basin of late Precambrian

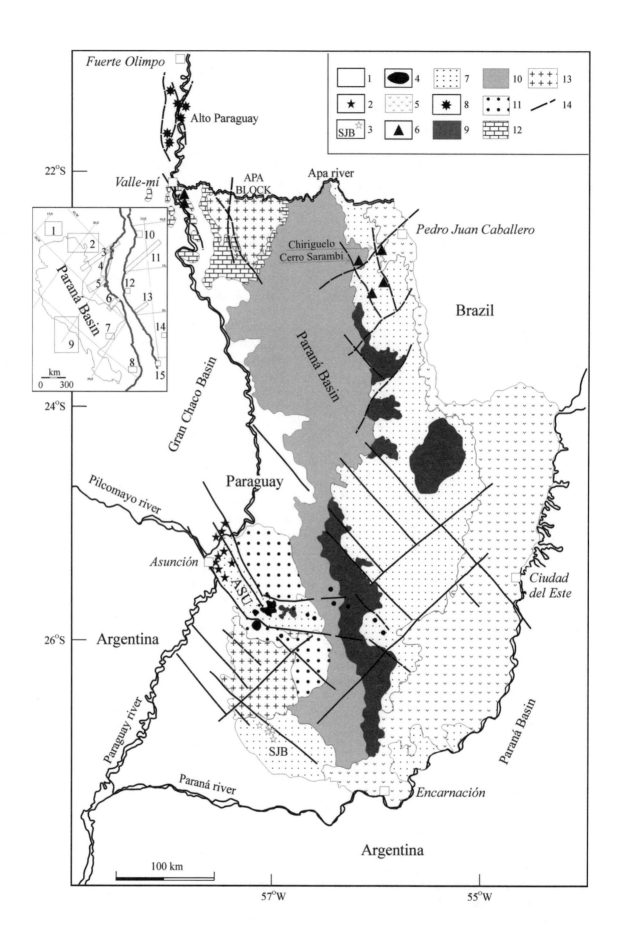

Legend items: 1, 2, 3 (SJB), 4, 5, 6, 7, 8, 9, 10, 11, 12, 13, 14

Fuerte Olimpo

Alto Paraguay

22°S

Valle-mí

APA BLOCK

Apa river

Pedro Juan Caballero

Chiriguelo
Cerro Sarambi

Brazil

Paraná Basin

Paraná Basin

Gran Chaco Basin

1

2

Paraná Basin

10

11

3

4

12

13

5

6

7

9

14

8

15

km
0   300

24°S

Pilcomayo river

Paraguay

Asunción

ASU

Ciudad
del Este

Argentina

26°S

Paraguay river

SJB

Paraná river

Encarnación

Argentina

57°W

55°W

100 km

to Early Cambrian age on the Pacific edge of the Brazilian Shield (Jezek et al., 1985; Mantovani et al., 2005).

Potassic magmatism from central–eastern Paraguay (and the Paraná flood tholeiites) is associated with the breakup of Gondwana and predates the separation of Africa and South America, while late sodic magmatism postdates the formation of the southern Atlantic Ocean (Comin-Chiaramonti and Gomes, 1995). Both potassic and sodic magmatic rocks were mainly emplaced along northwest-southeast-trending rift structures in eastern Paraguay, corresponding to a general northeast-southwest extensional event that started at least as early as Triassic times (Comin-Chiaramonti et al., 1996, 1999). Similar alkaline magmatic events—i.e., those in Early Cretaceous, later Cretaceous, and Paleogene time—were widespread everywhere in the Paraná-Angola-Etendeka system (Brazil, Angola, and Namibia; Hawkesworth et al., 1986; Comin-Chiaramonti and Gomes, 1995, 2005; Peate et al., 1999).

EASTERN PARAGUAY: PREVIOUS RADIOISOTOPIC AGES

The ages of the magmatic events in eastern Paraguay are poorly constrained by a large set of not very precise K/Ar, nonplateau Ar/Ar, Rb/Sr, and fission track data for both minerals and whole rocks (Table 1).

The Alto Paraguay rocks (sodic magmatism) and minerals (biotite, amphibole, and alkali feldspars) have an age span of 255–210 Ma (i.e., Upper Permian, Lopingian, to Upper Triassic, Norian; timescale according to Gradstein et al., 2004). On the whole, the large range of measured ages on amphiboles, alkali feldspars, and whole-rock samples indicates that the K/Ar and Rb/Sr isotopic data are affected by complexities linked earlier to subsolidus reactions and exsolutions, hydrothermal alteration, and weathering (cf. Velázquez et al., 1992, 1996). The smallest analytical errors and age range (i.e., 248–242 Ma) are yielded by K/Ar and Ar/Ar (nonplateau) analyses of biotite separates (cf. Table 1).

The age of the potassic alkaline magmatism from Amambay and Rio Apa (cf. Fig. 1) is poorly documented by previous K/Ar data on biotite and whole-rock samples and by fission-track data on titanite and apatite. The ages of basanites to trachytes span 147–128 Ma, i.e., from Upper Jurassic to Early Cretaceous. We note that ages younger than 133 Ma (cf. Eby and

Mariano, 1992) are not consistent with the field relationships (e.g., the Chiriguelo alkaline-carbonatitic complex; cf. Censi et al., 1989), which clearly show that the alkaline rocks and associated carbonatite predate the Paraná flow tholeiites (main peak at ca. 133 Ma).

The field relationships relative to the potassic alkaline rocks from the Asunción-Sapucai Graben (Comin-Chiaramonti et al., 1997) in the central provinces show that they represent a post-tholeiitic event. These rocks display a bimodal Early Cretaceous distribution (Barremian to Aptian), with the older averaged age at 129 ± 4 Ma (intrusive and effusive variants and dikes; mainly mineral K/Ar and Rb/Sr isochron ages) and the younger one represented only by potassic dikes at 119 ± 2 Ma (K/Ar whole-rock data).

The Misiones, San Juan Bautista, sodic magmatism is very poorly documented from a geochronological point of view. The only available K/Ar age (whole-rock, 120 Ma; cf. Comin-Chiaramonti et al., 1992), is similar to that of the younger Asunción-Sapucai-Villarrica potassic dikes.

Asunción sodic plugs, mainly in the northwestern Asunción-Sapucai-Villarrica region, have an age span of 61–39 Ma (Paleocene to Eocene), with an average of 50 ± 8 Ma (K/Ar ages on whole rocks; Comte and Hasui, 1971; Bitschene, 1987; Comin-Chiaramonti et al., 1991). A few sodic plugs and dikes outcropping in the central Asunción-Sapucai-Villarrica region yield similar K/Ar ages, ranging from 66 to 33 Ma. Despite the large interval of the measured ages, it is clear that Asunción-Sapucai-Villarrica sodic magmatism represents the youngest magmatic event in eastern Paraguay.

NEW ^{40}Ar/^{39}Ar AGES

Analytical Procedures

New ^{40}Ar/^{39}Ar analyses have been performed on seventy-six whole-rock samples and on mineral separates (alkali feldspar, plagioclase, amphibole, and biotite) from the five main alkaline magmatic suites of eastern Paraguay and Alto Paraguay at the Berkeley Geochronology Center (BGC, USA; nine analyses) and at the Geochronological Research laboratory of São Paulo University (USP, Brazil; sixty-seven analyses). Sample preparation, irradiation, and analyses followed methods and were performed in facilities described by Renne et al. (1996,

Figure 2. Geological map of eastern Paraguay (after Comin-Chiaramonti and Gomes, 1995). Color codes: 1—Neogene and Paleogene sedimentary cover (Gran Chaco; Argentina, Partim; eastern Paraguay); 2—Paleogene sodic alkaline rocks; 3—late Early Cretaceous sodic alkaline rocks (Misiones Province, San Juan Bautista [SJB]); 4—Early Cretaceous potassic alkaline rocks (post-tholeiites; ASU: Asunción-Sapucai-Villarica Graben); 5—Early Cretaceous tholeiites of the Paraná basin; 6—Early Cretaceous potassic alkaline rocks (pretholeiites); 7—Jurassic–Cretaceous sedimentary rocks (Misiones Formation); 8—Permo-Triassic alkaline rocks (Alto Paraguay province); 9—Permian sedimentary rocks (Independencia Group); 10—Permo-Carboniferous sedimentary rocks (Coronel Oviedo Group); 11—Ordovician–Silurian sedimentary rocks (Caacupé and Itacurubí groups); 12—Cambro-Ordovician platform carbonates (Itacupumí Group); 13—Archean and Neo-Proterozoic crystalline basement: high- to low-grade metasedimentary rocks, metarhyolites, and granitic intrusions; 14—faults. The inset indicates the distribution of the main areas occupied by alkaline magmatism in the Paraná-Angola-Etendeca system (cf. also Fig. 8). 1—Iporá; 2—Alto Paranaíba; 3—Cabo Frio; 4—Serra do Mar; 5—Ponta Grossa; 6—Anitápolis and Lages; 7—Piratiní; 8—Valle Chico; 9—Paraguay; 10—Kwanza basin, Angola; 11—Moçâmedes arch, Angola; 12—Virilundo, Angola; 13—Damara belt, Namibia; 14—Blue Hills, Namibia; 15—Dicker Willem Complex, Namibia.

TABLE 1. PREVIOUS RADIOISOTOPIC DATA FOR MESOZOIC–CENOZOIC MAGMATIC ROCKS OF EASTERN PARAGUAY

Region or locality	Occurrence	Rock type	Material	Method	Age (Ma)	Reference
Alto Paraguay:						
Sodic magmatism						
Cerro Siete Cabezas	Stock	Nepheline syenite	Am	K/Ar	227.9 ± 7.8	1
Cerro Siete Cabezas	Stock	Nepheline syenite	Bi	K/Ar	249.0 ± 3.0	1
Cerro Siete Cabezas	Stock	Syenite	Am	K/Ar	229.8 ± 8.3	1
Cerro Siete Cabezas	Stock	Syenite	Bi	K/Ar	244.4 ± 10.4	1
Cerro Siete Cabezas	Stock	Syenite–quartz syenite	AF-WR	Rb/Sr erorchron	255 ± 11	1
Cerro Siete Cabezas	Stock	Nepheline syenite	Am	Ar/Ar	236.0 ± 1.6	2
Cerrito	Stock	Nepheline syenite	Bi	K/Ar	253.2 ± 9.2	1
Fecho dos Morros	Stock	Nepheline syenite	Am	K/Ar	212.8 ± 14.8	1
Pão de Açucar	Stock	Nepheline syenite	Am	K/Ar	233.2 ± 7.2	1
Pão de Açucar	Stock	Nepheline syenite	Bi	K/Ar	248.3 ± 5.3	1
Pão de Açucar	Lava flow	Phonolite	WR	K/Ar	219.1 ± 13.3	3
Pão de Açucar	Lava flow	Trachyphonolite	Bi	Ar/Ar	242 ± 1.6	2
Pão de Açucar	Stock	Nepheline syenite	Bi	K/Ar	244.6	4
Pão de Açucar	Stock	Nepheline syenite	Bi	K/Ar	241.7	4
Pão de Açucar	Stock	Nepheline syenite	AF	K/Ar	211.3	4
Pão de Açucar	Stock	Nepheline syenite	Am	K/Ar	209.6	4
Cerro Boggiani	Stock	Nepheline syenite	Am	K/Ar	234.6 ± 13.7	1
Cerro Boggiani	Stock	Nepheline syenite	Am	K/Ar	234.0 ± 9.0	1
Cerro Boggiani	Lava flow	Peralkaline phonolite	WR	K/Ar	236.7 ± 10.9	1
Cerro Boggiani, Fecho dos Morros, Cerrito		Nepheline syenite Nepheline syenite Peralkaline phonolite	WR, Bi, Am, AF	Rb/Sr	255 ± 11	1
Rio Apa: Potassic magmatism						
Valle-Mí	Dike	Basanite	WR	K/Ar	142 ± 2	5
Amambay						
Cerro Chiriguelo	Stock	Ca-carbonatite	Bi	K/Ar	128 ± 5	6
Cerro Chiriguelo	Lava flow	Trachyte	Bi	K/Ar	146.7 ± 9.2	6a
Cerro Chiriguelo	Lava flow	Trachyte	WR	K/Ar	138.9 ± 9.2	6a
Cerro Sarambi	Dike	Syenite	WR	K/Ar	140 ± 1	5
Arroyo Gasory	Dike	Trachyte	Tit	Fission-track	137 ± 7	6
Arroyo Gasory	Dike	Trachyte	Ap	Fission-track	145 ± 8	6
Central Provinces (ASU):						
Potassic magmatism						
Cerro Km 23	Stock	Theralite	Bi	K/Ar	131.9 ± 5.0	7
Cerro Km 23	Dike	Basanite	WR	K/Ar	115.8 ± 4.2	7
Ybytyruzù (Cerro Acatí)	Lava flow	Trachyphonolite	Bi	K/Ar	125.9 ± 4.6	7
Ybytyruzù (Cerro Boni)	Lava flow	Latite	Bi	K/Ar	124.6 ± 4.2	7
Ybytyruzù (Cerro Itatí)	Lava flow	Phonotephrite	Bi	K/Ar	128.8 ± 4.6	7
Mbocayaty	Stock	Nepheline syenodiorite	AF	K/Ar	130.0 ± 3.4	8
Mbocayaty	Stock	Nepheline syenodiorite	Bi	K/Ar	129.2 ± 6.8	8
Mbocayaty	Stock	Essexite	Bi	K/Ar	128.2 ± 4.5	7
Mbocayaty	Stock	Essexite, nepheline syenodiorite	WR, Bi, AF	Rb/Sr isochron	126.5 ± 7.6	8
Aguapety Portón	Stock	Malignite	WR	K/Ar	138.1 ± 4.8	7
Aguapety Portón	Stock	Essexite	Bi	K/Ar	132.9 ± 5.5	7
Aguapety Portón	Stock	Essexite, malignite	WR, Bi, AF	Rb/Sr isochron	128.2 ± 4.5	8
Potrero Ybaté	Stock	Nepheline syenodiorite	WR, Bi, AF	Rb/Sr isochron	126.5 ± 7.6	8
Potrero Ybaté	Stock	Nepheline syenodiorite	AF	K/Ar	127.8 ± 5.6	8
Sapucai	Stock	Essexite	WR	K/Ar	131.0 ± 8.2	9
Sapucai	Lava flow	Phonolite	WR	K/Ar	136.4 ± 5.1	9
Sapucai	Dike	Basanite	WR	K/Ar	119.6 ± 7.2	9
Sapucai	Lava flow	Alkali basalt	WR	K/Ar	131.2 ± 5.1	9
Sapucai	Lava flow	Trachybasalt	WR	K/Ar	122.0 ± 4.0	2
Sapucai	Dike	Phonotephrite	WR	K/Ar	119.0 ± 4.0	2
Sapucai	Dike	Phonolite	WR	K/Ar	121.0 ± 4.0	2
Sapucai	Dike	Trachybasalt	WR	K/Ar	119.0 ± 4.0	2
Sapucai	Dike	Basanite	WR	K/Ar	118.0 ± 4.0	2
Sapucai	Dike	Phonotephrite	WR	K/Ar	119.0 ± 4.0	2
Sapucai	Stock	Theralite, syenogabbro	WR, Bi, AF	Rb/Sr isochron	126.5 ± 7.6	8

TABLE 1. *Continued*

Region or locality	Occurrence	Rock type	Material	Method	Age (Ma)	Reference
Cerro Santo Tomás	Stock	Syenogabbro	Bi	K/Ar	126.0 ± 4.5	7
Cerro Santo Tomás	Stock	Syenogabbro	FC	K/Ar	136.8 ± 5.0	7
Cerro Santo Tomás	Stock	Essexite	WR	K/Ar	136.5 ± 10.2	9
Cerro Santo Tomás	Dike	Basanite	Bi	K/Ar	130.1 ± 4.8	7
Cerro Santo Tomás	Dike	Basanite	Bi	K/Ar	127.9 ± 4.8	7
Cerro Santo Tomás	Stock	Nepheline syenodiorite	Bi	K/Ar	132.0 ± 11.5	3
Cerro Santo Tomás	Stock	Syenogabbro	WR, Bi, AF	Rb/Sr isochron	128.0 ± 8.0	7
Cerro Santo Tomás	Stock	Syenodiorite	WR, Bi, AF	Rb/Sr isochron	126.5 ± 7.6	8
Cerro Acahay	Dike	Trachybasalt	WR	K/Ar	118.0 ± 4.0	2
Cerro Arrúa-í	Stock	Nepheline syenodiorite	Bi	K/Ar	132.3 ± 8.4	8
Cerro Arrúa-í	Stock	Nepheline syenodiorite	WR, Bi, AF	Rb/Sr isochron	126.5 ± 7.6	8
Sapucai	Dike	Na-tephrite	WR	K/Ar	66.0 ± 2.0	2
Sapucai	Dike	Na-tephrite	WR	K/Ar	32.8 ± 0.9	2
Cerro Gimenez	Plug	Na-phonolite	WR	K/Ar	66.0 ± 4.6	8
Misiones: Sodic magmatism						
Estancia Guavira-y	Plug	Melanephelinite	WR	K/Ar	120 ± 5	10
Asunción: Sodic magmatism						
Cerro Patiño	Plug	Ankaratrite	WR	K/Ar	38.8 ± 2.3	7
Limpio	Plug	Melanephelinite	WR	K/Ar	50.2 ± 1.9	7
Cerro Verde	Plug	Ankaratrite	WR	K/Ar	57.0 ± 2.3	7
Villa Hayes	Plug	Ankaratrite	WR	K/Ar	58.4 ± 2.2	7
Cerro Ñemby	Plug	Melanephelinite	WR	K/Ar	45.7 ± 1.8	7
Remanso Castillo	Dike	Melanephelinite	WR	K/Ar	40.6 ± 1.7	7
Cerro Confuso	Plug	Phonolite	WR	K/Ar	55.3 ± 2.1	7
Cerro Confuso	Plug	Phonolite	WR	K/Ar	60.9 ± 4.4	7
Cerro Confuso	Plug	Phonolite	WR	K/Ar	59.3 ± 2.4	7
Nueva Teblada	Plug	Melanephelinite	WR	K/Ar	46.3 ± 2.0	7
Nueva Teblada	Plug	Melanephelinite	WR	K/Ar	56.7 ± 2.3	7
Cerro Lambaré	Plug	Melanephelinite	WR	K/Ar	48.9 ± 2.0	7
Cerro Lambaré	Plug	Melanephelinite	WR	K/Ar	48.9 ± 2.2	7
Cerro Tacumbú	Plug	Melanephelinite	WR	K/Ar	46.0 ± 7.0	3
Cerro Tacumbú	Plug	Melanephelinite	WR	K/Ar	41.3 ± 1.8	7

References: Previous radiometric ages for magmatic rocks from Alto Paraguay and eastern Paraguay. Abbreviations: WR, whole rock; Am, amphibole; Bi, biotite; FC, felsic concentrates; AF, alkali feldspar; Tit, titanite; Ap, apatite. Data source: 1. Velazquez et al., 1996; 2. Gomes et al., 1995; 3. Comte and Hasui, 1971; 4. Amaral et al., 1967; 5. Gibson et al., 1995a; 6. Eby and Mariano, 1992; 6a. Sonoki and Garda, 1988; 7. Bitschene, 1987; 8. Velazquez et al., 1992; 9. Palmieri and Arribas, 1975; 10. Comin-Chiaramonti et al., 1997.
Notes: ASU—Asunción-Sapucai-Villarica potassic alkaline central province; AF—alkali feldspar; Am—amphibole; Bi—biotite; WR—whole rock.

1998) and by Vasconcelos et al. (2002) for the argon-ion laser heating system at the BGC and USP, respectively. The mineral grains and rocks (16–25 mesh) were irradiated for 7 and 100 h respectively, along with Fish Canyon sanidine (28.02 Ma; Renne et al., 1998) neutron flux standards, at the TRIGA (Oregon State University) and at IPEN (Instituto de Pesquisas Energéticas/CNEN IEA-R1-USP) nuclear reactors, respectively. At both laboratories, the analyzed materials were step-heated using a fully automated noble gas extraction and purification system, and the Ar isotopic compositions were measured in static mode by a MAP-215-50 mass spectrometer. Isotopic run data were corrected for mass discrimination, radioactive decay, and nucleogenic interferences. Plateau and miniplateau ages are hereafter defined by at least three successive concordant steps and by at least 70 and 50%, respectively, of the total released ^{39}Ar. Errors are reported at the 2σ level. The decay constants recommended by Steiger and Jäger (1977) were used.

A critical evaluation and selection of the results obtained is based on the observed apparent age and Ar isotopic ratios ($^{38}Ar/^{39}Ar$ and $^{37}Ar/^{39}Ar$) and on X-ray diffractometry and electron microprobe analyses of the rocks and single mineral phases. In general, poor results (strongly discordant age spectra) were obtained for feldspars (systematically zoned with albite patches and/or exsolutions), leucite, and amphibole (spinoidal exsolution). Also, whole-rock, porphyritic samples and intrusive rock types are usually characterized by inhomogeneous and relatively high $^{37}Ar/^{39}Ar$ and nonconcordant apparent age spectra. In particular, whole-rock samples of potassic alkaline complexes, which are feldspars, nepheline- and leucite-rich, yielded systematically discordant age spectra, probably due to the sodium-potassium exchanges by intergranular fluids up to low hydrothermal temperatures (e.g., the leucite-nepheline analcimitization and K-Ca feldspar formation of Comin-Chiaramonti, 1979, and of Comin-Chiaramonti et al., 1979). In contrast, the best results (concordant age spectra, $^{37}Ar/^{39}Ar$ and $^{36}Ar/^{39}Ar$ isotopic spectra, consistent with the chemical analyses of the analyzed material) were obtained for lava whole rocks (mainly sodic alkaline rock types) and chiefly for nonaltered biotites

TABLE 2. Ar/Ar AGES OF PARAGUAY ALKALINE ROCKS

Region or locality	Laboratory	Rock type	Material	Plateau age (Ma)[1]	Isochron age (Ma)	MSWD	$^{40}Ar/^{36}Ar$ intercept
Alto Paraguay							
1. Stock 1	BGC	Syenite	Bi	240.6 ± 0.4	240 ± 2	1.3	299 ± 88
2. Cerro Siete Cabezas	BGC	Syenite	Bi	241.9 ± 0.4	239 ± 3	1.1	313 ± 82
3. Cerro Boggiani	USP	Phonolite	Bi	241.3 ± 0.7	241 ± 2	0.9	342 ± 74
Rio Apa							
1. Valle-Mí	BGC	Basanite (dike)	Bi	138.7 ± 0.2	139.0 ± 0.3	1.9	187 ± 123
Amambay							
1. Cerro Chiriguelo	BGC	Trachyte	Bi	137.6 ± 0.7	138.1 ± 1.7	0.6	184 ± 97
2. Cerro Sarambí	BGC	Glimmerite (vein)	Bi	139.3 ± 0.5	139.1 ± 0.9	0.2	214 ± 104
Central Provinces							
Ybytyruzú							
1. Cerro Km 23	USP	Essexitic gabbro	Bi	127.7 ± 0.1*			
2. Cerro San Benito	USP	Essexitic gabbro	Bi	127.4 ± 0.3	127.1 ± 1.9	1.3	394 ± 187
3. Cerro Santa Elena Villarrica	BGC	Essexite	Bi	125.6 ± 0.2	126.2 ± 0.5	1.2	194 ± 105
6. Mbocayaty	USP	Trachybasalt	Bi	126.4 ± 0.1	126.5 ± 0.9	0.5	264 ± 113
7. Aguapety Portón Serrania de Ybytymy	BGC	Essexite	Bi	126.2 ± 0.2	127.2 ± 0.7	0.35	204 ± 95
11. Cerro Cañada	USP	Alkali gabbro	Bi	126.1 ± 0.5	127.6 ± 1.4	1.7	114 ± 50
11A. Cerro Cañada	BGC	Ijolite	Bi	126.3 ± 0.2	126.2 ± 0.9	0.14	304 ± 110
11B. Cerro Cañada	USP	Alkali gabbro	Bi	127.5 ± 0.2*			
Cerro San José							
20. Cerro San José	USP	Syenogabbro	Bi	126.4 ± 0.4*	126.4 ± 3.9	0.81	306 ± 221
20A. Cerro San José	USP	Syenogabbro	Bi	128.5 ± 0.3[1]			
20B. Cerro San José Potrero	USP	Syenogabbro	Bi	128.5 ± 0.3*			
Ybaté							
23. Potrero Ybaté Sapucai	USP	Nepheline syenodiorite	Bi	124.1 ± 0.6			
25A. Sapucai (Cerro Verde)	USP	Essexite	Bi	124.6 ± 0.7	127.1 ± 1.4	0.76	198 ± 123
25B. Sapucai (Cerro Fidel)	USP	Trachyphonolite	Bi	126.4 ± 0.2	128.6 ± 2.9	2.76	138 ± 244
25C. Sapucai Paraguari	BGC	Trachyandesite (dike)	Bi	126.2 ± 0.1	126.2 ± 0.7	0.10	311 ± 52
27. Cerro Santo Tomás	USP	Nepheline syenodiorite	Bi	128.8 ± 0.4*	127.7 ± 1.1	2.1	383 ± 123
27A. Cerro Santo Tomás Acahay	USP	Basanite (dike)	Bi	127.2 ± 0.6	129.2 ± 3.1	1.0	226 ± 156
30. Acahay	USP	Trachybasalt (lava flow)	WR	127.0 ± 0.2	126.5 ± 1.5	0.10	294 ± 12
30A. Acahay Yaguarón	USP	Alkali gabbro	Bi	123.6 ± 0.5*			
33. Cerro Arrua-I	USP	Nepheline syenodiorite	WR	129.0 ± 0.2*	126.2 ± 1.7	2.1	411 ± 152
Misiones							
1. Estancia Guavira-y	USP	Melanephelinite (lava flow)	WR	119.8 ± 0.8*	118 ± 7	2.1	336 ± 77
2. Estancia Ramirez	USP	Melanephelinite (lava flow)	WR	118.3 ± 0.6*			
3. Cerro Caa Jhovy	USP	Peralkaline phonolite (lava dome)	WR	117.9 ± 0.9[1]			
4. Cerro Guayacan	USP	Tephrite (lava flow)	WR	117.2 ± 0.2	119.0 ± 1.3	1.9	257 ± 47
Asunción							
1. Cerrito	USP	Melanephelinite (lava flow)	WR	56.0 ± 0.5*			
2. Benjamin Aceval	USP	Melanephelinite (lava flow)	WR	56.4 ± 0.9*			
3. Cerro Verde	USP	Ankaratrite (lava flow)	WR	61.3 ± 0.3*			
4. San Jorge	USP	Ankaratrite (lava flow)	WR	57.0 ± 0.3			
5. Cerro Tacumbú	USP	Ankaratrite (lava flow)	WR	58.4 ± 0.4*			
6. Ñemby	USP	Ankaratrite (lava flow)	WR	60.7 ± 0.6	59 ± 2	0.7	434 ± 130

Notes: Summary of obtained plateau, mini-plateau, and integrated $^{40}Ar/^{39}Ar$ ages; plateau, defined by > 70% of total released gas, and mini-plateau, 50–70% of total released gas, are shown with asterisks *; (1) refers to mini-plateau age defined by 40% ^{39}Ar. BGC: Berkeley Geochronology Center, USA; USP: Geochronology Research Laboratory, São Paulo University, Brazil.

*Samples analyzed at the Berkeley Geochronology Center. The other samples were analyzed at the University of São Paulo.

BCG—Berkeley Geochronology Center; USP—Geochronological Research laboratory of São Paulo University.

MSWD—mean square of weighted deviates.

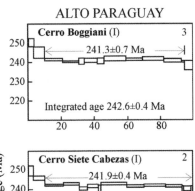

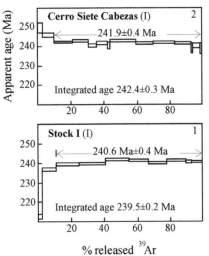

Figure 3. Sketch map showing the main outcrops along Alto Paraguay River (cf. Fig. 2 and Table 2) and the $^{40}Ar/^{39}Ar$ spectra. The arrows on the apparent age diagrams indicate steps included in the age plateau. 2 σ errors are shown.

(K_2O = 9–10.5 wt%; Comin-Chiaramonti and Gomes, 1995). Considering the generally low $^{40}Ar/^{39}Ar$, consistent with the high K/Ca of the microprobe analyses, biotite analyses particularly yield robust age data that are not affected by secondary alteration. A total of thirty-three samples yielded $^{40}Ar/^{39}Ar$ plateau and miniplateau ages and are discussed in this section. A summary of the geochronological results is presented in Table 2; the complete data set is available from the GSA Data Repository.[1]

Alto Paraguay

Biotite separates were analyzed for the evolved Na alkaline magmatic (intrusive) rocks from north to south: Cerro Boggiani, Cerro Siete Cabezas, and a small stock south of Cerro Siete Cabezas (Fig. 3). The samples yielded plateau ages that range from 240.6 ± 0.4 to 241.9 ± 0.4 Ma and are defined by eight to twelve successive heating steps and more than 85% of the total released gas. The plateau steps are characterized by relatively

[1]GSA Data Repository item 2007093, Ar/Ar isotopic data of selected samples (mineral separates and whole-rocks) of Paraguay alkaline magmatism, is available at www.geosociety.org/pubs/ft2007.htm, or on request from editing@geosociety.org, Documents Secretary, GSA, P.O. Box 9140, Boulder, CO 80301, USA.

homogeneous and low $^{37}Ar/^{39}Ar$, confirming that the analyzed material was "fresh" biotite. Inverse isochron ages ($^{40}Ar/^{39}Ar$ versus $^{36}Ar/^{39}Ar$) are indistinguishable from plateau ages and yield slightly larger errors and initial $^{40}Ar/^{36}Ar$ intercepts overlapping atmospheric values, suggesting the absence of excess ^{40}Ar component. Thus the plateau ages obtained indicate that Alto Paraguay sodic magmatism occurred at ca. 241 Ma during a short time interval (mean age 241.5 ± 1.3 Ma, corresponding to an Anisian age).

The integrated age (242.0 ± 1.6 Ma; no plateau age obtained) of the Pão de Açúcar trachyphonolite (Velázquez et al., 1996) is concordant with the plateau ages of the other samples.

Rio Apa and Amambay

Biotite separates of three samples (one basanite dike from Valle-Mí, one lava flow from Cerro Chiriguelo, and a glimmeritic vein from Cerro Sarambí) yielded plateau ages defined by 92–95% of the total gas released and range from 137.6 ± 0.7 to 139.5 ± 0.5 Ma (Fig. 4). The very low $^{37}Ar/^{39}Ar$ of plateau steps indicates that the analyzed materials were pure biotite. Isochron ages are concordant with plateau ages and yield at-

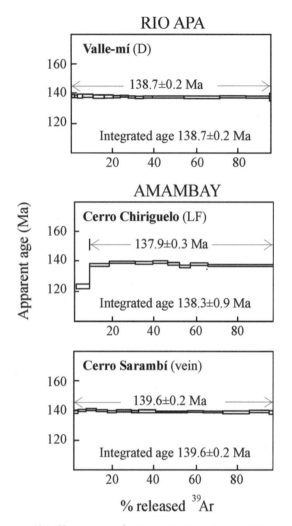

Figure 4. ^{40}Ar/^{39}Ar age spectra from the Amambay and Rio Apa regions of eastern Paraguay (cf. Fig. 2 and Table 2).

mospheric initial ^{40}Ar/^{36}Ar. Thus, the ages obtained (mean age 138.0 ± 1.6 Ma) are robust evidence of carbonatitic-potassic magmatism emplaced in the northern region of eastern Paraguay before the onset of Paraná-Angola-Etendeka flood volcanism. Notably, potassic alkaline magmatism of similar age (139 ± 3 Ma; Comin-Chiaramonti et al., 2005b) occurred in southeastern Bolivia (in the Velasco complexes).

Central Province (Asunción-Sapucai-Villarrica Region)

The largest data set concerns the post-tholeiitic magmatism from the Asunción-Sapucai-Villarrica region. ^{40}Ar/^{39}Ar analyses have been performed on a total of more than eighteen biotite separates from representative samples and on six whole-rock specimens. Plateau or miniplateau ages have been obtained on ten magmatic complexes, including the largest ones, i.e., the

Sapucai and Acahay alkaline complexes, including intrusives, as well as lava flows (Fig. 5). The ten plateau ages (obtained on nine biotite samples and one whole-rock sample) range from 124.6 ± 0.7 (Sapucai, Cerro Verde) to 128.5 ± 0.3 Ma (Cerro San José). Inverse isochron ages are concordant with plateau ages except for the youngest (Sapucai Serro Verde: isochron = 127.1 ± 1.4 Ma) and oldest samples (Cerro San José, no valid isochron age). Initial ^{40}Ar/^{36}Ar overlap atmospheric values (despite the large error on this initial value, due to the generally high ^{40}Ar/^{39}Ar) suggest a negligible role for excess ^{40}Ar. No significant age difference is apparent between intrusives and volcanics.

Thus, the highly concordant plateau ages suggest that the peak activity of the Asunción-Sapucai-Villarrica potassic magmatism occurred at 126.4 Ma (mean of the plateau ages = 126.4 ± 0.4 Ma) and was emplaced in a short time span, ~5–6 m.y. after the peak activity of the Paraná tholeiitic flood volcanism (133–132 Ma; Renne et al., 1992).

This Aptian-Barremian mean and peak activity age for the potassic Asunción-Sapucai-Villarrica magmatism is confirmed, but the total duration is slightly extended (123.6 ± 0.5 Ma to 129.0 ± 0.2 Ma) if miniplateau ages are also considered.

Misiones Province

Whole-rock samples from three lava flows and one lava dome of sodic affinity were analyzed (Fig. 6). The specimen from Cerro Guyacan yielded a plateau age of 117.2 ± 0.2 Ma, which is slightly lower than its isochron age (119.0 ± 1.3 Ma; atmospheric initial ^{40}Ar/^{36}Ar). Similar miniplateau ages have been obtained for the specimens from Estancia Guavira-y (119.8 ± 0.3 Ma) and Estancia Ramirez (118.3 ± 0.6). The mean of the plateau and miniplateau ages (118.3 ± 1.6 Ma) indicates that Misiones magmatic activity occurred at ca. 118 Ma (Aptian, late Early Cretaceous), roughly corresponding to the youngest age detected for the Florianópolis (119 Ma; Raposo et al., 1998), Ponta Grossa (120 Ma; Renne et al., 1996), and Santos–Rio de Janeiro tholeiitic dikes along the southeastern coast of Brazil (Renne et al., 1993; Turner et al., 1994).

Asunción Province

A total of one plateau age and five miniplateau ages have been obtained on lava flow whole-rock samples from the Na magmatic Asunción province (Fig. 7). The miniplateau ages range from 56.0 ± 0.5 to 61.3 ± 0.6 Ma, whereas the sample from the Ñemby ankaratrite yielded a plateau age at 60.7 ± 0.6 Ma. This latter sample yielded an isochron age (59 ± 2 Ma) with an atmospheric initial argon isotopic ratio. By contrast, no isochron ages were obtained for samples yielding miniplateau ages. Thus, a Paleocene age is suggested for the Asunción province (average of plateau and miniplateau ages = 58.7 ± 2.4 Ma), even though it should be noted that the age and duration of the sodic

Figure 5. Representative ^{40}Ar/^{39}Ar age spectra (best results) for the potassic Central province rock types and a sketch map showing the locations of the samples numbered as in Table 2.

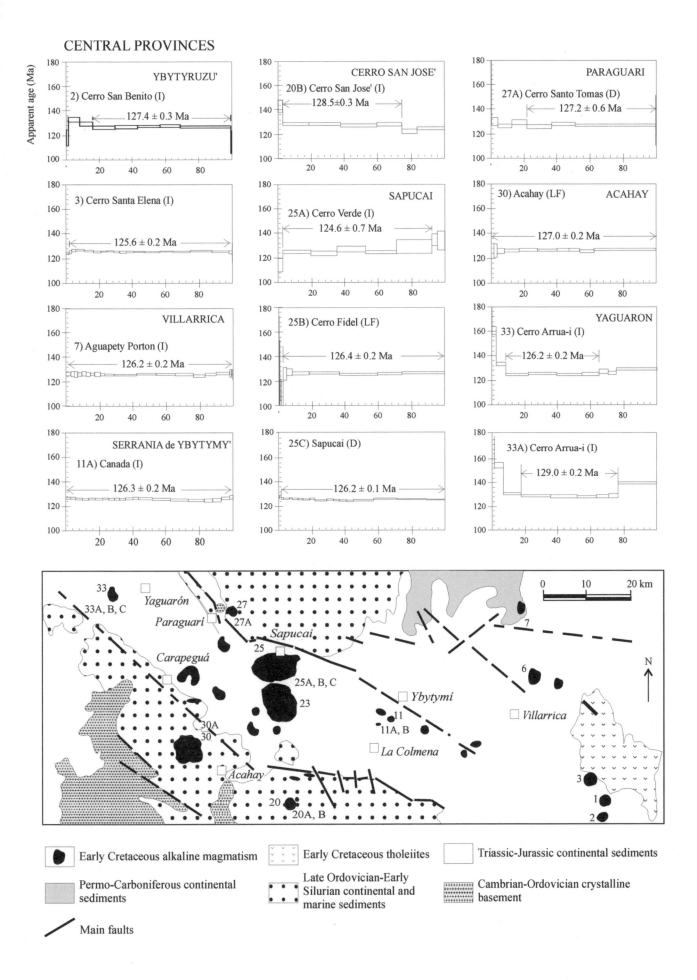

CENTRAL PROVINCES

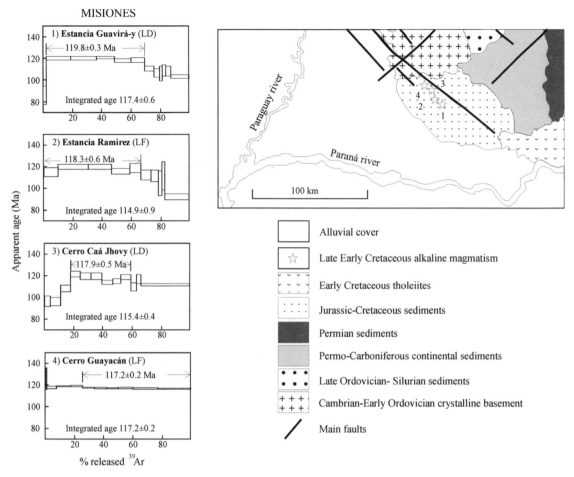

Figure 6. Representative ^{40}Ar/^{39}Ar age spectra for the Misiones province and sketch map with the locations of the samples numbered as in Table 2 (cf. Fig. 2). LD—lava dome; LF—lava flow.

alkaline events is generally less well constrained than those of the potassic alkaline events. These Early Cretaceous ages are comparable to those reported for some potassic alkaline complexes along the Cabo Frio Lineament (cf. Thompson et al., 1998).

COMPARISON WITH AGES OF VOLCANIC ROCKS FROM THE PARANÁ-ANGOLA-ETENDEKA SYSTEM

Continental flood basalts from the Paraná-Angola-Edendeka system cover an area of over 1.5×10^6 km^2. The Paraná-Angola-Edendeka province has two main associated tholeiitic dike swarms, one trending northwest-southeast (Ponta Grossa; cf. Fig. 8 and Piccirillo et al., 1990) and the other approximately paralleling the coast, i.e., trending north-south (Florianópolis and Etendeka; Erlank et al., 1984; Raposo et al., 1998) and with ~ENE-WSW attitude (Santos–Rio de Janeiro, Brazil, and Kwanza, Angola; Comin-Chiaramonti et al., 1983; Marzoli et al., 1999). The northwest-southeast-trending Ponta Grossa dikes are confined to South America and are parallel to the

Mesozoic rift-controlled basins, e.g., in Paraguay (Riccomini et al., 2005) and in Argentina (the Salado and Colorado basins of Hawkesworth et al., 1999, 2000). These basins reflect significant northeast-southwest extension in the Mesozoic that is not apparent in southern Africa (cf. Hawkesworth et al., 2000), although the Moçâmedes arch may be considered, to some extent, the symmetrical northeast-southwest expression of the Ponta Grossa arch (cf. Comin-Chiaramonti et al., 1991).

The ^{40}Ar/^{39}Ar ages of the Paraná-Angola-Edendeka basaltic flood tholeiites and Ponta Grossa dike swarm suggest that tholeiitic peak activity occurred at 133–130 Ma (Renne et al., 1992, 1993, 1996; Turner et al., 1994; Stewart et al., 1996; Ernesto et al., 1999; Hawkesworth et al., 1999; Marzoli et al., 1999; Kirstein et al., 2001). A slightly later generation of tholeiitic magmatism, i.e., 129–119 Ma, is represented by the coast-parallel dike swarms (Florianópolis, Santos–Rio de Janeiro, Kwanza, and Horingbaai; cf. inset of Fig. 8), indicating a tendency for younger tholeiitic dikes to concentrate toward the continental margins during the final stages of the main rifting immediately preceding the opening of the south Atlantic Ocean

ASUNCIÓN

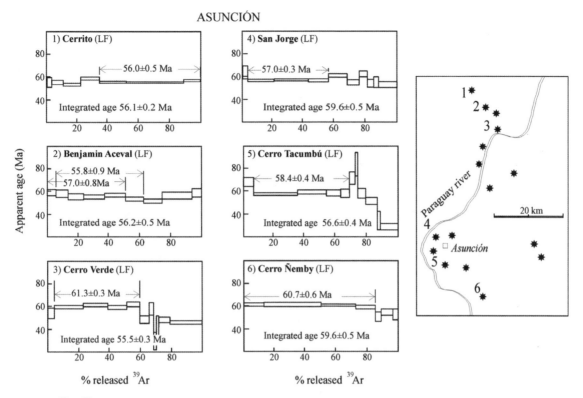

Figure 7. ^{40}Ar/^{39}Ar age spectra for the Asunción province and sketch map showing the locations of the samples numbered as in Table 2 (cf. Fig. 2). Note the prevailing mini-plateau (see text).

(cf. Raposo et al., 1998). Notably, the latter younger tholeiitic dikes have an age close to the age of the Asunción-Sapucai-Villarrica potassic magmatism (126.4 Ma).

On the other hand, distinct age intervals (mainly K/Ar data) are apparent in the whole Paraná-Angola-Edendeka system for the alkaline and alkaline-carbonatite complexes (Comin-Chiaramonti and Gomes, 2005, and references therein; cf. Fig. 8):

1. Pretholeiitic, i.e., 139–137 Ma (Early Cretaceous), potassic alkaline complexes occur in Angola (Moçâmedes arch; Alberti et al., 1999, and references therein), Namibia (Milner et al., 1995), and eastern Bolivia (Velasco complexes, Ar/Ar plateau age; Comin-Chiaramonti et al., 2005b).

2. Syntholeiitic, i.e., 129 ± 2 Ma (Early Cretaceous), potassic alkaline magmatism occurs in Brazil (Santa Catarina state, Ponta Grossa arch), Uruguay, and Angola (Alberti et al., 1999; Ruberti et al., 2005a,b). In Namibia the alkaline rock types slightly postdate (129–123 Ma) the main Paraná-Angola-Edendeka eruptive phase (Stewart et al., 1996; Peate, 1997; Hawkesworth et al., 1999). The first two events are generally represented by "low-Ti" potassic rocks (cf. Gibson et al., 1995a,b).

3. Post-tholeiitic, i.e., 109–106 Ma (late Early Cretaceous), potassic magmatism outcrops in Brazil (Ponta Grossa arch; Ruberti et al., 2005a), whereas Late Cretaceous (95 Ma;

Ar/Ar plateau age) sodic magmatism occurs in Angola (Kwanza basin; Marzoli et al., 1999).

4. Late Cretaceous, i.e., 81 ± 7 Ma, alkaline rocks of both potassic and sodic affinity outcrop in Brazil (Rio Grande do Sul and Santa Catarina states, Ponta Grossa arch, Serra do Mar, Alto Paranaíba, Goiás, and Mato Grosso states), Bolivia (Candelaria), and Namibia (Comin-Chiaramonti and Gomes, 2005). In particular, the alkaline magmatism from Alto Paranaíba and Santa Catarina districts (Brazil) is represented by "high-Ti" rock types of kamafugitic affinity (cf. Gibson et al., 1995a,b). The spatial distribution of Late Cretaceous alkaline magmatism suggests a close correlation between cratonic blocks (cb) and associated mobile belts (mb), e.g., São Francisco cb / Araçui mb, Congo cb / Damara mb, Luis Alves cb / Ribeira–Dom Feliciano mb (cf. Trompette, 1994; Kröner and Cordani, 2003).

5. Paleocene and Eocene alkaline magmatism is preserved in Brazil (i.e., 61 ± 4 Ma; Cabo Frio Lineament; Comin-Chiaramonti and Gomes, 1995, 2005) and in Namibia (49 Ma; Cooper and Reid, 1998), respectively.

If a comparison is made with the age of the alkaline rocks from eastern Paraguay, all the age intervals are present except for the Triassic and Late Cretaceous magmatism. In particular, the alkaline complexes from symmetrical lineaments of the African counterpart, compared with the Paraná basin (cf. Fig. 8),

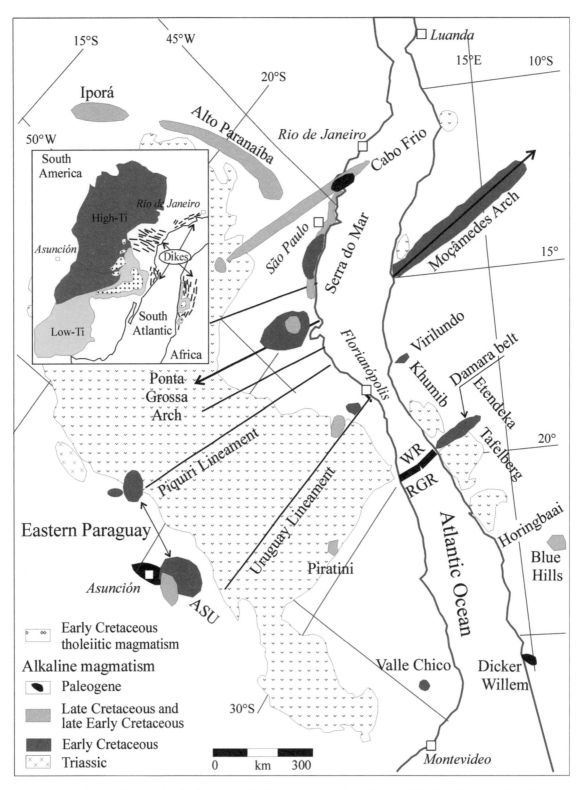

Figure 8. Sketch map showing the distribution of alkaline occurrences in and around the Paraná-Angola-Etendeka system (modified after Comin-Chiaramonti et al., 1997). ASU—Asunción-Sapucai-Villarica Graben; RGR—Rio Grande Rise; WR—Walvis Ridge. The arrows indicate the direction and dip of the main uplift of the Ponta Grossa and Moçâmedes arches. Inset: Sketch map of the Paraná-Angola-Etendeka system (cf. Piccirillo and Melfi, 1988); the arrows indicate the occurrences of the main dike swarms. The basaltic lavas are subdivided into broad high- and low-Ti groups and late-stage rhyolites.

e.g., Moçâmedes arch and Damara belt (northwest Namibia), display ages similar to the alkaline complexes from Brazil (Ponta Grossa arch) and Paraguay (Asunción-Sapucai-Villarrica Graben), respectively (Le Roex and Lanyon, 1998, and references therein; Alberti et al., 1999).

GEOCHEMISTRY OF MAGMATISM FROM EASTERN PARAGUAY AND FROM THE PARANÁ-ANGOLA-NAMIBIA SYSTEM

Incompatible element (IE) patterns, large-ion lithophile element (LILE) versus high field-strength element (HFSE) ratios, and Sr-Nd-Pb isotopic compositions indicate that the six magmatic events that occurred in eastern Paraguay were generated from geochemically distinct (enriched versus depleted) mantle sources. This section is based on previous work, namely that of Comin-Chiaramonti et al. (1991, 1997, 1999, 2001), Comin-Chiaramonti and Gomes (1995, 2005), Castorina et al. (1997), Alberti et al. (1999), and references in these papers. The geochemical data of the Paraguay rocks are compared with those of tholeiitic and alkaline rocks that occurred in the Paraná-Angola-Edendeka province and in a contiguous region of South America and southwestern Africa from the Early Cretaceous to the Paleogene (Fig. 8). Some chemical analyses representative of the most "primitive" alkaline rock types are reported in the appendix.

Incompatible Elements

The IE patterns of the Alto Paraguay sodic magmatic rocks (considering the less evolved rocks, i.e., nepheline syenites, SiO_2 = 55 wt% and MgO = 2.5 wt%) largely reflect their differentiated composition (e.g., negative Sr and Ti spikes), yet features such as positive Nb-Ta anomalies are most probably primary, that is, related to the mantle source of the parent magmas (Fig. 9).

The pre- and post-tholeiite potassic alkaline rocks (from Rio Apa-Amambay and Asunción-Sapucai-Villarrica, respectively; cf. Figs. 4 and 5) are low-Ti variants, according to Gibson et al. (1995a), and display quite similar IE patterns, in general characterized by LILE enrichment and HFSE depletion (Fig. 9A). Notably, high-Ti potassic rocks are typical of Late Cretaceous potassic magmatism in the Alto Paranaíba and Namibia Suites, which, in contrast, are characterized by HFSE enrichment (cf. Gibson et al., 1995a,b).

The Cretaceous low- and high-Ti Paraná flood tholeiites are distinct in terms of their relatively low elemental abundances and high LILE/HFSE ratios (Fig. 9B). In general, the marked Ta-Nb negative spike of the Paraná tholeiitic basalts is similar to that of the potassic alkaline magmas from eastern Paraguay, but marks a clear difference with the Mesozoic to Cenozoic sodic alkaline rocks from Paraguay and with the ocean island basalts (OIB) of the southern Atlantic islands of Tristan da Cunha and Trindade (Fig. 9C and D).

Specifically, similar to the whole Paraná basin, Paraguay has high-Ti tholeiites with IE abundances higher than those of the coeval low-Ti analogues and the negative Sr anomaly (cf. Comin-Chiaramonti et al., 1999, 2004). The upper Late Cretaceous (Misiones, San Juan Bautista) and Paleocene (Asunción) sodic alkaline rocks display almost identical IE patterns. With respect to the potassic alkaline magmatism, the Early Cretaceous and Paleocene sodic events differ, in general, by a marked negative K spike and positive HFSE spikes (Fig. 9C and F), with a general pattern similar to the Tristan da Cunha and Trindade ocean island magmas and, to some extent, to the Early Cretaceous potassic alkaline mafic rocks from Angola and Namibia (Fig. 9D).

In summary, the IE suggest a common signature for potassic alkaline magmatism from Paraguay and Paraná-Angola-Edendeka tholeiitic magmas, whereas Paraguay sodic alkaline rocks display IE patterns similar to those of Atlantic OIBs and African potassic alkaline basalts. Strongly light rare earth element (LREE)–enriched patterns suggest that Paraguayan potassic and sodic alkaline magmas issued from a garnet-bearing peridotite, yet their mantle source compositions were clearly distinct in terms of both IE composition and mineralogy. As proposed by Comin-Chiaramonti et al. (1997), the relative K enrichment of eastern Paraguay potassic rocks suggests that a K-bearing phase (e.g., phlogopite) was not a residual phase during partial melting of the mantle. Phlogopite, instead, was probably a residual phase in the mantle source for the sodic alkaline rocks, which is consistent with the lower melting degree inferred for the sodic magmatism compared to the potassic (e.g., 4–6% and 6–11% melting of a garnet mantle, respectively; Comin-Chiaramonti et al., 1997; Velázquez et al., 2006).

Considering "geodynamic indicators" (e.g., Pearce, 1983; Beccaluva et al., 1991), such as Nb/Zr versus Th/Zr and Ta/Yb versus Th/Yb (not shown; cf. Fig. 8 *in* Velázquez et al., 2006), the Misiones and Asunción-Sapucai-Villarrica sodic rocks fall in the array of basalts from nonsubduction settings, i.e., mid-ocean ridge basalts (MORB) +within-plate basalts (cf. Comin-Chiaramonti et al., 1997), along with the data from volcanic rocks of the Argentina central rift and the Bolivian Ayopaya Complex. On the other hand, both Serra Geral Formation high-and low-Ti tholeiites and the Asunción-Sapucai-Villarrica potassic rocks fall out of the field for non-subduction-related compositions in the same diagrams (cf. Comin-Chiaramonti et al., 1999). Using the ΔNb parameter of Fitton et al. (1997), the magmatic rock types from eastern Paraguay fit the field of the western Usa rift system (Basin and Range and lamproites: Fitton, this volume; Fig. 10; see also the appendix), for which it is suggested that the composition of the early Basin and Range basalts is best explained by mixing with an enriched component, represented by lamproites, in the subcontinental lithosphere.

Sr-Nd Isotopes

The investigated rocks from eastern Paraguay cover a wide range of Sr-Nd isotopic compositions (Fig. 10A) defining, on the

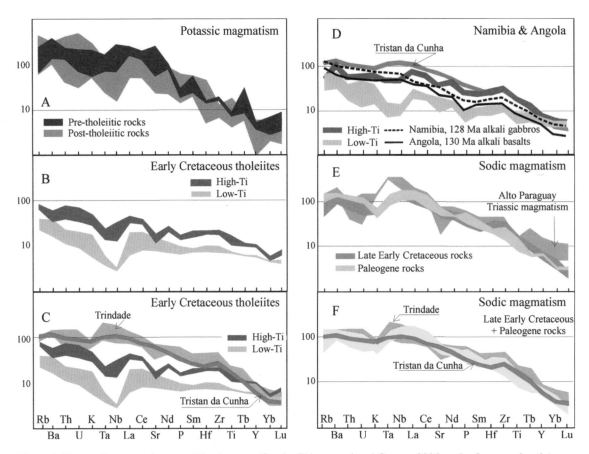

Figure 9. Eastern Paraguay: incompatible elements (Comin-Chiaramonti and Gomes, 2005, and references therein) normalized to the primitive mantle (Sun and McDonough, 1989) representative of the compositions of the mafic K-alkaline rocks (A) and tholeiites (B), compared with Trindade and Tristan da Cunha basanites (C) (Le Roex et al., 1990; Marques et al., 1999; Siebel et al., 2000; Ewart et al., 2004). (D) High- and low-Ti "uncontaminated" tholeiitic basalts from Angola and Namibia compared with Namibia alkaline gabbros and Angola alkali basalts (Alberti et al., 1999; Comin-Chiaramonti et al., 1999; Ewart et al., 2004) and with Tristan da Cunha basanites. (E) Triassic sodic rocks (Alto Paraguay: nepheline syenites with $SiO_2 = 55$ wt% and MgO = 2.5 wt%) compared with sodic alkaline rocks with upper Late Cretaceous and Paleogene ages from Paraguay (Comin-Chiaramonti and Gomes, 2005). (F) sodic mafic magmatism (late Early Cretaceous + Paleogene) compared with the field of Trindade and Tristan da Cunha basanites.

whole, a trend similar to the low-Nd array of Hart and Zindler (1989) (the "Paraguay array" of Comin-Chiaramonti et al., 1995a,b). Due to the high Sr and Nd content of the most "primitive" alkaline rocks (and associated carbonatites) from eastern Paraguay, Comin-Chiaramonti et al. (1997) suggested that the initial Sr-Nd isotopic ratios of such rocks can be considered crustally uncontaminated and, as a result, representative of the isotopic composition of the mantle source(s).

Initial $^{87}Sr/^{86}Sr$ (Sr_i) and $^{143}Nd/^{144}Nd$ (Nd_i) ratios range from the depleted quadrant to the enriched one. The potassic alkaline rocks, both pre- and post-tholeiites, have the highest initial Sr_i and the lowest Nd_i. Including the Cerro Chiriguelo and Sarambì carbonatites (Comin-Chiaramonti and Gomes, 1995), which occur associated with the pretholeiitic potassic rocks in northeastern Paraguay, Sr_i and Nd_i range from 0.70636 to 0.70721 and from 0.51194 to 0.51165, respectively. These values are quite distinct from those of the late Early Cretaceous

sodic rocks (Misiones: ca. 118 Ma; $Sr_i = 0.70486 \pm 0.00043$, $Nd_i = 0.51226 \pm 0.00015$) and the Paleocene sodic rocks (Asunción-Sapucai-Villarrica: ca. 60 Ma), which plot within the depleted quadrant ($Sr_i = 0.70369 \pm 0.00011$, $Nd_i = 0.51268 \pm 0.00006$) toward the high μ, i.e., high U/Pb mantle (HIMU)–depleted MORB mantle (DMM) components.

In any case, on the basis of the isotopic compositions it is possible to infer a time-related (Early Cretaceous to Paleogene) progressive increase in the role of the depleted mantle domain(s) in the genesis of alkaline magmatism in eastern Paraguay (Comin-Chiaramonti et al., 1999). On the other hand, the sodic magmatism, including the Alto Paraguay sodic magmatism (Triassic), tends to plot near the depleted quadrant. Notably, the Sr_i and Nd_i of the "uncontaminated" tholeiites (both high- and low-Ti) are intermediate between the potassic and sodic rocks (cf. Comin-Chiaramonti et al., 1999).

The Sr-Nd isotopic variation of the Paraguayan alkaline

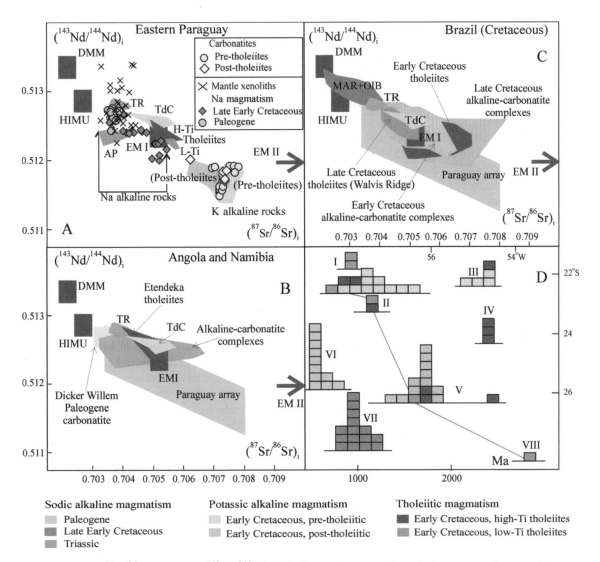

Figure 10. Initial $^{87}Sr/^{86}Sr$ (Sr_i) versus $^{143}Nd/^{144}Nd$ (Nd_i) diagram for magmatic rocks from eastern Paraguay (A) compared with Angola and Namibia (B) and with Brazil (C). AP—Alto Paraguay; TR—Trindade; TdC—Tristan da Cunha; DMM—depleted MORB (mid-ocean ridge basalt) mantle; EMI—enriched mantle, type I; EMII—enriched mantle, type II; HIMU—high μ (i.e., high U/Pb) mantle; H-Ti and L-Ti, high-Ti and low-Ti tholeiites; MAR—Mid-Atlantic Ridge; OIB—ocean island basalt. Data sources as in Figure 9 and Comin-Chiaramonti et al., 2001. Other symbols as in Figure 9. DMM, EMI, EMII, and HIMU fields after Zindler and Hart (1986), Hart and Zindler (1989), and Stracke et al., 2005. (D) distribution of T^{DM} ages (Ma) in eastern Paraguay: I—Ito Paraguay; II—Rio Apa; III—Amambay; IV—Carayó; V—Sapucai; VI—Asunción; VII—Misiones; VIII—Encarnación. Colors as in D. The red line joins "L-Ti" tholeiites. Data from Comin-Chiaramonti and Gomes, 2005.

rocks is larger than that of all other Paraná-Angola-Edendeka alkaline magmatism. In Angola and Namibia, at the easternmost fringe of the Paraná-Etendeka system, the Sr_i and Nd_i values for most Early Cretaceous Angolan K-alkaline-carbonatite complexes (Fig. 10B) vary between 0.70321 and 0.70466 and between 0.51273 and 0.51237, respectively, showing on the whole depleted characteristics relative to the bulk Earth (cf. Alberti et al., 1999). On the other hand, the Early Cretaceous alkaline-carbonatite complexes from Namibia have a similar Sr_i range (0.70351–0.70466), but almost constant Nd_i (0.51250–0.51244; Milner and Le Roex, 1996; Le Roex and Lanyon, 1998), because

the Sr-Nd isotopic compositions of these alkaline-carbonatitic magmas are insensitive to crustal assimilation (due to their high Sr and Nd contents; cf. Alberti et al., 1999). Notably, regarding the source region of carbonatites, Bell and Blenkinsop (1989) suggest that their Sr_i and Nd_i data may result from large-scale reservoirs, one corresponding to a lithosphere keel and the other to the asthenosphere.

The Sr_i (0.70425–0.70595; average 0.70527 ± 0.00034) and Nd_i (0.51213–0.51280; average 0.51224 ± 0.00011) of the Early Cretaceous Brazilian rocks (Fig. 10C) are generally higher and lower, respectively, than those of the coeval rocks (both sodic and potassic) of Angola and Namibia (cf. Comin-Chiaramonti et al., 1999; Comin-Chiaramonti and Gomes, 2005).

Late Cretaceous potassic alkaline-carbonatitic complexes have the following Sr_i and Nd_i mean values, respectively: Alto Paranaíba, $Sr_i = 0.70527 \pm 0.00036$ and $Nd_i = 0.51224 \pm 0.00006$ (Bizzi et al., 1994; Gibson et al., 1995b, and references therein); Taiúva-Cabo Frio and Serra do Mar, $Sr_i = 0.70447 \pm 0.00034$ and $Nd_i = 0.51252 \pm 0.00008$ (Thompson et al., 1998); Lajes, $Sr_i = 0.70485 \pm 0.00053$ and $Nd_i = 0.51218 \pm 0.00022$ (Comin-Chiaramonti et al., 2002). Note that the alkaline-carbonatite magmatism trends toward the Sr_i and Nd_i field delineated by the Late Cretaceous tholeiites from Walvis Ridge and Rio Grande Rise (cf. Richardson et al., 1982; Camboa and Rabinowitz, 1984).

Model Ages

Despite uncertainties related to the Sm/Nd fractionation (f) during melting and magma differentiation (cf. Arndt and Goldstein, 1987), Nd model ages (depleted mantle, T^{DM}; cf. De Paolo, 1988) may give a broad indication of the age of the main enrichment processes that may have affected the mantle source(s) of Paraguayan magmas. In general, Paraguayan potassic magmas display T^{DM} comparable to those of the Paraná-Angola-Edendeka tholeiites and higher than those of sodic magmas. The T^{DM} of the potassic alkaline rocks increases from the pretholeiitic rocks of northeastern Paraguay (peaks at 1.1–1.4 Ga, for $f \approx$ –0.5 to –0.7; Valle-Mí, Apa Block, and Amambay) to the post-tholeiitic Asunción-Sapucai-Villarrica potasssic alkaline complexes and dikes (1.7 Ga, $f \approx$ –0.4 to –0.5; cf. Fig. 10 D and the appendix; Comin-Chiaramonti et al., 1995a). The sodic alkaline rocks display Upper Proterozoic T^{DM} (0.9 Ga, Alto Paraguay; 0.6 Ga, Asunción-Sapucai-Villarrica; 1.0 Ga, Misiones, for $f \approx$ –0.4 to –0.5). The younger model ages parallel the Rio Paraguay Lineament and characterize the sodic magmatism.

The different geochemical behavior in the different sectors of the Paraná-Angola-Edendeka system also implies different sources. Using the T^{DM} (Nd) model ages on the whole Paraná-Angola-Etendeka system (cf. Comin-Chiaramonti and Gomes, 2005; Gastal et al., 2005), we observe that (1) the H-Ti flood tholeiites and dike swarms (cf. inset of Fig. 8) from the Paraná basin and the Early Cretaceous potassic rocks and carbonatites from eastern Paraguay range mainly from 0.9 to 1.7 Ga, whereas in Angola and Namibia the Early Cretaceous K-alkaline rocks range from 0.4 to 0.9 Ga; (2) the low-Ti tholeiites display a major T^{DM} variation, from 0.7 to 2.4 Ga (mean 1.6 ± 0.3), with an increase in the model ages from north to south; (3) Late Cretaceous alkaline rocks and associated carbonatites show model ages ranging from 0.6 to 1 Ga, similar to the model age shown by the Triassic to Paleogene sodic alkaline rock types along the Paraguay River. These model ages indicate that some notional distinct "metasomatic events" may have occurred during Paleoproterozoic to Neoproterozoic times as precursors to the alkaline and tholeiitic magmas in the Paraná-Angola-Etendeka system (cf. Comin-Chiaramonti et al., 1995a, 1997, 1999, 2004; Alberti et al., 1999).

The T^{DM} (Nd) model ages and mantle heterogeneity are also supported in the Paraná basin by (1) Re-Os isotope systematics for the potassic rocks of kamafugitic and kimberlitic affinity (Brazil Alto Paranaíba and Lages) where the Nd model ages cover an age range similar to the Os model ages and suggest different melting depths of heterogeneous lithospheric sources (Carlson et al., 1996; Araujo et al., 2001), and (2) the model of Meen et al. (1989; see also Castorina et al., 1996), a veined lithospheric mantle (amphibole/phlogopite-carbonate-lherzolite + CO_2 fluid type III and IV veins of Meen et al., 1989) of Proterozoic age, which may well account for the magmatism of the Paraná basin.

Pb Isotopes

The data available for the alkaline-carbonatite complexes and tholeiites from the Paraná-Angola-Etendeka system plot between HIMU and enriched mantle type I (EMI) end-members, and subordinately DMM and EMI, as well as crustal *sensu lato* components (e.g., EMII; Figs. 11 and 12). It should be noted that

---

Figure 11. $^{207}Pb/^{204}Pb$ and $^{208}Pb/^{204}Pb$ versus $^{206}Pb/^{204}Pb$ (initial ratios) for rock types from the Paraná-Angola-Edendeka system. Abbreviations: AP—Alto Paraguay; CR—"Central Rift" of northwest Argentina; DMM—depleted MORB (mid-ocean ridge basalt) mantle; EMI—enriched mantle, type I; EMII—enriched mantle, type II; HIMU—high μ (i.e., high U/Pb) mantle; MAR—Mid-Atlantic Ridge; NHRL—North Hemisphere Reference Line; OIB—ocean island basalt. Sources: NHRL after Hart (1984) and Hart et al. (1986, 1992); 132 Ma geochron after Ewart et al. (2004); CR after Lucassen et al. (2002). Other data sources: Eastern Paraguay—Comin-Chiaramonti et al. (1991, 1995a, 1996, 1997, 2001), Comin-Chiaramonti and Gomes (1995, 2005), Castorina et al. (1997), Marques et al. (1999); Angola and Namibia—Milner and Le Roex (1996), Cooper and Reid (1998), Le Roex and Lanyon (1998), Smithies and Marsh (1998), Alberti et al. (1999), Gibson et al. (1999), Harris et al. (1999), Kurszlaukis et al. (1999), Ewart et al. (2004, and references therein); Brazil (Lower Cretaceous)—Toyoda et al. (1994), Garda et al. (1995), Huang et al. (1995), Walter et al. (1995), Andrade et al. (1999), Gibson et al. (1999), Marques et al. (1999), Comin-Chiaramonti et al. (2002), Ruberti et al. (2002), Comin-Chiaramonti and Gomes (2005); Brazil (Upper Cretaceous)—Bellieni et al. (1990), Bizzi et al. (1994, 1995), Meyer et al. (1994), Toyoda et al. (1994), Gibson et al. (1995a,b, 1997, 1999), Carlson et al. (1996), Thompson et al. (1998); Atlantic Ocean, Walvis Ridge—Richardson et al. (1982); Rio Grande Rise—Camboa and Rabinowitz (1984); Mid-Atlantic Ridge—Hamelin et al. (1984), Ito et al. (1987), Fontignie and Schilling (1996); OIB—Halliday et al. (1988, 1995); Tristan da Cunha—Le Roex (1985), Le Roex et al. (1990), Ewart et al. (2004); Trindade—Marques et al., (1999), Siebel et al. (2000); Brazil (Paleogene), Serra do Mar—Thompson et al. (1998), Bennio et al. (2002); Abrolhos—Fodor et al. (1989); Northeastern Brazil—Fodor et al. (1998); Fernando de Noronha—Gerlach et al. (1987). DMM, EMI, EMII, and HIMU are approximations of mantle end-members taken from Hart and Zindler (1989) and Hart et al. (1992). In all the diagrams, the Rio Grande Rise basalts (not shown) plot in the same field as the Walvis Ridge samples.

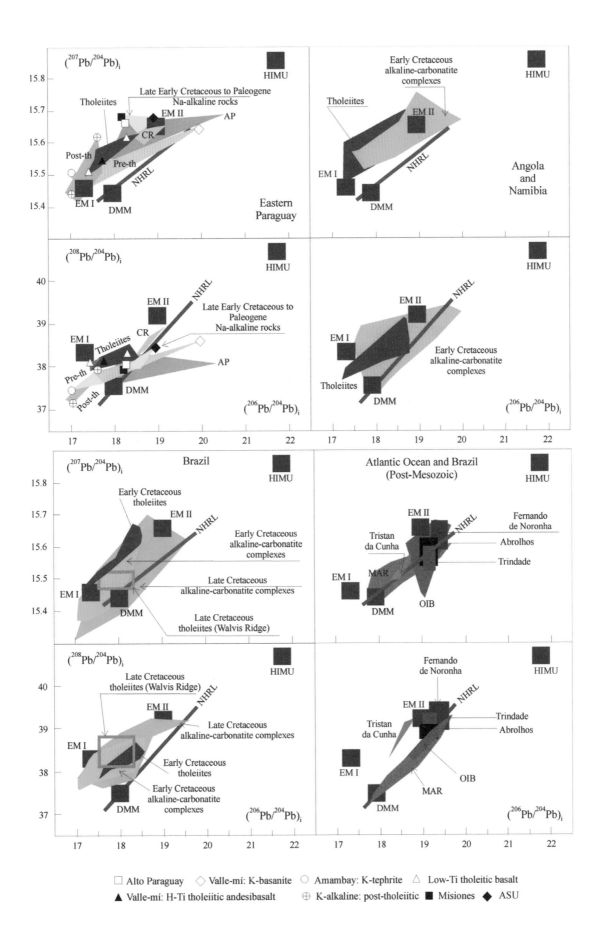

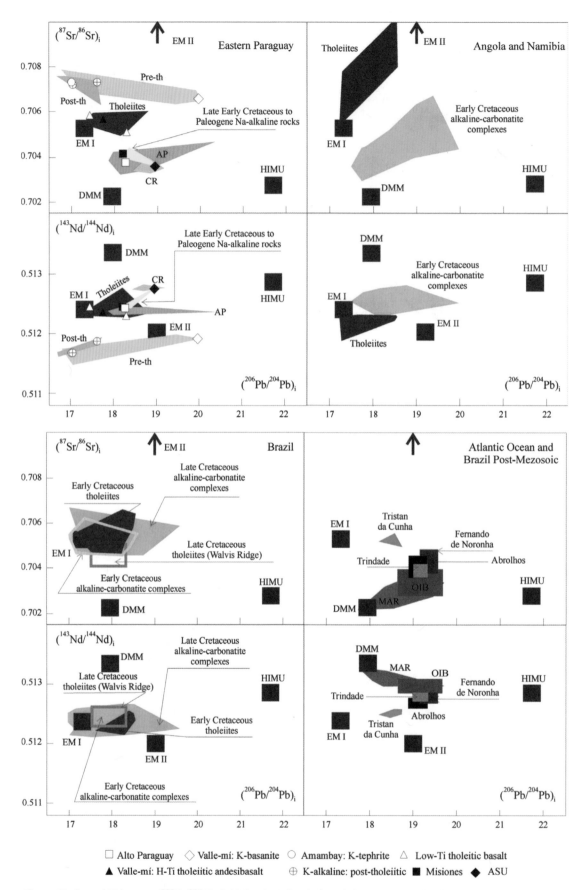

Figure 12. Sr_i and Nd_i versus $^{206}Pb/^{204}Pb$ (initial ratios). Symbols and data sources as in Figure 11.

the tholeiitic flood basalts from eastern Paraguay and from Angola-Namibia, at the westernmost and easternmost sides of the Paraná-Angola-Edendeka system, respectively, delineate different fields well (cf. Fig. 12).

The Mid-Atlantic Ridge and OIB delineate trends between the DMM and HIMU mantle components. In comparison, Paraná-Angola-Edendeka carbonatites plot close to the EMI/DMM–HIMU mixing lines for both Pb-Sr and Pb-Nd (cf. Comin-Chiaramonti and Gomes, 2005). This observation seems to confirm the advantages of using carbonatite over silicate rocks as indicators of mantle sources because of their rapid ascent to shallow depths or the surface and their buffering against crustal assimilation due to the high Sr, Nd, and Pb concentrations in the liquids.

GEOCHEMICAL IMPLICATIONS

The geochemical features of the Paraguay alkaline magmatism have been explained by enrichment processes (e.g., subduction—Hergt et al., 1991; Maury et al., 1992—or volatile-rich small-volume melts derived from the asthenosphere—McKenzie and O'Nions, 1995) occurring in a previously depleted mantle source. Comin-Chiaramonti et al. (1997) emphasize the lack of any geological evidence to support subduction processes in the Paraguayan region during Phanerozoic times. They suggest that the enrichment processes were related to small-volume melts in an old lithospheric mantle, possibly affected by a contribution from the previous Transamazonic to Brasiliano cycles.

Similarly, the main differences between low- and high-Ti flood tholeiites have been ascribed to different melting degrees (up to 5 and 20% for high- and low-Ti basalts, respectively) of a large-scale heterogeneous mantle source (Piccirillo and Melfi, 1988), possibly a veined peridotite where the distribution and frequency of the "metasomatizing" channels determine the different chemical signatures (cf. Comin-Chiaramonti and Gomes, 2005).

The Sr-Nd isotopic variations of the alkaline rocks from eastern Paraguay (Fig. 10A) may be explained (A) by generation from distinct portions of a large- and small-scale heterogeneous lithospheric mantle source, where small-scale heterogeneity is required by the variations in the Sr_i and Nd_i ratios in each magmatic event, or (B) in terms of mixing of magmas generated from an enriched mantle component (with an extreme EMI signature) and a depleted mantle component (DMM- or HIMU-like components).

Le Roex and Lanyon (1998), Thompson et al. (1998), and Ewart et al. (2004) postulated that Early Cretaceous alkaline-carbonatitic and tholeiitic magmatism from northwestern Namibia and Late Cretaceous alkaline and alkaline-carbonatitic magmatism from Alto Paranaíba–Serra do Mar (southern Brazil) would reflect variable contributions from the asthenospheric mantle components related to the Tristan da Cunha and Trindade plumes. On the contrary, Castorina et al. (1997),

Comin-Chiaramonti et al. (1997, 1999, 2004), and Alberti et al. (1999) suggested that the alkaline and alkaline-carbonatitic magmatism in the Paraná-Angola-Edendeka system originated from lithospheric mantle sources without appreciable participation of plume-derived materials.

On the basis of geochemical and geophysical data, Ernesto et al. (2000, 2002) proposed that the genesis of the Paraná-Angola-Edendeka tholeiites mainly reflects melting of heterogeneous subcontinental mantle reservoirs and that the geochemical and isotopic signatures of the Walvis Ridge and Rio Grande Rise basalts may be explained by contamination from detached continental lithospheric mantle left behind during the continental break-up processes (cf. also models after Foulger and Anderson, 2005; Foulger et al., 2005; and Lustrino, 2005; see detailed discussion in Anderson, 2006).

In the isotopic diagrams (cf. Figs. 11 and 12), Early Cretaceous potassic alkaline and tholeiitic magmatism from the Paraná-Angola-Edendeka system appears to be related to heterogeneous mantle sources from time-integrated HIMU to EM components. According to Tatsumi (2000), for example, relatively low $^{206}Pb/^{204}Pb$ and high $^{207}Pb/^{204}Pb$ compositions could be related to the delamination of pyroxenite restites formed by anatexis of the initial basaltic crust in Archean–Proterozoic times. We stress that in general, the enriched isotopic signatures of the Early Cretaceous alkaline magmatism decrease from west (Paraguay) to east (Brazil, southeast continental margin, and Angola and Namibia), reflecting the decrease in Nd model ages for potassic rocks from Paraguay to the east. These results suggest that Paraná-Angola-Edendeka system magmatism is related to both large- and small-scale heterogeneous mantle sources. Also, the isotopic signature of the Trindade and Abrolhos ocean islands (cf. Figs. 11 and 12) is similar to that of the alkaline-carbonatitic Early Cretaceous magmatism from Angola and Namibia, but quite different from that (EMI signature) of the Late Cretaceous–Paleogene analogue from the Alto Paranaíba, Ponta Grossa arch, and Cabo Frio–Taiúva–Serra do Mar areas (Comin-Chiaramonti and Gomes, 2005). According to Thompson et al. (1998), the Alto Paranaíba region would be the inland surface expression of the "dogleg" track left by the Trindade plume, but in terms of Sr-Nd-Pb isotopes, the contribution of components related to that plume, if any, is difficult to detect.

Hawkesworth et al. (1986) interpreted the Etendeka (Namibia) high-TiO_2 (HTZ) tholeiitic basalts as resulting from the melting of a Proterozoic lithospheric mantle, which, in the case of the Walvis Ridge basalts (cf. Richardson et al., 1982), was floating inside the oceanic asthenosphere during the opening of the south Atlantic. Alternatively, the elemental and isotopic signature of the HTZ basalts could be related to contamination of the oceanic mantle by ancient subcontinental lithospheric mantle. In summary, the isotopic signature of the Early and Late Cretaceous alkaline-carbonatite complexes of the Paraná-Angola-Edendeka system reflects ancient heterogeneities preserved in the subcontinental lithospheric mantle.

All the data indicate that they represent a thermally eroded metasomatic subcontinental upper mantle and/or delaminated lithospheric materials stored for a long time, e.g., toward the transition zone or deeper mantle (cf. Fig. 1) in Archean–Proterozoic times. In this context, considering the important differences in terms of trace elements patterns and Sr-Nd-Pb isotopic composition, the role of the Tristan da Cunha plume claimed by Ewart et al. (2004) is not apparent (cf. Figs. 9–12). Therefore, we believe, as documented by Ernesto et al. (1999, 2002), that the hypothesis of a mantle plume origin for Paraná-Angola-Edendeka magmatism is not compelling, with the caveat that mantle plumes may represent thermal perturbations.

GEODYNAMIC IMPLICATIONS

The geodynamic evolution of western Gondwana in the Early Cretaceous reflects the amalgamation processes that affected the region at least back to the time of the Brasiliano cycle, in both the Atlantic and the Pacific systems. The Brasiliano cycle developed between ca. 890 and 480 Ma in a diacronic way, creating the final arrangement of the basement of the South American Platform (Brito Neves et al., 1999; Mantovani et al., 2005, and references therein). During the Lower Ordovician a mosaic of lithospheric fragment linked by several (accretionary and collisional) Neo-Proterozoic mobile belts amalgamated to form Gondwana (Unrug, 1996). After the amalgamation, the Gondwana supercontinent accumulated Paleozoic and Mesozoic sediments. Concomitantly, it continuously laterally accreted on its western borders by means of successive orogenic belts, in the lower Paleozoic and in the Permian–Triassic, until the formation of Pangaea (Cordani et al., 2000, 2003). The main cratonic fragments, descending from ancestors of Pangaea, were reworked, such as the Amazonia, Rio Apa, Arequipa-Antofalla, and Rio de La Plata cratons. Smaller ancient crustal blocks at the present-day Paraguay boundaries were continuously reworked (Kröner and Cordani, 2003). In this context, the magmatism was driven by the relative extensional regimes derived by relative movements of the ancient blocks. For example, Alto Paraguay Middle Triassic alkaline magmatism occurred at the boundaries between the Rio Apa and Arequipa-Antofalla blocks and reveals extensional events at ca. 240 Ma, probably induced by counterclockwise and clockwise movements (north and south, respectively) hinged at a latitude of ~20°S (cf. Prezzi and Alonso, 2002).

The general geodynamic situation of Paraguay and neighboring countries can be seen from the present-day earthquakes typology combined with paleomagnetic and geological evidence. Earthquake mechanisms (Berrocal and Fernandes, 1995) highlight the distribution of the earthquakes with hypocenters > 500 km and < 70 km (Fig. 13). Deep earthquakes coincide with the inferred location under Paraguay of the subducting Nazca plate (cf. Fig. 2). In particular, the depths of lithospheric earthquakes, together with paleomagnetic results, delineate different

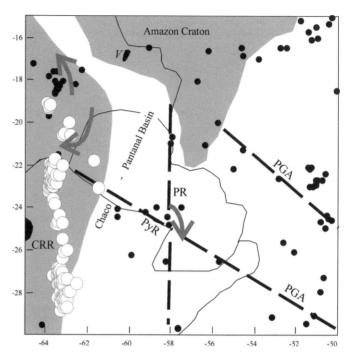

Figure 13. Earthquake distribution in Paraguay and neighboring regions. Open circles—earthquake hypocenters with depths >500 km; filled circles—earthquake hypocenters with depths <70 km (cf. Berrocal and Fernandes, 1995). The black lines indicate extensional lineaments: PGA—Ponta Grossa arch; PR—Paraguay Lineament; PyR—Pylcomaio Lineament. For reference, the Velasco (V) and Central rift (CRR) alkaline complexes of northwestern Argentina (140 and 90 Ma, respectively) are also shown. The dark gray lines with arrows delineate the rotational subplate trends (after Comin-Chiaramonti and Gomes, 1995, and Randall, 1998).

rotational paths at a latitude of ~18°–20°S, roughly corresponding to the Chaco-Pantanal basin, indicating extensional subplate tectonics in the Andean system (Randall, 1998).

Also crucial to the genesis of Paraná-Angola-Edendeka magma types is the link with the geodynamic processes that promoted the opening of the south Atlantic. According to Nürnberg and Müller (1991), seafloor spreading in the south Atlantic at Paraná-Angola-Edendeka latitude started at ca. 125–127 Ma (chron M4). North of the Walvis–Rio Grande ridges (latitude >28°), the onset of oceanic crust would be younger (ca. 113 Ma; Chang et al., 1988).

The Early Cretaceous alkaline and alkaline-carbonatitic complexes are subcoeval with the main flood tholeiites of the Paraná basin, and therefore occurred during the early stages of rifting, before the continental separation. On the other hand, the Late Cretaceous analogs emplaced during advanced stages of Africa–South America continental separation.

The origin of alkaline-carbonatitic magmatism in terms of plate tectonics is currently debated. Various models have been proposed involving deep mantle plumes or hotspots (up to sixteen; Stefanick and Jurdy, 1984) or shallow thermal anomalies

(Holbrook and Kelemen, 1993). Whatever the temperature, size, depth of origin, or number of hotspots, the plume model cannot account for the worldwide occurrence of alkaline-carbonatitic magmatism (see Smith and Lewis, 1999, for a discussion). Gibson et al. (1999) suggested that the Late Cretaceous alkaline-carbonatitic magmatism of Lages, located more than 1000 km south of the postulated head of the Trindade plume, was induced by "impressive" southward channeling of high-temperature melts away from the thick keels of the São Francisco craton. This model does not account for occurrences of Late Cretaceous potassic alkaline rocks from Rio Grande arch (e.g., Piratiní) or for the mechanical barriers formed, e.g., by the Ponta Grossa arch and by the Piquirí and Uruguay lineaments ("second-order plate boundaries," i.e., intraplate boundaries in line with the Paraná province, the Rio Grande, and the Walvis Ridge. Compare also the Ponta Grossa and Moçâmedes lineaments, according to Comin-Chiaramonti and Gomes, 1995. According to the interpretation of remote sensing data along the South American second-order boundary, Unternehr et al., 1988, suggest important dextral displacement between the two South American domains across this boundary).

A more convincing model for the south Atlantic is that proposed by Smith and Lewis (1999). The forces acting on plates that move at differential angular velocity and the presence of volatile-rich mantle sources ("wetspots") would drive the rifting to occur parallel to the preexisting (e.g., north-south) sutures corresponding to the "Adamastor Ocean" that separated the Rio de la Plata craton in South America from the Kalahari and Congo cratons in southern Africa ca. 580–550 Ma (cf. Frimmel and Fölling, 2004, for a discussion). Intraplate alkaline and alkaline-carbonatitic magmatism occurred where second-order plate boundaries (e.g., the Alto Paranaíba, the Ponta Grossa–Moçâmedes arches, the Uruguay Lineament, and the Damara belt; cf. Unternehr et al., 1988) intersect the axis of major rifting, possibly related to the erosion and cycling of continental lithospheric mantle toward the ridge axis.

In southern Brazil, alkaline and alkaline-carbonatitic magmatism is concentrated in regions with positive geoid anomalies (Molina and Ussami, 1999) that may be related to dense, very deep materials. Moreover, the different westward angular velocity of lithospheric fragments in the South America plate, as defined by second-order plate boundaries, as well as the different rotational trends at latitude 19°–20° (cf. Fig. 13), may favor decompression and melting at different times of variously metasomatized portions of the lithospheric mantle (wetspots) with variable isotopic signatures (cf. also Turner et al., 1994; Comin-Chiaramonti et al., 1999). The combined presence of even small amounts of water and carbon dioxide in the upper mantle may lower the melting temperature even by some hundred degrees (see Thybo, 2006, for a discussion).

This scenario could explain the presence of Late Cretaceous to Paleogene sodic magmatism in the Paraná-Angola-Edendeka system (cf. Fig. 8), even at the longitude of eastern Paraguay,

where there is evidence of active rifting structures (Comin-Chiaramonti et al., 1992, 1999). In this case, thermal perturbations may be channeled along the second-order plate boundaries, as also suggested by earthquake hypocenters in South America (cf. Fig. 13 and Berrocal and Fernandes, 1995).

CONCLUDING REMARKS

The new $^{40}Ar/^{39}Ar$ age data and the trace-element and Sr-Nd-Pb isotopic data for the Mesozoic magmatism from eastern Paraguay and from the Paraná-Angola-Namibia system lead to the following conclusions:

1. $^{40}Ar/^{39}Ar$ plateau ages show that distinct magmatic events have taken place in eastern Paraguay since the Middle Triassic in areas strongly characterized by extensional tectonics. The first event occurred during the Anisian (Middle Triassic) and was of sodic affinity. After a magmatic hiatus of ~100 m.y., during the Early Cretaceous, potassic alkaline magmatism pre- and postdated the emission of the Paraná basin flood tholeiites. Since the late Early Cretaceous–Paleogene, only sodic magmatism has occurred in Paraguay. The Sr-Nd-Pb isotopic data indicate that two main mantle components could have been involved in the Cretaceous to Paleogene magmatism in eastern Paraguay: an extreme and heterogeneous EMI component that is prevalent in the Cretaceous K-alkaline magmatism and a depleted component that appears to have been important in the sodic magmatism in the Middle Triassic, late Early Cretaceous, and Paleogene.

2. The potassic rocks form a compositional continuum, from moderately to strongly potassic and from alkali basalt or basanite to trachyte or phonolite. The sodic rocks include mainly ankaratrites, nephelinites, and phonolites.

3. The potassic suites (pre- and post-tholeiitic) are characterized by strongly fractionated REE "Ta-Nb-Ti anomalies." In contrast, slight positive anomalies for Ta and Nb are observed in the sodic rock.

4. Sr-Nd isotope data confirm the distinction of the potassic rocks, enriched in $^{87}Sr/^{86}Sr$ and low in $^{143}Nd/^{144}Nd$, from the sodic rocks, close to bulk Earth and transitional to the Paraná flood tholeiites. Crustal contamination does not appear to have been significant in the generation of the rocks investigated.

5. The Pb-Sr-Nd isotopic systematics show that the Early Cretaceous alkaline magmatism from the Paraná-Angola-Edendeka system appears to be related to heterogeneous mantle sources spanning DMM to HIMU and to time-integrated enriched mantle components. This alkaline magmatism mimics the coeval flood tholeiites in terms of isotopic compositions. The enriched isotopic signatures of the Early Cretaceous alkaline magmatism decrease from

west (Paraguay) to east (Brazil, the southeastern continental margin, and Angola and Namibia). A similar decreasing isotopic shift is also observed for the age of the magmatism in Paraguay and Brazil, i.e., Lower–Late Cretaceous to Paleogene. These results suggest that Paraná-Angola-Edendeka system magmatism is related to both large- and small-scale heterogeneities in the source mantle.

6. The source of potassic rocks is constrained by high levels of LILE, LREE, Th, U, and K relative to the composition of the primitive mantle.

7. The close association of potassic and sodic rock suites in eastern Paraguay demands that their parental magmas derived from small subcontinental mantle masses, vertically and laterally heterogeneous in composition and variously enriched in incompatible elements. Significant H-O-C and F are also expected in the mantle source from the occurrence of related carbonatites. These considerations may be extended to the whole Paraná-Angola-Etendeka system.

8. Any hypothesis involving mantle plume activity (Tristan da Cunha plume) at the margin of the Paraná basin is constrained by distinct lithospheric mantle characteristics and by paleomagnetic results (see Ernesto et al., 2002). Ernesto et al. (2002) demonstrate the following by paleomagnetic reconstruction: (A) The Paraná–Tristan da Cunha system, assuming this hotspot is a fixed point in the mantle, indicates that the TC plume was located ~800 km south of the Paraná magmatic province. Therefore, plume mobility would have been required in order to maintain the Paraná magmatic province–Tristan da Cunha relationships. (B) Assuming that Tristan da Cunha was located in the northern portion of the Paraná magmatic province (~20° from the present Tristan da Cunha position), the plume migrated southward from 133–132 Ma (the main volcanic phase in the area) to 80 Ma at a rate of ~40 mm/yr. From 80 Ma to the present, the plume remained virtually fixed, leaving a track compatible with the Africa plate movement. Notably, the southward migration of the plume is in opposition to the northward migration of the main Paraná magmatic phases (133 Ma in the south and 132 Ma in the north). (C) Regional thermal anomalies in the deep mantle, mapped by geoid and seismic tomography data, offer an alternative non-plume-related heat source for the generation of intracontinental magmatic provinces.

This does not preclude the possibility that thermal perturbations from the asthenosphere triggered magmatic activity in the lithospheric mantle in eastern Paraguay.

9. It is proposed that the variously isotopically enriched sources implied by the potassic magmas derived from a depleted lithospheric mantle pervasively invaded by IE-C-H-rich fluids. These are expected to have promoted the crystallization of K-rich phases (e.g., phlogopite) in a pristine peridotite, where they developed a veined network variously enriched in LILE and LREE under various redox condi-

tions. The newly formed veins ("enriched component") and peridotite matrix ("depleted component") underwent different isotopic evolutions with the time, depending on their parent/daughter ratio (cf. Comin-Chiaramonti et al., 1995a). This model may be extended to the Paraná flood tholeiites and to high- and low-Ti potassic magmatism from southeastern Brazil, Angola, and Namibia.

10. Isotopically distinct magmas were generated following two main "enrichment" events of the subcontinental upper mantle, estimated at 2.0–1.4 Ma and 1.0–0.5 Ga (Comin-Chiaramonti et al., 1997, 1999). This would have preserved isotopic heterogeneities over a long period of time, pointing to a nonconvective lithospheric mantle beneath different cratons or intercratonic regions.

11. The occurrence of sodic and potassic magmatism in the Paraná basin implies that there were also appropriate lithospheric sources to generate the flood tholeiites. Therefore, any hypothesis of an asthenospheric plume origin is not required, with the caveat that it might provide a thermal perturbation and/or a decompressional environment, as well as possible sources of Precambrian plume melts to contaminate the lithosphere.

12. The oversimplified model of mantle plumes is not satisfactory to explain most continental flood basalts and recurrent intraplate alkaline magmatism; therefore, following Ernesto et al. (2002), alternative thermal sources may be found in the mantle with no implication of material transfer from the core or the lower mantle to the lithosphere. Besides the indications from geoid anomalies, the existence of long-lived thermal or compositional anomalies in the mantle have already been demonstrated by velocity distribution models based on seismic tomography using both P- and S-waves (e.g., Zhang and Tanimoto, 1993; Li and Romanonicz, 1996; Van der Hilst et al., 1997; Liu et al., 2003).

On the whole, the geochemical results combined with the new $^{40}Ar/^{39}Ar$ ages for the magmatic events in eastern Paraguay indicate that any model proposed for the evolution of the Paraná-Angola-Edendeka system in terms of HIMU and EM end-members must be consistent with the following constraints: (A) HIMU and EMI-II are not restricted to the oceanic environment; (B) end-members are variously associated in space as a function of various protoliths; (C) mantle regions with HIMU and EMI isotope characteristics are capable of generating melts that can lead to the formation of a wide variety of silicate rocks, including melts enriched in CO_2 (cf. Bell, 1998); (D) the sodic alkaline rock types are systematically grouped together in fields quite distinct from the potassic alkaline fields in Paraguay, but fitting the fields of potassic alkaline-carbonatite rocks from Angola and Namibia; (E) even Na-alkaline rock types from the "Central Rift" of the sub-Andean system (Lucassen et al., 2002) fit the Triassic to Paleocene analogues from eastern Paraguay.

ACKNOWLEDGMENTS

The authors thank the Brazilian Agency Fundação de Amparo à Pesquisa do Estado de São Paulo, Projecto 97/01210–4 and 01/10714–3, for financial support. Technical support by T. Becker (Berkeley) and by the USA (the National Science Foundation) is kindly acknowledged. Highly constructive reviews from K. Bell (who has reservations about most of the ideas presented in this paper), A. Baksi, G. Foulger, and D. Jurdy are also kindly acknowledged.

APPENDIX

Table A-1 presents selected analyses of the main "primitive" alkaline rock types from eastern Paraguay.

TABLE A-1. SELECTED ANALYSES OF THE MAIN "PRIMITIVE" ALKALINE ROCK TYPES FROM EASTERN PARAGUAY

	Alto Paraguay	Valle-Mí	Amambay	Valle-Mí	Valle-Mí	Central region	Central K-ASU	Central K-ASU	Misiones	Asunción
Sample	RP-232	VM-3	SA-94	ST-1	VM-2	3006	77-PS245	D159-PS9	01A	3209
Rock type	Na nepheline syenite	K basanite	K tephrite	L-Ti tholeiitic basalt	H-Ti tholeiitic basalt	L-Ti tholeiitic basalt	K B-P ijolite	K AB-T trachy-basalt	Na ankara-mite	Na melane-phelinite
SiO_2 wt%	54.15	43.38	46.01	48.23	51.49	49.00	44.11	49.43	40.47	42.51
TiO_2	1.62	1.87	1.73	1.44	2.42	1.43	2.01	1.65	3.44	2.05
Al_2O_3	15.70	11.20	13.25	16.20	15.36	14.41	12.82	14.36	12.51	13.41
FeO_{tot}	8.35	10.17	8.31	12.33	12.22	12.28	9.16	8.37	11.64	10.10
MnO	0.43	0.20	0.12	0.17	0.20	0.16	0.19	0.16	0.19	0.20
MgO	2.75	12.08	5.70	7.51	4.13	7.12	9.53	7.61	10.06	9.54
CaO	3.96	15.66	8.46	8.59	6.80	10.92	10.32	7.55	11.27	10.45
Na_2O	6.42	1.38	2.92	2.52	3.04	2.19	3.34	2.29	4.03	5.80
K_2O	4.35	1.61	6.36	0.74	1.24	0.33	5.41	5.72	1.82	1.46
P_2O_5	1.13	0.54	0.59	0.20	0.37	0.14	0.75	0.34	0.71	1.18
LOI	0.38	1.95	6.04	1.58	2.48	1.53	1.23	1.07	1.82	2.19
Sum	99.24	99.94	99.49	99.51	99.75	99.49	98.87	98.55	97.96	98.89
Cr ppm	5	317	110	251	28	349	288	376	301	648
Ni	12	91	32	119	17	114	109	81	126	273
Rb	93	41	121	29.1	39	13.8	114	119	77.4	59
Ba	859	973	2465	301	463	160	1958	1334	953	980
Th	8.6	19.0	16.0	2.16	6.5	1.00	6.6	32.2	5.80	11.0
U	3.0	4.2	4.1	0.95	1.5	0.20	1.7	7.4	1.00	2.4
Ta	11.6	4.5	3.7	0.55	1.1	0.25	3.1	1.1	4.10	8.1
Nb	241	47.6	48.6	4.02	16	1.82	41	37	72	141
Sr	384	1151	2916	376	395	163	1624	1163	1001	1013
Hf	13.9	4.7	9.6	2.80	6.9	1.82			6.10	5.5
Zr	835	203	415	109	240	73	279	268	267	234
Y	71	23	31	31	52	27	19	19	23	33
La	136.00	156.6	162.6	12	23.1	5.50	108	81	63.21	119
Ce	273.01	309.2	268.7	39	53.1	15.01	204	119	117.98	186
Pr	28.50						23.8	13.0		
Nd	99.21	148.7	137.2	20	28.3	12.2	90.2	49.1	57.84	63.70
Sm	19.32	23.8	14.7	3.54	6.8	3.1	13.8	9.81	10.62	11.23
Eu	2.40	6.30	3.88	1.21	2.52	1.05	3.67	2.59	3.04	2.15
Gd	18.40	21.71			8.26		9.09	4.71	5.87	5.16
Tb	2.71	3.02	1.05	0.67	1.36	0.66	1.11	0.65	1.00	0.75
Dy	17.19	7.60			7.95		4.70	3.40	5.70	4.21
Ho	2.12	1.12	0.50		1.44		1.09	0.64		
Er	6.42	3.23	1.41		4.31		1.77	1.70	2.99	2.75
Tm	1.10	0.36			0.66		0.24	0.27		
Yb	5.80	3.20	1.30	2.11	4.48	2.10	1.36	1.32	1.87	1.79
Lu	0.80	0.61	0.21	0.39	0.63	0.27	0.24	0.19	0.27	0.27

(continued)

TABLE A-1. *Continued*

	Alto Paraguay	Valle-Mí	Amambay	Valle-Mí	Valle-Mí	Central region	Central K-ASU	Central K-ASU	Misiones	Asunción
$(La/Ta)_N$	0.70	2.08	2.16	1.30	1.25	1.31	2.02	4.39	0.92	0.88
Age (Ma)	241	139	139	133	133	133	126	126	118	60
Initial ratios										
$^{87}Sr/^{86}Sr$	0.703749	0.706618	0.707341	0.705862	0.705710	0.705157	0.707342	0.707249	0.704157	0.703596
$^{143}Nd/^{144}Nd$	0.512390	0.511868	0.511635	0.512420	0.512341	0.512266	0.511844	0.511650	0.512411	0.512717
$^{206}Pb/^{204}Pb$	18.278	19.968	17.033	17.432	17.752	18.300	17.624	17.040	18.197	18.935
$^{207}Pb/^{204}Pb$	15.662	15.641	15.506	15.511	15.546	15.618	15.620	15.439	15.682	15.677
$^{208}Pb/^{204}Pb$	38.062	38.589	37.465	38.103	38.145	38.315	37.915	37.156	37.936	38.432
$T^{DM}(Nd)$	867	1490	1440	870	1415	1751	1578	2323	925	533
$f_{Sm/Nd}$	−0.40	−0.51	−0.67	−0.46	−0.26	−0.22	−0.53	−0.39	−0.44	−0.46
μ	14.02	15.03	13.21	23.48	9.31	6.83	12.24	11.91	12.73	12.60
κ	3.34	3.50	1.51	2.35	4.87	5.17	4.01	6.76	5.56	5.62
Fe_2O_3/FeO	0.56	0.47	0.22	0.17	0.22	0.22	0.33	0.33	0.23	0.69
Mg#	0.51	0.72	0.59	0.56	0.42	0.55	0.71	0.68	0.66	0.69

Sources: After Comin-Chiaramonti and Gomes (1995, 2005) and Comin-Chiaramonti et al. (1999).

Notes: $(La/Ta)_N$ represents normalized ratios to the primitive mantle of Sun and McDonough (1989); T^{DM} is calculated assuming the following values for the depleted mantle: $^{143}Nd/^{144}Nd = 0.513151$ and $Sm/Nd = 0.2188$ (Faure, 1986); $f_{Sm/Nd} = [(^{147}Sm/^{144}Nd)_{Sample} − (^{147}Sm/^{144}Nd)_{CHUR}]/(^{147}Sm/^{144}Nd)_{CHUR}$ (DePaolo, 1988).

REFERENCES CITED

Alberti, A., Castorina, F., Censi, P., Comin-Chiaramonti, P., and Gomes, C.B., 1999, Geochemical characteristics of Cretaceous carbonatites from Angola: Journal of African Earth Sciences, v. 29, p. 735–759, doi: 10.1016/S0899-5362(99)00127-X.

Almeida, F.F.M., 1983, Relações tectônicas das rochas alcalinas mesozoicas da região meridional da plataforma Sul-Americana: Revista Brasileira de Geocioeencias, v. 13, p. 139–158.

Andrade, F.R.D., Möller, P., and Höhndorf, A., 1999, The effect of hydrothermal alteration on the Sr and Nd isotopic signatures of the Barra do Itapirapuã carbonatite, southern Brazil: Journal of Geology, v. 107, p. 177–191, doi: 10.1086/314339.

Araujo, A.L.N., Carlson, R.W., Gaspar, J.C., and Bizzi, L.A., 2001, Petrology of kamafugites and kimberlites from the Alto Paranaíba alkaline province, Minnas Gerais, Brazil: Contributions to Mineralogy and Petrology, v. 142, p. 163–177.

Arndt, N.T., and Goldstein, L.S., 1987, Use and abuse of crust-formation ages: Geology, v. 15, p. 893–895, doi: 10.1130/0091-7613(1987)15<893:UAAOCA>2.0.CO;2.

Beccaluva, L., Di Girolamo, P., and Serri, G., 1991, Petrogenesis and tectonic setting of the Roman volcanic province, Italy: Lithos, v. 26, p. 191–221.

Bell, K., 1998, Radiogenic isotope constraints on relationships between carbonatites and associated silicate rocks—A brief review: Journal of Petrology, v. 39, p. 99–102.

Bell, K., and Blenkinsop, J., 1989, Neodymium and strontium isotope geochemistry of carbonatites, in Bell, K., ed., Carbonatites, genesis and evolution: London, Unwin Hyman, p. 278–300.

Bellieni, G., Comin-Chiaramonti, P., Marques, L.S., Martinez, L.A., Melfi, A.J., Nardy, A.J.R., and Piccirillo, E.M., 1986, Continental flood basalts from the central-western regions of the Paraná plateau (Paraguay and Argentina): Petrology and petrogenetic aspects. Neues Jahrbuch: Mineralogische Abhandlungen, v. 154, p. 111–139.

Bellieni, G., Montes-Lauar, C.R., DeMin, A., Piccirillo, E.M., Cavazzini, G., Melfi, A.J., and Pacca, I.G., 1990, Early and Late Cretaceous magmatism from São Sebastião island (SE-Brazil): Geochemistry and petrology: Geochimica Brasiliensis, v. 4, p. 59–83.

Bennio, L., Brotzu, P., Gomes, C.B., D'Antonio, M., Lustrino, M., Melluso, L., Morbidelli, L., and Ruberti, E., 2002, Petrological, geochemical and Sr-Nd isotopic features of Eocene alkaline rocks of the Cabo Frio Peninsula (southern Brazil): Evidence of several magma series: Periodico di Mineralogia, v. 71, p. 137–158.

Berrocal, J., and Fernandes, C., 1995, Seismicity in Paraguay and neighbouring regions, in Comin-Chiaramonti, P., and Gomes, C.B., eds. Alkaline magmatism in Central–Eastern Paraguay: Relationships with coeval magmatism in Brazil: São Paulo, Edusp-Fapesp, p. 57–66.

Bitschene, P.R., 1987, Mesozoischer und Kanozoischer anorgener Magmatismus in Ostparaguay: Arbeiten zur Geologie und Petrologie zweier Alkaliprovinzen [Ph.D. dissertation]: Heidelberg University, 317 p.

Bizzi, L.A., Smith, B.C., De Wit, M.J., Macdonald, I., and Armstrong, R.A., 1994, Isotope characteristics of the lithospheric mantle underlying the SW São Francisco craton margin, Brazil, in International Symposium on the Physics and Chemistry of the Upper Mantle: Invited Lectures, São Paulo: Rio de Janeiro, Academia Brasileira de Ciencias, p. 227–256.

Bizzi, L.A., de Witt, M.J., Smith, C.B., McDonald, I., and Armstrong, R.A., 1995, Heterogeneous enriched mantle materials and Dupal-type magmatism along SW margin of the São Francisco craton, Brazil: Journal of Geodynamics, v. 20, p. 469–491, doi: 10.1016/0264-3707(95)00028-8.

Brito Neves, B.B., Costa Campos Neto, M., and Fuck, R.A., 1999, From Rodinia to Western Gondwana: An approach to the Brasiliano–Pan African cycle and orogenic collage: Episodes, v. 22, p. 155–166.

Camboa, L.A.P., and Rabinowitz, P.D., 1984, The evolution of the Rio Grande Rise in Southwest Atlantic: Marine Geology, v. 58, p. 35–58, doi: 10.1016/0025-3227(84)90115-4.

Carlson, R.W., Esperança, S., and Svisero, D.P., 1996, Chemical and Os isotopic study of Cretaceous potassic rocks from southern Brazil: Contributions to Mineralogy and Petrology, v. 125, p. 393–405, doi: 10.1007/s004100050230.

Castorina, F., Censi, P., Barbieri, M., Comin-Chiaramonti, P., Cundari, A., Gomes, C.B., and Pardini, G., 1996, Carbonatites from Eastern Paraguay: A comparison with coeval carbonatites from Brazil and Angola, in Comin-Chiaramonti, P., and Gomes, C.B., eds., Alkaline magmatism in Central–Eastern Paraguay: Relationships with coeval magmatism in Brazil: São Paulo, Edusp-Fapesp, p. 231–248.

Castorina, F., Censi, P., Comin-Chiaramonti, P., Gomes, C.B., Piccirillo, E.M., Alcover Neto, A., Almeida, R.T., Speziale, S., and Toledo, M.C., 1997, Geochemistry of carbonatites from Eastern Paraguay and genetic relationships with potassic magmatism: C, O, Sr and Nd isotopes: Mineralogy and Petrology, v. 61, p. 237–260, doi: 10.1007/BF01172487.

Censi, P., Comin-Chiaramonti, P., Demarchi, G., Longinelli, A., and Orué, D., 1989, Geochemistry and C-O isotopes of the Chiriguelo carbonatite, northeastern Paraguay: Journal of South American Earth Sciences, v. 2, p. 295–303, doi: 10.1016/0895-9811(89)90035-7.

Chang, H.K., Kowsmann, R.O., and de Figuereido, A.M.F., 1988, New concept on the development of east Brazilian margin basins: Episodes, v. 11, p. 194–102.

Comin-Chiaramonti, P., 1979, On the K-Ca feldspars. Neues Jahrbuch: Mineralogische Monatshefte, v. H11, p. 495–497.

Comin-Chiaramonti, P., and Gomes, C.B., eds., 1995, Alkaline magmatism in Central–Eastern Paraguay: Relationships with coeval magmatism in Brazil: São Paulo, Edusp-Fapesp, 464 p.

Comin-Chiaramonti, P., and Gomes, C.B., eds., 2005, Mesozoic to Cenozoic alkaline magmatism in the Brazilian Platform: São Paulo, Edusp-Fapesp, 752 p.

Comin-Chiaramonti, P., Gomes, C.B., Piccirillo, E.M., and Rivalenti, G., 1983, High-TiO2 basaltic dykes in the coastline of São Paulo and Rio de Janeiro States (Brazil): Neues Jahrbuch: Mineralogische Abandlungen, v. 146, p. 133–150.

Comin-Chiaramonti, P., Civetta, L., Petrini, R., Piccirillo, E.M., Bellieni, G., Censi, P., Bitschene, P., Demarchi, G., DeMin, A., Gomes, C.B., Castillo, A.M.C., and Velázquez, J.C., 1991, Paleogene nephelinitic magmatism in Eastern Paraguay: Petrology, Sr-Nd isotopes and genetic relationships with associated spinel-peridotite xenoliths: European Journal of Mineralogy, v. 3, p. 507–525.

Comin-Chiaramonti, P., Gomes, C.B., Petrini, R., DeMin, A., Velázquez, V.F., and Orué, D., 1992, A new area of alkaline rocks in Eastern Paraguay: Revista Brasileira de Geociências, v. 22, p. 500–506.

Comin-Chiaramonti, P., Castorina, F., Cundari, A., Petrini, R., and Gomes, C.B., 1995a, Dykes and sills from Eastern Paraguay: Sr and Nd isotope systematics, *in* Baer, G., and Heimann, A., eds., Physics and chemistry of dykes: Rotterdam, Balkema, p. 267–278.

Comin-Chiaramonti, P., Cundari, A., DeMin, A., Gomes, C.B., and Piccirillo, E.M., 1995b, Potassic magmatism from central–Eastern Paraguay: Petrogenesis and geodynamic inferences, *in* Comin-Chiaramonti, P., and Gomes, C.B., eds., Alkaline magmatism in Central-Eastern Paraguay: Relationships with coeval magmatism in Brazil: São Paulo, Edusp-Fapesp, p. 207–222.

Comin-Chiaramonti, P., DeGraff, J.M., Hegarty, K., and Petrini, R., 1996, Geological, geophysical and geochemical constraints on the Mesozoic–Paleogene magmatism in Eastern Paraguay: DICAMP, Quaderni di Mineralogia: Petrografia e Geochimica Applicate, v. 11, p. 1–19.

Comin-Chiaramonti, P., Cundari, A., Piccirillo, E.M., Gomes, C.B., Castorina, F., Censi, P., DeMin, A., Marzoli, A., Speziale, S., and Velázquez, V.F., 1997, Potassic and sodic igneous rocks from Eastern Paraguay: Their origin from the lithospheric mantle and genetic relationships with the associated Paraná flood tholeiites: Journal of Petrology, v. 38, p. 495–528, doi: 10.1093/petrology/38.4.495.

Comin-Chiaramonti, P., Cundari, A., DeGraff, J.M., Gomes, C.B., and Piccirillo, E.M., 1999, Early Cretaceous–Paleogene magmatism in Eastern Paraguay (western Paraná basin): Geological, geophysical and geochemical relationships: Journal of Geodynamics, v. 28, p. 375–391, doi: 10.1016/S0264-3707(99)00016-2.

Comin-Chiaramonti, P., Gomes, C.B., and Velázquez, V.F., 2000, The Mesoproterozoic rhyolite occurrences of Fuerte Olimpo and Fuerte San Carlos, northern Paraguay: Revista Brasileira de Geociências, v. 30, p. 785–788.

Comin-Chiaramonti, P., Princivalle, F., Girardi, V.A.A., Gomes, C.B., Laurora, A., and Zanetti, F., 2001, Mantle xenoliths from Ñemby, Eastern Paraguay: O-Sr-Nd isotopes, trace elements and crystal chemistry of hosted clinopyroxenes: Periodico di Mineralogia, v. 70, p. 205–230.

Comin-Chiaramonti, P., Gomes, C.B., Castorina, F., Censi, P., Antonini, P., Furtado, S., Ruberti, E., and Scheibe, L.F., 2002, Anitápolis and Lages alkaline-carbonatite complexes, Santa Catarina State, Brazil: Geochemistry and geodynamic implications: Revista Brasileira de Geociências, v. 32, p. 639–653.

Comin-Chiaramonti, P., Ernesto, M., Velázquez, V.F., and Gomes, C.B., 2004, Plumes beneath the Paraná Basin, Eastern Paraguay: Fact or fiction?: www.MantlePlumes.org/parana.html, p. 1–14.

Comin-Chiaramonti, P., Gomes, C.B., Censi, P., Gasparon, M., and Velázquez, V.F., 2005a, Alkaline complexes from the Alto Paraguay Province at the border of Brazil (Mato Grosso do Sul State) and Paraguay, *in* Comin-Chiaramonti, P., and Gomes, C.B., eds., Mesozoic to Cenozoic alkaline magmatism in the Brazilian Platform: São Paulo, Edusp-Fapesp, p. 71–148.

Comin-Chiaramonti, P., Gomes, C.B., Velázquez, V.F., Censi, P., Antonini, P., Comin-Chiaramonti, F., and Punturo, R., 2005b, Alkaline complexes from southeastern Bolivia, *in* Comin-Chiaramonti, P., and Gomes, C.B., eds., Mesozoic to Cenozoic alkaline magmatism in the Brazilian Platform: São Paulo, Edusp-Fapesp, p. 159–212.

Comte, D., and Hasui, Y., 1971, Geochronology of Eastern Paraguay by the potassium-argon method: Revista Brasileira de Geociências, v. 1, p. 33–43.

Cooper, A.F., and Reid, D.L., 1998, Nepheline sövites as parental magmas in carbonatite complexes: Evidence from Dicker Willem, southwest Namibia: Journal of Petrology, v. 39, p. 2123–2136, doi: 10.1093/petrology/39.11.2123.

Cordani, U.G., Sato, K., Teixeira, W., Tassinari, C.C.G., and Basei, M.A.S., 2000, Crustal evolution of the South American platform, *in* Cordani, U.G., et al., eds., Tectonic evolution of South America: 31st International Geological Congress, Rio de Janeiro, p. 19–40.

Cordani, U.G., Brito Neves, B.B., D'Agrella, M.S., and Trindade, R.I.F., 2003, From Rodinia to Gondwana: A review of the available evidence from South America: Gondwana Research, v. 6, p. 275–283, doi: 10.1016/S1342-937X(05)70976-X.

Cristiani, C., Matteini, M., Mazzuoli, R., Omarini, R., and Villa, I.M., 2005, Petrology of Late Jurassic–Early Cretaceous Tusaquillas and Abra Laite–Aguilar plutonic complexes (Central Andes, 23°05′S–66°05′W): A comparison with rift-related magmatism of NW Argentina and E Bolivia, *in* Comin-Chiaramonti, P., and Gomes, C.B., eds., Mesozoic to Cenozoic alkaline magmatism in the Brazilian Platform: São Paulo, Edusp-Fapesp, p. 213–240.

DePaolo, D.J., 1988, Neodymium isotope geochemistry: An introduction: Minerals and Rocks, v. 20, p. 1–87.

Eby, N.G., and Mariano, A.N., 1992, Geology and geochronology of carbonatites and associated alkaline rocks peripheral to the Paraná Basin, Brazil-Paraguay: Journal of South American Earth Sciences, v. 6, p. 207–216, doi: 10.1016/0895-9811(92)90009-N.

Erlank, A.J., Marsh, J.S., Duncan, A.R., Miller, R.McG., Hawkesworth, C.J., Betton, P.J., and Rex, D.C., 1984, Geochemistry and petrogenesis of the Etendeka volcanic rocks from South West Africa/Namibia: Johannesburg, Geological Society of South Africa, Special Publication 13, p. 195–245.

Ernesto, M., Raposo, M.I.B., Marques, L.S., Renne, P.R., Diogo, L.A., and DeMin, A., 1999, Paleomagnetism, geochemistry, [40]Ar-[39]Ar geochronology of the North-eastern Paraná magmatic province: Tectonic implications: Journal of Geodynamics, v. 28, p. 321–340, doi: 10.1016/S0264-3707(99)00013-7.

Ernesto, M., Marques, L.M., Piccirillo, E.M., Comin-Chiaramonti, P., and Bellieni, G., 2000, Paraná-Tristan da Cunha system: Plume mobility and petrogenetic implications: 31 International Gramsci Society, Rio de Janeiro, extended abstracts.

Ernesto, M., Marques, L.M., Piccirillo, E.M., Molina, E., Ussami, N., Comin-Chiaramonti, P., and Bellieni, G., 2002, Paraná Magmatic Province–Tristan da Cunha plume system: Fixed versus mobile plume, petrogenetic considerations and alternative heat sources: Journal of Volcanology and Geothermal Research, v. 118, p. 15–36, doi: 10.1016/S0377-0273(02)00248-2.

Ewart, A., Marsh, J.S., Milner, S.C., Duncan, A.R., Kamber, B.S., and Armstrong, R.A., 2004, Petrology and geochemistry of Early Cretaceous bimodal continental flood volcanism of the NW Etendeka, Namibia, Part 1, Introduction: Mafic lavas and re-evalutation of mantle source components: Journal of Petrology, v. 45, p. 59–105, doi: 10.1093/petrology/egg083.

Faure, G., 1986, Principles of isotope geology: New York, John Wiley and Sons, 589 p.

Fitton, J.C., 2007 (this volume), The OIB paradox, *in* Foulger, G.R., and Jurdy, D.M., eds., Plates, plumes, and planetary processes: Boulder, Colorado, Geological Society of America Special Paper 430, doi: 10.1130/2007.2430(20).

Fitton, J.C., Saunders, A.D., Norry, M.J., Hardarson, B.S., and Taylor, R.N., 1997, Thermal and chemical structure of the Iceland plume: Earth and Planetary Science Letters, v. 153, p. 197–208, doi: 10.1016/S0012-821X(97)00170-2.

Fodor, R.V., Mukasa, S.B., Gomes, C.B., and Cordani, U.G., 1989, Ti-rich Eocene basaltic rock, Abrolhos platform, offshore Brazil, 18°S: Petrology with respect to South Atlantic magmatism: Journal of Petrology, v. 30, p. 763–786.

Fodor, R.V., Mukasa, S.B., and Sial, A.N., 1998, Isotopic and trace-element indications of lithospheric and asthenospheric components in Paleogene alkalic basalts, northeastern Brazil: Lithos, v. 43, p. 197–217, doi: 10.1016/S0024-4937(98)00012-7.

Fontignie, D., and Schilling, J.G., 1996, Mantle heterogeneities beneath the South Atlantic: A Nd-Sr-Pb isotope study along the Mid-Atlantic Ridge (3°S–46°S): Earth and Planetary Science Letters, v. 146, p. 259–272.

Foulger, G.R., and Anderson, D.L., 2005, A cool model for the Iceland hotspot: Journal of Volcanology and Geothermal Research, v. 141, p. 1–22, doi: 10.1016/j.jvolgeores.2004.10.007.

Foulger, G.R., Natland, J.H., and Anderson, D.L., 2005, Genesis of the Iceland melt anomaly by plate tectonic processes, *in* Foulger, G.R., et al., eds., Plates, plumes, and paradigms: Boulder, Colorado, Geological Society of America Special Paper 388, p. 595–625.

Frimmel, H.E., and Fölling, P.G., 2004, Late Vendian closure of the Adamastor Ocean: Timing of tectonic inversion and syn-orogenic sedimentation in the Gariep Basin: Gondwana Research, v. 7, p. 685–699, doi: 10.1016/S1342-937X(05)71056-X.

Fulfaro, V.J., 1995, Geology of Eastern Paraguay, *in* Comin-Chiaramonti, P. and Gomes, C.B., eds., Alkaline magmatism in Central–Eastern Paraguay: Relationships with coeval magmatism in Brazil: São Paulo, Edusp-Fapesp, pp. 17–30.

Camboa, L.A.P., and Rabinowitz, P.D., 1984, The evolution of the Rio Grande Rise in Southwest Atlantic: Marine Geology, v. 58, p. 35–58, doi: 10.1016/0025-3227(84)90115-4.

Garda, G.M., Schorscher, H.D., Esperança, S., and Carlson, R.W., 1995, The petrology and geochemistry of coastal dikes from São Paulo State, Brazil: Implications for variable lithospheric contributions to alkaline magmas from the western margin of the South Atlantic: Anais da Academia Brasileira de Ciencias, v. 67, p. 191–216.

Gastal, M.P., Lafon, J.M., Hartmann, L.A., and Koester, E., 2005, Sm-Nd isotopic investigation of Neoproterozoic and Cretaceous igneous rocks from southern Brazil: A study of magmatic processes: Lithos, v. 82, p. 345–377, doi: 10.1016/j.lithos.2004.09.025.

Gerlach, D.C., Stormer, J.C., and Mueller, P.A., 1987, Isotope geochemistry of Fernando de Noronha: Earth and Planetary Science Letters, v. 85, p. 129–144, doi: 10.1016/0012-821X(87)90027-6.

Gibson, S.A., Thompson, R.N., Dickin, A.P., and Leonardos, O.H., 1995a, High-Ti and low-Ti mafic potassic magmas: Key to plume-lithosphere interactions and continental flood-basalt genesis: Earth and Planetary Science Letters, v. 136, p. 149–165, doi: 10.1016/0012-821X(95)00179-G.

Gibson, S.A., Thompson, R.N., Leonardos, O.H., Dickin, A.P., and Mitchell, J.G., 1995b, The Late Cretaceous impact of the Trindade mantle plume; Evidence from large-volume, mafic, potassic magmatism in SE Brazil: Journal of Petrology, v. 36, p. 189–229.

Gibson, S.A., Thompson, R.N., Weska, R.K., Dickin, A.P., and Leonardos, O.H.,

1997, Late-Cretaceous rift-related upwelling and melting of the Trindade starting mantle plume head beneath western Brazil: Contributions to Mineralogy and Petrology, v. 126, p. 303–314, doi: 10.1007/s004100050252.

Gibson, S.A., Thompson, R.N., Leonardos, O.H., Dickin, A.P., and Mitchell, J.G., 1999, The limited extent of plume-lithosphere interactions during continental flood-basalt genesis: Geochemical evidence from Cretaceous magmatism in southern Brazil: Contributions to Mineralogy and Petrology, v. 137, p. 147–169, doi: 10.1007/s004100050588.

Gomes, C.B., Laurenzi, M.A., Censi, P., DeMin, A., Velázquez, V.F., and Comin-Chiaramonti, P., 1995, Alkaline magmatism from northern Paraguay (Alto Paraguay): A Permo-Triassic province, *in* Comin-Chiaramonti, P., and Gomes, C.B., eds. Mesozoic to Cenozoic alkaline magmatism in the Brazilian Platform: São Paulo, Edusp-Fapesp, p. 223–230.

Gradstein, F.M., Ogg, J.G., Smith, A.G., Bleeker, W., and Lourens, L.J., 2004, A new geologic time scale with special reference to Precambrian and Neogene: Episodes, v. 27, p. 83–100.

Gudmundsson, O., and Sambridge, M., 1998, A regionalized upper mantle (RUM) seismic model: Journal of Geophysical Research, v. 103, p. B7121–B7136, doi: 10.1029/97JB02488.

Halliday, A.N., Dickin, A.P., Fallick, A.E., and Fitton, J.G., 1988, Mantle dynamics: A Nd, Sr, Pb and O isotopic study of the Cameroon Line volcanic chain: Journal of Petrology, v. 29, p. 181–211.

Halliday, A.N., Lee, D.C., Tommasini, S., Davies, G.R., Paslick, C.R., Fitton, G.J., and James, D.E., 1995, Incompatible trace elements in OIB and MORB and source enrichment in the sub-oceanic mantle: Earth and Planetary Science Letters, v. 133, p. 379–395, doi: 10.1016/0012-821X(95)00097-V.

Hamelin, B., Dupré, B., and Allègre, C.J., 1984, Lead-strontium isotopic variations along the East Pacific Rise and Mid-Atlantic Ridge: A comparative study: Earth and Planetary Science Letters, v. 67, p. 340–350, doi: 10.1016/0012-821X(84)90173-0.

Harris, C., Marsh, J.S., and Milner, S.C., 1999, Petrology of the alkaline core of the Messum igneous complex, Namibia: Evidence for the progressively decreasing effect of crustal contamination: Journal of Petrology, v. 40, p. 1377–1397, doi: 10.1093/petrology/40.9.1377.

Hart, S.R., 1984, A large-scale isotope anomaly in the southern hemisphere: Nature, v. 309, p. 753–757, doi: 10.1038/309753a0.

Hart, S.R., and Zindler, A., 1989, Constraints on the nature and the development of chemical heterogeneities in the mantle, *in* Peltier, W.R., ed., Mantle convection plate tectonics and global dynamics: New York, Gordon and Breach, p. 261–388.

Hart, S.R., Gerlach, D.C., and White, W.M., 1986, A possible new Sr-Nd-Pb mantle array and consequences for mantle mixing: Geochimica et Cosmochimica Acta, v. 50, p. 1551–1557, doi: 10.1016/0016-7037(86)90329-7.

Hart, S.R., Hauri, E.H., Oschmann, L.A., and Whitehead, J.A., 1992, Mantle plumes and entrainment: Isotopic evidence: Science, v. 256, p. 517–520, doi: 10.1126/science.256.5056.517.

Hawkesworth, C.J., Mantovani, M.S.M., Taylor, P.N., and Palacz, Z., 1986, Evidence from the Paraná of south Brazil for a continental contribution to Dupal basalts: Nature, v. 322, p. 356–359, doi: 10.1038/322356a0.

Hawkesworth, C.J., Kelley, S., Turner, S.P., Le Roex, A., and Storey, B., 1999, Mantle processes during Gondwana break-up and dispersal: Journal of African Earth Sciences, v. 28, p. 239–261, doi: 10.1016/S0899-5362(99)00026-3.

Hawkesworth, C.J., Gallagher, K., Kirstein, L., Mantovani, M.S.M., Peate, D.W., and Turner, S.P., 2000, Tectonic controls on magmatism associated with continental break-up: An example from the Paraná-Etendeka Province: Earth and Planetary Science Letters, v. 179, p. 335–349, doi: 10.1016/S0012-821X(00)00114-X.

Hegarty, K.A., Duddy, I.R., and Green, P.F., 1995, The thermal history in and around the Paraná Basin using apatite track analysis: Implications for hydrocarbon occurrences and basin formation, *in* Comin-Chiaramonti, P., and Gomes, C.B., eds., Alkaline magmatism in Paraguay and relationships with coeval magmatism in Brazil: São Paulo, Edusp-Fapesp, p. 41–50.

Hergt, J.M., Peate, D.W., and Hawkesworth, C.J., 1991, The petrogenesis of Mesozoic Gondwana low-Ti flood basalts: Earth and Planetary Science Letters, v. 105, p. 134–148, doi: 10.1016/0012-821X(91)90126-3.

Holbrook, W.S., and Kelemen, P.B., 1993, Large igneous province on the US Atlantic margin and implications for magmatism during continental breakup: Nature, v. 364, p. 433–436, doi: 10.1038/364433a0.

Huang, H.-M., Hawkesworth, C.J., Van Calsteren, P., and McDermott, F., 1995, Geochemical characteristics and origin of the Jacupiranga carbonatites, Brazil: Chemical Geology, v. 119, p. 79–99, doi: 10.1016/0009-2541(94)00093-N.

Iacumin, M., DeMin, A., Piccirillo, E.M., and Bellieni, G., 2003, Source mantle heterogeneity and its role in the genesis of Late Archean–Proterozoic (2.7–1.0 Ga) and Mesozoic (200 and 130 Ma) tholeiitic magmatism in the South American Platform: Earth-Science Reviews, v. 62, p. 365–397, doi: 10.1016/S0012-8252(02)00163-0.

Ito, E., White, W.M., and Göpel, C., 1987, The O, Sr, Nd and Pb isotope geochemistry of MORB: Chemical Geology, v. 62, p. 157–176, doi: 10.1016/0009-2541(87)90083-0.

Jezek, P., Wilner, A.P., Aceñolaza, F.G., and Miller, H., 1985, The Pungoviscana trough, a large basin of Late Precambrian to Early Cambrian age on the Pacific edge of the Brazilian shield: Geologische Rundschau, v. 74, p. 573–584, doi: 10.1007/BF01821213.

Kanzler, A., 1987, The southern Precambrian in Paraguay. Geological inventory and age relation: Zentralblatt Geologische und Paläontologische, v. 1, no. 7/8, p. 753–765.

Kirstein, L.A., Kelley, S., Hawkesworth, C., Turner, S., Mantovani, M., and Wijbrans, J., 2001, Protracted felsic magmatic activity associated with the opening of the South Atlantic: Journal of the Geological Society, London, v. 158, p. 583–592.

Kröner, A., and Cordani, U.G., 2003, African, southern Indian and South American cratons were not part of the Rondinia supercontinent: Evidence for field relationships and geochronology: Tectonophysics, v. 375, p. 325–352, doi: 10.1016/S0040-1951(03)00344-5.

Kurszlaukis, S., Franz, L., and Brey, G.P., 1999, The Blue Hill intrusive complex in southern Namibia: Relationships between carbonatites and monticellite picrites: Chemical Geology, v. 160, p. 1–18, doi: 10.1016/S0009-2541(99)00027-3.

Laux, J.H., Pimentel, M.M., Dantas, E.L., Armstrong, R., and Junges, S.L., 2005, Two Neoproterozoic crustal accretion events in the Brasilia belt, central Brazil: Journal of South American Earth Sciences, v. 18, p. 183–198, doi: 10.1016/j.jsames.2004.09.003.

Le Roex, A.P., 1985, Geochemistry, mineralogy and magmatic evolution of the basaltic and trachytic lavas from Gough Island, South Atlantic: Journal of Petrology, v. 26, p. 149–186.

Le Roex, A.P., and Lanyon, R., 1998, Isotope and trace element geochemistry of Cretaceous Damaraland lamprophyres and carbonatites, northwestern Namibia: Evidence for plume-lithosphere interaction: Journal of Petrology, v. 39, p. 1117–1146, doi: 10.1093/petrology/39.6.1117.

Le Roex, A.P., Cliff, R.A., and Adair, B.J.J., 1990, Tristan da Cunha, South Atlantic: Geochemistry and petrogenesis of a basanite-phonolite lava series: Journal of Petrology, v. 31, p. 779–812.

Li, X.D., and Romanonicz, B., 1996, Global mantle shear velocity model developed using nonlinear asymptotic coupling theory: Journal of Geophysical Research, v. 101, p. 22,245–22,272, doi: 10.1029/96JB01306.

Liu, H.K., Gao, S.S., Silver, P.G., and Zhang, Y., 2003, Mantle layering across central South America: Journal of Geophysical Research, v. 108, no. B11, 2510, doi: 10.1029/2002JB002208.

Lucassen, F., Escayola, M., Franz, G., Romer, R.L., and Koch, K., 2002, Isotopic composition of late Mesozoic basic and ultrabasic rocks from Andes (23–32° S)—Implications for the Andean mantle: Contributions to Mineralogy and Petrology, v. 143, p. 336–349.

Lustrino, M., 2005, How the delamination and detachment of lower crust can influence basaltic magmatism: Earth-Science Review, v. 72, p. 21–38, doi: 10.1016/j.earscirev.2005.03.004.

Mantovani, M.S.M., Quintas, M.C.L., Shukowsky, W., and Brito Neves, B.B., 2005, Delimitation of the Paranapanema Proterozoic block: A geophysical contribution: Episodes, v. 28, p. 18–22.

Marques, L.S., Ulbrich, M.N.C., Ruberti, E., and Tassinari, C.G., 1999, Petrology, geochemistry and Sr-Nd isotopes of the Trindade and Martin Vaz volcanic rocks (Southern Atlantic Ocean): Journal of Volcanology and Geothermal Research, v. 93, p. 191–216, doi: 10.1016/S0377-0273(99)00111-0.

Marzoli, A., Melluso, L., Morra, V., Renne, P.R., Sgrosso, I., D'Antonio, M., Duarte Morais, L., Morais, E.A.A., and Ricci, G., 1999, Geochronology and petrology of Cretaceous basaltic magmatism in the Kwanza basin (western Angola), and relationships with the Paraná-Etendeka continental flood basalt province: Journal of Geodynamics, v. 28, p. 341–356, doi: 10.1016/S0264-3707(99)00014-9.

Maury, R., Defant, C., and Joron, M.J., 1992, Metasomatism of the sub-arc mantle inferred from trace elements in Philippine xenoliths: Nature, v. 360, p. 661, doi: 10.1038/360661a0.

McKenzie, D.P., and O'Nions, M.J., 1995, The source regions of ocean island basalts: Journal of Petrology, v. 36, p. 133–159.

Meen, J.K., Ayers, J.C., and Fregeau, E.J., 1989, A model of mantle metasomatism by carbonated alkaline melts: Trace element and isotopic compositions of mantle source regions of carbonatite and other continental igneous rocks, *in* Bell, K., ed., Carbonatites, genesis and evolution: London, Unwin Hyman, p. 464–499.

Meyer, H.O.A., Blaine, L.G., Svisero, D.P., and Craig, B.S., 1994, Alkaline intrusions in western Minas Gerais, Brazil: Proceedings of the Fifth International Kimberlite Conference, Capetown, South Africa, p. 140–155.

Milner, S.C., and LeRoex, A.P., 1996, Isotope characteristics of the Okenyenya igneous complex, northwestern Namibia: Constraints on the composition of the early Tristan plume and the origin of the EM 1 mantle component: Earth and Planetary Science Letters, v. 141, p. 277–291, doi: 10.1016/0012-821X(96)00074-X.

Milner, S.C., Duncan, A.R., Whittingham, A.M., and Ewart, A., 1995, Trans-Atlantic correlation of eruptive sequences and individual silicic volcanics units within the Paraná-Etendeka igneous province: Journal of Volcanological and Geothermal Research, v. 69, p. 137–157, doi: 10.1016/0377-0273(95)00040-2.

Molina, E.C., and Ussami, N., 1999, The geoid in southern Brazil and adjacent regions: New constraints on density distribution and thermal state of lithosphere: Journal of Geodynamics, v. 28, p. 357–374, doi: 10.1016/S0264-3707(99)00015-0.

Nürnberg, D., and Müller, R.D., 1991, The tectonic evolution of South Atlantic from Late Jurassic to present: Tectonophysics, v. 191, p. 27–43, doi: 10.1016/0040-1951(91)90231-G.

Pearce, J.A., 1983, Role of the sub-continental lithosphere in magma genesis at active continental margins, *in* Hawkesworth, C.J., and Norry, M.J., eds., Continental basalts and mantle xenoliths: Nantwich, England, Shiva, p. 230–249.

Peate, D.W., 1997, The Paraná-Etendeka province, *in* Mahoney, J.J., and Coffin, M.F., eds., Large igneous provinces: Continental, oceanic and planetary flood volcanism: Washington, D.C., American Geophysical Union, Geophysical Monograph 100, p. 217–245.

Peate, D.W., Hawkesworth, C.J., Mantovani, M.S.M., Rogers, N.W., and Turner, S.P., 1999, Petrogenesis and stratigraphy of the high Ti/Y Urubici magma type in the Paraná flood basalt Province and implications for the nature of "Dupal"-type mantle in the South Atlantic Region: Journal of Petrology, v. 40, p. 451–473, doi: 10.1093/petrology/40.3.451.

Piccirillo, E.M., and Melfi, A.J., eds., 1988, The Mesozoic flood volcanism from the Paraná basin (Brazil): Petrogenetic and geophysical aspects: São Paulo, Iag-Usp, 600 p.

Piccirillo, E.M., Bellieni, G., Cavazzini, G., Comin-Chiaramonti, P., Petrini, R., Melfi, A.J., Pinese, J.P.P., Zantedeschi, P., and DeMin, A., 1990, Lower Cretaceous dyke swarms from the Ponta Grossa Arch: Petrology, Sm-Nd isotopes and genetic relationships with the Paraná flood volcanics: Chemical Geology, v. 89, p. 19–48, doi: 10.1016/0009-2541(90)90058-F.

Prezzi, C.B., and Alonso, R.N., 2002, New paleomagnetic data from the northern Argentina Puna: Central Andes rotation pattern reanalyzed: Journal of Geophysical Research, v. 107, no. B2, 2041, doi: 10.1029/2001JB000225.

Randall, D.R., 1998, A New Jurassic–Recent apparent polar wander path for South America and a review of central Andean tectonic models: Tectonophysics, v. 299, p. 49–74, doi: 10.1016/S0040-1951(98)00198-X.

Raposo, M.I.B., Ernesto, M., and Renne, P.R., 1998, Paleomagnetism and dating of the Early Cretaceous Florianópolis dike swarm (Santa Catarina Island), Southern Brazil: Physics of the Earth and Planetary Interiors, v. 108, p. 275–290, doi: 10.1016/S0031-9201(98)00102-2.

Renne, P.R., Ernesto, M., Pacca, I.G., Coe, R.S., Glen, J.M., Prévot, M., and Perrin, M., 1992, The age of Paraná flood volcanism, rifting of Gondwanaland, and Jurassic-Cretaceous boundary: Science, v. 258, p. 975–979, doi: 10.1126/science.258.5084.975.

Renne, P.R., Mertz, D.F., Teixeira, W., Ens, H., and Richards, M., 1993, Geochronologic constraints on magmatic and tectonic evolution of the Paraná Province [abs.]: Eos (Transactions, American Geophysical Union), v. 74, p. 553.

Renne, P.R., Deckart, K., Ernesto, M., Féraud, G., and Piccirillo, E.M., 1996, Age of the Ponta Grossa dike swarm (Brazil), and implications to Paraná flood volcanism: Earth and Planetary Science Letters, v. 144, p. 199–211, doi: 10.1016/0012-821X(96)00155-0.

Renne, P.R., Swisher, C.C., Deino, A.L., Karner, D.B., Owens, T., and DePaolo, D.J., 1998, Intercalibration of standards, absolute ages and uncertainties in ^{40}Ar/^{39}Ar dating: Chemical Geology: Isotope Geocience Section, v. 145, p. 117–152.

Riccomini, V., Velázquez, V.F., and Gomes, C.B., 2001, Cenozoic lithospheric faulting in the Asunción Rift, Eastern Paraguay: Journal of South American Earth Sciences, v. 14, p. 625–630, doi: 10.1016/S0895-9811(01)00037-2.

Riccomini, V., Velázquez, V.F., and Gomes, C.B., 2005, Tectonic controls of the Mesozoic and Cenozoic alkaline magmatism in central-southeastern Brazilian platform, in Comin-Chiaramonti, P., and Gomes, C.B., eds., Mesozoic to Cenozoic alkaline magmatism in the Brazilian Platform: São Paulo, Edusp-Fapesp, p. 31–56.

Richardson, S.H., Erlank, A.J., Duncan, A.R., and Reid, D.L., 1982, Correlated Nd, Sr and Pb isotope variations in Walvis Ridge basalts and implications for the evolution of their mantle source: Earth and Planetary Letters, v. 59, p. 327–342, doi: 10.1016/0012-821X(82)90135-2.

Ruberti, E., Castorina, F., Censi, P., Comin-Chiaramonti, P., Gomes, C.B., Antonini, P., and Andrade, F.R.D., 2002, The geochemistry of the Barra do Itapirapuã carbonatite (Ponta Grossa Arch, Brazil): A multiple stockwork: Journal of South American Earth Sciences, v. 15, p. 215–228, doi: 10.1016/S0895-9811(02)00031-7.

Ruberti, E., Gomes, C.B., and Comin-Chiaramonti, P., 2005a, The alkaline magmatism from the Ponta Grossa Arch, in Comin-Chiaramonti, P., and Gomes, C.B., eds., Mesozoic to Cenozoic alkaline magmatism in the Brazilian Platform: São Paulo, Edusp-Fapesp, p. 473–522.

Ruberti, E., Gomes, C.B., Tassinari, P., Antonini, P., and Comin-Chiaramonti, P., 2005b, The Early Cretaceous Valle Chico complex (Mariscala, SE Uruguay), in Comin-Chiaramonti, P., and Gomes, C.B., eds., Mesozoic to Cenozoic alkaline magmatism in the Brazilian Platform: São Paulo, Edusp-Fapesp, p. 591–628.

Siebel, W., Becchio, R., Volker, F., Hansen, M.A.F., Viramonte, J., Trumbull, R.B., Haase, G., and Zimmer, M., 2000, Trindade and Martin Vaz islands, South Atlantic: Isotopic (Sr, Nd, Pb) and trace element constraints on plume related magmatism: Journal of South American Earth Sciences, v. 13, p. 79–103, doi: 10.1016/S0895-9811(00)00015-8.

Smith, A.D., and Lewis, C., 1999, The planet beyond the plume hypothesis: Earth-Science Reviews, v. 48, p. 135–182, doi: 10.1016/S0012-8252(99)00049-5.

Smithies, R.H., and Marsh, J.S., 1998, The Marinkas Quellen carbonatite complex, southern Namibia: Carbonatite magmatism with an uncontaminated depleted mantle signature in a continental setting: Chemical Geology, v. 148, p. 201–212, doi: 10.1016/S0009-2541(98)00029-1.

Sonoki, I.K., and Garda, G.M., 1988, Idades K/Ar de rochas alcalinas do Brasil Meridional e Paraguai Oriental: Compilação e adaptação às novas constantes de decaimento: Boletim IG-USP, Série Cientifica, v. 19, p. 63–87.

Stefanick, M., and Jurdy, D.M., 1984, The distribution of hot spots: Journal of Geophysical Research, v. 98, p. 9919–9925.

Steiger, R.H., and Jäger, E., 1977, Subcommision on geochronology: Convention on the use of decay constants in geo- and cosmochronology: Earth and Planetary Science Letters, v. 36, p. 359–362, doi: 10.1016/0012-821X(77)90060-7.

Stewart, K., Turner, S., Kelley, S., Hawkesworth, C., Kirstein, L., and Mantovani, M., 1996, 3-D, 40Ar-39Ar geochronology in the Paraná continental flood basalt: Earth and Planetary Science Letters, v. 143, p. 95–109, doi: 10.1016/0012-821X(96)00132-X.

Stracke, A., Hofmann, A.W., and Hart, S.R., 2005, FOZO, HIMU, and rest of the mantle zoo: Geochemistry, Geophysics, Geosystems, v. 6, p. 1–20.

Sun, S.S., and McDonough, W.F., 1989, Chemical and isotopic systematics of oceanic basalts, in Saunders, D., and Norry, M.J., eds., Magmatism in the ocean basins: Geological Society of London Special Publication 42, p. 313–345.

Tatsumi, Y., 2000, Continental crust formation by crustal delamination in subduction zones and complementary accumulation of the enriched mantle I component in the mantle: Geochemistry, Geophysics, Geosystems, v. 1, 2000GC000094, 17 p.

Thompson, R.N., Gibson, S.A., Mitchell, J.G., Dickin, A.P., Leonardos, O.H., Brod, J.A., and Greenwood, J.C., 1998, Migrating Cretaceous–Eocene magmatism in the Serra do Mar alkaline Province, SE Brazil: Melts from the deflected Trindade mantle plume?: Journal of Petrology, v. 39, p. 1493–1526, doi: 10.1093/petrology/39.8.1493.

Thybo, H., 2006, The heterogeneous upper mantle velocity zone: Tectonophysics, v. 416, p. 53–79, doi: 10.1016/j.tecto.2005.11.021.

Toyoda, K., Horiuchi, H., and Tokonami, M., 1994, Dupal anomaly of the Brazilian carbonatites: Geochemical correlations with hotspots in the South Atlantic and implications for the mantle source: Earth and Planetary Science Letters, v. 126, p. 315–331, doi: 10.1016/0012-821X(94)90115-5.

Trompette, R., 1994, Geology of western Gondwana (2000–500 Ma) Pan African–Brasiliano: Aggregation of South America and Africa: Rotterdam, Balkema, 350 p.

Turner, S., Regelous, M., Kelley, S., Hawkesworth, C., and Mantovani, M., 1994, Magmatism and continental break-up in the South Atlantic: High precision ^{40}Ar-^{39}Ar geochronology: Earth and Planetary Science Letters, v. 121, p. 333–348, doi: 10.1016/0012-821X(94)90076-0.

Unrug, R., 1996, The assembly of Gondwanaland: Episodes, v. 19, p. 11–20.

Unternehr, P., Curie, D., Olivet, J.L., Goslin, J., and Beuzart, P., 1988, South Atlantic fits and intraplate boundaries in Africa and South America: Tectonophysics, v. 155, p. 169–179, doi: 10.1016/0040-1951(88)90264-8.

Ussami, N., Shiraiwa, S., and Dominguez, J.M.L., 1999, Basement reactivation in a sub-Andean foreland flexural bulge: The Pantanal wetland, SW Brazil: Tectonics, v. 18, p. 25–39, doi: 10.1029/1998TC900004.

VanDecar, J.C., James, D., and Assumpção, M., 1995, Seismic evidence for a fossil mantle plume beneath South America and implications for driving forces: Nature, v. 378, p. 25–31, doi: 10.1038/378025a0.

Van der Hilst, R.D., Widiyantoro, S., and Engdahal, E.R., 1997, Evidence for deep mantle circulation from global tomography: Nature, v. 386, p. 578–584, doi: 10.1038/386578a0.

Vasconcelos, P.M., Onoe, A.T., Kawashita, K., Soares, A.J., and Teixeira, W., 2002, ^{40}Ar/^{39}Ar geochronology at the Instituto de Geociências, USP: Instrumentation, analytical procedures and calibration: Anais da Academia Brasileira de Ciencias, v. 74, p. 297–342.

Velázquez, V.F., Gomes, C.B., Capaldi, G., Comin-Chiaramonti, P., Ernesto, M., Kawashita, K., Petrini, R., and Piccirillo, E.M., 1992, Magmatismo alcalino Mesozóico na porção centro-oriental do Paraguai: Aspectos geocronológicos: Geochimica Brasiliensis, v. 6, p. 23–35.

Velázquez, V.F., Gomes, C.B., Teixeira, W., and Comin-Chiaramonti, P., 1996, Contribution to the geochronology of the Permo-Triassic alkaline magmatism from the Alto Paraguay Province: Revista Brasileira de Geociências, v. 26, p. 103–108.

Velázquez, V.F., Comin-Chiaramonti, P., Cundari, A., Gomes, C.B., and Riccomini, C., 2006, Cretaceous Na-alkaline magmatism from Misiones province (Paraguay): Relationships with the Paleogene Na-alkaline analogue from Asunción and geodynamic significance: Journal of Geology, v. 114, p. 593–614, doi: 10.1086/506161.

Walter, A.V., Flicoteaux, R., Parron, C., Loubet, M., and Nahon, D., 1995, Rare-earth elements and isotopes (Sr, Nd, O, C) in minerals from the Juquiá carbonatite (Brazil): Tracers of a multistage evolution: Chemical Geology, v. 120, p. 27–44, doi: 10.1016/0009-2541(94)00101-D.

Wiens, F., 1986, Zur lithosgraphischen, petrographischen und structurellen Entwicklung des Rio Apa-Hochlandes, Nordost Paraguay [Ph.D. thesis]: Clausthal Universität, Clausthal-Zellerfeld, Germany, 280 p.

Zhang, Y.-S., and Tanimoto, T., 1993, High-resolution global upper mantle structure and plate tectonics: Journal of Geophysical Research, v. 98, p. 9793–9823.

Zindler, A., and Hart, S.R., 1986, Chemical geodynamics: Annual Review of Earth and Planetary Sciences, v. 14, p. 493–571, doi: 10.1146/annurev.ea .14.050186.002425.

MANUSCRIPT ACCEPTED BY THE SOCIETY JANUARY 31, 2007

The Geological Society of America
Special Paper 430
2007

The origin of the Columbia River flood basalt province: Plume versus nonplume models

Peter R. Hooper*

*Department of Geology, Washington State University, Pullman, Washington 99164, USA,
and Open University, Milton Keynes MK7 6AA, UK*

Victor E. Camp

Department of Geological Sciences, San Diego State University, San Diego, California 92182, USA

Stephen P. Reidel

Washington State University Tri-Cities, Richland, Washington 99352, USA

Martin E. Ross

*Department of Earth and Environmental Sciences, Northeastern University, 360 Huntington Avenue,
Boston, Massachusetts 02115, USA*

ABSTRACT

As a contribution to the plume-nonplume debate we review the tectonic setting in which huge volumes of monotonous tholeiite of the Columbia River flood basalt province of the Pacific Northwest, USA, were erupted. We record the timescale and the locations of these eruptions and estimates of individual eruption volumes, and we discuss the mechanisms of sheet-flow emplacement, all of which bear on the ultimate origin of the province. An exceptionally large chemical and isotopic database is used to identify the various mantle sources of the basalt and their subsequent evolution in large lower-crustal magma chambers. We conclude by discussing the available data in light of the various deep-mantle plume and shallow-mantle models recently advocated for the origin of this flood basalt province, and we argue that the mantle plume model best explains the eruption of such an exceptionally large volume of tholeiitic basalt within such an unusually short period and within such a restricted area.

Keywords: continental flood basalt, mantle plume, extension, tectonic setting, eclogite source

INTRODUCTION

Advocates of mantle plumes have long considered continental flood basalt provinces one of the most obvious expressions of plume activity (Richards et al., 1989; Campbell and Griffiths, 1990). As the youngest (Miocene) and smallest of the classic continental flood basalt provinces, the Columbia River province of the American northwest (Fig. 1) is particularly appropriate for pursuit of the plume versus nonplume debate. It is the most fully exposed and least altered of these provinces, and its easy accessibility has led to comprehensive studies providing detailed knowledge of the flow stratigraphy (Fig. 2) and the accumulation of an exceptionally large quantity of physical, chemical, and isotopic data bearing on its ultimate origin.

---

*E-mail: peterhooper@beeb.net.

Hooper, P.R., Camp, V.E., Reidel, S.P., and Ross, M.E., 2007, The origin of the Columbia River flood basalt province: Plume versus nonplume models, *in* Foulger, G.R., and Jurdy, D.M., eds., Plates, plumes, and planetary processes: Geological Society of America Special Paper 430, p. 635–668, doi: 10.1130/2007.2430(30). For permission to copy, contact editing@geosociety.org. ©2007 The Geological Society of America. All rights reserved.

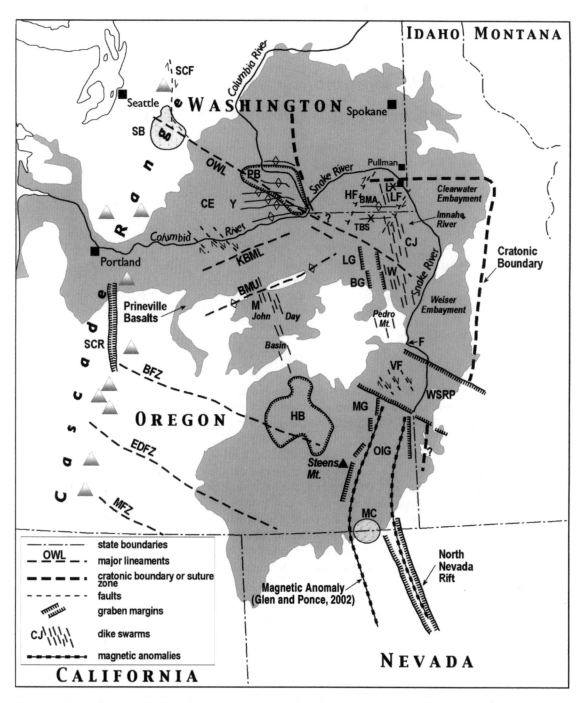

Figure 1. Map of the Columbia River flood basalt province (shaded), including the lower Steens basalt, interpreted as the oldest flood basalt unit, modified from Camp and Ross (2004). Note that in the graben of the Western Snake River Plain (WSRP), in east central Oregon, flood basalts are only present at depth along the northern margin, and, while flood basalts are shown beneath the Oregon-Idaho Graben (OIG), their presence there has not been proven. SCF—Straight Creek fault; SB—Snoqualmie Batholith; OWL—Olympic-Wallowa Lineament; CE—Columbia Embayment; PB—Pasco basin; Y—Yakima fold belt; HF—Hite fault (down to the west, but with minor postbasalt left lateral displacement); BMA—Blue Mountains anticline; L—Lewiston and the Lewiston basin syncline; LF—Limekiln fault (down to the west, but with minor postbasalt left lateral displacement); TBS—Troy Basin syncline; CJ—Chief Joseph Dike swarm; LG—La Grande Graben; BG—Baker Graben; W—Wallowa Mountains horst; KBML—Klamath–Blue Mountains Lineament (Riddihough et al., 1986); BMU—Blue Mountains uplift; M—Monument Dike swarm; F—Farewell Bend on the Snake River; VF—Vale fault zone; SCR—southern Cascade rift; BFZ—Brothers fault zone; EDFZ—Eugene-Denio fault zone; MFZ—McLaughlin fault zone; HB—Harvey basin; MG—Malheur Gorge; MC—McDermitt caldera.

Stratigraphy and Nomenclature of the Columbia River Flood Basalt Province

	Formation	Member	Other Units*	Isotopic Age (Ma)	Estimated Volume [kmv³]	Magnetic Polarity
Columbia River Basalt Group	Saddle Mountains Basalt	Lower Monumental		6[1]	15	N
		Ice Harbor		8.5[1]	75	N,R,N
		Buford	Swamp Creek		20	R
		Elephant Mountain	Craigmont	10.5[1]	440	R,T
		Pomona	Grangeville	12.0[1]	760	R
		Esquatzel	Icicle Flat		70	N
		Weissensels Ridge			20	N
		Asotin			220	N
		Wilbur Creek			70	N
		Umatilla			720	N
	Wanapum Basalt	Priest Rapids		14.5[1]	2,800	R
		Roza			1,300	T
		Shumaker Creek				N
		Frenchman Springs	Powatka	15.3[1]	6,410	N
		Lookingglass				
	Hiatus with saprolite horizon					
	Eckler Mountain Basalt	Dodge			170	N
		Robinette Mountain				N
	Hiatus with saprolite horizon					
Main Phase	Grande Ronde Basalt		Picture Gorge Basalt	15.0 [3]	GRB=148,600**	N2
						R2
					PGB= 2,400	N1
						R1
	Imnaha Basalt				10000**	N0
	Lower Steens Basalt			16.6 [2,3]	60000**	R0

* Isolated units whose stratigraphic position is only approximate
** Camp et al., 2003
Age sources
    [1]Tolan et al., 1989
    [2]Swisher et al., 1990
    [3]Hooper et al., 2002; Hooper, 2004

Figure 2. Stratigraphy and nomenclature of the Columbia River flood basalt province. Modified from Tolan et al. (1989).

The Columbia River Basalt Group forms a high plateau of stacked sheet flows between the Cascade Range to the west and the Rocky Mountains to the east. It covers large areas of southeast Washington state, northeast Oregon, and adjacent parts of western Idaho (Fig. 1). The voluminous tholeiitic basalt eruptions began ca. 16.6 Ma in and around Steens Mountain in east central Oregon (Swisher et al., 1990; Hooper et al., 2002a; Camp et al., 2003). The main focus of the eruptions then moved progressively north along the NNW-oriented Chief Joseph Dike swarm on the eastern borders of Oregon and Washington (Fig. 1) so that, by 6 Ma, the youngest flows were erupting in southeast and central Washington.

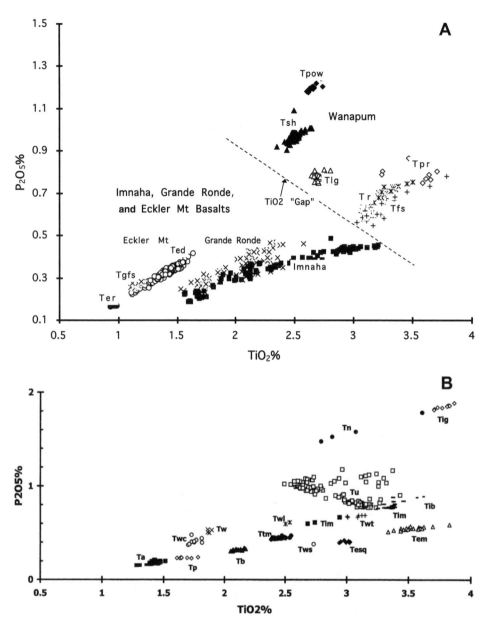

Figure 3. Chemical distinctions between Columbia River Basalt Group formations, members, and flows (after Hooper, 2000). (A) TiO$_2$ versus P$_2$O$_5$ plot of the Imnaha, Grande Ronde, Eckler Mountain, and Wanapum formations; the "Ti gap" separates all lower formations from the overlying Wanapum basalt. Ter—Robinette Mountain flow (Eckler Mt Formation); Tgfs—Field Springs flow (Grande Ronde Formation); Ted—Dodge flows (Eckler Mt Formation). Wanapum basalt flows: Tpow—Powatka flow; Tsh—Shumaker Creek flow; Tlg—Lookingglass flow; Tr—Roza Member; Tfs—Frenchman Springs Member; Tpr—Priest Rapids Member. (B) TiO$_2$ versus P$_2$O$_5$ plot of individual flows within the Saddle Mountains Basalt Formation. Note how multiple analyses of each flow form a short array at constant P$_2$O$_5$/TiO$_2$ ratios, suggesting gabbro fractionation. Ta—Asotin; Twc—Cloverland; Tp—Pomona; Tw—Wilbur Creek; Tb—Buford; Ttm—Tammany Creek; Twl—Lewiston Orchards; Tws—Slippery Creek; Tlm—Lower Monumental; Tn—basalt of Eden; Tu—Umatilla; Twt—Tenmile Creek; Tesq—Esquatzel; Tim—Martindale; Tig—Goose Island flow; Tib—Basin City; Tem—Elephant Mountain. The Umatilla Member includes two eruptions: the Umatilla flow followed by the Sillusi flow from the same vent, the younger invading and inflating the older in the central plateau, as shown in Figure 7.

Some fissure-dike systems are over 100 km long, feeding homogeneous basaltic sheet flows that were tens of meters thick and thousands of cubic meters in volume (Reidel et al., 1989; Tolan et al., 1989). Numerous flows have been correlated with their feeder dikes and traced over the Columbia plateau by their remarkably constant and distinct chemical compositions (Fig. 3; Hooper, 2000). Individual flows crossed the plateau from the feeder dikes on the eastern edge of the province to the Cascade Range, ponding in the Pasco basin on the way. Larger flows continued down the path of the ancestral Columbia River across the rising Cascade arch to the Pacific Ocean, a journey of over 600 km. The east-to-west transport was facilitated by an evolving east-to-west slope (1° to 2°) resulting both from the continued rise of the region to the east along the Idaho-Oregon border and from the continued deepening of the Pasco basin in the central plateau. Recent estimates suggest that an area greater than 200,000 km^2 was covered by 234,000 km^3 of basalt (Camp et al., 2003). Over 98% of this huge volume of basalt erupted in the first two million years (16.6 Ma to 14.5 Ma; Waters, 1961; Swanson et al., 1979; Tolan et al., 1989; Hooper et al., 2002a).

Most workers with direct experience of the Columbia River Basalt Group and associated magmatism have followed Brandon and Goles (1988) in embracing the plume model for the origin of the group (Draper, 1991; Pierce and Morgan, 1992; Hooper and Hawkesworth, 1993; Takahahshi et al., 1998; Hooper et al., 2002a; Camp and Ross, 2004). But alternative, nonplume, models have also been proposed (Carlson and Hart, 1987; Smith, 1992; King and Anderson, 1998; Humphreys et al., 2000; Christiansen et al., 2002; Hales et al., 2005). This contribution reviews the abundant physical, chemical, and isotopic data available for the Columbia River province and discusses the extent to which these data support either the mantle plume or the alternative shallow-mantle models.

TECTONIC SETTING OF THE COLUMBIA RIVER BASALT GROUP ERUPTIONS

The Columbia River Basalt Group was erupted in the intermontane zone between the Cascade Range and the Rocky Mountains. The basalts were extruded in huge volumes, primarily through the feeder dikes of the Chief Joseph Dike swarm, which parallels the edge of the North American craton within the relatively thin lithosphere of the accreted oceanic Blue Mountains terranes (Fig. 1). The large volumes, brief period, and restricted area of the tholeiitic flood basalt eruptions contrast with the more prolonged, partly contemporaneous but much smaller, calc-alkalic to alkalic eruptions localized along north-south grabens resulting from east-west extension (Fig. 4).

The boundary between the North American craton and the accreted terranes is marked by a complex suture zone that traces an erratic pattern across the intermontane zone. Running north from southeastern Oregon along the western edge of the Idaho Batholith, it parallels the Middle Cretaceous subduction zone (Armstrong et al., 1977; Fleck and Criss, 1985), which was later foreshortened and sheared by Late Cretaceous dextral transpression (Giorgis et al., 2005). Then, swinging abruptly through 90°, the boundary continues from east to west along the north-

ern margin of the accreted terranes beneath the Columbia River basalts on the north side of the Lewiston basin, tracked by its geophysical signature (Fig. 1; Mohl and Thiessen, 1995). Farther west, the suture again runs north-south to form the eastern margin of the Pasco basin (Fig. 1; Reidel, 1984; Sobczyk, 1994).

The Cretaceous subduction was responsible for the formation of the Idaho Batholith, the rise of the Rocky Mountains (in part), and the accretion of the oceanic island arc terranes from the west. When blocked by the accreted terranes at the end of the Cretaceous, subduction jumped to the western margin of the accreted terranes, there to initiate the Cascade volcanic arc during the Eocene. The North American craton, intruded by the Idaho Batholith, is presumably significantly thicker and more competent than the compilation of oceanic island arcs and associated sediments welded together by Jurassic and Cretaceous granitoids that make up the accreted terranes (Vallier and Brooks, 1987, 1995).

Where the basalts flowed northward from the Chief Joseph feeder dikes onto the craton of eastern Washington, they remain essentially flat-lying. Immediately south of the cratonic margin, however, where the same flows lie on the accreted terranes, the flows are deformed into east-west folds that parallel the cratonic margin along the Washington-Oregon border from Lewiston to the Yakima fold belt (Fig. 1). The east-west folds include the Blue Mountains anticline and adjacent Lewiston basin and Troy basin synclines (Ross, 1980; Hooper et al., 1995a) in addition to the "Lewiston structure," a steep anticline broken by a high-angle reverse fault in which flows dip up to 80° against the cratonic margin. This deformation culminates to the west at the "nose"

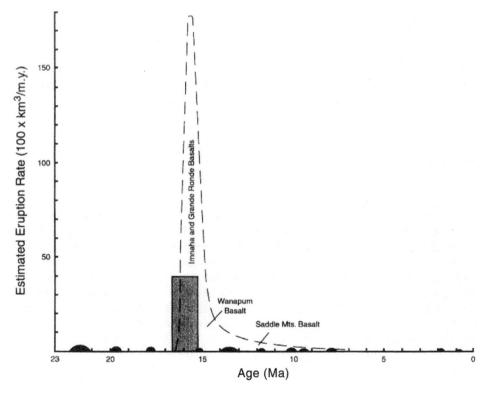

Figure 4. Eruption rates of the tholeiitic flood basalts (Steens [shaded], Imnaha, Grande Ronde, Wanapum, and Saddle Mountains basalts below the dashed line) contrasted with the eruption rates of the small-scale calc-alkalic to alkalic volcanicity (black) associated with active extension in east central Oregon (after Hooper et al., 2002a).

of the craton (Fig. 1) with the formation of the more complex Yakima fold belt.

The Yakima fold belt straddles the Pasco basin (Fig. 1), an actively subsiding basin where up to 20 km of Tertiary, including Eocene, sediment has accumulated. The sediment is topped by 4–5 km of ponded flood basalts. Using the thinning of individual flows over the anticlines, Reidel has shown that the Yakima folds formed during the Columbia River Basalt Group eruptions while the Pasco basin was actively subsiding (Reidel, 1984; Reidel et al., 1994). Similar evidence from the southwestern extension of the Yakima folds (Anderson, 1987) and the folds along the Washington-Oregon border (Ross, 1980) demonstrate that these structures grew during the basalt eruptions and continued later at a reduced rate (Reidel, 1984). The folds appear to result from north-south compression of the accreted terranes against the cratonic margin. This suggests that the northward component of the Late Cretaceous transpression documented by Giorgis et al. (2005) continued through the period of the flood basalt eruptions. The same north-south compressional strain remains detectable on the plateau today (Kim et al. 1986). Further evidence of the northward translation of the accreted terranes after the basalts erupted is seen in north-south vertical breccia zones with conspicuous horizontal slickensides that cut basalt flows in the Clearwater Embayment (P.R. Hooper, unpublished data) and in right lateral displacement farther south along vertical north-south faults that cross the Western Snake River Plain (Fig. 5A; Mabey, 1984; Ferns et al., 1993; and Hooper et al., 2002b). Given the probable 100 km shortening of the subduction zone along the western margin of the Idaho Batholith, into which the accreted terranes have been wedged, as proposed by Giorgis et al. (2005), the continued northward thrusting of the accreted terranes may have been the cause of the extraordinary east-west digression of the cratonic margin across southeast Washington (Fig. 1).

East-west extension has been a feature of the intermontane zone from British Columbia (Thorkelson, 1989; Breitsprecher et al., 2003) to Nevada (Fitton et al., 1988; Hawkesworth et al., 1995) since the Eocene. In northeast Washington the north-south-oriented Republic Graben and affiliated grabens of Eocene age are accompanied by calc-alkalic to alkalic magmatism (Holder et al., 1990; Hooper et al., 1995b; Morris and Hooper, 1997), while metamorphic core complexes resulting from east-west extension and lithospheric thinning developed in northern and southern Idaho (Coney, 1987). Local alkalic volcanicity in west central Idaho suggests that similar tectonic-magmatic activity, including east-west extension, may have continued through the Oligocene (Kauffman et al., 2003).

In the Miocene, the northern limit of obvious Basin and Range east-west extension was the northern edge of the Blue Mountains accreted terranes and farther east across western Idaho. Extensional features are minimal in the narrow zone forming the northern edge of the accreted terranes between the east-west-trending cratonic boundary in southeast Washington and the Olympic-Wallowa Lineament (Fig. 1; Taubeneck, 1970).

Extension becomes more obvious immediately south of the lineament with the development of the La Grande and Baker Grabens along its southern side. Here, graben formation was accompanied by the small calk-alkalic to alkalic eruptions of the Powder River volcanic field (Fig. 1; Gehrels et al., 1980; Bailey, 1990).

Farther south, in east central Oregon, the much wider (50 km) Oregon-Idaho Graben (OIG) developed at ca. 15.3 Ma, following the initial eruptions of the flood basalts (Lees, 1994). The Oregon-Idaho Graben is filled with small basalt to andesite to rhyolite eruptions and ash-flow tuffs of calc-alkalic to alkalic composition (Ferns et al., 1993; Lees, 1994; Binger, 1997; Cummings et al., 2000; Hooper et al., 2002a). In both the Eocene and the Miocene, this extension-related calc-alkalic magmatism has been attributed to decompressional melting below a locally extended and thinned lithosphere. In addition, both the Harney basin and the North Nevada rift have been interpreted as the result of east-west extension (Fig. 1; Walker, 1979; Zoback et al., 1994).

Lawrence (1976) suggested that a series of parallel WNW-ESE lineaments across northern California, Nevada, and Oregon were right lateral strike-slip faults that increased the degree of extension from north to south. Hooper and Conrey (1989) and Mann (1989) subsequently interpreted the lineaments as broader pull-apart structures or right lateral strike-slip extensional duplexes (Woodcock and Fischer, 1986), reemphasizing an increase in the cumulative amount of extension from north to south across eastern Oregon. The structural details of the right lateral strike-slip extensional duplexes are well displayed within the Oregon end of the Western Snake River Plain (WSRP, Fig. 5A).

The Western Snake River Plain is a wide graben across which basalt to rhyolite contacts are progressively displaced westward from north to south by a series of northwest-southeast right lateral faults, the Vale faults (Lawrence, 1976), that lie enechelon to the WNW-ESE trend of the Western Snake River Plain (Ferns et al., 1993). The amount of westward displacement to the south across the broad Western Snake River Plain Graben is recorded both by the displacement of the flood basalts and by the opening of the coeval, 50-km-wide Oregon-Idaho Graben along its southern side (the Adrian fault; Ferns et al., 1993). The parallel but much smaller Malheur Gorge half-graben, which lies south of the Western Snake River Plain Graben, has a similar structural pattern (Fig. 5A; Evans, 1990a,b). Rhyolite contacts are displaced right laterally across the half–graben, which is characterized by northwest-southeast faults enechelon to the WNW-ESE-trending graben. North and south of the narrow half-graben, north-south-trending listric faults are well developed (Fig. 5A), indicating that the regional direction of extension was east-west. Such east-west extension along the WNW-ESE-trending zones has resulted in the formation of narrow grabens or half-grabens, as observed in both the wider Western Snake River Plain and the much narrower Malheur Gorge half-graben (Evans, 1990a,b; Hooper et al., 2002b). Despite clear displacement of geologic contacts across the northwest-southeast faults

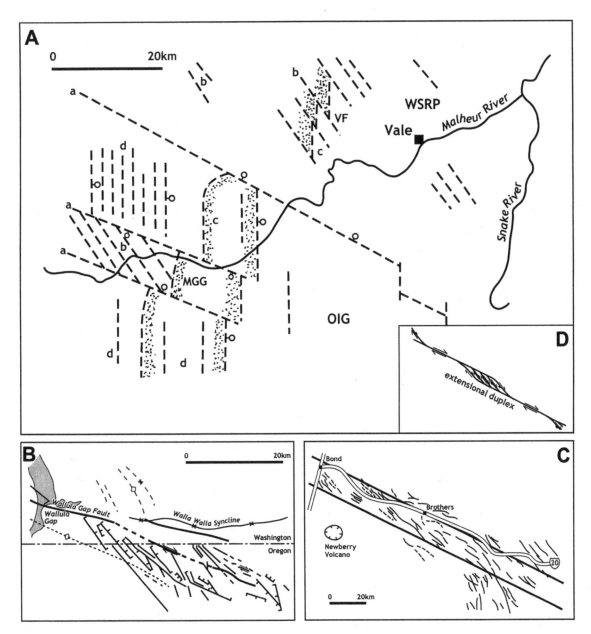

Figure 5. Right lateral extensional duplexes or "pull-apart" structures in eastern Oregon. (A) Cartoon of the structural elements of the Western Snake River Plain (WSRP, Fig. 1) Graben in east central Oregon, including the Vale faults (VF), the Malheur Gorge half-graben (MGG), and the western margin of the Oregon-Idaho Graben (OIG) defined by the rhyolite (stippled) intruded along the graben boundary faults. a—WNW-ESE faults bounding the two grabens: these faults displace rhyolite contacts right laterally both across the MGG and across the southern boundary fault (Adrian fault) of the WSRP Graben; b—northwest-southeast faults, within the grabens, which displace rhyolite right laterally across the Vale fault (VF) zone; c—displaced rhyolite contacts (stippled); d—north-south listric faults between the grabens and to the south of the MGG that define the direction of extension as east-west. Modified from Hooper et al. (2002b). (B) Fault pattern along part of the southern margin of the Olympic-Wallowa Lineament (OWL, Fig. 1) on the Washington-Oregon border as mapped by Swanson and Wright (Swanson et al., 1980, 1981), showing a similar structural geometry to Figure 5a. (C) Faults along the Brothers fault (Fig. 1) zone as mapped by Walker and Nolf (1981), showing a structural geometry similar to that of panel A. (D) Theoretical right lateral extensional duplex, after Woodcock and Fischer (1986).

in the Western Snake River Plain and the Malheur Gorge half-graben, slickensides are lacking, an absence we attribute to the extensional rather than the compressional nature of the duplexes. The repetition of the geometry of the structural elements of these duplexes, most obviously the enechelon arrangement of the northwest-southeast faults across the WNW-ESE trend of the structural zones, has led to the proposition that the Brothers fault zone (Walker and Nolf, 1981), the La Grande and Baker grabens (Gehrels et al., 1980), and the southern side of the Olympic-Wallowa Lineament (Swanson et al., 1980, 1981) are also extensional dextral strike-slip duplexes (Fig. 5; Hooper and Conrey, 1989; Hooper et al., 2002b).

In brief, the flood basalts erupted through the thinner accreted terranes along a line of fissures that parallel the cratonic margin immediately to the east. The tectonic setting was one of north-south compression, causing the accreted terranes to be pressed northward against the North American craton and east-west extension, which increased from north to south. Both the north-south compression and the increasing east-west extension southward during the Miocene are consistent with current models of oblique subduction of the Juan de Fuca and Gorda plates beneath North America, in which the increasing back-arc extension southward caused clockwise rotation of the Cascade arc (Magill et al., 1982; Wells and Heller, 1988; England and Wells, 1991; Wells et al., 1998; McCaffrey et al., 2000).

BASALT FLOW STRATIGRAPHY

Mapping of the Columbia River Basalt Group was initiated by Waters (1961) and later supported over the last thirty years by paleomagnetic measurements and increasingly precise major- and trace-elemental analyses. The thick stack of basalt flows can now be broken down into subgroups, formations, members, and individual flows that can be recognized and mapped across the province, as shown in Figures 2 and 3 (Swanson et al., 1979; Mangan et al., 1986; Bailey, 1989a; Reidel et al., 1989; Tolan et al., 1989; Hooper, 2000).

The main phase of eruptions, from ca. 16.6 to 15.0 Ma, began in eastern Oregon and continued unabated over six paleomagnetic intervals. Eruptions started with the reverse magnetic sequence R_0, best displayed in the lower flows at Steens Mountain (Fig. 1; Mankinen et al., 1987), and continued through to the normal magnetic sequence (N_2) that forms the upper part of the Grande Ronde basalt across the Columbia plateau (Fig. 2). This main phase is characterized by a lack of regional unconformities between successive flows, unlike the younger formations.

Steens Basalt

This article incorporates the lower Steens basalt into the Columbia River Basalt Group as the informal basal formation of the flood basalt stratigraphy (Fig. 2). Upgrading the lower Steens basalt to formational status must await more detailed investigations to clarify its distinction, chemically and geograph-

ically, from the upper Steens basalt. $^{40}Ar/^{39}Ar$ dates, combined with local mapping and petrochemical correlations, suggest that the Columbia River Basalt Group eruptions began at ca. 16.6 Ma in southeast Oregon in and around the Steens Mountain shield volcano (Fig. 1; Swisher et al., 1990; Binger, 1997; Hooper et al., 2002a; Camp et al., 2003). Johnson et al. (1998b) subdivided the 900 m Steens Mountain type section into the more primitive tholeiitic lower Steens basalt flows and the more evolved and mildly alkalic upper Steens basalt flows. This chemical break lies close to the reverse to normal paleomagnetic transition of Mankinen et al. (1987).

In well-mapped areas near Malheur Gorge, northeast of Steens Mountain, Oregon (Figs. 1 and 5A), the lowest unit (lower Pole Creek) of the basalt of Malheur Gorge is the chemical and petrographic equivalent of the lower Steens basalt. The conformably overlying upper Pole Creek and Birch Creek basalts are the chemical and petrographic equivalents of the Imnaha and Grande Ronde basalts of the Columbia River Basalt Group farther north (Fig. 6; Evans, 1990a,b; Ferns et al., 1993; Lees, 1994; Binger, 1997; Johnson et al., 1998a; Hooper et al., 2002a; Camp et al., 2003). Flows of lower Steens basalt composition extend over a large part of southeast Oregon (Fig. 1; Hart, 1982; Carlson and Hart, 1987, 1988). Flows similar to lower Steens basalt in magnetic polarity, petrography, and chemical composition conformably underlie Imnaha basalt as far north as the base of the Wallowa Mountains (C-1 sample of Carlson and Hart, 1988) and as far northeast as Squaw Butte, just north of Boise in Idaho (Martin, 1984; Hooper et al., 2002a).

Imnaha Basalt

While the lower Steens basalt is largely restricted to the Oregon plateau, the overlying Imnaha basalt is the oldest formation of the Columbia River Basalt Group in most of northeast Oregon and across the eastern Columbia plateau (Kleck, 1976; Hooper et al., 1984). It thickens northward from Malheur Gorge, where it lies conformably on the lower Steens basalt, across the northwest-trending Vale fault zone within the Western Snake River Plain Graben to Farewell Bend (Fig. 1; Lees, 1994; Hooper et al., 2002a). Farther north it fills deep canyons eroded in Pre-Tertiary rocks, finally creating a flat lava plateau dipping gently west to northwest. North from its type sections in the Imnaha Valley, Imnaha basalt is found in the Clearwater Embayment (Idaho), north to Pullman (Washington), and west to the eastern edge of the Pasco basin, where it occurs at the base of drill core (Tolan et al., 1989).

Feeder dikes of Imnaha basalt occur in the southern part of the Chief Joseph Dike swarm adjacent to the Imnaha Valley and south along the western side of the Snake River (Fig. 1). The latest estimates (Tolan et al., 1989; Camp et al., 2003) suggest that Imnaha basalt covered well over 50,000 km^2, with a volume exceeding 70,000 km^3. The most recent dates (Hooper et al., 2002a; Hooper, 2004) suggest ages for Imnaha basalt of between 16.5 and 15.0 Ma.

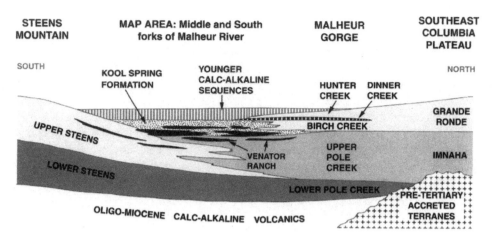

Figure 6. Schematic cross-section of volcanic units from Steens Mountain to Malheur Gorge, east central Oregon. Lower Steens basalt (lower Pole Creek) is conformably overlain by Imnaha basalt (upper Pole Creek) and Grande Ronde basalt (Birch Creek). After Camp et al. (2003).

Grande Ronde Basalt

Rocks of the Grande Ronde basalt dominate the Columbia River flood basalt province, forming over 60% of the total volume of the Columbia River Basalt Group (Camp et al., 2003). In contrast to the plagioclase-phyric Steens and Imnaha basalts, flows of the Grande Ronde are, with few exceptions, aphyric tholeiitic basaltic andesites (52–57% SiO_2). Grande Ronde basalt flows lie conformably on the Imnaha basalt, forming a thick and uniform sequence of typically flat-lying flows across the Columbia plateau. These are well exposed in the deep modern canyons of the Columbia and Snake rivers and their tributaries as they cross southeastern Washington and northeastern Oregon, and have been recovered from numerous exploratory boreholes in the search for gas and petroleum.

The thickness of the Grande Ronde basalt increases progressively northward from Malheur Gorge in east central Oregon to the Columbia plateau. Flows extend as far north as Spokane in eastern Washington, and they ponded in the actively deepening Pasco basin in central Washington, where they reach their maximum thickness of >4 km (Reidel et al., 1989, 1994). Many overflowed the basin and continued down the broad depression of the Columbia transarc lowlands (Beeson et al., 1989) across the rising volcanic arch of the Cascade Range to fill the Portland basin and continue west to the Pacific Ocean (Fig. 1; Wells et al., 1989). Individual Grande Ronde basalt eruptions are enormous, covering most of the Columbia plateau with volumes of up to 5000 km^3 (Reidel et al., 1989; Tolan et al., 1989). The composite Sentinel Bluffs Member, which may include more than one contemporaneous eruption, exceeds 10,000 km^3 (Reidel, 2005).

The $^{40}Ar/^{39}Ar$ ages are indistinguishable from those of the Imnaha basalt, between 15.0 and 16.5 Ma, but the Grande Ronde basalts are always stratigraphically the younger (Long and Duncan, 1982; Tolan et al., 1989; Hooper et al., 2002a; Hooper 2004). Feeder dikes of Grande Ronde basalt composition occur throughout the Chief Joseph Dike swarm, from Farewell Bend and Pedro Mountain in the south to the Washington-Idaho border (Fig. 1; H.C. Brooks, 2004, personal commun.; P.R. Hooper and S.P. Reidel, 2005, personal commun.).

The difficulty of subdividing this immense and compositionally uniform sequence of aphyric flows inhibited detailed work on the Columbia plateau for many years. The problem was partially overcome in the 1970s by the use of a portable fluxgate magnetometer. This recorded the basic magnetic polarity of each flow in the field, allowing the Grande Ronde basalt pile to be subdivided into four paleomagnetic units, from bottom to top: R_1, N_1, R_2, and N_2 (Fig. 2; Swanson et al., 1979), successive eruptions migrating northward with time (Camp, 1995). More detailed paleomagnetic measurements and mapping, combined with increasingly accurate chemical analyses (Fig. 3) for both major and trace elements, have subsequently made possible further subdivision of the Grande Ronde Basalt Formation into individual members and flows (Mangan et al., 1986; Reidel et al., 1989).

Picture Gorge Basalt

The relatively restricted Picture Gorge eruptions were contemporaneous with the Grande Ronde basalt eruptions. Picture Gorge basalt flows are confined to the John Day basin south of the pre-Miocene WSW-ENE-trending Blue Mountains uplift in north central Oregon. The uplift parallels the southern side of the Klamath–Blue Mountains Lineament (Riddihough et al., 1986), a gravity anomaly that marks the southeast margin of the Columbia Embayment (CE, Fig. 1). All Picture Gorge basalt flows appear to have been fed through the Monument Dike swarm in the John Day basin and across the Aldrich Mountains immediately south (Fig. 1; P.R. Hooper and S.P. Reidel, 2005, personal commun.). A few flows interfinger with N_1-R_2 Grande Ronde flows across the uplift, which have been dated by Baksi (1989) at 16.1 ± 0.2 Ma (Nathan and Fruchter, 1974; Bailey, 1989a; Hooper et al., 1993). The succession of Picture Gorge basalt flows is subdivided into three members and many individual flows (Bailey, 1989a). It covers 10,680 km^2 in the John Day basin, with a total volume of ~2400 km^3 (Tolan et al., 1989).

Eckler Mountain Basalt

The huge volumes of the lower Steens, Imnaha, Grande Ronde, and Picture Gorge basalts of the main eruptive phase display no evidence of significant time intervals between flows, as would be indicated by erosion of lower units, the development of soil horizons, or the presence of sedimentary interbeds. Flows of the main phase, therefore, appear to have poured out in rapid succession. A distinct hiatus, however, occurred in the eruptive record at the end of the Grande Ronde eruptions, across the whole Columbia plateau. Saprolite horizons are present in southeast Washington, where some lower flows of the next large formation, the Wanapum basalt, are absent. Farther west, this interval is occupied by sediments of the Ellensberg Formation eroded from the rising arch of the Cascade Range (Smith, 1988), and equivalent sediments occur locally around other margins of the plateau.

During this hiatus a few relatively small but distinctive flows of the Eckler Mountain basalt were erupted between saprolite horizons along the eastern end of the Oregon-Washington border, each flow with its own feeder dikes. The primitive Robinette Mountain flow was followed by four flows of the Dodge Member, whose distinctive plagioclase- and olivine-phyric petrography, weathering characteristics, and chemical composition provide a useful marker horizon in local mapping (Ross, 1989; Hooper et al., 1995a).

Wanapum Basalt

Eruption of larger flows resumed with the Wanapum basalt, which is dominated by flows of the Frenchman Springs, Roza, and Priest Rapids members (Fig. 2). Each of these members is composed of multiple flows with a rather uniform chemical composition, combining relatively low silica content with high abundances of TiO_2, P_2O_5, and other high field-strength trace elements (Fig. 3A). The many individual flows may be locally distinguished by characteristic phenocryst assemblages (Beeson et al., 1985; Martin, 1989). These large flows were interspersed with smaller, more evolved (higher levels of incompatible elements) flows: the Lookingglass, Shumaker Creek, and Powatka flows (Figs. 2 and 3A). Wanapum basalt flows are dated at 15.3–14.5 Ma (Tolan et al., 1989).

Saddle Mountains Basalt

The youngest eruptions of the Columbia River Basalt Group are the compositionally diverse flows of the Saddle Mountains basalt, which represent the waning period of the flood basalt eruptions. As a consequence of their relatively small volumes and of an eruption period extending over 8 m.y. (ca. 14.5–6.0 Ma; Tolan et al., 1989), many Saddle Mountains basalt flows fill river canyons cut through previous flows and are referred to as "intracanyon flows" (Waters, 1961; Swanson et al., 1979). Because of their compositional diversity, virtually all the Saddle

Mountains basalt flows can be distinguished from each other by chemical analyses (Fig. 3B) and correlated with their feeder dikes. Nearly all Saddle Mountains basalt dikes and flows are found along the northern end of the Chief Joseph Dike swarm in and around the Lewiston basin (P.R. Hooper and S.P. Reidel, 2005, personal commun.). The most northerly and youngest (8.5 Ma) known feeder dikes of the Columbia River Basalt Group form the Ice Harbor linear vent system (Swanson et al., 1975) along the eastern margin of the Pasco basin, although even younger flows of limited extent, the Tammany Creek and Lower Monumental flows (6 Ma), occur in the Lewiston basin syncline (Figs. 1 and 2).

MODE OF EMPLACEMENT

The formation of sheet flows, rather than more typical shield volcanoes, is the hallmark of a flood basalt province. Because the viscosities of the Columbia River Basalt Group flows and those of Hawaiian shield volcanoes are similar (Shaw and Swanson, 1970), the very different landscapes of the Columbia plateau and the Hawaiian Islands must result from different modes of eruption and emplacement. Sheet flows require large volumes of mafic magma to be extruded rapidly, typically from major fissures or feeder dikes, as in the case of the Columbia River basalts (Shaw and Swanson, 1970; Swanson et al., 1975). Such exceptionally voluminous and rapid eruptions require very large magma reservoirs that are probably confined to the lower crust because of the compositions of their mineral phases (see later) and the lack of surface collapse structures.

How the sheet flows were emplaced and the time required for the flows to cover such huge distances have proved controversial, with far-reaching implications for the effect on world climate and on mass biological extinctions (Saunders, 2005). Shaw and Swanson (1970) originally proposed that the lava flowed rapidly across the gently sloping plateau surface; each flow, they argued, covered hundreds of kilometers in weeks or even days, a process requiring turbulent flow. More recent studies have suggested that emplacement took several months (Reidel and Tolan, 1992; Reidel, 1998) or even longer (Long et al., 1991). Self and his associates (Finnemore et al., 1993; Self et al., 1996, 1998) have compared the detailed physical characteristics of Columbia River Basalt Group flows to those of other tholeiitic eruptions, particularly the much smaller flows of Hawaii. They concluded that laminar rather than turbulent flow dominated the Columbia River Basalt Group eruptions and that, like many smaller flows on Hawaii, the Columbia River flows were "emplaced as inflated compound pahoehoe flow fields via prolonged, episodic eruptions" (Self et al., 1998, p. 381).

Ongoing regional studies of compositional variation in vertical sections of individual sheet flows across the Columbia plateau (Reidel, 1998, 2005) demonstrate a recurring pattern of older eruptive units forming the top and bottom of a composite sheet flow, with younger eruptive units in the center. Examples (Fig. 7) include the Umatilla Member (Saddle Mountains basalt,

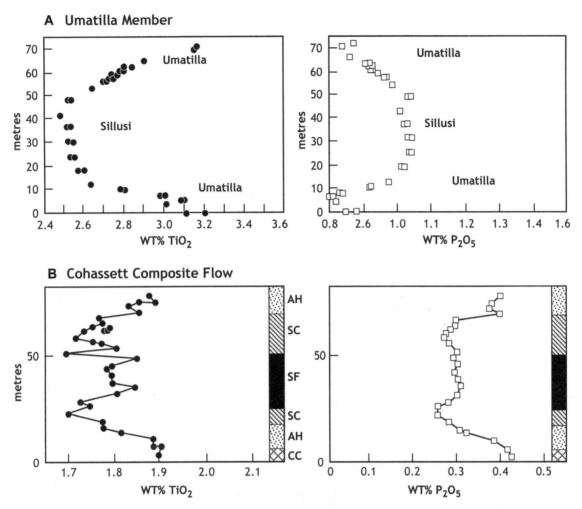

Figure 7. Chemical variation across vertical sections of composite sheet flows of the central Columbia plateau, illustrating invasive intrusion and inflation of older lavas by younger ones. (A) Umatilla Member, Saddle Mountains basalt. The Umatilla flow erupted before the Sillusi flow from the same volcanic fissure and vent in southeast Washington. (B) Cohassett composite flow, Grande Ronde basalt. The four chemical compositions were erupted from discrete fissures or vents in southeastern Washington. Chemical types from oldest to youngest: CC—California Creek; AH—Airway Heights; SC—Stember Creek; SF—Spokane Falls. After Reidel (1998, 2005).

Fig. 2). In the central plateau, the older Umatilla flow forms the top and bottom of the member, while the younger Sillusi flow of the same member occupies the center (Fig. 7A). Both of these flows erupted from the same volcano in southeast Washington around which they also formed individual surface lobes. A second, more complex, example is the composite Cohassett sheet flow of the Grande Ronde basalt in the central plateau. In the vertical section across the sheet flow (Fig. 7B), four individual compositional units are separated only by vesicular horizons, with the oldest units at the top and bottom and progressively younger units toward the center. Each compositional unit forms individual surface flows nearer their eruptive centers in eastern Washington (Reidel, 2005). These profiles confirm that the predominant mechanism for growth of the sheet flows in the central plateau is laminar flow with the progressive invasion and inflation of older lava units by younger units. This conclusion is compatible with the variations in composition across some dikes recorded by Ross (1983) and with the very small decrease in temperature (Ho and Cashman, 1997), the lack of significant crystallization, and the consequent lack of chemical variation from eruptive center to flow periphery, all of which properties have long been recognized as characteristics of Columbia River Basalt Group flows.

The time that these individual inflated sheet flows took to cover several hundred kilometers remains controversial. While Thordarson and Self (1998) suggested that it took as much as 10 years to emplace the 300-km-long Roza flow, Reidel (1998) calculated that the Umatilla Member, of similar length, could have been emplaced in weeks rather than years. Most recently, Keszthelyi et al. (2006) concluded that some of the Columbia River Basalt Group flows could have been emplaced in weeks or months.

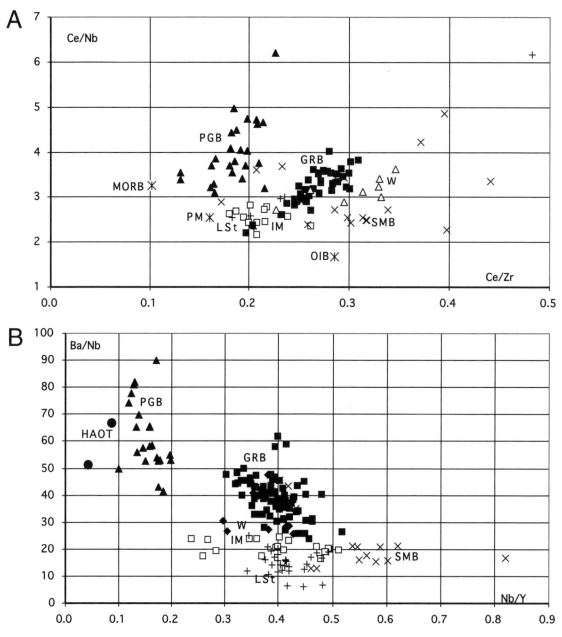

Figure 8. Trace element ratios used to discriminate primary and secondary sources for Columbia River Basalt Formations. (A) Ce/Zr versus Ce/Nb; (B) Nb/Y versus Ba/Nb. Stars—mantle sources; HAOT (filled circles, panel B only)—high-alumina olivine tholeiite; MORB—mid-ocean ridge basalt source; PM—primitive mantle; PGB (filled triangles)—Picture Gorge basalt; LSt (plus signs)—Lower Steens basalt; IM (open squares)—Imnaha basalt; GRB (filled squares)—Grande Ronde basalt; OIB—ocean island basalt source; W (open triangles in panel A, filled diamonds in panel B)—Wanapum basalt; SMB (crosses)—Saddle Mountains Basalt. Modified from Hooper and Hawkesworth (1993). The proximity of some formations (Lower Steens, Imnaha, Grande Ronde, and Wanapum) to OIB source compositions is used as evidence that these formations were derived from an enriched, OIB-like mantle source. In contrast, the proximity of the Picture Gorge basalt (and high-alumina olivine tholeiites and basalts from the Powder River volcanic field, not shown) to MORB source compositions is used as evidence of their derivation from a depleted MORB-like mantle. The increase in Ba/Nb ratios and arrays formed from positive correlations between Ce/Nb and Ce/Zr in the Picture Gorge and Grande Ronde basalts are interpreted as mixing arrays between the primary mantle source components and a secondary lithospheric source, while the lack of similar trends in the Imnaha and Lower Steens basalts implies that these formations were not significantly affected by the addition of such a lithospheric component, and their chemical variation can best be explained by variable degrees of partial melting and gabbro fractionation, as seen in Figure 10.

PETROGENESIS

Knowledge of the sequence and size of successive Columbia River basalt eruptions, and the abundance of chemical and isotopic data that is now available, places significant constraints on the sources and evolution of the lavas. The great volumes of basaltic lava that make up the Columbia River Basalt Group require partial melting events in the mantle as the primary magma source. Isotopic data require additional source components from the lithosphere. In addition, variations within and between flows require some crystal fractionation at crustal pressures. Finally, physical mixing of magmas, both by recharge of evolving magma in lower crustal reservoirs and by physical mixing during eruption, can be clearly demonstrated in a number of cases. The relative significance of these various processes differs for each Columbia River Basalt Formation.

Primary Source Component

Using trace-element and isotopic ratios, Hooper and Hawkesworth (1993), following Carlson (1984) and Church (1985) among others, showed that the Columbia River Basalt Group as a whole falls into three discrete subsets, reflecting three distinct original sources (viz., negative correlations between Ce/Nb and Ce/Zr and between Ba/Nb and Nb/Y; Hooper and Hawkesworth, 1993, Figs. 4–8). The first two subsets (Fig. 8) are assumed to mirror the same upper-mantle melting processes that formed mid-ocean ridge basalt (MORB) from depleted mantle and ocean island basalt (OIB) from enriched mantle. The largest Columbia River Basalt Group formations (the lower Steens, Imnaha, Grande Ronde, and Wanapum basalt formations) suggest a primary mantle source akin to that for OIBs, a view consistent with Hawaiian-like trace-element profiles (Hooper and Hawkesworth, 1993) and with the Sr, Nd, Pb, and He isotopic ratios of the relatively uncontaminated Imnaha basalt (Fig. 9; Dodson

et al., 1997; Bryce and DePaolo, 2004). The Picture Gorge subset (Bailey, 1989b; Brandon et al., 1993) has trace-element and isotopic ratios that suggest that its primary source was more akin to depleted (MORB-type) mantle, similar to the younger basalts of eastern Oregon, including the high-alumina olivine tholeiites (HAOTs) of Hart and Carlson (1987) and the Powder River volcanic field of Bailey (1990). The third subset, the Saddle Mountains basalt, has diverse chemical compositions and more evolved isotopic signatures (Fig. 9) that suggest an older, chemically variable mantle source from beneath the cratonic crust of the Precambrian North America plate enriched before 2000 Ma (Carlson, 1984).

Other Source Components

Figure 8 also demonstrates the difference between, on the one hand, the Imnaha basalt (and probably the lower Steens basalt), which displays no enrichment in a lithospheric component as indicated by an increase in the Ba/Nb ratio, and, on the other hand, both the Picture Gorge and Grande Ronde basalts, which do display clear enrichment in such lithospheric components (viz. positive correlations between Ce/Nb and Ce/Zr and variations of Ba/Nb without changes in Nb/Y [Fig. 8; Hooper and Hawkesworth, 1993]). As in the case of the isotopic differences (Fig. 9), the variability of ratios such as Ba/Nb is best explained by a mixing array between the original mantle component and a component with a lithospheric geochemical signature. The lithospheric component is characterized by conspicuous negative Ta-Nb anomalies (Hooper and Hawkesworth, 1993) and could represent either an enriched subcontinental lithospheric mantle (SCLM) or mafic lower crust. Hooper and Hawkesworth (1993) preferred the former explanation because of the limited fractionation of Sr/Zr observed in the silica-enriched Grande Ronde basalt. Sr/Zr fractionation would be required in a partial melting of crust in the presence of plagioclase

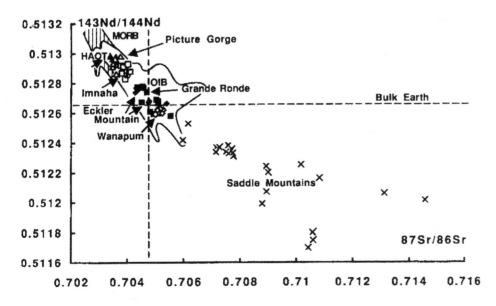

Figure 9. Sr and Nd isotope ratios for the Columbia River Basalt Group. HAOT (open triangles)—high-alumina olivine tholeiite; MORB (field with vertical lines)—mid-ocean ridge basalt; OIB (blank field)—ocean island basalt; closed triangles—Picture Gorge basalt; open squares—Imnaha basalt; closed squares —Grande Ronde basalt; closed diamonds—Eckler Mountain basalt; open circles—Wanapum basalt; crosses— Saddle Mountains basalt. From Hooper and Hawkesworth (1993).

feldspar. Recent work on Re/Os isotopic ratios (Hart et al., 1997; Chesley and Ruiz, 1998) rules out an SCLM component in the Imnaha basalt but suggests that a mafic component from the lower crust could be present. However, it is the Grande Ronde, not the Imnaha basalt, that contains the obvious lithospheric component (Fig. 8). Both SCLM and lower-crustal components may have been available, but the evidence at hand suggests that the lithospheric component forming the mixing arrays of the Grande Ronde and Picture Gorge basalts was an enriched SCLM.

Crustal Assimilation

The earlier work of Carlson et al. (1981), Carlson (1984), and Carlson and Hart (1987) argued that the Grande Ronde basalt was the combined result of crystal fractionation from different asthenospheric source magmas and assimilation of the upper crust. Assimilation of the upper crust was also suggested by Brandon et al. (1993) for the Picture Gorge basalt. Arguments against a significant upper-crustal component include the lack of any silicic magma that might be expected from such a process (DePaolo, 1983), while evidence from mineral phase–liquid investigations in the case of the Grande Ronde basalt has led Caprarelli and Reidel (2004) to conclude that (1) "very small pyroxene crystals preserving high P and T (i.e. lower crustal) information are an indication that magmas cannot have spent substantial amounts of time in shallow level (i.e. upper crustal) reservoirs" and (2) "our thermobarometric determinations indicate a major role for a deep magma reservoir in the lower crust." Crustal contamination has also been demonstrated by recent laser analyses of the isotopic zoning of plagioclase phenocrysts in Imnaha and Picture Gorge basalt magmas (Ramos et al., 2005). The combined results of these two studies suggest that both the pyroxene and the plagioclase phenocrysts crystallized at ~25 km below the surface, a depth consistent with magma reservoirs near the base of the relatively thin crust of the accreted Blue Mountains terranes.

Crystal Fractionation and Magma Mixing

The presence of phenocrysts in many Columbia River Basalt Group flows, even in a few of the lower normally aphyric Grande Ronde basalt flows, suggests that some crystal fractionation has occurred. This is confirmed by multiple chemical analyses of any one flow, which always show a small spread of incompatible-element abundances at constant ratios (Fig. 3). All phenocryst assemblages are dominated by labradorite plagioclase, indicating their formation in the crust. Clinopyroxenes are next in abundance, and, in the well-studied Grande Ronde basalt, these include both augite and pigeonite, while relatively Fe-rich olivine occurs occasionally in flows of other formations (Swanson et al., 1979; Reidel, 1983; Hooper et al., 1984; Bailey, 1989b). Caprarelli and Reidel (2004) demonstrate that the small pyroxene phenocrysts of the Grande Ronde basalt grew in the lower crust from a dry basaltic magma.

Wright and his co-workers (Wright et al., 1989) have long maintained from their field and experimental studies that crystal fractionation was not a major factor in the evolution of the Grande Ronde basalt and that the original magma was relatively Fe-rich. Hooper and Hawkesworth (1993) demonstrated that much of the variation within the Grande Ronde basalt included changes in such ratios as Rb/Zr, which cannot have been due to gabbro (crustal) fractionation, and they concluded that the maximum amount of gabbro fractionation observed within the Grande Ronde basalt was less than 10%, while the rest of the variation was due to mixing between the magma and the lithospheric components noted earlier. This restriction does not apply to the Imnaha basalt, where a convincing case can be made for three groups of Imnaha basalt flows derived from three different degrees of partial melting of their single enriched mantle source, followed in each case by crystal fractionation and recharge (Fig. 10; Hooper and Hawkesworth, 1993). Modeling of major and trace elements has shown that much of the crystal fractionation observed in the Imnaha basalt magmas included recharge of the magma reservoir by injection of new magma, a process that buffers the abundance of those major elements involved in the formation of the precipitated phenocrysts, while increasing the abundance of the incompatible elements (O'Hara and Mathews, 1981).

Finally, clear examples of the physical mixing of two chemically discrete magmas have been documented for Saddle Mountain basalt flows, both prior to eruption (Hooper, 1985) and after eruption (Reidel and Fecht, 1987). Physical mixing processes are best detected when the two end-member flows have distinctive compositions, but it could be anticipated that, during the large and rapid eruptions of the chemically similar Grande Ronde basalt flows, physical mixing of magmas occurred both before and after eruption.

DISCUSSION: PLUME VERSUS NONPLUME

The most fundamental problem confronting workers on the Columbia River basalts is whether this and other flood basalt provinces are derived from a deep-mantle plume or from some shallow-mantle process. Workers on the Oregon plateau have long advocated a model of volcanism associated with back-arc extension related to the Basin and Range province to account for the Steens and younger (HAOT) lavas (Eaton, 1984; Hart and Carlson, 1987; Carlson and Hart, 1988). In contrast, Draper (1991) was the first to suggest that these basalts have a mantle plume origin. Contrasting views are also apparent for eruptions along the Snake River Plain, which have been attributed both to a mantle plume process (Duncan, 1982; Draper, 1991; Pierce and Morgan, 1992) and to shallow-mantle nonplume processes (King and Anderson, 1998; Humphreys et al., 2000; Christiansen et al., 2002).

Scrutiny of Nonplume Interpretations and the Need for a Unified Model

Most workers with direct field experience of the Columbia River basalts have embraced a mantle plume origin as the most

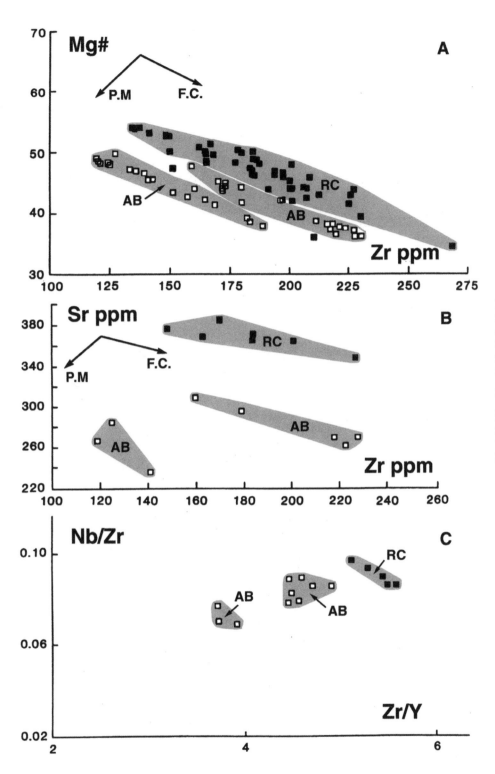

Figure 10. Evolution of Imnaha basalt flows. Panels A, B, and C illustrate three discreet groups of Imnaha basalt, and both A and B show how these may be formed by variable degrees of partial melting of an enriched mantle. Each group, then, has undergone crystal fractionation and recharge in lower crustal reservoirs. P.M.—partial melting trend; F.C.—fractional crystallization trend; RC—Rock Creek type of Imnaha basalt; AB—American Bar types of Imnaha basalt. After Hooper and Hawkesworth (1993).

plausible explanation for the exceptional volume of basalt erupted in less than 1.5 m.y. and over a very restricted area (Brandon and Goles, 1988; Hooper and Hawkesworth, 1993; Takahashi et al., 1998; Hooper et al., 2002a; Camp et al., 2003; Camp and Ross, 2004). Camp and Ross (2004) emphasize the need for a unifying model encompassing all contemporaneous magmatism on the Columbia plateau, the Oregon plateau, the Snake River Plain, and the northern Nevada rift, and they note that only the plume model achieves this. Nonplume interpretations appear to require explanations for the main-phase flood basalt eruptions independent of those for the eastern Snake River Plain and the Oregon plateau. Such interpretations describe long-lived tectonic regimes and slowly evolving processes that fail to account for the sudden outburst and short duration that typifies flood basalt volcanism.

It is not clear, for example, in the King and Anderson (1998) model, why the exceptionally large volumes of homogeneous Columbia River Basalt Group tholeiite should have erupted in only one area along the cratonic margin, which runs from Alaska to California, if the eruptions were controlled, as they argue, by the difference in thickness between the craton and the adjoining lithosphere. Christiansen et al. (2002) suggest that the eruptions may have been controlled by older structures beneath the Eastern Snake River Plain and the Brothers fault zone, which cross the cratonic margin close to the original center of the flood basalt eruptions. While the presence of these older structures and their ability to guide the location of later volcanicity are accepted, many equally significant structures meet the cratonic boundary to the north and south yet fail to generate huge volumes of flood basalt. The problem of explaining such sudden, short-lived, voluminous eruptions at one particular location remains, unless the additional presence of a thermal anomaly is invoked. Christiansen et al. (2002) also object to the plume model because it requires the coincidence of the hotspot track following an older tectonic boundary. The real coincidence, however, is between the older tectonic boundary and the westward migration of the North America plate, as derived independently from plate motions, regardless of whether a mantle plume is present.

Nonplume models that rely solely on back-arc extension (Carlson and Hart, 1988; Smith, 1992) fail to take into account the field, petrographic, and chemical evidence of the two distinct tectonomagmatic events that are particularly evident in east central Oregon (Fig. 4; Hooper et al., 2002a): (1) the long-lived, small-volume calc-alkaline to alkaline volcanicity of basalt through rhyolites and silicic tuffs that are specifically associated with graben formation (Hooper et al., 1995b) and (2) the huge and short-lived, relatively homogeneous, flood basalt eruptions. The former is consistent with a long-lived period of back-arc extension, but the latter is more reasonably attributed to the abrupt appearance of a thermal anomaly consistent with plume emplacement. These two events, which were in part contemporaneous, inevitably affected each other. Following the plume model, the thin lithosphere of the accreted terranes would have facilitated both a rapid rise of the plume head and decompressional partial melting (White and McKenzie, 1989, 1995). Surface eruption would have been aided by the ability of the magmas to find the weakest (thinned) zones (Thompson and Gibson, 1991) along highly oriented fissures aligned at right angles to the direction of long-lived back-arc extension above the plume head. Under this scenario, the elevated potential temperature of the plume head would have both weakened and uplifted the lithosphere, leading to increased east-west extension along the crest of the elevated region so forming the northern Nevada rift (Zoback et al., 1994; Glen and Ponce, 2002), the concentration of feeder dikes of the Chief Joseph Dike swarm

along the crest of the uplift, and the subsequent development of the unusually wide Oregon-Idaho Graben.

The derivation of the Columbia River Basalt Group lavas by extension alone is also in direct conflict with the experimental results of McKenzie and Bickle (1988) and White and McKenzie (1989, 1995). Their studies demonstrated that the surface eruption of continental flood basalts requires both a mantle source with abnormally high potential temperatures (T_p) and high stretching factors (β) typical of advanced stages of continental extension. Extension alone, no matter how advanced, cannot generate the melt volumes typical of continental flood basalt provinces. On the other hand, field evidence shows that at least some flood basalt provinces, such as the Deccan, have been generated prior to significant extension (Hooper, 1990), suggesting that high potential temperatures play the more critical role in melt generation on such a massive scale.

Merits of the Plume Model

Several lines of evidence support mantle plume genesis of the Columbia River Basalt Group eruptions. The earliest Columbia River basalt flows contain the chemical and isotopic signatures of ocean island basalts (OIBs) (Hooper and Hawkesworth, 1993; Bryce and DePaolo, 2004), a similarity required if a plume model is to be sustained. In common with magmas from the Snake River Plain, these early Columbia River basalt flows also have the high ^{3}He/^{4}He ratios (11.4 ± 0.7 Ra) expected of basalts derived from a deep mantle source (Dodson et al., 1997; D. Graham, 2003, personal commun.).

The recent identification of a plume tail beneath Yellowstone National Park to a depth of 500 km (Smith et al., 2005; Yuan and Dueker, 2005) appears to have verified a plume origin for Yellowstone and the Snake River Plain hotspot track. Plate reconstructions along this linear belt of age-progressive rhyolitic centers unambiguously place this plume tail near the McDermitt caldera along the Oregon-Nevada border at ca. 16–17 Ma, coincident in both time and space with the earliest Columbia River Basalt Group eruptions. Acceptance of a nonplume origin for the flood basalt event requires recognition that the Miocene location of the plume tail at the initiation site of flood basalt volcanism is a coincidence and incidental to a separate shallow-mantle process responsible for the flood basalt event. We suggest that it is more probable that the McDermitt region separates flood basalt volcanism to the north, underlain by the plume head, from the Yellowstone hotspot track to the east, underlain by the plume tail. The plume head / plume tail duality is consistent with the temporal and spatial relationships inherent in traditional plume models (Morgan, 1981; Richards et al., 1989; Campbell and Griffiths, 1990; Griffiths and Campbell, 1991; Hill et al., 1992; Weinberg, 1997).

The unusually high elevation of the extended terrain of eastern Oregon (approx. 1 km asl) is difficult to explain in the absence of elevated mantle temperatures. White and McKenzie

(1989, 1995) show that as β values (extension) increase in lithosphere underlain by normal mantle T_p, continental subsidence will increase accordingly. However, if the mantle T_p is elevated to 150 °C or more above normal, uplift will ensue, even at β values as high as 5 (White and McKenzie, 1995). Such uplift results from mantle melting and the addition of new crust, together with the reduced density provided by both the thermal anomaly and the residual mantle (Humphreys et al., 2000). A more buoyant lithosphere in the intermontane Pacific Northwest is consistent with findings of both Parsons et al. (1994) and Saltus and Thompson (1995), who modeled a residual mass deficit in the mantle beneath this uplifted region. Both studies concluded, independently, that the high altitude of this extended terrain must be isostatically supported by a broad mass of hot, low-density mantle that, they argue, is the Yellowstone mantle plume head.

One problem commonly cited that argues against a mantle plume genesis is the location of the main-phase eruptions from the Chief Joseph and Monument dike swarms up to 400 km north of the presumed impingement of the hotspot near the Oregon-Nevada border at ca. 16–17 Ma. This apparent inconsistency in the plume model has been addressed by both Geist and Richards (1993) and Camp (1995). Geist and Richards (1993) suggested that the plume head was initially deflected beneath the Chief Joseph Dike swarm by the northeast-subducting Farallon plate at ca. 17 Ma, followed by rapid recovery of the plume tail toward the western end of the Snake River Plain at ca. 14 Ma. Alternatively, Camp (1995) suggested that the spreading plume head was distorted against the thick cratonic margin of North America, deflecting it largely to the north beneath the progressively thinner lithosphere of the Blue Mountains accreted terranes. Northward flow of this hot mantle to shallower levels, it was argued, was accompanied by plume-generated uplift and decompressional melting.

Uplift above the plume head is one of the principal predictions of a plume model (White and McKenzie, 1989, 1995; Campbell and Griffiths, 1990), and such uplift immediately before and during the Columbia River Basalt Group eruptions is evident from the deep canyons in the prebasalt surface along the Idaho–northeast Oregon border, canyons that were filled by the initial flows of Imnaha basalt (Kleck, 1976; Hooper et al., 1984; but see Hales et al., 2005, for an alternative view). The continued development of a northward-dipping paleoslope during the main eruptive phase (Hooper and Camp, 1981; Camp, 1995; Camp and Ross, 2004) means that this uplift continued during the flood basalt eruptions. Major uplift above the erupting fissures is also consistent with both paleobotanical altitude estimates and the development of a broad precipitation shadow to the east that peaked at ca. 15 Ma (Pierce et al., 2002).

Plume genesis provides a unifying model capable of explaining all magmatic activity centered on east central Oregon and contemporaneous with the Columbia River Basalt Group eruptions from 17 to 6 Ma (Camp and Ross, 2004). The model

includes magmatism along the eastern Snake River Plain, the northern Nevada rift system, and the Brothers fault zone. Camp and Ross (2004) contend that the main-phase eruptions migrated progressively outward, from southeastern Oregon northward along the Chief Joseph and Monument dike swarms and southward into the eastern graben of the northern Nevada rift. Like spokes on a wheel, the focal point of these three outwardly radiating trends was close to the projected center of the mantle plume at ca. 16–17 Ma, as would be predicted from impingement and rapid spreading of a starting plume head (Campbell and Griffiths, 1990). The main-phase eruptions ceased at ca. 15 Ma when the plume head was sheared off by the thick westward-moving cratonic margin, thus severing the plume head from the feeding plume tail (Pierce and Morgan, 1992; Camp and Ross, 2004). At ca. 14 Ma, rhyolites began to erupt above the plume tail in southwestern Idaho, followed by the migration of bimodal eruptions along the Snake River Plain hotspot track (Pierce et al., 2002). Meanwhile the plume head, prevented from expanding east or farther north by the cratonic margin (Fig. 1), expanded west and south to form the westward-migrating bimodal eruptions along the Oregon High Lava Plains (Camp and Ross, 2004; Jordan et al., 2004) and the mixed eruptions associated with the northern Nevada rift (John et al., 2000), respectively.

The plume model we advocate is based on the experimental work of Campbell and Griffiths (1990), Cordery et al. (1997), Takahashi et al. (1998), and Yaxley (2000) among others, in which crustal basalt was subducted deep into the mantle, perhaps but not necessarily, to the core-mantle boundary, where it was reactivated to rise again as an eclogite-rich plume, the large head of which eventually spread out beneath the lithosphere. Experimental work indicates that such an iron-rich mantle source was capable of creating the large volumes of relatively evolved basaltic magma without significant quantities of silicic differentiates or mafic cumulates. Wright et al. (1979, 1989) have argued that the Grande Ronde basalt, in particular, required an unusually iron-rich mantle source such as would have been available in an eclogite-rich mantle plume (Takahashi et al., 1998; and Yaxley, 2000).

Rapid mantle flow, decompressional partial melting, and uplift all occurred in an environment of a relatively thin lithosphere and back-arc extension. This generated highly oriented north-south feeder dikes, with most of the basalt erupted from the Chief Joseph Dike swarm to produce the greatest volume of Columbia River Basalt Group magma in only ~1.5 m.y.

Our conclusion is that nonplume models, such as back-arc spreading, mantle upwelling adjacent to a cratonic margin, or generation of magmas along older tectonic boundaries, are not in themselves capable of explaining the ultimate cause of the Columbia River flood basalt eruptions. A thermal anomaly is required. We believe that an eclogite-rich plume emplaced into a tectonic environment of a thinner lithosphere and accompanied by active extension provides a model that more readily accounts for the salient features of this flood basalt province, including the unusually rapid extrusion of such exceptionally large volumes of evolved basaltic magma.

ACKNOWLEDGMENTS

We are grateful to the staff of the GeoAnalytical Laboratories of Washington State University, whose technical expertise and dedication over many years have facilitated work on the Columbia plateau. We thank the draftsmen at the Open University for their patience in drafting and redrafting many of the figures. Detailed reviews by Bob Christiansen, Tristram Hales, and Chris Hawkesworth greatly improved the original manuscript and are much appreciated.

REFERENCES CITED

Anderson, J.L., 1987, Structural geology and ages of deformation of a portion of the southwest Columbia Plateau, Oregon and Washington [Ph.D. thesis]: Los Angeles, University of Southern California, 283 p.

Armstrong, R.L., Taubeneck, W.H., and Hales, P.O., 1977, Rb/Sr and K/Ar geochronometry of the Mesozoic granite rocks and their Sr isotope composition, Oregon, Washington and Idaho: Geological Society of America Bulletin, v. 88, p. 397–411, doi: 10.1130/0016-7606(1977)88<397: RAKGOM>2.0.CO;2.

Bailey, D.G., 1990, Geochemistry and petrogenesis of Miocene volcanic rocks in the Powder River volcanic field, northeastern Oregon [Ph.D. thesis]: Pullman, Washington State University, 341 p.

Bailey, M.M., 1989a, Revisions to stratigraphic nomenclature of the Picture Gorge Basalt Subgroup, Columbia River Basalt group, *in* Reidel, S.P., and Hooper, P.R., eds., Volcanism and tectonism in the Columbia River flood-basalt province: Boulder, Colorado, Geological Society of America Special Paper 239, p. 67–84.

Bailey, M.M., 1989b, Evidence for magma recharge and assimilation in the Picture Gorge Basalt Subgroup, Columbia River Basalt Group, *in* Reidel, S.P., and Hooper, P.R., eds., Volcanism and tectonism in the Columbia River flood-basalt province: Boulder, Colorado, Geological Society of America Special Paper 239, p. 343–356.

Baksi, A.K., 1989, Re-evaluation of the timing and duration of extrusion of the Imnaha, Picture Gorge, and Grande Ronde Basalts, Columbia River Basalt Group, *in* Reidel, S.P., and Hooper, P.R., eds., Volcanism and tectonism in the Columbia River flood-basalt province: Boulder, Colorado, Geological Society of America Special Paper 239, p. 105–112.

Beeson, M.H., Fecht, K.R., Reidel, S.P., and Tolan, T.L., 1985, Regional correlations within the Frenchman Springs Member of the Columbia River Basalt Group: New insights into the middle Miocene tectonics of northwestern Oregon: Oregon Geology, v. 47, p. 87–96.

Beeson, M.H., Tolan, T.L., and Anderson, J.L., 1989, The Columbia River Basalt Group in western Oregon; Geologic structures and other factors that controlled flow and emplacement patterns, *in* Reidel, S.P., and Hooper, P.R., eds., Volcanism and tectonism in the Columbia River flood-basalt province: Boulder, Colorado, Geological Society of America Special Paper 239, p. 223–246.

Binger, G.B., 1997, The volcanic stratigraphy of the Juntura region, eastern Oregon [M.S. thesis]: Pullman, Washington State University, 206 p.

Brandon, A.D., and Goles, G.G., 1988, A Miocene subcontinental plume in the Pacific Northwest: Geochemical evidence: Earth and Planetary Science Letters, v. 88, p. 273–283, doi: 10.1016/0012-821X(88)90084-2.

Brandon, A.D., Hooper, P.R., Goles, G.G., and Lambert, R.St.J., 1993, Evaluating crustal contamination in crustal basalts: The isotopic composition of the Picture Gorge Basalt of the Columbia River Basalt Group: Contribu-

tions to Mineralogy and Petrology, v. 114, p. 452–464, doi: 10.1007/ BF00321750.

Breitsprecher, K., Thorkelson, D.J., Groome, D.J., and Dostal, J., 2003, Geochemical confirmation of the Kula-Farallon slab window beneath the Pacific Northwest in Eocene time: Geology, v. 31, p. 351–354.

Bryce, J.G., and DePaolo, D.J., 2004, Pb isotopic heterogeneity in basaltic phenocrysts: Geochimica et Cosmochimica Acta, v. 68, p. 4453–4468, doi: 10.1016/j.gca.2004.01.016.

Camp, V.E., 1995, Mid-Miocene propagation of the Yellowstone mantle plume head beneath the Columbia River basalt source region: Geology, v. 23, p. 435–438, doi: 10.1130/0091-7613(1995)023<0435:MMPOTY>2.3.CO;2.

Camp, V.E., and Ross, M.E., 2004, Mantle dynamics and genesis of mafic magmatism in the intermontane Pacific Northwest: Journal of Geophysical Research, v. 109, p. B08204, doi: 10.1029/2003JB002838.

Camp, V.E., Ross, M.E., and Hanson, W.L., 2003, Genesis of flood basalts and Basin and Range volcanic rocks from Steens Mountain to Malheur River Gorge, Oregon: Geological Society of America Bulletin, v. 115, p. 105–128, doi: 10.1130/0016-7606(2003)115<0105:GOFBAB>2.0.CO;2.

Campbell, I.H., and Griffiths, R.W., 1990, Implications of mantle plume structure for the evolution of flood basalts: Earth and Planetary Science Letters, v. 99, p. 79–93, doi: 10.1016/0012-821X(90)90072-6.

Caprarelli, G., and Reidel, S.P., 2004, Physical evolution of Grande Ronde Basalt magmas, Columbia River Basalt Group, north-western U.S.A: Mineralogy and Petrology, v. 80, p. 1–25, doi: 10.1007/s00710–003–0017–1.

Carlson, R.W., 1984, Isotopic constraints on Columbia River basalt genesis and the nature of the subcontinental mantle: Geochimica et Cosmochimica Acta, v. 48, p. 2357–2372, doi: 10.1016/0016-7037(84)90231-X.

Carlson, R.W., and Hart, W.K., 1987, Crustal genesis on the Oregon Plateau: Journal of Geophysical Research, v. 92, p. 6191–6206.

Carlson, R.W., and Hart, W.K., 1988, Flood basalt volcanism in the northwestern United States, *in* Macdougall, J.D., ed., Continental flood basalts: Dordrecht, Kluwer, p. 35–62.

Carlson, R.W., Lugmair, G.W., and Macdougall, J.D., 1981, Columbia River volcanism: The question of mantle heterogeneity or crustal contamination: Geochimica et Cosmochimica Acta, v. 45, p. 2483–2499, doi: 10.1016/ 0016-7037(81)90100-9.

Chesley, J.T., and Ruiz, J., 1998, Crust-mantle interaction in large igneous provinces: Implications from the Re-Os isotope systematics of the Columbia River flood basalts: Earth and Planetary Science Letters, v. 154, p. 1–11, doi: 10.1016/S0012-821X(97)00176-3.

Christiansen, R.L., Foulger, G.R., and Evans, J.R., 2002, Upper mantle origin of the Yellowstone hot spot: Geological Society of America Bulletin, v. 114, p. 1245–1256, doi: 10.1130/0016-7606(2002)114<1245:UMOOTY> 2.0.CO;2.

Church, S.E., 1985, Genetic interpretations of lead isotopic data from the Columbia River Basalt group, Oregon, Washington and Idaho: Geological Society of America Bulletin, v. 96, p. 676–690, doi: 10.1130/0016- 7606(1985)96<676:GIOLDF>2.0.CO;2.

Coney, P.J., 1987, The regional tectonic setting and possible causes of Cenozoic extension in the North American Cordillera, *in* Coward, M.P., et al., eds., Continental extensional tectonics: Geological Society, London, Special Publication 28, p. 177–186.

Cordery, M.J., Davies, G.F., and Campbell, I.H., 1997, Genesis of flood basalts from eclogite-bearing mantle plumes: Journal of Geophysical Research, v. 102, p. 20,179–20,197, doi: 10.1029/97JB00648.

Cummings, M.L., Evans, J.G., Ferns, M.L., and Lees, K.R., 2000, Stratigraphic and structural evolution of the middle Miocene syn-volcanic Oregon-Idaho graben: Geological Society of America Bulletin, v. 112, p. 668–682, doi: 10.1130/0016-7606(2000)112<0668:SASEOT>2.3.CO;2.

DePaolo, D.J., 1983, Comment on "Columbia River volcanism: The question of mantle heterogeneity or crustal contamination" by Carlson, R.W., Lugmair, G.W., and Macdougall, J.D: Geochimica et Cosmochimica Acta, v. 47, p. 841–844, doi: 10.1016/0016-7037(83)90117-5.

Dodson, A., Kennedy, B.M., and DePaolo, D.J., 1997, Helium and neon isotopes in the Imnaha Basalt, Columbia River Basalt Group: Evidence for a Yellowstone plume source: Earth and Planetary Science Letters, v. 150, p. 443–451, doi: 10.1016/S0012-821X(97)00090-3.

Draper, D.S., 1991, Late Cenozoic bimodal magmatism in the northern Basin and Range Province of southeastern Oregon: Journal of Volcanology and Geothermal Research, v. 47, p. 299–328, doi: 10.1016/0377-0273(91)90006-L.

Duncan, R.A., 1982, A captured island chain in the Coast Range of Oregon and Washington: Journal of Geophysical Research, v. 87, p. 10,827–10,837.

Eaton, G.P., 1984, The Miocene Great Basin of western North America as an extended back-arc region: Tectonophysics, v. 102, p. 275–295, doi: 10.1016/0040-1951(84)90017-9.

England, P., and Wells, R.E., 1991, Neogene rotations and quasicontinuous deformation of the Pacific Northwest continental margin: Geology, v. 19, p. 978–981, doi: 10.1130/0091-7613(1991)019<0978:NRAQDO>2.3.CO;2.

Evans, J.G., 1990a, Geology and mineral resources map of the Jonesboro Quadrangle, Malheur County, Oregon: Portland, Oregon, Department of Geology and Mineral Industries Geologic Map GMS-66, scale 1:24,000, 2 sheets.

Evans, J.G., 1990b, Geology and mineral resources map of the South Mountain Quadrangle, Malheur County, Oregon: Portland, Oregon, Department of Geology and Mineral Industries Geologic Map GMS-67, scale 1:24,000, 2 sheets.

Ferns, M.L., Brooks, H.C., Evans, J.G., and Cummings, M.L., 1993, Geologic map of the Vale 30 × 60 minute quadrangle, Malheur County, Oregon, and Owyhee County, Idaho: Portland, Oregon, Department of Geology and Mineral Industries Geologic Map GMS-78, scale 1:100,000, 12 p., 1 sheet.

Finnemore, S.L., Self, S., and Walker, G.P.L., 1993, Inflation features in lava of the Columbia River Basalts: Eos (Transactions, American Geophysical Union), v. 74, Fall Meeting Supplement, p. 555.

Fitton, J.G., James, D., Kempton, P.D., Omerod, D.S., and Leeman, W.P., 1988, The role of lithospheric mantle in the generation of late Cenozoic basic magmas in the western United States, *in* Menzies, I.A., and Cox, K.G., eds., Oceanic and continental lithosphere: Similarities and differences: Journal of Petrology, Special Volume, p. 331–350.

Fleck, R.J., and Criss, R.E., 1985, Strontium and oxygen isotope variations in Mesozoic and Tertiary plutons of central Idaho: Contributions to Mineralogy and Petrology, v. 90, p. 291–308, doi: 10.1007/BF00378269.

Gehrels, G.E., White, R.R., and Davis, G.A., 1980, The La Grande pull-apart basin, northeastern Oregon: Geological Society of America Abstracts with Programs, v. 12, p. 107.

Geist, D., and Richards, M.A., 1993, Origin of the Columbia Plateau and Snake River Plain: Deflection of the Yellowstone plume: Geology, v. 21, p. 789–792, doi: 10.1130/0091-7613(1993)021<0789:OOTCPA>2.3.CO;2.

Giorgis, S., Tikoff, B., and McClelland, W., 2005, Missing Idaho arc: Transpressional modification of the $^{87}Sr/^{86}Sr$ transition on the western edge of the Idaho batholith: Geology, v. 33, p. 469–472, doi: 10.1130/G20911.1.

Glen, J.M.G., and Ponce, D.A., 2002, Large-scale fractures related to inception of the Yellowstone hotspot: Geology, v. 30, p. 647–650, doi: 10.1130/0091-7613(2002)030<0647:LSFRTI>2.0.CO;2.

Griffiths, R.W., and Campbell, I.H., 1991, Interaction of mantle plume heads with the Earth's surface and onset of small-scale convection: Journal of Geophysical Research, v. 96, p. 18,295–18,310.

Hales, T.C., Abt, D.L., Humphreys, E.D., and Roering, J.J., 2005, Delamination origin for the Columbia River flood basalts and Wallowa Mountain uplift in NE Oregon, U.S.A.: Nature, v. 438, p. 842–845, doi: 10.1038/nature04313.

Hart, W.K., 1982, Chemical, chronological and isotopic significance of low potassium, high aluminum tholeiite in the north-western Great Basin [Ph.D. thesis]: Cleveland, Ohio, Case Western Reserve University, 410 p.

Hart, W.K., and Carlson, R.W., 1987, Tectonic controls on magma genesis and evolution in the northwestern United States: Journal of Volcanology and Geothermal Research, v. 32, p. 119–135, doi: 10.1016/0377-0273(87)90040-0.

Hart, W.K., Carlson, R.W., and Shirey, S.B., 1997, The role of lithosphere in continental basalt evolution: New Re-Os evidence from the northwestern United States: Earth and Planetary Science Letters, v. 150, p. 103–116, doi: 10.1016/S0012-821X(97)00075-7.

Hawkesworth, C., Turner, S., Gallagher, K., Hunter, A., Bradshaw, T., and Rogers, N., 1995, Calc-alkaline magmatism, lithospheric thinning and extension in the Basin and Range: Journal of Geophysical Research, v. 100, p. 10,271–10,286, doi: 10.1029/94JB02508.

Hill, R.I., Campbell, I.H., Davies, G.F., and Griffiths, R.W., 1992, Mantle plumes and continental tectonics: Science, v. 256, p. 186–193, doi: 10.1126/science.256.5054.186.

Ho, A.M., and Cashman, K.V., 1997, Temperature constraints on the Ginkgo flow of the Columbia River Basalt Group: Geology, v. 25, p. 403–406, doi: 10.1130/0091-7613(1997)025<0403:TCOTGF>2.3.CO;2.

Holder, G.M., Holder, R.W., and Carlson, D.H., 1990, Middle Eocene dike swarms and their relationship to contemporaneous plutonism, volcanism, core-complex mylonitization, and graben subsidence, Okanogan Highlands, Washington: Geology, v. 18, p. 1082–1085, doi: 10.1130/0091-7613(1990)018<1082:MEDSAT>2.3.CO;2.

Hooper, P.R., 1985, A case of simple magma mixing in the Columbia River Basalt Group: The Wilbur Creek, Lapwai and Asotin flows, Saddle Mountains Formation: Contributions to Mineralogy and Petrology, v. 91, p. 66–73, doi: 10.1007/BF00429428.

Hooper, P.R., 1990, The timing of crustal extension and the eruption of continental flood basalts: Nature, v. 345, p. 246–249, doi: 10.1038/345246a0.

Hooper, P.R., 2000, Chemical discrimination of Columbia River basalt flows: Geochemistry, Geophysics, Geosystems, v. 1, no. 6, doi: 10.1029/2000GC000040.

Hooper, P.R., 2004, Ages of the Steens and Columbia River flood basalts and their relationship to extension-related calc-alkalic volcanism in eastern Oregon: Reply: Geological Society of America Bulletin, v. 116, p. 249–250, doi: 10.1130/B25310R.1.

Hooper, P.R., and Camp, V.E., 1981, Deformation of the southeast part of the Columbia Plateau: Geology, v. 9, p. 323–328, doi: 10.1130/0091-7613(1981)9<323:DOTSPO>2.0.CO;2.

Hooper, P.R., and Conrey, R.M., 1989, A model for the tectonic setting of the Columbia River Basalt eruptions, *in* Reidel, S.P., and Hooper, P.R., eds., Volcanism and tectonism in the Columbia River flood-basalt province: Boulder, Colorado, Geological Society of America Special Paper 239, p. 293–306.

Hooper, P.R., and Hawkesworth, C.J., 1993, Isotopic and geochemical constraints on the origin and evolution of the Columbia River basalt: Journal of Petrology, v. 34, p. 1203–1246.

Hooper, P.R., Kleck, W.D., Knowles, C.R., Reidel, S.P., and Thiessen, R.L., 1984, Imnaha Basalt, Columbia River Basalt Group: Journal of Petrology, v. 25, p. 473–500.

Hooper, P.R., Steele, W.K., Conrey, R.M., Smith, G.A., Anderson, J.L., Bailey, D.G., Beeson, M.H., Tolan, T.L., and Urbanczyk, K.M., 1993, The Prineville Basalt, north-central Oregon: Oregon Geology, v. 55, p. 3–12.

Hooper, P.R., Gillespie, B.A., and Ross, M.E., 1995a, The Eckler Mountain basalts and associated flows, Columbia River Basalt Group: Canadian Journal of Earth Sciences, v. 32, p. 410–423.

Hooper, P.R., Bailey, D.G., and McCarley Holder, G.A., 1995b, Tertiary calc-alkaline magmatism associated with lithospheric extension in the Pacific Northwest: Journal of Geophysical Research, v. 100, p. 10,303–10,320, doi: 10.1029/94JB03328.

Hooper, P.R., Binger, G.B., and Lees, K.R., 2002a, Ages of the Steens and Columbia River flood basalts and their relationship to extension-related calc-alkalic volcanism in eastern Oregon: Geological Society of America Bulletin, v. 114, p. 43–50 (corrections p. 923–924).

Hooper, P.R., Johnson, J., and Hawkesworth, C., 2002b, A model for the origin of the western Snake River Plain as an extensional strike-slip duplex, Idaho and Oregon, *in* Bonnichsen, B., et al., eds., Tectonic and magmatic evolution of the Snake River Plain volcanic province: Idaho Geological Survey Bulletin 30, p. 59–67.

Humphreys, E.D., Dueker, D.L., Schutt, D.L., and Smith, R.B., 2000, Beneath Yellowstone: Evaluating plume and nonplume models using teleseismic images of the upper mantle: GSA Today, v. 10, p. 1–6.

John, D.A., Wallace, A.R., Ponce, D.A., Fleck, R.B., and Conrad, J.E., 2000, New perspectives on the geology and origin of the northern Nevada Rift, *in* Cluer, G.K., et al., eds., Geology and Ore Deposits 2000: The Great Basin and beyond: Geological Society of Nevada Symposium Proceedings, p. 127–154.

Johnson, J.A., Hooper, P.R., Hawkesworth, C.J., and Binger, G.B., 1998a, Geologic map of the Stemler Ridge Quadrangle, Malheur County, southeastern Oregon: U.S. Geological Survey Open-File Report OF 98-105, scale 1:24,000, 11 p., 1 sheet.

Johnson, J.A., Hawkesworth, C.J., Hooper, P.R., and Binger, G.B., 1998b, Major and trace element analyses of Steens basalt, southeastern Oregon: Reston, Virginia, U.S. Geological Survey Open-File Report 98-482, 26 p.

Jordan, B.T., Grunder, A.L., Duncan, R.A., and Deinio, A.L., 2004, Geochronology and age-progressive volcanism of the Oregon High Lava Plains: Implications for plume interpretation of Yellowstone: Journal of Geophysical Research, v. 109, p. B10202, doi:101029/2003JB002776.

Kauffman, J.D., Bush, J.H., and Lewis, R.S., 2003, Newly identified Oligocene alkali volcanics along the eastern margin of the Columbia Plateau, Latah and surrounding counties, Idaho: Geological Society of America Abstracts with Programs, v. 35, no. 6, poster 226-4.

Keszthelyi, L., Self, S., and Thordarson, T., 2006, Flood lavas on Earth: Io and Mars: Journal of the Geological Society, London, v. 163, p. 253–264.

Kim, K.S., Dischler, S.A., Aggson, J.R., and Hardy, M.P., 1986, State of in situ stresses determined by hydraulic fracturing at the Hanford site: Richland, Washington, Rockwell Hanford Operations Report RHO-BW-ST-73-P, 186 p.

King, S.D., and Anderson, D.L., 1998, Edge-driven convection: Earth and Planetary Science Letters, v. 160, p. 289–296, doi: 10.1016/S0012-821X(98)00089-2.

Kleck, W.D., 1976, Chemistry, petrography and stratigraphy of the Columbia River Basalt Group in the Imnaha River valley region, eastern Oregon and western Idaho [Ph.D. thesis]: Pullman, Washington State University, 203 p.

Lawrence, R.D., 1976, Strike-slip faulting terminates Basin and Range province in Oregon: Geological Society of America Bulletin, v. 87, p. 846–850, doi: 10.1130/0016-7606(1976)87<846:SFTTBA>2.0.CO;2.

Lees, K.R., 1994, Magmatic and tectonic changes through time in the Neogene volcanic rocks of the Vale area, Oregon [Ph.D. thesis]: Milton-Keynes, England, Open University, 283 p.

Long, P.E., and Duncan, R.A., 1982, $^{40}Ar/^{39}Ar$ ages of Columbia River basalt from deep bore holes in south-central Washington: Richland, Washington, Rockwell Hanford Operations Report RHO-BW-SA-233P, 13 p.

Long, P.E., Murphy, M.T., and Self, S., 1991, Time required to emplace the Pomona flow, Columbia River basalt: Eos (Transactions, American Geophysical Union), v. 72, Fall Meeting Supplement, p. 602.

Mabey, E.R., 1984, Geophysics and tectonics of the Snake River Plain, Idaho, *in* Bonnichsen, B., and Breckenridge, R.M., eds., Cenozoic geology of Idaho: Idaho Bureau of Mines and Geology Bulletin, v. 26, p. 139–153.

Magill, J.R., Wells, R.E., Simpson, R.W., and Cox, A.V., 1982, Post 12 m.y. rotation of southwest Washington: Journal of Geophysical Research, v. 87, p. 3761–3776.

Mangan, M.T., Wright, T.L., Swanson, D.A., and Byerly, G.R., 1986, Regional correlation of Grande Ronde Basalt flows, Columbia River Basalt Group, Washington, Oregon, and Idaho: Geological Society of America Bulletin, v. 97, p. 1300–1318, doi: 10.1130/0016-7606(1986)97<1300:RCOGRB>>2.0.CO;2.

Mankinen, E.A., Larson, E.E., Prevot, M., and Coe, R.S., 1987, The Steens Mountain (Oregon) geomagnetic polarity transition, part 3, Its regional significance: Journal of Geophysical Research, v. 92, p. 8057–8076.

Mann, G.M., 1989, Seismicity and late Cenozoic faulting in the Brownlee Dam area—Oregon-Idaho: A preliminary report: Reston, Virginia, U.S. Geological Survey Open-File Report 89-429, 46 p.

Martin, B.S., 1984, Paleomagnetism of basalts in northeast Oregon and west central Idaho [M.S. thesis]: Pullman, Washington State University, 133 p.

Martin, B.S., 1989, The Roza Member, Columbia River Basalt Group: Chemical stratigraphy and flow distribution, *in* Reidel, S.P., and Hooper, P.R., eds., Volcanism and tectonism in the Columbia River flood-basalt province: Boulder, Colorado, Geological Society of America Special Paper 239, p. 85–104.

McCaffrey, R., Long, M.D., Goldfinger, C., Zwick, P.C., Nabelek, J.L., Johnson, C.K., and Smith, C., 2000, Rotation and plate locking at the southern Cascadia subduction zone: Geophysical Research Letters, v. 27, p. 3117–3120, doi: 10.1029/2000GL011768.

McKenzie, D., and Bickle, M.J., 1988, The volume and composition of melt generated by extension of lithosphere: Journal of Petrology, v. 29, p. 625–679.

Mohl, G.B., and Thiessen, R.L., 1995, Gravity studies of an island arc–continent suture in west central Idaho and adjacent Washington, *in* Vallier, T.L., and Brooks, H.C., eds., The geology of the Blue Mountains Region of Oregon, Idaho and Washington: Petrology and tectonic evolution of Pre-Tertiary rocks of the Blue Mountains Region: Reston, Virginia, U.S. Geological Survey Professional Paper 1438, p. 497–516.

Morgan, W.J., 1981, Hotspot tracks and the opening of the Atlantic and Indian Oceans, *in* Emiliani, C., ed., The sea, v. 7, p. 443–487: New York, Wiley-Interscience.

Morris, G.A., and Hooper, P.R., 1997, Petrogenesis of the Colville Igneous Complex, northwest Washington: Implications for Eocene tectonics in the northern U.S. Cordillera: Geology, v. 25, p. 831–834, doi: 10.1130/0091-7613(1997)025<0831:POTCIC>2.3.CO;2.

Nathan, S., and Fruchter, J.S., 1974, Geochemical and paleomagnetic stratigraphy of the Picture Gorge and Yakima basalts (Columbia River Group) in central Oregon: Geological Society of America Bulletin, v. 85, p. 63–76, doi: 10.1130/0016-7606(1974)85<63:GAPSOT>2.0.CO;2.

O'Hara, M.J., and Mathews, R.E., 1981, Geochemical evolution in an advancing, periodically replenished, periodically tapped, continuously fractionated magma chamber: Journal of the Geological Society, London, v. 138, p. 237–277.

Parsons, T., Thompson, G.S., and Sleep, N.H., 1994, Mantle plume influence on the Neogene uplift and extension of the U.S. western Cordillera: Geology, v. 22, p. 83–86, doi: 10.1130/0091-7613(1994)022<0083:MPIOTN>2.3.CO;2.

Pierce, K.L., and Morgan, L.A., 1992, The track of the Yellowstone hot spot: Volcanism, faulting, and uplift, *in* Link, P.K., et al., eds., Regional geology of eastern Idaho and western Wyoming: Boulder, Colorado, Geological Society of America Memoir 179, p. 1–53.

Pierce, K.L., Morgan, L.A., and Saltus, R.W., 2002, Yellowstone plume head: Postulated tectonic relations to the Vancouver slab, continental boundaries, and climate, *in* Bonnichsen, B., et al., eds., Tectonic and magmatic evolution of the Snake River Plain Volcanic Province: Idaho Geological Survey Bulletin 30, p. 5–33.

Ramos, F.C., Wolff, J.A., and Tollstrup, D.L., 2005, Sr isotope disequilibrium in Columbia River flood basalts: Evidence for rapid, shallow-level open-system processes: Geology, v. 33, p. 457–460, doi: 10.1130/G21512.1.

Reidel, S.P., 1983, Stratigraphy and petrogenesis of the Grande Ronde Basalt from the deep canyon country of Washington, Oregon and Idaho: Geological Society of America Bulletin, v. 94, p. 519–542, doi: 10.1130/0016-7606(1983)94<519:SAPOTG>2.0.CO;2.

Reidel, S.P., 1984, The Saddle Mountains—The evolution of an anticline in the Yakima fold belt: American Journal of Science, v. 284, p. 942–978.

Reidel, S.P., 1998, Emplacement of Columbia River Basalt: Journal of Geophysical Research, v. 103, p. 27,393–27,410, doi: 10.1029/97JB03671.

Reidel, S.P., 2005, A lava flow without a source: The Cohassett flow and its compositional components, Sentinel Bluffs Member, Columbia River Basalt Group: Journal of Geology, v. 113, p. 1–21, doi: 10.1086/425966.

Reidel, S.P., and Fecht, K.R., 1987, The Huntzinger flow: Evidence of surface mixing of the Columbia River basalt and its petrogenetic implications: Geological Society of America Bulletin, v. 98, p. 664–677, doi: 10.1130/0016-7606(1987)98<664:THFEOS>2.0.CO;2.

Reidel, S.P., and Tolan, T.L., 1992, Eruption and emplacement of flood basalt: An example from the large-volume Teepee Butte Member, Columbia River Basalt Group: Geological Society of America Bulletin, v. 104, p. 1650–1671, doi: 10.1130/0016-7606(1992)104<1650:EAEOFB>2.3.CO;2.

Reidel, S.P., Tolan, T.L., Hooper, P.R., Beeson, M.H., Fecht, K.R., Bentley, R.D., and Anderson, J.L., 1989, The Grande Ronde Basalt, Columbia River Basalt Group: Stratigraphic descriptions and correlations in Washington, Oregon and Idaho, *in* Reidel, S.P., and Hooper, P.R., eds., Volcanism and tectonism in the Columbia River flood-basalt province: Boulder, Colorado, Geological Society of America Special Paper 239, p. 293–306.

Reidel, S.P., Campbell, N.P., Fecht, K.R., and Lindsey, K.A., 1994, Late Cenozoic structure and stratigraphy of south-central Washington, *in* Lasmanis, R., and Cheney, E.S., eds., Regional geology of Washington State: Washington Division of Geology and Earth Resources Bulletin, v. 80, p. 159–180.

Richards, M.A., Duncan, R.A., and Courtillot, V., 1989, Flood basalts and hotspot tracks: Plume heads and tails: Science, v. 246, p. 103–107, doi: 10.1126/science.246.4926.103.

Riddihough, R., Finn, C., and Couch, R., 1986, Klamath–Blue Mountain lineament, Oregon: Geology, v. 14, p. 528–531, doi: 10.1130/0091-7613(1986)14<528:KMLO>2.0.CO;2.

Ross, M.E., 1980, Tectonic controls of topographic development within Columbia River basalts in a portion of the Grande Ronde River–Blue Mountains region, Oregon and Washington: Oregon Geology, v. 42, p. 167–174.

Ross, M.E., 1983, Chemical and mineralogical variations within four dikes of the Columbia River Basalt Group, southeastern Columbia Plateau: Geological Society of America Bulletin, v. 94, p. 1117–1126, doi: 10.1130/0016-7606(1983)94<1117:CAMVWF>2.0.CO;2.

Ross, M.E., 1989, Stratigraphic relationships of subaerial, invasive, and intracanyon flows of Saddle Mountains Basalt in the Troy Basin, Oregon and Washington, *in* Reidel, S.P., and Hooper, P.R., eds., Volcanism and tectonism in the Columbia River flood-basalt province: Boulder, Colorado, Geological Society of America Special Paper 239, p. 131–142.

Saltus, R.W., and Thompson, G.A., 1995, Why is it downhill from Tonopah to Las Vegas?: A case for a mantle plume support of the high northern Basin and Range: Tectonics, v. 14, p. 1235–1244, doi: 10.1029/95TC02288.

Saunders, A.D., 2005, Large igneous provinces: Origin and environmental consequences: Elements, v. 1, p. 259–263.

Self, S., Thordarson, T., Keszthelyi, L., Walker, G.P.L., Hon, K., Murphy, M.T., Long, P.E., and Finnemore, S., 1996, A new model for the emplacement of Columbia River basalt as large inflated pahoehoe lava sheets: Geophysical Research Letters, v. 23, p. 2689–2692, doi: 10.1029/96GL02450.

Self, S., Keszthelyi, L., and Thordarson, T., 1998, The importance of pahoehoe: Annual Review of Earth and Planetary Sciences, v. 26, p. 81–110, doi: 10.1146/annurev.earth.26.1.81.

Shaw, H.R., and Swanson, D.A., 1970, Eruption and flow rates of flood basalts, *in* Gilmour, E.H., and Stradling, D.F., eds., Proceedings of the Second Columbia River Basalt Symposium: Cheney, Eastern Washington State College Press, p. 271–299.

Smith, A.D., 1992, Back-arc convection model for Columbia River basalt genesis: Tectonophysics, v. 207, p. 269–285, doi: 10.1016/0040-1951(92)90390-R.

Smith, G.A., 1988, Neogene syn-volcanic and syn-tectonic sedimentation in central Washington: Geological Society of America Bulletin, v. 100, p. 1479–1492, doi: 10.1130/0016-7606(1988)100<1479:NSASSI>2.3.CO;2.

Smith, R.B., Jordan, M., Puskas, C., Waite, G., and Farrell, J., 2005, Geodynamic models of the Yellowstone hotspot constrained by seismic tomography: Geological Society of America Abstracts with Programs, v. 37, p. 126.

Sobczyk, S.M., 1994, Crustal thickness and structures of the Columbia Plateau using geophysical methods [Ph.D. thesis]: Pullman, Washington State University, 208 p.

Swanson, D.A., Wright, T.L., and Helz, R.T., 1975, Linear vent systems and estimated rates of magma production and eruption for the Yakima basalt on the Columbia Plateau: American Journal of Science, v. 275, p. 877–905.

Swanson, D.A., Wright, T.L., Hooper, P.R., and Bentley, R.D., 1979, Revisions in stratigraphic nomenclature of the Columbia River Basalt Group: U.S. Geological Survey Bulletin 1457-G, 59 p.

Swanson, D.A., Wright, T.L., Camp, V.E., Gardner, J.N., Helz, R.T., Price, S.M., Reidel, S.P., and Ross, M.E., 1980, Reconnaissance geologic map of the Columbia River Basalt Group, Pullman and Walla Walla Quadrangles, southeast Washington and adjacent Idaho: Reston, Virginia, U.S. Geological Survey Miscellaneous Investigations Map I-1139, scale 1:250,000, 2 sheets.

Swanson, D.A., Anderson, J.L., Camp, V.E., Hooper, P.R., Taubeneck, W.H., and Wright, T.L., 1981, Reconnaissance geologic map of the Columbia River Basalt Group, northern Oregon and western Idaho: Reston, Virginia, U.S. Geological Survey Open-File Report 81-797, scale 1:250,000, 33 p., 6 sheets.

Swisher, C.C., Ach, J.A., and Hart, W.K., 1990, Laser fusion $^{40}Ar/^{39}Ar$ dating of the type Steens Mountain Basalt, southeastern Oregon, and the age of the Steens geomagnetic polarity transition: Eos (Transactions, American Geophysical Union), v. 71, Fall Meeting Supplement, p. 1296.

Takahashi, E., Nakajima, K., and Wright, T.L., 1998, Origin of the Columbia River basalts: Melting model of a heterogeneous plume head: Earth and Planetary Science Letters, v. 162, p. 63–80, doi: 10.1016/S0012-821X(98)00157-5.

Taubeneck, W.H., 1970, Dikes of Columbia River basalt in northeastern Oregon, western Idaho, and south-eastern Washington, *in* Gilmour, E.H., and Stradling, D., eds., Proceedings of the Second Columbia River Basalt Symposium: Cheney, Eastern Washington State College Press, p. 73–96.

Thompson, R.N., and Gibson, S.A., 1991, Subcontinental mantle plumes, hotspots, and pre-existing thinspots: Journal of the Geological Society, London, v. 148, p. 973–977.

Thordarson, T., and Self, S., 1998, The Roza Member, Columbia River Basalt Group: A gigantic pahoehoe lava flow field formed by endogenous processes?: Journal of Geophysical Research, v. 103, p. 27,411–27,445, doi: 10.1029/98JB01355.

Thorkelson, D.J., 1989, Eocene sedimentation and volcanism in the Fig Lake Graben, southwestern British Columbia: Canadian Journal of Earth Sciences, v. 26, p. 1368–1373.

Tolan, T.L., Reidel, S.P., Beeson, M.H., Anderson, J.L., Fecht, K.R., and Swanson, D.A., 1989, Revisions to the estimates of the areal extent and volume of the Columbia River Basalt Group, *in* Reidel, S.P., and Hooper, P.R., eds., Volcanism and tectonism in the Columbia River flood-basalt province: Boulder, Colorado, Geological Society of America Special Paper 239, p. 1–20.

Vallier, T.L., and Brooks, H.C., eds., 1987, Geology of the Blue Mountains Region of Oregon, Idaho and Washington: The Idaho Batholith and its border zone: Reston, Virginia, U.S. Geological Survey Professional Paper 1436, 196 p.

Vallier, T.L., and Brooks, H.C., eds., 1995, Geology of the Blue Mountains Region of Oregon, Idaho and Washington: Petrology and tectonic evolution of pre-Tertiary rocks of the Blue Mountains region: U.S. Geological Survey Professional Paper 1438, 540 p.

Walker, G.W., 1979, Revisions to the Cenozoic stratigraphy of Harney Basin, southeastern Oregon: U.S. Geological Survey Bulletin 1475, 35 p.

Walker, G.W., and Nolf, B., 1981, High Lava Plains, Brothers fault zone to Harney basin, Oregon, *in* Johnston, D.A., and Donnelly-Nolan, J., eds., Guides to some volcanic terranes in Washington, Idaho, Oregon and northern California: Reston, Virginia, U.S. Geological Survey Circular 838, p. 105–111.

Waters, A.C., 1961, Stratigraphic and lithologic variations in the Columbia River Basalt: American Journal of Science, v. 259, p. 583–611.

Weinberg, R.F., 1997, Rising of starting plumes through mantle of temperature-, pressure-, and stress-dependent viscosity: Journal of Geophysical Research, v. 102, p. 7613–7623, doi: 10.1029/97JB00266.

Wells, R.E., and Heller, P.L., 1988, The relative contribution of accretion, shear, and extension to Cenozoic tectonic rotation in the Pacific Northwest: Geological Society of America Bulletin, v. 100, p. 325–338, doi: 10.1130/0016-7606(1988)100<0325:TRCOAS>2.3.CO;2.

Wells, R.E., Simpson, R.W., Bentley, R.D., Beeson, M.H., Mangan, M.T., and Wright, T.L., 1989, Correlation of Miocene flows of the Columbia River Basalt Group from the central Columbia River Plateau to the coast of Oregon and Washington, in Reidel, S.P., and Hooper, P.R., eds., Volcanism and tectonism in the Columbia River flood-basalt province: Boulder, Colorado, Geological Society of America Special Paper 239, p. 113–130.

Wells, R.E., Weaver, C.S., and Blakely, R.J., 1998, Fore-arc migration in Cascadia and its neotectonic significance: Geology, v. 26, p. 759–762, doi: 10.1130/0091-7613(1998)026<0759:FAMICA>2.3.CO;2.

White, R.S., and McKenzie, D.P., 1989, Magmatism at rift zones: The generation of volcanic continental margins and flood basalts: Journal of Geophysical Research, v. 94, p. 7685–7729.

White, R.S., and McKenzie, D.P., 1995, Mantle plumes and flood basalts: Journal of Geophysical Research, v. 100, p. 17,543–17,586, doi: 10.1029/95JB01585.

Woodcock, N., and Fischer, M., 1986, Strike-slip duplexes: Journal of Structural Geology, v. 8, p. 725–735, doi: 10.1016/0191-8141(86)90021-0.

Wright, T.L., Swanson, D.A., Helz, R.T., and Byerly, G.R., 1979, Major oxide, trace element, and glass chemistry of Columbia River basalt samples collected between 1971 and 1977. Reston, Virginia, U.S. Geological Survey Open-File Report 79–711, 146 p.

Wright, T.L., Mangan, M., and Swanson, D.A., 1989, Chemical data for flows and feeder dikes of the Yakima basalt subgroup, Columbia River Basalt Group, Washington, Idaho and Oregon, and their bearing on a petrogenetic model: U.S. Geological Survey Bulletin 1821, 71 p.

Yaxley, G.M., 2000, Experimental study of the phase and melting relations of homogeneous basalt + peridotite mixtures and implications for the petrogenesis of flood basalts: Contributions to Mineralogy and Petrology, v. 139, p. 326–338, doi: 10.1007/s004100000134.

Yuan, H., and Dueker, K., 2005, Teleseismic P-wave tomogram of the Yellowstone plume: Geophysical Research Letters, v. 32, p. LO7304, doi;1029/2004GL022056.

Zoback, M.L., McKee, E.H., Blakely, R.J., and Thompson, G.A., 1994, The northern Nevada rift: Regional tectono-magmatic relations and middle Miocene stress direction: Geological Society of America Bulletin, v. 106, p. 371–382, doi: 10.1130/0016-7606(1994)106<0371:TNNRRT>2.3.CO;2.

MANUSCRIPT ACCEPTED BY THE SOCIETY JANUARY 31, 2007

DISCUSSION

7 January 2007, Alan D. Smith

I would contend that any unifying model for Late Cenozoic volcanism in the Pacific Northwest would be better based on a back-arc model, with the origin of the Columbia River Basalt Group (CRBG), Oregon plateau volcanism, and Chilcotin basalts in the Canadian Cordillera linked to asthenospheric upwelling behind the Cascade–Pemberton arc (Fig. D-1). The presence of OIB-like compositions in the CRBG does not indicate a plume origin: in nonplume models, subducted oceanic crust is remixed into the convecting mantle such that the recycled components otherwise attributed to plumes are ubiquitous in the asthenosphere (Meibom and Anderson, 2004). Contentions that helium isotopes, melt volumes, and eruption rates require a plume origin for large igneous provinces have been addressed by Anderson (2005a,b). Rather, the volume of the CRBG can be ascribed to the coincidence of back-arc upwelling with a region of thin lithosphere in an embayment in the Precambrian basement (Fig. D-1). Lithospheric thinning occurred mainly in the Eocene and Oligocene. Thus, convective upwelling, not extension, acts as the trigger for volcanism (Smith, 1992).

The eruptive centers of the Imnaha, Grande Ronde, and Wanapum formations occur in an area approximately 225 by 275 km (Fig. D-1). The volume of the CRBG could be met by 6–10% melting of an asthenosphere/continental mantle section 40–60 km thick under such an area in a back-arc model. Hooper et al. (this volume) dismiss continental mantle source components: "Recent work on Re/Os isotopic ratios (Hart et al., 1997; Chesley and Ruiz, 1998) rules out an SCLM component in the Imnaha basalt. . . ." The study by Hart et al. (1997) concerned the Oregon plateau and Snake River provinces, whereas Chesley

and Ruiz (1998) studied the CRBG. Both modeled generation of Os isotope ratios from subcontinental lithospheric mantle (SCLM) peridotite contaminated with metasomatic components including vein pyroxenites and subduction-derived fluids. However, the mixing lines are controlled not by the composition of the metasomatic components, which was the only parameter varied in both studies, but by the composition of the continental mantle on account of the high Os contents assumed therein (3 to 3.5 ppb Os).

Chesley and Ruiz (1998) also assumed $^{187}Os/^{188}Os = 0.110$ for the continental mantle, which would be appropriate for ancient cratonic mantle but not for young lithosphere beneath accreted terranes. Mantle xenoliths from further north in the Cordillera suggest $^{187}Os/^{188}Os = 0.129$, and 1.37 ppb Os would be more reasonable for continental mantle under the latter. Substituting the xenolith values into the calculation of Chesley and Ruiz (1998) allows generation of the Os isotopic composition of the Imnaha basalts from continental mantle containing 20–30% pyroxenite (Smith, 2003). Adopting the xenolith composition also relaxes the requirement for minimal involvement of continental mantle in the modeling of Hart et al. (1997). The Os isotope systematics of basalts from the Columbia River and Oregon plateau provinces are consistent with crustal contamination, but they do not preclude continental mantle sources.

The contemporaneous eruption of CRB and calc-alkaline volcanism in graben systems (presumably Powder River volcanism; Hooper et al., this volume) is readily explained as a result of compositional variations in the continental mantle section in a back-arc model. Rather, it is the plume model that would appear to encounter difficulties in explaining the eruption of different magma types from volcanic centers a few tens of

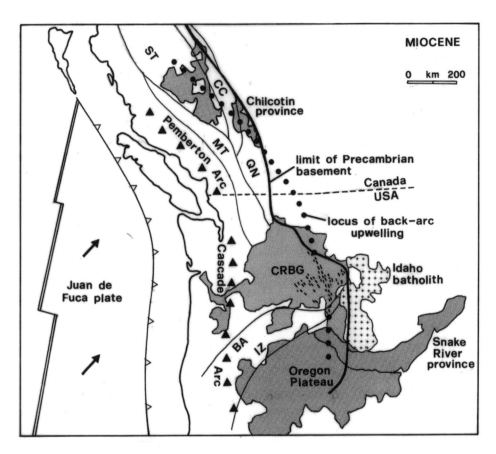

Figure D-1. Relationship of CRBG (eruptive centers of the Imnaha, Grande Ronde, and Wanapum formations indicated by dashed lines), Oregon plateau, and Chilcotin volcanism to the Pemberton and Cascade arcs in the Miocene. Note the absence of volcanism where back-arc upwelling occurred beneath Precambrian basement. Accreted terranes: BA, Baker; CC, Cache Creek; IZ, Izee; MT, Methow-Tyaughton; QN, Quesnel; ST, Stikine.

kilometers apart, when according to Camp and Ross (2004), the asthenosphere beneath the eastern half of the Columbia River province had been replaced with plume material by 15 Ma.

13 January 2007, Don L. Anderson

Before the mantle plume model was widely invoked, intraplate volcanism was attributed to shallow mantle and lithospheric processes. These included small-scale convection and propagating fractures (e.g., Smith, this volume) and a partially molten low-velocity zone. Continental basalts had a variety of shallow explanations involving the crust, lithosphere, and asthenosphere. Geologists sought to understand intraplate volcanism within the framework of geology and plate tectonics. With increasing popularity of the hotspot model, emphasis shifted to the core–mantle boundary and interpreting all intraplate volcanics as the result of mantle plumes. Other chapters and comments in this volume describe alternative models and give excellent descriptions of how normal plate tectonic processes (the plate hypothesis) can explain many features that are routinely attributed to plumes. Plate reorganizations, nonrigid plates, heterogeneous mantle, and abandonment of ridges, trenches, and back-arc basins are all involved. These processes are either ignored in the plume hypothesis or are attributed to plumes.

In the Pacific Northwest of North America we have ridges, ridge–trench collisions, young plate subduction, accreted ter-

ranes, sutures, slab windows, back-arc basins, batholiths, rifts, extension, and edges of cratons. We also have the possibilities and suggestions of delamination, fluids from an underlying slab, and melts from the mantle wedge and crust. With all of these processes and materials, why are deep mantle plumes also invoked? The reasons given are similarities to OIB, high $^3He/^4He$ ratios (R), short duration and uplift during volcanism, and the consensus of workers on the ground (Hooper et al., this volume). The plume model involves taking the material that is found in abundance around ridges, trenches, and collisional belts down to the core–mantle boundary and then bringing it back up to the same sorts of places, sometimes through an intervening slab. The plate model focuses on the upper boundary layer and processes in the crust, mantle wedge, and asthenosphere.

One has to be careful when interpreting continental flood basalts (CFBs) in terms of a deep plume, to avoid circular reasoning (http://www.mantleplumes.org/CRB.html). This is particularly true when chemical arguments are used, including R ($^3He/^4He$ ratios; Anderson, 2000a,b, 2001). Hooper et al. (this volume) argue in favor of a plume origin for the CRB using a series of oft-repeated assertions, some of them circular (see comments following chapter by Sheth, this volume). Arguments based on helium, large volumes, OIB-type chemistry, uplift concurrent with volcanism, and short duration of magmatism are particularly unfortunate examples. These arguments have been raised and refuted many times (see reviews by Anderson, 2005a,b;

Anderson and Natland, 2005; Smith, 1992; and http://www.mantleplumes.org/CRBDelam.html).

The arguments boil down to this: only plumes have rapid uplift, high R, and short durations. The CRB region has these characteristics and is therefore caused by a plume. The circularity of the short-duration argument is as follows. All plumes were assigned the property of high eruption rates only because CFBs were assumed to be plumes and these had high eruption rates (Richards et al., 1989). Hooper et al. (this volume) and many others now argue that because the CRBs are of short duration, they must therefore result from a plume.

OIB-like chemistry and high-R basalts have been found along mid-ocean ridges and in back-arc basins. In some cases, these have been attributed to hotspots thousands of kilometers away. Some basalts found along mid-ocean ridges (E-MORB) resemble OIB. The terms E-MORB, OIB, and OIB-like are used to describe alkalic olivine basalt on ridges, seamounts, islands, CFBs, and continental rifts (Natland comment 10 January 2007 on Fitton, this volume). OIB-like signatures occur at ridges and thousands of small seamounts (Natland and Winterer, 2005). Many oceanic hotspots are known to carry continental signatures that are plausibly related to delaminated continental crust or the continental fragment on which they erupted (Anderson, this volume; Natland, this volume). OIB-like chemistry cannot then be used to argue for a deep mantle plume, or even a hotspot, source.

High R was originally thought to be a proxy for high ^{3}He content and "therefore" for an undegassed primordial lower mantle origin; this is the helium paradox or fallacy. High-R basalts were attributed to plumes because Iceland, Yellowstone, and Hawaii had high-R magmas or gases, and these were "known to be from plumes." Most hotspots have helium isotope statistics that are indistinguishable from spreading ridges, and they have much lower ^{3}He contents than MORB (Anderson, 2001), consistent with a shallow origin in cumulates or restites.

The rates of CRB magmatism are comparable to the rates of arc magmatism and delamination (per kilometer of length) (Jicha et al., 2006). Asthenosphere from the mantle wedge upwells and displaces the dense foundering continental root at the same rate it is removed. This is in addition to the crustal space made available by spreading or stretching. Uplift accompanies delamination and asthenospheric upwelling. Foundered or delaminated material can be a local source for continental or "plume" signatures and of low-melting eclogite. Thus, underplating, uplift, large volumes, eclogite in the source, and the isotopic signatures of the CRB do not require deep mantle sources.

14 January 2007, James H. Natland

Accumulation of the great volume of CRB east of the active Cascades arc means that they erupted in some type of back-arc setting. Hooper et al. (this volume) are greatly knowledgeable about the CRB but seem unaware that western Pacific back-arc basins have many of the geochemical and other attributes that, for the CRB, they attribute to plumes. In my opinion, combining a plume with a back-arc setting simply beggars the imagination.

Figures in Hooper et al. (this volume) suggest that the average thickness of CRB before erosion was ~1.2 km. The Lau basin west of the Tonga arc covers roughly the same area (Taylor et al., 1996) and consists of ocean crust several kilometers thick that formed over the past ca. 4 Ma. Although only about a third of this is basalt flows and dikes, the full volume of ocean crust implies a rate of magmatism perhaps three times that of the CRB. The Lau basin, of course, is intraoceanic, and the basalts there erupted at spreading ridges, whereas the CRB erupted through continental crust. However, this does not negate comparisons of magmatic productivity; it simply says that the forms of magmatic activity, the tectonic controls on it, and probably the fertility of the source were different.

Lau basin basalt geochemistry indicates the influence of distinctive mantle components, including (1) a depleted MORB component, (2) an Indian Ocean component, (3) a Samoan component, and (4) a Louisville Ridge component (e.g., Volpe et al., 1988; Hickey-Vargas, 1998; Turner and Hawkesworth, 1998). Other back-arc basins are similarly complex (e.g., Stern et al., 1990; Taylor and Martinez, 2003). The Louisville component is attributed to subduction of volcaniclastic material on the Pacific plate, and the Samoan component to infiltration of Samoan mantle through a tear in the Pacific plate deep beneath the northern end of the Lau basin. But in neither case is a plume within the basin required. I also believe that Samoa results from lithospheric fracture, not a plume (Natland, 1980; http://www.mantleplumes.org/Samoa; Stuart et al., this volume; Natland and Winterer 12 January comment on Stuart et al., this volume). The Indian Ocean component is a regional signal originally produced by delamination of lower continental crust and dispersal of the Gondwana continents (e.g., Meyzen et al., 2005) but is present in the Lau basin because asthenosphere is drawn from the west into the convecting regime of the Tonga arc and Lau basin (Hickey-Vargas, 1998; Turner and Hawkesworth, 1998; Smith et al., 2001). Again, no plume is involved.

Although Hooper et al. (this volume) only briefly mention He isotopes, they assume the opinion of isotope geochemists that $R > 8$ requires a plume. One could presume that the geochemists are right but still ask whether a plume within the basin was necessary. After all, basalt with the Samoan signature in the Lau basin has such a signature (Poreda and Craig, 1992). But here I side with Anderson (1998; 13 January comment to this chapter) that sources with helium isolated from U and Th and with low intrinsic ^{3}He concentrations are likely to be important. This is because the sources of the CRB, with so many of its rocks having a continental isotopic signature, likely include cumulates produced during the ancient magmatism that created continental crust.

Cumulates, especially those with olivine, are mineral aggregates that contain volatiles originally trapped as bubbles during crystallization (e.g., Natland, 2003). Foundering or delamination of such cumulates (e.g., Daly, 1926; Jull and Kelemen, 2001; Anderson, this volume) beneath an oceanic island, an island arc, or an ancient batholith is sufficient to carry the volatiles into the lower lithosphere and convecting upper mantle, and any of these materials, transported into the melt domain by any mechanism, could be present beneath the CRB. The high-*R* source could even still be intact in the subcontinental lithosphere beneath Nevada near where the CRB originated. High *R* is simply an indication of the age of the original melting event, carried forward to the present day without affiliated Th and U to modify it in a cumulate time capsule. It is a simple matter of efficient separation of the vapor phase from the liquid and into a solid, having little to do, as Parman et al. (2005) have argued, with partitioning of U and Th into olivine from the melt. As such, it is far less a signature of a source in the lower mantle than of a place near the Earth's surface where volatile exsolution can occur. The CRB also includes both low-Ti and higher-Ti basalt (see Natland, this volume), with the latter likely requiring a titanian phase (ilmenite or rutile) in the melt source. That is, this component was not mantle peridotite but, more likely, eclogite (Takahashi et al., 1998), as agreed by Hooper et al. (this volume). A general link between high *R* in the CRB and eclogite in the source is indicated, but this is by no means proof of the existence of a mantle plume (Anderson, this volume; Natland, this volume).

17 January 2007, Alexei V. Ivanov

The Columbia River flood basalt province is located in a back-arc tectonic setting but is interpreted by Hooper et al. (this volume) among many others in the framework of the plume model. Two major plume-proponent arguments—the high ^{3}He/^{4}He and a large volume of lava erupted over a short period of time—have been suggested to be circular arguments rather than proplume evidence by Anderson (13 January) and Natland (14 January). I do not consider this point here.

Hooper et al. (this volume) write: "The largest CRBG formations (lower Steens, Imnaha, Grande Ronde, and Wanapum basalt formations) suggest a primary mantle source akin to that for OIBs (ocean island basalts), a view consistent with Hawaiian-like trace-element profiles (Hooper and Hawkesworth, 1993)." This is incorrect, at least in terms of Grande Ronde Formation and misleading for further evaluation of plume and nonplume models.

In Figure D-2, I plot trace-element data for 36 samples of the Grande Ronde basaltic andesites (analyses are taken from Hooper and Hawkesworth, 1993). They are characterized by relatively uniform trace-element patterns, which are significantly different from either modeled primitive OIB composition (Sun and McDonough, 1989) or from a typical Hawaiian basalt sample BHVO-1, which has served for many years as USGS reference material (http://minerals.cr.usgs.gov/geo_chem_stand/basaltbhvo1.html). Important features of OIB are Nb enrichment relative to K and La and Pb depletion relative to Ce and Pr on primitive mantle normalized diagrams (Fig. D-2A). Grande Ronde basaltic andesites show the exact opposite trace-element patterns. Thus, the sentence quoted above from Hooper et al. (this volume) expresses their interpretation that original OIB-like (plume) melts were contaminated by lithospheric material to produce the Grande Ronde basaltic andesites. This interpretation may or may not be correct. I do not evaluate the arguments used in support of this interpretation by Hooper et al. (this volume and earlier papers). The aim of my present comment is to show that the trace-element geochemistry of the voluminous Grande Ronde basaltic andesites can easily be explained by subduction-related processes and does not require a plume.

In Figure D-2B, the Grande Ronde basaltic andesites are compared with typical island arc basalts (IAB). Primitive IAB trace-element patterns are an average from high-Mg basalts of the most productive volcano of the northern hemisphere—the

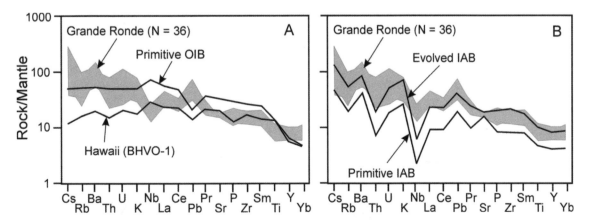

Figure D-2. Primitive-mantle-normalized diagram for Grande Ronde basaltic andesites compared with OIB (A) and IAB (B) reference compositions. Grande Ronde basaltic andesites are from Hooper and Hawkesworth (1993). Primitive OIB is from Sun and McDonough (1989). BHVO-1 is from Dulski (2001). Elements not analyzed by Dulski (2001) are from http://minerals.cr.usgs.gov/geo_chem_stand/basaltbhvo1.html. Primitive IAB is the average of high-Mg basalts of Klyuchevskoi volcano, Kamchatka (calculated from Dorendorf et al., 2000). Evolved IAB is a modeled composition calculated using the equilibrium crystallization equation of Shaw (1970), primitive IAB as starting melt composition, mineral/melt partition coefficients from Table D-1, and 65% of fractionated olivine, clinopyroxene, and plagioclase in the proportion 0.3:0.45:0.25, respectively.

Klyuchevskoi volcano, Kamchatka, Russia. It may be seen that practically all the troughs and peaks of the primitive IAB are represented in the Grande Ronde basaltic andesites with the exception of the Sr peak (Fig. D-2B). The Grande Ronde basaltic andesites have evolved SiO_2 from 53 wt% to above 57 wt%, and low MgO from 2.9 wt% to 5.9 wt% (e.g., Hooper and Hawkesworth, 1993), reflecting their fractionated nature. The major fractionating minerals in the Grande Ronde primary melts were olivine, clinopyroxene, and plagioclase (Fig. D-3). Subtracting these minerals from the primitive-IAB starting composition via equilibrium crystallization (Shaw, 1970) elevates concentrations of all trace elements except Sr, because Sr is compatible in plagioclase (Table D-1). Modeled evolved IAB composition is almost identical to the Grande Ronde basaltic andesites (Fig. D-2B).

It is possible to calculate the primary melt for the Grande Ronde basaltic andesites backward through addition of olivine, clinopyroxene, and plagioclase to the real Grande Ronde com-

TABLE D-1. MINERAL/MELT PARTITION COEFFICIENTS FOR SELECTED ELEMENTS USED FOR DERIVATION OF TRACE-ELEMENT PATTERNS OF EVOLVED ISLAND ARC BASALT MELT (SEE FIG. D-2B)

	Plagioclase	Clinopyroxene	Olivine
Cs	0.034	0.001	0.0001*
Rb	0.023	0.0047	0.0085
Ba	0.69	0.0007	0.001*
Th	0.064	0.012	0.01*
U	0.078	0.01	0.01*
K	0.05*	0.05*	0.003*
Nb	0.024	0.0077	0.0035
La	0.12	0.054	0.00001*
Ce	0.097	0.086	0.024
Pb	0.44	0.1*	0.0055
Pr	0.077	0.14	0.031
Sr	2.38	0.13	0.012
P	0.1*	0.125*	0.013
Zr	0.00018	0.12	0.015*
Sm	0.048	0.29	0.016
Ti	0.037	0.38	0.015*
Y	0.012	0.47	0.029
Yb	0.0098	0.43	0.053

Note: Values for plagioclase and olivine are based on Dunn and Sen (1994). Values for clinopyroxene are after the compilation of Zack et al. (1997). Asterisks mark extrapolated values.

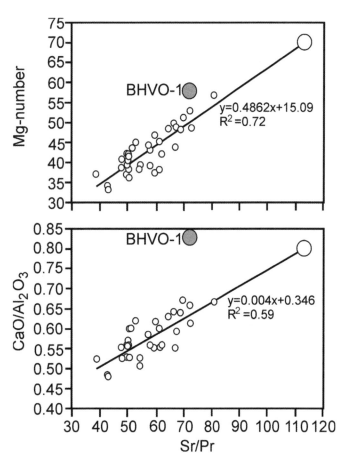

Figure D-3. Mg-number and CaO/Al_2O_3 versus Sr/Pr for Grande Ronde basaltic andesites (small open circles). Primitive mantle is after McDonough and Sun (1995). Mg-number, CaO/Al_2O_3, and Sr/Pr are used as proxies for olivine, clinopyroxene, and plagioclase fractionation, respectively. Large open circles represent plausible primary melt with high Sr/Pr, which reflects Sr enrichment of the mantle source by subduction-derived water fluid. Typical Hawaiian basalt (BHVO-1) is also marked as a large gray circle to show its off-trend position.

positions. This is not shown here, but the primary melt of the Grande Ronde basaltic andesites will be similar to that of the high-Mg basalts of the Klyuchevskoi volcano, differing by somewhat higher Th and Nb concentrations. Thus, the sentence from Hooper et al. (this volume) that I quote above should be modified to read: "The largest CRBG formations (e.g., Grande Ronde Formations) suggest a primary mantle source akin to that for IABs, a view consistent with island-arc trace element profiles."

The Columbia River flood basalt province is close to a subduction system, and thus IAB mantle source is expected, unlike the enigmatic plume source (Smith, 1992; comment by Smith, 7 January 2007). Subduction may be responsible not only for flood basalts situated near convergent boundaries but also for flood basalts at distances of 1000–2000 km away, such as the Siberian Traps (Ivanov, this volume).

20 January 2007, Ajoy K. Baksi

Hooper et al. (this volume), present a model arguing for a hotspot-related genesis for the CRB. Supporting evidence is based on geochronologic data and follows on earlier work (Hooper et al., 2002; Hooper, 2004). All pertinent radiometric data are critically reviewed for statistics and alteration of rock dates (Baksi, 1999, 2005, this volume, chapters 15 and 16). Errors are quoted at the 1σ level.

Two sets of data are cited by Hooper (2004), namely Duncan (pers. comm., 2003) and Hooper et al. (2002). For the former, no data tables are available for inspection. The ages cited appear to be valid from the statistical point of view. It is not possible to quantitatively assess the state of alteration of the rocks dated by Duncan. The very large error (±3.9 Ma) for the isochron age of BUK-5 suggests the rock contained large amounts of atmospheric argon and is altered. The ca. 15.5–16.0 Ma ages listed by Hooper (2004) for the Imnaha basalt cannot be critically evaluated for lack of isotopic data tables. These $^{40}Ar/^{39}Ar$ ages are in general agreement with the K-Ar dates determined in the 1970s, all of which were on altered whole-rock material (see Baksi, 1989). The ages determined by Duncan appear to be on altered material and do not serve as accurate measures of the time of crystallization.

Details of ages determined at the Open University were obtained (P.R. Hooper, pers. comm., 2003). Most of the these ages—listed in Hooper et al. (2002), Table 1—cannot be treated as accurate estimates of the time of crystallization of CRB rocks. First, the majority of the "ages" are replicate $^{40}Ar/^{39}Ar$ total fusion analyses. These are the equivalent of K-Ar dates, and do not address potential problems of partial $^{40}Ar*$ loss (by alteration?) and/or the presence of excess argon. The alteration index (A.I.) of these samples is three to ten times higher than the cutoff values for freshness (<0.0006 for basalts, <0.00006 for plagioclase). Seven samples were analyzed by the $^{40}Ar/^{39}Ar$ step-heating technique (Hooper et al., 2002). Four of these (SK078, KL046, MP028, and MP050) yield high A.I. values and are altered. Specimen KL333 appears to be relatively fresh, but its listed age (15.8 ± 2.8 Ma) is of very low precision and does not help narrow down the age of sections of the CRB. KL276 and SC054 have low A.I. values and gave plateau ages of 16.6 ± 0.1 and 15.9 ± 0.2 Ma, respectively. These two ages were rejected by Hooper et al. (2002), as "invalid because (they are) older than stratigraphic age." It is not clear what is meant by "stratigraphic age": the rest of the ages listed by these authors? The latter were determined on altered material and are underestimates of the correct age. The ages of KL276 and SC054 appear to be good estimates of the time of crystallization. There is no radiometric evidence that ">90% of the CRB (was) erupted between 16.1 and 15.0 Ma" (Hooper et al., 2002).

Hooper (2004) suggested that the ages of Baksi and Farrar (1990) are not in agreement with those determined by three other sets of scientists. This is not the case for the Berkeley Geochronology Center. Swisher et al. (1990) obtained ages of 16.6–16.5 Ma for the Steens basalt. The corresponding age of Baksi and Farrar (1990), when corrected to the calibrations preferred by Renne et al. (1998), is 16.5 Ma. As shown above, few if any of the Open University ages, as well as those determined by Duncan on the CRB, are valid estimates of the crystallization age. There are problems in the interpretation of the Baksi and Farrar (1990) ages with respect to the geomagnetic polarity time scale (see Baksi, 1993; Cande and Kent, 1995; Wilson and Gans, 2003). These may stem from incorrect ages and/or interpretation of the magnetostratigraphy of the CRB. For the latter, it was assumed that all reversals of the geomagnetic field in this time period (ca. 17–15 Ma) are trapped in CRB

lavas (see Baksi and Farrar, 1990; Baksi, 1993). The relevant $^{40}Ar/^{39}Ar$ analyses need to be critically evaluated by the A.I. technique. Unfortunately, the primary listing of the isotopic data sets was lost when (floppy) computer disks were damaged; the original charts related to the analyses at Queen's University in 1989 need to be reexamined. Some of the splits analyzed by Baksi and Farrar (1990) may have been altered because the powdered rocks were not washed in dilute nitric acid before analysis (see Baksi, this volume, chapters 15 and 16).

All data utilized to support hypotheses should be made available for critical examination. In particular, they should not be listed as "personal communication" or simply comprise unsupported numbers in abstracts.

28 January 2007, Peter Hooper, Vic Camp, Steve Reidel, and Marty Ross

We wish to thank Smith, Anderson, Natland, Ivanov, and Baksi for their comments on our chapter. All authors comment on the ambiguity of geochemical interpretations, asserting that OIB signatures and high $^{3}He/^{4}He$ ratios are the result of shallow-mantle processes unrelated to a plume origin. Smith and Anderson note that that the presence of OIB-like compositions does not indicate a plume origin. The largest CRBG formations do indeed have trace-element and isotopic ratios consistent with an original OIB-like source. However, we did not and do not claim that the chemical similarity of the Imnaha basalt to those of Hawaii proves a plume origin; only that such a similarity is necessary to such an origin.

We do not "dismiss continental mantle source components" as claimed by Smith. Rather we emphasize the role of a lithospheric component in mixing arrays for both the Picture Gorge and the Grande Ronde basalts and state "the variability of ratios such as Ba/Nb are best explained by a mixing array between the original mantle component and a component with a lithospheric geochemical signature. The lithospheric component is characterized by a conspicuous negative Nb-Ta anomaly (Hooper and Hawkesworth, 1993) and could represent either an enriched subcontinental lithospheric mantle (SCLM) or mafic lower crust" (see section "Other Source Components" in Hooper et al., this volume). We excepted the Imnaha basalt from having a lithospheric component, both because Chesley and Ruiz (1998) rightly or wrongly had used $^{187}Os/^{188}Os$ to discount SCLM as a component for these early flows and because Imnaha chemical trends lack the Ba/Nb enrichment and thus conform to normal partial melting and crystal fractionation trends. Hooper and Hawkesworth (1993) had demonstrated that the enrichment in the Grande Ronde basalt could not be a result of normal crystal fractionation, a conclusion previously reached by Wright et al. (1989) based on their chemical and experimental studies.

Ivanov argues that the Grande Ronde lavas could have been derived from subduction-related processes. Although the Imnaha basalt chemical trends conform to normal partial melting and crystal fractionation processes (Hooper et al., this volume,

Fig. 10), those of the Grande Ronde basalt do not (Hooper et al., this volume, Fig. 8A,B; Wright et al., 1989; Hooper and Hawkesworth, 1993). We have always been conscious of an important lithospheric component akin to calc-alkaline material probably derived from a subcontinental mantle enriched in the earlier subduction process (Hooper and Hawkesworth, 1993) and reactivated by the flood basalt event. To imply that we regard the Grande Ronde basalt as derived entirely from a single source similar to the OIBs is to misinterpret our present and previous papers.

28 January 2007, Vic Camp, Peter Hooper,
Steve Reidel, and Marty Ross

We thank Smith, Anderson, and Natland for their comments. Smith reiterates his back-arc model (Smith, 1992) for the genesis of the CRBG as an alternative to the plume model advocated in our chapter. A back-arc origin is also promoted by Natland who points out the similarity of the CRBG to oceanic basalts in western Pacific back-arc basins, which have high magma supply rates and similarly diverse geochemistry. Anderson advocates normal plate-tectonic and shallow-mantle processes for the generation of all continental flood basalts and cautions against the use of oft-repeated assertions and circular reasoning to support a plume genesis for the CRBG.

We maintain that the geological and field data for the CRBG are markedly consistent with the predictions inherent in models of plume impingement and just as strikingly inconsistent with all proposed alternative models. Here, we expand on some of the broader problems with the nonplume interpretations as supported by Anderson and some specific problems associated with the back-arc model advocated by Smith and Natland.

1. *Magma supply rates and duration of magmatism.* Smith, Anderson, and Natland suggest that the high eruption rates of the CRBG main-phase eruptions are no greater than those generated by normal plate-tectonic processes. We agree that the rates of magma accumulation at mid-oceanic ridges, island arcs, and in oceanic back-arc settings may also be considerable although relatively difficult to estimate with any accuracy. In the Aleutian arc, the eruption rates referred to by Anderson (Jicha et al., 2006) are 60–180 km^3/km/Ma. Equivalent calculations for the CRBG lavas erupted from the Chief Joseph Dike swarm are 500–600 km^3/km/Ma, significantly greater than the Aleutian-arc eruption rate. If Jarboe et al. (2006) are correct and the CRBG main phase was erupted in only 0.75 Ma, then the CRBG eruption rate is doubled. However, the real argument lies in high magma supply rates over exceedingly short periods of geological time. Ongoing back-arc extension, which has continued behind the Cascade arc since Eocene times, cannot by itself account for the dramatic, short-lived burst of basalt accumulation at ca.16 Ma.

2. *Flood-basalt volcanism in a continental back-arc setting.*

Volcanic arcs on continental margins are very well documented. It seems unlikely that the Cascade arc is the only one capable of generating a flood-basalt province behind it by back-arc extension. It is clear that the southern Cascade back-arc region was subjected to an unusual mantle-melting event, distinct in size, character, and short duration when compared to all other Phanerozoic examples in continental back-arc settings.

3. *The relationship between extension and magmatism.* The degree of basin-and-range extension behind the Cascade arc decreases from south to north, in marked contrast to the volume of the flood basalts, which increases in the same direction. This inverse relationship suggests that back-arc extension cannot have been the controlling factor to trigger flood-basalt volcanism. One must conclude that it was triggered by the abrupt arrival of a thermal and/or fertile anomaly at mantle depths.

4. *Concept of a unifying model.* Smith contends that any unifying model for Late Cenozoic volcanism in the Pacific Northwest would be better based on a back-arc model than on a plume model. How then does a back-arc model explain the extraordinary spatial and temporal connection of the CRBG flood-basalt province to the Snake River Plain (SRP) hotspot track (Fig. D-4)? There is no unifying mechanism in the back-arc model consistent with this relationship. Magmatism along the SRP emanates from the region of flood-basalt initiation in southeastern Oregon at ca. 16.6 Ma, along a series of calderas that progressively decrease in age from 16.5 Ma near the Nevada–Oregon border to 0.6 Ma at Yellowstone National Park. At Yellowstone, a mantle plume has been seismically resolved to a depth of at least 500 km (Yuan and Dueker, 2005; Waite et al., 2006). Projecting back in time places the Yellowstone plume in the vicinity of flood-basalt initiation at ca. 16.6 Ma. The geophysical data for the Yellowstone plume seem irrefutable and consistent with the geophysical data of both Parsons et al. (1994) and Saltus and Thompson (1995), demonstrating the existence of a broad mass of hot, low-density mantle in southeastern Oregon and northern Nevada, which, they argue, is the Yellowstone mantle-plume head. The field and geophysical data are consistent with the connection of a plume head/tail pair as predicted in traditional plume models. To believe that flood-basalt volcanism was generated by normal back-arc processes requires one to ignore this important relationship.

5. *The focal point of southeastern Oregon.* Plume skeptics also ignore the implications of the field data demonstrating the migration of flood-basalt eruptions along radial trends (Fig. D-4). CRBG volcanism began in southeastern Oregon and adjacent Nevada with the eruption of Steens basalts at ca. 16.6 Ma. These initial eruptions were soon followed by a short period of rhyolitic volcanism in this same localized region from ca. 16.5–15.5 Ma (Fig. D-4). This focused burst of bimodal magmatism occurs at the western edge of the SRP hotspot track. Stratigraphic relationships demonstrate

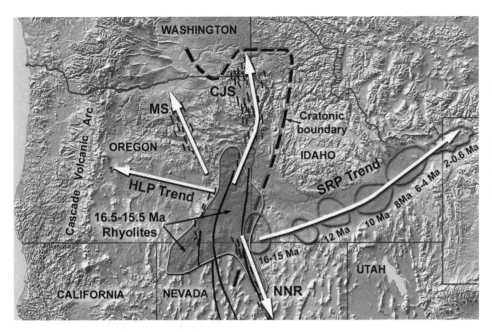

Figure D-4. Rhyolite magmatism and migrating volcanic trends associated with the CRBG-SRP magmatic system. HLP, High Lava Plains; SRP, Snake River Plain; CJS, Chief Joseph Dike swarm; MS, Monument Dike swarm; NNR, northern Nevada rift. Circular features along the SRP are age-progressive calderas and rhyolite centers from Pierce and Morgan (1992). Dark lines through the NNR and into southeastern Oregon are curvilinear magnetic anomalies (Glen and Ponce, 2002).

that volcanism rapidly migrated away from this focus along three radiating trends, forming the Chief Joseph and Monument dike swarms to the north and the NNR to the south (Fig. D-4). These radial trends are consistent with propagating volcanism above an expanding mantle-plume head (Camp and Ross, 2004) but are markedly inconsistent with the eruptive history expected from back-arc extension or any proposed nonplume model. Propagating trends along the SRP and the High Lava Plains (HLP) after ca. 15 Ma emanate from the same area of southeastern Oregon, as one would expect from the westward expansion of the plume head beneath weakened lithosphere (the Brothers fault zone) (e.g., Jordan et al., 2004), and from the east-northeast migration of the plume tail along the SRP hotspot track as a result of the relative motion of the North American plate (Pierce and Morgan, 1992).

Based on an unusually comprehensive interdisciplinary database, we maintain that the plume model for the CRBG-SRP magmatic system explains the available evidence better than any of the suggested alternative models.

28 January 2007, Peter R. Hooper

Baksi's remarks of 20 January appear to be aimed primarily at previous publications and are only marginally to do with the CRBG flood basalt as recounted in this volume. All but one of the dated samples he cites are of andesitic and rhyolitic rocks associated with the post-flood-basalt rifting in the Vale-Malheur Gorge area of eastern Oregon.

His criticisms of some of those dates must be answered by the chronological laboratories concerned. As nonspecialist geologists, we must accept what these labs provide. We did make

the effort to have some of the many Open University dates duplicated by the Corvallis laboratory, and these showed satisfactory agreement. These ages also tied in well with that of the Steens basalt from the Berkeley Geochronology Center, and, most importantly, were consistent with the stratigraphy as mapped independently in the field. That these ages are robust is implied by a more recent set of age data; Jarboe et al. (2006) pin down the Steens basalt magnetic reversal (R_0–N_0) to ca. 16.6 Ma and conclude that "the N_0 of the CRBG is the C5Cn.3n chron and that the bulk of the CRBG (the N_0–R_1–N_1–R_2–N_2 members) erupted in a short period of time: 0.75 Ma."

Thus, both the old and new dates appear to confirm that the main pulse of the CRBG on the Columbia plateau erupted between 16.1 and 15.0 Ma as stated in earlier papers (Hooper et al., 2002; Hooper, 2004), and when the earlier Steens basalt is added, the whole flood basalt eruption excepting the small volumes of late Saddle Mountains basalt occurred between 16.6 and 15.0 Ma as stated in this and previous papers.

28 January 2007, Don L. Anderson

Camp et al. (comment of 28 January 2007; hereafter CHRR) maintain that all the data for the CRBG are consistent with plume impingement and inconsistent with all proposed alternative models; all nonplume models are viewed as inadequate. But the plume model, taken alone, is also inadequate as well as implausible. Interpreting tomographic data is not as straightforward or unique as implied by CHRR (e.g., Yuan and Dueker, 2005). Crustal extension, delamination, hydration, eclogite components, mantle wind, focusing, deflection by a slab, edge effects, multiple plumes, and so on are required in order to make the plume model viable and to explain why similar tomographic structures elsewhere do not make Yellowstones. There is also the issue of why a deep mantle plume hit this place at this time, and similar questions regarding the other back-arc–plume coincidences. Although the association of plumes with continental breakup had some plausibility, the repeated attempted association of plumes with back-arcs, sutures, accretion, and collisions does not.

"The real argument," according to CHRR, lies in high magma rates over short periods of time. This is not a prediction of the plume model or of any thermal model; it is an after-the-fact rationalization of an observation. This argument is more consistent with a lithospheric stress-valve or extension mechanism coupled with underplating or ponding; these do not require a plume. The same also holds for progressive caldera development.

Many CFB provinces lie along convergent margins—presumably above slabs—and adjacent to cratons. Deccan, Karoo, Keweenawan, Emeishan, and Siberia are examples of CFB associated with accretion and convergence rather than continental breakup; these are all on mobile belts, adjacent to one or more cratons. In these areas, water from the underlying slab and delamination and fertilization from above are the most plausible explanations for low velocities and excess melting. Extensional stress, however, is a prerequisite for extrusion, as it is for the plume hypothesis.

CHRR argue that at Yellowstone a mantle plume has been seismically resolved to a depth of at least 500 km. This is not quite accurate (Fig. D-5). What has been mapped is a heterogeneous upper mantle, including low-velocity zones (LVZ) near Yellowstone and similar ones elsewhere. It is not known if these features are hot and upwelling. Yuan and Ducker (2005) conclude that the Yellowstone feature extends, at an angle, to 500 km depth (but no more). This rules out a conventional deep mantle thermal plume but is consistent with the fertile blob, top-down, and slab-dewatering models or with water solubility effects (Mierdel et al., 2007). The LVZ decreases in amplitude from 3.2% at 100 km to 0.9% at 450 km. The deep anomaly is offset by 200 km. Other hotspot-related anomalies also appear to terminate at 400 km or shallower (Anderson, this volume; Deuss, this volume), consistent with a shallow or top-down explanation.

LVZs elsewhere in western North America are similar in magnitude, volume, and depth extent to the Yellowstone anomaly; the volcanic output of these other places is very small, suggesting that the dominant control on volcanic output is strain. This probably also controls rates of magmatism.

The mapping of velocity into temperature is difficult, nonunique, and likely impossible; e.g., there is no one-to-one mapping or even necessarily a correlation among temperature, velocity, and density. Dense, cold eclogite sinkers can be LVZs (Anderson, this volume), as can hydrated peridotite. The tomography has been explained by small plumes rising beneath Yellowstone, leaving similar anomalies elsewhere unexplained. If the tomography is caused by excess temperature, a ΔT of 200–400 °C is required. Alternatively, fluids from the slab, or low-solidus crustal and lithosphere material, weakened and

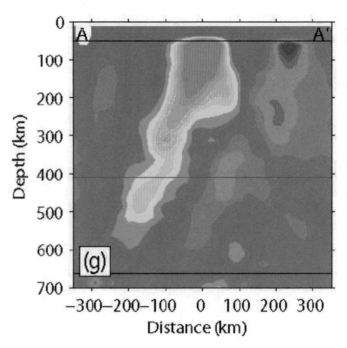

Figure D-5. Tomographic cross-section across Yellowstone. This is the most plume-like section (after Yuan and Dueker, 2005). An active plume upwelling would spread out laterally beneath the plate. The appearance of the feature depends on the orientation, cropping, and color scale. Red is low velocity, but one cannot infer temperature or upwelling from this kind of data alone. See also Hales et al. (2005), Dueker and Yuan (2004), and Google Images Yellowstone tomography.

eclogitized by Laramide hydration, promoted destabilization and removal, placing low-velocity fertile bodies into the mantle (Yuan and Ducker, 2005). This latter is the unified explanation for magmatism and LVZs in regions that involve sutures, accreted terranes, arcs, slab windows, overthickened crust, and fluids from underlying slabs. This explanation is not just a crack or a back-arc model; it is a general tectonic model that, by the way, explains the magmatism—or lack thereof—and uplift history (Hales et al., 2005).

28 January 2007, Alan D. Smith

In reply to the comment of Camp et al. of 28th January, I would like to clarify that it is convection, not extension, that serves as the trigger for CRBG volcanism in the back-arc model of Smith (1992). Back-arc upwelling is a consequence of plate interactions, and thus the interpretation offered for CRBG volcanism fits within the "plate model" for the origin of intraplate volcanism. The back-arc model is thus compatible with explanations (e.g., Christiansen and McKee, 1978; Christiansen and Lageson, 2003) that emphasize the role of large-scale plate interactions and lithospheric structure in generation of the Eastern Snake River Plain (ESRP).

Rhyolitic volcanism along the ESRP may show a linear age progression (Pierce and Morgan, 1992), but basaltic volcanism does not, and the plume tail model does not explain its persistence along the length of ESRP after migration of rhyolitic volcanism to the Yellowstone plateau (Christiansen et al., 2002). The low-velocity zone imaged by Yuan and Dueker (2005) also extends at least 350 km along the strike of the ESRP, such that the southwestern part of the anomaly currently underlies rhyolitic centers of 6 Ma age. If the low-velocity zone results from mantle upwelling, the length of this feature and the nonlinearity of ages in the basaltic volcanism would be more compatible with shallow nonplume convection mechanisms such as those proposed by Humphreys et al. (2000).

31 January 2007, Alexei Ivanov

In their comment of 28 January 2007, Hooper et al. did not reply to the key point of my comment of 17 January 2007, which is not ambiguity in the geochemical interpretation of OIB signatures. The key point is that the most voluminous Grande Ronde formation shows no sign of OIB signatures at all. Reference to Figure 8 of Hooper et al. (this volume) does not solve the problem. For example, the caption to Figure 8 states that "The proximity of some formations (. . . , Grande Ronde, . . .) to OIB source compositions is used as evidence that these formations were derived from an enriched, OIB-like, mantle source." Ironically, OIB is outside the limits of the Ba/Nb versus Nb/Y diagram (Fig. 8B of Hooper et al., this volume). Ba/Nb and Nb/Y in OIB are 7.3 and 1.7, respectively (Sun and McDonough, 1989).

5 February 2007, Peter Hooper, Vic Camp, Steve Reidel, and Marty Ross

Truth, like beauty we suppose, is in the eye of the beholder; so how best to interpret our Figure 8? Ivanov is correct in noting that the Ba/Nb and Nb/Y ratios of both the Imnaha and Grande Ronde basalts are outside the average values quoted by Sun and McDonough (1989) for OIBs. We would interpret this as evidence of some contamination of the early magmas with mafic components in the lower crust, as advocated by Chesley and Ruiz (1998) on the basis of their Re/Os isotope data.

The logic of our interpretation of Figure 8A,B (Hooper et al., this volume) is that:

1. The trace-element profiles of early Imnaha basalt, together with their isotopic ratios, closely match Hawaiian basalt and are distinct from subduction-related magmas (Hooper and Hawkesworth, 1993, Fig. 3).
2. The separate Imnaha and Grande basalt trends on Figure 8 overlap at their more primitive ends and thus are probably derived from a similar source.
3. The Imnaha trend may be explained by a combination of partial melting and crystal fractionation processes (Figs. 8 and 10).
4. The Grande basalt cannot be explained by crystal fraction-

ation (Rb/Zr rises with Rb/Sr) (Hooper and Hawkesworth, 1993, Fig. 9) and so is interpreted as a mixing array between the original OIB-like magma and a lithospheric component.

Smith has clarified his concept of the upwelling mantle in the back-arc area, which we appreciate, and we reemphasize the distinction between the magmatism directly associated with back-arc extension in eastern Oregon and the flood basalt eruption. This difference, so obvious in the field, we perceive as critical evidence against interpreting the flood basalt as a product of Smith's back-arc model.

Back-arc extension down the length of the "Inland Empire" of the northwestern United States, separating the Cascade and Idaho mountains, has been apparent since the subduction zone jumped west in the Eocene. It is seen in the Eocene Republic and related grabens of northeastern Washington state and has continued until the present. The extensional features are directly associated by calc-alkali to alkalic magmatism, which, erupting along the graben walls, fill the grabens (Hooper et al., 1995; Morris et al., 2000). In eastern Oregon this small-scale, very local, extension-related volcanism both pre- and postdates the sudden eruption of the huge volumes of monotonously tholeiitic basalt of Steens Mountain and Malheur Gorge (Hooper et al., 2002; Camp et al., 2003). The extension-related volcanism (basalt, andesite, rhyolite, and siliceous tuffs) erupted along the developing graben walls and fills the 50-km-wide Oregon–Idaho Graben. Thus, although back-arc extension was a factor before and after the flood basalt eruption, it created its own very distinctive type of volcanism and cannot easily be held responsible for the flood basalts for which, we maintain, a thermal anomaly at the base of the lithosphere is the most obvious explanation.

7 February 2007, Vic Camp, Peter Hooper, Steve Reidel, and Marty Ross

Anderson (comment of 28 January) states that "the plume model, taken alone, is also inadequate" in explaining the genesis of the CRBG. We do not argue that the plume model needs to be "taken alone." Numerous other factors are inevitably involved, as is apparent in our chapter (Hooper et al., this volume). Lithospheric thickness, for example, clearly influenced the location of volcanism, and extension provided a control on the orientation of feeder dikes. The highest eruption rates occurred in the region of least extension (the Chief Joseph Dike swarm), which lies contrary to Anderson's claim that the short duration of magmatism is "more consistent with a lithospheric stress-valve or extension mechanism, coupled with underplating or ponding." We agree instead with Smith (comment of 28 January) that mantle upwelling, not extension, was the trigger for CRBG volcanism, but we disagree with his assertion that the short-lived burst of mantle upwelling at ca. 16 Ma was a consequence of plate interactions. Extension and underplating may be factors, however vaguely expressed, but we assert once again that the eruption of such large volumes in such a brief period of time, during minimal extension, is surely better explained by the abrupt arrival of a large thermal and/or fertile anomaly. The persistence of this anomaly from southeastern Oregon to the present Yellowstone center, mirroring precisely the westward drift of the North American plate, is too obvious to be so lightly dismissed.

Anderson implies that our preferred model for CRBG genesis is similar to "other back-arc-plume coincidences" and to "the repeated attempted association of plumes with back-arcs." We agree with Anderson and other plume skeptics who see little rationale in applying a plume genesis to explain the style of volcanism common to most all areas of back-arc extension. However, flood-basalt magmatism is markedly uncommon in back-arc settings, the CRBG being the only obvious example. This unusual occurrence surely requires an additional mechanism distinct from the long-lived plate-tectonic processes inherent in all continental back-arc regions.

The failure of seismic studies to identify a mantle anomaly extending >200 km beneath Yellowstone has been the primary rationale for dismissing a plume genesis for the SRP and CRBG (e.g., Christiansen et al., 2002) until such an anomaly was resolved by Yuan and Dueker (2005) to a depth of ~500 km. We bow to Anderson's expertise in seismology and his argument that the low-velocity anomaly could also be interpreted as a fertile blob of cold eclogite or hydrated peridotite. We must point out, however, that many other geophysicists have interpreted the available data as supporting plume emplacement beneath southeastern Oregon and adjacent Nevada at ca. 16 Ma, and beneath the Yellowstone caldera today (Parsons et al., 1994; Saltus and Thompson, 1995; Bijwaard et al., 1998; Montelli et al., 2004; Yuan and Dueker, 2005; Waite et al., 2006). Anderson describes the deeper part of the Yellowstone anomaly as being offset by 200 km, and therefore, he relates it to "other hotspot-related anomalies [that] also appear to terminate at 400 km or shallower, . . . consistent with a shallow or top-down explanation." Instead, Yuan and Dueker (2005) state, "Our most important result is the resolution of a tilted low velocity plume that extends from beneath the Yellowstone caldera to 500 km depth. Further support for this conclusion derives from the observation of a localized 12-km depression in the depth of the 410-km discontinuity where the Yellowstone plume crosses this phase transition."

Much of this discussion on the chapter by Hooper et al. (this volume) revolves around a single question: What model can best explain all of the following:

1. The OIB-like end-member source and high ^{3}He/^{4}He ratios found in the CRBG lavas
2. The short duration and high eruption rates of CRBG volcanism
3. The radial, migrating trends of basalt eruption from a focused center
4. The subsequent age-propagation trend along the SRP hotspot track

5. The 500-km-deep low-velocity conduit currently centered beneath the Yellowstone caldera

All contributors to this discussion appear to agree that back-arc extension played a role in the evolution of the CRBG, but they disagree on the extent of that role. In our view, the burst of flood-basalt volcanism at ca. 16 Ma must have been triggered by an additional mechanism. Nonplume mechanisms suggested by the contributors include:

1. Mantle upwelling associated with plate interactions and exploitation of lithospheric structures (Smith, comment of 28 January; Christiansen and Lageson, 2003)
2. Mantle flow around a residuum body (Smith, comment of 28 January; Humphreys et al., 2000)
3. Subduction-related magmatism (Ivanov, comment of 17 January)
4. Top-down models involving mantle hydration, delamination, and/or fertilization (Anderson, comment of 28 January; Anderson, this volume)

Although these nonplume models have scientific merit, we contend that they cannot explain adequately all of the five constraints listed above. We maintain that a plume origin not only is consistent with each of these constraints but also is the only interpretation capable of combining the genesis of the CRBG flood-basalts and the SRP hotspot-track into a single unifying model.

REFERENCES CITED

Anderson, D.L., 1998, The helium paradoxes: Proceedings of the National Academy of Sciences of the USA, v. 95, p. 4822–4827.

Anderson, D.L., 2000a, The statistics of helium isotopes along the global spreading ridge system and the central limit theorem: Geophysical Research Letters, v. 27, p. 2401–2404.

Anderson, D.L., 2000b, The statistics and distribution of helium in the mantle: International Geology Review, v. 42, p. 289–311.

Anderson, D.L., 2001, A statistical test of the two reservoir model for helium: Earth and Planetary Science Letters, v. 193, p. 77–82.

Anderson, D.L. 2005a, Large igneous provinces, delamination, and fertile mantle: Elements, v. 1, p. 271–275.

Anderson, D.L., 2005b, The plume assumption: Frequently used arguments: www.mantleplumes.org.

Anderson, D.L., 2007 (this volume), The eclogite engine: Chemical geodynamics as a Galileo thermometer, in Foulger, G.R., and Jurdy, D.M., eds., Plates, plumes, and planetary processes: Boulder, Colorado, Geological Society of America Special Paper 430, doi: 10.1130/2007.2430(03).

Anderson, D.L., and Natland, J.H., 2005, A brief history of the plume hypothesis and its competitors: Concept and controversy, in Foulger, G.R., et al., eds., Plates, plumes, and paradigms, Geological Society of America Special Paper 388, p. 119–145.

Baksi, A.K., 1989, Reevaluation of the timing and duration of extrusion of the Imnaha, Picture Gorge and Grande Ronde Basalts, Columbia River basalt group, in Geological Society of America Special Paper 239, p. 105–112.

Baksi, A.K., 1993, A geomagnetic polarity time-scale for 17–0 Ma, based on ^{40}Ar/^{39}Ar plateau ages for selected geomagnetic reversals: Geophysical Research Letters, v. 20, p. 1617–1620.

Baksi, A.K., 1999, Reevaluation of plate motion models based on hotspot tracks in the Atlantic and Indian Oceans: Journal of Geology, v. 107, p. 13–26.

Baksi, A.K., 2005, Evaluation of radiometric ages pertaining to rocks hypothesized to have been derived by hotspot activity, in and around the Atlantic, Indian and Pacific Oceans, *in* Foulger, G.R., et al., eds., Plates, plumes, and paradigms, Geological Society of America Special Paper 388, p. 55–70.

Baksi, A.K., 2007 (this volume, chapter 15), A quantitative tool for detecting alteration in undisturbed rocks and minerals—I: Water, chemical weathering, and atmospheric argon, *in* Foulger, G.R., and Jurdy, D.M., eds., Plates, plumes, and planetary processes: Boulder, Colorado, Geological Society of America Special Paper 430, doi: 10.1130/2007.2430(15).

Baksi, A.K., 2007 (this volume, chapter 16), A quantitative tool for detecting alteration in undisturbed rocks and minerals—II: Application to argon ages related to hotspots, *in* Foulger, G.R., and Jurdy, D.M., eds., Plates, plumes, and planetary processes: Boulder, Colorado, Geological Society of America Special Paper 430, doi: 10.1130/2007.2430(16).

Baksi, A.K. and Farrar, E., 1990, Evidence for errors in the geomagnetic polarity time scale at 17–15 Ma: $^{40}Ar/^{39}Ar$ dating of basalts from the Pacific Northwest, USA: Geophysical Research Letters, v. 17, p. 1117–1120.

Bijwaard, H., Spakman, W., and Engdahl, E.R., 1998, Closing the gap between regional and global travel time tomography: Journal of Geophysical Research, v. 103, p. 30055–30078.

Camp, V.E., and Ross, M.E., 2004, Mantle dynamics and genesis of mafic magmatism in the Intermontane Pacific Northwest: Journal of Geophysical Research, v. 109, art. no. B08204, doi: 10.1029/2003JB002838.

Camp, V.E., Ross, M.E., and Hanson, W.L., 2003, Genesis of flood basalts and Basin and Range volcanic rocks from Steens Mountain to Malheur River Gorge, Oregon: Geological Society of America Bulletin, v. 115, p. 105–128.

Cande, S.C., and Kent, D.V., 1995, Revised calibration of the geomagnetic polarity timescale for the Late Cretaceous and Cenozoic: Journal of Geophysical Research, v. 100, p. 6093–6095.

Chesley, J.T., and Ruiz, J., 1998, Crust–mantle interaction in large igneous provinces: Implications from the re-Os isotope systematics of the Columbia River flood basalts: Earth and Planetary Science Letters, v. 154, p. 1–11.

Christiansen, R.L., Foulger, G.R., and Evans, J.R., 2002, Upper mantle origin of the Yellowstone hotspot: Geological Society of America Bulletin, v. 114, p. 1245–1256.

Christiansen, R.L., and Lageson, D.R., 2003, Structural control and plate-tectonic origin of the Yellowstone melting anomaly, *in* The hotspot handbook, Proceedings of Penrose Conference Plume IV: Beyond the plume hypothesis, Hveragerdi, Iceland (abstract).

Christiansen, R.L., and McKee, E.H., 1978, Late Cenozoic volcanic and tectonic evolution of the Great Basin and Columbia intermontane regions. Boulder, Colorado, Geological Society of America Memoir 152, p. 283–311.

Daly, R.A., 1926, Our Mobile Earth: New York, Charles Scribner's Sons, 342 p.

Deuss, A., 2007 (this volume), Seismic observations of transition-zone discontinuities beneath hotspot locations, *in* Foulger, G.R., and Jurdy, D.M., eds., Plates, plumes, and planetary processes: Boulder, Colorado, Geological Society of America Special Paper 430, doi: 10.1130/2007.2430(07).

Dorendorf, F., Wiechert, U., and Wörner, G., 2000, Hydrated sub-arc mantle: A source for the Klyuchevskoy volcano, Kamchatka/Russia: Earth and Planetary Science Letters, v. 175, p. 69–86.

Dueker, K., and Yuan, H., 2004, Upper mantle P-wave velocity structure from PASSCAL teleseismic transects across Idaho, Wyoming and Colorado: Geophysical Research Letters, v. 31, art. no. L08603, doi: 10.1029/2004GL019476.

Dulski, P., 2001, Reference materials for geochemical studies: New analytical data by ICP-MS and critical discussion of reference values: Geostandards Newsletter: The Journal of Geostandards and Geoanalysis, v. 25, p. 73–81.

Dunn, T., and Sen, C., 1994. Mineral/matrix partition coefficients for orthopyroxene, plagioclase, and olivine in basaltic to andesitic systems: A combined analytical and experimental study: Geochimica et Cosmochimica Acta, v. 58, p. 717–733.

Fitton, J.G., 2007 (this volume), The OIB paradox, *in* Foulger, G.R., and Jurdy,

D.M., eds., Plates, plumes, and planetary processes: Boulder, Colorado, Geological Society of America Special Paper 430, doi: 10.1130/2007.2430(20).

Glen, J.M.G., and Ponce, D.A., 2002, Large-scale fractures related to inception of the Yellowstone hotspot: Geology, v. 30, p. 647–650.

Hales, T.C., Abt, D.L., Humphreys, E.D., and Roering, J.J. 2005. Lithospheric instability origin for Columbia River flood basalts and Wallowa Mountains uplift in northeast Oregon: Nature, v. 438, p. 842–845. http://www.mantleplumes.org/CRBDelam.html.

Hart, W.K., Carlson, R.W., and Shirey, S.B., 1997, Radiogenic Os in primitive basalts from northwestern U.S.A.: Implications for petrogenesis: Earth and Planetary Science Letters, v. 150, p. 103–116.

Hickey-Vargas, R., 1998, Origin of the Indian Ocean-type isotopic signature in basalts from Philippine Sea plate spreading centers: An assessment of local versus large-scale processes: Journal of Geophysical Research, v. 103, p. 20,963–20,979.

Hooper, P.R., 2004, Ages of the Steens and Columbia River flood basalts and their relationship to extension-related calc-alkaline volcanism in eastern Oregon: Reply: Geological Society of America Bulletin, v. 116, p. 249–250.

Hooper, P.R., Bailey, D.G., and McCarley Holder, G.A., 1995, Tertiary calc-alkaline magmatism associated with lithospheric extension in the Pacific Northwest: Journal of Geophysical Research, v. 100, p. 303–319.

Hooper, P.R., Binger, G.B., and Lees, K.R., 2002, Ages of the Steens and Columbia River flood basalts and their relationship to extension-related calc-alkaline volcanism in eastern Oregon: Geological Society of America Bulletin, v. 114, p. 43–50.

Hooper, P.R., Camp, V.E., Reidel, S.P., and Ross, M.E. 2007 (this volume), The origin of the Columbia River flood basalt province: Plume versus nonplume models, *in* Foulger, G.R., and Jurdy, D.M., eds., Plates, plumes, and planetary processes: Boulder, Colorado, Geological Society of America Special Paper 430, doi: 10.1130/2007.2430(30).

Hooper, P.R., and Hawkesworth, C.J., 1993, Isotopic and geochemical constraints on the origin and evolution of the Columbia River basalts: Journal of Petrology, v. 34, p. 1203–1246.

Humphreys, E.D., Dueker, K.G., Schutt, D.L., and Smith, R.B., 2000, Beneath Yellowstone: Evaluating plume and nonplume models using teleseismic images of the upper mantle: GSA Today, v. 10, p. 1–7.

Ivanov, A., 2007 (this volume), Evaluation of different models for the origin of the Siberian traps, *in* Foulger, G.R., and Jurdy, D.M., eds., Plates, plumes, and planetary processes: Boulder, Colorado, Geological Society of America Special Paper 430, doi: 10.1130/2007.2430(31).

Jarboe, N.A., Coe, R.S., Renne, P.R., and Glen, J.M., 2006, $^{40}Ar/^{39}Ar$ ages of the Early Columbia River Basalt Group: Determining the Steens Mountain geomagnetic polarity reversal (R_0N_0) as the top of the C5R chron and the Imnaha normal (N_0) as the C5Cn.3n chron: Washington, D.C., American Geophysical Union Abstract V51D-1702.

Jicha, B.R., Scholl, D.W., Singer, B.S., Yogodzinski, G.M., and Kay, S.M., 2006, Revised age of Aleutian Island Arc formation implies high rate of magma production: Geology v. 34, p. 661–664.

Jordan, B.T., Grunder, A.L., Duncan, R.A., and Deino, A.L., 2004, Geochronology of age-progressive volcanism of the Oregon High Lava Plains: Implications for the plume interpretation of Yellowstone: Journal of Geophysical Research, v. 109, art. no. B10202, doi: 10.1029/2003JB002776.

Jull, M., and Kelemen, P.B., 2001, On the conditions for lower crustal convective instability: Journal of Geophysical Research, v. 106, p. 6423–6446.

McDonough, W.F., and Sun, S.-S., 1995. The composition of the Earth: Chemical Geology, v. 120, p. 223–253.

Meibom, A., and Anderson, D.L., 2004, The statistical upper mantle assemblage: Earth and Planetary Science Letters, v. 217, p. 123–139.

Meyzen, C.M., Ludden, J.N., Humler, E., Luais, R., Toplis, M.J., Mével, C., and Storey, M., 2005, New insights into the origin and distribution of the Dupal isotope anomaly in the Indian Ocean mantle from MORB of the Southwest Indian Ridge: Geochemistry, Geophysics, Geosystems, v. 6, p. 1–34, doi: 10.1029/2005GC000979.

Mierdel, K., Keppler, H., Smyth, J.R., and Langenhorst, F., 2007, Water solubility in aluminous orthopyroxene and the origin of Earth's asthenosphere: Science, v. 315, p. 364.

Montelli, R., Nolet, G., Dahlen, F.A., Masters, G., Engdahl, E.R., and Hung, S.-H., 2004, Finite-frequency tomography reveals a variety of plumes in the mantle: Science, v. 303, p. 338–343, doi: 10.1126/science.1092485.

Morris, G.A., Larson, P.B., and Hooper, P.R., 2000, "Subduction style" magmatism in a non-subduction setting: The Colville Igneous Complex, NE Washington State, USA: Journal of Petrology, v. 41, p. 43–67.

Natland, J.H., 1980. The progression of volcanism in the Samoan linear volcanic chain: American Journal of Science, v. 280A, p. 709–735.

Natland, J.H., 2003, Capture of mantle helium by growing olivine phenocrysts in picritic basalts from the Juan Fernandez Islands, SE Pacific: Journal of Petrology, v. 44, p. 421–456.

Natland, J.H., 2007 (this volume), ΔNb and the role of magma mixing at the East Pacific Rise and Iceland, *in* Foulger, G.R., and Jurdy, D.M., eds., Plates, plumes, and planetary processes: Boulder, Colorado, Geological Society of America Special Paper 430, doi: 10.1130/2007.2430(21).

Natland, J.H., and Winterer, E.L., 2005. Fissure control on volcanic action in the Pacific, *in* Foulger, G.R., et al., eds., Plates, plumes, and paradigms: Boulder, Colorado, Geological Society of America Special Paper 388, p. 687–710, doi: 10.1130/2005.2388(39).

Parman, S.W., Kurz, M.D., Hart, S.R., and Grove, T.L., 2005, Helium solubility in olivine and implications for high ^{3}He/^{4}He in ocean island basalts: Nature, v. 437, p. 1140–1143.

Parsons, T., Thompson, G.S., and Sleep, N.H., 1994, Mantle plume influence on the Neogene uplift and extension of the U.S. western Cordillera: Geology, v. 22, p. 83–86.

Pierce, K.L., and Morgan, L.A., 1992, The track of the Yellowstone hotspot: Volcanism, faulting, and uplift, *in* Link, P., ed., Regional geology of eastern Idaho and western Wyoming: Boulder, Colorado, Geological Society of America Memoir 179, p. 1–53.

Poreda, R., and Craig, H., 1992, He and Sr isotopes in the Lau Basin mantle: Depleted and primitive mantle components: Earth and Planetary Science Letters, v. 113, p. 487–493.

Renne, P.R., Swisher, C.C., Deino, A.L., Karner, D.B., Owens, T.L., and DePaolo, D.J., 1998, Intercalibration of standards, absolute ages and uncertainties in 40Ar-39Ar dating: Chemical Geology, v. 145, p. 117-152, doi: 10.1016/S0009-2541(97)00159-9.

Richards, M.A., Duncan, R.A., and Courtillot, V.E., 1989. Flood basalts and hotspot tracks; Plume heads and tails: Science, v. 246, p. 103–107.

Saltus, R.W., and Thompson, G.A., 1995, Why is it downhill from Tonopah to Las Vegas?: A case for a mantle plume support of the high northern Basin and Range: Tectonics, v. 14, p. 1235–1244.

Shaw, D.M., 1970. Trace element fractionation during anatexis: Geochimica et Cosmochimica Acta, v. 34, p. 237–243.

Sheth, H.C., 2007 (this volume), Plume-related regional prevolcanic uplift in the Deccan Traps: Absence of evidence, evidence of absence, *in* Foulger, G.R., and Jurdy, D.M., eds., Plates, plumes, and planetary processes: Boulder, Colorado, Geological Society of America Special Paper 430, doi: 10.1130/2007.2430(36).

Smith, A.D., 1992, Back-arc convection model for Columbia River Basalt genesis: Tectonophysics, v. 207, p. 269–285.

Smith, A.D., 2003, Critical evaluation of Re-Os and Pt-Os isotopic evidence on the origin of intraplate volcanism: Journal of Geodynamics, v. 36, p. 469–484.

Smith, A.D., 2007 (this volume), A plate model for Jurassic to recent intraplate volcanism in the Pacific Ocean basin, *in* Foulger, G.R., and Jurdy, D.M., eds., Plates, plumes, and planetary processes: Boulder, Colorado, Geological Society of America Special Paper 430, doi: 10.1130/2007.2430(23).

Smith, G.P., Wiens, D.A., Fischer, K.M., Dorman, L.M., Webb, S.P., and Hildebrand, J.A., 2001, A complex pattern of mantle flow in the Lau backarc: Science, v. 292, p. 713–716.

Stern, R.J., Lin, P.N., and Morris, J.D., 1990, Enriched back-arc basin basalts from the northern Mariana Trough: Implications for the magmatic evolution of back-arc basins: Earth and Planetary Science Letters, v. 100, p. 210–225.

Stuart, W.D., Foulger, G.R., and Barall, M., 2007 (this volume), Propagation of the Hawaiian-Emperor volcano chain by Pacific plate cooling stress, *in* Foulger, G.R., and Jurdy, D.M., eds., Plates, plumes, and planetary processes: Boulder, Colorado, Geological Society of America Special Paper 430, doi: 10.1130/2007.2430(24).

Sun, S.-S., and McDonough, W.F., 1989. Chemical and isotopic systematics of oceanic basalts: Implications for mantle composition and process, *in* Sounders, A.D., and Norry, M.J., eds., Magmatism in the oceanic basins: Geological Society of London Special Publication 42, p. 313–345.

Swisher, C.C., Ach, J.A., and Hart, W.K., 1990, Laser fusion ^{40}Ar/^{39}Ar dating of the type Steens Mountain basalt, south-eastern Oregon and the age of the Steens geomagnetic polarity transition [abs.]: Eos (Transactions, American Geophysical Union), v. 71, p. 1296.

Takahashi, E., Nakajima, K., and Wright, T.L., 1998, Origin of the Columbia River basalts: Melting model of a heterogeneous plume head: Earth and Planetary Science Letters, v. 162, p. 63–80.

Taylor, B., and Martinez, F., 2003, Back-arc basalt systematics: Earth and Planetary Science Letters, v. 210, p. 481–497.

Taylor, B., Zellmer, K., Martinez, F., and Goodliffe, A., 1996, Sea-floor spreading in the Lau back-arc basin: Earth and Planetary Science Letters, v. 144, p. 35–40.

Turner, S., and Hawkesworth, C., 1998, Using geochemistry to map mantle flow beneath the Lau Basin: Geology, v. 26, p. 1019–1022.

Volpe, A.M., Macdougall, J.D., and Hawkins, J.W., 1988, Lau Basin basalts (LBB): Trace element and Sr-Nd isotopic evidence for heterogeneity in backarc basin mantle: Earth and Planetary Science Letters, v. 90, p. 174–186.

Waite, G.P., Smith, R.B., and Allen, R.M., 2006, Vp and Vs structure of the Yellowstone hot spot from teleseismic tomography: Evidence for an upper mantle plume: Journal of Geophysical Research, v. 111, art. no. B04303, doi: 10.1029/2005JB003867.

Wilson, D.S., and Gans, P.B., 2003, Magnetostratigraphy of the Eldorado Mountains volcanic complex and the calibration of the early to middle Miocene polarity time scale: Geophysical Research Letters, v. 30, art. no. 1635, doi: 10.1029/2003GL017085.

Wright, T.L., Mangan, M., and Swanson, D.A., 1989, Chemical data for flows and feeder dikes of the Yakima basalt subgroup, Columbia River Basalt Group, Washington, Idaho and Oregon, and their bearing on a petrogenetic model: U.S. Geological Survey Bulletin 1821, 71 p.

Yuan, H., and Dueker, K., 2005, Teleseismic P-wave tomogram of the Yellowstone plume: Geophysical Research Letters, v. 32, L07304, doi: 10.1029/2004GL022056.

Zack, T., Foley, S.F., and Jenner, G.A., 1997, A consistent partition coefficient set for clinopyroxene, amphibole and garnet from laser ablation microprobe analysis of garnet pyroxenites from Kakanui, New Zealand: Neues Jahrbuch für Mineralogie-Abhandlungen, v. 172, p. 23–41.

The Geological Society of America
Special Paper 430
2007

Evaluation of different models for the origin of the Siberian Traps

Alexei V. Ivanov*

Institute of the Earth's Crust, Siberian Branch, Russian Academy of Sciences, Lermontov Street, 128, 664033, Irkutsk, Russia

ABSTRACT

Various types of evidence, including the size and volume of the Siberian Traps, the timing and duration of eruptions, paleotectonic and paleogeographic reconstructions, lithospheric structure, heatflow, and the trace-element and radiogenic isotope compositions of lava, are reviewed in this chapter. The major evidence may be summarized as follows. The Siberian Traps erupted in a number of brief volcanic events from the Late Permian until the end of the Middle Triassic. They occupied a vast region ($\sim$7 $\times$ 10^6 km^2) in a back-arc tectonic setting. The overall volume of erupted rocks was as much as $\sim$4 $\times$ 10^6 km^3, with most of the volume erupted within the Tunguska syncline. This syncline experienced long-term subsidence before initiation of the volcanism, and the region is now underlain by a relatively thin lithosphere, which is $\sim$180 km thick. Two types of trace-element patterns are observed in the Siberian Traps: subordinate high-Ti ocean island basalt–like patterns and dominant low-Ti island arc basalt–like patterns. In radiogenic isotope and trace-element coordinates, mixing trends between these two types of magma are absent, or at least not evident. Some volcanic rocks contain primary magmatic mica. These are considered in light of different models. Each model can explain, or was thought to explain, particular observations. However, some evidence can be fatal for some models. For example, the enormous size and volume of the Siberian Traps cannot be explained in the framework of impact and edge-driven convection models and are problematic for lithospheric delamination models. Plume models face problems in explaining the uplift and subsidence pattern and the absence of mixing curves between expected high-Ti primary plume melts and contaminated low-Ti melts. Therefore, a model that relates Siberian Trap magmatism and subduction is suggested. In this model, subducting slabs brought significant amounts of water into the mantle transition zone. Consequent release of water from the transition zone lowered the solidus of the upper mantle, leading to voluminous melting. Major supporting observations for this model include (1) the tectonic position of the Siberian Traps in a back-arc setting of Permian subduction systems, (2) island arc basalt–like trace-element patterns for the majority of the erupted basalts, (3) primary mica found in volcanic rocks, and (4) experimental data on the high water capacity of the mantle transition zone and its recharging via the subduction process.

Keywords: Siberian Traps, subduction, plume, delamination, convection, impact

---

*E-mail: aivanov@crust.irk.ru.

Ivanov, A.V., 2007, Evaluation of different models for the origin of the Siberian Traps, *in* Foulger, G.R., and Jurdy, D.M., eds., Plates, plumes, and planetary processes: Geological Society of America Special Paper 430, p. 669–691, doi: 10.1130/2007.2430(31). For permission to copy, contact editing@geosociety.org.

INTRODUCTION

A number of completely different models have been discussed in the scientific literature to explain the origin of the Siberian Traps. These include the extraterrestrial bolide impact model (Jones et al., 2002) and terrestrial models, which consider the Siberian Traps to result from a large mantle plume from the core-mantle boundary (Campbell and Griffiths, 1990), lithospheric delamination involving a weaker plume (Elkins-Tanton, 2005), redistribution of heat in the upper mantle without a plume (King and Anderson, 1998; Puffer, 2001), or saturation of the upper mantle with water following prolonged subduction beneath Siberia (Ivanov and Balyshev, 2005). The plume model is the most popular, and thus it is considered the conventional model. Other models are viewed as alternatives. Each model employs its own line of arguments. Useful evidence includes magma volumes, timing of magmatism, uplift history, geophysical data, and trace-element and isotope geochemistry. This chapter evaluates the conventional and alternative models from the viewpoint of the evidence.

TERMINOLOGY

Large Igneous Province and Flood Basalt Province

The Siberian Traps contain mafic, ultramafic, and silicic rocks, both intrusive and effusive. In this sense the Siberian Traps is a large igneous province (LIP) because it is large and igneous (for discussion of the definition and classification of LIPs, see http://www.mantleplumes.org/LIPClass.html#Discussion). The Siberian Traps were built from one or more volcanic events involving the outpouring of large volumes of mainly basaltic magma. The large volumes distinguish the Siberian Traps, which are, in this sense, a flood basalt province (FBP). Therefore, in this chapter I use two terms: (1) the Siberian Traps LIP, meaning the spatially and temporally related rocks of different composition and probably variable origin, and (2) the Siberian Traps FBP, meaning the large volume of mainly basaltic volcanic units. The majority of the Siberian Traps FBP units have distinct subduction-like major- and trace-element features such as low concentrations of high field-strength elements (HFSE) relative to large-ion lithophile elements (LILE) (Fedorenko et al., 1996; Puffer, 2001; Ivanov and Balyshev, 2005).

Plumes and Plate Tectonic Processes

Morgan (1971) suggested that plumes are localized upwelling convective currents originating in the lower mantle, probably at the core-mantle boundary. He suggested that plumes are the driving force for plate tectonics. As noted by Korenaga (2005), in the fluid dynamic literature a plume is any self-buoyancy-driven flow. In this sense, all subducting slabs, sinking delaminated lower crust, and lower-mantle upwellings are plumes. Such usage of the term however, is unacceptable to Earth sci-

entists, who distinguish between subduction-related volcanism occurring at plate boundaries and intraplate volcanism occurring far from plate boundaries. Many different, sometimes contradictory, definitions of plumes can be found in the literature. A complete survey of plume definitions is beyond the scope of this chapter, but the following quote from Campbell (2005, p. 265) expresses a commonly held view of what a "mantle plume" is:

High-pressure experimental studies of the melting point of iron-nickel alloys show that the core is several hundred degrees hotter than the overlying mantle. A temperature difference of this magnitude is expected to produce an unstable boundary layer above the core which, in turn, should produce plumes of hot, solid material that rise through the mantle, driven by their thermal buoyancy. Therefore, from theoretical considerations, mantle plumes are the inevitable consequence of a hot core.

Such plumes, referred to as Morganian by Courtillot et al. (2003), are often viewed as phenomena unrelated to plate tectonics. Plates can be driven by cooling from above without internal heating (Anderson, 2001), whereas Morganian plumes are driven by heating from the core (Courtillot et al., 2003; Campbell, 2005). According to the conventional view, Morganian plumes are necessary to form LIPs. They may or may not lead to breakup of a continent (e.g., the central Atlantic magmatic province and the Siberian Traps, respectively). Therefore, in evaluating plume models in this chapter, I use the term *plume* only for first-order upwelling currents, which are either unrelated to plate tectonics or play a role as its driving force. Any upper-mantle processes (e.g., shallow recycling of subducted material, upper-mantle convection) are considered nonplume plate tectonic processes. Lithospheric delamination may be related either to plume or to plate tectonic processes depending on the original causes of the delamination.

EVIDENCE

Size and Volume

The Siberian Traps LIP is one of the largest in size and volume on Earth, though estimates vary significantly. For example, Fedorenko et al. (1996) referred to the work of Milanovskiy (1976), whose estimates of its size and volume were $\sim 4 \times 10^6$ km^2 and $> 2 \times 10^6$ km^3. Fedorenko et al. (1996) wrote, "We believe that even this volume may be underestimated." Reichow et al. (2002) estimated separately the sizes of the Siberian Traps LIP located on the Siberian craton and in the west Siberian basin, as 2.6×10^6 km^2 (1×10^6 km^3) and 1.3×10^6 km^2 (1.3×10^6 km^3), respectively. The distribution of igneous rocks that can be attributed to the Siberian Traps LIP is shown in Figure 1, from Masaitis (1983). According to this author, the area of the Siberian Traps LIP is $\sim 7 \times 10^6$ km^2, and the volume could be as much as 4×10^6 km^3. This volume is not the largest reported in the literature. Dobretsov (2003) estimated that it is over 16×10^6 km^3, including the Kara and Barents undersea areas.

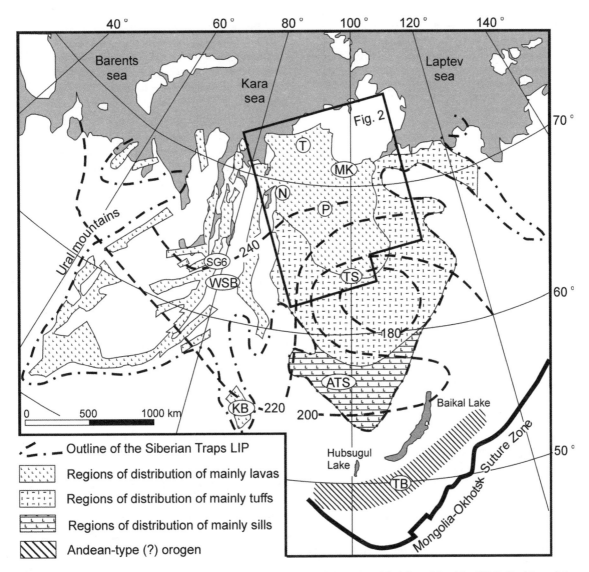

Figure 1. The Siberian Traps LIP (large igneous province; simplified and modified from Masaitis, 1983). Position of the Mongolia-Okhotsk Suture Zone and the Permo-Triassic volcano-plutonic complexes of Mongolia and Transbaikalia (Russia) are shown after Zorin (1999) and Yarmolyuk et al. (2001), respectively. The bold dashed lines with numbers represent depths in km of the L-boundary (the base of the lithosphere), detected by nuclear explosion seismic data, which probably marks the lithosphere-asthenosphere boundary (Pavlenkova and Pavlenkova, 2006). SG6—superdeep drillhole 6; WSB— west Siberian basin; KB—Kuznetsk basin; N—Noril'sk; T—Tajmyr; MK—Maimecha-Kotui; P—Putorana; TS— Tunguska syncline; ATS—Angara-Taseevskaya syncline; TB—Transbaikalian belt. The rectangle shows the location of the area shown in Figure 2. The Angara-Taseevskay syncline is after Malitch (1999).

The various estimates of the size of the Siberian Traps LIP reported in the literature are summarized in Table 1. Most of these, however, are rough guesses from the size and average thickness of volcanogenic deposits. Precise calculation of the volume has been performed only by Vasil'ev et al. (2000). These authors focused on the cratonic part and calculated separately the present-day preserved volume of lava and volcanoclastic and intrusive rocks using geological survey data. Note that the size of the Siberian Traps erupted on the platform after Vasil'ev et al. (2000) is on the same order as the size assumed by Mi-

lanovskiy (1976) and Reichow et al. (2002) for the whole LIP. This is partly because a large part of the Siberian Platform experienced only intrusive magmatism with extensive but relatively low-volume sills that are hardly exposed on the surface and known mostly from drilling. The volume estimated by Vasil'ev et al. (2000) is 1.752×10^6 km^3, over half of which is located in the Tunguska syncline. Tuffs are abundant within the syncline (Fig. 1).

The estimates discussed earlier form three general groups: (1) a size of $\sim 4 \times 10^6$ km^2 and a volume of $\sim 2 \times 10^6$ km^3

TABLE 1. ESTIMATES OF THE SIZE AND VOLUME OF THE SIBERIAN TRAPS LARGE IGNEOUS PROVINCE

Papers (in chronological order)	Size (10^6 km^2)	Volume (10^6 km^3)
Milanovskiy (1976)	~4	≥2
Masaitis (1983)	~7 × 10^6	≤4
Fedorenko et al. (1996)	n.r.	>2
Vasil'ev et al. (2000)[t]	4.3[t] × 2 ~ 8.6	1.75[t, φ] × 2 ~ 3.5
Reichow et al. (2002)	3.9	2.3
Dobretsov (2003)	n.r.	≥16

Notes: n.r.—not reported.

[t]The calculations of Vasil'ev et al. (2000) were only for the Siberian craton, so those calculations multiplied by 2 to estimate the total.

[φ]The present-day preserved value.

(Milanovskiy, 1976; Reichow et al., 2002), (2) a size of ~7 × 10^6 km^2 and a volume of ~4 × 10^6 km^3 (Masaitis, 1983; Vasil'ev et al., 2000), and (3) a size of over $1 × 10^7$ km^2 and a volume of $1.6 × 10^7$ km^3 (Dobretsov, 2003). In my view, the first group underestimates by a factor of about 2, whereas the third group may be an overestimation. A volume on the order of $4 × 10^6$ km^3 is probably close to the true value. It is worth mentioning that this includes various types of rocks from ultrabasic to acidic, with basalts the major rock type. Ultrabasic magmas (meimechites) erupted in large volumes in the Meimecha-Kotui region. Rhyolites and dacites are found in subordinate amounts in the west Siberian basin (Masaitis, 1983; Medvedev et al., 2003). The volume of the Siberian Traps FBP is hard to estimate precisely, because it requires additional constraints from timing. However, it should be only slightly below the total LIP value.

Timing

Comments on the Use of Different Isotopic Systems. The question of timing is discussed here on the basis of U-Pb and ^{40}Ar/^{39}Ar dating. Both radioisotopic systems may yield erroneous ages due to loss of radiogenic lead and argon (giving younger apparent ages) and inherited lead and extraneous argon (giving older apparent ages). In general, particular criteria such as concordance in the U-Pb system and plateaus in the ^{40}Ar/^{39}Ar system have been developed to separate the true crystallization ages from apparent ages. However, when dealing with a short time duration, an error of a few million years may bias the conclusions (e.g., see Renne, 1995, for ^{40}Ar/^{39}Ar reconsideration of the Noril'sk-I intrusion and Mundil et al., 2004, for debate on the U-Pb age of the Permo-Triassic boundary). Zircon and baddeleyite U-Pb ages are usually considered more reliable than ^{40}Ar/^{39}Ar ages. However, zircon and baddeleyite are rare minerals in basaltic melts, and thus there are few published U-Pb ages for the Siberian Traps FBP. ^{40}Ar/^{39}Ar ages are more abundant. They are reported (recalculated) here relative to the same age of 98.79 Ma for the GA1550 standard (Renne et al., 1998) unless otherwise stated. The choice of this age merely reflects

the fact that it was used in the two most recent reviews of the Siberian Traps (Reichow et al., 2002; Ivanov et al., 2005). Accepting this age for the GA1550 standard makes the ^{40}Ar/^{39}Ar ages ~0.7% younger than the U-Pb ages. The reason for the difference is most likely due to biased ^{40}K decay constants (e.g., Min et al., 2000; Ivanov, 2006). Using a more recent value of 98.5 Ma (Spell and McDougall, 2003) for the GA1550 standard makes the ^{40}Ar/^{39}Ar ages ~0.9% younger than the U-Pb ages (Ivanov, 2006). The errors in the ^{40}Ar/^{39}Ar ages shown here are at the 2σ level and include reported analytical errors on the J-factor, but not errors that would result from inhomogeneities of the standards (because this information is rarely reported in original publications) and errors due to intercalibration of the standards.

Duration of Siberian Traps FBP Magmatism. A short duration for Siberian Traps FBP magmatism was suggested on the basis of two ^{40}Ar/^{39}Ar ages (Renne and Basu, 1991), for the stratigraphically oldest and youngest lava units in the northern part of the FBP (for locations of dated samples, see Fig. 2). Later it was shown that these ages are statistically indistinguishable from the ^{40}Ar/^{39}Ar age for the Permo-Triassic boundary dated from the Meishan section in China (Renne et al., 1995). Since then, the Siberian Traps FBP has been widely considered to have formed during 1 m.y. or less despite the fact that older and younger ages have been published (e.g., Dalrymple et al., 1995; Reichow et al., 2002). For example, Walderhaug et al. (2005) obtained Middle–Late Triassic ^{40}Ar/^{39}Ar ages and paleomagnetic poles for dolerite sills emplaced within Taimyr peninsula and concluded that these rocks are unrelated to the Siberian Traps FBP because of the assumption that the Siberian Traps FBP was a Permo-Triassic boundary event. However, to be correct, it is unknown when there were major (by volume) events within the Siberian Traps. Figure 2 summarizes dated localities within approximately one-fourth of the Siberian Traps LIP. It clearly shows that most Permo-Triassic ages come from a narrow strip from the Noril'sk-Kharaelkah to the Maimecha-Kotui subprovince. Huge regions have not been studied by precise radioisotopic age dating!

A basis for suggesting that the Siberian Traps FBP formed over a prolonged interval of time up to the Middle–Late Triassic, and probably in multiple volcanic events, is the following: (1) Tunguska subprovince, with the most voluminous volcanic deposits (see the earlier estimations of the volume) is characterized by Middle Triassic and Middle–Late Triassic ^{40}Ar/^{39}Ar ages (240.7 ± 2.8 Ma for Korvuchana tuffs and 232.1 ± 4.6 Ma for an uppermost lava unit; Baksi and Farrar, 1991); (2) Usol'skii sill, emplaced within Angara-Taseevskaya syncline (Fig. 1), yielded an Early Triassic ^{40}Ar/^{39}Ar age (243.9 ± 1.4 Ma; Ivanov et al., 2005); (3) combined paleomagnetic and ^{40}Ar/^{39}Ar studies show that doleritic sills from Taimyr peninsula were emplaced in the Middle–Late Triassic (230.2 ± 14, 230.7 ± 2.5, and 232.5 ± 8 Ma; Walderhaug et al., 2005); (4) Daldykan dolerite sill emplaced within the Noril'sk-Kharaelakh subprovince yielded

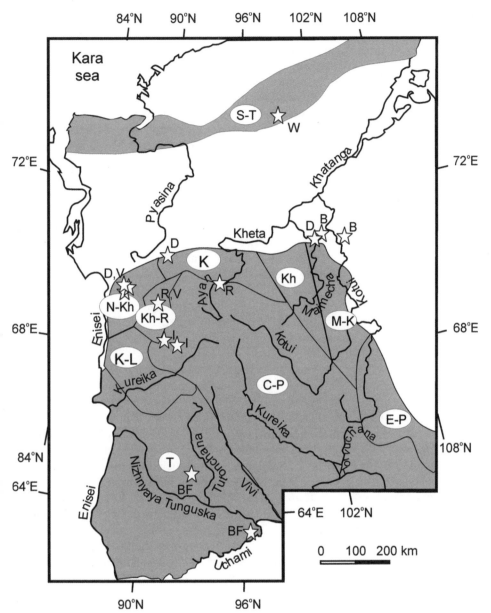

Figure 2. Volcanic subprovinces of the central and northern part of the Siberian Traps LIP (large igneous province; after Zolotuhin et al., 1984). N-Kh—Noril'sk-Kharaelakh; K-L—Kureika-Letninsk; Kh-R—Khantaisk-Rybninsk; T—Tunguska; K—Kamensk; S-T—south Taimyr; Kh—Kheta; C-P—central Putorana; M-K—Maimecha-kotui; E-P—eastern Putorana. Stars mark localities of $^{40}Ar/^{39}Ar$ dated samples: D—Dalrymple et al. (1995); V—Venkatesan et al. (1997); R—Renne and Basu (1991); I—Ivanov et al. (2006); BF—Baksi and Farrar (1991); W—Walderhaug et al. (2005); B—Basu et al. (1995). U-Pb-dated samples are located in the Noril'sk-Kharaelakh and Maimecha-Kotui subprovinces, practically at the places marked D and B (Kamo et al., 1996, 2003). One sample dated by Renne and Basu (1991) was taken from a submeridian part of the Ayan River near the junction of the C-P and K subprovinces, not in the central part of the C-P as shown in Renne and Basu (1991). The localities of samples marked BF are restored from regions studied by Dmitriev (1973), who donated the samples for the $^{40}Ar/^{39}Ar$ study of Baksi and Farrar (1991), and may be not absolutely precise. Volcanic rocks between the south Taimyr and Noril'sk-Kharaelakh, Kamensk, and Kheta subprovinces and to the west of the River Enisei are covered by Mesozoic and Cenozoic sediments. The area south to the River Uchami was not studied at the same level in the 1980s and therefore is not marked as a separate subprovince. Each subprovince is characterized by its own stratigraphical units, which were correlated with each other on the basis of petrography and chemistry.

Middle Triassic $^{40}Ar/^{39}Ar$ ages 239.5 ± 0.8 as the average of the two ages; Dalrymple et al., 1995); (5) lamproite dikes in the southern mountains framing the Kuznetsk basin yielded Early and Middle Triassic $^{40}Ar/^{39}Ar$ ages (the original $^{40}Ar/^{39}Ar$ values relative to the MCA-11 standard with no reported age were 244.0 ± 0.8, 244.4 ± 0.8, 245.7 ± 0.7, and 236.5 ± 3.8 Ma; Vrublevskii et al., 2004); (6) the Nadezhdinsky and Hona-Makitsky suites within the central Putorana subprovince yielded $^{40}Ar/^{39}Ar$ ages of 246.6 ± 2.4 and 241.0 ± 2.5 Ma, respectively (the absolute values may be slightly biased by the J-factor, but the age difference of 5.6 ± 3.5 Ma will remain unchanged for these two suites, as discussed by Ivanov et al., 2006); (7) combined paleomagnetic and biostratigraphic study of the SG-6 superdeep drillhole revealed that volcanic rocks in the deep rift structures of the

west Siberian basin were formed up to the Early–Middle Triassic (Olenekian-Anisian) (Kazanskii et al., 2000).

Many of the volcanic and intrusive rocks mentioned earlier have traditionally been considered part of the Siberian Traps FBP. In the section on geochemistry I shall show that Usol'skii sill dolerites are practically indistinguishable on the basis of trace elements from geochemically uniform upper lava units in the Noril'sk-Kharaelakh subprovince. Therefore, magmas of similar composition were erupted in different regions at different times. A detailed $^{40}Ar/^{39}Ar$ study of the Ethiopian Traps FBP revealed the same feature: chemically similar magmas erupted during different pulses of volcanism (Kieffer et al., 2004).

It is worth repeating that thick lavas in the north Siberian Traps were formed briefly at the Permo-Triassic boundary, but

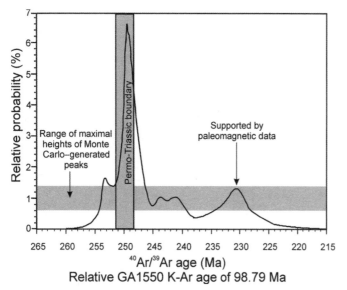

Figure 3. Relative probability distribution of $^{40}Ar/^{39}Ar$ ages (after Ivanov et al., 2006) and original ages (from Baksi and Farrar, 1991; Renne and Basu, 1991; Basu et al., 1995; Dalrymple et al., 1995; Venkatesan et al., 1997; Reichow et al., 2002; Ivanov et al., 2005, 2006; Walderhaug et al., 2005). The Permo-Triassic boundary is from Renne et al. (1995).

in other regions voluminous eruptions occurred during other, probably also brief, periods of time (Fig. 3).

Timing of Initial Eruptions within the Siberian Traps FBP. Basu et al. (1995) reported a Late Permian $^{40}Ar/^{39}Ar$ age of 253.3 ± 2.6 Ma for phlogopite from olivine nephelinite in the Arydjansky Suite, which represents the initial phase of magmatism in the northern part of the Maimecha-Kotui subprovince. This age is significantly older than the Permo-Triassic age of the initial lava suites of the Khantaisk-Rybninsk subprovince dated in the same laboratory (Renne and Basu, 1991). However, for the Arydjansky Suite, Kamo et al. (2003) reported the U-Pb perovskite age of 251.7 ± 0.4 Ma, which is statistically indistinguishable from the Permo-Triassic U-Pb zircon and baddeleyite age 251.2 ± 0.3 Ma for the Noril'sk-I ore-bearing intrusion of the Noril'sk-Kharaelakh subprovince (Kamo et al., 1996).

Ivanov (2006) compared $^{40}Ar/^{39}Ar$ and U-Pb ages obtained for the same intrusions and lava units from the Siberian Traps LIP, among other regions, and showed that the $^{40}Ar/^{39}Ar$ ages are systematically younger than the U-Pb ages. The exception is the $^{40}Ar/^{39}Ar$–U-Pb pair age for the Arydjansky Suite. This inconsistency may be explained either by excess argon in the phlogopite, which leads to an older apparent $^{40}Ar/^{39}Ar$ age, or by an incorrect correction for the initial Pb composition in the perovskite, which would lead to a younger U-Pb age.

Reichow et al. (2002) obtained Late Permian $^{40}Ar/^{39}Ar$ ages for biotites from olivine gabbros of the Van Eganskaya borehole within the west Siberian basin (253.4 ± 0.8 and 252.5 ± 1.5 Ma). These ages and a $^{40}Ar/^{39}Ar$ age for the Arydjansky Suite are statistically indistinguishable. To verify this finding, Ivanov et al.

(2006) performed a Monte Carlo test as follows. First, it was assumed that all Permo-Triassic $^{40}Ar/^{39}Ar$ ages are true crystallization ages. Second, a data set of ages in the range 260–220 Ma with errors similar to true analytical errors was stochastically created. Third, the Permo-Triassic and stochastic ages were plotted as a probability distribution. The second and third steps were repeated about one hundred times. The highest stochastic peak heights are lower than that of the late Permian $^{40}Ar/^{39}Ar$ peak, showing that the coincidence between west Siberian basin gabbro and Arydjansky Suite ages is not due to stochastic reasons (Fig. 3).

In the Kuznetsk basin there are mafic rocks of both Late Permian and Early Triassic age (e.g., Dobretsov et al., 2005). The reported $^{40}Ar/^{39}Ar$ ages (Dobretsov et al., 2005) are, however, not supported by analytical details, whereas all age spectra shown by Fedoseev et al. (2005) revealed disturbed K-Ar isotope systems in the samples studied (though the authors, unaware of this disturbance, interpreted the spectra as true crystallization ages). Paleomagnetic study of the SG-6 superdeep drillhole has shown that the initial volcanic rocks are as old as latest Late Permian (upper Late Tatatrian) (Kazanskii et al., 2000).

In summarizing the data presented in this section I should mention that the question of timing of the initial eruptions within the Siberian Traps FBP requires further geochronological investigation. At present, it is obvious that in the west Siberian basin the oldest mafic magmas erupted at the end of the Permian. Probably simultaneous eruptions of mafic magmas took place in the Kuznetsk basin and in the northern part of the Siberian craton (in the Maimecha-Kotui subprovince).

Timing of the Siberian Traps LIP. As mentioned earlier, the Siberian Traps LIP is a broad term that includes, in addition to mafic rocks, intrusive rocks of acidic composition. So-called anorogenic granitoids (mainly granodiorites and granosyenites) are abundant in the Kuznetsk basin and its mountain borders and on the Taimyr peninsula (e.g., Vernikovsky et al., 2003; Dobretsov et al., 2005). The Bolgokhtokh granodiorite intrusion is known in the Noril'sk-Kharaelakh subprovince (e.g., Kamo et al., 2003). An important feature of these anorogenic granitoids is that the oldest are characterized by zircon U-Pb ages that are statistically indistinguishable from those of the Permo-Triassic boundary (256 ± 8, 253 ± 4, and 245 ± 7 Ma for different massifs of the late Kolba Complex within the mountain border of the Kuznetsk basin (Vladimirov et al., 2001) and 249.0 ± 5.2 Ma for the Taimyr (Vernikovsky et al., 2003); it is debated whether the U-Pb age of the Permo-Triassic boundary is 251.4 ± 0.3 Ma (Bowring et al., 1998) or 252.6 ± 0.3 Ma (Mundil et al., 2004). The youngest anorogenic granitoids yielded Middle–Late Triassic ages (231 ± 11 and 225 ± 4 Ma for different massifs of the Monastyrskii Complex within the mountains bordering the Kuznetsk basin (Vladimirov et al., 2001) and 229.0 ± 0.4 Ma for the Bolgokhtokh granodiorite intrusion (Kamo et al., 2003). Therefore, the mafic and acidic magmatism of the Siberian Traps was, in general, coeval.

Uplift History

There is debate about the subsidence-uplift history in the region of the Siberian Traps LIP. Czamanske et al. (1998) noted that volcanic rocks on the Siberian craton erupted within a subsided Tunguska syncline and that these rocks were not preceded by any uplift, as would be expected from a plume model (e.g., Campbell and Griffiths, 1990; Campbell, 2005). Saunders et al. (2005) claimed that there was plume-related uplift in the area of the west Siberian basin and that this uplift is now hidden beneath Mesozoic sedimentary cover.

In Figure 4 a paleogeographic map of Siberia is shown for the Early Permian (Vinogradov, 1968). The subsidence-uplift pattern shown in Figure 4 remained virtually unchanged until the Permo-Triassic boundary, as reflected by paleogeographic maps reconstructed through different stratigraphic ages (Vinogradov, 1968). Deep rifts in the west Siberian basin (the long, linear south-north structures in Fig. 1) developed mainly in the Triassic along a preexisting long north-south linear uplift. Uplift commenced at least 25–30 m.y. before initiation of Siberian Traps magmatism. Tectonic development of this region may be interpreted in the context of collision between Euro-American (the left part of Fig. 4), Kazakh (bottom left part of Fig. 4), and Siberian (central part of the Fig. 4) paleocontinents (Podurushin, 2002). Saunders et al. (2005), however, argued for a relationship between the uplift and a plume. The question as to whether the subsidence-uplift history recorded for Siberia is associated with a plume is considered in more detail below.

Tectonic Setting

In the Permian and the Permo-Triassic, Siberia was part of Pangaea. It was surrounded by convergent plate boundaries with subduction of oceanic plates beneath the continent (Fig. 5) (Nikishin et al., 2002). In present-day coordinates, the southern remnant of the convergent plate boundary is represented by the Mongolia-Okhotsk suture zone (e.g., Zonenshain et al., 1990; Zorin, 1999). It is ~800 km south of the southeastern end of the Siberian Traps LIP (Fig. 1). Noril'sk is ~2500 km from the Mongolia-Okhotsk suture zone. According to paleotectonic reconstructions in the east and southwest (in present-day coordinates) there should be other Permian subduction systems (Fig. 5).

The Transbaikalian volcano-plutonic belt was located between the Mongolia-Okhotsk suture zone and the Siberian Traps LIP (Fig. 1). This belt developed in the Permian and the Triassic and is composed of granitic and syenitic batolites, basaltic and rhyolitic volcanic deposits, and dikes (e.g., Yarmolyuk et al., 2001). The origin of the belt is debated. Some authors attribute it to the Siberian Traps LIP (Dobretsov, 2003, 2005), but others consider it an independent intraplate rift phenomenon (Yarmolyuk et al., 2001) or a continental margin Andean-type orogen (Zorin et al., 1998; Zorin, 1999; Nikishin et al., 2002).

Geophysics

Lithospheric Structure. The upper mantle structure beneath Siberia has been investigated in a number of studies using nuclear explosion data (Thybo and Perchuc, 1997; Morozov et al., 1999; Egorkin, 2001, 2004; Pavlenkova et al., 2002; Pavlenkova, 2006; Pavlenkova and Pavlenkova, 2006). Notable features are alternating low- and high-velocity zones in the upper mantle, despite some mismatch between published results on their depths. It is not clear from the seismic data at what depth the lithosphere-asthenosphere boundary lies. Most likely the base of the lithosphere is at the so-called L-boundary (Pavlenkova et al., 2002; Pavlenkova, 2006; Pavlenkova and Pavlenkova,

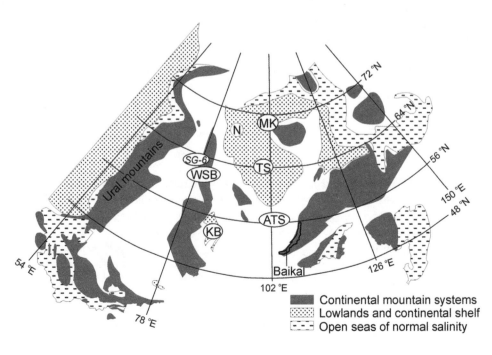

Figure 4. Paleogeographic reconstruction sketch for the Asselian-Sakmarian (Early Permian, ca. 280–290 Ma) of Siberia (simplified after Vinogradov, 1968). The locations of major Permo-Triassic units presented in Figure 1 and modern geographic coordinates are shown for orientation. SG6—superdeep drillhole 6; WSB—west Siberian basin; KB—Kuznetsk basin; N—Noril'sk; MK—Maimecha-Kotui; TS—Tunguska syncline; ATS—Angara-Taseevskaya syncline. The southern distribution of major elements is limited by the border of the former Soviet Union.

■ Continental mountain systems
▦ Lowlands and continental shelf
⌐--⌐ Open seas of normal salinity

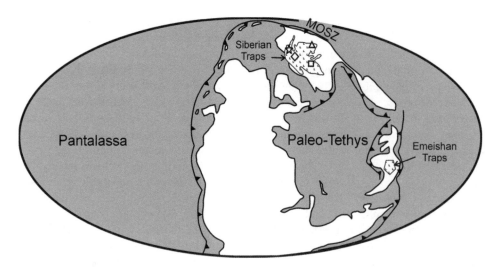

Figure 5. Paleotectonic map of Pangaea in the Late Permian (ca. 255 Ma). The white and gray fields represent paleolandmasses and paleo-oceans, respectively. The solid lines with black triangles mark convergent plate boundaries (simplified from http://www.scotese.com). The approximate positions of the Siberian Traps large igneous province and the Mongolia-Okhotsk suture zone are shown after Figure 1. The triangle, square, diamond, and star mark the approximate locations of samples from the Angara-Taseevskaya syncline, the West Siberian basin, and Noril'sk-Kharaelakh and Maimecha-Kotui subprovinces, respectively, discussed in the text.

2006). Lithospheric low-velocity zones have been interpreted as thin fluid-rich or partial melt–bearing layers (e.g., Thybo and Perchuc, 1997; Pavlenkova and Pavlenkova, 2006; Pontevivo and Thybo, 2006). A 3-D model of the L-boundary is shown in Figure 1 (from Pavlenkova and Pavlenkova, 2006). The thinnest rounded part corresponds to the position of the Tunguska syncline, and the thickest part underlies the west Siberian basin. The seismic structure disagrees with a thermal model that places the hottest, thinnest parts and the coldest, thickest parts of the lithosphere beneath the west Siberian basin and the Siberian craton, respectively (Artemieva, 2006).

Heatflow. The Siberian craton is characterized by a low mean heatflow of ~40 mW/m², which is typical for cratons (Lysak, 1984). According to thermal modeling (Zorin and Vladimirov, 1989; Hyndman et al., 2005) such a low heatflow implies a very long (billions of years) history of stability of the lithosphere. In other words, if the lithosphere was significantly thinned at ca. 250 Ma, elevated heatflow would be expected. But this is not observed. However, the calculated thermal thicknesses of the lithosphere vary widely, from 180 to 350 km (Artemieva, 2006). The thermal thickness of the lithosphere beneath the west Siberian basin is ~100 km, similar to that of many mobile belts (Artemieva, 2006). As noted earlier, these results disagree with seismic data (Fig. 1).

Geochemistry

Noril'sk-Kharaelakh Chemostratigraphic Subdivisions. The Noril'sk-Kharaelakh subprovince is characterized by the presence of economically important Cu-Ni-Pt deposits (e.g., Naldrett et al., 1992). Due to this, more than 3 km thickness of lava was drilled out, and samples recovered horizon-by-horizon were analyzed in great detail (e.g., Lightfoot et al., 1993; Wooden et al., 1993).

Lava strata in the Noril'sk-Kharaelakh subprovince were subdivided into several suites and subsuites (formations and subformations; Fedorenko, 1981; Zolotukhin et al., 1984, 1986)

(Fig. 6). These suites and subsuites were grouped into early, middle, and late assemblages, which in practice correspond to three major chemical units: high-Ti, low-Ti–2, and low-Ti–1 (Fedorenko, 1981; Fedorenko et al., 1996). High- and low-Ti series were distinguished by Lightfoot et al. (1993) on the basis of a TiO_2-Mg# diagram. Ivanov and Balyshev (2005) suggested using a discrimination line calculated as $TiO_2 = 3.45 - 0.0317 \times$ Mg#, where Mg# is $Mg/(Mg + 0.85Fe_{total})$ (elements are presented in atomic units, and Fe_{total} is calculated from FeO_{total}). High-Ti rocks are characterized by higher ratios of light and middle rare earth elements (REE) to heavy REE compared to the low-Ti rocks. For example, Fedorenko et al. (1996) separated rocks of the high- and low-Ti series by a divider of Gd/Yb = 2. These authors used the Th/U ratio to distinguish two subseries of the low-Ti rocks; a lower Th/U, generally below 3.5, characterized the low-Ti–1 subseries, and a higher Th/U, above 3.2, characterized the low-Ti–2 subseries. In the Δ8/4–Th/U diagram, basalts of the Noril'sk-Kharaelakh subprovince are separated into two groups (Fig. 7); a group of DUPAL basalts with Δ8/4 > ~50, Th/U > ~2.5, and with systematically more radiogenic Sr and Nd isotopes and a group of non-DUPAL basalts with Δ8/4 < ~70, Th/U < ~3.0, and nonradiogenic Sr and Nd isotopes (Δ8/4 is calculated as $100 \times [(^{208}Pb/^{204}Pb) / (1.209 \times {}^{206}Pb/^{204}Pb) + 15.627)]$ (Hart, 1984), reflecting a shift in the $^{208}Pb/^{204}Pb$– $^{206}Pb/^{204}Pb$ diagram from a linear regression line through north Atlantic MORBs. *DUPAL* is an acronym for the names of Dupré and Allègre (1985). It appears that DUPAL and non-DUPAL basalts alternate in the continuous lava pile (Fig. 6). Changes between different chemostratigraphic subdivisions are very sharp, without any mixing trends on isotopic and trace-element diagrams (Fig. 7).

DUPAL basalts are systematically depleted in Pt and Pd (usually at the detection limit of neutron activation, with the exception of part of the Tuklonsky Suite; Wooden et al., 1993). Non-DUPAL basalts yield higher concentrations of Pt and Pd distributed in a Gaussian fashion (Fig. 8). DUPAL rocks are generally more enriched in the radiogenic isotopes Sr and Nd,

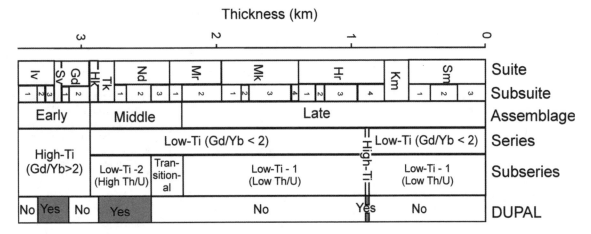

Figure 6. Chemostratigraphic subdivision of lavas at Noril'sk (modified from Lightfoot et al., 1993; Wooden et al., 1993; Fedorenko et al., 1996; Al'mukhamedov et al., 2004). *DUPAL* is an acronym of the names of Dupré and Allègre (1985). DUPAL is defined on the basis of the Δ8/4-Th/U diagram (Fig. 7), where Δ8/4 is the vertical deviation from linear regression for north Atlantic mid-ocean ridge basalts on the $^{208}Pb/^{204}Pb-^{206}Pb/^{204}Pb$ diagram (Hart, 1984). Suite names: Ivakinsky (Iv), Syverminsky (Sv), Gudchikhinsky (Gd), Hakanchansky (Hk), Tuklonsky (Tk), Nadezhdinsky (Nd), Morongovsky (Mr), Mokulaevsky (Mk), Haraelakhsky (Hr), Kumginsky (Km), and Samoedsky (Sm). Low-Ti lavas make up 80% of the lava sequence. DUPAL-type lavas make up up to ~20% of the lava sequence in both the high- and low-Ti units.

though there is a large overlap. Initial values at 250 Ma are as follows: $^{87}Sr/^{86}Sr$ from 0.7053 to 0.7088 and εNd from 2 to –11 for DUPAL and $^{87}Sr/^{86}Sr$ from 0.7041 to 0.7072 and εNd from 4.2 to –2.5 for non-DUPAL rocks, where εNd ≡ ($[^{143}Nd/^{144}Nd]_{sample}$ / $[^{143}Nd/^{144}Nd]_{chondrite}$ – 1) × 10,000 at 250 Ma. Limited data for osmium isotopes show that Tuklonsky DUPAL rocks and Gudchikhinsky$_2$ non-DUPAL rocks have similar positive γOs values of 3.4–6.5 and 5.3–6.1, respectively, where γOs ≡ ($[^{187}Os/^{188}Os]_{sample}$ / $[^{187}Os/^{188}Os]_{chondrite}$ – 1) × 100 at 250 Ma; Horan et al., 1995).

Such distribution of Pt and Pd together with radiogenic isotopes and trace-element data can be interpreted as evidence for an eclogitic source of the DUPAL basalts of the Noril'sk-Kharaelakh subprovince. It is worth mentioning that DUPAL basalts are interpreted as derivates of an eclogitic (recycled crustal) source everywhere in the world (see the discussion in Ivanov and Balyshev, 2005), including the classical DUPAL of the Indian Ocean (e.g., Escrig et al., 2004; Meyzen et al., 2005).

Figure 9 gives an overview of trace-element patterns in different chemostratigraphic units of the Noril'sk-Kharaelakh subprovince. It may be seen that there are two major types, with trace-element concentrations (1) at a level of ocean island basalts (OIB) and (2) at a level of enriched mid-ocean ridge basalts (E-MORB). The Nadezhdinsky Suite is enriched in the most incompatible elements (Rb to K), like OIB, but depleted in more compatible elements (Sr to Yb), like E-MORB. Despite the similar levels of concentration, the trace-element patterns exhibit more irregular patterns than do OIB and E-MORB. A general feature of most suites is depletion of Ta relative to La and K. The Gudchikhinsky Suite is an exception. However, it differs from E-MORB in depletion of K and Ba and absence of

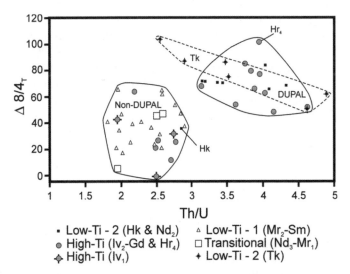

Figure 7. Subdivision of basaltic lavas at Noril'sk into DUPAL and non-DUPAL subsets on the basis of Th/U and Δ8/4 ratios, where Δ8/4 is calculated as $100 × [(^{208}Pb/^{204}Pb)/(1.209 × [^{206}Pb/^{204}Pb] + 15.627)]$ (Hart, 1984). The original data for the analysis are taken from the most complete source of data (Wooden et al., 1993). The subscript $_T$ refers to initial values at 250 Ma. Suite abbreviations: Iv—Ivakinsky; Sv—Syverminsky; Gd—Gudchikhinsky; Hk—Hakanchansky; Tk—Tuklonsky; Nd—Nadezhdinsky; Mr—Morongovsky; Mk—Mokulaevsky; Hr—Haraelakhsky; Km—Kumginsky; Sm—Samoedsky. The subscripted numbers added to suite abbreviations refer to subsuite numbers.

depletion in Pb relative to Ce and Pr. The Ivakinsky$_1$ Subsuite is similar to OIB, but differs from it in depletion of Sr and enrichment of Ba, Th, U, and P relative to neighboring elements.

Heavy REE are compatible (e.g., Green, 1994) in garnet,

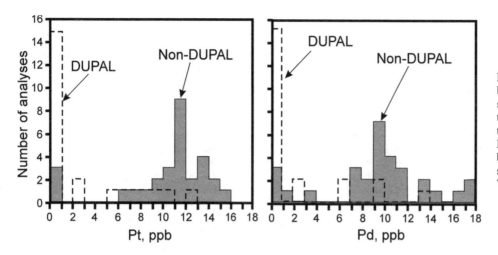

Figure 8. Distribution of Pt and Pd in basalts from the Noril'sk-Kharaelakh subprovince. All low Pt and Pd concentrations in non-DUPAL basalts belong to the Ivakinsky$_1$ subsuite. Most of the high Pt and Pd concentrations in DUPAL basalts are from samples of the Tuklonsky Suite. The original data are from Wooden et al. (1993).

and therefore melting of garnet-bearing mantle will produce particular REE patterns in the derivative melts with high ratios of light to heavy REE at low concentrations of heavy REE. In spinel, heavy REE are incompatible, producing low ratios of light to heavy REE with highly variable concentrations of the heavy REE (compare Models 1 and 2 in Fig. 10). This reasoning would lead to estimation that the source mineralogy of Siberian Traps basalts lies within the shallow (less than 50–60 km depth) spinel stability field of the mantle. The lithosphere is at least 180 km thick beneath the Tunguska syncline and more than 240 km thick beneath Noril'sk. The lithosphere could not be thinned to 60 km and restored to the normal cratonic value in only 250 m.y. (Zorin and Vladimirov, 1989; Hyndman et al., 2005). Therefore, either voluminous melting appeared solely within cold cratonic lithosphere, which is unlikely from the thermal point of view, or geochemical data require another explanation. From Mg# variations, only a few stratigraphic units at Noril'sk can be regarded as containing primary mantle-derived melts. These are the non-DUPAL Gudchikhinsky$_2$ Subsuite and the DUPAL Tuklonsky Suite (Figs. 6 and 10). Their compositions can be modeled as ~18% and ~45–70% partial melting from garnet lherzolite and eclogite, respectively (Fig. 10). All other stratigraphic units contain variably differentiated melts. Due to the differentiation, these melts are characterized by higher Yb content. It should be noted that the modeling presented in Figure 10 shows in principle the possibility of derivation of the Siberian Traps basalts from garnet-bearing sources with consequent differentiation rather than finding the best-fitting solution for each suite.

The Maimecha-Kotui Subprovince. In the Meimecha-Kotui subprovince, a lava stratum ~4 km thick formed. Initial eruptions of the Arydzhansky Suite high-Ti lava took place probably as early as the Late Permian, as shown by $^{40}Ar/^{39}Ar$ dating (Basu et al., 1995), but this was questioned by a Permo-Triassic U-Pb perovskite age (Kamo et al., 2003; see the earlier section on dating). A large part of the stratum consists of low-Ti basalts similar to the Noril'sk-Kharaelakh low-Ti basalts. The uppermost Maymechinsky Suite was preceded by the high-Ti Del-

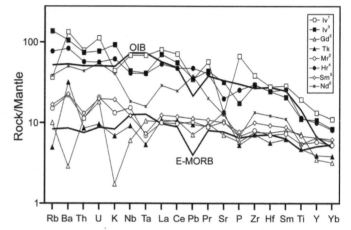

Figure 9. Primitive mantle-normalized trace-element diagrams for representative samples of different chemostratigraphic units of the Noril'sk-Kharaelakh subprovince. The data are from Wooden et al. (1993). Hypothetical pure compositions of ocean island basalt (OIB) and enriched mid-ocean ridge basalt (E-MORB) are shown after Sun and McDonough (1989). The primitive mantle composition is after McDonough and Sun (1995).

kansky Suite (Fedorenko and Czamanske, 1997). High-Ti lavas including unusual picritic lavas, meimechites, comprise ~50% by volume. The Maymechinsky Suite makes up no more than one-third of the total lava sequence. This suite is intruded by carbonatites with a U-Pb baddeleyite age of 250.2 ± 0.3 Ma (Kamo et al., 2003).

Meimechites are of particular interest. These are unusual high-Mg (MgO > 18 wt%), high-Ti (TiO$_2$ > 1 wt%) mafic and ultramafic rocks (SiO$_2$ < 52 wt%) (Le Bas and Streckeisen, 1991). They are characterized by porphiric texture, with olivine and serpentine phenocrysts in a dark feldspar-free groundmass. Primary biotite also occurs occasionally (e.g., Fedorenko and Czamanske, 1997). These rocks exhibit enriched OIB-type trace-element patterns without Ta-Nb depletion and depleted Sr and Nd isotopes ($^{87}Sr/^{86}Sr$ from 0.7030 to 0.7034 and εNd from 2.8

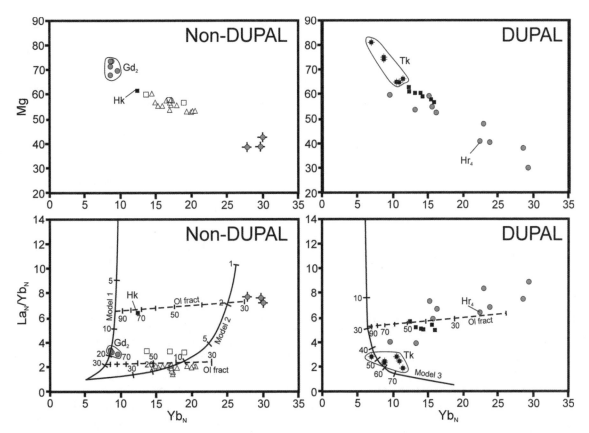

Figure 10. La_N/Yb_N and Mg# [mole MgO / (MgO + FeO$_t$)] versus Yb_N. The subscript $_N$ refers to chondrite-normalized values (McDonough and Sun, 1995). Models 1, 2, and 3 show equilibrium partial melting of garnet lherzolite, spinel lherzolite, and eclogite, respectively. The numbers close to the curves represent the percentage of partial melting. The dashed lines represent the equilibrium crystallization of olivine. The numbers close to the lines represent the percentage of remaining magma. Suite abbreviations: Iv—Ivakinsky; Sv—Syverminsky; Gd—Gudchikhinsky; Hk—Hakanchansky; Tk—Tuklonsky; Nd—Nadezhdinsky; Mr—Morongovsky; Mk—Mokulaevsky; Hr—Haraelakhsky; Km—Kumginsky; Sm—Samoedsky. The subscripted numbers added to suite abbreviations refer to subsuite numbers. The equation of Shaw (1970) was used for the calculations. The compositions of the modeled sources of melting were as follows: (1) garnet lherzolite—52% of Ol, 30% of Opx, 15% of Cpx, 3% of Gt; (2) spinel lherzolite—55% of Ol, 15% of Opx, 15% of Cpx, 15% of Spl; (3) eclogite—55% of Cpx, 45% of Grt; (where Ol, Opx, Cpx, Grt, and Spl are olivine, orthopyroxene, clinopyroxene, garnet, and spinel, respectively). The concentrations of La and Yb in lherzolites and eclogite were accepted as five times the chondritic values and as in normalized mid-ocean ridge basalt, respectively (Sun and McDonough, 1989; McDonough and Sun, 1995). The sources of the mineral/melt distribution coefficients are as follows: Ol and Opx—Dunn and Sen (1994); Cpx—Hart and Dunn (1993); Grt—Zack et al. (1997); Spl—Kelemen et al. (1990). The compositions of the Siberian Traps at Noril'sk are after Wooden et al. (1993). Symbols as in Figure 7.

to 5.9; Arndt et al., 1995). Arndt et al. (1995) concluded that it is not evident from what source (lithospheric or sublithospheric) the meimechites originated. Kogarko and Ryabchikov (1995) suggested they formed from interaction of sublithospheric melts with lithospheric refractory harzburgites. In both cases a low degree of melting for the primary meimechite melts was accepted (Arndt et al., 1995; Kogarko and Ryabchikov, 1995).

A high Mg content is usually considered evidence of high temperature. On the basis of analysis of melt inclusions in the meimechites and numerical calculations, Sobolev et al. (1991) suggested that meimechites were derived from a depth of 230–300 km at a temperature of 1800–1900 °C due to mantle upwelling of a plume from greater depths (~700 km). Elkins-Tanton

et al. (2007) performed an experimental study of the meimechites and concluded that their primary melts were derived from a source with as much as 1 wt% of H_2O at ~180 km depth and a temperature of 1700 °C. They noted that incorporation of additional H_2O and CO_2 would lower the temperature to 1550 °C.

The West Siberian Basin and Angara-Taseevskaya Syncline. Medvedev et al. (2003) and Reichow et al. (2005) provided analytical data on almost the same samples recovered from a number of drillholes in the west Siberian basin. The volcanic rock compositions vary from mafic to acidic. If only mafic rocks are considered, according to these authors' data and the classification scheme of Fedorenko et al. (1996), all but one sample belongs to the low-Ti basalts (tholeiites and shoshonites).

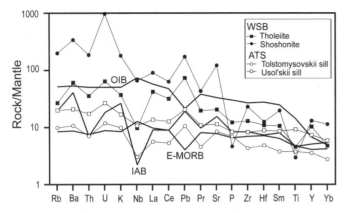

Figure 11. Primitive mantle-normalized trace-element diagrams for representative samples from the west Siberian basin (WSB) and the Angara-Taseevskaya syncline (ATS). Data are after Medvedev et al. (2003) and Ivanov et al. (2003), respectively. Hypothetical pure compositions of ocean island basalt (OIB) and enriched mid-ocean ridge basalt (E-MORB) are shown after Sun and McDonough (1989). The average of high-Mg basalts from Klyuchevskoi volcano, Kamchatka, is shown as representative of island-arc basalt (IAB) (Dorendorf et al., 2000). The primitive mantle composition is after McDonough and Sun (1995).

Shoshonites and tholeiites exhibit similar island arc basalt (IAB)–like trace-element patterns, with Ba, U, Pb, and Sr relative enrichment and Nb relative depletion, though shoshonites are characterized by higher concentrations of most trace elements (Fig. 11). The dolerite sills of the Angara-Taseevskaya syncline are characterized by lower concentrations of the trace element at the level of E-MORB. But the IAB-like trace-element pattern is also typical of the dolerite sills. Usol'skii and Tolstomysovskii sill dolerites are geochemical analogues of the low-Ti–1 subseries and the low-Ti–2 subseries (Tuklonsky Suite) of the Noril'sk-Kharaelakh subprovince, respectively. This is reflected in their almost identical trace-element patterns and Sr isotopes (Ivanov et al., 2003).

Petrographic examination of the dolerite sills revealed that some samples contain primary magmatic mica, probably biotite (Fig. 12).

DISCUSSION: EVIDENCE AND MODELS

Impact Model

Interest in the impact model for the origin of the Siberian Traps FBP is mainly due to the idea that collisions of large extraterrestrial bodies with the Earth cause biotic mass extinctions, an idea that was originally suggested for the Cretaceous-Tertiary boundary (Alvarez et al., 1980) and also debated over years for the Permo-Triassic boundary (see Erwin et al., 2002, for a review). Two varieties of the impact model can be discussed in relation to the Siberian Traps FBP: One proposes that the impact site was somewhere within the Siberian Traps FBP (Jones et al., 2002), and the other proposes that the impact site was antipodal to the Siberian Traps FBP (Hagstrum, 2005). The former model has related Cu-Ni-Pt mineralization in the Siberian Traps and impactor material. The latter model could explain the absence of a Permo-Triassic impact crater because of its disappearance due to subduction of the Pantalassa oceanic flow (see Fig. 5). In any case, the sizes of the impactor and the impact crater must be

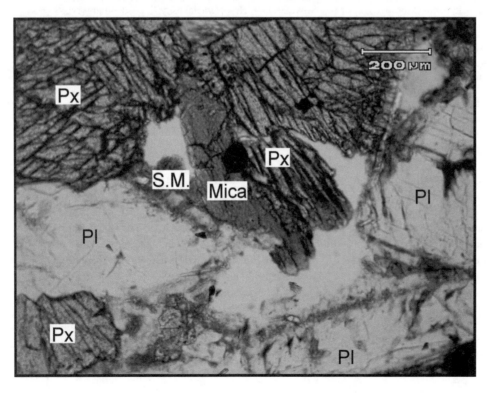

Figure 12. A view of a dolerite thin section (sample ST-21-05 from Tolstomysovskii sill) under a polarized microscope (parallel nicols). Px—pyroxene; Pl—plagioclase; S.M.—secondary mineral developed on plagioclase. The small black dots indicate ore minerals. Olivine is rare (out of the image). Mica crystals show intergrowths with, mainly, pyroxene. The dolerite sill geology is described in detail by Feoktistov (1979) and briefly by Ivanov et al. (2005).

on the order of the Cretaceous-Tertiary meteorite and the Chicxulub Crater or even larger (≥10–15 km for the impactor and 200–300 km for the crater). However, each of these two impact models meets irresolvable problems in explaining the volume of the outpoured basaltic magma and the size of the Siberian Traps LIP (Fig. 1; Table 1). Various calculations suggest that an impact of such a size might have initiated magma generation on the order of 10^5 km^3 (Glikson, 1999; Ivanov and Melosh, 2003). If an initial hot geotherm and thin oceanic lithosphere is considered, an increase in magma volume of up to 2.5×10^6 km^3 would be expected due to the additional effect of decompression (Jones et al., 2005). As Jones et al. (2005) noted, the largest impact would have generated a footprint up to 1000 km in diameter if it occurred on oceanic lithosphere, but much smaller if it occurred on continental lithosphere. Therefore, the extent of the Siberian Traps LIP (Fig. 1) is too large even for the most favored conditions of the impact model.

Besides the previously mentioned problems, a large impact would have left geochemical fingerprints in Permo-Triassic boundary sedimentary records. All earlier findings of such fingerprints have either been disproved or at least engendered doubts (see the critiques by Erwin et al., 2002, and by Koeberl et al., 2002, for reviews). The apparently prolonged volcanism of the Siberian Traps FBP, starting in the Late Permian and continuing until the end of the Middle Triassic, is also against the impact model.

Plume Model

To explain the large size of LIPs, Campbell and Griffiths (1990) suggested a starting plume head model in which a plume has a particular flattened head and a thin tail structure. The head is responsible for the voluminous eruptions. As Campbell (2005) recently pointed out, this model makes testable predictions: (1) new plumes consist of a large head and a thin tail; (2) flattened plume heads should be 2000–2500 km in diameter; (3) plumes must originate from a hot boundary layer, probably the core-mantle boundary; (4) both head and tail should erupt high-temperature picrites; (5) the temperature excess of a plume head is highest at the center of the head and decreases toward the margin; (6) picrites should erupt early during flood volcanism and be most abundant near the center of the plume head and less abundant toward the margin; and (7) flood basalt should be preceded by domal uplift of 500–1000 m at the center of the dome. Not all of the listed predictions can be tested for an ancient event such as that which produced the Siberian Traps FBP without additional assumptions. The following assumptions are numbered to correspond to the related predictions.

1. For example, we do not know whether a Siberian plume, if one existed, had a head-and-tail structure. According to the formulated plume model, the head and tail should yield a LIP (FBP) and a continuous chain of volcanoes, respectively (Campbell, 2005). A continuous volcanic chain from the Siberian Traps LIP to any active or more recent hotspot is unknown, though some authors speculate that there is a link between Iceland and the Siberian Traps and that the volcanic track between them is hidden beneath the polar ocean (e.g., Burke and Torsvik, 2004; Chernysheva et al., 2005).

2. The Siberian Traps LIP is larger than the expected 2000–2500 km circular flattened head (Fig. 1). Actually, the Siberian Traps do not exhibit a rounded structure, but rather their volcanic rocks filled preexisting elongated depressions (compare Figs. 1 and 4).

3. The depth of origin of the supposed plume cannot be inferred by any means, though Sobolev et al. (1991) interpreted their data on meimechites to suggest that these melts came from as deep as 700 km. On the basis of these extrapolations, the depth of origin and how to find it are both unclear.

4. Picrites are known from many parts of the Siberian Traps LIP. However, at least some are cumulates (for example, picrites within sills of the Angara-Taseevskaya syncline). Detailed petrographic studies are required from all over the Siberian Traps FBP to test a primary versus cumulative origin of the picrites.

5. The temperature profile of the mantle at 250 Ma is not known. However, we may expect the greatest erosion of the lithosphere where the plume is hottest, i.e., at its center. The geographical center of the Siberian Traps LIP is the western part of the Tunguska syncline (Fig. 1). The lithosphere there is significantly thinned (Fig. 1). This is in agreement with the prediction. However, because of the absence of uplift in the Tunguska syncline (Fig. 4), Saunders et al. (2005) placed the plume center beneath the West Siberian basin. The lithospheric thickness is great there (Fig. 1).

6. In the Noril'sk-Kharaelakh subprovince, picrites are recorded within the Gudchikhinsky and Tuklonsky suites (see Fig. 6 for the stratigraphic locations of the suites). These are not the earliest suites. In the Maimecha-Kotui subprovince, the most magnesium, and expectedly the highest-temperature melts were erupted at the very end of the Permo-Triassic volcanism pulse, once more the opposite of what is predicted. If we locate the plume center in the geographical center of the Siberian Traps LIP, the meimechites appear to be at the margin of the supposed plume head. If instead we assume that the meimechites mark the position of the plume center, the plume would have been highly asymmetric even if the FPB extended into the Kara and Laptev undersea areas (Fig. 1).

7. There was no domal uplift at the center of the suggested plume head. Both the Tunguska and the Maimecha-Kotui subprovinces are located in subsided areas (Fig. 4). Burov and Guillou-Frottier (2005) calculated that a plume beneath a continent may result in a complex pattern of uplifts and sedimentary basins. They even showed that a subsidence may overlie the center of the plume head. One may then question the testability of the domal uplift prediction, when

the presence of uplift prior to emplacement of the Emeishan Traps (He et al., 2003) is taken as classical evidence of a plume (Campbell, 2005), but the presence of subsidence prior to the eruption of the Siberian Traps is also taken as evidence of a plume. However, even incorporation of a "realistic lithosphere model" cannot explain the observed subsidence-uplift pattern (Fig. 4) in the context of the plume model. According to the calculations of Burov and Guillou-Frottier (2005), while a plume is entering the upper mantle (crossing the 660 km discontinuity) a small to moderate uplift occurs above the plume center. After a few m.y. the uplift may be converted to deep subsidence. The subsidences in the Tunguska and Maimecha-Kotui subprovinces originated long before the initiation of the volcanism (at least 25–30 m.y. before), and these subsidences remained virtually unchanged until the Permo-Triassic volcanism (Vinogradov, 1968).

There are a number of possible explanations as to why the evidence expected on the basis of the plume model is not observed in the Siberian Traps (for example, picrite melts could have ponded in crustal magmatic chambers and not reached the surface). However, a successful theory cannot be built on non-observed evidence. In my view, much available evidence contradicts the expectations of the plume model, while data on paleogeography argue strongly against the plume model. The only supporting evidence is the high temperatures calculated for the meimechite melts (Sobolev et al., 1991;), which are hard to explain in terms of conventional thinking regarding a "dry" peridotitic source of melt. The addition of volatiles lowers the melting temperature required to produce meimechites (Elkins-Tanton et al., 2007). The effect of volatile addition thus requires further thorough investigation.

Wooden et al. (1993) considered geochemical aspects of the plume model for the Noril'sk-Kharaelakh subprovince. These authors suggested a model in which a deep crustal reservoir is periodically replenished with sublithospheric magma (the Ivakinsky$_1$ Subsuite type of magma), and tapped. Fractionation and assimilation of wallrocks leads to subduction-like magmas (e.g., the Tuklonsky and Nadezhdinsky Suite types of magma; see Fig. 9 and 11). Such a model would explain the alternating non-DUPAL and DUPAL lavas (Fig. 6), which would represent sublithospheric melts and crustal contaminated melts, respectively. However, this model is unlikely to pertain for several reasons. First, there are no mixing trends between "uncontaminated" (Ivakinsky$_1$) and "contaminated" (e.g., Tuklonsky and Nadezhdinsky Suite) magmas, as would be expected (Fig. 7). Second, it does not explain why the non-DUPAL lavas of the late assemblage bear subductionlike signatures that make them remarkably different in trace-element geochemistry from the Ivakinsky$_1$ lavas (Fig. 9). Involvement of lithospheric mantle instead of lower crust to explain the subductionlike trace-element patterns would also require mixing trends between uncontaminated and contaminated melts that are not observed (Fig. 7).

The subductionlike trace-element pattern of the low-Ti basalts has usually been considered a fingerprint of lithospheric contamination (e.g., Lightfoot et al., 1993; Reichow et al., 2005, for the Siberian Traps). Recently, Kieffer et al. (2004) studied the question of lithospheric sources for Oligocene Ethiopian traps, which also contain typical high- and low-Ti basalt series, and concluded that "the lithospheric mantle did not contribute significantly to the formation of any of the Oligocene lavas from northern Ethiopia." This shows that the interpretation of subduction-like trace-element patterns in continental basalts as lithospheric melts is at least not unique. Reviewing lithospheric mantle melting literature, Kieffer et al. (2004) posed the question "Is the label 'lithosphere' just given to the source of any magma whose composition is thought to be inconsistent with that of an asthenosphere or plume source?"

The timing of volcanism (Fig. 3) can be explained by a starting plume model if dense eclogite patches (e.g., pieces of subducted oceanic crust) were incorporated into the plume (Lin and van Keken, 2005). However, the plume model is not a unique explanation for the pulsing volcanism (see later discussion).

Edge-Driven Convection Model

The edge-driven convection model was suggested by King and Anderson (1998) as an alternative to the plume model. Its basic idea is that at a boundary separating thick cratonic lithosphere and thinner lithosphere, upper-mantle convective flow may be controlled by the lithospheric structure. Puffer (2001) combined this model with evidence of IAB-like trace-element features in Siberian Traps basalts. According to Puffer (2001), lithospheric mantle of the Siberian craton attained subduction-like features (trace elements characteristic of the upper mantle wedge) long before the Permo-Triassic magmatism as a result of paleosubductions. This source, which contained a fusible water-rich mineral assemblage, melted, producing voluminous but brief volcanism as a result of redistribution of heat in the upper mantle due to craton-induced convection. However, this model cannot explain the size of the Siberian Traps LIP (Fig. 1) and the subduction-like features in the volcanic rocks both on and off the craton (Fig. 11). Nor can it explain the alternating pattern of DUPAL and non-DUPAL basalts in the Noril'sk-Kharaelakh subprovince unless the model is coupled with a lithospheric delamination model (see later discussion). It does not resolve the problem of the high temperatures of the meimechite melts.

Lithospheric Delamination Model

Elkins-Tanton (2005) suggested a model in which a weak plume caused a low degree of partial melting. Melt crystallized within the lithosphere in the form of dense eclogites and the lithospheric root, subsequently delaminated via a Rayleigh-Taylor instability. After the delamination, deeper sublithospheric material upwelled into the place of the delaminated lithospheric root and melted to a high degree. Combining this model and the observation of alternating DUPAL and non-DUPAL rocks (Fig. 6), I suggest the following scenario. The sublithospheric melts are represented by Ivakinsky$_1$ Subsuite lavas. The delaminated lithospheric root is represented by an eclogite layer, which would be stable at ~185 km depth (Anderson, 2006). After delamination the eclogitic material crossed the solidus and almost completely melted. These melts are represented by the Ivakinsky$_2$ and Ivakinsky$_3$ Subsuites. Sublithospheric material rose and produced non-DUPAL Gudchikhinsky melts. Another portion of the lithospheric eclogite delaminated and, after crossing the eclogite solidus, melted again, producing the Tuklonsky and Nadezhdinsky Suite DUPAL basalts. Next, a portion of the sublithospheric material rose to a shallow level, where it melted to a high degree, producing monotonous non-DUPAL lavas of the late assemblage.

Lithospheric thinning occurred beneath the Tunguska syncline (Fig. 1), which is filled by the most voluminous volcanics (Vasil'ev et al., 2000), in agreement with the delamination model. The number of volcanic pulses (Fig. 3) could be explained as the number of delamination events that occurred in different parts of the Siberian Traps FBP.

However, geochemically similar low-Ti basalts were erupted in different parts of the Siberian Traps FBP (see the section headed "Geochemistry"). Their similar trace-element compositions require similar mechanisms of origin. Therefore, a single weak mantle plume could not have produced the whole FBP via the process of delamination. If several plumes were involved, this model would inherit all the problems of the classical mantle plume model discussed earlier. In other words, the delamination model requires a cause for the delamination on the scale of the whole Siberian Traps FBP. In addition, it offers no solution for coeval basaltic effusive and acidic intrusive volcanism.

Subduction-Related Model

Low-Ti volcanic rock is the dominant rock type among the mafic rocks of the Siberian Traps FBP. In the Noril'sk-Kharaelakh subprovince it comprises up to 80% of the basalts by volume, and in the west Siberian basin and the Angara-Taseevskaya syncline practically all the rocks belong to this type. In the Maimecha-Kotui subprovince low-Ti rocks are the lowest percentage, but they still make up to half of the lava sequence. The important feature of the low-Ti rocks is their IAB-like trace-element pattern (Figs. 9 and 11). Interestingly, the percentage of low-Ti rocks seems to be higher in the southern part of the Siberian Traps FBP and lower in the northern part. If this feature is not an artifact of incomplete sampling, it must correspond to the position of the erupted rocks relative to the paleosubduction systems (Fig. 5); the closer to the subduction system, the more prominent the IAB-like fingerprint would be expected to be. This geochemical evidence, coupled with location of the Siberian Traps FBP in the back-arc tectonic setting (Fig. 5), led to the suggestion that subduction beneath the Siberian part of Pangaea and voluminous eruptions of the Siberian Traps FBP are two

related phenomena (Ivanov and Balyshev, 2005). A similar conclusion was reached by Zhu et al. (2005) for the Emeishan Traps FBP, which formed in the Late Permian in a similar tectonic setting (Fig. 5). These authors suggested interaction between a lower-mantle plume, depleted upper mantle, and a recycled subduction component. The question, however, is this: "Is the lower-mantle plume necessary?" In the case of the Siberian Traps FPB, the lower-mantle plume is necessary only to explain the meimechite petrology, which requires too high a temperature for a likely origin in upper-mantle plate tectonic processes. However, the meicmechite source could be wet (>1 wt% of water), and if so, a high-temperature model is not necessary. This question is thus of particular interest for future investigation.

Ivanov and Balyshev (2005) suggested that the mantle transition zone beneath the Siberian part of Pangaea was saturated by water from subduction during the Permian. Dewatering of the transition zone during the latest Permian created a wet upper-mantle source that melted to a high degree. Silver et al. (2006) argued that rapid eruption of voluminous basalts within cratons could be related to tectonic events that were preceded by relatively long-term (>1 m.y.) supersolidus conditions. They discussed several possible causes of the supersolidus conditions maintained beneath the cratons, one of which is subduction-related increase in volatile components.

This model can explain the large volume of eruptions. For example, in Kamchatka the two most productive volcanoes, neighboring Klyuchevskoi and Tolbachik volcanoes, lie above the edge of a modern subducting slab in an area where lateral flow of normal asthenospheric mantle may influence the hydrated mantle-wedge source (Portnyagin et al., 2005). These volcanoes are characterized by averaged Holocene eruption rates of ~0.02 and 0.01 km^3 per year of essentially basaltic magma (Fedotov, 1984). In 1975–1976, the large Tolbachick fissure erupted 2.2 km^3 of lava and pyroclastics, which is comparable to the volume of the largest fissure eruptions in Iceland and Hawaii. The total volume of Holocene volcanic rocks in the vicinity of Tolbachik volcano is ~10^2 km^3 (Fedotov, 1984). The so-called zone of the Klyuchevskoi group of volcanoes is ~110 km long and up to 70 km wide (Fedotov, 1984). That is not much smaller than the size of the northern subprovinces of the Siberian Traps (e.g., Noril'sk-Kharaelakh, ~330 × 110 km; Khantaisk-Rybninsk, ~250 × 110 km; Kamensk, ~390 × 110 km; Kheta, ~390 × 130 km; Kureika-Letninsk, ~330 × 150 km; Fig. 2). Extrapolating these values to the size of the whole Siberian Traps FBP easily makes the total volume on the order of $4 × 10^6$ km^3 (Table 1). A direct comparison of island arc and Siberian Trap volcanism is not necessary because of the differences in lithospheric thinning, depth of melting, water recycling style, and so on.

This model may explain the geochemistry of the Siberian Traps FBP and was inspired by the IAB-like geochemistry. Alternating DUPAL and non-DUPAL rocks in the Noril'sk-Kharaelakh subprovince can be explained by recycling of the oceanic crust (eclogites). Patches of eclogite material could have been brought to sublithospheric depths by composite diapirs such as those modeled by Yasuda and Fujii (1998). Eclogite has a much lower solidus than lherzolite and will melt at high (50–70%) degrees of partial melting, while lherzolitic ambient rocks remain solid (Anderson, 2005). This could explain the practical absence of mixing between DUPAL and non-DUPAL melts (Fig. 7).

This model probably can explain the eruptions in a several short pulses during the 20–25 m.y. overall duration of the magmatism and the temporal and spatial coincidence of basaltic and acidic magmatism (granitic batholites are usual in subduction setting). The model does not require uplift, as the plume model does. Because the essence of the model is chemical modification of the upper mantle with lowering of its solidus, it does not require thinning of the lithosphere, though preexisting thinning beneath the Tunguska syncline would have increased melt productivity. The primary mica in typical dolerite samples of the Angara-Taseevskaya syncline (Fig. 12) and the meimechites of the Maimecha-Kotui subprovince (Fedorenko and Czamanske, 1997), as well as the shoshonites in the west Siberian basin (Medvedev et al., 2003), can be regarded as evidence for water in the source of melting.

Saturation of the transition zone with water is a popular model (Bercovici and Karato, 2003; Ohtani, 2005). The transition zone has a high water capacity due to the high water capacity of its major minerals, wadsleyite and ringwoodite (Ohtani, 2005). Superhydrous phases can survive under fast subduction

conditions in a subducting slab (Litasov and Ohtani, 2003) (Fig. 13). In the anhydrous mineral stability field at depths of 200–250 km, the water could be transported by nominally anhydrous minerals such as omphacite, garnet, and rutile (Katayama et al., 2006). Due to the metastable mineral assemblage, the subducting slab tends to deflect horizontally while crossing the 410 and 660 km discontinuities (Bina et al., 2001). Figure 13 shows that subducting slabs should have recharged the transition zone with water throughout geological history. Increasing the amount of water may lead to partial melting within the transition zone (Huang et al., 2005). The mantle transition zone is denser than silicate melt, but in the bottom part of the upper mantle, silicate melt is, conversely, denser than ambient peridotitic mantle (Matsukage et al., 2005; Sakamaki et al., 2006). However, increasing the amount of water leads to a decrease in melt density. Current experimental studies, which have large uncertainties, suggest that the melt will rise from the bottom of the upper mantle due to its own buoyancy at ~6 wt% of saturation without any additional source of heating. Dense hydrous magnesian silicate E, which is stable at these depths (Fig. 13), contains 11.4 wt% H_2O (Ohtani, 2005). With conductive heating, the phase E will decompose, releasing H_2O into the silicate melt. Therefore, the creation of high water concentrations in silicate melts in the bottom part of the upper mantle is not impossible in principle. The process of saturation of the transition zone with water will be inevitably reflected on the Earth's surface as volcanism. Here I suggest that Siberian Traps vol-

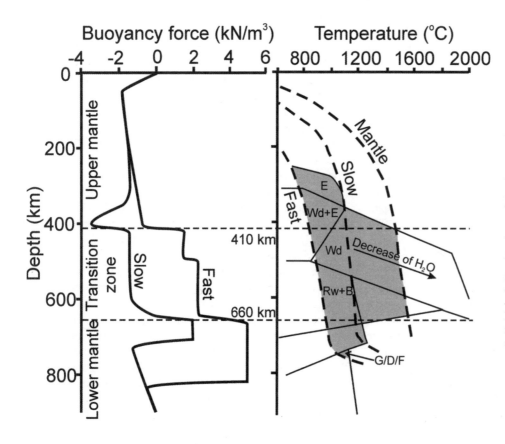

Figure 13. Saturation of the transition zone with water via subduction. The left panel shows that a fast subducting slab attains positive buoyancy while crossing the 410 km discontinuity. Both fast and slow subducting slabs attain positive buoyancy while crossing the 660 km discontinuity (after Bina et al., 2001). The right panel shows the stability fields (shadowed) of deep water-bearing minerals in the CaO-MgO-Al_2O_3-SiO_2-H_2O system (after Litasov and Ohtani, 2003). Water-bearing minerals: Wd—wadsleyite; Rw—ringwoodite; E—dense hydrous magnesian silicate; B and G/D/F —superhydrous phases. The straight lines show boundaries between different mineral assemblages (not included for simplicity).

canism and other similar trap volcanism is likely to have been formed by this process.

Subduction-Related Model and Other Traps

Almost thirty years ago, Cox (1978) noted that the Parana, Karoo, and Beacon (Ferrar and Kirkpatrick) traps erupted within the southern part of Gondwana in back-arc tectonic settings, and because of this coincidence he speculated that their origin was linked to subduction. Other traps, not investigated but mentioned in conjunction with the subduction-related model, were the Deccan Traps, the Siberian Traps, the North Atlantic igneous province, and the Columbia River plateau. The detailed investigation of the Siberian Traps presented here shows that the subduction-related model is viable. The origin of the Columbia River plateau as a back-arc phenomenon was considered by Smith (1992). In addition to these examples, the Emeishan Traps were considered in relation to subduction by Zhu et al. (2005). All these examples have abundant or dominant low-Ti basalts. The Oligocene Ethiopian province is one of the few examples that cannot be directly related to subduction. Low-Ti basalts are also abundant there (Kieffer et al., 2004). If all low-Ti basalts originated as a result of hydration of the sublithospheric mantle, as suggested here, this implies that dehydration of transition zones may appear not only in areas of nearly coeval subduction but also in regions of paleosubductions. A relationship between Cenozoic African volcanism and ancient paleosubducted slab was suggested by Balyshev and Ivanov (2001).

CONCLUSIONS

If the correct origin of the Siberian Traps is to be known, a number of important pieces of evidence must be explained:

1. The Siberian flood basalts erupted in a number of brief volcanic events from the Late Permian until the end of the Middle Triassic. Acidic intrusive magmatism was almost coeval with the basaltic volcanism.
2. The Siberian Traps erupted within the Pangaean supercontinent in a back-arc tectonic setting.
3. The size and volume of the Siberian Traps flood basalts were ~7×10^6 km^2 and 4×10^6 km^3. Most of the erupted rocks are within the Tunguska syncline. The abundant tuff units correspond to the syncline.
4. The Tunguska syncline experienced long-term subsidence prior to the initiation of the volcanism. The present-day lithosphere is thinned beneath the same region.
5. Low-Ti volcanic rocks with prominent subduction-like trace elements are the dominant rock type within the Siberian Traps. High-Ti rocks are subordinate. In the Noril'sk area, DUPAL and non-DUPAL types of basalts alternate. There is no evidence for mixing between the different rock types.
6. Petrographic examination of volcanic rocks reveals primary magmatic mica in some samples (though the num-

ber of samples with primary water-bearing minerals is not known).
7. The inferred temperature of meimechite melts in dry (or semidry) conditions is too high for the upper mantle, and the volatile contents of the source is unknown.

Various models for the origin of the Siberian Traps discussed in the literature were tested against the stated evidence. The models that fit the observations most poorly seem to be the impact and edge-driven convection models. The plume model can explain the evidence if some expected evidence has not yet been (or cannot in principle be) observed. The high inferred temperature of the meimechites is the only real supporting evidence. However, the addition of volatiles lowers the temperature of the meimechite primary melts.

The lithospheric delamination model is viable, though it has problems with explaining the size of the Siberian Traps and the causes of delamination on the scale of the Siberian Traps FBP. It offers no solution for coeval effusive basaltic and intrusive acidic magmatism. It explains the meimechite origin as a volatile-rich source of melting.

A subduction-related model is proposed in this chapter. This model can explain the enormous volume and size of the Siberian Traps, their tectonic setting, the coeval basaltic and acidic magmatism, the geochemistry of the dominant low-Ti basalt type, and the presence of water-bearing minerals in rocks. This model can explain meimechite origin without a thermal (plume) anomaly by the addition of large quantities of volatiles. The various pieces of evidence, coupled with assessment of existing models, enable us to draw the conclusion that the Siberian Traps phenomenon was related to an upper-mantle plate tectonics process and not a lower-mantle plume.

ACKNOWLEDGMENTS

Gillian R. Foulger motivated me to write this chapter. Linda Elkins-Tanton and Andrew D. Saunders made valuable comments on the manuscript and offered sensible criticism of the subduction-related model. Gillian R. Foulger and Donna M. Jurdy provided helpful editorial support. Viktor V. Rybov and Artem Y. Shevko helped with locating samples provided by Russian geologists for the ^{40}Ar/^{39}Ar study of Renne and Basu (1991). The work was supported by Russian Foundation for Basic Research Projects 05-05-64477 and 05-05-64281.

REFERENCES CITED

Al'mukhamedov, A.I., Medvedev, A.Ya., and Zolotukhin, V.V., 2004, Chemical evolution of the Permian–Triassic basalts of the Siberian Platform in space and time: Petrology, v. 12, p. 297–311.

Alvarez, L.W., Alwarez, W., Asaro, F., and Michel, H.V., 1980, Extraterrestrial cause for the Cretaceous–Tertiary extinction: Science, v. 208, p. 1095–1108, doi: 10.1126/science.208.4448.1095.

Anderson, D.L., 2001, Top-down tectonics: Science, v. 293, p. 2016–2018, doi: 10.1126/science.1065448.

Anderson, D.L., 2005, Large igneous provinces, delamination, and fertile mantle: Elements, v. 1, p. 271–275.

Anderson, D.L., 2006, Speculations on the nature and cause of mantle heterogeneity: Tectonophysics, v. 416, p. 7–22, doi: 10.1016/j.tecto.2005.07.011.

Arndt, N., Lehnert, K., and Vasil'ev, Y., 1995, Meimechites: Highly magnesian, contaminated alkaline magmas from the subcontinental lithosphere: Lithos, v. 34, p. 41–59, doi: 10.1016/0024-4937(94)00032-W.

Artemieva, I.M., 2006, Global 1 deg × 1 deg thermal model TC1 for the continental lithosphere: Implications for lithosphere secular evolution: Tectonophysics, v. 416, p. 245–277.

Baksi, A.K., and Farrar, E., 1991, ^{40}Ar/^{39}Ar dating of the Siberian Traps, USSR: Evaluation of the ages of the two major extinction events relative to episodes of flood-basalt volcanism in USS and the Deccan Traps, India: Geology, v. 19, p. 461–464, doi: 10.1130/0091-7613(1991)019<0461:ADOTST>2.3.CO;2.

Balyshev, S.O., and Ivanov, A.V., 2001, Low-density anomalies in the mantle: Ascending plumes and/or heated fossil lithospheric plates: Doklady Earth Sciences, v. 380, p. 858–862.

Basu, A.R., Poreda, R.J., Renne, P.R., Teichmann, F., Vasiliev, Yu.R., Sobolev, N.V., and Turrin, B.D., 1995, High-^{3}He plume origin and temporal-spatial evolution of the Siberian flood basalts: Science, v. 269, p. 822–825.

Bercovici, D., and Karato, S.-I., 2003, Whole-mantle convection and the transition-zone water filter: Nature, v. 425, p. 39–44, doi: 10.1038/nature01918.

Bina, C.R., Stein, S., Marton, F.C., and Van Ark, E.M., 2001, Implications for slab mineralogy for subduction dynamics: Physics of the Earth and Planetary Interiors, v. 127, p. 51–66, doi: 10.1016/S0031-9201(01)00221-7.

Bowring, S.A., Erwin, D.H., Jin, Y.G., Martin, M.W., Davidek, K., and Wang, W., 1998, U/Pb zircon geochronology of the end-Permian mass extinction: Science, v. 280, p. 1039–1045, doi: 10.1126/science.280.5366.1039.

Burke, K., and Torsvik, T.H., 2004, Derivation of large igneous provinces of the past 200 million years from long-term heterogeneities in the deep mantle: Earth and Planetary Science Letters, v. 227, p. 531–538, doi: 10.1016/j.epsl.2004.09.015.

Burov, E., and Guillou-Frottier, L., 2005, The plume head–continental lithosphere interaction using a tectonically realistic formulation for the lithosphere: Geophysical Journal International, v. 161, p. 469–490, doi: 10.1111/j.1365-246X.2005.02588.x.

Campbell, I.H., 2005, Large igneous provinces and the mantle plume hypothesis: Elements, v. 1, p. 265–270.

Campbell, I.H., and Griffiths, R.W., 1990, Implications of mantle plume structure for the evolution of flood basalts: Earth and Planetary Science Letters, v. 99, p. 79–93, doi: 10.1016/0012-821X(90)90072-6.

Chernysheva, E.A., Kharin, G.S., and Stolbov, N.M., 2005, Basaltic magmatism in Arctic seas related to the Mesozoic activity of the Iceland plume: Petrology, v. 13, p. 289–304.

Courtillot, V., Davaille, A., Besse, J., and Stock, J., 2003, Three distinct types of hotspots in the Earth's mantle: Earth and Planetary Science Letters, v. 205, p. 295–308.

Cox., K.G., 1978, Flood basalts, subduction and the break-up of Gondwanaland: Nature, v. 274, p. 47–49.

Czamanske, G.K., Gurevich, A.B., Fedorenko, V., and Simonov, O., 1998, Demise of the Siberian plume: Paleogeographic and paleotectonic reconstruction from the prevolcanic and volcanic records, North-Central Siberia: International Geology Review, v. 40, p. 95–115.

Dalrymple, G.B., Czamanske, G.K., Fedorenko, V.A., Simonov, O.N., Lanphere, M.A., and Likhachev, A.P., 1995, A reconnaissance ^{40}Ar/^{39}Ar study of ore-bearing and related rocks, Siberian Russia: Geochimica et Cosmochimica Acta, v. 59, p. 2071–2083, doi: 10.1016/0016-7037(95)00127-1.

Dmitriev, Yu.I., 1973, Mesozoic trap volcanism in the centre and periphery of the Tunguska syncline: Izvestiya AN USSR, Geological series, no. 10, p. 58–67 (in Russian).

Dobretsov, N.L., 2003, Mantle plumes and their role in the formation of anorogenic granitoids: Geologiya i Geofizika, v. 44, p. 1243–1261.

Dobretsov, N.L., 2005, Large igneous provinces of Asia (250 Ma): Siberian and Emeishan traps (plateau basalts) and associated granitoids: Geologiya i Geofizika, v. 46, p. 870–890.

Dobretsov, N.L., Vladimirov, A.G., and Kruk, N.N., 2005, Permo-Triassic magmatism in the Altai-Sayan fold system as a reflection of the Siberian superplume: Doklady Earth Sciences, v. 400, p. 40–43.

Dorendorf, F., Wiechert, U., and Wörner, G., 2000, Hydrated sub-arc mantle: A source for the Klyuchevskoy volcano, Kamchatka/Russia: Earth and Planetary Science Letters, v. 175, p. 69–86, doi: 10.1016/S0012-821X(99)00288-5.

Dunn, T., and Sen, C., 1994, Mineral/matrix partition coefficients for orthopyroxene, plagioclase, and olivine in basaltic to andesitic systems: A combined analytical and experimental study: Geochimica et Cosmochimica Acta, v. 58, p. 717–733, doi: 10.1016/0016-7037(94)90501-0.

Dupré, H.B., and Allègre, C.J., 1985, Pb-Sr-Nd isotope data of Indian ocean ridges: New evidence of large-scale mapping of mantle heterogeneities: Earth and Planetary Science Letters, v. 76, p. 288–298.

Egorkin, A.V., 2001, Upper mantle structure below the Daldyn-Alakitsk kimberlite field by nuclear explosion seismograms: Geology of Ore Deposits, v. 43, p. 19–32.

Egorkin, A.V., 2004, Mantle structure of the Siberian platform: Izvestiya–Physics of the Solid Earth, v. 40, p. 385–394.

Elkins-Tanton, L.T., 2005, Continental magmatism caused by lithospheric delamination, *in* Foulger, G.R., et al., Plates, plumes, and paradigms: Boulder, Colorado, Geological Society of America Special Paper 388, p. 449–462.

Elkins-Tanton, L.T., Draper, D.S., Agee, C.B., Jewell, J., Thorpe, A., and Hess, P.S., 2007, The last lavas erupted during the main phase of the Siberian flood volcanic province: Results from experimental petrology: Contributions to Mineralogy and Petrology, v. 153, p. 191–209, doi: 10.1007/x00410-006-0140-1.

Erwin, D.H., Bowring, S.A., and Yugan, J., 2002. End-Permian mass extinctions: A review, *in* Koeberl, C., and MacLeod, K.G., Catastrophic events and mass extinctions: Impacts and beyond: Boulder, Colorado, Geological Society of America Special Paper 356, p. 363–383.

Escrig, S., Capmas, F., Dupré, B., and Allègre, C.J., 2004, Osmium isotopic constraints on the nature of the DUPAL anomaly from Indian mid-oceanic-ridge-basalts: Nature, v. 431, p. 59–63, doi: 10.1038/nature02904.

Fedorenko, V.A., 1981, Petrochemical series of effusive rocks in Noril'sk region: Geologiya i Geofizika, v. 6, p. 77–88 (in Russian).

Federenko, V.A., and Czamanske, G.K., 1997, Results of new field and geochemical studies of the volcanic and intrusive rocks of the Maymecha-Kotuy area, Siberian flood-basalt province, Russia: International Geology Review, v. 39, p. 479–531.

Fedorenko, V.I., Lightfoot, P.C., Naldrett, A.J., Czamanske, G.K., Hawkesworth, C.J., Wooden, J.L., and Ebel, D.S., 1996, Petrogenesis of the flood-basalt sequence at Noril'sk, North Central Siberia: International Geology Review, v. 38, p. 99–135.

Fedoseev, G.S., Sotnikov, V.I., and Rikhvanov, L.P., 2005, Geochemistry and geochronology of Permo-Triassic basites in the northwestern Altai-Sayan folded area: Geologiya i Geofizika, v. 46, p. 289–302.

Fedotov, S.A., editor-in-chief, 1984, Large Tolbachik fissure eruption: Kamchatka 1975–1976: Moscow, Nauka (in Russian).

Feoktistov, G.D., 1979, Petrology and conditions for formation of trap sills: Nauka, Novosibirsk (in Russian).

Glikson, A.Y., 1999, Oceanic mega-impacts and crustal evolution: Geology, v. 27, p. 387–390, doi: 10.1130/0091-7613(1999)027<0387:OMIACE>2.3.CO;2.

Green, T.H., 1994, Experimental studies of trace-element partitioning applicable to igneous petrogenesis: Sedona 16 years later: Chemical Geology, v. 117, p. 1–36, doi: 10.1016/0009-2541(94)90119-8.

Hagstrum, J.T., 2005, Antipodal hotspots and bipolar catastrophes: Were large oceanic large-body impacts the cause?: Earth and Planetary Science Letters, v. 236, p. 13–27, doi: 10.1016/j.epsl.2005.02.020.

Hart, S., 1984, A large-scale isotope anomaly in the Southern Hemisphere mantle: Nature, v. 309, p. 753–757, doi: 10.1038/309753a0.

Hart, S.R., and Dunn, T., 1993, Experimental cpx/melt partitioning of 24 trace elements: Contributions to Mineralogy and Petrology, v. 113, p. 1–8, doi: 10.1007/BF00320827.

He, B., Xu, Y.-G., Chung, S.-L., Xiao, L., and Wang, Y., 2003, Sedimentary evidence for a rapid kilometer-scale crustal doming prior to the eruption of the Emeishan flood basalts: Earth and Planetary Science Letters, v. 213, p. 391–405, doi: 10.1016/S0012-821X(03)00323-6.

Horan, M.F., Walker, R.J., Fedorenko, V.A., and Czamanske, G.K., 1995, Osmium and neodymium isotopic constraints on the temporal and spatial evolution of the Siberian flood basalt sources: Geochimica et Cosmochimica Acta, v. 59, p. 5159–5168, doi: 10.1016/0016-7037(96)89674-8.

Huang, X.G., Xu, Y.S., and Karato, S.I., 2005, Water content in the transition zone from electrical conductivity of wadsleyite and ringwoodite: Nature, v. 434, p. 746–749, doi: 10.1038/nature03426.

Hyndman, R.D., Currie, C.A., and Mazzotti, S.P., 2005, Subduction zone backarcs, mobile belts, and orogenic heat: GSA Today, v. 15, p. 4–11.

Ivanov, A.V., 2006, Systematic difference between U-Pb and $^{40}Ar/^{39}Ar$ dates: Reason and evolution techniques: Geochemistry International, v. 44, 1041–1047, doi: 10.1134/50016702906100090.

Ivanov, A.V., and Balyshev, S.V., 2005, Mass flux across the lower-upper mantle boundary: Vigorous, absent, or limited? *in* Foulger, G.R., et al., Plates, plumes, and paradigms: Boulder, Colorado, Geological Society of America Special Paper 388, p. 327–346.

Ivanov, A.V., Feoktistov, G.D., Rasskazov, S.V., Yasnygina, T.A., Markova, M.E., He, H., and Boven, A., 2003, Geochemical characteristics and $^{40}Ar/^{39}Ar$ age of sills from the Kansk-Taseevskaya basin: To a question of duration of Permo-Triassic traps magmatism of the Siberian Platform, *in* Second Russian conference on isotopic geochronology, "Isotopic Geochronology for Solving Problems of Geodynamics and Ore Formations": St. Petersburg, Center of Informational Culture, p. 179–183.

Ivanov, A.V., Rasskazov, S.V., Feoktistov, G.D., He, H., and Boven, A., 2005, $^{40}Ar/^{39}Ar$ dating of Usol'skii sill in the southeastern Siberian Traps Large Igneous Province: Evidence for long-lived magmatism: Terra Nova, v. 17, p. 203–208, doi: 10.1111/j.1365-3121.2004.00588.x.

Ivanov, A.V., Ryabov, V.V., Shevko, A.Y., and He, H., 2006, Single short episode or multiple non-coeval episodes of voluminous basaltic volcanism of the Siberian Traps?: Data of $^{40}Ar/^{39}Ar$ dating, *in* Proceedings of III Russian conference on isotopic geochronology, Moscow, IGEM RAS, p. 278–282 (in Russian).

Ivanov, B.A., and Melosh, H.J., 2003, Impacts do not initiate volcanic eruptions: Eruptions close to the crater: Geology, v. 31, p. 869–872, doi: 10.1130/G19669.1.

Jones, A.P., Price, G.D., Price, N.J., DeCarli, P.S., and Clegg, R., 2002, Impact induced melting and the development of large igneous provinces: Earth and Planetary Science Letters, v. 202, p. 551–561, doi: 10.1016/S0012-821X(02)00824-5.

Jones, A.P., Wünemann, K., and Price, G.D., 2005, Modeling impact volcanism as a possible origin for the Ontong Java Plateau, *in* Foulger, G.R., et al., Plates, plumes, and paradigms: Boulder, Colorado, Geological Society of America Special Paper 388, p. 711–720.

Kamo, S.L., Czamanske, G.K., and Krogh, T.E., 1996, A minimum U-Pb age for Siberian flood-basalt volcanism: Geochimica et Cosmochimica Acta, v. 60, p. 3505–3511, doi: 10.1016/0016-7037(96)00173-1.

Kamo, S.L., Czamanske, G.K., Amelin, Yu., Fedorenko, V.A., Davis, D.W., and Trofimov, V.R., 2003, Rapid eruption of Siberian flood-volcanic rocks and evidence for coincidence with the Permian-Triassic boundary and mass extinction at 251 Ma: Earth and Planetary Science Letters, v. 214, p. 75–91, doi: 10.1016/S0012-821X(03)00347-9.

Katayama, I., Nakashima, S., and Yurimoto, H., 2006, Water content in natural eclogite and implication for water transport into the deep upper mantle: Lithos, v. 86, p. 245–259, doi: 10.1016/j.lithos.2005.06.006.

Kazanskii, A.Yu., Kazanskii, Yu.P., Saraev, S.V., and Moskvin, V.I., 2000, The Permo-Triassic boundary in volcanosedimentary section of the West-Siberian plate according to paleomagnetic data (from studies of the core from the Tyumenskaya superdeep borehole SD-6): Geologiya i Geofizika, v. 41, p. 327–339.

Kelemen, P.B., Joyce, D.B., Webster, J.D., and Holloway, J.R., 1990, Reaction between ultramafic wall rock and fractionating basaltic magma, Part 2, Experimental investigation of reaction between olivine tholeiite and harzburgite at 1150 and 1050°C and 5 kbar: Journal of Petrology, v. 31, p. 99–134.

Kieffer, B., Arndt, N., Lapierre, H., Bastien, F., Bosch, D., Pecher, A., Yirgu, G., Ayalew, D., Weis, D., Jerram, D.A., Keller, F., and Meugniot, C., 2004, Flood and shield basalts from Ethiopia: Magmas from the African superswell: Journal of Petrology, v. 45, p. 793–834, doi: 10.1093/petrology/egg112.

King, S.D., and Anderson, D.L., 1998, Edge-driven convection: Earth and Planetary Science Letters, v. 160, p. 289–296, doi: 10.1016/S0012-821X(98)00089-2.

Koeberl, C., Gilmour, L., Reimold, W.U., Claeys, P., and Ivanov, B., 2002, End-Permian catastrophe by bolide impact: Evidence of a gigantic release of sulfur from the mantle: Comment: Geology, v. 30, p. 855–856, doi: 10.1130/0091-7613(2002)030<0855:EPCBBI>2.0.CO;2.

Kogarko, L.N., and Ryabchikov, I.D., 1995, Conditions of meimechite magmas formation (polar Siberia) based on the geochemical data: Geokhimia, v. 12, p. 1699–1709.

Korenaga, J., 2005, Why did not the Ontong Java Plateau form subaerially?: Earth and Planetary Science Letters, v. 234, p. 385–399, doi: 10.1016/j.epsl.2005.03.011.

Le Bas, M.J., and Streckeisen, A.L., 1991, The UIGS systematics of igneous rocks: Journal of the Geological Society, London, v. 148, p. 825–833.

Lightfoot, P.C., Hawkesworth, C.J., Hergt, J., Naldrett, A.J., Gorbachev, N.S., Fedorenko, V.A., and Doherty, W., 1993, Remobilisation of the continental lithosphere by a mantle plume: Major-, trace-element, and Sr-, Nf-, and Pb-isotope evidence from picritic and tholeiitic lavas of the Noril'sk District, Siberian Trap, Russia: Contributions to Mineralogy and Petrology, v. 114, p. 171–188, doi: 10.1007/BF00307754.

Lin, S.C., and van Keken, P.E., 2005, Multiple volcanic episodes of flood basalts caused by thermochemical mantle plumes: Nature, v. 436, p. 250–252, doi: 10.1038/nature03697.

Litasov, K., and Ohtani, E., 2003, Stability of various hydrous phases in CMAS pyrolite-H_2O system up to 25 GPa: Physics and Chemistry of Minerals, v. 30, p. 147–156, doi: 10.1007/s00269-003-0301-y.

Lysak, S.V., 1984, Terrestrial heat flow in the south of East Siberia: Tectonophysics, v. 103, p. 205–215, doi: 10.1016/0040-1951(84)90084-2.

Malitch, N.S., editor-in-chief, 1999, Geological map of Siberian platform and adjoining areas: St. Petersburg, Vserissuusjuu Geologicheskii Institute, scale 1:1,500,000.

Masaitis, V.L., 1983, Permian and Triassic volcanism of Siberia: Zapiski Vserossiiskogo Mineralogicheskogo Obshestva, v. 4, p. 412–425 (in Russian).

Matsukage, K.N., Jing, Z.C., and Karato, S., 2005, Density of hydrous silicate melt at the conditions of Earth's deep upper mantle: Nature, v. 438, p. 488–491, doi: 10.1038/nature04241.

McDonough, W.F., and Sun, S.-S., 1995, The composition of the Earth: Chemical Geology, v. 120, p. 223–253, doi: 10.1016/0009-2541(94)00140-4.

Medvedev, A.Ya., Al'mukhamedov, A.I., and Kirda, N.P., 2003, Geochemistry of Permo-Triassic volcanic rocks of West Siberia: Geologiya i Geofizika, v. 44, p. 86–100.

Meyzen, C.M., Ludden, J.N., Humler, E., Luais, B., Toplis, M.J., Mevel, C., and Storey, M., 2005, New insights into the origin and distribution of the DUPAL isotope anomaly in the Indian Ocean mantle from MORB of the Southwest Indian Ridge: Geochemistry, Geophysics, Geosystems, v. 6, Q11K11.

Milanovskiy, Y.Y., 1976, Rift zones of the geologic past and their associated formations: Report 2: International Geology Review, v. 18, p. 619–639.

Min, K., Mundil, R., Renne, P.R., and Ludwig, K.R., 2000, A test for systematic errors in $^{40}Ar/^{39}Ar$ geochronology through comparison with U/Pb analysis of a 1.1-Ga rhyolite: Geochimica et Cosmochimica Acta, v. 64, p. 73–98, doi: 10.1016/S0016-7037(99)00204-5.

Morgan, W.J., 1971, Convection plumes in the lower mantle: Nature, v. 230, p. 42–43, doi: 10.1038/230042a0.

Morozova, E.A., Morozov, I.B., Smithson, S.B., and Solodilov, L., 1999, Heterogeneity of the uppermost mantle beneath Russian Eurasia from the ultralong-range profile quartz: Journal of Geophysical Research, v. 104, p. 20,329–20,348, doi: 10.1029/1999JB900142.

Mundil, R., Ludwig, K.R., Metcalfe, I., and Renne, P.R., 2004, Age and timing of the Permian mass extinctions: U/Pb dating of closed-system zircons: Science, v. 305, p. 1760–1763, doi: 10.1126/science.1101012.

Naldrett, A.J., Lightfoot, P.C., Fedorenko, F.A., Gorbachev, N.S., and Doherty, W., 1992, Geology and geochemistry of intrusions and flood basalts of the Noril'sk region, USSR, with implications for the origin of the Ni-Cu ores: Economic Geology and the Bulletin of the Society of Economic Geologists, v. 87, p. 975–1004.

Nikishin, A.M., Ziegler, P.A., Abbott, D., Brunet, M.-F., and Cloetingh, S., 2002, Permo-Triassic intraplate magmatism and rifting in Eurasia: Implications for mantle plumes and mantle dynamics: Tectonophysics, v. 351, p. 3–39, doi: 10.1016/S0040-1951(02)00123-3.

Ohtani, E., 2005, Water in the mantle: Elements, v. 1, no. 1, p. 25–30.

Pavlenkova, N.I., 2006, Long-range profile data on the upper-mantle structure in the Siberian platform: Russian Geology and Geophysics, v. 47, p. 626–641.

Pavlenkova, G.A., and Pavlenkova, N.I., 2006, Upper mantle structure of the Northern Eurasia from peaceful nuclear explosion data: Tectonophysics, v. 416, p. 33–52, doi: 10.1016/j.tecto.2005.11.010.

Pavlenkova, G.A., Priestley, K., and Cipar, J., 2002, 2D model of the crust and uppermost mantle along rift profile, Siberian craton: Tectonophysics, v. 355, p. 171–186, doi: 10.1016/S0040-1951(02)00140-3.

Podurushin, V.F., 2002, Geodynamics of the West-Siberian platform and its margins: Izvestiya Vysshikh Uchebnykh Zavedeniy. Geologiya i Razvedka, v. 1, p. 30–37 (in Russian).

Pontevivo, A., and Thybo, H., 2006, Test of the upper mantle low velocity layer in Siberia with surface waves: Tectonophysics, v. 416, p. 113–131, doi: 10.1016/j.tecto.2005.11.015.

Portnyagin, M., Hoernle, K., Avdeiko, G., Hauff, F., Werner, R., Bindeman, I., Uspensky, V., and Garbe-Schönberg, D., 2005, Transition from arc to oceanic magmatism at the Kamchatka-Aleutian junction: Geology, v. 33, p. 25–28, doi: 10.1130/G20853.1.

Puffer, J.H., 2001, Contrasting high field strength element content of continental flood basalts from plume versus reactivated-arc sources: Geology, v. 29, p. 675–678, doi: 10.1130/0091-7613(2001)029<0675:CHFSEC>2.0.CO;2.

Reichow, M.K., Saunders, A.D., White, R.V., Pringle, M.S., Al'mukhamedov, A.I., Medvedev, A.I., and Kirda, N.P., 2002, $^{40}Ar/^{39}Ar$ dates from the West Siberian Basin: Siberian flood basalt province doubled: Science, v. 296, p. 1846–1849, doi: 10.1126/science.1071671.

Reichow, M.K., Saunders, A.D., White, R.V., Al'mukhamedov, A.I., and Medvedev, A.Ya., 2005, Geochemistry and petrogenesis of basalts from the West Siberian Basin: An extension of the Permo-Triassic Traps, Russia: Lithos, v. 79, p. 425–452, doi: 10.1016/j.lithos.2004.09.011.

Renne, P.R., 1995, Excess ^{40}Ar in biotite and hornblende from the Norilsk 1 intrusion, Siberia: Implication for the age of Siberian Traps: Earth and Planetary Science Letters, v. 131, p. 165–176, doi: 10.1016/0012-821X(95)00015-5.

Renne, P.R., and Basu, A.R., 1991, Rapid eruption of the Siberian Traps flood basalts at the Permo-Triassic boundary: Science, v. 253, p. 176–179, doi: 10.1126/science.253.5016.176.

Renne, P.R., Zichao, Z., Richards, M.A., Black, M.T., and Basu, A.R., 1995, Synchrony and casual relations between Permian-Triassic boundary crises and Siberian flood volcanism: Science, v. 269, p. 1413–1416, doi: 10.1126/science.269.5229.1413.

Renne, P.R., Swisher, C.C., Deino, A.L., Karner, D.B., Owens, T.L., and DePaolo, D.J., 1998, Intercalibration of standards, absolute ages and uncertainties in $^{40}Ar/^{39}Ar$ dating: Chemical Geology, v. 145, p. 117–152, doi: 10.1016/S0009-2541(97)00159-9.

Sakamaki, T., Suzuki, A., and Ohtani, E., 2006, Stability of hydrous melt at the base of the Earth's upper mantle: Nature, v. 439, p. 192–194, doi: 10.1038/nature04352.

Saunders, A.D., England, R.W., Reichow, M.K., and White, R.V., 2005, A mantle plume origin for the Siberian traps: Uplift and extension in the West Siberian Basin, Russia: Lithos, v. 79, p. 407–424, doi: 10.1016/j.lithos.2004.09.010.

Shaw, D.M., 1970, Trace element fractionation during anatexis: Geochimica et Cosmochimica Acta, v. 34, p. 237–243, doi: 10.1016/0016-7037(70)90009-8.

Silver, P.G., Behn, M.D., Kelley, K., Schmitz, M., and Savage, B., 2006, Understanding cratonic flood basalts: Earth and Planetary Science Letters, v. 245, p. 190–201, doi: 10.1016/j.epsl.2006.01.050.

Smith, A.D., 1992, Back-arc convection model for Columbia River basalt genesis: Tectonophysics, v. 207, p. 269–285, doi: 10.1016/0040-1951(92)90390-R.

Sobolev, A.V., Kamenetsky, V.S., and Kononkova, N.N., 1991, New data on petrology of meymechites: Geochimiya, v. 8, p. 1084–1095.

Spell, T.L., and McDougall, I., 2003, Characterization and calibration of $^{40}Ar/^{39}Ar$ dating standards: Chemical Geology, v. 198, p. 189–211, doi: 10.1016/S0009-2541(03)00005-6.

Sun, S.-S., and McDonough, W.F., 1989, Chemical and isotopic systematics of oceanic basalts: Implications for mantle composition and process, *in* Sounders, A.D., and Norry, M.J., Magmatism in the oceanic basins: Geological Society, London, Special Publication 42, p. 313–345.

Thybo, H., and Perchuc, E., 1997, The seismic 8° discontinuity and partial melting in continental mantle: Science, v. 275, p. 1626–1629, doi: 10.1126/science.275.5306.1626.

Vasil'ev, Yu.R., Zolotukhin, V.V., Feoktistov, G.D., and Prusskaya, S.N., 2000, Evaluation of the volumes and genesis of Permo-Triassic Trap magmatism of the Siberian Platform: Geologiya i Geofizika, v. 41, p. 1696–1705 (in Russian).

Venkatesan, T.R., Kumar, A., Gopalan, K., and Al'mukhamedov, A.I., 1997, ^{40}Ar-^{39}Ar age of Siberian basaltic volcanism: Chemical Geology, v. 138, p. 303–310, doi: 10.1016/S0009-2541(97)00006-5.

Vernikovsky, V.A., Pease, V.L., Vernikovskaya, A.E., Romanov, A.P., Gee, D.G., and Travin, A.V., 2003, First report of early Triassic A-type granite and syenite intrusions from Taimyr: Product of the northern Eurasian superplume?: Lithos, v. 66, p. 23–36, doi: 10.1016/S0024-4937(02)00192-5.

Vinogradov, A.P., editor-in-chief, 1968, The lithologo-paleogeographic maps of USSR, v. 2: Moscow, Ministry of Geology of the USSR and USSR Academy of Sciences.

Vladimirov, A.G., Kozlov, M.S., Shokal'skii, S.P., Khalilov, V.A., Rudnev, S.N., Kruk, N.N., Vystavnoi, S.A., Borisov, S.M., Berezhikov, Y.K., Metsner, A.N., Babin, G.A., Mamlin, A.N., Murzin, O.M., Nazarov, G.V., and Makarov, V.A., 2001, Major epochs of the intrusive magmatism of Kuznetsk Alatau, Altai, and Kalba (from U-Pb isotope dates): Geologiya i Geofizika, v. 42, p. 1157–1178.

Vrublevskii, V.V., Gertner, I.F., Polyakov, G.V., Izokh, A.E., Krupchatnikov, V.I., Travin, A.V., and Voitenko, N.N., 2004, Ar-Ar isotopic age of lamproite dikes of the Chua complex, Gornyi Altai: Doklady Earth Sciences, v. 399A, p. 1252–1255.

Walderhaug, H.J., Eide, E.A., Scott, R.A., Inger, S., and Golionko, E.G., 2005, Paleomagnetism and $^{40}Ar/^{39}Ar$ geochronology from the South Taimyr igneous complex, Arctiv Russia: A Middle–Late Triassic magmatic pulse after Siberian flood-basalt volcanism: Geophysical Journal International, v. 163, p. 501–517, doi: 10.1111/j.1365-246X.2005.02741.x.

Wooden, J.L., Czamanske, G.K., Fedorenlo, V.A., Arndt, N.T., Chauvel, C., Bouse, R.M., King, B.-S.W., Knight, R.J., and Siems, D.F., 1993, Isotopic and trace-element constraints on mantle and crustal contributions to characterization of Siberian continental flood basalts, Noril'sk area, Siberia: Geochimica et Cosmochimica Acta, v. 57, p. 3677–3704, doi: 10.1016/0016-7037(93)90149-Q.

Yarmolyuk, V.V., Litvinovsky, B.A., Kovalenko, V.I., Jahn, B.M., Zanvilevich, A.N., Vorontsov, A.A., Zhuravlev, D.Z., Posokhov, V.F., Kuz'min, D.V., and Sandimirova, G.P., 2001, Formation stages and sources of the peralkaline granitoid magmatism of the Northern Mongolia–Transbaikalia rift belt during the Permian and Triassic: Petrology, v. 9, p. 302–328.

Yasuda, A., and Fujii, T., 1998, Ascending subducted oceanic crust entrained within mantle plumes: Geophysical Research Letters, v. 25, p. 1561–1564, doi: 10.1029/98GL01230.

Zack, T., Foley, S.F., and Jenner, G.A., 1997, A consistent partitioning coefficient set for clinopyroxene, amphibole and garnet from laser ablation microprobe analyses of garnet pyroxenite from Kakanui, New Zealand: Neues Jahrbuch für Mineralogie-Abhandlungen, v. 172, p. 23–41.

Zhu, B.-G., Hu, Y.-G., Chang, X.-Y., Xie, J., and Zhang, Z.-W., 2005, The Emeishan large igneous province originated from magmatism of a primitive mantle plus subducted slab: Russian Geology and Geophysics, v. 46, p. 904–921.

Zolotukhin, V.V., Vilenskii, A.M., Vasil'ev, Yu.P., Mezhvilk, A.A., Ryabov, V.V., and Shcherbakova, Z.V., 1984, Magnesium basic rocks of the western Siberian platform and questions of nickel-bearing: Novosibirsk, Nauka (in Russian).

Zolotukhin, V.V., Vilenskii, A.M., and Dyuzhikov, O.A., 1986, Basalts of Siberian platform: Novosibirsk, Nauka (in Russian).

Zonenshain, L.P., Kuzmin, M.I., and Natapov, L.M., 1990. Geology of the USSR: Plate tectonic synthesis: Washington, D.C., American Geophysical Union, Geodynamic Series 21.

Zorin, Y.A., 1999, Geodynamics of the western part of the Mongolia-Okhotsk collisional belt, Trans-Baikal region (Russia) and Mongolia: Tectonophysics, v. 306, p. 33–56, doi: 10.1016/S0040-1951(99)00042-6.

Zorin, Yu.A., and Vladimirov, B.M., 1989, On the genesis of trap magmatism of the Siberian platform: Earth and Planetary Science Letters, v. 93, p. 109–112, doi: 10.1016/0012-821X(89)90188-X.

Zorin, Yu.A., Belichenko, V.G., Turutanov, E.K., Kozhevnikov, V.M., Sklyarov, E.V., Tomurtogoo, O., Khosbayar, P., Arvisbaatar, N., and Biambaa, C., 1998, Terranes of East Mongolia and Central Trans-Baikal region and evolution of the Mongolia-Okhotsk fold belt: Geologiya i Geofizika, v. 39, p. 11–25.

MANUSCRIPT ACCEPTED BY THE SOCIETY JANUARY 31, 2007

DISCUSSION

15 January 2007, Ajoy K. Baksi

Ivanov's assessment of aspects of Siberian Trap volcanism serves as a useful adjunct to our knowledge about flood basalt volcanism. However, numerous ages used therein do not qualify as proper estimates of the crystallization ages; that is, $^{40}Ar/^{39}Ar$ ages need to be evaluated for both statistical validity and freshness of samples dated (Baksi, this volume, chapters 15 and 16).

First, the ages reported in Baksi and Farrar (1991a) are known to be incorrect on the basis of both mass spectrometric problems and ages of standards used (Baksi and Farrar, 1991b; Baksi, 2005). These analyses are now evaluated for freshness (Baksi, 2007a). The plateau steps for ST-154 ($K = 0.60\%$) and ST-524 ($K = 0.13\%$) give alteration index (A.I.) values in the range 0.044–0.0030 and 0.020–0.0015, respectively. The cutoff for freshness is <0.0006; both rocks are altered and cannot give proper estimates of the time of crystallization. ST-563 is a dike, and its A.I. cannot be unequivocally used for a freshness test (see discussion of Baksi, this volume, chapter 16). The ages of Walderhaug et al. (2005) must all be discounted because they fail statistical tests for proper plateaus and/or are altered as based on the A.I. test. Earlier efforts (Renne and Basu, 1991; Dalrymple et al., 1995; Venkatesan et al., 1997; Reichow et al., 2002) must all be critically evaluated, in particular by the A.I. technique. This is not always possible because some of the data sets are currently not available for examination. Some of the analyses fail the freshness test; the major portions of mafic rocks appear to have been extruded in ca. 1 Ma around 250 Ma. The best (U–Pb) ages are those of Kamo et al. (2003) at 252–251 Ma, and the screened $^{40}Ar/^{39}Ar$ ages are in general agreement with these.

The total duration of volcanism remains unknown. All relevant $^{40}Ar/^{39}Ar$ ages must be critically examined by the techniques set out earlier (see Baksi, 2007). I strongly urge scientists to carry out such critical examination for themselves before advancing hypotheses related to the (temporal) formation of this immense province and its possible environmental effects.

19 January 2007, Alexei Ivanov

Baksi (15 January) raises two important questions about (1) duration of the Siberian Traps and (2) reliability of published $^{40}Ar/^{39}Ar$ ages.

The U–Pb ages of Kamo et al. (2003) were obtained on samples from a localized area compared to the whole Siberian Traps, and assigning these ages to other remote regions is questionable. A paleomagnetic study in the Noril'sk area shows that volcanism started there at the end of a reversed-polarity chron (Latest Permian) and continued during a normal-polarity chron (earliest Lower Triassic) (e.g., Heunemann et al., 2004). A paleomagnetic study of volcanic rocks recovered from the superdeep drill hole No. 6 (SG-6) within the west Siberian basin revealed five normal- and four reversed-polarity chrons, suggesting a duration of volcanism from the Tatarian to the end of the Olenekian (end of the Lower Triassic; roughly 9 Ma duration). As for the Walderhaug et al. (2005) article, whose Middle Triassic $^{40}Ar/^{39}Ar$ ages on Taimyr dolerite sills were questioned by Baksi (15 January), these $^{40}Ar/^{39}Ar$ ages were complemented by a paleomagnetic study that revealed paleomagnetic poles in agreement with sill emplacement in the Middle Triassic. U–Pb and $^{40}Ar/^{39}Ar$ ages obtained on the same lava units and intrusions appeared to be in good agreement, taking into account a slight systematic difference of <1% (Ivanov, 2006). Among these, the Bolgokhtokh granodiorite intrusion is dated as old as 228.9 ± 0.3 Ma and 226.8 ± 0.8 Ma by U–Pb and ^{40}Ar/^{39}Ar, respectively. (The original ^{40}Ar/^{39}Ar age [Dalrymple et al., 1995] is recalculated relative to 98.5 Ma for GA1515 standard.) Similar types of intrusions are found elsewhere within the Siberian Traps and dated by U–Pb from ca. 250 to ca. 225 Ma (see references in Ivanov, this volume). Thus, long-lived magmatism of the Siberian Traps is expected, even if no ^{40}Ar/^{39}Ar ages had been published.

As mentioned by Baksi (15 January), most published ^{40}Ar/^{39}Ar ages for the Siberian Traps are presented without full analytical details. So A.I., suggested as a measure of ^{40}Ar/^{39}Ar age reliability by Baksi (this volume, chapters 15 and 16), can be calculated only by the authors of those papers. Plateau steps from ^{40}Ar/^{39}Ar stepwise-heating dating of plagioclases from Ivanov et al. (2005) yielded A.I. values of 0.00067 ± 0.00036 (error is one standard deviation) for plateau steps of two dated samples. This is about ten times higher than the cutoff value for plagioclases suggested by Baksi (this volume). The low-temperature steps with too low apparent ages (Fig. 2 in Ivanov et al., 2005) are characterized by high A.I. up to 0.5, showing alteration of plagioclases at crystal edges and within cracks. The weighted mean of the two plateau ages is 243.9 ± 1.4 Ma (relative to 129.4 Ma for LP6) or 243.9 ± 5.8 Ma if possible subsampling inhomogeneity of LP6 is accounted for (Ivanov et al., 2005).

Baksi (15 January) reports that the A.I. for his samples (Baksi and Farrar, 1991a) varies between 0.0015 and 0.04, and he rejects these ages as altered on the basis of the cutoff value of 0.0006 for whole-rock samples. Ivanov et al. (2006) dated three samples using the ^{40}Ar/^{39}Ar stepwise-heating technique. The samples yielded concordant plateau and isochron ages. Application of the method of Baksi (this volume, chapters 15 and 16) shows that for plateaus of these samples, A.I. is between 0.002 and 0.05. According to Baksi (this volume, chapters 15 and 16), these ages should be rejected. However, there are some arguments to keep these ages as true crystallization ages. First of all, the age for the Nadezhdinsky suite agrees within analytical uncertainty with those published by Venkatesan et al. (1997), whose ages are in agreement with the U-Pb ages of Kamo et al. (2003). Second, there is a difference between the ages obtained for samples from the Nadezhdinsky and Hona-Makitsky suites, whereas there is no difference in their A.I. indices. Third, Baksi (this volume, chapters 15 and 16) acknowledges that the A.I. will be higher than the cutoff value in fresh samples of volcanic rocks if they were derived from a sub-duction-derived water-bearing mantle source. I argue that the Siberian Traps (Ivanov, this volume) and probably other flood basalt provinces (see Ivanov, 17 January comment to discussion of Hooper et al., this volume) were derived from such mantle

sources. Therefore, suggested A.I. cutoff values may be not applicable for the Siberian Traps.

As to the rest, I fully agree with Baksi (15 January) that careful evaluation of ^{40}Ar/^{39}Ar data is essential for correct interpretation of geological models.

REFERENCES CITED

Baksi, A.K., 2005, Evaluation of radiometric ages pertaining to rocks hypothesized to have been formed by hotspot activity, in and around the Atlantic, Indian and Pacific oceans, *in* Foulger, G.R., et al., eds., Plates, plumes, and paradigms: Boulder, Colorado, Geological Society of America Special Paper 388, p. 55–70, doi: 10.1130/2005.2388(05).

Baksi, A.K., 2007a (this volume, chapter 15), A quantitative tool for detecting alteration in undisturbed rocks and minerals—I: Water, chemical weathering, and atmospheric argon, *in* Foulger, G.R., and Jurdy, D.M., eds., Plates, plumes, and planetary processes: Boulder, Colorado, Geological Society of America Special Paper 430, doi: 10.1130/2007.2430(15).

Baksi, A.K., 2007b (this volume, chapter 16), A quantitative tool for detecting alteration in undisturbed rocks and minerals—II: Application to argon ages related to hotspots, *in* Foulger, G.R., and Jurdy, D.M., eds., Plates, plumes, and planetary processes: Boulder, Colorado, Geological Society of America Special Paper 430, doi: 10.1130/2007.2430(16).

Baksi, A.K., and Farrar, E., 1991a, ^{40}Ar/^{39}Ar dating of the Siberian Traps, USSR: Evaluation of the ages of two major extinction events relative to flood-basalt volcanism in the USSR and the Deccan Traps, India: Geology, v. 19, p. 461–464.

Baksi, A.K., and Farrar, E., 1991b, ^{40}Ar/^{39}Ar ages of whole-rock basalts (Siberian Traps) in the Tunguska and Noril'sk regions, USSR: Eos (Transactions, American Geophysical Union), v. 72, p. 570.

Dalrymple, G.B., Czamanske, G.K., Fedorenko, V.A., Simonov, O.N., Lanphere, M.A., and Likhachev, A.P., 1995, A reconaissance ^{40}Ar/^{39}Ar geochronologic study of ore-bearing and related rocks, Siberian Russia: Geochimica et Cosmochimica Acta, v. 59, p. 2071–2083.

Heunemann, C., Krása, D., Soffel, H.C., Gurevitch, E., and Bachtadse, V., 2004, Directions and intensities of the Earth's magnetic field during a reversal: Results from the Permo-Triassic Siberian Trap basalts, Russia: Earth and Planetary Science Letters, v. 218, p. 197–213.

Hooper, P.R., Camp, V.E., Reidel, S.P., and Ross, M.E. 2007 (this volume), The origin of the Columbia River flood basalt province: Plume versus non-plume models, *in* Foulger, G.R., and Jurdy, D.M., eds., Plates, plumes, and planetary processes: Boulder, Colorado, Geological Society of America Special Paper 430, doi: 10.1130/2007.2430(30).

Ivanov, A.V., 2006, Systematic difference between U–Pb and ^{40}Ar/^{39}Ar dates: Reasons and evaluation techniques: Geochemistry International, v. 44, p. 1041–1047.

Ivanov, A., 2007 (this volume), Evaluation of different models for the origin of the Siberian traps, *in* Foulger, G.R., and Jurdy, D.M., eds., Plates, plumes, and planetary processes: Boulder, Colorado, Geological Society of America Special Paper 430, doi: 10.1130/2007.2430(31).

Ivanov, A.V., Rasskazov, S.V., Feoktistov, G.D., He, H., and Boven, A., 2005, ^{40}Ar/^{39}Ar dating of Usol'skii sill in the southeastern Siberian Traps Large Igneous Province: Evidence for long-lived magmatism: Terra Nova, v. 17, p. 203–208.

Ivanov, A.V., Ryabov, V.V., Shevko, A.Y., and He., H., 2006, Single short episode or multiple non-coeval episodes of voluminous basaltic volcanism of the Siberian Traps? Data of ^{40}Ar/^{39}Ar dating, *in* Proceedings of III Russian conference on isotopic geochronology, Moscow, IGEM RAS, p. 278–282 (in Russian).

Kamo, S.L., Czamanske, G.K., Amelin, Y., Fedorenko, V.A., Davis, D.W., and Trofimov, V.R., 2003, Rapid eruption of Siberian flood-volcanic rocks and evidence for coincidence with the Permo-Triassic boundary and mass extinction at 251 Ma: Earth and Planetary Science Letters, v. 214, p. 75–91.

Reichow, M.K., Saunders, A.D., White, R.V., Pringle, M.S., Al'Mukhamedov, A.I., Medvedev, A.I., and Kirda, N.P., 2002, ^{40}Ar/^{39}Ar dates from the West Siberian Basin; Siberian flood basalt province doubled: Science, v. 296, p. 1846–1849.

Renne, P.R., and Basu, A.R., 1991, Rapid eruption of the Siberian Trap flood basalts at the Permo-Triassic boundary: Science, v. 253, p. 176–179.

Venkatesan, T.R., Kumar, A., Gopalan, K., and Al'Mukhamedov, A.I., 1997, ^{40}Ar/^{39}Ar age of Siberian basaltic volcanism: Chemical Geology, v. 138, p. 303–310.

Walderhaug, H.J., Eide, E.A., Scott, R.A., Inger, S., and Golionko, E.G., 2005, Paleomagnetism and ^{40}Ar-^{39}Ar geochronology from the South Taimyr igneous complex, Arctic Russia: A Middle–Late Triassic magnetic pulse after Siberian flood-basalt volcanism: Geophysical Journal International, v. 163, p. 501–517.

The Geological Society of America
Special Paper 430
2007

Eastern Anatolia: A hotspot in a collision zone without a mantle plume

Mehmet Keskin*

Istanbul University, Faculty of Engineering, Department of Geological Engineering, 34320 Avcılar, Istanbul, Turkey

ABSTRACT

Eastern Anatolia is one of the best examples of an active continental collision zone in the world. It comprises one of the high plateaus of the Alpine-Himalaya mountain belt, with an average elevation of ~2 km above sea level. Almost two-thirds of this plateau is covered by young volcanic units related to collision. They range in age from 11 Ma to Recent and have a thickness of up to 1 km in places. The collision-related volcanic province is not confined to Eastern Anatolia, but extends across much of the Caucasus in the east, including Eastern Turkey, Armenia, Azerbaijan, Georgia, and Southern Russia, spanning a distance of some 1000 km. The region covered by the collision-related volcanic sequences comprises a regional domal shape (~1000 km in diameter), and this unique morphology is comparable to that of the Ethiopian high plateau except for its north-south shortened asymmetrical shape. Recent geophysical data reveal that the lithospheric mantle is exceptionally thin or absent beneath this regional dome, indicating that the dome is currently supported by the asthenospheric mantle. Because of these features, the Eastern Anatolia–Iranian plateau and the Lesser Caucasus region as a whole can be regarded as the site of a "melting anomaly" or "hotspot" closely resembling the setting proposed for mantle plumes. However, geologic and geochemical data provide evidence against a plume origin. Instead, the results of recent geophysical studies, coupled with geologic, geochemical, and experimental findings, support the view that both domal uplift and extensive magma generation can be linked to the mechanical removal of a portion or the whole thickness of the mantle lithosphere, accompanied by passive upwelling of normal-temperature asthenospheric mantle to a depth as shallow as 40–50 km. Mechanical removal of the mantle lithosphere might be controlled by delamination in the north beneath the Erzurum-Kars plateau, while it might be linked to slab steepening and break-off in the south. Therefore, magma generation beneath Eastern Anatolia may have been controlled by adiabatic decompression of the asthenosphere. The Eastern Anatolian example is important in showing that not only plumes but also shallow plate tectonic processes have the potential to generate regional domal structures in the Earth's lithosphere as well as large volumes of magma in continental intraplate settings.

Keywords: Eastern Anatolia, collision, volcanism, domal uplift, slab break-off, steepening, melting anomaly, hotspot, mantle plume, fertile mantle, intraplate, subduction component

*E-mail: keskin@istanbul.edu.tr.

Keskin, M., 2007, Eastern Anatolia: A hotspot in a collision zone without a mantle plume, *in* Foulger, G.R., and Jurdy, D.M., eds., Plates, plumes, and planetary processes: Geological Society of America Special Paper 430, p. 693–722, doi: 10.1130/2007.2430(32). For permission to copy, contact editing@geosociety.org. ©2007 The Geological Society of America. All rights reserved.

INTRODUCTION

Orogenic belts formed by collisions between continents contain invaluable records of the geological history of the Earth and therefore have always attracted the attention of Earth scientists. The Anatolian-Iranian plateau is one of two regions where active continent-continent collision is currently taking place, the other being the Tibetan plateau (see Fig. 1 *in* Dewey et al., 1986). Therefore, Eastern Anatolia, which is the western part of the Anatolian-Iranian plateau, can be regarded as a spectacular natural laboratory where the early stages of continent-continent collision and their effects can be thoroughly studied.

Previous studies to date (e.g., Şengör and Kidd, 1979; Dewey et al., 1986) have shown that collision occurred between the Eurasian and Arabian continents, resulting in the formation of an extensive (~150,000 km^2) high plateau with an average elevation of 2 km above sea level. These studies also revealed that the region reached this elevation as a block after the Serravalian (ca. 13–11 Ma; Gelati, 1975), when the terminal collision of Arabia with Eurasia started (Şengör and Kidd, 1979).

Volcanic activity initiated immediately after rapid block uplift of Eastern Anatolia and became widespread all over the region, producing subaerial lava flows and pyroclastic products that are very variable in their composition (i.e., from calc-alkaline to alkaline, from basalts to high silica rhyolites) and eruptive style (i.e., from Hawaiian to Plinian) (Pearce et al., 1990; Keskin et al., 1998; Yılmaz et al., 1998). The volcanic activity initiated in the north around the Erzurum-Kars plateau with calc-alkaline lavas and migrated to the SSE, becoming more alkaline (i.e., basically sodic alkaline) in character (Keskin, 2003). A vast volume of volcanic material was produced by this activity, covering almost two-thirds of the region (i.e., ~43,000 km^2 in Eastern Anatolia alone) and reaching over 1 km in thickness in some localities. Although fissure eruptions dominated the volcanic activity, there are more than twenty major volcanic centers (e.g., Mt. Nemrut, Mt. Ararat, Mt. Tendürek) and numerous small ones in Eastern Anatolia, corresponding basically to central eruption sites (Figs. 1 and 2). Although it is difficult to calculate the total volume of the volcanic material produced because of variations in the thickness of the volcanic succession and erosion in the region, the estimated total volume is a minimum of 15,000 km^3 in Eastern Anatolia alone (assuming an average volcanic thickness of 300–350 m). The erupted volume may represent only a small fraction of the melt generated beneath the region, because a greater proportion presumably was emplaced deeper in the crust as plutonic intrusions. Thus, there must have been enormous magma generation beneath the whole region related to the collision of Arabia with Eurasia. As a result, the Anatolian-Iranian plateau can be regarded as one of the Earth's "hotspots" or "melting anomalies."

The Eastern Anatolian topographic uplift resembles the Tibetan plateau and has been viewed as a younger version of it in many studies (e.g., Şengör and Kidd, 1979; Dewey et al., 1986; Barazangi, 1989). In these studies, the Eastern Anatolian lithosphere is considered to have doubled in thickness (to ~250–300 km) as a result of collision. However, recent geophysical studies (Gök et al., 2000, 2003; Al-Lazki et al., 2003, 2004; Piromallo and Morelli, 2003; Al-Damegh et al., 2004; Maggi and Priestley, 2005) have revealed that the mantle lithosphere is almost completely absent beneath a greater portion of the region (Figs. 3 and 4).

On the basis of these results and the geology of the region, Şengör et al. (2003) proposed that the East Anatolian high plateau is a mantle-supported, north-south-shortened domal structure whose east-west topographic profile along the 40°N parallel is very similar to that of the Ethiopian high plateau (Fig. 4A). At present, it is difficult to recognize the dome in topographic maps because the topography of the region has been strongly modified by volcanoes and river drainage systems.

In what follows, I deal with a number of problems, including

- How and why the region gained its elevation and the aforementioned domal shape
- How great volumes of collision-related magma were generated in the region
- What tectonic processes are responsible for both magma generation and the regional uplift

The organization of the article is as follows: the first section focuses on the geology of the region, the second summarizes new geophysical findings about the lithospheric structure of the region, the third deals with the geochemical characteristics of the collision-related volcanic units, and the fourth discusses competing geodynamic models proposed for the region, with an emphasis on the inherent discrepancies in each model. The final two sections present a discussion and conclusion.

Figure 1. (A) Simplified geological map of Eastern Anatolia showing tectonic units, collision-related volcanic products, and volcanic centers. E-K-P—Erzurum-Kars plateau; NATF and EATF—North and East Anatolian transform faults. Volcanic centers: Ag—Mt. Ağrı (Ararat); Al^1—Mt. Aladağ (southeast of Ağrı); Al^2—Mt. Aladağ (NW of Horasan); Bi—Mt. Bingöl; Bl—Mt. Bilicandaği; D—Mt. Dumanlidağ; E—Mt. Etrüsk; H—Mt. Hamadağ; K—Mt. Karatepe; Ki—Mt. Kisirdağ; M—Mt. Meydandağ; N—Mt. Nemrut; S—Mt. Süphan; T—Mt. Tendürek; Y—Mt. Yaglicadağ; Z—Mt. Ziyaretdağ. (B) Major tectonic blocks of Eastern Anatolia. The borders are modified from Şengör et al. (2003). I—Rhodope-Pontide fragment; II—Northwest Iranian fragment; III—Eastern Anatolian Accretionary Complex (EAAC); IV—Bitlis-Pötürge Massif; V—Arabian foreland. Dark green areas—outcrops of ophiolitic mélange; pink and red areas—collision-related volcanic units; white areas—undifferentiated units or young cover formations. (C) Distribution of the oldest radiometric ages of the volcanic units. Ages are from Ercan et al. (1990), Pearce et al. (1990), and Keskin et al. (1998). Initiation ages of the volcanism are contoured in 1 m.y. intervals. PS—Pontide suture; BPS—Bitlis-Pötürge suture; CS—inferred cryptic suture between the Eastern Anatolian Accretionary Complex and the Bitlis-Pötürge suture (BPS).

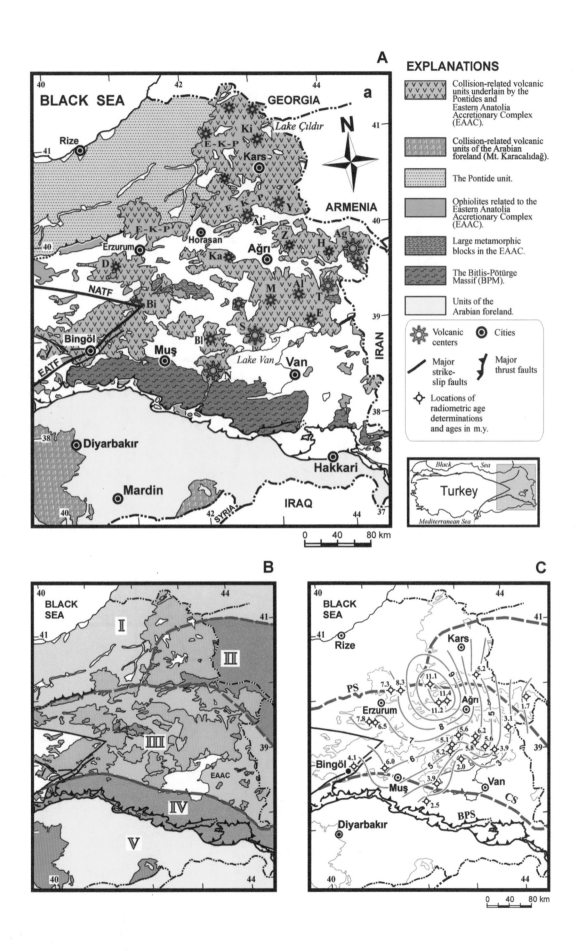

EXPLANATIONS

Collision-related volcanic units underlain by the Pontides and Eastern Anatolia Accretionary Complex (EAAC).

Collision-related volcanic units of the Arabian foreland (Mt. Karacalıdağ).

The Pontide unit.

Ophiolites related to the Eastern Anatolia Accretionary Complex (EAAC).

Large metamorphic blocks in the EAAC.

The Bitlis-Pötürge Massif (BPM).

Units of the Arabian foreland.

Volcanic centers

Cities

Major strike-slip faults

Major thrust faults

Locations of radiometric age determinations and ages in m.y.

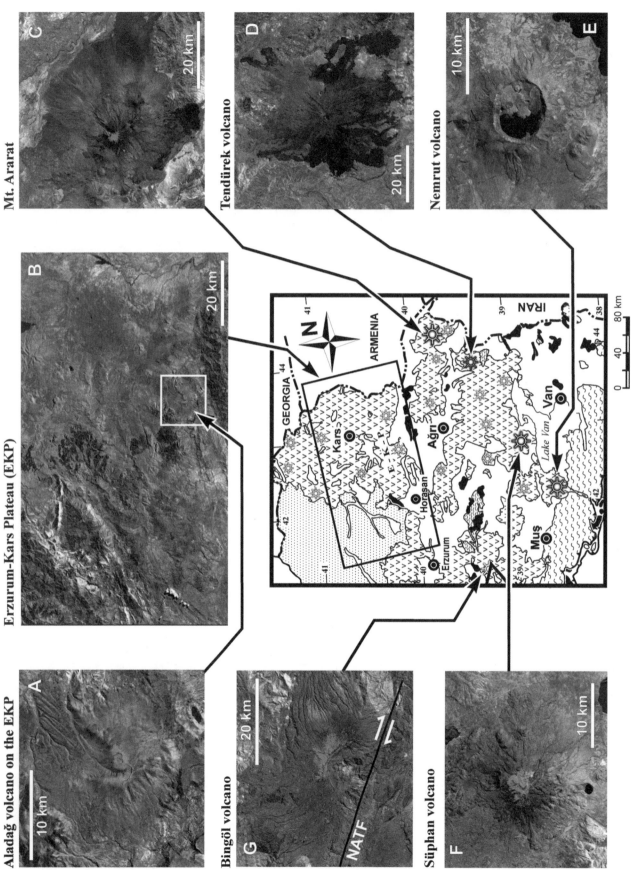

Figure 2. MrSID satellite view of major volcanic centers of Eastern Anatolia. (A) Mt. Aladağ volcano on the Erzurum-Kars plateau in the northernmost part of Eastern Anatolia; (B) a general view of the Erzurum-Kars plateau in the northeast; (C) Mt. Ararat, a double-peaked stratovolcano; (D) Tendürek, a shield volcano in the south; (E) Mt. Nemrut volcano in the northeast; (F) Süphan stratovolcano in the northern part of Lake Van; and (G) Bingöl volcano, a truncated volcano near the North Anatolian transform fault (NATF). The reddish-brownish-colored areas correspond to volcanic units, while the purple to pinkish areas are either basement units (e.g., areas in the northwest) or young sedimentary cover formations. In general

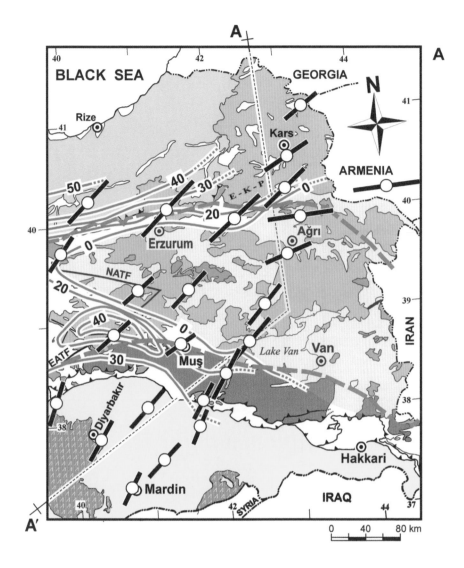

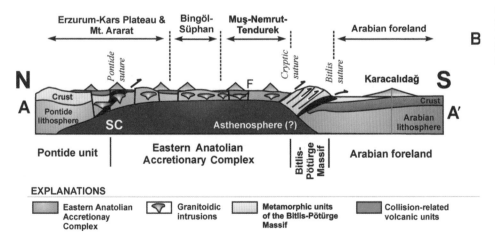

Figure 3. (A) Map showing collision-related volcanic and tectonic units, mantle lid thicknesses, and shear wave–splitting fast polarization directions (from Fig. 1 *in* Sandvol et al., 2003b) in Eastern Anatolia. E-K-P—Erzurum-Kars plateau; NATF and EATF—North and East Anatolian transform faults. The contours (red) indicate the mantle lid (i.e., lithospheric mantle) thicknesses in km (contours are from Fig. 2 *in* Şengör et al., 2003). The light bluish–colored triangular area surrounded by the cities of Ağrı, Erzurum, Muş, and Van in the center of the figure represents the area with no mantle lid. The thick, dashed dark blue lines represent the northern and southern borders of the Eastern Anatolian Accretionary Complex (see also Fig. 1B). Note that areas of inferred complete lithospheric detachment almost exactly coincide with the extent of the Eastern Anatolian Accretionary Complex. (B) Cross-section summarizing the lithospheric structure across Eastern Anatolia (not to scale). The crustal and lithospheric thicknesses are from Şengör et al. (2003) and Zor et al. (2003). The direction of the cross-section (A–A′) is shown in panel A. Sources of the geochemical data: Ercan et al. (1990), Pearce et al. (1990), Keskin et al. (1998). SC—subduction component; F—strike-slip faults.

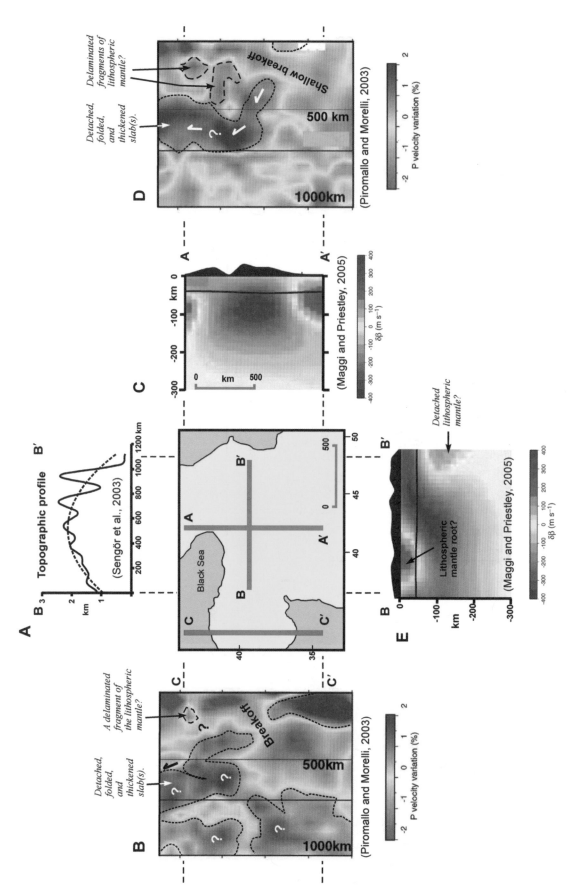

Figure 4. (A) East-west topographic profile along the 40°N parallel (Şengör et al., 2003). The smooth line represents least-squares simplifications of the topography. (B–E) Tomographic cross-sections showing the lithospheric structure of Eastern Anatolia (Piromallo and Morelli, 2003; Maggi and Priestley, 2005).

GEOLOGY

The basement of the Anatolian-Iranian plateau is made up of microcontinents accreted to one another during the Late Cretaceous to Early Tertiary (Şengör, 1990). These microcontinents are separated from one another by ophiolite belts and accretionary complexes. Five different tectonic blocks are recognized in Eastern Anatolia (Fig. 1B): (1) The Eastern Rhodope–Pontide fragment in the northwestern part of the region (I in Fig. 1B), which underlies the southwestern and northeastern parts of the Erzurum-Kars plateau (i.e., EKP in Fig. 1B); (2) the Northwest Iranian fragment (II in Fig. 1B), a tectonic block overlain by the eastern part of the Erzurum-Kars plateau (i.e., the Horasan, Aladağ, Kağizman, and Kars areas and Mt. Ararat) (Keskin et al., 2006); (3) the Eastern Anatolian Accretionary Complex in the middle of the region located between the Aras River and the Bitlis-Pötürge Massif (III in Fig. 1B); (4) the Bitlis-Pötürge unit, which is exposed along the Taurus belt (IV in Fig. 1B); and (5) autochthonous units of the Arabian continent or foreland (V in Fig. 1B). Except for the Eastern Anatolian Accretionary Complex, all the tectonic blocks correspond to the aforementioned microcontinents.

The Eastern Rhodope–Pontide unit is located in the northernmost part of the region. Its basement is represented by a metamorphic massif named the Pulur Complex (Topuz et al., 2004). The Pulur Complex is composed of a heterogeneous set of granulite-facies rocks ranging from quartz-rich mesocratic gneisses to silica- and alkali-deficient, Fe-, Mg-, and Al-rich melanocratic rocks (Topuz et al., 2004). A thick volcano-sedimentary arc sequence overlies this metamorphic basement. This sequence is regarded as an ensialic, south-facing magmatic arc formed by north-dipping subduction under the Eurasian continental margin (Yılmaz et al., 1997) in a period between the Albian and the Oligocene (Şengör et al., 2003).

The Northwest Iranian fragment is masked by collision-related volcanic units in Eastern Anatolia. It is exposed in Armenia around the Tsakhkuniats basement outcrop and the Hankavan-Takarly and Agveran massifs (Karapetian et al., 2001). The unit is composed of a heterogeneous rock sequence consisting of trondhjemitic, phyillitic, albite-plagiogranitic, and plagiogranite- and granite-migmatitic lithologies (Karapetian et al., 2001).

The Eastern Anatolian Accretionary Complex forms a 150 to 180-km-wide, northwest-southeast-extending belt in the middle of the region. It represents the remnant of a huge subduction-accretion complex formed between the Late Cretaceous and the Oligocene on a north-dipping subduction zone located between the Rhodope-Pontide microcontinent in the north and the Bitlis-Pötürge microcontinent in the south (Şengör et al., 2003). It consists of two contrasting rock units: an ophiolitic mélange of Late Cretaceous age and Paleocene to Late Oligocene flysch sequences incorporated into the ophiolitic mélange as north-dipping tectonic slices. These flysch slices become younger from north to south and shallower from the Cretaceous to the Oligocene (Şengör et al., 2003). This observation is consistent with the polarity of the subduction zone thought to have created the Eastern Anatolian accretionary prism by underthrusting.

The Bitlis-Pötürge Massif is exposed in a northwest-southeast-extending belt along the Eastern Taurus mountain range. It is regarded as the easternmost extremity of the Menderes-Taurus block. It consists of moderately to highly metamorphosed sedimentary and igneous units.

Shallow marine deposits of Oligocene to Middle Miocene age unconformably overlie these tectonic blocks in some places (not shown in Figs. 1 and 3). Collision-related subaerial volcanic units, on the other hand, unconformably overlie both these five tectonic blocks and the aforementioned marine deposits, masking the basement units over great distances (Fig. 1A and B). These volcanic units become younger to the SSE (Keskin, 2003) (Fig. 1C).

LITHOSPHERIC AND MANTLE STRUCTURE BENEATH THE REGION

Results from the Eastern Turkey Seismic Experiment (ETSE) project (Gök et al., 2000, 2003; Al-Lazki et al., 2003; Sandvol et al., 2003a; Angus et al., 2006) indicate that the mantle lithosphere is either very thin or absent beneath a considerable portion of the region between the Aras River (broadly corresponding to the southern border of the Erzurum-Kars plateau) in the north and the Bitlis-Pötürge Massif in the south (Figs. 3 and 4). Barazangi et al. (2006) point out that the uppermost mantle beneath this crustal block strongly attenuates Sn waves and has one of the lowest Pn velocities on Earth (~7.6 km/s). Furthermore, the crustal thicknesses obtained from receiver function studies reveal a gradual change from <38 km in the southeast around the southern part of the Bitlis Suture Zone to 50 km in the north beneath the Erzurum-Kars plateau (Fig. 4 *in* Zor et al., 2003; see also Çakır et al., 2000, and Çakır and Erduran, 2004), averaging some 45 km. This indicates that a crust of almost normal thickness overlies an extremely thin mantle lithosphere, or perhaps it is almost directly underlain by the asthenosphere (see also the cross-section in Fig. 3B). These results are also confirmed by the studies of Hearn and Ni (1994), Maggi et al. (2002), Sandvol and Zor (2004), and Maggi and Priestley (2005), suggesting that the temperature of the mantle significantly increased beneath this area. Moreover, high Bouguer gravity anomalies also suggest that the Moho is almost directly underlain by hot asthenospheric material beneath Eastern Anatolia (Ateş et al., 1999; Barazangi et al., 2006). This interpretation is also supported by the results of a recent magnetotelluric study conducted in the region by Türkoğlu et al. (2006). On the basis of their geoelectric images, these researchers argue that the upper mantle beneath Eastern Anatolia has a very low resistivity, and this is consistent with the presence of shallow asthenosphere beneath the region.

When all the geophysical findings are taken into consideration, a reasonable interpretation is that Eastern Anatolia's crust is hot and weak (Reilinger et al., 1997) and is made up of crustal slivers that are in motion relative to one another (Angus et al., 2006). A lithospheric thickness of ~45 km is normal in extensional areas, such as Iceland, but unusual in a continental collision setting with a compressional tectonic regime. What all these findings may imply is that a huge portion of the mantle lithosphere was lost from beneath Eastern Anatolia.

Shear-wave-splitting fast polarization directions (Fig. 3A) are quite uniform, exhibiting northeast-southwest orientations beneath the region (Sandvol et al., 2003b). Sandvol et al. (2003b) argue that there is a fundamental difference between the "present-day mantle flow directions" and surface deformation across the Arabia and Anatolia plates. Therefore, they suggest that the observed mantle flow directions are asthenospheric and not litho-

spheric. These findings are also consistent with the tomographic results and imply that most of the Eastern Anatolian mantle lithosphere has been removed.

High-resolution deeper tomographic images obtained by inversion of P-wave delay times beneath the region (Piromallo and Morelli, 2003) also support the aforementioned detachment model. Tomographic sections from Piromallo and Morelli (2003) indicate that there is a positive anomaly around the transition zone beneath Eastern Anatolia (at ~300–500 km depth; Fig. 4B and D), extending laterally to the west and merging with one beneath the Aegean Sea (Faccenna et al., 2006). These images might be interpreted as evidence for a continuous subducting slab beneath the region, extending from the Hellenic subduction

TABLE 1. MAJOR-ELEMENT, TRACE-ELEMENT, AND ISOTOPIC DATA ON REPRESENTATIVE VOLCANIC ROCK TYPES ACROSS EASTERN ANATOLIA

	Area	Source	Stage	Sample	SiO_2	TiO_2	Al_2O_3	Fe_2O_3	MnO	MgO	CaO	Na_2O	K_2O
North	EKP	Keskin	Early	MK289	49.09	1.26	16.92	10.56	0.15	7.02	10.13	3.11	0.51
	EKP	Keskin	Early	MK281	56.99	1.83	16.12	8.81	0.11	2.23	5.73	4.68	2.00
	EKP	Keskin	Middle	MK237	56.47	0.99	18.43	6.63	0.10	4.16	7.33	3.91	1.35
	EKP	Keskin	Middle	MK277	60.58	0.80	15.85	5.22	0.08	4.08	6.10	4.11	2.18
	EKP	Pearce	Late	3011	46.00	2.42	16.90	13.90	0.21	5.81	8.49	4.04	0.61
	EKP	Keskin	Late	MK144	48.76	1.34	15.18	9.10	0.33	8.40	11.40	3.29	1.08
	EKP	Keskin	Late	MK139	49.67	1.47	17.03	9.77	0.15	7.06	9.78	4.06	0.73
	EKP	Keskin	Late	MK130	54.36	1.90	15.81	9.71	0.18	4.56	7.90	4.87	0.99
	EKP	Pearce	Late	PL2/21	64.40	0.99	17.12	4.64	0.03	0.26	4.66	4.60	1.94
	Mt. Ararat	Pearce		3132	51.40	2.08	17.54	9.88	0.15	6.68	8.42	4.12	0.86
	Mt. Ararat	Pearce		3031	56.60	1.35	16.94	7.60	0.11	3.16	6.08	4.25	1.82
	Mt. Ararat	Pearce		3041	58.20	0.97	17.50	6.37	0.09	4.08	6.57	4.22	1.80
	Mt. Ararat	Pearce		3131	62.60	0.79	18.00	5.40	0.09	1.42	4.59	4.67	1.81
Middle													
	Bingöl	Pearce		2212	58.20	1.34	17.90	7.69	0.17	1.37	5.94	4.32	2.36
	Bingöl	Pearce		2214	70.60	0.36	15.30	2.30	0.04	0.24	1.56	4.22	4.84
	Süphan	Pearce		2521	62.80	1.01	16.10	6.38	0.11	1.45	3.94	4.79	2.71
	Süphan	Pearce		2531	63.90	0.87	17.30	5.90	0.11	1.33	3.80	5.71	2.58
South	Mus	Buk-Tem		4	45.65	2.72	17.46	14.11	0.19	4.58	9.06	4.58	1.16
	Mus	Pearce		2112	50.10	2.86	15.56	12.60	0.20	4.60	8.00	3.92	1.32
	Mus	Pearce		2111	53.70	1.45	18.20	9.53	0.18	2.21	6.38	4.90	2.21
	Mus	Buk-Tem		1	56.35	1.55	17.54	7.94	0.12	3.36	6.28	4.40	1.95
	Mus	Buk-Tem		21	61.47	1.33	17.24	6.31	0.04	0.81	3.75	5.33	3.27
	Mus	Buk-Tem		16	63.01	0.81	16.65	5.08	0.09	2.42	5.09	3.80	2.78
	Mus	Pearce		2141	66.20	0.50	18.00	2.54	0.00	0.00	2.01	5.52	4.32
	Nemrut	Pearce		2362	47.70	3.08	16.66	14.10	0.20	6.46	9.48	3.54	0.68
	Nemrut	Özdemir		Yoz 81	51.75	2.69	15.98	12.47	0.18	3.20	7.01	4.19	1.80
	Nemrut	Özdemir		Z-12	58.58	1.25	16.87	7.71	0.18	1.72	4.06	5.40	3.75
	Nemrut	Pearce		2022	64.30	0.55	15.70	5.54	0.19	0.26	1.41	6.02	5.58
	Nemrut	Pearce		2421	66.00	0.42	16.20	4.35	0.12	0.00	1.20	6.14	4.81
	Nemrut	Özdemir		Z-11	68.45	0.49	11.88	7.53	0.21	0.03	0.53	5.64	5.11
	Nemrut	Özdemir		Cu-11	70.15	0.38	10.07	6.72	0.17	0.02	0.37	5.52	4.42
	Tendürek	Pearce		3121	48.90	2.35	17.00	11.70	0.20	4.10	6.53	5.38	1.56
	Tendürek	Sen		31	51.00	1.87	17.14	11.20	0.19	4.10	7.35	5.22	1.65
	Tendürek	Sen		28	53.50	1.52	17.20	9.96	0.19	2.88	6.24	5.31	2.13
	Tendürek	Pearce		3111	58.40	1.24	18.10	5.98	0.14	1.34	3.56	6.56	3.73

zone to Central and Eastern Anatolia (Faccenna et al., 2006), where it seems to have been detached (Fig. 4B and D). Note that beneath both Eastern and Western Anatolia, slabs appear to have significantly thickened by a factor of two to three. This is not unusual; it is now well understood that slabs can significantly deform during their descent into the more viscous lower mantle by means of folding and thickening (Lay, 1994; Hafkenscheid et al., 2006). Fast and steeply subducting slabs can fold and thicken by a factor of two to three (e.g., Gaherty and Hager, 1994; Christensen, 1996), whereas slowly subducting slabs at small angles can thicken to twice their original thickness when they enter the more viscous lower mantle (e.g., Gaherty and Hager, 1994; Becker et al., 1999).

GEOCHEMICAL CHARACTERISTICS OF THE COLLISION-RELATED VOLCANIC UNITS

Two of the most striking aspects of Eastern Anatolia are the volume and compositional variability of collision-related volcanic products erupted during the Neogene and the Quaternary. Over half the region is covered with young volcanic units (Figs. 1 and 3), ranging in age from 11.4 Ma to present (Fig. 1C). In this section a short description of the geochemical characteristics of the volcanic units and their spatial changes are presented. Major-element, trace–element, and isotopic data from representative lava types across Eastern Anatolia are given in Table 1.

P_2O_5	L.O.I.	Total	Sc	V	Cr	Co	Ni	Cu	Zn	Ga	Rb	Sr	Y	Zr	Nb	Ba
0.26	1.47	99.01	28	163	261	47	150	96	104	20	11	455	25	115	8	143
0.45	0.97	99.04	22	183	3	11	3	20	83	18	52	417	31	213	14	335
0.30	1.40	99.66	18	132	51	20	22	16	69	20	20	564	17	158	12	495
0.28	0.84	99.28	11	99	116	21	95	32	55	20	62	489	16	159	14	450
0.58	1.13	100.09		120	22		43		89		5	518		275	10	
0.54	2.47	99.41	24	201	264	62	212	66	85	16	18	867	26	119	15	533
0.31	0.30	100.03	29	183	162	39	85	41	75	22	9	507	25	147	11	461
0.33	0.14	100.61	26	163	11	35	32	20	75	19	15	415	33	214	7	175
0.35	1.17	100.16		126	8		6	26	73		50	554	22	225	13	701
0.33	0.17	101.63	28	213	298	32	100	43	89		11	531	26	187	10	221
0.31	0.82	99.07	16	191	81	19	25	26	84		47	463	28	229	12	461
0.30	0.29	100.39		111	103		67		69		46	532	22	203	13	
0.29	0.47	100.13		56	5		10		73		41	379		207	13	
0.32	2.12	101.73		108	24		33		73		75	433		306	20	
0.11	1.89	101.46		13	2		6		41		155	155	19	307	23	
0.32	1.05	100.65		46	3		6		76		82	207	45	364	13	
0.24	0.46	102.20		42	6		10		70		80	200		338	10	392
0.72		100.23	8	115	30	37	35		81	23	19	1082	30	339	22	155
0.49	1.24	100.89	26	253	99	30	23	40	127		28	393	39	313	21	180
0.44	1.68	100.88		67	12		21		100		63	422	40	430	28	184
0.52		100.01	15	133	8	26	19		100	25	54	492	32	30	35	335
0.37		99.92	9	99	10	18	10		77	28	140	340	41	517	36	436
0.25		99.98	9	77	24	29	18		41	19	95	399	28	269	24	693
0.19	1.11	100.39					11		83		138	242	56	647	39	409
0.37	−0.02	102.25	24	250	97	44	44	40	116		7	401	31	200	11	125
0.53		99.80	22	203		26		6	75	24	39	361	43	269	18	387
0.44	0.10	100.06	9	41		11		6	74	26	91	340	55	425	29	566
0.10	0.42	100.07			1		9		103		93	19		466	33	245
0.07	0.37	99.68			8				53		98	94	51	531	34	
0.01	0.10	99.98	8			1		3	175	37	211	2	64		68	
0.02	1.70	99.54	1			1		13	15	37	249	2	150		74	8
1.02	0.24	98.98		106			12		114		22	695	42	389	34	329
	0.08	99.80									23	504	28	246	26	455
	0.33	99.26									37	525	34	342	33	546
0.42	0.39	99.86		36	11		7		114		95	286	54	475	42	538

(continued)

TABLE 1. *Continued*

	Area	Source	Sample	La	Ce	Pr	Nd	Sm	Eu	Gd	Tb	Dy	Ho	Er	Tm
North	EKP	Keskin	MK289	12.3	26.3	3.4	15.3	3.64	1.34	3.8	0.73	4.32	0.88	2.62	0.40
	EKP	Keskin	MK281	24.7	48.9	5.9	23.6	5.12	1.79	5.4	0.81	4.90	1.03	2.92	0.44
	EKP	Keskin	MK237	30.1	48.9	5.8	23.0	3.90	1.28	3.6	0.55	3.06	0.58	1.62	0.24
	EKP	Keskin	MK277	28.1	48.6	5.3	19.3	3.48	0.99	3.0	0.47	2.50	0.51	1.48	0.24
	EKP	Pearce	3011		46.4		27.5	6.17	2.05	5.6		5.99		3.50	
	EKP	Keskin	MK144	30.6	68.9	8.3	33.7	6.14	1.74	5.7	0.71	4.05	0.83	2.20	0.32
	EKP	Keskin	MK139	13.5	29.6	3.8	15.8	3.64	1.33	4.1	0.65	4.14	0.80	2.46	0.35
	EKP	Keskin	MK130	12.8	30.2	4.0	18.9	4.48	1.56	4.9	0.90	5.15	1.04	3.08	0.47
	EKP	Pearce	PL2/21											8.0	
	Mt. Ararat	Pearce	3132	18.9	52.7		26.6	5.17	1.63		0.69		0.83		
	Mt. Ararat	Pearce	3031	26.9	67.8		29.1	5.50	1.62		0.53				
	Mt. Ararat	Pearce	3041	29.6	64.1		25.8	4.41	1.36		0.51				
	Mt. Ararat	Pearce	3131		47.3		20.6	3.96	1.22	3.5		3.58		1.96	
Middle	Bingöl	Pearce	2212		65.1		26.7	5.04	1.43	4.3		4.53		2.55	
	Bingöl	Pearce	2214												
	Süphan	Pearce	2521		65.1		31.0	6.77	1.98	7.2		7.69		4.63	
	Süphan	Pearce	2531		86.3		33.1	5.57	1.33	6.1		6.67		4.03	
South	Muş	Buk-Tem	4	25.0	50.0		31.6	6.60	1.90		0.78				
	Muş	Pearce	2112	28.6	77.7		44.1	8.45	2.40		1.31		1.57		
	Muş	Pearce	2111	35.4	90.7		45.8	8.92	2.33		1.07				
	Muş	Buk-Tem	1	44.0	69.0		29.2	5.80	1.80		0.00				
	Muş	Buk-Tem	21	47.0	83.0		35.8	6.90	1.50		0.97				
	Muş	Buk-Tem	16	41.0	64.0		26.0	4.70	1.30		0.93				
	Muş	Pearce	2141	41.9	113.1		57.1	11.06	1.63		1.16				
	Nemrut	Pearce	2362	14.4	44.4		23.6	5.56	4.41		0.92		1.28		
	Nemrut	Özdemir	Yoz 81	31.3	64.1	8.4	35.3	7.60	2.53	7.5	1.29	7.33	1.64	4.15	0.60
	Nemrut	Özdemir	Z-12	48.1	90.8	11.3	47.5	10.10	2.80	9.0	1.58	9.27	1.93	5.56	0.87
	Nemrut	Pearce	2022	43.3	100.9		47.3	10.33	5.99		1.09				
	Nemrut	Pearce	2421												
	Nemrut	Özdemir	Z-11	62.0	178.5	16.6	63.9	14.80	1.03	13.0	2.47	15.67	3.36	10.67	1.74
	Nemrut	Özdemir	Cu-11	109.6	239.8	27.3	111.7	23.20	1.53	22.5	4.21	25.27	4.88	15.44	2.35
	Tendürek	Pearce	3121	50.6	126.3		58.5	10.34	2.88		1.35				
	Tendürek	Sen	31	44.2	90.2	10.5	37.7	7.37	2.54	9.1		6.06	1.10	3.11	0.48
	Tendürek	Sen	28	57.0	115.0	13.2	46.6	8.92	3.08	11.5		7.74	1.40	3.86	0.62
	Tendürek	Pearce	3111	58.6	128.4		50.9	8.53	2.41		1.34				

Sources: Buk-Tem—Buket and Temel (1998); Keskin—Keskin et al. (1998, 2006); Özdemir—Özdemir et al. (2006); Pearce—Pearce et al. (1990); Sen-Sen et al. (2004).

Note: EKP—Erzurum-Kars plateau.

Classification

Collision-related volcanic rocks across the region span the whole compositional range, from basalts to rhyolites. There is significant variation in lava chemistry in the north-south direction between the Erzurum-Kars plateau in the north and the Muş-Nemrut-Tendürek volcanoes in the south (Figs. 5A–F). The volcanic units of the Erzurum-Kars plateau are calc-alkaline (they follow a calc-alkaline trend on the $K_2O + Na_2O$, Fe_2O_3, MgO diagram, which is not shown here), while those of the Muş-Nemrut-Tendürek volcanoes are alkaline to mildly alkaline in character. Lavas of the Bingöl and Süphan volcanoes display transitional chemical characteristics (Pearce et al., 1990) (Fig. 5B and E).

Spatial Variations in Magmatism and Source Compositions

Multielement Patterns. In order to highlight spatial variations in subduction and intraplate components in collision-related magmatism across the region, incompatible multi-element patterns normalized to N-type mid-ocean ridge basalt (MORB) composition are presented in Figure 5G–I. The elements are arranged in the order suggested by Pearce (1983). In these di-

Yb	Lu	Hf	Ta	Pb	Th	U	$^{87}Sr/^{86}Sr_{(i)}$	$^{143}Nd/^{144}Nd_{(i)}$	$^{206}Pb/^{204}Pb_{(i)}$	$^{207}Pb/^{204}Pb_{(i)}$	$^{208}Pb/^{204}Pb_{(i)}$	$\delta^{18}O$
2.61	0.41	3.1	0.47	21.2	1.30	0.41	0.703904	0.512847	17.663	15.553	37.493	
2.67	0.41	4.3	0.89	9.0	6.55	1.90	0.703931	0.512898	18.734	15.603	38.748	
1.55	0.25	4.1	0.70	12.6	6.67	1.11						
1.33	0.21	3.8	0.88	11.5	7.50	2.06	0.704322	0.512781	18.682	15.605	38.721	7.2
3.29							0.703800					
2.18	0.34	2.8	0.71	5.6	3.34	1.03	0.704570	0.512791	19.022	15.664	39.053	
2.37	0.35	2.6	0.53	2.6	1.65	0.70	0.703705	0.512931	18.933	15.667	39.031	
3.05	0.48	4.7	0.45	5.1	2.52	0.67	0.703390	0.512930	18.939	15.623	38.895	
							0.704180					
2.46	0.41	4.2	0.43	3.0	2.07	0.30	0.703890					
2.74	0.46	5.4	0.67	10.0	5.78	1.99	0.704170					
1.88	0.34	4.1	0.68	8.0	6.29	2.20	0.704160					
1.95							0.704410					
2.53							0.704730	0.512745				
				21.0	22.00		0.705060					
4.40				12.0	10.00		0.705050	0.512842				
3.97							0.704660					
2.80	0.50	5.4			2.20	14.00	0.704660	0.512860				
4.00	0.61	8.4	1.51	6.0	3.95	0.73	0.704320					
4.41	0.61	8.8	1.67	7.0	8.81	1.71	0.704430	0.512793				
3.00	0.70	6.2			7.50	6.00	0.704160	0.512800				
3.40	0.60	9.4			19.00	17.00	0.704820	0.512760				
2.40	0.50	5.9			12.00	15.00	0.705000	0.512710				
4.47	0.80	10.7	3.03	20.0	17.28	7.68	0.705440					
3.04	0.45	4.5	0.70		1.66	0.23	0.703570					
4.04	0.60	6.7	1.20	1.3	6.30	1.30						
5.94	0.89	9.9	2.20	2.3	12.70	3.00						
5.44	0.71	12.2	1.70		8.97	1.84	0.706080					
				11.0	12.00		0.705050					
12.14	1.99	25.6	4.70	23.3	25.00	3.70						
15.03	2.02	32.4	4.80	13.0	34.20	10.10						
4.75	0.76	7.6	1.47	12.0	3.79	0.96	0.705630					
2.94	0.49	5.0	1.10	11.3	5.13	1.63	0.705743	0.512676				
3.91	0.61	7.0	1.51	15.4	6.45	1.95	0.705889	0.512676				
5.28	0.93	9.8	1.63	17.0	12.20	3.73	0.705340	0.512816				

agrams, the incompatibility of mobile elements increases from Sr to Ba, while that of immobile elements increase from Yb to Th during lherzolite melting. Only the samples with $SiO_2 < 60$ (wt%) have been plotted on these diagrams, because fractional crystallization and crustal assimilation mask the ability of these patterns to reveal mantle sources. On these diagrams, calc-alkaline volcanic units on the Erzurum-Kars plateau and Mt. Ararat display patterns typical of continental arc volcanics. They are likely to have been derived from an enriched mantle source containing a distinct subduction signature (Fig. 5G). This signature decreases to the south and diminishes around the Muş-Nemrut-Tendürek volcanoes (Fig. 5I), where the lavas are alkaline and display an intraplate signature (Fig. 5H) (Pearce et al., 1990).

Ta/Yb versus Th/Yb Plots. On a Ta/Yb versus Th/Yb dia-

gram, the calc-alkaline lavas of the Erzurum-Kars plateau display a consistent displacement from the mantle metasomatism array toward higher Th/Yb ratios, forming a trend subparallel to the main mantle metasomatism array (MM; Fig. 6A). The aforementioned displacement suggests that the Erzurum-Kars plateau mantle source region had a distinct subduction component. On the other hand, the presence of the trend subparallel to the main MM may be linked to magma chamber processes such as fractional crystallization (FC) or assimilation and fractional crystallization (AFC). It should be noted that this diagram is not suitable for differentiating between FC and AFC processes, because the modeled trajectories for AFC and FC processes are almost parallel. The alkaline basic lavas of the Muş-Nemrut-Tendürek volcanoes displayed a progressive shift from the MM with

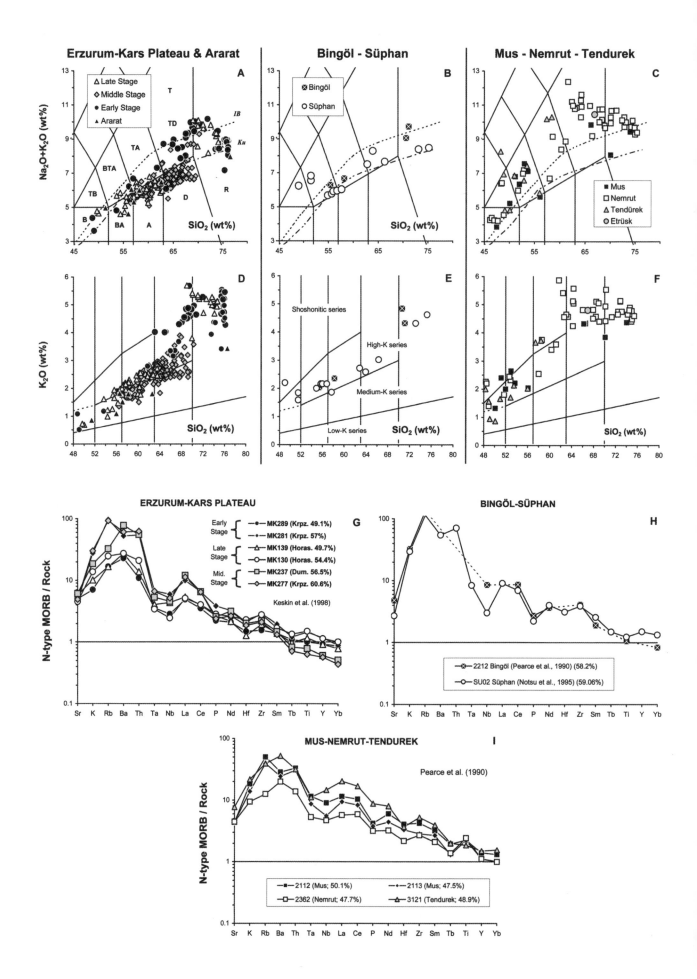

increasing SiO_2 (Fig. 6C). Pearce et al. (1990) argued that the lavas following this trend might have been derived from an enriched source with or without a slight subduction signature and then evolved through combined AFC.

Th/Ta versus MgO and Ta Plots. In order to highlight compositional variations in magmatism across the region and their possible links with mantle source regions, Th/Ta ratios of basic samples (with MgO > 3 wt% and SiO_2 < 52 wt%) have been plotted against their MgO and Ta values in Figure 6D and E. This ratio is specifically selected because it can be used to differentiate between lavas with subduction and those with intraplate signatures. Note that both Th and Ta are highly incompatible with one another during melting and with most minerals crystallizing from mafic to intermediate liquids. The Th/Ta ratio is therefore independent of partial melting and fractional crystallization, providing that anhydrous phases (i.e., plagioclase, olivine, augite, and magnetite) are the dominant crystallizing or residual assemblages. On these two diagrams, data from lavas from northern areas (i.e., the Erzurum-Kars plateau and Mt. Ararat) have consistently higher Th/Ta ratios and, in general, lower Ta concentrations compared to those from the southern areas (i.e., the Muş-Nemrut-Tendürek volcanoes). When interpreted with the findings from multielement diagrams, this relationship indicates that lavas on the Erzurum-Kars plateau and Mt. Ararat in the north were possibly derived from a mantle source containing a distinct subduction signature, in contrast to the lavas of the southern areas (i.e., the Muş-Nemrut-Tendürek volcanoes), which were derived from a source displaying an intraplate signature with or without a slight subduction component. These observations imply that there is a north-south variation in source composition, with a southward increase in intraplate signature.

Fractional Crystallization

Crystallization assemblages in the collision-related lavas of Eastern Anatolia also display variations across the region. Lavas in the north contain hydrous assemblages (e.g., amphibole) as well as anhydrous minerals, whereas those in the south are generally dominated by anhydrous minerals (Pearce et al., 1990). This indicates that lavas are richer in water in the north than in the south, consistent with their subduction signature. Geochemical data are also consistent with these petrographic observations: the lavas containing hydrous minerals (e.g., amphiboles) display distinct depletion in Y with increasing Rb (Low-Y series trend in Fig. 6F) in contrast to the lavas of the southern areas (i.e., Muş-Nemrut-Tendürek; High-Y trend in Fig. 6H) which contain anhydrous minerals that exhibit positive to flat gradients.

Summary of the Geochemical Findings

The geochemical evidence presented so far indicates that volcanic products in the north around the Erzurum-Kars plateau and Mt. Ararat are calc-alkaline in character and likely to have been derived from a slightly enriched mantle source containing a distinct subduction signature (Figs. 5G–I; also see Fig. 2 *in* Keskin, 2003). This signature decreases to the south and diminishes around the Muş-Nemrut-Tendürek volcanoes, where the lavas are alkaline and display an intraplate signature. Radiometric dating results published to date indicate that volcanic activity began earlier in the north than in the south and migrated south over time (Fig. 1C). However, it should be noted that there are few good dates from the older lavas in the southern part of Eastern Anatolia. Therefore, further research is needed to confirm this trend.

The striking results of recent geophysical studies (Gök et al., 2000, 2003; Al-Lazki et al., 2003, 2004; Piromallo and Morelli, 2003; Maggi and Priestley, 2005; Angus et al., 2006), along with the geochemical findings discussed earlier, lead us to question the validity of geodynamic models proposed for the Eastern Anatolian Collision Zone in a number of studies reported in the literature. Therefore, prior to focusing on the issue of what process was responsible for the loss of mantle lithosphere, I first review the competing geodynamic models and their discrepancies.

Figure 5. Major-oxide and trace-element diagrams for classifying the lavas of Eastern Anatolia. (A–C) Classification of volcanic units of Eastern Anatolia on the total alkali versus silica diagram of Le Bas et al. (1986). Data for the Erzurum-Kars plateau are from Keskin et al. (1998) and Pearce et al. (1990). Data for Bingöl-Süphan are from Pearce et al. (1990) and Notsu et al. (1995), while the data for Muş-Nemrut-Tendürek are taken from Pearce et al. (1990), Buket and Temel (1998) (only for Muş), Şen et al. (2004) (only for Tendürek), and Özdemir et al. (2006) (only for Nemrut). The diagrams are arranged from north to south: the Erzurum-Kars plateau in the north, Bingöl-Süphan areas in the central-west, Muş-Nemrut-Tendürek areas in the south. Abbreviations: B—basalt; BA—basaltic andesite; TB—trachybasalt; BTA—basaltic trachyandesite; A—andesite; TA—trachyandesite; D—dacite; TD—trachydacite; T—trachite; R—rhyolite; IB—alkaline-subalkaline divide of Irvine and Baragar (1971); Ku—alkaline-subalkaline divide of Kuno (1966). Alkalinity increases from north to south. (D–F) Classification of the volcanic units of Eastern Anatolia on the K_2O versus silica diagram of Peccerillo and Taylor (1976). The data sources are the same as described for panels A–C. (G–I) N-type mid-ocean ridge basalt (MORB)–normalized patterns for volcanic samples from the Eastern Anatolian collision zone. The normalization values are from Sun and McDonough (1989). The sources of the data are presented in the diagrams. The numbers in brackets in the legend of panel G are SiO_2 wt% values. The elements are arranged in the order suggested by Pearce (1983). Note that the samples from the Erzurum-Kars plateau in the north contain a distinct subduction signature, while lavas of the Muş-Nemrut-Tendürek areas display an intraplate signature with or without a slight subduction signature. Samples from the Bingöl-Süphan area display intermediate characteristics between those of the Erzurum-Kars plateau and the Muş-Nermrut-Tendürek areas.

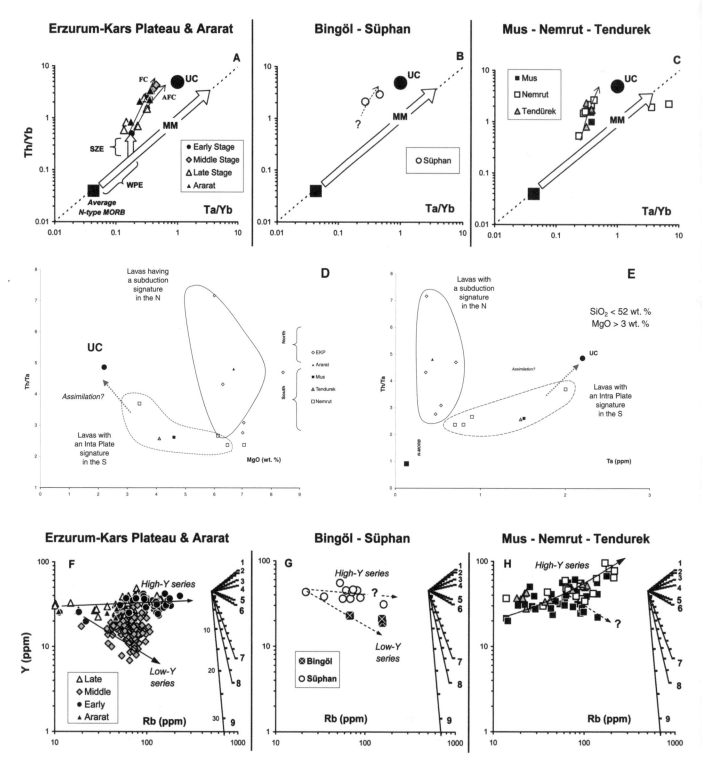

Figure 6. Trace-element diagrams used to study spatial variations in the petrogenesis and geochemistry of the lavas of Eastern Anatolia. (A–C) Th/Yb versus Ta/Yb diagrams (after Pearce, 1983) for basic and intermediate lavas (SiO_2 < 60%) from the Eastern Anatolian Collision Zone. Data for the Erzurum-Kars plateau are from Pearce et al. (1990) and Keskin et al. (1998). Data for Bingöl-Süphan are from Pearce et al. (1990) and Notsu et al. (1995), while the data for Muş-Nemrut-Tendürek are taken from Pearce et al. (1990), Buket and Temel (1998) (only for Muş), Şen et al. (2004) (only for Tendürek), and Özdemir et al. (2006) (only for Nemrut). MM—mantle metasomatism array; SZE—subduction zone enrichment; WPE—within-plate enrichment; UC—upper-crustal composition of Taylor and McLennan (1985); FC—fractional crystallization vector; AFC—assimilation combined with fractional crystallization curve. The FC vector has been modeled for 50% crystallization of an assemblage consisting of 50% plagioclase and 50% amphibole from a basic magma. The AFC curve has been drawn for an *r* value of 0.3. Note that

COMPETING GEODYNAMIC MODELS
AND THEIR DISCREPANCIES

Ten different geodynamic models have been proposed for the genesis of collision-related magmatism beneath the Eastern Anatolian Collision Zone (Fig. 7). Some of the earlier studies (e.g., the tectonic escape model of McKenzie, 1972) did not address the problem of why and how huge volumes of magmas were generated beneath the region. Any geodynamic model proposed for the Eastern Anatolian Collision Zone should, however, answer this critical question, because the topographic expression, tectonic elements, and magma generation are clearly all associated with the same mechanism. Each model is now discussed thoroughly, along with its weaknesses and strengths.

The Tectonic Escape of Microplates
to the East and West (McKenzie, 1972)

Discrepancies: A close examination of the model of McKenzie (1972) reveals that it does not entirely account for the strain induced by the 2.5 cm/yr convergence of the Arabia and Eurasia plates (Dewey et al., 1986). In addition, this model cannot explain why and how huge volumes of magma were generated beneath the region and how the region was elevated to form an extensive plateau now 2 km above sea level. It also does not provide an answer to the question of why the lithospheric mantle is absent beneath a greater portion of Eastern Anatolia (Fig. 7, Model 1).

Renewed Continental Subduction of the Arabia Plate
beneath Eastern Anatolia (Rotstein and Kafka, 1982)

Discrepancies: This model is not supported by any seismic evidence (Fig. 7, Model 2). There are no seismic data that suggest a currently subducting oceanic or continental lithospheric plate beneath the Bitlis-Pötürge Massif and Eastern Anatolia, attached to the Arabian plate.

Detachment and Northward Movement of a Subducting Slab
beneath Eastern Anatolia (Innocenti et al., 1982a,b)

On the basis of available radiometric dating results and data on chemical zonation in volcanic units across the collision zone,

Innocenti et al. (1982a,b) suggested that the andesitic volcanic front migrated northward by 150–200 km during the Pliocene. According to them, this is evidence for detachment of the subducted slab immediately after continental collision (Fig. 7, Model 3). According to their model, the detached slab moved northward while it was sinking in the asthenosphere. They suggest that this movement generated progressively lower-intensity magmatism from south to north. In their view, the volcanism becomes younger from south to north. In this model, calc-alkaline magmas that formed the Plio-Quaternary volcanic belt in the north were generated above the subducting slab, while the alkaline magmas representing the Miocene volcanic belt in the south were derived from the asthenosphere's upwelling through the gap (i.e., slab window) behind the detached subducting slab.

Although the model of Innocenti et al. (1982a,b) is one of the earliest, it is remarkable in that the possibility of slab detachment and the consequent effects on magma genesis in the Eastern Anatolian Collision Zone were envisaged thirteen years earlier than the "slab break-off model" was proposed by Davies and von Blanckenburg (1995). The latest geodynamic model, "slab steepening and break-off beneath a large subduction-accretion complex," (Şengör et al., 2003), also proposes a similar slab detachment process for magma genesis, although the slab in the models of Şengör et al. (2003) and Keskin (2003) did not move northward after break-off but instead steepened beneath a large subduction-accretion complex until it broke off, creating a gradually widening mantle wedge beneath the region.

Discrepancies: A more detailed study of collision-related volcanism on the Erzurum-Kars plateau (Keskin, 1994), which comprises the northernmost part of the Eastern Anatolian volcanic province, has shown that volcanism initiated at ca. 11 Ma in the north (Keskin et al., 1998), then migrated south over time (Keskin, 2003). These findings are the opposite of what is expected by the model of Innocenti et al. (1982a, b).

Rifting along East-West-Oriented Late
Miocene–Pliocene Basins (Tokel, 1985)

Tokel (1985) cited data from drill cores gathered from east-west-oriented Upper Miocene–Pliocene basins in Eastern Anatolia. He argued that these basins are bounded by gravity faults and are filled with at least 2000 m of limnic and fluvial deposits intercalated with voluminous "tholeiitic" and "alkaline" volcanic

---

this kind of diagram is not useful for differentiating the FC process from the AFC process. Lavas of the Erzurum-Kars plateau contain a distinct subduction zone enrichment (SZE) signature. (D–E) Th/Ta versus MgO and Ta diagrams highlighting the spatial variations in the geochemistry of lavas across the region. The data sources are the same as for panels A–C. For an explanation, see the text. (F–H) Rb versus Y diagrams displaying theoretical Rayleigh fractionation vectors for 50% crystallization of the phase combinations (given below) from a common magma composition. The tick marks on each vector correspond to 5% crystallization intervals. The data sources are the same as for panels A–C. The bulk partition coefficient values used in the modeling are from Table 2 *in* Keskin et al. (1998). The FC vectors have been modeled using the "FC-Modeler program" of Keskin (2002). Phase combinations for the vectors (see below for meanings of abbreviations): (1) $plg_{.5}$ + $cpx_{.3}$ + $olv_{.2}$ (B); (2) $plg_{.5}$ + $cpx_{.5}$ (B) or ~$plg_{.5}$ + $cpx_{.3}$ + $olv_{.2}$ (I); (3) $plg_{.5}$ + $amp_{.5}$ (B) or $plg_{.5}$ + $cpx_{.5}$ (I); (4) $plg_{.2}$ + $opx_{.1}$ + $cpx_{.6}$ + $olv_{.1}$ (I); (5) $plg_{.5}$ + $cpx_{.5}$ (A); (6) $plg_{.5}$ + $amp_{.5}$ (I); (7) $plg_{.4}$ + $amp_{.4}$ + $gt_{.2}$ (I); (8) $plg_{.5}$ + $amp_{.5}$ (A); (9) $plg_{.4}$ + $amp_{.4}$ + $gt_{.2}$ (A). Abbreviations: plg—plagioclase; cpx—clinopyroxene; opx—orthopyroxene; olv—olivine; amp—amphibole; gt—garnet. B—basic; I—intermediate; and A—acid magma compositions.

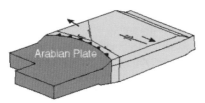

Model 1: The tectonic escape of microplates to the east and west (*McKenzie*, 1972).

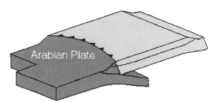

Model 2: Renewed continental, subduction of the Arabian plate beneath Eastern Anatolia (*Rotstein & Kafka*, 1982).

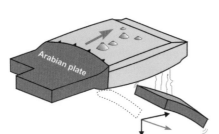

Model 3: Detachment and northward movement of a subducting slab beneath Eastern Anatolia (*Innocenti et al.*, 1982a,b).

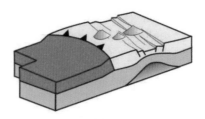

Model 4: Lithosphere extension following collision. Rifting along E-W oriented Late Miocene-Pliocene basins (*Tokel*, 1985) possibly accompanied by decompression melting of "normal asthenosphere" due to extension (*McKenzie & Bickle*, 1988).

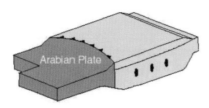

Model 5: Continental collision and subsequent thickening of the Anatolian crust/lithosphere (*Dewey et al.*, 1986).

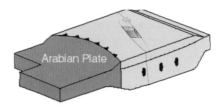

Model 6: Localized extension associated with pull-apart basins in strike-slip systems (*Dewey et al.*, 1986; *Pearce et al.*, 1990; *Keskin et al.*, 1998).

Model 7: Mantle plume impacts following collision (tested by *Pearce et al.*, 1990; proposed as a model by *Ershov and Nikishin*, 2004).

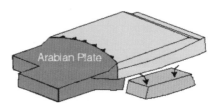

Model 8: Delamination of mantle lithosphere beneath the region (proposed by *Pearce et al.*, 1990; refined by *Keskin et al.*, 1998).

Figure 7. Competing geodynamic models proposed for Eastern Anatolia.

products. He suggested that recent tectonics in Eastern Anatolia were dominated by an extensional stress regime. On the basis of the mathematical model of Turcotte (1983), he proposed that these depressions and the sediments deposited therein were related to a "rifting event" in the region (Fig. 7, Model 4).

Discrepancies: The fault plane solutions of earthquakes in the region indicate that the faults are either strike-slip or reverse, which is inconsistent with extension (i.e., a rift setting). A close examination of the east-west-oriented basins in the region reveals that they are not rift-related, but are instead dominantly pull-apart basins related to strike-slip fault systems.

Decompression melting of normal asthenosphere as a result

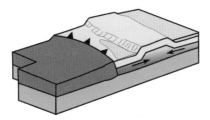

Model 9: Inflow of lower crust driven by the isostatic response to denudation and sedimentation in surrounding areas (*Mitchell & Westaway*, 1999).

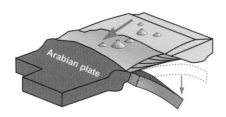

Model 10: Slab-steepening and breakoff beneath a subduction-accretion complex (proposed by *Sengor et al.*, 2003; supported by *Keskin*, 2003).

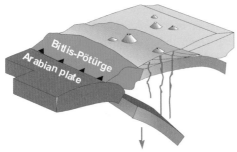

Modified slab-breakoff model involving two slabs subducting to the north (*Barazangi et al.*, 2006).

Figure 7. *Continued*

of regional extension (McKenzie and Bickle, 1988) requires that a stretching factor of ~2.5 generated melts in dry asthenosphere at a depth of 50 km and a temperature of around 1280 °C. As is well known, the region is not being stretched, so this is not a likely scenario.

Continental Collision and Subsequent Thickening of the Anatolian Lithosphere (Dewey et al., 1986)

Dewey et al. (1986) argued that Eastern Anatolia owes its high elevation to a doubled (~300 km) lithospheric thickness (Fig. 7, Model 5). They believed this thickening occurred as a result of continental collision between the Arabian and Eurasian continents. They also point out that the lavas were erupted both through north-south cracks that extend into the Arabian foreland and through transcurrent pull-aparts (Fig. 7, Model 6).

Following the model of Dewey et al. (1986), Yılmaz et al. (1987) suggested that the young volcanism in Eastern Anatolia could be linked to heating of the lower continental crust and mantle lithosphere that had been subjected to lithospheric thickening. Similarly, on the basis of their geochemical data, Koronovskiy and Demina (1996) argued that heating due to crustal thickening may explain the young volcanism of the Lesser Caucasus, adjacent to Eastern Anatolia.

Discrepancies: It is now well understood that the region would not have been isostatically elevated to ~2 km if a 250- to 300-km-thick, dense (3.2–3.3 g/cm³) mantle lithosphere had

been attached to the base of a lighter (2.7–2.8 gr/cm³) crust (Şengör, 2002; Şengör et al., 2003). This model is also not supported by the recent tomographic data (e.g., Al-Lazki et al., 2003; Gök et al., 2003; Maggie and Priestley, 2005) presented earlier.

Pearce et al. (1990) discuss the point that a 50% increase in thickness of the metasomatized mantle lithosphere lowers a significant portion of this layer to a depth below that of amphibole breakdown, forming garnet and releasing water. This may initiate localized melting, but it also lowers the geotherm. When this happens, most of the metasomatized layer remains significantly below the solidus and thus does not produce magma (Pearce et al., 1990). Therefore, it is difficult to explain the huge volumes of magma generated in the region by the models of Yılmaz et al. (1987) and Koronovskiy and Demina (1996).

Localized Extension Associated with Pull-Apart Basins in Strike-Slip Systems (Dewey et al., 1986; Pearce et al., 1990; Keskin et al., 1998)

In their pioneering study, Dewey et al. (1986) highlighted the connection between the formation of pull-apart basins and volcanism (Fig. 7, Model 6). They pointed out that there are two different neotectonic magmatic suites in the region: the nepheline-hypersthene normative alkaline basalts of mantle origin and the silicic to mafic calc-alkaline suite. They suggested that both suites occur in pull-apart basins in strike-slip regimes and north-south extensional fissures. They argued that the position and shape of magmatic intrusions might have been controlled by "flaking of the elastic lid," particularly beneath the pull-apart basins. In this model, magma generation is linked to local extension and small-scale delamination events beneath these basins (e.g.,. the Erzincan, Karasu-Pasinler-Horasan, and Muş basins). These authors also argued that rapid lithospheric stretching and small-scale delamination beneath pull-apart basins can generate melting in the mantle.

Although Pearce et al. (1990) considered delamination the dominant process that caused voluminous magma generation beneath the region, they also argued that it might have been accompanied by other stretching mechanisms, such as pull-apart basins. They also suggested that deviatoric stress perpendicular to the principal direction of compression might also have had some effect.

Keskin et al. (1998) emphasized the role of strike-slip faulting in pull-apart basins in focusing magmas on the Erzurum-Kars plateau, north of the region. They pointed out that, compared to nearby areas, a much thicker (2–4 km) sequence of volcanic and volcano-clastic rocks was deposited in these gradually subsiding basins. However, it is not clear whether these faults simply provided fractures that enabled magma to reach the surface or whether the associated localized extension in pull-apart basins also encouraged melting in the mantle. Discrepancies: Collision-related volcanic units are not confined to pull-apart basins. Instead, they cover a much greater area away from these basins. Therefore, it is doubtful that a pull-apart model can explain the genesis of all the collision-related magmatism in the region.

Hotspot Activity Related to a Mantle Plume (Discussed by Pearce et al., 1990; Proposed as a Model by Ershov and Nikishin, 2004)

The possibility of plume-related "hotspot" activity in Eastern Anatolia was first discussed by Pearce et al. (1990) and recently proposed as a model for the Anatolian-Iranian plateau by Ershov and Nikishin (2004) (see Fig. 7, Model 7). Pearce et al. (1990) point out the remarkable correlation between topographic and volcanic expressions in Eastern Anatolia. The Eastern Anatolian topographic uplift has an asymmetric (i.e., deformed) dome shape (Şengör et al., 2003) whose long axis aligns approximately east-west. The overall volcanic expression is also asymmetric, extending ~300 km in the direction of compression but 900 km perpendicular to it (Pearce et al., 1990). This remarkable correlation between the topography and volcanic expression brings into question whether there is a mantle plume beneath the Eastern Anatolian Collision Zone. Note that a plume is defined here as a passive, diapiric upwelling of material from the deep mantle.

As previously stated, Şengör et al. (2003) argued that the cause of the domal uplift in both Eastern Anatolia and the Ethiopian high plateau was the same: hot, rising asthenosphere beneath crust bereft of underlying mantle lithosphere. Although

domal uplift related to a mantle plume is expected to have a symmetrical shape, it may acquire an asymmetrical shape in a collision setting due to compression. However, there is no modern or ancient example anywhere in the world of a plume-related dome structure deformed by shortening in a collision zone.

Ershov and Nikishin (2004) propose that volcanism on the Eastern Anatolian plateau and in Armenia can be explained by extraordinary lateral spreading of the African "superplume." They argue that the lithosphere is relatively thin along a south-north line extending from Afar to Anatolia due to a previous orogenic collapse event that occurred at around 550 Ma (i.e., an Africa-Mozambique-Arabia orogen). In their view, lateral asthenospheric mantle flow from the African superplume flowed along a lithospheric channel and moved to the north, resulting in the migration of the volcanism and uplift along a north-south belt. They claim that volcanism migrated to the north from Kenya (37–45 Ma) to Syria (9–13 Ma), Anatolia (11 Ma), and finally Armenia (11–2.8 Ma) (see Figs. 2 and 3 *in* Ershov and Nikishin, 2004). They argue that a slab break-off event took place around 11 Ma along the Eastern Taurus belt and produced a slab window through which the asthenospheric flow passed and reached the Anatolian and Armenian plateaus, creating uplift and extensive volcanism in the north.

Discrepancies: Most domal structures thought to be formed by plumes are expected to contain fault systems and dike swarms distributed radially. Such faults and dikes are absent in Eastern Anatolia. Fault plane solutions of earthquakes imply that the faults are either transform or reverse, not normal as would be expected in a plume-related domal structure. A plume model cannot explain why volcanic units contain a distinct subduction component in the northern part of Eastern Anatolia and why this component gradually diminishes to the south. It is also difficult to explain by a plume model why volcanism migrated south with time and why there is a gradual change in magma chemistry from calc-alkaline in the north to alkaline in the south. As pointed out by Pearce et al. (1990), volcanic activity over the last six million years demonstrates a temporal change from more regional-scale activity to localized activity on a set of aligned central volcanoes. Such an evolutionary sequence is the reverse of what is expected in plume-related volcanic activity. On the basis of these discrepancies, I argue that a plume is not a viable model for the Eastern Anatolian Collision Zone. However, it should be noted that some of the characteristic features discussed earlier may not be observed in every hotspot setting, as in the case of the Ethiopian Rift reported by Peccerillo et al. (2003).

Lateral spreading of the African superplume over great distances through a north-south channel along a previously thinned lithospheric domain (Ershov and Nikishin, 2004) does not seem to be a viable model, because not only does it involve an unrealistic scenario that involves the lateral migration of plume-related material for unreasonably great distances, but also both magma generation and domal uplift can theoretically be generated by slab break-off alone without a need for a plume-related hot mantle flow, as will be thoroughly covered in the discussion section.

Delamination of Mantle Lithosphere beneath the Region (Proposed by Pearce et al., 1990; Refined by Keskin et al., 1998)

Delamination of a thickened thermal boundary layer is plausible, because such a layer is colder and thus denser than the underlying asthenosphere (Fig. 7, Model 8). It could therefore be convectively replaced by asthenosphere (Houseman et al., 1981; England and Houseman, 1988). Platt and England (1993) argued that magmatism in mountain belts could be evidence of delamination of the lower part of the thickened mantle lithosphere. Figure 8A and B illustrates the delamination model in a 3-D block diagram for Eastern Anatolia (modified from Keskin, 1994). This process is likely to be an effective mechanism for generating large volumes of collision-related magma across the region, because asthenosphere is brought into close contact with a thickened layer of metasomatized lithosphere (Pearce et al., 1990). When delamination occurs, it causes a perturbation in what is left of the mantle lithosphere, raising some parts of it above its solidus. While sinking into the asthenosphere, the delaminated block of the mantle lithosphere may release water that also promotes melting (Elkins-Tanton, 2004). These two mechanisms play an important role in the generation of extensive partial melting in the mantle and can produce widespread volcanism in the region (Fig. 8B).

Pearce et al. (1990) argued that the region is characterized by a set of mantle domains that run parallel to the collision zone. They suggest that each domain has yielded magmas of particular composition since the beginning of the magmatism in the region. This may also be regarded as supporting evidence for the delamination model.

On the basis of estimates of the active slip rates, total convergence, and timing of collision-related deformation across the Arabia-Eurasia Collision Zone, coupled with the interpretation of a cross-section produced by the National Iranian Oil Company (1977), Allen et al. (2004) suggest that the collision-related magmatism, which initiated at ca. 11 Ma (Keskin et al., 1998), predated shortening of the crust in the region. Therefore, they argued, a sudden and regional delamination event is not a viable model. However, results obtained from three independent seismic studies presented earlier reveal that most of the Eastern Anatolian Collision Zone is devoid of a mantle lithosphere. Therefore, geophysical findings support a major lithospheric detachment beneath the region and contradict the interpretation of Allen et al. (2004).

Discrepancies: As discussed earlier, new geophysical data indicate that there appears to be no lithospheric mantle over a greater portion of the area beneath the region. If this is the case, the delamination must have been a shallow event involving the whole lithospheric mantle and perhaps even the lower crust (e.g., as exemplified by Lustrino, 2005). In the absence of metasomatized lithospheric mantle, the source region of the volcanism would be asthenospheric mantle, because all lavas across Eastern Anatolia have mantle signatures.

Because the basement of a great portion of Eastern Anatolia, between the Aras River in the north and Lake Van in the south, is represented by a subduction-accretion complex (i.e., EAAC in Fig. 1B), and such large subduction-accretion complexes are devoid of lithospheric roots because they are produced on and supported by subducting oceanic slabs, the delamination model cannot be a viable one for the areas covered by the Eastern Anatolian Accretionary Complex as discussed later.

Inflow of Lower Crust Driven by the Isostatic Response to Denudation and Sedimentation in Surrounding Areas (Mitchell and Westaway, 1999)

On the basis of their study of Neogene–Quaternary uplift and magmatism in the Greater Caucasus, Mitchell and Westaway (1999) proposed an alternative model to explain the formation of high mountain ranges and plateaus such as the Greater and Lesser Caucasus, including the Armenian highlands adjacent to Northeastern Anatolia. They argued that the rate and spatial scale of uplift of the Caucasus are too great to be the result of plate convergence alone; therefore, some other processes must have been operational.

Mitchell and Westaway (1999) argued that when crustal material is hotter than 300 °C, it starts to behave in a ductile way, deforming plastically. The depth at which this temperature is reached (~15–20 km) broadly corresponds to the boundary between the plastic lower crust and the brittle upper crust. In the lower crust, the direction of movement (i.e., direction of flow) is determined by pressure gradients caused by lateral variations in the depth of the base of the brittle layer (Mitchell and Westaway, 1999). In this model, most of the crustal deformation occurs in the lower crust in an atectonic fashion (e.g., Kaufman and Royden, 1994).

The model of Mitchell and Westaway (1999) is dramatically different from the other competing models in that crustal thickening is not caused directly by plate motions. Their model involves lateral inflow of ductile lower crust, driven by the isostatic response to denudation of a mountain range and sedimentation in its surroundings. According to these authors, the start of the uplift of the Caucasus and surrounding areas was related to changes in environmental conditions in the Late Miocene. The Messinian draw-down of sea level in the Mediterranean region resulted in the complete desiccation of the Black Sea (Hsu and Giovanoli, 1979). This was accompanied by draw-down of the Caspian Sea level. This resulted not in an increase in subaerial relief, but also in an increase in the denudation rate of the Greater Caucasus. Coupled denudation and sedimentation (Fig. 7, Model 9) caused lateral inflow into the lower crust toward the base of the mountain range, resulting in uplift along the length of the Caucasus.

Mitchell and Westaway (1999) suggested that atectonic thickening of the continental crust keeping mantle lithosphere thickness constant would have raised the temperature in the mantle lithosphere, resulting in melting and magma generation

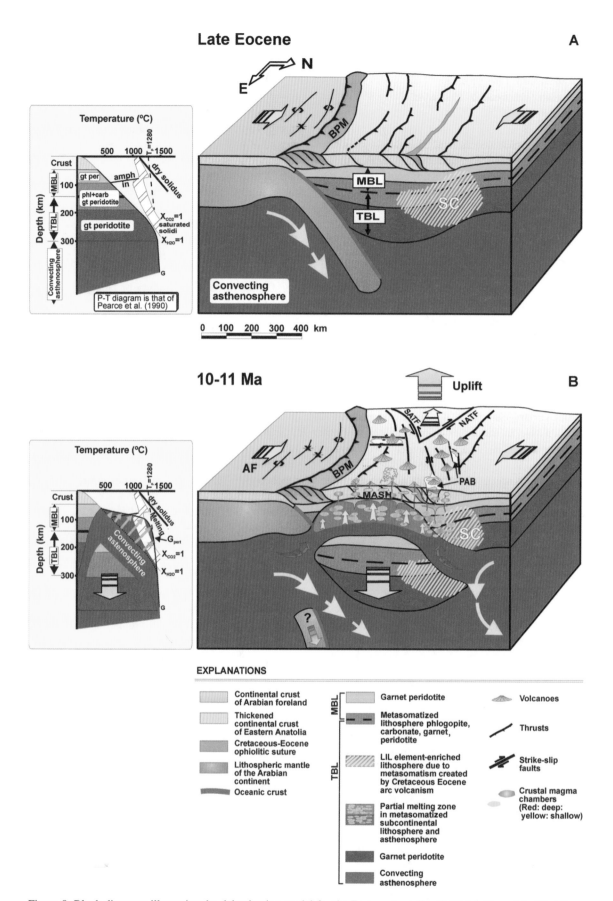

Figure 8. Block diagrams illustrating the delamination model for the Eastern Anatolian Collision Zone. Modified from Keskin (1994). MBL—mechanical boundary layer; TBL—thermal boundary layer; SC—subduction component; SATF and NATF—South and North Anatolian transform faults; BPM—Bitlis Pötürge Massif; MASH—melting, assimilation, storage, and homogenization; PAB—pull-apart basins.

as suggested by Koronovskiy and Demina (1996). They argued that this process was responsible for both uplift and volcanism in the Lesser Caucasus, including Armenia, adjacent to Eastern Anatolia. They also suggested that this process could be a viable model for Eastern Anatolia (R. Westaway, personal commun., 2002).

Discrepancies: In this model, thickening occurred only in the lower crust by means of lateral flow driven by plastic deformation (Fig. 7, Model 9). In such a case, a normal thickness of lithospheric mantle is still expected beneath the thickened crust, because there is no reason why it should have been detached from the base of the crust or along the thermal boundary layer. However, there is strong seismic evidence for a major lithospheric detachment event beneath the region, as discussed earlier. Moreover, as previously discussed, an increase in the thickness of the lithosphere could not have generated a significant amount of magma, because it would remain well below its solidus (Pearce et al., 1990). Therefore, the model of Mitchell and Westaway (1999) is not consistent with new geophysical findings and fails to explain the volume and variability of magmatic products across the region.

Slab Steepening and Break-off beneath a Subduction-Accretion Complex (Proposed by Şengör et al., 2003; Supported by Keskin, 2003)

Şengör et al. (2003) pointed out that areas with no mantle lithosphere, located in the southern part of the Erzurum-Kars plateau, coincide broadly with the Eastern Anatolian Accretionary Complex, of Late Cretaceous to earliest Oligocene age. The basement of a great portion of Eastern Anatolia, between the Aras River in the north and Lake Van in the south, is represented by the Eastern Anatolian Accretionary Complex (Figs. 1B). Following the subduction-accretion hypothesis (Şengör and Yılmaz, 1981; Şengör and Natal'in, 1996b), Şengör et al. (2003) argue that the Eastern Anatolian Accretionary Complex can be regarded as a remnant of a large accretionary prism located between the Pontides and the Bitlis-Pötürge Massif that formed on northward-subducting oceanic lithosphere. In contrast to continental blocks, large subduction-accretion complexes do not have their own lithospheric roots, because they are produced on and supported by subducting oceanic slabs. In theory, this area should have been underlain by a subducting slab, not by subcontinental mantle lithosphere, before the lithospheric detachment event. Therefore, what took place beneath the region could not have been a shallow lithospheric delamination event (Keskin, 2003; Şengör et al., 2003). Because tomography provides no evidence for a mantle lid beneath the region, the underlying slab must have detached and sunk into the asthenosphere, possibly immediately prior to the domal uplift of the region at ca. 11–13 Ma. Therefore, this event can be ascribed to the past break-off of the inferred slab beneath the Eastern Anatolian Accretionary Complex (Fig. 7, Model 10). Deep tomographic sections of Piromallo and Morelli (2003) illustrating a

detached slab at around 300–500 km depth (Fig. 4B and D) support this interpretation.

A modified version of the slab break-off model was recently proposed by Barazangi et al. (2006). The authors argue that there were two northward-subducting slabs in close proximity beneath the region during the Late Miocene: the one in the north subducted beneath the Eastern Anatolian Accretionary Complex, while the other shallowly descended beneath the Bitlis-Pötürge Massif, i.e., the oceanic segment of the Arabian lithosphere. In their view, the slab in the south (i.e., the Arabian slab) beneath the Bitlis-Pötürge Massif broke off ca. 11 Ma, producing widespread volcanism and resulting in regional uplift across the Eastern Anatolian plateau.

There seems to be a discrepancy in this model. Although I agree with the possibility of Arabian slab break-off in the south, the timing of this event does not seem to be reasonable. Arabian slab break-off beneath the Bitlis-Pötürge Massif ca. 11 Ma would have generated widespread volcanism on the massif, because it would have opened up a slab window right below this continental sliver (See Fig. 7, the last model; see also Fig. 7C in Barazangi et al., 2006). However, such young and widespread volcanism (≤11 Ma) does not exist along the Bitlis-Pötürge Massif, and this contradicts the model of Barazangi et al. (2006). Therefore, Arabian slab break-off in the south might have been an older event (e.g., Eocene–Oligocene).

There appear to be inconsistencies in all models except for the delamination and slab steepening and break-off models. In view of these arguments, a model involving steepening and break-off of a subducting slab beneath a huge subduction-accretion complex can better explain the geodynamic evolution of the Eastern Anatolian Collision Zone (Şengör et al., 2003) and widespread magmatism (Keskin, 2003) (Fig. 7, Model 10), as will be discussed in the next section.

DISCUSSION

Keskin (2003) showed that volcanic activity in Eastern Anatolia began earlier in the north (i.e., almost coeval with the rapid regional block uplift at ca. 11–13 Ma) than in the south, migrating south with time (Fig. 1C). This migration was accompanied by significant variation in lava chemistry in the north-south direction between the Erzurum-Kars plateau in the north and the Muş-Nemrut-Tendürek volcanoes in the south (Figs. 5 and 6). As presented earlier, volcanic products erupted in the north around the Erzurum-Kars plateau are calc-alkaline in character with a distinct subduction signature, in contrast to the ones in the south around the Muş-Nemrut-Tendürek volcanoes, which are alkaline with an intraplate signature (Pearce et al., 1990). The volcanic units of the Bingöl and Süphan volcanoes display transitional chemical characteristics (Fig. 5).

Keskin (2003) pointed out that these spatial and temporal variations in magma genesis, coupled with the uplift history of the region, can be explained by a model involving steepening of a northward-subducting slab beneath a large subduction-accretion complex, namely the Eastern Anatolian Accretionary Complex, followed by break-off (Şengör at al., 2003) at ca. 10–11 Ma. Keskin (2003) points out that the slab, whose subduction was generating the Pontide arc in the north, was not attached to the Arabia plate. Instead, it was possibly attached to the Bitlis-Pötürge block before break-off (Fig. 9). The oceanic realm between the Bitlis-Pötürge Massif and the Arabia plate had been closed much earlier (i.e., in the Late Eocene: Pearce et al., 1990; Şengör et al., 2003). Therefore, it is not surprising that researchers failed to reach a consensus regarding the timing of the collision event in the region.

According to Şengör et al. (2003), the oceanic realm between the Pontides and the Bitlis-Pötürge Massif was completely closed in the Oligocene (see Fig. 3 *in* Şengör et al., 2003, and Fig. 3C *in* Keskin, 2003). After a period between the Oligocene and the Serravalian, during which the Eastern Anatolian Accretionary Complex was shortened and thus thickened over the slab, the hidden subduction possibly stopped (Fig. 10A; see also Fig. 3D *in* Keskin, 2003). As a result, left unsupported by subduction, the dense oceanic lithospheric slab may have steepened and finally detached from the Eastern Anatolian Accretionary Complex, opening an asthenospheric mantle wedge that gradually widened to the south (Figs. 9 and 10B and C). This possibly created suction on the asthenosphere, generating mantle flow to the south (Figs. 9B and 10B and C).

Emplacement of the asthenospheric mantle with a subduction component (especially water) and a potential temperature of 1280 °C at shallow depths (~40–50 km) beneath the Eastern Anatolian Accretionary Complex would have generated extensive adiabatic decompression melting in the mantle wedge (Keskin, 2003). It also probably generated regional block uplift, producing the regional domelike structure (Şengör et al., 2003) (Figs. 9B and 10C).

The presence of such asthenospheric flow related to the opening of an asthenospheric mantle wedge may provide an answer to the question of why the volcanic activity initiated much earlier in the north on the Erzurum-Kars plateau and migrated to the south over time. Similarly, it better explains why the volcanic products are calc-alkaline, with a distinct subduction signature in the north (Figs. 9B and 10B and C).

Deep tomographic images of the region (Piromallo and Morelli, 2003) provide evidence for a lithospheric mantle fragment, possibly a slab, at a depth of around 300–500 km, currently sinking into the asthenosphere beneath Eastern Anatolia (Fig. 4B and D). What this may indicate is that the detachment of the oceanic lithosphere of the Arabia plate took place in the past, perhaps millions of years ago (i.e., ca. 11 Ma; Figs. 9 and 10C).

I suggest that the mantle source region owed its exceptional fertility to a subduction component inherited from a previous subduction event (i.e., the subduction beneath the Pontides during the Eocene and Oligocene), to the oceanic crustal material previously subducted beneath the region (see inferred detached lithospheric fragments in Fig. 4D), or to a combination of both.

A process similar to the latter has recently been proposed by a number of researchers (e.g., Gasparik, 1997; Anderson, 2000, 2004a, and this volume; Balyshev and Ivanov, 2001; Foulger, 2002; Foulger and Natland, 2003; Ivanov, 2003; Foulger et al., 2005) to explain low-velocity anomalies in the mantle as well as the genesis of magmatism in exceptionally fertile mantle domains (e.g., the Icelandic hotspot; Foulger et al., 2005). Alternatively, delaminated lithospheric blocks beneath the northern part of the region (i.e., the Erzurum-Kars plateau), where a lithospheric mantle root is thin but still exists, might have contributed to the magma generation by dewatering themselves as they sank (a process described by Elkins-Tanton, 2004). Such detached blocks seem to exist in the mantle wedge beneath the Erzurum-Kars plateau, although they are not clear in the tomographic images of Piromallo and Morelli (2003) (Fig. 4D). As pointed out by Anderson (2004a,b), melting anomalies can result from fertile patches or regions of shallow mantle with a low melting point, and this seems to be the case for Eastern Anatolia.

Recent experimental studies by Regard et al. (2005) and Faccenna et al. (2006) and the deep tomographic images of Piromallo and Morelli (2003) support the slab break-off model for Eastern Anatolia. Faccenna et al. (2006) argue that deep deformation of the Bitlis-Hellenic slab by means of slab break-off beneath Eastern Anatolia and its lateral effect in the western part of the subduction system (i.e., Hellenic arc and Aegean region) in the form of slab rollback might be responsible for the fundamental plate tectonic reorganization during the Late Miocene–Early Pliocene period in Anatolia and the surrounding areas. They claim that the aforementioned reorganization in the slab geometry beneath the Eastern Mediterranean region might have created the North and South Anatolian transform fault systems that currently control the dynamics of the whole Neotectonic system. Results of a recent GPS study by Reilinger et al. (2006) support the interpretation of Faccenna et al. (2006) and further indicate that slab rollback, possibly driven by slab break-off in the east, might be responsible for both the westward motion of Anatolia and the counterclockwise rotation of the whole of Arabia and Anatolia.

In order to understand how a wide hot orogen with a relatively thin lithosphere (e.g., the Tibetan and Anatolian-Iranian plateaus) is deformed during collision, Cruden et al. (2006) constructed a set of analogue vice models and conducted 2-D numerical experiments. Results of their experimental studies revealed that ductile lower crust and mantle in the weak lithosphere could flow laterally parallel to the orogen, producing upright folding in the upper crust and decoupled horizontal strain in the lower crust. One of their experiments (experiment 32) produced a result impressively similar to what is seen in the deformation style of the Eastern Anatolian–Iranian plateau (i.e., in terms of its fold and fault geometry), supporting the presence of an exceptionally thin lithosphere over a hot and relatively buoyant mantle beneath Eastern Anatolia.

In another recent study, Hafkenscheid et al. (2006) investigated the Mesozoic–Cenozoic subduction history of the Tethyan

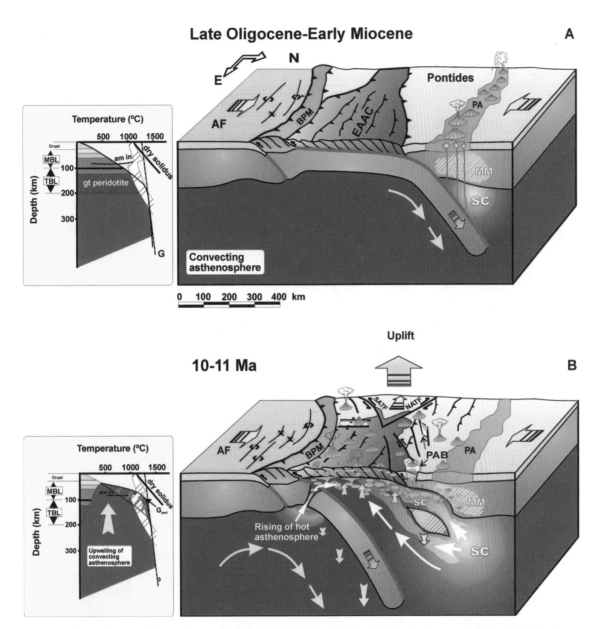

Figure 9. Block diagrams illustrating the slab-steepening and breakoff model for the Eastern Anatolian Collision Zone. MBL—mechanical boundary layer; TBL—thermal boundary layer; AF—Arabian foreland; BPM—Bitlis Pötürge Massif; EAAC—Eastern Anatolian Accretionary Complex; PA—Pontide arc; MM—mantle metasomatism; SC—subduction component; SATF and NATF—South and North Anatolian transform faults. For explanations of colors, see Figure 8. The white arrows indicate the possible flow direction of the asthenosphere. Modified from Şengör et al. (2003). See also Figure 3 *in* Keskin (2003).

region by integrating independent information from mantle tomography (Bijwaard at al., 1998) and tectonic reconstructions. Their aim was to test three different tectonic reconstructions proposed by Dercourt et al. (1993), Şengör and Natal'in (1996a), Norton (1999), and Stampfli and Borel (2002, 2004) by comparing the predicted thermal signature of the subducted lithosphere to the tomographic mantle structure underneath the Tethyan region. They argue that the sizes and positions of their analyzed tomographic volumes can be best explained by a slab break-off event ca. 12 Ma beneath Eastern Anatolia and 30 Ma

farther to the east (i.e., in the Northern Zagros Suture Zone). They point out that the slab break-off might have initiated ca. 30 Ma beneath the Northern Zagros Suture Zone and then propagated both eastward and westward along the suture zone, reaching Eastern Anatolia ca. 12 Ma.

An interesting feature of the Eastern Anatolian collision zone is the gradual weakening of collision-related volcanism across the region. On the basis of their numerical model, Gerbi et al. (2006) argue that when wholesale lithospheric delamination occurs, the result is dramatic heating of the lower crust by

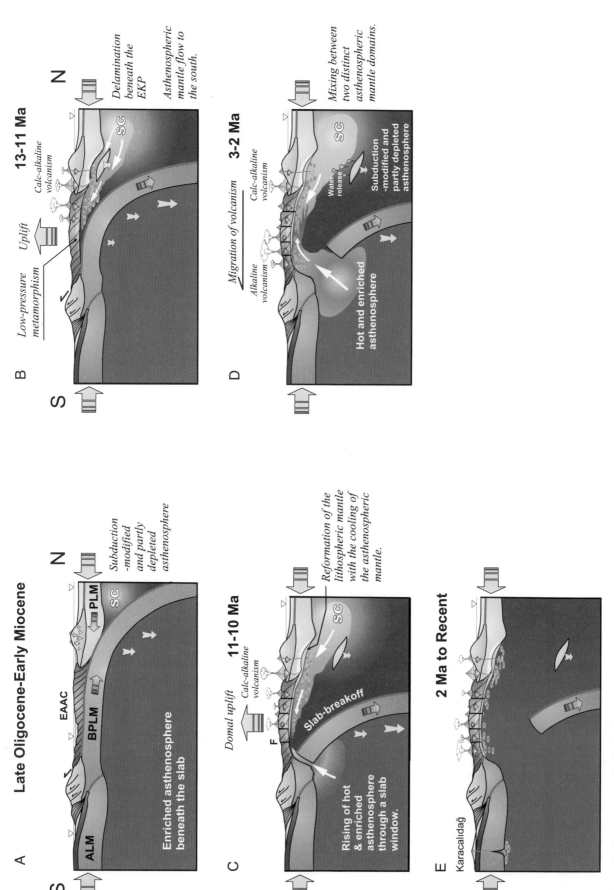

Figure 10. Cross-sections displaying the evolution of the Eastern Anatolian collision zone in time. EAAC—Eastern Anatolian Accretionary Complex; ALM—Arabian lithospheric mantle; BPLM—lithospheric mantle of the Bitlis-Pötürge Massif; PLM—lithospheric mantle of the Pontides; SC—asthenospheric mantle containing a subduction component; EKP—Erzurum-Kars plateau.

the asthenosphere. This process causes low-pressure metamorphism while the hot asthenosphere cools, turning into the lithospheric mantle beneath the crust (Fig. 10C). Thickening of the lithosphere by reformation of the lithospheric mantle in this way (i.e., via conductive cooling of the asthenospheric mantle) might have been a process important for the weakening of volcanic activity during the course of time across Eastern Anatolia.

Şengör (2006) supports this interpretation and further argues that the increasing alkalinity in volcanism (e.g., on the Erzurum-Kars plateau) can be explained by thickening of the lithosphere beneath the region. He points out that thickening of lithosphere with the cooling of mantle resulted in deepening of the foci of melting, and this increased the alkalinity of the volcanic rocks with time. Although in theory deepening of the foci of melting controls the alkalinity of melts in the nature, the magma composition depends more on the composition of the source material. Provided that the source is enriched, deepening of the foci of melting may increase the alkalinity. However, if the source was previously depleted or modified by subducted material, the composition of lavas generated would be dramatically different from that of lavas derived from an enriched source. For example, magmas from the deep Hawaiian source are mostly tholeiitic. Similarly, magma generation occurs deep in the mantle wedge beneath island arcs, deeper than most rift settings, but produces calc-alkaline lavas reflecting the source chemistry. In the case of the Eastern Anatolian mantle, the source is fairly depleted and appears to have been strongly modified by a subduction component. Therefore, I argue that the increasing alkalinity of lavas in time can be more simply explained by lateral flow of enriched asthenospheric mantle beneath the region. It seems likely that the asthenospheric flow was from the north to the south during the early stages of volcanism in response to the suction effect created by slab steepening (i.e., an asthenospheric flow from north to south; Figs. 10B and C). This might have carried the subduction-modified and partly depleted mantle from the north, producing magmas with a subduction signature (Keskin, 2003).

After a certain degree of steepening, the slab appears to have broken off, creating a "slab window" (Fig. 10C). Hotter asthenosphere once located beneath the slab might have filled this window, generating extensive melting beneath the collision zone along a linear belt. In response to this radical change beneath Eastern Anatolia, asthenospheric mantle flow possibly changed its direction from south to north, bringing hot, enriched, fertile asthenospheric material once located beneath the slab (i.e., underneath the Arabian continent) to shallow depths (Fig. 10D). Shear wave–splitting fast polarization directions (Sandvol et al., 2003b), which display quite a uniform distribution with northeast-southwest orientations (Fig. 3A) are consistent with the inferred mantle flow direction. This process might have contributed to magma generation and the resultant volcanism in the south around the Muş-Nemrut-Tendürek volcanoes, aligning as a southwest-northeast-trending belt that is probably subparallel to the aforementioned slab window (Figs. 1 and 2). This

may also explain why these volcanoes produced lavas with variable degrees of within-plate signature. Mixing between these two distinct sources (i.e., the partly depleted one containing a subduction component in the north and the enriched one in the south) combined with AFC in crustal chambers may be responsible for generating the great variety of volcanic material across the region (Fig. 10D).

In their recent experimental study, Kincaid and Griffiths (2004) showed that steepening of a slab in a subduction system can promote melting both in the mantle wedge and in the slab itself. The results of Kincaid and Griffiths (2004) indicate that a slab-steepening mode of rollback favors steeper flow trajectories into the wedge apex, enhancing decompression melting within the wedge. Flow velocities toward the slab center line immediately increase by a factor of three to five, and slab surface temperatures rapidly increase by 100–200 °C in this part of the slab. Therefore, slab steepening not only results in melting in the overlying mantle wedge, but also promotes melting in the slab itself. This process might have been important for magma generation beneath the Eastern Anatolian Collision Zone, possibly during the early stages of the collision-related magmatism.

Another interesting feature of the collision-related units in Eastern Anatolia is that the primitive lavas across the region have trace-element and isotopic signatures reflective of a mantle origin, not a crustal origin. Because the accretionary prism had directly overlain hot asthenospheric mantle, one would expect widespread crustal anatexis beneath the region and eruption of lavas with crustal signatures, which does not seem to have been the case for Eastern Anatolian lavas. The numerical modeling studies by Bodorkos et al. (2002) and Gerbi et al. (2006) can provide a solution to this dilemma. Both groups of authors argue that when the asthenosphere rises to the base of crust in response to the delamination, it significantly heats the crust but cannot cause anatexis. According to Gerbi et al. (2006), the thermal anomaly propagates upward to the middle and upper crusts, generating low-pressure metamorphism (550–600 °C at 17 km) within 35 km of the crust, but "not anatectic conditions" in it. If erosion accompanies the lithospheric delamination, the warming effect of the shallow asthenosphere remains limited to the base of the crust. Because the asthenosphere loses heat, it changes into lithospheric mantle (in a process called the reformation of lithospheric mantle), isolating the crust from the hot asthenosphere (Figs. 10B and C). These findings provide an answer to the question of why collision-related basaltic lavas of Eastern Anatolia have mantle signatures.

It should be noted that the slab steepening and break-off model is viable only if the basement of a greater part of Eastern Anatolia is represented by an accretionary complex (i.e., the Eastern Anatolian Accretionary Complex), as proposed by Şengör et al. (2003), and if only one north-dipping subducting slab was forming this accretionary prism. Because the collision-related volcanic sequence masks the basement units over great distances, it is difficult to find evidence that sheds light on whether a great portion of the basement of Eastern Anatolia is represented by the Eastern Anatolian Accretionary Complex. Şengör's (2006) interpretation of current surface motion vectors seems to provide an answer to this question. He showed that the deformation style of the area that is presumably covered by the Eastern Anatolian Accretionary Complex (he names this area "the squashy zone") is unique for mélange material, because it accommodates most of the deformation in the form of thrusts and strike-slip faults, transmitting a relatively smaller fraction of surface movement to rigid continental units in the north (i.e., the Pontides and the Northwest Iranian fragment). This finding supports the idea that the area masked by volcanic successions is indeed underlain by accretionary prism material.

The slab steepening and break-off model proposed for the genesis of the collision-related volcanism in Eastern Anatolia differs from the original model of Davies and von Blanckenburg (1995) because it involves a large accretionary complex and the steepening of the slab beneath it. A number of recent studies have addressed the importance of the slab break-off process for collision zones (e.g., Nemcok et al., 1998; Chemenda et al., 2000; Maury et al., 2000; Coulon et al., 2002; Haschke et al., 2002; Maheo et al., 2002; Ferrari, 2004; Williams et al., 2004; Knapp et al., 2005; Molinaro et al., 2005; Koulakov, and Sobolev, 2006). The slab steepening and break-off process beneath large subduction-accretion complexes, accompanied by magma generation and the emplacement of magmas, may be a process very important in the making of continental crust in the "Turkic-type" (Şengör and Natal'in, 1996b) orogenic belts that comprise a large part of the Asian continent (Şengör et al., 2003).

In addition to various processes discussed so far, strike-slip faulting might have played an important role in focusing magmas by generating localized extension and volcanism in associated pull-apart basins (Dewey et al., 1986; Pearce et al., 1990; Keskin et al., 1998). In a recent study, Cooper et al. (2002) supported this view and suggested that the mafic magmas beneath Northwestern Tibet might have been created by a mantle upwelling beneath the releasing bends of the strike-slip fault systems. They also presented a model for magma generation in such systems. Therefore, as in the Tibetan plateau (e.g., Williams et al., 2004), the uplift and magmatism history of Eastern Anatolia may be related to more than one geodynamic process.

CONCLUSION

The Eastern Anatolian–Iranian high plateau can be regarded as a hotspot or "melting anomaly" coinciding with a regional domal structure that is squeezed in a collision zone in the north-south direction. By virtue of these features, the region closely resembles a mantle plume setting. However, the Eastern Anatolian domal uplift lies in a collision zone, in contrast to plume-related hotspots located in intraplate settings (e.g., the Ethiopian high plateau).

The Eastern Anatolian lithosphere is at present bereft of its mantle component (Şengör et al., 2003). This indicates that a huge piece (perhaps almost the whole thickness) of the mantle

lithosphere was detached from the overlying crust in the past. If this removal of the denser mantle material is responsible for both the regional uplift and coeval volcanism, the detachment must have occurred at ca. 11–13 Ma, at the same time as the onset of those events. The volume opened up by the removal of the mantle lithosphere would have been filled by hot, fertile (i.e., containing a subduction component) asthenospheric upwelling, which would have resulted in both the formation of the regional domal structure (Şengör et al., 2003) and extensive magma generation and volcanism due to adiabatic decompression melting (Keskin, 2003).

On the basis of combined geologic, geophysical, and geochemical data, it can thus be argued that the Eastern Anatolian domal uplift is not related to a mantle plume; instead its formation is linked to shallow plate tectonic processes. Temporal and spatial variations in lava chemistry, coupled with the uplift history and age relationships of the volcanic products in the Eastern Anatolian Collision Zone, may be linked to slab steepening and break-off beneath a subduction-accretion complex (Keskin, 2003; Şengör et al., 2003) in the south, where the mantle lid is absent (Fig. 9B). Slab steepening was possibly associated with asthenospheric flow that resulted in gradual change in the geochemical character of the volcanics erupted. I argue that lithospheric delamination might still be a more viable model for the northern areas where a lithospheric mantle root, although thinned, still exists (e.g., the Erzurum-Kars Plateau; Fig. 9B). The tomographic sections of Piromallo and Morelli (2003) support this view (Fig. 4D). These two processes explain the voluminous magma generation and resultant volcanism in addition to the formation of the domal uplift across the region better than other competing geodynamic models.

The Eastern Anatolian example is particularly important, because it shows that shallow plate tectonic processes can generate both regional lithospheric domal structures and great volumes of magma in the absence of a mantle plume (see also Keskin, 2005). This observation contradicts the proposal of Şengör (2001), who argues that all hotspots and long-wavelength domes on the Earth's surface are related to mantle plumes.

Further research is needed for a better understanding of collision-related magma genesis in Eastern Anatolia and its connection with slab break-off and other alternative processes. Issues regarding source characteristics, melting mechanisms, the mode and extent of magma-crust interaction, and crustal melting also need further investigation.

ACKNOWLEDGMENTS

I thank Professor Gillian R. Foulger for inviting me to contribute to this volume. An earlier version of this paper was presented as a keynote talk at the Chapman Conference, The Great Plume Debate: The Origin of LIPs and Hotspots, at Ft. William, Scotland, in 2005. I am grateful to the Research Fund of Istanbul University for their support (Project UDP-589/12072005). I particularly thank Ercan Aldanmaz and Angelo Peccerillo for their helpful comments on an earlier version of this paper. My thanks also go to staff members of the Department of Earth Sciences, Durham University, England, for kindly making their research facilities available for me during my sabbatical leave and providing me with a stimulating research environment. I must also give special thanks to the people at St. Chad's College, Durham University, especially to Principal Joseph P. Cassidy and Senior Tutor Margaret Masson for their kindness and generosity during my sabbatical studies. I also thank the people at Ustinov College (Durham University) for their valuable help.

REFERENCES CITED

Al-Damegh, K., Sandvol, E., Al-Lazki, A., and Barazangi, M., 2004, Regional seismic wave propagation (Lg and Sn) and Pn attenuations in the Arabian Plate and surrounding regions: Geophysical Journal International, v. 157, p. 775–795, doi: 10.1111/j.1365-246X.2004.02246.x.

Al-Lazki, A., Seber, D., Sandvol, E., Türkelli, N., Mohamad, R., and Barazangi, M., 2003, Tomographic Pn velocity and anisotropy structure beneath the Anatolian plateau (eastern Turkey) and the surrounding regions: Geophysical Research Letters, v. 30, no. 24, p. 8043, doi: 10.1029/2003GL017391.

Al-Lazki, A., Sandvol, E., Seber, D., Barazangi, M., Türkelli, N., and Mohamad, R., 2004, Pn tomographic imaging of mantle lid velocity and anisotropy at the junction of the Arabian, Eurasian and African plates: Geophysical Journal International, v. 158, p. 1024–1040, doi: 10.1111/j.1365-246X.2004.02355.x.

Allen, M., Jackson, J., and Walker, R., 2004, Late Cenozoic reorganization of the Arabia-Eurasia collision and the comparison of short-term and long-term deformation rates: Tectonics, v. 23, p. TC2008, doi: 10.1029/2003TC001530.

Anderson, D.L., 2000, The thermal state of the upper mantle: No role for mantle plumes: Geophysical Research Letters, v. 27, p. 3623–3626, doi: 10.1029/2000GL011533.

Anderson, D.L., 2004a, Reheating slabs by thermal conduction in the upper mantle: http://www.mantleplumes.org/HotSlabs2.html.

Anderson, D.L., 2004b, What is a plume?: http://www.mantleplumes.org/PlumeDLA.html.

Anderson, D.L., 2007 (this volume), The eclogite engine: Chemical geodynamics as a Galileo thermometer, *in* Foulger, G.R., and Jurdy, D.M., eds., Plates, plumes, and planetary processes: Boulder, Colorado, Geological Society of America Special Paper 430, doi: 10.1130/2007.2430(03).

Angus, D.A., Wilson, D.C., Sandvol, E., and Ni, J.F., 2006, Lithospheric structure of the Arabian and Eurasian collision zone in Eastern Turkey from S-wave receiver functions: Geophysical Journal International, v. 166, no. 3, p. 1335–1346.

Ateş, A., Kearey, P., and Tufan, S., 1999, New gravity and magnetic anomaly maps of Turkey: Geophysical Journal International, v. 136, p. 499–502, doi: 10.1046/j.1365-246X.1999.00732.x.

Balyshev, S.O., and Ivanov, A.V., 2001, Low-density anomalies in the mantle: ascending plumes and/or heated fossil lithospheric plates?: Doklady Earth Sciences, v. 380, no. 7, p. 858–862.

Barazangi, M., 1989, Continental collision zones: Seismotectonics and crustal structure, *in* James, D., ed., Encyclopedia of solid Earth geophysics: New York, Van Nostrand Reinhold, p. 58–75.

Barazangi, M., Sandvol, E., and Seber, D., 2006, Structure and tectonic evolution of the Anatolian plateau in eastern Turkey, *in* Dilek, Y., and Pavlides, S., eds., Postcollisional tectonics and magmatism in the Mediterranean region and Asia: Boulder, Colorado, Geological Society of America Special Paper 409, p. 463–474, doi: 10.1130/2006.2409(22).

Becker, T.W., Faccenna, C., O'Connell, R.J., and Giardini, D., 1999, The development of slabs in the upper mantle: Insights from numerical and laboratory experiments: Journal of Geophysical Research, v. 104, p. 15,207–15,226, doi: 10.1029/1999JB900140.

Bijwaard, H., Spakman, W., and Engdahl, E.R., 1998, Closing the gap between regional and global travel time tomography: Journal of Geophysical Research, v. 103, p. 30,055–30,078, doi: 10.1029/98JB02467.

Bodorkos, S., Sandiford, M., Oliver, N.H.S., and Cawood, P.A., 2002, High-T, low-P metamorphism in the Palaeoproterozoic Hall's Creek Orogen, northern Australia: The middle crustal response to a mantle-related thermal pulse: Journal of Metamorphic Geology, v. 20, p. 217–237, doi: 10.1046/j.1525-1314.2002.00339.x.

Buket, E., and Temel, A., 1998, Major-element, trace-element, and Sr-Nd isotopic geochemistry and genesis of Varto Muş volcanic rocks, Eastern Turkey: Journal of Volcanology and Geothermal Research, v. 85, p. 405–422, doi: 10.1016/S0377-0273(98)00064-X.

Çakır, Ö., and Erduran, M., 2004, Constraining crustal and uppermost mantle structure beneath station TBZ (Trabzon, Turkey) by receiver function and dispersion analyses: Geophysical Journal International, v. 158, p. 955–971, doi: 10.1111/j.1365-246X.2004.02345.x.

Çakır, Ö., Erduran, M., Çınar, H., and Yılmaztürk, A., 2000, Forward modelling receiver functions for crustal structure beneath station TBZ (Trabzon, Turkey): Geophysical Journal International, v. 140, p. 341–356, doi: 10.1046/j.1365-246x.2000.00023.x.

Chemenda, A.I., Burg, J.P., and Mattauer, M., 2000, Evolutionary model of the Himalaya-Tibet system: Geopoem based on new modelling, geological and geophysical data: Earth and Planetary Science Letters, v. 174, p. 397–409, doi: 10.1016/S0012-821X(99)00277-0.

Christensen, U.R., 1996, The influence of trench migration on slab penetration into the lower mantle: Earth and Planetary Science Letters, v. 140, p. 27–39, doi: 10.1016/0012-821X(96)00023-4.

Cooper, K.M., Reid, M.R., Dunbar, N.W., and McIntosh, W.C., 2002, Origin of mafic magmas beneath northwestern Tibet: Constraints from [230]Th-[238]U disequilibria, Geochemistry, Geophysics, Geosystems, v. 3, no. 1, art. no. 1065.

Coulon, C., Megartsi, M., Fourcade, S., Maury, R.C., Bellon, H., Louni-Hacini, A., Cotten, J., Coutelle, A., and Hermitte, D., 2002, Post-collisional transition from calc-alkaline to alkaline volcanism during the Neogene in Oranie (Algeria): Magmatic expression of a slab breakoff: Lithos, v. 62, p. 87–110, doi: 10.1016/S0024-4937(02)00109-3.

Cruden, A.R., Nasseri, M.H.B., and Pysklywec, R., 2006, Surface topography and internal strain variation in wide hot orogens from three-dimensional analogue and two-dimensional numerical vice models, *in* Buiter, S.J.H., and Schreurs, G., eds., Analogue and numerical modelling of crustal-scale processes: Geological Society, London, Special Publication 253, p. 79–104.

Davies, J.H., and von Blanckenburg, F., 1995, Slab breakoff: A model of lithosphere detachment and its test in the magmatism and deformation of collisional orogens: Earth and Planetary Science Letters, v. 129, p. 85–102, doi: 10.1016/0012-821X(94)00237-S.

Dercourt, J., Ricou, L.E., and Vrielynck, B., eds., 1993, Atlas Tethys: Palaeoenvironmental Maps. New York, Elsevier.

Dewey, J.F., Hempton, M.R., Kidd, W.S.F., Şaroğlu, F., and Şengör, A.M.C., 1986, Shortening of continental lithosphere: The neotectonics of Eastern Anatolia—A young collision zone, *in* Coward, M.P., and Ries, A.C., Collision tectonics: Geological Society of London Special Publication 19, p. 3–36.

Elkins-Tanton, L.T., 2004, Continental magmatism caused by lithospheric delamination: http://www.mantleplumes.org/LithDelam.html.

England, P.C., and Houseman, G.A., 1988, The mechanics of the Tibetan Plateau: Philosophical Transactions of the Royal Society, London, v. A326, p. 301–320.

Ercan, T., Fujitani, T., Madsuda, J.-I., Notsu, K., Tokel, S., and Tadahide, U.I., 1990, Doğu ve güneydoğu Anadolu Neojen-Kuvaterner volkanitlerine ilişkin yeni jeokimyasal, radyometrik ve izotopik verilerin yorumu: M.T.A Dergisi, v. 110, p. 143–164.

Ershov, A.V., and Nikishin, A.M., 2004, Recent geodynamics of the Caucasus–Arabia–East Africa Region: Geotectonics, v. 38, no. 2, p. 123–136.

Faccenna, C., Bellier, O., Martinod, J., Piromallo, C., and Regard, V., 2006, Slab detachment beneath eastern Anatolia: A possible cause for the formation of the North Anatolian fault: Earth and Planetary Science Letters, v. 242, p. 85–97, doi: 10.1016/j.epsl.2005.11.046.

Ferrari, L., 2004, Slab detachment control on mafic volcanic pulse and mantle heterogeneity in central Mexico: Geology, v. 32, no. 1, p. 77–80, doi: 10.1130/G19887.1.

Foulger, G.R., 2002, Plumes or plate tectonic processes?: Astronomy and Geophysics, v. 43, p. 6.19–6.23.

Foulger, G.R., and Natland, J.H., 2003, Is "hotspot" volcanism a consequence of plate tectonics?: Science, v. 300, p. 921, doi: 10.1126/science.1083376.

Foulger, G.R., Natland, J.H., and Anderson, D.L., 2005, A source for Icelandic magmas in remelted Iapetus crust: Journal of Volcanology and Geothermal Research, v. 141, p. 23–44, doi: 10.1016/j.jvolgeores.2004.10.006.

Gaherty, J.B., and Hager, B.H., 1994, Compositional vs. thermal buoyancy and the evolution of subducted lithosphere: Geophysical Research Letters, v. 21, p. 141–144, doi: 10.1029/93GL03466.

Gasparik, T., 1997, A model for the layered upper mantle: Physics of the Earth and Planetary Interiors, v. 100, p. 197–212, doi: 10.1016/S0031-9201(96)03240-2.

Gelati, R., 1975, Miocene marine sequence from Lake Van, Eastern Turkey: Rivista Italiana di Paleontologia e Stratigrafia, v. 81, p. 477–490.

Gerbi, C.C., Johnson, S.E., and Koons, P.O., 2006, Controls on low-pressure anatexis: Journal of Metamorphic Geology, v. 24, p. 107–118, doi: 10.1111/j.1525-1314.2005.00628.x.

Gök, R., Türkelli, N., Sandvol, E., Seber, D., and Barazangi, M., 2000, Regional wave propagation in Turkey and surrounding regions: Geophysical Research Letters, v. 27, no. 3, p. 429–432, doi: 10.1029/1999GL008375.

Gök, R., Sandvol, E., Türkelli, N., Seber, D., and Barazangi, M., 2003, Sn attenuation in the Anatolian and Iranian plateau and surrounding regions: Geophysical Research Letters, v. 30, no. 24, p. 8042, doi: 10.1029/2003GL018020.

Hafkenscheid, E., Wortel, M.J.R., and Spakman, W., 2006, Subduction history of the Tethyan region derived from seismic tomography and tectonic reconstructions, Journal of Geophysical Research, Solid Earth, v. 111, no. B8, article no. B08401.

Haschke, M.R., Scheuber, E., Gunther, A., and Reutter, K.-J., 2002, Evolutionary cycles during the Andean orogeny: Repeated slab breakoff and flat subduction?: Terra Nova, v. 14, p. 49–55, doi: 10.1046/j.1365-3121.2002.00387.x.

Hearn, T.N., and Ni, J.F., 1994, Pn velocities beneath continental collision zones: The Turkish-Iranian Plateau: Geophysical Journal International, v. 117, p. 273–283.

Houseman, G.A., McKenzie, D.P., and Molnar, P., 1981, Convective instability of a thickened boundary layer and its relevance for the thermal evolution of continental collision belts: Journal of Geophysical Research, v. 86, p. 6115–6132.

Hsu, K.J., and Giovanoli, F., 1979, Messinian event in the Black Sea: Palaeogeography, Palaeoclimatology, Palaeoecology, v. 29, nos. 1–2, p. 75–93.

Innocenti, F., Manetti, P., Mazzuoli, R., Pasquaré, G., and Villari, L., 1982a, Anatolia and north-western Iran, in Thorpe, R.S., ed., Andesites: New York, John Wiley & Sons.

Innocenti, F., Mazzuoli, R., Pasquaré, G., Radicati di Brozolo, F., and Villari, L., 1982b, Tertiary and Quaternary volcanism of the Erzurum-Kars area (Eastern Turkey): Geochronological data and geodynamic evolution: Journal of Volcanology and Geothermal Research, v. 13, p. 223–240, doi: 10.1016/0377-0273(82)90052-X.

Irvine, T.N., and Baragar, W.R.A., 1971, A guide to the chemical classification of the common volcanic rocks: Canadian Journal of Earth Sciences, v. 8, p. 523–548.

Ivanov, A.V., 2003, Plumes or reheated slabs?: http://www.mantleplumes.org/HotSlabs1.html.

Karapetian, S.G., Jrbashian, R.T., and Mnatsakanian, A.K., 2001, Late collision rhyolitic volcanism in the north-eastern part of the Armenian Highland:

Journal of Volcanology and Geothermal Research, v. 112, no. 1–4, p. 189–220, doi: 10.1016/S0377-0273(01)00241-4.

Kaufman, P.S., and Royden, L.H., 1994, Lower crustal flow in an extensional setting: Constraints from the Halloran Hills region, eastern Mojave Desert, California: Journal of Geophysical Research, v. 99, p. 15,723–15,739, doi: 10.1029/94JB00727.

Keskin, M., 1994, Genesis of collision-related volcanism on the Erzurum-Kars Plateau, Northeastern Turkey [Ph.D. thesis]: University of Durham, England.

Keskin, M., 2002, FC-Modeler: a Microsoft[R] Excel[c] spreadsheet program for modeling Rayleigh fractionation vectors in closed magmatic systems: Computers and Geosciences, v. 28, no. 8, p. 919–928, doi: 10.1016/S0098-3004(02)00010-9.

Keskin, M., 2003, Magma generation by slab steepening and breakoff beneath a subduction-accretion complex: An alternative model for collision-related volcanism in Eastern Anatolia, Turkey: Geophysical Research Letters, v. 30, no. 24, p. 8046, doi:10.1029/ 2003GL018019.

Keskin, M., 2005, Domal uplift and volcanism in a collision zone without a mantle plume: Evidence from Eastern Anatolia: http://www.mantleplumes.org/Anatolia.html.

Keskin, M., Pearce, J.A., and Mitchell, J.G., 1998, Volcano-stratigraphy and geochemistry of collision-related volcanism on the Erzurum-Kars Plateau, North Eastern Turkey: Journal of Volcanology and Geothermal Research, v. 85, nos. 1–4, p. 355–404, doi: 10.1016/S0377-0273(98)00063-8.

Keskin, M., Pearce, J.A., Kempton, P.D., and Greenwood, P., 2006, Magmacrust interactions and magma plumbing in a post-collision setting: Geochemical evidence from the Erzurum-Kars Volcanic Plateau, Eastern Turkey, in Dilek, Y., and Pavlides, S., eds., Postcollisional tectonics and magmatism in the Mediterranean region and Asia: Boulder, Colorado, Geological Society of America Special Paper 409, p. 475–505, doi: 10.1130/2006.2409(23).

Kincaid, C., and Griffiths, R.W., 2004, Variability in flow and temperatures within mantle subduction zones, Geochemistry, Geophysics, Geosystems, v. 5, article no. Q06002.

Knapp, J.H., Knapp, C.C., Raileanu, V., Matenco, L., Mocanu, V., and Dinu, C., 2005, Crustal constraints on the origin of mantle seismicity in the Vrancea Zone, Romania: The case for active continental lithospheric delamination: Tectonophysics, v. 410, p. 311–323, doi: 10.1016/j.tecto.2005.02.020.

Koronovskiy, N.V., and Demina, L.I., 1996, A model for the collision volcanism of the Caucasian segment of the Alpine Fold Belt: Doklady Akademii Nauk Rossi, v. 350, p. 519–522.

Koulakov, I., and Sobolev, S.V., 2006, A tomographic image of Indian lithosphere break-off beneath the Pamir-Hindukush region: Geophysical Journal International, v. 164, p. 425–440, doi: 10.1111/j.1365-246X.2005.02841.x.

Kuno, H., 1966, Lateral variation of basalt magma types across continental margins and island arcs: Bulletin of Volcanology, v. 29, p. 195–222.

Lay, T., 1994, The fate of descending slabs: Annual Review of Earth and Planetary Sciences, v. 22, p. 33–61, doi: 10.1146/annurev.ea.22.050194.000341.

Le Bas, M.J., Le Maitre, R.W., Streckeisen, A., and Zanettin, B., 1986, A chemical classification of volcanic rocks based on the total alkali-silica diagram: Journal of Petrology, v. 27, p. 745–750.

Lustrino, M., 2005, How the delamination and detachment of lower crust can influence basaltic magmatism: Earth-Science Reviews, v. 72, nos. 1–2, p. 21–38, doi: 10.1016/j.earscirev.2005.03.004.

Maggi, A., and Priestley, K., 2005, Surface waveform tomography of the Turkish-Iranian plateau: Geophysical Journal International, v. 160, no. 3, p. 1068–1080, doi: 10.1111/j.1365-246X.2005.02505.x.

Maggi, A., Priestley, K., and McKenzie, D., 2002, Seismic structure of the Middle East: Eos (Transactions, American Geophysical Union), v. 83, no. 47 (fall meeting supplement) [abs.], p. S51B–1041.

Maheo, G., Guillot, S., Blichert-Toft, J., Rolland, Y., and Pecher, A., 2002, A slab breakoff model for the Neogene thermal evolution of South Karakorum and South Tibet: Earth and Planetary Science Letters, v. 195, no. 1–2, p. 45–58, doi: 10.1016/S0012-821X(01)00578-7.

Maury, R.C., Fourcade, S., Coulon, C., El Azzouzi, M., Bellon, H., Coutelle, A.,

Oubadi, A., Semroud, B., Megartsi, M., Cotton, J., Belanteur, Q., Louni-Hacini, A., Pique, A., Capdevila, R., Hernandez, J. and Rehaulp, J.P., 2000, Post-collisional Neogene magmatism of the Mediterranean Maghreb margin: A consequence of slab breakoff: Comptes Rendus de l'Académie des Sciences, Série 2, Fascicule A: Sciences de la Terre et des Planets, v. 331, no. 3, p. 159–173.

McKenzie, D.P., 1972, Active tectonics of the Mediterranean: Geophysical Journal of the Royal Astronomical Society, v. 30, no. 2, p. 109–185.

McKenzie, D.P., and Bickle, M.J., 1988, The volume and composition of melt generated by extension of the lithosphere: Journal of Petrology, v. 29, p. 625–679.

Mitchell, J., and Westaway, R., 1999, Chronology of Neogene and Quaternary uplift and magmatism in the Caucasus: Constraints from K-Ar dating of volcanism in Armenia: Tectonophysics, v. 304, p. 157–186, doi: 10.1016/S0040-1951(99)00027-X.

Molinaro, M., Zeyen, H., and Laurencin, X., 2005, Lithospheric structure beneath the south-eastern Zagros Mountains, Iran: Recent slab break-off?: Terra Nova, v. 17, p. 1–6, doi: 10.1111/j.1365–3121.2004.00575.x.

National Iranian Oil Company, 1977, Geological cross-sections north-west Iran: Tehran, National Iranian Oil Company.

Nemcok, M., Pospisil, L., Lexa, J., and Donelick, R.A., 1998, Tertiary subduction and slab break-off model of the Carpathian-Pannonian region: Tectonophysics, v. 295, p. 307–340, doi: 10.1016/S0040-1951(98)00092-4.

Norton, I.O., 1999, Global plate reconstruction model, technical report: Houston, Texas, ExxonMobil.

Notsu, K., Fujitani, T., Ui, T., Matsuda, J., and Ercan, T., 1995, Geochemical features of collision-related volcanic rocks in central and eastern Anatolia, Turkey: Journal of Volcanology and Geothermal Research, v. 64, p. 171–192, doi: 10.1016/0377-0273(94)00077-T.

Özdemir, Y., Karaoğlu, O., Tolluoğlu, A.U., and Güleç, N., 2006, Volcanostratigraphy and petrogenesis of the Nemrut stratovolcano (East Anatolian High Plateau): The most recent post collisional volcanism in Turkey: Chemical Geology, v. 226, no. 3–4, p. 189–211.

Pearce, J.A., 1983, Role of the sub-continental lithosphere in magma genesis at active continental margins, *in* Hawkesworth, C.J., and Norry, M.J., eds., Continental basalts and mantle xenolites: Nantwich, Shiva, p. 230–249.

Pearce, J.A., Bender, J.F., De Long, S.E., Kidd, W.S.F., Low, P.J., Güner, Y., Şaroğlu, F., Yılmaz, Y., Moorbath, S., and Mitchell, J.G., 1990, Genesis of collision volcanism in Eastern Anatolia, Turkey: Journal of Volcanology and Geothermal Research, v. 44, p. 189–229, doi: 10.1016/0377-0273(90)90018-B.

Peccerillo, A., and Taylor, S.R., 1976, Geochemistry of Eocene calc-alkaline volcanic rocks from the Kastamonu area, northern Turkey: Contributions to Mineralogy and Petrology, v. 58, p. 63–81, doi: 10.1007/BF00384745.

Peccerillo, A., Barberio, M.R., Yirgu, G., Ayalew, D., Barbieri, M., and Wu, T.W., 2003, Relationship between mafic and peralkaline silicic magmatism in continental rift settings: A petrological, geochemical and isotopic study of the Gedemsa volcano, Central Ethiopian Rift: Journal of Petrology, v. 44, no. 11, p. 2003–2032, doi: 10.1093/petrology/egg068.

Piromallo, A., and Morelli, P., 2003, P wave tomography of the mantle under the Alpine-Mediterranean area: Journal of Geophysical Research, v. 108, doi: 10.1029/2002JB001757.

Platt, J.P., and England, P.C., 1993, Convective removal of lithosphere beneath mountain belts: Thermal and mechanical consequences: American Journal of Science, v. 293, p. 307–336.

Regard, V., Bellier, O., Martinod, J., and Faccenna, C., 2006, Analogue experiments of subduction vs. collision processes: Insight for the Iranian tectonics: JSEE, Fall 2005, v. 7, no. 3, p. 1–9.

Reilinger, R., McClusky, S., Oral, B., King, R., Toksoz, N., Barka, A., Kınık, I., Lenk, O., and Şanlı, I., 1997, Global positioning system measurements of present-day crustal movements in the Arabia–Africa–Eurasia plate collision zone: Journal of Geophysical Research, v. 102, no. B5, p. 9983–9999, doi: 10.1029/96JB03736.

Reilinger, R., McClusky, S., Vernant, P., Lawrence, S., Ergintav, S., Çakmak, R.,

Özener, H., Kadirov, F., Guliev, I., Stepanyan, R., Nadariya, M., Hahubia, G., Mahmoud, S., Sakr, K., ArRajehi, A., Paradissis, D., Al-Aydrus, A., Prilepin, M., Guseva, T., Evren, E., Dmitrotsa, A., Filikov, S.V., Gomez, F., Al-Ghazzi, R., and Karam, G., 2006, GPS constraints on continental deformation in the Africa-Arabia-Eurasia continental collision zone and implications for the dynamics of plate interactions: Journal of Geophysical Research, v. 111, p. B05411, doi: 10.1029/2005JB004051.

Rotstein, Y., and Kafka, A.L., 1982, Seismotectonics of the southern boundary of Anatolia, eastern Mediterranean region: Subduction, collision and arc jumping: Journal of Geophysical Research, v. 87, p. 7694–7706.

Sandvol, E., and Zor, E., 2004, Upper mantle P- and S-wave velocity structure beneath eastern Anatolian plateau: Eos (Transactions, American Geophysical Union), AGU Fall Meeting 2004, p. S13B–1056.

Sandvol, E., Türkelli, N., and Barazangi, M., 2003a, The Eastern Turkey Seismic Experiment: The study of a young continent-continent collision: Geophysical Research Letters, v. 30, no. 24, p. 8038, doi: 10.1029/2003GL018912.

Sandvol, E., Türkelli, N., Zor, E., Gök, R., Bekler, T., Gürbüz, C., Seber, D., and Barazangi, M., 2003b, Shear wave splitting in a young continent-continent collision: An example from Eastern Turkey: Geophysical Research Letters, v. 30, no. 24, p. 8041, doi: 10.1029/2003GL017390.

Şen, P.A., Temel, A., and Gourgaud, A., 2004, Petrogenetic modelling of Quaternary post-collisional volcanism: A case study of central and eastern Anatolia: Geological Magazine, v. 141, no. 1, p. 81–98, doi: 10.1017/S0016756803008550.

Şengör, A.M.C., 1990, A new model for the late Palaeozoic–Mesozoic tectonic evolution of Iran and implications for Oman, *in* Robertson, A.H.F., et al., eds., The geology and tectonics of the Oman region: Geological Society of London Special Publication 49, p. 797–831.

Şengör, A.M.C., 2001, Elevation as indicator of mantle-plume activity: Boulder, Colorado, Geological Society of America Special Paper 352, p. 183–225.

Şengör, A.M.C., 2002, The Turkish-Iranian High Plateau as a falcogenetic structure: Workshop on the Tectonics of Eastern Turkey and the Northern Anatolian plate, Erzurum, Turkey, p. 28.

Şengör, A.M.C., 2006, Tectonics of the Turkish-Iranian High Plateau: Lake Van Drilling Project (ICDP), PaleoVan Workshop in Van, Turkey, June 6–9.

Şengör, A.M.C., and Kidd, W.S.F., 1979, Post-collisional tectonics of the Turkish-Iranian Plateau and a comparison with Tibet: Tectonophysics, v. 55, p. 361–376, doi: 10.1016/0040-1951(79)90184-7.

Şengör, A.M.C., and Natal'in, B.A., 1996a, Paleotectonics of Asia: Fragments of a synthesis, *in* Yin, A., and Harrison, T.M., eds., The tectonic evolution of Asia: New York, Cambridge University Press, p. 486–640.

Şengör, A.M.C., and Natal'in, B.A., 1996b, Turkic-type orogeny and its role in the making of the continental crust: Annual Review of Earth and Planetary Sciences, v. 24, p. 263–337, doi: 10.1146/annurev.earth.24.1.263.

Şengör, A.M.C., and Yılmaz, Y., 1981, Tethyan evolution of Turkey: A plate tectonic approach: Tectonophysics, v. 75, p. 181–241, doi: 10.1016/0040-1951(81)90275-4.

Şengör, A.M.C., Özeren, S., Zor, E., and Genç, T., 2003, East Anatolian high plateau as a mantle-supported, N-S shortened domal structure: Geophysical Research Letters, v. 30, no. 24, p. 8045, doi: 10.1029/2003GL017858.

Stampfli, G.M., and Borel, G.D., 2002, A plate tectonic model for the Paleozoic and Mesozoic constrained by dynamic plate boundaries and restored synthetic oceanic isochrons: Earth and Planetary Science Letters, v. 196, p. 17–33, doi: 10.1016/S0012-821X(01)00588-X.

Stampfli, G.M., and Borel, G.D., 2004, The TRANSMED transects in space and time: Constraints on the paleotectonic evolution of the Mediterranean domain, *in* Cavazza, W., et al., eds., The TRANSMED atlas: The Mediterranean region from crust to mantle, p. 53–80: New York, Springer.

Sun, S.S., and McDonough, W.F., 1989, Chemical and isotopic systematics of oceanic basalts: Implications for mantle composition and processes, *in* Saunders, A.D., and Norry, M.J., eds., Magmatism in ocean basins: Geological Society, London, Special Publication 42, p. 313–345.

Taylor, S.R., and McLennan, S.M., 1985, The continental crust: Its composition and evolution: Geoscience texts: London, Blackwell Scientific, 312 p.

Tokel, S., 1985, Mechanism of crustal deformation and petrogenesis of the Neogene volcanics in East Anatolia: Special Publication of Turkiye Jeoloji Kurumu, Ketin Symposium, ed. T. Ercan and M.A. Çağlayan, p. 121–130.

Topuz, G., Altherr, R., Kalt, A., Satır, M., Werner, O., and Schwarz, W.H., 2004, Aluminous granulites from the Pulur complex, NE Turkey: A case of partial melting, efficient melt extraction and crystallization: Lithos, v. 72, p. 183–207, doi: 10.1016/j.lithos.2003.10.002.

Turcotte, D.L., 1983, Mechanism of crustal deformation: Journal of the Geological Society of London, v. 140, p. 701–724.

Türkoğlu, E., Unsworth, M., Çağlar, I., Tuncer, V., Avşar, U., and Tank, B., 2006, Magnetotelluric imaging of the Eurasian-Arabian collision in Eastern Anatolia: International Association of Geomagnetism and Aeronomy Working Group 1.2 on Electromagnetic Induction in the Earth, Extended Abstract 18, Workshop, El Vendrell, Spain, September, 17–23.

Williams, H.M., Turner, S.P., Pearce, J.A., Kelley, S.P., and Harris, N.B.W., 2004, Nature of the source regions for post-collisional, potassic magmatism in Southern and Northern Tibet from geochemical variations and inverse trace element modeling: Journal of Petrology, v. 45, no. 3, p. 555–607, doi: 10.1093/petrology/egg094.

Yılmaz, Y., Şaroğlu, F., and Güner, Y., 1987, Initiation of the Neomagmatism in East Anatolia: Tectonophysics, v. 134, nos. 1–3, p. 177–199.

Yılmaz, Y., Tüysüz, O., Yiğitbaş, E., Genç, S.C., and Şengör, A.M.C., 1997, Geology and tectonic evolution of the Pontides, *in* Robinson, A.G., ed., Regional and petroleum geology of the Black Sea and surrounding region: Tulsa, Oklahoma, American Association of Petroleum Geologists Memoir 68, p. 183–226.

Yılmaz, Y., Güner, Y., and Şaroğlu, F., 1998, Geology of the quaternary volcanic centers of the east Anatolia: Journal of Volcanology and Geothermal Research, v. 85, nos. 1–4, p. 173–210, doi: 10.1016/S0377-0273(98)00055-9.

Zor, E., Gürbüz, C., Türkelli, N., Sandvol, E., Seber, D., and Barazangi, M., 2003, The crustal structure of the East Anatolian Plateau from receiver functions: Geophysical Research Letters, v. 30, no. 24, p. 8044, doi: 10.1029/2003GL018192, doi: 10.1029/2003GL018192.

MANUSCRIPT ACCEPTED BY THE SOCIETY JANUARY 31, 2007

The Geological Society of America
Special Paper 430
2007

Phantom plumes in Europe and the circum-Mediterranean region

Michele Lustrino*
Eugenio Carminati
Dipartimento di Scienze della Terra, Università degli Studi di Roma La Sapienza, P.le A. Moro, 5, 00185 Rome, Italy,
and Istituto di Geologia Ambientale e Geoingegneria (IGAG) CNR, c/o Dipartimento di Scienze della Terra,
Università degli Studi di Roma La Sapienza, P.le A. Moro, 5, 00185, Rome, Italy

ABSTRACT

Anorogenic magmatism of the circum-Mediterranean area (the Tyrrhenian Sea, Sardinia, Sicily Channel, and the Middle East) and of continental Europe (the French Massif Central, Eifel, the Bohemian Massif, and the Pannonian basin) has been proposed to be related to the presence of one or more mantle plumes. Such conclusions based on geochemical data and seismic tomography are not fully justified because (1) a given chemical and isotopic composition of a magma can be explained by different petrogenetic models, (2) a given petrogenetic process can produce magmas with different chemical and isotopic composition, (3) tomographic studies do not furnish unique results (i.e., different models give different results), and (4) the commonly adopted interpretation of seismic wave velocity anomalies exclusively in terms of temperature is not unique; velocities are also dependent on other parameters, such as composition, melting, anisotropy, and anelasticity. Tomography and geochemistry are powerful tools but must be used in an interdisciplinary way, in combination with geodynamics and structural geology. Alone they cannot provide conclusive evidence for or against the existence of mantle plumes.

The existence of large and/or extensive thermal anomalies under Europe is here considered unnecessary, because other models, based on the existence of upper-mantle heterogeneity, can explain the major-element, trace-element, and isotopic variability of the magmas. Volcanism in central Europe (the French Massif Central, Germany, and the Bohemian Massif) is concentrated in Cenozoic rifted areas and is here interpreted as the result of passive asthenosphere upwelling driven by decompression. Similarly, anorogenic magmatism in Sardinia, the Tyrrhenian Sea, and the Pannonian basin is explained as the result of lithospheric stretching in a back-arc geodynamic setting. The most important factors determining the locus and, in part, the geochemical characteristics of magmatic activity are the Moho and the lithosphere-asthenosphere boundary depths. Where both are shallowed by tectonic processes (e.g., in rift zones or back-arc basins), passive upwelling of asthenospheric mantle can explain the magmatic activity.

Keywords: petrology, geodynamics, mantle plume, lithosphere, Mediterranean

---

*E-mail: michele.lustrino@uniroma1.it.

Lustrino, M., and Carminati, E., 2007, Phantom plumes in Europe and the circum-Mediterranean region, *in* Foulger, G.R., and Jurdy, D.M., eds., Plates, plumes, and planetary processes: Geological Society of America Special Paper 430, p. 723–745, doi: 10.1130/2007.2430(33). For permission to copy, contact editing@geosociety.org. ©2007 The Geological Society of America. All rights reserved.

INTRODUCTION

The most popular model for the Earth's mantle proposed in the last few decades requires a subcontinental upper-mantle structure with a variably enriched lithospheric mantle and a relatively depleted asthenospheric mantle. On the other hand, the oceanic lithospheric and asthenospheric mantles are considered to be more similar in chemical composition and to have a generally depleted character. The depleted character of the uppermost oceanic mantle was deduced indirectly from the composition of magmas emplaced along mid-ocean ridges. Some "anomalous" compositions, mostly recorded on punctiform magmatic manifestations and known as ocean island basalts (OIB), are considered to be related not to the upper mantle but rather to deep-mantle sources, possibly ascending from the core-mantle boundary or the upper-lower mantle transition zone (e.g., Wilson, 1963; Olson et al., 1987; Courtillot et al., 2003). The difference between mid-ocean ridge basalts (MORB) and OIB has often been related to different compositions of the two mantle sources: relatively depleted for the first and relatively enriched (or even pristine) for the second. This dichotomy was based on the assumption of a nearly total separation between upper and lower mantle and the presence of two separate systems of mantle convection divided by the 660 km seismic discontinuity (e.g., Karato, 1997; Tackley, 2000).

Since the 1990s, global tomographic models (e.g., van der Hilst et al., 1997; Bijwaard et al., 1998) have provided insights into the distribution of seismic wave velocities in the whole mantle. Fast and slow velocity anomalies have been interpreted as evidence for cold and hot bodies in the mantle. Although partial evidence for this interpretation is available for the upper mantle, no clear proof is available for the lower mantle. A major inference from the results of global tomography analyses is that a total separation between upper and lower mantle reservoirs (one of the major requirements to invoke the peculiar geochemical signature for OIB magmatism) is no longer acceptable a priori. In fact, on the basis of seismic tomography models it has been proposed that subducting slabs cross the 660 km discontinuity (e.g., Grand et al., 1997; Grand, 2002) and may bend and accumulate within the D″ layer, at the core-mantle boundary (Hutko et al., 2006). Nonetheless, the plume model is still based on such geochemical distinctions.

In order to explain OIB magmatism with the mantle plume model, geochemists propose a contrasting model: from one side, plume advocates invoke isolated sources (i.e., a closed system that has never contributed to magma production and consequently is considered to be undegassed and with high $^3He/^4He$ ratios). But, at the same time, such a reservoir cannot be considered a closed system because it must allow the entrance of subducted oceanic crust where it should be stored for at least 2 b.y. This requirement is necessary to explain the high uranogenic Pb isotopic ratios observed (e.g., $^{206}Pb/^{204}Pb > 21$; e.g., Hofmann and White, 1982).

In this chapter we discuss the main problems arising from an uncritical use of geochemical data and tomography images

to propose the existence of hot mantle upwellings from the core-mantle boundary or shallower depths (e.g., the 660 km discontinuity) under "hotspots." An upwelling of hot mantle is commonly called a mantle plume. The difference between the potential temperature (T_p; i.e., the temperature that a volume of the mantle would have if brought to the surface adiabatically without melting; McKenzie and Bickle, 1988) of "normal" asthenosphere (with $T_p \sim 1280\ °C$) and mantle plume material can be as high as 300 °C (e.g., Richards et al., 1989; Griffiths and Campbell, 1990), even if its ΔT_p can be reduced to as low as 100 °C (e.g., McKenzie and Bickle, 1988; White and McKenzie, 1989). The requirement of such a large ΔT_p (i.e., the existence of extremely hot upwelling volumes of mantle) has been proposed to explain the huge volumes of magma emplaced in continental provinces such as the Deccan, Paranà-Etendeka, and Siberian traps, as well as Kerguelen and Ontong Java and others (e.g., Saunders et al., 1992; Mahoney and Coffin, 1997). These geochemical and geophysical models are based on the assumption that the source regions of large igneous provinces (volumes of magmas on the order of some million km^3 produced in a relatively short time, ~ 1–2 m.y.) are entirely peridotitic. However, during the last decade, new models have suggested the presence of lithologies (eclogites, pyroxenites, garnet granulites, etc.) with solidus temperatures several hundred degrees lower than those of peridotic mantle (e.g., Cordery et al., 1997; Hirschmann, 2000; Yaxley, 2000; Kogiso et al., 2004). At least in some cases, enhanced melt productivity could be the consequence of chemical anomalies (e.g., the presence of low-temperature melting point assemblages) rather than thermal anomalies (as proposed in mantle plume models). As an example of questionable plumes, defined on the basis of geochemical and tomography analyses, we discuss some volcanic areas in the European and circum-Mediterranean area.

The main conclusions of this review can be summarized as follows:

1. Geochemical data must be used with care, because (a) a given chemical and isotopic composition can be explained by different models and (b) a single petrogenetic model can produce magmas with different chemical and isotopic compositions.
2. Tomographic studies do not furnish unique results, and different studies give different results.
3. The presence of deeply rooted mantle plumes or anomalously hot upper mantle is unlikely and unnecessary to propose in the circum-Mediterranean area.
4. Most Cenozoic magmatism in the circum-Mediterranean area can be considered related to passive upwelling of asthenospheric material as a consequence of lithospheric thinning in rift or back-arc areas.

GEODYNAMIC SETTING

European continental structure is the result of several orogenic cycles dating back at least to the Precambrian. Cenozoic evolution of the western Mediterranean area occurred in the framework of convergence between Africa and Europe that led, during the lower Cretaceous–Paleocene, to the consumption of the Tethyan ocean(s) (see Polino et al., 1990, and Schmid et al., 1996, for a discussion on Cretaceous paleogeography) along an ESE-dipping subduction zone (with European-Iberian lithosphere sinking below the Africa plate) running continuously from the Alps to the Betics via Corsica and the Balearics (Fig. 1). There, European-Iberian lithosphere sank below the Africa plate. The Tertiary–present diachronous collisional stage was possibly associated with slab detachment events and the presence of continental microplates between Africa and Europe (Doglioni, 1991; von Blanckenburg and Davies, 1995; Schmid et al., 1996; Carminati et al., 1998).

During the Cenozoic, a rift system (the Rhine, Rhône, and Bresse grabens) developed from the North Sea to the Mediterranean (Ziegler, 1992) contemporaneously with the development of the Alpine orogeny (Fig. 2). Since the Oligocene, the onset of west-dipping subduction of the Adriatic plate beneath the European plate drove the development of the Apennine and Maghrebide belts. The fast radial rollback of the subducting plate (up to five times faster than Africa-Europe convergence) resulted in the opening of two diachronous back-arc basins, the Lower Miocene–Langhian Algerian-Provençal basin and the Langhian–present Tyrrhenian Sea. The rollback of the subduction zones also caused partial disruption of the Alpine-Betic chain due to back-arc extension and the migration and rotation of segments of this chain (e.g., Corsica, Calabria, and the Kabilies). These are now entrained within the Apennines-Maghrebides belt or located in the western Mediterranean back-arc basin (Fig. 1).

The Balkan, Aegean, and Anatolian areas have been characterized, since the Mesozoic, by the development of a polyphased double-vergence orogenic belt (the Dinarides, Hellenides, and Taurides). This orogen is the result of at least two or three subduction zones, as shown by the occurrence of two distinct oceanic sutures (the Vardar-Izmir-Ankara and Sub-Pelagonian ophiolites). These represent one or two (depending on the reconstruction) branches of the Mesozoic Tethyan Ocean and the present oceanic subduction of the Ionian and Aegean seas (e.g., Carminati and Doglioni, 2004; Fig. 1). The east-dipping subduction of the Adriatic lithosphere beneath the Dinarides-

Hellenides started as early as the Cretaceous and in the Tertiary became the main process acting in the eastern Mediterranean area. In the Tertiary, widespread extension overprinted compressional structures in the Dinarides-Hellenides-Taurides orogen. Much as in the western Mediterranean, boudinage of the preexisting Alps and Dinarides orogens also occurred in the Pannonian basin, which is a back-arc basin related to the eastward-retreating, westward-dipping Carpathian Subduction Zone, which has been active from the Oligocene to the present (Fig. 1).

KEY PARAMETERS OF BASIC MAGMAS

Widespread volcanic activity accompanied the Cenozoic evolution of continental Europe, northern Africa, Mashrek (the Middle East), and the Mediterranean Sea. Cenozoic igneous rocks emplaced within this large region show an extremely wide range of chemical compositions that can be grouped into (1) sodic, mildly alkaline compositions, often associated with tholeiitic rocks with geochemical characteristics resembling magmas emplaced in oceanic intraplate tectonic settings (e.g., OIB-like compositions); (2) ocean-floor rocks, whose compositions vary from N (normal)-MORB to E (enriched)-MORB and low-K calc-alkaline basalts and andesites; (3) calc-alkaline rocks with geochemical characteristics resembling those of magmas emplaced in subduction-related settings; (4) potassic to ultrapotassic alkaline rocks with mildly to strongly SiO_2-undersaturated compositions; and (5) rare, exotic compositions such as lamproites, lamprophyres, and carbonatites (Wilson and Downes, 1991; Wilson and Bianchini, 1999; Conticelli and Melluso, 2004; Peccerillo, 2005; Beccaluva et al., 2007; Lustrino and Wilson, 2007). The Sr-Nd-Pb isotopic compositions of these products comprise virtually all the known worldwide reservoirs, with the extreme heterogeneous compositions of the mantle sources of this sector of the European-Mediterranean lithosphere-asthenosphere system reflecting its complex geodynamic history (e.g., Lustrino and Wilson, 2007).

This review focuses mainly on the first type of product (i.e., Na-alkaline and tholeiitic magmas), because in many cases these have been interpreted as the result of partial melting of an anomalously hot mantle. The origin of this type of igneous activity (called "anorogenic") has often been related to the presence

---

Figure 1. Evolution of the Mediterranean and adjacent areas since 45 Ma. (A) At ca. 45 Ma the Alps were linked to the Betics down to the Gibraltar Strait. The Alps-Betics chain developed on top of a southeastward-dipping subduction zone that consumed an ocean to the northwest of the belt. Along the Dinarides and the Hellenides, an east-northeast-dipping subduction zone was active. Corsica and Sardinia were attached to the Iberian peninsula. (B) At ca. 30 Ma the Apennines-Maghrebides subduction developed along the Alps-Betics retrobelt, where oceanic or thinned preexisting continental lithosphere was present to the east. The subduction zone was characterized by a fast, mainly southeastward rollback. Similarly, the Carpathians started to develop along the Dinarides retrobelt (i.e., the Balkans). The fronts of the Alps-Betics orogen are crosscut by the Apennines-related subduction back-arc extension, which drove the opening of the Liguro-Provençal basin, the Valencia Trough, and the north Algerian basin. Between 19 and 15 Ma Corsica and Sardinia were separated from Iberia and rotated counterclockwise to their present-day position. (C) At 15 Ma the Liguro-Provençal basin, the Valencia Trough, and the north Algerian basin were almost completely opened, and active extension jumped to the east of Corsica and Sardinia, leading to the opening of the Tyrrhenian back-arc basin. Similarly, the Carpathians migrated eastward due to the rollback of the subduction zone, and the Pannonian back-arc basin was generated. The Dinarides subduction slowed down due to the presence of the thick Adriatic continental lithosphere to the west, whereas to the south, the Hellenic subduction was very active due to the presence in the footwall plate of Ionian oceanic lithosphere. (D) At present four subduction zones are active in the Mediterranean: the west-directed Apennines-Maghrebides, the west-directed Carpathians, the northeast-directed Dinarides-Hellenides-Taurides, and the southeast-directed Alps. Modified from Carminati and Doglioni (2004).

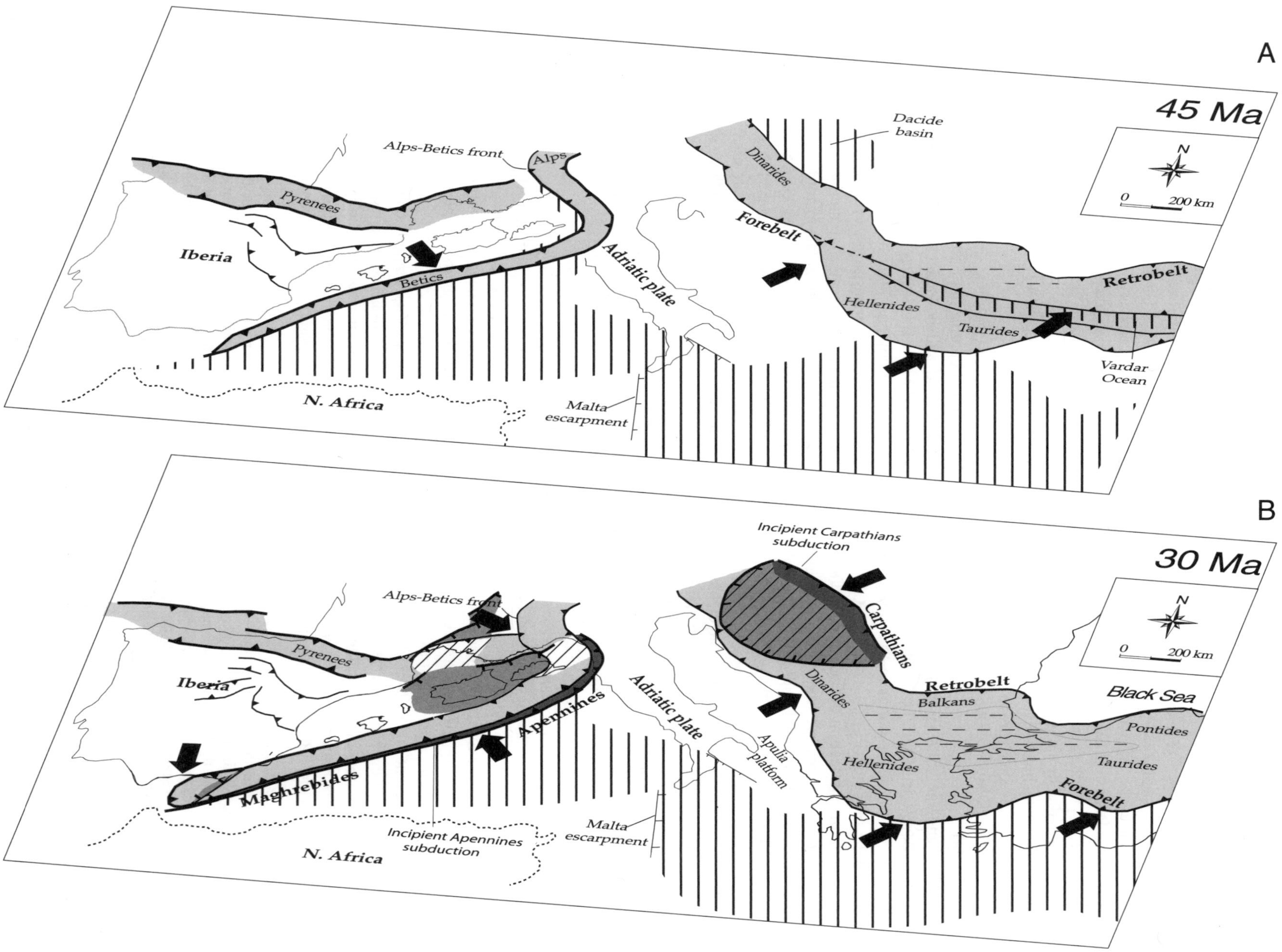

A

45 Ma

N

0 200 km

Retrobelt

Vardar
Ocean

Dacide
basin

Taurides

Hellenides

Dinarides

Forebelt

Adriatic plate

Malta
escarpment

Alps

Alps-Betics front

Pyrenees

Betics

Iberia

N. Africa

B

30 Ma

N

0 200 km

Black Sea

Pontides

Taurides

Forebelt

Retrobelt

Balkans

Hellenides

Carpathians

Incipient Carpathians
subduction

Dinarides

Apulia
platform

Adriatic plate

Malta
escarpment

Apennines

Incipient Apennines
subduction

Alps-Betics front

Pyrenees

Maghrebides

Iberia

N. Africa

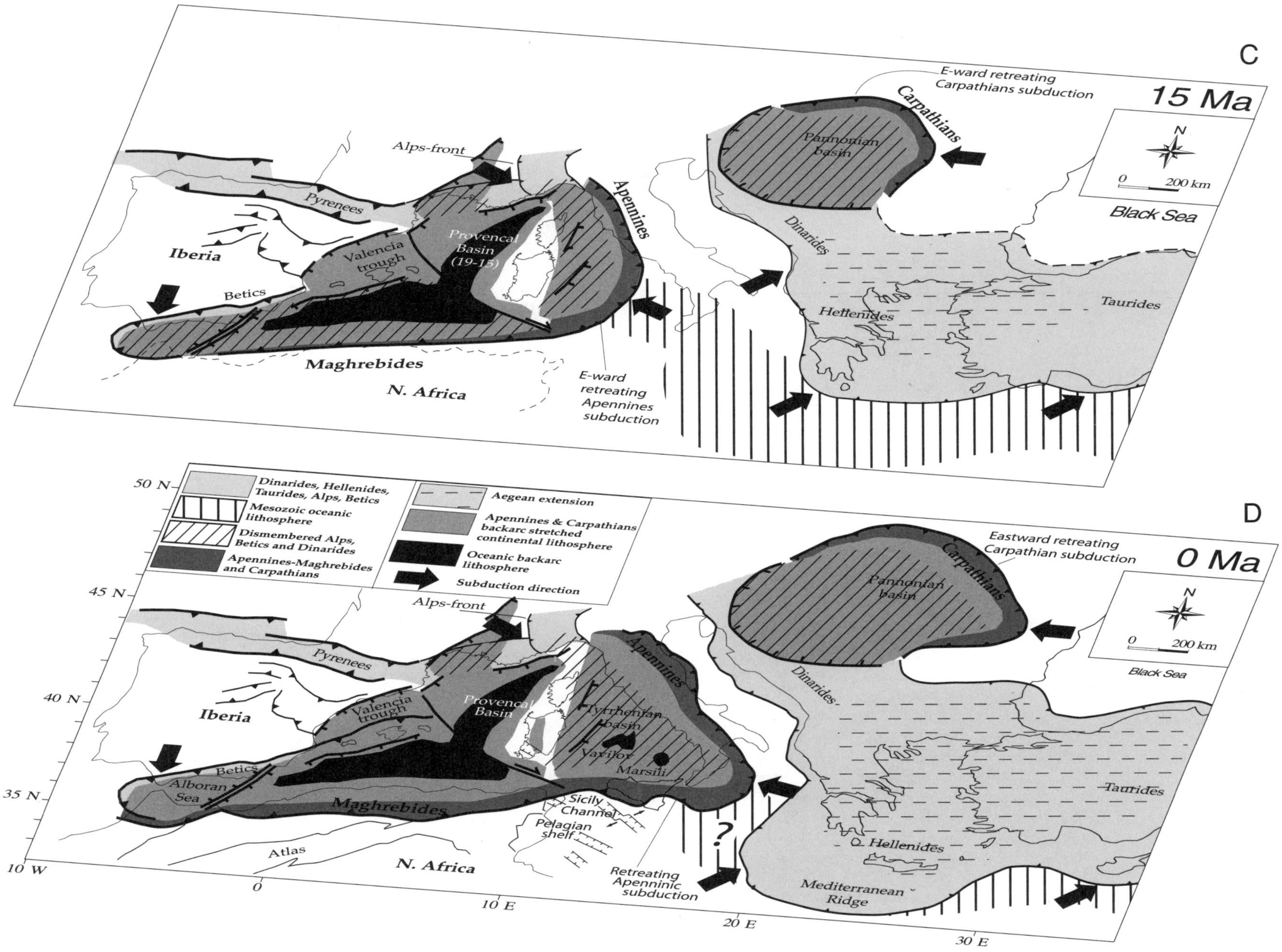

Figure 1. Continued

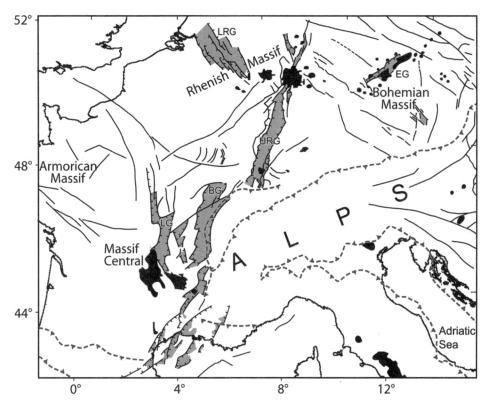

Figure 2. Map of the main features of the European Cenozoic rift system in the Alpine and Pyrenean foreland, showing Cenozoic faults (black lines), rift-related sedimentary basins (light gray), and volcanic fields (black). The dashed barbed lines indicate the Alpine deformation front. BG—Bresse Graben; EG—Eger Graben; LG—Limagne Graben; LRG—Lower Rhine Graben; URG—Upper Rhine Graben. Modified from Dèzes et al. (2004).

of deep mantle plumes that interact with upper mantle in several ways. Many of the petrological studies proposing the existence of a single giant mantle plume or the presence of several smaller plumes have been based on tomographic images that depict anomalously low-velocity (interpreted as hot) regions at variable depths.

The first problem is to try to define what an "anorogenic" magma is. This topic has been approached by several authors (e.g., Wilson and Bianchini, 1999; Wilson and Downes, 2006; Lustrino and Wilson, 2007). Proposals are mostly based on interelemental incompatible trace-element ratios (e.g., Ba/Nb, La/Nb, Th/Yb, etc.), major-element ratios (e.g., Na_2O/K_2O), and isotopic systematics (mostly $^{87}Sr/^{86}Sr$, $^{143}Nd/^{144}Nd$, and, to lesser extent, $^{207}Pb/^{204}Pb$). However, it must be noted that no clear geochemical distinction exists between "anorogenic" and "orogenic" magmas because of the existence of common transitional or "hybrid" compositions. Following Lustrino and Wilson (2007), the mafic anorogenic volcanic rocks considered in this review have the following geochemical characteristics:

1. Most of the anorogenic rocks are mildly to strongly alkaline (basanites, nephelinites, hawaiites, and alkali basalts) plus rarer tholeiitic basalts and basaltic andesites. The SiO_2 content of the most primitive mafic rocks is 40–52 wt%, and the sum of $Na_2O + K_2O$ is mostly 3–10 wt%. In general, these rocks are sodic, with Na_2O/K_2O in the range 1.4–5.0. The TiO_2 contents of the most mafic samples (MgO >7 wt%) range from ~1 to ~6 wt%.

2. Most of the primitive anorogenic rocks have primitive mantle-normalized trace-element patterns resembling those of a HIMU (high μ, i.e., high U/Pb mantle) OIB end-member. The most common features of the trace-element patterns are peaks at the high field-strength elements (Nb-Ta-Hf-Zr) and negative K and Pb anomalies.

3. The Sr and Nd isotopic compositions are mostly confined to the depleted quadrant, with $^{87}Sr/^{86}Sr$ lower than present-day bulk silicate earth estimates (BSE = 0.70445) and $^{143}Nd/^{144}Nd$ higher than chondritic uniform reservoir values (ChUR = 0.51264). The $^{206}Pb/^{204}Pb$ ratios are in general relatively radiogenic (18.8–20.4), whereas $^{207}Pb/^{204}Pb$ and $^{208}Pb/^{204}Pb$ are highly variable (Lustrino and Wilson, 2007).

GEOCHEMISTRY-TOMOGRAPHY-GEODYNAMICS: IS THERE AN OBVIOUS LINK?

The classification of a rock as "anorogenic" is particularly important in very complex areas such as the circum-Mediterranean area. Indeed, if we want to investigate the composition of the upper mantle and its response to tectonics, we must exclude magmas that show clear effects of crustal contamination at shallow levels or that have suffered metasomatic modifications along subduction margins. Crustal contamination of melts in magma chambers at crustal levels is relatively easy to detect on the basis of petrographic observations (e.g., the presence of quartz xenocrysts or crustal/mantle xenoliths) and on the basis of some key geochemical parameters (e.g., compatible and incompatible trace elements and Sr-Nd isotopic variation with MgO, SiO_2 etc.). More complex is the identification of crustal contamination at deeper levels in the mantle.

What is important to stress here is that virtually all the igneous rocks reflect in their chemistry the effects of interaction between mantle (i.e., peridotitic) and recycled crustal (i.e., pyroxenitic or eclogitic) lithologies (e.g., Hofmann and White, 1982; Hirschmann and Stolper, 1996; Kogiso et al., 2003, 2004; Meibom and Anderson, 2003; Lustrino, 2005, 2006; Stracke et al., 2005).

The presence of crustal lithologies at mantle depths obviously influences the composition of the partial melts of a mixed source. Crustal lithologies have a solidus temperature ~50–300 °C lower than peridotitic mantle at similar pressures. This means that, in an adiabatically upwelling mantle, crustal lithologies may melt several tens of km deeper than ambient mantle. No consensus has been so far reached about the effective role of pyroxenitic/eclogitic material in basalt petrogenesis (e.g., Hirschmann et al., 2003; Kogiso et al., 2003; Keshav et al., 2004; Sobolev et al., 2005; Huang and Frey, 2005; Lustrino, 2006).

Neither mixing between peridotite and bulk clinopyroxenite nor mixing between peridotite and clinopyroxenite partial melts provides a satisfactory explanation for the range of chemical compositions observed in basaltic magma. A complex style of interaction between peridotite and clinopyroxenite partial melts involving assimilation of some phases and stabilization of other minerals is more likely (e.g., Arai and Abe, 1995; Yaxley, 2000; Lustrino, 2005). The effects of such chromatographic styles of metasomatism are hard to model geochemically and are still not understood in detail.

Regarding geophysics, the results of tomographic models are regularly proposed as compelling evidence for the existence of mantle plumes. The basic idea is that low-velocity anomalies in tomographic images reflect relatively high-temperature zones. There are two main limitations in the use of tomography to detect plumes. The first and more obvious is that different studies frequently yield different results and lead different conclusions to be drawn. The case of the so-called Iceland plume is emblematic. Bijwaard and Spakman (1999) and Zhao (2001) used global tomography to image a narrow low-velocity anomaly below Iceland extending down to the core-mantle boundary. Alternative tomographic results provided by Ritsema et al. (1999), Foulger et al. (2000), and Montelli et al. (2004) showed that below Iceland a low-velocity anomaly interpretable as a plume can be resolved clearly only in the upper mantle. Discrepancies are related to the different analysis approaches and, more important, to the different graphical presentations and interpretations of the results. Particularly important is the type of color used in cross-sections (Anderson, 1999). For example, Bijwaard and Spakman (1999) and Zhao (2001) were able to illustrate a continuous low-velocity anomaly extending down to the lower mantle by saturating their color scales at the red end at a velocity anomaly of ~0.5%. Anomalies of ~5% in the upper mantle and 0.5% in the lower mantle were all shown as the same color. On the other hand, if the color scale is saturated at ~2%, as, for example, was done by Ritsema et al. (1999), the strong upper mantle anomalies appear red but the weak lower mantle anomalies only yellow or orange, so there is no visual impression of a continuous structure. In fact, most tomographic models for the mantle beneath Iceland basically agree that there are strong anomalies in the upper mantle and only weak ones in the lower mantle. The weak lower-mantle anomalies vary from model to model because they are close to the noise level and not robust enough to be easily repeatable between different studies.

Surface waves are well suited for investigating the uppermost mantle on a global scale. Although narrow plume stems may be theoretically too narrow to be imaged by global tomography, large ponds of plume material (on the order of 1000 km in width) beneath the lithosphere are detectable. Body waves are regularly used to investigate the whole mantle, although teleseismic tomography may be used for detailed analyses of the upper few hundred km of the mantle beneath particular areas. Whole-mantle (or global) tomography uses large global data sets to derive 3-D models of the entire mantle. The main pitfalls of such models derive from errors due to uneven distribution of rays from earthquakes and seismometers throughout the world and from weak resolving power (on the order of a few hundred km). Such a low resolution should encourage caution when interpreting low-velocity bodies in the lower mantle as plumes (Foulger and Natland, 2003).

The second limitation is more subtle. As discussed, Earth scientists commonly interpret low-velocity bodies in tomographic images as hot zones and high-velocity bodies as cold zones. However, they often neglect to account for the fact that other factors also affect the velocity of seismic waves. These include pressure, rock composition, melting, anisotropy, and anelasticity. Consequently, anomalously low-velocity bodies in the mantle may not automatically be interpreted as hot material and assumed to indicate the presence of a plume. For example, Anderson (2006) argues that high-density eclogitic material may be characterized by lower shear wave velocities compared to peridotitic material at the same depth, thus resulting in "red zones" in tomographic images and giving the impression of the existence of hot mantle. High-velocity bodies are generally interpreted as indicating a subducting slab that is colder than the adjoining lithosphere due to the fact that thermal reequilibration is slower than slab sinking. Although such an interpretation may not be absolutely certain, the correspondence, worldwide, of high-velocity bodies with subcrustal earthquake alignments down to the bottom of the transition zone (Wadati-Benioff planes) provides independent supporting evidence for this interpretation. Such independent evidence, however, is missing for the low-velocity bodies beneath presumed hotspots and elsewhere. It may be concluded that at the moment there is insufficient evidence for mantle plumes in seismology alone.

Caution is thus needed when interpreting geochemical data and tomographic results in terms of geodynamic features such as hotspots. A further problem with such an interpretation derives from the fact that in the recent literature a clear and unique definition of a hotspot is not available. In his seminal work Morgan (1971) defined a hotspot as a zone of anomalous intraplate volcanism underlain by a narrow upwelling plume originating from the deep mantle. Such a definition has been progressively modified to account for unexpected observations, and presently the hotspot family consists of a wide variety of processes and geometries, generating (in the better cases only semantic) confusion. Concepts like fossil plumes (e.g., Stein and Hofmann, 1992; Anguita and Hernàn, 2000; Rotolo et al., 2006), dying plumes (Davaille and Vatteville, 2005), recycled plume heads (Gasperini et al., 2000), tabular plumes (Hoernle et al., 1995), fingerlike plumes (e.g., Granet et al., 1995), depleted plumes (Bernard-Griffiths et al., 1991), baby plumes (Ritter, 2007), channeled plumes (e.g., Oyarzun et al., 1997), plumes with thoroidal chemical compositions (Mahoney et al., 1992), head-free plumes (e.g., Ritter, 2007), cold plumes (Garfunkel, 1989; Anguita and Hernàn, 2000; Jung et al., 2006), and others have been proposed in an attempt to explain geophysical and geochemical features not compatible with the original definition of plumes. Courtillot et al. (2003) tried to eliminate such problems and divided hotspots into three main categories, depending on whether their origin is at the core-mantle boundary, at the base of the upper mantle, or in the lithosphere. It is evident that the original model of Morgan (1971) has been so enlarged that the use of the terms "hotspot" and "mantle plume" may have lost significance because they now comprise processes from lithospheric plate tectonics to deep-mantle dynamics. Anderson (2005), scoring proposed mantle plumes worldwide, concluded that there is no place on Earth that fulfils all the five characteristics required by Courtillot et al. (2003) to identify a mantle plume (i.e., the presence of a hotspot track, a large igneous province at the start of this track, high buoyancy flux, high ^{3}He/^{4}He ratios, and low seismic shear wave velocity at 500 km depth).

In particular, plume-supporting geochemists have attributed the high ^{3}He/^{4}He ratios of some OIB to the involvement of an undegassed (i.e., primitive) lower-mantle source component. However, ^{3}He/^{4}He isotope systematics cannot be considered clear evidence of derivation from a primitive (lower-mantle) source, as shown by several studies (e.g., Anderson, 1998; Meibom et al., 2003, 2005; Gautheron et al., 2005; Parman et al., 2005). The most important criticisms of ^{3}He/^{4}He systematic use as "proof" of lower-mantle derivation are as follows: (1) often MORB show higher ^{3}He than OIB, an effect opposite to that expected because the sources of MORB are much more depleted (and therefore degassed) than the proposed lower-mantle sources, HIMU OIB; (2) high ^{3}He could be related not only to undegassed sources but also to recycled oceanic sediments where cosmic dust (rich in ^{3}He) accumulated; (3) an undegassed reservoir would have high ^{3}He but also high ^{4}He (derived from U and Th decay), and therefore it is not obvious that an undegassed (i.e., primitive) mantle reservoir should be characterized by high ^{3}He/^{4}He; (4) the bulk partition coefficient for He is not known in detail, and recent experimental studies show that it is lower than that of U + Th (the parent isotopes of ^{4}He). This means that depleted (i.e., degassed) mantle sources evolve with higher He/(U + Th) compared with primitive (i.e., undegassed) mantle, therefore leading to higher ^{3}He/^{4}He, which is associated with low total helium contents.

MANTLE PLUMES IMAGED IN THE CIRCUM-MEDITERRANEAN AREA

Regarding the circum-Mediterranean area, the presence of a mantle plume has been suggested based on some geochemical characteristics of mafic magmas (e.g., unradiogenic ^{87}Sr/^{86}Sr, radiogenic ^{143}Nd/^{144}Nd, radiogenic ^{206}Pb/^{204}Pb, low ratios of large-ion lithophile elements [LILE] to high field-strength elements [HFSE], etc.) and on tomographic images. We suggest that this approach may lead to wrong conclusions because comparable geochemical features also characterize magmas derived from passive upwelling of upper mantle and tomographic images can be interpreted in different ways. In this section we review a series of case studies of mantle plumes whose existence has been postulated on geochemical and tomographic grounds.

Tyrrhenian Sea

A classical example is from the circum-Tyrrhenian area, where the presence of a deep-mantle plume has been hypothesized mostly on geochemical (e.g., Bell et al., 2004) and geophysical or geological bases (Locardi and Nicolich, 2005). Bell et al. (2004) proposed the existence of a mantle plume on the basis of a pronounced isotopic polarity along the length of Italy (i.e., a northward increase in ^{87}Sr/^{86}Sr, δ^{18}O, and δ^{13}C and a decrease in ^{143}Nd/^{144}Nd, ^{206}Pb/^{204}Pb, and ^{3}He/^{4}He isotopic ratios). They interpreted this as mixing between two end-members, one depleted and the other enriched. The enriched end-member (called Italian enriched mantle, or ITEM; Bell et al., 2004) is considered to represent a deep-mantle reservoir sited at the D″ core-mantle boundary layer and isolated from mantle convection. These authors discussed the opening of the Mediterranean region along the southwestward continuation of the Rhine-Rhône rift system. They concluded that the geochemical signature of the circum-Tyrrhenian lavas (i.e., the presence of hypothetical geochemical end-members like EM-I [enriched mantle type I], FOZO [focal zone] or ITEM) is the expression of melts originating in the deep mantle.

According to Locardi and Nicolich (2005), an eastward-migrating deep-seated thermal plume would be responsible for (1) the opening of several basins and the reduction of lithospheric thickness, (2) fragmentation of the western Mediterranean, (3) subsequent rotations and collision of microplates, and (4) the production of magmas with an extremely wide range of chemical compositions over a short time interval. Bell et al. (2004) and Locardi and Nicolich (2005) seem to ignore the existence of subduction beneath Italy. In particular, Locardi and Nicolich (2005) interpreted the indisputable seismically active belt in southern Italy as the effect of a convective cell associated with hot asthenosphere inducing stress and seismic activity at the interface with the neighboring cooler mantle.

In drawing their conclusions, Bell et al. (2004) and Locardi and Nicolich (2005) did not take into consideration some basic features of Cenozoic Italian magmatism, in particular: (1) from Oligocene until Recent times, volcanic activity with a subduction-like geochemical signature has occurred from the northwest (Sardinia) through the central Tyrrhenian Sea (the Magnaghi and Vavilov seamounts) to the southeast (the Aeolian Archipelago; e.g., Lustrino, 2000, Savelli, 2002, Peccerillo, 2005); (2) the chemical compositions of Late Miocene–Quaternary igneous rocks emplaced along the western (Sardinia) and the eastern (Italian peninsula) branch of the Tyrrhenian Sea are completely different (Peccerillo, 2005; Lustrino et al., 2007); (3) the composition of Italian peninsula volcanic rocks (mostly potassic to ultrapotassic) has never been found in oceanic intraplate tectonic settings (i.e., no OIB show potassic to ultrapotassic compositions); (4) the Tyrrhenian Sea crust is very deep compared to the depth of oceanic crust of a similar age (Malinverno, 1981; Spadini et al., 1995), in disagreement with the hypothesis that doming would be induced by a mantle heat anomaly; (5) the calculated T_p of the Tyrrhenian Sea exceeds only by a few tens of °C that of a mantle under normal conditions (i.e., T_p ~1320 °C versus 1280 °C, respectively; Zito et al., 2003; Cella et al., 2006); (6) numerical modeling suggests that the development of the Apennine belt cannot be explained by the occurrence of a mantle plume but requires tectonic forces such as those in subduction settings (e.g., Carminati et al., 1999); (7) subcrustal earthquakes indicate the existence of a slab at least below the northern Apennines and the Calabrian arc down to a depth of 500 km (Amato and Selvaggi, 1991; Giardini and Velonà, 1991; Selvaggi and Chiarabba, 1995; Carminati et al., 2002); and (8) the off-scraping of sediments previously deposited on continental lithosphere and their accretion in the Apennines thin-skinned accretionary wedge suggests the subduction of continental crust for some 170 km in the northern Apennines (Carminati et al., 2005) and some 280 km in the southern Apennines (Scrocca et al., 2005). These values may be added to previous oceanic lithosphere subduction to obtain a total subduction of more than 200 km and 500 km in the two areas, whereas below Calabria, some 770 km of oceanic lithosphere have been subducted since 23 Ma (Gueguen et al., 1998). These estimates are consistent with the results of tomographic models showing high-velocity anomalies beneath Italy with comparable lengths (Piromallo and Morelli, 2003).

This evidence suggests that the opening of the Tyrrhenian Sea was the result of a southeastward radial rollback of the northwest-directed Apennines-Maghrebides Subduction Zone that started ca. 30 Ma (e.g., Carminati et al., 1998; Gueguen et al., 1998). This is shown in Figure 1.

Middle East and Sicily Channel

In other cases the presence of "fossil" plume heads has been proposed on the basis of the relative constancy of the $^{87}Sr/^{86}Sr$ and $^{143}Nd/^{144}Nd$ isotopic ratios of Israeli basalts erupted during the Meso-Cenozoic (e.g., Stein and Hofmann, 1992) or subma-rine volcanism in the Sicily Channel rift (Rotolo et al., 2006). In particular, Stein and Hofmann (1992) proposed the existence of a fossil plume head stagnating beneath the lithospheric mantle at least since the Proterozoic because Mesozoic/Cenozoic Israeli lavas show a relatively homogeneous Sr-Nd isotopic composition. However, this model is not able to explain the strong isotopic difference between Miocene–Pliocene Israeli lavas and coeval igneous rocks from neighboring areas (e.g., Lebanon, Syria, and Jordan; Shaw et al., 2003; Krienitz et al., 2006; Lustrino and Sharkov, 2006). The differences in composition between lavas emplaced a few tens of km away from each other can be more easily explained with shallower (lithospheric) mantle sources activated by decompression along the Dead Sea transform fault. Rotolo et al. (2006) proposed the occurrence of a fossil plume head beneath an area at least as wide as Sicily on the basis of a common HIMU-FOZO component in the volcanic rocks of Pantelleria and Linosa islands in the Sicily Channel, Mt. Etna and the Hyblean Mountains in Sicily, Ustica Island in southwestern Tyrhhenian Sea, and Alicudi, one of the Aeolian Archipelago islands.

The relatively constant HIMU-FOZO-like trace-element and Sr-Nd-Pb isotopic character of the Sicily Channel lavas falls within the chemical range of the common mantle reservoir (CMR) as defined by Lustrino and Wilson (2007). The CMR is believed to represent the "average composition" of the upper asthenospheric mantle in the entire circum-Mediterranean area, without any relationship to deep-seated mantle sources (Lustrino and Wilson, 2007).

Sardinia

Concerning the magmatism of Sardinia, the presence of a mantle plume head recycled back into the mantle has been hypothesized. Gasperini et al. (2000) proposed for the Pleistocene volcanic rocks of northern Sardinia (the Logudoro area) an origin from a mantle source represented by recycled oceanic plateaus. Because these authors related the existence of oceanic plateaus (unusually thick regions of ocean floor) to anomalous production of basaltic magma as a consequence of an anomalously hot mantle, they concluded that these lavas reflect the existence of plume heads recycled back into the mantle. The timing, location, and mechanism are unspecified. This model has been strongly criticized on geochemical, petrological, and geological grounds by Lustrino et al. (2002, 2004, 2007). In particular, the geochemical peculiarity of igneous rocks of Sardinia (e.g., relatively high SiO_2, low CaO and CaO/Al_2O_3, relatively high Ni, relatively low HFSE, low HREE, high Ba/Nb and La/Nb, slightly high $^{87}Sr/^{86}Sr$, and unradiogenic $^{143}Nd/^{144}Nd$ and $^{206}Pb/^{204}Pb$ ratios) is not explainable by anomalously hot mantle but requires the presence of heterogeneous, metasomatized shallow (lithospheric) mantle sources. In particular, this geochemical peculiarity is thought to be related to an orthopyroxene-rich lithospheric mantle source. The origin of this enrichment in orthopyroxene would be a consequence of SiO_2-rich melt derived from delaminated and detached ancient lower continental crust reacting with mantle peridotite (for more details, see Lustrino, 2005, and Lustrino et al., 2007).

Mediterranean Sea and Central–Western Europe

An extreme example of a mantle plume model is the work of Hoernle et al. (1995), who, on the basis of a geochemical review coupled with a tomographic investigation, proposed the existence beneath the entire circum-Mediterranean area of a large tabular-shaped mantle plume whose stem would be located beneath the Canary Archipelago. Concerning the geochemical database, Hoernle et al. (1995) used not only the composition of typical "anorogenic" magmas (such as those from the French Massif Central, the Rhenish Massif, the Rhine and Eger grabens, and the Pannonian basin, Sicily) but also the composition of potassic/ultrapotassic lavas from the Roman comagmatic province and the calc-alkaline to ultrapotassic lavas from Aeolian lavas. This despite the fact that those compositions (never recorded among OIB) are classically considered to be generated in subduction-related settings (e.g., Conticelli et al., 2002, 2004; Peccerillo, 2005; Peccerillo and Lustrino, 2005).

Other authors proposed models resembling that of Hoernle et al. (1995). Among these, the papers of Oyarzun et al. (1997) and Macera et al. (2003) should be mentioned. These authors proposed a process of plume channeling with the stem of the hypothetical thermal anomaly centered beneath the Canary Archipelago or the Cape Verde Islands, then moving northeastward through Portugal, central-southeastern Spain, central-southern France, and central-northern Germany up to the Bohemian Massif, and eventually ending in northeastern Italy (the Veneto volcanic province). This model is based on two considerations: (1) most of the mafic rocks from these areas show relatively uniform and common geochemical compositions, and (2) these compositions resemble HIMU-FOZO end-members. Because these end-members have commonly been associated with deep-mantle sources, the reasoning "uniform HIMU-FOZO composition = origin from a common mantle source = origin from a deep mantle plume" is at work. The hypothesis of plume channeling cannot be considered reliable for at least four reasons: (1) there is no age progression of the volcanic rocks from the southeast (Cape Verde–Canaries) to the northwest (Bohemian Massif and Veneto province), as would be expected for a northwest-moving channeled plume; (2) tomographic investigations fail to clearly image a low-velocity zone connecting the Canaries to the Bohemian Massif (e.g., Piromallo and Morelli, 2003); (3) the channeling of hot mantle material beneath the lithospheric thermal boundary layer would necessarily imply thermal erosion of the base of the lithosphere, resulting in the development of geochemical heterogeneities, in contrast with the overall homogeneous compositions of the Cenozoic lavas emplaced between the Canaries and the Bohemian Massif; and (4) the HIMU-FOZO-CMR–like chemical composition of the Cenozoic anorogenic volcanic rocks from this area (and worldwide in general) can also be explained by invoking much shallower mantle sources (e.g., Stracke et al., 2005; Lustrino and Wilson, 2007).

French Massif Central

The Paleocene–Quaternary volcanic activity recorded in the French Massif Central, partially associated with the Limagne-Forez Graben that developed in the Oligocene, has often been related to the presence of a mantle plume (e.g., Froidevaux et al., 1974; Granet et al., 1995; Sobolev et al., 1997; Wilson and Patterson, 2001). The strongest evidence supporting a mantle plume origin for the circum-Mediterranean anorogenic Cenozoic igneous province magmas is provided by the correlation, in some areas, between the incompatible trace-element content and the Sr-Nd-Pb isotope characteristics of the most primitive mafic magmas and those of HIMU-OIB. There is also evidence for contemporaneous regional basement uplift and lithospheric thinning, and seismic tomography images show fingerlike upper-mantle low-velocity anomalies beneath the volcanic fields.

On the other hand, contrasting interpretations of the evolution of the European Cenozoic rift system (Fig. 2) have emphasized the role of passive rifting in response to the buildup of Pyrenean and Alpine collision-related compressional stresses in magma generation processes (Dèzes et al., 2004). Such stresses, upon reaching a critical value, may have caused rifting in the European foreland. In this scenario, anomalous uplift of the Massif Central area may be interpreted as shoulder uplift related to the Rhône and Bresse rifts. Rifting in the Limagne-Forez Graben may have further contributed to flexural uplift of the adjacent areas. This scenario is consistent with the Neogene–Recent uplift of the Massif Central (Granet et al., 1995). Minor Cretaceous uplift may simply be related to foreland propagation of compressional stresses related to the Alpine and Pyrenean orogeny. Within this rift-related hypothesis, the magmatism may be considered the result of passive upwelling of asthenospheric material and consequent partial melting. This interpretation is consistent with the tomographic results that show that the low-velocity anomaly is limited to the upper mantle. Moreover, if the nature of the magmatism is rifting-related, this may explain the absence of the hotspot traces that should be evident, given the non-negligible relative motion of the Europe plate with respect to the underlying mantle since the Paleocene (Fekiakova et al., 2006). The oldest magmatic activity that occurred before the main formation of the Limagne-Forez Graben cannot easily be explained by this rift-related hypothesis. However, the very small volume of magma produced during this phase is also contrary to what is expected from a mantle plume.

Eifel and Neighboring Areas

The Rhenish Massif (located at the junction of the Rhine, Leine, and Ruhr Grabens, close to the Eifel area) suffered a domal uplift of the Hercynian basement of ~300 m after the early Miocene, i.e., 20–40 m.y. after the onset of Rhine Graben rifting. The uplift was associated with volcanism that shifted to the Eifel volcanic fields west of the Rhine in the Quaternary. The last eruption occurred ca. 11 Ka (Lippolt 1983).

The existence of a mantle plume beneath northern Germany

(the Rhenish Massif) has been inferred by analogy with the Cenozoic igneous activity of the French Massif Central, where tomographic studies have indicated the presence of a fingerlike low-velocity anomaly extending down to transition zone depths (410–660 km; Granet et al., 1995; Sobolev et al., 1997; Wilson and Patterson, 2001). Additional supporting evidence for an Eifel plume is provided by ~250 m of uplift during the Quaternary. Ritter et al. (2001) provided the first detailed images of the mantle structure beneath Eifel based on high-resolution teleseismic tomography. The top of the velocity anomaly is well constrained at a depth of 50–60 km (Ritter, 2007), corresponding to the base of the lithosphere. Pilidou et al. (2005) obtained high-resolution Rayleigh wave tomography images showing low-velocity anomalies beneath the Eifel volcanoes down to ~400 km. The low-velocity anomalies imaged are, however, smaller in magnitude than those observed in other potential hotspot areas (e.g., the Azores and Iceland). In their global tomography model, Montelli et al. (2004) imaged a low-velocity anomaly only down to 650 km. The upper-mantle character of the low-velocity anomaly, if interpreted as hot, may be consistent with mantle upwelling related to intraplate extensional stresses. This hypothesis is consistent with the correlation argued to exist between the timing of magmatic activity and changes in the regional stress field (Wilson and Bianchini, 1999). According to this scenario, the main volcanic phases may have been associated with periods of compressional stress relaxation in the foreland of the Alpine orogenic belt.

Ritter (2007) has attributed the magnitude of the low-velocity anomaly to a mantle excess temperature of ~100–200 °C combined with ~1% partial melt. An apparent "hole" in the shear wave velocity anomaly at ~200 km depth could have been caused by the onset of partial melting (Ritter, 2007). A shear wave–splitting analysis of the Eifel area (Walker et al., 2005) shows a first-order parabolic pattern in a fast polarization azimuth around the hotspot. This feature suggests a lattice preferred orientation of olivine fast axes in the asthenosphere. The preferred orientation was interpreted as a result of the interaction between the slow WSW absolute motion of the Eurasia plate and mantle upwelling beneath the Eifel volcanic fields. The parabolic pattern is similar to that observed beneath Hawaii. It has been concluded, however, that the anisotropy beneath the Eifel hotspot and the surrounding Rhenish Massif is mostly contained in the asthenosphere. Walker et al. (2005) suggest that the Eifel upwelling is sporadic. In our opinion, this is more likely to be the result of varying crustal intraplate stresses that periodically change and facilitate sporadic eruption than to be the result of varying a low excess upwelling temperature.

According to Goes et al. (1999), a mantle plume beneath central Europe can be traced in tomographic images down to lower mantle depths (~2000 km). In any case, the absence of an east-west age-progressive hotspot track with the oldest magmatism confined to the easternmost sectors (Heldburg and Rhön) and the youngest to the west (Eifel; Wedepohl and Baumann, 1999) is inconsistent with the kinematics of the Europe plate (which moved eastward as a consequence of the opening of the Atlantic Ocean). If we consider the anorogenic volcanic activity in central Europe as extending eastward up to the Bohemian Massif (where volcanic rocks of the Eger Graben show an age interval of ca. 77 to ca. 0.26 Ma), the absence of the hotspot track is even harder to explain in the context of a mantle plume model. Additionally, Fekiacova et al. (2007) estimate that at ca. 40 Ma the Tertiary Hocheifel volcanic field was ~1000 km southwest of its current geographic position, making a geodynamic link with the present-day seismic structure beneath the Eifel impossible.

Several authors have drawn conclusions against the hypothesis of a deep-mantle plume or small plumes or diapirs in this area. Bogaard and Wörner (2003) and Jung et al. (2005) favored passive upwelling of the asthenosphere as the main cause of magma generation. Meyer et al. (2002) proposed as a potential mantle magma source of the Rhön Massif the metasomatically overprinted subcontinental lithosphere, forced to partially melt by crustal extensional processes. Fekiacova et al. (2007) exclude any genetic relationship between the Eocene (ca. 44–35 Ma) and Quaternary volcanic activity in the Eifel area and relate the origin of the Tertiary Hocheifel volcanic field to Rhine Graben formation. Haase et al. (2004) modeled the petrogenesis of the Westerwald magmas in terms of partial melting of the thermal boundary layer at the base of the lithospheric mantle as a consequence of adiabatic decompression during lithospheric thinning.

Pannonian Basin

The Pannonian basin and the Carpathian arc show a geodynamic evolution similar to that of the Tyrrhenian basin and the Apenninic arc (Fig. 1). In both areas a retreating subduction system developed during the Cenozoic, roughly coeval with two geochemically, petrographically, and volcanologically contrasting types of igneous activity: an older event with orogenic geochemical characteristics and a younger with more or less anorogenic geochemical characteristics (e.g., Seghedi et al., 2005; Harangi et al., 2006; Harangi and Lenkey, 2007). Also, the age and shape of the Carpathian belt can be compared to those of the Apennine chain in Italy (e.g., Carminati and Doglioni, 2004; Fig. 1).

The Neogene anorogenic magmatic products of the Pannonian basin have been explained as the result of a mantle plume finger impingement beneath the lithosphere (e.g., Embey-Isztin and Dobosi, 1995; Wilson and Patterson, 2001; Seghedi et al., 2004). In particular, despite the relatively small volume of basaltic magma, Embey-Isztin and Dobosi (1995) proposed the involvement of a mantle plume to explain the origin of the Neogene volcanic rocks of the Pannonian basin, linking the extension of the lithosphere to an active rifting process (i.e., extension of the lithosphere initiated by a mantle plume). Seghedi et al. (2004) proposed that mantle partial melting was triggered by the upwelling of fingerlike upper-mantle plumes similar to those proposed beneath the Massif Central in France and the Eifel district in Germany.

Peak anorogenic volcanism took place at ca. 6 Ma in the Pannonian basin, ~5 m.y. after the end of the main extensional phases (Harangi and Lenkey, 2007, and references therein), although the first anorogenic volcanic rocks were emplaced at ca. 11 Ma, coeval with the last extensional phases (Harangi, 2001). Moreover, most of the alkaline anorogenic volcanism in the Pannonian basin is offset from the areas with major lithospheric thinning (i.e., they are localized in areas with relatively thick lithosphere). These two observations were taken as evidence against any role for lithospheric extension in the genesis of anorogenic magmatism in the Pannonian basin. Notwithstanding these two observations, Harangi and Lenkey (2007) argue that the presence of a mantle plume beneath this area is highly unlikely. According to these authors, the magmatic activity is related to the presence of a relatively high temperature regime, a consequence of the ca. 17–11 Ma lithospheric thinning. The time lapse between extension and the peak of magmatism is readily explained by the thermal inertia of lithosphere,

TABLE 1. AVERAGE COMPOSITION OF THE LESS DIFFERENTIATED SAMPLES (MgO > 7 wt%)

		SiO_2	TiO_2	Al_2O_3	Fe_2O_3	MnO	MgO	CaO	Na_2O	K_2O
Spain	Average	44.47	2.57	13.98	11.96	0.18	9.80	10.88	3.66	1.74
	St. dev.	2.61	0.38	2.03	1.07	0.02	2.27	1.71	0.68	0.56
	n	74	74	74	74	74	74	74	74	74
Maghrebian Africa	Average	41.87	3.08	12.26	12.58	0.20	11.47	12.61	3.22	1.7
	St. dev.	3.05	0.61	2.52	1.60	0.05	3.52	1.92	1.03	0.9
	n	117	117	117	117	117	117	117	117	117
France	Average	44.84	2.87	13.61	12.28	0.19	9.84	10.52	3.40	1.7
	St. dev.	2.30	0.36	1.18	0.70	0.02	1.94	1.25	0.60	0.5
	n	259	259	259	259	259	259	259	259	259
Italy (Sardinia UPV)	Average	50.20	2.06	15.60	10.07	0.14	8.14	7.77	3.73	1.8
	St. dev.	1.79	0.37	0.82	0.58	0.03	0.86	0.80	0.71	0.5
	n	109	109	109	109	109	109	109	109	109
Italy (Sardinia RPV)	1 sample	45.78	3.13	15.21	11.69	0.16	7.55	10.37	3.49	2.23
Italy (Ustica and Sicily Channel)	Average	47.01	2.19	16.00	11.24	0.17	8.97	9.32	3.42	1.11
	St. dev.	1.48	0.56	1.36	1.06	0.03	2.06	1.07	0.44	0.36
	n	49	49	49	49	49	49	49	49	49
Italy (Mt. Etna)	Average	48.25	1.51	16.26	10.90	0.16	8.33	10.12	3.19	0.9
	St. dev.	1.32	0.13	1.84	0.96	0.02	1.50	1.02	0.44	0.4
	n	20	20	20	20	20	20	20	20	20
Italy (Hyblean Mountains)	Average	47.17	2.15	15.38	11.28	0.17	8.65	10.03	3.58	0.87
	St. dev.	3.06	0.51	0.91	1.94	0.02	1.37	1.34	1.27	0.68
	n	77	77	77	77	77	77	77	77	77
Italy (Veneto province)	Average	46.03	2.54	13.85	12.17	0.16	9.94	10.16	2.93	1.34
	St. dev.	3.00	0.44	1.08	0.96	0.03	1.93	1.63	0.73	0.48
	n	128	128	128	128	128	128	128	128	128
Germany (Eifel)	Average	42.63	2.67	13.28	11.22	0.20	9.81	12.74	3.32	3.03
	St. dev.	1.68	0.23	1.19	0.52	0.01	1.72	1.47	0.56	0.77
	n	64	64	64	64	64	64	64	64	64
Germany (Rhön)	Average	43.30	2.94	12.92	12.61	0.18	10.92	12.10	3.01	1.25
	St. dev.	1.56	0.46	0.89	0.91	0.02	1.23	0.72	0.62	0.53
	n	8	8	8	8	8	8	8	8	8
Germany (Vogelsberg)	Average	45.08	2.56	12.88	12.15	0.18	11.61	10.80	2.96	1.14
	St. dev.	3.47	0.38	0.95	0.79	0.02	2.19	1.43	0.39	0.46
	n	43	43	43	43	43	43	43	43	43
Germany (Westerwald)	Average	44.64	2.63	12.92	12.36	0.19	11.30	10.86	3.05	1.39
	St. dev.	1.73	0.38	0.75	0.71	0.01	1.65	0.93	0.64	0.41
	n	55	55	55	55	55	55	55	55	55
Bohemian Massif	Average	41.86	3.24	11.77	13.19	0.20	11.12	13.62	2.98	1.19
	St. dev.	2.49	0.87	1.67	1.34	0.03	2.97	2.05	0.69	0.47
	n	254	254	254	254	254	254	254	254	254
Pannonian basin	Average	46.60	2.25	15.42	10.46	0.16	9.30	9.32	3.88	1.92
	St. dev.	1.72	0.24	1.00	0.95	0.02	1.54	0.90	0.68	0.55
	n	37	37	37	37	37	37	37	37	37
Turkey	Average	46.22	2.52	14.48	12.21	0.15	8.90	9.93	3.45	1.50
	St. dev.	2.00	0.48	1.14	1.61	0.04	1.44	0.91	0.76	0.54
	n	99	99	99	99	99	99	99	99	99
Mashrek	Average	45.70	2.47	14.57	12.83	0.17	8.84	9.68	3.68	1.26
	St. dev.	1.81	0.46	1.04	1.05	0.03	1.18	1.19	0.67	0.62
	n	235	231	231	231	231	231	231	231	231

which remains close to the solidus temperature as a consequence of local perturbation caused by mantle flow. On the basis of geophysical, geological, and petrological observations (e.g., the presence of a high-velocity anomaly at ~500 km depth, the absence of regional doming, the ongoing subsidence in many parts of the Pannonian basin, and the low volume of magma produced), Harangi and Lenkey (2007) excluded the possibility of the existence of a mantle plume beneath the Pannonian basin.

DISCUSSION AND CONCLUDING REMARKS

Notwithstanding the extremely wide range of composition of the Cenozoic anorogenic products emplaced in the circum-Mediterranean area, some common geochemical key parameters can be highlighted (Table 1 and Fig. 3). In particular, Lustrino and Wilson (2007) proposed the existence of a CMR characterized by a restricted range of major- and trace-element compositions

P_2O_5	Na_2O/K_2O	$^{87}Sr/^{86}Sr$	$\varepsilon\Sigma\rho$	$^{143}Nd/^{144}Nd$	$\varepsilon N\delta$	$^{206}Pb/^{204}Pb$	$^{207}Pb/^{204}Pb$	$^{208}Pb/^{204}Pb$	Rb
0.76	2.45	0.703971	−6.8	0.512772	2.6	19.13	15.65	38.98	48
0.32	1.33	0.000395	5.6	0.000060	1.2	0.11	0.03	0.18	42
74	74	11	11	5	5	2	2	2	69
1.0	2.50	0.703586	−12.3	0.512852	4.1	19.69	15.63	39.51	44
0.5	1.53	0.000444	6.3	0.000097	1.9	0.69	0.04	0.51	25
114	117	42	42	39	39	17	17	17	91
0.8	2.25	0.703573	−12.5	0.512887	4.9	19.43	15.62	39.21	54
0.2	0.89	0.000285	4.0	0.000054	0.9	0.27	0.02	0.21	23
259	259	96	96	77	77	29	29	29	230
0.5	2.39	0.704463	0.2	0.512515	−2.4	17.89	15.59	37.98	38
0.2	1.33	0.000116	1.6	0.000059	1.2	0.11	0.01	0.15	16
109	109	12	12	7	7	4	4	4	93
0.39	1.56	0.704010	−6.2	0.512850	4.1	19.23	15.64	39.10	49
0.58	3.49	0.703302	−16.3	0.513002	7.1	19.47	15.65	39.12	22
0.23	1.34	0.000415	5.9	0.000052	1.0	0.17	0.03	0.18	8
49	49	16	16	10	10	9	9	9	48
0.4	4.74	0.703444	−14.2	0.512875	4.6	19.83	15.64	39.46	36
0.1	2.67	0.000161	2.3	0.000049	1.0	0.17	0.01	0.14	16
20	20	98	98	38	38	8	8	8	88
0.73	7.67	0.703068	−19.6	0.513021	7.4	19.70	15.64	39.25	16
0.39	6.57	0.000194	2.8	0.000069	1.4	0.17	0.03	0.21	11
77	77	40	40	29	29	24	24	24	84
0.88	2.47	0.703341	−15.7	0.512912	5.3	19.30	15.63	39.09	30
0.32	1.11	0.000167	2.4	0.000036	0.7	0.23	0.02	0.18	12
128	128	25	25	25	25	18	18	18	117
0.65	1.23	0.704339	−1.6	0.512744	2.0	19.17	15.63	39.41	70
0.19	0.66	0.000353	5.0	0.000080	1.6	0.26	0.01	0.24	22
64	64	26	26	19	19	6	6	6	34
0.75	2.82	0.703557	−12.7	0.512825	3.6	19.40	15.60	39.15	52
0.22	1.20	0.000250	3.6	0.000021	0.4	0.09	0.02	0.09	18
8	8	45	45	36	36	6	6	6	14
0.65	3.13	0.703482	−13.7	0.512797	3.1	19.21	15.62	38.97	66
0.18	1.55	0.000269	3.8	0.000081	1.6	0.19	0.03	0.25	54
43	43	35	35	32	32	10	10	10	39
0.67	2.40	0.703590	−12.2	0.512832	3.7	19.42	15.60	39.15	45
0.25	1.00	0.000226	3.2	0.000025	0.5	0.20	0.02	0.28	20
55	55	17	17	13	13	9	9	9	18
0.82	2.92	0.703396	−15.0	0.512863	4.3	19.66	15.62	39.32	42
0.27	1.32	0.000155	2.2	0.000053	1.0	0.14	0.10	0.18	21
254	254	79	79	75	75	18	18	18	173
0.68	2.25	0.703748	−10.0	0.512809	3.3	19.03	15.62	38.91	58
0.18	0.95	0.000648	9.2	0.000101	2.0	0.24	0.02	0.16	18
37	37	19	19	18	18	15	15	15	33
0.63	2.55	0.703515	−13.3	0.512914	5.3	19.25	15.66	39.16	25
0.34	0.93	0.000384	5.4	0.000069	1.3	0.06	0.02	0.05	15
99	99	21	21	17	17	3	3	3	76
0.80	3.23	0.703299	−16.3	0.512884	4.8	19.10	15.61	38.87	17
0.48	0.91	0.000230	3.3	0.000048	0.9	0.14	0.03	0.16	9
231	231	56	56	56	56	38	38	38	146

(continued)

TABLE 1. *Continued*

		Sr	Ba	Sc	V	Cr	Co	Ni	Y	Zr	Nb	La	Ce
Spain	Average	937	801	22	233	298	58	185	28	222	78	61.72	118.44
	St. dev.	295	194	3	29	158	35	95	5	75	22	23.48	40.00
	n	69	69	4	54	67	64	69	68	68	40	69	70
Maghrebian Africa	Average	1333	984	24	253	373	52	191	29	276	109	65.59	128.68
	St. dev.	459	343	6	44	220	14	100	6	83	35	26.47	48.26
	n	94	90	23	82	87	30	90	82	84	83	39	36
France	Average	892	666	20	208	269	48	192	28	287	93	59.79	115.19
	St. dev.	240	205	5	55	140	5	92	4	67	23	19.45	36.37
	n	240	235	186	180	221	187	223	205	212	212	135	135
Italy (Sardinia UPV)	Average	794	959	17	181	297	46	210	21	203	42	40.52	78.79
	St. dev.	184	366	3	21	77	6	54	3	52	14	13.17	27.97
	n	93	81	34	80	81	61	81	90	80	92	81	81
Italy (Sardinia RPV)	1 sample	970	528	20	254	212	44	126	28	223	70	47.10	96.40
Italy (Ustica and Sicily Channel)	Average	609	319	27	216	300	47	182	25	200	45	35.24	63.24
	St. dev.	195	125	3	37	90	14	83	4	59	16	13.01	21.67
	n	48	33	26	30	33	32	30	44	43	33	36	33
Italy (Mt. Etna)	Average	1039	574	27	250	97	41	55	22	195	36	52.11	101.13
	St. dev.	285	171	3	60	116	6	68	1	43	15	13.62	32.07
	n	88	65	61	7	62	62	63	6	63	5	65	83
Italy (Hyblean Mountains)	Average	757	423	29	214	309	47	218	30	198	58	47.85	96.19
	St. dev.	434	249	5	38	56	3	28	5	83	35	29.79	60.63
	n	84	83	15	73	73	73	73	73	73	73	73	78
Italy (Veneto province)	Average	829	486	—	217	290	46	190	25	238	54	43.01	89.36
	St. dev.	245	200	—	35	118	6	61	4	88	21	14.59	30.85
	n	117	114	—	40	116	40	116	117	116	116	114	115
Germany (Eifel)	Average	991	1102	28	319	369	48	189	28	239	100	83.10	153.33
	St. dev.	193	197	5	14	153	14	86	3	30	22	33.90	37.25
	n	34	14	2	2	14	14	13	12	13	12	2	3
Germany (Rhön)	Average	893	733	31	298	365	56	211	27	263	84	60.03	118.63
	St. dev.	121	119	4	60	169	3	102	5	62	17	15.56	29.44
	n	14	8	4	8	8	4	8	8	8	8	8	8
Germany (Vogelsberg)	Average	841	683	24	234	425	52	271	24	207	63	47.17	96.87
	St. dev.	224	185	4	38	114	5	78	4	47	18	14.71	29.24
	n	39	36	36	36	36	36	36	36	36	36	35	35
Germany (Westerwald)	Average	789	599	23	241	320	50	209	24	260	79	50.52	101.58
	St. dev.	167	169	3	36	126	6	77	3	58	24	15.24	28.32
	n	18	14	14	14	14	14	14	14	14	14	14	14
Bohemian Massif	Average	998	799	26	245	328	51	205	26	267	102	77.25	141.63
	St. dev.	328	260	7	78	234	9	125	6	106	35	30.12	50.95
	n	173	168	124	108	152	120	158	148	151	148	87	89
Pannonian basin	Average	927	848	21	209	242	41	180	23	237	75	57.40	104.17
	St. dev.	256	268	5	21	87	6	75	4	82	26	22.61	35.72
	n	31	34	14	30	34	20	29	33	34	30	34	30
Turkey	Average	826	426	23	208	288	53	163	24	209	60	33.22	67.67
	St. dev.	245	220	3	24	150	13	72	4	53	25	9.61	17.62
	n	87	85	56	83	85	55	84	84	85	85	50	50
Mashrek	Average	953	427	20	192	253	52	196	27	199	47	41.19	82.72
	St. dev.	435	272	3	21	74	7	64	27	66	26	27.20	44.59
	n	155	160	89	130	169	95	170	157	151	121	133	117

Sources: References in Lustrino and Wilson (2007).
Notes: n—number of samples; UPV—unradiogenic Pb volcanics; RPV—radiogenic Pb volcanics.

as well as Sr-Nd-Pb isotopic ratios. However, the existence of Cenozoic anorogenic volcanic rocks with relatively common geochemical features (resembling CMR) cannot be directly interpreted as requiring a physically continuous mantle source region. Rather, the presence of homogeneous geochemical compositions suggests the existence of common petrogenetic processes. In-

deed, a continuous (upper) mantle source able to feed all Cenozoic volcanic activity, from the Canaries to the west to Turkey to the east and from Libya–Maghrebian Africa to the south to the Bohemian Massif and Germany to the north, is hardly imaginable.

If one postulates the existence of different regions in the upper mantle able to produce melts with relatively homogeneous

Pr	Nd	Sm	Eu	Gd	Tb	Dy	Ho	Er	Tm	Yb	Lu	Th	Pb	U
15.80	53.13	12.11	3.24	10.39	1.04	6.96	1.13	3.36	0.33	2.10	0.28	6.23	6.00	1.45
—	23.49	3.85	1.04	2.27	0.72	1.42	—	0.63	—	0.47	0.03	1.98	—	0.65
1	43	23	37	20	16	20	1	2	1	36	22	23	1	4
14.90	54.83	9.19	3.44	7.28	1.10	4.93	0.90	2.29	0.28	1.70	0.27	7.06	2.79	2.15
5.15	19.06	3.14	3.70	2.49	0.30	1.07	0.18	0.41	0.05	0.37	0.21	2.91	1.04	0.86
10	39	31	38	14	21	27	10	26	10	38	22	28	6	12
13.79	50.85	9.48	2.95	8.58	1.13	6.01	1.10	2.72	0.39	2.16	0.32	10.58	4.11	2.14
3.43	13.98	2.09	0.58	1.70	0.19	0.90	0.16	0.40	0.06	0.32	0.05	3.48	1.04	0.55
80	130	136	136	99	82	109	81	98	56	136	118	221	71	33
9.10	34.25	6.68	2.11	6.11	0.79	4.13	0.76	1.80	0.22	1.34	0.18	5.11	4.60	0.99
3.36	8.27	1.16	0.31	1.67	0.15	0.54	0.09	0.23	0.04	0.17	0.03	1.88	1.13	0.53
9	75	10	11	10	7	10	9	10	4	10	9	11	11	5
12.10	46.60	8.27	2.53	7.02	1.01	5.66	0.88	2.26	0.35	1.97	0.30	5.90	4.00	1.30
9.23	32.68	6.33	2.20	7.13	0.93	5.20	0.90	2.33	0.51	1.98	0.32	4.50	1.00	1.24
4.60	13.81	2.68	0.82	2.63	0.36	1.58	0.24	0.54	0.22	0.36	0.07	1.89	0.00	0.49
4	8	9	9	4	5	4	4	4	8	9	9	11	3	5
12.70	40.23	8.48	2.66	6.60	0.88	4.80	0.83	2.11	0.31	2.15	0.26	7.32	—	2.15
—	16.20	1.01	0.48	—	0.10	—	—	—	—	0.23	0.05	2.13	—	0.60
1	26	52	62	1	62	1	1	1	1	51	6	61	—	61
12.32	46.19	8.81	3.06	8.07	2.13	6.11	1.51	2.81	0.82	2.26	0.35	5.27	3.21	1.29
7.87	27.73	3.42	1.25	2.07	1.94	1.55	0.68	0.58	0.75	0.52	0.05	3.20	2.03	0.82
20	30	43	43	18	37	26	27	26	27	36	32	73	22	30
10.59	37.57	8.39	2.73	7.91	1.04	5.49	0.94	2.31	0.31	1.72	0.26	5.61	3.11	1.37
3.26	13.84	2.26	0.73	2.29	0.22	0.98	0.13	0.38	0.05	0.23	0.05	1.64	0.57	0.48
42	108	54	64	45	33	54	42	55	33	54	54	49	5	40
—	64.36	9.84	2.72	7.49	1.00	4.71	0.79	2.05	—	1.81	0.35	8.95	3.00	2.36
—	10.57	2.72	0.26	—	0.10	—	—	—	—	0.14	0.07	3.25	—	0.85
—	17	17	3	1	2	1	1	1	—	3	2	2	1	2
—	52.68	10.53	3.08	8.03	—	5.64	—	2.19	—	1.76	0.20	—	3.88	—
—	13.78	2.82	0.87	1.11	—	0.64	—	0.58	—	0.21	0.01	—	1.05	—
—	8	8	8	8	—	8	—	8	—	8	4	—	8	—
10.19	40.56	8.68	2.52	6.47	0.95	4.83	0.86	2.22	0.29	1.71	0.24	5.56	2.89	1.33
3.52	10.33	1.51	0.33	0.92	0.12	0.44	0.07	0.26	0.03	0.28	0.05	1.66	0.59	0.42
23	35	33	35	35	23	35	23	35	23	35	35	22	22	22
12.71	47.57	9.02	2.72	7.78	1.04	5.40	0.95	2.33	0.30	1.80	0.25	5.91	3.41	1.75
3.24	11.09	1.65	0.45	1.20	0.14	0.65	0.11	0.27	0.03	0.19	0.03	1.90	0.84	0.54
13	14	14	14	13	14	13	13	13	13	14	14	14	14	14
14.36	61.91	10.77	3.17	9.16	1.34	6.43	1.00	2.39	0.83	2.14	0.61	10.50	3.08	1.17
3.49	19.19	2.99	0.85	2.07	0.60	1.62	0.23	0.36	0.33	0.85	2.78	3.83	2.13	1.08
39	94	95	78	71	44	52	46	45	4	89	78	125	63	115
9.70	42.31	7.76	2.39	6.01	0.83	4.63	0.86	2.18	—	2.14	0.35	8.31	3.68	1.07
—	14.69	2.59	0.60	0.03	0.21	0.28	—	0.10	—	0.40	0.10	3.30	1.86	0.49
1	19	17	17	2	9	2	1	2	—	16	12	21	8	4
8.92	34.09	7.30	2.34	6.70	1.00	4.85	0.91	2.08	0.34	1.62	0.23	4.63	2.22	1.38
2.13	8.16	1.41	0.45	1.03	0.56	0.87	0.21	0.36	0.28	0.31	0.03	1.37	1.51	0.55
30	51	51	50	47	45	47	45	47	29	50	49	48	46	45
9.71	35.91	8.21	2.64	6.67	0.90	4.37	0.78	1.84	0.24	1.46	0.20	3.22	2.01	0.97
3.54	16.98	2.47	0.73	1.87	0.25	0.68	0.15	0.23	0.04	0.21	0.03	1.79	1.25	0.49
32	87	49	89	44	57	37	35	36	31	49	45	101	72	58

geochemical characteristics (Table 1 and Fig. 3), the ultimate origin of magmatism should be addressed. Is it related to the presence of anomalously hot mantle rising from deep (e.g., the upper-lower mantle boundary) or very deep (e.g., lower mantle-core boundary) sources? Or is it related to shallow-mantle processes mostly linked with lithospheric extension causing asthenospheric passive upwelling and consequential adiabatic partial melting? We favor the second hypothesis, where magmatism is unrelated to any heat excess of the mantle sources but is rather controlled by plate kinematics (e.g., Anderson, 2001). The volcanism in central Europe (the French Massif Central, Germany, and the Bohemian Massif) is concentrated in Cenozoic

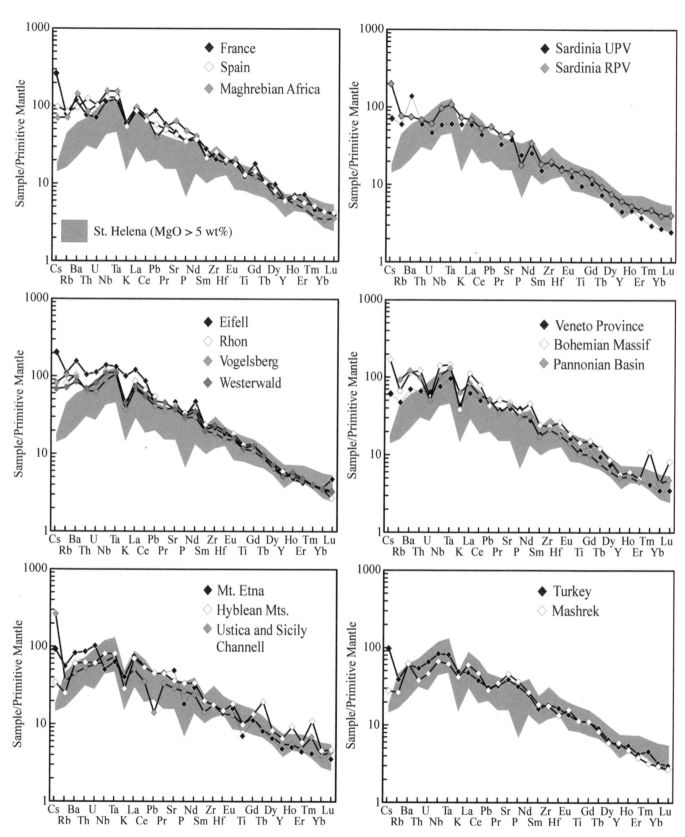

Figure 3. Primitive mantle-normalized diagrams (normalizing factors after Sun and McDonough, 1989) of several primitive (MgO > 8 wt%) anorogenic volcanic districts of the circum-Mediterranean area. References in Lustrino and Wilson (2007). The field of the least differentiated (MgO > 5 wt%) volcanic rocks from St. Helena Island (considered to represent typical high μ, i.e., high U/Pb mantle–ocean island basalts) has been plotted for comparison (references in Lustrino and Sharkov, 2006). UPV—unradiogenic Pb volcanics; RPV—radiogenic Pb volcanics.

rift areas and can be easily interpreted (as we so interpret it) as the result of passive asthenosphere upwelling. Similarly, the anorogenic magmatism in Sardinia, the Tyrrhenian Sea, and the Pannonian basin may be explained as the result of lithospheric stretching in a back-arc geodynamic setting.

Plume advocates may argue that most of the European Cenozoic rift system is amagmatic (i.e., lacking significant volcanic activity; Fig. 2). Consequently, a magmatogenetic process related only to lithospheric thinning and passive upwelling of asthenospheric mantle is unreliable. Nevertheless, we can explain this vast amagmatic zone in the central portion of the European Cenozoic rift system using a plume-free model. According to McKenzie and Bickle (1988) and White and McKenzie (1989), huge volumes of magmas derived from partial melting of upper mantle without excess heat (e.g., with $T_p \sim 1280\ °C$) can develop only under thin lithosphere (~70 km thick) in association with large stretching factors (i.e., $\beta >5$). Unfortunately, little is known about the origin of igneous activity characterized by much lower volumes of magmas. The European Cenozoic rift system is characterized by $\beta < 5$ (values >5 are associated with continental rifts evolving to pure oceanic crust), and in this case the melt thickness generated by adiabatic decompression of asthenospheric mantle with normal potential temperature ($T_p \sim 1280\ °C$) is very limited (White and McKenzie, 1989). Anyway, the volumes of igneous rocks in the volcanic districts considered here are at least two to four orders of magnitude smaller than the volumes of magmas emplaced in continental (e.g., Paraná-Etendeka, Deccan, Siberian Traps) and oceanic large igneous provinces (e.g., Kerguelen, Ontong Java). This means that lithospheric extension alone cannot explain the extreme melt productivity of mantle sources corresponding to large igneous provinces, but it can be considered a potential mechanism for feeding the much less voluminous Cenozoic anorogenic volcanic activity of the circum-Mediterranean area. It is also worth noting that most of the products of this area are sodic alkaline lavas (Lustrino and Wilson, 2007), considered to have arisen from low degrees of partial melting of mantle sources (generally <5%), opposite to the bulk of the products of large igneous provinces (tholeiitic basalts and basaltic andesites), considered to be the products of larger degrees of melting (generally ~10% or more; Jaques and Green, 1980; Falloon et al., 1988, 1999; Schwab and Johnston, 2001).

For this reason, we suggest that the key factor determining partial melting in rift zones and its upwelling to the Earth's surface is lithospheric thickness. From Figure 4 it is clear that igneous activity in the European Cenozoic rift system is concentrated where lithospheric thickness is less (around 60–80 km) and is virtually absent where it reaches maximum values (up to 140 km). This is evidence for an effective role of lithospheric stretching in promoting adiabatic decompression of asthenospheric mantle where the lithosphere is relatively thin. It could be argued that the shallow lithospheric-asthenospheric boundary below the Massif Central and Eifel could be due to thermal thinning induced by a rising plume rather than to stretching. However, the correspondence between minimum lithospheric

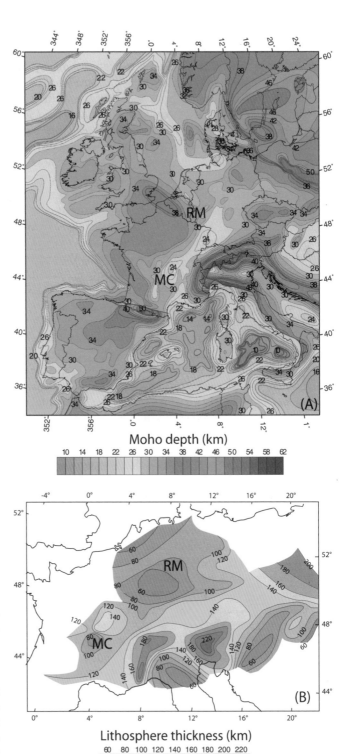

Figure 4. (A) Moho depth below central western Europe and the central Mediterranean (modified from Dèzes and Ziegler, 2002). (B) Lithospheric thickness below central western Europe (modified from Babuska and Plomerová, 1992). MC—Massif Central; RM—Rhenish Massif.

and crustal thicknesses (Fig. 4) in these areas excludes such an interpretation. The fact that relatively large lithospheric thicknesses occur below the upper Rhine Graben in central Germany and small thicknesses below the Rhenish Massif and the Massif Central is, in our opinion, related to the interaction between rifting and post-Variscan lithospheric geometry. Central Germany coincided with internal sectors of the Variscan orogeny (e.g., the Saxo-Thuringian zone) and was likely characterized by post-Variscan lithospheric thicknesses significantly greater than at the northern and southern boundaries of the chain. The superposition of a similar amount of Cenozoic stretching in internal and external sectors of the Variscan chain may explain the resulting differences in the lithospheric thickness. Accordingly, igneous activity in central Europe is mostly confined to the northern (Rhenish Massif) and the southern (French Massif Central) borders of the Variscan belt.

We conclude that in strongly rifted areas (e.g., the Massif Central, Eifel, and the Eger Graben in the Bohemian Massif), lithospheric stretching can induce passive partial melting of asthenospheric mantle. According to this process, the sublithospheric mantle is sucked upward as a consequence of rift development. Following the conclusions of Wilson and Patterson (2001) and Lustrino and Wilson (2007), in order to avoid confusion we suggest calling such passive upwellings not "mantle plumes" but "upper-mantle diapiric instabilities." According to McKenzie and Bickle (1988) and White and McKenzie (1989), large volumes of lava can be produced in the presence of limited excess heat related to thermal plumes. We propose that the large volume of magmas produced in the French Massif Central and in the Eifel-Rhön-Vogelsberg areas (though still much smaller than the volume produced in large igneous provinces) is instead related to low-temperature melting-point lithologies in the upper mantle and not to anomalously hot mantle sources. Indeed, the melt productivity of a mantle reservoir can be enhanced not only by 100–200 °C hotter sources, but also by crustal lithologies with low solidus temperatures (generally subducted and/or delaminated crustal slices that have solidus temperatures up to 300 °C lower than peridotitic assemblages at similar pressures; e.g., Hirschmann, 2000).

Petrological and tomographic methods and interpretations normally used to suggest the existence of mantle plumes are not conclusive. Moreover, frequently local or regional geodynamic and tectonic evolution can explain anorogenic magmatism without the involvement of deep-seated plumes. Our final conclusion is that a detailed knowledge of the regional geology, petrology, geophysics, and geodynamics and a fully interdisciplinary approach are necessary to propose large-scale processes on the basis of scant and unconstrained assumptions.

ACKNOWLEDGMENTS

The authors wish to express their warmest thanks to (in order of seniority) Enrica Mascia, Raffaella Bigoni, Bianca Lustrino, Alessio Carminati, Laura Lustrino, and Silvia Carminati for their patience during the writing of this manuscript. M.L. thanks Marge Wilson, Uli Achauer, and the PLUME (Plume-Like Upper Mantle instabilities beneath Europe) research group for vibrant discussions on the origin of Cenozoic anorogenic magmatism in the circum-Mediterranean region. M.L. also thanks Pino Daniele, Rino Zurzolo, Tullio De Piscopo, Joe Amoruso, Jamese Senese, Tony Esposito, and Fabio Forte. Official reviews by Romain Meyer and two anonymous reviewers plus the editorial handling of Gillian Foulger have been greatly appreciated.

REFERENCES CITED

Amato, A., and Selvaggi, G., 1991, Terremoti crostali e subcrostali nell'Appennino settentrionale: Studi Geologici Camerti, v. 1991, no. 1 (special issue), p. 75–82.

Anderson, D.L., 1998, The helium paradoxes: Proceedings of the National Academy of Sciences of the United States of America, v. 95, p. 4822–4827, doi: 10.1073/pnas.95.9.4822.

Anderson, D.L., 1999, Are color cross sections really Rorschach tests?: Eos (Transactions, American Geophysical Union), v. 80, p. F719.

Anderson, D.L., 2001, Topside tectonics: Science, v. 293, p. 2016–2018, doi: 10.1126/science.1065448.

Anderson, D.L., 2005, Scoring hot spots: The plume and plate paradigms, *in* Foulger, G.R., et al., eds., Plates, plumes, and paradigms: Boulder, Colorado, Geological Society of America Special Paper 388, p. 31–54.

Anderson, D.L., 2006, Speculations on the nature and cause of mantle heterogeneity: Tectonophysics, v. 416, p. 7–22, doi: 10.1016/j.tecto.2005.07.011.

Anguita, F., and Hernàn, F., 2000, The Canary Islands origin: A unifying model: Journal of Volcanology and Geothermal Research, v. 103, p. 1–26, doi: 10.1016/S0377-0273(00)00195-5.

Arai, S., and Abe, N., 1995, Reaction of orthopyroxene in peridotite xenoliths with alkali-basalt melt and its implication for genesis of Alpine-type chromitite: American Mineralogist, v. 80, p. 1041–1047.

Babuska, V., and Plomerová, J., 1992, The lithosphere in central Europe—Seismological and petrological aspects: Tectonophysics, v. 207, p. 141–163, doi: 10.1016/0040-1951(92)90475-L.

Beccaluva, L., Bianchini, G., and Wilson, M., eds., 2007, Cenozoic volcanism in the Mediterranean area: Boulder, Colorado, Geological Society of America Special Paper 418, 358 p.

Bell, K., Castorina, F., Lavecchia, G., Rosatelli, G., and Stoppa, F., 2004, Is there a mantle plume below Italy? Eos (Transactions, American Geophysical Union), v. 85, p. 541–547.

Bernard-Griffiths, J., Fourcade, S., and Dupuy, C., 1991, Isotopic study (Sr, Nd, O and C) of lamprophyres and associated dykes from Tamazert (Morocco): Crustal contamination processes and source characteristics: Earth and Planetary Science Letters, v. 103, p. 190–199, doi: 10.1016/0012-821X(91)90160-J.

Bijwaard, H., and Spakman, W., 1999, Tomographic evidence for a narrow whole mantle plume below Iceland: Earth and Planetary Science Letters, v. 166, p. 121–126, doi: 10.1016/S0012-821X(99)00004-7.

Bijwaard, H., Spakman, W., and Engdahl, E.R., 1998, Closing the gap between regional and global travel time tomography: Journal of Geophysical Research, v. 103, p. 30,055–30,078, doi: 10.1029/98JB02467.

Boggard, P.J.F., and Worner, G., 2003, Petrogenesis of basanitic to tholeiitic volcanic rocks from the Miocene Vogelsberg, Central Germany: Journal of Petrology, v. 44, p. 569–602.

Carminati, E., and Doglioni, C., 2004, Mediterranean geodynamics, *in* Encyclopedia of Geology: Amsterdam, Elsevier, p. 135–146.

Carminati, E., Wortel, M.J.R., Spakman, W., and Sabadini, R., 1998, The role of slab detachment processes in the opening of the western-central Mediterranean basins: Some geological and geophysical evidence: Earth and Planetary Science Letters, v. 160, p. 651–665, doi: 10.1016/S0012-821X(98)00118-6.

Carminati, E., Giunchi, C., Argnani, A., Saladini, R., and Fernandez, M., 1999, Dynamic modelling of the Northern Apennines: Implications for Plio-

Quaternary vertical motions: Tectonics, v. 18, p. 703–718, doi: 10.1029/1999TC900015.

Carminati, E., Giardina, F., and Doglioni, C., 2002, Rheological control of subcrustal seismicity in the Apennines subduction (Italy): Geophysical Research Letters, v. 2002, doi: 10.1029/2001GL014084.

Carminati, E., Negredo, A.M., Valera, J.L., and Doglioni, C., 2005, Subduction-related intermediate-depth and deep seismicity in Italy: Insights from thermal and rheological modelling: Physics of the Earth and Planetary Interiors, v. 149, p. 65–79, doi: 10.1016/j.pepi.2004.04.006.

Cella, F., de Lorenzo, S., Fedi, M., Loddo, M., Mongelli, F., Rapolla, A., and Zito, G., 2006, Temperature and density of the Tyrrhenian lithosphere and slab and new interpretation of gravity field in the Tyrrhenian Basin: Tectonophysics, v. 412, p. 27–47, doi: 10.1016/j.tecto.2005.08.025.

Conticelli, S., and Melluso, L., eds., 2004, A showcase of the Italian research in petrology: Magmatism in Italy: Periodico di Mineralogia, v. 73 (special issue), 251 p.

Conticelli, S., D'Antonio, M., Pinarelli, L., and Civetta, L., 2002, Source contamination and mantle heterogeneity in the genesis of Italian potassic and ultrapotassic volcanic rocks: Sr-Nd-Pb isotope data from Roman province and southern Tuscany: Mineralogy and Petrology, v. 74, p. 189–222, doi: 10.1007/s007100200004.

Conticelli, S., Melluso, L., Perini, G., Avanzinelli, R., and Boari, E., 2004, Petrologic, geochemical and isotopic characteristics of potassic and ultrapotassic magmatism in central-southern Italy: Inferences on its genesis and on the nature of mantle sources, *in* Conticelli, S., and Melluso, L., eds., A showcase of the Italian research in petrology: Magmatism in Italy: Periodico di Mineralogia, v. 73 (special issue), p. 135–164.

Cordery, M.J., Davies, G.F., and Campbell, I.H., 1997, Genesis of flood basalts from eclogite-bearing mantle plumes: Journal of Geophysical Research, v. 102, p. 20,179–20,197, doi: 10.1029/97JB00648.

Courtillot, V., Davaille, A., Besse, J., and Stock, J., 2003, Three distinct types of hot spots in the Earth's mantle: Earth and Planetary Science Letters, v. 205, p. 295–308, doi: 10.1016/S0012-821X(02)01048-8.

Davaille, A., and Vatteville, J., 2005, On the transient nature of mantle plumes: Geophysical Research Letters, v. 32, p. L14309, doi: 10.1029/2005GL023029.

Dèzes, P., and Ziegler, P.A., 2002, Moho depth map of Western and Central Europe: http://www.unibas.ch/eucor-urgent.

Dèzes, P., Schmid, S.M., and Ziegler, P.A., 2004, Evolution of the European Cenozoic Rift System: Interaction of the Pyrenean and Alpine orogens with the foreland lithosphere: Tectonophysics, v. 389, p. 1–33, doi: 10.1016/j.tecto.2004.06.011.

Doglioni, C., 1991, Una interpretazione della tettonica globale: Le Scienze, v. 270, p. 32–42.

Embey-Isztin, A., and Dobosi, G., 1995, Mantle source characteristics for Miocene–Pleistocene alkali basalts, Carpathian-Pannonian Region: A review of trace elements and isotopic composition: Acta Vulcanologica, v. 7, p. 155–166.

Falloon, T.J., Green, D.H., Hatton, C.J., and Harris, K.L., 1988, Anhydrous partial melting of a fertile and depleted peridotite from 2 to 30 kb and application to basalt: Journal of Petrology, v. 29, p. 1257–1282.

Falloon, T.J., Green, D.H., Danyushevsky, L.V., and Faul, U.H., 1999, Peridotite melting at 1.0 and 1.5 GPa: An experimental evaluation of techniques using diamond aggregates and mineral mixes for determination of near-solidus melts: Journal of Petrology, v. 40, p. 1343–1375, doi: 10.1093/petrology/40.9.1343.

Fekiacova, Z., Mertz, D.F., and Hofmann, A.W., 2007, Geodynamic setting of the Tertiary Hocheifel volcanism (Germany), Part 2, Geochemistry and Sr, Nd and Pb isotopic compositions, *in* Ritter, J.R.R., and Christensen, U.R., eds., Mantle plumes—An interdisciplinary approach: Heidelberg, Springer-Verlag, p. 207–239.

Foulger, G.R., and Natland, J.H., 2003, Is "Hot spot" volcanism a consequence of plate tectonics?: Science, v. 300, p. 921–922, doi: 10.1126/science.1083376.

Foulger, G.R., Pritchard, M.J., Julian, B.R., Evans, J.R., Allen, R.M., Nolet, G.,

Morgan, W.J., Bergsson, B.H., Erlendsson, P., Jakobsdottir, S., Ragnarsson, S., Stefansson, R., and Vogfjord, K., 2000, The seismic anomaly beneath Iceland extends down to the mantle transition zone and no deeper: Geophysical Journal International, v. 142, p. F1–F5, doi: 10.1046/j.1365-246x.2000.00245.x.

Froidevaux, C., Brousse, R., and Bellon, H., 1974, Hot spot in France?: Nature, v. 248, p. 749–751, doi: 10.1038/248749a0.

Garfunkel, Z., 1989, Tectonic setting of Phanerozoic magmatism in Israel: Israel Journal of Earth Sciences, v. 38, p. 51–74.

Gasperini, D., Blichert-Toft, J., Bosch, D., del Moro, A., Macera, P., Telouk, P., and Albarede, F., 2000, Evidence from Sardinian basalt geochemistry for recycling of plume heads into the Earth's mantle: Nature, v. 408, p. 701–704, doi: 10.1038/35047049.

Gautheron, C., Moreira, M., and Allègre, C., 2005, He, Ne and Ar composition of the European lithospheric mantle: Chemical Geology, v. 217, p. 97–112, doi: 10.1016/j.chemgeo.2004.12.009.

Giardini, D., and Velonà, M., 1991, The deep seismicity of the Tyrrhenian Sea: Terra Nova, v. 3, p. 57–64.

Goes, S., Spakman, W., and Bijwaard, H., 1999, A lower mantle source for Central European volcanism: Science, v. 286, p. 1928–1931, doi: 10.1126/science.286.5446.1928.

Grand, S.P., 2002, Mantle shear-wave tomography and the fate of subducted slabs: Philosophical Transactions of the Royal Society of London, v. A 360, p. 2475–2491.

Grand, S.P., van der Hilst, R.D., and Widiyantoro, S., 1997, Global seismic tomography: A snapshot of convection in the Earth: GSA Today, v. 7, p. 1–7.

Granet, M., Wilson, M., and Achauer, U., 1995, Imaging a mantle plume beneath the French Massif Central: Earth and Planetary Science Letters, v. 136, p. 281–296, doi: 10.1016/0012-821X(95)00174-B.

Griffiths, R.W., and Campbell, I.H., 1990, Stirring and structure in mantle starting plumes: Earth and Planetary Science Letters, v. 99, p. 66–78, doi: 10.1016/0012-821X(90)90071-5.

Gueguen, E., Doglioni, C., and Fernandez, M., 1998, On the post 25 Ma geodynamic evolution of the western Mediterranean: Tectonophysics, v. 298, p. 259–269, doi: 10.1016/S0040-1951(98)00189-9.

Haase, K.M., Goldschmidt, B., and Garbe-Schonberg, C.-D., 2004, Petrogenesis of Tertiary continental intra-plate lavas from the Westerwald region, Germany: Journal of Petrology, v. 45, p. 883–905, doi: 10.1093/petrology/egg115.

Harangi, S., 2001, Neogene magmatism in the Alpine-Pannonian Transition Zone—A model for melt generation in a complex geodynamic setting: Acta Vulcanologica, v. 13, p. 25–39.

Harangi, S., and Lenkey, L., 2007, Genesis of the Neogene to Quaternary volcanism in the Carpathian-Pannonian region: Role of subduction, extension and mantle plume, *in* Beccaluva, L., et al., eds., Cenozoic volcanism in the Mediterranean area: Boulder, Colorado, Geological Society of America Special Paper 418, p. 67–92.

Harangi, S., Downes, H., and Seghedi, I., 2006, Tertiary–Quaternary subduction processes and related magmatism in the Alpine-Mediterranean region, *in* Gee, D., and Stephenson, R., eds., European lithosphere dynamics: Geological Society of London Memoir 32, p. 167–190.

Hirschmann, M.M., 2000, Mantle solidus: Experimental constraints and the effects of peridotite composition: Geochemistry, Geophysics, Geosystems, http://146.201.254.53/publicationsfinal/articles/2000gc000070/fs2000gc000070.html.

Hirschmann, M.M., and Stolper, E.M., 1996, A possible role for garnet pyroxenite in the origin of the "garnet signature" in MORB: Contributions to Mineralogy and Petrology, v. 124, p. 185–208, doi: 10.1007/s004100050184.

Hirschmann, M.M., Kogiso, T., Baker, M.B., and Stolper, E.M., 2003, Alkalic magmas generated by partial melting of garnet pyroxenite: Geology, v. 31, p. 481–484, doi: 10.1130/0091-7613(2003)031<0481:AMGBPM>2.0.CO;2.

Hoernle, K., Zhang, Y.S., and Graham, D., 1995, Seismic and geochemical evidence for large-scale mantle upwelling beneath the eastern Atlantic and western and central Europe: Nature, v. 374, p. 34–39, doi: 10.1038/374034a0.

Hofmann, A., and White, W., 1982, Mantle plumes from ancient oceanic crust: Earth and Planetary Science Letters, v. 57, p. 421–436, doi: 10.1016/0012-821X(82)90161-3.

Huang, S., and Frey, F.A., 2005, Recycled oceanic crust in the Hawaiian Plume: Evidence from temporal geochemical variations within the Koolau Shield: Contributions to Mineralogy and Petrology, v. 149, p. 556–575, doi: 10.1007/s00410-005-0664-9.

Hutko, A.R., Lay, T., Garnero, E.J., and Revenaugh, J., 2006, Seismic detection of folded, subducted lithosphere at the core-mantle boundary: Nature, v. 441, doi: 10.1038/nature04757.

Jacques, A.L., and Green, D.H., 1980, Anhydrous melting of peridotite at 0–15 Kb pressure and the genesis of tholeiitic basalts: Contributions to Mineralogy and Petrology, v. 73, p. 287–310, doi: 10.1007/BF00381447.

Jung, C., Jung, S., Hoffer, E., and Berndt, J., 2006, Petrogenesis of Tertiary mafic alkaline magmas in the Hocheifel, Germany: Journal of Petrology, v. 47, p. 1637–1671.

Jung, S., Pfander, J.A., Brugmann, G., and Stracke, A., 2005, Sources of primitive alkaline volcanic rocks from the Central European Volcanic Province (Rhön, Germany) inferred from Hf, Os and Pb isotopes: Contributions to Mineralogy and Petrology, doi 10.1007s00410-005-0029-4.

Karato, S.I., 1997, On the separation of crustal component from subducted oceanic lithosphere near the 660 km discontinuity: Physics of the Earth and Planetary Interiors, v. 99, p. 103–111, doi: 10.1016/S0031-9201(96)03198-6.

Keshav, S., Gudfinnsson, G.H., Sen, G., and Fei, Y., 2004, High-pressure melting experiments on garnet clinopyroxenite and the alkalic to tholeiitic transition in ocean-island basalts: Earth and Planetary Science Letters, v. 223, p. 365–379, doi: 10.1016/j.epsl.2004.04.029.

Kogiso, T., Hirschmann, M.M., and Frost, D.J., 2003, High-pressure partial melting of garnet pyroxenite: Possible mafic lithologies in the source of ocean island basalts: Earth and Planetary Science Letters, v. 216, p. 603–617, doi: 10.1016/S0012-821X(03)00538-7.

Kogiso, T., Hirschmann, M.M., and Petermann, M., 2004, High-pressure partial melting of mafic lithologies in the mantle: Journal of Petrology, v. 45, p. 2407–2422, doi: 10.1093/petrology/egh057.

Krienitz, M.S., Haase, K.M., Mezger, K., Eckardt, V., and Shaikh-Mashail, M.A., 2006, Magma genesis and crustal contamination of continental intraplate lavas in northwestern Syria: Contributions to Mineralogy and Petrology, v. 151, p. 698–716, doi: 10.1007/s00410-006-0088-1.

Lippolt, H.J., 1983, Distribution of volcanic activity in space and time, *in* Fuchs, K., et al., eds., Plateau uplift: The Rhenish Shield—A case history: Berlin, Springer, p. 112–120.

Locardi, E., and Nicolich, R., 2005, Crust-mantle structures and Neogene–Quaternary magmatism in Italy: Bollettino di Geofisica Teorica ed Applicata, v. 46, p. 169–180.

Lustrino, M., 2000, Volcanic activity during the Neogene to Present evolution of the western Mediterranean area: A review: Ofioliti, v. 25, p. 87–101.

Lustrino, M., 2005, How the delamination and detachment of lower crust can influence basaltic magmatism: Earth-Science Review, v. 72, p. 21–38, doi: 10.1016/j.earscirev.2005.03.004.

Lustrino, M., 2006, Comment on "High-pressure melting experiments on garnet clinopyroxenite and the alkalic to tholeiitic transition in ocean-island basalts" by Keshav et al.: Earth and Planetary Science Letters, v. 241, p. 993–996, doi: 10.1016/j.epsl.2005.10.024.

Lustrino, M., and Sharkov, E., 2006, Neogene volcanic activity of western Syria and its relationship with Arabian plate kinematics: Journal of Geodynamics, v. 42, p. 115–139.

Lustrino, M., and Wilson, M., 2007, The Circum-Mediterranean anorogenic Cenozoic Igneous Province: Earth-Science Reviews, v. 81, p. 1–65.

Lustrino, M., Melluso, L., and Morra, V., 2002, The transition from alkaline to tholeiitic magmas: A case study from the Orosei-Dorgali Pliocene volcanic district (NE Sardinia, Italy): Lithos, v. 63, p. 83–113, doi: 10.1016/S0024-4937(02)00113-5.

Lustrino, M., Morra, V., Melluso, L., Brotzu, P., d'Amelio, F., Fedele, L., Franciosi, L., Lonis, R., and Petteruti Liebercknecht, A.M., 2004, The Cenozoic igneous activity of Sardinia, *in* Conticelli, S., and Melluso, L., eds., A

showcase of the Italian research in petrology: Magmatism in Italy: Periodico di Mineralogia, v. 73 (special issue), p. 105–134.

Lustrino, M., Melluso, L., and Morra, V., 2007, The geochemical peculiarity of "Plio-Quaternary" volcanic rocks of Sardinia in the circum-Mediterranean Cenozoic Igneous Province, *in* Beccaluva L., et al., eds., Cenozoic volcanism in the Mediterranean area: Boulder, Colorado, Geological Society of America Special Paper 418, p. 277–301.

Macera, P., Gasperini, D., Piromallo, C., Blichert-Toft, J., Bosch, D., del Moro, A., and Martin, S., 2003, Geodynamic implications of deep mantle upwelling in the source of Tertiary volcanics from the Veneto region (southeastern Alps): Journal of Geodynamics, v. 36, p. 563–590, doi: 10.1016/j.jog.2003.08.004.

Mahoney, J.J., and Coffin, M.F., eds., 1997, Large igneous provinces: Continental, oceanic and planetary flood volcanism: Washington, D.C., American Geophysical Union, Geophysical Monograph 100, 438 p.

Mahoney, J.J., le Roex, A.P., Peng, Z., Fisher, R.L., and Natland, J.H., 1992, Southwestern limits of Indian Ocean ridge mantle and the origin of low $^{206}Pb/^{204}Pb$ Mid-Ocean Ridge Basalt: Isotope systematics of the central southwest Indian Ridge (17°–50°E): Journal of Geophysical Research, v. 97, p. 19,771–19,790.

Malinverno, A., 1981, Quantitative estimates of age and Messinian paleobathymetry of the Tyrrhenian Sea after seismic reflection, heat flow and geophysical models: Bollettino di Geofisica Teorica ed Applicata, v. 23, p. 159–171.

McKenzie, D., and Bickle, M.J., 1988, The volume and composition of melt generated by extension of the lithosphere: Journal of Petrology, v. 29, p. 625–679.

Meibom, A., and Anderson, D.L., 2003, The statistical upper mantle assemblage: Earth and Planetary Science Letters, v. 217, p. 123–139, doi: 10.1016/S0012-821X(03)00573-9.

Meibom, A., Anderson, D.L., Sleep, N.H., Frei, R., Chamberlain, C.P., Hren, M.T., and Wooden, J.L., 2003, Are high $^3He/^4He$ ratios in oceanic basalts an indicator of deep-mantle plume components?: Earth and Planetary Science Letters, v. 208, p. 197–204, doi: 10.1016/S0012-821X(03)00038-4.

Meibom, A., Sleep, N.H., Zahnle, K., and Anderson, D.L., 2005, Models for noble gases in mantle geochemistry: Some observations and alternatives, *in* Foulger, G.R., et al., eds., Plates, plumes, and paradigms: Boulder, Colorado, Geological Society of America Special Paper 388, p. 347–363.

Meyer, R., Abratis, M., Viereck-Gotte, L., Madler, J., Hertogen, J., and Romer, R.L., 2002, Mantelquellen des Vulkanismus in der Thüringischen Rhön: Beiträge zur Geologie von Thüringen, v. 9, p. 75–105.

Montelli, R., Nolet, G., Dahlen, F.A., Masters, G., Engdahl, E.R., and Hung, S.-H., 2004, Finite-frequency tomography reveals a variety of plumes in the mantle: Science, v. 303, p. 338–343, doi: 10.1126/science.1092485.

Morgan, W.J., 1971, Convective plumes in the lower mantle: Nature, v. 230, p. 42–43, doi: 10.1038/230042a0.

Olson, P., Schubert, G., and Anderson, C., 1987, Plume formation in the D″ layer and the roughness of the core-mantle boundary: Nature, v. 327, p. 409–413, doi: 10.1038/327409a0.

Oyarzun, R., Doblas, M., Lòpez-Ruiz, J., and Cebrià, J.M., 1997, Opening of the Central Atlantic and asymmetric mantle upwelling phenomena: Implications for long-lived magmatism in western north Africa and Europe: Geology, v. 25, p. 727–730, doi: 10.1130/0091-7613(1997)025<0727:OOTCAA>2.3.CO;2.

Parman, S.W., Kurz, M.D., Hart, S.R., and Grove, T.L., 2005, Helium solubility in olivine and implications for high $^3He/^4He$ in ocean island basalts: Nature, v. 437, p. 1140–1143, doi: 10.1038/nature04215.

Peccerillo, A., 2005, Plio-Quaternary volcanism in Italy: Berlin, Springer, 365 p.

Peccerillo, A., and Lustrino, M., 2005, Compositional variations of the Plio-Quaternary magmatism in the circum-Tyrrhenian area: Deep- vs. shallow-mantle processes, *in* Foulger, G.R., et al., eds., Plates, plumes, and paradigms: Boulder, Colorado, Geological Society of America Special Paper 388, p. 421–434.

Pilidou, S., Priestley, K., Debayle, E., and Gudmundsson, O., 2005, Rayleigh wave tomography in the North Atlantic: High resolution images of the Iceland, Azores and Eifel mantle plumes: Lithos, v. 79, p. 453–474, doi: 10.1016/j.lithos.2004.09.012.

Piromallo, C., and Morelli, A., 2003, P wave tomography of the mantle under the Alpine-Mediterranean area: Journal of Geophysical Research, v. 108, no. B2, p. 2065, doi: 10.1029/2002JB001757.

Polino, R., Dal Piaz, G.V., and Gosso, G., 1990, Tectonic erosion at the Adria margin and accretionary processes for the Cretaceous orogeny of the Alps: Mémoire Société Géologique France, v. 156, p. 345–367.

Richards, M.A., Duncan, R.A., and Courtillot, V.E., 1989, Flood basalts and hot spot tracks: Plume heads and tails: Science, v. 246, p. 103–107, doi: 10.1126/science.246.4926.103.

Ritsema, J., van Heijst, H.J., and Woodhouse, J.H., 1999, Complex shear wave velocity structure imaged beneath Africa and Iceland: Science, v. 286, p. 1925–1928, doi: 10.1126/science.286.5446.1925.

Ritter, J.R.R., 2007, The seismic signature of the Eifel plume, *in* Ritter, J.R.R., and Christensen, U.R., eds., Mantle plumes—An interdisciplinary approach: Heidelberg, Springer-Verlag, p. 379–404.

Ritter, J.R.R., Jordan, M., Christensen, U.R., and Achauer, U., 2001, A mantle plume below the Eifel volcanic fields, Germany: Earth and Planetary Science Letters, v. 186, p. 7–14, doi: 10.1016/S0012-821X(01)00226-6.

Rotolo, S.G., Castorina, F., Cellura, D., and Pompilio, M., 2006, Petrology and geochemistry of submarine volcanism in the Sicily Channel Rift: Journal of Geology, v. 114, p. 355–365, doi: 10.1086/501223.

Saunders, A.D., Storey, M., Kent, R.W., and Norry, M.J., 1992, Consequences of plume-lithosphere interactions, *in* Storey, B.C., et al., eds., Magmatism and the causes of continental break-up: Geological Society of London Special Publication 68, p. 31–39.

Savelli, C., 2002, Time-space distribution of magmatic activity in the western Mediterranean and peripheral orogens during the past 30 Ma (a stimulus to geodynamic considerations): Journal of Geodynamics, v. 34, p. 99–126, doi: 10.1016/S0264-3707(02)00026-1.

Schmid, S.M., Pfiffner, O.A., Froitzheim, N., Schönborn, G., and Kissling, E., 1996, Geophysical-geological transect and tectonic evolution of the Swiss-Italian Alps: Tectonics, v. 15, p. 1036–1064, doi: 10.1029/96TC00433.

Schwab, B.E., and Johnston, D.A., 2001, Melting systematics of modally variable, compositionally intermediate peridotites and the effects of mineral fertility: Journal of Petrology, v. 42, p. 1789–1811, doi: 10.1093/petrology/42.10.1789.

Scrocca, D., Carminati, E., and Doglioni, C., 2005, Deep structure of the Southern Apennines (Italy): Constraints vs. speculations: Tectonics, v. 24, TC3005, doi: 10.1029/2004TC001634.

Seghedi, I., Downes, H., Vaselli, O., Szakacs, A., Balogh, K., and Pécskay, Z., 2004, Post-collisional Tertiary–Quaternary mafic alkaline magmatism in the Carpathian-Pannonian region: A review: Tectonophysics, v. 393, p. 43–62, doi: 10.1016/j.tecto.2004.07.051.

Seghedi, I., Downes, H., Harangi, S., Mason, P.R.D., and Pécskay, Z., 2005, Geochemical response of magmas to Neogene–Quaternary continental collision in the Carpathian-Pannonian region: A review: Tectonophysics, v. 410, p. 485–499, doi: 10.1016/j.tecto.2004.09.015.

Selvaggi, G., and Chiarabba, C., 1995, Seismicity and P-wave velocity image of the Southern Tyrrhenian subduction zone: Geophysical Journal International, v. 121, p. 818–826.

Shaw, J.E., Baker, J.A., Menzies, M.A., Thirlwall, M.F., and Ibrahim, K.M., 2003, Petrogenesis of the largest intraplate volcanic field on the Arabian plate (Jordan): A mixed lithosphere-asthenosphere source activated by lithospheric extension: Journal of Petrology, v. 44, p. 1657–1679, doi: 10.1093/petrology/egg052.

Sobolev, S.V., Zeyen, H., Granet, M., Achauer, U., Bauer, C., Werling, F., Altherr, R., and Fuchs, K., 1997, Upper mantle temperatures and lithosphere-asthenosphere system beneath the French Massif Central constrained by seismic, gravity, petrologic and thermal observations: Tectonophysics, v. 275, p. 143–164, doi: 10.1016/S0040-1951(97)00019-X.

Sobolev, A.V., Hofmann, A.W., Sobolev, S.V., and Nikogosian, I.K., 2005, An olivine-free mantle source of Hawaiian shield basalts: Nature, v. 434, p. 590–597, doi: 10.1038/nature03411.

Spadini, G., Cloetingh, S., and Bertotti, G., 1995, Thermo-mechanical modelling of the Tyrrhenian Sea: Lithospheric necking and kinematics of rifting: Tectonics, v. 14, p. 629–644, doi: 10.1029/95TC00207.

Stein, M., and Hofmann, A.W., 1992, Fossil plume head beneath the Arabian lithosphere?: Earth and Planetary Science Letters, v. 114, p. 193–209, doi: 10.1016/0012-821X(92)90161-N.

Stracke, A., Hofmann, A.H., and Hart, S.R., 2005, FOZO, HIMU, and the rest of the mantle zoo: Geochemistry, Geophysics, Geosystems, v. 6, p. Q05007, doi:10.1029/2004GC000824.

Sun, S.S., and McDonough, W.F., 1989, Chemical and isotopic systematics of oceanic basalts: Implications for mantle compositions and processes, *in* Saunders, A.D., and Norry, M.J., eds.: Magmatism in the ocean basins: Geological Society of London Special Publication 42, p. 313–345.

Tackley, P.J., 2000, Mantle convection and plate tectonics: Toward and integrated physical and chemical theory: Science, v. 288, p. 2002–2007, doi: 10.1126/science.288.5473.2002.

van der Hilst, R.D., Widiyantoro, S., and Engdahl, E.R., 1997, Evidence for deep mantle circulation from global tomography: Nature, v. 386, p. 578–584, doi: 10.1038/386578a0.

von Blanckenburg, F., and Davies, J.H., 1995, Slab breakoff: A model for syncollisional magmatism and tectonics in the Alps: Tectonics, v. 14, p. 120–131, doi: 10.1029/94TC02051.

Walker, K.T., Bokelmann, G.H.R., Klemperer, S.L., and Bock, G., 2005, Shear-wave splitting around the Eifel hot spot: Evidence for a mantle upwelling: Geophysical Journal International, v. 163, p. 962–980.

Wedepohl, K.H., and Baumann, A., 1999, Central European Cenozoic plume volcanism with OIB characteristics and indication of a lower mantle source: Contributions to Mineralogy and Petrology, v. 136, p. 225–239.

White, R., and McKenzie, D., 1989, Magmatism at rift zones: The generation of volcanic continental margins and flood basalts: Journal of Geophysical Research, v. 94, p. 7685–7729.

Wilson, J.T., 1963, A possible origin of the Hawaiian Islands: Canadian Journal of Physics, v. 41, p. 863–870.

Wilson, M., and Bianchini, G., 1999, Tertiary–Quaternary magmatism within the Mediterranean and surrounding regions, *in* Durand, B., et al., eds., The Mediterranean basins: Tertiary extension within the Alpine Orogen: Geological Society of London Special Publication 156, p. 141–168.

Wilson, M., and Downes, H., 1991, Tertiary–Quaternary extension-related alkaline magmatism in western and central Europe: Journal of Petrology, v. 32, p. 811–849.

Wilson, M., and Downes, H., 2006, Tertiary–Quaternary intra-plate magmatism in Europe and its relationship to mantle dynamics, *in* Gee, D., and Stephenson, R., eds., European lithosphere dynamics: Geological Society of London Memoir 32, p. 147–166.

Wilson, M., and Patterson, R., 2001, Intraplate magmatism related to short-wavelength convective instabilities in the upper mantle: Evidence from the Tertiary–Quaternary volcanic province of western and central Europe, *in* Ernst, R.E., and Buchan, K.L., eds., Mantle plumes: Their identification through time: Boulder, Colorado, Geological Society of America Special Paper 352, p. 37–58.

Yaxley, G.M., 2000, Experimental study of the phase and melting relations of homogeneous basalt + peridotite mixtures and implications for the petrogenesis of flood basalts: Contributions to Mineralogy and Petrology, v. 139, p. 326–338, doi: 10.1007/s004100000134.

Zhao, D., 2001, Seismic structure and origin of hot spots and mantle plumes: Earth and Planetary Science Letters, v. 192, p. 251–265, doi: 10.1016/S0012-821X(01)00465-4.

Ziegler, P.A., 1992, European Cenozoic rift system: Tectonophysics, v. 208, p. 91–111, doi: 10.1016/0040-1951(92)90338-7.

Zito, G., Mongelli, F., de Lorenzo, S., and Doglioni, C., 2003, Geodynamical interpretation of the heat flow in the Tyrrhenian Sea: Terra Nova, v. 15, p. 425–432, doi: 10.1046/j.1365-3121.2003.00507.x.

MANUSCRIPT ACCEPTED BY THE SOCIETY JANUARY 31, 2007

DISCUSSION

22 January 2007, Françoise Chalot-Prat

The article by Lustrino and Carminati (this volume) presents an exhaustive review of the different hypotheses for the nature of mantle sources of Cenozoic to Quaternary Na-alkaline mafic volcanics from the circum-Mediterranean and European region and the causes of mantle melting within the geodynamic setting of eruptions.

This magmatism is often proposed, based on geochemistry and tomography, to be related to mantle plumes coming from the core–mantle boundary. Nevertheless, the authors show that, considering not only major-element and isotope geochemistry and tomographic models but also the geodynamic evolution of both regions, the mantle sources are likely to be much shallower and mostly in a heterogeneous metasomatized lithospheric mantle. Mantle melting is checked by plate tectonic processes. In these regions where lithospheric extension was and is always the prevailing mechanism during eruptions, mantle melting occurs by decompression during rifting-related uplift and is enhanced by the low melting points of crust-contaminated lithologies.

Furthermore, I think that the main problem concerning the "plume or no-plume" debate comes from petrologists themselves. They often include plate tectonic concepts in their models without taking into account the tectonic constraints they imply. Besides, a number of petrologists exclude a priori the mantle lithosphere as a possible source of some basalts, whereas experimental work (dating from the 1970s and now largely confirmed) and numerous studies on mantle xenoliths support it. I will try to clarify my comments.

A magma should never be defined from a geodynamic point of view, such as orogenic (subduction-related) or anorogenic (not subduction-related). This last term not only is too vague but is meaningless in terms of plate tectonics.

A magma is defined in terms of major- and trace-element composition. On the whole its composition is a function of the composition of its solid source (if it is a partial melt) or liquid source (if it derives from a parent magma), and also the P-T-f parameters that prevailed during both its formation and ascent up to its emplacement at the surface or at depth.

Magmatism, extrusive or/and intrusive, is associated with all the stages of the Wilson orogenic cycle. The absence of volcanics does not mean the absence of magmatism, as suggested by some petrologists. Crustal under- and intraplating are common processes at every stage of the cycle. Also, and it is a major point, volcanism always implies that melt genesis is synchronous with opening of fractures up to the surface of the continental or oceanic lithosphere. Whatever the geodynamic setting, the feeder dike network of any eruptive area is dependent on regional tectonic constraints and thus on the plate-tectonic context. The same applies to plutonism, except that this results from the closure of the fractures upward at depth. In any case, there

is always a link between magmatism and tectonics, which cannot be clarified by geochemistry and tomography and is ignored in the plume hypothesis.

The fact that a type of mantle magma (the subject of Lustrino and Carminati, this volume) occurs, either always, frequently, or even never at some stages, depends only on the composition of the mantle involved.

If more than one type occurs during the same stage, sometimes at the same site, and even synchronously (geologically speaking), this means that more than one mantle source is involved. So either the tapped mantle source is heterogeneous or distinct mantle sources at different depths are tapped simultaneously.

Then the fact that one (or several) type(s) of mantle source(s) is (are), either always, frequently, or even never involved in some stages, depends on the plate-tectonic setting (including lithosphere type and thickness) at the time of the eruptions. The plate-tectonic processes involved generate tectonic constraints (decoupling or/and shearing between rheologically different zones) within the lithospheric or/and asthenospheric mantles, or even the lower–upper mantle transition zone, that trigger melting. These processes must work with brittle behavior of overlying crustal or lithospheric layers, leading to the formation of feeder dikes.

The ultimate question concerns the location of the mantle sources accounting for their composition deduced from the mafic eruptives. The necessary involvement of plate-tectonic processes leads to consideration of three possible locations: the asthenospheric mantle, the lithospheric mantle, and the upper–lower mantle transition zone. As a well-accepted hypothesis, the asthenosphere would correspond to a "depleted" mantle (primitive mantle residue) source of N-MOR basalts, the only type of basalt occurring in only one type of plate-tectonic context (oceanic spreading). This fact is at odds with the hypothesis that the asthenosphere includes blobs of enriched mantle. The lithospheric mantle and the upper–lower mantle transition zone would correspond to an "enriched" mantle source of both other types of basalts—alkaline and calc-alkaline. From continental mantle xenoliths or ridge oceanic mantle sample studies, this enriched mantle, always heterogeneous on a regional scale, was initially a depleted mantle that later underwent melt percolation and metasomatism. It reflects the mixing between an "asthenospheric" component and melts coming either from the underlying mantle (asthenosphere) or from eclogitic oceanic and/or continental subducted lithospheres. It is even probable that portions of eclogitic slabs, tracers of previous sutures, are stored in the lithospheric mantle for a long time after subduction. Besides, any enriched mantle may have been involved in several successive orogenic cycles with several episodes of melting and metasomatism, which means that the effects of continental crust recycling (including old oceanic crust) likely dominate over those of recent oceanic crust.

29 January 2007, Romain Meyer

Françoise Chalot-Prat makes an important point about the geochemical and petrological location of the mantle source for Central European primary magmas. Petrological and geochemical investigation of igneous rocks from the Rhoen Mountains and the Grabfeld (Heldburg Dike Swarm), integral areas of the Central European Volcanic Province (CEVP), have revealed the metasomatically overprinted, heterogeneous, subcontinental lithosphere to be a potential mantle source (Meyer et al., 2002).

The age spectrum of these rocks, according to new ^{40}Ar/^{39}Ar data, is clearly divided into two distinct subsets, with volcanic rocks of the Rhoen dated at 20–18 Ma and those of the Grabfeld area in the SE dated at 16–14 Ma (Abratis et al., in press). This clearly indicates regionally and temporally distinct evolution for the Miocene volcanism. The composition of the volcanic rocks in the two volcanic fields is remarkably diverse and includes alkali basalts, tholeiites and minor basanites, and nephelinites. Comparable Cenozoic igneous suites are reported from the nearby Vogelsberg and the northern Hessian Depression.

The geochemical and isotopic characteristics of the different magma types correlate strongly for every rock type in both areas, e.g., ^{87}Sr/^{86}Sri = 0.7034 to 0.7040; εNd = 3.9 to 4.7; and ^{206}Pb/^{207}Pbi = 19.0 to 19.3. REE patterns can be interpreted as reflecting melts derived by variable degrees of partial melting within the transitional zone between a garnet–peridotite and spinel–peridotite mantle facies, close to the base of the lithosphere (Meyer et al., 2002). Similar radiogenic isotope signatures (Sr, Nd, Pb) for the same magma types in both subareas of the CEVP point to a chemically uniform source that is identical to the source of most of the other, western to mid-European magma fields. Minor differences in the magma composition result either from decreasing degrees of melting of the subcontinental mantle or from increasing depths of melting. A major question for the understanding of rift-related magmatism is why the volcanic activity ceased (18 Ma) in the Rhoen area, whereas it continued in the Vogelsberg and the northern Hessian Depression in the west to northwest and started in the Grabfeld (16 Ma) farther to the southeast.

Partial melting of the mantle source was induced by adiabatic decompression in both areas. A lack of magmatic activity between 18 and 16 Ma in combination with the transition and onset of magmatism from the Rhoen Mountains into the Grabfeld during this volcanic period indicates a geochemically similar, fertile mantle package at subcontinental levels below both areas. In the Rhoen, this fertile mantle domain acted as a magma source over ca. 2 Ma (20–18 Ma) before tapping a virtually identical reservoir in the Grabfeld area with a comparable lifetime between 16 and 14 Ma.

REFERENCES CITED

Abratis, M., Mädler, J., Hautmann, S., Leyk, H.-J., Meyer, R., Lippolt, H.J., and Viereck-Götte, L., In press, Two distinct Miocene age ranges of basaltic rocks from the Rhön and Heldburg areas (Germany) based on ^{40}Ar/^{39}Ar step heating data: Chemie der Erde—Geochemistry, doi: 10.1016/j.chemer.2006.03.003. Available online at: www.sciencedirect.com/science/journal/00092819.

Lustrino, M., and Carminati, E., 2007 (this volume), Phantom plumes in Europe and the circum-Mediterranean region, *in* Foulger, G.R., and Jurdy, D.M., eds., Plates, plumes, and planetary processes: Boulder, Colorado, Geological Society of America Special Paper 430, doi: 10.1130/2007.2430(33).

Meyer, R., Abratis, M., Viereck-Götte, L., Mädler, J., Hertogen, J., and Romer, R.L., 2002, Mantelquellen des Vulkanismus in der thüringischen Rhön: Beiträge zur Geologie von Thüringen, v. 9, p. 75–105.

The Geological Society of America
Special Paper 430
2007

Mechanisms of crustal growth in large igneous provinces: The north Atlantic province as a case study

Laurent Geoffroy*
Université du Maine, EA 3264, Ave O. Messiaen, 72085 Le Mans Cedex 9, France
Charles Aubourg
Université de Cergy-Pontoise, CNRS UMR 7072, 95031 Cergy-Pontoise Cedex, France
Jean-Paul Callot
IFP, 1–4 Avenue de Bois Préau, 92852 Rueil Malmaison Cedex, France
Jean-Alix Barrat
Université de Bretagne Occidentale, IUEM, CNRS UMR 6538, Technopole Brest-Iroise, 29280 Plouzané, France

ABSTRACT

The mechanisms of magma crust accretion at large igneous provinces (LIPs) are questioned using arguments based on the north Atlantic case. Published and new data on the calculated flow vectors within dike swarms feeding the early traps and subsequent seaward-dipping reflector lavas suggest that most of the mafic magmas forming the north Atlantic LIP transited through a small number of igneous centers. The magma was injected centrifugally in dike swarms at some distance away from individual igneous centers along the trend of the maximum horizontal stress acting in the crust, feeding lava piles via dikes intersecting the ground surface. This mechanism is similar to that observed in present-day Iceland and, more generally, in mafic volcano-tectonic systems. The absence of generalized vertical magma transit in a LIP has major geodynamic consequences. We cannot link the surface extent of LIP magmas to the dimensions of the mantle melting zone as proposed in former plume head models. The distribution of LIP magmas at the surface is primarily controlled by the regional stress field acting within the upper crust, but is also affected by magma viscosity. The igneous centers feeding LIPs most likely represent the crustal expression of small-scale convective cells of the buoyant mantle naturally located beneath the mechanical lithosphere.

Keywords: traps, volcanic margin, dike swarm, small-scale convection, AMS

INTRODUCTION

This contribution focuses on the localization of upper-mantle melting at the origin of large igneous provinces (LIPs). We need first to summarize the different views on LIP origin, which is a subject of great debate (see the Web site www.MantlePlumes.org).

Two distinctive stages of development are recognized in LIPs: the widespread emplacement of flood basalts (the "trap stage") and a (possible) consecutive "volcanic margin stage" during which synextension magmatism is concentrated along the breakup zone (e.g., White and McKenzie, 1989; Eldholm et al., 1995; Courtillot et al., 1999; Geoffroy, 2005). It is important to

*E-mail: laurent.geoffroy@univ-lemans.fr.

Geoffroy, L., Aubourg, C., Callot, J.-P., Barrat, J.-A., 2007, Mechanisms of crustal growth in large igneous provinces: The north Atlantic province as a case study, *in* Foulger, G.R., and Jurdy, D.M., eds., Plates, plumes, and planetary processes: Geological Society of America Special Paper 430, p. 747–774, doi: 10.1130/2007.2430(34). For permission to copy, contact editing@geosociety.org. ©2007 The Geological Society of America. All rights reserved.

bear in mind the existence of these two developmental stages insofar as the geographic distribution and volumes of magma are different in each of them.

Trap Stage

Traps or plateau basalts are flat-lying accumulations of mafic lavas that are emplaced during a relatively short time span. Some authors have used sedimentary records (White and Lovell, 1997) and wide-angle seismic surveys (Al-Kindi et al., 2003) to argue that magma underplating may also occur during this early stage. During trap emplacement, the amount of tectonic extension is usually small, and dilatation through dike injection seems to predominate over extension associated with normal faults (e.g., Doubre and Geoffroy, 2003).

Many authors have explained the large uplifted oceanic and/or continental areas covered with plateau basalts, as well as oceanic hotspots located at the crest of broad seafloor swells, by the presence of more or less axisymmetric hot mantle plume heads beneath the lithosphere (e.g., Morgan, 1971; Courtney and White, 1986; Olson and Nam, 1986). White and McKenzie (1989) summarize the geological and geophysical features that are generally linked with postulated mantle plumes beneath the lithosphere. Mantle plumes are primarily thought to represent hot gravitational instabilities formed at the core-mantle boundary due to a core-mantle thermal boundary layer (e.g., Anderson, 2004). In parallel, the enriched trace-element geochemistry of early traps at LIPs and oceanic islands basalts (OIB), as well as their rare gas isotopic ratios, is commonly explained by the postulated primitive composition of a lower-mantle reservoir (e.g., Courtillot et al., 2003). Many models of plume head–lithosphere interaction have been discussed, such as: (1) sudden impact of a very hot plume (e.g., Richards et al., 1989), (2) progressive thermal erosion of the basal lithosphere by a long-lived incubating plume head (e.g., Kent et al., 1992), (3) interaction of a plume head with a lithosphere of variable thickness (e.g., White and McKenzie, 1989; Thompson and Gibson, 1991), (4) small-scale convection within the plume head (e.g., Fleitout et al., 1986).

However, a range of new concepts and experiments has challenged the mantle plume theory (see the extensive references at www.MantlePlumes.org). The reinterpretation of mantle seismic tomography raises questions about deep-seated mantle plumes, as exemplified by the Icelandic case (e.g., Foulger, 2002). The chemistry of early flood basalts and OIB could also be explained by melting of a much shallower compositionally heterogeneous mantle (e.g., Gallagher and Hawkesworth, 1992; Anderson, 1994). Notably, melting of eclogites (old subducted slabs) is proposed as a possible component in igneous provinces developed along ancient orogenic sutures (e.g., Foulger and Anderson, 2005). Hotter than normal mantle is also debated as a cause of LIP magmatism (e.g., Green et al., 2001). Different top-to-down processes have been proposed as an alternative to hot plumes, most of them invoking, although at quite different scales, upward counterflow processes in the mantle due to the gravitational instability of cold lithospheric roots (e.g., King and Anderson, 1998; Lustrino, 2005).

VPM Stage

Most volcanic passive margins (VPMs, Fig. 1) are consecutive to the emplacement of traps and associated with anomalously thick oceanic crust following continental breakup. The

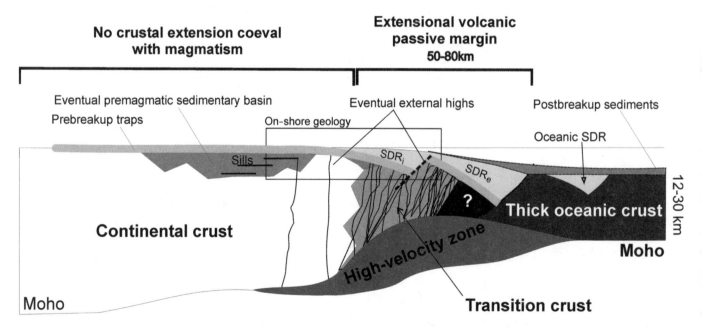

Figure 1. Elements of a volcanic passive margin (from Geoffroy, 2005). Note the emplacement of traps before the synmagmatic stretching and thinning of the continental crust. SDR$_i$—inner seaward-dipping reflectors; SDR$_e$—external seaward-dipping reflectors.

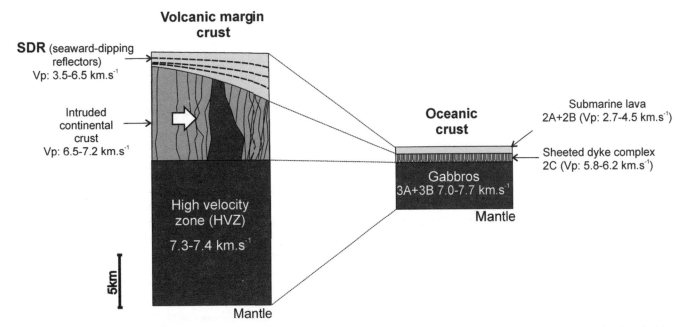

Figure 2. Normal oceanic crust and volcanic passive margin igneous crust: comparison of layering and compressional wave seismic velocities (V_p). White arrow—seaward gradient of mafic intrusions in the crust. Data from White et al. (1987) and Juteau and Maury (1999).

main crustal characteristics of VPMs are listed in White et al. (1987), Eldholm et al. (1995), Bauer et al. (2000), and Geoffroy (2005) and include data from both geophysical and geological surveys (Fig. 1). These margins are associated with significant magma accretion during lithosphere extension or rifting and subsequent breakup. Interestingly, the top-to-down trilogy of basalts, sheeted complex, and gabbros postulated for the VPM crust is strongly analogous with the structure of oceanic crust (Fig. 2). From the surface down to the Moho, the VPM crust (Fig. 1) is composed of (1) several wedges of seaward-dipping volcanic rocks (otherwise known as seismic SDR), (2) intensively intruded and stretched continental crust, and (3) large volumes of material of high compressional wave velocity (V_p) usually interpreted as underplated high-Mg gabbros. SDR are made up of aerial or subaerial lavas, but also contain volcanic ejecta (e.g., hyaloclastites, tuffs, etc.). The tectonic significance of SDR has been discussed, principally in Eldholm et al. (1995) and Geoffroy (2005). It is noteworthy that an inner SDR prism is usually located immediately above the stretched and intruded continental crust (e.g., Roberts et al., 1979; Planke et al., 2000; Geoffroy, 2005; Fig. 1).

Mantle plume specialists believe that VPMs originate from continental breakup over the hot plume head or residual tail that remains after trap emplacement (e.g., White et al., 1987; White and McKenzie, 1989; Courtillot et al., 1999). However, small-scale convection due to the rifting process itself and/or pre-existing lateral variations in lithosphere thickness (e.g., Mutter et al., 1988; Keen and Boutilier, 1995; King and Anderson, 1998) could also account for the huge volumes of magma associated with VPMs without invoking any excess in mantle temperature.

From numerical models of rapidly stretched continental lithosphere, Van Wijk et al. (2001) point out that the characteristic melt thickness of VPMs may be obtained solely through adiabatic melting of the sublithospheric buoyant mantle immediately before plate breakup. Anderson (1994 and 1995) proposed that the huge volumes produced at VPMs are the result of pull-apart processes over the fertile part of the mantle located beneath the mechanical lithosphere (called the "perisphere" or thermal boundary layer in the present article). All these non-plume models assume that plate tectonics and plate geometry (craton edges) are sufficient by themselves to account for the origin of VPMs and LIPs as a whole.

Melting Localization at LIPs

A fundamental aspect of LIP buildup concerns the localization of melting areas beneath or within the lithosphere. This problem has commonly been addressed through magma geochemistry. However, as outlined earlier, it is not always straightforward to characterize the mantle reservoirs involved (i.e., lithospheric or asthenospheric mantle) because of the contamination of the melts by various crustal lithologies as illustrated by the Scottish Tertiary volcanics (e.g., Dickin et al., 1987, among others). Another approach to discussing the origin of the igneous activity is to establish the pattern of magma flow within the oceanic or continental crust. This was performed at the scale of a Proterozoic LIP by Ernst and Baragar (1992). From an AMS (anisotropy of magnetic susceptibility) study in dikes, these authors argued that the McKenzie giant dike swarm (and related LIP) was fed centrifugally and horizontally in the continental crust from a central source (toward which the dikes converge) that was itself fed vertically from a mantle plume. Later, in the section on flow vectors, we return to the AMS methodology followed by these authors to infer the flow pattern in dikes. We should note that Phanerozoic LIPs are generally not associated with giant dikes (see the section on dike swarms and stress fields), and that no studies on the more recent LIP are specifically concerned with magma transfer from the mantle to the upper crust. Instead, in most LIP plume models, it is assumed either implicitly or explicitly that "the extent of [flood basalts] gives a good indication of the area underlain by the mushroom head of hot mantle carried up by the plume" (White and Mc-Kenzie, 1989). The proposed diameters for melting plume heads could thus reach values of up to 2000 km (White and McKenzie, 1989; Hill et al., 1992).

The question of LIP feeding is also addressed by authors defending models that do not involve mantle plumes. Denying the existence of dike swarms that radiate from a single point, they propose that traps at LIPs are fed upward through regional dike swarms whose location is controlled solely by plate-related lithospheric stresses (e.g., Favela and Anderson, 1999; McHone et al., 2004), not by a plume-related stress field (i.e., radial due to a lithospheric swell, as in Ernst et al., 1995).

Thus, the most commonly held view of LIP formation involves a vertical transfer of magma from extensive zones of melting in the mantle to the upper crust (intrusions) or the Earth's surface (lavas, volcanic ejecta). This transfer is assumed to be direct (primary magmas) or indirect (magmas differentiated in crustal reservoirs). Inferring that the area covered by flood basalts corresponds more or less to the extent of mantle melting at depth is an important assumption, because it determines the whole geodynamic model for the origin of LIPs. In the present study, we make use of the north Atlantic case to question this assumption by establishing the actual mechanisms of magma accretion during LIP growth.

THE NORTH ATLANTIC PROVINCE: WHAT KIND OF MAGMA FEEDING SYSTEM?

Volume Estimates for the North Atlantic LIP

In the north Atlantic volcanic province (Figs. 3 and 4), the earliest oceanic magnetoanomaly is dated as C24r, i.e., ca. 56 to 54 Ma according to the Berggren et al. (1995) timescale. The emplacement of Paleocene flood basalts spanned ~6 m.y., from 62 to 57 or 56 Ma (Hitchen and Richtie, 1993). Ar-Ar ages from the East Greenland Coastal Dike swarm (Lenoir et al., 2003) indicate that the breakup is bracketed between 54 and 51 Ma, which is consistent with a sudden Eocene breakup event, coeval with the earliest oceanic accretion.

Coffin and Eldholm (1994) and Eldholm and Grue (1994) estimated the minimum volumes of magma in the north Atlantic volcanic province (including initial traps and VPMs) at no more than 6×10^6 km^3, with rates of magma production reaching 2 km^3/yr. However, such an estimate is difficult to establish and should be considered only as a maximum value. For instance, most authors favoring the plume hypothesis unhesitatingly assume that west Greenland, east Greenland, the Faeroes, and the British Tertiary igneous province are parts of the same flood basalt province (e.g., Saunders et al., 1997; Larsen et al., 1999). This shortcut hypothesis is debatable, because there is no continuity of outcrop between west and east Greenland. Estimating the volume of underplated mafic magmas from the high-velocity zone (HVZ; Figs. 1 and 2) may also lead to unreliable results. Most authors agree that the HVZ represents substantial amounts of magma accreted ("underplated") at the Moho (e.g., Eldholm

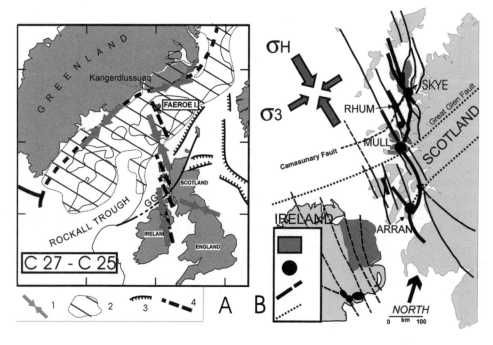

Figure 3. (A) The north Atlantic volcanic province during the C27–C25 trap stage. GG—Great Glen fault; 1—trend of the maximum horizontal stress inferred from dike swarms and fault tectonics (data from Geoffroy et al., 1993 and 1994); 2—approximate distribution of Paleocene traps and sill swarms; 3—faults; 4—dike swarms. Note that the existence of Paleocene dikes along the southeast Greenland coast is only hypothetical (see Lenoir et al., 2003). (B) The British Tertiary igneous province, mainly after Speight et al. (1982). σ_H—maximum horizontal stress; σ_3—minimum principal stress; 1—traps; 2—igneous center; 3—dike swarms; 4—main crustal faults.

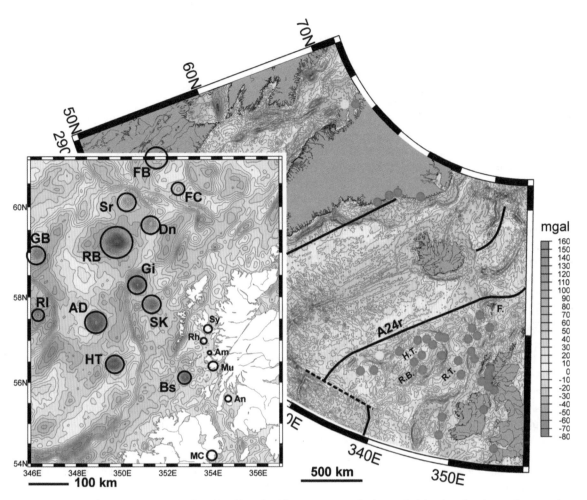

Figure 4. Recognized (red dots) and inferred (yellow dots) igneous centers in the north Atlantic volcanic province, not including those located at the continent-ocean transition (free-air gravity map; Smith and Sandwell, 1997). Magnetoanomaly 24 (A24r) is approximately located. This compilation is not exhaustive, because a number of new igneous centers probably remain to be discovered. The West Erlend, Erlend, and Brendan igneous centers (Smythe et al., 1983; Hitchen and Richtie, 1993), north of the Shetland Islands, lie outside the map area. Most igneous centers are recognized as Paleocene (also younger in some cases), with the possible exception of Rosemary Bank and Anton Dohrn, which are probably Maastrichtian (see Hitchen and Richtie, 1993, and references therein). Insert abbreviations: AD—Anton Dohrn; Am—Ardnamurchan; An—Arran; Bs—Blackstone; Dn—Darwin; FB—Faeroe Bank; FC—Faeroe Channel; GB—George Blight Bank; Gi—Geikie; HT—Hebrides Terrace; MC—Mourne-Carlingford; Mu—Mull; Rh—Rhum; RB—Rosemary Bank; RI—Rockall Island; Sr—Sigmundur; SK—St. Kilda; Sy—Skye. Abbreviations for the map at right: F.—Faeroe Islands; H.T.—Hatton Trough; R.B.—Rockall Bank; R.T.—Rockall Trough.

and Grue, 1994; Holbrook et al., 2001). However, Gernigon et al. (2004) have challenged the HVZ magma interpretation beneath the Voring Mesozoic basin. Should their observations be correct and applicable to other sedimentary basins, this would considerably decrease the estimated magma volume in the north Atlantic volcanic province.

Igneous Center Distribution in the North Atlantic Igneous Province

An important characteristic of LIPs is the ubiquity of igneous centers punctuating the nonrifted and rifted continental or transitional crust. In the uppermost crust, igneous centers are represented by magma chambers and overlying hypovolcanic

intrusions, making up the roots of large polygenic volcanoes (e.g., Vann, 1978; Irvine et al., 1998; Bauer et al., 2000; Chandrasekhar et al., 2002). In the north Atlantic, these igneous centers are well known from direct observation and/or potential field data (Fig. 4). To a first approximation, the large subcircular or elliptical magnetic and gravity anomalies associated with the igneous centers can be modeled by cylinder-shaped bodies of mafic to ultramafic rocks extending down to the Moho (e.g., Bott and Tuson, 1973; Bott and Tantrigoda, 1987).

The internal structure of these bodies is unknown. For instance, Bauer et al. (2000) proposed a crustal-scale interconnected network of mafic planar intrusions for the Messum Igneous Complex in Namibia. In the north Atlantic volcanic province, at least thirty-eight igneous centers were probably active during the trap stage (Fig. 4). Although not all igneous centers have yet been recognized and/or dated, their distribution appears to be (1) 2-D in map view and (2) unrelated to the thickness of the crust (Fig. 4). According to Callot (2002), the spacing between off-shore trap-stage igneous centers would vary from ~75 ± 30 km in the Hatton area to 100 ± 40 km in the Rockall–Faeroe Bank area (Fig. 4). This spacing decreases locally and significantly in the British Tertiary igneous province (35 ± 3 km for the Sy-Mu-Am-Mu group (Callot, 2002). In the latter case, the distribution of igneous centers is evidently controlled by the location of the main Caledonian-inherited discontinuities that were reactivated during the Tertiary (Fig. 4B; e.g., Roberts, 1974).

Many igneous centers are also associated with the north Atlantic volcanic province breakup process, so they are seen to punctuate the volcanic margins (e.g., Barton and White, 1997; Korenaga et al., 2000; Callot, 2002; Callot et al., 2002; Callot and Geoffroy, 2004). Apart from the eroded along-strike exposures of the innermost parts of a VPM, as observed on the southeast Greenland coast (Figs. 4 and 5; Myers, 1980; Bromann-Klausen and Larsen, 2002), the igneous centers associated with the breakup stage are less easy to distinguish physically due to their lower gravity and magnetic contrasts with the enclosing transitional or igneous crust (Fig. 5). However, the presence of relative gravity highs (Fig. 5) and magnetic anomalies (Gac and Geoffroy, 2005) suggest that igneous centers in the VPM transitional crust display an aligned or zigzag 1-D pattern or else a 2-D arrangement within a narrow band (Callot et al., 2002). Callot (2002) measured a 155 ± 7 km spacing of igneous centers in on-shore east Greenland (an area of low to moderate crustal thinning), decreasing to 58 ± 12 km off-shore, where the litho-spheric thinning was greatest (Fig. 5 and related caption). This latter value compares well with the wavelength of gravity and magnetic segmentation observed along the U.S. East Coast VPM at the continent-ocean transition (Behn and Lin, 2000).

It is noteworthy that, for at least two north Atlantic volcanic province igneous centers (Rosemary Bank and Anton Dohrn; Fig. 4), the magmatic activity is likely to be Late Cretaceous in age (e.g., Jones et al., 1974; Hitchen and Richtie, 1993), thus predating the postulated Paleocene emplacement of the so-called Icelandic mantle plume. In addition, magmatic activity at igneous centers is often a persistent phenomenon. In some centers that were active during trap emplacement, highly differentiated magma continued to be intruded during and even after the breakup process, sometimes at great distances from the volcanic margins (for example, the end-Eocene granites of the Mourne, Skye, and Lundy igneous centers; Fig. 4) (Hitchen and Richtie, 1993; Saunders et al., 1997).

There is general agreement that most LIP volcanism is of subaerial and fissural type. Low-viscosity tholeiitic or intermediate lava flows were fed by dikes intersecting the ground surface as in present-day Hawaii, Afar, or Icelandic volcanic systems (e.g., Self et al., 1997). Basaltic tuffs result from the dikes themselves (monogenic cones along fissures) or can be produced by ash eruptions from igneous centers related to polygenic volcanoes. It is thus evident that igneous centers and dike swarms play an essential role in distributing magmas within LIPs, not only at the ground surface (lavas and ejecta), but also within the intruded continental crust (magma crystallizing within the dikes themselves and in the igneous centers, hypovolcanic complexes, and magma chambers).

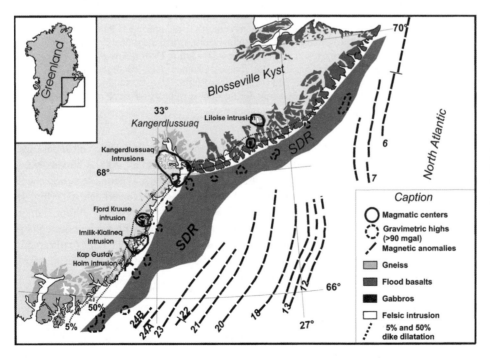

Figure 5. Geological map of eastern Greenland showing the locations of exposed igneous centers and off-shore gravity highs. Modified from Esher and Pulvertaft (1995).

Dike Swarms and Stress Fields

In the north Atlantic volcanic province continental domain, only a small number of off-shore and on-shore dike swarms have been identified from aeromagnetic surveys and direct observations, respectively (Fig. 3) (e.g., Larsen, 1978; Myers, 1980; Speight et al., 1982; Kirton and Donato, 1985; Bromann-Klausen and Larsen, 2002). For some off-shore areas, however, high-resolution aeromagnetic data may be missing or remain untreated. None of the recognized dike swarms correspond to giant dike swarms (i.e., swarms of dikes exceeding ~30 m in thickness): in the whole north Atlantic volcanic province area, outcropping dikes (even through eroded basement) exhibit an average thickness rarely exceeding 2 m during the trap stage (e.g., Speight et al., 1982) and ranging from 3 m (highly stretched crust) to 8 m (weakly stretched crust) along the VPM (Bromann-Klausen and Larsen, 2002). Notable exceptions include individual intrusions such as the ~20-m-thick Cleveland Dike (Scotland; McDonald et al., 1988) and the ~600-m-thick Kraemer Island dikelike intrusion (east Greenland, Kangerlussuaq area).

Dike swarms trend parallel or subparallel to the maximum horizontal stress (e.g., Anderson, 1951), so the general pattern of the north Atlantic volcanic province dike swarms may reflect the stress field in this area during the Paleogene (Fig. 3A). However, this stress field was not uniform. A regional northwest-southeast-trending maximum stress during the Paleocene was associated with prebreakup trap emplacement in the British Tertiary igneous province (e.g., Vann, 1978) and in the Faeroes area (Geoffroy et al., 1994; Fig. 3B). At the scale of the north Atlantic volcanic province, the maximum horizontal stress was apparently radial and focused on the Kangerdlussuaq area, which contains outcrops of a large system of igneous intrusions (Fig. 3A). It should be noted that most of the observed dike swarms in the north Atlantic volcanic province are centered on individual igneous centers (Fig. 3B). This is well established both in the trap area (e.g., Vann, 1978; Speight et al., 1982) and on the exposed parts of the VPM (Myers, 1980; Bromann-Klausen and Larsen, 2002). In all cases studied, the finite horizontal magma dilatation associated with these swarms increases toward the igneous centers.

Therefore, in the north Atlantic volcanic province (and generally in LIPs), magma transport in the brittle crust follows specific flow paths. The covering of vast continental areas by repetitive lava flows and volcanic products coming from a small number of fissure systems is not a hypothesis but a fact. The formation of dikes also plays a significant role in the magmatic accretion of the transitional crust (Figs. 1 and 2). Although this mechanism is now well accepted, it still needs to be firmly integrated into a mantle or lithosphere model for Phanerozoic LIPs, because the LIP magma feeding system is evidently connected with the distribution of mantle melting at depth. Some authors have proposed that dikes are fed vertically from mantle ridge structures (or linear thinned zones) that are undergoing partial melting and follow the same trend as the swarms (e.g.,

Speight et al., 1982). Such a view is also implicit in the work of Al-Kindi et al. (2003). Others, such as White and McKenzie (1989), suggest that the magma migrates upward from a more or less homogeneously subcircular melting mantle plume head (see a variation of this plume model by introducing preexisting lithospheric thin spots into Thompson and Gibson, 1991, and Nielsen et al., 2002).

In the present study, we discuss these views using a statistical analysis of the flow vectors in selected dike swarms emplaced during both the trap and the SDR stages of north Atlantic volcanic province evolution.

MEASURING FLOW VECTORS IN DIKES AND DIKE SWARMS

Magma Flow Vectors Estimated by AMS

In the field, it is often difficult to determine with precision the fossilized flow vector in dikes. This is due to the scarcity of observable flow indicators both inside (e.g., oriented phenocrysts, elongated gas vesicles) and along the walls of the intrusions (e.g., mechanical lineations; Baer and Reches, 1987). Our method for studying magma flow in the north Atlantic volcanic province dikes is based on AMS. The AMS technique consists of determining the maximum, intermediate, and minimum principal axes (K1, K2, and K3, respectively) of the magnetic susceptibility ellipsoid of a rock sample submitted to a weak magnetic field (for an explanation of the technique, see Rochette et al., 1991). The application of AMS to the petrofabric study of basaltic dikes has been extensively discussed (e.g., Ellwood, 1978; Knight and Walker, 1988; Hargraves et al., 1991; Rochette et al., 1991; Ernst and Baragar, 1992; Staudigel et al., 1992; Baer, 1995; Varga et al., 1998; Aubourg et al., 2002). Briefly, the magnetic foliation in basalts (i.e., the plane containing axes K1 and K2, perpendicular to K3) is considered to reflect the distribution of ferromagnetic oxides in the rock mass. Depending notably on the time of appearance of these grains during differentiation of the magma, their distribution is thought to correspond directly (e.g., Borradaile, 1988) or indirectly (Hargraves et al., 1991) to the fossilized magmatic foliation (i.e., a flow plane; see Nicolas, 1992). A number of authors assume that the AMS axis K1 (i.e., the magnetic lineation) yields the orientation of the long axes of multidomain magnetic grains (i.e., Borradaile, 1988). In dikes, K1 would thus indicate the trend (but not the absolute direction) of the magma flow, i.e., the magmatic lineation (e.g., Staudigel et al., 1992; Varga et al., 1998). The commonly observed obliquity of K1 axes relative to the walls of a dike was termed "imbrication fabric" by Knight and Walker (1988). This imbrication has been used by some authors to determine the absolute direction of flow (e.g., Blanchard et al.,1979; Knight and Walker, 1988; Staudigel et al., 1992; Baer, 1995). This fabric would result from the downstream and oblique distribution of phenocrysts due to the strong velocity gradients existing in the magma near the walls of a dike (Fig. 6A).

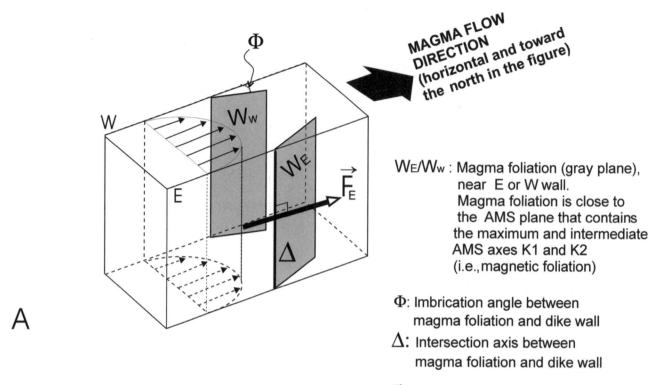

WE/WW : Magma foliation (gray plane),
near E or W wall.
Magma foliation is close to
the AMS plane that contains
the maximum and intermediate
AMS axes K1 and K2
(i.e., magnetic foliation)

Φ: Imbrication angle between
magma foliation and dike wall

Δ: Intersection axis between
magma foliation and dike wall

$\vec{F}_E$: Flow vector, eastern wall of the dike

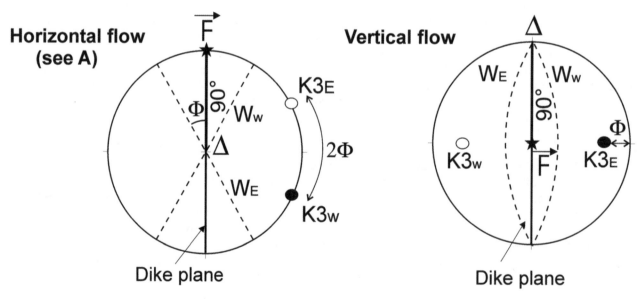

F: Flow vector (merged vectors from the eastern and western walls)
Note: With the above geometric method the orientation of the magnetic axes
K1 and K2 is not taken into account any more)

Figure 6. The AMS (anisotropy of magnetic susceptibility) "magmatic foliation" method (Geoffroy et al., 2002). (A) Velocity fluid within a dike (Newtonian laminar flow) and related orientation of the imbricate magmatic foliation. The flow vector at each wall F is considered as the axis, on the dike wall, perpendicular to the intersection axis Δ between the magnetic foliation, whose pole is the AMS axis K3, and the wall. (B) Example of flow vector determination in the case of horizontal (left) or subvertical (right) downward flow. All projections are lower hemisphere.

Flow Vectors Determined Solely from K3 Axes

It is difficult to use the AMS technique in basaltic rocks because independent measurements of the magma flow direction from thin sections in the (K1, K2) planes demonstrate that the K2 axis may also be subparallel to the alignment of phenocrysts in the rock mass (Ellwood, 1978; Moreira et al., 1999; Geoffroy et al., 2002; Callot and Geoffroy, 2004). This observation implies that K1 cannot be used indiscriminately as an indicator of the flow lineation in basaltic intrusions or lavas. In some cases, K1 would represent the intersection axis between shear-type and foliation-type magma planes (Callot and Guichet, 2003). Geoffroy et al. (2002) have proposed avoiding the incorrect use of K1 axes as flow indicators by considering just the angle—when it can be distinguished—between magnetic *planes* with respect to each wall of the dikes (Fig. 6). In other words, we should use solely the K3 axes (poles of magnetic foliation) and the poles of the dike walls to determine a mean flow vector for the intrusion in order to avoid any misinterpretation of K1 or K2 as flow axes (Fig. 6). For example, Ernst and Baragar (1992) conclude that there was a centrifugal flow of magma within the Proterozoic McKenzie giant dike swarm away from the area of convergence of the dike trends (where they infer vertical flow). Although this is a major result, it is based solely on K1 statistics. Their data should be reworked using the K3 methodology described earlier.

In addition, it may be appropriate to consider not just individual dikes but sets of parallel intrusions of similar thickness and composition, assumed (or demonstrated) to be of the same age (e.g., Callot et al., 2001). In such cases, the working hypothesis is that the set of intrusions were (1) emplaced at the same depth, (2) derived from the same reservoir, and (3) governed by the same dynamics. We can then analyze the statistical grouping of the K3 axes from the whole-core data obtained from the two opposite walls of the dikes (which yield statistical imbrications at the walls), with all data represented in terms of "dike coordinates" (see Rochette et al., 1991). We follow the same reasoning and computation to determine the mean flow orientation within the swarm as applied in the case of a single intrusion.

MAGMA FLOW DURING THE TRAP EMPLACEMENT STAGE: THE CASE OF THE ISLE OF SKYE

Dike Swarms on the Isle of Skye

The Paleocene tholeiitic dike swarms of the British Tertiary igneous province follow a northwest-southeast to NNW-SSE trend and are related to Paleocene to Eocene igneous centers (e.g., Vann, 1978; Speight et al., 1982) (Figs. 3 and 7). Two sub-parallel sets of Paleocene dikes are known on the Isle of Skye (Mattey et al., 1977). Alkaline dikes seem uniformly distributed over the island and are associated with a small finite dilatation (not exceeding 1%). They probably fed the Skye Main Lava Series (Mattey et al., 1977). These dikes are postdated by a prominent tholeiitic swarm, focused on the Skye igneous center, that is associated with significant northeast-southwest-trending magma dilatation (up to 20%) displaying a dual positive gradient (Fig. 7) not only toward the symmetry axis of the swarm but also toward the igneous center (Speight et al., 1982). According to Bell (1976) and Mattey et al. (1977), this major swarm fed the (nowadays, mostly eroded) tholeiitic traps (the "Preshal Mohr lavas").

AMS studies have already been conducted in the Skye acid ring–dikes (Geoffroy et al., 1997) and the mafic cone sheet (Herrero-Bervera et al., 2001). Both of these studies concluded that magma flow within the annular intrusions of the igneous center was probably subvertical and governed by bottom-to-top pressure gradients from a central crustal magma reservoir. Herrero-Bervera et al. (2001) also investigated the flow pattern within nine intrusions belonging to the regional dike swarm, concluding that there had been some lateral magma flow within the mafic dikes. However, both their methodology and their AMS interpretation were questioned by Aubourg and Geoffroy (2003).

AMS Study of the Tholeiitic Swarm: A Technical Approach

We present here, for the first time, a study carried out in 1995 on magma flow in the Isle of Skye dike swarm (Geoffroy and Aubourg, 1997). We sampled 522 samples in the walls of 30 basaltic dikes (1J to 30J). To avoid turbulent flow, we cored dikes with a thickness not exceeding 1.65 m (average thickness 0.9 m). We preferentially cored dike margins, where cooling had been more rapid, to obtain the largest flow velocity gradients and avoid postinjection rearrangement of the flow fabric. We selected only the basal 2.2 cm of the cores for measurement to minimize any effects due to weathering. The dikes were sampled from six sites (Fig. 7A–F) located at different distances from the igneous center along the general northwest-southeast trend of the swarm. Although all the dikes are tholeiitic, two of them (5J and 17J) nevertheless display high K_2O contents (>1 wt%) (Table 1). The tholeiitic dikes clearly belong to the Preshal Mohr type of basalts (Mattey et al., 1977; Kent and Fitton, 2000). The well-defined trends in Figure 8 suggest a single parental melt composition, with magmatic processes dominated by crystal fractionation, possibly within the same reservoir. The magnetic susceptibilities range from 10^{-4} to 10^{-1} SI, which clearly indicates the predominance of magnetite in the rock mass. The rock magnetic fabric is dominantly planar, with an average magnetic foliation ratio exceeding the average magnetic lineation in 74% of the dike walls. In most dikes, there is a closer clustering of the K3 axes (Table 2) compared to the much more scattered K1 axes at both walls of the intrusions. Oriented thin sections made from five key samples demonstrate that both K1 and K2 could represent the flow lineation (Geoffroy and Aubourg, 1997), which justifies our choice in using only the magnetic foliation to determine the flow vector orientation.

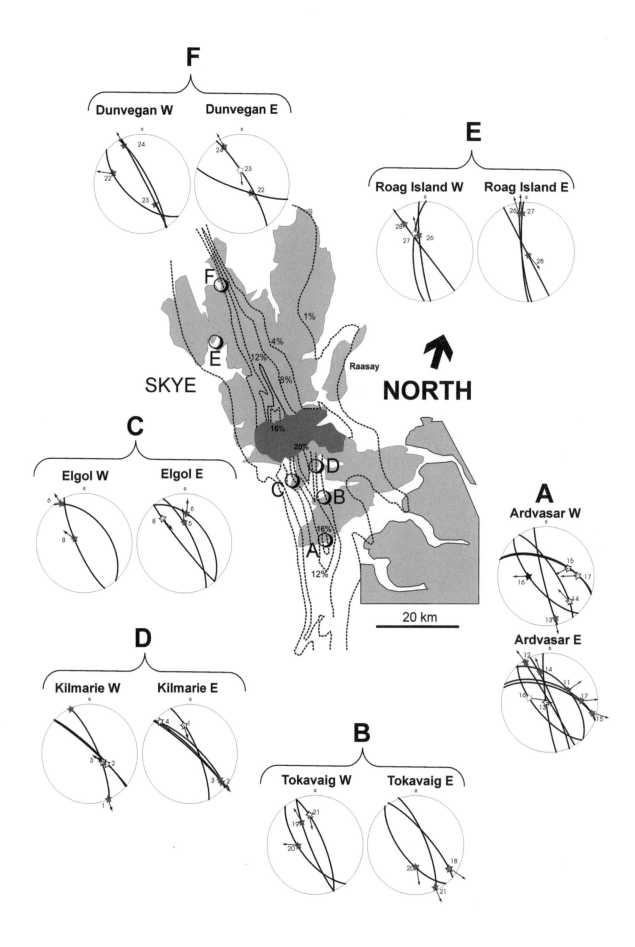

F

Dunvegan W   Dunvegan E

E

Roag Island W   Roag Island E

C

Elgol W   Elgol E

D

Kilmarie W   Kilmarie E

A

Ardvasar W

Ardvasar E

B

Tokavaig W   Tokavaig E

SKYE

Raasay

NORTH

20 km

1%
4%
12%
8%
16%
20%
16%
12%

TABLE 1. GEOCHEMISTRY OF DIKES FROM THE ISLE OF SKYE

Dike	SiO_2	TiO_2	Al_2O_3	Fe_2O_3	MnO	MgO	CaO	Na_2O	K_2O	P_2O_5	LOI	Total	Sr
8J	46.25	1.63	13.20	15.45	0.25	6.10	11.40	2.12	0.08	0.15	3.09	99.72	135
9J	45.50	1.46	14.60	13.70	0.19	6.07	10.35	1.65	0.08	0.13	5.4	99.13	181
11J	44.50	2.27	14.15	14.50	0.19	4.62	12.30	3.00	0.28	0.20	3.77	99.78	211
4J	47.20	2.13	12.67	17.00	0.26	4.88	9.60	2.44	0.43	0.22	2.49	99.32	139
5J	48.10	1.72	13.92	14.12	0.23	4.41	8.42	2.84	1.11	0.32	4.39	99.58	341
7J	46.25	0.79	17.89	10.05	0.17	6.95	13.60	1.75	0.05	0.07	2.21	99.78	121
1J	46.40	1.23	13.84	13.94	0.23	7.30	12.35	1.99	0.15	0.10	1.91	99.44	116
2J	44.10	0.84	14.25	11.20	0.19	8.65	14.20	1.47	0.15	0.07	4.35	99.47	143
3J	44.25	1.16	14.75	11.90	0.25	6.74	12.60	1.90	0.12	0.10	5.82	99.59	147
18J	46.45	1.11	14.25	12.30	0.23	6.74	13.70	1.96	0.17	0.10	2.32	99.33	130
19J	47.50	1.05	14.25	12.70	0.21	7.85	13.05	1.88	0.04	0.09	1.3	99.92	105
20J	46.90	1.10	13.80	12.68	0.25	7.14	13.12	1.96	0.06	0.10	2.73	99.84	121
15J	47.00	0.95	15.80	11.40	0.19	7.01	13.50	1.95	0.11	0.10	1.94	99.95	141
16J	46.25	0.79	18.60	9.60	0.15	6.74	13.25	1.81	0.15	0.08	2.08	99.5	142
17J	51.35	1.76	14.05	13.10	0.19	3.45	6.65	3.10	1.94	0.37	3	98.96	420
12J	46.40	1.59	14.10	14.10	0.23	6.44	12.10	2.30	0.18	0.15	2.39	99.98	161
13J	47.90	1.16	14.13	13.10	0.22	7.03	12.50	1.94	0.19	0.11	1.59	99.87	131
14J	46.90	1.11	14.50	12.95	0.22	6.93	13.10	1.95	0.11	0.10	2.23	100.10	141
28J	46.60	0.84	15.76	11.28	0.18	9.38	12.18	2.05	0.06	0.07	1.49	99.89	90
29J	45.75	0.85	15.24	11.52	0.18	8.34	11.65	2.17	0.18	0.07	3.66	99.61	173
30J	46.70	1.00	14.60	12.35	0.20	8.05	12.40	2.07	0.07	0.08	2.43	99.95	99
25J	46.75	1.07	13.95	12.50	0.20	8.00	12.05	2.33	0.16	0.09	2.19	99.29	117
26J	47.15	1.08	14.75	12.80	0.20	7.14	12.65	2.08	0.09	0.09	1.37	99.40	100
27J	47.40	1.06	14.35	13.06	0.21	7.21	12.70	2.21	0.12	0.09	1.3	99.71	111
21J	45.90	2.02	13.10	16.25	0.23	5.48	10.55	2.68	0.26	0.18	2.9	99.55	162
23J	46.75	1.07	13.90	12.50	0.20	7.82	12.16	2.41	0.14	0.09	2.70	99.74	114
24J	47.05	1.08	14.00	12.70	0.20	7.64	12.35	2.07	0.19	0.09	2.32	99.69	102

Because extreme caution is needed in applying AMS to mafic rocks, we used a very critical approach to interpret our results (see details in the notes to Table 2). In our study, twenty-four dikes out of a total of thirty provided results that could be interpreted in terms of orientation of magnetic foliation relative to at least one of the dike walls (Table 2 and Fig. 7). However, there is frequently a large discrepancy between the flow vectors computed from walls on either side of a dike, at the worst yielding nearly opposite directions (Table 2 and Fig. 7). We were able to determine consistent flow vectors (see notes to Table 2) for both of the walls in only eight dikes (1J, 6J, 20J, 21J, 22J, 24J, 26J, and 28J), while considering as (possibly) valid the single-wall data from 5J, 11J, 12J, and 18J (Table 2).

Figure 7 presents all the interpretable data for both walls of each dike at all the studied sites. Figure 9 presents a statistical analysis of K3 and K1 axes grouped by area: all data from the northwest or southeast of Skye are pooled (Groups 1 and 2, respectively) with the exception of data from site Ardvasar (A). In Figure 9, K1 and K3 statistics are expressed in dike coordinates (Rochette et al., 1991; this means than all AMS data are rotated with the dike plane oriented vertically, assuming an arbitrary north-south trend).

Analysis and Interpretation of Results

We summarize the Skye data as follows (Table 2; Figs. 7, 9, and 10). Analysis of the individual data reveals that, apart from dikes 1J and 21J, the flow vectors are all downward-plunging on the Isle of Skye. Southeast of the igneous center (sites A–D), the flow pattern is complex. The flow vector is oriented outward from the igneous center for two dikes (1J and 18J), while it is directed inward in three cases (5J, 6J, and 12J) and downward in two others (11J and 20J). Northwest of the igneous center, the flow vectors are plunging both downward and outward (24J, 26J, and 28J), with the exception of the inward- and downward-plunging 22J (Figs. 7, 9, and 10).

The statistical study of data for the area northwest of the igneous center (Fig. 9, Group 1) shows that (in dike coordinates) the magnetic foliation (K3) yields a very clear imbrication at the "eastern" walls (Φ angle −7° with respect to the "north"), but this becomes less well defined at the western walls (Φ angle +4° with respect to the "north"). This indicates that the dominant flow in this area is lateral and directed toward the northwest (see Fig. 6). A clear imbrication is also encountered southeast of the igneous center (Fig. 9, Group. 2) at the eastern walls (Φ angle

Figure 7. Flow vectors calculated from the dike walls on the Isle of Skye (see method in Fig. 6). Black stars and outward-directed arrows—downward vectors; white stars and inward-directed arrows—upward vectors. Isodilatation curves (in %) are calculated from the dike thicknesses (in Speight et al., 1982).

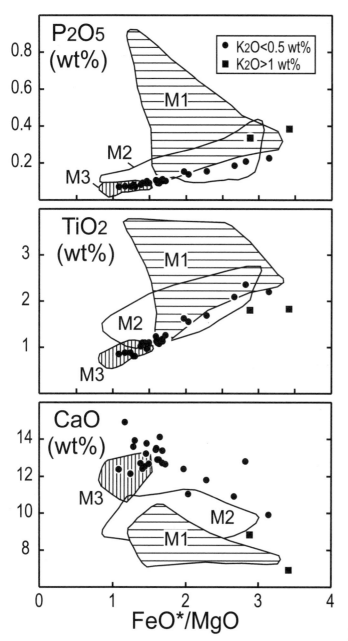

Figure 8. P$_2$O$_5$, TiO$_2$, and CaO contents plotted against FeO*/MgO. The contents are recalculated to 100% on an H$_2$O-free basis. The fields for the Mull Plateau Group / Skye Main Lava Series (M1), Coire Gorm (M2), Central Mull tholeiites / Preshal More (M3) basalt types are drawn using the analyses of dikes obtained by Kent and Fitton (2000).

+18° with respect to "north" on the diagram), with the statistical foliation parallel to the dike at the western walls (Φ angle 0°). This indicates a general southeastward lateral magma flow, i.e., away from the igneous center. Finally, at the Ardvasar site (Fig. 7) the statistical orientation of K3 at the dike walls is clouded by the coexistence of lateral magma flows directed (geographically) toward both the northwest and the southeast (two clear maxima are observed for K3 at each wall, apart from the east-west trend in Fig. 9).

To summarize, the majority of flow vectors in basaltic dikes from Skye are subhorizontal to downward plunging, which argues strongly in favor of lateral feeding from one or several high-level magma chambers (Fig. 10). The generalized northwestward and southeastward flows in areas northwest and southeast of the Skye igneous center, respectively, can be interpreted in terms of centrifugal lateral feeding of dikes (belonging to the major tholeiitic swarm) from a magma chamber located at the Skye igneous center (Fig. 10). Because both northwestward and southeastward lateral flows are encountered south of Skye, we propose to interpret this result as indicating either (1) a double-feeding source or (2) the existence of postinjection back-flow fabrics in these dikes, as reported elsewhere (Philpotts and Asher, 1994). According to the first hypothesis, and because the Skye regional swarm is connected to the Mull igneous center via a sigmoidal but continuous dilatation axis (Fig. 3B), we tentatively propose that the northwestward flows recorded on southern Skye could have originated from the contemporaneous Mull igneous center. Alternatively, we could also invoke the existence of a small magma chamber south of Skye (see the high level of finite dilatation calculated from dikes south of Skye in Fig. 7).

More specifically, we should note that the steepest plunges of flow vectors are often observed on dikes with the shallowest dips (e.g., dike 11J). In addition, most dikes that yield opposite directions of flow from one wall to the other (i.e., "class 4" dikes; see Table 2) also exhibit the shallowest dips (dikes 16J and 17J, for example). These particular cases could well be explained by a normal shear transposition of the flow-related fabric during solidification of the magma. Also, some of the results from northwest Skye (e.g., 27J and 28J) probably represent the effect of a lateral intrusive flow combined with a lateral Couette-type displacement, respectively sinistral and dextral, in excellent agreement with the NNW-SSE orientation of the maximum principal horizontal stress during the Tertiary in this region (England, 1988; Geoffroy et al., 1993; see Fig. 3). Another interesting point is the clear downward plunge of many flow vectors at some distance from the proposed feeding center (Figs. 7 and 10). The occurrence of a downward flow component in individual dikes has already been suggested from the analysis of flow markers (Baer and Reches, 1987) and petrofabrics (Shelley, 1985; Aubourg et al., 2002). Northwest of the Skye igneous center, ~50 km away from the magma chamber, our data suggest that the magma flow vector could have been systematically downward in most of the dike swarm at depths of probably less than 3 km below the Paleocene topographic surface. Because magma is injected from chambers at a level of neutral buoyancy (Rhyan, 1987), such a pattern could reflect the increase in magma density due to cooling within dikes farther away from their feeder source. Decreasing both lateral pressure gradients and negative buoyancy of the magma with respect to its host rock would promote convection of the magma within the dike fissure. Another hypothesis would involve levels of neutral buoyancy inclined away from the summit of the Skye polygenic volcano, but this

TABLE 2. CALCULATION OF FLOW VECTORS CARRIED OUT INDEPENDENTLY AT EACH DIKE WALL

Dike	Site (Fig. 7)	West dike wall — Flow west Strike	Plunge	Dynamic	δK3	Φ	Class	East dike wall — Flow east Strike	Plunge	Dynamic	δK3	Φ	Class	γ	Class	Dike (both walls) — Flow Strike	Plunge	Dynamic
1J	Kilmarie	156	0	Horizontal	9	5	B	348	40	Upward	42	19	B	41	BB2	341	20	Up
2J	Kilmarie	119	63	Upward	10	21	A	129	17	Down	38	22	B	132	AB4			Down
3J	Kilmarie	121	55	Upward	22	18	B	130	11	Down	11	18	A	136	BA4			Down
4J	Kilmarie	339	60	Down	11	4	NO	315	13	Upward	34	25	B	130	OB			
5J	Elgol	Undefined						358	57	Down	20	23	A		OA	20	23	Down
6J	Elgol	328	9	Down	10	23	A	3	39	Down	3	18	A	43	AA2	343	25	Down
8J	Elgol	294	71	Down	9	7	B	317	32	Upward	8	27	A	139	BA4			
11J	Ardvasar	Undefined						60	52	Down	21	54	A		OA	60	52	Down
12J	Ardvasar	328	27	Down	29	2	NO	328	4	Down	7	18	A		OA	328	16	Down
13J	Ardvasar	168	16		15	18	A	262	84	Upward	3	5	A	105	AA4			
14J	Ardvasar	136	31	Upward	33	21	B	340	36	Down	5	26	A	134	BA4			
15J	Ardvasar	69	49	Upward	28	26	B	104	12	Down	10	45	A	133	BA4			
16J	Ardvasar	268	61	Down	9	8	B	283	54	Upward	29	12	B	169	BB4			
17J	Ardvasar	88	34	Upward	20	16	B	89	33	Downward	32	10	B	179	BB4			
18J	Tokavaig	326	32	Upward	16	4	NO	154	19	Down	12	43	A		OA	154	19	Down
19J	Tokavaig	328	46	Down	19	13	B	159	15	Down	14	4	NO		BO			
20J	Tokavaig	266	61	Down	5	30	A	180	53	Down	15	7	B	45	AB2	217	64	Down
21J	Tokavaig	349	38	Upward	22	8	B	155	5	Down	11	16	A	45	BA2	341	17	Up
22J	Tokavaig	288	35	Down	27	39	A	134	68	Down	10	12	A	75	AA3	269	71	Down
23J	Tokavaig	150	48	Down	7	20	A	353	67	Upward	21	19	B	116	AB4			
24J	Tokavaig	335	17	Down	15	10	B	332	19	Down	20	23	A	3	BA1	334	18	Down
26J	Roag Island	340	62	Down	12	18	A	350	20	Down	30	30	A	41	AA2	347	41	Down
27J	Roag Island	328	56	Upward	11	13	A	354	22	Down	19	31	A	141	AA4			
28J	Roag Island	322	31	Down	23	26	A	152	84	Down	29	16	B	65	AB3	321	63	Down

Notes: Because dikes trend mostly northwest-southeast to NNW-SSE, we refer to the dike wall facing the southwest or northeast as the west or east wall, respectively. We systematically used the imbrication angle φ between the local orientation of the wall and the mean minimum susceptibility (K3) axes to calculate the flow vector (see Fig. 6). The angular uncertainty of K3, with 95% confidence, is δK3. The following quality criteria were used. The fabric is thought to be interpretable when φ > 5°. If φ ≤ 5° (NO entry in the table), the result is considered questionable given the uncertainty of the orientation of the dike wall (±1° resolution using a Topochaix™ magnetic compass, due to irregularities on the dike wall, etc.). When δK3 > φ, this corresponds to class A results. In such cases, the imbrication angle is truly determined given the uncertainty of K3. When δK3 ≤ φ, this corresponds to class B results. In such cases, there is an overlap between the two confidence area and the pole of the wall. The quality of the data at the scale of a dike depends on the consistency between the flow vectors calculated at each wall. The angle γ between the two flow vectors defines the following subclasses i in Table 2: i = 1 for γ < 30°, i = 2 for 30° ≤ γ < 60°, i = 3 for 60° ≤ γ < 90°, and i = 4 for γ ≥ 90°. Quality AB3 means that the W wall of a dike is of A quality, the E wall of B quality, and that the angular separation between the W and E flow vectors is comprised between 30° and 60°. We are confident only of the results from qualities AA1 to BB3; the latter type is the least reliable. We exclude any interpretation using data of quality i = 4. We selected dikes for the final interpretation even if only one wall was determined, provided their quality was A (e.g., dike 18J, where only the eastern wall provides flow). All detailed Skye data, individual core and averaged dike AMS (anisotropy of magnetic susceptibility) and magnetic susceptibility data, are available on simple request to the first or second author (laurent.geoffroy@univ-lemans.fr or aubourg@geol.u-cergy.fr).

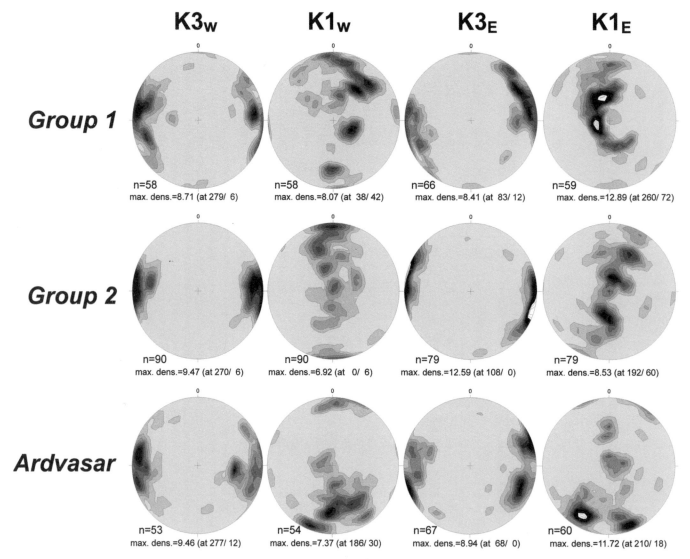

Figure 9. Plot of the K3 and K1 axes (poles of magnetic foliation) from the western and eastern margins of the dikes in "dike coordinates." Group 1—E and F together; Group 2—B, C, and D. Ardvasar is A in Figure 7.

seems to conflict with the strong plunges of flow vectors northwest of Skye.

MAGMA FLOWS AT THE VPM STAGE: THE EAST GREENLAND CASE

Between ~66°N and 68°N, the southeast Greenland coast partly exposes the western VPM that formed during the Eocene when Greenland and Europe split away to form the Reykjanes basin (Figs. 3A and 5). Three fieldwork campaigns in 1998, 1999, and 2000 were chiefly aimed at establishing the mean magma flows within a mafic dike swarm that crosscuts the transitional crust. This major dike swarm trends northeast to NNE, with a clear dilatation gradient across-strike of the margin (northwest to southeast; Fig. 5). The gradient also increases toward the coastal igneous centers that punctuate the margin (Myers, 1980;

Bromann-Klausen and Larsen, 2002; Callot, 2002). The coastal outcrop area represents the flexed transitional crust located beneath an inner SDR wedge, which is nowadays eroded (Geoffroy, 2005). Many dikes were passively tilted during SDR formation. While some of them were injected during the flexing, another set of vertical intrusions postdates the crustal flexing (e.g., Karson and Brooks, 1999; Bromann-Klausen and Larsen, 2002; Lenoir et al., 2003).

We focused especially on the dike swarm centered on the Imlik-Kialineq igneous center (Fig. 5). This swarm is located at the southwest edge of the intrusive complex. A total of 44 dikes were sampled over a distance of 125 km, representing a total of 1172 drilling cores, making this analysis one of the most extensive ever carried out at the scale of a dike swarm. Based on a quantitative comparison between K1 and the observed textural fabric (i.e., paramagnetic phenocrysts) in thin sections from 52

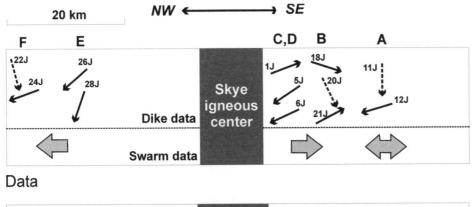

Data

Figure 10. Summary of flow vector data and interpretation for the Isle of Skye.

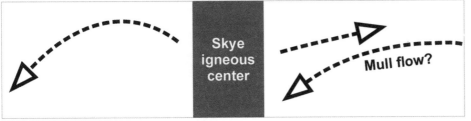

Interpretation

cores, we concluded that neither K1 nor K2 could represent a valid estimate of the flow vector orientation (Geoffroy et al., 2002; Callot and Geoffroy, 2004). We thus drilled specifically chilled margins, considering only the imbricate foliation fabrics reliable for inferring flow vectors.

The results of this study have already been published (Callot et al., 2001; Callot and Geoffroy, 2004) and are only summarized here (Fig. 11):

1. Dike flow vectors could be interpreted for 24 of the 44 intrusions studied (Fig. 11). In all cases but two, the individual flow vector is directed to the southwest.
2. Magma flow vectors at the scale of the dike swarm studied are remarkably consistent with a subhorizontal magma flow toward the southwest.

We thus have little doubt that the overlying SDR (volcanic formations) along this VPM were fed laterally (i.e., along strike) from the upper-crustal igneous centers, not vertically as initially thought.

MAGMA FEEDING MODEL FOR THE NORTH ATLANTIC LIP

The Accretion Center Model

Thus, in the north Atlantic volcanic province, we can infer that both traps and SDR are fed laterally by magmas collected in central crustal reservoirs. By itself, this result is not surprising (but needed to be confirmed), because this type of lateral feeding mechanism has long been established at slow-spreading or moderate-spreading oceanic accretion axes (e.g., Staudigel et al., 1992), in Iceland (e.g., Sigurdsson, 1987) or in Hawaii (e.g., Fiske and Jackson, 1972; Knight and Walker, 1988; Tilling and Dvorak, 1993; Parfitt et al., 2002). Such a mechanism seems to be predominant in mafic volcano tectonic systems (Parfitt et al., 2002). The lateral feeding model is also in good agreement with the general observation that lavas forming traps and SDR are differentiated by fractional crystallization in high-level crustal magma chambers (e.g., Cox, 1980; Andreasen et al., 2004). The mechanisms that control dike nucleation in magma chambers have been thoroughly investigated and do not need to be further discussed here (e.g., McLeod and Tait, 1999). Figure 12 presents a horizontal plan illustrating the concept of a single LIP "accretion center" at the depth of the magma chamber. This accretion center model defines the elementary volcano tectonic segmentation in LIP-related volcanic rifts and margins (Geoffroy, 2005; see also Ebinger and Casey, 2001). One may debate the importance of lateral transport of low-viscosity magmas along cracks in controlling the regional distribution of traps and SDR. It is not easy to determine the along-strike length of individual dikes because dikes, like any tabular intrusion, are segmented in 3-D. Nevertheless, magma has been shown to flow laterally as far as 100 km in the Hawaii dikes (e.g., Parfitt et al., 2002). We suggested earlier (Fig. 10) that some of the dikes on the Isle of Skye are fed by the Mull igneous center, corresponding to ~200 km of lateral flow (Fig. 3B). McDonald et al. (1988) concluded from geochemical evidence that the Cleveland Dike (Fig. 3B) was fed laterally in a single pulse from the Mull igneous center, which would represent up to 430 km of subhori-

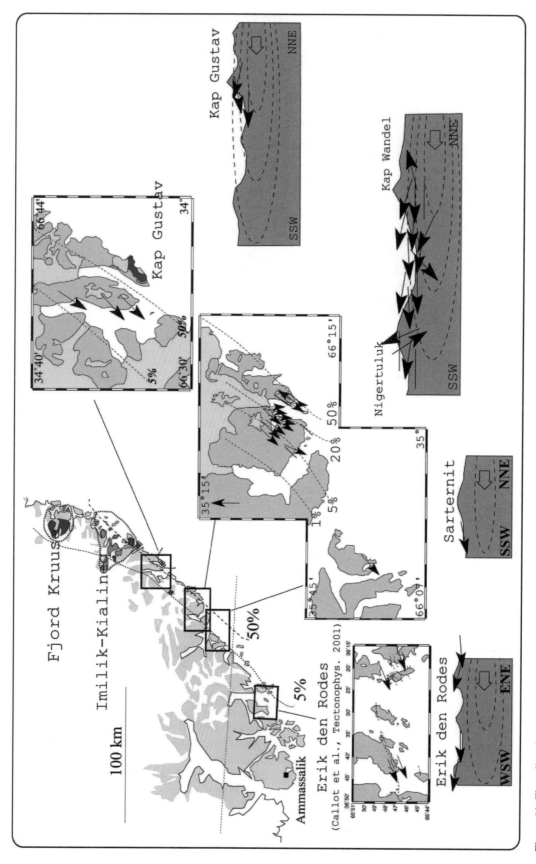

Figure 11. Flow directions obtained for the east Greenland margin, on map and vertical cross-sections along strike (from Callot and Geoffroy, 2004).

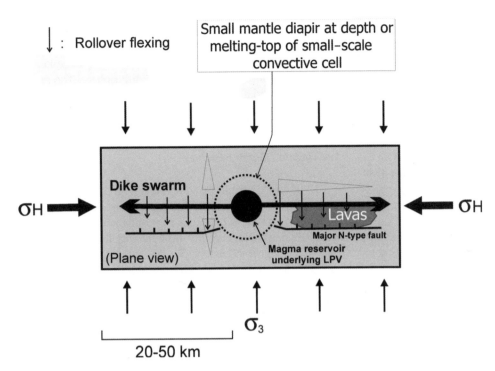

Figure 12. Concept of a large igneous province accretion center, from Geoffroy (2005). Dikes inject from igneous centers in the trend of the maximum horizontal stress (σ_H). The figure illustrates the idea that tectonic stresses within the crust indirectly control the distribution of lavas. This control is twofold: (1) magma erupts along dikes after lateral transport from the central magma reservoir; (2) the flow and distribution of lavas erupting from the feeder dikes is primarily controlled by their intrinsic viscosity, but also, in many cases, by the hanging-wall flexural topography of active normal faults that develop parallel to the dikes, following the trend of σ_H. This tectonic control of lava flow takes place both during trap formation (e.g., Doubre and Geoffroy, 2003) and during the volcanic passive margin breakup stage (seaward-dipping reflector development). The seaward-dipping reflectors are analogous to fault-controlled rollover structures that develop during the volcanic activity (Geoffroy, 2005). σ_3—minimum principal stress; LPV—large polygenic volcano.

zontal flow. Because low-viscosity lavas may flow over great distances from their eruptive fissures (Self et al., 1997; see also Fig. 12 for the role of faults), the observations made earlier imply that the areal extent of LIP lavas is controlled by crustal processes rather than the mantle.

Could the Whole of the North Atlantic LIP Magmas Be Drained through Individual Magma Centers?

Igneous centers are thus the key to understanding the distribution of melting in the mantle underlying LIPs. Although the approach is highly speculative, it is possible to estimate the volume of magma that has transited through individual north Atlantic igneous centers. We estimate that a minimum of thirty-eight igneous centers were active during the trap stage in the north Atlantic (Fig. 4). The number of igneous centers active during the Eocene breakup stage is presently unknown (see Chapter 1). Basing our estimate solely on the trap stage, we obtain a volume of 10^6 km^3, which seems a very large upper-limit value for the magma extruded and intruded during the period from 62 to 58 Ma (Paleocene–Earliest Eocene) (see White and McKenzie, 1989; Eldholm and Grue, 1994). The average upper-limit output rate at an individual igneous center would thus be ~7 × 10^{-3} km^3/yr. This value could be compared with the volume of magma represented by a dike intrusion phase to estimate an average eruption rate. However, as is explained further later, such an exercise may be meaningless.

One of the best-documented cases is from Hawaii, which provides an output of 3.3 × 10^{-3} km^3 (Cervelli et al., 2002). A

similar value (7.5 × 10^{-3} km^3) is obtained from GPS measurement in the Galapagos, following the 1995 Fernandina flank eruption (Jónsson et al., 1999). The former estimates relate to both thin (<1 m) and nonfeeder intrusions. Although in a different geodynamic context, the lateral dike intrusion event monitored in the Izu Islands during the year 2000 corresponds to an estimated magma volume of ~1 km^3 (Nishimura et al., 2001). The volume of the Cleveland Dike (thickness ~20 m) appears to attain 85 km^3 according to McDonald et al. (1988). Self et al. (1997) report a volume of 1300 km^3 for a single lava flow in the Columbia River LIP, thus implicitly setting a lower-limit value for the associated feeder dike. With such a range of values (over six orders of magnitude), it is not possible to evaluate the average volume of magma coming from a single igneous center. However, to obtain a gross estimate of the dike intrusion frequency and related magma volumes, we can tentatively refer to the two most intensively investigated mafic central volcanoes, i.e., Kilauea in Hawaii (e.g., Tilling and Dvorak, 1993) and Krafla in Iceland (e.g., Sigurdsson, 1987; Hofton and Foulger, 1996), which are situated in intraplate and plate boundary settings, respectively. In both cases, the magma supply at igneous centers seems highly dependent on the progressive buildup of stress within the surrounding crust, irrespective of whether these stresses are due to gravity (Hawaii) and/or plate tectonics (Iceland).

The most recent intrusive activity on Kilauea (since 1956) appears to fit with at least one dike intrusion every 4 yr, with periods of much higher activity (see Tilling and Dvorak, 1993). In the case of Krafla, it seems that periods of quiescence lasting 100–150 yr (periods of tectonic stress concentration) alternate with episodic faulting or diking events (the last one spanning 6 yr from 1975), during which about twenty dikes were injected laterally (Sigurdsson, 1987) parallel to the trend of the maximum horizontal stress. We should note that, in both Hawaii and Iceland, the volume of magma intrusion in dikes during an intrusive or eruptive event largely exceeds the volume represented by magma chamber deflation. In this way, diking events reflect the continuous feeding of the upper-crustal reservoirs by the mantle. The total flow out the Krafla reservoir during the last period of activity was ~1.08 km^3, which corresponds to an average of 8×10^{-3} km^3/yr over a period of 125 yr. The mean output rate of Hawaii gives a strikingly similar value when averaged since 1840 (Tilling and Dvorak, 1993). These values compare well with the estimated maximum output rate of ~7×10^{-3} km^3/yr for an individual north Atlantic volcanic province igneous center at the trap stage, suggesting that these igneous centers could be good candidates for the feeding of the whole LIP.

IMPLICATIONS FOR MANTLE MELTING

In this section we discuss the implications of our results on mantle models. We use the term *lithosphere* to refer to the mechanical entity, including the crust and part of the upper mantle, that is able to sustain stress over geological periods (e.g., Anderson, 1995). This lithosphere is thermally conductive. It is separated from the large-scale convecting mantle by a thermal boundary layer (TBL) in which temperature tends asymptotically to a convective-type gradient. This boundary layer is considered either stable or convective on a small scale (e.g., Jaupart and Mareschal, 1999; Morency et al., 2002).

Although other melting materials could be involved (see the first section of this article), the adiabatic decompressive melting of rising mantle is generally acknowledged as the primary source of magmas at LIPs and oceanic ridges. In such cases, the area of mantle melting is evidently primarily controlled by the effective mantle temperature, pressure, and volatile contents.

Our data suggest that no generalized vertical magma transfer occurs in LIPs (apart from beneath the crustal igneous centers themselves). This leads to major geodynamic consequences. The question we then need to address is the meaning of igneous centers in relation to the pattern of mantle melting. With such a localized distribution of feeders for magma crust accretion in LIPs, how could the mantle be homogeneously melting (as in the initial plume head model)?

We should bear in mind that, at the trap stage (see the section on igneous center distribution), the distribution of igneous centers seems partly independent of the premagmatic rift zones in the north Atlantic (Fig. 4). However, major discontinuities (e.g., Late Caledonian strike-slip faults, but also Mesozoic normal faults reactivating Caledonian thrusts) clearly have some influence on the igneous centers' distribution. During the VPM stage, this cannot be the case, because the igneous centers are regularly distributed within the necked and segmented crust (the area associated with SDR development; see Geoffroy, 2005;

Fig. 5). In this latter context, their spacing seems related to the amount of lithosphere necking associated with the breakup. At both stages of LIP evolution, the nonrandom distribution of igneous centers strongly suggests the existence of some kind of small-scale fluid instability within the lithosphere. The related "fluidlike" material could be present as melts (hypothesis 1: low-viscosity magma diapirs) or, as discussed further later, in the solid state (hypothesis 2: small-scale mantle diapirs).

Are Magma Diapirs Possible?

Hypothesis 1 could be compatible with the following scenario: mafic to ultramafic magma rises homogeneously through the mantle lithosphere, collecting as a continuous sill-like layer(s) at the Moho, where it partially differentiates, then ascends as diapirs to form igneous centers (Fig. 13A). It is likely that magma collects at levels of neutral buoyancy, thus explaining the presence of the HVZ at Moho depth under LIPs (Figs. 1 and 2; e.g., Fyfe, 1992; Holbrook et al., 2001). However, it is extremely improbable that Rayleigh-Taylor instabilities could develop in a basaltic layer (sill-like?), and indeed this should be ruled out. First, this would imply that the bulk density of the magma decreases more rapidly through fractional crystallization than the density increases due to cooling. Second, the fluid behavior of a magma, whether of Newtonian or power-law type, depends on several factors, including its temperature and crystal content (e.g., Weinberg and Podlachikov, 1995). A mafic magma extracted from a reservoir is more likely to behave as a Newtonian fluid, with a viscosity not exceeding 10^2 Pa/s (e.g., Spera, 1980). On the other hand, the lower-crust viscosity, even in high heatflow areas, is not expected to be lower than 10^{18} Pa/s. This constitutes a very strong obstacle for the development of Rayleigh-Taylor instabilities for a low-viscosity fluid. Nevertheless, some authors accept that a high-temperature diapir may decrease the host-rock viscosity at its edges (Spera, 1980; Rubin, 1993). This phenomenon could be enhanced by partial fusion of the country rock. In addition, if the magma diapir behaves as a power-law fluid enclosed in a power-law "ductile" crust, the buoyant stress of the diapir may also decrease the wallrock viscosity (Weinberg and Podlachikov, 1995). However, whatever the true fluid behavior of the magma, the viscosity of the Q-rich lower crust would probably not fall beneath 10^{16} Pa/s (see, for example, Weinberg and Podlachidov, 1995). Such a high viscosity ratio between the magma and the host rock suggests that this latter behaves elastically and would fracture (Rubin, 1993). This may be related to the accepted geological observation that the emplacement of mafic plutons is always fracture-associated and never involves processes of diapiric intrusion (e.g., Shaw, 1980). Moreover, we note that even the existence of acid diapirs can be questioned (e.g., Clemens and Mawer, 1992). Finally, we should add that buoyant rising magma diapirs are expected to slow down, cool, and finally solidify beneath the brittle-ductile transition in the crust. Such a scenario would completely contradict the postulated existence of large mafic magma chambers in the LIP upper crust.

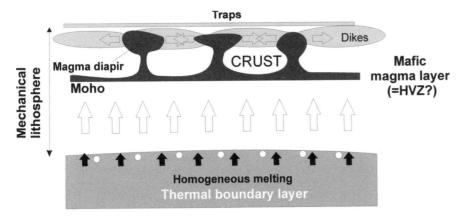

A Traps origin: magma diapir hypothesis (rejected)

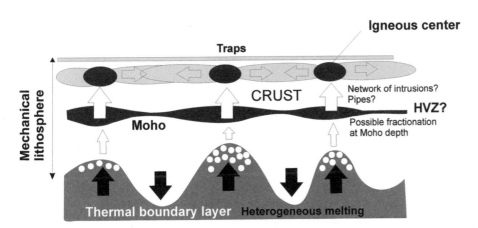

Figure 13. Two interpretations of the relationship between thermal boundary layer mantle and igneous centers: (A) homogeneous melting and magma diapirism (ruled out in the discussion) and (B) small-scale convection model (favored). HVZ—high-velocity zone.

B Traps origin: small-scale convection hypothesis (favored)

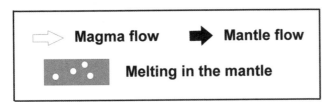

We thus believe that the only plausible explanation for the distribution and role of LIP igneous centers as magma feeders is that melting is focused within the TBL mantle itself (hypothesis 2). This may occur in two cases: (A) if the TBL is a nonconvective steady-state hot layer, but melt products or melting are localized in specific areas, or (B) if the TBL exhibits small-scale 3-D convection, and melting occurs specifically at the top of the uprising cells.

Steady-State TBL Hypothesis

We first discuss the concept of a steady-state buoyant TBL with inhomogeneous melting.

Many authors have discussed the general issue of melt extraction and migration in an adiabatically flowing melting mantle, especially at spreading ridges (e.g., Spiegelman and Reynolds, 1999). Both magma percolation through a solid compacting matrix and magma hydrofracturing have been investigated, with the latter mechanism favored (e.g., Shaw, 1980; Spera, 1980; Nicolas, 1990). Most authors acknowledge that, overall, melt buoyancy contributes in a major way to magma flow in the mantle compared with compaction and dynamic pressures (e.g., Schmeling, 2000). Therefore, we do not expect large amounts of lateral magma flow within the mantle. However, provided the magma is extracted, a highly viscous dehydrated layer could be developed at the top of the melting zone that may act as a barrier channeling the magma flow from below (e.g., Morgan, 1987; Choblet and Parmentier, 2001). This led Madge et al. (1997) to explain along-axis variations in the thickness of igneous crust at slowly spreading ridges by the lateral upslope migration of melts. Such a process is assumed to occur at the top of large-scale undulations in a passive buoyant mantle whose topology is inherited from the 3-D lithosphere structure. In particular, areas with excess conductive cooling would act as melt deflectors. While this model contradicts the diapiric model of oceanic accretion proposed by Lin et al. (1990), it might be compatible with the existence of a regional and continuous low-resistivity layer in northeast Iceland, as shown from both magnetotelluric and electrical measurements, with a minimum depth beneath the Krafla and Grimsvotn igneous centers (e.g., Bjornssön, 1985). This layer was interpreted as corresponding to a partially molten mantle (5–20% partial melt) at the roof of uprising asthenosphere domes. However, the model of Madge et al. (1997) can hardly be applied to the north Atlantic continental lithosphere case, at least at the trap stage. Indeed, there appears to be no clear correlation at this stage between the thickness of basaltic products and the initial (Cretaceous, i.e., premagmatic) thermal state of the very heterogeneous lithosphere.

Melting instabilities could also develop laterally within a thermally destabilized subhorizontal asthenosphere that is close to melting or partially molten. This scenario has been discussed by Tackley and Stevenson (1993) and Schmeling (2000), who investigated the lateral development of such instabilities using Newtonian and non-Newtonian viscosities, respectively. Although such instability propagation could explain the alignment of volcanoes with a progressive variation of ages, it cannot account for the observed 2-D distribution of igneous centers of similar ages within the north Atlantic LIP (Fig. 4).

Convective Destabilization of the TBL

We now explore the most plausible mechanism, i.e., partial melting of the TBL at the top of buoyant small-scale convection cells (Fig. 13). The spacing of igneous centers beneath the thinned north Atlantic volcanic province lithosphere at the VPM is shorter than for igneous centers punctuating the thicker trap lithosphere, thus suggesting a relationship between igneous centers and mantle dynamics. The TBL is increasingly considered to be undergoing natural small-scale convection, even in the absence of any additional heat supply (i.e., without invoking a mantle plume). This has been highlighted using different mantle rheologies and boundary conditions in a large number of experimental (e.g., Davaille and Jaupart, 1993, 1994) and numerical studies (e.g., Dumoulin et al., 1999; Callot, 2002; Morency et al., 2002).

During the trap stage, we should point out that the distribution of igneous centers is more or less homogeneous in 2-D map view (Fig. 4), and not specifically associated with, for example, thinned Mesozoic crust. Therefore, at this stage the upwelling of the melting mantle does not appear to be primarily dependent on previous tectonic stretching and thinning of the lithosphere. Such a conclusion has also been drawn from the time evolution of igneous geochemistry in the British Tertiary igneous province by Thompson and Morrison (1988) and Kerr (1994). These authors proposed a progressive and localized penetration of melting mantle into the continental lithosphere beneath the Skye and Mull igneous centers to explain the chemistry of the successive magma series.

To test the hypothesis that igneous center spacing is correlated with small-scale 3-D convection in the TBL, we would need to compare the characteristic wavelength λ of this small-scale convection with the average spacing of igneous centers in the north Atlantic (Callot, 2002). Theoretically, λ should be close to the thickness of the convective layer itself, because the convective cell aspect ratio generally lies between 1.00 and 1.35 (e.g., Houseman et al., 1981). Therefore, to resolve this issue, we need to evaluate the thickness of the convective TBL beneath a 60 Ma, relatively heterogeneous lithosphere. Although this thickness cannot be accessed directly, e.g., from geophysical data, it could be determined indirectly from experimental data. For example, Davaille and Jaupart (1993, 1994) propose an equation in which λ is inversely proportional to the surface heat flux Qs (at the time of the convection). From the data of Morency et al. (2002), we can also derive different relations (depending on the type of mantle viscosity) between λ and Qm, the mantle heatflow beneath the conductive lithosphere.

Theoretically, we could also estimate (1) the crustal heat production (and its time evolution; see Artemieva and Mooney, 2002) and (2) the cooling and recovery of the continental lithosphere. These estimates could be used to correct present-day surface heatflow (or lithosphere thickness) in the north Atlantic volcanic province and correlate it with the past wavelengths of small-scale convection cells (Callot, 2002). The final step in testing the hypothesis would be to compare the theoretical wavelength with the actual spacing of igneous centers. While this point is not fully investigated here, we nevertheless give some first-order idea of the validity of small-scale convection at LIPs. It is clear that the previous reasoning should be primarily applied to comparing trap areas that undergo little lithosphere thinning before the onset of trap formation. It is indeed difficult to estimate the true premagmatic lithosphere thickness if strong extension occurred (this depends notably on the rate of lateral cooling during extension). This should rule out any direct application to the north Atlantic volcanic province because of its relatively complex Mesozoic evolution (e.g., Van Wijk and Cloetingh, 2002; Scheck-Wenderoth et al., 2006). However, according to the previous argument, and as a first approximation, the thicker the present-day lithosphere (or the lower the surface heat flux), the longer the average spacing between igneous centers. Callot (2002) tested this hypothesis on a range of trap areas worldwide (Deccan, Siberia, Parana-Etendeka, etc.) by making use of available geological, seismological, and heatflow data, but without correcting for the cooling and evolution of the lithosphere since the associated trap emplacement (however, in many cases we may consider the error in thickness as falling within the uncertainties of estimation of present-day lithosphere thickness).

Taking account of the evident serious limitations previously outlined, Figure 14 suggests a positive correlation between lithosphere thickness (and indirectly TBL small-scale convection) and the spacing of igneous centers, even when considering the off-shore Hebrides data (heterogeneous in thickness lithosphere). Note that the main discrepancy with the general correlation shown in Figure 14 concerns the British Tertiary igneous province igneous centers, where the close spacing is fault controlled. It is noteworthy that many igneous centers lie along inherited lithospheric-scale or crustal-scale discontinuities (especially Late Caledonian subvertical shear zones) (Fig. 4). At the same time, we should also take into account the existence of a transient Paleocene lithospheric stress field with the maximum horizontal stress converging toward a single point (Fig. 3). This could suggest a relationship during trap formation between a sudden change in regional-scale stress field and the enhancement of TBL destabilization, especially along the Caledonian discontinuities that were slightly reactivated during the Paleocene.

The distribution of igneous centers during the breakup stage seems much more focalized along the thinned and stretched Eocene VPM than at the trap stage (Fig. 5). We note a strong decrease in the spacing of igneous centers (or igneous centers postulated from off-shore gravity and magnetism) from the onshore internal margin (thicker crust and lithosphere) and the off-shore or distal margin (thinnest crust or lithosphere with outer SDR). The data presented here strongly support a mechanism of VPM accretion similar to that at slowly spreading ridges, which partly explains the strong analogy in structure between the two crusts (Fig. 2). We point out that the wavelength of magma segmentation along the east Greenland VPM (located off-shore) is very similar to that observed along the adjacent Reykjanes Ridge (Gac and Geoffroy, 2005). Similar observations have been made elsewhere (Behn and Lin, 2000). Some authors have proposed that magma may be focused at the top of small-scale convective cells, not only at slowly spreading ridges (Lin et al., 1990), but also at volcanic passive margins (e.g., Mutter et al., 1988; Keen and Boutilier, 1999). Kelemen and Holbrook (1995) and Holbrook et al. (2001) present arguments in favor of a strongly active upwelling mantle at VPMs, with active rates up to four times faster than passive rates (stretching-related). According to Huismans et al. (2001) as well as Van Wijk et al. (2001), a significant component of active mantle upwelling (and consecutive melting) may naturally occur beneath rifts at the end of passive stretching. However, Nielsen et al. (2002) argue for a slight temperature excess in the mantle. In any case, considering the true 3-D architecture of a VPM, the active mantle upwelling should be regarded as small-scale 3-D (channeled along the breakup zone) and certainly not 2-D axisymmetric (see Geoffroy, 2005).

CONCLUSION

We show in this article that magma distribution at the LIP ground surface has little to do with the extent of mantle melting at depth. At both stages of LIP evolution, the magma is channeled

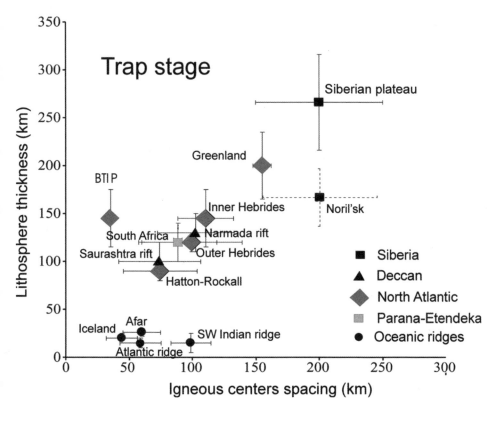

Figure 14. Estimated thickness (in km) of the seismological lithosphere against igneous centers spacing. Data from large igneous provinces (trap stage only) but also from several oceanic ridges (in km; from Callot, 2002).

through pinpoint crustal pathways that extend downward to localized melting zones in the mantle. We thus propose a magma feeding model for LIPs that is quite distinct from previous views of homogeneous mantle melting over plume heads (Fig. 15A) or homogeneous melting over deep "mantle ridges" (Fig. 15B). At both LIP stages, the best model describing the pattern described during traps emplacement invokes the existence of small-scale convection within the mantle (Fig. 15C). Small-scale convection in the TBL is not a specific LIP-related phenomena and may correspond to a generalized process beneath the mechanically rigid lithosphere (e.g., Korenaga and Jordan, 2002; Morency et al., 2002). It is probably a natural consequence of the negative buoyancy of the bottom of the lithosphere. However, such small-scale convection would have to be enhanced to explain the quite sudden mantle melting during the trap stage in LIPs (and not elsewhere or at any other time). A transient excess in TBL temperature (e.g., plume head emplacement) could cause enhanced convection. However, we suggest that other controls, such as a transient Paleocene compressive stress field (see Doubre and Geoffroy, 2003) acting in a lithosphere of highly variable thickness, could indirectly trigger mantle melting. This could be tentatively explained by diapiric destabilization at the top of the existing buoyant small-scale convecting cells, especially along reactivated lithospheric sutures. This latter explanation probably fits best with the available observations as well as the time and space constraints.

There is a strong structural analogy between VPMs and oceanic crust, which is based on their layering (Fig. 2), along-axis segmentation (Lin et al., 1990; Behn and Lin, 2000; Callot et al., 2002; Gac and Geoffroy, 2005), and crustal growth mechanisms (this study; for spreading ridges see Staudigel et al., 1992; Madge et al., 1997). This is probably due to similar mantle and crustal growth processes (Geoffroy, 2005). To improve our understanding of VPMs, we need to ask why some continental rifts function as oceanic rifts, although with enhanced magmatism (Fig. 2), while other rift systems do not. Both types of rift system (i.e., amagmatic and magmatic) are formed under extensional stress regimes and may develop at the margins of cratonic areas; the east Greenland VPM is an example of a magmatic system, whereas the Baikal rift is an amagmatic system (e.g., Pavlenkova et al., 2002). This suggests that lithosphere edge effects (e.g., Anderson, 1994; Sheth, 1999) are not the sole indirect cause of magmatism at VPMs. We argued elsewhere that the pattern of lithospheric deformation at VPMs during breakup is quite dependent on the weakening of the upper-lithosphere mantle by low-viscosity anomalies located within the mantle lithosphere (the soft-point model of continental breakup; see, for example, Callot et al., 2002, and Geoffroy, 2005). These low-viscosity anomalies explain both the 3-D strain localization and the rift propagation at VPMs (Callot et al., 2002; Geoffroy, 2005). They fit very well with the postulated zones of mantle melting at depth presented here, thus giving a consistent model of combined magma, rheologic, and tectonic evolution at VPMs

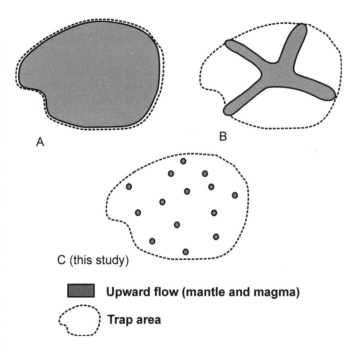

Figure 15. Mantle models for trap provinces (plane view, outcrop area delimited by dashed line). The shaded areas are sublithospheric areas associated with upward mantle flow and adiabatic melting. (A) Plume head model. (B) Mantle ridge model. (C) Small-scale convection/diapirism model (favored).

(Geoffroy, 2005). Here again, the 3-D sublithospheric convection pattern provides a key to understanding the origin and evolution of VPMs. Probably all rift systems are associated with small-scale convection in the TBL (see Korenaga and Jordan, 2002). The differences in along-strike segmentation between nonvolcanic (e.g., Rhine Graben) and volcanic rifts (e.g., the Ethiopian rift) suggest that the wavelength of small-scale convection is different in the two cases, thus also implying differences in mantle viscosity (see the discussion of the 3-D pattern of oceanic accretion in Choblet and Parmentier, 2001). Our study does not provide a solution for this particular issue. However, we suggest that during the latest stages of continental breakup in a LIP the small-scale convection suddenly turns from a poorly organized sublithospheric pattern (large wavelengths) to a more regular spreading-type regime (smaller wavelengths; see Fig. 14).

ACKNOWLEDGMENTS

This article is contribution no. 2173 of the GDR Marges. The research was co-supported by Euromargin (LEC-01) and Institut Paul-Emile Victor (IPEV) (Program 290). Don Anderson, Gillian Foulger, Donna Jurdy, and Jolante Van Wijk are warmly thanked for their constructive and interesting comments. M. Carpenter is thanked for his efficient correction of the English language.

REFERENCES CITED

Al-Kindi, S., White, N., Sinha, M., England, R., and Tiley, R., 2003, Crustal trace of a hot convective sheet: Geology, v. 31, p. 207–210, doi: 10.1130/ 0091-7613(2003)031<0207:CTOAHC>2.0.CO;2.

Anderson, D.L., 1994, The sublithospheric mantle as the source of continental flood basalts: The case against the continental lithosphere and plume head reservoir: Earth and Planetary Science Letters, v. 123, p. 269–280, doi: 10.1016/0012-821X(94)90273-9.

Anderson, D.L., 1995, Lithosphere, asthenosphere and perisphere: Review of Geophysics, v. 33, p. 125–149, doi: 10.1029/94RG02785.

Anderssön, D.L., 2004, What is a plume?: http://www.mantleplumes.org/ PlumeDLA.html.

Anderson, E.M., 1951, The dynamics of faulting and dyke formation with applications to Britain: Edinburgh, Oliver and Boyd, 206 p.

Andreasen, R., Peate, D.W., and Brooks, C.K., 2004, Magma plumbing systems in large igneous provinces: Inferences from cyclical variations in Palaeogene East Greenland basalts: Contributions to Mineralogy and Petrology, v. 147, p. 438–452, doi: 10.1007/s00410-004-0566-2.

Artemieva, I.M., and Mooney, W.D., 2002, On the relation between cratonic lithosphere thickness, plate motions, and basal drag: Tectonophysics, v. 358, p. 211–231, doi: 10.1016/S0040-1951(02)00425-0.

Aubourg, C., and Geoffroy, L., 2003, Comment on paper: Magnetic fabric and inferred flow direction in dikes, conesheets and sill swarms, Isle of Skye, Scotland: Journal of Volcanology and Geothermal Research, v. 122, p. 143–144, doi: 10.1016/S0377-0273(02)00467-5.

Aubourg, C., Giordano, G., Mattei, M., and Speranza, F., 2002, Magma flow in sub-aqueous rhyolitic dikes inferred from magnetic fabric analysis (Ponza Island, W. Italy): Physics and Chemistry of the Earth, v. 27, p. 1263–1272.

Baer, G., 1995, Fracture propagation and magma flow in segmented dikes: Field evidence and fabric analyses, Makhtesh Ramon, Israel, *in* G. Baer and A. Heimann, eds., Physics and chemistry of dikes: Rotterdam, Balkema, p. 125–140.

Baer, G., and Reches, Z., 1987, Flow patterns of magma in dikes, Makhtesh Ramon, Israel: Geology, v. 15, p. 569–572, doi: 10.1130/0091-7613(1987) 15<569:FPOMID>2.0.CO;2.

Barton, A.J., and White, R.S., 1997, Volcanism on the Rockall continental margin: Journal of the Geological Society of London, v. 154, p. 531–536.

Bauer, K., Neben, K.S., Schreckenberger, S., Emmermann, B., Hinz, R., Fechner, K., Gohl, N., Schulze, K., Trumbull, R.B., and Weber, K., 2000, Deep structure of the Namibia continental margin as derived from integrated geophysical studies: Journal of Geophysical Research, v. 105, p. 28,829– 28,853.

Behn, M.D., and Lin, J. 2000, Segmentation in gravity and magnetic anomalies along the U.S. East Coast passive margin: Implications for incipient structure of the oceanic lithosphere: Journal of Geophysical Research, v. 105, p. 25,769–25,790.

Bell, J.D., 1976, The Tertiary intrusive complex on the Isle of Skye: Proceedings of the Geological Association, v. 87, p. 247–271.

Berggren, W.A., Kent, D.V., Swisher, C.C., and Aubry, M., 1995, A revised Cenozoic geochronology and chronostratigraphy, in Berggren, W.A., et al., eds. Geochronology, time scales and global stratigraphic correlation: Society of Economic Paleontologists and Mineralogists Special Publication 54, p.129–212.

Bjornsson, A., 1985, Dynamics of crustal rifting in NE Iceland: Journal of Geophysical Research, v. 90, p. 10,151–10,162.

Blanchard, J.P., Boyer, P., and Gagny, C., 1979, Un nouveau critère de sens de mise en place dans une caisse filonienne: Le "pincement" des minéraux aux épontes: Tectonophysics, v. 53, p. 1–25, doi: 10.1016/0040-1951(79) 90352-4.

Borradaile, G.H., 1988, Magnetic susceptibility, petrofabric and strain: Tectonophysics, v. 156, p. 1–20, doi: 10.1016/0040-1951(88)90279-X.

Bott, M.H.P., and Tantrigoda, A.D.A., 1987, Interpretation of the gravity and magnetic anomalies over the Mull tertiary intrusive complex, NW Scotland: Journal of the Geological Society, London, v. 144, p. 17–28.

Bott, M.H.P., and Tuson, J., 1973, Deep structure beneath the Tertiary volcanic regions of Skye, Mull and Ardnamurchan, North-west Scotland: Nature Phys-Sci, v. 242, p. 114–116.

Bromann-Klausen, M., and Larsen, H.C., 2002, East-Greenland coast-parallel dike swarm and its role in continental break-up, *in* Menzies, M.A., et al., eds., Volcanic rifted margins: Boulder, Colorado, Geological Society of America Special Paper 362, p. 133–158.

Callot, J.P., 2002, Origine, structure et développement des marges volcaniques: L'exemple du Groenland [Ph.D. thesis]: Université Paris VI, France, 584 p.

Callot, J.P., and Geoffroy, L., 2004, Magma flow in the East Greenland dyke swarm inferred from study of anisotropy of magnetic susceptibility: Magmatic growth of a volcanic margin: Geophysical Journal International, v. 159, p. 816–830, doi: 10.1111/j.1365-246X.2004.02426.x.

Callot, J.P., and Guichet, X., 2003, Rock texture and magnetic lineation in dykes: A simple analytical model: Tectonophysics, v. 366, p. 207–222, doi: 10.1016/S0040-1951(03)00096-9.

Callot, J.P., Geoffroy, L., Aubourg, C., Pozzi, J.P., and Mege, D., 2001, Magma flow direction of shallow dykes from the East Greenland margin inferred from magnetic fabric studies: Tectonophysics, v. 335, p. 313–329, doi: 10.1016/S0040-1951(01)00060-9.

Callot, J.-P., Geoffroy, L., and Brun, J.-P., 2002, Development of volcanic passive margins: Three-dimensional laboratory models: Tectonics, v. 21, p. 25–37.

Cervelli, P., Segall, P., Amelung, F., Garbeil, H., Meertens, C., Owen, S., Miklius, A., and Lisowski, M., 2002, The 12 September 1999 Upper East Rift Zone dike intrusion at Kilauea Volcano, Hawaii: Journal of Geophysical Research, v. 107, doi: 10.1029/2001JB000602.

Chandrasekhar, D.V., Mishra, D.C., Poornachandra Rao, G.V.S., and Mallikharjuna Rao, J., 2002, Gravity and magnetic signatures of volcanic plugs related to Deccan volcanism in Saurashtra, India, and their physical and geochemical properties: Earth and Planetary Science Letters, v. 201, p. 277–292, doi: 10.1016/S0012-821X(02)00712-4.

Choblet, G., and Parmentier, E.M., 2001, Mantle upwelling and melting beneath slow spreading ridges: Earth and Planetary Science Letters, v. 184, p. 589–604.

Clemens, J.D., and Mawer, C.K., 1992, Granitic magma transport by fracture: Tectonophysics, v. 204, p. 339–360, doi: 10.1016/0040-1951(92)90316-X.

Coffin, M.F., and Eldholm, O., 1994, Large igneous provinces: Crustal structure, dimensions, and external consequences: Review of Geophysics, v. 32, p. 1–36, doi: 10.1029/93RG02508.

Courtillot, V., Jaupart, C., Manighetti, I., Tapponnier, P., and Besse, J., 1999, On causal links between flood basalts and continental break-up: Earth and Planetary Science Letters, v. 166, p. 177–195, doi: 10.1016/S0012-821X(98) 00282-9.

Courtillot, V., Davaile, A., Besse, J., and Stock, J., 2003, Three distinct types of hotspots in the Earth's mantle: Earth and Planetary Science Letters, v. 205, p. 295–308, doi: 10.1016/S0012-821X(02)01048-8.

Courtney, R.C., and White, R.S., 1986, Anomalous heat flow and geoid across the Cape Verde Rise: Evidence for dynamic support from a thermal plume in the mantle: Geophysical Journal of the Royal Astronomical Society of London, v. 87, p. 815–867.

Cox, K.G., 1980, A model for flood basalt volcanism: Journal of Petrology, v. 21, p. 629–650.

Davaille, A., and Jaupart, C., 1993, Transient high Rayleigh number thermal convection with large viscosity variations: Journal of Fluid Mechanics, v. 253, p. 141–166, doi: 10.1017/S0022112093001740.

Davaille, A., and Jaupart, C., 1994, Onset of thermal convection in fluids with temperature-dependent viscosity: Application to the oceanic mantle: Journal of Geophysical Research, v. 99, p. 19,853–19,866, doi: 10.1029/ 94JB01405.

Dickin, A.P., Jones, N.W., Thirlwall, M.F., and Thompson, R.N., 1987, A Ce/Nd isotope study of crustal contamination processes affecting Palaeocene magmas in Skye, Northwest Scotland: Contributions to Mineralogy and Petrology, v. 96, p. 455–464, doi: 10.1007/BF01166690.

Doubre, C., and Geoffroy, L., 2003, Rift-zone development around a plume-related magma centre on the Isle of Skye (Scotland): A model for stress inversions: Terra Nova, v. 15, p. 230–237, doi: 10.1046/j.1365-3121.2003.00494.x.

Dumoulin, C., Doin, M.P., and Fleitout, L., 1999, Heat transport in stagnant lid convection with temperature and pressure dependent Newtonian or non-Newtonian rheology: Journal of Geophysical Research, v. 104, p. 12,759–12,777.

Ebinger, C.J., and Casey, M., 2001, Continental break-up in magmatic provinces: An Ethiopian example: Geology, v. 29, p. 527–530, doi: 10.1130/0091-7613(2001)029<0527:CBIMPA>2.0.CO;2.

Eldholm, O., and Grue, K., 1994, North Atlantic volcanic margins: Dimensions and production rates: Journal of Geophysical Research, v. 99, p. 2955–2968, doi: 10.1029/93JB02879.

Eldholm, O., Skogseid, J., Planke, S., and Gladczenko, P., 1995, Volcanic margin concept, *in* Banda, E., ed., Rifted ocean-continent boundaries: Kluwer Academic, p. 1–16.

Ellwood, B.B., 1978, Flow and emplacement direction determined from selected basaltic bodies using magnetic anisotropy measurements: Earth and Planetary Science Letters, v. 41, p. 254–264, doi: 10.1016/0012-821X(78)90182-6.

England, R.W., 1988, The early Tertiary stress regime in NW Britain: Evidence from the pattern of volcanic activity, *in* Norton, A.C., and Parson, L.M., eds., Early Tertiary volcanism and the opening of the NE Atlantic: Geological Society of London Special Publication 39, p. 381–389.

Ernst, R.E., and Baragar, W.R.A., 1992, Evidence from magnetic fabric for the flow pattern of magma in the Mackenzie giant radiating dyke swarm: Nature, v. 356, p. 511–513, doi: 10.1038/356511a0.

Ernst, R.E., Head, W.J., Parfitt, E., Grosfils, E., and Wilson, L., 1995, Giant radiating dike swarms on Earth and Venus: Earth-Science Reviews, v. 39, p. 1–58, doi: 10.1016/0012-8252(95)00017-5.

Esher, J.C., and Pulvertaft, T.C.R., 1995, Geological map of Greenland: Copenhagen, Geological Survey of Greenland, scale 1:2,500,000.

Favela, J., and Anderson, D.L., 1999, Extensional tectonics and global volcanism, *in* Boschi, E., Ekstrom, G., and Morelli, A., eds., Problems in geophysics for the new millennium: Bologna, Italy, Editrice Compositori, p. 463–498.

Fiske, R.F., and Jackson, E.D., 1972, Orientation and growth of Hawaiian volcanic rifts: The effect of regional structure and gravitational stresses: Proceedings of the Royal Society, London, v. 329, p. 299–326.

Fleitout, L., Froidevaux, C., and Yuen, D., 1986, Active lithosphere thinning: Tectonophysics, v. 132, p. 271–278, doi: 10.1016/0040-1951(86)90037-5.

Foulger, G.R., 2002, Plumes, or plate tectonic processes?: Astronomy and Geophysics, v. 43, p. 6.19–6.23.

Foulger, G.R., and Anderson, D.L., 2005, A cool model for the Iceland hotspot: Journal of Volcanology and Geothermal Research, v. 141, p. 1–22, doi: 10.1016/j.jvolgeores.2004.10.007.

Fyfe, W.S., 1992, Magma underplating of continental crust: Journal of Volcanology and Geothermal Research, v. 50, p. 33–40, doi: 10.1016/0377-0273(92)90035-C.

Gac, S., and Geoffroy, L., 2005, Axial magnetic anomaly segmentation along the East-Greenland volcanic passive margin compared to the magnetic segmentation of the Mid-Atlantic Ridge: A similar origin?: Geophysical Research Abstracts, v. 7, p. 7049.

Gallagher, K., and Hawkesworth, C., 1992, Dehydration melting and the generation of continental flood basalts: Nature, v., 358, p. 57–59.

Geoffroy, L., 2005, Volcanic passive margins: Comptes Rendus Géosciences, v. 337, p. 1395–1408, doi: 10.1016/j.crte.2005.10.006.

Geoffroy, L., and Aubourg, C., 1997, Mantle cylinders as rift-zone initiators in continental crust: The British Tertiary Volcanic Province as a case study: European Union of Geosciences, Strasbourg, France: Terra Nova, v. 9, Abstract Supplement no. 1, p. 75. IX, Strasbourg.

Geoffroy, L., Bergerat, F., and Angelier, J., 1993, Modification d'un champ de contrainte régional par un champ de contraintes magmatiques local: Exemple de l'Ile de Skye (Ecosse) au Paléocène: Bulletin de la Société Géologique de France, v. 164, p. 541–552.

Geoffroy, L., Bergerat, F., and Angelier, J., 1994, Tectonic evolution of the Greenland-Scotland ridge during the Paleogene: New constraints: Geology, v. 22, p. 653–656, doi: 10.1130/0091-7613(1994)022<0653:TEOTGS>2.3.CO;2.

Geoffroy, L., Olivier, P., and Rochette, P., 1997, Structure of a hypovolcanic acid complex inferred from magnetic susceptibility anisotropy measurement: The Western Red Hills granites (Skye, Scotland, Thulean Igneous Province): Bulletin of Volcanology, v. 59, p. 147–159, doi: 10.1007/s004450050182.

Geoffroy, L., Callot, J.P., Aubourg, C., and Moreira, M., 2002, Magnetic and plagioclase linear fabric discrepancy in dykes: A new way to define the flow vector using magnetic foliation: Terra Nova, v. 14, p. 183–190, doi: 10.1046/j.1365-3121.2002.00412.x.

Gernigon, L., Ringenbach, J.C., Planke, S., and Le Gall, B., 2004, Deep structures and breakup along volcanic rifted margins: Insights from integrated studies along the outer Vøring Basin (Norway): Marine and Petroleum Geology, v. 21, p. 363–372, doi: 10.1016/j.marpetgeo.2004.01.005.

Green, D.H., Falloon, T.J., Eggins, S.M., and Yaxley, G.M., 2001, Primary magmas and mantle temperatures: European Journal of Mineralogy, v. 13, p. 437–451, doi: 10.1127/0935-1221/2001/0013-0437.

Hargraves, R.B., Johnson, D., and Chan, C.Y., 1991, Distribution anisotropy: The cause of AMS in igneous rocks?: Geophysical Research Letters, v. 18, p. 2193–2196.

Herrero-Bervera, E., Walker, G.P.L., Canon-Tapia, E., and Garcia, M.O., 2001, Magnetic fabric and inferred flow direction in dikes, conesheets and sill swarms, Isle of Skye, Scotland: Journal of Volcanology and Geothermal Research, v. 106, p. 195–210, doi: 10.1016/S0377-0273(00)00293-6.

Hill, R.I., Campbell, I.H., Davies, G.F., and Griffiths, R.W., 1992, Mantle plumes and continental tectonics: Science, v. 256, p. 186–193, doi: 10.1126/science.256.5054.186.

Hitchen, K., and Richtie, J.D., 1993, New K-Ar ages and a provisional chronology for the offshore part of the British Tertiary Igneous Province: Scottish Journal of Geology, v. 29, p. 73–85.

Hofton, M.A., and Foulger, G.R., 1996, Post-rifting anelastic deformation around the spreading plate boundary, north Iceland, 2, Implications of the model derived from the 1987–1992 deformation field: Journal of Geophysical Research, v. 101, p. 25,423–25,436, doi: 10.1029/96JB02465.

Holbrook, W.S., Larsen, H.C., Korenaga, J., Dahl-Jensen, T., Reid, I.D., Kelemen, P.B., Hopper, J.R., Kent, G.M., Lizarralde, D., Bernstein, S., and Detrick, R.S., 2001, Mantle thermal structure and active upwelling during continental breakup in the North Atlantic: Earth and Planetary Science Letters, v. 190, p. 251–266, doi: 10.1016/S0012-821X(01)00392-2.

Houseman, G.A., McKenzie, D.P., and Molnar, P., 1981, Convective instability of a thickened boundary layer and its relevance for the thermal evolution of continental convergent belts: Journal of Geophysical Research, v. 86, p. 6115–6132.

Huismans, R.S., Podlachikov, Y.Y., and Cloetingh, S., 2001, Transition from passive to active rifting: Relative importance of asthenospheric doming and passive extension of lithosphere: Journal of Geophysical Research, v. 106, p 11,271–11,291.

Irvine, T.N., Andersen, J.C., and Brooks, C.K., 1998, Included blocks (and blocks within blocks) in the Skaergaard intrusion: Geologic relations and the origins of rhythmic modally graded layers: Geological Society of America Bulletin, v. 110, p. 1398–1447, doi: 10.1130/0016-7606(1998)110<1398:IBABWB>2.3.CO;2.

Jaupart, C., and Mareschal, J.C., 1999, The thermal structure and thickness of continental roots: Lithos, v. 48, p. 93–114, doi: 10.1016/S0024-4937(99)00023-7.

Jones, E.J.W., Ramsay, A.T.S., Preston, N.J., and Smith, A.C.S., 1974, A Cretaceous guyot in the Rockall Trough: Nature, v. 251, p. 129–131, doi: 10.1038/251129a0.

Jónsson, S., Zebker, H., Cervelli, P., Segall, P., Garbeil, H., Mouginis-Mark, P., and Rowland, S., 1999, A shallow-dipping dike fed the 1995 flank eruption at Fernandina volcano, Galápagos, observed by satellite radar interferometry: Geophysical Research Letters, v. 26, no. 8, p. 1077–1080, doi: 10.1029/1999GL900108.

Juteau, T., and Maury, R., 1999, The oceanic crust: From accretion to mantle recycling: Springer-Praxis Series in Geophysics, ed. Springer-Verlag, 390 p.

Karson, J.A., and Brooks, C.K., 1999, Structural and magmatic segmentation of the Tertiary east Greenland volcanic rifted margin, *in* Ryan, P., and McNiocaill, C., eds., J.F. Dewey volume on continental tectonics: Geological Society of London Special Publication 164, p. 313–318.

Keen, C.E., and Boutilier, R.R., 1995, Lithosphere-asthenosphere interactions below rifts, *in* Banda, E., ed., Rifted ocean-continent boundaries: Dordrecht, Holland, Kluwer Academic, p. 17–30.

Keen, C.E., and Boutilier, R.R., 1999, Small-scale convection and divergent plate boundaries: Journal of Geophysical Research, v. 104, p. 7389–7606.

Kelemen, P.B., and Holbrook, W.S., 1995, Origin of thick, high-velocity igneous crust along the U.S. East Coast Margin: Journal of Geophysical Research, v. 100, p. 10,077–10,094, doi: 10.1029/95JB00924.

Kent, R.W., and Fitton, J.G., 2000, Mantle sources and melting dynamics in the British Palaeogene Igneous Province: Journal of Petrology, v. 41, p. 1023–1040, doi: 10.1093/petrology/41.7.1023.

Kent, R.W., Storey, M., and Saunders, A.D., 1992, Large igneous provinces: Sites of plume impact or plume incubation?: Geology, v. 20, p. 891–894, doi: 10.1130/0091-7613(1992)020<0891:LIPSOP>2.3.CO;2.

Kerr, A.C., 1994, Lithospheric thinning during the evolution of continental large igneous provinces: A case study from the North Atlantic Tertiary Province: Geology, v. 22, p. 1027–1030, doi: 10.1130/0091-7613(1994)022<1027:LTDTEO>2.3.CO;2.

King, S.D., and Anderson, D.L., 1998, Edge-driven convection: Earth and Planetary Science Letters, v. 160, p. 289–296, doi: 10.1016/S0012-821X(98)00089-2.

Kirton, S.R., and Donato, J.A., 1985, Some buried Tertiary dykes of Britain and surrounding waters deduced by magnetic modelling and seismic reflection methods: Journal of the Geological Society of London, v. 142, p. 1047–1057.

Knight, M.D., and Walker, G.P.L., 1988, Magma flow directions in dikes of the Koolau complex, Oahu, determined from magnetic fabric studies: Journal of Geophysical Research, v. 93, p. 4301–4319.

Korenaga, J., and Jordan, T.H., 2002, On the state of sublithospheric upper mantle beneath a supercontinent: Geophysical Journal International, v. 169, p. 179–189.

Korenaga, J., Holbrook, W.S., Kent, D.-V., Kelemen, P.B., Detrick, R.S., Larsen, H.C., Hopper, J.R., and Dahl-Jensen, T., 2000, Crustal structure of the southeast Greenland margin from joint refraction and reflection seismic tomography: Journal of Geophysical Research, v. 105, p. 21,591–21,614, doi: 10.1029/2000JB900188.

Larsen, H.C., 1978, Offshore continuation of the East Greenland dyke swarm and North Atlantic Ocean formation: Nature, v. 274, p. 220–223, doi: 10.1038/274220a0.

Larsen, T.B., Yuen, D.A., and Storey, M., 1999, Ultrafast mantle plumes and implications for flood basalt volcanism in the Northern Atlantic region: Tectonophysics, v. 311, p. 31–43, doi: 10.1016/S0040-1951(99)00163-8.

Lenoir, X., Feraud, G., and Geoffroy, L., 2003, High-rate flexure of the East Greenland volcanic margin: Constraints from ^{40}Ar/^{39}Ar dating of basaltic dykes: Earth and Planetary Science Letters, v. 214, p. 515–528.

Lin, J., Purdy, G.M., Schouten, H., Sempere, J.C., and Zervas, C., 1990, Evidence from gravity data for focused magmatic accretion along the Mid-Atlantic Ridge: Nature, v. 344, p. 627–632, doi: 10.1038/344627a0.

Lustrino, M., 2005, How the delamination and detachment of lower crust can influence basaltic magmatism: Earth-Science Reviews, v. 72, p. 21–38, doi: 10.1016/j.earscirev.2005.03.004.

Madge, L.S., Sparks, D.W., and Detrick, R.S., 1997, The relationship between buoyant mantle flow, melt migration, and gravity bull's eyes at the Mid-Atlantic Ridge between 33°N and 35°N: Earth and Planetary Science Letters, v. 148, p. 59–67, doi: 10.1016/S0012-821X(97)00039-3.

Mattey, D.P., Gibson, J.L., Marriner, G.F., and Thompson, R.N., 1977, The diagnostic geochemistry, relative abundance and spatial distribution of high-calcium, low-alkali olivine tholeiitic dykes in the Lower Tertiary regional swarm of the Isle of Skye, NW Scotland: Mineralogical Magazine, v. 41, p. 273–285, doi: 10.1180/minmag.1977.041.318.16.

Mcdonald, R., Wilson, L., Thorpe, R.S., and Martin, A., 1988, Emplacement of the Cleveland Dyke: Evidence from geochemistry, mineralogy, and physical modelling: Journal of Petrology, v. 29, p. 559–583.

McHone, J.G., Anderson, D.L., and Fialko, Y.A., 2004, Giant dikes: Patterns and plate tectonics, *in* Do plumes exist?: http://www.mantleplumes.org/GiantDikePatterns.html.

McLeod, P., and Tait, S., 1999, The growth of dykes from magma chambers: Journal of Volcanology and Geothermal Research, v. 92, p. 231–245, doi: 10.1016/S0377-0273(99)00053-0.

Moreira, M., Geoffroy, L., and Pozzi, J.P., 1999, Ecoulement magmatique dans les dykes du point chaud des Açores: Étude par anisotropie de susceptibilité magnétique (Asn) dans l'île de San Jorge: Comptes Rendus de l'Académie des Sciences de Paris, v. 329, p. 15–22.

Morency, C., Doin, C., and Dumoulin, C., 2002, Convective destabilization of a thickened continental lithosphere: Earth and Planetary Science Letters, v. 202, p. 303–320, doi: 10.1016/S0012-821X(02)00753-7.

Morgan, J.P., 1987, Melt migration beneath mid-ocean spreading centers: Geophysical Research Letters, v. 14, p. 1238–1241.

Morgan, W.J., 1971, Convection plumes in the lower mantle: Nature, v. 230, p. 42–43, doi: 10.1038/230042a0.

Mutter, J.C., Buck, W.R., and Zehnder, C.M., 1988, Convective partial melting, 1: A model for the formation of thick basaltic sequences during the initiation of spreading: Journal of Geophysical Research, v. 93, p. 1031–1048.

Myers, J.S., 1980, Structure of the coastal dyke swarm and associated plutonic intrusions of East Greenland: Earth and Planetary Science Letters, v. 46, p. 407–418, doi: 10.1016/0012-821X(80)90054-0.

Nicolas, A., 1990, Melt extraction from mantle peridotites: Hydrofracturing and porous flow, with consequences for oceanic ridge activity, *in* Rhyan, M.P., ed., Magma transport and storage: New York, John Wiley and Sons, p.159–173.

Nicolas, A., 1992, Kinematics in magmatic rocks with special reference to gabbros: Journal of Petrology, v. 33, p. 891–915.

Nielsen, T.K., Larsen, H.C., and Hopper, J.R., 2002, Contrasting rifted margin styles south of Greenland: Implications for mantle plume dynamics: Earth and Planetary Science Letters, v., 200, p. 271–286.

Nishimura, T., Ozawa, S., Murakami, M., Sagiya, T., Tada, T., Kaidzu, M., and Ukawa, M., 2001, Crustal deformation caused by magma migration in the northern lzu Islands, Japan: Geophysical Research Letters, v. 28, p. 3745–3748, doi: 10.1029/2001GL013051.

Olson, P., and Nam, I.S., 1986, Formation of seafloor swells by mantle plumes: Journal of Geophysical Research, v. 91, p. 7181–7192.

Parfitt, E.A., Gregg, T.K.P., and Smith, D.K., 2002, A comparison between subaerial and submarine eruptions at Kilauea Volcano, Hawaii: Implications for the thermal viability of lateral feeder dikes: Journal of Volcanology and Geothermal Research, v. 113, p. 213–242, doi: 10.1016/S0377-0273(01)00259-1.

Pavlenkova, G.A., Priestley, K., and Cipar, J., 2002, 2D model of the crust and uppermost mantle along rift profile, Siberian Craton: Tectonophysics, v. 355, p. 171–186, doi: 10.1016/S0040-1951(02)00140-3.

Philpotts, A.R., and Asher, P.M., 1994, Magmatic flow-direction indicators in a giant diabase feeder dike: Geology, v. 22, p. 363–366, doi: 10.1130/0091-7613(1994) 022<0363:MFDIIA>2.3.CO;2.

Planke, S., Symonds, P.A., Alvestad, E., and Skogseid, J., 2000, Seismic volcanostratigraphy of large-volume basaltic extrusive complexes on rifted margins: Journal of Geophysical Research, v. 105, p. 19,335–19,352, doi: 10.1029/1999JB900005.

Rhyan, M.P., 1987, Neutral buoyancy and the mechanical evolution of magmatic systems, *in* Mysen, B.O., ed., Magmatic processes: Physicochemical principles: Geochemistry Society Special Publication 1, p. 259–287.

Richards, M.A., Duncan, R.A., and Courtillot, V., 1989, Flood basalts and hotspot tracks: Plume heads and tails: Science, v. 246, p. 103–107, doi: 10.1126/science.246.4926.103.

Roberts, D.G., 1974, Structural development of the British Isles, the continental margin, and the Rockall Plateau, *in* Burk, C.A., and Drake, C.L., eds., The geology of continental margins: New York, Springer, p. 343–359.

Roberts, D.G., Morton, A.C., Murray, J.W., and Keene, J.B., 1979, Evolution of volcanic rifted margins: Synthesis of leg 81 results on the west margin of Rockall plateau: Initial Reports, DSDP, v. 81, p. 883–911.

Rochette, P., Jenatton, L., Dupuy, C., Boudier, F., and Reuber, I., 1991, Diabase dikes emplacement in the Oman ophiolite: A magnetic fabric study with reference to geochemistry, *in* Peters, T.J., et al., eds., Ophiolite genesis and evolution of the oceanic lithosphere: Dordrecht, Holland, Kluwer Academic, p. 55–82.

Rubin, A.M., 1993, Dikes vs. diapirs in visco-elastic rock: Earth and Planetary Science Letters, v. 119, p. 641–659, doi: 10.1016/0012-821X(93)90069-L.

Saunders, A.D., Fitton, J.G., Kerr, A.C., Norry, M.J., and Kent, R.W., 1997, The North Atlantic igneous province, *in* Mahoney, J.J., and Coffin, M.F., eds., Large igneous provinces: Continental, oceanic, and planetary flood volcanism: Washington, D.C., American Geophysical Union, Geophysical Monograph 100, p. 45–93.

Scheck-Wenderoth, M., Faleide, J.I., Raum, T., Mjelde, R., Di Primo, R., and Horsfield, B., 2006, Results from 2D/3D structural modeling of the Norwegian margin, *in* ILP Task Force on Sedimentary Basin, Québec, September 18–22, p. 58.

Schmeling, H., 2000, Partial melting and melt segregation in a convecting mantle, *in* Bagdassarov, N., and Laporte, D., eds., Physics and chemistry of partially molten rocks: Dordrecht, Holland, Kluwer Academic, p. 141–178.

Self, S., Thordarson, T., and Keszthelyi, L., 1997, Emplacement of continental flood basalt lava flows, *in* Mahoney, J.J., and Coffin, M.F., eds., Large igneous provinces: Continental, oceanic, and planetary flood volcanism: Washington, D.C., American Geophysical Union, Geophysical Monograph 100, p. 381–410.

Shaw, H.R., 1980, The fracture mechanism of magma transport from the mantle to the surface, *in* Hargraves, R.B., ed., Physics of magmatic processes: Princeton, New Jersey, Princeton University Press, p. 201–264.

Shelley, D., 1985, Determining paleo-flow directions from groundmass fabrics in the Lyttelton radial dykes: Journal of Volcanological and Geothermal Research, v. 25, p. 69–79, doi: 10.1016/0377-0273(85)90005-8.

Sheth, H.C., 1999, A historical approach to continental flood basalt volcanism: Insights into pre-volcanic rifting, sedimentation, and early alkaline magmatism: Earth and Planetary Science Letters, v. 168, p. 19–26, doi: 10.1016/S0012-821X(99)00045-X.

Sigurdsson, H., 1987, Dike injection in Iceland: A review: St. John's, Newfoundland, Geological Association of Canada Special Paper 34, p. 55–64.

Smith, W.H.F., and Sandwell, D.T., 1997, Global seafloor topography from satellite altimetry and ship depth soundings: Science, v. 277, 1957–1962.

Smythe, D.K., Chalmers, J.A., Skuce, A.G., Dobinson, A., and Mould, A.S., 1983, Early opening history of the North Atlantic, 1, Structure and origin of the Faeroe-Shetland Escarpment: Geophysical Journal of the Royal Astronomical Society, v. 72, p. 373–398.

Speight, J.M., Skelhorn, R.R., Sloan, T., and Knapp, R.J., 1982, The dike swarms of Scotland, *in* Sutherland, D.S., ed., Igneous rocks of the British Isles: New York, John Wiley and Sons, p. 449–459.

Spera, F.J., 1980, Aspects of magma transport, *in* Hargraves, R.B., ed., Physics of magmatic processes: Princeton, New Jersey, Princeton University Press, p. 265–318.

Spiegelman, M., and Reynolds, J.R., 1999, Combined dynamic and geochemical evidences for convergent melt flow beneath the east Pacific Rise: Nature, v. 402, p. 282–285, doi: 10.1038/46260.

Staudigel, H., Gee, J., Tauxe, L., and Varga, R.J., 1992, Shallow intrusive directions of sheeted dikes in the Troodos ophiolite: Anisotropy of magnetic susceptibility and structural data: Geology, v. 20, p. 841–844, doi: 10.1130/0091-7613(1992)020<0841:SIDOSD>2.3.CO;2.

Tackley, P.J., And Stevenson, D.J., 1993, A mechanism for self-perpetuating volcanism on the terrestrial planets, *in* Stone, D.B., and Runcorn, S.K., eds., Flow and creep in the solar system: Observations, modeling and theory: Dordrecht, Holland, Kluwer, p. 307–322.

Thompson, R.N., and Gibson, S.A., 1991, Subcontinental mantle plumes, hotspots and preexisting thinspots: Journal of the Geological Society of London, v. 148, p. 973–977.

Thompson, R.N., and Morrison, M.A., 1988, Asthenospheric and lower lithospheric mantle contributions to continental extensional magmatism: An example from the British Tertiary Province: Chemical Geology, v. 68, p. 1–15, doi: 10.1016/0009-2541(88)90082-4.

Tilling, R.I., and Dvorak, J.J., 1993, Anatomy of a basaltic volcano: Nature, v. 363, p. 125–133, doi: 10.1038/363125a0.

Vann, I.R., 1978, The siting of Tertiary volcanicity, *in* Bowes, D.R., and Leake, B.E., eds., Crustal evolution in northwestern Britain and adjacent regions: Journal of the Geological Society of London, special issue 10, p. 393–414.

Van Wijk, J.W., and Cloetingh, S.A.P.L., 2002, Basin migration caused by slow lithospheric extension: Earth and Planetary Science Letters, v. 198, p. 275–288, doi: 10.1016/S0012-821X(02)00560-5.

Van Wijk, J.W., Huismans, R.S., Ter Voorde, M., and Cloetingh, S.A.P.L., 2001, Melt generation at volcanic continental margins: No need for a mantle plume?: Geophysical Research Letters, v. 28, p. 3995–3998, doi: 10.1029/2000GL012848.

Varga, R.J., Gee, J.S., Staudigel, H., and Tauxe, L., 1998, Dike surface lineations as magma flow indicators within the sheeted dike complex of the Troodos Ophiolite, Cyprus: Journal of Geophysical Research, v. 103, p. 5241–5256, doi: 10.1029/97JB02717.

Weinberg, R.F., and Podlachikov, Y.Y., 1995, The rise of solid-state diapirs: Journal of Structural Geology, v. 17, p. 1183–1195, doi: 10.1016/0191-8141(95)00004-W.

White, N., and Lovell, B., 1997, Measuring the pulse of a plume with the sedimentary record: Nature, v. 387, p. 888–891, doi: 10.1038/43151.

White, R.S., and McKenzie, D., 1989, Magmatism at rift zones: The generation of volcanic continental margins and flood basalts: Journal of Geophysical Research, v. 94, p. 7685–7729.

White, R.S., Spence, G.D., Fowler, S.R., McKenzie, D.P., Westbrook, G.K., and Bowen, A.N., 1987, Magmatism at rifted continental margins: Nature, v. 330, p. 439–444, doi: 10.1038/330439a0.

MANUSCRIPT ACCEPTED BY THE SOCIETY JANUARY 31, 2007

DISCUSSION

22 January 2007, Jolante W. van Wijk and Laurent Gernigon

The data presented in the chapter by Geoffroy et al. (this volume) suggest that a significant part of the magmas forming the central North Atlantic Igneous Province (NAIP) were injected in a limited number of igneous centers, feeding dikes (subhorizontally) over (sometimes) large distances. More work is needed, but if the suggested pattern of the calculated dike flow vectors as presented in this study is correct, existing models for emplacement processes may need to be reevaluated.

Prior work on emplacement processes often assumed that magma is transferred vertically and that, therefore, the surface extent of flood basalts gives a good indication of the extent of mantle melting at depth. Geoffroy and co-workers, however, find that magma feeds a limited number of igneous centers and then travels from these centers outward, subhorizontally. They propose, for example, that seaward-dipping magmatic reflector sequences could be fed laterally from central crustal reservoirs instead of vertically as initially thought. This conclusion confirms an earlier discussion advanced by Klausen and Larsen (2002) on the east Greenland coast parallel dike swarm. Geoffroy et al. (this volume) provide new quantitative results and show that this lateral feeding model proves to be valid for other areas in the north Atlantic as well.

The authors propose that no direct relationship needs to exist between the magma distribution at the surface of the Earth and the extent of mantle melting at depth. The distribution of igneous centers in the crust is more indicative of the extent of mantle melting. Several lines of arguments exist on how melt accumulates in igneous centers, and the authors favor a process in which melt migration horizontally in the mantle over large distances is unlikely, such that the igneous centers thus form above or close to the location of mantle melting. To explain the distribution of igneous centers, small-scale convection-induced magmatism just below the lithosphere is proposed. Geoffroy et al. (this volume) do not support diapiric structures inside the continental lithosphere to explain the igneous feeding systems. Nevertheless, other studies have supported such a mechanism for different study areas and under specific conditions; Gerya et al. (2004), for example, describe and model diapiric features in the Bushveld Complex explained by the temporal inversion of a vertical temperature gradient as a result of the emplacement of large quantities of hot and mafic magma onto cold material. Drury et al. (2001) simulate diapiric intrusions inside a cold cratonic area.

If the melting region in the mantle is indeed reflected by the distribution of igneous centers, a large, quite uniform thermal plume head is not supported by these observations, according to the authors.

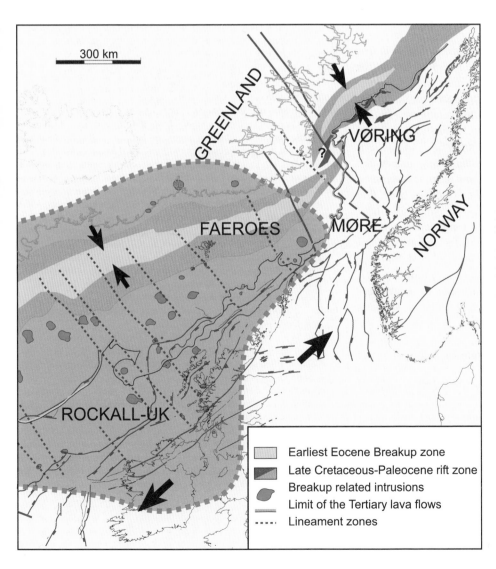

Figure D-1. Distribution of breakup-related intrusions, major fracture systems, and shear zones in the North Atlantic Igneous Province (NAIP). The regional pink surface represents the NAIP area with atypical(?) concentration of igneous centers. The major Paleocene intrusive complexes may be associated with old inherited structures and mostly focus in the Rockall–UK area. Map modified after Doré et al. (1999), Gernigon (2002), and Hitchen (2004).

The suggested link between these sublithospheric instabilities (small-scale convection) and the igneous centers is clearly at a developing stage at this moment, and no geophysical evidence exists yet that allows us to prove the presence of these mantle instabilities during Paleocene time. Emplacement and formation of huge igneous complexes at depth is far from being well understood. Recent potential-field investigations in the Rockall–UK area (Edwards, 2002) for example, suggest that massive high-density igneous complexes vary in shape and can be present deep in the crust or restricted to only the upper part of the crust. We cannot exclude that different modes of emplacement may exist (Edwards, 2002). More geophysical data are needed to detect and investigate igneous intrusions in other parts of the NAIP as well.

The authors discuss about three dozen igneous centers located in the central NAIP. Most of them are Paleocene in age, but some remain undated, and some datings are not well constrained (e.g., Hitchen, 2004; Meyer et al., this volume). To our knowledge, these magmatic features far from the breakup axis have not been recognized in the rest of the NAIP area to the same extent (Fig. D-1). This could change in the future when more datasets are processed, but for now it is unknown whether the Rockall–UK area style of massive igneous accumulation is representative of the entire NAIP. On the Norwegian margin, for example, Berndt et al. (2000) identified the Hel Graben Sill Complex, but this feeder system does not seem similar to the feeders investigated by Geoffroy et al. (this volume).

Along-margin variations in melt accumulation could be expected, as the tectonic evolution (and stress field) of different margin segments has varied during the long rifting history that ultimately resulted in breakup. Lithosphere-scale processes (such as the regional stress field, as mentioned by Geoffroy et al., this volume) affected by inherited structures play a role in how magma transfer and accumulation will develop. Deep shear zones such as the Great Glen fault can probably affect the upper mantle as suggested by deep seismic data cutting the Moho (e.g., Klemperer and Hobbs, 1991) or/and act as conduits for melt transport. Figure D-1 illustrates the distribution of major (recognized) fracture and shear zones in the north Atlantic, suggesting a geographical and structural relation with breakup-related intrusions.

At this point, many open questions remain, including these:

1. How has melt accumulated in the NAIP? Are there regional variations?
2. Does the suggested subhorizontal melt transfer mechanism proposed by Geoffroy et al. (this volume) apply to the entire NAIP?
3. If there indeed exists a dense concentration of igneous feeding centers in the Rockall–UK area, how can this be explained? Why are similar features not clearly observed north of the Faeroes-Shetland basin?
4. In general, how do melt migration and accumulation work in continental lithosphere, and what is the relationship/connection with mantle melting?

It is clear that more research is needed to establish the extent, style, and distribution of igneous feeding centers along the margins of the NAIP. A more complete picture would enable testing the model proposed by Geoffroy et al. (this volume) for magma transfer in large igneous provinces.

REFERENCES CITED

Berndt, C., Skogly, O.P., Planke, S., and Eldholm, O., 2000, High-velocity breakup-related sills in the Vøring basin, off Norway: Journal of Geophysical Research, v. 105, p. 28,443–28,454.

Doré, A.G., Lundin, E.R., Jensen, L.N., Birkeland, Ø., Eliassen, P.E., and Fichler, C., 1999, Principal tectonic events in the northwest European Atlantic margin, *in* Fleet, A.J., and Boldy, S.A.R., eds., Petroleum geology of northwest Europe: Proceedings of the 5th Conference: London, Geological Society of London, p. 41–61.

Drury, M.R., van Roermund, H.L.M., Carswell, D.A., de Smet, J.H., van den Bergh, A.P., and Vlaar, N., 2001, Emplacement of deep upper-mantle rocks into cratonic lithosphere by convection and diapiric upwelling: Journal of Petrology, v. 42, p. 131–140.

Edwards, J.W.F., 2002, Development of the Hatton-Rockall basin, north-east Atlantic Ocean: Marine and Petroleum Geology, v. 19, p. 193–205.

Geoffroy, L., Aubourg, C., Callot, J.-P., and Barrat, J.A., 2007 (this volume), Mechanisms of crustal growth in large igneous provinces: The north Atlantic province as a case study, *in* Foulger, G.R., and Jurdy, D.M., eds., Plates, plumes, and planetary processes: Boulder, Colorado, Geological Society of America Special Paper 430, doi: 10.1130/2007.2430(34).

Gernigon, L., 2002, Extension et magmatisme en contexte de marge passive volcanique: Déformation et structure crustale de la marge norvégienne externe (Domaine Nord-Est Atlantique) [Extension and magmatism in a volcanic margin context: deformation and structure of the Norwegian margin] (Ph.D. thesis): Brest, University of Brest, "Domaines Océaniques," Institut Universitaire Européen de la Mer, Université de Bretgagne Occidentale, 300 p.

Gerya, T., Uken, R., Reinhardt, J., Watkeys, M.K., Maresch, W.V., and Clarke, B.M., 2004, "Cold" diapirs triggered by intrusion of the Bushveld Complex: Insight from two-dimensional numerical modeling, *in* Whitney, D.L., et al., eds., Gneiss domes in Orogeny: Boulder, Colorado, Geological Society of America Special Paper 380, p. 117–127.

Hitchen K., 2004, The geology of the UK Hatton–Rockall margin: Marine and Petroleum Geology, v. 21, p. 993–1012.

Klausen, M.B., and Larsen, H.C., 2002, East Greenland coast-parallel dike swarm and its role in continental breakup, *in* Menzies, M.A., et al., eds., Volcanic rifted margins: Boulder, Colorado, Geological Society of America Special Paper 362, p. 133–158.

Klemperer, S.L., and Hobbs, R.W., 1991, The BIRPS atlas of deep seismic reflection profiles around the British Isles: Cambridge, Cambridge University Press, Cambridge, 128 p. + 99 sections.

Meyer, R., van Wijk, J., and Gernigon, L., 2007 (this volume), The North Atlantic igneous province: A review of models for its formation, *in* Foulger, G.R., and Jurdy, D.M., eds., Plates, plumes, and planetary processes: Boulder, Colorado, Geological Society of America Special Paper 430, doi: 10.1130/2007.2430(26).

The Geological Society of America
Special Paper 430
2007

K-T magmatism and basin tectonism in western Rajasthan, India, results from extensional tectonics and not from Réunion plume activity

Kamal K. Sharma*

Government Postgraduate College, Sirohi (Rajasthan) 307001, India

ABSTRACT

Evolution of sedimentary basins took place in the Barmer, Jaisalmer, and Bikaner regions during K-T (Cretaceous–Tertiary) time in western Rajasthan, India. These intracratonic rift basins developed under an extensional tectonic regime from Early Jurassic to Tertiary time. Rift evolution resulted in alkaline magmatism at the rift margins. This magmatism is dated at 68.5 Ma and has been considered an early phase of Deccan volcanism. Deccan volcanism, sedimentary basin development, and the alkaline magmatism of western Rajasthan have thus been considered the products of Réunion plume activity. However, sedimentary basin evolution began in western Rajasthan prior to Deccan volcanism and K-T alkaline magmatism. Gondwanaland fragmentation during the Mesozoic caused the development of the rift basins of Gujarat and western Rajasthan. This resulted in the opening of the Jurassic rift system and mildly alkaline magmatism at ca. 120 Ma in western India. This event was pre-K-T, and plume activity has not hypothesized for it. Continental fragmentation under an extensional tectonic regime during K-T time resulted in the magmatism and basin tectonism in western Rajasthan. Crustal development during the K-T period in western Rajasthan resulted from an extensional tectonic regime and is not the manifestation of Réunion plume activity.

Keywords: K-T magmatism, mantle plume, western Rajasthan, extensional tectonics, K-T basins

INTRODUCTION

The northwestern Indian shield has a unique evolutionary history extending from Precambrian to Tertiary time (Figs. 1–3). The crust in the region evolved through orogenic, anorogenic, magmatic, and granulite exhumation phases during Precambrian time. The Aravalli, Delhi, and Sirohi orogenies (Fig. 3), which date from ca. 1850 Ma, 1400 Ma, and 830 Ma, respec- tively, testify to the compressional tectonic regime that existed in the shield at those times. Rodinia fragmentation (at ca. 750 Ma) resulted in a change from compressional to extensional lithospheric tectonism, and this in turn initiated development of the Malani silicic large igneous province (SLIP) in an intraplate rift setting in northwest India, Pakistan, and the Seychelles during the Neoproterozoic (750-680 Ma; Sharma, 2004b, 2005).

*E-mail: sharmasirohi@yahoo.com.

Sharma, K.K., 2007, K-T magmatism and basin tectonism in western Rajasthan, India, results from extensional tectonics and not from Réunion plume activity, *in* Foulger, G.R., and Jurdy, D.M., eds., Plates, plumes, and planetary processes: Geological Society of America Special Paper 430, p. 775–784, doi: 10.1130/ 2007.2430(35). For permission to copy, contact editing@geosociety.org. ©2007 The Geological Society of America. All rights reserved.

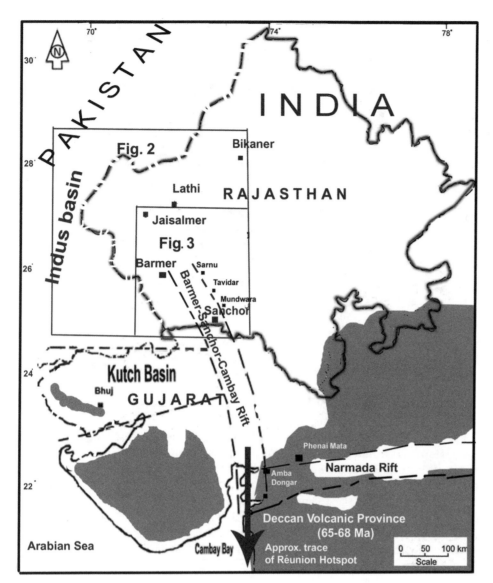

Figure 1. Map of northwest–central India showing the extent of the Deccan igneous province (shaded area), the Narmada Rift Valley, and the Barmer-Cambay rift and alkaline complexes. The arrow marks the approximate path of the postulated Réunion plume in the Late Cretaceous (after Campbell and Griffiths, 1990).

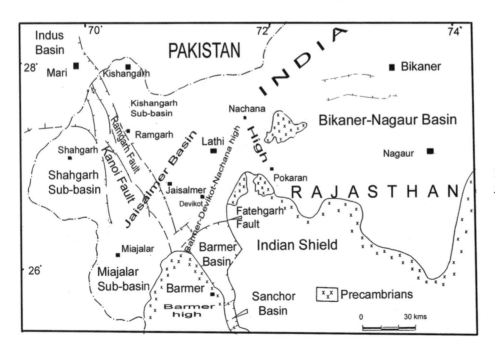

Figure 2. Structural map of western Rajasthan (after Misra et al. 1993).

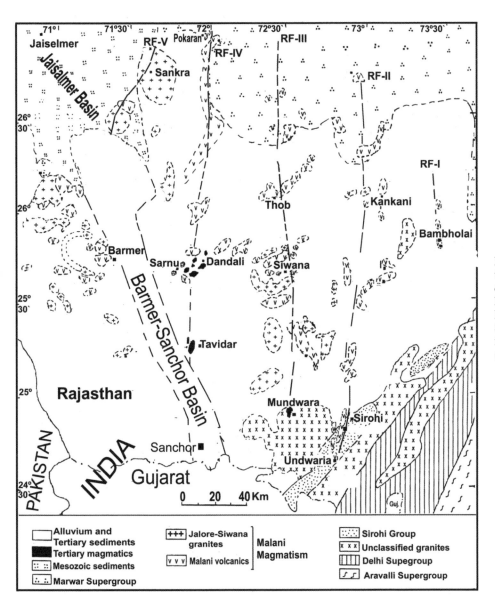

Figure 3. Interpretive geological map of the Malani large igneous province indicating the roughly north-south-trending Neoproterozoic rift fractures (RF-I to RF-V) and the Cretceous–Tertiary Barmer-Sanchor basin and magmatism. Map modified from Bhushan (2000) and Sharma (2005).

The Phanerozoic geological evolution of this region has largely centered around the formation of rift basins in response to global tectonic processes that resulted in separation of the India plate from Gondwana during Jurassic and Cretaceous times (Table 1, Figs. 2 and 3). The growth and geometry of Mesozoic to Tertiary basins have largely been controlled by northwest-southeast- and northeast-southwest-trending fracture systems (Fig. 2). The Phanerozoic crustal evolution of western Rajasthan was characterized by the development of three sedimentary basins, the Jaisalmer basin (Figs. 2 and 3), the Barmer-Sanchor basin (Fig. 2), and the Bikaner-Nagaur basin (Fig. 2), along with alkali magmatism at Mundwara, Sarnu, and Tavidar (Fig. 3).

This chapter summarizes K-T magmatism and basin evolution in western Rajasthan. The Barmer-Sanchor rift (Figs. 2 and 3) and associated alkaline magmatism resembles the Cambay and Narmada rift settings of the Deccan volcanic province (Fig. 1).

The former has been interpreted as a pre-outburst phase of a plume currently under Réunion (Basu et al., 1993; Raval and Veeraswamy, 2000, 2003; Raval, 2001, 2003; Roy and Jakhar, 2002; Roy, 2003, 2004). In this chapter I reappraise the tectonics commonly attributed to a postulated Réunion plume during the K-T period in western Rajasthan. I include a brief discussion of relevant lithologies and tectonic implications, and I present a viable working model for the tectonics of K-T magmatism and basin evolution in western Rajasthan (Fig. 1).

K-T MAGMATISM IN WESTERN RAJASTHAN

Mundwara Igneous Complex

Coulson (1933) reported a suite of igneous rocks popularly called the Mundwara igneous complex west and northwest of

TABLE 1. PHANEROZOIC STRATIGRAPHY OF WESTERN RAJASTHAN

Age	Barmer basin	Jaisalmer basin	Bikaner-Nagaur basin
Cenozoic			
Holocene	Dune sands, alluvium gravels Utarlai Formation	Shumar Formation	Dune sands Mar Formation
Pleistocene			
Pliocene			
Miocene			
Oligocene			
Eocene			
Priabonian		Bandah Formation	
Bartonian			
Lutenian			
Ypresian	Kapurdi Formation Mataji ka Dungar Formation Akli Formation	Khuiala Formation	Jogira Formation
Paleocene			Marh Formation
Thanetian	Barmer Formation Fatehgarh Formation	Sanu Formation	
Montain			
Danian			Palana Formation
Mesozoic			
Cretaceous			
Santonian			
Coniacian			
Turonian			
Cenomanian			
Albian			
Aptian		Habur Formation	
Neocomian	Sarnu Formation	Pariwar Formation	
Jurassic			
Tithonian		Bhadasar Formation	
Kimmeridgian		Baisakhi Formation	
Oxfordian			
Callovian	Jaisalmer Formation	Jaisalmer Formation	
Bathonian			
	Lathi Formation	Lathi Formation	
Lias			
Triassic			
Paleozoic			
Permian			Badhaura Formation
Carboniferous			Bap boulder bed
Devonian to Ordovician			
Cambrian			
	Birmania Formation	Birmania Formation	Nagaur Formation Bilara limestone
Neoproterozoic	Randha Formation Malani rocks/Basement Complex	Randha Formation Malani rocks/Basement Complex	Jodhpur sandstone Malani rocks

Source: After Bhandari (1999).

the village of Mundwara in the Sirohi region (Fig. 3). This complex contains plutonic, hypabyssal, and volcanic members ranging from ultrabasic to alkaline in composition. On the basis of the intrusive nature of the rocks at Sarnu and Tavidar (Fig. 3) into Cretaceous sandstones, Srivastava (1988) proposed a Paleocene age for these rocks. Basu et al. (1993) reported an age of 68.5 Ma for the alkaline olivine gabbro. Using the $^{40}Ar/^{39}Ar$ method, Rathore et al. (1996) analyzed several samples from the Mundwara region and reported an age of 70 Ma for the mafic rocks and 64 Ma for the syenite. The reported ages for the Mund-

wara rocks are not coherent, and more precise geochronology is needed. The isotope age of 68.5 Ma for the Mundwara rocks is interpreted as an early phase of the Réunion hotspot activity in the region (Basu et al., 1993; Raval and Veeraswamy, 2000; Roy and Jakhar, 2002; Roy, 2003).

Tavidar Volcanic Suite

Agarwal (1984) described the petrography and geochemistry of the Tavidar volcanics (Fig. 3). On the basis of chemical discrimination, Upadhyaya et al. (1988) divided the Tavidar volcanics into two groups: (1) rhyolite, quartz-trachyte, and trachyte and (2) basalts, including hawaiite and mugerite. The distribution behavior of major and trace elements and their field relationships indicate that the intermediate and acidic volcanics in the area are cogenetic, whereas the basalts are younger. Rathore (1995) reported a $^{40}Ar/^{39}Ar$ age of 64–66 Ma for the felsic rocks.

Alkaline Complex of Sarnu-Dandali

The Sarnu-Dandali region is located at the eastern margin of the Cambay-Barmer basin in the Barmer region of Rajasthan (Fig. 3). The mildly alkaline igneous rocks (ca. 120 Ma) occur below Lower Cretaceous Sarnu sandstone in the Sarnu region (Chandrasekaran, 1987). Later a variety of acidic, intermediate, and alkaline magmatism took place during K-T time. Basu et al. (1993) reported a mean age for the Sarnu-Dandali alkali pyroxenite of 68.57 ± 0.08 Ma. The most common plutonic rocks are alkali pyroxenite, micromelteigite, melanephelinite, ijolite, and feldspathic ijolite. The dikes are carbonatites, foidal syenites, phonolites, and phonolite porphyries (Chandrasekaran, 1987).

K-T Sedimentary Basins

The evolution of sedimentary basins in western Rajasthan began in the Mesozoic and continued into the Tertiary (Tables 1–4). The most important K-T basins are the Jaisalmer, Barmer, and Bikaner-Nagaur basins (Fig. 2).

TABLE 2. LITHOSTRATIGRAPHY OF THE JAISALMER BASIN AND TECTONOMAGMATIC EVENTS RELATING TO INDIA AND ADJACENT CONTINENTS

Era	Period	Epoch	Age	Ma	Formation	Tectonomagmatic events
Cenozoic	Quaternary	Holocene		0.01	Shumar	
		Pleistocene	Calabrian	1.81		
	Neogene	Pliocene	Gelsian	2.58		
			Piacenzian	3.60		
			Zanclean	5.32		
		Miocene	Messinian	7.12		
			Tortonian	11.2	Hiatus	
			Serrvallian	14.8		
			Langhian	16.4		
			Burdigallian	20.5		
			Acquitanian	23.8		
	Paleogene	Oligocene	Chattian	28.5		
			Rupelian	33.7	Bandah	Collision-uplift
		Eocene	Priabonian	37.0		
			Bartonian	41.3	Marine	
			Lutenian	49.0		
			Ypresian	55.0	Khuiala	
		Paleocene	Thanetian	57.9	Sanu	
			Selandian	61.0		
			Danian	65.5		India-Seychelles separation (Deccan and K-T magmatism in Rajasthan)
Mesozoic	Cretaceous	Late	Maastrichtian	71.3	Hiatus	
			Campanian	83.5		
			Santonian	85.8		Madagascar-India separation
			Coniacian	89.0		
			Turonian	93.5		
			Cenomanian	98.9		
		Early	Albian	112.2		
			Aptian	121.0		India-Australia-East Antarctica separation (Rajmahal)
			Barremian	127.0	Pariwar	
			Hauterivian	132.0		
			Valanginian	136.5		
			Barriasian	142.0	Marine	
	Jurassic	Late	Tithonian	150.7	Bhadasar	
			Kimmeridglan	154.1	Baisakhi	
			Oxfordian	159.4	Jaisalmer	
		Middle	Callovian	164.4		Opening of the Mozambique Ocean
			Bathonian	169.2	Lathi	

Basement: Precambrian Malani igneous rocks and Marwar Supergroup rocks

Sources: Lithostratigraphy modified from Misra et al. (1993); tectonomagmatic events from Torsvik et al. (2005).

TABLE 3. TERTIARY SEQUENCE OF THE BIKANER BASIN (KOLAYAT) REGION

Age	Formation	Lithology
Pleistocene to recent	Kolayat Formation (Continental fluvial, arid, and eolian facies)	Conglomerates, ferruginous sandstone, ironstone nodule, calcareous grot kankar, gypsite, sand and sandy alluvium
Lower to Middle Eocene	Jogira Formation (Calcareous facies)	Foraminiferal limestone, shale, marl, and fuller's earth
Upper Paleocene	Marh Formation (Arenaceous facies)	Ferruginous sandstone, clay, and shale
Lower Paleocene	Palana Formation (Carbonaceous facies)	Sandstone, carbonaceous shale, and lignite

Source: After Ghosh (1983).

Jaisalmer Basin

The Jaisalmer basin extends over 30,000 km² in southwest Rajasthan (Fig. 2). It extends to the Mari region of Pakistan and forms part of the Indus basin (Fig. 2). The basin is controlled by wrench-fault tectonics (Misra et al., 1993) and is divided into four tectonic elements: the Mari-Jaisalmer high shown by the Kanoi and Ramgarh faults (Fig. 2), the Shahgarh sub-basin (Fig. 2), the Miajalar sub-basin, and the monoclinal Kishangarh sub-basin (Fig. 2).

The lowermost horizon in the Jaisalmer region is the Lathi Formation, which is of Jurassic age and was named after Lathi village (Fig. 1) on the Pokaran-Jaisalmer road (Fig. 2, Tables 1 and 2). This horizon comprises a fluvial sequence of conglomerate and sandstone. The sandstone comprises several silicified land plant fossils (Das Gupta, 1975). After deposition of the Lathi Formation, a marine transgression occurred and the Jaisalmer basin formed. Marine sedimentation began with the Jaisalmer Formation, followed by the Baisakhi, Bhadasar, Pariwar, and Habur formations (Tables 1 and 2). The Tertiary sequence in Jaisalmer basin is represented by the Sanu, Khuiala, Bandah, and Shumar formations (Das Gupta, 1975).

Bikaner-Nagaur Basin

The Bikaner-Nagaur basin (Fig. 2) is situated in the northern part of Rajasthan and comprises the Bikaner and Nagaur districts. Ghosh (1983) presented the Tertiary stratigraphy of the Bikaner basin (Table 3). Tertiary sedimentation began with the Palana Formation, which was deposited during the Paleocene in

TABLE 4. GENERALIZED STRATIGRAPHY OF THE BARMER BASIN

Age	Formation	Lithology
Recent	Dune sand and sediments	Alluvium sands, river alluvium, and gravel wash
Subrecent and (?) older	Uttaralai Formation (3–4 m)	Thin gypseous limestone and salt sequence with unconsolidated sands, kankar, and gravel beds
		———Unconformity———
Lower Eocene	Kapuradi Formation (30 m)	Lacustrine fuller's earth deposits(?) interbeded with marine bioclastic limestone
	Mataji ka Dungar Formation (180 m)	Shallow marine orthoquartzite and hard sandstone with pisolite and ball clay and impure bentonitic clay bands at the base
		———Unconformity———
	Akli Formation (280 m) Akli Bentonite Member Thumbali Member	Volcanogenic bentonite sequence at the top and sandstone lignite sequence in the basal part
		———Unconformity———
	Barmer Formation (520 m) Madai Member Barmer Hill Member	Shallow marine sandstone with rare plant fossils and orthoquartzite bands grading into conglomerate, sandstone with plant fossils, and volcanogenic clays
Paleocene	Fatehgarh Formation (520 m) Fatehgarh Scarp Member Vinjori Member	Sunstone layer with mixed bivalve and gastropod casts at the top. Dominantly of ochreous clay bands, variegated sandy siltstone, and sandstone sequence with coquina beds.
		———Unconformity———
	Volcanic formation (?)	Acid to basic volcanic rocks mainly in the form of sills and dikes and local intrusive porphyries
		———Intrusive contact———
Cretaceous	Sarnu Formation (80 m)	Indurated, terrestrial sandstone and siltstone with plant fossils
		———Unconformity———
Callovo-Oxfordian	Jaisalmer Formation (15 m)	Marine, fossiliferous, arenaceous limestone
Bathonian to Lias	Lathi Formation (?)	Terrestrial arenaceous sequence with wood plant fossils and fossilferous tree trunks
Precambrian	Malani magmatism	Rhyolites, basalts, and granite

Source: After Das Gupta (1974).

subtropical swampy conditions. The Marh Formation indicates encroachment of the sea during the Upper Paleocene–Lower Eocene. The Jogira Formation, of Lower to Middle Eocene age, succeeds this. There was complete withdrawal of the sea after the Eocene in this region, and the Kolayat Formation was laid down under fluvial, arid, and eolian conditions from the Pleistocene to recent times.

Barmer Basin

The Barmer basin lies in the southeast part of the Jaisalmer basin, and it forms a north-south graben over 100 km in length (Fig. 2). The Cambay rift basin extends northward through the Sanchor and Barmer basins (Fig. 1). The presence of Cretaceous–Paleocene volcanogenic sediments indicates that the Barmer basin developed as a composite, second-order graben. It contains predominantly terrestrial sediments ranging from Jurassic to Cretaceous in age. The Paleocene–Eocene period was marked by deposition of coastal, marine, shallow-water sediments.

The basal unit of the Barmer basin is the Fatehgarh Formation (Table 4). This formation comprises mixed siliciclastic, carbonate, and phosphoritic rocks of Cretaceous age. Mathur et al. (2005) reported spherules, glassy balls, highly magnetic fine dust, and microbrecciated matrix and suggested that the spherules are related to the volcanic source or to K-T boundary impact ejecta from the Chicxulub impact in the Gulf of Mexico. However, more evidence is required to support this suggestion.

Sisodia et al. (2005) reported a loose fragmental layer 3–5 cm thick in siliceous earth from the Barmer basin, comprising volcanic debris such as glass shards, agglutinates, hollow spheroids, kinked biotites, feldspars, olivines, ilmenite, and native iron. They suggested that the bands of siliceous earth were volcanic ash deposited during the Upper Cretaceous–Lower Paleocene. This suggests that the development of the Barmer basin took place contemporaneously with Deccan volcanism.

RÉUNION PLUME MODEL

Alkaline magmatism at Mundwara (Fig. 3) in north Cambay is interpreted as representing the earliest manifestation of the Réunion mantle plume in northwest India (Basu et al., 1993). Mahoney et al. (2002) postulated a pre-Deccan marine phase of early Réunion hotspot activity after studying Cretaceous volcanic rocks in the South Tethyan Suture Zone of Pakistan.

Raval and Veeraswamy (2000) presented a detailed account of interaction between the Réunion plume and continental lithosphere in western India. The dynamics of this interaction were grouped into pre-, syn-, and post-outburst phases; with the main plume outburst phase represented by the Deccan volcanism at 65.5 ± 0.5 Ma (Basu et al., 1993; Baksi, 1994). K-T magmatism and sedimentary basin development in western Rajasthan were considered pre-outburst plume activity and included (1) the formation of the Mundwara and Sarnu igneous complexes, dated at 68.53 ± 0.16 Ma (Basu et al., 1993); (2) a event involving Pb loss from the Delhi-Aravalli region, dated at ca. 70 Ma (Fig. 3)

(Sivaraman and Raval, 1995); and (3) the intraplate volcanism of Parh (Pakistan) at ca. 70 Ma (Mahoney et al., 2002). During this pre-outburst phase, the Indian plate was moving northward at the high velocity of ~16–19 cm/yr (Klootwijk et al., 1992), resulting in the development of a linear corridor from the Parh Group (Pakistan) through Sarnu-Dandali and Mundwara to the Réunion plume outburst around Cambay (Fig. 1).

Roy and Jakhar (2002) and Roy (2003, 2004) endorsed the Réunion plume model (Raval and Veeraswamy, 2000, 2003) and suggested that Deccan volcanism; the Cambay-Barmer, Jaisalmer, and Bikaner-Nagaur basins; and alkaline magmatism in Sarnu-Dandali, Tavidar, and Mer-Mundwara represent manifestations of plume–continental lithosphere interaction. The presence of high-gravity anomalies, high heatflow, and a seismic low-velocity zone below the Cambay-Barmer region were considered the geophysical expressions of Réunion plume impingement on the base of the Indian lithosphere. It was suggested that the bolide impact at the K-T boundary intensified already ongoing plume volcanism in northwest India, though the mechanism of this is not clear (Roy and Jakhar, 2002; Roy, 2003).

AN ALTERNATIVE VIEW

The basement rocks outcropping in western Rajasthan are dominantly Neoproterozoic Malani rocks (750–770 Ma; Torsvik et al., 2001). Following the anorogenic Malani magmatism, the crust subsided and an intracratonic sag basin developed. This basin contains unfossiliferous, siliciclastic, carbonate, and evaporite facies commonly known as the Marwar Supergroup (Fig. 3). The sedimentation in the Marwar basin continued up to the Cambrian-Precambrian boundary.

Western Rajasthan remained almost tectonically dormant during the Paleozoic. With the exception of some glacial deposits laid down in the Permo-Carboniferous, no major sedimentary units were deposited in the region. The K-T geodynamics in northwest India were complex and involved several major phenomena, including the development of the Cambay-Barmer rift and sedimentary basins in the Jaisalmer, Barmer, and Bikaner regions and alkali magmatism on the rift shoulders at Barmer, Tavidar, and Mundwara. The mantle plume model has been invoked to explain various observations, but important factors such as continental fragmentation, the origin of the Arabian Sea, plate tectonics, the K-T mass biological extinction, and the nature of postulated mass transfer from the core-mantle boundary to the surface have not been adequately taken into consideration.

Sedimentary basin evolution began in western Rajasthan prior to Deccan volcanism and K-T boundary alkaline magmatism. The Jurassic basin of Kutch (Fig. 1) and Jaisalmer in Rajasthan resulted from separation of the Indian continent from eastern Gondwana (Naqvi, 2005). The extensional regime resulted in opening of the Mozambique Ocean and development of the Jurassic rift system in western India. This is marked by mildly alkaline igneous rocks below the Lower Cretaceous Sarnu sandstone in the Barmer region (Fig. 3). Sedimentation ceased in the Barmer and Jaisalmer basins during the Upper Cretaceous (ca. 85 Ma). The marine regression from western Rajasthan during this period correlates with the breakaway of Madagascar from India (Torsvik et al., 2000). The separation of India and Madagascar resulted in tectonic changes that caused the sea to withdraw from western Rajasthan (Table 2).

The next phase of basin rejuvenation in western Rajasthan occurred during the K-T boundary period. This coincided with the separation of the Seychelles from India, large-scale Deccan volcanism, development of the Cambay rift, and alkali magmatism in western Rajasthan. Crustal rifting during the K-T boundary period resulted in large-scale Deccan volcanism (Sheth, 2005a,b) and development of the Cambay-Barmer rift system. The extensional tectonics resulted in deep fractures manifested by the development of rift basins, alkali magmatism, and lamprophyre-carbonatite dikes. Mitra et al. (2006) described the Cambay basin as a classic example of an extension basin that subsequently remained tectonically active and continued to exhibit graben subsidence and horst- and fault-bounded features. The development of the Cambay-Sanchor-Barmer rift caused reactivation of Precambrian fractures and resulted in magmatism at the basin margin at Mundwara, Tavidar, and Sarnu (Sharma, 2004a).

NONPLUME TECTONISM

Tectonically, the Jaisalmer basin comprises part of the shelf portion of the Indus geosyncline (Pareek, 1984). It is separated from the Bikaner-Nagaur (Fig. 2) basin by the Pokaran-Nachana high (Fig. 2) to the northwest and from the Barmer basin by the Barmer-Devikot-Nachana high in the south (Fig. 2). A pronounced northwest-southeast-trending regional step-faulted Jaisalmer-Mari high zone traverses the center of the basin. This high divides the basin into three parts. To the northeast is the monoclinal Kishangarh sub-basin (Fig. 2), to the south Miajalar basin, and to the southwest Shahgarh sub-basin (Fig. 2). Zaigham and Mallick (2000) identified the Indus basin as an extension basin resulting from an inferred fossil rift crustal feature overlain by a thick sedimentary sequence. The extensional regime that resulted in the Indus basin initiated in the region as a result of divergence of the Indo-Pakistan subcontinent from Gondwanaland. The Indus basin (Fig. 1) is a fossil-rift feature and is characterized by horst and graben structures together with a system of transcurrent faults. The origin of Jaisalmer basin is related to Indus basin evolution at the beginning of the Triassic (Pareek, 1984) and not to "Réunion plume" activity.

The development of Bikaner-Nagaur basin in northwest Rajasthan is attributed to a plume by Roy and Jakhar (2002) and Roy (2003). The basin developed along east-west-striking fault blocks in the Bikaner-Nagaur region, which extends westward into Pakistan. The Palana Formation (Table 3), which constitutes the basal unit, was deposited under continental conditions and contains important deposits of lignite. Volcanogenic sediments and igneous activity are totally absent on the basin margins. Mafic and alkaline magmatism and other plume manifestations (Campbell, 2005) are also absent, arguing against plume-lithosphere interaction there. The Bikaner-Nagaur basin formed in relation to collision between the India plate and Tibet ca. 55–50 Ma (Naqvi, 2005).

Raval (2001, 2003) attributed the current seismicity in northwestern India to Réunion plume activity. In his model, plume-lithosphere interaction below western Rajasthan, the Cambay rift, and further south caused mantle upwelling, lithospheric thinning, and a low–seismic velocity zone, weakening the crust and resulting in the present seismic activity. Sharma (2006) argued against this unlikely scenario. He showed that the distribution of earthquake epicenters correlates with lineaments related to the paleoaccretionary corridors, rift margins, and fault zones. These preexisting weak zones are vulnerable to reactivation and remobilize episodically. The northwestern Indian Shield exhibits evidence of several geomorphological changes during the Quaternary. For example, western Rajasthan and Gujarat experienced rapid tectonic upheaval subsequent to the Vedic period (ca. 5,000–10,000 yr B.C.). The reason for the apparently constant reactivation of the region throughout geological history might be the weakened lithosphere, which is criss-crossed with faults, rather than plume–Indian lithosphere interaction. Courtillot et al. (1986) presume that the India plate drifted over a Réunion plume of deep-mantle origin, causing a large plume head to upwell below the India plate as well as K-T volcanism. However, mantle imaging shows the absence of evidence in the continental lithosphere supporting plume impact or incubation models beneath the Deccan volcanic province (Ravi Kumar and Mohan, 2005).

The Barmer basin and associated alkali igneous complexes developed at basin margins during the K-T period. The Mundwara, Tavidar, and Sarnu igneous complexes are interpreted as evidence of Réunion plume activity in western Rajasthan (Basu, et al., 1993; Raval and Veeraswamy, 2000; Roy, 2003, 2004). Sharma (2004a, 2005), however, reports that these Tertiary alkaline complexes resulted from reactivation of the Precambrian Malani fracture system during the development of the Cambay-Barmer rift under an extensional tectonic regime (Fig. 3). This extensional tectonic regime developed during Mesozoic–Tertiary time as a result of Gondwana fragmentation (Sharma, 2004a).

SUMMARY

Gondwanaland fragmentation during the Mesozoic resulted in extensional tectonics in the northwestern Indian Shield. This led to the development of rift basins in Gujarat and western Rajasthan. Deccan volcanism, separation of the Seychelles microcontinent from India, sedimentary basin development in western Rajasthan, and alkaline magmatism in Mundwara, Sarnu-Dandali, and elsewhere have been postulated to be the products of Réunion plume activity in western India. However, basin development began in western Rajasthan during the Jurassic, and such a plume cannot account for this. Continuation of the extensional tectonic regime resulted in deep fractures in the continental and oceanic lithosphere. The Cambay-Sanchor-Barmer rift developed in the continental lithosphere. Mundwara, Sarnu-

Dandali, and Barmer magmatism, with nephelinite-carbonatite affinities at the basin margin, occurred in a typical rift-tectonic setting. Jaisalmer basin evolution is related to the Indus basin of Pakistan and does not have alkaline magmatism, volcanogenic sediments, or other plume manifestations. The Barmer basin is a typical rift setting with high rift shoulders and alkaline-carbonatite magmatism. The tectonic setting and crustal development of western Rajasthan during the K-T period thus occurred in a typical extensional tectonic regime and does not fit a model of Réunion plume activity.

ACKNOWLEDGMENTS

I express a deep sense of gratitude to Professor Gillian R. Foulger and Dr. Hetu Sheth for assisting and encouraging me in various ways in preparing this manuscript. I am thankful to Dr. Sheth for regularly providing me with important information. Dr. Manoj Pandit and an anonymous reviewer provided helpful reviews that improved the manuscript.

REFERENCES CITED

Agarwal, V., 1984, Geochemistry of the volcanic rocks around Tavidar, District Jalore, Rajasthan [Ph.D. thesis]: Jaipur, University of Rajasthan, 110 p.

Baksi, A.K., 1994, Geochronological studies on whole-rock basalts, Deccan Traps, India: Evolution of the timing of volcanism relative to the K-T boundary: Earth and Planetary Science Letters, v. 121, p. 43–56, doi: 10.1016/0012-821X(94)90030-2.

Basu, A.R., Renne, P.R., Das Gupta, D.K., Teichmann, F., and Poreda, R.J., 1993, Early and late alkali igneous pulses and a high-^{3}He plume origin for the Deccan flood basalts: Science, v. 261, p. 902–906, doi: 10.1126/science .261.5123.902.

Bhandari, A., 1999, Phanerozoic stratigraphy of western India: A review, *in* Kataria, P., ed., Geology of Rajasthan: Status and perspective: Udaipur, Mohan Lal Sukhadia University Press, p. 126–174.

Bhushan, S.K., 2000, Malani rhyolite—A review: Gondwana Research, v. 3, no. 1, p. 65–77, doi: 10.1016/S1342-937X(05)70058-7.

Campbell, I.H., 2005, Large igneous provinces and the mantle plume hypothesis: Elements, v. 1, p. 265–269.

Campbell, I.H., and Griffiths, R.W., 1990, Implications of mantle plume structure for the evolution of flood basalts: Earth and Planetary Science Letters, v. 99, p. 79–93, doi: 10.1016/0012-821X(90)90072-6.

Chandrasekaran, 1987, Geochemistry of the basic, acid and alkaline intrusive rocks of Sarnu-Dandali area, Barmer District, Rajasthan [Ph.D. thesis]: Jaipur, University of Rajasthan, 108 p.

Coulson, A.L., 1933, The geology of the Sirohi State, Rajasthan: Bangalore, Geological Survey of India Memoir 63, 166 p.

Courtillot, V., Besse, J., Vandamme, D., Montigny, R., Jaeger, J.J., and Cappetta, H., 1986, Deccan flood basalts at the Cretaceous/Tertiary boundary?: Earth and Planetary Science Letters, v. 80, p. 361–374, doi: 10.1016/0012-821X(86)90118-4.

Das Gupta, S.K., 1974, Stratigraphy of the Rajasthan Shelf, *in* Proceedings of the IV Indian Colloquium on Micropaleontology Stratigraphy, India, p. 219–233.

Das Gupta, S.K., 1975, Revision of the Mesozoic–Tertiary stratigraphy of the Jaisalmer Basin, Rajasthan: Indian Journal of Earth Science, v. 2, p. 77–94.

Ghosh, R.N., 1983, Tertiary clay deposits of Kolayat and adjacent areas in Bikaner district, Rajasthan: Indian Minerals, v. 37, no. 4, p. 56–69.

Klootwijk, C.T., Gee, J.S., Peirce, J.W., Smith, G.M., and Mcfadden, P.L., 1992, An early India-Asia contact: Palaeomagnetic constraints from Ninety East Ridge, ODP leg 121: Geology, v. 20, p. 395–398, doi: 10.1130/0091-7613 (1992)020<0395:AEIACP>2.3.CO;2.

Mahoney, J.J., Duncan, R.A., Khan, W., Gnos, E., and McCormick, G.R., 2002, Cretaceous volcanic rocks of the South Tethyan Suture Zone, Pakistan: Implications for the Réunion hotspot and Deccan Traps: Earth and Planetary Science Letters, v. 203, p. 295–310, doi: 10.1016/S0012-821X(02)00840-3.

Mathur, S.C., Gour, S.D., Loyal, R.S., Tripathi, A., Tripathi, R.P., and Gupta, A., 2005, Occurrence of magnetic spherules in the Maastrichtian bone bed of the Fatehgarh Formation, Barmer Basin, India: Current Science, v. 89, p, 1259–1268.

Misra, P.C., Singh, N.P., Sharma, D.C., Upadhyay, H., Kakroo, A.K., and Saini, M.L., 1993. Lithostratigraphy of west Rajasthan basins: Dehradun, Oil and Natural Gas Commission report.

Mitra, T., Maitra, A., and Misra, R.S., 2006, Syn-rift tectonics and depositional pattern at North Cambay Basin, India: Interplay of basin tectonics, structural style and sedimentation: GEO 2006, Seventh Middle East Geosciences Conference and Exhibition, Manama, Bahrain, March 27–29.

Naqvi, S.M., 2005, Geology and evolution of the Indian plate (From Hadean to Holocene—4 Ga to 4 Ka): New Delhi, Capital Publishing, 448 p.

Pareek, H.S., 1984, Pre-Quaternary geology and mineral resources of Northwestern Rajasthan: Calcutta, Geological Survey of India Memoir 115, 95 p.

Rathore, S.S., 1995, Geochronological studies of Malani volcanic and associated igneous rocks of southwest Rajasthan, India: Implications to crustal evolution [Ph.D. thesis]: Baroda, Maharaja Sayajirao University, 175 p.

Rathore, S.S., Venkatesan, T.R., and Srivastva, R.K., 1996, Mundwara alkali igneous complex, Rajasthan: Chronology and Sr isotope systematic: Journal of the Geological Society of India, v. 48, no. 5, p. 517–528.

Raval, U., 2001, Earthquakes over Kutch: A region of "trident" space-time geodynamics: Current Science, v. 8, no. 97, p. 809–815.

Raval, U., 2003, Interaction of mantle plume with Indian continental lithosphere since Cretaceous: Bangalore, Geological Society of India Memoir 53, p. 449–479.

Raval, U., and Veeraswamy, K., 2000, The radial and linear modes of interaction between mantle plume and continental lithosphere: A case study from western India: Journal of the Geological Society of India, v. 56, no. 5, p. 525–536.

Raval, U., and Veeraswamy, K., 2003, India-Madagascar separation: Breakup along a pre-existing mobile belt and chipping of the craton: Gondwana Research, v. 6, p. 467–485, doi: 10.1016/S1342-937X(05)70999-0.

Ravi Kumar, M., and Mohan, G., 2005, Mantle discontinuities beneath the Deccan volcanic province: Earth and Planetary Science Letters, v. 237, p, 252–263.

Roy, A.B., 2003, Geological and geophysical manifestations of the Réunion plume–Indian lithosphere interactions—Evidence from Northwest India: Gondwana Research, v. 6, no. 3, p. 487–500.

Roy, A.B., 2004, The Phanerozoic reconstitution of Indian Shield as the aftermath of break-up of the Gondwanaland: Gondwana Research, v. 7, no. 2, p. 387–406.

Roy, A.B., and Jakhar, S.R., 2002, Geology of Rajasthan (Northwest India), Precambrian to recent: Jodhpur, Scientific Publishers, 421 p.

Sharma, K.K., 2004a, K-T magmatism of western Rajasthan, India: Manifestation of Réunion plume activity or extensional lithospheric tectonics?: American Geophysical Union, Fall Meeting [abs. V51B–0560].

Sharma, K.K., 2004b, The Neoproterozoic Malani magmatism of the northwestern Indian shield: Implications for crust-building processes: Proceedings of the Indian Academy of Science (Earth Planetary Science), v. 113, no. 4, p. 795–807.

Sharma, K.K., 2005, Malani magmatism: An extensional lithospheric tectonic origin, *in* Foulger, G.R., et al., eds., Plates, plumes, and paradigms: Boulder, Colorado, Geological Society of America Special Paper 388, p. 463–476.

Sharma, K.K., 2006, Intraplate seismicity of the northwestern Indian shield: Implication for the reactivation of palaeo-tectonic elements: Geophysical Research Abstracts, v. 8, 03253.

Sheth, H., 2005a, From Deccan to Réunion: No trace of a mantle plume, *in* Foulger, G.R., et al., eds., Plates, plumes, and paradigms: Boulder, Colorado, Geological Society of America Special Paper 388, p. 477–502.

Sheth, H.C., 2005b, Were the Deccan flood basalts derived in part from ancient oceanic crust within the Indian continental lithosphere?: Gondwana Research, v. 8, p.109–127.

Sisodia, M.S., Singh, U.K., Lashkari, G., Shukla, P.N., Shukla, A.D., and Bhandari, N., 2005, Mineralogy and trace element chemistry of the siliceous Earth of Barmer basin, Rajasthan: Evidence for a volcanic origin: Journal Earth System Science, v. 114, no. 2, p. 111–124.

Sivaraman, T.V., and Raval, U., 1995, U-Pb isotope study of zircons from a few granitoids of Delhi-Aravalli belt: Journal of the Geological Society of India, v. 46, p. 461–475.

Srivastava, R.K., 1988, Magmatism in the Aravalli mountain range and its environs: Bangalore, Geological Society of India Memoir 7, p. 77–94.

Torsvik, T.H., Tucker, R.D., Ashwal, L.D., Carter, L.M., Jamtveit, B., Vidyadharan, K.T., and Venkataramana, P., 2000, Late Cretaceous India-Madagascar fit and timing of break-up related magmatism: Terra Nova, v. 12, p. 220–224, doi: 10.1046/j.1365-3121.2000.00300.x.

Torsvik, T.H., Carter, L.M., Ashwal, L.D., Bhushan, S.K., Pandit, M.K., and

Jamtvit, B., 2001, Rodinia redefined or obscured: Palaeomagnetism of the Malani Igneous Suite (NW India): Precambrian Research, v. 108, p. 319–333, doi: 10.1016/S0301-9268(01)00139-5.

Torsvik, T.H., Pandit, M.K., Redfield, T.F., Ashwal, L.D., and Webb, S.J., 2005, Remagnetization of Mesozoic limestones from the Jaisalmer basin, NW India: Geophysical Journal International, v. 161, p. 57–64, doi: 10.1111/j.1365-246X.2005.02503.x.

Upadhyaya, R., Srivastava, R.K., and Agarwal, V., 1988, A statistical approach to the study of an igneous suite—A case history of Tavidar volcanics: Neues Jahrbuch für Mineralogie Abhandlungen, v. 159, no. 3, p. 311–324.

Zaigham, N.A., and Mallick, K., 2000, A prospect of hydrocarbon associated with fossil-rift structures of the Southern Indus Basin, Pakistan: American Association of Petroleum Geologists Bulletin, v. 84, p. 11.

Manuscript Accepted by the Society January 31, 2007

DISCUSSION

19 January 2007, Rajat Mazumder

Sharma (this volume) summarizes some aspects of Cretaceous-Tertiary (K-T) magmatism and basin tectonism in western Rajasthan. He criticizes the existing idea that K-T magmatism in western Rajasthan is a consequence of the Réunion plume and provides an alternative model, that crustal development during the K-T period in western Rajasthan is a manifestation of a plume-unrelated extensional tectonic regime.

Sharma (this volume) presents a broad lithological description of the stratigraphic units and reviews existing geochronological data from the magmatic rocks that will be useful to readers. However, the geochemical aspects of the magmatic rocks, particularly those of the Sarnu-Dandali Complex, are underemphasized. Sharma (this volume) refers to a Ph.D. dissertation that is largely inaccessible to interested readers. Uplift by a mantle plume in a sedimentary basin has many significant consequences (such as palaeogeographic shallowing, diversion in sediment dispersal pattern, occurrences of penecontemporaneous deformation structures, etc.) for the sedimentation system that, in turn, can help geologists to identify the influence of plume-induced uplift on Earth's surface processes from stratigraphic records (cf. Rainbird and Ernst, 2001, and references therein) if there are any. This requires detailed sedimentary facies analysis, determination of palaeogeographic setting, and mode of sequence building in terms of sea level change and contemporary basinal tectonics. From the sedimentological and stratigraphic account of the K-T basins of western Rajasthan as presented by Sharma (this volume), however, it is unclear whether such effort has yet been made. If not, detailed sedimentological analysis is essential to resolve many key issues and, thus, of paramount importance to prove or disprove the plume model.

Needless to say, researchers will gain valuable insights from Sharma's review in this volume to argue for plume-induced or plume-unrelated extensional basin tectonics and K-T magmatism in western Rajasthan. I congratulate the author on his effort.

30 January 2007, Kamal Kant Sharma

Mazumder points out that uplift by a mantle plume in a sedimentary basin has many significant consequences on the sedimentation system. Basin development and magmatism occurred during K-T time in western Rajasthan and are considered to be the Réunion-plume pre-outburst phase. It is thought that the main outburst phase resulted in uplift and volcanism in the Deccan province. K-T basin development was initiated in western Rajasthan before Réunion plume interactions and continued to the end of the Tertiary. Surprisingly, the Bikaner Tertiary basin, which developed under continental conditions with lignite deposits, is also related to plume activity. K-T sedimentary basin development has a complex evolutionary history. The Jaisalmer basin is a part of Indus tectonism and represents horst and graben features such as the Shahgarh and Kishangarh subbasins. The Barmer–Sanchor rift evolved along with the Cambay basin.

The K-T basins of western Rajasthan are engulfed in vast desert sands, and much facies analysis work has not yet been done. In my article (Sharma, this volume) I referred to an unpublished thesis in order to communicate the original source of the work. However, it has also been published in the form of different papers and as a book section, and citing all these would have been a repetition. I thank Mazumder for his comments on my article.

REFERENCES CITED

Rainbird, R.H., Ernst, R.E., 2001, The sedimentary record of mantle plume uplift, *in* Ernst, R.E., and Buchan, K.L., eds., Mantle plumes: Their identification through time: Boulder, Colorado, Geological Society of America Special Paper 352, p. 227–245.

Sharma, K.K., 2007 (this volume), K-T magmatism and basin tectonism in western Rajasthan, India, results from extensional tectonics and not from Réunion plume activity, *in* Foulger, G.R., and Jurdy, D.M., eds., Plates, plumes, and planetary processes: Boulder, Colorado, Geological Society of America Special Paper 430, doi: 10.1130/2007.2430(35).

The Geological Society of America
Special Paper 430
2007

Plume-related regional prevolcanic uplift in the Deccan Traps: Absence of evidence, evidence of absence

H.C. Sheth*

Department of Earth Sciences, Indian Institute of Technology Bombay, Powai, Mumbai 400076, India

ABSTRACT

From the mantle plume model it would be expected that one to a few kilometers of regional, domal lithospheric uplift occurred 5–20 m.y. before the onset of flood basalt volcanism. This uplift resulted from heat conduction out of and dynamic support by the hot, buoyant, rising plume head. Field evidence for such uplift would comprise sedimentary sequences that reflect progressive basin shallowing before volcanism or (in the case of differential uplift along faults) widespread conglomerates derived from the basement rocks and underlying the first lavas. Local uplifts and subsidences cannot be used to invoke or rule out plume-caused uplift. Over large areas of the Late Cretaceous Deccan flood basalt province, the base of the lava pile is in the subsurface. Basalt-basement contacts are observed along the periphery of the province and in central India (the Satpura and Vindhya ranges), where substantial post-Deccan uplift is evident. Here, extensive horizontal Deccan basalt flows directly overlie extensive low-relief planation surfaces cut on various older rocks (Archean through Mesozoic) with different internal structures. Locally, thin, patchy Late Cretaceous clays and limestones (the Lameta Formation) separate the basalts and basement, but some Lameta sediments are known to have been derived from already erupted Deccan basalt flows in nearby areas. Thus, the eruption and flowage of the earliest Deccan basalt lava flows onto extensive flat planation surfaces developed on varied bedrock, and the nearly total absence of basement-derived conglomerates at the base of the lava pile throughout the province, are evidence against prevolcanic lithospheric uplift (both regional and local), and thereby the plume head model. There has been major (~1 km) post-Deccan, Neogene uplift of the Indian peninsula and the Sahyadri (Western Ghats) Range, which runs along the entire western Indian rifted margin, well beyond the Deccan basalt cover. This uplift has raised the regional Late Cretaceous lateritized surface developed on the Deccan lava pile to a high elevation. This uplift cannot reflect Deccan-related magmatic underplating, but is partly denudational, is aided by a compressive stress regime throughout India since the India-Asia collision, and is possibly also related to active eastward flow of the sublithospheric mantle. The easterly drainage of the Indian peninsula, speculated to be dome-flank drainage caused by the plume head, predates the uplift. Field evidence from the Deccan and India is in conflict with a model of plume-caused regional uplift a few million years before the onset of volcanism.

---

*E-mail: hcsheth@iitb.ac.in.

Sheth, H.C., 2007, Plume-related regional prevolcanic uplift in the Deccan Traps: Absence of evidence, evidence of absence, *in* Foulger, G.R., and Jurdy, D.M., eds., Plates, plumes, and planetary processes: Geological Society of America Special Paper 430, p. 785–813, doi: 10.1130/2007.2430(36). For permission to copy, contact editing@geosociety.org. ©2007 The Geological Society of America. All rights reserved.

Keywords: mantle plume, flood basalt, tectonics, uplift, planation surface, volcanism, Deccan

INTRODUCTION: FLOOD BASALTS, PLUME HEADS, AND LITHOSPHERIC UPLIFT

The deep mantle plume model of Morgan (1972, 1981) for flood basalt volcanism was further developed from fluid dynamical experiments and numerical modeling (Richards et al., 1989; Campbell and Griffiths, 1990; Hill, 1991; Farnetani and Richards, 1994; and several others). Such modeling indicated that thermal plumes rising buoyantly from the core-mantle boundary should develop large bulbous "heads" about 1000 km in diameter by entrainment of surrounding mantle, and that the heads should remain connected to the source region by narrow "tails." A key premise is that mantle plumes are hotter than ambient mantle, and the large volumes of magma erupted in flood basalt provinces are due to the high-temperature plumes undergoing extensive decompressional melting.

Campbell and Griffiths (1990) calculated that a new ("starting") plume head, 1000 km in diameter, flattens to a disc twice as wide after impinging on the base of the lithosphere (Fig. 1). One key predicted consequence of plume upwelling is that the plume head will raise the lithosphere over a broad area (~1000 km or more across) before flood basalt volcanism begins. This is a consequence of heat conduction out of and dynamic support by the hot plume head. Significant domal uplift would be expected to occur 10–20 m.y. before flood volcanism, when the top of the plume head is still well below the lithosphere (Fig. 1). The uplift is followed by subsidence as the plume head begins to melt extensively and the magmas are transported toward and erupted at the surface. The amplitude of the surface uplift has been calculated at 1–4 km (Campbell and Griffiths, 1990; Farnetani and Richards, 1994), depending on model parameters such as plume excess temperature.

It has been claimed that geological and geomorphological evidence from most flood basalt provinces of the world fulfills the predictions and patterns of the model just described (Cox, 1989; Kent, 1991; Campbell, 2005). Here, I examine in detail the evidence from the Deccan flood basalt province and India and conclude that the model is inconsistent with the field evidence, which rules out regional pre-eruption domal uplift in the province. The objective of this chapter is to systematically and clearly lay out the field evidence and not to formulate or present an alternative model to the plume model, because such models, e.g., those involving continental breakup and rift-related convection and mantle melting, have been presented elsewhere (e.g., Sheth, 2005a).

EVIDENCE REGARDING UPLIFT IN FLOOD BASALTS

Geological evidence for prevolcanic uplift can take the form of pre-eruption sedimentary sequences that reflect progressive

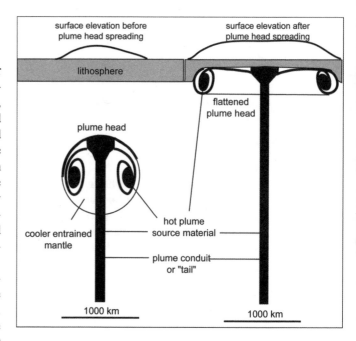

Figure 1. Basic structure of a starting mantle plume and the patterns of surface uplift predicted a few million years before and after plume head impingement on the lithosphere. Modified from Hill (1991).

basin shallowing before volcanism (e.g., Rainbird and Ernst, 2001; Mazumder, 2005, 2006). Alternatively, if there is differential tectonic uplift along faults, rapid erosion of the basement rocks can be expected to produce beds of conglomerate or coarse clastics below the first lavas. If the uplift were regional, such sediments would be expected to have a regional distribution. In the Permo-Triassic Emeishan flood basalt province of China, He et al. (2003) have argued for domal, kilometer-scale regional prevolcanic uplift based on paleogeographic shallowing recorded in sediments and conglomerate horizons underlying the initial basalt lavas. They support the plume head model and reject nonplume mechanisms. Prevolcanic uplift has also been identified in the east and west Greenland flood basalt provinces of Paleocene age (e.g., Dam et al., 1998).

However, evidence for prevolcanic uplift is absent, or contentious, beneath several other flood basalts. In Russia, the Siberian flood basalts of Permo-Triassic age (ca. 250 Ma) overlie the coal-bearing Tungusskaya Series (320–250 Ma) throughout most of their great areal extent (Czamanske et al., 1998). The Tungusskaya Series includes the Tungusska basin, the largest coal-bearing basin in the world. This suggests continuous crustal subsidence before and during the flood eruptions, which Czamanske et al. (1998) considered incompatible with dynamic support from a plume head beneath the Siberian lithosphere.

Subsequently, Reichow et al. (2002) reported and dated extensive subsurface basalts from the west Siberian basin, considerably enlarging the area of the Siberian flood basalt province. On this basis, Saunders et al. (2005) argue that prevolcanic uplift did occur in the Siberian Traps, not over the present outcrop area of the traps, but over the west Siberian basin. They note that this uplift had previously been ascribed to a plate tectonic cause—a Hercynian tectonic event (Peterson and Clarke, 1991)—but prefer a plume head as the explanation for the uplift. However, according to the plume head model, lithospheric uplift should commence 10–20 m.y. before the flood basalt eruptions and reach a maximum 5–10 m.y. before the eruptions (Campbell and Griffiths, 1990; Hill, 1991). If the west Siberian basin had been uplifted for tens of m.y. before the Siberian Trap eruptions, as has been suggested (A. Ivanov, personal commun., 2005), it would be inconsistent with the plume head model.

In the Miocene Columbia River flood basalt province, the absence of prevolcanic uplift has been noted by Hales et al. (2005), who observe that mild pre-eruptive subsidence was followed by syneruptive uplift of a few hundred meters and a long-term uplift of 2 km. In the 120 Ma Ontong Java oceanic plateau, the world's largest, crustal uplift during or immediately preceding the eruptions was negligible or much less than expected. The entire plateau was constructed below sea level with few subaerial eruptions, and this observation and several others have prompted reconsideration of the plume model and exploration of alternative models (Ingle and Coffin, 2004; Tejada et al., 2004; Clift, 2005; Korenaga, 2005).

The absence of uplift before volcanism in some flood basalt provinces has been explained within the framework of the plume model by proposing flattening and spreading of the plume head against a thick, cold, strong lithosphere or lateral migration of magma into the crust (Campbell and Griffiths, 1990). These variations make the plume model impossible to falsify or test. It is argued, for example, that uplift patterns can be complex depending on factors such as lithospheric strength and plume excess temperature, buoyancy, and shape. It is thus argued that not all plumes cause uplift, and flood basalt provinces *without* prevolcanic uplift may well be plume generated (Burov and Guillou-Frottier, 2005.) If so, the plume head model makes no specific and testable predictions for field geologists and cannot be evaluated against their observations. The plume head model would become untenable, however, when a flood basalt province showed not merely an *absence of evidence* for prevolcanic uplift, but also *evidence for absence* of such uplift. This is not only possible, I argue, but is in fact the scenario in the Deccan flood basalt province, India.

INDIAN AND DECCAN GEOLOGY IN A NUTSHELL

The Deccan province, of Late Cretaceous age (ca. 65 Ma), was associated with the break-off of the Seychelles microcontinent from the Indian subcontinent along India's western margin (Fig. 2). Prior to this, at ca. 88 Ma, Greater India (India plus the Seychelles) broke off from Madagascar, an event that was associated with massive flood basalt volcanism on Madagascar and relatively minor volcanic-intrusive activity in India (Storey et al., 1995; Pande et al., 2001). The following description of the main features of Indian and Deccan geology (Figs. 2 and 3) is taken from Sheth (2005a), and many relevant references can be found therein. The Indian subcontinent has a rich rock record from Early Archean up to Recent times, with at least six Archean or Early Proterozoic cratonic nuclei recognized. These are the Aravalli, Bundelkhand, Singbhum, Bastar (Bhandara), and Dharwar cratons and the high-grade granulite terrain in the south (Fig. 2). Several major rift zones traverse the subcontinent: the Godavari and Mahanadi rifts in the east, the Cambay rift in the NNW, and the Kachchh rift in the northwest. The ENE-trending Satpura Mountain Range, considered by many a horst block, separates the Narmada and Tapi rifts, and this zone is a major 1600-km-long, long-active tectonic zone in the central part of India. The Indian rifts are known to run along major Precambrian tectonic trends. The Narmada rift is considered by some to be a Proterozoic protocontinental suture. The western Indian coast and the Cambay rift parallel the NNW-SSE, Proterozoic Dharwar orogenic trend. Another major Precambrian orogenic trend, the northeast-southwest Aravalli trend, splays out into two at its southern end: the east-west Delhi trend (along which the Mesozoic Kachchh rift has developed) and the main northeast-southwest Aravalli trend, which continues into the Saurashtra peninsula.

Figure 3 shows the main rock formations that make up the Indian Shield: a large portion of the shield is made up of Archean and Proterozoic crystalline rocks, and there are many Proterozoic and Phanerozoic sedimentary basins on them. The Deccan basalt pile, which obscures the basement from observation over 0.5 million km^2, is thickest (~2,000 m) along the Sahyadri Range (better known as the western Ghats) near the west coast and thins progressively eastward and southeastward, such that along the eastern fringes of the province the lava pile is only ~200 m thick or less. Whereas the lava pile in the western Ghats region and in the interior areas of the province is made up almost completely of fairly evolved subalkaline tholeiitic basalts (Sheth, 2005b), felsic and alkaline magma types are also prominent along the rift zones and along the west coast. Considerable volumes of felsic and mafic tuffs and rhyolite and trachyte lavas crop out along the coast, e.g., at Bombay (Sheth et al., 2001; Sheth and Ray, 2002). The west coast and the rift zones are also where significant tectonic-structural deformation has affected the lava pile, such as a pronounced seaward-dipping monoclinal flexure (the Panvel flexure) along the west coast (Sheth, 1998). Regional, dominantly north-south-oriented dike swarms of basalts, dolerites, lamprophyres, basanites, and allied alkaline rocks crop out along the coast. Significant volumes of felsic rocks and many alkaline complexes (several of which include carbonatites) are also found along the Narmada rift and the Cambay rift. The Narmada-Satpura-Tapi zone also contains major linear doleritic and alkaline dike swarms that trend ~ENE-WSW.

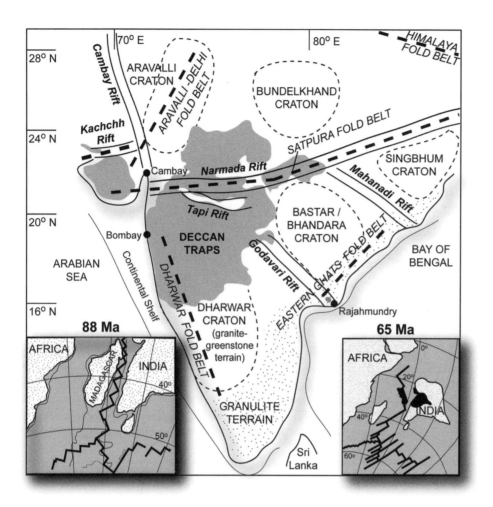

Figure 2. Map showing the basic structural framework of the Indian subcontinent, including Precambrian cratons (boundaries approximate), structural trends defined by Precambrian fold belts (primarily Proterozoic), the rift zones (primarily Phanerozoic), and the present outcrop area of the Deccan flood basalt province (shaded). Modified from Sheth (2005a,b). The inset figures show the breakup of Greater India from Madagascar at ca. 88 Ma (Storey et al., 1995; Pande et al., 2001) and the break-off of the Seychelles microcontinent (located at the northern tip of the Mascarene plateau, black) from India at ca. 65 Ma (Norton and Sclater, 1979).

The pronounced linearity of the west coast and the continental margin suggest structural control. The newly formed western Indian continental margin and the rift zones may have constituted major vent areas for the Deccan lavas, as inferred from abundant mafic dike swarms and intrusions, high heatflow, and aligned thermal springs (Sheth, 2000). The Deccan basalts continue beyond the west coast and onto the continental shelf. The Cambay rift and the region offshore of the west coast have productive oil and gas fields. Much of the Cambay region is covered by Tertiary and Quaternary sediments (up to 5 km thick), and at places the underlying basalts are known, from seismic data, to be over 4 km thick.

Over a large part of the province, the contact of the Deccan lavas with the prevolcanic basement is not exposed. The Deccan lavas overlie a complex Archean and Proterozoic basement along the southern and southeastern periphery of the province. In the northern and northeastern parts of the province, i.e., central India, they overlie diverse geological formations: the large Vindhyan sedimentary basin (Middle–Late Proterozoic), the large Gondwana sedimentary basin (Carboniferous to Early Cretaceous), the Late Cretaceous Bagh and Lameta sediments, and also Archean and Early Proterozoic crystalline rocks (granites,

gneisses, and metasediments). I will now consider the main geological features of the Deccan proper and southern India that have key significance with regard to the plume-related uplift issue.

THE SAHYADRI RANGE (WESTERN GHATS) AND INDIAN PENINSULAR DRAINAGE

Many major Indian rivers originate in the Sahyadri Range (Western Ghats), not far from the west coast, and instead of draining into the Arabian Sea a few tens of kilometers to the west, they flow for hundreds of kilometers eastward to the Bay of Bengal (Fig. 4). Cox (1989) speculated that this pronounced easterly drainage was a consequence of regional domal uplift caused by the Deccan plume head. He noted that the Narmada and the Tapi, two major Indian rivers, flow westward. He ascribed this to their exploiting a rift system in the dome. Why such a rift system would produce a westerly drainage (*toward* where the topographically high center of the uplifted dome would be) is unclear. He also presented drainage maps from the Karoo (South Africa) and the Parana flood basalt provinces and considered each example as the preserved half of an originally com-

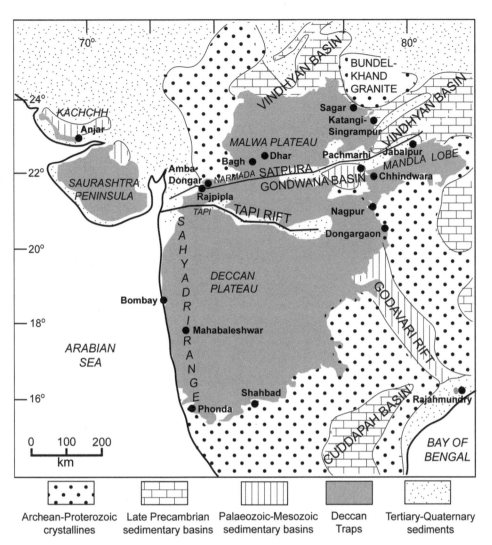

Figure 3. Sketch map of western and central India showing its main geological features, main geographic-physiographic features (italicized), and the outcrop of the Deccan flood basalts (shaded). The Bagh and Lameta sediments and the Bijawar metamorphics are too small to show at the scale of the map, but the locations of the Vindhyan and Satpura Gondwana basins are indicated. Also shown are localities mentioned in the text. Modified from Sheth (2005a).

plete dome-flank pattern produced by a plume head. Summerfield (1990) pointed out that the Cox (1989) model ignored aspects such as the Cenozoic drainage development in Africa. It is important to recognize that no quantitative data (such as apatite fission-track ages) have ever been offered in support of the plume-caused doming postulated by Cox (1989), and as I discuss here, the model ignores and is incompatible with well-known geological facts about India.

It should be noted that though the Sahyadri Range is where the Deccan basalt pile is best exposed, the Sahyadri is by no means confined to the Deccan lava field. It constitutes a NNW-SSE-trending and 1500-km-long "Great Escarpment" extending to the southern tip of India, well beyond the Deccan basalt cover (Radhakrishna, 1952/2001; Ollier, 1990; Gunnell and Radhakrishna, 2001). The escarpment parallels the western rifted margin of India and has been retreating eastward due to erosion since its formation along the line of rifting at ca. 65 Ma (e.g., Widdowson, 1997a). There is only one break in the escarpment throughout its great length, namely the "Palghat Gap" (Fig. 4).

The gap has a maximum elevation of 300 m above mean sea level and an average width of only 13 km, and is entirely rock-floored without any alluvium cover (Subramanian and Muraleedharan, 1985; Gunnell and Radhakrishna, 2001). Its origin has been controversial. On both sides of the gap, the Ghats reach great heights of >2.5 km amsl. The Nilgiri charnockite massif to the north of the Palghat Gap and the Palni-Kodaikanal Massif to the south of it are made up of Precambrian hypersthene-bearing granites and granulites (the so-called "charnockites") (Rajesh and Santosh, 2004). These mountains are the highest in peninsular India and rank among the highest mountains in shield areas anywhere in the world (Gunnell and Louchet, 2000). Summits of the western Ghats and the Karnataka plateau, between the Deccan basalt outcrop and the southern charnockite massifs, also approach 2 km in height (e.g., Bababudan, Fig. 4) and are built of Precambrian metamorphic rocks such as quartzites and gneisses. In comparison, the highest peak of the western Ghats in the Deccan basalt region, Kalsubai, stands at 1646 m (Fig. 4).

The substantially greater heights of the charnockite massifs

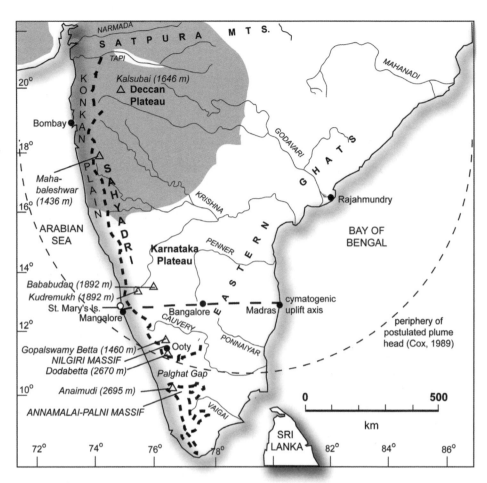

Figure 4. The main elements of the physiography of the Indian peninsula, based on Ollier and Powar (1985) and Sheth (2005a). Note the pronounced easterly drainage. The outcrop area of the Deccan flood basalts is shaded. The Sahyadri (western Ghats) escarpment is shown by the heavy broken line, and some major summits of the Ghats are indicated, as are some localities cited in the text. The circumference of the plume head proposed by Cox (1989) and the axis of cymatogenic uplift at 13°S latitude proposed by Subrahmanya (1994) are also shown. The St. Mary's Islands rhyodacite lavas (ca. 85 Ma) along the west coast are older than the Deccan Traps and related to the India-Madagascar break-up event (Pande et al., 2001).

may well reflect both the original topography and the greater resistance of charnockite to erosion than basalt, because these charnockites are among the hardest rocks known, with an extremely low fracture density (Gunnell and Louchet, 2000). What is most important, however, is that the entire Sahyadri is the precipitous western edge of an uplifted plateau that has been tilted eastward, and the plateau surface has an aged character; it is an ancient flat land surface that has been rejuvenated in relatively recent (Neogene) times by major tectonic uplift (Radhakrishna, 1952/2001, 1993; Vaidyanadhan, 1977), and the uplift has continued during the Quaternary (e.g., Powar, 1993; Valdiya, 2001). The Sahyadri, e.g., at Mahabaleshwar (1436 m, Fig. 4), has a very youthful aspect today, with stupendous west-facing scarps and many great waterfalls. Mahabaleshwar sits atop a spectacular ~1200-m-thick exposed Deccan basalt sequence (Fig. 5), and yet the top of the Mahabaleshwar plateau represents a regional low-relief, Late Cretaceous paleosurface developed on the uppermost basalts after the cessation of the eruptions, represented by laterite profiles 25–50 m thick (Widdowson, 1997a). In southern India, laterites or bauxites cap the high-elevation summits built of Precambrian rocks. The western Ghats escarpment has much the same form whether it is developed on the Deccan basalts or on the massive and structureless charnockites

in southern India (see Figures 10.10 and 10.11 *in* Ollier and Pain, 2000). Sahyadri uplift is evidently not related to Deccan volcanism, but is more appropriately designated as "rift-shoulder uplift." Along with the Sahyadri, the Konkan Plain to the west of it was also rising during the Tertiary (e.g., Powar, 1993), and places on the west coast, such as Mangalore (Fig. 4), are currently rising (e.g., Subrahmanya, 1996).

As regards the easterly drainage pattern of the peninsula, Widdowson and Cox (1996) showed that in the Mahabaleshwar area the easterly drainage cuts through the axis of a north-south-aligned, south-plunging anticlinal structure identified from regional dips of the basalt lavas and the laterite cap. The anticlinal structure, as they pointed out, developed subsequent to the drainage and had no effect on the drainage lines. Previously, Ollier and Powar (1985) observed that the drainage pattern of peninsular India is dendritic over both the region of the Deccan lavas and the older basement. They therefore suggested that the drainage developed subsequent to the eruption of the Deccan lavas. The newly formed Deccan lava field would have provided a regional slope to the east. However, they also noted that the drainage was antecedent to the uplift of the Sahyadri Range, i.e., the easterly drainage existed prior to the rise of the Ghats. The eastward-draining Godavari-Krishna River system also was

Figure 5. View of the Sahyadri (western Ghats) escarpment at Mahabaleshwar, showing the 1200-m-thick, horizontally disposed Deccan flood basalt lava pile. The top of the lava pile is a heavily lateritized Late Cretaceous surface. Photo by H.C. Sheth.

already in existence by the time the Deccan lavas were in eruption. A thin (tens of meters), 64 Ma (Deccan-age) basalt flow sequence at Rajahmundry near the southeastern coast of India (Figs. 3 and 4) has strong chemical similarities with some of the southernmost Deccan basalts, and Baksi et al. (1994) have proposed that the Rajahmundry basalts were intracanyon flows (akin to many in the Columbia River basalt province) that erupted in the southern Deccan and flowed to the southeastern coast of India along existing river systems. Knight et al. (2003) consider this scenario of long-distance surface transport and that of local eruptions at Rajahmundry equally possible, though no dikes or intrusions that are potential feeders to the local lava pile have been reported around Rajahmundry.

The easterly drainage of the Indian peninsula was also antecedent to the uplift of the eastern Ghats, according to Ollier and Powar (1985). The eastern Ghats, made up primarily of Precambrian charnockites, schists, and other metamorphic rocks, form disconnected hill ranges in eastern India and are thus unlike the western Ghats, though in southeastern India the eastern Ghats have an aspect similar to that of the western Ghats, and major rivers such as the Cauvery were antecedent to them and have cut steep gorges through them (Ollier and Powar, 1985). However, even the easterly direction of drainage in southern India may be an oversimplification, because there appears to be a distinct east-west axis of cymatogenic uplift (crustal arching) at 13°N latitude, to the north of which the rivers flow northeast, and to the south of which the drainage is to the southeast (Subrahmanya, 1994; Fig. 4). Mangalore on the west coast and Madras on the east coast are both located on this axis, and both

are actively rising relative to the sea (Bendick and Bilham, 1999). The uplift of the Sahyadri (western Ghats) and the attendant uplift of the Konkan Plain to the west of it are definitely post-Deccan (Radhakrishna, 1952/2001, 1993; Widdowson and Cox, 1996; Widdowson, 1997a), and both offshore sedimentary evidence and apatite fission-track data that have recently become available are consistent with this (Gunnell and Gallagher, 2001; Gunnell et al., 2003). The easterly drainage of the Indian peninsula is older than the uplift of the Ghats (both the western and the eastern), and the uplift of the Ghats is much younger than Deccan volcanism. The proposition of Cox (1989), that the easterly drainage is a *result* of a plume-generated lithospheric dome, is therefore without basis.

DECCAN BASALT–BASEMENT CONTACTS: CLUES TO PREVOLCANIC UPLIFT, OR THE LACK THEREOF

Over most of the Deccan province today, the lava-basement contact is in the subsurface, often at considerable depth. There is a huge thickness (~1700 m exposed) of Deccan basalts in the Sahyadri Range and an additional ~500 m in the subsurface, as identified from seismic data (Kaila et al., 1981). Seismic data also suggest two linear, Mesozoic sedimentary basins in the subsurface below the Deccan basalts in the Narmada-Tapi region (Kaila, 1988; Sridhar and Tewari, 2001). The northern Narmada basin is 1000 m thick, and the southern Tapi basin is 1800 m thick. To the north of the Narmada River, Deccan basalts overlie the Middle–Late Proterozoic sediments of the Vindhyan basin, or the older crystalline basement, directly. The Vindhyan basin does not extend to the south of the Narmada River. To the south of the river, in the Satpura Range, Deccan lavas overlie the thick Gondwana sedimentary basin (Carboniferous to Early Cretaceous), or the Precambrian basement. The Gondwana basin does not extend to the north of the Narmada River and is particularly well exposed in the Pachmarhi area (Fig. 3) due to substantial post-Deccan uplift along basin boundary faults.

Two other sedimentary rock formations, the Bagh and Lameta formations, underlie the Deccan basalts locally in central India and the Satpura region (e.g., Mohabey, 1996; Sahni et al., 1996; Tandon, 2002; Khosla and Sahni, 2003). Both are of Late Cretaceous age and are much smaller than the Vindhyan and Gondwana basins. Both are also known in Indian literature as "infra-trappeans" and locally separate the Deccan basalts from the Vindhyans, the Gondwanas, or the Archean-Proterozoic crystalline basement, as the case may be. The Bagh Formation consists of sandstones and limestones formed during a Late Cretaceous marine *transgression* in the western Narmada Valley (Sheth, 2005a), the region where one would expect maximum uplift from a putative Deccan plume head at this exact time. The fluvial and lacustrine Lameta Formation is also of Late Cretaceous age and consists of thin (~20 m) limestones and clays. Its maximum thickness is 40 m at the type locality near Jabalpur (Fig. 3) (Tandon, 2002). The Lametas overlie the Gondwana sediments or Precambrian rocks and are themselves overlain by the Deccan basalt flows in some sections, based on which they were considered an older formation than the basalts. More recent field, geochemical, and clay mineralogical studies have indicated, however, that the Lameta clays in the Jabalpur region were derived from the Deccan basalts themselves, and thus basalts were already erupting in nearby areas and supplied material for these sediments, of restricted lateral and vertical extent (Prasad and Khajuria, 1995; Salil et al., 1997; Tandon, 2002; Shrivastava and Ahmad, 2005). Clearly, the Lameta beds in some sections predate the Deccan basalts, and the Lameta beds in other sections are derived from (and so younger than) other Deccan basalts.

Pre-eruption uplift is recorded in the Lameta sediments of the Dongargaon basin of the Nagpur region (Fig. 3), where Tandon (2002) recorded a clear "shallowing up" trend from shallow lake deposits to a paleosol before the terrain was buried by the first Deccan basalt flow. He ascribed this, however, to prevolcanic surface uplift of the area on the order of *meters only* and possibly also to "mock aridity." Mock aridity (Harris and van Couvering, 1995) refers to extreme local aridity resulting from active volcanism (e.g., Kilauea) while the region as a whole may be under humid tropical conditions (Hawaii). Evidence for mock aridity has been found in "intertrappean" sediments interlayered with (not below) the Deccan basalt flows at Anjar, in the northwestern part of the province (Khadkikar et al., 1999; Fig. 3). If mock aridity was already at work when the thin, upward-shallowing Lameta sequence at Dongargaon was forming (Tandon, 2002), volcanism was already active nearby.

To conclude, there were both local uplifts and subsidences just before volcanism in parts of central India, but local and vertically restricted uplift (as recorded in the Lameta beds of Dongargaon basin) cannot be used to support the plume model, and local Late Cretaceous subsidence and marine invasion (as in the Bagh area, Fig. 3) cannot be used to refute it. Such local tectonics are easily related to local processes such as the filling and emptying of magma chambers, emplacement of plutons and sills, and faulting. Relevant in this connection is a basalt- and basement-derived conglomerate bed reported in the Rajpipla area (Fig. 3) that unconformably overlies tilted Late Cretaceous sediments and is overlain by the Deccan basalts (Widdowson, 2005). Rajpipla is located near the edge of the Deccan basalt outcrop, and the basement is exposed in the vicinity. This outcrop suggests that uplift and/or tilting and erosion of the Late Cretaceous sediments occurred in quick succession just before volcanism. A basalt flow, or possibly several flows, was then erupted nearby, and continuing uplift and erosion produced the conglomerate, which was in turn covered by the subsequent lava flows. This appears to be a case of local uplift such as can be produced by moderately large intrusions or faulting. Abundant dikes, intrusions, and plugs have been known from Rajpipla (e.g., Krishnamurthy, 1971), and only 40 km to the east is the Amba Dongar carbonatite-alkaline complex, where the Late Cretaceous marine Bagh sandstones have been domed up by an intrusive basalt plug (Ray et al., 2003). The Rajpipla conglomerate outcrop appears very similar to conglomerates interlayered with some Paleogene basalt flows on the Isle of Skye, Scotland. At Preshal Beg on Skye, a spectacularly columnar-jointed olivine tholeiite lava flow (of maximum thickness 120 m) of the Talisker Bay Group is underlain by a conglomerate of the Preshal Beg Conglomerate Formation (Fig. 6). The conglomerate is a very poorly sorted and chaotic deposit that was derived locally from the nearby volcanics by local uplift and probably slope instabilities (Williamson and Bell, 1994; Emeleus and Bell, 2005).

My own field work in the Pachmarhi-Chhindwara-Nagpur region (Fig. 3) has revealed that the Deccan lavas directly overlie a varied basement without any intervening coarse clastics of Late Cretaceous age. Kale et al. (1992), with field data from the southernmost fringes of the Deccan (e.g., around Phonda and Shahbad, Fig. 3), have reported a considerable variation (~200 m) in the elevation (amsl) of the basalt-basement contact, based on which they have argued that the distribution of the lava flows was strongly controlled by preexisting topography. They have not reported any basement-derived conglomerates under the lavas, however, and this indicates the absence of local, immediately pre-eruption uplift. Nor have Choubey (1971) or Dixey (1970/1999) reported such clastics from central India as a whole. Sections in the Deccan province with strata that reflect prevolcanic uplift are thus minuscule in number, and the uplift here does not exceed a few meters. What, then, is the evidence in central India for or against regional pre-eruption uplift? To answer this question, we must first understand the geomorphological concept of a planation surface and its tectonic underpinnings.

Figure 6. Chaotic, completely unsorted Preshal Beg Formation conglomerate under a spectacularly columnar-jointed olivine tholeiite lava flow of the Talisker Bay Group, Isle of Skye, Scotland. The person near the lower right corner of the photo provides a scale. Photo by H.C. Sheth.

PLANATION SURFACES: A KEY CONCEPT IN TECTONIC GEOMORPHOLOGY

Planation surfaces, discussed at length by Ollier and Pain (2000), are regionally extensive, flat or nearly flat surfaces produced by advanced erosion (usually fluvial) of any earlier topography to the existing base level of erosion, which is in most cases sea level. (Large and high inland plateaus like the Tibetan plateau have their own base level, and a large river may act as a subregional base level.) Planation surfaces are also called *erosion surfaces* or *peneplains,* and these terms will be used here interchangeably. *Paleosurfaces* (Widdowson, 1997b) constitute a broader category of ancient land surfaces that can be exogenic (caused by erosion or weathering) or endogenic (e.g., lava plains). A planation surface is therefore a kind of exogenic paleosurface.

On a peneplain, relief is minor to absent, and erosion is therefore negligible, but deep in situ weathering and lateritization are typical. Summerfield (1991) states that low relief, absence of erosion, and sediment starvation are vital prerequisites for the development of indurated paleosols of the nature of ferricretes or laterites. A planation surface thus indicates long-term tectonic stability and the absence of uplift and erosion. If a low-lying peneplain is uplifted to form a plateau, the dissection of the plateau by further erosion works toward forming a younger planation surface at a lower level, broken only by some surviving remnants ("monadnocks") of the higher surface (Fig. 7A). Also, because there is no process that can produce a flat erosional surface out of a jagged terrain at a high elevation, the presence

of a geologically young planation surface at high elevation today suggests rapid and recent uplift (Ollier and Pain, 2000).

It is important to be able to recognize planation surfaces correctly. First, planation surfaces are products of bedrock erosion, not sediment deposition. A vast, flat alluvial plain is not a planation surface. A planation surface is difficult to prove on horizontally bedded rocks, because these tend to produce flat surfaces on erosion anyway. These are purely structural surfaces. A planation surface is most easily recognized when it cuts across, or bevels, diverse rocks with varied internal structures (Fig. 7B). The presence of beveled cuestas over a wide area of folded or tilted rocks is a good indicator of a former planation surface, because in the absence of a former planation surface, a cuesta, which is a structurally controlled landform on dipping strata, would show a sharp ridge crest, and never a level top (Fig. 7C and D). However, even in the absence of beveled cuestas, the phenomenon of "accordant summit levels"—complexly deformed rocks of widely varying types with their summits at more or less the same general level—suggests the existence of a former extensive land surface that lay somewhat higher and is now being dissected again (Fig. 7E). Planation surfaces are of great significance in tectonics and are a key concept in tectonic geomorphology, as Ollier and Pain (2000) show with many examples worldwide. For example, on the basis of the strikingly accordant summit levels seen in many major fold-thrust mountain belts (the Himalayas, Alps, and Andes) and identifiable planation surfaces and other evidence, they argue that fold-thrust mountains are essentially uplifted peneplains or plateaus subsequently dissected into the present rugged topography, and the rock

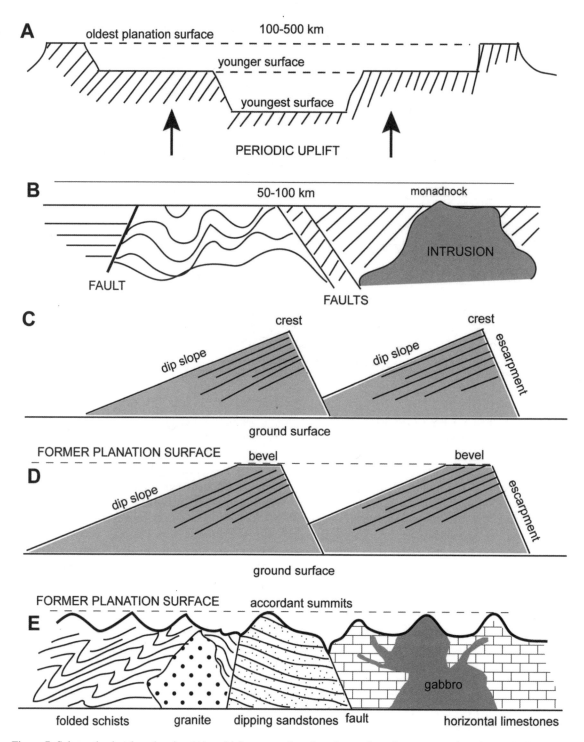

Figure 7. Schematic sketches showing (A) multiple, successive planation surfaces in an area undergoing periodic uplift, i.e., a "polycyclic" landscape; (B) a planation surface cut on complex rocks with a resistant monadnock left behind; (C) cuestas on dipping-resistant rocks, which normally have sharp crests (view is along strike direction); (D) beveled cuestas representing a former planation surface; and (E) accordance of summit levels in a lithologically diverse and complexly deformed terrain suggesting a former planation surface. Based on Ollier and Pain (2000). The vertical scale is greatly exaggerated in panels A, B, and C.

deformation reflected in their complex internal structure has nothing to do with the present mountainous topography. Planation surfaces at rifted continental margins have been used to track uplift. For example, Bonow et al. (2006a,b) show that the west Greenland rifted margin contains successive planation surfaces ~2 km above present sea level, and that this margin has experienced multiple uplift and erosion events since the culmination of rifting in the Labrador Sea during the Middle Eocene. They conclude that the present relief was formed during the late Neogene. Similarities with the Sahyadri, which has experienced major uplift in the Neogene, are striking.

Figure 8 shows a schematic NNW-SSE profile (after Gunnell, 1998) through southern India showing the multiple planation surfaces that bevel varied bedrock types and structures (see Figure 4 for locations of the summits). S_o (~2500 m elevation) is the original, Gondwanic planation surface (Late Jurassic or Early Cretaceous), S_1 (~2200 m) a Late Cretaceous surface well seen in the Nilgiri massif (around Ooty, Fig. 4) and surrounding highlands, S_2 probably an Early or Middle Tertiary surface (identified partly based on bauxite occurrences and suggested to be the top of the Deccan basalt pile), and S_3 a Late Tertiary surface. S_4 is the lowest and youngest surface, Mio-Pliocene in age. Note that postdenudational upwarping has produced a fanning of the surfaces S_1 to S_3, which in the opinion of Gunnell (1998) is probably unrelated to the rift-flank uplift of the western Ghats, but is related to crustal loading by the Deccan basalt pile and late Neogene intraplate crustal deformation in the Indian Ocean basin that has been affecting peninsular India as well (Subrahmanya, 1996).

PLANATION SURFACES UNDER THE DECCAN BASALTS

As noted earlier, the top of the Deccan basalt pile in the western Ghats, such as at Mahabaleshwar (1436 m), is a heavily lateritized surface of Late Cretaceous to early Tertiary age (Fig. 5) developed on the uppermost basalts after the eruptions ceased (Widdowson and Cox, 1996; Widdowson, 1997a; Widdowson and Gunnell, 1999). The lateritization is consistent with observations of lateritized planation surfaces worldwide (Ollier and Pain, 2000), and the surface indicates a stable tectonic regime not punctuated by uplift and erosion but marked by advanced rock weathering. Its present high elevation reflects purely post-Deccan uplift, amounting at least to several hundred meters if it originally lay a few hundred meters above what was then sea level (Widdowson, 1997a), and probably to more than a kilometer. This surface reflects stable tectonic conditions after the eruptions ceased. If a planation surface can also be recognized *below* the Deccan basalt pile in a particular region, however, long-term tectonic stability *before* flood volcanism will be indicated.

This scenario does appear to obtain. Widdowson and Gunnell (1999) note that, with the exception of the thin infratrappean Lameta and Bagh beds that indicate fluvial and shallow marine conditions at some localities in the province during the Late

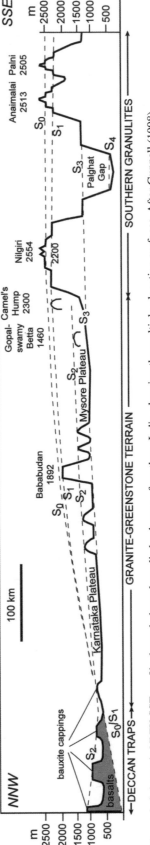

Figure 8. Schematic NNW-SSE profile through the polycyclic landscape of southern India, showing the multiple planation surfaces. After Gunnell (1998).

Cretaceous, the basalts were erupted onto the pre-Deccan (Gondwanan?) land surface. The surface had lain exposed, over very large areas, to continental weathering for a long time. Evidence for pre-Deccan planation surfaces is best sought in central India, where the basalt-basement contacts are exposed over a large region, and the basement comprises a very varied assemblage of Archean granites and gneisses, the Proterozoic Bijawar metamorphics (dominantly quartzites), the Middle to Late Proterozoic Vindhyan sequence (dominantly sandstone-shale-limestone), the Carboniferous to Jurassic Gondwana Supergroup, and the Late Cretaceous Bagh and Lameta beds (Fig. 3). Dixey (1970/1999) and Choubey (1971) identified erosion surfaces under the Deccan lavas in central India, and Dixey (1970/1999) commented on their usefulness in deciphering subsequent tectonic movements. Choubey (1971) noted that a significant feature of the landscape of central India is that most Vindhyan and Bijawar hills stand approximately at a level of 590 m, with a remarkable summit accordance. He suggested that these are the remnants of

a vast Cretaceous peneplain. The Deccan basalts cap this surface. At lower elevations (440–360 m), the Deccan basalts cap another planation surface developed on the softer Gondwana sediments. Both surfaces existed before the Deccan eruptions and represented a long interval of weathering with a deep lateritic saprolith. In this section, I list the main observations from specific areas (Fig. 3).

Katangi-Singrampur

Deccan basalt flows cap Vindhyan rocks near Katangi and Singrampur (Figs. 3 and 9), and five different flows with a total thickness of 150 m were recognized by Choubey (1971). A thin Lameta limestone bed separates the lowermost flow from the Vindhyan basement. Choubey (1971) observed that nowhere do the basalts occupy valleys in the Vindhyans—they would be expected to if any topographic relief existed before the eruptions—and the base of the basalts is remarkably flat, always at 590 m,

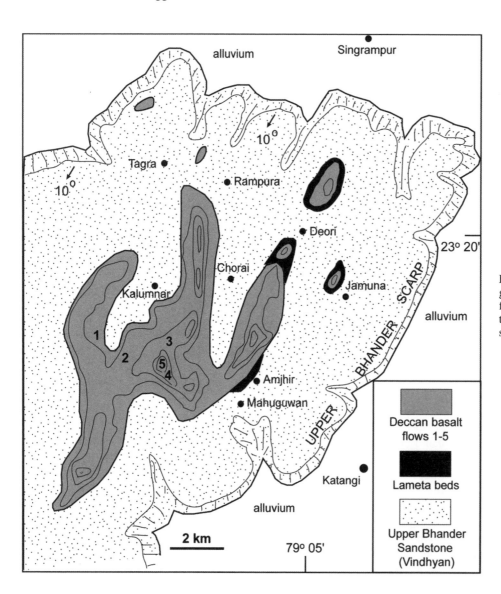

Figure 9. Geological map of the Singrampur-Katangi area (Fig. 3), simplified after Choubey (1971). Note that the top of the gently dipping Vindhyan sandstone is a flat planation surface.

irrespective of whether the basalts lie directly over the Lameta bed or the Vindhyans. These observations led him to suggest that the Vindhyans were completely peneplained before the eruptions of the Deccan lavas, and he considered this a spectacular example of a peneplain.

Sagar

Sagar (Fig. 3) is located near the northern margin of the Deccan outcrop, and the area exposes both Lameta beds and Vindhyans under the basalts (West, 1959; West and Choubey, 1964). Ten lava flows were recognized here by West (1959) and West and Choubey (1964), with a total thickness of 180 m. The base of the flows is at 425 m, with either the Lametas or the Vindhyans directly below. In the area some 25–35 km SSE of Sagar, Choubey (1971) mentioned exposures of ferruginous laterite ~2 m thick, overlying the Vindhyans and overlain by the Deccan basalts. He interpreted the laterite to represent the sub-basaltic erosion surface or peneplain.

Dhura-Jamunia

In this area 40 km east of Sagar (Fig. 3), Choubey (1971) recorded three Deccan basalt flows that he could correlate with the flows in the Katangi and Sagar areas. The basalts are horizontal and overlie the Vindhyan rocks.

Jabalpur

In the Jabalpur area (Fig. 3), Choubey (1971) noted a Late Cretaceous erosion surface extending over Archeans, Bijawars, and Gondwanas at 440–425 m, over which the (thin) Lametas were deposited, followed by the Deccan basalts. Flat-topped ridges of the Bijawar quartzites here are capped by 7–8 m of lateritic duricrust, which he reasonably interpreted as the remnants of the planation surface.

Dhar Forest and Bari

In this area (Fig. 3), Choubey (1971) noted a remarkable summit accordance at 300 m on Archean gneisses, Bijawar quartzites, and the Vindhyan sandstones, and he remarked that these hard rocks, though considerably disturbed, form a flat terrace, most likely the remnant of an uplifted peneplain. He considered this area a fine one for examining an exhumed pre-volcanic landscape.

Monadnocks

Some 100 km south of Sagar, at the village of Bitli (440 m), Choubey (1971) noted a conspicuous isolated hill of nearly vertical beds of Bijawar quartzites, with a very flat top. He suggested that this is an isolated monadnock (surviving remnant of an eroded peneplain). It is tempting to see a similarity between this and Ayers Rock, a huge monadnock made up of very steeply dipping feldspathic sandstone strata in a vast, flat, pre-Pliocene (>5 Ma) peneplain in central Australia (Ollier and Pain, 2000, p. 28–30). Together, the regional planation surfaces, the laterite that caps them, and the isolated monadnock constitute very convincing evidence for long-term tectonic stability before the Deccan eruptions in central India. Yet more critical evidence comes from the Pachmarhi area of the Satpura dome.

SATPURA DOME

The sub–Deccan Trap Gondwana sedimentary basin is spectacularly exposed in the Satpura Mountain Range due to post-Deccan cymatogenic uplift, and the region of uplift mimics an ENE-WSW-elongated dome (Venkatakrishnan, 1984, 1987). Pachmarhi town, located on the dome (Fig. 10A and B), sits on a thick (~1000 m) lens of Early Triassic (Upper Gondwana) Pachmarhi sandstone containing subordinate shales and conglomerates and capped by extensive ferricrete. It is arguably the doming that has exhumed these sandstones from beneath the Deccan basalt cover, and the most spectacular feature of the dome is the ENE-WSW-trending, 160-km-long, south-facing, free-face scarp that has been carved into the Pachmarhi sandstone, with a maximum height of 280 m (Fig. 10A and B). Many outliers of this sandstone lie scattered south of the present position of the scarp, which has been retreating northward.

The Pachmarhi sandstone strata dip north by 5°–10° and are underlain by the Bijori Formation (Upper Permian, Lower Gondwana), dominated by shales (Fig. 10B). Many large Deccan Trap dolerite dikes and intrusions can be found intruding the Pachmarhis and Bijoris at lower elevations, particularly in the gorges cut by rivers such as the Denwa. The dipping strata are beveled by planation surfaces. Venkatakrishnan (1984, 1987) recognized three distinct planation surfaces in the region (Fig. 10A and B). The lowest and youngest is the Bijori Surface (640 m), which is currently forming and slopes northward toward Pachmarhi from the basalt cliffs to the south. Older and higher than the Bijori Surface is the Pachmarhi Surface (920 m), and still older and higher is a planation surface (~1300 m) that is preserved in three accordant sandstone summits of Dhupgarh (1352 m), Mahadeo (1330 m), and Chauragarh (1308 m) that rise over Pachmarhi within a few kilometers of it. Venkatakrishnan (1984) named this the Dhupgarh Surface and suggested that the Narmada River Valley to the north of Pachmarhi has been recurrently acting as a local base level, periodically rising and falling, for streams eroding the plateau.

On Dhupgarh, which is the highest peak in the Satpura Range, a Deccan basalt flow unconformably overlies the north-dipping Pachmarhi sandstone (Venkatakrishnan, 1984), and Venkatakrishnan (1984) considered the base of the basalt flow—the Dhupgarh Surface—the same as the Late Cretaceous peneplain described by Dixey (1970/1999), the sub–Deccan Trap erosion surface of Choubey (1971), and probably also the intercontinental Gondwana or African Surface of King (1953).

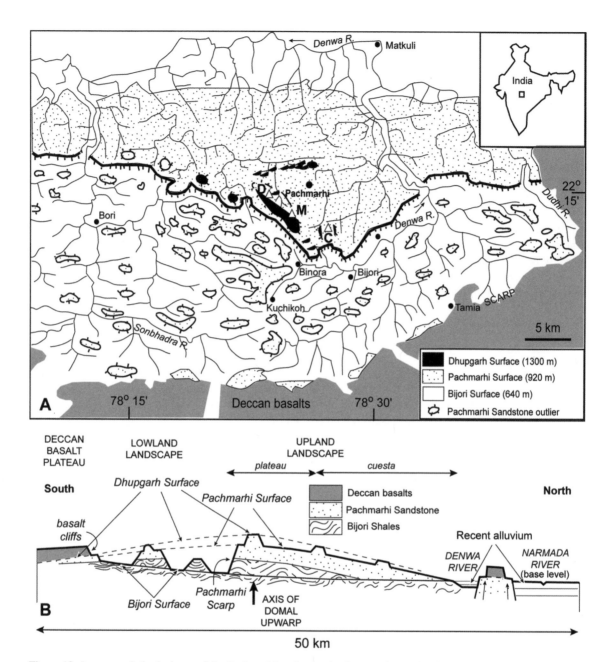

Figure 10. A geomorphological map of the Pachmarhi region on the Satpura dome, showing the Pachmarhi scarp, planation surfaces, structural elements, Pachmarhi sandstone outliers, drainage lines, and some localities. Innumerable Deccan Trap dikes and intrusions in the sandstones are not shown for clarity. D—Dhupgarh (1352 m); M—Mahadeo (1330 m); C—Chauragarh (1308 m). Based on Venkatakrishnan (1984). (B) Generalized geological cross-section of the Pachmarhi area showing the main landscape elements, slightly modified from Venkatakrishnan (1987). Note the upwarped planation surfaces, and that the surfaces bevel the internal structure of the Pachmarhi sandstones and the underlying Bijori shales. The surfaces are also characterized by laterite or ferricrete development. The highest and oldest planation surface is the Dhupgarh surface (only approximate in panel A), and it is the pre–Deccan Trap erosion surface of Choubey (1971), and the Late Cretaceous peneplain of Dixey (1970/1999).

Although the sandstone-basalt contact at Dhupgarh is an unconformity, the unconformity does not mean there were uplift and erosion right before the Deccan eruptions, because the entire rock record from the Middle Triassic to the Late Cretaceous is missing here. Also, because the unconformity is a regional planation surface, indicating long-term tectonic stability, the missing rock record cannot be ascribed to rapid erosion just before volcanism.

Choubey (1971) regarded the Satpura doming as one of the finest and most spectacular examples of cymatogenic uplift in the world, and Venkatakrishnan (1984, 1987) observed that the multiple planation surfaces here were warped as a result of this post-Deccan uplift. He identified the doming and warping in the Satpura region on the basis of evidence such as the warped contact between the Pachmarhi sandstones and the Bijori shales, reversals of structural dips, warped planation surfaces, cave levels, and the drainage pattern. Major rivers in the region, like the Denwa and the Dudhi (Fig. 10A), maintained their preexisting courses across the rising plateau, and in the process cut profound gorges. These rivers are therefore good examples of antecedent drainage (Choubey, 1971; Venkatakrishnan, 1984). The smaller rivers, however, were defeated by the uplift, and therefore occupied lithostructural breaks in the Pachmarhi sandstones, producing the unusual rectangular and barbed drainage pattern found there (Fig. 10A). A key observation by Venkatakrishnan (1984) is that *rivers such as Denwa and Dudhi originated well to the south of the present-day scarp, several hundred meters below the Pachmarhi Surface, and have cut their way northward through the evolving Satpura dome in spectacular gorges.* He suggests that the most reasonable explanation of this is drainage antecedence; the rivers were in existence before the Satpura uplift and kept cutting the domal upwarp that was forming in their path. To summarize, the uplift and warping of the Pachmarhi block, with its planation surfaces, is post-Deccan (just as the uplift of the Sahyadri and the Indian peninsula is post-Deccan).

What was the cause of this post-Deccan uplift? The Sahyadri uplift (all along the rifted margin) cannot be related to any magmatic underplating that may have occurred during Deccan volcanism, because the essentially thermal uplift associated with magmatic underplating (Cox, 1980; McKenzie, 1984) decays rapidly after the magmatism (e.g., Clift, 2005). The uplift and deformation of large regions of peninsular India, including the noticeable buckling and upwarping at 13°N latitude and around Pachmarhi, may be related at least in part to the dominantly compressional stress regime throughout peninsular India since the suturing of India and Asia along the Himalayas at ca. 55 Ma (e.g., Klootwijk et al., 1992; Gordon et al., 1998; Valdiya, 1998), which has caused reactivation of ancient and older weak zones and much recent intraplate seismicity (e.g., Subrahmanya, 1994; Valdiya, 2001; Vita-Finzi, 2002). The Sahyadri uplift is denudational in part (Widdowson and Cox, 1996; Widdowson and Gunnell, 1999) and may also be related to active eastward flow of the sublithospheric mantle relative to the lithosphere as hypothesized by Doglioni et al. (2003) based on several lines of evidence.

DISCUSSION AND CONCLUSIONS

Campbell and Griffiths (1990) cited Pachmarhi as the center of a broad, uplifted dome produced by their conjectured Deccan plume head, quoting Choubey (1971) that the lava-basement contact at Pachmarhi is over 1 km above sea level. But as many lines of evidence discussed earlier show unambiguously, the doming of the Pachmarhi region (notably not in a circular pattern but elongated ENE-WSW along the Satpura tectonic trend), and the entire Satpura Range of which it is a part, is not prevolcanic but rather postvolcanic. Casshyap and Khan (2000) argued for "pre–Deccan Trap doming" of the Indian landmass based on sedimentological field studies in the Pachmarhi region, but the timing and location of this doming cannot be related to a putative Deccan plume by any means. They identified three separate uplift events, the youngest of which resulted in Late Jurassic to earliest Cretaceous sediments that were derived from a source in northwestern India. An uplift event centered on northwestern India, and preceding the Late Cretaceous–Paleocene Deccan volcanism by ~70 m.y., cannot be considered prevolcanic uplift from the Deccan plume.

I have presented and reviewed the evidence for planation surfaces below the Deccan basalts in central India. The flatness of the pre-Deccan landscape constructed on various older rocks, the remarkable horizontality of the Deccan basalt flows over long distances, and laterites found on the pre-Deccan landscape together form quite compelling evidence for pre-Deccan planation surfaces. This evidence, along with a nearly universal absence of indicators of pre-eruption uplift throughout the extent of the province, runs counter to the idea that a large plume head upwelled beneath India in the Late Cretaceous, producing regional domal uplift, the current drainage pattern, and Deccan flood basalt volcanism. Prevolcanic regional lithospheric uplift is not to be found in the Deccan, and real field evidence (the planation surfaces) indicates its actual absence. There has been major kilometer-scale post-Deccan uplift, however, which has brought the pre-Deccan planation surfaces from their originally low elevations (relative to the base level at that time) to their present high elevations. A very similar scenario has been noted in the flood basalts of Yemen by Menzies et al. (1997), who noted that these voluminous basalt lavas overlie thick paleosols developed on underlying fluvial and marine sediments. The paleosols are widespread, 5–70 m thick, and ferruginous. Following Summerfield (1991), they suggested that these paleosols indicate both low relief and a lack of intense denudation and sediment starvation. They noted ongoing sediment starvation during the eruption of some 3000 m of volcanics in 6 m.y. with few intervening sediments.

How are we to explain the lack of pre-eruption uplift in flood basalts, as described here? Campbell and Griffiths (1990) argued that prevolcanic regional domal uplift due to a plume head may not be significant due to lateral migration of magma in the crust, or difficult to recognize because early uplift may be overprinted by later subsidence. These are not satisfactory explanations, because uplift is expected to have begun 10–20 m.y. before volcanism, and maximum uplift is expected to have occurred 5–10 m.y. before volcanism (Campbell and Griffiths, 1990; Fig. 1), when the plume head was still well below the lithosphere and initial low-degree melts may barely have begun to form, much less have been injected into the crust. Later subsidence that overprinted and erased evidence of early pre-eruption uplift cannot be invoked, because such subsidence is simply not seen in flood basalts like the Deccan and Columbia River basalts, which show major postvolcanic uplift instead (as described earlier in this article and in Hales et al., 2005). And this uplift should expose evidence of early pre-eruption uplift, if there ever was any.

There is a serious need to reconsider the plume head model for the Deccan Traps because few if any aspects of the plume model are compatible with geological and geophysical data from the Deccan province (Sheth, 2005a,b; see also Ravi Kumar and Mohan, 2005, and Senthil Kumar et al., this volume). The plume model is also under reconsideration for many other flood basalts (e.g., Ingle and Coffin, 2004; Hales et al., 2005; and several articles in Foulger et al., 2005). In a very recent summary of the plume model, Campbell (2005) claims that all the events expected on the basis of the model, including prevolcanic regional uplift, are supported by field data from flood basalt provinces of the world, such as the Emeishan and the Deccan. But prevolcanic regional uplift in the Emeishan province (He et al., 2003) and in the west and east Greenland provinces (Dam et al., 1998) are more exceptions than the rule, and even in these provinces it remains to be seen to what extent the prevolcanic uplift reflects magmatic underplating as opposed to dynamic support from a thermal anomaly in the mantle.

In conclusion, the uplift of the Sahyadri Range (western Ghats) of India is not related to a putative Deccan mantle plume: it is not domal; it occurred all along the western Indian rifted margin, well beyond the Deccan lava cover, and is most appropriately considered a rift-shoulder uplift affecting the Deccan basalt pile and the basement rocks equally, maintained at least in part by denudational unloading (Widdowson, 1997a). Planation surfaces below the basalts reflect long-term tectonic stability prior to volcanism, and their significant elevations today reflect post-Deccan uplift, major uplift having occurred in the Neogene (i.e., <23 Ma). The uplift has been aided by a compressional regime throughout the Indian Shield since ca. 55 Ma. The easterly drainage of the Indian peninsula is not dome-flank drainage produced by a plume head but was antecedent to the uplift. The uplift of both the western and the eastern Ghats postdates Deccan volcanism and the easterly drainage. There is thus not only an *absence of evidence* for prevolcanic regional uplift in the Deccan flood basalt province, but actual *evidence for absence* thereof. The abundant plume conjecture in the voluminous Dec-

can literature has no factual basis (see also Sheth, 2005a,b); the plume model is untenable there.

ACKNOWLEDGMENTS

Debate and interaction with colleagues at the 2003 GSA Penrose Conference Plume IV: Beyond the Plume Hypothesis (Iceland), and the 2005 American Geophysical Union Chapman Conference The Great Plume Debate (Scotland) were invaluable, as were the field trips associated with both. I particularly thank Gillian Foulger for our extensive interaction over mantle plumes during the last five years. The first version of this article benefited considerably from very helpful reviews by Richard Ernst, Mike Widdowson, and an anonymous reviewer. Yanni Gunnell and Colin Pain provided valuable, constructive reviews of the revised manuscript, and Rajat Mazumder and Peter Japsen also offered helpful unofficial comments. Volume editors Gillian Foulger and Donna Jurdy offered editorial comments at both stages that helped improve the presentation.

REFERENCES CITED

Baksi, A.K., Byerly, G.R., Chan, L.-H., and Farrar, E., 1994, Intracanyon flows in the Deccan province, India?: Case history of the Rajahmundry Traps: Geology, v. 22, p. 605–608, doi: 10.1130/0091-7613(1994)022<0605: IFITDP>2.3.CO;2.

Bendick, R., and Bilham, R., 1999, Search for buckling of the southwest Indian coast related to Himalayan collision, *in* McFarlane, A., et al., eds., Himalaya and Tibet: Mountain roots to mountain tops: Boulder, Colorado, Geological Society of America Special Paper 328, p. 313–321.

Bonow, J.M., Lidmar-Bergström, K., and Japsen, P., 2006a, Palaeosurfaces in central West Greenland as reference for identification of tectonic movements and estimation of erosion: Global and Planetary Change, v. 50, p. 161–183, doi: 10.1016/j.gloplacha.2005.12.011.

Bonow, J.M., Japsen, P., Green, P.F., Chalmers, J.A., and Lidmar-Bergström, K., 2006b, Significance of elevated planation surfaces for conclusions of uplift and tectonic events on passive margins: Geophysical Research Abstracts, v. 8, p. 09869.

Burov, E., and Guillou-Frottier, L., 2005, The plume head–continental lithosphere interaction using a tectonically realistic formulation for the lithosphere: Geophysical Journal International, v. 161, p. 469–490, doi: 10.1111/j.1365-246X.2005.02588.x.

Campbell, I.H., 2005, Large igneous provinces and the mantle plume hypothesis: Elements, v. 1, p. 265–269.

Campbell, I.H., and Griffiths, R.W., 1990, Implications of mantle plume structure for the evolution of flood basalts: Earth and Planetary Science Letters, v. 99, p. 79–93, doi: 10.1016/0012-821X(90)90072-6.

Casshyap, S.M., and Khan, A., 2000, Tectono-sedimentary evolution of the Gondwanan Satpura Basin of central India: Evidence of pre-trap doming, rifting and palaeoslope reversal: Journal of African Earth Sciences, v. 31, p. 65–76, doi: 10.1016/S0899-5362(00)00073-7.

Choubey, V.D., 1971, Pre–Deccan Trap topography in central India and crustal warping in relation to Narmada rift structure and volcanic activity: Bulletin of Volcanology, v. 35, p. 660–685.

Clift, P.D., 2005, Sedimentary evidence for modest mantle temperature anomalies associated with hotspot volcanism, *in* Foulger, G.R., et al., eds., Plates, plumes, and paradigms: Boulder, Colorado, Geological Society of America Special Paper 388, p. 279–287.

Cox, K.G., 1980, A model for flood basalt vulcanism: Journal of Petrology, v. 21, p. 629–650.

Cox, K.G., 1989, The role of mantle plumes in the development of continental drainage patterns: Nature, v. 342, p. 873–877, doi: 10.1038/342873a0.

Czamanske, G.K., Gurevitch, A.B., Fedorenko, V., and Simonov, O., 1998, Demise of the Siberian plume: Palaeogeographic and palaeotectonic reconstruction from the prevolcanic and volcanic record, north-central Siberia: International Geology Review, v. 40, p. 95–115.

Dam, G., Larsen, M., and Sonderholm, M., 1998, Sedimentary response to mantle plumes: Implications from Palaeocene onshore successions: West and East Greenland: Geology, v. 26, p. 207–210.

Dixey, F., 1970/1999, The geomorphology of Madhya Pradesh, India, *in* Murthy, T.V.V.G.R.K., ed., Studies in earth sciences: W.D. West Commemoration Volume: New Delhi, Today and Tomorrow Publishers, p. 195–224 (1970); reprinted in Subbarao, K.V., ed., 1999, Deccan volcanic province, v. 1–2: Bangalore, Geological Society of India Memoir 43, p. 575–590(1999).

Doglioni, C., Carminati, E., and Bonatti, E., 2003, Rift asymmetry and continental uplift: Tectonics, v. 22, doi: 10.1029/2002TC001459.

Emeleus, C.H., and Bell, B.R., 2005, The Palaeogene volcanic districts of Scotland (4th edition): Nottingham, British Geological Survey, 212 p.

Farnetani, C., and Richards, M.A., 1994, Numerical investigations of the mantle plume initiation model for flood basalt events: Journal of Geophysical Research, v. 99, p. 13,813–13,833, doi: 10.1029/94JB00649.

Foulger, G.R., Natland, J.H., Presnall, D.C., and Anderson, D.L., 2005, eds., Plates, plumes, and paradigms: Boulder, Colorado, Geological Society of America Special Paper 388, 881 p.

Gordon, R.G., DeMets, C., and Royer, J.-Y., 1998, Evidence for long-term diffuse deformation of the lithosphere of the equatorial Indian Ocean: Nature, v. 395, p. 370–374, doi: 10.1038/26463.

Gunnell, Y., 1998, The interaction between geological structure and global tectonics in multistoreyed landscape development: A denudation model for the South Indian shield: Basin Research, v. 10, p. 281–310, doi: 10.1046/j.1365-2117.1998.00072.x.

Gunnell, Y., and Gallagher, K., 2001, Short- and long-term denudation rates in cratonic environments: Estimates for Precambrian Karnataka (South India), *in* Gunnell, Y., and Radhakrishna, B.P., eds., 2001, Sahyadri: The great escarpment of the Indian subcontinent: Bangalore, Geological Society of India Memoir 47 (nos. 1–2), p. 445–461.

Gunnell, Y., and Louchet, A., 2000, Rock hardness and long-term divergent weathering as factors of relief development in the charnockite massifs of South Asia: Zeitschrift für Geomorphologie, v. 44, p. 33–57.

Gunnell, Y., and Radhakrishna, B.P., eds., 2001, Sahyadri: The great escarpment of the Indian subcontinent: Bangalore, Geological Society of India Memoir 47 (nos. 1–2), 1054 p.

Gunnell, Y., Gallagher, K., Carter, A., Widdowson, M., and Hurford, A.J., 2003, Denudation history of the continental margin of western peninsular India since the early Mesozoic—Reconciling apatite fission-track data with geomorphology: Earth and Planetary Science Letters, v. 215, p. 187–201, doi: 10.1016/S0012-821X(03)00380-7.

Hales, T.C., Abt, D.L., Humphreys, E.D., and Roering, J.J., 2005, A lithospheric instability origin for Columbia River flood basalts and Wallowa Mountains uplift in northeast Oregon: Nature, v. 438, p. 842–845, doi: 10.1038/nature04313.

Harris, J., and van Couvering, J., 1995, Mock aridity and the palaeoecology of volcanically influenced ecosystems: Geology, v. 23, p. 593–596, doi: 10.1130/0091-7613(1995)023<0593:MAATPO>2.3.CO;2.

He, B., Xu, Y.-G., Chung, S.-L., Xiao, L., and Wang, Y., 2003, Sedimentary evidence for a rapid, kilometer-scale crustal doming prior to the eruption of the Emeishan flood basalts: Earth and Planetary Science Letters, v. 213, p. 391–405, doi: 10.1016/S0012-821X(03)00323-6.

Hill, R.I., 1991, Starting plumes and continental break-up: Earth and Planetary Science Letters, v. 104, p. 398–416, doi: 10.1016/0012-821X(91)90218-7.

Ingle, S.P., and Coffin, M.F., 2004, Impact origin for the greater Ontong Java Plateau?: Earth and Planetary Science Letters, v. 218, p. 123–134, doi: 10.1016/S0012-821X(03)00629-0.

Kaila, K.L., 1988, Mapping the thickness of Deccan Trap flows in India from DSS studies and inferences about a hidden Mesozoic basin in the Narmada-Tapti region, *in* Subbarao, K.V., ed., Deccan flood basalts: Bangalore, Geological Society of India Memoir 10, p. 91–116.

Kaila, K.L., Murthy, P.R.K., Dixit, M.M., and Lazarenko, M.A., 1981, Crustal structure from the seismic soundings along the Koyna II (Kelsi-Loni) profile in the Deccan Trap area, India: Tectonophysics, v. 73, p. 365–384, doi: 10.1016/0040-1951(81)90223-7.

Kale, V.S., Kulkarni, H.C., and Peshwa, V.V., 1992, Discussion on a geological map of the southern Deccan Traps, India, and its structural implications: Journal of the Geological Society of London, v. 149, p. 473–475.

Kent, R.W., 1991, Lithospheric uplift in eastern Gondwana: Evidence for a long-lived mantle plume system?: Geology, v. 19, p. 19–23, doi: 10.1130/0091-7613(1991)019<0019:LUIEGE>2.3.CO;2.

Khadkikar, A.S., Sant, D.A., Gogte, V., and Karanth, R.V., 1999, The influence of Deccan volcanism on climate: Insights from lacustrine inter-trappean deposits, Anjar, western India: Palaeogeography, Palaeoclimatology, Palaeoecology, v. 147, p. 141–149, doi: 10.1016/S0031-0182(98)00156-4.

Khosla, A., and Sahni, A., 2003, Biodiversity during the Deccan volcanic eruptive episode: Journal of Asian Earth Sciences, v. 21, p. 895–908, doi: 10.1016/S1367-9120(02)00092-5.

King, L.C., 1953, Canons of landscape evolution: Geological Society of America Bulletin, v. 64, p. 721–752.

Klootwijk, C.T., Gee, J.S., Pierce, J.W., Smith, G.M., and McFadden, P.L., 1992, An early India-Asia contact: Palaeomagnetic constraints from Ninety-East Ridge, ODP Leg 121: Geology, v. 20, p. 395–398, doi: 10.1130/0091-7613(1992)020<0395:AEIACP>2.3.CO;2.

Knight, K.B., Renne, P.R., Halkett, A., and White, N., 2003, $^{40}Ar/^{39}Ar$ dating of the Rajahmundry Traps, eastern India, and their relationship to the Deccan Traps: Earth and Planetary Science Letters, v. 208, p. 85–99, doi: 10.1016/S0012-821X(02)01154-8.

Korenaga, J., 2005, Why did not the Ontong Java Plateau form subaerially?: Earth and Planetary Science Letters, v. 234, p. 385–399, doi: 10.1016/j.epsl.2005.03.011.

Krishnamurthy, P., 1971, Petrology of the dyke rocks of the western portions of Rajpipla hills, Broach district, Gujarat, India: Bulletin of Volcanology, v. 35, p. 930–946.

Mazumder, R., 2005, Proterozoic sedimentation and volcanism in the Singbhum crustal province, India, and their implications: Sedimentary Geology, v. 176, p. 167–193, doi: 10.1016/j.sedgeo.2004.12.011.

Mazumder, R., 2006, Large igneous provinces, mantle plumes and uplift: A case study from the Dhanjori Formation, India: www.mantleplumes.org/Dhanjori.html.

McKenzie, D.P., 1984, A possible mechanism for epeirogenic uplift: Nature, v. 307, p. 616–618, doi: 10.1038/307616a0.

Menzies, M.A., Baker, J., Chazot, G., and Al'Kadasi, M., 1997, Evolution of the Red Sea volcanic margin, western Yemen, *in* Mahoney, J.J., and Coffin, M.F., eds., Large igneous provinces: Oceanic, continental, and planetary flood volcanism: Washington, D.C., American Geophysical Union, Geophysical Monograph 100, p. 29–43.

Mohabey, D., 1996, Depositional environment of Lameta Formation (Late Cretaceous) of Nand-Dongargaon inland basin, Maharashtra: The fossil and lithological evidences, *in* Sahni, A., ed., Cretaceous stratigraphy and palaeoenvironments: Bangalore, Geological Society of India Memoir 37, p. 363–386.

Morgan, W.J., 1972, Deep mantle convection plumes and plate motions: Bulletin of the American Association of Petroleum Geologists, v. 20, p. 203–213.

Morgan, W.J., 1981, Hotspot tracks and the opening of the Atlantic and Indian Oceans, *in* Emliani, C., ed., The sea: New York, Wiley, p. 443–487.

Norton, I.O., and Sclater, J.G., 1979, A model for the evolution of the Indian Ocean and the breakup of Gondwanaland: Journal of Geophysical Research, v. 84, p. 6803–6830.

Ollier, C.D., 1990, Mountains, *in* Barto-Kyriakidis, A., ed., Critical aspects of the plate tectonic theory, v. 2: Athens, Theophrastus Publications, p. 211–236.

Ollier, C.D., and Pain, C.F., 2000, The origin of mountains: London, Routledge, 368 p.

Ollier, C.D., and Powar, K.B., 1985, The Western Ghats and the morphotectonics of peninsular India: Zeitschrift für Geomorphologie: Suppl. Bd., v. 54, p. 57–69.

Pande, K., Sheth, H.C., and Bhutani, R., 2001, ^{40}Ar-^{39}Ar age of the St. Mary's Islands volcanics, southern India: Record of India-Madagascar break-up on the Indian subcontinent: Earth and Planetary Science Letters, v. 193, p. 39–46, doi: 10.1016/S0012-821X(01)00495-2.

Peterson, J.A., and Clarke, J.W., 1991, Geology and hydrocarbon habitat of the West Siberian Basin, v. 32: Tulsa, Oklahoma, American Association of Petroleum Geologists, 93 p.

Powar, K.B., 1993, Geomorphological evolution of Konkan coastal belt and adjoining Sahyadri uplands with reference to Quaternary uplift: Current Science, v. 64, p. 793–796.

Prasad, G.V.R., and Khajuria, C.K., 1995, Implications of the infra- and intertrappean biota from the Deccan, India, and the role of volcanism in Cretaceous-Tertiary boundary extinction events: Journal of the Geological Society of London, v. 152, p. 289–296.

Radhakrishna, B.P., 1952/2001, The Mysore plateau: Its structural and physiographical evolution: Bulletin of the Mysore Geologists Association, v. 3, p. 1–53 (1954); reprinted in Gunnell, Y., and Radhakrishna, B.P., eds., Sahyadri: The great escarpment of the Indian subcontinent: Bangalore, Geological Society of India Memoir 47 (nos. 1–2) (2001), p. 71–82.

Radhakrishna, B.P., 1993, Neogene uplift and geomorphic rejuvenation of the Indian peninsula: Current Science, v. 64, p. 787–793.

Rainbird, R.H., and Ernst, R.E., 2001, The sedimentary record of mantle plume uplift, in Ernst, R.E., and Buchan, K.L., eds., Mantle plumes: Their identification through time: Boulder, Colorado, Geological Society of America Special Paper 352, p. 227–245.

Rajesh, H.M., and Santosh, M., 2004, Charnockitic magmatism in southern India, in Sheth, H.C., and Pande, K., eds., Magmatism in India through time: Proceedings of the Indian Academy of Sciences (Earth and Planetary Sciences), v. 113, p. 565–585.

Ravi Kumar, M., and Mohan, G., 2005, Mantle discontinuities beneath the Deccan volcanic province: Earth and Planetary Science Letters, v. 237, p. 252–263, doi: 10.1016/j.epsl.2005.06.034.

Ray, J.S., Pande, P., and Pattanayak, S.K., 2003, Evolution of the Amba Dongar carbonatite complex: Constraints from ^{40}Ar/^{39}Ar chronologies of the inner basalt and an alkaline plug: International Geology Review, v. 45, p. 857–862.

Reichow, M., Saunders, A.D., White, R.V., Pringle, M.S., Al'mukhamedov, A.I., Medvedev, A., and Korda, N., 2002, New ^{40}Ar-^{39}Ar data on basalts from the West Siberian Basin: Extent of the Siberian flood basalt province doubled: Science, v. 296, p. 1846–1849, doi: 10.1126/science.1071671.

Richards, M.A., Duncan, R.A., and Courtillot, V.E., 1989, Flood basalts and hotspot tracks: Plume heads and tails: Science, v. 246, p. 103–108, doi: 10.1126/science.246.4926.103.

Sahni, A., Venkatachala, B.S., Kar, R.K., Rajanikanth, A., Prakash, T., Prasad, G.V.R., and Singh, R.Y., 1996, New palynological data from the Deccan intertrappean beds: Implications for the latest record of dinosaurs and synchronous initiation of the volcanic activity in India, in Sahni, A., ed., Cretaceous stratigraphy and palaeoenvironments: Bangalore, Geological Society of India Memoir 37, p. 267–284.

Salil, M.S., Shrivastava, J.P., and Pattanayak, S.K., 1997, Similarities in the mineralogical and geochemical attributes of detrital clays of Maastrichtian Lameta Beds and weathered Deccan basalt, central India: Chemical Geology, v. 136, p. 25–32, doi: 10.1016/S0009-2541(96)00128-3.

Saunders, A.D., England, R.W., Reichow, M.K., and White, R.V., 2005, A mantle plume origin for the Siberian Traps: Uplift and extension in the West Siberian Basin, Russia: Lithos, v. 79, p. 407–424, doi: 10.1016/j.lithos.2004.09.010.

Senthil Kumar, P., Menon, R., and Koti Reddy, G., 2007 (this volume), Crustal geotherm in southern Deccan basalt province, India: The Moho is as cold as adjoining cratons, in Foulger, G.R., and Jurdy, D.M., eds., Plates, plumes,

and planetary processes: Geological Society of America Special Paper 430, p. 275–284, doi: 10.1130/2007.2430(14).

Sheth, H.C., 1998, A reappraisal of the coastal Panvel flexure, Deccan Traps, as a listric-fault-controlled reverse drag structure: Tectonophysics, v. 294, p. 143–149, doi: 10.1016/S0040-1951(98)00148-6.

Sheth, H.C., 2000, The timing of crustal extension, dyking, and the eruption of the Deccan flood basalts: International Geology Review, v. 42, p. 1007–1016.

Sheth, H.C., 2005a, From Deccan to Réunion: No trace of a mantle plume, in Foulger, G.R., et al., eds., Plates, plumes, and paradigms: Boulder, Colorado, Geological Society of America Special Paper 388, p. 477–501.

Sheth, H.C., 2005b, Were the Deccan flood basalts derived in part from ancient oceanic crust within the Indian continental lithosphere?: Gondwana Research, v. 8, p. 109–127, doi: 10.1016/S1342-937X(05)71112-6.

Sheth, H.C., and Ray, J.S., 2002, Rb/Sr-^{87}Sr/^{86}Sr variations in Bombay trachytes and rhyolites (Deccan Traps): Rb-Sr isochron, or AFC process?: International Geology Review, v. 44, p. 624–638.

Sheth, H.C., Pande, K., and Bhutani, R., 2001, ^{40}Ar-^{39}Ar ages of Bombay trachytes: Evidence for a Palaeocene phase of Deccan volcanism: Geophysical Research Letters, v. 28, p. 3513–3516, doi: 10.1029/2001GL012921.

Shrivastava, J.P., and Ahmad, M., 2005, A review of research on late Cretaceous volcanic-sedimentary sequences of the Mandla Lobe: Implications for Deccan volcanism and the Cretaceous/Palaeogene boundary: Cretaceous Research, v. 26, p. 145–156, doi: 10.1016/j.cretres.2004.11.009.

Sridhar, A.R., and Tewari, H.C., 2001, Existence of a sedimentary graben in the western part of the Narmada zone: Seismic evidence: Journal of Geodynamics, v. 31, p. 19–31, doi: 10.1016/S0264-3707(00)00013-2.

Storey, M., Mahoney, J.J., Saunders, A.D., Duncan, R.A., Kelley, S.P., and Coffin, M.F., 1995, Timing of hotspot-related volcanism and the breakup of Madagascar and India: Science, v. 267, p. 852–855, doi: 10.1126/science.267.5199.852.

Subbarao, K.V., ed., 1999, Deccan volcanic province, v. 1–2: Bangalore, Geological Society of India Memoir 43, 947 p.

Subrahmanya, K.R., 1994, Post-Gondwana tectonics of peninsular India: Current Science, v. 67, p. 527–530.

Subrahmanya, K.R., 1996, Active intraplate deformation in South India: Tectonophysics, v. 262, p. 231–241, doi: 10.1016/0040-1951(96)00005-4.

Subramanian, K.S., and Muraleedharan, M.P., 1985, Origin of the Palghat Gap in South India—A synthesis: Journal of the Geological Society of India, v. 26, p. 28–37.

Summerfield, M.A., 1990, Geomorphology and mantle plumes: Nature, v. 344, p. 387–388, doi: 10.1038/344387c0.

Summerfield, M.A., 1991, Sub-aerial denudation of passive margins: Regional elevation versus local relief models: Earth and Planetary Science Letters, v. 102, p. 460–469, doi: 10.1016/0012-821X(91)90036-H.

Tandon, S.K., 2002, Records of the influence of Deccan volcanism on contemporary sedimentary environments in central India: Sedimentary Geology, v. 147, p. 177–192, doi: 10.1016/S0037-0738(01)00196-8.

Tejada, M.L.G., Mahoney, J.J., Castillo, P.R., Ingle, S.P., Sheth, H.C., and Weis, D., 2004, Pin-pricking the elephant: Evidence on the origin of the Ontong Java Plateau from Pb-Sr-Hf-Nd isotopic characteristics of ODP Leg 192 basalts, in Fitton, J.G., et al., eds., Origin and evolution of the Ontong Java Plateau: London, Geological Society Special Publication 229, p. 133–150.

Vaidyanadhan, R., 1977, Recent advances in geomorphic studies of peninsular India: A review: Indian Journal of Earth Sciences, S. Ray Volume, p. 13–35.

Valdiya, K.S., 1998, Dynamic Himalaya: Bangalore, Universities Press, 198 pp.

Valdiya, K.S., 2001, Tectonic resurgence of the Mysore plateau and surrounding regions in cratonic southern India: Current Science, v. 81, p. 1068–1089.

Venkatakrishnan, R., 1984, Parallel scarp retreat and drainage evolution, Pachmarhi area, Madhya Pradesh, central India: Journal of the Geological Society of India, v. 25, p. 401–413.

Venkatakrishnan, R., 1987, Correlation of cave levels and planation surfaces in the Pachmarhi area, Madhya Pradesh: A case for base level control: Journal of the Geological Society of India, v. 29, p. 240–249.

Vita-Finzi, C., 2002, Intraplate neotectonics in India: Current Science, v. 82, p. 400–402.

West, W.D., 1959, The source of the Deccan Trap flows: Journal of the Geological Society of India, v. 1, p. 44–57.

West, W.D., and Choubey, V.D., 1964, The geomorphology of the country around Sagar and Katangi, M.P.: An example of superimposed drainage: Journal of the Geological Society of India, v. 5, p. 41–55.

Widdowson, M., 1997a, Tertiary palaeosurfaces of the SW Deccan, western India: Implications for passive margin uplift, *in* Widdowson, M., ed., Palaeosurfaces: Recognition, reconstruction, and palaeoenvironmental interpretation: London, Geological Society Special Publication 120, p. 221–248.

Widdowson, M., 1997b, The geomorphological and geological importance of palaeosurfaces, *in* Widdowson, M., ed., Palaeosurfaces: Recognition, reconstruction, and palaeoenvironmental interpretation: London, Geological Society Special Publication 120, p. 1–12.

Widdowson, M., 2005, The Deccan basalt-basement contact: Evidence for a plume-head generated CFBP? [abs.]: American Geophysical Union Chapman Conference, The Great Plume Debate, Scotland, p. 69–70.

Widdowson, M., and Cox, K.G., 1996, Uplift and erosional history of the Deccan Traps, India: Evidence from laterites and drainage patterns of the Western Ghats and Konkan coast: Earth and Planetary Science Letters, v. 137, p. 57–69, doi: 10.1016/0012-821X(95)00211-T.

Widdowson, M., and Gunnell, Y., 1999, Lateritization, geomorphology and geodynamics of a passive continental margin: The Konkan and Kanara coastal lowlands of western peninsular India: Kingston upon Thames, England, International Association of Sedimentologists Special Publication 27, p. 245–274.

Williamson, I.T., and Bell, B.R., 1994, The Palaeocene lava field of west-central Skye, Scotland: Stratigraphy, palaeogeography and structure: Transactions of the Royal Society of Edinburgh: Earth Sciences, v. 85, p. 39–75.

MANUSCRIPT ACCEPTED BY THE SOCIETY JANUARY 31, 2007

DISCUSSION

5 December 2006, Peter R. Hooper

In his introductory remarks, Sheth (this volume) quotes a recent article by Hales et al. (2005) in support of the absence of prevolcanic uplift in the Miocene Columbia River flood basalt province. This is an unfortunate example of a single and unfounded statement being repeated in the literature to support the author's thesis despite clear and overwhelming evidence, geological, biological, and geophysical, provided by numerous authors that exactly the opposite is true and very easily demonstrated. This evidence is discussed, and the relevant authors referenced, in Hooper et al. (this volume). Here I simply repeat the most obvious geological evidence for pre- and postvolcanic uplift of the area underlying the main feeder dikes of the Columbia River basalts.

The Chief Joseph Dike swarm, the dominant feeder system for the Columbia River basalts, runs NNW through and immediately to the east of the Wallowa Mountains, Oregon. In the Imnaha and all the adjacent Snake and Salmon river valleys to the east of the Wallowa Mountains, the earliest flood basalt eruptions fill very steep-sided canyons cut into the older rocks. These areas have been mapped in detail over a large area (e.g., Kleck, 1976; Hooper et al., 1984; see references in Hooper et al., this volume). One does not have to be a rocket scientist, or even a basalt petrologist, to realize that these steep, deep canyons were formed by river erosion of a rapidly rising landmass and that this uplift must have occurred immediately before the basalt eruptions. For Hales et al. (2005) to interpret these deep canyons as ponding in incipient basins and hence suggestive of prebasalt subsidence is disingenuous and clearly not tenable. Nor is subsidence to the east of the Wallowa Mountains necessary to those authors' interesting thesis that underplating beneath the dramatic uplift of the Wallowa block was a potential factor in the formation of the voluminous Grande Ronde basalts.

The subsequent development of equivalently shaped canyons through these early basalts, including Hells Canyon, the deepest in the United States, dictates equally clearly that this same area continued to rise rapidly after the basalt eruptions. That this rise was specifically along the line of the Chief Joseph Dike swarm is illustrated by the steady rise in the Imnaha–Grande Ronde basalt contact, west to east across this zone from Lewiston (Idaho) to tens of kilometers up the Clearwater River. Although the cause of uplift may be open to debate, the occurrence of uplift, both immediately before and after the eruption of the Columbia River basalts, is clearly demonstrated and beyond reasonable dispute.

I hope that this simple direct evidence of prebasalt uplift beneath the Columbia River basalt feeder system will prevent Sheth's statement being repeated through the literature in the future.

6 December 2006, H.C. Sheth

Hooper argues that there was significant prevolcanic uplift in the Columbia River basalt (CRB) province, reflected in the deep fluvial canyons developed in the prebasaltic rocks that the basalt lavas filled up. He also states that Hales et al. (2005), whom I have cited as arguing for lack of such uplift, are wrong, and therefore my statement about the lack of such uplift in the CRB province is wrong as well.

This may well be. Having no experience of CRB geology, I reserve judgment on whether Hales et al. (2005) are correct in relating the canyons to incipient basins and pre-eruption subsidence, or whether Hooper (5 December 2006) and Hooper et al. (this volume) are correct in ascribing them to pre-eruption uplift and fluvial erosion. While working on my article, I cited the existing, published literature. In the absence of any published criticism of the work of Hales et al. (2005), I could only quote them as the most recent study I knew of then, relating to the pre-volcanic vertical motion of the CRB.

I thus reserve further judgment regarding my statement about the CRB. In any case, I barely mention the CRB, keeping discussion of other flood basalts to a minimum and focusing on the Deccan, where I claim sufficient experience. At the same time, it helps to have a brief comparative mention of other flood basalts in one's chapter.

It is possible, as Hooper says, that the CRB example does not support my thesis. However, I stand by the main thrust of my chapter, that there was no pre-eruption uplift (not just absent evidence, but evident absence) in the Deccan.

19 December 2006, T.C. Hales, J.J. Roering, and D.L. Abt

Hooper raises a number of interesting questions regarding the nature and timing of uplift in the area surrounding the source of the Columbia River basalts. His comments highlight the difficulty in determining where (or even if) surface uplift or subsidence occurred before eruption of the Columbia River basalts. Despite these challenges, Hooper's suggestion that the presence of pre-Imnaha canyons provides evidence of significant uplift in the region is simply not supported by the literature. Geoscientists have recently made significant strides in deciphering the coupling between topography and tectonic forcing (e.g., Burbank et al., 1996; Whipple, 2004), and importantly, none of this work suggests that the mere presence of deep canyons can be interpreted as evidence for surface uplift.

Steep and deep canyons form as the result of changes in base level (as first described by Powell, 1875; Gilbert, 1877), which can be controlled by local rock uplift or other processes such as basin reorganization. In some cases, such as the Southern Alps of New Zealand and Taiwan, rapid rock and surface uplift cause rapid base level lowering (e.g., Whitehouse, 1988; Dadson et al., 2003). In contrast, numerous examples demonstrate rapid river incision caused by other factors. Among them is another canyon of significance in North America, the Grand Canyon, which cuts through the Colorado plateau as a result of drainage capture or lake overflow (Spencer and Pearthree, 2003, and references therein), not rapid uplift. In fact, we view the Colorado Plateau as a useful analogy for pre-Imnaha northeastern Oregon (Goles, 1986). In our case, Hells Canyon likely formed from the Late Neogene capture of the Snake River by the Columbia River (Alpha and Vallier, 1994; Van Tassel et al., 2001). As a result, Hells Canyon is a wonderful example of a canyon of history and circumstance; capture of the Snake River basin increased the Columbia River's drainage area and effectively drained Neogene Lake Idaho, resulting in increased local stream power and rapid incision (Van Tassel et al., 2001). Most importantly, this suggests that using canyon incision as evidence for rapid uplift is a fundamentally flawed endeavor that ignores the principles of landscape evolution (Powell, 1875; Gilbert, 1877).

So how can we determine the pre-eruptive history of the area? As described in Hales et al. (2005), the best markers of uplift are the Columbia River basalts themselves because they are well dated and were emplaced as relatively flat layers (Hooper and Camp, 1981; Tolan et al., 1989; Hooper, 1997). The impressive magnitude and detailed nature of lava flow mapping in the area allows us to use them as deformation markers. Furthermore, we argue that the relatively flat orientation of the early Grande Ronde basalt flows lends strong support for the absence of rapid pre-eruptive uplift. If significant pre-eruptive uplift occurred, we would expect significant thinning of the R1 Grande Ronde lava flows from the topographic high created by the uplift. As pointed out by Hooper in his comment, significant thinning of the lava flows does not occur until the last phases of the Grande Ronde eruptions, a point discussed by Hales et al. (2005) as relating to syneruptive uplift caused by changes in mantle buoyancy. Another reason we are inclined toward minimal pre-Imnaha topography is the presence of the lava flows themselves. Columbia River basalt flows have been characterized as inflated flows, which can form only on low topographic slopes (typically less than 5°) (Self et al., 1997). If significant pre-eruptive uplift had occurred in the area, one would expect to see channelization, levee development, and a range of lava flow forms consistent with basalt emplacement on steep slopes (Hon et al., 1994).

Based on these lines of evidence, we suggest that rapid pre-eruptive uplift of the magnitude proposed by Hooper is unlikely. Instead, we favor minimal pre-eruptive uplift and possibly pre-eruptive subsidence (Elkins-Tanton, 2005). Granted, the physical evidence for subsidence is relatively poor, mostly because we lack the ability to resolve paleotopography before emplacement of basalt flows. However, we are able to easily discount rapid pre-eruptive uplift for the reasons outlined above. Determining the pre-eruptive uplift history of this region may be possible via thermochronometry, and we welcome any comments that may shed light on this compelling question.

12 January 2007, Don L. Anderson

There are various mechanisms for having large and rapid uplifts concurrent with magmatism. The plume hypothesis is unique in predicting that uplift should precede magmatism by 5 to 20 Ma. The delamination hypothesis predicts a period of subsidence preceding uplift and magmatism. A critical test of the plume versus alternative mechanisms involving delamination, asthenospheric upwelling, slab windows, and slab detachment is the timing between subsidence, uplift, and magmatism. If large igneous provinces are emplaced on deep sedimentary basins or at sea level (or below sea level if emplaced on oceanic ridges or triple junctions, e.g., Ontong Java plateau), then this would argue against the plume hypothesis, particularly if there is no heatflow anomaly. Uplift is not the issue; it is the timing of the uplift. Numerical models of plumes do not predict the sequencing of observed uplifts in spite of repeated claims to the contrary.

20 January 2007, Peter R. Hooper

I reply to the comment of Hales and others, 19 December 2006.

Whipple (2004) and subsequent work indeed illustrate the significance of factors other than rapid uplift for the formation of deep canyons, but that hardly precludes the established view that uplift triggers river erosion. River capture, the most obvious alternative to rapid uplift, also requires the captured system to be at a much higher elevation than the capturing system. Although the present canyon system, including Hells Canyon, may well be caused in significant part by the capture of the upper Snake River by the Salmon-Columbia River system, that capture occurred long after the Columbia River basalt group (CRBG) eruptions and was clearly aided by the continued uplift of the area along the Idaho-Oregon border as evidenced by the rise of the Imnaha Grande Ronde basalt contact from west to east as noted earlier. There is no evidence of an earlier river-capturing event capable of forming the pre-Imnaha basalt steep-sided canyons. I re-emphasize the very steep-sided nature of the canyons filled by the first CRBG eruptions (Imnaha basalt), the lack of evidence for any river-capturing event before those eruptions, and consider the rapid rise of the area above the CRBG feeder system as the only plausible explanation for the formation of the steep-sided Imnaha basalt-filled canyons.

Hales et al. (comment, 19 December 2006) are incorrect in stating that significant thickening of Grande Ronde flows from east to west is confined to the last phases of the Grande Ronde eruptions. It is true that the sudden uplift along the northwest-trending Limekiln fault prevented anything but small volcanic cones of Grande Ronde N2 lava developing to the southeast of that structure while large volumes of N2 lava erupted on the northwestern down throw side. But it is equally true that all Grande Ronde flows (R1 through N2) thicken westward. As recorded many years ago, this creates an off-lap effect, reflecting the continuing rise of the area beneath the CRBG feeder system. The thickness of CRBG flows in the central Columbia plateau is almost three times the thickness of the basalt that accumulated around the feeder dikes (Hooper and Camp, 1981; Tolan et al., 1989; Hooper, 1997).

The argument of Hales et al. (comment, 19 December 2006) that doming before eruption would cause channeling rather than growth by inflation is equally unrealistic. First, the Imnaha flows and Grande Ronde flows on the east side of the Wallowa uplift have strikingly different physical properties to the inflated flows of the central Columbia plateau; their mode of emplacement has yet to be investigated. Second, the implication that doming over the plume head, if it occurred, would have caused flows to be channeled and not fed by inflation fails to recognize the likely slope angle that would result from doming. For instance, if the radius of the dome were 500 km and the central elevation 1–2 km, then the dip on the dome flank would be 0.250°. Even assuming that the surfaces were perfectly smooth, the yield strength of erupted basalt is such that this low slope angle would be unlikely to modify the evolving flow patterns or methods of flow growth. These low angles are likely to be the underlying cause of the massive inflated pahoehoe sheets that are so characteristic of the central Columbia plateau (Thordarson and Self, 1998) and that are consistent features of many other continental flood basalts (Jerram and Widdowson, 2004).

I conclude by reiterating the belief that the evidence for rapid rise of the area underlying the main feeder dikes for the CRBG along the eastern margin of the Columbia plateau is compelling and that the arguments advocated by Hales et al. (comment, 19 December 2006) are, at best, equivocal. To this must be added the geophysical and botanical evidence for doming (Parsons et al., 1994; Saltus and Thompson, 1995; Pierce et al., 2002).

2 February 2007, Mike Widdowson

Sheth (this volume) offers an interesting, albeit controversial, interpretation to some of the geological and geomorphological data from the Deccan volcanic province (DVP). He continues an essentially plume-skeptic theme adopted in previous articles (e.g., Sheth, 1999, 2005). Here, he challenges the "plume model" by attempting to demonstrate a "lack of regional uplift" before the Deccan eruptions. There are some excellent background sections, but it would be wrong to conclude that this chapter offers either a comprehensive or a balanced review of the available geophysical, geochemical, and geomorphological data. Moreover, the issues regarding plume-related uplift and associated geophysical data concerning evidence for a Deccan plume are actually far more complex than admitted (e.g., Kennet and Widyantoro, 1999; Burov and Guillou-Frottier, 2005).

Sheth correctly focuses on the volcano–sedimentary evolution of the DVP in seeking evidence for, or against, pre-eruptive doming of the region but provides only those interpretations consistent with a particular viewpoint. Alternative interpretations do exist, and these are more widely accepted. Of more immediate concern is that this article fails to present us with a viable alternative to the currently preferred plume-based view for the DVP; this is a serious omission. Sheth's assertion that the Deccan volcanic province (DVP) is not the result of a Late Cretaceous mantle plume might be made more credible if it were tested and explored through application of the available, alternative models (e.g., edge-driven convection, EDC). Instead, only a critique of previous work is offered.

It is apparent that Sheth favors the EDC family of models (e.g., Sheth, 1999, 2005) but fails to consider that these might present even greater inconsistencies if adopted as an alternate explanation of the DVP and the wider pre-, syn-, and postrift evolution of the western Indian margin. For any model to be credible for the DVP, it must adequately explain all of the following points:

1. Independent of global eustatic changes, offshore sedimentation rates increase significantly around peninsula India

during the Late Cretaceous and then wane again during early Palaegene (most scholars would conclude that this indicates regional uplift of the continental hinterland).

2. The fact that rift-related diking, the Seychelles rifting event, and associated sea-floor spreading clearly postdate the bulk of the DVP tholeiitic eruptions (e.g., Hooper, 1990; Devey and Stephens, 1991; Widdowson et al., 2000).

3. The major DVP eruptive units young (both stratigraphically and chronologically) southward (Cox, 1983; Devey and Lightfoot, 1986; Mitchell and Widdowson, 1991; Vandamme et al., 1991).

Most contend that this DVP structural asymmetry is consistent with the rapid drift of peninsular India over a mantle-melting anomaly. Without resorting to the petrogenetic arguments rehearsed elsewhere, it remains an effective argument that these stratigraphical, volcanological, geochronological, and paleomagnetic data are more consistent with the plume model of DVP evolution than with any of the available alternatives (Jerram and Widdowson, 2005).

Like many geoscientists, I am not particularly committed to any model of CFBP evolution. Even so, most would agree that any geological/geophysical model is simply a conceptual framework that aids an understanding of a wide range of observations; typically, it is usually a simplification of the natural world and thus often imperfect. However, we should at least demand equal testing of all available models and that this testing should first require determining the model outcome and then its comparison with actual observational data.

A vast body of recent geological and geomorphological data now exists for the DVP, including examples by Sheth et al. (2001a,b), which might allow the first attempts at this approach. However, much of the observation presented here by Sheth regarding uplift, or lack thereof, is either equivocal or else can equally well be explained in a manner consistent with the very plume-rift interaction model he seeks to contend (e.g., Cox, 1978; White et al., 1987; Courtillot et al., 1999; Ernst and Buchan, 2001).

Another serious omission is the lack of reference to Courtillot et al. (2003) (or, indeed, to any of the Deccan research published by Courtillot and colleagues). These authors conclude, reasonably, that in order to demonstrate the presence of a plume, five specific and stringent observational criteria must be met. Unsurprisingly, most of the "mantle plume" sites around the globe fail these tests. However, of those examples that do apparently fulfill all the criteria of plume-related volcanism, the Deccan is given as a prime example. Interestingly, Courtillot et al. (2003) do not consider pre-DVP uplift as being a critical factor, and whether such uplift can actually prove that mantle plumes exist, or not, still remains a polemic issue (see response by Anderson, 12 January).

To the credit of those who initiated this debate, we can now accept that the wider plume model needs to be scrutinized and perhaps used with more circumspection. However, it is neither logical nor scientific to conclude that simply because not every fact or interpretation fits a particular geological model, the model then becomes redundant or disproved. For any (CFBP) model to be acceptable, it must be capable of explaining most of the collected observations, and also be viable under most of those comparable geological situations found elsewhere (Jerram and Widdowson, 2005). Accordingly, we should now focus on which model best fits the available DVP data and thereby seek to develop and refine a wider understanding of the geologically complex phenomena that gave rise to this CFBP.

5 February 2007, H.C. Sheth

My chapter was meant to present the geological and geomorphological evidence from the Deccan Traps that has a bearing on the issue of plume-related prevolcanic uplift. Any model should address this evidence. Widdowson misses the point. I never intended a geophysical-geochemical-geomorphological synthesis or the discussion of a whole family of genetic models that he expected simply because the Sheth (2005) article that he quotes offers these, at some length, for the Deccan. Many workers (e.g., chapters in this volume and Foulger et al., 2005) evaluate the plume and other models for other specific areas or in general. I have not invoked the EDC model previously or here.

I believe in field data and logic. Widdowson does not call my discussion logical, though it is not logical to continue to defend a plume model for the Deccan Traps that has failed all predictions and tests and has no physical basis. He does not question the field facts given here such as the Indian peninsular drainage being antecedent to the western Ghats uplift and not related to a hypothetical plume-generated dome, or the planation surfaces in central India over which the Deccan lavas erupted. These observations bear greatly on the plume issue. The Courtillot et al. (2003) article is shown by Anderson (2005) to use highly subjective criteria and even circular reasoning for picking plumes, but Widdowson considers the former authoritative and a standard. That article is irrelevant here. Previously, Widdowson (2005) suggested a single conglomerate at Rajpipla to potentially reflect plume-caused uplift. My much more down-to-earth interpretation of this conglomerate, supported by comparable examples in Skye, does not evoke comment by him. His new suggestion that prevolcanic uplift is irrelevant because Courtillot et al. (2003) think so well illustrates the moving goalposts that plume-skeptics must wrestle with.

Widdowson considers my chapter unbalanced in interpretation. The plume view has been abundantly promoted in the existing literature in a way that is far from balanced by criticism. (Anderson and Natland, 2005, estimate that some 500 articles published in the last 10 years simply assume the existence of plumes.) More seriously, much of the plume-advocate literature distorts or ignores well-documented facts. The high-level contact between the Deccan lavas and the Gondwana sedimentary rocks in the Pachmarhi area cannot be evidence for plume uplift as was used. A plume-generated dome with rivers flowing away from it simply does not exist; the drainage antecedence was known well before that model was proposed but was ignored. Widdowson asserts that other interpretations of the data I have presented are possible. He does not itemize them, but I assume flattening of a plume head, a strong lithosphere, or some similar ad hoc embellishment of the plume hypothesis would do. These arguments I noted, and this is not my approach.

I answer the three specific points Widdowson raises:

1. Some references would help. There is a huge Tertiary sedimentary pile along the western Indian rifted margin, and several articles in Gunnell and Radhakrishna (2001) discuss major Neogene uplift of the western Ghats and the peninsula. Widdowson reports major uplift in the late Cretaceous (90–65 Ma?). In fact, much Deccan volcanism had already occurred by the KTB, and any uplift can be well explained by magmatic underplating or the buoyancy of melt-depleted residues. See also comment by Anderson (12 January 2007).

2. True, Seychelles–India separation and sea-floor spreading began after much (not all) Deccan volcanism had occurred. This does not preclude rifting (not an instantaneous process) occurring before or with volcanism. Dikes were arguably being emplaced along the future coastline and, in the Narmada-Tapi region and elsewhere, were feeding a lava pile that was growing laterally and vertically, and at some point breakup occurred. This is fully consistent with a rift-related convective melting model (see Fig. 13 of Sheth, 2005).

3. A half-truth. There is a southward younging of stratigraphic formations in the western Ghats. But Widdowson does not cite new knowledge from geochemical-stratigraphic work (Peng et al., 1998; Mahoney et al., 2000; Sheth et al., 2004) that the northern and northeastern Deccan areas expose formations very similar to (or the same as) the younger formations to the south. Because lavas can flow great distances, it is the feeder dikes that really matter. We are only beginning to locate these (Bondre et al., 2006; Vanderkluysen et al., 2006; Ray et al., 2007), and this argument is lost. A simple southward age progression does not exist and is not even expected when a putative plume head 1000 km across melts simultaneously over large areas.

Widdowson does not raise one point on which I may be mistaken. This is my usage of "planation surface" for the heavily lateritized surface preserved as erosional remnants at the top of the Deccan Traps in the western Ghats (Widdowson and Cox, 1996; Widdowson, 1997). These authors consider it a constructional surface and the weathered, lateritized top of the basalt pile, implying little erosion subsequent to the eruptions. Planation surfaces are by definition produced by extensive, advanced erosion (and do often undergo lateritization subsequently). I thank Yanni Gunnell for recently pointing out to me this possible error. Recent e-mail discussions with him and Cliff Ollier are much appreciated. Note that Ollier and Pain (2000) do not believe the high lateritized surface to be the top of the lava pile; they suggest relief inversion of a laterite-floored river system. This is another issue.

I do not reject the plume model for the Deccan because, as Widdowson puts it, "just not every fact is consistent with it. . . ." The facts quite inconsistent with the plume model, here and worldwide, now number hundreds, like swarms of midges, flying in the face.

5 February 2007, Peter Hooper and Mike Widdowson

The assertions by both Sheth and Anderson (in the comments on Hooper et al., this volume) that, because any one suggested result of the plume model is not obvious, the plume model is therefore void and to be abandoned, are surely unacceptable. What geological model, even plate tectonics itself, can meet such demands? We are looking for the model that best explains flood basalt volcanism. The melt production models of White and McKenzie (1989) and White (1993) indicate that neither elevated mantle temperatures nor lithospheric extension and associated decompressive melting of the mantle (McKenzie, 1978) would alone be sufficient to generate the volume of magma observed in the large CFBs. These authors argue that it is the combination of both the arrival of a plume head and the associated thermal modification of the overlying lithosphere, with its attendant onset of extension and mantle decompression, that together generate massive mantle melting. It is, therefore, unrealistic to criticize only one part of this plume-rift scenario and so conclude that plumes are an inadequate explanation of CFBs. If we accept these models as a basis for understanding mantle melting and CFB evolution, then lithospheric extension, underplating, and ponding (Cox, 1980, 1993) are all explicable consequences of the plume-rift scenario. Most importantly, such models remain consistent with our current knowledge of many CFBs.

For Anderson (comments on Hooper et al., this volume) to describe the western margin of India as a "convergent" margin runs entirely counter to current and conventional geological understanding of the region. Both the structure and the geomorphology of the western margin are unequivocally that of extensional passive margin, alternatively described as a "volcanic rifted margin." If, and it is not clear, Anderson is alluding to much more ancient mobile zones between Archean cratons (see Sheth, this volume, Fig. 2), then such an association appears vague in the extreme. In the Karoo, although the northern part of the province appears to have erupted over such a zone, the bulk of the Karoo flood basalts erupted over Lesotho. On the Deccan, the center of the main eruption is Kalsubai Mountain, well within a stable craton.

Perhaps uniquely, the Indian margin has been the site of several subparallel rifting episodes over a considerable period of geological time, all associated with the Mesozoic dismemberment of the Gondwana supercontinent: the separation of Madagascar and India from Africa at ca. 160–170 Ma, the separation of Madagascar and India at ca. 88–90 Ma, and the separation of India from the Seychelles–Mascarene plateau at ca. 62–64 Ma. Of interest here is the fact that all these separations apparently followed an inherited structural fabric, which is often described within the India landmass as the "Dharwar trend," which currently runs NNW–SSE in present-day peninsular India. Indeed, modern earthquake epicenters are often traced to lineaments and structural heterogeneities that together form this "fabric" (Widdowson and Mitchell, 1999). Although it may be true that development and spreading along the Carlsberg Ridge has more recently (Paleogene–Neogene?) altered the general stress field in the Arabian Sea to that now comprising a component of a compressive rather than extensional effects (Whiting et al., 1994), there is no indication that western India ever has developed, or is even likely to develop into, a convergent margin.

By contrast, it is true that the northern part of the Indian plate now constitutes part of a major convergent margin, but even northern India records a change from an essentially Mesozoic extensional regime resulting from the northerly subduction of Tethyan ocean crust beneath Eurasia (during which the east-west–trending Narmada and Tapti rifts formed, deepened, and were filled with Jurassic and Cretaceous marine and terrestrial clastic sediments; see Sheth, this volume, Fig. 2), to a Paleogene compressive regime when the Indian and Eurasian plates actually collided and the Himalayas began to develop. However, even if this essentially north–south oriented convergent tectonics represents the convergent margin that is being described by Anderson, then the timing of the India–Asia collision, and its effect in northern India, both significantly predate and are entirely spatially disparate with the eruption of the Deccan Traps.

Of critical importance to the evolution of the Deccan Traps CFB is the timing of the separation of the Seychelles–Mascarene plateau. Hooper (1990) has argued that rifting largely postdates eruption, and current paleomagnetic data (e.g., Todal and Eldholm, 1998) reveal that the earliest sea floor spreading between the two continental fragments occurred during Chron 28 (i.e., 62.5–64 Ma), further supporting this view. Thus, the timing of Deccan eruption versus Seychelles–Mascarene separation not only places important constraints on the models promulgated by White (1993) but also presents a serious challenge to the alternative EDC model favored by Sheth.

7 February 2007, Mike Widdowson

Sheth misreads key aspects of my earlier comment. For the record, I consider "uplift" an issue central to the plume debate. I reiterate, uplift "remains a polemic issue" (i.e., worthy of discussion). Word limits preclude the desired detailed dialogue, but I raise some important issues below.

I entirely agree that concerted fieldwork is necessary, but both this chapter and Sheth (2005) offer little new field data and instead rely largely on an interpretation of previous authors' information. Such retrospectives do not permit the reader to eval-

uate the relative merits of plume versus nonplume models and so cannot materially progress debate.

The argument regarding southward (not eastward, as suggested by Sheth) younging of the main Deccan edifice remains robust. Three independent lines of geological evidence support this interpretation; two independent lines are usually deemed sufficient to indicate a scientific "truth."

Sheth concludes, logically, that the western Ghats is the product of a post-Deccan denudational processes. This particular interpretation has long been available (e.g., Widdowson and Cox, 1996; Widdowson, 1997; Gunnell and Fleitout, 1998; Widdowson and Mitchell, 1999). But this issue is not in dispute, so why raise it here?

Sheth asserts that the plume-head drainage idea of Cox (1989) is problematic. Perhaps, but the fact that radial drainage patterns do occur in key CFBPs remains a valid, if inexplicable (?), observation. Cox's idea was superseded by arguments provided in Widdowson and Cox (1996), Widdowson (1997; see Fig. 14), and apatite fission track analysis data (Gunnell et al., 2003) and so becomes irrelevant for contending pre-eruptive uplift.

Sheth argues, correctly, that the nature of the pre-Deccan paleosurface holds important clues regarding pre-eruptive uplift in the DVP (Jerram and Widdowson, 2005). Much of this surface remains buried by the Deccan lavas and is both inaccessible and unknowable. It only becomes exposed around the northern and eastern periphery of the main lava pile. Such peripheral localities, including many of those described by Sheth, were hundreds of kilometers from the Deccan eruptive loci. If any uplift did occur here, it would have been minimal at such large distances from the focus of putative plume-head uplift and thus consistent with that affecting the Dongargaon basin, for example (Tandon, 2002; Samant and Mohabey, 2005).

The pre-eruptive paleosurface has been significantly modified by the crustal loading of the Deccan edifice and, in its western extensions, suppressed far below datum. Thus, the gross form and elevation of this basement–basalt contact are largely artifacts of posteruptive flexural adjustment. Nevertheless, Sheth argues that this highly modified surface reveals a "peneplain" and that its preservation as such precludes significant fluvial incision. Possibly, but peneplains are the consequence of erosion, and the classical, albeit obsolete, Davisian model requires regional uplift as a trigger for peneplanation to proceed. Etchplanation is more appropriate to the development of the pre-Deccan surface (e.g., Büdel, 1982). Here, thick alteration mantles accumulate through tropical weathering of surfaces exposed during prolonged periods of tectonic stability. If, as Sheth requires, such conditions had characterized the pre-Deccan land surface, then the widespread absence of deep weathering mantle preserved beneath the lava units may instead indicate that this landscape had been thoroughly stripped before DVP eruptions. Etchplain stripping may be achieved through widespread fluvial erosion induced by regional uplift (Borger and Widdowson, 2001).

Offshore sedimentary records in the Krishna, Godavari, and Narmada-Tapti basins all reveal significant increases in Late Cretaceous depositional flux (Halkett et al., 2001): These data are consistent with pre-eruptive regional erosion of peninsular India, perhaps starting with the stripping of an easily erodable weathering mantle.

If pre-eruptive (plume-driven?) uplift had occurred in pre-Deccan peninsular India, what might then be recorded in the erosional and sedimentary chronologies of the DVP peripheral regions? Removal of any easily erodable weathering mantle, perhaps; minimal changes in elevation, possibly; development of shallow basins receiving fine clastic input from the plume-uplift effects hundreds of kilometers away, maybe. This interpretation of the available infra- and intratrappean sedimentary (i.e., Lameta Beds) data is equally plausible using the same compendium of field evidence provided by Sheth. Accordingly, I offer a modified, précis version of Sheth's own summary:

Any original flatness and elevation of the pre-Deccan landscape has been significantly modified by syn- and post-eruptive isostatic adjustment deriving, initially, from the loading of the DVP edifice, and subsequently by denudational unloading. The occurrence of a stripped, pre-eruptive etchplain, together with associated offshore sedimentological data, are consistent with those phenomena predicted had a large plume head upwelled beneath India during the Late Cretaceous.

Post-Deccan uplift has elevated both the pre-Deccan, and post-Deccan surfaces. This uplift of the western Ghats is not related to a putative Deccan plume: it is not domal, occurs beyond the limits of the Deccan lava cover, and represents a later, denudationally driven, uplift (Widdowson, 1997). Thus, the easterly drainage of the Indian peninsula is not plume-related dome flank drainage and is largely antecedent to denudational uplift effects.

To summarize, of those observations described by Sheth, most, if not all, can equally and adequately be explained by the passage of India over a static, spatially restricted, mantle-melting anomaly during the Late Cretaceous. For want of a better term, and until consensus offers me a better alternative, I will continue to call this anomaly, *sensu lato,* a "mantle plume." I end by reiterating the rationale to my initial comment: The challenge to Sheth remains to deliver us an alternative, "nonplume" model that can better explain the Deccan CFBP.

9 February 2007, H.C. Sheth

I reiterate that I do not reject the plume model for the Deccan just because only one observation (on uplift) is in conflict with it. I have mentioned many others (Sheth, 2005), and Kumar et al. (this volume) give us yet another. I am also *not* proposing the EDC model (King and Anderson, 1995, 1998); the repeated assertion that I am is mystifying. Plate tectonics is surely not a hypothesis at the same level as the plume hypothesis.

I propose that the Deccan is related to continental rifting and invoke a simple, rifting-related convective melting model

(Sheth, 2005) that, unlike the plume model, does not require or predict precursory uplift or high temperatures. These are unique to the plume hypothesis. I refer Hooper and Widdowson to recent work (e.g., Van Wijk et al., 2001) showing that mantle plumes (or, rather, anomalously high temperatures) are not required for volcanism at rifting continental margins. This work, developed for the north Atlantic, fits the Deccan well.

I completely agree that western India has been an extensional-rifting region for a long time. I presume that Anderson (comment of 28 January 2007, on Hooper et al., this volume) was referring to early phases of the Greater India–Asia collision and Indian plate flexure that, with approximately north–south compression, might have caused the ENE–WSW-trending Narmada-Tapi tectonic zone to experience extension.

I disagree with Hooper (1990) that there was no prevolcanic extension in the Deccan (Sheth, 2000; Ray et al., 2007). Prevolcanic, and certainly prebreakup, extension produced many preferentially oriented dike swarms in the Deccan, many of which I take as feeders to the lavas. Much of the lava pile erupted, and then the breakup occurred, during which younger dike swarms were emplaced. Thus, continental breakup certainly followed the main phase of volcanism. There is nothing inconsistent here with the rifting-related melting model I endorse.

In response to the comment of Widdowson of 7 February 2007, I concede that many observations in my chapter (Sheth, this volume) are neither new nor my own. My objective was to compile as many of these as possible that are relevant to the plume debate in one place and to give due credit to past works. I do not see this as a shortcoming of my chapter.

I am not proposing an eastward age progression for the Deccan in place of southward. I said that no age progression is evident and is not even expected if there were a large plume head supplying melt over extensive areas.

The post-Deccan uplift of the western Ghats and the drainage issue, which Widdowson says are not new, are discussed at length because some of the greatest champions of the plume model (e.g., Campbell, 2005; Campbell and Davies, 2006) still vigorously defend the plume model, citing the drainage pattern of India as evidence for a Deccan plume.

Widdowson feels that "the fact that radial drainage patterns do occur in key CFBPs remains a valid, if inexplicable (?) observation." This observation is neither valid nor inexplicable. The drainage patterns are not radial, just as the uplifts, which are substantially younger, are not domal. The key CFBPs have experienced major Neogene uplift (e.g., Ollier and Pain, 2000; Bonow et al., 2006) whose cause(s) we do not fully understand. Doglioni et al. (2003), who propose an active eastward-directed asthenospheric mantle flow since at least 40 Ma, as the cause of ridge depth asymmetries and continental margin uplifts, have an interesting hypothesis. Both the western Indian rifted margin and the west Greenland margin, which have experienced significant Neogene uplift, directly block this proposed eastward mantle flow.

The pre-Deccan planation surfaces around Pachmarhi are not hundreds of kilometers from the Deccan eruptive loci as Widdowson asserts. Huge dikes and intrusions are known from the area, and inferred to be the feeders of lavas nearby (Sen, 1980). Work is currently ongoing on these intrusions and their chemical-isotopic correlations to the lava flows.

The field evidence from Dongargaon shows that the uplift was purely local and of a few meters (Tandon, 2002), which is not evidence for a plume. I have not claimed that local, pre-eruptive subsidence refutes the plume model. Nor have I stated that the pre-Deccan eruptive surface in the western Ghats, which is not visible at all, was a peneplain. Nothing can be said about it.

Widdowson states that "the obsolete Davisian model requires regional uplift as a trigger for peneplanation to proceed." On the contrary, Ollier and Pain (2000) write that peneplains necessarily develop at or near the base level of erosion, and because there is no geomorphic process capable of producing a peneplain at high elevation, high-level peneplains indicate relatively recent uplift.

Widespread lateritization of the sub-Trap planation surfaces around Pachmarhi and in central India is indeed apparent (Venkatakrishnan, 1984, 1987).

The point about the Lameta sediments, which Widdowson considers consistent with fine clastic input from a far-away plume-uplifted area, was that several of these sediments contain volcanic detritus, and thus eruptions were already underway in the source areas. This proves the uplift to be syn- or posteruption, perfectly consistent with emplacement of hot intrusions in the crust or buoyancy of the melt-depleted residue in the shallow mantle.

Widdowson offers a rewritten summary of the thesis of my paper. I agree with the second half in full but with the first only in part. Lastly, he states that "of those observations described by Sheth, most, if not all, can equally and adequately be explained by the passage of India over a static, spatially restricted, mantle-melting anomaly during the Late Cretaceous." My point is that they are equally well, if not better, explained without such an anomaly. Widdowson now redefines as a plume any static, spatially restricted, mantle-melting anomaly. He thus drops uplift, helium, even temperature as criteria for plumes. Others argue that even fixity is not a criterion. Frequent redefining of "plume" in this way also does not materially progress debate.

There is an observation in my chapter that, for the sake of accuracy, I wish to set right. Venkatakrishnan (1987) correlates a lava flow at the top of Mt. Dhupgarh, Pachmarhi, with lava flows capping the Gondwanas at Tamia (Fig. 10b) and considers the base of the Dhupgarh flow as the sub-trap planation surface. I have visited the Pachmarhi–Tamia area (Fig. 10a). The Dhupgarh "lava flow" apparently refers to the spheroidally weathered basaltic/doleritic rock outcropping at the western edge of the mountain, at Sunset Point, some 75 m below the summit itself. This is a north–south-trending doleritic dike that is seen clearly in the road cut about 50 m below Sunset Point. The dike does not quite reach the summit, which shows no lava cover, and the

whole mountain comprises Gondwana sandstone intruded by dikes. Three lava flows occur at Tamia (Sen, 1980; Venkatakrishnan, 1987), and it remains to be seen whether their base corresponds to the Dhupgarh Surface or the Pachmarhi Surface. The "basalt cliffs" shown in the section of Venkatakrishnan (1987) (Fig. 10b) are not cliffs: the section greatly exaggerates the basalt thickness. The imposing near-vertical Tamia scarp comprises the Gondwanas intruded by a sill complex seen at lower elevations (Sen, 1980), and the three basalt lava flows simply cap it. "Basalt cliffs" in Figure 10b should thus read "basalt-capped cliffs." However, this correction does not affect the validity of the main observations and interpretations, including those by Venkatakrishnan (1984, 1987), about planation surfaces and relatively recent uplift.

I conclude with some general remarks. The uplift issue is very relevant to the plume debate; the prediction of prevolcanic regional uplift is unique to the plume hypothesis. Field data are unambiguous and should be utilized. They were not available to the originators of the plume hypothesis nearly 40 years ago. The originators and early enthusiasts had little direct experience with particular provinces, which resulted in a large number of claims about things that do not exist (e.g., the Cambay triple junction story, see Sheth, 2005). Critical tests of the Deccan plume should have come from regional experts, either from India or abroad. The purported prevolcanic doming in the Emeishan province has been made much of, e.g., by Campbell (2005) and Campbell and Davies (2006), who even confidently cite such doming in the Deccan as fact. Even the claimed Emeishan case is suspect (see comments by me and by Hamilton on the chapter by Xu and He, this volume).

Finally, to call a plume viable, one should appeal to evidence, not uncertainties. To say that plumes are possible because there are too many unknowns is not a logical approach. However, that it is one used by many plume theoreticians is clear from several chapters in this volume and the discussion following the chapter by Garnero et al. (this volume). Dark matter, baby universes, superstrings, and eleven dimensions are also all theoretically possible, without physical evidence, and some would like to have them. Plumes are similar. A plume proponent a couple of years ago compared the plume to an electron, by nature invisible but beyond doubt. The more meaningful questions to me are: Do plumes have to exist? Is there evidence for them? And can known, well-understood, nonplume upper mantle processes explain the observations instead?

I thank editor Gillian Foulger for ably coordinating this discussion and for the opportunity to clarify my objectives and position.

REFERENCES CITED

Alpha, T.R., and Vallier, T.L., 1994, Physiography of the Seven Devils Mountains and adjacent Hells Canyon of the Snake River, Idaho and Oregon: Reston, Virginia, United States Geological Survey Professional Paper 1439, p. 91–100.

Anderson, D.L., 2005, Scoring hotspots: The plume and plate paradigms, *in*

Foulger, G.R., et al., eds., Plates, plumes, and paradigms: Boulder, Colorado, Geological Society of America Special Paper 388, p. 31–54, doi: 10.1130/2005.2388(04).

Anderson, D.L., and Natland, J.H., 2005, A brief history of the plume hypothesis and its competitors: Concept and controversy, *in* Foulger, G.R., et al., eds., Plates, plumes, and paradigms: Boulder, Colorado, Geological Society of America Special Paper 388, p. 119–145, doi: 10.1130/2005.2388(08).

Bondre, N.R., Hart, W.K., and Sheth, H.C., 2006, Geology and geochemistry of the Sangamner mafic dike swarm, western Deccan volcanic province, India: Implications for regional stratigraphy: Journal of Geology, v. 114, p. 155–170.

Bonow, J.M., Lidmar-Bergström, K., and Japsen, P., 2006, Palaeosurfaces in central West Greenland as reference for identification of tectonic movements and estimation of erosion: Global and Planetary Change, v. 50, p. 161–183.

Borger, H., and Widdowson, M., 2001, Indian laterites, and lateritious residues of southern Germany: A petrographic, mineralogical, and geochemical comparison: Zeitschrift für Geomorphologie N. F., v. 45, p. 177–200.

Büdel, J., 1982, Climatic Geomorphology (transl. L. Fischer and D. Busche): Princeton, New Jersey, Princeton University Press, 443 p.

Burbank, D.W., Leland, J., Fielding, E., Anderson, R.S., Brozovic, N., Reid, M.R., and Duncan, C., 1996, Bedrock incision, rock uplift and threshold hillslopes in the northwestern Himalayas: Nature, v. 379, p. 505–510.

Burov, E., and Guillou-Frottier, L., 2005, The plume head–continental lithosphere interaction using a tectonically realistic formulation for the lithosphere: Geophysical Journal International, v. 161, p. 469–490.

Campbell, I.H., 2005, Large igneous provinces and the mantle plume hypothesis: Elements, v. 1, p. 265–269.

Campbell, I.H., and Davies, G.F., 2006, Do mantle plumes exist?: Episodes, v. 29, p. 162–168.

Courtillot, V., Jaupart, C., Manighetti, I., Tapponnier, P., and Besse, J., 1999, On causal links between flood basalts and continental breakup: Earth and Planetary Science Letters, v. 166, p. 177–195.

Courtillot, V., Davaille, A., Besse, J., and Stock, J., 2003, Three distinct types of hotspots in the Earth's mantle: Earth and Planetary Science Letters, v. 205, p. 295–308.

Cox, K.G., 1978, Flood basalts and the breakup of Gondwanaland: Nature, v. 274, p. 47–49.

Cox, K.G., 1980, A model for flood basalt volcanism: Journal of Petrology, v. 21, p. 629–650.

Cox, K.G., 1983, The Deccan Traps and the Karoo: Stratigraphic implications of possible hot-spot origins: IAVCEI, programme and abstracts of XVIII General Assembly, Hamburg, Germany, p. 96.

Cox, K.G., 1989, The role of mantle plumes in the development of continental drainage patterns: Nature, v. 342, p. 873–877.

Cox, K.G., 1993, Continental magmatic underplating: Philosophical Transactions of the Royal Society of London (A), v. 342, p. 155–166.

Dadson, S.J., Hovius, N., Chen, H., Dade, B., Hsieh, M.-L., Willett, S.D., Hu, J.-C., Horng, M.-J., Chen, M.-C., Stark, C.P., Lague, D., and Lin, J.-C., 2003, Links between erosion, runoff variability and seismicity in the Taiwan orogen: Nature, v. 426, p. 648–651.

Devey, C.W., and Lightfoot, P.C., 1986, Volcanology and tectonic control of stratigraphy and structure in the western Deccan Traps: Bulletin of Volcanology, v. 48, p. 195–207.

Devey, C.W., and Stephens, W.E., 1991, Tholeiitic dikes in the Seychelles and the original extent of the Deccan: Journal of the Geological Society of London, v. 148, p. 979–983.

Doglioni, C., Carminati, E., and Bonatti, E., 2003, Rift asymmetry and continental uplift: Tectonics, v. 22, doi: 10.1029/2002TC001459.

Elkins-Tanton, L.T., 2005, Continental magmatism caused by lithospheric delamination, in Foulger, G.R., et al., eds., Plates, plumes, and paradigms: Boulder, Colorado, Geological Society of America Special Paper 388, p. 449–462.

Ernst, R.E., and Buchan, K.L., 2001, Large mafic events through time and links

to mantle-plume heads, *in* Ernst, R.E., and Buchan, K.L., eds., Mantle plumes: Their identification through time: Boulder, Colorado, Geological Society of America Special Paper 352, p. 483–575.

Garnero, E.J., Lay, T., and McNamara, A., 2007 (this volume), Implications of lower-mantle structural heterogeneity for existence and nature of whole-mantle plumes, *in* Foulger, G.R., and Jurdy, D.M., eds., Plates, plumes, and planetary processes: Boulder, Colorado, Geological Society of America Special Paper 430, doi: 10.1130/2007.2430(05).

Gilbert, G.K., 1877, Geology of the Henry Mountains (Utah): Washington, D.C., United States Government Printing Office, 172 p.

Goles, G.G., 1986, Miocene basalts of the Blue Mountains province in Oregon. I: Compositional types and their geological settings: Journal of Petrology, v. 27, p. 495–520.

Gunnell, Y., and Fleitout, L., 1998, Shoulder uplift of the Western Ghats passive margin, India: A denudational model: Earth Surface Processes and Landforms, v. 23, p. 391–404.

Gunnell, Y., and Radhakrishna, B.P., eds., 2001, Sahyadri: The great escarpment of the Indian subcontinent: Bangalore, Geological Society of India Memoir 47(1–2), 1054 p.

Gunnell, Y., Gallagher, K., Carter, A., Widdowson, M., and Hurford, A.J., 2003, Denudation history of the continental margin of western peninsular India since the early Mesozoic—Reconciling apatite fission track data with geomorphology: Earth and Planetary Science Letters, v. 215, p. 187–201.

Hales, T.C., Abt, D.L., Humphreys, E.D., and Roering, J.J., 2005, Lithospheric instability as an origin for Columbia River Flood Basalts and Wallowa Mountains uplift in NE Oregon, USA: Nature, v. 432, p. 842–845.

Halkett, A., White, N., Chandra, K., and Lal, N.K., 2001, Dynamic uplift of the Indian Peninsula and the Réunion Plume: Abstract, American Geophysical Union, Fall Meeting, T11A-0845.

Hon, K., Kauahikaua, J., Denlinger, R.P., and Mackay, K., 1994, Emplacement and inflation of pahoehoe sheet flows: Observations and measurements of active lava flows on Kilauea Volcano, Hawaii: Geological Society of America Bulletin, v. 106, p. 351–370.

Hooper, P.R., 1990, The timing of crustal extension and the eruption of continental flood basalts: Nature, v. 349, p. 246–249.

Hooper, P.R., 1997, The Columbia River flood basalt province: Current status, *in* Mahoney, J.J., and Coffin, M.F., eds., Large igneous provinces: Continental, oceanic, and planetetary volcanism: Washington, D.C., American Geophysical Union Geophysical Monograph 100, p. 1–27.

Hooper, P.R., and Camp, V.E., 1981, Deformation of the southeast part of the Columbia Plateau: Geology, v. 9, p. 323–328.

Hooper, P.R., Kleck, W.D., Knowles, C.R., Reidel, S.P., and Thiessen, R.L., 1984, Imnaha Basalt, Columbia River Basalt Group: Journal of Petrology, v. 25, p. 473–500.

Hooper, P.R., Camp, V.E., Reidel, S.P., and Ross, M.E., 2007 (this volume), The origin of the Columbia River flood basalt province: Plume versus non-plume models, *in* Foulger, G.R., and Jurdy, D.M., eds., Plates, plumes, and planetary processes: Boulder, Colorado, Geological Society of America Special Paper 430, doi: 10.1130/2007.2430(30).

Jerram, D., and Widdowson, M., 2004, The anatomy of continental flood basalt provinces: Geological constraints on the processes and products of flood volcanism. Lithos, v. 79, p. 355–366.

Jerram, D.W., and Widdowson, M., 2005, The anatomy of Continental Flood Basalt Provinces: geological constraints on the processes and products of flood volcanism: Lithos, v. 79, p. 385–405.

Kennett, B.L.N., and Widiyantoro, S., 1999, A low seismic wavespeed anomaly beneath northwestern India: A seismic signature of the Deccan plume?: Earth and Planetary Science Letters, v. 165, p. 145–155.

King, S.D., and Anderson, D.L., 1995, An alternative mechanism of flood basalt formation: Earth and Planetary Science Letters, v. 136, p. 269–279.

King, S.D., and Anderson, D.L., 1998, Edge-driven convection: Earth and Planetary Science Letters, v. 160, p. 289–296.

Kleck, W.D., 1976, Chemistry, petrography and stratigraphy of the Columbia River Basalt Group in the Imnaha River valley region, eastern Oregon and western Idaho [Ph.D. thesis]: Pullman, Washington State University, 203 p.

Kumar, P.S., Menon, R., and Reddy, G.K., 2007 (this volume), Crustal geotherm in southern Deccan basalt province, India: The Moho is as cold as adjoining cratons, *in* Foulger, G.R., and Jurdy, D.M., eds., Plates, plumes, and planetary processes: Boulder, Colorado, Geological Society of America Special Paper 430, doi: 10.1130/2007.2430(14).

Mahoney, J.J., Sheth, H.C., Chandrasekharam, D., and Peng, Z.X., 2000, Geochemistry of flood basalts of the Toranmal section, northern Deccan Traps, India: Implications for regional Deccan stratigraphy: Journal of Petrology, v. 41, p. 1099–1120.

McKenzie, D.P., 1978, Some remarks on the development of sedimentary basins: Earth and Planetary Science Letters, v. 40, p. 25–32.

Mitchell, C., and Widdowson, M., 1991, A geological map of the Southern Deccan Traps, India and its structural implications: Journal of the Geological Society of London, v. 148, p. 495–505.

Ollier, C.D., and Pain, C.F., 2000, The origin of mountains: London, Routledge, 368 p.

Parsons, T., Thompson, G.S., and Sleep, N.H., 1994, Mantle plume influence on the Neogene uplift and extension of the U.S. western Cordillera: Geology, v. 22, p. 83–86.

Peng, Z.X., Mahoney, J.J., Hooper, P.R., Macdougall, J.D., and Krishnamurthy, P., 1998, Basalts of the northeastern Deccan Traps, India: Isotopic and elemental geochemistry and relation to southwestern Deccan stratigraphy: Journal of Geophysical Research, v. 103, p. 29,843–29,865.

Pierce, K.L., and Morgan, L.A., 1992, The track of the Yellowstone hot spot: Volcanism, faulting, and uplift, *in* Link, P.K., et al., eds., Regional geology of eastern Idaho and western Wyoming: Geological Society of America Memoir 179, p. 1–53.

Pierce, K.L., Morgan, L.A., and Saltus, R.W., 2002, Yellowstone plume head: Postulated tectonic relations to the Vancouver slab, continental boundaries, and climate, *in* Bonnichsen, B., et al., eds., Tectonic and magmatic evolution of the Snake River Plain Volcanic Province: Idaho Geological Survey Bulletin 30, p. 5–33.

Powell, J.W., 1875, Exploration of the Colorado River of the West, and its tributaries: Washington, D.C., United States Government Printing Office, 302 p.

Ray, R., Sheth, H.C., and Mallik, J., 2007, Structure and emplacement of the Nandurbar–Dhule mafic dyke swarm, Deccan Traps, and the tectonomagmatic evolution of flood basalt, Bulletin of Volcanology, v. 69, p. 537–551, doi: 10.1007/s00445-006-0089-y.

Saltus, R.W., and Thompson, G.A., 1995, Why is it downhill from Tonopah to Las Vegas?: A case for mantle plume support of the high northern Basin and Range: Tectonics, v. 14, p. 1235–1244.

Samant, B., and Mohabey, D.M., 2005, Response of flora to Deccan volcanism: A case study from Nand–Dongargaon basin of Maharashtra, implications to environment and climate: Gondwana Geological Magazine Special Volume, v. 8, p. 151–164.

Self, S., Thordarson, T., and Keszthelyi, L., 1997, Emplacement of continental flood basalt lava flows, *in* Mahoney, J.J., and Coffin, M.F., eds., Large igneous provinces: Continental, oceanic, and planetary volcanism: Washington, D.C., American Geophysical Union Geophysical Monograph 100, p. 381–409.

Sen, G., 1980, Mineralogical variations in the Delakhari sill, Deccan Trap intrusion, central India: Contributions to Mineralogy and Petrology, v. 75, p. 71–78.

Sheth, H.C., 1999, Flood basalts and large igneous provinces from deep mantle plumes: Fact, fiction and fallacy: Tectonophysics, v. 311, p. 1–29.

Sheth, H.C., 2000, The timing of crustal extension, dyking, and the eruption of the Deccan flood basalts: International Geology Review, v. 42, p. 1007–1016.

Sheth, H.C., 2005, From Deccan to Réunion: No trace of a mantle plume, *in* Foulger, G.R., et al., eds., Plates, plumes, and paradigms: Boulder, Colorado, Geological Society of America Special Paper 388, p. 477–501, doi: 10.1130/2005.2388(29).

Sheth, H.C., 2007 (this volume), Plume-related regional prevolcanic uplift in the Deccan Traps: Absence of evidence, evidence of absence, *in* Foulger, G.R., and Jurdy, D.M., eds., Plates, plumes, and planetary processes: Boulder, Colorado, Geological Society of America Special Paper 430, doi: 10.1130/2007.2430(36).

Sheth, H.C., Pande, K., and Bhutani, R., 2001a, ^{40}Ar-^{39}Ar ages of Bombay trachytes: Evidence for a Palaeocene phase of Deccan volcanism: Geophysical Research Letters, v. 28, p. 3513–3516.

Sheth, H.C., Pande, K., and Bhutani, R., 2001b, ^{40}Ar-^{39}Ar age of a national geological monument: The Gilbert Hill basalt, Deccan Traps, Bombay: Current Science, v. 80, p. 1437–1440.

Sheth, H.C., Mahoney, J.J., and Chandrasekharam, D., 2004, Geochemical stratigraphy of flood basalts of the Bijasan Ghat section, Satpura Range, India: Journal of Asian Earth Sciences, v. 23, p. 127–139.

Spencer, J.E., and Pearthree, P.A., 2003, Headward erosion versus closed-basin spillover as alternative causes of Neogene capture of the ancestral Colorado River by the Gulf of California, *in* Young, R.A., and Spamer, E.E., eds., Colorado River: Origin and Evolution: Grand Canyon, Arizona, Grand Canyon Association, p. 215–219.

Tandon, S.K., 2002, Records of the influence of Deccan volcanism on contemporary sedimentary environments in central India: Sedimentary Geology, v. 147, p. 177–192.

Thordarson, T., and Self, S., 1998, The Roza Member, Columbia River Basalt Group: A gigantic pahoehoe lava flow field formed by endogenous processes? Journal of Geophysical Research, v. 103, p. 27,411–27,445.

Todal, A., and Eldholm, O., 1998, Continental margin of western India and Deccan large igneous province: Marine Geophysical Research, v. 20, p. 273–291.

Tolan, T.L., Reidel, S.P., Beeson, M.V., Anderson, J.L., Fecht, K.R., and Swanson, F.J., 1989, Revisions to the estimates of the areal extent and volume of the Columbia River Basalt Group: Boulder, Colorado, Geological Society of America Special Paper 239, p. 1–20.

Vandamme, D., Courtillot, V., Besse, J., and Montigny, R., 1991. Palaeomagnetism and age determination of the Deccan Traps (India): Results of a Nagpur-Bombay traverse and review of earlier work: Reviews of Geophysics, v. 29, p. 159–190.

Vanderkluysen, L., Mahoney, J.J., Hooper, P.R., and Sheth, H.C., 2006, Location and geometry of the Deccan Traps feeder system inferred from dike geochemistry: Eos (Transactions, American Geophysical Union), v. 87, no. 52, fall meeting supplement, abstract V13B-0681.

Van Tassel, J., Ferns, M.L., McConnell, V., and Smith, G.R., 2001, The mid-Pliocene Imbler fish fossils, Grande Ronde Valley, Union County, Oregon, and the connection between Lake Idaho and the Columbia River: Oregon Geology, v. 63, p. 77–96.

Van Wijk, J.W., Huismans, R.S., Ter Voorde, M., and Cloetingh, S., 2001, Melt generation at volcanic continental margins: No need for a mantle plume? Geophysical Research Letters, v. 28, p. 3995–3998.

Venkatakrishnan, R., 1984, Parallel scarp retreat and drainage evolution, Pachmarhi area, Madhya Pradesh, central India: Journal of Geological Society of India, v. 25, p. 401–413.

Venkatakrishnan, R., 1987, Correlation of cave levels and planation surfaces in the Pachmarhi area, Madhya Pradesh: A case for base level control: Journal of Geological Society of India, v. 29, p. 240–249.

Whipple, K.X., 2004, Bedrock rivers and the geomorphology of active orogens: Annual Review of Earth and Planetary Science, v. 32, p. 151–185.

White, R.S. 1993, Melt production rates in mantle plumes: Philosophical Transactions of the Royal Society of London (A), v. 342, p. 137–153.

White, R.S., Spence, D., Fowler, S.R., Mckenzie, D.P., Westbrook, G.K., and Bowen, E.N., 1987, Magmatism at rifted continental margins: Nature, v. 330, p. 439–444.

White, R.S., and McKenzie, D.P., 1989, Magmatism at rift zones: The generation of volcanic continental margins and flood basalts: Journal of Geophysical Research, v. 94, p. 7685–7729.

Whitehouse, I.E., 1988, Geomorphology of the central Southern Alps, New

Zealand: The interaction of plate collision and atmospheric circulation: Zeitschrift für Geomorphologie Supplementeband, v. 69, p. 105–116.

Whiting, B.M., Karner, G.D., and Driscoll, N.W., 1994, Flexural and stratigraphic development of the West Indian continental margin: Journal of Geophysical Research, v. 99, p. 13,791–13,811.

Widdowson, M., 1997, Tertiary palaeosurfaces of the SW Deccan, western India: Implications for passive margin uplift, *in* Widdowson, M., ed., Palaeosurfaces: Recognition, reconstruction, and palaeoenvironmental interpretation: London, Geological Society of London Special Publication 120, p. 221–248.

Widdowson, M., 2005, The Deccan basalt–basement contact: Evidence for a plume-head generated CFBP? American Geophysical Union Chapman Conference, "The Great Plume Debate," Ft. William, Scotland, 28 August–1 September 2005, p. 69–70 (abstract).

Widdowson, M., and Cox, K.G., 1996, Uplift and erosional history of the Deccan Traps, India: Evidence from laterites and drainage patterns of the West-ern Ghats and Konkan coast: Earth and Planetary Science Letters, v. 137, p. 57–69.

Widdowson, M., and Mitchell, C., 1999, Large-scale stratigraphical, structural and geomorphological constraints for earthquakes in the southern Deccan Traps, India: The case for denudationally-driven seismicity, *in* Subbarao, K.V., ed., The Deccan Province: Geological Society of India Memoir, v. 43, p. 219–232.

Widdowson, M., Pringle, M.S., and Fernandez, O.A., 2000, A post K-T boundary (Early Palaeocene) age for Deccan-type feeder dikes, Goa, India: Journal of Petrology, v. 41, p. 1177–1194.

Xu, Y.-G., and He, B., 2007 (this volume), Thick, high-velocity crust in the Emeishan large igneous province, southwestern China: Evidence for crustal growth by magmatic underplating or intraplating, *in* Foulger, G.R., and Jurdy, D.M., eds., Plates, plumes, and planetary processes: Boulder, Colorado, Geological Society of America Special Paper 430, doi: 10.1130/2007.2430(03).

The Geological Society of America
Special Paper 430
2007

Nd and Sr isotope systematics and geochemistry of a plume-related Early Cretaceous alkaline-mafic-ultramafic igneous complex from Jasra, Shillong plateau, northeastern India

Rajesh K. Srivastava*
Anup K. Sinha
Igneous Petrology Laboratory, Department of Geology, Banaras Hindu University, Varanasi 221 005, India

ABSTRACT

This article describes an Early Cretaceous Jasra alkaline-mafic-ultramafic igneous complex related to the Kerguelen hotspot–mantle plume system of the Indian Ocean. This complex, emplaced in the Shillong plateau, consists mainly of pyroxenite, gabbro, and nepheline syenite and is closely associated with the Barapani-Tyrsad shear zone, Kopali faults, and Um Ngot Lineaments. Pyroxenite and gabbro occur as separate plutons, whereas nepheline syenites occur either in the form of small dikes in pyroxenites or as differentiated bodies in the gabbros. A few mafic dikes, contemporaneous with gabbro, cut pyroxenite and granite bodies. Mineral compositions classify the pyroxenites into pyroxenite and alkali pyroxenite and the gabbros into essexite and olivine gabbro. The chemical and normative compositions of mafic-ultramafic rocks also show their alkaline nature. Nepheline syenite samples are miaskitic in character (agpaitic index < 1), suggesting the involvement of a CO_2-related phase in their genesis. Chemical data do not support any simple genetic relationship between the different rock units of the complex. Field data also support this conclusion and suggest that these rock units have different genetic histories. Major-element-based discrimination function diagrams suggest the ocean island basalt (OIB) affinity of the Jasra samples. The OIB nature of these samples is further corroborated by their radiogenic isotopic compositions.

Sr and Nd isotopic compositions ($^{87}Sr/^{86}Sr_{initial}$ between 0.706523 and 0.708891 and $^{143}Nd/^{144}Nd_{initial}$ between 0.512258 and 0.512464) suggest that these rocks were derived from a mixing of mantle components such as HIMU (mantle with a high U/Pb ratio) and EM (enriched mantle) components, which show an isotopic composition similar to FOZO (focal zone) composition. Lherzolite mantle was metasomatized into an alkaline wehrlite by CO_2, released by low-degree melting of a carbonated mantle peridotite. Melting of such a metasomatized mantle source may produce ultrabasic alkaline silicate magma, from which the different rock units of the Jasra Complex were crystallized. The geological, geochemical, geochronological, and isotopic data also suggest a spatial and temporal association with the Kerguelen plume activity.

*E-mail: rajeshgeolbhu@yahoo.com.

Srivastava, R.K., and Sinha, A.K., 2007, Nd and Sr isotope systematics and geochemistry of a plume-related Early Cretaceous alkaline-mafic-ultramafic igneous complex from Jasra, Shillong plateau, northeastern India, *in* Foulger, G.R., and Jurdy, D.M., eds., Plates, plumes, and planetary processes: Geological Society of America Special Paper 430, p. 815–830, doi: 10.1130/2007.2430(37). For permission to copy, contact editing@geosociety.org. ©2007 The Geological Society of America. All rights reserved.

Keywords: Alkaline-ultramafic-mafic complex, radiogenic isotopes, geochemistry, Jasra, Shillong plateau, Kerguelen plume, CO_2 metasomatism

INTRODUCTION

A large igneous province (LIP) located in the southern Indian Ocean covers a massive region consisting of the Kerguelen plateau, Broken Ridge, Bunbury, and northeastern India (see Fig. 1), which represents the volcanic outpouring from a Kerguelen hotspot–mantle plume system since ca. 132 Ma (Kent et al., 1997, 2002; Coffin et al., 2002). The Kerguelen LIP also marks the breakup of India and Australia and is usually believed to be related to hotspot activity over a period of up to 130 m.y. (Frey et al., 2000; Kent et al., 2002).

An association between alkaline magmatism and mantle plumes has frequently been proposed (Franz et al., 1999; Bell, 2001, 2002). This is more realistic if alkaline magmatism is found associated with carbonatite. There are some other models, such as rift-setting or subduction-related magmatism (Luhr, 1997; Sheth et al., 2000; Verma, 2006) that may also explain the genesis of alkaline magmas (Wilson, 1989). Alkaline magma may be generated by low-degree melting of an enriched mantle (EM) source, by pronounced differentiation of a mafic magma, or even by crustal contamination of mantle-derived magma; volatile contents and mantle mineralogy also affect magma compositions (Mahoney et al., 1985; Wilson, 1989; Winter, 2001). It is also observed that the latest phases of continental flood basalts (CFB) are spatially and temporally associated with alkaline and carbonatite magmatism (Bell, 2001, 2002; Heaman et al., 2002). One reason for relating such alkaline magmatism to plume activity is the isotopic similarity between these rocks and ocean island basalts (OIB). To establish such relationships, radiogenic isotope data are helpful to identify components such as depleted mid-ocean ridge basalt (MORB) mantle (DMM); mantle with a high U/Pb ratio (HIMU); enriched mantle with high Rb/Sr, low U/Pb, and Sm/Nd ratios (EM I); enriched mantle with high Rb/Sr and U/Pb and low Sm/Nd ratios (EM II); and focal zone mantle component (FOZO) (Zindler and Hart, 1986; Hart, 1988; Hart et al., 1992; Bell, 2002).

The Shillong plateau of northeastern India underwent Early Cretaceous basic and alkaline carbonatite activity. Many earlier workers proposed spatial and temporal association of these magmatisms with a plume currently beneath Kerguelen (Storey et al., 1992; Kent et al., 1997, 2002; Ray et al., 1999; Srivastava et al., 2005). It is suggested that Kerguelen hotspot was situated very close to the eastern Indian margin between 100 and 120 Ma (cf. Fig. 5 *in* Coffin et al., 2002; Fig. 4 *in* Kent et al., 2002) and responsible for the Rajmahal basalts exposed over the Gondwana Supergroup and for the Sylhet basalts and the alkaline-carbonatite magmatism of the Shillong plateau. The Jasra alkaline-mafic-ultramafic complex (Fig. 2A) selected for this study is situated in the Karbi Anglong district of Assam in the eastern part of Shillong plateau. Mamallan et al. (1994) were

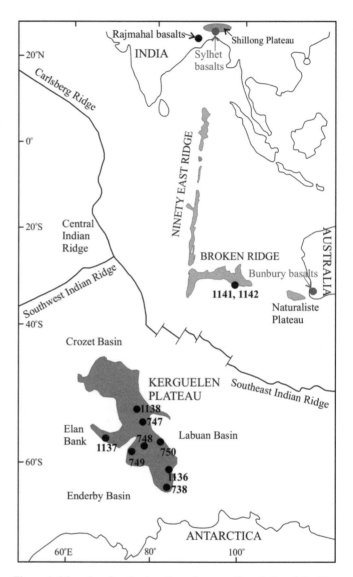

Figure 1. Map showing the location of magmatic activity of the Kerguelen plateau, Broken Ridge, Naturaliste plateau, Bunbury, Rajmahal-Sylhet, and associated ultramafic-alkaline-carbonatite complexes of Shillong plateau (modified from Kent et al., 2002). Age data for these magmatic rocks are presented in Table 1.

the first to document this complex, and Srivastava and Sinha (2004) subsequently presented detailed petrological and whole-rock geochemical data for these rocks. Heaman et al. (2002) dated this complex applying the U-Pb method to zircon and baddeleyite, which placed it at 105.2 ± 0.5 Ma, roughly contemporaneous with other nearby complexes (Table 1). Because there are no isotopic data available that allow us to discuss the petrogenesis of different rock types associated with the Jasra Com-

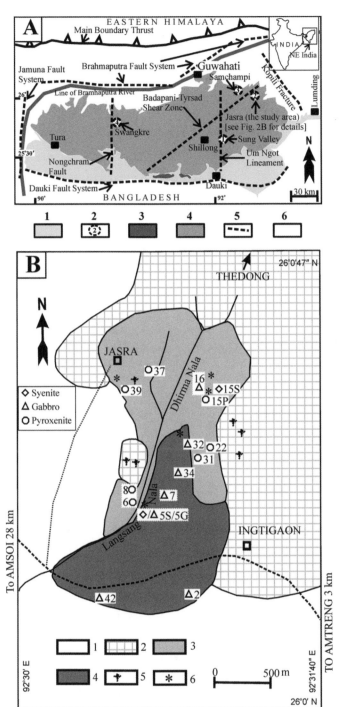

Figure 2. (A) Regional geological and tectonic setup of the Shillong plateau (modified from Srivastava and Sinha, 2004). 1—Cretaceous–Tertiary sediments; 2—ultramafic-alkaline-carbonatite complexes; 3—Sylhet basalts; 4—Archean gneissic complex, Shillong Group rocks, mafic igneous rocks, and Proterozoic granites; 5—major fault systems; 6—alluvium and recent sediments. (B) Geological map of the Jasra alkaline ultramafic-mafic complex (after Srivastava and Sinha, 2004). 1—Shillong Group rocks; 2—Neoproterozoic granite; 3—pyroxenite; 4—gabbro; 5—mafic dike exposures; 6—nepheline syenite exposures. Locations of geochemically analyzed samples are marked on the map.

plex, this article presents the first radiogenic isotopic data from the Jasra alkaline-mafic-ultramafic complex. Using other petrological and geochemical data, we attempt to evaluate the nature of the source, whether derived from mantle components or with some other genetic history, for the genesis of this complex.

GEOLOGICAL SETTING

The Shillong plateau, northeastern India, underwent extensive magmatic activity that ranged from granitic to alkaline composition. Most of the rocks were emplaced into the Proterozoic Shillong Group, consisting of orthoquartzite and phyllite, within the basement Archean gneissic complex (Desikachar, 1974; Mazumdar, 1976). In addition, Neoproterozoic granite plutons, small bodies of metamorphosed mafic igneous rocks, many exposures of early Cretaceous basaltic flows (Sylhet Traps), and alkaline-(carbonatite)-mafic-ultramafic complexes are emplaced in the Shillong Group (Fig. 2A; Mazumdar, 1976; Ghosh et al., 1994; Kumar et al., 1996; Das Gupta and Biswas, 2000).

The Shillong plateau (Fig. 2A) is a rectangular-shaped uplifted horstlike feature bounded on all sides by deep fault systems (Evans, 1964; Desikachar, 1974; Nandy, 1980; Gupta and Sen, 1988; Golani, 1991). The Jasra alkaline-mafic-ultramafic igneous complex is situated within the prominent fracture system of the Barapani-Tyrsad Shear Zone, Kopili fault, and Um Ngot Lineaments (Fig. 2A). No geological or geophysical feature observed or reported from the Shillong plateau indicates extensional or subduction tectonics (Evans, 1964; Desikachar, 1974; Nandy, 1980; Gupta and Sen, 1988; Das Gupta and Biswas, 2000; Kayal et al., 2006). It is believed that the set of north-south-trending faults and lineaments of the Shillong plateau were initially developed during Late Jurassic–Early Cretaceous times and closely associated with the early Cretaceous igneous intrusions, including alkaline and carbonatite complexes (Gupta and Sen, 1988).

Mamallan et al. (1994) and Srivastava and Sinha (2004) have described the geology of the Jasra alkaline-mafic-ultramafic igneous complex. This complex consists of pyroxenite, gabbro, a few mafic dikes, and nepheline syenite (Fig. 2B). Coarse-grained pyroxenite and gabbro form separate plutons, although it is difficult to observe any field relationship between the pyroxenite and gabbro plutons. Srivastava and Sinha (2004) reported that a few mafic dikes, contemporaneous with the gabbro plutons, sharply cut the pyroxenites as well as the granitic rocks, but not the gabbroic plutons. This relationship suggests that the mafic gabbro and dikes are younger than the pyroxenite. Small dikes and dikelets of nepheline syenite cut all rock types and form the youngest unit of this complex. In places, a few thin sections exhibit accumulation of nepheline and alkali pyroxene grains and show that these rocks have an ijolitic nature. This is important because ijolite-nephelinite is thought to be often associated with the alkaline-carbonatite magmatism (Le Bas, 1987, 1989). Calcite grains are also observed in the pyroxenite samples, but no separate exposure of ijolite or carbonatite has yet been found.

Srivastava and Sinha (2004) have already described, in detail, the petrography of major rock types associated with the Jasra

TABLE 1. AGE DATA ON BASALTS FROM THE KERGUELEN PLATEAU, BROKEN RIDGE, NATURALISTE PLATEAU, BUNBURY, AND RAJMAHAL-SYLHET IGNEOUS PROVINCE AND ASSOCIATED ULTRAMAFIC-ALKALINE-CARBONATITE COMPLEXES OF SHILLONG PLATEAU

Method	Material	Age (Ma)	References
Kerguelen plateau			
Ar-Ar	Basalt from ODP site 1136	118–119	Coffin et al. (2002); Duncan
	Basalt from ODP site 1137	107–108	(2002)
	Basalt from ODP sites 749 and 750	110–112	
	Basalt from ODP site 1138	~100	
Broken Ridge			
Ar-Ar	Basalt from ODP sites 1141 and 1142	~100	Coffin et al. (2002); Duncan (2002)
Naturaliste plateau			
Ar-Ar	Basalt	100.6 ± 1.2	Pyle et al. (1995)
Bunbury, western Australia			
Ar-Ar	Basalt lava	123–130	Frey et al. (1996); Coffin et al. (2002)
Rajmahal-Sylhet flood basalts province			
Ar-Ar	Basalts	105–118	Baksi (1995); Coffin et al. (2002); Kent et al. (2002)
Ultramafic-alkaline-carbonatite complexes (Shillong plateau)			
1. Sung Valley			
Ar-Ar	Pyroxenite (WR) and phlogopite from carbonatite	107.2 ± 0.8	Ray et al. (1999)
Rb-Sr	Carbonatite (WR), pyroxenite (WR), and phlogopite from carbonatite	106 ± 11	Ray et al. (2000)
U-Pb	Perovskite from ijolite	115.1 ± 5.1	Srivastava et al. (2005)
2. Jasra			
U-Pb	Zircon and baddeleyite from differentiated gabbro	105.2 ± 0.5	Heaman et al. (2002)
3. Samchampi			
Fission track	Apatite	~105	Acharya et al. (1986)
4. Swangkre			
K-Ar	Lamprophyre	107 ± 3	Sarkar et al. (1996)

Note: See Figure 6 for locations.

Complex (Table 2). Based on a study of about seventy samples, mineral compositions along with textures classify the pyroxenites as clinopyroxenite, olivine clinopyroxenite, and alkali pyroxenite, whereas the gabbro samples show variations from monzogabbro (essexite) to olivine gabbro. The syenite samples contain nepheline.

ANALYTICAL TECHNIQUES

Whole-rock major elements (for sixteen samples) and trace (+ rare earth) elements (for ten samples) were analyzed at Activation Laboratories Ltd. in Ancaster, Canada. An ICP-OES (inductively coupled plasma optical emission spectrometer, model Thermo-Jarret Ash ENVIRO II) was used to analyze major elements, whereas an ICP-MS (inductively coupled plasma mass spectrometer, model Perkin Elmer Sciex ELAN 6000) was used

to determine trace-element concentrations. Several international geochemical reference material samples (STM1, MRG1, DNC1, SY3, and W2) were run along with Jasra samples in order to assess the accuracy of our result. Precision and accuracy were ~5% and 5%–10% when reported at 100X, the detection limit, for major oxides and trace or rare earth elements, respectively. Whole-rock major oxide and CIPW-normative compositions are presented in Table 3. The CIPW norms, rock types, and Mg# for all samples were automatically computed using the SINCLAS Computer Program (Verma et al., 2002). In the middlemost option of this program the iron-ratio used for CIPW norm calculation is taken from Middlemost (1989). Table 4 presents whole-rock trace-element and rare earth element compositions. The locations of the samples analyzed are marked on the geological map (Fig. 2B).

Radiogenic isotopic compositions were measured at the Re-

TABLE 2. PETROGRAPHIC CHARACTERS OF DIFFERENT IGNEOUS ROCK UNITS ASSOCIATED WITH THE JASRA ALKALINE-MAFIC-ULTRAMAFIC COMPLEX

Rock type (oldest to youngest)	Mineral compositions (in order of abundance)	Texture	Color index
Pyroxenite (oldest unit)	Clinopyroxenes (68–74%; diopside, aegirine-augite, aegirine, augite, and titan-augite), opaques (~12%; magnetite and ilmenite), olivine, biotite, orthoclase, plagioclase, apatite, sphene, and calcite	Hypidiomorphic and porphyritic	~94
Alkaline pyroxenite	Clinopyroxenes (54–57%; diopside, aegirine-augite, aegirine, augite, and titan-augite), olivine (10–14%), nepheline (5–10%), opaques (magnetite and ilmenite), biotite, orthoclase, plagioclase, apatite, sphene, and calcite	Hypidiomorphic and porphyritic	80–85
Essexite	Clinopyroxenes (44–60%; augite, titan-augite, aegirine-augite, aegirine, and diopside), feldspars (24–35%; plagioclase and orthoclase), olivine, nepheline (5–9%), opaques (magnetite and ilmenite), apatite, sphene, biotite, zircon, and rutile	Hypidiomorphic and panidiomorphic	57–66
Olivine gabbro and mafic dikes	Clinopyroxenes (~36%; augite, titan-augite, aegirine-augite, aegirine, and diopside), plagioclase (~38%), olivine (~7%), opaques (magnetite and ilmenite), orthoclase (4%), nepheline (4%), apatite, sphene, biotite, zircon, and rutile	Hypidiomorphic and panidiomorphic textures in gabbro samples; intergranular or ophitic textures in dike samples	~50
Nepheline syenite (differentiated gabbro)	Orthoclase (~46%), plagioclase (~15%), clinopyroxenes (~13%; aegirine, aegirinie-augite, and augite), nepheline (~8%), biotite, opaques (magnetite and ilmenite), calcite, zircon, sphene, apatite, and olivine	Hypidiomorphic, porphyritic, and poikilitic	~25
Nepheline syenite dikes (youngest unit)	Orthoclase (~60%), plagioclase (~10%), clinopyroxenes (~9%; aegirine, aegirinie-augite, and augite), nepheline (~6%), biotite, opaques (magnetite and ilmenite), calcite, zircon, sphene, apatite, and olivine	Hypidiomorphic, porphyritic, and poikilitic	~19

search Center for Mineral Exploration (RCMRE), Beijing, China. A Thermo Finnigan MAT 262 multicollector mass spectrometer was used for the radiogenic isotope (Sm-Nd and Rb-Sr) analyses. International standards La Jolla and BCR-1 were used for the Nd isotope analyses, and NBS987 and NBS607 were used for the Sr isotope analyses. The two standard deviation uncertainties on the measured $^{87}Sr/^{86}Sr$ and $^{143}Nd/^{144}Nd$ isotope ratios are considered better than ± 0.002% and ± 0.001%, respectively. Analyzed isotope data are presented in Table 5. Initial isotopic ratios were calculated assuming an emplacement age of 105 Ma for the Jasra Complex determined by a precise U-Pb method using zircon or baddeleyite from a differentiated gabbro, which placed it at 105.2 ± 0.5 Ma (Heaman et al., 2002). We do not have ages for other units of the complex, but it is presumed that older and younger phases of this complex may have been emplaced at an interval of ~5 m.y. Initial isotopic ratios were also calculated assuming ages of 100 Ma and 110 Ma and found to be within the analytical error bars.

GEOCHEMISTRY

Whole-Rock Geochemistry

Classification based on the total alkali and silica contents (TAS) is presented in Table 3. This classification, based on International Union of Geological Sciences recommendations (Le Maitre, 2002), was obtained with the help of the SINCLAS computer program (Verma et al., 2002). Pyroxenite samples were classified as melanephelinite foidite to melanephelinite basanite. The gabbroic samples indicate melanephelinite tephrite, potassic trachybasalt, alkali basalt, and melanephelinite basanite compositions. The syenite samples are tephriphonolite in nature. Thus, we have concluded that most samples show melanephelinitic compositions. CIPW-normative compositions (Table 4) also corroborate the alkaline nature of these rocks. All of the pyroxenite samples contain alkaline-normative minerals such as nepheline, leucite, and dicalcium silicate. Similarly, the gabbro samples have orthoclase, albite, and nepheline in their normative compositions. The agpaitic indices [$(Na_2O+K_2O)/Al_2O_3$ ratio or $(Na^+K)/Al$] for both the nepheline syenite samples are <1, suggesting a miaskitic affinity (Heinrich, 1966; Bates and Jackson, 1980). This is an important feature; miaskitic nepheline syenites have a very close relationship with carbonatite magmatism, but it is also realized that more analyses of nepheline syenite samples is required for a strong conclusion to be reached. Such syenitic rock is also found to be enriched in CO_2, leading to crystallization of cancrinite and calcite (Heinrich, 1966). A high Mg number ($\cong 62$) and MgO > 6% are indicative of primary magma (Green, 1971; Luhr, 1997; Velasco-Tapia and Verma, 2001). Some samples used in the present study have similar Mg numbers, and most have similar MgO contents. Other samples, including syenites, are probably differentiated products of a

TABLE 3. WHOLE-ROCK MAJOR-ELEMENT (wt% OXIDES) AND CIPW NORM COMPOSITIONS OF DIFFERENT ROCK UNITS OF THE JASRA ALKALINE-MAFIC-ULTRAMAFIC COMPLEX

Sample no. →	Pyroxenites							Gabbros							Ne syenites	
	JS/6	JS/8	JS/15P	JS/22	JS/31	JS/37	JS/39	JS/2	JS/5G	JS/7	JS/16	JS/32	JS/34	JS/42	JS/5S	JS/15S
SiO_2	43.29	39.85	41.79	41.67	39.18	36.84	38.40	41.71	45.68	44.98	49.62	41.48	45.32	41.79	51.40	53.24
TiO_2	3.81	4.76	4.62	2.54	6.14	5.94	5.67	5.19	4.24	4.65	2.52	5.84	5.05	4.87	2.60	0.76
Al_2O_3	5.72	7.72	5.55	8.64	5.99	8.87	9.82	9.10	11.08	11.01	5.94	9.58	11.34	9.18	16.29	19.18
Fe_2O_3	12.37	14.60	13.35	13.98	15.86	16.08	15.36	16.00	14.21	14.93	9.50	13.80	13.68	15.51	10.07	4.78
MnO	0.16	0.21	0.14	0.24	0.23	0.23	0.22	0.23	0.22	0.22	0.19	0.23	0.16	0.22	0.15	0.07
MgO	14.36	11.26	11.62	9.41	10.26	7.16	6.98	8.40	6.54	6.41	8.74	6.55	8.33	8.58	2.33	0.83
CaO	16.44	15.31	19.31	14.70	17.23	16.34	14.56	14.21	10.88	10.82	17.69	13.65	11.39	13.58	4.81	4.39
Na_2O	1.02	1.62	0.78	2.97	1.03	2.92	3.49	2.33	3.01	3.03	1.49	2.75	2.32	2.33	4.61	3.22
K_2O	1.80	1.51	0.99	2.04	1.57	2.02	2.21	1.36	2.34	2.52	2.36	2.82	1.16	1.55	5.12	8.40
P_2O_5	0.31	0.76	0.34	0.35	0.94	1.08	1.07	1.33	0.94	1.00	0.36	0.98	0.26	1.08	0.78	0.06
LOI	1.08	1.33	1.22	2.14	0.59	1.27	1.22	–	–	–	0.99	1.08	–	–	0.70	3.90
Total	100.36	98.93	99.71	98.68	99.02	98.75	99.00	99.86	99.14	99.57	99.40	98.76	99.01	98.69	98.86	98.83
Mg#	72.30	64.32	66.18	61.14	59.26	51.00	51.51	55.10	53.66	50.09	68.26	52.60	58.74	56.39	38.40	31.88
Al															8.62	10.15
Na															3.42	2.39
K															4.25	6.97
A.I.*															0.60	0.40
A.I.**															0.89	0.92
CIPW norm																
Or	5.82	9.69						8.16	14.11	15.15	14.29	8.21	7.01	9.41	31.06	52.49
Ab								3.98	16.29	12.78	4.23		18.81	3.50	28.55	11.67
An			8.96	4.42	7.30	5.33	4.77	10.51	10.01	9.15	2.62	5.67	17.48	10.28	8.86	13.81
Lc	8.49	7.26	4.71	9.91	7.50	9.74	10.62									
Ne	4.76	7.71	3.68	14.28	4.87	13.93	16.58	8.69	5.25	9.20	4.70	13.06	0.68	9.07	6.22	9.29
Di	49.46	43.46	52.42	45.44	46.17	33.39	38.45	42.49	31.54	31.79	66.25	46.89	31.11	41.98	8.61	9.42
Ol	17.07	14.24	11.25	12.51	11.72	10.07	7.49	9.45	7.90	9.22		2.07	11.41	10.17	5.79	1.71
Mt	2.17	3.35	2.37	3.24	2.82	3.70	3.52	3.59	4.47	3.36	2.15	3.16	3.09	3.52	3.97	1.94
Il	7.37	9.38	9.02	5.06	12.02	11.74	11.16	10.01	8.22	8.98	4.90	11.49	9.80	9.50	5.07	1.53
Ap	0.73	1.83	0.81	0.85	2.25	2.60	2.57	3.13	2.22	2.36	0.86	2.35	0.62	2.57	1.86	0.15
Cs	4.12	3.09	6.80	4.28	5.36	9.50	4.84									
Rock type	PIC	BSN (mnp)	PB	BSN (mnp)	FOI (mnp)	FOI (mnp)	FOI (mnp)	TEP (mnp)	TB (pot)	TEP (bsn)	B (alk)	TEP (mnp)	B (alk)	BSN (mnp)	TPH	TPH

Notes: – indicates either zero or below zero; A.I.—agpaitic index; A.I.* $[(Na_2O+K_2O)/Al_2O_3]$, ** $[(Na+K)/Al]$; Or—orthoclase; Ab—albite; An—andesite; Lc—leucite; Ne—nepheline; Di—diopside; Ol—olivine; Mt—magnetite; Il—ilmenite; Ap—apatite; Cs—dicalcium silicate; PIC—picrite; PB—picrobasalt; BSN (mnp)—basanite (melanephelinite); FOI (mnp)—foidite (melanephelinite); TEP (mnp)—tephrite (melanephelinite); TB (pot)—potassic trachybasalt; TEP (bsn)—tephrite (basanite); B (alk)—alkali basalt; TPH—tephriphonolite.

TABLE 4. TRACE-ELEMENT AND RARE EARTH ELEMENT ANALYSES (ppm) OF DIFFERENT ROCK UNITS OF THE JASRA ALKALINE-MAFIC-ULTRAMAFIC COMPLEX

Sample no. →	Pyroxenites				Gabbros				Ne syenites	
	JS/6	JS/15P	JS/22	JS/37	JS/2	JS/5G	JS/32	JS/34	JS/5S	JS/15S
Cr	1080	104	504	21	178	b.d.l.	126	91	b.d.l.	b.d.l.
Ni	194	179	166	147	106	b.d.l.	151	158	b.d.l.	b.d.l.
Sc	37	42	21	23	25	20	20	28	6	2
V	286	308	414	465	409	337	439	583	98	181
Rb	48	37	73	36	23	129	80	27	129	141
Ba	595	384	605	746	549	809	1225	385	1540	3542
Sr	781	727	954	2052	1666	1690	2473	1252	2021	5312
Ga	14	15	16	23	20	25	20	18	25	24
Ta	2.3	3.8	5.1	9.1	5.5	6.7	12.7	3.7	9.0	1.2
Nb	33	46	82	122	82	133	172	47	133	55
Hf	6.1	7.6	6.6	13.3	9.1	11.2	14.7	6.6	15.4	5.2
Zr	197	209	285	516	341	452	584	228	741	280
Y	16	16	28	32	37	40	48	22	39	10
Th	3.2	6.5	11.9	15.0	7.3	13.2	14.5	4.0	29.1	4.7
U	1.1	2.4	2.7	4.5	1.7	3.3	3.0	0.8	7.5	1.5
La	36.2	55.8	74.1	113.0	94.8	99.0	138.0	35.0	125.0	17.0
Ce	71.9	107.0	137.0	238.0	199.0	204.0	283.0	72.5	239.0	29.8
Pr	9.79	14.20	15.40	30.10	25.60	25.10	34.60	9.64	27.60	3.44
Nd	41.1	59.9	58.7	123.0	100.0	95.1	136.0	41.9	98.1	12.2
Sm	7.5	10.4	9.7	20.0	16.7	15.7	22.4	7.9	14.7	2.2
Eu	2.50	3.32	2.90	6.31	5.22	4.97	6.93	2.81	5.36	0.53
Gd	5.9	7.7	7.4	14.2	12.3	11.5	16.0	6.7	9.7	1.8
Tb	0.8	1.0	1.1	1.8	1.7	1.6	2.2	1.0	1.5	0.3
Dy	3.6	4.5	5.7	7.9	7.9	8.0	10.6	4.9	7.5	1.7
Ho	0.6	0.7	1.0	1.2	1.3	1.4	1.7	0.8	1.3	0.3
Er	1.4	1.7	2.9	2.8	3.4	3.7	4.6	2.3	3.6	0.9
Tm	0.17	0.18	0.39	0.29	0.39	0.43	0.54	0.29	0.45	0.13
Yb	1.0	1.2	2.6	1.9	2.5	2.9	3.6	1.9	3.1	0.9
Lu	0.14	0.16	0.37	0.25	0.31	0.39	0.47	0.27	0.41	0.14

Notes: b.d.l.—below detection limit; detection limit for Cr and Ni is 20 ppm.

primary magma. Thus, we believe that the Jasra samples studied either are derived from a primary magma or are differentiated products of primary magma.

With reference to Harker and other variation diagrams, in an earlier paper (Srivastava and Sinha, 2004) we suggested that none of the three major rock types shows any co-genetic relationship with the others and that they probably have different genetic histories. A few new variation diagrams (not presented in Srivastava and Sinha, 2004) are shown in Figure 3. Important characteristics observed from these plots are as follows:

1. The pyroxenite and gabbro samples show similar chemical compositions and linear trends, but this probably does not suggest any simple differentiation process.
2. This observation is more evident from the plots of Mg# versus SiO_2 (Fig. 3A) and Zr versus SiO_2 (Fig. 3D) plots. On these plots all three rock types plot separately and show different crystallization trends. More significant, at a given Zr content different rock types have different silica contents

(Fig. 3D), which is difficult to explain by simple differentiation of a single batch of magma.

3. One syenite sample (JS/5S), collected from the differentiated part of a gabbro, indicates differentiation; it has a high concentration of high field-strength elements (HFSE).

Thus, it is apparent that different rock types of the Jasra Complex probably have different evolutionary histories.

These geochemical characteristics are further evident from the primordial mantle-normalized multielement and chondrite-normalized rare earth element (REE) patterns for the Jasra samples (Fig. 4). Multielement plots of pyroxenite and gabbro samples show similar patterns, quite unlike those from the syenite samples. However, the REE patterns of all the samples are similar. Again, it is difficult to explain the relationships between different rock types of the Jasra Complex by a differentiation process; if gabbro is a differentiated product of pyroxenite, it should have higher concentrations of REE than the pyroxenite, but this is not so. The multielement pattern of one syenite sample

TABLE 5. Nd AND Sr ISOTOPIC COMPOSITIONS IN SAMPLES FROM THE JASRA ALKALINE-MAFIC-ULTRAMAFIC COMPLEX

Sm-Nd compositions

Sample no. ↓	Rock type	Sm (ppm)	Nd (ppm)	$^{147}Sm/^{144}Nd$	$^{143}Nd/^{144}Nd$ (2σ)	$^{143}Nd/^{144}Nd_{initial}$	εNd
JS/5S	Ne syenite	16.92	111.00	0.09221	0.512450 ± 11	0.512387	−2.26
JS/15A	Ne syenite	2.54	13.94	0.11030	0.512334 ± 6	0.512258	−4.78
JS/22	Pyroxenite	49.73	346.60	0.08677	0.512487 ± 6	0.512427	−1.48
JS/37	Pyroxenite	21.76	133.90	0.09828	0.512532 ± 11	0.512464	−0.76
JS/34	Gabbro	8.68	45.18	0.11623	0.512397 ± 17	0.512317	−3.63

Rb-Sr compositions

Sample no. ↓	Rock type	Rb (ppm)	Sr (ppm)	$^{87}Rb/^{86}Sr$	$^{87}Sr/^{86}Sr$ (2σ)	$^{87}Sr/^{86}Sr_{initial}$	εSr
JS/5S	Ne syenite	141.30	2087.0	0.19600	0.707949 ± 10	0.707657	46.6
JS/15A	Ne syenite	155.10	5320.0	0.08438	0.707060 ± 10	0.706934	36.3
JS/22	Pyroxenite	8.31	960.8	0.03481	0.706575 ± 10	0.706523	30.5
JS/37	Pyroxenite	39.49	2062.0	0.05542	0.706757 ± 11	0.706674	32.6
JS/34	Gabbro	29.68	1248.0	0.06885	0.708994 ± 13	0.708891	64.1

Notes: Initial Nd and Sr isotope ratios along with epsilon values were calculated assuming an emplacement age of 105 Ma. Epsilon values are calculated using present-day ratios of $^{87}Sr/^{86}Sr = 0.7045$ (DePaolo, 1988) and $^{87}Rb/^{86}Sr = 0.0827$ (DePaolo, 1988) for bulk silicate earth (BSE) and $^{143}Nd/^{144}Nd = 0.512638$ (Goldstein et al., 1984) and $^{147}Sm/^{144}Nd = 0.1967$ (Jacobsen and Wasserburg, 1980) for chondritic uniform reservoir (CHUR). The internal precision for Sr and Nd isotope ratio measurement is ±0.0006 and ±0.0007, respectively, whereas the external precision for both isotopes is ±0.0009. Two standards, La Jolla and BCR-1, were run with the Jasra samples for the measurement of $^{143}Nd/^{144}Nd$ ratios; measured values were 0.511854 ± 7 and 0.512613 ± 8, respectively. Similarly, NBS987 and NBS607 standards were run during $^{87}Sr/^{86}Sr$ analyses; measured values were 0.710239 ± 10 and 1.20054 ± 12, respectively.

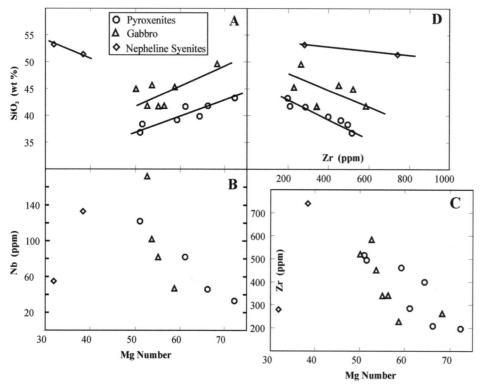

Figure 3. Variation diagram plotted between Mg number and (A) SiO₂, (B) Nb, and (C) Zr and between (D) SiO₂ and Zr. IAB—island arc basalts; CRB—continental rift basalts; OIB—ocean island basalts; MORB—mid-ocean ridge basalts.

(JS/5S) is similar to the patterns of gabbro samples, but the other sample (JS/15S) shows an entirely different pattern, suggesting that the two samples have different crystallization histories. These characteristic HFSE patterns also illustrate no or positive Nb, Ta (with respect to K and La), and Ti (with respect to Sm and Tb) anomalies. Such geochemical characteristics clearly preclude a role for crust in the genesis of Jasra samples; crustal contaminated samples should have negative Nb, Ta, and Ti anomalies (Wilson, 1989; Carlson, 1991; Saunders et al., 1992; Kent, 1995; Winter, 2001). Similar geochemical characteristics are also observed for some modern OIB (Saunders et al., 1992; Kent, 1995). This feature of the Jasra samples is very clearly observed in Figure 5, from Agrawal et al. (2004) and Verma et al. (2006). These authors have presented a set of new discrimination function diagrams based on the major-element data so as to interpret the nature of magma type and the emplacement tectonic environment. This set of plots can discriminate among island arc, continental rift, ocean island, and mid-ocean ridge tectonic settings. On these diagrams the Jasra samples proved to be of OIB affinity, because a high percentage (~86–100%) of the Jasra samples plot in the OIB field in these diagrams. Patterns of light REE (LREE) enrichment are observed either in the subduction-related magmatism (i.e., that involving crustal material) or that derived from low-degree partial mantle melting or through some metasomatic process (Cullers and Graf, 1984). It is correct that samples derived from a crust-contaminated melt show LREE-enriched patterns, but such samples should also have negative Nb, Ta, and Ti anomalies on multielement patterns, which is different from the Jasra samples. A small degree of melting (<10%) of mantle peridotite may show an enriched LREE pattern (Cullers and Graf, 1984). Thus, low-degree partial melting of mantle together with a metasomatic process looks more appropriate for the Jasra rocks. These observed geochemical characteristics most likely preclude the possibility of crustal contamination and suggest that the Jasra samples are derived from a mantle-derived melt of OIB affinity.

Radiogenic Isotope Geochemistry

The $^{87}Sr/^{86}Sr_{initial}$ and $^{143}Nd/^{144}Nd_{initial}$ ratios in the Jasra samples show variation between 0.706523–0.708891 and 0.512258–0.512464, respectively. All samples have positive εSr values (+30.5 to +64.1) and negative εNd values (–0.76 to –4.78). Isotopic data for all the samples are plotted in the $^{87}Sr/^{86}Sr_{initial}$ versus $^{143}Nd/^{144}Nd_{initial}$ diagram (Fig. 6A). All samples plot close to the EM field. The pyroxenite samples have the lowest $^{87}Sr/^{86}Sr_{initial}$ ratios, and the gabbro sample has the highest $^{143}Nd/^{144}Nd_{initial}$ ratios, whereas the $^{143}Nd/^{144}Nd_{initial}$ ratios show very limited variation. The high $^{87}Sr/^{86}Sr$ values observed in these samples are probably the result of metasomatism by fluids or melts with an EM signature.

Zindler and Hart (1986) and Hart (1988) defined different mantle reservoirs on the basis of their isotopic taxonomy and pro-

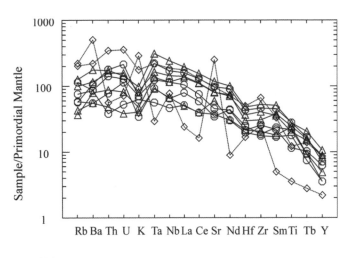

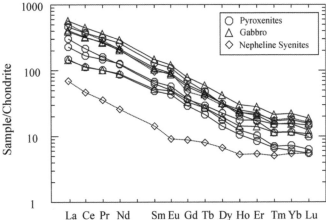

Figure 4. Primordial mantle–normalized multielement and chondrite-normalized rare earth element patterns for different samples from the Jasra Complex. The primordial mantle and chondrite values are taken from McDonough et al. (1992) and Evensen et al. (1978), respectively.

posed at least four mantle components: DMM, HIMU, EM I, and EM II. In addition to these four components, Hart et al. (1992) proposed another mantle component named FOZOs. This component has been interpreted using two different models, but in both, FOZO mixes in plumes containing EM and HIMU components, one from the core-mantle boundary and the other from the lower mantle. The major difference between these two models is that in the first model all components are derived from below 670 km, and in the other model they are derived from near the 670 km discontinuity itself. From the isotopic data available for the Jasra samples, it is difficult to distinguish between these two models without some more isotopic data, particularly $^{206}Pb/^{204}Pb$, $^{207}Pb/^{204}Pb$, $^{208}Pb/^{204}Pb$, $^{3}He/^{4}He$, and noble gas data (Hart et al., 1992; Bell, 2002). The EM component may have been derived either from subducted crustal material in the mantle or from mantle metasomatism. But, as mentioned earlier, subduction-derived rocks should have negative Nb, Ta, and Ti anomalies, which are not observed in these rocks (Fig. 4). Thus, we prefer

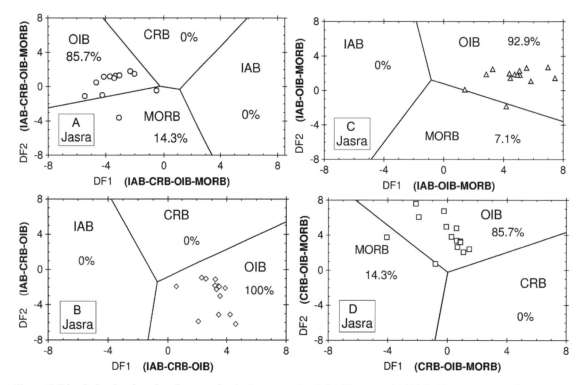

Figure 5. Discrimination function diagrams for the Jasra samples (after Verma et al., 2006). The percentage given next to the tectonic setting name represents the percentage of success obtained for the Jasra samples. IAB—island arc basalts; CRB—continental rift basalts; OIB—ocean island basalts; MORB—mid-ocean ridge basalts.

the mantle metasomatism model. Bell et al. (1998) pointed out that metasomatism is restricted mainly to the lithosphere, which can trap melt inclusions generated from the convecting asthenosphere. Bell (2001, 2002) suggested plume involvement in the genesis of carbonatitic, alkalic, and kimberlitic magmatism.

It is not easy to estimate the exact proportion of the different mantle components involved in generating the range of isotopic compositions preserved in the samples from the Jasra Complex, but possibly mixing between the HIMU and EM components, similar to the FOZO model, could explain the range in Sr and Nd isotopic composition (Hart et al., 1992). The isotope data are also consistent with mixing of other end-members such as DMM or prevalent mantle with EM I and EM II or with the involvement of continental crust. Pb isotope data are probably required to distinguish between these possibilities. Mixing between HIMU and EM is noticed for many alkaline and carbonatite complexes (Kramm and Kogarko, 1994; Tilton and Bell, 1994; Simonetti et al., 1995, 1998; Bell and Simonetti, 1996; Kalt et al., 1997; Srivastava et al., 2005). Of course, to corroborate this observation we require more isotopic data, particularly $^{206}Pb/^{204}Pb$, $^{207}Pb/^{204}Pb$, $^{208}Pb/^{204}Pb$, and $^3He/^4He$ data.

The Sung Valley alkaline carbonatite complex is thought to be contemporaneous with the Jasra and other alkaline-carbonatite complexes of Shillong plateau (Kumar et al., 1996; Ray et al., 1999; Heaman et al., 2002; Srivastava et al., 2005). The $^{87}Sr/^{86}Sr_{initial}$ and $^{143}Nd/^{144}Nd_{initial}$ ratios obtained for the Sung Valley Com-

plex are also plotted for comparison (Fig. 6A). Close isotopic similarities are observed between the silicate samples from both the complexes; the gabbro sample from Jasra shows exactly the same isotopic composition observed for an ijolitic sample from Sung Valley.

DISCUSSION

The main inferences from the field, petrological, geochemical, and radiogenic isotope data for samples from the Jasra Complex presented in this article include the following:

1. This Early Cretaceous (U-Pb age 105.2 ± 0.5 Ma) alkaline-mafic-ultramafic complex consists of pyroxenite, gabbro, mafic dikes, and nepheline syenite emplaced within the Proterozoic Shillong Group of the Shillong plateau.
2. The pyroxenites and gabbros do not show any direct field relationships, while the mafic dikes, evidently contemporaneous with the gabbro, intrude the granites as well as the pyroxenites. These dikes do not cut the gabbro bodies. Nepheline syenite is the youngest unit of this complex and occurs as differentiates of gabbro or as small dikes or dikelets intruded into the pyroxenite with sharp edge contacts.
3. Primary calcite grains occur in the pyroxenite samples. Some thin sections show ijolitic composition.
4. The agpaitic indices of the nepheline syenites are <1, indi-

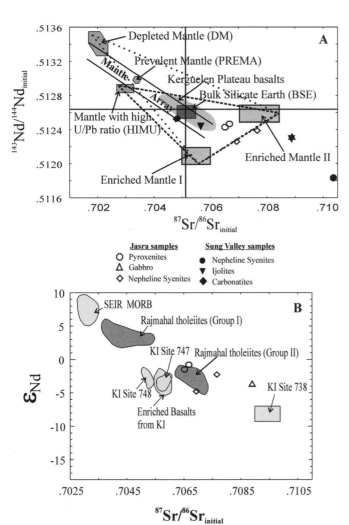

Figure 6. (A) $^{143}Nd/^{144}Nd$ and $^{87}Sr/^{86}Sr$ isotope correlation plot showing the main oceanic mantle components of Zindler and Hart (1986). The mantle array is defined by many mid-ocean ridge basalts and ocean island basalts, and the bulk earth values of $^{143}Nd/^{144}Nd$ and $^{87}Sr/^{86}Sr$ can be observed from this trend (Faure, 1986, 2001; Dickin, 1995). The isotopic data of the Sung Valley samples are taken from Srivastava et al. (2005). The Kerguelen plateau basalt field is based on data from Michard et al. (1986), Dosso et al. (1988), Storey et al. (1992), and Mahoney et al. (1995). (B) $^{87}Sr/^{86}Sr_{initial}$ and εNd plot for the studied samples and the Kerguelen plateau basalts. The source for the Kerguelen data is the same as for panel A. SEIR—Southeast Indian Ridge.

cating the miaskitic nature of these rocks and suggesting the involvement of CO_2-related activity in their genetic history.

5. The geochemical compositions of different rock units do not support any direct genetic relationship between the different rock types of the complex.

6. Major-element-based discrimination function diagrams suggest OIB affinity.

7. Nd-Sr isotopic compositions suggest the involvement of an EM component in the genesis of the rocks studied.

From these observations, it appears that the different rock types of the Jasra Complex do not have any co-genetic relationships. The overall alkaline nature of the different rock types and the miaskitic nature of the nepheline syenite samples of the Jasra Complex indicate the involvement of CO_2 in their genesis. Harmer (1999) has presented a model for the genesis of such silicate rocks involving a primitive carbonate liquid generated by low-degree melting of carbonated mantle peridotite. This model explains why such a melt contains appreciable alkalis and dissolves an adequate amount of olivine and pyroxene to provide the Al, Fe, and Si necessary for the crystallization of the silicate minerals. The other possibility may be a liquid immiscibility process, but the relation of Mg# to Ba and La ratios precludes this possibility (Hamilton et al., 1989; Srivastava and Sinha, 2004; Srivastava et al., 2005). On the basis of experimental work on the immiscible carbonate liquid and associated silicate liquid under any temperature and pressure and irrespective of whether the parental silicate melt was nephelinite or phonolite, Hamilton et al. (1989) have shown that the Ba/La ratio should be higher in an immiscible carbonate melt than in an associated silicate melt.

Because a similar picture is observed in the present study, we favor Harmer's (1999) model for the genesis of the different rock types of the Jasra Complex. Upton (1967) also pointed out that pyroxenite in alkalic environments may result from a reaction between silicate rock and carbonatite magma. This model is also supported by the experimental work of Wallace and Green (1988), Lee and Wyllie (1997), and Wyllie and Lee (1998). These workers showed that carbonate melts can be generated by direct melting of carbonated peridotite at depths equivalent to ~2–3.5 GPa, that they are Mg-rich and contain significant (up to 7%) amounts of alkalis, and that they may, depending on the fertility of the peridotite and the Na/K balance, be in equilibrium with phlogophite ± paragasite-richterite amphibole (Wallace and Green, 1988; Sweeney, 1994). This deep-seated melt may consume orthopyroxene (lherzolite) and transform it into metasomatic clinopyroxene and olivine. During this process, CO_2 is released and progressively metasomatizes the lherzolite to an alkaline wehrlite. With continued metasomatic enrichment, the alkali wehrlite melts to form an ultrabasic alkaline silicate magma that may crystallize to alkaline ultramafic-mafic rocks.

The isotopic data of the Jasra samples are plotted together with data for Kerguelen OIB and Rajmahal tholeiites (Michard et al., 1986; Dosso et al., 1988; Storey et al., 1992; Mahoney et al., 1995) in Figure 6B. The data from the Jasra samples are similar to the isotopic compositions of the Kerguelen OIB and Rajmahal Group II tholeiites. This observation is well supported by the geochemical composition of these rocks (Fig. 5). A Kerguelen plume origin has been proposed for these basalts (Mahoney et al., 1992; Coffin et al., 2002; Duncan, 2002; Kent et al., 2002). On the basis of the temporal and spatial relationships of the alkaline and carbonatite complexes of the Shillong plateau and radiogenic isotope compositions, it is suggested that a plume currently beneath Kerguelen was also responsible for these complexes (Ray et al., 1999; Heaman et al., 2002; Srivastava et al.,

2005). The spatial and temporal relationship is clearly evident from Table 1 and Figure 1, which implies a genetic link between the ca. 107–115 Ma alkaline-carbonatite complexes in northeastern India and the Kerguelen OIB.

Alkaline magma can also be generated in continental rift–like extensional tectonic environments (Wilson, 1989; Sheth et al., 2000; Winter, 2001; Verma, 2006). Although many features, such as OIB-like magma generated from an EM source, are similar in both plume and extensional tectonics, but none of the geological or geophysical features noticed from the Shillong plateau suggest any possibility of extensional rift tectonics in the region. Thus, a plume model for the genesis of the Jasra samples seems more practical.

CONCLUSIONS

Field relationships, geochemistry, and the radiogenic compositions of the early Cretaceous Jasra alkaline-mafic-ultramafic complex suggest that the different rock units (pyroxenite, gabbro, and nepheline syenite) do not have any direct genetic relationship with one another and probably have different genetic histories. Discrimination function diagrams based on the major-element data and radiogenic isotope data clearly suggest an OIB-like origin for the Jasra samples. The radiogenic isotopic compositions suggest that these rocks are derived from the mixing of mantle components such as HIMU and EM. Low-degree melting of a metasomatized mantle peridotite produces carbonatite melts that might dissolve an adequate amount of olivine and pyroxene to provide the Al, Fe, and Si necessary for crystallization of silicate minerals. The CO_2 released by this process progressively metasomatizes the lherzolite to an alkaline wehrlite, and melts derived from alkaline wehrlite (ultrabasic alkaline silicate magma) may be responsible for the crystallization of different rocks of the Jasra Complex. This genetic history suggests that the Jasra alkaline-mafic-ultramafic complex is an integrated part of the alkaline-carbonatite magmatism of the Shillong plateau and has been closely associated with Kerguelen plume activity.

ACKNOWLEDGMENTS

The authors thank Yu Jie and Sun Shihua (Research Center for Mineral and Resource Exploration, Beijing, China) for analyzing the radiogenic isotopic compositions of the Jasra samples, Keith Bell for his constructive comments on an earlier version of this manuscript, and Gareth Davies for pointing out important errors in an earlier version. We are grateful to Surendra P. Verma for his detailed, constructive comments, which were very helpful during the revision of an earlier version. We also sincerely thank Gillian R. Foulgar and Donna Jurdy (editors of this volume) and an anonymous reviewer for their comments, which were valuable in improving this manuscript. Council of Scientific and Industrial Research, New Delhi, provided financial assistance for this work (Scheme no. 24 (0251)/01/EMR-II).

REFERENCES CITED

Acharya, S.K., Mitra, N.D., and Nandy, D.R., 1986, Regional geology and tectonic setting of northeast India and adjoining region: Kolkota, Geological Survey of India Memoir 119, p. 6–12.

Agrawal, S., Guevara, M., and Verma, S.P., 2004, Discriminant analysis applied to establish major-element field boundaries for tectonic varieties of basic rocks: International Geology Review, v. 46, p. 575–594.

Baksi, A.K., 1995, Petrogenesis and timing of volcanism in the Rajmahal flood basalt province, Northeastern India: Chemical Geology, v. 121, p. 73–90, doi: 10.1016/0009-2541(94)00124-Q.

Bates, R.L., and Jackson, J.A., 1980, Glossary of geology (2nd edition): Falls Church, Virginia, American Geological Institute, 751 p.

Bell, K., 2001, Carbonatites: Relationships to mantle-plume activity, in Ernst, R.E., and Buchan, K.L., eds., Mantle plumes: Their identification through time: Boulder, Colorado, Geological Society of America Special Paper 353, p. 267–290.

Bell, K., 2002, Probing the mantle: The story from carbonatites: Eos (Transactions, American Geophysical Union), v. 83, p. 273–277, doi: 10.1029/2002EO000190.

Bell, K., and Simonetti, A., 1996, Carbonatite magmatism and plume activity: Implications from the Nd, Pb, and Sr isotope systematics of Oldoinyo Lengai: Journal of Petrology, v. 37, p. 1321–1339, doi: 10.1093/petrology/37.6.1321.

Bell, K., Kjarsgaard, B.A., and Simonetti, A., 1998, Carbonatites—Into the twenty-first century: Journal of Petrology, v. 39, p. 1839–1845, doi: 10.1093/petrology/39.11.1839.

Carlson, R.W., 1991, Physical and chemical evidence on the cause of source characteristics of flood basalt volcanism: Australian Journal of Earth Sciences, v. 38, p. 525–544.

Coffin, M.F., Pringle, M.S., Duncan, R.A., Gladezenko, T.P., Storey, M., Müller, R.D., and Gahagan, L.A., 2002, Kerguelen hotspot magma output since 130 Ma: Journal of Petrology, v. 43, p. 1121–1139, doi: 10.1093/petrology/43.7.1121.

Cullers, R.L., and Graf, J.L., 1984, Rare-earth elements in igneous rocks of the continental crust: Predominantly basic and ultrabasic rocks, in Henderson, P., ed., Rare earth element geochemistry: Amsterdam, Elsevier, p. 237–274.

Das Gupta, A.B., and Biswas, A.K., 2000, Geology of Assam: Bangalore, Geological Society of India, 169 p.

DePaolo, D.J., 1988, Neodymium isotope geochemistry: An introduction: New York, Springer.

Desikachar, S.V., 1974, A review of the tectonic and geological history of eastern India in terms of plate tectonic theory: Journal of the Geological Society of India, v. 15, p. 137–149.

Dickin, A.P., 1995, Radiogenic isotope geology: Cambridge, Cambridge University Press, 490 p.

Dosso, L., Bougault, H., Beuzart, P., Calvez, J.-Y., and Joron, J.-L., 1988, The geochemical structure of the South East Indian Ridge: Earth and Planetary Science Letters, v. 88, p. 47–59, doi: 10.1016/0012-821X(88)90045-3.

Duncan, R.A., 2002, A time frame for construction of the Kerguelen Plateau and Broken Ridge: Journal of Petrology, v. 43, p. 1109–1119, doi: 10.1093/petrology/43.7.1109.

Evans, P., 1964, The tectonic framework of Assam: Journal of the Geological Society of India, v. 5, p. 80–96.

Evensen, N.M., Hamilton, P.J., and O'Nion, R.K., 1978, Rare earth abundances in chondritic meteorites: Geochimica et Cosmochimica Acta, v. 42, p. 1199–1212, doi: 10.1016/0016-7037(78)90114-X.

Faure, G., 1986, Principles of isotope geology: New York, Wiley, 653 p.

Faure, G., 2001, Origin of igneous rocks—The isotopic evidence: Berlin, Springer, 496 p.

Franz, G., Steiner, G., Volker, F., Pudlo, D., and Hammerschmidt, K., 1999, Plume related alkaline magmatism in central Africa—The Meidob Hills (W Sudan): Chemical Geology, v. 157, p. 27–47, doi: 10.1016/S0009-2541(98)00195-8.

Frey, F.A., McNaughton, N.J., Nelson, D.R., de Laeter, J.R., and Duncan, R.A.,

1996, Petrogenesis of the Bunbury basalt, Western Australia: Interaction between the Kerguelen plume and Gondwana lithosphere?: Earth and Planetary Science Letters, v. 144, p. 163–183, doi: 10.1016/0012-821X(96)00150-1.

Frey, F.A., Coffin, M.F., Wallace, P.J., Weis, D., and ODP Leg 183 Shipboard Scientific Party, 2000, Origin and evolution of a submarine large igneous province: The Kerguelen plateau and Broken Ridge, southern Indian Ocean: Earth and Planetary Science Letters, v. 176, p. 73–89, doi: 10.1016/S0012-821X(99)00315-5.

Ghosh, S., Chakrabarty, S., Paul, D.K., Bhalla, J.K., Bishui, P.K., and Gupta, S.N., 1994, New Rb-Sr isotopic ages and geochemistry of granitoids from Meghalaya and their significance in middle to late Proterozoic crustal evolution: Indian Minerals, v. 48, p. 33–44.

Gołani, P.R., 1991, Nangcharam fault: A major dislocation zone from western Meghalaya: Journal of the Geological Society of India, v. 37, p. 31–38.

Goldstein, S.L., O'Nions, R.K., and Hamilton, P.J., 1984, A Sm-Nd study of atmospheric dusts and particles from major river systems: Earth and Planetary Science Letters, v. 70, p. 221–236, doi: 10.1016/0012-821X(84)90007-4.

Green, D.H., 1971, Composition of basaltic magmas as indicators of conditions of origin: Application to oceanic volcanism: Philosophical Transactions of the Royal Society of London, v. 268A, p. 707–725.

Gupta, R.P., and Sen, A.K., 1988, Imprints of Ninety-East Ridge in the Shillong Plateau: Indian Shield: Tectonophysics, v. 154, p. 335–341, doi: 10.1016/0040-1951(88)90111-4.

Hamilton, D.L., Bedson, P., and Esson, J., 1989, The behavior of trace elements in the evolution of carbonatites, *in* Bell, K., ed., Carbonatites—Genesis and evolution: London, Unwin Hyman, p. 405–427.

Harmer, R.E., 1999, The petrogenetic association of carbonatite and alkaline magamatism: Constraints from the Spitskop Complex, South Africa: Journal of Petrology, v. 40, p. 525–548, doi: 10.1093/petrology/40.4.525.

Hart, S.R., 1988, Heterogeneous mantle domains: Signatures, genesis and mixing chronologies: Earth and Planetary Science Letters, v. 90, p. 273–296, doi: 10.1016/0012-821X(88)90131-8.

Hart, S.R., Hauri, E.H., Oschmann, L.A., and Whitehead, J.A., 1992, Mantle plumes and entrainment: Isotopic evidence: Science, v. 256, p. 517–520, doi: 10.1126/science.256.5056.517.

Heaman, L.M., Srivastava, R.K., and Sinha, A.K., 2002, A precise U-Pb zircon/baddeleyite age for the Jasra igneous complex, Karb-Analong district, Assam, NE India: Current Science, v. 82, p. 744–748.

Heinrich, E.W., 1966, The geology of carbonatites: Chicago, Rand McNally, 555 p.

Jacobsen, S.B., and Wasserburg, G.J., 1980, Sm-Nd isotopic evolution of chondrites: Earth and Planetary Science Letters, v. 50, p. 139–155, doi: 10.1016/0012-821X(80)90125-9.

Kalt, A., Hegner, E., and Satir, M., 1997, Nd, Sr and Pb isotopic evidence for diverse mantle sources of East Africa Rift carbonatites: Tectonophysics, v. 278, p. 31–45, doi: 10.1016/S0040-1951(97)00093-0.

Kayal, J.R., Arefiev, S.S., Barua, S., Hazarika, D., Gogoi, N., Kumar, A., Chowdhury, S.N., and Kalita, S., 2006, Shillong plateau earthquakes in northeast India region: Complex tectonic model: Current Science, v. 91, p. 109–114.

Kent, R.W., 1995, Continental and oceanic flood basalt provinces: Current and future perspectives, *in* Srivastava, R.K., and Chandra, R., eds., Magmatism in relation to diverse tectonic settings: Rotterdam, A.A. Balkema, p. 17–42.

Kent, R.W., Saunders, A.D., Kempton, P.D., and Ghose, N.C., 1997, Rajmahal basalts, eastern India: Mantle sources and melt distribution at a volcanic rifted margin, *in* Mahoney, J.J., and Coffin, M.F., eds., Large igneous provinces: Continental, oceanic and planetary flood volcanism: Washington, D.C., American Geophysical Union, Geophysical Monograph 100, p. 145–182.

Kent, R.W., Pringle, M.S., Müller, R.D., Saunders, A.D., and Ghose, N.C., 2002, $^{40}Ar/^{39}Ar$ geochronology of the Rajmahal basalts, India, and their relationship to the Kerguelen Plateau: Journal of Petrology, v. 43, p. 1141–1153, doi: 10.1093/petrology/43.7.1141.

Kramm, U., and Kogarko, L.N., 1994, Nd and Sr isotope signatures of the Khibina and Lovozero agpaitic centers, Kola alkaline province, Russia: Lithos, v. 32, p. 225–242, doi: 10.1016/0024-4937(94)90041-8.

Kumar, D., Mamallan, R., and Diwedy, K.K., 1996, Carbonatite magmatism in northeast India: Journal of Southeast Asian Earth Sciences, v. 13, p. 145–158, doi: 10.1016/0743-9547(96)00016-5.

Le Bas, M.J., 1987, Nephilinites and carbonatites, *in* Fitton, J.G., and Upton, B.G.J., eds., Alkaline igneous rocks: Geological Society of London Special Publication 30, p. 53–83.

Le Bas, M.J., 1989, Diversification of carbonatite, *in* Bell, K., ed., Carbonatites—Genesis and evolution: London, Unwin Hyman, p. 428–447.

Lee, W.J., and Wyllie, P.J., 1997, Liquid immiscibility in the join $NaAlSiO_4$-$NaAlSiO_8$-$CaCO_3$ at 1 Gpa: Implications for crustal carbonatites: Journal of Petrology, v. 38, p. 1113–1135, doi: 10.1093/petrology/38.9.1113.

Le Maitre, R.W., 2002, Igneous rocks: A classification and glossary of terms (2nd edition): Cambridge, Cambridge University Press, 236 p.

Luhr, J.F., 1997, Extensional tectonics and the diverse primitive volcanic rocks in the western Mexican volcanic belt: Canadian Mineralogist, v. 35, p. 473–500.

Mahoney, J.J., Macdougall, J.D., Lugmair, G.W., Gopalan, K., and Krishnamurthy, P., 1985, Origin of contemporaneous tholeiitic and K-rich alkalic lavas: A case study from the northern Deccan plateau, India: Earth and Planetary Science Letters, v. 72, p. 39–53.

Mahoney, J.J., le Roex, A.P., Peng, Z., Fisher, R.L., and Natland, J.H., 1992, Southwestern limits of Indian Ocean Ridge mantle and the origin of low $^{206}Pb/^{204}Pb$ mid-ocean ridge basalt: Isotope systematics of the Central Southwest Indian Ridge (17°–50°E): Journal of Geophysical Research, v. 97, p. 19,771–19,790.

Mahoney, J.J., Jones, W.B., Frey, F.A., Salters, V.J.M., Pyle, D.G., and Davies, H.L., 1995, Geochemical characteristics of lavas from Broken Ridge, the Naturaliste Plateau and southernmost Kerguelen Plateau: Cretaceous plateau volcanism in the southeast Indian Ocean: Chemical Geology, v. 120, p. 315–345, doi: 10.1016/0009-2541(94)00144-W.

Mamallan, R., Kumar, D., and Bajpai, R.K., 1994, Jasra ultramafic-mafic-alkaline complex: A new find in the Shillong plateau, northeastern India: Current Science, v. 66, p. 64–65.

Mazumdar, S.K., 1976, A summary of the Precambrian geology of the Khasi Hills, Meghalaya: Geological Survey of India Miscellaneous Publication, v. 23, no. 2, p. 311–334.

McDonough, W.F., Sun, S.-S., Ringwood, A.E., Jagoutz, E., and Hofmann, A.W., 1992, K, Rb and Cs in the earth and moon and the evolution of the earth's mantle: Geochimica et Cosmochimica Acta, v. 56, p. 1001–1012, doi: 10.1016/0016-7037(92)90043-I.

Michard, A., Montigny, R., and Schlich, R., 1986, Geochemistry of the mantle beneath the Rodriguez triple junction and the South-East Indian Ridge: Earth and Planetary Science Letters, v. 78, p. 104–114, doi: 10.1016/0012-821X(86)90176-7.

Middlemost, E.A.K., 1989, Iron oxidation ratios, norms and the classification of volcanic rocks: Chemical Geology, v. 77, p. 19–26, doi: 10.1016/0009-2541(89)90011-9.

Nandy, D.R., 1980, Tectonic patterns in northeastern India: Indian Journal of Earth Sciences, v. 7, p. 103–107.

Pyle, D.G., Christie, D.M., Mahoney, J.J., and Duncan, R.A., 1995, Geochemistry and geochronology of ancient southeast Indian Ocean and southwest Pacific Ocean seafloor: Journal of Geophysical Research, v. B100, p. 22,261–22,282, doi: 10.1029/95JB01424.

Ray, J.S., Ramesh, R., and Pande, K., 1999, Carbon isotopes in Kerguelen plume-derived carbonatites: Evidence for recycled inorganic carbon: Earth and Planetary Science Letters, v. 170, p. 205–214, doi: 10.1016/S0012-821X(99)00112-0.

Ray, J.S., Trivedi, J.R., and Dayal, A.M., 2000, Strontium isotope systematics of Amba Dongar and Sung Valley carbonatite-alkaline complexes, India: Evidence for liquid immiscibility, crustal contamination and long-lived Rb/Sr enriched mantle source: Journal of Asian Earth Sciences, v. 18, p. 585–594, doi: 10.1016/S1367-9120(99)00072-3.

Sarkar, A., Datta, A.K., Poddar, B.K., Bhattacharyya, B.K., Kollapuri, V.K., and Sanwal, R., 1996, Geochronological studies of Mesozoic igneous rocks from eastern India: Journal of Southeast Asian Earth Sciences, v. 13, p. 77–81, doi: 10.1016/0743-9547(96)00009-8.

Saunders, A.D., Storey, M., Kent, R.W., and Norry, M.J., 1992, Consequences of plume-lithosphere interactions, *in* Storey, B.C., et al., eds., Magmatism and the causes of continental break-up: Geological Society of London Special Publication 68, p. 41–60.

Sheth, H.C., Torres-Alvarado, I.S., and Verma, S.P., 2000, Beyond subduction and plumes: A unified tectono-petrogenetic model for the Mexican volcanic belt: International Geology Review, v. 42, p. 1116–1132.

Simonetti, A., Bell, K., and Viladkar, S.G., 1995, Isotopic data from the Amba Dongar Carbonatite Complex, west-central India: Evidence for an enriched mantle source: Chemical Geology, v. 122, p. 185–198, doi: 10.1016/0009-2541(95)00004-6.

Simonetti, A., Goldstein, S.L., Schmidberger, S.S., and Viladkar, S.G., 1998, Geochemical and Nd, Pb and Sr isotope data from Deccan alkaline complexes—Inferences for mantle sources and plume-lithosphere interaction: Journal of Petrology, v. 39, p. 1847–1864, doi: 10.1093/petrology/39.11.1847.

Srivastava, R.K., and Sinha, A.K., 2004, Geochemistry of early Cretaceous alkaline ultramafic-mafic complex from Jasra, Karbi Anglong, Shillong plateau, Northeastern India: Gondwana Research, v. 7, p. 549–561, doi: 10.1016/S1342-937X(05)70805-4.

Srivastava, R.K., Heaman, L.M., Sinha, A.K., and Shihua, S., 2005, Emplacement age and isotope geochemistry of Sung Valley Alkaline-Carbonatite Complex, Shillong Plateau, Northeastern India: Implications for primary carbonate melt and genesis of the associated silicate rocks: Lithos, v. 81, p. 33–54, doi: 10.1016/j.lithos.2004.09.017.

Storey, M., Kent, R.W., Saunders, A.D., Hergt, J., Salters, V.J.M., Whitechurch, H., Sevigny, J.H., Thirlwall, M.F., Leat, P., Ghose, N.C., and Gifford, M., 1992, Lower Cretaceous volcanic rocks on continental margins and their relationship to the Kerguelen Plateau: Proceedings of the Ocean Drilling Program: Scientific Results, v. 120, p. 33–53.

Sweeney, R.J., 1994, Carbonatite melt compositions in the earth's mantle: Earth and Planetary Science Letters, v. 128, p. 259–270, doi: 10.1016/0012-821X(94)90149-X.

Tilton, G.R., and Bell, K., 1994, Sr-Nd Pb isotope relationships in Late Archean carbonatites and alkaline complexes: Applications to the geochemical evolution of the Archean mantle: Geochimica et Cosmochimica Acta, v. 58, p. 3145–3154, doi: 10.1016/0016-7037(94)90042-6.

Upton, B.G.J., 1967, Alkaline pyroxenites, *in* Wyllie, P.J., ed., Ultramafics and related rocks: New York, John Wiley and Sons, p. 281–288.

Velasco-Tapia, F., and Verma, S.P., 2001, First partial melting inversion model for a rift-related origin of the Sierra de Chichinautzin volcanic field, central Mexican volcanic belt: International Geology Review, v. 43, p. 788–817.

Verma, S.P., 2006, Extension-related origin of magmas from a garnet-bearing source in the Los Tuxtlas volcanic field, Mexico: International Journal of Earth Sciences, v. 95, p. 871–901, doi: 10.1007/s00531-006-0072-z.

Verma, S.P., Torres-Alvarado, I.S., and Sotelo-Rodríguez, Z.T., 2002, SINCLAS: Standard igneous norm and volcanic rock classification system: Computers and Geosciences, v. 28, p. 711–715, doi: 10.1016/S0098-3004(01)00087-5.

Verma, S.P., Guevara, M., and Agrawal, S., 2006, Discriminating four tectonic settings: Five new geochemical diagrams for basic and ultrabasic volcanic rocks based on log-ratio transformation of major-element data: Journal of Earth System Science, v. 115, p. 485–528.

Wallace, M.E., and Green, D.H., 1988, An experimental determination of primary carbonatite magma composition: Nature, v. 335, p. 343–346, doi: 10.1038/335343a0.

Wilson, M., 1989, Igneous petrogenesis: London, Chapman and Hall, 466 p.

Winter, J.D., 2001, An introduction to igneous and metamorphic petrology: Englewood Cliffs, New Jersey, Prentice Hall, 697 p.

Wyllie, P.J., and Lee, W.-J., 1998, Model system controls on conditions for formation of magnesiocarbonatite and calciocarbonatite magmas from the mantle: Journal of Petrology, v. 39, p. 1885–1893, doi: 10.1093/petrology/39.11.1885.

Zindler, A., and Hart, S.R., 1986, Chemical geodynamics: Annual Review of Earth and Planetary Sciences, v. 14, p. 493–571, doi: 10.1146/annurev.ea.14.050186.002425.

MANUSCRIPT ACCEPTED BY THE SOCIETY JANUARY 31, 2007

DISCUSSION

22 December 2006, Surendra P. Verma

Srivastava and Sinha present new data on major and trace elements as well as Sr and Nd isotopes for ca. 105 Ma rocks from the poorly studied Shillong plateau of northeastern India and interpret them in terms of a Kerguelen hotspot-mantle plume system proposed to have been established since ca. 132 Ma. Briefly, this "plume relationship" is mainly based on (1) the fact that Shillong plateau geochemistry does not suggest any direct genetic relationship among the different rock types; (2) low (<1) values of the agpaitic index of differentiated nepheline syenite magmas, which suggest CO_2-related activity in their genesis; (3) major element chemistry showing an OIB affinity; and (4) inferred initial $^{87}Sr/^{86}Sr$ and $^{143}Nd/^{144}Nd$ data that suggest mixing of HIMU, EMI, and EMII mantle components.

None of the above-mentioned features actually demands (or even supports) a plume relationship. The lack of a genetic relationship between different rock types may have to be explained in terms of heterogeneous mantle sources and crustal processes as well as more complex petrogenetic models than attempted by the authors. The agpaitic index based on the major elements Na, K, and Al, without any reference to the actual CO_2 contents, can hardly be used to suggest involvement of CO_2-related activity. The OIB affinity per se cannot be interpreted as exclusively plume-related (see, for example, Fitton, this volume). The interpretation of Sr and Nd isotope data solely in terms of mantle components is based on the assumption that crustal contamination was insignificant, which may not be the case. The thickness and type of continental crust underlying the Shillong plateau may have to be investigated and taken into account. Furthermore, their interpretation of Sr and Nd isotope data is not unique in terms of these three mantle components (HIMU, EMI, and EMII in their Fig. 6A); other combinations, such as the one formed by BSE, EMI, and EMII, are equally feasible. Finally, the spatial relationship of Shillong plateau magmatism with the Kerguelen plume is not clear from their Figure 1; for clarification, plate reconstructions for the past and their uncertainties will have to be carefully analyzed.

The phrase "plume related" in the title of the chapter by Srivastava and Sinha therefore reflects the authors' preference for the plume hypothesis rather than an evaluation of all plausible models.

1 January 2007, Rajesh K. Srivastava and Anup K. Sinha

First we thank Verma for his thoughtful interest in our work. Verma raises a question on the genetic association of the studied early Cretaceous Jasra alkaline complex with the Kerguelen hotspot-mantle plume system of the Indian Ocean. This association has already been established by many previous workers (Mahoney et al., 1992; Storey et al., 1992; Kumar et al., 1996; Kent et al., 1997; Ray et al., 1999; Coffin et al., 2002; Duncan, 2002; Kent et al., 2002; Srivastava et al., 2005; and many more). In our article we present additional data that further support this association. The studied complex comes from a region where many magmatic rocks are reported, and all these are thought to be associated with the Kerguelen plume. Most of these rocks were emplaced in the period 105–115 Ma. Kerguelen-plume-derived rocks are found not only on the Shillong plateau but also in adjacent western regions over and within Gondwana basins. The rocks of the present study show a similar emplacement age and similar geochemical and isotopic compositions. Thus, we favor the plume model. Another point is that there is no geological or geophysical evidence available that supports a subduction or rift setting for the Shillong plateau (Evans, 1964; Desikachar, 1974; Nandy, 1980; Gupta and Sen, 1988; Das Gupta and Biswas, 2000; Kayal et al., 2006).

5 January 2007, Don L. Anderson

A plume origin of continental basalts (CFB) is often based on similarities with ocean island basalts (Hooper et al., this volume; Srivastava and Sinha, this volume). The continental origin of some ocean island basalts is based on the reverse argument (see references in Anderson, this volume). The delamination of lower continental crust can explain the similarity in composition of these paired igneous provinces (Anderson, this volume) plus the asthenospheric signature in CFB. Asthenosphere upwells as the eclogitic lower crust sinks. At the same time, the absence of precursory uplift, heatflow anomalies, and high magma temperatures are also explained. The continental affinity of Kerguelen is well known.

30 January 2007, Kamal K. Sharma

Srivastava and Sinha (this volume) endorse spatial and temporal association of the Jasra igneous complex, Shillong plateau, northeastern India, with Kerguelen plume activity. They also state that the absence of evidence for extensional tectonics supports the plume model for the genesis of Jasra igneous rocks. The magmatic activity of the Kerguelen plateau, Broken Ridge, Naturaliste plateau, Bunbury, Rajmahal-Sylhet, and Ninety-East

Ridge are interpreted as Kerguelen hotspot activity. I make the following comments:

1. The origin of the Rajmahal-Sylhet Trap, Bengal basin, Eighty- and Ninety-East Degree ridges, Andaman-Nicobar Ridge, and eastward subduction are important tectonic features of the southeastern Indian plate, which evolved through the prolonged period of time of approximately 120 Ma. The tectonic evolution of these features cannot be explained as Kerguelen hotspot activity. The hotspot model for the evolution of the Rajmahal Traps and/or the break-up of Gondwanaland is a matter of conjecture and dispute. The controversy still exists as to whether the supposed plume is at Crozet Rise or Kerguelen and whether the path was associated with Eighty East Degrees or Ninety East Degrees, respectively (Baksi, 1994; Subrahmanyam et al., 1999; Singh et al., 2004). In view of this, it is not feasible to explain the Jasra igneous complex with the Kerguelen hotspot model.

2. Mukhopadhyay et al. (1986) and Mukhopadhyay (2000) suggested that the structural style, Gondwana sedimentation, and volcanism imply that the Rajmahal Traps evolved through decompressional melting subsequent to rifting of Gondwanaland. Intertrappean shales and sandstones are reported along with traps from Rajmahal (Naqvi, 2005), which indicates subsidence of a basin and an extensional tectonic regime during the early Cetaceous. Correlation of the Jasra igneous complex with the Rajmahal Traps by Srivastava and Sinha (this volume) accords a similar extensional tectonic setting there. The association of the Jurassic–Early Cretaceous north-south trending faults and lineaments of the Shillong plateau with Early Cretaceous igneous intrusions, including alkaline and carbonatite complexes, is clearly indicative of extensional tectonism rather than plume–lithosphere interaction.

3. There is debate regarding the continental or oceanic origin of the Kerguelen plateau. The presence of granitic and syenitic plutons suggests a continental origin. Thus, Kerguelen represents a continental mass at the Antarctic margin during early opening of the Indian Ocean (Naqvi, 2005, and references therein). Figure 1 in that article shows the Southwest Indian Ridge as divider between Kerguelen and the Ninety Degree Ridge and other tectonic elements. For all these reasons, associating Kerguelen hotspot activity with the Jasra igneous complex seems inappropriate. Almost similar Ar-Ar age data from Jasra, Rajmahal, Broken Ridge, and Bunbury, western Australia, compel one to consider alternative, nonplume mechanisms for their evolution.

4 February 2007, Rajesh K. Srivastava and Anup K. Sinha

We agree with the comments made by Anderson (5 January 2007).

Sharma (comment 30 January 2007) doubts the relationship between the Jasra igneous complex and the Kerguelen hotspot–mantle plume system. We appreciate his interest in our work, but his comments are exclusively based on the Rajmahal Traps and Gondwana sedimentation. We do not compare our data with Andaman-Nicobar Ridge volcanism. He ignores other geological facts (mentioned in our chapter and others) regarding igneous activity associated with the Kerguelen hotspot–mantle plume system. Even for the Rajmahal-Sylhet Traps, Kent et al. (1997, 2002) clearly associated the basalts with the Kerguelen plume. There are many articles (e.g., in the Special Volume of Journal of Petrology, v. 43, 2002) related to magmatism associated with the Kerguelen hotspot–mantle plume system. The other igneous complexes (including the famous Sung Valley igneous complex) reported from the Shillong plateau also clearly show their association with the Kerguelen hotspot–mantle plume system. These observations are based on detailed geological and geochemical (including isotopic) data.

We, on the basis of data from the Jasra igneous complex and also data available from other, similar complexes of the Shillong plateau, reached the conclusion that the Jasra complex has very similar geological and geochemical signatures to those observed in other magmatic rocks associated with the Kerguelen hotspot–mantle plume system of the Indian Ocean. Available data on the Jasra igneous complex are closer to "plume-related" magmatism than to extensional tectonism. Of course, further detailed work is required before an unquestionable conclusion can be reached.

REFERENCES CITED

Anderson, D.L., 2007 (this volume), The eclogite engine: Chemical geodynamics as a Galileo thermometer, *in* Foulger, G.R., and Jurdy, D.M., eds., Plates, plumes, and planetary processes: Boulder, Colorado, Geological Society of America Special Paper 430, doi: 10.1130/2007.2430(03).

Baksi, A.K., 1994, Revised plate motions relative to the hotspots from combined Atlantic and Indian Ocean hotspot tracks: Comment and reply: Geology, v. 22, p. 276–277.

Coffin, M.F., Pringle, M.S., Duncan, R.A., Gladezenko, T.P., Storey, M., Müller, R.D., and Gahagan, L.A., 2002, Kerguelen hotspot magma output since 130 Ma: Journal of Petrology, v. 43, p. 1121–1139.

Das Gupta, A.B., and Biswas, A.K., 2000, Geology of Assam: Bangalore, Geological Society of India, 169 p.

Desikachar, S.V., 1974, A review of the tectonic and geological history of eastern India in terms of plate tectonic theory: Journal of the Geological Society of India, v. 15, p. 137–149.

Duncan, R.A., 2002, A time frame for construction of the Kerguelen Plateau and Broken Ridge: Journal of Petrology, v. 43, p. 1109–1119.

Evans, P., 1964, The tectonic framework of Assam: Journal of the Geological Society of India, v. 5, p. 80–96.

Fitton, J.G., 2007 (this volume), The OIB paradox, *in* Foulger, G.R, and Jurdy, D.M., eds., Plates, plumes, and planetary processes: Boulder, Colorado, Geological Society of America Special Paper 430, doi: 10.1130/2007 .2430(20).

Gupta, R.P., and Sen, A.K., 1988, Imprints of Ninety-East Ridge in the Shillong Plateau, Indian Shield: Tectonophysics, v. 154, p. 335–341.

Hooper, P.R., Camp, V.E., Reidel, S.P., and Ross, M.E., 2007 (this volume), The origin of the Columbia River flood basalt province: Plume versus non-

plume models, *in* Foulger, G.R., and Jurdy, D.M., eds., Plates, plumes, and planetary processes: Boulder, Colorado, Geological Society of America Special Paper 430, doi: 10.1130/2007.2430(30).

Kayal, J.R., Arefiev, S.S., Barua, S., Hazarika, D., Gogoi, N., Kumar, A., Chowdhury, S.N., and Kalita, S., 2006, Shillong plateau earthquakes in northeast India region: Complex tectonic model: Current Science, v. 91, p. 109–114.

Kent, R.W., Saunders, A.D., Kempton, P.D., and Ghose, N.C., 1997, Rajmahal basalts, eastern India: Mantle sources and melt distribution at a volcanic rifted margin, *in* Mahoney, J.J., and Coffin, M.F., eds., Large igneous provinces: Continental, oceanic and planetary flood volcanism: Washington, D.C., American Geophysical Union Geophysical Monograph 100, p. 145–182.

Kent, R.W., Pringle, M.S., Müller, R.D., Saunders, A.D., and Ghose, N.C., 2002, $^{40}Ar/^{39}Ar$ geochronology of the Rajmahal basalts, India, and their relationship to the Kerguelen Plateau: Journal of Petrology, v. 43, p. 1141–1153.

Kumar, D., Mamallan, R., and Diwedy, K.K., 1996, Carbonatite magmatism in northeast India: Journal of Southeast Asian Earth Sciences, v. 13, p. 145–158.

Mahoney, J.J., le Roex, A.P., Peng, Z., Fisher, R.L., and Natland, J.H., 1992, Southwestern limits of Indian Ocean Ridge mantle and the origin of low $^{206}Pb/^{204}Pb$ mid-ocean ridge basalt: Isotope systematics of the Central Southwest Indian Ridge (17°–50°E): Journal of Geophysical Research, v. 97, p. 19,771–19,790.

Mukhopadhyay, M., 2000, Deep crustal structure of the West Bengal basin deduced from gravity and DSS data: Journal of the Geological Society of India, v. 56, p. 351–364.

Mukhopadhyay, M., Verma, R.K., and Ashraf, M.H., 1986, Gravity field and structures of Rajmahal hills: Examples of the Palaeo-Mesozoic continental margin in eastern India: Tectonophysics, v. 131, p. 353–367.

Nandy, D.R., 1980, Tectonic patterns in northeastern India: Indian Journal of Earth Sciences, v. 7, p. 103–107.

Naqvi, S.M., 2005, Geology and evolution of the Indian plate: New Delhi, Capital Publishing, 448 p.

Ray, J.S., Ramesh, R., and Pande, K., 1999, Carbon isotopes in Kerguelen plume-derived carbonatites: Evidence for recycled inorganic carbon: Earth and Planetary Science Letters, v. 170, p. 205–214.

Singh, A.P., Kumar, N., and Singh, B., 2004, Magmatic underplating beneath the Rajmahal Traps: Gravity signature and derived 3-D configuration: Proceedings of the Indian Academy of Sciences (Earth and Planetary Science), v. 113, p. 759–769.

Srivastava, R.K., and Sinha, A.K., 2007 (this volume), Nd and Sr isotope systematics and geochemistry of a plume-related Early Cretaceous alkaline-mafic-ultramafic igneous complex from Jasra, Shillong plateau, northeastern India, *in* Foulger, G.R., and Jurdy, D.M., eds., Plates, plumes, and planetary processes: Boulder, Colorado, Geological Society of America Special Paper 430, doi: 10.1130/2007.2430(37).

Srivastava, R.K., Heaman, L.M., Sinha, A.K., and Shihua, S., 2005, Emplacement age and isotope geochemistry of Sung Valley alkaline-carbonatite complex, Shillong Plateau, Northeastern India: Implications for primary carbonate melt and genesis of the associated silicate rocks: Lithos, v. 81, p. 33–54.

Storey, M., Kent, R.W., Saunders, A.D., Salters, V.J., Hergt, J., Whitechurch, H., Sevigny, J.H., Thirlwall, M.F., Leat, P., Ghose, N.C., and Gifford, M., 1992, Lower Cretaceous volcanic rocks on continental margins and their relationship to the Kerguelen Plateau, *in* Wise, S.W., and Schlich, R., Jr., eds., Proceedings of the Ocean Drilling Program, scientific results, v. 120: College Station, Texas Ocean Drilling Program, p. 33–53.

Subrahmanyam, C., Thakur, N.K., Gangadhara Rao, T., Khanna. R., Ramana. M.V., and Subrahmanyam, V., 1999, Tectonics of the Bay of Bengal: New insights from satellite gravity and ship borne geophysical data: Earth and Planetary Sciences Letters, v. 171, p. 237–251.

The Geological Society of America
Special Paper 430
2007

A *bimodal large igneous province and the plume debate: The Paleoproterozoic Dongargarh Group, central India*

Sarajit Sensarma*

National Facility for Geochemical Research, School of Environmental Sciences, Jawaharlal Nehru University, New Delhi 110 067, India

ABSTRACT

A bimodal large igneous province (LIP) with subequal volumes of nearly coeval felsic-mafic volcanic rocks occurs in the ca. 2500 Ma Dongargarh Group of central India, perhaps the only LIP of this kind known to date. Although some of its features match the expectations of the plume model, the longer eruption times ($\sim\leq$30–73 Ma) and bimodal distribution of lava types do not. Melting of crust and mantle, driven by a common thermal perturbation in an extensional tectonic setting, and interactions of the crustal and mantle melts gave rise to the province. This contrasts with contemporary mantle-melting or crust-melting models for LIP genesis.

Keywords: rhyolite, high-Mg basalts, crust-mantle, large igneous province (LIP), plume, Paleoproterozoic, India

INTRODUCTION

Large igneous provinces (LIPs) are generally considered to be of two types:

1. Mafic LIPs (MLIPs; Bryan and Ernst, 2006), also called large basaltic provinces (LBPs; Sheth, 2006), as typified by continental flood basalts (CFB), which originated during short-duration eruptive events of mantle melts with volumetrically insignificant felsic lavas erupting late in the history of volcanism (e.g., the Deccan Traps; Melluso et al., 2006, and references therein).

2. Silicic-dominated igneous provinces (SLIPs) with <10% basalts (e.g., the Early Cretaceous volcanic rifted margin in eastern Australia; e.g., Bryan et al., 2002), the Jurassic Chon Aike province in South America, the Antarctic peninsula (e.g., Pankhurst et al., 1998; Riley and Leat, 1999), and the Neoproterozoic Malani igneous province in northwestern

India (e.g., Sharma, 2004). These are thought to be the products of crustal melting, with the mantle providing the heat. SLIPs may have comparable eruptive volumes to CFBs, but their duration of emplacement may be up to 40–60 m.y. (e.g., Ernst and Buchan, 2001; Bryan et al., 2002).

Here I describe a third type of LIP from central India—a bimodal LIP with subequal proportions of felsic and mafic volcanic rocks, the Paleoproterozoic Dongargarh province (Sensarma et al., 2004; Sensarma, 2005). It is perhaps the only bimodal LIP known to date (e.g., Foulger, this volume), hitherto considered absent from the geological record (Bryan et al., 2002). This province has felsic rocks early in the sequence, something not commonly reported.

Contemporary models for LIP genesis consider thermal anomalies in the form of mantle plumes rising from deep within the mantle (e.g., Morgan, 1971; Courtillot et al., 1999; Campbell, 2005; Ernst et al., 2005), compositional heterogeneity and

---

*E-mail: sensarma2002@yahoo.co.in.

enhanced fertility in the upper mantle coupled with processes consequential to plate tectonics (e.g., Foulger and Anderson, 2005; Foulger et al., 2005), and lower-crustal delamination that recycles continental mantle lithosphere into the asthenosphere and triggers eruptions of flood basalts (e.g., Tanton and Hager, 2000; Lustrino, 2005). The purpose of the present contribution is to highlight the uniqueness of the Dongargarh province, where substantial melting of both the crust and mantle was apparently driven by a common thermal perturbation.

I first critically appraise stratigraphic, geochronological, petrographic, and geochemical constraints to illustrate how different source regions were involved and interacted over a period of ~30–73 m.y. at ca. 2500 Ma. I then evaluate features of the Dongargarh LIP in terms of the predictions of the plume model and briefly discuss the implications. The major conclusions are that LIPs are more diverse than previously thought and that large-scale crust-mantle interactions could be a plausible mechanism for LIP genesis that has not, to date, been considered in contemporary models.

GEOLOGIC SETTING

The Dongargarh volcanic-sedimentary succession in the central Indian craton called the Dongargarh Group (Fig. 1) is 10–12 km thick and 80–100 km wide and extends southerly for ~300 km into the Kotri area (Bastar). The Dongargarh Group is folded into the regional "Sitagota syncline" and is metamorphosed to low-grade greenschist-facies. Magmatism in the belt took place in an extensional setting (e.g., Krishnamurthy et al., 1990; Sensarma et al., 2000, 2004), and is presumably related to global Neoarchean–Paleoproterozoic rifting (e.g., Blake and Groves, 1987).

The lithostratigraphic units within the Dongargarh Group (Fig. 2) form a structurally concordant sequence without any regional unconformity (Sensarma and Mukhopadhyay, 2003). The volcanics (Fig. 3) comprise subequal volumes of felsic pyroclastic rocks (average SiO_2 ~75 wt%, ≤4 km thick) and mafic lava flows that together constitute about two-thirds of the preserved stratigraphy. The basal felsic volcanic rocks (the Bijli rhyolite) are followed by three mafic volcanic rock formations interspersed with volcanogenic wacke and quartz arenite, respectively. Two mafic volcanic formations, the Pitepani volcanics (~1 km thick) and the more voluminous younger Sitagota volcanics (>3 km thick) (Fig. 2), have interlayered high-Mg basalts (MgO ~ 7.5–12 wt%) and low-Mg basalts (MgO ~6 wt%). Volumetrically minor andesite (Sensarma, 2001) to basaltic andesite (Neogi et al., 1996) (<1 km thick) constitutes the youngest unit. Granitic plutons (the Dongargarh granite) intrude the Dongargarh Group.

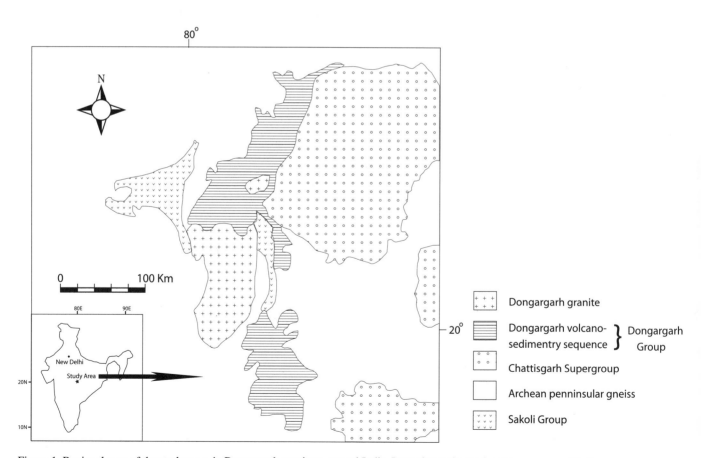

Figure 1. Regional map of the study area, in Dongargarh province, central India. Inset shows the study area on a map of India.

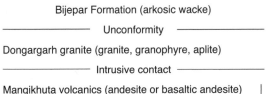

Bijepar Formation (arkosic wacke)

——————— Unconformity ———————

Dongargarh granite (granite, granophyre, aplite)

——————— Intrusive contact ———————

Mangikhuta volcanics (andesite or basaltic andesite)
Karutola Formation (quartz arenite)
Sitagota volcanics (high-Mg basalts–low-Mg basalts)   **Dongargarh**
Chandsuraj Formation (pebbly volcanogenic wacke)   **Group**
Pitepani volcanics (high-Mg basalts–low-Mg basalts)
Bijli rhyolite (felsic pyroclastic rocks)
    (with Halbitola rhyolitic breccioconglomerate member)

Basement?

Figure 2. Lithostratigraphic succession in the Dongargarh province, central India. The principal lithology of each of the stratigraphic units is given in parentheses. From Sensarma and Mukhopadhyay (2003).

The felsic volcanic rocks underlie the mafic lava flows in the Kotri area also. The Amgaon Gneissic Complex, with vestiges of tonalite-trondhjemite-granodiorite suite of rocks (U-Pb zircon date 3562 ± 2 Ma), is unconformably overlain by the volcanics in the Kotri area, and thus possibly constitutes the Archean basement to the Kotri-Dongargarh rocks (Ghosh, 2004). However, our field mapping in the northern part of the belt could not confirm this observation (Sensarma, 2001; Sensarma and Mukhopadhyay, 2003).

Although the Dongargarh granite intruded the Dongargarh Group, the rhyolites and the Dongargarh granite have close geochemical similarities and are products of the same tectonothermal event (see Roy et al., 2000, and references therein). Hence, the felsic volcanic-plutonic activity in the belt is approximately coeval; the Dongargarh granite represents a continuation of the magmatic event that initially produced the rhyolites. The high- and low-Mg basalts are compositionally distinct (e.g., Fig. 4A and C). A genetic connection exists between the felsic volcanic rocks and the basalts in both the Pitepani volcanics and the Sitagota volcanics, because the compositions of all of these either

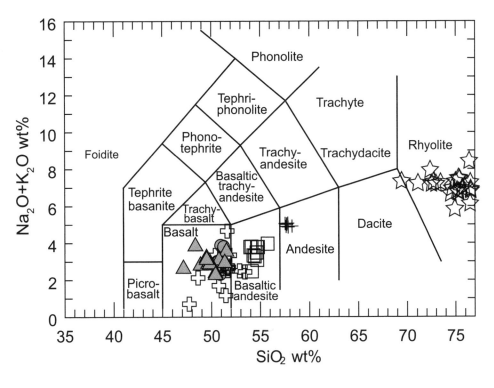

Figure 3. The bimodal compositions of the volcanic rocks in the Dongargarh province. PV—Pitepani volcanics; SHMB—siliceous high-magnesian basalts; SV—Sitagota volcanics; MV—Mangikhuta volcanics. The stratigraphic positions of the volcanic formations are given in Figure 2. After Le Bas et al. (1986).

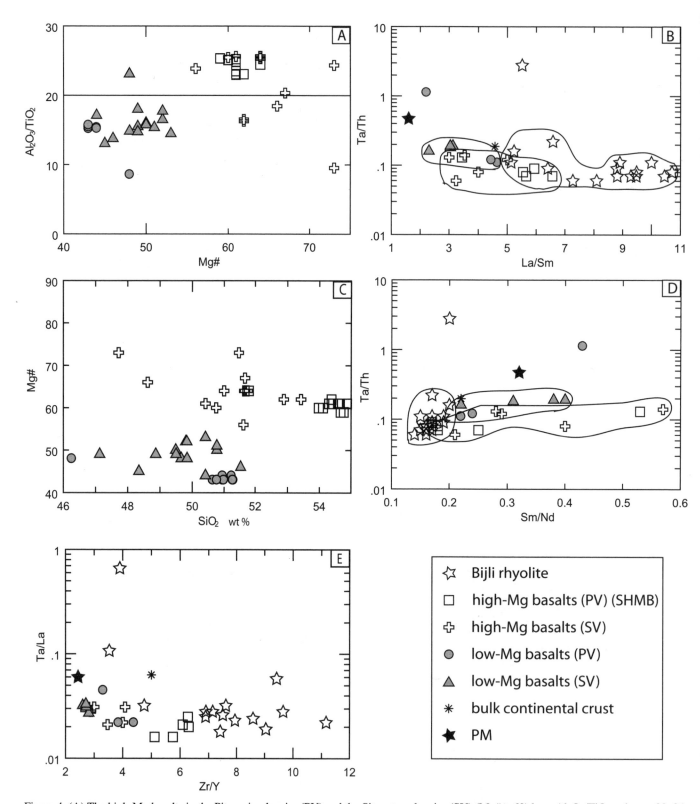

Figure 4. (A) The high-Mg basalts in the Pitepani volcanics (PV) and the Sitagota volcanics (SV) (Mg# ≥ 60) have Al_2O_3/TiO_2 values ~ 20–24, whereas low-Mg basalts (Mg# ≤ 52) have Al_2O_3/TiO_2 ratios ≤ 16. Mg# = 100 × (Mg/[Mg + Fe^{2+}]). (B) A La/Sm versus Ta/Th plot illustrating overlapping La/Sm values for the Bijli rhyolites and high-Mg basalts, particularly for high-Mg basalts (the Pitepani volcanics), at Ta/Th (0.06–0.08), common in granulites. Low-Mg basalts, however, have lower La/Sm values and higher Ta/Th (0.1–0.2) values. (C) The high- and low-Mg basalts are compositionally distinct on a plot of SiO_2 wt% versus Mg#. The high-Mg basalts in the Pitepani volcanics have higher SiO_2 concentrations (~54 wt%) compared to those of high-Mg basalts (SiO_2 ~ 50–52 wt%) in the Sitagota volcanics at comparable Mg# (60–64). Low-

vary within close range or show smooth trends on immobile refractory element ratio plots such as, e.g., Ta/Th, La/Sm, Sm/Nd, Zr/Y, and Ta/La plots (Fig. 4B, D, and E). The mafic rocks are more tholeiitic than komatiitic, and the relationship between the Pitepani volcanics and the Sitagota volcanics is discussed in the next section.

COMPOSITIONS OF THE VOLCANIC ROCKS

The geochemical data discussed here include those of Asthana et al. (1996), Neogi et al. (1996), and Sensarma (2001). In the last of these, concentrations of major oxides and trace elements such as Rb, Ba, Sr, Zr, Y, Ni, and Cr were determined using XRF (X-ray fluorescence spectroscopy), whereas other trace and rare earth elements were determined using INAA (instrumental neutron activation analysis) at the Institut für Mineralogie und Geochemie, Cologne, Germany.

The geochemical details of high-silica rhyolite are provided in Sensarma (2001) and Sensarma et al. (2004). These are characterized by high contents of silica ($SiO_2 \geq 74$ wt%), large-ion lithosphere elements (LILE) (Rb > 100 ppm, Cs > 1 ppm, K > 2.4 wt%), high field-strength elements (HFSE) (Zr > 270 ppm, Nb > 18 ppm, Ta > 1.5 ppm, Hf > 4.1 ppm, Y > 28 ppm), and rare earth elements (REE) (La > 31.9 ppm, Sm > 6.1 ppm, Eu > 0.17 ppm, Lu > 0.36 ppm). In addition, they have $Na_2O + K_2O$ > 6.8 wt%, FeO_t/MgO > 8. Combined with low contents of CaO (≤ 1 wt%), Ba (<1000 ppm), and Sr (<80 ppm), these characteristics point to an "A-type" granitic composition. A high eruption temperature (900–950 °C) is indicated for the melts (Sensarma et al., 2004). The Bijli rhyolite is the only preserved felsic volcanic unit in the area.

The mafic volcanics in the Pitepani volcanics and the Sitagota volcanics, separated by volcanogenic wacke of the Chandsuraj Formation (Fig. 2), are intrinsically similar in composition. The high-Mg basalts are generally characterized by MgO ~ 7.5-12 wt% at SiO_2 ~ 50–52 wt%; the maximum MgO measured is 15.3 wt% (sample no. 8, Neogi et al., 1996). These rocks have $Na_2O + K_2O$ > 2 wt%, FeOt ~ 9.5 wt%, TiO_2 ~ 0.5 wt%, Al_2O_3/TiO_2 ~ 20–24 wt% (Fig. 4A), CaO/Al_2O_3 ~ 0.8, Mg# $\geq$ 60, and Ni up to 250 ppm and are subalkaline basalts to basaltic

andesites (Fig. 3). On the basis of International Union of Geological Sciences classification for high-Mg volcanic rocks, the rocks can be classified as basalts, with a few samples as picrite (samples with MgO > 12 wt%).

The high-Mg basalts in the Pitepani volcanics have higher SiO_2 (~54 wt%; Fig. 4C), LILE (e.g., Rb, Ba, Sr), and light REE (LREE) contents at a given Mg#. These rocks differ from other common volcanic rocks, including boninite, and are designated siliceous high-magnesian basalts (SHMB) following Sensarma et al., (2002, and references therein). The mantle-derived Koolau and Lanai tholeiitic basalts in Hawaii are known to have high SiO_2 contents (SiO_2 > 51 wt%). When compared with high-silica Koolau (Frey et al., 1994) and Lanai (West et al., 1992) lavas, the SHMB in the Dongargarh province have higher SiO_2 ($\geq$54 wt% versus 51–53 wt%), higher alkali ($Na_2O + K_2O$, ~3.5 wt% versus $\leq$3.5 wt%), but lower FeO (~9.5 wt% versus 10–12 wt%) and Al_2O_3 ($\leq$13 wt% versus $\geq$13 wt%) and significantly low TiO_2 (~0.5 wt% versus ~2 wt%) values at comparable MgO contents. Generally lower SiO_2, LILE, and LREE contents, but higher MgO, Mg#, Ni, and Cr values, indicate a more primitive character for the Sitagota volcanics high-Mg basalts.

The low-Mg basalts in the Pitepani volcanics and the Sitagota volcanics have MgO 6–8 wt% at SiO_2 ~ 48–50 wt%, FeOt ~ 12 wt%, TiO_2 ~ 1 wt%, Mg# $\leq$ 52, Ni < 150 ppm, Al_2O_3/TiO_2 $\leq$16 (Fig. 4A), and CaO/Al_2O_3 $\leq$0.7 and are compositionally similar to CFB. Thus, it seems that two batches of basalt, one with high Mg and the other with low Mg, erupted in pulses, to form first the Pitepani volcanics and then the thicker Sitagota volcanics within the Dongargarh Group. The larger volume in the younger pulse (the Sitagota volcanics) may be attributed to a higher magma supply rate with increasing extension of the lithosphere.

The high-Mg basalts are related to a parental Mg-richer magma by olivine fractionation (Sensarma et al., 2002). On the basis of the thermometer of Niu (2005; T °C $= 1026e^{0.01894 \times MgO\ wt\%}$), and taking the picrite sample so far known to have maximum MgO in the province (MgO = 15.3 wt%; Neogi et al., 1996) as parental melt, a temperature of 1370 °C is estimated. The primary liquid with higher MgO would have a higher temperature. The ol-normative low-Mg tholeiites (MgO ~ 6–8 wt%),

Mg basalts are compositionally similar to continental flood basalts. (D) Several high-Mg basalt (Pitepani volcanics) samples cluster with rhyolites (Sm/Nd ~ 0.15), implying close genetic relations. Sm/Nd (0.2–0.4) in low-Mg basalts is comparable to primitive mantle (PM; 0.32); for high-Mg basalts, Sm/Nd (0.2–0.6) shows a larger spread. (E) Zr/Y values in rhyolites, high-Mg basalts, low-Mg basalts, and PM plot along a smooth curve, suggesting genetic relationships. Note that high-Mg basalts (Pitepani volcanics) plot closer to rhyolites, whereas low-Mg basalts (Sitagota volcanics) lie at the other end, closer to PM. PM from Palme and O'Neill (2003) and bulk continental crust from McLennan (2001). SHMB—siliceous high-magnesian basalts.

Note: The concentrations of major and trace elements (e.g., Rb, Sr, Zr, Y, Ni, and Cr) were determined by XRF (X-ray fluorescence spectroscopy) using fused glass discs. The analyses were made with a Philips PW 2400 XRF spectrometer equipped with an automatic sample changer. Based on extensive comparison with standard rocks, XRF data accuracy is estimated to be <3%. For INAA (instrumental neutron activation analysis) data, ~100 mg powdered samples were irradiated in the carousel of the TRIGA reactor of the Institut für Kernchemie, Universität Mainz, Germany, and subsequently counted on large Ge detectors at the Institut für Geologie und Mineralogie, Universität zu Köln, Germany. INAA data precision is within 10% of measurement errors. Details of analytical techniques and accuracies obtained from interlaboratory comparison are given in Sensarma et al. (2002) and references therein. The Ta values are calculated from the given Nb values for a few selected mafic volcanic rock samples in Neogi et al. (1996), taking chondritic Nb/Ta = 19.9 (Münker et al., 2003).

on the other hand, may have been crystallized from a lower-temperature (1149-1193 °C; Niu, 2005) magma.

SIZE OF THE PROVINCE

The volumes of erupted melts and their original areal extent are difficult to estimate because the rocks are deformed and have been subjected to prolonged erosion. The present areal extent of the felsic and mafic volcanic rocks is roughly estimated at ~30,000 km². The size of the nearly coeval Dongargarh granite further adds to the volume, perhaps to a major extent. The relationship between the mafic volcanics of comparable age in the southern part of the craton (i.e., the subalkaline basalts, basaltic andesite, and high-Mg basalts; Srivastava et al., 2004, and references therein) and the Dongargarh mafic volcanic rocks in the northern part is unknown. Thus, although the estimated area of the Dongargarh Group is smaller than the minimum size suggested for a LIP (≥50,000 km²; Coffin and Eldholm, 1994; Sheth, 2006), the original size of the province may have been much larger. In any case, the appropriate lower size limit for a LIP is controversial, and the size of igneous provinces may form a continuous spectrum (see http://www.mantleplumes.org/TopPages/LIPClassTop.html).

DURATION OF MAGMATIC ACTIVITY

The age data for the province are sparse, and mainly from the felsic rocks. Nevertheless, some important constraints may be derived from the available U-Pb data. The U-Pb single-crystal zircon dating of the oldest and youngest rhyolite yielded emplacement ages of 2525 ± 15 Ma and 2506 ± 4 Ma, respectively (Ghosh, 2004), indicating a span of ≤ 38 m.y. for felsic volcanism in the area. The intrusive Dongargarh granite is correlated to the well-known Cu-Mo-Au-bearing Malanjkhand granite, located ~120 km north of the Dongargarh Group in the craton. The U-Pb zircon dating of the Malanjkhand granite yields ages of 2478 ± 9 Ma and 2477 ± 10 Ma (Panigrahi et al., 2002). An age of ca. 2500 Ma is also obtained by Re-Os geochronology in molybdenite from the Malanjkhand deposits (Stein et al., 2004). So the Dongargarh granite is possibly of similar age. Although there are no isotopic age data for the mafic rocks, the U-Pb zircon age constraints for the rhyolites and the granitic activity in the province together suggest that the Dongargarh Group (including sedimentary components) may have been formed between ca. 2540 Ma and ca. 2467 Ma, i.e., within a period of ~73 m.y., but surely longer than ~38 m.y. The total duration of formation of the Dongargarh Group is thus reasonably similar to the age range (40–60 m.y.) suggested for SLIPs (e.g., Ernst and Buchan, 2001; Bryan et al., 2002).

CONTEMPORANEITY OF FELSIC AND MAFIC VOLCANISM

The Dongargarh belt is characterized by eruptions of separate, but in part coeval, felsic and mafic magmas. The evidence for nearly contemporaneous rhyolitic and basaltic magmas is as follows.

Petrographic Evidence

Many rounded enclaves of plagioclase-phyric basalt are found in the Bijli rhyolite. Several basaltic enclaves contain quartz xenocrysts with rounded embayed margins and sharp extinction, similar to those present in rhyolites. Also, pumice in felsic rocks from several localities has vesicles filled with devitrified basaltic glass. The textural relations, including the rounded margins of the enclaves, imply that the basaltic melts were incorporated into rhyolite while both were in a molten state. This is suggestive of the mingling of felsic-mafic melts in the area and argues for the near-contemporaneity of both melts.

Chemical Evidence

In the basaltic fractionation model, a large amount of basalt is needed, and it would produce only a small amount of derivative felsic melt. Also, continuous ranges of differentiated product of a basaltic magmalike andesite, dacite, and rhyodacite should occur in spatial proximity. The presence of a "Daly gap" (Fig. 3) and a large volume of rhyolite in the province therefore do not support such an origin for the Bijli rocks. The spread in the concentration of compatible elements in feldspar and Fe-Ti oxides in the samples (e.g., Sr, Ba, Eu, and Ti) (Fig. 5) may then indicate fractional crystallization of feldspar and Fe-Ti oxides, presumably from crust-derived melts.

Evidence for crustal involvement in rhyolite genesis includes the presence of a silica gap (Fig. 3), the similarity of initial $^{87}Sr/^{86}Sr$ ratios of the rhyolites to the crustal value (~0.703) at ca. 2500 Ma (0.7057 ± 0.0015, Sarkar et al., 1981; 0.70305 ± 0.0017, Krishnamurthy et al., 1990), and high alkali and HFSE contents. The low Ta/Th (0.06–0.08) obtained in many Bijli samples (Fig. 4B and D), in comparison to bulk continental crust (BCC: Ta/Th = 0.19; McLennan, 2001), are encountered in granulites (Rudnick and Presper, 1990). Combined with high eruption temperature (900–950 °C) and lower than expected values of LILE (e.g., Rb and U) in some samples (Fig. 5), these observations could indicate the contribution of a deeper crustal material in the Bijli rhyolite. On the other hand, shallow melting of a possible calc-alkaline granitic source may also be responsible for the generation of these rhyolites with depressed Al, Ca, and Sr contents, as demonstrated by Patiño Douce (1997) for the generation of metaluminous A-type liquids. Large variations in incompatible-element concentrations in high-silica rhyolites are often caused by crystallization of accessory phases with high concentrations in these elements. This may also partly be attributed to source heterogeneity at different crustal depths. So partial melting of the regional crust at various depths seems reasonable for the origin of the Bijli rhyolite. Subsequent fractional crystallization may have magnified the variation, particularly in trace-element chemistry.

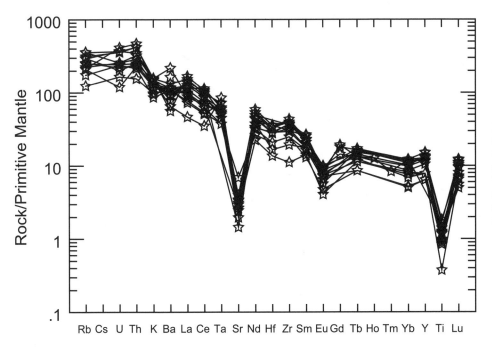

Figure 5. Primitive mantle (PM)–normalized incompatible-elements plots of the Bijli rhyolite. Negative Sr, Eu, and Ti anomalies are characteristic of the rocks. The plots show lower than expected Rb and U values in a few samples. Primitive mantle data are taken from Palme and O'Neill (2003).

A La/Sm diagram is an efficient method for constraining magmatic processes as well as crustal- and mantle-melts interactions. On a La/Sm versus Ta/Th plot (Fig. 4B), La/Sm (~3–6.5) ratios in the high-Mg basalts (the Pitepani volcanics and Sitagota volcanics), particularly in the Pitepani volcanics, overlap some of the rhyolites (5–11) at similar Ta/Th values. The Sm/Nd and Zr/Y values in the Pitepani volcanics (Fig. 4D and E) also overlap or plot close to those of rhyolites.

Olivine fractionation, by which the high-Mg basalts are related to a more MgO-rich parental picritic melt, cannot increase the La/Sm ratios in the melts, because both elements are highly incompatible in olivine. I have shown earlier that an enriched, depleted, or hydrous mantle source is unsuitable, because such melts are not consistent with high-Mg basalts in the Pitepani volcanics, i.e., those of SHMB composition (see Sensarma et al., 2002). The comparison of these SHMB with high-SiO$_2$ Hawaiian tholeiites, considered to have recycled subducted materials in their source (West et al., 1992), provides further evidence that an enriched mantle source cannot give rise to SHMB composition. So the elevated La/Sm in high-Mg basalts in the Pitepani volcanics (Fig. 4B) must have resulted from the assimilation of rhyolite compositions. The overlapping Sm/Nd and Zr/Y values in the rhyolites, and the high-Mg basalts in the Pitepani volcanics (Fig. 4D and E), support this model. The higher SiO$_2$ contents in high-Mg basalts in the Pitepani volcanics compared to many high-Mg basalt samples in the Sitagota volcanics at comparable Mg# (Fig. 4C) further suggests additional silica supply into high-Mg basalts in the Pitepani volcanics. A 15–20% input of rhyolitic melts into a parental high-Mg melt is suggested for the origin of the high-Mg basalts in the Pitepani volcanics (Sensarma et al., 2002). A lesser extent of assimilation may explain

the lower La/Sm ratios (~3–4) and lower SiO$_2$ contents in the younger high-Mg basalts. The increase in high-Mg basalt volumes concomitant with decline in rhyolite input upward in the Sitagota volcanics may suggest higher rates of magma supply with drop in pressure in a widening rift, as mentioned earlier.

The low-Mg basalts, on the other hand, maintain typical tholeiitic basalt compositions and have La/Sm values (~2.0–4.5) much lower than the rhyolites and not far from primitive mantle (PM) (~2; Fig. 4B). In other plots also (Fig. 4D and E), the low-Mg basalts move away from rhyolites and plot closer to the PM, signifying that they may be relatively uncontaminated. The inability of low-Mg basalts to assimilate rhyolite may be attributed to lower temperatures (~1150 °C), i.e., lower heat content.

In summary, crustal melts and mantle-derived high-Mg basalts interacted closely to form geochemically hybrid rhyolites and basalts, especially in the older mafic volcanic pulse (the Pitepani volcanics). This undoubtedly suggests a coeval nature for the rhyolites and basalts.

TESTING A PLUME ORIGIN

The expectations of the plume model include eruption of high-temperature magmas with high MgO content in the range of 18–22 wt%, a high rate of melt production, and an increase in the eruptive volume in later syn- or postrift pulses. There may be precursory or syneruption uplift and emplacement of lavas typically within 1–5 m.y. (Campbell, 2005; Ernst et al., 2005).

The picture that emerges from the previous discussion above is that (1) major felsic-mafic magmatism in Dongargarh is related in time and space, and emplacement of the lavas took ~30–73 m.y., reasonably comparable to the durations found for silicic LIPs (40–60 m.y.; e.g., Ernst and Buchan, 2001; Bryan et al., 2002); (2) eruptions of ~1350 °C picritic magma erupted as part of the igneous activity; (3) pulsatory mafic volcanism occurred, with melt volumes in the second pulse (the Sitagota volcanics) exceeding those of the first (the Pitepani volcanics); (4) felsic-mafic melt interactions probably decreased with time in the Sitagota volcanics. Some of these features match the expectations of the plume model.

However, there are also certain discrepancies, particularly concerning the duration of emplacement of the lavas. Total emplacement duration was not as short as 1–5 m.y., as suggested by Campbell (2005). On the other hand, though SLIPs have a comparable emplacement duration to that of the Dongargarh volcanics, they do not characteristically include the large volumes of mafic lavas that occur in the Dongargarh province. The MgO content (15.3 wt%) of the presumed parental picritic magma is less than the suggested MgO values for parental melts in the plume model.

Thus, the Dongargarh LIP does not unequivocally exhibit all the characteristics expected by the plume model. At the same time, large-scale nearly coeval genesis of mafic-felsic melts and their interactions, leading to the generation of a wide range of melt compositions, are not part of contemporary nonplume models either.

IMPLICATIONS AND CONCLUSIONS

Studies of LIP genesis have considered either mantle melting or crustal anatexis. It is of great significance that a common process of crust-mantle interaction is primarily responsible for building the bimodal LIP in Dongargarh. The major conclusions of this study are as follows:

1. Bimodal LIPs exist, in addition to mafic and silicic LIPs.
2. Melting of both crust and mantle below Dongargarh, driven by a common thermal perturbation along with interactions of the melts, gave rise to the province. The thermal perturbation lasted for longer (~60 m.y.) than the few million years generally reported for LIPs.
3. Such close interactions between crustal and mantle melts over a longer duration contrasts with contemporary models for LIP genesis.

ACKNOWLEDGMENTS

I am grateful to Gillian Foulger, Dhruba Mukhopadhyay, Sumit Chakraborty, Françoise Chalot-Prat, Hetu Sheth, Tom Sisson, and two anonymous reviewers for their incisive and helpful comments. Godfrey Fitton provided invaluable support and advice, clarifying many points. His continuous help during the revision has greatly improved the presentation of the manuscript. A Deutscher Akademischer Austausch Dienst fellowship supported analytical work at Cologne, Germany.

REFERENCES CITED

Asthana, D., Dash, M.R., Pophare, A.N., and Khare, S.K., 1996, Interstratified low-Ti and high-Ti volcanics in arc-related Khairagarh Group of Central India: Current Science, v. 71, no. 4, p. 304–306.

Blake, T.S., and Groves, D.I., 1987, Continental rifting and the Archean-Proterozoic transition: Geology, v. 15, p. 229–232, doi: 10.1130/0091-7613 (1987)15<229:CRATAT>2.0.CO;2.

Bryan, S.E., and Ernst, R.E., 2006, Proposed revision to large igneous province classification: http://www.mantleplumes.org/LIPClass2.html.

Bryan, S.E., Riley, T.R., Jerram, D.A., Stephens, C.J., and Leat, P.T., 2002, Silicic volcanism: An undervalued component of large igneous provinces and volcanic rifted margin, *in* Menzies, M.A., et al., eds., Volcanic rifted margins: Boulder, Colorado, Geological Society of America Special Paper 362, p. 99–120.

Campbell, I.H., 2005, Large igneous provinces and the mantle plume hypothesis: Elements, v. 1, p. 265–269.

Coffin, M.F., and Eldholm, O., 1994, Large igneous provinces: Crustal structure, dimensions, and external consequences: Review of Geophysics, v. 32, p. 1–36, doi: 10.1029/93RG02508.

Courtillot, V., Jaupart, C., Manighetti, I., Tapponnier, P., and Besse, J., 1999, On causal links between flood basalts and continental breakup: Earth and Planetary Science Letters, v. 166, p. 177–195, doi: 10.1016/S0012-821X(98) 00282-9.

Ernst, R.E., and Buchan, K.L., 2001, Large mafic magmatic events through time and links to mantle-plume heads, *in* Ernst, R.E., and Buchan, K.L., eds., Mantle plumes: Their identification through time: Boulder, Colorado, Geological Society of America Special Paper 352, p. 483–575.

Ernst, R.E., Buchan, K., and Campbell, I.H., 2005, Frontiers in large igneous province research: Lithos, v. 79, p. 271–297, doi: 10.1016/j.lithos .2004.09.004.

Foulger, G.R., 2007 (this volume), The "plate" model for the genesis of melting anomalies, *in* Foulger, G.R., and Jurdy, D.M., eds., Plates, plumes, and planetary processes: Boulder, Colorado, Geological Society of America Special Paper 430, doi: 10.1130/2007.2430(01).

Foulger, G.R., and Anderson, D.L., 2005, A cool model for the Iceland hotspot: Journal of Volcanology and Geothermal Research, v. 141, p. 1–22, doi: 10.1016/j.jvolgeores.2004.10.007.

Foulger, G.R., Natland, J.H., and Anderson, D.L., 2005, A source for Icelandic magmas in remelted Iapetus crust: Journal of Volcanology and Geothermal Research, v. 141, p. 23–44, doi: 10.1016/j.jvolgeores.2004.10.006.

Frey, F.A., Garcia, M.O., and Roden, M.F., 1994, Geochemical characteristics of Koolau volcanics: Implications of intershield geochemical differences among Hawaiian volcanoes: Geochimica et Cosmochimica Acta, v. 58, no. 5, p. 1441–1462, doi: 10.1016/0016-7037(94)90548-7.

Ghosh, J.G., 2004, Geochronological constraints on the evolution of the Kotri linear belt and its basement: Records of the Geological Survey of India, v. 136, part 2, p. 24–26.

Krishnamurthy, P., Sinha, D.K., Rai, A.K., Seth, D.K., and Singh, S.N., 1990, Magmatic rocks of the Dongargarh Supergroup, Central India—Their petrological evolution and implications on metallogeny: Nagpur, Geological Survey of India Special Publication 28, p. 303–319.

Le Bas, M.J., Le Maitre, R.W., Streckeisen, A., and Zanettin, B., 1986, A chemical classification of volcanic rocks based on total alkali-silica diagram: Journal of Petrology, v. 27, p. 747–750.

Lustrino, M., 2005, How the delamination and detachment of lower crust can influence basaltic magmatism: Earth-Science Reviews, v. 72, p. 21–38, doi: 10.1016/j.earscirev.2005.03.004.

McLennan, S.M., 2001, Relationships between the trace element composition of sedimentary rocks and upper continental crust: Geochemistry, Geophysics, Geosystems, v. 2, p. 2000GC000109, ISSN 1525–2027.

Melluso, L., Mahoney, J.J., and Dallai, L., 2006, Mantle sources and crustal input as recorded in high-Mg Deccan Traps basalts of Gujrat (India): Lithos, v. 89, p. 259–274, doi: 10.1016/j.lithos.2005.12.007.

Morgan, W.J., 1971, Convection plume in the lower mantle: Nature, v. 230, p. 42–43, doi: 10.1038/230042a0.

Münker, C., Pfänder, J.A., Weyer, S., Büchl, A., Kleine, T., and Mezger, K., 2003, Evolution of planetary cores and the Earth-Moon system from Nb/Ta systematics: Science, v. 301, p. 84–87, doi: 10.1126/science.1084662.

Neogi, S., Miura, H., and Hariya, Y., 1996, Geochemistry of Dongargarh volcanic rocks, Central India: Implications for the Precambrian mantle: Precambrian Research, v. 76, p. 77–91, doi: 10.1016/0301-9268(95)00025-9.

Niu, Y., 2005, Generation and evolution of basaltic magma: Some basic concepts and new views on the origin of Mesozoic–Cenozoic basaltic volcanism in eastern China: Geological Journal of China Universities, v. 11, p. 9–46.

Palme, H., and O'Neill, H. St. C., 2003, Cosmochemical estimates of mantle composition, *in* Treatise on geochemistry, v. 2: Oxford, Elsevier, p. 1–38.

Panigrahi, M.K., Misra, K.C., Brteam, B., and Naik, R.K., 2002, Genesis of the granitoid affiliated copper-molybdenum mineralization at Malajkhand, central India: Facts and problems [extended abs], *in* Proceedings of the 11th Quadrennial International Association on the Genesis of Ore Deposits Symposium and Geocongress, Windhoek, Namibia, July 22–26, n.p.

Pankhurst, R.J., Leat, P.T., Sruoga, P., Rapela, C.W., Márquez, M., Storey, B.C., and Riley, T.R., 1998, The Chon Aike silicic igneous province of Patagonia and related rocks in Antarctica: A silicic large igneous province: Journal of Volcanology and Geothermal Research, v. 81, p. 113–136, doi: 10.1016/S0377-0273(97)00070-X.

Patiño Douce, A.E., 1997, Generation of metaluminous A-type granites by low-pressure melting of calc-alkaline granitoids: Geology, v. 25, no. 8, p. 743–746, doi: 10.1130/0091-7613(1997)025<0743:GOMATG>2.3.CO;2.

Riley, T.R., and Leat, P.T., 1999, Large volume silicic volcanism along the proto-Pacific margin of Gondwana: Lithological and stratigraphical investigations from the Antarctic Peninsula: Geological Magazine, v. 136, p. 1–16, doi: 10.1017/S0016756899002265.

Roy, A., Ramachandra, H.M., and Bandopadhyay, B.K., 2000, Supracrustal belts and their significance in the crustal evolution of central India: Calcutta, Geological Survey of India Special Publication 55, p. 361–380.

Rudnick, R.L., and Presper, T., 1990, Geochemistry of intermediate- to high-pressure granulites, *in* Vielzeuf, D., and Vidal, Ph., eds., Granulites and crustal evolution: Dordrecht, Kluwer.

Sarkar, S.N., Gopalan, K., and Trivedi, J.R., 1981, New data on the geochronology of the Precambrians of Bhandara-Durg, Central India: Indian Journal of Earth Science, v. 8, p. 131–151.

Sensarma, S., 2001, Volcanic stratigraphy and petrology of the Dongargarh belt near Salekasa-Darekasa, Bhandara district, Maharashtra [Ph.D. thesis]: Calcutta, Jadavpur University.

Sensarma, S., 2005, The Dongargarh Group: A large igneous province at the Archean-Proterozoic transition in India, *in* AGU Chapman Conference: The Great Plume Debate: The origin and impact of LIPs and hot spots: Washington, D.C., American Geophysical Union, p. 87.

Sensarma, S., and Mukhopadhyay, D., 2003, New insight on the stratigraphy and volcanic history of the Dongargarh Belt, central India: Gondwana Geological Magazine Special Volume 7, p. 129–136.

Sensarma, S., Palme, H., Deloule, E., and Mukhopadhyay, D., 2000, Evidence of liquid immiscibility in the Early Proterozoic andesitic rocks, Dongargarh Supergroup, Central India, and possible tectonic implication, *in* Goldschmidt Conference, Oxford, England, September 3–8, p.101–102.

Sensarma, S., Palme, H., and Mukhopadhyay, D., 2002, Crust-mantle interaction in the genesis of siliceous high magnesian basalts (SHMB): Evidence from the Early Proterozoic Dongargarh Supergroup, India: Chemical Geology, v. 187, p. 21–37, doi: 10.1016/S0009-2541(02)00020-7.

Sensarma, S., Hoernes, S., and Mukhopadhyay, D., 2004, Relative contributions of crust and mantle to the origin of Bijli Rhyolite in a palaeoproterozoic bimodal volcanic sequence (Dongargarh Group), central India: Proceedings of the Indian Academy of Science (Earth and Planetary Science), v. 113, no. 4, p. 619–648.

Sharma, K.K., 2004, The Neoproterozoic Malani magmatism of the northwestern Indian Shield: Implications for crust-building processes: Proceedings of the Indian Academy of Science (Earth and Planetary Science), v. 113, no. 4, p. 795–807.

Sheth, H.C., 2006, Large igneous provinces (LIPs): Definition and recommended terminology, and a hierarchical classification: http://www.MantlePlumes.org/LIPclass.html.

Srivastava, R.K., Singh, R.K., and Verma, S.P., 2004, Neoarchean mafic volcanic rocks from the southern Bastar greenstone belt, Central India: Petrological and tectonic significance: Precambrian Research, v. 131, p. 305–322, doi: 10.1016/j.precamres.2003.12.013.

Stein, H.J., Hannah, J.L., Zimmerman, A., Markey, R.J., Sarkar, S.C., and Pal, A.B., 2004, A 2.5 Ga porphyry Cu-Mo-Au deposit at Malanjkhand, Central India: Implications for Late Archaean continental assembly: Precambrian Research, v. 134, p. 189–226, doi: 10.1016/j.precamres.2004.05.012.

Tanton, L.T.E., and Hager, B.H., 2000, Melt intrusion as a trigger for lithospheric foundering and the eruption of the Siberian flood basalts: Geophysical Research Letters, v. 27, p. 3937–3940, doi: 10.1029/2000GL011751.

West, H.B., Garcia, M.O., Gerlach, D.C., and Romano, J., 1992, Geochemistry of tholeiites from Lanai, Hawaii: Contributions to Mineralogy and Petrology, v. 112, p. 520–542, doi: 10.1007/BF00310782.

Manuscript Accepted by the Society January 31, 2007

The Geological Society of America
Special Paper 430
2007

Thick, high-velocity crust in the Emeishan large igneous province, southwestern China: Evidence for crustal growth by magmatic underplating or intraplating

Yi-Gang Xu*
Bin He
*Key Laboratory of Isotope Geochronology and Geochemistry, Guangzhou Institute of Geochemistry,
Chinese Academy of Sciences, 510640 Wushan, Guangzhou, China*

ABSTRACT

Geophysical, geological, and petrologic data in southwestern China have been integrated in order to characterize magmatic underplating associated with the Late Permian Emeishan large igneous province (LIP; ca. 260 Ma). Seismic reflection and refraction reveals a heterogeneous crustal structure with high-velocity layers or bodies in the upper crust (6.0–6.6 km/s), lower crust (7.1–7.8 km/s), and upper mantle (8.3–8.6 km/s). These seismically anomalous bodies are all confined in the inner zone of the prevolcanic domal structure, but are generally absent in the intermediate and outer zones. There is a decreasing trend in crustal thickness from the inner zone (>60 km, with a ~20-km-thick high-velocity lower crust, or HVLC) via the intermediate zone (~45 km) to the outer zone (<40 km). Because the domal uplift immediately preceding eruption of the Emeishan basalts was unambiguously related to a mantle plume, such a configuration highlights a genetic relationship between the formation of the high-velocity crust and the mantle plume that led to the eruption of the Emeishan basalts. It is proposed that the HVLC may have resulted from magmatic underplating associated with the Emeishan volcanism, whereby the fast mantle represents the residues left after extensive melt extraction from the plume head. Magmatic underplating can also account for the prolonged crustal uplift that formed the Chuandian "old land" in southwestern China. Petrologic modeling further suggests that the HVLC may represent fractionated cumulates from picritic melts and that the Emeishan basalts represent residual melts after polybaric fractionations. This relationship allows a reestimation of the volume of Emeishan magmas, which is as much as 3.8×10^6 km^3, typical of plume-generated LIPs in the world.

Keywords: mantle plumes, uplift, underplating, petrologic modeling, Emeishan

*E-mail: yigangxu@gig.ac.cn.

Xu, Y.-G, and He, B., 2007, Thick, high-velocity crust in the Emeishan large igneous province, SW China: Evidence for crustal growth by magmatic underplating or intraplating, *in* Foulger, G.R., and Jurdy, D.M., eds., Plates, plumes, and planetary processes: Geological Society of America Special Paper 430, p. 841–858, doi: 10.1130/2007.2430(39). For permission to copy, contact editing@geosociety.org. ©2007 The Geological Society of America. All rights reserved.

INTRODUCTION

Magmatic underplating is an important feature of large igneous provinces (LIPs). On the basis of basaltic compositions and mantle melts, Cox (1980, 1993) suggested that ponding and polybaric fractionation of primary picritic magmas at the base of crust may be an important process in the generation of continental flood basalts, and magmatic underplating is potentially a large contribution to the crustal accretion. This hypothesis has received supporting evidence in geophysical, geological, and geochemical or petrologic observations for over twenty years.

First, there is a growing body of seismic data that allows imaging of the crust-mantle structure of hotspots (e.g., Watts and ten Brink, 1989; Caress et al., 1995), of volcanic rifted margins (e.g., White and McKenzie, 1989; Barton and White, 1997; Bauer et al., 2000; Menzies et al., 2002), and of oceanic plateaus (e.g., Furumoto et al., 1976; Mauffret and Leroy, 1997). Seismic data consistently indicate crustal thickening and the existence of high-velocity layers at the base of the crust in LIPs. The thickness of these layers ranges from a few kilometers to more than 20 km, and their seismic velocities are intermediate between those of the mantle (>8 km/s) and those typical of the lower crust (<7 km/s). These high-velocity layers are commonly interpreted as results of magmatic underplating (Kelemen and Holbrook, 1995; Farnetani et al., 1996; Trumbull et al., 2002).

Second, subsidence and exhumation history in basins of volcanic rifted margins and permanent uplift in the continental LIPs are surface manifestations of the geologic processes at depth. As the lithosphere is thinned by uniform stretching, rifts and basins subside to maintain isostatic equilibrium, while in volcanic rifted margins this subsidence is counterbalanced by uplift resulting from the addition of new igneous material to the crust (White and McKenzie, 1989). Many basins in north Atlantic volcanic rifted margins underwent permanent uplift of several hundred meters, which was likely caused by magmatic underplating (Saunders et al., 2007). After having studied the drainage patterns of three continental LIPs, Cox (1989) suggested that topographic doming associated with plume activity can be preserved after ~200 m.y. and that crustal thickening by magmatic underplating is the most likely cause for the persistence of such features. Although the work by Cox (1989) on the Deccan Traps receives some criticism (Sheth, this volume), the link between underplating and persistent uplift remains valid.

Finally, petrologic modeling provides supporting evidence for the hypothesis of magmatic underplating beneath the LIPs. On the basis of thermal and petrophysical modeling, Furlong and Fountain (1986) suggested that an underplating layer more than 10 km thick (with a compressional wave velocity of $V_p = 7.0$–7.8 km/s) has been added to the base of the continental crust. Farnetani et al. (1996) used the petrological code MELTS (Ghiorso and Sack, 1995) to model the compositional evolution of primary picritic magmas, assuming that crystal fractionation is the dominant mechanism through which picritic liquids evolve toward a basaltic composition. They argued that high-velocity layers

(7.4–7.9 km/s) detected at the base of the crust beneath oceanic and hotspot tracks may represent fractionated cumulates from picritic magmas.

Magmatic underplating is therefore an integral part of volcanism and can be constrained in three independent ways, namely, through geophysical, geologic, and petrologic approaches. Although an individual approach has been used to characterize magmatic underplating in LIPs, so far no multidisciplinary integration has been performed in a single continental LIP. The Late Permian Emeishan flood basalts in southwestern China are ideal for such an exercise, because evidence from all three aspects is present: (1) A lot of geophysical experiments (e.g., deep seismic sounding, seismic tomography, and gravity survey) have been performed in this region since the 1980s, thus providing insights into the crust-mantle structure; (2) sedimentary evidence has been preserved for the crustal uplift before and after the Emeishan volcanism, which is likely related to deep processes (He et al., 2003, 2006; Xu et al., 2004); and (3) the Emeishan basalts and coeval layered mafic and ultramafic intrusions are well exposed in this region and have stimulated many petrologic and geochemical analyses (Chung and Jahn, 1995; Xu et al., 2001; Zhou et al., 2002, 2005; Zhong et al., 2004). This makes it possible to determine the nature of cumulate products left at depth by fractionation of the Emeishan basalts.

In this contribution we review the geophysical data available in this area and highlight the main characteristics of the crust and mantle structure in the Emeishan LIP. In particular, we view the spatial variation in crustal thickness in relation to the domal structure defined by sedimentary data (He et al., 2003), which has been interpreted as plume-induced uplift. We attribute the unusually thick, high-velocity crust in the central part of the Emeishan LIP to magmatic underplating and/or intraplating. The nature of the underplated materials is constrained by relating petrologic modeling results to the observed seismic velocity.

GEOLOGICAL BACKGROUND AND PREVIOUS STUDIES

The Late Permian Emeishan basalts are erosional remnants of the voluminous mafic volcanic successions occurring in the western margin of the Yangtze craton, southwestern China. They are exposed in a rhombic area of 250,000 km² (Xu et al., 2001) bounded by the Longmenshan thrust fault in the northwest and the Ailaoshan–Red River slip fault in the southwest (Fig. 1). However, some basalts and mafic complexes exposed in the Simao basin, northern Vietnam (west of the Ailaoshan fault), and in Qiangtang terrain (northwest of the Longmenshan fault) make possible an extension of the Emeishan LIP (Chung et al., 1998; Xiao et al., 2003; Hanski et al., 2004). The Red River fault is also considered a suture zone along which the Yangtze craton was subducted by at least 150 km underneath the Sanjiang tectonic zone (Liu et al., 2000). Therefore, besides translational disruption by the Red River fault, the Emeishan LIP was also affected by reworking during the Mesozoic closure

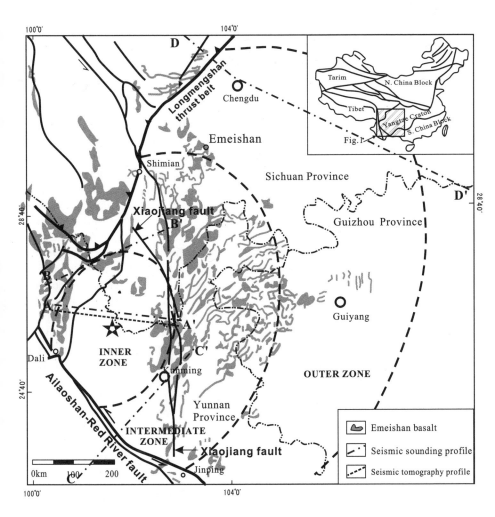

Figure 1. Map showing the geology of the Emeishan large igneous province and the distribution of geophysical experiments (the inset map of China shows the area depicted). The dashed circular curves separate the inner, intermediate, and outer zones, which are defined in terms of extent of erosion of the Maokou Formation (He et al., 2003). The dash-dotted lines (A–A′, B–B′, C–C′, and D–D′) indicate the profiles along which the seismic sounding results shown in Figure 2 were obtained.

of the Tethys Ocean and the Cenozoic collision between the Indian and Asian continents.

The Emeishan basalts uncomfortably overlie the late Middle Permian carbonate formation (i.e., the Maokou limestone) and are, in turn, covered by the uppermost Permian in the east and west and by the Upper Triassic sediments in the central part. A number of observations indicate that the central part of the Emeishan LIP has experienced uplift:

1. Most of the flood basalts were eroded away in the inner zone except for a remnant in a few localities. Most of remnant flood basalts are distributed in the intermediate zone (Fig. 1).
2. The moderate and high-grade metamorphic Mesoproterozoic Kangding Complex, a Neoproterozoic spilite-keratophyre sequence with greenschist-facies metamorphism, and epimetamorphic sedimentary rock sequences are exposed in the inner zone (Precambrian metamorphic crystalline basement is exposed). However, mainly Paleozoic sedimentary sequences are exposed in the intermediate zone.
3. Numerous layered mafic intrusions and ring felsic complexes are exposed only in the central Emeishan LIP. To further characterize this area, He et al. (2003) correlated and compared the biostratigraphic unit of the Maokou Formation and showed a systematic thinning of the strata beneath the Emeishan basalts. The surface of thinned carbonates is an unconformity, with karst paleotopography and local basal conglomerates, the clasts of which were derived from the uppermost Maokou Formation. This suggests that stratigraphic thinning likely resulted from differential erosion due to regional uplift. Isopachs of the Maokou Formation further delineate a circular uplifted area (He et al., 2003) very similar to the crustal doming above an upwelling mantle plume predicted by laboratory experiments (Campbell and Griffiths, 1990) and by numerical simulations (Farnetani and Richards, 1994). Prevolcanic crustal doming in the Emeishan LIP led to a new division of the province, i.e., inner, intermediate, and outer zones (Fig. 1) that have different extents of erosion of the Maokou Formation (He et al., 2003).

The Emeishan flood volcanism succession comprises predominantly basaltic flows and pyroclastic deposits, with minor amounts of picrites and basaltic andesites, with a total thickness ranging from several hundred meters up to 5 km (Xu et al., 2001). The Emeishan basalts have been divided into two major magma types: high-Ti (Ti/Y > 500) and low-Ti (Ti/Y < 500) basalts (Xu et al., 2001). Data compilation reveals a systematic change in basalt type from the inner to the intermediate zones (Xu et al., 2004). In general, the domed region comprises thick (2000–5000 m) sequences of dominant low-Ti volcanic rocks and subordinate picrites (Chung and Jahn, 1995) and high-Ti and alkaline lavas (Xu et al., 2001, 2003; Xiao et al., 2003). In contrast, thin sequences (<500 m) of high-Ti volcanic rocks mainly occur on the periphery of the domal structure. It has been shown that the high-Ti and low-Ti lavas require very different mantle conditions (Xu et al., 2001). Therefore, this spatial variation reflects the thermal gradient of the mantle from which the Emeishan basalts were derived (Xu et al., 2004).

GEOPHYSICAL EVIDENCE FOR A THICK AND HETEROGENEOUS CRUST IN THE EMEISHAN LIP

Geophysical investigation has been carried out in this region since the 1980s with the purpose of mineral exploration and of characterizing the crust-mantle structure under the Emeishan LIP (Tan, 1987; Zhang et al., 1988; Teng, 1994). The experiments have included six deep seismic sounding (DSS) profiles (Yan et al., 1985; Xiong et al., 1986; Cui et al., 1987; Tan, 1987), several seismic tomography experiments (Zhang and Ma, 1985; Sun et al., 1991; Liu et al., 2000), and gravity and magnetic measurements (Liu et al., 1987; Teng, 1994). The resulting data are of critical importance in understanding the crust-mantle structure in this area. Unfortunately, most of these data were published in Chinese literature that is not accessible to foreign researchers. In this section we summarize the main features of these geophysical data with special attention to spatial variation of the crust-mantle structure in relation to the domal structure.

Deep Seismic Sounding (DSS)

General Description of Methods. During long-range seismic wide-angle profile refraction or reflection investigations, large (1–38 t) shots fired in deep (20–30 m) holes were used to probe deeply and obtain clear seismic arrival times from the crust and upper mantle. Some 196 sets of analog magnetic tape recorders (GPS, France; 7030G, Germany; DZSM-1, China) were employed along the profiles to record timing and shot-instant pulses (detector distance: 1.5–3 km). Four to eight shots were fired for each profile. In the field, a split spread configuration was used for each shot point, with seismographs deployed on both sides or one side of each shot point. Altogether, up to 240 recording sites were occupied along the profiles (two deployments of 120 instruments) by the portable FM cassette-tape recording seismographs. The usability of the seismic data was 86.4–73.2%. Seismic phases were identified by their kinematic and dynamic characteristics on the record section graphs.

In order to build a reasonable initial crust model, different methods (e.g., the common depth point, t2-x2, time term, and different time-distance methods) were used. Both forward and inverse velocity travel-time modeling were employed in establishing the crustal models. Details of the experiments and data treatment can be found in Teng (1994). Here we briefly summarize the main features of four DSS profiles that traverse different parts of the Emeishan LIP (Fig. 1).

Lijiang-Zhehai Profile (A–A'). This east-west-trending profile is 350 km long and runs over the inner zone of the Emeishan LIP (Fig. 2A). It is centered on the Panzhihua site, the second-largest V-Ti magmatic ore deposit in the world. The crust under this profile can be divided into three layers. The upper crust is ~25 km thick, with a V_p ranging from 4.0 to 6.6 km/s. The velocity of the uppermost crust (<5 km) in the middle part of the profile (5.8–6.0 km/s) is higher than those at the western and eastern ends (4.2–5.5 km/s). This might indicate the presence of mafic sills and intrusions in the center part of the Emeishan LIP. The velocity increases to 6.3–6.6 km/s in the lower part of the upper crust, significantly higher than the average velocity of the continental upper crust (<6 km/s; Rudnick and Jackson, 1995). It then decreases to 5.6–6.0 km/s in the depth range of 26–40 km (middle crust). The lower crust extends from 40 to 55 km and is characterized by a V_p of 6.5–6.8 km/s. A higher velocity (6.8–7.0 km/s) has been observed locally (Tan, 1987). This layer is sitting above a layer of 7.6–7.9 km/s. A layer of similar velocity is also detected in the Lijiang-Xinshi profile (see the following sections). However, the higher velocity (>8.0 km/s) observed in the Lijiang-Xinshi profile is not detected along this profile. While the interpretation of the layer of 7.6–7.9 km/s remains uncertain, the crust beneath the inner zone is at least 55 km, which is thicker than those in other parts of the Yangtze craton (~40 km).

Lijiang-Xinshi Profile (B–B'). This northeast-trending profile is 407 km long and runs across the inner zone and intermediate zone of the Emeishan LIP. The gross crustal structure along this profile is shown in Figure 2A. The most salient feature is that the crust is segregated by Xiaojiang fault, Anninghe fault, and Jinhe fault, which cut through the lower crust and reach the upper mantle. The crustal thickness under the center of the profile could be >70 km if the layer of 7.6–7.8 km/s is taken as a part of the lower crust. The high-velocity rock body (~6.0 km/s) near the surface in the center part of the Emeishan LIP is probably related to mafic sills and intrusions.

Simao-Malong Profile (C–C'). This profile is situated outside of the Emeishan LIP and crosses the Red River fault (Fig. 1). The crustal thickness in this region ranges from 37 to 45 km (Fig. 2C), significantly less than that in the inner zone of the Emeishan LIP. Another salient feature is that the layer with velocity intermediate between crust and mantle (7.6–7.8 km/) is not present under this region. The P-wave velocity of the upper mantle is 8.06–8.10 km/s.

Heishui-Wulong Profile (D–D'). This profile is located in the northern part of the outer zone of the Emeishan LIP. The crustal thickness in this region ranges from 40 to 45 km. A thin

A

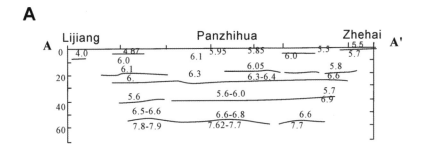

B

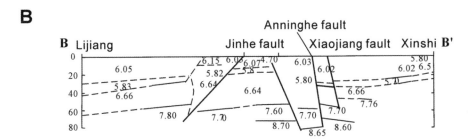

C

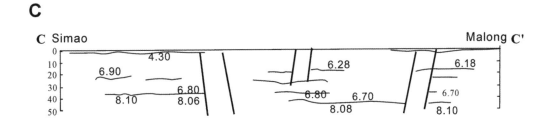

D

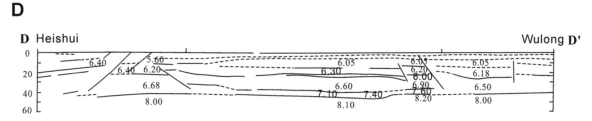

Figure 2. Seismic sounding profiles (modified from Yan et al., 1985; Xiong et al., 1986; Cui et al., 1987; Tan, 1987; Teng, 1994). The locations of the profiles (lines A–A′, B–B′, C–C′, and D–D′) are shown in Figure 1.

(<5 km) layer with a V_p of 7.1–7.4 km/s is detected at the middle of the profile.

Seismic Tomography

The seismic tomography along the Lijiang-Zhehai profile (the same as the DSS profile A–A′) is reproduced in Figure 3. The best horizontal resolution for this velocity image is 20–30 km, and the best vertical resolution is 10–15 km in the crust and ~30 km in the upper mantle (Liu et al., 2000). Three main features are noted from Figure 3:

1. A heterogeneous velocity structure is observed in both vertical and horizontal directions. Vertically, continuous and discontinuous alternating high- and low-velocity strip-shaped zones are observed at different depths. Horizontally, high- and low-velocity strips are obviously divided into several sections, and the strongest heterogeneity of the crustal structure is near the central part of the Emeishan LIP, in which 6.0 km/s is found at the depth of 5 km. A high-velocity (6.1–6.4 km/s), lens-shaped body 60 km long and 10 km thick is present in the upper crust at a depth of 10–20 km.

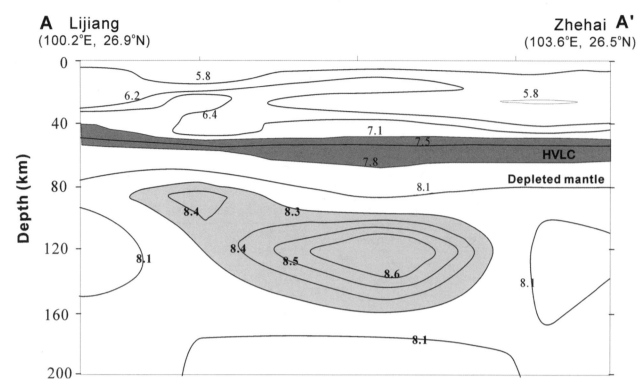

Figure 3. The seismic tomographic velocity structure of the crust and upper mantle beneath west Yangtze craton (modified from Liu et al., 2001). The location of the seismic profile from Lijiang to Zhehai (line A–A′) is shown in Figure 1. HVLC—high-velocity lower crust.

2. A layer with high seismic P-wave velocities (7.1–7.8 km/s) is present in the crust and mantle transition zone. Its velocity increases rapidly from 6.8 km/s to 7.1–7.5 km/s from a depth of 40 km to 55 km and then slowly to 7.8 km/s to a depth of 60–70 km. This layer is relatively thicker in the central part of the profile than at the ends of the profile. The average thickness of this layer is 20 km, with a maximum of 25 km at the center of the profile. It decreases rapidly to 10 km near Lijiang (to the west) and decreases gradually to 15 km in Zhehai (to the east).

3. There is a high-velocity, lens-shaped body (8.3–8.6 km/s) in the upper mantle at a depth of 110–160 km. Surrounded by normal-velocity mantle (~8.1 km/s), this high-velocity body is ~200 km long in the east-west direction and ~50 km thick. It is important to note that the location of this fast zone matches well that of the thick lower crust layer. In addition, the geometric coincidence of this high-V_p body and the inner zone of the doming structure is also remarkable (Fig. 1).

Deep Seismic Reflection

Deep seismic reflection has been employed to determine the crustal thickness in the Emeishan LIP, which, together with the domal structure, is shown in Figure 4. It is noted that there is a gradual decrease in crustal thickness from the center to the margin of the Emeishan LIP. The crustal thickness in the central part ranges from 55 to 64 km (average 61.5 km). The crust in the eastern part is also relatively thick, ranging from 38 to 54 (average 45 km), but thinner than in the central part. The data beyond the Emeishan LIP are sparse but define a range of 35–45 km.

Summary

Various geophysical approaches reveal some consistent features regarding the crust-mantle structure in the Emeishan LIP. The crustal structure and crust-mantle transition in the central part of the Emeishan LIP are different from those in other areas. Specifically, three anomalously high-velocity bodies (compared to their respective environs) have been detected in the upper crust (6.0–6.6 km/s), the crust-mantle transition zone (7.1–7.8 km/s), and the upper mantle (8.3–8.6 km/s). All these seismically anomalous bodies are confined to the inner zone, but are generally absent or not obvious in the intermediate and outer zones (e.g., Cui et al., 1987).

The layer with a velocity > 7.0 km/s at the crust-mantle transition zone and underlying fast mantle (>8.3 km/s) has been observed in many LIPs, for example, the Columbia plateau (8.4 km/s; Catchings and Mooney, 1988), Rhine Graben (8.3–8.4 km/s; Zucca, 1984), and Ontong Java plateau (8.6 km/s; Furumoto et al., 1976). Therefore, the crust-mantle velocity structure in the Emeishan case shares features observed in other large

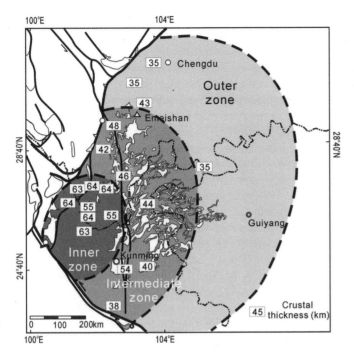

Figure 4. Crustal thickness data plotted over the domal area of the Emeishan large igneous province. Data sources: Zhang et al. (1988); Yuan (1995).

igneous provinces. In the following sections we will evaluate the cause(s) of crustal thickening in this region and the formation of the high-velocity crust.

CRUSTAL GROWTH BY MAGMATIC UNDERPLATING DURING EMEISHAN VOLCANISM

High-Velocity Layer at the Crust-Mantle Transition: Upwelling Mantle or Lower Crust?

The layer of 7.1–7.8 km/s is ubiquitously present in the crust-mantle transitional zone in the Emeishan LIP. Similar velocity has been observed in the east African rift (Davis and Slack, 2002) and Baikal rift (Brazier and Nyblade, 2003) and is considered evidence for upwelling mantle. Tan (1987) and Zhang et al. (1988) have adopted this interpretation and regard it as evidence for upwelling asthenosphere in the Panxi (i.e., the Panzhihua-Xichang) rift zone. However, the mantle upwelling model is inconsistent with the fast upper mantle ($V_p > 8.3$ km/s; Liu et al., 2000) rather than the low-velocity mantle expected in extensional region. On the other hand, the so-called Panxi rift is Permian–Triassic in age (Cong, 1988; Zhang et al., 1988). Consequently, the upwelling mantle is not absolutely expected at the present time. Other arguments against the low-velocity mantle model include thick crust along the Panxi paleorift rather than the thin crust expected for the extended region. Consequently, we interpret the layer of 7.1–7.8 km/s as a part of the lower crust.

In fact, high-velocity lower crust (HVLC) is a characteristic of LIPs (Coffin and Eldholm, 1994; Menzies et al., 2002).

Formation of the HVLC by Magmatic Underplating Associated with the Emeishan Plume

Two distinct dynamic models have been proposed to explain the formation of the HVLC (see the review of Kelemen and Holbrook, 1995). The plume model invokes a deep-seated thermal anomaly in upwelling mantle (White and McKenzie, 1989; Farnetani et al., 1996), whereas "secondary convection models" (Mutter et al., 1984) call upon rapid upwelling of a mantle source without a thermal anomaly. A possible scenario in the second case is lower-crustal (eclogitic?) detachment and subsequent uplift as mantle flows into the new space. However, this model encounters several problems:

1. If crustal delamination induced mantle melting that gave rise to the Emeishan flood volcanism, a thinner crust would be expected underneath the center of the Emeishan LIP. As stated in the previous section, a layer with a seismic velocity of 7.1–7.8 km/s is part of the lower crust rather than the upper mantle. Even if this high-velocity layer is not taken into account, the crust beneath the inner zone is thicker than that under the intermediate zone.
2. Crustal rebound subsequent to crustal delamination would be expected to be an instantaneous effect, i.e., to occur over a short time span (a few million years). This is inconsistent with persistent uplift that lasted for >40 m.y. in the Emeishan case.
3. Figure 4 shows a gradual decrease in crustal thickness from the center to the margin of the Emeishan LIP, with thicker crust confined to the inner zone. This spatial variation cannot be explained by the delamination model, because there is no reason to expect the delaminated crust to have a circular shape.
4. Finally, crustal delamination has not yet been physically and mechanically demonstrated. The lower crust and convective mantle are separated by buoyant subcontinental lithospheric mantle. Whether delaminated crust could penetrate the rigid lithospheric mantle is unclear.

The common practice to constrain the genesis of the HVLC is to use the geochemistry of a suite of lava samples to estimate the physical properties of the cumulates and then compare these properties with those determined by the geophysical survey (Farnetani et al., 1996; MacLennan et al., 2001) or to combine modeling results and experimental data to draw the conditions under which the HVLC was formed (Kelemen and Holbrook, 1995; Trumbull et al., 2002). Both approaches require knowledge of a number of assumptions and parameters (e.g., about the nature of parental magmas and physical conditions under which fractionation takes place) that are not easy to determine. On the other hand, because the underplated igneous crust has never been directly sampled, it is impossible to date its formation age, leaving the geodynamic setting under which the HVLC was formed poorly constrained. The observation that the crust-mantle structure in the Emeishan LIP varies systematically in relation to the domal structure provides another insight into the origin of the HVLC.

In the Emeishan case, the HVLC is overlain by a high-velocity upper crust (V_p = 6.0–6.6 km/s) and a lens-shaped, fast upper mantle (V_p = 8.1–8.6 km/s). Farnetani et al. (1996) suggested that the coexistence of HVLC with high upper crustal velocity is diagnostic of deep melting and that HVLC layers beneath oceanic plateaus and hotspots are an integral part of plume volcanism. A similar interpretation can thus be applied in the Emeishan case. More important, the distribution of these seismically anomalous bodies at different levels is well within the inner zone of the Emeishan LIP. For instance, high-velocity upper crust is not observed in the intermediate and outer zones. The western and eastern margins of this seismically anomalous body correspond to longitudes of 100.8°E and 102.8°E, respectively, which agree well with the geographic location of the inner zone (Xu et al., 2004). These observations are indicative of a common factor that governed the formation of these seismically anomalous bodies. As discussed by He et al. (2003), the crustal domal uplift immediately preceding eruption of the basalts most likely resulted from thermal or dynamic doming by a mantle plume that generated the Emeishan basalts. Because the domal structure and the crust-mantle structure were obtained by independent methods, this correlation strongly suggests the generation of the thick HVLC by an Emeishan plume. The fast upper mantle might represent the residues left after extensive melt extraction from the plume head. Strongly refractory (olivine-rich) mantle is low in density, which translates to high compressional wave velocities. The high-velocity upper crust and the HVLC represent intraplating and underplating of plume-derived melts at different levels of the crust.

Experiments show that primary melts derived by partial melting of mantle materials at depth are likely picritic, with MgO >16 wt%. This contrasts with the predominant evolved nature (MgO < 8%) of the erupted lava. Crystal fractionation has been invoked to explain this discrepancy (Cox, 1980), because picritic magma will pond at the crust-mantle boundary due to density contrast (Sparks et al., 1980). Crystal fractionation produces cumulates, causing underplating and thickening of the crust (i.e., formation of the HVLC). According to the plume hypothesis (Campbell and Griffiths, 1990), the melt production at the central part of the plume head would be more important than at the plume head periphery due to the temperature gradient across the plume head. Another reason may be related to the uplift of mantle and crust above the plume head. The space created by lithospheric uplift will be instantaneously filled by upwelling asthenosphere or a plume. This induces more decompression melting in the region above the plume head compared to the plume periphery, where the lithospheric uplift is limited. To a first-order approximation, the amount of melt or cumulates trapped

at the crust-mantle boundary could be proportional to the expected melt production. This explains the observed variation in crustal thickness across the doming structure of the Emeishan LIP (Fig. 4; Xu et al., 2004). This interpretation is supported by recent geochronologic studies that show the synchronism between the emplacement of mafic and ultramafic intrusions in the inner zone of the Emeishan LIP and the eruption of the Emeishan basalts. SHRIMP U-Pb zircon dating reveals that mafic and ultramafic intrusions were emplaced at 259–261 Ma (Zhou et al., 2002, 2004), virtually identical to the eruption age of the Emeishan basalts inferred from the stratigraphic constraints (Courtillot et al., 1999), if a new geologic timescale (Gradstein et al., 2004) is applied.

Other evidence supporting magmatic underplating or intraplating as the origin of high-velocity crust comes from the Late Permian–Middle Triassic sedimentary record (Fig. 5). Stratigraphic units below the Emeishan basalts are similar to other parts of the Yangtze craton (Zhang et al., 1988; Feng et al., 1997), indicating that this region experienced a geological history similar to that of other parts of Yangtze craton before the Emeishan volcanism (Fig. 5A). However, there was a dramatic change in the sedimentary environment after the Emeishan volcanism in southwestern China (Wang et al., 1994; Xu et al., 2004). The most striking change was the appearance of an elliptical basement core (known as the Chuandian "old land") in the center of the Emeishan LIP (Fig. 5B). Depositional patterns from Upper Permian to Middle Triassic suggest that crustal uplift persisted to the Middle Triassic (Fig. 5C). Such a prolonged uplift (>45 m.y.) may have been due to magmatic underplating (McKenzie, 1984; White and McKenzie, 1989; Brodie and White, 1994), because in some instances underplating-related uplift has lasted for >200 m.y. (Cox, 1989). It is shown in Figure 5B that the "Chuandian old land" is surrounded by terrestrial (Xuanwei Formation) and marine clastic (Longtan Formation) rocks. This sedimentologic configuration is indicative of extensive erosion of the Emeishan basalts in the inner zone, which is consistent with scarce Emeishan basalts in the inner zone compared to those in intermediate zone (Fig. 1). The Xuanwei clasts represent eroded products from Emeishan basalts in the inner zone (He et al., 2007). Depositional patterns from Upper Permian to Middle Triassic show transgression, because marine clastic rocks and limestones progressively overlapped the Chuandian old land (Fig. 5D).

Possible Components of the HVLC

Before adopting a petrologic approach to constrain the composition of the HVLC based on the observed velocity values (Kelemen and Holbrook, 1995; Farnetani et al., 1996; Trumbull et al., 2002), it is important to determine (1) whether garnet plays a role in the observed high-velocity layer and (2) whether this thick crust is composed of solidified liquids or of fractionated cumulate derived by crystal fractionation from an evolving liquid that later migrated upward to form the shallow crust.

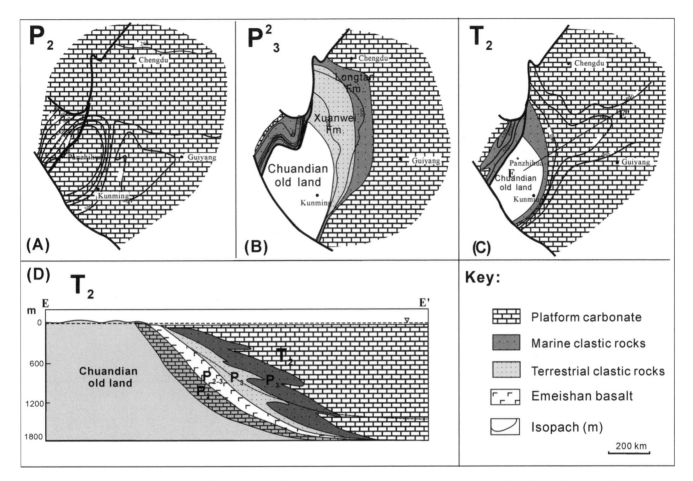

Figure 5. Diagram showing changes in sedimentation and lightfaces paleogeography before and after the Emeishan volcanism. (A) A stable, homogeneous carbonate platform in the Maokou Stage, which is much bigger than the Emeishan large igneous province. The isopach contours of remnant Maokou Formation indicates a domal thinning. (B) Paleogeography in the Wujiapingian. (C) Lithofacies paleogeography in the Middle Triassic. (D) A cross-section along the profile E–E′ (see panel C for the location of the profile). The numbers indicate thickness in meters. Data compiled BGMRGP (1987, 1991); Wang et al. (1994); from Feng et al. (1997). Modified from He et al. (2006).

Garnet has potentially important effects on the seismic crustal structure. For instance, 20% volume of garnet in a MORB-like bulk composition would increase V_p from 7.1 km/s to 7.4–7.5 km/s (Trumbull et al., 2002). This effect should be taken into account in southwestern China, where the crust is as thick as 65 km, a condition favorable for the stability of garnet (>30 km). Nevertheless, the following considerations argue against a significant role for garnet at the base of crust of the Emeishan LIP:

1. The high-velocity layer in southwestern China does not occur as a horizontal layer, as would be expected if the velocity of the lower crust were controlled by plagioclase-garnet phase transformation, because the latter takes place at fairly constant pressure. For instance, a V_p of 7.1 km/s in the Panxi area occurs at a depth of ~50 km; in contrast, similar velocity has been detected at a shallow depth (<40 km) near Lijiang.

2. The plagioclase-garnet phase transformation occurs at a depth of ~30 km. However, in southwestern China, the velocity at this depth varies between 6.2 km/s and 6.4 km/s, significantly lower than that for garnet granulites (>7.0 km/s, Rudnick, 1995).

3. As will be demonstrated later, the HVLC is likely composed of mafic-ultramafic components that have less aluminous contents than the overlying gabbros. The lack of a significant amount of plagioclase may hamper the garnet-formation reaction (Farnetani et al., 1996).

4. Finally, if the HVLC is due to the presence of garnet, the high density of this material would result in mechanical instability of the mafic crust. Delamination may put fundamental limitations on the thickness of the crust (~30 km; Sobolev and Babeyko, 1989). Trumbull et al. (2002) suggested that the crust in the Namibian margin (ca. 130 Ma) may have lost its lower part. This clearly did not happen in southwestern China, although the HVLC may be much older (ca. 260 Ma).

The HVLC may be composed of cumulates crystallized from picritic liquids derived by partial melting at relatively high pressure. Alternatively, the picritic melts formed some extensive and horizontal sills that crystallized as a nearly closed system (Bedard et al., 1988; Dick et al., 1991). In this scenario, little crustal differentiation occurred at the site, and some picritic magmas may have formed sills in the accreting lower crust, crystallizing in their entirety at depth. In the case of the Emeishan LIP, the HVLC may be predominantly composed of cumulates of pyroxene and olivine (Xu et al., 2003), although entrapment of some residual melts among cumulus minerals cannot be fully ruled out (e.g., Zhu et al., 2003). This argument is made based on the following lines:

1. The wide range in V_p of 7.1–7.8 km/s observed in southwestern China is not consistent with the total melt crystallization model, in which no variation in seismic velocity with depth is expected.
2. As argued by Kelemen and Holbrook (1995), the formation of sills likely occurred within 10 km of the surface, where picritic magma can cool rapidly. In a thick crust (>20 km), magmas underplated at the base of crust can undergo extensive crystal fractionation as a result of slow cooling. This is the situation in southwestern China, where the original crust thickness was ~35–40 km.
3. Solidified melts cannot account for the observed V_p, as high as 7.8 km/s. As demonstrated by Farnetani et al. (1996) and later by this study, high velocity (7.8 km/s) can be readily explained only by cumulate derived by fractionation from picritic melts.

Relationship between the Emeishan Basalts and the HVLC: Petrologic Modeling

In another section of this article, a genetic link between the Emeishan basalts and the high-velocity crust is established on the basis of the variation in crustal thickness across the domal structure and the synchronism between the emplacement of mafic or ultramafic intrusions and the eruption of the Emeishan basalts. Here we follow the approach of Farnetani et al. (1996) and use the observed V_p to constrain the composition of the HVLC. Low MgO (mostly < 7%) and Ni contents suggest that the Emeishan basalts underwent extensive crystal fractionation from parental magmas either in magma chambers or en route to the surface (Xu et al., 2001). We assume that fractionation of picritic magma produces ultramafic cumulates (i.e., HVLC) and residual low-MgO basaltic magma (the Emeishan basalts). This hypothesis is apparently at odds with a petrographic feature that suggests plagioclase and clinopyroxene rather than olivine as dominant fractionated minerals. Yet the fractionation of plagioclase and clinopyroxene likely took place under relatively low pressure, whereas cumulates in the HVLC in southwestern China were settled at ~15 kilobars.

To test this model, we used the MELTS program of Ghioso

and Sack (1995) to model the compositional evolution of deep-mantle plume melts, assuming that crystal fractionation is the dominant mechanism through which parental melts evolve toward basaltic magmas. The oxygen fugacity used in the calculation is the quartz-fayalite-magnetite (QFM) buffer. We used recent estimates of primary magmas (Xu and Chung, 2001) as initial melts and calculated the fractionated mineral assemblage as under 10 and 15 kilobars. These pressures are equivalent to the thickness of the crust prior to basaltic underplating in south-

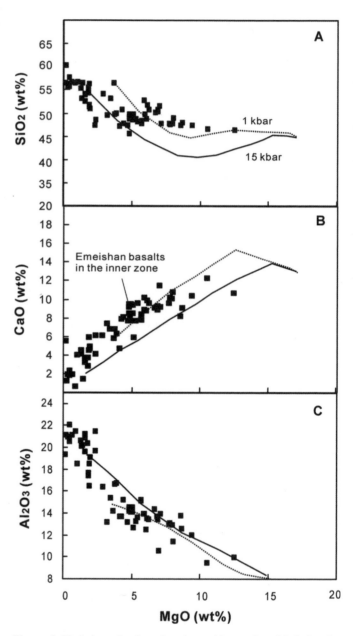

Figure 6. Variation of selected major oxides against MgO for the Emeishan basalts in the inner zone (Xu et al., 2003, and unpublished data). Also shown are fractionation trends at 1 kilobar (kbar) and 15 kilobars, calculated using MELTS by Ghioso and Sack (1995). The composition of the primary melt is after Xu and Chung (2001).

TABLE 1. CALCULATION OF FRACTIONATION AT DIFFERENT PRESSURES

	SiO_2	TiO_2	Al_2O_3	FeO	MgO	CaO	Na_2O	K_2O	Sp	Cpx	Opx	Ol	F	Calculated V_p
									\multicolumn{4}{c}{Proportion of fractionated minerals}		
Parental melt	44.75	2.55	7.27	12.50	16.77	12.78	1.31	0.19						
Residual melt														
15 kilobars	39.98	4.53	9.56	17.23	13.01	11.1	2.41	0.40	2.5	93.7	2.4	1.5	54.0	7.5–7.8 km/s
10 kilobars	41.01	4.86	10.89	16.44	10.59	10.45	2.72		2.0	77.3	8.5	12.1	59.0	7.5–7.8 km/s
5 kilobars	43.13	4.67	11.58	14.40	9.35	10.92	2.70	0.89	2.5	70.3	0.0	27.2	56.3	7.6–7.9 km/s
3 kilobars	44.05	4.45	11.5	13.49	9.2	11.52	2.56	0.82	2.4	58.6	0.0	39.0	52.4	7.6–8.0 km/s
1 kilobars	44.76	4.22	11.23	12.8	9.3	12.3	2.40	0.36	2.5	53.0	0.0	44.5	48.2	7.8–8.0 km/s

Note: Calculation is performed using petrologic code MELTS (Ghioso and Sack, 1995). Sp—spinel; Cpx—clinopyroxene; Opx—orthopyroxene; Ol—olivine; V_p—compressional wave velocity.

Source: Composition of parental magmas is from Xu and Chung (2001).

western China (~40 km, i.e., the crustal thickness outside of the Emeishan LIP). Plagioclase does not appear as a liquidus phase during fractionation under these pressures, consistent with the absence of Eu anomalies and increase in Al_2O_3 with decreasing MgO in the Emeishan basalts (Cong, 1988; Xu et al., 2001). The main fractionating phases include spinel (Sp), orthopyroxene (Opx), clinopyroxene (Cpx), and olivine (Ol). Opx generally crystallizes at the beginning of fractionation, which is then dominated by Cpx. Figure 6 shows that fractionation of picrites can produce residual melts that are compositionally similar to the Emeishan basalts. A 54% fractionation of picritic melts at 15 kilobars generates an ultramafic cumulus with a volumetric proportion of Sp : Cpx : Opx : Ol of 2.5 : 93.7 : 2.4 : 1.5 (Table 1). The calculated V_p for this fractionated assemblage is 7.5–7.8 km/s. A similar result is obtained for the fractionation at 10 kilobars (Table 1). These calculated V_p are at the high end of the observed seismic velocities (7.1–7.8 km/s) in the Emeishan LIP.

Zhu et al. (2003) argued that entrapment of a few percentage points of residual melts in cumulus minerals can lower the calculated velocity to the observed values (7.1 km/s). Alternatively, the cumulate layer with lower V_p (~7.1 km/s) may be derived from a fractionating melt that lost mafic minerals in an early stage of fractionation. For instance, if we chose experimental melts produced at 10 kilobars (Hirose and Kushiro, 1993) as initial melts, fractionation of this melt at 10 kilobars would produce an assemblage of Sp, Cpx, and plagioclase with a volume proportion of 0.8:58.2:41. The corresponding V_p ranges between 7.0 and 7.3 km/s, which is very similar to the low end of the observed seismic velocity. Another possibility to lower V_p involves the hybridization of underplated magmas with the preexisting crustal material. At this stage, it is not possible to make a distinction between these alternatives.

The high-velocity (6.0–6.6 km/s) upper crust in the inner zone of the Emeishan LIP may have resulted from magmatic intraplating. To test this model, we performed calculations at lower pressures (5, 3, and 1 kilobars). Because olivine enters its liquidus phase earlier at lower pressures than at high pressure, the involvement of olivine (a high-velocity mineral) in fraction-

ated assemblage (olivine gabbros) yields a high-velocity range of 7.6–8.0 km/s (Table 1), significantly exceeding the observed values. Even if a two-stage fractionation is considered (i.e., residual melt after fractionation at Moho depth, fractionation again at shallow crustal levels), a velocity greater than 7.0 km/s is obtained for cumulates (see Farnetani et al., 1996). The observed high-velocity upper crust in the Emeishan case is therefore unlikely related to mineral segregation processes of plume-derived melts. It is proposed here that it may represent a mush of mineral cumulates and solidified residual melts.

IMPLICATIONS FOR THE TOTAL VOLUME OF THE EMEISHAN MAGMAS

While the mantle plume model has been increasingly adopted for the generation of the Emeishan basalts (Chung and Jahn, 1995; Xu et al., 2001), some doubts have been cast on the viability of this model because of the relatively small dimension of this igneous province compared with typical LIPs (e.g., Thompson et al., 2001). The Emeishan basalts underwent extensive erosion (Xu et al., 2004; He et al., 2006) and tectonic disruption (Xiao et al., 2003). Thus the estimates of their total volumes remain problematic. Lin (1985) estimated the average lava thickness of the Emeishan basalts to be ~700 m. Taking the exposure surface of 2.5×10^5 km^2, the entire volume of the Emeishan basalts is estimated to be ~0.2×10^6 km^3. This represents a minimum estimate (Xu et al., 2001) because (1) complicated tectonic movements in Meso-Cenozoic eras in this region cut off the western extension of the LIP (Chung et al., 1998; Xiao et al., 2003), (2) erosion must have removed a significant portion of the eruptive sequences, and (3) the associated intrusives are not taken into account.

As discussed earlier, the emplacement of igneous materials in the lower crust was genetically related to the erupted basalts; therefore, this high-velocity layer can be considered an integral part of the Emeishan LIP. It is possible that the HVLC occurs predominantly within the inner zone of the dome, given the contrast in the crust-mantle structure between the inner zone and other zones. In this sense, a minimum volume (2.5×10^6 km^3) of igneous materials accreted to preexisting crust can be estimated. A higher estimate could be expected if intraplating (i.e., magmas trapped at different crustal level) is taken into account. This estimate, in turn, provides constraints on the volume of the erupted lavas, because cumulates and erupted magmas can be related by mass balance (Cox, 1989): $C_P = C_L X_L + C_C X_C$, where C_P, C_L, and C_C are the concentrations of an element in parental magma, erupted liquid, and cumulate, respectively, and X_L and X_C are the respective mass fractions of erupted liquid and cumulates. Using this equation and the elemental concentrations in parents, erupted basalts, and cumulate, the ratio X_C/X_L of 0.65 is obtained. Accordingly, the minimum volume of erupted magmas is estimated to be 3.8×10^6 km^3, equivalent to the volume of the typical LIPs on the Earth (Coffin and Eldholm, 1994). As a consequence, the volume of igneous materials is not at odds with the plume model proposed for the formation of the Emeishan LIP.

CONCLUSIONS

1. Geophysical survey (deep seismic refraction or reflection and seismic tomography) reveals a thick and seismically heterogeneous crust beneath the Emeishan LIP. The crustal thickness decreases gradually from the inner zone (~60 km) via the intermediate zone (<45 km) to the outer zone (< 40 km). The crust in the inner zone is composed of a relatively high-velocity (6.0–6.6 km/s) upper crust and a 20-km-thick high-velocity (7.1–7.8 km/s) lower crust. These high-velocity bodies are not apparent in the intermediate and outer zones. This, together with the synchronism of the emplacement of mafic or ultramafic intrusions and the eruption of Emeishan basalts, strongly suggests a genetic link between the plume activity and the formation of the high-velocity crust if the crustal doming preceding the volcanism resulted from dynamic uplift induced by the plume that generated the Emeishan basalts.

2. Considerable crustal thickening in the inner zone may have resulted from magmatic underplating associated with the Emeishan volcanism. Crustal growth by basaltic accretion to the preexisting crust finds its supporting evidence in systematic stratigraphic study and paleogeographic reconstruction that reveal a prolonged (~45 m.y.) crustal elevation (i.e., the Chuandian old land) in the Emeishan LIP. The dimension of this domal permanent uplift is roughly the same as that of HVLC detected by seismic imaging. Moreover, the extent of permanent uplift (~2 km) estimated in terms of geologic basis (He et al., 2006) suggests a 20-km-thick underplated layer if the isostatic theory is applied (MacLennan and Lovell, 2002). This inferred volume of underplated materials is similar to the average thickness of the HVLC observed in the Emeishan LIP.

3. Petrologic modeling further suggests that the HVLC is composed of cumulates (pyroxenite and olivine pyroxenites) crystallized from picritic melts. The data presented in this study therefore lend support to the conceptual model proposed by Farnetani et al. (1996) for the formation of the HVLC in LIPs.

4. The volume of the erupted lavas can be estimated because cumulates and erupted magmas can be related by mass balance (Cox, 1989). This yields a minimum volume of erupted magmas of 3.8×10^6 km^3, equivalent to the volume of the typical LIPs on the Earth (Coffin and Eldholm, 1994).

ACKNOWLEDGMENTS

We thank C. Farnetani and an anonymous referee for constructive reviews. We also thank Gillian Foulger and Donna Jurdy for their editorial comments and patience. This study is jointly supported by the Natural Science Foundation of China (40421303 and 40234046), the Chinese Academy of Sciences, and the Guangzhou Institute of Geochemistry (GIG-CX-04-06).

REFERENCES CITED

Barton, A.J., and White, R.S., 1997, Crustal structure of the Edoras Bank continental margin and mantle thermal anomalies beneath the North Atlantic: Journal of Geophysical Research, v. 105, p. 25,829–25,853.

Bauer, A.J., Neben, S., Schreckenberger, B., Emmermann, R., Hinz, K., Fechner, N., Gohl, K., Schulze, A., Trumbull, R.B., and Weber, K., 2000, Deep structure of the Namibia continental margin as derived from integrated geophysical studies: The MAMBA experiment: Journal of Geophysical Research, v. 105, p. 25,829–25,853, doi: 10.1029/2000JB900227.

Bedard, J.H., Sparks, R.S.J., Renner, R., Cheadle, M.J., and Hallworth, M.A., 1988, Peridotite sills and metasomatic gabbros in the eastern layered series of the Rhum complex: Journal of the Geological Society of London, v. 145, p. 207–224.

Brazier, R.A., and Nyblade, A.A., 2003, Upper mantle P velocity structure beneath the Baikal Rift from modeling regional seismic data: Geophysical Research Letters, v. 30, no. 4, art. no. 1153.

Brodie, J., and White, N., 1994, Sedimentary basin inversion caused by igneous underplating: Northwest European continental shelf: Geology, v. 22, p. 147–150, doi: 10.1130/0091-7613(1994)022<0147:SBICBI>2.3.CO;2.

Bureau of Geology and Mineral Resources of Guizhou Province (BGMRGP), 1987, Regional geology of Guizhou province: Beijing, Geological Publishing House, 698 p. (in Chinese with English abstract).

Bureau of Geology and Mineral Resources of Sichuan Province (BGMRSP), 1991, Regional geology of Sichuan province: Beijing, Geological Publishing House, 745 p. (in Chinese with English abstract).

Campbell, I.H., and Griffiths, R.W., 1990, Implications of mantle plume structure for the evolution of flood basalts: Earth and Planetary Science Letters, v. 99, p. 79–93, doi: 10.1016/0012-821X(90)90072-6.

Caress, D.V., McNutt, M.K., Detrick, R.S., and Mutter, J.C., 1995, Seismic imaging of hotspot-related crustal underplating beneath the Marquesas Islands: Nature, v. 373, p. 600–603, doi: 10.1038/373600a0.

Catchings, R.D., and Mooney, W.D., 1988, Crustal structure of the Columbia Plateau: Evidence for continental rifting: Journal of Geophysical Research, v. 93, p. 459–474.

Chung, S.L., and Jahn, B.M., 1995, Plume-lithosphere interaction in generation of the Emeishan flood basalts at the Permian-Triassic boundary: Geology, v. 23, p. 889–892, doi: 10.1130/0091-7613(1995)023<0889:PLIIGO>2.3.CO;2.

Chung, S.L., Jahn, B.M., Wu, G.Y., Lo, C.H., and Cong, B.L., 1998, The Emeishan flood basalt in SW China: A mantle plume initiation model and its connection with continental break-up and mass extinction at the Permian-Triassic boundary, *in* Flower, M.F.J., et al., eds., Mantle dynamics and plate Interaction in East Asia: Washington, D.C., American Geophysical Union, Geodynamics Series, v. 27, p. 47–58.

Coffin, M.F., and Eldholm, O., 1994, Large igneous provinces: Crustal structure, dimension, and external consequences: Review of Geophysics, v. 32, p. 1–36, doi: 10.1029/93RG02508.

Cong, B., 1988, The formation and evolution of the Panxi paleo-rift system: Beijing, Scientific Publishing House (in Chinese).

Courtillot, V.E., Jaupart, C., Manighetti, I., Tapponnier, P., and Besse, J., 1999, On causal links between flood basalts and continental breakup: Earth and Planetary Science Letters, v. 166, p. 177–195, doi: 10.1016/S0012-821X(98)00282-9.

Cox, K.G., 1980, A model for flood basalt volcanism: Journal of Petrology, v. 21, p. 629–650.

Cox, K.G., 1989, The role of mantle plumes in the development of continental drainage patterns: Nature, v. 342, p. 873–877, doi: 10.1038/342873a0.

Cox, K.G., 1993, Continental magmatic underplating: Royal Society of London Philosophical Transactions (Series A), v. 342, p. 155–166.

Cui, Z.Z., Luo, D.Y., Chen, J.P., Zhang, Z.Y., and Huang, L.Y., 1987, Deep crust structure and tectonics in the Panxi area: Acta Geophysica Sinica, v. 30, p. 566–579.

Davis, P.M., and Slack, P.D., 2002. The uppermost mantle beneath the Kenya dome and relation to melting, rifting and uplift in East Africa: Geophysical Research Letters, v. 29, no. 7, art. no. 1117.

Dick, H.J.B., Meyer, P.S., Bloomer, S., Kirby, S., Stakes, D., and Mawer, C., 1991, Lithostratigraphic evolution of an in-situ section of oceanic layer 3: Proceedings of the Ocean Drilling Program, Scientific Results, v. 118, p. 439–538.

Farnetani, C.G., and Richards, M.A., 1994, Numerical investigation of the mantle plume initiation model for flood basalt event: Journal of Geophysical Research, v. 99, p. 13,813–13,883, doi: 10.1029/94JB00649.

Farnetani, C.G., Richards, M.A., and Ghiorso, M.S., 1996, Petrological models of magma evolution and deep crustal structure beneath hotspots and flood basalts: Earth and Planetary Science Letters, v. 143, p. 81–94, doi: 10.1016/0012-821X(96)00138-0.

Feng, Z.Z., Yang, Y.Q., and Jin, Z.K., 1997, Lithofacies paleography of Permian of south China: Beijing, Petroleum University Press, 242 p. (in Chinese with English abstract).

Furlong, K.P., and Fountain, D.M., 1986, Continental crustal underplating: Thermal considerations and seismic-petrological consequence: Journal of Geophysical Research, v. 91, p. 8285–8294.

Furumoto, A.S., Webb, J.P., Odegard, M.E., and Husong, D.M., 1976, Seismic studies on the Ontong Java Plateau: Tectonophysics, v. 34, p. 71–90.

Ghioso, M.S., and Sack, R.O., 1995, Chemical mass transfer in magmatic processes, IV, A revised and internally consistent thermodynamic model for the interpretation and extrapolation of liquid-solid equilibria in magmatic systems at elevated temperatures and pressure: Contributions to Mineralogy and Petrology, v. 119, p. 197–212.

Gradstein, F.M., Ogg, J.G., Smith, A.G., Bleeker, W., and Lourens, L.J., 2004, A new geologic time scale, with special reference to Precambrian and Neogene: Episodes, v. 27, p. 83–100.

Hanski, E., Walker, R.J., Huhma, H., Polyakov, G.V., Balykin, H.T.T., and Phuong, N.T., 2004, Origin of the Permian-Triassic komatiites, northwestern Vietnam: Contributions to Mineralogy and Petrology, v. 147, p. 453–469, doi: 10.1007/s00410-004-0567-1.

He, B., Xu, Y.G., Chung, S.L., Xiao, L., and Wang, Y.M., 2003, Sedimentary evidence for a rapid, kilometer scale crustal doming prior to the eruption of the Emeishan flood basalts: Earth and Planetary Science Letters, v. 213, p. 391–405, doi: 10.1016/S0012-821X(03)00323-6.

He, B., Xu, Y.G., Wang, Y.-M., and Luo, Z.Y., 2006, Sedimentation and lithofacies paleogeography in SW China before and after the Emeishan flood volcanism: New insights into surface response to mantle plume activity: Journal of Geology, v. 114, p. 117–132, doi: 10.1086/498103.

He, B., Xu, Y.G., Huang, X.L., Luo, Z.Y., Shi, Y.R., Yang, Q.J., and Yu, S.Y., 2007, Age and duration of the Emeishan flood volcanism, SW China: Geochemistry and SHRIMP zircon U-Pb dating of silicic ignimbrites, post-volcanic Xuanwei Formation and clay tuff at the Chaotian section: Earth and Planetary Science letters, v. 255, p. 306–323.

Hirose, K., and Kushiro, I., 1993, Partial melting of dry peridotites at high pressure: Determination of compositions of melts segregated from peridotite using aggregates of diamond: Earth and Planetary Science Letters, v. 114, p. 477–489, doi: 10.1016/0012-821X(93)90077-M.

Kelemen, P., and Holbrook, S., 1995, Origin of thick, high-velocity igneous crust along the U.S. East Coast margin: Journal of Geophysical Research, v. 100, p. 10,077–10,094, doi: 10.1029/95JB00924.

Lin, J.Y., 1985, Spatial and temporal distribution of Emeishan basaltic rocks in three southwestern provinces (Sichuan, Yunnan and Guizhou) of China: Chinese Science Bulletin, v. 12, p. 929–932.

Liu, J., Liu, F., He, J., Chen, H., and You, Q., 2001, Study of seismic tomography in Panxi paleorift area of southwestern China—Structural features of crust and mantle and their evolution: Science in China (Series D), v. 44, p. 277–288.

Liu, Y.L., Wu, C.Z., Chen, J.C., Liu, H.C., Wu, L.G., and Li, S.Y., 1987, Gravity anomaly features of Dukou-Xizang area and a study of rift problems, *in* Yuan, X.C., ed., Contribution to the Panzhihua-Xichang Rift, China, III: Beijing, Geological Publishing House, p. 90–98 (in Chinese).

Liu, F.T., Liu, J.H., Zhong, D.L., He, J.K., and Yu, Q., 2000, The subducted slab of Yangtze continental block beneath the Tethyan orogen in western Yunnan: Chinese Science Bulletin, v. 45, p. 466.

MacLennan, J., and Lovell, B., 2002, Control of regional sea level by surface uplift and subsidence caused by magmatic underplating of Earth's crust: Geology, v. 30, p. 675–678, doi: 10.1130/0091-7613(2002)030<0675: CORSLB>2.0.CO;2.

MacLennan, J., McKenzie, D., Gronvold, K., and Slater, L., 2001, Crustal accretion under northern Iceland: Earth and Planetary Science Letters, v. 191, p. 295–310, doi: 10.1016/S0012-821X(01)00420-4.

Mauffret, A., and Leroy, S., 1997, Seismic stratigraphy and structure of the Caribbean igneous province: Tectonophysics, v. 283, p. 61–104, doi: 10.1016/S0040-1951(97)00103-0.

McKenzie, D., 1984, A possible mechanism for epeirogenic uplift: Nature, v. 307, p. 616–618, doi: 10.1038/307616a0.

Menzies, M.A., Klemperer, S.L., Ebinger, C.J., and Baker, J., 2002, Characteristics of volcanic rifted margins, *in* Menzies, M.A., et al., eds., Volcanic rifted margins,: Boulder, Colorado, Geological Society of America Special Paper 362, p. 1–14.

Mutter, J.C., Talwani, M., and Stoff, P.L., 1984, Evidence for a thick oceanic crust adjacent to the Norwegian margin: Journal of Geophysical Research, v. 89, p. 483–502.

Rudnick, R.L., 1995, Making continental crust: Nature, v. 278, p. 571–578.

Rudnick, R.L., and Jackson, I., 1995, Measured and calculated elastic wave speeds in partially equilibrated mafic granulite xenoliths: Implications for the properties of an underplated lower continental crust: Journal of Geophysical Research, v. 100, p. 10,211–10,218, doi: 10.1029/94JB03261.

Saunders, A.D., Jones, S.M., Morgan, L.A., Pierce, K.L., Widdowson, M., and Xu, Y.G., 2007, The role of mantle plumes in the formation of continental large igneous provinces: Field evidence used to constrain the effects of regional uplift: Chemical Geology, v. 241, p. 282–318.

Sheth, H.C., 2007 (this volume), Plume-related regional prevolcanic uplift in the Deccan Traps: Absence of evidence, evidence of absence, *in* Foulger, G.R., and Jurdy, D.M., eds., Plates, plumes, and planetary processes: Boulder, Colorado, Geological Society of America Special Paper 430, doi: 10.1130/2007.2430(36).

Sobolev, S.V., and Babeyko, A.Y., 1989, Phase transformations in the lower continental crust and its seismic structure, *in* Mereu, R.S., et al., eds., Properties and processes of Earth's lower crust: Washington, D.C., American Geological Union, Geophysical Monograph 51, p. 311–320.

Sparks, R.S.J., Meyer, P., and Sigurdsson, H., 1980, Density variation amongst mid-ocean ridge basalts: Implications for magma mixing and the scarcity of primitive lava: Earth and Planetary Science Letters, v. 46, p. 419–430, doi: 10.1016/0012-821X(80)90055-2.

Sun, R., Liu, F., and Liu, J., 1991, Seismic tomography of Sichuan: Acta Geophysica Sinica, v. 34, p. 708–716.

Tan, T.K., 1987, Geodynamics and tectonic evolution of the Panxi rift: Tectonophysics, v. 133, p. 287–304, doi: 10.1016/0040-1951(87)90271-X.

Teng, W.J., 1994. Physics and dynamics of the lithosphere in the Kangdian tectonic belt: Beijing, Scientific Press, 256 p. (in Chinese).

Thompson, G.M., Ali, J.R., Song, X., and Jolley, D.W., 2001, Emeishan basalts, SW China: Reappraisal of the formation's type area stratigraphy and a discussion of its significance as a large igneous province: Journal of the Geological Society of London, v. 158, p. 593–599.

Trumbull, R.B., Sobolev, S.V., and Bauer, K., 2002. Petrophysical modeling of high seismic velocity crust at the Namibian volcanic margin, *in* Menzies, M.A., et al., eds., Volcanic rifted margins: Boulder, Colorado, Geological Society of America Special Paper 362, p. 225–234.

Wang, L.T., Lu, Y.B., Zhao, S.J., and Luo, J.H., 1994. Permian lithofacies paleogeography and mineralization in south China: Beijing, Geological Publishing House, 147 p. (in Chinese with an English abstract).

Watts, A.B., and ten Brink, U.S., 1989, Crustal structure, flexure and subsidence history of the Hawaiian Islands: Journal of Geophysical Research, v. 94, p. 10,473–10,500.

White, R., and McKenzie, D., 1989, Magmatism at rift zones: The generation of volcanic continental margins and flood basalts: Journal of Geophysical Research, v. 94, p. 7685–7729.

Xiao, L., Xu, Y.-G., Chung, S.-L., He, B., and Mei, H.J., 2003, Chemostratigraphic correlation of Upper Permian lava succession from Yunnan Province, China: Extent of the Emeishan Large Igneous Province: International Geologic Review, v. 45, p. 753–766.

Xiong, S., Teng, J., Yin, Z., Lai, M., and Huang, Y., 1986, Explosion seismological study of the structure of the upper mantle at the southern part of the Panxi tectonic belt: Acta Geophysica Sinica, v. 29, p. 235–244.

Xu, Y.G., and Chung, S.L., 2001, The Emeishan large igneous province: Evidence for mantle plume activity and melting conditions: Geochimica, v. 30, p. 1–9.

Xu, Y.G., Chung, S.L., Jahn, B.M., and Wu, G.Y., 2001, Petrologic and geochemical constraints on the petrogenesis of Permian–Triassic Emeishan flood basalts in southwestern China: Lithos, v. 58, p. 145–168, doi: 10.1016/S0024-4937(01)00055-X.

Xu, Y.G., Mei, H.J., Xu, J.F., Huang, X.L., Wang, Y.J., and Chung, S.L., 2003, Origins of two differentiation trends in the Emeishan flood basalts: Chinese Science Bulletin, v. 48, p. 390–394, doi: 10.1360/03tb9082.

Xu, Y.G., He, B., Chung, S.L., Menzies, M.A., and Frey, F.A., 2004, The geologic, geochemical and geophysical consequences of plume involvement in the Emeishan flood basalt province: Geology, v. 32, no. 10, p. 917–920, doi: 10.1130/G20602.1.

Yan, Q., Zhang, G., Kan, R., and Hu, H., 1985, The crust structure of Simao to Malong profile, Yunnan Province, China: Journal of Seismological Research, v. 8, no. 2, p. 249–280.

Yuan, X.C., 1995, Geophysical maps of China: Beijing, Geological Publishing House, 200 p.

Zhang, Y.G., and Ma, D.B., 1985, The investigation of seismic wave velocity and crustal structure in Sichuan region by use of mine explosions: Acta Geophysica Sinica, v. 28, no. 4, p. 377–388.

Zhang, Y.X., Luo, Y.N., and Yang, Z.X., 1988, Panxi Rift: Beijing, Geological Publishing House, 466 p. (in Chinese with English abstract).

Zhong, H., Yao, H., and Prevec, S.A., 2004, Trace-element and Sr-Nd isotopic geochemistry of the PGE-bearing Xinjie layered intrusion in SW China: Chemical Geology, v. 203, p. 237–252, doi: 10.1016/j.chemgeo.2003.10.008.

Zhou, M., Malpas, J., Song, X., Robinson, P.T., Sun, M., Kennedy, A.K., Lesher, C.M., and Keays, R.R., 2002, A temporal link between the Emeishan large igneous province (SW China) and the end-Guadalupian mass extinction: Earth and Planetary Science Letters, v. 196, p. 113–122, doi: 10.1016/S0012-821X(01)00608-2.

Zhou, M.F., Robinson, P.T., Lesher, C.M., Keays, R.R., Zhang, C.J., and Malpas, J., 2005, Geochemistry, petrogenesis and metallogenesis of the Panzhihua gabbroic layered intrusion and associated Fe-Ti-V oxide deposits, Sichuan Province, SW China: Journal of Petrology, v. 46, p. 2253–2280, doi: 10.1093/petrology/egi054.

Zhu, D., Luo, T.Y., Gao, Z.M., and Zhu, C.M., 2003, Differentiation of Emeishan flood basalts at the base of the crust and throughout the crust of Southwest China: International Geology Review, v. 45, p. 471–477.

Zucca, J.J., 1984, The crustal structure of the southern Rhinegraben from reinterpretation of seismic refraction data: Journal of Geophysics, v. 55, p. 13–22.

MANUSCRIPT ACCEPTED BY THE SOCIETY JANUARY 31, 2007

DISCUSSION

21 December 2006, H.C. Sheth

Xu and He (this volume) write:

After having studied drainage patterns of three continental LIPs, Cox (1989) suggested topographic doming associated with plume activity can be preserved after ca. 200 Ma, and crustal thickening by magmatic underplating is the most likely cause for the persistence of such features. Although the work by Cox (1989) on the Deccan Traps receives some criticism (Sheth, this volume), the link between underplating and persistent uplift remains valid.

One of the objectives of my article (Sheth, this volume) was to show that major uplift in many flood basalts is significantly younger (tens of millions of years) than the flood volcanism and associated underplating. It is important to get the facts right. We are simply not seeing uplift *commencing* at or just before flood basalt volcanism and *persisting* for tens of hundreds of millions of years, as Cox (1989) imagined and Xu and He have quoted. Rather, we are seeing uplift *commencing* tens of millions of years after the flood volcanism and all underplating have ended. See my chapter in this volume for studies that have argued for major Neogene uplift of the Indian peninsula and western Indian rifted margin.

In this connection, Xu and He may be interested in reading three articles (Bonow et al., 2006a,b; Japsen et al., 2006) published on west Greenland this year. These nicely complement my chapter. The authors' main argument is that major *Neogene* uplift events (>2 km) raising flat planation surfaces have shaped the present topography of west Greenland, and the youngest is as young as 7–2 Ma. The flood volcanism occurred in the Paleogene, of course. Is it really Paleogene underplating, then, that has caused such major uplift?

It is well to acknowledge that we do not know. But keeping the facts right and looking for a solution is a better practice, to my mind, than lumping every uplift event of any age whatsoever with magmatic underplating and hypothetical plume heads that simply won't die.

See also the Discussion of Sheth (this volume).

5 January 2007, Warren B. Hamilton

Many papers on the Emeishan igneous province have appeared in the Western literature in the past decade, but mostly they present only local trace elements or other details or, like this article by Xu and He (this volume) and many others of which one or both are co-authors, they consist of internally and externally inconsistent plumological conjectures. The papers have been uncritically cited by many proponents of hypothetical plumes elsewhere. Supporting data for the syntheses are in the Chinese literature, where they cannot be evaluated by Western readers—but the data base obviously is weak.

My comments here express some of the frustration of a reader trying to comprehend the Emeishan. I have been in the area briefly (but have not looked at the Emeishan) and appreciate the difficulties of working in a region of deep weathering and mostly poor exposures.

The schematic regional geologic map by Xu and He (this volume, Fig. 1; essentially the same as He et al., 2003, Fig. 1) shows a quite different distribution of basalt and altogether different regional structure than does the map by Lo et al. (2002, Fig. 1), even though the two papers share as an author Chung, who has been senior author on a number of Emeishan papers and must have access to the published data.

[Note added 16 February 2007: After receiving this Discussion, Xu and He (see their following response) substituted a new geologic map for their initial Figure 1. My comment above applies to the initial Figure 1. The new map uses the strikingly different Emeishan-basalt distribution of Lo et al. (2002, Fig. 1B) while retaining the fault pattern of the initial Figure 1 of Xu and He (this volume). On the new figure, great regional post-Emeishan faults cross small to huge outcrops of Emeishan without offsetting them; and the purported isopachs of the Maokou Formation cross through younger major faults without being offset. Adding these new conflicts to the numerous weaknesses and inconsistencies in this and other Emeishan reports reinforces my conviction that the geology and geophysics are too poorly known to provide support for the unique and elaborate geodynamic and crustal-evolution speculations of Xu and He (this volume) and other plumologists.]

A province of igneous rocks exists, but is it properly a "large igneous province" of "flood basalts"? Correlative A- and I-type granites occur in the heart of the area (Zhong et al., 2007). The areal extent of rocks of a likely unified province is small, ~0.3 × 10^6 km^2 (Alt et al., 2005). Preserved sections apparently are mostly thin, and the 5-km local section schematized by Lo et al. (2002, inset in Fig. 1) is of rocks (including kilometer-thick pyroclastics) suggestive to me of a volcano rather than flood basalts. (Or perhaps much of the volcanic section was shallow-submarine, and the purported great dome never existed, which I infer from other data.) What are the igneous rock types (not just their Ti/Y ratios), and do they have subregional stratigraphy? The assignment to a single age and petrographic province of basalts on all sides of the major faults (Xu and He, this volume, Fig. 1) in this complex region is implausible.

The primary basis for the assertion of Xu and He (this volume) that the Emeishan basalt "is unambiguously related to a mantle plume" (i.e., it can be forced into a speculative plume model) appears to be the prebasalt doming deduced by He et al. (2003). The latter work, alone and in comparison with Xu and He (this volume), contains so many inconsistencies that it is not at all clear that the purported dome exists. Thus, He et al. (2003) show the Permian limestones to be syndepositionally faulted along a regional prebasalt fault, with a great change in thickness across the fault (their Fig. 4), yet their isopachs (Fig. 1) all pass undisturbed across this purported fault. They define doming (Fig. 3) by zones of fusulinids (all rocks were deposited in the photic zone) in unscaled stratigraphic columns of Middle Permian limestone that show basalt as lying on progressively older zones in the center of the dome; but their geologic map (Fig. 1) shows most of the truncated limestone sections as 20 to 50 km from the nearest outcrops of basalt. Further, most of the fusulinid identifications in the domal region are of a few genera only, not of the preferable species, so stratigraphic control on the truncations may be poor.

In Xu and He (this volume, Fig. 5), the fault noted above has been replaced by a concordant section of, from lowest upward, Middle Permian limestone, Emeishan basalt, Upper Permian terrestrial and marine clastic rocks, and Lower Triassic limestone. The limestone varies little in thickness: the stratigraphic basis for the flank of the dome has disappeared. The great doming, the basis for the hypothetical plume, nevertheless is still postulated; the dome produced no unconformity but rose high from just below sea level immediately before the basalt was deposited, then was promptly flattened and ended up slightly below sea level again.

Xu and He (this volume, Fig. 5) present sketches of interpretations of velocity boundaries in reconnaissance "seismic sounding" lines. However, their description of the seismic experimental details is too cursory (the details are again in a Chinese publication) for the reader to judge the quality and reliability of the results. The individual lines were apparently interpreted by different interpreters with quite different methods and philosophies, and Xu and He (this volume) attach geodynamic significance to the differences. One line shows steep faults continuous to depths of 80 km (!), one shows a midcrustal low-velocity zone continuous for 200 km laterally, and one shows only monotonic velocity increases. I infer that the data are of poor quality and that the lines have been seriously overinterpreted.

Further, the basal-crustal layer of $V_p > 7$ km/s, to which Xu and He (this volume) attribute great local plumological significance, is common in many regions. Xu and He (this volume) give the reader no basis for judging the viability of the tomography (Fig. 3), which also is important to their model.

I do not share the authors' faith in chemical calculations, particularly in the absence of mineralogical data, to determine the precise sources and complex fractionation histories of the rising melts that produced the evolved rocks erupted at the surface.

I may be wrongly maligning a good data base. If so, I urge the authors to make more of that base accessible to non-Chinese readers and to clarify the major inconsistencies among their various papers.

4 February 2007, Yi-Gang Xu and Bin He

In our chapter we refer to the relationship between permanent uplift and underplating in LIPs. This is mentioned by Cox (1989) and other groups using numerical modeling and field geology (e.g., McKenzie, 1984; White et al., 1987; MacLennan and Lovell, 2002). The link between underplating and persistent uplift is valid.

Sheth argues (his comment of 21 December 2006) that uplift at Deccan commencing at or just before flood basalt volcanism and persisting for tens or hundreds of millions of years is absent. Without field experience on the Deccan Traps, we cannot comment except to say that we were inspired by the approach of Cox (1989) to investigate the Emeishan LIP by relating geomorphology and deep mantle processes. Widdowson arrived at different conclusions from Sheth (Saunders et al., 2007) and proposes three episodes of crustal uplift at the Deccan. Prevolcanism uplift is most likely in the Kutch–Cambay–western Narmada region, though sedimentary evidence is equivocal. This may be because of the short-lived thermal and dynamic uplift in the north, because the northward movement of India decoupled the initial plume-head-impact site from the plume center. The pattern of uplift preserved in the sedimentary sequences surrounding the Deccan and the distribution of lavas are consistent with northward movement of the continent over a stationary hotspot (Saunders et al., 2007). Permanent uplift occurs along the western Ghats as a result of magmatic underplating, isostatic adjustment caused by scarp recession and erosion, and deposition across the thinned continental margin (Widdowson and Cox, 1996; Widdowson, 1997). It seems clear that transient and permanent uplift occurred, features shared with the Emeishan case (He et al., 2003, 2006; Xu and He, this volume).

We regret that the schematic geologic map by Xu and He (this volume, Fig. 1) shows a different distribution of basalt from that by Lo et al. (2002). We took the distribution from a Memoir of the Chinese Geological Survey (Regional Geology of Sichuan, Yunnan and Guizhou provinces, Geological Publishing House, Beijing). Lo et al. (2002) based their map on the Ph.D. thesis of Huang (1986). Discrepancies might reflect mapping difficulties where erosion is intensive and volcanic sequences are tilted.

We see no problem with the occurrence of correlative A- and I-type granites in the area (Zhong et al., 2007, i.e., in the inner zone in terms of our subdivision scheme). When plume–derived basalts pond in the crust, they melt roof rocks, provided their melting temperature is sufficiently low (Campbell, 2001). The resultant products vary widely in composition. Thus, the coexistence of A- and I-type granites is a consequence of plume-related melting events and has been documented at many LIPs. The exposure of granites in the inner zone is the result of intensive erosion, as demonstrated by He et al. (2003).

Basalt is not abundant on the western Yangtze craton (~0.3 $\times 10^6$ km^2; Xu et al., 2001), but this does not argue against "the Emeishan LIP." The Emeishan province is a remnant of deep erosion and is also disrupted tectonically (Courtillot et al., 1999; Xu et al., 2004; He et al., 2006). Tectonic disruption accounts for the late Permian mafic-ultramafic bodies in southeast Yunnan (Xiao et al., 2003) and northern Vietnam (Hanski et al., 2004). Coeval late Permian mafic magmatism also occurs in Guangxi province and Songpan-Ganze region (see also Xiao et al., 2004a; Zhou et al., 2006), suggesting a much larger exposure of the original basalts.

The absence of basalts in the uplifted area (i.e., the inner zone) is caused by enhanced erosion in the uplifted area. Nevertheless, lava section thicknesses in the inner zone are 1–5 km (Xu et al., 2004), with most sections >2 km thick. The volcanic succession comprises predominantly basaltic flows and subordinate pyroclastics (Ross et al., 2005), a feature most evident in the Yongsheng section.

A wide range of rocks occur in the province, including picrite, basalts (tholeiitic and alkali), basaltic andesites, hawaiites, mugearite to benmoreite, phonolite, and rhyolite. Because Ti/Y is an effective petrogenetic indicator (Peate et al., 1992; Xu et al., 2001) and helps decipher the thermal structure of the postulated mantle plume (Xu et al., 2004), a chemical classification of rock type is adopted. Although low-Ti and high-Ti lavas do not necessarily correspond to specific rock types, most low-Ti lavas are tholeiitic and basaltic andesite, whereas most high-Ti lavas are transitional. The temporal and spatial relationship between low- and high-Ti basalts (see Xu et al., 2001, 2004; Xiao et al., 2004b) reveals the stratigraphy. Over 350 analyses throughout ten entire sections provide an excellent geochemical dataset.

Our age assessment does not rely on one age, but from stratigraphic correlation, Ar-Ar dating on basalts, and zircon U-Pb dating on mafic intrusions, alkaline rocks, and silic ignimbrite (Zhou et al., 2002, 2005; Zhong and Zhu, 2006; Zhong et al., 2007; He et al., 2007).

Isopachs (Xu and He, this volume, Fig. 1) are from interpolating the thicknesses of 67 sections of the Maokou Formation. The possible influence of long-term faulting (active from Late Permian to present) was not included. Indeed, it is difficult to determine the exact location of the faults of a zone tens of kilometers wide because a limited number of stratigraphic columns were investigated, and the basalts and underlying strata are strongly tilted. Thus, the isopachs are shown undisturbed across the purported fault, though they do reflect a syndoming fault across which the Maokou Formation thickens from ~50 m to >300 m. Stratigraphic columns (except section A) of Middle Permian limestone are scaled in Figure 3, with the thickness of the Maokou Formation marked at lower left in every section. The Maokou limestone in the Inner zone is incomplete because of erosion and thermally-derived metamorphism, and as a result, only a few fusilinids have been found in the domal region.

We emphasize that all the basalts overlie the Maokou Formation (Fig. 3 in He et al., 2003). Hamilton's comment that limestone sections in Figure 1 (He et al., 2003) are separated from the nearest basalt outcrops by 20–50 km is incorrect. Some outcrops of basalts are simply too small to be clear on the map.

Figure 5 shows changes in sedimentation and paleogeography before and after volcanism, so faults were not included. It is clear that the fault lies between the Chuandian old land (i.e., the center of the LIP) and the east of the LIP. Only in the east does a concordant section occur, of (in upward order) Middle Permian limestone, Emeishan basalt, Upper Permian terrestrial and marine clastic rocks, and Lower Triassic limestone. Two unconformities occur between the Maokou limestone and Emeishan basalts and between basalts and Upper clastic rocks (He et al., 2003). The unconformities are distinct in the inner zone, where the Maokou limestone is variably thinned and capped by Triassic sediments. Figure 5 shows clearly that the center of the Emeishan LIP was domally uplifted during the Late Permian to Triassic (He et al., 2006).

Hamilton is correct that the basal-crustal layer of $Vp > 7$ km/s is common and widespread. Such a high-velocity structure can be explained by different mechanisms. Hamilton ignores the reasoning we used to develop the plume-related model. We reiterate: The seismically anomalous bodies at different levels are all distributed within the inner zone of the Emeishan LIP, e.g., high-velocity upper crust is absent in the intermediate and outer zones. The western and eastern margins of this seismically anomalous body correspond to longitudes 100.8°E and 102.8°E, agreeing well with the location of the inner zone (Xu et al., 2004). These observations indicate a common factor governing the formation of these seismically anomalous bodies and crustal doming.

As petrologist/geologists we cannot assess the quality of the tomography (Fig. 3) and the experimental results. Figure 5 compiles velocity data from different groups, and overinterpretation might have occurred. In our chapter, we focused on the DDP results and velocity structure, not the fault penetration depths. We were encouraged by the first-order features of these geophysical data because different approaches showed similar crust–mantle structure (e.g., thick and high velocity in the inner zone). Also, the crust–mantle structure inferred from velocity profiles correlates with domal structure inferred from sedimentary records. We believe this correlation is not just a coincidence.

Two independent data types constrain the genesis of the HVLC; the crust–mantle structure in the Emeishan LIP varies systematically with the domal structure and the petrologic calculations. We accord equal importance to these two arguments. The latter approach has been widely employed, and thermodynamic modeling of magmatic fractionation at variable P-T conditions is well established (Ghioso and Sack, 1995). This approach has been justified by Farnetani et al. (1996), MacLennan et al. (2001), and Trumbull et al. (2002).

REFERENCES CITED

Alt, J.R., Thompson, G.M., Zhou, M.-F., and Song, X., 2005, Emeishan large igneous province, SW China: Lithos, v. 79, p. 475–489.

Bonow, J.M., Japsen, P., and Lidmar-Bergstrom, K., 2006a, Palaeosurfaces in central West Greenland as reference for identification of tectonic movements and estimation of erosion: Global and Planetary Change, v. 50, p. 161–183.

Bonow, J.M., Japsen, P., Lidmar-Bergstrom, K., Chalmers, J.A., and Pedersen, A.K., 2006b, Cenozoic uplift of Nuusuuaq and Disko, West Greenland—Elevated erosion surfaces as uplift markers of a passive margin: Geomorphology, v. 80, p. 325–337.

Campbell, I.H., 2001. Identification of ancient mantle plumes, *in* Ernst, R.E., and Buchan, K.L., eds., Mantle plumes: Their identification through time: Boulder, Colorado, Geological Society of America Special Paper 352, p. 5–22.

Courtillot, V.E., Jaupart, C., Manighetti, I., Tapponnier, P., and Besse, J., 1999. On causal links between flood basalts and continental breakup: Earth and Planetary Science Letters, v. 166, p. 177–195.

Cox, K. G., 1989, The role of mantle plumes in the development of continental drainage patterns: Nature, v. 342, p. 873–877.

Farnetani, C.G., Richards, M.A., and Ghiorso, M.S., 1996, Petrological models of magma evolution and deep crustal structure beneath hotspots and flood basalts: Earth and Planetary Science Letters, v. 143, p. 81–94.

Ghioso, M.S., and Sack, R.O., 1995, Chemical mass transfer in magmatic processes, IV. A revised and internally consistent thermodynamic model for the interpretation and extrapolation of liquid-solid equilibria in magmatic systems at elevated temperatures and pressure: Contributions to Mineralogy and Petrology, v. 119, p. 197–212.

Hanski, E., Walker, R.J., Huhma, H., Polyakov, G.V., Balykin, P.A., Hoa, T.T., and Phuong, N.T., 2004. Origin of the Permian-Triassic komatiites, northwestern Vietnam: Contributions to Mineralogy and Petrology, v. 147, p. 453–469.

He, B., Xu, Y.-G., Chung, S.-L., Xiao, L., and Wang, Y., 2003, Sedimentary evidence for a rapid, kilometer-scale crustal doming prior to the eruption of the Emeishan flood basalts: Earth and Planetary Science Letters, v. 213, p. 391–405.

He, B., Xu, Y.G., Wang, Y.M., and Zhen, Y.L., 2006, Sedimentation and lithofacies paleogeography in SW China before and after the Emeishan flood volcanism: New insights into surface response to mantle plume activity: Journal of Geology, v. 114, p. 117–132.

He, B., Xu, Y.-G., Huang, X.-L., Luo, Z.-Y., Shi, Y.-R., Yang, Q.-J., and Yu, S.-Y., Age and duration of the Emeishan flood volcanism, SW China: Geochemistry and SHRIMP zircon U–Pb dating of silicic ignimbrites, postvolcanic Xuanwei Formation and clay tuff at the Chaotian section, Earth and Planetary Sciences Letters, v. 255, p. 306–323.

Huang, K.N., 1986, The petrological and geochemical characteristics of the Emeishan basalts from SW China and the tectonic setting of their formation [Ph.D. thesis]: Beijing, Institute of Geology, Academy Sinica, 140 p. (in Chinese).

Japsen, P., Bonow, J.M., Green, P.F., Chalmers, J.A., and Lidmar-Bergstrom, K., 2006, Elevated, passive continental margins: Long-term highs or Neogene uplifts? New evidence from West Greenland: Earth and Planetary Science Letters, v. 248, p. 330–339.

Lo, C.-H., Chung, S.-L., Lee, T.-Yi, and Wu, G., 2002, Age of the Emeishan flood magmatism and relations to Permian-Triassic boundary events: Earth and Planetary Science Letters, v. 198, p. 449–458.

MacLennan, J., and Lovell, B., 2002, Control of regional sea level by surface uplift and subsidence caused by magmatic underplating of Earth's crust: Geology, v. 30, p. 675–678.

MacLennan, J., McKenzie, D., Gronvold, K., and Slater, L., 2001, Crustal accretion under northern Iceland: Earth and Planetary Science Letters, v. 191, p. 295–310.

McKenzie, D., 1984, A possible mechanism for epeirogenic uplift: Nature, v. 307, p. 616–618.

Peate, D.W., Hawkesworth, C.J., and Mantovani, M.S.M., 1992, Chemical stratigraphy of the Parana lavas (South America): Classification of magma-types and their spatial distribution: Bulletin of Volcanology, v. 55, p. 119–139.

Ross, P.-S., Ukstins Peate, I., McClintock, M.K., Xu, Y.G., Skilling, I.P., White, J.D.L., and Houghton, B.F., 2005, Mafic volcaniclastic deposits in flood basalt provinces: A review: Journal of Volcanology and Geothermal Research, v. 145, p. 281–314.

Saunders, A.D., Jones, S.M., Morgan, L.A., Pierce, K.L., Widdowson, M., and Xu, Y.G., 2007, The role of mantle plumes in the formation of continental LIPs: Field evidence used to constrain the effects of regional uplift: Chemical Geology, v. 241, p. 282–318.

Sheth, H.C., 2007 (this volume), Plume-related regional prevolcanic uplift in the Deccan Traps: Absence of evidence, evidence of absence, *in* Foulger, G.R., and Jurdy, D.M., eds., Plates, plumes, and planetary processes: Boulder, Colorado, Geological Society of America Special Paper 430, doi: 10.1130/2007.2430(36).

Trumbull, R.B., Sobolev, S.V., and Bauer, K., 2002, Petrophysical modeling of high seismic velocity crust at the Namibian volcanic margin, *in* Menzies, M.A., et al., eds. Volcanic rifted margins: Boulder, Colorado, Geological Society of America Special Paper 362, p. 225–234.

White, R.S., Spence, G.D., Fowler, S.R., McKenzie, D.P., Westbrook, G.K., and Bowen, A.N., 1987, Magmatism at rifted continental margins: Nature, v. 330, p. 439–444.

Widdowson, M., 1997, Tertiary palaeosurfaces of the SW Deccan, Western India: Implications for passive margin uplift, *in* Widdowson, M., ed., Palaeosurfaces: Recognition, reconstruction and palaeoenvironmental interpretation: London, Geological Society of London Special Publication 120, p. 221–248.

Widdowson, M., and Cox, K.G., 1996, Uplift and erosional history of the Deccan Traps, India: Evidence from laterites and drainage patterns of the Western Ghats and Konkan Coast: Earth and Planetary Science Letters, v. 137, p. 57–69.

Xiao, L., Xu, Y.-G., Chung, S.-L., He, B., and Mei, H.J., 2003, Chemostratigraphic correlation of Upper Permian lava succession from Yunnan Province, China: Extent of the Emeishan LIP: International Geologic Review, v. 45, p. 753–766.

Xiao, L., Xu, Y.-G., Xu, J.-F., Bin He, B., and Pirajno, F., 2004a, Chemostratigraphy of flood basalts in the Garze-Litang Region and Zangza Block: Iimplications for western extension of the Emeishan LIP, SW China:. Acta Geologica Sinica, v. 78, p. 61–67.

Xiao, L., Xu, Y.-G., Mei, H.J., Zheng, Y.F., He, B., and Pirajno, F., 2004b, Distinct mantle sources of low-Ti and high-Ti basalts from the western Emeishan LIP, SW China: Implications for plume-lithosphere interaction: Earth and Planetary Science Letters, v. 228, p. 525–546.

Xu, Y.-G., Chung, S.L., Jahn, B.M., and Wu, G.Y., 2001, Petrologic and geochemical constraints on the petrogenesis of Permian–Triassic Emeishan flood basalts in SW China: Lithos, v. 58, p. 145–168.

Xu, Y.-G., He, B., Chung, S.L., Menzies, M.A., and Frey, F.A., 2004, The geologic, geochemical and geophysical consequences of plume involvement in the Emeishan flood basalt province: Geology, v. 30, p. 917–920.

Xu, Y.-G., and He, B., 2007 (this volume), Thick, high-velocity crust in the Emeishan large igneous province, southwestern China: Evidence for crustal growth by magmatic underplating or intraplating, *in* Foulger, G.R., and Jurdy, D.M., eds., Plates, plumes, and planetary processes: Boulder, Colorado, Geological Society of America Special Paper 430, doi: 10.1130/2007.2430(39).

Zhong, H., and Zhu, W.-G., 2006, Geochronology of layered mafic intrusions from the Pan-Xi area in the Emeishan LIP, SW China: Mineral Deposit, v. 41, p. 599–606.

Zhong, H., Zhu, W.-G., Chu, Z.Y., He, D.F., and Song, X.Y., 2007, SHRIMP U-Pb zircon geochronology, geochemistry, and Nd-Sr isotopic study of contrasting granites in the Emeishan LIP, SW China: Chemical Geology, v. 236, p. 112–133.

Zhou, M.F., Malpas, J., Song, X.Y., Robinson, P.T., Sun, M., Kennedy, A.K., Lesher, C.M., and Keays, R.R., 2002, A temporal link between the Emeishan LIP (SW China) and the end-Guadalupian mass extinction: Earth and Planetary Science Letters, v. 196, p. 113–122.

Zhou, M.F., Robinson, P.T., Lesher, C.M., Keays, R.R., Zhang, C.J., and Malpas, J., 2005, Geochemistry, petrogenesis, and metallogenesis of the Panzihua gabbroic layered intrusion and associated Fe-Ti-V-oxide deposits, Sichuan Province, SW China: Journal of Petrology, v. 46, p. 2253–2280.

Zhou, M.-F., Zhao, J.-H., Qi, L., Su, W.H., and Hu, R.Z., 2006, Zircon U-Pb geochronology and elemental and Sr-Nd isotopic geochemistry of Permian mafic rocks in the Funing area, SW China: Contributions to Mineralogy and Petrology, v. 151, p. 1–19.

The Geological Society of America
Special Paper 430
2007

The coronae of Venus: Impact, plume, or other origin?

Donna M. Jurdy*

Department of Earth and Planetary Sciences, Northwestern University, Evanston, Illinois 60208, USA

Paul R. Stoddard*

Department of Geology and Environmental Geosciences, Northern Illinois University, DeKalb, Illinois 60115, USA

ABSTRACT

The surface of Venus hosts hundreds of circular to elongate features, ranging from 60 to 2600 km, and averaging somewhat over 200 km, in diameter. These enigmatic structures have been termed "coronae" and attributed to either tectonovolcanic or impact-related mechanisms. A quantitative analysis of symmetry and topography is applied to coronae and similarly sized craters to evaluate the hypothesized impact origin of these features. Based on the morphology and global distribution of coronae, as well as crater density within and near coronae, we reject the impact origin for most coronae. The high level of modification of craters within coronae supports their tectonic nature. The relatively young Beta-Atla-Themis region has a high coronal concentration, and within this region individual coronae are closely associated with the chasmata system. Models for coronae as diapirs show evolution through a sequence of stages, starting with uplift, followed by volcanism and development of annuli, and ending with collapse. With the assumption of this model, a classification of coronae is developed based merely on their interior topography. This classification yields corona types corresponding to stages that have a systematic variation of characteristics. We find that younger coronae tend toward being larger, more eccentric, and flatter than older ones, and generally occur at higher geoid and topography levels.

Keywords: Venus, coronae, impact craters, volcanism, tectonism

INTRODUCTION

Of the terrestrial planets, Venus most resembles Earth, but with key differences. The two planets have similar size, density, and surface basalt composition; however, Venus, unlike Earth, undergoes retrograde rotation, has no magnetic field, lacks water, and has a very dense atmosphere. Although Venus does not display Earthlike plate tectonics, it does show evidence of both volcanic and tectonic activity. Here we will argue that numerous large circular features on Venus, called coronae, are a manifestation of these processes.

Venus's surface hosts nearly 1000 unambiguous impact craters, ranging in diameter from 1.5 to 280 km. The planetary crater density has been used to infer a relatively young surface age for Venus, in the range of 750–300 Ma (Phillips et al., 1992; Schaber et al., 1992). A more recent estimate (McKinnon et al., 1997) widens that range, to 1000–300 Ma. To a first order, the crater distribution approximates a random distribution (Phillips et al., 1992); however, terranes may differ in crater density (Ivanov and Basilevsky, 1993; Price et al., 1996). More specifically, a slight deficit, of about twenty craters, has been documented, statistically, near Venus's chasmata (Stefanick and Jurdy,

*E-mails: donna@earth.northwestern.edu, prs@geol.niu.edu.

Jurdy, D.M., and Stoddard, P.R., 2007, The coronae of Venus: Impact, plume, or other origin? *in* Foulger, G.R., and Jurdy, D.M., eds., Plates, plumes, and planetary processes: Geological Society of America Special Paper 430, p. 859–878, doi: 10.1130/2007.2430(40). For permission to copy, contact editing@geosociety.org. ©2007 The Geological Society of America. All rights reserved.

1996). The majority of the craters appear pristine in radar images, although slightly fewer than two hundred display clear modification by either volcanic or tectonic activity or both. Craters, when viewed at the highest available resolution, however, often reveal evidence of subtle modification (Herrick, 2006). In addition, some craters show enigmatic, parabolic halos: impact-related phenomena unique to Venus. These parabolic halo-associated craters preserve impact debris that has settled in the presence of zonal winds and may represent the most recent 10% of Venus's history (Basilevsky and Head, 2002).

Veneras 15 and 16 mapped Venus's surface with radar that Barsukov et al. (1986) used to identify ringlike, uplifted features, named "coronae" or "ovoids." Pronin and Stofan (1990), using corona morphology, further classified 32 features that had been identified on ~20% of the planet, as imaged by Venera radar. With the Magellan probe's improved resolution and radar coverage exceeding 90%, Stofan et al. (1992) were able to catalogue and characterize 362 structures that have an "annulus of concentric tectonic features." These coronae and coronalike features were classified according to morphology; individual features ranged from 60 to 2000+ km in diameter, and many of them displayed circular to elliptical annuli and raised interiors.

Coronae on Venus have been attributed to a variety of mechanisms. When coronae were first identified on Venus's surface, they were considered "volcano-tectonic" features because of associated deformation and lava flows (Barsukov et al., 1986). However, in noting five hundred additional circular features of "unclear origin," the authors speculated whether these could have resulted from the "reworking" of ancient impact basins (Barsukov et al., 1986). In their analysis of Venera data, however, Pronin and Stofan (1990) selected twenty-one features for which they identify corona characteristics. From these examples they documented an evolutionary sequence for coronae with initial uplift and volcanism and later annulus development. On the basis of raised topography and associated volcanism, coronae were attributed to diapirs or hotspots, and their clustering and location at tectonic sites further suggested that coronae were related to Venus's global tectonics.

Noting, from Magellan images, the nonrandom distribution of coronae and the age progression for overlapping coronae, Stofan et al. (1992) attributed coronae to the effects of mantle plumes beneath a stationary venusian surface. Later, Stofan and Smrekar (2005) attributed Venus's large topographic rises ("regiones") to mantle plumes. They inferred multiple scales of upwelling on Venus, with coronae operating at an intermediate scale between volcanoes and larger volcanic rises. Furthermore, Stofan and Smrekar (2005) postulated that due to the lack of plate tectonics, Venus may release heat via a larger number of secondary upwellings, coming from a shallower level, which generate plumes. Koch and Manga (1996) replicated the raised rims of coronae with a model for a rising diapir that spreads at the level where it reaches neutral buoyancy. Based on the diapir model, DeLaughter and Jurdy (1999) reclassified coronae according to the extent of interior uplift of each structure, which they interpreted as a measure of stage, or degree of maturity, of

individual features. Noting the selective location of coronae, Johnson and Richards (2003) argued that coronae could be due to small-scale transient effects coexisting with larger-scale upwellings, such as those that produce major highland provinces. A more complete history and description of the variety of proposed models for coronae formation are provided by Herrick et al. (2005).

Recently, Venus's coronae have once again been interpreted as impact-related. Vita-Finzi et al. (2005) analyzed an expanded database of 514 coronae, of which 362 had been catalogued by Stofan et al. (1992); the remainder, which they termed "stealth" coronae, were features with incomplete annuli from the catalogue of Tapper et al. (1998). Based on comparison of the morphology and distribution of Venus's coronae with those of lunar craters, Vita-Finzi et al. (2005) argued that coronae are impact features and that the variation in corona form results from the location of the impact and subsequent modification. Hamilton (2005) asserted that, on Venus, a gradation exists between pristine, generally accepted craters and much older, highly deformed features, and that circular features classified as coronae are in fact the consequence of ancient impacts. As noted in both these studies, a reinterpretation of coronae as impact features gives a much-expanded catalogue of craters. This would result in a much older estimate of the age of Venus's surface and bring into question the proposed resurfacing event at ca. 1000–300 Ma. In addition, the identification of Venus's coronae as impact features would require revisiting models for the evolution and heat loss of our sister planet.

In this study, we evaluate a variety of mechanisms that have been proposed to form coronae. Starting with the impact hypothesis, we assess evidence for the age and activity of coronae and compare their distribution with that of known impact craters. We develop and apply a quantitative approach to compare the topographic symmetry of selected, similarly sized features classified as either coronae or craters. Next, we investigate whether the proposed evolutionary model for coronae results in stages that have systematic characteristics, such as size, shape, and dip. We end by proposing a model for coronae that best explains the observations and noting remaining questions.

DATA SETS

The Magellan mission, 1990–1994, provided nearly full coverage of Venus. Because of the thick atmosphere, radar was used, operating in three modes: nadir-directed altimetry, synthetic-aperture-radar (SAR) imaging, and thermal emission radiometry. The altimetry footprint was dependent on direction and latitude, but generally ranged between 10 and 30 km, and the vertical resolution was typically 5–50 m. Over its three cycles, the Magellan radar succeeded in imaging 98% of the planet's surface at high resolution (~100 m), changing the angle of incidence between cycles (Pettengill et al., 1991; Saunders et al., 1991). Consequently, ~10% of the areas were imaged two or three times with different incident angles, allowing very high-resolution topographic analysis using stereo imaging, with op-

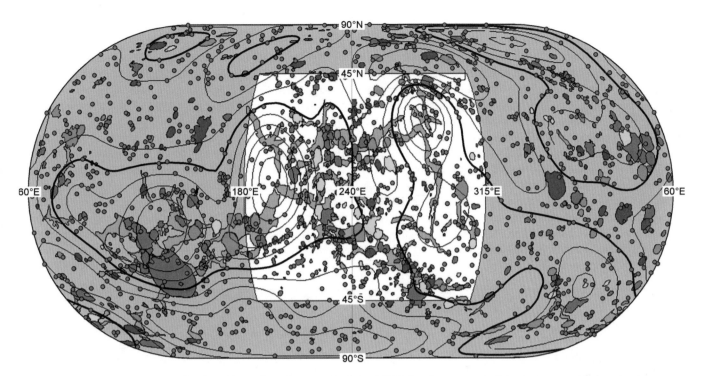

Figure 1. Eckert IV (equal area) projection of Venus's surface, centered on 120°W, showing craters (red dots), chasmata (pink regions), coronae (gray, yellow, green, and blue regions, as defined in the text), and geoid (order 10) 30 m contours. The unshaded area represents the region depicted in Figure 2.

timal lateral resolution exceeding that of the altimetry by a factor of ~100 (Plaut, 1993). The nearly complete coverage provided by Magellan and Pioneer Venus allowed the gravity field, and the corresponding potential field (the geoid), to be determined at a relatively high resolution (order 90) (Sjogren et al., 1997). For our study, we use the geoid field, as well as global altimetry data and the SAR images collected by Magellan.

Classification of coronae remains subjective, as reflected in the numerous published corona catalogues. Herrick et al. (2005), for example, addressed the issue of differentiating volcanoes from coronae, noting the problem that some features are in catalogues of both types. For our analysis we draw our data set from the intersection of the Price and Suppe (1995) and the DeLaughter and Jurdy (1999) catalogues (Fig. 1). Price and Suppe (1995) mapped 669 distinct features as coronae, defined as "circular to irregular volcanic-tectonic features characterized by an annulus of concentric deformation," and ranked according to the increasing proportion of "new volcanic flows" associated with each. A set of 335 coronae that DeLaughter and Jurdy (1999) were able to classify derives from their analysis of a total of 394 features from three sources (Schaber et al., 1992; Stofan et al., 1992; Magee Roberts and Head, 1993). We further discuss this scheme in a later section.

Impact crater distribution and morphology are the primary tools used to analyze planetary surface ages and processes. Here we use the 940-crater catalogue of Phillips and Izenberg (personal commun., 1994; Phillips et al., 1992), as shown in Figure 1. As previously mentioned, these craters have a first-order ran-

dom global distribution, indicating a nearly uniform surface age for Venus. More detailed analysis (Price and Suppe, 1995) suggested a terrane-based density structure, with plains the most heavily cratered (and thus oldest) terrane. Morphologically, Phillips et al. (1992), using radar imagery, identified the minority of craters that have been obviously modified: 158 tectonized and 55 embayed, with 19 craters showing clear evidence of being both tectonized and embayed. However, close analysis of craters at very high resolution (<100 m), made possible with stereo imaging, reveals tectonic and volcanic activity for numerous craters, suggesting that perhaps most of the craters have been modified (Herrick and Sharpton, 2000; Herrick, 2006). Both studies thus concluded that volcanic activity on Venus may be more widespread than initially believed. It is important to remember, however, that stereo imaging is possible for only a small percentage of the surface, so any such studies are necessarily very limited in their scope. For example, in their study of the Beta-Atla-Themis region, Matias and Jurdy (2005) found only 13 out of 153 craters with the necessary double coverage and only two with triple; thus in that region fewer than 10% of the craters are candidates for stereo imaging. We use the global crater data set of Phillips and Izenberg in our analyses, as well as their assessment of modification. We argue that those craters obviously modified, as judged directly from radar images, have suffered more significant alteration than the subtle modifications that can be discerned only with stereo imaging, and thus are more indicative of major alteration processes.

Like the Earth, Venus has a global rift system. The 1978

Pioneer missions to Venus provided radar and gravity data that enabled Schaber (1982) to identify a global system of extensional features on Venus, which he cited as evidence of tectonic activity, despite the apparent lack of Earth-style plate tectonics. Schaber attributed the extension to upwelling-related processes, such as at Earth's continental rift zones, but noted the global scope of these extension zones, similar in scale to Earth's mid-ocean ridge system. These rift zones can be fit by four great circle arcs (Schaber, 1982). Using the nearly global coverage provided by Magellan, Solomon et al. (1992) characterized these rift zones, termed "chasmata," as rugged regions with some of Venus's deepest troughs, extending thousands of kilometers. They noted the extreme relief, with elevation changing as much as 7 km in just 30 km distance. The 54,464-km-long Venus chasmata system, as defined in greater detail by Magellan, can be fit by great circle arcs at the 89.6% level, and when corrected for the smaller size of the planet, the total length of the chasmata system measures (Jurdy and Stefanick, 1999) within 2.7% of the 59,200 km length of the spreading ridges determined for Earth by Parsons (1981). The chasmata with the greatest relief on Venus experienced linear rifting during the latest stage of tectonic deformation (Head and Basilevsky, 1998). The chasmata shown in Figure 1 were derived from mapping by Price and Suppe (1995). The geoid of Venus, as determined from Magellan data, is superposed on the other features in Figure 1. We display a smooth geoid field (order 10), similar in scale to the features of interest.

Venus's regiones, broad areas of relief, number around ten. Stofan and Smrekar (2005) noted that regiones range in diameter from 1000 to 2700 km, rise between 0.5 and 2.5 km above the surrounding terranes, and have positive gravity anomalies. A regio might be dominated by rifts, as are Atla and Beta, discussed here (Fig. 2), or by volcanism, as are Imdr, Bell, and

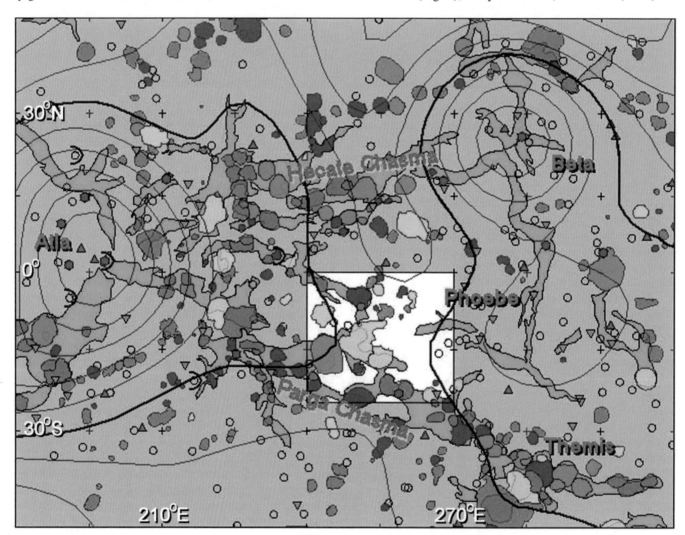

Figure 2. The "BAT" region—the area between Beta, Atla, and Themis regiones. Pink—chasmata; yellow—domal coronae; green—circular coronae; blue—calderic coronae (classifications from DeLaughter and Jurdy, 1999). Dark gray—unclassified coronae; open circles—pristine craters; red upward-pointing triangles—embayed craters; light blue downward-pointing triangles—tectonized craters; purple circles—craters that have been both tectonized and embayed. Craters with dark halos are indicated by black arcs. Contour lines are for geoid (order 10) 30 m contours. The unshaded area is shown in detail in Figure 4.

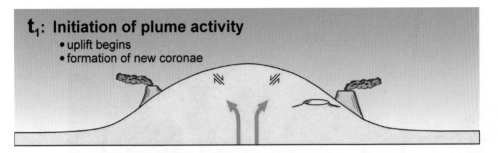

t_1: **Initiation of plume activity**
- uplift begins
- formation of new coronae

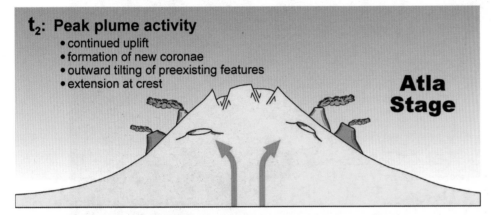

t_2: **Peak plume activity**
- continued uplift
- formation of new coronae
- outward tilting of preexisting features
- extension at crest

Atla Stage

Figure 3. Model for the evolution of regio features, such as Atla and Beta.

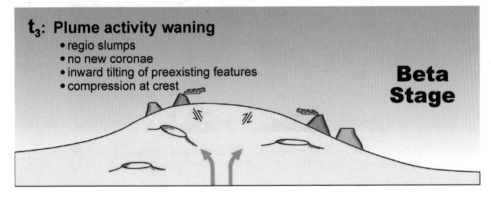

t_3: **Plume activity waning**
- regio slumps
- no new coronae
- inward tilting of preexisting features
- compression at crest

Beta Stage

Dionne, or be dominated by coronae, as is Themis (Stofan and Smrekar, 2005). The rift-dominated regiones, Atla and Beta, have the greatest topographic expression. Atla and Beta are the sites of several rift intersections and the two major geoid highs on Venus. Curiously, the geoid "bulls' eyes" also coincide with the intersections of arcs fitting the chasmata. The current deformation of Venus's surface has been described as being caused by a swell-push force, the result of a steep gradient of the geoid height (Sandwell et al., 1997). Thus, these areas may be experiencing the most intense deformation on the planet, and the network of rifts may have formed in response to this deformation.

Coronae occur in many rift segments, yet none actually occurs at these intersection points. Perhaps just as remarkable, Atla has a partial ring of four domal coronae, all between four and five geoid contours from the crest, while Beta has a partial ring of six or so calderic coronae between three and four contours from its crest. Possibly, thicker crust at the regiones in-

hibits the formation of coronae in association with chasmata (Bleamaster and Hansen, 2004). On the other hand, using geoid-topography ratios, models for these highlands suggest thinning of a thick lithosphere (150–350 km) to as little as 100 km over an anomalously hot (by as much as 400–1000K) asthenosphere (Moore and Schubert, 1997). Such an analysis (which requires wavelengths greater than 600 km) is not appropriate for coronae, unfortunately, given their smaller size as well as their close proximity to each other. The observed distribution of coronae relative to the regiones led Stoddard and Jurdy (2003) to hypothesize that Atla represents a younger phase of large-scale upwelling than Beta (Fig. 3). Craters, initially formed flat, show some tilting around Atla and Beta (Jurdy et al., 2003; Matias et al., 2004) consistent with an active uplift of Atla and a recent slumping of Beta. Additionally, coronae often intertwine with chasmata or are contained by the chasmata walls (Fig. 4).

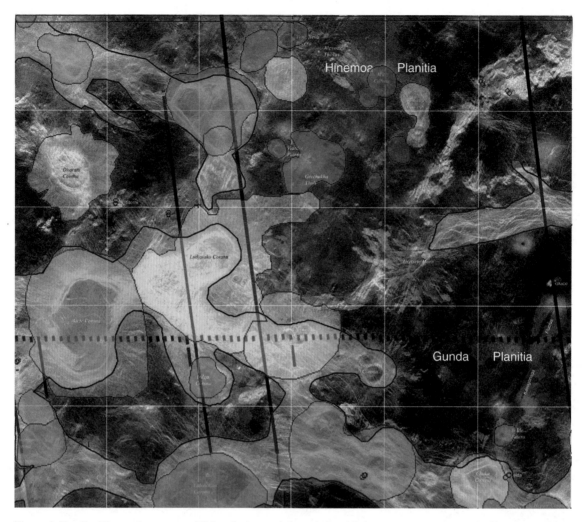

Figure 4. Detail of Parga chasm exemplifying the intertwining relationship between many coronae (yellow, green, blue, green, and gray) and chasmata (pink). The region extends from 240°E to 270°E, 0°S to 25°S. 5° grid. The radar image is from http://planetarynames.wr.usgs.gov/images/v40_comp.pdf.

TESTING THE IMPACT ORIGIN HYPOTHESIS

Crater Statistics

An ancient impact origin for coronae, as opposed to recent endogenic activity, makes several testable predictions. If, as suggested by Hamilton (2005, 2007) and Vita-Finzi et al. (2005), coronae are ancient impact features, with ages of up to 3.9 Ga, they should be significantly more heavily cratered than the younger, average-aged surface. Previously, Namiki and Solomon (1994) evaluated impact crater density within coronae interiors, finding evidence for significantly lower densities within late-stage coronae, which they defined as being dominated by volcanism. Price and Suppe (1995) evaluated crater density on various terranes and found that coronae, as a terrane type, have low crater densities. Using the corona data set of Stofan et al. (1992), DeLaughter and Jurdy (1997) found low crater density out to four corona radii near the uplifted coronae, whereas the

density was about normal for those coronae with collapsed interiors. In our analysis, we found that although the 669 coronae from the map of Price and Suppe (1995) occupy 10.6% of Venus's surface, they host only 7.0% (66 of 940) of the crater population (Table 1), suggesting that the coronae are, as a whole, younger than the average surface age. On the other hand, an excess (again, when compared to surface area) of Phillips et al.'s (1992) unambiguously tectonized craters (24, or 15.2%, of 158 total) are found within coronae, indicating that coronae are more tectonically active than the average surface region. Somewhat surprisingly, embayed craters are under-represented within coronae (5.5% of the planetary population on 10.6% of the planet's surface). Of course, with smaller populations, interpretations from these statistics become less certain. For a random distribution, the standard deviation is the square root of the number counted (Fisher, 1973, sect. 15). Thus, for a count of 4 items, one standard deviation would include counts of 2–6 (4 ± 2). Obviously, for such small sets, it would not be possible to achieve counts that

TABLE 1. CRATER DENSITY IN CORONAE, RIFTS, AND THE BAT REGION

	Number of features	Coverage (% of Venus's surface area)	Craters on feature type		Tectonized craters on feature type		Embayed craters on feature type		Tectonized (T) and embayed (E) craters on feature type	
			No.	Percent of total craters	No.	Percent of tectonized craters	No.	Percent of embayed craters	No.	Percent of T and E craters
All coronae	669	10.6	66	7.0	24	15.2	3	5.5	2	10.5
Domal (Y)	39	0.9	3	0.3	1	0.6	0	0.0	0	0.0
Circular (G)	83	2.0	14	1.5	5	3.2	1	1.8	1	5.3
Calderic (B)	164	2.2	15	1.6	5	3.2	0	0.0	0	0.0
Rift segments	57	8.3	59	6.3	32	20.3	8	14.5	7	36.8
BAT region	1	26.5	224	23.9	44	27.8	27	49.1	11	57.9

Notes: "BAT" region—the area between Beta, Atla, and Themis regiones; number of features—total number of each feature on Venus's surface; coverage—total percentage of Venus's surface area covered by each type of feature. For crater columns, the total number of craters found on each type of feature is given, as well as the percentage of the total number of that type of crater found on Venus. For example, 24 tectonized craters were found on all coronae, which represents 15.2% of the 158 tectonized craters found on Venus.

deviate enough from the average to achieve two full standard deviations.

Another consequence of the impact origin hypothesis is that coronae, as ancient impact sites, should be concentrated on the oldest areas of the planet's surface. However, we claim that the opposite is true, that coronae are actually concentrated in the youngest region of Venus's surface. Coronae are most heavily concentrated in the so-called BAT region, the area between Atla, Beta, and Themis regiones (Fig. 2). We define this region roughly as between 45°N and 45°S, 180°E and 315°E. Using these boundaries, we find that 292 (43.6%) of all coronae lie within the BAT region, which itself comprises only 26.4% of the planet's surface. Unambiguous BAT region impact craters total 224 (23.9%), but if we assume that coronae are additional impact sites, 32% of all impact features are found here, which then indicates a somewhat older than average surface age. However, by most accounts the BAT region represents the youngest, most active region on Venus (e.g., Head and Basilevsky, 1998).

Several lines of evidence support this claim. Examining the set of young, commonly agreed-upon craters shows that the BAT region is somewhat deficient, arguing for a younger, not older, region. Also, the BAT region contains nearly two-thirds (by area) of all rifts as identified by Price and Suppe (1995). In their global sequence of tectonic deformation, Head and Basilevsky (1998) found that linear rifting prevailed in the latest stage of events. That rifts are among the most active (or most recently active) features on Venus can be further demonstrated by their relative dearth of craters and the plethora of tectonized and embayed craters (Table 1). Crater Uvaysi (2.3°N, 198.2°E) provides additional support of our conclusion that the BAT region has experienced very recent activity. This crater, at the intersection of three chasmata and nearly at Atla Regio's crest, has been classified as both tectonized and embayed. Opportunely, the clear evidence of modification is coupled with the presence of a radar-dark parabola with Uvaysi, near the apex of Atla Regio. As ar-

gued by Matias and Jurdy (2005), these two occurrences constrain the volcanism and tectonism of the crater as recent, because parabolic haloes remain from only the last 10% of Venus's surface history. Uvaysi is one of eleven of the planet's nineteen craters both tectonized and embayed (nearly 60%) contained within the BAT region. This nearly 60% measures more than double what would be expected based solely on area. Looking at cratering and stratigraphy, Vezolainen et al. (2004) suggested that Beta uplift began after the average age of the surface (T) and has continued until after 0.5 T. Basilevsky and Head (2007), on the basis of stratigraphic relations to neighboring terranes, also suggested recent or current uplift of Beta Regio. Taken in its entirety, the BAT region itself also shows the relative lack of craters and the excess of modified craters seen by the rifts. Furthermore, the two largest geoid highs coincident with Atla and Beta regiones may indicate a dynamic nature for these features, and thus provide additional support for a young age for the BAT region.

Coronae, if they indeed are the results of ancient impacts, should predate active rifts. Coronae and chasmata, however, are intimately related, as can be seen in Figure 2. Even a cursory inspection of Hecate and Parga chasmata (extending between Atla and Beta and between Atla and Themis, respectively), depicted in Figures 2 and 4, shows this relation. In many cases, corona boundaries seem constrained by the rift walls (Fig. 4). A more quantitative analysis of corona and rift orientation was also undertaken (Stoddard and Jurdy, 2004). This comparison shows that although there is no apparent relation for coronae outside rifts, coronae within rifts tend to parallel the rift axis (Fig. 5). Given their locations and orientations, we argue that coronae within rifts must develop as part of the rifting process and/or continue forming postrifting. If, on the other hand, coronae were ancient, predating the rifts, we would find them not in the rifts, but bisected by the rifts. On the basis of the previous analyses, we here conclude that coronae cannot be due to ancient impacts.

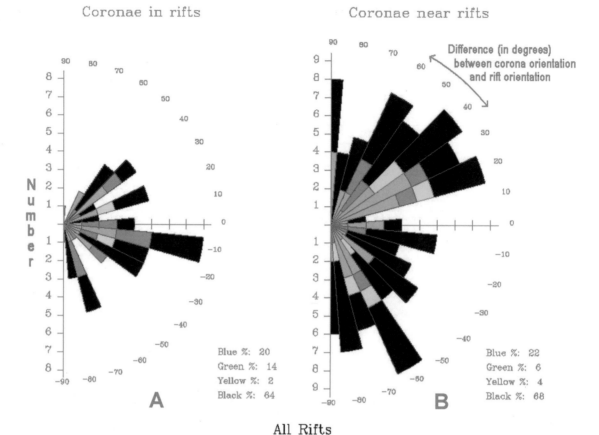

Figure 5. Corona orientation relative to rifts. The orientation of the long axis of the best-fitting ellipse to each corona is compared to the orientation of the nearest rift segment. (A) For coronae within the central rift graben, there is a preferred corona orientation subparallel to the rift axis. (B) Coronae near, but not in, rifts do not show this behavior. Color scheme (yellow, green, blue) matches the classification scheme (domal, circular, calderic, respectively) of DeLaughter and Jurdy (1999).

A model of coronae as ancient impacts needs to address the lack of cratering between the hypothesized ancient corona impacts and young crater impacts and the lack of transitional features between coronae and craters. In positing an old "impact age" for coronae, neither Hamilton (2005) nor Vita-Finzi et al. (2005) addressed in detail transitional craters, i.e., those younger than ancient "corona" impacts but older than the more commonly accepted impact set. The size distributions of craters and coronae (Fig. 6 *in* Vita-Finzi et al., 2005) clearly show two well-defined and distinct populations.

Spatial distribution has been proposed as a means for determining the origin of coronae. Vita-Finzi et al. (2005), in their argument for an impact origin for coronae, claimed that the corona distribution on Venus resembles global impact distributions on both Venus and the Moon. We challenge their conclusion on two grounds. First, we assert that the lunar comparison is flawed, because the catalogue of 1562 lunar craters used, while the best currently available, consists of named features only, and thus has a strong near-side bias, as well as a bias against high-latitude features (Deborah Lee Soltesz, USGS, 2006, personal commun.). Correspondingly, their lunar "crater density traces," based on that catalogue, peak at 0°N, 0°E (Fig. 10 *in*

Vita-Finzi et al., 2005). Second, we observe that the venusian crater distributions were incorrectly displayed by Vita-Finzi et al. (2005). When corrected for decreasing area with latitude, the venusian distribution appears random in both latitude (with the exception of the drop-off toward the south pole due to gaps in satellite coverage) and longitude, unlike the corona distribution (Fig. 6). We therefore argue that the corona distribution on Venus differs significantly from the crater distribution and cannot be used to argue for similar origins. Furthermore, we attribute the complementary distribution of craters and coronae with longitude to crater removal by corona-related volcanotectonic activity.

Quantitative Analysis of Circular Symmetry

Craters by their nature are circular. They are excavated by a roughly hemispherical shock wave, and thus, almost regardless of impact angle, will be round rim-and-basin structures (Melosh, 1989). Underlying structural features, such as faults, and later tectonic deformation can modify crater shape. Perhaps, therefore, the strongest test of an impact origin for coronae is the circularity of these features.

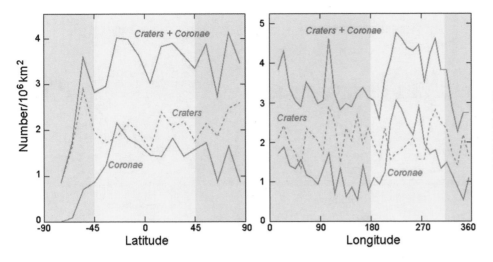

Figure 6. Coronae (red, solid lines), craters (blue, dashed lines), and combined (purple, solid lines) densities with latitude and longitude for Venus (10° bins). The lighter background indicates BAT region limits.

Here we introduce an approach for the assessment of a feature's circular symmetry. Using altimetry data, we compared, by cross-correlation, multiple profiles across a single feature. Jurdy and Stoddard (2005, http://jove.geol.niu.edu/faculty/stoddard/05ChapmanPoster.pdf) provided an example in which Mead crater and two coronae, all measuring ~280 km across, were analyzed. They found that for each corona, profiles cross-correlated at only 25–30% of perfect cross-correlation. Profiles for Mead crater, however, correlated at a much higher level, 80%. Here we report an expanded study for five features generally classified as craters and six whose classification as coronae has been questioned by Hamilton (this volume), the results of which are summarized in Figure 7. We chose only the largest craters, because altimetry data are too coarse to allow enough data points for analyses of smaller features, and also because they are of similar size to the coronae in our study. For each feature, thirty-six profiles (taken every 10°) were extracted from the altimetry data. The average slope was removed from each profile (to nullify the effects of any postemplacement tilting), and the results were aligned and then averaged together. For each feature, each profile was then correlated against the average, and the correlations themselves were averaged to give an assessment of circular symmetry. A perfectly circular feature would have a correlation average of 100%, indicating that each profile was identical to the average profile.

Figure 7A–E shows the results for five craters. Note that for Mead, Cleopatra, Meitner, and Isabella the profiles display the typical rim and basin structure expected for craters, but for Klenova (E) the average profile is more domal, with only a few of the individual profiles looking craterlike. The "contested" coronae are shown in Figure 7F–K. The average profiles for Eurynome (F), Maya (H), and C21 (I) appear craterlike, albeit with more variation among the individual profiles than seen in the generally agreed-upon craters. Anquet (G) has a rim-and-basin structure, but unlike typical craters, has a basin elevated above the surrounding plains. Acrea (K) appears to be a small hill in a large depression, again with a high degree of variation. Ninhursag (J) is clearly domal, and cannot be viewed as a crater.

The variability of the profiles, and thus the circularity of each feature, is summarized in Figure 7L. Those features universally agreed upon as craters (in yellow) have the highest correlation percentages, all at or above 80%, with the exception of Klenova. The disputed features (Fig. 7F–K) are not as circular, although C21 is close. Based on this analysis, we conclude that Klenova has been mischaracterized as an impact crater, and also that C21, a feature previously classified as a corona, may indeed be of impact origin (Table 2). The cases for Maya and Eurynome are more ambiguous. We propose that this type of correlation analysis can be used in an objective assessment of the circularity, and therefore the origin, of the remaining catalogue of similar features.

To address the noncircularity of coronae, Vita-Finzi et al. (2005) and Hamilton (2007) suggested deformation of these features by postimpact tectonic activity. Such activity must be local rather than regional; otherwise, a preferred orientation of the long axes of coronae, reflecting the tectonic stress regime, should be apparent. This is not the case, in relation either to the major tectonic features or to the chasmata (Fig. 5). We have found no correlation between the long axis of individual coronae and their dip direction, as might be expected if coronae were initially circular and their ellipticity and orientation were both related to later deformation.

EVOLUTIONARY MODEL

Here we consider an evolutionary model for coronae, based on rising diapirs, as an alternative to the impact hypothesis. Coronae were assigned to three distinct morphological groups using Magellan altimetry (DeLaughter and Jurdy, 1999). In this classification, *domal* coronae (numbering 54 of the features classified by DeLaughter and Jurdy, 1999) are distinguished by a central uplift with no surrounding moat, and may have associated radial fracturing, often visible only in the SAR images. A flattened interior and an annular moat characterize 93 *circular* coronae; portions of their interiors may be lower than the surrounding plains. *Calderic* coronae, with more than 50% of the

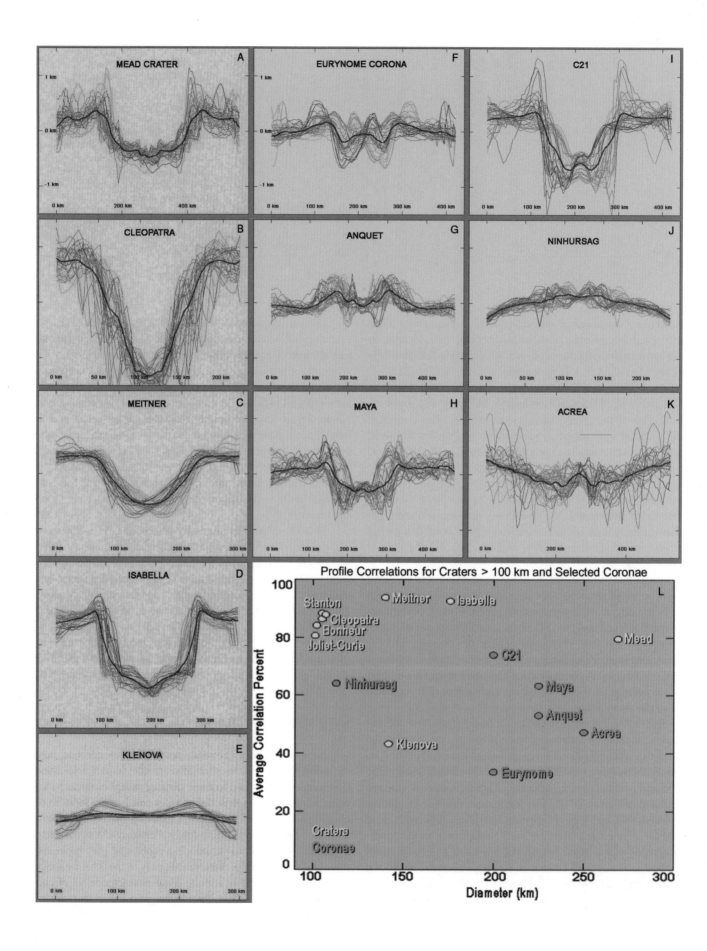

MEAD CRATER — A

CLEOPATRA — B

MEITNER — C

ISABELLA — D

KLENOVA — E

EURYNOME CORONA — F

ANQUET — G

MAYA — H

C21 — I

NINHURSAG — J

ACREA — K

Profile Correlations for Craters > 100 km and Selected Coronae — L

TABLE 2. CHARACTERISTICS OF CRATERS LARGER THAN 100 KM IN DIAMETER AND SELECTED CORONAE

Feature	Latitude (°N)	Longitude (°E)	Diameter (km)	Correlation %	Common classification	New Classification
Joliet-Curie	−1.7	62.4	100.9	81	Crater	Crater
Bonheur	9.7	288.8	102.2	84	Crater	Crater
Cleopatra	65.9	7.0	105.0	86	Crater	Crater
Stanton	−23.2	199.3	107.0	88	Crater	Crater
Meitner	−55.5	321.7	140.0	94	Crater	Crater
Klenova	78.2	104.7	141.9	43	Crater	Corona? (domal)
Isabella	−29.2	204.2	176.0	93	Crater	Crater
Mead	12.5	57.0	268.7	80	Crater	Crater
Ninhursag	−38.0	23.5	113.0	64	Corona	Corona (domal)
C21	29.0	243.0	200.0	74	Corona	Crater
Maya	23.0	98.0	225	63	Corona	Crater?
Eurynome	26.5	94.5	200	34	Corona	Corona? (circular)
Anquet	26.5	98.0	225	53	Corona	Corona (circular)
Acrea	24.0	243.5	250	47	Corona	Corona (calderic)

interior lower than the surrounding plains, constitute the majority (188), and display raised rims and annular moats. The three groups are gradational; consequently, boundaries in this classification are arbitrary. In Figure 8, we show the classification along with radar images of representative coronae corresponding to the stages. The attraction of this scheme is the simplicity of application: one needs to establish only the elevation of the corona interior relative to its surroundings. A further appeal of the approach is the possibility that the three groups may represent evolutionary stages of corona development, from initial diapir uplift to ultimate collapse.

Corona Characteristics by Type

For our analysis of classified coronae, we used the subset of DeLaughter and Jurdy's (1999) catalogue that corresponds to those 669 distinct features mapped by Price and Suppe (1995) as coronae. A total of 287 were matched to the morphologically classified coronae (DeLaughter and Jurdy, 1999). This correlation yielded a smaller set: 39 domal coronae, 83 circular, and 165 calderic, with 382 features remaining unclassified. A more sophisticated scheme could be devised to incorporate more features, but in our study we used this subset.

Next, we investigated the consequences of this classification scheme: Do the three groups of coronae—domal, circular, and calderic—represent stages of corona evolution? If so, an age progression should be evident from the density of impact craters and their modification. As noted, the distribution of impact craters on Venus very nearly approaches random. In Table 1, the crater counts are documented for all coronae, as well as the mor-

phological subgroups. Some intriguing patterns emerge. Coronae cover 10% of the surface of Venus, but contain only 7% of the craters, indicating a younger than average age. Likewise, although domal coronae occupy 0.9% of the total surface area, they contain only 0.3% of the craters; thus the crater density on these coronae is about one-third of what would be expected for average-aged features, and is also less than that for all coronae as a group. Additionally, the circular and calderic coronae have only three-fourths of the number of craters expected for their areas. These crater densities are consistent with the inferred stages, i.e., with the domal the youngest. The circular and calderic coronae, however, have an overpopulation, by 50%, of tectonically modified craters. This analysis (Table 1) shows that coronae, as a set, stand out as younger features, ones with lower crater density. Similarly, Price and Suppe (1995), in their terrane-based study, found that coronae and coronalike features are second only to large volcanoes in having the lowest crater density. Furthermore, our study provides evidence of tectonic modification, as would be expected for the proposed older coronae. In addition, crater densities and modification are consistent with the classification of coronae by evolutionary stage. Alternatively, if coronae were of ancient impact origin, we would expect to find them more heavily cratered, not less, than the average terrane.

Do Coronae Evolve in Size, Shape, and Orientation through Their Lifetime?

If these coronae do, in fact, represent a diapir life cycle, we would expect to see systematic variations in some coronal attributes, such as size, shape, dip, topographic and geoidal

---

Figure 7. Comparison of topographic profiles across craters and coronae on Venus. For each of panels A through K, thirty-six individual profiles are shown, in blue through green, based on the orientation of the profile. Blue is west-east, proceeding clockwise through south-north, east-west, north-south, and back to west-east. The average profile is depicted by the bold black line. (A–E) Profiles for five craters. (F–K) Profiles for six coronae. (L) Summary of a circularity study for these eleven features plus four others, based on the average correlation among the individual profiles. A 100% average correlation would indicate thirty-six identical profiles and perfect circular symmetry. The features in panels A–E, commonly accepted as craters, are depicted in yellow; green indicates the features in panels F–K, commonly accepted as coronae.

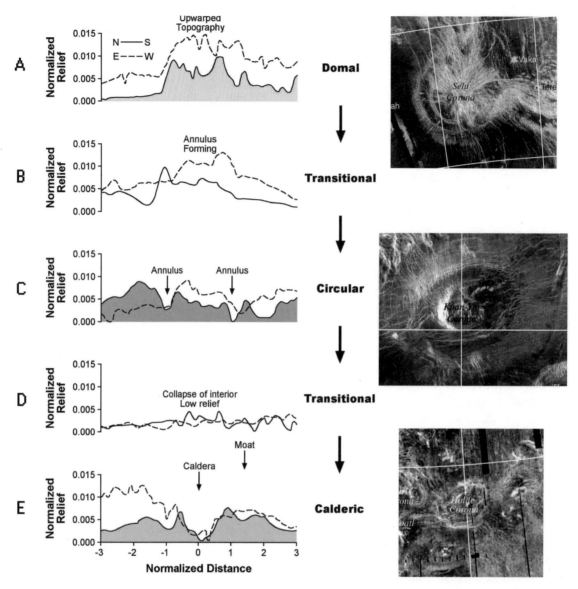

Figure 8. Coronae classification scheme, with example profiles. (A) Domal corona Selu, centered at 42.5°S, 6°E, diameter = 150 km. (B) Domal or circular transitional corona Earhart, 71°N, 136°E, 185.5 km. (C) Circular corona Kuan-Yin, 4.3°S, 10°E, 125 km. (D) Circular or calderic transitional corona Demeter, 55°N, 295°E, 333.5 km. (E) Calderic corona Holde, 53.5°N, 155°E, 100 km. Profiles after DeLaughter and Jurdy (1999). The radar images for Selu, Kuan-Yin, and Holde are from http://planetarynames.wr.usgs.gov/vgrid.html.

elevations, and so on, independent of the morphological criteria by which the coronae were classified. Figure 9A shows length versus dip, and Figure 9B shows the eccentricity versus dip for all coronae, comparing the whole set. The geoid versus topography is shown in Figure 9C. The domal coronae are shown as yellow, the circular as green, and the calderic as blue, with the remaining unclassified ones as black. Although the data show considerable scatter, analysis reveals some interesting patterns.

Quartile analysis provides a useful characterization of the range of values for a set. In quartile analysis the values range from q0 to q4. The lowest fourth of the values range from q0 to q1, the next fourth of the values range from q1 to q2; similarly,

the third quarter of the values range from q2 to q3, and the final, top quarter of the values range from q3 to q4. The numbers q1 and q3 are often referred to as the first and third quartiles, and q2 is usually referred to as the median. The numbers q0 and q4 are the minimum and maximum values. For a set with a statistically defined "normal distribution," the quartiles can be related to the standard deviation: For a normal (or Gaussian) distribution, 68.3% of the values lie within one standard deviation of the mean. Alternatively, the range between the first and third quartiles contains 50% of the values, and the points are within 0.675 standard deviation of the mean (Fisher, 1973). We apply this simple, yet informative analysis to the sets of coronae.

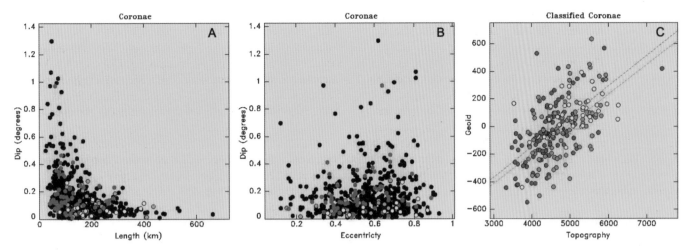

Figure 9. Comparison of various coronal parameters, by corona classification. Yellow—domal coronae; green—circular coronae; blue—calderic coronae; black—unclassified coronae. (A) Length versus dip for coronae. (B) Eccentricity versus dip for coronae. (C) Topography versus geoid; crosshairs indicate average values and standard deviations for each set.

The quartile analysis documents a distinct separation by stage (Table 3). Size strongly depends on the stage: three-fourths of the domal coronae are larger than three-fourths of the calderic ones, with circular coronae intermediate in size. Why are domal coronae (yellow) bigger and more eccentric? Perhaps the initial corona eruption corresponds to an active diapir that later withdraws. The ellipticity is also a function of stage: more than half the circular and calderic coronae have eccentricities of less than 0.50, while more than half the domal ones have eccentricities of over 0.70. A tilt was determined for each feature by determining the dip of a best-fitting plane through the region. The tilt or dip determined for coronae also seems related to the stage: three-fourths of the domal coronae dip less than three-fourths of the calderic, and the circular are intermediate. Although a continuum exists between corona stages, some characteristics distinguish uplifted coronae from largely collapsed ones: the domal coronae are larger, more eccentric, but flatter than the calderic coronae. Systematically, the circular coronae lie between the domal and the calderic for almost all parameters we defined. These patterns further support the morphologic classification of coronae (DeLaughter and Jurdy, 1999) as a simply determined, but useful, indication of stage or degree of maturity of individual features. Thus, the consistent continuum implies an evolutionary sequence, and we infer that the morphology of coronae indicates the stage, with the domal youngest, the circular intermediate, and the calderic the oldest. On the basis of these observations, we suggest that an objectively defined algorithm could be universally applied to the entire catalogue, allowing classification of many, if not most, of these features.

DISCUSSION

In our study we examined the cratering record, distribution, and morphology of coronae. We argued that these preclude an

TABLE 3. QUARTILES FOR CORONA CHARACTERISTICS

	Eccentricity		
	q1	q2	q3
Domal (Y)	.505	.700	.735
Circular (G)	.410	.490	.610
Calderic (B)	.410	.510	.640
	Length (km)		
	q1	q2	q3
Domal (Y)	142	188	281
Circular (G)	129	167	225
Calderic (B)	75	102	150
	Dip (degrees)		
	q1	q2	q3
Domal (Y)	.051	.062	.082
Circular (G)	.057	.101	.169
Calderic (B)	.086	.128	.219

Note: Eccentricity, length, and dip given for each corona type with median (q2) and top (q3) and bottom (q1) quartiles.

impact origin for these features. A tectonovolcanic origin better fits our observations. We then presented and evaluated a model that considers coronae as manifestations of diapir evolution and found it consistent with characteristics of coronae.

What Mechanism Formed Venus's Coronae?

Crater density inside and near coronae argues for their being young and therefore volcanically and tectonically active features. Impact crater density within coronae lies below the planetary average; yet even with this lower density, the proportion of obviously tectonized craters considerably exceeds the number expected based on the total coronal area. Also, coronal distribution is more consistent with a volcanotectonic than with an impact origin. Coronae are strongly concentrated in the young BAT region and are closely associated with rifts, themselves sites of recent, if not ongoing, activity. We also show that longitudinal corona distribution, a pulse centered in the BAT region, clearly differs from the more random arrangement of impact craters. Even if we were to assume that all coronae are impact features, the combined corona and crater distribution is still not random in longitude (Fig. 6). Taken together, these observations are more consistent with coronae being young, active features, rather than ancient impact sites.

Are Coronae Venusian Analogs of Earth's Plumes?

Although there are hundreds of features that have been identified as coronae, only fifty-four of these have more than 50% of their interiors raised above the exterior (here classified as domal and potentially active). These correspond to 0.9% of the surface of the planet. Are these plumes on Venus? Here we define a "plume" as a deep-seated, long-lived thermal perturbation with a volcanic and tectonic surface expression. On Earth, the combination of plume and plate tectonic activity leads to Hawaiian-style island chains. On Mars, where no large-scale lateral tectonic activity has been documented, plume activity remains localized, resulting in the largest known shield volcanoes, such as Olympus Mons. On Venus, if coronae are in fact caused by plumes, there could be more than fifty. In comparison, Earth's currently active plumes have been variously numbered from a mere handful to well over one hundred. No unanimity exists on a catalogue of hotspots for Earth. For example, two analyses that correlated hotspot locations with geoid highs to infer the dynamic link (Chase, 1979; Crough and Jurdy, 1980) employed twenty-four and forty-two hotspots, respectively. So, comparing corona analyses, if all fifty-four domal coronae were active plumes, this would correspond to Venus's having an excess of 30% to more than 100% when compared with its sister planet, Earth. This number, though high, may not be unreasonable for a planet lacking plate tectonics to transport heat from the interior. Nonetheless, the presence of over six hundred coronae, implying that there was the same number of plumes in the last 1000–350 m.y. of Venus, does seem inordinately high, and therefore argues against a distinct plume source for each of these. We agree with Stofan and Smrekar (2005) that the regiones, such as Atla, Beta, Themis, and Phoebe, would better correspond to planetary plumes, i.e., large-scale uplifts. The smaller number of regiones (about ten), as discussed earlier, agrees more with our understanding about the capability of a planet's core to generate plumes that support uplifts. Ideally, the size and strength of individual plumes should be considered in this comparison, but these characteristics exceed the scope of this article.

The areal pattern of the domal coronae presents another argument against their having deep sources: in Figure 2, for example, note the four closely spaced domal coronae (yellow) between two contour lines around Atla's peak. That there are active

plumes beneath each of these locations does not seem reasonable. In the terrestrial studies cited, hotspots are generally restricted to positive "residual" geoid regions, once the dominating effect of subduction has been removed. However, on Venus, coronae are found to correlate with midgeoid levels (Jurdy and Stefanick, 1999). This effect is apparent in Figure 9C, as almost a topographic limit, which the domal coronae—the ones we infer as active—cannot exceed.

Shallow diapirs may offer a more reasonable explanation for coronae. Admittance studies of several coronae (Hoogenboom et al., 2004) showed coronae as active upwellings, a result that has been interpreted to indicate isostatic compensation (Stofan and Smrekar, 2005). Based on fluid dynamic models, Hansen (2003) argued that coronae could be attributed to compositional diapirs, as opposed to the large rises and regiones, like Atla and Beta, for which thermal plumes are often invoked. We find that the observations presented in this article are consistent with a model of coronae as diapirs, either thermal or compositional, evolving through a sequence of stages, starting with uplift, followed by volcanism and development of annuli, and ending with collapse. As we have shown, a classification of coronae based merely on their interior topography leads to stages with a systematic set of characteristics. Younger coronae are larger, more eccentric, and flatter, and generally occur at higher geoid and topography levels.

A fuller understanding of coronae, and therefore of their underlying causes, could be achieved by extending the approaches we have taken to the entire set of 669 features mapped as "coronae" by Price and Suppe (1995). The study we presented in this article was more limited in scope and applied to the subset of coronae that had been categorized by DeLaughter and Jurdy (1999) by inspection. Any classification by visual examination is open to interpretation; thus we propose using an objective algorithm to assign each feature to a class based on its topography —whether uplifted, flat, or collapsed. The characteristics of each feature, such as tilt, size, ellipticity, altitude, geoid, and particularly circularity, can then be determined quantitatively, as we demonstrated here. We suspect that the group of coronae showing collapsed interiors may harbor a few impact craters, such as C21, as described in our analysis here. A complete and systematic examination of all features on Venus classified as coronae should allow further evaluation of the diapir versus impact models for origin of these enigmatic circular features.

The relationships among coronae, regiones, and chasmata are complex, but in all likelihood hold the key to understanding the resurfacing processes for Venus, and in turn, the global heat dissipation mechanisms. In the absence of Earthlike plate tectonics, plumes, whether on the scale of regiones or of coronae, must play an important role in both phenomena.

ACKNOWLEDGMENTS

We thank reviewers Alexander Basilevsky, Claudio Vita-Finzi, and editor Gillian Foulger for their careful and thoughtful reviews. Warren Hamilton's comments on the first version of this article helped us in developing an approach to assess circularity of features. We thank Michael Stefanick for his assistance with statistics and interpretation of data.

REFERENCES CITED

Barsukov, V.L., Basilevsky, A.T., Burba, G.A., Bobinna, N.N., Kryuchkov, V.P., Kuzmin, R.O., Nikolaeva, O.V., Pronin, A.A., Ronca, L.B., Chernaya, I.M., Shashkina, V.P., Garanin, A.V., Kushky, E.R., Markov, M.S., Sukhanov, A.L., Kotelnikov, V.A., Rzhiga, O.N., Petrov, G.M., Alexandrov, Y.N., Sidorenko, A.I., Bogomolov, A.F., Skrypnik, G.I., Bergman, M.Y., Kudrin, L.V., Bokshtein, I.M., Kronrod, M.A., Chochia, P.A., Tyuflin, Y.S., Kadnichansky, S.A., and Akim, E.L., 1986, The geology and geomorphology of the Venus surface as revealed by the radar images obtained by Venera 15 and 16: Journal of Geophysical Research, v. 91, p. D378–D398.

Basilevsky, A.T., and Head, J.W., 2002, Venus: Timing and rates of geologic activity: Geology, v. 30, p. 1015–1018, doi: 10.1130/0091-7613(2002)030 <1015:VTAROG>2.0.CO;2.

Bleamaster, L.F., and Hansen, V.L., 2004, Effects of crustal heterogeneity on the morphology of chasmata, Venus: Journal of Geophysical Research, v. 109, E02004, doi: 10.1029/2003JE002193.

Chase, C.G., 1979, Subduction, the geoid, and lower mantle convection: Nature, v. 282, p. 464–468, doi: 10.1038/282464a0.

Crough, S.T., and Jurdy, D.M., 1980, Subducted lithosphere, hotspots, and the geoid: Earth and Planetary Science Letters, v. 48, p. 15–22, doi: 10.1016/0012-821X(80)90165-X.

DeLaughter, J.E., and Jurdy, D.M., 1997, Venus resurfacing by coronae: Implications from impact craters: Geophysical Research Letters, v. 24, p. 815–818, doi: 10.1029/97GL00687.

DeLaughter, J.E., and Jurdy, D.M., 1999, Corona classification by evolutionary stage: Icarus, v. 139, p. 81–92, doi: 10.1006/icar.1999.6087.

Fisher, R.A., 1973, Statistical methods for research workers: New York, Hafner, 362 p.

Hamilton, W.B., 2005, Plumeless Venus preserves an ancient impact-accretionary surface, *in* Foulger, G.R., et al., eds., Plates, plumes, and paradigms: Boulder, Colorado, Geological Society of America Special Paper 388, p. 781–814.

Hamilton, W.B., 2007 (this volume), An alternative Venus, *in* Foulger, G.R., and Jurdy, D.M., eds., Plates, plumes, and planetary processes: Boulder, Colorado, Geological Society of America Special Paper 430, doi: 10.1130/2007.2430(41).

Hansen, V.L., 2003, Venus diapirs: Thermal or compositional?: Geological Society of America Bulletin, v. 115, p. 1040–1052, doi: 10.1130/B25155.1.

Head, J.W., and Basilevsky, A.T., 1998, Sequence of tectonic deformation in the history of Venus: Evidence from global stratigraphic relationships: Geology, v. 26, p. 35–38, doi: 10.1130/0091-7613(1998)026<0035:SOTDIT>2.3.CO;2.

Herrick, R.R., 2006, Updates regarding the resurfacing of Venusian impact craters: Lunar and Planetary Science Conference 37 [abs.].

Herrick, R.R., and Sharpton, V.L., 2000, Implications from stereo-derived topography of Venusian impact craters: Journal of Geophysical Research, v. 105, E8, doi: 10.1029/1999JE001225.

Herrick, R.R., Dufek, J., and McGovern, P.J., 2005, Evolution of large shield volcanoes on Venus: Journal of Geophysical Research, v. 110, doi: 10.1029/2004JE002283.

Hoogenboom, T., Smrekar, S.E., Anderson, F.S., and Houseman, G., 2004, Admittance survey of type 1 coronae on Venus: Journal of Geophysical Research, v. 109, doi: 10.1029/2003JE002171.

Ivanov, M.A., and Basilevsky, A.T., 1993, Density and morphology of impact craters on tessera terrain: Venus Geophysical Research Letters, v. 20, p. 2579–2582.

Johnson, C.L., and Richards, M.A., 2003, A conceptual model for the relationship between coronae and large-scale mantle dynamics on Venus: Journal of Geophysical Research, v. 108, E6, p. 5058, doi: 10.1029/2002JE001962.

Jurdy, D.M., and Stefanick, M., 1999, Correlation of Venus surface features and geoid: Icarus, v. 139, p. 93–99, doi: 10.1006/icar.1999.6089.

Jurdy, D.M., and Stoddard, P.R., 2005, Uplift and rifting on Venus: Role of plumes [abs.]: American Geophysical Union Chapman Conference: The Great Plume Debate, Fort William, Scotland, August 28–September 1, p. 54.

Jurdy, D.M., Stoddard, P.R., and Matias, A., 2003, Upwellings on Venus: Evidence from coronae and craters [abs.]: Geological Society of America Penrose Conference: Plume IV: Beyond the Plume Hypothesis, Hveragerdi, Iceland, August 25–29, n.p.

Koch, D.M., and Manga, M., 1996, Neutrally buoyant diapirs: A model for Venus coronae: Geophysical Research Letters, v. 23, p. 225–228, doi: 10.1029/95GL03776.

Magee Roberts, K., and Head, J.W., 1993, Large-scale volcanism associated with coronae on Venus: Implications for formation and evolution: Geophysical Research Letters, v. 20, p. 1111–1114.

Matias, A., and Jurdy, D.M., 2005, Impact craters as indicators of tectonic and volcanic activity in the Beta-Atla-Themis region, Venus, *in* Foulger, G.R., et al., eds., Plates, plumes, and paradigms: Boulder, Colorado, Geological Society of America Special Paper 388, p. 825–839.

Matias, A., Jurdy, D.M., and Stoddard, P.R., 2004, Stereo imaging of impact craters in the Beta-Atla-Themis (BAT) region, Venus [abs.]: 35th Lunar and Planetary Science Conference, Houston, Texas, abstract 1383.

McKinnon, W.B., Zahnle, K.J., Ivanov, B.A., and Melosh, H.J., 1997, Cratering on Venus: Models and observations, *in* Bougher, S.W., et al., eds., Venus II: Geology, geophysics, atmosphere, and solar wind environment: Tucson, University of Arizona Press, 969–1014.

Melosh, H.J., 1989, Impact cratering: New York, Oxford University Press.

Moore, W.B., and Schubert, G., 1997, Venusian crustal and lithospheric properties from nonlinear regressions of highland geoid and topography: Icarus, v. 128, p. 415–428, doi: 10.1006/icar.1997.5750.

Namiki, N., and Solomon, S.C., 1994, Impact crater densities on volcanoes and coronae on Venus: Implications for volcanic resurfacing: Science, v. 265, p. 929–933, doi: 10.1126/science.265.5174.929.

Parsons, B., 1981, The rates of plate creation and consumption: Geophysical Journal of the Royal Astronomical Society of London, v. 67, p. 437–448.

Pettengill, G.H., Ford, P.G., Johnson, W.T.K., Raney, R.K., and Soderblom, A., 1991, Magellan: Radar performance and data products: Science, v. 252, p. 260–265, doi: 10.1126/science.252.5003.260.

Phillips, R.J., Raubertas, R.F., Arvidson, R.E., Sarkar, I.C., Herrick, R.R., Izenberg, N., and Grimm, R.E., 1992, Impact craters and Venus resurfacing history: Journal of Geophysical Research, v. 97, p. 15,923–15,948.

Plaut, J.J., 1993, Stereo imaging, *in* Guide to Magellan Image Interpretation: Pasadena, California, California Institute of Technology, Jet Propulsion Laboratory Publication 93–24, p. 33–43.

Price, M., and Suppe, J., 1995, Constraints on the resurfacing history of Venus from the hypsometry and distribution of tectonism, volcanism, and impact craters: Earth, Moon, and Planets, v. 71, p. 99–145, doi: 10.1007/BF00612873.

Price, M.H., Watson, G., Suppe, J., and Brankman, C., 1996, Dating volcanism and rifting on Venus using impact crater densities: Journal of Geophysical Research, v. 101, p. 4657–4671, doi: 10.1029/95JE03017.

Pronin, A.A., and Stofan, E.R., 1990, Coronae on Venus: Morphology, classification, and distribution: Icarus, v. 87, p. 452–474, doi: 10.1016/0019-1035(90)90148-3.

Sandwell, D.T., Johnson, C.L., Bilotti, F., and Suppe, J., 1997, Driving forces for limited tectonics on Venus: Icarus, v. 129, p. 232–244, doi: 10.1006/icar.1997.5721.

Saunders, R.S., Spear, A.J., Allin, P.C., Austin, R.S., Berman, A.L., Chandlee, R.C., Clark, J., DeCharon, A.V., De Jong, E.M., Griffith, D.G., Gunn, J.M., Hensley, S., Johnson, W.T.K., Kirby, C.E., Leung, K.S., Lyons, D.T., Michaels, G.A., Miller, J., Morris, R.B., Morrison, A.D., Piereson, R.G., Scott, J.F., Shaffer, S.J., Slonski, J.P., Stofan, E.R., Thompson, T.W., and Wall, S.D., 1991. Magellan mission summary: Journal of Geophysical Research, v. 97, p. 13,067–13,090.

Schaber, G.G., 1982, Venus: Limited extension and volcanism along zones of lithospheric weakness: Geophysical Research Letters, v. 9, p. 499–502.

Schaber, G.G., Strom, R.G., Moore, H.J., Soderblom, L.A., Kirk, R.L., Chadwick, D.J., Dawson, D.D., Gaddis, L.R., Boyce, J.M., and Russell, J., 1992,

Geology and distribution of impact craters on Venus: What are they telling us?: Journal of Geophysical Research, v. 97, p. 13,257–13,301.

Sjogren, W.L., Banerdt, W.R., Chodas, P.W., Konopliv, A.S., Balmino, G., Barriot, P., Arkani-Hamed, J., Colvin, T.R., and Davies, M.E., 1997, The Venus gravity field and other geodetic parameters, *in* Bougher, S.W., et al., eds., Venus II: Geology, geophysics, atmosphere, and solar wind environment: Tucson, University of Arizona Press, p. 1125–1161.

Solomon, S.C., Smrekar, S.E., Bindschadler, D.L., Grimm, R.E., Kaula, W.M., McGill, G.E., Phillips, R.J., Saunders, R.S., Schubert, G., Squyres, S.W., and Stofan, E.R., 1992, Venus tectonics: An overview of Magellan observations: Journal of Geophysical Research, v. 97, p. 13,199–13,255.

Stefanick, M., and Jurdy, D.M., 1996, Venus coronae, craters and chasmata: Journal of Geophysical Research, v. 101, p. 4637–4643, doi: 10.1029/95JE02709.

Stoddard, P.R., and Jurdy, D.M., 2003, Uplift of Venus geoid highs: Timing from coronae and craters [abs.]: 34th Lunar and Planetary Science Conference, Houston, Texas, abstract 2129.

Stoddard, P.R., and Jurdy, D.M., 2004, Venus' Atla and Beta Regiones: Formation of chasmata and coronae [abs.]: Eos (Transactions, American Geophysical Union), v. 85.

Stofan, E.R., Sharpton, V.L., Schubert, G., Baer, G., Bindschadler, D.L., Janes, D.M., and Squyres, S.W., 1992, Global distribution and characteristics of coronae and related features on Venus: Implications for origin and relation to mantle processes: Journal of Geophysical Research, v. 97, p. 13,347–13,378.

Stofan, E.R., and Smrekar, S.E., 2005, Large topographic rises, coronae, large flow fields, and large volcanoes on Venus: Evidence for mantle plumes? *in* Foulger, G.R., et al., eds., Plates, plumes, and paradigms: Boulder, Colorado, Geological Society of America Special Paper 388, p. 841–861.

Tapper, S.W., Stofan, E.R., and Guest, J.E., 1998, Preliminary analysis of an expanded corona database [abs.]: 29th Lunar and Planetary Science Conference, Houston, Texas, abstract 1104.

Vezolainen, V., Solomatav, V.S., Basilevsky, A.T., and Head, J.W., 2004, Uplift of Beta Regio: Three-dimensional models: Journal of Geophysical Research, v. 109, doi:10.1029/2004E002259.

Vita-Finzi, C., Howarth, R.J., Tapper, S.W., and Robinson, C.A., 2005, Venusian craters, size distribution, and the origin of craters, *in* Foulger, G.R., et al., eds., Plates, plumes, and paradigms: Boulder, Colorado, Geological Society of America Special Paper 388, p. 815–823.

MANUSCRIPT ACCEPTED BY THE SOCIETY JANUARY 31, 2007

DISCUSSION

3 December 2006, Warren B. Hamilton

Thousands of old, mostly circular structures with impact-compatible morphology and rim diameters from 3 to 2500 km saturate large tracts of both uplands and lowlands of Venus. Most of these old structures are ignored in conventional work, including that by Jurdy and Stoddard (this volume), and the small fraction that are considered at all are deemed endogenic and young. The well-preserved structures typically are circular-rimmed shallow basins with gentle debris aprons, often with external gentle circular moats, and often with internal central peaks or peak rings. Basins are superimposed with the cookie-cutter bites required by impacts but incompatible with endogenic origins. Doublets and composite shapes attest to disruption of many bolides by gravity and dense atmosphere. Superpositions and analogy with the dated youngest large lunar impact basin indicate most of these structures to be older than 3.85 Ga if they record impacts: the venusian landscape is relict from late-stage planetary accretion. See Hamilton (2005, this volume) for evidence, illustrations, and discussions. Superimposed on the old structures are ~1000 small craters, rim diameters of all but a few <100 km, their scarceness reflecting atmospheric destruction of most small bolides, acknowledged by all as impact structures and very conservatively designated.

Jurdy and Stoddard (this volume) minimally address these matters in their presentation of numerical arguments for the conventional assumption that the old circular structures formed during the last billion years by endogenic processes. From the thousands of these structures, they consider only a small, arbitrary subset, particularly conspicuous in radar backscatter imagery of uplands, of "coronae." These mostly are circular rimmed depressions with rimcrest diameters between ~70 and 500 km.

They include fewer than half of the large quasicircular structures and fragments thereof visible in uplands; perhaps one-twentieth of the large variably filled structures, seen primarily in radar altimetry, of the twice-as-extensive lowlands; and none of the several thousand additional small, old circular structures in both uplands and lowlands. Jurdy and Stoddard (this volume) select "coronae" by arbitrary criteria that limit them mostly to uplands and hence err in reasoning that uplands and circular structures must be products of plumes and diapirs. Further, they wrongly deny the existence of structures transitional between the ~1000 small young impact structures and their selected structures and of old, small structures (see my papers for illustrations of both).

Jurdy and Stoddard (this volume) depict coronae as elongate and aligned within rift zones, and therefore endogenic. The conflict between these conjectural elongate structures and the actual mostly circular ones is clarified by their Figure 4. Their "coronae" are elongate blobs drawn arbitrarily outside those circular structures most obvious in low-resolution reflectivity, or even drawn to enclose several such structures each. Their analysis of these blobs is irrelevant. They dismiss, without discussion, the circular and composite structures actually at issue as "calderic depressions," despite lack of dimensional and geometric similarity to any modern terrestrial magmatic features.

Jurdy and Stoddard (this volume) present stacked radial topographic profiles through five young impact craters accepted by all, and through six coronae. They acknowledge that four (I say five) of these six coronae have impact-compatible topography but argue that because five of the six profiles have lower circular symmetry than do those of the accepted impact structures, all but one are endogenic. (The one they accept as an impact is indistinguishable visually from hundreds of other "coronae.") The perceived distinction reflects flawed methodology and the greater age and modification of coronae. The Magellan Stereo Toolkit software, presumed source of the profiles, adds artifacts and digital noise, which cannot be interactively edited, where radar brightness contrast is low and features are large, the common case with old structures (C.G. Cochrane, 2005, and written communications, 2006). Jurdy and Stoddard (this volume) magnify these defects with vertical exaggerations up to 160:1. The extreme exaggerations of their Figures 3 and 8 negate conclusions derived from those also.

The Jurdy and Stoddard (this volume) statements regarding distribution of accepted "young" impact craters are based on an early tabulation that undercounted craters in bedrock terrains. Their confident statements regarding ages and origins of surfaces, landforms, and coronae are merely intertwined assumptions that are false if the structures at issue record impacts. Their deductions regarding geoid and uplands are false if uplands include impact-melt constructs, as I advocate. Their references to my work and predictions are miscitations.

22 December 2006, Claudio Vita-Finzi

To the outsider it must seem that we are going round in circles, or perhaps ovoids. For example, Jurdy and Stoddard (this volume) suggest that interpreting coronae as impact craters would result in a crater density for the BAT region of Venus indicative of a "somewhat older than average surface age," whereas they know it is relatively young because it lacks impact craters and boasts many rifts that are also poor in impact craters. Such arguments cut no ice with those who draw no distinction between coronae and craters. Even the statistical dialogue is not conducted on a single wavelength: measures of circularity are neither here nor there if you have decided that craters may be deformed by erosion or tectonics.

Perhaps we should move on to more productive matters. Three come immediately to mind. First, we need to know the varying atmospheric and geological conditions under which venusian impact craters formed, and this we can do only by dating them using criteria other than crater density. Second, the isotopic evidence points to a wetter past. When was it, how wet was it, and did impact history have anything to do with its demise? Third, if some, several, or all the coronae originated in plumes or diapirs, can we construct a model for the interior of Venus that can support them simultaneously or serially? But the first step is surely to follow Hamilton in looking at all quasicircular structures on Venus and not just those we call coronae, whereupon—witness the fine image by S.W. Tapper in Vita-Finzi et al. (2005), Figure 11, p. 821 (reproduced here as Fig. D-1)—we find a landscape as lunar as the Moon's.

27 December 2006, Richard J. Howarth

At the end of the initial paragraph of their section on testing the impact hypothesis, Jurdy and Stoddard's quotation from Fisher is outdated: the approximation to a normal distribution is quite inadequate for small samples. The authors do not even say what their "count" of four is supposed to be—it must be the

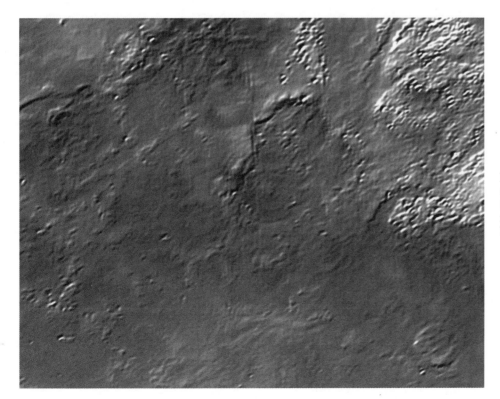

Figure D-1. Shaded relief image generated from Magellan altimetry of part of Scarpellini quadrangle illustrates subtle multicratered topography. The crater at bottom right has a diameter of ~200 km.

number of occurrences in so many sampled units. Even if these are implicit, and they ought to be explicit, e.g., in *n* cells of *X* by *X* km on the surface of Venus you only find four instances, etc. (see Howarth, 1998, for a survey of recent results on this topic).

Jurdy and Stoddard (this volume) talk about a "random" distribution without specifying what they mean by it. In fact, they are almost certainly talking about a Poisson distribution. In the second paragraph of their section on circular symmetry, they write of "profiles cross-correlated at only 25–30% of perfect cross-correlation." This kind of statement is inadequate. The sample size (i.e., number of data points), the type of metric used to assess the "cross-correlation," and the results of a statistical test of significance should all be given.

Even had these topics been adequately addressed, Jurdy and Stoddard's statistical assertions seem to me to add little weight to the key arguments, which are essentially geological and have been addressed, e.g., by Vita-Finzi et al. (2005).

9 January 2007, Ellen R. Stofan

Jurdy and Stoddard (this volume) describe coronae, a feature type that has been recognized in the peer-reviewed literature since their first identification in Venera radar images of the Venus surface (e.g., Barsukov et al., 1986; Basilevsky et al., 1986; Pronin and Stofan, 1990; Stofan et al., 1991; Head et al., 1992; Janes et al., 1992; Squyres et al., 1992; Namiki and Solomon, 1994; Stefanick and Jurdy, 1996; Hansen, 2003; Johnson and Richards, 2003; Hoogenboom et al., 2004). The majority of the community working with Venus data clearly differentiate between impact craters and coronae, as each has distinct morphologies and distributions. Jurdy and Stoddard (this volume) provide a clear and concise overview of the characteristics of coronae and why an impact origin is not consistent with these characteristics.

Their article provides clear evidence as to why their origin is related to endogenic processes. Endogenic processes are capable of producing depressions, and several models have been discussed in the peer-reviewed literature that are consistent with the observed range of corona topography (<1.0 km to >2.0 km) (e.g., Koch and Manga, 1996; Smrekar and Stofan, 1997). The impact cratering process, on the other hand, dominantly produces highly circular depressions. Clearly some coronae are depressions and some coronae are circular, but it is important to understand that both exogenic and endogenic processes can produce circular depressions and to focus on the evidence that allows one to determine which cause is more likely. As Jurdy and Stoddard (this volume) conclude, the data clearly support an endogenic origin.

18 January 2007, Warren B. Hamilton

Just the place for a Snark! I have said it twice:
That alone should encourage the crew.
Just the place for a Snark! I have said it thrice:
What I tell you three times is true.

                                                                    Lewis Carroll

Vita-Finzi et al. (2005) and I (Hamilton, 2005, and this volume) argue that the thousands of circular venusian structures, mostly rimmed depressions that reach giant sizes and that predate obvious small, young impact structures, are products of ancient impacts. Although early interpreters of low-resolution venusian radar imagery recognized this possibility, Stofan (e.g., Stofan et al., 1985) and a few others speculated, years before detailed imagery became available, that Venus is too earthlike to preserve a primordial surface; hence, the circular structures must be young and endogenic. This conjecture soon became dogma, and the impact option has seldom been mentioned since. Broadly conflicting young-endogenic classifications and genetic rationales for some of the circular structures have been presented in hundreds of papers by Stofan and by many others (e.g., Jurdy and Stoddard, this volume), but the great majority of circular structures that are candidates for old impacts have been ignored by Stofan and most other venusian specialists. Many of these structures are huge and, if products of impacts, must, by analogy with dated lunar structures, be older than 3.8 Ga. The structures are mostly or entirely older than the small impact structures, as accepted by all observers, that conventionally have been assigned maximum ages of somewhere between 0.3 and 1.5 Ga on the basis of poorly constrained calculations of the effect on fragile bolides of the extremely dense venusian atmosphere; ages reach 3.8 Ga in my terms.

Few post-1988 mainline papers mention the impact option, and none has rigorously evaluated it, but Stofan here implies that because a number of papers, including five of her own and the chapter in this volume by Jurdy and Stoddard, agree that the circular structures are endogenic, and because the manuscripts were peer-reviewed (by other specialists who also assume endogenic origins), endogenic origins should be accepted. That the authority-by-repetition thus appealed to is poorly supported by evidence is shown by Stofan's many papers, and that the implied consensus does not exist is shown by the mutual incompatibility of her evolving speculations with those by others including Jurdy and Stoddard (this volume).

Stofan's papers deal with various venusian features that she regards as products of endogenic magmatism, mostly intrusive, despite their complete dissimilarity to any modern terrestrial structures. Her papers have concentrated on 400 or so "coronae," an arbitrary subset of conspicuous midsize circular structures. She has presented conceptual and numerical models wherein assumed parameters and unearthly processes enable endogenic results, but the models do not explain the characteristic circularity of the rimmed depressions except as local coincidences. Stofan's statement here that "endogenic processes can produce circular depressions" is but wishful thinking with regard to the thousands of examples with rim diameters up to 2000 km. Her papers have not addressed the common impact-compatible morphology, whereas even the anti-impact Jurdy and Stoddard chapter in this volume acknowledges that three of the six coronae whose morphology it considers in detail "appear crater-like" and that one of those three likely is an impact structure and the other two may be. The great majority of the old circular structures in the plains, which comprise two-thirds of Venus, are excluded from Stofan's studies, so her published statements about restricted distribution of coronae are invalid with regard to the broad population of structures at issue, and her global-dynamic speculations cantilevered from those statements also are invalid.

Stofan here notes, correctly, that "The impact cratering process . . . dominantly produces highly circular depressions" but then implies ("some coronae are circular and some coronae are depressions") that such structures are but a minor venusian type. The reader need look no further than the plains image in Figure D-1 to see that circular depressions are the rule. Stofan and Smrekar (2005) themselves stated that "the most typical shape for a corona is a depression or rimmed depression." The several papers by Stofan that mention the impact option dismiss it with arguments that I refuted in my papers as irrelevant because most of them (like most arguments in the Jurdy and Stoddard chapter in this volume) address only a biased subset of the circular structures, chosen by criteria that limit their geographic and size distributions, and further deal not with the circular rimmed depressions at issue but with imagined boundaries far outside them or even outside groups of the structures.

10 February 2007, Donna M. Jurdy and Paul R. Stoddard

In his discussion, Hamilton criticizes our data set and definition of the outlines of the features we analyze as arbitrary. Both statements are inaccurate. As we stated in our chapter, we used the data sets of Stofan et al. (1992), Magee Roberts and Head (1993), and Price and Suppe (1995), which in turn were not picked arbitrarily but based on a defined set of observed topographic, structural, and volcanic features. We used outlines that were mapped by Price and Suppe (1995) based on volcanic flows related to the structures, as interpreted from the radar imagery. We documented our use of the subset of the classified corona catalogue that could be matched with those 669 distinct features mapped by Price and Suppe (1995) as "coronae," which they defined in their catalogue as "circular to irregular volcanic-tectonic features characterized by an annulus of concentric deformation."

Hamilton suggests that if we were to pick the outlines of the features "correctly," we would find most to be circular to near-circular and cites our Figure 4 as evidence. Even if we were to accept this criticism, many of the internal structures in Figure 4 are still clearly noncircular. Dhorani and Ludjatako are clearly oblong, and Atete, Krumine (the northern part, as this is one that may have had "several such structures"), and Javine are all irregularly shaped. Others, such as Dilga, may be more circular, although Dilga was fairly round even as mapped by Price. Furthermore, our quantitative analysis assesses circularity not based on the mapping by Price but on Magellan topography. Finally, it must be pointed out, again, that circularity does not necessitate impact origin, but noncircularity requires a much more complex history, with many other implications, than a simple impact.

The remaining major concern Hamilton raises is the inappropriateness of the Magellan Stereo Toolkit software for features of the size we analyzed. We agree, which is why we did not use it for the study we report in this volume. (However, we did reference other studies that employed stereo imaging; this may be where the confusion arises.) Stereo imaging for high-resolution topography can be attempted for only about 10% of Venus's surface with multiple radar coverage. For our topographic analysis, we used the nearly global Magellan altimetry data.

In his discussion, Vita-Finzi suggests that we engage in circular reasoning. Although we admit that many of our arguments are based on the circularity of craters and lack thereof of many coronae, we disagree that our BAT region argument is circular. The BAT region has many indications of being more active, and therefore younger, than the average age of the venusian surface—uplifts with associated high geoid values, extensive rift systems, and the high concentration of tectonized, embayed craters as well as volcanic activity—all independent of raw crater counts. Crater Uvaysi retains its parabolic halo, indicative of the most recent 10% of craters. The severe modification of this parabola-associated crater dates tectonic activity as having occurred recently. Furthermore, independent stratigraphic study of the BAT region to neighboring areas also shows it to be young (Basilevsky et al., 1997).

Vita-Finzi correctly points out that tectonics can change the shape of craters, and erosion can modify, to some degree, the circularity as measured in our chapter. However, particularly for tectonized features, one would expect all features in a region to be elongated in more or less the same direction. Even with multiple tectonic events, the latest will be imprinted on all features. Analysis of corona orientation does not support this.

The correlation percentage referred to by Howarth is the standard cross correlation between the test and average profiles, divided by the autocorrelation of the average profile. The number of data points per profile was uniform for each feature, ranging from 41 for the smallest to 108 for the largest features.

Howarth has commented that our use of Fisher's description of sampling errors is outdated. However, Fisher's discussion of sampling errors in counting objects (Fisher, 1973, sect. 15, p. 57) is both clear and understandable as well as familiar to Earth scientists. It is, of course, based on assuming a Poisson distribution. We had noted the limitations of small sample size for some counts. We maintain that the statistical analysis of craters has utility in establishing the relative ages of planetary surfaces.

Stofan in her discussion gives a historical perspective on studies of Venus's coronae. We thank her for comments and appreciate her insightful summary of past work.

None of the criticisms has seriously addressed our distribution arguments: Why are coronae, on average, less heavily cratered if they are indeed older features? Why do coronae tend to be more densely distributed in and near rift zones? If coronae were classified as craters, the BAT region with its surplus of coronae would then appear older than the average surface of Venus. What independent evidence exists for this?

We have presented a technique designed to quantitatively analyze features for crater-like morphology. Certainly this technique could be refined to consider better such factors as diameter/depth ratios, but we feel that even this rudimentary approach is much better than the eyeball method—with its associated subjectivity—for determining features' origins. Finally, we reiterate that only a complete, quantitative analysis of all mapped features identified as coronae would most conclusively address their origin.

REFERENCES CITED

Barsukov, V.L., Basilevsky, A.T., Burba, G.A., Bobinna, N.N., Kryuchkov, V.P., Kuz'min, R.O., Nikolaeva, O.V., Pronin, A.A., Ronca, L.B., Chernaya, I.M., Shashkina, V.P., Garanin, A.V., Kushky, E.R., Markov, M.S., Sukhanov, A.L., Kotel'nikov, V.A., Rzhiga, O.N., Petrov, G.M., Alexandrov, Yu.N., Sidorenko, A.I., Bogomolov, A.F., Skrypnik, G.I., Bergman, M.Yu., Kudrin, L.V., Bokshtein, I.M., Kronrod, M.A., Chochia, P.A., Tyuflin, Yu.S., Kadnichansky, S.A., and Akim, E.L., 1986, The geology and geomorphology of the Venus surface as revealed by the radar images obtained by Venera 15 and 16: Journal of Geophysical Research, v. 91, p. D378–398.

Basilevsky, A.T., Pronin, A.A., Ronca, L.B., Kryuchkov, V.P., Sukhanov, A.L., and Markov, M.S., 1986, Styles of tectonic deformation on Venus: Analysis of Veneras 15 and 16 data: Journal of Geophysical Research, v. 91, p. 399–411.

Basilevsky, A.T., Head, J.W., Schaber, G.G., and Strom, R.G., 1997, The resurfacing history of Venus, *in* Bougher, D.M., et al., eds., Venus II, geology, geophysics, atmosphere, and solar wind environment: Tucson, Arizona, University of Arizona Press, p. 1047–1084.

Cochrane, C.G., 2005, Topographic modelling from SAR imagery of impact craters on Venus [Ph.D. thesis]: London, Imperial College, University of London, 230 p.

Fisher, R.A., 1973, Statistical methods for research workers: New York, Hafner, 362 p.

Hamilton, W.B., 2005, Plumeless Venus preserves ancient impact-accretionary surface, *in* Foulger, G.R., et al., ed., Plates, plumes, and paradigms: Boulder, Colorado, Geological Society of America Special Paper 388, p. 781–814, doi: 10.1130/2005.2388(44).

Hamilton, W.B., 2007 (this volume), An alternative Venus, *in* Foulger, G.R., and Jurdy, D.M., eds., Plates, plumes, and planetary processes: Boulder, Colorado, Geological Society of America Special Paper 430, doi: 10.1130/2007.2430(41).

Hansen, V.L., 2003, Venus diapirs: Thermal or compositional?: Geological Society of America Bulletin, v. 115, p. 1040–1052.

Head, J.W., Crumpler, L.S., Aubele, J.C., Guest, J.E., and Saunders, R.S., 1992, Venus volcanism: Classifications of volcanic features and structures, associations, and global distribution from Magellan data: Journal of Geophysical Research, v. 97, p. 13153–13198.

Hoogenboom, T., Smrekar, S.E., Anderson, F.S., and Houseman, G., 2004, Admittance survey of type 1 coronae on Venus: Journal of Geophysical Research, v. 109, doi: 10.1029/2003JE002171.

Howarth, R.J., 1998, Improved estimators of uncertainty in proportions, point-counting and pass-fail test results: American Journal of Science, v. 298, p. 594–607.

Janes, D.M., Squyres, S.W., Bindschadler, D.L., Baer, G., Schubert, G., Sharpton, V.L., and Stofan, E.R., 1992, Geophysical models for the formation and evolution of coronae on Venus: Journal of Geophysical Research, v. 97, p. 16,055–16,067.

Johnson, C.L., and Richards, M.A., 2003, A conceptual model for the relationship between coronae and large-scale mantle dynamics on Venus: Journal of Geophysical Research, v. 108(E6), p. 5058, doi: 10.1029/2002JE001962.

Jurdy, D.M., and Stoddard, P.R., 2007 (this volume), The coronae of Venus: Impact, plume, or other origin?, *in* Foulger, G.R., and Jurdy, D.M., eds., Plates, plumes, and planetary processes: Boulder, Colorado, Geological Society of America Special Paper 430, doi: 10.1130/2007.2430(40).

Koch, D.M., and Manga, M., 1996, Neutrally buoyant diapirs: A model for Venus coronae: Geophysical Research Letters, v. 23, p. 225–228.

Magee Roberts, K., and Head, J.W., 1993, Large-scale volcanism associated with coronae on Venus: Implications for formation and evolution: Geophysical Research Letters, v. 20, p. 1111–1114.

Namiki, N., and Solomon, S.C., 1994, Impact crater densities on volcanoes and coronae on Venus: Implications for volcanic resurfacing: Science, v. 265, p. 929–933.

Price, M., and Suppe, J., 1995, Constraints on the resurfacing history of Venus from the hypsometry and distribution of tectonism, volcanism, and impact craters: Earth, Moon, and Planets, v. 71, p. 99–145.

Pronin, A.A., and Stofan, E.R., 1990, Coronae on Venus: Morphology, classification, and distribution: Icarus, v. 87, p. 452–474.

Smrekar, S.E., and Stofan, E.R., 1997, Coupled upwelling and delamination: A new mechanism for coronae formation and heat loss on Venus: Science, v. 277, p. 1289–1294.

Squyres, S.W., Janes, D.M., Baer, G., Bindschadler, D.L., Schubert, G., Sharpton, V.L., and Stofan, E.R., 1992, The morphology and evolution of coronae on Venus: Journal of Geophysical Research, v. 97, p. 13,611–13,634.

Stefanick, M., and Jurdy, D.M., 1996, Venus coronae, craters and chasmata: Journal of Geophysical Research, v. 101, p. 4637–4643.

Stofan, E.R., Bindschadler, D.L., Head, J.W., and Parmentier, E.M., 1991, Corona structures on Venus: Models of origin: Journal of Geophysical Research, v. 96, p. 20,933–20,946.

Stofan, E.R., and Smrekar, S.E., 2005, Large topographic rises, coronae, large flow fields, and large volcanoes on Venus: Evidence for mantle plumes?, *in* Foulger, G.R., et al., eds., Plates, plumes, and paradigms: Boulder, Colorado, Geological Society of America Special Paper 388, p. 841–861, doi: 10.1130/2005.2388(47).

Stofan, E.R., Head, J.W., and Grieve, R.A.F., 1985, Classification of circular features on Venus: 1984 Report on Planetary Geology and Geophysics Program, Houston, Texas, National Aeronautics and Space Administration, p. 103–104.

Stofan, E.R., Sharpton, V.L., Schubert, G., Baer, G., Bindschadler, D.L., Janes, D.M., and Squyres, S.W., 1992, Global distribution and characteristics of coronae and related features on Venus: Implications for origin and relation to mantle processes: Journal of Geophysical Research, v. 97, p. 13,347–13,378.

Vita-Finzi, C., Howarth, R.J., Tapper, S.W., and Robinson, C.A., 2005, Venusian craters, size distribution, and the origin of coronae, *in* Foulger, G.R., et al., eds., Plates, plumes, and paradigms: Boulder, Colorado, Geological Society of America Special Paper 388, p. 815–823, doi: 10.1130/2005.2388(45).

The Geological Society of America
Special Paper 430
2007

An alternative Venus

Warren B. Hamilton*

Department of Geophysics, Colorado School of Mines, Golden, Colorado 80401, USA

ABSTRACT

Conventional interpretations assign Venus a volcanotectonic surface, younger than 1 Ga, pocked only by 1000 small impact craters. These craters, however, are superimposed on a landscape widely saturated with thousands of older, and variably modified, small to giant circular structures, which typically are rimmed depressions with the morphology expected for impact origins. Conventional analyses assign to a fraction of the most distinct old structures origins by plumes, diapirs, and other endogenic processes, and ignore the rest. The old structures have no analogues, in their venusian consensus endogenic terms, on Earth or elsewhere in the solar system, and are here argued to be of impact origin instead. The 1000 undisputed young "pristine" craters (a misnomer, for more than half of them are substantially modified) share with many of the old structures impact-diagnostic circular rims that enclose basins and that are surrounded by radial aprons of debris-flow ejecta, but conventional analyses explain the impact-compatible morphology of the old structures as coincidental products of endogenic uplifts complicated by magmatism. A continuum of increasing degradation, burial, and superposition connects the younger and truly pristine young impact structures with the most modified of the ancient structures. Younger craters of the ancient family are superimposed on older ones in impact-definitive cookie-cutter bites and are not deflected as required by endogenic conjectures. Four of the best-preserved of the pre-"pristine" circular structures are huge, with rimcrests 800–2000 km in diameter, and if indeed of impact origin, must have formed, by analogy with lunar dating, no later than 3.8 Ga. Much of the venusian plains is seen in topography to be saturated with overlapping 100–600 km circular structures, almost all of which are disregarded in conventional accounts. Several dozen larger ancient plains basins reach 2500 km in diameter, are themselves saturated with midsize impact structures, and may date back even to 4.4 Ga. Giant viscously spread "tessera plateaus" of impact melt also reach 2500 km in diameter; the youngest are little modified and are comparable in age, as calibrated by superimposed "pristine" impact structures, to the least modified of the giant impact basins, but the oldest are greatly modified and bombarded. The broad, low "volcanoes" of Venus formed within some of the larger of the ancient rimmed structures, resemble no modern volcanic complexes on Earth, and may be products of the collapse and spread of impact-fluidized central uplifts. Venusian plains are saturated with impact structures formed as transient-ocean sediments were deposited. The variable burial of, and compaction into, old craters by plains fill is incompatible with the popular contrary inference of flood basalt plains. Early "pristine" craters were formed in water-saturated sediments, subsequent greenhouse desiccation of which

*E-mail: whamilto@mines.edu.

Hamilton, W.B., 2007, An alternative Venus, *in* Foulger, G.R., and Jurdy, D.M., eds., Plates, plumes, and planetary processes: Geological Society of America Special Paper 430, p. 879–911, doi: 10.1130/2007.2430(41). For permission to copy, contact editing@geosociety.org. ©2007 The Geological Society of America. All rights reserved.

produced regional cracking and wrinkling of the plains and superabundant mud volcanoes ("shields"). The minimal internal planetary mobility indicated by this analysis is compatible with geophysical evidence. The history of the surface of Venus resembles that of Mars, not Earth.

Keywords: impact cratering, planetary accretion, plumes, planetary evolution

INTRODUCTION

Venus displays clear evidence of prolonged surface stability. It has unimodal topography and no plate tectonics. Its rift systems record only minor extension (Connors and Suppe, 2001). Venus nevertheless is widely assumed to be about as active internally as Earth. This paradox is popularly resolved by assigning a young maximum age to the 1000 most obvious impact craters, by rationalizing that the planet was wholly resurfaced endogenically before those craters began forming, and by assuming that the thousands of older circular structures have non-impact origins. I argue here that the morphology of those older structures instead requires that they be variably degraded and buried impact structures. The modified structures include many far larger than any young ones, and must be at least as old as 3.8 Ga if they indeed record impacts. The history of Venus has been profoundly misunderstood in conventional literature. Vita-Finzi et al. (2004, 2005) are among the very few researchers who have argued since ca. 1990 for the presence of old impact structures on Venus.

At least half of the 1000 accepted "pristine" craters are in fact modified, many severely, and there is a morphologic continuum from these, via increasing superposition, erosion, and burial, into and through the thousands of older structures. Many regions, both uplands and plains, are saturated with the old structures. There is much evidence for an ancient hydrosphere and for thick plains sedimentation. This article proceeds from a broad exploration of venusian problems (Hamilton, 2005). Some of the present conclusions (including those regarding preserved marine features and the nonmagmatic nature of "volcanoes") differ markedly from those in my 2005 paper, but the two papers are complementary.

IMAGERY

The surface of Venus is known primarily from radar imagery obtained by the Venus-orbiting spacecraft Magellan during 1990–1994. The most used of this imagery is mosaicked synthetic-aperture-radar (SAR) backscatter (reflectivity), plotted as grayscale brightness draped on generalized topography. Recorded Magellan SAR resolution cells, before resampling to reoriented 75 × 75 m pixels, vary from 120 × 120 m to 120 × 280 m. The imagery differs profoundly from the photographic and optical scanner imagery that it superficially resembles. The brightness of radar reflectivity is a function of slope (greatest from slopes facing the satellite), surface roughness at centimeter and decimeter scale near the radar wavelength (the rougher, the brighter), and the dielectric constant of surface material (the higher, the brighter). Substantial variations in topography often cannot be seen on the radar imagery, important features can be invisible, and intersecting structures appear to run together. The appearance of features varies greatly with their orientation and with the direction and inclination of the radar beam. Slopes facing the satellite are shortened and slopes away lengthened, so symmetrical ridges appear to be hogbacks of dipping strata.

East-looking imagery with latitudinally varying inclination provides the most extensive coverage. East-looking imagery with greater inclinations from the vertical is available for about one-sixth of the planet. Paired images with different inclinations can be viewed in optical or computerized stereoscopy; however, the apparent lengths of slopes, and the vertical and horizontal positions of the ridges and valleys defined by their intersections, vary with steepness and trends of those slopes, and confusing illusions abound in steep topography. Computerized stereoscopy relies primarily on matching brightness transitions, which can vary greatly, and migrate or disappear, with different look angles; mismatches are common, and false topography, with convincing visual aspect, can be generated (Cochrane, 2005). The MST (Magellan Stereo Toolkit) computer program, used widely to generate perspective images and topographic profiles, is very unreliable where brightness contrasts are low and features are large, as commonly is the case with old structures; it can, on the one hand, mismatch wholly different features, and, on the other hand, add topographic noise to featureless areas (C.G. Cochrane, 2005, and written communications, 2006). Investigators who rely on it (e.g., Matias and Jurdy, 2005) can be seriously misled. West-looking imagery is available for part of the planet, but the incompatible brightnesses of the opposite look directions commonly preclude satisfactory stereoscopy. Venusian global topography comes primarily from Magellan nadir measurements, which have low resolution and contain artifacts and noise. Detailed profiles derived from this topography, and given huge vertical exaggeration, also mislead investigators (such as Jurdy and Stoddard, this volume) who rely on the apparent details. Most backscatter and altimetric imagery shown here was downloaded from the Web site http://astrogeology.usgs.gov/Projects/Map-a-Planet, which almost instantly provides a seamless mosaic of any specified area on its own sinusoidal projection.

Hundreds of pseudoperspective diagrams (e.g., Grindrod and Hoogenboom, 2006; Grindrod et al., 2006; Krassilnikov and

Head, 2003) of the circular structures have been published. The impact rim, basin, and apron morphology of most of the structures illustrated commonly are displayed but are obfuscated by the invariably great vertical exaggerations (often unmentioned in captions or texts) of at least 10×, typically 20 or 30×, and sometimes even 50×, and by altimetric artifacts that also are exaggerated. (The one pseudoperspective plot in this article has an exaggeration of only 3×.) Also common in the venusian literature are topographic profiles with even greater vertical exaggerations. Herrick et al. (2005, Fig. 1D) showed profiles across Chloris Mons (which is shown in Fig. 8 of this article, which appears later) with a vertical exaggeration of ~150:1, rendering the 1° slopes of the very low central dome within the crater as gigantic 70° slopes. Grindrod et al. (2006, their Fig. 3) similarly depicted a broad, low mound (Atai Mons) that rises within a circular rim 150 km in diameter, with a vertical exaggeration of 60:1. Many venusian interpretations (e.g., Jurdy and Stoddard, this volume) are based on illusions in such exaggerated and artifact-ridden illustrations.

Circular Structures

The surface of Venus displays thousands of circular structures that obviously are mostly older than the 1000 minimally modified craters agreed by all observers to be of impact origin, and much of the surface is saturated with the old structures. Fewer than 1000 of the old structures are recognized in the conventional literature—most of the large ones, and nearly all of the small ones, are ignored—and are presumed to be endogenic. However, the better preserved of these ignored old circular structures share with the young impact structures the basin, rim, and apron morphology expected of impact origins and have the requisite cookie-cutter superpositions. From these there are all gradations of greater modification, overprinting, and burial. Several thousand of the old structures have circular rims 200–2500 km in inner diameter, and if they are indeed of impact origin, even the youngest must, by analogy with lunar dating, be at least as old as 3.85 Ga.

Impact structures begin as transient excavated and imploded cavities, but the floors of all but the smallest immediately rebound and, simultaneously, the walls cave in. The final rims are far beyond the initial large excavations and are not perfect circles in anisotropic targets. Shock melting, decompression and delayed melting in very large excavations, and granular fluidization by shock without melting, may be extensive.

Conflicting Endogenic Conjectures. The endogenic explanations for the old circular structures have many imaginative variants constrained primarily by the assumption that plumes, diapirs, and other upwellings and downwellings must be responsible. Grindrod and Hoogenboom (2006), Herrick et al. (2005), Jurdy and Stoddard (this volume), Krassilnikov and Head (2003), and Stofan and Smrekar (2005) recently provided additions to the conflicting speculations. Here is the list by Herrick et al. (2005, p. 1; I omit their citations of supporting publica-

tions) of some of the mutually incompatible conjectures for the origin of coronae, the most common label given large, old venusian structures dominated by circular rimmed depressions: "coronae are caused by plumes originating from a midmantle layer; coronae are caused by breakup of a mantle plume head; coronae form from plumes interacting with thin lithosphere . . . ; coronae are caused by detached diapirs, perhaps followed by retrograde subduction and or delamination; and coronae are formed by small, long-lived plumes from midmantle depths that evolve to delaminate the lithosphere." These hypothetical processes have no known terrestrial analogues. The Herrick et al. paper, like most others, does not even mention impact origins as an option.

Nomenclature. Venusian literature is rich in unfamiliar terms that render papers opaque to nonspecialists. Some terms are shared with other planets, but the most-used are unique to Venus. The fewer than 1000 old circular structures that are acknowledged in the conventional literature, arbitrarily selected from the thousands that exist, are assigned labels that include *arachnoid, astrum, corona, stealth corona, crustal plateau, mons, nova, patera, tessera, tick,* and *volcano,* plus various hybrids. These terms are all used with exclusively endogenic connotations. The only common term, *volcano,* conveys a misleading impression of morphology and may be incorrect in implying a magmatic construct.

Most abundant of the large circular structures are coronae (which mostly have raised rims enclosing basins), "volcanotectonic features that are apparently unique to Venus" in the solar system (Grindrod et al., 2006, p. 265). The implausibility of such uniqueness vanishes if coronae, and the other circular structures, instead record a ubiquitous exogenic planetary process.

Numerical Modeling. Endogenic speculations often are accompanied by circular-rationalization numerical modeling based on values and concepts assumed because they enable calculation of the desired results. Kaula (1995) recognized that such modeling, including his own, can amount to "wish fulfillment" and emphasized that all explanations of mantle-circulation causes of hypothetical resurfacing are contrived rationalizations. Herrick et al. (2005, p. 11) conceded that their own modeling was based on "debatable and largely unconstrained" assumptions. Nevertheless, most venusian modelers (e.g., Stofan and Smrekar, 2005) do not acknowledge that they are merely extrapolating their speculations. Anderson (2007) demonstrated that the common exclusion of pressure-varying parameters from such modeling renders it essentially useless.

None of the qualitative or quantitative endogenic conjectures account for the characteristic circularity and impactlike morphology and superpositions of the structures at issue.

Planetology and Plumology

The Moon's small to huge rimmed circular structures were widely regarded as products of endogenic processes until the 1960s. A few astute observers, from Grove Karl Gilbert to Robert

Dietz, argued for impact origins and for preservation of a landscape dating back to the era of planetary accretion, but this view did not prevail until space-age evidence made it inescapable. Similarly, terrestrial circular structures now proved to record ancient impacts were commonly assumed, until even later, to be "volcanotectonic" or "cryptovolcanic."

Consensus regarding the large rimmed circles of Venus went in the opposite direction, from impact to volcanotectonic—and for the opposite reasons, with conjecture trumping evidence. The mist-shrouded surface of Venus was seen first, in the 1970s and 1980s, with low-resolution radar imagery, and many large rimmed circular structures were then recognized. These have the size distribution and rim, basin (often with central peaks or peak rings), and apron morphology expected of impact structures and were early regarded as of possible impact origin, even though this required that they record very ancient accretion of the planet (e.g., Basilevsky et al., 1987; Grieve and Head, 1981; Masursky et al., 1980).

Theorists (e.g., Stofan et al., 1985), however, rationalized that preservation of ancient surfaces and structures on "Earth's twin" was impossible, and exported terrestrial plume conjecture to Venus to explain the planet's circular structures. Speculation regarding terrestrial plumes, hypothetical columns of hot material rising from deep in the mantle, became popular during the 1970s and 1980s. The present volume contains many articles expressing terrestrial pro- and anti-plume arguments; see also Foulger et al. (2005) and G.R. Foulger's Web site, http://www .mantleplumes.org, for many more papers on both sides of the debate. The physical and geochemical rationales presented for plumes are contradicted by most available information from the systems at issue, which instead indicate processes limited, or almost limited, to the upper mantle (Anderson, 2007). The geologic rationales for hotspot tracks and the like have been disproved. Nevertheless, venusian plume conjectures quickly became entrenched. Impact explanations were discarded, years before high-resolution imagery became available, in favor of a chain of suppositions linked with a few facts: Venus is almost as large and dense as Earth; Venus is mostly unfractionated, is as hot, volatile-rich, and mobile internally as Earth, so no ancient surface can be preserved; Venus lacks plate tectonics, but Earth has plumes, so Venus must lose heat mostly by plumes; the venusian atmosphere is now hot and dry, so the surface was never appreciably modified by atmospheric or hydrospheric processes. Although only the statements regarding size, density, lack of plate tectonics, and present atmosphere are known to be correct (the other statements being conjectures, most of which are falsified by Anderson, 2007), this chain promptly became dogma. The chain was lengthened, not evaluated, when Magellan imagery was obtained: venusian structures resemble nothing attributed to plumes on Earth [true], but whatever is observed must be due to endogenic processes (conjecture), so plumes operate uniquely on Venus [!]. This composite chain is accepted in nearly all mainline geologic and geodynamic venusian literature (e.g., Stofan and Smrekar, 2005; Turcotte, 1995). Nikolaeva

(1993) was among the very few who continued to argue for impacts after plumology became fashionable.

Atmosphere and Hydrosphere

A recurring theme in this article is that Venus had oceans during much of the early period recorded by its older impact structures. The present highly evolved greenhouse atmosphere is dense (~90 bars), hot (surface T ~450°C), and dry (96.5% CO_2, 3.5% N_2, ~250 ppm H_2O, traces of many other gases). So what was the composition and state of the early atmosphere, and could there indeed have been a hydrosphere?

The mainline venusian literature assumes that Venus had a hot, anhydrous atmosphere and surface throughout the era recorded by everything now exposed. Venus nevertheless must have accreted with abundant water. Although accretion began from local feeding zones in the protoplanetary disk, by ~10^7 years from the beginning of accretion, when the planets had something like half their final masses, eccentricities of orbits of protoplanets in the asteroid belt (components of which are graded Jupiterward in general composition, from metallic and stony through hydrous carbonaceous masses to dirty ices) had been so pumped up by Jupiter and protoplanets that they were abundantly accreted to the inner planets until those reached essentially their final sizes before something like 10^8 years had elapsed (Raymond et al., 2006). Much water may also have come in with dust particles (Kulikov et al., 2006). Much water would have been lost to space from impact erosion, from the magma oceans resulting from accretion, and from erosion by violent radiation by the infant sun, and the uncertainties are huge. Among the options are early oceans, which permit explanation of the present atmospheric deuterium/hydrogen ratio of Venus, which is ~120 times that of Earth, as contrasted with the lack of mass fractionation of oxygen and nitrogen isotopes (Kulikov et al., 2006).

Lowlands dominate Venus. The plains are radar-dark, hence smooth-surfaced, and conventionally are assumed to be flood basalts, but lava flow speculation is countered by the almost total lack of possible sources—volcanoes, dikes, rifts—for eruption of the hypothetical lavas. Evidence that the plains instead are floored by consolidated sediments, thermally metamorphosed by the subsequent greenhouse atmosphere, was presented by Jones and Pickering (2003) and Hamilton (2005). This article emphasizes that thick fill progressively buried old impact structures formed synchronously with deposition. The fill is compacted into many old structures, but the youngest impact structures of the ancient family postdate most of the fill and likely include correlatives of the older "pristine" craters. The surface smoothness of the radar-dark plains, the horizontally laminated character of their material (now hard rock) as imaged by Soviet landers, the compaction of plains materials into and above craters, and the fluidized impact ejecta from many small "pristine" craters in the plains all accord with sedimentary fill. Lowlands are crossed by meandering channels, up to thousands of km long and marked by cut-off meanders, point bars, and deltas, that have semiconstant widths of 1–3 km and depths of ~50 m. The geometry of channels and distributaries resembles that of Earth's submarine turbidite-feeding channels, and turbidites, rather than the conventionally assumed lava flows, may account for lobate patterns on plains surfaces (Jones and Pickering, 2003). However, only magmatic explanations for the channels are advocated in the conventional venusian literature. For example, Oshigami and Namiki (2005) argued for erosion by lavas, and Lang and Hansen (2006) proposed continuous linear sapping from below by lava flowing thousands of km beneath uniformly thin cover—even though these implausible processes cannot account for the continuity and constant dimensions of the channels.

Large tracts ("shield fields") of the plains, including the filled interiors of many old circular structures, are randomly pimpled by perhaps a million small smooth-surfaced cones, typically several km in diameter, <200 m high, and often bearing small crestal craters. These cones commonly are assumed to be of basalt, but they nowhere define rifts or other suggestions of magmatic sources. The cones may be mud volcanoes developed from wet sediments overpressured by top-down heating by the developing greenhouse atmosphere into which the ocean evaporated (Hamilton, 2005). Terrestrial mud volcanoes, the most important pathway for degassing deeply buried sediments (Dimitrov, 2002), are similar in morphology and variety to the small venusian shields. The slight deformation recorded by plains wrinkle ridges and reticulate fracture systems ends abruptly against bedrock uplands, and also may be due to atmospheric thermal effects.

Old impact structures are variably smoothed and degraded by erosion, much of it likely submarine, and many impact structures may have formed underwater. Much lowland sediment may have been recycled from local comminuted impact debris, but erosion of uplands presumably also provided much sediment. The general aspect of venusian uplands is of scoured bedrock and etched landforms, with accumulations of sediments in small and large low areas and in perched plains. Integrated upland dendritic drainage systems, and other evidence of major fluvial erosion, have been suggested locally but are not present on a large scale, although poorly integrated stream valleys may be widespread, and braided streams (appropriate for the gentle gradients and large sediment supplies) appear to be present. Perhaps wind erosion was a major denudation agent when the planet rotated faster. Given strong winds and extensive impact pulverization, the dense, corrosive, and perhaps supercritical atmosphere would have been a powerful transport medium, and planar sedimentation would have been favored, whether or not eolian sediment was dumped into standing water. There is no high stillstand shoreline apparent in radar imagery, but possible recessional shorelines are preserved in some areas.

PROPERTIES OF VENUS

Venusian and solar system data, independent of interpretation of circular structures, provide no support for the chain of conjectures on which consensus plumological speculations are based. Orbital modeling requires that terrestrial planets formed rapidly (Chambers, 2004; Raymond et al., 2006). Abundant isotopic (e.g., Kleine et al., 2004) and other evidence from the sampled parts—Earth, the Moon, and meteorites representing the Moon, Mars, and asteroids—of the inner solar system requires that planets fractionated as they accreted. Earth, the Moon, and Mars, and hence presumably Venus, were near their final sizes, and internally fractionated, before 4.45 Ga., only ~10^8 years after condensation of the protoplanetary disk began. They have not remained mostly unfractionated as assumed by many geochemists since the 1950s and by modern plumologists and venusian specialists. Venus has no magnetic field: the core does not convect, and may be solid. Most of the ^{40}Ar ever generated from Earth's ^{40}K is now in the atmosphere (Anderson, 2007). Venus has only one-fourth as much absolute ^{40}Ar in its atmosphere as does Earth; the simplest explanation is that Venus has only a similar fraction as much heat-generating potassium, as is expected from its assembly primarily from a feeding zone condensed at higher temperature than was Earth's. Radioactive ^{40}K has a relatively short half-life and is now only a minor contributor to terrestrial heat, but it was a major contributor in the young Earth, and Earth's huge heat content is mostly residual from its early history. Current terrestrial dynamic mobility is greatly enhanced by water and CO_2 cycled back into the mantle by plate tectonics. Venus lacks plate tectonics and likely has a volatile-poor mantle. Venus is far stiffer than Earth because it is both colder and lower in weakening and melt-enhancing volatiles.

Lithosphere Strength

The great contrast between terrestrial and venusian geoids highlights the dissimilarity between internal properties of the two planets. Earth's topography is supported isostatically by lateral variations in density, mostly at lower-crustal and uppermost-mantle depths; topography is almost invisible in the geoid, which reflects density variations, including subducted slabs, deeper in the mantle (Fig. 1A). Venusian topography and geoid (Fig. 1B and C) correlate directly over a broad dimensional range. Venusian topography may be supported by lithosphere far stronger, even over very long periods, than that of Earth (Kaula, 1994), despite the high surface temperature imposed by the greenhouse atmosphere. If the thousands of large circular structures of Venus are indeed primarily of impact origins, as I argue here, the high-strength option is required.

Popular conjecture nevertheless presumes that Venus is so hot and active that the lithosphere must be weak, and the correlation between geoid and topography is commonly assumed to require very thin, even zero thickness, elastic lithosphere, and shallow isostatic compensation (e.g., Anderson and Smrekar, 2006), or else dynamic disequilibrium, whereby rising mantle currents push up the uplands, and sinking currents pull down the lowlands, or, oppositely, in an illustration of the lack of constraints in endogenic conjectures, whereby rising currents thin

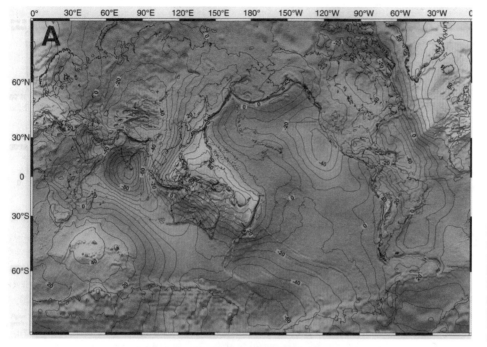

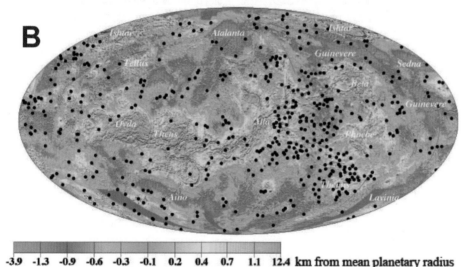

-3.9 -1.3 -0.9 -0.6 -0.3 -0.1 0.2 0.4 0.7 1.1 12.4 **km from mean planetary radius**

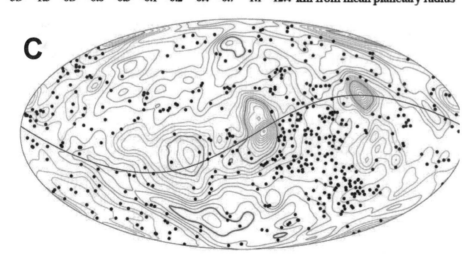

Figure 1. Contrasted geoids show Earth to have thin, weak lithosphere, and Venus to have strong, thick lithosphere and hence far less internal mobility despite high surface temperature. (A) Terrestrial geoid, 10 m contours, blue and pink low, green-yellow-brown high. Topography is mostly compensated isostatically at shallow depths, and shows almost no correlation with geoid. (B) Venusian topography and (C) geoid correlate directly over a broad range of wavelengths because the lithosphere is far stiffer than Earth's and supports topography with strength rather than isostasy. Geoid height (C) shown by 10 m contours, 0 not shown, red positive, blue negative; spherical harmonic degrees 2–30 are fully retained, but, because reliability falls off at higher degrees, a Gaussian roll-off is applied to degrees 30–60. Venus maps are centered on the equator and longitude 180°. Centers of arbitrarily selected backscatter-identified coronae, which do not include most large lowland circular structures here regarded as of impact origin, are marked by dots in (B) and (C). Panel A provided by David Sandwell, panels B and C by Catherine Johnson, who supplied the topographic scale, corrected from that published by Johnson and Richards (2003).

the crust under lowlands, and sinking ones thicken it under uplands (e.g., Johnson and Richards, 2003; Vezolainen et al., 2004). Anderson and Smrekar (2006) derived their postulate of extremely variable venusian lithosphere primarily from the spherical-harmonic gravity field for wavelengths <~700 km, although gravity is very poorly determined at those short wavelengths (Wieczorek, 2007): the Anderson and Smrekar modeling is based on artifacts. Most of the topography is commonly regarded as perhaps 0.5 Ga in age, so an implausible corollary of these postulates is that the thermal and dynamic imbalances have been maintained for hundreds of millions of years.

Jurdy and Stoddard (this volume) attributed tectonic significance to the specific location of selected venusian coronae (which I regard as impact structures) on a geoid truncated at degree and order 10. Because this truncation both extremely generalizes the spherical-harmonic series and distorts it with artifacts, such attribution is unwarranted; further, there are abundant old circular structures both higher and lower on that invalid geoid than those selected by Jurdy and Stoddard.

Mass Imbalance

The integrated sum of venusian excess-mass anomalies is expressed by the great-circle heavy line, calculated by C.L. Johnson from the geoid, across Figure 1C. That this is approximately equatorial is unlikely to be a coincidence. I attribute the topographically high areas to changes in density as a consequence of melting by giant impacts, for which an equatorial bias is not expected, and suggest that the mass imbalance reoriented the planet with regard to its spin axis. (Anderson, 2007, proposed such an explanation for the nearly equatorial position of the Tharsis upland of Mars.) Braking by the shift may have contributed to slowing of venusian rotation. Perhaps the minor surface deformation recorded by venusian rift zones and plains undulations includes byproducts of the shift. (On Earth, spin imbalances are readily accommodated by shallow-plate motions.) The most conspicuous rift intersection on Venus is at Ozza Mons, an uncommonly large "volcano" (to me, an impact-fluidization construct) on the equator, where three rifts meet at angles near 120°, so lithosphere weakening as a byproduct of exothermic events may be indicated. Less regular but similar is the intersection at Theia Mons, which rises from the low, much-modified tessera upland of Beta Regio. (Venusian papers cite these locations to opposite effect, as evidence for control of volcanoes and rifts by plumes.)

Earth's Properties

Although Earth is far more mobile than Venus, widely assumed terrestrial properties and evolution also are overstated in the direction of excess mobility. The global heat loss commonly assumed for Earth, ~44 TW, is ~40% above the measured value integrated for seafloor age, and is based on models that incorporate an erroneously high thermal conductivity of hot oceanic

lithosphere (Hofmeister and Criss, 2005) and that lack a sound physical basis (Anderson, 2007). Among other properties that require the lower mantle to be vastly less mobile than postulated by plumists are the great decrease of thermal expansivity, and the increase of viscosity, with increasing pressure; the high thermal conductivity of the deep mantle [which may even preclude convection]; the probability of irreversible layering; and the likelihood that the cause of the deepest-mantle low–seismic velocity regions is higher iron contents, not higher temperatures (Anderson, 2007). Numerical modelers of plumes do not properly incorporate these parameters.

CIRCLES, CIRCLES, CIRCLES: IMPACT STRUCTURES YOUNG AND OLD

The unearthly landscapes of Venus (Fig. 2) show unimodal topography—no plate tectonics—and thousands of circular structures that typically have raised rims that range from a few km to 2000 km in inner diameter. The circular structures generally are well exposed in uplands but variably buried in lowlands by fill that accumulated as the structures developed. On Mercury, the Moon, the south half of Mars, and some satellites of Jupiter and Saturn, such rimmed circular structures are known to record bolide impacts. Only on Venus are they widely presumed to be, with the exception of 1000 minimally modified small craters, of endogenic origins of types unique in the solar system. *None* of the diverse conjectures of endogenic origins in the mainline literature address the consistent impact-compatible morphology and superpositions of these structures.

Many papers convey the erroneous impression that such structures are sparse. The geologic maps by Ivanov and Head (2001, their Plates 8 and 9) of the north half of the area of Figure 2A left the many circular structures undiscriminated within hypothetical, and implausible, circumglobal stratigraphic volcanic units such as "densely fractured plains material." On the other hand, Stofan and Smrekar (2005) recognized about three-fourths of the obvious large structures in this same view (but none of the many small old structures, and not the dozen or so more that can be inferred additionally from topography), and classed them mostly as endogenic coronae. Stofan and Smrekar, like Jurdy and Stoddard (this volume), drew corona boundaries far outside the rimmed circles, lumped overlapping or adjacent circles as single coronae, described features in terms that obscure circularity and impact-compatible morphology, and based endogenic conjectures on isolated examples. To me, the circular rims, the basins they commonly enclose, and the cookie-cutter superpositions are the features that should be addressed. Impacts by bolides pancaked or fragmented in the dense atmosphere (see Cochrane and Ghail, 2006, for examples among "pristine" craters) and superimposed impacts provide the general explanation for the many structures that diverge significantly from circularity and typically show, where well constrained, rims composited of circular arcs.

Some of the many small old circular structures, with rim

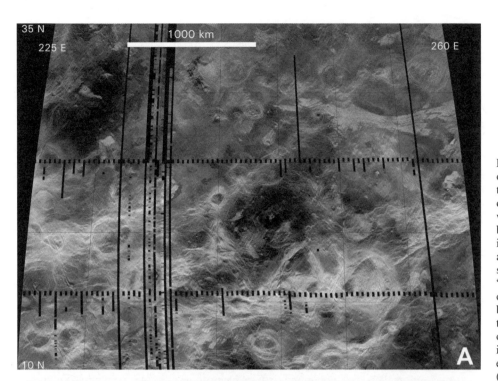

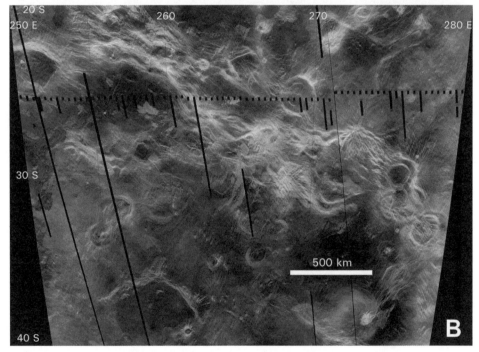

Figure 2. Regions mostly saturated with old circular impact structures that, in these radar-backscatter images, are more obvious in highlands than in lowlands, wherein they are variably buried. The better-preserved structures have impact-indicative raised rims that enclose basins and are surrounded by debris aprons; some are multiring. (A) Of eleven small "pristine" craters in this area, the only one obvious at this scale is the radar-bright tailed object in the central part of the dark area near the northwest corner of the area. All other circular structures in this panel are commonly considered endogenic, formed by plumes and diapirs. (B) About half of the large structures visible are conventionally termed coronae and paterae, either individually or as arbitrary composites, and regarded as endogenic, whereas the other half are overlooked. Of the many small circular structures within the view, but mostly inconspicuous at this scale, only about one-third (e.g., bottom edge, left of center) are classed as "pristine" craters and are commonly accepted as products of impacts. Some of the small circular structures here regarded as due to older impacts are shown in Figure 3. Although the Parga rift zone crosses the area from upper left to right center, most structures within it retain circular shapes.

diameters from ~5 to 100 km, that are within the view of Figure 2B are shown in Figure 3. Most such small structures are unmentioned and unmapped in conventional venusian literature. Some of these small structures arguably should be on the "pristine" list because the arbitrary criteria for identification of modified structures as "pristine" admit fewer structures in uplands, where reflectivity contrasts are low, than in lowlands, where they are high.

Size and Frequency Evidence for Impact Origins

The abundance of bolides wandering about the inner solar system falls off exponentially with increasing size, and the abundances and sizes of impact craters on the airless Moon (Stöffler et al., 2006) and thin-atmosphere Mars (Frey, 2006a) define almost straight lines on log-log plots. Most incoming small venusian bolides are destroyed in the 90 bar atmosphere, and even

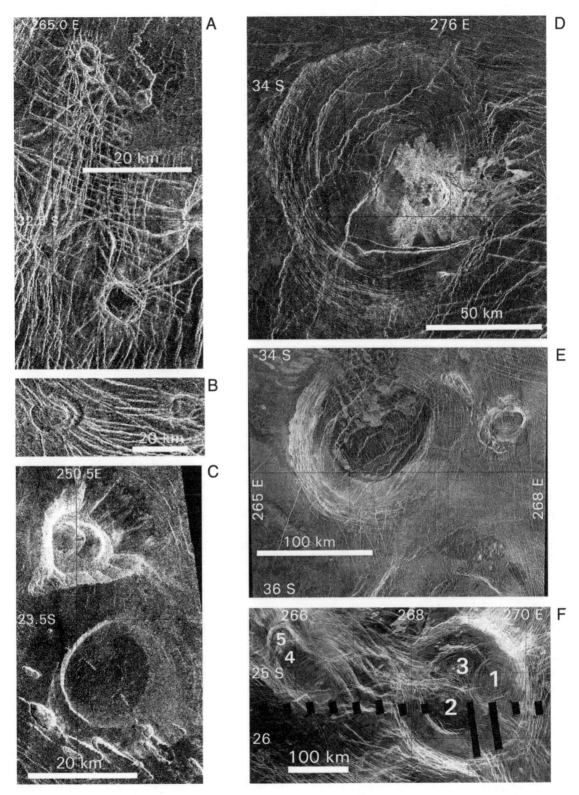

Figure 3. Small, old probable-impact circular structures within the area of Figure 2B. (A) Two of these small structures retain exposed rims; the northern one appears to be a "tick," with thrust-faulted margins indicative of oblique impact into consolidated sediments. (B) Two small structures retain exposed rims; the west one is a doublet. (C) The northern structure retains an exposed rim and some of the apron. The rim of the southern structure is exposed in the west but likely mostly buried by sediment in the east. Both structures are within an older impact structure obvious in Figure 2B. (D) Pristine Elza Crater is superimposed on a mostly buried old crater into which covering sediments are compacted. (E) A small compound crater (in the northeast quadrant) has an exposed rim surrounding a closed depression. The large crater (Nordenflycht Patera, left) is mostly buried, but compaction of sediments leaves a large closed depression. The north part of a more deeply buried large crater shows in the southeast. (F) A large, mostly buried crater (Hervor Corona, right) is cut by three nested craters (from almost simultaneous impacts by fragments of a single bolide?), of which 1 cuts 2, which cuts 3; in the northwest, 4 cuts 5, with the rims of both partly exposed. Only Elza Crater among all of these structures is conventionally assigned an impact origin.

the largest of the 1000 venusian craters universally accepted as of impact origins do not clearly define such a line (cf. McKinnon et al., 1997). Most bolides capable of making craters even 100–200 km in diameter apparently are destroyed in the dense venusian atmosphere.

The old circular structures that I see as of impact origin do fit straight log size–log frequency lines. Analysis is complicated by the arbitrary omission of the many small old structures, and the relatively few giant structures, from conventional databases, and by the custom of measuring the diameters of larger old structures not to their topographic rims but to arbitrarily greater diameters, but even these biased measurements define straight log-log lines. Stofan et al. (1992) defined a log-log straight line from their early coronal data, derived mostly from structures with rim diameters of ~100–600 km, but dismissed its impact significance, as a coincidence, in favor of a plume conjecture with which they could not explain the correlation. Glaze et al. (2002) confirmed the log-log relationship with a larger coronal data set but did not even mention its possible impact significance. Vita-Finzi et al. (2005) used a still larger data set, also showed the log-log size distribution to be as required by impact origins, and emphasized, correctly, that this is powerful evidence for such origins.

Young "Pristine" Craters

About 1000 unmodified to moderately modified small rimmed circular structures on Venus are universally accepted as impact craters. The definitive list is maintained by Robert Herrick at http://www.lpi.usra.edu/research/vc/vchome.html. Only nine of these craters have rimcrest diameters > 100 km, and only one, at 270 km, is > 200 km. Their geographic distribution is random (Vita-Finzi et al., 2005, their Fig. 1; Matias and Jurdy, 2005, and Jurdy and Stoddard, this volume, attach significance to local divergence from randomness of nonrobust small samples). The criterion for inclusion is that there can be no doubt as to impact origin. All other circular features are considered endogenic only by default, yet the list is widely accepted (but not by Vita-Finzi et al.) as including all possible impact structures on the planet. Although often termed "pristine," fewer than half of them actually fit that description and preserve sharp topography, radar-bright breccia floors, and unmodified lobate flow-breccia aprons (Fig. 4A). Venusian aprons are dominated by lobate debris, rather than by ballistic ejecta as on the airless Moon, because of the dense atmosphere and high gravity.

Herrick (2006) emphasized that ~60% of the "pristine" craters that can be measured accurately are in fact much modified. Many are partly infilled, and their ejecta blankets partly covered, by younger materials; many are breached, some tilted, and a few rifted. Herrick assumed that the fill and cover are volcanic, while puzzling over the lack of eruption sites to feed either inside or outside lavas, and did not consider processes of erosion and sedimentation. I see the crater fills as sedimentary, in part

because most of the rims are unbreached and the fills apparently were derived from the crater walls.

Many "pristine" craters may have formed in shallow water, and many more may have formed while the plains sediments were still water-rich. Many craters of the older family may be submarine.

Central-peak Lachappelle Crater and peak-ring Barton Crater show two degrees of erosion and of sedimentary burial of interiors, rims, and aprons (Fig. 4B). Like hundreds of the "pristine" craters, they predate thermal wrinkling of plains, and hence, in my terms, predate greenhouse top-down metamorphism of plains sediments and formed when the sediments were wet. Doublet Heloise Crater (Fig. 4C) is more deeply buried by thermally wrinkled plains sediments. A tilted, breached, and much modified crater that is accorded conventional "pristine" status was shown by Hamilton (2005, Figure 3C). Herrick (2006) and Matias and Jurdy (2005) referred to other tilted and modified craters of the young family.

Age of "Pristine" Craters

The "pristine" craters commonly are assumed to have formed within the past 0.3, 0.5, 0.7, or 1.5 billion years, the date varying with the assumptions used by the modelers, on a planetary surface broadly resurfaced at about that limiting time by endogenic processes that have no terrestrial analogues. This age modeling, at its best, integrates estimated numbers and sizes of captured bolides of different types (metallic and stony [but not weak carbonaceous] asteroid fragments, and comets) with estimates for ablation, fragmentation, pancaking, dispersal, and retardation in the dense atmosphere and for crater dimensions produced by surviving bolides and fragments, then seeks fits to the size-abundance distribution of the observed craters. Korycansky and Zahnle (2005) thus calculated that the observed craters formed within the last 0.73 b.y., and suggested that uncertainties in their assumptions limited the true maximum age to within a factor of two of this calculated value, say, 0.35–1.5 Ga; but uncertainties are large and multiplicative, and that small error limit is optimistic. Prior modeling by others (e.g., McKinnon et al., 1997) yielded preferred maximum ages of between 0.3 and 0.7 Ga.

As Schultz (1993) recognized, dating ambiguities permit the "pristine" venusian craters to go back to the early history of the planet. Models and observations agree that most small bolides are destroyed in the venusian atmosphere, and a critical factor for modeling is the survival of larger bolides. Almost all objects that now have orbits reaching inside Earth's orbit and that are capable of generating craters larger than 100 km in diameter on an airless terrestrial planet are comets, dirty iceballs prone to atmospheric destruction (Shoemaker, 1994, 1998; McKinnon et al., 1997). Earth's thin atmosphere probably produces far more disruption even of stony meteorites than generally appreciated (Bland and Artemieva, 2003), and the venusian atmosphere is

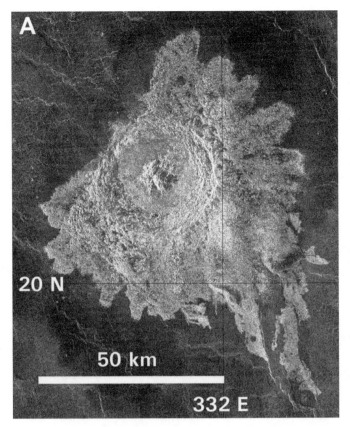

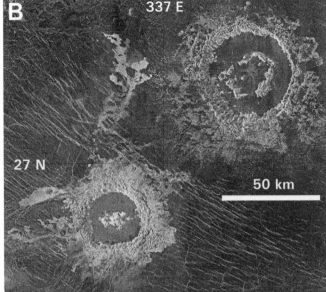

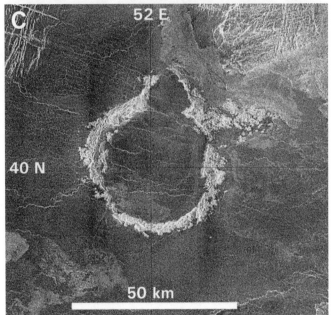

Figure 4. Modified structures universally accepted as produced by impacts. (A) Aurelie Crater is one of only ~40% of the accepted structures that warrant the designation "pristine." It has sharp topography, a radar-bright breccia floor, a central peak, and an apron of lobate debris-flow breccias. (B) Central-peak Lachapelle Crater (lower left) is partly filled by radar-dark sediments derived from the walls, and its ejecta apron was partly eroded and partly buried by plains sediments before those sediments were thermally wrinkled. Peak-ring Barton Crater is more degraded, its interior more deeply filled, and its apron more buried by plains material. (C) Double-impact Heloise Crater is breached and almost buried by plains sediments and predates their thermal wrinkling.

almost 100 times as dense as Earth's. There are so few large "pristine" craters on Venus that no statistical use of them is robust, and no straight-line log-log size-frequency relationship can be established to suggest that a size is represented beyond which all bolides survived to generate craters. Nevertheless, the young model ages incorporate the assumption that such a limiting size is well within the observed population.

Only craters of the "pristine" family are superimposed on many of the large circular venusian structures for which lunar analogy indicates a minimum age of 3.8 Ga. More modified, hence still older, large impact structures are pocked by variably more, up to full saturation, of the older family of structures.

"Pristine" and Older Craters

Many hundreds of examples provide a continuum between the small misnamed "pristine" impact craters and the more degraded structures commonly presumed to be endogenic. This gradation is denied in mainline venusian literature, which is emphatic (e.g., Strom et al., 2005) that the 1000 "pristine" craters are strikingly different from older rimmed circular structures and are the only impact structures on the planet. Such assertions notwithstanding, several thousand circular structures similar in size to, or modestly larger than, the 1000 commonly accepted impact structures are present on Venus, but show variably more modification. Some of these are shown in Figure 3, and more are shown now, before proceeding on to the thousands of larger old circular structures. Small and large old structures are of similar limiting ages, as calibrated by degrees of modifications and by superpositions. Nearly all small, old structures, and perhaps three-fourths of the visible large, old structures, are ignored in conventional venusian work.

Ten small craters, only two of which are widely accepted as of impact origins, are shown in Figure 5. The three same-size craters of Figure 5B are similar in aspect except for their quite different erosional smearing and sedimentary cover, and perhaps formed simultaneously from three fragments of a bolide disrupted in the atmosphere, with 1 having impacted on land, 2 in shallow water, and 3 in deeper water. (Water, erosion, and sedimentation are not considered possible in conventional venusian work.) Figure 5C shows four of the older family of structures that are largely buried by, and predate the thermal wrinkling of,

plains material. The "tick," the venusian term for a minor structural type commonly assumed to be endogenic, by contrast postdates plains deformation and likely was produced by a bolide that struck solidly lithified subhorizontal strata at a low angle and generated in them shallow thrust faults, convex downrange. The Spider (Shoemaker and Shoemaker, 1996) and Gosses Bluff (Milton et al., 1996) impact structures of Australia, and Upheaval Dome of Utah (Scherler et al., 2006), have similar thrust patterns and record such impacts and targets.

The plains craters of Figure 6 show an age sequence by their superpositions and modifications. The outer part of the apron of large Isabella Crater, an accepted impact structure, is subdued, and this modification, and also the extremely long runout of its southeast lobe, may be products of an impact that was either shallow submarine or occurred while the sediments were still saturated with water. Such long runout lobes are seen only about plains craters, not upland ones. Acoustic fluidization may gen-

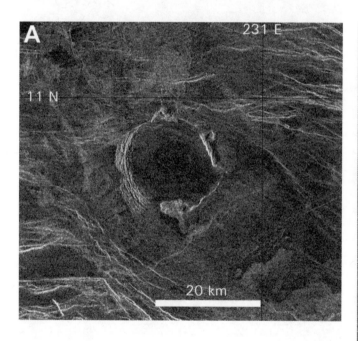

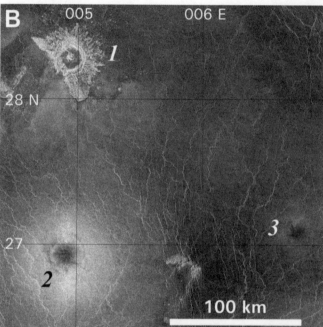

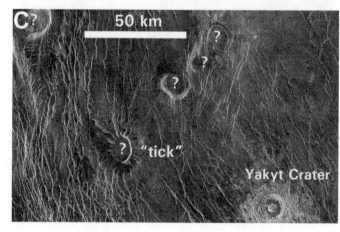

Figure 5. Small, circular structures, old and new, among which only Browning and Yakyt craters are commonly accepted as products of impacts. (A) Small, breached crater, mostly buried by plains sediments, that is conventionally classed as endogenic. Compare with "pristine" Heloise Crater, Figure 4C. (B) Variable degradation of three initially similar craters. 1—Sediment partly fills, and laps onto apron of, Browning Crater. 2—Erosion-smeared impact structure that was partly buried by plains sediments before thermal wrinkling and retains a closed crater. 3—Similar, but still more subdued, structure largely buried by plains sediments. Impact, erosion, and burial of 2 and 3 may have been submarine. (C) Five small likely impact structures are marked by question marks. The northern four of these are largely buried by, and predate the thermal deformation of, plains sediments, and may record submarine impacts. See text for discussion of the "tick." The area extends from 2°N to 3°N, and from 169.0°E to 170.5°E.

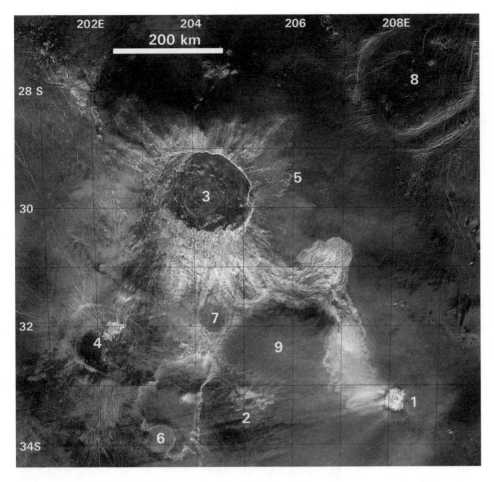

Figure 6. Variably preserved impact craters, numbered in inferred order of increasing age. 1—Tiny Cohn Crater. 2—Cluster of four tiny, likely simultaneously formed, craters. 3—Isabella Crater, the second-largest generally accepted impact crater on Venus, rimcrest diameter 175 km (note imperfect circularity), filled by dark sediment; the outer parts of the lobate debris apron and of long runout to the southeast are smeared and subdued, so may have been shallow marine. 4—A deep crater, with topographic rim 60 km in diameter, preserves its debris apron; partly covered by Isabella apron. 5—Tiny Alimat Crater (left of number), buried except for its rim. 6 and 7—Craters with rims ~50 km in diameter. 8—Mostly buried crater, rim diameter ~200 km. 9—Rimmed depression compacted above 200 km crater; Isabella debris flows were deflected by the rim. 1, 3, and 5 are commonly accepted as "pristine"; 4 and 8 are classed as endogenic coronae; and the others are overlooked. Other more modified craters can be inferred on detailed imagery.

erally account for the runouts (cf. Collins and Melosh, 2003), but with an assist from contained water.

The abundant small likely impact structures on Venus that are overlooked in conventional analysis probably greatly outnumber the 1000 accepted "pristine" structures, so several thousand of those old, small structures are exposed. The distribution of the old structures is not random about the planet, as is that of "pristine" impact structures. The small old structures commonly are present where coronae are exposed, but are mostly lacking on the youngest of the huge circular structures and on the youngest tessera terrains, and also are unseen in many plains areas. This accords with the obviously large age range of tessera, and with the variable burial of plains structures. Large impact structures in the plains are much more likely to print through to the surface, as illustrated here and by Vita-Finzi et al. (2005, their Fig. 5).

A pseudoperspective view of Aramaiti corona displays, with modest vertical exaggeration, little-modified impact morphology (Fig. 7). Aramaiti is the same size as Mead, the largest venusian crater commonly accepted as of impact origin. Aramaiti's circular rim, steep on the inside and apparently stepped down by concentric collapsed terraces, encloses a basin with a peakring uplift. The outside of the rim is a gentle ejecta apron with a broad, shallow ring syncline, a low outer rise beyond that, and,

as seen in detailed imagery, radial debris-flow lobes. The structure formed late in the depositional history of the surrounding plains, and presumably smoothing of the structure was submarine. Three probable impact craters that bracket Aramaiti in size are shown and described in Figure 8. McDaniel and Hansen (2005) accepted my recognition of Aramaiti as an impact structure, and noted others of similar morphology; Hansen had long regarded all coronae as endogenic.

Despite its obviously impact-compatible morphology, Aramaiti has been singled out for endogenic conjectures by plumists. Stofan and Smrekar (2005) speculated that Aramaiti formed by delamination of an inward-migrating rim atop a plume. Grindrod and Hoogenboom (2006) attributed Aramaiti to sinking of the surface into the center of a cylindrical upwelling that resulted from a density inversion of dense lithosphere overlying light asthenosphere. Neither conjecture addressed the crater's remarkable circularity and impact-compatible morphology, or considered an impact origin. Such speculations might reasonably be invoked to explain an isolated example of this type, but not the hundreds with similar morphology.

Grindrod and Hoogenboom (2006, their Fig. 1B–D) showed backscatter images and highly exaggerated pseudoperspective views of three of the many Aramaitilike coronae with rim

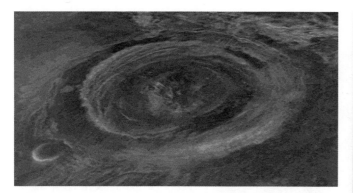

Figure 7. Northeastward pseudoperspective view of Aramaiti Corona (26°S, 82°E), showing an impact morphology typical of well-preserved circular structures conventionally classed as endogenic. The rimcrest is 270 km in diameter, and its steep inner slope appears to be terraced by slumps. The crater has a central peak-ring uplift. A smoothed and darkened conical ejecta apron slopes into the outer ring syncline. A broad outer rise, with outside diameter 400 km, preserves faint debris lobes. The impact and softening of morphology may have been submarine. Figure prepared by Trent Hare, U.S. Geological Survey, by draping a radar-brightness image on a topographic model with vertical exaggeration of 3:1.

diameters near 200 km. All have impact-morphology basins, rims, and gentle aprons of radial debris flows, and are seen on inspection to owe their slight elongations and irregularities to superimposed circles and circular arcs: they are products of impacts by pancaked or fragmented bolides. A number of other small craters with impact-compatible morphology can be seen in each of their images, but Grindrod and Hoogenboom considered only endogenic origins.

If the slight deficit of "pristine" craters on "coronae" calculated by Jurdy and Stoddard (this volume, their Table 1) is real, and not an illusion due to the ambiguous statistics of small samples, likely explanations lie in the under-reporting of young impacts in bedrock areas, discussed elsewhere in this article, and perhaps also may indicate that submarine impacts like that of Aramaiti Corona continued well into the era of formation of "pristine" craters.

Superimposed Old Impact Structures

Venus displays thousands of old superimposed rimmed circular structures. The morphologically younger take cookie-cutter bites from the older, as required by impact origins. Examples are illustrated by figures in this article and in Hamilton (2005). Were the structures endogenic, related to subsurface intrusions in accord with any of the diverse plume and diapir conjectures, the younger would be deformed against the older, which is not observed. Plume advocates evade this powerful evidence by lumping superimposed or neighboring rimmed basins as single coronae. Törmänen et al. (2005) termed seventy examples of superimposed structures "multiple coronae," and, without mention of their sequences, geometries, and impact morpholo-

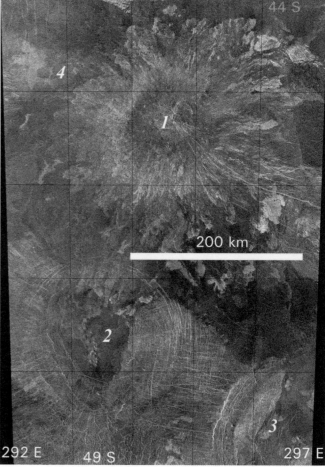

Figure 8. Three modified probable impact craters. 1 has a subdued rim, central peak, and well-preserved lobate ejecta apron that flowed into 2, the floor and rim of which are nearly circular despite the lopsided backscatter appearance. Most of the rim of large crater 3 is outside this view and is buried. Conventional interpretation: 1 is an endogenic volcano, Chloris Mons; 2 and 3 are endogenic coronae. Small feature 4 does not appear at high resolution to be a crater.

gies, assumed them to be endogenic. Further, they considered only isolated clusters, and did not discuss regions saturated with such structures.

A crater-saturated plains region, wherein at least fifteen craters were superimposed while being progressively buried by accumulating plains material, is shown in Figure 9. All structures, except for two tiny "pristine" craters, are mostly buried, and some are completely buried and are visible only as depressions produced by compaction into them of plains material. This burial and compaction provide strong evidence that venusian lowlands are filled by sediments, not by the basalts of popular assumption. Such burial is well known on Mars (Buczkowski et al., 2005; Frey, 2006b) but has not been considered in the conventional venusian literature, even though many "pristine" craters also are partly buried (Fig. 4B and C; Herrick, 2006).

Complex superpositions obvious in radar reflectivity are

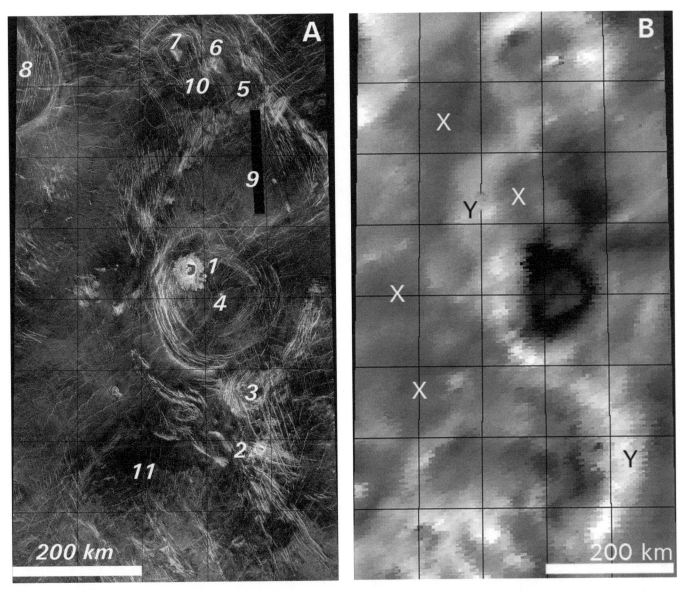

Figure 9. Progressive burial, by compacted plains sediments, of impact structures that saturate the region from 19°N to 27°N and from 95°E to 100°E. (A) Radar backscatter. (B) Altimetry, with relief ~1 km; white is high, black low. The impact structures visible in panel A are numbered in inferred order of age. 1 and 2 (right of number) are small "pristine" Horner and Criss craters. 3 is a crater with rim diameter 70 km. 4 is a subdued crater with a 180 km rim. 5, 6, and 7 are probable craters, each ~50 km in diameter. 8 and 9 are 200 km craters, 10 a 180 km crater; 9 and 10 are mostly buried. 11 is a compaction depression over a buried crater ~200 km in diameter. Additional buried structures revealed by compaction can be inferred from (B); Xs mark shallow depressions over possible buried craters, and Ys mark possible buried rims. In all, 13 impact craters older than "pristine" 1 and 2 are inferred. Among all of these, 4, 8, and 10 are conventionally classed as endogenic coronae, and the others are disregarded.

illustrated in Figures 10, 11, and 12. The great craters in the southwest-younging chain in Figure 12 are pocked by an approximately global-average array of "pristine" craters but by little else, and hence formed late in the pre-"pristine" sequences. A progressive chain of possible nearly simultaneous small impact features is shown in Figure D-1A in the discussion that follows this article. Other regions saturated with mostly buried and variably bombarded craters are illustrated next.

Large Ancient Impact Structures

The old structures of probable impact origin discussed to this point, like those illustrated by Hamilton (2005), are shown primarily by radar-brightness imagery, although Figure 9 includes an altimetric image. Structures obvious in reflectivity are sparser in lowlands than in uplands, which has led to many erroneous statements about abundances (e.g., Jurdy and Stoddard,

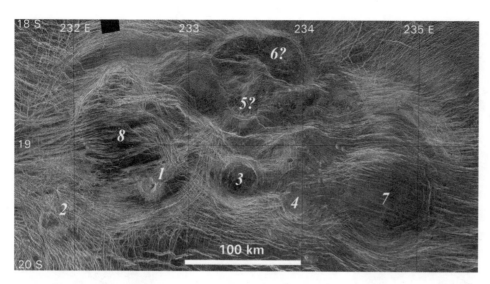

Figure 10. Terrain saturated with eroded superimposed impact craters, numbered in inferred order of increasing age. 1—Darline Crater, with rim diameter 13 km. 2—Slightly degraded crater with 15 km rim. 3—Subdued 30 km crater. 4—More subdued 22 km crater. 5 and 6, and several other circular arcs not marked, may be remnants of other craters. 7—Crater with rim diameter ~80 km. 8—Crater with rimcrest ~130 km. Conventional analysis: 1 is "pristine," 2 is ignored, and 3–8 are lumped together as Beruth Corona and assumed to have a unified endogenic origin. (3–8 may have formed in rapid succession from fragments of a large bolide.)

this volume). Fewer than 1000 of the large old circular structures, with rimcrests ~100–2000 km in diameter where preserved, are recognized, and assigned to unique-to-Venus structural types, in the conventional literature, and these are disproportionately in uplands. Many reports bundle clusters as single features. Several thousand more large structures, in addition to the several thousand overlooked small, old structures discussed previously, are apparent in radar reflectivity and altimetry and yet are disregarded in the conventional literature.

Both conventionally listed and conventionally ignored structures are approximately circular, or are fragments of circles preserved in composites of superimposed circles. Many hundreds of the structures retain raised topographic rims enclosing

Figure 12. Southwestward "view" over large superimposed impact basins toward the horizon at 90°E of a venusian hemisphere. The largest well-preserved circular basin on Venus is Artemis (top center). The topographic rim, which is gentle outside and steeper inside, is at the inner edge of the radar-bright ring and is 2000 km in diameter. The axis of the broad, shallow ring syncline is near the outer part of the bright ring; a low outer rise and broad apron extend far to the left from there. This chain of very large young structures is unique on Venus. Superpositions show that the structures become progressively younger southwestward. Nearly simultaneous impacts of fragments of a giant bolide are inferred. Everything visible at this scale, including the large and small circles, is commonly considered endogenic.

Figure 11. Cookie-cutter superpositions of impact craters. Crater 1 cuts larger 2, which cuts much larger 3. A possible 30 km crater that also cuts 3 is marked by x. Conventional interpretation: the craters collectively comprise Acrea Corona, and nothing in the view relates to impact.

basins, and many of these also retain flanking aprons. Even plume advocates Stofan and Smrekar (2005, p. 850) emphasized that the "typical shape for a corona is a depression or rimmed depression"—which is to be expected for impact structures but not for products of hypothetical plumes. Erratic chains of volcanoes, such as those cited (I think wrongly) as evidence for terrestrial plumes, are lacking on Venus. What I see as ejecta aprons of classic impact structures—the lobate outer slopes of rimmed structures, as in many of the accompanying figures—commonly are regarded as lava flows erupted from the flanks of volcanoes, despite the lack of rift zone sources.

Saturation of Lowlands with Large Impact Structures

Much of the venusian lowlands is saturated with thousands of overlapping apparent impact basins, mostly 100–2000 km in rim diameter, that are variably superimposed, degraded, and buried. The large old structures that I attribute to impact commonly are well shown by reflectivity in venusian uplands. But uplands comprise only ~30% of the planet, and much of those uplands consists of tessera plateaus younger than most old impact structures, and so lacks their imprint. Only relatively sparse old circular structures commonly are obvious in radar brightness alone in the lowlands that comprise the rest of the surface, and the common misconception that the structures are sparse in the lowlands reflects underuse of the topographic information. The topographic imagery shows that large parts of these lowlands are in fact saturated with overlapping circular structures. Most of these lowland structures are invisible in reflectivity alone, or show only as discontinuous arcuate rims projecting through the surface, whereas their shapes are revealed in topography because of compaction of plains fill into variably buried basins formed during the era of sedimentation. My previous report (Hamilton, 2005) said little about these mostly buried plains structures because I was unaware of how well they often show in altimetry until I read Vita-Finzi et al. (2005).

Paired images of reflectivity and altimetry are shown for a small region in Figure 9, for large regions in Figures 13 and 14, and for a huge region in Figure 15. Most papers accept nothing shown in these figures, save sparse small "pristine" craters that are invisible in Figure 15, and almost so in Figures 13 and 14, as products of impacts, whereas I see these areas as mostly saturated with large impact structures. Most of the venusian lowlands resembles Figures 13–15, although the clarity of the circular structures in altimetry varies widely with superpositions and obliteration of older structures by younger ones and with masking by variable thicknesses of sediments. The basinal structures are particularly obscure or sparse near some uplands, which I attribute to thick sedimentation.

Plains saturation by overlapping circular basins, with rims 150–600 km in diameter, is strikingly shown by the topographic image of Figure 13B. Many distinct circular basins, mostly rimmed, of diverse ages, and cookie-cutter superpositions of younger on older, are displayed by the different depths of sur-

face offset by sediment compaction. The topographic expression apparently is due to compaction of sediments into impact basins that formed intermittently during the era of sedimentation but before its end. The crosses mark centers, to an arbitrary lower level of confidence, of what I see as likely old impact structures. Most of the circular structures are invisible in the backscatter image (Fig. 13A), which shows one corona, Ma, with likely impact morphology, plus several other named coronae of indistinct morphology, all of which appear in altimetry to be composite-impact structures. Small old impact structures are more obscured by burial, and relatively few of them can be seen in this low-resolution altimetry. The tessera remnants here predate some of these plains-saturating structures but are unsaturated by them, hence postdate many.

Simon Tapper first recognized the spectacular basin-revealing topography of the area of Figure 13. His shaded-relief image was published by Vita-Finzi et al. (2005, their Fig. 11), who emphasized that the circular structures record impacts. Earlier, Tapper (in Tapper et al., 1998) had gone along with the conventional assumption that these structures are endogenic. Tapper et al. (1998) stated that they had identified 228 new coronae from a global survey of "altimetry and synthetic stereo" images; they did not explain how they had selected these structures from the thousands displayed, did not mention the impact-indicative overprinting of older by younger structures, and concentrated on fitting the structures into Stofan's imaginative endogenic classification of circular features.

The paired images of Figure 14 show a more subdued impact-saturated plains region. The incomplete network of low arcuate ridges apparent in reflectivity (A) is seen in altimetry (B) to mark parts of the rims of overlapping probable impact structures. At least fifty variably overlapping and overprinted rimmed circular basins, and remnants thereof, mostly buried by plains sediments, are apparent, and have rim diameters > 100 km. The areally varying clarity of the structures may reflect the thicknesses of overlying sediments. Within this area of ~6 × 10^6 km^2, there are only two small named coronae, Ituana and Clonia, each of which is a doublet crater consisting of two overlapping rimmed-basin circles. Both Ituana basins have central peaks.

Few of the thousands of lowland circular basins, and their remnants, are recorded on standard lists or maps of venusian structures. This is a function of the restrictive definitions of *corona* and other venusian terms, which arbitrarily exclude most circular structures actually present, and of the underuse of topographic information. The often published claim that circular structures are sparse in the lowlands is false, and the derivative endogenic conjectures (e.g., Jurdy and Stoddard, this volume) that relate uplands and old circular structures are invalid.

Giant Impact Basins

Four well-preserved giant circular rim-and-basin structures were illustrated, and described in impact terms, by Hamilton

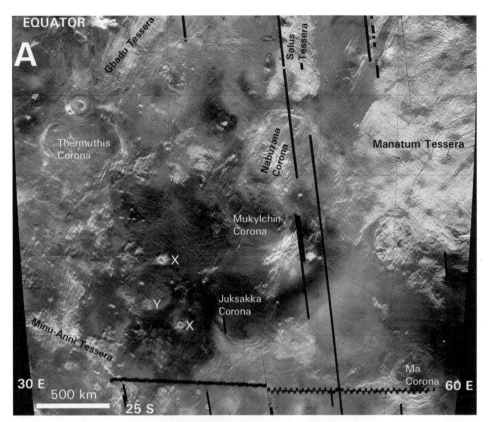

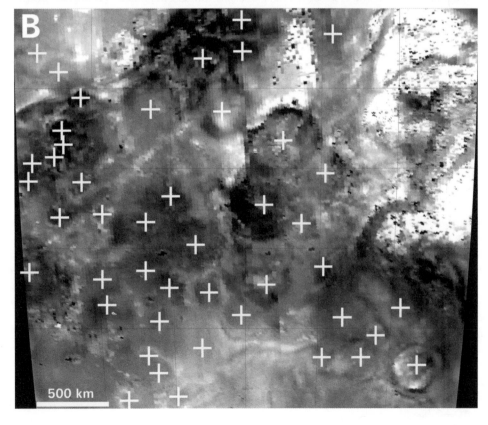

Figure 13. Paired (A) radar reflectivity and (B) altimetric images of a crater-saturated region of plains and tessera remnants. In panel A, remnants of tessera are seen to project through plains sediments, which host a few coronae that mostly have indistinct morphology. Xs mark the two largest and most conspicuous of the fourteen small, young craters conventionally recognized as impact structures in this area. The altimetric grayscale grades from white, the highest parts of the map area, to black, the lowest; total relief is ~4 km, but most of the region is within a 2 km range. The large and variably superimposed circular depressions (crosses) are inferred to mark impact basins, formed while plains sediments were being deposited, that were buried to different depths, and into which the sediments were compacted.

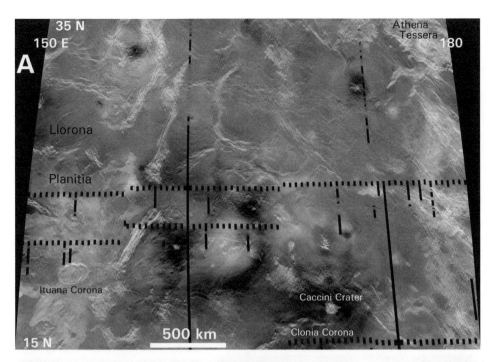

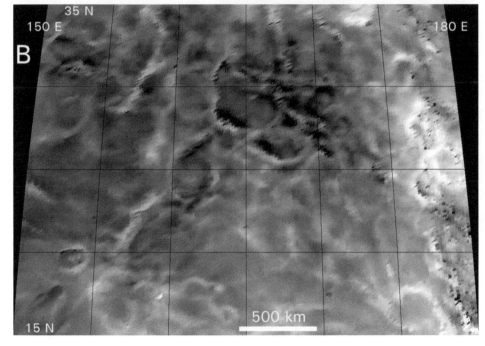

Figure 14. (A) Radar reflectivity and (B) altimetric images of a plains region saturated with large, old, mostly buried impact structures. Parts of rims stand above plains sediments as low arcuate ridges. The topography shows compaction of sediments into basins and a complex array of superimposed likely impact structures, of which those apparent at this scale mostly have rims 100–400 km in diameter. Almost all of these structures are overlooked in conventional work, and only two are named. Ituana Corona (above label) and small Clonia Corona (left of label but inconspicuous at this scale) both appear on detailed imagery to be doublet impact craters. Caccini Crater (below label; rim 38 km in diameter) is the largest of the seven small accepted "pristine" impact craters in this view.

(2005). These structures are Artemis (shown at a small scale here at the southwest end of the chain of nearly simultaneous large basins in Fig. 12), Lakshmi, Heng-O, and Quetzalpetlatl, which have rim-crest diameters of, respectively, 2000, 1400, 900, and 800 km. All but Heng-O preserve great debris aprons. These huge structures are assigned widely conflicting speculative endogenic origins, with mantle upwellings or plumes commonly invoked (e.g., Bannister and Hansen, 2006), but with downwellings invoked in other variants, in the mainline literature.

Many older and more modified giant circular structures also are likely of impact origins. A huge region is shown in Figure 15. At this small scale, the midsize impact basins, 100–600 km in rim diameters, such as are shown in Figure 14 (it and Fig. 9 are within this region), are visible primarily as reticulate patterns in the altimetry. The topographic image (Fig. 15B) also shows giant quasicircular depressions, 800–2500 km in diameter, whose floors lie 1 km or more below the general level of surrounding lowlands. These, and the similar giant basins throughout other

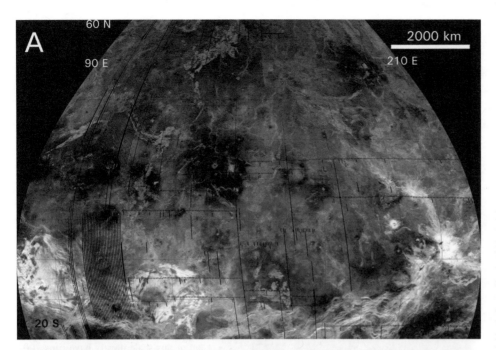

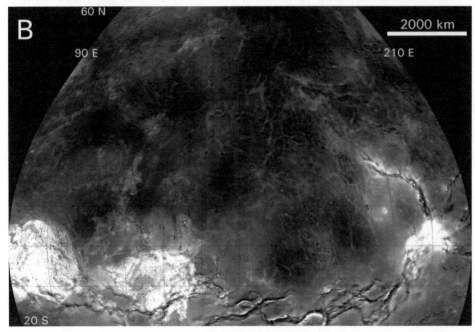

Figure 15. (A) Radar reflectivity and (B) altimetric images of a huge region, much of which is saturated with probable impact structures, with rim diameters up to ~600 km, that show primarily as reticulations in the altimetry. Giant quasicircular depressions, 800–2000 km in diameter, are emphasized (dark) by the topography, whereas their discontinuous rims are shown as arcuate remnants in the reflectivity, and are inferred to be ancient impact basins. The two large quasicircular plateaus in the southwest, bright in both images, are tessera plateaus. Ganis rift and ridge, trending southeast and south below right center and shown best in the topography, are superimposed on an impact-saturated landscape. The figure does not present a global view, but is truncated at the labeled meridians.

plains regions, comprise the lowest parts of the planetary surface, and are much less obvious in reflectivity (Fig. 15A).

I presume that these ancient giant basins are also products of impacts. The structures dimensionally resemble the Aitken–South Pole impact basin (rim diameter 2700 km) on the Moon, and Hellas (1800 km) on Mars, though preservation of the venusian structures is poorer, akin to that of the still older, and mostly buried, giant impact basins on Mars described by Frey (2006a). Midsize basins, like those of Figure 14, saturate much of the area of these great venusian basins.

Conventional venusian literature commonly regards the giant basins as products of circular downwellings (antiplumes), but the unconstrained conjectures vary widely. One of the best defined of these basins is Atalanta, 2300 km in diameter and just north of the area shown in Figure 15, whose gentle floor lies 1 or 2 km below the broad topographic rim that encloses most of it. Gauthier and Arkani-Hamed (2000, p. 1) summarized three of the published downwelling conjectures for Atalanta. It "is the result of a young mantle coldspot or an immature mantle downwelling and the lowland is the surface expression prior to thickening of the crust." Or "the central depression [is due to] crustal thinning and the surrounding positive topography to crustal thickening" above the downwelling mantle. Or, their own preference, the crust was thickened over the downwelling and initially stood high, but subsequent cooling was accompanied by transformation of the thickest deep crust to dense phases, causing subsidence. Like the diverse ad hoc speculations regarding endogenic origins for coronae and other venusian circular structures, none of these mutually contradictory mechanisms has either local constraints or any analogues elsewhere in the solar system—yet an impact explanation, which encounters no such obstacles, is not considered.

Large Impact Melt and Impact Fluidization Constructs

Broad pancakelike "tessera" plateaus and the (misnamed?) "volcanoes" of Venus probably formed from impact melts and fluidized debris, although, like everything else on the planet older than the 1000 "pristine" craters, they conventionally are assigned young endogenic origins. Elkins-Tanton and Hager (2005) modeled extensive magma generation by large impacts. Shock melt can be much augmented by decompression and delayed melts. Neither plateaus nor "volcanoes" have modern terrestrial analogs, but presumably their equivalents were plentiful in the bolide-bombarded early Earth. Perhaps the terrestrial Neoarchean Stillwater and Paleoproterozoic Bushveld and Sudbury complexes are eroded remnants of analogous magmatic masses. Among these, only Sudbury has yet been proved to be of impact origin, but it shows that impacts can produce voluminous magmatism. Transitions link venusian tessera plateaus and "volcanoes" to impact basins and indicate the huge constructs to also be products of impacts. The best-preserved plateaus and "volcanoes" almost lack superimposed ancient midsize impact structures (coronae, etc.), whereas small remnants, largely buried by plains sediments, are dismembered by large impact structures of the old family. Tesserae and "volcanoes" thus span much of the visible pre–"pristine" crater history of Venus.

Tessera Plateaus. The least buried and least bombarded tessera (or "crustal") plateaus are quasicircular, have diameters of 1000–2500 km, stand several km above surrounding plains, and display abundant structural evidence of thin-skinned spreading. The distinctive surface structures of these giant pancakes record radial outflow, with radial shortening by folding and with surfaces whose slopes decrease exponentially outward, combined with circumferential extension displayed by radial graben (e.g., Hansen et al., 2000). The features commonly are attributed to plumes, although, as for all other venusian structures thought endogenic, specific speculations vary greatly, and even antiplumes are invoked. I (Hamilton, 2005) proposed an impact-melt interpretation: tessera plateaus are igneous complexes crystallized from giant impact-melt lakes that spread sluggishly outward like continental ice sheets. Hansen (2005) also adopted this explanation.

Lakshmi Planum and Ishtar Terra show that a tessera plateau

formed from a giant impact basin (Hamilton, 2005). Lakshmi is a smooth-floored basin with a raised circular rim 1400 km in diameter, pocked only by "pristine" craters. A huge debris apron is preserved about much of its perimeter, and it appears to be a magma-filled impact structure formed at about the end of the pre–"pristine" crater history. A great tessera pancake spread into the lowlands from broad ruptures in the northeast half of Lakshmi's rim. This pancake is the westernmost of the three tessera plateaus (products of simultaneous great impacts?) that partly flowed together to make the compound Ishtar Terra upland.

All tesserae have their approximate quotas of "pristine" craters, but they vary widely in age relative to accumulation of impact structures of the ancient family. The great icesheetlike plateaus that are preserved whole, with only their outer edges apparently buried beneath plains sediments, are pocked primarily by "pristine" craters, so are similar in age to the youngest of the great rimmed impact basins, including Artemis. Both hugebasin Lakshmi and the western Ishtar tessera impact melt that broke away from it are similarly pocked primarily by "pristine" craters. From these well-preserved giant pancakes, there are all gradations to small, isolated remnants of older complexes, scattered about the planet, mostly buried by plains sediments and variably disrupted by old impact structures. These remnants can be recognized by their distinctive structure. Increments of the gradation in morphology and cratering are shown in figures in this article and in Hamilton (2005). The two plateaus shown at very small scale in Figure 15 carry very few old-family structures. On the other hand, the small, low, irregular Saluf and Manatum remnants of a tessera plateau are older than a trio of particularly conspicuous, hence relatively young, nested large impact basins of the old family (Fig. 13B), but postdate the early craters of that region. Tesserae thus formed during at least the latter half or so of the era recorded by the landscape-saturating midsize craters. This era may have lasted hundreds of millions of years; see the later discussion. Perhaps much of the crystalline crust of venusian uplands was formed of products of early impact-melt lakes.

"Volcanoes." Broad, low quasicircular domes and cones— up to many hundreds of km in diameter yet commonly only 1 or 2 km high, mostly with single rather than multiple peaks (there are a few doublets), with large, irregular sags, not calderas or vents, at their crests, and lacking rift zones and associated cones and flows—are strewn randomly about the planet. The edifices are popularly termed "volcanoes" and deemed endogenic, but the characteristics just noted distinguish them from modern terrestrial volcanoes and lava fields. Their occurrence relates them to impacts, and I am dubious of the voluminous magmatism implied by the term *volcano*. The constructs typically have outer slopes of <0.5°, and upper slopes of only 1 or 2°. These extremely gentle slopes are shown in the conventional literature with extremely exaggerated topographic profiles and pseudoperspective images that make them appear to be great conical or domiform volcanoes. Thus, Basilevsky and Head (2000) presented "perspective views" of a number of "volcanoes" with unmentioned extreme vertical exaggerations such that low edifices,

200–300 km in diameter and only 1 or 2 km high, rising mostly from rimmed basins, appear to be gigantic steep rounded cones 50 or 70 km high.

The occurrence of the "volcanoes" indicates impact origins, whether or not the constructs are properly magmatic. Most are superimposed on large rimmed depressions (and so often are termed "volcano-corona hybrids"). "Volcanoes" and basins obviously are linked, and to me the links are bolide impacts. Some of the "volcanoes" partly overflow their rims. Some of the largest reveal no enclosing rims, and for these I infer overfilling of impact basins. The large, well-preserved "volcanoes" are pocked only by sparse small "pristine" impact craters, but elsewhere tattered remnants of older "volcanoes" protrude through younger cover and are also overprinted by large and small ancient impact structures (Hamilton, 2005).

Also conventionally classed as "volcanoes" or "mons" are subdued structures (e.g., 1 in Fig. 8) that have rims, basins, and aprons—to me, again, impact structures. Herrick et al. (2005) explained that particular rim-and-basin structure as resulting from sagging of the broad top of a volcano as a causative mantle plume "died out."

These edifices may be products of impact-shock fluidization rather than of impact melting. Their lobate outer flanks commonly are inferred to be lava flows, but these resemble the debris aprons of the young "pristine" craters agreed by all to record impacts (e.g., Figs. 4–6). The formation of large impact structures can neither be tested experimentally nor extrapolated from small-scale experiments, but powerful computers enable modeling. Key to comprehension of complex craters is the duration of shock fluidization and the amount of uplift of temporarily fluidized crater floors, for the fluidized central uplift can shoot far above the pre-impact ground surface and then collapse downward and outward, even over-riding the collapsed transient crater rim (Pierazzo and Collins, 2003, their Fig. 6). The formation of a venusian "volcano" can be visualized as an extremely rapid accompaniment of the impact event, and the broad crestal sags, which do not resemble calderas, as products of rapid spreading of the collapsing debris pile. See Collins and Melosh (2003) for analysis of the mechanism of sustained fluidal behavior of nonmolten material during long-distance motion on gentle slopes. Or perhaps the venusian "volcanoes" formed from mixed fluidized debris and impact melt. The Sudbury igneous complex of Ontario, with its impact breccias atop a fractionated magma lake, might have formed as a Venus-type "volcano."

Concentric and Radial Features

Many small and midsize venusian circular structures display impact-type central peaks or peak rings within their basins, as shown by preceding figures. Additionally, many structures show concentric and radial structures outside their rims, and these may record both impact-shock and surficial processes.

Multiring Structures. Many venusian circular structures are multiring. Circular waveforms, outside the rims of the basins,

define broad, shallow ring synforms, beyond which are broad, low rises. Such structures characterize many terrestrial impact structures. By contrast, craters on Mars and the Moon, where targets consisted of deep impact rubble within which there was little to hold structure, mostly lack such rings.

Concentric deformation related to a large terrestrial impact is shown by the Vredefort structure of South Africa, which has been eroded 7 or 10 km since it formed 2.02 Ga (e.g., Brink et al., 1997; Grieve and Therriault, 2000; Reimold and Gibson, 1996). Small basal remnants of, and top-down contact metamorphism by, what may have been a huge impact-melt lake are preserved. The target was a thick platform section of Paleoproterozoic and Neoarchean sedimentary and volcanic rocks and underlying older Archean basement. Concentric structures are shown by geologic mapping, and much of the concentricity is apparent on satellite imagery despite the generally low relief, surficial cover, and weathering (Fig. 16). Gravity and magnetic maps show low-resolution concentricity. The initial central peak within the crater is now represented by a core, 20 km in radius, of basement rocks, of which the inner part was raised from the deep crust, inside a collar, to an outer radius of 35 km, of vertical to outward-overturned strata. The collar displays close-spaced but irregular concentric structure. Beyond the collar are minor arcuate anticlines, synclines, and faults, and a broad, deep synclinorium with

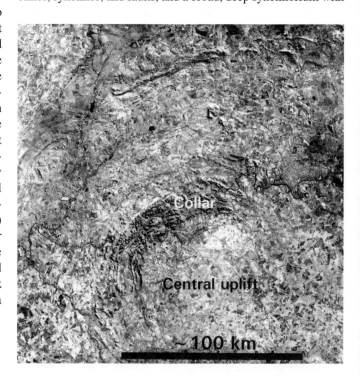

Figure 16. Northwest part of the deeply eroded Paleoproterozoic Vredefort impact structure, South Africa, which displays much concentric structure despite deep erosion and widespread surficial deposits. Around the central uplift of basement rocks is a collar of steep to overturned Neoarchean and Paleoproterozoic stratified rocks. A broad syncline within the major synclinorium is visible at left center, and other concentric elements are visible far beyond the collar. Landsat image provided by Frank Wieland.

an axis at a radius of ~55 km. Concentricity is incomplete, and relation to the impact structure uncertain, beyond the outer flank, near 75 km, of that synclinorium. No basin rim is preserved. Identified radial structures are irregular.

The crater Chicxulub, the Cretaceous-Tertiary-boundary impact structure in Yucatan, is well preserved beneath a cover of Cenozoic sediments, and is known from drilling and geophysical data (Pope et al., 2004). The crater has a peak ring ~80 km in diameter, a crater rim at ~145 km, an outer trough at ~200 km, and a deformation limit at a diameter of ~250 km. Many old venusian circular structures have such dimensions and proportions. The Triassic Manicouagan impact structure of Quebec was developed in Proterozoic gneisses covered thinly by Paleozoic strata, the latter now preserved only (as displaced blocks?) within a circular moat, 60 km in diameter (Grieve and Head, 1983). Lower-crustal rocks are exposed in the central uplift. A sheet of impact melt, with an initial volume likely of >500 km³, covered the central region, except for the central peak, but not the moat. The original crater rim is not preserved. Other terrestrial impact structures also have circular basins outside their rims. Upheaval Dome, Utah (Scherler et al., 2006), is a small-crater example.

Fine-Scale Concentric Structure. No terrestrial craters with which I am familiar display on optical imagery the close-spaced concentric fracturing that characterizes parts of some venusian structures, e.g., the inner slope of the broad, shallow ring synform just outside the rim of the giant Artemis structure (Hamilton, 2005, Fig. 14). Such venusian occurrences are dominantly in scoured bedrock. In the discussion following this article, I show two images of a wind-scoured impact crater in the Sahara Desert. Far more concentric structure is apparent on the radar image (analogous to venusian imagery) than on the optical image. I presume the venusian fracturing to be explicable in impact-compatible terms. Anti-impact venusian papers cite circular fracturing as evidence against impacts, but because they present no plausible endogenic explanation, this argument is unconvincing. That no such fracturing is visible in craters formed in impact-rubble regoliths on Mercury, Mars, and the Moon indicates that targets of bedrock, not rubble, are required.

Some of the concentric lines about old venusian impact structures may record surficial processes. Many venusian plains structures (e.g., Fig. 17) that have impact morphology softened by what was suggested previously to be submarine erosion display abundant concentric features that differ markedly from the Artemis-style circular grooving and perhaps mark wave-cut shorelines in seas shrinking by evaporation. These concentric structures postdate the formation of lobate aprons of impactite. Grindrod et al. (2006) noted this age relationship and assumed endogenic origins, for which, however, they had no plausible explanation.

Radial Structure. Many circular plains structures with general impact morphology show conspicuous radial radar-bright grooves that also may be of surficial origin. The grooves are the subjects of many papers that interpret them as dikes or graben

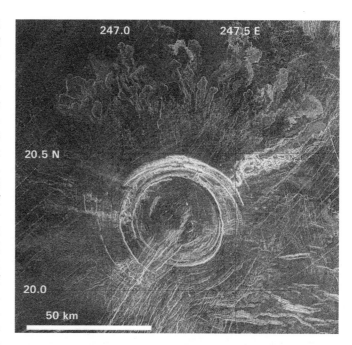

Figure 17. Plains impact structure (Serova Patera) with concentric rings that may be wave-cut shorelines from an evaporating sea. The structure has impact morphology—a circular rim enclosing a crater, and a flanking lobate apron. The nonintersecting radar-bright lines that form concentric arcs about, particularly, the higher part of the structure appear to be following contours. The arcs are most conspicuous on the north-facing parts of both outer and inner slopes. Conventional explanation: endogenic structure, which was never in contact with water.

on endogenic volcanoes (e.g., Aittola and Kostama, 2002; Grindrod et al., 2005; Wilson and Head, 2006). Most radial systems are centered on broad, low, gentle mounds that are enclosed within, or overflow, large circular rims that I regard as impact structures, and have radial aprons of lobate impactite. As emphasized earlier, these "volcanoes" are broad, low, and gentle; lack calderas, cones, rifts, and composites of superimposed volcanoes, and resemble no modern earthly volcanic edifices that would indicate magmatic origins. The grooves may be products of collapsing impact-fluidized central peaks rather than of magmatism. The grooves lack associated cones and flows.

Grindrod et al. (2005) analyzed the radial grooves of four of these broad, low structures. In their upper reaches, the grooves are valleys, typically several kilometers wide and hundreds of meters deep, with angle-of-repose sides. Both widths and depths decrease downslope. The valleys cannot mark graben, because, as Grindrod et al. emphasized, the required extensions would be on the order of 100 times too great to be explicable by the very slight doming that could be inferred from inflation models for the very low rises. (Many papers do explain such radial features with doming, likely because authors are misled by their own greatly exaggerated depictions of topography.) Grindrod et al. explained the features as thick dikes, injected from a central intrusion, although they recognized that no igneous features are associated with the postulated dikes. The hypothetical dike

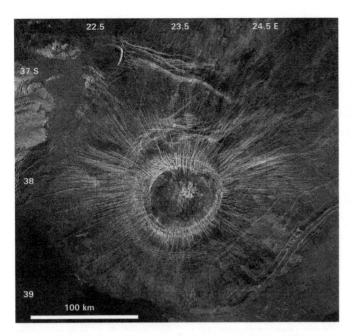

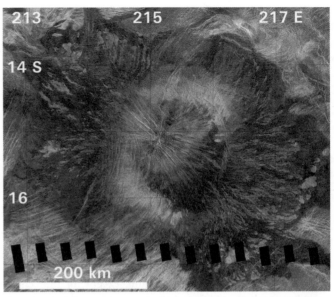

Figure 18. Ninhursag Corona shows impact morphology. The radar-bright rim stands < 2 km above the lowest part of the surrounding plains and encloses a basin, partly filled by radar-dark sediment, from which rises a small peak that is lower than the rim. The radial structure may relate to impact-debris flow, as discussed in the text. Conventional explanation: plume plus volcanism.

Figure 19. Mbokumu Mons has a broad, low, radially streaked mound (dark) within a subdued impact-crater rim (light) that is surrounded by an apron of lobate debris. The relief from the top of the mound to the outer edge of the apron is only 1.0–1.5 km, and the overall slope is ~0.5°. Radial lines (light) extend from the crest of the mound, cross much of the rim, and give way to debris lobes. Suggestion: part of the debris apron was sloughed off the collapsing impact-fluidized central peak, almost instantaneously after impact. Conventional explanation: endogenic volcanism.

systems resemble nothing on modern Earth: they are not sources of cones and flows; they are vastly too numerous and thick, and require impossibly large circumferential extension given the lack of doming; they neither anastomose nor step en echelon, as do dikes; and they do not extend beyond the lobate aprons of their host structures. The lobate outer aprons commonly are explained as lava flows from the postulated dikes in those specific places where the grooves reach the lobes, although such lobes are indistinguishable from those elsewhere on the same circular structures where no grooves are present, and from the common lobate aprons that lack such radial lines. Wilson and Head (2006) explained the lack of surface magmatism along purported dikes in terms of magma density that varied precisely with radial distance to allow them to reach the surface but not erupt—an implausible coincidence. Ernst et al. (2001) speculated that such venusian "dikes" require plumes, a non sequitur even if dikes are present.

Ninhursag Corona (Fig. 18) displays characteristic circular impact morphology on which are superimposed radial grooves. Its rim encloses a partly filled crater from which protrudes a central peak, and a large apron of radial debris slopes very gently outward from the rim. The radial grooves decrease in abundance outward almost to the limit of that apron but nowhere extend beyond it. Although many of the grooves look on the backscatter image as though they cross the rim from the enclosed depression, the position of the inner edge of the rim appears in altimetry to be inside the innermost concentric features on the backscatter

image. (See profile of Ninhursag, vertical exaggeration ~70:1, by Jurdy and Stoddard, this volume, their Fig. 7; the slopes appear gentle even displayed with this great distortion.) The grooves are oriented down the extremely gentle slope, show no dendritic pattern suggestive of stream gullies, and lack features suggestive of igneous dikes. The grooves may be flow lines and channels left by catastrophically flowing fluidized impact debris. The subdued character of the impactite lobes and the concentric etching on the upper outer slopes accord with a submarine origin for the complex.

Radial systems of radar-bright lines are common on venusian "volcanoes," although most are less regular than that of Figure 18. An example, with a suggested mode of impact debris-flow formation in the caption, is shown in Figure 19. For an endogenic explanation of the same structure, illustrated with topographic profiles in which the slopes of <1° are exaggerated 50:1 and 100:1, see Aittola and Kostama (2002) and Kostama and Aittola (2003).

Anti-Impact Arguments

Thousands of old venusian circular structures have the general morphology expected of impact craters—basins enclosed by circular rims that, in the better preserved examples, are still surrounded by lobate debris aprons, and that display cookie-cutter superpositions. None of the hundreds of papers advocating endogenic origins address these features. Conventional papers assume that the structures are endogenic, and seek hypothetical mechanisms that might produce isolated examples as cross-sections, not as circles. As noted previously, the spectacular impact morphology shown in Figure 7 was assigned two convoluted and incompatible endogenic origins in two recent papers, neither of which mentioned the impact option. Papers argue unconstrained endogenic speculations at length, and the few that mention the impact option dismiss it without meaningful discussion. Among the rare discussions is that by Jurdy and Stoddard (this volume).

The vast and continuous size range of old circular structures, which display impact-compatible morphology throughout a range of topographic rim diameters from ~5 to 2500 km, defies explanation by endogenic mechanisms. The statement by Stofan and Smrekar (2005, p. 850) that "the narrow size range and distribution of coronae are inconsistent with an impact origin," and similar statements by Jurdy and Stoddard (this volume), reflect selection, as "coronae," of an arbitrary small sample of circular structures that omits the thousands of small structures (e.g., Fig. 3), the dozens of giant ones (e.g., Fig. 12), and even most of the midsize structures (e.g., Figs. 13 and 14), because they do not fit assumed endogenic concepts. Further, Stofan and Smrekar, and Jurdy and Stoddard, measure diameters to the outermost recognized concentric features, or to irregular boundaries somewhere else outside the rims, or to boundaries outside groups of circular structures, rather than to circular topographic rims, as is done for "pristine" structures, and thus approximately double the diameters of the relatively few structures they do consider. Even as thus limited in size range and inconsistently measured, the log size–log frequency distribution of coronae yields approximately a straight line, as expected for impact explanations (see references cited earlier).

A misunderstood uneven areal distribution of the old structures was cited by Jurdy and Stoddard (this volume) as evidence against impact. This misconception is another expression of inclusion of only a small, biased sample of perhaps 5% of the structures at issue, selected because they fit an arbitrary standard of conspicuous visibility that excludes the great majority of lowland structures, half the upland structures, and nearly all small old structures in both settings. Their elaborate statistical manipulation of this nonrepresentative sample is irrelevant to the debate. Truly uneven distribution does characterize uplands, much of which consists of huge impact-melt constructs, tesserae, that postdate most of the older impact structures, but old uplands and lowlands are both saturated with circular structures.

Jurdy and Stoddard (this volume) cite the lack of structures transitional in size and character between young and old families as evidence against impact origins, but precisely such transitional structures are in fact conspicuous throughout Venus, as this article shows. At least half of the "pristine" craters are substantially to severely modified, and the modification series continues throughout the older structures. The conspicuous old structures are larger to vastly larger than the young ones, but there are thou-

sands of small old structures also. I looked at about one-tenth, the Beta Regio area from 10°N to 40°N and 265°E to 305°E, of the Jurdy and Stoddard BAT study region, in detailed imagery, and found, in addition to the standard-list "pristine" craters, about twenty small structures of transitional aspect, plus, in addition to the coronae they showed, another seventy more modified small and midsize structures that I see as probable to possible impact features. This subregion is dominated by middle-aged tesserae on which relatively few pre-pristine structures are to be expected.

Jurdy and Stoddard (this volume) draw elongate, irregular blobs far outside the rimmed circular structures at issue, or outside groups of neighboring structures; term the blobs "coronae"; and claim the elongations of the blobs to be incompatible with impacts. Their own Figure 4 shows that the relevant structures are dominantly circular, so the exercise has no bearing on the discussion. They also miscite the irregularities in noisy, artifact-ridden low-resolution altimetric profiles, magnified with erratic vertical exaggerations of 65:1 to 160:1, as evidence against impacts.

The arbitrary criteria for identification of modified structures as "pristine" admit fewer structures in uplands, where reflectivity contrasts are low, than in lowlands, where they are high. This subjective factor can account for the slight deficit (if it be statistically valid) near coronae of standard-list "pristine" craters claimed by Matias and Jurdy (2005).

Many dismissals of impact origins are in the form of assertions of the assumptions required by venusian plumology: Venus is too active internally to preserve ancient features, it was resurfaced recently by plumes, its landscape has never been modified by erosion and sedimentation, and so on. These assumptions are at issue and are not proper arguments. Conversely, if the old circular structures are indeed of impact origins, these assumptions are all disproved.

If viable anti-impact arguments exist, I am unaware of them.

EVOLUTION OF VENUS

There are several thousand large circular structures on Venus for which the obvious explanation is formation by bolide impacts, and yet the reader will find no objective discussion of this in the conventional venusian literature. If my analysis is correct, that the abundant old rimmed circular structures on Venus are indeed impact structures, analogy with dated lunar materials requires that they mostly date from early planetary history. At issue is not merely pushing venusian "resurfacing" a bit further back in time. Venus retains a landscape shaped by impacts recording the tail end of main planetary accretion, complicated by effects of a transient hydrosphere and a runaway greenhouse atmosphere.

Age of Surface

The oldest radar-visible features in venusian uplands may be as old as 4.4 Ga. Venus has been too immobile internally—

too cold, too low in volatiles, or both—for its inner workings to have much affected its visible surface since. Even the minor endogenic rifting and warping it does display may be due to repositioning stresses. Venus is more a big sister to Mars than the twin to Earth of popular analogy.

The thousands of impact structures on Venus that are older than the 1000 accepted "pristine" craters can be assigned a general minimum age by reference to Imbrium, the youngest large impact basin on the nearside of the Moon, which formed at 3.85 Ga (Stöffler et al., 2006) or 3.91 Ga (Gnos et al., 2004). Among other large lunar impact structures, only undated Orientale basin, mostly on the lunar far side, may be younger. Imbrium and Orientale are pocked only by small impact craters, analogous to venusian "pristine" craters although these are vastly more numerous on the airless Moon. Imbrium has a rim diameter of 1200 km, Orientale 900 km, but their bolides would have made smaller basins on higher-gravity Venus. Dated rocks from the Moon go back to 4.45 or 4.50 Ga, so the Moon had most of its present size by that very early time. So did Earth, for terrestrial crustal zircons, recycled into younger rocks, as old as 4.43 Ga are known. Subsequent bolides gardened the surfaces but added relatively little mass.

A great bombardment of large bolides, from ca. 3.95–3.85 Ga, is widely assumed to have affected the Moon, hence necessarily also Earth and Venus. The bombardment was inferred by Dalrymple and Ryder (1993) from the Gaussian scatter of Ar/Ar dates of many shock-melted glasses in impact breccias collected on Apollo missions. The postulate of a late heavy bombardment suffers from the implausibility of parking numerous large bolides somewhere in the inner solar system for hundreds of millions of years until they were released at ca. 3.9 Ga, or otherwise suddenly deriving them, and it is incompatible with photogeologic evidence (Baldwin, 2006). Haskin et al. (1998) showed that the glasses used by Dalrymple and Ryder could have come from the vast Imbrium ejecta blanket, and thus that the dates may record only the Imbrium event, their spread representing diffusion and analytical scatter. Stöffler et al. (2006) agreed that evidence of a late bombardment now appears to be lacking, whereas Norman et al. (2006) disagreed. Much smaller bolides have since continued to impact the Moon, but there may have been no concentrated terminal bombardment. Lunar zircon dates, from granophyres and gabbros, decrease exponentially in abundance with decreasing age from 4.3 to 3.9 Ga (Meyer et al., 1996). I suggest that these granophyres and gabbros were produced in impact-melt lakes and represent the exponential decline of accretion intensity after the Moon reached essentially its final size, as expected from orbital considerations. In these terms, 3.9 Ga approximately dates the end of the tail of accretion of large bolides in the inner solar system, including Venus.

The largest well-preserved impact basins on Venus (Artemis, Fig. 12, and Lakshmi, Heng-O, and Quetzalpetlatl, all illustrated and discussed by Hamilton, 2005) are pocked mostly by craters of the "pristine" family, hence formed late in the era recorded by the old impact structures. I presume that these large structures

(and also the similarly pocked youngest of the tessera plateaus) correlate approximately with Imbrium, ca. 3.9 Ga. The equally large quasicircular depressions of Figure 15 are markedly older, for the fill in at least some of them is saturated by old midsize craters. Frey (2006b) evaluated the size-frequency distribution of analogous craters superimposed on similar giant impact basins on Mars in terms of an accretionary model, and deduced a minimum age of 4.26 Ga for those basins. Because Earth, the Moon, Mars, and presumably Venus had almost their present sizes before 4.4 Ga, I presume that the venusian impact landscape may well reach back to 4.4 Ga.

Crust of Venus during Late Accretion

Mars and the Moon had mostly rubble surfaces, gardened by impact upon impact, during late main accretion. Many more venusian craters of the old family may have formed on solid crust rather than regolith. This would account for the bedrock-target aspect of upland craters, and the clean aspect of the best-displayed old lowland-saturating ones (e.g., Fig. 13). Small remnants of tessera plateaus widely show as basement, so the oldest visible crust might have consisted of the products of impact melts, rather than of still older magma-ocean fractionates. The surface of Venus was largely stabilized before 3.8 Ga, judging from the saturation of much of its surface by midsize and giant impact structures, and likely long before. Venus nevertheless must have retained plentiful internal heat far longer than did small, fast-cooling Mars and the Moon, and its thermal history awaits evaluation in terms of new concepts.

ACKNOWLEDGMENTS

Trent Hare, Catherine Johnson, David Sandwell, and Frank Wieland provided some of the illustrations used here. Comments on a short preliminary manuscript by Tracy Gregg, Keith Howard, Eugene Smith, and two anonymous, hostile reviewers, and on a subsequent full version by Gillian Foulger, Donna Jurdy, G.J.H. McCall, Claudio Vita-Finzi, and Howard Wilshire, resulted in many improvements in content and presentation.

REFERENCES CITED

Aittola, M., and Kostama, V.-P., 2002, Chronology of the formation process of Venusian novae and the associated coronae: Journal of Geophysical Research, v. 107, no. E11, 26 p., doi:10.1029/2001JE001528.

Anderson, D.L., 2007, The new theory of the Earth: Cambridge, England, Cambridge University Press, 408 p.

Anderson, F.S., and Smrekar, S.E., 2006, Global mapping of crustal and lithospheric thickness on Venus: Journal of Geophysical Research, v. 111, no. E8, doi: 10.1029/2004JE002395.

Baldwin, R.B., 2006, Was there ever a terminal lunar cataclysm? With lunar viscosity arguments: Icarus, v. 184, p. 308–318, doi: 10.1016/j.icarus.2006 .05.004.

Bannister, R.A., and Hansen, V.I., 2006, Geologic analysis of deformation in the interior region of Artemis (Venus, 34°S 132°E): Lunar and Planetary Science 37, paper 1370, 2 p. (CD-ROM).

Basilevsky, A.T., and Head, J.W., 2000, Rifts and large volcanoes on Venus— Global assessment of their age relations with regional plains: Journal of Geophysical Research, v. 105, p. 24,583–24,611, doi: 10.1029/ 2000JE001260.

Basilevsky, A.T., Ivanov, B.A., Burba, G.A., Chernaya, I.A., Kryuchkov, V.P., Nikolaeva, O.V., Campbell, D.B., and Ronca, L.B., 1987, Impact craters of Venus—A continuation of the analysis of data from the Venera 15 and 16 spacecraft: Journal of Geophysical Research, v. 92, p. 12,869–12,901.

Bland, P.A., and Artemieva, N.A., 2003, Efficient disruption of small asteroids in Earth's atmosphere: Nature, v. 424, p. 288–291, doi: 10.1038/nature01757.

Brink, M.C., Waanders, F.B., and Bischoff, A.A., 1997, Vredefort—A model for the anatomy of an astrobleme: Tectonophysics, v. 270, p. 83–114, doi: 10.1016/S0040-1951(96)00175-8.

Buczkowski, D.L., Frey, H.V., Roark, J.H., and McGill, G.E., 2005, Buried impact craters—A topographic analysis of quasi-circular depressions, Utopia Basin, Mars: Journal of Geophysical Research, v. 110, no. E3, 8 p., doi: 10.1029/2004JE002324.

Chambers, J.E., 2004, Planetary accretion in the inner Solar System: Earth and Planetary Science Letters, v. 223, p. 241–252, doi: 10.1016/j.epsl.2004 .04.031.

Cochrane, C.G., 2005, Topographic modelling from SAR imagery of impact craters on Venus [Ph.D. thesis]: Imperial College, University of London, 230 p.

Cochrane, C.G., and Ghail, R.G., 2006, Topographic constraints on impact crater morphology on Venus from high-resolution stereo synthetic aperture radar digital elevation models: Journal of Geophysical Research, v. 111, no. E4, 11 p., doi:10.1029/2005JE002570.

Collins, G.S., and Melosh, H.J., 2003, Acoustic fluidization and the extraordinary mobility of sturzstroms: Journal of Geophysical Research, v. 108, no. B10, 14 p., doi 10:1029/2003JB002465.

Connors, C., and Suppe, J., 2001, Constraints on magnitude of extension on Venus from slope measurements: Journal of Geophysical Research, v. 106, p. 3237–3260, doi: 10.1029/2000JE001256.

Dalrymple, G.B., and Ryder, G., 1992, ^{40}Ar/^{39}Ar age spectra of Apollo 15 impact melt rocks by laser step-heating and their bearing on the history of lunar basin formation: Journal of Geophysical Research, v. 98, p. 13,085–13,095.

Dimitrov, L.I., 2002, Mud volcanoes—The most important pathway for degassing deeply buried sediments: Earth-Science Reviews, v. 59, p. 49–76, doi: 10.1016/S0012-8252(02)00069-7.

Elkins-Tanton, L.R., and Hager, B.H., 2005, Giant meteoroid impacts can cause volcanism: Earth and Planetary Science Letters, v. 239, p. 219–232, doi: 10.1016/j.epsl.2005.07.029.

Ernst, R.E., Grosfils, E.B., and Mège, D., 2001, Giant dike swarms—Earth, Venus and Mars: Annual Review of Earth and Planetary Sciences, v. 29, p. 489–534, doi: 10.1146/annurev.earth.29.1.489.

Foulger, G.R., et al., eds., 2005, Plates, plumes, and paradigms: Boulder, Colorado, Geological Society of America Special Paper 388, 881 p.

Frey, H.V., 2006a, Impact constraints on, and a chronology for, major events in early Mars history: Journal of Geophysical Research, v. 111, no. E8, 11 p., doi:10.1029/2005JE002449.

Frey, H.V., 2006b, Impact constraints on the age and origin of the lowlands of Mars: Geophysical Research Letters, v. 33, no. 8, 4 p., doi:10.1029/ 2005GL0244.

Gauthier, M., and Arkani-Hamed, J., 2000, Formation of Atalanta Planitia, Venus: Lunar and Planetary Science 31, paper 1721, 2 p. (CD-ROM).

Glaze, L.F., Stofan, E.R., Smrekar, S.E., and Baloga, S.M., 2002, Insights into corona formation through statistical analyses: Journal of Geophysical Research, v. 107, no. E12, doi: 10:1029/2002JE001904.

Gnos, E., Hofmann, B.A., Al Kathiri, A., Lorenzetti, S., Eugster, O., Whitehouse, M.J., Villa, I.M., Jull, A.J.T., Eikenberg, J., Spettel, B., Krähenbühl, U., Franchi, I.A., and Greenwood, R.C., 2004, Pinpointing the source of a lunar meteorite—Implications for the evolution of the Moon: Science, v. 305, p. 657–659, doi: 10.1126/science.1099397.

Grieve, R.A.F., and Head, J.W., 1981, Impact cratering, a geological process on the planets: Episodes, v. 4, no. 2, p. 3–9.

Grieve, R.A.F., and Head, J.W., 1983, The Manicouagan impact structure— An analysis of its original dimensions and form: Journal of Geophysical Research, v. 88, supplement, p. A807–A818.

Grieve, R.A.F., and Therriault, A., 2000, Vredefort, Sudbury, Chicxulub—Three of a kind?: Annual Review of Earth and Planetary Sciences, v. 28, p. 305–338, doi: 10.1146/annurev.earth.28.1.305.

Grindrod, P.M., and Hoogenboom, T., 2006, Venus—The corona conundrum: Astronomy & Geophysics, v. 47, no. 3, p. 3.16–3.21, doi: 10.1111/j.1468-4004.2006.47316.x.

Grindrod, P.M., Nimmo, F., Stofan, E.R., and Guest, J.E., 2005, Strain at radially fractured centers on Venus: Journal of Geophysical Research, v. 110, no. E12, 13 p., doi:10.1029/2005JE002416.

Grindrod, P.M., Stofan, E.R., Brian, A.W., and Guest, J.E., 2006, The geological evolution of Atai Mons, Venus—A volcano-corona "hybrid": Journal of the Geological Society of London, v. 163, p. 265–275, doi: 10.1144/ 0016-764904-502.

Hamilton, W.B., 2005, Plumeless Venus preserves ancient impact-accretionary surface, *in* Foulger, G.R., et al., eds., Plates, plumes, and paradigms: Boulder, Colorado, Geological Society of America Special Paper 388, p. 781–814.

Hansen, V.L., 2005, Crustal plateaus as ancient large impact features—A hypothesis: Lunar and Planetary Science 36, paper 2251, 2 p. (CD-ROM).

Hansen, V.L., Phillips, R.J., Willis, J.J., and Ghent, R.R., 2000, Structure of tessera terrain, Venus—Issues and answers: Journal of Geophysical Research, v. 105, p. 4135–4152, doi: 10.1029/1999JE001137.

Haskin, L.A., Korotev, R.L., Rockow, K.M., and Joliff, B.L., 1998, The case for an Imbrium origin of the Apollo thorium-rich impact-melt breccias: Meteoritics & Planetary Science, v. 33, p. 959–975.

Herrick, R.R., 2006, Updates regarding the resurfacing of Venusian impact craters: Lunar and Planetary Science 37, paper 1588, 2 p. (CD-ROM).

Herrick, R.R., Dufek, J., and McGovern, P.J., 2005, Evolution of large shield volcanoes on Venus: Journal of Geophysical Research, v. 110, no. 1, 19 p., doi:10.1029/2004JE002283.

Hofmeister, A.M., and Criss, R.E., 2005, Earth's heat flux revised and linked to chemistry: Tectonophysics, v. 395, p. 159–177, doi: 10.1016/j.tecto.2004 .09.006.

Ivanov, M.A., and Head, J.W., 2001, Geology of Venus—Mapping of a global geotraverse at 30°N latitude: Journal of Geophysical Research, v. 106, p. 17,515–17,566, doi: 10.1029/2000JE001265.

Johnson, C.L., and Richards, M.A., 2003, A conceptual model for the relationship between coronae and large-scale mantle dynamics on Venus: Journal of Geophysical Research, v. 108, no. E6, 18 p., doi:10.1029/ 2002JE001962.

Jones, A.P., and Pickering, K.T., 2003, Evidence for aqueous fluid-sediment transport and erosional processes on Venus: Journal of the Geological Society of London, v. 160, p. 319–327.

Jurdy, D.M., and Stoddard, P.R., 2007 (this volume), Venus's coronae: Impact, plume, or other origin? *in* Foulger, G.R., and Jurdy, D.M., eds., Plates, plumes, and planetary processes: Boulder, Colorado, Geological Society of America Special Paper 430, doi: 10.1130/2007.2430(40).

Kaula, W.M., 1994, The tectonics of Venus: Royal Society of London Philosophical Transactions, v. A-349, p. 345–355.

Kaula, W.M., 1995, Venus reconsidered: Science, v. 270, p. 1460–1464, doi: 10.1126/science.270.5241.1460.

Kleine, T., Mezger, K., Palme, H., and Münker, C., 2004, The W isotope evolution of the bulk silicate Earth—Constraints on the timing and mechanisms of core formation and accretion: Earth and Planetary Science Letters, v. 228, p. 109–123, doi: 10.1016/j.epsl.2004.09.023.

Korycansky, D.G., and Zahnle, K.J., 2005, Modeling crater populations on Venus and Titan: Planetary and Space Science, v. 53, p. 695–710, doi: 10.1016/j.pss.2005.03.002.

Kostama, V.-P., and Aittola, M., 2003, The arched graben of Venusian coronanovae: Lunar and Planetary Science 34, paper 1144, 2 p. (CD-ROM).

Krassilnikov, A.S., and Head, J.W., 2003, Novae on Venus—Geology, classification, and evolution: Journal of Geophysical Research, v. 108, no. E9, paper 12, 48 p., doi: 10.1029/2002JE0011983.

Kulikov, Yu.N., Lammer, H., Lichtenegger, H.I.M., Terada, N., Ribas, I., Kolb, C., Langmayr, D., Lundin, R., Guinan, E.F., Barabash, S., and Biernat, H.K., 2006, Atmospheric and water loss from early Venus: Planetary and Space Science, v. 54, p. 1425–1444, doi: 10.1016/j.pss.2006.04.021.

Lang, N.P., and Hansen, V.I., 2006, Venusian channel formation as a subsurface process: Journal of Geophysical Research, v. 111, no. E4, 15 p., doi: 10.1029/2005JE002629.

Masursky, H., Eliason, E., Ford, P.G., McGill, G.E., Pettengill, G.H., Schaber, G.G., and Schubert, G., 1980, Pioneer Venus radar results—Geology from images and altimetry: Journal of Geophysical Research, v. 85, p. 8232–8260.

Matias, A., and Jurdy, D.M., 2005, Impact craters as indicators of tectonic and volcanic activity in the Beta-Atla-Themis region, Venus, *in* Foulger, G.R., et al., eds., Plates, plumes, and paradigms: Boulder, Colorado, Geological Society of America Special Paper 388, p. 825–839.

McDaniel, K., and Hansen, V.L., 2005, Circular lows, a genetically distinct subset of coronae?: Lunar and Planetary Science 36, paper 2367, 2 p. (CD-ROM).

McKinnon, W.B., Zahnle, K.J., Ivanov, B.A., and Melosh, H.J., 1997, Cratering on Venus—Models and observations, *in* Bougher, S.W., et al., eds., Venus II: Tucson, University of Arizona Press, p. 969–1014.

Meyer, C., Williams, I.S., and Compston, W., 1996, Uranium-lead ages for lunar zircons—Evidence for a prolonged period of granophyre formation from 4.32 to 3.88 Ga: Meteoritics & Planetary Science, v. 31, p. 370–387.

Milton, D.J., Glikson, A.Y., and Brett, R., 1996, Gosses Bluff—A latest Jurassic impact structure, central Australia, Part 1, Geological structure, stratigraphy, and origin: AGSO Journal of Australian Geology & Geophysics, v. 16, p. 453–486.

Nikolaeva, O.V., 1993, Largest impact features on Venus—Non-preserved or non-recognizable?: Lunar and Planetary Science Conference 24, p. 1083–1084.

Norman, M.D., Duncan, R.A., and Huard, J.J., 2006, Identifying impact events within the lunar cataclysm from ^{40}Ar-^{39}Ar ages and compositions of Apollo 16 impact melt rocks: Geochimica et Cosmochimica Acta, v. 7, p. 6032–6049.

Oshigami, S., and Namiki, N., 2005, Cross-sectional profile of Baltis Vallis channel on Venus—Reconstruction from Magellan SAR brightness data: Lunar and Planetary Science 36, paper 1555, 2 p. (CD-ROM).

Pierazzo, E., and Collins, G., 2003, A brief introduction to hydrocode modeling of impact cratering, *in* Claeys, P., and Henning, D., eds., Submarine craters and ejecta-crater correlation: New York, Springer, p. 323–340.

Pope, K.O., Kieffer, S.W., and Ames, D.E., 2004, Empirical and theoretical comparisons of the Chicxulub and Sudbury impact structures: Meteoritics & Planetary Science, v. 39, p. 97–116.

Raymond, S.N., Quinn, T., and Lunine, J.I., 2006, High-resolution simulations of the final assembly of Earth-like planets, 1, Terrestrial accretion and dynamics: Icarus, v. 183, p. 265–282, doi: 10.1016/j.icarus.2006.03.011.

Reimold, W.U., and Gibson, R.L., 1996, Geology and evolution of the Vredefort impact structure, South Africa: Journal of African Earth Sciences, v. 23, p. 125–162, doi: 10.1016/S0899-5362(96)00059-0.

Scherler, D., Kenkmann, T., and Jahn, A., 2006, Structural record of an oblique impact: Earth and Planetary Science Letters, v. 248, p. 28–38, doi: 10.1016/j.epsl.2006.05.002.

Schultz, P.H., 1993, Searching for ancient Venus: Lunar and Planetary Science Conference 24, p. 1255–1256.

Shoemaker, E.M., 1994, Late impact history of the solar system: EOS (Transactions, American Geophysical Union), v. 75, no. 16, supplement, p. 50.

Shoemaker, E.M., 1998, Long-time variations in the impact cratering rate on Earth: Geological Society of London Special Publication 140, p. 6–10.

Shoemaker, E.M., and Shoemaker, C.S., 1996, The Proterozoic impact record of Australia: AGSO Journal of Australian Geology & Geophysics, v. 16, p. 379–396.

Stofan, E.R., and Smrekar, S.E., 2005, Large topographic rises, coronae, flow fields, and volcanoes on Venus—Evidence for mantle plumes? *in* Foulger, G.R., et al., eds., Plates, plumes, and paradigms: Boulder, Colorado, Geological Society of America Special Paper 388, p. 841–861.

Stofan, E.R., Head, J.W., and Grieve, R.A.F., 1985, Classification of circular features on Venus: 1984 Report on Planetary Geology and Geophysics Program: Washington, D.C., National Aeronautics and Space Administration, p. 103–104.

Stofan, E.R., Sharpton, V.L., Schubert, G., Baer, G., Bindschadler, D.L., Janes, D.M., and Squyres, S.W., 1992, Global distribution and characteristics of coronae and related features on Venus—Implications for origin and relation to mantle processes: Journal of Geophysical Research, v. 97, p. 13,347–13,378.

Stöffler, D., Ryder, G., Ivanov, B.A., Artemieva, N.A., Cintala, M.J., and Grieve, R.A.F., 2006, Cratering history and lunar chronology: Reviews in Mineralogy and Geochemistry, v. 60, p. 519–596, doi: 10.2138/rmg.2006.60.05.

Strom, R.G., Malhotra, R., Ito, T., Yoshida, F., and Kring, D.A., 2005, The origin of planetary impactors in the inner Solar System: Science, v. 309, p. 1847–1850, doi: 10.1126/science.1113544.

Tapper, S.W., Stofan, E.R., and Guest, J.E., 1998, Preliminary analysis of an expanded corona database: Lunar and Planetary Science 29, paper 1104, 2 p. (CD-ROM).

Törmänen, T., Aittola, M., Kostama, V.-P., and Raitala, J., 2005, Distribution and classification of multiple coronae on Venus: Lunar and Planetary Science 36, Paper 1640, 2 p. (CD-ROM).

Turcotte, D.L., 1995, How does Venus lose heat?: Journal of Geophysical Research, v. 100, no. E8, p. 16,931–16,940, doi: 10.1029/95JE01621.

Vezolainen, A.V., Solomatov, V.S., Basilevsky, A.T., and Head, J.W., 2004, Uplift of Beta Regio—Three-dimensional models: Journal of Geophysical Research, v. 109, no. E8, 8 p., doi:10.1029/2004JE002259.

Vita-Finzi, C., Howarth, R.J., Tapper, S., and Robinson, C., 2004, Venusian craters and the origin of coronae: Lunar and Planetary Science 35, paper 1564, 2 p. (CD-ROM).

Vita-Finzi, C., Howarth, R., Tapper, S., and Robinson, C., 2005, Venusian craters, size distributions and the origin of coronae, *in* Foulger, G.R., et al., eds., Plates, plumes, and paradigms: Boulder, Colorado, Geological Society of America Special Paper 388, p. 815–823.

Wieczorek, M.A., 2007, The gravity and topography of the terrestrial planets, *in* Treatise on Geophysics (in press; available at http://www.ipgp.jussieu.fr.)

Wilson, L., and Head, J.W., 2006, Lateral dike injection and magma eruption around novae and coronae on Venus: Lunar and Planetary Science 37, paper 1125, 2 p. (CD-ROM).

MANUSCRIPT ACCEPTED BY THE SOCIETY JANUARY 31, 2007

DISCUSSION

24 December 2006, G.J.H. McCall

Warren Hamilton's article is clearly the result of very detailed study of venusian surface features. Venus is the only planet in our solar system for which we have to rely on radar images, and this in itself provides additional constraints on interpretation. Hamilton has carefully considered the likely radar image obtained from different geological materials, e.g., basalt, sediments, breccias. He readily admits that his interpretations conflict with the interpretations of the conventional school of J.W. Head and others. It is a remarkable fact that, whereas for the moon, Mercury, Mars, and the many planetary satellites the conventional wisdom is virtually to apply Impact (rather than volcanism) as an a priori or default explanation for any craters or circular structures of disputed origin, the reverse is the case for Venus. Volcanic processes are invoked except for about a thousand "undisputed young pristine" venusian craters. Hamilton rightly suggests that "pristine" has been in many cases incorrectly used, and many are by no means pristine.

In this space age, conventional interpretations tend with the passage of years to be taken as "graven in stone," but of course they are not. I remember well a meeting in London at which Dan McKenzie put forward a quite different interpretation of Venus, rejecting the conventional resurfacing model and calling for anomalies in crater counts.

I feel that so little is really known for certain about Venus that no careful interpretations such as that of Hamilton should be rejected out of hand. As Lichtenberg has said, "question everything once." I myself am an independent thinker, like Hamilton. Indeed, I think that there is yet room for such unconventional statements about Mars, even though we know much more for certain about Mars than about Venus. Take, for example, the venusian craters Aurelia, Lachapelle, and Barton, illustrated by Hamilton. They are surely the radar-image analogues of the type of Martian crater, including the delightfully named Tooting Crater, illustrated and discussed by Barlow (2006) and Hartmann and Barlow (2006). Such Martian craters are surrounded by what appear to be successive thin flows, possibly of slurry, which conventionally are interpreted as "ejecta blankets." But are they? "Spirit" and "Opportunity" have completely revolutionized Martian "geology" by recognizing wholesale alteration of early Martian volcanic rocks to carbonates and sulfates under an early hydrous regime in which the water likely occurred as transient emissions. These craters could equally well be volcanic. There is absolutely no certainty that they are impact craters, and there must have been early volcanism that produced the basaltic parents of these altered rocks and the clasts in the aeolian sediments imaged by "Opportunity." The associated heat could have mobilized the alteration products into several generations of slurries, for instance, in Tooting Crater (see cover illustration, Meteoritics Planetary Science 41[10]). Impact origin is only an a priori assumption. Less is known for certain about Mars than is popularly supposed; it is early days yet, and our knowledge of Venus is in its infancy. Meanwhile let us allow divergent interpretations of both to see the light of day.

I am not myself in any way versed in interpretation of the surface of Venus. I do not know whether Hamilton's interpretations are correct or not. That is not the point. I do have greater interest in Mars, especially from the meteoritic and eruptive viewpoints, and there can be no doubt that Olympus Mons is an ultrasized shield volcano with a caldera (McCall, 2006). Experience of mapping Menengai, Suswa, Kilombe, Silali, and Ambrym caldera volcanoes in Africa and Vanuatu led me to recognize many of the classic features of such volcanoes, albeit highly magnified. In general also, my wide experience of volcanic terrains and extensive examination of Mars images suggest that there will prove to be far more volcanic craters on Mars than is now conventionally supposed. There must be an earlier generation of volcanic rocks to have been altered to carbonate and sulfate minerals, and many simple younger craters may have been misidentified.

On Venus there are some quite unarguable volcanic structures (e.g., the bunlike tholoids), and, considering the nature of its atmosphere, Venus must have had a history of major volcanism. Nevertheless, Hamilton's arguments are convincing and cannot be dismissed out of hand, and impact may have been much more important in shaping that planet's surface than is conventionally supposed. The state of knowledge is that many of the venusian craters and circular structures are ambiguous, and our knowledge of the geochronology of the planet is at present comparable with that of pre-Curie Victorian terrestrial geology: sequence known but no vestige of knowledge of a time scale.

Hamilton has identified sediments and mud volcanoes. In doing this he is relying on radar contrasts and other observed relationships, but, drawing on imagery from other planets and the moon, I suspect that the plains are composed of volcanic material. I personally doubt whether Venus ever had transient oceans. My experience of mud volcanoes related to hydrocarbons in the Makran of Iran suggests that what Hamilton is envisaging is something quite different from those small features, which would defy detection in the radar imagery of Venus.

3 January 2007, Warren B. Hamilton

Joe McCall provided helpful comments on the manuscript for this article, and I thank him for his continuing interest. We disagree on the nature of the venusian plains. He shares the overwhelming-majority view that the plains are volcanic, whereas I regard them as formed of sediments metamorphosed thermally by a runaway-greenhouse atmosphere. McCall makes few specific statements for discussion, so I list some of my reasons; see my present book chapter and Hamilton (2005) for elaboration.

1. The plains are radar-dark, hence smooth-surfaced at centimeter and decimeter scale.
2. Soviet-lander images show thinly slabby laminated material.
3. Turbidite channels and lobes can be inferred from radar imagery.
4. The plains show no fissures or other obvious lava sources.
5. Old circular-rimmed basins (the older of the "pristine impact structures" agreed on by all, plus my superabundant ancient impact structures) formed concurrently with deposition of plains materials, which progressively flooded the fretted topography of the structures, buried them, and compacted into and over them.
6. Many wet-sediment and underwater impacts can be inferred.
7. Thermal wrinkling and contraction structures, imposed only on plains material, are explicable by top-down heating and desiccation.
8. The million or so smooth small, low, gentle-sided "shields" strewn about the plains resemble terrestrial mud volcanoes, not rough, steep-sided, fissure-following lava cones, and plausibly relate to top-down heating.

McCall terms the "bun-like tholoids" "unarguable volcanic structures"; but they are arguable. "Tholus," defined vaguely as "small domical mountain or hill," is applied to 55 venusian structures from 15 to 300 km in diameter in the U.S. Geological Survey Gazetteer of Planetary Nomenclature (http://planetary-names.wr.usgs.gov). Most have raised circular rims, many of which, like Lama Tholus (Fig. D-1B), enclose broad, shallow basins. I regard most of them, like venusian circular-rimmed structures given other designations, as ancient impact structures. Figure D-1B shows two other circular rim-and-basin structures that I also see as impacts. Lama's radial debris-flow lobes are deflected against the aprons of the other two and also against the lobate apron (in the northwest corner) of a large out-of-view doublet crater, Aruru Corona, so Lama is youngest. Everything in view is conventionally deemed endogenic.

Or perhaps McCall means the "farra" (singular, farrum) of specialist terminology: rare low, flat-topped circular "pancake domes," typically several tens of kilometers in diameter and several hundred meters high, mostly in elongate clusters. The domes have been casually likened to terrestrial silicic lava domes, but their "morphology and dimensions . . . make them unlike any type of terrestrial subaerial volcano" (Bridges, 1995), and their radar response is utterly unlike that of terrestrial lava domes (Plaut et al., 2004). Note the circularity of the Seoritsu Farra and the superposition of younger domes on older, without deflection required by magmatic origins (Fig. D-1A). Perhaps these are an eastward-younging chain of almost simultaneous impacts of fragments of a large bolide, disrupted by gravity and superdense atmosphere, into soft, water-saturated sediments.

31 January 2007, Suzanne E. Smrekar and Ellen Stofan

The study of impact craters is now 50 years old and was developed in support of the Apollo program.

Impact craters are defined by their shape and ejecta blanket. They are depressions; the shape follows a transition from smaller, bowl-shaped craters to those with smaller depth-to-diameter ratios and central peaks, to larger multiring basins with even smaller depth-to-diameter ratios. The transition between these shapes is a function of planetary gravity. The ejecta blankets result from the shock wave that hurls the fragment surface onto the surrounding area. The effects of atmospheres and target properties on crater morphology are well understood through studies of craters on Earth, the moon, Mars, Mercury, and the icy satellites of the outer solar system. The physics of the processes that form these shapes has been examined via fieldwork on terrestrial craters, morphologic studies of craters on other planets, experimental studies, and theoretical modeling. Impact cratering is a well-understood process (e.g., Melosh, 1989).

Coronae are defined by their concentric fracture rings (Stofan et al., 2001). They typically also have radial fractures. They have a wide range of topography morphologies, with similar numbers of topographically high and topographically low features (Stofan

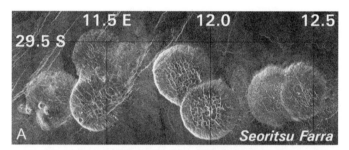

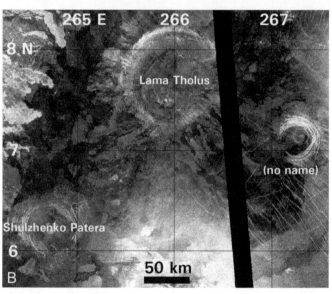

Figure D-1. Plains features commonly regarded as volcanoes that appear instead to be of impact origins. (A) Chain of low pancake domes which show eastward-younging cookie-cutter superpositions, not magmatic-interference patterns, and that may be constructs of impacts into soft sediments by fragments of a disrupted large bolide. (B) Lama Tholus has impact-compatible basin, rim, and lobate debris apron, and lacks morphologic terrestrial volcanic analogues.

et al., 2001; Glaze et al., 2002). All coronae have at least minor associated volcanism, and many have extensive volcanism. They are commonly interpreted to form above either mantle upwellings or downwellings because of the radial and concentric fracturing and range of topographic morphologies (e.g., Smrekar and Stofan, 1997; Hoogenboom and Houseman, 2006). They are enigmatic in that they are unique to Venus. One explanation is that either the thick continental lithosphere or the low-viscosity asthenosphere prevents the formation of coronae on Earth (Smrekar and Stofan, 1997). Alternatively, interior thermal boundary layers are affected by a stagnant lid as compared to plate tectonics and may generate more small-scale plumes (Parmentier and Sotin, 2000; Jellinek et al., 2002). Arguments as to why an impact origin for coronae is unlikely are well laid out in Jurdy and Stoddard (this volume) and generally ignored in the article by Hamilton (this volume).

Throughout his article in this volume, Hamilton ignores the physics of impact cratering in attempting to explain characteristics of coronae; in fact, he cites little of the relevant literature on the topic. We do not intend to refute Hamilton's paper point by point but simply identify examples of major inconsistencies. First, Hamilton completely ignores the issue of transition of shape as a function of size. As on all planets, venusian craters transition to multiring basins at a given size (e.g., McKinnon et al., 1997). All coronae have concentric rings. To suggest that the concentric fractures seen at coronae are, in fact, multiring basin structures is to completely ignore both the morphology and physics of impact basin rings, which form via slumping of the crater walls and thus never exceed three rings. In contrast, corona concentric rings are narrow, closely spaced fractures, with typically a dozen or more at a given corona. To suggest that the radial fractures seen at most coronae can be formed through debris flow processes (flow lines or channels) has no basis. The appearance of tectonic features in radar images is well understood through decades of radar studies of Earth. The radial features at most coronae are grabens, with steep walls and flat floors. Their location and morphology are completely consistent with a tectonic origin, and completely inconsistent with formation by flow processes.

Additionally, Hamilton applies his explanations for corona characteristics selectively, not to mention with the use of phrases such as "I think . . ." and "I presume . . ." rather than offering sound evidence. For example, he states "Venusian aprons are dominated by ground-hugging lobate debris flows, rather than by ballistic ejecta as on the airless Moon, because of the dense atmosphere and high gravity." This statement ignores both the observed ejecta blankets on Venus, which bear characteristics of both ballistic and flow emplacement. The physics of how the atmosphere affects ejecta on Venus has been analyzed and modeled, using actual examples of venusian impact craters (Schultz, 1992). The ejecta around craters bears no resemblance to the flow aprons around coronae, which are lava flows that can be traced to the fractures and edifices from which they originated. Hamilton (this volume) rejects a volcanic origin for nearly every type of feature despite the morphological evidence. Again, we refer

him to the extensive literature on terrestrial radar studies (e.g., Ford et al., 1989, 1993), which provide excellent analogues for the volcanic edifices and lava flows seen in Magellan images.

Controversial ideas serve a key role in the advancement of science through challenging our assumptions and paradigms. Hamilton's proposal goes too far in that he ignores basic physics and makes selective observations in the service of his hypothesis. The details of corona formation are still debated in the literature, but the impact hypothesis was rejected long ago on the basis of sound, scientific, unbiased research. For an informed discussion of the origin of coronae, see Jurdy and Stoddard (this volume).

4 February 2007, Warren B. Hamilton

The scores of papers published by Smrekar and Stofan, with various co-authors, over the past 20 years have been dominated by speculations regarding the origin, always assumed to be endogenic, of the fraction of old circular venusian structures that they term "coronae." Their comment above, of 31 January 2007 (henceforth, S&S07) continues their consideration of these structures in isolation from the thousands of other old structures, mostly ignored, from which they arbitrarily discriminate coronae. As I show in my article in this volume (henceforth, H07) and in Hamilton (2005), the infrequent mentions, and prompt dismissals, of exogenic origins by Smrekar and Stofan in their papers prove only that their assumptions of endogenesis are incompatible with impact origins. S&S07 approvingly cite the anti-impact article by Jurdy and Stoddard (this volume); but, as I show (see the discussion following that article, and H07), the arguments of Jurdy and Stoddard (this volume) also are irrelevant because, among other disqualifications, they address hypothetical endogenic blobs rather than the rimmed circular structures abundant in their own study area. The statement by S&S07 that coronae "form either above mantle upwellings or downwellings" —diametrically opposed speculations—illuminates the lack of constraints on endogenic conjectures.

The ejecta aprons of small, young venusian impact craters (agreed on by all) are distinctively lobate. Diverse ground-hugging and atmospheric-interaction mechanisms have been proposed (e.g., Herrick et al., 1997; Barnouin-Jha and Schultz, 1998), although long-runout fluidized flows can only be ground-hugging (e.g., Purdie and Petford, 2005). These lobate aprons much resemble the lobate aprons of coronae (compare Figs. 4–6 of H07 with Figs. 7 and 18), with appropriate scaling of lobes for diameters of structures (cf. Barnouin-Jha and Schultz, 1998). The assertion in S&S07 that "the ejecta around craters bear no resemblance to the flow aprons around coronae" is false. The assertion that corona aprons "are lava flows that can be traced to the fractures and edifices from which they originated" is based only on interpretations of rare and ambiguous relationships; dike eruptions, for example, have nowhere been documented. The physics of large impacts cannot be extrapolated from small ones and can only be modeled. Although shock fluidization probably is a major process (e.g., Pierazzo and Collins, 2003), S&S07

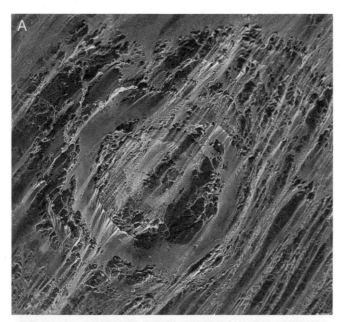

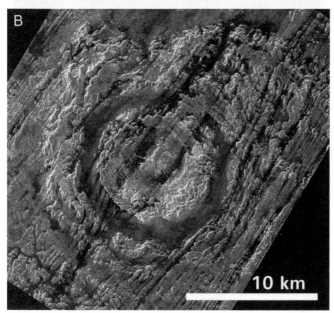

Figure D-2. (A) Optical image of Aorounga Crater, northern Chad, courtesy of NASA. Structure is centered near 19.2°E, 19.1°N. North at top. Wind-sculpted ridges and narrow sand sheets are aligned with prevailing wind, and minor concentric structures are obscure. (B) Radar image of Aorounga Crater, courtesy of NASA. Smooth sand too thick to see through is radar-dark. Corona-like concentric fracturing is obvious. Illuminated from northwest; apparent small triangular facets are caused by slant-radar hogback illusion. Black corners are areas with no data.

state that "radial features [for which I infer a surficial origin consequent on shock fluidization] at most coronae are grabens"—an assertion that is disproved because the required extensions along concentric circles about the structures are 100 times too great to be explicable by the very slight doming possible with inflation models (Grindrod et al., 2005).

S&S07 assert that analogies with modern terrestrial features require endogenic volcanic interpretations for many features of coronae. In fact, no modern terrestrial features resemble venusian "volcanoes," which are hundreds to many hundreds of kilometers in diameter but mostly only 1 or 2 km high, typically rise from rimmed circles, and mostly have only single central peaks, either broad plateaus or exceedingly gentle summits, with broad sags but not calderas. These venusian constructs (of shock melt and shock-fluidized material?) may be analogous to those once present atop the terrestrial Precambrian proved-impact structures of Vredefort and Sudbury, and possible-impact Bushveld and Stillwater, and such features may have been abundant on the pre-3.8 Ga Earth. That Ford et al. (1989, p. 27–54) presented radar images of various small young terrestrial volcanic features does not confirm the interpretation by Ford et al. (1993, p. 109–134) that the very different features of Venus also are volcanic. Thus, the great 800 × 300 km stream of rough-surfaced lobate flows (termed lava by Ford et al., 1993, although in scale, source, and character they resemble nothing on modern Earth) of Mylitta Fluctus originates in what appears to me to be a complex of simultaneous(?) impact structures (Jord Corona, Tarbell Patera, and Alcott Crater) and thus is analogous to, but appropriately larger than, the fluidized-ejecta stream of similar appearance from Isabella Crater (H07, Fig. 6).

I agree with S&S07 that the concentric fractures in many of their coronae lack common conspicuous analogues in known impact structures elsewhere, but I have not confused these with multiring craters. That no endogenic process yet conceived could produce the concentric structures has not influenced conjectures by Smrekar and Stofan. The impact regime, by contrast, makes it easy to visualize formation of shock-induced concentric features in bedrock targets, whereas they could not form in the deep-rubble targets of Mars and the moon, and I suspect that such structures are abundant but commonly unseen in terrestrial bedrock structures. In Figure D-2, I show two images of Aorounga Crater, the impact origin of which is confirmed by shatter cones and shock metamorphism (McHone et al., 2002), in the Sahara Desert. The optical image (Fig. D-2A) is dominated by eolian features, whereas the satellite synthetic-aperture-radar image (Fig. D-2B; analogous to venusian imagery) sees through the thin cover and shows abundant concentric fracturing of venusian-corona type.

Coronae, as hypothetical endogenic structures, "are unique to Venus" (S&S07), whereas impact structures are ubiquitous on solid bodies in the Solar System. This contrast justifies skepticism of endogenic-corona conjecture and consideration of the obvious option that the thousands of small to huge old circular structures of Venus are ancient impact structures.

REFERENCES CITED

Barlow, S.G., 2006, Impact craters in the northern hemisphere of Mars. Layered ejecta and central pit characteristics: Meteoritics and Planetary Science, v. 41, p. 1425–1436.

Barnouin-Jha, O.S., and Schultz, P.H., 1998, Lobateness of impact ejecta deposits from atmospheric interactions: Journal of Geophysical Research, v. 103, p. 25,739–25,756.

Bridges, N.T., 1995, Submarine analogs to venusian pancake domes: Geophysical Research Letters, v. 22, p. 2781–2784.

Ford, J.P., Blom, R.G., Crisp, J.A., Elachi, C., Farr, T.G., Saunders, R.S., Thellig, E.E., Wall, S.D., and Yewell, S.B., 1989, Spaceborne radar observations—A guide for Magellan radar-image analysis: Pasadena, California, Jet Propulsion Laboratory Publication 89-41, 126 p.

Ford, J.P., Plaut, J.J., Weitz, C.M., Farr, T.G., Senske, D.A., Stofan, E.R., Michaels, G., and Parker, T.J., 1993, Guide to Magellan image interpretation: Pasadena, California, Jet Propulsion Laboratory Publication 93-24, 148 p.

Glaze, L.S., Stofan, E.R., Smrekar, S.E., and Bologa, S.M., 2002, Insights into corona formation through statistical analyses: Journal of Geophysical Research, v. 107, doi: 10.1029/2002JE001904.

Grindrod, P.M., Nimmo, F., Stofan, E.R., and Guest, J.E., 2005, Strain at radially fractured centers on Venus: Journal of Geophysical Research, v. 110, no. E12, paper 2, 13 p.

Hamilton, W.B., 2005, Plumeless Venus preserves an ancient impact-accretionary surface, *in* Foulger, G.R., et al., eds., Plates, plumes, and paradigms: Boulder, Colorado, Geological Society of America Special Paper 388, p. 781–814, doi: 10.1130/2005.2388(44).

Hamilton, W.B., 2007 (this volume), An alternative Venus, *in* Foulger, G.R., and Jurdy, D.M., eds., Plates, plumes, and planetary processes: Boulder, Colorado, Geological Society of America Special Paper 430, doi: 10.1130/2007.2430(41).

Hartmann, W.K., and Barlow, S.G., 2006, Nature of Martian uplands: Effect on Martian meteorite age distribution and secondary cratering: Meteoritics and Planetary Science, v. 41, p. 1453–1467.

Herrick, R.L., Sharpton, V.L., Malin, M.C., Lyons, S.E., and Feely, K., 1997, Morphology and morphometry of impact craters, *in* Bougher, S.W., et al., eds., Venus II: Tucson, University of Arizona Press, p. 1015–1046.

Hoogenboom, T., and Houseman, G.A., 2006, Rayleigh-Taylor instability as a mechanism for corona formation on Venus: Icarus, v. 180, p. 292–307.

Jellinek, A.M., Lenardic, A., and Manga, M., 2002, The influence of interior mantle temperature on the structure of plumes: Heads for Venus, tails for the Earth: Geophysical Research Letters, v. 29, p. 15–32, doi: 1029/2001GL014624.

Jurdy, D.M., and Stoddard, P.R., 2007 (this volume), Venus's coronae: Impact, plume, or other origin?, *in* Foulger, G.R., and Jurdy, D.M., eds., Plates, plumes, and planetary processes: Boulder, Colorado, Geological Society of America Special Paper 430, doi: 10.1130/2007.2430(40).

McCall, G.J.H., 2006, A caldera volcano of Brobdingnagian scale: Geoscientist, v. 16, p. 28–30.

McHone, J.F., Greeley, R., Williams, K.K., Blumberg, D.G., and Kuzmin, R.O., 2002, Space shuttle observations of terrestrial SIR-C and X-SAR radars: Meteoritics and Planetary Science, v. 37, p. 407–420.

McKinnon, W.B., Zahnle, K.J., Ivanov, B.A., and Melosh, H.J., 1997, Cratering on Venus: Models and observations, *in* Brougher, S.W., et al., eds., Venus II: Tucson, University of Arizona Press, p. 969–1014.

Melosh, H.J., 1989, Impact cratering geology: Tucson, University of Arizona Press, 245 p.

Parmentier, E.M., and Sotin, C., 2000, Three-dimensional numerical experiments on thermal convection in a very viscous fluid: Implications for the dynamics of a thermal boundary layer at high Rayleigh number: Physics of Fluids, v. 12, p. 609–617.

Pierazzo, E., and Collins, G., 2003, A brief introduction to hydrocode modeling of impact cratering, *in* Claeys, P., and Henning, D., eds., Submarine craters and ejecta-crater correlation: New York, Springer, p. 323–340.

Plaut, J.J., Anderson, S.W., Crown, D.A., Stofan, E.R., and Jakob, J.v.Z., 2004, The unique radar properties of silicic lava domes: Journal of Geophysical Research, v. 109, no. E3, paper 1, 12 p.

Purdie, P., and Petford, N., 2005, Addams Crater, Venus—Outflow analogous with a submarine debris flow?: Lunar and Planetary Science, v. 36, paper 1044, 2 p. (CD-ROM).

Schultz, P.H., Atmospheric effects on ejecta emplacement and crater formation on Venus from Magellan: Journal of Geophysical Research, v. 97, p. 16,183–16,248.

Smrekar, S.E., and Stofan, E.R., 1997, Coupled upwelling and delamination: A new mechanism for coronae formation and heat loss on Venus, Science, v. 277, p. 1289–1294.

Stofan, E.R., Tapper, S.W., Guest, J.E., Grindrod, P., and Smrekar, S.E., 2001, Preliminary analysis of an expanded corona database for Venus: Geophysics Research Letters, v. 28, p. 4267–4270.

Geological Society of America
Special Paper 430
2007

Interaction between local magma ocean evolution and mantle dynamics on Mars

Chris C. Reese*
Viatcheslav S. Solomatov
Christopher P. Orth
Department of Earth and Planetary Sciences, Washington University in St. Louis, St. Louis, Missouri 63130, USA

ABSTRACT

Large-scale features of the crustal structure on Mars, including the hemispherical dichotomy and Tharsis, were established very early in planetary history. Geodynamical models for the origin of the dichotomy and Tharsis, such as lithospheric recycling and a plume from the core-mantle boundary, respectively, involve solid-state mantle flow and are difficult to reconcile with timing constraints. An alternative point of view is that the martian crustal asymmetry and Tharsis can be associated with the upwelling and spreading of large, impact-induced melt regions, i.e., local magma oceans.

While the local magma ocean–induced upwelling model satisfies timing constraints on dichotomy and Tharsis formation, it neglects any interaction with longer-timescale mantle dynamics and cannot explain recent volcanic activity at Tharsis. In this study, fully 3-D, spherical shell simulations are used to investigate coupling between local magma oceans and mantle dynamics with radiogenic heating and core heatflow. For low core heat flux, it is found that upwellings driven by local magma ocean buoyancy are transient features of planetary evolution that is dominated by sublithospheric instabilities. With increasing core heat flux, local magma ocean–induced upwellings strongly influence the pattern of thermal plumes from the core-mantle boundary, which can remain stable for ~4.5 b.y. The predicted melt volumes, present-day melting rate, and crustal structure are compared to observational constraints.

Keywords: Mars, magma ocean, mantle convection

INTRODUCTION

Geological and geochemical observations indicate rapid development of large-scale crustal structure early in martian history. Both the hemispherical crustal dichotomy (Zuber, 2001; Frey et al., 2002; Nimmo and Tanaka, 2005) and Tharsis province (Banerdt and Golombek, 2000; Phillips et al., 2001; Johnson and Phillips, 2005; Solomon et al., 2005) were emplaced within ~1 b.y. of planet formation. Geochemical analyses of martian meteorites are also consistent with early silicate differentiation (Harper et al., 1995; Lee and Halliday, 1997; McLennan, 2001).

The martian crustal hemispherical dichotomy is expressed topographically as a north-south slope of ~0.036° such that the south pole is ~6 km higher than the north pole (Smith et al., 1999), with a complex boundary zone (Frey et al., 1998), and

*E-mail: creese@levee.wustl.edu.

Reese, C.C., Solomatov, V.S., and Orth, C.P., 2007, Interaction between local magma ocean evolution and mantle dynamics on Mars, *in* Foulger, G.R., and Jurdy, D.M., eds., Plates, plumes, and planetary processes: Geological Society of America Special Paper 430, p. 913–932, doi: 10.1130/2007.2430(42). For permission to copy, contact editing@geosociety.org. ©2007 The Geological Society of America. All rights reserved.

geologically as a contrast between heavily cratered southern highlands and smoother northern lowlands. The size-frequency distribution of circular depressions in the resurfaced northern plains is similar to the crater distribution in the south, implying that southern basement crust is of approximately the same age as northern crust, i.e., early Noachian (Frey et al., 2002), consistent with geochemical constraints on differentiation. Mars's crustal structure is constrained by topography and gravity data (Lemoine et al., 2001; Smith et al., 2001) from Mars Global Surveyor. Models with uniform crustal density and varying crustal thickness (Zuber et al., 2000; Zuber, 2001; Neumann et al., 2004), consistent with the data, predict a lower bound on the mean crustal thickness of ~50 km and a general crustal thinning trend from south to north. The crustal structure in the dichotomy boundary zone is complex, exhibiting a smooth transition in Arabia Terra but more scarplike regional offsets elsewhere.

Several hypotheses have been suggested for hemispherical dichotomy formation. Excavation by one or more large impacts (Wilhelms and Squyres, 1984; Frey and Schultz, 1988) is inconsistent with the lack of correlation between crustal structure and the boundary surface expression, the noncircular shape and complex boundary topography, and the lack of filled basin gravity signatures such as Utopia for the northern lowlands as a whole (Zuber et al., 2000; Zuber, 2001). The late stages of a transient episode of plate tectonics has been suggested (Sleep, 1994), although subduction of 35-km-thick crust (the mean lowland thickness) may be difficult due to the deeper basalt-eclogite transition on Mars. Long-wavelength (spherical harmonic degree 1) mantle convection is possible for Mars but requires a layered viscosity structure (Zhong and Zuber, 2001). Such a convective planform could produce a dichotomy by enhanced melting and crustal production at the upwelling or crustal erosion by the downwelling. Timescales for surface recycling or mantle flow are difficult to reconcile with early formation of the dichotomy (Solomon et al., 2005), and preservation of Utopia crustal structure makes northern crustal thinning unlikely (Neumann et al., 2004). Gravitational instability due to crystallization of a global magma ocean can lead to mantle overturn that may have a strong degree 1 component and might provide a mechanism for dichotomy formation (Hess and Parmentier, 2001; Elkins-Tanton et al., 2005a,b).

Later in the Noachian, magmatism and tectonism were focused at Tharsis province, a broad topographic rise comprised of regional centers of volcanism and deformation (e.g., Anderson et al., 2001). Large crustal structure variations (Zuber et al., 2000) and layered volcanics in Valles Marineris (McEwen et al., 1999) support the notion of constructional magmatism as the primary contributor to Tharsis elevation (Solomon and Head, 1982). Localized magmatism resulted in a lithospheric load with an estimated volume of ~3×10^8 km^3 (Phillips et al., 2001). Faulting patterns dating to the late Noachian are consistent with a flexural loading model based on present-day topography and gravity, suggesting that the load was emplaced by that time

(Banerdt and Golombek, 2000). Volcanism at Tharsis continued throughout martian history but at much lower levels (Hartmann and Neukum, 2001). Tharsis formation probably postdates development of the hemispherical crustal dichotomy. Crustal thickening is superimposed on the global south-north trend (Zuber et al., 2000; Zuber, 2001; Neumann et al., 2004), and Tharsis volcanic units are emplaced on older magnetized crust (Johnson and Phillips, 2005).

Hypotheses for Tharsis formation have focused on a planform of mantle convection dominated by a single thermal plume as an explanation for the monopolar magmatic and tectonic activity. Deep-mantle mineralogical phase changes are capable of stabilizing a one-plume pattern of convection (Weinstein, 1995; Harder and Christensen, 1996; Breuer et al., 1998; Spohn et al., 1998; Harder, 2000), but no plume model has reproduced development of Tharsis development on a timescale consistent with observation. Alternatively, if the martian mantle is layered and thick southern crust acts as an insulating layer, hot plumes can develop beneath the insulating layer (Wenzel et al., 2004), although this hypothesis does not account for the location of Tharsis near the dichotomy boundary. The spatial correlation of the Tharsis rise and dichotomy boundary motivates the hypothesis of edge-driven convection as the mechanism for Tharsis formation (King and Redmond, 2005). This hypothesis requires a deep, cold, stable layer beneath southern hemisphere crust in order to initiate small-scale convection, and if edge driven convection occurs, it should be prevalent everywhere along the dichotomy boundary rather than focused at Tharsis.

An alternative possibility is that planetary formation processes and initial thermal and compositional conditions play an important role. Large impacts occurring during late-stage accretion (e.g., Agnor et al., 1999) can produce localized, intact melt regions (Tonks and Melosh, 1992, 1993), i.e., local magma oceans. A local magma ocean evolves to a uniform global layer, or global magma ocean, only if buoyant melt is rapidly redistributed prior to any significant solidification (Tonks and Melosh, 1992, 1993). A simple fluid dynamical model for local magma ocean evolution (Reese and Solomatov, 2006) suggests that if cooling and crystallization are fast compared to melt redistribution, the driving force for isostatic adjustment diminishes, melt region viscosity increases significantly, and restoration of a globally symmetric, fully molten layer can be difficult. Instead, a localized mantle upwelling develops in response to the buoyant, partially molten, local magma ocean. Two preliminary studies of this mechanism have produced results that are in qualitative agreement with early crustal structure and timing constraints for dichotomy and Tharsis formation (Reese et al., 2004; Reese and Solomatov, 2006). However, both studies failed to address interaction between local magma ocean–induced upwellings and longer-timescale mantle dynamics. The study of Reese et al. (2004) was designed to isolate local magma ocean upwelling effects and neglected radiogenic mantle heating and core heatflow, which are crucial for continued melting throughout planetary history (Hartmann and Neukum, 2001; Neukum et al., 2004). Reese and Solomatov (2006) did not consider local magma ocean–mantle convection interaction at all.

In this article, the coupling of local magma ocean evolution and mantle dynamics with internal heating and core heat flux is investigated. The study is organized as follows. The fluid dynamical model for the evolution of a local magma ocean and hypothesis for origin of the dichotomy and Tharsis (Reese and Solomatov, 2006) are reviewed. The effects of local magma ocean evolution on martian mantle convection and magmatic evolution are explored using fully 3-D, spherical shell geometry simulations. Results are compared with observational constraints, and future work is suggested.

LOCAL MAGMA OCEAN EVOLUTION

The intact melt region generated by a large impact can form a global magma ocean only if the ratio of the isostatic adjustment time to the melt cooling time is small (Tonks and Melosh, 1992, 1993). Convection in the local magma ocean and the radiative boundary condition at the surface control the crystallization time. Complete isostatic adjustment involves both radial relaxation as the buoyant melt region moves upward and a lateral component as melt is redistributed around the planet to a spherically symmetric state. In this section, a model for local magma ocean evolution is reviewed (Reese and Solomatov, 2006).

Formation of an Impact-Induced Local Magma Ocean

The thermal state of proto-Mars prior to large impacts is not well constrained. A study that incorporates an improved planetesimal size distribution (but neglects planetesimal heating due to decay of short-lived radioisotopes) suggests that the interior can avoid widespread and deep melting during early accretion (Senshu et al., 2002). Mars could be a remnant planetary embryo from the runaway growth stage of planet formation (Chambers and Wetherill, 1998; Chambers 2004). After runaway growth, the planetesimal mass spectrum is characterized by Mars mass embryos separated by a factor of ~10 from a continuous power law distribution (Kokubo and Ida, 1996, 1998, 2000). Impacting bodies with masses ~0.1–1.0 times the lunar mass seem likely. Large impacts are sufficiently energetic to produce melting and generate a localized magma ocean.

High-velocity impacts generate a spherical shock wave that propagates into the target. Near the impact site, there is an isobaric core where shock pressures are approximately uniform. Outside the isobaric core, shock pressures decay with distance. For a hemispherical model of the shock pressure distribution (Fig. 1), vertical incidence, target to projectile density ratio of 1, and pressure drop off outside the isobaric core, varying like r^{-2}, the radius of the region shocked to pressure P (Tonks and Melosh, 1992, 1993), may be expressed as

$$r = 2^{1/3} \left(\frac{v_i}{v_i^m} \right)^{1/2} a, \tag{1}$$

where a is the impactor radius, v_i is the impact velocity, and

$$v_i^m = \frac{C}{S} \left[\left(1 + \frac{rSP}{\rho C^2} \right)^{1/2} - 1 \right]. \tag{2}$$

(See Table 1 for the symbols for and values of physical parameters referred to in this article.) Implicit in this expression is the assumption of a linear shock velocity–particle velocity relationship (Melosh, 1989),

$$U = C + Su_p, \tag{3}$$

where U is the shock velocity and u_p is the particle velocity. For dunite at shock pressures $P = \geq 73$ GPa, $C = 4.4$ km/s and $S = 1.5$ (Kieffer, 1977). Oblique impacts complicate this simple geometry, concentrating more energy downrange (Pierazzo and Melosh, 2000). The shock pressure corresponding to complete melting of solidus dunite is $P = 115$ GPa (Tonks and Melosh, 1993). The radius of a region shocked to a melt fraction φ can be calculated from the near-solidus Hugoniot in pressure-entropy space (Tonks and Melosh, 1993) and the relationship between melt fraction and entropy. The critical melt fraction $\varphi_{cr} \sim 0.4$ corresponds to a rheological transition from low-viscosity liquid to high-viscosity partially molten solid (Solomatov, 2000). For linear melt fraction dependence on entropy, this corresponds to a shock pressure of ~70 GPa. For an impact velocity of $v_i \sim 7$ km/s, the radius of the region shocked to φ_{cr} may be expressed as

$$R \approx 1.7a. \tag{4}$$

The volume of the region with $\varphi \geq \varphi_{cr}$ may be expressed as

$$V = \frac{2}{3} \pi R^3 \left(1 - \frac{3}{8} \frac{R}{R_p} \right), \tag{5}$$

where R_p is the planet radius and the term in parentheses corresponds to a geometrical correction giving the volume between the spherical surface of radius R_p and a hemisphere of radius R centered on the surface (Tonks and Melosh, 1992).

While some of the melt produced by a large impact gets redistributed about the planet by the excavation flow, sufficiently large impacts generate melt deep in a planet where it remains subsequent to crater excavation. The transient crater radius is given by the π-scaling law for the gravity regime (Schmidt and Housen, 1987),

$$r_c = \frac{1}{2} \left(\frac{4}{3} \pi \right)^{1/3} C_D \left(3.22 \frac{g}{v_i^2} \right)^{-\beta} a^{1-\beta}, \tag{6}$$

where $C_D = 1.6$ and $\beta = 0.22$ are empirical scaling parameters (Melosh, 1989). For an impact velocity $v_i \sim 7$ km/s,

$$r_c \approx 300 \text{ km} \left(\frac{a}{100 \text{ km}} \right)^{0.78}. \tag{7}$$

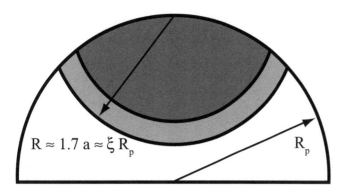

Figure 1. Hemispherical model of impact melt distribution. The radius of the region with melt fraction $\varphi > \varphi_{cr}$ is R, the impactor radius is a, and the relationship between the two is $R = 1.7a$ (see text).

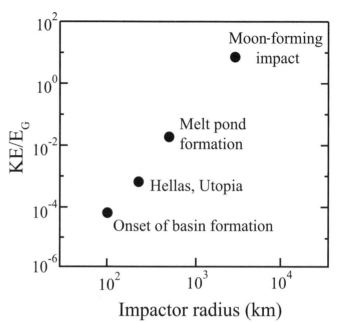

The scaling laws $R \sim a$ and $r_c \sim a^{0.78}$ imply that as impactor size increases, the melt region radius increases faster than the crater radius. Thus, the retained melt fraction increases with increasing a, and eventually a large volume of melt can be retained (Fig. 2).

While an exact criterion for local magma ocean formation is not easily defined, Tonks and Melosh (1992) suggest equal melt and transient crater volume. For a crater with a parabolic

TABLE 1. SYMBOLS FOR AND VALUES OF PHYSICAL PARAMETERS

Parameter	Symbol	Value
Thermal capacity	c_p	1200 J/(kg K)
Shock velocity parameter	C	4.4 km/s
Cratering scaling parameter	C_D	1.6
Melt activation energy	E_m	250 kJ/mole
Gravitational acceleration	g	3.7 m/s²
Specific heat production rate	H_0	2.1×10^{-11} W/kg
Thermal conductivity	k	4 W/(m K)
Silicate latent heat of fusion	L	500 kJ/kg
Mars mass	M	6.4×10^{23} kg
Mars radius	R_p	3398 km
Gas constant	R^*	8.314 J/(mole K)
Shock velocity parameter	S	1.5
Surface temperature	T_s	220 K
Initial mantle temperature	$T_{i,0}$	1700 K
Impact velocity	v_i	7 km/s
Thermal expansion	α	2×10^{-5} K⁻¹
Cratering scaling parameter	β	0.22
Thermal diffusivity	κ	10^{-6} m²/s
Decay constant	λ	3.3×10^{-10} yr⁻¹
Viscosity (solid)	η_s	10^{21} Pa s
Viscosity (partial melt)	η_m	10^{17} Pa s
Viscosity (liquid)	η_l	0.1 Pa s
Viscosity (melt near solidus)	η_f	100 Pa s
Solid silicate density	ρ	3500 kg/m³
Silicate density (@ ϕ_{cr})	ρ_{cr}	3200 kg/m³
Liquid silicate density	ρ_l	2800 kg/m³
Stefan-Boltzmann constant	σ_B	5.67×10^{-8} W/(m² K⁴)
Yield stress	τ_y	10^7 Pa
Critical melt fraction	ϕ_{cr}	0.4
Crustal spreading parameter	ζ_{cr}	0.3

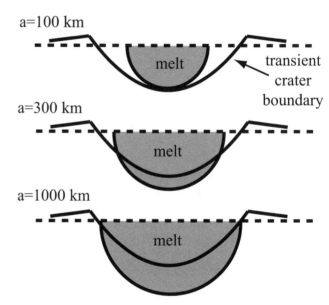

Figure 2. (Top) Comparison of impactor kinetic energies (KE) scaled by the martian gravitational binding energy E_G for different-sized events. Formation of a Hellas- or Utopia-sized basin is based on an ~200 km radius impactor with impact velocity v_i (Table 1). The energy of a moon-forming impact on the proto-Earth is based on an impactor to target mass ratio ~0.1 and an impact velocity equal to the terrestrial escape velocity. (Bottom) The effect of increasing impactor size (a) on melt retained below the excavation zone (after Tonks and Melosh, 1992).

cross-section (Melosh, 1989), the crater volume $V_c = \gamma_c/(2\pi r_c^3)$, where the typical depth to radius ratio $\gamma_c \approx 1/2 - 2/3$. Equating volumes V and V_c gives the critical impactor radius for local magma ocean formation, $a_{cr} \approx 700 - 1500$ km. Such impactor size can be expected based on improved models of planetary formation (e.g., Agnor et al., 1999). The lower bound of this range corresponds to a local magma ocean, with $R \sim 1000$ km. Magma pond–forming impacts are significantly larger and more energetic than basin-forming impacts (Fig. 2).

Local Magma Ocean Crystallization Timescale

For ultramafic silicates at low pressure and large melt fraction, the viscosity $\eta_l \sim 0.1$ Pa s (Bottinga and Weill, 1972; Shaw, 1972; Persikov et al., 1990; Bottinga et al., 1995). At near-liquidus conditions, komatiite viscosities range from 0.1 to 10.0 Pa s depending on MgO content (Huppert and Sparks, 1985). For low-viscosity liquids, a power law governs viscosity variations with temperature (Bottinga et al., 1995), implying a minor effect over the melting range. Variations due to pressure are also small for a completely depolymerized melt (Andrade, 1952; Gans, 1972). The most significant rheological variation occurs at the critical melt fraction φ_{cr}, where the viscosity increases to a value near that of partially molten solids (e.g., Solomatov and Stevenson, 1993). Thus, the viscosity of the magma ocean between $\varphi = 1$ and $\varphi = \varphi_{cr}$ is assumed to be $\eta_l \sim 0.1$ Pa s with an uncertainty factor of ~ 10.

Vigorous convection in the low-viscosity melt region is the primary mode of cooling and crystallization. As the region cools, a partially solid layer develops at the bottom (Fig. 3). The magma ocean heat flux is determined by convection in the magma and the radiative boundary condition at the surface. In the blackbody approximation, the surface heat flux may be expressed as

$$F = \sigma_B T_s^4, \tag{8}$$

where σ_B is the Stefan-Boltzmann constant and T_s is the surface temperature.

For low-viscosity magma, convection is in a turbulent regime (Kraichnan, 1962; Shraiman and Siggia, 1990; Siggia, 1994). Neglecting the effect of rheological variations near the surface, the heat flux in the soft turbulence regime may be expressed as

$$F = 0.089k(T - T_s)^{4/3}\left(\frac{\rho_l a g}{\eta_l \kappa}\right)^{1/3}, \tag{9}$$

where T is the potential temperature. The convective heat flux must match the radiative heat flux at the surface. In the blackbody approximation, at a potential temperature of 2000 K, the surface temperature and heat flux are ~ 1400 K and 3×10^5 W/m^2, respectively.

The timescale for crystallization to the critical melt fraction may be expressed as

$$t_{crys} \approx \frac{[L(1 - \varphi_{cr}) + c_p \Delta T]M}{FS}, \tag{10}$$

where the mass $M = \rho_l V$ and $S = \pi R^2$. For the variation of melt fraction between liquidus and solidus suggested by the experi-

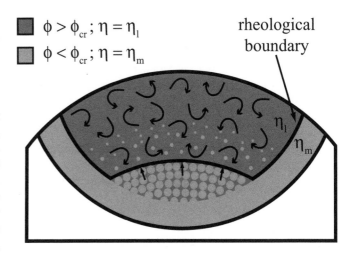

Figure 3. Melt region crystallization. As the magma pond cools via convection, the temperature drops below the liquidus at the base of the melt region and crystallization proceeds from the bottom up. When the temperature decreases to the rheological transition temperature, a partially solid region develops, and with further cooling, a rheological front propagates through the melt region. φ—melt fraction; φ_{cr}—critical melt fraction; η—viscosity; η_m—viscosity (partial melt).

mental data of McKenzie and Bickle (1988), the temperature change corresponding to $\Delta\varphi = 1 - \varphi_{cr}$ is $\Delta T \approx 0.4(T_l - T_s) \approx 0.4(2000-1400)$ K ≈ 240 K. For local magma ocean size $R \sim 1000$ km, the crystallization time $t_{crys} \sim 300$ yr.

Development of a stable, stagnant lid (Davaille and Jaupart, 1993) reduces the convective driving temperature. The rheological temperature scale that drives convection beneath the lid may be expressed as

$$\Delta T_{d,m} \sim \left[\frac{\partial \ln\eta}{\partial T}\right]^{-1} \sim \left[\frac{E_m}{R^* T^2}\right]^{-1}, \tag{11}$$

where E_m and R^* are the activation energy and gas constant, respectively. The activation energy of olivine tholeiite melt at 1 Gpa, $E_m \sim 250$ kJ/mol (Kushiro, 1986). The convective heat flux in the stagnant lid regime (Solomatov and Moresi, 2000) may be expressed as

$$F = 0.53k\Delta T_{d,m}^{4/3}\left(\frac{\rho_l a g}{\eta_l \kappa}\right)^{1/3}. \tag{12}$$

At a temperature $T \sim 2000$ K, $\Delta T_{d,m} \sim 100$ K and $F \sim 10^5$ W/m^2, a factor of three smaller than the surface recycling case. However, the stability of a viscous lid with a thickness on the order of centimeters (the thickness $\sim k\Delta T/F$ with $\Delta T \sim 1000$ K) seems unlikely due to continuing impact disruption and convective stress-induced recycling.

Impacts of the sizes considered here distribute vaporized and molten ejecta around the planet and inject water into the atmosphere (Sleep and Zahnle, 1998; Segura et al., 2002), which acts as a greenhouse gas. A steam atmosphere significantly reduces the ability of the atmosphere to radiate in the infrared (Matsui and Abe, 1986; Kasting, 1988; Zahnle et al., 1988). For large amounts of water, an abrupt atmospheric transition from low to near–melting surface temperature occurs for an outgoing radiative flux of ~300 W/m². The present-day net solar flux at Mars is ~100 W/m². If a steam atmosphere limits surface heat flux to 200 W/m², the magma ocean crystallization time increases by a factor of ~1000 compared to the blackbody model. On the other hand, at even higher surface temperatures, the atmosphere can radiate at a much higher rate due to increased transparency in the visible and the near-infrared (Zahnle et al., 1988). The effect of a thick atmosphere and high surface temperature on local magma ocean evolution requires further study.

Isostatic Adjustment to a Local Magma Ocean

Because the melt region density is less than the density of solid silicates, the melt region is positively buoyant and the planet deforms in order to restore isostatic equilibrium. The isostatic adjustment timescale depends on the melt region size and the effective viscosity of the solid part of the planet. A crude, zero-order estimate of the radial relaxation timescale can be made based on Stokes flow of an inviscid sphere of radius R through a distance R,

$$t_{iso} \sim 3 \frac{\eta_s}{(\rho - \rho_l)gR} \tag{13}$$

with solid planet viscosity η_s (e.g., Karato and Wu, 1993), liquid silicate density ρ_p and solid silicate density ρ. For nominal parameters (Table 1) and $R \sim 1000$ km, $t_{iso} \sim 3 \times 10^4$ yr.

For $t_{crys} \gg t_{iso}$, the melt region crystallizes to the critical crystal fraction before melt is redistributed by isostatic adjustment of the planet. For the blackbody model and $R \sim 1000$ km, convection with surface recycling or a stagnant lid yields an isostatic adjustment time—approximately thirty to one hundred times larger than the crystallization time. A critical parameter controlling whether a melt region evolves as a global magma ocean or a local partial melt pond is the viscosity of the solid part of the planet, η_s. Strongly temperature-dependent mantle viscosity implies that the thermal state of Mars at the time of the impact plays a crucial role. Additionally, non-Newtonian rheology may dominate at higher stress levels, i.e., larger local magma ocean sizes. If a thick atmosphere significantly reduces the surface heat flux, the resulting long crystallization time facilitates development of a global magma ocean. For sufficiently large local magma oceans, the isostatic adjustment time could be greatly reduced. For a local magma ocean size $R \sim 1000$ km, the differential stress $(\rho - \rho_l)gR \sim 1$ GPa, which is close to the power law creep breakdown stress (Tsenn and Carter, 1987) of $10^2 \mu$ with shear modulus μ. Furthermore, this differential stress approaches the ultimate strength of rocks ~1–2 GPa (Kinsland, 1978; Davies, 1982). In these cases, isostatic adjustment could be ex-

tremely fast. The viscous isostatic adjustment model probably holds only below an upper limit of $R/R_p \sim 0.5$.

Isostatic Adjustment to a Partially Molten Local Magma Ocean

Subject to the caveats of the previous section and within the uncertainties of various parameters, it is possible for a local magma ocean to crystallize in place prior to development of a global layer. After local magma ocean crystallization to the critical crystal fraction, several changes occur that affect the evolution. Crystallization reduces the driving density difference for radial relaxation. A more significant effect is the rheological transition from a liquid to a partially molten solid, which greatly increases the viscosity. Recent experiments (Hirth and Kohlstedt, 1995a,b, 2003; Kohlstedt and Zimmerman, 1996; Scott and Kohlstedt, 2004a,b) suggest that the viscosity $\eta_m \sim \eta_s \exp(-\alpha_\eta \varphi)$ where $\alpha_\eta \sim 25$ (Scott and Kohlstedt, 2004b). The viscosity drops by one to two orders of magnitude for $\varphi \sim 0.1$–0.2. For this study, the viscosity of partially molten material $\eta_m \sim 10^{17}$ Pa s. Additionally, at sufficiently high stresses, the material can fail. The low-pressure yield strength of rocks (coherence strength) $\tau y \sim$ 10–100 MPa and increases with pressure (frictional law).

For the much higher viscosity, η_m, the convective cooling time increases dramatically. Although melt migration in the partially molten local magma ocean can occur before any further convective cooling, partial melt region buoyancy drives isostatic adjustment on a timescale that is small compared to melt migration. Isostatic adjustment involves both a radial and a lateral component, i.e., the partial melt region moves radially outward and spreads laterally until spherical symmetry is restored. While the timescale for radial relaxation is controlled by the rheology of the solid mantle, the lateral component of isostatic adjustment involves deformation of the melt region itself.

The percolation velocity of melt through the solid matrix may be expressed as

$$v_{perc} = \frac{k_\varphi(\rho - \rho_l)g}{\eta_f} \frac{(1 - \varphi)}{\varphi}. \tag{14}$$

In the low–Reynolds number limit, the Ergun equation for k_φ (Ergun, 1952) reduces to the Blake-Carman-Kozeny equation (e.g., Bird et al., 1960),

$$k_\varphi = \frac{\varphi^3 D_p^2}{150(1 - \varphi)^2}, \tag{15}$$

where D_p is the grain size. For $D_p \sim 3$ mm and $\varphi \sim 3\%$, $k_\varphi \sim 10^{12}$ m². Near the solidus, the partial melt viscosity is higher due to lower temperatures. At low pressures, the basaltic melt viscosity formed near the anhydrous solidus $\eta_f \sim 100$ Pa s (Kushiro, 1980, 1986) depending on composition and, in particular, silica content, which affects the degree of polymerization. For this value of η_f and a local magma ocean size of $R \sim$ 1000 km, the timescale for percolation to reduce the melt fraction to 3% is $t_{perc} \sim 10^8$ yr.

At the critical melt fraction, φ_{cr}, the density difference that drives isostatic adjustment $\rho - \rho_{cr} = (\rho - \rho_l)\varphi_{cr} \sim 300$ kg/m³. The Stokes flow estimation for the radial relaxation time (equation 10) is increased by a factor of $(\rho - \rho_l) / (\rho - \rho_{cr}) \sim \varphi_{cr}^{-1} \sim 3$. For nominal parameters (Table 1) and $R \sim 1000$ km, $t_{iso} \sim 10^5$ yr. Thus, isostatic adjustment is fast compared to melt percolation.

Partially molten material extruded onto the surface spreads laterally to complete the process of isostatic adjustment. The lateral spreading time is estimated from an axisymmetric gravity current model. Depending on local magma ocean size and radial relaxation rate, the gravity current can be in a viscous regime or a yield stress regime (Fig. 4).

For nominal parameters (Table 1) and a sufficiently large volume flux, $Q \sim V/t_{iso}$, basal shear stresses can exceed the yield stress. In this case, the gravity current radius (Nye, 1951; Blake, 1990) may be expressed as

$$S \sim \left(\frac{\rho_{cr}gQ^2}{\tau_y}\right)^{1/5} t^{2/5}, \qquad (16)$$

where ρ_{cr} is the density at the critical melt fraction and τy is the yield stress. The flow thickness may be expressed as

$$Z \sim (\tau_y Q^{1/2}/\rho_{cr}g)^{2/5}t^{1/5}. \qquad (17)$$

The criterion that the flow area equals the planetary surface area $S^2 \sim 2 R_p{}^2$ gives the spreading time.

For the driving force–yield stress balance to dominate throughout spreading, $\eta_m U/Z \gg \tau_y$ with characteristic spreading velocity $U \sim R_p/t_{sprd}$. The criterion for plastic spreading gives a critical volume flux of

$$Q_{cr} \gg 4\pi^{2/5}R_p^2\tau_y^2/(\rho_{cv}g\eta_m). \qquad (18)$$

For nominal parameters (Table 1), $Q_{cr} \sim 6 \times 10^6$ m³/s, corresponding to $R_{cr} \sim 1700$ km. The spreading time at this transition $t_{sprd} \sim 10^3$ yr, implying that radial relaxation limits isostatic adjustment.

Below the transition local magma ocean size, R_{cr}, the volume flux drops below the critical one for basal yielding, $Q < Q_{cr}$, and spreading proceeds in a viscous regime. Another implication is that the lateral spreading time becomes larger than the radial relaxation time, implying that a constant volume flow condition is more appropriate. Surface cooling results in large rheological variations within the flow. For sufficiently great crustal viscosity, flow can be controlled by crustal deformation at the flow periphery (Griffiths, 2000). However, for the large-scale flows considered here, the crust yields and dynamics are controlled by deformation of the bulk. Thus, we consider a rheologically uniform flow and bulk cooling as the flow-limiting

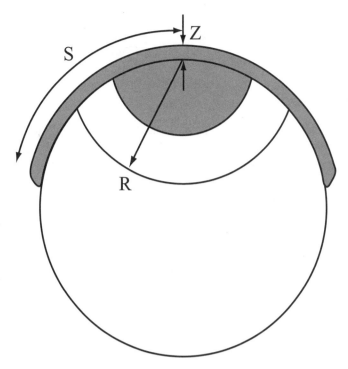

Figure 4. Isostatic adjustment subsequent to initial crystallization. The initial distribution of the buoyant partial melt region of size R is shown with a line. Partially molten material is shaded gray. The timescale for radial relaxation is estimated from a simple Stokes flow model. As buoyant material rises up, it flows out over the surface. The lateral spreading is estimated from an axisymmetric gravity current model. The height and radius of the lateral flow are Z and S, respectively. In the model, it is assumed that all material is extruded and flows over a rigid, undeformable surface. In reality, some material may flow beneath the lithosphere. Additionally, a deformable substratum would limit flow (coefficients in lateral spreading equations would change).

process. For an axisymmetric, viscous, constant-volume, divergent gravity current, the radius (Huppert, 1982) may be expressed as

$$S \sim \pi^{-3/8}\left(\frac{\rho_{cv}gV^3}{\eta_m}\right)^{1/8} t^{1/8}, \qquad (19)$$

and the thickness as

$$Z \sim (\eta_m V/(\rho_{cr}g))^{1/4}t^{-1/4}. \qquad (20)$$

The spreading timescale in this regime increases from $t_{sprd} \sim 10^5$ yr for $R \sim R_{cr}$ to 10^7 yr at $R \sim$1000 km.

The bulk viscosity of the flow increases with time due to cooling, which slows and eventually stops the flow (Sakimoto and Zuber, 1995; Bercovici and Lin, 1996). Spreading should essentially stop after approximately one diffusion time, for cooling from above and below this time satisfies $Z/2 \sim 2(\kappa t_{cool})^{1/2}$.

For $t_{sprd} \gg t_{cool}$, the partial melt cools and spreading stops before a global layer forms. For $R \sim 1000$ km, $t_{sprd} \sim 10^7$ yr, while the cooling time is an order of magnitude smaller, $t_{cool} \sim 10^6$ yr.

Hypothesis for the Origin of the Hemispherical Dichotomy and Tharsis

A hypothesis for dichotomy and Tharsis formation in the context of local magma ocean evolution has recently been suggested (Reese and Solomatov, 2006). In this scenario, an impact-induced local magma ocean rapidly crystallizes and isostatic adjustment produces a global layer of partially molten silicates with a degree-1 asymmetry (Fig. 5). A second, smaller impact occurs after isostatic adjustment to the first partially molten magma ocean is complete. This local magma ocean crystallizes, and its size is such that the lateral spreading component of isostatic adjustment proceeds in a viscous regime. In this case, bulk cooling results in cessation of spreading and formation of a large axisymmetric dome of material (Fig. 5). Both the dichotomy and Tharsis are the result of similar processes of local magma ocean evolution; the only difference is the size of the initial impact-induced melt regions.

In the context of this hypothesis, the earliest formed martian crust is simply the solidified partial melt layer produced by the earlier and larger of the two impacts. A mean crustal thickness of 50 km, corresponding to 4.4% of the planetary volume (Zuber, 2001), requires a local magma ocean size of $R \sim 1700$ km (equation 5). This value of R coincides with R_{cr}, in which case lateral spreading occurs in a yield stress–controlled regime and takes $\sim 10^4$ yr. The layer thickness decreases from $Z \sim (4\,\tau_y R_p / (\rho_{cr}\,g))^{1/2} \sim 100$ km on axis to near zero at the antipode (Fig. 5). Of course, while this layer continues to cool and solidify it can also relax viscously. Long-term preservation of the thickness variation depends on the mean thickness and viscosity structure (Zuber et al., 2000; Nimmo and Stevenson, 2001).

The formation of Tharsis could be due to the evolution of a smaller impact-induced melt region. At the low end of the R range, the retained, complete melt volume is $\sim 5 \times 10^{17}$ m³. This is of the order of the volume estimated for magmatic material comprising Tharsis (Phillips et al., 2001). The distribution of this material after isostatic adjustment can be estimated from the viscous lateral spreading model (equation 19). For a cooling time of $\sim 10^6$ yr, the radius $S \sim 3000$ km, corresponding to an area of $\sim 25\%$ of the planetary surface area. The central thickness at the cessation of spreading is $Z \sim (\eta_m\,V / (\rho_{cr}\,g))^{1/4}\,t_{cool}^{-1/4} \sim 20$ km. The crustal thickness variability within Tharsis (Zuber et al., 2000; Zuber, 2001) may be related to regionally enhanced melting.

NUMERICAL SIMULATIONS OF THE INTERACTION BETWEEN LOCAL MAGMA OCEAN EVOLUTION AND MANTLE DYNAMICS

Although a fully coupled simulation of magma ocean processes and mantle convection is beyond the scope of this study,

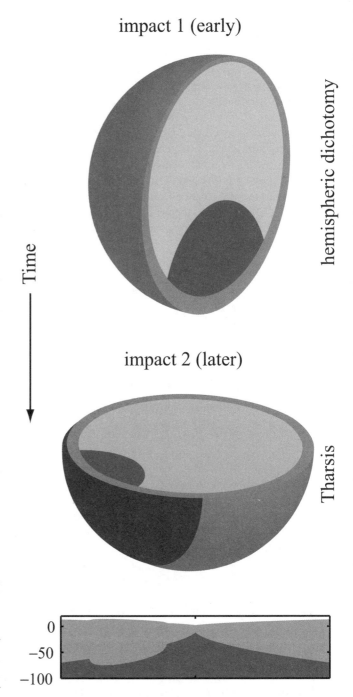

Figure 5. (Top) The hypothesis for hemispherical crustal dichotomy and Tharsis formation. The dichotomy is associated with an early Noachian local magma ocean that quickly ($\sim 10^4$–10^5 yrs) evolves to a partially molten layer with a degree-1 asymmetry. During the middle–late Noachian, a smaller local magma ocean is formed that evolves to an axisymmetric dome associated with the Tharsis rise. (Bottom) Crustal structure after isostatic adjustment can be estimated from the lateral spreading model. The figures shows a circum-Mars profile of crustal thickness through Tharsis. The vertical scale is in kilometers. Light and dark gray represent crust and mantle, respectively. The variable crustal structure within Tharsis may be due to the merger of late-stage magma ocean evolution with mantle dynamics, which can lead to subsequent localized melting (cf. Figs. 6–8).

some important features of the problem can be addressed. As cooling and crystallization proceed, the melt fraction and driving density difference for isostatic adjustment decrease, the viscosity increases, and the magma ocean processes discussed in the previous section slow down. At some point the dynamics merge with longer-timescale convective processes in the mantle.

Initial Temperatures

The initial mantle temperature is not well constrained. While early core formation implies high temperatures (Stevenson, 2001), no thermal model of Mars formation produces high temperatures in the deep interior (Senshu et al., 2002). Here an initial mantle temperature of $T_{i,0} = 1700$ K is assumed. The surface temperature $T_s = 220$ K. The initial cold boundary layer thickness is controlled by the accretion timescale $\delta_0 \sim 2\,(\kappa\,t_{acc})^{1/2}$. Due to the viscosity gradient limitations of the numerical method, a thicker boundary layer must be considered, $\delta_0 \sim 300$ km. The mantle is near the solidus at the base of the cold boundary layer.

The initial temperature difference between the core and mantle, $\Delta T_{cmb,0} = T_{cmb,0} - T_{i,0}$, with initial core-mantle boundary temperature $T_{cmb,0}$, depends on partitioning of the gravitational energy released during core formation (Flasar and Birch, 1973). The existence of an early dynamo (Stevenson, 2001) provides constraints. In the absence of surface recycling, $\Delta T_{cmb,0} \sim$ 150–200 K is sufficient to sustain a transient dynamo (Williams and Nimmo, 2004). For the purposes of this numerical model, thermal evolution of the core is not addressed, i.e., T_{cmb} is treated as an adjustable parameter that is constant in time.

Viscosity

An exponential temperature-dependent viscosity law is used,

$$\eta = b \exp(-\gamma T), \qquad (21)$$

where b and γ are constant and chosen to fix the initial interior viscosity $\eta_{i,0} = \eta(T_{i,0})$ and the initial viscosity contrast between the surface and mantle $\Delta\eta_0 = \exp(\gamma(T_{i,0} - T_s)) = 10^2$. A rigid upper-surface boundary condition results in a relatively immobile lithosphere.

Internal Heating

Geochemical analyses suggest early separation of a radiogenic isotope-enriched crust (McLennan, 2001). The internal heating rate may be expressed as

$$H(t) = C_d H_0 \exp(-\lambda t), \qquad (22)$$

with depletion factor C_d, initial specific heat production H_0, and decay constant λ (Table 1).

Melting

The mantle solidus is parameterized according to experimental data for peridotite (Scarfe and Takahashi, 1986; Ito and Takahashi, 1987; McKenzie and Bickle, 1988; Zhang and Herzberg, 1994),

$$T_m = 1374 + 130p - 5.6p^2, \qquad (23)$$

where T_m and p are in Kelvin and GPa, respectively. After the slope reaches $dT_m / dp = 10$ K/GPa, T_m increases linearly with pressure. Only dry melting at pressures below ~ 8 GPa is considered. The melt fraction may be expressed as

$$f = \frac{c_p}{L}(T - T_m). \qquad (24)$$

The model parameters (Table 1) imply a melting rate of $\sim 0.2\%$/K above solidus. All melt is assumed to be immediately extracted from the mantle, resulting in crustal growth through intrusive and extrusive magmatism.

Crustal Growth and Spreading

Crustal thickness variations generate lateral pressure gradients that drive crustal deformation, resulting in relaxation of topographic relief. Crustal spreading can be addressed using a viscous (Nimmo and Stevenson, 2001) or viscoelastic (Zuber et al., 2000) rheology. The spatially localized and high melting rates for the models considered here generate very large lateral pressure gradients, in excess of the cohesive strength of rocks. This motivates an alternative crustal spreading model based on a yield strength criterion.

When the product of the topographic relief and slope, $\zeta = z(\theta,\varphi)|\nabla z(\theta,\varphi)|$, reaches some critical value, ζ_{cr}, material is rapidly redistributed. Here θ and φ are the spherical coordinate polar and azimuthal angles, respectively. This idea is implemented by requiring z to satisfy a nonlinear diffusion equation with a transport coefficient that increases rapidly as $\zeta \to \zeta_{cr}$. During crustal growth, the conditions of isostacy and $\zeta \le \zeta_{cr}$ are maintained. The crustal structure predicted by this model would be subject to further viscous or viscoelastic relaxation (Zuber et al., 2000; Nimmo and Stevenson, 2001).

Partially Molten Local Magma Oceans

Local magma ocean conditions can be mimicked by including a partially molten, buoyant region in the simulation. This condition is approximated by choosing a suitably large magma ocean temperature, T_{ocean}. In the numerical model, initial magma ocean buoyancy is due solely to thermal expansion. In contrast, the driving buoyancy for isostatic relaxation in the scaling analysis is due to the melt-solid density difference. Production

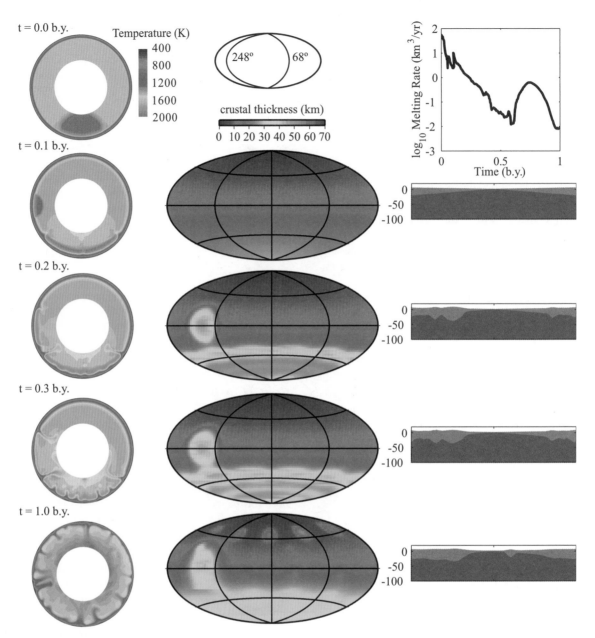

Figure 6. Model A Noachian evolution (Table 3). (Left) Evolution of the temperature field for a slice along longitude lines of 68°E to 248°E. t—time. (Center) Global map of the crustal thickness evolution. (Right, top) The total melting rate as a function of time. (Right, bottom) Circum-Mars profiles of crustal thickness along longitudes 68°E to 248°E. Light and dark gray represent crust and mantle, respectively.

of melt associated with isostatic adjustment of the partial melt region is governed by equations (23) and (24).

With this approach, the extrusion volume V can be calibrated with the scaling analysis result (equation 5) by adjusting T_{ocean}. For the cases considered here, $T_{ocean}(p) = T_m(p)$ and the driving density difference $\Delta\rho \sim \rho\alpha(T_m - T_{i,0}) \sim 30$ kg/m³. This is about an order of magnitude smaller than $\Delta\rho$ due to the partial melt–solid density difference, and thus the isostatic adjustment time increases by an order of magnitude.

Two local magma oceans are included in the numerical models. The larger and earlier magma ocean has an axis of symmetry about the south pole, size $R \sim 1700$ km and time of formation $t = 0.0$ b.y. (Figs. 6–8). The axis of symmetry of the second, smaller magma ocean passes through 0°N, 248°E, with the size $R \sim 1000$ km and the time of formation $t = 0.1$ b.y. (Figs. 6–8).

To model the initial rapid redistribution of partial melt associated with the larger magma ocean, all melt produced during

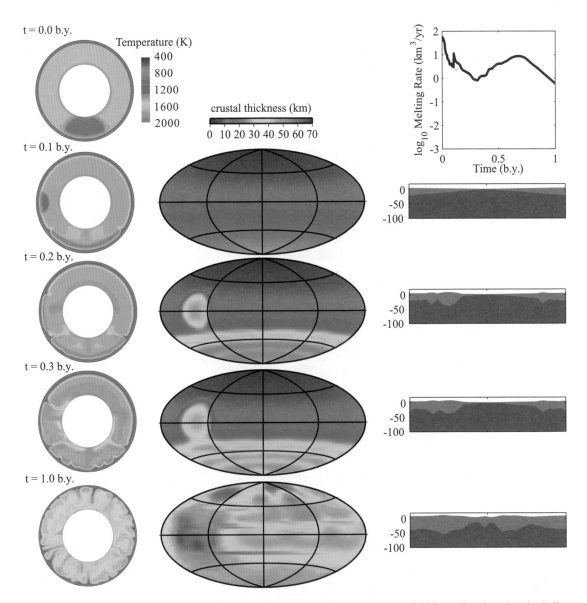

Figure 7. Model B Noachian evolution (Table 3). (Left) Evolution of the temperature field for a slice along longitude lines of 68°E to 248°E. t—time. (Center) Global map of the crustal thickness evolution. (Right, top) The total melting rate as a function of time. (Right, bottom) Circum-Mars profiles of crustal thickness along longitudes 68°E to 248°E. Light and dark gray represent crust and mantle, respectively.

the first 0.1 b.y. is redistributed axisymmetrically about the planet according to $z(\theta,\varphi) = V/(2\pi R_p^2)\, \theta/\pi$, with θ in radians. All subsequent melt is superposed on this initial distribution and subject to the crustal spreading model.

Convection Equations and Numerical Method

In the large Prandtl number limit, which neglects inertial terms in the momentum balance, and the Boussinesq approximation, where density variations are neglected everywhere except in the buoyancy term, the equations of convection expressing conservation of mass, momentum, and energy are

$$\frac{\partial u_i}{\partial x_i} = 0, \tag{25}$$

$$\frac{\partial \tau_{ij}}{\partial x_j} = \frac{-\partial p}{\partial x_i} + \alpha \rho g_i, \tag{26}$$

$$\rho c_p \left(\frac{\partial T}{\partial t} + u_i \frac{\partial T}{\partial x_i} \right) = k \frac{\partial^2 T}{\partial x_j \partial x_j} + \rho H - \rho L \frac{\partial f}{\partial t}, \tag{27}$$

with spatial coordinates x_i, time t, velocity u_i, dynamic pressure p, coefficient of thermal expansion α, gravitational acceleration g_i, temperature T, specific heat at constant pressure c_p, thermal

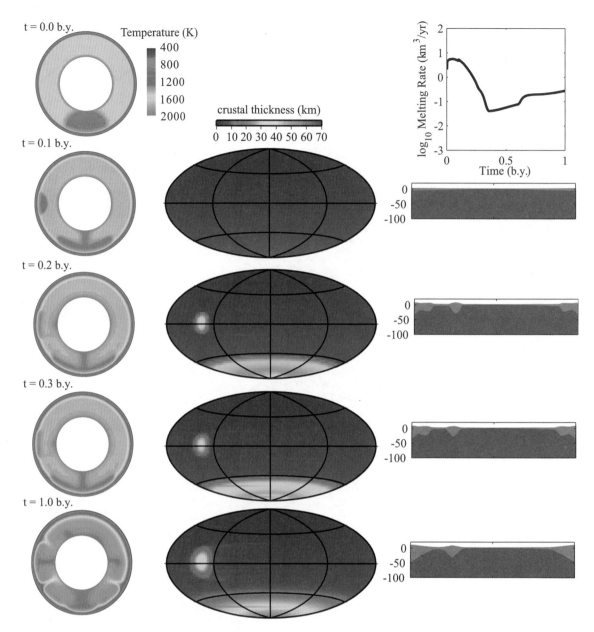

Figure 8. Model C Noachian evolution (Table 3). (Left) Evolution of the temperature field for a slice along longitude lines of 68°E to 248°E. t—time. (Center) Global map of the crustal thickness evolution. (Right, top) The total melting rate as a function of time. (Right, bottom) Circum-Mars profiles of crustal thickness along longitudes 68°E to 248°E. Light and dark gray represent crust and mantle, respectively.

conductivity k, specific heat production rate H, latent heat of fusion L, and melt fraction f. The deviatoric stress tensor may be expressed as

$$\tau_{ij} = \eta\left(\frac{\partial u_i}{\partial x_j} + \frac{\partial u_j}{\partial x_i}\right). \tag{28}$$

The fully 3-D spherical shell geometry code TERRA is used to study the effect of local magma oceans on martian mantle con-

vection and magmatic evolution (Baumgardner, 1985; Bunge and Baumgardner, 1995; Reese et al., 2004).

NUMERICAL SIMULATION RESULTS

In this section, numerical simulation results are summarized. Particular attention is focused on early magmatic evolution (Table 2), crustal structure (Figs. 6–8), and conditions for which melting continues at lower levels to the present day (Fig. 9).

TABLE 2. MARTIAN EPOCHS AND AGES	
Epoch	Age (Ga)
Noachian	4.5–3.5
Hepserian	3.5–1.8
Amazonian	1.8–0.0

TABLE 3. ADJUSTABLE MODEL PARAMETERS

Parameter	Model A	Model B	Model C
T_{cmb} (K)	§	1800	2000
C_d	0	0.3	0.3
$\eta_{i,0}$ (Pa s)	10^{21}	10^{21}	10^{22}

Note: §—insulating bottom boundary condition

Three models are considered in which the core-mantle boundary temperature, depletion factor, and initial interior viscosity were varied (Table 3).

Noachian Evolution

The thermal and magmatic evolution for the first 1 b.y. of martian evolution for these cases are summarized in Figures 6–8. To isolate the effect of the local magma oceans, a model with no internal or bottom heating was considered (Model A, cf. Reese et al., 2004). For this case, the initial magma ocean results in a crustal thickness that decreases from ~30 km to 0 km from the south to the north pole. Small-scale lithospheric instabilities are developed in the southern hemisphere by ~0.3 b.y., and no signature of the initial upwellings associated with the local magma oceans remains after this time. The melting rate decreases by four orders of magnitude in ~1 b.y. with a pulse at ~0.7 corresponding to onset of small-scale convection in the northern hemisphere. The thickest crust is in Tharsis province. In the absence of any source of internal heat, mantle cooling results in unrealistically low interior temperatures (a midshell temperature of ~1400 K at 1 b.y.).

Model B has modest internal and bottom heating (Table 2) and the same initial interior viscosity as Model A. The initial

crustal structure associated with the first magma ocean is primarily determined by T_{ocean} and $\eta_{i,0}$ and thus is approximately the same as that of Model A. Vigorous plumes and lithospheric instabilities result in efficient thermal homogenization that removes any signature of the initial upwellings. While the maximum crustal thickness of ~70 km again occurs in Tharsis province (Fig. 7), there is also significant melting in the northern hemisphere. The melting rate decreases by two orders of magnitude over the Noachian epoch.

Model C has stronger bottom heating and an initially more viscous mantle (Table 2). The initial crustal thickness is much less due to the fact that the extrusion rate ~t_{iso}^{-1} ~ $\eta_{i,0}$. Subsequent magmatism and crustal growth is confined to the southern hemisphere and the Tharsis region. In contrast to what is seen in Models A and B, the local magma oceans produce mantle plumes that remain stable throughout the Noachian. In fact, these plumes remain stable throughout planetary evolution. The thickest crust occurs at the south pole. The melting rate decreases from the initial value but is actually increasing at the end of the Noachian (Fig. 8).

Hesperian-Amazonian Evolution

The magmatic evolution of the models after 1 b.y. depends strongly on the value of T_{cmb} (Table 2, Fig. 9) and the interior viscosity. For Model A, the magmatic rate is zero after ~1 b.y. In the case of Model B, the magmatic rate decays gradually to zero by ~3 b.y., followed by episodes of melting, with rates peaking at ~0.003 km³/yr and continuing until the present day. Model C exhibits a pulse in the melt production rate just after 1 b.y. associated with plume head melting and a relatively constant rate of ~0.1 km³/yr from ~2 b.y. to the present day.

The existence of young volcanism on Mars (Hartmann et al., 1999) must be addressed by any mantle convection model (see, e.g., Kiefer, 2003). Constraints on the present-day melt production rate are provided by cratering studies. An estimate for the resurfacing rate during the middle to late Amazonian is ~0.01 km²/yr (Hartmann and Neukum, 2001). Assuming that the resurfacing of small craters on young surface flows requires a depth of ~10 m and an extrusive to intrusive efficiency of ~10% implies that a lower bound on the present-day melting rate is 10^3 km³/yr. Model B just satisfies this requirement, while Model C produces significantly more melt (Fig. 9).

The present-day model topography can be compared to Mars Global Surveyor topography (Fig. 10). While no single model

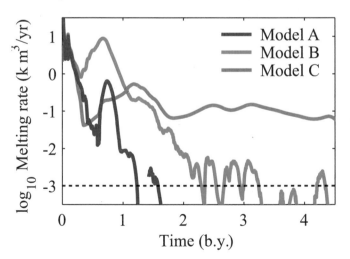

Figure 9. The melt volume production rate as a function of time for Models A, B, and C. For the middle to late Amazonian epoch, Model B exhibits transient episodes of melting while Model C produces an approximately constant rate of 0.1 km³/yr after ~2 b.y. The melting rate required to bury 10 m topography at the observed resurfacing rate (Hartmann and Neukum, 2001) is shown with a dashed line.

produces a significant quantitative correlation with observation, individual features of the models are intriguing. The long-wavelength patterns of topography for Models A and B agree qualitatively with observation. The western hemisphere topography of Model B reproduces the approximate size, location, and scale of Tharsis province. In both Model A and Model B, superimposed on the north-south topographic slope are small-scale features generated by melting associated with lithospheric instabilities. Model C produces upwelling plumes from the core-mantle boundary that remain stable throughout planetary evolution. There are four strong plumes, one beneath the southern hemisphere and three in the Tharsis region. The topography is the surface expression of local magma ocean evolution and continued melting in the plume heads. One interesting possibility is that the Elysium volcanic province corresponds to a stable plume, as in Model C, or to melting induced by lithospheric instability, as in Models A and B.

DISCUSSION AND CONCLUSIONS

The formation of local magma oceans is one process that shaped the initial thermal and compositional state of Mars. The effects of local magma ocean evolution on crust formation, core formation, and development of geochemical reservoirs in the mantle can provide a fundamental framework for understanding the early evolution of the planet. In this study, the implications of local magma oceans for the rapid development of large-scale crustal structure on early Mars were addressed. In particular, a hypothesis for the formation of the hemispherical crustal dichotomy and Tharsis based on local magma ocean evolution (Reese and Solomatov, 2006) was further explored using 3-D spherical shell simulations.

The models considered produce crustal volumes that are of the same order of magnitude as those based on gravity and topography data that predict a lower bound on the mean crustal thickness of ~50 km (Zuber et al., 2000; Zuber, 2001; Neumann et al., 2004). The average crustal thicknesses for Models A, B, and C are ~20, 40, and 10 km, respectively. Crustal structure varies between models. In general, all models exhibit a south-north thinning trend and thickening in the Tharsis region. Models A and B predict small-scale variations attributed to melting associated with lithospheric instabilities. Model C produces plume head melting beneath the southern hemisphere and for three plumes in the Tharsis region. The present-day melting rate is another observational constraint. Melting ceases in Model A at ~1 b.y., episodes of melting with peak rates $\sim 3 \times 10^3$ km^3/yr continue to the present day in Model B, and Model C has an approximately constant melting rate of 0.1 km^3/yr after ~2 b.y. To resurface 10 m topography, the present-day melting rate must be ~$10^{?3}$ km^3/yr.

This preliminary study has several topics that can be addressed in future work:

1. Young volcanism on Mars seems to require some degree of internal and/or bottom heating (e.g., Models B and C and

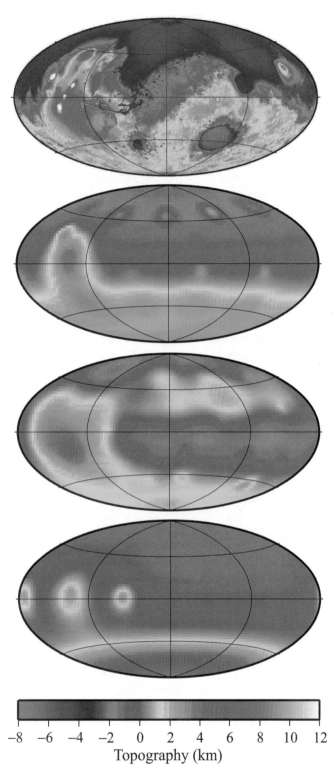

Figure 10. (Top diagram) Mars Global Surveyor topography. (Bottom three diagrams) present-day topography for Models A, B, and C.

Topography (km)

Kiefer, 2003). In this study, T_{cmb} was fixed, but in reality it depends on the coupled thermal evolution of the mantle and the core. The existence of a transient dynamo (Stevenson, 2001) is a strong constraint on the early core heat flux (Nimmo and Stevenson, 2000; Williams and Nimmo, 2004). Thermal evolution of the core can be included in the models to address dynamo constraints.

2. The stability of a plume or plumes beneath Tharsis throughout martian evolution remains an outstanding question. While Model C produces three stable plumes in the western hemisphere, it also predicts a large plume beneath the southern hemisphere. This disagrees with the observation of middle–late Noachian volcanism focused at Tharsis and of geologically recent volcanism limited to Tharsis and Elysium. The long duration of the Tharsis plume or plumes relative to the initial upwelling beneath the southern hemisphere must be addressed by future work.

3. No crust-mantle feedback was considered in the models. Enrichment of the crust in heat-producing elements can lead to higher Moho temperatures in the southern hemisphere, where the crust is thick. This asymmetry in the mantle surface temperature can have important implications for mantle dynamics (Lenardic and Moresi, 2003; Wenzel et al., 2004).

4. Preservation of early-formed geochemical reservoirs for 4.5 b.y. (e.g., Kleine et al., 2004) is difficult to reconcile with vigorous mantle mixing. The characteristic interior velocities for Models B and C correspond to approximately ten and three mantle overturns, respectively. The mixing efficiency also depends on the planform of convection, which in Model C remains relatively constant throughout evolution. Preservation of mantle compositional heterogeneity can be addressed in future work.

5. In this study, the sizes and locations of large impacts were specified. A preferable approach would be to generate impactor sizes from an appropriate distribution and specify impact locations randomly. One possible result is a global, laterally heterogeneous magma ocean produced by overlapping local magma oceans. Another possibility is that only local magma oceans are generated due to the small number of sufficiently large impacts during late-stage accretion.

REFERENCES CITED

Agnor, C.B., Canup, R.M., and Levison, H.F., 1999, On the character and consequences of large impacts in the late stage of terrestrial planet formation: Icarus, v. 142, p. 219–237, doi: 10.1006/icar.1999.6201.

Anderson, R.C., Dohm, J.M., Golombek, M.P., Haldemann, A.F.C., Franklin, B.J., Tanaka, K.L., Lias, J., and Peer, B., 2001, Primary centers and secondary concentrations of tectonic activity through time in the western hemisphere of Mars: Journal of Geophysical Research, v. 106, p. 20,563–20,586, doi: 10.1029/2000JE001278.

Andrade, E.N.C., 1952, Viscosity of liquids: Proceedings of the Royal Society, London, v. A215, p. 36–43.

Banerdt, W.B., and Golombek, M.P., 2000, Tectonics of the Tharsis region of Mars: Insights from MGS topography and gravity: Lunar Planetary Science Conference, v. 31, no. 2038.

Baumgardner, J.R., 1985, Three dimensional treatment of convective flow in the Earth's mantle: Journal of Statistical Physics, v. 39, p. 501–511, doi: 10.1007/BF01008348.

Bercovici, D., and Lin, J., 1996, A gravity current model of cooling mantle plume heads with temperature dependent buoyancy and viscosity: Journal of Geophysical Research, v. 101, p. 3291–3309, doi: 10.1029/95JB03538.

Bird, R.B., Stewart, W.E., and Lightfoot, E.N., 1960, Transport phenomena: New York, Wiley.

Blake, S., 1990, Viscoplastic models of lava domes, *in* Fink, J.H., ed., Lava flows and domes: Emplacement mechanisms and hazard implications: New York, Springer, p. 88–128.

Bottinga, Y., and Weill, D.F., 1972, The viscosity of magmatic silicate liquids: A model for calculation: American Journal of Science, v. 272, p. 438–475.

Bottinga, Y., Richet, P., and Sipp, A., 1995, Viscosity regimes of homogeneous silicate melts: American Mineralogist, v. 80, p. 305–318.

Breuer, D., Yuen, D.A., Spohn, T., and Zhang, S., 1998, Three dimensional models of Martian mantle convection with phase transitions: Geophysical Research Letters, v. 25, p. 229–232, doi: 10.1029/97GL03767.

Bunge, H.-P., and Baumgardner, J.R., 1995, Mantle convection modeling on parallel virtual machines: Computers in Physics, v. 9, p. 207–215, doi: 10.1063/1.168525.

Chambers, J.E., 2004, Planetary accretion in the inner solar system: Earth and Planetary Science Letters, v. 223, p. 241–252, doi: 10.1016/j.epsl.2004.04.031.

Chambers, J.E., and Wetherill, G.W., 1998, Making the terrestrial planets: N-body integrations of planetary embryos in three dimensions: Icarus, v. 136, p. 304–327, doi: 10.1006/icar.1998.6007.

Davaille, A., and Jaupart, C., 1993, Thermal convection in lava lakes: Geophysical Research Letters, v. 20, p. 1827–1830.

Davies, G.F., 1982, Ultimate strength of solids and formation of planetary cores: Geophysical Research Letters, v. 9, p. 1267–1270.

Elkins-Tanton, L.T., Zaranek, S.E., Parmentier, E.M., and Hess, P.C., 2005a, Early magnetic field and crust on Mars from magma ocean cumulate overturn: Earth and Planetary Science Letters, v. 236, p. 1–12, doi: 10.1016/j.epsl.2005.04.044.

Elkins-Tanton, L.T., Hess, P.C., and Parmentier, E.M., 2005b, Possible formation of ancient crust on Mars through magma ocean processes: Journal of Geophysical Research, v. 110, E12S01, doi: 10.1029/2005JE002480.

Ergun, S., 1952, Fluid flow through packed columns: Chemical Engineering Progress, v. 48, p. 89–94.

Flasar, F.M., and Birch, F., 1973, Energetics of core formation: A correction: Journal of Geophysical Research, v. 78, p. 6101–6103.

Frey, H., and Schultz, R.A., 1988, Large impact basins and the mega-impact origin for the crustal dichotomy on Mars: Geophysical Research Letters, v. 15, p. 229–232.

Frey, H., Sakimoto, S.E., and Roark, J., 1998, The MOLA topographic signature at the crustal dichotomy boundary zone on Mars: Geophysical Research Letters, v. 25, p. 4409–4412, doi: 10.1029/1998GL900095.

Frey, H.V., Roark, J.H., Shockey, K.M., Frey, E.L., and Sakimoto, S.E.H., 2002, Ancient lowlands on Mars: Geophysical Research Letters, v. 29, doi: 10.1029/2001GL013832.

Gans, R.F., 1972, Viscosity of the Earth's core: Journal of Geophysical Research, v. 77, p. 360–366.

Griffiths, R.W., 2000, The dynamics of lava flows: Annual Review of Fluid Mechanics, v. 32, p. 477–518, doi: 10.1146/annurev.fluid.32.1.477.

Harder, H., 2000, Mantle convection and the dynamic geoid of Mars: Geophysical Research Letters, v. 27, p. 301–304, doi: 10.1029/1999GL008418.

Harder, H., and Christensen, U.R., 1996, A one-plume model of Martian mantle convection: Nature, v. 380, p. 507–509, doi: 10.1038/380507a0.

Harper, C.L., Nyquist, L.E., Bansal, B., Wiesmann, H., and Shih, C.Y., 1995, Rapid accretion of early differentiation of Mars indicated by Nd/Nd in SNC meteorites: Science, v. 267, p. 213–217, doi: 10.1126/science.7809625.

Hartmann, W.K., and Neukum, G., 2001, Cratering chronology and the evolution of Mars: Space Science Reviews, v. 96, p. 165–194, doi: 10.1023/A:1011945222010.

Hartmann, W.K., Malin, M., McEwen, A., Carr, M., Soderblom, L., Thomas, P., Danielson, E., James, P., and Ververka, J., 1999, Evidence for recent volcanism on Mars from crater counts: Nature, v. 397, p. 586–589, doi: 10.1038/17545.

Hess, P.C., and Parmentier, E.M., 2001, Implications of magma ocean cumulate overturn for Mars: Lunar Planetary Science Conference, v. 32, no. 1319.

Hirth, G., and Kohlstedt, D.L., 1995a, Experimental constraints on the dynamics of the partially molten upper mantle: Deformation in the diffusion creep regime: Journal of Geophysical Research, v. 100, p. 1981–2001, doi: 10.1029/94JB02128.

Hirth, G., and Kohlstedt, D.L., 1995b, Experimental constraints on the dynamics of the partially molten upper mantle: Deformation in the dislocation creep regime: Journal of Geophysical Research, v. 100, p. 15,441–15,449, doi: 10.1029/95JB01292.

Hirth, G., and Kohlstedt, D.L., 2003, Rheology of the mantle wedge, in, Eiler, J., ed., Inside the subduction factory: Washington, D.C., American Geophysical Union, p. 83–106.

Huppert, H.E., 1982, The propagation of two-dimensional and axisymmetric viscous gravity currents over a rigid horizontal surface: Journal of Fluid Mechanics, v. 121, p. 43–58, doi: 10.1017/S0022112082001797.

Huppert, H.E., and Sparks, R.S.J., 1985, Komatiites 1, Eruption and flow: Journal of Petrology, v. 26, p. 694–725.

Ito, E., and Takahashi, E., 1987, Melting of peridotite at uppermost lower mantle conditions: Nature, v. 328, p. 514–517, doi: 10.1038/328514a0.

Johnson, C.L., and Phillips, R.J., 2005, Evolution of the Tharsis region of Mars: Insights from magnetic field observations: Earth and Planetary Science Letters, v. 230, p. 241–254, doi: 10.1016/j.epsl.2004.10.038.

Karato, S.-I., and Wu, P., 1993, Rheology of the upper mantle: A synthesis: Science, v. 260, p. 771–778, doi: 10.1126/science.260.5109.771.

Kasting, J.F., 1988, Runaway and moist greenhouse atmospheres and the evolution of Earth and Venus: Icarus, v. 74, p. 472–494, doi: 10.1016/0019-1035(88)90116-9.

Kiefer, W.S., 2003, Melting in the martian mantle: Shergottite formation and implications for present-day mantle convection on Mars: Meteorites and Planetary Science, v. 39, p. 1815–1832.

Kieffer, S.W., 1977, Impact conditions required for formation of melt by jetting in silicates, in Roddy, D.J., et al., eds., Impact and explosion cratering: New York, Pergamon, p. 751–769.

King, S.D., and Redmond, H.L., 2005, The crustal dichotomy and edge driven convection: A mechanism for Tharsis rise volcanism?: Lunar Planetary Science Conference, v. 36, no. 1960 [abs.].

Kinsland, G.L., 1978, The effect of the strength of materials on the interpretation of data from opposed anvil high pressure devices: High Temperatures–High Pressures, v. 10, p. 627–639.

Kleine, T., Mezger, K., Münker, C., Palme, H., and Bischoff, A., 2004, Hf-W isotope systematics of chondrites, eucrites, and martian meteorites: Chronology of core formation and early mantle differentiation in Vesta and Mars: Geochimica et Cosmochimica Acta, v. 68, p. 2935–2946, doi: 10.1016/j.gca.2004.01.009.

Kohlstedt, D.L., and Zimmerman, M.E., 1996, Rheology of partially molten mantle rocks: Annual Review of Earth and Planetary Sciences, v. 24, p. 41–62, doi: 10.1146/annurev.earth.24.1.41.

Kokubo, E., and Ida, S., 1996, On runaway growth of planetesimals: Icarus, v. 123, p. 180–191.

Kokubo, E., and Ida, S., 1998, Oligarchic growth of protoplanets: Icarus, v. 131, p. 171–178.

Kokubo, E., and Ida, S., 2000, Formation of protoplanets from planetesimals in the solar nebula: Icarus, v. 143, p. 15–27.

Kraichnan, R.H., 1962, Turbulent thermal convection at arbitrary Prandtl number: Physics of Fluids, v. 5, p. 1374–1389, doi: 10.1063/1.1706533.

Kushiro, I., 1980, Viscosity, density, and structure of silicate melts at high pressures, and their petrological applications, in Hargraves, R.B., ed., Physics of magmatic processes: Princeton, New Jersey, Princeton University Press, p. 93–121.

Kushiro, I., 1986, Viscosity of partial melts in the upper mantle: Journal of Geophysical Research, v. 91, p. 9343–9350.

Lee, D.C., and Halliday, A.N., 1997, Core formation on Mars and differentiated asteroids: Nature, v. 388, p. 854–857, doi: 10.1038/42206.

Lemoine, F.G., Smith, D.E., Rowlands, D.D., Zuber, M.T., Neumann, G.A., Chinn, D.S., and Pavlis, D.E., 2001, An improved solution of the gravity field of Mars (GMM-2B) from Mars Global Surveyor: Journal of Geophysical Research, v. 106, p. 23,359–23,376, doi: 10.1029/2000JE001426.

Lenardic, A., and Moresi, L., 2003, Thermal convection below a conducting lid of variable extent: Heat flow scalings and two dimensional, infinite Prandtl number numerical simulations: Physics of Fluids, v. 15, p. 455–466, doi: 10.1063/1.1533755.

Matsui, T., and Abe, Y., 1986, Evolution of an impact induced atmosphere and magma ocean on the accreting Earth: Nature, v. 319, p. 303–305, doi: 10.1038/319303a0.

McEwen, A.S., Malin, M.C., Carr, M.H., and Hartmann, W.K., 1999, Voluminous volcanism on early Mars revealed in Valles Marineris: Nature, v. 397, p. 584–586, doi: 10.1038/17539.

McKenzie, D., and Bickle, M.J., 1988, The volume and composition of melt generated by extension of the lithosphere: Journal of Petrology, v. 29, p. 625–679.

McLennan, S.M., 2001, Crustal heat production and the thermal evolution of Mars: Geophysical Research Letters, v. 28, p. 4019–4022, doi: 10.1029/2001GL013743.

Melosh, H.J., 1989, Impact cratering: A geological process: New York, Oxford University Press.

Neukum, G., Jaumann, R., Hoffmann, H., Hauber, E., Head, J.W., Basilevsky, A.T., Ivanov, B.A., Werner, S.C., van Gasselt, S., Murray, J.B., McCord, T., and the HRSC co-investigator team, 2004, Recent and episodic volcanic and glacial activity on Mars revealed by the High Resolution Stereo Camera: Nature, v. 432, p. 971–979, doi: 10.1038/nature03231.

Neumann, G.A., Zuber, M.T., Wieczorek, M.T., McGovern, M.T., Lemoine, F.G., and Smith, F.G., 2004, Crustal structure of Mars from gravity and topography: Journal of Geophysical Research, v. 109, doi: 10.1029/2004JE002262.

Nimmo, F., and Stevenson, D., 2000, The influence of early plate tectonics on the thermal evolution and magnetic field of Mars: Journal of Geophysical Research, v. 105, p. 11,969–11,979, doi: 10.1029/1999JE001216.

Nimmo, F., and Stevenson, D.J., 2001, Estimates of Martian crustal thickness from viscous relaxation of topography: Journal of Geophysical Research, v. 106, p. 5085–5098, doi: 10.1029/2000JE001331.

Nimmo, F., and Tanaka, K., 2005, Early crustal evolution of Mars: Annual Review of Earth and Planetary Sciences, v. 33, p. 133–161, doi: 10.1146/annurev.earth.33.092203.122637.

Nye, J.F., 1951, Mechanics of glacier flow: Journal of Glaciology, v. 2, p. 82–93.

Persikov, E.S., Zharkov, V.A., Bukhtiyarov, P.G., and Polskoy, S.F., 1990, The effect of volatiles on the properties of magmatic melts: European Journal of Mineralogy, v. 2, p. 621–642.

Phillips, R.J., Zuber, M.T., Solomon, S.C., Golombek, M.P., Jakosky, B.M., Banerdt, W.B., Smith, D.E., Williams, R.M.E., Hynek, B.M., Aharonson, O., and. Hauck, S.A., II, 2001, Ancient geodynamics and global scale hydrology on Mars: Science, v. 291, p. 2587–2591, doi: 10.1126/science.1058701.

Pierazzo, E., and Melosh, H.J., 2000, Melt production in oblique impacts: Icarus, v. 145, p. 252–261, doi: 10.1006/icar.1999.6332.

Reese, C.C., and Solomatov, V.S., 2006, Fluid dynamics of local martian magma oceans: Icarus, v. 184, p. 102–120, doi: 10.1016/j.icarus.2006.04.008.

Reese, C.C., Solomatov, V.S., Baumgardner, J.R., Stegman, D.R., and Vezolainen, A.V., 2004, Magmatic evolution of impact-induced Martian mantle plumes and the origin of Tharsis: Journal of Geophysical Research, v. 109, doi: 10.1029/2003JE002222.

Sakimoto, S.E.H., and Zuber, M.T., 1995, The spreading of variable viscosity axisymmetric radial gravity currents: Applications to the emplacement of

Venus pancake domes: Journal of Fluid Mechanics, v. 301, p. 65–77, doi: 10.1017/S0022112095003806.

Scarfe, C.M., and Takahashi, E., 1986, Melting of garnet peridotite to 13 GPa and the early history of the upper mantle: Nature, v. 322, p. 354–356, doi: 10.1038/322354a0.

Schmidt, R.M., and Housen, K.R., 1987, Some recent advances in the scaling of impact and explosion cratering: International Journal of Impact Engineering, v. 5, p. 543–560, doi: 10.1016/0734-743X(87)90069-8.

Scott, T., and Kohlstedt, D.L., 2004a, The effect of large melt fraction on the deformation behavior of peridotite: Implications for the rheology of Io's mantle: Lunar Planetary Science Conference, v. 35, no. 1304.

Scott, T.J., and Kohlstedt, D.L., 2004b, The effect of large melt fraction on the deformation behavior of peridotite: Implications for the viscosity of Io's mantle and the rheologically critical melt fraction: Eos (Transactions, American Geophysical Union), v. 85, no. 4582.

Segura, T.L., Toon, O.B., Colaprete, A., and Zahnle, K., 2002, Environmental effects of large impacts on Mars: Science, v. 298, p. 1977–1980, doi: 10.1126/science.1073586.

Senshu, H., Kuramoto, K., and Matsui, T., 2002, Thermal evolution of a growing Mars: Journal of Geophysical Research, v. 107, doi: 10.1029/2001JE001819.

Shaw, H.R., 1972, Viscosities of magmatic silicate liquids: An empirical method of prediction: American Journal of Science, v. 272, p. 870–893.

Shraiman, B.I., and Siggia, E.D., 1990, Heat transport in high Rayleigh number convection: Physical Review A, v. 42, p. 3650–3653, doi: 10.1103/PhysRevA.42.3650.

Siggia, E.D., 1994, High Rayleigh number convection: Annual Review of Fluid Mechanics, v. 26, p. 137–168, doi: 10.1146/annurev.fl.26.010194.001033.

Sleep, N.H., 1994, Martian plate tectonics: Journal of Geophysical Research, v. 99, p. 5639–5655, doi: 10.1029/94JE00216.

Sleep, N.H., and Zahnle, K., 1998, Refugia from asteroid impacts on early Mars and early Earth: Journal of Geophysical Research, v. 103, p. 28,529–28,544, doi: 10.1029/98JE01809.

Smith, D.E., Zuber, M.T., Solomon, S.C., Phillips, R.J., Head, J.W., Garvin, J.B., Banerdt, W.B., Muhleman, D.O., Pettengill, G.H., Neumann, G.A., Lemoine, F.G., Abshire, J.B., Aharonson, O., Brown, C.D., Hauck, S.A., Ivanov, A.B., McGovern, P.J., Zwally, H.J., and Duxbury, T.C, 1999, The global topography of Mars and implications for surface evolution: Science, v. 284, p. 1495–1503, doi: 10.1126/science.284.5419.1495.

Smith, D.E., Zuber, M.T., Frey, H.V., Garvin, J.B., Head, J.W., Muhleman, D.O., Pettengill, G.H., Phillips, R.J., Solomon, S.C., Zwally, H.J., Banerdt, W.B., Duxbury, T.C., Golombek, M.P., Lemoine, F.G., Neumann, G.A., Rowlands, D.D., Aharonson, O., Ford, P.G., Ivanov, A.B., Johnson, C.L., McGovern, P.J., Abshire, J.B., Afzal, R.S., and Sun, X., 2001, Orbiter Laser Altimeter: Experiment summary after the first year of global mapping of Mars: Journal of Geophysical Research, v. 106, p. 23,689–23,722, doi: 10.1029/2000JE001364.

Solomatov, V.S., 2000, Fluid dynamics of magma oceans, *in* Canup, R., and Righter, K., eds., Origin of the Earth and Moon: Tucson, University of Arizona Press, p. 323–338.

Solomatov, V.S., and Moresi, L.-N., 2000, Time dependent stagnant lid convection on the Earth and other terrestrial planets: Journal of Geophysical Research, v. 105, p. 21,795–21,817.

Solomatov, V.S., and Stevenson, D.J., 1993, Nonfractional crystallization of a terrestrial magma ocean: Journal of Geophysical Research, v. 98, p. 5391–5406.

Solomon, S.C., and Head, J.W., 1982, Evolution of the Tharsis province of Mars: The importance of heterogeneous lithospheric thickness and volcanic construction: Journal of Geophysical Research, v. 82, p. 9755–9774.

Solomon, S.C., Aharonson, O., Aurnou, J.M., Banerdt, W.B., Carr, M.H., Dombard, A.J., Frey, H.V., Golombek, M.P., Hauck, S.A., II, Head, J.W., III, Jakosky, B.M., Johnson, C.L., McGovern, P.J., Neumann, G.A., Phillips, R.J., Smith, D.E., and Zuber, M.T., 2005, New perspectives on ancient Mars: Science, v. 307, p. 1214–1220, doi: 10.1126/science.1101812.

Spohn, T., Sohl, F., and Breuer, D., 1998, Mars: Astronomy and Astrophysics Review, v. 8, p. 181–235, doi: 10.1007/s001590050010.

Stevenson, D.J., 2001, Mars' core and magnetism: Nature, v. 412, p. 214–219, doi: 10.1038/35084155.

Tonks, W.B., and Melosh, H.J., 1992, Core formation by giant impacts: Icarus, v. 100, p. 326–346, doi: 10.1016/0019-1035(92)90104-F.

Tonks, W.B., and Melosh, H.J., 1993, Magma ocean formation due to giant impacts: Journal of Geophysical Research, v. 98, p. 5319–5333.

Tsenn, M.C., and Carter, N.L., 1987, Upper limits of power law creep of rocks: Tectonophysics, v. 136, p. 1–26.

Weinstein, S.A., 1995, The effects of a deep mantle endothermic phase change on the structure of thermal convection in silicate planets: Journal of Geophysical Research, v. 100, p. 11,719–11,728, doi: 10.1029/95JE00710.

Wenzel, M.J., Manga, M., and Jellinek, A.M., 2004, Tharsis as a consequence of Mars' dichotomy and layered mantle: Geophysical Research Letters, v. 31, doi: 10.1029/2003GL019306.

Wilhelms, D.E., and Squyres, S.W., 1984, The martian hemispheric dichotomy may be due to a giant impact: Nature, v. 309, p. 138–140, doi: 10.1038/309138a0.

Williams, J.-P., and Nimmo, F., 2004, Thermal evolution of the Martian core: Implications for an early dynamo: Geology, v. 32, p. 97–100, doi: 10.1130/G19975.1.

Zahnle, K.J., Kasting, J.F., and Pollack, J.B., 1988, Evolution of a steam atmosphere during Earth's accretion: Icarus, v. 74, p. 62–97, doi: 10.1016/0019-1035(88)90031-0.

Zhang, J.Z., and Herzberg, C., 1994, Melting experiments on anhydrous peridotite KLB-1 from 5.0 to 22.5 Gpa: Journal of Geophysical Research, v. 99, p. 17,729–17,742, doi: 10.1029/94JB01406.

Zhong, S., and Zuber, M.T., 2001, Degree-1 mantle convection and the crustal dichotomy on Mars: Earth and Planetary Science Letters, v. 189, p. 75–84, doi: 10.1016/S0012-821X(01)00345-4.

Zuber, M.T., 2001, The crust and mantle of Mars: Nature, v. 412, p. 220–227, doi: 10.1038/35084163.

Zuber, M.T., , Solomon, S.C., Phillips, R.J., Smith, D.E., Tyler, G.L., Aharonson, O., Balmino, G., Banerdt, W.B., Head, J.W., Johnson, C.L., Lemoine, F.G., McGovern, P.J., Neumann, G.A., Rowlands, D.D., and Zhong, S., 2000, Internal structure and early thermal evolution of Mars from Mars Global Surveyor topography and gravity: Science, v. 287, p. 1788–1793, doi: 10.1126/science.287.5459.1788.

MANUSCRIPT ACCEPTED BY THE SOCIETY JANUARY 31, 2007

DISCUSSION

27 December 2006, Joe McCall

This is a most interesting article, which requires a very deep knowledge of mathematics and physics (which I do not have) to fully comprehend. It illustrates the two totally contrasting methods of working, both acceptable. As a terrestrial geologist, I have always worked from the observed surface geological features, extending what I have observed to conjecture about the deeper origins. There is a contrasting methodology that involves modeling processes at mantle and even core-boundary levels, using seismology, geochemistry, and evidence from kimberlites. This is epitomized by "plumologists" such as my former student Ian Campbell at Australian National University, Canberra.

The work described by Reese et al. (this volume), comparable with the second approach but applied to Mars, really in no way impinges on my simple comparison between Olympus Mons and terrestrial caldera volcanoes (McCall, 2006a). Olympus Mons shows all the features of a shield volcano with a summit caldera, and a recent discussion with Prof. Lionel Wilson suggests to me that, as in many terrestrial shield volcanoes, the huge dome was erected by fissure eruptions; there may have been no summit crater before caldera formation. An important point I made is that there is an active analogue on Io. Whether impact was involved in the early mantle-level processes, which ultimately gave rise to this activity, must remain a matter of conjecture, and indeed, we may never know for certain. At this level of modeling, it seems possible that diverse models will be advanced, all credible. However, the fact that a model is credible does not mean that this is what actually happened.

4 January 2007, Lionel Wilson

Reese et al. (this volume) mention as issues requiring future work the temporal stability of plumes. There is some evidence on timescales stemming from the observation by my colleague Eve Scott (Wilson et al., 2001) that the presence of cross-cutting but interlocking calderas on the summits of most martian shield volcanoes implies that, in general, any shallow magma reservoir involved in such a caldera-forming event must have a finite life and must then cease to receive any significant magma supply from the mantle for long enough (which we calculate to mean tens of millions of years) for it to cool below the solidus. Only in this way is a new magma reservoir likely to form partly (or completely) offset from the previous one. But if a new shallow magma reservoir is to start growing, each newly emplaced near-surface early intrusion must be fed by extra magma faster than it cools. The details depend on the size and geometry of the first intrusion, but there must be an initial magma flux pulse estimated to be ~100 to 200 m^3 s^{-1} for at least a few weeks. This is followed by a period during which the minimum required flux decreases with time as the initial intrusion swells and becomes more resistant to cooling (because its cooling outer boundary layer only thickens in proportion to the square root of time).

There are uncertainties in timescales that arise from a range of options in how the geometry of the initial intrusion will develop, but simply to maintain fully formed reservoirs with the sizes implied by the visible martian calderas requires magma fluxes of ~1 to 10 m^3 s^{-1} (these values correspond to 0.03 to 0.3 km^3 a^{-1}, similar to their mean flux for the last 2 Ga in model C). Overall, the implications of our analysis were that, for each volcano, there has to be (1) a way of periodically generating an initial large pulse of magma and (2) a way of turning off the magma supply for long periods. To satisfy all the constraints, including mean volcano growth rates, averaged over the entire volcano lifetime, estimated to be within a factor of two to three of ~0.05 m^3 s^{-1}, the simplest solution is for there to be a potentially active phase lasting for ca. 1 Ma alternating with a quiet phase lasting for ca. 100 Ma, the Tharsis volcanoes each having experienced ten to twenty such cycles.

Considerable variation around these values is possible, but not by orders of magnitude. An explanation for this long-term episodicity is needed. I can imagine that if such a timescale were established, other consequences would follow: Perhaps the large initial magma pulse of a new active phase might be related to there being a threshold for the first extraction of melt from new partial melt zones.

6 January 2007, James H. Natland

The plume hypothesis is often justified by manifestations of large-scale volcanism. The usual refrain for Earth is, "What about Hawaii?" For Mars, it is, "What about Olympus Mons?" The Hawaiian Ridge culminates most actively at two large shield volcanoes, Mauna Loa and Kilauea. Olympus Mons is one of the several Tharsis shield volcanoes of Mars and is by far the largest shield volcano in the solar system. The rationale for plumes at these places is that both Hawaii and Olympus Mons are huge and seemingly require some type of conveyor belt of mantle material to explain them. The conveyor belt has to be hot, and its flow rapid, in order to explain eruption of great thicknesses of basalt in short periods of time. This in turn is presumed to require a hot source, arguably a thermal boundary layer at great depth, to trigger plumes. However, consider another type of conveyor belt.

Hamilton (2002) and Natland and Winterer (2005) emphasized that plates considered in their entirety influence asthenospheric convection, particularly counterflow away from subduction zones. If the top of the asthenosphere is generally above its solidus (Presnall and Gudfinsson, 2005), buoyant partial melts will tend to collect in general beneath the entire base of the lithosphere. Evidently even very strange, low-degree partial melts, with compositions of potassic olivine nephelinite, are still aggregating today beneath the oldest and coldest portions of the Pacific plate now entering the Japan Trench. These erupt at small volcanoes where the plate begins its flexure (Hirano et al., 2006). Thus, picture the Pacific plate slipping to its destiny on a film of lubricious partial melt where the effective stress is reduced to nil; this is essentially overthrust faulting on an enormous scale. The slippery layer will not be even because the melt fraction and quantity should be greater over fertile spots (patches, schlieren, blobs of eclogite, etc.), over somewhat warmer spots, and where changes in plate thickness are fairly abrupt (i.e., adjacent to fracture zones). I contend that this obviates the need for deep plume conduits.

With the trench geometry in the western Pacific, the strongest counterflow working its way back toward the East Pacific Rise should lie approximately beneath the Hawaiian Ridge and parallel to it. This is necessarily coincident with the most favorable location of propagation of a tensional fracture based on thermoelastic cooling of the Pacific plate (see Natland and Winterer comment on Stuart et al., this volume). The "bow" on the thermoelastic stress field of the cooling plate of Stuart et al. (this volume) produces a line of weakness that coincides precisely with an upward bow in the pattern of asthenospheric counterflow. The conveyor belt is not a plume but is turned on its side in such a way that large amounts of melt aggregate continually from a large volume of mantle to supply Hawaii. No other volcanic chains are active in the vicinity to dissipate or weaken this flow. We may even surmise that the melt layer is squeezed or overpressured by the action of the plate sliding above it, as are the fluids that lubricate thrust faults and decollements at accretionary prisms, and that this contributes to the volume and elevation of Mauna Loa. Of course this is supremely speculative, but then, I think we should admit, so are plumes.

Reese et al. (this volume) argue that formation of Olympus Mons and the other large Tharsis volcanoes early in the history of the planet is incompatible with an origin by means of mantle plumes. They advocate instead that large-scale impact melting produced local magma oceans, from which these volcanoes derive. They develop an evolutionary model to show how long-term thermal effects from a deep thermal boundary layer at the Martian core–mantle transition may have permitted types of follow-up plumes to develop and persist, allowing volcanism to cover up crater fields for long times thereafter.

But was an impact-produced local magma ocean even necessary? Suppose that the Martian surface once was effectively a single lithospheric plate capped with a thick primordial crust that, perhaps because of the relatively small size and fairly rapid cooling of the surface of the planet, was never dynamically unstable. The crust never foundered back into the planetary interior, like the presumed protocrust of Earth, and it never split up into separate plates. But when the whole planet was hotter, that one large surface plate must at some time have overlain a mantle that was at or above its solidus temperature regionally all around the planet. These are the same conditions I have postulated above for beneath the Pacific plate: an asthenosphere refulgent with partial melt, concentrated near the top, but in this case global; an asthenosphere weakest and with least viscosity near the top (Zhong and Zuber, 2001); and eruption occurring only where concatenation of stresses on the lithosphere permitted. I see no reason why a widespread layer, perhaps the entire shallow asthenosphere, could not have provided the magma. I cannot say whether forces acting within the lithosphere produced the stress points or whether they were punctures produced by impacts. But it seems to me that a broad comparison can be drawn between the mechanisms of large-scale "midplate" shield volcanism on planets Earth and Mars that is tied to the general direction necessarily taken by cooling of both bodies, but without mantle plumes.

8 January 2007, Joe McCall

The magnetic stripes recorded by Mars Global Surveyor from ~400 km altitude (Bowler, 2005) are a major surprise. It is not the intention here to question the geophysical statement or the detailed relationships with Mars surface features. However, as a geologist, involved for many years with terrestrial mapping in the east African rift valleys and a major plate convergence zone in Iran as well as long concern with meteorites and planetology, I question the attribution to early plate tectonics on Mars for the pattern revealed.

The pattern displayed is of a panglobal striping traversing the planet's surface from east to west. There are swirls, widenings, constrictions to cutting out, and even circular patterns, but one cannot escape the conclusion that this patttern is panglobal and almost certainly a single impress on the planet's surface and not a conglomeration of separate spreading patterns imposed at different times comparable with the patterns of magnetic stripes imposed by spreading on the Earth's oceans (of which the latest cycle only is now preserved).

It is important to consider exactly what the terrestrial magnetic stripes are. They are a feature of oceanic crust only. Each successive stripe represents a newer age of formation (spreading), recording a magnetic polar reversal. The oceans are megarifts, opening up from a median volcanic ridge. The east African rift is in effect an aborted ocean, where eruptive distension occurred but because of its unique situation between the passive margined Atlantic and Indian ocean systems, arose in a situation where compression was acting on both sides of the continent, and it could not spread out to form an ocean (McCall, 2006b). The rifts associated with ocean spreading initially developed on Earth amid supercontinents, breaking them up (e.g., Gondwana).

The problem with equating the Mars magnetic field with terrestrial plate tectonics is that we have apparently a panglobal assemblage of stripes affecting only the Martian analogue of "oceanic crust" and later erasures of the stripes where volcanic provinces and inferred impacts have been superimposed. What is lacking is the "foreland" continental marginal masses. A resolution of this problem bordering on the absurd would be the dictum that "on Mars continents vanish; on Earth oceans vanish." However, the pattern has a regularity of average stripe width, and magnetic reversals on Earth vary widely in duration and so width. A relationship to a number of successive openings at different magnetic polarities can thus surely be ruled out. Magnetic stripes on Earth also have a wide variation in trend within the single Jurassic–Present cycle, because the source ridges of the striped oceanic crust develop initially with different generalized trends and are also to a degree sinuous. The pattern on Mars suggests an early panplanetary crust, with this early crust splitting along linear zones and a new crust forming between the separations in a time of reversed polarity throughout the planet.

What seems evident is that this is not analogous to terrestrial plate tectonics; it cannot relate to a system of rifting of continents and spreading provinces—the surface areas (continent analogues) on which the initial spreading occurred are absent, and so the comparison with terrestrial magnetic stripes is far fetched. Plate tectonics involves a movement of lithospheric plates geographically, that is, across the planetary surface. Where is the evidence of this?

This is, surely, a surprise peculiarity of Mars—sui generis? The final question that comes to mind is "should we look at other planets and satellites for evidence of similar stripes?"

REFERENCES CITED

Bowler, S., 2005, Mars on a plate: Geoscientist, v. 15, no. 12, p. 8.

Hamilton, W.B., 2002, The closed upper-mantle circulation of plate tectonics, *in* Stein, S., and Freymueller, J.T., eds., Plate boundary zones: Washington, D.C., American Geophysical Union Geodynamics Monograph 30, p. 359–410.

Hirano, N., Takahashi, E., Yamamoto, J., Abe, N., Ingle, S.P., Kaneoka, I., Hirata, T., Kimura, J.-I., Ishii, T., Ogawa, Y., Machnida, S., and Suyehiro, S., 2006, Volcanism in response to plate flexure: Science, v. 313, p. 1426–1428.

McCall, G.J.H., 2006a, A caldera volcano of Brobdingnagian scale: Geoscientist, v. 16, p. 28–30.

McCall, G.J.H., 2006b, Tales from Afar: Geoscientist, v. 16, no, 7, p. 8–9.

Natland, J.H., and Winterer, E.L., 2005, Fissure control on volcanic action in the Pacific, *in* Foulger, G.R., et al., eds., Plates, plumes, and paradigms: Boulder, Colorado, Geological Society of America Special Paper 388, p. 687–710, doi: 10.1130/2005.2388(39).

Presnall, D.C., and Gudfinnsson, G., 2005, Carbonate-rich melts in the oceanic low-velocity zone and deep mantle, *in* Foulger, G.R., et al., eds., Plates, plumes, and paradigms: Boulder, Colorado, Geological Society of America Special Paper 388, p. 207–216, doi: 10.1130/2005.2388(13).

Reese, C., Solomatov, V.S., and Orth, C.P., 2007 (this volume), Interaction between local magma ocean evolution and mantle dynamics on Mars, *in* Foulger, G.R., and Jurdy, D.M., eds., Plates, plumes, and planetary processes: Boulder, Colorado, Geological Society of America Special Paper 430, doi: 10.1130/2007.2430(42).

Stuart, W.D., Foulger, G.R., and Barall, M., 2007 (this volume), Propagation of the Hawaiian-Emperor volcano chain by Pacific plate cooling stress, *in* Foulger, G.R., and Jurdy, D.M., eds., Plates, plumes, and planetary processes: Boulder, Colorado, Geological Society of America Special Paper 430, doi: 10.1130/2007.2430(24).

Wilson, L., Scott, E.D., and Head, J.W., 2001, Evidence for episodicity in the magma supply to the large Tharsis volcanoes: Journal of Geophysical Research, v. 106(E1), p. 1423–1433.

Zhong, S., and Zuber, M.T., 2001, Degree-1 mantle convection and the crustal structure of Mars: Earth and Planetary Science Letters, v. 189, p. 75–84.

The Geological Society of America
Special Paper 430
2007

The mantle plume debate in undergraduate geoscience education: Overview, history, and recommendations

Brennan T. Jordan*

Department of Earth Sciences, University of South Dakota, Vermillion, South Dakota 57069, USA

ABSTRACT

Mantle plume theory has been widely, but not universally, accepted in the geosciences for several decades, but recent critical evaluation has led to an intense debate regarding the existence of mantle plumes. I provide an overview of mantle plume theory and the current skepticism. The results of a poll taken after the 2005 American Geophysical Union Chapman Conference, The Great Plume Debate, are presented to give general readers a sense of which arguments specialists consider strongest in favor of, or opposing, mantle plume theory. Mantle plume theory first appeared in introductory textbooks in the late 1970s, and the theory was presented in most introductory textbooks by the end of the 1980s. In light of the current debate, most recent editions have introduced language indicting debate and uncertainty about mantle plume theory. However, none of these textbooks offers any alternative hypotheses; advanced textbooks also give little attention to alternative hypotheses. I assert that, without the presentation of alternative theories, students will simply accept the one presented as fact. The current debate should be seen as a teaching opportunity. It should be conveyed to students that this debate reflects the fact that geology is a very dynamic science and that first-order problems in the geosciences remain to be addressed by current and future generations.

Keywords: mantle plume, hotspot, education, textbook

INTRODUCTION

In most introductory geoscience textbooks, mantle plume theory is presented as the explanation for age-progressive intraplate volcanism (e.g., Hawaii) and anomalous plate margin volcanism (e.g., Iceland). Mantle plume theory is also presented in textbooks for relevant upper-division courses, such as igneous petrology and structural geology or tectonics. While mantle plume theory has been widely accepted in the geoscience community for several decades, there have been skeptics from the beginning, and the last decade has seen revitalization of crit-

ical evaluation of mantle plume theory, the development of alternative theories, and vigorous debate of the subject. The purpose of this article is to examine the current, and historical, presentation of mantle plume theory in undergraduate geoscience courses and to make recommendations to undergraduate educators and textbook authors for treatment of the mantle plume debate at the undergraduate level. Because the target audience for this paper includes nonspecialists, I will open with a brief introduction to mantle plume theory and the current debate; for more extensive reviews from both sides of the debate, see articles from a plume advocate perspective by Davies (1999, chap. 13; 2005), Campbell

*E-mail: brennan.jordan@usd.edu.

Jordan, B.T., 2007, The mantle plume debate in undergraduate geoscience education: Overview, history, and recommendations, *in* Foulger, G.R., and Jurdy, D.M., eds., Plates, plumes, and planetary processes: Geological Society of America Special Paper 430, p. 933–944, doi: 10.1130/2007.2430(43). For permission to copy, contact editing@geosociety.org. ©2007 The Geological Society of America. All rights reserved.

(2006), and Sleep (2006 and this volume), or from the plume-skeptic perspective by Anderson and Natland (2005) and Foulger (2005 and this volume).

Mantle Plume Theory

In the midst of the plate tectonic revolution, J. Tuzo Wilson (1963, 1965) introduced the concept of hotspots as an explanation for intraplate volcanism (e.g., Hawaii) and anomalous plate margin volcanism (e.g., Iceland), phenomena that were generally considered without simple explanation in the evolving plate tectonic model. The hotspot model postulated the presence of unusually hot mantle underlying these areas of persistent anomalous volcanism but did not propose a mechanism for their origin. The term *hotspot* is now used generically to refer to areas of anomalous volcanism without implying a process; the term *melting anomaly* is also used in this way and has the advantage of not implying anomalously high temperature, a contested point in the debate. Jason Morgan (1971, 1972a,b) proposed mantle plume theory as an explanation for hotspots. Morgan postulated that hotspots could be surface manifestations of thermal plumes in the mantle, and that such plumes would arise from the core-mantle boundary, the most significant thermal boundary layer in the interior of the Earth.

The plumes that Morgan envisioned were essentially vertical features, fixed with respect to one another. This original mantle plume theory has been added to and modified in a number of ways. Many modifications have been criticized as being ad hoc (e.g., Anderson, 2000); the fact that the theory is susceptible to ad hoc modification is recognized by many of its proponents (e.g., Sleep, this volume). However, many modifications are based on reasonable, though not definitive, models of the properties of the interior of the Earth. Inasmuch as these models constitute significant changes from the Morgan (1971, 1972a,b) model, they should properly be considered distinct alternative theories, testable aspects should be identified, and the theories should be evaluated. This article addresses mantle plume theory in a broad sense as consisting of a family of theories spawned from Morgan's original model.

Mantle plume theory, restricted to the original definition of Morgan (1971, 1972a,b), has already been disproved by the demonstration that hotspots are not fixed with respect to one another (e.g., Burke et al., 1973; Molnar and Stock, 1987). A model has been proposed by Steinberger and O'Connell (1998, 2000) that is not ad hoc and may account for nonfixity of hotspots. They impose a mantle structure based on seismic tomography and allow that structure to dictate flow, then model the effect of this flow on rising plumes. The parameters of the model may be contested, but it is predictive, and not ad hoc. Two other important addenda or modifications are these:

1. Mantle plumes originating with plume heads give rise to large igneous provinces (flood basalts) (e.g., Morgan, 1981; Richards et al., 1989; Campbell and Griffiths, 1990; Griffiths and Campbell, 1990).

2. Thermochemical plumes, in which buoyancy may arise from compositional heterogeneity in addition to temperature, have been modeled by Farnetani and Samuel (2005). The results suggest a more diverse array of plume forms. Testable aspects of this model should be identified and considered.

Beyond age-progressive volcanism, evidence that plume advocates commonly consider in support of mantle plume theory includes common association with flood basalts, distinctive isotopic signatures (including helium), topographic swells and geoid highs, seismic tomography, and the results of analog and numerical modeling. It is important to note that all of these lines of argument have either been used by plume skeptics to argue against a plume origin or addressed in their arguments (e.g., Anderson and Natland, 2005; Foulger, 2005 and this volume).

Plume Skepticism and Alternative Hypotheses

According to Anderson and Natland (2005), mantle plume theory was received with considerable skepticism when it was proposed in the early 1970s. Early criticism of mantle plume theory was based on a variety of arguments and lines of evidence, including fluid dynamics, geophysical data, and other observations. The subject was controversial enough that it was one of several topics chosen for selected readings coverage in the Cowen and Lipps (1975) volume *Controversies in the Earth Sciences*.

For most of the 1980s and 1990s, as mantle plume theory was reinvigorated in the wake of publication of analog and numerical modeling results and application of these models (e.g., Richards et al., 1989), critical evaluation diminished. Several leading researchers, notably Don Anderson (e.g., Anderson et al., 1992), continued to critique the theory and seek alternatives. Skeptical analysis of mantle plume theory expanded at the beginning of the twenty-first century. Through several conferences and the catalyzing efforts of Gillian Foulger in the development of the www.mantleplumes.org Web site, a cohesive and active community developed to focus on the critical evaluation of mantle plume theory and the development and advancement of alternative theories.

The criticisms of mantle plume theory are founded on a number of points, including the following (a list not intended to be exhaustive):

1. That hotspots are not hot, with no abnormal heatflow (e.g., Stein and Stein, 2003) or indication of high magmatic temperatures (e.g., Green and Falloon, 2005).
2. That helium isotope ratios considered characteristic of mantle plumes are not due to high ^3He, and could be generated at shallow depths (e.g., Anderson, 2000; Meibom and Anderson, 2004); other isotopic characteristics also can be explained by shallow processes (e.g., Anderson and Natland, 2005).
3. Most hotspots do not meet the standard criteria for mantle plumes, which commonly include age-progressive volcanic

track originating with flood basalt, high buoyancy flux, high maximum ^{3}He/^{4}He ratios, and a low-velocity anomaly in tomographic images (e.g., Courtillot et al., 2003; Anderson, 2005).

4. Seismic tomography indicates no plume in the lower mantle beneath many postulated plumes (e.g., Foulger et al., 2001), and many seismic tomography images showing lower-mantle plumes saturate the images, creating misleading results.

5. Durations of volcanism and the relationship between large igneous province development and extension favor lithospheric processes for the development of these provinces (e.g., Sheth, 2000).

6. When a locale fails to conform to the standard criteria for mantle plumes (e.g., the criteria of Courtillot et al., 2003, and Anderson, 2005), ad hoc modifications to the theory are made, rendering the theory untestable (e.g., Foulger, this volume).

7. Simpler models can be developed based on shallow plate tectonic processes (e.g., Anderson, 2001; Hamilton, 2003; Foulger, this volume).

Many of these points are contested by plume advocates. To read more thorough recent accounts from both sides, see Davies (2005) and Foulger (2005).

Alternative hypotheses for hotspots had been around prior to the proposal of mantle plume theory, and more were suggested in the 1970s. The early history of alternative theories includes propagating cracks, shear melting, self-perpetuating volcanic chains, membrane stresses, and gravitational anchors; see Anderson and Natland (2005) and references therein for more detailed discussion of these developments. Along with the renewed critical evaluation of mantle plume theory that arose at the beginning of the twenty-first century have come a number of new or revived alternative hypotheses for the origin of hotspot volcanism. These models have in common the view that relatively shallow (lithospheric or upper-mantle) processes are responsible for hotspots. The alternative models can be grouped in four categories: increased fertility for melting (e.g., Foulger and Anderson, 2005; Foulger et al., 2005a for Iceland); fractures tapping magmas already present at depth (e.g., Hamilton, 2003); shallow convection (e.g., the edge-driven convection of King and Anderson, 1998); and extraterrestrial mechanisms (e.g., Hagstrum, 2005). The first three mechanisms share the interpretation of hotspots as a second-order effect of plate tectonics, the "plate" model of Foulger (this volume).

Current Status of the Debate

Following the 2005 American Geophysical Union Chapman Conference, The Great Plume Debate, I conducted an online poll of participants, as well as those who had expressed interest but not attended. Some of the results may be interesting for non-specialists trying to understand the issues involved in the debate. Before presenting the results, two caveats should be noted. This conference brought together people from both sides of the debate; thus, the unfiltered results are not reflective of overall opinion in the relevant fields of the geosciences. Second, the total number of respondents is not large, and the filtered results have a smaller n; this was not a scientific poll, though I attempted to conduct it without bias. Poll participation requests were sent to 107 people, and 66 responded (62%). There were twelve questions in the poll, five of which are discussed in this article, in the following paragraphs and the recommendations section.

The first question was a multiple-choice question intended to ascertain the participants' positions in the plume debate. The question, choices offered, and results are presented in Table 1. The five available choices were written to offer strong and moderate pro-plume and plume-skeptical responses as well as a middle-of-the-road response. The results of this question were used to filter all other results into three groups for further analysis, plume advocates (first and second responses), middle (third response), plume skeptics (fourth and fifth responses). It is a credit to the conveners (who represented both sides) that the responses to this question were so balanced, with 29% pro-plume, 24% skeptical, and the remainder in the middle. Another poll result that speaks in favor of the kind of dialog represented by this conference is that 30% of respondents who attended the conference reported that their answer to this question changed from what it would have been before the conference because of something they had heard at the conference.

TABLE 1. POLL ON VALIDITY OF PLUME THEORY

Question and responses	Votes	Percent
Question: Which of the following best describes your perspective on the debate regarding mantle plume theory?		
Most intraplate volcanism and anomalous plate margin volcanism can be explained by mantle plume theory; mantle plume theory is well supported by observations, data, and modeling.	0	0
Most large systems of intraplate volcanism and anomalous plate margin volcanism can be explained by mantle plume theory; mantle plume theory is well supported by observations, data, and modeling at these sites. Smaller systems may be explained by other processes.	19	29
Mantle plume theory is a generally viable model for most large systems of intraplate volcanism and anomalous plate margin volcanism, but there are unresolved inconsistencies in observations, data, and modeling; alternative models should be developed and tested.	30	46
Mantle plume theory has many inconsistencies with data and observations and is probably not a viable explanation for intraplate and anomalous plate margin volcanism; most data and observations are more consistent with shallow (lithospheric and upper-mantle) processes.	12	18
Mantle plume theory has been disproved; intraplate volcanism and anomalous plate margin volcanism are best understood as the result of shallow (lithospheric and upper-mantle) processes.	4	6

TABLE 2. POLL ON ARGUMENTS FOR AND AGAINST MANTLE PLUME THEORY

Questions and responses	All respondents (%)	Plume advocates (%)	Middle (%)	Plume skeptics (%)
Question: Which of the following do you feel most strongly argue for mantle plume theory (you may select more than one)?	*n* = 61	*n* = 18	*n* = 28	*n* = 15
Age-progressive volcanic chains	**72**	**89**	**79**	40
Seismic tomography	48	67	57	7
Large igneous provinces	43	67	50	0
Numerical and analog models show plume formation	43	56	46	20
Evidence for high magmatic temperatures	30	44	36	0
Isotopic (other than helium) and trace-element signatures	25	44	25	0
Helium isotope ratios	21	39	21	0
No other viable explanation	5	6	4	7
None of the above. I believe that mantle plume theory is not supported by observations, data, and modeling.	15	0	4	**53**
Question: Which of the following do you feel most strongly argue against mantle plume theory (you may select more than one)?	*n* = 60	*n* = 16	*n* = 28	*n* = 16
Many plumes do not meet standard criteria for plumes	**47**	**31**	**54**	50
Inconsistencies or problems (other than nonfixity) with hotspot tracks	37	19	36	56
Lack of evidence for high temperatures	33	0	32	**69**
Geochemical signatures are more easily explained by other processes	27	0	32	44
Relationship between large igneous province development and extension	23	6	14	56
Hotspots are not fixed with respect to one another	20	0	18	44
Poor application of the scientific method in support of mantle plume theory	20	0	14	50
Seismic tomography	17	6	18	25
None of the above. I believe that mantle plume theory is strongly supported by observations, data, and modeling.	22	56	14	0

Note: Bold type indicates the leading answer to each question for each group.

Two questions were asked to identify which observations, data, or lines of argument the participants felt most strongly supported or opposed mantle plume theory. Eight choices were provided for each question, encompassing a selection of the main arguments commonly offered by both sides, as well as a none-of-the-above response. Participants could select more than one choice. These questions, responses, and results are summarized in Table 2 and Figure 1. These results are presented in full and filtered; the filtered responses are particularly interesting.

Age-progressive volcanic chains are identified as the strongest line of evidence supporting mantle plume theory in the unfiltered results and in the results of each of the three filtered groups, including 40% of the plume skeptics. This is not a surprise, because this was one of the primary observations leading to the concept of hotspots (Wilson, 1963) and the development of mantle plume theory (Morgan, 1971, 1972a,b). Results of seismic tomography, the relationship with large igneous provinces (flood basalts), and the results of modeling were also indicated by >40% of responses in the unfiltered results and those of the plume advocates and the middle group. Interestingly, the order of the top seven responses was the same in the unfiltered, plume advocate, and middle groups. The relatively unsatisfying response, no other viable explanation, received low support, but it was approximately equal in all groups.

In the unfiltered results the argument that respondents indicated they felt most strongly argued against mantle plume theory was that many proposed plumes do not meet the standard criteria for plumes; this was also the leading selection of the plume advocates and the middle group. Inconsistencies or problems (other than nonfixity, which was another choice) with hotspot tracks was the only other response given by a significant number of plume supporters. The middle group also considered significant the lack of evidence of high temperature and geochemical signatures more easily explained by other processes. Plume skeptics rated all of the arguments against mantle plume theory, other than seismic tomography, pretty highly. The single response that received the most support from this group was the lack of evidence for high temperatures.

An interesting result of the poll, considering the unfiltered data, is that the lines of evidence are seen as significant arguing both ways. Evidence of high magmatic temperatures rated 30% in support of mantle plume theory, and lack of evidence for high temperatures rated 33% against plume theory. These mutually exclusive responses offer the promise of resolution, because apparently both sides have made compelling, but not convincing, arguments with regard to temperature. Geochemical arguments also went both ways in approximately equal proportions (supporting plumes: helium isotopes 21%, other signatures 25%; against plumes: geochemical signatures more easily explained by other processes 27%). This area offers less short-term promise of resolution, because the chemical arguments are largely model-driven on both sides.

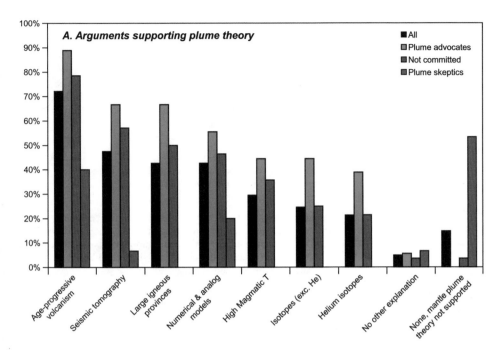

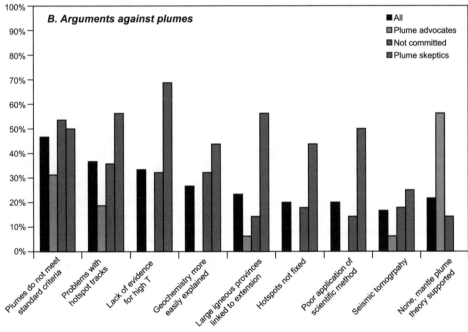

Figure 1. Histograms of poll data presented in Table 2, indicating lines of evidence that participants in the 2005 American Geophysical Union Chapman Conference, The Great Plume Debate, identified as most strongly supporting (A) or arguing against (B) mantle plume theory. The data are presented both unfiltered and filtered by leaning in the plume debate as discussed in the text.

Hierarchy of Theories and the Standing of Mantle Plume Theory

Theories originate as hypotheses, logical explanations of observations, subject to subsequent testing. The boundary between hypothesis and theory is not clearly defined, but generally, when a hypothesis is supported by subsequent observations and becomes widely accepted, it is elevated to the level of a theory. In developing a framework within which to discuss the standing of mantle plume theory, it is useful to have a defined

hierarchy of theories. A useful classification scheme is that of Dutch (1980), who described a three-tiered hierarchical classification of theories: central, frontier, and fringe. Dutch (1980) classifies hypotheses as fringe theories; most do not stand the test of time and will be discarded. Some theories are supported but have unresolved inconsistencies and/or serious alternatives; these are frontier theories, which constitute the mainstream thinking of a scientific discipline. Central theories are no longer seriously disputed and form the foundation of a discipline.

Twenty-five years ago, Dutch (1980) considered plate

tectonic theory at the central-frontier transition. At present, plate tectonics is a central theory in geology. Mantle plume theory is commonly regarded as a fundamental addendum to plate tectonic theory, and therefore might also seem to hold the status of a central theory. However, the current vigorous debate about the existence of plumes suggests that mantle plume theory is more properly regarded as a frontier theory. This classification is also consistent with the fact that even among plume advocates there are differing views about the nature and origin of plumes (e.g., Courtillot et al., 2003; Montelli et al., 2004, 2006; Farnetani and Samuel, 2005). Many plume skeptics would consider it a disproved hypothesis or fringe theory.

PRESENTATION OF MANTLE PLUME THEORY IN UNDERGRADUATE TEXTBOOKS

Hotspots are a first-order planetary-scale phenomena of the Earth, and most textbooks dealing with the Earth on a large scale, or addressing topics directly relating to hotspots, must address their origin. The following questions are addressed here. How are hotspots explained in undergraduate geoscience textbooks? Does the explanation reflect the current debate about the origin of hotspots? What is the history of presentation of mantle plume theory in introductory textbooks?

Mantle Plume Theory in Current Introductory Geology Textbooks

Following are quotations from several commonly used introductory textbooks regarding hotspots and mantle plumes. These brief quotations have been chosen in an attempt to reflect the various presentations of mantle plumes and, particularly, the degree of uncertainty in mantle plume theory that is expressed in the recent texts. A general discussion follows the quotations.

Almost all geologists accept the notion that hot-spot volcanism is caused by some type of upwelling in the mantle beneath the plates. However, the mantle plume hypothesis—that these upwellings are narrow conduits of material rising from the deep mantle—remains controversial.

And later, discussing plume heads as the cause of flood basalts,

Others dispute this hypothesis, pointing out that continental flood basalts often seem to be associated with preexisting zones of weakness in plates—suggesting that the magmas are generated by convective processes in the upper mantle. (Press et al., 2004)

We are not completely sure, but the simple idea of mantle plumes rising as narrow columns from the deep interior explains many such features [hotspots] and has become a generally accepted part of plate tectonic theory. (Hamblin and Christiansen, 2004)

We now recognize that most intraplate volcanism occurs where a mass of hotter than normal mantle material called a mantle plume ascends toward the surface. Although the depth at which (at least some) of these

mantle plumes originate is still hotly debated, many appear to form deep within the Earth at the core-mantle boundary. (Tarbuck and Lutgens, 2005)

Most mantle plumes are thought to originate just above the core mantle boundary, where heat from the Earth's core warms the base of the mantle. (Recent evidence, however, suggests that some may originate in the upper mantle.) (Marshak, 2005)

The plume hypothesis for explaining hot spots is not yet adequately tested and is very controversial. It is possible that not all hot spots have the same origin. (Smith and Pun, 2006)

Some geologists believe that these partially melted zones may be the sources of great plumes of rising hot mantle that may be driving the motion of the Earth's plates and fueling the volcanism that occurs at intraplate hot spots such as the Hawaiian Islands and Yellowstone National Park. (Chernicoff and Whitney, 2007)

"Hot spots" in the crust (where the geothermal gradient is locally very high) have been hypothesized to be due to hot mantle plumes, which are narrow upwellings of hot material within the mantle.

And later,

However, researchers using new data have questioned the extent, and even the existence, of deep mantle plumes. (For an in-depth discussion on the debate about mantle plumes, go to www.mantleplumes.org/.) (Plummer et al., 2007)

These quotations do express some uncertainty about the uniformity of plumes and their depth of origin, and some convey a sense of the scientific debate surrounding the issue. The treatment in most of these texts suggests a status of frontier theory, in the usage of Dutch (1980), for mantle plume theory. However, a critical point is that all of these recent texts go on to explain mantle plume theory in detail, and none offers an alternative hypothesis.

History of the Presentation of Mantle Plume Theory in Introductory Textbooks

Mantle plume theory began to appear in introductory textbooks in the late 1970s. Press and Sevier (1978) included an introduction to mantle plumes, which they appropriately described as a "controversial concept," in their second edition; the first edition, published four years earlier, did not include a reference to mantle plumes. Several current introductory textbooks have editions dating back to the 1970s and early 1980s. Hamblin and Christiansen (2004) dates back to 1975. The first edition (Hamblin, 1975) uses the term "plume" to describe hotspots, but does not indicate a lower-mantle origin. By the fifth edition, Hamblin and Christiansen (1989), they identify plumes as arising from >700 km depth and state that "the simple idea has become a generally accepted part of plate tectonic theory."

The text by Plummer et al. (2007) can be traced back to

1979. By the second edition (Plummer and McGeary, 1982; a first edition was not available for review), mantle plume theory was described with frequent references to Morgan, and plumes were depicted as originating at the core-mantle boundary. All this was followed by a note of skepticism: "Deep-sea drilling has shown, however, that not all aseismic ridges increase in age along their lengths. This evidence has led to alternate hypotheses for the origin of aseismic ridges. It may pose difficulties for the plume hypothesis itself." Interestingly, that exact text remains in the current edition.

The Tarbuck and Lutgens (2005) text was first published in 1985. In the first edition, Tarbuck and Lutgens (1985) explained hotspots as "plumes of molten rock that rise from deep within the mantle," but they were not explicit about a depth of origin for these plumes.

So mantle plumes have been offered as an explanation for hotspot volcanism in introductory textbooks for two to three decades. Press and Sevier (1978) treated mantle plume theory as a fringe theory, in the usage of Dutch (1980), a treatment appropriate for the time. Other authors treated mantle plume theory as a frontier or central theory. Anderson (2005) describes interest in plume-related research as "steady but low" through the 1980s; ironically, it was in that decade that mantle plume theory became a fixture in most introductory textbooks. Most authors have made changes in the most recent editions of their textbooks to reflect better the current debate over mantle plumes. But, again, none offers any mention of proposed alternatives.

Other Introductory Geoscience Courses

Many students take other closely related introductory geoscience courses, such as environmental geology, which is often an accepted substitute for physical geology in American geology departments, or oceanography. Quotations from texts in these subjects follow.

Environmental geology:
In contrast to the huge curtain-shaped mass of mantle that rises beneath a spreading center, a mantle plume is a relatively small rising column of hot, plastic mantle rock. Many plumes rise from great depths in the mantle, probably because small zones of rock near the core-mantle boundary become hotter and more buoyant than surrounding regions of the deep mantle. Others may form as a result of heating in shallower portions of the mantle. (Thompson and Turk, 2005)

Oceanography:
These volcanic mountains [the Hawaiian Islands and seamounts] are created over a deep-seated "hot spot" called a mantle plume. Mantle plumes are places where molten rock originates deep below the asthenosphere. (Pinet, 2003)

Both these quotations refer to mantle plumes as facts, though Thompson and Turk (2005) do indicate that some may be shallow features. These texts clearly treat mantle plume theory as a central theory in the hierarchy of Dutch (1980). Some oceanography and environmental geology textbooks simply do not take the discussion any further than hotspots. It is inevitable that textbooks for geoscience subjects that are less focused on internal Earth processes, such as these, give briefer coverage to subjects like hotspots and mantle plumes. It is commonly necessary to simplify the treatment of complex or controversial issues in these texts. However, changes in language to indicate that mantle plumes are hypothesized or theorized would require only minor modifications.

Upper-Division Courses

Explanations for the underlying causes of hotspot volcanism are generally presented in petrology textbooks. Mantle plume theory is also touched on in some textbooks used for other upper-division courses in subjects such as structural geology, geophysics, and volcanology. Following are quotations from two leading current petrology textbooks for consideration.

Island chains in the same plate follow subparallel paths and progress in the same direction, leading Wilson (1963) to conclude these chains were the result of volcanism generated by hot spots, or rising plumes generated by thermal anomalies originating in the deep mantle (or even the core), and thus anchored in place beneath the moving plates.

And later,

Explanations for hot spots include thermal perturbations in the deep mantle or core, compositional plumes of less dense material (Anderson, 1975), and volatile influx causing melting point reduction and rise of buoyant melts. (Winter, 2001)

First proposed by Morgan (1971), mantle plumes had been confirmed to exist by the end of the century through seismic tomography imaging (see, for example, articles in the March 19, 1999, and May 14, 1999, issues of Science).

And later,

The proposal of Wilson (1963) that the [Hawaiian Island–Emperor Seamount] chain originated by the movement (at that rate [8–9 cm/y]) of the oceanic lithosphere over an essentially stationary mantle plume is now accepted by most geologists. (Best, 2003)

Winter (2001) has been cautious to be descriptive of the literature on the subject rather than stating his own conclusions, and in doing so he even mentions alternative hypotheses, though he does not go into them in any depth. The first quotation from Best (2003) is one that would be contested by most plume skeptics. The second *Science* article referred to by Best (2003) summarizes the tomography of Bijwaard and Spakman (1999) for Iceland. Subsequent tomographic studies suggest that the low-velocity anomaly under Iceland is either entirely limited to the upper mantle, arguing against a mantle plume (e.g., Foulger et al., 2001), or only weak and discontinuous in the lower mantle, suggesting a "pulsating plume" (Montelli et al., 2006). The statement made in the second quotation is true, though the citation of Wilson (1963) is incorrect, because he did not propose a mantle plume.

RECOMMENDATIONS

Mantle plume theory is a frontier theory, in the usage of Dutch (1980). It should be treated as such in undergraduate textbooks, and taught that way in undergraduate classrooms. Through either direct reference to debate or the use of terms such as *hypothesized,* the current field of introductory texts does a pretty good job of reflecting the present debate regarding mantle plumes. The fact of the matter is that even among the plume advocate community there are divergent views about the nature of plumes, so an expression of uncertainty about the theory is appropriate. However, in the thirty-five years since the proposal of mantle plume theory, no introductory textbook that I have found has ever discussed an alternative hypothesis.

When people are first exposed to the debate over mantle plumes as an explanation for hotspots, the first question most ask is "What else could they be?" I believe that this is because, even if textbooks express some uncertainty regarding a theory, if no alternative theory is proffered, the reader has little option other than to accept the presented theory. This is especially true in introductory courses, which generally emphasize transmission of facts and concepts rather than critical thinking; after all, there is a tremendous amount of material to be covered in these courses.

I strongly recommend that textbook authors complement the current, often detailed, description of mantle plume theory with at least a paragraph introducing the fact that there are alternative hypotheses, and explore at least one in some depth. The more detailed description of an alternative hypothesis could be in the form of considering an alternative model applied to a specific location, such as the model for Iceland proposed by Foulger and Anderson (2005) and by Foulger et al. (2005a). Even in the absence of the presentation of such a model in a textbook, instructors could introduce an alternative hypothesis in lectures. My argument is that this would lend legitimacy to what I hope all, no matter which side of the debate they are on, would agree is appropriate language in these textbooks, treating mantle plume theory as a frontier theory, a work in progress, contested by some.

Some may feel that the debate over mantle plume theory is too complex to explain to introductory students. However, introductory students are taught that melting within the Earth can result from heating, decompression, or adding water and decreasing the melting point, so the idea that hotspot magmatism could result from something other than unusually hot mantle should be within their grasp. The poll conducted in the wake of The Great Plume Debate conference asked participants if they considered the debate over mantle plume theory too complex to explain to introductory students. The question, choices, and results, are presented in Table 3. Only 10% of respondents, overall, considered the mantle plume debate too complex for the introductory classroom. Interestingly, it was the plume skeptics who had the greatest concern that the subject may be too complex. The leading response (70% overall) was that the issue was not too complex and should be introduced in introductory classes; this was the leading choice of each group, though among plume advocates it was tied with the response that introducing the debate was unnecessary because the theory is sound.

In related geoscience courses, such as oceanography and environmental geology, the approach taken by current textbooks is understandable, but could be improved. Current texts either skip the issue by leaving anomalous volcanism as the result of hotspots, or present mantle plume theory as fact, as a central theory. All that would be required to more accurately reflect the status of the debate would be to change a word or two to include language such as *theory* or *hypothesis.*

Upper-division courses are actually in the best position to benefit from more thorough discussion of the mantle plume debate. Courses in subjects such as petrology, tectonics, and geophysics could explore the debate by considering the relevant data and observations deriving from these fields and exploring how they bear on the debate. From this perspective, the current treatment of the mantle plume debate in some petrology textbooks is disappointing. I would encourage textbook authors to present alternative hypotheses, though because mantle plume theory is favored in the discipline, an emphasis on this theory is still appropriate. As for introductory classes, I recommend first covering the alternative hypotheses broadly, then exploring at least one in depth as applied to a specific locale.

TABLE 3. POLL ON TEACHING THE MANTLE PLUME DEBATE

Question and responses	All respondents (%)	Plume advocates (%)	Middle (%)	Plume skeptics (%)
Do you consider the content of the debate regarding mantle plume theory too complex for, or difficult to explain to, introductory geoscience students?	*n* = 61	*n* = 15	*n* = 30	*n* = 16
Yes	10	7	7	19
No, but mantle plume theory is sound enough that presenting the debate is unnecessary.	16	47	10	0
No, the debate and some alternatives can be taught to introductory students.	70	47	83	69
No, and as alternatives are more consistent with observations, they should be presented instead.	3	0	0	13

I would encourage educators and textbook authors to remain abreast of developments in this debate and the rapidly evolving ideas on both sides. However, I would caution them to approach the literature seeking balance. This can be difficult. Much of the mainstream literature on locations where plumes have been invoked ignores the debate and alternative hypotheses. On the other side, Foulger et al. (2005b) and the Web site www.mantleplumes .org are resources mostly dedicated to the exploration of alternative hypotheses. Because neither the mainstream literature nor these other resources alone provide a balanced presentation of the debate, it may take some effort on the part of the reader or researcher to find balance. The mantle plume theory, its modifications, and alternative hypotheses should all be subject to the same high standard of critical evaluation. If the scientific method works properly, the current debate will either result in disproving mantle plume theory or the evolution of a stronger, more robust, mantle plume theory.

Mantle plume theory, as generally taught in the classroom, is a simple, elegant theory that explains the magmatism that plate tectonics does not. It is therefore appealing to treat it as an addendum to plate tectonics and give it the status of a central theory in the classroom. However, we are doing the students a disservice by doing so. And, by not encouraging open minds and critical thinking, we are perhaps doing damage to the future of our own science. The mantle plume debate offers an opportunity to demonstrate to students, at all levels, that geology is a dynamic science still addressing first-order questions about how the Earth works through research that they can be involved in if they choose.

ACKNOWLEDGMENTS

I am grateful for funding from the American Geophysical Union, which supported my attendance at the 2005 Chapman Conference, The Great Plume Debate. I thank the conveners and participants in that conference for a lively and productive discussion, and in particular I appreciate the time many took to complete my online poll, for which results are presented here. I thank Robert Tilling, Stephen Marshak, and Gillian Foulger for constructive reviews of this manuscript.

REFERENCES CITED

Anderson, D.L., 1975, Chemical plumes in the mantle: Geological Society of America Bulletin, v. 86, p. 1593–1600, doi: 10.1130/0016-7606(1975)86 <1593:CPITM>2.0.CO;2.

Anderson, D.L., 2000, The statistics of helium isotopes along the global spreading ridge system: Geophysical Research Letters, v. 27, p. 2401–2404, doi: 10.1029/1999GL008476.

Anderson, D.L., 2001, Top-down tectonics: Science, v. 293, p. 2016–2018, doi: 10.1126/science.1065448.

Anderson, D.L., 2005, Scoring hotspots: The plume and plate paradigms, *in* Foulger, G.R., et al., eds., Plates, plumes, and paradigms: Boulder, Colorado, Geological Society of America Special Paper 388, p. 119–145.

Anderson, D.L., and Natland, J.H., 2005, A brief history of the plume hypothesis and its competitors: Concept and controversy, in Foulger, G.R., et al.,

eds., Plates, plumes, and paradigms: Boulder, Colorado, Geological Society of America Special Paper 388, p. 119–145.

Anderson, D.L., Tanimoto, T., and Zhang, Y.-S., 1992, Plate tectonics and hotspots: The third dimension: Science, v. 256, p. 1645–1650, doi: 10.1126/science.256.5064.1645.

Best, M.G., 2003, Igneous and metamorphic petrology (2nd edition): Malden, Massachusetts, Blackwell, 729 p.

Bijwaard, H., and Spakman, W., 1999, Tomographic evidence for a narrow whole mantle plume below Iceland: Earth and Planetary Science Letters, v. 166, p. 121–126, doi: 10.1016/S0012-821X(99)00004-7.

Burke, K., Kidd, W.S.F., and Wilson, J.T., 1973, Relative and latitudinal motion of Atlantic hotspots: Nature, v. 245, p. 133–137.

Campbell, I.H., 2006, Large Igneous Provinces and the mantle plume hypothesis: Elements, v. 1, p. 265–269.

Campbell, I.H., and Griffiths, R.W., 1990, Implications of mantle plume structure for the evolution of flood basalts: Earth and Planetary Science Letters, v. 99, p. 79–93, doi: 10.1016/0012-821X(90)90072-6.

Chernicoff, S., and Whitney, D., 2007, Geology (4th edition): Upper Saddle River, New Jersey, Prentice Hall, 679 p.

Courtillot, V., Davaille, A., Besse, J., and Stock, J.M., 2003, Three distinct types of hotspots in the Earth's mantle: Earth and Planetary Science Letters, v. 205, p. 295–308, doi: 10.1016/S0012-821X(02)01048-8.

Cowen, R., and Lipps, J.H., 1975, Controversies in the Earth sciences: San Francisco, West, 437 p.

Davies, G.F., 1999, Dynamic Earth: Plates, plumes and mantle convection: Cambridge, England, Cambridge University Press, 458 p.

Davies, G., 2005, A case for mantle plumes: Chinese Science Bulletin, v. 50, p. 1541–1554, doi: 10.1360/982005-918.

Dutch, S.I., 1980, Notes on the nature of fringe science: Journal of Geological Education, v. 30, p. 6–13.

Farnetani, C.G., and Samuel, H., 2005, Beyond the thermal plume paradigm: Geophysical Research Letters, v. 32, p. L07311, doi: 10.1029/2005GL022360.

Foulger, G., 2005, Mantle plumes: Why the current scepticism?: Chinese Science Bulletin, v. 50, p. 1555–1560, doi: 10.1360/982005-919.

Foulger, G.R., 2007 (this volume), The "plate" model for the genesis of melting anomalies, *in* Foulger, G.R., and Jurdy, D.M., eds., Plates, plumes, and planetary processes: Boulder, Colorado, Geological Society of America Special Paper 430, doi: 10.1130/2007.2430(01).

Foulger, G.R., and Anderson, D.L., 2005, A cool model for the Iceland hotspot: Journal of Volcanology and Geothermal Research, v. 141, p. 1–22, doi: 10.1016/j.jvolgeores.2004.10.007.

Foulger, G.R., Pritchard, M.J., Julian, B.R., Evans, J.R., Allen, R.M., Nolet, G., Morgan, W.J., Bergsson, B.H., Erlendsson, P., Jakobsdottir, S., Ragnarsson, S., Stefansson, R., and Vogfjord, K., 2001, Seismic tomography shows that upwelling beneath Iceland is confined to the upper mantle: Geophysical Journal International, v. 146, p. 504–530, doi: 10.1046/j.0956-540x .2001.01470.x.

Foulger, G.R., Natland, J.H., Presnall, D.C., and Anderson, D.L., 2005a, Genesis of the Iceland melt anomaly by plate tectonic processes, *in* Foulger, G.R., et al., Plates, plumes, and paradigms: Boulder, Colorado, Geological Society of America Special Paper 388, p. 595–625.

Foulger, G.R., Natland, J.H., Presnall, D.C., and Anderson, D.L., eds., 2005b, Plates, plumes, and paradigms: Boulder, Colorado, Geological Society of America Special Paper 388, 881 p.

Green, D.H., and Falloon, T.J., 2005, Primary magmas at mid-ocean ridges, "hot spots" and other intraplate settings: Constraints on mantle potential temperature, *in* Foulger, G.R., et al., eds., Plates, plumes, and paradigms: Boulder, Colorado, Geological Society of America Special Paper 388, p. 217–248.

Griffiths, R.W., and Campbell, I.H., 1990, Stirring and structure in mantle plumes: Earth and Planetary Science Letters, v. 99, p. 66–78, doi: 10.1016/0012-821X(90)90071-5.

Hagstrum, J.T., 2005, Antipodal hotspots and bipolar catastrophes: Were oceanic large-body impacts the cause?: Earth and Planetary Science Letters, v. 236, p. 13–27, doi: 10.1016/j.epsl.2005.02.020.

Hamblin, W.K., 1975, Earth's dynamic systems (1st edition): Minneapolis, Minnesota, Burgess, 578 p.

Hamblin, W.K., and Christiansen, E.H., 1989, Earth's dynamic systems (5th edition): New York, Macmillan, 576 p.

Hamblin, W.K., and Christiansen, E.H., 2004, Earth's dynamic systems (10th edition): Upper Saddle River, New Jersey, Prentice Hall, 790 p.

Hamilton, W.B., 2003, An alternative Earth: GSA Today, v. 13, p. 4–12, doi: 10.1130/1052-5173(2003)013<0004:AAE>2.0.CO;2.

King, S.D., and Anderson, D.L., 1998, Edge-driven convection: Earth and Planetary Science Letters, v. 160, p. 289–296, doi: 10.1016/S0012-821X(98)00089-2.

Marshak, S., 2005, Earth: Portrait of a planet (2nd edition): New York, W.W. Norton, 844 p.

Meibom, A., and Anderson, D.L., 2004, The statistical upper mantle assemblage: Earth and Planetary Science Letters, v. 217, p. 123–139, doi: 10.1016/S0012-821X(03)00573-9.

Molnar, P., and Stock, J.M., 1987, Relative motions of hotspots in the Pacific, Atlantic, and Indian oceans since late Cretaceous time: Nature, v. 327, p. 587–591, doi: 10.1038/327587a0.

Montelli, R., Nolet, G., Dahlen, F.A., Masters, G., Engdahl, E.R., and Hung, S.H., 2004, Finite-frequency tomography reveals a variety of plumes in the mantle: Science, v. 303, p. 338–343, doi: 10.1126/science.1092485.

Montelli, R., Nolet, G., Dahlen, F.A., and Masters, G., 2006, A catalogue of deep mantle plumes: New results from finite-frequency tomography: Geochemistry, Geophysics, Geosystems, v. 7, p. Q1107, doi: 10.1029/2006GC001248.

Morgan, W.J., 1971, Convection plumes in the lower mantle: Nature, v. 230, p. 42–43, doi: 10.1038/230042a0.

Morgan, W.J., 1972a, Deep mantle convection plumes and plate motions: Bulletin of the American Association of Petroleum Geologists, v. 56, p. 203–213.

Morgan, W.J., 1972b, Plate motions and deep mantle convection: Geological Society of America Bulletin, v. 132, p. 7–22.

Morgan, W.J., 1981, Hot spot tracks and the opening of the Atlantic and Indian Oceans, *in* Emiliani, C., ed., The sea: New York, Wiley, p. 443–487.

Pinet, P.R., 2003, Invitation to oceanography (3rd edition): Sudbury, Massachusetts, Jones and Bartlett, 556 p.

Plummer, C.C., and McGeary, D., 1982, Physical geology (2nd edition), Dubuque, Iowa, William C. Brown, 513 p.

Plummer, C.C., Carlson, D.H., and McGeary, D., 2007, Physical geology (11th edition): New York, McGraw Hill Higher Education, 617 p.

Press, F., and Sevier, R., 1978, Earth (2nd edition): San Francisco, W.H. Freeman, 649 p.

Press, F., Sevier, R., Grotzinger, J., and Jordan, T.H., 2004, Understanding Earth (4th edition): San Francisco, W.H. Freeman, 618 p.

Richards, M.A., Duncan, R.A., and Courtillot, V.E., 1989, Flood basalts and hot-spot tracks: Plume heads and tails: Science, v. 246, p. 103–107, doi: 10.1126/science.246.4926.103.

Sheth, H.C., 2000, The timing of crustal extension, diking, and the eruption of the Deccan flood basalts: International Geologic Review, v. 42, p. 1007–1016.

Sleep, N.H., 2006, Mantle plumes from top to bottom: Earth-Science Reviews, v. 77, p. 231–271, doi: 10.1016/j.earscirev.2006.03.007.

Sleep, N.H., 2007 (this volume), Origins of the plume hypothesis and some of its implications, *in* Foulger, G.R., and Jurdy, D.M., eds., Plates, plumes, and planetary processes: Boulder, Colorado, Geological Society of America Special Paper 430, doi: 10.1130/2007.2430(02).

Smith, G.A., and Pun, A., 2006, How does Earth work? Physical geology and the process of science: Upper Saddle River, New Jersey, Prentice Hall, 674 p.

Stein, C., and Stein, S., 2003, Sea floor heat flow near Iceland and implications for a mantle plume: Astronomy and Geophysics, v. 44, p. 1.8–1.10.

Steinberger, B., and O'Connell, R.J., 1998, Advection of plumes in mantle flow; implications on hotspot motion, mantle viscosity and plume distribution: Geophysical Journal International, v. 132, p. 412–434, doi: 10.1046/j.1365-246x.1998.00447.x.

Steinberger, B., and O'Connell, R.J., 2000, Effects of mantle flow on hotspot motion, *in* Richards, M.A., et al., eds., The history and dynamics of global plate motions: Washington, D.C., American Geophysical Union, Geophysical Monograph 120, p. 377–398.

Tarbuck, E.J., and Lutgens, F.K., 1985, The Earth: An introduction to physical geology: Columbus, Ohio, Charles E. Merrill, 594 p.

Tarbuck, E.J., and Lutgens, F.K., 2005, The Earth: An introduction to physical geology (8th edition): Upper Saddle River, New Jersey, Prentice Hall, 711 p.

Thompson, G.R., and Turk, J., 2005, Earth science and the environment (3rd edition): Belmont, California, Thomson Brooks/Cole, 656 p.

Wilson, J.T., 1963, A possible origin of the Hawaiian Islands: Canadian Journal of Physics, v. 41, p. 863–870.

Wilson, J.T., 1965, Evidence from ocean islands suggesting movement in the earth, *in* Blackett, P.M.S., et al., eds., A symposium on continental drift: Philosophical Transactions of the Royal Society of London, series A, v. 258, p. 145–167.

Winter, J.D., 2001, An introduction to igneous and metamorphic petrology: Upper Saddle River, New Jersey, Prentice Hall, 697 p.

MANUSCRIPT ACCEPTED BY THE SOCIETY JANUARY 31, 2007

DISCUSSION

27 December 2006, Robert I. Tilling

With this chapter, Brennan Jordan has rendered a great service not only for geoscience educators but also for geoscientists (like me) whose research does not directly bear on plate-tectonics phenomena. Even though not personally involved in the current debate, we nonspecialists still want to know its ultimate outcome and are curious whether the debate protagonists can reach some common ground, if not consensus. Toward this end, Jordan provides a highly readable historical overview and nontechnical distillation of the basic arguments that the specialists —advocate or skeptic—regard as the most diagnostic and persuasive in debating the mantle-plume theory.

For added perspective in gauging the present status of the debate, he conducted an online poll of specialists following the 2005 AGU Chapman Conference on the "The Great Plume Debate," involving mostly conference participants but also some experts not at the meeting. The number of poll participants was relatively small (66 responded to the 107 requests sent), and Jordan admits that the poll is not "scientific." Yet, even though the poll results are not statistically robust, I agree with Jordan that they may be useful to "non-specialists in trying to understand the issues involved in the debate." Particularly interesting, though hardly surprising, was the finding that age-progressive volcanic chains were considered by the majority of specialists (even 40% of the plume skeptics) to constitute the strongest line of evidence in support of the mantle-plume theory. Somewhat unexpected, however, was the poll result that some lines of evidence were viewed by many specialists as supporting or arguing against mantle plumes. This dichotomy in part reflects, in my view, the existing inconsistencies and variability in the definitions and perceptions of the elements used in framing the debate.

To evaluate the presentation of mantle-plume theory to students, Jordan surveyed the latest editions of several widely used undergraduate geoscience textbooks. Encouragingly, he found that most of these acknowledge some degree of uncertainty regarding the origin and nature of hotspots and mantle plumes, thus reflecting the recent resurgence in critically reevaluating earlier held ideas. He also worried that none of the undergraduate textbooks presented alternatives to the mantle-plume theory, and advanced textbooks also gave only scant consideration to alternative hypotheses. Jordan argues that the insufficient presentation of alternative theories in college-level textbooks might compel students to accept the mantle-plume theory presented as fact.

Although I share his concern, I suspect that, at the introductory (Geology 101) level, the presentation of multiple theories might tend to confuse rather than enlighten the average undergraduate. However, I fully concur with Jordan in advocating increased focus on alternative theories in textbooks targeted for above-average students pursuing graduate studies in any of the geosciences or related disciplines. Yet interestingly, very few respondents (~10%) in Jordan's poll considered the mantle-plume debate to be "too complex for the introductory classroom." Of course, we must bear in mind that the poll respondents are research-focused specialists. I wonder how the results might have turned out had the same poll question been posed to nonspecialists, especially teachers of geosciences.

Overall, I found Jordan's study to be a well-balanced, unbiased treatment of the relevant issues on both sides of the current debate about mantle-plume theory. The results of his informal poll conducted among specialists were especially interesting, and they beg to be followed up by more broadly based, scientifically rigorous polls, involving not only plate-tectonics specialists but also rank-and-file practitioners of the geosciences in academia, government, and industry. Perhaps these larger polls—ideally with hundreds or thousands of participants—could be conducted under the auspices of some geoscience professional organization (e.g., American Geophysical Union, American Geological Institute, Geological Society of America) or the National Science Foundation. I wager that the results of a more comprehensive poll would afford some instructive and surprising comparisons with those from Jordan's limited poll.

28 December 2006, Stephen Marshak

In response to a request from the editor, Gillian Foulger, I offer the following comments on Brennan Jordan's chapter from the perspective of a textbook author. Jordan suggests that introductory texts, such my *Earth: Portrait of a Planet* and *Essentials of Geology,* do not provide sufficient discussion of alternatives to plumes as an explanation for hot-spot volcanism.

The choice of deciding what to include and what not to include in an introductory text proves to be a delicate balancing act. Jordan wants to see more coverage of plume alternatives, but reviewers argue that alternatives are too hard to explain using requisite simplified terminology, so students become frustrated when reading about them. Further, reviewers emphasize that 99% of the students in an introductory geology course will not become geologists, so an extended discussion of plume alternatives goes beyond the needs of an introductory course. Such comments highlight the fact that, in the case of introductory texts, too much detail may be worse than too little detail if the book is to appeal to the broadest audience.

The debate about plume coverage reflects the broader dilemma: that adopters have conflicting desires when choosing texts. They want to see more topics covered and more detail provided (especially as regards their personal area of expertise) but at the same time want the books to be shorter and simpler because the bulk of students are nonmajors and do not have the patience to read long texts. For this reason, it is not surprising that authors tend to wait for the dust to settle before committing precious page space to ideas that have not yet stood the test of time. Also, because publishers do not want the page count of a book to become too large, one can ask: In the space available, is it more important to have an extended discussion of Hurricane Katrina and the Indian Ocean tsunami or an extended discussion of plume alternatives?

Adopters generally indicate a preference for more discussion of issues that have societal impact than of ones that are primarily of academic concern. That said, introductory books should reflect the latest discoveries in geoscience and should convey a sense that active research continues to take place. In the case of the plume debate, the growth of literature about alternatives implies that the subject has matured sufficiently for a brief mention of alternatives to be appropriate. Thus, the third edition of my books will include one.

7 February 2007, Brennan T. Jordan

I thank Tilling and Marshak for their comments on my manuscript and have several comments to offer in reply. Tilling's comments generally reflect the perspective with which I hope this contribution is received. I aimed to provide an unbiased overview of mantle plume theory and the history and current state of presentation of this subject matter in undergraduate textbooks. I believe that although the poll I conducted in the wake of the 2005 Chapman Conference is a nonscientific poll of a very narrow subset of the geoscience community, it provides insights into the debate that can be considered with value to both specialist and nonspecialist. Perhaps the most important result from this poll is that most lines of evidence considered in the plume debate are used to argue both ways, and apparently effectively, as people with strong feelings on either side find these lines of evidence compelling in their favor. I believe that what one should take away from this observation is that we need new data, better data, or new perspectives to resolve this debate; arguing about the existing data set has created a deadlock.

There is one additional poll result that I will report in this reply. Tilling points out that, in regard to the low proportion of poll respondents (10%) who believe that the mantle plume debate is too complex for introductory students, "we must bear in mind that the poll respondents are research-focused specialists." Certainly, all of the participants in the conference are researchers, but a significant number are also educators; 57% of the poll respondents indicated that they teach an introductory geoscience course, and 60% indicated that they teach upper-division undergraduate courses. Tilling's point is still well taken: The respondents are not representative of the overall geoscience community, and, given their interests, they are more likely to conceive of approaches to distill the complicated aspects of this issue for introductory students.

Both Tilling and Marshak comment on issues and limitations of presenting controversies in introductory textbooks. Because Marshak is the author of one of the leading physical geology textbooks, I am most grateful for his perspective. I am sensitive to the delicate balance that authors of introductory textbooks must try

to strike, and I do not want this chapter to be read as a critique of their coverage. Authors are tugged in opposing directions by publishers, adopters, and reviewers. In that sense, this chapter, recommending the inclusion of an alternative hypothesis to add depth to the now commonly portrayed uncertainty in the plume hypothesis, only contributes to the authors' struggle. I know that textbook authors cannot thoroughly convey the standing of every theory in their field or present both sides of every controversy; to do so would produce a long, baffling text that portrayed a science in disarray. But I do think that it is healthy to present at least a couple of controversies to demonstrate that the science is still actively addressing fundamental issues. This should be exciting to the more interested students. If controversies are to be presented, students will appreciate that both sides have merit only if alternative hypotheses are presented. The essence of my recommendations is that I believe that the mantle plume debate is a worthy candidate for this kind of coverage.

REFERENCES CITED

Marshak, S., 2005, Earth: Portrait of a planet (2nd edition): New York, W.W. Norton, 748 p.

Marshak, S., 2007, Essentials of geology (2nd edition): New York, W.W. Norton, 545 p.

The Geological Society of America
Special Paper 430
2007

Graphic solutions to problems of plumacy

John C. Holden
Plumatic Asylum, P.O. Box 853, Omak, Washington 98841-0853, USA
Peter R. Vogt*
Plumatic Asylum–East, 3555 Alder Road, Port Republic, Maryland 20676, USA

PREFACE

The mantle plume is just the youngest member of a big and colorful family of geological fads and fashions: diluvialisms and catastrophisms, earths expanding and contracting, global tectonics new and old. Some of these fads have become bandwagons rolling from theory to fact. Others are intellectual white elephants gathering library dust. We do not know yet how the mantle plume will fare; certainly it has not quite attracted the bandwagon that the new global tectonics did.

Since plumes are better hidden from observation than plates, it may take years to prove or disprove their existence. This is just as well; one cannot write a research proposal to prove that the Earth is round or that the continents drift.

In this article we hope to cut through the hullabaloo surrounding mantle plumes by offering a graphic commentary on the "hot" topic. Any resemblance between persons or deities depicted here and those living or dead may or may not be coincidental. If kings and statesmen fall to the cartoonist's pen, why not scientists, their students, instruments, and Mother Earth herself?

INTRODUCTION

It was Wilson (1963) who first suggested that hotspots were fixed in the upper mantle and created aseismic ridges as crustal plates moved over them. This idea was expanded to include a mechanism that would also account not only for hotspots but also for the causal forces of plate tectonics and continental drift. In short, it was proposed that ascending convection plumes exist in the deep mantle below active hotspot volcanism (Morgan, 1971, 1972). Geometrically, these features are toroidal cells with narrow (200–300-km-wide) vertical axes through which hot material is transferred from the lower to the higher regions in the mantle.

An entire science has developed around the study of plumes, albeit perhaps no more scientific than social, spiritual, or political science. If the concept is valid, most first-order features of the Earth's surface may be attributed directly or indirectly to plumes. We term this discipline "plumacy," its practitioners "plumatics" (also "plume freaks"). Recognizing the parallels between religious and geological faiths, we also propose the term *aplumatics* for those who do not believe that mantle plumes exist; we carry over the term *agnostic* to apply to all those fainthearted Earth scientists who refuse to debate the issue on grounds that the existence or nonexistence of mantle plumes cannot be proved since the terms are not defined. This position is untenable, since the definition and etymology of the word *plume* have been published in the geological literature, by Anderson (1975). Because he is a little-known author (just one of the numerous Andersons in earth science) publishing in an obscure journal, it is worthwhile to reproduce his definition here.

Tozer (1973) has objected to the use of the word *plume* because of prior usage and connotations. However, the various definitions of *plume* and its antecedents in French, Latin, and German seem to provide enough flexibility to describe the phenomenon, its implications, and its raison d'être on the one hand and its inventors, supporters, and detractors on the other: *Plume* (English, from French and Latin, pluma)—a feather, a long handsome feature, a token of honor or prowess; a prize, to pride or congratulate; to preen. *Plombe* (Germanic)—a plug. *Plombe* (Old English from West Germanic)—something especially desirable, as a good position. *Panache* (French)—trail, stripe, swagger; *fumée* (French)—smoke, fumes, steam; *fumer* (verb)—to fume, to dung, to manure; *plumitif* (familiar)—scribbler, pen-pusher.

We leave the definition of hotspots (also called melting spots and melting anomalies) as an exercise for the reader.

*E-mail: ptr_vogt@yahoo.com.

Holden, J.C., and Vogt, P.R., 1977, Graphic solutions to problems of plumacy: Eos (Transactions, American Geophysical Union), v. 58, no. 7, p. 573–580. ©1977 American Geophysical Union. Reproduced by permission of American Geophysical Union for inclusion in Foulger, G.R., and Jurdy, D.M., eds., Plates, plumes, and planetary processes: Geological Society of America Special Paper 430, p. 945–953, doi: 10.1130/2007.2430(44).

PLUMES, ANTIPLUMES, AND OTHER EXPLANATIONS FOR HOTSPOTS

Plumes are often invoked to explain the source of the force causing seafloor spreading and plate motion. Unfortunately, no airtight case has been made for the cause of the plumes themselves, and the original problem of where it all begins is not solved but merely pushed deeper into the Earth's interior. Therefore, we must in all fairness discuss not only mantle plumes but alternate theories put forth to explain hotspots.

One wonders what was being smoked when the plume concept was formed (Fig. 1). It is a well-known dictum in geopolitics that for popular acceptance the terms one chooses are as important as the ideas themselves. Therefore, when the plume concept was under construction, the stem of the plume was cleverly dubbed a pipe, which immediately brings to mind kimberlite pipes of mantle origin. . . . Bingo, a winner! After all, it is much more propitious for material to move "up the pipes" than to go "down the tubes." However, to suggest that the term *pipe* was chosen only for its public relations value is not entirely fair. After all, when geophysicists conceptualize, they habitually use the stems of their pipes as a scale for relative proportions. So it is only logical that the stem of a plume should be called a pipe; Q.E.D.

A few years after the plume idea was brought out, Shaw and Jackson (1973) proposed a slight modification of it in their gravitational anchor theory (Fig. 2). According to this scheme, dense crustal residues are sinking beneath Pacific hotspots at the eastern ends of the Hawaiian, Tuamoto, and Austral chains as low-density basaltic magmas are distilled out and up to form the spot on the hotspot. The geometry of this specialized convection cell is that of an upside-down plume, or antiplume as it were. Shaw and Jackson are absolutely correct in assuming that these anchors do not drive the plates but rather act as pinning points. We presume that the chain (not the volcanic chain) attached to the anchor (and nowhere discussed by the original authors) extends out of the orifice of the volcano. Where it goes after that is uncertain; perhaps it is attached to the nearest continent, and if it does not drive the oceanic plate, it pulls the continent over it. Because this theory accounts for the large shield volcanoes at Hawaii and elsewhere, gravitational anchors are associated with an abundance of tephra-laden hot air that often tends to obfuscate a clear understanding of these features.

On occasions when the clouds do clear somewhat, some authors can identify stress fields in the Hawaiian chain (not the anchor chain) indicating the stress of the anchor (Jackson and Shaw, 1975). They define two stress fields, a Polynesian and a Hawaiian field. We see that a third pattern and a reassessment of all three fields yields a remarkable phenomenon (Fig. 3). Carefully plotting the three lineation sets actually spells out this fact, and in Latin, no less. We misinterpret these data to read, "We think; therefore, they exist." The mouse diagram showing trend bearings about an imaginary center marks the overall trend of the Hawaiian chain (the volcanic chain, not the anchor chain).

Figure 1. Conception of the mantle plume theory, adapted liberally from W.J. Morgan (unpuffed data, 1977).

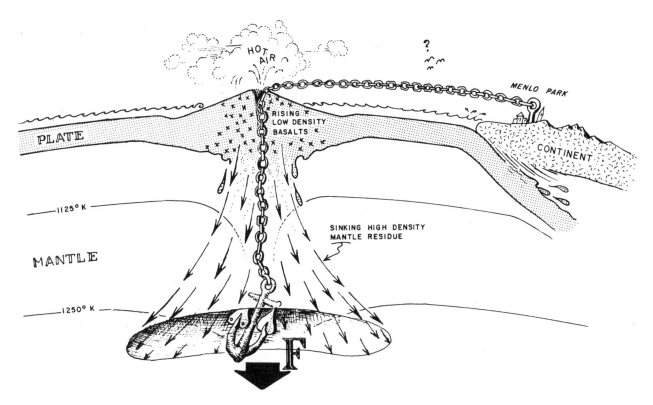

Figure 2. Gravitational anchor theory, showing the origin of Hawaii (Shaw and Jackson, 1973). According to this hypothesis, even the motion of California can be explained (although the motions of Californians remain as mysterious as ever).

Plumes, antiplumes, and propagating fractures are certainly not the only possible explanations for features such as the Hawaiian chain. Ancient legend has it that the island of Japan is situated atop a giant carp and that every time the beast shifts position Mount Fuji erupts. As most legends have some basis in fact, we propose that this creature also finds its home throughout the viscous asthenosphere (Fig. 4). Could it be that as it swims, at a rate of only a few centimeters per year within the mantle, it leaves behind a buoyant trail of tholeiitic bubbles that rise ponderously to create aseismic ridges? We name this fish *Asthenicthes aseismotathes,* or the asthenospheric fish that makes aseismic ridges. No doubt there are some who would question our taxonomy and would want to call the species a form of crappie. Readers may find something fishy about this theory, but there is at least something fishy about the other theories as well.

How plumes are oriented, orifice up or orifice down, awaits future adjudication. On the other hand, we can make some definite statements about the distributions of plumes. There are, for example, two plumes on the Arctic and Antarctic circles, namely Iceland and the Balleny Islands, respectively. If the Balleny Islands are, in fact, a plume, as Morgan (1972) predicted, this plume is nearly perfectly antipodal to the Iceland plume (Holden, 1976b). If these two hotspots are taken as midpoints of the two world rift systems, the rifts have an interesting relationship to each other, as shown in Figure 5. One supposes that this proves the athletic excellence of the Creator, for the game is certainly horseshoes, and He has scored two perfect ringers. It would seem that this particular game has been going on for at least 200 m.y. if the rift margins of Antarctica are any indication. This continent has been located at the South Pole since Pangaea broke apart, and its rift margins are all close to the Antarctic Circle (Holden, 1976a). Unfortunately, this accounts for only two plumes; we leave it up to the reader to devise explanations for the distribution of the remaining 120.

HOW MANY PLUMES?

Mantle plumes are in the midst of a population explosion that threatens to engulf the earth in a volcanic catastrophe (Vogt, 1972). The facts speak eloquently for themselves—no need to consult the Club of Rome. In 1971 there were 20 plumes (Morgan, 1971, 1972); in a mere half decade the population has risen to no fewer than 122 (Burke and Wilson, 1976) . Our extrapolations from these data show that there will be one million hotspots by the year 2000. We hope someone proves that hotspots do not exist, before it is too late.

CRACKED SEWER PIPES

If mantle material rises below Iceland, it must somehow spread out from the top (head) of the plume. Presumably the flow occurs primarily in the asthenosphere below the plate. Vogt

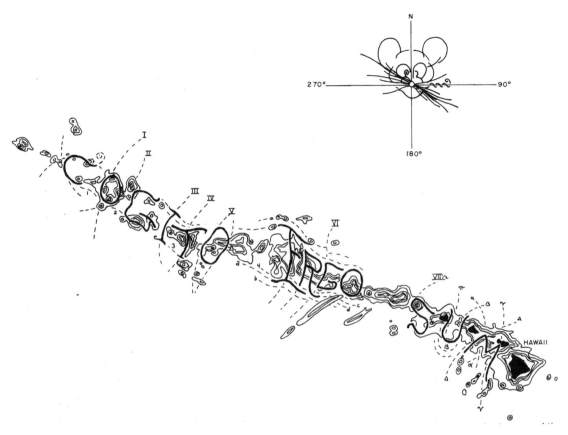

Figure 3. Stress field lineations plotted from volcano distribution along the Hawaiian chain. The geophysical relevance of the message spelled out is a matter of intense debate.

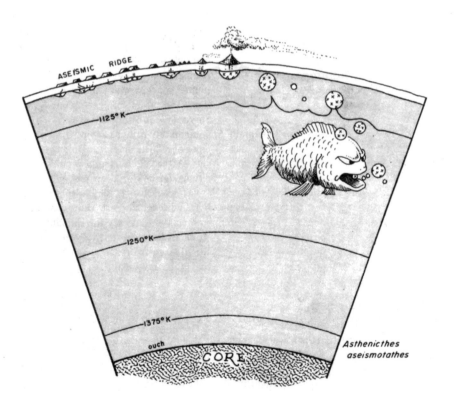

Figure 4. Alternative to the mantle plume theory (based on an ancient Japanese legend).

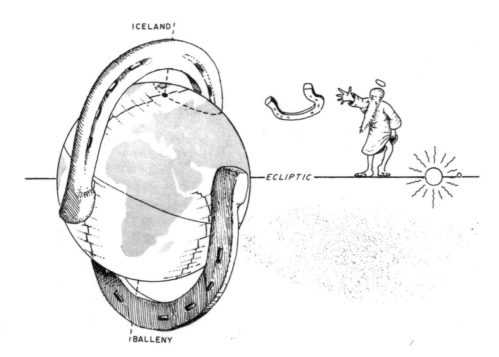

Figure 5. Schematic of a theory for the origin of mid-oceanic rifts, hotspots, the solar system, creation, and God.

(1971, 1974) and Schilling (1973) have suggested that this flow is concentrated in the bidirectional pipe formed below the spreading axis as a result of the extensive partial melting and lowered viscosity there (Fig. 6). This slightly tilted "cracked sewer pipe" flushes away the material rising in the stem of the vertical pipe.

The sewer pipe is cracked (or, more exactly, half-cracked like its author) in the sense that tholeiitic magmas must rise vertically to feed the constantly accreting oceanic crust. To further complicate matters, transform faults offsetting the spreading axis also offset the subaxial conduit, thus creating transform dams (Vogt and Johnson, 1975). Despite all this faulty plumbing, the conduit can do it, beHolden though it is to numerous mantle demons (Fig. 6). Fluctuations in plume discharge down the cracked sewer pipes are thought to leave V-shaped ridges imprinted on the ocean crust (Vogt, 1971). Vogt is currently searching bathymetric charts to see whether there is a good spot for a *P* on it.

The extreme case of a fluctuating plume is the mantle blob (Schilling, 1975), a concept possibly suggested to Schilling by the behavior of a lava lamp. We expect future hotspot authors to embellish this already descriptive terminology with, for example, mantle drops and dabs and drips, mantle splats and burps and belches, spurts and. . . .

Before leaving the intriguing though semiscatological subject of tectonic toiletimetry, we offer an interesting historical note. The famous early geophysicist Jules Verne had his team descend the vertical pipe beneath Iceland in their Journey to the Center of the Earth and ascend through a volcano in the Mediterranean. This path no doubt accounts for the great difficulties they encountered, since mantle material rises at Iceland and sinks in the Mediterranean subduction zone. They were fighting against strong currents all the way. (Soon after this ordeal, Verne spent eighty days in a balloon floating around the world to recuperate.)

PLUME CHEMISTRY

It now seems that the final answers concerning the Earth's largest features will be found by those dedicated research teams studying the rarest elements and smallest subdivisions of matter.

Hotspot basalts have long been known to differ from ocean floor basalts. Even beginning Geology 1 students could take some Icelandic basalts for granite, since some of them are. Sunken continental crust was once thought to exist at depth but has since dissolved into mantle. More recently (Schilling, 1973, 1975), systematic variations in some isotope ratios and LIL (large ionic lithophile) elements along the mid-ocean ridge away from hotspots have suggested unique PHMP (primary hot mantle plume) chemistry distinct from the DLVL (depleted low-velocity layer). Although the various PHMPs differ among themselves in LILs, such as La, K, P, and Ti, the DLVL always has fewer LILs than the PHMP.

We see that hotspot basalts are really a GAS (geochemical alphameric soup) that seems diabolically difficult to understand, probably because it is the work of the UMDs (upper-mantle demons; see Fig. 7). The decoding of this GAS clearly requires elaborate technologies, such as mass spectrometers, neutron activators, eutecticized equilibrating quasi–partial melting fractionators, and multiphased polyglazing computers. To operate this fancy gadgetry and produce significant results, a LAGS (large army of graduate students) is also required (Fig. 8).

To make chemistry of the ocean crust more accessible to the

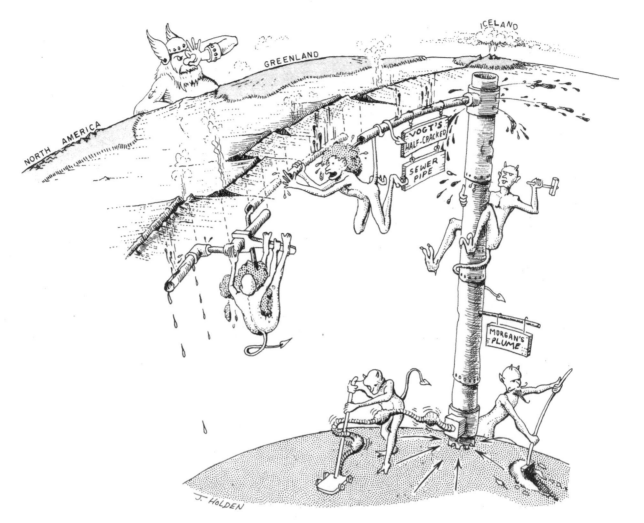

Figure 6. Mantle plume materials transported by faulty plumbing system from the lower regions to the mid-oceanic ridge. (Devils not to scale.)

simpleminded (and penniless) layman, Vogt and Johnson (1973) have invented "magnetic telechemistry." In this modern version of water-witching, all that is needed is a $2 million aircraft or a $15 million research ship outfitted with a magnetometer. Iron-rich basalts associated with mantle plumes can be charted once the measurements are corrected for sunken submarines and schools of magnetic fish. The concentrations of other elements (gold!?) can be predicted accurately, at least to within a few orders of magnitude, over at least 0.1% of the ocean floor.

GLOBAL SYNCHRONISM

For many years now, dashed correlation lines have connected short, fat magnetic anomalies with tall, skinny magnetic anomalies and even with featureless magnetic plains. But such correlation is simply a corollary of the Geologic Correlation Axiom. According to this rule, any variable can be made to correlate with any other variable once the vertical and horizontal scales are suitably adjusted, the eyeball suitably trained, and the correlation lines suitably slanted to the author's prejudices.

Vogt (1972) has applied this axiom to show that hotspot volcanoes, magnetic reversals, dinosaur populations, the stock market, and the annual number of Norman Watkins papers all follow the same global rhythm. Although further proofs of global synchronism are really unnecessary, Kennett and Thunell (1975) have shown that volcanic activity along the world's island arcs has indeed fluctuated in sympathy with the hotspot discharge curve inferred by Vogt (1972). But why should volcanoes in the Marianas or the Caribbean "know" what the Hawaiian volcanoes are doing? One answer is that mantle plumes speed up plate motions and thereby increase subduction rates and related volcanic activity. However, since many scientists still do not accept mantle plumes, we offer a yet more plausible mechanism in Figure 9. (To the reader who might question who would fill the plume-juice bag, we would answer, "Nitpicker!")

LOGJAM TECTONICS

Having dwelt too long on the generation of aseismic ridges, we close by considering their destruction. Basalts deposited on

Figure 7. Strange chemistry of ocean island tholeiites, hawaiites, balonites, and similar hotspot-generated rocks, attributed to the culinary habits of X. From Vulcan et al. (regurgitated material, in preparation, 1977).

Figure 8. Geochemical research on mantle plumes, as conducted at a modern university. The significance of the Azores, Iceland, and Mickey Mouse in this scheme is still under discussion.

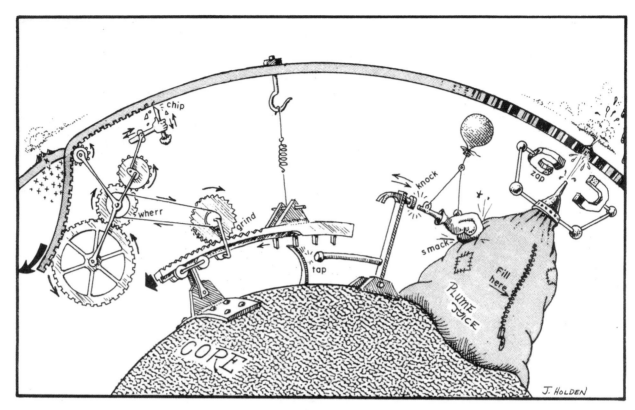

Figure 9. Diagram hypothesizing the correlation between discharge of island arc and hotspot volcanoes. Mantle viscosity is assumed to be zero as a first approximation (with apologies to S.K. Runcorn).

continental crust will tend to be eroded off, along with some of the underlying intrusives. Aseismic ridges formed on ocean crust will probably eventually be subducted into the mantle (The Gospel According to Le Pichon, Francheteau, and Bonnin, aka *Plate Tectonics,* 1973). However, the bulky, buoyant aseismic ridge poses a problem to plate digestion, and subduction is postponed as long as possible. In the meantime, the island arc and trench migrate toward the downgoing plate as a result of back-arc spreading. This migration is slowed or stopped where the ridge is being subducted, and (Eureka!) a cusp is formed (Vogt, 1973). Isn't that clever? Figure 10 shows the battering ramifications of the pushy process postponing plate plunging.

The only other theory for island arc formation besides log-jam tectonics is ping-pong ball tectonics. This theory holds that the Earth is a giant ping-pong ball, its surface dented inward in numerous places to produce island arcs (Frank, 1968). Frankly, we find any theory that likens the Earth to a wet, dirty, dented ping-pong ball too ludicrous to debate.

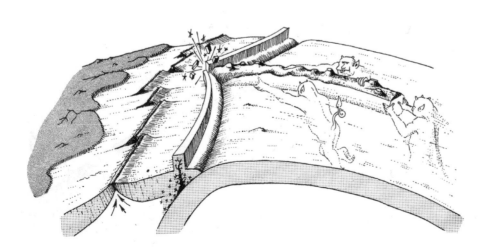

Figure 10. Illustration of logjam tectonics (see text for explanation of symbols).

ACKNOWLEDGMENTS

The authors completed this article while on sabbatical at the Plumatic Asylum, Lower Slobbovian Institute of Advanced Geoquackery. P.R.V. wishes to thank the anonymous reviewers for censoring the original manuscript and the original copyeditor, Joan Welsh of the American Geophysical Union, for humor above and beyond the call of duty. He also thanks the lampooned individuals and institutions not to press libel charges. This paper does not necessarily reflect the views of the Plumatic Asylum, the Institute of Geotectonic Redundancies, the International Stop Continental Drift Society, the U.S. Government, or the authors. Plumatic Asylum contribution 0000001.

REFERENCES CITED

Anderson, D.L., 1975, Chemical plumes in the mantle: Geological Society of America Bulletin, v. 86, p. 1593–1600, doi: 10.1130/0016-7606(1975) 86<1593:CPITM>2.0.CO;2.

Burke, K.C., and Wilson, J.T., 1976, Hot spots on the earth's surface: Scientific American, v. 235, p. 46–57.

Frank, F.C., 1968, Curvature of island arcs: Nature, v. 220, p. 363, doi: 10.1038/220363a0.

Holden, J.C., 1976a, Permian–Triassic continental configurations and the origin of the Gulf of Mexico: Comment: Geology, v. 4, p. 324–325, doi: 10.1130/0091-7613(1976)4<324:PCCATO>2.0.CO;2.

Holden, J.C., 1976b, Present and past symmetry of rifts [abs.]: Eos (Transactions, American Geophysical Union), v. 57, p. 89.

Jackson, E.D., and Shaw, H.R., 1975, Stress fields in central portions of the Pacific plate: Delineated in time by linear volcanic chains: Journal of Geophysical Research, v. 80, p. 1861–1874.

Kennett, J.P., and Thunell, R.C., 1975, Global increase in Quaternary explosive volcanism: Science, v. 187, p. 497–503, doi: 10.1126/science.187.4176.497.

Le Pichon, X., Francheteau, J., and Bonnin, J., 1973, Plate tectonics: New York, Elsevier, 300 p.

Morgan, W.J., 1971, Convection plumes in the lower mantle: Nature, v. 230, p. 42–43, doi: 10.1038/230042a0.

Morgan, W.J., 1972, Deep mantle convection plumes and plate motions: American Association of Petroleum Geologists Bulletin, v. 56, p. 202–213.

Schilling, J.-G., 1973, Iceland mantle plume: Geochemical study of Reykjanes ridge: Nature, v. 242, p. 565–571, doi: 10.1038/242565a0.

Schilling, J.-G., 1975, Azores mantle blob: Rare-earth evidence: Earth and Planetary Science Letters, v. 25, p. 103–115, doi: 10.1016/0012-821X(75)90186-7.

Shaw, H.R., and Jackson, E.D., 1973, Linear island chains in the Pacific: Results of thermal plumes or gravitational anchors?: Journal of Geophysical Research, v. 78, p. 8634–8652.

Tozer, D.C., 1973, Thermal plumes in the earth's mantle: Nature, v. 244, p. 398–400, doi: 10.1038/244398a0.

Vogt, P.R., 1971, Asthenosphere motion recorded by the ocean floor south of Iceland: Earth and Planetary Science Letters, v. 13, p. 153–160, doi: 10.1016/0012-821X(71)90118-X.

Vogt, P.R., 1972, Evidence for global synchronism in mantle plume convection and possible significance for geology: Nature, v. 240, p. 338–342, doi: 10.1038/240338a0.

Vogt, P.R., 1973, Subduction and aseismic ridges: Nature, v. 241, p. 189–191, doi: 10.1038/241189a0.

Vogt, P.R., 1974, The Iceland phenomenon: Imprints of a hot spot on the ocean crust, and implications for flow below the plates, in Kristjansson, L., ed., Geodynamics of Iceland and the North Atlantic area: Dordrecht, Holland, Reidel, p. 105–126.

Vogt, P.R., and Johnson, G.H., 1973, Magnetic telechemistry of oceanic crust?: Nature, v. 245, p. 373–375, doi: 10.1038/245373a0.

Vogt, P.R., and Johnson, G.L., 1975, Transform faults and longitudinal flow below the midoceanic ridge: Journal of Geophysical Research, v. 80, p. 1399–1428.

Wilson, J.T., 1963, Continental drift: Scientific American, v. 208, p. 86–100.

The Geological Society of America
Special Paper 430
2007

Plumacy reprise

Peter R. Vogt*
Plumatic Asylum–East, 3555 Alder Road, Port Republic, Maryland 20676, USA
John C. Holden
Plumatic Asylum, P.O. Box 853, Omak, Washington 98841-0853, USA

Keywords: adhocity, allwetspot, aplumatic, auto-obliteration, Fertile Earth Mother (or re-gurgitation) hypothesis, geobleme, geohyperbolae, geophallic, Great Aplumatic Challenge, Hannukah oil lamp effect, hypothesis radiation, Loch Ness Monster principle, neoastro-blemicist, notsofixed, notsohotspot, numeroplume, platonics, platonizer, ploomz, plumacy, plumatic, plumology, plumotaxonomist, protoplume, superduperplume, tomograph, tomo-plume, Wundergeologist

Key Acronyms: AGU, B&W, BS, DSDP, DVD, FNA, FORTRAN, GAC, GSA, HOLE, IELW, JCH, LAGS, LIP, LNMP, MOR, MWHM, NOAA, NSF, OHLP, OJP, SUV, UK, US, VSR

INTRODUCTION AND HISTORY

We offer a long-overdue reprise (a cognate of reprisal) of Holden and Vogt (1977; reproduced in this book in all its corny details). We try to offend as many authors as possible, including ourselves, but don't take too many sides between the plumatics (Believers) and the aplumatics (Nonbelievers), who, as this volume and P³ (Foulger et al., 2005b) demonstrate, are mounting the Great Aplumatic Challenge (GAC; pronounced "gack!"). We try to navigate our article (the raft in Fig. 1) through this swamp of controversy. Meanwhile, even plumes are no longer sure of their existence (Fig. 2).

The term *plumacy* covers not only the "science" (plumol-ogy; Hamilton, 2005), but also the psychology and scientology of plumatics, and the pseudoscience, Christian Science, and social science of plumes (sometimes misspelled "ploomz" by orthographically challenged aplumatics). However, we hope this new article will finally lay to rest any doubts about the existence or nonexistence of mantle plumes, just as Dietz and Holden (1987) finally laid to rest any doubts about Creationism. May doubts and doubters rest in peace!

In this article we follow the usual suspects and distinguish between the observable surface manifestations, traditionally called "hotspots," and the conjectured deep mantle plumes that Morgan (1971, 1972) proposed underlie sites of hotspot igneous activity or swell development. However, given the very, very slowly accumulating heatflow evidence (e.g., McNutt, 2002; DeLaughter et al., 2005) that hotspots (unlike their students) are really not so hot, we call them, in the new tradition of rediscov-ering Old English / Proto-Germanic agglutination (hot spot = hotspot), "notsohotspots." Hot lavas, after all, can be formed and erupted simply by depressurization, and some mantle melts more easily than other. For example, a small amount of water re-duces melting points. Some notsohotspots may really be all-wetspots. Or are those postulating this idea all wet?

Long after the last yellowed, bescribbled xeroxes of Holden-Vogt masterpieces were finally peeled off the office door of a paleo–graduate student long moved on to a tenured professorship at McDonald's Big Mac University, some may believe we are long deceased. Why else would we delay this reprise for thirty years—because of repeated rejections by narrow-minded jour-nal editors? Not so; we never gave them the pleasure! We have

*E-mail: ptr_vogt@yahoo.com.

Vogt, P.R., and Holden, J.C., 2007, Plumacy reprise, *in* Foulger, G.R., and Jurdy, D.M., eds., Plates, plumes, and planetary processes: Geological Society of Amer-ica Special Paper 430, p. 955–974, doi: 10.1130/2007.2430(45). For permission to copy, contact editing@geosociety.org. ©2007 The Geological Society of America. All rights reserved.

Figure 1. Authors Holden and Vogt as they try to steer their raft (this article) through the swamp of controversy over deep mantle plumes.

simply been monitoring the rate at which plumacy (oxymoronically dubbed "plumology" by Hamilton, 2005) is advancing as a science, and decided to revisit plumacy only every thirty years. Even so, readers will immediately note that most of our 1977 figures (not to be dismissed as "cartoons") remain perfectly valid today.

While the various mantle devils (a Holden specialty, perhaps reflecting his ancestry) depicted in Holden and Vogt's Figures 6, 7, and 10; God in their Figure 5; and Mickey Mouse in their Figures 3 and 6 do not age, the authors—as well as "plume-pop" Jason Morgan (Fig. 1 of that article and Figs. 1 and 3 of this article)—have aged and spread a bit. Morgan equated plumes to pipes, so back then we showed him smoking a 1000-km-long plume or pipe. But was Jason just blowing smoke, his mantle

plume concept a pipe dream? Or will aplumatics one day have to put plumes in *their* pipes and smoke them? In any case, Professor Morgan has stuck to his guns—or plumes—and still has a large following of plumatics, although the heatflow community has cooled to piping hot plumes. Yes, plumes have cooled down since 1977. But their friends and foes are hotter than ever—at least under their collars.

Why did Holden and Vogt dare to enter the ploomz swamp in the first place—as we do again (Fig. 1)? For one thing, because we both had jumped on the mantle plume (Morgan, 1971) bandwagon early—in the case of Holden, even before there was a band or even a wagon. We modestly failed to note this historical fact in Holden and Vogt (1977), but now the truth can be told. (If it's published, it must be true!) Probably unknown to

Figure 2. "Do we exist or not?" The deep mantle plume's view of his or her predicament.

modern researchers who don't cite, let alone read, papers from the last millennium, Jack (back then called John) Holden costarred with our late friend and mentor, Bob Dietz (1914–1995). With smug modesty we quote from geohistorian William Glen (2005, 99): "The influence of Dietz and Holden's 1970 Pangaea reconstruction paper on bolstering the theory of fixed mantle plumes is difficult to exaggerate."

Dietz and Holden (1970a,b) had based their Pangaea reconstructions directly on Wilson (1963)—as if they had referred to Moses's rock tablet instead of the Holy Bible. The pair had not yet heard of mantle plumes—because mantle plumes would be proposed only later (Morgan, 1971). In fact, who knows but that the father of mantle plumes, Morgan (1971), was in part led to ploomz by the "fixed thermal centers" of Dietz and Holden (1970 a,b)? That makes one of us (J.C.H.) a possible grandfather of mantle plumes—or at least a disreputable stepgranduncle.

The late Bob Dietz lived on to see the fixity of his fixed thermal centers challenged, so we imagine his ghost is still trying to

nail down the Tristan da Cunha hotspot (Fig. 4), a fixed point in the Dietz-Holden reconstructions.

Actually, as Wikipedia reminds us, Dietz began to think about these things much earlier—ten years before the Wilson (1963) paper. According to a Dietz biography, this *Wundergeologist* spied through the welter of Pacific seamounts "the Emperor chain of seamounts that extended from the end of the Hawaiian Island–Midway chain" and "speculated over lunch with Robert Fisher in 1953 that something must be carrying these old volcanic mountains northward like a conveyor belt." Many ideas are born over lunch, and we suspect the two had a few drinks. Geohistorians have failed to note that this epic Dietz-Fisher lunch gave birth to seafloor spreading / plate tectonics and hotspots / mantle plumes!

For the record, J. Holden still retains–out of reach of marauding mountain lions and eager collectors–and regularly polishes and strokes the globes used in those 1970 Eulerian analog reconstructions of Pangaea. Back then, *mirabile dictu,* geoscientists could move plates around without digital computers! When the *Antiques Roadshow* finally comes to Omak, those globes will be there!

And, only a year later (before Morgan, 1972), Vogt (1971) published unambiguous (to him, but ambiguous to others) seafloor evidence in the form of "V-shaped ridges (VSRs)" that hot upper mantle flows under the Reykjanes Ridge "pipe" away from Iceland. Thus, VSRs supported Morgan's (1971) plume hypothesis (see Jones et al., 2002, and Vogt and Jung, 2005, for basically unchanged interpretations).

Unlike the hotspot fixity ideas (Fig. 4) of Dietz and Holden (1970a,b) (which survived for a full three years, not refuted until Molnar and Atwater, 1973), VSRs were ignored or dismissed for many years. At one unforgettable early 1970s Princeton seminar, an impudent, "brilliant," but here unnamed Princeton graduate student even asked Vogt if the V pattern was not rather an artifact of plotting ridge-axis-parallel features on a Mercator map! (Finally, years later, the Reykjanes VSRs were rediscovered by Cambridge University, where R. White militarily renamed them "chevrons"; e.g., White and Morton, 1995).

Considering Figure 4 of Holden and Vogt (1977) and the abundance of fish around Iceland, we suggest that the pattern of nested VSRs better resembles the ribs of an Icelandic cod skeleton than spineless chevrons, with the Reykjanes Ridge spreading axis as its backbone. To this day, the VSRs south of Iceland stick in the craws of aplumatics, who have few alternative interpretations. Are these plume skeptics being asked to swallow a fish or just a big fish story?

We begin by identifying a few of the trends that have developed in the three decades since Holden and Vogt (1977).

THE RADIATION OF PLUMES: STILL MORE EVIDENCE FOR EVOLUTION

The most obvious plumatic development has been what Anderson and Natland (2005, p. 120–121) lament as "a large

Figure 3. Iceland mantle plume, as modified from Figure 6 *in* Holden and Vogt (1977). Resemblance of humans (not to scale) to D.L. Anderson, G.R. Foulger, and W.J. Morgan not coincidental.

number of ad hoc modifications, such as incubating plumes, fossil plumes, plumelets, superplumes [more about those below], lateral plumes, secondary plumes, chemical plumes, radially zoned plumes, and so on." The "so on" includes wandering plumes (Fig. 5). Seismic tomography, if interpreted in terms of plumes (seek and ye shall find), has further multiplied the possible plume species (what we here call "tomoplumes"; e.g., Bijwaard and Spakman, 1999; Montelli et al., 2004; Dziewonski, 2005), and numerical modeling (e.g., Phillips and Bunge, 2005) shows a great variety of plumelike upwellings (what we here call "numeroplumes") from a thermal boundary layer in the mantle, especially the core-mantle boundary. However, numeroplumes apparently cannot arise by internal mantle heating.

Radiation (the evolutionary kind) has created many species, subspecies, and races of tomoplumes and numeroplumes—a challenge to plumotaxonomists. The various types of plumes are multiplying and evolving even as we write, but the process (sexual reproduction? dividing like amoebae? virgin birth?) is not understood. Differences in size, shape, and color help the layman distinguish the species. However, readers should remember that numeroplumes, although interesting and beautiful, are only virtual plumes, like the belly dancers and gunships in videogames. However, both tomoplumes—which, as they are based on real data, must be real—and numeroplumes owe their evolution and radiation since Holden and Vogt (1977) to the evolution of computers with ever more memory and computation speed.

Figure 4. The ghost of Robert S. Dietz (1914–1995) is still trying to fix the Tristan da Cunha hotspot, a key assumption in the continental reassemblies of Dietz and Holden (1970a,b) and later fixed-hotspot-based models.

Human fantasy and the enormous volume (920,000,-000,000,000,000 m^3) of Earth's mantle allows room for almost every imaginable mantle structure, composition, and process, including plumes (e.g., Fig. 9 of Holden and Vogt, 1977; Fig. 1 of Anderson, 2005; Fig. 8 of Ivanov and Balyshev, 2005; Fig. 1 of Lay, 2005; Fig. 8 of Smith, 2005). Figure 1 of Anderson (2005)

made the "cover our butts—they can't both be wrong" cover of P^3, an astonishing accomplishment given that Anderson was only the fourth editor. The controversial P^3 tome may someday rank with Darwin's 1859 *Origin of Species,* Morgan's *Plate Motions and Deep Mantle Convection,* and the even more controversial but less well-known opus published by Morgan a decade later:

Figure 5. Professor A. Peccerillo flunks a hapless wandering Italian plume (see also Peccerillo and Lustrino, 2005).

The Aquatic Ape (Morgan, 1982). Lay's Figure 1 attempts to find a middle ground in the split lobotomy shown on the P³ cover. Some of the Layman's plumes rise from the core-mantle boundary, while others emerge like large hatched worms from the tops or flanks of giant manure piles in the lower mantle, which resemble that strange object (a giant sea cucumber with flattened ears?) labeled "Ancient" in the "Plate Model" of Anderson (2005). This object was supposed to be a superplume (G.R. Foulger, 2007, personal commun.), but deep-mantle superplumes are actually acceptable to Professor Anderson so long as they don't crash the plate tectonic party.

The aplumatic community dismisses as largely untestable adhocity—as a bunch of moving targets (and moving plumes)—the wealth of plume variants we have described. These all evolved in a mere thirty-five years from the simple protoplume, which, like Minerva/Athena, "leapt forth from [her father's] brain fully formed" (from the Ancient Greek, by way of Wikipedia)—in this case, from the mind of Morgan (1971, 1972). We don't know whether, like Zeus/Jupiter, W. Jason suffered from headaches during his mental pregnancy, but we can imagine that the Great Aplumatic Challenge must be a real headache for the aging professor.

However, skeptics fail to understand that Darwin also spoke to mantle plumes! We have simply experienced a Darwinian "hypothesis radiation" from simple, geophallic Morganian protoplumes. The "drivers" in this evolutionary radiation have been new and different data, good data and bad data, as well as good and bad ideas. After all, we humans now enjoy and try to preserve thousands of species of birds, all presumably descended from an Archeopteryxlike protobird. (They may deny it, but plumologists/plumatics and plume skeptics/aplumatics both descended from the same early Mesozoic mammals in a series of evolutionary spurts called radiations, one after the Cretaceous/Tertiary extinction—perhaps caused by the deniable [Sheth, 2005] Deccan plume head—and another during the Late Paleocene thermal maximum.)

Charles Darwin will ultimately decide which of the many plume and nonplume hypotheses, tomoplumes and numeroplumes, will survive to be reproduced in future textbooks. In the meantime, as we preserve biodiversity, shouldn't we also preserve plumodiversity? The difficulty or impossibility of disproving some plume variants (e.g., Anderson and Natland, 2005) simply helps maintain plumodiversity and keeps our planet interesting, our journals full, our conferences attended, and our grants funded! As long as debate continues, our planet is really many planets—one real and many solar systems of virtual ones. It will be a boring world indeed when the one real planet stands up.

As for semantic confusion about terms such as *plume* and *hotspot* (Anderson and Natland, 2005), that's no big deal either. It's simply "fuzzy semantics," and, like "fuzzy math," should not be confused with "fuzzy thinking." After all, a plume by any other name is still a plume!

Age Progressions and Younging and Aging along Hotspot Tracks—Do We Suffer from Too Many Data?

In a field famous for researching stuff that happened unimaginably long ago, plumatics have refreshed the lexicon by speaking of "younging directions" and "younging rates" along hotspot tracks, as if they had reversed the very arrow of time. Most of us would be happy to young in any direction! Perhaps the younging of volcanoes leads us to the mantle plume as the fountain of youth, its overhead lava fountains giving birth to reddish infant rocks ever so briefly of zero age.

The Morgan mantle plume hypothesis involves plume fixity—already downgraded to relative fixity by Molnar and Atwater (1973)—and then an age progression along the supposed track. The original plume hypothesis made some specific and testable predictions, and early data seemed to pass the test. However, many a great theory has been destroyed by too many data, and age-dating volcanoes has become as troublesome and error-prone as age-dating certain ladies. In the good old days, two age dates along a chain sufficed and had a 50% chance of capturing the right younging direction. The odds were even pretty good, with the large error bars of whole-rock K-Ar dates on ancient rotten, worm-infested basalt, that those dates would fit predictions of fixed hotspot models.

Then came more and more age dates, filling the scatter plots with clouds of data, like swarms of killer bees, comprehensible only if hotspot volcanoes erupt any time and any place along the track—or anywhere else, for that matter. Inconsistencies began to arise, even for the most "Hawaiian-looking" seamount chain in the Atlantic (the New England seamounts; McHone, 1996). Things went downhill for the chain, plumologically speaking, after the first (1979) age versus distance scatter plot was later shown to be based on data fabricated by a brilliant but overenthusiastic Woods Hole / MIT plumatic graduate student.

The biggest blow to hotspot fixity was the failure of age dates along the Emperor seamounts to conform to formation at the present site of Hawaii. After we learned that "HowEYE" was actually "HaWAH'ee," a pillar in the Temple of Plumacy began to crack! Now the truth is out: plumes have been liberated to wander (e.g., the downgraded Italian plume; Fig. 5), or even streak along singlemindedly in the Earth's mantle, perhaps floating balloonlike in the mantle wind, before settling down at one spot, carefully chosen to overlie a lower-mantle seismic wavespeed anomaly (see under "Intelligent Plumes" later). Even the most famous bend on the planet, the Hawaii-Emperor Bend, became, like its students, "slower and older" (Sharp and Clague, 2006). The bend had begun to move in time, after sitting peacefully at 43 Ma sharp for many years. Just recently it has begun to age—at least a million years per year—and is now a grizzled 50 Ma.

Plumatics should be relieved, however, that Ajoy Baksi (this volume) is discarding one isotope age date after another, reducing bloated data sets to very small sizes or nil. This reverse trend

Figure 6. Professor E. Garnero (Garnero et al., this volume) diagnoses Earth with a case of deep-mantle superpiles.

can bring us back to the early golden age of plumacy, before the data explosion began to misfit predictions and thereby undermine theories plumatics know to be right.

Superduperplumes and other Geohyperbolae

Yet another trend of the past thirty years has been increasing use of geohyperbolae. It seems that normal terms with adjectives (or letting the numbers speak for themselves) no longer suffice. For example, topographic swells (rises) were known (about as well as today) at the time of Holden and Vogt (1977), but then came the snap-off-your-tongue "superswells" (e.g., the South Pacific superswell, McNutt and Fisher, 1987). A few years later, superswell scientist Marcia McNutt et al. (1990) dubbed the inferred, albeit nonexistent, Cretaceous Darwin Rise an "ancient superswell." Of course, superswells require "superplumes," also called "megaplumes" (e.g., Anderson and Natland, 2005; Dziewonski, 2005), and the superplume below Africa is also called the "Great African Plume," but may recently have been downgraded to a mere superpile (Fig. 6; Garnero et al., this volume). Superplumes, like Superman (created in 1932 and still wearing blue and red, like political maps of the United States or tomographic maps of seismic wavespeed in the mantle), can accomplish the seemingly impossible, as illustrated by Cox and Van Arsdale (2002), who attribute the very Mississippi Embayment to a "Cretaceous superplume event." And what on Earth besides superplumes could "control magnetic reversal frequency" (Larson and Olson, 1991)?

Meanwhile, the mid-oceanic ridge (MOR) community was not to be outdone, hyperbolically speaking. Early papers had distinguished between slow- and fast-spreading ridges, but then the East Pacific Rise, ~10°S–20°S, became a "superfast" spreading center, even as the long Cretaceous period of normal polarity became—you guessed it!—the "Cretaceous superchron." And although marine scientists have known since the late 1960s that the Southwest Indian and Gakkel (Mid-Arctic or Nansen) ridges were spreading slowly (7–15 mm/yr total opening rates), more recent authors decided this qualified as "ultraslow" (e.g., Michael et al., 2003) or "superslow" (Mendel et al., 1997). This forced the senior author to find another term for the Terceira rift, opening at only 4 mm/yr, still slower than ultraslow/superslow. Vogt and Jung (2004) thus introduced "hyperslow" to the plate kinematic lexicon. About the same time, long MOR segments—e.g., along the Southwest Indian Ridge—became "supersegments."

We don't yet know what plumatics will call the small end of the mantle plume spectrum (mini-, micro-, or even nanoplumes?), although minihotspots have already been discovered and named "petit spots" (Hirano et al., 2006). However, given the difficulty of finding or disproving even macroplumes, wouldn't disproving miniplumes be as challenging as looking for midget Loch Ness Monsters (discussed further later)?

Will the plume-slayers soon follow with their own hyperjargon? Iceland is by any measure a globally unique superhotspot (even if it's not superhot, or even hot). If Iceland's raison d'être is instead the regurgitation of subducted Caledonian crust (Foulger et al., 2005a), it must have involved supersubducted supercrust; otherwise why aren't there more Icelands on our planet, given the plethora of plates subducted over the eons? However,

Figure 7. Two contrasting Venutian views: Hasn't Professor Hamilton made enough of an impact already? Is Professor Jurdy seeing things?

plumatics lag behind MORmen like Ken Macdonald, who compared the MOR to an "infinite onion." Although the infinite plume has not been postulated, we may be dealing with an eternal plume controversy.

HOT PLUMES TAKING THE HEAT (BUT MAYBE NOT GIVING ANY)

It's an understated understatement that mantle plumes, especially the deep ones, have taken a beating in recent years: if people like Don Anderson and Gillian Foulger keep chopping away (Fig. 3), will even the Iceland pipe eventually lose its footing? Meanwhile, papers are beginning to appear with negatives brazenly placed in their very titles. (Like most scientists, we have time to read only the titles of papers we cite.) Gillian Foulger innocently asks, "Why the current skepticism?" (when we know Foulger herself is largely responsible). Hetu Sheth (2005) finds "no trace of a [Réunion / Deccan Traps] mantle plume." Rocchi et al. (2005) find "no plume, no rift magmatism in the West Antarctic Rift." In that deceptively named Internet tabloid www.MantlePlumes.org, a Web page headline screams "STILL NO ITALY PLUME!" while author A. Peccerillo says, "No

thanks" to the Italian plume (Peccerillo and Lustrino, 2005). Aplumatic Anderson (2000) finds "no role for mantle plumes" in the upper mantle's thermal state. Fairhead and Wilson (2005) ask: "Do we need mantle plumes?" ("Not really," we knew their answer would be.) Ritsema and Allen (2003) find plumes "elusive." Heatflow experts found that hotspots weren't so hot (e.g., De Laughter et al., 2005), and even the heatflow variations over that hotspot's hotspot, Hawaii, were found by McNutt (2002) to reflect mere porewater movement, "not plume properties."

Hamilton (2005), following the declaration of airports and restaurants smokeless, examined 21 Galileo mission radar images of the Venusian surface and declared the entire planet (Venus) "plumeless" (Fig. 7), although he hedged this (p. 789) with "no need for plumes." Nitpickers point out that there was no need for Warren Hamilton to exist either, but he does exist. (On some days, Venutian plumatic Donna Jurdy [Fig. 7; Matias and Jurdy, 2005] and others [e.g., Kiefer and Hager, 1991] probably wish he didn't.) Even on home planet Earth, Hamilton had announced only three years earlier, to a shocked and awed audience of geophysicists, "Plumes do not exist" (Hamilton, 2002). Meanwhile, ignoring Mark Twain's warning about premature obituaries, McNutt (2006) is pounding "another nail in the

Figure 8. The explosion and crash of mantle plume populations.

plume coffin." Glen (2005) even went so far as to downgrade plumacy from paradigm to "quasi-paradigm." Even notsohotspots have been questioned: Once a confirmed plumatic himself, waffler Vogt (1991) asked if Bermuda was not instead formed of "non-hotspot processes." And (as befits his usual guarded ambivalence in discussing mantle plumes), Anderson (2005) carefully disguised his skepticism in a double entendre: His paper's title begins with "Scoring hotspots," and we note that definitions of *scoring* include "to cancel by drawing a line through, to cut so as to mark with scratches," and "to berate, excoriate, castigate. . . ." Thirty years before (Anderson, 1975), he had already castigated plumatics as "fumers, smokers, swaggerers, scribblers, pen-pushers, dungers, and manurers" (reworded from Holden and Vogt, 1977).

We doubt, however, that true plumatics will be deterred by all this naysaying; indeed, they will only dismiss the naysayers as former U.S. vice president Spiro Agnew (or, more likely, his speechwriter) once famously dismissed his critics—as "nattering nabobs of negativism."

PLUME POPULATION EXPLOSION AND CRASH

Holden and Vogt (1977) first noted with alarm the burgeoning plume population (from 20 in 1972 to 122 by 1976). Using state-of-art mathematics and card-punched FORTRAN II, they predicted that nearly one million would exist by the year 2000. Vogt and Holden (in the present article) are happy to report that we overestimated, albeit only by two to three orders of magnitude (not too shabby in fields such as astronomy, astrology, social science, and geoscience). However, the actual plume population did reach "about 5200" at the end of the millennium (Malamud and Turcotte, 1999). Plumes had evidently reached an unsustainable population (like lemmings in the Arctic, rabbits in Australia, and humans on Earth), and turf wars among plumes began at depth, thus explaining the increased volcanic, seismic, and tsunamigenic activity in and on the overlying plates in the last quarter century.

However, as sometimes happens with stock markets, oil supplies, real estate, and the price of tulip bulbs, the plume population crashed (Fig. 8), as already noted by geohistorian Glen (2005). Some would call it a "bubble effect"; bubbles bear an uncanny resemblance to the heads of mantle plumes (e.g., Griffiths and Campbell, 1991; Campbell, 2006). At present, the mantle plume population may be as small as five or six (Courtillot et al., 2003). Alarmed, the Plumatic Asylum recently sued the U.S. government to force inclusion of mantle plumes as "federally endangered species." However, plumatics can hope that Stephen J. Gould's "punctuated equilibrium" theory—small populations are able to evolve quite rapidly—will apply to mantle plumes. In fact, we see evidence of this rapid evolution in the present volume.

While plumes have declined in number, plume (and no-plume) publications continue to explode. P[3] and the present volume have added yet more to the throng. Just from 1970 to 1984, the annual number of papers (N_p) with "hotspot" or "mantle plume" as keywords had risen from zero to approximately seventy-five (Fig. 5 of Tucholke and Vogt, 1986). Anderson and Natland (2005) performed a search similar to that of Tucholke and Vogt, showing an early 1970s peak and a sharp increase in plume publication rate after Griffiths and Campbell (1991) revealed to all who would read their paper that plumes had heads and tails (the "heads I win, tails you lose" phenomenon). A further ballooning of mantle plume publications was inspired by Kellogg and Wasserburg (1990), who showed how ^{3}He could be transported, like Josephinite, from the lower mantle by rising plumes, and why mantle plume–derived helium, which is even lighter than light, also makes balloons rise faster and higher.

We dare not extrapolate plume publication rates into the future; with electronic storage and plenty of room on neighboring planets, there seems to be no physical limit to the number of papers that can be published about mantle plumes or hotspots! As t approaches eternity, so will

$$\sum_{1970}^{\infty} N_p(t)\Delta t, \qquad (1)$$

the cumulative number of plume/nonplume publications.

We attribute the remarkable opposing trends of "plume" versus "plume paper" populations to the Loch Ness Monster principle (LNMP), which we define here for the first time. Very few research or other papers are written about the salmon in Loch Ness; those salmon, although faster, smaller, and easier to digest, are like the now familiar tectonic plates: Why study and issue papers or press releases about the known or obvious? Earth is round, and plates exist, at least for the great majority of us (although some antique cars still sport Holden's "Stop Continental Drift" bumper stickers). Meanwhile, innumerable papers and books, not to speak of news reports and films, have appeared and continue to appear on the search for the Loch Ness Monster, which, perhaps like the elusive mantle plume (Ritsema and Allen, 2003), lives unseen at depth, like the giant mantle carp in Figure 4 of Holden and Vogt (1977). We define the LNMP thus: "The number of studies conducted and articles published about a concept is inversely proportional to its plausibility." However, we predict that even LNM researchers will at some point give up their search and simply declare the creature forever real, as in "Yes, Virginia, there are mantle plumes—and one lit up briefly under your namesake state's bottom in Middle Eocene times."

EVIDENCE FOR THE SHAPE, COLOR, BEHAVIOR, AND INTELLIGENCE OF MANTLE PLUMES

Holden and Vogt (1977; their Fig. 3) already demonstrated that mantle plumes can speak to literate geologists (at least in Latin), using fracture patterns (see also under "Crack Hypothe-

ses" later). If so, plumes must be alive. However, the mainstream geoscience community failed to recognize this, and only in later years even discovered that plumes have heads and tails (Griffiths and Campbell, 1991), and that they can move about and change their shapes and colors at will. Seeming to defy Reynolds numbers and thermal diffusivities, mantle plumes are able to change really fast—from one paper to the next. To escape detection, they often masquerade as food items, making the mantle wrongly appear, to those fruity nutcakes who smoke what they shouldn't, as marble cake or plum pudding (e.g., Anderson and Natland, 2005), made visible only when mathematically mixed with banana doughnut kernels. Unfortunately, some plumes, barring unlikely errors in the superscience of seismic tomography, seem miserably deformed, staggering, tottering, twisted, Quasimodolike creatures (e.g., Bijwaard and Spakman, 1999; Zhao, 2004; Ritsema, 2005; Van der Hilst and de Hoop, 2005).

Computer coloring has also allowed advanced seismotomographers to elucidate plume color (though color has so far eluded paleontologists studying dinosaurs). Typically, tomoplumes are recorded and digitally mastered on tomographs, which resemble paint-by-number paintings, as each color equates to a given numerical range of sound or shear wave speed. Such tomographs are then published in coloring books and discussed in color bars. Unfortunately, these books are precolored, preventing readers from exercising their own imaginations and coloring pencils to find their favorite plumes (as well as Mickey Mouse and Pluto) in the crazy patterns.

Most virile plumes turn out to be saturated red, but weaker, perhaps diseased or juvenile plumes can be yellow or even sickly greenish-yellow. This conveys the basic idea that plumes cause fluids to ooze out or be vomited from the Earth's interior.

Negative plumes may also exist, and are blue (e.g., Phillips and Bunge, 2005). Blue negative plumes have their heads down (versus looking up at the Heavens) as some politically blue Americans allegedly do.

Plumatics found it especially demeaning for Anderson (1999) to dismiss colored tomoplumes as mere Rorschach (pronounced "raw-shock" according to Wikipedia) inkblots. After all, modern psychiatrists score the inkblot tests of hapless patients according to their "level of vagueness" and take note of "any illogical, incongruous, or incoherent aspects of the responses" (Wikipedia). Endless graduate student hours spent fine-tuning computer codes, hours of supercomputer time used inverting uncountable numbers of seismic recordings, and the result is just a colored inkblot? Next thing, unimaginative parade-rainers will be telling us the stars in the Big Dipper do not resemble a large bear! Fortunately for plumatics, Anderson's (1999) paper appears to have been rejected by peer reviewers, perhaps because he failed to realize that *Rorschach* was just the nom de plume of an early plumatic. *Ror* is misspelled German for *Rohr* (pipe), and *Schach* means "chess," so Freudian psychiatrist Hermann Rorschach (1884–1922) was playing games with pipes, as many still do today.

Returning to the question of plume color, we note that

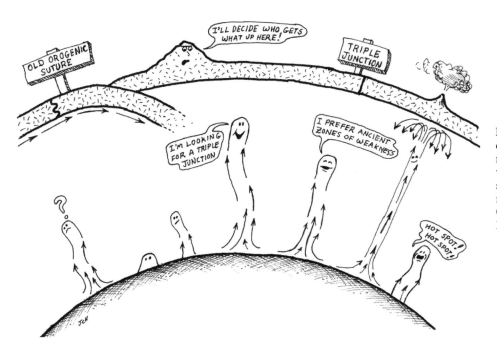

Figure 9. Mantle plumes (actually colored red or yellow) are born in the lower mantle, but then rise in this world, always on the lookout for lithospheric weak spots such as triple junctions, ancient igneous centers, and intracontinental rifts. Only dumb plumes bump their heads on Precambrian cratons.

Dziewonski (2005) demonstrated with statistics (which Mark Twain taught are *not lies*), that most plumes are "at home" in a red district of the lower mantle. Blue districts, he found, are pretty plumeless. Unfortunately, GSA's editorial policy prohibits our drawing political inferences from this observation. However, as we tend to associate red, not blue, with higher temperature (with apologies to Max Planck and astrophysics), the red color of plumes must mean they are warmer than the surrounding mantle, so we consider further debate over plume temperatures pointless.

In this article, we further assume that plumes are living beings, and represent them as such in Figures 1, 2, 5, and 9. Older readers will note a certain resemblance (Fig. 9) to the "schmoos" of the late Al Capp's comic strip *Li'l Abner* or to Halloween ghosts or Klansmen in white bed sheets. Granted, but such resemblances are purely coincidental. Readers should not assume from our B&W renditions, whether arrow-headed or not, that plumes actually resemble ghosts, but rather that we are too cheap to pay for color illustrations. We also disavow any racial overtones in our depiction of plumes, shown by seismologists to be red, as white and black.

Given that some aplumatic readers still doubt the intelligence of plumes, we offer Figure 9 as proof. Young, growing plumes seek adventure and want to grow into the upper mantle, where they can speak up and be heard (at least by seismologists). However, since they are intelligent, they scan what's overhead and naturally plan to seek out weak spots like rifts and previously active volcanic areas. An example of a teenage rebel plume is the Oregon High Lava Plains plume (OHLP), a ca. 15 Ma offspring of the Yellowstone plume (Jordan, 2005). The OHLP decided not to sit still, but swam west, leaving a track 120° from the predicted one. In the last 5 m.y. it has slowed down,

however, and may be deciding to turn around and come back home. Only real superdumb plumes would aim for and bump their heads against the bottoms of Precambrian shields, and we may never know who they are!

If really lucky, a smart young plume sees a triple junction overhead and heads for that, never mind the angle. Plumes *must* have some intelligence; otherwise why would the Shatsky Rise large igneous province be a triple junction trace (Sager, 2005), or the Azores plateau (e.g., Vogt and Jung, 2004) the site of a modern triple junction? Why else would African igneous activity recur intermittently at the same sites over hundreds of millions of years (Bailey and Woodley, 2005)? New plumes below Africa evidently like to follow in the mantle footsteps of their granddaddies and great-great-granddaddies, an observation that appears to put an upper limit on plume intelligence. Finally, it has already been shown by Vogt (1972) that plumes are able to communicate with each other over tens of thousands of kilometers. How else could they plan to rise in unison against those overbearing plates?

To be fair, we note that mantle plumes were first shown to be living, intelligent beings by Professor Erin Beutel at the 2005 Fort William Chapman Conference. In one of the stimulating evening sessions, held in the hotel's (licensed) dining salon, Dr. Beutel represented herself as a mantle plume, and to musical accompaniment by a Scottish Highland band, offered the premier performance of The Dance of the Mantle Plumes. Did she learn this from Venutian plumes (Fig. 7)? A bit vertically challenged, however, Dr. Beutel could represent only a scaled-down (1:1,000,000) plume. We assume this was one of her NSF or NOAA-funded "professional outreach" activities. We hope Dr. Beutel's Dance of the Mantle Plumes, which was immortalized on videotape by Dr. Bruce Julian, the official conference

Figure 10. Grading (scoring) deep mantle plumes.

filmer-of-unseemly-behavior, will soon be released by the GSA and/or AGU as a DVD.

Although we are not certain, deep mantle plumes may also have feelings, and we should give them the benefit of the doubt (as some conservatively assume human cells are equivalent to human beings). Plumes probably are not at all pleased being represented as deep-mantle schmoos (Fig. 9) or given low grades (e.g., Courtillot et al., 2003; Anderson, 2005; Figs. 5 and 10), although they most probably prefer that to other representations (elsewhere or here and now) as objects such as giant toadstools, water towers, pipes, thunderheads, umbrellas, mushroom clouds, piles (Fig. 6), or phalluses. (The ghost of Sigmund Freud smiles knowingly at the many geophallic depictions of salt domes, magma chambers, and mantle plumes—probably reflecting the excessive amounts of time geoscientists spend in the field or at sea on research cruises or glued to their computers.) However, our mantle plume friends will surely agree that even depiction as toadstools, piles, or phalluses is preferable to outright nonexistence.

HAS THE "MULTIPLE WORKING HYPOTHESIS METHOD" GONE MAD?

In our poll on the Cretaceous-Tertiary boundary (Vogt and Holden, 1981), we noted the large number and wild variety of hypotheses that had been advanced for the extinction of di-

nosaurs. (Poll results remain unpublished because we still await more replies, although none have arrived in the past twenty years.) Evidently the Multiple Working Hypothesis method (MWHM) had gone mad. However, shortly before we had finished our draft, the Alvarez et al. (1980) iridium spike had speared this hypothesis zoo—which had begun to resemble the extragalactic truck stop in *Star Wars*—into extinction. As is often the case in the Earth and other environmental sciences, the magic iridium spike was not found by testing the asteroid impact hypothesis, but was tripped over in the process of an unrelated study. Similarly, magnetic anomalies were not mapped in the 1950s and early 1960s to test seafloor spreading. As plumatics trip the light fantastic, they have perhaps just not yet tripped over their equivalent of the iridium spike.

Can Asteroid Impacts Melt Rocks, Cause LIPs, and Give Rise to Plumes?

It's no secret that impacts convert kinetic energy into heat, and very large impacts on the moon melted some of its green cheese, flooding impact sites with oceans of magma, the marea. However, geologists were slow to learn from lunologists. One down-to-Earth man stood out, however, Robert S. Dietz. This giant discovered most of what we know about the Earth today, including seafloor spreading, fixed hotspots (Fig. 4; Dietz and Holden, 1970b; Bob Fisher, 1953, lunch discussion), the fact

Figure 11. Scientists searching for proof that impacts might cause LIPs or hotspot trailheads never seem to catch this secretive pair together.

that Earth is more than 6010 years old at this time (Dietz and Holden, 1987), and, yes, that giant impacts occur and can melt the Earth's crust (Dietz, 1961). Dietz called such large impact craters "astroblemes" (star wounds), e.g., that nickel mine in Canada that earlier geologists had called the Sudbury Lopolith. Dietz (1986) topped it all when he showed ignorant geologists that a bolide triggered the Bahama Nexxus (whatever that is), the Newark Series, and related igneous activity along the eastern U.S. margin (McHone, 1996), while splitting Pangaea and triggering seafloor spreading! How much else *is* there to explain?

Eventually other geologists began to realize that impactors could have melted Earth rocks (e.g., Green, 1972). Moreover, Alvarez et al. (1980) later exhumed the bones of terminal-Maastrichtian dinosaurs and foram tests, washed off the stardust, and showed they all had died of astroblemes. In recent years, the idea of large bolides melting large volumes of rock to form large igneous provinces (LIPs) or even give rise to plumes has been revisited, if not yet rehabilitated (e.g., Jones et al., 2005). One problem faced by LIP neoastroblemicists is that the LIPs and other hotspots seem never to be in the same place as the astroblemes, and contrariwise (Fig. 11). This problem is being overcome with auto-obliteration, which is similar to what often happens to furry critters on roads and ultimately to SUVs once oil begins to run out. Jones et al. (2005) suggest the Ontong-Java plateau, the largest heap of lava on the present Earth, could have

been formed by an asteroid striking young oceanic crust. However, most investigators, while scratching their heads over the Ontong-Java, consider it only a geobleme, and so far Jones seems not to have made a major impact.

Driving Mantle Convection Models over the Edge

While some authors have gone over the edge in their attempts to explain notsohotspots, King (2005), building on King and Anderson (1998), decided to go under it. Modeling the thermal convection that is generated in the asthenosphere by the thermal contrast between thick, cold cratonic lithosphere and thin new oceanic lithosphere, he predicts downflow under the continental edge and upflow under the middle of the adjacent oceanic plate. Once an ocean has widened to 2000 km or more, this upflow (which to us looks more like sideways flow) generates off-ridge hotspots, King suggests. However, the author admits that this model is no better than the plume model at explaining the North Atlantic / Icelandic notsohotspot. We are suspicious of papers in which the author(s) does (do) not claim to have explained everything with their model.

Can Platonics Explain Notsohotspots?

Most other nonplume explanations for notsohotspots attribute them to processes or properties involving plates and their

Figure 12. Geo-zippergate: An otherwise reputable geoscientist still looking for the legendary Great Hawaiian Propagating Crack southeast of the Big Island and Loihi, unzipping the Pacific plate to let Hawaiian lavas flow out.

interactions (Anderson, 2005). For example, lithospheric stress is blamed for cracking the Earth and forming "hotspot tracks"; localized volcanism is attributed to fertile patches in the upper mantle and to zones of plate weakness. Globally synchronized volcanism (Vogt, 1972) is attributed to episodic and ongoing reorganizations of plate motions, causing episodes of increased midplate stress. Following Anderson (2005) and Plato, we call the "plate model" or "plate paradigm" platonics and its practitioners and supporters platonizers, who shave with Ockham's Razor and dream of an idealized world where Newtonian physics explains everything. However, plumatics point to one definition of *platonic* as "of an academic nature, devoid of substantiality."

Cracked Models

Explaining midplate igneous activity in terms of cracks is not a new idea. Some otherwise respectable scientists continue to believe that the Hawaiian chain is being formed by a crack forever propagating, in response to stresses within the Pacific plate, at a few cm/yr toward the East Pacific Rise (Stuart et al., this volume). Finding the actual crack, however, has proved challenging to the stressed-out experts (Fig. 12). Other geologists, traditionally skilled at connecting two volcanoes with a straight or curved line (e.g., Vogt, 1974) to define the fracture required to support

their pet structural model, have interpreted the pattern of Hawaiian volcanic centers as a response to regional stress fields (Jackson and Shaw, 1975). Holden and Vogt (1977, their Fig. 3) played this same game, showing that plumes do "talk" to anyone willing to read the message in the cracks. After all, are volcano distributions that different from tealeaf patterns? A much more recent crack paper by Natland and Winterer (2005) waffles (Neogenically) in the plume versus crack debate, but reads a different message in the Cretaceous "tealeaves," that is to say, volcanoes. Those authors posit that large numbers of western Pacific underwater volcanoes, now guyots and seamounts, erupted through diverse cracks (versus from mantle plumes) in the young Pacific plate.

We can't offer more insight, and we resist the temptation to put into print puerile puns such as "The crack hypothesis, advanced by half-cracked crackpots, may not be all it's cracked up to be, although it cracks up plumatics," not to speak of unprintable Freudian allusions.

The Regurgitation or "Fertile Earth Mother" Hypothesis (Iceland)

Foulger et al. (2005a; hereafter FNA) offer a different platonic, aplumatic model for Iceland, which has stubbornly clung to the MOR axis even as the latter is trying to get away from its embrace (e.g., Lundin and Doré, 2005). This (both the papers and the observation) has annoyed plumatics, who expected a bit more model compliance from their star ridge-centered hotspot. The FNA model explains why this notsohotspot is also notsofixed.

FNA attribute the "large melt volume to the remelting of subducted oceanic crust trapped in the Caledonian suture in the form of eclogite or mantle peridotite fertilized by resorbed eclogite." This concept of a Fertile Earth Mother was first recognized by the Norwegian Vikings who settled Iceland and discovered its fertile basaltic soil (which, however, soon was washed into the Atlantic because of fertile overgrazing sheep, even after the Lord Himself became their Shepherd in A.D. 1000). *MOR* is, after all, the Norwegian word for "mother," so the Vikings also were first to speculate that life itself may have originated in ancient hot, wet cracks.

We find it miraculous that subducted Caledonian crust was quietly carted along over the Earth in the cargo hold of the giant Pangaean plate—at least 600 km thick (depth to base of present seismic anomaly!)—for the 300+ m.y. between Caledonian subduction/suturing and the first Paleocene melting, even before Greenland and Eurasia had begun to pull apart. Another miracle is the seemingly inexhaustible supply of basalt magma; is the Iceland melt source a Hannukah-type oil lamp that miraculously burns for seven days (whereas for God, of course, one day may be at least ten million years)? The Foulger et al. (2005a) "Hannukah oil lamp effect" (HOLE) is explained by those authors as the result of the Caledonian suture's "X" relation to the Mid-Atlantic rift, allowing the MOR to overlie subducted crust for ages . . . and ages. . . . However, X may also represent the unknown substance FNA were smoking when they came up with this idea.

FNA have gingerly stepped away from the dangerously but not anomalously hot partially molten mantle that many (e.g., Vogt, 1971; Jones et al., 2002; Sleep, 2002) think they know must flow along and under the MOR, driven by the along-ridge topographic head (as the "ridge-push" force acting on the plate at right angles to the MOR). FNA were not sure what to say about those pesky V-shaped ridges, long touted as evidence for such flow, so they didn't say anything, because in fact their hypothesis predicts only the flow of mantle's milk, not a stampede of its cows! But maybe it's unfair to expect any theory to explain *all* observations!

A notoriously deferential scientist by nature, Foulger (2006) has also recently rehabilitated the brilliant but unfairly discredited ideas of geopioneers Alfred Wegener and V.V. Beloussov, neither of whom was fooled by Iceland's surficial oddities such as active volcanoes or Pleistocene–Holocene lava flows. They—and Foulger—knew, deep down, Iceland's underlying truth: that continental crust must lie cleverly concealed under all those lava flows. Wegener also realized that places like Iceland are "in the way" of good pre-drift reconstructions, just as Iceland still squats darkly in the way of our understanding deep-Earth pro-

cesses, taunting us all—platonizers, plumatics, astroblemicists, or whatever.

CONCLUSION: WHAT ENDGAME PLUMACY?

The plume debate is starkly defined: Deep mantle plumes (like the Loch Ness Monster or God) either exist or they don't exist—just as one is either pregnant or not pregnant. Discovering that second plume or second Loch Ness Monster wouldn't make the front page. If plumes exist, their exact number and locations are not quite so interesting. In fact, there would probably have to be a reproducing population.

During the Fort William Chapman Conference discussions, D. Zhao was asked if seismic tomography would eventually resolve the issue. Zhao speculated that this could well happen within the next decade, when every last seismogram wiggle has been dialed into the inversion. We call this the "IELW" (invert every last wiggle) prediction (Fig. 13) and have buried it in a time capsule.

We suspect that the deep mantle plume problem can't be solved (or resolved) with certainty except by detonating the world's entire nuclear stockpile and recording the signals around the world. As a relatively minor benefit to mankind, to provide "added value" besides settling the War over Plumes, the threat of nuclear war and more mass extinctions would be simultaneously eliminated. Of course this would cause massive unemployment among highly trained scientists, engineers, technicians, program managers, and nuclear test ban treaty junkies, not to speak of scientifically illiterate politicians. However, we offer the idea of this experiment in a spirit of cautious optimism that it won't actually happen.

It's also possible that a volcano somewhere has already coughed up a xenolith of a rock that once actually saw and felt the core-mantle boundary. Who's to say it *isn't* possible? A rock named Josephinite once claimed to have been carried up from the very core by a deep mantle plume, but Miss Josephine was exposed as a pretender. Short of discovering such a plumatic's Philosopher's Stone, we doubt if geochemistry will ever solve the problem. Hordes of elements and isotopes have been shouting their messages at us for decades, but geochemists still haven't deciphered what exactly they are saying. Of course, when we hear geochemists like Dr. Dean Presnall speak of "model-system phase relations in the $CaO-MgO-Al_2O_2-SiO_2-Na_2O-FeO$ (CMASNF) systems," do we know what exactly they are saying? (Readers will note that we had two geochemical cartoons in Holden and Vogt (1977), but have added no new ones; we are just too confused.) High hopes had been pinned on helium-3, but even that light and noble isotope is sending ambiguous messages. No one seems to be able to agree about who is wearing the trousers in its curious matrimony with helium-4, or are "San Francisco values" at work here?

Modeling mantle convection has become ever more sophisticated (e.g., Phillips and Bunge, 2005) and includes ever more

Figure 13. Seismotomographers looking for the plumatic's Grail: tiny wiggles they can extract from the noise and invert into tomoplumes.

parameters, physical laws, possible effects, and geometries, with ever smaller time steps and cell sizes in finite-element calculations. All this has to happen with ever less research funding and a much smaller LAGS (large army of graduate students; a SAGS?) than when the term was defined by Holden and Vogt (1977). The modelers (i.e., mostly the LAGgards working for them) are exploring the multidimensional jungle of parameter space: will the *real* parameter values (e.g., viscosity, thermal expansivity, radiation, density, composition, etc.) please come out of the jungle and identify yourselves to the explorers? But no matter how convincing numeroplumes may be, we still have to find their likenesses as tomoplumes. We predict that will probably happen sometime between the next AGU and never in a million years.

Of course, the search for plumes may have a surprise or two; that "iridium spike" might be out there somewhere, waiting to be tripped over.

In the meantime, wanting to "prebut" any "impartially moderated" "discussion" of this article, in addition to completing it so late that no such discussion could take place, we decided to end on a polite, conciliatory, and even politically correct note. We celebrate plumodiversity by adapting multiculturalism and rainbow coalitions to the Earth's mantle, which, as we noted earlier, has enough room for all contenders (Fig. 14). We apologize to anyone whose pet theory or plume has not been bashed; there is room aplenty to add more (Figure 14 is only one of an infinite number of different slices through a sphere).

ACKNOWLEDGMENTS

The authors—armed with pepper and mosquito spray to ward off marauding bears, mountain lions, and mosquitoes, not to speak of black widow spiders lurking under outhouse seats—completed much of this work under the supervision of Professor Gillian Foulger, Dr. Bruce Julian, the co-author's wife, and ten cats at the secret Plumatic Asylum headquarters, hidden in an abandoned hippie commune outside Omak, Washington.

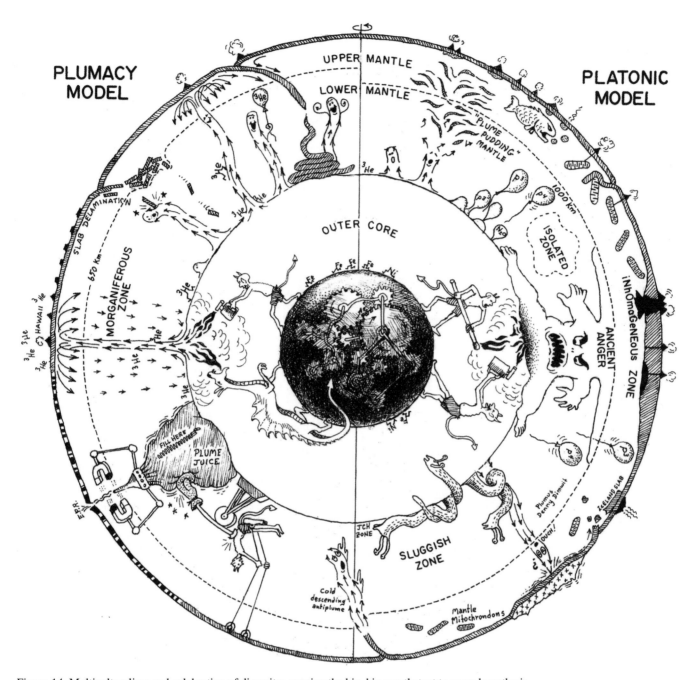

Figure 14. Multiculturalism and celebration of diversity: opening the big, big mantle tent to every hypothesis.

Foulger conceived the ideas for Figures 5 and 6, besides smuggling police mugshots of herself and other suspects to J. Holden. Recognizing the many contributions by Professor Foulger, the Plumatic Asylum has conferred on her its highest award: Honorary Senior Aplumatic Fellow. Foulger thus becomes the Nancy Pelosi of the previously male-dominated Plumatic Asylum.

This article does not necessarily reflect the views of the Geological Society of America, the editors of this volume, the International Stop Continental Drift Society, the Naval Research Laboratory, the U.S. Geological Survey, the Flat Earth Society, Princeton University, Cambridge University, or the authors. Plumatic Asylum Contribution 0000002. Planning has already begun for Contribution 0000003, due to be published in 2037.

REFERENCES CITED

Alvarez, W., Alvarez, L.W., Asaro, F., and Michel, H.V., 1980, Extraterrestrial cause for the Cretaceous–Tertiary extinction: Science, v. 208, p. 1095–1108.

Anderson, D.L., 1975, Chemical plumes in the mantle: Bulletin of the Geological Society of America, v. 86, p. 1593–1600, doi: 10.1130/0016-7606 (1975)86<1593:CPITM>2.0.CO;2.

Anderson, D.L., 1999, Are color cross sections really Rorschach tests?: Eos (Transactions, American Geophysical Union), v. 80, p. F719.

Anderson, D.L., 2000, Thermal state of the upper mantle: No role for mantle plumes: Geophysical Research Letters, v. 27, p. 2623–2626.

Anderson, D.L., 2005, Scoring hotspots: The plume and plate paradigms, *in* Foulger, G.R., et al., eds., Plates, plumes, and paradigms: Boulder, Colorado, Geological Society of America Special Paper 388, p. 31–54, doi: 10.2230/2005.2388(04).

Anderson, D.L., and Natland, J.H., 2005, A brief history of the plume hypothesis and its competitors: Concept and controversy, *in* Foulger, G.R., et al., eds., Plates, plumes, and paradigms: Boulder, Colorado, Geological Society of America Special Paper 388, p. 119–145, doi:10.1130/2005.2388(08).

Bailey, D.K., and Woodley, A.R., 2005, Repeated, synchronous magmatism within Africa: Timing, magnetic reversals, and global tectonics, *in* Foulger, G.R., et al., eds., Plates, plumes, and paradigms: Boulder, Colorado, Geological Society of America Special Paper 388, p. 365–377, doi:10.1130/2005.2388(22).

Baksi, A.K., 2007 (this volume), A quantitative tool for detecting alteration in undisturbed rocks and minerals—II: Application to argon ages related to hotspots, *in* Foulger, G.R., and Jurdy, D.M., eds., Plates, plumes, and planetary processes: Boulder, Colorado, Geological Society of America Special Paper 430, doi: 10.1130/2007.2430(16).

Bijwaard, H., and Spakman, W., 1999, Tomographic evidence for a narrow whole mantle plume below Iceland: Earth and Planetary Science Letters, v. 166, p. 121–126, doi: 10.1016/S0012-821X(99)00004-7.

Campbell, I.H., 2006, Large Igneous Provinces and the mantle plume hypothesis: Elements, v. 1, p. 265–269.

Cox, R.T., and Van Arsdale, R.B., 2002, The Mississippi Embayment, North America: A first-order continental structure generated by the Cretaceous superplume event: Journal of Geodynamics, v. 34, p. 163–176, doi: 10.1016/S0264-3707(02)00019-4.

Courtillot, V., Davaille, A., Besse, J., and Stock, J., 2003, Three distinct types of hotspots on the Earth's mantle: Earth and Planetary Science Letters, v. 205, p. 295–308, doi: 10.1016/S0012-821X(02)01048-8.

DeLaughter, J.E., Stein, C.A., and Stein, S., 2005, Hotspots: A view from the swells, in Foulger, G.R., et al., eds., Plates, plumes, and paradigms: Boulder, Colorado, Geological Society of America Special Paper 388, p. 257–278, doi: 10.1130/2005.2388(16).

Dietz, R.S., 1961, Astroblemes: Scientific American, August, p. 51–58.

Dietz, R.S., 1986, Triassic–Jurassic event, Newark basalts, and impact-generated Bahama Nexxus: Meteoritics, v. 21, no. 4, p. 355–356.

Dietz, R.S., and Holden, J.C., 1970a, The breakup of Pangaea: Scientific American, October, p. 30–41.

Dietz, R.S., and Holden, J.C., 1970b, Reconstruction of Pangaea: Breakup and dispersion of continents, Permian to present: Journal of Geophysical Research, v. 75, p. 4939–4955.

Dietz, R.S., and Holden, J.C., 1987, Creation/evolution satiricon: Winthrop, Washington, The Bookmaker, 140 p.

Dziewonski, A.M., 2005, The robust aspects of global seismic tomography, *in* Foulger, G.R., et al., eds., Plates, plumes, and paradigms: Boulder, Colorado, Geological Society of America Special Paper 388, p. 147–154, doi: 10.1130/2005.2388(09).

Fairhead, J.D., and Wilson, M., 2005, Plate tectonic processes in the South Atlantic Ocean: Do we need mantle plumes? *in* Foulger, G.R., et al., eds., Plates, plumes, and paradigms: Boulder, Colorado, Geological Society of America Special Paper 388, p. 537–553, doi:10.1130/2005.2388 (32).

Foulger, G.R., 2006, Older crust underlies Iceland: Geophysical Journal International, v. 165, p. 672–676, doi: 10.1111/j.1365-246X.2006.02941.x.

Foulger, G.R., Natland, J.H., and Anderson, D.L., 2005a, Genesis of the Iceland melt anomaly by plate tectonic processes, *in* Foulger, G.R., et al., eds., Plates, plumes, and paradigms: Boulder, Colorado, Geological Society of America Special Paper 388, p. 595–625, doi:10.1130/2005 .2388(35).

Foulger, G.R., Natland, J.H., Presnall, D.C., and Anderson, D.L., eds., 2005b, Plates, plumes, and paradigms: Boulder, Colorado, Geological Society of America Special Paper 388, 881 p.

Garnero, E.J., Lay, T., and McNamara, A., 2007 (this volume), Implications of lower mantle structural heterogeneity for existence and nature of whole mantle plumes, *in* Foulger, G.R., and Jurdy, D.M., eds., Plates, plumes, and planetary processes: Boulder, Colorado, Geological Society of America Special Paper 430, doi: 10.1130/2007.2430(05).

Glen, W., 2005, The origins and early trajectory of the mantle plume quasi-paradigm, *in* Foulger, G.R., et al., eds., Plates, plumes, and paradigms: Boulder, Colorado, Geological Society of America Special Paper 388, p. 91–117, doi: 10.1130/2005.2388(07).

Green, D.H., 1972, Archaean greenstone belts may include terrestrial equivalents of lunar marea?: Earth and Planetary Science Letters, v. 15, p. 263–270, doi: 10.1016/0012-821X(72)90172-0.

Griffiths, R.W., and Campbell, I., 1991, Interaction of mantle plume heads with the Earth's surface and onset of small-scale convection: Journal of Geophysical Research, v. 96, p. 18,295–18,310.

Hamilton, W.B., 2002, The closed upper-mantle circulation of plate tectonics: Washington, D.C., American Geophysical Union Monograph 30, p. 359–410.

Hamilton, W.R., 2005, Plumeless Venus preserves an ancient impact-accretionary surface, *in* Foulger, G.R., et al., eds., Plates, plumes, and paradigms: Boulder, Colorado, Geological Society of America Special Paper 388, p. 781–814, doi: 10:1130/2005.2388(44).

Hirano, N., Takahashi, E., Yamamoto, J., Abe, N., Ingle, S., Kaneoka, I., Hirata, T., Kimura, J.-I., Ishii, T., Ogawa, Y., Mashida, S., and Suyehiro, K., 2006, Volcanism in response to plate flexure: Science, v. 313, p. 1426–1428, doi: 10.1126/science.1128235.

Holden, J.C., and Vogt, P.R., 1977, Graphic solutions of plumacy: Eos (Transactions, American Geophysical Union), v. 58, p. 573–580; reprinted in this volume.

Ivanov, A.V., and Balyshev, S.O., 2005, Mass flux across the lower-upper mantle boundary: Vigorous, absent, or limited? *in* Foulger, G.R., et al., eds., Plates, plumes, and paradigms: Boulder, Colorado, Geological Society of America Special Paper 388, p. 327–346, doi:10.1130/2005.2388(20).

Jackson, E.D., and Shaw, H.R., 1975, Stress fields in central portions of the Pacific plate: Delineated in time by linear volcanic chains: Journal of Geophysical Research, v. 80, p. 1861–1974.

Jones, A.P., Wünemann, K., and Price, G.D., 2005, Modeling impact volcanism as a possible origin for the Ontong Java Plateau, *in* Foulger, G.R., et al., eds., Plates, plumes, and paradigms: Boulder, Colorado, Geological Society of America Special Paper 388, p. 711–720, doi:10.1130/2005.2388(40).

Jones, S.M., White, N., and McLennan, J., 2002, V-shaped ridges around Iceland: Implications for spatial and temporal patterns of mantle convection: Geochemistry, Geosystems, Geophysics, v. 3, no. 10, p. 1–23, doi: 10.1029/2002GC000361.

Jordan, B.T., 2005, Age-progressive volcanism of the Oregon High Lava Plains: Overview and evaluation of tectonic models, *in* Foulger, G.R., et al., eds., Plates, plumes, and paradigms: Boulder, Colorado, Geological Society of America Special Paper 388, p. 503–515, doi:10.1130/2005.2388(30).

Kellogg, L.H., and Wasserburg, G.J., 1990, The role of plumes in mantle helium fluxes: Earth and Planetary Science Letters, v. 99, p. 276–289, doi: 10.1016/0012-821X(90)90116-F.

Kiefer, W.S., and Hager, B.H., 1991, A mantle plume model for the equatorial highlands of Venus: Journal of Geophysical Research, v. 96, p. 20,947–20,966.

King, S.D., 2005, North Atlantic topographic and geoid anomalies: The result of a narrow ocean basin and cratonic roots? *in* Foulger, G.R., et al., eds.,

Plates, plumes, and paradigms: Boulder, Colorado, Geological Society of America Special Paper 388, p. 653–664, doi:10.1130/2005.2388 (37).

King, S.D., and Anderson, D.L., 1998, Edge-driven convection: Earth and Planetary Science Letters, v. 160, p. 289–296, doi: 10.1016/S0012-821X (98)00089-2.

Larson, R.L., and Olson, P., 1991, Mantle plumes control magnetic reversal frequency: Earth and Planetary Science Letters, v. 107, p. 437–447, doi: 10.1016/0012-821X(91)90091-U.

Lay, T., 2005, The deep mantle thermo-chemical boundary layer: The putative mantle plume source, *in* Foulger, G.R., et al., eds., Plates, plumes, and paradigms: Boulder, Colorado, Geological Society of America Special Paper 388, p. 193–205, doi:10.1130/2005.2388(12).

Lundin, E.R., and Doré, A.G., 2005, Fixity of the Iceland "hotspot" on the Mid-Atlantic Ridge: Observational evidence, mechanisms, and implications for Atlantic volcanic margins, *in* Foulger, G.R., et al., eds., Plates, plumes, and paradigms: Boulder, Colorado, Geological Society of America Special Paper 388, p. 627–652.

Malamud, B.D., and Turcotte, D.L., 1999, How many plumes are there?: Earth and Planetary Science Letters, v. 174, p. 113–124, doi: 10.1016/S0012-821X(99)00257-5.

Matias, A., and Jurdy, D., 2005, Impact craters as indicators of tectonic and volcanic activity in the Beta-Atla-Themis region, Venus, *in* Foulger, G.R., et al., eds., Plates, plumes, and paradigms: Boulder, Colorado, Geological Society of America Special Paper 388, p. 825–839, doi: 10.1130/2005.2388 (46).

McHone, J.G., 1996, Constraints on the mantle plume model for Mesozoic alkaline intrusions in northeastern North America: Canadian Mineralogist, v. 34, p. 325–334.

McNutt, M.K., 2002, Heat-flow variations over the Hawaiian Swell controlled by near-surface processes, not plume properties, *in* Hawaiian volcanoes: Deep underwater perspectives: Washington, D.C., American Geophysical Union, Geophysical Monograph 128, p. 355–364.

McNutt, M., 2006, Another nail in the plume coffin?: Science, v. 315, p. 39–40.

McNutt, M.K., and Fisher, K.M., 1987, The South Pacific superswell, *in* Keating, B.H., et al., eds. Seamounts, islands, and atolls: Washington, D.C., American Geophysical Union, Geophysical Monograph 43, p. 25–34.

McNutt, M.K., Winterer, E.L., Sager, W.W., and Natland, J.H., 1990, The Darwin Rise: A Cretaceous superswell?: Geophysical Research Letters, v. 17, p. 1101–1104.

Mendel, V., Sauter, D., Parson, L., and Vanney, J.-R., 1997, Segmentation and morphotectonic variations along a super slow spreading center: The Southwest Indian Ridge: Marine Geophysical Research, v. 19, p. 505–533, doi: 10.1023/A:1004232506333.

Michael, P.J., Langmuir, C.H., Dick, H.J.B., Snow, J.E., Goldstein, S.L., Graham, D.W., Lehnert, K., Kurras, G., Jokat, W., Muehe, R., and Edmonds, H.N., 2003, Magmatic and amagmatic seafloor generation at the ultraslow spreading Gakkel Ridge, Arctic Ocean: Nature, v. 423, p. 956–961, doi: 10.1038/nature01704.

Molnar, P., and Atwater, T., 1973, Relative motion of hotspots on the mantle: Nature, v. 246, p. 288–291, doi: 10.1038/246288a0.

Montelli, R., Nolet, G., Dahlen, F.A., Masers, G., Engdahl, E.R., and Hung, S.-H., 2004, Finite-frequency tomography reveals a variety of plumes in the mantle: Science, v. 303, p. 338–343, doi: 10.1126/science.1092485.

Morgan, E., 1982, The aquatic ape: New York, Stein and Day, 170 p.

Morgan, W.J., 1971, Convection plumes in the lower mantle: Nature, v. 230, p. 42–43, doi: 10.1038/230042a0.

Morgan, W.J., 1972, Plate motions and deep mantle convection: Boulder, Colorado, Geological Society of America Memoir 132, p. 7–22.

Natland, J.H., and Winterer, E.L., 2005, Fissure control on volcanic action in the Pacific, *in* Foulger, G.R., et al., eds., Plates, plumes, and paradigms: Boulder, Colorado, Geological Society of America Special Paper 388, p. 687–710, doi: 10.1130/2005.2388(39).

Peccerillo, A., and Lustrino, M., 2005, Compositional variations of Plio-

Quaternary magmatism in the circum-Tyrrhenian area: Deep versus shallow mantle processes, *in* Foulger, G.R., et al., eds., Plates, plumes, and paradigms: Boulder, Colorado, Geological Society of America Special Paper 388, p. 421–434, doi:10.1130/2005.2388(25).

Phillips, B.R., and Bunge, H.-P., 2005, Heterogeneity and time dependence in 3D spherical mantle convection models with continental drift: Earth and Planetary Science Letters, v. 233, p. 121–135, doi: 10.1016/j.epsl.2005 .01.041.

Ritsema, J., 2005, Global seismic structure maps, *in* Foulger, G.R., et al., eds., Plates, plumes, and paradigms: Boulder, Colorado, Geological Society of America Special Paper 388, p. 11–18, doi:10.1130/2005.2388(02).

Ritsema, J., and Allen, R.M., 2003, The elusive mantle plume: Earth and Planetary Science Letters, v. 207, p. 1–12, doi: 10.1016/S0012–821X(02) 01093–2.

Rocchi, S., Armienti, P., and Di Vincenzo, G., 2005, No plume, no rift magmatism in the West Antarctic Rift, *in* Foulger, G.R., et al., eds., Plates, plumes, and paradigms: Boulder, Colorado, Geological Society of America Special Paper 388, p. 435–447, doi:10.1130/2005.2338(26).

Sager, W.W., 2005, What built Shatsky Rise, a mantle plume or ridge tectonics? *in* Foulger, G.R., et al., eds., Plates, plumes, and paradigms: Boulder, Colorado, Geological Society of America Special Paper 388, p. 721–733, doi: 1130/2005.2388(41).

Sharp, W.D., and Clague, D.A., 2006, 50-Ma initiation of Hawaiian-Emperor Bend records major change in Pacific plate motion: Science, v. 313, p. 1281–1284, doi: 10.1126/science.1128489.

Sheth, H.C., 2005, From Deccan to Réunion: No trace of a mantle plume, *in* Foulger, G.R., et al., eds., Plates, plumes, and paradigms: Boulder, Colorado, Geological Society of America Special Paper 388, p. 477–501, doi:10.1130/2005.2388(29).

Sleep, N.H., 2002, Ridge-crossing mantle plumes and gaps in tracks: Geochemistry, Geophysics, Geosystems, v. 3, doi:10:10.1029/2001/GC000290 (33).

Smith, A.D., 2005, The streaky-mantle alternative to mantle plumes and its bearing on bulk-Earth geochemical properties, *in* Foulger, G.R., et al., eds., Plates, plumes, and paradigms: Boulder, Colorado, Geological Society of America Special Paper 388, p. 303–325, doi: 10.1130/2005 .2388(19).

Stuart, W.D., Foulger, G.R., and Barall, M., 2007 (this volume), Propagation of the Hawaiian-Emperor volcano chain by Pacific plate cooling stress, *in* Foulger, G.R., and Jurdy, D.M., eds., Plates, plumes, and planetary processes: Boulder, Colorado, Geological Society of America Special Paper 430, p. doi: 10.1130/2007.2430(24).

Tucholke, B.E., and Vogt, P.R., 1986, Perspectives on the geology of the North Atlantic Ocean, *in* Vogt, P.R., and Tucholke, B.E., eds., The Western North Atlantic region, v. M of The Geology of North America: Boulder, Colorado, Geological Society of America, p. 1–18.

Van der Hilst, R.D., and de Hoop, M.V., 2005, Banana-doughnut kernels and mantle tomography: Geophysical Journal International, v. 163, p. 956–961.

Vogt, P.R., 1971, Asthenosphere motion recorded by the ocean floor south of Iceland: Earth and Planetary Science Letters, v. 13, p. 153–160, doi: 10.1016/0012-821X(71)90118-X.

Vogt, P.R., 1972, Evidence for global synchronism in mantle plume convection, and possible significance for geology: Nature, v. 240, p. 338–342, doi: 10.1038/240338a0.

Vogt, P.R., 1974, Volcano spacing, fractures, and thickness of the lithosphere: Earth and Planetary Science Letters, v. 21, p. 235–252, doi: 10.1016/0012-821X(74)90159-9.

Vogt, P.R., 1991, Bermuda and Appalachian-Labrador rise: Common non-hotspot processes?: Geology, v. 19, p. 41–44, doi: 10.1130/0091-7613(1991)019 <0041:BAALRC>2.3.CO;2.

Vogt, P.R., and Holden, J.C., 1981, Extinctions: The democratic solution: Eos (Transactions, American Geophysical Union), v. 62, p. 537–538.

Vogt, P.R., and Jung, W.-Y., 2004, The Terceira Rift as hyper-slow, hotspot-dominated oblique spreading axis: A comparison with other slow-spreading plate boundaries: Earth and Planetary Science Letters, v. 218, p. 77–90, doi: 10.1016/S0012-821X(03)00627-7.

Vogt, P.R., and Jung, W.-Y., 2005, Paired basement ridges: Spreading axis migration across mantle heterogeneities? *in* Foulger, G.R., et al., eds., Plates, plumes, and paradigms: Boulder, Colorado, Geological Society of America Special Paper 388, p. 555–579, doi: 10.1130/2005.2388(33).

White, R.S., and Morton, A.C., eds., 1995, The Iceland plume and its influence on the evolution of the NE Atlantic: Journal of the Geological Society of London, v. 152, p. 933–1052.

Wilson, J.T., 1963, A possible origin of the Hawaiian Islands: Canadian Journal of Physics, v. 41, p. 863–870.

Zhao, D., 2004, Global tomographic images of mantle plumes and subducting slabs: Insight into deep Earth dynamics: Physics of the Earth and Planetary Interiors, v. 146, p. 3–34, doi: 10.1016/j.pepi.2003.07.032.

MANUSCRIPT ACCEPTED BY THE SOCIETY JANUARY 31, 2007

Index

A

absolute motion. *See* hotspot reference frames; plate motions

Academy of Sciences Rise–Kvakhon arc, 477

adiabatic melting and expansion
dry melting of asthenosphere during lithosphere detachment, 6
fraction of partial melting, 30–31
melt thickness, 31
melting depth, 30–31
petrological grids, 30
runaway adiabatic decompression melting, 13, 53, 55

Afar
asthenosphere flow from hotspot, 192
East African Rift, 5, 402, 405
lower mantle source, 11, 129
transition-zone observations, 129, 132
triple ridge intersection, 3, 5, 19

African magmatism. *See also* Afar
East African Rift, 5, 402–403, 405
enhanced midplate igneous activity (40 Ma), 581
superswell model for Africa, 74, 76–77

age progression
Hawaiian chain, 17, 336
Hawaiian-Emperor chain, 17, 336, 465
Hess Rise, 456
High Lava Plains, 18, 663
hotspot tracks, 38–39
importance in mantle plume debate, 936–937, 942–943, 960–961
Louisville, 356
New England Seamounts, North Atlantic Ocean, 38–39
Pacific Ocean melting anomalies, 18
Shatsky Rise, 456
Snake River Plain, 650
St. Helena Seamount chain, 16
tensional cracks, 355–356

A.I. *See* alteration index

Alder Creek rhyolite sanidine, 306–307

Aleutian trench, 344, 350, 464, 466

alkali basalts
alteration index, 290
association of alkaline magmatism with plumes, 816
Cameroon line, 400, 402
Columbia River Basalt Group, 650, 656, 662
composition of alkali basalt from rift systems, 402–404
isotopic similarity to OIBs, 816, 825
ΔNb values, 397, 402
sources of alkaline magmatism, 816, 826

alkaline rocks. *See* alkali basalts; anorogenic rocks; Cameroon line; eastern Paraguay; Mediterranean region; Paraná-Angola-Edendeka system

Alpine orogeny, 725, 728

alteration (chemical weathering). *See also* argon ages; dating methods
Alder Creek rhyolite sanidine, 306–307

alteration tests and [36]Ar content, 286–288, 300, 328
Atlantic Ocean rocks, 314–316, 317
Bedout Crater, 296
Caribbean rocks, 311, 313–314
Central European volcanic province, 326–327
Chagos-Laccadive Ridge, 314, 315
climate effects, 297–298
Columbia River Basalts, 295, 329, 331
cryptomelane, 296–297
Dabi volcanics, 309, 311, 313
Darwin rise, 330, 331
Deccan Traps dikes, 327
Deccan Traps plagioclase feldspar, 307, 309, 311–312
Deccan Traps whole-rock basalt, 290, 296, 297, 307–309, 310
Deep Sea Drilling Project (DSDP), 313, 319
Easter chain, 322–323
effect on K-Ar dates, 286–287
episodic alteration, 286, 298, 328, 329
Ferrar magmatic province, 295–296, 297, 298
flood basalts, 307–309, 310, 328, 329, 331
Galapagos, 319
Gettysburg Sill, 289–290
Gilbert Ridge, 320–322
Guyots (northwest Pacific), 318, 319, 331–332
Hawaiian-Emperor bend, 320
Indian Ocean rocks, 314, 315
Izu-Bonin-Marianas trench, 329–330
K-Ar method, underestimation of ages, 286–288, 299–300, 306–307
Kirkpatrick Basalt, 297
Line volcanic chain, 318, 319
melt–crust interactions, 326–327
mobility of large-ion lithophile elements (LILE), 414
Musicians seamounts, 316, 318
Newark Trend Basins, 290
Nicoya peninsula, Costa Rica, 313, 315
Ninetyeast Ridge, 314, 315
oceanfloor rocks, 309, 311, 313
oceanic plateau rocks, 311, 313–314, 315
Pacific Ocean rocks, 316, 318–323
Palisade rhyolite, 292–294
plagioclase from central Atlantic magmatic province, 290–292
pyroxenite analysis and paleoatmosphere sampling, 297, 298
Rajahmundry Traps, 290, 291, 329
Rajmahal Traps, 290, 313, 314
seafloor andesitic basalts, 309, 313
seafloor rocks linked to hotspot tracks, 323–324
Siberian Traps, 329
Singhbum craton, 296, 297
St. Helena, 316, 317
subduction zone–related studies, 299, 328, 329
Tokelau seamounts, 320, 322
Tristan da Cunha, 315–316
validity of hotspot-related ages, summary, 323–324

water, role in raising [36]Ar content, 287–288, 300
alteration index (A.I.)
alkali basalts, 290
biotite, 294, 328, 329
hornblende, 294, 295, 329
K-feldspar, 292–294, 298–299, 300–301
plagioclase feldspar, 290–292, 296, 297, 298, 301, 328–329
Siberian Traps, 689–690
Western Ghats (Sahyadri Range), 309
whole-rock basalts, 288–290, 295–296, 300, 307

AMS. *See* anisotropy of magnetic susceptibility

Amsterdam–St. Paul hotspot, 190, 197–198, 200

Anatolian-Iranian plateau, 694, 699, 710, 714. *See also* Eastern Anatolia

andesite, 415, 416, 418–420, 421

Angola, 619

anisotropy of magnetic susceptibility (AMS), 749–750, 753–761

anorogenic rocks
anorogenic granitoids in the Siberian Traps LIP, 674
definition and characteristics, 725, 728
Malani magmatism, 781
Mediterranean region, 728, 732–739, 739–740

Antarctica. *See also* West Antarctic Rift
Antarctic apparent polar wander path, 351–352
chemical weathering, 295–296, 297–298, 300
and Pangaea, 947
plate motions in hotspot fixed reference frame, 72, 181, 363
uncertainties about rigidity of Antarctic plate, 27, 339

Aorounga Crater, 910

apparent polar wander path (APWP). *See also* fixed hotspot model; paleomagnetism
Antarctic apparent polar wander path, 351–352
comparison of paleomagnetic and hotspot models, 343–346
crack model explanation of hotspot drift, 351, 355–356
drilling core data, uses, 339–341
fishhook shape of Pacific APWP, 339–340, 341, 343, 345
geocentric axial dipole assumption, 336–337
hotspot motion as explanation, 349–351
Jurassic and Early Cretaceous wander, 339, 341, 343–344, 347
Late Cretaceous and early Cenozoic wander, 339, 341–342, 343–344, 347–351
magnetic anomaly studies, 339–340, 347, 355
nondipole fields, 337, 338–339, 347, 348
northward motion implied by paleomagnetic data versus age, 347, 349
overview and definitions, 336–339, 355
Pacific apparent polar wander path studies, 339–342
Pacific paleomagnetic poles, dates, and locations, 339–342

apparent polar wander path (*continued*)
  paleolatitude implied by paleomagnetic data versus age, 347, 349
  polar wander in the hotspot reference frame, 343–346, 347
  polar wandering, causes, 338–339
  statistical significance and uncertainty, 346
  true polar wander (TPW)
    definition, 338
    effect on paleolatitude, 347, 348
    explanation for Late Cretaceous APWP, 347
    relationship to hotspot drift, 338–339, 349–351
APWP. *See* apparent polar wander path
argon
  alteration tests and ^{36}Ar content, 286–288, 300, 328
  atmospheric values for argon, 287, 328
  in paleoatmosphere, 297, 298, 299
argon ages. *See also* alteration (chemical weathering); alteration index; dating methods
  ^{40}Ar/^{39}Ar step-heating technique, 286, 287, 306, 609–610
  atmospheric argon (^{36}Ar) contents of silicates, 288, 300, 306
  diffusion loss of ^{40}Ar*, 286, 287
  K-Ar method, reliability and sources of error, 286–288, 299–300, 306–307
  leaching of basalt to remove alteration products, 288, 289, 324
  leaching of plagioclase to remove alteration products, 288, 290–292, 301, 322–324, 328–329
  leaching of seafloor rocks to remove alteration products, 324
  Paraguay alkaline rocks, 610–614
  recommendations for dating of seafloor rocks, 324–325
  statistical tests, 287, 296, 306
  subduction zone–related studies, 299, 328
  validity of hotspot-related ages, summary, 323–324
  water, role in raising ^{36}Ar content, 287–288, 300
Arperos Basin, 473, 475, 477, 478, 491
Ascension, 192, 195, 196, 362
Ascension-Cameroon system, 16
aseismic ridges
  Galápagos volcanic province, 511
  plumacy, 947, 950, 952
  subduction, 44, 204–205, 206, 952
  time-progressive ridges linked with LIPs, 397
  Walvis Ridge, 16, 316, 616, 624
asthenosphere. *See also* asthenosphere flow; mantle
  chemical heterogeneities, 175–177
  compositional lithosphere, 169, 171, 178, 179–180
  D″ layer as source of asthenosphere material, 73
  decoupling zone between lithosphere and lower mantle, 66, 73, 361
  enrichment of OIB, 410–411
  flow direction and He isotopic anomalies, 200–201

geochemical plume provinces along mid-ocean ridges, 175–177, 191–194, 191–199
geochemistry, 174–177
mass balance, 50, 167
melt-extraction–induced mantle strengthening, 169, 178
partially molten asthenosphere model, 56, 60, 63, 120, 355
plum-pudding model, 166, 172, 174–177, 205–206
plumes as source of asthenosphere material, 73, 77
potential density, 169
potential temperature, 167, 168, 173
progressive melt-extraction, buoyancy changes, 171–172
progressive melt-extraction, differences between MORBs and OIBs, 174–178
seismic anisotropies, 361–362
seismic low-velocity zone (LVZ), 167
seismic shear-wave velocities, 167–168
shear heating, kinematic model, 361–362
sinks at mid-ocean ridges, 73, 167, 169, 178, 180
temperature elevation relative to mantle below, 68, 73, 74
thickness, 73
viscosity variations, 73, 361–362
asthenosphere flow. *See also* asthenosphere; mantle convection
asthenosphere consumption
  dragdown or entrainment by subducting slabs, 166–167, 173–174, 178, 180, 190
  lithosphere growth by plate cooling, 167, 178, 180, 190
  sinks at mid-ocean ridges, 73, 167, 169, 178, 180, 190
  total consumption, 180, 182
asthenospheric sink at southern margin of Australia, 192–194, 193, 199–200
compositional lithosphere, 169, 171, 178, 179–180
Easter Island vortex, 184–185, 193
flow direction and He isotopic anomalies, 200–201
global plume-fed asthenosphere flow model
  buoyancy effects of progressive melt-extraction, 171–172
  comparison to conventional whole mantle flow model, 166
  comparison to counterflow models, 166–167, 205, 493–494
  critiques and alternatives, 204–207
  dynamic isostacy, 183–184, 184, 186
  limitations of model, 185–186
  lithosphere thickness, 179, 180
  map, 190
  overview, 166–168, 189–190
  plate velocities, 177, 179, 181
  plume strengths, 179, 181, 183–184
  potential density, 169
  potential temperature, 167, 168, 173
  predicted pattern of asthenosphere flow, 185, 190–191

restite roots to hotspot swells, 169–171
subcontinental plumes, 185, 191
temperature of asthenosphere, 168–169
thermal buoyancy effects, 172
thin-spherical-shell finite element model equations, 178–179
thin-spherical-shell finite element model solution method, 179–182
horizontal flow in plume hypothesis, 77
mass balance, 50, 167
plate-driven counterflow, 51, 58–59, 60, 930–931
seismic wavespeed maps compared to flow, 193–194
shear drag by moving lithosphere, 169–170, 178, 190
Atlantic-African geoid anomaly, 594, 598–599
Atlantic Ocean. *See also* Mid-Atlantic Ridge; North Atlantic Igneous Province; *specific locations*
  alteration (chemical weathering) of rocks, 314–316, 317
  asthenosphere flow, 190, 191–192
  average depth anomaly in western Atlantic, 566
  bathymetric and geochemical variations along Mid-Atlantic Ridge, 194–195
  Cameroon line, 16, 191–192, 400, 402, 411–412
  currently active hotspots, 508
  large igneous provinces, 508
  maps, 508
  melting anomalies, plate model, 14–16
  plume provinces, 191, 192
  South Atlantic, formation, 625
  tectonic history of North Atlantic Igneous Province, 528
  V-shaped volcano trails
    hotspot and plate motion interpretation, 532
    Kolbeinsey Ridge, 532
    lateral and vertical transport models, 375–376, 550, 949, 957
    Reykjanes Ridge, 532, 957
    Tristan da Cunha, 375, 376
    V-shaped ridges or escarpments (VSRs), 957, 969
Australia
  asthenospheric sink at southern margin of Australia, 192–194, 199–200
  rapid northern motion of craton, 193, 199, 202, 583
Australian-Antarctic discordance, 193, 194, 197, 199–200, 202
axisymmetric calculations for thermal plume structure
  with coefficient of thermal expansion as a function of depth, 110, 111–113, 114–115, 116
  constant-viscosity convection calculations, 108–109
  with internal heating, 109–110, 111, 115
  limitations of method, 116
  methods, 106–108, 116
  with phase transformations, 110, 113–114, 115
  Rayleigh number effect, 109, 110

with stable layer at base of mantle, 110, 112–114, 115, 116
with temperature-dependent viscosity, 109–111, 112
Azores, 4, 16, 191–192, 194–195

B

Baffin Island. *See also* North Atlantic Igneous Province
^{3}He/^{4}He ratios, 13
geochemistry, 13
magmatism, 447, 529
map, 527
seaward-dipping reflector sequences (SDRSs), 533, 549–550
Baja, 124, 125
Balleny plume, 193, 194, 947, 949
basalts. *See also* flood basalts; mid-ocean ridge basalts; ocean island basalts
alteration index for whole-rock basalts, 288–290, 295–296, 300, 307
columnar basalt, 595, 792, 793
composition of basaltic continental magmas, 18
DUPAL (Dupré and Allègre) basalts, 676–678, 682–684
Basin and Range province. *See also* Yellowstone hotspot
composition of lamproites, 402, 403
extension, 640, 648, 662
magmatic history, 18, 402, 581
bathymetric maps
bathymetric and geochemical variations along Mid-Atlantic Ridge, 194–195
Bermuda, 264
Cape Verde, 264
Crozet, 264
Galápagos volcanic province, 512
Hawaii, 264
Iceland, 9
Mid-Atlantic Ridge, 195
Pacific Ocean, 499, 505
Réunion, 264
Reykjanes Ridge, 9
bathymetry. *See also* swells
age relationship to depth, 37
bathymetric and geochemical variations along Mid-Atlantic Ridge, 194–195
Hawaiian swell, 224
Seismic Wave Exploration in the Lower Lithosphere (SWELL), 212, 224–225
South Pacific isotopic and thermal anomaly (SOPITA), 505
Bedout Crater, 296
Bermuda
absolute velocity in hotspot reference frame, 266
age, 558
bathymetric maps, 264, 555
Bermudite sheet igneous episodes, 559, 570, 584, 585
boreholes on the Bermuda Islands, 559–560
boreholes on the ocean floor, 559–560, 560–561
calcareous dune deposits (aeolianites), 557

comparison of Bermuda cluster with Japanese seamounts, 561–562
erosion of original islands, 561
geoid anomaly, 566
geophysics and seismology, 558–559
gravity anomaly, 559, 566
hydrothermal circulation, 270, 272, 568
intrusive sheets (dikes), 559–561, 570–571, 579, 581, 584–586
magnetic anomaly, 554–555, 558, 564
setting and morphology, 557–558
two-stage igneous history, 584–585
volcanic history, 557–561
Bermuda hotspot
arguments for and against hotspot explanation, 571–574
buoyancy flux, 584
heatflow anomaly, 266–267, 554–555, 567–569
heatflow measurement sites, 264, 555
hotspot track, 39, 76
intermittent blobs of hot mantle (lava lamp), 554, 574, 584
plume score, 584
relationship to Mississippi Embayment area magmatism, 557, 573–574, 579, 584
Bermuda Pedestal, 552, 556–557, 560, 561, 578
Bermuda Platform, 556–557, 558–559, 560, 561
Bermuda Rise
basement topography, 564–565, 566, 568, 576, 580
bathymetric maps, 264, 555, 572
boreholes on the ocean floor, 560–561
Bowditch seamount, 557, 561, 585
Challenger Bank, 557, 558, 561, 585
compensation depth, 566, 578
crust age, 565, 579
crust thickness, 565
depth anomalies, 555, 565–566, 570–571, 577–578
dimensions, 562
earthquakes, 564, 566–567, 572, 585
edge-driven convection, 579, 580, 584
geoid anomalies, 554–555, 565, 566, 572, 576–578
geologic map, 563, 564
global plate kinematic reorganization and Bermuda event, 580–583
gravity anomaly, 566
heatflow anomaly measurements, 266–267, 554–555, 567–569
heatflow measurement sites, 264, 555
J Anomaly Ridge, 564, 579
Kane fracture zone, 554–555, 564–565, 569–570, 571
lithosphere motion, 554, 572–573
lithosphere preconditioning, 579–580, 584
lithosphere stresses, 567
magnetic anomaly, 564
northeast elongation, 557, 562, 579–580, 584
overview, 553–557
Plantagenet Bank (Argos), 557, 561, 586
sediment thickness, 562–564
seismic reflecting horizon A^v, 553–554, 560, 561, 568, 570

seismic reflector A^t, 564, 569
seismic tomography, 569
seismicity, 566–567, 585–586
sheet (dike) intrusion, 559–561
Southeast Bermuda Deep, 570–571
two-stage igneous history, 584–585
uplift, 569–570, 574–579, 581
bolide impacts. *See* impact craters; meteorite-impact hypothesis
Boussinesq approximation
classical Boussinesq approximation, 140
extended Boussinesq approximation, 139–140, 143
for mantle convection at core-mantle boundary, 83, 89, 93, 95, 99
Bouvet, 3, 19–20, 184, 192–193
Bowditch seamount, 557, 561, 585
Bowers Ridge, 466, 467
Bowie
asthenosphere flow from hotspot, 193, 194
depth of source, 11, 129
hotspot track, 67, 70
transition-zone observations, 124–125, 129
Brazil. *See also* Paraná-Angola-Edendeka system
alkaline magmatism, 615, 625
Brazilian shield, 604, 605, 607
Sr-Nd isotopic compositions of magmatic rocks, 619–620
Bridge River Ocean, 474, 477, 479
British Igneous Province, 529, 531, 541, 550
British Tertiary igneous province, 750, 752, 753, 755, 766–767. *See also* Isle of Skye
Broken Ridge, 383, 816, 818, 829
Bunbury basalt, 816, 818, 829
buoyancy number (B), 57
Bushveld, 899, 910
Bushveld complex, 447, 773, 899, 910

C

Caledonian orogeny, 526, 534, 535
Caledonian suture, 15–16, 69, 389, 537, 969
Cameroon line, 16, 191–192, 400, 402, 411–412
Campanian lineations, 452, 454
Canary, 191–192, 194–195, 267, 510, 560
Cape Verde
absolute velocity in hotspot reference frame, 265–266
asthenosphere flow from hotspot, 192
bathymetry and heatflow measurement sites, 264
Cape Verde Rise, 566, 567, 570–571
geochemistry, 194–195
heatflow anomaly measurements, 269–270
plume model, 579, 580, 584
plume score, 584
preconditioning of lithosphere, 579–580
seismicity, 585
uplift, 560
volcanoes, 265, 577, 579
Caribbean Great arc, 474–475, 482, 489–490
Caribbean plate, 473–474, 482, 489–490, 491
Carnegie Ridge, 396, 398, 399, 511
Caroline Islands, 193, 194

Cayman Trough, 581, 582
Central Atlantic Magmatic Province (CAMP), 5, 290–292. *See also* dikes; Pangaea
Central European volcanic province, 326–327, 745
Central Indian Ridge
  asthenosphere flow, 192
  dehydration front at 60-km depth, 171
  geochemical relationships to hotspots, 197
  map, 392
  ΔNb for basalts, 391–392, 393
Chagos-Laccadive Ridge, 314, 315, 417
Challenger Bank, 557, 558, 561, 585
chemical weathering. *See* alteration (chemical weathering)
Chicxulub, 681, 781, 901
Chinook plate, 456
Chinook Trough. *See also* Pacific plate
  description, 455
  formation during ridge reorganization, 461, 467
  Kula plate initiation, 455, 466
  map, 453
  as Pacific-Farallon transform fault, 455
  as rift, 456, 457
Circe–Ascension–Saint Helena–Cameroon region, 16
Closepet Granite batholith, 276, 277, 278, 279–281
CMB. *See* core-mantle boundary
Coast Range (Pacific Northwest), 316
Cobb, 68, 181, 184, 193, 194
Cocos plate, 396, 473, 483, 491
Cocos Ridge, 373, 396, 398, 399, 473
Columbia River Basalt Group (CRBG). *See also* Snake River Plain
  ^{3}He/^{4}He ratios, 650, 657–658, 659
  alteration (chemical weathering), 295, 329, 331
  ^{40}Ar/^{39} ages, 642–643, 661, 663
  area covered, 638
  back-arc extension, 648, 650, 662–663, 665–666
  Blue Mountain terranes, 639, 643, 648, 651
  Brothers fault zone, 642, 650, 651, 663
  calc-alkaline to alkaline magmatism, 650, 656, 662
  Cascade arch, 638, 639, 642, 656–657, 662
  Cascade Range, 637, 643
  chemical distinctions between formations, members, and flows, 638
  Chief Joseph dike swarm, 639, 642–643, 651–652, 663, 665
  crustal assimilation, 648
  crystal fractionation, 647, 648, 649, 660, 661–662, 665
  east-west extension, 639–642, 650–651, 664–666
  Eckler Mountain Basalt, 644
  eclogite-rich plume, 651, 658–659
  enriched subcontinental lithospheric mantle (SCLM) component, 647–648, 656, 661
  enrichment in lithosphere components, 646–648
  evolution of flood basalts, 647, 648–649, 651–652
  high alumina olivine tholeiite (HAOT), 646, 647, 648

Imnaha Basalt, 642–643, 647–649, 661, 665
  isotope ratios, 647, 661
  Lewiston basin, 639, 644
  magmatic rate, 638
  map, 636
  migration of eruptions, 643, 651, 662–663, 665, 667
  nonplume explanations, 648, 650, 656–659, 663–665
  overview, 635–639, 656–657
  Pasco basin, 636, 638, 640, 643
  petrogenesis, 646–648
  Picture Gorge Basalt, 643, 647–648
  plume head of Yellowstone plume, 650
  plume model, 648, 650–652, 661–663, 665–666
  primary magma source, 646–647
  Saddle Mountains Basalt, 644–645
  sheet flow emplacement and composition, 644–645
  Steens basalt, 636, 639, 642–643, 643, 662–663
  stratigraphy, 637, 642–645
  stretching factors (β, 650, 651
  tectonic setting of eruptions, 639–642
  trace element patterns, 638, 646–647, 659–660, 661, 665
  uplift and subsidence, 6, 650–651, 657–658, 787, 803–805
  volume estimates, 98, 638
  Wanapum Basalt, 638, 639, 644
  Yakima fold belt, 639–640
Comores
  asthenosphere flow from hotspot, 192, 193
  hotspot track, 67, 70
  transition-zone observations, 124–125, 133
continental collision
  Eurasian-Arabian collision, 694, 707–719
  Euro-American and Siberian paleocontinents, 675
  India-Eurasia collision, 356, 580, 581, 808, 809
continental crust. *See also* oceanic crust
  delaminated continental crust as source of Icelandic basalts, 437–439, 447–448
  density and shear-wave seismic velocity of minerals, 51
  lower continental crust as source of Icelandic basalts, 434–437, 446–448
  Nb deficiency, 390
  recycling into mantle, 5–6, 13, 60
convection models of multiscale plumes. *See also* mantle convection; plume hypothesis; plumes; superplumes
  anelastic approximation of laws of conservation, 139–140
  boundary conditions, 140
  Cartesian geometry, 140–141
  classical Boussinesq approximation, 140
  complex models for multiscale plumes, 143–158
  conservation of mass equation, 138
  conservation of moment of momentum equation, 138
  depth-dependent thermal expansivity, 142–144, 145, 150–158, 159
  equation of state, 139

extended Boussinesq approximation, 139–140, 143
  heat equation, 139, 140, 141
  heat transfer equation for homogeneous material, 139
  model description and background, 138–143
  momentum equation, 138, 141
  multiscale nature of mantle plumes, overview, 158–160
  phase buoyancy parameter, 141–142, 150
  post-perovskite phase transitions, 159–160
  Prandtl number, 140
  radiative thermal conductivity effects, 144–145, 150–151, 158–159, 163
  Rayleigh number, 140, 141, 143–144, 160, 163
  residual temperature fields, 143, 146, 150, 152
  rheological relationship, 139
  secondary plumes, 138, 145, 148–150, 156–159
  shear heating in convecting layers, 145, 149–150
  simple models of depth-dependent properties, 143, 144
  stream functions, 141, 143, 147, 150, 153
  stress tensor equation, 139
  temperature- and depth-dependent viscosity, 142, 143–145
  thermal conductivity equation, 143
  thermal convection attributes, 138
  thermal expansion coefficient, 139
  upper mantle plume creation, 145, 148–150, 156–159
  viscosities above D", 145, 148
Cordillera, 477, 482, 491, 581–582
core
  cooling history, 94
  core-reflected S-waves (ScS), 230
  formation after moon-forming impact, 94
  thermal history, 35, 163
core-mantle boundary (CMB). *See also* D" layer; large low–shear velocity provinces; mantle; ultra-low-velocity zones
  Boussinesq approximation for mantle convection, 83, 89, 93, 95, 99
  chemical stratification, 90, 94–95, 97, 99
  controversy and critiques of models, 89–99
  density of low-velocity anomalies, 11
  density stratification, 89, 92, 94–95
  high shear velocities near subduction zones, 80–81
  narrow whole-mantle plumes, 89–91, 92, 97
  partial melting, 82, 84, 85
  perovskite (Pv) phase transformations, 80–82
  post-perovskite structure (pPv), 80–81, 94
  realistic mantle simulations (RMS), 90, 92–93, 94, 95–97
  seismic wave anisotropy, 80, 82
  seismic wave velocity variations (V_s and V_p), 80–83
  slab accumulation at core-mantle boundary, 724
  subadiabatic temperature gradients in deep mantle, 53, 63, 90–91, 104–105
  subducted MORB slab penetration into lower mantle, 85

superadiabatic temperature region, 63, 104, 105
themochemically dense piles, 83–85, 93, 98–99
thermal boundary layer (TBL)
  coefficient of thermal expansion, 89, 92,
    93–95, 105
  formation, 104, 108
  as plume source, 80, 89, 90–91, 115, 120
  pressure effects, 89, 90, 93
  Rayleigh number effect on temperature,
    89–90, 96
  temperature changes, 80, 89–91, 96
  temperature dependence of viscosity, 83,
    89–90, 95–96, 99
  thickness, 90
  ultra-low-velocity zones (ULVZs), 80, 82,
    84–85
  uniform bottom heating, 90, 91, 92–93
Corner seamounts, 38
counterflow
  comparison to eclogite engine, 59
  comparison to global plume-fed asthenosphere
    flow model, 166–167, 205, 493–494
  global plume-fed asthenosphere flow model,
    166–167, 493–494
  plate-driven counterflow, 51, 58–59, 60,
    493–494, 930–931
crack model
  age progression of tensional cracks, 355–356
  alternative crack theories for hotspots, 342
  delamination, 36
  explanation of hotspot drift, 351, 355–356
  lithospheric delamination, 36
  overview, 17, 18, 497–498, 503
  plume ascent through cracks, 36
  propagating crack model of swell relief,
    210–211, 230, 231
  thermoelastic crack model, 498, 501, 503, 505
  time progression, 17, 355
CRBG. *See* Columbia River Basalt Group
Crozet
  absolute velocity in hotspot reference frame,
    266
  asthenosphere flow from hotspot, 192, 193,
    200–201
  bathymetry and heatflow measurement sites,
    264
  description of islands, 264–265
  heatflow anomaly and distance from hotspot,
    266
  heatflow anomaly measurements, 265, 266,
    269–270
  volcanic history, 264–265
crust. *See* continental crust; oceanic crust

D

D″ layer. *See also* core-mantle boundary
  creation of superplumes, 150, 158–159
  displacement of material by slabs, 49, 54, 57,
    724
  Fe-enrichment, 84, 92
  high wavespeeds under North America plate,
    569
  as hotspot source, 11

influence of increasing radiative transfer of heat,
    144, 146
residual temperature, 146, 150, 152
slab accumulation, 724
as source of asthenosphere material, 73
stabilization by excess density, 89, 92
stratification, 113–114, 115, 116
temperature field, 143–145, 150, 151
thermal conductivity, 150, 158
thermal-electrical coupling between core and
  lower mantle, 140
viscosities above D″ layer, 145, 148
Dabi volcanics, 309, 311, 313
damp mantle melting, 509, 516–520, 521
Darwin rise, 330, 331
dating methods. *See also* alteration (chemical
    weathering); argon ages
  $^{40}Ar/^{39}Ar$ step-heating technique, 286, 287, 306
  geomagnetic polarity time scale (GPTS), 286,
    661
  K-Ar method, reliability and sources of error,
    286–288, 299–300, 306–307
  recommendations for argon dating of seafloor
    rocks, 324–325
  statistical tests, 287, 296, 306
Deccan basalt province. *See also* Deccan Traps;
    Dharwar craton; India
  Closepet Granite batholith, 276, 277, 278,
    279–281
  conglomerate bed at Rajpipla, 792, 806
  contact with prevolcanic basement, 788,
    791–792, 796, 799, 808
  crust and lithosphere thickness, 278, 279–280
  crustal contribution to heatflow, 279–281, 282
  crustal geotherm modeling, 276, 281–283
  crystallization differentiation, 446
  Dhar Forest and Bari area, 797
  Dhura-Jamunia area, 797
  eruption history, 278
  geology, 278–279, 787–788, 789
  heat production, 279, 280
  heatflow at surface, 276, 278, 279, 281
  Jabalpur area, 797
  Katangi-Singrampur area, 796–797
  magma mixing, 417, 421
  maps, 276, 776
  Moho heatflow, 281, 282–283
  Moho temperature, 281
  Monadnocks area, 797
  Pachmarhi sandstone, 792, 797–799, 806, 810
  planation surfaces under Deccan basalts,
    795–797, 799–800, 806, 807, 810
  and Réunion, 275–276
  rivers and drainage, 788–791, 806–807, 809
  Sagar area, 797
  sedimentary basin relationships with basalt
    Bagh sedimentary basin, 788–789, 792,
      795–796
    Godavari basin, 809
    Gondwana sedimentary basin, 788–789, 792,
      797, 806–807, 810
    Krishna basin, 809
    Lameta Formation, 788–789, 792, 795–797,
      809, 810

mock aridity, 792
Narmada basin, 791, 792, 809
Vindhyan sedimentary basin, 788, 789,
  791–792, 796–797
uplift and subsidence
  absent or minimal pre-eruption uplift, 792,
    800, 805–810
  cymatogenic uplift, 790, 791, 797, 799
  drainage patterns and uplift, 788–791, 800,
    806–807, 809
  lithospheric uplift predicted over plume
    heads, 786, 792, 800, 805–810
  Sahyadri (Western Ghats) uplift, 790–791,
    800
  Satpura doming and post-Deccan uplift,
    797–799
Deccan Traps. *See also* Deccan basalt province;
    Western Ghats (Sahyadri Range)
  age, 7, 307, 309
  alteration of dikes, 327
  alteration of plagioclase feldspar, 307, 309,
    311–312
  alteration of whole-rock basalt, 290, 296, 297,
    307–309, 310
  contact with prevolcanic basement, 788,
    791–792, 796, 799, 808
  eruption during continental breakup, 5
  geology, 787–788, 789
  K-T boundary, 307
  Re-Os whole-rock isochron, 296, 297
  volume estimates, 7
decompression melting
  asthenospheric consumption, 166, 169
  buoyant decompression melting instability, 53,
    55, 56
  in delamination, 60
  and Galileo thermometer, 52–53
  and lithosphere creation, 166, 169
  modeling, in plumes, 19–20
  peridotite conversion to harzburgite, 171–172
  secondary decompression melting, 77
  suppression by thick lithosphere, 201
deep mantle. *See* core-mantle boundary; D″ layer
Deep Sea Drilling Project (DSDP)
  alteration of rocks, 313, 318–319
  Bermuda Rise, 560–561, 564, 569, 579, 586
  Hess Rise, 456
  Irish volcanic margin, 531
  Meiji seamount, 343, 465
  Obruchev Rise, 464
  OIB and N-MORB studies, 404–405
  paleomagnetic data, 336, 337, 342, 343, 349
delamination. *See also* lithosphere; lithospheric
    detachment
  definitions, 437
  delaminated continental crust as source of
    Icelandic basalts, 437–439, 447–448
  fertile blob production, 44, 98
  Hawaiian chain, 36
  Iceland, 437–439
  Norway, 534
  stoping, 48, 49, 437, 438
  uplift, 437–438
  volcano formation, 92, 98

depleted mantle. *See also* mantle
  $^3He/^4He$ ratios in shallow mantle, 64
  Nd model ages in eastern Paraguay, 620, 627–628
  production rate, 35
  progressive melt-extraction, differences between MORBs and OIBs, 174–178
  slab dehydration and melting, 5–6
depth measurements. *See* bathymetry
Detroit seamount
  age, 320, 343, 465, 480
  geochemistry, 466
  maps, 464, 467
  paleomagnetic position, 480
Dharwar craton
  crustal structure, 277–279
  Dharwar trend, 787, 808
  heatflow, 276, 282–283
  maps, 277, 788
  radiogenic heat production, 276, 278, 279, 280
diapirs. *See* fertile blobs; mantle convection
Dietz, Robert S., plumacy, 956–957, 959, 966–967
dikes
  accretion center model, 761, 763
  Bermuda Rise, 559–561, 570–571, 579, 581, 584–586
  Brazil, 612
  British Tertiary igneous province, 755
  Chief Joseph dike swarm, 639, 642–643, 651–652, 663, 665
  Cleveland Dike, 753, 761, 763
  Deccan Traps dikes, alteration, 327
  East Pacific Rise, 417–418, 419
  eastern Paraguay, 604, 607, 614–615, 620
  effects of lithospheric stress, 39, 48, 376–377, 750, 753
  giant dike swarms, 749–750, 753, 755
  Iceland, 438
  lithosphere rejuvenation and underplating, 229
  magma flow vector determined from K3 axes, 754–755, 759, 760
  magma flow vector estimation by anisotropy of magnetic susceptibility (AMS), 749–750, 753–755, 757
  magmatic volume estimates, 763–764
  mantle ridge model, 753, 768
  melt transport at ridge-crossing seamount chains, 376–377
  radial dike formation, 34, 750
  regional tectonic constraints, 744
  tensional stress fields, 376–377
discontinuities. *See also* D″ layer; mantle; transition zone
  410-km velocity discontinuity
    olivine phase transitions, 122, 126–128, 131–134
    phase change, 10
    temperature effect on depth, 122, 126–128, 130
  520-km velocity discontinuity, 126
  660-km velocity discontinuity
    garnet phase transitions, 122, 126–128, 132–133, 134
    olivine phase transitions, 122, 126–128, 131–134

temperature and composition effects on depth, 122
    temperature effect on depth, 126–128, 130
  effects of temperature and composition, 122, 126–128
  topography, 10–11
Discovery, 67, 124–125, 190, 192, 193
Dongargarh Group (Dongargarh LIP)
  bimodal (felsic and mafic) compositions of rocks, 832–833, 836–837, 838
  contemporaneous rhyolitic and basaltic magmatism, 836–837
  duration of magmatism, 836
  geologic setting, 832–835
  incompatible element composition, 836, 837
  lithostratigraphic succession, 832–833
  major-element and trace-element composition, 832–837
  map, 832
  plume model, 837–838
  size of province, 836
DSDP. *See* Deep Sea Drilling Project
dunites, density estimates, 54–55
dunk tectonics, 57
Dupal anomaly, 195, 411
DUPAL (Dupré and Allègre) basalts, 676–678, 682–684

E

earthquakes
  Bermuda Rise, 564, 566–567, 572, 585
  eastern Paraguay, 624
  off the coast of Chile, 213–214
  off the coast of Guatemala, 214–215
  Rat Island event (Hawaiian swell), 212–213
East African Rift, 5, 402–403, 405. *See also* Afar; African magmatism
East Pacific Rise
  $^3He/^4He$ ratios, 201
  asthenosphere flow, 193, 194, 199, 930–931
  axial melt lenses, 417–418, 419, 421
  effects of magma mixing, 416–417, 417, 418–421
  geochemical relationships to hotspots, 198–199
  geochemistry of seamount OIB and N-MORB, 397, 400
  isotope signature along Pacific-Antarctic Ridge, 198–199
  magmatism and geology, overview, 417–418
  mantle heterogeneity, 420–421
  maps, 400, 401
  mixing of truly depleted MORB, E-MORB, and andesite, 416, 418–421
  ΔNb for basalt, 416
  ΔNb of OIB and N-MORB, 396, 397, 400, 401, 402
  Pacific-Antarctic East Pacific Rise, 416
  rare earth element mixing models, 416, 419
  rare earth element (REE) concentrations in basalt, 416, 418–419
  spreading rate, 364
  thermoelastic stress, 501, 931

Easter Island and Easter chain
  alteration (chemical weathering), 322–323
  Easter microplate, 185, 193
  location on moving ridge, 362
  lower mantle source, 129
  as primary hotspot, 132
  seismic observations of transition zone, 124–125, 127, 128–129, 129, 131–132
  triple ridge intersection, 4
  vortex in asthenosphere flow, 184–185, 193
Eastern Anatolia. *See also* Anatolian-Iranian plateau
  Bitlis-Pötürge Massif, overview, 695, 699
  crustal thickness, 699
  domal uplift, 694, 710, 713, 718
  Eastern Anatolian Accretionary Complex, overview, 695, 697, 699
  Eastern Anatolian Collision Zone, 705–707, 710–713, 715–718
  Eastern Rhodope–Pontide fragment, 695, 699
  Eurasian-Arabian collision, 694, 707–719
  fractional crystallization, 703, 705, 706–707
  geochemistry, 701–706
  geodynamic models
    continental collision with crust/lithosphere thickening, 708, 709
    delamination of mantle lithosphere, 708, 711, 712
    detachment and northward movement of subducting slab, 707, 708
    hotspot activity related to mantle plume, 708, 710
    inflow of lower crust in response to denudation and sedimentation, 709, 711, 713
    localized extension of pull-apart basins in strike-slip systems, 708, 709–710
    renewed continental subduction of the Arabian plate, 707, 708
    rifting along east-west-oriented Late Miocene-Pliocene basins, 707–709
    slab-steepening and breakoff beneath subduction-accretion complex, 709, 713–718
    tectonic escape of microplates, 707, 708
  incompatible element patterns of volcanic rock types, 702–703, 704
  isotopic data for volcanic rock types, 700–703
  lithospheric and mantle structure, 697–701
  major-element composition of volcanic rock types, 700–703, 704
  maps, 695, 696, 697
  migration of volcanic activity, 694, 705, 707, 710, 713–714
  Northwest Iranian fragment, 695, 699, 718
  overview, 694
  satellite view of major volcanic centers, 696
  Ta/Yb versus Th/Yb diagrams, 703, 705, 706
  Th/Ta versus MgO and Ta diagrams, 705, 706
  topographic profile, 698
  trace-element composition of volcanic rock types, 700–703, 704

variation and classification of volcanic rock
types, 702
volume estimates, 694
eastern Paraguay. *See also* Paraná-Angola-
Edendeka system
ages
^{40}Ar/^{39}Ar ages, analytical procedures, 607,
609–610
argon ages of alkaline rocks, 610–614
Asunción sodic alkaline province, 607, 609,
612, 614, 615
Nd model ages of depleted mantle, 619, 620,
627–628
radioisotopic ages (previous ages), 607,
608–609
Alto Paraguay sodic alkaline province,
604–606, 607, 608, 611
analyses of main alkaline rock types, 627–628
Apa and Amambay potassic alkaline province,
604–605, 606, 607, 608
Asunción-Sapucai-Villarica potassic alkaline
central province (ASU), 604, 605, 606,
607, 608–609
Asunción sodic alkaline province
^{40}Ar/^{39}Ar age (new age), 612, 614, 615
map, 605
previous age, 607, 609
earthquakes, 624
fractionation of rare earth elements, 617, 625–626
geochemical implications related to plume
hypothesis, 623–624, 626
geodynamic significance, 624–625
geological–geophysical background, 604,
605–607
Gran Chaco basin, 604, 605, 606
incompatible elements, 617, 618
large-ion lithophile elements (LILE), 617, 626
low-velocity anomaly in upper mantle, 605
magmatic events, 625
maps, 605, 606
Misiones sodic alkaline province, 605, 607, 609,
612, 614
Nazca slab, 605, 624, 626
overview, 604
Paraná basin, 604–605, 606, 620, 624, 625–626
Paraná Serra Geral flood tholeiites, 604, 605,
611–612
Pb isotopes, 620–623, 625–626
Rio Apa and Amambay potassic alkaline
province, 611–612
seismic tomography data, 605, 626
Sr-Nd isotopic compositions of magmatic rocks,
617–619, 620, 625
eclogite engine. *See also* Galileo thermometer;
mantle convection
assumptions, 49, 57
basalt-eclogite transition conditions, 49
comparison to other mantle convection models
counterflow, 59
fertile blobs versus hot plumes, 42, 43–44,
60–61
geochemical models, 60
overview, 56–58
plate model, 58–59

compressibility of magmas, 52
cooling from above as driving force, 51
and decompression melting, 52–53
deep mantle displacement by subducted
material, 49, 54, 57, 724
delaminated material, changes in mantle, 49,
51–52
density variations from phase changes, 49,
54–55, 57
density variations from pressure changes, 49
as Galileo thermometer, 50, 52–53, 54–57
and ground rules of mantle dynamics, 59–60
high-resolution seismology, 55
mantle convection as an engine, 48–49
overview of eclogite cycle, 48–49, 52–53
partially molten asthenosphere model, 56, 60,
63, 120, 355
and plate model, 58–59
problems and critiques of eclogite engine
model, 51–52
radioactive heating, 49, 51, 52–53, 55, 63
runaway adiabatic decompression melting, 13,
53, 55
thermal implications, 53–54, 63–64
top-down convection, 50, 51, 54, 58, 59
underplating and ponding, 49
yo-yo tectonics, 49, 57, 58, 63
eclogites. *See also* eclogite engine; oceanic crust
buoyancy variation with temperature, 51, 411
definition, 48
density estimates, 54
density variation with depth, 50–51, 56
density variations from phase changes, 49,
54–55
eclogite from subducted oceanic crust, 5–6, 14,
49, 52, 95
fate in mantle, 54, 55, 58, 133–134
formation by partial melting, 48
high density of MORB-eclogite, 50, 54
large igneous provinces from eclogite melting,
60–61
low seismic velocities, 50, 51
melt-retention buoyancy, 52
secondary eclogite formation, 48, 54
as tholeiite source, 228, 253, 258
volumes of melt plotted against lithosphere age,
12
Eifel hotspot, 10, 67, 69, 732–733, 739–740
Eltanin fracture zone
geochemistry, 198
and Louisville chain, 481–482, 483, 504, 505
maps, 473, 476, 482
Emeishan large igneous province
crustal heterogeneity, 844–847
crustal thickness measurements, 845–846, 847,
848, 849–852
deep seismic reflection, 846, 847
deep seismic sounding (DSS) profiles, 844–845
geological background, 842–844, 851, 855, 856
geophysical measurements, 844–847
high-velocity lower crust (HVLC)
calculation of fractionation ad different
pressures, 851
as characteristic of LIPs, 847, 851–852

formation by magmatic underplating,
847–848, 852
geophysical measurements, 844–846, 852,
855–856
lower crust versus upwelling, 847
petrologic modeling, 842, 850–851, 852
possible components, 848–850
seismic tomographic velocity structure, 846
variation of major oxides against MgO, 851
Maokou limestone thickness, 855, 856–857
map, 843
pre- and postvolcanic sedimentation and
paleogeography, 848, 849, 857
prevolcanic domal uplift, 7, 786, 843, 854–855,
856
seismic tomography, 845–846
total thickness, 844–845
volume estimates, 851–852
Emperor seamounts. *See also* Hawaiian-Emperor
chain
age, 343, 465–466, 480
Deep Sea Drilling Project and Ocean Drilling
Program, 336, 337, 343
lack of LIP, 8, 17
magmatic rate, 8
ΔNb for Emperor seamount basalt, 393, 395
paleomagnetic studies, 336, 347, 348–349,
355–356, 480
satellite gravity map, 453, 467
thermoelastic stress, 503
Emperor Trough
age estimates, 463–464, 466
description, 455
formation as rift, 456
as fracture zone, 467
map, 453
as sinistral transform fault between old and new
triple junctions, 461, 462, 480
Tenji seamount, 464
as transform fault linking areas of spreading,
456, 459
Ethiopian Traps, 673

F

Faeroe Islands, 531
Farallon plate
boundaries, 459–460, 461–462, 463
breakup, 27, 483
configuration, Late-Jurassic to Early Cretaceous,
474
configuration, Middle Jurassic, 472, 474
motion, 452, 453, 461
origin, 482
subduction, 477, 479
Farallon slab, 566, 569
Fernando da Noronha, 191, 192
Ferrar magmatic province, 295–296, 297, 298
fertile blobs
comparison to plume hypothesis, 42–44, 60–61,
98, 206, 411
in Galileo thermometer model, 52–53, 411
hotspot motions, 56
large igneous province formation, 537–538

fertile blobs (*continued*)
  melting anomaly sources, 56, 493
  OIB formation hypothesis, 388, 404–405,
      406–407
  piclogite packages, 48, 52
  plate-driven counterflow, 51, 58–59, 60, 493–494
  production by delamination, 44
  radioactivity in, 63
  reflectors and scatterers in mantle, 50, 51, 55, 57
fixed hotspot model. *See also* apparent polar
      wander path; hotspot reference frames;
      hotspots; plume hypothesis
  asthenosphere counterflow, 60
  Hawaiian-Emperor bend and changes in plate
      motion, 27
  history and development of model, 336,
      342–343
  linear seamount chain studies, 336
  plate velocity in hotspot reference frame, 65–66,
      68–72, 76–77, 181, 356
  true polar wander, effect on paleolatitudes,
      338–339
fixity. *See also* apparent polar wander path; hotspot
      reference frames; hotspots; plume
      hypothesis
  hotspot pairs, 72–73
  nonfixity prediction from fluid dynamics, 43
  Pacific hotspots, 343, 360
  plate-driven counterflow and hotspot motion,
      51, 58, 60
  plume hypothesis predictions, 42, 73
flood basalts. *See also* basalts; Deccan basalt
      province; mid-ocean ridge basalts;
      ocean island basalts
  alteration (chemical weathering), 307–309, 310,
      328, 329, 331
  lithospheric uplift predicted over plume heads,
      786, 792, 800, 805–810
  in plume hypothesis, 786, 816, 829
  prevolcanic uplift, 786–787
Foundation seamounts
  age, 491
  asthenosphere flow from hotspot, 193, 194
  effects of intraplate stress, 491, 493
  initiation, 483, 491
  map, 473
  ridge-approaching hotspot, 38
FOZO (focal zone, in mantle)
  isotope signatures, 425, 431, 433, 440, 823–824
  noble gas trapping mechanism, 64
  previously melted (depleted) source, 64
  recycled subducted ocean crust, 393, 395
  rhyolite similarity to FOZO, 425, 431, 433, 440
French Massif Central, 732, 733, 740

G

Galápagos Archipelago
  alteration (chemical weathering), 319
  geochemistry, 396, 398–399
  low velocity zone in upper mantle, 509
  map, 399
  ΔNb for basalt, 396, 398
Galápagos plume, 193, 194, 362, 396

Galápagos volcanic province
  aseismic ridges, 511
  bathymetric map, 512
  characteristics of crust, 509, 510
  crustal compensation model, 512, 514
  damp melting, 509, 516
  gravity anomaly, 512, 514, 521
  gravity modeling and compensation of
      tomography, 512, 513–514
  seismic tomography model, 512
  seismic velocity structure, 511, 512
  V-shaped volcano trails, 515
  vertical density (Vp) gradients, 511, 512, 513,
      514
Galileo thermometer (GT), 47, 50, 52–53, 54–57,
      411. *See also* eclogite engine
Garnero, E., 961
garnet clinopyroxenites, 48
garnet phase transitions at 660-km discontinuity,
      122, 126–128, 132–133, 134
geoid
  Atlantic-African geoid anomaly, 594, 598–599
  Bermuda Rise geoid anomalies, 554–555, 565,
      566, 572, 576–578
  effect of Rayleigh number (Ra), 109, 110
  geodesy plate motion models, 364, 366, 371,
      372
  Hawaiian swell geoid anomalies, 224–225
  hotspot swells, geoid anomalies, 106–108
  map, 884
  scaling of geoid anomaly, formula, 107
  Seismic Wave Exploration in the Lower
      Lithosphere (SWELL), 224–225
  Venus, 861, 862–863, 883–885
geomagnetic polarity time scale (GPTS), 286, 583,
      661
geothermometry. *See* olivine geothermometer
Gettysburg Sill, 289–290
Gilbert Ridge, 320–322
global seismic tomography. *See* tomographic
      mantle models; tomography
Gondwana
  Atlantic-African geoid anomaly, 594, 598–599
  fracture tessellation model
    crack propagation in spherical shells, 597
    hexagonal fracture systems, 595
    icosadeltahedral stress tessellation, 594–597
    overview, 594–595, 599, 600–601
    tensile stress fields, 594, 595–597
  hotspots associated with tessellation, 597–598
  Indian separation from Gondwana, 777, 781,
      782, 789
  LIP eruption in icosahedral rifts, 595, 596, 597,
      601
  lithospheric weak belts, 596, 600–601
  map at 200 Ma, 594–595
  rift triple junctions, 594, 595, 601
  rifting of Zealandia in Late Cretaceous, 479
  source of Alexander terrane, 477
Gough
  alteration of rocks, 316, 317
  asthenosphere flow from hotspot, 192, 193
  trace elements injected into Mid-Atlantic Ridge,
      195, 196

Gravina-Nutzotin-Gambier arc, 477, 478
Great Meteor seamount, 38, 191, 192, 195
Greenland. *See also* North Atlantic Igneous
      Province
  geological map, 752
  high-velocity lower-crustal bodies (LCBs), 533
  magma flows at volcanic passive margin (VPM)
      stage, 760–761, 762
  magmatism, 529
  planation surfaces, 795
  seaward-dipping reflector sequences (SDRSs),
      529, 760–761
  seismic velocity structure, 511
  uplift history, 529, 535, 786
Greenland-Iceland Ridge
  magmatism, 15, 532, 534, 539
  maps, 9, 527
Guadalupe, 68, 181, 184, 193–194
gumbo, 52
Guyots (northwest Pacific), 318, 319, 331–332

H

Hawaii. *See also* Hawaiian swell; Seismic Wave
      Exploration in the Lower Lithosphere
  [3]He/[4]He ratios, 17
  absolute velocity in hotspot reference frame,
      263, 267
  age of hotspot track, 263
  age progression of Hawaiian chain, 17, 336
  area and volume, 9, 19
  asthenosphere flow from hotspot, 190, 192, 193,
      194
  bathymetry and heatflow measurement sites,
      263, 264
  buoyancy flux, 33
  crustal age, 263
  damp melting, 509
  differences from other Pacific volcanoes, 17
  finite-frequency tomography, 17
  heatflow, 262–263, 265, 266, 269–270
  isotope ratios of Hawaiian basalts, 175
  Kilauea eruption of 1959
    magma composition, 237–239, 240, 243–245,
        255
    parental crystallization temperature, 240,
        243–245, 246–248, 253
  kinematics of shallow Hawaiian plume model,
      364–365
  large igneous province formation, 9, 17, 19
  lava geochemistry variability, 17, 175
  lithosphere delamination, 36
  lithospheric thinning, 210–211
  lower mantle source, 121, 129
  magmatic rates, 8–9, 170, 267, 763
  mantle equilibration depths and pressures,
      245–246
  Mauna Loa volcano
    parental crystallization temperature, 240,
        247–248, 253
    parental melt composition, 240, 246, 253,
        255–256
  mock aridity, 792
  ΔNb for basalts, 393, 395

Nuuanu glasses, 255–256
physical properties of cooling magma, 269
picrites, 252–253, 255–256
plate model, 18
possible hydrothermal circulation masking of anomalies, 17, 106, 267
progressive melt-extraction of basalts, 175
propagation rate of volcanic track, 362
Puna Ridge glasses, 237, 238–240, 245–246, 253
pyrolite, partial melting trends, 245
receiver functions and hotspot topography, 129–130, 134
as ridge-leaving hotspot, 38
seismic tomography, 17, 211–212
sill model, 268–269
temperature excess of source region, 31–32, 49
temperature of hotspot source, 12, 17
tholeiite, 228–229
total thermal power from magma, 267, 269
transition zone thickness, 17
Hawaiian-Emperor bend
  alteration (chemical weathering), 320
  chronology, 8, 25–27, 286, 343
  and events around the Pacific basin margin, 481
  and global plate kinematic reorganization, 581–583
  hotspot motion within mantle as explanation, 27, 349–351, 356
  and India-Eurasia collision, 356
  lithosphere rejuvenation, 37
  plate propagation direction, 17, 26–27
  thermoelastic stress, 503
Hawaiian-Emperor chain. *See also* Emperor seamounts
  age progression of Hawaiian-Emperor chain, 17, 336, 465
  characteristics of crust, 510
  crack model
    age progression of tensional cracks, 355–356
    alternative crack theories for hotspots, 342
    delamination, 36
    explanation of hotspot drift, 351, 355–356
    lithospheric delamination, 36
    overview, 17, 18, 497–498, 503
    plume ascent through cracks, 36
    propagating crack model of swell relief, 210–211, 230, 231
    thermoelastic crack model, 498, 501, 503, 505
    time progression, 17, 355
  hotspot motion within mantle as explanation for hotspot drift, 27, 349–351, 356
  hotspot origin in abandoned triple junction, 456, 461, 462, 466–468
  lava geochemistry variability, 17
  magmatic rates, 8–9, 17
  mantle plume explanation, 336, 497
  Pacific hotspot models, 342–343
  satellite gravity map, 453, 467
  thermoelastic stress, 498, 501
  true polar wander (TPW), relationship to hotspot drift, 338–339, 349–351
Hawaiian lineations, 453–454, 463

Hawaiian swell. *See also* Hawaii; Seismic Wave Exploration in the Lower Lithosphere; swells
  alternative causes of swell, 210–211, 228–232
  bathymetry, 224
  buoyancy flux, 33, 170
  compensation depth, 210–211, 224–225
  compositional buoyancy model, 210, 211, 225, 228–229
  dynamic model, 34
  dynamic support model, 210–211, 212, 224, 225
  fracture zones, 230
  geoid anomalies, 224–225
  heatflow measurements, 17, 106
  hybrid thermal rejuvenation–dynamic support/ thinning model, 225–226, 228–232
  kinematic model, 33–34
  lithospheric erosion, 170, 171
  lithospheric rejuvenation, 219–220, 222, 223, 225
  low-velocity body in mantle, 215, 220, 222–223, 224
  magnetotelluric (MT) data, 223–224, 225
  petrology, 228
  propagating crack model of swell relief, 210–211, 230, 231
  restite swell root, 169–171
  size, 18, 33, 36
  thermal rejuvenation model, 210, 225
  thermoelastic stress, 503, 931
  topographic map, 33
  volume flux, 33, 34, 170
Heard, 192, 193, 200–201
heatflow. *See also* hydrothermal circulation; mantle convection
  Bermuda, 264, 266–267, 554–555, 567–569
  Cape Verde, 269–270
  conductive heatflow near swells, 267
  contribution of cooling Hawaiian magma, 267–270
  convective heatflow near swells, 267
  cooling history of Earth, 104
  Crozet, 265, 266, 269–270
  Deccan basalt province
    crustal contribution to heatflow, 279–281, 282
    heatflow, surface, 276, 278, 279, 281
    Moho heatflow, 281, 282–283
  Dharwar craton, 276, 282–283
  Hawaii, 262–263, 265, 266, 269–270
  Hawaiian swell, 17, 106
  heat flux of Earth, estimation and measurement, 35, 40, 885
  heatflow anomalies at hotspot swells, calculation method, 106–108
  heatflow anomalies at hotspot swells, measurements, 106, 281–282
  heatflow anomalies predicted by reheating and dynamic plume models, 262–263, 270
  heatflow in hotspots, estimated total, 106
  lithosphere reheating model, 261, 262–263, 264–265, 270
  mantle heat flux, 35
  pogo measurements, 263, 266, 267, 271, 272
  Réunion, 263–264, 265, 276, 281–282

helium (He)
  concentrations in OIB and MORB, 2, 542
  ^{3}He/^{4}He ratios for plume and plate models, 13–14, 205
  lack of correlation with depth of mantle plumes, 201, 205, 542, 730, 934
Hess Deep, 417, 421, 434
Hess Rise
  age estimates, 456, 478, 479
  age progression, 456
  association with Pacific-Farallon-Izanagi triple junction, 456, 480
  association with Pacific-Farallon spreading center, 456, 459, 478
  maps, 453, 473–474
Hessian Depression, 745
HFSE. *See* high field strength elements
high field strength elements (HFSE)
  crystallization differentiation, 415
  Dongargarh Group, 835, 836
  eastern Paraguay, 617
  enrichment in oxide gabbro and silicic veins, 436
  high ionic potential, 414
  Jasra alkaline-mafic-ultramafic complex, 821, 823
  Mediterranean area, 730, 731
  ΔNb for Pacific-Antarctic East Pacific Rise basalt, 415, 416
  Pacific-Antarctic East Pacific Rise basalt, 416
  partitioning into melts, 414–415, 419
  Siberian Traps, 670
  similarity to rare earth elements, 415
High Lava Plains (HLP)
  age, 663, 965
  age progression, 18, 663
  magmatism, 651
  maps, 4, 663
  teenage rebel plume, 965
  time progressive trail, 18
Hikurangi plateau, 5, 473–475, 478, 479, 481. *See also* Ontong Java–Manihiki–Hikurangi composite plateau
Hokkaido Trough, 453, 455, 459, 463
hotspot reference frames (HSRF). *See also* fixed hotspot model; fixity; hotspots
  absolute velocity in hotspot reference frame
    Bermuda, 266
    Cape Verde, 265–266
    Crozet, 266
    Hawaii, 263, 267
    plate velocities, 65–66, 68–72, 76–77, 181, 356
    Réunion, 263
    ridge axes in Pacific and Atlantic, 39
  apparent polar wander path (APWP), 343–346, 347
  azimuth and rates of hotspot tracks, 67–68, 70–71
  azimuth-only method for plate velocity, 66, 68–72, 76–77
  comparison of paleomagnetic and hotspot models, 343–346
  geodesy plate motion models, 364, 366, 371, 372

hotspot reference frames (*continued*)
   global plate motions in deep hotspot reference frame, 363–364
   global plate motions in shallow hotspot reference frame, 363–364, 365–366
   hotspots on plate boundaries, exclusion from reference frame, 360
   HS3–NUVEL1A plate kinematic model, 361, 362–364, 366
   International Terrestrial Reference Frame (ITRF2000), 366
   kinematic consequences of shallow hotspot reference frame, 365, 366–369, 371–373
   kinematics of shallow Hawaiian plume model, 364–365
   no-net-rotation (NNR) reference frame, 360, 361, 366, 371
   overview, 360–361
   westward drift of lithosphere, 360, 366–367
hotspot tracks. *See also* age progression; hotspot reference frames; ridge-crossing seamount chains
   age progression of hotspot tracks, 38–39
   azimuth and rate of hotspot tracks, 67–68, 70–71
   azimuth-only method for plate velocity, 66, 68–72, 76–77
   fertile blobs, 56
   Monteregian hotspot track, 38
   plate-driven counterflow and hotspot motion, 51, 58, 60
hotspots. *See also* fixed hotspot model; hotspot reference frames; melting anomalies; plume hypothesis; plumes; swells
   Atlantic Ocean, currently active hotspots, 508
   buoyancy flux (BF), 33
   correlation with large low–shear velocity provinces, 82
   criteria used to characterize plumes and hotspots, 122
   crustal thickness increase near hotspots, 267
   definitions, 388, 730
   depth of sources, 10–11
   deviation from plume model predictions, 2, 509
   geoid anomalies at hotspot swells, 106–108
   global heat flux, 35, 40
   global volume flux, 35, 40
   heatflow anomalies, estimated, 262–263
   heatflow anomalies at hotspot swells, calculation method, 106–108
   heatflow anomalies at hotspot swells, measurements, 106, 281–282
   heatflow in hotspots, estimated total, 106
   history of hotspot concept, 2, 508–509
   hotspot pairs, 72–73
   hotspot seamount chains on the Pacific plate, 337
   hydrothermal circulation, possible masking of anomalies, 17, 106, 262, 267
   kinematic consequences of shallow hotspot reference frame, 367–369, 371–373
   locations, 3, 184, 584
   magmatic volume, 48
   map of global hotspots, 3

Pacific hotspot models, 342–343
   pairs of hotspots, 72–73
   plate-driven counterflow and hotspot motion, 51, 58–59, 60
   and plate motion, 25–27
   plume strengths, 179, 181, 183–184
   potential temperatures of melting sources and mantle, 104
   shallow sources of hotspots, 361
   subsidence and uplift chronology, 6–7, 36–37
   topographic anomalies at hotspot swells, 106–108
   validity of hotspot-related ages, summary, 323–324
hydrothermal circulation
   Bermuda, 270, 272, 568
   Hawaii, 17, 106, 267
   Juan de Fuca, 267
   overview, 267
   possible masking of heatflow at hotspots, 17, 106, 262, 267

I

Iceland. *See also* North Atlantic Igneous Province
   ^{3}He/^{4}He ratios, 532
   area and volume, 9, 14, 19
   asthenosphere flow of plume material, 190, 191
   buoyancy flux, 584
   Caledonian suture, 15–16, 69, 389, 537, 969
   damp melting, 509, 516
   delaminated continental crust as source of Icelandic basalts, 437–439, 447–448
   gabbro as source of basalt, 433–434, 436
   geochemistry, 389–390, 395, 421–437, 440, 532
   heterogeneity of basalt source, 437
   large igneous province formation, 15, 19
   location on moving ridge, 362
   low velocity zone in upper mantle, 10, 509, 729
   lower continental crust as source of basalts, 434–437, 446–448
   lower mantle source, 129
   magma mixing model
      critiques and discussion, 444–448
      delaminated continental crust as source of Icelandic basalts, 437–439, 447–448
      gabbro as source of basalt, 433–434, 436
      isotope ratios, 425
      lower continental crust as source of basalts, 434–437, 446–448
      picrites, 253, 433–434, 436, 444
      plume model as alternative, 445
      rare earth elements, 426–429, 430
      summary, 439–440
      trace elements, 429, 431–433
   magmatic rates, 9, 14, 32
   map, 9
   ΔNb of basalts, 389–390, 395, 422–423, 439–440, 444–445
   ΔNb variation for Icelandic basalt, 416
   Nb/Y versus Zr/Y variation for Icelandic basalt, 389–390, 395
   picrites, 253, 433–434, 436, 444
   plate tectonics model, 15–16

plume score, 584
   projection of partial melting trends for Iceland pyrolite, 246
   rhyolite-basalt mixing, 421–426, 422
   rhyolite similarity to FOZO, 425, 431, 433, 440
   temperature estimates for anomaly, 15, 31, 32
   Theistareykir volcanic system, 240, 241
   thin transition zone under Iceland, 130
   TiO content of basalts, 421–423, 434–437
   topographic map, 33
   trace elements injected into Mid-Atlantic Ridge, 194–195
Iceland-Faeroe Ridge, 9, 14–15, 32, 527, 534
Iceland-Greenland Ridge, 32
Icelandic Volcanic plateau, 9
icosahedrons
   crack propagation in spherical shells, 597
   fracture tessellation model of Gondwana breakup, overview, 594–595, 599, 600–601
   hexagonal fracture systems, 595
   icosadeltahedral stress tessellation, 594–597
   LIP eruption in icosahedral rifts (Gondwana), 595, 596, 597, 601
   natural examples of icosahedral arrangements, 594, 599
   tensile stress fields, 594, 595–597
impact craters. *See also* meteorite-impact hypothesis; Venus
   Aorounga Crater, 910
   Bushveld, 899, 910
   Chicxulub, 681, 781, 901
   Gosses Bluff impact structure, Australia, 890
   Manicouagan impact structure, 901
   overview and description, 908
   Spider impact structure, Australia, 890
   Stillwater, 899, 910
   Sudbury structure, 899, 900, 910, 967
   Upheaval Dome, Utah, 890, 901
   Vredefort structure, South Africa, 900, 910
incompatible elements. *See also* large-ion lithophile elements; light rare earth elements
   depletion in MORB, 174, 176
   eclogite plums, 177, 514
   enrichment in E-MORB, 174
   enrichment in ocean-island basalts, 2, 13, 388, 410
   enrichment in shallow mantle, 388
   Icelandic basalt, 389–391, 432, 436
   Paraguay rocks, 617, 618
   Scottish Midland Valley, 403
India. *See also* Deccan basalt province; Dharwar craton; Malani large igneous province; Western Ghats (Sahyadri Range); western Rajasthan
   Cambay rift, 787–788, 810
   Eastern Ghats, 791, 800
   geology, overview, 787, 788
   India-Eurasia collision, 356, 580, 581, 808, 809
   Indian shield, geological features and evolution, 775, 782, 787, 789, 800
   planation surfaces in southern India, 795
   rivers and drainage, 788–791, 806–807, 809

separation from Gondwana, 777, 781, 782, 789
separation from Madagascar, 787, 788, 807–808
separation from the Seychelles, 787, 788, 806, 807, 808
Indian Ocean, 190–191, 192–193, 314, 315. *See also specific locations*
Insular terrane, 474, 475, 477, 478
intraplate volcanism
alkali basalts, 397
buoyant decompression melting instability, 56
East African rift system, 405–406
Late Cretaceous to Early Eocene volcanism, 475–476, 479–481
oceanic plateaus and diffuse island chains in the Middle Jurassic to Early Cretaceous, 474, 477–478
overview, 472
plate configuration in the Late Jurassic to Early Cretaceous, 474, 477
plate configuration in the Middle Jurassic, 472, 474, 477
plate reorganization in the Early Eocene, 476, 482–483
plate reorganization in the Late Oligocene, Miocene, and Pliocene, 483–484
plate reorganization in the Middle Cretaceous, 475, 478–479
plum-pudding plume model, 205
shear heating and Pacific magmatism, 362
subduction of Pacific plate in Early Eocene, 476, 480
subduction of Pacific plate in Late Cretaceous, 475, 479–480
whole-mantle recycling of material, 205
Irish volcanic margin, 531
Isle of Skye
British Igneous Province, 529, 531, 541, 550
dike swarms, overview, 755
geochemistry of dikes, 755, 757, 758
magma flow during trap emplacement, 755–760, 761
magma flow vector estimation by anisotropy of magnetic susceptibility (AMS), 755–761
maps, 750, 756
Preshal Beg conglomerate and columnar-jointed olivine tholeiite, 792, 793
Izanagi plate. *See also* Kula plate
boundaries, 458–461, 467
breakup, 454, 476, 480–481
configuration in the Late-Jurassic to Early Cretaceous, 474
configuration in the Middle Jurassic, 472, 474
motion, 453, 454, 458, 461, 462
single Kula/Izanagi plate, 454, 456, 461–462, 480
subduction, 476, 480
Izu-Bonin-Mariana arc, 475, 479, 482, 483, 485
Izu-Bonin-Marianas trench, 329–330

J

Japanese lineations, 453–454, 455, 456, 463
Japanese seamounts, 561–562

Jasra alkaline-mafic-ultramafic complex
age, 816, 818
CO_2 metasomatism, 823–824, 825–826
discrimination function diagrams, 823, 824
geological setting, 817–818
and Kerguelen hotspot, 816, 825–826, 828–830
lack of co-genetic relationships of rocks, 824–825, 826
light rare earth elements (LREE), 823
major-element and CIPW composition of rocks, 818, 820
maps, 817
metasomatized lithosphere, 823–824, 825–826
overview, 816–817
petrography of major rock types, 817, 818, 819
radiogenic isotope ratios, 823–824, 825
similarity of tholeiites to Kerguelen OIB, 825
trace-element and rare earth element analyses, 818, 821, 823
variation diagrams, 821, 822
whole-rock geochemistry, 819–823, 824
Juan de Fuca
hotspot motions, 363
hydrothermal circulation, 267
plate motions, 27, 72, 181, 366
plate subduction, 642
thermoelastic stress, 501
thermoelastic stress near Juan de Fuca Ridge, 499, 501, 505
transition-zone observations, 124–125, 129
Juan Fernandez hotspot
asthenosphere flow from hotspot, 193, 194, 199
geochemistry, 255–256
hotspot location and strength, 184
hotspot track, 67, 491
Juan Fernandez microplate, 185, 193, 198
Juan Fernandez Ridge, 473, 483–484, 491, 492

K

K-T boundary, 307, 309, 781, 901
K-T magmatism, 777–779, 781
K-T sedimentary basins
Barmer basin, 778, 781
Bikaner-Nagaur basin, 778, 780–781
geological evolution, 777, 779, 781–783
Jaisalmer basin, 778, 779, 780, 781
map, 776
stratigraphy, 778
Kamchatka trench, 456, 464, 466
Kamchatkan arc, 483
Kamuikotan Complex, 478, 482
Karmutsen-Nikolai Formation, 477, 482
Karoo flood basalt and dikes
crystallization differentiation, 446
drainage, 788
magma mixing, 417, 421
Paleozoic and Permian rifts, 595
stratigraphy, 807
triple junction, 601
Kerguelen hotspot
$^3He/^4He$ ratios, 200
asthenosphere flow from hotspot, 190, 192, 193, 200–201

characteristics of crust, 510
and Jasra alkaline-mafic-ultramafic complex, 816, 825–826, 828–830
separation from Ninetyeast Ridge and Rajmahal basalts, 383
Kerguelen plateau, 383, 816, 818, 825, 829
Kirkpatrick Basalt, 297
Kolbeinsey Ridge. *See also* Iceland; North Atlantic Igneous Province
crust thickness, 532
geochemistry, 422–427, 429–431, 433–435, 437–440, 444–447
V-shaped volcano trail, 532
Komandorsky basin, 466, 467
komatiites, high-temperature source, 52
Kruzenstern fracture zone, 453, 455, 459, 463–464, 466
Kula lineations, 453, 455, 456, 457
Kula plate. *See also* Izanagi plate
motion, 462, 466
Pacific-Kula boundary, 461, 463
plate initiation, 453–454, 456, 461, 466
single Kula/Izanagi plate, 454, 456, 461–462, 480
splitting off from Farallon North plate, 480
subduction, 453, 476, 480

L

Labrador Sea, 581, 582
large igneous province (LIP) formation. *See also* large igneous provinces; traps; volcanic passive margins
accretion center model, 761, 763
compensation depth, 513, 531
dynamic rifting model, 536, 538–539
edge-driven convection model, 536, 538–539, 551
fertile mantle model, 536, 537–538, 541, 551
igneous center distributions, 751–752, 763–764, 766–767, 773–774
LIP eruption in icosahedral rifts (Gondwana), 595, 596, 597, 601
lithospheric delamination, 535–537, 550–551
localization of melting areas, 749–750
magma diapir model, 764–765
magma flow during trap emplacement at Isle of Skye, 755–760, 761
magma flows at volcanic passive margin (VPM) stage in East Greenland, 760–761, 762
mantle plume theory, 526, 536, 539–540, 551, 670
mantle ridge model, 753, 768
meteorite impact theory, 537
nonplume alternative models, overview, 540–542
overview, 747–748, 831–832
plume hypothesis, arguments against, 526
plume model, 748–750, 753, 768
small-scale convection model, 536, 538–539, 551, 765–768, 772–774
trap stage of LIP formation, 748
volcanic passive margin (VPM) stage of LIP formation, 748–749

large igneous provinces (LIPs). *See also* large igneous province (LIP) formation
  Atlantic Ocean, 508
  bimodal (felsic and mafic) LIPs, 832–833, 838
  comparison of off-ridge and on-ridge LIPs, 509–510
  continental LIP eruption during continental rifting, 18–19
  crust thickness, 509–510, 510
  damp mantle melting, 516–520, 521
  from eclogite melting, 60–61
  geophysical constraints on mantle melting
    crustal thickness versus mantle potential temperature, 517–521
    overview, 514–515, 521
    parameters that govern mantle melting, 516–518, 520–521
    seismic velocity as a function of mean pressure, 517
    vertical density (Vp) gradients versus mantle potential temperature, 517–521
    volumetric melt flux and its temporal variations, 515–516
  gravity modeling and compensation of tomography, 511–514
  mafic LIPs (MLIPs), 831
  magmatic rates, 8
  map, 4
  meteorite-impact hypothesis, 537
  and ridge-crossing seamount chains, 383
  seismic modeling and uncertainty analysis of oceanic crust, 510–511, 512
  seismic observations of LIPs, 509, 510
  seismic velocity structure, 511
  silicic-dominated igneous provinces (SLIPs), 775, 831, 836, 838
  structure and physical properties, 508, 509–514
  subsidence and uplift chronology, 6–7, 18
  underplating, 842
  vertical density (Vp) gradients, 509–510, 511, 514, 517–520
  volume and area criteria, 8, 509
large-ion lithophile elements (LILE)
  Dongargarh Group, 835, 836–837
  eastern Paraguay, 617, 626
  lower mantle source, 55
  Mediterranean area, 730
  mobility during alteration, 414
  Siberian Traps, 670
large low–shear velocity provinces (LLSVPs). *See also* core-mantle boundary; superplumes
  bulk sound velocity variations, 81
  chemical anomalies, 81–82, 85, 93
  correlation with surface hotspots, 82
  density increase, 81, 84
  high temperature, 84
  and hotspots, 82
  lack of post-perovskite structure (pPv), 81
  partial melting, 85
  perovskite (Pv) phase transformations, 80–82
  seismic evidence for sharp edges, 82, 84, 93, 99
  superplume hypothesis, 83
  themochemically dense piles, 83–85, 93, 98–99
  thermochemical models, 83–85

Lau Basin, 658
Leonardo da Vinci, 285, 305
light rare earth elements (LREE). *See also* rare earth elements
  crystallization differentiation, 415
  depletion in N-MORB compared to E-MORB, 391, 393
  Dongargarh Group, 835
  East Pacific Rise, 415–416, 418–419
  Iceland, 426, 429, 432, 434
  Jasra Complex, 823
  North Atlantic Igneous Province magmas, 531
  Paraguayan magmas, 617, 626
  Paraná-Angola-Edendeka system, 617, 626
LILE. *See* large-ion lithophile elements
Liliuokalani Ridge, 453, 459
Line Islands
  alteration (chemical weathering), 318, 319
  asthenosphere flow from hotspot, 193, 194
  lack of age progression, 355, 484
  Late Cretaceous ocean island volcanism, 480, 484
  maps, 473–474
LIP. *See* large igneous provinces
lithosphere. *See also* delamination; metasomatized lithosphere
  compositional lithosphere, 169, 171, 178, 179–180
  kinematic consequences of shallow hotspot reference frame, 365, 366–369, 371–372
  lithosphere reheating model, 261, 262–263, 264–265, 270
  lithospheric thickening with age, 5, 178, 190
  rejuvenation, 36–37
  thickness, 179, 180
  westward drift of lithosphere, 360, 366–367
lithospheric delamination. *See* delamination
lithospheric detachment. *See also* delamination
  Emeishan large igneous province, 847
  magma production due to gravitational instability, 6
  recycling of material into mantle, 5–6, 13, 60
  slab detachment in Eastern Anatolia, 697, 700, 707–708, 711, 713–714
  subsidence before LIP volcanism, 6, 19
  uplift, 437–438
LLSVPs. *See* large low–shear velocity provinces
Louisville chain
  age progression, 356
  asthenosphere flow from hotspot, 190, 193, 194
  and Eltanin fracture zone, 481–482, 483, 504, 505
  eruption rates, 8
  lower mantle source, 129
  maps, 473–474, 482
  and Osbourn Trough, 481, 484
  ridge-approaching hotspot, 38
  thermoelastic stress, 501
LREE. *See* light rare earth elements

M

Macdonald seamount, 124–125, 129, 193, 194
Madagascar, breakup from India, 787, 788, 807–808

Madeira, 191–192, 194–195
Magellan seamounts, 473–474, 477–478, 480, 484
Magellan spacecraft data, 860–862, 867, 875, 881–882, 909
Magellan Stereo Toolkit, 875, 877, 880, 881
magmas. *See also* hotspots; mid-ocean ridge basalts; ocean island basalts
  compressibility of eclogite magmas, 52
  crustal contamination of melts, 728–729, 749
  global volume flux, 35, 40
  volume of midplate volcanism, 48, 50
  volume of volcanism at plate margins, 48
magnetic anomalies
  apparent polar wander path (APWP), 339–340, 347, 355
  Campanian lineations, 452, 454
  Hawaiian lineations, 453–454, 463
  Japanese lineations, 453–454, 455, 456, 463
  Kula lineations, 453, 455, 456, 457
  M-sequence lineations, Pacific plate, 453, 455, 457
  northern Pacific magnetic lineations, overview, 452–453, 454
Malani large igneous province, 775, 777, 781, 782, 831
Manahiki plateau, 474–475, 478, 479, 481. *See also* Ontong Java–Manihiki–Hikurangi composite plateau
mantle. *See also* asthenosphere; core-mantle boundary; D″ layer; mantle convection; mantle geochemistry; mantle plume theory; tomographic mantle models; transition zone
  chemical heterogeneities, 5–6, 175–177
  chemical stratification, 50–51, 56, 63
  cooling by slabs, 63–64, 73
  damp melting, 509, 516–520, 521
  deep mantle displacement by subducted material, 49, 54, 57, 724
  density and phase changes at 400 km, 55, 58
  density and shear-wave seismic velocity of minerals, 51
  density estimates, 54, 105
  depleted MORB-source mantle (DMM), 393
  eclogite fate in mantle, 54, 55, 58, 133–134
  enriched mantle (EM-1 and EM-2), 393, 395–396
  fertility, 3, 5
  focus zone (FOZO)
    isotope signatures, 425, 431, 433, 440, 823–824
    previously melted (depleted) source, 64
    recycled subducted ocean crust, 393, 395
  geophysical constraints on mantle melting
    crustal thickness versus mantle potential temperature, 517–521
    overview, 514–515, 521
    parameters that govern mantle melting, 516–518, 520–521
    seismic velocity as a function of mean pressure, 517
    vertical density (Vp) gradients versus mantle potential temperature, 517–521
    volumetric melt flux and its temporal variations, 515–516

global shear wavespeed variations in shallow mantle, 167–168
heat flux, 35
high U/Pb mantle (HIMU), 393, 395–396, 731–732
kinematic consequences of shallow hotspot reference frame, 368–369
mantle plumes, definitions, 64, 97, 670
mass flux at hotspots, 35
multiscale nature of mantle plumes, overview, 158–160
narrow whole-mantle plumes, 89–91, 92, 97
oceanic mantle structure, 724
outer boundary layer thickness, 50
plum-pudding model, 166, 172, 174–177, 205–206
primitive mantle (PM) composition, 390
reflectors and scatterers, 50, 51, 55, 57
slab dehydration and melting, 5–6, 206–207
slab penetration into lower mantle, 72, 85, 167, 569, 580
statistical upper mantle assemblage (SUMA) model, 52, 60
subadiabatic temperature gradients in deep mantle, 53, 63, 90–91, 104–105
subcontinental mantle structure, 724
superadiabatic temperature region, 63, 104, 105
temperature variation, 63, 90–91, 104
thermal boundary layer (TBL) at base of lithosphere, 90–91, 96, 764, 765–766, 766–767
thermal boundary layer (TBL) at core-mantle boundary
    formation, 104, 108
    as plume source, 80, 89, 90–91, 115, 120
    pressure effects, 89, 90
    Rayleigh number effect on temperature, 89–90, 96
    temperature changes, 80, 89–90, 90–91, 96
    temperature dependence of viscosity, 83, 89–90, 95, 99
    thickness, 90
    transfer of heat from core, 89
thermal conductivity, 105
thermal diffusivity, 105
uniform bottom heating, 90, 91, 92–93
upper mantle plume creation, 145, 148–150, 156–159
viscosity variations, 73, 74, 105
whole-mantle-convection and superswell model, 65–66, 68, 74, 76–77
mantle convection. *See also* asthenosphere flow; convection models of multiscale plumes; eclogite engine; heatflow; mantle plume theory
    chaos theory calculations, 92, 93
    convective overturn and asthenosphere heating, 68, 73, 74
    edge-driven convection
        Bermuda Rise, 579, 580, 584
        North Atlantic Igneous Province (NAIP), 536, 538–539, 551
        Siberian Traps, 682, 685
    ground rules of mantle dynamics, 59–60

kinematic consequences of shallow hotspot reference frame, 368–369
the mantle as an engine, 48–49
narrow whole-mantle plumes, 89–91, 92, 97
realistic mantle simulations (RMS), 90, 92–93, 94, 95–97
shear heating in convecting layers, 145, 149–150
small-scale convection, 536, 538–539, 551, 765–768, 772–774
thermal convection, 49, 138
thermodynamic properties, 83, 89
top-down tectonics models, 50–51, 54, 58–59, 93, 205
whole-mantle-convection and superswell model, 65–66, 68, 74, 76–77
mantle depletion. *See* depleted mantle
Mantle Electromagnetic and Tomography (MELT) experiments, 174, 211, 214
mantle geochemistry. *See also* eclogite engine; mantle; olivine
    bathymetric and geochemical variations along Mid-Atlantic Ridge, 194–195
    chemical heterogeneities, 175–177
    chemical stratification, 50–51, 56, 63
    compositional lithosphere, 169, 171, 178, 179–180
    flow direction and He isotopic anomalies, 200–201
    geochemically distinct provinces along mid-ocean ridges, 175–177, 191–199, 201–202
    high U/Pb mantle (HIMU), 393, 395–396, 731–732
    melt-extraction–induced mantle strengthening, 169, 178
    primitive mantle (PM) composition, 390
    progressive melt-extraction, buoyancy changes, 171–172
    progressive melt-extraction, differences between MORBs and OIBs, 174–178
mantle plume theory. *See also* mantle; mantle convection; plume hypothesis; plumes
    age progression, importance in mantle plume debate, 936–937, 942–943, 960–961
    alternatives and critiques, 35–39, 42–44, 89–99, 934–938
    definitions, 64, 97, 670, 934
    evolution of mantle plume theory, 934, 957–960
    hierarchy of theories and mantle plume theory, 937–938, 939, 940
    large igneous province (LIP) formation, 526, 536, 539–540, 551, 670
    mantle plumes, definitions, 64, 97, 670
    mantle plumes postulated in the Mediterranean area, 730–735
    Morganian plumes, 670
    multiscale nature of mantle plumes, overview, 158–160
    narrow whole-mantle plumes, 89–91, 92, 97
    overview, 80, 104–105, 119–120, 526, 539–540
    poll (following American Geophysical Union Chapman Conference, The Great Plume Debate)

on arguments for and against mantle plume theory, 936–937, 942–943
number of educators participating in poll, 944
poll description, 935, 942, 943
on teaching the mantle plume debate, 940, 943
on validity of plume theory, 935
supercontinent breakup, 593–594
in undergraduate science education
    lack of alternative hypothesis discussion, 938, 939–941, 943, 944
    presentation in introductory geology textbooks, 938–939, 940, 943, 944
    presentation in other geoscience courses, 939, 940
    presentation in upper-division courses, 939–940
    recommendations, 940–941
    teaching the mantle plume debate, 940, 943
    upper mantle plume creation, 145, 148–150, 156–159
Marcus Wake seamounts, 453, 473–474, 480, 484
Marion
    asthenosphere flow from hotspot, 192, 193, 197
    geochemistry, 197
    hotspot location and strength, 184
    hotspot track, 67, 70
Marquesas
    asthenosphere flow from hotspot, 193, 194
    characteristics of crust, 510
    crustal compensation model, 512
    eruption rates, 8
    thermoelastic stress, 501
    transition-zone observations, 124–125
Mars
    Hellas, 916
    hemispherical crustal dichotomy, 913–914, 920, 926
    hydrous alteration of craters, Spirit and Opportunity data, 907
    lack of tectonic activity, 872
    local magma oceans
        crystallization timescale, 917–918, 930
        impact-induced formation, 914, 915–917, 931
        interaction between local magma ocean evolution and mantle dynamics, 920–926, 929–932
        isostatic adjustment to partially molten local magma ocean, 918–920
        magma ocean–induced upwelling model for dichotomy and Tharsis formation, 920, 925–927, 930–931
    magnetic stripes, 931–932
    Mars Global Surveyor, 914, 925–926, 931
    martian meteorites, 913
    Olympus Mons, 872, 907, 930–931
    physical parameters, 916
    Tharsis
        alternative explanations for formation, 914, 930
        magma ocean–induced upwelling model for formation, 920, 925–927, 930–931
        magmatic volume estimates, 914
        planetary mass imbalance, 885

Mars (*continued*)
　Tooting Crater, 907
　Utopia, 914
Martin Vaz, 67, 69, 72, 192
Mauritius, 383
McKenzie giant dike swarm, 749–750, 755
Median Tectonic arc, 475, 477
Mediterranean region
　anorogenic magmatism, source, 739–740
　anorogenic rocks, geochemical characteristics,
　　728, 732–739
　Central European Volcanic Province (CEVP),
　　745
　Eifel, 10, 67, 69, 732–733, 739–740
　European Cenozoic rift system, 725, 728
　evolution of region since 45 Ma, 725, 726–727
　French Massif Central, 732, 733, 740
　geodynamic setting, 724–725
　heterogeneous metasomatized lithospheric
　　mantle, 728–729, 731, 733, 744–745
　HIMU-FOZO composition of circum-
　　Mediterranean rocks, 731–732
　lithosphere thickness, 730, 739–740, 744
　major-element composition of magmas,
　　734–737
　mantle plumes postulated in the Mediterranean
　　area, 730–735
　Middle East, 731
　Moho depth, 739
　Pannonian basin, 733–735
　passive asthenosphere upwelling, 724, 730, 732,
　　733, 739–740
　plume explanation of Central European
　　magmatism, 327, 729–730, 732–733,
　　739, 744–745
　Rhenish Massif, 732–733, 740
　Sardinia, 731–732
　Sicily Channel, 731
　trace-element composition of magmas,
　　734–737, 738
　Tyrrhenian Sea, 730–731
Meiji
　abandoned spreading center, 461, 481
　age, 343, 480
　junction of the Kamchatka and Aleutian
　　Trenches, 456
　maps, 464, 467
　traveltime of the triple junction to Meiji, 463
　volcanism interrupted by plate reorganization,
　　480
melting anomalies. *See also* hotspots; large
　　igneous provinces; plate model
　alternative explanations, 1–2, 121, 508–509
　association with triple junctions, 4
　definition, 508
　extensional stress, 4
　fertile blob and plume hypotheses, comparison,
　　42, 43–44, 60–61, 98, 206
　fertile blobs as sources, 56, 493
　heat, influence of, 12–13
　lithosphere, influence of, 12–13
　mantle fertility and, 3, 42, 48
　mantle temperatures, 19
　OIB geochemistry, 13, 19

plate model, 18–19
subsidence and uplift chronology, 6–7
thermal mantle plume explanations, 508–509
topography, 10–11
variability, 1, 19
Mendocino fracture zone, 452–453, 454, 456, 459,
　　475
Mercury, 885, 901, 907, 908
metasomatized lithosphere. *See also* lithosphere
　circum-Mediterranean region, 728–729, 731,
　　733, 744–745
　continental crust recycling into mantle, 5–6, 13,
　　60
　eastern Paraguay, 620, 623, 624, 625
　interaction between peridotite and
　　clinopyroxenite partial melts, 729
　Jasra alkaline-mafic-ultramafic complex,
　　823–824, 825–826
　mantle metasomatism array, 703, 706
　plate model, 5–6
　subcontinental lithospheric mantle peridotite,
　　656
　at subduction margins, 60, 728–729
　thickening in Eastern Anatolia, 709, 711, 715
meteorite-impact hypothesis. *See also* impact
　　craters
　K-T boundary, 307, 309, 781, 901
　large igneous province formation, 537, 966–967
　Siberian Traps, 680–681, 685
Mid-Atlantic Ridge (MAR). *See also* Atlantic
　　Ocean
　^{3}He/^{4}He ratios, 194, 195, 196
　asthenosphere flow, 191–192, 194–195, 949
　bathymetric variations along ridge, 195
　geochemical variations along ridge, 194–195,
　　196
　map of south Mid-Atlantic Ridge, 392
　trace elements injected into Mid-Atlantic Ridge,
　　195–197
mid-ocean ridge basalts (MORB). *See also* basalts;
　　depleted mantle; ocean island basalts
　comparison to OIB, 2, 388
　enriched MORB (E-MORB)
　　enrichment of MORB by plume-fed
　　　asthenosphere, 411
　　geochemistry, 390–391, 404–405
　　incompatible element enrichment, 174
　　mixing with truly depleted MORB and
　　　andesite, 416, 418–420
　　progressive melt-extraction, 174–175, 177
　　rare earth element composition, 416
　estimation of temperature by petrology, 44
　FeO^T and MgO contents of whole rocks and
　　glasses, 239, 241–244, 246, 252–253,
　　258
　incompatible element depletion, 174, 176
　isotopic evolution of recycled oceanic crust,
　　206–207
　mantle heterogeneity and diversity of MORBs, 49
　mixing between truly depleted MORB, E-
　　MORB, and andesite, 416, 418–421
　normal MORB (N-MORB)
　　comparison to enriched MORB (E-MORB),
　　　391

depletion, compared to primitive mantle, 388
MgNo versus Nb/Y in Pacific-Antarctic East
　　Pacific Rise basalt, 416
Nb deficiency, 390
Nb/U value, 390
ΔNb values for discrimination between OIB
　　and N-MORB, 391, 395, 404
rare earth element composition, 416
upper-mantle source, 388, 404
progressive melt-extraction, differences between
　　MORBs and OIBs, 174–178
Siqueiros glasses, composition, 242, 246, 252,
　　254, 256–258
temperature variations, causes, 254–255
temperatures of hotspot and MORB tholeiites,
　　241–243
thermal variability in MORB source, 63–64
truly depleted MORB, 416, 418
Mid-Pacific Mountains, 343, 473–475, 477–478,
　　480
midplate volcanism. *See* hotspots
model S20RTS, 123, 124, 128–129. *See also*
　　tomographic mantle models
Moho
　depth in Iceland, 532
　depth in Mediterranean region, 739
　depth on east flank of Bermuda Rise, 558–559
　Eastern Anatolia, 699
　geometry, 511, 512–513, 515
　heatflow in Deccan basalt province, 281,
　　282–283
　North Atlantic Igneous Province, 532, 533,
　　748–749, 750, 764
　temperature in Deccan basalt province, 281
Monteregian hills, 38
Moon
　accretion, 94, 883, 904
　age of rocks, 904
　Aitken–South Pole impact basin, 898
　ballistic ejecta, 888, 900, 901, 909
　craters, proof of impact origin, 881–882, 885
　dates of bolide bombardment, 904
　Imbrium, 904
　size distribution of craters, 886
MORB. *See* mid-ocean ridge basalts
Morgan, W.J., plumacy, 946, 956, 957, 958,
　　959–960
mud volcanoes derived from wet sediments, 883,
　　907–908
Muir seamount, 555, 562
Murray fracture zone, 453, 483
Musicians seamounts
　alteration (chemical weathering), 316, 318
　alteration of rocks, 316, 318

N

NAIP. *See* North Atlantic Igneous Province
Namibia, 619
Nazca plate
　age, 473, 483
　creation from Farallon plate, 483, 491
　hotspots beneath plate, 193, 492
　intraplate stresses, 492

maps, 473
Nazca slab, 605, 624
plate velocity, 364–365, 369
subduction, 364, 484, 624
velocity, 364–365
Nazca Ridge
age, 484
characteristics of crust, 510
Easter hotspot, 25, 27
end of volcanism, 483
maps, 473, 484
neon (Ne) in OIB and MORB, 14
New England Seamounts, North Atlantic Ocean
age progression, 38–39
buoyancy flux, 38
hotspot critiques and alternatives, 38–39
hotspot track, 38
topographic map, 38
Newark Trend Basins, 290
Nicoya peninsula, Costa Rica, 313, 315
Ninetyeast Ridge
alteration (chemical weathering), 314, 315
fossil transform fault, 383
geochemical connections to Atlantic and Indian
oceans, 195
ridge-leaving hotspot, 38
separation from Kerguelen plateau, 383
niobium. *See also* high field strength elements
deficiency in continental crust, 390
deficiency in N-MORB, 390
Icelandic basalt, 416
Nb/U values for basalts, 390
ΔNb values
alkali basalts, 397, 402, 403
basalts from ocean islands, 391, 393, 395–396
definition and background, 390, 414, 415
discrimination between OIB and N-MORB,
391, 395, 404
East Pacific Rise OIC and N-MORB, 396,
397, 400, 401, 402
effects of basalt-rhyolite magma mixing, 417
effects of rare earth elements, 415–417
Emperor seamount basalt, 393, 395
Galápagos Archipelago basalt, 396, 398–399
Hawaiian basalt, 393, 395
HIMU, EM-1, and EM-2, 396
Icelandic basalt, 389–390, 395, 422–423,
439–440, 444–445
mixing between truly depleted MORB,
E-MORB, and andesite, 416, 418–420,
421
N-MORB, 391, 393, 395
Pacific-Antarctic East Pacific Rise basalt,
415, 416
and source heterogeneity, 415
southern Mid-Atlantic and southwest and
central Indian ridge basalts, 391–392,
393, 394
noble gases. *See* argon; helium; neon
North Atlantic Igneous Province (NAIP). *See also*
Atlantic Ocean; Iceland
^{3}He/^{4}He ratios, 532, 537, 542
accretion center model, 761, 763
ages of rocks, 528, 532–533, 550, 750

area, 526
characteristics of crust, 510
convective destabilization of the TBL, 764,
766–767
dike swarms and stress fields, 753
distribution and timing of volcanism, 550,
751–752, 763–764
dynamic rifting model, 536, 538–539
edge-driven convection model, 536, 538–539,
551
eruption during continental breakup, 5
fertile mantle model, 536, 537–538, 541, 551
geochemistry, 529–531, 542
high-velocity lower-crustal bodies (LCBs), 533,
539, 540, 542
high-velocity zone (HVZ), 748, 749, 750–751,
764, 765
igneous center distributions, 751–752, 763–764,
766–767, 773–774
lithospheric delamination, 535–537, 550–551
localization of melting areas, 749–750
magma flow during trap emplacement at Isle of
Skye, 755–760, 761
magma flows at volcanic passive margin (VPM)
stage in East Greenland, 760–761, 762
magmatic rate, 750, 763–764
mantle plume theory, 526, 536, 539–542, 551
mantle seismic velocity structure, 534
maps, 527, 750, 773
nonplume alternative models, overview,
540–542
Ocean Drilling Program (ODP), 529
picrites, 529, 541, 542
plume hypothesis, arguments against, 526, 542
plume model, 526, 748–750, 753, 768
postbreakup magmatism, 532, 773–774
prebreakup and breakup-related magmatism,
529–532, 550
rifting episodes, 526–529, 536
seaward-dipping reflector sequences (SDRs or
SDRSs)
Baffin Island, 533, 549–550
Greenland, 529, 760–761
Irish volcanic margin, 531
North Atlantic Igneous Province (NAIP), 29,
529, 531, 533, 549–550
Norway, 29, 533
in volcanic passive margin crust structure,
748–749
Vøring Plateau, 533
small-scale convection model, 536, 538–539,
551, 765–768, 772–774
steady-state thermal boundary layer hypothesis,
764, 765–766
subsidence and uplift chronology, 6–7
tectonic history, overview, 528
uplift history, 527–528, 534–535
volume estimates, 526, 532–534, 750–751, 763
North Atlantic volcanic province. *See* North
Atlantic Igneous Province
North Sea basin, 403, 406
Norway
Aegir rift, 581
high-velocity lower-crustal bodies (LCBs), 533

lithospheric delamination, 534
magmatism at Norwegian volcanic margins,
529, 532–533
seaward-dipping reflector sequences (SDRSs),
29, 533
small-scale convection, 539
uplift at Norwegian volcanic margins, 534–535

O

Ob-Lena, 192, 193
Obruchev Rise
age, 465
formation at ridge axis, 465–466
Kamchatka–Aleutian trench intersection, 466
maps, 464
size and description, 463, 464–465
Ocean Drilling Program (ODP)
comparison to SWELL data, 212–213
Detroit seamount, 343, 465
North Atlantic Igneous Province (NAIP), 529
Obruchev Rise, 464
OIB and N-MORB studies, 404–405
paleomagnetic data, 336, 340, 342, 343
ocean island basalts (OIB). *See also* basalts;
hotspots; mid-ocean ridge basalts
^{3}He/^{4}He ratios, 2
blob or streak hypothesis, 388, 404–405,
406–407
comparison to MORB, 2, 388
enrichment of OIB by global plume-fed
asthenosphere, 410–411
FeO^T and MgO contents of whole rocks and
glasses, 239–240, 241–244, 246,
252–253, 256
geochemistry, overview, 13–14, 19, 388–389,
404–407
HIMU OIB, 393, 395–396
hypotheses of formation, 388
incompatible element enrichment, 2, 13, 388,
410
isotopic evolution of recycled oceanic crust,
206–207
ΔNb for basalts from ocean islands, 391, 393,
395–396
Nb/U value, 390
ΔNb values for discrimination between OIB and
N-MORB, 391, 395, 404
OIB paradox, 387–388, 406–407, 410–412
perisphere hypothesis, 388, 406, 407
plume hypothesis, 388, 397, 724
progressive melt-extraction, differences between
MORBs and OIBs, 174–178
Ocean Seismic Network (OSN), 211–214, 212,
221–222
oceanic crust. *See also* continental crust
creation at mid-ocean ridges, 388
crustal compensation model, 512
dehydration and melting of subducted oceanic
crust, 5–6, 206–207
density and shear-wave seismic velocity of
minerals, 51
eclogite from subducted oceanic crust, 5–6
layering and seismic wave velocities, 749

oceanic crust (*continued*)
 mass balance, 50
 petrology of lower ocean crust, 438
 production rates, 35
 seismic modeling and uncertainty analysis, 510–511, 512
 vertical density (*Vp*) gradients, 509
ODP. *See* Ocean Drilling Program
OIB. *See* ocean island basalts
Oku-Niikappu Complex, 475, 477, 478
olivine. *See also* olivine tholeiite magmas
 Anderson-Grüneisen parameter, 142
 phase transitions in transition zone, 58, 122, 126–128, 131–134
olivine geothermometer
 definition, 236
 H_2O versus K_2O contents of glasses and melts, 239, 240, 241, 242, 246
 Hawaii temperature relative to mid-ocean ridges, 12
 Kilauea volcano parental crystallization temperature, 240, 243–245, 247–248, 253
 mantle potential temperature (T_p) calculations assumptions, 243–245, 252–254, 258
 critiques and discussion, 252–259
 mantle equilibration depths and pressures, 245–248
 mantle potential temperature results, 247–248, 254
 potential temperature variations, 254–255
 projection of partial melting trends for pyrolite, 245, 246–247
 Mauna Loa volcano, magma composition, 240, 246, 253, 255–256
 Mauna Loa volcano, parental crystallization temperature, 240, 247–248, 253
 olivine phenocrysts, significance, 236–237, 254, 257
 overview, 11–12, 236–237
 oxidation state effect, 237
 parental liquid calculation methods, 11–12, 236–237
 Puna Ridge glasses, 237, 238–240, 245–246, 253
 Réunion, 240, 241
 Theistareykir volcanic system, Iceland, crystallization temperature, 240, 241
 Theistareykir volcanic system, Iceland, parental crystallization temperature, 240
 water, effect on liquidus temperature, 237
 xenocrysts, 237
olivine tholeiite magmas. *See also* olivine; tholeiite
 FeO^T and MgO contents of whole rocks and glasses, 239–240, 241–244, 246, 252–253, 256
 forsterite (Fo) content, 252, 254, 255–256, 257
 glass compositions, 238
 high alumina olivine tholeiite (HAOT) in Columbia River Basalts, 646, 647, 648
 Kilauea eruption of 1959, magma composition, 237–239, 240, 243–245, 255

Kilauea eruption of 1959, parental crystallization temperature, 240, 243–245, 247–248, 253
Mauna Loa volcano, magma composition, 240, 246, 253, 255–256
Mauna Loa volcano, parental crystallization temperature, 240, 247–248, 253
picrites, 252–253, 255–256, 433–434, 444
Preshal Beg conglomerate and columnar-jointed olivine tholeiite, 792
Puna Ridge glasses, 237, 238–240, 245–246, 253
Réunion, magma composition, 240, 241
Réunion, parental crystallization temperature, 240, 241
temperatures of hotspot and MORB tholeiites, 241–243
Theistareykir volcanic system, Iceland, parental crystallization temperature, 240, 241
Ontong Java plateau
 age, 313, 314, 479
 characteristics of crust, 510
 creation at ridge, 5, 478, 490
 creation at triple junction, 383, 481
 magma volume, 384
 magmatic rate, 8
 maps, 337, 340, 458, 473–475
 paleomagnetic data, 341–342
 subsidence and uplift chronology, 6, 787
Ontong Java–Manihiki–Hikurangi composite plateau, 1, 5, 6
Oregon High Lava Plains. *See* High Lava Plains
orthopyroxene, density and phase changes at 400 km, 55, 58
Osbourn Trough
 abandonment, 481
 and Louisville chain, 481, 484
 maps, 473–474, 476, 484
 rifting, 478, 479
Osmium (Os)
 in DUPAL basalts, 677
 Os isotope systematics in Columbia River Basalt Group, 648, 656, 665
 Os isotope systematics in plum-pudding plume model, 75–76, 77
 Re-Os isochron age for Deccan Traps, 296, 297, 309
 Re-Os isochron age for Dongargarh Group, 836

P

Pacific-Antarctic East Pacific Rise. *See* East Pacific Rise
Pacific-Farallon-Izanagi triple junction. *See also* triple junctions
 age of seafloor at ends of triple junction jump, 462–465
 association with Hess Rise, 456
 association with Shatsky Rise, 454–455, 458
 Hawaiian hotspot origin in abandoned triple junction, 456, 461, 462, 466–468
 jump of 800 km at M21 time (147 Ma), 454, 458, 462
 jump of 2000 km (ca. 90 Ma), 456, 459, 461

motion of triple junction over time, 453, 454, 456
Pacific-Farallon Ridge, 452, 480, 482, 484, 492, 493
Pacific Ocean. *See also specific locations*
 alteration (chemical weathering) of rocks, 316, 318–323
 asthenosphere flow, 190, 192–193, 194, 199
 bathymetry and topography of the Pacific seafloor, 499, 504–505
 bootstrap method for Pacific plate motion, 69–70
 isotope signature along Pacific-Antarctic Ridge, 198–199
 lack of evidence for numerous plumes, 39
 map, 473
 melting anomalies, age progression, 18
 melting anomalies, plate model, 17–18
 Mid-Pacific Mountains, 343, 473–475, 477–478, 480
 plume provinces, 194
 shear-wave velocity structures of Pacific seafloor, 167
 South Pacific isotopic and thermal anomaly (SOPITA), 505
Pacific-Phoenix Ridge, 478, 479–480, 484
Pacific plate. *See also* plate motions; Pukapuka volcanic ridges; Shatsky Rise; *specific locations*
 bathymetry and topography of the Pacific seafloor, 499, 504–505
 bootstrap method for Pacific plate motion, 69–70
 comparison of Pacific and Antarctic apparent polar wander paths, 351–352
 configuration
  71 Ma (chron 32), 461, 463
  80 Ma, 458
  84 Ma (chron 34), 461, 462
  ca. 90 Ma (quiet zone), 459, 460
  ca. 90 Ma (after triple junction jump), 459, 461
  ca. 110 Ma (end of M sequence), 459, 460
  125 Ma (M0 time), 458–459
  Late Jurassic to Early Cretaceous, 474, 477
  Middle Jurassic, 472, 474, 477
 decoupling zone between lithosphere and lower mantle, 66, 73
 Euler poles, 363, 371
 formation date, 453
 hotspot seamount chains on the Pacific plate, 337
 Jurassic and Early Cretaceous motion, 339, 341, 343–344, 347
 Late Cretaceous and early Cenozoic motion, 339, 341–342, 343–344, 347–351
 Late Cretaceous to Early Eocene volcanism, 475–476, 479–481
 M-sequence lineations, 453, 455, 457
 map of current volcanism and oceanic plateaus, 473
 northern Pacific magnetic lineations, overview, 452–453, 454
 northern Pacific satellite gravity map, 452, 453

northward motion implied by paleomagnetic data versus age, 347, 349
oceanic plateaus and diffuse island chains in the Middle Jurassic to Early Cretaceous, 474, 477–478
Pacific hotspot models, 342–343
plate boundaries in the past, 343–344
plate motions, history, 344
plate reorganization
  early Campanian, 475, 480–481
  Early Eocene, 476, 482–483
  Late Oligocene, Miocene, and Pliocene, 483–484
  Middle Cretaceous, 475, 478–479
rotation, 363
stresses acting on the Pacific plate
  and intraplate volcanism, 472, 484–485
  microplates at triple junctions, 5, 185, 193
  overview, 472, 484–485, 491–493
  plate reorganization in the Early Eocine, 483
  plate reorganization in the Late Oligocene, 483–484, 491
  plate reorganization in the Middle Cretaceous, 479
  thermoelastic stress field, 18, 498, 500–503, 505, 931
subduction, 475, 476, 479–480, 930–931
tectonic evolution of the northwest Pacific, overview, 457–458, 466–468
velocity, 362–364, 554
paleomagnetism. *See also* apparent polar wander path; magnetic anomalies
Deep Sea Drilling Project (DSDP), paleomagnetic data, 336, 337, 342, 343, 349
magnetic anomaly studies, 339–340, 347, 355
Ocean Drilling Program (ODP), paleomagnetic data, 336, 340, 342, 343
overview, 336
Palisade rhyolite, alteration (chemical weathering), 292–294
Pangaea
  and Antarctica, 947
  and Atlantic-African geoid anomaly, 594, 598–599
  formation, 624
  map, 676
  reconstruction, 957, 967, 969
  and Siberia, 675, 676, 683, 685
Pannonian basin, 733–735
Paraguay. *See* eastern Paraguay; Paraná-Angola-Edendeka system
Paraná-Angola-Edendeka system. *See also* eastern Paraguay
  ages of rocks, 616–617, 619
  drainage, 788
  geochemical implications related to plume hypothesis, 623–624
  geodynamic significance, 624–625
  incompatible elements, 618
  maps, 616
  Nd model ages of depleted mantle, 620
  overview, 614
  Pb isotopes, 620–623

Sr-Nd isotopic compositions of magmatic rocks, 619–620
tholeiitic dike swarms, 614–615
partial melting
  asthenosphere mass balance, 50
  core-mantle boundary (CMB), 82, 84, 85
  density reduction, 49, 53, 54
  eclogite formation, 48
  fraction of partial melting in adiabatic melting and expansion, 30–31
  large low–shear velocity provinces (LLSVPs), 85
  partial melting trends for pyrolite, 246–247
  projection of partial melting trends for Hawaiian pyrolite, 245
  Reykjanes Ridge, 32
  role in chemical stratification, 13
Peccerillo, A., plumacy, 959
peneplains. *See* planation surfaces
peridotite
  buoyancy changes from progressive melt-extraction, 172
  melting experiments, 246, 248–249
  refertilized peridotite, 52
  variation of density with depth, 50–51, 56
petrological grids
  assumptions, 32
  calculations, 30–31
  fraction of partial melting, 30
  Hawaii hotspot temperature estimates, 31–32
  Iceland hotspot temperature estimates, 31, 32
  MORB temperature estimates, 44
Phoenix plate
  configuration in the Late-Jurassic to Early Cretaceous, 474
  configuration in the Middle Jurassic, 472, 474
  configuration in the Paleocene, 476
  North New Guinea plate, 480, 484
  subduction, 479, 484
piclogite, characteristics, 48, 52
picrites
  definition, 421
  Hawaii, 252–253, 255–256
  high temperature source, 52
  Iceland, 253, 433–434, 436, 444
  North Atlantic Igneous Province (NAIP), 529, 541, 542
  olivine tholeiite magmas, 252–253, 255–256, 433–434, 444
  phenocrysts of plagioclase and clinopyroxene in Iceland, 253
Pioneer fracture zone, 452–453
Pioneer missions to Venus, 861, 862
Pitcairn Islands
  asthenosphere flow from hotspot, 193, 194
  eruption rates, 8
  hotspot track, 68, 70
  thermoelastic stress, 501
  transition-zone observations, 124–125
planation surfaces
  under Deccan basalts, 795–797, 799–800, 806, 807, 810
  in Greenland, 795
  overview, 793–795, 808
  in southern India, 795

Plantagenet Bank (Argos), 557, 561, 586
plate model
  chronology of volcanism, 8–10
  and eclogite engine, 58–59
  extensional stress, 3, 5
  $^3He/^4He$ ratios for plume and plate models, 13–14
  heat requirements for melting, 12–13
  magmatic rates, 8–9
  magmatic volume, factors affecting, 3
  mantle fertility, 3, 5
  mantle inhomogeneity, 5–6
  melting anomalies, overview, 3–6, 19–20
  overview, 58–59, 472
  partially molten asthenosphere model, 56, 60, 63, 120, 355
  PLATE (plate, lithosphere, and asthenosphere tectonics and eclogite) paradigm, 59
  subsidence and uplift chronology, 6–7
  temperature difference measurements, 11–12
  top-down convection, 50, 51, 54, 58, 59
  vertical motion associated with volcanism, 6–7
  volume of volcanism, 8–9
plate motions. *See also* apparent polar wander path; Pacific plate; plate tectonic theory
  azimuth-only method for plate velocity, 66, 68–72, 76–77
  bootstrap method for Pacific plate motion, 69–70
  decoupling of plates from mantle, 367
  geodesy plate motion models, 364, 366, 371, 372
  global plate motions in hotspot reference frames, 72, 181, 363–364, 365–366
  no-net-rotation (NNR) reference frame, 360, 361, 366, 371
  North American plate model, 70–71, 76
  NUVEL-1A model relative plate motion, 68, 72, 362–365
  plate velocity, absolute motion model, 177, 179, 181
  plate velocity in hotspot reference frame, 65–66, 68–72, 76–77, 181, 356
  resistance from asthenospheric drag, 171
  sinusoidal global flow field, 366
  westward drift of lithosphere, 360, 366–367
PLATE (plate, lithosphere, and asthenosphere tectonics and eclogite) paradigm. *See* plate model
plate tectonic theory. *See also* plate model; plate motions
  early development, 2
  and hotspots, 2
  and mantle plume theory, 938
  plate tectonic cycle relationship to plume rates, 73–74
  top-down tectonics models of mantle convection, 50–51, 54, 58–59, 93, 205
plateaus. *See also* Ontong Java Plateau
  Anatolian-Iranian plateau, 694, 699, 710, 714. *See also* Eastern Anatolia
  Hikurangi plateau, 5, 473–475, 478, 479, 481
  Icelandic Volcanic plateau, 9

plateaus (*continued*)
Kerguelen plateau, 383, 816, 818, 825, 829
Manahiki plateau, 474–475, 478, 479, 481
Mascarene plateau, 314, 324, 417, 788, 808
Pacific plate, current volcanism and oceanic plateaus, 473
Pacific plate, Middle Jurassic to Early Cretaceous, 474, 477–478
Shillong plateau, 816, 817
tessera plateaus, Venus
age range, 891
Beta Regio, 885
characteristics, 899
formation from impact melts and fluidized debris, 899, 903
images, 896, 898
oldest tessera plateaus as basement, 904
sparsity of old impact structures, 891, 895, 903
Tibetan plateau, 694, 718, 793
Vøring Plateau, 529, 532, 533, 538
plumacy. *See also* mantle plume theory; plume hypothesis
age progression, 960–961
antipodal plumes, 947, 949
aseismic ridges, 947, 950, 952
asthenosphere flow from mid-ocean rifts, 947, 949
cracked models, 968
definitions, 945, 955
evolution of mantle plume theory, 957–960
fixity, 957, 959, 960
global synchronism, 950
grading (scoring) of mantle plumes, 959, 966
Iceland plume, 947, 949–950, 958
impact theory of LIP formation, 966–967
invert every last wiggle (IELW) prediction, 969–970
Loch Ness Monster principle (LNMP), 961, 964, 969
logjam tectonics, 950, 952
mantle carp (*Asthenicthes aseismotathes*), 947, 948, 957, 964
mantle devils, 950, 951, 952, 956
mantle plume theory overview, 945–948, 955–957, 962–963, 971
multiculturalism and celebration of diversity, 970–971
notsohotspots, 955, 963, 967–968, 969
numbers of plumes, 947, 963–964
Pangaea, 957, 967, 969
platonics, 967–968
plume chemistry, 949–950, 951, 969
regurgitation or "Fertile Earth Mother" hypothesis (Iceland), 968–969
shape, color, and intelligence of mantle plumes, 959, 964–966
stress fields, 946, 948
superduperplumes and geohyperbole, 961–963
superpiles, 961
thermal models of mantle convection, 967
V-shaped ridges or escarpments (VSRs), 957
Venus, 962

plume hypothesis. *See also* convection models of multiscale plumes; fixity; mantle plume theory; superplumes; thermal plumes
age progression of hotspot tracks, 38–39
alternatives and critiques, 35–39, 42–44, 89–99, 934–938
buoyancy flux, 33
comparison of plume and fertile blob hypotheses, 42, 43–44, 60–61, 98, 206, 411
comparison of plume and plate models, 508–509
delamination alternative, 36
global heat flux, 35, 40
global volume flux, 35, 40
gravitational anchor hypothesis, 228, 342, 946–947
^{3}He/^{4}He ratios for plume and plate models, 13–14
history of hypothesis
classical plume hypothesis, assumptions, 2–3, 19, 42
contemporary plume hypothesis, assumptions, 42–43
mantle homogeneity and petrology, 49, 51–52
hotspots, deviation from predicted behavior, 2, 509
injection experiments, 89, 92
lithospheric uplift predicted over plume heads, 786, 792, 800, 805–810
mantle plumes, definitions, 64, 97, 670
narrow whole-mantle plumes, 89–91, 92, 97
overview, 39–40
plum-pudding plume model, 75–76, 77, 166, 172, 174–177, 205–206
plume pipes, 72, 73, 74, 946–947, 956
poll (following American Geophysical Union Chapman Conference, The Great Plume Debate)
on arguments for and against mantle plume theory, 936–937, 942–943
number of educators participating in poll, 944
poll description, 935, 942, 943
on teaching the mantle plume debate, 940, 943
on validity of plume theory, 935
subsidence and uplift chronology, 6–7, 36–37, 786–787
on Venus
evolutionary diapir model, 860, 867–871, 876–878
mantle upwelling models, 860
overview and background, 860, 881–882
volume of starting plume heads, 34–35
plumes. *See also* convection models of multiscale plumes; fixity; hotspots; mantle plume theory; plume hypothesis; superplumes; thermal plumes
as dynamic features, 36
geochemically distinct provinces along mid-ocean ridges, 175–177, 191–199, 201–202
locations, 181, 184
low-velocity anomalies and, 729

mantle plumes, definitions, 64, 97, 670
Morganian plumes, 670
multiscale nature of mantle plumes, overview, 158–160
narrow whole-mantle plumes, 89–91
plate tectonic cycle relationship to plume rates, 73–74
plume clusters, 83
plume strengths, 179, 181, 183–184
ponding of plume material in transition zone, 126, 135, 166
source of asthenosphere material, 73, 77
upper mantle plume creation, 145, 148–150, 156–159
polar wander. *See* apparent polar wander path
ponding, 49, 126, 135, 166. *See also* underplating
post-perovskite phase transitions, 80–81, 94, 159–160
potential temperature, definition, 724
Prandtl number, 140
Preliminary Reference Earth Model (PREM)
densities of mafic and ultramafic melts, 51
depth-variable density model, 139
seismic profiles, 55, 215, 218–220, 223, 225
Pukapuka volcanic ridges, 364

R

Rajahmundry Traps, 290, 291, 292, 328–329, 329
Rajasthan. *See* western Rajasthan
Rajmahal-Sylhet igneous province, 818, 825, 829–830
Rajmahal Traps, 290, 313, 314, 829–830
Rano Rahi seamounts, 397, 401
rare earth elements (REE). *See also* light rare earth elements
chondrite-normalized concentrations in basalt, 416
concentrations in East Pacific Rise basalt, 416, 418–419
effects of magma mixing on rare earth elements in Icelandic basalt, 426–429, 430
similarity to high field strength elements (HFSE), 415
Raton, 67, 69, 124–125
Rayleigh number (Ra)
convection models of multiscale plumes, 140, 141, 143–144, 160, 163
definition, 104–105
effect on geoid, heatflow, and topographic anomalies, 109, 110
effect on temperature of lower mantle thermal boundary layer, 89–90, 96
mantle stratification and, 50, 57
surface Rayleigh number, 140
Rayleigh-Taylor instabilities, 6, 50, 52, 56–57, 138
receiver functions. *See also* seismology; tomography
deepening of 410-km discontinuity below Yellowstone hotspot, 129
Hawaiian hotspot topography, 129–130, 134
lack of global coverage, 123, 130, 135
sensitivity to narrow features, 123, 130, 131, 133

thin transition zone under Iceland, 130
topography on the 410-km and 660-km
    discontinuities, 10
recycling (crust)
    continental crust, recycling into mantle, 5–6,
        13, 60
    isotopic evolution of recycled oceanic crust,
        206–207
    mantle mass flux, 35
    mass balance, 50
    role in OIB geochemistry, 13–14, 728–729
    role of recycled near-surface materials in OIB,
        728–729
    slabs, 6, 14
    subducted ocean crust in FOZO, 393, 395
    whole-mantle recycling of material, 205
REE. *See* rare earth elements
reference frames. *See* hotspot reference frames
region D″. *See* D″ layer
rejuvenation models
    hybrid thermal rejuvenation-dynamic
        support/thinning model, 228
    hybrid thermal rejuvenation–dynamic
        support/thinning model, 225–226,
        228–232
    overview, 36–37, 229
    thermal rejuvenation model, 210, 225
Réunion
    absolute velocity in hotspot reference frame,
        263
    age of hotspot track, 263, 275
    asthenosphere flow from hotspot, 192
    bathymetry and heatflow measurement sites,
        263–264
    characteristics of crust, 510
    crustal age, 263–264
    and Deccan volcanism, 275–276, 781
    geochemical influence on ridges, 197
    heatflow anomaly, 263–264, 265, 276,
        281–282
    heatflow measurements, 264, 265, 276
    and K-T magmatism, 781
    location on abandoned ridge and fracture zone,
        368
    location on transform fault, 362
    magma composition, 240, 241
    parental crystallization temperature, 240, 241
    partial melting trends for olivine glasses, 245
    Piton de la Fournaise eruption of 1939, 241
    relationship with western Rajasthan, India, 777,
        778, 781, 782–783
    ridge-crossing hotspot, 38
Reykjanes Ridge. *See also* Iceland; North Atlantic
        Igneous Province
    asthenosphere flow, 32, 171, 189–190
    bathymetric map, 9
    crust thickness, 9, 532
    dehydration front at 60-km depth, 171
    geochemistry of basalt, 422–427, 429–431,
        433–435, 437–440, 444–447
    map, 9
    partial melting, 32
    V-shaped volcano trails, 532, 957
Rhenish Massif, 732–733, 740

rhyolite
    mixing with basalt, 421, 422
    occurrence in Iceland basalts, 421
    similarity to FOZO, 425, 431, 433, 440
    source, 444, 446–447
ridge-crossing seamount chains. *See also* hotspot
        tracks; ridge-transform intersections;
        seamounts
    fertile blob model, 377, 380–383, 384
    fixed plume model, 375, 377, 384
    horizontal tensional stress ahead of RTIs, 377,
        384
    and large igneous provinces, 383, 384
    melt source, 376
    melt transport, 376–377
    nonplume mechanism for formation, 375–377,
        383–384
    other seamount geometry's, 383
    ridge reorganization model, 377–380, 383,
        384–385
    South Atlantic plate motions and ridge location
        over time, 380–383
    two-dimensional plane-strain elastic finite
        element model, 384
ridge-transform intersections (RTIs), 375, 377,
        384. *See also* transform faults
rifts
    Aegir rift, 581
    Afar triple junction, 3, 5, 19
    Barmer-Sanchor rift, 776–777, 784
    Cambay rift, 787–788, 810
    Chinook Trough, 456, 457
    composition of alkali basalt, 402–404
    composition of alkali basalt from rift systems,
        402–404
    continental LIP eruption during continental
        rifting, 18–19
    dynamic rifting model of LIP formation, 536,
        538–539
    East African Rift, 5, 402–403, 405
    Emperor Trough, 456
    European Cenozoic rift system, 725, 728
    Gondwana
        LIP eruption in icosahedral rifts, 595, 596,
            597, 601
        rift triple junctions, 594, 595, 601
        rifting of Zealandia in Late Cretaceous, 479
    Karoo flood basalt, 595
    North Sea basin, 403, 406
    Norway, Middle Jurassic rifting, 527–528
    Osbourn Trough, 478, 479
    rift development and breakup of
        supercontinents, 593, 931
    rift-to-drift transition, 404, 411
    Scottish Midland Valley, 402–403, 404, 406
    Venus
        corona orientation relative to rifts, 865, 866
        map, 861
        overview, 862, 880
        relationship between coronae and chasmata,
            863–864, 865
        scarcity of craters near chasmata, 859
    West Antarctic Rift, 583, 601, 962
Rodinia fragmentation, 775

Rodrigues, 192, 193, 195
RTI. *See* ridge-transform intersections

S

S20RTS, 123, 124, 128–129. *See also* tomographic
        mantle models
Sahyadri Range. *See* India; Western Ghats
        (Sahyadri Range)
Sala y Gomez Ridge
    age, 484
    Easter hotspot, 25, 27
    initiation, 484, 491, 492
    magmatism, 482
    map, 473, 484
Samoa Islands
    age, 483
    asthenosphere flow from hotspot, 193, 194
    depth of source, 11, 129
    eruption rates, 8
    extensional stress, 5, 491
    hotspot track, 68, 70
    large melting anomaly, 18
    map, 473
    thermoelastic stress, 501
    transition-zone observations, 124–125, 129
San Felix, 67, 124, 184, 193–194
Sardinia, 731–732
Scott plume, 193, 194
Scottish Midland Valley, 13, 402–403, 404, 406
SDR. *See* seaward-dipping reflector sequences
SDRS. *See* seaward-dipping reflector sequences
seamounts. *See also* Detroit seamount; Emperor
        seamounts; Foundation seamounts;
        ocean island basalts; ridge-crossing
        seamount chains
    Bowditch seamount, 557, 561, 585
    comparison of Bermuda cluster with Japanese
        seamounts, 561–562
    Corner seamounts, 38
    geochemistry of East Pacific Rise seamounts,
        397, 400
    Great Meteor seamount, 38, 191, 192, 195
    hotspot seamount chains on the Pacific plate, 337
    Japanese seamounts, 561–562
    Macdonald seamount, 124–125, 129, 193, 194
    Magellan seamounts, 473–474, 477–478, 480,
        484
    Marcus Wake seamounts, 453, 473–474, 480,
        484
    Meiji seamount, 343, 465
    melt transport at ridge-crossing seamount
        chains, 376–377
    Muir seamount, 555, 562
    Musicians seamounts, 316, 318
    New England Seamounts, 38–39
    New England Seamounts, North Atlantic Ocean,
        38–39
    numbers of, 388
    Rano Rahi seamounts, 397, 401
    St. Helena Seamount chain, 16
    Tenji seamount, 464
    Tokelau seamounts, 320, 322
    Wentworth seamounts, 456

seaward-dipping reflector sequences (SDRSs or SDRs). *See also* large igneous province (LIP) formation; volcanic passive margins
    Baffin Island, 533, 549–550
    Greenland, 529, 760–761
    Irish volcanic margin, 531
    North Atlantic Igneous Province (NAIP), 29, 529, 531, 533, 549–550
    Norway, 29, 533
    in volcanic passive margin crust structure, 748–749
    Vøring Plateau, 533
seismic tomography. *See* receiver functions; Seismic Wave Exploration in the Lower Lithosphere; seismology; tomographic mantle models
Seismic Wave Exploration in the Lower Lithosphere (SWELL). *See also* Hawaii; Hawaiian swell; seismology; swells
    bathymetry, 224
    comparison of model with triangle approaches, 220–222
    compensation depth, 210–211, 224–225
    critiques and alternative causes of swell, 228–232
    differential pressure gauge (DPG), 212, 213–214
    dispersion measurement, 211–212, 216, 218, 220, 223
    earthquake amplitude and noise spectra for Rat Island event, 212–213
    earthquakes off the coast of Chile, 213–214
    earthquakes off the coast of Guatemala, 214–215
    frequency-dependent phase measurements, 215
    geoid anomalies, 224–225
    hybrid thermal rejuvenation-dynamic support/thinning model, 225–226, 228–232
    lateral refraction, 215–216, 221
    lateral variations across swell, 216–217, 225
    lithospheric rejuvenation, 219–220, 222, 223, 225
    Low-Cost Hardware for Earth Applications and Physical Oceanography (L-CHEAPO), 211
    low-velocity body in mantle, 215, 220, 222–223, 224
    magnetotelluric (MT) data, 223, 225
    overview, 211–212, 225–226
    phase measurements across the pilot array, 215–216
    phase velocity curve determinations, 215–217
    phase velocity inversion for structure at depth, 217–220
    resolution limits, 216, 217, 220, 222–223
    shear velocity models, 218–219, 218–220
    signal coherence assessment, 214–215
    signal-to-noise assessment, 212–214
    site location map, 212
    tradeoff between prediction error and smoothness, 217–218

seismology. *See also* receiver functions; tomography
    body waves for whole-mantle tomography, 158, 729
    core-reflected S-waves (ScS), 84, 230
    effects of temperature, 11
    Fresnel zones, 123, 130, 133, 134
    global shear wavespeed variations in shallow mantle, 167–168
    ocean bottom seismometers (OBSs), 510, 512, 513
    seismic modeling and uncertainty analysis of oceanic crust, 510–511, 512
    SS precursors and transition zone discontinuities, 122–126, 130, 131–135
    SS shear waves, 123
    surface waves
        lithospheric plume detection, 131, 158, 167–168, 211
        sensitivity to azimuthal anisotropy, 182
        sensitivity to shear velocities, 167–168, 217
        surface wave tomography, 182
    temperature, effect on seismic wavespeeds, 11
    velocity of seismic waves, factors affecting, 729
    wide-angle reflection and refraction seismics (WAS), 510, 513–514, 515, 521
Seychelles, breakup from India, 787, 788, 806, 807, 808
Shatsky Rise
    age progression, 456
    geochemistry, 477
    large igneous province, 5
    mantle plume model, 454
    maps, 453, 473, 474, 475
    MORB-like origin, 458
    Pacific-Farallon-Izanagi triple junction, 5, 383, 454–456, 458, 477–478
Shenandoah Igneous Province, 581
Shillong plateau, 816, 817
Shona, 190
Siberian Traps
    age, 6, 18, 674, 689
    alteration index (A.I.), 689–690
    Angara-Taseevskaya syncline geochemistry, 680
    chemistratigraphic subdivisions of lavas, 676–678
    dolerite sills, 672, 680, 689
    DUPAL (Dupré and Allègre) and non-DUPAL basalts, 676–678, 682–684
    duration of Siberian Traps FBP magmatism, 672–674, 689
    edge-driven convection model, 682, 685
    geochemistry, 676–680, 683–684
    heatflow, 676
    high-Ti basalts, 676, 678, 685
    impact model, 680–681, 685
    lithospheric delamination model, 682–683, 685
    lithospheric structure, 675–676
    low-Ti basalts, 676–677, 678, 679–680, 682–683, 685
    magmatic mica, 680, 684, 685
    Maimecha-Kotui subprovince, 676–678, 678–679
    maps, 671, 673

    Noril'sk-Kharaelakh subprovince, 676–678, 681, 682, 683
    paleogeographic reconstruction, 675, 676
    plume model, 681–682, 685
    rare earth element patterns, 676, 677–678
    Siberian Traps flood basalt province (FBP), definition, 670
    Siberian Traps large igneous province (LIP), definition, 670
    size and volume estimates, 6, 18, 98, 670–672
    subduction-related model, 683–685
    subprovinces and dates localities, map, 673
    subsidence and uplift chronology, 6, 18, 98, 675, 681–682, 786–787
    tectonic setting, 675
    timing of initial eruptions, 674
    timing of Siberian Traps LIP, 674
    trace element patterns, 677–678, 680, 682–683, 685
    Tunguska syncline
        lithospheric thinning, 676, 678, 681, 683–684
        map, 671
        subsidence, 675, 681–682, 685, 786
        volume, 671, 685
    West Siberian basin geochemistry, 679–680
Sicily Channel, 731
Sierra Leone hotspot, 194–195
Singhbum craton, 296, 297
slabs. *See also* subducted oceanic crust
    accumulation at core-mantle boundary, 724
    accumulation in D" region, 724
    asthenosphere entrainment by subducting slabs, 167, 173–174
    cooling of mantle, 63–64, 73
    deep mantle displacement by subducted material, 49, 54, 57, 724
    deep mantle penetration, 85
    detachment and movement of subducting slab, 707, 708
    Farallon slab, 566, 569
    kinematic consequences of shallow hotspot reference frame, 367–369, 371–373
    penetration into lower mantle, 72, 85, 167, 569, 580
    recycling, 6, 14
    slab dehydration and melting, 5–6, 206–207
    slab dip, 372–373, 483, 491
    slab-steepening and breakoff beneath subduction-accretion complex, 709, 713–718
    stagnation in transition zone, 85, 569, 580, 584
    thickening of subducting slabs in mantle, 701
    water in subducted slabs, 299–300, 329, 670, 683–684, 690
Snake River capture by Salmon-Columbia River, 804, 805
Snake River Plain. *See also* Columbia River Basalt Group
    age progression, 650, 664
    geochemistry, 650
    hotspot track, 650, 651, 662, 663
    plate model, 664–665
    plume model, 650–651
    tectonic setting, 640–642

Society Islands, 193–194, 501
Sorachi terrane, 474, 475, 478, 482
Southeast Indian Ridge
    ^{3}He/^{4}He ratios, 198, 200
    asthenosphere flow, 190, 192–193
    geochemical relationships to hotspots, 197
    geochemical variations along ridge, 195,
        197–198
    maps, 392, 394
    ΔNb for basalts, 391–392, 393, 394
Southwest Indian Ridge, 192–193, 197, 391–392,
        393
St. Helena island
    alteration of rocks, 316, 317
    asthenosphere flow from hotspot, 192
    trace elements injected into Mid-Atlantic Ridge,
        195, 196
St. Helena Seamount chain
    age progression, 16
Stalemate fracture zone, 455, 461, 464, 466
statistical upper mantle assemblage (SUMA)
        model, 52, 60
Stillwater, 899, 910
stoping, 48, 49, 437, 438. *See also* delamination
stretching factors (β, 650, 651, 709, 739
subducted oceanic crust. *See also* slabs
    deep mantle displacement by subducted
        material, 49, 54, 57, 724
    eclogite from subducted oceanic crust, 5–6, 14,
        49, 52, 95
    slab dehydration and melting, 5–6, 206–207
    statistical upper mantle assemblage (SUMA)
        model, 52, 60
    transition zone as barrier to subduction, 55, 95
subsidence. *See also* Deccan basalt province
    Columbia River Basalt Group (CRBG), 6,
        650–651, 657–658, 787, 803–805
    large igneous provinces (LIPs), 6–7, 18
    North Atlantic Igneous Province (NAIP), 6–7
    at Siberian Traps, chronology, 6, 18, 98, 675,
        681–682, 786–787
    Siberian Traps subsidence and uplift
        chronology, 6, 18, 98, 675, 681–682,
        786–787
    Tunguska syncline, 675, 681–682, 685, 786
    uplift and subsidence from hotspots, 6–7, 36–37
Sudbury structure, 899, 900, 910, 967
supercontinents, 593, 594, 931. *See also*
        Gondwana; Pangaea
superplumes. *See also* convection models of
        multiscale plumes; large low–shear
        velocity provinces
    as alternative to thermal plume hypothesis, 104,
        105
    compositional origin, 11, 17
    creation from D" layer, 158–159
    definition, 138
    density, 11
    multiscale nature of mantle plumes, 158–160
    secondary plumes, 138, 145, 148–150, 156–159
superswells, 65–66, 68, 74, 76–77
surface waves
    lithospheric plume detection, 131, 158,
        167–168, 211

sensitivity to azimuthal anisotropy, 182
sensitivity to shear velocities, 167–168, 217
surface wave tomography, 729
swells. *See also* bathymetry; Hawaiian swell;
        hotspots; Seismic Wave Exploration in
        the Lower Lithosphere
    geoid anomalies at hotspot swells, 106–108
    Hawaiian swell
        buoyancy flux, 33
        dynamic model, 34
        heatflow measurements, 17
        kinematic model, 33–34
        size, 18, 33, 36
        topographic map, 33
        volume flux, 33, 34
    heatflow anomalies at hotspot swells,
        calculation method, 106–108
    heatflow anomalies at hotspot swells,
        measurements, 106, 281–282
    superswells, 65–66, 68, 74, 76–77
    topographic anomalies at hotspot swells,
        106–108

T

Tahiti, 8, 11, 124–125, 129
Tasman Ridge, 474, 476, 481
tessellation. *See* icosahedrons
Tethys Ocean closure, 580, 581–582, 584, 842
thermal boundary layer (TBL) at base of
        lithosphere, 90–91, 96, 764, 765–766,
        766–767
thermal boundary layer (TBL) at core-mantle
        boundary
    formation, 104, 108
    as plume source, 80, 89, 90–91, 115, 120
    pressure effects, 89, 90
    Rayleigh number effect on temperature, 89–90,
        96
    temperature changes, 80, 89–91, 96
    temperature dependence of viscosity, 83, 89–90,
        95, 99
    thickness, 90
    transfer of heat from core, 89
thermal models and thermal reference models.
        See heatflow; thermal plumes
thermal plumes. *See also* plume hypothesis;
        plumes
    axisymmetric calculations
        coefficient of thermal expansion as a function
            of depth, 110, 111–113, 114–115, 116
        constant-viscosity convection calculations,
            108–109
        with internal heating, 109–110, 111, 115
        limitations of method, 116
        methods, 106–108, 116
        with phase transformations, 110, 113–114,
            115
        Rayleigh number effect, 109, 110
        with stable layer at base of mantle, 110,
            112–114, 115, 116
        with temperature-dependent viscosity,
            109–111, 112
    temperature dependence of viscosity, 107

thermal conductivity effects, 115–116
thermal plume hypothesis, overview, 29–32, 80,
        103–106, 116, 508–509
thermochemical piles, 83–85, 93, 98–99. *See also*
        large low–shear velocity provinces
tholeiite. *See also* olivine tholeiite magmas
    differentiation of Hawaiian tholeiite, 228–229
    eclogite as source, 228, 253, 258
    Iceland magmatic volumes, 14
    Rajmahal-Sylhet igneous province, 825
    similarity to Kerguelen OIB, 825
    temperatures of hotspot and MORB tholeiites,
        241–243
Tibetan plateau, 694, 718, 793
titanium
    high-Ti basalts in Siberian Traps, 676, 678, 685
    low-Ti basalts in Siberian Traps, 676–677, 678,
        679–680, 682–683, 685
    TiO content of Iceland basalts, 421–423,
        434–437
Tokelau seamounts, 320, 322
tomographic mantle models
    comparison of V_s and V_p models, 80–81
    finite-frequency model PRI-S05, 123, 124,
        129–130
    S20RTS, 123, 124, 128–129
    SS–SdS differential times, 122–123
    whole-mantle-convection and superswell model,
        65–66, 68, 74, 76–77
tomography. *See also* receiver functions;
        seismology; tomographic mantle models
    body waves for whole-mantle tomography, 158,
        729
    finite-frequency tomography, 10
    global tomography, 10–11, 729
    low velocity zones in melting anomalies, 10,
        729
    mantle structure from seismology, 10–11
    surface wave tomography, 729
    teleseismic tomography, 10
Tonga trench, 18, 483, 505, 658
topography of the Earth, map, 15
transform faults
    Chinook Trough, 455
    Emperor Trough, 456, 459, 461, 462, 480
    horizontal tensional stress ahead of ridge-
        transform intersections (RTIs), 377, 384
    Ninetyeast Ridge fossil transform fault, 383
    Réunion, 362
    ridge-transform intersections (RTIs), 375, 377,
        384
transition zone. *See also* discontinuities; mantle;
        receiver functions
    410-km velocity discontinuity
        olivine phase transitions, 122, 126–128,
            131–134
        phase change, 10
        temperature effect on depth, 122, 126–128,
            130
    660-km velocity discontinuity
        garnet phase transitions, 122, 126–128,
            132–133, 134
        olivine phase transitions, 122, 126–128,
            131–134

transition zone (*continued*)
  slab penetration into lower mantle, 72, 85, 167, 569, 580
  temperature and composition effects on depth, 122, 126–128, 130
  as barrier to subduction, 55, 95
  eclogite fate in transition zone, 54, 55, 133–134
  olivine phase transitions, 122
  ponding of plume material in transition zone, 126, 134, 166
  seismic tomography models, 128–130
  SS precursor seismic data, 122–126, 130, 131–135
  thickness, global average, 124, 132, 134–135
  thickness, plume average, 124, 131–133
  topography, 10–11
  water-rich transition zone, 683–684
traps. *See also* Deccan Traps; large igneous province (LIP) formation; Siberian Traps
  British Tertiary igneous province, 753
  definition, 748
  Ethiopian Traps, 673
  magma flow during trap emplacement at Isle of Skye, 755–760, 761
  mantle plume explanation, 748
  nonplume explanation, 748
  Rajahmundry Traps, 290, 291, 292, 328–329, 329
  Rajmahal Traps, 290, 313, 314, 829–830
  trap stage of LIP formation, 748
triple junctions. *See also* Pacific-Farallon-Izanagi triple junction
  Afar, 3, 5, 19
  association with melting anomalies, 4
  Azores, 4
  Bouvet, 4
  Cambay rift, 810
  Easter Island, 4
  Emperor Trough as sinistral transform fault between old and new triple junctions, 480
  Gondwana, 594, 595, 601
  Hawaiian-Emperor hotspot origin in abandoned triple junction, 456, 461, 462, 466–468
  Karoo, 601
  microplates at triple junctions, 5, 185, 193
  Ontong Java plateau creation at triple junction, 383, 481
  Rodrigues, 192, 193, 195
  Shatsky Rise creation at triple junction, 383
  traveltime of the triple junction to Meiji, 463
Tristan da Cunha
  alteration (chemical weathering), 315–316
  asthenosphere flow from hotspot, 192, 193
  asymmetrical V-shaped array of seamounts, 375, 376
  geochemistry, 617, 618–619, 620
  hotspot location and strength, 184
  hotspot reference frame and fixity, 361, 957, 959
  hotspot track, 67
  map, 392
  and Paraná-Angola-Etendeka magmatism, 623–624, 626

plate model for melting anomaly, 16
trace elements injected into Mid-Atlantic Ridge, 195
Tristan-Gough, 38. *See also* Gough
Tuamotu Ridge, 25, 27, 473–474, 491, 492–493
Tyrrhenian Sea, 730–731

U

ultra-low-velocity zones (ULVZs), 80, 82, 84–85. *See also* core-mantle boundary
undergraduate education and mantle plume theory
  lack of alternative hypothesis discussion, 938, 939–941, 943, 944
  presentation in introductory geology textbooks, 938–939, 940, 943, 944
  presentation in other geoscience courses, 939, 940
  presentation in upper-division courses, 939–940
  recommendations, 940–941
  teaching the mantle plume debate, 940, 943
underplating. *See also* ponding
  eclogite engine, 49
  high-velocity lower-crustal body (LCB) in North Atlantic Igneous Province, 533
  in large igneous provinces, overview, 842
  lithosphere rejuvenation and underplating by dikes, 229
  Marquesas swell, buoyancy of underplating material, 514
  melting anomalies, 510
  petrologic modeling, 842
  and uplift, 842
uplift. *See also* Deccan basalt province; swells
  Bermuda Rise, 569–570, 574–579, 581
  Cape Verde, 560
  cymatogenic uplift, 790, 791, 797, 799
  delamination, 437–438
  Eastern Anatolia, 694, 710, 713, 718
  Emeishan large igneous province, 7, 786, 843, 854–855, 856
  flood basalts, prevolcanic uplift, 786–787
  Greenland, 529, 535, 786
  lithospheric detachment, 437–438
  lithospheric uplift predicted over plume heads, 786, 792, 800, 805–810
  at Norwegian volcanic margins, 534–535
  subsidence and uplift chronology
    Columbia River Basalt Group (CRBG), 6, 650–651, 657–658, 787, 803–805
    large igneous provinces (LIPs), 6–7, 18
    melting anomalies, 6–7
    North Atlantic Igneous Province (NAIP), 6–7, 527–528, 534–535
    Ontong Java plateau, 6, 787
    plate model predictions, 6–7
    plume hypothesis, 6–7, 36–37, 786–787
    plume hypothesis predictions, 6–7, 36–37, 786–787
    Siberian Traps, 6, 18, 98, 675, 681–682, 786–787
  and underplating, 842
uplift and subsidence from hotspots, 6–7, 36–37
Western Ghats (Sahyadri Range), 790–791

V

V-shaped volcano trails
  Atlantic Ocean
    hotspot and plate motion interpretation, 532
    Kolbeinsey Ridge, 532
    Reykjanes Ridge, 532, 957
    Tristan da Cunha, 375, 376
    V-shaped ridges or escarpments (VSRs), 957, 969
  Galápagos volcanic province, 515
  hotspot and plate motion interpretation, 532, 957
  lateral and vertical transport models, 375–376, 550, 949, 957
velocity discontinuities. *See* discontinuities
Vema
  asthenosphere flow from hotspot, 192–193
  geochemistry, 238
  hotspot location and strength, 184
  hotspot track, 67
  seismic observations of transition zone, 124–125
  transition-zone observations, 124–125
Venera 15 and 16 radar mapping, 860, 876
Venus
  Acrea, 867, 868, 869, 895
  age of surface, 859, 903–904
  Anquet, 867, 868, 869
  Aramaiti Corona, 891–892
  atmosphere, 882–883
  BAT region (area between Beta, Atla, and Themis regiones)
    activity, 865, 877
    age estimates, 865, 875
    craters and coronae, 865, 867, 872, 877, 903
    map, 862
  C21, 867, 868, 869, 872
  Chloris Mons, 881, 892
  Cleopatra, 867, 868, 869
  coronae
    age estimates, 866, 876, 880, 881
    assessment of circular symmetry, 866–867, 868, 874
    background, 860, 875–877, 882, 908–909
    classification by type, 869–871
    conflicting endogenic models, 881
    crater statistics, 864–866
    definitions, 861
    distribution, 863, 866–867, 875–876
    evolutionary diapir model, 860, 867–871, 876–878
    impact origin hypothesis, 860, 864–867, 874–877, 908–909
    mantle upwelling models, 860
    map, 861
    mechanism of formation, 871–872
    morphology, 860
    nomenclature, 881
    numbers of, 860, 861
    numerical modeling, 881
    orientation relative to rifts, 865, 866
    relationship between coronae and chasmata, 863–864, 865

shallow source, 872
spatial distribution, 860
topography, 867, 868, 874–875
crust during late accretion, 904
Eurynome, 867, 868, 869
evolution of surface, 903–904, 907–908
geoid, 861, 862–863, 883–885
hydrosphere, 882–883
imagery and data sets, 860–863, 880–881
impact craters and basins
   age estimates, 874, 880, 888–889
   background, 859, 908–909
   crater density in coronae, rifts, and the BAT
      region, 865
   distribution by longitude and latitude, 867
   giant impact basins, 894, 895, 897–899
   halos, 860
   map, 861
   modification and alteration of "pristine"
      craters, 861, 874, 880, 888–889
   overview and description, 859–860, 908,
      909–910
   relationship between "pristine" and older
      craters, 889–892
   scarcity of craters near chasmata, 859
   topography, 867, 868, 875
Isabella, 867, 868, 869, 890–891, 910
Klenova, 867, 868, 869
lack of plate tectonics (unimodal topography),
   859, 872, 880, 882–883, 885
lithosphere strength, 883–885
Magellan spacecraft data, 860–862, 867, 875,
   881–882, 909
Magellan Stereo Toolkit, 875, 877, 881
mass imbalance, 885
Maya, 867, 868, 869
Mead, 867, 868, 869, 891
Meitner, 867, 868, 869
Ninhursag, 867, 868, 869, 902
old circular structures and impact hypothesis
   age estimates, 880, 881
   endogenic explanations, 881, 902–903
   exclusion from conventional literature, 881,
      885–888, 889–892, 909
   fine-scale concentric structure, 901
   giant impact basins, 894, 895, 897–899
   mud volcanoes derived from wet sediments,
      883, 907–908
   multiring structures, 900–901, 910
   nomenclature for circular structures, 881
   radial structures, 901–902
   relationship between "pristine" and older
      craters, 889–892
   size and frequency relationships, 886, 888
   small ancient impact structures, 893–895,
      896–898, 908
   small old structures, 886, 887, 890–892

submarine formation of, 883
superpositions, 885, 889–890, 892–895
volcanoes as impact-fluidization constructs,
   885, 899–900, 901, 902, 910
Pioneer missions to Venus, 861, 862
plume hypothesis, 860, 881–882, 962
properties of Venus, 883–885
regiones, 862–863, 872
resurfacing, 860, 880
rifts (chasmata)
   corona orientation relative to rifts, 865, 866
   map, 861
   overview, 862, 880
   relationship between coronae and chasmata,
      863–864, 865
   scarcity of craters near chasmata, 859
sedimentation, 882–883, 888–891, 892–893,
   895–897, 907–908
tessera plateaus
   age range, 891
   Beta Regio, 885
   characteristics, 899
   formation from impact melts and fluidized
      debris, 899, 903
   images, 896, 898
   oldest tessera plateaus as basement, 904
   sparsity of old impact structures, 891, 895,
      903
   Venera 15 and 16 radar mapping, 860, 876
   volcanoes as impact-fluidization constructs, 885,
      899–900, 901, 902, 910
Virginia, igneous activity, 573–574, 580, 581–582,
   584
Vizcaino-Guerrero arc, 477
volcanic chains. *See* hotspot tracks
volcanic passive margins (VPMs). *See also* large
      igneous province (LIP) formation;
      seaward-dipping reflector sequences
   crust layering and seismic wave velocities,
      749
   magma flows at VPM stage in East Greenland,
      760–761, 762
   plume hypothesis, 749
   prebreakup and breakup-related magmatism in
      North Atlantic, 529–532
   tectonic explanations, 749
   volcanic passive margin stage of LIP formation,
      748–749
volcanoes. *See also* V-shaped volcano trails
   correlation of volcanism and extensional stress,
      5, 48
   delamination, 92, 98
   mud volcanoes derived from wet sediments,
      883, 907–908
Vøring Basin, 534, 535, 751
Vøring Plateau, 529, 532, 533, 538
Vredefort structure, South Africa, 900, 910

W

Walvis Ridge, 16, 316, 616, 624
water in subducted slabs, 299–300, 329, 670,
   683–684, 690
Wentworth seamounts, 456
West Antarctic Rift, 583, 601, 962
West Virginia, igneous activity, 573–574, 581–582
Western Cordillera, 477, 482, 581
Western Ghats (Sahyadri Range). *See also* Deccan
      Traps; India
   age, 307, 309
   alteration index, 309
   charnockites, 789–790
   geology, 787, 789–791
   Palghat Gap, 789
   rivers and drainage, 788–791, 806–807
   uplift, 790–791, 800
western Rajasthan. *See also* Deccan basalt
      province; Malani large igneous province
   Barmer-Sanchor rift, 776–777, 784
   extensional tectonics, 775, 781–783, 784
   geological evolution, overview, 775, 777, 778,
      781–782
   K-T magmatism, 777–779, 781
   K-T sedimentary basins
      Barmer basin, 778, 781
      Bikaner-Nagaur basin, 778, 780–781
      geological evolution, 777, 779, 781–783
      Jaisalmer basin, 778, 779, 780, 781
      map, 776
      stratigraphy, 778
   maps, 776
   Marwar Supergroup, 781
   Mundwara igneous complex, 777–778, 781–782
   Phanerozoic stratigraphy, 778
   and Réunion plume, 777, 778, 781, 782–783
   Sarnu-Dandali alkaline complex, 779, 781, 784
   Tavidar volcanics, 778–779
wet melting, 6. *See also* damp mantle melting
wetspots, 35, 360, 361, 625. *See also* hotspots

Y

Yellowstone hotspot
   deepening of 410-km discontinuity, 129
   and formation of Coastal Range, 316
   hotspot track, 67, 69
   low velocity zone in upper mantle, 10, 664
   plume tail at 500-km depth, 650, 664
   seismic transition-zone observations, 124–125,
      129, 131
   shallow source, 76
   tomographic cross-section, 664
Yellowstone–Snake River Plain, 18, 650. *See also*
      High Lava Plains; Snake River Plain
yo-yo tectonics, 49, 57, 58, 63